ADVANCED INORGANIC CHEMISTRY

Sixth Edition

Sixth Edition

ADVANCED INORGANIC CHEMISTRY

F. Albert Cotton

W. T. Doherty-Welch Foundation Distinguished Professor of Chemistry
Texas A&M University
College Station, Texas, USA

Geoffrey Wilkinson

Sir Edward Frankland Professor of Inorganic Chemistry
Imperial College of Science and Technology
University of London
London, England

Carlos A. Murillo

Professor
University of Costa Rica
Ciudad Universitaria, Costa Rica
Adjunct Professor
Texas A&M University
College Station, Texas, USA

Manfred Bochmann

Professor
School of Chemistry
University of Leeds
Leeds, England

A WILEY-INTERSCIENCE PUBLICATION
JOHN WILEY & SONS, INC.
New York • Chichester • Weinheim • Brisbane • Singapore • Toronto

Library of Congress Cataloging in Publication Data:

Advanced inorganic chemistry. --6th ed. / by F. Albert Cotton, Geoffrey Wilkinson, Carlos A. Murillo, and Manfred Bochmann; with a chapter on boron by Russell Grimes.
    p. cm.
  Prev. ed. entered under Cotton.
  "A Wiley-Interscience publication."
  Includes index.
  ISBN 0-471-19957-5 (alk. paper)
  1. Chemistry, Inorganic. I. Cotton, F. Albert (Frank Albert),
  1930-   . II. Wilkinson, Geoffrey, III. Murillo, Carlos A.,
  IV. Bochmann, Manfred, IV. Grimes, Russell N.
  GD151.2.C68   1999
  546--dc21

                                98-8020

Printed in the United States of America.

10 9 8 7 6 5 4 3 2 1

Preface

Readers of this book will, if they are familiar with the fifth edition, immediately recognize that there are several major changes. Most obvious is the fact that there are now four authors. This change was planned and implemented by Geoffrey Wilkinson and me before his death. We had also decided to have the chapter on boron chemistry done by an expert on boron chemistry since neither of us had ever felt sufficiently conversant with that vast and complex subject. We were very pleased when Russell Grimes consented to accept the task. With regard to the rest of the book, the general division of labor was to be such that each author undertook about one-fourth of the book. As fate would have it, Geoffrey Wilkinson was able to finish written drafts of all his chapters just days before he so unexpectedly passed away. Thus his deft hand will be very much evident here. Beyond that, however, I have assumed the responsibility for this new edition.

The other major change that Geoffrey Wilkinson and I have made is to redistribute the content of the book. It has always been our desire to keep the physical size of the book to about that of the fifth edition, or, if possible, a little less. Another point that influenced our thinking was the increased appearance of books that deal with the broad principles of fields like organometallic chemistry, reaction mechanisms, bioinorganic chemistry and catalysis. We thus decided to eliminate some of the broad topical chapters previously included (e.g., Bioinorganic Chemistry) and enlarge the coverage of the individual elements (or groups) to deal with these types of chemistry on an element by element basis. This does not necessarily correspond to a reduction in space devoted to important topics. For example, more, not less, space is now devoted to the biochemistry of iron. In its arrangement of material, this edition is now more like the third edition than the fourth or fifth, though of course the amount of material is much greater.

We have continued to handle other aspects of the book in the same way as in the earlier editions. We have adhered to the Periodic Table as the organizing principle. We have also continued our practice of including references to the research literature following the earlier rules, as follows. Only references that have appeared since the previous edition are normally given; documentation of many statements may be found in earlier editions. In citing references we mention only the name of the corresponding author (or the first, if there are three or more) except when there are only two authors, when both names are given.

F. Albert Cotton
College Station, Texas

Preface to the First Edition

It is now a truism that, in recent years, inorganic chemistry has experienced an impressive renaissance. Academic and industrial research in inorganic chemistry is flourishing, and the output of research papers and reviews is growing exponentially.

In spite of this interest, however, there has been no comprehensive textbook on inorganic chemistry at an advanced level incorporating the many new chemical developments, particularly the more recent theoretical advances in the interpretation of bonding and reactivity in inorganic compounds. It is the aim of this book, which is based on courses given by the authors over the past five to ten years, to fill this need. It is our hope that it will provide a sound basis in contemporary inorganic chemistry for the new generation of students and will stimulate their interest in a field in which trained personnel are still exceedingly scarce in both academic and industrial laboratories.

The content of this book, which encompasses the chemistry of all of the chemical elements and their compounds, including interpretative discussion in the light of the latest advances in structural chemistry, general valence theory, and, particularly, ligand field theory, provides a reasonable achievement for students at the B.Sc. honors level in British universities and at the senior year graduate level in American universities. Our experience is that a course of about eighty lectures is desirable as a guide to the study of this material.

We are indebted to several of our colleagues, who have read sections of the manuscript, for their suggestions and criticism. It is, of course, the authors alone who are responsible for any errors or omissions in the final draft. We also thank the various authors and editors who have so kindly given us permission to reproduce diagrams from their papers: specific acknowledgements are made in the text. We sincerely appreciate the secretarial assistance of Miss C. M. Ross and Mrs. A. B. Blake in the preparation of the manuscript.

F. A. COTTON
Cambridge, Massachusetts

G. WILKINSON
London, England

Contents

PART 4   THE ROLE OF ORGANOMETALLIC CHEMISTRY IN CATALYSIS

Appendices

Index

Abbreviations in Common Use

1. Chemicals, Ligands, Radicals, etc.

Ac	acetyl, CH_3CO
acac	acetylacetonate anion
acacH	acetylacetone
acacen	bisacetylacetoneethylenediimine
AIBN	azoisobutyronitrile
am	ammonia (or occasionally an amine)
An	any actinide element
n[Ane]X_m	cyclic polydentate ligand; n = ring size; X = O, S, NR
Ar	aryl or arene (ArH)
aq	aquated, H_2O
ATP	adenosine triphosphate
9-BBN	9-borabicyclo[3,3,1]nonane
bdt	the anion of 1,2-benzenedithiol (H_2bdt)
BINAP	2,2'-bis(diphenylphosphino)-1,1'-binaphthyl
bipy	2,2'-dipyridine, or bipyridine
Bu	butyl (Bu^n, normal-; Bu^i, iso-; Bu^s, secondary-; or Bu^t, tertiary-butyl)
Bz	benzyl
C,lmn	cryptate ligand with l-, m-, and n-membered rings (also lmn-crypt)
cat	catechol (o-dihydroxybenzene)
COD or cod	cycloocta-1,5-diene
COT or cot	cyclooctatetraene
Cp, Cp', Cp*	cyclopentadienyl, C_5H_5; MeC_5H_4; C_5Me_5
cy	cyclohexyl
dab or dad	1,4-diaza-1,3-butadienes, $RN{=}CH{-}CH{=}NR$
dba	*trans, trans*-dibenzylideneacetone
depe	1,2-bis(diethylphosphino)ethane
depm	1,2-bis(diethylphosphino)methane
diars	o-phenylenebisdimethylarsine, o-$C_6H_4(AsMe_2)_2$
dien	diethylenetriamine, $H_2NCH_2CH_2NHCH_2CH_2NH_2$
diglyme	diethylene glycol dimethyl ether, $CH_3O(CH_2CH_2O)_2CH_3$
dike	a diketonate anion, such as acetylacetonate
diop	{[2,2-dimethyl-1,3-dioxolan-4,5-diyl)bis(methylene)]-bis(diphenylphosphine)}
diphos	any chelating diphosphine, but usually 1,2-bis(diphenyl-phosphino)ethane, dppe
DME	dimethoxyethane (also glyme)
DMF or dmf	N,N'-dimethylformamide, $HCONMe_2$
dmg	the anion of dimethylglyoxime ($dmgH_2$)

dmpe	1,2-bis(dimethylphosphino)ethane
dmpm	1,2-bis(dimethylphosphino)methane
DMSO or dmso	dimethyl sulfoxide, Me_2SO
DPhF, DTolF	N,N'-diphenylformamidinate, N,N'-di-p-tolylformamidinate
dpma	bis(diphenylphosphinomethyl)phenylarsine
dppe	1,2-bis(diphenylphosphino)ethane
DPPH	diphenylpicrylhydrazyl
dppm	bis(diphenylphosphino)methane
dppp	bis(diphenylphosphino)propene
E	electrophile or element
$EDTAH_4$	ethylenediaminetetraacetic acid
$EDTAH_{4-n}^{n-}$	anions of $EDTAH_4$
en	ethylenediamine, $H_2NCH_2CH_2NH_2$
Et	ethyl
Fc	ferrocenyl (Fc' for substituted Fc)
form	a formamidinate anion
Fp	$Fe(CO)_2Cp$
gly	glycinate anion
glyme (=DME)	ethylene glycol dimethyl ether, $CH_3OCH_2CH_2OCH_3$
heidi	the anion of $N(CH_2COOOH)_2(CH_2CH_2OH)$ (H_3heidi)
hfa	hexafluoroacetylacetonate anion
HMPA	hexamethylphosphoric triamide, $OP(NMe_2)_3$
hpp	the anion of 1,3,4,6,7,8-hexahydro-2H-pyrimido-[1,2-a]-pyrimidine (Hhpp)
ind	indenyl
L	ligand
Ln	any lanthanide element
M	central (usually metal) atom in compound
MAO	methylalumoxane
m-C-n	macrocyclic polyether with m-membered ring and n oxygen atoms
Me	methyl
mes	mesityl
mes*	2,4,6-$C_6H_2R_3$, where R is usually a bulky group such as Bu^t
Me_6tren	tris(2-dimethylaminoethyl)amine, $N(CH_2CH_2NMe_2)_3$
mnt	maleonitriledithiolate anion
MTO	methyltrioxorhenium(VII), CH_3ReO_3
NBD or nbd	norbornadiene
NBS	N-bromosuccinimide
np^2 (=PNP)	bis(2-diphenylphosphinoethyl)amine, $HN(CH_2CH_2PPh_2)_2$
np^3	tris(2-diphenylphosphinoethyl)amine, $N(CH_2CH_2PPh_2)_3$
$NTAH_3$	nitrilotriacetic acid, $N(CH_2COOH)_3$
OAc	acetate anion
oep	octaethylporphyrin
OTf	teflate anion, $OTeF_5^-$
ox	oxalate anion, $C_2O_4^{2-}$
Pc	phthalocyanine
Ph	phenyl, C_6H_5

phen	1,10-phenanthroline
pic	picolinate anion
pip	piperidine
pn	propylenediamine (1,2-diaminopropane)
PNP ($=np^2$)	bis(2-diphenylphosphinoethyl)amine, $HN(CH_2CH_2PPh_2)_2$
porph	porphyrin (or any porphyrin)
pp^3	tris(2-diphenylphosphinoethyl)phosphine, $P(CH_2CH_2PPh_2)_3$
PPN^+ ($=PNP^+$)	$[(Ph_3P)_2N]^+$
Pr	propyl (Pr^n or Pr^i)
pts	p-toluenesulfonate (also tos)
py	pyridine
pz or pyr	pyrazolyl
QAS	tris(2-diphenylarsinophenyl)arsine, $As(o\text{-}C_6H_4AsPH_2)_3$
QP	tris(2-diphenylphosphinophenyl)phosphine, $P(o\text{-}C_6H_4PPh_2)_3$
R	alkyl (preferably) or aryl group
R_F	perfluoro alkyl group
S	solvent
Sacac	thioacetylacetonate anion
sal	salicylaldehyde
salen	bissalicylaldehydeethylenediimine
saloph	the dianion of disalicylidene-o-phenylenediamine
silox	the anion of Bu^t_3SiOH
tacn	1,4,7-trimethyl-1,4,7-triazacyclononane
TAN	tris(2-diphenylarsinoethyl)amine, $N(CH_2CH_2AsPh_2)_3$
TAP	tris(3-dimethylarsinopropyl)phosphine, $P(CH_2CH_2CH_2AsMe_2)_3$
TAS	bis(3-dimethylarsinopropyl)methylarsine, $MeAs(CH_2CH_2CH_2AsMe_2)_2$
TCNE	tetracyanoethylene
TCNQ	7,7,8,8-tetracyanoquinodimethane
teen	N,N,N',N'-tetraethylethylenediamine
tempo	2,2,6,6-tetramethylpiperidine-1-oxyl
terpy	terpyridine
TFA	trifluoroacetic acid
THF or thf	tetrahydrofuran
THT or tht	tetrahydrothiophene
tmen	N,N,N',N'-tetramethylethylenediamine (also TMEDA, tmeda)
TMPA	tris(2-pyridylmethyl)amine
tmtaa	dibenzotetramethyltetraaza[14]annulene
tn	1,3-diaminopropane(trimethylenediamine)
tol	tolyl ($CH_3C_6H_4$)
tos	p-toluenesulfonate (also pts)
Tp, Tp*	hydridotrispyrazoylborate; substituted Tp
TPA	1,3,5-triaza-7-phosphaadamantane
TPN ($= np^3$)	tris(2-diphenylphosphinoethyl)amine, $N(CH_2CH_2PPh_2)_3$
TPP	meso-tetraphenylporphyrin
tren	tris(2-aminoethyl)amine, $N(CH_2CH_2NH_2)_3$
trien	triethylenetetraamine, $(-CH_2NHCH_2CH_2NH_2)_2$
triflate (ion)	$CF_3SO_3^-$

triphos	any chelating triphosphine
trop	anion of tropolone
TSN	tris(2-methylthiomethyl)amine, $N(CH_2CH_2SMe)_3$
TSP	tris(2-methylthiophenyl)phosphine, $P(o\text{-}C_6H_4SMe)_3$
TSeP	tris(2-methylselenophenyl)phosphine, $P(o\text{-}C_6H_4SeMe)_3$
TTA	2-thenoyltrifluoroacetone, $C_4H_3SCOCH_2COCF_3$
tu	thiourea
X	halogen or pseudohalogen
Xy	xylol, 2,6-dimethylphenyl

2. Miscellaneous

Å	angstrom unit, 10^{-10} m
asym	asymmetric or antisymmetric
bcc	body centered cubic
BM	Bohr magneton
bp	boiling point
ccp	cubic close packed
CFSE	crystal field stabilization energy
CFT	crystal field theory
CIDNP	chemically induced dynamic nuclear polarization
cm^{-1}	wavenumber
CT	charge transfer
CVD	chemical vapor deposition
dec	decomposes
d-	dextorotatory
endor	electron nuclear double resonance
ESCA	electron spectroscopy for chemical analysis (= XPE, X-ray photoelectron spectroscopy)
esr or epr	electron spin (or paramagnetic) resonance
eV	electron volt
EXAFS	extended X-ray absorption fine structure
FT	Fourier transform (for nmr or ir)
g	*g*-value
(g)	gaseous state
glc	gas-liquid chromatography
h	Planck's constant
hcp	hexagonal close packed
HOMO	highest occupied molecular orbital
Hz	hertz, s^{-1}
ir	infrared
IUPAC	International Union of Pure and Applied Chemistry
l-	levorotatory
(l)	liquid state
LCAO	linear combination of atomic orbitals
LFSE	ligand field stabilization energy
LFT	ligand field theory
LUMO	lowest unoccupied molecular orbital

MAS nmr	magic angle spinning nmr
MO	molecular orbital
mp	melting point
NOE	nuclear Overhauser effect
nmr	nuclear magnetic resonance
PE	photoelectron (spectroscopy)
R	gas constant
(s)	solid state
SCE	saturated calomel electrode
SCF	self-consistent field
SCF-Xα-SW	self-consistent field, Xα, scattered wave (form of MO theory)
sp	square pyramid(al)
str	vibrational stretching mode
sub	sublimes
sym	symmetrical
tbp	trigonal bipyramid(al)
U	lattice energy
uv	ultraviolet
VB	valence bond
Z	atomic number
ε	molar extinction coefficient
ν	frequency (cm^{-1} or Hz)
μ	magnetic moment in Bohr magnetons
χ	magnetic susceptibility
θ	Weiss constant

ADVANCED INORGANIC CHEMISTRY

Sixth Edition

Part 1

SURVEY OF PRINCIPLES

Chapter 1

SOME CROSS-CUTTING TOPICS

1-1 Scope and Purpose

As explained in the Preface, the organization of this sixth edition does not include chapters on broad classes of compounds (e.g., organometallic compounds) nor on broad topics (e.g., bioinorganic chemistry). Much material that previously appeared in such chapters is now distributed in the present rewritten chapters on the chemistry of individual elements or groups of elements. However, there are still cross-cutting concepts that are best treated generically. That will be done partly in this chapter, but also, for a few topics that are mainly relevant to transition metal chemistry, in Chapter 16.

1-2 Polyhedra for Coordination and Cluster Compounds

One way to describe the structure of a coordination compound, or any molecule, AB_n, wherein a central atom A is linked to n peripheral atoms, B, is to state the polyhedron whose vertices correspond to the positions of the B atoms. Thus, we describe $TiCl_4$ as tetrahedral and PF_5 as trigonal bipyramidal. For cluster compounds, with or without a central atom, it is obvious that the polyhedron defined is a useful description of the structure. In this section the major coordination and cluster polyhedra will be reviewed.

STRUCTURES OF COORDINATION COMPOUNDS

Coordination Number 2

There are two geometric possibilities, linear and bent. If the two ligands are identical, the general types and their symmetries are linear, L—M—L, $D_{\infty h}$; bent, L—M—L, C_{2v}. This coordination number is, of course, found in numerous molecular compounds of divalent elements, but is relatively uncommon otherwise. In many cases where stoichiometry might imply its occurrence, a higher coordination number actually occurs because some ligands form "bridges" between two central atoms. In terms of the more conventional types of coordination compound—those with a rather metallic element at the center—it is restricted mainly to some complexes of

Cu^I, Ag^I, Au^I, and Hg^{II}. Such complexes have linear arrangements of the metal ion and the two ligand atoms, and typical ones are $[ClCuCl]^-$, $[H_3NAgNH_3]^+$, and NCHgCN. The gold(I) halides present good examples of linear two coordination; they consist of zigzag chains of the type (1-I). The metal atoms in cations such as $[UO_2]^{2+}$, $[UO_2]^+$, and $[PuO_2]^{2+}$, which are linear, may also be said to have coordination number 2, but these oxo cations interact fairly strongly with additional ligands and their actual coordination numbers are much higher; it is true, however, that the central atoms have a specially strong affinity for the two oxygen atoms. Linear coordination also occurs in the several trihalide ions, such as I_3^- and $ClBrCl^-$.

(1-I)

Coordination Number 3

The two most symmetrical arrangements are planar and pyramidal with D_{3h} and C_{3v} symmetry, respectively. Both these arrangements are found often among molecules formed by trivalent central elements. Among complexes of the metallic elements this is a rare coordination number; nearly all compounds or complexes of metal cations with stoichiometry MX_3 have structures in which sharing of ligands leads to a coordination number for M that exceeds 3. There are, however, a few exceptions, such as the planar HgI_3^- ion that occurs in $[(CH_3)_3S^+][HgI_3^-]$, the MN_3 groups that occur in $Cr(NR_2)_3$ and $Fe(NR_2)_3$, where R = $(CH_3)_3Si$, and various gold(I) complexes.

In a few cases (e.g., ClF_3 and BrF_3), a T-shaped form of three coordination (symmetry C_{2v}) is found.

Coordination Number 4

This is a highly important coordination number, occurring in hundreds of thousands of compounds, including, *inter alia*, most of those formed by the element carbon, essentially all those formed by silicon, germanium, and tin, and many compounds and complexes of other elements. There are three principal geometries. By far the most prevalent is tetrahedral geometry which has symmetry T_d when ideal. Tetrahedral complexes or molecules are almost the only kind of 4-coordinate ones formed by nontransition elements; whenever the central atom has no electrons in its valence shell orbitals except the four pairs forming the σ bonds to ligands, these bonds are disposed in a tetrahedral fashion. With many transition metal complexes, square geometry occurs because of the presence of additional valence shell electrons and orbitals (i.e., partially filled d orbitals), although there are also many tetrahedral complexes formed by the transition metals. In some cases (e.g., with Ni^{II}, Co^{II}, and Cu^{II} in particular), there may be only a small difference in stability between the tetrahedral and the square arrangement and rapid interconversions may occur.

Square complexes are also found with nontransition central atoms when there are two electron pairs present beyond the four used in bonding; these two pairs lie above and below the plane of the molecule. Examples are XeF_4 and $(ICl_3)_2$. Similarly, when there is one extra electron pair, as in SF_4, an irregular arrangement of symmetry C_{2v} is adopted. This can be regarded as what remains of a trigonal bipyramid (e.g., PF_5) when one equatorial atom is replaced by a lone pair of electrons.

Coordination Number 5

Though less common than numbers 4 and 6, coordination number 5 is still very important. There are two principal geometries, and these may be conveniently designated by stating the polyhedra that are defined by the set of ligand atoms. In one case the ligand atoms lie at the vertices of a trigonal bipyramid (*tbp*) (1-II), and in the other at the vertices of a square pyramid (*sp*) (1-III). The *tbp* belongs to the symmetry group D_{3h}; the *sp* belongs to the group C_{4v}. It is interesting and highly important that these two structures are similar enough to be interconverted without great difficulty. Moreover, many 5-coordinate complexes have structures that are intermediate between these two prototype structures. This ready deformability and interconvertibility gives rise to one of the most important types of fluxionality (Section 1-3).

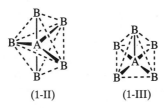

(1-II)   (1-III)

While the *tbp* seems to be somewhat more common than the *sp*, there is no general predictive rule. For example, the $[MCl_5]^{3-}$ ions (M = Cu, Cd, Hg) are *tbp*, but $[InCl_5]^{2-}$ and $[TlCl_5]^{2-}$ are *sp*, and there are compounds that contain both *tbp* and *sp* ions in the same crystal (e.g., $[Ni(CN)_5]^{3-}$).

Pentagonal planar coordination, as in $[Te(S_2COEt)_3]^-$, where two ligands are bidentate and one monodentate, is very unusual. It seems to be due to the presence of two stereochemically active lone pairs.

Coordination Number 6

This is perhaps the most common coordination number, and the six ligands usually lie at the vertices of an *octahedron* or a distorted octahedron.

There are three principal forms of distortion of the octahedron. One is tetragonal, elongation or contraction along a single C_4 axis; the resultant symmetry is only D_{4h}. Another is rhombic, changes in the lengths of two of the C_4 axes so that no two are equal; the symmetry is then only D_{2h}. The third is a trigonal distortion, elongation or contraction along one of the C_3 axes so that the symmetry is reduced to D_{3d}. These three distortions are illustrated in Fig. 1-1.

The tetragonal distortion most commonly involves an elongation of one C_4 axis and, in the limit, two trans ligands are lost completely, leaving a square, 4-coordinate complex. The trigonal distortion transforms the octahedron into a trigonal antiprism.

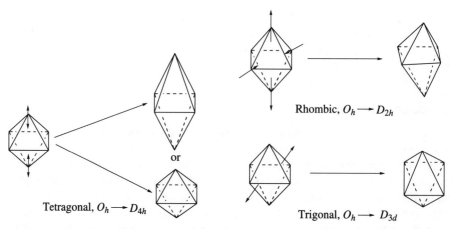

Rhombic, $O_h \longrightarrow D_{2h}$

Tetragonal, $O_h \longrightarrow D_{4h}$

Trigonal, $O_h \longrightarrow D_{3d}$

Figure 1-1 The three principal types of distortion found in real octahedral complexes.

Another type of 6-coordinate geometry, much rarer but nonetheless important, is that in which the ligands lie at the vertices of a *trigonal prism*; the ideal symmetry is D_{3h}. This arrangement has often been observed in complexes with chelating ligands and in a few metal sulfides, namely, MoS_2 and WS_2, where it was first seen many years ago, and more recently in MM'_3S_6 (M = Mn, Fe, Co, Ni; M' = Nb, Ta). The chelate complexes that best exemplify this type of coordination contain the 1,2-dithiolene or 1,2-diselenolene type ligands, $RC(S)-C(S)R$, $RC(Se)-C(Se)R$.

Structures lying between the extremes of trigonal prismatic and antiprismatic are sometimes found. As shown in Fig. 1-2, we may define the range of structures according to a twist angle ϕ, which is 0° for the prism and 60° for the antiprism. Ligands such as the dithiolenes (1-IV) and the tropolonato anion (1-V), which are somewhat inflexible and have too short a "bite" (the distance between the two atoms bonded to the metal atom) to reach across the distance between vertices of an octahedron, sometimes dictate a ϕ angle of <60°.

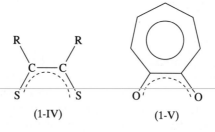

(1-IV)   (1-V)

Rather recently it has been found that the trigonal prism can occur in some discrete ML_6 species. The first examples were $ZrMe_6^{2-}$ and $HfMe_6^{2-}$, which are discussed under the chemistry of those elements.

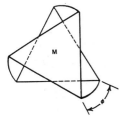

Figure 1-2 A 6-coordinate structure intermediate between the trigonal prism and antiprism projected down the threefold axis. The twist angle ϕ is measured in the plane of projection.

Coordination Number 7

There are three important geometric arrangements, as shown in Fig. 1-3. Both experimental data and theory indicate that, except where a bias might arise from the requirements of a particular polydentate ligand, these three structures are of similar stability. Moreover, interconversions are not likely to be seriously hindered, so that 7-coordinate complexes should be prone to fluxionality, as is often observed.

Coordination Number 8

There are three especially important idealized structures: the cube (O_h), the square antiprism (D_{4d}), and the triangulated dodecahedron (D_{2d}). All three are depicted in Fig. 1-4, which also shows how each of the latter two can be obtained by distortions of the cube. The cube rarely occurs in discrete complexes, although it is found in various solid arrays (e.g., the CsCl structure). Since each of the other two structures, which can be so easily obtained from it, allow the same close metal–ligand contacts while alleviating the ligand–ligand repulsions, their energetic superiority over the cube is understandable.

The dodecahedron can be viewed as a pair of interpenetrating tetrahedra: a flattened one defined by the B vertices and an elongated one defined by the A vertices. There are also three nonequivalent sets of edges, one set being those marked m in Fig. 1-4. The m edges are generally those spanned when there are four bidentate ligands with a short bite. Detailed analysis of the energetics of $M-X$ and $X-X$ interactions suggests that there will in general be little difference between the energies of the square antiprism and the dodecahedral arrangement, unless other factors, such as the existence of chelate rings, energies of partially filled inner shells, exceptional opportunities for orbital hydridization, or the like, come into play. Both arrangements occur quite commonly, and in some cases [e.g., the $M(CN)_8^{n-}$ (M = Mo or W; n = 3 or 4) ions] the geometry varies from one kind

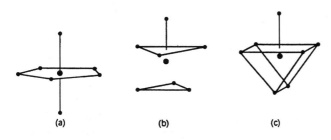

Figure 1-3 The three important geometries for seven coordination: (a) Pentagonal bipyramid (D_{5h}); (b) Capped octahedron (C_{3v}); (c) Capped trigonal prism (C_{2v}).

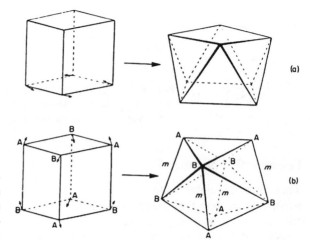

(a)

(b)

Figure 1-4 The two most important ways of distorting the cube; (a) to produce a square antiprism; (b) to produce a dodecahedron.

to the other with changes in the counterion in crystalline salts, on changing from crystalline to solution phases, and on changing the oxidation state of the metal atom [e.g., $TaCl_4(dmpe)_2$ is approximately square antiprismatic while $TaCl_4(dmpe)_2^+$ is more nearly dodecahedral].

A form of eight coordination, which is a variant of the dodecahedral arrangement, is found in several compounds containing bidentate ligands in which the two coordinated atoms are very close together (ligands said to have a small "bite"), such as NO_3^- and O_2^{2-}. In these, the close pairs of ligand atoms lie on the m edges of the dodecahedron (see Fig. 1-4b); these edges are then very short. Examples of this are the $Cr(O_2)_4^{3-}$ and $Co(NO_3)_4^{2-}$ ions and the $Ti(NO_3)_4$ molecule.

Three other forms of octacoordination, which occur less often and are essentially restricted to actinide and lanthanide compounds, are the hexagonal bipyramid (D_{6h}) (1-VI), the bicapped trigonal prism (D_{3h}) (1-VII) and the bicapped trigonal antiprism (D_{3d}) (1-VIII). The hexagonal bipyramid is restricted almost entirely to the oxo ions, where an OMO group defines the axis of the bipyramid, though it is occasionally found elsewhere.

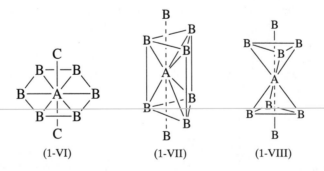

(1-VI)   (1-VII)   (1-VIII)

Higher Coordination Numbers

Of these, only nine displays appreciable regularity of form. The tricapped trigonal prism, shown in Fig. 1-5, is rather common, being found, for example, in the

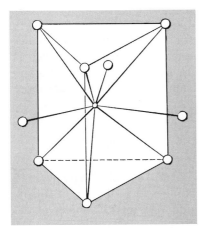

Figure 1-5 The tricapped trigonal prism structure found in many 9-coordinate complexes.

$[M(H_2O)_9]^{3+}$ ions of the lanthanides and $[ReH_9]^{2-}$. Another idealized structure, which is rarer, is that of a square antiprism capped on one rectangular face. Even higher coordination numbers, 10 to 12, are sometimes found for the largest metal ions. In general, these do not conform to any regular geometry, although for 10 coordination a bicapped square antiprism is sometimes found, for example, in $K_4[Th(O_2CCO_2)_4(H_2O)_2]\cdot 2H_2O$. A distorted icosahedral arrangement for 12 coordination is found in $[Ce(NO_3)_6]^{2-}$ and $[Pr(naph)_6]^{3+}$, where naph is 1,8-naphthyridine. The small bite of these bidentate ligands makes possible the high coordination number.

STRUCTURES OF CAGE AND CLUSTER COMPOUNDS

The formation of polyhedral cages and clusters is now recognized as an important and widespread phenomenon, and examples may be found in nearly all parts of the periodic table. A cage or cluster is in a certain sense the antithesis of a complex; yet there are many similarities due to common symmetry properties. In each type of structure a set of atoms defines the vertices of a polyhedron, but in a complex these atoms are each bound to one central atom and not to each other, whereas in a cage or cluster there need not be a central atom and the essential feature is a system of bonds connecting each atom directly to its neighbors in the polyhedron.

To a considerable extent the polyhedra found in cages and clusters are the same as those adopted by coordination compounds (e.g., the tetrahedron, trigonal bipyramid, and octahedron), but there are also others (see especially the polyhedra with six vertices), and cages with more than six vertices are far more common than coordination numbers >6. It should be noted that triangular clusters, as in $[Re_3Cl_{12}]^{3-}$ or $Os_3(CO)_{12}$, though not literally polyhedra, are not essentially different from polyhedral species such as $Mo_6Cl_8^{4+}$ or $Ir_4(CO)_{12}$.

Just as all ligand atoms in a set need not be identical, so the atoms making up a cage or cluster may be different; indeed, to exclude species made up of more than one type of atom would be to exclude the majority of cages and clusters, including some of the most interesting and important ones.

Four Vertices

Tetrahedral cages or clusters have long been known for the P_4, As_4, and Sb_4 molecules and in more recent years have been found in polynuclear metal carbonyls such as $Co_4(CO)_{12}$, $Ir_4(CO)_{12}$, $[\eta^5\text{-}C_5H_5Fe(CO)]_4$, $RSiCo_3(CO)_9$, $Fe_4(CO)_{13}^{2-}$, $Re_4(CO)_{12}H_4$, and a number of others; B_4Cl_4 is another well-known example and doubtless many more will be encountered.

Five Vertices

Polyhedra with five vertices are the trigonal bipyramid (*tbp*) and the square pyramid (*sp*). Both are found among the boranes and carboranes (e.g., the *tbp* in $B_3C_2H_5$ and the *sp* in B_5H_9), as well as among the transition elements. Examples of the latter are $Os_5(CO)_{16}$ (*tbp*) and $Fe_3S_2(CO)_9$ (*sp*).

Six Vertices

Octahedral cages and clusters are numerous, especially among the transition metals. Examples are $Rh_6(CO)_{16}$ and $[Co_6(CO)_{14}]^{4-}$, as well as the metal halide type clusters, M_6X_8 and M_6X_{12}, for M = Zr, Nb, Ta, Mo, W, and Re. The $B_6H_6^{2-}$ and $B_4C_2H_6$ species are also octahedral, although B_6H_{10} is a pentagonal pyramid. Less regular geometries are also known such as the bicapped tetrahedron in $Os_6(CO)_{18}$ and capped square pyramid in $H_2Os_6(CO)_{18}$.

Seven Vertices

Such polyhedra are relatively rare. The isoelectronic $B_7H_7^{2-}$ and $B_5C_2H_7$ species have pentagonal bipyramidal (D_{5h}) structures. The $Os_7(CO)_{21}$ molecule has a capped octahedron of Os atoms with three CO groups on each Os.

Eight Vertices

Eight-atom polyhedral structures are numerous, and a common polyhedron is the *cube*; this is in direct contrast to the situation with eightfold coordination, where a cubic arrangement of ligands is rare because it is disfavored relative to the square antiprism and the triangulated dodecahedron in which ligand–ligand contacts are reduced. In the case of a cage compound, of course, a structure in which contacts between atoms are maximized will tend to be favored (provided good bond angles can be maintained), since bonding rather than repulsive interactions exist between neighboring atoms.

Relatively few cases with eight like atoms in a cubic array are known. α-Polonium is the only element known to have simple cubic close packing. Among the main group elements cubic C_8R_8, Si_8R_8, Ge_8R_8, and Sn_8R_8 molecules have been reported[1] and among the transition metals there have been several reports, some

[1]A. Sekiguchi *et al.*, *J. Am. Chem. Soc.* **1992,** *114,* 6260.

with empty cubes ($[Ni_8Se_6(PPr^i_3)_6]$,[2] $[Ni_8Cl_4(PPh)_6(PPh_3)_4]$) and some with an atom at the cube center ($Cu_8S[S_2P(OR)_2]_6$[3]).

The other cubic systems all involve two different species of atom that alternate as shown in (1-IX). In all cases either the A atoms or the B atoms or both have appended atoms or groups. The following list collects some of the many cube species, the elements at the alternate vertices of the cube being given in bold type.

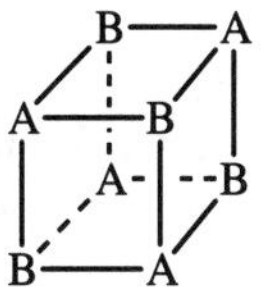

A (and appended groups)	B (and appended groups)
Mn(CO)$_3$	SEt
Os(CO)$_3$	O
PtMe$_3$ or **PtEt**$_3$	Cl, **Br**, I, OH
CH$_3$**Zn**	OCH$_3$
Tl	OCH$_3$
η^5-C$_5$H$_5$**Fe**	S
Me$_3$As**Cu**	I
Ph**Al**	NPh
Co(CO)$_3$	Sb
FeSR	S
Mo(H$_2$O)$_3$	S

(1-IX)

Although the polyhedron in cubane, or in a similar molecule, may have the full O_h symmetry of a cube, the A_4B_4-type structures can have at best tetrahedral, T_d, symmetry since they consist of two interpenetrating tetrahedra.

It must also be noted that only when the two interpenetrating tetrahedra happen to be exactly the same size will all the ABA and BAB angles be equal to 90°. Since the A and the B atoms differ, it is not in general to be expected that this will occur. In fact, there is, in principle, a whole range of bonding possibilities. At one extreme, represented by $[(\eta^5$-C$_5$H$_5)$Fe(CO)]$_4$, the members of one set are so close together that they must be considered to be directly bonded, whereas the other set (the C atoms of the CO groups) are not at all bonded among themselves but only to those in the first set. In this extreme, it might be better to classify the system as having a tetrahedral cluster (of Fe atoms) supplemented by bridging CO groups.

At the other extreme are the A_4B_4 systems in which all A—A and B—B distances are too long to admit of significant A—A or B—B bonding; thus the system can be regarded as genuinely cubic (even if the angles differ somewhat from 90°). This is true of most of the systems listed previously. The atoms, however, in the smaller of the two tetrahedra tend to have some amount of direct interaction

[2]D. Fenske *et al.*, *Angew. Chem. Int. Ed. Engl.* **1992**, *31*, 321.
[3]J. P. Fackler, Jr. *et al.*, *J. Am. Chem. Soc.* **1995**, *117*, 9778.

with one another, thus blurring the line of demarcation between the *cluster* and *cage* types.

A relatively few species are known in which the polyhedron is, at least approximately, a triangulated dodecahedron (Fig. l-4b). These are the boron species $B_8H_8^{2-}$, $B_6C_2H_8$, and B_8Cl_8.

Nine Vertices

Cages with nine vertices are rare. Representative ones are Bi_9^{5+} (in $Bi_{24}Cl_{26}$), $B_9H_9^{2-}$, and $B_7C_2H_9$, all of which have the tricapped trigonal prism structure (Fig. 1-5), and Sn_9^{4-}, which is a square antiprism capped on one square face.

Ten Vertices

Species with 10 vertices are well known. In $B_{10}H_{10}^{2-}$ and $B_8C_2H_{10}$ the polyhedron (1-X) is a square antiprism capped on the square faces (symmetry D_{4d}). But there is a far commoner structure for 10-atom cages, commonly called the adamantane structure after the hydrocarbon *adamantane* ($C_{10}H_{16}$), which has this structure; it is depicted in 1-XI and consists of two subsets of atoms: a set of four (A) that lie at the vertices of a tetrahedron and a set of six (B) that lie at the vertices of an octahedron. The entire assemblage has the T_d symmetry of the tetrahedron. From other points of view it may be regarded as a tetrahedron with a bridging atom over each edge or as an octahedron with a triply bridging atom over an alternating set of four of the eight triangular faces.

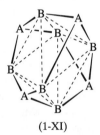

(1-X)   (1-XI)

The adamantane structure is found in dozens of A_4B_6-type cage compounds formed mainly by the main group elements. The oldest recognized examples of this structure are probably the phosphorus(III) and phosphorus(V) oxides, in which we have P_4O_6 and $(OP)_4O_6$, respectively. Other representative examples include $P_4(NCH_3)_6$, $(OP)_4(NCH_3)_6$, $As_4(NCH_3)_6$, and $(MeSi)_4S_6$.

Eleven Vertices

Perhaps the only known eleven-atom cages are $B_{11}H_{11}^{2-}$ and $B_9C_2H_{11}$.

Twelve Vertices

Twelve-atom cages are not widespread but play a dominant role in boron chemistry. The most highly symmetrical arrangement is the icosahedron (1-XII), which has

12 equivalent vertices and I_h symmetry. Icosahedra of boron atoms occur in all forms of elemental boron, in $B_{12}H_{12}^{2-}$, and in the numerous carboranes of the $B_{10}C_2H_{12}$ type. A related polyhedron, the cuboctahedron (1-XIII) is found in several borides of stoichiometry MB_{12}.

(1-XII)   (1-XIII)

1-3   Fluxionality (Stereochemical Nonrigidity)

Most molecules have a single, well-defined nuclear configuration. The atoms execute approximately harmonic vibrations about their equilibrium positions, but in other respects the structures may be considered rigid. There are, however, many cases in which molecular vibrations or intramolecular rearrangements carry a molecule from one nuclear configuration into another. When such processes occur at a rate permitting detection by some physical or chemical method, the molecules are designated as *stereochemically nonrigid*. In some cases, the two or more configurations are not chemically equivalent and the process of interconversion is called *isomerization* or *tautomerization*. In other cases, the two or more configurations are chemically equivalent, and this type of stereochemically nonrigid molecule is called *fluxional*. These will be our main concern here.

The rearrangement processes involved in stereochemically nonrigid molecules are of particular interest when they take place rapidly, although there is a continuous gradation of rates and no uniquely defined line of demarcation can be said to exist between "fast" and "slow" processes. The question of the speed of rearrangement most often derives its significance when considered in relation to the time scale of the various physical methods of studying molecular structure. In some of these methods, such as electronic and vibrational spectroscopy and gas phase electron diffraction, the act of observation of a given molecule is completed in such an extremely short time ($<10^{-11}$ s) that processes of rearrangement may seldom if ever be fast enough to influence the results. Thus for a fluxional molecule, where all configurations are equivalent, there will be nothing in the observations to indicate the fluxional character. For interconverting tautomers, the two (or more) tautomers will each be registered independently, and there will be nothing in the observations to show that they are interconverting.

It is the technique of nmr spectroscopy that most commonly reveals the occurrence of stereochemical nonrigidity, since its time scale is typically in the range 10^{-2} to 10^{-5} s. The rearrangements involved in stereochemically nonrigid behavior are rate processes with activation energies. When these activation energies are in the range 25 to 100 kJ mol^{-1} the rates of the rearrangements can be brought into the range of 10^2 to 10^5 s^{-1} at temperatures between +150 and −150°C. Thus by proper choice of temperature, many such rearrangements can be controlled so that they are slow enough at lower temperatures to allow detection of individual

Figure 1-6   The inversion of a pyramidal molecule WXYZ. Note that if X, Y, and Z are all different, the *invertomers* are enantiomorphous.

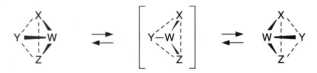

molecules, or environments within the molecules, and rapid enough at higher temperatures for the signals from the different molecules or environments to be averaged into a single line at the mean position. Thus by studying nmr spectra over a suitable temperature range, the rearrangement processes can be examined in much detail. Only one study by any other spectroscopy, infrared, has been reported.[4]

Fluxional Coordination Compounds

Coordination polyhedra are usually thought of in essentially static terms, that is, as if there are no intramolecular interchanges of ligands. In many cases, especially for octahedral complexes, this is valid, but there is a growing body of evidence that nonrigidity, particularly fluxionality, is not uncommon. In fact, for 5-coordinate complexes and most of those with coordination numbers of 7 or higher, nonrigidity is the rule rather than the exception.

A common type of fluxional behavior is the inversion of pyramidal molecules (Fig. 1-6). In the case of NH_3 and simple non-cyclic amines the activation energies, which are equal to the difference between the energies of the pyramidal ground configurations and the planar transition states, are quite low (24–30 kJ mol^{-1}) and the rates of inversion extremely high (e.g., 2.4×10^{10} s^{-1} for NH_3). Actually, in the case of NH_3, the inversion occurs mainly by quantum mechanical tunnelling through the barrier rather than by passage over it. In most cases, however, passage over a barrier (i.e., a normal activated rate process) is operative. With phosphines, arsines, R_3S^+, and R_2SO species the barriers are much higher (>100 kJ mol^{-1}), and inversions are slow enough to allow isolation of enantiomers in cases such as RR'R''P and RR'SO.

Among 4-coordinate transition metal complexes fluxional behavior based on planar/tetrahedral interconversions is of considerable importance. This is especially true of nickel(II) complexes, where planar complexes of the type $Ni(R_3P)_2X_2$ have been shown to undergo planar $\rightleftharpoons$ tetrahedral rearrangements with activation energies of about 45 kJ mol^{-1} and rates of $\sim 10^5$ s^{-1} at about room temperature.

Trigonal Bipyramidal Molecules

A class of fluxional molecules of great importance are those with a *tbp* configuration. When all five appended groups are identical single atoms, as in AB_5, the symmetry of the molecule is D_{3h}. The two apical atoms B_1 and B_2 (Fig. 1-7) are equivalent but distinct from the three equatorial atoms B_3, B_4, B_5, which are equivalent among themselves. In general, experiments such as measuring nmr spectra of B nuclei, which can sense directly the kind of environmental difference represented by B_1, B_2 versus B_3, B_4, B_5, should indicate the presence of two sorts of B nuclei in *tbp*

[4]J. J. Turner *et al.*, *J. Phys. Chem.* **1995**, *99*, 17532.

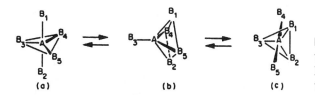

Figure 1-7 The *tbp–sp–tbp* interconversion, the so-called Berry mechanism or pseudorotation for 5-coordinate molecules.

molecules. In many cases, for example, the ^{13}C spectrum of $Fe(CO)_5$, and the ^{19}F spectrum of PF_5 (to name the two cases where such observations were first made), all five B nuclei appear to be equivalent in the nmr spectrum, even though other experimental data with a shorter time scale, such as diffraction experiments and vibrational spectroscopy, confirm the *tbp* structure.

All the ligands in the nmr spectrum in these cases appear to be equivalent because they pass rapidly between the axial and equatorial sites. Theory shows that if two nuclei occupying sites whose resonance frequencies ν_1 and ν_2 differ by $\Delta\nu$ s^{-1} change places at a frequency greater than $\Delta\nu$ s^{-1}, only one resonance at $\frac{1}{2}(\nu_1 + \nu_2)$ will be observed. Since a ligand can move from an axial to an equatorial site only if there is a simultaneous shift of a ligand from an equatorial site to an axial one, it is clear that only two types of intramolecular* exchange processes are possible: (1) those in which each step involves one axial and one equatorial ligand, and (2) those in which both axial ligands simultaneously exchange with two equatorial ones. In cases for which direct evidence has been obtained, the second type of process (2-for-2 exchange) is indicated.

It is important to realize that the nmr experiment can never do more than distinguish between two algebraically different permutations (i.e., 1-for-1, or 2-for-2, as above); it can never reveal the detailed pathways of the atoms. Two plausible, idealized pathways have been suggested for the 2-for-2 rearrangement of a *tbp* molecule. One of them, first suggested by R. S. Berry in 1960, is shown in Fig. 1-7. Not only do the *tbp* and *sp* configurations of an AB_5 molecule tend to differ little in energy, but, as Berry pointed out, they can also be interconverted by relatively small and simple angle deformation motions and in this way axial and equatorial vertices of the *tbp* may be interchanged. As shown in Fig. 1-7, the *sp* intermediate (b) is reached by simultaneous closing of the B_1AB_2 angle from 180° and opening of the B_4AB_5 angle from 120° so that both attain the same intermediate value, thus giving a square set of atoms, B_1, B_2, B_4, B_5, all equivalent to each other. This *sp* configuration may then return to a *tbp* configuration in either of two ways, one of which simply recovers the original while the other, as shown, places the erstwhile axial atoms B_1, B_2 in equatorial positions and the erstwhile equatorial atoms B_4, B_5 in the axial positions. Note that B_3 remains an equatorial atom and also that the molecule after the interchange is, effectively, rotated by 90° about the $A–B_3$ axis. Because of this apparent, but not real rotation, the Berry mechanism is often called a pseudorotation and the atom B_3 is called the pivot atom. Of course, the process can be repeated with B_4 or B_5 as the pivot atom, so that B_3 too will change to an axial position.

---

*In both PF_5 and $Fe(CO)_5$ the persistence of $^{31}P–^{19}F$ and $^{57}Fe–^{13}C$ coupling rules out dissociative or bimolecular processes, and there is no reason to doubt that the overwhelming majority if not all fluxional *tbp* molecules rearrange intramolecularly.

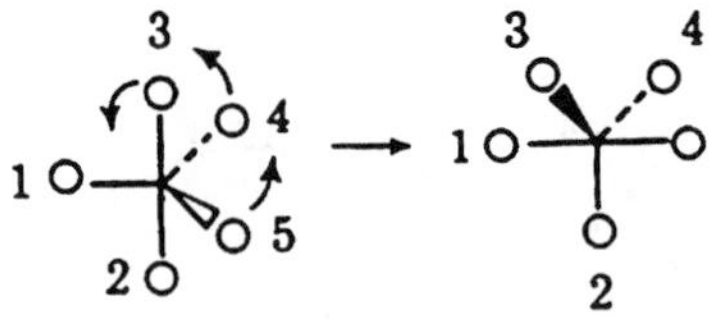

Figure 1-8 The turnstile rotation.

A second process that also results in a 2-for-2 exchange, called the "turnstile rotation" for obvious reasons, is shown in Fig. 1-8. As already noted, no choice between these is possible on the basis of the nmr spectra themselves for an AB_5 molecule, but theoretical work on PF_5 and other species favors the Berry process.

Coordination Number 6 or More

The octahedron is usually rather rigid, and fluxional or rapid tautomeric re-arrangements generally do not occur in octahedral complexes unless metal-ligand bond breaking is involved. Among the few exceptions are certain iron and ruthenium complexes of the type $M(PR_3)_4H_2$. The cis and trans isomers of $Fe[PPh(OEt)_2]_4H_2$, for example, have separate, well-resolved signals at $-50°C$ that broaden and collapse as the temperature is raised until at $60°C$ there is a single sharp multiplet indicative of rapid interconversion of the two isomeric structures. The preservation of the $^{31}P-^1H$ couplings affords proof that the rearrangement process is nondissociative. The distortion modes postulated to account for the interconversions are shown in Fig. 1-9. The rearrangement of "octahedral" bis and tris chelate complexes is considered later.

Stereochemical nonrigidity, especially if it is fluxional, seems likely to be consis-tently characteristic of complexes with coordination numbers of 7 or greater. All 7-coordinate complexes so far investigated by nmr techniques have shown ligand-atom equivalence even though there is no plausible structure for a 7-coordinate complex that would give static or instantaneous equivalence.

Eight-coordinate structures are usually fluxional. The dodecahedral structure, which is one of the commonest, has two distinct subsets of ligands, but these can easily interchange by rearrangement of the dodecahedron to a square antiprism, and then back, as shown in Fig. 1-10. The fundamental feature of this process is the opening of one or more edges shared by adjacent triangular faces to generate one or more square faces, followed by reclosing of edges in a different way. In this case, two such triangle–square–triangle transformations occur; the Berry process can be viewed as entailing only one such process for each step. There are other

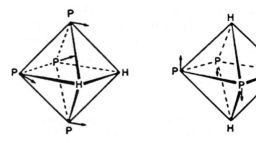

Figure 1-9 The types of distor-tion postulated to lead to intercon-version of cis and trans isomers of $Fe[PPh(OEt)_2]_4H_2$.

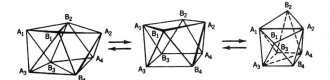

Figure 1-10 Interconversion of A and B vertices of a dodecahedron by way of a square antiprism intermediate or transition state.

systems (e.g., icosahedral carboranes) in which this simple step may provide a basis for polytopal rearrangements.

In the case of 9-coordinate species, where the ligands adopt the D_{3h} capped trigonal prism arrangement shown in Fig. 1-11, there is also an easy pathway for interchanging ligands of the two sets, and for species such as ReH_9^{2-}, $ReH_8PR_3^-$, and $ReH_7(PR_3)_2$, attempts to detect the presence of hydrogen atoms in two different environments by nmr have failed. Figure 1-11 shows the probable form of the rearrangement that causes the rapid exchange.

Trischelate Complexes

Trischelate complexes exist in enantiomeric configurations Λ and Δ about the metal atom and the process of inversion (interconversion of enantiomers) is of considerable interest. When the metal ions are of the inert type, it is often possible to resolve the complex; then the process of racemization can be followed by measurement of optical rotation as a function of time. Possible pathways for racemization fall into two broad classes: those without bond rupture and those with bond rupture.

There are two pathways without bond rupture that have been widely discussed. One is the trigonal, or Bailar, twist and the other is the rhombic, or Ray-Dutt, twist, shown in Fig. 1-12 (a) and (b), respectively. Twist processes are, of course, not confined to chelate complexes.

There are several plausible dissociative pathways, in which a 5-coordinate intermediate of either *tbp* or *sp* geometry is formed. One which seems likely in some cases is shown in Fig. 1-12(c). While associative pathways entailing a 7-coordinate intermediate with a coordinated solvent molecule have been hypothesized, there is no evidence for their actual occurrence.

While unequivocal determination of rearrangement pathways is usually impossible, experimental data can often exclude certain possibilities. For example, in the racemization of $[Cr(C_2O_4)_3]^{3-}$ a ring-opening path is virtually demanded since all oxalate oxygen atoms exchange with solvent water at a rate faster than that for oxalate exchange but almost equal to that of racemization.

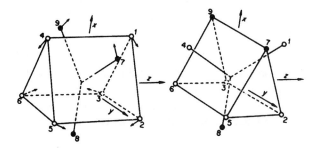

Figure 1-11 The postulated pathway by which the ligands may pass from one type of vertex to the other in the D_{3h} tricapped trigonal prism.

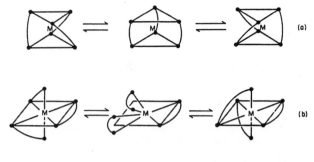

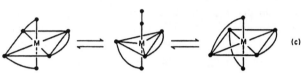

Figure 1-12 Three possible modes of intramolecular racemization of a trischelate complex, (a) trigonal twist, (b) rhombic twist, (c) one of several ring-opening paths.

A considerable amount of effort has been devoted to M(dike)₃ complexes because by using unsymmetrical diketonate ligands, the processes of isomerization and racemization can be studied simultaneously. Since isomerization can occur *only* by a dissociative pathway, it is often possible to exploit well-designed experiments to yield information on the pathways for both isomerization and racemization.

To illustrate the approach, let us consider some of the data and deductions for the system $Co[CH_3COCHCOCH(CH_3)_2]_3$, measured in C_6H_5Cl. It was found that both the isomerization and the racemization are intramolecular processes, which occur at approximately the same rate and with activation energies that are identical within experimental error. It thus appears likely that the two processes have the same transition state. This excludes a twist mechanism as the principal pathway for racemization. Moreover, it was found that isomerization occurs mainly with inversion of configuration. This imposes a considerable restriction on the acceptable pathways. Detailed consideration of the stereochemical consequences of the various dissociative pathways, and combinations thereof, leads to the conclusion that for this system the major pathway is through a *tbp* intermediate with the dangling ligand in an axial position as in Fig. 1-12(c).

Evidence for the trigonal twist mechanism has been obtained in a few complexes in which the "bite" of the ligand is small, thus causing the ground state configuration to have a small twist angle ϕ, as defined in Fig. 1-2. Complexes in which there are three chelating ligands with small bite typically contain ligands such as $R_2NCS_2^-$ or $RC(NR')_2^-$. Since their structures are already distorted considerably from octahedral towards trigonal prismatic, a fully trigonal prismatic transition state is energetically rather accessible.

Metal Carbonyl Scrambling

Di- and polynuclear metal carbonyl compounds have a general tendency to engage in a type of fluxional behavior called carbonyl scrambling. This type of behavior arises because of some of the inherent properties of metal to CO bonding. As shown in Fig. 1-13, the energy of a binuclear system consisting of two metal atoms

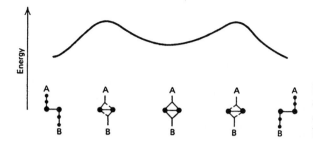

Figure 1-13 Potential energy curve for the concerted exchange of two CO ligand groups via a bridging intermediate.

and two CO groups does not in general vary a great deal (<30 kJ mol^{-1}) over the entire range of configurations from that in which there is one terminal CO ligand on each metal atom to that in which both CO groups are forming symmetrical bridges. In many cases the terminal arrangement is more stable than the bridging one, and the overall process allows CO (A) to pass from the left metal atom to the one on the right while CO (B) is simultaneously making the opposite journey. This sort of process may also occur around a three-membered ring (or even a larger one). The converse case, where a bridged arrangement is more stable than the terminal one, also occurs. Concerted processes of this type account for most of the known cases in which CO groups are scrambled over a skeleton of two or more metal atoms. Let us examine a few illustrations.

The $Cp_2Fe_2(CO)_4$ molecule exists in solution as a mixture of cis and trans isomers with bridging CO groups, as follows.

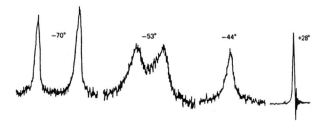

The ^1H nmr resonances for the rings should appear at different positions and, as shown in Fig. 1-14, this is observed at −70°C. However, at +28°C only a single sharp signal at the intermediate position is seen. Clearly, between −70°C and room temperature some process by which the cis and trans isomers are interconverted becomes very rapid. This process cannot be a simple rotation because of the central rigid ring system. The nmr spectrum of the ^{13}C atoms of the CO groups shows that

Figure 1-14 The ^1H nmr spectra of the *cis-* and *trans-*[$(\eta^5\text{-}C_5H_5)Fe(CO)_2$]$_2$ at several temperatures.

the cis–trans interconversion is accompanied by interchange of the CO groups between bridging and terminal positions. The explanation for both these processes is that the CO bridges open in a concerted way to give a nonbridged $Cp(OC)_2Fe-Fe(CO)_2Cp$ intermediate in which rotation about the Fe—Fe bond takes place. This rotation will be followed by a reclosing of bridges, but that may produce either a cis or a trans isomer, regardless of which one was present before bridge opening. In addition, the CO groups that swing into bridge positions need not be the same ones that were there at the outset, so that bridge/terminal interchange will also result.

Another example of CO scrambling is provided by the molecule shown in Fig. 1-15, which has four different types of CO group, *a–d*, as indicated. In a ^{13}C nmr spectrum, only at $-139°C$ or lower are all the different resonances seen. When the temperature is raised to about $-60°C$ only two resonances, in an intensity ratio of $5:2$, are seen. This is because the five approximately coplanar CO groups (types *a*, *b*, and *c*) are rapidly cycling around over the five available sites, as in Fig. 1-15(a). Between $-60°C$ and room temperature, this two-line spectrum collapses to a one-line spectrum as the five in-plane CO groups come into rapid exchange with the other two, probably by a process in which each set of three *a*, *b*, and *c* carbonyl groups on each iron atom rapidly rotates as in Fig. 1-15(b). The combined effect of both rapid processes is to move all CO groups over all seven sites rapidly, thus making them all appear equivalent in the nmr spectrum, even though there are four distinct types at any instant.

This example introduces another rather common type of CO scrambling process, namely, localized rotation of the CO groups in an $M(CO)_3$ unit that is bound to some other portion of a large molecule. In virtually every known case, this rotation will occur rapidly below the decomposition temperature of the compound, regardless of whether there is any internuclear scrambling of the CO groups. However, the activation energies, and hence the temperatures of onset, for these localized rotations vary considerably. A rather nice example is provided by $Os_6(CO)_{18}$ (a bicapped tetrahedron), in which the three distinct types of $Os(CO)_3$ unit exhibit coalescence temperatures of *ca.* -100, -10, and $+40°C$. It has not been possible to determine which signal is associated with each type of $Os(CO)_3$ however.

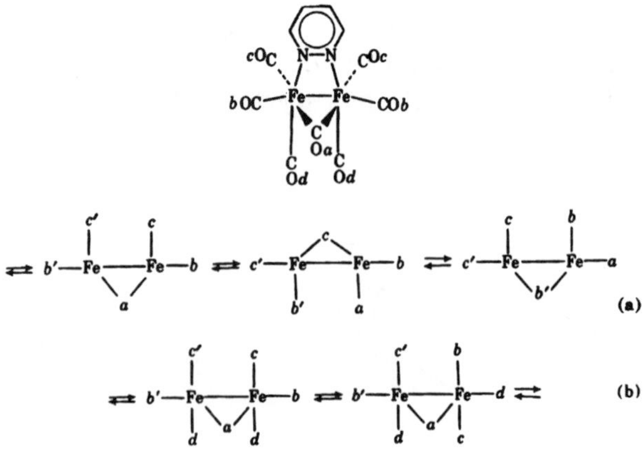

Figure 1-15 A molecule that displays: (a) a cyclic mode of rearrangement of the five coplanar CO groups; (b) a mode of rearrangement of three terminal CO groups on the same iron atom.

Cluster Rotation within CO Shells

It is well known that in molecules such as $Fe_3(CO)_{12}$ and $Rh_4(CO)_{12}$, where there are several structurally nonequivalent types of CO groups, the ^{13}C nmr spectrum at higher temperatures shows only one signal. This indicates that all CO groups pass through all of the different types of sites rapidly. In the case of $Rh_4(CO)_{12}$, the signal in the fast exchange regime also shows unequivocally that these CO groups "see" each of the four Rh atoms equally. It is easily possible to explain these results by the repeated occurrence of the sorts of processes just discussed. For each of these molecules the key step would be the concerted opening of CO bridges to give intermediates that contain only terminal groups, followed by the reclosing of bridges using other CO groups and spanning other edges of the Fe_3 or Rh_4 cluster. It has been pointed out, however, that for these two cases, and some others, a different mechanism could also account for the ^{13}C nmr results.[5]

This alternative mechanism is based on the notion that the set of 12 CO groups is packed around the central cluster of metal atoms in such a way as to define a distorted icosahedron. Since the central cluster has only a few of the symmetry elements of a regular icosahedron, its presence inside the icosahedron leads to small, symmetry-lowering distortions. However, if we imagine that the cluster as a whole can reorient within the shell of the CO groups, accompanied by small readjustments of the distortions from full icosahedral symmetry, we have a mechanism for rendering all of the CO groups equivalent on a time-average basis. At the present time, no experiment has yet been devised to distinguish between this mechanism and the more conventional ones for $M_n(CO)_m$ molecules in solution.

In the crystalline state, however, there is experimental evidence showing that at least limited reorientations of metal clusters within shells of CO groups do occur. The clearest example is provided by $Fe_3(CO)_{12}$, crystals of which are disordered as shown in Fig. 1-16. Magic-angle spinning ^{13}C nmr spectra between 31 and $-121°C$ show that the Fe_3 cluster flips between these two orientations at room temperature but is frozen into one or the other at lower temperatures. This has the effect of averaging the bridge CO resonances with a pair of the terminal ones, as well as averaging the other terminal resonances in pairs. In crystalline $Co_4(CO)_{12}$ the situation is a little more complicated, but some form of rotation of the Co_4 cluster within the shell of CO groups occurs. In both of these cases, the proposal of cluster rotation within a fixed shell of CO groups is based on the reasonable assumption that because of intermolecular forces the CO shell itself does not rotate.

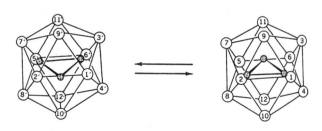

Figure 1-16 An idealized view of the structure of $Fe_3(CO_{12})$, where the set of CO groups is represented as a regular icosahedron and the two orientations for the Fe_3 cluster are shown.

[5]B. F. G. Johnson *et al.*, *Polyhedron* **1993**, *12*, 977.

Fluxional Organometallic Compounds

Fluxionality is characteristic of certain classes of organometallic compounds and is found occasionally in others. The phenomenon is seen characteristically in compounds containing conjugated cyclic polyolefins such as cyclopentadienyl, cycloheptatrienyl, or cyclooctatetraene, to name the three most common ones, attached to a metal atom through at least one but less than all of their carbon atoms, as illustrated by the following partial structures.

In each of these structures there are several structurally different ring carbon atoms and hydrogen atoms; thus complex nmr spectra would be expected. For example, the $(\eta^1\text{-}C_5H_5)M$ system should have a complex downfield multiplet of the AA'BB' type and relative intensity 4 for the four olefinic H atoms and an upfield multiplet of relative intensity 1 for the H atom attached to the same carbon atom as the metal atom. In fact, at room temperature, or not far above, most compounds containing moieties of these kinds exhibit only one sharp singlet in the ^1H or ^{13}C spectrum for the entire organic ligand. The explanation for this is that at higher temperatures the place of attachment of the metal atom to the ring is shifting rapidly over some or all members of the set of such equivalent points, thus conferring time-average equivalence on the C and H atoms. Two such hopping processes for the $(\eta^1\text{-}C_5H_5)M$ case are illustrated in Fig. 1-17. Because of the nature of the motion of the ring relative to the metal atom, these systems have been called "ring whizzers."

In each of the ring whizzer systems full characterization requires experimental information on three points: (1) structure of the configuration in which the molecule rests between hops; (2) the rate of hopping at various temperatures, from which activation parameters can be obtained; (3) the pathway used in the hopping process.

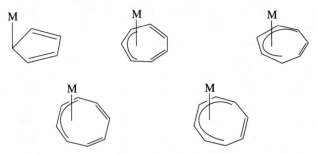

Figure 1-17 Two rearrangement pathways for a $(\eta^1\text{-}C_5H_5)M$ moiety. (a) 1,2 shifts. (b) 1,3 shifts.

(a)

(b)

As the $(\eta^1\text{-}C_5H_5)M$ case shows, there is usually more than one set of jumps that can be considered.

On all these points information is provided by nmr spectra recorded at lower temperatures. When a temperature is reached such that the spectrum remains the same at still lower temperatures, it is generally safe to assume that this spectrum (the slow-exchange spectrum) indicates the structure of the molecule between jumps. In dozens of cases structures of the above types were expected, and their presence was confirmed in this way. Of course X-ray study of the crystalline compounds can usually also be carried out, and in every case the X-ray and low-temperature nmr results have been in agreement. If nmr spectra are then recorded at 10 to 20°C temperature intervals between the slow-exchange limit and the higher temperatures and each one is matched to a computer-calculated spectrum for a given rate of rearrangement, the activation parameters can be evaluated by plotting the rates and temperatures in the usual ways.

In quite a few cases it can be shown that the appearance of the *intermediate* spectra would be different for different hopping patterns (such as the 1,2 shifts and 1,3 shifts in Fig. 1-17), and a decision can therefore be made between them. In all cases of $(\eta^1\text{-}C_5H_5)M$ systems, observations of this kind coupled with other data have favored the 1,2-shift pathway. For the homologous $(\eta^1\text{-}C_7H_7)M$ systems more diversity has been found. When M = $Re(CO)_5$, only 1,2 shifts are observed, but when M = $(\eta^5\text{-}C_5H_5)Ru(CO)_2$ both 1,2 and 1,3 shifts occur, although the former predominate. In the case of $Ph_3Sn(\eta^1\text{-}C_7H_7)$ the fluxional pathway entails 1,5 shifts.

For the $(\eta^6\text{-}1,3,5,7\text{-}Me_4C_8H_4)M(CO)_3$ molecules, with M = Cr, Mo, and W, only 1,2 shifts have been detected, but in the $(\eta^6\text{-}C_8H_8)M(CO)_3$ molecules, the predominant pathway is by 1,3 shifts, with 1,2 shifts accounting for only about 30% of the jumps. For systems of the other three types only 1,2 shifts have been observed.

Allyl complexes are characteristically fluxional, the principal pathway being the $\eta^3\text{-}\eta^1\text{-}\eta^3$ process shown below.

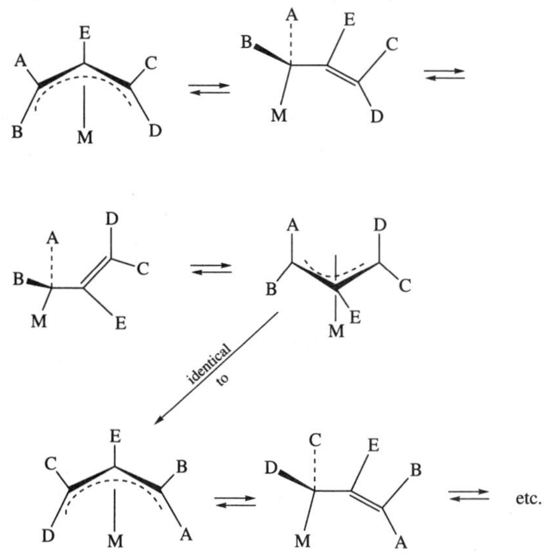

There are also compounds in which ligands exchange their modes of bonding. This is illustrated, for example, by Ti(η^1-C_5H_5)$_2$(η^5-C_5H_5)$_2$ as follows.

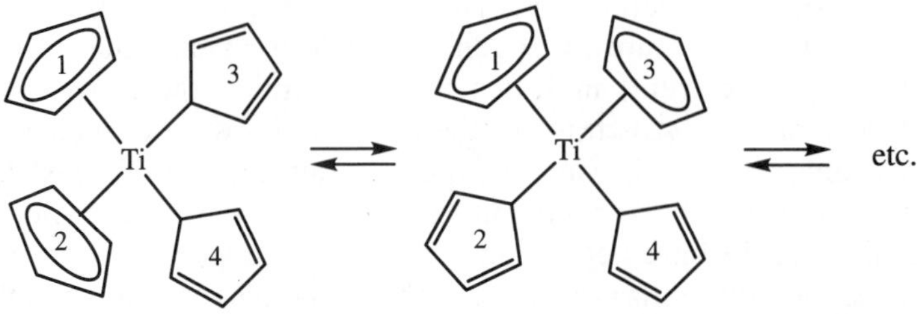

1-4 The Use of Ligand Bulk and Other Properties to Enhance Stability

The term "stable" is commonly used in a loose operational sense. If a compound persists under certain conditions (specified or unspecified) it is called stable. This is confusing, because the real questions have to do with inherent thermodynamic stability, kinetic control of reactivity, the nature of the available reactants (e.g., O_2, H_2O) and, finally, the thermodynamic stability of the reaction products.

Many "unstable" substances are simply reactive, i.e., *kinetically* unstable. In recent years a realization of this has led to the design of ligands or substituents that diminish reactivity. Most commonly, this is done simply by placing so much steric hindrance in the potential reaction path that an otherwise thermodynamically favored reaction cannot get underway. In other cases, it is a question of choosing ligands or substituents that are themselves unable to engage in a type of reaction that could result in decomposition. In this section examples of both of these approaches will be presented.

Multiple Bond Formation by Heavier Main Group Elements

It used to be ($\geq$30 years ago) a truism that while C, O, and N readily formed stable compounds in which there are double and triple bonds to these atoms, their heavier congeners (Si, Ge, Sn, S, Se, P, As) did not form such bonds. Explanations tended to imply a thermodynamic cause, that is, they suggested reasons why π bonds between the heavier elements would be too weak to exist. Inner shell repulsions and poor pπ–pπ overlap were usually blamed.

Today it is clear that the problem was kinetic and by sterically encumbering the multiple bonds they can be protected from electrophilic attack. It may be noted that the π bonds between atoms of the heavier elements are indeed weaker than those in the first row, but this alone is not the reason why, until recently, compounds could not be isolated. For example, it is believed, based partly on cis/trans isomerization studies, that while the C—C π bond has a strength of about 272 kJ mol^{-1}, the Si—Si π bond strength is about 170 kJ mol^{-1}. Thus a Si=Si is thermodynamically stable, but weak enough to be much more reactive than a C=C bond.

As first shown in 1981, a molecule containing an Si=Si bond can be stabilized by a simple steric "fix," that is, using the bulky mesityl groups:

$$R_2Si=SiR_2 \qquad R =$$

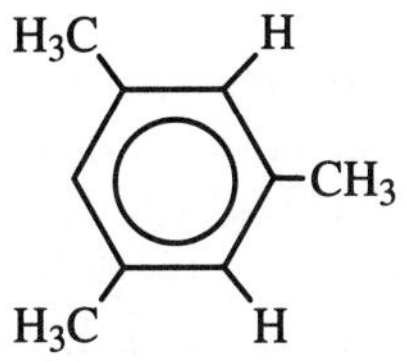

Many isolable $R_2Si=SiR_2$ compounds are now known, most containing even larger R groups, such as 2-adamantyl, Me_3C and Tip, which is a sort of supermesityl, 1-XIV.

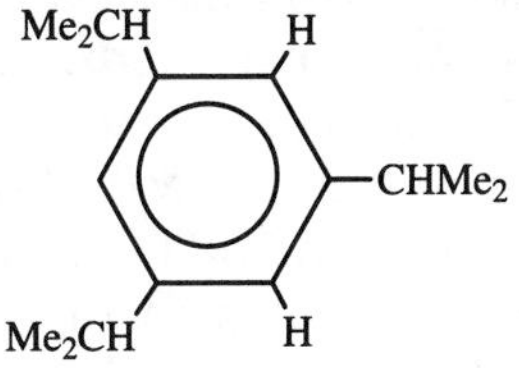

(1-XIV)

The first "stabilized" Si=Si bond was soon followed by a P=P bond in $(Me_3Si)_3$ $CP=PC(SiMe_3)_3$.

Very bulky groups such as those just mentioned, and even larger ones such as 1-XV, have now been employed to prepare stable compounds containing, in addition to Si=Si and P=P bonds, also Si=C, Si=P, Si=As, Ge=Ge, Ge=C, Sn=Sn, and Sn=C bonds.[6]

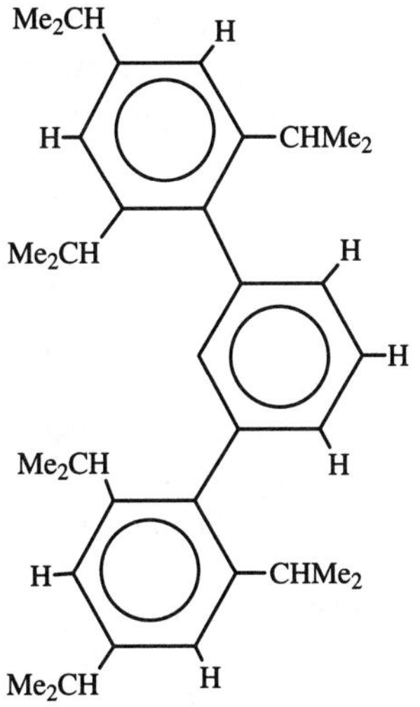

(1-XV)

[6]A. Sekiguchi et al., J. Am. Chem. Soc. **1992**, 114, 6260.

Steric Protection Against Oligomerization

The employment of bulky ligands to prevent dimerization or other oligomerization process is a well-known but important practice. In this way compounds or complexes with lower than usual coordination numbers are obtained. It was shown many years ago that several of the transition metals will form mononuclear tris-amido complexes when the amido group is $(Me_3Si)_2N$. It is also possible to suppress or entirely prevent the tendency of alkoxides to form polynuclear oligomers by employing bulky alkyl or aryl groups. For example, $Ti(OEt)_4$ forms a tetramer in which each Ti atom achieves octahedral coordination, but when very bulky R groups are substituted for the ethyl groups, dimers or even monomers can be stabilized.

Recently, an interesting application of this principle has been made in the case of molybdenum(III) dialkyl amides. With the small amido groups Me_2N and Et_2N these dimerize, though not by the common process of forming an edge-sharing bioctahedron, with bridging NR_2 groups, but, as discussed in Section 18-C-10, by forming an ethane-like molecule with a triple $Mo{\equiv}Mo$ bond. When the R_2N ligands are $(Me_3C)(3,5\text{-}Me_2C_6H_3)N$, the two $(R_2N)_3Mo$ units cannot approach each other to form a $Mo{\equiv}Mo$ triple bond, but the $(R_2N)_3Mo$ molecule can react avidly with N_2O to form $(R_2N)_3MoN$ and $(R_2N)_3MNO$.[7]

Another recent example of stabilizing a very low coordination number is provided by the R_2Fe compound in which $R = 2,4,6\text{-}(Me_3C)_3C_6H_2$, whereby a 2-coordinate iron(II) compound is formed.[8]

Forestalling β-Hydrogen Transfer

Another class of supposedly "non-existent compounds," were alkyls of transition metals. It was tacitly assumed, years ago, that $M{-}C$ bonds, where M is a transition metal, were very weak and thus simply by breakage of such bonds decomposition was inevitable. It is now recognized that this is much too simple an idea and that in many cases the decomposition is not initiated by $M{-}C$ bond breaking but by a process called β-hydrogen transfer:

$$M{-}\underset{H_2}{C}{-}\underset{H_2}{C}{-}R \longrightarrow M{\leftarrow}\overset{\overset{H}{|}}{\underset{\underset{CH_2}{\|}}{CR}}^{\,H} \longrightarrow \begin{array}{l}\text{other mostly}\\ \text{irreversible}\\ \text{steps}\end{array}$$

The clear inference to be drawn is that the use of alkyl groups that have no β-hydrogen atoms, such as methyl, benzyl ($C_6H_5CH_2^-$), neopentyl ($Me_3CCH_2^-$) and trimethylsilylmethyl ($Me_3SiCH_2^-$), to cite the most commonly used ones, might allow the isolation of compounds of types where the ethyl or propyl, etc. homologs decompose rapidly. This has proven to be correct. For example, $TiEt_4$ decomposes at $-80°C$, while $Ti(CH_2CMe_3)_4$ is stable at its melting point of $90°C$. It must be noted that to some extent steric crowding may also contribute to stability since $Ti(CH_3)_4$ decomposes at about $-40°C$.

It is advisable to conclude with a word of caution. The decomposition of metal alkyls may be governed by many factors, of which the possibility of β-elimination

[7]C. C. Cummins *et al.*, *J. Am. Chem. Soc.* **1996**, *118*, 709.
[8]W. Siedel *et al.*, *Angew. Chem. Int. Ed. Engl.* **1995**, *34*, 325.

is only one. However, the use of an alkyl that cannot do so is often sufficient to suppress rapid decomposition.

1-5   Design of Specialized Ligands

Classical coordination chemistry employing simple, or relatively simple, ligands (e.g., NH_3, Cl^-, H_2O) is presented in introductory textbooks and will not be reviewed here. However, the design and use of a variety of complex, polydentate ligands that are designed to achieve specific purposes is continuing to be an important frontier of research in inorganic chemistry. In this section, several aspects of this subject will be summarized. We begin with some basic concepts, but assume that the reader has some familiarity with the fundamentals.[9]

The Chelate Effect[10]

The term chelate effect refers to the enhanced stability of a complex system containing chelate rings as compared to the stability of a system that is as similar as possible but contains none or fewer rings. As an example, consider the following equilibrium constants:

$$Ni^{2+}(aq) + 6NH_3(aq) = [Ni(NH_3)_6]^{2+}(aq) \qquad \log \beta = 8.61$$

$$Ni^{2+}(aq) + 3en(aq) = [Ni\ en_3]^{2+}(aq) \qquad \log \beta = 18.28$$

The system $[Ni\ en_3]^{2+}$ in which three chelate rings are formed is nearly 10^{10} times as stable as that in which no such ring is formed. Although the effect is not always so pronounced, such a chelate effect is a very general one.

To understand this effect, we must invoke the thermodynamic relationships:

$$\Delta G^\circ = -RT \ln \beta$$

$$\Delta G^\circ = \Delta H^\circ - T\Delta S^\circ$$

Thus β increases as ΔG° becomes more negative. A more negative ΔG° can result from making ΔH° more negative or from making ΔS° more positive.

As a very simple case, consider the reactions, and the pertinent thermodynamic data for them, given in Table 1-1. In this case the enthalpy difference is well within experimental error; the chelate effect can thus be traced entirely to the entropy difference.

In the example first cited, the enthalpies make a slight favorable contribution, but the main source of the chelate effect is still to be found in the entropies. We

[9]Recent monographs: A. von Zelewsky, *Stereochemistry of Coordination Compounds,* Wiley, 1996; A. E. Martell and R.D. Hancock, *Metal Complexes in Aqueous Solutions,* Plenum, 1996; A. E. Martell and R. J. Motekaitis, *The Determination and Use of Stability Constants,* VCH Publishers, 1989.

[10]A. E. Martell *et al., Coord. Chem. Rev.* **1994,** *133,* 39.

Table 1-1   Two Reactions Illustrating a Purely Entropy-Based Chelate Effect

$$Cd^{2+}(aq) + 4CH_3NH_2(aq) = [Cd(NH_2CH_3)_4]^{2+}(aq) \qquad \log \beta = 6.52$$
$$Cd^{2+}(aq) + 2H_2NCH_2CH_2NH_2(aq) = [Cd(en)_2]^{2+}(aq) \qquad \log \beta = 10.6$$

Ligands	$\Delta H°$ (kJ mol^{-1})	$\Delta S°$ (J mol^{-1} deg^{-1})	$-T\Delta S°$ (kJ mol^{-1})	$\Delta G°$ (kJ mol^{-1})
4CH$_3$NH$_2$	−57.3	−67.3	20.1	−37.2
2 en	−56.5	+14.1	−4.2	−60.7

may look at this case in terms of the following metathesis:

$$[Ni(NH_3)_6]^{2+}(aq) + 3\ en(aq) = [Ni\ en_3]^{2+}(aq) + 6NH_3(aq) \qquad \log \beta = 9.67$$

for which the enthalpy change is −12.1 kJ mol^{-1}, whereas $-T\Delta S° = -43.0$ kJ mol^{-1}. The enthalpy change corresponds very closely to that expected from the increased crystal field stabilization energy of [Ni en$_3$]$^{2+}$, which is estimated from spectral data to be −11.5 kJ mol^{-1} and can presumably be so explained.

As a final example, which illustrates the existence of a chelate effect despite an unfavorable enthalpy term, we may use the reaction

$$[Ni\ en_2(H_2O)_2]^{2+}(aq) + tren(aq) = [Ni\ tren(H_2O)_2]^{2+}(aq) + 2\ en(aq)$$

$$\log \beta = 1.88 \qquad\qquad [tren = N(CH_2CH_2NH_2)_3]$$

For this reaction we have $\Delta H° = +13.0$, $-T\Delta S° = -23.7$, and $\Delta G° = -10.7$ (all in kJ mol^{-1}). The positive enthalpy change can be attributed both to greater steric strain resulting from the presence of three fused chelate rings in Ni tren, and to the inherently weaker M—N bond when N is a tertiary rather than a primary nitrogen atom. Nevertheless, the greater number of chelate rings (3 *vs* 2) leads to greater stability, owing to an entropy effect that is only partially canceled by the unfavorable enthalpy change.

Probably the main cause of the large entropy increase in each of the three cases we have been considering is the net increase in the number of unbound molecules—ligands *per se* or water molecules. Thus, although 6 NH$_3$ displace 6 H$_2$O, making no net change in the number of independent molecules, it takes only 3 en molecules to displace 6 H$_2$O.

It should be pointed out, however, that the thermodynamic explanation of the chelate effect, in particular the contribution of entropy as presented above, is actually not as straightforward as it might appear. The entropy change for a reaction depends on the standard state chosen for reference and for very concentrated solutions one might chose unit mole fraction instead of one molal and the chelate effect would disappear. However, this is not realistic and for solutions one molal (or less) there is a *real* chelate effect. In very dilute solutions (0.1 M or less) where complexation of metal ions is generally most important, the chelate effect is of major importance and is properly understood as entropically driven.

Another more pictorial way to look at the problem is to visualize a chelate ligand with one end attached to the metal ion. The other end cannot then get very far away, and the probability of it, too, becoming attached to the metal atom is greater than if this other end were instead another independent molecule, which would have access to a much larger volume of the solution. This view provides an explanation for the decreasing magnitude of the chelate effect with increasing ring size, as illustrated by data such as those shown below for copper complexes of $H_2N(CH_2)_2NH_2$ (en) and $H_2N(CH_2)_3NH_2$ (tn):

$$[Cu\ en_2]^{2+}(aq) + 2tn(aq) = [Cu\ tn_2]^{2+}(aq) + 2\ en(aq) \qquad \log \beta = -2.86$$

Of course, when the ring that must be formed becomes sufficiently large (seven membered or more), it becomes more probable that the other end of the chelate molecule will contact another metal ion than that it will come around to the first one and complete the ring. It is important to note that for the comparison of five- and six-membered rings, the simple size effect that is generally valid may (and does) fail for the smaller metal ions such as Ni^{II} and Cu^{II}. This is because of strain energy effects. To minimize strain energy in a five-membered ring with a small metal ion, the M—L distances must be long and, hence, the metal–ligand bond energy is diminished.

The Macrocyclic Effect

This term refers to the greater thermodynamic stability of a complex with a cyclic polydentate ligand when compared to the complex formed by a comparable noncyclic ligand. A representative comparison would be between the following pair:

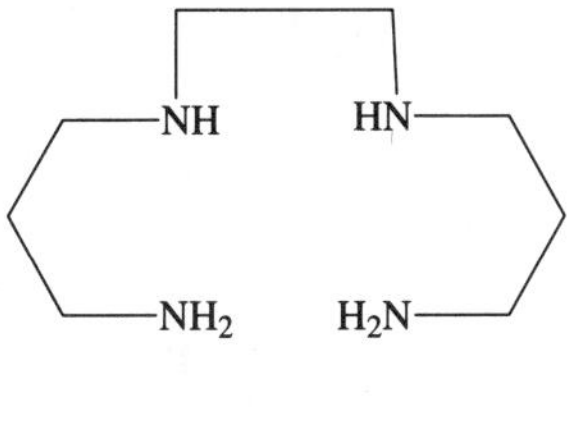

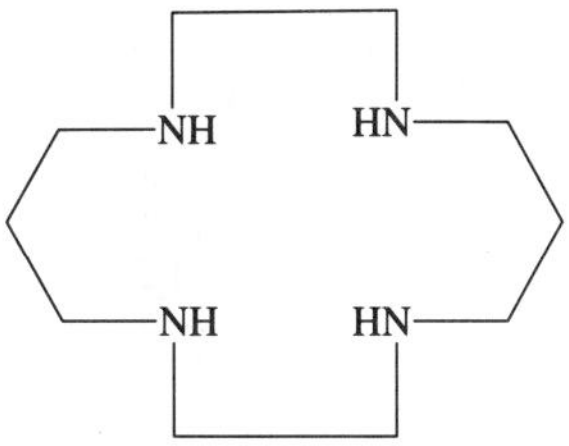

The formation of Zn (II) complexes by these two ligands, that is, the reactions

$$Zn^{2+}(aq) + L_i(aq) \longrightarrow ZnL_i^{2+}(aq)$$

have been shown to have the following thermodynamic parameters:

	L_1	L_2
Log K	11.25	15.34
$-\Delta H°$ (kJ mol^{-1})	44.4	61.9
$\Delta S°$ (J deg^{-1} mol^{-1})	66.5	85.8

The macrocyclic effect is evident in the increase of 4.09 units in log K. It can be seen that the overall effect is of both enthalpic and entropic origin. The relative importance of these two contributions varies from case to case.

Cyclic Polydentate Ligands

The macrocyclic effect is now widely exploited. Several important ligands are shown as 1-XVI to 1-XXI. The first one is porphyrin, which is usually found as one of its numerous derivatives, including those occurring in hemoglobin, myoglobin, cytochromes, etc. The trinitrogen ligands shown as 1-XIX have, over the past decade, assumed tremendous importance.[11] No common metal ion will fit inside so as to give a planar MN_3 core, but that is not their role. What they do, to perfection, is to occupy three mutually cis positions in an octahedral complex.

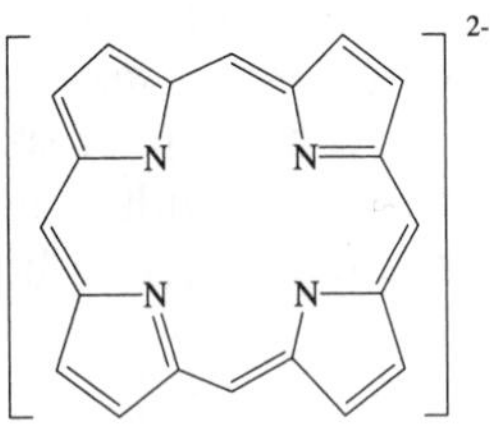

(1-XVI)	(1-XVII)	(1-XVIII)
(1-XIX)	(1-XX)	(1-XXI)

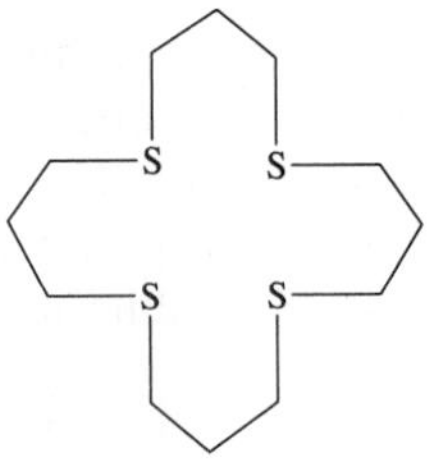

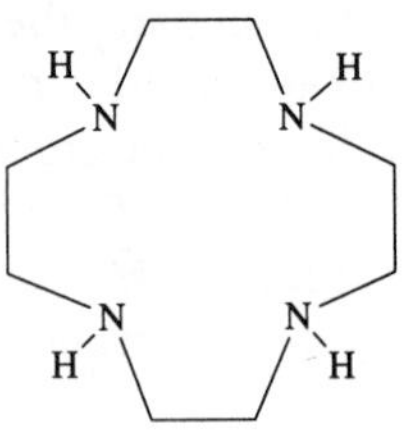

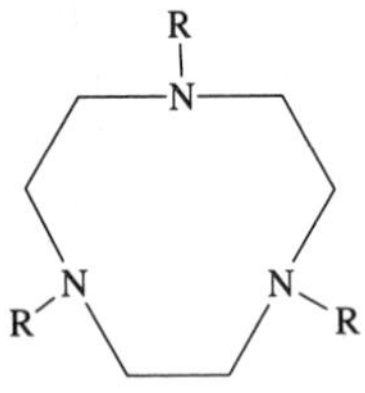

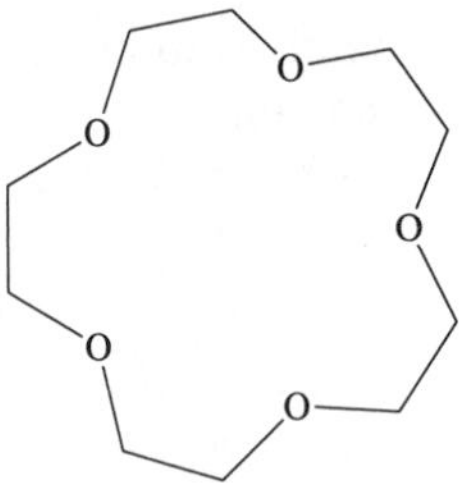

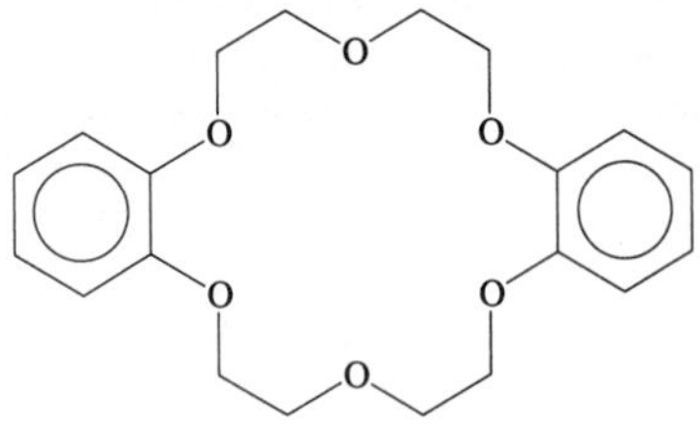

Crown Ethers

These are a subclass of cyclic polydentate ligands that complex alkali and alkaline earth metal ions particularly well. Two examples are shown as 1-XX and 1-XXI. Since systematic names for these are unwieldy, a handy notation in which 1-XX and 1-XXI are called, respectively, 15-crown-5 (even more compactly, 15-C-5) and dibenzo-18-crown-6. Crown ethers with as many as ten oxygen atoms are known and several are commercially available. For more details and references see Section 11-14.

Cryptands and Sepulchrates

These are not merely cyclic but polycyclic polydentate ligands. Most cryptands have the general formula 1-XXII. Again, a simplified code for naming them is a practical

[11]K. Wieghardt *et al., Inorg. Chem.* **1995,** *34,* 6440.

necessity. They are called "cryptand-*mmn*," where *m* and *n* are as defined in (1-XXII). One of the commonest is cryptand-222.

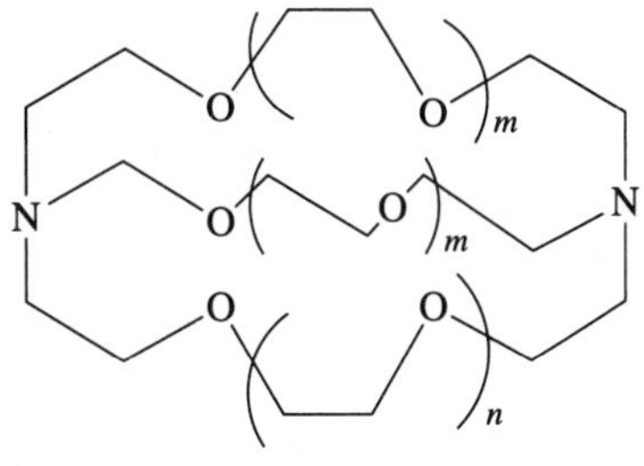

(1-XXII)

These ligands have two characteristics that make them unusually interesting. Because they are chelating ligands of high denticity, they give very high formation constants, and since the size of ion that will best fit the cavities can be predetermined by changing the ring size, these ligands can be designed to be selective.

In addition to the cryptates, which are synthesized apart from metal ions and then used to form complexes, there are other types of multicyclic ligands called *encapsulating ligands*, which are synthesized around the metal ion and cannot release it. Complexes of this sort are sometimes called sepulchrates. Two of these are (1-XXIII) and (1-XXIV). An encapsulation complex allows studies to be carried out under extremely acidic or basic conditions since the metal ion, though it cannot be removed, can be oxidized or reduced. Such ligands also can enforce unusual coordination geometries; in the examples shown the coordination is much closer to trigonal prismatic than to octahedral.

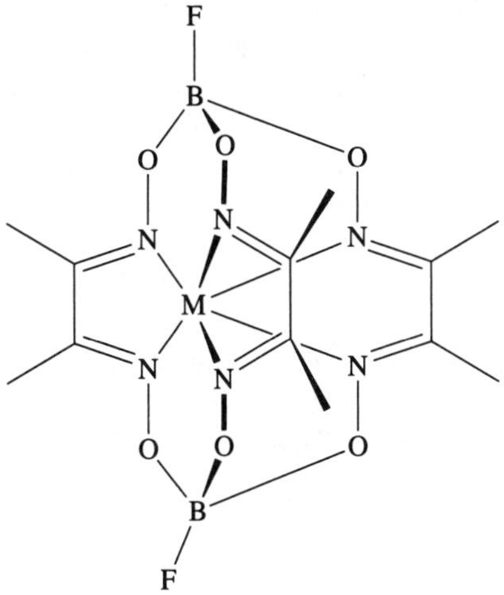

(1-XXIII)

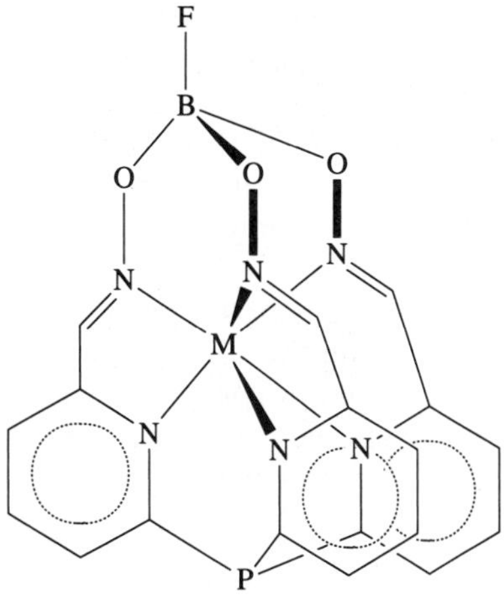

(1-XXIV)

Ligands of Unusual Reach

Ordinarily bidentate ligands occupy cis positions around a metal ion. This is because two potential donor atoms separated by a chain long enough to be able to span two trans positions would have a very low probability of actually doing so. It would be more likely to form only one bond to a given metal atom, while using its second donor atom to coordinate to a different metal atom, or not at all. This is simply an inorganic example of the well-known problem in organic chemistry of synthesizing very large rings. By appropriately designing the connection between the donor atoms, however, ligands that span trans positions in a square complex or the two sites in a linear LML complex can be made. An example is (1-XXV). Large chelate ring compounds with 12 to 72 membered rings can sometimes be made from flexible bidentate ligands [e.g., $Me_2N(CH_2)_nNMe_2$ or $R_2P(CH_2)_nPR_2$].

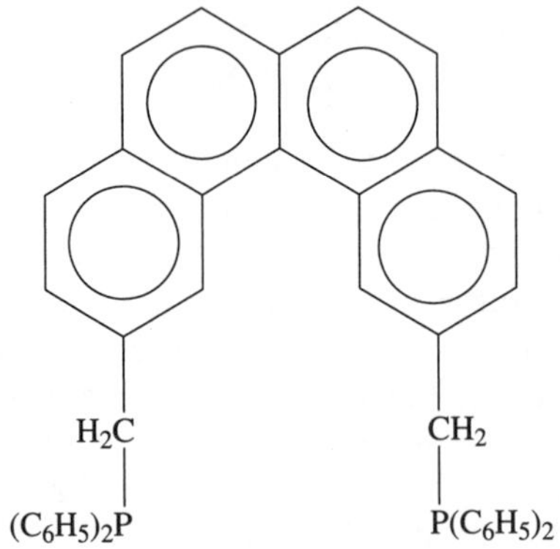

H_2C CH_2

$(C_6H_5)_2P$ $P(C_6H_5)_2$

(1-XXV)

Conformations of Chelate Rings

Simple diagrams of chelate rings in which the ring conformation is ignored are adequate for many purposes. Indeed, in some cases, such as β-diketonate complexes, the rings are planar and no problem arises. The relative stabilities and certain spectroscopic properties of many chelate complexes, however, can be understood only by considering the effects of the ring conformations, as in the important case of five-membered rings such as those formed by ethylenediamine.

Figure 1-18 shows three ways of viewing the puckered rings, and identifies the absolute configurations in the λ,δ notation. As indicated clearly in the figure, the chelate ring has as its only symmetry element a C_2 axis. It must therefore be chiral, and the two forms of a given ring are enantiomorphs. When this source of enantiomorphism is combined with the two enantiomorphous ways, Λ and Δ, of orienting the chelate rings about the metal atom (Fig. 1-19), a number of diastereomeric molecules become possible, specifically, the following eight:

$$\Lambda(\delta\delta\delta) \qquad \Delta(\lambda\lambda\lambda)$$
$$\Lambda(\delta\delta\lambda) \qquad \Delta(\lambda\lambda\delta)$$
$$\Lambda(\delta\lambda\lambda) \qquad \Delta(\lambda\delta\delta)$$
$$\Lambda(\lambda\lambda\lambda) \qquad \Delta(\delta\delta\delta)$$

The two columns are here arranged so as to place an enantiomorphous pair on each line. In the following discussion we shall mention only members of the Λ series; analogous energy relationships must of course exist among corresponding members of the Δ series.

(a)

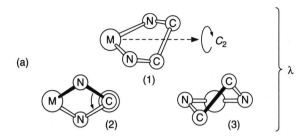

(1)   (2)   (3)   } λ

(b)

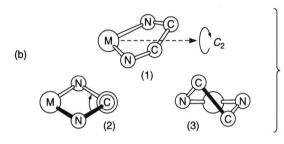

(1)   (2)   (3)

Figure 1-18 Different ways of viewing the puckering of ethyl-enediamine chelate rings. The absolute configurations λ and δ are defined. [Reproduced by permission from C. J. Hawkins, Absolute Configurations of Metal Complexes, Wiley, New York, 1971.]

The relative stabilities of the four diastereomers have been extensively investigated. First, it can easily be shown that the diastereomers must, in principle, differ in stability because there are different nonbonded (repulsive) interactions between the rings in each case. Figure 1-20 shows these differences for any two rings in the complex. When any reasonable potential function is used to estimate the magnitudes of the repulsive energies, it is concluded that the order of decreasing stability is

$$\Lambda(\delta\delta\delta) > \Lambda(\delta\delta\lambda) > \Lambda(\delta\lambda\lambda) > \Lambda(\lambda\lambda\lambda)$$

This is not the actual order, however, because enthalpy differences between diastereomers are rather small ($2-3$ kJ mol^{-1}), and an entropy factor must also be considered. Entropy favors the $\delta\delta\lambda$ and $\delta\lambda\lambda$ species because they are three times as probable as the $\delta\delta\delta$ and $\lambda\lambda\lambda$ ones. Hence the best estimate of relative stabilities, which in fact agrees with all experimental data, becomes

$$\Lambda(\delta\delta\lambda) > \Lambda(\delta\delta\delta) \approx \Lambda(\delta\lambda\lambda) \gg \Lambda(\lambda\lambda\lambda)$$

In crystalline compounds, the $\Delta(\delta\delta\delta)$ isomer (or its enantiomorph) has been found most often, but the other three have also been found. These crystallographic

Δ

Δ

Figure 1-19 Trischelate octahedral complexes (actual symmetry: D_3) showing how the absolute configurations Λ and Δ are defined according to the translation (twist) of the helices.

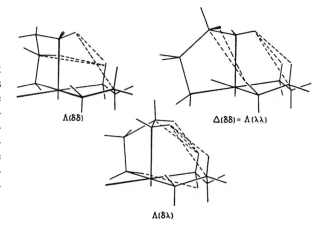

Figure 1-20 The different sets of repulsive interactions that exist between the three different pairs of ring conformations in octahedral ethylenediamine complexes; broken lines represent the significant repulsive interactions. [Reproduced by permission from C. J. Hawkins, *loc. cit.*]

results probably prove nothing about the intrinsic relative stabilities, since hydrogen bonding and other intermolecular interactions can easily outweigh the small intrinsic energy differences.

Nuclear magnetic resonance studies of solutions of Ru^{II}, Pt^{IV}, Ni^{II}, Rh^{III}, Ir^{III}, and Co^{III} [M en$_3$]$^{n+}$ complexes have yielded the most useful data, and the general conclusions seem to be that the order of stability suggested here is correct and that ring inversions are very rapid. Both experiment and theory suggest that the barrier to ring inversion is only ~25 kJ mol^{-1}. Thus the four diastereomers of each overall form (Λ or Δ) are in labile equilibrium.

One of the interesting and important applications of the foregoing type of analysis is to the determination of absolute Δ or Λ configurations by using substituted ethylenediamine ligands of known absolute configuration. This is nicely illustrated by the [Co(*l*-pn)$_3$]$^{3+}$ isomers. The absolute configuration of *l*-pn [pn = 1,2-diaminopropane, NH$_2$CH(CH$_3$)—CH$_2$NH$_2$] is known. It would also be expected from consideration of repulsions between rings in the tris complex (as indicated in Fig. 1-20) that pn chelate rings would always take a conformation that puts the CH$_3$ group in an equatorial position. Hence, an *l*-pn ring can be confidently expected to have the δ conformation shown in Fig. 1-21. Note that because of the extreme unfavorability of having axial CH$_3$ groups, only two tris complexes are expected to occur, namely, $\Lambda(\delta\delta\delta)$ and $\Delta(\delta\delta\delta)$. But by the arguments already advanced for en rings, the Λ isomer should be the more stable of these two, by 5 to 10 kJ mol^{-1}. Thus we predict that the most stable [Co(*l*-pn)$_3$]$^{3+}$ isomer must have the absolute configuration Λ about the metal.

In fact, the most stable [Co(*l*-pn)$_3$]$^{3+}$ isomer is the one with + rotation at the sodium-D line, and it has the same circular dichroism spectrum, hence the same absolute configuration as (+)-[Co en$_3$]$^{3+}$. The absolute configuration of the latter

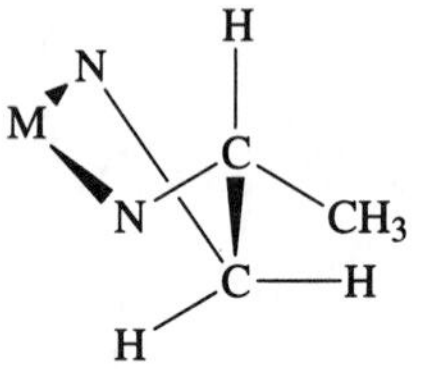

Figure 1-21 The absolute configuration and expected conformation (i.e., with an equatorial CH$_3$ group) for an M(*l*-pn) chelate ring.

has been determined, and it is indeed Λ. Thus the argument based on conformational analysis is validated.

Biochemical Applications

In many instances ligands have been designed to support one or more metal ions in an environment believed to be similar to that in a metalloenzyme, or class of metalloenzymes. It is not intended that such ligands have any overall similarity to the overall natural environment, but only that they provide a *local* environment around the metal ion or ions that might be structurally—and if possible functionally—similar to the local environment of the metal ion(s) in the natural system. There are many examples that might be chosen to illustrate this. We give here only one, which, while not necessarily better than others, is recent and representative of this type of activity.

There are numerous copper-containing enzymes, e.g., tyrosinase, that use a two-copper active site to bind and activate O_2 so it can oxygenate hydrocarbons, especially aryl groups. To model this process, the ligand **1** in Fig. 1-22 was designed and it was shown[12] that after it had bound two copper(I) ions, **2**, it could be exposed to O_2 and the remaining steps shown would occur. This ligand was especially designed to make this sequence of reactions possible.

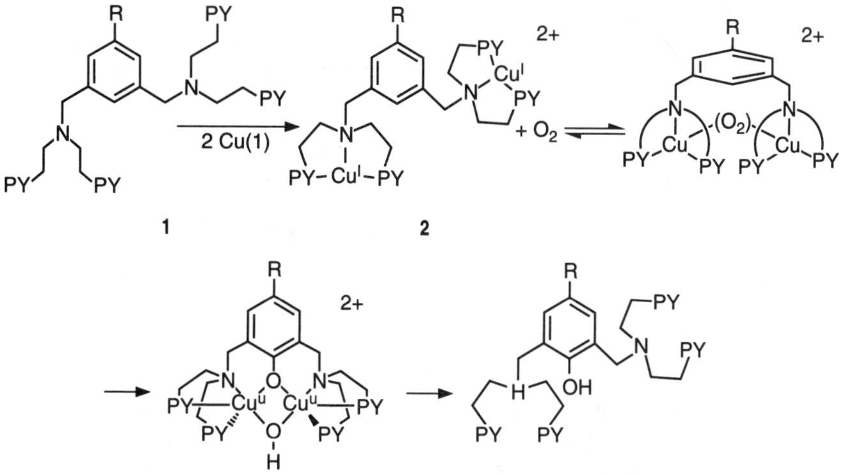

Figure 1-22 An example of how a specially designed ligand, **1**, and its Cu^I complex, **2**, can mimic the behavior of an enzyme that catalyzes the oxidation by O_2 of an aryl group to a phenol.

1-6 Isoelectronic and Isolobal Relationships

Seemingly diverse metal–ligand fragments, L_xM and L'_yM', may have fundamental similarities because they are either isoelectronic, isolobal, or both. An appreciation of these relationships and how they can serve to predict chemical and structural similarities is also of importance in organometallic chemistry, generally.

[12]K. D. Karlin *et al.*, *J. Am. Chem. Soc.* **1994,** *116,* 1324.

Isoelectronic Relationships

Since this concept is a relatively familiar one, let us simply provide a few examples that are pertinent to metal atom cluster chemistry and organometallic chemistry. Some of them derive directly from the isoelectronic relationships already noted in connection with simple metal carbonyl molecules.

1. Species of the same or similar composition containing metal atoms from the same group are isoelectronic:

   $Mn(CO)_5/Re(CO)_5$ $CpMo(CO)_3/CpW(CO)_3$ $Co(CO)_3/Ir(CO)_3$

2. Species with the same or similar ligands containing metal atoms from different groups with a suitable change in net charge are isoelectronic

   $Fe(CO)_3^-/Co(CO)_3$

3. Species in which NO is substituted for CO, with proper charge adjustment, if necessary, are isoelectronic

   $NiNO/Co(CO)_2$ $CpFe(NO)R/CpCo(CO)R$

4. The replacement of one CO ligand by two electrons or two H atoms gives isoelectronic species

   $Fe_5C(CO)_{15}/FeC(CO)_{14}]^{2-}$ $[Re_6C(CO)_{19}]^{2-}/[Re_6C(CO)_{18}H_2]^{2-}$

5. Based upon the fact that arene ligands such as C_6H_6 and $C_5H_5^-$ are six-electron donors, we have sets of isoelectronic species such as

   $C_6H_6Cr/C_5H_5Mn/Cr(CO)_3$ or C_5H_5Ni/C_6H_6Co

Isolobal Relationships

Two molecular fragments are isolobal if the number, symmetry properties, shapes, and approximate energies of their frontier orbitals are the same. They may or may not also be isoelectronic. For example, the HB and HC fragments are isolobal (but not isoelectronic), whereas the H_2C^- and H_2N moieties are both isolobal and isoelectronic. These relationships are illustrated in Fig. 1-23. As shown there, the symbol ⟷ is used to express isolobality. Also shown in Fig. 1-23 is the idea that we may choose to picture the isolobality in more than one way. Thus, for the HB and HC fragments, we can either envision one *sp* hybrid orbital whose axis is colinear with the H—B or H—C bond, plus two *p* orbitals perpendicular to this axis, or we may envision a state of full hybridization, where the frontier orbitals in each case are three of the four in a set of sp^3 tetrahedral hybrids. The orbitals whose similarity is critical in determining isolobality are called the *frontier orbitals*.

The illustrations of isolobality given in Fig. 1-23 may seem so obvious as to be trivial, but the concept gains power when generalized to more disparate systems.

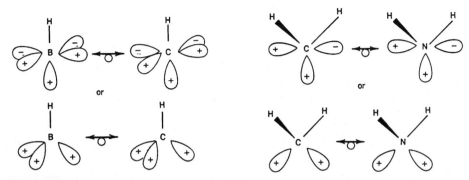

Figure 1-23 Some simple examples of isolobal species. Each pair is represented in two ways: above, with least hybridization; below, with fullest hybridization.

Consider, for example, the species shown in Fig. 1-24. The practical importance of these isolobalities is that, in general we may expect the existence of stable, isostructural molecules in which one component is replaced by another that is isolobal to it. As an illustration of this, all of the following clusters may be regarded as related by isolobal replacements; in each case, there is a tetrahedral core:

$$Co_4(CO)_{12} \qquad Co_3FeH(CO)_{12} \qquad C_5H_5NiCo_3(CO)_9$$
$$C_6H_6Co_4(CO)_9 \qquad HCCo_3(CO)_9 \qquad PCo_3(CO)_9$$

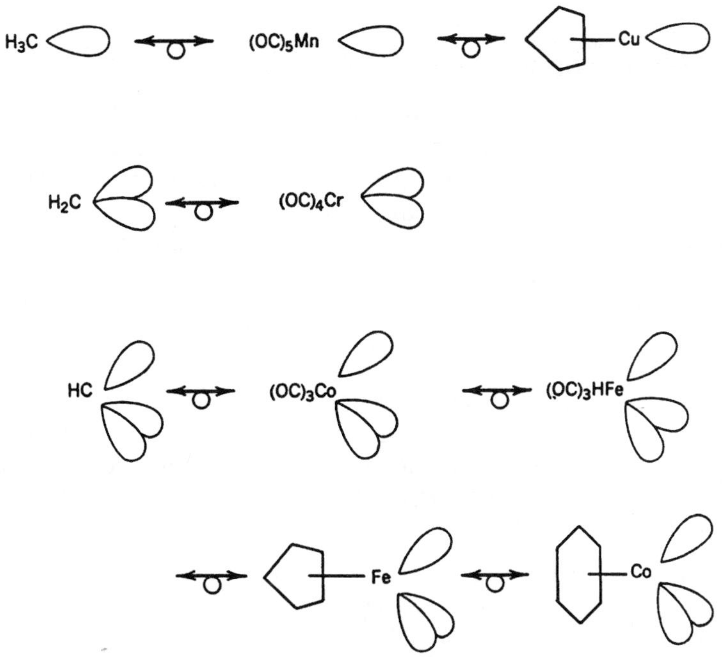

Figure 1-24 Some more general examples of isolobal species.

1-7   Bond Stretch (or Distortional) Isomerism[13]

This is (or was) a subject that waxed and waned between the writing of the fifth edition and this one. However, it is still mentioned in recent and contemporary literature and therefore, in keeping with the purpose of this book (to prepare the student to read the chemical literature), it will be briefly discussed. The episode began with the proposal that one geometrical isomer of $(PMe_2Ph)Cl_2MoO$ existed in two forms, green and blue, in which the Mo to O distances were 1.80 and 1.68 Å, respectively, where all other molecular dimensions were virtually the same. Other supposedly similar pairs of compounds were subsequently claimed and a theoretical study appeared to offer support for the phenomenon.

However, experimental studies soon showed that in the original case and apparently in all similar ones, as well as more complicated ones,[14] no such bond stretch isomers exist. The isomer(s) with the longer bond are simply the normal (short-bonded) molecule co-crystallized with an impurity so that at some crystal sites a M—Cl bond occurs where there would be an M=O bond in the pure oxo compound. The refinement of X-ray data from such a mixed crystal leads to the appearance of a longer M=O bond. Also, the contamination of a blue compound by a yellow impurity gives the appearance that there is a green compound. From the theoretical side the concept seems counter-intuitive (reminiscent of "polywater"?) and, indeed, a high-quality molecular quantum study[15] failed to confirm the earlier theoretical support.

It has been pointed out that the well-known and genuine phenomenon of spin-state isomerism, where molecules of the same composition and general structure will often show significant metrical differences in different spin states, should not be confused with the alleged bond-stretch isomerism.[16]

1-8   Relativistic Effects[17]

As commonly employed, atomic and molecular quantum mechanical calculations do not entail relativistic considerations, but over the last few decades it has become clear that many facets of the chemistry of the heaviest elements, certainly from about hafnium on, are significantly affected.

The basis of these effects is found in the relativistic mass increase for a moving particle. Its mass, m, increases with its velocity, v, according to the relation

$$m = m_0[1-(v/c)^2]^{-1/2}$$

where m_0 and c are the rest mass and the speed of light, respectively. For the heavier elements, with their high nuclear charges, electrons near the nucleus (viewed classically as particles revolving about the nucleus) move at speeds that are greater than 10% of the velocity of light. For gold (Z = 79) we have for the $1s$ electrons $v/c \approx 0.6$ and thus $m = 1.23\ m_0$.

[13]G. Parkin, *Accts. Chem. Res.* **1992,** *25,* 455; *Chem. Rev.* **1993,** *93,* 887.
[14]J. H. Enemark *et al., Inorg. Chem.* **1994,** *33,* 15.
[15]J. Song and M. B. Hall, *Inorg. Chem.* **1991,** *30,* 4433.
[16]G. Parkin and R. Hoffmann, *Angew. Chem. Int. Ed. Engl.* **1994,** *33,* 1462.
[17]N. Kaltsoyannis, *J. Chem. Soc. Dalton Trans.* **1997,** 1 and earlier references cited therein.

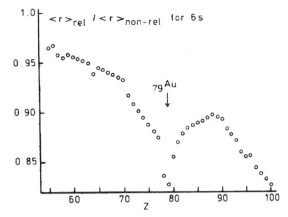

Figure 1-25 The radii of the $6s$ orbitals of atoms from atomic numbers 55 (Cs) to 100 (Fm), expressed as the ratio of the actual (i.e., relativistic) radius to that expected in the absence of relativistic effects.

Because the Bohr radius of an orbit depends on m^{-1},

$$a_0 = 4\pi\epsilon_0 h^2/mZe^2$$

the mass increase causes significant orbital contractions, as shown in Fig. 1-25. For Pt and Au the contractions of the $6s$ orbital are nearly 20%.

Rather than dealing in any more detail with the theoretical aspects of this subject, we shall list some of the phenomena of direct, practical interest to chemists that result from relativistic effects on electronic structure.

1. The 2-coordinate radius of Au^I is 0.08 Å smaller than that of Ag^I.[18]
2. While Zn_2^{2+} and Cd_2^{2+} are not important, Hg_2^{2+} is very important.
3. The so-called inert-pair effect for Tl, Pb, and Bi as well as the low cohesive energy of elemental mercury arise from relativistic contraction of the $6s$ orbitals.
4. The lanthanide contraction, classically ascribed entirely to the shielding effect of the $4f$ electrons, is also partly (*ca* 15%) due to relativistic effects.
5. Spin-orbit coupling is of relativistic origin and becomes so large in the heavier elements that the electronic states of the atoms are better classified by the values of J (total angular momentum) than by S and L separately. This has profound effects on spectroscopic and magnetic properties. It also affects properties that are more properly called chemical, such as the anomalously large electron affinities of the Hg and Tl atoms. The very large spin-orbit coupling splits the $6p$ shell into two shells $6p_{1/2}$ and $6p_{3/2}$, with the former relativistically contracted and hence more stable.
6. The electronic structures and bonding in actinide compounds cannot be reliably treated without the inclusion of relativistic effects. A specific example is provided by uranocene, $(C_8H_8)_2U$,[19] but there are many others.[20,21]

[18]H. Schmidbauer *et al.*, *J. Am. Chem. Soc.* **1996**, *118*, 7006.
[19]A. H. H. Chang and R. M. Pitzer, *J. Am. Chem. Soc.* **1989**, *111*, 2500.
[20]M. Pepper and B. E. Bursten, *Chem. Rev.* **1991**, *91*, 719.
[21]N. Kaltsoyannis and B. E. Bursten, *Inorg. Chem.* **1995**, *34*, 2735.

1-9   Zintl Compounds[22]

These compounds are named for the German chemist Eduard Zintl who pioneered their study during the 1920s and 1930s. It is instructive to survey the main classes of compounds that are *not* Zintl compounds in order to show why they are appropriately considered as a separate class.

When *very* electropositive elements, the alkali, alkaline earth and lanthanide metals, combine with *very* electronegative elements, oxygen, nitrogen, and the halogens as well as combinations of these such as the anions NO_3^-, ClO_4^-, SO_4^{2-}, relatively simple salt-like compounds which are very ionic in character and typically have an extensive aqueous chemistry are formed. The anions may often be large ($S_2O_7^{2-}$, $P_3O_9^{3-}$) or even infinite polymers, as in the chains, sheets, and networks formed by silicates, but we still regard these as ionic salts held together by essentially electrostatic forces.

When metallic elements combine with each other the products are called alloys and usually have metallic properties.

When non-metallic compounds combine with one another they form compounds that are non-ionic and non-metallic, sometimes molecular (As_4N_4, P_4S_{10}) and sometimes infinite polymers (SiS_2, As_2S_3, SiO_2).

However, in addition to these familiar classes, there remains a large number of important compounds, most of which fall into the class of Zintl compounds. They are formed when highly electropositive metals combine with the moderately electronegative (P, As, Sb, Se), or metalloidal (Ge, Ga), or even metallic (Tl, Sn, Pb) main group elements. In these compounds there is an essentially complete transfer of valence electrons from the very electropositive metal atoms to the main group elements which then form polyhedra, bands, chains, or sheets that are negatively charged. Within these anionic structures there is usually a network of electron-pair bonds, with each atom achieving a closed octet, and the compounds are diamagnetic semiconductors. There are also borderline cases where all these requirements may not be entirely met, but most often the key feature is a combination of charge and structure that allows each atom to have an octet of valence shell electrons.

These compounds are discussed here as a class because they are formed by a large number of elements and they cannot, therefore, be described satisfactorily under the heading of one element or even one group of elements, since many Zintl compounds contain elements from two groups, e.g., $Na_2Ga_3Sb_3$.

Some representative compounds will now be presented. Some contain discrete Zintl anions while others have extended one-, two-, and three-dimensional structures. It is also possible, as just mentioned, to have Zintl anions consisting of more than one main group element.

Examples of small, structurally simple Zintl anions are Si_4^{4-} and Si_4^{6-}, 1-XXVI and 1-XXVII. In these it is evident that the octet rule is obeyed. Each Si atom in Si_4^{4-} is surrounded by three electron pairs in the Si—Si bonds plus another lone pair. When one Si—Si bond is broken, two more electrons must be added, one to each of the outer Si atoms and thus we go from Si_4^{4-} to Si_4^{6-}.

[22]S. M. Kauzlarich, ed., *Chemistry, Structure and Bonding of Zintl Phases and Ions,* VCH Publishers, 1996.

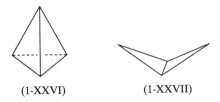

(1-XXVI)   (1-XXVII)

Among the larger discrete anions are the well known P_7^{3-} (also As_7^{3-}) that has the structure 1-XXVIII. This structure is similar to that of P_4S_3, a neutral molecule, but since the mere replacement of three S atoms by three P atoms would leave it short three electrons, it is a 3− anion. The $SnSb_6^{4-}$ ion is isostructural, with Sn at the cap position. The $Sn_2P_6^{12-}$ anion, 1-XXIX, is structurally analogous to Si_2Cl_6, but since each P atom has two electrons fewer than Cl, two electrons are added to each P atom so that each atom has an octet and the overall charge is 12−. This and $SnSb_6^{4-}$ are examples of hetero Zintl anions. There are isoelectronic and isostructural analogs of $Sn_2P_6^{12-}$, *viz.*, $Si_2Te_6^{6-}$ and $Ga_2Sb_6^{12-}$.

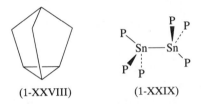

(1-XXVIII)   (1-XXIX)

Another hetero Zintl anion is the $Si_2P_6^{10-}$ ion, 1-XXX, where the manner in which all bonds and octets are formed should be obvious. Silicon and phosphorus also combine to form an infinite chain anion, $(SiP_2)_x^{2x-}$, 1-XXXI.

(1-XXX)   (1-XXXI)

It should be noted that all the E_n^{2-} ions with E = S, Si, Te are genuine Zintl anions, although this is not usually emphasized. A recent example is the Te_2^{2-} ion.[23]

There are also some discrete Zintl anions that obey more complex valence rules such as In_{11}^{8-}, where the structure is that of a trigonal prism capped on all five faces and there is an "extra" electron which is probably expelled into a conduction band.[24] Other examples are provided by ions such as Sn_9^{4-}, which can be obtained by reduction of Sn^{2+} in liquid ammonia containing dissolved sodium.

In general, discrete Zintl anions, whether they obey the simple octet rule or are electronically more complex, can often be obtained intact from the initial Zintl phase by substituting tetraalkylammonium ions for the alkali metal cations or by encapsulating the latter in a cryptand such as cryptate-222 (see 1-XXII).[25]

[23]R. C. Haushalter *et al.*, *J. Alloy Comp.* **1995**, *229*, 175.
[24]R. Nesper *et al.*, *Z. Naturforsch.* **1993**, *486*, 754.
[25]R. C. Haushalter *et al.*, *Inorg. Chem.* **1983**, *22*, 1809.

An infinite sheet structure is found in $Sr[Sn_2As_2]$, where cyclohexane-like Sn_3As_3 6-rings are fused together on all edges. A completely 3-dimensional array is found in $Na_2Ga_3Sb_3$.

Zintl anions are capable of acting as ligands, as shown, for example, by the compound $P_7Cr(CO)_3^{3-}$, 1-XXXII. One of the basal P—P bonds of P_7^{3-} (1-XXVIII) has opened to allow the P_7^{3-} unit to serve as a 6-electron donor to $Cr(CO)_3$.

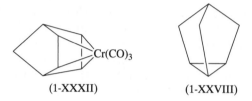

(1-XXXII)   (1-XXVIII)

1-10 Chemical Vapor Deposition and Inorganic Materials

Inorganic compounds are uniquely suited to the production of materials with well-defined characteristics and advanced design. A notable example is the deposition of thin films of high purity and controlled structure, as part of semiconducting or opto-electronic devices or as surface coatings, using volatile precursor compounds. The growth of films by the thermal decomposition of compounds or compound mixtures from the gas phase is known as *Chemical Vapor Deposition* (CVD)[26] and relies on the control of physical properties such as volatility through appropriate ligand design. In most cases the deposition process involves the saturation of an inert gas stream with the vapor of a volatile precursor compound which then passes over a heated substrate where thermal decomposition of the precursor leads to the deposition of a solid film. The advantage of the CVD technique is uniformity of coverage, film thickness and composition. Most films are polycrystalline, with either random or strongly oriented crystallite orientations. In the case of semiconductor devices, the substrates tend to be single crystals of silicon, gallium arsenide, etc. Provided the film to be deposited consists of a compound with a related crystal structure and closely matched unit cell parameters, the structure and orientation of the substrate is continued into the deposited film, a process known as *epitaxial growth*. In such a way devices with a complex sequence of very thin films of different compositions and precisely defined electronic characteristics can be grown epitaxially.

Numerous types of materials have been made by CVD techniques, such as metal carbides, borides, silicides, oxides, nitrides, and chalcogenides, as well as films of pure metals.[27] For example, thermolysis of $Cp*PtMe_3$ deposits films of platinum on glass or silicon wafer substrates,[28] while $Pd(hfacac)_2$ (hfacac = 1,1,1,5,5,5-

[26]W. S. Rees (ed.), *CVD of Nonmetals,* VCH, Weinheim, **1996;** J. T. Spencer, *Prog. Inorg. Chem.* **1994,** *41,* 145.

[27]Overview of metal film deposition: M. J. Hampden-Smith and T. T. Kodos, *Chem. Vap. Deposition* **1995,** *1,* 8 and 39; H. D. Kaesz, *Coordination Compounds in New Materials and in Materials Processing,* in: *ACS Symp. Ser.* **1994,** *565,* 444.

[28]H. D. Kaesz *et al., J. Am. Chem. Soc.* **1989,** *111,* 8779.

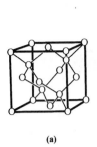 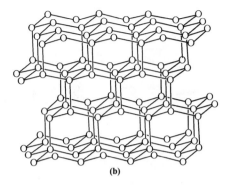

Figure 1-26 The diamond structure seen from two points of view. (a) The conventional cubic unit cell. (b) A view showing how layers are stacked; these layers run perpendicular to the body diagonals of the cube.

(a)　(b)

hexafluoropentanedionato anion) and related Cu complexes give palladium or copper films.[29] Metal alkoxides $M(OR)_n$ are precursors for metal oxide films.[30] Polydentate ligands have proved useful to ensure the volatility of divalent metal compounds; e.g., the tridentate alkoxide -OCBu^t(CH$_2$OR)$_2$ gives volatile Ca and Ba compounds.[31] Polyether complexes of alkaline earth carboxylates $M(O_2CCF_3)_2(L)$ are precursors for optical materials such as barium titanate,[32] and mixtures of alkoxides $M(OR)_n$ and carboxylates $M'(O_2CR)_m$ give mixed-metal oxides.[33] Thermolysis of the amino-borane $[Et_2NBH_2]_2$ deposits boron nitride films on silica,[34] and heating titanium thiol complexes, $TiCl_4(HSR)_2$, leads to films of TiS_2, a cathode material in lithium batteries.[35]

A particularly important area is the design and production of semiconductor devices. Semiconductors are crystalline materials differing from metals in having a much lower concentration of current carriers and lacking freely delocalized electrons. Instead, all valence electrons are tied up in a valence band, while the unoccupied orbitals form a potential conduction band. The difference between insulators and semiconductors is the size of the energy gap between valence and conduction band: whereas in insulators this energy gap is too large for electrons to be promoted from the valence to the conduction band, in semiconductors this can be achieved for example on exposure to light.

A typical semiconducting material is silicon. It crystallises in the diamond lattice (Fig. 1-26) but unlike diamond its band gap is narrow enough (1.12 eV) to make it semiconducting.

Since the electrical properties of semiconductors are strongly influenced by the incorporation of heteroelements, ultrapure semiconductor grade silicon is obtained by the reduction of $SiCl_4$ or $SiHCl_3$ with very pure Mg or Zn, followed by the growth from the molten silicon of a columnar single crystal which is further purified

[29]M. J. Hampden-Smith *et al., Chem. Mater.* **1996,** *8,* 1119; *Chem. Vap. Deposition* **1997,** *3,* 85.
[30]D. C. Bradley, *Polyhedron* **1994,** *13,* 1111.
[31]W. A. Herrmann *et al., Angew. Chem. Int. Ed. Engl.* **1994,** *33,* 105.
[32]M. J. Hampden-Smith *et al., Inorg. Chem.* **1996,** *35,* 6995.
[33]M. J. Hampden-Smith *et al., J. Sol-Gel Science Technol.* **1997,** *8,* 35.
[34]A. R. Phani *et al., J. Chem. Soc., Chem. Commun.* **1993,** 684.
[35]C. H. Winter *et al., Inorg. Chem.* **1993,** *32,* 3807.

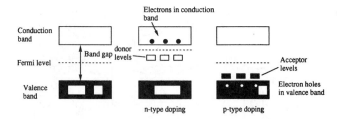

Figure 1-27 Schematic representation of doping leading to n-type and p-type semiconductors.

by zone refining. Single crystal films of Si are deposited by the thermal decomposition of SiH_4. Silicon wafers for the production of silicon chips are made by slicing the single-crystal block at a fixed orientation relative to the unit cell axes. High-purity silicon is probably the purest substance ever made. Purity is often expressed by the carrier concentration. A carrier concentration of 10^{14} cm^{-3} corresponds to only 1 carrier atom per 5×10^8 silicon atoms.

The inclusion of low concentrations of elements such as P, As, or Sb adds electrons to the conduction band (n-type doping), while replacement of Si atoms by B, Al, or Ga produces electron holes in the valence band (p-type Si) (Fig. 1-27). The majority of electronic devices employ p-n junctions.

A series of isoelectronic compounds with identical crystal structures but different band gaps can be envisaged if half the silicon atoms in Fig. 1-26 are replaced by elements with fewer valence electrons, e.g., aluminum or gallium, and the other half with elements that make up the electron deficit, i.e., Group 15 atoms. This leads to the so-called III-V semiconductors, important examples being GaAs and InP. Similarly, the chalcogenides of divalent metals give II-VI materials, such as ZnSe or CdTe. All these materials crystallize in the cubic zinc blende (ZnS) structure. The closely related hexagonal wurtzite lattice is also often found (Fig. 1-28).

The band gap decreases with increasing atomic number of the elements, so that materials such as cadmium mercury telluride $Cd_{1-x}Hg_xTe$, in which a small proportion of the cadmium positions are taken by mercury, become conducting on exposure to infrared light and are used in heat seeking and night vision devices.

Organometallic compounds are particularly important for the production of III-V and II-VI semiconductors, in a process known as *metal-organic chemical vapor*

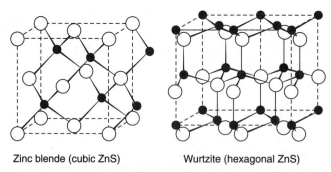

Zinc blende (cubic ZnS)   Wurtzite (hexagonal ZnS)

Figure 1-28 Unit cells of the zinc blende and wurtzite structures. Black circles indicate metal cations, open circles anions. Note the similarity to the diamond structure, Fig. 1-26.

deposition (MOCVD) or metal-organic vapor phase epitaxy (MOVPE).[36,37] A gas stream of metal alkyls with sufficient volatility, typically $GaMe_3$ or $ZnMe_2$, is mixed with a volatile reaction partner, such as AsH_3 for III-V or H_2Se or Pr^i_2Te for II-VI materials. Films are deposited by passing the gas mixture over a heated single-crystal substrate. Multiple layers can be grown in this way. Very controlled film growth can be achieved at low pressures (metal-organic molecular beam epitaxy, MOMBE).

$$GaMe_3 + AsH_3 \xrightarrow[-CH_4]{\Delta} GaAs$$

$$ZnMe_2 + H_2Se \xrightarrow[-CH_4]{\Delta} ZnSe$$

"Single-source" precursors which contain both the metal and the heteroelement in one molecule at a defined stoichiometric ratio have also been successful in some cases, although their generally lower volatility and the possibility of incorporating carbon impurities have limited their application.[38] Examples are $[Me_2Ga(\mu\text{-}AsBu^t_2)]_2$ to give GaAs, and bulky chalcogenolates $[M(ER)_2]_2$ [R = $C_6H_2Bu^t_3$ or $Si(SiMe_3)_3$] and $M[Bu^t_2P(E)NR]_2$ (M = Zn, Cd; E = S, Se, Te) for II-VI compounds.[39]

1-11   Bioinorganic Chemistry[40]

Biochemistry is not merely an elaboration of organic chemistry. The chemistry of life involves, in essential and indispensable ways, at least 25 elements. In addition to the "organic" elements C, H, N, and O, there are 9 other elements that are required in relatively large quantities, and called, therefore, *macronutrients*. These elements are Na, K, Mg, Ca, S, P, Cl, Si, and Fe. There are also many other elements, *micronutrients*, that are required in small amounts by at least some forms of life: V, Cr, Mn, Co, Ni, Cu, Zn, Mo, W, Se, F, and I. As research activity intensifies, and as instrumental methods of analysis and detection become more sophisticated and sensitive, it is likely that other elements will be added to the list of micronutrients. The elements Cr, Ni, W, and Se have been added only within the last few years.

The metallic elements play a variety of roles in biochemistry. Several of the most important roles are the following:

1. *Regulatory action* is exercised by Na^+, K^+, Mg^{2+}, and Ca^{2+}. The flux of these ions through cell membranes and other boundary layers sends signals that turn metabolic reactions on and off.

[36]G. B. Stringfellow, *Organometallic Vapor-Phase Epitaxy: Theory and Practice*, Academic Press, New York, 1989.
[37]P. O'Brien in: D. W. Bruce and D. O'Hare (eds.), *Inorganic Materials*, Wiley, New York, 1992, p. 491; F. Maury, *Chem. Vap. Deposition* **1996**, *2*, 113
[38]A. H. Cowley and R. A. Jones, *Polyhedron* **1994**, *13*, 1149.
[39]M. Bochmann *et al.*, *Chem. Vap. Deposition* **1995**, *1*, 78; **1996**, *2*, 85.
[40]S. J. Lippard and J. M. Berg, *Principles of Bioinorganic Chemistry*, University Science Books, Mill Valley, CA, 1994.

2. *The structural role* of calcium in bones and teeth is well known, but many proteins owe their structural integrity to the presence of metal ions that tie together and make rigid certain portions of these large molecules, portions that would otherwise be only loosely linked. Metal ions particularly known to do this are Ca^{2+} and Zn^{2+}.

3. An enormous amount of *electron-transfer* chemistry goes on in biological systems, and nearly all of it critically depends on metal-containing electron-transfer agents. These include cytochromes (Fe), ferredoxins (Fe), and a number of copper-containing "blue proteins," such as azurin, plastocyanin, and stellacyanin.

4. *Metalloenzymes or metallocoenzymes* are involved in a great deal of enzymatic activity, which depends on the presence of metal ions at the active site of the enzyme or in a key coenzyme. Of the latter, the best known is vitamin B_{12}, which contains cobalt. Important metalloenzymes include carboxypeptidase (Zn), alcohol dehydrogenase (Zn), superoxide dismutase (Cu, Zn), urease (Ni), and cytochrome P-450 (Fe).

5. All aerobic forms of life depend on *oxygen carriers*, molecules that carry oxygen from the point of intake (such as the lungs) to tissues where O_2 is used in oxidative processes that generate energy. Examples of the most important oxygen carriers and the metals they contain are:

Hemoglobins (Fe), found in all mammals.

Hemerythrins (Fe), found in marine invertebrates.

Hemocyanins (Cu), found in arthropods and molluscs.

For many of the more prominent biochemically active elements details will be given in the pertinent chapters.

1-12   The Reference Literature of Inorganic Chemistry

In this book the literature is cited in a selective way. Only those articles in the primary literature (i.e., research journals) that have appeared since the preceding edition of the book was completed, 1988, and prior to the end of 1996 are cited. (A few references in 1997 are also included.) Even in this, there is no intention to be complete, but only to be reasonably helpful. References are also given to the secondary literature, that is, review journals. Primary literature references, even within the period 1988–1996 are sometimes intentionally omitted when there is a reference to the secondary literature that contains them.

There is also a tertiary literature, namely, certain monographs on particular elements or classes of compounds and, more important, encyclopedic works. Monographs, and some reviews, are cited in the Additional References lists in each chapter (or, in Chapters 17 and 18, each sub-chapter).

The encyclopedic tertiary literature is not explicitly cited, but it is very valuable to readers seeking further references and older references. The following works are the major ones.

R. B. King, Ed., *Encyclopedia of Inorganic Chemistry*, John Wiley & Sons, New York, 1994.
J. E. Macintyre, *Exec. Ed.*, *Dictionary of Inorganic Compounds*, Chapman and Hall, London, 1992. Since 1992 there have been four supplements.

Comprehensive Coordination Chemistry, G. Wilkinson, R. D. Gillard and J. A. McCleverty, Eds., Pergamon Press, Oxford, 1987.

J. Buckingham, Ed., *Dictionary of Organometallic Compounds*, Chapman and Hall, London, 1984.

The organometallic chemistry of all the elements is covered systematically in E. W. Abel, F. G. A. Stone, and G. Wilkinson, Eds., *Comprehensive Organometallic Chemistry*, Elsevier, Amsterdam, 1982, covers the subject up to 1981 and the supplementary edition II covers the literature from 1982 to 1994.

Finally, we note *Gmelin's Handbuch der anorganische Chemie*, published since 1974 by Springer-Verlag, Berlin-New York. These volumes, of which there are many, cover their subjects in minute detail. The older volumes (the series began in 1927) are in German, but for the last dozen years or so they have been appearing in English. Unfortunately, many elements have not been updated for many years.

Part 2

THE CHEMISTRY OF THE MAIN GROUP ELEMENTS

Chapter 2

HYDROGEN

GENERAL REMARKS

2-1 Introduction

Three isotopes of hydrogen are known: ^1H, ^2H (deuterium or D), and ^3H (tritium or T). Although isotope effects are greatest for hydrogen, justifying the use of distinctive names for the two heavier isotopes, the chemical properties of H, D, and T are essentially identical except in matters such as rates and equilibrium constants of reactions; in addition, diverse methods of isotope separation are known. The normal form of the element is the diatomic molecule: the various possibilities are H_2, D_2, T_2, HD, HT, and DT.[1] Naturally occurring hydrogen contains 0.0156% deuterium, whereas tritium occurs naturally in amounts of the order of 1 in 10^{17}.

Deuterium as D_2O is separated from water by fractional distillation or electrolysis and by utilization of very small differences in the free energies of the H and D forms of different compounds, the H_2O—H_2S system being particularly favorable in large-scale use:

$$HOH(l) + HSD(g) = HOD(l) + HSH(g) \qquad K \sim 1.01$$

Deuterium oxide is available in ton quantities. It is used in certain nuclear reactions, to synthesize deuterated compounds as nmr solvents, and for labelled compounds in reaction mechanism and spectroscopic studies.

Tritium, which is radioactive (β^-, $t_{1/2} = 12.4$ y), is made by the reaction $^6Li(n,\alpha)^3H$ in nuclear reactors. It is also formed in plasmas[2] as $^3H^+$ and by cosmic ray induced nuclear reactions in the upper atmosphere. The decay of ^3H probably accounts for traces of ^3He in the atmosphere.

Dihydrogen (bp 20.28 K) is a colorless, odorless gas virtually insoluble in water. It is made industrially by the steam reforming of hydrocarbons, notably methane

---

[1]Molecular H_2 (and D_2) have ortho and para forms in which the nuclear spins are aligned or opposed, respectively. This leads to very slight differences in bulk physical properties, and the forms can be separated by gas chromatography.
[2]A. Carrington and J. R. McNab, *Acc. Chem. Res.* **1989,** *22,* 218.

(reaction 2a), or from coke (reaction 2b), followed by the water-gas shift reaction (reaction 2c):

$$CH_4 + H_2O = CO + 3H_2 \qquad \Delta H = +206 \text{ kJ mol}^{-1} \qquad \text{(a)}$$

$$C + H_2O = CO + H_2 \qquad \text{(b)}$$

$$CO + H_2O = CO_2 + H_2 \qquad \Delta H = -46 \text{ kJ mol}^{-1} \qquad \text{(c)}$$

Reaction 2b is strongly endothermic, and the process is therefore run at a high temperature (*ca.* 900–1100°C). Steam reforming is carried out over nickel or Ni/Al$_2$O$_3$ catalysts, although Rh/Al$_2$O$_3$ catalysts and higher hydrocarbon feedstocks have also been used. Most of the hydrogen thus produced is consumed directly in large-scale processes such as the production of methanol, using Cu/ZnO catalysts:

$$CO + 2H_2 = CH_3OH$$

In this reaction the residual CO_2 content can be tolerated. Another major H$_2$ consuming process is the manufacture of ammonia. This requires pure H$_2$; carbon oxides are poisons for the ammonia catalyst and have to be removed, CO_2 by scrubbing, and residual CO (as well as traces of CO_2) by catalytic conversion to CH$_4$ (methanation) which is recycled.

Hydrogen gas is sometimes naturally found in geological structures; recently, crushed basalt and H$_2$O in the dark under anaerobic conditions was found to generate H$_2$. This would account for natural emissions and explain the energy source for bacteria growing in underground systems.[3]

Although efforts have been made to achieve photochemical or metal-catalyzed cleavage of water to H$_2$ and O$_2$, none is yet practical. However, hydrogen has been obtained from OH groups[4] in the reaction [Cp$'$ = 1,3-(Me$_3$Si)$_2$C$_5$H$_3$], which is an oxidative elimination:

$$Cp'_4U^{III}_2(\mu\text{-}OH)_2 \longrightarrow Cp'_4U^{IV}_2(\mu\text{-}O)_2 + H_2$$

Hydrogen is not exceptionally reactive, in part due to the highly endothermic dissociation:

$$H_2 = 2H \qquad \Delta H_0^0 = 434.1 \text{ kJ mol}^{-1}$$

Hydrogen burns in air to form water and will react with oxygen and the halogens explosively under certain conditions. At higher temperatures the gas will reduce many oxides either to lower oxides or to the metal. In the presence of suitable catalysts and above room temperature it reacts with N$_2$ to form NH$_3$. With electropositive metals and most non-metals it forms hydrides.

Despite the high bond energy of H$_2$, the molecule is readily cleaved at low temperature and pressure by many transition metal complexes, when two metal-

[3]T. O. Stevens and J. P. McKinley, *Science* **1995,** *270,* 450.
[4]R. A. Andersen *et al., J. Am. Chem. Soc.* **1996,** *118,* 901.

hydrogen bonds are formed, for example,

$$RhCl(PPh_3)_3 + H_2 \rightleftharpoons RhH_2Cl(PPh_3)_3$$

These reactions involve initial coordination of molecular hydrogen (see Sect. 2-16).

Atomic hydrogen ($t_{1/2}$ *ca.* 0.3 s) can be obtained by UV radiation, in discharge tubes or electric arcs. The recombination produces high temperatures. Hydrogen atoms produced by mercury sensitization[5] can be used at 1 atm and 0–150°C for organic syntheses on a preparative scale. The hydrogen atoms are generated as follows:

$$Hg + h\nu = Hg*$$
$$H_2 + Hg* = 2H + Hg$$

2-2 The Bonding of Hydrogen

The chemistry of hydrogen depends mainly on three electronic processes:

1. *Loss of the Valence Electron.* The $1s$ valence electron may be lost to give the hydrogen ion H^+ which is merely the proton. Its small size ($r \sim 1.5 \times 10^{-13}$ cm) relative to atomic sizes ($r \sim 10^{-8}$ cm) and its charge result in a unique ability to distort the electron clouds surrounding other atoms; the proton accordingly never exists as such, except in gaseous ion beams; in condensed phases it is always closely associated with other atoms or molecules.

2. *Acquisition of an Electron.* The hydrogen atom can acquire an electron, attaining the $1s^2$ structure of He, to form the hydride ion H^-. This ion exists as such essentially only in the saline hydrides formed by electropositive metals (Section 2-13).

3. *Formation of an Electron-Pair Bond.* The majority of hydrogen compounds contain an electron-pair bond. The number of carbon compounds of hydrogen is legion, and most of the less metallic elements form numerous hydrogen derivatives.

The chemistry of many of these compounds is highly dependent on the nature of the element (or the element plus its other ligands) to which hydrogen is bound. Particularly dependent is the degree to which compounds undergo dissociation in polar solvents and act as acids:

$$HX \rightleftharpoons H^+ + X^-$$

Also important for chemical behavior is the electronic structure and coordination number of the molecule as a whole. This is readily appreciated by considering the covalent hydrides BH_3, CH_4, OH_2, and FH. The first not only dimerizes (Chap. 5) but is a Lewis acid, methane is chemically unreactive and neutral, ammonia has a lone pair and is a base, water can act as a base or as a very weak acid, and FH is appreciably acidic in water.

[5]R. H. Crabtree *et al., J. Am. Chem. Soc.* **1991**, *113*, 2233.

Except in H_2 itself, where the bond is nonpolar, all other H—X bonds possess polar character to some extent. The dipole may be oriented either way, and important chemical differences arise accordingly. Although the term "hydride" might be considered appropriate only for compounds with H negative, many compounds that act as acids in polar solvents are properly termed covalent hydrides. Thus although HCl and $HCo(CO)_4$ behave as strong acids in aqueous solution, they are gases at room temperature and are undissociated in non-polar solvents.

Since water is the most important hydrogen compound, we consider first hydrogen bonds. These dominate all the chemistry of water, aqueous solutions, hydroxylic solvents, hydroxo species and so on. They are of crucial importance in all biological systems, being responsible *inter alia* for base pairing in nucleic acids and in the linking of polypeptide chains in proteins. Hydrogen-bonding also has immense importance for molecular recognition and in defining crystal structures.

THE CLASSICAL HYDROGEN BOND; WATER; HYDRATES; HYDROGEN IONS; ACIDS AND BASES

2-3   The Classical Hydrogen Bond[6]

The literature on this topic is so extensive that only a brief discussion as it involves inorganic chemistry can be given here. The concept of the hydrogen bond was due to Huggins, Latimer, and Rodebush at the University of California, Berkeley in 1920. It is the term given to the relatively weak interaction between a hydrogen atom bound to an electronegative atom and another atom that is also generally electronegative and has one or more lone pairs enabling it to act as a base. We can thus refer to proton donors XH and proton acceptors Y and can give the following generalized representation of a hydrogen bond:

$$\overset{\delta-}{X}\text{——}\overset{\delta+}{H}\text{- - - -}Y$$

Such interaction is strongest when both X and Y are first-row elements. It might be assumed that H-bonding occurs in the direction of the lone pairs and this was confirmed for the NH···O=C bond in a survey of many structures. However, this is not generally so, H bonds often preferring linearity or near linearity. Lone pair directionality is more important in organic compounds for sp^2 rather than sp^3 lone pairs. Although O—H···O and N—H···O are the most common H bonds, others include N—H···S,[7] N—H···N,[8] S—H···S,[9] S—H···O,[9] and C—H···O.[10] Although O—H···O bonds have been intensively studied in organic chemistry, they are also

[6]D. A. Smith, ed., *Modeling the Hydrogen Bond*, A.C.S. Symposium Ser No. 569, A.C.S. Washington, DC, 1994; F. Hibbert and J. Emsley, *Adv. Phys. Org. Chem.* **1990**, *26*, 255; C. B. Aakeroy and K. R. Seddon, *Chem. Soc. Rev.* **1993**, 397.
[7]M. A. Walters *et al., Inorg. Chem.* **1995**, *34*, 1090.
[8]T. Steiner, *J. Chem. Soc., Chem. Commun.* **1995**, 1331.
[9]P. M. Boorman *et al., J. Chem. Soc., Chem. Commun.* **1992**, 1656.
[10]F. Steiner, *J. Chem. Soc., Chem. Commun.* **1994**, 2341; M. G. Davidson, *J. Chem. Soc., Chem. Commun.* **1995**, 919; P. Behrens *et al., Angew. Chem. Int. Ed. Engl.* **1995**, *34*, 2680; J. Janiak *et al., Chem. Eur. J.* **1995**, *1*, 637.

Table 2-1 Some Parameters of Hydrogen Bonds

$X-H\cdots Y$ Bond	Compound	Bond Energy (kJ mol^{-1})	Depression of Stretching Frequency (cm^{-1})	$X\cdots Y$ Distance (Å)	$X-H$ Distance (Å)
$F\cdots H\cdots F$	$KHF_2{}^a$	~212	~2450	2.26	1.13
$F-H\cdots F$	HF(g)	~28.6	700	2.55	
$O-H\cdots O$	$(HCO_2H)_2$	29.8	~460	2.67	
$O-H\cdots O$	$H_2O(s)$	~21	~430	2.75	1.01
$O-H\cdots O$	$B(OH)_3$			2.74	1.03
$N-H\cdots N$	Melamine	~25	~120	3.00	
$N-H\cdots N$	N_2H_5Cl		~460	3.12	
$C-H\cdots N$	$(HCN)_n$		180	3.2	

^{a}For $KF\cdot 2H_2O$ see K. Schwartz *et al.*, *Inorg. Chem.* **1994**, *33*, 4774.

important in inorganic chemistry, for example, in the association of alcohols with metal alkoxides.[11]

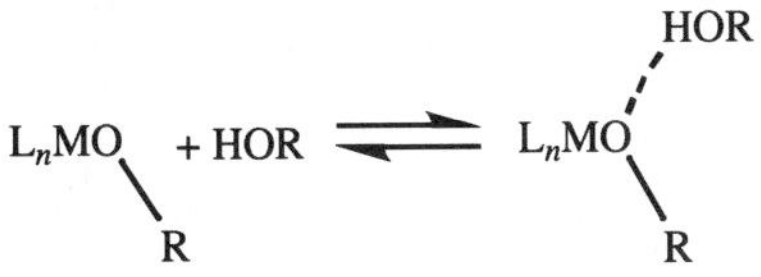

$$L_nMO\diagdown_R + HOR \;\rightleftharpoons\; L_nMO\diagup^{HOR}_{\diagdown R} \tag{1}$$

Groups such as PH, ClH, and BrH, as well as transition metal hydrides MH_n that will be discussed later, can act as donors, while acceptor atoms can be N, O, F, Cl, Br, Se, or P.

Early evidence for H-bonding arose from comparisons of physical properties of hydrogen compounds, the classic cases being the abnormally high boiling points of NH_3, H_2O, and HF compared to those of PH_3, H_2S, and HCl. Detailed X-ray diffraction and neutron diffraction data are now available, and ion cyclotron resonance and other techniques allow estimates of bond energies.

The strength of an H-bond may be defined as the enthalpy of the process:

$$X\!-\!\!-\!\!-H\!-\!-\!-Y \longrightarrow X\!-\!\!-\!\!-H + Y$$

There are three main types: weak, with enthalpies 10–50 kJ mol^{-1}; strong, 50–100 kJ mol^{-1}; very strong, >100 kJ mol^{-1}, the best example of which is $F-H\cdots F^-$.

Examples of H-bonds and the range of certain properties are given in Table 2–1.

Very *short, strong H-bonds* are found in HF_2^-, $H(O_2CR)_2^-$, $H(ONO_2)_2^-$, and in various other species.[12] In some cases such as $O\cdots H\cdots O$ and $F\cdots H\cdots F$, the bonds are symmetrical. There appears to be no unique correlation between $X\cdots Y$ and

[11]G. van Koten *et al.*, *J. Am. Chem. Soc.* **1995**, *117*, 10939.
[12]G. Gilli *et al.*, *J. Am. Chem. Soc.* **1994**, *116*, 909.

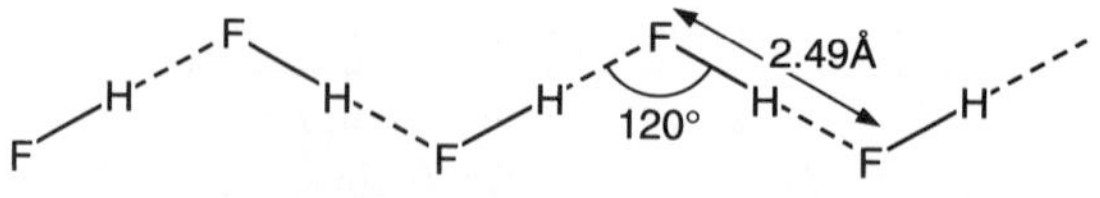

Figure 2-1   The structure of crystalline hydrogen fluoride.

$X-H$ distances and the entire environment has to be considered. Figure 2-1 shows an example of $F-H-F$ bonding in the $(HF)_n$ polymer. Since the structure might be expected to be linear, that finding lends strong support for preferred directions in some cases. By contrast, HCN polymers are linear.

Multicenter H-bonds[13] are known in a number of compounds, the most common being 3-center bonds of the type (2-IA); four-center bonds are rare. H-bonds of type (2-IB) are also known.

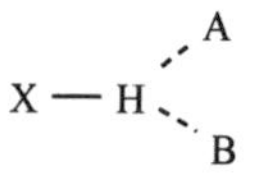

(2-IA) 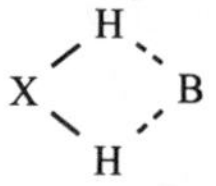   (2-IB)

Hydrogen bonds with $C-H\cdots X$ *interactions*[14] can be formed if the $C-H$ bond is relatively polar as it is when C is bonded to electronegative groups as in $HCCl_3$. A *bis*(carbene)-H^+ complex has been shown to have the structure (2-IIA)[15] with a linear $C-H-C$ bond.

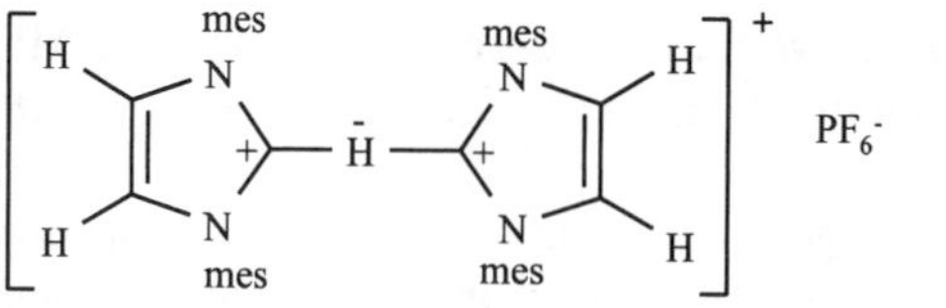

(2-IIA) 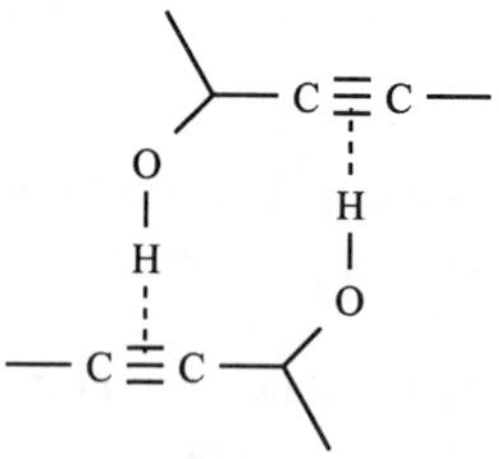   (2-IIB)

Finally, there are some π interactions[16] with OH, NH, and $C\equiv C$, an example being (2-IIB).

2-4   Ice and Water

The structural nature of ice and, *a fortiori*, of water are very complex matters that can be treated but briefly here.

There are nine known modifications of ice, each stable over a certain range of temperature and pressure. Ordinary ice, ice-I, which forms from liquid water at

[13]For references see R. H. Crabtree *et al.*, *Inorg. Chem.* **1995**, *34*, 3474.
[14]T. Steiner *et al.*, *Chem. Commun.* **1996**, 1277.
[15]A. A. Arduengo *et al.*, *J. Am. Chem. Soc.* **1995**, *117*, 572.
[16]J. A. K. Howard *et al.*, *J. Am. Chem. Soc.* **1996**, *116*, 4081; C. A. Hunter *et al.*, *Chem. Commun.* **1996**, 2529.

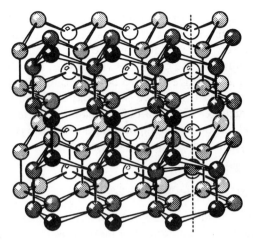

Figure 2-2 The structure of ice-I. Only the oxygen atoms are shown.

0°C and 1 atm, has a rather open structure built of puckered, six-membered rings (Fig. 2-2). Each H_2O is tetrahedrally surrounded by the oxygen atoms of four neighboring molecules, and the whole array is linked by unsymmetrical hydrogen bonds. The $O-H\cdots O$ distance is 2.75 Å (at 100 K), and the H atom lies 1.01 Å from one oxygen atom and 1.74 Å from the other. Each oxygen atom has two near and two far hydrogen atoms, but there are six distinct arrangements, two being illustrated in Fig. 2-3, all equally probable. However, the existing arrangement at any one oxygen atom eliminates certain of these at its neighbors. A rigorous analysis of the probability of any given arrangement in an entire crystal leads to the conclusion that at the absolute zero ice-I should have a disordered structure with a zero-point entropy of 3.4 J mol⁻¹ deg⁻¹, in excellent agreement with experiment. This result in itself constitutes a good proof that the hydrogen bonds are unsymmetrical; if they were symmetrical, there would be a unique, ordered structure, hence no zero-point entropy. Entirely similar considerations confirm the presence of a network of unsymmetrical H bonds in KH_2PO_4 and $Ag_2H_3IO_6$, whereas the absence of zero-point entropy in K[FHF] accords with the symmetrical structure of the FHF⁻ ion.

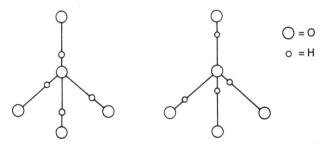

$\bigcirc = O$

$\circ = H$

Figure 2-3 Two of the six possible configurations of the hydrogen atoms about an oxygen atom in ice.

The structure of liquid water has been the subject of much study, experimentally and theoretically. In addition to water dimers[17] there are trimers[18] and tetramers that have ring structures. The pentamer is also cyclic while the heptamer prefers a 3-dimensional structure. In the complex $C_6H_6(H_2O)_6$, the water appears to have a structure containing four single and two double donor water molecules.[19]

Water becomes supercritical above 374°C and 220 atm, becoming less polar. This allows homogenization with non-polar organic materials, thus making them available for chemical reactions. An important potential use is the rapid destruction of toxic organic compounds and waste by oxygen.[20]

2-5   Hydrates and Aqua Ions[21]

Crystalline hydrates of metal ions and of organic substances, especially those with N—H and O—H bonds, are numerous. For metal ions, the oxygen is always bound to the metal and the lone pairs on it can be directed toward the metal and involved in bonding but can, however, also form H bonds. There is hence flexibility, allowing stabilization in lattices of many different types of hydrated structure.

It is often of practical importance to be able to distinguish between coordinated water, $M(OH_2)$, and non-coordinated lattice water. Infrared spectra can be diagnostic, a key feature being the appearance around 1650 cm^{-1} of a δ_{HOH} absorption for coordinated water.[22]

The configuration at the oxygen atom of a coordinated water molecule may be pyramidal or planar.[23] There appears to be no significant energy difference between these two.

The structures and dynamics of hydrates have been studied by both *ab initio* calculations[24] and by a variety of experimental techniques. Hydration numbers for the most common cations and anions are known.[25]

The rates of exchange of coordinated water molecules with bulk solvent vary over nearly 20 orders of magnitude. The $[Ir(H_2O)_6]^{3+}$ ion is the slowest and $[Cs(H_2O)_n]^+$ is the fastest. The rates for all common cations are presented in Fig. 2-4.

Gas hydrates[26] are a particular class of inclusion compounds where, generally, molecules of a gas can be trapped in a lattice having receptor holes of the right size.

[17]J. Bertran *et al., J. Am. Chem. Soc.* **1994,** *116,* 8249.
[18]J. E. Fowler and H. F. Schaefer, III, *J. Am. Chem. Soc.* **1995,** *117,* 446.
[19]K. D. Jordan *et al., J. Am. Chem. Soc.* **1994,** *116,* 11568.
[20]T. Clifford and K. Bartle, *Chem. Brit.* **1993,** 499.
[21]H. Ohtaki and T. Radnai, *Chem. Rev.* **1993,** *93,* 1157 (417 refs); G. W. Nelson and J. E. Enderby, *Adv. Inorg. Chem.* **1989,** *34,* 195; H. D. Lutz, *Bonding and Structures of Solid Hydrates* in *Structure and Bonding* Vol. 69, M. J. Clarke, ed., Springer, New York, 1988.
[22]G. J. Kubas *et al., Organometallics* **1992,** *11,* 3396.
[23]S. P. Best *et al., J. Chem. Soc. Dalton Trans.* **1993,** 2711.
[24]L. G. M. Petterson *et al., J. Am. Chem. Soc.* **1994,** *116,* 8705; U. Koelle *et al., Inorg. Chem.* **1995,** *34,* 306.
[25]H. Bertagnolli and T. S. Ertel, *Angew. Chem. Int. Ed. Engl.* **1994,** *33,* 45.
[26]R. L. Christiansen and D. E. Sloan Jr., *Ann. N.Y. Acad. Sci.* **1994,** *715,* 283; E. D. Sloan, *Clathrate Hydrates of Natural Gases,* Dekker, New York, 1990; A. Müller *et al., Angew. Chem. Int. Ed. Engl.* **1995,** *34,* 2328; J. L. Attwood *et al.,* eds. *Inclusion Compounds,* Academic Press, London, Vol. I, 1984 *et seq.*

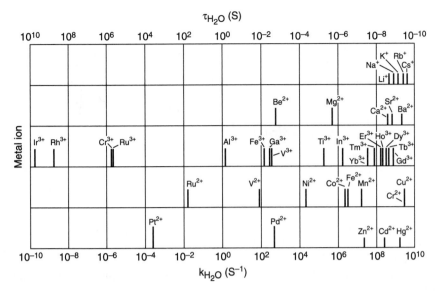

Figure 2-4 Exchange rate constants (k) and mean residence times (τ) for coordinated H_2O molecules. $\tau = k^{-1}$. Tall bars are directly measured values while short bars give values derived from complex formation rates. Reproduced from A. E. Merbach *et al.*, *J. Am. Chem. Soc.* **1996,** *118,* 5265.

The first example, discovered by Humphrey Davy in 1811, was $Cl_2 \cdot 7H_2O$. Hydrates are commonly obtained when water is frozen in presence of a gas such as Ar, Kr, Xe,[27] Cl_2, CO_2, and so on. *Methane hydrate*[28] occurs as a solid under permafrost or on the deep sea bed in various parts of the world in vast quantities. Utilization presents problems and is not likely to be necessary for some time when conventional natural gas deposits run out.

Gas hydrates have two common structures. One has 46 molecules of water that form 6 medium size and 2 small cages; this is for small molecules. Complete cage filling to give the composition $X \cdot 5.6H_2O$ is rare. Medium cage filling gives $X \cdot 7.67H_2O$ and chlorine hydrate $Cl_2 \cdot 7.3H_2O$ approaches this. The second structure formed with larger molecules such as $CHCl_3$ or $EtCl$ has 8 large and 16 smaller cages.

Salt hydrates. These may be formed when R_4N^+ or R_3S^+ salts crystallize from aqueous solution; examples are $Bu_4^tN^+PhCO_2^- \cdot 39.5H_2O$ and $(Bu_3S)F \cdot 20H_2O$.

The frameworks here are H-bonded water molecules, some H bonded to F^- or the oxygen of anions. The cations and parts of the anions (e.g., PhC of benzoate) occupy cavities randomly and incompletely. Hydrates of strong acids, for example, $HPF_6 \cdot 7.67H_2O$, $HBF_4 \cdot 5.75H_2O$, and $HClO_4 \cdot 5.5H_2O$ are also known.

It may be noted that there is some structural analogy between this type of hydrate and zeolites. For example, $[Me_4N]OH \cdot 5H_2O$ is isostructural with the silicate

[27]T. Pietrass *et al.*, *J. Am. Chem. Soc.* **1995,** *117,* 7520.
[28]R. H. Crabtree, *Chem. Rev.* **1995,** *95,* 987; K. Lekvam and P. Ruoff, *J. Am. Chem. Soc.* **1993,** *115,* 8565.

sodalite and there are parallels between 12- and 17-Å gas hydrates and other zeolite structures.

When salt hydrates are melted under CO_2 pressure, the gas is reversibly absorbed. Thus $(Me_4N)F \cdot 4H_2O$ at 50°C and 100 kPa absorbs CO_2 to give a concentration *ca* 1.9 *M*. This may lead to new methods for removal of CO_2 from gas streams.[29]

2-6   Hydroxonium Cations

Since the reaction

$$H(g) = H^+(g) + e \qquad \Delta H = 569 \text{ kJ mol}^{-1}$$

is more endothermic than that for Xe, one consequence is that the hydrogen ion can be found *only* when its compounds such as $HCl(g)$ are dissolved in media that can solvate the protons. The solvation process provides the energy for the H—X bond rupture, as indicated by the water solvation reaction estimated from thermodynamic cycles:

$$H^+(g) + nH_2O = H^+(aq) \qquad \Delta H = -1091 \text{ kJ mol}^{-1}$$

Although the hydrogen ion is commonly written as H^+, solvation is *always* assumed, as is the case for other ions such as Na^+ or Fe^{2+}.

The reduction of solvated protons to H_2, e.g., by reaction of Zn or Fe with dilute acids, is one of the most fundamental redox reactions. The reaction is not simple as it depends upon the nature of the solvent, concentrations and temperature. Reduction can be achieved, of course, electrolytically and by a variety of other reactions, some being catalytic.[30]

In crystalline compounds the hydroxonium ions H_3O^+, $H_5O_2^+$, $H_7O_3^+$, $H_9O_4^+$, and $H_{14}O_6^{2+}$ [31] have been characterized.[32] In compounds commonly written as $H_2PtCl_6 \cdot 2H_2O$ or $HBF_4 \cdot nH_2O$, the acidic proton is always present as the oxonium ion.

In some acids, such as H_2SO_4, the protons can be bound to the O atoms, while for other acids like $H_4Fe(CN)_6$, the anhydrous form has FeCN$\cdots$H$\cdots$NCFe hydrogen bonds; no oxonium ion is present in $2PO(OH)_3 \cdot H_2O$ and $(CO_2H)_2 \cdot 2H_2O$ that has a 3-dimensional H-bonded structure.

The structural role of H_3O^+ in a crystal often resembles that of NH_4^+; thus $H_3O^+ClO_4^-$ and $NH_4^+ClO_4^-$ are isomorphous. The important difference is that compounds of H_3O^+ and similar ions generally have much lower melting points than have NH_4^+ salts.

The structure of H_3O^+ is that of a flat pyramid (2-III) while that of $H_5O_2^+$ is shown in (2-IV); the other ions are more complicated and O$\cdots$H—O distances may

[29]R. Quinn *et al., J. Am. Chem. Soc.* **1995,** *117,* 329.
[30]U. Koelle, *New J. Chem.* **1992,** *16,* 157.
[31]H. Henke and W. H. Kuks, *Z. Krist.* **1987,** *181,* 113.
[32]P. C. Junk and J. L. Attwood, *J. Chem. Soc., Chem. Commun.* **1995,** 1551; C. A. Reed *et al., Inorg. Chem.* **1995,** *34,* 5403.

differ considerably. Alcohols can form similar ions with H^+. Although the H_4O^{2+} ion is unknown, many stable 4-coordinate oxygen compounds, such as those with the basic beryllium acetate structure, have been made and structurally characterized. A recent new example is $[(Ph_3PAu)_4O][BF_4]_2$.[33]

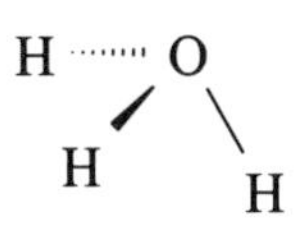

(2-III) 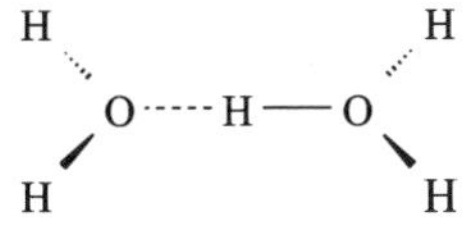   (2-IV)

2-7 Anionic Species

The "solvation" of H^+ by anions can form species such as $[H(NO_3)_2]^-$, $[H(SO_3F)_2]^-$, and $[H(HCO_2)_2]^-$ that are commonly found in crystals. The anions HF_2^- and $[Cl-H-Cl]^-$,[34] which can be isolated as stable salts of $[K(18-C-6)]^+$, can be included in the same category.

The hydrated hydroxyl ion $H_3O_2^-$ has been recognized in transition metal hydrated hydroxo species such as $[Cr(bipy)_2(H_2O)OH]^{2+}$.[35] The formation of these compounds occurs as follows (2-V).

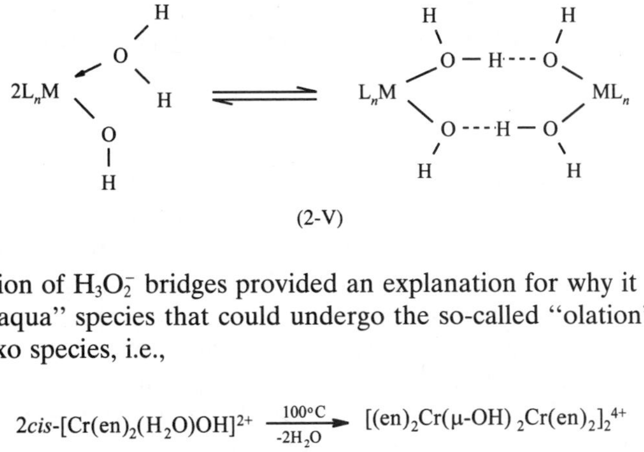

(2-V)

The recognition of $H_3O_2^-$ bridges provided an explanation for why it was the only "*cis*-hydroxoaqua" species that could undergo the so-called "olation" reaction to give dihydroxo species, i.e.,

$$2cis\text{-}[Cr(en)_2(H_2O)OH]^{2+} \xrightarrow[-2H_2O]{100°C} [(en)_2Cr(\mu\text{-}OH)_2Cr(en)_2]_2^{4+} \qquad (2)$$

There are of course some genuine hydroxoaqua complexes without $H_3O_2^-$ bridges. Single bridged compounds are also known.[36] The bridges can be *gauche* (2-VI) or *anti* (2-VII); the O—O distances are in the range 2.41–2.54 Å. Finally it may be noted that free hydroxide-water clusters have been studied both experimentally and theoretically. They are of the type $OH^-(OH_2)_n$ where n can be from 1 to *ca* 30.[37]

[33] H. Schmidbauer *et al., Nature (London)* **1995**, *377*, 503.
[34] J. L. Attwood *et al., Inorg. Chem.* **1990**, *29*, 467.
[35] M. Ardon and A. Bino, *Struct. Bonding (Berlin)* **1987**, *65*, 1.
[36] K. Wieghardt *et al., J. Chem. Soc. Dalton Trans.* **1994**, 2041.
[37] S. S. Xantheas, *J. Am. Chem. Soc.* **1995**, *117*, 10373.

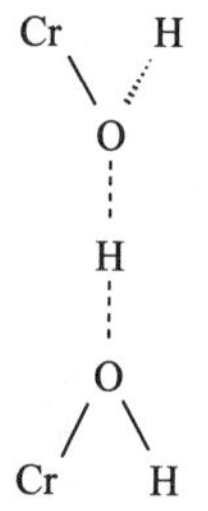

(2-VI) 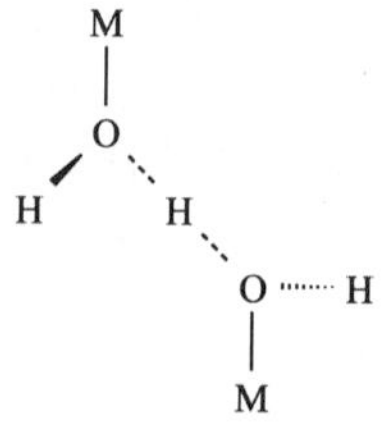   (2-VII)

STRENGTHS OF PROTONIC ACIDS

An important characteristic of hydrogen compounds (HX) is the extent to which they ionize in water or other solvents, that is, the extent to which they act as acids. The strength of an acid depends not only on the nature of the acid itself but very much on the medium in which it is dissolved. Thus CF_3CO_2H and $HClO_4$ are strong acids in water, whereas in 100% H_2SO_4 the former is nonacidic and the latter only a very weak acid. Similarly, H_3PO_4 is a base in 100% H_2SO_4. Although acidity can be measured in a wide variety of solvents, the most important is water.

It is convenient to note here that 1,8-bis(dimethylamino)naphthalene and related compounds are very strong bases and act as a *proton sponge*, removing H^+ and forming $N{\cdots}H{\cdots}N$ H-bonds that are symmetrical and nearly linear.

2-8   Binary Acids

Although the intrinsic strength of H—X bonds is one factor, other factors are involved, as the following consideration of the appropriate thermodynamic cycles for a solvent system indicates. The intrinsic strength of H—X bonds and the thermal stability of covalent hydrides seem to depend on the electronegativity and size of the element X. The variation in bond strength in some binary hydrides is shown in Fig. 2-5. There is a fairly smooth *decrease* in bond strength with *increasing Z* in a periodic group and a general *increase across* any period.

For HX dissolved in water we may normally (but not always—see later) assume that dissociation occurs according to the equation:

$$HX(aq) = H^+(aq) + X^-(aq)$$

The dissociation constant K is related to the change in Gibbs free energy by the relation

$$\Delta G^0 = -RT \ln K$$

and the free energy change is in turn related to the changes in enthalpy and entropy *via* the relation

$$\Delta G = \Delta H - T\Delta S$$

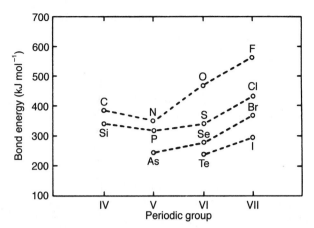

Figure 2-5 Variations in mean H—X bond energies.

in which R is the gas constant and T is the absolute temperature, which we shall take to be 298 K in the following discussion. The dissociation process may be considered to be the sum of several other reactions (i.e., as one step in a thermodynamic cycle). Table 2-2 summarizes the Gibbs free energy changes for these several steps.

Hydrogen fluoride in *dilute* aqueous solution is a very weak acid; the pK values for the halogen acids change only slightly (3 units overall) from HI to HBr to HCl, but there is a precipitous change, by 10.2 units, on going to HF. The classic thermodynamic explanation for this is that the HF bond is so much stronger than those of the other HX molecules that even an unusually high free energy of solvation of F^- cannot compensate for it; thus most dissolved HF molecules remain undissociated. However, this is not correct. From ir studies of the four aqueous HX acids it was inferred that the H_3O^+ ion in aqueous HF is indeed present in large quantity but with a perturbed spectrum indicative of strong H-bonding to F^-. In other words, HF *is* dissociated, but tight ion pairs, $F^-\cdots H^+-OH_2$, unique to F^-, which is a far better participant in H-bonding than Cl^-, Br^-, or I^-, reduce the thermodynamic activity coefficient of H_3O^+.

Table 2-2 Free Energy Changes (kJ mol^{-1}) for Dissociation of HX Molecules in Water at 298 K

Process	HF	HCl	HBr	HI
HX(aq) = HX(g)	23.9	−4.2	−4.2	−4.2
HX(g) = H(g) + X(g)	535.1	404.5	339.1	272.2
H(g) = H$^+$(g) + e	1320.2	1320.2	1320.2	1320.2
X(g) + e = X$^-$(g)	−347.5	−366.8	−345.4	−315.3
H$^+$(g) + X$^-$(g) = H$^+$(aq) + X$^-$(aq)	−1513.6	−1393.4	−1363.7	−1330.2
HX(aq) = H$^+$(aq) + X$^-$(aq)	18.1	−39.7	−54.0	−57.3
pK_A(= ΔG^0/5.71)	3.2	−7.0	−9.5	−10

On increasing the HF concentration new equilibria become important.[38] The ion pair $H_3O^+ F^-$ increasingly dissociates with the formation of HF_2^-:

$$H_3O^+F^- + HF \rightleftharpoons H_3O^+ + HF_2^-$$

X-ray study of the crystalline hydrates $H_2O \cdot nHF$, $n = 1, 2$, and 4, shows that these are $H_3O^+F^-$, $H_3O^+HF_2^-$, and $H_3O^+H_3F_4^-$.

2-9 Oxo Acids

The second main class of acids are compounds with $X-OH$ groups of the type H_nXO_m, for example, H_3PO_4, or better $O=P(OH)_3$.

For oxo acids certain generalizations may be made concerning (a) the magnitude of K_1, and (b) the ratios of successive constants, K_1/K_2, K_2/K_3, and so on. The value of K_1 seems to depend on the charge on the central atom. Qualitatively, it is reasonable to suppose that the greater the positive charge, the more the process of proton loss will be favored on electrostatic grounds. If this positive charge is taken to be the so-called formal charge, semiquantitative correlations are possible. The formal charge in an oxo acid H_nXO_m is computed assuming the structure of the acid to be $O_{m-n}X(OH)_n$. Each $X-(OH)$ bond is formed by sharing one X electron and one OH electron and is thus *formally* nonpolar. Each $X-O$ bond is formed by using two X electrons and thus represents a net loss of one electron by X. Therefore the formal positive charge on X is equal to the number of $X-O$ bonds, hence equal to $(m - n)$. The data in Table 2-3 show that with the exception of the acids listed in brackets, which are special cases to be discussed presently, the following relations between $m - n$ (or formal positive charge on X) and the values of K_1 hold:

$$
\begin{array}{lll}
\text{For } m - n = 0 & pK_1 \sim 8.5 \pm 1.0 & (K \sim 10^{-8} \text{ to } 10^{-9}) \\
\text{For } m - n = 1 & pK_1 \sim 2.8 \pm 0.9 & (K \sim 10^{-2} \text{ to } 10^{-4}) \\
\text{For } m - n \geq 2 & pK_1 \ll 0 \text{ (the acid is very strong)}
\end{array}
$$

With very few exceptions, the difference between successive pK's is 4 to 5. Phosphorous acid (H_3PO_3) obviously is out of line with the other acids having $m - n = 0$ and seems to fit fairly well in the group with $m - n = 1$. This is, in fact, where it belongs, since there is independent evidence (Chapter 10) that its structure is $OPH(OH)_2$ with H bonded directly to P. Similarly, H_3PO_2 has a pK_1 that would class it with the $m - n = 1$ acids where it, too, belongs, since its structure is $OP(H)_2(OH)$, with the two H hydrogen atoms directly bound to P.

"Carbonic acid" is exceptional in that the directly measured pK_1, 6.38, does not refer to the process

$$H_2CO_3 = H^+ + HCO_3^-$$

since CO_2 in solution is only partly in the form of H_2CO_3, but largely present as

---

[38]R. Braddy *et al.*, *J. Fluorine Chem.* **1994,** *66,* 63.

Table 2-3   Strengths of Oxo Acids, H_nXO_m, in Water K_2 (pK_2)

($m - n$)	Examples	$-\log K_1$ (pK_1)	$-\log K_2$ (pK_2)	$-\log K_3$ (pK_3)
0	HClO	7.5		
	HBrO	8.68		
	H_3AsO_3	9.22	?	?
	H_4GeO_4	8.59	13	?
	H_6TeO_6	8.80	?	?
	$[H_3PO_3]$	1.8	6.15	
	H_3BO_3	9.22	?	?
1	H_3PO_4	2.12	7.2	12
	H_3AsO_4	3.5	7.2	12.5
	H_5IO_6	3.29	6.7	~15
	H_2SO_3	1.90	7.25	
	H_2SeO_3	2.57	6.6	
	$HClO_2$	1.94		
	HNO_2	3.3		
	$[H_2CO_3]$	6.38 (3.58)	10.32	
2	HNO_3	Large neg. value		
	H_2SO_4	Large neg. value	1.92	
	H_2SeO_4	Large neg. value	2.05	
3	$HClO_4$	Very large neg. value		
	$HMnO_4$	Very large neg. value		
−1(?)	H_3PO_2	2	?	?

more loosely hydrated species $CO_2(aq)$. When a correction is made for the equilibrium

$$CO_2(aq) + H_2O = H_2CO_3(aq)$$

the pK_1 value of 3.58 is obtained, which falls in the range for other $m - n = 1$ acids.

Many metal ions whose solutions are acidic can be regarded as oxo acids. Thus although the hydrolysis of metal ions is often written as shown here for Fe^{3+}:

$$Fe^{3+} + H_2O = Fe(OH)^{2+} + H^+$$

it is just as valid thermodynamically and much nearer to physical reality to recognize that the Fe^{3+} ion is coordinated by six water molecules and to write:

$$[Fe(H_2O)_6]^{3+} = [Fe(H_2O)_5OH]^{2+} + H^+ \qquad K_{Fe}^{3+} \approx 10^{-3}$$

From this formulation it becomes clear why the Fe^{2+} ion, with a lower positive charge, is less acidic or, in alternative terms, less hydrolyzed than the Fe^{3+} ion:

$$[Fe(H_2O)_6]^{2+} = [Fe(H_2O)_5OH]^+ + H^+ \qquad K_{Fe^{2+}} \ll K_{Fe^{3+}}$$

It should be noted that one cannot necessarily compare the acidity of the bivalent ion of one metal with that of the trivalent ion of *another* metal in this way, however. There appears to be no good general rule concerning the acidities of hydrated metal ions at the present time, although some attempts have been made at correlations.

Table 2-4 K_1/K_2 Ratio for Dicarboxylic Acids, $HO_2C(CH_2)_nCO_2H$

n	1	2	3	4	5	6	7	8
K_1/K_2	1120	29.5	17.4	12.3	11.2	10.0	9.5	9.3

2-10   Theory of Ratios of Successive Constants

It was shown many years ago by Niels Bjerrum that the ratios of successive acid dissociation constants could be accounted for in a nearly quantitative way by electrostatic considerations. Consider any bifunctional acid HXH:

$$HXH = HX^- + H^+ \qquad K_1$$

$$HX^- = X^{2-} + H^+ \qquad K_2$$

There is a purely statistical effect that can be considered in the following way. For the first process, dissociation can occur in two ways (i.e., there are two protons, either of which may dissociate), but recombination in only one; whereas in the second process dissociation can occur in only one way, but recombination in two (i.e., the proton has two sites to which it may return, hence twice the probability of recombining). Thus on purely statistical grounds one would expect $K_1 = 4 K_2$. Bjerrum observed that for the dicarboxylic acids $HO_2C(CH_2)_nCO_2H$, the ratio K_1/K_2 was always >4, but decreased rapidly as n increased (see Table 2-4). He suggested the following explanation. When the two points of attachment of protons are close together in the molecule, the negative charge left at one site when the first proton leaves strongly restrains the second one from leaving by electrostatic attraction. As the separation between the sites increases, this interaction should diminish.

By making calculations using Coulomb's law,[39] Bjerrum was able to obtain rough agreement with experimental data. The principal difficulty in obtaining quantitative agreement lies in a choice of dielectric constant, since some of the lines of electrostatic force run through the molecule ($D \sim 1$–10), others through neighboring water molecules (D uncertain), and still others through water having the dielectric constant (~ 82) of pure bulk water. Nearly quantitative agreement is obtained using more elaborate models that take into account the variability of the dielectric constant. The important point here for our purposes is to recognize the physical principles involved without necessarily trying to obtain quantitative results.

Thus the large separation in successive pK's for the oxo acids is attributable to the electrostatic effects of the negative charge left by the dissociation of one proton on the remaining ones. In bifunctional binary acids, where the negative charge due to the removal of one proton is concentrated on the very atom to which the second proton is bound, the separation of the constants is extraordinarily great: K_1 and K_2 for H_2S are $\sim 10^{-7}$ and 10^{-14}, respectively, whereas for water we have

$$H_2O = H^+ + OH^- \qquad K_1 = 10^{-14}$$

$$OH^- = H^+ + O^{2-} \qquad K_2 < 10^{-36} \quad \text{estimated}$$

---

[39]$F \propto q_1q_2/Dr$, where F is the force; q_1 and q_2 the charges separated by r; and D the dielectric constant of the medium between them.

Table 2-5 The Hammett Acidity Function H_0 for Several Acids

Acid	$-H_0$	Acid	$-H_0$
$HSO_3F + SbF_5$ (14.1 mol%)	26.5	HF (100%)	15.1
$HF + SbF_5$ (0.6 mol%)	21.1	$HF + NaF$ (1 M)	8.4
HSO_3F	15	H_3PO_4	5
$H_2S_2O_7$	15	H_2SO_4 (63% in H_2O)	4.9
CF_3SO_3H	13.8	HCO_2H	2.2
H_2SO_4	11.9		

2-11 Pure Acids and Relative Acidities; Superacids[40]

The concepts of hydrogen ion concentration and pH discussed previously are mean-ingful only for dilute aqueous solutions of acids. A widely used means of gauging acidity in other media and at high concentrations is the Hammett acidity function H_0, which is defined in terms of the behavior of one or more indicator bases B, for which there is the protonation equilibrium

$$B + H^+ = BH^+$$

The acidity function is defined as

$$H_0 = pK_{BH^+} - \log \frac{[BH^+]}{[BH]}$$

In very dilute solutions

$$K_{BH^+} = \frac{[B][H^+]}{[BH^+]}$$

so that in water, H_0 becomes synonymous with pH. By using suitable organic bases (e.g., *p*-nitroaniline) and suitable indicators over various ranges of concentration and acidities or by nmr methods, it is possible to interrelate values of H_0 for strong acids extending from dilute solutions to the pure acid.

For a number of strong acids in aqueous solution up to concentrations about 8 M, the values of H_0 are very similar. This suggests that the acidity is independent of the anion. The rise in acidity with increasing concentration can be fairly well predicted by assuming that the hydrogen ion is present as $H_9O_4^+$, so that protonation can be represented as

$$H_9O_4^+ + B \rightleftharpoons BH^+ + 4H_2O$$

Values of H_0 for some pure liquid acids are given in Table 2-5. It is to be noted particularly that for HF the acidity can be very substantially increased by the

[40]G. A. Olah, *Angew. Chem. Int. Ed. Engl.* **1995,** *34,* 1393; **1993,** *32,* 767; G. A. Olah, G. K. S. Prakash and J. Sommer, *Superacids,* Wiley, 1985; T. A. O'Donnell, *Superacids and Acid Melts as Inorganic Reaction Media,* VCH, New York, 1993; F. Arata, *Adv. Catal.* **1990,** *37,* 165 (solid superacids); F. Aubke *et al.,* in *Synth. Fluorine Chem.,* G. A. Olah *et al.,* eds., Wiley, New York, 1992, pp. 43–86 (transition metal derivatives of strong protic acids and superacids).

additions of a Lewis acid or fluoride ion acceptor, for example:

$$2HF + SbF_5 \rightleftharpoons H_2F^+ + SbF_6^-$$

but its acidity is decreased by addition of NaF owing to formation of the HF_2^- ion. Antimony pentafluoride is commonly used as the Lewis acid, since it is comparatively easy to handle, being a liquid, and is commercially available. However, other fluorides such as BF_3, NbF_5, and TaF_5 behave in a similar way. Acid media with $-H_0$ values above about six are referred to as *superacids*, since they are upward of 10^6 times as strong as a *1 M* aqueous solution of a strong acid. The addition of SbF_5 to HSO_3F dramatically raises the $-H_0$ value from 15 for 0% SbF_5 to ~17 at 0.4 mol % SbF_5 and finally to 26.5 at 90 mol %, the latter being the highest $-H_0$ value known. The SbF_5–FSO_3H system, which is very complicated, has been thoroughly investigated by nmr and Raman spectra. The acidity is due to the formation of the $H_2SO_3F^+$ ion. The equilibria depend on the ratios of the components; with low ratios of SbF_5 to FSO_3H the main ones are the following:

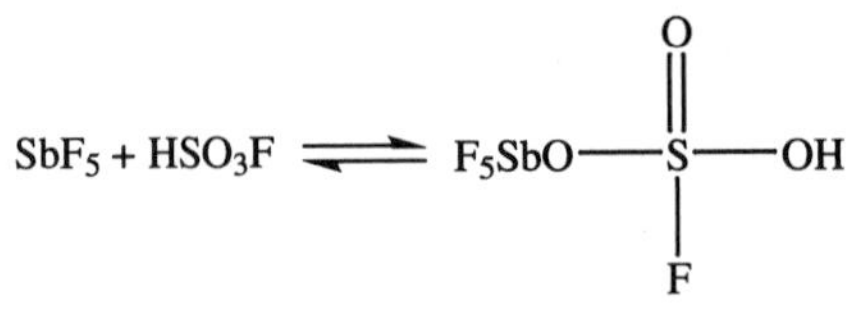

$$SbF_5 + HSO_3F \rightleftharpoons F_5SbO\!-\!\!\overset{\displaystyle O}{\underset{\displaystyle F}{\overset{\|}{\underset{|}{S}}}}\!\!-\!OH$$

$$H(F_5SbSO_3F) + HSO_3F \rightleftharpoons [F_5SbOSO_2F]^- + H_2SO_3F^+$$

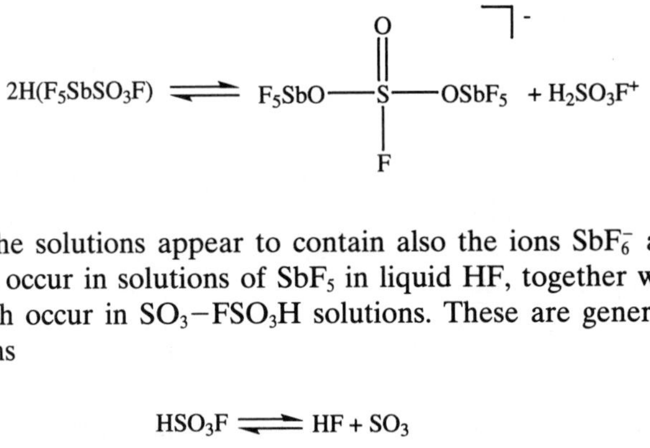

$$2H(F_5SbSO_3F) \rightleftharpoons \left[F_5SbO\!-\!\!\overset{\displaystyle O}{\underset{\displaystyle F}{\overset{\|}{\underset{|}{S}}}}\!\!-\!OSbF_5\right]^- + H_2SO_3F^+$$

At higher ratios the solutions appear to contain also the ions SbF_6^- and $[F_5Sb\!-\!F\!-\!SbF_5]^-$, which occur in solutions of SbF_5 in liquid HF, together with HS_2O_6F and HS_3O_9F, which occur in SO_3–FSO_3H solutions. These are generated by the additional reactions

$$HSO_3F \rightleftharpoons HF + SO_3$$

$$3SbF_5 + 2HF \rightleftharpoons HSbF_6 + HSb_2F_{11}$$

$$2HSO_3F + 3SO_3 \rightleftharpoons HS_2O_6F + HS_3O_9F$$

The SbF_5–HSO_3F solutions are very viscous and are normally diluted with liquid sulfur dioxide so that better resolution of nmr spectra is obtained. Although the equilibria appear not to be appreciably altered for molecular ratios $SbF_5 \cdot HSO_3F$ <0.4, in more concentrated SbF_5 solutions the additional equilibria noted here are shifted to the left by removal of SbF_5 as the stable complex $SbF_5 \cdot SO_3$, which can be obtained crystalline.

Table 2-6 Properties of Some Strong Acids in the Pure State

Acid	mp (°C)	bp (°C)	κ^a (temperature, °C)	ε^b (temperature, °C)
HF	−83.36	19.74	1.6×10^{-6} (0)	84 (0)
HCl	−114.3	−85.09	3.5×10^{-9} (−85)	14.3 (−114)
HBr	−86.92	−66.78	1.4×10^{-10} (−84)	7.33 (−86)
HI	−50.85	−35.41	8.5×10^{-10} (−45)	3.57 (−45)
HNO_3	−41.59	82.6	3.77×10^{-2} (25)	
$HClO_4$	−112	(109 extrap.)	1.085×10^{-4} (25)	
HSO_3F	−88.98	162.7	1.044×10^{-2} (25)	~120 (25)
H_2SO_4	10.377	~270 d^c		110 (20)

^{a}Specific conductance (ohm^{-1} cm^{-1}). Values are often very sensitive to impurities.
^{b}Dielectric constant divided by that of a vacuum.
^{c}Constant-boiling mixture (338°C) contains 98.33% of H_2SO_4; d = with decomposition.

Extensive studies have also been made on the $HSO_3F-Sb(SO_3F)_5$ conjugate superacid system[41]; this may be more acidic than the two strongest known HSO_3F compounds, namely, $HSO_3F-SbF_2(SO_3F)_3$ and $HSO_3F-Ta(SO_3F)_5$.

2-12 Properties of Some Common Strong Acids

A collection of some properties of the more common and useful pure strong acids is found in Table 2-6.

Hydrogen Fluoride[42]

The acid is made by the action of concentrated H_2SO_4 on CaF_2 and is the principal source of fluoride compounds.

It is available in steel cylinders, with purity approximately 99.5%; it can be purified further by distillation. Although liquid HF attacks glass rapidly, it can be handled conveniently in apparatus constructed either of copper or Monel metal or of materials such as polytetrafluoroethylene (Teflon or PTFE) and Kel–F (a chlorofluoro polymer).

The high dielectric constant is characteristic of hydrogen-bonded liquids. Since HF forms only a two-dimensional polymer, it is less viscous than water. In the vapor, HF is monomeric above 80°C, but at lower temperatures the physical properties are best accounted for by an equilibrium between HF and a hexamer, $(HF)_6$, which has a puckered ring structure. Crystalline $(HF)_n$ has zigzag chains (Fig. 2-1).

After water, liquid HF is one of the most generally useful solvents. Indeed in some respects it surpasses water as a solvent for both inorganic and organic compounds, which often give conducting solutions as noted previously; it can also be used for cryoscopic measurements.

The self-ionization equilibria in liquid HF are

$$2HF \rightleftharpoons H_2F^+ + F^- \qquad K \sim 10^{-10}$$

$$F^- + HF \rightleftharpoons HF_2^- \xrightarrow{HF} H_2F_3^- \quad \text{and so on}$$

[41]F. Aubke *et al., Inorg. Chem.* **1995,** *34,* 2269.
[42]W. D. Chandler *et al., Inorg. Chem.* **1995,** *34,* 4943.

The structures of fluoronium ions,[43] H_2F^+, $H_3F_2^+$, and $H_7F_6^+$ as SbF_6^- and $Sb_2F_{11}^-$ salts have been determined as well as those of the anions $H_2F_3^-$, $H_3F_4^-$, $H_4F_5^-$, and $H_5F_6^-$ as potassium salts.

The formation of the stable hydrogen-bonded anions accounts in part for the extreme acidity. In the liquid acid the fluoride ion is the conjugate base, and ionic fluorides behave as bases. Fluorides of M^+ and M^{2+} are often appreciably soluble in HF, and some such as TlF are very soluble.

The only substances that function as "acids" in liquid HF are those such as SbF_5, previously noted, which increase the concentration of H_2F^+. The latter ion appears to have an abnormally high mobility in such solutions.

Reactions in liquid HF are known that also illustrate amphoteric behavior, solvolysis, or complex formation. Although HF is waterlike, it is not easy, because of the reactivity, to establish an emf series, but a partial one is known.

In addition to its utility as a solvent system, HF as either liquid or gas is a useful fluorinating agent, converting many oxides and other halides into fluorides.

Hydrogen Chloride, Bromide, and Iodide[42]

These acids are quite similar but differ from HF; they are normally pungent gases; in the solid state they have hydrogen-bonded zigzag chains and there is probably some hydrogen bonding in the liquid. Hydrogen chloride is made by the action of concentrated H_2SO_4 on concentrated aqueous HCl or NaCl; HBr and HI may be made by catalytic reaction of $H_2 + X_2$ over platinized silica gel or, for HI, by interaction of iodine and boiling tetrahydronaphthalene. The gases are soluble in a variety of solvents, especially polar ones. The solubility in water is not exceptional; in moles of HX per mole of solvent at 0°C and 1 atm, the solubilities in water, 1-octanol, and benzene, respectively, are HCl, 0.409, 0.48, 0.39; HBr, 1.00, 1.30, 1.39; HI, 0.065, 0.173, 0.42.

The self-ionization is very small:

$$3HX \rightleftharpoons H_2X^+ + HX_2^-$$

Liquid HCl has been fairly extensively studied as a solvent, and many organic and some inorganic compounds that can be protonated dissolve giving conducting solutions:

$$B + 2HCl \rightleftharpoons BH^+ + HCl_2^-$$

The low temperatures required and the short liquid range are limitations, but conductimetric titrations are readily made.

Salts of the ion H_2Cl^+ have not been isolated, but crown ether salts of the HCl_2^- and HBr_2^- ions have been structurally characterized. These are angular with strong H-bonds, though the parameters vary with the cation.[44]

Nitric Acid

Nitric acid is made industrially by oxidation of ammonia with air over platinum catalysts. The resulting nitric oxide is absorbed in water in the presence of air to

[43]D. Mootz and K. Bartmann, *Z. Naturforsch.* **1991,** *46B,* 1659.
[44]J. L. Attwood *et al., Inorg. Chem.* **1990,** *29,* 467.

form NO_2, which is then hydrated. The normal concentrated aqueous acid ($\sim$70% by weight) is colorless but often becomes yellow as a result of photochemical decomposition, which gives NO_2:

$$2HNO_3 \xrightarrow{\;h\nu\;} 2NO_2 + H_2O + {}^1\!/_2O_2$$

The so-called fuming nitric acid contains dissolved NO_2 in excess of the amount that can be hydrated to $HNO_3 + NO$.

Pure nitric acid can be obtained by treating KNO_3 with 100% H_2SO_4 at 0°C and removing the HNO_3 by vacuum distillation. The pure acid is a colorless liquid or white crystalline solid; the latter decomposes above its melting point according to the previous equation for the photochemical decomposition, hence must be stored below 0°C.

The acid has the highest self-ionization of the pure liquid acids. The initial protolysis

$$2HNO_3 \rightleftharpoons H_2NO_3^+ + NO_3^-$$

is followed by rapid loss of water:

$$H_2NO_3^+ \rightleftharpoons H_2O + NO_2^+$$

so that the overall self-dissociation is

$$2HNO_3 \rightleftharpoons NO_2^+ + NO_3^- + H_2O$$

Pure nitric acid is a good ionizing solvent for electrolytes, but unless they produce the NO_2^+ or NO_3^- ions, salts are sparingly soluble.

In dilute aqueous solution, nitric acid is approximately 93% dissociated at 0.1 M concentration. Nitric acid of concentration below 2 M has little oxidizing power. The concentrated acid is a powerful oxidizing agent and, of the metals, only Au, Pt, Ir, and Re are unattacked, although a few others such as Al, Fe, Cu are rendered *passive*, probably owing to formation of an oxide film; magnesium alone can liberate hydrogen and then only initially from dilute acid. The attack on metals generally involves reduction of nitrate. *Aqua regia* ($\sim$3 vol. of conc. HCl + 1 vol. of conc. HNO_3) contains free chlorine and ClNO, and it attacks gold and platinum metals, its action being more effective than that of HNO_3 mainly because of the complexing function of chloride ion. Red fuming nitric acid contains N_2O_4. Some metals, notably tantalum, are quite resistant to HNO_3 but dissolve with extreme vigor if HF is added, to give TaF_6^- or similar ions. Nonmetals are usually oxidized by HNO_3 to oxo acids or oxides. The ability of HNO_3, especially in the presence of concentrated H_2SO_4, to nitrate many organic compounds is attributable to the formation of the nitronium ion, NO_2^+.

Gaseous HNO_3 has a planar structure (Fig. 2-6), although hindered rotation of OH relative to NO_2 probably occurs.

Perchloric Acid

This is available in concentrations 70 to 72% by weight. The water azeotrope with 72.5% of $HClO_4$ boils at 203°C, and although some Cl_2 is produced, which

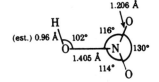

Figure 2-6   The structure of the nitric acid molecule in the vapor.

can be swept out by air, there is no hazard involved. The anhydrous acid is best prepared by vacuum distillation of the concentrated acid in the presence of the dehydrating agent $Mg(ClO_4)_2$; it reacts explosively with organic material. The pure acid, mp $-112°C$, has been structurally characterized.[45] It is nearly identical with the structure of ClO_3F (Chapter 13) and has three $Cl-O$ bonds 1.42 Å and one 1.61 Å, the latter being involved in connecting $HClO_4$ molecules into chains *via* weak $O\cdots H$ bridges. It decomposes at $-101.4°C$ to give $HClO_4\cdot0.25H_2O$ and Cl_2O_7. Numerous crystalline hydrates have been structurally characterized as well as anhydrous perchlorates such as $Cu(ClO_4)_2$.[46] The hot concentrated acid oxidizes organic materials vigorously or even explosively; it is a useful reagent for the destruction of organic matter, especially after pretreatment with, or in the presence of, H_2SO_4 or HNO_3. The addition of concentrated $HClO_4$ to organic solvents such as ethanol should be avoided, even if the solutions are chilled.

Perchlorate salts of organic or organometallic cations or complexes with organic ligands must be handled with *extreme* caution; substitution of ClO_4^- by $CF_3SO_3^-$ or other non-coordinating anions is *always* advisable.

Sulfuric Acid

This is prepared on an enormous scale by the lead chamber and contact processes. In the former, SO_2 oxidation is catalyzed by oxides of nitrogen (by intermediate formation of nitrosylsulfuric acid, $HOSO_2ONO$); in the latter, heterogeneous catalysts such as Pt are used for the oxidation. Pure H_2SO_4 is a colorless liquid that is obtained from the commercial 98% acid by addition first of SO_3 or oleum and then titration with water until the correct specific conductance or melting point is achieved.

The phase diagram of the $H_2SO_4-H_2O$ system is complicated, and eutectic hydrates such as $H_2SO_4\cdot H_2O$ (mp 8.5°C) and $H_2SO_4\cdot2H_2O$ (mp $-38°C$) occur. In pure crystalline H_2SO_4 there are SO_4 tetrahedra with $S-O$ distances 1.42, 1.43, 1.52, and 1.55 Å, linked by strong H-bonds. There is also extensive H-bonding in the concentrated acid. In the gas phase the structure is $O_2S(OH)_2$, which is essentially that in the liquid although the latter is extensively H-bonded.

Pure H_2SO_4 shows extensive self-ionization resulting in high conductivity. The equilibrium

$$2H_2SO_4 \rightleftharpoons H_3SO_4^+ + HSO_4^- \qquad K_{10°C} = 1.7 \times 10^{-4} \text{ mol}^2 \text{ kg}^{-2}$$

[45]A. Simon and H. Borrman, *Angew. Chem. Int. Ed. Engl.* **1988**, *27*, 1339.
[46]F. Favier *et al., J. Chem. Soc. Dalton Trans.* **1994**, 3119.

is only one factor, since there are additional equilibria due to dehydration:

$$2H_2SO_4 \rightleftharpoons H_3O^+ + HS_2O_7^- \qquad K_{10°C} = 3.5 \times 10^{-5} \text{ mol}^2 \text{ kg}^{-2}$$

$$H_2O + H_2SO_4 \rightleftharpoons H_3O^+ + HSO_4^- \qquad K_{10°C} = 1 \text{ mol}^2 \text{ kg}^{-2}$$

$$H_2S_2O_7 + H_2SO_4 \rightleftharpoons H_3SO_4^+ + HS_2O_7^- \qquad K_{10°C} = 7 \times 10^{-2} \text{ mol}^2 \text{ kg}^{-2}$$

Estimates of the concentrations in 100% H_2SO_4 of the other species present, namely, H_3O^+, HSO_4^-, $H_3SO_4^+$, $HS_2O_7^-$, and $H_2S_2O_7$ can be made; for example, at 25°C, HSO_4^- is 0.023 M.

Pure H_2SO_4 and dilute oleums have been greatly studied as solvent systems, but interpretation of the cryoscopic and other data is often complicated. Sulfuric acid is not a very strong oxidizing agent, although the 98% acid has some oxidizing ability when hot. The concentrated acid reacts with many organic materials, removing the elements of water and sometimes causing charring, for example, of carbohydrates. Many substances dissolve in the 100% acid, often undergoing protonation. Alkali metal sulfates and water also act as bases. Organic compounds may also undergo further dehydration reactions, for example:

$$C_2H_5OH \xrightarrow{H_2SO_4} C_2H_5OH_2^+ + HSO_4^- \xrightarrow{H_2SO_4} C_2H_5HSO_4 + H_3O^+ + HSO_4^-$$

Because of the strength of H_2SO_4, salts of other acids may undergo solvolysis, for example:

$$NH_4ClO_4 + H_2SO_4 \rightleftharpoons NH_4^+ + HSO_4^- + HClO_4$$

There are also examples of acid behavior. Thus H_3BO_3 reacts to give quite a strong acid:

$$H_3BO_3 + 6H_2SO_4 \rightleftharpoons B(HSO_4)_3 + 3H_3O^+ + 3HSO_4^-$$

$$B(HSO_4)_3 + HSO_4^- \rightleftharpoons B(HSO_4)_4^-$$

The addition of SO_3 to H_2SO_4 gives what is known as *oleum* or fuming sulfuric acid $(SO_3)_n \cdot H_2O$. The constitution of concentrated oleums is controversial, but with equimolar ratios the major constituent is pyrosulfuric (disulfuric) acid $(H_2S_2O_7)$. At higher concentrations of SO_3, Raman spectra indicate the formation of $H_2S_3O_{10}$ and $H_2S_4O_{13}$. Pyrosulfuric acid has higher acidity than H_2SO_4 and ionizes thus:

$$2H_2S_2O_7 \rightleftharpoons H_2S_3O_{10} + H_2SO_4 \rightleftharpoons H_3SO_4^+ + HS_3O_{10}^-$$

The acid protonates many materials; $HClO_4$ behaves as a weak base, and CF_3CO_2H is a nonelectrolyte in oleum.

Fluorosulfuric Acid

Fluorosulfuric acid is made by the reaction:

$$SO_3 + HF = FSO_3H$$

or by treating KHF_2 or CaF_2 with oleum at ~250°C. When freed from HF by sweeping with an inert gas, it can be distilled in glass apparatus. Unlike $ClSO_3H$, which is explosively hydrolyzed by water, FSO_3H is relatively slowly hydrolyzed. The acid is one of the strongest pure liquid acids and, as noted in Sect. 2-11, is used in superacid systems. An advantage over other acids is its ease of removal by distillation in vacuum. The extent of self-ionization

$$2FSO_3H \rightleftharpoons FSO_3H_2^+ + FSO_3^-$$

is much lower than for H_2SO_4 and consequently interpretation of cryoscopic and conductometric measurements is fairly straightforward.

In addition to its solvent properties, FSO_3H is a convenient laboratory fluorinating agent. It reacts readily with oxides and salts of oxo acids at room temperature. For example, K_2CrO_4 and $KClO_4$ give CrO_2F_2 and ClO_3F, respectively.

Trifluoromethanesulfonic Acid[47]

This very strong ($-H_o = 15.1$), useful acid (CF_3SO_3H, bp 162°C) is often given the trivial name "triflic" acid and its salts called "triflates." It is very hygroscopic and forms the monohydrate $CF_3SO_3H \cdot H_2O$, mp 34°C. Its salts are similar to perchlorates but non-explosive, and the $CF_3SO_3^-$ ion is less likely than ClO_4^- to be found disordered in crystal structures. The ion is also a useful leaving group in many organic and inorganic syntheses.

Related and useful protonating fluorocarbon acids are $(CF_3SO_2)_2CH_2$, $(CF_3SO_2)_2CHPh$, and $(CF_3SO_2)_2NH$ that also give non-coordinating anions like $CF_3SO_3^-$.[48]

Other Acids

Although "tetrafluoroboric" and "hexafluorophosphoric" acids do not exist as such, but only in the form of oxonium salts, for example, $H_5O_2^+BF_4^-$, in nonaqueous solvents such as carboxylic anhydrides or ethers there can be strong acid behavior. Tetrafluoroboric acid in diethyl ether is a useful strong acid, $[Et_2OH]^+BF_4^-$.

Trifluoroacetic acid,[49] $K_D = 0.66$ mol L^{-1}, forms a variety of hydrates such as $[H_5O_2]^+[H(O_2CCF_3)_2]^- \cdot 6H_2O$. Another useful strong acid is the solid Nafion H^+,[50] where Nafion is a perfluorinated ion exchange polymer.

Before leaving the topic of acids, the use of very strong bases to remove H^+ from very weak acids can be noted. These compounds are called *proton sponges*.[51]

[47]P. J. Stang and M. R. White, *Triflic Acid and its Derivatives, Aldrichchimica Acta* **1983,** *16,* 15.
[48]A. R. Siedle and J. C. Huffman, *Inorg. Chem.* **1990,** *29,* 3131.
[49]D. Mootz and M. Schilling, *J. Am. Chem. Soc.* **1992,** *114,* 7435.
[50]K. J. Cavell *et al., J. Chem. Soc., Dalton Trans.* **1992,** 1381.
[51]R. W. Alder, *Chem. Rev.* **1989,** *89,* 1215; B. Brzezinski *et al., J. Mol. Struct.* **1992,** *274,* 75.

One example is 1,8-bis(dimethylamino)naphthalene, which forms $N \cdots H^+ \cdots N$ H-bonds that are symmetrical and nearly linear. Non-ionic cage compounds such as $MeN{=}P(MeNCH_2CH_2)_3N^{52}$ are also very strong bases.

METAL HYDRIDES

Although virtually all the elements form hydrides, those of non-metallic elements such as boron hydrides, NH_3, PH_3, etc., are discussed in the appropriate chapters. Only metal hydrides are discussed in this section.

2-13 The Hydride Ion; Saline Hydrides

The tendency of the hydrogen atom to form the negative ion is much lower than for the more electronegative halogen elements. This may be seen by comparing the energetics of the formation-reactions:

$$\tfrac{1}{2}H_2(g) \longrightarrow H(g) \qquad \Delta H = 218 \text{ kJ mol}^{-1}$$

$$\tfrac{1}{2}Br_2(g) \longrightarrow Br(g) \qquad \Delta H = 113 \text{ kJ mol}^{-1}$$

$$H(g) + e \longrightarrow H^-(g) \qquad \Delta H = \text{-}67 \text{ kJ mol}^{-1}$$

$$Br(g) + e \longrightarrow Br^-(g) \qquad \Delta H = \text{-}345 \text{ kJ mol}^{-1}$$

$$\tfrac{1}{2}H_2(g) + e \longrightarrow H^-(g) \qquad \Delta H = 151 \text{ kJ mol}^{-1}$$

$$\tfrac{1}{2}Br_2(g) + e \longrightarrow Br^-(g) \qquad \Delta H = \text{-}232 \text{ kJ mol}^{-1}$$

Thus, owing to the endothermic character of the H^- ion, only the most electropositive metals — the alkalis and the alkaline earths — form saline or saltlike hydrides, such as NaH and CaH_2. The ionic nature of the compounds is shown by their high conductivities just below or at the melting point and by the fact that on electrolysis of solutions in molten alkali halides hydrogen is liberated at the *anode*.

X-ray and neutron diffraction studies show that in these hydrides the H^- ion has a crystallographic radius between those of F^- and Cl^-. Thus the electrostatic lattice energies of the hydride and the fluoride and chloride of a given metal will be similar. These facts and a consideration of the Born-Haber cycles lead us to conclude that *only* the most electropositive metals *can* form ionic hydrides, since in these cases relatively little energy is required to form the metal ion.

The hydrides and some of their physical properties are given in Table 2-7. The heats of formation compared with those of the alkali halides, which are about 420 kJ mol^{-1}, reflect the inherently small stability of the hydride ion.

For the relatively simple two-electron system in the H^- ion, it is possible to calculate an effective radius for the free ion, 2.08 Å. It is of interest to compare

[52]J. G. Verkade *et al.*, *J. Am. Chem. Soc.* **1993**, *115*, 5015.

Table 2-7   The Salt-like Hydrides and Some of Their Properties

Salt	Structure	Heat of Formation $-\Delta H_f(298\ K)$ (kJ mol^{-1})	M—H Distance (Å)	Apparent Radius of H$^-$ (Å)^a
LiH	NaCl type	91.0	2.04	1.36
NaH	NaCl type	56.6	2.44	1.47
KH	NaCl type	57.9	2.85	1.52
RbH	NaCl type	47.4	3.02	1.54
CsH	NaCl type	49.9	3.19	1.52
CaH$_2$	Slightly distorted hcp	174.5	2.33^b	1.35
SrH$_2$	Slightly distorted hcp	177.5	2.50	1.36
BaH$_2$	Slightly distorted hcp	171.5	2.67	1.34
MgH$_2$	Rutile type	74.5		1.30

^{a}See text.
^{b}Although half the H$^-$ ions are surrounded by four Ca^{2+} and half by three Ca^{2+}, the Ca—H distances are the same.

this with some other values, specifically, 0.93 Å for the He atom, ~0.5 Å for the H atom, 1.81 Å for the crystallographic radius of Cl$^-$, and 0.30 Å for the covalent radius of hydrogen, as well as with the values of the "apparent" crystallographic radius of H$^-$ given in Table 2-7.

The values in the table are obtained by subtracting the Goldschmidt radii of the metal ions from the experimental M—H distances. The value 2.08 Å for the radius of free H$^-$ is at first sight surprisingly large, being more than twice that for He. This results because the H$^-$ nuclear charge is only half that in He and the electrons repel each other and screen each other (~30%) from the pull of the nucleus. Table 2-7 indicates that the apparent radius of H$^-$ in the alkali hydrides never attains the value 2.08 Å and also that it decreases markedly with decreasing electropositive character of the metal. The generally small size is probably attributable in part to the easy compressibility of the rather diffuse H$^-$ ion and partly to a certain degree of covalence in the bonds.

Preparation and Chemical Properties

The saline hydrides are prepared by direct interaction at 300 to 700°C. The rates are in the order Li > Cs > K > Na, largely because of the activation energy term in the Arrhenius equation. Extremely active hydrides of Li, Na, and K can be made by interaction of hydrogen with BuLi + TMEDA in hexane (LiH) or of BuLi + M(OBu^t) + TMEDA in hexane (NaH and KH).

The saline hydrides are crystalline solids, white when pure but usually gray owing to traces of metal. They can be dissolved in molten alkali halides and on electrolysis of such a solution, for example, CaH$_2$ in LiCl + KCl at 360°C, hydrogen is released at the anode. They react instantly and completely with even the weakest acids, such as water, according to the reaction

$$MH + H^+ = M^+ + H_2$$

The standard potential of the H_2/H^- couple has been estimated to be -2.25 V, making H^- one of the most powerful reducing agents known.

The hydrides of Rb, Cs, and Ba may ignite spontaneously in moist air. Thermal decomposition at high temperatures gives the metal and hydrogen. Lithium hydride alone can be melted (mp 688°C) and it is unaffected by oxygen below red heat or by chlorine or dry HCl; it is seldom used except for the preparation of the more useful complex hydride $LiAlH_4$ discussed in Chapter 6. Sodium hydride is available as a dispersion in mineral oil; although the solid reacts violently with water, the reaction of the dispersion is less violent. It is used extensively in organic synthesis and for the preparation of $NaBH_4$. Sodium hydride acts as a strong base for hydrogen abstractions, but as a reducing agent it is slow and inefficient though improved by addition of nickel acetate as catalyst or by sodium *t*-amyloxide. It is soluble in aqueous hexamethylphosphoramide giving a blue solution (cf. Chap. 3). Calcium hydride reacts smoothly with water and is a useful source of H_2 (1000 cm^3/g) as well as a convenient drying agent for gases and organic solvents. Finally, it may be noted that the compound 1,8-naphthalenediylbis(dimethylborane) (2-VIII) reacts with KH in THF to give a bridged monohydride. The compound also removes H^- from borohydride and other hydridic compounds and has been called a "hydride sponge" by analogy with a "proton sponge" noted earlier, which is similar but with NMe_2 instead of BMe_2 groups. It also gives a bridged anion with F^-.

(2-VIII)

The metal hydrides of a more covalent nature such as those of Mg, Al, and Ga are discussed under the chemistry of those elements.

Transition Metal Hydrides

There is a vast and rapidly increasing literature on transition metal hydrides, their chemical reactions and their role in hydrogen transfers, as in catalytic reactions of alkenes and other substrates. We can only outline the present knowledge which includes also what are called "agostic H atoms" and new types of H-bonding as well as compounds with $M—H$ and $M(H_2)$ bonds.

2-14 Classical Hydrides with M—H Bonds[53]

The first complexes to be made, by W. Hieber in the 1930s, were $HCo(CO)_4$ and $H_2Fe(CO)_4$, but their structures and the nature of the $M—H$ bond were not known

[53]J. A. Martinho Simões and J. L. Beauchamp, *Chem. Rev.* **1990**, *90*, 629; P. Dedieu, ed., *Transition Metal Hydrides*, VCH, New York, 1992.

until much later. The discovery of Cp_2ReH by G. Wilkinson in 1955 initiated modern studies as this molecule allowed the $M-H$ stretching region in the ir spectrum to be identified, at ~ 2000 cm^{-1}, while the ^1H nmr spectrum showed a unique line on the high field side of Me_4Si. The molecule also demonstrated for the first time that metal atoms in complexes could be protonated as the compound is as basic as ammonia:

$$Cp_2ReH + H^+ \rightleftharpoons Cp_2ReH_2^+$$

There are now hundreds if not thousands of compounds of the d group elements, lanthanides and actinides. All manner of ligands may be present such as CO, PR_3, η^5-$C_5H_5^-$, arenes, CN^-, halides, pyrazolylborates, and so on. Some compounds will be noted under the chemistry of the particular element involved.

Metal-hydride compounds having also NO as ligand are fairly common,[54] an example being $Cp^*W(NO)(CH_2SiMe_3)H$ which also illustrates the possibility of $MH(CR_3)$ complexes. Rather uncommon are peroxo compounds like $[OsH(\eta^2-O_2)L]$-BPh_4, $L = C_6H_{11}PCH_2CH_2PC_6H_{11}$,[55] and alkoxo species[56] such as $[W_2(\mu-H)(OPr^i)_7]_2$, $W_4(H_2)_2(OPr^i)_{14}$, and $W_6(\mu-H)_4H(\mu-CPr^i)(\mu-OPr^i)_7(OPr^i)_5$. Complexes with simple amines can be stable, an early example being $[RhH(NH_3)_5]^{2+}$; a similar aqua complex $[CrH(OH_2)_5]^{2+}$ has only a short lifetime. A few paramagnetic hydrides are known, e.g., $Ir(H_2)_2(Cl)_2(PPr_3^i)_2$ and $Ta(H_2)_2(Cl)_2(dmpe)_2$.

Synthetic Methods

The synthesis of hydrido species can involve interactions with molecular hydrogen, hydride sources such as $NaBH_4$ or $LiAlH_4$, alcohols and other organic sources. Some representative syntheses are as follows.

$$\textit{trans-}PtCl_2(PEt_3)_2 \xrightarrow{\text{KOH-EtOH}} \textit{trans-}PtClH(PEt_3)_2 \qquad (3)$$

$$FeI_2(CO)_4 \xrightarrow[\text{THF}]{\text{NaBH}_4} FeH_2(CO)_4 \qquad (4)$$

$$Rh^{3+}(aq) + SO_4^{2-} + NH_3(aq) + Zn \longrightarrow [RhH(NH_3)_5]SO_4 \qquad (5)$$

Methods that include direct reaction with dihydrogen are illustrated by the following:

$$RuCl_2(PPh_3)_3 + H_2 + Et_3N \longrightarrow RuCl(H)(PPh_3)_3 + Et_3NHCl \qquad (6)$$

$$Cp_2ZrMe_2 + 2H_2 \xrightarrow{60\,°C/60\ atm.} Cp_2Zr(H)_2 + 2CH_4 \qquad (7)$$

$$WMe_6 + 3PMe_3 + 5H_2 \xrightarrow{1\ atm.} WH_4(PMe_3)_3 + 6CH_4 \qquad (8)$$

[54] P. Legzdins *et al., Organometallics* **1995,** *14,* 2543.
[55] A. Mezzatti *et al., J. Chem. Soc., Chem. Commun.* **1994,** 1597.
[56] M. H. Chisholm *et al., Chem. Commun.* **1996,** 1331; *J. Am. Chem. Soc.* **1995,** *117,* 1974; M. H. Chisholm, *Chem. Soc. Rev.* **1995,** *24,* 79.

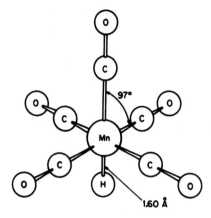

Figure 2-7   The structure of HMn(CO)$_5$.

$$(Ph_3P)_3Rh[N(SiMe_3)_2] + H_2 \xrightarrow{\text{1 atm.}} RhH(PPh_3)_3 + (Me_3Si)_2NH \tag{9}$$

$$Ru(COD)(COT) + H_2 \xrightarrow{PCy_3} trans\text{-}Ru(H)_2(H_2)_2(PCy_3)_2 + C_8H_{16} \tag{10}$$

Structures and Spectra

Many structures having terminal and/or bridging H atoms have been determined by X-ray diffraction. Although neutron diffraction is preferable, large, well formed crystals are usually required and far fewer structures have been determined this way. An early structure is shown in Fig. 2-7.

Spectroscopic data are useful. In the infrared MH stretches are in the region 2300–1600 cm^{-1} with bends appearing from 900–600 cm^{-1}. Assignments can be confirmed by deuterium substitution. Nuclear magnetic resonance data can be obtained in a variety of solvents including strong acids like H$_2$SO$_4$. The lines are usually up-field from SiMe$_4$ from -1 to -35 ppm although in some cases lines have been seen at -60 ppm and also, much less commonly, down-field as far as 22.87 ppm in Cp*TaCl$_2$(CH$_2$CMe$_3$)H.[57] So far there appears to be little correlation between spectroscopic data and M—H bond lengths; the shifts doubtless depend on the nature of the whole molecule.

The nmr measurement of dipolar relaxation (T_1) can give data that can be correlated with M—H bond distances, however, and has been useful in distinguishing between MH and M(H$_2$) species as discussed later. Enhancement of nmr signals by parahydrogen induced polarization can allow detection of isomers in low concentrations.[58]

In phosphorus ligand complexes coupling of ^{1}H and ^{31}P often gives useful structural information but in 5 and 7 or higher coordinate species nonrigidity is commonly observed, for example, in Cr(H)$_2$[P(OMe)$_3$]$_5$.

[57] R. R. Schrock *et al.*, *J. Am. Chem. Soc.* **1995**, *117*, 6609.
[58] S. B. Duckett *et al.*, *Chem. Commun.* **1996**, 383.

Chemically, M—H bonds can be detected by the reaction with CCl_4 that gives $CHCl_3$ and M—Cl. The hydride ligand has both a very high *trans* influence (shown by long M—L distances in *trans* H—M—L compounds) and also a strong *trans* effect, which means that ligands in the *trans* positions to H commonly undergo facile substitution. The factors that affect the protonic or hydridic behavior of compounds with M—H bonds are generally similar to those determining this behavior in non-transition metal compounds as exemplified by BH_3, CH_4, NH_3, OH_2, and FH. The most important factors in complexes are the strength of the M—H bond and the nature of the ligands. For a reaction

$$L_nMH \rightleftharpoons L_nM^- + H^+$$

the key is the stability of the conjugate base, that is, the electronic structure of the anion. Strong π bonding of the ligand L will stabilize ML_n^- by delocalization of negative charge and this is why the most acidic hydrides are those with CO or PF_3 as ligands. Substitution of CO by a poorer π acid can dramatically reduce the pK_A as the following examples in H_2O show:

$HCo(CO)_4$	Strong acid	$HV(CO)_6$	Strong acid
$HCo(CO)_3(PPh_3)$	$pK_A = 7.0$	$HV(CO)_5(PPh_3)$	$pK_A = 6.8$

There are, of course, hydrides that are essentially neutral, e.g., $HRuCl(CO)_2(PPh_3)_2$, while the basic hydrides either have lone pairs, as in Cp_2ReH mentioned earlier, or are electron rich as in d^8 and d^{10} complexes, such as those of Ru^0, Rh^I, or Pt^{II}, that can undergo not only protonations but oxidative-additions of a variety of molecules.

It may also be noted that in all discussions of acidity, it must always be remembered that the acidity depends on the solvent and the extent, consequently, of dissociation. The hydrides of early transition metals tend to be hydridic; 2nd and 3rd row d metal compounds are usually less acidic than their 1st row analogues.

Hydrogen Bridges

Early evidence for bridging H atoms came from nmr spectra of the cation $[Cp(CO)_3W]_2H^+$, which showed satellites due to the coupling of H with both tungsten atoms symmetrically (^{184}W, spin ½). Many bridged species, both homo- and heterometallic, have been confirmed by spectroscopic and diffraction methods. The main types are given in the following:

More complicated bridge systems have been found in yttrium and lanthanide compounds, for example, (2-IX) and (2-X).

(2-IX)   (2-X)

The bridges $(\mu$-H$)_4$ and $(\mu_4$-H$)$ occur in $(Cp*Ru)_2(\mu$-H$)_4$ and $[Mo_4(\mu_4$-H$)(OCH_2$-Bu$^t)_{12}]^-$,[59] respectively, and $(\mu_6$-H$)$ is known in $[Co_6(\mu_6$-H$)(CO)_{15}]^-$. Many other alkoxide hydrides are known.[60]

All single bridges are *bent* with angles ranging from *ca.* 78° to 124°; the M—H—M asymmetric stretches are around 1700 cm^{-1}.

A representative complex, studied by neutron diffraction, is $(\eta^5$-Cp*Co$)_2$-$(\mu$-H$)_3$ which has a short Co—Co bond (2.253 Å).[61] A few cases of bridges *unsupported* by M—M bonds are known, e.g., $[(\mu$-H$)W_2(CO)_{10}]^-$ and $[(\mu$-H$)_2W_3(CO)_{13}NO]^-$.[62]

Some reactions of M—H bonds are shown in Fig. 2-8. Most involve some type of cleavage to give H·, H$^+$, or H$^-$. The energies for homolytic and heterolytic cleavages[63] have been determined by a variety of methods and much data is available.[64]

Hydrogen transfers can be induced thermally or photochemically and the kinetics of transfers from compounds such as HMn(CO)$_5$ and Cp*Mo(CO)$_3$H have been determined.[65]

Polyhydrides[66]

These are metal compounds that have 3 to 9 hydrogen atoms which may be neutral, cationic, or anionic. They usually have tertiary phosphines or η^5-C$_5$H$_5$ as ligands but some homoleptic species are known. Some examples are given in Table 2-8.

Whether or not a given compound has only M—H bonds or whether some of the hydrogen atoms are present as M(H$_2$) groups has caused much study and reformulation in a number of cases; an example is a "tetrahydride" shown to be Ir(H$_2$)$(\eta^2$-H$_2$)Cl(PPr$_3^i$).

Polyhydrides can be obtained in different oxidation states up to the highest for a given element; many, according to ^{1}H nmr spectra, are non-rigid. Reductive elimination of H$_2$ can give lower oxidation state compounds of greater reactivity due to coordinative unsaturation.

[59]M. H. Chisholm *et al., J. Am. Chem. Soc.* **1994,** *116,* 389.
[60]M. H. Chisholm, *Chem. Soc. Rev.* **1995,** *24,* 79.
[61]T. F. Koetzle *et al., Inorg. Chem.* **1996,** *35,* 2698 (lists data for other M(μ-H)$_3$M species).
[62]J. T. Lin *et al., Inorg. Chem.* **1994,** *33,* 1948.
[63]R. M. Bullock, *Comments Inorg. Chem.* **1991,** *12,* 10.
[64]D. Wang and R. J. Angelici, *J. Am. Chem. Soc.* **1996,** *118,* 935.
[65]T-Y. Cheng and R. M. Bullock, *Organometallics* **1995,** *14,* 4032.
[66]Z. Lin and M. B. Hall, *Coord. Chem. Rev.* **1994,** *135/136,* 845 (classical and non-classical polyhydrides, 60 refs).

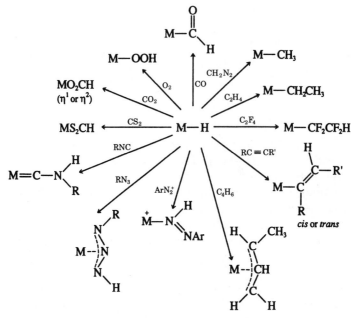

Figure 2-8 Some intramolecular hydrogen transfer or insertion reactions.

Polyhydrido Anions

There is now an extensive array of solids of the general composition A_nMH_m where A is an alkali cation usually Li^+ or Na^+. The first well characterized anion ReH_9^{2-} is shown in Fig. 2-9.

The solid compounds are commonly made by high temperature interaction of the metal powders with saltlike hydrides or with Mg or alkali metals under high pressures of H_2 (or D_2).[67] Typical of the ions so produced are the planar $[PtH_4]^{2-}$ and $[RhH_4]^{2-}$, linear $[PtH_2]^{2-}$, the octahedral $[RuH_6]^{2-}$ and $[RhH_6]^{2-}$. The electronic structures and bonding in compounds such as Mg_2RuH_4 have been considered in

Table 2-8 Examples of Polyhydrides Having M—H Bonds

Compound	Geometry
$WH_6(PR_3)_3$	Tricapped trigonal prism
$ReH_4(PPh)_2$ [a]	Monocapped square antiprism
$ReH_5(PMe_2Ph)_3$	Dodecahedral
$OsH_5(PMe_2Ph)_3^+$ [b]	″
$OsH_6(PR_3)_2$	″

[a] J. L. Spencer *et al., Organometallics* **1995,** *14,* 568; R. H. Crabtree *et al., Inorg. Chem.* **1996,** *35,* 695.
[b] K. G. Caulton *et al., Inorg. Chem.* **1994,** *33,* 4996.

[67] W. Bronger *et al., Z. anorg. allg. Chem.* **1996,** *622,* 462; *Angew. Chem. Int. Ed. Engl.* **1994,** *33,* 1112; *J. Alloys Compd.* **1995,** *229,* 1 (Review on synthesis and structure); K. Yvon *et al., J. Alloys Compd.* **1995,** *204,* L5; R. O. Moyer, Jr. *et al., J. Solid State Chem.* **1996,** *121,* 56.

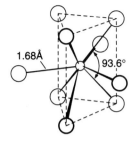

Figure 2-9 The structure of the ReH_9^{2-} ion. The hydrogen atoms define a trigonal prism centered on all rectangular faces.

detail.[68] K_3MnH_5 [69] contains both H^- ions and tetrahedral MnH_4^{2-} ions. Finally, it is well known that palladium in particular and some other metals can absorb considerable quantities of hydrogen.[70]

Carbonyl Hydrides and Hydrido Anions

As noted earlier $HCo(CO)_4$ and $H_2Fe(CO)_4$ were the first known hydrides. Carbonyl hydrides and carbonylate anions have been intensively studied in part because they are intermediates in many metal catalyzed reactions involving CO and H_2. The structure of $HMn(CO)_5$ is shown in Fig. 2-7.

Some compounds and their properties are shown in Table 2-9. The iron and cobalt carbonyl hydrides form pale yellow solids or liquids at low temperatures and in the liquid phase begin to decompose above ~ -10 and $-20°C$, respectively; they are relatively more stable in the gas phase, however, particularly when diluted with carbon monoxide. They have revolting odors and are readily oxidized by air. The carbonyl hydride, $HMn(CO)_5$, is appreciably more stable.

The carbonyl hydrides are often acidic, but pK_a values are solvent dependent. For example, while for $HCo(CO)_4$ in H_2O the pK_a is < 2, in MeCN it is 8, comparable to that of HCl in MeCN.

The hydrides are usually obtained by acidification of carbonylate or substituted carbonylate anions. In the case of highly reduced anions like $Na_3Ta(CO)_5$, interaction with BH_4^- in EtOH-THF may be adequate to give $[HM(CO)_5]^{2-}$.

Polynuclear cluster hydrides can be made similarly, but special methods may be used, for example,

$$Fe(CO)_5 + Et_3N \xrightarrow{H_2O, 80°C} [Et_3NH][HFe_3(CO)_{11}]$$

$$[HFe_3(CO)_{11}]^- + Fe(CO)_5 \longrightarrow [HFe_4(CO)_{13}]^- + 3CO$$

$$[HRu(CO)_4]^- \xrightarrow{CO, H_2O} [HRu_3(CO)_{11}]^-$$

$$[HFe(CO)_4]^- + (THF)Mo(CO)_5 \longrightarrow [HFeMo(CO)_9]^-$$

[68] G. J. Miller *et al.*, *Inorg. Chem.* **1994**, *33*, 1330.
[69] W. Bronger *et al.*, *Z. anorg. allg. Chem.* **1996**, *22*, 1145.
[70] T. B. Flanagan and Y. Sakamoto, *Platinum Metals Rev.* **1993**, *37*, 26. Y. Fukai, *The Metal-Hydrogen System*, *Springer Series* in *Materials Science, 21*, Springer, New York, 1993.

Table 2-9 Some Carbonyl Hydrides and Their Properties[a]

Compound	Form	Mp (°C)	M−H Stretching Frequency (cm^{-1})	^1H nmr δ (ppm)	Comment
$HMn(CO)_5$	Colorless liquid	−25	1783	−7.5	Stable liquid at 25°C
$H_2Fe(CO)_4$	Yellow liquid, colorless gas	−70		−11.1	Decomposes at −10°C giving H_2 + (red)$H_2Fe_2(CO)_8$
$H_2Fe_3(CO)_{11}$	Dark red liquid			−15	
$HCo(CO)_4$	Yellow liquid, colorless gas	−26	~1934	−10	Decomposes above mp giving H_2 + $CO_2(CO)_8$
$HW(CO)_3(\eta\text{-}C_5H_5)$	Yellow crystals	69	1854	−7.5	Stable for a short time in air
$[HFe(CO)_3(PPh_3)_2]^+$	Yellow			−7.6	Formed from $Fe(CO)_3(PPh_3)_2$ in conc H_2SO_4

[a]For pK_a values see U. Koelle, *New J. Chem.* **1992**, *16*, 157.

The anionic hydrido carbonyls have been much studied because of their intermediacy in catalytic processes like the water gas shift reaction, reduction of organic substances, and other reactions. Thus chromium group hydrides $[HM(CO)_4PR_3]^-$ are highly efficient H-transfer agents and for reactions with alkyl halides a reactivity order was established: cis-$[HW(CO)_4PR_3]^-$ > cis-$[HCr(CO)_4PR_3]^-$ > $[HW(CO)_5]^-$ > $[CpV(CO)_3H]^-$ > $[HCr(CO)_5]^-$ > $[HRu(CO)_4]^-$ > trans-$HFe(CO)_3PR_3$ >> $[HFe(CO)_4]^-$. Cluster anions, most of which are carbonyl or substituted carbonyl compounds, may have both terminal and bridging groups and in some cases, such as the ion $[HCo_6(CO)_{15}]^-$ made by protonation of $[Co_6(CO)_{15}]^{2-}$, a neutron diffraction study indicates that the H atom is in the center of a Co_6 unit. Infrared and inelastic neutron scattering spectra suggest that the μ_6-H coordination is highly sensitive to the surroundings such as the nature of the cations and that some of the H atoms may be μ_2-H or μ_3-H inside or outside the Co cluster.[71]

In rhodium clusters, e.g., $[Rh_{13}(CO)_{14}H_{5-n}]^{n-}$ ($n = 2$ and 3), only one of the H atoms is encapsulated and ^1H nmr spectra show that this atom interacts with all 13 Rh atoms (^{103}Rh, spin ½) and is thus migrating within the cluster. Proton transfer reactions of interstitial hydrides have been studied[72] and more complex heteronuclear species[73] can be made by reactions such as

$$[Ir(H)_2(Me_2CO)_2(PPh_3)_2]SbF_6 + Re(H)_7(PPh_3)_2 \xrightarrow[-Me_2CO]{-H_2} [(PPh_3)_2(H)_3Re(\mu-H)_3Ir(H)(PPh_3)_2]SbF_6$$

(11)

2-15 η^2-Dihydrogen Complexes[74,75]

The possibility of the bonding of the hydrogen molecule to transition metals was considered many years ago by E. Singleton in connection with the nature of "$RuH_4(PPh_3)_2$," which readily lost H_2 in solution. The compound "$FeH_4(PR_3)_2$" acted similarly and showed an unusual ir band at ca. 2400 cm^{-1}. In addition the oxidative-addition reactions of many d^8 complexes, reactions such as

$$Ir^ICl(CO)(PPh_3)_2 + H_2 \rightleftharpoons Ir^{III}(H)_2Cl(CO)(PPh_3)_2$$

implied the close approach or actual bonding of H_2 before H—H cleavage could occur, as in the general case

$$L_nM + H_2 \rightleftharpoons L_nM(H_2) \rightleftharpoons L_nM(H)_2$$

In 1984, G. J. Kubas isolated the tungsten(0) complex $W(\eta^2\text{-}H_2)(CO)_3(PPr^i_3)_2$ by reduction of a higher oxidation state halide under hydrogen. The core structure of the compound is shown in Fig. 2-10.

[71]J. Eckert et al., Inorg. Chem. 1989, 28, 4055.
[72]R. J. Weberg and J. R. Norton, J. Am. Chem. Soc. 1990, 112, 1105.
[73]X-L. Luo et al., Inorg. Chem. 1993, 32, 2118 and references therein.
[74]R. H. Crabtree, Angew. Chem. Int. Ed. Engl. 1993, 32, 789.
[75]D. M. Heinekey and W. Oldham Jr., Chem. Rev. 1993, 93, 913 (153 references); P. G. Jessop and E. Morris, Coord. Chem. Rev. 1992, 121, 155; R. H. Crabtree, Acc. Chem. Res. 1990, 23, 95; K. W. Zilm and J. M. Millar, Adv. Magn. Opt. Reson. 1990, 15, 163; G. J. Kubas, Acc. Chem. Res. 1988, 21, 120.

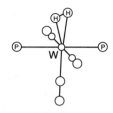

Figure 2-10   The core structure of $W(\eta^2\text{-}H_2)(CO)_3(PPr^i_3)_2$.

The subsequent discovery of more $\eta^2\text{-}H_2$ compounds has led to reexamination of many compounds previously described as polyhydrido species, and some have been reformulated as containing at least some of the hydrogen as $\eta^2\text{-}H_2$. Methods of distinguishing between $M(\eta^2\text{-}H_2)$ and $M(H)_2$ are the following:

1. X-ray crystallography, but this is often unreliable because of the very large uncertainties in locating hydrogen atoms.

2. Neutron diffraction, which is limited by the usual requirement that rather large single crystals be grown.

3. The measurement of longitudinal relaxation times (T_1) in the ^1H nmr spectra.[76] The basic idea is that the closer two protons are together the shorter T_1. However, T_1 depends on magnetic field strength and on temperature as well, so that it is important to make comparisons at a specified magnetic field strength and to determine the minimum value of T_1 as a function of temperature. It has been proposed that the $M(\eta^2\text{-}H_2)$ description is appropriate when $T_1(\text{min})$ is <40 ms and the $M(H)_2$ description when $T_1(\text{min})$ is >100 ms, both at 250 MHz. Between these values there is presumably an intermediate degree of dissociation of the H_2 molecule.

4. The strength of H to D coupling can give very useful information on the H to D distance.[77]

5. The rotation of H_2 about the $M\cdots H_2$ axis occurs as for C_2H_4 in $M(\mu\text{-}C_2H_4)$ complexes, and these barriers can be measured by inelastic neutron scattering[78] or by nmr at low temperatures. The barriers are typically in the range of 2–15 kJ mol^{-1} although slightly lower as well as higher ones (e.g., 46 kJ mol^{-1} in $[Cp_2NbPR_3(\eta^2\text{-}H_2)]^+$) are also found.

6. Infrared spectra for hydrido complexes usually show M—H stretches in the 1600–2000 cm^{-1} range, whereas in some $\eta^2\text{-}H_2$ complexes H—H stretching vibrations may be seen in the range 2500–3100 cm^{-1} (cf. 4000 cm^{-1} for H_2 itself).

The bonding in an $M(\eta^2\text{-}H_2)$ unit is believed to have a strong formal resemblence to the μ-bonding of metal olefin complexes.[79] In general, there is a combination of donation from the σ bonding orbital of H_2 and back-donation from a suitable metal

[76]H. Berke *et al., Inorg. Chem.* **1993,** *32,* 3270, 3628.
[77]N. S. Hush *et al., J. Am. Chem. Soc.* **1996,** *118,* 3753.
[78]F. A. Jalon *et al., J. Am. Chem. Soc.* **1995,** *117,* 10123; J. Eckert *et al., Inorg. Chem.* **1996,** *35,* 1292.
[79]H. F. Schaefer III *et al., J. Am. Chem. Soc.* **1996,** *118,* 870; T. Ziegler *et al., J. Am. Chem. Soc.* **1995,** *117,* 11482.

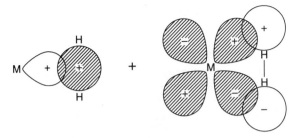

Figure 2-11   The two components in the bonding of an M(η^2-H$_2$) unit.

d orbital to the σ antibonding orbital of H$_2$. This picture is shown in Fig. 2-11. Some direct evidence for the back-bonding is provided by Mössbauer spectra for compounds of iron.[80] The greater the contribution of either component (or of both) the more the H—H distance should increase relative to that in H$_2$ itself (0.75 Å). Beyond a certain point the system will be better described in terms of the other limit, M(H)$_2$.

Table 2-10 lists some M(η-H$_2$) complexes in which the H—H distances remain rather short. There are also "stretched" species, with considerably longer distances, but still correctly considered to be M(η-H$_2$) complexes. Consider, for example, the comparison between ReH$_9^{2-}$ and ReH$_7$[P(p-C$_6$H$_4$Me)$_3$]$_2$. In the latter there is one H to H distance of 1.36 Å and thus these two hydrogen atoms are best regarded as a η^2 ligand, giving the Re atom a coordination number of 8. In the ReH$_9^{2-}$ ion, however, all H to H distances are >2 Å and this is a 9-coordinate nonahydride.

There is a series of osmium complexes in which there is a wide range of J$_{HD}$ values,[81] indicative of considerable variation in the H—H distance. In general the shorter H—H distances seem to be found in complexes with more strongly π-acidic ligands. A very short distance is found in Cr(CO)$_3$(PPr$_3^i$)$_2$(H$_2$), which loses H$_2$ very readily.[82] Another example is the original Kubas complex (Fig. 2-10); in which, again there is a very short H—H distance. This compound effervesces on contact with water[83]:

$$W(\eta^2\text{-}H_2)(CO)_3(PPr^i_3)_2 \xrightarrow[\text{THF}]{+\,H_2O} W(OH_2)(CO)_3(PPr^i_3)_2 + H_2 \qquad (12)$$

In general the tendency to loss of H$_2$ is correlated with the shortness of the H—H distance and this in turn depends on the other ligands present.[84] Cationic species seem to be more resistant to loss of H$_2$ than neutral ones. For example [CpRu (dmpe)(η^2-H$_2$)]$^+$ loses H$_2$ only slowly on refluxing in MeCN.

Protonation of M—H by acids[85] often leads to the prompt evolution of H$_2$, but,

[80]R. H. Morris and M. Schlaf, *Inorg. Chem.* **1994**, *33*, 1725; T. F. Koetzle *et al.*, *J. Am. Chem. Soc.* **1994**, *116*, 7677.

[81]H. Taube *et al.*, *J. Am. Chem. Soc.* **1994**, *116*, 9506, 4352, 2874; **1993**, *115*, 2545.

[82]G. J. Kubas *et al.*, *Inorg. Chem.* **1994**, *33*, 2954.

[83]G. J. Kubas *et al.*, *Organometallics* **1992**, *11*, 3390.

[84]T. Le-Husebo and C. M. Jensen, *Inorg. Chem.* **1993**, *32*, 3797.

[85]R. A. Henderson and K. E. Oglieve, *J. Chem. Soc., Dalton Trans.* **1993**, 3431; K. G. Caulton *et al.*, *Inorg. Chem.* **1995**, *34*, 2894, 1788; *J. Am. Chem. Soc.* **1995**, *117*, 9473; R. M. Bullock *et al.*, *Organometallics* **1996**, *15*, 2504.

Table 2-10   Representative H—H Distances (Å) Using Different Methods[a]

	H—H Distance		
Complex	X-ray Diffraction	Neutron Diffraction	NMR
$Cr(H_2)(CO)_3(PR_3)_2$	0.67^b		0.85^c
$Mo(H_2)(CO)(dppe)_2$		0.74	0.88
$W(H_2)(CO)_3(PPr^i_3)_2$	0.75	0.82	0.89
$[FeH(H_2)(dppe)_2]^+$	0.87	0.82	0.90
$[Cp(H_2)Ru(dppe)]^+$			1.02

[a]Adapted from data kindly provided by Dr. G. J. Kubas, Los Alamos and from G. J. Kubas *et al., Inorg. Chem.* **1994**, *33*, 2954, where an additional table gives IR frequencies (ν_{H-H} range 2690–3080 cm^{-1}), rotational barriers (kJ mol^{-1}, e.g., $Mo(H_2)(CO)_3(PCy_3)$, 5.52; $Mo(H_2)(CO)(dppe)_2$, 2.92), enthalpy of H—H bonding [ΔH, kJ mol^{-1}, e.g., $Mo(H_2)(CO)_3(PCy_3)$, −27.2; $W(H_2)(CO)_3(PCy_3)$, −39.3] and values for J_{HD} (Hz range 21–35).
[b]$R = Pr^i$.
[c]$R = Cy$.

in principle, intermediates should be considered:

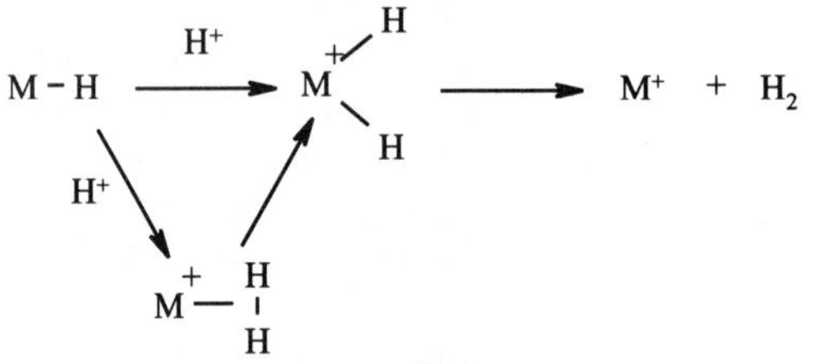

(2-XI)

In general, protonation of M—H is a way to synthesize $M(H_2)$ compounds. In the case of $Cp*Os(CO)_2H$, protonation was actually shown to give an equilibrium mixture of the $Os(H)_2^+$ and $Os(H_2)^+$ species. Protonation can even occur with an acid as weak as methanol.[86]

$$Ru(H)_2(dmpe)_2 \xrightarrow{MeOH} trans\text{-}[RuH(\eta^2\text{-}H_2)(dmpe)_2]^+ + MeO^- \qquad (13)$$

Of course, direct introduction of H_2 into a suitable receptor is a way, though not a general one, to prepare η^1-H_2 compounds. The following reaction[87] is a case in point:

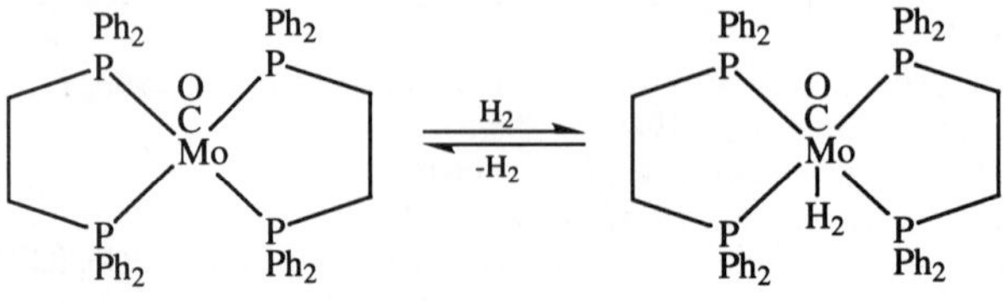

[86]L. D. Field *et al., Inorg. Chem.* **1994**, *33*, 2009.
[87]G. J. Kubas *et al., J. Am. Chem. Soc.* **1993**, *115*, 569.

It is a striking indication of the sensitivity of the mode of binding of hydrogen to metal atoms that if reactions similar to the above are attempted with ligands having Et or Bu^i groups instead of Ph on the phosphorus atoms (meaning that the diphosphine ligands are more basic) the products are 7-coordinate molecules with two hydrido ligands, e.g., $Mo(depe)_2(CO)(H)_2$.

More commonly, the H_2 receptor has to be generated *in situ* by reduction as in the following examples:

$$TcCl_4(PPh_3)_4 \xrightarrow{\text{Zn, dppe}} TcCl(dppe)_2 \underset{}{\overset{H_2}{\rightleftharpoons}} TcCl(\eta^2\text{-}H_2)(dppe)_2 \qquad (14)$$

$$ReCl_5 + 4Na + 2dppe \xrightarrow[H_2]{THF} ReCl(\eta^2\text{-}H_2)(dppe)_2 \qquad (15)$$

Finally, it may be noted that η^2-H_2 complexes can display acidity.[88] While the pK of H_2 itself is estimated to be *ca.* 35 in THF, coordination can drastically lower this. For example, *trans*-$[Os(H_2)(dppe)_2(MeCN)](BF_4)_2$ is markedly acidic (pK_a = -2), dissociating to give H^+ and $[OsH(dppe)_2(MeCN)]^+$.[89]

2-16 Agostic C—H···M Interactions and Others

It was first shown in the early 1970s that whenever metal atoms having fewer than the number (usually 18) of valence shell electrons normally required could do so, they would interact with a nearby C—H bond to obtain a share in the C—H bonding pair.[90] Schematically, this is shown in 2-XII. Later workers, in the 1980s, discovered additional examples and proposed the term *agostic* (from Greek, drawing towards) to describe such bonding.[91] It is now widely recognized to be important,[92] and two classic examples are shown in Fig. 2-12.

(2-XII)

It is to be emphasized that the nature of an agostic C—H···M interaction is different from that of a hydrogen bond. In the agostic case the C—H bond *supplies electrons* to the metal atom, whereas in a hydrogen bond, X—H···Y, the X—H partner *seeks electrons* on the responding atom Y. Put differently, agostic interactions are 3c-2e bonds, whereas hydrogen bonds are 3c-4e bonds.

The energies of agostic interactions vary, and many are weak and dynamic, i.e., they dissociate and reform rapidly at room temperature, although some are

[88]R. H. Morris *et al.*, *J. Am. Chem. Soc.* **1994**, *116*, 3375; *Organometallics* **1993**, *12*, 3808.
[89]R. H. Morris *et al.*, *Organometallics* **1996**, *15*, 2270.
[90]F. A. Cotton and R. L. Luck, *Inorg. Chem.* **1989**, *28*, 3210.
[91]M. Brookhart *et al.*, *Prog. Inorg. Chem.* **1988**, *36*, 1.
[92]M. Etienne *et al.*, *J. Chem. Soc., Chem. Commun.* **1994**, 1661; *Organometallics* **1994**, *13*, 410; M. L. H. Green *et al.*, *J. Chem. Soc., Dalton Trans.* **1992**, 3077; H. E. Selman and J .S. Merola, *J. Am. Chem. Soc.* **1991**, *113*, 4008; J. B. Sheridan *et al.*, *J. Chem. Soc., Dalton Trans.* **1995**, 931; X-L. Luo *et al.*, *Inorg. Chem.* **1995**, *34*, 6538.

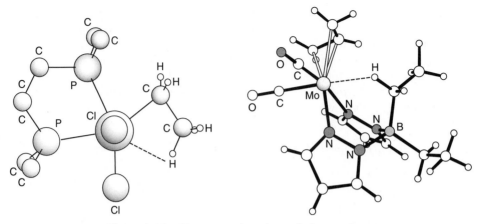

Figure 2-12 Two examples of agostic interactions.

strong enough to be slow on the nmr time scale.[93] Nuclear magnetic resonance evidence for agostic interactions is sometimes provided by a low value for the $^1J_{CH}$ coupling constant. While most examples of agostic interactions involve *d*-block metal atoms, the *lanthanide elements* can also show $M{\cdots}H{-}C$[94] and $M{\cdots}H{-}Si$[95] agostic bonding. Extensive studies have been made on many other $X{-}H{-}M$ interactions, where X can be Si,[96] Sn, B,[97] P, S, and there are also $C{-}H{-}M$, $M{-}H{-}M$, and $M{-}H{-}M'$ interactions.

Si$-$H bonds represent a particularly interesting case because the interaction can be strong enough for the existence of η^2-SiH$_4$ complexes,[98] as in 2-XIII.

$$R_2P\cdots\underset{\underset{PR_2}{|}}{\overset{\overset{PR_2}{|}}{Mo}}\cdots CO$$

(2-XIII)

2-17 H···H Bonds

Recently it has been proposed that non-covalent bonds can be formed between $X{-}H$ groups with partially positive hydrogen atoms and the hydrogen atoms of some $M{-}H$ moieties, in particular, for the $O{-}H{\cdots}H{-}M$ and $N{-}H{\cdots}H{-}M$ pairs. Specific examples[99] are provided by 2-XIV, 2-XV, and 2-XVI. These are all *intramo-*

[93]G. S. Girolami *et al., Organometallics* **1994,** *16,* 1646.
[94]D. L. Clark *et al., J. Am. Chem. Soc.* **1993,** *115,* 8461.
[95]W. S. Rees *et al., Angew. Chem. Int. Ed. Engl.* **1996,** *35,* 419.
[96]U. Schubert, *Adv. Organomet. Chem.* **1990,** *30,* 151; U. Rosenthal *et al., J. Am. Chem. Soc.* **1995,** *117,* 10399.
[97]F. Teixidor *et al., Organometallics* **1995,** *14,* 3952.
[98]X.-L. Luo *et al., J. Am. Chem. Soc.* **1995,** *17,* 1159.
[99]W. Yao and R. H. Crabtree, *Inorg. Chem.* **1996,** *35,* 3008; for intermolecular NH···H$-$M bonding see A. L. Rheingold *et al., Chem. Commun.* **1996,** 991.

lecular cases, but *inter*molecular interactions of the same types have been proposed, e.g., phenol with $WH(CO)_2(NO)PMe_3$, on the basis of infrared spectra and other indirect evidence.[100] A more complex interaction, 2-XVII, has also been reported.[101]

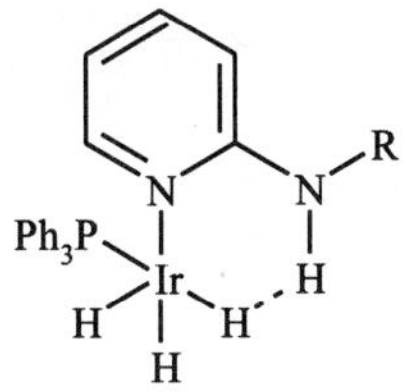

(2-XIV) 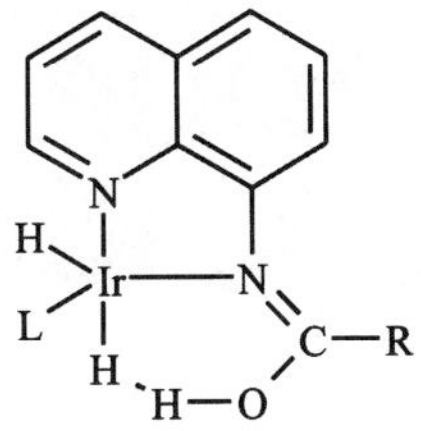   (2-XV)

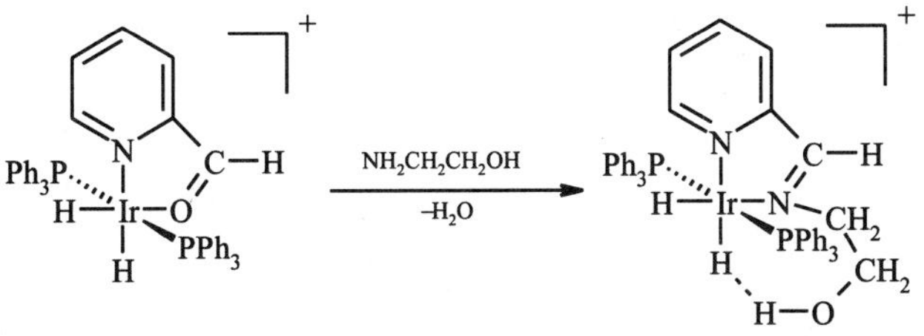

(2-XVI) 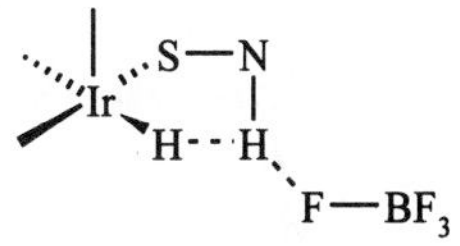   (2-XVII)

Additional References

D. Braga, *et al., J. Am. Chem. Soc.* **1995,** *117,* 5156 (C—H⋯OCM H-bonding in first row metal carbonyls); *Organometallics* **1996,** *15,* 2692 (M—H⋯O H-bonding in organometallic complexes and clusters; Data from Cambridge Crystallographic Data Centre).

J. P. Collman, *et al., Angew. Chem. Int. Ed. Engl.* **1994,** *33,* 1537 (Reduction of protons and oxidation by hydrogen, electrochemically by porphyrins; reduction of N_2 and catalysis).

R. H. Crabtree, *Chem. Rev.* **1995,** *95,* 987 (Aspects of methane chemistry includes hydrates, H atom abstraction, etc.).

M. Y. Darensbourg, *Adv. Organomet. Chem.* **1987,** *27,* 1 (anionic MH species, 129 refs.).

R. A. Henderson, *Angew. Chem. Int. Ed. Engl.* **1996,** *35,* 946 (Protonation mechanisms for organometallics, unsaturated C ligands and bioinorganics).

F. Hensel and P. P. Edwards, *Science (Washington, DC)* **1996,** *271,* 1692 (Hydrogen: the first metallic element).

G. van Koten, *et al., Chem. Commun.* **1996,** 1309; *Inorg. Chem.* **1996,** *35,* 526, 534 (C≡CH⋯ClPt; NH⋯S and OH⋯O H-bonds in Pd^{II} complexes).

D. Philip and J. F. Stoddard, *Angew. Chem. Int. Ed. Engl.* **1996,** *35,* 1154 (Classical H-bonds in self assembly E⋯X, E = F, Cl, O, S, N, π-systems, X = F, Cl, O, S, N, C).

J. J. Schneider, *Angew. Chem. Int. Ed. Engl.* **1996,** *35,* 1069 (Si—H and C—H activation by transition metal complexes: a step towards isolable alkane complexes?).

---

[100]R. H. Crabtree *et al., J. Chem. Soc., Chem. Commun.* **1995,** 2175; H. Berke *et al., J. Am. Chem. Soc.* **1996,** *118,* 1105.

[101]R. H. Morris *et al., Inorg. Chem.* **1996,** *35,* 3001.

Chapter 3

THE GROUP 1 ELEMENTS: Li, Na, K, Rb, Cs, Fr

GENERAL REMARKS

3-1 Introduction

The closely related elements lithium, sodium, potassium, rubidium, and cesium, often termed the alkali metals, have a single s electron outside a noble gas core. Some relevant data are listed in Table 3-1.

As a result of the low ionization enthalpies for the outer electrons and the sphericity and low polarizability of the resulting M^+ ions, the chemistry of these elements is principally that of their $+1$ ions. No other cations are known or, in view of the values of the second ionization enthalpies, expected.* The ions M^-, where the s shell is filled, are discussed in Section 3-4.

The chemistry of the elements is mainly that of ionic salts in the solid state and solvated cations. Although some lithium and even sodium compounds are soluble in organic solvents, such compounds as $(LiCH_3)_4$ have essentially ionic Li^+; for sodium and potassium compounds, close ion pairing can occur, as discussed in later sections.

The gaseous diatomic molecules (M_2) are covalently bonded.

The element *francium* is formed in the natural radioactive decay series and in nuclear reactions. All its isotopes are radioactive with short half-lives. The ion behaves as would be expected from its position in the group.

Of all the groups in the Periodic Table, the Group 1 metals show most clearly and with least complication the effect of increasing size and mass on chemical and physical properties. Thus all of the following *decrease* through the series: (*a*) melting points and heats of sublimation of the metals; (*b*) lattice energies of all salts except those with the very smallest anions (because of irregular radius ratio effects); (*c*) effective hydrated radii and the hydration energies (see Table 3-2); (*d*) ease of thermal decomposition of nitrates and carbonates; (*e*) strength of the covalent bonds in the M_2 molecules; (*f*) heats of formation of fluorides, hydrides, oxides, and carbides (because of higher lattice energies with the smaller cations). Other trends also can readily be found.

---

*Electrochemical oxidation of Cs^+ at $-35°C$ in acetonitrile has been claimed (K. Moock and K. Seppelt, *Angew. Chem. Int. Ed. Engl.* **1989**, *28*, 1676).

Table 3-1   Some Properties of Group 1 Metals

Element	Electronic Configuration	Metal Radius (Å)	Ionization Enthalpies (kJ mol^{-1})		mp (°C)	bp (°C)	E^{0a} (V)	$-\Delta H_{diss}{}^{b}$ (kJ mol^{-1})
			1st	2nd $\times 10^{-3}$				
Li	[He]2s	1.52	520.1	7.296	180.5	1326	−3.02	108.0
Na	[Ne]3s	1.86	495.7	4.563	97.8	883	−2.71	73.3
K	[Ar]4s	2.27	418.7	3.069	63.7	756	−2.92	49.9
Rb	[Kr]5s	2.48	402.9	2.640	38.98	688	−2.99	47.3
Cs	[Xe]6s	2.65	375.6	2.26	28.59	690	−3.02	43.6
Fr	[Rn]7s							

^{a}For $M^{+}(aq) + e = M(s)$.
^{b}Energy of dissociation of the diatomic molecule M_2.

The Li^{+} ion is exceptionally small and hence has an exceptionally high charge-radius ratio, comparable to that of Mg^{2+}. The properties of a number of lithium compounds are therefore anomalous (in relation to the other Group 1 elements) but resemble those of magnesium compounds. Many of the anomalous properties arise because the salts of Li^{+} with small anions are exceptionally stable owing to their very high lattice energies, whereas salts with large anions are relatively unstable owing to poor packing of very large with very small ions. Lithium hydride is stable to approximately 900°C, but NaH decomposes at 350°C. Lithium nitride is stable, whereas Na_3N does not exist at 25°C. Lithium hydroxide decomposes at red heat to Li_2O, whereas the other hydroxides MOH sublime unchanged; LiOH is also considerably less soluble than the other hydroxides. The carbonate (Li_2CO_3) is thermally much less stable relative to Li_2O and CO_2 than are other alkali metal carbonates (M_2CO_3). The solubilities of Li^{+} salts resemble those of Mg^{2+}. Thus LiF is sparingly soluble (0.27 g/100 g H_2O at 18°C) and can be precipitated from ammoniacal NH_4F solutions; LiCl, LiBr, LiI, and especially $LiClO_4$ are soluble in solvents such as ethanol, acetone, and ethyl acetate, and LiCl is soluble in pyridine. Sodium perchlorate ($NaClO_4$) is less soluble than $LiClO_4$ in various solvents by factors of 3 to 12, whereas $KClO_4$, $RbClO_4$, and $CsClO_4$ have solubilities only 10^{-3} of that of $LiClO_4$. Since the spherical ClO_4^{-} ion is virtually non-polarizable and the alkali metal perchlorates form ionic crystals, the high solubility of $LiClO_4$ is mainly attributable to strong solvation of the Li^{+} ion. Lithium bromide in hot concentrated solution has the unusual property of dissolving cellulose. Lithium sulfate does *not* form alums and is not isomorphous with the other sulfates.

Other ions that have chemical behavior closely resembling that of the Group 1 ions are:

1. Ammonium ions, NH_4^{+}, RNH_3^{+}, . . . , R_4N^{+}. Salts of NH_4^{+} generally resemble those of K^{+} quite closely in their solubilities and crystal structures.

2. The thallium(I) ion Tl^{+} behaves in certain respects as an alkali metal ion (although in others more like Ag^{+}). Its ionic radius (1.54 Å) is comparable to that of Rb^{+}, although it is more polarizable. Thus TlOH is a water-soluble, strong base, which absorbs CO_2 from the air to form the

carbonate. The sulfate and some other salts are isomorphous with the alkali metal salts.

3. Other unipositive, essentially spherical cations often behave like alkali ions of comparable size. For example, the very stable di(η-cyclopentadienyl)cobalt(III) ion and its analogues with similar "sandwich" structures have precipitation reactions similar to those of Cs^+, and $[(\eta-C_5H_5)_2Co]OH$ is a strong base that absorbs CO_2 from the air and forms insoluble salts with large anions.

3-2 The Elements

Sodium and K have high abundances (2.6 and 2.4%) in the lithosphere and occur in large deposits of rock salt, NaCl, and carnallite, $KCl \cdot MgCl_2 \cdot 6H_2O$. Lithium, Rb, and Cs have much lower abundances and occur mainly in a few silicate minerals.

Lithium and Na are obtained by electrolysis of fused salts or of low melting eutectics such as $CaCl_2 + NaCl$. Potassium, Rb, and Cs are made by treating molten MCl with Na vapor in a countercurrent fractionating tower; the metals are best purified by distillation. At the boiling point, the vapors have *ca.* 1% of M_2.

Lithium, Na, K, and Rb are silvery, but Cs has a golden-yellow appearance. In air, Li, Na, and K rapidly tarnish; Rb and Cs must be handled in argon.

Because there is only one valence electron per metal atom, the binding energies in the close-packed lattices are relatively weak and the metals are soft with low melting points. Liquid sodium has been much studied in view of its use as a coolant in nuclear reactors and many reactions occur in the melt.[1] The most important alloys are the liquid Na/K ones where the eutectic with 72.2% K has mp $-12.3°C$. The liquid metals, due to their low melting points, are effective solvents for other metals. Thus barium dissolves in liquid Li and Na. These solutions absorb N_2 to give Li_3N or LiBaN, depending on the conditions. For Na, compounds such as NaBaN, $NaBa_3N$ and Na_5Ba_3N, can be obtained.[2]

The metals, usually Na or K, dissolve in mercury with considerable vigor. Sodium amalgam (Na/Hg) is liquid when sodium-poor ($\leq 7\%$), but solid when rich; the most sodium-rich crystalline solid is Na_3Hg.[3] Sodium amalgams are useful reducing agents in non-protic solvents such as ethers or aromatics. Potassium graphite, KC_8, made by heating the components at 150–200°C for about 30 min under an inert atmosphere, is a useful reducing agent in tetrahydrofuran or THF-pyridine.[4] The interaction of Na vapor with anhydrous sodalite, $Na_6[Si_6Al_6O_{24}]$, leads to color changes from white to blue, purple, and finally to "black sodalite." Solid state nmr studies indicate cages containing Na_3^{3+} and paramagnetic Na_4^{3+} clusters. These materials are of interest because of their optical, electronic, and magnetic properties.[5]

Lithium is relatively light (density 0.53 g/cm^3) and has the highest melting and boiling points and also the longest liquid range of all the alkali metals; it has also an extraordinarily high specific heat. These properties should make it an excellent

[1]C. C. Addison, *The Chemistry of Liquid Alkali Metals,* Wiley, New York, 1984.
[2]G. J. Snyder and A. Simon, *J. Am. Chem. Soc.* **1995,** *117,* 1996; *J. Chem. Soc., Dalton Trans.* **1994,** 1159; P. Hubberstay and P. G. Roberts, *J. Chem. Soc., Dalton Trans.* **1994,** 667.
[3]H-J. Dieseroth and M. Rochnia, *Angew. Chem. Int. Ed. Engl.* **1993,** *32,* 1494.
[4]F. A. Cotton *et al., Inorg. Chem.* **1995,** *34,* 5424.
[5]G. Engelhart *et al., Chem. Commun.* **1996,** 729.

coolant in heat exchangers, but it is also very corrosive—more so than other liquid metals—which is a great practical disadvantage; it is used to deoxidize, desulfurize, and generally degas copper and copper alloys.

Sodium metal may be dispersed by melting on various supporting solids (sodium carbonate, kieselguhr, etc.) or by high-speed stirring of a suspension of the metal in various hydrocarbon solvents held just above the melting point of the metal. Dispersions of the latter type may be poured in air, and they react with water only with effervescence. They are often used synthetically where sodium shot or lumps would react too slowly. Sodium and potassium, when dispersed on supports such as carbon, alumina, or silica are often more reactive than the metals.

The high electrode potentials of the metals suggest potential use in batteries and indeed several are known, for example, one with a Li anode, a polyvinylpyridine-I_2 cathode, and LiI as solid electrolyte; another with a liquid Na anode separated from a sulfur cathode by a β-Al_2O_3 solid electrolyte operates at 300°C.

The chemical reactivity of the metals toward all chemical reactants, except N_2, increases from Li to Cs. Usually the least reactive, lithium is only rather slowly attacked by water at 25°C, whereas sodium reacts vigorously, potassium inflames, and rubidium and cesium react explosively. With liquid Br_2, Li and Na barely react, whereas the others do so violently. Lithium does not replace the weakly acidic hydrogen in $C_6H_5C\equiv CH$, whereas the other alkali metals do so, yielding hydrogen gas. With N_2, however, Li gives a ruby-red crystalline nitride Li_3N (Mg also reacts to give Mg_3N_2); at 25°C this reaction is slow, but it is quite rapid at 400°C. Both Li and Mg can be used to remove nitrogen from other gases. When heated with carbon, both Li and Na react to form the *acetylides* Li_2C_2 and Na_2C_2. The heavier alkali metals also react with carbon, but give nonstoichiometric interstitial compounds where the metal atoms enter between the planes of carbon atoms in the lamellar graphite structure. This difference may be attributed to size requirements for the metal, both in the ionic acetylides ($M_2^+C_2^{2-}$) and in the penetration of the graphite.

A particularly fundamental chemical difference between lithium and its congeners, attributable to cation size, is the reaction with oxygen. When the metals are burnt in air or oxygen at 1 atm, lithium forms the oxide Li_2O, with only a trace of Li_2O_2, whereas the other alkali oxides (M_2O) react further, giving as principal products the peroxides M_2O_2 and (for K, Rb, and Cs) the superoxides MO_2.

3-3 Alkali Metals in Liquid Ammonia and Other Solvents

The metals, and to a lesser extent Ca, Sr, Ba, Eu, and Yb, are soluble in liquid ammonia and certain other solvents, giving solutions that are blue when dilute. These solutions conduct electricity electrolytically and measurements of transport numbers suggest that the main current carrier, which has an extraordinarily high mobility, is the solvated electron. Solvated electrons are also formed in aqueous or other polar media by photolysis, radiolysis with ionizing radiations such as X rays, electrolysis, and probably some chemical reactions. The high reactivity of the electron and its short lifetime (in 0.75 M $HClO_4$, 6×10^{-11} s; in neutral water, $t_{1/2}$ *ca.* 10^{-4} s) make detection of such low concentrations difficult. Electrons can also be trapped in ionic lattices or in frozen water or alcohol when irradiated and again blue colors are observed. In very pure liquid ammonia, the lifetime of the solvated electron may be quite long (1% decomposition per day), but under ordinary conditions initial rapid decomposition occurs with water present.

In *dilute ammonia solutions*, the metal is dissociated into solvated metal ions (M^+) and electrons.

$$M(s)(\text{dispersed}) \rightleftharpoons M(\text{in solution}) \rightleftharpoons M^+ + e$$

$$2e \rightleftharpoons e_2$$

The broad absorption around 15,000 Å accounts for the common blue color; since the metal ions are colorless, this absorption must be associated with the solvated electrons. Magnetic and esr studies show the presence of "free" electrons, but the decrease in paramagnetism with increasing concentration suggests that the ammoniated electrons can associate to form diamagnetic species containing electron pairs. The electron is considered to be "smeared out" over a large volume so that the surrounding solvent molecules experience electronic and orientational polarization. The electron is trapped in the resultant polarization field, and repulsion between the electron and the electrons of the solvent molecules leads to the formation of a cavity within which the electron has the highest probability of being found. In ammonia this is estimated to be approximately 3 to 3.4 Å in diameter. This cavity concept is based on the fact that solutions are of much lower density than the pure solvent; that is, they occupy far greater volume than that expected from the sum of the volumes of metal and solvent. In methylamine, however, nmr studies suggest that most of the electron density is located on the N atoms of the solvent cavity. With increasing metal concentration, the solutions become copper colored, have a metallic luster and high conductivity, and contain M^- ions as discussed below.

Comparative *ab initio* MO studies of $Na(NH_3)_n$ and their ions and those of $Na(H_2O)_n$ species[6] have been reported.

In the Li/NH_3 system a golden yellow solid, $Li(NH_3)_4$, is formed. X-ray and neutron diffraction studies of the deuterated compound,[7] $Li(ND_3)_4$ (mp 89 K), show that one Li—N distance is much longer than the other three, suggesting the formulation $Li(ND_3)_3 \cdot ND_3$. The methylamine compound, $Li(MeNH_2)_4$ (mp 155 K), is also known,[8] and methylamine can assist in stabilizing Li and Na in amines or ethers in which they normally are not soluble. In the liquid state $Li(MeNH_2)_4$ has electrical and magnetic properties resembling those of a metal but it is a semiconductor in the solid state.

The metals are also soluble in a variety of other solvents such as $OP(NMe_2)_3$, amines, and ethers, to a greater or lesser extent depending on the dielectric constant and complexing ability of the solvent. The solubility in Me_2O or THF can be substantially increased by addition of cryptands, or crown ethers, and neat liquid crown ethers, for example, 12-C-4 (mp 16°C) can be used.

The ammonia and amine solutions of alkali metals are useful for preparing both organic and inorganic compounds. Thus Li in methylamine shows great selectivity in its reducing properties, but both this reagent and Li in ethylenediamine are quite

[6]K. Hashimoto and K. Morokuma, *J. Am. Chem. Soc.* **1995**, *117*, 4151.
[7]W. S. Glaunsinger *et al.*, *J. Am. Chem. Soc.* **1989**, *111*, 9260.
[8]J. L. Dye *et al.*, *J. Am. Chem. Soc.* **1989**, *111*, 5957.

powerful and can reduce aromatic rings to cyclic monoolefins. Sodium in liquid ammonia is probably the most widely used system for preparative purposes; the solution is moderately stable, but the decomposition reaction

$$Na + NH_3(l) = NaNH_2 + \tfrac{1}{2}H_2$$

can occur photochemically and is catalyzed by transition metal salts. Sodium amide can be conveniently prepared by treatment of Na with liquid ammonia in the presence of a trace of $FeCl_3$. Amines react similarly.

3-4 Alkalides and Electrides[9]

Solutions of the metals (except Li) in ethers may contain not only M^+_{solv} and e^-_{solv} but also the anion formed in a disproportionation reaction, which is driven by the solvation energy of the cation:

$$2M(s) \rightleftharpoons M^+ + M^-$$

When cryptands or crown ethers are added to stabilize the cations, crystalline solids may be isolated, some containing the M^- anions, *alkalides*, and others containing trapped electrons, *electrides*.

The alkalides are the more kinetically and thermally stable. The crystalline solids are yellow-bronze in color and can be studied by nmr (^{23}Na and ^{133}Cs) in both the solid state and solution since they are diamagnetic. Examples are $[Na(C222)]^+Na^-$ and $[Rb(15\text{-}C\text{-}5)_2]^+Na^-$. In the latter the cation is sandwiched between two crown ethers and it has been shown that the presence of two different crown ethers is favorable,[10] as in $[K(18\text{-}C\text{-}6)(12\text{-}C\text{-}4)]^+Na^-$.

The isolation of crystalline electrides is much more difficult and they are favored by the formation of the ML_2^+ cations. Examples are black, paramagnetic $[Cs(18\text{-}C\text{-}6)_2]^+e^-$ and $[Cs(15\text{-}C\text{-}5)_2]^+e^-$, which becomes antiferromagnetic below 4.6 K.[11]

Alkalides or electrides in Me_2O or thf can be used to reduce salts of metals such as Au, Pt, or Cu to give very finely divided metals.[12] Solutions of mixed alkali metals in amines can also be made and have been much studied.[13] The compounds $Li(MeNH_2)_n^+Na^-$ and $LiNa(EtNH_2)_n$ have Li^+ and Na^- in both solid and liquid phases, and have a bronze color.

COMPOUNDS OF THE GROUP 1 ELEMENTS

3-5 Binary Compounds

The metals react directly with most nonmetals to give one or more binary compounds; they also form alloys and compounds with other metals such as Pb and Sn.

[9]J. L. Dye, *Chem. Tracts. Inorg. Chem.* **1993,** *5,* 243 (119 refs); J. L. Dye and R. H. Huang, *Chem. Brit.* **1990,** 239.
[10]J. L. Dye *et al., J. Am. Chem. Soc.* **1993,** *115,* 9542.
[11]J. L. Dye *et al., J. Am. Chem. Soc.* **1991,** *113,* 1605.
[12]K-L. Tsai and J. L. Dye, *J. Am. Chem. Soc.* **1991,** *113,* 1650.
[13]M. G. DeBacker *et al., J. Am. Chem. Soc.* **1996,** *118,* 1997.

Oxides are obtained by combustion of the metals. Although Na normally gives Na_2O_2, it will take up further oxygen at elevated pressures and temperatures to form NaO_2. The per- and superoxides of the heavier alkalis can also be prepared by passing stoichiometric amounts of oxygen into their solutions in liquid ammonia; ozonides (MO_3) are also known. The structures of the ions O_2^{2-}, O_2^-, and O_3^- and of their alkali salts are discussed in Sections 11-5 and 11-6. The increasing stability of the per- and superoxides as the size of the alkali ions increases is noteworthy and is a typical example of the stabilization of larger anions by larger cations through lattice energy effects.

Owing to the highly electropositive character of the metals, the various oxides (and also sulfides and similar compounds) are readily hydrolyzed by water according to the following equations:

$$M_2O + H_2O = 2M^+ + 2OH^-$$

$$M_2O_2 + 2H_2O = 2M^+ + 2OH^- + H_2O_2$$

$$2MO_2 + 2H_2O = O_2 + 2M^+ + 2OH^- + H_2O_2$$

The oxide Cs_2O has the *anti*-$CdCl_2$ structure and is the only known oxide with this type of lattice. An abnormally long $Cs-Cs$ distance and a short $Cs-O$ distance imply considerable polarization of the Cs^+ ion. Rubidium and Cs form *suboxides* such as Rb_9O_2, $Rb_{12}O_2$, and $Cs_{11}O_3$ that are highly colored and often metallic in appearance. Their structures have metal clusters with $M-M$ bonds; for example, in Rb_9O_2 there is a confacial bioctahedron of Rb atoms with an O atom in the center of each.

The *hydroxides*, MOH, are white crystalline solids soluble in water and alcohols. They sublime unchanged at 350–400°C and the vapors contain mainly dimers. There is great diversity in the structures of the solid hydroxides depending on the metal radii and orientation of the OH groups leading to H-bonds of different strengths. Depending upon the temperature of crystallization, for example, NaOH can have three different forms.[14]

Measurements of the proton affinities of MOH in the gas phase show that the base strength increases from lithium to cesium, but this order need not be observed in aqueous or alcoholic solutions where the base strength of the hydroxide is reduced by solvent effects and hydrogen bonding. In suspension in nonhydroxylic solvents such as 1,2-dimethoxyethane, the hydroxides are exceedingly strong bases and can conveniently be used to deprotonate a wide variety of weak acids such as PH_3 ($pK \sim 27$) or C_5H_6 ($pK \sim 16$). The driving force for the reaction is provided by the formation of the stable hydrate:

$$2KOH(s) + HA = K^+A^- + KOH \cdot H_2O(s)$$

The alkali metals form a multitude of compounds with the elements of Groups 13–16, only a few of which can be thought of in *simple* ionic terms (i.e., in terms of M^+ ions and anions with complete octets). The vast majority are far richer in

[14]H. P. Beck and G. Lederer, *Angew. Chem. Int. Ed. Engl.* **1993**, *32*, 271.

the metalloidal element (e.g., NaP_7, $SrSi_2$, and $LiGe$) and are Zintl compounds containing complex polynuclear anionic structures. These materials are structurally and electronically transitional between ionic compounds and alloys. Most of them can be made either by direct reaction of the elements or by reaction of a liquid ammonia solution of the alkali metals with compounds of the metalloidal components. Examples of ionic sulfides[15] are Na_2S, NaS_4, K_2S_5, and Cs_2S_6, while a more complicated sulfide is $Li_2S_6(tmen)_2$, which has a μ-η^2-S_6^{2-} unit as in (3-I).

(3-I)

Zintl compounds to be mentioned in other chapters are compounds or phases that have anions such as Sn_9^{4-}, $Sn_2Bi_2^{2-}$ or Pb_5^{2-}. The alkali metal salts can often be isolated crystalline by complexation of the cation with crown ethers or cryptands. Recent examples are Na_4Sn[16] and $Li_2Ba_4Si_6$, the latter having a Si_6^{10-} ring.[17]

Saline hydrides were discussed in Chapter 2.

3-6 Other Compounds

Alkoxides. The metals dissolve readily in alcohols with evolution of hydrogen to give this important class of compounds. The solutions of Na or K in C_2H_5OH or Bu^tOH are commonly used in organic chemistry as a source of nucleophilic OR^- ions or as reducing agents. Lithium alkoxides have been particularly well studied structurally. Most are of the type $[LiOR]_n$ where n can be 2–6.[18] The Li atoms are commonly bridged by alkoxide oxygen atoms (3-II).

(3-II)

The only alkali metal *organo peroxide* that has been structurally characterized, $[Li\eta^2\text{-}O_2Bu^t]_{12}$, has a complex aggregate with Li^+ bridges between the $O-O$ bonds.[19]

Amido Complexes. The compounds of the general type MNHR and MNR_2, especially those of Li, are important reagents in both organic and organometallic chemistry as well as starting materials for the synthesis of other metallic amido and

[15]J. Cusick and I. Dance, *Polyhedron* **1991**, *10*, 2629 (sulfides and selenides); K. Tatsumi *et al.*, *Inorg. Chem.* **1993**, *32*, 4317.
[16]F. Guerin and D. Richeson *et al.*, *J. Chem. Soc., Chem. Commun.* **1993**, 2213.
[17]H. G. von Schnering *et al.*, *Angew. Chem. Int. Ed. Engl.* **1996**, *35*, 984.
[18]See e.g., L. M. Jackman *et al.*, *J. Am. Chem. Soc.* **1993**, *115*, 6267; L. Brandsma *et al.*, *Organometallics* **1991**, *10*, 1623.
[19]G. Boche *et al.*, *Chem. Eur. J.* **1996**, *2*, 604.

imido compounds. The compounds are obtained by reaction of amines with LiBu^n, NaH or the metals K, Rb, Cs.

Although many structures have been determined by X-ray study (*vide infra*), the nature of the species in solution is not well known. The degree of aggregation may be similar to that in the solid but in many cases is solvent dependent: species may be mono- or dimeric or equilibrium mixtures thereof. However, spectroscopic methods for the determination of aggregation constants in dilute solutions have been developed.[20] The data should lead to a better understanding of reactivities and mechanisms.

The structures of crystalline materials can be relatively simple, such as that of $[C_6F_5N(H)Li(thf)_2]_2$, which has a Li_2N_2 ring with planar N^{21}, or more complicated, such as that of $Li_3(RNCH_2CH_2)_3N$, $Li_3(Me_3SiNCH_2CH_2)_3N(thf)_2$,[22] or the tetrameric $[CsNH(SiMe_3)]_4$, which has a Cs_4 tetrahedron capped on each face by μ_3-amido groups.[23]

Many of the compounds are solvates with Et_2O, dioxane and other donors.[24] A simple example is that in (3-III).

$$(Me_3Si)_2N \underset{\underset{(OEt_2)_n}{Li}}{\overset{\overset{(OEt_2)_n}{Li}}{<\quad>}} N(SiMe_3)_2$$

(3-III)

Perhaps the most complex structure is that of an amide $\{(c\text{-}C_5H_9)NH\}_{12}OLi_{14}$ that is a cage molecule with a salt-like distorted face-centered cube of lithium cations, with a central μ_6-O atom that arises from water.[25]

Phosphido compounds can be made similarly by reactions such as

$$\text{BuLi} + \text{HP}\left(\underset{OMe}{\bigcirc}\right)_2 \longrightarrow \text{LiP}\left(\underset{OMe}{\bigcirc}\right)_2 + \text{BuH}$$

and corresponding AsR_2^-, SbR_2^- compounds have also been made as well as SR^- derivatives. Examples are $[LiP(SiMe_3)_2]_6$ and $[LiSC_6H_3\text{-}2,6mes_2]_2(Et_2O)_2$. The structural types include monomers, dimers, ladders, and other polymers.[26]

Dilithium derivatives of primary amines (also phosphines and arsines) are also known, usually as amorphous solids, but $[(RNLi_2)_{10}(Et_2O)_6]\cdot Et_2O$, R = α-naphthyl and a $RPLi_2$ species have been structurally characterized since being soluble in organic solvents they can be crystallized.[27]

[20]A. Streitweiser *et al.*, *J. Am. Chem. Soc.* **1993**, *115*, 8024; B. L. Lucht and D. B. Collum, *J. Am. Chem. Soc.* **1994**, *116*, 6009.

[21]D. Stalke *et al.*, *Chem. Commun.* **1996**, 1639.

[22]J. G. Verkade *et al.*, *Inorg. Chem.* **1995**, *34*, 2179.

[23]T. P. Hanusa *et al.*, *J. Am. Chem. Soc.* **1992**, *114*, 6590.

[24]For detailed studies, theory, and references see papers by D. B. Collum *et al.*, e.g., B. L. Lucht and D. B. Collum, *J. Am. Chem. Soc.* **1996**, *118*, 2217; **1995**, *117*, 9863.

[25]R. E. Mulvey *et al.*, *Chem. Commun.* **1996**, 1065.

[26]See H. C. Aspinall and M. R. Tillotson, *Inorg. Chem.* **1996**, *35*, 5; C. L. Raston *et al.*, *J. Chem. Soc., Chem. Commun.* **1995**, 47; M. Driess *et al.*, *Angew. Chem. Int. Ed. Engl.* **1995**, *34*, 316; J. J. Ellison and P. P. Power, *Inorg. Chem.* **1994**, *33*, 4231.

[27]For references see M. Driess *et al.*, *Angew. Chem. Int. Ed. Engl.* **1996**, *35*, 986.

3-7 Ionic Salts and M⁺ Ions in Solution

Salts of the bases MOH are crystalline, ionic solids, colorless except where the anion is colored. For the alkali metal ions the energies required to excite electrons to the lowest available empty orbitals could be supplied only by quanta far out in the vacuum ultraviolet (the transition $5p^6 \rightarrow 5p^56s$ in Cs⁺ occurs at ~1000 Å). However, colored crystals of compounds such as NaCl are sometimes encountered. Color arises from the presence in the lattice of holes and free electrons, called color centers, and such chromophoric disturbances can be produced by irradiation of the crystals with X rays and nuclear radiation. The color results from transitions of the electrons between energy levels in the holes in which they are trapped. These electrons behave in principle similarly to those in solvent cages in the liquid ammonia solutions, but the energy levels are differently spaced and consequently the colors are different and variable. Small excesses of metal atoms produce similar effects, since these atoms form M⁺ ions and electrons that occupy holes where anions would be in a perfect crystal.

The structures and stabilities of the ionic salts are determined in part by the lattice energies and by radius ratio effects. Thus the Li⁺ ion is usually tetrahedrally surrounded by water molecules or negative ions, although $\text{Li(H}_2\text{O)}_6^+$ has also been found. On the other hand, the large Cs⁺ ion can accommodate eight near-neighbor Cl⁻ ions, and its structure is different from that of NaCl, where the smaller cation Na⁺ can accommodate only six near-neighbors. The Na⁺ ion appears to be 6-coordinate in some nonaqueous solvents.

The salts generally have high melting points, electrical conductivity in melts, and ready solubility in water. They are seldom hydrated when the anions are small, as in the halides, because the hydration energies of the ions are insufficient to compensate for the energy required to expand the lattice. Owing to its small size, the Li⁺ ion has a large hydration energy, and it is often hydrated in its solid salts when the same salts of other alkalis are unhydrated (namely, $\text{LiClO}_4\cdot 3\text{H}_2\text{O}$). For *strong* acids, the lithium salt is usually the *most* soluble in water of the alkali metal salts, whereas for *weak* acids the lithium salts are usually *less* soluble than those of the other alkalis.

The large size of the Cs⁺ and Rb⁺ ions frequently allows them to form ionic salts with rather unstable anions, such as various polyhalide anions and the superoxides already mentioned.

Since few salts are sparingly soluble in water, there are few *precipitation reactions* of the aqueous ions. Generally the larger the M⁺ ion, the more numerous are its insoluble salts. Thus sodium has very few; the mixed Na—Zn and Na—Mg uranyl acetates [e.g., $\text{NaZn(UO}_2\text{)}_3(\text{CH}_3\text{CO}_2)_9\cdot 6\text{H}_2\text{O}$], may be precipitated almost quantitatively under carefully controlled conditions from dilute acetic acid solutions. The perchlorates and hexachloroplatinates of K, Rb, and Cs are rather insoluble in water and virtually insoluble in 90% ethanol. These heavier ions may also be precipitated by the ion $[\text{Co(NO}_2)_6]^{3-}$ and various other large anions. Sodium tetraphenylborate $\text{NaB(C}_6\text{H}_5)_4$, which is moderately soluble in water, precipitates the tetraphenylborates of K, Rb, and Cs from neutral or faintly acid aqueous solutions.

Lithium halides when solvated show remarkable structural diversity.[28] A simple one is $[\text{LiBr}\cdot\text{Et}_2\text{O}]_4$, which has a cubic Li_4Br_4 core with OEt₂ bound to the Li atoms, whereas $[(\text{LiCl})_4\cdot 3.5\text{tmeda}]_2$ crystallizes as a bicyclic system of fused 6- and 4-

[28]P. von R. Schleyer *et al.*, *Inorg. Chem.* **1995,** *34,* 6553 and references therein.

Table 3-2 Data on Hydration of Aqueous Group 1 Ions

	Li^+	Na^+	K^+	Rb^+	Cs^+
Crystal radii[a] (Å)	0.86	1.12	1.44	1.58	1.84
Approximate hydrated radii (Å)	3.40	2.76	2.32	2.28	2.28
Approximate hydration numbers[b]	25.3	16.6	10.5		9.9
Hydration energies (kJ mol^{-1})	519	406	322	293	264
Ionic mobilities (at ∞ dil., 18°C)	33.5	43.5	64.6	67.5	68

[a]Ladd radii.
[b]From transference data.

membered rings. Bromine and CsF combine slowly to produce $CsF \cdot Br_2$ and $2CsF \cdot Br_2$, in which layers of vertical Br_2 molecules lie between sheets (or double sheets) of CsF. The Br—Br distance is 2.313 Å compared to 2.280 Å in gaseous Br_2. There appears to be a small degree of charge transfer from the Br_2 sheets to the CsF sheets.[29] Square planar alkali compounds such as lithium porphyrins and related compounds[30] are also known.

The M$^+$ Ions in Aqueous Solution

The primary hydration shell for Li^+ in aqueous solution is tetrahedral.[31] Only tetrahydrated salts are formed except when there is also hydration of the anions. X-ray scattering studies show that the primary hydration number of K^+ is 4, and since Na^+ forms the stable $Na(NH_3)_4^+$ ion in liquid ammonia, it too presumably has a first coordination sphere of four water molecules. There is no direct evidence regarding Rb^+ or Cs^+, but a higher number, probably 6, seems likely.

In all cases electrostatic forces operate beyond the first shell, and additional water molecules are bound in layers of decreasing firmness. The larger the cation itself, the less it binds outer layers. Thus although the crystallographic radii *increase* down the group, the hydrated radii *decrease* (Table 3-2). The hydration number of Li^+ is very large and Li^+ salt solutions generally deviate markedly from ideal solution behavior, showing abnormal colligative properties such as very low vapor pressures and freezing points. Also, hydration energies of the gaseous ions decrease. The decrease in size of the hydrated ions is manifested in various ways. The mobility of the ions in electrolytic conduction increases, and so generally does the strength of binding to ion-exchange resins.

The equilibria

$$A^+(aq) + [B^+R^-](s) = B^+(aq) + [A^+R^-](s)$$

where R represents the resin and A^+ and B^+ the cations, have been measured and the order of preference is usually $Li^+ < Na^+ < K^+ < Rb^+ < Cs^+$, although irregular behavior does occur in some cases. The usual order may be explained if we assume that the binding force is essentially electrostatic and that under ordinary conditions the ions within the water-logged resin are hydrated approximately as they are outside it. Then the ion with the smallest hydrated radius (which is the one with

[29]K. Seppelt *et al.*, *Chem. Eur. J.* **1996,** *2,* 1303.
[30]M. Albrecht and S. Kotula, *Angew. Chem. Int. Ed. Engl.* **1996,** *35,* 1291.
[31]R. G. Keil *et al.*, *Inorg. Chem.* **1989,** *28,* 2764.

the largest "naked" radius) will be able to approach most closely to the negative site of attachment and will hence be held most strongly according to Coulomb's law. The efficiency of separating alkali metal ions on cation exchangers can be increased significantly by adding chelating agents like EDTA to the eluting solution. These agents bind more strongly to the ions that are less strongly held to the resin, thus enhancing the separation factors.

3-8 Alkali Metal Complexes[32]

The M^+ ions are only very weakly complexed by simple anions, for example, F^- in 1 M fluoride solution, where the order falls $Li \rightarrow Cs$. Although lithium is an exception, as discussed later, chelation is usually essential to complex formation. Complexes are formed with β-diketones, nitrophenols, and, of course, cryptates and macrocyclic polyethers. Some complexes, such as those with hexafluoroacetyl-acetone, are sublimable at 200°C, though the metal—ligand bonds are doubtless quite polar. The anhydrous β-diketonates are usually insoluble in organic solvents, indicating an ionic nature, but in the presence of additional coordinating ligands, including water, they may become soluble even in hydrocarbons; for example, sodium benzoylacetonate dihydrate is soluble in toluene and tetramethyleth-ylenediaminelithium hexafluoroacetylacetonate in benzene.

This behavior has allowed the development of solvent-extraction procedures for alkali metal ions. Thus not only can the trioctylphosphine oxide adduct in $Li[PhC(O)CHC(O)Ph][OP(octyl)_3]_2$ be extracted from aqueous solutions into p-xylene, but this process can also be used to separate lithium from other alkali metal ions. Even Cs^+ can be extracted from aqueous solutions by 1,1,1-trifluoro-3-(2' thenoyl) acetone (TTA) in $MeNO_2$—hydrocarbons, and all of the M^+ ions can be extracted into chloroform by crown ethers, in a manner that depends upon the associated anions.[33]

Lithium forms a very wide range of complexes[34] with amines, ethers, carboxylates, alkoxides, dialkylamides, and many other ligands whose structures are usually quite different from those of the other M^+ ions. Some of these are shown schematically in 3-IV. However, we may note that in many of these compounds lithium can have coordination numbers 3–7. Some ionophores with high preference for Li^+ are crown ethers with long aliphatic chains and other macrocycles with functional groups may have utility for therapy in brain disorders for which Li^+ salts are used.

(3-IV)

[32]For references and examples see H. Bock *et al.*, *Angew. Chem. Int. Ed. Engl.* **1994**, *33*, 875; D. N. Reinhoudt *et al.*, *J. Am. Chem. Soc.* **1994**, *116*, 123; S. F. Lincoln *et al.*, *Inorg. Chem.* **1993**, *32*, 2195.

[33]R. A. Bartsch *et al.*, *J. Am. Chem. Soc.* **1993**, *115*, 3370.

[34]P. von R. Schleyer *et al.*, *Adv. Inorg. Chem.* **1991**, *37*, 47; R. E. Mulvey, *Chem. Soc. Rev.* **1991**, *20*, 167; H. Olsher *et al.*, *Chem. Rev.* **1991**, *91*, 137; N. S. Poonia and A. V. Bajaj, *Coord. Chem. Rev.* **1988**, *87*, 55; P. G. Willard *et al.*, *Angew. Chem. Int. Ed. Engl.* **1996**, *35*, 1322.

Simple examples are Li(16-C-4)NCS, which is tetragonal pyramidal, while LiI·en$_3$ has Li$^+$ in a distorted octahedron of N atoms. Chloride complex ions such as (3-V) have been characterized; LiBF$_4$[OP(NMe$_2$)$_3$] is soluble in benzene and has strong Li—F interactions and LiNO$_3$(diacetamide) is 5-coordinate with η^1-NO$_3$.

$$\left[\begin{array}{c} \text{Li(OEt}_2)_2 \\ | \\ \text{Cl} \\ \text{(Et}_2\text{O)}_2\text{Li} \diagdown \diagup \text{Li(OEt}_2)_2 \\ \text{Cl} \\ | \\ \text{Li(OEt}_2)_2 \end{array} \right]^{2+}$$

(3-V)

The effectiveness of THF and dimethylethers of ethylene- and diethyleneglycols as media for reactions involving sodium metal may be due in part to slight solubility of Na, but the driving force is undoubtedly the complexation of Na$^+$ by the chelate ethers. The most important ethers are the crowns and cryptates (O, N) which bind M$^+$ strongly and often with high selectivity. The affinity of such a ligand for an ion is strongly dependent on how well the ion fits into the cavity that the ligand can provide for it. At the same time, the strength of complexation and to some extent the selectivity also depend on the solvent. Illustrations of selectivity, provided by cryptate-221 and -222, are shown in Fig. 3-1.

For cryptate-222, for example, K$^+$ fits the cavity very well, but Li$^+$ and Na$^+$ are too small to make good contacts with the oxygen atoms and Rb$^+$ and Cs$^+$ are too large to enter without appreciable steric strain. An example of an alkali cation (Rb$^+$) occupying the cavity in cryptate-222, and coordinated by the six oxygen and two nitrogen atoms, is presented in Fig. 3-2.

An 18-C-6 crown ether with a side chain terminating in an NH$_2$ group is selective for Na$^+$/K$^+$ in the transport of ions across a liquid membrane, thus mimicking natural ion-selective transport agents such as monemsin. Many of the natural agents

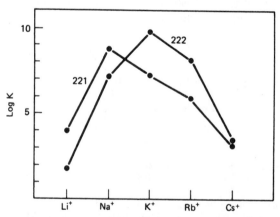

Figure 3-1   Stability constants for cryptate-221 and -222 versus alkali ion in MeOH/H$_2$O (95:5).

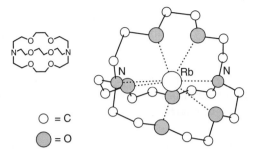

○ = C

● = O

Figure 3-2 The structure of the cation in the salt [RbC$_{18}$H$_{36}$N$_2$O$_6$]SCN·H$_2$O.

are small polypeptides of which valinomycin (3-VI) is another. The structures of two such complexes are shown in Fig. 3-3.

(3-VI)

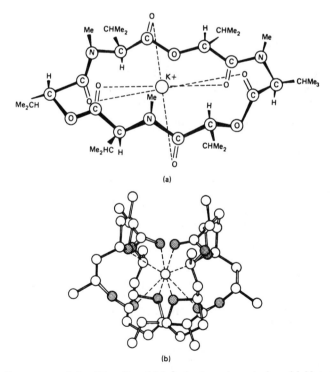

(a)

(b)

Figure 3-3 Structures of the K$^+$ salts of (a) [D-hydroxyisovaleric acid-N-methyl-L-valine]$_3$ or enniatin B and (b) nonactin.

Some new calixene crown ionophores show even better K^+/Na^+ selectivity than does valinomycin,[35] while a substituted diaza-18-crown is highly selective for K^+ and Ba^{2+}.[36]

Although few complex ions with only N donor ligands are stable for Na^+-Cs^+, there are many of these for Li^+. Thus the effectiveness of $Me_2N(CH_2)_2NMe_2$ (tmen) in increasing the effectiveness of lithium alkyls is due to formation of $[Li(tmen)_2]^+$ with concomitant increase in nucleophilicity of the anion.

Porphyrin Complexes Complexes of octaethylporphyrin and other porphyrins can be obtained[37] by the reaction

$$(porph)H_2 + 2MN(SiMe_3)_2 \xrightarrow[\text{reflux}]{\text{thf or dme}} M_2(porph)(solv)_n + 2HN(SiMe_3)_2$$

The lithium compounds, which, as noted earlier have square planar Li, are very soluble in polar solvents, are useful reagents for the synthesis of transition metal porphyrin complexes. There are also ionic species of the type $[Li\ solv_n]^+[Li\ porph]^-$.

3-9 Organometallic Compounds[38]

Many compounds with M—C bonds are known for all the alkali metals, but those of lithium are doubtless the most important because they are utilized in both organic and inorganic syntheses.

Lithium Alkyls and Aryls

One of the largest uses for lithium metal is in the synthesis of organolithium compounds which are of immense utility, generally resembling Grignard reagents, but more reactive.[39] They are commonly made by direct interaction of Li with an alkyl or aryl chloride in ether, benzene, or alkanes; ethers are slowly attacked, however, by LiR.

$$RCl + 2Li = RLi + LiCl$$

In ether, MeLi gives complexes such as Li_4Me_3Br and Li_4Me_3I. A common use for *n*- or *t*-butyllithium in hexane, benzene, or ethers is to carry out metal-hydrogen

[35]R. Ungaro *et al.*, *Chem. Eur. J.* **1996,** *4,* 436.
[36]R. M. Izatt *et al.*, *J. Am. Chem. Soc.* **1995,** *117,* 11507.
[37]J. Arnold *et al.*, *Inorg. Chem.* **1994,** *33,* 4332.
[38]E. Weiss, *Angew. Chem. Int. Ed. Engl.* **1993,** *32,* 1501. A review (220 references); L. Brandsma and H. D. Verkrvijsse, *Preparative Polar Organometallic Chemistry,* B. M. Trost, ed., Vols. 1 and 2, 1987, 1990, Springer, New York.
[39]B. J. Wakefield, *Organolithium Methods,* Academic Press, London, 1990; C. Lambert and D. B. Collum, *Acc. Chem. Res.* **1992,** *25,* 448; P. von R. Schleyer, *Angew. Chem. Int. Ed. Engl.* **1994,** *33,* 1129.

or metal-halogen exchanges, for example,

$$2Bu^nLi + (\eta^5\text{-}C_5H_5)_2Fe \longrightarrow (\eta^5\text{-}C_5H_4Li)_2Fe + 2n\text{-}C_4H_{10}$$

$$Bu^nLi + o\text{-}Br(NO_2)C_6H_4 \longrightarrow o\text{-}Li(NO_2)C_6H_4 + n\text{-}C_4H_9Br$$

Organolithium reagents are very reactive to water and to air, often being spontaneously flammable. They are soluble in hydrocarbons, have high volatility and may be sublimed in vacuum. Lithium alkyls are often considered to be carbanionic in reactions (although radicals are formed in some cases). However, although some solvated lithium compounds, for example, $Li(crown)_2CHPh_2$, do have isolated planar carbanions, almost all lithium alkyls and aryls are associated in both the solid state and in solutions. There is accordingly a wide variation in reactivities depending on differences in aggregation and ion-pair interactions, all of which are solvent dependent. Thus $LiCH_2Ph$, which is monomeric in THF, reacts with a given substrate more than 10^4 times as fast as does $(LiMe)_4$. In noncoordinating solvents Bu^tLi is hexameric, while in THF there is an equilibrium.

$$(Bu^tLi)_4 \rightleftharpoons 2(Bu^tLi)_2$$

The addition of tmen, $Me_2N(CH_2)_2NMe_2$, to lithium reagents increases the nucleophilicity remarkably by complexing the Li^+ ion.

The aryls and alkyls of main group and transition metals that are coordinately unsaturated have Lewis acid character. In alkylations of halides by LiR, in contrast to the use of MgR_2, lithium *alkylate anions* are often formed, for example,

$$Mo_2(O_2CMe)_4 + 8MeLi \xrightarrow{\text{THF}} [Li(THF)]_4[Mo_2Me_8]$$

$$MgPh_2 + 2PhLi \longrightarrow Li_2[MgPh_4]$$

$$WMe_6 + 2MeLi \longrightarrow Li_2[WMe_8]$$

The lithium alkylates of copper are of great utility in organic syntheses.

Structures

The structures of organolithiums, like those of $LiNR_2$, LiOR, $LiPR_2$, LiSR, and so on noted above, have generated much literature.[40]

Quite generally (*a*) the Li^+ ion is bound to the principal donor atom, for example, C in the alkyls or N in $LiNR_2$ by multicenter bonds and may also interact with other atoms such as H in the organic moiety; (*b*) lithium may have coordination numbers from 2 to 6 though four is most common; (*c*) lithium commonly is found in tetrahedra of lithium atoms, that is, Li_4 but other polyhedra occur. It is also found in rings such as $(LiN)_3$ that may, in certain cases, have hydrogen bridges between lithium and a metal $Li(\mu\text{-}H)_nM$.

[40]W. N. Setzer and P. von R. Schleyer, *Adv. Organomet. Chem.* **1985**, *24*, 353 (extensive review); P. von R. Schleyer, *Chem. Rev.* **1990**, *90*, 1061 (The Li bond); F. M. Bickelhaupt *et al., Organometallics* **1996**, *15*, 2923 [Bonding in $(MeLi)_n$].

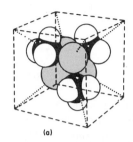
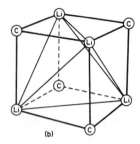

Figure 3-4   The structure of $(CH_3Li)_4$ (a) showing the tetrahedral Li_4 unit with the CH_3 groups located symmetrically above each face of the tetrahedron. The structure can also be regarded as derived from a cube (b).

The structure of the simplest, methyllithium, is shown in Fig. 3-4; that of $(EtLi)_4$ is similar. In both compounds the Li atoms are at the corners of a tetrahedron with the alkyl groups (μ_3) centered over the facial planes. Although the CH_3 group is symmetrically bound to three Li atoms, in the ethyl, the α-carbon of CH_2CH_3 is closer to one Li atom than the other two.

Some other structures of LiR compounds include:

$LiCH(SiMe_3)_2$	polymeric chains
$LiC(SiMe_3)_3$	dimer
$LiBu^t$	tetramer
$LiPr^i$	hexamer
$Li(c\text{-}C_6H_{11})$	hexamer

The structures of many compounds solvated by thf, dme, tmen, etc., are also known.[41] Thus $[LiBu^t(OEt_2)]_2$ is a dimer similar to (3-II), $[LiBu^n(tmeda)]_n$ has a zig-zag chain, while $[LiMe(thf)]_4$ has a structure similar to that in Fig. 3-4 but with THF bound to each Li atom.

Lithium aryls have been less studied but several structures are known, e.g., $[Li(2,4,6\text{-}Pr^iC_6H_2)]_4$[42] and $Li(2,4,6\text{-}Ph_2C_6H_2)(OEt_2)_2$,[43] the latter being monomeric with planar 3-coordination of Li.

Although most alkali metal compounds have M—C σ bonds, cyclopentadienyl π-complexes are known for Li, Na, and K. These are commonly polymeric but solvation can break the chains and even result in cation-anion separation.[44]

Examples of structures are the following where S is a coordinated solvent such as THF, pyridine, or tmen.

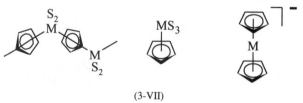

(3-VII)

[41]See, e.g., T. Koffke and D. Stalke, *Angew. Chem. Int. Ed. Engl.* **1993**, *37*, 580; C. A. Ogle *et al.*, *Organometallics* **1993**, *12*, 1960; R. E. Mulvey *et al.*, *J. Am. Chem. Soc.* **1993**, *115*, 1573.
[42]P. P. Power *et al.*, *J. Am. Chem. Soc.* **1993**, *115*, 11353 and references therein.
[43]G. S. Girolami *et al.*, *Organometallics* **1992**, *11*, 3907.
[44]S. Harder *et al.*, *Organometallics* **1996**, *16*, 118.

The simplest sandwich, Cp_2Li^-, is made by interaction of LiCp and Ph_4PCl in thf and isolated as the PPh_4^+ salt. The structure is similar to those of other Cp_2M compounds[45] with staggered parallel rings. Cesium gives a 3-decker, $[Cp_3Cs_2]^-$, that is bent like $[Cp_3Tl_2]^-$.

Polylithium Compounds

These are compounds having two or more Li atoms bound to C or on adjacent carbons.[46] Compounds such as $(CH_2Li_2)_n$, Li_4C_3, or Li_4C can be made by interaction of $LiBu^n$ with MeCN, $MeC\equiv CH$ and so on, or for $(CH_2Li_2)_n$ by thermal decomposition of MeLi. Calculations predicting structures of a variety of CLi_n species have been done.[47]

Simple compounds like dilithioacetylene $LiC\equiv CLi$ and dilithiodiacetylene, known much earlier, can be included as polylithiums.

The highest polylithiums known so far (usually detected by mass spectra) are CLi_6, OLi_4, and SLi_4.

Other Alkali Metal Compounds[48]

Compounds of Na-Cs are essentially ionic, usually sparingly soluble, if at all, in hydrocarbon solvents and exceedingly air and moisture sensitive. Some syntheses include:

$$Me_2Hg + K/Na \longrightarrow MeK + Na/Hg$$

$$KOBu^t \xrightarrow[- LiOBu^t]{MeLi} MeK \xrightarrow{(Me_3Si)_3CH} (Me_3Si)_3CK$$

Although $(MeNa)_4$ is similar structurally to $(MeLi)_4$, the methyls of K, Rb, and Cs have a NiAs type structure. In $(PhMe_2Si)_3CK$ there are linear chains of alternate K^+ and planar $(PhMe_2Si)_3C^-$ anions. Some compounds with bulky ligands, e.g., $RbC(SiMe_3)_3$, are soluble in hydrocarbons.

More important are the compounds formed by acidic hydrocarbons such as cyclopentadiene, indene, and acetylenes. These are obtained by reaction with sodium in liquid ammonia or, more conveniently, sodium dispersed in THF, glyme, diglyme, or dimethylformamide (DMF).

$$3C_5H_6 + 2Na \longrightarrow 2C_5H_5^-Na^+ + C_5H_8$$

$$RC\equiv CH + Na \longrightarrow RC\equiv C^-Na^+ + \tfrac{1}{2}H_2$$

[45]S. Harder and M. H. Prosenc et al., Angew. Chem. Int. Ed. Engl. 1996, 35, 97; D. Stalke, Angew. Chem. Int. Ed. Engl. 1994, 33, 2169 (a review of alkali and alkaline earth metallocenes).
[46]A. Maercker, Angew. Chem. Int. Ed. Engl. 1992, 31, 584.
[47]J. Ivanic and C. J. Marsden, J. Am. Chem. Soc. 1993, 115, 7503.
[48]C. Schade and P. von R. Schleyer, Adv. Organomet. Chem. 1987, 27, 169 (extensive review); P. von R. Schleyer et al., Organometallics 1993, 12, 1193; J. D. Smith et al., Organometallics 1994, 13, 753; Angew. Chem. Int. Ed. Engl. 1995, 34, 687.

Many more hydrocarbons, as well as aromatic ketones, triphenylphosphine oxide, triphenylarsine, azobenzene, and so on, can form highly colored *radical anions* when treated at low temperatures with sodium or potassium in solvents such as THF. For the formation of such anions it must be possible to delocalize the negative charge over an aromatic system. Species such as the benzanide ($C_6H_6^-$), naphthalenide, or anthracenide ions can be detected and characterized spectroscopically and by esr. The sodium−naphthalene system $Na^+[C_{10}H_8]^-$ in an ether is widely used as a powerful reducing agent (e.g., in nitrogen fixing systems employing titanium catalysts) and for the production of complexes in low oxidation states. The blue solution of sodium and benzophenone in THF, which contains the *ketyl* or radical ion, is a useful and rapid reagent for the removal of traces of oxygen from nitrogen.

Additional References

H. V. Borgstedt and C. K. Matthews, *Applied Chemistry of Alkali Metals,* Plenum, New York, 1987.

L. Brandsma and H. Verkruijsse, *Preparative Organometallic Chemistry* 1, Springer, Berlin, 1987.

L. Brandsma, *Preparative Organometallic Chemistry* 2, Springer, Berlin, 1990.

J. L. Dye, *Electrides: from 1D Heisenberg Chains to 2D Pseudo-Metals*, Inorg. Chem. **1997,** *36,* 3816.

P. D. Lickiss and C. M. Smith, *Coord. Chem. Rev.* **1995,** *145,* 75 (Alkali and alkaline earth silyl compounds, e.g., Ph₃SiLi).

A. M. Sapse and P. von R. Schleyer, *Lithium Chemistry: a Theoretical and Experimental Overview*, Wiley, New York, 1995.

G. N. Schrauzer and K. Klippel, Eds., *Lithium in Biology and Medicine*, VCH, Weinheim, 1991.

Chapter 4

THE GROUP 2 ELEMENTS: Be, Mg, Ca, Sr, Ba, Ra

4-1 General Remarks

Some pertinent data for the elements are given in Table 4-1. Beryllium has unique chemical behavior with a predominantly covalent chemistry, although it forms an aqua ion $[Be(H_2O)_4]^{2+}$. Magnesium has a chemistry intermediate between that of Be and the heavier elements, but it does not stand in as close relationship with the predominantly ionic heavier members as might have been expected from the similarity of Na, K, Rb, and Cs. It has considerable tendency to covalent bond formation, consistent with the high charge/radius ratio. For instance, like beryllium, its hydroxide can be precipitated from aqueous solutions, whereas hydroxides of the other elements are all moderately soluble, and it readily forms bonds to carbon.

The metal atomic radii are smaller than those of the Group 1 metals owing to the increased nuclear charge; the number of bonding electrons in the metals is twice as great, so that the metals have higher melting and boiling points and greater densities.

All are highly electropositive metals, however, as is shown by their high chemical reactivities, ionization enthalpies, standard electrode potentials and, for the heavier ones, the ionic nature of their compounds. Although the energies required to vaporize and ionize these atoms to the M^{2+} ions are considerably greater than those required to produce the M^+ ions of the Group 1 elements, the high lattice energies in the solid salts and the high hydration energies of the $M^{2+}(aq)$ ions compensate for this, with the result that the standard potentials are similar to those of the Li—Cs group.

The potential E^0 of beryllium is considerably lower than those of the other elements, indicating a greater divergence in compensation by the hydration energy, the high heat of sublimation, and the ionization enthalpy. As in Group 1, the smallest ion crystallographically (i.e., Be^{2+}) has the largest hydrated ionic radius.

All the M^{2+} ions are smaller and considerably less polarizable than the isoelectronic M^+ ions. Thus deviations from complete ionicity in their salts due to polarization of the cations are even less important. However, for Mg^{2+} and, to an exceptional degree for Be^{2+}, polarization of anions by the cations does produce a degree of covalence for compounds of Mg and makes covalence characteristic for Be. Accordingly, only an estimated ionic radius can be given for Be^{2+}; the charge/radius ratio

Table 4-1   Some Physical Parameters for the Group 2 Elements

Element	Electronic Configuration	mp (°C)	Ionization Enthalpies (kJ mol^{-1}) 1st	2nd	E^0 for M^{2+}(aq) + 2e = M(s) (V)	Ionic Radius (Å)^a	Charge —— Radius
Be	[He]$2s^2$	1278	899	1757	-1.70^b	0.31^b	6.5
Mg	[Ne]$3s^2$	651	737	1450	-2.37	0.78	3.1
Ca	[Ar]$4s^2$	843	590	1146	-2.87	1.06	2.0
Sr	[Kr]$5s^2$	769	549	1064	-2.89	1.27	1.8
Ba	[Xe]$6s^2$	725	503	965	-2.90	1.43	1.5
Ra	[Rn]$7s^2$	700	509	979	-2.92	1.57	1.3

^{a}Ladd radii.
^{b}Estimated.

is greater than for any other cation except H^+ and B^{3+}, which again do not occur as such in crystals. The closest ratio is that for Al^{3+} and some similarities between the chemistries of Be and Al exist. Examples are the resistance of the metal to attack by acids owing to formation of an impervious oxide film on the surface, the amphoteric nature of the oxide and hydroxide, and Lewis acid behavior of the chlorides. However, Be shows just as many similarities to zinc, especially in the structures of its binary compounds and in the chemistry of its organic compounds. Thus BeS is insoluble in water, although Al_2S_3, CaS, and so on are rapidly hydrolyzed.

Calcium, Sr, Ba, and Ra form a closely allied series in which the chemical and physical properties of the elements and their compounds vary systematically with increasing size in much the same manner as in Group 1, the ionic and electropositive nature being greatest for radium. Again the larger ions can stabilize certain large anions, e.g., the peroxide and superoxide ions, polyhalide ions, and so on. Some examples of systematic group trends in the series Ca–Ra are (*a*) hydration tendencies of the crystalline salts increase; (*b*) solubilities of sulfates, nitrates, chlorides, and so on (fluorides are an exception) decrease; (*c*) solubilities of halides in ethanol decrease; *(d)* thermal stabilities of carbonates, nitrates, and peroxides increase; (*e*) rates of reaction of the metals with hydrogen increase. Other similar trends can be found.

All isotopes of *radium* are radioactive, the longest-lived isotope being ^{226}Ra (α; ~1600 years). This isotope is formed in the natural decay series of ^{238}U and was first isolated by Pierre and Marie Curie from pitchblende. Once widely used in radiotherapy, it has largely been supplanted by radioisotopes made in nuclear reactors.

The elements zinc, cadmium, and mercury, which have two electrons outside filled penultimate *d* shells, are classed in Group 12. Although the difference between the calcium and zinc subgroups is marked, zinc, and to a lesser extent cadmium, show some resemblance to beryllium or magnesium in their chemistry. We discuss these elements separately (Chapter 15), but note here that zinc, which has the lowest second ionization enthalpy in the Zn, Cd, Hg group, still has a value (1726 kJ mol^{-1}) similar to that of beryllium (1757 kJ mol^{-1}), and its standard potential (-0.76 V) is considerably less negative than that of magnesium.

There are a few ions with ionic radii and chemical properties similar to those of Sr^{2+} or Ba^{2+}, notably those of the $+2$ lanthanides (Section 19-13), especially the europium ion Eu^{2+}, and its more readily oxidized analogues, Sm^{2+} and Yb^{2+}. Because of this fortuitous chemical similarity, europium is frequently found in Nature in Group 2 minerals, and this is a good example of the geochemical importance of such chemical similarity.

Although the differences between the first and second ionization enthalpies, especially for beryllium, might suggest the possibility of a stable $+1$ state, there is no evidence to support this. Calculations using Born-Haber cycles show that owing to the much greater lattice energies of MX_2 compounds, MX compounds would be unstable and disproportionate:

$$2MX \longrightarrow M + MX_2$$

It may be noted that ^{90}Sr is one of the most toxic radionuclides formed in nuclear fission of ^{235}U and ^{239}Pu. It has been widely distributed by nuclear weapons testing and marginally increased by the Chernobyl accident. Analytical problems have been studied in detail[1] as have attempts to find extraction procedures using crown and other polyethers, polyethyleneglycols, etc.[2]

As a result of the small size, high ionization enthalpy, and high sublimation energy of beryllium, its lattice and hydration energies are insufficient to provide complete charge separation and the formation of simple Be^{2+} ions. In fact, in all compounds whose structures have been determined, even those of the most electronegative elements (i.e., BeO and BeF_2), there appears to be substantial covalent character in the bonding.

To allow the formation of two covalent bonds, unpairing of the $2s$ electrons is required. Although linear 2-coordination is confirmed for molecules in the vapor phase, examples being $BeCl_2$, $Be[N(SiMe_3)_2]_2$ and $BeBu_2^t$, and possibly in solution, it is not known so far in solids. The most common coordination number is 4, followed by 3. In many cases the higher coordination is achieved by dimerization or polymerization; when bulky ligands are used, monomeric 3-coordinate species can be obtained. Beryllium compounds can also achieve maximum coordination *via* Lewis acid behavior as in etherates, $BeCl_2(OR_2)_2$, or complex ions such as $[BeF_4]^{2-}$ and $[Be(H_2O)_4]^{2+}$.

It is to be noted that beryllium compounds are *exceedingly toxic*, particularly if inhaled, and great care and precautions must be taken in handling them. Chelating hydroxy- and polyaminocarboxylic acids have been studied[3] as sequestering agents for Be^{2+}.

4-2 Elemental Beryllium

The major ores are *beryl*, $Be_3Al_2(SiO_3)_6$, and *bertrandite*, a hydrated silicate; extraction from the ores is complicated. The element is obtained by electrolysis of molten $BeCl_2$ to which a small amount of NaCl is added to increase the electrical conductivity.

[1]K. Wendt *et al.*, *Angew. Chem. Int. Ed. Engl.* **1995**, *34*, 181; E. Finckh *et al.*, *Angew. Chem. Int. Ed. Engl.* **1995**, *34*, 183; A. Knöckel *et al.*, *Angew. Chem. Int. Ed. Engl.* **1995**, *34*, 186.
[2]R. D. Rogers *et al.*, *Inorg. Chem.* **1994**, *33*, 5682.
[3]A. Mederos *et al.*, *Inorg. Chem.* **1995**, *34*, 1579.

The gray metal is the second lightest after Li (1.86 g/cm^3) and quite hard and brittle. Since the absorption of electromagnetic radiation depends on the electron density in matter, beryllium has the lowest stopping power per unit mass thickness of all suitable construction materials. It is used for "windows" in X-ray apparatus and has other special applications in nuclear technology. Like aluminum, metallic beryllium is rather resistant to acids unless finely divided or amalgamated, owing to the formation of an inert and impervious oxide film on the surface. Thus although the standard potential (-1.70 V) would indicate rapid reaction with dilute acids (and even H$_2$O), the rate of attack depends greatly on the source and fabrication of the metal. For the very pure metal the relative dissolution rates are HF > H$_2$SO$_4 \sim$ HCl > HNO$_3$. The metal dissolves rapidly in 3M H$_2$SO$_4$ and in 5M NH$_4$F, but very slowly in HNO$_3$. Like aluminum, it also dissolves in strong bases, forming what is called the beryllate ion.

4-3 Binary Compounds

The white crystalline *oxide* BeO is obtained on ignition of beryllium or its compounds in air. It resembles Al$_2$O$_3$ in being highly refractory (mp 2570°C) and in having polymorphs; the high-temperature form (>800°C) is exceedingly inert and dissolves readily only in a hot syrup of concentrated H$_2$SO$_4$ and (NH$_4$)$_2$SO$_4$. The more reactive forms dissolve in hot alkali hydroxide solutions or fused KHSO$_4$.

Addition of the OH$^-$ ion to BeCl$_2$ or other beryllium solutions gives the *hydroxide*. This is amphoteric, and in alkali solution the *beryllate* ion, probably [Be(OH)$_4$]$^{2-}$, is obtained. When these solutions are boiled, the most stable of several polymorphs of the hydroxide can be crystallized.

In binary compounds generally, the structures are often those of the corresponding zinc analogs. Thus the low temperature form of BeO has the wurtzite structure while the most stable Be(OH)$_2$ polymorph has the Zn(OH)$_2$ structure.

Beryllium halides, all four of which are known, are deliquescent and cannot be obtained from their hydrates by heating, since HX is lost as well as H$_2$O. The fluoride is obtained as a glassy hygroscopic mass by heating (NH$_4$)$_2$BeF$_4$. The glassy form has randomly oriented chains of $\cdots$F$_2$BeF$_2$Be$\cdots$ similar to those in BeCl$_2$ and BeBr$_2$ but disordered. Two crystalline modifications are known, which appear to be structurally analogous to the quartz and cristobalite modifications of SiO$_2$. At 555°C BeF$_2$ melts to a viscous liquid that has low electrical conductivity. The polymerization in the liquid may be lowered by addition of LiF, which forms the [BeF$_4$]$^{2-}$ ion.

Beryllium chloride is prepared by passing CCl$_4$ over BeO at 800°C. On a small scale the pure chloride and bromide are best prepared by direct interaction in a hot tube. The white crystalline chloride (mp 405°C) dissolves exothermically in water; from HCl solutions the salt [Be(H$_2$O)$_4$]Cl$_2$ can be obtained. Beryllium chloride is readily soluble in oxygenated solvents such as ethers. In melts with alkali halides, chloroberyllate ions [BeCl$_4$]$^{2-}$ may be formed, but this ion does not exist in aqueous solution.

The structure of (BeCl$_2$)$_n$ is shown in Fig. 4-1. The coordination of Be is not exactly tetrahedral, since the ClBeCl angles are only 98.2° which means that the BeCl$_2$Be units are somewhat elongated in the direction of the chain axis. In (BeMe$_2$)$_n$ the CBeC angle is 114°. These distortions from the ideal tetrahedral angle are

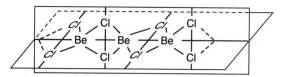

Figure 4-1 The structure of polymeric $BeCl_2$ in the crystal. The structure of $Be(CH_3)_2$ is similar.

related to the presence (or absence) of the lone pairs on the bridge group or atom that are available for bonding to the metal.

Beryllium hydride is difficult to obtain pure; it is doubtless polymeric with bridging hydrides. The *nitride*, Be_3N_2, is obtained by interaction of Be with NH_3 or N_2 at *ca.* 1000°C; the colorless crystals are hydrolyzed by water. The metal also reacts on heating with sulfur, and BeS has the ZnS type structure; with ethylene at *ca.* 450°C, BeC_2 is formed.

4-4 Coordination Complexes

Oxygen Ligands

In strongly acid solutions the aqua ion $[Be(H_2O)_4]^{2+}$ occurs, and crystalline salts with various anions can be obtained. The water in such salts is more firmly retained than is usual for aquates, indicating strong binding. Thus the sulfate is dehydrated to $BeSO_4$ only on strong heating, and $[Be(H_2O)_4]Cl_2$ loses no water over P_2O_5. Solutions of beryllium salts are acidic; this may be ascribed to the acidity of the aqua ion, the initial dissociation being

$$[Be(H_2O)_4]^{2+} \rightleftharpoons [Be(H_2O)_3(OH)]^+ + H^+$$

The addition of soluble carbonates to beryllium salt solutions gives only basic carbonates. Beryllium salt solutions also have the property of dissolving additional amounts of the oxide or hydroxide. This behavior is attributable to the formation of complex species with Be—OH—Be or Be—O—Be bridges. The rapidly established equilibria involved in the hydrolysis of the $[Be(H_2O)_4]^{2+}$ ion are very complicated and depend on the anion, the concentration, the temperature, and the pH. The main species, which will achieve four-coordination by additional water molecules, are considered to be $Be_2(OH)^{3+}$ and $Be_3(OH)_3^{3+}$ (probably cyclic).

The $[Be_3(OH)_3]^{3+}$ ion predominates at pH 5.5 in perchlorate solution. Various crystalline hydroxo complexes have been isolated. In concentrated alkaline solution the main species is $[Be(OH)_4]^{2-}$.

Other important beryllium oxygen complexes are the alkoxides, a typical example being the *tert*-butoxide (4-I); other, 3-coordinate alkoxides, e.g., $Be\{O(2,4,6-Bu_3^tC_6H_2)\}_2(OEt_2)$, are known and readily made by interaction of LiOR and $BeCl_2(OEt_2)_2$.[4] The tetrahedral structure of the latter was also confirmed, being similar to that in other solvates.

[4]P. P. Power *et al.*, *Inorg. Chem.* **1993**, *32*, 1724.

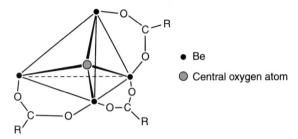

● Be
◉ Central oxygen atom

Figure 4-2   The structure of the basic carboxylate complexes $Be_4O(O_2CR)_6$. Only three RCO_2 groups are shown.

$$Bu'O—Be\underset{\underset{Bu'}{O}}{\overset{\overset{Bu'}{O}}{<}}Be\underset{\underset{Bu'}{O}}{\overset{\overset{Bu'}{O}}{>}}Be—OBu'$$

(4-I)

Some important complexes have the formula $Be_4O(O_2CR)_6$ and are formed by refluxing the hydroxide with carboxylic acids. These white crystalline compounds are soluble in organic solvents, even alkanes, but are insoluble in water and lower alcohols. They are inert to water but are hydrolyzed by dilute acids; in solution they are un-ionized. They have the structures illustrated in Fig. 4-2. The central oxygen atom is tetrahedrally surrounded by the four beryllium atoms and each beryllium atom is tetrahedrally surrounded by four oxygen atoms. Zinc and Co also form such complexes, as does the ZrO^{2+} ion, with benzoic acid. The zinc complexes are rapidly hydrolyzed by water, in contrast to those of beryllium. The acetate complex has been utilized as a means of purifying beryllium by solvent extraction from an aqueous solution into an organic layer.

There are many other carboxylic and oxyacid complexes that have been studied in solution. The salicylate, obtained by interaction of the acid with $BeSO_4·4H_2O$ and $Ba(OH)_2·8H_2O$ in water, has a relatively simple structure; $Be[C_6H_4(O)CO_2](OH_2)_2$ has tetrahedral Be with the chelate salicylate bound through the phenolic O atom and the η^1-OC(O) group.[5]

Nitrogen Ligands

Although simple amine complexes such as $[Be(NH_3)_4]Cl_2$ are known, they are sensitive to hydrolysis.

Dialkylamides, $[Be(NR_2)_2]_n$, are made from $BeCl_2$ and $LiNR_2$. A representative example, (4-II), has both 3- and 4-coordinate Be.

$$Me_2NBe\underset{\underset{Me_2}{N}}{\overset{\overset{Me_2}{N}}{<}}Be\underset{\underset{Me_2}{N}}{\overset{\overset{Me_2}{N}}{>}}BeNMe_2$$

(4-II)

[5]H. Schmidbauer *et al., Inorg. Chem.* **1991**, *30*, 3101.

Pyrazolylborates, whose steric demands can be readily altered by substitution, have been studied both structurally and by solvent extraction procedures. The compound [η^3-HB(3-Bu^{t}pz)$_3$]BeMe was made by interaction of the thallium salt of HTp* and (BeMe$_2$)$_n$ with elimination of TlMe; it has the structure shown in (4-III).[6]

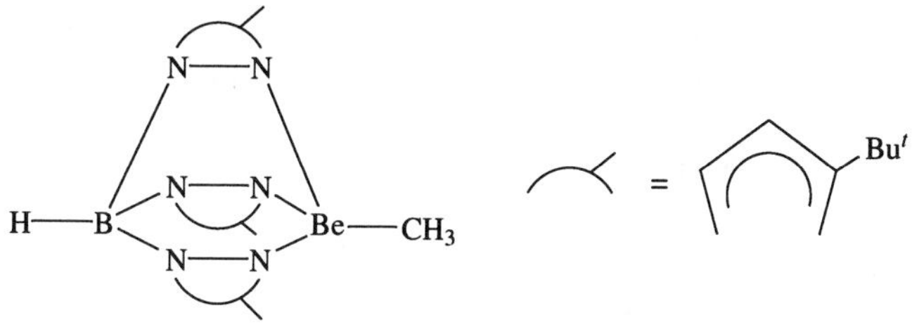

(4-III)

Detailed studies have been made using [H$_2$B(pz)$_2$]$^-$, [HB(pz)$_3$]$^-$, and [B(pz)$_4$]$^-$ for the extraction of BeCl$_2$ (and also Mg^{2+}, Ca^{2+}, and Sr^{2+}) from aqueous solutions into CH$_2$Cl$_2$.[7] The extractions can be selective and the crystal structures of Be[B(pz)$_4$]$_2$ and [Be(HB(pz)$_3$)OH]$_3$ were determined; in the former the Be is tetrahedral and bound to 4 N atoms while in the latter there is a cyclic (BeOH)$_3^{3+}$ moiety.

The *phthalocyanine* complex is unique in that the rigid configuration of the macrocycle leads to planar Be bound to four N atoms.

Sulfur and Phosphorus Ligands. Only a few beryllium compounds of second row elements are known. Examples are the sulfide[8] Be[S(2,4,6-Bu$^t_3$C$_6$H$_2$)]$_2$thf·toluene, Cp*Be(PBu$^t_2$)[9] and the pyrazolylborate (L) thiolate, LBeSH.

Halide Complexes. The *tetrafluoroberyllates*, which are obtained by dissolving BeO or Be(OH)$_2$ in concentrated solutions or melts of acid fluorides such as NH$_4$HF$_2$. The tetrahedral ion has a crystal chemistry similar to that of SO$_4^{2-}$, and corresponding salts (e.g., PbBeF$_4$ and PbSO$_4$) usually have similar structures and solubility properties. Beryllium fluoride (BeF$_2$) readily dissolves in water to give mainly BeF$_2$(H$_2$O)$_2$ according to ^{9}Be nmr spectra. In 1 *M* solutions of (NH$_4$)$_2$BeF$_4$ the ion BeF$_3^-$ occurs from 15 to 20%.

There are also chloro complexes M^{I}Be$_2$Cl$_5$, M = K, Rb, NH$_4$, that have μ-Cl atoms in a chain structure quite unlike the M^{I}Be$_2$F$_5$ analogs that have infinite sheet anions in which there are hexagonal rings of BeF$_4$ tetrahedra sharing corners (*cf.* silicate anions Si$_2$O$_5^{2-}$, Section 8-7).

4-5   Organoberyllium Compounds

Although beryllium alkyls can be obtained by the interaction of BeCl$_2$ with lithium alkyls or Grignard reagents, they are best made in a pure state by heating the metal and a mercury dialkyl, for example:

$$HgMe_2 + Be \xrightarrow{110°C} BeMe_2 + Hg$$

[6]R. Han and G. Parkin, *Inorg. Chem.* **1993**, *32*, 4968.
[7]Y. Sohrin *et al.*, *J. Am. Chem. Soc.* **1993**, *115*, 4128.
[8]P. P. Power *et al.*, *Inorg. Chem.* **1993**, *32*, 1724.
[9]J. L. Atwood *et al.*, *J. Chem. Soc., Chem. Commun.* **1990**, 692.

The alkyl can be collected by sublimation or distillation in a vacuum. On the other hand, the aryls are made by reaction of a lithium aryl in a hydrocarbon with $BeCl_2$ in diethyl ether:

$$2LiC_6H_5 + BeCl_2 \longrightarrow 2LiCl + Be(C_6H_5)_2$$

The alkyls are liquids or solids of high reactivity, spontaneously flammable in air and violently hydrolyzed by water. The bonding in polymers like $(BeMe_2)_n$ is of the 3c-2e type.

The aryls can also be made by the reaction:

$$3BeEt_2 + 2B(aryl)_3 \longrightarrow 3Be(aryl)_2 + 2BEt_3$$

The *o*- and *m*-tolyl compounds are dimers, presumed to have structure (4-IV).

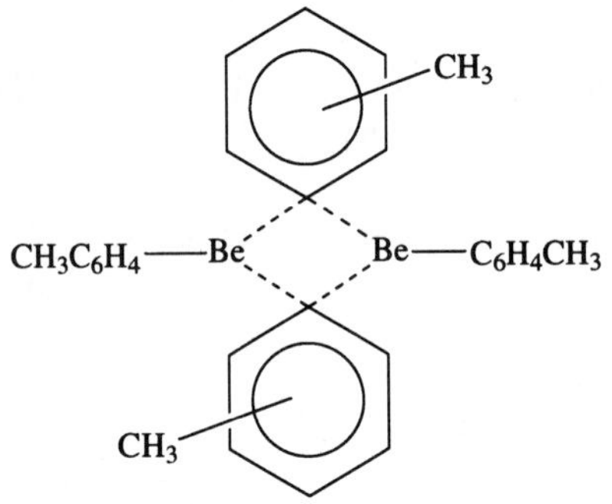

(4-IV)

The higher alkyls are progressively less highly polymerized; diethyl- and diisopropylberyllium are dimeric in benzene, but the *tert*-butyl is monomeric; the same trend is found in aluminum alkyls.

As with several other elements, notably Mg and Al, there are close similarities between the alkyls and hydrides, especially in the complexes with donor ligands. For the polymeric alkyls, especially $BeMe_2$, strong donors such as Et_2O, Me_3N, or Me_2S are required to break down the polymeric structure. Mixed hydrido alkyls are known: thus pyrolysis of diisopropylberyllium gives a colorless, nonvolatile polymer:

$$x(iso\text{-}C_3H_7)_2Be \xrightarrow{200°C} [(iso\text{-}C_3H_7)BeH]_x + xC_3H_6$$

However, above 100°C the *tert*-butyl analogue gives pure BeH_2. With tertiary amines, reactions of the following types may occur:

$$BeMe_2 + Me_3N \longrightarrow Me_3NBeMe_2$$

$$2BeH_2 + 2R_3N \longrightarrow [R_3NBeH_2]_2$$

The trimethylamine hydrido complex appears to have structure (4-V).

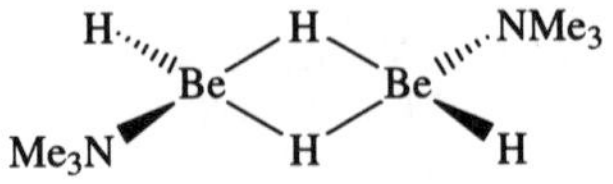

(4-V)

Beryllium alkyls give colored complexes with 2,2'-bipyridine [e.g., bipy Be(C$_2$H$_5$)$_2$, which is bright red]; the colors of these and similar complexes with aromatic amines given by beryllium, zinc, cadmium, aluminum, and gallium alkyls are believed to be due to electron transfer from the M—C bond to the lowest unoccupied orbital of the amine. The pyrazolylborate (L), LBeMe, mentioned earlier, Li$_2$BeMe$_4$, and Me$_2$Be(quinuclidine) are also known.

Although the cyclopentadienyls (η^5-C$_5$H$_5$)BeX, X = Cl, Br, Me have a normal Cp ring, the very air sensitive Cp$_2$Be has rather weakly bound rings with one η^1 and one η^5 ring. As a consequence, Cp$_2$Be undergoes two types of very rapid fluxional processes on heating[10] according to ^{1}H nmr spectra. These can be formulated as η^2, η^5, gear wheel mechanism, and η^3, η^3, transitions associated with molecular inversion.

MAGNESIUM, CALCIUM, STRONTIUM, BARIUM, AND RADIUM

4-6 Occurrence; The Elements

Except for radium, the elements are widely distributed in minerals and in the sea. They occur in substantial deposits such as *dolomite* (CaCO$_3$·MgCO$_3$), *carnallite* (MgCl$_2$·KCl·6H$_2$O), and *barytes* (BaSO$_4$). Calcium is the third most abundant metal terrestrially.

Magnesium is produced in several ways. An important source is dolomite from which, after calcination, the calcium is removed by ion exchange using seawater. The equilibrium is favorable because the solubility of Mg(OH)$_2$ is lower than that of Ca(OH)$_2$:

$$Ca(OH)_2 \cdot Mg(OH)_2 + Mg^{2+} \longrightarrow 2Mg(OH)_2 + Ca^{2+}$$

The most important processes for preparation of magnesium are (*a*) the electrolysis of fused halide mixtures (e.g., MgCl$_2$ + CaCl$_2$ + NaCl) from which the least electropositive metal Mg is deposited, and (*b*) the reduction of MgO or of calcined dolomite (MgO·CaO). The latter is heated with ferrosilicon:

$$CaO \cdot MgO + FeSi = Mg + silicates\ of\ Ca\ and\ Fe$$

and the magnesium is distilled out. Magnesium oxide can be heated with coke at 2000°C and the metal deposited by rapid quenching of the high-temperature equilibrium, which lies well to the right:

$$MgO + C \rightleftharpoons Mg + CO$$

Magnesium may in the long run replace aluminum in many applications because the supply available in seawater is virtually unlimited.

Calcium and the other metals are made only on a relatively small scale, by the reaction:

$$6CaO + 2Al \xrightarrow{\ 1200°C\ } 3Ca + Ca_3Al_2O_6$$

or reduction of the halides with sodium.

[10]P. E. Blöchl *et al.*, *J. Am. Chem. Soc.* **1994,** *116,* 11177.

Radium is isolated in the processing of uranium ores; after coprecipitation with barium sulfate, it can be obtained by fractional crystallization of a soluble salt.

Magnesium is a greyish-white metal with a surface oxide film that protects it to some extent chemically—thus it is not attacked by water, despite the favorable potential, unless amalgamated. It is readily soluble in dilute acids and is attacked by most alkyl and aryl halides in ether solution to give Grignard reagents.

Highly reactive Mg can be made in various ways[11] for use in reductions, for example, by reduction of the halide with molten Na or K or more conveniently by interaction of the metal with anthracene and MeI in THF, which gives an orange "adduct":

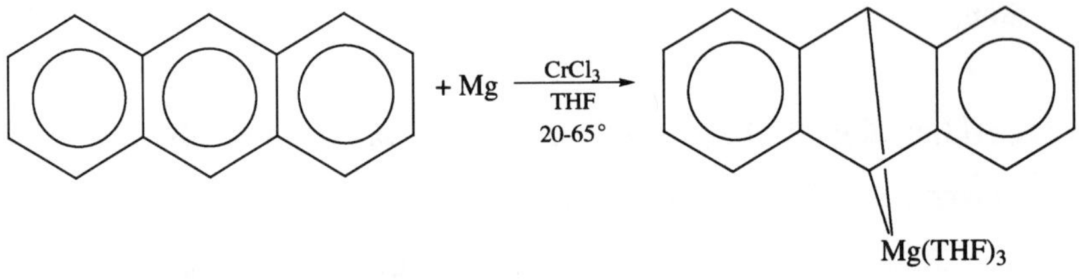

or by decomposition of MgH_2 at low pressure and 250°C.

Calcium and the other metals are soft and silvery, resembling sodium in their chemical reactivities, although somewhat less reactive. These metals are also soluble, though less readily and to a lesser extent than sodium, in liquid ammonia, giving blue solutions similar to those of the Group 1 metals. These blue solutions are also susceptible to decomposition (with the formation of the amides) and have other chemical reactions similar to those of the Group 1 metal solutions. They differ, however, in that moderately stable metal ammines such as $Ca(NH_3)_6^{2+}$ can be isolated on removal of solvent at the boiling point.

Calcium, Sr, and Ba metals are also activated in thf and other ethers or toluene when saturated with NH_3. Syntheses of organo compounds, alkoxides, amides, etc., can be achieved readily from the metals by reaction with compounds having acidic H-atoms.[12]

4-7 Binary Compounds

Oxides

These are white, high melting crystalline solids with NaCl type lattices. Calcium oxide, mp 2570°C, is made on a vast scale:

$$CaCO_3 \rightleftharpoons CaO + CO_2(g) \quad \Delta H^0_{298} = 178.1 \text{ kJ mol}^{-1}$$

Magnesium oxide is relatively inert, especially after ignition at high temperatures, but the other oxides react with water, evolving heat, to form the hydroxides. They also absorb carbon dioxide from the air. Magnesium hydroxide is insoluble in water

[11]A. Fürstner, *Angew. Chem. Int. Ed. Engl.* **1993**, *32*, 164.
[12]S. R. Drake and D. J. Otway, *Polyhedron* **1992**, *11*, 745.

($\sim 1 \times 10^{-4}$ g/L at 20°C) and can be precipitated from Mg^{2+} solutions; it is a much weaker base than the Ca-Ra hydroxides, although it has no acidic properties and unlike $Be(OH)_2$ is insoluble in an excess of hydroxide. The Ca-Ra hydroxides are all soluble in water, increasingly so with increasing atomic number [$Ca(OH)_2$, ~2g/L; $Ba(OH)_2$, ~60 g/L at ~20°C], and all are strong bases.

There is no optical transition in the electronic spectra of the M^{2+} ions, and they are all colorless. Colors of salts are thus due only to colors of the anions or to lattice defects. The oxides may also be obtained with defects, and BaO crystals with ~0.1% excess of metal in the lattice are deep red.

Halides

The anhydrous halides can be made by dehydration of the hydrated salts. For rigorous studies, however, magnesium halides are best made by removal of solvent from solvates made by reactions such as:

$$Mg + HgX_2 \xrightarrow{\text{boiling ether}} MgX_2(solv) + Hg$$

$$2Mg + SnCl_4 \xrightarrow{\text{THF}} 2MgCl_2(THF)_2 + Sn$$

Magnesium and calcium halides readily absorb water. The tendency to form hydrates, as well as the solubilities in water, decrease with increasing size, and Sr, Ba, and Ra halides are normally anhydrous. This is attributed to the fact that the hydration energies decrease more rapidly than the lattice energies with the increasing size of M^{2+}.

The fluorides vary in solubility in the reverse order (i.e., Mg < Ca < Sr < Ba) because of the small size of the F^- relative to the M^{2+} ion. The lattice energies decrease unusually rapidly because the large cations make contact with one another without at the same time making contact with the F^- ions.

The halides are all typically ionic solids, but can be vaporized as molecules, the structures of which are not all linear. On account of its dispersion and transparency properties, CaF_2 is used for prisms in spectrometers and for cell windows (especially for aqueous solutions). It is also used to provide a stabilizing lattice for trapping lanthanide +2 ions.

Hydrides

Magnesium hydride is obtained by interaction of Mg and H_2 at *ca.* 500°C. An active form is readily made by interaction of $PhSiH_3$ and Bu_2^nMg in ether-heptane solvents in presence of tmen. The solvated pyrophoric powders are used as reducing agents and hydride sources.[13] The hydrides of Ca, Sr, and Ba are also made by direct interaction (Section 2-13). For Ca, Sr, and Ba the structure is of the $PbCl_2$ type, whereas MgH_2 has the rutile structure.[14]

---

[13]M. J. Michalczyk, *Organometallics* **1992**, *11*, 2307.
[14]For refs see M. Kaupp and P. von R. Schleyer, *J. Am. Chem. Soc.* **1993**, *115*, 11202.

Carbides

All the metals or their oxides react with carbon at high temperatures to give carbides MC_2.[15] These are ionic acetylides whose properties are discussed in Chapter 7. The carbide Mg_2C_3[16] has been obtained relatively pure by reacting Mg dust with pentane at *ca.* 680°C. This has Mg atoms coordinated by four linear C_3 groups in a tetrahedral arrangement, three *via* terminal C atoms and one by a

$$C-\!\!-C\overset{\displaystyle Mg}{\overset{\diagup\diagdown}{-\!\!-}}C$$

unit.

Other Compounds

Direct reaction of the metals with other elements can lead to binary compounds such as borides, silicides, arsenides, and sulfides. Many of these are ionic and are rapidly hydrolyzed by water or dilute acids. At ~300°C, Mg reacts with N_2 to give colorless, crystalline Mg_3N_2 (resembling Li and Be in this respect). The other metals also react normally to form M_3N_2, but other stoichiometries are known. One such compound is Ca_2N, which has an *anti*-$CdCl_2$ type of layer structure, as does Cs_2O. In Ca_2N, however, there is one "excess" electron per formula unit. These excess electrons evidently occupy delocalized energy bands within metal atom layers, causing a lustrous, graphitic appearance. Ba_2N is similar, while another compound, $NaBa_3N_7$ is obtained on slow cooling of Ba and N_2 in liquid sodium. It has a chain structure with Ba^{2+} and N_3^- units and there is a close structural and bonding relationship to the alkali metal suboxides[17] (Section 3-5).

Zintl phases[18] such as $Ca_{31}Sn_{20}$ and $Sr_{31}Pb_{20}$ are known; they are made by high temperature interaction of the elements. Other phases such as Ca_5Si_3, CaSn, and Ca_2Sn are characterized.

4-8   The Ions: Salts and Complexes[19]

Apart from complexes such as the alkoxides, dialkylamides, and organo compounds discussed in later sections, all of the Mg—Ra elements give ionic compounds in the solid state and in solutions. They form salts with common anions like the halides, NO_3^-, RCO_2^-, ClO_4^-, SO_4^{2-}, and so on. Because of its smaller size Mg^{2+} is the hardest of the ions. While Mg^{2+} and Ca^{2+} salts are usually hydrated, those of Sr, Ba, and Ra are commonly anhydrous. The coordination number for Mg^{2+} is almost invariably 6 but a few cases of 5, 7, and 8 coordination are known.[20] The

[15]R. Hoffmann *et al., Inorg. Chem.* **1992,** *31,* 1734 and references therein.

[16]H. Fjellvåg and P. Karen, *Inorg. Chem.* **1992,** *31,* 3260.

[17]P. E. Rauch and A. Simon, *Angew. Chem. Int. Ed. Engl.* **1992,** *31,* 1520.

[18]J. D. Corbett *et al., Inorg. Chem.* **1993,** *32,* 4349.

[19]N. S. Poonia and A. V. Bajoj, *Chem. Rev.* **1979,** *79,* 291; *Coord. Chem. Rev.* **1988,** *87,* 55 (971 refs).

[20]O. Carugo *et al., J. Chem. Soc., Dalton Trans.* **1993,** 2127 (comparisons of Mg^{2+} and Ca^{2+} coordination from crystallographic data); J. P. Glusker *et al., J. Am. Chem. Soc.* **1996,** *118,* 5752 (Ca^{2+} coordination and comparison with Be, Mg, and Zn); *Inorg. Chem.* **1994,** *33,* 419 ($[Mg(H_2O)_6]^{2+}$ coordination from crystallographic data).

heavier elements commonly have coordination numbers of 7 and 8 and some examples of 9 and 10 coordination are known. For all the elements, when forming covalent compounds such as alkoxides or dialkylamides, 3- and 4-coordination is very common.

The cations Mg^{2+} and Ca^{2+} can be solvated by acetone, methanol, MeCN, HMPA, THF, and other ethers. In aqueous solutions the formation constants of complexes vary enormously and depending on the choice of ligand high selectivity for a particular ion can be obtained. These are the main trends:

1. For small or highly charged anions and certain uni- and bidentate ligands, the constants decrease with increasing crystal radii, Mg > Ca > Sr > Ba.
2. For oxoanions like NO_3^-, SO_4^{2-}, and IO_4^-, the order accords with the hydrated radii, Mg < Ca < Sr < Ba.
3. For hydroxycarboxylic, polycarboxylic, and polyaminocarboxylic acid ligands, the order is Mg < Ca > Sr > Ba.

The formation constants for cryptates, which show greatly enhanced stability and selectivity compared to crown ethers, can be $\sim 10^6$ times greater than those of the ethers. As for Na^+ and K^+, there are both synthetic and natural ligands with selective affinity for Ca^{2+}, the latter exercising a controlling effect in Ca^{2+} metabolism. One example is the antibiotic A_{23187}, a monocarboxylic acid that binds and transports Ca^{2+} across membranes. It is tridentate and forms a 7-coordinate complex $Ca(N,O,O)_2(H_2O)$.

Oxygen chelate compounds, among the most important being those of the ethylenediaminetetraacetate (EDTA) type, readily form complexes in alkaline aqueous solution, for example:

$$Ca^{2+} + EDTA^{4-} = [Ca(EDTA)]^{2-}$$

The complexing of calcium by EDTA and also by polyphosphates is of importance, not only for removal of calcium ions from water, but also for the volumetric estimation of calcium. In $Ca[CaEDTA]\cdot 7H_2O$ the coordination of the Ca ion is 8. Calcium ions are involved in many natural processes such as bones $(Ca_5(PO_4)_3OH)$, blood clotting, transmission of nerve impulses, and so on as well as in selective binding to various proteins with a range of functions.

4-9 Alkoxides and Related Compounds[21]

Alkoxides and aryloxides have been intensively studied because of their interest, especially for Ba, as precursors to oxide materials having desirable electronic properties; this development was stimulated by the discovery of superconducting materials such as $YBa_2Cu_3O_{7-x}$. The same interest applies to other volatile complexes like β-diketonates. The alkyl or aryl oxides (also siloxides) are readily obtained by a variety of methods, notably: (a) direct interaction of the metals with ROH in THF, toluene saturated with NH_3,[22] or other ether solvents; (b) the interaction of

[21]D. C. Bradley, *Chem. Rev.* **1989**, *89*, 1317.
[22]S. R. Drake *et al.*, *Inorg. Chem.* **1993**, *32*, 5704.

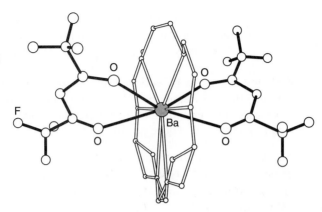

Figure 4-3   Molecular structure of $Ba(hfa)_2(18\text{-crown-}6)$. Reproduced by permission from *J. Chem. Soc., Chem. Commun.* **1991,** 971.

$MI_2(THF)_4$ in THF with KOR^{23}; (c) the interaction of metal alkyls or amides[24] with ROH.

The structural types range from discrete monomeric species through relatively simple dimers and trimers to large complicated aggregates such as $Ca_9(OCH_2CH_2OMe)_{18}(HOCH_2CH_2OMe)_2$ or $Ba_6O(OC_2H_4OMe)_{10}(HOC_2H_4OMe)_4$. The complexity can arise because of conformational and bonding flexibility for RO ligands and, as well, not only η^1 but μ_1-, μ_2-, and μ_3-bonding modes. Additional complexity can be introduced also by having different metals present as in $Mg[Al(OBu^s)_4]_2$ and $(THF)_2Mg[(\mu\text{-}OPh)_2Al(OPh)_2]_2{}^{25}$ and the ethyleneglycol complex $BaCu(C_2H_6O_2)_n(C_2H_4O_2)_2$ $n = 3, 6.^{26}$ Particularly for the largest ion, Ba^{2+}, the X-ray crystal structures are very complicated with the metal often having different coordination numbers in the same compound.[27]

A relatively simple structure is that of $Ca(OAr)_2(THF)_3$, Ar = $2,6\text{-}Bu_2^t\text{-}4\text{-}MeC_6H_2$, which has a distorted *tbp* coordination for Ca. Other types are the alkane soluble Sr and Ba species made by the reaction

$$Ba(OBu^t)_2 + [Sn(OB\mu^t)_2]_2 \longrightarrow Ba[(OBu^t)_3Sn]_2$$

in which the Ba atom is in the middle of a "sandwich" type structure with octahedral BaO_6 and trigonal planar O atoms.

β-diketonates. These compounds[28] are made by methods similar to those for the alkoxides. Many of them are volatile, examples being the Ca, Sr, and Ba hexafluoroacetylacetonates having coordinated crown ether 18-C-6. The structure of the Ba compound,[29] Fig. 4-3, has the chelate hfa ligands above and below the plane of the crown ring.

[23]T. P. Hanusa *et al., J. Am. Chem. Soc.* **1994,** *116,* 2409.
[24]K. G. Caulton *et al., Inorg. Chem.* **1994,** *33,* 994.
[25]J. W. Gilje *et al., Polyhedron* **1994,** *13,* 1045.
[26]C. J. Page *et al., Inorg. Chem.* **1992,** *31,* 1784.
[27]H. Bock *et al., Angew. Chem. Int. Ed. Engl.* **1995,** *34,* 1353.
[28]S. R. Drake *et al., Inorg. Chem.* **1995,** *34,* 5295; **1993,** *32,* 4464; *J. Chem. Soc., Dalton Trans.* **1993,** 2883; R. E. Sievers *et al., Inorg. Chem.* **1994,** *33,* 798; D. J. Williams *et al., Inorg. Chem.* **1993,** *32,* 3233.
[29]J. A. T. Norman and G. P. Pez, *J. Chem. Soc., Chem. Commun.* **1991,** 971.

4-10   Complexes of Nitrogen and Phosphorus Ligands

Nitrogen Ligands

Ammonia and alkyl amines do not form complexes in aqueous solution, but the solid halides can give adducts of the type $MCl_2 \cdot nNH_3$ and magnesium halides give complexes such as $MgBr_2py_4$.

Perhaps the most important of magnesium complexes are the chlorophylls that are involved in photosynthesis in plants; the structure of one of these is shown in diagram (4-VI).

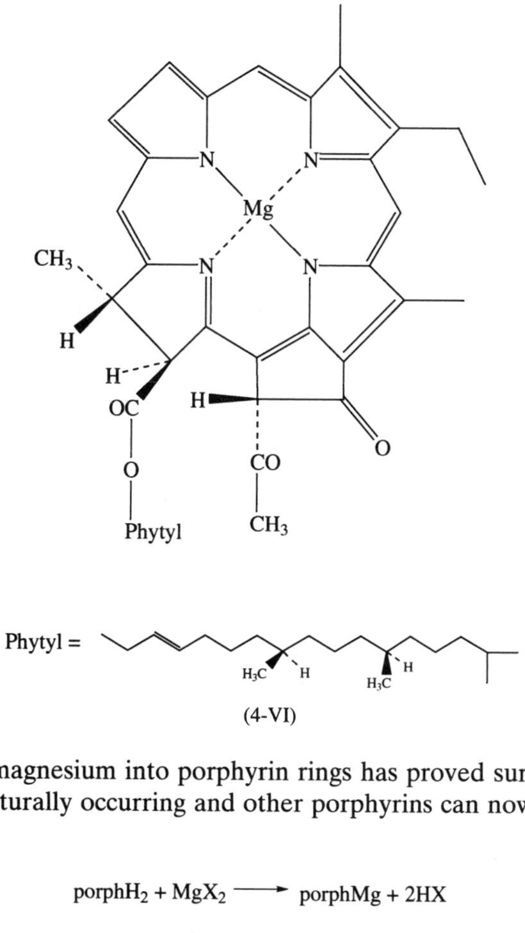

(4-VI)

The insertion of magnesium into porphyrin rings has proved surprisingly difficult. Compounds of naturally occurring and other porphyrins can now be made readily by the reaction

$$porphH_2 + MgX_2 \longrightarrow porphMg + 2HX$$

using non-coordinating and nonionizing solvents such as toluene with a nucleophilic base like NEt_3 and magnesium compounds like $MgBr_2(OEt)_2$ or MgI_2.[30] An Mg porph complex linked to Zn and free base porphyrins in a "molecular wire" that transmits light energy can act as a fluorescence switch.[31]

It may be noted that Mg^{2+} is too large to give a planar complex and is normally out of plane with a square pyramidal geometry. This exposure makes the magnesium easy to remove from the porphyrin ring.

[30]J. S. Lindsey and J. N. Woodford, *Inorg. Chem.* **1995,** *34,* 1063.
[31]D. F. Bocian *et al., J. Am. Chem. Soc.* **1996,** *118,* 3996.

Amides. These are a major class of compounds readily obtained *via* reactions such as

$$Ba + HN(SiMe_3)_2 \xrightarrow{\text{THF/NH}_3} Ba[N(SiMe_3)_2]_2(THF)_2$$

$$BaCl_2 + 2NaN(SiMe_3)_2 \xrightarrow{\text{Et}_2O} Ba[N(SiMe_3)_2]_2(OEt_2)_2$$

The desolvated amides can be obtained by heating the solvates in vacuum.

The structures[32] of $M[N(SiMe_3)_2]_2$, M = Mg—Ba, have 3-coordinate metal atoms. The Mg compound is planar, the Ca and Sr compounds show slight pyramidalization, while the Ba compound is pyramidal (see 4-VIIA,B).

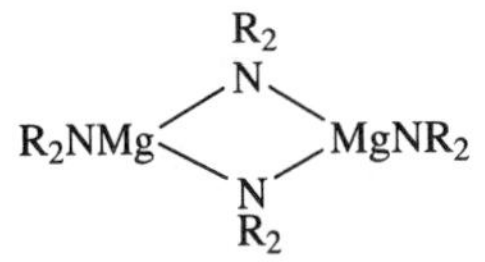

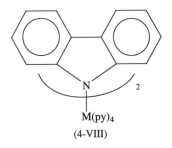

(4-VIIA) 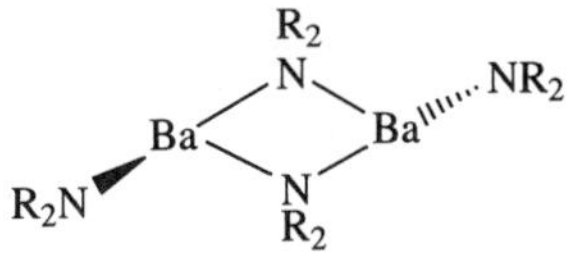 (4-VIIB)

Other amides include the benzyls $[Mg(NBz_2)]_2$ and $Li_2Mg(NBz_2)_4$[33] and the monomeric, non-linear, 2-coordinate complex, $Mg[N(SiMePh_2)_2]$[34]; there are many more complicated ones. Mixed Al—Mg compounds such as $[Me_2Al(\mu\text{-}Et_2N)_2Mg\mu\text{-}Me]_2$ can undergo insertions of carbodiimides, $R'NCNR'$, into the Mg—C bond,[35] a reaction well known for transition metal-to-carbon bonds as in WMe_6. The unique *o*-phenylenediaminomagnesium cluster[36] is shown in Fig. 4-4; the nitride $Mg_6(N)(BuNH)_9$ and organometallic clusters are also known.

Different types of amides[37] are those of Ca, Sr, and Ba made by interaction of the metal with carbazole in the presence of NH_3 and pyridine (4-VIII).

M(py)$_4$

(4-VIII)

Imido compounds have been made by interaction of amines with Grignard or dimagnesium compounds. The reaction with Grignards gives what are termed "magnesylamines."

[32]M. Kraupp and P. von R. Schleyer, *J. Am. Chem. Soc.* **1993**, *115*, 11202 and references therein.
[33]W. Clegg *et al.*, *J. Chem. Soc., Chem. Commun.* **1994**, 769.
[34]P. P. Power *et al.*, *Inorg. Chem.* **1994**, *33*, 4800 (gives many references).
[35]C-C. Chang *et al.*, *Organometallics* **1996**, *15*, 2571.
[36]W. Clegg *et al.*, *J. Chem. Soc., Chem. Commun.* **1994**, 97.
[37]P. von R. Schleyer *et al.*, *J. Am. Chem. Soc.* **1992**, *114*, 10880.

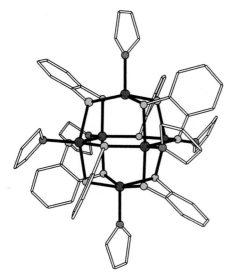

Figure 4-4 The molecular structure of *o*-phenylenediamido magnesium tetrahydrofuran solvate (hydrogen atoms omitted). Reproduced by permission from *J. Chem. Soc., Chem. Commun.* **1994**, 97.

$$RNH_2 + 2R'MgX \longrightarrow \text{"}RN(MgX)_2\text{"} + 2R'H$$

The actual product from interaction of EtMgBr and PhNH$_2$ has the structure [(Et$_2$O)Mg]$_6$(NPh)$_4$Br$_4$. The similar reaction with MgEt$_2$ gives the imido cluster [Mg(NPh)THF]$_6$ which can act as a NPh transfer agent, e.g., to Ph$_2$CO giving Ph$_2$CNPh.[38]

Pyrazolylborates. A number of compounds of these ligands have been structurally characterized.[39,40,41] Derivatives of the type (η^3-Tp*)MgR, R = Me, Ph, CH$_2$SiMe$_3$, OR, SR, Cl, etc., are 4-coordinate. They can undergo a disproportionation reaction to give 6-coordinate "sandwich" complexes (η^3-Tp*)$_2$Mg.[39] Sandwich complexes of Ca, Sr, and Ba, MTp$_2^*$, have now been made directly by interaction of MI$_2$ with KTp* or KBp*, Bp* = [H$_2$B(3,5-Me$_2$pz)$_2$].[41]

Pyrazolylborates have now been also used to extract Mg, Ca, Sr, and Be from aqueous solutions into CH$_2$Cl$_2$: selectivity depends on the nature of the ligand. The structures of [HBpz$_3$]$_2$M, M = Mg, Ca have distorted octahedral coordination with tridentate N bonding; the beryllium compounds discussed earlier were tetrahedral.

Phosphides[42] similar to the amides have been made for Ca and Ba by interaction of M[N(SiMe$_3$)$_2$]$_2$ with (Me$_3$Si)$_2$PH, an example being Ba[P(SiMe$_3$)$_2$]$_2$(THF)$_4$. Towards tertiary phosphines, the Group 2 elements are notoriously weak acceptors, but a complex of Mg that also has Mg—C bonds (4-IX) is known.[43]

[38]P. P. Power *et al.*, *Inorg. Chem.* **1996**, *35*, 3254; **1994**, *33*, 4860.

[39]R. Haq and G. Parkin, *J. Am. Chem. Soc.* **1992**, *114*, 748.

[40]E. Carmona *et al.*, *Polyhedron* **1996**, *15*, 3453.

[41]M. Ruf and H. Vakrenkamp, *Inorg. Chem.*, **1996**, *35*, 6571.

[42]S. R. Drake *et al.*, *Polyhedron* **1993**, *12*, 2307; M. Westerhausen and W. Schwartz, *J. Organomet. Chem.* **1993**, *463*, 51.

[43]G. van Koten *et al.*, *Inorg. Chem.* **1996**, *35*, 3436.

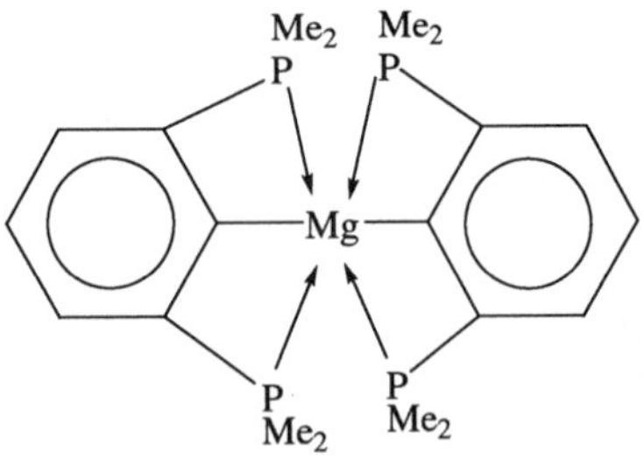

(4-IX)

A *stannide*[44] was made by the reaction:

$$Ca + Me_3Sn\!-\!SnMe_3 \xrightarrow{\text{THF}} Ca(SnMe_3)_2(THF)_4$$

4-11   Other Complexes

The -SR, -SeR, and -TeR compounds[45] can be obtained by addition, e.g., of mes*SH to $[Mg\{N(SiMe_3)_2\}_2]_2$, mes* = $2,4,6\text{-Bu}_3^t C_6 H_2$. Similar complexes of Ca, Sr, and Ba have also been made. Typical examples are the 3-coordinate $[Mg(Smes^*)_2]_2$, the 4-coordinate $Mg(Semes^*)_2(THF)_2$ and $Mg[TlSi(SiMe_3)_3]_2$.

4-12   Organometallic Compounds[46]

Magnesium compounds are the most important for both organic and inorganic syntheses; the most widely used reagents are Grignards, RMgX, and dialkyl or diaryl magnesiums. The Grignards are made by interaction of Mg with an organic halide in Et_2O or other ethers. The reaction is fastest with iodides and I_2 can be used as an initiator with RCl or RBr.

The nature of Grignard reagents *in solution* has led to much confusion in part due to impurities, to traces of oxygen or water, and to differing conditions and methods of study. The initial, rate-determining step in their formation is that of electron transfer from the Mg surface to the σ^* anti-bonding orbital of the C—X bond. This produces a radical anion–radical cation pair; further reactions involving radicals lead to the reagents.[47] It is further well established that the following equilibria are involved between *solvated* species:

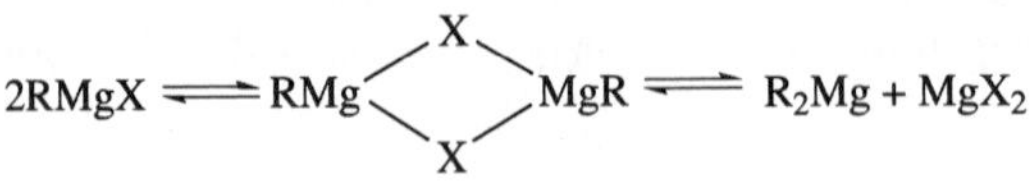

$$2RMgX \rightleftharpoons RMg\underset{X}{\overset{X}{\diamondsuit}}MgR \rightleftharpoons R_2Mg + MgX_2$$

[44]M. Westerhausen, *et al., Angew. Chem. Int. Ed. Engl.* **1994,** *33,* 1493.

[45]K. Ruhlandt-Senge *et al., Inorg. Chem.* **1995,** *34,* 2587, 3499; D. Grindelberger and J. Arnold, *Inorg. Chem.* **1994,** *33,* 6293.

[46]C. E. Holloway and M. Melnik, *J. Organomet. Chem.* **1994,** *465,* 1 (structural data to end 1991); E. Weiss, *Angew. Chem. Int. Ed. Engl.* **1993,** *32,* 1523 (Mg).

[47]H. M. Walborsky and M. Topolski, *J. Am. Chem. Soc.* **1992,** *114,* 3455 and references therein.

In dilute solutions and in the more strongly donor solvents, monomeric species predominate but in Et_2O, at high Mg concentrations, polymeric species are present. The equilibria have been studied by nmr and by EXAFS[48] spectra.

The isolation and structural determination of several crystalline RMgX compounds shows that the essential structure is $RMgX(solvent)_n$. Specific examples are $PhMgBr(OEt_2)_2$ and $MeMgBr(THF)_3$, which are respectively 4- and 5-coordinate.[49] One of the most unusual Grignard reagents is the completely substituted ruthenocene $[C_5(MgCl)_5]_2Ru$ made from the corresponding $[C_5(HgCl)_5]_2Ru$ which provides a route to other substituted compounds.[50] The MgR_2 compounds can be obtained either by dry reactions

$$HgR_2 + Mg \longrightarrow MgR_2 + Hg$$

or by addition of dioxane to Et_2O solutions of Grignards which precipitates $MgCl_2$. For the smaller alkyls and aryls, polymerization occurs via μ_2-C bridges as in $(MgR_2)_n$, R = Me, Et, Ph; these compounds are not very soluble in ethers. Use of bulky groups can give linear, 2-coordinate species such as $Mg[C(SiMe_3)_3]_2$ and $Mg(2,4,6-Bu^tC_6H_2)_2$. There are also magnesiato species such as $[Li\ tmen]_2[MgMe_4]$.[51]

More complicated structures are formed by bifunctional ligands,[52] as in $[o$-phenyleneMgTHF$]_4$ where the Mg atoms form a tetrahedron. The compound $[CpMgOEt]_4$ has a cuboidal structure with Mg coordinated to Cp and three O atoms.

Interaction of RMgX or R_2Mg with ethers and amines can lead to disproportionation[53] resulting in a cation and a more reactive magnesiate anion, e.g., where L = a 2,1,1-cryptand:

$$2R_2Mg + L \longrightarrow RMgL^+ + R_3Mg^-$$

The reaction of $(MgMe_2)_n$ with triazacyclononane(L) in benzene[54] gives $[LMg(\mu\text{-Me})_3MgL]^+_2[Mg_3Me_8]^{2-}\cdot1.5C_6H_6$. This magnesiate anion can be regarded structurally as a segment of the $(MgMe_2)_n$ chain.

Mixed metal compounds can also be made, examples being $Mg(AlMe_4)_2$, $Mg[Al(OMe)_2Me_2]_2$ and $[MeMg(\mu\text{-NPr}^i)AlMe_2]_4$.[55] Finally, silicon analogues of Grignard reagents stabilized by amine coordination have been isolated, e.g., $MgBr(SiMe_3)$tmeda.[56]

Calcium, Strontium, and Barium.[57] The organic compounds of these metals are of little synthetic use compared to the magnesium compounds. The most interesting and best studied are the cyclopentadienyls and related compounds.[58] These can be

[48]H. Bertagnoli and T. S. Ertel, *Angew. Chem. Int. Ed. Engl.* **1994**, *33*, 45.
[49]F. Bickelhaupt et al., *Organometallics* **1991**, *10*, 1531.
[50]C. H. Winter et al., *J. Am. Chem. Soc.* **1996**, *118*, 5506.
[51]R. J. Wehmschulte and P. P. Power, *Organometallics* **1995**, *14*, 3264.
[52]F. Bickelhaupt et al., *J. Am. Chem. Soc.* **1993**, *115*, 2808.
[53]H. G. Richey et al., *J. Am. Chem. Soc.* **1993**, *115*, 9333.
[54]E. Weiss et al., *Angew. Chem. Int. Ed. Engl.* **1994**, *33*, 1257.
[55]C-C. Cheng et al., *J. Chem. Soc., Dalton Trans.* **1994**, 315.
[56]A. Ritter et al., *Angew. Chem. Int. Ed. Engl.* **1995**, *34*, 1030.
[57]T. P. Hanusa, *Polyhedron* **1990**, *9*, 1345 (review).
[58]D. Stalke, *Angew. Chem. Int. Ed. Engl.* **1994**, *33*, 2168 (review of Group I and II compounds).

non-solvated or solvated. The latter have bent structures[59] as have the solvated indenyl complexes.[60] Some substituted butadiene complexes of Ca and Sr are of the type (diene)M(THF)$_4$.[61]

Additional References

J. A. Cowan, Ed., *The Biological Chemistry of Magnesium,* VCH, New York, 1995.

H. G. Emblem and K. Hargreaves, *Rev. Inorg. Chem.* **1995,** *15,* 109 (barium, properties, uses, 90 references).

C. E. Holloway and M. Melnik, *J. Organomet. Chem.* **1994,** *465,* 1 (structural data for organo compounds to end of 1991).

C. Röhr, *Angew. Chem. Int. Ed. Engl.* **1996,** *35,* 1199. Structural chemistry of Gp 1/Gp 2 sub-oxo and nitrido compounds, e.g., $Na_{14}Ba_{14}CaN_6$.

B. J. Wakefield, *Organomagnesium Methods in Organic Syntheses,* Academic, London, 1995.

---

[59]N. H. Allinger *et al., J. Am. Chem. Soc.* **1995,** *117,* 7452.
[60]J. S. Overby and T. P. Hanusa, *Organometallics* **1996,** *15,* 2205.
[61]K. Mashima *et al., J. Am. Chem. Soc.* **1994,** *116,* 6977.

Chapter 5

BORON

By Russell N. Grimes

5-1 Introduction

Boron is in many ways unlike any other element. Although it shares with carbon the very rare ability to form stable covalently bonded molecular frameworks by bonding to itself, its propensity to adopt polyhedral cluster structures has generated a quite distinct stereochemistry that has no counterpart elsewhere in the Periodic Table (although, as will be seen, the boranes furnish important models for clusters in general). Just as the unique properties of carbon are a consequence of its combination of 4 valence bonding orbitals, 4 valence electrons, and moderately high electronegativity, those of boron are directly traceable to its having only 3 valence electrons and lower electronegativity. Thus, while carbon readily forms rings and chains based on two-center covalent bonds (electron pairs shared by two atoms), boron is more versatile; not only does it form ring structures with electron-donor atoms such as N, P, O, and S, but in addition boron can compensate for its "electron deficiency" through multicenter bonds in which three or more atoms are linked by a single electron pair (see below), a mode that encourages cluster formation. The tendency for boron to cluster is so pervasive that atoms of many other elements readily combine with it to form stable polyhedral molecules, called heteroboranes. This effectively broadens boron chemistry to include most of the Periodic Table, and extends even to elements, such as Li, N, and O, that are otherwise rarely found in electron-delocalized polyhedral frameworks.

The uniqueness of boron is highlighted by the sharp contrasts between its properties and those of the other Group 13 elements—all of which have much lower ionization energies and are distinctly metallic—as well as those of its diagonal neighbor, silicon. While boron has some silicon-like characteristics, e.g., both are semiconducting as pure elements and are found in Nature as oxygenated minerals, there are sharp differences in their chemistries; for example, silicon hydrides (silanes) adopt hydrocarbon-like chain structures, quite unlike the boranes. Although boron is not an abundant element, comprising about 8 ppm of the earth's crust (about a third that of lithium and slightly less than lead), boron-containing minerals, described in the following section, are concentrated in arid regions such as Turkey and California's Death Valley, and can be mined cheaply and converted into commercially important boron compounds including $NaBH_4$, an inexpensive bulk chemical.

Neutral monoboron species are invariably trigonal planar molecules of the type BX_3 involving trivalent boron, a consequence of the easy promotion of the s^2p

valence shell ground state to a hybridized sp^2 state having 3 singly occupied valence orbitals in a trigonal planar arrangement. Since this promotion requires little energy in comparison to that released by the formation of 3 covalent B—X bonds, monovalent boron compounds (BX) are unknown in isolable form. Addition of X:$^-$ electron-pair donors to BX_3 (for example, when X is H or halogen) generates stable tetrahedral BX_4^- anions in which boron, though sp^3 hybridized and *tetracoordinate*, remains *trivalent*, i.e., formally B^{III}. The BX_3 species in which X is a halogen are further stabilized by (1) donation of electron density from lone pairs on X into the "vacant" p_z orbital of boron, creating some multiple-bond character and shortening the B—X bond distance, and (2) ionic B$^+$X$^-$ resonance contributions arising from electronegativity differences between B and X. The latter effect is particularly important for B—F and B—O bonds, as can be seen by the bond-shortening that is observed despite the apparent absence of $p\pi$-$p\pi$ bonding. However, the role of ionic bonding in boron chemistry is limited to partial contributions of this kind, since the energy required to fully ionize a boron atom is so high that the B^{3+} ion is never generated under ordinary conditions. This underlines a principal difference between boron and its Group 13 congeners Al, Ga, In, and Tl, whose cations play a major role in their chemistry.

5-2 The Element

Occurrence and Isolation

Boron in Nature consists of the isotopes ^{10}B (19.6%) and ^{11}B (80.4%) and is one of the few elements whose isotopic distribution, hence its atomic weight, varies measurably in samples originating in different geographical locations; for this reason, the atomic weight can be given only to two decimal places (10.81). The ^{10}B isotope has a special property that has long been exploited in nuclear reactors, and currently is of intense interest in the field of cancer therapy, namely, its very high neutron absorption cross section (3835 barns). Boron-containing control rods have been used to moderate the neutron flux in reactors, and the reaction of ^{10}B with slow (thermal) neutrons generates ^{7}Li and α particles, the latter having energies that make them lethal to nearby cells. In boron neutron capture therapy (BNCT),[1] practiced in Japan for some time and now in clinical trial in the United States, tumor tissue having a high ^{10}B concentration, administered in the form of stable compounds of low toxicity, is exposed to a flux of low-energy neutrons. Neither the boron nor the neutrons alone have an appreciable effect on the patient, but in combination they offer a potentially effective noninvasive treatment for certain types of cancers. Derivatives of the icosahedral $B_{12}H_{12}^{2-}$ ion and the carborane $C_2B_{10}H_{12}$ (Sections 5-4 and 5-5) bound to tumor-specific antigens are the main focus of interest in this area.

   With very minor exceptions, boron in Nature is bonded to oxygen, and is never found as the free element. The important sources are *borax* and *kernite*, $Na_2B_4O_5(OH)_4 \cdot nH_2O$ ($n = 8$ for borax and 2 for kernite), both of which are mined from large deposits in desert regions as mentioned above. The tendency of elemental

[1]M. F. Hawthorne, *Angew. Chem. Int. Ed. Engl.* **1993**, *32*, 950; J. H. Morris, *Chem. Brit.* **1991**, *27*, 331.

boron to acquire electron-rich atoms such as C, N, or O is so strong that it is extremely difficult to isolate the element in high purity. Consequently, while 97–98% boron can be obtained easily by reduction of its oxides (see first equation below), samples of the very pure element must be prepared by other means, notably the reduction of boron trihalides (second equation) or the decomposition of BI_3 or boron hydrides at high temperature. In general, pure boron is obtainable only on a limited scale (kilograms).

$$B_2O_3 + 3Mg \longrightarrow 2B \text{ (96-98\%)} + 3MgO$$

$$BBr_3 + \tfrac{3}{2}H_2 \xrightarrow[\text{1000-1200°C}]{\text{Ti}} B \text{ (>99.9\%)} + 3HBr$$

Other methods have also been used to produce high-purity boron, including the electrolytic reduction of KBF_4 in molten $KCl-KF$ mixtures.

Elemental crystalline boron is nearly inert—for example, it reacts with hot concentrated nitric acid only in finely powdered form, and not at all with boiling HF. The bulk solid is extremely hard and is a refractory material, the β-rhombohedral form melting at 2180°C.

Structure

The uniqueness of boron is clearly seen in its elemental forms, the number and structural complexity of which exceed those of any other element. At least five distinct allotropes are known, all of which contain icosahedral B_{12} cluster units that in most cases are accompanied by other boron atoms lying outside the icosahedral cages. The most thermodynamically stable form, β-rhombohedral boron, has 105 B atoms in its unit cell, while the β-tetragonal phase has 192 atoms and is still not completely elucidated despite years of study!

This complexity is a consequence of the fundamental fact, already noted, that the boron atom has more valence orbitals than valence electrons. Such a situation, sometimes referred to as electron deficiency, is actually quite common among metallic elements, and results in delocalization of some of the valence electrons in energy levels (bands) extending throughout the bulk structure. This is not possible for boron, whose high ionization energy restricts it to covalent electron-pair sharing in stable arrangements that allow each boron atom to fully utilize its 4 orbitals and to acquire an octet of electrons. The extremely stable B_{12} icosahedron basically solves this problem, since 26 of its 36 valence electrons suffice to fill the 13 bonding MOs in the cage framework (see the discussion of skeletal electron-counting rules later in this chapter), with the 10 "extra" electrons serving to link the icosahedra together in the solid-state structure. However, there are a number of ways to accomplish this with reasonable thermodynamic stability, some of which incorporate additional boron atoms as mentioned above—hence the variety of allotropes found in Nature. The simplest form, α-rhombohedral boron, contains B_{12} icosahedra in a nearly close-packed arrangement having 12 atoms in the unit cell. In the much more complex β-rhombohedral phase, the structural motif consists of a central B_{12} icosahedron surrounded by 12 pyramidal B_6 units, each centered over a vertex of the icosahedron to form an 84-boron "supercluster."

The extraordinary experimental difficulties that these structures present is illustrated by the case of "α-tetragonal boron." This material was prepared in 1943 and was long thought to have 50 B atoms per unit cell, but has since been found to contain carbon or nitrogen (depending on conditions during synthesis), and is now assigned the composition $B_{50}X_2$, where X is C or N—an example of the previously mentioned propensity of elemental boron to associate with electron-rich atoms.

5-3 Borides

Compounds of boron and the less electronegative elements (metals and metalloids) are loosely referred to as borides, a very large class encompassing hundreds of known compounds. Well-characterized binary transition metal borides number over 120, and include compounds of all of the d-block elements except Zn, Ag, Cd, Au, and Hg (Cr alone forms nine known borides).

The compositions and structures of the metal borides are exceedingly varied and range from metal-rich (e.g., M_4B, M_3B) to very boron-rich compounds (MB_{50}, MB_{66}), with the latter group predominant. As a class, these materials exhibit intriguing, often unique, solid-state architectures, and their physical properties—extreme hardness, high melting points, inertness to air and mineral acids, and high thermal conductivity—make them attractive to modern technology. Some metal borides (e.g., TiB_2) have electrical conductivities far exceeding those of the metal itself, and melting points above 3000°C are not uncommon.

Table 5-1 summarizes the main trends in metal boride structures. As can be

Table 5-1 Examples of Metal Borides and Their Structures[a]

Formula	Examples	B—B bonds	Boron arrangements
M_4B	Mn_4B	Isolated atoms	
M_3B	Ni_3B		
M_2B	Be_2B		
M_3B_2	V_3B_2	Pairs	
MB	NIB	Zigzag chains	
	FeB		
$M_{11}B_8$	$Ru_{11}B_8$	Branched chains	
M_3B_4	Ta_3B_4	Double chains	
	Cr_3B_4		
MB_2	CrB_2	Two-dimensional	
	TiB_2	network	
	MgB_2		
	ZrB_2		
	GdB_2		
MB_4	LaB_4	Three-dimensional	B_6 octahedral
MB_6	LaB_6	network	
MB_{12}	YB_{12}		B_{12} cubooctahedron
MB_{15}	NaB_{15}		B_{12} icosahedron
M_3B_{12}	B_4C		
MB_{66}	YB_{66}		$B_{12}(B_{12})_{12}$ giant icosahedron

[a]Adapted from J. Etourneau and P. Hagenmuller, *Phil. Mag. B,* **1985,** *52,* 589.

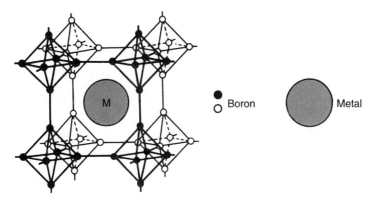

Figure 5-1 Structure of MB_6 borides containing B_6 octahedral clusters.

seen, the extent of boron-boron bonding varies from none at all (i.e., isolated B atoms) in metal-rich systems to extensive (e.g., B_6 and B_{12} polyhedra) in very boron-rich compounds. Intermediate compositions often contain isolated B_2 pairs or two-dimensional B_n chains or sheets, as shown.

Other boron-rich borides adopt varying geometries, some of them extraordinarily complex; for example, YB_{66} has a cubic unit cell containing 1584 B atoms and 24 Y atoms! (The structure is related to the β-rhombohedral form of elemental boron, mentioned earlier.)

The MB_6 borides, formed by a number of metals including Ca, Sr, Ba, Sc, Cr, Y, Th, and several lanthanides (Eu, Yb), adopt the structure shown in Fig. 5-1 consisting of interconnected B_6 octahedra with metal atoms in interstitial locations.

Related structures are found in MB_{12} borides (M = Sc, Ni, Y, Zr, Hf, W) having B_{12} cuboctahedra (5-I) and metal atoms in cubic lattices of the NaCl type.

(5-I)

Another important group of metal borides has the composition M_3B_4 and features the geometry of Ta_3B_4 shown in Fig. 5-2, with boron "double chains" sandwiched between layers of metal atoms.[2]

Of the non-metal borides, by far the most important are those of carbon, usually referred to as boron carbides since carbon is the more electronegative element. These materials, of interest for their potential in high-temperature thermoelectric energy conversion, have compositions varying from $B_{13}C_2$ to $B_{24}C$ and are narrow-band semiconductors. Boron carbides are prepared by reduction of B_2O_3 with C in an electric furnace or by reaction of BCl_3-H_2 mixtures with carbon fibers. The most thoroughly studied of these compounds, so-called "boron carbide," is used industrially as an abrasive (it is nearly as hard as diamond), in bulletproof vests, in armor for aircraft, and as a neutron absorber in nuclear reactors (see above);

[2]R. M. Minyaev and R. Hoffmann, *Chem. Mater.* **1991,** *3,* 547.

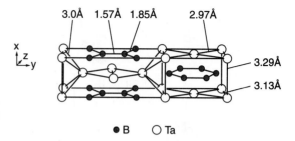

Figure 5-2 Unit cell of Ta_3B_4 (ref. 2); reproduced by permission of the American Chemical Society.

boron carbide fibers are exceedingly strong and chemically inert. This compound was long formulated as B_4C, but is now known to more closely approximate $B_{13}C_2$, although its composition can vary depending on the method of preparation and treatment, in some cases approaching $B_{12}C_3$. The solid consists of a network of icosahedral B_{12} units linked by linear $C-B-C$ chains that lie outside the icosahedra; since there is one such chain per B_{12} cluster, the $B_{13}C_2$ formula is accounted for. Facile replacement of one boron atom by carbon in some of the icosahedral units (but *not* in the $C-B-C$ chains) leads to the observed variation in boron-carbon ratio.

5-4   Boron Hydrides

Background

The binary boron-hydrogen compounds, or boranes, are at the core of a huge, exceedingly rich area of boron-based cluster chemistry, and indeed of molecular clusters generally. Experimental and theoretical studies of the boranes have had a fundamental impact on all of chemistry, organic as well as inorganic, in part because of whole new fields such as metallaboranes and carboranes that have emerged. Even more importantly, the understanding of borane structural principles has created a revolution in the way that the covalent bond is perceived. For example, the fields of metal cluster chemistry and nonclassical carbocations both owe much to theoretical and structural studies of the boranes, notably by W. N. Lipscomb (who was awarded a Nobel prize for this work in 1976) and his students and coworkers. The molecular structures of several binary hydrides are depicted in Fig. 5-3.

The synthesis of the boranes, elucidation of their structural and bonding principles, and development of their chemistry has been a continuing saga that began in 1910 with the work of the great German chemist Alfred Stock, a true experimental genius. In work extending over 25 years in an area in which virtually all prior reports turned out to be erroneous, he and his coworkers isolated and characterized the lower boranes B_2H_6, B_4H_{10}, B_5H_9, B_5H_{11}, B_6H_{10}, and $B_{10}H_{14}$, of which all but the last are volatile liquids or gases that are readily oxidized on contact with air, in some circumstances explosively. These properties, which had defeated earlier attempts by others to properly study the boranes (and had led to such incorrect claims as the existence of "BH_3" and "B_3H_3" as isolable compounds), require that the lighter members be handled in an air-free apparatus; to deal with this problem, Stock invented the chemical glass vacuum line equipped with a pumping system

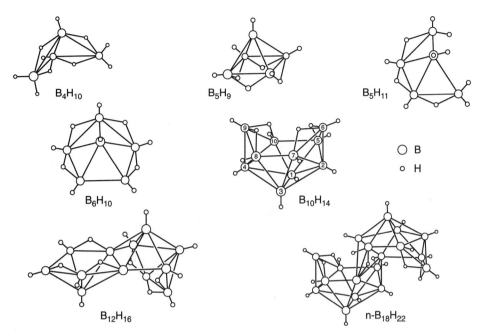

Figure 5-3 Structures of selected boranes determined by X-ray crystallography.

and U-traps for separating volatile materials *via* fractional distillation. Among the many experimental difficulties faced by these workers was the reactivity of boranes toward the unsaturated hydrocarbon lubricants available at that time, preventing the use of greased stopcocks; this was circumvented by another Stock invention, the mercury float valve, a greaseless device used in place of stopcocks. Remarkably, all operations including weighing and analyses of the purified boranes were conducted inside the vacuum apparatus.

Except for B_2H_6, which was generated *via* decomposition of B_4H_{10}, the lower boron hydrides investigated in Stock's laboratory were generated from reactions of crude magnesium boride (very impure Mg_3B_2) with mineral acids; because of contamination by magnesium silicide, the volatile products included lower silanes as well as boranes. Characteristically, Stock took up the challenge and became an early pioneer of silane chemistry in addition to his seminal work on boron hydrides (in yet another major contribution, Stock investigated adverse physiological symptoms that he suffered, recognized the cause as mercury poisoning caused by exposure to that element in his laboratory, and wrote the first scientific papers on that subject).

Subsequent to Stock's work, many additional hydrides have been prepared and characterized, including a number by R. Schaeffer and associates, and today about 35 neutral binary boranes (including isomers) are known (Table 5-2). Many derivatives containing organic or halo substituents have been synthesized, and thousands of other species have been generated *via* insertion of metal or nonmetal heteroatoms into borane frameworks as described in Section 5-6. The polyhedral borane anions (see below) add yet another dimension to the remarkably rich area of polyborane chemistry.

Table 5-2 The Neutral Boron Hydrides[a]

Formula	Structure	mp, °C	bp, °C	Thermal stability	Stability to air
B_2H_6	Ethylene-like (D_{2h})	−164.85	−92.59	Stable at 25°C	Explosively reactive
B_4H_{10}	Butterfly	−120	18	Unstable at 25°C	Reacts slowly, if pure
B_5H_9	Square pyramid	−46.8	60	Stable at 25°C; dec slowly at 150°C	Spontaneously flammable
B_5H_{11}	Icosahedral fragment	−122	65	Unstable at 25°C	Spontaneously flammable
B_6H_{10}	Pentagonal pyramid	−62.3	108	Stable at 25°C	Reacts very slowly
B_6H_{12}	Icosahedral fragment[b]	−82.3	80–90	Stable at 25°C	Reacts rapidly
B_8H_{12}	Icosahedral fragment			Unstable at 25°C	Reacts rapidly
B_8H_{14}	Unknown			Dec above −30°C	Reacts rapidly
B_8H_{16}	Unknown			Moderately unstable	
B_8H_{18}	Conjuncto-2,2′-$(B_4H_9)_2$[b]				
n-B_9H_{15}	Icosahedral fragment	2.6		Stable	Stable
i-B_9H_{15}	Icosahedral fragment[b]			Dec above −35°C	Stable
$B_{10}H_{14}$	Isocahedral fragment	99.5	213[c]	Stable at 150°C	Stable
$B_{10}H_{16}$	Conjuncto-1,1′-$(B_5H_8)_2$[d]			Stable	Stable
$B_{10}H_{16}$	Conjuncto-1,2′-$(B_5H_8)_2$	18.4		Stable	Reacts rapidly
$B_{10}H_{16}$	Conjuncto-2,2′-$(B_5H_8)_2$	−23–−21		Stable	Stable over short periods
$B_{10}H_{18}$	Unknown			Unstable at 25°C	
$B_{11}H_{15}$	Icosahedral fragment	Solid		Dec above 0°C	

Formula	Structure	mp (°C)		
$B_{12}H_{16}$	Edge-fused B_8 and B_6 cages	64–66	Slow dec at 25°C	Stable
$B_{13}H_{19}$	Edge-fused B_9 and B_6 cages	44	Stable	Stable
$B_{14}H_{18}$	Edge-fused B_{10} and B_6 cages[b]	Liquid	Dec above 100°C	Stable
$B_{14}H_{20}$	Edge-fused B_8 cages	Solid	Stable	
$B_{14}H_{22}$	Unknown	Liquid	Unstable at 25°C	
$B_{15}H_{23}$	Vertex-fused B_9 and B_7 cages	Solid	Moderately stable	Reacts
$B_{16}H_{20}$	Edge-fused B_{10} and B_8 cages	108–112	Stable	Stable
$n\text{-}B_{18}H_{22}$	Edge-fused B_{10} cages	179–180	Stable	Stable
$i\text{-}B_{18}H_{22}$	Edge-fused B_{10} cages	128–129	Stable	Stable
$B_{20}H_{16}$	Face-fused B_{10} cages	196–199	Stable	Hygroscopic
$B_{20}H_{26}$	Conjuncto-1,5'-$(B_{10}H_{13})_2$	114–115	Stable	Stable
$B_{20}H_{26}$	Conjuncto-2,2'-$(B_{10}H_{13})_2$	178–179	Stable	Stable
$B_{20}H_{26}$	Conjuncto-2,6'-$(B_{10}H_{13})_2$	154–155	Stable	Stable
$B_{20}H_{26}$	Conjuncto-6,6'-$(B_{10}H_{13})_2$	198–199	Stable	Stable
$B_{20}H_{26}$	Conjuncto-1,2'-$(B_{10}H_{13})_2$	139–142	Stable	Stable
$B_{20}H_{26}$	Conjuncto-2,5'-$(B_{10}H_{13})_2$	109–111	Stable	Stable
$B_{20}H_{26}$	Conjuncto-5,5'-$(B_{10}H_{13})_2$	97–98	Stable	Stable

[a] Isolated and characterized species.
[b] Proposed structure.
[c] Extrapolated value.
[d] Volatile solid.

Preparation of Neutral Boron Hydrides

Until recent years, the standard method of synthesis for most polyboranes (molecules larger than diborane) was pyrolysis of B_2H_6, a comparatively inefficient approach that generates mixtures of volatile boranes, solid polymeric hydrides of varying composition, and hydrogen gas. This method was applied on a very large (multiton) scale in the 1950s to prepare B_5H_9 and $B_{10}H_{14}$ for use in the development of new rocket and jet fuels by the U.S. military, a project later abandoned. Today, non-pyrolytic methods are available for preparing essentially all of the known boron hydrides, many of them taking advantage of the cheap availability of metal borohydrides, MBH_4. These compounds, important industrial and laboratory reagents that are widely used as reducing agents in organic synthesis, in metal plating, and in the bleaching of wood pulp, are prepared in reactions of metal hydrides with diborane or boron trihalides:

$$B_2H_6 + 2LiH \longrightarrow 2LiBH_4$$

$$BCl_3 + 4NaH \xrightarrow{Al_2Cl_6} NaBH_4 + 3NaCl$$

Diborane itself can be obtained from $NaBH_4$ and boron trifluoride etherate, from $LiAlH_4$ and boron trichloride, or *via* reduction of boron trihalides with hydrogen:

$$3NaBH_4 + 4BF_3 \longrightarrow 3NaBF_4 + 2B_2H_6$$

$$3LiAlH_4 + 4BCl_3 \longrightarrow 3LiAlCl_4 + 2B_2H_6$$

$$BCl_3 + 3H_2 \longrightarrow \tfrac{1}{2}B_2H_6 + 3HCl$$

Perhaps the simplest and best laboratory route is hydride abstraction from BH_4^- ion:

$$BH_4^- + BX_3 \xrightarrow[25°C]{CH_2Cl_2} \tfrac{1}{2}B_2H_6 + HBX_3^- \quad (X = Cl, Br)$$

Hydride abstraction, an approach developed by S. G. Shore and his coworkers, can also be applied to larger anions to generate very pure higher boranes, in most cases as the only volatile product. Thus, treatment of $B_3H_8^-$ salts (readily obtained in reactions of $NaBH_4$) with boron trihalides provides the best available synthesis of tetraborane(10):

$$B_3H_8^- + BX_3 \longrightarrow B_4H_{10} + HBX_3^- + (BH)_x \text{ solid} \quad (X = F, Cl, Br)$$

In a similar fashion, B_5H_{11} can be obtained from the $B_4H_9^-$ ion.

Pentaborane(9), generated some time ago on a pilot plant scale *via* diborane pyrolysis as part of the fuel development program mentioned above, is currently stockpiled in large quantity by the U.S. Government. A convenient laboratory-scale preparation utilizes the bromotriborohydride anion:

$$NBu_4^+B_3H_8^- \xrightarrow{HBr} NBu_4^+B_3H_7Br^- \xrightarrow{95\text{-}100°C} B_5H_9 + NBu_4^+Br^- + B_2H_6 + H_2$$

Hexaborane(10), another of the original Stock boranes, can be synthesized *via* insertion of boron into a bromopentaborane framework:

$$1\text{-}BrB_5H_8 \xrightarrow[\text{-}H_2]{KH\,/\,\text{-}78\,°C} K^+[BrB_5H_7]^- \xrightarrow[\text{-}78\,°C]{B_2H_6} K^+[BrB_6H_{10}]^- \xrightarrow[\text{-}KBr]{\text{-}78\,°C} B_6H_{10}$$

Decaborane(14), the smallest solid borane and an important precursor to higher hydrides as well as carboranes (see below), was formerly prepared *via* diborane pyrolysis; however, this approach has been rendered obsolete by the development of several high-yield routes based on the BH_4^- or $B_5H_8^-$ anions (the latter species is easily obtained *via* deprotonation of B_5H_9).

$$NaBH_4 \xrightarrow[\text{alkyl halide}]{\text{acid or}} NaB_{11}H_{14} \xrightarrow{\text{oxidant}} B_{10}H_{14}$$

$$B_5H_8^- \xrightarrow[\text{24h, 25°C}]{CH_2Cl_2} B_9H_{14}^- \xrightarrow{BBr_3} B_{10}H_{14} \text{ (hydride abstraction)}$$

$$B_5H_8^- \xrightarrow[\text{25°C}]{FeCl_2/FeCl_3 \text{ or } RuCl_3} B_{10}H_{14} \text{ (metal-promoted fusion)}$$

Of the characterized higher boranes larger than $B_{10}H_{14}$, all except $B_{11}H_{15}$ are of the *conjuncto* class, consisting of two or more cluster units joined in some manner (for two examples see Fig. 5-3). A variety of synthetic approaches has been employed to prepare these hydrides, in most cases generating products serendipitously with no real control over the product compositions or structures. Typical examples are as follows:

$$B_5H_9 + NaH \longrightarrow B_{11}H_{14}^- + \xrightarrow[\text{HCl}]{\text{-}78\,°C} B_{11}H_{15}$$

$$B_5H_9 + H_2 \xrightarrow[\text{discharge}]{\text{electric}} 1,1\text{-}(B_5H_8)_2$$

$$B_6H_{10} + i\text{-}B_9H_{15} \xrightarrow{0\,°C} B_{15}H_{23}$$

$$B_{10}H_{14} \xrightarrow[\text{electron radiolysis}]{hv,\ heat,\ or} (B_{10}H_{13})_2 \text{ isomers}$$

$$closo\text{-}B_{10}H_{10}^{2-} \xrightarrow{Fe^{3+}} B_{20}H_{18}^{2-} \xrightarrow[\text{EtOH}]{H^+} n\text{-} + i\text{-}B_{18}H_{22}$$

$$Me_2SB_9H_{13} \xrightarrow{heat} i\text{-}B_{18}H_{22}$$

An unusual example of a "designed" synthesis of a previously unknown binary boron hydride is the formation of $B_{12}H_{16}$ (Fig. 5-3), the only known neutral dodecaborane, *via* metal-promoted oxidative fusion of hexaborane(9) anions:

$$B_6H_9^- \xrightarrow{Fe^{2+}/Fe^{3+}} B_{12}H_{16} + B_6H_{10}$$

This approach has also been used to prepare $n\text{-}B_{18}H_{22}$ from $B_9H_{12}^-$ and $HgBr_2$, and as a route to $B_{10}H_{14}$ from $B_5H_8^-$ and metal ions (*vide supra*).

Preparation of Polyhedral Borane Anions

The species $B_nH_n^{2-}$ are known for $n = 6$ to 12, and all are deltahedral clusters (polyhedra having only triangular faces) as shown in Fig. 5-4. As a group these anions are far more stable, thermally and hydrolytically, than the neutral boranes; this is particularly true of $B_{10}H_{10}^{2-}$ and $B_{12}H_{12}^{2-}$ (Figs. 5-4e and g), whose salts can be heated to several hundred degrees C without decomposition. In sharp contrast to the neutral hydrides, the $B_nH_n^{2-}$ dianions are highly soluble in water and form stable aqueous solutions. Synthetic routes to these anions vary, but most follow a general pattern of condensation of small borane fragments *via* processes whose mechanistic details are still largely obscure.

$$2NaBH_4 + 5B_2H_6 \xrightarrow[100\text{-}180°C]{Et_3N} Na_2B_{12}H_{12} + 13H_2$$

$$2LiAlH_4 + 5B_2H_6 \xrightarrow{160°C} Li_2B_{10}H_{10} + \text{other } B_nH_n^{2-} \text{ salts}$$

The facile synthesis of large clusters such as $B_{10}H_{10}^{2-}$ and $B_{12}H_{12}^{2-}$ from $NaBH_4$ and diborane is especially remarkable, and evidently involves the $B_3H_8^-$ ion as a major building-block intermediate. However, the most common route to $B_{10}H_{10}^{2-}$ is the reaction of $B_{10}H_{14}$ with triethylamine in boiling xylene:

$$B_{10}H_{14} + Et_3N \xrightarrow{-H_2} B_{10}H_{12}(NEt_3)_2 \longrightarrow (Et_3NH^+)_2B_{10}H_{10}{}^{2-}$$

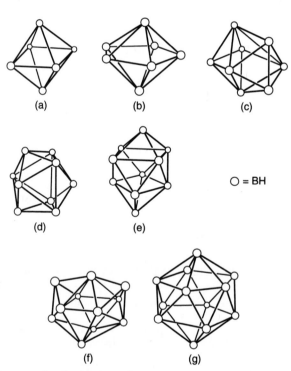

$O = BH$

Figure 5-4 Structures of polyhedral $B_nH_n^{2-}$ dianions. (a) $B_6H_6^{2-}$. (b) $B_7H_7^{2-}$. (c) $B_8H_8^{2-}$. (d) $B_9H_9^{2-}$. (e) $B_{10}H_{10}^{2-}$. (f) $B_{11}H_{11}^{2-}$. (g) $B_{12}H_{12}^{2-}$.

Salts of this dianion can also be prepared *via* the pyrolysis of $Et_4N^+BH_4^-$ and the treatment of $Et_3N \cdot BH_3$ with diborane at 170°C.

Structures and Bonding

The closed polyhedral (deltahedral) geometries adopted by the $B_nH_n^{2-}$ dianions are designated by the prefix *closo*. Most neutral boranes, in contrast, have open structures that can be viewed as fragments of closo polyhedra as follows: formal removal of one vertex from a closo parent cluster generates a *nido* cage (from the Latin word for nest), while removal of two and three vertices give *arachno* (Greek "web") and *hypho* ("net") frameworks, respectively. This principle is illustrated in Fig. 5-5, which depicts the boron frameworks in *closo*-$B_6H_6^{2-}$, *nido*-B_5H_9, and *arachno*-B_4H_{10}.

The unique molecular geometries of the boron hydrides, unsuspected in Stock's time (he assumed hydrocarbon-like ring or chain structures) were gradually revealed through X-ray diffraction studies of the lower boranes during the 1950s and 1960s. The structures of B_4H_{10}, B_5H_9, B_5H_{11}, B_6H_{10}, B_8H_{12}, and B_9H_{15}, all volatile liquids at room temperature, were elucidated by Lipscomb and his associates from X-ray data collected on crystals mounted in glass capillaries and cooled in a stream of cold N_2; that of B_2H_6 (m.p. $-165°C$) required cooling with liquid helium! These remarkable achievements, which were preceded by the structure determination of $B_{10}H_{14}$ by Kasper, Lucht, and Harker in 1948 (a notable accomplishment in the precomputer era), were followed by elucidation of numerous other borane structures including most of the hydrides listed in Table 5-2. The boron frameworks in most of the smaller neutral hydrides are icosahedral fragments (reflecting the icosahedral B_{12} units in elemental boron as discussed above), the notable exceptions being B_5H_9 and the $(B_5H_8)_2$ isomers.

The early efforts to apply hydrocarbon-based bonding models to the boranes were frustrated by the so-called "electron deficiency" of boron, with its three valence electrons *vs.* four for carbon. Long before the structure of diborane (B_2H_6) was deduced from infrared spectroscopy (later confirmed by X-ray crystallography and electron diffraction), it was clear that its 12 valence electrons are insufficient for an ethane-like $H_3B{-}BH_3$ structure, which has 7 bonds and would require 14 electrons. The actual geometry, shown in Fig. 5-6, contains 4 B–H terminal and two 3-center B–H–B bridge bonds, *each of the latter type using a single electron pair to bind 3 atoms together*, thus accounting for the 12 valence electrons and utilizing 4 valence orbitals (approximately sp^3 hybridized) on each boron atom.

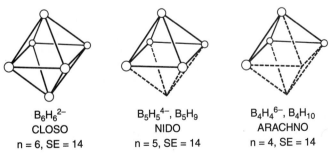

$B_6H_6^{2-}$	$B_5H_5^{4-}$, B_5H_9	$B_4H_4^{6-}$, B_4H_{10}
CLOSO	NIDO	ARACHNO
n = 6, SE = 14	n = 5, SE = 14	n = 4, SE = 14

Figure 5-5 Boron frameworks based on a 6-vertex, 14-skeletal electron octahedral cluster.

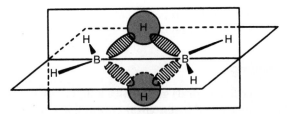

Figure 5-6 Structure of B_2H_6 depicted as two BH_2 units joined by two B—H—B 3-center, 2-electron bridge bonds.

This structure is analogous not to ethane, but rather to ethylene (C_2H_4), with which it is isoelectronic; indeed the pair of B—H—B bridges can be regarded as "protonated double bonds." Since all bonding orbitals are occupied in both B_2H_6 and C_2H_4, there is in fact no "deficiency" of electrons in either molecule. Nonetheless, the notion of electron deficiency as an intrinsic property of the boranes became entrenched over the years. Although this traditional label still has its defenders, the boron hydrides as a class are intrinsically stable, filled-shell molecular systems (all bonding orbitals filled) and are *not in any true respect deficient in electrons*, certainly not in the sense that most chemists use the word. However, the term is quite applicable to fragments such as BH_3 and its carbon analogue CH_3^+ which possess vacant valence orbitals and are strong Lewis acids.

The neutral boranes can be described in valence-bond language utilizing localized two-center (B—H, B—B) and 3-center (B—H—B or B—B—B) bonds. This approach, developed primarily by Lipscomb and his associates, solved the classical problem of accounting for the boron hydride structures, which differ so sharply from those of the hydrocarbons. Figure 5-7 illustrates its application to B_6H_{10}, whose pentagonal- pyramidal geometric structure is depicted on the left and one representation of its valence-bond structure on the right. The important thing to note is that the total number of bond pairs in this description is 14, comprised of six 3-center bonds (2 B—B—B and 4 B—H—B) plus eight 2-center bonds (2 B—B and 6 B—H terminal), thus accounting for all 28 valence electrons in the molecule. Moreover, each boron atom utilizes its 4 valence orbitals.

The complete theory takes into account valence bond angles and other considerations, and invokes partial 3-center bonding and resonance forms to provide detailed

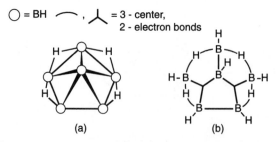

Figure 5-7 (a) Geometric structure of B_6H_{10} showing atom connectivities. (b) Valence-bond structure of B_6H_{10} represented as a combination of 2-center and 3-center electron-pair bonds.

bonding descriptions. More generally, and particularly when dealing with the polyhedral borane anions and other closo systems, molecular orbital theory is applied to borane cage structures.

A particularly useful tool is the Polyhedral Skeletal Electron Pair Theory (PSEPT),[3] originated by K. Wade, elaborated by D. M. P. Mingos and others, and known colloquially as "Wade's Rules." This approach and its applications to carbocations, metal clusters, and other non-boron species, are described in numerous books and reviews. Here we simply outline the basic rules.

(1) Polyhedral clusters are considered to be assemblies of building-block units such as BH, CH, $Co(\eta^5\text{-}C_5H_5)$, $Fe(CO)_3$, etc., each of which contributes to skeletal bonding a characteristic number of electrons (those available after the bonding and nonbonding electrons within the unit are accounted for).

(2) The total number of skeletal bonding pairs from all of the individual building-block units determines the cluster geometry, starting with the assumption that each unit supplies 3 orbitals for construction of skeletal MOs; hence there will be $3n$ MOs in an n-vertex closo polyhedron, and theory shows that $n+1$ of these will be bonding (comprised of n MOs tangential to the surface and a unique MO inside the polyhedron that is formed by overlap of inward-pointing orbitals of the n skeletal units).

The tetrahedron constitutes a special case in that the number of skeletal bonding MOs is usually 6, accommodating 6 electron pairs; since this is also the number of edges, most tetrahedral clusters can also be viewed as "classical" species having one electron pair for each bonding interaction.

(3) The removal of one or more vertices from a closo polyhedral system leaves the number of skeletal bonding MOs unchanged; thus, a 6-vertex closo cluster (octahedral, as in Fig. 5-5), a 5-vertex nido cluster (square pyramidal), and a 4-vertex arachno cluster (butterfly or square planar, depending on whether adjacent or nonadjacent vertices are removed from the octahedron) all have 6 bonding MOs and hence require 7 bonding electron pairs. Expressed in terms of the number of vertices n, this equals $n+1$, $n+2$, $n+3$, and $n+4$ pairs for closo, nido, arachno, and hypho clusters, respectively.

(4) The *capping principle* states that if a triangular face of a closo cluster is capped (for example, by adding a seventh vertex to an octahedron) the number of skeletal bonding MOs remains unchanged. Thus, a 7-vertex capped octahedron will have 7 bonding pairs, or in general terms, an n-vertex capped-closo cluster will have n bonding pairs.

The skeletal electron contribution of a BH unit is clearly 2, since 2 of the 4 valence electrons are required for the B—H bond. Similarly, CH groups (as in carboranes, see below) are 3-electron donors. B—H—B bridging hydrogen atoms and "extra" terminal hydrogen atoms (as in BH_2 units) contribute 1 electron each.

As an illustration of these principles, consider B_5H_9 (Figs. 5-3 and 5-5). The 5 BH units and 4 bridging hydrogen atoms supply $10 + 4 = 14$ electrons or 7 pairs total. Since $n = 5$, this is an $n+2$ electron pair system and hence is predicted to be nido, i.e., a square pyramid, in accordance with its known structure. The $B_nH_n^{2-}$

[3]D. M. P. Mingos and D. J. Wales, *Introduction to Cluster Chemistry;* Prentice Hall, Englewood Cliffs, N.J., 1990.

dianions each have *n* BH units, providing *n* electron pairs, plus another pair from the dinegative charge; they are thus *n*+1 electron pair systems and are expected to be closo, as indeed they are (Fig. 5-4). One last example: B_4H_{10}, with 4 BH groups and 6 extra hydrogen atoms, has 8 + 6 = 14 electrons or 7 pairs, and therefore is an *n* + 3 electron pair system (*n* = 4), corresponding to an arachno cluster and consistent with its established butterfly geometry (Figs. 5-3 and 5-5). The PSEPT approach is particularly useful when applied to boron clusters containing heteroatoms, as will be seen in discussions of carboranes, metallaboranes, and metallacarboranes to follow.

Reactions and Properties: General Comments

The neutral boranes are colorless, diamagnetic compounds, the lower members of which are highly reactive toward oxygen, typically generating solid B_2O_3 and water in very exothermic processes. The considerable heat generated in these combustions (e.g., ΔH^0 for B_2H_6 = −2137.7 kJ mol^{-1}) far exceeds that obtained by burning hydrocarbons, and led to exploratory use of boranes as fuels or fuel additives several decades ago as mentioned earlier. The hydrides B_2H_6, B_4H_{10}, B_5H_{11}, and B_6H_{12} (among others) are thermally unstable, forming higher hydrides and solid polymers *via* very complex mechanisms on standing at room temperature. In contrast, B_5H_9 and B_6H_{10} are thermally stable but are readily oxidized on contact with air. The higher molecular weight boranes tend to be less reactive; for example, $B_{10}H_{14}$ is a white solid with an unpleasant odor (toxic, as are all of the volatile boranes) that can be handled in fume hoods, is easily sublimed on mild heating, and only slowly hydrolyzes on exposure to moist air, while the $B_{18}H_{22}$ isomers are nonvolatile at room temperature and essentially nonreactive toward air and water.

The chemistry of three boron hydrides—B_2H_6, B_5H_9, and $B_{10}H_{14}$—has been extensively documented, and only an abbreviated summary of some of the main findings can be given here. *Diborane* is a versatile reagent, especially in its application to organic chemistry *via* the *hydroboration reaction* that was developed by H. C. Brown and his students into a versatile synthetic tool. This reaction, normally conducted either with B_2H_6 generated *in situ* or with convenient derivatives such as 9-borabicyclo[3.3.1]nonane or 9-BBN, shown below, results in the anti-Markovnikov, cis addition of H and BH_2 units to C=C or C≡C bonds. The alkylboranes so generated are easily converted to a variety of products including ketones, alcohols, and many other species.

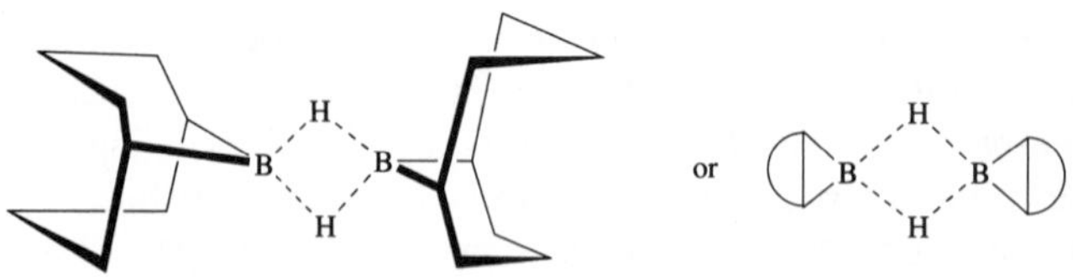

Because it affords a specific route to compounds that have anti-Markovnikov, cis stereochemistry (often difficult to achieve by other means), the hydroboration reaction has assumed considerable importance in organic synthesis. Also important is the use of diborane as a reducing agent, in which it functions as an electrophile, for example, readily reducing carboxylic acids and esters; BH_4^-, in contrast, is a nucleophile and attacks electrophilic centers.

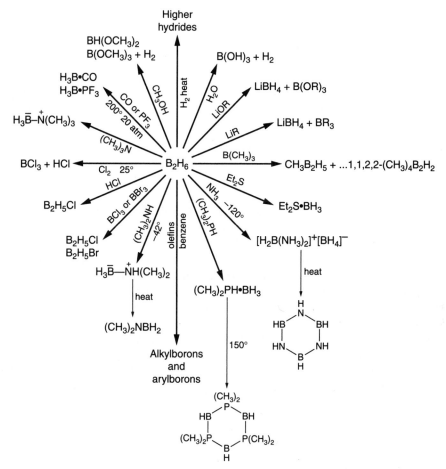

Figure 5-8 Characteristic reactions of diborane.

Other reactions typical of diborane are shown in Fig. 5-8. A prominent feature of its chemistry is the formation of B_2H_6L or $H_3B \cdot L$ adducts with Lewis bases. The BH_3 fragment itself (a truly electron-deficient species) is an extremely strong Lewis acid, indeed stronger even than BF_3. In certain adducts such as H_3BCO, there is evidence of π back-bonding from the B—H bonds to acceptor orbitals on the non-borane fragment, somewhat analogous to that between transition metals and CO groups in metal carbonyls. Ammonia and some amines cleave diborane asymmetrically, forming salts of the type $[L_2BH_2]^+BH_4^-$ that can be converted into cyclic products on heating.

Pentaborane(9) is the smallest thermally stable polyborane, and it is an important precursor to many small carboranes, metallaboranes, and other classes of boron clusters (discussed below). Its bridging hydrogen atoms are protonic, and treatment of B_5H_9 with metal hydrides or alkyllithium reagents in polar solvents generates the $B_5H_8^-$ anion, whose unprotonated B—B bond reacts with electrophiles to generate bridged complexes as shown in Fig. 5-9.

Reactions of $B_5H_8^-$ with transition metal reagents can also lead to full incorporation of the metal into the cage skeleton, as discussed in Section 5-6. Among the well-

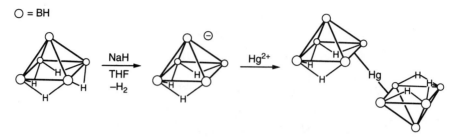

Figure 5-9 Deprotonation of B_5H_9 to form the $B_5H_8^-$ ion and complexation with Hg^{2+} to generate $\mu, \mu\text{-}Hg(B_5H_8)_2$.

studied reactions of B_5H_9 are its interactions with halogens: the apically substituted 1-XB_5H_8 derivative is produced on treatment with Br_2, I_2, or Cl_2 (in the last case $AlCl_3$ is required). Basally halogenated and polyhalogenated derivatives have also been prepared. Pentaborane is readily alkylated *via* reactions with alkyl halides under Friedel-Crafts conditions; as expected, substitution occurs preferentially at the negative apex boron. Reactions of B_5H_9 with alkynes in the presence of Lewis bases take a very different course, producing insertion of carbon into the borane framework to form carboranes, as outlined in Section 5-5.

 Decaborane(14), $B_{10}H_{14}$, is a hydride of considerable synthetic importance as a precursor to large metallaboranes and other heteroborane clusters, including the icosahedral $RR'C_2B_{10}H_{10}$ carboranes. For example, $B_{10}H_{12}L_2$ adducts are readily obtained on direct treatment of $B_{10}H_{14}$ with Lewis bases, the latter reagents attacking at the electropositive 6,9 boron atoms (see Fig. 5-3):

$$B_{10}H_{14} + 2MeCN \longrightarrow \textit{arachno-}B_{10}H_{12}(NCMe)_2 + H_2$$

Adducts of this type react with alkynes to generate carboranes (see below) and can also undergo cage closure on heating to form salts of the *closo-*$B_{10}H_{10}^{2-}$ dianion, as in the treatment of $B_{10}H_{14}$ with Et_3N in boiling toluene mentioned earlier. Addition of one or two boron atoms to the B_{10} skeleton can be effected with $LiBH_4$ or other reagents to produce the $B_{11}H_{14}^-$ anion and the icosahedral *closo-*$B_{12}H_{12}^{2-}$ dianion, respectively.

 Other well-studied reactions of decaborane(14) include reduction to the dianion, deprotonation, and electrophilic substitution:

$$B_{10}H_{14} + 2Na \xrightarrow{NH_3(l)} 2Na^+ + B_{10}H_{14}{}^{2-}$$

$$B_{10}H_{14} + OH^- \longrightarrow B_{10}H_{13}^- + H_2O$$

$$B_{10}H_{14} + X_2 \xrightarrow{AlCl_3} 1\text{- and } 2\text{-}B_{10}H_{13}X + 2,4\text{-}B_{10}H_{12}X_2 \quad (X = Cl, Br, I)$$

$$B_{10}H_{14} + MeX \xrightarrow{AlCl_3} Me, Me_2, Me_3, Me_4 \text{ derivatives (substitution at borons 1-4)}$$

In some cases, the product species have different arrangements of hydrogen on the open face in comparison to $B_{10}H_{14}$ itself; thus, the $B_{10}H_{14}^{2-}$ dianion and the 6,9-

$B_{10}H_{12}L_2$ adducts have B–H–B bridges on the 7,8 and 5,10 edges as well as extra terminal hydrogen atoms on the 6,9 boron atoms.

Polyhedral $B_nH_n^{2-}$ Dianions

In striking contrast to the neutral boranes, salts of these ions are highly soluble in water and form stable aqueous solutions (at least in neutral and basic conditions). In general, stability increases with size; $B_{12}H_{12}^{2-}$ is very likely the most robust of all known molecular clusters, having a decomposition temperature above 700°C (K^+ salt) and resisting cage degradation even in boiling nitric acid. The $B_{10}H_{10}^{2-}$ ion, whose apex BH units are negatively charged relative to the 8 equatorial vertices, is somewhat more reactive, but is almost as thermally stable as $B_{12}H_{12}^{2-}$. Both of these ions are essentially inert to biological systems, with low toxicity comparable to that of NaCl, and cage-substituted derivatives of them have been employed in boron neutron capture therapy (BNCT) studies mentioned earlier.

On oxidation, $B_{10}H_{10}^{2-}$ undergoes loss of hydrogen and formation of dimeric species such as $B_{20}H_{18}^{2-}$ and $B_{20}H_{19}^{-}$, which consist of discrete B_{10} polyhedra linked by B—H—B or B—B bonds.

5-5 Carboranes

One of the most intensively studied classes of boron clusters is the carboranes (more formally, carbaboranes), whose defining feature is the presence of one or more carbon atoms in a borane cluster framework. Since their discovery in the late 1950s, the study of carboranes has evolved into a field of enormous scope and versatility, with thousands of characterized compounds (including metal and hetero-atom derivatives) and an almost bewildering variety of structural types and composi-tions involving most of the natural elements in the Periodic Table. The majority of these species contain two skeletal carbon atoms per cage (primarily reflecting the synthetic routes employed) but 1-, 3-, 4-, and even 5-carbon carboranes are also known. Inevitably, the field overlaps closely related areas, including the cyclic organoboranes. Even some planar heterocycles have isomers with carborane-type geometries; for example, while 1,4-difluoro-1,4-diborabenzene (FBC_4H_4BF) has a classical cyclic planar structure, its dihydrogen counterpart, $H_4C_4B_2H_2$, is a pentago-nal pyramidal *nido*-carborane (see below, Fig. 5-13e). Borderline cases can arise involving bridging carbon atoms that one may or may not regard as part of the cage skeleton. In general, however, the carboranes form a distinct, recognizable class of "nonclassical" molecules in which the cage carbon atoms participate in delocalized bonding and often achieve coordination numbers of 5 or 6.

Closo-Carboranes

The $C_2B_{n-2}H_n$ and $CB_{n-1}H_n^-$ families (listed in Table 5-3, with the structures of the former group shown in Fig. 5-10) are isoelectronic analogues of the polyhedral $B_nH_n^{2-}$ dianions discussed above.

Each of these *n*-vertex carboranes has the same cage geometry as its $B_nH_n^{2-}$ counterpart, except that $C_2B_3H_5$ has no known $B_5H_5^{2-}$ analogue. Application of the PSEPT rules (see p 145) is straightforward, with contributions of 3 electrons from

Table 5-3   Closo-Carboranes[a]

Vertices	Dicarbon series	Monocarbon series	Cage geometry
5	$1,5\text{-}C_2B_3H_5$		Trigonal bipyramid
6	$1,2\text{-}C_2B_4H_6$	$CB_5H_6^-$, CB_5H_7	Octahedron
6	$1,6\text{-}C_2B_4H_6$		Octahedron
7	$2,4\text{-}C_2B_5H_7$		Pentagonal bipyramid
7	$2,3\text{-}C_2B_5H_7$		Pentagonal bipyramid
8	$1,7\text{-}C_2B_6H_8$	$CB_7H_8^-$	Triangulated dodecahedron
9	$1,7\text{-}C_2B_7H_9$		Tricapped trigonal prism
10	$1,6\text{-}C_2B_8H_{10}$	$CB_9H_{10}^-$	Bicapped square antiprism
10	$1,10\text{-}C_2B_8H_{10}$		Bicapped square antiprism
11	$1,8\text{-}C_2B_9H_{11}$	$CB_{10}H_{11}^-$	Octadecahedron
12	$1,2\text{-}C_2B_{10}H_{12}$	$CB_{11}H_{12}^-$	Icosahehedron
12	$1,7\text{-}C_2B_{10}H_{12}$		Icosahehedron
12	$1,12\text{-}C_2B_{10}H_{12}$		Icosahehedron

[a] Isolated and characterized species.

each CH unit and 2 electrons from each BH producing in each case a total electron count of $2n+2$ ($n+1$ pairs), thus favoring closo structures. The closed-shell electronic systems of these clusters, with extensive delocalization of the skeletal electrons, is reflected in their high thermal and oxidative stability. Especially stable are the pentagonal bipyramidal $2,4\text{-}C_2B_5H_7$ cluster (Fig. 5-10d) and the icosahedral 1,2-, 1,7-, and $1,12\text{-}C_2B_{10}H_{12}$ isomers (Fig. 5-10j,k,l) and their derivatives. The $C_2B_{10}H_{12}$ compounds, with 26 skeletal electrons, are in effect 3-dimensional analogues of benzene and have been labeled "superaromatic."

As Table 5-3 indicates, isomers exist for several of the closo systems (although not all of the possible isomers have been found), and thermal rearrangements of the cage skeleton are common in these species, as described below. The lower members are colorless volatile liquids, while the 1,2-, 1,7-, and $1,12\text{-}C_2B_{10}H_{12}$ isomers (known informally as ortho-, meta-, and para-carborane, respectively) are white nonvolatile solids.

Preparative routes to these compounds vary. The smaller parent species ($C_2B_3H_5$, $C_2B_4H_6$, $C_2B_5H_7$, and alkyl derivatives of these) are formed *via* high-energy combination of volatile boranes and alkynes in electric discharges or flash reactions; alkylcarboranes can also be obtained from alkylboranes, as in the formation of $1,5\text{-}C_2B_3Et_5$ *via* condensation of $(Et_2B)_3CEt$.[4] The intermediate members ($C_2B_6H_8$, $C_2B_7H_9$, $C_2B_8H_{10}$) are prepared *via* degradation of icosahedral carboranes as discussed below. The important $1,2\text{-}C_2B_{10}H_{12}$ carborane and many C,C'-substituted derivatives are synthesized in reactions of $B_{10}H_{12}L_2$ Lewis base adducts with alkynes (Fig. 5-11).

This method is applicable to a wide variety of alkynes (e.g., R,R' = H, alkyl, aryl, alkoxy, alkenyl, alkynyl), but reagents containing OH or COOH moieties degrade the boron cage and cannot be employed. However, carboranes containing these functional groups can be obtained *via* electrophilic substitution on the parent compound as outlined below.

[4] W. Wrackmeyer *et al., J. Chem. Soc., Chem. Commun.* **1995,** 1691.

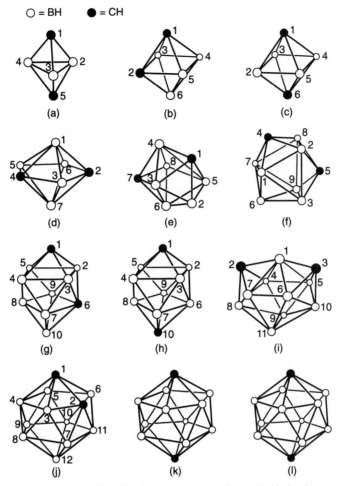

Figure 5-10 Structures of $C_2B_{n-2}H_n$ *closo*-carboranes (not all of the known isomers are shown). (a) $1,5\text{-}C_2B_3H_5$. (b) $1,2\text{-}C_2B_4H_6$. (c) $1,6\text{-}C_2B_4H_6$. (d) $2,4\text{-}C_2B_5H_7$. (e) $1,7\text{-}C_2B_6H_8$. (f) $4,5\text{-}C_2B_7H_9$. (g) $1,6\text{-}C_2B_8H_{10}$. (h) $1,10\text{-}C_2B_8H_{10}$. (i) $2,3\text{-}C_2B_9H_{11}$. (j) $1,2\text{-}C_2B_{10}H_{12}$. (k) $1,7\text{-}C_2B_{10}H_{12}$. (l) $1,12\text{-}C_2B_{10}H_{12}$.

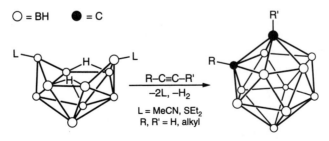

Figure 5-11 Insertion of an alkyne into the decaborane cage to form an icosahedral $RR'C_2B_{10}H_{10}$ carborane.

The reaction chemistry of the $RR'C_2B_{10}H_{10}$ isomers has been actively investigated since their discovery in the mid-1950s (first reported in the open literature in 1963). Primarily, this effort has been directed toward the synthesis of exceptionally stable polymers that exploit the remarkable thermal stability and "electron-sink" properties of the carborane units. A number of plastics, elastomers, and oils that incorporate 1,7- or $1,12-C_2B_{10}H_{10}$ cages linked by SiR_2O, Hg, S, or other groups have been characterized, and some have been produced commercially as tubing, coatings, gaskets, and liquid phases for high-temperature gas chromatography, among other uses. In a different application, organosubstituted derivatives of $1,2-C_2B_{10}H_{12}$ are currently in use in clinical trials of the previously mentioned BNCT treatment of inoperable brain tumors.

Four types of reactions of the closo-carboranes are particularly important: deprotonation and introduction of functional groups at the cage CH groups; electrophilic substitution at the boron vertices; base-promoted removal of BH units to form nido cages; and polyhedral rearrangement. Owing to commercial availability, this chemistry is far more developed in the $C_2B_{10}H_{12}$ isomers than in the other closo species. As a consequence of the relatively electronegative carbon atoms, the CH protons in these cages are slightly acidic. In the icosahedral $C_2B_{10}H_{12}$ system, the CH acid strength decreases in the three isomers in the order 1,2 > 1,7 > 1,12, and these protons can be removed by BuLi, Grignards, or other nucleophiles. The resulting C-mono- or C,C'-dilithio derivatives can be converted to a host of C-organosubstituted products; for example, treatment of 1,2-, 1,7-, or $1,12-Li_2C_2B_{10}H_{10}$ with CO_2 and subsequent acidification affords the corresponding $(HOOC)_2C_2B_{10}H_{10}$ dicarboxylic acids. Similarly, C- and C,C'-alkyl and alkenyl derivatives can be obtained in reactions with organo halides.

$$1,2\text{- or }1,7\text{-}C_2B_{10}H_{12} + 2BuLi \xrightarrow[60-80°C]{C_6H_6} Li_2C_2B_{10}H_{10} \xrightarrow{2RX} R_2C_2B_{10}H_{10} + 2LiX$$

Although the $1,2-C_2B_{10}H_{12}$ cage is not degraded by acids, strong bases remove the B(6)—H or B(3)—H vertex (the only BH units adjacent to two carbon atoms, and the most electropositive boron atoms in the cage), forming the $nido\text{-}C_2B_9H_{12}^-$ monoanion; this species in turn can be deprotonated to give the $nido\text{-}C_2B_9H_{11}^{2-}$ dianion.

$$C_2B_{10}H_{12} + EtO^- + 2EtOH \longrightarrow 7,8\text{-}C_2B_9H_{12}^- + H_2 + B(OEt)_3$$

$$C_2B_9H_{12}^- + H^- \longrightarrow C_2B_9H_{11}^{2-} + H_2$$

The 11-vertex dianion, named "dicarbollide" (from the Spanish *olla*, pot) has a pentagonal C_2B_3 open face that is analogous to $C_5H_5^-$, a fact that led directly to the synthesis of metallacarboranes (following section).

A separate series of closo-carboranes, isoelectronic with the $B_nH_n^{2-}$ dianions and the neutral $C_2B_{n-2}H_n$ compounds, consists of the monocarbon anions $CB_{n-1}H_n^-$ (Table 5-3). The preparative routes to these species vary widely, but a commonly used method for incorporating a single carbon atom into a borane cage involves the conversion of cyanoboranes to amino derivatives; the cyano carbon atom satisfies its valence requirements *via* absorption into the cluster framework.

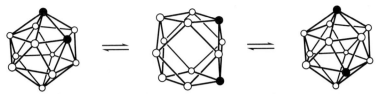

Figure 5-12 Proposed diamond-square-diamond rearrangement of $1,2\text{-}C_2B_{10}H_{12}$ to $1,7\text{-}C_2B_{10}H_{12}$ *via* a cubooctahedral intermediate.

One example is the following:

$$B_{10}H_{14} + RN \equiv C \longrightarrow (RH_2N)CB_{10}H_{12} \xrightarrow[\text{2) Me}_2\text{SO}_4]{\text{1) H}^+} (RMe_2N)CB_{10}H_{12} \xrightarrow{\text{Na}} CB_{10}H_{11}^-$$

The icosahedral $CB_{11}H_{12}^-$ ion and its derivatives have attracted attention in recent years as "least-coordinating" anions, much valued in studies of highly reactive cations such as the long-sought silylium ions SiR_3^+.[5]

Polyhedral Rearrangements

A characteristic feature of closo-carboranes and other polyhedral boron clusters is their thermal isomerization involving intramolecular rearrangement of the cage skeleton. The major driving force is the separation of the relatively positive cage carbon nuclei, which tends to favor isomers having maximum C—C separation; thus, octahedral $1,2\text{-}C_2B_4H_6$ converts to the 1,6 isomer at 250°C. Similarly, icosahedral $1,2\text{-}C_2B_{10}H_{12}$ (*o*-carborane, Fig. 5-10j) undergoes rearrangement above 425°C to the 1,7-isomer (*m*-carborane, Fig. 5-10k), and at still higher temperatures decomposes with partial conversion to the 1,12-isomer (*p*-carborane, Fig. 5-10l); this last product is probably the most stable covalent neutral molecule known, with thermal stability extending well above 700°C. More than 30 years of experimental and theoretical studies of the 1,2 to 1,7 rearrangement have still not fully resolved the mechanism; the current evidence[6] suggests that more than one pathway is involved, with "diamond- square-diamond" (dsd) processes of the kind originally proposed by Lipscomb (Fig. 5-12) playing a role, although modified by rotations of faces in the cuboctahedral intermediate.

Nido- and Arachno-Carboranes

Carboranes having open-cage geometry form a much larger, and structurally far more diverse, group than the closo series. As indicated in Table 5-4 and the selected examples in Fig. 5-13, they include 1- to 4-carbon systems exhibiting widely varying geometry.

The smallest of these, *nido*-$1,2\text{-}C_2B_3H_7$ (Fig. 5-13a), an isoelectronic and isostructural analogue of B_5H_9, is unstable, polymerizing in the liquid state to form a white solid. Parent *nido*-$C_2B_4H_8$ (Fig. 5-13c), the first carborane reported in the

[5] C. A. Reed *et al.*, *Phosphorus, Sulfur, and Silicon* **1994**, *93*, 485.
[6] V. Roberts and B. F. G. Johnson, *J. Chem. Soc., Dalton Trans.* **1994**, 759.

Table 5-4   Open-Cage Carboranes[a,b]

Vertices	Cages				
	1-Carbon	2-Carbon	3-Carbon	4-Carbon	5-Carbon
5	a-CB$_4$H$_{10}$[c]	n-C$_2$B$_3$H$_7$			
6	n-CB$_5$H$_9$	n-2,3-C$_2$B$_4$H$_8$	n-2,3,4-C$_3$B$_3$H$_7$	n-C$_4$B$_2$H$_6$	n-C$_5$BH$_6^{+c}$
		n-2,4-C$_2$B$_4$H$_7^-$	n-2,3,5-C$_3$B$_3$H$_7$[c]		
7		n-C$_2$B$_5$H$_9$[c]			
8	a-CB$_7$H$_{13}$	n-C$_2$B$_6$H$_{10}$[d]		n-C$_4$B$_4$H$_8$[c,d]	
		a-C$_2$B$_6$H$_{12}$[d]			
		h-C$_2$B$_6$H$_{13}$			
9	n-CB$_8$H$_{12}$	n-C$_2$B$_7$H$_{11}$	h-C$_3$B$_6$H$_{13}^{-c}$		
	a-CB$_8$H$_{14}$	a-C$_2$B$_7$H$_{13}$[d]			
10	n-CB$_9$H$_{12}^-$	n-C$_2$B$_8$H$_{12}$[d]	n-C$_3$B$_7$H$_{11}$[c]	n-C$_4$B$_6$H$_{10}$[d]	
	a-CB$_9$H$_{14}^-$	a-C$_2$B$_8$H$_{14}$	a-C$_3$B$_7$H$_{13}$[c]		
11	n-CB$_{10}$H$_{13}^-$	n-C$_2$B$_9$H$_{13}$[d]	n-C$_3$B$_8$H$_{12}$	n-C$_4$B$_7$H$_{11}$[d]	
				a-C$_4$B$_7$H$_{13}$[c,d]	
12		n-C$_2$B$_{10}$H$_{13}^{-d}$		n-C$_4$B$_8$H$_{12}$[c]	
		a-C$_2$B$_{10}$H$_{15}^{-c}$		a-C$_4$B$_8$H$_{14}$[c]	
13				n-C$_4$B$_9$H$_{13}$[c,e]	
15				n-C$_4$B$_{11}$H$_{15}$[c,d,e]	

[a] Isolated and characterized species.
[b] n = nido, a = arachno, h = hypho.
[c] Known only as C- or B-substituted derivatives.
[d] Two or more isomers known.
[e] Proposed structure.

literature, was originally isolated from thermal or base-promoted reactions of acetylene with B$_5$H$_9$ that also generate other products, including derivatives of the *nido*-CB$_5$H$_9$ monocarbon carborane (Fig. 5-13b). Tricarbon carboranes include *nido*-C$_3$B$_3$H$_7$ and its alkyl derivatives and, more recently, C$_3$B$_n$ species in which n = 6-8.[7]

The open-cage carboranes and their deprotonated anions are in general quite reactive, and the more accessible of them are excellent ligands for sandwich-binding to metal ions as described below. Especially widely used are the 6-vertex *nido*-RR′C$_2$B$_4$H$_4^{2-}$ and 11-vertex *nido*-RR′C$_2$B$_9$H$_9^{2-}$ ions, which are generated by deprotonation of the neutral carboranes in Figs. 5-13c and 5-13g, respectively, and have planar C$_2$B$_3$ open faces. The 12-vertex *nido*-RR′C$_2$B$_{10}$H$_{10}^{2-}$ dianions, with C$_2$B$_4$ open faces, can form 13- and 14-vertex metallacarboranes on coordination to metal ions.

The larger nido and arachno cages are usually obtained *via* degradation (in one or more steps) of icosahedral carboranes, as previously mentioned, but the smaller members are typically prepared in reactions of alkynes with neutral boranes, as in the synthesis of 2,3-RR′C$_2$B$_4$H$_8$ (Fig. 5-14). However, the *nido*-2,4-C$_2$B$_4$H$_7^-$ "carbons apart" ion can be obtained by removal of an apex boron from *closo*-2,4-C$_2$B$_5$H$_7$ (Fig. 5-10d).[8]

[7] L. G. Sneddon *et al.*, *J. Am. Chem. Soc.* **1993,** *115,* 10004; B. Štíbr *et al.*, *J. Chem. Soc., Chem. Commun.* **1995,** 795.
[8] T. Onak *et al.*, *Inorg. Chem.* **1988,** *27,* 3679.

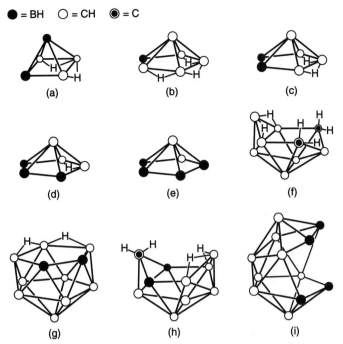

● = BH   ○ = CH   ◉ = C

(a)   (b)   (c)

(d)   (e)   (f)

(g)   (h)   (i)

Figure 5-13 A selection of *nido-* and *arachno*-carborane structures. (a) *nido*-1,2-$C_2B_3H_7$. (b) *nido*-2-CB_5H_9. (c) *nido*-2,3-$C_2B_4H_8$. (d) *nido*-2,3,4-$C_3B_3H_7$. (e) *nido*-2,3,4,5-$C_4B_2H_6$. (f) *arachno*-$C_2B_7H_{13}$. (g) *nido*-$C_2B_9H_{13}$. (h) *arachno*-$C_3B_7H_{13}$. (i) *nido*-$C_4B_8H_{12}$. Compounds (h) and (i) are known only as C-substituted derivatives.

Transition metal-promoted reactions are important in the synthesis of some nido- and arachno-carboranes, as in the preparation of a C_4B_4 cage *via* alkyne incorporation into $Et_2C_2B_4H_6$ (Fig. 5-15a).[9] Metal-promoted oxidative fusion[10] of $R_2C_2B_4H_4^{2-}$ ligands in bis(carboranyl) metal sandwich complexes (a type of reaction that can also be applied to borane, metallaborane, and metallacarborane synthesis) forms 12-vertex C_4B_8 carboranes (Fig. 5-15b). Other carborane fusion reactions have generated products proposed to have 15-vertex C_4B_{11} cage frameworks.

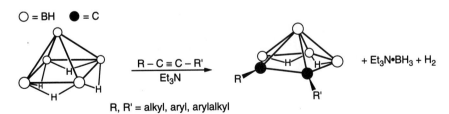

○ = BH   ● = C

$$R-C\equiv C-R' \xrightarrow{\quad Et_3N \quad}$$

+ Et_3N•BH_3 + H_2

R, R' = alkyl, aryl, arylalkyl

Figure 5-14 Base-promoted synthesis of *nido*-2,3-RR'$C_2B_4H_6$ carboranes.

[9] L. G. Sneddon *et al.*, *J. Am. Chem. Soc.* **1989**, *111*, 592.
[10] R. N. Grimes, *Coord. Chem. Rev.* **1995**, *143*, 71.

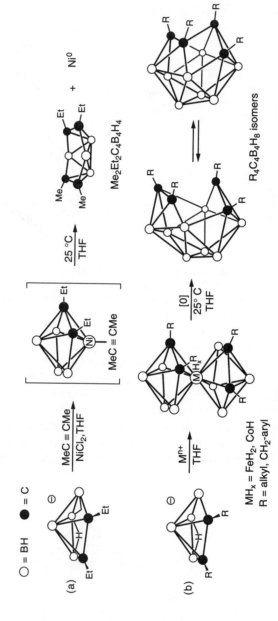

Figure 5-15 Transition metal-promoted syntheses of tetracarbon carboranes.

As shown, the $R_4C_4B_8H_8$ compounds in which R is a small alkyl group are fluxional in solution; for larger substituents (e.g., benzyl), only the "open" structure is seen.[10] Both the "open" and "closed" isomers are 28-electron cage species that show marked distortion from regular icosahedral geometry (which accommodates only 26 electrons); however, neither form adopts the nido geometry expected from PSEPT analysis as described earlier, i.e., a 13-vertex polyhedron with one missing vertex. Reduction of the neutral species affords the 30-electron *arachno*-$R_4C_4B_8H_8^{2-}$, whose structure is a fragment of a 14-vertex closo polyhedron (bicapped hexagonal antiprism), as predicted by the PSEPT rules.

The open-cage carboranes exhibit a very rich chemistry whose exploration is still in an early stage. In general, the carbon atoms prefer low-coordinate vertices on the open face of the cage, and the insertion of metal and nonmetal heteroatoms into these faces provides one of the main themes of carborane chemistry. Although the open-cage compounds are in general more reactive than the closo-carboranes, they are far less so than the lower boranes, and hence their interactions with metals are more easily controlled—a point that will be apparent in comparing the metallaboranes with the metallacarboranes, discussed in the following section.

5-6 Metallaboranes and Metallacarboranes

In 1965 M. F. Hawthorne and his students, having found that the icosahedral 1,2-$C_2B_{10}H_{12}$ carborane cage can be converted to the *nido*-$C_2B_9H_{12}^{2-}$ "dicarbollide" ion *via* base-degradation and deprotonation as described above, noted an apparent similarity between the C_2B_3 open face in this species and the $C_5H_5^-$ ion. This implied that transition metal ions should form η^5-coordinated sandwich complexes analogous to the metallocenes, an idea that was soon verified by experiment and launched the new field of metallacarborane chemistry. In the three decades since this momentous discovery, the general area of metallaboron clusters—which now includes not only metallacarboranes per se but also the polyhedral metallaboranes, metal-borane and metal-carborane σ- and μ-complexes, and involves most of the main-group and transition metals—has grown to enormous size, encompassing thousands of compounds and a rich structural chemistry. The following discussion presents only a very rudimentary outline of the main classes, with a few examples selected to illustrate the wide range of cage geometries.

Metallaboranes are borane cages containing one or more metal atoms, but no carbon atoms, in the skeletal framework. (The "metalla" usage is derived from the "oxa/aza" convention of organic chemistry and denotes a metal replacing a skeletal boron atom.) Reactions between boron hydrides and neutral metal reagents, or more commonly, between boron hydride anions and cationic metals or metal-ligand units, have been employed to generate a plethora of metallaboranes ranging from very small 4- and 5-vertex species to multi-cage "macropolyhedral" systems (Figs. 5-16 and 5-17). With few exceptions, these syntheses are uncontrolled, often affording products of novel structure but in widely varying yields. Consequently, synthetic routes having broad applicability are almost nonexistent, but a recent review by L. Barton in *Comprehensive Organometallic Chemistry II* on this subject provides a useful guide to this area.

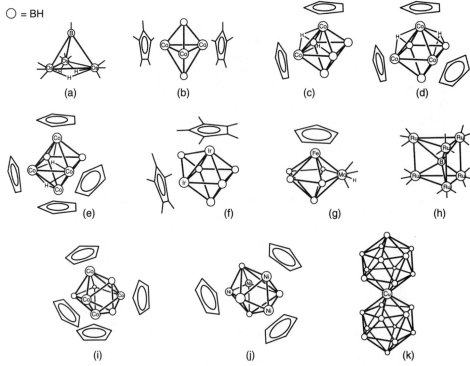

Figure 5-16 Some closo-metallaboranes (in a few cases Cp or Cp* ligands are omitted for clarity). (a) $H_3(CO)_9Os_3BCO$. (b)$Cp*_3Co_3B_2H_4$. (c) $Cp_2Co_2B_4H_6$. (d) $Cp_3Co_3B_3H_5$. (e) $Cp_4Co_4B_2H_4$. (f) $Cp*_2Ir_2B_5H_5$. (g) $CpFe(CO)_3HMoB_5H_5$. (h) $H_2(CO)_{18}Ru_6B^-$. (i) $Cp_4Co_4B_4H_4$. (j) $Cp_4Ni_4B_4H_4$. (k) $Cu(B_{11}H_{11})_2^{3-}$.

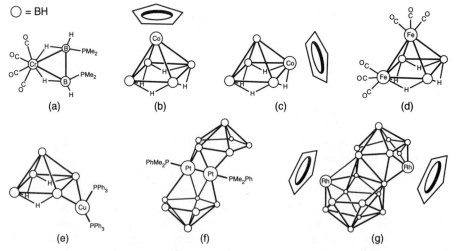

Figure 5-17 Selected open-cage metallaboranes. (a) $(CO)_4CrB_2H_4(PMe_3)_2$. (b) 1-$CpCoB_4H_8$. (c) 2-$CpCoB_4H_8$. (d) 1,2-$(CO)_6Fe_2B_3H_7$. (e) μ-$(Ph_3P)_2CuB_4H_8$. (f) $(Me_2PPh)_2Pt_2B_{12}H_{18}$. (g) $Cp_2Rh_2B_{17}H_{19}$.

Since borane substrates, particularly the smaller ones, are quite reactive toward metal reagents, it is frequently the case that attack of the metal can occur at more than one site; moreover, the incorporation of the first metal atom into the cage may promote further reactions leading to di- or even trimetallic products, as well as isomers. Consequently, metallaboranes are often formed as complex mixtures of isolable species: typically, the first closo-metallaborane clusters were obtained from the reaction of the *nido*-$B_5H_8^-$ ion with $CoCl_2$ and $NaCp$, which generated both the nido complex 2-$CpCoB_4H_8$ (Fig. 5-17c) and a family of Co_2, Co_3, and Co_4 closo-metallaboranes. In a different approach, reactions of metal clusters with monoboron reagents such as BH_4^- or BH_3·THF have been employed to generate metal-rich complexes (e.g., $Cp_4Co_4B_2H_4$, Fig. 5-16e) and "metal boride" clusters that contain encapsulated boron atoms, e.g., octahedral $H(CO)_{17}Ru_6B^{11}$ and trigonal-prismatic $H_2(CO)_{18}Ru_6B^-$, Fig. 5-16h.[12]

Structural patterns in the metallaboranes can be interpreted *via* the PSEPT (Wade's Rules described earlier). In applying this approach to metal-containing clusters, it is assumed that each metal atom occupying a vertex in the cage utilizes exactly 3 orbitals for binding to its neighbors in the framework. For main-group metals, which in most cases have 4 valence orbitals, this leaves one orbital for bonding to an external atom or group such as H, alkyl, halogen, etc. However, transition metals, with 9 valence orbitals, have 6 orbitals remaining after allocating 3 for cage bonding; these are used to bond to external groups and to store nonbonding electrons. The skeletal electron contribution that a metal-ligand unit makes to the cage framework is then readily calculated: thus, in a neutral CpCo fragment, which has 14 valence electrons, the metal atom uses 3 orbitals to bind to the Cp ring, 3 orbitals for bonding to the carborane ligand, and 3 orbitals for nonbonding electron storage. Since the binding of Cp to Co requires 6 electrons, and 6 electrons are nonbonding, 2 electrons are available for cage bonding and hence CoCp is *isolobal* with BH:

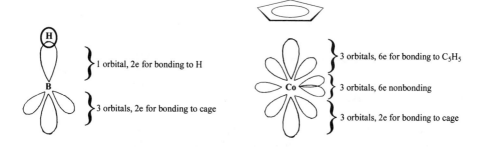

Similarly, the commonly encountered $Fe(CO)_3$ fragment is also a 2-electron donor while CpNi supplies 3 electrons (analogous to CH). These relationships are illustrated by the complexes 1- and 2-$CpCoB_4H_8$ (Fig. 5-17b and c) and 1,2-$(CO)_6Fe_2B_3H_7$ (Fig. 5-17d), each of which has a nido 5-vertex structure analogous to B_5H_9 (Fig. 5-3b) with CpCo or $(CO)_3Fe$ units replacing BH. In all of these clusters, there are 14 skeletal electrons (10 from the 5 BH or metal units and 4

[11]S. G. Shore *et al.*, *Inorg. Chem.* **1989,** *28,* 3284.
[12]C. E. Housecroft *et al.*, *J. Chem. Soc., Chem. Commun.* **1992,** 842.

from the bridging hydrogen atoms), corresponding to 7 pairs; since $n = 5$, these are $(n+2)$-electron-pair systems in accord with their nido geometries. As another example, icosahedral $Cp_2Ni_2B_{10}H_{10}$ is related to $C_2B_{10}H_{12}$ and $B_{12}H_{12}^{2-}$ *via* formal replacement of two CH or two BH$^-$ groups with two CpNi units.

Most metallaboranes obey the PSEPT paradigm, with closo species adopting the polyhedral structures of their isoelectronic $B_nH_n^{2-}$ analogues (Fig. 5-4) and the open-cage complexes having the geometries of their neutral borane counterparts (Fig. 5-3). However, there are notable exceptions. For example, the $Cp_4M_4B_4H_4$ clusters (M = Co, Ni), shown in Fig. 5-16 (i and j) have, respectively, 16 and 20 skeletal electrons, but adopt closo 8-vertex polyhedral structures like that of $B_8H_8^{2-}$, an 18 electron system (Fig. 5-4c). The failure of these two metallaboranes to adopt the PSEPT-predicted geometries (capped-closo and nido, respectively) has been explored theoretically. A special situation (not a PSEPT "violation" *per se*) arises in the case of bis(dicarbollyl) complexes of the type $M(C_2B_9H_{12})_2$ when M is a metal ion having more than 6 d electrons. This produces an excess of electrons beyond the 13 bonding pairs per cage that can be accommodated in an icosahedron and causes slip-distortion (movement of the metal to a position away from the center of the C_2B_3 faces) rather than the usual closo-to-nido cage-opening.

Metal-borane and metal-carborane σ- and μ-complexes are boranes or carboranes having metal substituents external to the cage and bound to it by single (σ) or bridging (μ) bonds, as in μ-[Cu(PPh$_3$)$_2$]B$_4$H$_8$ (Fig. 5-17e).

Metallacarboranes have both metal and carbon atoms in the cage skeleton. In contrast to the metallaboranes, syntheses of metallacarboranes *via* low- or room-temperature metal insertion into carborane anions in solution are more controllable, usually occurring at a well-defined C_2B_n open face and forming a single isomer. Other preparative routes, such as metal insertion into neutral carboranes at high temperature, involve more complex processes and tend to give less clean-cut results, and such approaches are rarely employed today. Typical syntheses that are based on readily available 11-vertex and 7-vertex *nido*-carborane ligands are as shown:

$$C_2B_9H_{11}^{2-} \xrightarrow[\text{THF}]{Ni^{2+}} \xrightarrow{\text{air}} Ni(C_2B_9H_{11})_2$$

$$C_2B_9H_{11}^{2-} + Fe^{2+} \xrightarrow[\text{THF}]{C_5H_5^-} \xrightarrow{\text{air}} CpFe(C_2B_9H_{11})$$

$$Et_2C_2B_4H_4^{2-} + Co^{2+} \xrightarrow[\text{THF}]{C_5H_5^-} \xrightarrow{\text{air}} CpCo(Et_2C_2B_4H_4)$$

Metallacarborane structures are mostly consistent with the PSEPT approach. It is impossible to convey the richness of this field in a short chapter, but a few representative examples of the many hundreds of structurally characterized species are shown in Fig. 5-18.

The large majority of the known compounds have two carbon atoms per cage, but there are now many examples of 1-, 3-, and 4-carbon metallacarboranes. Complexes are known involving most transition metals, many lanthanides, and numerous main-group elements in groups 1, 2, 13, 14, 15, and 16; in addition, combinations of these metals may occur in the same cage framework. Given this scope and the availability of many different carborane ligands, it is safe to say that the metallacarboranes form by far the largest and most diverse class of cluster compounds known.

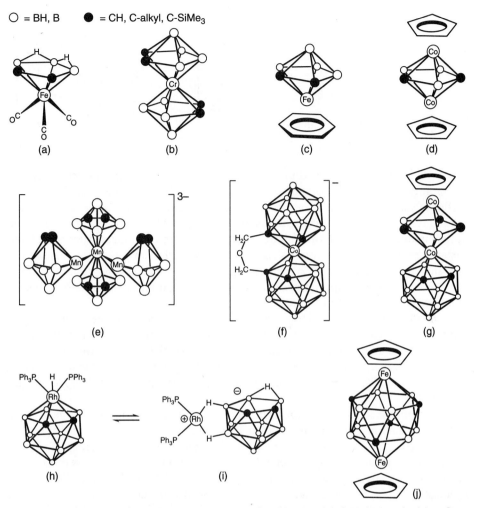

Figure 5-18 Examples of metallacarboranes. (a) *nido*-1,2,3-(CO)$_3$Fe(C$_2$B$_3$H$_7$). (b) Cr[2,3-(Me$_3$Si)$_2$C$_2$B$_4$H$_4$]$_2$. (c) 1,2,3-(C$_6$H$_6$)Fe(Et$_2$C$_2$B$_4$H$_4$). (d) 1,7,2,4-Cp$_2$Co$_2$(C$_2$B$_3$H$_5$). (e) Mn$_3$[2,3-(Me$_3$Si)$_2$C$_2$B$_4$H$_4$]$_4^{3-}$. (f) Co(1,2-C$_2$B$_9$H$_{10}$)$_2$-μ-O(CH$_2$)$_2$. (g) CpCo(MeEt$_2$C$_3$B$_2$Et$_2$)Co(C$_2$B$_9$H$_{11}$). (h) (Ph$_3$P)$_2$HRh(C$_2$B$_9$H$_{11}$). (i) *exo-nido*-[(Ph$_3$P)$_2$RhH$_2$]$^+$[C$_2$B$_9$H$_{10}$]$^-$. (j) Cp$_2$Fe$_2$(Me$_4$C$_4$B$_8$H$_8$).

As a group, metallacarboranes are more robust than metallaboranes and their reactivity is easier to control; consequently their chemistry has developed more systematically. The closo-metallacarboranes exhibit reactions typical of carboranes, including substitution at carbon *via* C-lithiated derivatives, cage expansion and degradation, and electrophilic halogenation, while the *nido*-metallacarboranes undergo bridge-deprotonation just as their borane analogues do (Section 5-4). However, the most intensively studied chemistry of metallacarboranes involves the metal center(s), and much attention has been directed in recent years to exploiting the properties of these compounds in practical ways. Earlier work on the development of hydrogenation catalysts based on rhodium- and iridium-carborane complexes, in which the catalytically active species are of the exo-nido type shown in

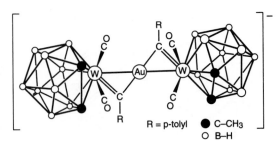

Figure   5-19 μ-Au(C-C$_6$H$_4$Me)$_2$-
[(CO)$_2$W(Me$_2$C$_2$B$_9$H$_9$)]$_2^-$.

R = p-tolyl   ● C–CH$_3$
                      ○ B–H

Fig. 5-18i,[13] has been followed in recent years by studies of early transition metal-carboranes as catalysts for olefin polymerization.[14] In these applications, the metalla-carborane species resemble their metallocene-based analogues; however, there are other areas, such as the construction of new types of electronic, magnetic, and optical materials, in which the unique structural and other characteristics of metalla-carboranes may be put to use. "Venus flytrap" linked-dicarbollide ligands such as that shown encapsulating a Co^{3+} ion in Fig. 5-18f have been prepared as a possible way to bind radiotransition metals to tumor-associated monoclonal antibodies for purposes of diagnosis and therapy.[15]

Metallacarborane chemistry overlaps extensively with studies of metal clusters, as in the characterization of many complexes containing both metal-metal bonded arrays and carborane ligands. Figure 5-19 depicts a gold-bridged tungsten carbene complex[16] that is representative of a very large and diverse family of alkylidene and alkylidyne heavy metal compounds incorporating MC$_2$B$_9$ or MC$_2$B$_{10}$ cage units.[17]

Heteroboranes (or heteroboron clusters) are boron cages containing one or more nonmetal atoms in the skeletal framework; if skeletal carbon atoms are also present, these species are called *heterocarboranes*. Examples involving nearly all of the elements in Groups 13, 14, 15, and 16 are now known, including even atoms as electropositive as Na and as electronegative as O. As an illustration, Group 15 atoms such as N, P, and As are isoelectronic with CH and are 3-electron donors to the cage (with a lone pair of electrons directed outward); accordingly, icosahedral species such as PCB$_{10}$H$_{11}$, P$_2$B$_{10}$H$_{10}$, N$_2$B$_{10}$H$_{10}$, and As$_2$B$_{10}$H$_{10}$ (all of which are isoelec-tronic with C$_2$B$_{10}$H$_{12}$) are well known. Other extensively studied classes of hetero-boron clusters include the thia- and selenaboranes, alumina- and gallaboranes, and heterocarborane analogues of all these groups. Still further variation may be achieved by introducing transition metals to heteroboron cage frameworks to give *heterometallaboranes* and *heterometallacarboranes*. In virtually all cases, the ob-served structures of heteroboron clusters adhere to the PSEPT rules, although certain elements create ambiguous situations; e.g, S can function as either a 4-electron or 2-electron donor depending on whether it has one or two exo-polyhe-dral lone pairs.

[13]J. A. Belmont *et al., J. Am. Chem. Soc.* **1989,** *111,* 7475.
[14]D. J. Crowther and R. F. Jordan, *Makromol. Chem., Macromol. Symp.* **1993,** *66,* 121.
[15]D. E. Harwell *et al., Can. J. Chem.* **1995,** *73,* 1044.
[16]J. E. Goldberg *et al., J. Chem. Soc., Dalton Trans.* **1992,** 2495.
[17]F. G. A. Stone, *Adv. Organomet. Chem.* **1990,** *31,* 53.

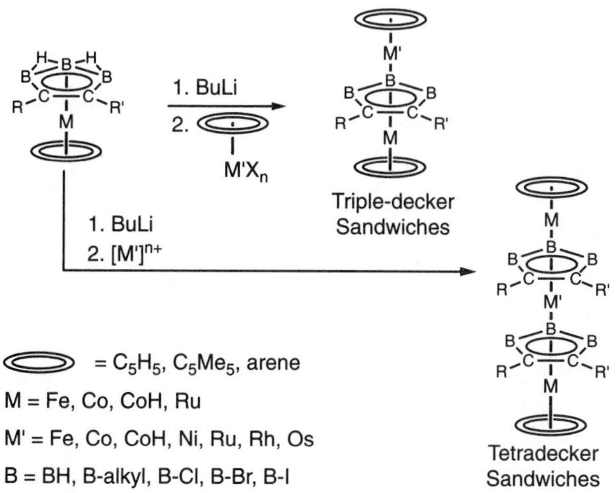
1. BuLi
2.
M'Xn
Triple-decker Sandwiches
1. BuLi
2. [M']n+
= C5H5, C5Me5, arene
M = Fe, Co, CoH, Ru
M' = Fe, Co, CoH, Ni, Ru, Rh, Os
B = BH, B-alkyl, B-Cl, B-Br, B-I
Tetradecker Sandwiches

Figure 5-20 Synthesis of multidecker sandwiches *via* metal stacking reactions.

Multidecker metal sandwich complexes[18,19] incorporate carborane (C_2B_3) or organoborane (e.g., C_3B_2, C_2SB_2, C_4B, C_4B_2) rings that are facially bound to metal atoms on both sides of the ring plane. These are metallacarboranes by definition, but are also *bona fide* organometallic complexes, since the planar boron-containing rings are isoelectronic counterparts of hydrocarbon ligands; for example, $CpCo(C_2B_3H_5)CoCp$ (Fig. 5-18d) contains a formal $C_2B_3H_5^{4-}$ ring ligand which is a 6π-electron analogue of $C_5H_5^-$.

The synthesis in the 1970s of C_2B_3-bridged triple-decker sandwiches (the first stable, neutral multidecker complexes), as depicted in Fig. 5-20, has been extended to larger molecular systems having 4-6 decks,[18,19,20] to linked-sandwich chains, and to C_3B_2-bridged polydecker sandwiches (Fig. 5-21); the Ni polymer is an electrical semiconductor.[19] The linked-sandwich oligomers[20] are remarkably stable, soluble in organic solvents, and in many cases can undergo reversible oxidation and reduction at their metal centers; the paramagnetic species tend to show extensive electron delocalization in the chain.

The construction of such compounds *via* stepwise building-block approaches, as in the synthesis of the hexadecker sandwiches[21] in Fig. 5-22, lends order and controllability to this area, and may allow the rational development of new materials having desired combinations of electronic, magnetic, optical, or other properties.

5-7 Boron Halides

Trihalides

In contrast to the aluminum trihalides, which form Al_2X_6 dimers and thereby complete the valence octet on Al, the boron trihalides are strictly monomeric, trigonal

[18]R. N. Grimes, *Chem. Rev.* **1992,** *92,* 251; *Applied Organometal. Chem.* **1996,** *10,* 209.
[19]W. Siebert, *Adv. Organometal. Chem.* **1993,** *35,* 187–210.
[20]X. Meng *et al., J. Am. Chem. Soc.* **1993,** *115,* 6143.
[21]X. Wang *et al., J. Am. Chem. Soc.* **1995,** *117,* 12227.

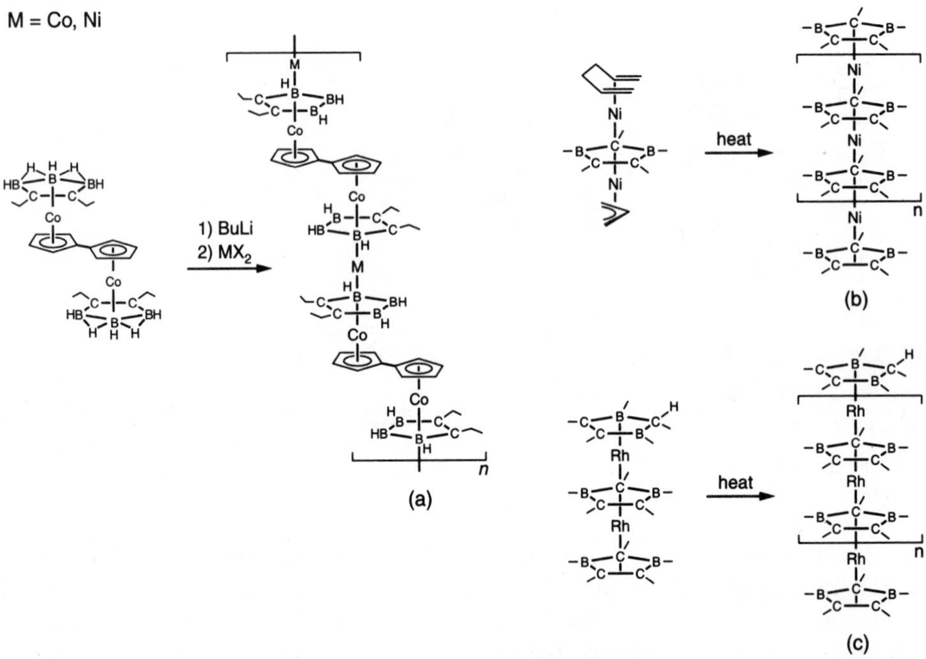

M = Co, Ni

Figure 5-21   Preparation of fulvalene-linked carborane sandwich oligomers (a) and diboro-lyl sandwich polymers (b and c).

planar molecules. The difference is attributed to back-donation from "nonbonding" electrons on the halogen atoms into the "empty" p_z orbital of boron, which lends some double-bond character to the $B-X$ bonding (manifested by bond shortening) and stabilizes the monomer. (In the heavier Group 13 elements, which have larger, more diffuse p_z orbitals, such back-bonding is unfavorable relative to dimerization). With the exception of BI_3, the boron trihalides can be prepared directly from the

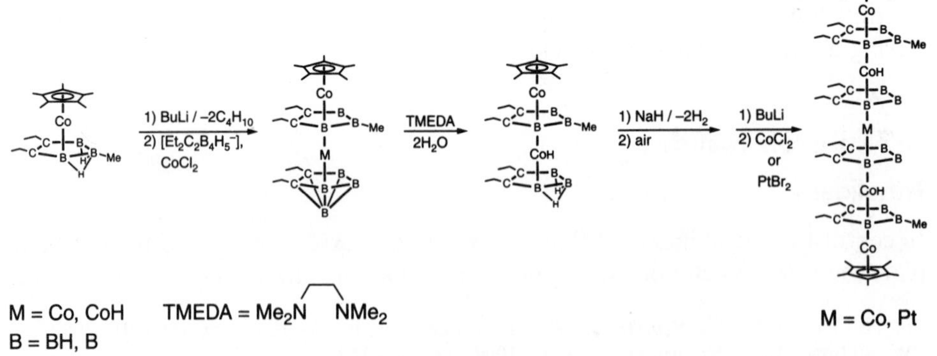

M = Co, CoH   TMEDA = Me_2N NMe_2

B = BH, B

M = Co, Pt

Figure 5-22   Designed synthesis of hexadecker sandwich complexes from small building-block units.

Table 5-5 Boron Halides[a]

Compound	mp, °C	bp, °C	Compound	dec, °C $(t_{1/2})$[b]
BF_3	−127.1	−99.9	B_8Cl_8	200 (18 h)
BCl_3	−107	12.5	B_9Cl_9	400 (85 h)
BBr_3	−46	91.3	$B_{10}Cl_{10}$	300 (100 h)
BI_3	49.9	210	$B_{11}H_{11}$	300 (3 h)
B_2F_4	−56.0	−34	$B_{12}Cl_{12}$	200 (5 h)
B_2Cl_4	−92.6	65.5	B_7Br_7	200 (18 h)
B_2Br_4	—	22.5[c]	B_8Br_8	200 (18 h)
B_2I_4	125	60[d]	B_9Br_9	200 (7 d)
B_4Cl_4	95	—	$B_{10}Br_{10}$	200 (24 h)
B_4Br_4	108	—	B_8I_8	—
			B_9I_9	—

[a] Isolated and characterized species.
[b] Approximate half-life at the temperature shown (J. A. Morrison, *Chem Rev.* **1991**, *91*, 35).
[c] 5.5 mm Hg.
[d] 10^{-4} mm Hg.

elements, but BF_3 is normally obtained *via* treatment of B_2O_3 with CaF_2 in H_2SO_4. As can be seen in Table 5-5, BI_3 is a solid, but the others are volatile liquids or gases at room temperature. All four compounds are Lewis acids, but the acid strength decreases markedly in going from BBr_3 to BF_3, evidently reflecting increasing importance of internal $B-X$ π-bonding. On mixing two different trihalides, rapid halogen exchange takes place to form equilibrium mixtures of mixed-halogen species, e.g., BCl_3 and BBr_3 generate BCl_2Br and $BClBr_2$. Figure 5-23 summarizes some typical reactions of the lower boron halides.

Boron trifluoride, a colorless gas, forms adducts with a wide variety of electron-pair donors including water, alcohols, ethers, and amines; a well-known reagent is the $BF_3 \cdot OEt_2$ adduct. With water below 20°C it forms the hydrates $BF_3 \cdot H_2O$ and $BF_3 \cdot 2H_2O$, which in the liquid state are partially dissociated, e.g., $BF_3 \cdot 2H_2O = BF_3OH^- + H_3O^+$. In dilute solution, BF_3 forms the tetrafluoroborate anion, whose acid solutions are known as fluoroboric acid:

$$4BF_3 + 6H_2O \longrightarrow 3H_3O^+ + 3BF_4^- + B(OH)_3$$

Owing to its powerful Lewis acidity, BF_3 is an effective reagent in organic synthesis, for example, promoting the conversion of alcohols and acids to esters, the polymerization of olefins and olefin oxides, and acylations and alkylations (in a manner similar to Friedel-Crafts processes). Mechanistic studies of some reactions of the latter type, such as the ethylation of benzene by C_2H_5F, have shown that the BF_3 functions as a scavenger for HF *via* the formation of HBF_4 and thus participates stoichiometrically rather than catalytically.

Although HBF_4 (fluoroboric acid) cannot be isolated per se, various salts and solvates, e.g., $NaBF_4$, NH_4BF_4, and $HBF_4 \cdot Et_2O$ are well known, and some solvates

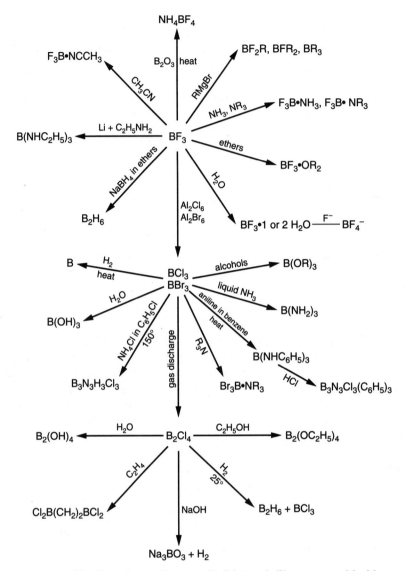

Figure 5-23   Reactions of boron trihalides and diboron tetrachloride.

contain oxonium ions; an example is $[CH_3OH_2]^+[BF_4]^-$, mp $-41°C$. The tetrahedral BF_4^- ion, an isoelectronic analogue of CF_4, is similarly tetrahedral, and is known as a relatively (but not always) noncoordinating anion.

   Boron trichloride, boron tribromide, and boron triiodide, in contrast to the trifluoride, are all rapidly hydrolyzed in water (explosively so in the case of BI_3) to generate $B(OH)_3$ and HX; as indicated above, all are stronger Lewis acids than BF_3. While BCl_3 and BBr_3 are volatile liquids, BI_3 is an unstable white solid that polymerizes on standing. The BX_4^- anions are all known, but the larger ones —especially BI_4^-—require stabilization by a large cation.

Subhalides

Boron-halogen compounds containing B$-$B bonds, and in which the halogen:boron ratio is less than 3, were once restricted largely to the diboron tetrahalides and a few other species, but this field has grown considerably in recent years and now comprises a wide variety of structure types. The tetrahalides, B_2X_4, are known for all four halogens, and all consist of two BX_2 groups joined by a B$-$B single bond. Curiously, in the case of B_2Cl_4 and B_2Br_4, the structure is staggered (nonplanar) in the gas phase but planar in the solid state, which is a strong indication of facile rotation (barrier < 10 kJ) around the B$-$B bond and is consistent with single-bond (σ) character.

Diboron tetrafluoride can be prepared from BF_3 and elemental boron at 1850°C, a reaction that first produces BF (a compound that can be isolated by condensation at -196°C), which reacts with BF_3 to give B_2F_4 and unstable higher boron fluorides such as B_8F_{12}.

Diboron tetrachloride is obtained *via* electric discharge through BCl_3 in the presence of copper or mercury. In an important reaction, B_2Cl_4 readily adds across C$=$C double bonds to form *bis*(dichloroboryl) products that can in turn be used to synthesize a variety of species (Fig. 5-23).

Polyhedral boron halides[22] have B:X ratios of 1:1 and feature closed-cage (closo) structures like those of the $B_nH_n^{2-}$ anions (Fig. 5-4). However, unlike the latter group, the neutral B_nX_n clusters are stable, despite having *only n skeletal electron pairs*, and thus do not obey the PSEPT rules. This has been explained in terms of unfavorable cage-to-halogen orbital overlap which causes the $(n+1)$th framework orbitals in B_nX_n clusters to be essentially nonbonding, so that the HOMOs are now the nth orbitals.[22] As a consequence, 10-vertex perchloro and perbromo clusters are isolable both as neutral $B_{10}X_{10}$ and anionic $B_{10}X_{10}^{2-}$ species.

The characterized compounds are listed in Table 5-5. Except for B_4Cl_4, the polyhedral halides are prepared *via* thermal disproportionation of the tetrahalides; thus B_2Cl_4 forms BCl_3, B_8Cl_8, B_9Cl_9, $B_{10}Cl_{10}$, $B_{11}Cl_{11}$, and $B_{12}Cl_{12}$. The $B_{10}Cl_{10}$ and $B_{10}Br_{10}$ species can also be obtained by oxidation of the corresponding dianions or thermolysis of their salts. The polyhedral halides are intensely colored and can be sublimed under vacuum; their stability varies, the most robust being B_9Cl_9 and B_9Br_9, both of which survive at 300°C.

The tetrahalide B_4Cl_4 is something of a special case. This yellow sublimable solid, known for decades, can be prepared in small amounts by radio-frequency discharge through BCl_3 over Hg, and has long been of interest from a theoretical viewpoint because of its regular tetrahedral geometry and its *formal* skeletal electron count of 8, and the fact that the corresponding B_4H_4 borane (or anions thereof) are nonexistent. Photoelectron spectroscopy confirms that the molecule is stabilized *via* extensive σ and π interaction of Cl bonding electrons with the skeletal bonding molecular orbitals.[22] Treatment of B_4Cl_4 with a fivefold excess of BBr_3 generates B_4Br_4 and several mixed-ligand $B_4Cl_nBr_{n-4}$ species.[23] The chemistry of B_4Cl_4 has been extensively studied, and among its interesting aspects are its ability to undergo

[22]J. A. Morrison, *Chem Rev.* **1991**, *91*, 35.
[23]L. Ahmed *et al.*, *Inorg. Chem.* **1992**, *31*, 1858.

cluster expansion and fusion, forming chlorinated derivatives of B_5H_9, B_6H_{10}, and $B_{10}H_{14}$.

5-8 Boron-Nitrogen Compounds

Azaboranes and Boron Nitrides

Much of B-N chemistry relates to the fact that a BN unit is isoelectronic with CC and that the electronegativity and atomic radius of carbon are almost exactly intermediate between the values for boron and nitrogen. Consequently, there are many azaborane analogues of hydrocarbons in which B and N replace a pair of carbon atoms, as in 5-II to 5-VIII:

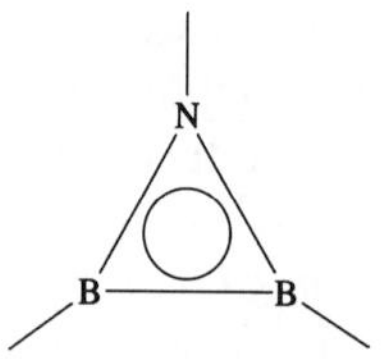

(5-II) 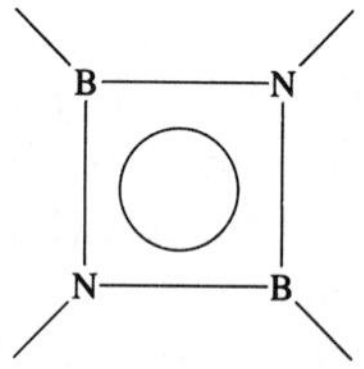   (5-III)

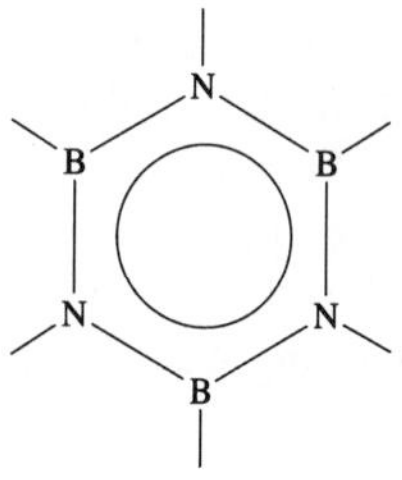

(5-IV) 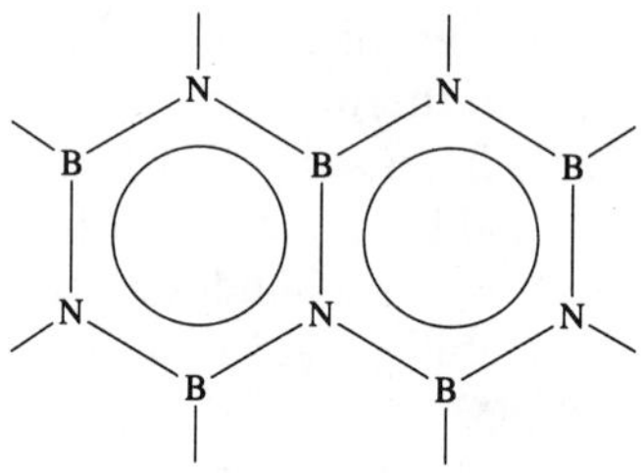   (5-V) 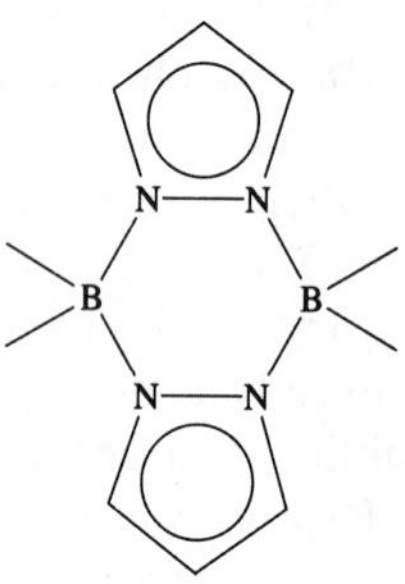   (5-VI)

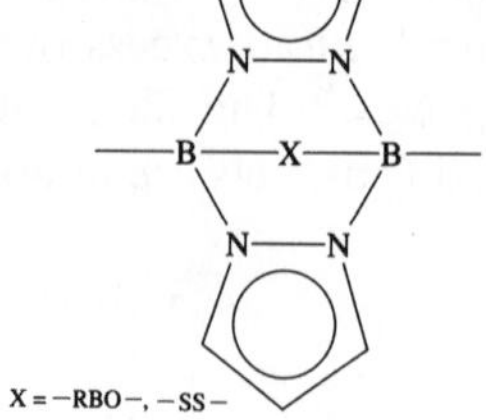

X = −RBO−, −SS−

(5-VII) 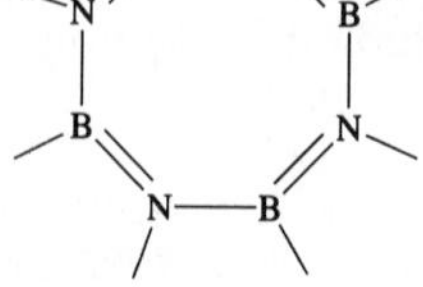   (5-VIII)

Types 5-VI and 5-VII are *pyrazaboles*, a widely investigated class that incorporates 5-membered pyrazole (C_3N_2) rings.[24] The best-known and most important boron-nitrogen ring compound is *borazine*, $B_3N_3H_6$, a benzene-like regular hexagon (5-IV). Like benzene, borazine is stabilized by π-delocalization that lends it a degree of aromatic character, evidenced by the ability of its alkyl derivatives to form sandwich complexes such as $(\eta^6\text{-}B_3N_3R_6)Cr(CO)_3$. However, the resemblance largely ends here, because borazine is far more reactive than benzene, owing to partial negative and positive charges on the N and B atoms, respectively. Thus, borazine readily adds HX to form cyclic $(H_2NBHX)_3$ species where X = halogen, OH, OR, or other groups. At room temperature borazine slowly decomposes, and it reacts with water on heating to form NH_3 and $B(OH)_3$.

Borazine can be prepared in a variety of ways, including the reaction of NH_3 with BCl_3 to form B,B',B''-trichloroborazine, $Cl_3B_3N_3H_3$, which is then treated with $NaBH_4$ to yield $B_3H_3H_6$. However, these and other literature methods are hampered by limitations of scale and other problems. Improved routes that utilize reactions of ammonia-borane $(H_3N\cdot BH_3)$ at 140–160°C or alternatively, $(NH_4)_2SO_4$ and $NaBH_4$ at 120–140°C, to afford very pure borazine on a multigram scale have been reported recently.[25]

A solid state compound closely related to borazine is *hexagonal boron nitride* (BN), an increasingly important ceramic material that is used as a lubricant and for other purposes, and to which considerable research effort is currently directed. This compound is isoelectronic with graphite and has a similar sheet-like structure except that the hexagonal rings are aligned in an eclipsed fashion (each atom lying directly above and below atoms in neighboring sheets) rather than staggered as in graphite. The B and N atoms alternate vertically in successive layers. As with borazine and benzene, the structural similarity to graphite is deceiving; hexagonal BN is colorless, electrically nonconducting, and is more resistant to chemical attack than graphite. The usual synthetic routes involve high-temperature reactions of cheap nitrogen compounds such as ammonia or urea with borates or boric oxide, but obtaining pure, carbon-free BN is difficult. Some promising new approaches include the use of polyborazylene (a linear polymer of borazine) as a precursor to BN[26] and reactions of B-haloborazines with alkali metals; the latter route forms an amorphous BN product that can be converted to tubular BN by thermal annealing.[27] There is much current interest in BN thin films produced by CVD.[28]

At high pressure and temperature (e.g., 2000°C and 60 kbar), hexagonal BN converts to *cubic boron nitride*, a material that has a tetrahedral structure analogous to diamond and is almost as hard; it is employed as a high-temperature abrasive in situations where diamond forms carbides and hence cannot be used.

Other Boron-Nitrogen Inorganic Compounds

Hydrogenation of borazine yields cyclic nonplanar $B_3N_3H_{12}$, a cyclohexane analogue. Indeed, many if not most small hydrocarbons have their B-N counterparts, e.g.,

[24]C. Haben *et al.*, *Inorg. Chem.* **1989,** *28,* 2659.
[25]T. Wideman and L. G. Sneddon, *Inorg. Chem.* **1995,** *34,* 1002.
[26]F. J. Fazen *et al.*, *Chem. Mater.* **1995,** *7,* 1942.
[27]E. J. M. Hamilton *et al.*, *Chem. Mater.* **1995,** *7,* 111.
[28]R. T. Paine and C. K. Narula, *Chem. Rev.* **1990,** *90,* 73, and references therein.

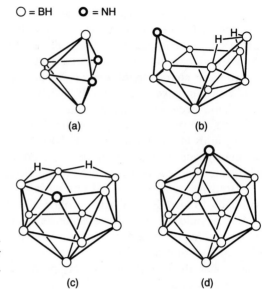

O = BH   ⬤ = NH

(a)   (b)

(c)   (d)

Figure 5-24 Boron-nitrogen clusters.
(a) *nido*-$N_2B_4H_6$ (known as alkyl or
$SiMe_3$ derivatives). (b) *nido*-NB_9H_{12}.
(c) *nido*-$NB_{10}H_{13}$. (d) $NB_{11}H_{12}$.

H_3BNH_3, which is isoelectronic and isostructural with ethane. However, there is also a rapidly developing interest in *boron-nitrogen clusters* that are structural and electronic counterparts of boranes rather than classical organic molecules.[29] Figure 5-24 presents a few examples, all of whose structures are in accord with PSEPT predictions (NH is a 4-electron donor and can formally replace a BH_3 unit). The 6-vertex nido-$N_2B_4H_6$ cage in Fig. 5-24a (known in the form of substituted derivatives) is noteworthy: although it is isoelectronic with B_6H_{10}, its geometry is that of a pentagonal bipyramid with an equatorial vertex missing rather than an apical vertex as in B_6H_{10}.

5-9   Boron-Phosphorus Compounds

To a point, the structures and reactivities of B-N compounds are reflected in B-P and B-As chemistry. For example, phosphaboranes of the type R_2BPR_2', like their aminoborane analogues, cyclize to form puckered rings such as $(R_2BPR_2')_3$; however, monomers having bulky or electron-releasing substituents on boron are stable, e.g., Et_2PBPh_2. There is a fairly extensive boron-phosphorus ring chemistry that largely resembles the azaboranes, as well as a family of phosphaborane and phosphacarborane cluster compounds (mentioned earlier) in which one or more P atoms formally replace CH or BH_2 units in the cage framework. Analogous compounds of As, and to a very limited degree Sb, are also known.

   Boron phosphide occurs in two forms, one of which, *cubic BP,* has a diamond-like structure analogous to cubic boron nitride (see above). The other variety, $B_{12}P_{1.8}$, has a partially disordered crystal structure that contains icosahedral B_{12} units, as found in many metal borides (Section 5-3). Cubic BP is extremely inert, resisting attack by boiling concentrated acids or bases, is not oxidized in air below

[29]P. Paetzold, *Pure Appl. Chem.* **1991,** *63,* 345; *Phosphorus, Sulfur, and Silicon* **1994,** *93–94,* 39.

1000°C, and can be heated to 2500°C without decomposition. Cubic BAs is structurally analogous and also relatively unreactive, although somewhat less so than BP.

5-10 Boron-Oxygen Compounds

Oxides and Acids

Boron oxides and their derivatives are technologically important and are relatively inexpensive to produce since, as noted earlier, essentially all boron in Nature is in oxygenated form. Boron-oxygen compounds contain predominantly trigonal planar BO_3, and to a lesser degree tetrahedral BO_4 units, as in the borate anions in Fig. 5-25.

The B—O bond is quite strong (560-790 kJ mol^{-1}), and boranes and related compounds are thermodynamically unstable relative to B_2O_3 and $B(OH)_3$. The principal oxide, B_2O_3, is very difficult to crystallize and normally exists in a glassy state ($d = 1.83$ g cm^{-3}) composed of randomly oriented B_3O_3 rings with bridging O atoms. In the normal crystalline form, ($d = 2.56$ g cm^{-3}) trigonal BO_3 units are linked through their oxygens; at high pressures and temperatures a high-density ($d = 3.11$ g cm^{-3}) phase having BO_4 tetrahedra is produced. Molten B_2O_3 (m.p. 450°C) readily dissolves metal oxides to form colored *borate glasses*, and this is one of the main commercial uses of boric oxide. In particular, *borosilicate glasses* find wide application in glassblowing and the production of glass objects because of their small coefficient of thermal expansion.

Hydrolysis of B_2O_3 generates *orthoboric acid* (boric acid), $B(OH)_3$, another major commercial product that is manufactured on a scale of hundreds of thousands of tons per year by acidification of aqueous solutions of *borax*, a naturally occurring mineral (see Section 5-2). Crystalline $B(OH)_3$ contains planar BO_3^{3-} units (Fig.

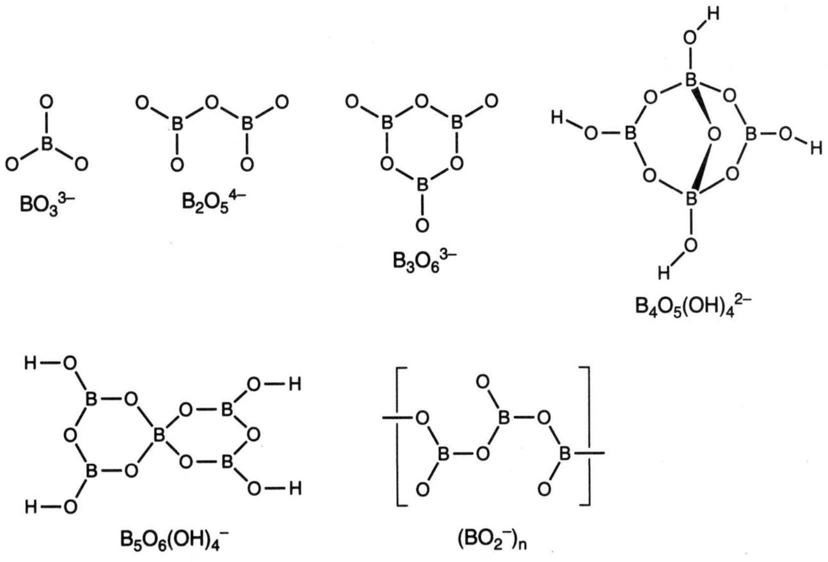

Figure 5-25 Some important borate anions.

5-25) linked *via* asymmetrical hydrogen bonds, and the layers are well separated (3.18 Å) in a manner reminiscent of graphite. The sheetlike structure of boric acid is manifested in its waxy, flaky properties. The compound is very weakly acidic in water ($pK = 9.3$), forming $H_3O^+B(OH)_4^-$; thus it functions not as a proton donor but rather as an OH^- acceptor, i.e., as a Lewis acid. In concentrated solution, species such as $B_3O_3(OH)_4^-$ are produced, and as the pH is increased, still other ions, e.g., $B_4O_5(OH)_4^{2-}$ (tetraborate) and $B_5O_6(OH)_4^-$ (pentaborate), both shown in Fig. 5-25, are formed *via* attack of OH^- on neutral trigonal boron. In dilute solution, the favored product is tetrahedral $B(OH)_4^-$; thus, when borax, which contains the $B_4O_5(OH)_4^{2-}$ ion, is dissolved in water, the tetraborate ion disintegrates to form $B(OH)_4^-$.

Some other important aspects of boric acid chemistry are summarized in Fig. 5-26. Among these is the formation of *borate esters* [$B(OR)_3$, R = alkyl or aryl], usually obtained as colorless liquids, on treatment with alcohols and H_2SO_4. The vast literature on these compounds falls generally within the purview of organic chemistry, and will not be developed here; however, it will be noted that a well-known qualitative test for boron involves treatment of the sample with methanol to form $B(OMe)_3$, which produces a bright green color in a Bunsen burner flame. A major early discovery in this area was the synthesis of *boronic acids* by E. Frankland in 1862, *via* partial oxidation of trialkylboranes, with subsequent hydrolysis of the ester:

$$BR_3 + O_2 \longrightarrow RB(OR)_2 \xrightarrow{2H_2O} RB(OH)_2$$

Heating boronic acids or treatment with P_4O_{10} at room temperature results in loss of H_2O to form *boroxines*, a much-studied class of cyclic organoboranes.

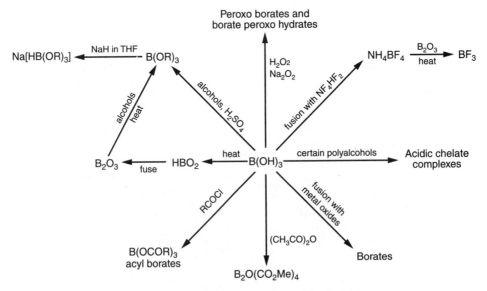

Figure 5-26 Selected reactions of boric acid.

Boric acid itself is reversibly dehydrated on heating, generating first HBO_2 (*metaboric acid*) and finally boric oxide:

$$B(OH)_3 \xrightarrow[\text{-H}_2\text{O}]{\text{heat}} HBO_2 \xrightarrow[\text{-H}_2\text{O}]{\text{heat}} B_2O_3$$

Depending on the temperature, three different structural modifications of metaboric acid are formed. From 100–130°C, one obtains the orthorhombic phase containing sheets of $B_3O_3(OH)_3$ rings linked by hydrogen bonds; at 130–150°C, the monoclinic form, which has a complex structure containing both 3- and 4-coordinate boron, is generated, while heating above 150°C produces cubic HBO_2, which features tetrahedral BO_4 units linked by $O-H-O$ bonds.

Borates

Metal borates occur in Nature in a variety of minerals, most of which are hydrated and consist of discrete polyanions with some oxygen atoms having attached protons, often accompanied in the crystal lattice by H_2O molecules. Anhydrous borates, readily generated by heating the hydrated compounds, have polymeric assemblies of planar BO_3 and/or tetrahedral BO_4 units that are linked *via* shared oxygen atoms; the more important of these are depicted in Fig. 5-25. Not all of these anions are found in the hydrated borates: a general principle appears to be that only those borate polyanions having BO_4 units are stable in aqueous solution, because 3-coordinate boron is readily attacked by water, causing the structure to disintegrate. The main structural feature in hydrated borates is a 6-membered B_3O_3 ring in which there is at least one BO_4 unit, as in the $B_4O_5(OH)_4^{2-}$ ion found in borax, and $B_5O_6(OH)_4^-$, which occurs in potassium pentaborate (Fig. 5-25). Borax itself is a major industrial chemical, produced annually in the millions of tons and employed in the manufacture of fiberglass, borosilicate glass, and fire retardants among other products.

Certain borates that have the capability of generating hydrogen peroxide (H_2O_2) in aqueous solution are formulated as *peroxyborates*. The well-known compound sodium perborate, which is represented both as $NaBO_3 \cdot 4H_2O$ and as $NaBO_2$. $H_2O_2 \cdot 3H_2O$, contains the peroxoanion $B_2(O_2)_2(OH)_4^{2-}$; this species consists of two tetrahedral $BO_2(OH)^{2-}$ groups joined *via* a pair of $O-O$ linkages. Peroxyborates are marketed as household bleaching agents.

5-11 Boron-Sulfur Compounds

Boron and sulfur combine to form an extensive series of molecular and solid-state substances, but there is little resemblance to boron-oxygen chemistry. *Diboron trisulfide*, B_2S_3, is a yellow solid with a layer structure that consists of hexagonal BSBSBS rings connected by additional sulfur atoms that form inter-ring $B-S-B$ bridges, quite unlike the 3-dimensional B_2O_3 arrangement (but reminiscent of hexagonal BN). A different boron sulfide, B_8S_{16}, is a molecular compound with a porphine-like structure containing four pentagonal $BSBS_2$ rings connected by $B-S-B$ bridges. Numerous boron-sulfur ring compounds having alternating S and B(SH) groups are known, and *thioborates* such as $B(SR)_3$ have been thoroughly studied. Many of these same themes are found as well in boron-selenium chemistry.

Sulfur, selenium, and tellurium are also capable of participating in electron-delocalized cluster bonding, and there is an extensive chemistry of polyhedral thiaboranes, thiacarboranes, and thiametallacarboranes and their Se and Te analogues, in which the group 16 atom (as a 4-electron donor to skeletal bonding with an exo-polyhedral lone pair) formally replaces NH, CH_2, or BH_3 in a cage framework. Typical examples are *nido*-$S_2B_9H_9$, *nido*-$Se_2B_9H_9$, and *closo*-$(PR_3)_2PtSeB_{10}H_{10}$.

Additional Recent References

E. Abel, F. G. A. Stone, and G. Wilkinson, Eds., *Comprehensive Organometallic Chemistry II*, Pergamon Press, Oxford, 1995, Vol. 1.

D. Emin *et al.*, Eds., *Boron-Rich Solids (AIP Conference Proceedings No. 140)*, American Institute of Physics, New York, 1986.

J. F. Liebman, A. Greenberg, and R. E. Williams, Eds., *Advances in Boron and the Boranes*, VCH Publishers, Inc., New York, 1988.

D. M. P. Mingos and D. J. Wales, *Introduction to Cluster Chemistry*, Prentice Hall: Englewood Cliffs, N.J., 1990.

W. Siebert, Ed., *Advances in Boron Chemistry*, Royal Society of Chemistry, Cambridge, 1997.

Chapter 6

THE GROUP 13 ELEMENTS: Al, Ga, In, Tl

GENERAL REMARKS

6-1 Electronic Structures and Valences

The electronic structures and some other fundamental properties of the Group 13 elements are listed in Table 6-1.

Whereas for the smallest member of Group 13, boron, cationic species of the type $[B(H_2O)_6]^{3+}$ are unknown, the remaining four elements have a rich cationic chemistry, in line with their metallic character. They also form many molecular compounds that are covalent.

While the main oxidation state is III, there are certain compounds in which the oxidation state is formally II. Some of these are $M^I M^{III}$ mixed-valent compounds, while others have metal-metal bonds as in $[Cl_3Ga-GaCl_3]^{2-}$. The univalent state becomes progressively more stable as the group is descended, and for thallium the Tl^I-Tl^{III} relationship is a dominant feature of the chemistry. This occurrence of an oxidation state two units below the group valence is sometimes called the inert pair effect, which first makes itself evident here, although it is adumbrated in the low reactivity of mercury in Group 12, and is much more pronounced in Groups 14 and 15. The term refers to the resistance of a pair of s electrons to participate in covalent bond formation. Thus mercury is difficult to oxidize, allegedly because it contains only an inert pair ($6s^2$), Tl readily forms Tl^I rather than Tl^{III} because of the inert pair in its valence shell ($6s^2 6p$), and so on. The concept of the inert pair does not actually tell us anything about the ultimate reasons for the stability of certain lower valence states, but it is a useful label. The phenomenon is not due to intrinsic inertness, that is, unusually high ionization potential of the pair of s electrons, but rather to the decreasing strength of the bonds as a group is descended. Thus, for example, the sum of the second and third ionization enthalpies (kJ mol^{-1}) is lower for In (4501) than for Ga (4916), with Tl (4820) intermediate. There is, however, a steady decrease in the mean thermochemical bond energies, for example, among the trichlorides: Ga, 242; In, 206; Tl, 153 kJ mol^{-1}. The relative stabilities of oxidation states differing in the presence or absence of the inert pair are further discussed in connection with the Group 14 elements (Chapter 7). Only where it can be shown that the electron pair has no stereochemical consequences is it reasonable to assume an s^2 pair. Recent theoretical work shows that relativistic effects make an important contribution to the inert pair effect.

Table 6-1   Some Properties of the Group 13 Elements

Property	Al	Ga	In	Tl
Electronic configuration	$[Ne]3s^2 3p$	$[Ar]3d^{10}4s^2 4p$	$[Kr]4d^{10}5s^2 5p$	$[Xe]4f^{14}5d^{10}6s^2 6p$
Ionization enthalpies (kJ mol^{-1})				
1st	576.4	578.3	558.1	589.0
2nd	1814.1	1969.3	1811.2	2862.8
3rd	2741.4	2950.0	2689.3	2862.8
4th	11563.0	6149.7	5571.4	4867.7
mp (°C)	660	29.8	157	303
bp (°C)	2327	~2250	2070	1553
E^0 for $M^{3+} + 3e = M(s)^a$	−1.66	−0.35	−0.34	0.72
Radius M^{3+} (Pauling) (Å)	0.50	0.62	0.81	0.95
Radius M^+ (Å)		1.13	1.32	1.40

^{a}In$^+ + e =$ In, $E^0 = -0.178$ V; Tl$^+ + e =$ Tl, $E^0 = -0.3363$ V.

In the trihalide, trialkyl, and trihydride compounds there are some resemblances to the corresponding boron chemistry. Thus MX_3 compounds behave as Lewis acids and can accept other neutral donor molecules or anions to give tetrahedral species; the acceptor ability generally decreases in the order Al > Ga > In, with the position of Tl uncertain. There are, however, notable distinctions from boron. These are in part due to the reduced ability to form multiple bonds and to the ability of the heavier elements to have coordination numbers exceeding four. Thus although boron gives $Me_2B=NMe_2$, Al, Ga, and In give dimeric species, for example $[Me_2AlNMe_2]_2$, in which there are NMe_2 bridging groups and both the metal and nitrogen atoms are 4-coordinate. Similarly, the boron halides are all monomeric, whereas those of Al, Ga, and In are all dimeric or polymeric, with bridging halides. The formation of four-membered rings, as in $X_2Al(\mu\text{-}X)_2AlX_2$, is the most common way of achieving coordinative saturation. Secondly, 5-coordinate compounds such as $(Me_3N)_2AlH_3$ have trigonal bipyramidal (*tbp*) structures, which of course are impossible for boron adducts.

The stereochemistries and oxidation states of Group 13 are summarized in Table 6-2.

THE ELEMENTS

6-2   Occurrence, Isolation, and Properties

Aluminum, the commonest metallic element in the Earth's crust (8.8 mass %) occurs widely in Nature in silicates such as micas and feldspars, as the hydroxo oxide (*bauxite*), and as *cryolite* (Na_3AlF_6). In spite of the abundance, aluminum does not appear to play a role in biology, although it has been linked to Alzheimer's disease, a degenerative brain disorder.

The other elements are found only in trace quantities. Gallium and indium occur in aluminum and zinc ores, for example in bauxite, but the richest sources contain <1% of Ga and still less In. Thallium is widely distributed and is usually recovered from flue dusts from the roasting of pyrites.

Table 6-2 Oxidation States and Stereochemistries of Group 13 Elements

Oxidation state	Coordination number	Geometry	Examples
+1	6	Distorted octahedral	TlF
	4	Tetrahedral	$[AlBr(NEt_3)]_4$, $[Ga\{CH(SiMe_3)_2\}]_4$
+2	4	Tetrahedral	$[Ga_2Cl_6]^{2-}$, $[In_2Cl_6]^{2-}$, $[AlBr_2(PhOMe)]_2$
	3	Trigonal	$Al_2\{CH(SiMe_3)_2\}_4$
+3[a]	3	Planar	$In[Co(CO)_4]_3$, $Al(mes)_3$
	4[a]	Tetrahedral	$[AlCl_4]^-$, $[GaH_4]^-$, $Al_2(CH_3)_6$, Ga_2Cl_6
	5	*tbp*	$AlCl_3(NMe_3)_2$, $[In(NCS)_5]^{2-}$, $Me_2GaCl(phen)$, $InCl_3(PPh_3)_2$
		sp	$[InCl_5]^{2-}$, $EtAl(salen)$, *p*-tolylTl$(S_2CNEt_2)_2$
	6	Octahedral	$[Al(H_2O)_6]^{3+}$, $Ga(acac)_3$, $[AlF_6]^{3-}$, $[GaCl_2(phen)_2]^+$
	8	Dodecahedral	$[Al(BH_4)_4]^{-b}$

[a]Most common states.
[b]S. G. Shore *et al.*, *Inorg. Chem.* **1994,** *33,* 5443.

Aluminum is prepared on a vast scale. Bauxite is purified by dissolution in sodium hydroxide, filtration over drum filters to remove impurities such as iron oxides, precipitation, and calcining to Al_2O_3. This is then dissolved in cryolite at 800 to 1000°C and the melt is electrolyzed. The production is highly energy-intensive and requires cheap sources of electricity. Aluminum is a hard, strong, white metal. Although highly electropositive, it is resistant to corrosion because a hard, tough film of oxide is formed on the surface. Thick oxide films, some with the proper porosity when fresh to trap particles of pigment, are often electrically applied to Al. Aluminum is soluble in dilute mineral acids, but is passivated by concentrated nitric acid. If the protective film of the oxide is overcome, for example, by scratching or by amalgamation, rapid attack even by water can occur. The metal is dissolved in aqueous alkali hydroxide with liberation of hydrogen to give alkali aluminate. It is also attacked by halogens and various nonmetals. Highly purified Al is quite resistant to acids and is best attacked by hydrochloric acid either containing a little $CuCl_2$, or in contact with Pt, some H_2O_2 also being added during the dissolution.

Gallium, indium, and *thallium* are usually obtained by electrolysis of aqueous solutions of their salts; for Ga and In this possibility arises because of large overvoltages for hydrogen evolution on these metals. They are soft, white, comparatively reactive metals, dissolving readily in acids; however, thallium dissolves only slowly in sulphuric or hydrochloric acid, since the Tl^I salts formed are only sparingly soluble. Gallium, like Al, is soluble in sodium hydroxide. The elements react rapidly at room temperature, or on warming, with the halogens and nonmetals such as sulphur. Gallium of extremely high purity for use in the semiconductor industry is produced by zone-refining.

The increased first ionization enthalpy of Ga compared to Al reflects the fact that the filled $3d^{10}$ shell of Ga shields the increased nuclear charge only incompletely. The exceptionally low melting point of Ga has no simple explanation. Since its boiling point (2070°C) is not abnormal, Ga has the longest liquid range of any known substance and finds use as a thermometer liquid.

CHEMISTRY OF THE TRIVALENT STATE BINARY COMPOUNDS

6-3   Oxygen Compounds

Aluminum oxides rank among the most important large-scale industrial chemicals. Stoichiometrically there is only one oxide of aluminum, namely, *alumina* (Al_2O_3). This simplicity, however, is compensated for by the occurrence of various polymorphs, hydrated species, and so on, the formation of which depends on the conditions of preparation. There are several forms of anhydrous Al_2O_3, most importantly α-Al_2O_3 and γ-Al_2O_3. Other trivalent metals (Ga, Fe) form oxides that crystallize in the same two structures. In α-Al_2O_3 the oxide ions form a hexagonal close-packed array and the aluminum ions are distributed symmetrically among the octahedral interstices, so that each oxygen atom is surrounded by four Al atoms. The γ-Al_2O_3 structure is sometimes regarded as a "defect" spinel structure, that is, as having the structure of spinel with a deficit of cations (see below). Another modification, κ-Al_2O_3, has a layer structure built up of sheets and ribbons made up of AlO_6 octahedra connected by AlO_4 tetrahedra.[1]

Alpha-Al_2O_3 is stable at high temperatures and also indefinitely metastable at low temperatures. It occurs in Nature as *corundum* and may be prepared by heating γ-Al_2O_3 or any hydrous oxide above 1000°C. It is very hard and resistant to hydration and attack by acids. Gamma-Al_2O_3 is obtained by dehydration of hydrous oxides at low temperatures ($\sim$450°C); it takes up water readily and dissolves in acids. The Al_2O_3 that is formed on the surface of the metal has still another structure, namely, a defect rock salt structure; there is an arrangement of Al and O ions in the rock salt ordering with every third Al ion missing.

There are several important hydrated forms of alumina corresponding to the stoichiometries Al(O)OH and Al(OH)$_3$. Addition of ammonia to a boiling solution of an aluminum salt produces a form of Al(O)OH known as *boehmite*, which may be prepared in other ways also. A second form of Al(O)OH occurs in Nature as the mineral *diaspore*. The true hydroxide Al(OH)$_3$ is obtained as a crystalline white precipitate when carbon dioxide is passed into alkaline "aluminate" solutions. It occurs in Nature as the mineral *gibbsite*. Materials sometimes referred to as β-aluminas have other ions such as Na^+ and Mg^{2+} present. They possess the idealized composition $Na_2O \cdot 11Al_2O_3$. They can act as ion exchangers, have high electrical conductivity, and are potential solid state electrolytes for batteries.

Aluminas used in chromatography or as catalyst supports are prepared by heating hydrated oxide to various temperatures so that the surfaces may be partially or wholly dehydrated. The activity of the alumina depends critically on the treatment, subsequent exposure to moist air, and other factors. Concentrated aqueous $AlCl_3$ solutions can be spun into fibers which on drying produce filaments of Al_2O_3. They are non-toxic, unlike asbestos, possess excellent heat insulation properties, and are stable up to 1400°C.

Gallium oxides form a similar system, affording a high-temperature α- and a low-temperature γ-Ga_2O_3 phase. The trioxide is formed by heating the nitrate, the sulfate, or the hydrous oxides that are precipitated from Ga^{III} solutions by the action of ammonia. Beta-Ga_2O_3 contains both tetrahedrally and octahedrally coordinated

---

[1]H. L. Gross and W. Mader, *Chem. Commun.* **1997**, 55.

gallium with Ga—O distances of 1.83 and 2.00 Å, respectively. The hydrous oxides GaO(OH) and Ga(OH)$_3$ are similar to their Al analogues.

Indium gives yellow In$_2$O$_3$, which is known in only one form, and a hydrated oxide In(OH)$_3$. Thallium has only the brown-black Tl$_2$O$_3$, which begins to lose oxygen at about 100°C to give Tl$_2$O. The action of NaOH on Tl^{III} salts gives what appears to be the oxide, whereas with Al, Ga, and In the initial products are basic salts.

Aluminum, gallium, and thallium form mixed oxides with other metals. First, there are aluminum oxides containing only traces of other metal ions. These include ruby (Cr^{3+}) and blue sapphire (Fe^{2+}, Fe^{3+}, and Ti^{4+}). Synthetic ruby, blue sapphire, and gem-quality corundum are now produced synthetically in large quantities. Second, there are mixed oxides containing macroscopic proportions of other elements, such as *spinel* (MgAl$_2$O$_4$) and *crysoberyl* (BeAl$_2$O$_4$).

Alkali metal compounds can be made by heating Al$_2$O$_3$ or Ga$_2$O$_3$ with an alkali oxide at 1000°C. Aluminates, like silicates, tend to give lattices with anions composed of AlO$_4$ tetrahedra sharing edges, for example, Na$_{14}$[Al$_4$O$_{13}$] and Na$_{17}$Al$_5$O$_{16}$ that has chains, although edge-sharing structures may occur, as in K$_6$Ga$_2$O$_6$. Calcium aluminate, Ca$_3$Al$_2$O$_6$, is an important constituent of cement, besides Ca silicates. It contains cyclic anions [Al$_6$O$_{18}$]$^{18-}$ (6-I) formed by vertex-sharing tetrahedra.

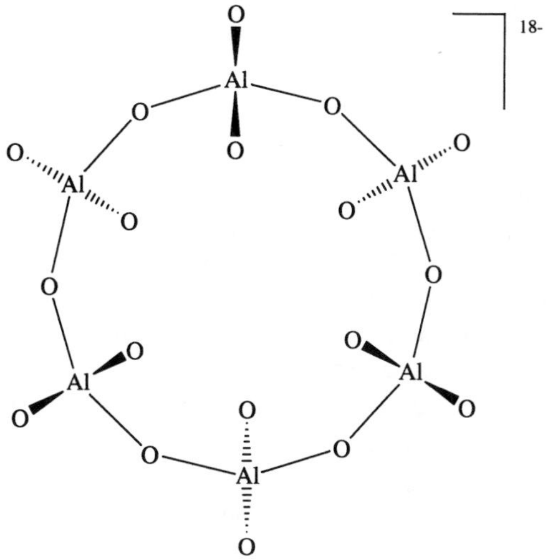

(6-I)

Aluminum phosphate, AlPO$_4$ is isoelectronic with SiO$_2$, and like quartz (*cf.* Section 8-7) it passes through a number of phases, from α- and β-AlPO$_4$ through *berlinite-*, *tridymite-*, and *crystobalite*-like structures before it melts at >1600°C. The hydrothermal synthesis of AlPO$_4$ in the presence of organic bases such as amines leads to cage structures similar to zeolites (see Section 8-7). Unlike zeolites, however, these AlPO$_4$-phases are electroneutral and do not act as ion exchangers. They are used as molecular sieves; some have larger pores than zeolites. The aluminophosphate VPI-5 has ring channels of 18 tetrahedral units, with a free diameter of 12-13 Å, while AlPO$_4$-8 has 14-ring channels. Gallium phosphate made in the presence

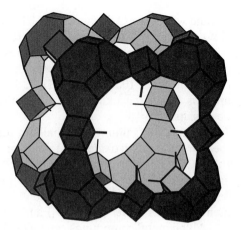

Figure 6-1 Supercage at the intersection of the 13 Å channels in a GaPO₄ molecular sieve. Each corner is the center of an MO₄ tetrahedron. (Reproduced with permission from M. Eskermann *et al., Nature* **1991,** *352,* 320.)

of quinuclidinium fluoride has an even more open structure. The intersections of its 13 Å channels form exceptionally large supercages of 29-30 Å free diameter (Fig 6-1).[2] Above 700°C the structure collapses to tridymite-GaPO₄. There are also Al and Ga phosphates with polyanionic sheet structures, again very reminiscent of silicates.[3]

6-4   Halides

All four trihalides of each element are known, with one exception. The compound TlI₃, obtained by adding iodine to thallous iodide, is not thallium(III) iodide, but rather thallium(I) triiodide [Tlᴵ(I₃)]. This situation may be compared with the non-existence of iodides of other oxidizing cations such as Cu²⁺.

The coordination numbers found in the crystalline halides are shown in Table 6-3. The fluorides of Al, Ga, and In are all high melting [1290, 950 (subl.), 1170°C, respectively] and form lattices with MF₆ octahedra. The chlorides, bromides, and iodides have lower melting points which correlate with the coordination number:

Table 6-3 Coordination Numbers of Metal Atoms in Group 13 Halides in the Solid State

	F	Cl	Br	I
Al	6	6	4	4
Ga	6	4	4	4
In	6	6	6	4
Tl	6	6	4	4

[2]M. Eskermann *et al., Nature* **1991,** *352,* 320.
[3]J. M. Thomas *et al., J. Chem. Soc., Chem. Commun.* **1991,** 1520; **1994,** 565.

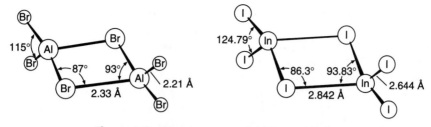

Figure 6-2   The structures of Al_2Br_6 and In_2I_6.

$AlCl_3$, 193°C (at 1700 mm); $GaCl_3$, 78°C; $InCl_3$, 586°C. The structure of $AlCl_3$ changes at the melting point from a 6-coordinate polymer to a 4-coordinate dimer which also exists in the gas phase.

The dimeric trihalides form discrete molecules (Fig. 6-2). The halides dissolve readily in aromatic solvents such as benzene, in which they are dimeric. As Fig. 6-2 shows, the configuration of halogen atoms about each metal atom is far from ideally tetrahedral. The formation of such dimers is attributable to the tendency of the metal atoms to complete their octets.

At high temperatures the dimers $M_2(\mu\text{-}X)_2X_4$ dissociate into monomers. The dissociation enthalpies of $Al_2X_6(g) \rightarrow 2\,AlX_3(g)$ (X = Cl, Br, I) are 46–63 kJ mol^{-1}. The monomers are planar like BX_3. The D_{3h} symmetry of MX_3 has been confirmed by matrix isolation and gas phase electron diffraction studies at >380°C.[4]

The interaction of trichlorides of Al, Ga, or In with other metal chlorides such as $CaCl_2$, $CrCl_3$, $AuCl_3$, or UCl_5 leads to mixed halides with halogen bridges, for example, $Cl_4Nb(\mu\text{-}Cl)_2AlCl_2$, many of which are quite volatile. Thus the vapor pressure of $NbCl_3$ is increased by a factor of 10^{13} at 600°C.

Thallium(III) Halides

The chloride, which is most commonly used, can be prepared by the sequence:

$$\text{Tl, TlCl, or } Tl_2CO_3 \xrightarrow{\text{ClNO}} TlCl_3 \cdot NOCl \xrightarrow{\text{heat}} TlCl_3$$

Solutions of $TlCl_3$ and $TlBr_3$ in CH_3CN, which are useful for preparative work, are conveniently obtained by treating solutions of the monohalides with Cl_2 or Br_2. Solid $TlCl_3$ loses chlorine at about 40°C and above, to give the monochloride, and the tribromide loses bromine at even lower temperatures to give first $TlBr_2$, which is actually $Tl^I[Tl^{III}Br_4]$. The fluoride is stable to about 500°C.

6-5 Other Binary Compounds

Binary compounds such as carbides, nitrides, phosphides, and sulfides are commonly made by direct interaction of the elements.

[4]A. Haaland *et al.*, *J. Chem. Soc., Dalton Trans.* **1992**, 2209.

Aluminum carbide (Al_4C_3) is formed at temperatures of 1000 to 2000°C. It reacts instantly with water to produce methane, and since there are discrete carbon atoms (C—C = 3.16 Å) in the structure, it may be considered as a "methanide" with a C^{4-} ion, but this is doubtless an oversimplification.

The *nitrides* AlN, GaN, and InN are known. Only aluminum reacts directly with nitrogen; GaN is obtained on reaction of Ga or Ga_2O_3 at 600 to 1000°C with NH_3, and InN by pyrolysis of $(NH_4)_3InF_6$. All have a wurtzite structure (Fig. 1-28). They are fairly hard and stable, as might be expected from their close structural relationship to diamond and diamond-like BN. Because of its hardness and resistance to chemical attack, AlN has found numerous materials applications. Films of AlN can be produced from $AlCl_3$ and NH_3 at high temperatures, or better from volatile organometallic precursors such as $AlMe_3$ and NH_3.

$$Me_3Al\cdot NH_3 \xrightarrow[-CH_4]{\Delta} [Me_2AlNH_2]_3 \xrightarrow[-CH_4]{\Delta} [MeAlNH]_n \xrightarrow[-CH_4]{\Delta} AlN$$

The chemical vapor deposition (CVD) of metal amides $M_2(NMe_2)_6$ can be used similarly for MN (M = Al, Ga) film growth.[5]

Binary compounds between Group 13 and the heavy Group 15 elements, such as GaAs and InP, are known as III-V compounds, based on the old III, V numbering of groups. They have a smaller band gap and are important as semiconductors; like other materials of this kind, such as Si and Ge, they possess zincblende (diamond) structures. Single crystal films are most commonly grown using volatile organometallic precursors (*cf.* Section 1-10):

$$GaMe_3 + AsH_3 \longrightarrow GaAs + 3CH_4$$

GaAs is used in light emitting diodes (LEDs).

Aluminum forms a series of chalcogenides Al_2E_3 (E = S, Se, Te) which hydrolyze readily. The structures of the chalcogenides of Ga, In, and Tl are more varied and include cubic and hexagonal Ga_2S_3 polymorphs and hexagonal chalcogenides of divalent metals such as GaS, GaSe, GaTe, InS, and InSe. The thermolysis of $[Bu^tGaS]_4$ leads to the metastable cubic phase of GaS. The compounds TlS and TlSe are mixed-valence species, $Tl^I[Tl^{III}E_2]$ (E = S, Se).

Indium and thallium form a number of binary compounds with alkali metals in which the group 13 elements form well-defined anionic clusters (Zintl ions, see Section 1-9). Examples are K_8In_{11} (Fig. 6-3) which has considerably fewer (2n-4) electrons than the minimum described by Wade's rules (2n + 2), $K_8In_{10}Zn$ and $K_{10}In_{10}M$ (M = Ni, Pd, Pt).[6] Closo-In_{16} and nido-In_{11} clusters have also been found. Thallium, too, forms Zintl clusters; Na_2Tl contains Tl_4^{8-} tetrahedra, while K_8Tl_{11} is similar to In, and KTl contains Tl_6^{6-} octahedra.[7]

[5]R. G. Gordon *et al., Mat. Res. Symp. Proc.* **1991,** *204,* 95; L. V. Interrante *et al., Mat. Res. Symp. Proc.* **1991,** *204,* 135.
[6]J. D. Corbett *et al., Inorg. Chem.* **1993,** *32,* 1059; *J. Am. Chem. Soc.* **1993,** *115,* 9089; *Inorg. Chem.* **1992,** *32,* 1895.
[7]Z. Dong and J. D. Corbett, *J. Am. Chem. Soc.* **1993,** *115,* 11299.

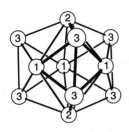

Figure 6-3 Structure of $[In_{11}]^{8-}$ in K_8In_{11}. The cluster is a penta-capped trigonal prism. (Reproduced with permission from J. D. Corbett and S. C. Sevov, *Inorg. Chem.* **1991,** *30,* 4875.)

COMPLEX COMPOUNDS

6-6   The Aqua Ions; Oxo Salts, Aqueous Chemistry

The elements form a wide variety of salts including hydrated chlorides, nitrates, sulfates, and perchlorates, as well as sparingly soluble phosphates.

The *aqua ions,* $[M(H_2O)_6]^{3+}$, exist in both aqueous solutions and crystalline salts; $[Tl(H_2O)_6]^{3+}$ has two trans H_2O molecules that are more strongly bound than the others. Aqueous solutions of thallium halides also contain $[TlX(H_2O)_5]^{2+}$ and *trans*-$[TlX_2(H_2O)_4]^{+}$.[8] The ions are acidic,

$$[M(H_2O)_6]^{3+} \rightleftharpoons [M(H_2O)_5OH]^{2+} + H^+$$

and measured in the presence of noncomplexing anions, the constants K_A for Al, Ga, In and Tl, respectively, are $\sim 10^{-5}$, 10^{-3}, 10^{-4}, and 10^{-1}. Salts of weak acids are extensively hydrolyzed by water. Indeed, $[Al(H_2O)_6]^{3+}$ will protonate the acetate ion:

$$[Al(H_2O)_6]^{3+} + MeCO_2^- \longrightarrow [Al(H_2O)_5OH]^{2+} + MeCO_2H$$

The $[Al(H_2O)_5OH]^{2+}$ ion has been detected in dilute solution but there is a very high constant for the dimerization:

$$2[Al(H_2O)_5OH]^{2+} \rightleftharpoons \left[(H_2O)_4Al \underset{\underset{H}{O}}{\overset{\overset{H}{O}}{<}} \hspace{-0.3em} \rangle Al(H_2O)_4 \right]^{4+}$$

Over a wide pH range under physiological conditions in chloride solution the species appear to be $[Al(H_2O)_5OH]^{2+}$, $Al(OH)_3$, $[Al(OH)_4]^-$, $[Al_3(OH)_{11}]^{2-}$, $[Al_6(OH)_{15}]^{3+}$, and $[Al_8(OH)_{22}]^{2+}$.

The hydrolyzed species depend markedly on the exact conditions. Addition of Na_2CO_3 gives further condensation of the dimer to $[AlO_4Al_{12}(OH)_{24}(H_2O)_{12}]^{7+}$. The analogous Ga_{13} ion is also known. These Keggin-type ions consist of twelve MO_6 octahedra surrounding a tetrahedral MO_4 unit.[9]

Study of the substitution of the aluminum aqua ion by ligands such as SO_4^{2-},

[8]J. Glaser *et al., J. Am. Chem. Soc.* **1995,** *117,* 5089.
[9]J. W. Akitt *et al., J. Chem. Soc., Dalton Trans.* **1988,** 1347.

citrate, and EDTA in dilute solutions, using ^{27}Al nmr spectroscopy, has been stimulated by the recognition of (a) the key role of Al^{3+} in acid lakes due to leaching
from silicates by acid rain, (b) the toxicity of Al to aquatic life and, (c) the increased
concentration of aluminosilicates in human brains in senile dementia (Alzheimer's
disease). Aluminum ions bind strongly to physiologically important compounds
such as the phosphate group of adenosine triphosphate (ATP) (4000 times stronger
than Mg^{2+}),[10] and to blood serum proteins.

Alums, $M^IAl(SO_4)_2 \cdot 12H_2O$, where M^+ is almost any ion except Li^+, which is
too small to be accommodated without loss in stability of the structure, give their
name to a large number of analogous salts of +3 ions including those of Ti, V, Cr,
Mn, Fe, Co, Ir, Ga, and In. There are three cubic structures consisting of $[M(H_2O)_6]^+$,
$[M(H_2O)_6]^{3+}$, and two SO_4^{2-} ions, differing slightly in detail depending on the size
of M^+.

Aluminates and Gallates

The hydroxides are amphoteric:

$$Al(OH)_3(s) = Al^{3+} + 3OH^- \qquad K \approx 5 \times 10^{-33}$$

$$Al(OH)_3(s) = AlO_2^- + H^+ + H_2O \qquad K \approx 4 \times 10^{-13}$$

$$Ga(OH)_3(s) = Ga^{3+} + 3OH^- \qquad K \approx 5 \times 10^{-37}$$

$$Ga(OH)_3(s) = GaO_2^- + H^+ + H_2O \qquad K \approx 10^{-15}$$

The oxides and the metals also dissolve in alkali bases as well as in acids.
The oxides and hydroxides of In and Tl are, by contrast, basic; hydrated Tl_2O_3 is
precipitated from solution even at pH 1 to 2.5. The nature of the aluminate and
gallate solutions has been much studied by Raman and nmr spectra. The principal
species in solution for Al is $[Al(OH)_4]^-$; only where Na^+ is the cation is there
evidence for species in solution that may be related to solids that crystallize from
aluminate solutions. The latter contain dimeric anions with 4-coordinate Al,
$[(HO)_3AlOAl(OH)_3]^{2-}$, and others with octahedral Al^{3+}.

Nuclear magnetic resonance spectroscopy of the ^{27}Al nucleus, both in solution
and the solid state, has proved useful for the determination of the coordination
numbers and, to some extent, the stereochemistry of Al. Four-, 5-, and 6-coordinate
geometries can be differentiated on the basis of their chemical shifts.

6-7   Halide Complexes and Adducts

Fluorides

The hydrated fluorides $AlF_3 \cdot nH_2O$ (n = 3 or 9) can be obtained by dissolving
Al in aqueous HF. The nonahydrate is very soluble in water, and ^{19}F nmr spectra

---

[10]T. Kiss *et al., Inorg. Chem.* **1991,** *30,* 2130.

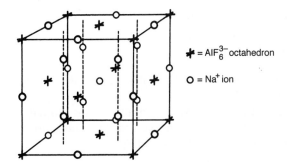

$\maltese$ = AlF_6^{3-} octahedron

$\bigcirc$ = Na^+ ion

Figure 6-4 The cubic structure of *cryolite*, Na_3AlF_6.

show the presence of $AlF_3(H_2O)_3$ as well as the ions AlF_4^-, $[AlF_2(H_2O)_4]^+$, and $[AlF(H_2O)_5]^{2+}$. The existence of discrete AlF_4^- anions has long been uncertain but is now confirmed,[11] although octahedral fluoro anions such as $[Al_2(\mu\text{-}F)_2F_6(OH_2)_2]^{2-}$ are also known. At high fluoride concentration and in crystalline solids the AlF_6^{3-} ion may be found. This anion is present in the most important fluoro salt Na_3AlF_6, *cryolite*. The cryolite structure (Fig. 6-4) is adopted by many other salts containing small cations and large octahedral anions.

Other Halides

The dimeric trihalides M_2X_6 are split by donor molecules to give mononuclear adducts. The halides dissolve in water with partial hydrolysis to give acidic solutions from which hydrates may be obtained. With amines tetrahedral complexes such as $R_3N \cdot AlCl_3$ are formed, while in acetonitrile $AlCl_3$ dissociates into $[Al(NCMe)_5Cl]^{2+}$ and $AlCl_4^-$. Similarly, Al_2Cl_6 in THF forms $[AlCl_2(THF)_4]AlCl_4$, and the ammonia complex $AlCl_3(NH_3)_2$ has the structure $[AlCl_2(NH_3)_4]^+[AlCl_4]^-$. Oxochlorides of sulfur form adducts of the type $SOCl_2 \cdot nAlCl_3 (n = 1 - 3)$ which act as aprotic superacids and are able to crack alkanes under ambient conditions.[12]

The addition of Cl^- to Al_2Cl_6 gives the simplest of the chloro anions, the tetrahedral $AlCl_4^-$. More complex anions are formed at higher Cl^- concentrations, e.g., in $AlCl_3/NaCl$ molten salt media which are sometimes employed as electrolytes and highly polar high-temperature solvents. The equilibria involved at 175 to 300°C are

$$2AlCl_4^- \rightleftharpoons Al_2Cl_7^- + Cl^-$$

$$3Al_2Cl_7^- \rightleftharpoons 2Al_3Cl_{10}^- + Cl^-$$

$$2Al_3Cl_{10}^- \rightleftharpoons 3Al_2Cl_6 + 2Cl^-$$

Interaction of $AlCl_3$ with *N*-butylpyridinium chloride or 1-methyl-3-ethyl-imidazolium chloride gives conducting *liquids* at room temperature that are good solvents

[11]N. Herron *et al.*, *J. Am. Chem. Soc.* **1993**, *115*, 3028; *Inorg. Chem.* **1993**, *32*, 2985.
[12]I. Akhrem *et al.*, *Chem. Eur. J.* **1996**, *2*, 812.

for both organic and inorganic compounds. The principal equilibrium is

$$2AlCl_4^- \rightleftharpoons Al_2Cl_7^- + Cl^-$$

but $Al_3Cl_{10}^-$ and Al_2Cl_6 may also be present, as in the $AlCl_3/NaCl$ systems, depending on $AlCl_3$ concentration.

The tetrahaloaluminates are hydrolyzed by water, but gallium can be extracted from 8 M HCl solutions into ethers, where the ether phase contains $GaCl_4^-$ ions, so that at some point there is a change in coordination number:

$$[Ga(H_2O)_6]^{3+} \xrightarrow{\text{HCl}} [GaCl_n(H_2O)_m]^{(3-n)+} \xrightarrow{\text{HCl}} GaCl_4^-$$

Oxonium salts, for example, $[(Et_2O)_n H]^+ MCl_4^-$ from the reaction of MCl_3 with HCl in ether are viscous oils.

The formation of $AlCl_4^-$ and $AlBr_4^-$ ions is essential to the functioning of Al_2Cl_6 and Al_2Br_6 as Friedel-Crafts catalysts, since in this way the necessary carbonium ions are formed. The spectra and structures of intermediates of the Friedel-Crafts reaction have been studied. There are two types, one molecular (6-II) and the other ionic (6-III), depending on the solvent. In CH_2Cl_2 only (6-II) is formed.

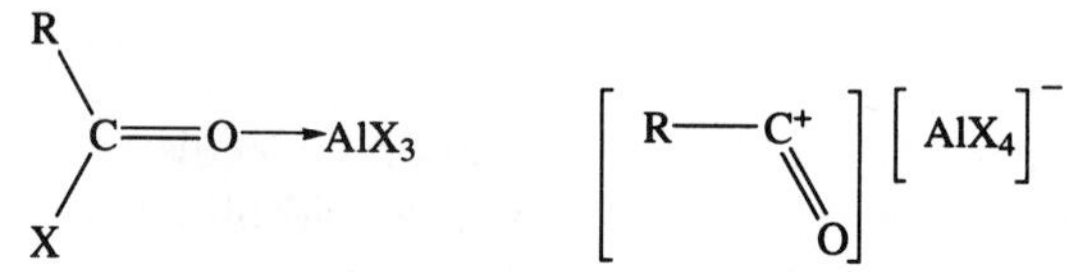

(6-II)   (6-III)

The oxocarbenium ion can then react further, for example,

$$RCO^+ + C_6H_6 \longrightarrow [RCOC_6H_6]^+ \longrightarrow RCOC_6H_5 + H^+$$

Thallium(III) halides form the hydrates $TlX_3(H_2O)_2$ in solution with *tbp* structure.

In addition to MCl_4^- there are 5- or 6-coordinate species, $InCl_5^{2-}$, InF_6^{3-}, $[TlCl_5(OH_2)]^{2-}$, and $TlCl_6^{3-}$. Similar pseudohalide complexes, for example, $[In(NCS)_5]^{2-}$ and $[In(NCS)_6]^{3-}$, may also be obtained, depending on the size of the cation.

For the TlI_4^- ion,[13] the stability of the iodide in contact with Tl^{III} is a result of the stability of the ion, since TlI_3 is itself unstable relative to $Tl^I(I_3)$. Thallium also forms the ion $Tl_2Cl_9^{3-}$, which has the confacial bioctahedral structure.

Cationic Complexes

Apart from the aqua ions and partially substituted species such as $[GaCl(H_2O)_5]^{2+}$ noted previously, complexes with pyridine, bipyridine, and phenanthroline are known, for example, $[GaCl_2(phen)_2]^+$ and $[Ga(phen)_3]^{3+}$. For thallium(III) com-

[13]R. Burns *et al., Z. Krist.* **1994,** *209,* 686.

plexes, the best route appears to be oxidation of thallium(I) halide in acetonitrile solution by halogen, followed by addition of ligands to get complex ions.

Neutral Adducts

The trihalides (except the fluorides), and other R_3M compounds such as the trialkyls, triaryls, mixed R_2MX compounds, and AlH_3, all function as Lewis acids, forming 1:1 adducts with a great variety of Lewis bases. This is one of the most important aspects of the chemistry of the Group 13 elements. The Lewis acidity of the AlX_3 groups (where $X = Cl$, CH_3, etc.) has been extensively studied thermodynamically, and basicity sequences for a variety of donors have been established.

While tetrahedral 4-coordinate Lewis base adducts MX_3L are the most common, particularly for Al and Ga, 5- and 6-coordinate adducts are also known, mainly for the heavier members of the group. Indium trichloride forms *tbp* complexes $InCl_3L_2$ ($L = THF$, PPh_3) and octahedral *fac*-$InCl_3L_3$ ($L = THF$, Me_3PO, dimethylform-amide, etc.).[14]

6-8   Chelate Complexes

The most important octahedral complexes of the Group 13 elements are those containing chelate rings. Typical are those of β-diketones, pyrocatechol (6-IV), dicarboxylic acids (6-V), and 8-quinolinol (6-VI). The neutral complexes dissolve readily in organic solvents, but are insoluble in water. The acetylacetonates have low melting points (<200°C) and vaporize without decomposition. The anionic complexes are isolated as the salts of large univalent cations. The 8-quinolinolates are used for analytical purposes. Tropone (T) gives an 8-coordinate anion of indium in $Na[InT_4]$.

(6-IV) 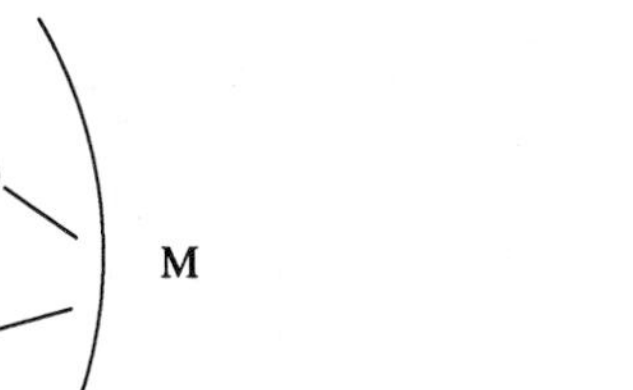   (6-V)

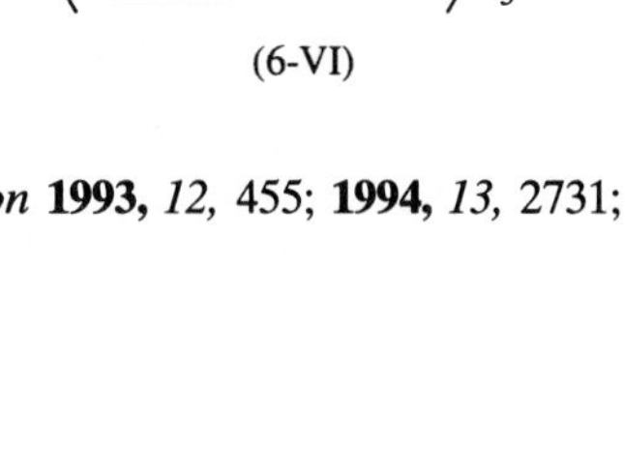

(6-VI)

[14]R. L. Wells *et al.*, *Polyhedron* **1993**, *12*, 455; **1994**, *13*, 2731; C. J. Wilkins, *J. Chem. Soc., Dalton Trans.* **1993**, 3111.

Aluminum β-diketonates have been much studied by nmr methods because of their stereochemical nonrigidity. Aluminum(III), Ga^{III}, and In^{III} form stable complexes with di-, tri-, and hexadentate chelating ligands in which the metal is octahedrally coordinated. These complexes are stable under physiological conditions. Complexes of the radioisotopes ^{67}Ga (γ, $t_{1/2}$ = 3.25 d), ^{68}Ga (β^+, $t_{1/2}$ = 68 min), and ^{111}In (γ, $t_{1/2}$ = 2.80 d), are used as radiopharmaceutical imaging agents because they have appropriate energies and half-lives for γ-ray imaging or positron emission tomography. In these applications it is the function of the polydentate ligand to ensure that the complex is selectively transported to the target organ.[15]

The *carboxylates* of indium and thallium are obtained by dissolving the oxides in acid. Acetate and trifluoroacetate salts are used extensively as reagents in organic synthesis. Certain other thallium compounds have been used also. The trifluoroacetate, $Tl(O_2CCF_3)_3$, will directly "thallate" aromatic compounds to give arylthallium species, for example, $C_6H_5Tl(O_2CCF_3)_2$ (*cf.* aromatic mercuration, Section 15-15) and oxidize arenes to biaryls.

Sulfur complexes of Al, Ga, and In are less common, but the tris(dithiocarbamates), $M(dtc)_3$, and thiolato anions, $Ga(SR)_4^-$, are known. The gallium and indium compounds differ structurally from most other tris(dithiocarbamates) in their close approach to a prismatic arrangement of S atoms.

6-9 Alkoxides

The OR group usually acts as a bridging ligand, although the degree of association of Group 13 alkoxides is controlled by the size of R. Only the alkoxides of aluminum are generally important, particularly the isopropoxide which is used in organic synthesis for the reduction of ketones. Aluminum alkoxides can be made by the reactions

$$Al + 3ROH \xrightarrow[\text{catalyst, warm}]{1\% \ HgCl_2 \ as} Al(OR)_3 + {}^3/_2H_2$$

$$AlCl_3 + 3RONa \longrightarrow Al(OR)_3 + 3NaCl$$

The alkoxides are readily hydrolyzed in water. $Al(OPr^i)_3$ is oligomeric and crystallizes as a tetramer with three tetrahedral Al centers surrounding a central AlO_6 unit (6-VII).[16] The bulkier *t*-butoxide is a dimer (6-VIII) both in solution and as a solid, as are the siloxides $[Al(OSiR_3)_3]_2$ (R = Me, Et).[17] The sterically highly congested compound $Al(OAr^*)_3$ is monomeric, with a trigonal-planar AlO_3 core (6-IX) (Ar* = 2,6-Bu$_2^i$-4-MeC$_6$H$_2$).[18] The related alkyls $AlR(OAr^*)_2$ (R = Me, Et) are also monomeric. These coordinatively unsaturated compounds have wide Al—O—C angles (>140°) and very short Al—O distances, possibly indicative of a partial Al—O double-bond character.

[15]D. Parker *et al., Polyhedron* **1991**, *10*, 1951; F. Nepveu *et al., Inorg. Chim. Acta* **1993**, *211*, 141; D. M. Roundhill *et al., Inorg. Chem.* **1994**, *33*, 1241.
[16]K. G. Caulton *et al., Polyhedron* **1991**, *10*, 1639.
[17]M. H. Chisholm *et al., Polyhedron* **1991**, *10*, 1367.
[18]A. R. Barron and M. D. Healy, *Angew. Chem. Int. Ed. Engl.* **1992**, *31*, 921.

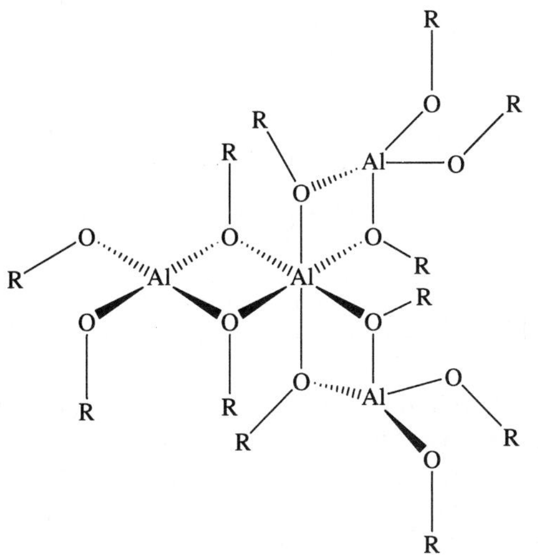

(6-VII)

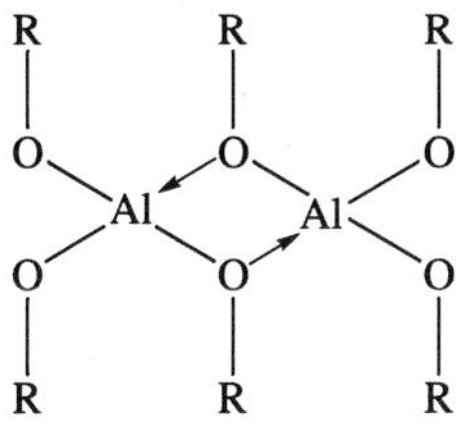

(6-VIII) 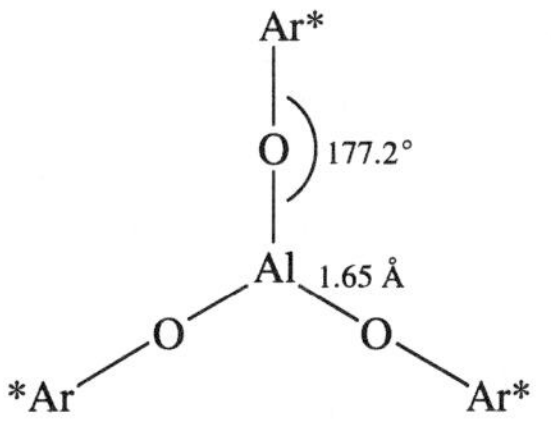   (6-IX)

177.2°

Al 1.65 Å

Nitrato Complexes

Gallium and thallium form *nitrato* complexes by the reaction of N_2O_5 with NO_2^+ [$GaCl_4$]$^-$ or $TlNO_3$, respectively. The gallium ion in NO_2^+[$Ga(NO_3)_4$]$^-$ appears to have unidentate NO_3 groups, by contrast with [$Fe(NO_3)_4$]$^-$, which has bidentate groups and is 8-coordinate, although the radius of Ga^{III} (0.62 Å) is only slightly smaller than that of Fe^{III} (0.64 Å).

6-10 Hydrides and Complex Hydrides[19]

Aluminum hydride is obtained by interaction of $LiAlH_4$ with 100% H_2SO_4 in THF:

$$2LiAlH_4 + H_2SO_4 = 2AlH_3 + 2H_2 + Li_2SO_4$$

In the interaction of $LiAlH_4$ with $AlCl_3$, intermediates like $AlHCl_2$ are formed.

The white hydride is thermally unstable. Unlike boranes, [AlH_3]$_\infty$ is more ionic and forms a three-dimensional lattice isostructural with AlF_3. With donor ligands, however, a range of molecular complexes AlH_3L and AlH_3L_2 are formed, indicative of the Lewis acidic behavior of AlH_3.

[19]A. H. Cowley *et al., J. Organomet. Chem.* **1995,** *500,* 81.

Gallium hydride had long remained elusive. The authentic compound is accessible *via* halide exchange:

$$Ga_2Cl_6 + 4Me_3SiH \xrightarrow[-20°C]{} [H_2GaCl]_2 + 4Me_3SiCl$$

$$[H_2GaCl]_2 + 2LiGaH_4 \xrightarrow[-30°C]{} [GaH_3]_2 + 2Li[GaH_3Cl]$$

Unlike $[AlH_3]_\infty$, gallane is a molecular compound with a structure like borane, $H_2Ga(\mu\text{-}H)_2GaH_2$. It is volatile and thermally and hydrolytically very sensitive (decomp $> 30°C$).[20]

The relationship of Al and Ga hydrides to boranes is underlined by the formation of mixed hydrides such as $HM\{(\mu\text{-}H)_2BH_2\}_2$ (M = Al, Ga) and $Al\{(\mu\text{-}H)_2BH_2\}_3$. The volatile, thermally unstable $GaBH_6$ has a $[H_2Ga(\mu\text{-}H)BH_2(\mu\text{-}H)]_\infty$ infinite chain structure.[21] Aluminum and Ga hydrides and their adducts are of interest as precursors to metal films and III-V semiconductors.

Hydride Anions

Both Al and Ga hydride anions are obtained by the reaction

$$4LiH + MCl_3 \xrightarrow{Et_2O} LiMH_4 + 3LiCl$$

However, for AlH_4^- the sodium salt can be obtained by direct interaction:

$$Na + Al + 2H_2 \xrightarrow[150°C/2000\ psi/24h]{THF} NaAlH_4$$

The salt is obtained by precipitation with toluene and can be converted efficiently to the lithium salt:

$$NaAlH_4 + LiCl \xrightarrow{Et_2O} LiAlH_4 + NaCl(s)$$

The most important compound is *lithium aluminum hydride*, $LiAlH_4$, a nonvolatile crystalline solid, stable below 120°C, that is explosively hydrolyzed by water. In the crystal there are tetrahedral AlH_4^- ions with an average Al—H distance of 1.55 Å. The Li^+ ions each have four near hydrogen neighbors (1.88 − 2.00 Å) and a fifth that is more remote (2.16 Å). Lithium aluminum hydride is soluble in diethyl and other ethers and can be solubilized in benzene by crown ethers. In ethers, the Li^+, Na^+, and R_4N^+ salts of AlH_4^- and GaH_4^- tend to form three types of species depending on the concentration and on the solvent, namely, either loosely or tightly bound aggregates or ion pairs. Thus $LiAlH_4$ is extensively associated in diethyl ether, but at low concentrations in THF there are ion pairs. Sodium aluminum hydride ($NaAlH_4$) is insoluble in diethyl ether.

[20]A. J. Downs and C. R. Pulham, *Adv. Inorg. Chem.* **1994,** *41,* 171.
[21]A. J. Downs *et al., Angew. Chem. Int. Ed. Engl.* **1997,** *36,* 890.

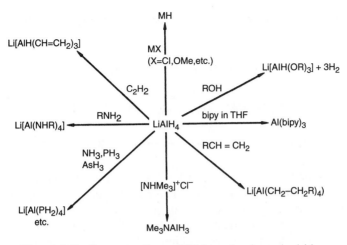

Figure 6-5 Some reactions of lithium aluminum hydride.

Some reactions of $LiAlH_4$ are given in Fig. 6-5. It is a significantly stronger reducing agent than BH_4^- and is widely used as a reductant in organic and inorganic chemistry. Related hydrido anions that are selective reducing agents are the benzene soluble "Red-Al," $Na[AlH_2(OCH_2CH_2OMe)_2]$, and $Li[AlH(OBu^t)_3]$.

Lithium and sodium salts of AlH_6^{3-} are best made by direct interaction of the metals under H_2 pressure, but can be made by the reaction

$$2NaH + NaAlH_4 \longrightarrow Na_3AlH_6$$

Lithium gallium hydride decomposes slowly even at 25°C to give LiH, Ga, and H_2. Comparing the MH_4^- ions of B, Al, and Ga, the thermal and chemical stabilities vary according to the ability of MH_3 to act as an acceptor for H^-. The order, and also the M—H force constants of MH_4^-, follow the sequence B > Al >> Ga.

Aluminum forms a number of borohydride anions with 3c-2e bonds, e.g., $[AlH_n(BH_4)_{4-n}]^-$ (n = 0, 1, 2, 3). The anion $[Al(\eta^2\text{-}BH_4)_4]^-$ is a rare example of 8-coordinate Al^{III}.[22] Lithium aluminum hydride reacts with ammonium salts and amines to give Al amides:[23]

$$LiAlH_4 + [Me_2NH_2]Cl \longrightarrow \text{\textonethird}[H_2Al(\mu\text{-}NMe_2)]_3 + LiCl + 2H_2$$

$$LiAlH_4 + 2R_2NH \xrightarrow{Et_2O} (R_2N)_2Al(\mu\text{-}H)_2Li(OEt_2)_2 + 2H_2 \quad (R = SiMe_3)$$

Donor Adducts

These are similar to borane adducts, the stability order being B > Al > Ga, and also similar to adducts of the halides and alkyls, where the stability order is

[22]S. G. Shore *et al., Inorg Chem.* **1994,** *33,* 5443.
[23]A. J. Downs *et al., Polyhedron* **1992,** *11,* 1295; A. Heine and D. Stalke, *Angew. Chem. Int. Ed. Engl.* **1992,** *31,* 854.

halides > alkyls > hydrides. The most studied adducts are the trialkylamine alanes (alane = AlH_3):

$$Me_3NAlCl_3 + 3LiH \xrightarrow{Et_2O} Me_3NAlH_3 + 3LiCl$$

$$Me_3NH^+Cl^- + LiAlH_4 \xrightarrow[-60°C]{Et_2O} Me_3NAlH_3 + LiCl + H_2$$

$$3LiAlH_4 + AlCl_3 + 4NMe_3 \longrightarrow 4Me_3NAlH_3 + 3LiCl$$

Trimethylamine gives both 1:1 and 2:1 adducts, but the latter are stable only in the presence of an excess of amine. The monoamine adduct, a white, volatile, crystalline solid (mp 75°C), which is readily hydrolyzed by water and slowly decomposes to $(AlH_3)_n$, is monomeric and tetrahedral. The bis(amine) is *tbp* with axial N atoms. Tetrahydrofuran also gives 1:1 and 2:1 adducts, but diethyl ether, presumably for steric reasons, gives only the 1:1 compound, although a mixed THF-Et_2O adduct exists. Typical reactions of AlH_3 and its adducts are shown in Fig. 6-6.

Adducts of GaH_3 with amines are accessible from tetrahydrogallate:

$$LiGaH_4 + [Me_3NH]Cl \longrightarrow Me_3NGaH_3 + H_2 + LiCl$$

$GaH_3(NMe_3)$ is thermally stable (mp 70.5°C). The stability of the adducts $GaH_3(L)$ decreases for L = Me_2NH > Me_3N > PMe_3 > C_5H_5N > R_2O. Preparation of

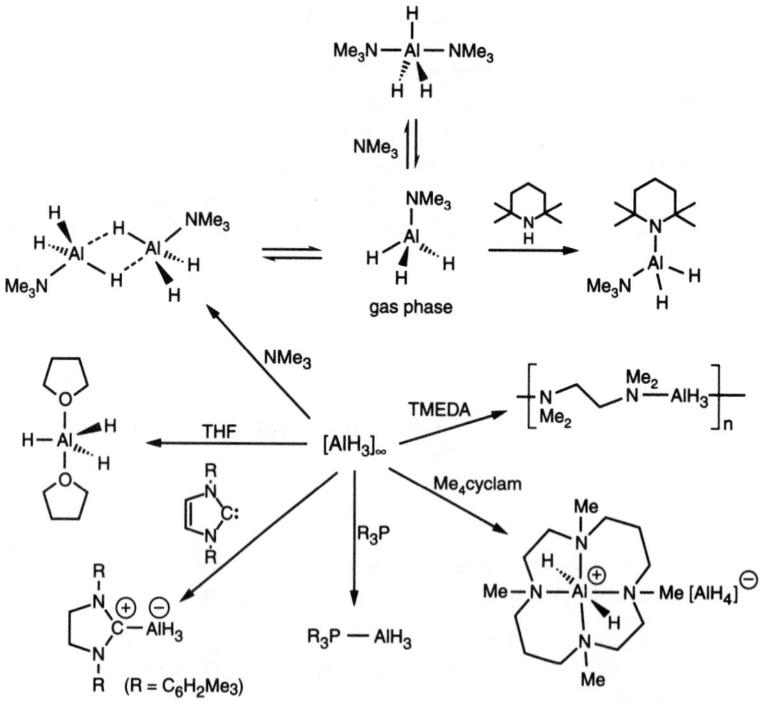

Figure 6-6   Reactions of $[AlH_3]_\infty$ and its donor adducts.

the gallanes illustrates a useful principle regarding the use of a weak donor as solvent:

$$Me_3NGaH_3 + BF_3 \xrightarrow[\text{weak}]{Me_2S} Me_2SGaH_3 + Me_3NBF_3$$

strong-weak   strong   weak-weak   strong-strong

Because the weak-weak, strong-strong combination is favored over two weak-strong adducts, the net effect is to displace the strong donor Me_3N by the weaker Me_2S.

6-11 Organometallic Compounds

Organoaluminum compounds, particularly $AlEt_3$, are commercially very important as activators for olefin polymerization catalysts and are produced on a large scale, in spite of their pyrophoric nature and violent reaction with water. In the laboratory organoaluminum compounds can be made from mercury alkyls by transmetallation:

$$2Al + 3R_2Hg \longrightarrow 2R_3Al + 3Hg$$

or by reaction of Grignard reagents with $AlCl_3$:

$$RMgCl + AlCl_3 \longrightarrow RAlCl_2, R_2AlCl, R_3Al$$

The lower alkyls are spontaneously flammable and extremely reactive liquids and all organoaluminums are sensitive to air and to water, alcohols, halocarbons, and other compounds.

Direct methods for large-scale synthesis stemmed from studies by K. Ziegler that showed that aluminum hydride or $LiAlH_4$ reacts with olefins to give alkyls or alkyl anions—a reaction specific for B and Al hydrides:

$$AlH_3 + 3C_nH_{2n} \longrightarrow Al(C_nH_{2n+1})_3$$

$$LiAlH_4 + 4C_nH_{2n} \longrightarrow Li[Al(C_nH_{2n+1})_4]$$

Although $(AlH_3)_n$ cannot be made by direct interaction of Al and H_2, in the presence of aluminum alkyl the following reaction to give the dialkyl hydride can occur:

$$Al + {}^3/_2 H_2 + 2AlR_3 \longrightarrow 3AlR_2H$$

This hydride will then react with olefins:

$$AlR_2H + C_nH_{2n} \longrightarrow AlR_2(C_nH_{2n+1})$$

The direct interaction of Al, H_2, and olefin (hydroalumination) is used to give either the dialkyl hydrides or the trialkyls. These are also accessible by reduction of organoaluminum halides and exist as dimers or trimers, depending on the steric requirements of the alkyl or aryl group. Examples are $[Bu_2^iAlH]_3$, $[mes*AlH_2]_2$, and $mes*GaH_2$.[24]

The degree of association of aluminum alkyls $[AlR_3]_n$ is dependent on the steric bulk of R. Many are dimeric, with μ-R 3c-2e bonds, but bridging and terminal alkyl ligands exchange rapidly. For example, at $-75°C$ the ^1H nmr spectrum of trimethylaluminum exhibits separate resonances for the terminal and bridging methyl groups, but on warming, these begin to coalesce, and at room temperature only one sharp peak is observed. The exchange process in Al_2Me_6 involves dissociation to the monomer and reformation of the dimer:

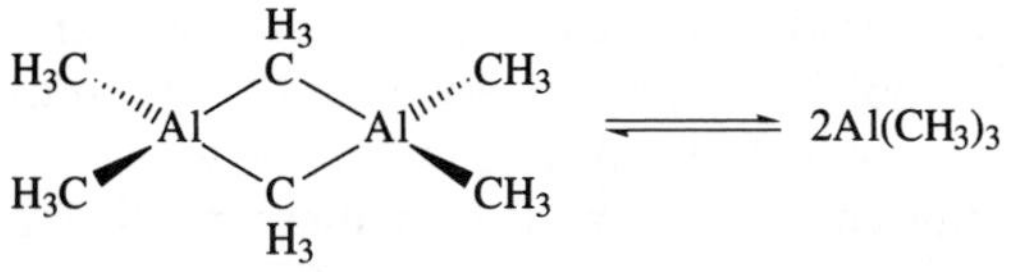

$$\rightleftharpoons \quad 2Al(CH_3)_3$$

The extent of the dissociation of Al_2Me_6 is very small, 0.0047% at 20°C, even though the exchange process is fast. Triethylaluminum is also a dimer, while $AlBu_3^i$ and $AlBu_3^t$ are monomeric.

Aluminum phenyls R_2AlPh (R = alkyl or Ph) are dimeric, with bridging phenyl groups (6-X). Bulkier aryls such as trimesitylaluminum are trigonal-planar monomers. Tribenzylaluminum is unique in having a chain structure with π-coordination between a vacant p-orbital of Al and the *ortho*-carbon of the phenyl ring of the neighboring AlR_3 unit (6-XI). π-Coordination is also found in Al cyclopentadienyl compounds. The compound Cp_2AlMe has two η^2-bonded Cp rings in the solid state (6-XII);[25] it reacts with $B(C_6F_5)_3$ to give a salt of the aluminocenium cation, $[Al(\eta^5$-$Cp)_2]^+[MeB(C_6F_5)_3]^-$,[26] which, like its decamethyl analogue,[26] $Cp_2^*Al^+$, possesses a sandwich structure (6-XIII).

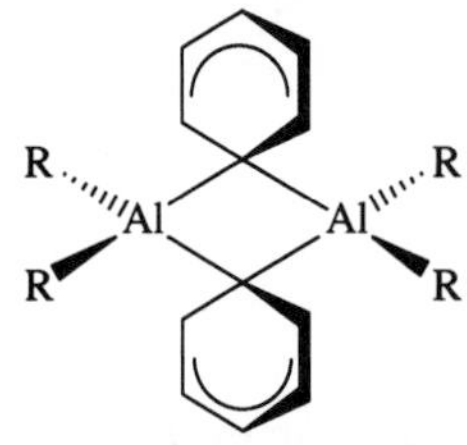

(6-X) 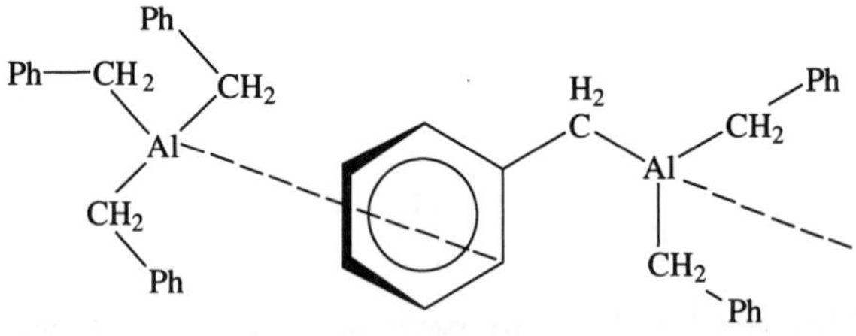   (6-XI)

[24]W. Uhl, *Z. anorg. allg. Chem.* **1989,** *570,* 37; P. P. Power *et al., Inorg. Chem.* **1994,** *33,* 5611 and 6300; *J. Am. Chem. Soc.* **1996,** *118,* 791; A. H. Cowley *et al., Angew. Chem. Int. Ed. Engl.* **1994,** *33,* 1253.

[25]P. J. Shapiro *et al., Organometallics* **1994,** *13,* 3324.

[26]M. Bochmann and D. M. Dawson, *Angew. Chem. Int. Ed. Engl.* **1996,** *35,* 2226; H. Schnöckel *et al., Angew. Chem. Int. Ed. Engl.* **1993, 32,** 1655.

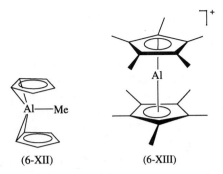

(6-XII)   (6-XIII)

Aluminum alkyls are Lewis acids. They form adducts with amines and ethers, $AlR_3 \cdot L$, and react with the anions of salts. The reaction of K_2SO_4 and $AlMe_3$ gives $K_2[SO_4(AlMe_3)_4]$. They also form tetrahedral complex anions:

$$3K + 4AlMe_3 \longrightarrow 3K[AlMe_4] + Al$$

$$Na[Et_3AlOEt] + AlEt_3 \longrightarrow Na[AlEt_4] + \tfrac{1}{2}[Et_2AlOEt]_2$$

The *trialkyls of Ga, In, and Tl* are less stable than those of Al, and, like those of boron, do *not* dimerize. In the crystal Me_3In and Me_3Tl are tetramers, with only weak bonds between the individual MMe_3 molecules. Trimethylgallium, $GaEt_3$, and $InMe_3$ are used for the production of III-V semiconductors (see Section 1-10). Trimethylgallium can be made electrolytically using a gallium anode and Grignard reagents:

$$6MeMgX + 2Ga \xrightarrow{\ Et_2O\ } 2GaMe_3OEt_2 + 3Mg + 3MgX_2$$

Completely ether- and halide-free high-purity $GaMe_3$ for semiconductor applications is difficult to obtain. However, the compound forms crystalline adducts with non-volatile phosphines such as $Ph_2PC_2H_4PPh_2$, which dissociate on heating to give ultra-pure $GaMe_3$. This method of purifying volatile metal alkyls *via* the reversible formation of crystalline adducts is quite general and has also been applied to other semiconductor precursors such as $ZnMe_2$ and $CdMe_2$.

Alkyl Halide Compounds

Aluminum alkyl halides of the types R_2AlX and $RAlX_2$ are halide-bridged dimers in which, unlike Al_2Me_6, Al is surrounded by an electron octet. They are stronger Lewis acids than the trialkyls. The sterically very crowded compounds mes*MX_2 are monomeric; they fail to give complexes with Et_2O (M = Al, Ga, In; X = Cl, Br).[27]

Other technically important compounds are the *sesquichlorides* such as $Me_3Al_2Cl_3$ and $Et_3Al_2Cl_3$. These compounds can be made by direct interaction of

[27]P. P. Power *et al.*, *Organometallics* **1993**, *12*, 1086; H. W. Roesky *et al.*, *Inorg. Chem.* **1993**, *32*, 3343.

Al or Mg—Al alloy with the alkyl chloride. This reaction fails for propyl and higher alkyls since the alkyl halides decompose in the presence of alkyl aluminum halides to give alkenes and HCl.

Whereas Al alkyls and alkyl halides are completely hydrolyzed in water, dialkyl compounds of Ga, In, and Tl give well-characterized compounds in aqueous solution. Thallium gives very stable ionic derivatives of the type R_2TlX (X = halide, SO_4^{2-}, CN^-, NO_3^-, etc.) which resemble the isoelectronic mercury dialkyls R_2Hg in being unaffected by air and water. The $[TlMe_2]^+$ ion is linear, but does not form an aquo complex like $Me_2Ga(H_2O)_2^+$.

In general, primary amines may form neutral or ionic adducts with aluminum halides, with ionization being favored for the more labile bromides and iodides.[28] Ammonia forms an ionic adduct with Me_2GaCl, $[Me_2Ga(NH_3)_2]^+Cl^-$.

Triethylaluminum, the sesquichloride $(C_2H_5)_3Al_2Cl_3$, and alkyl hydrides are used together with transition metal halides, alkoxides, or organometallic complexes as catalysts (e.g., Ziegler catalysts) for the polymerization of alkenes (Chapter 22). They are also used widely as reducing and alkylating agents for transition metal complexes.

In alkylations of metal halides with AlR_3 usually only one alkyl group is transferred, since the dialkyl aluminum halides are much less powerful alkylating agents than the trialkyls. The alkylaluminums react with compounds having acidic hydrogen atoms to give the alkane; in some cases alkylation may occur, for example,

$$R_3COH \xrightarrow[\substack{\text{toluene}\\100\text{-}120\,°C}]{AlMe_3} R_3CMe$$

Compounds with Bonds to N, P, and As

The alkyls are Lewis acids, combining with donors such as amines, phosphines, ethers, and thioethers to give tetrahedral, 4-coordinate species. Thus Me_3NAlMe_3 in the gas phase has C_{3v} symmetry with staggered methyl groups. With tetramethylhydrazine and $(CH_3)_2NCH_2N(CH_3)_2$, 5-coordinate species that appear to be of the kind shown in (6-XIV) are obtained, although at room temperature exchange processes cause all methyl groups and all ethyl groups to appear equivalent in the proton nmr spectrum.

(6-XIV)

With $(CH_3)_2NCH_2CH_2N(CH_3)_2$ a complex is formed that has an AlR_3 group bound to each nitrogen atom. The interaction of the alkylaluminums or mixed alkyls such as $AlClEt_2$ with primary or secondary amines gives adducts initially.

[28]D. Atwood and J. Jeglier., *Inorg. Chem.* **1996,** *35,* 4277.

On heating, these adducts lose alkane to give oligomers that have Al—N—Al bonds, each Al becoming 4-coordinate by interaction with a nitrogen atom lone pair.

$$2Et_2BrAlNH_2Bu^i \xrightarrow{120°C} (EtBrAlNHBu^i)_2 + 2C_2H_6$$

$$nMe_3Al + nNH_2Me \longrightarrow (MeAlNMe)_n + 2nCH_4$$

The dimers have a four-membered Al_2N_2 ring (6-XV) which may have isomers if the groups are different. The aluminum imides $[RAl(NR')]_n$ usually have cubane or larger cage structures (6-XVI); trimers (e.g., the "alumazene" 6-XVII) and heptamers have also been found, depending on the size of R and R′ and the reaction conditions.[29]

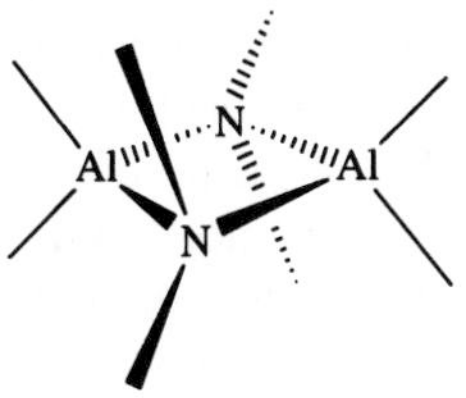

O Al

● NR, O

(6-XV) 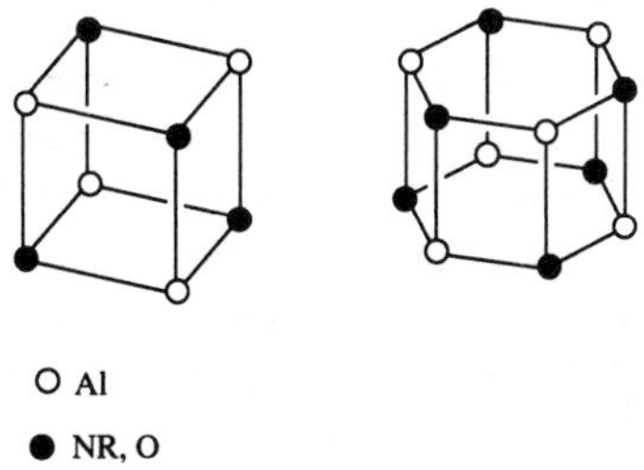   (6-XVI)

$R = C_6H_2Pr^i_3$

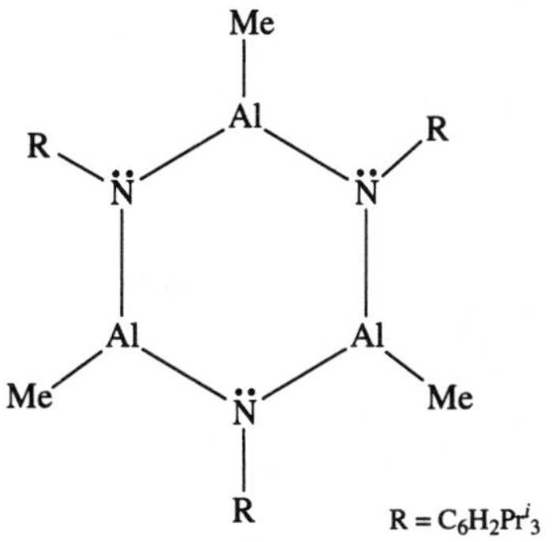

(6-XVII)

With suitably bulky ligands monomeric amido complexes are obtained. All possess relatively short Al—N bonds, typically 1.8 − 1.9 Å. Examples are $Bu^t_2AlNR_2$ (R = mes, $SiPh_3$), (mes)$Al(NR_2)_2$, $ClGa(NR_2)_2$, and $M(NR_2)_3$ (R = $SiMe_3$; M = Al, Ga, In, Tl).[30] Aluminum, Ga, and In compounds with M—P and M—As bonds are of interest as "single source" precursors to III-V semiconductors and as models for the gas phase reactions during the deposition of films, e.g., GaAs from $GaMe_3$ and AsH_3. The structural chemistry encountered closely resembles that of the Al—N systems. Compounds of the composition $[R_2M(\mu-ER'_2)]_n$ are usually cyclic

[29]K. M. Waggoner and P. P. Power, *J. Am. Chem. Soc.* **1991,** *113,* 3385.
[30]P. P. Power *et al., Organometallics* **1994,** *13,* 2792.

dimers (6-XV) or trimers. If P—H or As—H bonds are present, controlled thermolysis leads to alkane elimination and formation of higher cluster species:

$$2[R_2M(\mu\text{-PHR}')]_2 \xrightarrow{\ \Delta\ } [RM(\mu_3\text{-PR}')]_4 + 4RH$$

Some compounds, such as the unsubstituted arsenide derivative $[Bu_2^tGa(\mu\text{-AsH}_2)]_3$, give GaAs at temperatures as low as 110°C.[31]

Whereas, as mentioned above, aluminum trialkyls oligomerize ethene under pressure, cationic compounds of the type $[L_2AlMe]^+X^-$ and $[(L_2AlMe)_2(\mu\text{-Me})]^+X^-$ [L = benzamidinate anion, X = $MeB(C_6F_5)_3$] give high molecular weight polyethene under mild conditions.[32]

Compounds with M—Chalcogenide Bonds

The compound Al_2Me_6 reacts violently with water to give methane and $Al(OH)_3$. However, if the hydrolysis is carried out under controlled conditions, e.g., by reacting Al_2Me_6 with $Al_2(SO_4)_3 \cdot 16H_2O$ in toluene, a highly viscous amorphous product of the approximate composition $[Al(Me)O]_n$, called methylalumoxane (MAO), is formed. This is a mixture of polymers of varying structures, with a molecular weight of *ca.* 900 − 1500. Methylalumoxane is usually rich in methyl groups, e.g., $[Al(Me)_{1.3-1.4}O]_n$, and contains varying amounts of Al_2Me_6 (up to 30–40% in the case of some commercial products). Although cyclic components such as $[Al(Me)O]_3$ have been isolated from MAO solutions, the low activation barrier of Al towards ligand exchange implies that well-defined components rapidly rearrange in solution to give an equilibrium mixture of oligomers, including cage compounds of the types (6-XVI). Some light was shed on the reactions involved in MAO formation by the low-temperature hydrolysis of $AlMe_3$ in THF solutions containing controlled amounts of water which leads to $(Me_2AlOH)_4 \cdot n$THF, the first stage in the build-up of cage structures. The protolysis reaction can be stopped at this stage by the addition of methyllithium to give isolable $(Me_2AlOLi)_4 \cdot 7$THF· LiCl, which contains a slightly puckered 8-membered Al_4O_4 ring structure.[33]

In spite of its ill-defined nature, MAO has become one of the most important aluminum alkyls; it is produced commercially on a large scale to be an activator for zirconocene-based olefin polymerization catalysts (see Chapter 22).

The hydrolysis of $AlBu_3^i$ is easier to control and gives a number of well-defined products which illustrate the structural principles in alumoxanes. At −78°C the hydroxide, $[Bu_2^iAl(\mu\text{-OH})]_3$, is the major product, while under more vigorous conditions oligomers such as (6-XVIII) and the clusters $[Bu^tAl(\mu_3\text{-O})]_n$ (n = 6, 8, 9) result. Their structures are based on the Al_4O_4 cube and the Al_6O_6 hexagonal prism, *cf.* (6-XVI). It is likely that MAO consists of frameworks constructed along similar principles. The cage compounds $[Bu^tAlO]_n$ (n = 6, 8, 9) react, similar to MAO, with Cp_2ZrMe_2 to give polymerization catalysts. The reaction is initiated by the cleavage of an Al—O bond and methyl anion transfer to Al, to give, e.g., (6-XIX).[34]

[31]A. H. Cowley *et al., Organometallics* **1991,** *10,* 652. For a review on organometallic III-V precursors see: A. H. Cowley and R. A. Jones, *Angew. Chem. Int. Ed.* **1989,** *28,* 1208.
[32]M. P. Coles and R. F. Jordan, *J. Am. Chem. Soc.* **1997,** *119,* 8125.
[33]H. W. Roesky *et al., J. Am. Chem. Soc.* **1997,** *119,* 7505.
[34]A. R. Barron *et al., J. Am. Chem. Soc.* **1995,** *117,* 6465.

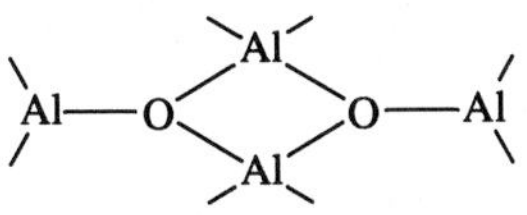

(6-XVIII) 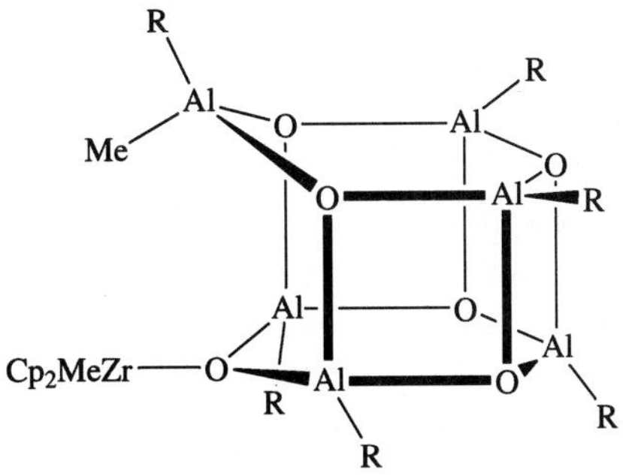   (6-XIX)

The hydrolysis of bulky aryls of Al and Ga leads similarly to compounds of the type $(Ar_2MOH)_n$ ($n = 2, 3$); in this case the water complex $(mes)_3Ga \cdot OH_2 \cdot THF$ can also be isolated.[35] Cyclic compounds $(ArAlO)_4$ react with AlR_3 under alkyl ligand redistribution to give compounds based on ribbons of Al_2O_2 four-membered rings.[36]

The reaction of AlR_3 with alcohols and phenols gives alkyl alkoxides which are generally dimeric, $[R_2Al(\mu\text{-}OR')]_2$, although $[Me_2Al(OMe)]_3$ is a cyclic trimer. The monoalkoxides are significantly less reactive than the trialkyls. With an excess of alcohol, the products $[RAl(OR')_2]_n$ and $[Al(OR')_3]_n$ result. The reaction between Al_2Me_6 and the highly hindered phenols $HOAr^*$ ($Ar^* = 2,6\text{-}Bu^t_2\text{-}4\text{-}MeC_6H_2$, $2,6\text{-}Ph_2C_6H_3$) give the monomeric 3-coordinate compounds $MeAl(OAr^*)_2$. They act as Lewis acids and form tetrahedral adducts not only with amines and PMe_3 but also with aldehydes. In contrast to the reactivity of most Al alkyls, these aldehyde complexes are resistant to $C=O$ insertion into the $Al-Me$ bond. The formaldehyde complex $MeAl(OAr^*)_2(CH_2=O)$ acts as a source of formaldehyde for a variety of organic reactions.[37] Other examples illustrating steric control of the coordination geometry are the trigonal-planar monomers $Bu^t_2M(OAr^*)$ ($M = Al, Ga$).

The reactions of Al, Ga, and In trialkyls with hydroperoxides or O_2 give isolable hydroperoxide complexes (6-XX) which act as oxidizing agents.[38]

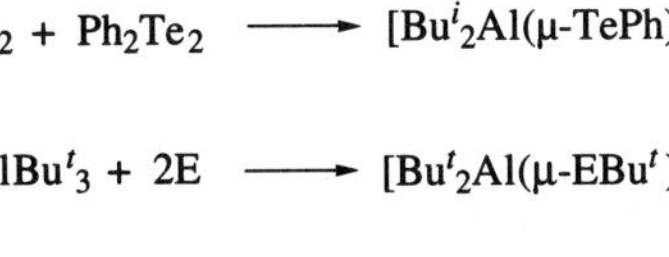

(6-XX)

Complexes of the heavier chalcogenides can be made by a variety of routes; here too the ER ligands are bridging:

$$[Bu^i_2AlH]_2 + Ph_2Te_2 \longrightarrow [Bu^i_2Al(\mu\text{-}TePh)]_2 + H_2$$

$$2AlBu^t_3 + 2E \longrightarrow [Bu^t_2Al(\mu\text{-}EBu^t)]_2$$

$$2In(SePh)_3 + 4InR_3 \longrightarrow 3[R_2In(\mu\text{-}SePh)]_2$$

[35]H. W. Roesky et al., J. Am. Chem. Soc. **1996**, 118, 1380.
[36]R. J. Wehmschulte and P. P. Power, J. Am. Chem. Soc. **1997**, 119, 8387.
[37]A. R. Barron et al., Organometallics **1990**, 9, 3086; H. Yamamoto et al., J. Am. Chem. Soc. **1990**, 112, 7422.
[38]Y. A. Alexandrov et al., J. Organomet. Chem. **1991**, 418, 1; A. R. Barron et al., Organometallics **1993**, 12, 4908; J. Am. Chem. Soc. **1989**, 111, 8966.

Tris(thiolato) complexes are made from MCl_3 and NaSR. While Al and Ga are insufficiently reactive, metallic In is oxidized by dichalcogenides E_2R_2 to give $[In(ER)_3]_n$. The compound $In(SePh)_3$ has a chain structure with octahedral indium, $[In(\mu_2\text{-SePh})_3]_\infty$.[39]

Alkyl metal thiolates and metal trithiolates are, like the alkoxides, generally dimeric or polymeric. Exceptions are the bulky monomeric compounds $M(SC_6H_2Bu_3^t)_3$ (M = Al, Ga, In). Compounds with SH ligands are only rarely isolable, an example being $[Bu_2^tGa(\mu\text{-SH})]_2$ obtained by the treatment of $GaBu_3^t$ with H_2S at ambient temperature. The reactions of MBu_3^t with E = S, Se, or Te give the cubane compounds $[Bu^tM(\mu_3\text{-E})]_4$ (M = Al, Ga). The gas phase thermolysis of $[Bu^tGaS]_4$ affords films of the metastable cubic phase of GaS.[40]

6-12 Transition Metal Complexes

The interaction of sodium salts of metal carbonyl anions, for example, $[Co(CO)_4]^-$, with Group 13 metal chlorides commonly gives metal—metal bonded compounds that are readily soluble in organic solvents. Examples are $AlCo_3(CO)_9$, $In[Mn(CO)_5]_3$, and $Tl[Co(CO)_4]_3$. Aluminum, Ga, and In alkyl halides react with anionic metal carbonyl complexes to give, e.g., (6-XXI). These complexes act as precursors for the gas phase epitaxial growth of bimetallic films, e.g., β-CoGa.[41] Complex (6-XXII) contains formal Tl=Cr double bonds and is one of a family of metal carbonyl complexes with linear M=E=M moieties (E = main group element, e.g., Ge, Pb).[42]

$$L(CO)_nM\!\!-\!\!E\overset{\displaystyle Do}{\underset{\displaystyle R}{\overset{\displaystyle R}{\Big\langle}}}$$

E = Al, Ga; M = Fe, Co, Ni. $[(OC)_5Cr\!=\!Tl\!=\!Cr(CO)_5]^-$

(6-XXI)   (6-XXII)

THE CHEMISTRY OF OXIDATION STATES I AND II[43]

Since the elements in Group 13 have the outer electron configurations ns^2np, it is natural to consider whether monovalent ions might exist.

[39]D. G. Tuck *et al.*, *J. Chem. Soc., Dalton Trans.* **1988,** 1045; *Polyhedron* **1989,** *8,* 865; T. Beachley *et al.*, *Organometallics* **1992,** *11,* 3144; K. Sasaki *et al.*, *Chem. Lett.* **1991,** 415; A. H. Cowley *et al.*, *Angew. Chem. Int. Ed. Engl.* **1991,** *30,* 1143.
[40]M. B. Power and A. R. Barron, *J. Chem. Soc., Chem. Commun.* **1991,** 1315; *Chem. Mater.* **1993,** *5,* 1344; A. H. Cowley *et al.*, *Angew. Chem. Int. Ed. Engl.* **1991,** *30,* 1143; K. Ruhlandt-Senge and P. P. Power, *Inorg. Chem.* **1991,** *30,* 2633.
[41]R. A. Fischer *et al.*, *Angew Chem. Int. Ed. Engl.* **1993,** *32,* 746 and 748; *Inorg. Chem.* **1994,** *33,* 934.
[42]B. Schiemenz and G. Huttner, *Angew Chem. Int. Ed. Engl.* **1993,** *32,* 1772.
[43]C. Dohmeier *et al.*, *Angew. Chem. Int. Ed. Engl.* **1996,** *35,* 129.

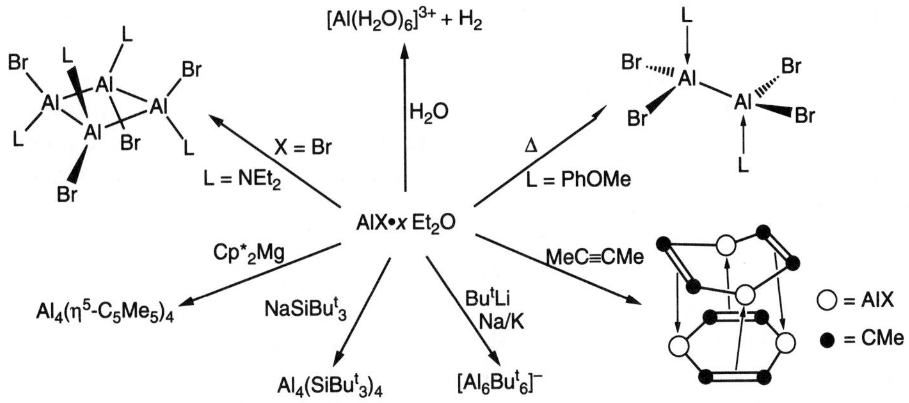

[Al(H₂O)₆]³⁺ + H₂ → $[Al(H_2O)_6]^{3+} + H_2$

H_2O

$X = Br$
$L = NEt_2$

Δ
$L = PhOMe$

$AlX \cdot x\,Et_2O$

Cp^*_2Mg

$NaSiBu^t_3$

Bu^tLi
Na/K

$MeC{\equiv}CMe$

$Al_4(\eta^5\text{-}C_5Me_5)_4$

$Al_4(SiBu^t_3)_4$

$[Al_6Bu^t_6]^-$

○ = AlX
● = CMe

Figure 6-7 Reactions of Al(I) halides.

6-13 Aluminum and Gallium (I, II)

Aluminum monochloride and GaCl are high temperature species. The former is made from Al vapor and HCl gas at 1200 K in high yield:

$$Al(g) + HCl(g) \xrightarrow{<0.2\ mbar} AlCl + \tfrac{1}{2}H_2$$

It can be trapped as a dark red solid on cold ($-196°C$) surfaces and disproportionates above $-90°C$:

$$3AlCl \rightleftharpoons AlCl_3 + 2Al$$

Aluminum(I) bromide, GaCl, and GaBr are similarly accessible. The position of the thermodynamic equilibrium has long precluded the development of the chemistry of Al^I. However, the co-condensation of AlCl with Et$_2$O and toluene gives metastable red solutions of AlX·OEt$_2$ (X = Cl, Br) which can be kept for several hours. These solutions are the starting point for a diversified chemistry; typical reactions are shown in Fig. 6-7.

The co-condensation of AlBr and NEt$_3$ gives crystals of the tetramer [AlBr(NEt$_3$)]$_4$ which contains an Al$_4$ ring (Al—Al 2.64 Å). If anisole is used as the donor ligand, slow disproportionation takes place leading to Al$^{II}_2$Br$_4$(anisole)$_2$, with an Al—Al bond length of 2.53 Å. Hydrolysis of AlX gives [Al(H$_2$O)$_6$]$^{3+}$ and hydrogen.

A number of organometallic Al^I compounds are known. Aluminum(I) chloride reacts with Cp*$_2$Mg to give tetrahedral [Al(η^5-Cp*)]$_4$; this is also accessible from the reduction of [Cp*AlCl$_2$]$_2$ with potassium metal.[44] The compound is structurally reminiscent of the boron halide B$_4$Cl$_4$. The gallium analogue [GaCp*]$_6$ is octahedral, with very weak Ga—Ga interactions (*ca.* 4.1 Å.).[45] The compound AlCl·Et$_2$O reacts with LiBu^t in the presence of Na/K alloy to give the anion [Al$_6$Bu$^t_6$]$^-$, which

[44]H. Schnöckel *et al., Angew. Chem. Int. Ed. Engl.* **1994,** *33,* 862; 1754; **1991,** *30,* 564.
[45]H. Schnöckel *et al., Angew. Chem. Int. Ed. Engl.* **1997,** *36,* 860.

is thought to contain an Al_6 octahedron.[46] The reduction of $[Bu_2^iAlCl]_2$ with potassium in hexane gives $K_2[Al_{12}Bu_{12}^i]$ with an icosahedral Al_{12} core (Al—Al 2.68 – 2.70 Å).[47] The reaction of Al^I iodide with $LiNR_2$ gives the unusual anion $[Al_{77}(NR_2)_{20}]^{2-}$ (R = $SiMe_3$), one of the largest structurally characterized clusters.[48]

The reactions of Al^I compounds indicate the tendency to reach the oxidation state III. $[AlCp^*]_4$ reacts with $AlCl_3$ with disproportionation to give the aluminocenium cation:[49]

$$[AlCp^*]_4 + AlCl_3 \longrightarrow 2Al + [(\eta^5\text{-}Cp^*)_2Al]^+[(\eta^1\text{-}Cp^*)AlCl_3]^-$$

Gallium(I) chloride reacts with H_2 to give GaH_2Cl, which can be trapped in a matrix at low temperatures.[50] Ga_2O and the non-stoichiometric Ga_2S are also known; GaCp is a highly volatile unstable compound and unlike InCp and TlCp is probably monomeric.

A number of Al and Ga dialkyls exist with M—M bonds:

$$2R_2AlCl + 2K \xrightarrow[-KCl]{} \begin{array}{c} R \quad\quad R \\ \backslash \quad\quad \backslash\backslash\backslash \\ Al \text{———} Al \\ / \quad\quad \blacktriangle \\ R \quad\quad R \end{array}$$

$$R = CH(SiMe_3)_2, C_6H_2Pr^i_3$$

Similar In^{II} and Tl^{II} have also been made. These compounds can be isolated only with very bulky R groups.[51] These M^{II} compounds are reduced by alkali metals to give stable radical anions:

$$\begin{array}{c} R \quad\quad R \\ \backslash \quad\quad \backslash\backslash\backslash \\ Al \text{—} Al \\ / \quad\quad \blacktriangle \\ R \quad\quad R \end{array} \xrightarrow{Li} \left[\begin{array}{c} R \quad\quad R \\ \backslash \quad\quad \backslash\backslash\backslash \\ Al \text{====} Al \\ / \quad\quad \blacktriangle \\ R \quad\quad R \end{array} \right]^{\cdot -}$$

Whereas the R_2M moieties in M_2R_4 are twisted with respect to each other ($\theta \approx 45°$), the radical anions are almost planar, with short M—M distances, indicative of the partial π character of the M—M bond.[52]

Solid Ga_2X_4 is prepared by comproportionation of $GaCl_3$ and Ga. It has the structure $Ga^I[Ga^{III}X_4]$ (X = Cl, Br), whereas in the presence of donor ligands the molecular adducts $Ga^{II}_2X_4·2L$ are formed, with Ga—Ga bonds. A Ga—Ga distance

[46]H. Schnöckel *et al.*, *Angew. Chem. Int. Ed. Engl.* **1993,** *32,* 1428.
[47]W. Uhl *et al.*, *Angew. Chem. Int. Ed. Engl.* **1991,** *30,* 179.
[48]H. Schnöckel *et al.*, *Nature* **1997,** *387,* 379.
[49]H. Schnöckel *et al.*, *Angew. Chem. Int. Ed. Engl.* **1993,** *32,* 1655.
[50]H. Schnöckel *et al.*, *J. Chem. Soc. Dalton Trans.* **1992,** 3393.
[51]N. Wiberg *et al.*, *Angew. Chem. Int. Ed. Engl.* **1996,** *35,* 65.
[52]W. Uhl, *Angew. Chem. Int. Ed. Engl.* **1993,** *32,* 1386; P. P. Power *et al.*, *Inorg. Chem.* **1993,** *32,* 2983.

of 2.406 Å is found in $Ga_2Cl_4(dioxane)_2$. The reaction of $Ga[GaCl_4]$ with sulfur and pyridine gives $[GaSClpy]_3$, whereas treatment with aromatic solvents leads to stable arene complexes of Ga^I of which (6-XXIII) is an example.[53] Thallium(I) forms analogous complexes.

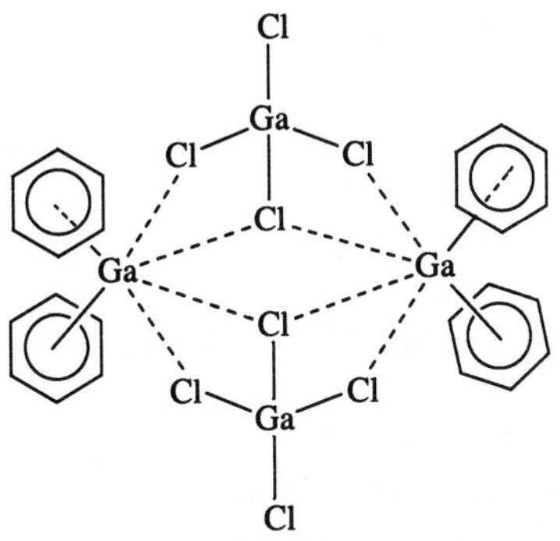

(6-XXIII)

The alkylation of $Ga_2Br_4(dioxane)_2$ with $LiC(SiMe_3)_3$ gives the tetrahedral $[Ga^I\{C(SiMe_3)_3\}]_4$.[54] The chalcogenides GaS, GaSe, and GaTe can be made directly from the elements. The anodic dissolution of Ga in $6M$ HCl or HBr at 0°C followed by addition of Me_4NX precipitates white crystalline salts of the ion $[Ga_2X_6]^{2-}$ that are stable and diamagnetic. In the staggered ethane-like ions there is a Ga—Ga bond of 2.39 Å for X = Cl and 2.41 Å when X = Br, so that formally, Ga is in the II oxidation state. Reduction of the bulky aryl derivative $ArGaCl_2$ with potassium metal gives the unusual Ga^0/Ga^I mixed valence compound $K_2[Ga_3Ar_3]$ (6-XXIV), with short (2.42 Å) Ga—Ga bonds in the triangular Ga_3 unit.[55]

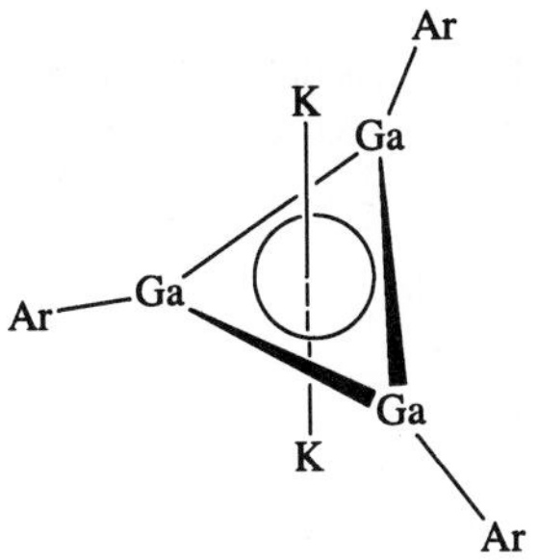

(6-XXIV)

6-14 Indium (I, II)

The indium(I) cation is unstable in aqueous solution and can be obtained only in low concentration. Stable solutions in acetonitrile are generated by treating silver

[53]H. Schmidbaur, *Angew. Chem. Int. Ed. Engl.* **1985**, *24*, 893.
[54]W. Uhl *et al.*, *Angew. Chem. Int. Ed. Engl.* **1992**, *31*, 1364.
[55]G. H. Robinson *et al.*, *Organometallics* **1996**, *15*, 3798.

triflate, AgO_3SCF_3, with indium amalgam, and aqueous solutions which are stable for several hours[56] are obtained by subsequent dilution with oxygen-free water. The In^+ ion is rapidly oxidized by both H^+ ion and by air, and is also unstable to disproportionation, as can be shown from the following potentials:

$$In^{3+} + 3e = In \qquad E^0 = -0.343 \text{ V}$$

$$In^{3+} + 2e = In^+ \qquad E^0 = -0.426 \text{ V}$$

$$In^+ + e = In \qquad E^0 = -0.178 \text{ V}$$

The extremely reactive In^{2+} aquated ion is formed in the reduction of In^{3+} with hydrated electrons.

The halides InX (X = Cl, Br, and I) as well as In_2O, In_2S, and In_2Se are obtained by solid state reactions; they are unstable in water. The dihalides, that is, $In^I[InX_4]$, are made by refluxing indium with dihalogens in xylene:

$$2In + 3X_2 = 2InX_3$$

$$2InX_3 + In = 3InX_2$$

A number of mixed valence halides exist, such as In_2Cl_3 ($In_3^I In^{III}Cl_6$), In_5Cl_9 ($In_3^I In_2^{II}Cl_9$), and In_7Cl_9 ($In_6^I In^{III}Cl_9$); In^{II} complexes can be made by the reaction

$$2[R_4N]^+X^- + 2InX_2 \xrightarrow{\text{xylene}} [R_4N]_2[In_2X_6]$$

Unlike $[Ga_2X_6]^{2-}$, the $[In_2X_6]^{2-}$ ions disproportionate in nonaqueous solvents to give $In^I X_2^-$ and $In^{III}X_4^-$.

The cyclopentadienyls $[In(C_5H_4R)]_n$ (R = H, Me) form infinite zig-zag chains; $[InCp^*]_6$ is a hexamer with an octahedral In_6 core. The compound $In_4Ar^*_6$ ($Ar^* = 2,4,6\text{-}Pr_3^iC_6H_2$) is composed of three Ar^*_2In units bound (In—In = 2.70 Å) to a central In atom.[57]

6-15 Thallium (I, II)

The unipositive state is quite stable, and in aqueous solution it is distinctly more stable than Tl^{III}:

$$Tl^{3+} + 2e = Tl^+$$

$$E^0 = +1.25 \text{ V } [E^0 = +0.770, 1M \text{ HCl}; +1.26, 1M \text{ HClO}_4]$$

The Tl^+ ion is not very sensitive to pH, although the Tl^{3+} ion is extensively hydrolyzed to $TlOH^{2+}$ and the colloidal oxide even at pH 1 to 2.5; the redox potential

[56]S. K. Chandra and E. S. Gould, *Inorg. Chem.* **1996**, *35*, 3881.
[57]P. J. Brothers *et al., Angew. Chem. Int. Ed. Engl.* **1996**, *35*, 2555.

is hence very dependent on pH as well as on the presence of complexing anions. Thus, as indicated by the previously listed potentials, the presence of Cl^- stabilizes Tl^{3+} more (by formation of complexes) than Tl^+ and the potential is thereby lowered.

The colorless Tl^+ ion has a radius of 1.54 Å, which can be compared to those of K^+, Rb^+, and Ag^+ (1.44, 1.58, and 1.27 Å). Its chemistry thus resembles that of both the alkali and silver(I) ions.

The Tl^+ ion has been proposed as a probe for the behavior of K^+ in biological systems. The two isotopes ^{203}Tl and ^{205}Tl (70.48%) have a nuclear spin, and nmr signals are readily detected both in solutions and in solids; also the Tl^I (and Tl^{III}) resonances are very sensitive to the environment and have large solvent-dependent shifts. For Tl^+ it is possible to correlate shifts with solvating ability, hence the utility as a probe in biological systems.

In crystalline salts, the Tl^+ ion is usually 6- or 8-coordinate. The yellow hydroxide is thermally unstable, giving the black oxide Tl_2O at about 100°C. The latter and the hydroxide are readily soluble in water to give strongly basic solutions that absorb carbon dioxide from the air; TlOH is a weaker base than KOH, however. Many Tl^+ salts have solubilities somewhat lower than those of the corresponding alkali salts, but otherwise are similar to and quite often isomorphous with them. Examples of such salts are the cyanide, nitrate, carbonate, sulfate, phosphate, perchlorate, and alum.

Thallium(I) sulfate, nitrate, and acetate are moderately soluble in water, but—except for the very soluble TlF—the halides are sparingly soluble. The chromate and the black sulfide Tl_2S, which can be precipitated by hydrogen sulfide from weakly acidic solutions, are also insoluble. Thallium(I) chloride also resembles silver chloride in being photosensitive; it darkens on exposure to light. Incorporation of Tl^I halides into alkali halides gives rise to new absorption and emission bands, because complexes of the type that exist also in solutions (most notably TlX_2^- and TlX_4^{3-}) are formed; such thallium-activated alkali halide crystals are used as phosphors (e.g., for scintillation radiation detectors). Thallium(I) chloride is insoluble in ammonia, unlike AgCl.

Other than those with halide, oxygen, and sulfur ligands, Tl^I gives rather few complexes. The dithiocarbamates $Tl(S_2CNR_2)$, made from aqueous Tl_2SO_4 and the sodium dithiocarbamates, are useful reagents for the synthesis of other metal dithiocarbamates from the chlorides in organic solvents, since the Tl dithiocarbamate is soluble and TlCl is precipitated on reaction. The structure of the *n*-propyl complex shows that it is polymeric, with $[TlS_2CNPr_2]_2$ dimeric units linked by Tl—S bonds.

Electron-exchange reactions in the Tl^I—Tl^{III} system have been intensively studied and appear to be two-electron transfer processes; various Tl^{III} complexes participate under appropriate conditions.

When oxygen, nitrogen, and sulfur bound to organic groups are also bound to Tl^I, the Tl—X bond appears to be more covalent than the bond to alkali metal ions in similar compounds. Thallium compounds tend to be polymeric rather than ionic. Thus the acetylacetonate is a linear polymer with 4-coordinate Tl.

Thallium alkoxides are obtained by reactions such as:

$$4Tl + O_2 + 4EtOH \longrightarrow [TlOEt]_4 + 2H_2O$$

The ethoxide, a useful reagent for the synthesis of thallium organometallics, is a liquid, whereas the methoxide is crystalline. All are tetramers and the methoxide has a cuboidal structure. Vibrational studies indicate that there are weak Tl··Tl interactions in these molecules.

Thallium(I) forms compounds with transition metals, as in $TlCo(CO)_4$, that are mainly "salts" of carbonylate anions and tend to be ionic. Metal-metal bonds may be cleaved, for example,

$$[Cp(CO)_3Mo]_2 + 2Tl \longrightarrow 2TlMo(CO)_3Cp$$

Thallium(I) thiolates form polymers, often with complex structures. Thus TlSPh is ionic and best described as $[Tl_7(SPh)_6]^+[Tl_5(SPh)_6]^-$, while TlSBu^t is an octameric cage.[58] Numerous anionic polychalcogenide complexes are known, such as $[Tl_2Te_2]^{2-}$ and $[Tl_2(S_4)_2]^{2-}$ (6-XXV).[59] As expected for a "soft" metal, Tl—chalcogenide bonds are strong and highly covalent.

(6-XXV)

The addition of KOH and cyclopentadiene to an aqueous solution of Tl_2SO_4 gives the air-stable TlCp. The compound sublimes readily but forms infinite zig-zag chains in the solid state, ···Cp···Tl···Cp···Tl···, similar to InCp. TlCp is a useful reagent for transferring Cp ligands to other metals. By contrast, $(C_5R_5)Tl$ (R = CH_2Ph) is a dimer, with a weak Tl^I···Tl^I interaction.[60] The anion $[TlCp_2]^-$ has a bent metallocene structure, reminiscent of stannocene $SnCp_2$ with which it is isoelectronic.[61]

Tl^{II} compounds are hardly known. A well-authenticated example is the silyl derivative Tl_2R_4 [R = $Si(SiMe_3)_3$], which resembles the Al^{II} and Ga^{II} alkyls discussed above and has a Tl—Tl bond length of 2.91 Å. It is unstable in solution. The Tl^{2+} ion has been proposed to occur in reactions involving Tl^+ with one-electron oxidants and Tl^{3+} with reductants, and has been detected in flash photolysis where it has a half life of only 0.5 ms:

$$H_2O + Tl^{3+} \xrightarrow{\ h\nu\ } Tl^{2+} + OH^{\bullet} + H^+$$

Redox and disproportionation reactions have been studied and estimates made for the couples Tl^{3+}/Tl^{2+}, $E = 0.33$ V and Tl^{2+}/Tl^+, $E = 2.2$ V.

All thallium compounds are *exceedingly* poisonous.

[58]B. Krebs *et. al.*, *Angew. Chem. Int. Ed. Engl.* **1989,** *28,* 1682.
[59]S. S. Dhingra and M. G. Kanatzidis, *Inorg. Chem.* **1993,** *32,* 2298.
[60]H. Schumann *et al.*, *J. Organomet. Chem.* **1988,** *357,* 7.
[61]D. Stalke *et al.*, *Angew. Chem. Int. Ed. Engl.* **1993,** *32,* 1774.

Additional References

A. R. Barron, Oxide, Chalcogenide and Related Clusters of Aluminum, Gallium, and Indium. *Comments Inorg. Chem.* **1993,** *14,* 123.

A. J. Downs, Ed., *The Chemistry of Aluminum, Gallium, Indium, and Thallium,* Blackie, Glasgow, 1993.

G. H. Robinson, Ed., *Coordination Chemistry of Aluminum,* VCH Publishers, New York, 1993.

Chapter 7

CARBON

GENERAL REMARKS

There are more compounds of carbon than of any other element except hydrogen, and most of them are best regarded as organic chemicals.

The electronic structure of the C atom in its ground state is $1s^2 2s^2 2p^2$, with the two $2p$ electrons unpaired, following Hund's rule. To account for the normal four-covalence, we must consider that it is promoted to a valence state based on the configuration $2s2p_x2p_y2p_z$. The ion C^{4+} does not arise in any normal chemical process, but something approximating the C^{4-} ion may possibly exist in some carbides. In general, however, carbon forms covalent bonds.

In organic reactions there is abundant evidence for transient carbonium ions (R_3C^+), carbanions (R_3C^-), and carbenes $(:CR_2)$. Some stable carbonium ions like Ph_3C^+ and carbanions like $C(CN)_3^-$ can be isolated as well as radicals like $Ph_3C^{\cdot}$. In most of these cases the charge on the electron must be delocalized over the entire system for stability. Transition metal complexes with carbene or carbyne ligands, $L_nM{=}CR_2$ and $L_nM{\equiv}CR$, are discussed in Chapters 16 and 21.

Coordination Numbers

In virtually all its stable compounds carbon forms four bonds and has coordination numbers of 2 ($\equiv$C$-$ or $=$C$=$), 3 ($=$C$\lessdot$), or 4, with linear, triangular (planar), and tetrahedral geometries, respectively; CO has coordination number 1. In interstitial carbides (Section 7-3), certain metal cluster compounds[1] (Section 7-9), and very stable trigonal bipyramidal and octahedral penta- and hexa(aurio)methanium cations of the type $(LAu)_5C^+$ and $(LAu)_6C^{2+}$, where L is a phosphine,[2] carbon atoms are found with coordination numbers of 4, 5, or 6. Coordination number 5 is also found in compounds with bridging alkyls such as Al_2Me_6, in some carboranes (Section 5-12), and in reactive carbocations.[3]

[1]B. F. G. Johnson *et al.*, *J. Chem. Soc. Dalton Trans.* **1996,** 2395; M. Akita *et al.*, *Chem. Commun.* **1997,** 1557.

[2]H. Schmidbaur *et al.*, *Angew. Chem. Int. Ed. Engl.* **1990,** *29,* 1399; **1988,** *27,* 1544; *Interdisc. Sci. Rev.* **1992,** *17,* 213.

[3]G. K. Surya Prakash and P. v. R. Schleyer, *Stable Carbocation Chemistry,* John Wiley and Sons, Inc., New York, 1997; G. A. Olah, *Angew. Chem. Int. Ed. Engl.* **1995,** *34,* 1393.

Table 7-1 Some Bond Energies Involving Carbon, Silicon, and Sulfur

Bond	Energy (kJ mol^{-1})	Bond	Energy (kJ mol^{-1})
C$-$C	356	C$-$O	336
Si$-$Si	226	Si$-$O	368
S$-$S	226	S$-$O	$\sim$330

Catenation

The unusual stability of catenated carbon compounds, compared with those of silicon and sulfur, can be appreciated by considering the bond-energy data shown in Table 7-1. Thus the simple *thermal* stability of $-C_n-$ chains is high because of the intrinsic strength of C$-$C bonds. The relative stabilities toward oxidation follow from the fact that C$-$C and C$-$O bonds are of comparable stability, whereas for Si, and probably also for S, the bond to oxygen is considerably stronger. Thus, given the necessary activation energy, compounds with a number of Si$-$Si links are converted very exothermically into compounds with Si$-$O bonds.

THE ELEMENT

Carbon has the isotopic composition ^{12}C 98.89% and ^{13}C 1.11%. Only ^{13}C has nuclear spin (S = ½); with Fourier transform nmr, spectroscopic measurements can routinely be made using natural abundance.

The radioisotope ^{14}C (β^-, 5570 years), which is widely used as a tracer, is made by thermal neutron irradiation of Li or Al nitride, ^{14}N$(n,p)^{14}$C. It is available not only as CO_2 or carbonates, but also in numerous labeled organic compounds. Its formation in the atmosphere and absorption of CO_2 by living organisms provide the basis of radiocarbon dating.

7-1 Allotropy of Carbon: Diamond, Graphite, and Fullerenes

Diamond, graphite, and the fullerenes differ in their physical and chemical properties because of differences in the arrangement and bonding of the carbon atoms. Diamond is the densest (3.51 *vs* 2.22 and 1.72 g cm^{-3} for graphite and C_{60}, respectively), but graphite is more stable than diamond, by 2.9 kJ mol^{-1} at 300 K and 1 atm pressure; it is considerably more stable than the fullerenes (see later). From the densities it follows that to transform graphite into diamond, pressure must be applied, and from the thermodynamic properties of the two allotropes it can be estimated that they would be in equilibrium at 300 K under a pressure of $\sim$15,000 atm. Of course, equilibrium is attained extremely slowly at this temperature, and this property allows the diamond structure to persist under ordinary conditions.

The energy required to vaporize graphite to a monoatomic gas is an important quantity, since it enters into the estimation of the energies of all bonds involving carbon. It is not easy to measure directly because even at very high temperatures, the vapor contains appreciable fractions of C_2, C_3, and so on. However, the value is known to be 716.9 kJ mol^{-1} at 300 K; in using older tables of bond energies, the heat of sublimation of graphite may have been taken as $\sim$520 or 574 kJ mol^{-1}.

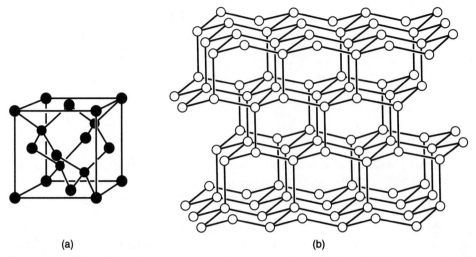

Figure 7-1 The diamond structure seen from two points of view. (a) The conventional cubic unit cell. (b) A view showing how two layers are stacked; these layers run perpendicular to the diagonals of the cube.

Diamond[4]

This form of carbon is almost invariably found with the cubic structure shown in Fig. 7-1. There is also a hexagonal form (lonsdaleite),[5] found in certain meteorites and also available synthetically, in which the puckered layers are stacked in an ABAB· · · pattern instead of the ABCABC· · · pattern. The hexagonal form is probably unstable toward the cubic, since unlike the cubic, it contains some eclipsed bonds.

Diamond is the hardest solid known. It has a high density and index of refraction, the highest mp (~4000°C), thermal conductivity (25 W/cm °C at room temperature, more than six times that of copper), and lowest molar entropy (2.4 J mol⁻¹K⁻¹) of any element.

Although graphite can be directly converted into diamond at temperatures of ~1400°C and pressures above 5 GPa, in order to obtain useful rates of conversion, an alloy containing transition metal catalysts such as Co, Fe, and Ni is used. It appears that a thin film of molten alloy forms on the graphite, dissolving some and reprecipitating it as diamond, which is less soluble; about 75% of the world's supply of industrial-quality diamonds are now synthetic. By accidental or deliberate introduction of trace impurities, colored stones can be produced—nitrogen causes a yellow coloration, and boron gives blue. It is also possible to form diamond at low and medium pressures by the transformation of simple carbon-containing materials.[6] The chemical reactivity of diamond is much lower than that of carbon in the form

[4]E. Wilks and J. Wilks, *Properties and Applications of Diamond,* Butterworth Heinemann, Oxford, 1991.
[5]R. F. Davis, Ed., *Diamond Films and Coatings: Development, Properties, and Applications,* Noyes Publications, New Jersey, 1993.
[6]A. P. Rudenko *et al., Russ. Chem. Rev. (Engl. Transl.)* **1993,** *62,* 87.

of macrocrystalline graphite or the various amorphous forms. Diamond can be made to burn in air by heating it to 600 to 800°C.

Graphite

Hexagonal graphite has a layer structure as shown in Fig. 7-2. The separation of the layers is 3.35 Å, which is about equal to the sum of van der Waals radii and indicates that the forces between layers should be relatively small. Thus the observed softness and particularly the lubricity of graphite can be attributed to the easy slippage of these layers over one another. It will be noted that within each layer each carbon atom is surrounded by only three others. After forming one σ bond with each neighbor, each carbon atom would still have one electron and these are paired up into a system of π bonds (7-I). Resonance with other structures having different but equivalent arrangements of the double bonds makes all C—C distances equal at 1.415 Å. This is a little longer than the C—C distance in benzene, where the bond order is 1.5, and agrees with the assumption that the bond order in graphite is ~1.33.

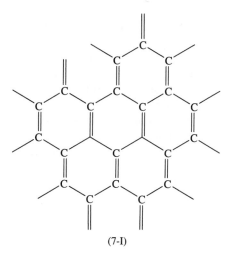

(7-I)

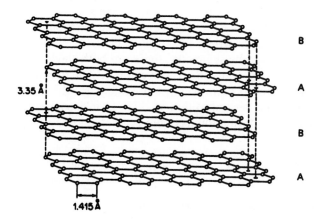

Figure 7-2 The normal structure of graphite.

Table 7-2 Selected Characteristics of Some Isolated Fullerenes

[n]fullerene	Symmetry of some known isomers	Color		Hydrocarbon solvent
		Thin film	Bulk	
60	$I_h{}^a$	Mustard yellow	Dark red to black	Magenta[b]
70	$D_{5h}{}^a$	Red	Black	Port-wine
76	$D_2{}^a$			Yellow-green
78	C_{2v}(I), C_{2v}(II), D_3			Golden chestnut brown
82	C_2, C_{2v}, C_{3v}, D_2, D_{2d}			Greenish-yellow
84[c]	D_2, D_{2d}			Yellow-green

[a]Only possible geometric isomer; $C_{76}(D_2)$ can have optical isomers.
[b]Brown in π-donor solvents due to charge-transfer interactions.
[c]One of the most abundant higher fullerenes.

Actually two modifications of graphite exist, differing in the order of the layers. In no case do all the carbon atoms of one layer lie directly over those in the next layer, but, in the structure shown in Fig. 7-2, carbon atoms in every other layer are superposed. This type of stacking, which may be designated (ABAB· · ·), is apparently the most stable and exists in the commonly occurring hexagonal form of graphite. There is also a rhombohedral form, frequently present in naturally occurring graphite, in which the stacking order is (ABCABC· · ·); that is, every third layer is superposed. It seems that local areas of rhombohedral structure can be formed by mechanical deformation of hexagonal crystals and can be removed by heat treatment.

Fullerenes[7]

These are a family of polyhedral carbon allotropes, C_{2n}; they have been isolated for n values in the range 30–48; larger giant fullerenes have been suggested as components of soot[8] and interstellar matter. Some characteristics are given in Table 7-2. The most common of these roughly spherical molecules, C_{60}, consists of arrays of 60 atoms with the geometry of a truncated icosahedron, having 20 hexagons and 12 pentagons, a geometry typical of a soccer ball, as shown in Fig. 7-3. While all of the carbon atoms of C_{60} are equivalent by symmetry, there are both single and double bonds with C—C distances of 1.453 and 1.383 Å, respectively.[9] The fullerene C_{60} is one of the most strained molecules known so far, but exhibits great kinetic stability. However, it shows measurable decomposition at ~750°C; it decomposes faster in the presence of O_2. The heat of formation has been estimated to be ~42.5

[7]W. Krätschmer, *Synth. Metals* **1995,** *70,* 1309; R. Taylor, ed., *The Chemistry of Fullerenes,* World Scientific, New Jersey, 1995; H. W. Kroto, *Angew. Chem. Int. Ed. Engl.* **1992,** *31,* 111.
[8]R. Taylor, *J. Chem. Soc. Chem. Commun.* **1994,** 1629.; A. Gügel *et al., J. Mater. Chem.* **1997,** *7,* 1327.
[9]See for example: M. M. Olmstead *et al., Inorg. Chem.* **1995,** *34,* 390.

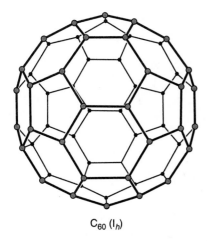

C_{60} (I_h)

Figure 7-3   The idealized structure of [60]fullerene; there are twelve pentagons and twenty hexagons.

kJ mol^{-1}, which makes it thermodynamically much less stable than either diamond or graphite; the heat of formation of C_{70} indicates that it is slightly more stable than C_{60}[10] (40.4 *vs* 42.5 kJ mol^{-1}, respectively).

In C_{70}, whose structure is shown in Fig. 7-4, there are five types of carbon atoms and thus eight distinct types of C—C bonds. Four of these constitute the common sides of hexagons, while the other four are formed when five- and six-membered rings fuse; the C—C distances vary from 1.39 to 1.54 Å.[11] Structures of other fullerenes are also shown in Fig. 7-4. It should be noted that many higher fullerenes are chiral, e.g., D_2-C_{76}, D_3-C_{78}, and D_2-C_{84}.[12]

The simplest dimer of a fullerene, C_{120} (7-II), can be synthesized in *ca.* 18% yield by vigorously vibrating a solid mixture of C_{60} and 20 molar equivalents of KCN powder;[13] it has a dumbbell shape with a nearly square C_4 ring connecting the cages.

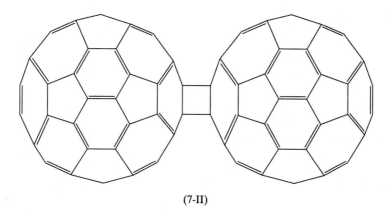

(7-II)

[10]C. Rüchardt, *Angew. Chem. Int. Ed. Engl.* **1994,** *33,* 996.
[11]K. Hedberg *et al., J. Am. Chem. Soc.* **1997,** *119,* 5314.
[12]See for example, F. Diederich *et al., Angew. Chem. Int. Ed. Engl.* **1997,** *36,* 2268.
[13]K. Komatsu *et al., Nature* **1997,** *387,* 583.

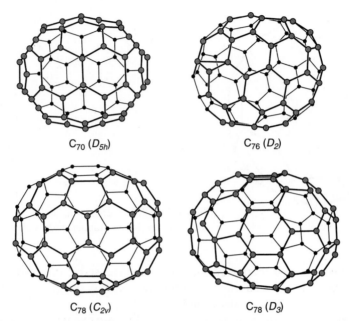

C$_{70}$ (D$_{5h}$)   C$_{76}$ (D$_{2}$)

C$_{78}$ (C$_{2v}$)   C$_{78}$ (D$_{3}$)

Figure 7-4 The idealized structures of some fullerenes, including two geometric isomers of C$_{78}$. In the larger fullerenes the hexagons are predominantly around the "waist" while the combination of hexagons-pentagons is prevalent in the more spherical surfaces.

Fullerene production is a simple process.[14] A large electric current is passed through graphite rods in a quenching atmosphere of inert gas (He), thus evaporating the rods to produce a light, fluffy condensate called "fullerene soot." Because of the relatively high solubility in common organic solvents such as benzene, methods for chromatographic separation and others[15] have been capable of isolating gram quantities of pure C$_{60}$ and C$_{70}$[16] and lesser amounts of C$_{76}$, C$_{78}$, C$_{80}$, C$_{82}$, C$_{84}$, C$_{96}$, and other fullerenes. The soluble fullerene fraction is composed of ~80% C$_{60}$, ~20% C$_{70}$, and ~1% higher fullerenes. Upon concentration of the solutions, crystals are formed which normally contain significant amounts of solvent and other guest molecules in the intersticies, e.g., C$_{70}$(S$_8$)$_6$ and C$_{76}$(S$_8$)$_6$.[17] Solvent-free crystals can be produced by sublimation under vacuum.

Amorphous carbons, carbon black, soot, charcoals, and so on, are forms of graphite or fullerenes. The physical properties depend on the nature and magnitude of the surface area. They show electrical conductivity, have high chemical reactivity due to oxygenated groups on the surface, and readily intercalate other molecules (see later). Graphite and amorphous carbons as supports for Pd, Pt, and other metals are widely used in catalysis and for the preparation of diamond films.[18]

[14]R. E. Smalley, *Acc. Chem. Res.* **1992**, *25*, 98; W. Krätschmer *et al.*, *Nature* **1990**, *347*, 354.
[15]E. C. Constable, *Angew. Chem. Int. Ed. Engl.* **1994**, *33*, 2269.
[16]J. M. Tour *et al.*, *J. Am. Chem. Soc.* **1994**, *116*, 6939.
[17]G. Roth *et al.*, *Angew. Chem. Int. Ed. Engl.* **1994**, *33*, 1651.
[18]I. J. Ford, *J. Appl. Phys.* **1995**, *78*, 510; Z. Feng and K. Komvopoulos, *J. Appl. Phys.* **1995**, *78*, 2720; Y. Nakamura *et al.*, *J. Mater. Sci.* **1992**, *27*, 6437.

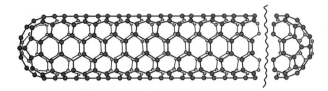

Figure 7-5 A schematic representation of a carbon nanotube showing a broken cap. Note that the pentagons are necessary only on the caps.

Strong graphite fibers are made by pyrolysis, at 1500°C or above, of oriented organic polymer fibers (e.g., those of polyacrylonitrile, polyacrylate esters, or cellulose). When incorporated into plastics the reinforced materials are light and very strong. Other forms of graphite such as foams, foils, or whiskers can also be made.

An interesting group of materials are the *carbon nanotubes*[19] prepared by arc-evaporation of graphite. They are needle-like cylindrical tubes of graphitic carbon capped by fullerene-like hemispheres, as shown schematically in Fig. 7-5. They can be opened by nitric acid;[20] the opened tubes can then be filled with various metal oxides.[21] The nanotubes are metallic, semiconducting, or insulating depending on the preparation method.[22] There are potential applications in many fields but none is yet operational.

Other Forms

Chaoite is a very rare carbon mineral. Other cubic and hexagonal forms, all rare and poorly understood—are also known. Small *polycarbon molecules*,[23] C_n, $n \sim 2$–10 have been described but they are not stable under normal conditions. The diatomic molecule, C_2, occurs in a variety of chemical reactions, and in a wide range of astrophysical objects. Energetic treatment of hydrocarbons, either by irradiation, by highly exothermic stripping reactions, or by exposure to high temperature, invariably produces C_2. Dicarbon appears in the atmospheres of carbon stars, the sun, comets, and diffuse interstellar clouds. In the ground state $X^1\Sigma_g^+$, the electronic configuration is $KK(\sigma_g 2s)^2(\sigma_u 2s)^2(\pi_u 2p)^4$. The triatomic molecule, C_3, has also been detected and is presumed to be linear. The C_4 to C_{10} molecules constitute a very minor fraction of the saturated vapor over graphite up to 3000 K and none of them has been detected in the gas phase except *via* mass spectrometry. Their calculated geometries are shown in Fig. 7-6.

Larger rings of acetylenic cyclo[*n*]carbons (7-III) have been reported to be formed in the gas from some stable precursors such as cyclobutenodedehydro[*n*] annulenes (*n* =18,24,30).[24] Although they have not been isolated as pure solids,

[19]B. I. Yakobson and R. E. Smalley, *American Scientist* **1997**, *85*, 324.
[20]M. L. H. Green *et al.*, *Nature* **1994**, *372*, 159.
[21]M. L. H. Green *et al.*, *Chem. Commun.* **1996**, 2489.
[22]See for example H. W. Kroto *et al.*, *Nature* **1997**, *388*, 52; P. L. McEuen *et al.*, *Science* **1997**, *275*, 1922.
[23]W. Weltner, Jr. and R. J. van Zee, *Chem. Rev.* **1989**, *89*, 1713.
[24]F. Diederich *et al.*, *J. Am. Chem. Soc.* **1990**, *112*, 1607.

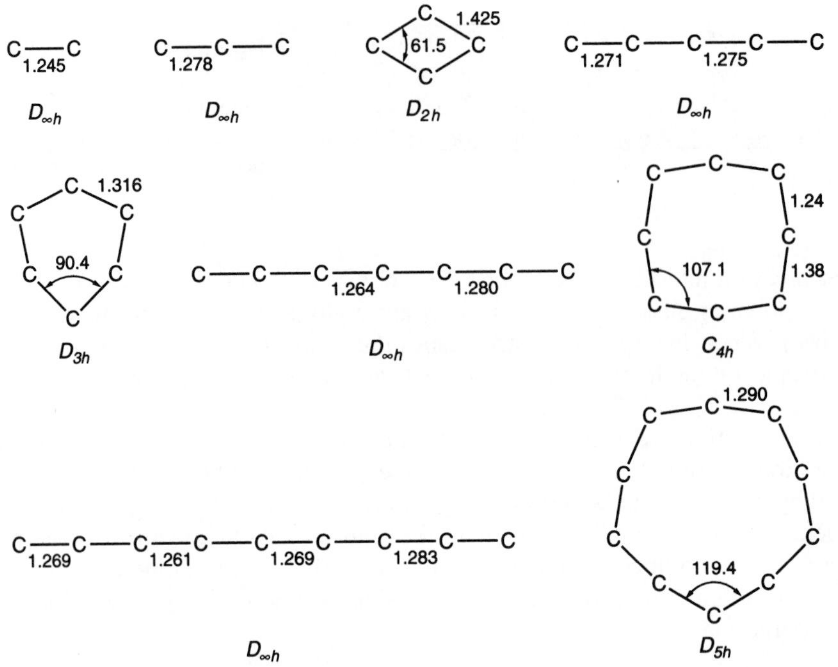

Figure 7-6 Calculated ground-state geometries of the C_n, $(n = 2\text{-}10)$ molecules. Bond lengths are shown in angstroms and bond angles in degrees. Note that there are also low-lying linear structures that are close in energy to the cyclic form in the case of C_4, C_6, and C_8.

complexes of C_{18} and C_{24} have been made, for example, by the oxidative cyclization:

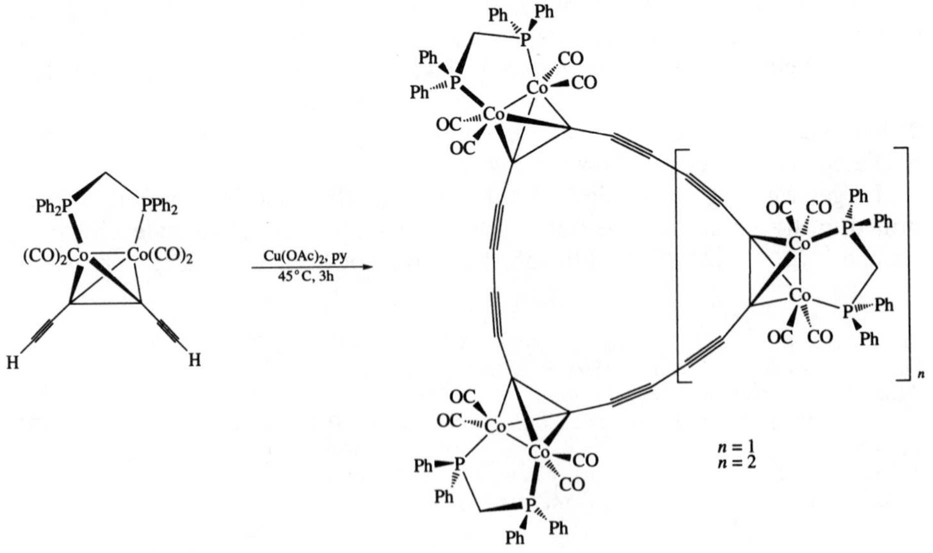

These are isolated as air-stable shiny black needles. The cyclo[18]carbon-containing complex has been characterized by X-ray crystallography.[25]

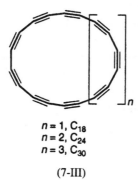

$$n = 1, C_{18}$$
$$n = 2, C_{24}$$
$$n = 3, C_{30}$$

(7-III)

7-2   Intercalation Compounds of Graphite

The layered structure of graphite allows molecules and ions to penetrate between the layers, forming interstitial or lamellar compounds. There are two basic types: those in which the graphite becomes nonconducting and those in which high electrical conductivity remains or is enhanced. Only two substances of the first type are known, namely, graphite oxide and graphite fluoride. *Graphite "oxide"* is obtained by treating graphite with strong aqueous oxidizing agents such as fuming nitric acid or $HNO_3 + KClO_3$. The idealized stoichiometry is C_8O_2OH and there are $-C=O$ and $-C-OH$ groups in the material, which is fairly acidic.

Graphite fluoride[26] is obtained by fluorination. At 2 atm pressure in liquid HF at 20°C a material C_xF, $5 > x > 2$ is first obtained, which then forms $CF_{12}{}^+HF_2{}^-$; electrolysis of alkali fluoride melts with C anodes also gives C_xF. At high temperatures, 400–600°C, fluorination gives white $(CF)_n$.

Electrically conducting intercalation compounds,[27] also called lamellar compounds, are formed by insertion of various atoms, molecules, or ions between the layers of graphite. The intercalates can be made by spontaneous reaction with alkali metals, or halides like $FeCl_3$, ReF_6, or AsF_5.[28] The latter possess high electrical conductivities. It has been proposed that the interaction of graphite with AsF_5, AsF_5/F_2, and O_2AsF_6 give materials of composition $C_{14}AsF_6$ in which the fluoride ligands of the $AsF_6{}^-$ anion are nestled in contiguous threefold sets of carbon atom hexagons of the graphite.

Electrochemical intercalation is cleaner and controllable:

$$C_n + M^+ + e \rightleftharpoons M^+ + C_n{}^-$$

$$C_n + X^- - e \rightleftharpoons C_n{}^+ + X^-$$

[25]F. Diederich *et al.*, *J. Am. Chem. Soc.* **1990**, *112*, 4966.
[26]N. Kumagal et al., *J. Appl. Electrochem.* **1995**, *25*, 869.
[27]P. P. Trzaskoma, *J. Electrochem. Soc.* **1993**, *140*, L103.
[28]F. Okino and N. Bartlett, *J. Chem. Soc. Dalton Trans.* **1993**, 2081.

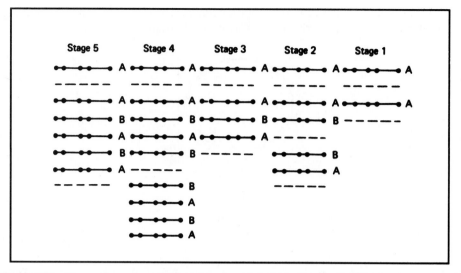

Figure 7-7   The various stages of graphite intercalation compounds. The letters A and B refer to the stacking pattern of carbon layers, and the two carbon layers flanking a guest layer are always equivalent, that is, have their carbon atoms superposed.

The guest ions penetrate during cathodic reduction or anodic oxidation of graphite. Solvent may also be incorporated into the layers as in $K(THF)C_{24}$ or $C_n(MeNO_2)_2$-PF_6. Among the earliest lamellar compounds were those of alkali metals of initial composition C_8M (Cs, Rb, K) but C_6Li; also known are $C_{24}M$ and $C_{60}M$. All these materials ignite in air and react explosively with water. They are widely used in organic and inorganic syntheses[29] and have high electrical conductivity.

Generally, for any given guest species, there is a whole series of stoichiometric compositions obtainable, each corresponding to a "stage." This term refers to the frequency with which the layers of the graphite host are invaded, as shown in Fig. 7-7. For a stage n compound every nth layer contains guest species, so that the highest concentration of guest occurs in the stage 1 compound.

NONMOLECULAR COMPOUNDS

7-3   Carbides

Only compounds in which carbon is combined with elements of similar or lower electronegativity are called carbides (see also Section 7-9).

Preparative methods for carbides of all types include: (*a*) direct union of the elements at high temperature (2200°C and above); (*b*) heating a compound of the metal, particularly the oxide, with carbon; and (*c*) heating the metal in the vapor of a suitable hydrocarbon. Carbides of Cu, Ag, Au, Zn, and Cd, also commonly called *acetylides*, are prepared by passing acetylene into solutions of the metal salts; with Cu, Ag, and Au, ammoniacal solutions of salts of the unipositive ions are used

---

[29]V. V. Aksënov *et al.*, *Uspekhi Khimi* **1990**, *59*, 1267; A. Fürstner, *Angew. Chem. Int. Ed. Engl.* **1993**, *32*, 64.

to obtain Cu_2C_2, Ag_2C_2, and Au_2C_2 (uncertain), whereas for Zn and Cd the acetylides ZnC_2 and CdC_2 are obtained by passing acetylene into alkane solutions of dialkyl compounds. The Cu and Ag acetylides are explosive, being sensitive to both heat and mechanical shock.

1. *Saltlike Carbides.*[30] The most electropositive metals form carbides having physical and chemical properties indicating that they are essentially ionic. The colorless crystals are hydrolyzed by water or dilute acids at ordinary temperatures, and hydrocarbons corresponding to the anions $C^{4-}(CH_4)$, C_2^{2-} (C_2H_2) and C_3^{4-} (C_3H_4) are formed.

Examples of *methides* are Be_2C and Al_4C_3; the former having an antifluorite structure. Aluminum carbide, Al_4C_3, hydrolyzes according to

$$Al_4C_3 + 12H_2O \longrightarrow 4Al(OH)_3 + 3CH_4$$

Other methides are the cubic ternary carbides of the type A_3MC[31] where A is mostly a rare earth or transition element (e.g., Sc, Y, La—Na, Gd—Lu) and M is a metallic or semimetallic main group element (e.g., Al, Ge, In, Tl, Sn, Pb). These perovskite carbides are typically hydrolyzed by dilute HCl to give ~84-97 (wt-%) methane and 3-16% saturated and unsaturated higher hydrocarbons.

Besides those mentioned above, there are many carbides that contain C_2^{2-} ions, or anions that can be so written to a first approximation. For the M_2^I compounds, where M^I may be one of the alkali metals or one of the coinage metals, and for the $M^{II}C_2$ compounds, where M^{II} may be an alkaline earth metal, and for $M_2^{III}(C_2)_3$ compounds in which M^{III} is Al, La, Pr, or Tb, this description is probably a very good approximation. In these, the postulation of C_2^{2-} ions requires that the metal ions be in their normal oxidation states. In those instances where accurate structural parameters are known, the C—C distances lie in the range 1.19 to 1.24 Å. The compounds react with water and the C_2^{2-} ions are hydrolyzed to give acetylene only, for example,

$$Ca^{2+} + C_2^{2-} + 2H_2O \longrightarrow HCCH + Ca(OH)_2$$

Most of the MC_2 acetylides have the CaC_2 structure commonly described as derived from the NaCl structure with the $[C—C]^{2-}$ ions lying lengthwise in the same direction along one of the cell axes, thus causing a distortion from cubic symmetry to tetragonal symmetry with one axis longer than the other two. However, the fine details of the validity of this description have recently come into question.[32] There is some doubt as to whether the C_2 units are parallel to the *c*-axis since there is spectral broadening in the ^{13}C nmr spectra; thus it is possible that the C_2 units are only statistically aligned along the axis.

[30]V. A. Gubanov *et al.*, *Electronic Structure of Refractory Carbides and Nitrides,* Cambridge University Press, Cambridge, 1994.
[31]W. Jeitschko *et al.*, *Z. Naturforsch.* **1997,** *52b,* 176.
[32]See for example R. Hoffmann *et al.*, *Inorg. Chem.* **1992,** *31,* 1734 and references therein.

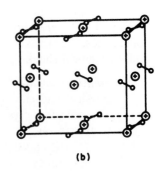

Figure 7-8 The structures of (a) CaC_2 (the C_2 units are possibly only statistically aligned along the *c*-axis) and (b) ThC_2 (the latter somewhat simplified).

(a)   **(b)**

In thorium carbide the C_2^{2-} ions are lying flat in parallel planes in such a way that two axes are equally lengthened with respect to the third. These structures are shown in Fig. 7-8. Lithium carbide, Li_2C_2, (Section 3-2) has a structure similar to that of CaC_2.

There are, however, a number of carbides that have structures similar to those discussed previously, meaning that the carbon atoms occur in discrete pairs, but they cannot be satisfactorily described as C_2^{2-} compounds[33]; these include YC_2, TbC_2, YbC_2, LuC_2, UC_2, Ce_2C_3, Pr_2C_3, and Tb_2C_3. For these MC_2 compounds, neutron-scattering experiments show that (*a*) the metal atoms are essentially trivalent and (*b*) the C—C distances are 1.28 to 1.30 Å for the lanthanide compounds and 1.34 Å for UC_2. These facts and other details of the structures are consistent with the view that the metal atoms lose not only the electrons necessary to produce C_2^{2-} ions (which would make them M^{2+} ions), but also a third electron, mainly to the antibonding orbitals of the C_2^{2-} groups, thus lengthening the C—C bonds (*cf.* C—C = 1.19 Å in CaC_2). There are actually other, more delocalized, interactions among the cations and anions in these compounds, since they have metallic properties. The M_2C_3 compounds have the metals in their trivalent states, C—C distances of 1.24 to 1.28 Å, and also direct metal-metal interactions. These carbides, which cannot be represented simply as aggregates of C_2^{2-} ions and metal atoms in their normal oxidation states, are hydrolyzed by water to give only 50 to 70% of HCCH, while C_2H_4, CH_4, and H_2 are also produced. There is no detailed understanding of these hydrolytic processes.

An interesting *sesquicarbide*, Mg_2C_3, can be prepared relatively pure by reaction of Mg dust with *n*-pentane at ~680°C. In the solid state, it contains C_3^{4-} units; these species are linear, isoelectronic with CO_2, and have C—C distances of 1.332(2) Å.[34] Other known sesquicarbides are Sc_3C_4 (which contains a portion of C_3 groups next to C_2 and C_1 (units) and a relatively stable lithium derivative, Li_4C_3. The hydrolysis of this type of carbide proceeds according to:

$$C_3^{4-} + 4HCl \xrightarrow{-4Cl^-} \underset{H}{\overset{H}{\diagdown}} C = C = C \underset{H}{\overset{H}{\diagup}}$$

[33]J. K. Gibson, *J. Phys. Chem.* **1998,** *102,* 4501.
[34]H. Fjellvåg and P. Karen, *Inorg. Chem.* **1992,** *31,* 3260.

Finally, numerous ternary metal carbides containing Ln metals, transition metals, and discrete C_2 units are also well established.[35] In many such carbides the C—C distance are in the range 1.32 to 1.47 Å and have been described as double bonds. Thus, they are formally derivatives of the ethylenic tetraanion C_2^{4-}, e.g., $Ln_3M(C_2)_2$ (M = Fe, Co, Ni, Ru, Rh, Os, and Ir). The C_2 units can bind to the transition metal atoms in a $1,1-\mu_2-C_2^{4-}$ (7-IVa), $1,2-\mu_2-C_2^{4-}$ (7-IVb), $1,2-\eta^2-\mu_3-C_2^{4-}$ (7-IVc), or $1,1,2-\mu_3-C_2^{4-}$ (7-IVd) manner. Other ternary carbides such as $Dy_{12}Mn_5C_{15}$ and $Ln_{3.67}FeC_6$ have M—C_2 units (7-IVe) which can be interpreted as the deprotonation product of vinylidene groups, e.g., M=C=CH$_2$. Such ligands are isoelectronic and isostructural with terminal CO groups. In $Ln_3Mn(C_2)_3$ (Ln = Gd and Tb), the vinylidene ligands, C_2^{2-}, are bound as in (7-IVf). In the carbide $ScCrC_2$, the C—C separation of 1.60 Å is suggestive of a single bond, thus formally containing the C_2^{6-} anion; the core is shown in (7-IVg).

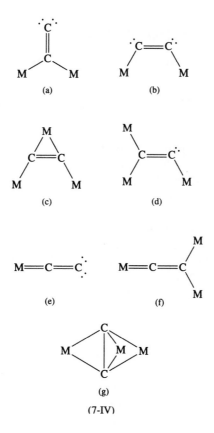

(7-IV)

2. *Interstitial Carbides.*[30] These have characteristically very high melting points, great hardness, and metallic conductivity.

3. *Covalent Carbides.* Although other carbides (e.g., Be$_2$C) are at least partially covalent, the two elements that approach carbon closely in size and electro-

[35]R. B. King, *J. Organomet. Chem.* **1997**, *536-537*, 7.

negativity, namely, silicon and boron, give completely covalent compounds. Silicon carbide[36] (SiC), known as carborundum, is an extremely hard, infusible, and chemically stable material made by reducing SiO_2 with carbon in an electric furnace. It has three structural modifications, in all of which there are infinite three-dimensional arrays of Si and C atoms, each tetrahedrally surrounded by four of the other kind.

Boron carbide (B_4C) is also an extremely hard, infusible, and inert substance, made by reduction of B_2O_3 with carbon in an electric furnace at 2500°C, and has a very unusual structure. The C atoms occur in linear chains of 3, and the boron atoms in icosahedral groups of 12 (as in crystalline boron itself). These two units are then packed together in a sodium chloride-like array. There are, of course, covalent bonds between C and B atoms as well as between B atoms in different icosahedra. A graphite-like boron carbide (BC_3) has been made by interaction of benzene and BCl_3 at 800°C.

SIMPLE MOLECULAR COMPOUNDS

Some of the more important inorganic carbon compounds and their properties are listed in Table 7-3.

7-4   Carbon Halides

Carbon tetrafluoride is an extraordinarily stable compound. It is the end product in the fluorination of any carbon-containing compound. Laboratory preparations involve the fluorination of silicon carbide at high temperature. The SiF_4 also formed is removed easily by passing the mixture through a 20% NaOH solution. The CF_4 is unaffected, whereas the SiF_4 is immediately hydrolyzed; the difference exists because in CF_4 carbon is coordinately saturated, whereas silicon in SiF_4 has $3d$ orbitals available for coordination of OH^- ions in the first step of the hydrolysis reaction. A more convenient one-step and high-yield synthesis involves the room-temperature reaction of graphite-free calcium cyanamide (Section 7-7) and F_2 in the presence of trace amounts of CsF^{37} to suppress the formation of NF_3:

$$CaNCN + 3F_2 \xrightarrow[\text{12h}]{25\,°C} CaF_2 + N_2 + CF_4$$

Carbon tetrachloride is a common solvent; it is fairly readily decomposed photochemically[38] and also quite often readily transfers chlorine to various substrates, CCl_3 radicals often being formed simultaneously at high temperatures (300–500°C). It is often used to convert oxides into chlorides. Although it is thermodynamically unstable with respect to hydrolysis, the absence of acceptor orbitals on carbon makes the attack very difficult.

Carbon tetrabromide, a pale yellow solid at room temperature, is insoluble in water and other polar solvents but soluble in some nonpolar solvents such as

[36]See for example: *Diam. Relat. Mater.* **1997**, *6,* 1243-1571 (an issue dedicated to SiC characterization, crystal structure, defects, applications, bulk and epithaxial growth, and other related materials).
[37]T. M. Klapötke *et al.*, *Inorg. Chem.* **1992**, *31*, 3864.
[38]X. L. Zhou and J. P. Cowin, *J. Phys. Chem.* **1996**, *100,* 1055.

Table 7-3 Some Simple Compounds of Carbon

Compound	mp (°C)	bp (°C)	Remarks
CH_4	−183	−161	Odorless and flammable
CF_4	−184	−130	Very stable
CCl_4	−23	77	Moderately stable; non-flammable; high liver toxicity
CBr_4	88–90	190	Decomposes slightly on boiling
CI_4	168	130–140 (1–2 torr)	Decomposes before boiling; can be sublimed under low pressure; red crystals
COF_2	−114	−83	Easily decomposed by H_2O but stable otherwise
$COCl_2$	−118	8	"Phosgene"; highly toxic; used in the manufacture of polyurethane plastics
$COBr_2$		65	Fumes in air; $COBr_2 + H_2O \rightarrow CO_2 + 2HBr$
$CO(NH_2)_2$	132		Isomerized by heat to NH_4NCO; used as a fertilizer
CO	−205	−191	Odorless and toxic
CO_2	−57 (5.2 atm)	−79	Odorless; non-combustible; sublimes at −78.5°C (1 atm)
C_3O_2	−111	6.8	Evil-smelling gas; easily polymerizable
COS	−138	−50	Flammable; slowly decomposed by H_2O
CS_2	−109	46	Flammable and toxic
$(CN)_2$	−28	−21	Very toxic; colorless; water soluble
HCN	−13.4	25.6	Very toxic; high dielectric constant (116 at 20°C) for the associated liquid; weak acid

benzene; it can form clathrates with several solvents such as hexa(phenylthio)benzene.[39]

Carbon tetraiodide is a bright red, crystalline material, possessing an odor like that of iodine. Both heat and light cause decomposition to iodine and tetraiodoethylene. The tetraiodide is insoluble in water and alcohol, though attacked by both at elevated temperatures, and soluble in benzene and chloroform from which it can be recrystallized. It may be prepared by the reaction:

$$CCl_4 + 4C_2H_5I \xrightarrow{AlCl_3} CI_4 + 4C_2H_5Cl$$

The increasing instability, both thermal and photochemical, of the carbon tetrahalides with increasing weight of the halogen correlate with a steady decrease in the $C-X$ bond energies:

$$C-F \ 485 \qquad C-Cl \ 327 \qquad C-Br \ 285 \qquad C-I \ 213 \qquad kJ \ mol^{-1}$$

Many of the mixed halides $CX_{4-n}X'_n$ are also known.[40] There are also derivatives with three halide atoms, e.g., $CClF_2I$ or $CBrF_2I$. However, the prototype of the chiral compounds, $CBrClFI$, has not been prepared yet but $CHBrClF$ has.

[39] H. Kiefte, *J. Phys. Chem.* **1993,** 97, 12949.
[40] A. H. Gouliaev and A. Senning in *Comprehensive Organic Functional Group Transformations,* T. L. Gilchrist, ed., Vol. 6, Chapter 6.07, Pergamon, 1995.

The *carbenoid* LiCHCl$_2$·3py[41] has been prepared at temperatures below $-78°C$ according to:

$$CH_2Cl_2 \xrightarrow[\text{py}]{\text{Bu}^n\text{Li}} LiCHCl_2 \cdot 3py$$

The crystals are stable only at low temperature; a crystallographic analysis revealed that the geometry at both the Li and C atoms is tetrahedral.

The *carbonyl halides* (X$_2$CO: X = F, Cl, Br), or mixed (e.g., ClBrCO) are hydrolytically unstable compounds. Urea, (NH$_2$)$_2$CO, is related but more stable. They all have longer CO bonds than do simple ketones because of partial double bonding of X to C, which weakens the carbon-to-oxygen π bonding. The fluoride F$_2$CO is a useful reagent for displacing either hydrogen by fluorine from P—H, N—H, and C—H bonds or oxygen by fluorine from oxides such as those of B, Si, Ge, Sn, P, Se, Te, N, Cl, I, V, Nb, Ta, Cr, Mo, W, and U.[42]

7-5 Carbon Oxides

There are four stable oxides of carbon: CO, CO$_2$, C$_3$O$_2$, and C$_{12}$O$_9$. The last is the anhydride of mellitic acid (7-V) and not discussed, nor are unstable oxides such as CO$_3$, C$_2$O, C$_2$O$_3$, C$_3$O, C$_4$O, and C$_6$O.[43]

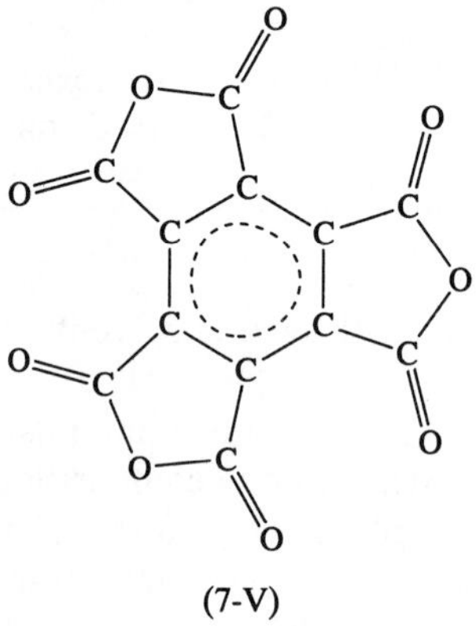

(7-V)

Carbon monoxide is formed when carbon is burned with a deficiency of oxygen. The following equilibrium exists at all temperatures, but is not rapidly attained at ordinary temperatures:

$$2CO(g) \rightleftharpoons C(s) + CO_2(g)$$

Carbon monoxide is made industrially on a huge scale, together with hydrogen as "synthesis gas," and is used for a variety of large-scale organic syntheses (Section 22-5). Although CO is an exceedingly weak Lewis base, one of its most important properties is the ability to act as a donor ligand toward transition metals. Nickel

[41]G. Boche, *Angew. Chem. Int. Ed. Engl.* **1996,** *35,* 1518.
[42]C. J. Schack and K. O. Christe, *Inorg. Chem.* **1988,** *27,* 4771.
[43]R. J. Van Zee *et al., J. Am. Chem. Soc.* **1988,** *110,* 609.

metal reacts with CO to form $Ni(CO)_4$ and Fe reacts under more forcing conditions to give $Fe(CO)_5$. Metal carbonyls are discussed in detail in Section 16-3.

Carbon monoxide is very toxic, rapidly giving a bright red complex with the hemoglobin of blood. Carbon monoxide reacts with alkali metals in liquid ammonia to give the alkali metal *carbonyls;* these white solids contain the $[OCCO]^{2-}$ ion.

Carbon dioxide,[44] an important green house gas,[45] is obtained in combustion of carbon and hydrocarbons, calcination of $CaCO_3$, and so on. It forms complexes with transition metals (Section 7-14) and inserts into MH and other bonds (Section 21-3). The gas is very soluble in ethanolamines, which are used to scrub CO_2 from gas streams. Liquid CO_2 at pressures up to 400 bar is a solvent for some organic compounds and is used to extract caffeine from coffee beans; many studies of other applications of supercritical CO_2 have been conducted.[46]

Numerous chemical, electrochemical, and photochemical processes have been used to reduce CO_2, the most abundant C_1 feedstock on Earth, to CO or to organic compounds using transition metal complexes as catalysts, but no large-scale production method has been developed yet.

One of the most studied reactions is the hydrogenation of CO_2 to produce formic acid.[47]

$$CO_2(g) + H_2(g) \longrightarrow HCO_2H(l) \qquad \Delta G^0 = 32.9 \text{ kJ/mol};$$

$$\Delta H^0 = -31.2 \text{ kJ/mol}; \Delta S^0 = -215 \text{ J/(mol K)}$$

This reaction is catalyzed by many complexes including metal hydrides as shown schematically:

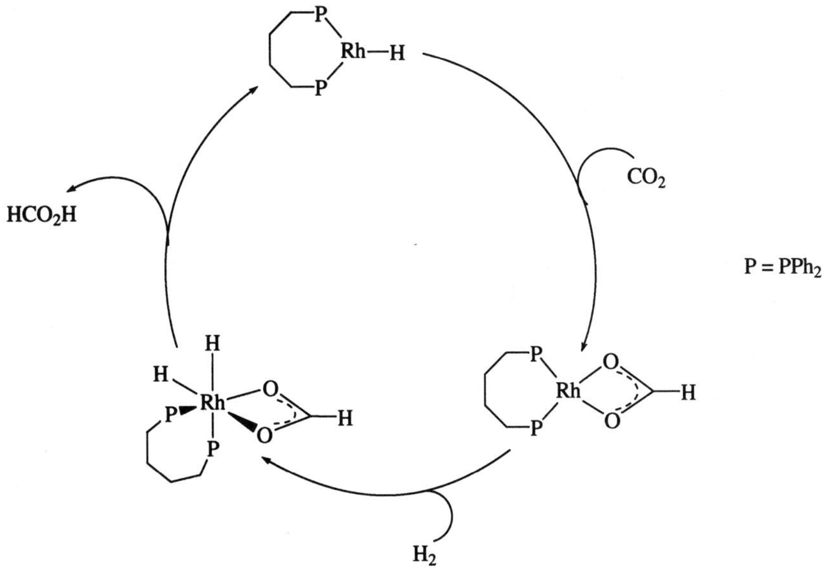

[44]D. H. Gibson, *Chem. Rev.* **1996,** *96,* 2063.
[45]U. Siegenthaler and J. L. Sarmiento, *Nature* **1993,** *365,* 119.
[46]See for example: S. M. Howdle *et al., J. Am. Chem. Soc.* **1997,** *119,* 6399 and references therein.
[47]W. Leitner, *Angew. Chem. Int. Ed. Engl.* **1995,** *34,* 2207; R. Noyori *et al., Chem. Rev.* **1995,** *95,* 259.

Contrary to the efficient manner in which green plants consume CO_2 in the photosynthetic process, direct photochemical reduction is difficult. However, CO_2 can be reduced to its radical anion, $CO_2^{\bar{}}$ in the presence of $[Ni_3(\mu_3\text{-}I)_2(dppm)_3]$.[48]

Oxygen-atom transfer is rare for CO_2, but a metathetical exchange in the presence of Sn or Ge diamides readily produces isocyanates and carbodiimides:[49]

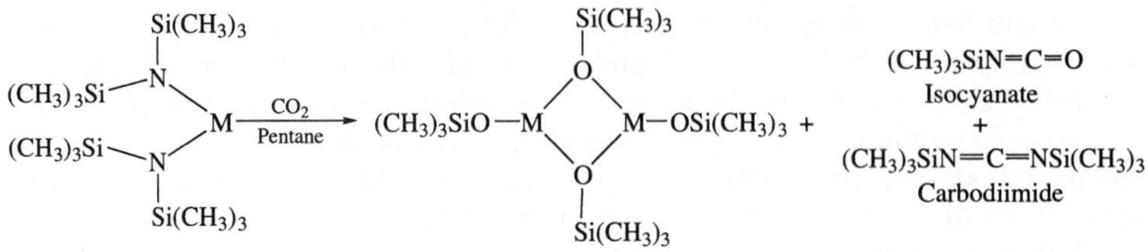

Carbon dioxide has also been reduced to methanol, formate, oxalate, methane, and carbon monoxide.

Carbon suboxide (C_3O_2),[50] an evil-smelling gas, is formed by dehydrating malonic acid with P_2O_5 in vacuum at 140 to 150°C, or better, by thermolysis of diacetyltartaric anhydride. The molecule is linear and can be represented by the structural formula $O{=}C{=}C{=}C{=}O$. It is stable at -78°C, but at 25°C it polymerizes, forming yellow to violet products. Photolysis of C_3O_2 gives C_2O, which will react with olefins:

$$C_2O + CH_2{=}CH_2 \longrightarrow CH_2CCH_2 + CO$$

It reacts but slowly with water[51] to give malonic acid, of which it is the anhydride, but more rapidly with stronger nucleophiles:

$$C_3O_2 + 2H_2O \longrightarrow HO_2CCH_2CO_2H$$

$$C_3O_2 + 2NHR_2 \longrightarrow R_2NCOCH_2CONR_2$$

Carbonic Acid

Though CO is formally the anhydride of formic acid, its solubility in water and bases is slight. It will give formates when heated with alkalis, however. As just noted, C_3O_2 gives malonic acid. However, CO_2 is by far the most important carbonic acid anhydride, combining with water to give *carbonic acid,* for which the following equilibrium constants are conventionally written:

$$\frac{[H^+][HCO_3^-]}{[H_2CO_3]} = 4.16 \times 10^{-7}$$

$$\frac{[H^+][CO_3^{2-}]}{[HCO_3^-]} = 4.84 \times 10^{-11}$$

[48]C. P. Kubiak *et al., J. Am. Chem. Soc.* **1993,** *115,* 6470.
[49]L. R. Sita *et al., J. Am. Chem. Soc.* **1996,** *118,* 10912.
[50]Z. Pei and S. D. Worley, *J. Phys. Chem.* **1994,** *98,* 5135.
[51]T. T. Tidwell *et al., J. Chem. Soc., Chem. Commun.* **1995,** 2547.

The equilibrium quotient in the first equation is not really correct. It assumes that all CO_2 dissolved and undissociated is present as H_2CO_3, which is not true. In fact, the greater part of the dissolved CO_2 is only loosely hydrated, so that the correct first dissociation constant, using the "true" activity of H_2CO_3, has a value of about 2×10^{-4}, which is more nearly in agreement with expectation for an acid with the structure $(HO)_2CO$.

The rate at which CO_2 comes into equilibrium with H_2CO_3 and its dissociation products when passed into water is measurably slow, and this indeed is what has made possible an analytical distinction between H_2CO_3 and the loosely hydrated CO_2 (aq). This slowness is of great importance physiologically and in biological (e.g., carbonic anhydrase catalyzed hydration), analytical, and industrial chemistry.

The neutralization of CO_2 occurs by two paths. For example, for pH < 8 the principal mechanism is direct hydration of CO_2:

$$CO_2 + H_2O = H_2CO_3 \qquad \text{(slow)}$$

$$H_2CO_3 + OH^- = HCO_3^- + H_2O \qquad \text{(instantaneous)}$$

An etherate of H_2CO_3 is obtained by interaction of HCl with Na_2CO_3 at low temperature in dimethyl ether. The resultant white crystalline solid (mp $-47°C$), which decomposes at about 5°C, is probably $OC(OH)_2 \cdot O(CH_3)_2$.

Thermal decomposition of NH_4HCO_3 gives H_2CO_3 in the gas phase.[52]

7-6 Oxocarbon Anions[53]

Oxo carbon anions that have aromatic character are derived from the hydroxy acids (7-VIa-d). They are made relatively easily. Electrochemical reduction of CO, which proceeds *via* radical ions gives $C_4O_4^{2-}$; interaction of K and CO at 170°C gives $K_6C_6O_6$. Cyclization reactions of acetylenic ethers, for example, $Bu^tOC \equiv COBu^t$ by $Co_2(CO)_8$, and treatment of the resulting butoxides with CF_3CO_2H also gives the anions.

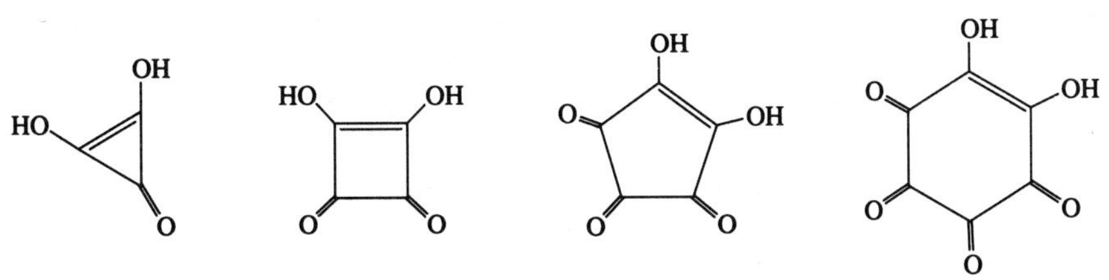

(a)   (b)   (c)   (d)

(7-VI)

[52]H. Schwarz, *Angew. Chem. Int. Ed. Engl.* **1987**, *26*, 354.
[53]R. Gleiter *et al.*, *J. Org. Chem.* **1995**, *60*, 5878.

7-7 Compounds with C—N Bonds; Cyanides and Related Compounds

An important area of "inorganic" carbon chemistry is that of compounds with C—N bonds. The most important species are the cyanide, cyanate, and thiocyanate ions and their derivatives. We can regard many of these compounds as being pseudohalogens or pseudohalides, but the analogies, although reasonably apt for cyanogen, $(CN)_2$, are not especially valid in other cases.

1. *Cyanogen.* There are three known isomers of composition C_2N_2:

$$N{\equiv}C{-}C{\equiv}N \qquad C{=}N{-}C{\equiv}N \qquad C{=}N{-}N{=}C$$
$$\mathbf{1} \qquad\qquad\quad \mathbf{2} \qquad\qquad\quad \mathbf{3}$$

Isomer **2**, isocyanogen,[54] and isomer **3**, diisocyanogen,[55] have been detected by nmr and other spectroscopies; isocyanogen is extremely unstable and polymerizes above $-80°C$. Isomer **1**, cyanogen, is a flammable gas (Table 7-3) which is stable even though it is unusually endothermic ($\Delta Hf^0_{298} = 297$ kJ mol^{-1}). It can be prepared by oxidation of HCN using (*a*) O_2 with a silver catalyst, (*b*) Cl_2 over activated carbon or silica, or (*c*) NO_2 over calcium oxide-glass; the last reaction allows the NO produced to be recycled:

$$2HCN + NO_2 \longrightarrow (CN)_2 + NO + H_2O$$

Cyanogen can also be obtained from the cyanide ion by aqueous oxidation using Cu^{2+} (*cf.* the $Cu^{2+}{-}I^-$ reaction):

$$Cu^{2+} + 2CN^- \longrightarrow CuCN + \tfrac{1}{2}(CN)_2$$

or acidified peroxodisulfate. A better procedure for dry $(CN)_2$ employs the reaction

$$Hg(CN)_2 + HgCl_2 \longrightarrow Hg_2Cl_2 + (CN)_2$$

The cyanogen molecule, $N{\equiv}C{-}C{\equiv}N$, is linear. It dissociates into CN radicals, and, like RX and X_2 compounds, it can oxidatively add to lower-valent metal atoms giving dicyano complexes, for example,

$$(Ph_3P)_4Pd + (CN)_2 \longrightarrow (Ph_3P)_2Pd(CN)_2 + 2PPh_3$$

A further resemblance to the halogens is the disproportionation in basic solution:

$$(CN)_2 + 2OH^- \longrightarrow CN^- + OCN^- + H_2O$$

[54]F. Bickelhaupt *et al.*, *J. Am. Chem. Soc.* **1991**, *113*, 6104.
[55]F. Stroh *et al.*, *Chem. Phys. Letts.* **1989**, *160*, 105.

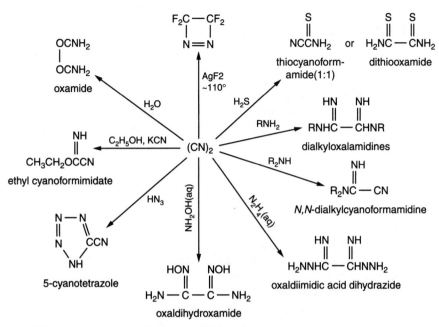

Figure 7-9 Some reactions of cyanogen. Other products may also be obtained by fluorination (e.g., $CF_3N=NCF_3$, NCCFNCl).

Thermodynamically this reaction can occur in acid solution, but it is rapid only in base. Cyanogen has a large number of reactions, some of which are shown in Fig. 7-9. A stoichiometric mixture of O_2 and $(CN)_2$ burns, producing one of the hottest flames (~5050 K) known from a chemical reaction.

   Impure $(CN)_2$ can polymerize on heating to give a polymer, paracyanogen (7-VIIa), which will depolymerize above ~850°C. Interaction of KCN and ICN gives a linear polycyanogen (7-VIIb).

(a)

(b)

(7-VII)

Isocyanogen reacts with HCl and Br_2 to give $NCN=CHCl$ and $NCN=CBr_2$, respectively.[56]

2. *Hydrogen Cyanide.* Like the hydrogen halides, HCN is a covalent, molecular substance, but is a weak acid in aqueous solution ($pK = 9.0$).

[56]F. Bickelhaupt *et al.*, *J. Org. Chem.* **1993**, *58*, 6930.

Proton transfer studies, however, show that as with normal protonic acids, direct proton transfer to base B

$$B: + HCN \longrightarrow BH^+ + CN^-$$

occurs without participation of water.

The colorless gas is extremely toxic (though much less so than H_2S); it is formed on addition of strong acids to cyanides and on a large scale industrially by the reaction:

$$CH_4 + NH_3 \xrightarrow[\text{Pt}]{1200\,°C} HCN + 3H_2 \qquad \Delta H = -247 \text{ kJ mol}^{-1}$$

Hydrogen cyanide condenses at 25.6°C to a liquid with a very high dielectric constant (107 at 25°C). Here, as in similar cases, such as water, the high dielectric constant is due to association of intrinsically very polar molecules by hydrogen bonding. Liquid HCN is unstable and can polymerize violently in the absence of stabilizers: in aqueous solutions polymerization is induced by ultraviolet light.

Hydrogen cyanide is thought to have been one of the small molecules in the Earth's primeval atmosphere and to have been an important source or intermediate in the formation of biologically important chemicals. Among the many polymerized products of HCN are the trimer, aminomalononitrile, $HC(NH_2)(CN)_2$, the tetramer, diaminomalononitrile, and polymers of high molecular weight. Furthermore, under pressure with traces of water and ammonia, HCN pentamerizes to adenine, and HCN can also act as a condensing agent for amino acids to give polypeptides.

Industrial uses of HCN are for synthesis of methyl methacrylate and to form adiponitrile (for adipic acid and nylon) by addition to 1,3-butadiene in the presence of nickel(0) phosphite complexes. Waste HCN is also oxidatively hydrolyzed to give oxamide for use as fertilizer.

3. *Cyanides.* Sodium cyanide is made by absorbing gaseous HCN in NaOH or Na_2CO_3 solution. It used to be made by the reaction of molten sodium with ammonia first to give $NaNH_2$, which reacts with carbon to give sodium cyanamide Na_2NCN and finally NaCN according to the stoichiometry

$$NaNH_2 + C \Longleftrightarrow NaCN + H_2$$

In crystalline alkali cyanides at normal temperatures, the CN^- ion is rotationally disordered and is thus effectively spherical with a radius of 1.92 Å. Hence NaCN has the NaCl structure.

The main use of NaCN is in the extraction of gold and silver from their ores by the formation of cyano complexes (Section 7-12). The ions Ag^+, Hg_2^{2+}, and Pb^{2+} give insoluble cyanides.

Calcium cyanamide (CaNCN) is made in an impure form, largely for fertilizer use, by the reaction

$$CaC + N_2 \xrightarrow{\sim 1000\,°C} CaNCN + C \qquad \Delta H = -297 \text{ kJ mol}^{-1}$$

The cyanamide ion is linear and is isostructural and isoelectronic with CO_2.

Cyanamide (H_2NCN), a very irritating and caustic crystalline solid (mp 45°C), is prepared by hydrolysis of CaNCN:

$$CaNCN + H_2O + CO_2 \rightleftharpoons CaCO_3 + H_2NCN$$

In alkaline solution at 80°C cyanamide dimerizes to dicyandiamide,

$$2H_2NCN \rightleftharpoons (H_2N)_2C=NCN$$

and this in turn may be converted to *melamine* (7-VIII), the cyclic trimer of cyanamide, by heating in NH_3. Melamine is more easily made from urea,

$$6NH_2CONH_2 \xrightarrow{\text{100 atm, 300°C}} C_3N_3(NH_2)_3 + 6NH_3 + 3CO_2$$

and the CO_2 and NH_3 formed can be recycled to give urea. Melamine is used for polymers and plastics.

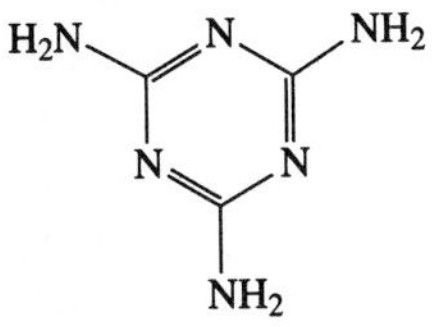

(7-VIII)

4. *Cyanogen Halides.* The most important compound is *cyanogen chloride* (bp 13°C), which is obtained by the action of Cl_2 on HCN, by electrolysis of aqueous solutions of HCN and NH_4Cl, and in other ways. It may be polymerized thermally to *cyanuric chloride,* which has the cyclic triazine structure (7-IX) similar to that of melamine. The chlorine atoms in $C_3N_3Cl_3$ are labile and there is an extensive organic chemistry of triazines, since these compounds are widely used in herbicides and dye stuffs. Compounds of the type $(XCN)_3$, X = F or Cl, react with stoichiometric amounts of F_2/AsF_5 in $CFCl_3$ to give $[X_3C_3N_3F][AsF_6]$.[57]

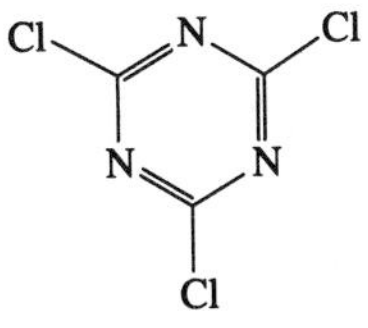

(7-IX)

Fluorination of $C_3N_3Cl_3$ gives $C_3N_3F_3$, which can be cracked to give FCN. Although this is stable as a gas (bp −46°C), it polymerizes at 25°C. Cyanogen bromide is similar to ClCN; the vapors are highly irritant and poisonous. Cyanogen iodide is made by treating $Hg(CN)_2$ with I_2. A linear cation of composition $[FCNF]^+$ is also known.[58]

[57]P. v. R. Schleyer *et al., Inorg. Chem.* **1993,** *32,* 1523.
[58]T. M. Klapötke, *Angew. Chem. Int. Ed. Engl.* **1991,** *30,* 1485.

The cyanogen halides generally behave like other halogenoids.

Compounds between CN and other halogenoid radicals are known, such as NCN_3 formed by the reaction

$$BrCN + NaN_3 \rightleftharpoons NaBr + NCN_3$$

5. *Cyanate and its Analogous S, Se, and Te Ions.* The linear ion OCN^- is obtained by mild oxidation of aqueous CN^-, for example:

$$PbO(s) + KCN(aq) \longrightarrow Pb(s) + KOCN(aq)$$

The free acid, $K = 1.2 \times 10^{-4}$, decomposes in solution to NH_3, H_2O, and CO_2. Compounds of the type $(OCN)_2$ have been observed as transient species. Stable covalent compounds such as $P(NCO)_3$ and some metal complexes are known. The compounds are usually prepared from halides by interaction with $AgNCO$ in benzene or NH_4OCN in acetonitrile or liquid SO_2. In such compounds or complexes, either the O or N atoms of OCN can be bound to other atoms[59] and this possibility exists also for SCN. In general, most nonmetallic elements seem to be N-bonded.

Thiocyanates are obtained by fusing alkali cyanides with sulfur; the reaction of S with KCN is rapid and quantitative, and S in benzene or acetone can be titrated with KCN in 2-propanol with bromthymol blue as indicator. Thiocyanate is the product of detoxification of CN^- in living systems.

Thiocyanogen is obtained by MnO_2 oxidation

$$2SCN^- \rightleftharpoons (SCN)_2 + 2e \quad E^0 = -0.77 \text{ V}$$

but since it is rapidly decomposed by water, it is best made by action of Br_2 on AgSCN in an inert solvent. In the free state $(SCN)_2$ rapidly and irreversibly polymerizes to brick red polythiocyanogen, but it is most stable in CCl_4 or CH_3CO_2H solution, where it exists as NCSSCN.

6. *Dicyanopolyynes.*[60] These contain linear chains of "naked" carbon atoms with alternating single and triple bonds; the ends are capped with $C\equiv N$ groups as in 7-X.

$$NC \equiv\!\!\!\!\!=\!\!\!\!\!\equiv\!\!\!\!\!=\!\!\!\!\!\equiv\!\!\!\!\!\Big]_n\!\!\!\!\!=\!\!\!\!\!\equiv CN$$

(7-X)

The synthesis of the dicyanopolyynes can be achieved by vaporization of graphite in the presence of cyanogen. The smaller chains (8-18 C atoms) can be separated by extraction with toluene and then chromatography. The materials with longer chains (500 or more C atoms) are also soluble in most organic solvents.

The CF_3-capped analogues are also known.

---

[59]S. D. Worley *et al.*, *J. Phys. Chem.* **1989**, *93*, 4598.
[60]R. J. Lagow *et al.*, *Science* **1995**, *267*, 362; A. Hirsch *et al.*, *Chem. Eur. J.* **1997**, *3*, 1105.

7. *Nitrosyl Cyanide and Its Isomers.* Various isomers of elemental composition CN_2O are known. The most stable is nitrosyl cyanide (**A**) prepared according to:

$$NOCl + AgCN \xrightarrow[\text{-30°C}]{\text{CHCl}_3/\text{EtOH}} ONCN + AgCl$$

The blue-green gas is stable in glass vessels for several hours. Upon irradiation ($\lambda = 193$ nm) it forms isonitrosylcyanide (**B**). However, at wavelengths greater than 570 nm, another isomer also appears, namely, nitrosyl isocyanide (**C**); they can be trapped in argon materials at 15 K and interconvert photochemically[61] according to:

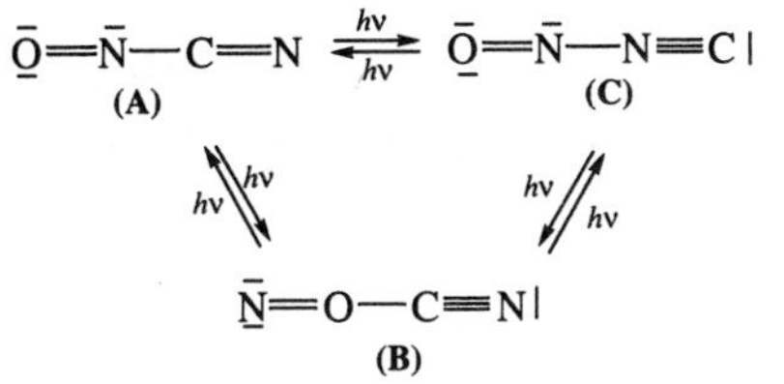

$$\bar{O}{=}\bar{N}{-}C{\equiv}N \underset{h\nu}{\overset{h\nu}{\rightleftharpoons}} \bar{O}{=}\bar{N}{-}N{\equiv}C\,|$$
$$\quad\quad\text{(A)}\quad\quad\quad\quad\quad\text{(C)}$$

$$h\nu \updownarrow h\nu \quad\quad\quad h\nu \mathbin{/\!/} h\nu$$

$$\bar{N}{=}O{-}C{\equiv}N\,|$$
$$\text{(B)}$$

Isonitrosyl isocyanide, NONC, has not been detected yet.

7-8 Compounds with C—S Bonds

Binary Carbon Sulfides[62]

Several compounds with the formula C_xS_y are known (Fig. 7-10). Most of them are molecular compounds, but a few polymeric materials such as $(CS_2)_n$, $(CS)_n$, and $(C_3S_5)_n$ have also been described. Polycarbon sulfide displays high electrical conductivity.

These compounds are prepared by a variety of methods, some of which are described by the following reactions:

$$2C_3S_5H_2 \xrightarrow{\Delta} C_6S_8 + 2H_2S$$

$$C_2S_6 \xrightarrow{2\text{LiBHEt}_3} \xrightarrow{\text{CSCl}_2} C_5S_7$$

$$[Zn(C_3S_5)_2]^{2-} + 2S_2Cl_2 \longrightarrow C_3S_8 + {}^{1}\!/{}_{2}C_6S_{12} + ZnCl_4^{2-}$$

$$[Zn(C_3S_5)_2]^{2-} \xrightarrow[\substack{-2SO_2 \\ -ZnCl_4^{2-}}]{+2SO_2Cl_2} [\beta\text{-}C_3S_5]_n$$

Carbon Disulfide

This compound is produced by the direct interaction of C and S at high temperatures. A similar yellow liquid CSe_2 is made by the action of CH_2Cl_2 on molten selenium;

[61]G. Maier *et al.*, *Angew. Chem. Int. Ed. Engl.* **1997**, *36*, 1707.
[62]T. B. Rauchfuss *et al.*, *Inorg. Chem.* **1994**, *33*, 4537; **1993**, *32*, 5467; *J. Am. Chem. Soc.* **1989**, *111*, 3463.

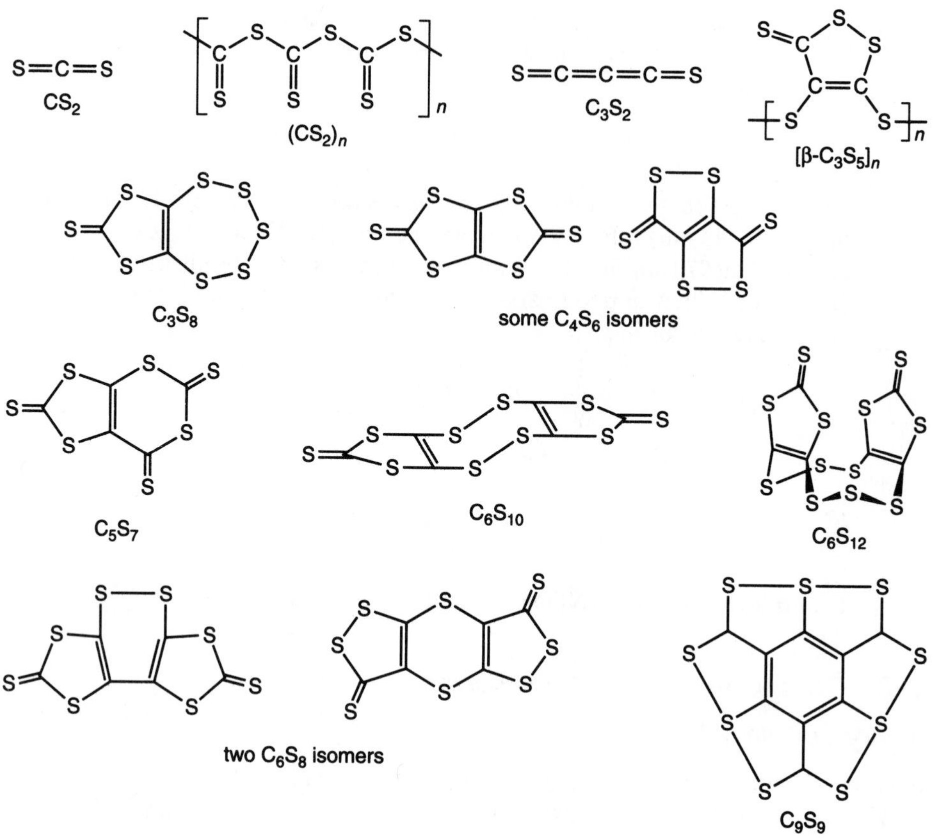

Figure 7-10   Some C_xS_y compounds.

it has a worse smell than CS_2 but, unlike it, is nonflammable. The selenide slowly polymerizes spontaneously, but CS_2 does so only under high pressures, to give a black solid.

In addition to its high flammability in air, CS_2 is a very reactive molecule and has an extensive chemistry. It is used to prepare carbon tetrachloride industrially:

$$CS_2 + 3Cl_2 \longrightarrow CCl_4 + S_2Cl_2$$

Important reactions of CS_2 involve nucleophilic attacks on carbon by the ions SH^- and OR^- and by primary or secondary amines, which lead, respectively, to thiocarbonates, xanthates, and dithiocarbamates, for example,

$$SCS + :SH^- \longrightarrow S_2CSH^- \xrightarrow{OH^-} CS_3^{2-}$$

$$SCS + :OCH_3^- \longrightarrow CH_3OCS_2^-$$

$$SCS + :NHR_2 \xrightarrow{OH^-} R_2NCS_2^-$$

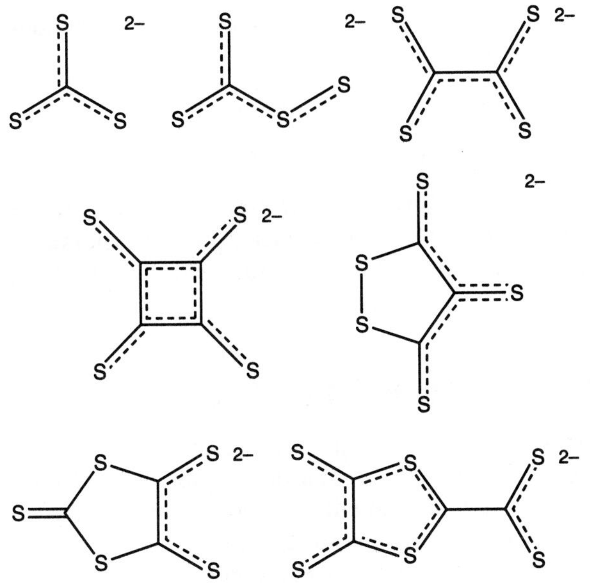

Figure 7-11 Some thiocarbon anions.

The molecule readily enters into insertion reactions, for example,

$$M(NR_2)_4 + 4CS_2 \longrightarrow M(S_2CNR_2)_4$$

The monosulfide, COS, is generally similar in its reactions.

Thiocarbonates

Thiocarbonates are readily formed by the action of SH^- on CS_2 in alkaline solution, and numerous yellow salts containing the planar ion are known. Heating CS_3^{2-} with S affords orange tetrathiocarbonates, which have the structure $[S_3C—S—S]^{2-}$. The free acids can be obtained from both these ions as red oils, stable at low temperatures. Other thiocarbon ligands are shown in Fig. 7-11. They have an important place in coordination chemistry, where they are found bound to many metal atoms in species such as $Ni(C_3S_5)_2^-$, $CpNi(C_3S_5)$,[63] and $(p\text{-cymene})_2Ru_2Cl_2(C_4S_6)$.[64]

Dithiocarbamates; Thiuram Disulfides

Dithiocarbamates are normally prepared as alkali metal salts by the action of primary or secondary amines on CS_2 in the presence of, say, NaOH. The zinc, manganese, and iron dithiocarbamates are extensively used as agricultural fungicides, and zinc salts as accelerators in the vulcanization of rubber. Alkali metal dithiocarbamates are usually hydrated and are dissociated in aqueous solution. When anhydrous, they are soluble in organic solvents in which they are associated.

[63]C. Faulmann et al., *J. Chem. Soc., Dalton Trans.* **1996**, 2261.
[64]T. B. Rauchfuss et al., *Angew. Chem. Int. Ed. Engl.* **1995**, *34*, 1890.

On oxidation of aqueous solutions by H_2O_2, Cl_2, or $S_2O_8^{2-}$ thiuram disulfides, of which the tetramethyl is the commonest, are obtained:

$$Cl_2 + 2Me_2NCS_2^- \longrightarrow Me_2NC\underset{\underset{S}{\|}}{-}S-S-\underset{\underset{S}{\|}}{C}NMe_2 + 2Cl^-$$

Thiuram disulfides, which are strong oxidants, are also used as polymerization initiators (for, when heated, they give radicals) and as vulcanization accelerators.

Tetraethylthiuram disulfide is "Antabuse," the agent for rendering the body allergic to ethanol.

CARBON AND CARBON COMPOUNDS AS LIGANDS

The most important complexes involving carbon-containing ligands in which carbon is bound to the metal are discussed separately; those of carbon monoxide in Chapter 16, organometallic compounds of transition elements, and organometallic compounds of the main group elements in the corresponding chapters.

In this chapter we discuss carbon and other C-bonded ligands.

7-9   Carbon

Except for the fullerenes (see later), carbon atoms seldom coordinate a metal atom; a well-characterized example of a naked carbon atom is a molybdenum carbide[65] prepared, in a multistep procedure, by removing a terminal proton from a corresponding methylidyne species:

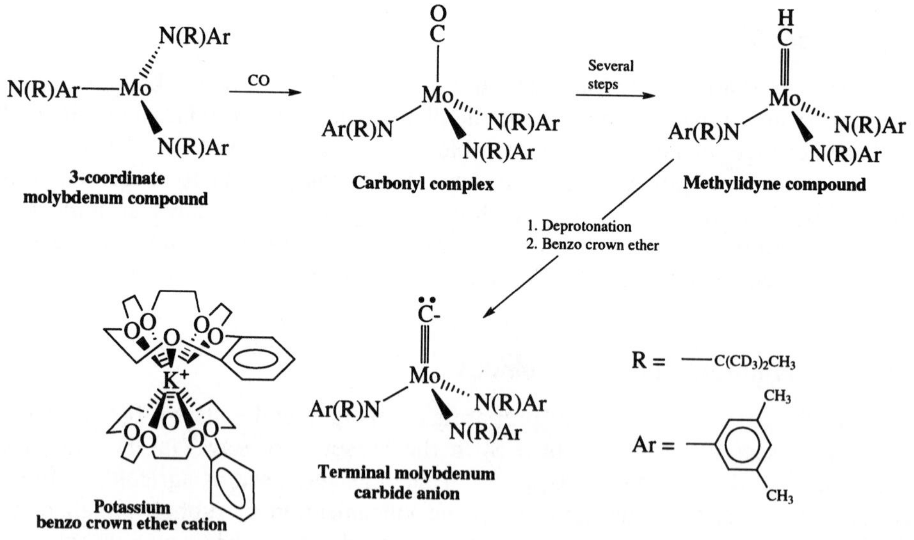

The carbon atom has a lone pair of electrons and is attached to the molybdenum atom by a triple bond. The carbanion has an electronic structure similar to that of the equivalent nitride compound containing a molybdenum-nitrogen triple bond.

[65]C. C. Cummins *et al.*, *Chem. Commun.* **1997**, 1995.

There are also a number of complexes with M—C—M' bridges. For example, reduction of the tetraphenylporphyrin (tpp)FeCl by $NaBH_4$ in CH_2Cl_2 gives (tpp)Fe, which in turn, with CI_4, gives $[(tpp)Fe]_2C$ as dark purple crystals. Mössbauer spectra indicate that this has Fe^{IV} and thus (tpp)Fe=C=Fe(tpp).

The 2-coordinate carbon atom may be part of a 1,3-dimetallallene system (7-XIa) or part of a C-metalated carbyne complex (7-XIb).

$$(tpp)Fe{=}C{=}ML_n \qquad L_nM{\equiv}C{-}M'L_n$$
$$\text{(a)} \qquad\qquad \text{(b)}$$

$$(7\text{-XI})$$

The ML_n group in 7-XIa can be $Re_2(CO)_9$, $Mn_2(CO)_9$, $Cr(CO)_5$, or $Fe(CO)_4$. An example of the second type is $HB(pz)_3(CO)_2 Mo{\equiv}C{-}FeCp(CO)_2$,[66] made by reacting $HB(pz)_3(CO)_2Mo{\equiv}C{-}Cl$ and $KFe(CO)_2Cp$. Among the *carbide-containing clusters*, interstitial carbon can be found in clusters such as $Gd_{10}Cl_{18}C_4$, but the majority are cluster compounds derived from π-acid ligands,[67] especially carbon monoxide (see also Section 7-3). The carbide is frequently derived by cleavage of the C—O bond. Examples of reactions where both the C and O atoms are trapped in the same molecule are those of metal—metal bonded ditungsten or tetratungsten alkoxides:[68]

$$\left.\begin{array}{l} W_2(\mu\text{-CO})(OR)_6 + W_2(OR)_6 \\[2em] W_4(OR)_{12} + CO \end{array}\right\} \longrightarrow W_4(\mu_4\text{-C})(O)(OR)_{12}$$

The skeleton of $W_4(\mu_4\text{-C})(O)(OH_2Bu^t)_{12}$ is given in 7-XII. The carbon atom is frequently found as a discrete C^{4-} unit encapsulated in scores of clusters of metals such as iron, ruthenium, osmium, cobalt, rhodium, nickel, and rhenium but a few contain encapsulated C_2 units (see later).

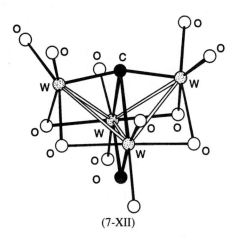

$$(7\text{-XII})$$

[66]J. L. Templeton *et al.*, *J. Am. Chem. Soc.* **1991**, *113*, 2324.
[67]Y. Chi *et al.*, *J. Am. Chem. Soc.* **1996**, *118*, 3289 and references therein; B. F. G. Johnson *et al.*, *J. Chem. Soc. Dalton Trans.* **1996**, 2395; **1997**, 3251.
[68]M. H. Chisholm *et al.*, *J. Am. Chem. Soc.* **1992**, *114*, 7056.

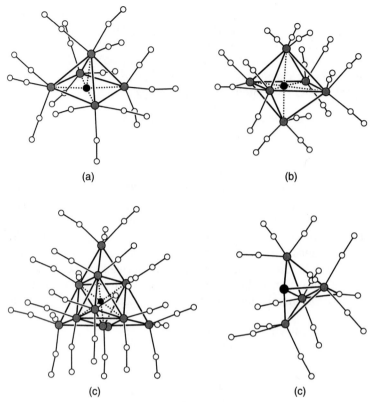

(a)   (b)

(c)   (c)

Figure 7-12 The structures of some representative carbido carbonyl cluster species. (a) $Fe_5C(CO)_{15}$, (b) $Ru_6C(CO)_{17}$, (c) $[Os_{10}C(CO)_{24}]^{2-}$, and (d) $Fe_4C(CO)_{13}$.

The earliest carbido cluster to be recognized was $Fe_5C(CO)_{15}$, shown in Fig. 7-12(a); the Ru and Os analogues with similar structures are also known. Octahedral species with a carbon atom in the center were next to be recognized, the first having been $Ru_6C(CO)_{17}$, Fig. 7-12(b), and a number of larger ones are now well characterized, for example, $[Os_{10}C(CO)_{24}]^{2-}$, Fig. 7-12(c). The truly encapsulated carbon atoms are relatively unreactive and it is the smaller cluster, $Fe_4C(CO)_{13}$, Fig. 7-12(d), that has shown the greatest chemical activity, as summarized in Fig. 7-13. The exposed carbon atom in the Fe_4C system can be thought of as a possible model for a surface carbon atom in heterogeneous catalytic processes.

The synthetic chemistry of carbido metal carbonyl clusters is not very systematic. Many of them are obtained by pyrolysis of precursors or by refluxing precursors for long periods in high boiling solvents. Among the earliest organometallic complexes known to contain the C_2 unit are anions of the type $[Co_3Ni_7(CO)_xC_2]^{n-1}$ with (x,n) combinations of (16,2), (16,3), and (15,3). These contain scrambled sets of 10 metal atoms arranged in a distorted form of cubic close packing and C—C distances of ~1.48 Å. An interesting tantalum complex is formed by reaction of a Ta^{III} alkoxide with CO:[69]

$$2Ta(OSiBu^t_3)_3 + CO \xrightarrow[\text{benzene}]{25\,°C} (Bu^t_3SiO)_3Ta=O + \frac{1}{2}[(Bu^t_3SiO)_3Ta]_2(\mu\text{-}C_2)$$

[69]P. T. Wolczanski *et al.*, *J. Am. Chem. Soc.* **1989,** *111,* 9056.

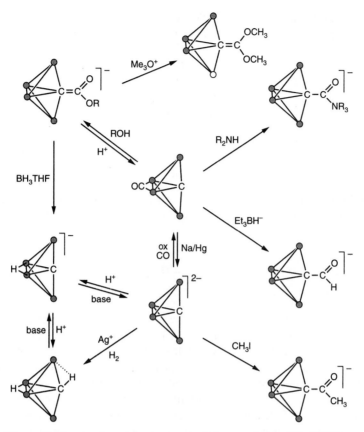

Figure 7-13 A schematic summary of the reactions of Fe$_4$C(CO)$_{13}$.

The latter can be hydrolyzed to produce ethene:

$$[(\text{Bu}^t{}_3\text{SiO})_3\text{Ta}]_2(\mu\text{-C}_2) \xrightarrow[\text{THF, }110^\circ\text{C, }15\text{h}]{\text{H}_2\text{O (excess)}} \text{C}_2\text{H}_4 + 6\text{Bu}^t{}_3\text{SiOH} + \text{Ta}_2\text{O}_5 \cdot n\text{H}_2\text{O}$$

Thus it can formally be considered as a dicarbene (7-XIII(a)). However, structural and spectroscopic data suggest that the C$_2$ units in [(Bu$^t_3$O)W](μ-C$_2$) and [PtCl(PR$_3$)$_2$](μ-C$_2$)[70] are best described as dicarbyne (7-XIII(b)) and yndiyl (7-XIII(c)) groups, respectively.

$$\begin{array}{ccc} \text{M}=\text{C}=\text{C}=\text{M} & \text{M}\equiv\text{C}-\text{C}\equiv\text{M} & \text{M}-\text{C}\equiv\text{C}-\text{M} \\ \textbf{(a)} & \textbf{(b)} & \textbf{(c)} \end{array}$$

(7-XIII)

The C$_2$ unit has also been found to bind in various other ways, including a fully encapsulated manner, as in (7-XIV), which represents the core of the [Co$_6$Ni$_2$C$_2$(CO)$_{16}$]$^{2-}$ anion. Other clusters, such as Ru$_5$(μ_5-C$_2$)(μ-SMe)$_2$ (μ-PPh$_2$)$_2$(CO)$_{12}$, have also been prepared.[71] Theoretical studies of such clusters[72] indicate that the C$_2$ unit interacts strongly with the metal *via* electron donation from

[70]S. Takahashi *et al.*, *Organometallics* **1988,** *11,* 2257.
[71]C. J. Adams *et al.*, *J. Chem. Soc. Dalton Trans.* **1997,** 2937.
[72]J.-F. Halet *et al.*, *Organometallics* **1997,** *16,* 2590.

the $C_2\sigma(p)-$ and filled π-bonding MOs into vacant metallic MOs, supplemented by back-donation from filled metallic MOs to the $C_2\pi^*$ orbitals.

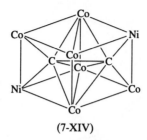

(7-XIV)

A cyclopropenium complex[73] consisting of a nearly equilaterial C_3 ring with an iron center bonded to each vertex can be prepared as follows:

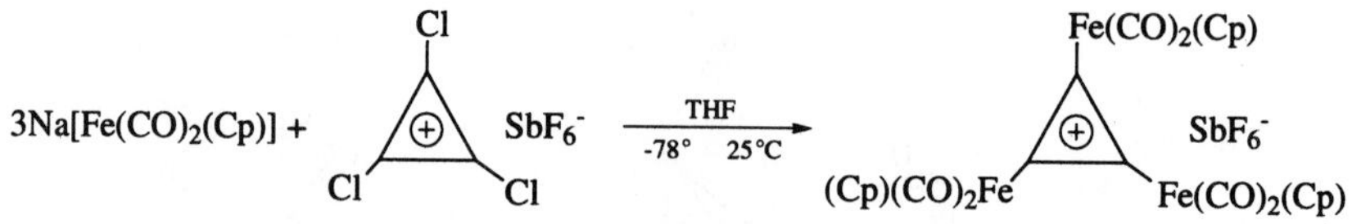

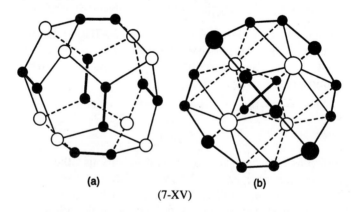

Other metal complexes containing all carbon ligands are illustrated in Fig. 7-14.[74]

Other carbide compounds which have recently received a lot of attention are the so-called *metallocarbohedrenes*[75] or *Met-Cars*. These are robust clusters of the general formula M_8C_{12} (M = Ti, V, Zr, Hf, Cr, Fe, Mo, etc.) as well as binary metal combinations such as $Ti_xM_yC_{12}$ ($x + y = 8$).[76] These clusters are produced in laser-induced plasma reactors or by similar high energy processes, but they have not been isolated so far. It has been proposed that they contain metal-carbon bonds distributed in a symmetrical, cage-like structure with the geometry of a pentagonal dodecahedron, a polygon with twelve pentagonal faces (7-XV(a)) and overall point symmetry T_h. However, recent spectroscopic[77] and theoretical studies suggest that lower symmetry structures are perhaps more stable, e.g., the tetracapped tetrahedral T_d structure (7-XV(b)).

(a)   **(b)**

(7-XV)

[73]M. S. Morton and J. P. Selegue, *J. Am. Chem. Soc.* **1995**, *117*, 7005.
[74]M. I. Bruce *et al.*, *Chem. Commun.* **1996**, 2405.
[75]A. W. Castleman, Jr., *Science* **1992**, *255*, 1411; M. T. Bowers, *Acc. Chem. Res.* **1994**, *27*, 324.
[76]A. W. Castleman, Jr., *J. Am. Chem. Soc.* **1994**, *116*, 5295; **1993**, *115*, 7415.
[77]See for example, L.-S. Wang *et al.*, *J. Am. Chem. Soc.* **1997**, *119*, 7417.

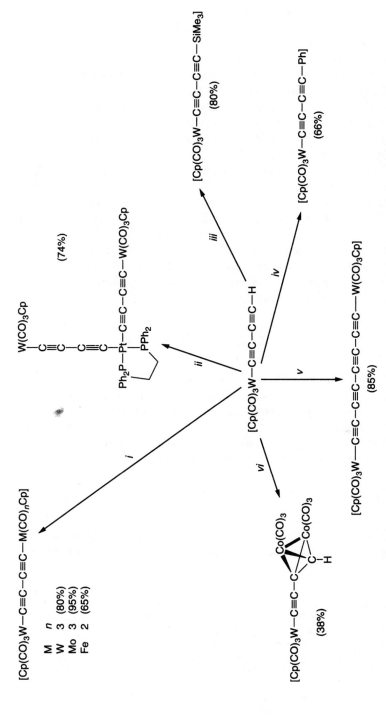

Figure 7-14 Preparation of some metal complexes containing all-carbon ligands. Reagents: *i*, CuI, NHEt₂, [ML_n]Cl; *ii*, CuI, NHEt₂, [PtCl₂(dppe)]; *iii*, LDA, SiClMe₃; *iv*, C₆H₅I, [Pd(PPh₃)₄], CuI; *v*, [Cu(tmeda)]Cl, O₂; *vi*, [Co₂(CO)₈]. Yields are given in parentheses.

241

7-10   Fullerenes

Fullerenes have a very rich chemistry; they react with both organic and inorganic compounds. Some of the simpler compounds known are the so-called *endohedral fullerenes* which contain encapsulated atoms;[78] examples are Ln@C_{2n} (Ln = Sc, Y, La, Ce, Nd, Sm, Eu, Gd; n = 37-48). They are formulated as $Ln^{3+}@C_{2n}^{3-}$. Larger fullerenes can encapsulate two metal atoms; examples of such dimetallofullerenes are $Ln_2@C_{80}$ (Ln = Sc, La), $Sc_2@C_{82}$, and $Sc_2@C_{84}$. Both ^{13}C and ^{139}La nmr studies of $La_2@C_{80}$ show a singlet at 290 K, giving an indication that the metal atoms are delocalized in the cage,[79] but the signal broadening as temperature changes suggests that the two La atoms can stop in certain positions. Smaller fullerenes can encapsulate certain gases, e.g., M@C_{60} and M@C_{70}, M = He, Ne, Ar, Kr, or Xe.[80] There are also a series of alkali metal doped fullerenes. In solution, the highest C_{60}^{n-} fullerides known are n = 6 on the electrochemical timescale and n = 5 for low-temperature spectroscopic identification. All others have been isolated in the solid state; for example, the salt of the tetraanionic fulleride $[Na(2.2.2crypt)^+]_4[C_{60}^{4-}]$.[81] The C_{60}^{n-} fullerides with n = 1, 2, 3, and 4 are all paramagnetic. While the A_3C_{60}, A = alkali metal, are metallic and superconducting,[82] the A_4C_{60} salts are insulating; this is not well understood yet. In their crystals, the alkali ions typically occupy interstices between the C_{60}^{n-} anions. A layered structure is present in the crystals of $[K(2.2.2crypt)]_2[C_{60}]$;[83] it is prepared according to:

$$C_{60} \xrightarrow[\text{toluene/(2.2.2.crypt)}]{\text{DMF/K}} [K(2.2.2.crypt)]_2[C_{60}]$$

Other fullerides such as $[PPN^+]_2[C_{60}^{2-}]$ are also known.[84]

   In contrast to the relative ease of the preparation of fullerides, the oxidation of fullerenes is not straightforward. It has been determined that the ease of oxidation follows the pattern $C_{60} < C_{70} < C_{78}(C_{2v}) < C_{76} < C_{84} < C_{78}(D_3)$.[85] Thus it was no surprise that the first isolated fullerene carbocation contains one of the bigger fullerenes, namely C_{76}; it can be made according to:

$$C_{76} + [Ar_3N^{\cdot +}][CB_{11}H_6Br_6^-] \xrightarrow{\text{o-dichlorobenzene}} [C_{76}^+][CB_{11}H_6Br_6^-] + Ar_3N$$

where $[Ar_3N^{\cdot +}]$ is the radical cation of tris(2,4-dibromophenyl)amine;[86] this is the first example of an all-carbon carbocation. The chemistry of the fullerenes is domi-

---

[78]F. T. Edelmann, *Angew. Chem. Int. Ed. Engl.* **1995**, *34*, 981.; H. Shinohara *et al.*, *Fullerene Sci. Technol.* **1997**, *5*, 829 (a review on the structural properties).
[79]T. Akasaka *et al.*, *Angew. Chem. Int. Ed. Engl.* **1997**, *36*, 1643.
[80]M. Saunders *et al.*, *J. Am. Chem. Soc.* **1994**, *116*, 2193.
[81]Y. Sun and C. A. Reed, *Chem. Commun.* **1997**, 747 and references therein.
[82]R. C. Haddon, *Acc. Chem. Res.* **1992**, *25*, 127; W. E. Broderick *et al.*, *J. Am. Chem. Soc.* **1994**, *116*, 5849.
[83]T. F. Fässler *et al.*, *Angew. Chem. Int. Ed. Engl.* **1997**, *36*, 486.
[84]C. A. Reed, *J. Am. Chem. Soc.* **1994**, *116*, 4145.
[85]L. Echegoyen *et al.*, *J. Am. Chem. Soc.* **1995**, *117*, 7801.
[86]C. A. Reed *et al.*, *J. Am. Chem. Soc.* **1996**, *118*, 13093.

nated by reactions of the $C=C$ double bonds; nucleophilic additions and substitutions are legion.[87] Some examples are:

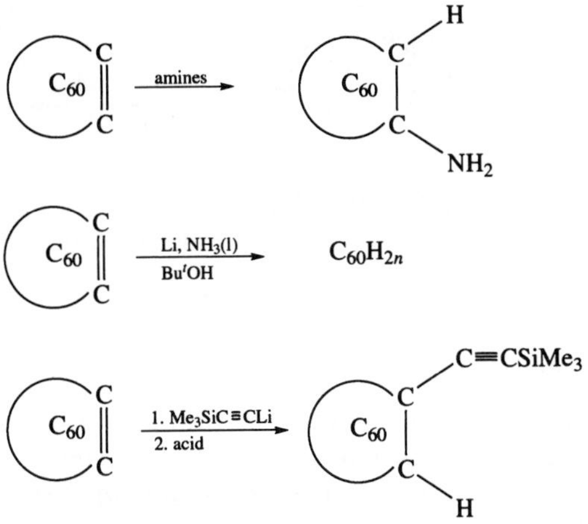

A similar situation is encountered in reactions with transition metal complexes where the fullerenes behave much like large olefins. For example, the addition of Vaska-type iridium complexes, such as $Ir(CO)Cl(PR_3)_2$, to fullerenes which has proven to be an effective way of obtaining materials that are suitable for crystallographic studies:[88]

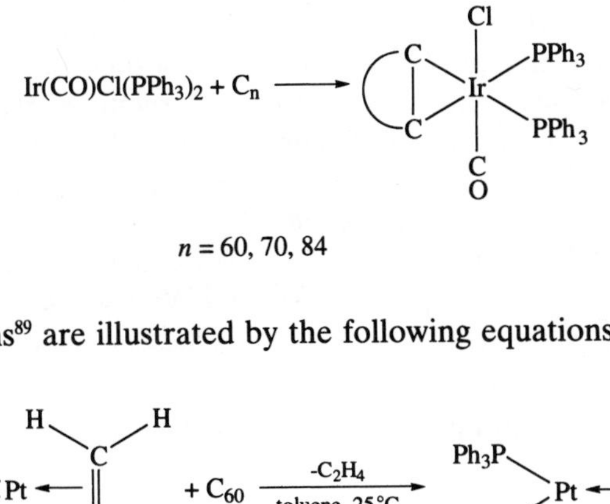

$$n = 60, 70, 84$$

Other reactions[89] are illustrated by the following equations:

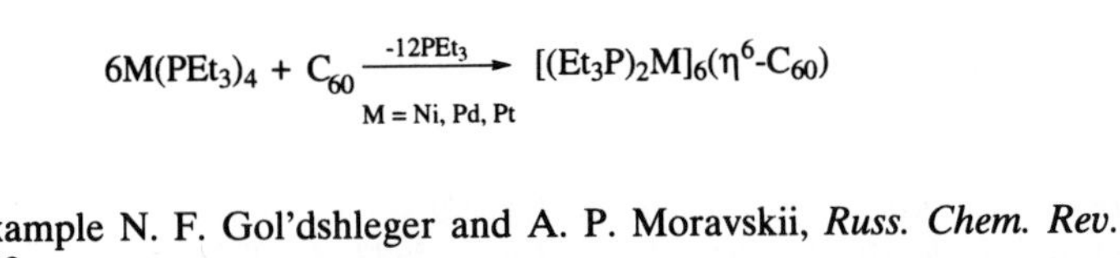

$$6M(PEt_3)_4 + C_{60} \xrightarrow[M = Ni, Pd, Pt]{-12PEt_3} [(Et_3P)_2M]_6(\eta^6\text{-}C_{60})$$

[87]See for example N. F. Gol'dshleger and A. P. Moravskii, *Russ. Chem. Rev. (Engl. Ed.)* **1997,** *66,* 323.
[88]A. L. Balch *et al., J. Am. Chem. Soc.* **1994,** *116,* 2227; *Inorg. Chem.* **1994,** *33,* 5238.
[89]P. J. Fagan *et al., Acc. Chem. Res.* **1992,** *25,* 134.

Cyclopentadienyl analogues can be made,[90] for example:

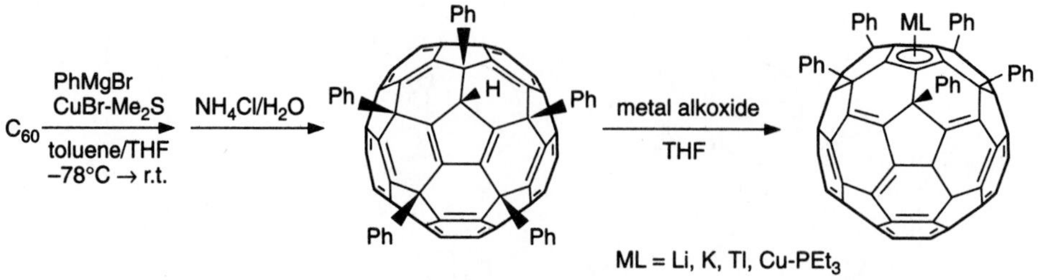

ML = Li, K, Tl, Cu-PEt₃

7-11 Cyanide Ion[91]

The CN^- ion can form complexes in aqueous solution with transition metal ions and particularly strong complexes with Zn^{2+}, Cd^{2+}, and Hg^{2+}.

Complexes can be formed in both low and high oxidative states, for example, $K_4[Ni^0(CN)_4]$, $K_3[Fe^{III}(CN)_6]$. Probably only in d metal complexes such as the Ni^0 is there any appreciable π bonding, and since CN^- is a strong nucleophile, π bonding need not be invoked to explain the stability of complexes in high oxidation states and the negative charge on the ion would decrease π acceptor behavior compared to say CO. Cyanide ion occupies a high position in the spectrochemical series, gives rise to a large nephelauxetic effect, and produces a large trans effect.

Unidentate CN^- ions, which generally exhibit sharp, intense bands between 2000 and 2200 cm^{-1}, in the ir region always bind through C, but since the N atom has a lone pair (see later), cyanide can act as a bridge giving M—C—N—M links. Such bridges are found in many crystalline cyanides such as AuCN, $Zn(CN)_2$, or $Cd(CN)_2$, all of which are polymeric with chain structures. Many complex compounds, such as the Prussian blues (Section 17-E-6), have linear MCNM units; mixed metal bridges M'CNM'' can also be formed, examples are the heptanuclear species $Fe\{(CN)Cu(trispyridylmethylamine)\}_6^{8-}$ and $(NEt_4)_2[M\{NCCr(CO)_5\}_6]^{92}$ (7-XVI), where M = Si^{IV}, Ge^{IV}, Sn^{IV} and the corresponding species where M = Cr^{III}, Mn^{II}, Fe^{II}, Co^{II}, Ni^{II}, Cu^I, Zn^{II}, Co^0, Mo^0, and W^0. Cyanide-bridged complexes can also be made by reactions of cyano complex anions, for example,

$$Cp_2ZrI_2 + K_2Pt(CN)_4 \longrightarrow \{Cp_2ZrPt(CN)_4\}_n$$

Sometimes isomers are possible. In $KFeCr(CN)_6$, the green isomer has Fe^{II}—CN—Cr^{III} bridges and the red isomer has Cr^{III}—CN—Fe^{II} linkages.

[90]M. Sawamura *et al.*, *J. Am. Chem. Soc.* **1996**, *118*, 12850.
[91]K. R. Dunbar and R. A. Heintz, *Prog. Inorg. Chem.* **1997**, *45*, 283.
[92]W. P. Fehlhammer *et al.*, *Inorg. Chim. Acta* **1992**, *198-200*, 513.

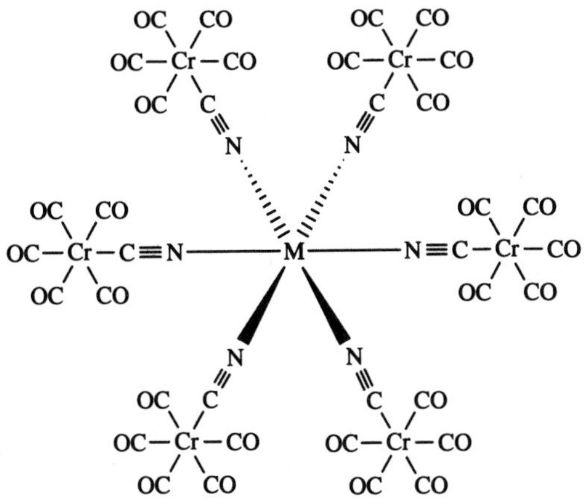

(7-XVI)

Observed binding modes are:[93]

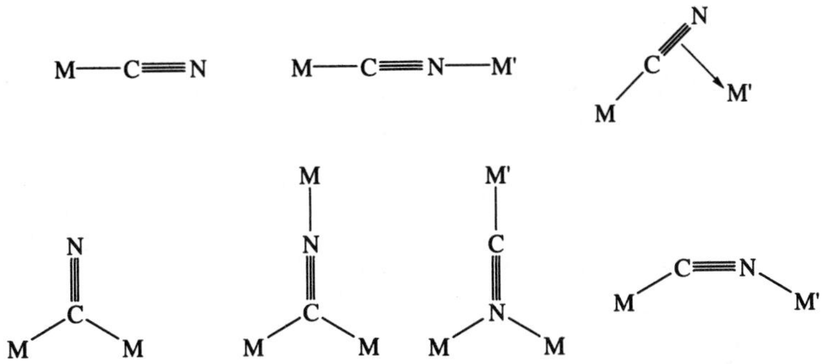

Most cyano complexes have the general formula $[M^{n+}(CN)_x]^{(x-n)-}$ and are anionic, but mixed complexes such as $[M(CN)_5X]^{n-}$, $X = H_2O$, NH_3, CO, NO, H, or halogen are common. Certain cyano complexes can be obtained in different oxidation states with the same geometry $M(CN)_x^{n-}$, $M(CN)_x^{(n+1)-}$, $M(CN)_x^{(n+2)-}$.

The *anhydrous acids* can often be isolated, for example, $H_3[Rh(CN)_6]$ and $H_4[Fe(CN)_6]$. These differ from those of other complex ions such as $[PtCl_6]^{2-}$or $[SiF_6]^{2-}$ that are invariably hydrated and are H_3O^+ salts.

The hydrogen atoms in the cyanides are H-bonded $M-CN{\cdots}H{\cdots}NC-M$ and different types of structure arise depending on the stoichiometry. For example, in $H[Au(CN)_4]$ there are sheets. For octahedral anions there is a difference in structure depending on whether the number of protons equals half the number of cyanide ions. For $H_3[M(CN)_6]$ compounds an infinite, regular three-dimensional array is formed in which the hydrogen bonds are perhaps symmetrical, whereas in other cases the structures appear to be more complicated.

The formation of linear bridges implies basicity on the N atom of a MCN group and indeed methylation or protonation can occur to give isocyano groups, MCNMe,

[93]I. Dance *et al.*, *Angew. Chem. Int. Ed. Engl.* **1995**, *34*, 314.

or MCNH as in $W(CO)_5CNH$; bridging cyanides of the (7-XVII) type can also be protonated to μ,η^2-HNC groups (7-XVIII).

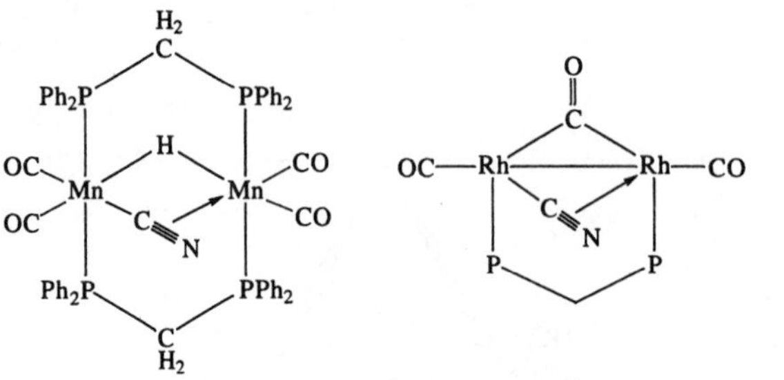

(7-XVII)

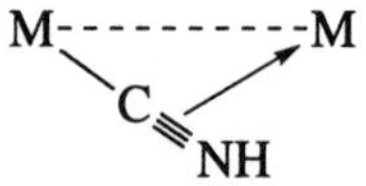

(7-XVIII)

The formation of complexes with Lewis bases, especially BPh_3, that is, $MCNBPh_3$, is of importance in the hydrocyanation of alkenes by nickel complexes (Section 22-4).

Finally, *cyanogen* can act as a ligand:

$$(AgAsF_6)_n + n(CN)_2 \longrightarrow \{Ag[(CN)_2]_2\}_n(AsF_6)_n$$

where there is an AgNC—CNAg group.

7-12 Isocyanides[94]

Isocyanides (RNC) are better σ donors and poorer π acceptors than CO, as indicated by the observation that typical homoleptic isonitrile complexes of many metals are in higher oxidation states than the typical carbonyl complexes of the same metal. Some metal isocyanide complexes are given in Table 7-4. It should be pointed out that there are no carbonyl analogues of those compounds in the higher oxidation

Table 7-4 Metal Isocyanide Complexes of 3*d* Transition Metals

		$[Co(CNR)_4]^-$		
$[Cr(CNR)_6]$	$[Fe(CNR)_5]$	$[Co_2(CNR)_8]$	$[Ni(CNR)_4]$	$[Cu(CNR)_4]^+$
$[Cr(CNR)_6]^+$	$[Fe_2(CNR)_9]$	$[Co(CNR)_5]^+$	$[Ni_4(CNR)_7]$	
$[Cr(CNR)_6]^{2+}$	$[Fe(CNR)_6]^{2+}$	$[Co(CNR)_5]^{2+}$	$[Ni(CNR)_4]^{2+}$	
$[Cr(CNR)_7]^{2+}$		$[Co_2(CNR)_{10}]^{4+}$		

[94]H. Werner, *Angew. Chem. Int. Ed. Engl.* **1990,** *29,* 1077.

Table 7-5 Lowering of CO and CN Frequencies in Analogous Compounds, Relative to Values for Free CO and CNAr[a]

Molecule	Δv (cm^{-1}) for each fundamental mode		
Cr(CO)$_6$	43	123	160
Cr(CNAr)$_6$	68	140	185
Ni(CO)$_4$	15	106	
Ni(CNAr)$_4$	70	125	

[a]Ar represents C$_6$H$_5$ and p-CH$_3$OC$_6$H$_4$.

states for which π bonding is of little significance. There are also a few homoleptic isonitrilates; an example is [Co(2,6-Me$_2$C$_6$H$_3$NC)$_4$]$^-$ which is isoelectronic with [Co(CO)$_4$]$^-$.

Table 7-5 shows that the extent to which the CN stretching frequencies in Cr(CNAr)$_6$ and Ni(CNAr)$_4$ molecules are lowered relative to the frequencies of the free CNAr molecules exceeds that by which the CO modes of the corresponding carbonyls lie below the frequency of CO.

Although terminal MCNR groups are normally close to linear with CNC angles of 165 to 179°, *bent isocyanides* with angles ~130 to 135° are known. This situation is thus similar to other linear/bent systems like M—NO or M—NNR (Section 9-20). The bending can be accounted for by a contribution of (7-XIXa) to (7-XIXb), both representing extreme forms:

$$\text{M}\!=\!\!=\!\text{C}\!=\!\!=\!\text{N} \qquad\qquad \text{M}\!-\!\!-\!\text{C}\!\equiv\!\text{NR}$$

(a) R (b)

(7-XIX)

In Ru(CNBu^t)$_4$PPh$_3$, for example, the PPh$_3$ and bent Bu^{t}NC groups are equatorial with the almost linear Bu^{t}NC groups in axial positions. The ir bands for bent groups are in the 1830 to 1870 cm^{-1} region compared to bands ~1960 cm^{-1} for linear RNC groups.

They are prepared by ligand substitution reactions,[95] sometimes using a heterogeneous catalyst such as CoCl$_2$, activated carbon, or metallic platinum on an oxide support as in

$$\text{Fe(CO)}_5 + 5\text{-}n\ \text{CNR} \longrightarrow \text{Fe(CNR)}_{5-n}(\text{CO})_n + n\text{CO}$$

$$\text{Cr}(\eta^6\text{-C}_{10}\text{H}_8)_2 \xrightarrow[-2\text{C}_{10}\text{H}_8]{\text{CNR}} \text{Cr(CNR)}_6$$

$$\text{M(N}_2)_2(\text{PR}_3)_4 \xrightarrow{+\,2\text{CNR}'} \text{M(CNR')}_2(\text{PR}_3)_4 + 2\text{N}_2$$

M = Mo, W

[95]A. J. L. Pombeiro and R. L. Richards, *Coord. Chem. Rev.* **1990,** *104,* 13.

They can also be made from halides and by electrophilic attacks on cyanide complexes as in

$$[L_nMCN]^{n+} \xrightarrow{\text{RX}} [L_nMCNR]^{(n-1)+} + X^-$$

Diisocyanides[96] such as $CN(CH_2)_nNC$ and $CNC(Me)_2(CH_2)_2C(Me)_2NC$ give chelate octahedral complexes of Cr, Mn, Fe, and Co in 0 to + 3 oxidation states and square Rh^I cations. They can also act as bridges between metal atoms in Ni, Pd, Pt, or Rh complexes such as $[Rh_2(\mu\text{-}L)_4]^{2+}$. Other isocyanides are

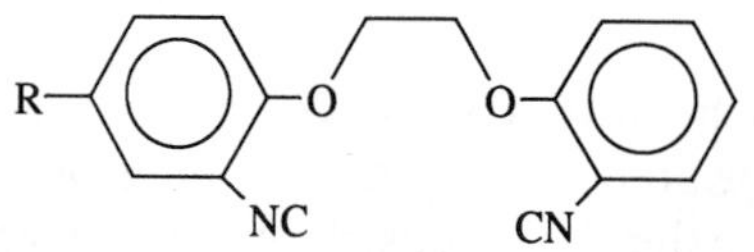

and

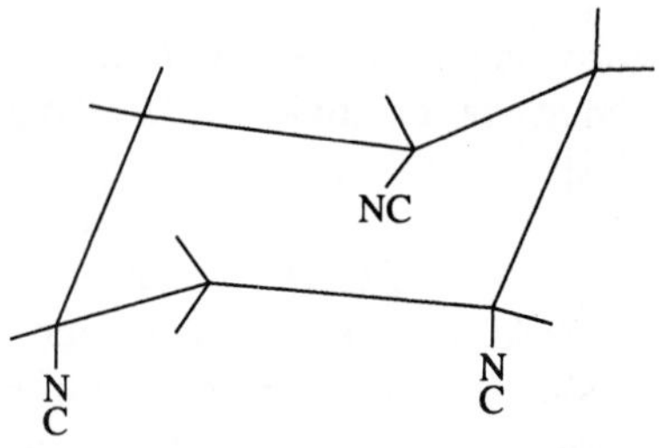

but the *cis,cis*,1,3,5-tricyanocyclohexane does not act as a tridentate to a triangle of metals.

Bridging isocyanides[97] are of several types:

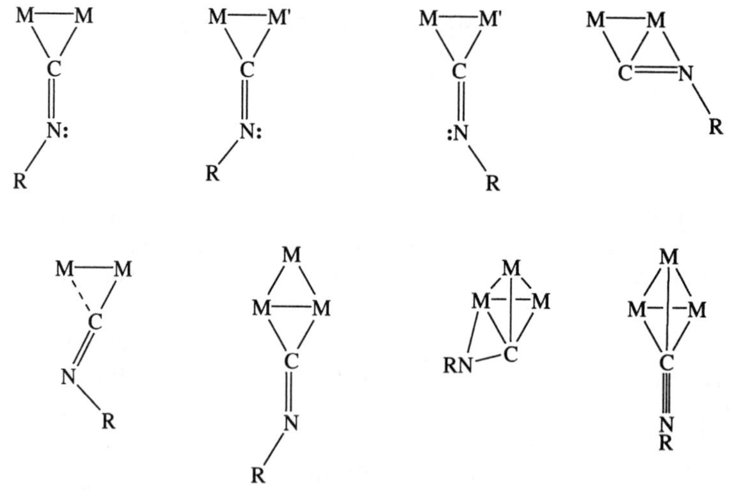

The CNC bonds are greatly bent with angles ~120°C. There may also be exchange between terminal and bridge groups, as in $Fe_2(\mu\text{-}CNEt)_3(CNEt)_6$, possibly *via* semibridged species. One example of triply bridging isocyanide is

[96]R. J. Puddephatt *et al.*, *Inorg. Chem.* **1994**, *33*, 2355.
[97]P. Royo *et al.*, *Organometallics* **1992**, *11*, 1229; J. L. Haggitt and D. M. P. Mingos, *J. Chem. Soc. Dalton Trans.* **1994**, 1013.

$Pd_3(\mu_3\text{-}\eta^1\text{-CNXy})(\mu\text{-dppm})_3^{2+},{}^{98}$ made according to:

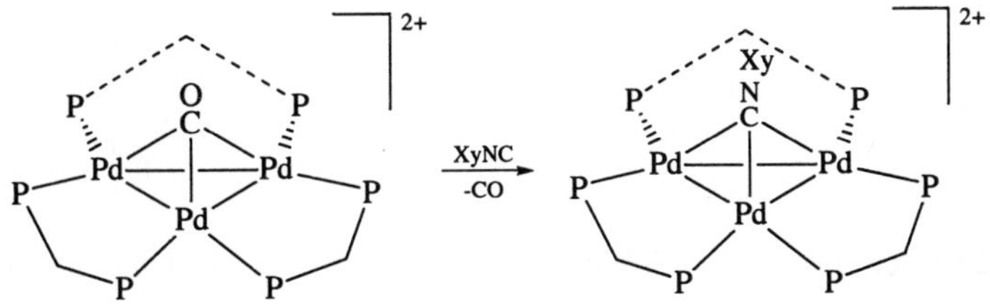

Reactions of Isocyanide Complexes[99]

Isocyanide complexes undergo many types of reactions. Some of the more important are the following:

1. *Coupling reactions* which lead to C—C bond formation. Some examples are

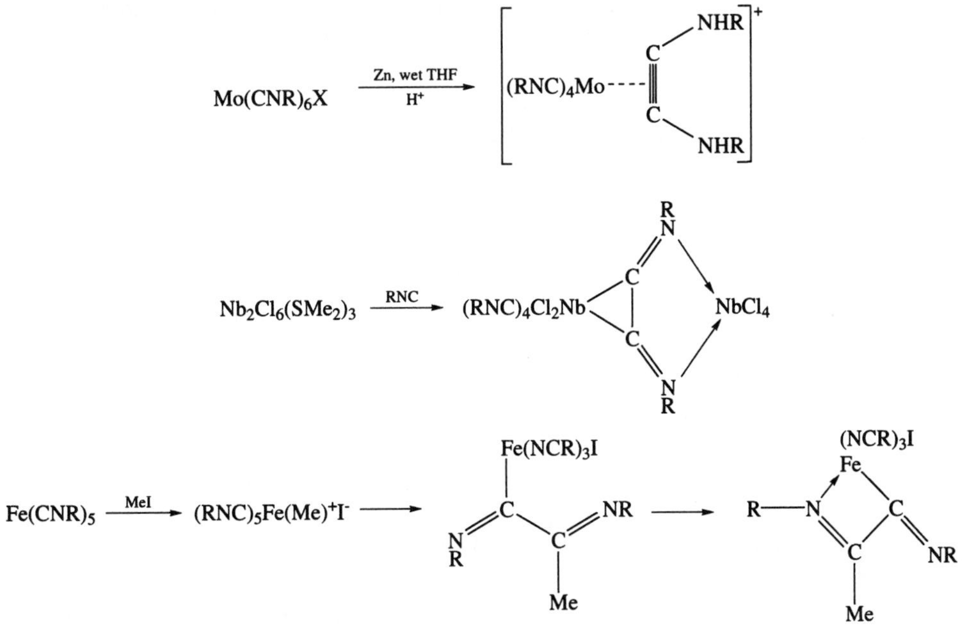

2. *Electrophilic attacks* can give alkylidyne compounds (Chapter 16)

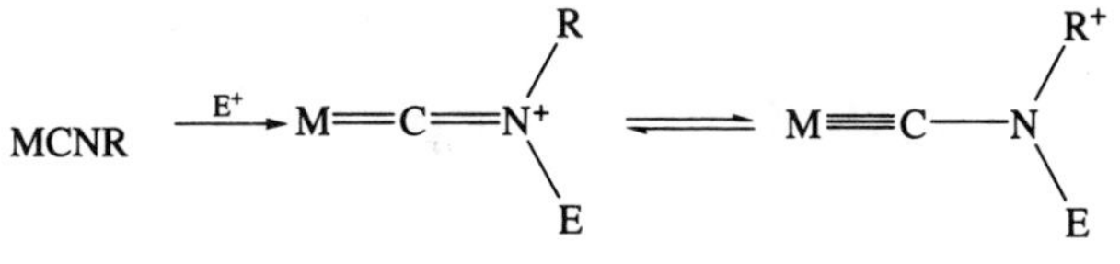

[98]R. J. Puddephatt et al., Inorg. Chem. **1994**, 33, 1497.

[99]A. J. L. Pombeiro and R. L. Richards, Coord. Chem. Rev. **1990**, 104, 13 (a review on reactions of alkynes, isocyanides, and cyanides at dinitrogen-binding, transition metal centers).

A specific example being

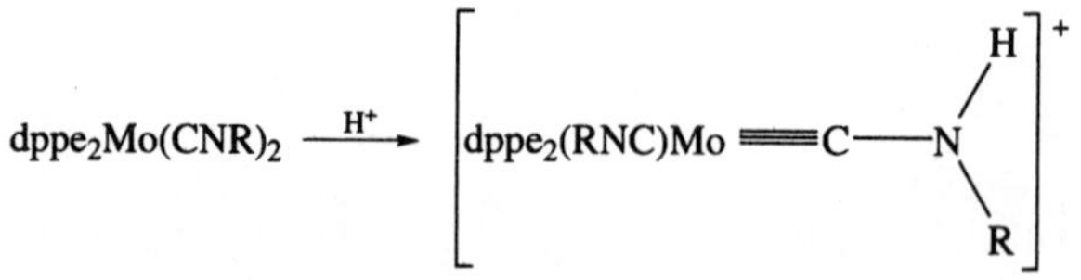

$$\text{dppe}_2\text{Mo(CNR)}_2 \xrightarrow{\text{H}^+} \left[\text{dppe}_2\text{(RNC)Mo} \equiv\!\!\!\equiv\!\!\!\equiv C\!\!-\!\!N \begin{matrix} H \\ \\ R \end{matrix} \right]^+$$

while bridging isocyanides have high nucleophilicity and are readily attacked, for example,

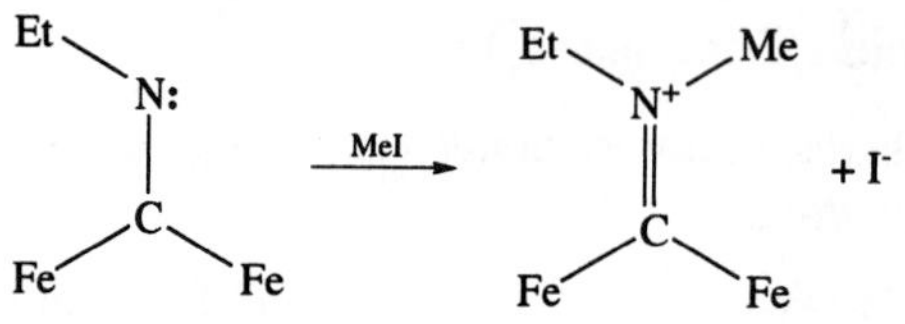

3. *Nucleophilic attacks* either at C or at N can give alkylidene complexes, for example,

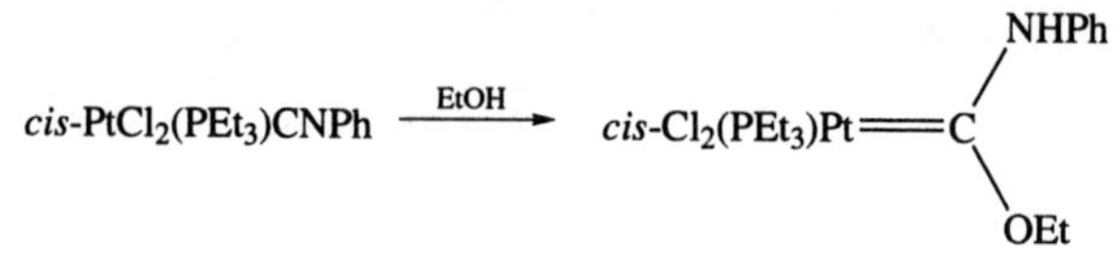

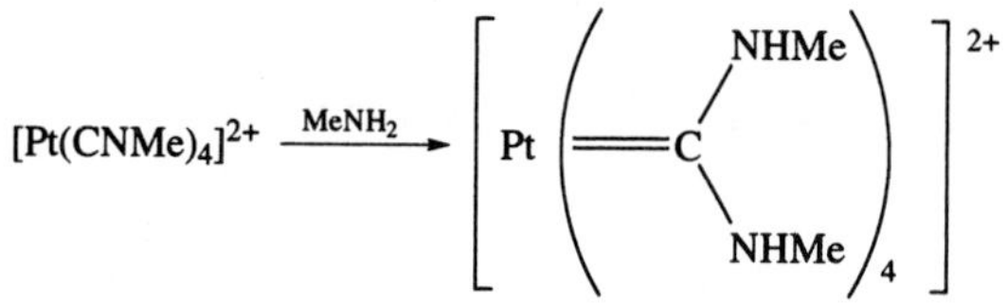

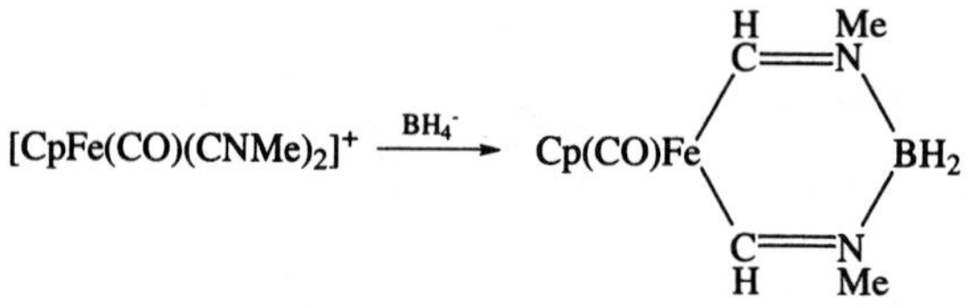

$$\textit{cis-}\text{PtCl}_2(\text{PEt}_3)\text{CNPh} \xrightarrow{\text{EtOH}} \textit{cis-}\text{Cl}_2(\text{PEt}_3)\text{Pt}\!=\!\!C \begin{matrix} \text{NHPh} \\ \\ \text{OEt} \end{matrix}$$

$$[\text{Pt(CNMe)}_4]^{2+} \xrightarrow{\text{MeNH}_2} \left[\text{Pt} \left(\!=\!\!C \begin{matrix} \text{NHMe} \\ \\ \text{NHMe} \end{matrix} \right)_4 \right]^{2+}$$

$$[\text{CpFe(CO)(CNMe)}_2]^+ \xrightarrow{\text{BH}_4^-} \text{Cp(CO)Fe} \begin{matrix} \text{H}\quad\text{Me} \\ \text{C}\!=\!\!\text{N} \\ \qquad\qquad\text{BH}_2 \\ \text{C}\!=\!\!\text{N} \\ \text{H}\quad\text{Me} \end{matrix}$$

4. *Dealkylation* also can occur in some reactions to give a cyano complex:

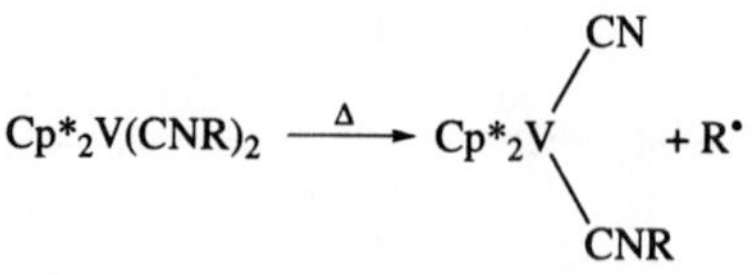

$$\text{Cp*}_2\text{V(CNR)}_2 \xrightarrow{\Delta} \text{Cp*}_2\text{V} \begin{matrix} \text{CN} \\ \\ \text{CNR} \end{matrix} + \text{R}^{\bullet}$$

Finally, although not a reaction of isocyanide complexes as such, except as labile intermediates, it can be noted that metal hydrido or alkyl compounds can undergo *insertion reactions* to give acylimidoyl complexes, which can react further:

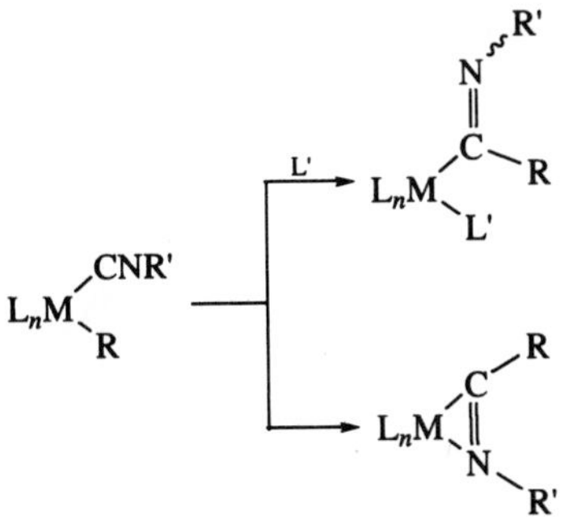

Isocyanide complexes are involved in catalytic hydrogenations (Chapter 22) of isocyanides.

7-13 Carbon Dioxide, Carbon Disulfide

Carbon Dioxide[100]

The three cumulenes, CO_2, CS_2, and COS, react in rather similar ways to give complexes, but those of CO_2 have been best studied because of the hope of direct conversion of CO_2 into organic chemicals.

Although CO_2 is a linear molecule, all structurally characterized complexes contain $M-C$ bonds and possess a bent OCO unit. Compounds of the following types are known:[101]

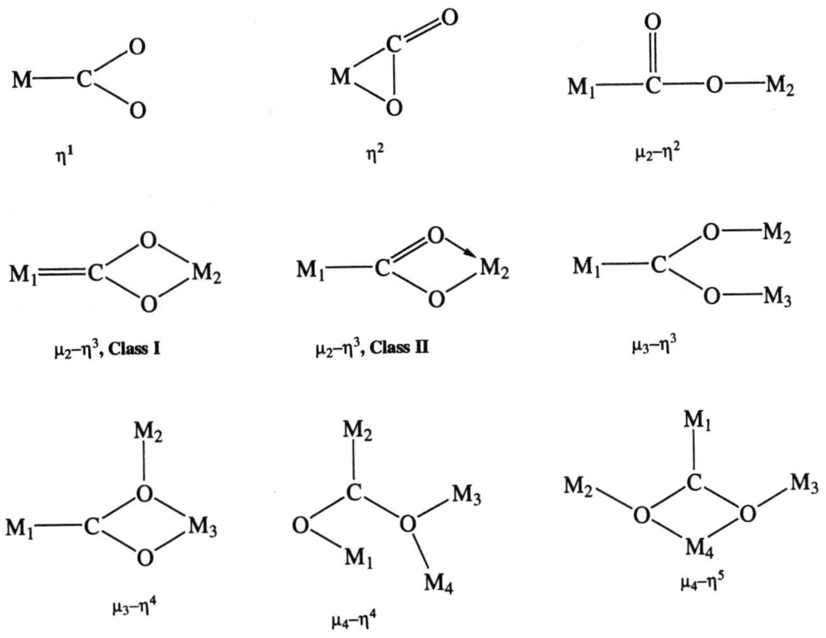

[100]D. H. Gibson, *Chem. Rev.* **1996**, *96*, 2063.
[101]See for example: M. Orchin *et al.*, *Polyhedron* **1993**, *12*, 1423.

Many complexes are prepared by direct carbonation as in

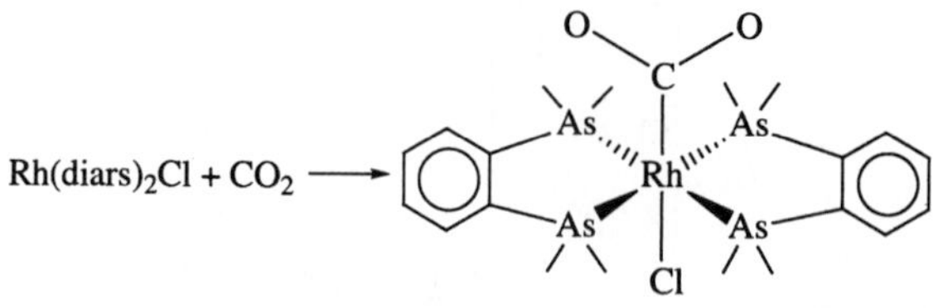

$$Rh(diars)_2Cl + CO_2 \longrightarrow$$

but other methods such as aerobic oxidation[102] also work for some metal carbonyl complexes as in

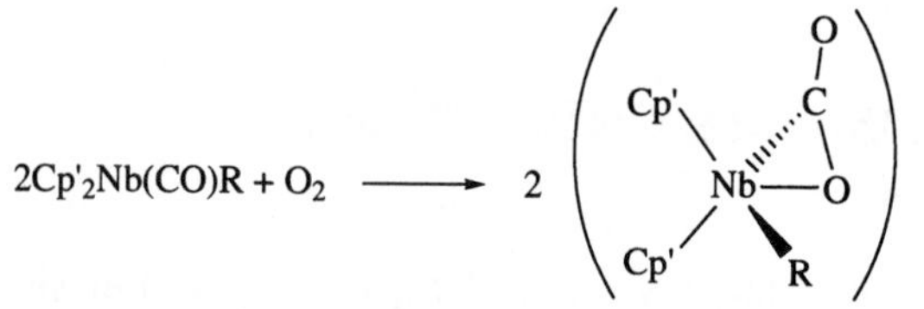

$$2Cp'_2Nb(CO)R + O_2 \longrightarrow 2$$

Cp' = η^5-C$_5$H$_4$Me; R = CH$_2$SiMe$_3$, CH$_2$CMe$_3$, CH$_3$

A coupling reaction of $(Ph)(PEt_3)_2Pt(OH)$ and $(HOCO)Pt(PEt_3)_2Ph$ gives

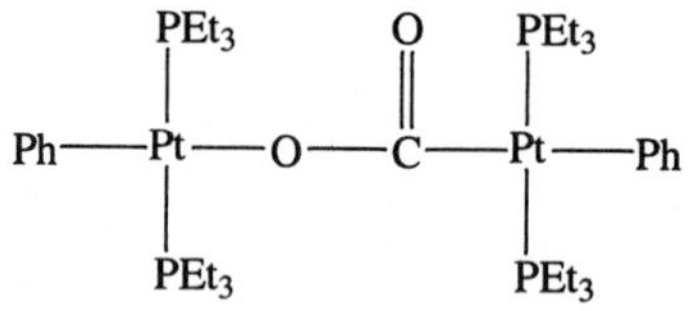

$$Ph-Pt-O-C-Pt-Ph$$

The η^2 compounds are best characterized and a good example is *trans*-Mo(CO$_2$)$_2$(PMe$_3$)$_4$. These have ir bands at ~1660 and 1630 cm^{-1}, whereas η^1 compounds absorb at ~1550 and 1220 cm^{-1}.

Carbon dioxide readily undergoes insertion reactions probably *via* initial CO$_2$ complexing, with metal alkoxides, hydroxides, oxides, dialkylamides, and metal hydrides and alkyls:[103]

[102]K. M. Nicholas *et al.*, *J. Am. Chem. Soc.* **1992**, *114*, 6579.
[103]See for example: S. Sakaki and Y. Musashi, *Inorg. Chem.* **1995**, *34*, 1914 and references therein.

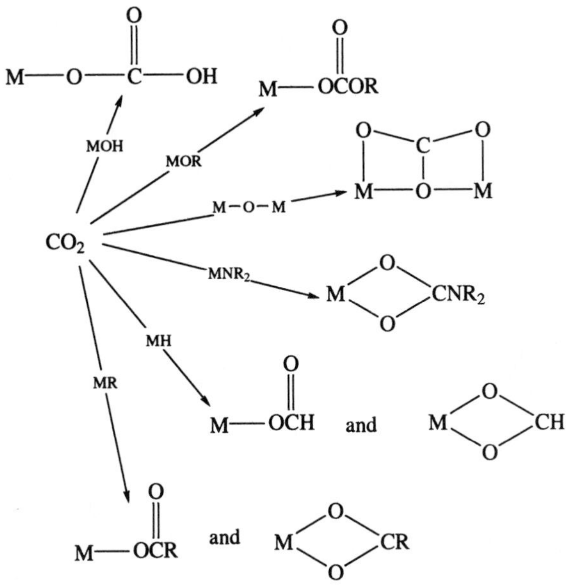

$$M-O-\overset{\overset{\displaystyle O}{\|}}{C}-OH$$

$$M-O\overset{\overset{\displaystyle O}{\|}}{C}OR$$

MOH

MOR

$$M-OCOR$$

$$M-O-M$$

$$CO_2$$

MNR$_2$

$$M\overset{O}{\underset{O}{\big\langle}}\overset{}{\big\rangle}C\overset{O}{\underset{O}{\big|}}\overset{}{\big|}M$$

$$M\overset{O}{\underset{O}{\big\langle}}CNR_2$$

MH

MR

$$M-O\overset{\overset{\displaystyle O}{\|}}{C}H \quad and \quad M\overset{O}{\underset{O}{\big\langle}}CH$$

$$M-O\overset{\overset{\displaystyle O}{\|}}{C}R \quad and \quad M\overset{O}{\underset{O}{\big\langle}}CR$$

There are a number of reactions of CO_2 with metal complexes in which prior coordination is likely. For example, the anion $W(CO)_5OH^-$ reacts with CO_2, COS, or CS_2 to give the corresponding bicarbonate or thiocarbonate complexes.[104] With complexes of oxophilic metals like Ti or Zr, deoxygenation to CO may occur while in others disproportionation to give CO and CO_3^{2-} occurs; an example of the latter reaction is

$$2Ru(CO)_4^{2-} + 2CO_2 \longrightarrow Ru_2(CO)_8^{2-} + CO + CO_3^{2-}$$

A "Witting Reaction" on a coordinated CO_2 unit forms a metal ketene complex[105] according to:

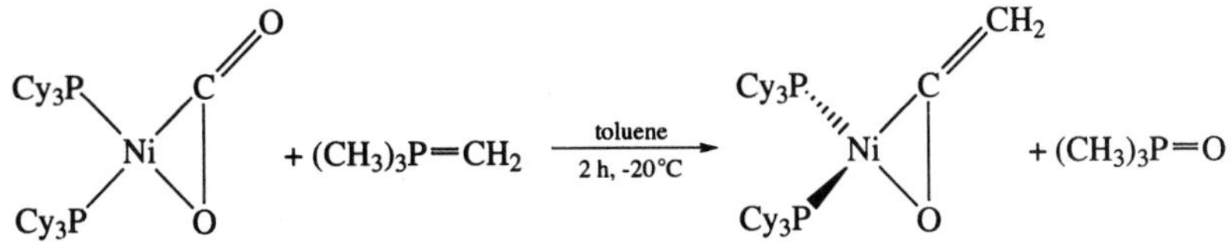

$$\begin{array}{c} Cy_3P \\ \diagdown \\ Ni \\ \diagup \\ Cy_3P \end{array} \overset{O}{\underset{O}{\big\langle}}C \quad + (CH_3)_3P=CH_2 \quad \xrightarrow[\text{2 h, -20°C}]{\text{toluene}} \quad \begin{array}{c} Cy_3P \\ \diagdown \\ Ni \\ \diagup \\ Cy_3P \end{array} \overset{CH_2}{\underset{O}{\big\langle}}C \quad + (CH_3)_3P=O$$

Fixation of atmospheric CO_2 by the dinuclear di-μ-hydroxo copper(II) complex $\{[CuHB(3,5-Pr_2^i pz)_3]_2(\mu\text{-}OH)_2\}$ gives the corresponding carbonato complex[106] 7-XX.

[104]D. J. Darensbourg *et al.*, *Inorg. Chem.* **1996**, *35*, 4406.
[105]J. K. Gong *et al.*, *J. Am. Chem. Soc.* **1996**, *118*, 10305.
[106]N. Kitajima *et al.*, *J. Chem. Soc., Chem. Commun.* **1990**, 1357.

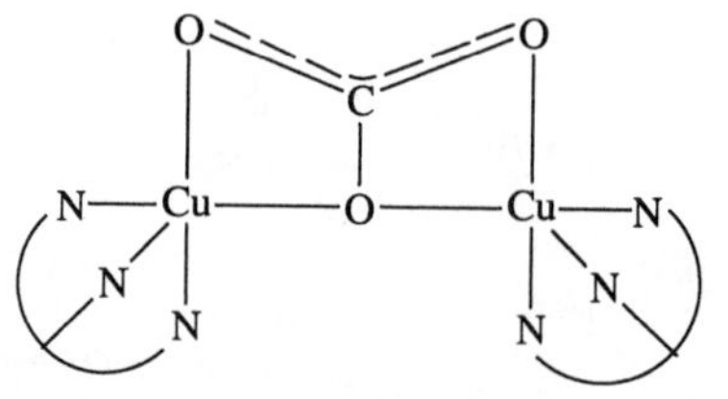

(7-XX)

Carbon Disulfide[107]

Carbon disulfide is a very versatile ligand that gives a wide variety of complexes which, apart from η^1-S end-on coordination, all involve bonding to carbon:

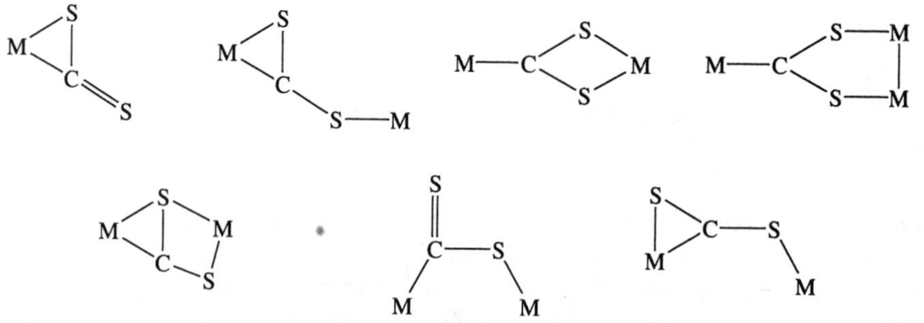

The CS_2 complexes are prepared by the reaction of CS_2 with transition metal complexes. The ligand in the M-η^2-CS_2 and M-η^1-CS_2 coordination modes acts as a good π-acceptor and a poor σ-donor as shown by the lower CS stretching frequencies found in the complexes relative to that of free CS_2.

Not only COS but CSSe and CSe_2 as well as thiocyanates, cyanates, Ph_2CS, and so on, give similar species,

[107]K. K. Pandey, *Coord. Chem. Rev.* **1995,** *140,* 37.

The coordinated CS_2 is very reactive and a large number of reactions are known. A few are the following:

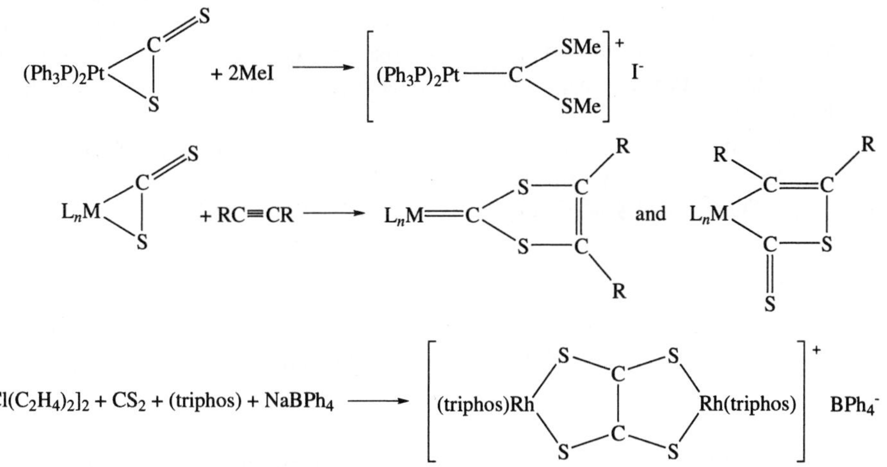

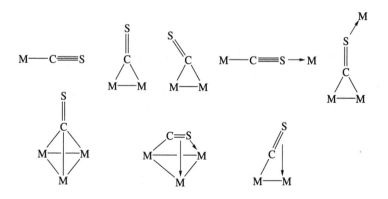

Like CO_2, CS_2 undergoes insertion reactions with M—H, M—R, and so on, bonds, e.g.,

$$CS_2 + M-H \longrightarrow MS_2CH \text{ or } MSC(S)H$$

7-14 Thiocarbonyl Compounds[108]

Although unlike CO, CS is not stable, polymerizing above −160°C, many MCS compounds can be made. The types of bonding are similar to those of CO:

[108]K. J. Klabunde *et al.*, *Chem. Rev.* **1988,** *88,* 391; see also H. Werner, *Angew. Chem. Int. Ed. Engl.* **1990,** *29,* 1077.

In general, linear CS is a stronger σ-donor and π-acceptor than CO, but this depends on the electron richness of the metal and in some cases the reverse is true.

Few binary metal thiocarbonyls are known. Nickel tetrathiocarbonyl, $Ni(CS)_4$, is unstable at room temperature, but many mixed carbonyl-thiocarbonyl complexes are stable and have properties similar to those of the parent carbonyl compounds. Examples are $Cr(CO)_5(CS)$, $(\eta^6\text{-}C_6H_6)Cr(CO)_2(CS)$,[109] $Cr(CO)_3(CS)[P(OPh)_3]_2$,[110] and $Fe(CO)_4(CS)$.

The CS ligand has a greater preference for the formation of bridges than CO. Examples of this behavior are $(dppe)_2(CO)W-C\equiv S-W(CO)_5$ and $[HB(pz)_3](CO)_2W(\mu\text{-}CS)Mo(CO)_2(\eta^5\text{-}C_9H_7)$[111] (7-XXI).

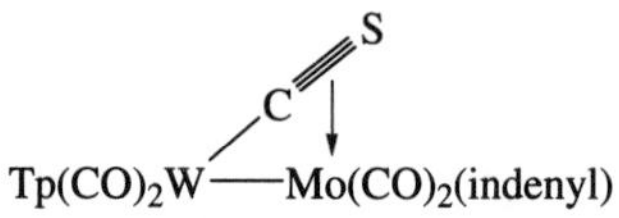

$$Tp(CO)_2W-\!\!\!-Mo(CO)_2(\text{indenyl})$$
(7-XXI)

The compounds are often made by desulfurization of CS_2 complexes, e.g.:

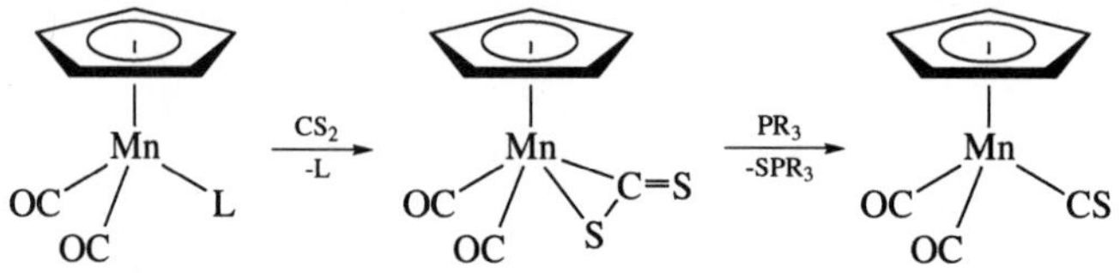

but they can also be made by interaction of $CSCl_2$:

$$[Fe(CO)_4]^{2-} \xrightarrow{\text{SCCl}_2} [Fe(CO)_4CS]$$

and in other ways such as:

$$(Me_3P)_2(CO)_2Fe(\eta^2\text{-}CS_2Me) \xrightarrow{\text{Na/Hg}} Fe(CS)(CO)_2(PMe_3)_2$$

The CS group can undergo a wide variety of reactions, including electrophilic attack at sulfur and nucleophilic attack at C, for example,

$$L_nMCS + MeI \longrightarrow [L_nMCSMe]^+I^-$$

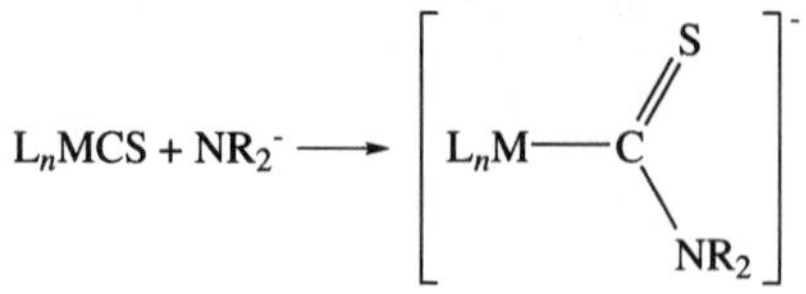

[109]D. F. R. Gilson *et al.*, *Inorg. Chem.* **1992**, *31*, 322.
[110]D. F. R. Gilson *et al.*, *Inorg. Chem.* **1994**, *33*, 804.
[111]R. J. Angelici *et al.*, *J. Am. Chem. Soc.* **1989**, *111*, 4995.

and intramolecular transfers:

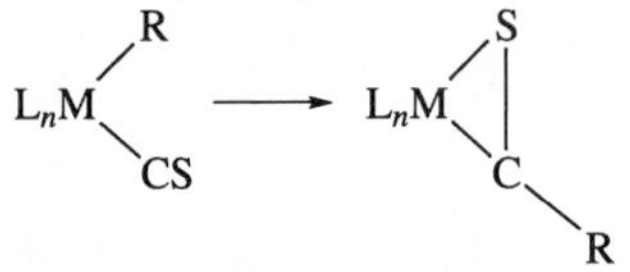

Compounds with other chalcogenides, CE (E = Se and Te) are also known; an example is the series of complexes $OsCl_2(CO)(CE)(PPh_3)_2$.[112]

7-15 Other C-Bonded Ligands

Acetonitrile (Section 9-14), although usually bonded through nitrogen, is acidic and can give species with $(H)M—CH_2CN$ groups while CH_3NO_2 similarly can give $(H)M—CH_2NO_2$ groups. Carbon bonded *acetylacetonates* are noted in Section 11-15; C-bonded dipyridyls and other cyclometallated *N*-compounds can be formed by cleavage of $C—H$ bonds, for example, in orthometallation reactions:

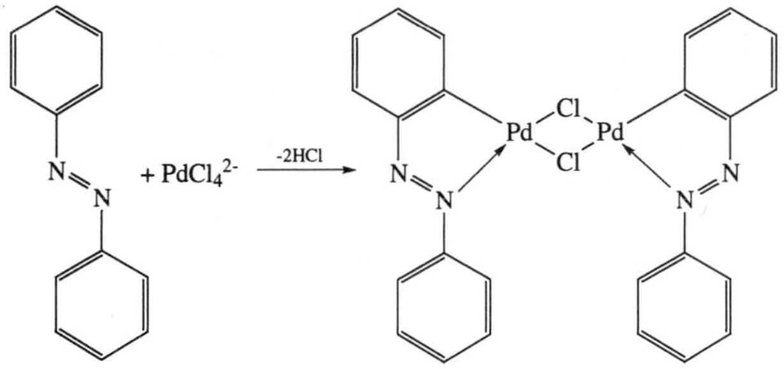

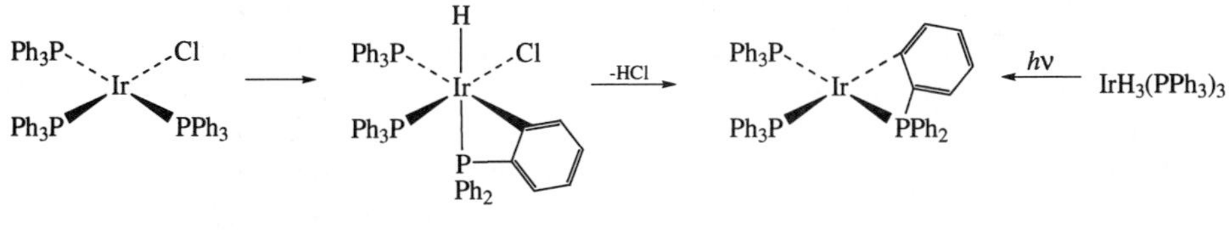

Additional References

W. E. Billups and M. A. Ciufolini, Eds., *Buckminsterfullerenes,* VCH Publishers, New York, 1993.

R. F. Curl and R. E. Smalley, *Scientific American* **1991**, *October,* 54.

R. F. Curl, *Angew. Chem. Int. Ed. Engl.* **1997**, *36,* 1566 (Nobel Lecture).

G. Davies, Ed., *Properties and Growth of Diamond,* Short Run Press, Exeter, 1994.

M. S. Dresselhaus *et al., Science of Fullerenes and Carbon Nanotubes,* Academic Press, San Diego, 1996.

P. W. Fowler *et al., An Atlas of Fullerenes,* Clarendon, Oxford, 1995.

H. Kroto, *Angew. Chem. Int. Ed. Engl.* **1997**, *36,* 1578 (Nobel Lecture).

C. O'Driscoll, *Designs on C_{60}, Chemistry in Britain* **1996**, *September,* 32.

R. E. Smalley, *Angew. Chem. Int. Ed. Engl.* **1997**, *36,* 1594 (Nobel Lecture).

N. E. Tolbert and J. Preiss, Eds., *Regulation of Atmospheric CO_2 and O_2 by Photosynthetic Carbon Metabolism,* Oxford University Press, Oxford, 1994.

[112]W. R. Roper *et al., J. Organomet. Chem.* **1988**, *338,* 393.

Chapter 8

THE GROUP 14 ELEMENTS: Si, Ge, Sn, Pb

GENERAL REMARKS

8-1 Group Trends

There is no more striking example of an enormous discontinuity in general properties between the first- and the second-row elements followed by a relatively smooth change toward more metallic character thereafter than in Group 14. Little of the chemistry of silicon can be inferred from that of carbon. Carbon is strictly nonmetallic; silicon is essentially nonmetallic; germanium is metalloid; tin and especially lead are metallic. Some properties of the elements are given in Table 8-1.

Catenation

Though not as extensive as in carbon chemistry, catenation is an important feature of Group 14 chemistry in certain types of compounds. Chains occur in Si and Ge hydrides (up to Si_6H_{14} and Ge_9H_{20}) and in Si halides (only Ge_2Cl_6 is known); very long chains are present in organic polymers of silicon,[1] germanium, and tin[2] (Section 8-12). All elements of the group form compounds that contain polyhedral clusters of metal atoms, e.g., Na_4Pb_9, Rb_2Sn_3, and $Na_4Ge_9(en)_5$ (Section 8-4).

However, there is a general if not entirely smooth decrease in the tendency to catenation in the order $C \gg Si > Ge \approx Sn \gg Pb$, which may be ascribed partly to diminishing strength of the C—C, Si—Si, Ge—Ge, Sn—Sn, and Pb—Pb bonds (Table 8-2).

Bond Strengths

The strengths of single covalent bonds between Group 14 atoms and other atoms (Table 8-2) generally decrease from Si to Pb. In some cases there is an initial rise from C to Si followed by a decrease. These energies do not, of course, reflect the ease of heterolysis of bonds, which is the usual way in chemical reactions; thus, for example, in spite of the high Si—Cl or Si—F bond energies, compounds containing

---

[1] T. D. Tilley, *Acc. Chem. Res.* **1993,** *26,* 22; L. Hevesi *et al., J. Chem. Soc., Chem. Commun.* **1995,** 769; L. V. Interrante *et al., J. Am. Chem. Soc.* **1994,** *116,* 12085.
[2] N. Devylder *et al., Chem. Commun.* **1996,** 711.

Table 8-1 Some Properties of the Group 14 Elements

Element	Electronic structure	mp (°C)	bp (°C)	Ionization enthalpies (kJ mol)$^{-1}$				Electro-negativity	Covalent radius^a (Å)
				1st	2nd	3rd	4th		
C	[He]$2s^22p^2$	>3550^b	4827	1086	2353	4618	6512	2.5–2.6	0.77
Si	[Ne]$3s^23p^2$	1410	2355	786.3	1577	3228	4355	1.8–1.9	1.17
Ge	[Ar]$3d^{10}4s^24p^2$	937	2830	760	1537	3301	4410	1.8–1.9	1.22
Sn	[Kr]$4d^{10}5s^25p^2$	231.9	2260	708.2	1411	2942	3928	1.8–1.9	1.40^c
Pb	[Xe]$4f^{14}5d^{10}6s^26p^2$	327.5	1744	715.3	1450	3080	4082	1.8	1.44^d

^{a}Tetrahedral (i.e., sp^3 radii).
^{b}Diamond.
^{c}Covalent radius of Sn^{II}, 1.63 Å.
^{d}Ionic radius of Pb^{2+}, 1.33 (CN 6); of Pb^{4+}, 0.775 Å.

these bonds are highly reactive. Since the charge separation in a bond is a critical factor, the bond ionicities must also be considered when interpreting the reactivities toward nucleophilic reagents. Thus Si—Cl bonds are much more reactive than Si—C bonds because, though stronger, they are more polar, Si$^{\delta+}$—Cl$^{\delta-}$, rendering the silicon more susceptible to attack by a nucleophile such as OH$^-$.

It may be noted also that (a) there is a steady decrease in E—C and E—H bond energies and (b) E—H bonds are stronger than E—C bonds.

Electronegativities

The electronegativities of the Group 14 elements have been a contentious matter.[3] Although C is generally agreed to be the most electronegative element, certain evidence, some of it suspect, has been interpreted as indicating that Ge is more electronegative than Si or Sn. It is to be remembered that electronegativity is a very qualitative matter, and it seems most reasonable to accept a slight progressive decrease Si → Pb.

Table 8-2 Approximate Average Bond Energies^a

Group 14 element	Energy of bond (kJ mol^{-1}) with							
	Self	H	C	F	Cl	Br	I	O
C	356	416		485	327	285	213	336
Si	210–250	323	250–335	582	391	310	234	368
Ge	190–210	290	255	465	356	276	213	
Sn	105–145	252	193		344	272	187	

^{a}For thermodynamic data, bond distances, and some other properties of Si compounds see M. J. S. Dewar *et al., Organometallics* **1986**, *5*, 375.

[3]For recent papers see: L. C. Allen, *J. Am. Chem. Soc.* **1989**, *111*, 9003; M. Dräger *et al., Organometallics*, **1994**, *13*, 2733.

It can be noted that Zn in HCl reduces only germanium halides to the hydrides, which suggests a higher electronegativity for Ge than for Si or Sn. Also, dilute aqueous NaOH does not affect GeH_4 or SnH_4, but SiH_4 is rapidly hydrolyzed by water containing a trace of OH^-. This is consistent with, though not necessarily indicative of, the Ge—H or Sn—H bonds either being nonpolar or having the positive charge on hydrogen. Finally, germanium halides are hydrolyzed in water only slowly and reversibly.

The Divalent State

The term "lower valence" indicates the use of fewer than four electrons in bonding. Thus although the *oxidation state* of carbon in CO is usually formally taken to be 2, this is only a formalism, and carbon uses more than two valence electrons in bonding. True divalence is found in carbenes and the Si, Ge, and Sn analogues (MR_2) discussed later; the high reactivity of carbenes may result from the greater accessibility of the sp^2 hybridized lone pair in the smaller carbon atom. The stable divalent compounds of the other elements can be regarded as carbenelike in the sense that they are bent with a lone pair and undergo the general type of carbene reactions to give new bonds to the element, that is,

$$R_2C: \longrightarrow R_2C\overset{\displaystyle /}{\underset{\displaystyle \backslash}{}}$$

However, the divalent state becomes increasingly stable down the group and is dominant for lead.

Inspection of Table 8-1 clearly shows that this trend cannot be explained exclusively in terms of ionization enthalpies, since these are essentially the same for all the elements Si—Pb. The "inert pair" concept probably holds only for the lead ion Pb_{aq}^{2+} where there could be a $6s^2$ configuration. In more covalent Pb^{II} compounds and most Sn^{II} compounds there are stereochemically active lone pairs—indeed some EX_2 and ER_2 compounds of Ge and Sn can act as donor ligands.

Relativistic effects (Section 1-8) due to an increase in the *sp*-promotion gap from Sn to Pb are also well established.[4] Other factors that undoubtedly govern the relative stabilities of the oxidation states are promotion energies, bond strengths for covalent compounds, and lattice energies for ionic compounds. Taking first the promotion energies, it is rather easy to see why the divalent state becomes stable if we remember that the M—X bond energies generally decrease in the order Si—X, Ge—X, Sn—X, Pb—X. (The factor that stabilizes CH_4 relative to $CH_2 + H_2$ despite the much higher promotional energy required in forming CH_4, is the great strength of the C—H bonds and the fact that two more of these are formed in CH_4 than in CH_2.) Thus if we have a series of reactions $MX_2 + X_2 = MX_4$ in which the M—X bond energies are decreasing, it is obviously possible that this energy may eventually become too small to compensate for the $M^{II} \rightarrow M^{IV}$ promotion energy and the MX_2 compound becomes the more stable. The progression is illustrated by ease of

---

[4]W. H. E. Schwarz, in *Theoretical Models of Chemical Bonding*, B. Maksic, Ed., Springer: Berlin, 1990; vol. 2, p. 593.

addition of chlorine to the dichlorides:

$$GeCl_2 + Cl_2 \longrightarrow GeCl_4 \qquad \text{(very rapid at 25 °C)}$$

$$SnCl_2 + Cl_2 \longrightarrow SnCl_4 \qquad \text{(slow at 25 °C)}$$

$$PbCl_2 + Cl_2 \longrightarrow PbCl_4 \qquad \text{(only under forcing conditions)}$$

Note that even $PbCl_4$ decomposes except at low temperatures, while $PbBr_4$ and PbI_4 do not exist, probably owing to the reducing power of Br^- and I^-. For ionic compounds matters are not so simple but, since the sizes of the (real or hypothetical) ions, M^{2+} and M^{4+}, increase down the group, it is possible that lattice energy differences no longer favor the M^{4+} compound relative to the M^{2+} compound in view of the considerable energy expenditure required for the process:

$$M^{2+} \longrightarrow M^{4+} + 2e$$

Of course, there are few compounds of the types MX_2 or MX_4 that are entirely covalent or ionic (almost certainly no ionic MX_4 compounds), so that the previous arguments are oversimplifications, but they indicate roughly the factors involved. For solutions no simple argument can be given, since Sn^{4+} and Pb^{4+} probably have no real existence.

It should be noted that compounds appearing from their stoichiometries to contain Group 14 elements in formal oxidation states between II and IV are generally mixed valence compounds. For example, Ge_5F_{12} consists of four Ge^{II} and one Ge^{IV} suitably coordinated and linked by fluorine atoms. Similarly Sn_3F_8 is built of octahedral $Sn^{IV}F_6$ and pyramidally coordinated $Sn^{II}F_3$ in $1:2$ ratio, linked by shared F atoms.

The only authentic stable compounds of Group 14 elements with formal oxidation number III are of the types (8-I) and (8-II). Type (8-Ia) for Si has a half-life of only ~10 min, but the others have lives of 3 to 12 months. The stability of these radicals is due to the great steric hindrance to attack resulting from the bulk of the ligands.

$E = Si, Ge, and Sn$

(8-I)

$E = Ge, Sn$

(8-II)

Tin, like iron, is an element for which Mössbauer spectroscopy can be broadly useful.[5] Using the Sn^{119m} nucleus, isomer shift (IS) measurements allow one to determine oxidation number and to some extent to estimate structural and bonding features. The IS values are usually in the order $Sn^{II} > Sn^0 > Sn^{IV}$, though this cannot be trusted absolutely.

[5]See for example: C. Carini *et al., J. Chem. Soc. Dalton Trans.* **1989,** 289; R. Barbieri *et al., J. Chem. Soc. Dalton Trans.* **1989,** 519.

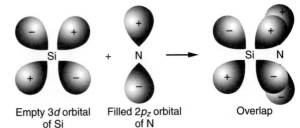

Figure 8-1 Formation of a $d\pi$-$p\pi$ bond between Si and N in trisilylamine.

Empty $3d$ orbital of Si   Filled $2p_z$ orbital of N   Overlap

Multiple Bonding

In the early 1960s, unstable transient intermediates with $Si=C$, $p\pi-p\pi$ bonding were discovered in reactions such as

$$H_2Si\diagup\!\!\diagdown \xrightarrow{\ 560°C\ } H_2Si=CH_2 + CH_2=CH_2$$

Compounds $R_2Si=CR'_2$ and $R_2M=MR'_2$ for Si, Ge, and Sn have now been isolated provided that bulky groups are used as discussed in Section 8-13.

Although stoichiometric similarities occur, for example, CO_2 and SiO_2, Me_2CO, and Me_2SiO, there is no structural or chemical similarity. Also, reactions that might have been expected to yield a carbonlike product do not do so—thus dehydration of silanols, $R_2Si(OH)_2$, produces $(R_2SiO)_n$ and $R_2(OH)SiOSi(OH)R_2$.

Multiple bonding of the $d\pi-p\pi$ type for Si is quite well established, especially in bonds to O, S, N, and P. It is important to note, however, that this does not necessarily lead to conjugation in the sense usual for carbon multiple bonded systems. Observations of the following types provide evidence for $d\pi-p\pi$ bonding.

1. Trisilylamines such as $(H_3Si)_3N$ differ from $(H_3C)_3N$ in being planar rather than pyramidal[6] and in being very weak Lewis bases; $(H_3Si)_2NH$ is also planar while H_3SiNCO has a linear $Si-N-C-O$ chain in the vapor phase, although there is some bending with $\angle Si-N-C = 158°$ and $\angle N-C-O = 176°$ in the crystalline solid. These observations can be explained by supposing that nitrogen forms dative π bonds to the Si atoms. In the planar state of $(H_3Si)_3N$, the nonbonding electrons of N would occupy the $2p_z$ orbital, if we assume that the N—Si bonds are formed using $sp_x p_y$ trigonal hybrid orbitals of nitrogen. Silicon has empty $3d$ orbitals, which are of low enough energy to be able to interact appreciably with the nitrogen $2p_z$ orbital. Thus the N—Si π bonding is due to the kind of overlap indicated in Fig. 8-1. It is the additional bond strength to be gained by this $p\pi-d\pi$ bonding that causes the NSi_3 skeleton to adopt a planar configuration, whereas with $N(CH_3)_3$, where the carbon has no low-energy d orbitals, the σ bonding alone determines the configuration, which is pyramidal as expected.

   The molecule $(H_3Si)_3P$ is pyramidal, however, indicating that the second-row atom P is less able to contribute a p orbital to $p\pi-d\pi$ bonding than is nitrogen.

[6]D. W. H. Rankin *et al.*, *J. Chem. Soc. Dalton Trans.* **1990**, 161.

Similarly, H_3GeNCO is *not* linear in the gas phase. Evidently effective bonding in the linear structure occurs only for Si—N and not Ge—N bonds.

2. The disilyl ethers $(R_3Si)_2O$ all have large angles at oxygen (140-180°), and both electronic and steric explanations have been suggested. Electronically, overlap between filled oxygen $p\pi$ orbitals and silicon $d\pi$ orbitals would improve with increasing angle and in the limit might favor linearity. When $R = C_6H_5$ the angle at oxygen is 180°, but there may also be a strong steric factor here as well.

3. Silanols, such as Me_3SiOH, are stronger acids than the carbon analogues and form stronger hydrogen bonds;[7] for Ph_3EOH the acidities are in the order $C \approx Si \gg Sn$. The hydrogen bonding can be ascribed to Si—O π bonding involving one of the two unshared pairs of the silanol oxygen and the $3d$ orbital of Si to give a situation somewhat similar electronically to that of the nitrogen atom in an imine $R_2C=NH$. One unshared pair still remains on the oxygen; it can be protonated in superacid media.[8]

A similar situation arises with the acid strength of R_3ECO_2H, where the order is $Si \geq Ge > C$; in this case the $d\pi-p\pi$ bonding probably acts to stabilize the anion. The order of π bonding, $C > Si > Ge \geq Sn > Pb$, can be obtained from hydrogen bonding and nmr studies on amines. Thus $(Me_3Si)_3N$ is virtually nonbasic, the germanium compound is about as basic as a tertiary amine, and the tin compound is more basic than any organic amine. The same order is found in REX_3 when X = alkyl, but when X = halogen it is $C < Si < Ge < Sn$.

Silicon-29 nmr spectra in both solids and solutions can be correlated with bonding types, namely, all σ bonds as in Me_4Si, $\sigma + \pi$ types as in $Me\overset{\frown}{O}$—$Si(OMe)_3$ and those where there are both types as in $H_2Si\overset{\frown}{—}\ddot{F}_2$.

Stereochemistry[9]

The stereochemistries of the Group 14 compounds are given in Table 8-3.

IV Oxidation State

All the elements form tetrahedral compounds and some chiral compounds such as $GeHMePh(\alpha\text{-napthyl})$ are known. It was first believed that planar SiO_4 groups occurred in complexes of the type (8-III) but this is not so, although distorted SiO_4 tetrahedra do occur.

(8-III)

[7] D. C. Hrncir *et al.*, *J. Chem. Soc., Chem. Commun.* **1990,** 306.
[8] C. A. Reed *et al.*, *J. Chem. Soc., Chem. Commun.* **1994,** 2519.
[9] R. R. Holmes, *Chem. Rev.* **1996,** *96,* 927; K. M. Baines and W. G. Stibbs, *Coord. Chem. Rev.* **1995,** *145,* 157.

Table 8-3 Valence and Stereochemistry of Group 14 Elements

Valence	Coordination number	Geometry	Examples
II	2	ψ-Trigonal (angular)	$Ge(NBu^i_2)_2$, $Pb(C_5H_5)_2$
	3	Trigonal	$[Ir_2(PbI)(CO)_2I(\mu\text{-dpma})]^{+a}$
	3	Pyramidal	$SnCl_2 \cdot 2H_2O$, $SnCl_3^-$, $Sn[CH(PPh_2)_2]_2$, $Pb(SPh)_3^-$
	4	Square planar	$Sn\{N(SePPh_2)_2\}_2^b$ ($SnSe_4$ core)
	4	ψ-*tbp*	Pb^{II} in Pb_3O_4, $Sn(S_2CNR_2)_2$, $Sn(\beta\text{-dike})_2$
	4	ψ-*sp*	$M(tmtaa)$, $M = Ge$, Sn, Pb^c
	5	ψ-Octahedral	SnO (blue-black form)
	6	Octahedral	PbS (NaCl type), GeI_2 (GaI_2 type)
	7	Complex	$[SC(NH_2)_2]_2PbCl^+$, $(18\text{-C-}6)SnCl^+$
	6, 7	ψ-Pentagonal bipyramidal + complex ψ-8-coordination	$Sn^{II}[Cn(EDTA)H_2O \cdot H_2O$
	8	Distorted square antiprism	$Pb(C_4O_3HNH_2)_2(H_2O)_2^d$
	9, 10	Complex	$Pb(NO_3)_2$(disemicarbazone), $Pb(O_2CMe)_2 \cdot 3H_2O$
	12	Cubooctahedron	$[Pb(NO_3)_6]^{4-e}$
IV	2	Linear	$Cp(CO)_2Mo\equiv GeC_6H_3\text{-}2,6,$ mes_2^f
	3	Trigonal	Ph_3Si^+; $(mes^*)_2Ge=Se^g$
	4	Tetrahedral	SiO_2, $SiCl_4$, GeH_4, $PbMe_4$
	5	*tbp*	$Me_3SnCl(py)$, $SnCl_5^-$, SiF_5^-, $RSiF_4^-$
	5^h	*sp*	$[XSi(O_2C_6H_4)_2]^-$
	6	Octahedral	SiF_6^{2-}, $[Si(acac)_3]^+$, $[Si(ox)_3]^{2-}$, GeO_2, $PbCl_6^{2-}$, *trans-*$GeCl_4(py)_2$, $Sn(S_2CNEt_2)_4$
	7	Pentagonal bipyramidal	$Ph_2Sn(NO_3)_2(OPPh_3)$
	8	Dodecahedral	$Sn(NO_3)_4$, $Pb(O_2CMe)_4$

[a]A. L. Balch *et al., J. Am. Chem. Soc.* **1991,** *113,* 1252.
[b]R. Cea-Olivares *et al., Chem. Commun.* **1996,** 519.
[c]A. H. Cowley *et al., Inorg. Chem.* **1992,** *31,* 3871.
[d]L. A. Hall *et al., Polyhedron* **1994,** *13,* 45.
[e]R. E. Cramer *et al., J. Chem. Soc. Dalton Trans.* **1994,** 563.
[f]R. S. Simons and P. P. Power, *J. Am. Chem. Soc.* **1996,** *118,* 11966.
[g]R. Okazaki *et al., Angew. Chem. Int. Ed. Engl.* **1994,** *33,* 2316.
[h]See also: M. Ye and J. G. Verkade, *Inorg. Chem.* **1993,** *32,* 2796.

Since valence shell expansion by utilization of outer *d* orbitals can occur, 5- and 6-coordinate species are common.

Pentacoordination is found mainly (a) in ions MX_5^- and $R_nMX_{5-n}^-$ stabilized in lattices by large cations, that are usually trigonal bipyramidal; (b) spiro compounds with O, S, and N chelates such as 8-IV and 8-V that are usually square pyramidal. The distortion to square pyramidal structures in the solid tends to increase with decreasing electronegativity of the central atom; (c) adducts of halides or substituted halides with donor ligands, for example, MX_4L; (d) for Sn, polymeric compounds R_3SnX where X acts as a bridge.

(8-IV)   (8-V)

Octahedral coordination is common for all the elements, although for ions and adducts, whether a compound will be 5- or 6-coordinate depends on delicate energy balances and cannot be predicted.

II Oxidation State

As noted previously, in most divalent compounds there is a lone rather than an *inert* pair. Thus in Ge_5F_{12} the Ge^{II} atoms are square pyramidal with the lone pair occupying the sixth position and the same is true in the blue-black form of SnO and of PbO where there are MO_5 ψ-octahedra.

In many Sn^{II} compounds there are atoms at three corners of a tetrahedron and a lone pair of electrons at the fourth. Thus $SnCl_2 \cdot 2H_2O$ has a pyramidal $SnCl_2OH_2$ molecule, the second H_2O not being coordinated (it is readily lost at 80°C), while $K_2SnCl_4 \cdot H_2O$ consists of ψ-tetrahedral $SnCl_3^-$ ions and Cl^- ions. The ψ-tetrahedral SnF_3^- ion is also known, and the $Sn_2F_5^-$ ion consists of two SnF_3^- ions sharing a fluorine atom. Other Sn^{II} compounds such as $SnCl_2$ or SnS similarly involve three coordination, but with a bridging group between the metal atoms.

For lead with its relatively large radius, higher coordination numbers have been established, but in some of these there is no evidence for stereochemically active lone pairs as in the 10-coordinate $Pb(\eta^2\text{-}NO_3)_2$(2,9-diformyl-1,10-phenanthrolinedi-semicarbazone). However, in the 1,4,7-triazacyclononane complex, $tacnPb(NO_3)_2$, the electron pair *is* stereochemically active.

THE ELEMENTS

8-2   Occurrence, Isolation, and Properties

Silicon is second only to oxygen in weight percentage of the earth's crust ($\sim$28%) and is found in an enormous diversity of silicate minerals. Germanium, Sn, and Pb are relatively rare elements ($\sim$10^{-3} wt%), but they are well known because of their technical importance and the relative ease with which Sn and Pb are obtained from natural sources.

Silicon is obtained in the ordinary commercial form by reduction of SiO_2 with carbon or CaC_2 in an electric furnace. Similarly, Ge is prepared by reduction of GeO_2 with C or H_2. Silicon and Ge are used as semiconductors in transistors and for microcircuitry and for this reason their properties, including surface chemistry, have been studied extensively.[10] Exceedingly high purity is essential, and special methods are required to obtain usable materials; these include fractional distillation

[10]H. N. Waltenburg and J. T. Yates, Jr., *Chem. Rev.* **1995,** *95,* 1589.

of halides, $SiHCl_3$, and so on, reduction of halides by H_2 or thermal decomposition of SiH_4 to give the element, then zone refining of the latter. Thin films of semi-insulating polycrystalline Si on Si single crystals for integrated circuits can be made by interaction of SiH_4 and N_2O.

Tin and Pb are obtained from the ores in various ways, commonly by reduction of their oxides with carbon. Further purification is usually effected by dissolving the metals in acid and depositing the pure metals electrolytically.

Silicon is rather unreactive. It is attacked by halogens giving tetrahalides, and by alkalis giving solutions of silicates. It is not attacked by acids except HF; presumably the stability of SiF_6^{2-} provides the driving force here. Highly reactive Si, prepared by the reaction,

$$3CaSi_2 + 2SbCl_3 \longrightarrow 6Si + 2Sb + 3CaCl_2$$

reacts with water to give SiO_2 and hydrogen.

Germanium is somewhat more reactive than Si and dissolves in concentrated H_2SO_4 and HNO_3. Tin and Pb dissolve in several acids and are rapidly attacked by halogens. They are attacked slowly by cold alkali, rapidly by hot, to form stannates and plumbites. Lead often appears to be much more noble and unreactive than would be indicated by its standard potential of -0.13 V. This low reactivity can be attributed to a high overvoltage for hydrogen and also in some cases to insoluble surface coatings. Thus Pb is not dissolved by dilute H_2SO_4 and concentrated HCl.

8-3   Allotropic Forms

Silicon and Ge are normally isostructural with diamond and the covalent bonds are 2.35 Å for Si. By use of very high pressures, denser forms with distorted tetrahedra have been produced. The graphite structure (Section 7-1), which requires the formation of $p\pi{-}p\pi$ bonds, is unique to carbon.

Tin has two crystalline modifications, with the equilibria

$$\alpha\text{-Sn} \underset{\text{"gray"}}{\overset{18°C}{\rightleftharpoons}} \beta\text{-Sn} \underset{\text{"white"}}{\overset{232°C}{\rightleftharpoons}} Sn(l)$$

α-Tin, or gray Sn (density at 20°C = 5.75), has the diamond structure. The metallic form, β or white Sn (density at 20°C = 7.31), has a distorted close-packed lattice. The approach to ideal close packing accounts for the considerably greater density of the β metal compared with the diamond form.

Lead exists only in a *ccp,* metallic form. This is a reflection both of its preference for divalence rather than tetravalence and of the relatively low stability of the Pb—Pb bond.

COMPOUNDS OF THE GROUP 14 ELEMENTS

8-4   Anions; Binary and Ternary Compounds

The reduction by alkali metals of Ge, Sn, and Pb salts in liquid ammonia or in the presence of hexamethylphosphoramide or cryptands has allowed the isolation of many stable Zintl anions such as Ge_9^{2-},[11] Ge_9^{4-},[12] Sn_9^{4-}, Sn_9^{3-}, Sn_4^{4-}, Pb_9^{3-},[13] and E_5^{2-} (E = Ge, Sn, and Pb).[14] It may be noted that the E_4^{4-} ions are $20e$ species, isoelectronic with P_4, As_4, or Sb_4 and similarly tetrahedral and diamagnetic; E_5^{2-} species have trigonal bipyramidal geometry ($\sim D_{3h}$). The anion in (K crypt)$_3$Sn$_9$ has a tricapped trigonal prismatic structure and is paramagnetic with 21 skeletal electrons. Various anions have a large formal negative charge, e.g., Sn_5^{12-} and Sn_6^{14-}; they have complex structures. A diamagnetic Na_5Sn_{13} species forms in the sodium–tin system in high yield following prolonged reaction of a quenched mixture at 280°C in a tantalum container; it has infinite chains of interconnected clusters built of tin pentagons.[15] There are also mixed ions of various sorts such as $Sn_5S_{12}^{4-}$,[16] $Sn_4Se_{10}^{4-}$,[17] $Sn_3S_7^{2-}$,[18] $Sn_2Bi_2^{2-}$, SnS_6^{4-},[19] $Ge_2Se_8^{4-}$,[20] $Pb_2Se_3^{2-}$, and $SnTe_4^{4-}$.

While interactions of alkali metals or magnesium with elemental silicon and germanium (or their oxides) can give various silicides or germanides, most of these products do not contain true anions and some are semiconductors. However, in $Li_{12}Si_7$ there are Si_5 rings and trigonal planar Si-centered Si_4 units in the lattice.

Dibarium silicide (Ba_2Si) has an *anti*-$PbCl_2$ structure with Si^{4-} octahedrally coordinated by Ba atoms; Si_4 tetrahedra also occur in K_3LiSi_4.

There are binary compounds such as SnX_2, X = S or Se,[21] SiS_2, and GeS_2. The latter have chains of tetrahedral MS_4 units linked by S bridges. Tin disulfide has a CaI_2 lattice, each Sn having six S neighbors; $SnS_{2-x}Se_x$ forms intercalation compounds.[22] Silicon nitride (Si_3N_4) can be prepared by many of the standard powder technology methods[23] or by regular synthetic methods, for example:

$$H_2SiCl_2 + NH_3 \longrightarrow \left(-SiH_2-NH-\right)_x \longrightarrow Si_3N_4$$

$$SiCl_4 + 8MeNH_2 \xrightarrow{-4MeNH_3Cl} Si(NHMe)_4$$

$$3Si(NHMe)_4 + 4NH_3 \longrightarrow Si_3N_4 + 12MeNH_2$$

It is a refractory compound of high technological importance.[24]

There are various polymorphs of silicon carbide made by high temperature interaction; some have wurtzite (ZnS) or diamond structures. It is exceedingly hard and inert; it finds uses in polishing products, furnace linings, and semiconductor technology.

[11]V. Queneau and S. C. Sevov, *Angew. Chem. Int. Ed. Engl.* **1997**, *36*, 1754.
[12]H. G. von Schnering *et al.*, *Z. anorg. allg. Chem.* **1997**, *623*, 1037.
[13]T. F. Fässler and M. Hunziker, *Inorg. Chem.* **1994**, *33*, 5380.
[14]J. Campbell and G. J. Schrobilgen, *Inorg. Chem.* **1997**, *36*, 4078.
[15]J. T. Vaughey and J. D. Corbett, *Inorg. Chem.* **1997**, *36*, 4316.
[16]J. B. Parise *et al.*, *J. Chem. Soc., Chem. Commun.* **1994**, 69.
[17]G. J. Schrobilgen *et al.*, *Inorg. Chem.* **1995**, *34*, 6265.
[18]J. B. Parise, *J. Chem. Soc., Chem. Commun.* **1994**, 527.
[19]H. Kessler *et al.*, *Inorg. Chem.* **1997**, *36*, 4697.
[20]W. S. Sheldrick and B. Schaaf, *Z. Naturforsch.*, **1994**, *B49*, 655.
[21]S. R. Bahr and P. Boudjouk, *Inorg. Synth.* **1997**, *31*, 86.
[22]D. O'Hare, *Chem. Soc. Rev.* **1992**, 121.
[23]R. A. Andrievskii, *Russ. Chem. Rev. (Engl. Trans.)* **1995**, *64*, 291.
[24]H.-P. Baldus and M. Jansen, *Angew. Chem. Int. Ed. Engl.* **1997**, *36*, 329.

Table 8-4   Tetravalent Hydrides and Halides of Group 14 Elements

Hydrides[a]		Fluorides and Chlorides[b]		
MH₄	Other	MF₄	MCl₄	Other
SiH_4 bp −112°C	$Si_2H_6 \rightarrow Si_6H_{14}$ bp −145°C	SiF_4 subl −95.7°C	$SiCl_4$ bp 57.6°C mp −70°C	$Si_2Cl_6 \rightarrow Si_6Cl_{14}$ bp −145°C $Si_2F_6 \rightarrow Si_{16}F_{34}$ bp −18.5°C Si_2Br_6 mp 90.8°C
GeH_4 bp −88°C	$Ge_2H_6 \rightarrow Ge_9H_{20}$ bp −29°C	GeF_4 subl −37°C	$GeCl_4$ bp 83°C mp −49.5°C	Ge_2Cl_6 mp 40°C Ge_5F_{12} $Ge + GeF_4 \rightarrow Ge_7F_{16}$[c]
SnH_4 bp −52.5°C PbH_4(?)	Sn_2H_6	SnF_4 subl 704°C PbF_4	$SnCl_4$ bp 114.1°C $PbCl_4$ d. 105°C	

[a]There are mixed Si—Ge hydrides and *cyclo* silanes as well as isomers that may be separable by g.l.c.
[b]All MX₄ compounds except PbBr₄ and PbI₄ are known, as well as mixed halides of Si (e.g., SiF₃I and SiFCl₂Br) and even SiFClBrI.
[c]J. Köhler and J.-H. Chang, *Z. anorg. allg. Chem.* **1997,** *623,* 596.

Finally, ceramic composites belonging to the ternary Si—C—N system are of high technical relevance. Crystalline solids of composition SiC_2N_4[25] have a polymeric network structure; they are made according to:

$$nSiCl_4 + 2nMe_3Si\text{—}N\text{=}C\text{=}N\text{—}SiMe_3 \xrightarrow[100°C]{\text{py}} [Si(N\text{=}C\text{=}N)_2]_n + 4nMe_3SiCl$$

THE TETRAVALENT STATE

8-5   Hydrides

All the hydrides, Table 8-4, are colorless.

Silanes

Monosilane (SiH_4) is best prepared on a small scale by heating SiO_2 and $LiAlH_4$ at 150 to 170°C. On a larger scale SiO_2 or alkali silicates are reduced by a NaCl—$AlCl_3$ eutectic (mp 120°C) containing Al metal, or with hydrogen at 400 atm and 175°C. The original Stock procedure of acid hydrolysis of magnesium silicide gives a mixture of silanes. Chlorosilanes may also be reduced by $LiAlH_4$. Higher silanes can also be made by photolysis of SiH_4—H_2 mixtures.

Only SiH_4 and Si_2H_6 are indefinitely stable at 25°C; the higher silanes decompose giving H_2 and mono- and disilane, possibly indicating SiH_2 as an intermediate.

[25]R. Riedel *et al., Angew. Chem. Int. Ed. Engl.* **1997,** *36,* 603.

The hydridic reactivity of the Si—H bond in silanes and substituted silanes may be attributed to charge separation $Si^{\delta+}$—$H^{\delta-}$ that results from the greater electronegativity of H than of Si. Silanes are spontaneously flammable in air, for example,

$$2Si_4H_{10} + 13O_2 \longrightarrow 8SiO_2 + 10H_2O$$

Although silanes are stable to water and dilute mineral acids, rapid hydrolysis occurs with bases;

$$Si_2H_6 + (4+n)H_2O \longrightarrow 2SiO_2 \cdot nH_2O + 7H_2$$

The silanes are strong reducing agents. With halogens they react explosively at 25°C, but controlled replacement of H by Cl or Br may be effected in the presence of AlX_3 to give halogenosilanes such as SiH_3Cl.

Transition metal complexes in which SiH_4 is coordinated in a η^2-fashion via a Si—H σ bond have been prepared according to

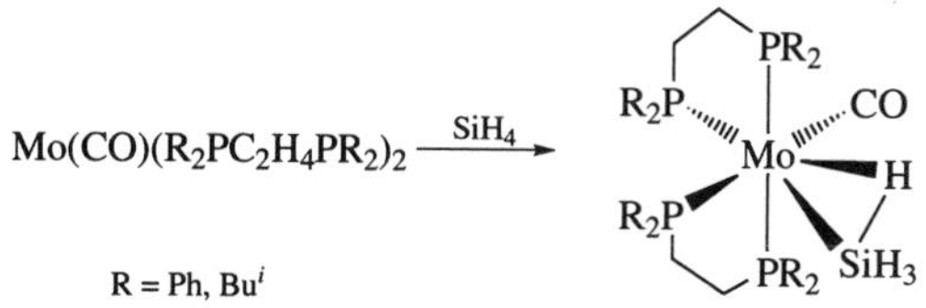

$$Mo(CO)(R_2PC_2H_4PR_2)_2 \xrightarrow{\text{SiH}_4}$$

R = Ph, Bu^i

Those compounds show a tautomeric equilibrium between the η^2-SiH_4 complex and the hydridosilyl species:[26]

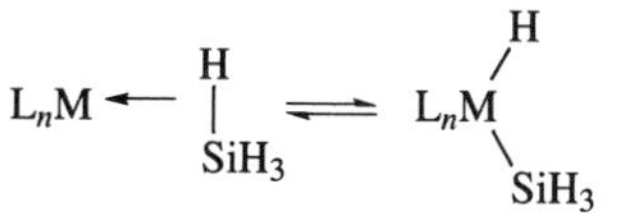

Monogermane together with Ge_2H_6 and Ge_3H_8 can be made by heating GeO_2 and $LiAlH_4$ or by addition of $NaBH_4$ to GeO_2 in acid solution. Higher germanes are made by electric discharge in GeH_4. Germanes are less flammable than silanes, although still rapidly oxidized in air, and the higher germanes increasingly so. The reaction of GeH_4 with ozone in solid argon produces a variety of species such as the transient germanone (H_2GeO), a germylene-water complex (H_2O—GeH_2), hydroxygermylene, germanic acid, and germanol.[27] However, the germanes are resistant to hydrolysis, and GeH_4 is unaffected by even 30% NaOH.

Stannane (SnH_4) is obtained by interaction of $SnCl_4$ and $LiAlH_4$ in ether at −30°C. It decomposes rapidly when heated and yields β-tin at 0°C. Although it is stable to dilute acids and bases, 2.5 M NaOH causes decomposition to Sn and some stannate. Stannane is easily oxidized and can be used to reduce organic compounds (e.g., C_6H_5CHO to $C_6H_5CH_2OH$, and $C_6H_5NO_2$ to $C_6H_5NH_2$). With concentrated

[26]X. L. Luo et al., J. Am. Chem. Soc. **1995**, 117, 1159.
[27]R. Withnall and L. Andrews, J. Phys. Chem. **1990**, 94, 2351.

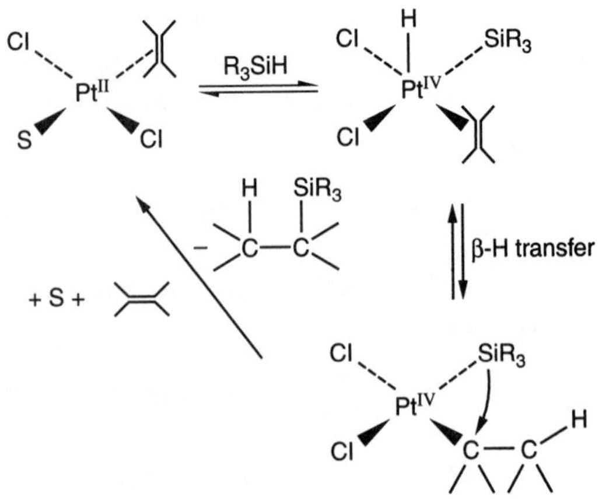

Figure 8-2 Simplified cycle for hydrosilation of alkenes by platinum complexes (S = solvent).

acids at low temperatures, the solvated stannonium ion is formed by the reaction

$$SnH_4 + H^+ \longrightarrow SnH_3^+ + H_2$$

Plumbane (PbH_4) is said to be formed in traces when Mg—Pb alloys are hydrolyzed by acid or when Pb^{2+} salts are reduced cathodically, but its existence is doubtful.[28]

All the elements form stable organohydrides (R_nMH_{4-n}); they are readily made by reduction of the corresponding chlorides with $LiAlH_4$. There are also a number of compounds of transition metals with silyl groups [e.g., $H_3SiCo(CO)_4$].

Perhaps the most important reaction of compounds with an Si—H bond, such as Cl_3SiH or Me_3SiH, and one that is of commercial importance, is the Speier or *hydrosilation*[29] reaction of alkenes, for example:

$$RCH{=\!=}CH_2 + HSiCl_3 \longrightarrow RCH_2CH_2SiCl_3$$

Olefins usually give the terminal product. The reaction is catalyzed by transition metal complexes, e.g., H_2PtCl_6 or $RhCl(PPh_3)_3$; it may involve the oxidative addition of Si—H bonds across a transition metal as shown in Fig. 8-2.

Silicon tetrahydride and GeH_4 react with K, Rb, and Cs to form H_3SiM^I and H_3GeM^I, which have NaCl type structures at room temperature; more complex structures are known for other cations.[30]

The unusual reducing properties of $SiHCl_3$ are discussed in Section 8-6.

8-6   Halides

The more important halides are given in Table 8-4.

[28]See for example: W. Thiel *et al., J. Phys. Chem.* **1993,** *97,* 4381 and references therein.
[29]M. Brookhart and B. E. Grant, *J. Am. Chem. Soc.* **1993,** *115,* 2151; A. R. Cutler *et al., J. Am. Chem. Soc.* **1995,** *117,* 10139.
[30]W. Sundermeyer *et al., Angew. Chem. Int. Ed. Engl.* **1994,** *33,* 216.

Fluorides

These are obtained by fluorination of the other halides or by direct interaction; GeF_4 is best made by heating $BaGeF_6$. Tetrafluorides of Si and Ge are hydrolyzed by an excess of water to the hydrous oxides; the main product from SiF_4 and H_2O in the gas phase is $F_3SiOSiF_3$. In an excess of aqueous HF, the hexafluoro anions (MF_6^{2-}) are formed. These anions are also found with certain trivalent cations, e.g., $SmSnF_7$ and $TlPbF_7$.[31] Similarly, a solid state reaction of tin metal and $(NH_4)HF_2$ gives:

$$2Sn + 7(NH_4)HF_2 \xrightarrow{300°C} 2(NH_4)_3SnF_7 + 4H_2 + NH_3$$

In these EF_7^{3-} species the stiochiometry might be misleading in regard to their structure because they all contain $[SnF_6]^{2-}$ octahedra and F^- anions surrounded by the corresponding cations.[32] Germanium tetrafluoride can be reduced to give hygroscopic mixed-valent species such as Ge_5F_{12} and Ge_7F_{16}.[33] Tin tetrafluoride is polymeric, with Sn octahedrally coordinated by four bridging and two nonbridging F atoms. Non-stoichiometric lead tetrafluoride is made by the action of F_2 on PbF_2, but the pure material can be prepared by high-pressure fluorination of "PbF_{4-x}"; it is isostructural with SnF_4.[34]

Silicon Halides

Silicon tetrachloride is made by chlorination of Si at red heat. Hexachlorodisilane (Si_2Cl_6) can be obtained by interaction of $SiCl_4$ and Si at high temperatures or, along with $SiCl_4$ and higher chlorides, by chlorination of a silicide such as that of calcium. The higher members, which have highly branched structures, can also be obtained by amine-catalyzed reactions such as

$$5Si_2Cl_6 \longrightarrow Si_6Cl_{14} + 4SiCl_4$$

$$3Si_3Cl_8 \longrightarrow Si_5Cl_{12} + 2Si_2Cl_6$$

and by photolysis of $SiHCl_3$. The products are separated by fractional distillation.

All the chlorides are immediately and completely hydrolyzed by water, but careful hydrolysis of $SiCl_4$ gives $Cl_3SiOSiCl_3$ and $(Cl_3SiO)_2SiCl_2$.

Hexachlorodisilane is a useful reducing agent for compounds with oxygen bound to S, N, or P; under mild conditions, at 25°C in $CHCl_3$ chlorooxosilanes are produced. It is particularly useful for converting optically active phosphine oxides $R^1R^2R^3PO$ into the corresponding phosphine. Since the reduction is accompanied by configurational inversion, the intermediacy of a highly nucleophilic $SiCl_3^-$ ion (*cf.* PCl_3) has

[31] O. Graudejus and B. G. Müller, *Z. anorg. allg. Chem.* **1996**, *622*, 1601.
[32] C. Plitzko and G. Meyer, *Z. anorg. allg. Chem.* **1997**, *623*, 1347.
[33] J. Köhler and J.-H. Chang, *Z. anorg. allg. Chem.* **1997**, *623*, 596.
[34] M. Bork and R. Hoppe, *Z. anorg. allg. Chem.* **1996**, *622*, 1557.

been proposed:

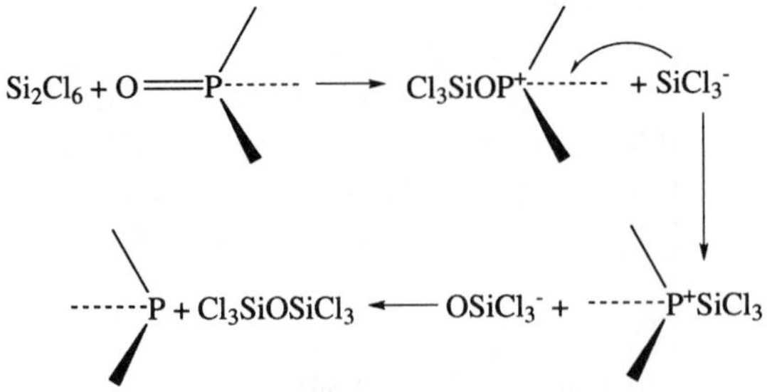

$$Si_2Cl_6 + O{=\!\!=}P \cdots \longrightarrow Cl_3SiOP^+ \cdots + SiCl_3^-$$

$$\cdots P + Cl_3SiOSiCl_3 \longleftarrow OSiCl_3^- + \cdots P^+SiCl_3$$

The postulation of $SiCl_3^-$ can also accommodate the equally useful, clean, selective reductions by trichlorosilane (bp 33°C) and also the formation of C—C and Si—C bonds by reaction of $SiHCl_3$ with CCl_4, RX, RCOCl, and other halogen compounds in the presence of amines. In these cases the hypothetical $SiCl_3^-$ could be generated by the reaction

$$HSiCl_3 + R_3N \rightleftharpoons R_3NH^+ + SiCl_3^-$$

followed by

$$SiCl_3^- + Cl_3C{-\!-}CCl_3 \longrightarrow SiCl_4 + Cl^- + Cl_2C{=\!\!=}CCl_2$$

$$SiCl_3^- + RX \longrightarrow [R^- + XSiCl_3] \longrightarrow RSiCl_3 + X^-$$

There is precedent for the postulation of the $SiCl_3^-$ ion, since trisubstituted organosilanes (R_3SiH), react with bases to give silyl ions (R_3Si^-).

Silicon tetrabromide[35] can be prepared by a multi-step procedure by reaction of $C_6H_5SiCl_3$ with $LiAlH_4$ then Br_2 and finally HBr according to:

$$4C_6H_5SiCl_3 + 3LiAlH_4 \xrightarrow[3h/50°C]{Et_2O} 4C_6H_5SiH_3 + 3AlCl_3 + 3LiCl$$

$$C_6H_5SiH_3 + 3Br_2 \xrightarrow[0°C\ to\ 80°C]{CCl_4/N_2} C_6H_5SiBr_3 + 3HBr$$

$$C_6H_5SiBr_3 + HBr \xrightarrow[70°C/5-6h]{C_6H_6/AlBr_3} SiBr_4 + C_6H_6$$

It can also be prepared from the elements. The fuming liquid is decomposed by water into silicic acid and HBr with great evolution of heat; it reacts violently with potassium.

Chloride Oxides

A variety of chlorooxosilanes, both linear and cyclic, is known. Thus controlled hydrolysis of $SiCl_4$ with moist ether, or interaction of Cl_2 and O_2 on hot silicon, gives $Cl_3SiO(SiOCl_2)_nSiCl_3$, where $n = 1$ to 4.

[35]H.-G. Horn and D. Kuczkowiak, *Z. anorg. allg. Chem.* **1996,** *622,* 1083.

Germanium Tetrachloride

This differs from $SiCl_4$ in that only partial hydrolysis occurs in aqueous 6 to 9 M HCl and there are equilibria involving species of the type $[Ge(OH)_nCl_{6-n}]^{2-}$; the tetrachloride can be distilled and separated from concentrated HCl solutions of GeO_2.

Tin and Lead Halides

The tetrachlorides are also hydrolyzed completely only in water; in the presence of an excess of acid they form chloroanions, as discussed in Section 8-9. All SnX_4 (X = Cl, Br, and I) are mononuclear with tetrahedral structures, but SnF_4[36] is polymeric unlike SiF_4, and can be made according to:

$$2SnF_2 + 2I_2 + 2MeCN \longrightarrow SnF_4(MeCN)_2 + SnI_4$$

$$SnF_4(MeCN)_2 \xrightarrow{\Delta} SnF_4 + 2MeCN$$

8-7 Oxygen Compounds of Silicon

Silica[37]

Pure SiO_2 occurs in only two forms, *quartz* and *cristobalite*. The silicon atom is always tetrahedrally bound to four oxygen atoms, but the bonds have considerable ionic character. In *cristobalite* the Si atoms are placed as are the C atoms in diamond with the O atoms midway between each pair. In *quartz* there are helices, so that enantiomorphic crystals occur, and these may be easily recognized and separated mechanically.

The interconversion of *quartz* and *cristobalite* on heating requires breaking and re-forming of bonds, and the activation energy is high. However, the rates of conversion are profoundly affected by the presence of impurities, or by the introduction of alkali metal oxides or other "mineralizers," and amorphous silica can also be converted to crystalline quartz in the presence of Mg^{2+} or Ca^{2+}.

What was believed to be a third form of SiO_2 "tridymite" is probably a solid solution of mineralizer and silica.

Slow cooling of molten silica or heating any form of solid silica to the softening temperature gives an amorphous material that is glassy in appearance and is indeed a glass in the general sense, that is, a material with no long-range order but rather a disordered array of polymeric chains, sheets, or three-dimensional units.

Dense forms of SiO_2, called coesite and stishovite, were first made under drastic conditions (250–1300°C at 35–120 katm), but they were subsequently identified in meteor craters where the impact conditions were presumably similar; stishovite has the rutile structure. Both are chemically more inert than normal SiO_2 to which they revert on heating.

---

[36]D. Tudela and J. A. Patron, *Inorg. Synth.* **1997**, *31*, 92.
[37]C. R. Helms and B. E. Deal, eds., *The Physics and Chemistry of SiO_2 and $Si-SiO_2$ Interface*, 2nd ed., Plenum Press, New York, 1993; E. Philippot *et al.*, *J. Solid State Chem.* **1996**, *123*, 1.

Silica is relatively unreactive towards Cl_2, H_2, acids, and most metals at ordinary or slightly elevated temperatures, but it is attacked by fluorine, aqueous HF, alkali hydroxides, fused carbonates, and so on.

Amorphous silica, silica "gel," can be made by hydrolysis of alkoxides such as $Si(OEt)_4$; it is used, when dehydrated, as a drying agent, and chromatographic and catalyst support material. It appears to contain $Si(OSi\equiv)_4$, $Si(OSi\equiv)_3OH$, and $Si(OSi\equiv)_2(OH)_2$ groups. The nmr studies on ^{29}Si indicate that silica found in plants, flagellates, and other biological systems has the same type of structure as silica gel.

Silicates

There is an enormous variety of silicates, including rock forming and synthetic minerals. Only the briefest survey can be given. The basic unit in most silicates is the SiO_4 tetrahedron; these occur singly or by sharing O atoms in small groups that may be linear or cyclic, in infinite chains and sheets. There are also a few silicates which contain 5-coordinate SiO_5 frameworks;[38] a well established example is that of $CaSi_2O_5$.[39]

Although silicate structures have been determined mainly by X-ray diffraction, Raman and magic-angle spinning nmr using ^{29}Si and ^{27}Al are useful for microcrystalline and amorphous solids, glasses and solutions.[40] In aluminosilicates such spectra can distinguish between tetrahedral and octahedral Al.

Alkali Silicates; Silicate Solutions

Sodium silicates are made on a large scale by fusion of SiO_2 and Na_2CO_3 at ~1500°C for detergent and other uses.

The free SiO_4^{4-} ion has a lifetime of only milliseconds, and silicate anions are polymeric. Many species have been identified by trimethylsilation of hydroxy groups and glc analysis of the products. The solutions contain linear, cyclic, and cage polymeric ions in proportions depending on the pH, concentration, and temperature. For alkali ions, most of the Si is in ions with charge $>10^-$ but for Me_4N^+ in solutions below 50°C the major species is the $[Si_8O_{20}]^{8-}$ anion,[41] which in the crystal has a double ring (8-VI). Silicon-29 nmr of Ge substituted species confirm that in solution there are both double-four and double-three rings (8-VI and 8-VII). The sort of species actually isolated after silation are of the type $Si_6O_{15}(SiMe_3)_6$, $Si_6O_{18}(SiMe_3)_{12}$, and $Si_8O_{12}(OMe)_8$.

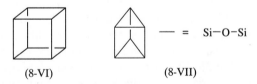

(8-VI)   (8-VII)

The alkaline silicate species are commonly in dynamic equilibrium and on undergoing pH changes or dilution, equilibria may be quite rapidly established,

[38]R. Tacke and M. Mühleisen, *Inorg. Chem.* **1994,** *33,* 4191; R. Tacke *et al., Angew. Chem. Int. Ed. Engl.* **1994,** *33,* 1186.

[39]R. J. Angel *et al., Nature* **1996,** *384,* 441.

[40]See for example: R. K. Harris *et al., J. Chem. Soc. Dalton Trans.* **1997,** 2533.

[41]R. K. Harris *et al., J. Chem. Soc. Dalton Trans.* **1996,** 3349.

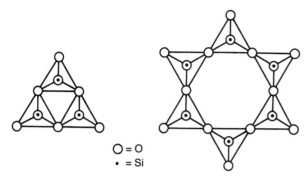

$O = O$
$\bullet = Si$

Figure 8-3 Examples of cyclic silicate anions; $Si_3O_9^{6-}$ and $Si_6O_{18}^{12-}$.

within hours. However, for Me_4N^+ solutions, which are important in zeolite synthesis (see next section), approaches to equilibrium can take days or weeks. Minerals containing alkaline-earth cations dissolve at rates that correlate with ionic size, whereas minerals containing first-row transition metals dissolve at rates that vary with the number of cation d-electrons.[42]

Simple Orthosilicates[43]

Some crystalline silicates contain discrete SiO_4^{4-} ions; since the cations are coordinated by the O atoms the nature of the structure depends on the coordination number of the cation. In *phenacite* (Be_2SiO_4) and *willemite* (Zn_2SiO_4) the cations are surrounded by a tetrahedrally arranged set of four oxygen atoms. There are a number of compounds of the type M_2SiO_4, where M^{2+} is Mg^{2+}, Fe^{2+}, Mn^{2+}, or some other cation with a preferred coordination number of 6, in which the SiO_4^{4-} anions are so arranged as to provide interstices with six oxygen atoms at the apices of an octahedron in which the cations are found. In *zircon* ($ZrSiO_4$) the Zr^{4+} ion is 8-coordinate, although not all $Zr-O$ distances are equal. It may be noted that, although the $M-O$ bonds are probably more ionic than the $Si-O$ bonds, there is doubtless some covalent character to them, and these substances should not be regarded as literally ionic in the sense $[M^{2+}]_2[SiO_4^{4-}]$ but rather as somewhere between this extreme and the opposite one of giant molecules. There are also other silicates containing discrete SiO_4 tetrahedra.

Other Discrete, Noncyclic Silicate Anions[44]

The simplest of the condensed silicate anions—that is, those formed by combining two or more SiO_4 tetrahedra by sharing of oxygen atoms—is the pyrosilicate ion ($Si_2O_7^{6-}$). This ion occurs in *thortveitite* ($Sc_2Si_2O_7$), *hemimorphite* [$Zn_4(OH)_2Si_2O_7$], and in other minerals. It is interesting that the $Si-O-Si$ angle varies from 131 to 180° in these substances. In the lanthanide silicates ($Ln_2O_3 \cdot 2SiO_2$) there are $O_3SiOSi(O)_2OSiO_3^{8-}$ ions.

Cyclic Silicate Anions

The structures of two such cyclic ions are shown schematically in Fig. 8-3. It should be clear that the general formula for any such ion must be $Si_nO_{3n}^{2n-}$. The ion

[42]W. H. Casey and H. R. Westrich, *Nature* **1992**, *355*, 157.
[43]A. Akella and D. A. Keszler, *Inorg. Chem.* **1995**, *34*, 1308.
[44]P. J. Grandinetti *et al.*, *J. Am. Chem. Soc.* **1996**, *118*, 3493.

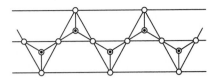

Figure 8-4 A linear chain silicate anion.

$Si_3O_9^{6-}$ occurs in *benitoite* ($BaTiSi_3O_9$) and probably in $Ca_2BaSi_3O_9$; the ion $Si_6O_{18}^{12-}$ occurs in *beryl* ($Be_3Al_2Si_6O_{18}$).

Infinite Chain Anions

The two main types of infinite chain anions are the *pyroxenes,* which contain single-strand chains of composition $(SiO_3^{2-})_n$ (Fig. 8-4) and the *amphiboles,* which contain double-strand, cross-linked chains or bands of composition $(Si_4O_{11}^{6-})_n$. Note that the general formula of the anion in a pyroxene is the same as in a silicate with a cyclic anion. Silicates with this general stoichiometry are often called "metasilicates," especially in older literature. There is actually neither metasilicic acid nor any discrete metasilicate anion. With the exception of the few "metasilicates" with cyclic anions, such compounds contain infinite chain anions.

Examples of pyroxenes are *enstatite* ($MgSiO_3$), *diopside* [$CaMg(SiO_3)_2$], and *spodumene* [$LiAl(SiO_3)_2$], the last being an important lithium ore. In the latter there is one unipositive and one tripositive cation instead of two dipositive cations. Indeed, the three compounds cited illustrate very well the important principle that within rather wide limits, *the specific cations or even their charges are unimportant as long as the total positive charge is sufficient to produce electroneutrality.* This may be easily understood in terms of the structure of the pyroxenes in which the $(SiO_3)_n$ chains lie parallel and are held together by the cations that lie between them. Obviously the exact identity of the individual cations is of minor importance in such a structure.

A typical amphibole is *tremolite,* $Ca_2Mg_5(Si_4O_{11})_2(OH)_2$. Although it would not seem to be absolutely necessary, amphiboles apparently always contain some hydroxyl groups attached to the cations. Aside from this, however, they are structurally similar to the pyroxenes, in that the $(Si_4O_{11}^{6-})_n$ bands lie parallel and are held together by the metal ions lying between them. Like the pyroxenes and for the same reason, they are subject to some variability in the particular cations incorporated.

Because of the strength of the $(SiO_3)_n$ and $(Si_4O_{11})_n$ chains in the pyroxenes and amphiboles, and also because of the relative weakness and lack of strong directional properties in the essentially electrostatic forces between them *via* the metal ions, we might expect such substances to cleave most readily in directions parallel to the chains. This is in fact the case, dramatically so in the various *asbestos* minerals. Asbestos is an imprecise commercial term referring to fibrous silicates such as *chrysotile,* white serpentine, $Mg_6Si_4O_{10}(OH)_8$, and *crocidolite,* blue amphibole, $Na_2Fe^{II}_3Fe^{III}_2Si_8O_{22}(OH)_2$. These materials, though frequently used in the past, are now considered dangerous and exposure to fibers in air can initiate a rare form of lung cancer.[45]

[45]J. A. Hardy and A. E. Aust, *Chem. Rev.* **1995,** *95,* 97.

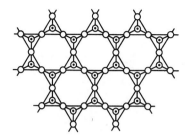

Figure 8-5 Idealized structure of a sheet silicate anion.

Infinite Sheet Anions[46]

When SiO_4 tetrahedra are linked into infinite two-dimensional networks as shown in Fig. 8-5, the empirical formula for the anion is $(Si_2O_5^{2-})_n$. Many silicates have such sheet structures with the sheets bound together by the cations that lie between them. Such substances might thus be expected to cleave readily into thin sheets, and this expectation is confirmed in the micas, which are silicates of this type.

Framework Minerals

The next logical extension in this progression from simple SiO_4^{4-} ions to larger and more complex structures would be to three-dimensional structures in which every oxygen is shared between two tetrahedra. The empirical formula for such a substance would be simply $(SiO_2)_n$; that is, we should have silica. However, if some silicon atoms in such a three-dimensional framework structure are replaced by aluminum, the framework must be negatively charged and there must be other cations uniformly distributed through it.

Aluminosilicates[47] of this type are the feldspars, zeolites, and ultramarines, which (except for the last) are among the most widespread, diverse, and useful silicate minerals in Nature. Moreover, many synthetic zeolites have been made in the laboratory, and have important uses. The feldspars are the major constituents of igneous rocks and include such minerals as *orthoclase* ($KAlSi_3O_8$), which may be written $K[(AlO_2)(SiO_2)_3]$ to indicate that one fourth of the oxygen tetrahedra are occupied by Al atoms, and *anorthite* $\{CaAl_2Si_2O_8$ or $Ca[(AlO_2)_2(SiO_2)_2]\}$, in which half the tetrahedra are AlO_4 and half are SiO_4.

A semiprecious deep-blue gem *lapis lazuli* has been known from ancient times and is available in synthetic forms under the name *ultramarine*. These are aluminosilicates of the sodalite type that contain sulfur in the form of the radical anions S_3^- and S_2^-. The former, always present, causes a deep-blue color, and when S_2^- is also present a green hue is produced.

Zeolites[48]

Zeolites, sometimes called *porotectosilicates,* are the most important framework silicates. A zeolite may be defined as an aluminosilicate with a framework structure

[46]See for example: K. Kosuge and A. Tsunashima, *J. Chem. Soc., Chem. Commun.* **1995**, 2427.
[47]T. W. Swaddle, *Chem. Soc. Rev.* **1994**, 319.
[48]S. L. Suib, *Chem. Rev.* **1993**, *93,* 803; P. K. Dutta and M. Ledney, *Prog. Inorg. Chem.* **1997**, *44,* 209.

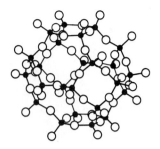

Figure 8-6 The arrangement of AlO_4 and SiO_4 tetrahedra that gives the cubooctahedral cavity in some zeolites and felspathoids; dot represents Si or Al.

enclosing cavities occupied by large ions and water molecules, both of which have considerable freedom of movement, permitting ion exchange and reversible dehydration. The framework consists of an open arrangement of corner-sharing tetrahedra where SiO_4 are partially replaced by AlO_4 tetrahedra, which requires sufficient cations to achieve electroneutrality. The cavities are occupied by H_2O molecules. An idealized formula is $M_{x/n}^{n+}[(AlO_2)_x(SiO_2)_y]\cdot nH_2O$. Magic-angle nmr spectra show that there are five distinct $Si(OAl)_n(OSi)_{4-n}$ structures, $n = 0$–4 corresponding to the tetrahedral SiO_4. There is a rule (Loewenstein's) that there is an alternation of Al and Si on the tetrahedral sites, that is, two Al atoms cannot be adjacent and there are no $Al-O-Al$ links. Some typical cavities occurring in zeolites are shown in Figs. 8-6 and 8-7. A representative formula for a naturally occurring zeolite is that of *faujasite,* $Na_{13}Ca_{11}Mg_9KAl_{55}Si_{137}O_{384}\cdot235H_2O$.

There are more than 50 natural and well over 100 synthetic zeolites, the latter all made by hydrothermal synthesis. The main uses are as molecular sieves and catalyst supports for platinum group and other metals.[49]

Molecular sieves[50] is the term first given to zeolites dehydrated by heating *in vacuum* at about 350°C. However, there are now other materials, such as phosphorus substituted zeolites, microporous silicas, and $AlPO_4$ frameworks; the latter are made hydrothermally in the presence of quaternary ammonium salts, like zeolites.

Dehydration of the synthetic zeolite Linde *A,* $Na_{12}[(AlO_2)_{12}(SiO_2)_{12}]\cdot27H_2O$, leaves cubic microcrystals in which the AlO_4 and SiO_4 tetrahedra are linked together

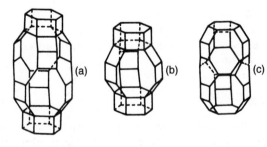

Figure 8-7 Cavities of different dimensions in (a) *chabazite* ($Ca_6Al_{12}Si_{24}O_{72}\cdot H_2O$), (b) *gmelinite* [$(Na_2Ca)_4Al_8Si_{16}O_{48}\cdot24H_2O$], and (c) *erionite* ($Ca_{4.5}Al_9Si_{27}O_{72}\cdot27H_2O$).

[49]M. E. Davis, *Acc. Chem. Res.* **1993,** *26,* 111; W. M. H. Sachtler, *Acc. Chem. Res.* **1993,** *26,* 383.
[50]See for example: A. Corma, *Chem. Rev.* **1997,** *97,* 2373 (microporous and mesoporous molecular sieves and their use in catalysis).

to form a ring of eight O atoms on each face of the unit cube and an irregular ring of six O atoms across each corner. In the center of the unit cell is a large cavity about 11.4 Å in diameter, which is connected to six identical cavities in adjacent unit cells by the eight-membered rings, which have inner diameters of ~4.2 Å. In addition, the large cavity is connected to eight smaller cavities, ~6.6 Å in diameter, by the six-membered rings, which provide openings ~2.0 Å in diameter. In the hydrated form all the cavities contain water molecules. In the anhydrous state the same cavities may be occupied by other molecules brought into contact with the zeolite, provided such molecules are able to pass through the apertures connecting cavities. Molecules within the cavities then tend to be held there by attractive forces of electrostatic and van der Waals types. Thus the zeolite will be able to absorb and strongly retain molecules just small enough to enter the cavities. It will not absorb at all those too big to enter, and it will absorb weakly very small molecules or atoms that can enter but also leave easily. For example, zeolite A will absorb straight chain but not branched chain or aromatic hydrocarbons.

High silica zeolites are exemplified by ZSM-5 (Mobil Oil) and silicalite (Union Carbide); these zeolites are also made hydrothermally using large tetraalkyl ammonium ions as templates for crystal growth; soluble double five-membered ring silicates are possible intermediates. Subsequent heating of the microcrystalline product removes organic matter and water. Magic-angle spinning nmr spectra show that there are AlO_4 tetrahedra and the general formula is $H_x[(AlO_2)_x(SiO_2)_{96-x}] \cdot 16H_2O$, where x is normally ~3 and the pore diameters about 5.5 Å. The main feature of these zeolites and other zeolites is that due to the dimensions and geometry of the channels, only certain reactants can enter and diffuse, leading to shape selectivity in catalytic reactions.

Since the catalytic, sorption, and ion-exchange properties are strongly dependent on the Al—Si ratio it is important to be able to vary this. The Al content can be decreased, for example, by treatment with $SiCl_4$, which removes Al as $AlCl_3$. It can be increased in high silica zeolites by $AlCl_3$ vapor:

$$(SiO_2)_x + 4AlCl_3 \longrightarrow Al^{3+}[(AlO_2)_3(SiO_2)_{x-3}] + 3SiCl_4$$

or aqueous KOH treatment. Use of sodium gallate also allows replacement of Si by Ga. Much information on zeolite and other silicate structures can be obtained by MAS nmr of ^{27}Al, ^{29}Si, or ^{17}O.[51]

Titanosilicates[52]

There are various Ti-substituted molecular sieves, e.g., TS-1 which is a ZSM-5 type molecular sieve.[53] They are characterized by having isomorphous substitution of some Si^{IV} by Ti^{IV} ions in the zeolite framework. These materials find wide use in synthesis and as oxidation catalysts.

[51]G. J. Kennedy *et al., J. Am. Chem. Soc.* **1994**, *116*, 11000.
[52]R. Murugavel and H. W. Roesky, *Angew. Chem. Int. Ed. Engl.* **1997**, *36*, 477.
[53]For other titanosilicates, see T. Maschmeyer *et al., Chem. Commun.* **1997**, 1847 and references therein.

Clays[54]

Among the infinite sheets previously noted are clay minerals such as *montmorillonite, kaolin,*[55] and *talc* that are ubiquitous in Nature. All are hydrous with layered structure and can accordingly intercalate molecules and ions. Like zeolites they have ion-exchange properties, often with high selectivity. They have a wide range of uses including catalytic ones where transition metal ions may be intercalated. *Pillared clays* are clays in which very large cations such as $Nb_6Cl_{12}^{2+}$ or $[Al_{13}O_4(OH)_{24}(H_2O)_{12}]^{7+}$ are intercalated to form "pillars" (8-VIII); this leads to expanded layers, improved diffusion, sorption, and catalytic properties. After heating, such pillared clays have holes or cavities and zeolite-like behavior; an advantage is their cheapness.

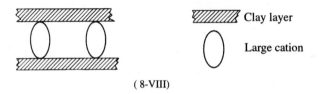

Clay layer

Large cation

(8-VIII)

8-8   Oxygen Compounds of Germanium, Tin, and Lead

Oxides and Hydroxides

Germanium dioxide is known in the quartz, stishovite, and crystobalite structures analogous to those of SiO_2 and transformations are catalyzed by ions such as Na^+. Tin dioxide exists in three different modifications of which the rutile form (in the mineral *cassiterite)* is most common; PbO_2 shows only the rutile structure. The basicity of the dioxides appears to increase from Si to Pb. Silicon dioxide is purely acidic; GeO_2 is less so and in concentrated HCl gives $GeCl_4$; SnO_2 is amphoteric, though when made at high temperatures, or by dissolving Sn in hot concentrated HNO_3 it is, like PbO_2, remarkably inert to chemical attack.

There is little evidence that there are true tetrahydroxides, $M(OH)_4$, and the products obtained by hydrolysis of hydrides, halides, alkoxides, and so on, are best regarded as hydrous oxides. Thus the addition of OH^- to Sn^{IV} solutions gives a white gelatinous precipitate that when heated is dehydrated through various intermediates and gives SnO_2 at 600°C. However, true hydroxides can be made from organometallic compounds; for example, $[(SnMe_2)_2(OH)_3]ClO_4$ is a polymer which consists of 5-coordinate Sn^{IV} units with di- and monohydroxo bridging.[56] Many examples are also known of dinuclear compounds containing the $[(SnR_2)_2(\mu\text{-}OH)_2]^{2+}$ unit.[57] Oxo-containing organometallic species such as *stannoxanes* of the type $[\{SnR_2(\mu\text{-}O)\}_n]$ are often insoluble polymers, but if bulky R groups are

[54]N. Clauer and S. Chaudhuri, *Clays in Crustal Environments,* Springer-Verlag, Berlin, 1995.
[55]H. H. Murray *et al.,* eds., *Kaolin Genesis and Utilization,* The Clay Minerals Society, Boulder, 1993.
[56]S. Funahashi *et al., J. Chem. Soc. Dalton Trans.* **1994,** 2749.
[57]A. Castineiras *et al., Inorg. Chim. Acta* **1994,** *216,* 257.

used they can be solubilized:[58]

$$Sn_2R_4 \xrightarrow[0°C/2h]{Me_3NO/hexane} [\{SnR_2(\mu\text{-}O)\}_2] \xrightarrow[\text{toluene, reflux, 5 min}]{H_2O,\ THF} [\{SnR_2(OH)\}_2(\mu\text{-}O)]$$

Oxo Anions

The known chemistry of germanates is much less extensive than that of silicates. Although there are many similarities, there are some structural differences because Ge more readily accepts higher coordination numbers. In $K_2Ge_8O_{17}$, for example, six of the Ge atoms are in GeO_4 tetrahedra but two are 5-coordinate, as silicon seldom is under normal conditions. Germanates containing the $Ge(OH)_6^{2-}$, $[Ge_3O_{10}]^{8-}$,[59] and $[Ge_6O_{18}]^{12-}$ [60] ions are also known.

In dilute aqueous solutions the major germanate ions appear to be $[GeO(OH)_3]^-$, $[GeO_2(OH)_2]^{2-}$, and $\{[Ge(OH)_4]_8(OH)_3\}^{3-}$. Fusion of SnO_2 or PbO_2 with K_2O gives K_2EO_3, which has chains of edge-sharing EO_5 square pyramids. Crystalline alkali metal stannates and plumbates can be obtained as trihydrates, for instance, $K_2SnO_3 \cdot 3H_2O$; they contain octahedral $E(OH)_6^{2-}$ anions.

8-9 Complexes of Group 14 Elements

Most of the complexes of germanium, tin, and lead in the IV oxidation state contain halide ions or donor ligands that are oxygen, nitrogen, sulfur, or phosphorus compounds.

Anionic Species

In aqueous solution the most stable fluoroanion is SiF_6^{2-}, whose high stability accounts for the incomplete hydrolysis of SiF_4 in water:

$$2SiF_4 + 2H_2O \longrightarrow SiO_2 + SiF_6^{2-} + 2H^+ + 2HF$$

The ion is usually made by attack of HF on hydrous silica and is stable even in basic solution. Although the salts that crystallize are normally those of the SiF_6^{2-} ion, the pentafluorosilicate ion is found in compounds such as $[Ph_4As][SiF_5]$. Typical reactions producing SiF_5^- salts are

$$SiO_2 + 5HF(aq) + [R_4N]Cl \xrightarrow{CH_3OH} [R_4N]SiF_5 + HCl + 2H_2O$$

$$SiF_4 + [R_4N]F \xrightarrow{CH_3OH} [R_4N]SiF_5$$

The nmr data for the SiF_5^- ion and also for the similar species, $RSiF_4^-$ and $R_2SiF_3^-$, indicate *tbp* structures,[61] but above $-60°C$ exchange processes do occur; in phenyl

[58]M. A. Edelman *et al.*, *J. Chem. Soc., Chem. Commun.* **1990,** 1116.
[59]C. Linke and M. Jansen, *Z. Naturforsch.* **1996,** *51b*, 1591.
[60]H.-J. Brandt and H. H. Otto, *Z. Kristallogr.* **1997,** *212*, 34.
[61]R. R. Holmes, *J. Am. Chem. Soc.* **1989,** *111*, 3250.

compounds, the Ph groups are equatorial. Six-coordinate $NH_4[Si(NH_3)F_5]$ can be prepared by reaction of silicon powder with NH_4HF_2 in sealed Monel ampoules at 400°C.[62] Several GeF_5^- salts are also known.

Germanium,[63] tin, and lead also form hexafluoro anions; for example, dissolution of GeO_2 in aqueous HF followed by the addition of KF at 0°C gives crystals of K_2GeF_6. The Ge and Sn anions are hydrolyzed by bases, but most Pb salts are hydrolyzed even by water. Many tin species, $SnF_{6-n}X_n^{2-}$, X = OH^-, Cl^-, Br^-, and so on have been studied by nmr. Anhydrous hexafluorostannates can be made by dry fluorination of the stannates ($M_2^ISnO_3\cdot3H_2O$).

The hexachloro ions of Ge and Sn are normally made by the action of HCl or M^ICl on MCl_4. The thermally unstable yellow salts of $PbCl_6^{2-}$ are obtained by the action of HCl and Cl_2 on $PbCl_2$. Under certain conditions, pentachloro complexes of Ge and Sn may be stabilized, for example, by the use of $(C_6H_5)_3C^+$ as the cation or by the interaction of MCl_4 and $Bu_4N^+Cl^-$ in $SOCl_2$ solution.

Other anionic species include the ions $[Ge(SO_3F)_6]^{2-}$ and $[Sn(SO_3F)_5]^-$ formed by oxidation of the element with $S_2O_6F_2$ in HSO_3F, the nitrate $[Sn(NO_3)_6]^{2-}$, and $SnCl_5(THF)^-$.[64] There are also distorted *tbp* spirocyclic oxygen or sulfur chelated anions (e.g., 8-IV and 8-V) that are nonrigid in solution, oxalates $[M(ox)_3]^{2-}$, and other carboxylates.

Cationic Species[65]

There are comparatively few cationic complexes, the most important being the octahedral β-diketonates and tropolonates of Si and Ge such as $[Ge(acac)_3]^+$ and Si $trop_3^+$. "Siliconium" ions can also be formed by reactions such as

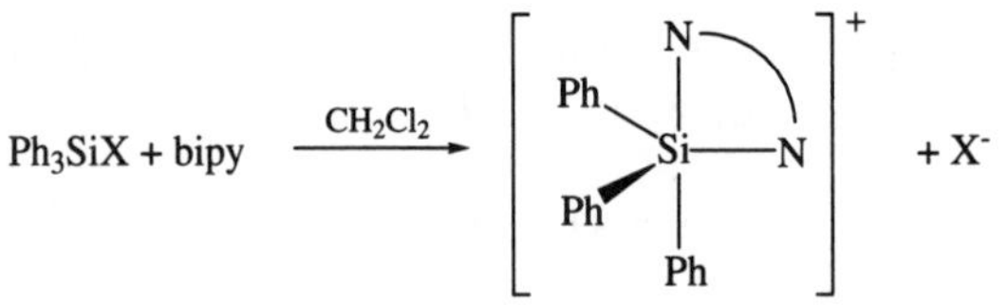

or by oxidation of $SiCl_2(bipy)_2$ to *cis*-$[SiCl_2(bipy)_2]^{2+}$.

An interesting cyclotrigermenium cation[66] with a 2π system can be isolated in high yield as air- and moisture-sensitive yellow crystals:

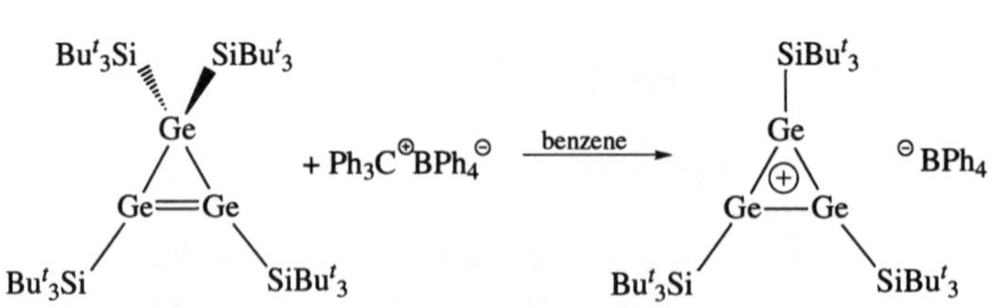

[62]C. Plitzko and G. Meyer, *Z. anorg. allg. Chem.* **1996,** *622,* 1646.
[63]J. Stepien-Damm *et al., Z. Kristallogr.* **1996,** *211,* 936.
[64]G. R. Willey *et al., J. Chem. Soc. Dalton Trans.* **1997,** 2677.
[65]R. J. P. Corriu *et al., Angew. Chem. Int. Ed. Engl.* **1994,** *33,* 1097.
[66]P. R. Schleyer, *Science* **1997,** *275,* 39.

The three germanium atoms form an equilateral triangle similar to that of the carbon analogue, the cyclopropenium ion;[67] the cation has Ge—Ge bonds (Section 8-12).

Neutral Species[68]

These are numerous and quite varied in type. Some are 4-coordinate such as SnX_4 and $X_nSn(NR_2)_{4-n}$[69] but the majority are 6-coordinate, examples being *trans*-$SnCl_2$ (β-dike)$_2$, $SnCl_2(S_2CNEt_2)_2$, $Sn[(OC_2H_4)_2N(C_2H_4OH)]_2$, and porph SnX_2.[70] Both lower and higher coordination numbers also occur, examples being five in PhSi(*o*-$C_6H_4O_2$)$_2$ and presumably seven or eight in $Sn(S_2CNEt_2)_4$.

Adducts

The tetrahalides are prone to add neutral ligands to form adducts that are usually 6-coordinate. Typical examples are *trans*-$SiF_4(py)_2$, *cis*-$SiF_4(bipy)$, $SiCl_4L_2$ (L = py, PMe_3), and numerous *cis*-$SnX_4(L—L)$ and *cis* or *trans*-SnX_4L_2 compounds.

The Lewis acid order is $SnCl_4 > SnBr_4 > SnI_4$.

8-10 Alkoxides, Carboxylates, and Oxo Salts

All four elements form *alkoxides*, but those of silicon, for example, $Si(OC_2H_5)_4$, often called *silicate esters*, are the most important; the surface of glass or silica can also be alkoxylated. They can be used in the synthesis of ceramic materials.[71] Alkoxides are normally obtained by the standard method:

$$MCl_4 + 4ROH + 4\ \text{amine} \longrightarrow M(OR)_4 + 4\ \text{amine·HCl}$$

Silicon alkoxides are rapidly hydrolyzed by water, eventually to hydrous silica, but polymeric hydroxo alkoxo intermediates occur. Organo alkoxides such as the *silyl ester* (MeO)$_3$SiMe and its derivatives are widely known.[72]

Of the *carboxylates*, *lead tetraacetate* is the most important because it is used in organic chemistry as a strong but selective oxidizing agent.[73] It is made by dissolving Pb_3O_4 in hot glacial acetic acid or by electrolytic oxidation of Pb^{II} in acetic acid. In oxidations the attacking species is generally considered to be $Pb(O_2CMe)_3^+$, which is isoelectronic with the similar oxidant $Tl(O_2CMe)_3$, but this is not always so, and some oxidations are known to be free radical in nature. The trifluoroacetate is a white solid, which will oxidize even heptane to give the CF_3CO_2R species, from which the alcohol ROH is obtained by hydrolysis; benzene similarly gives phenol.

The tetraacetates of Si, Ge, Sn, and Pb also form complex anions such as $[Pb(O_2CMe)_6]^{2-}$ or $[Sn(O_2CMe)_5]^-$. For $M(O_2CMe)_4$, Si and Ge are 4-coordinate

[67]A. Sekiguchi *et al.*, *Science* **1997**, *275*, 60.
[68]C. Y. Wong and J. D. Woollins, *Coord. Chem. Rev.* **1994**, *130*, 175.
[69]See for example: H. Schmidbaur *et al.*, *Chem. Ber./Recueil* **1997**, *130*, 1159; 1167.
[70]D. P. Arnold and J. P. Bartley, *Inorg. Chem.* **1994**, *33*, 1486.
[71]T. J. Boyle and R. W. Schwartz, *Comments Inorg. Chem.* **1994,** *16*, 243.
[72]J. G. Verkade *et al.*, *Inorg. Chem.* **1990**, *29*, 1065.
[73]T. L. Holton and H. Shechter, *J. Org. Chem.* **1995**, *60*, 4725.

with unidentate acetate; Pb has only bidentate acetates, whereas the smaller Sn has a very distorted dodecahedron.

Tin(IV) sulfate, $Sn(SO_4)_2 \cdot 2H_2O$, can be crystallized from solutions obtained by oxidation of Sn^{II} sulfate; it is extensively hydrolyzed in water.

Tin(IV) nitrate is obtained as a colorless volatile solid by interaction of N_2O_5 and $SnCl_4$; it contains bidentate NO_3^- groups giving dodecahedral coordination. The compound reacts with organic matter.

8-11   Organo Compounds

The general formula is $R_{4-n}EX_n(n = 0$ to 3), where R is alkyl or aryl and X is any of a wide variety of atoms or groups (H, halogen, OR', NR_2', SR', $Mn(CO)_5$, etc.). The elements may also form part of heterocyclic rings or cages,[74] for example $(R_2EO)_3$ or the silsesquioxanes $[RSiO_{3/2}]_n$[75] such as $R_8Si_8O_{12}$ (8-IX).

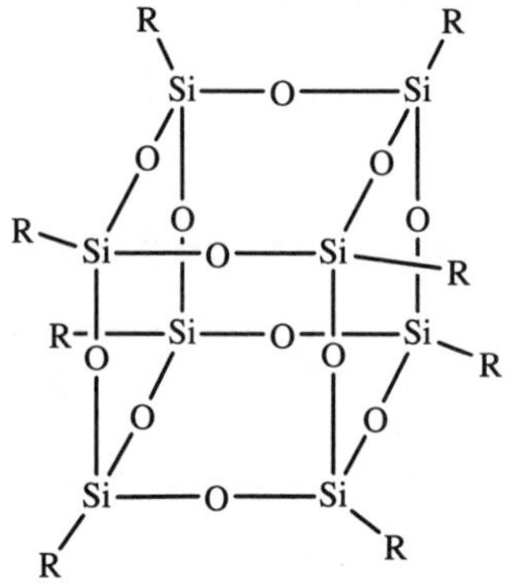

(8-IX)

For a given class of compounds, members with C—Si and C—Ge bonds have higher thermal stability and lower reactivity than those with bonds to Sn and Pb. In catenated compounds similarly, Si—Si and Ge—Ge bonds are more stable and less reactive than Sn—Sn and Pb—Pb bonds; for example, Si_2Me_6 is very stable, but Pb_2Me_6 blackens in air and decomposes rapidly in CCl_4, although it is fairly stable in benzene.

The bonds to carbon are usually made *via* interaction of lithium, mercury, or aluminum alkyls or RMgX and the Group 14 halide, but there are many special synthetic methods, some of which are noted later.

Silicon and Germanium

The organo compounds of Si and Ge are very similar in their properties. We discuss only Si compounds.

[74]See for example: P. G. Harrison, *J. Organometal. Chem.* **1997,** *542,* 141 (a review on silicate cages, 221 references).
[75]F. J. Feher *et al., Chem. Commun.* **1997,** 1185.

Silicon-carbon bond dissociation energies are less than those of C—C bonds but are still quite high, in the region 250 to 335 kJ mol^{-1}. The tetraalkyls and -aryls are hence thermally quite stable; $Si(C_6H_5)_4$, for example, boils unchanged at 530°C.

The chemical reactivity of Si—C bonds is generally greater than that of C—C bonds because (a) the greater polarity of the bond $Si^{\delta+}$—$C^{\delta-}$ allows easier nucleophilic attack on Si and electrophilic attack on C than for C—C compounds, and (b) displacement reactions at silicon are facilitated by its ability to expand the coordination number above 4 by utilization of d orbitals.

The reactions of Si compounds have no mechanism analogous to S_N1 reactions at carbon and are generally complicated. Substitution reactions at 4-coordinate silicon characteristically proceed *via* an associative mechanism involving 5-coordinate transition states. Retention or inversion of stereochemistry may occur depending on the nature of the entering or leaving groups, namely,

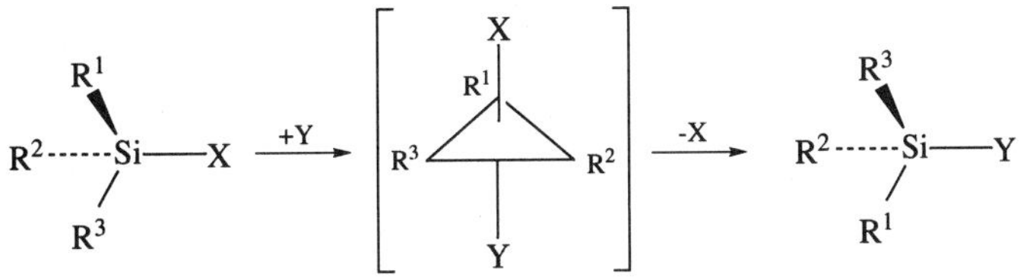

With the same leaving group, both retention and inversion can be observed; hard nucleophiles tend to attack equatorially to give retention, soft ones apically leading to inversion. Mechanisms depend crucially on the solvent used. If this has donor ability like DMF or DMSO it may attack first to form the 5-coordinate species and is then displaced by an incoming nucleophile.

A characteristic feature of organosilicon (and -germanium) chemistry, setting it strikingly apart from carbon chemistry, is the great ease with which R_3Si (and R_3Ge) groups migrate; a factor of up to 10^{12} as compared to analogous carbon compounds is typical. Among the best studied migration reactions are anionic 1,2-shifts, represented generally by the equation:

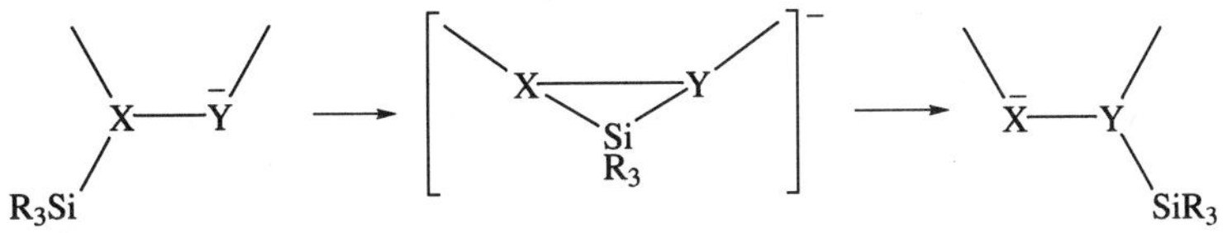

where X—Y may be N—N, O—N, or S—C. As indicated, a transition state involving 5-coordinate Si is postulated; since carbon has no valence shell d orbitals, it cannot form such a transition state easily, and such 1,2-shifts are "forbidden" in the Woodward-Hoffmann sense.

Radicals are less important in silicon than in carbon chemistry. However, silicon radicals have been detected in solution by esr and have been isolated in matrices. They are made by hydrogen abstraction with t-butoxy and other radicals generated photochemically, for example,

$$R_3SiH + Me_3CO^{\bullet} \longrightarrow R_3Si^{\bullet} + Me_3COH$$

The stable $R_3Si^{\cdot}$ radicals with extremely bulky R groups have been isolated. A comparison of the rates of reactions such as

$$p\text{-}XC_6H_4ER_3 + H_2O \longrightarrow R_3EOH + C_6H_5X$$

in aqueous-methanolic $HClO_4$ gives the order $Si(1) < Ge(36) \ll \ll Sn(3 \times 10^5) \ll Pb(2 \times 10^8)$, which suggests that with increasing size, there is increased availability of outer orbitals, which allows more rapid initial solvent coordination to give the 5-coordinated transition state.

Another stable radical is obtained by treatment of GeClR, R = 2,6-$C_6H_3mes_2$, with 1 equivalent of KC_8.[76] The dark crystalline $Ge_3L_3^{\cdot}$ has a triangular cyclogermenyl core (see Section 8-12) of the type (8-X); the core is also similar to that of the cyclotrigermanium cation mentioned in Section 8-9.

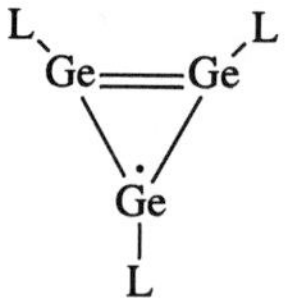

(8-X)

Alkyl- and Arylsilicon Halides

These compounds are of special importance because of their hydrolytic reactions. They may be obtained by normal Grignard procedures from $SiCl_4$, or, in the case of the methyl derivatives, by the Rochow process, in which methyl chloride is passed over a heated, copper-activated silicon:

$$CH_3Cl + Si(Cu) \longrightarrow (CH_3)_nSiCl_{4-n}$$

The halides are liquids that are readily hydrolyzed by water, usually in an inert solvent. In certain cases, the silanol intermediates R_3SiOH, $R_2Si(OH)_2$, and $RSi(OH)_3$[77] can be isolated, especially if they are stabilized by large ligands, for example:

$$RSiCl_3 \xrightarrow{\text{H}_2\text{O/aniline}} RSi(OH)_3$$

R = Bu^t, $Co_3(CO)_9C$, and Bu^t

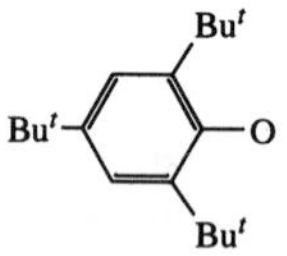

However, with smaller substituents the diols and triols frequently condense, under the hydrolysis conditions, to siloxanes that have $Si-O-Si$ bonds.[78] The exact nature

---

[76]P. P. Power *et al., Chem. Commun.* **1997**, 1595.
[77]H. W. Roesky *et al, Acc. Chem. Res.* **1996**, *29*, 183.
[78]See for example: R. H. Baney *et al., Chem. Rev.* **1995**, *95*, 1409; D. A. Loy and K. J. Shea, *Chem. Rev.* **1995**, *95*, 1431; P. A. Agaskar and W. G. Klemperer, *Inorg. Chim. Acta* **1995**, *229*, 355.

of the products depends on the hydrolysis conditions and linear, cyclic, and complex cross-linked polymers of varying molecular weights can be obtained. They are often referred to as silicones; the commercial polymers usually have R = CH_3, but other groups may be incorporated for special purposes.

Controlled hydrolysis of the alkyl halides in suitable ratios can give products of particular physical characteristics. The polymers may be liquids, rubbers, or solids, which have in general high thermal stability, high dielectric strength, and resistance to oxidation and chemical attack.

Examples of simple siloxanes are $Ph_3SiOSiPh_3$ and the cyclic trimer or tetramer $(Et_2SiO)_{3(or\ 4)}$; linear polymers contain $-SiR_2-O-SiR_2-O-$ chains, whereas the cross-linked sheets have the basic unit $RSi(O-)_3$.

A very large number of hetero- and metallasiloxanes derived from silanediols, disilanols, silanetriols, and trisilanols have been prepared[79] and their chemistries have been reviewed.[80] An example is:

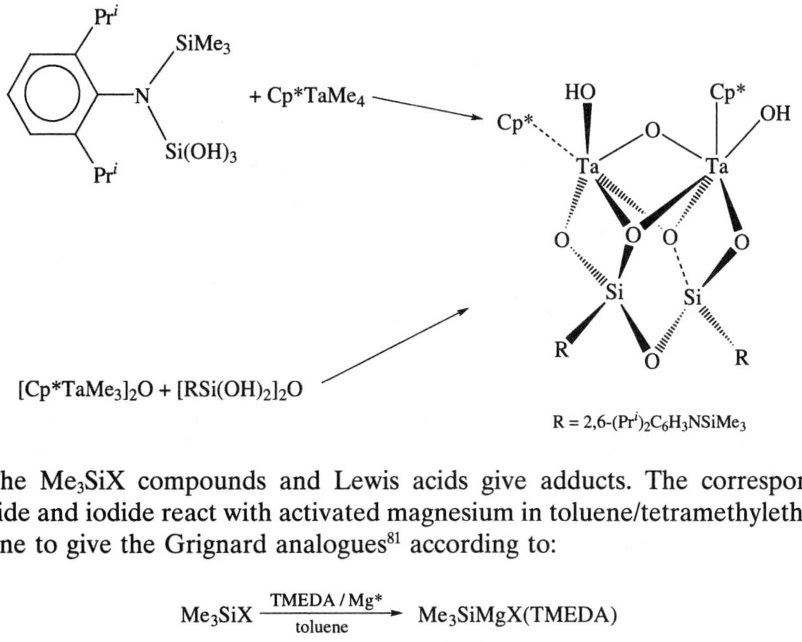

R = 2,6-$(Pr^i)_2C_6H_3NSiMe_3$

The Me_3SiX compounds and Lewis acids give adducts. The corresponding bromide and iodide react with activated magnesium in toluene/tetramethylethylen-diamine to give the Grignard analogues[81] according to:

$$Me_3SiX \xrightarrow[\text{toluene}]{\text{TMEDA}/Mg^*} Me_3SiMgX(TMEDA)$$

Tin

Where the compounds of tin differ from those of Si and Ge they do so mainly because of a greater tendency of Sn^{IV} to show coordination numbers >4. The *tetraalkyl complexes* such as Me_4Sn are essentially tetrahedral with Sn$-$C distances in the range 2.10 $-$ 2.14 Å;[82] most tetraaryl compounds have idealized S_4 symmetry. They react with bases to give compounds such as Li$[SnMe_5]$ and $R_4Sn(bipy)$.

[79]See for example: W. A. Hermann *et al.*, *Angew. Chem. Int. Ed. Engl.* **1994**, *33*, 1285; H. W. Roesky *et al.*, *Inorg. Chem.* **1997**, *36*, 3392.
[80]H. W. Roesky *et al.*, *Chem. Rev.* **1996**, *96*, 2205.
[81]A. Ritter *et al.*, *Angew. Chem. Int. Ed. Engl.* **1995**, *34*, 1030
[82]B. Krebs *et al.*, *Acta Crystallogr.*, **1989**, *C49*, 1066.

Trialkyltin compounds of the type R_3SnX, of which the best studied are the CH_3 compounds ($X = ClO_4$, F, NO_3, etc.), are of interest in that they are always associated in the solid by anion bridging (8-XI and 8-XII); the coordination of the tin atom is close to *tbp* with planar $SnMe_3$ groups. When X is RCO_2 the compounds may in addition be mononuclear with unidentate or bidentate carboxylate groups.

(8-XI)   (8-XII)

The R_3SnX (and also R_3PbX) compounds also form 1:1 and 1:2 adducts with Lewis bases, and these also generally appear to contain 5-coordinate Sn, with the alkyl groups axial. The highly specific action of R_3SnX compounds in biological systems where toxicity is independent of X is probably due to dissociation. In water the trimethyl perchlorate gives $[Me_3Sn(H_2O)_2]^+$.

Dialkyltin compounds (R_2SnX_2) have behavior similar to that of the trialkyl compounds. Thus the fluoride Me_2SnF_2 is again polymeric, with bridging F atoms, but Sn is octahedral and the Me—Sn—Me group is linear. However, the chloride and bromide have low melting points (90 and 74°C) and are essentially molecular compounds, only weakly linked by halogen bridges. The nitrate $Me_2Sn(NO_3)_2$ is strictly molecular with bidentate nitrate groups.

The halides also give conducting solutions in water and the aqua ion has the linear C—Sn—C group characteristic of the dialkyl species (*cf.* the linear species Me_2Hg, Me_2Tl^+, Me_2Cd, and Me_2Pb^{2+}), probably with four water molecules completing octahedral coordination. The linearity in these species appears to result from maximizing of *s* character in the bonding orbitals of the metal atoms. The ions Me_2SnCl^+ and Me_2SnOH^+ also exist, and in alkaline solution *trans*-$[Me_2Sn(OH)_4]^{2-}$ occurs.

Catenated organotin compounds are discussed in Section 8-12.

Organotin hydrides (R_3SnH), which can be made by $LiAlH_4$ reduction of the halide or in other ways, are useful reducing agents in organic and inorganic chemistry;[83] some of the reactions are known to proceed by free-radical pathways. The hydrides undergo additional reactions with alkenes or alkynes similar to the hydrosilation reaction, which provides a useful synthetic method for organotin compounds containing functional groups.

$$R_3SnH + R_2C{=}CR'_2 \longrightarrow R_3SnCR_2CR'_2H$$

Unlike hydrosilation (Section 8-5), hydrostannation proceeds *via* radical chains propagated by $R_3Sn^{\cdot}$ radicals; intermediate radicals $R_3SnCR_2CR_2^{\cdot}$ have been characterized. The rates for hydrostannation reactions are greatly increased (100–600 times) under sonication conditions.[84]

[83]F. A. Cotton *et al., Inorg. Chim. Acta* **1996,** *252,* 239.
[84]E. Nakamura *et al., J. Am. Chem. Soc.* **1989,** *111,* 6849.

Organotin compounds are widely used as marine antifouling compounds, wood preservatives, stabilizers for poly(vinyl chloride), antihelmintics, and so on.

Tin sulfide in marine sediments can react with CH_3I—a ubiquitous biogenic molecule in sea water—to give tin methyls:

$$SnS + 4MeI \longrightarrow MeSnI_3 + Me_3SI$$

Finally, oxygen-containing organometallic species such as insoluble stannoxanes have been mentioned (Section 8-8). A soluble cyclodistannoxane is $[SnR_2(\mu\text{-O})]_2$, $R = CH(SiMe_3)_2$; it has chalcogenide analogues $(R = Bu^t)$.[85]

Lead[86]

There is an extensive organolead chemistry, but the best known compounds are $PbMe_4$ and $PbEt_4$. These compounds were made in large quantities for use as antiknock agents in gasoline,[87] although their use has been essentially eliminated in favor of nonpolluting or less polluting oxygenates such as methyl t-butyl ether, dimethyl carbonate, and MeOH.

The major commercial synthesis is by the interaction of a sodium-lead alloy with CH_3Cl or C_2H_5Cl in an autoclave at 80 to 100°C, without solvent for C_2H_5Cl but in toluene at a higher temperature for CH_3Cl. The reaction is complicated and not fully understood, and only a quarter of the lead appears in the desired product:

$$4NaPb + 4RCl \longrightarrow R_4Pb + 3Pb + 4NaCl$$

The required recycling of the lead is disadvantageous, and electrolytic procedures have been developed. One process involves electrolysis of $NaAlEt_4$ with a lead anode and mercury cathode; the sodium formed can be converted into NaH and the electrolyte regenerated:

$$4NaAlEt_4 + Pb \longrightarrow 4Na + PbEt_4 + 4AlEt_3$$

$$4Na + 2H_2 \longrightarrow 4NaH$$

$$4NaH + 4AlEt_3 + 4C_2H_4 \longrightarrow 4NaAlEt_4$$

Tetraethyllead is conveniently prepared on a laboratory scale by reacting $Pb(O_2CCH_3)_4$ and EtMgCl in THF.

The alkyls are nonpolar, highly toxic liquids. The tetramethyl begins to decompose around 200°C and the tetraethyl around 110°C, by free-radical mechanisms.

The coordination numbers vary over a wide range for Pb^{IV} in organometallic compounds. Four-coordinate acyl complexes[88] can be made by reaction of Pb_2mes_6

[85]M. F. Lappert et al., J. Chem. Soc., Chem. Commun. **1990**, 1116.
[86]See for example: K. H. Pannell et al., Organometallics **1993**, 12, 4278; K. Dehnicke et al., Z. Naturforsch. **1997**, 52b, 149.
[87]S. W. Benson, J. Phys. Chem. **1988**, 92, 1531.
[88]K. H. Pannell et al., Organometallics **1993**, 12, 4278.

and Li followed by the addition of acyl chloride:

$$mes_3Pb-Pbmes_3 + 2Li \longrightarrow 2[mes_3Pb^-]Li^+ \xrightarrow{2RCOCl} 2mes_3PbCOR + 2LiCl; R = Me, Ph$$

In the crystal, $PbMe_3I$ forms polymeric chains (8-XIII) *via* bent μ_2-I bridges and almost linear $I-Pb-I$ units.[89]

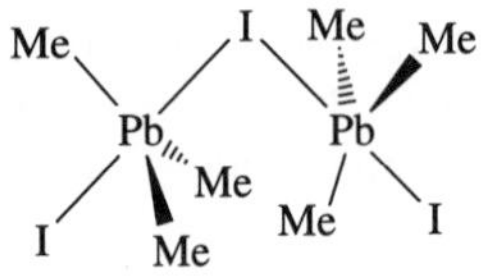

(8-XIII)

Six-coordination is observed in $[PhPbCl_5]^{2-}$[90] while 7-coordination is seen in $(Ph)_2Pb(2,6\text{-pyridinedicarboxylate})(H_2O)_2$.[91]

ELEMENT TO ELEMENT AND MULTIPLE BONDING

8-12 Silanes, Germanes, Stannanes, Plumbanes, and Oligomers

Catenation is widespread among the elements of Group 14. As indicated earlier, the tendency to catenation decreases in the order $C \gg Si > Ge \approx Sn \gg Pb$.

Small Chains

Organocompounds with bonds between two group 14 elements are legion. Those of the type $R_3E-E'R_3'$[92] (E and E' can be any element of the group: C, Si, Ge, Sn, and Pb; where R and R' can be organic groups such as methyl or phenyl) are known for essentially all combinations of E and E'.

They can be prepared in various ways which involve the use of halide compounds and lithium salts[93] or magnesium metal,[94] for example:

$$PbCl_2 + 2(THF)_3Li[Si(SiMe_3)_3] \xrightarrow[OEt_2]{-78 \text{ to } 25\,°C} [(Me_3Si)_3Si]_2 + Pb + 2LiCl$$

$$2\{(Me_3Si)_3C\}Me_2PbBr + Mg \xrightarrow{THF} [\{(Me_3Si)_3C\}Me_2Pb]_2 + MgBr_2$$

There are numerous compounds with longer chains that are relatively stable. For example, the reaction of sodium in liquid ammonia with $Sn(CH_3)_2Cl_2$ gives $[Sn(CH_3)_2]_n$, which consists mainly of linear molecules with chain lengths of 12 to 20 (and perhaps more), as well as at least one cyclic compound, $[Sn(CH_3)_2]_6$. Linear

[89]K. Dehnicke *et al., Z. Naturforsch.* **1997,** *52b,* 149.
[90]A. V. Yaysenko *et al., Zh. Strukt. Khim.* **1989,** *30,* 1192.
[91]H. Preut *et al., Acta Crystallogr.* **1988,** *C44,* 755.
[92]L. Párkányi *et al., Inorg. Chem.* **1996,** *35,* 6622; **1994,** *33,* 180; M. Dräger *et al., Organometallics* **1994,** *13,* 2733; K. H. Pannell *et al., Inorg. Chem.* **1992,** *31,* 522.
[93]R. A. Geanangel *et al., Inorg. Chem.* **1992,** *31,* 1626; **1994,** *33,* 6357; **1993,** *32,* 602.
[94]K. H. Pannell *et al., Inorg. Chem.* **1994,** *33,* 6406.

polystannanes[95] can be prepared in a more straightforward manner by employing the hydrostannolysis reaction and diorganotin hydrides, for example:

$$2R'_3Sn\text{—}NMe_2 + H\text{—}(R_2Sn)_n\text{—}H \longrightarrow R'_3Sn\text{—}(R_2Sn)_n\text{—}SnR'_3 + 2HNMe_2$$

Similarly, branched polystannanes can be made according to:

$$RSnH_3 + 3Me_2N\text{—}SnMe_3 \longrightarrow (Me_3Sn)_3SnR + 3HNMe_2$$

There is also a series of anions of the type ER_3^-. The best studied are the *silyl anions*,[96] especially the silyllithiums. For example, $Si(SiMe_3)_3^-$ is a versatile reagent for the synthesis of a great variety of polysilyl derivatives. It can be prepared by the reaction of $Si(SiMe_3)_4$ and methyllithium[97] according to

$$Si(SiMe_3)_4 + MeLi \xrightarrow[\text{4 days}]{\text{THF}} (Me_3Si)_3SiLi(THF)_3 + Me_4Si$$

Polymers

Long chains of silicon,[98] germanium,[99] and tin[100] atoms are found in polymers which have average molecular weights as high as *ca.* 1×10^6. They find use as moulding materials, rubbers, and ceramic precursors.

The polymers show extensive σ-delocalization and they are of various types,[101] such as

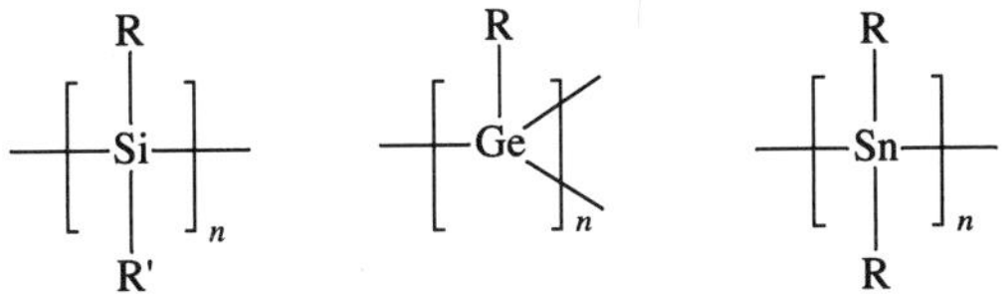

They can be made in various ways, for example:

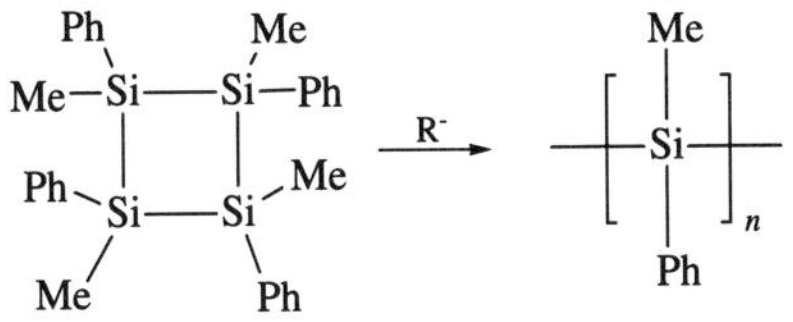

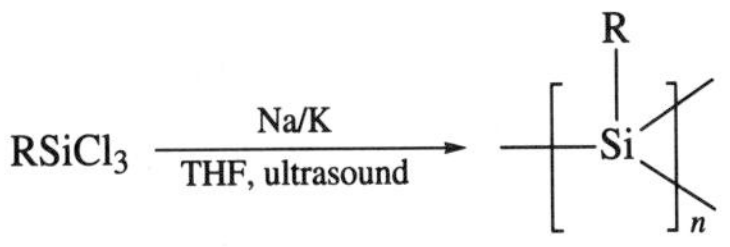

[95]See for example: L. R. Sita, *Adv. Organomet. Chem.* **1995,** *38,* 189 (a review on the structure and property relationships of polystannanes).
[96]K. Tamao and A. Kawachi, *Adv. Organomet. Chem.* **1995,** *38,* 1.
[97]G. M. Sheldrick *et al., Inorg. Chem.* **1993,** *32,* 2694.
[98]E. Fossum and K. Matyjaszewski, *Macromolecules* **1995,** *28,* 1618.
[99]P. A. Bianconi *et al., Macromolecules* **1993,** *26,* 869.
[100]J. R. Babcock and L. R. Sita, *J. Am. Chem. Soc.* **1996,** *118,* 12481.
[101]I. Manners, *Angew. Chem. Int. Ed. Engl.* **1996, ** *35,* 1602.

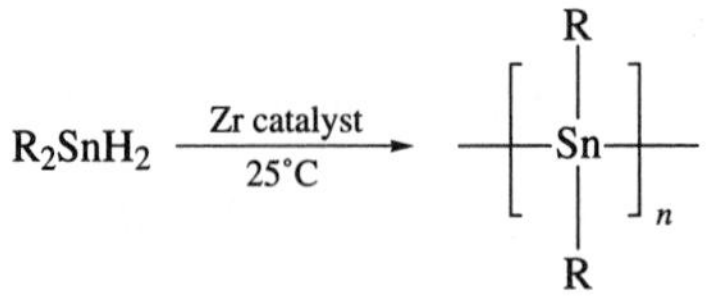

$$R_2SnH_2 \xrightarrow[25°C]{\text{Zr catalyst}} \left[\begin{array}{c} R \\ | \\ Sn \\ | \\ R \end{array} \right]_n$$

Rings and Cages[102]

The smallest and perhaps the most important ones are the cyclotrisilanes,[103] -germanes,[104] -stannanes,[105] and their mixed compounds.[106] These are normally prepared by reaction of halides of the type R_2EX_2 with sodium or lithium naphthalenide, as in

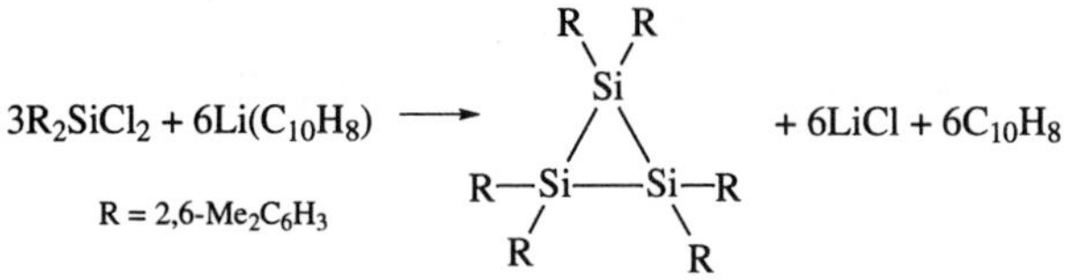

$$3R_2SiCl_2 + 6Li(C_{10}H_8) \longrightarrow \begin{array}{c} R \quad R \\ \diagdown / \\ Si \\ / \quad \diagdown \\ R{-}Si{-}{-}Si{-}R \\ / \qquad \diagdown \\ R \qquad\quad R \end{array} + 6LiCl + 6C_{10}H_8$$

$$R = 2,6\text{-Me}_2C_6H_3$$

$$3(Et_3Si)_2SiBr_2 + 6Na \longrightarrow [(Et_3Si)_2Si]_3 + 6NaCl$$

They are very reactive; an important synthetic reaction for which they are used is the formation of silylenes and disilenes:

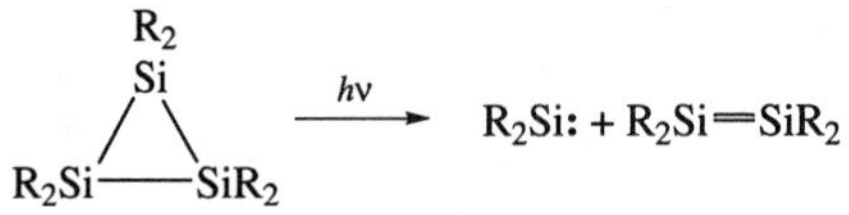

$$\begin{array}{c} R_2 \\ Si \\ \diagup \diagdown \\ R_2Si{-}{-}{-}SiR_2 \end{array} \xrightarrow{h\nu} R_2Si: + R_2Si{=}SiR_2$$

If the size of the R group is small, then the R_2Si: species react further to give more $R_2Si{=}SiR_2$.

Other reactions involve the cleavage of all Si—Si bonds as in

$$[(2\text{-}(Me_2NCH_2C_6H_4)_2Si]_3 \xrightarrow{\Delta} 3R_2Si:$$

or the cleavage of only one Si—Si bond[107] as in

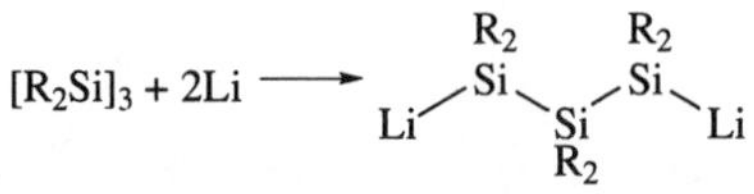

$$[R_2Si]_3 + 2Li \longrightarrow \begin{array}{c} R_2 \qquad R_2 \\ Si \qquad Si \\ \diagup \diagdown \diagup \diagdown \\ Li \qquad Si \qquad Li \\ R_2 \end{array}$$

[102]A. Sekiguchi and K. Sakurai, *Adv. Organomet. Chem.* **1995,** *37,* 1.
[103]M. Weidenbruch, *Chem. Rev.* **1995,** *95,* 1479.
[104]M. Weidenbruch *et al., Angew. Chem. Int. Ed. Engl.* **1995,** *34,* 1085.
[105]M. Weidenbruch *et al., Angew. Chem. Int. Ed. Engl.* **1994,** *33,* 1846.
[106]A. Heine and D. Stalke, *Angew. Chem. Int. Ed. Engl.* **1994,** *33,* 113; W. Ando *et al., Angew. Chem. Int. Ed. Engl.* **1994,** *33,* 659; *Organometallics* **1993,** *12,* 803; G. Maier *et al., Angew. Chem. Int. Ed. Engl.* **1995,** *34,* 1439.
[107]J. Belzner *et al., Angew. Chem. Int. Ed. Engl.* **1994,** *33,* 2450.

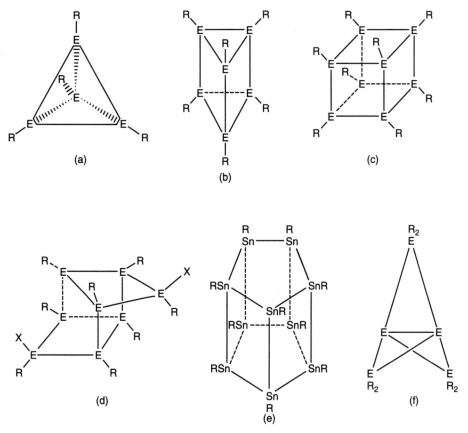

Figure 8-8 Some cluster structures for silicon, germanium, and tin.

Cyclotrigermanium cations and radicals have been mentioned in Sections 8-9 and 8-11, respectively.

There are also polycyclic compounds such as cyclotetra-, cyclopenta-, and cyclo-hexasilanes[108] as well as Ge and Sn derivatives[109] and a large number of clusters such as tetrahedral $(ER)_4$ compounds,[110] $(ER)_6$ trigonal prisms,[111] $(ER)_8$ cubanes,[112] propellanes,[113] and other types of structures.[114] A few cluster structures are shown in Fig. 8-8.

[108]E. Hengge and R. Janoschek, *Chem. Rev.* **1995,** *95,* 1495.
[109]S. P. Mallela and R. A. Geangel, *Inorg. Chem.* **1994,** *33,* 1115.
[110]N. Wiberg *et al., Angew. Chem. Int. Ed. Engl.* **1996,** *35,* 1333.
[111]H. Sakurai *et al., Angew. Chem. Int. Ed. Engl.* **1989,** *28,* 55.
[112]A. Sekiguchi *et al., J. Am. Chem. Soc.* **1992,** *114,* 6260.
[113]L. R. Sita and I. Kinoshita, *J. Am. Chem. Soc.* **1992,** *114,* 7024.
[114]S. Nagase, *Acc. Chem. Res.* **1995,** *28,* 469; L. R. Sita and R. D. Bickerstaff, *J. Am. Chem. Soc.* **1989,** *111,* 3769.

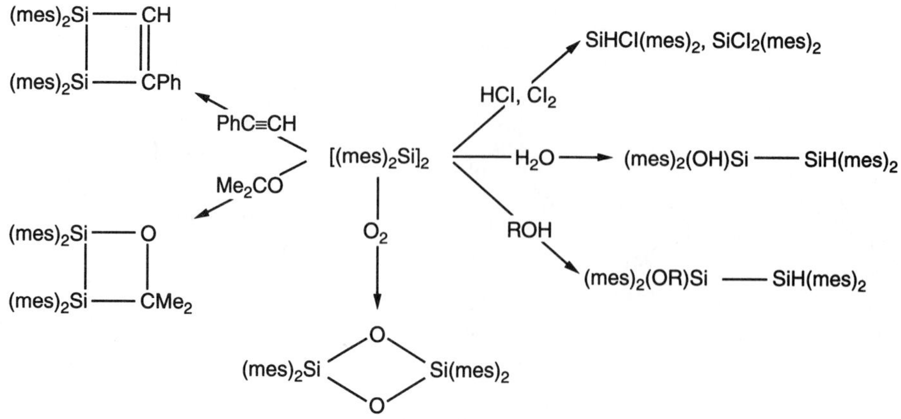

Figure 8-9 Reaction of tetramesitylsilene (mes = mesityl).

8-13 Silenes,[115] Germenes,[116] and Stannenes[117]

The first isolable disilene compound, a yellow tetramesityl disilene, was made by R. West *via* the reaction

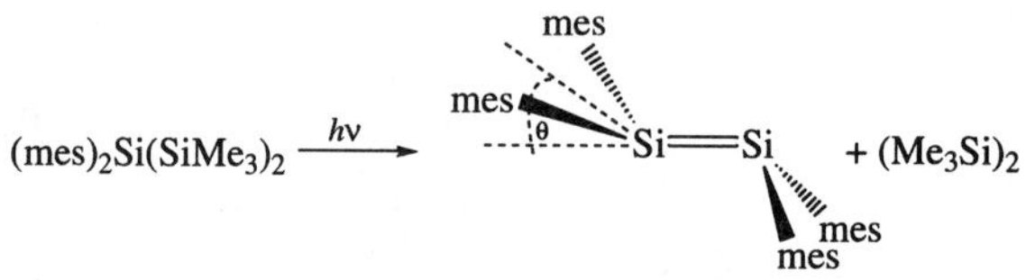

It is an air sensitive but reasonably thermally stable solid with a slightly bent, trans structure ($\theta = 18°$). In most other disilenes the θ angle is smaller (*ca.* 0°) and the Si atoms and the atoms to which they are bonded are nearly coplanar. The typical Si=Si bond (2.15 ± 0.01 Å) is ~10% shorter than a typical Si—Si bond. Other factors consistent with double bond formulation for disilenes are the deshielded ^{29}Si nmr chemical shifts and large ^1J (SiSi) coupling constants.

The key to the synthesis of stable disilenes has been the protection of the Si=Si bond by sterically encumbering substituents. Other disilenes are made by reduction of R_2SiCl_2 with Li, Na,[118] or Li naphthalenide. The 1,2-di(1-adamantyl)dimesitylsilene analogue has cis and trans isomers for which activation energies of isomerization are 120 kJ mol^{-1}.[119]

Some reactions of the mesityl are given in Fig. 8-9. The oxo compound (8-XIV) has a nearly planar Si(μ-O)$_2$Si ring, and the Si—Si distance is also shorter

[115]M. Weidenbruch, *Coord. Chem. Rev.* **1994**, *130*, 275; R. Okazaki and R. West, *Adv. Organomet. Chem.* **1996**, *39*, 231.
[116]J. Escudié *et al.*, *Coord. Chem. Rev.* **1994**, *130*, 427.
[117]K. M. Baines and W. G. Stibbs, *Adv. Organomet. Chem.* **1996**, *39*, 275 (a review on stable doubly bonded compounds of germanium and tin).
[118]M. Kira *et al.*, *Angew. Chem. Int. Ed. Engl.* **1994**, *33*, 1489.
[119]R. West *et al.*, *Organometallics* **1989**, *8*, 2664.

than normal in $(\mu\text{-O})_2$ compounds, suggesting that the oxygen breaks only the π and not the σ component of the bond.

Among the digermenes which have been structurally characterized are {(2,6-$C_6H_3Et_2)_2Ge\}_2$ (8-XV), {2,6-$C_6H_3Et_2)(mes)Ge\}_2$,[120] and {[(Me$_3$Si)$_2$CH]$_2$Ge}$_2$; a tin analogue of the latter is also known. Another distannene which has been structurally characterized is {[(Me$_3$Si)$_3$Si]$_2$Sn}$_2$.[121] The core of the digermenes and distannenes is typically more distorted from planar geometry; thus the standard bond description of a σ and a π bond applicable to olefins and silenes *is not* as appropriate for most digermenes and distannenes. Theoretical studies indicate that π bonding does not occur and that tin is pyramidal sp^3 with conjugative interaction leading to some sort of Sn—Sn bond or possibly

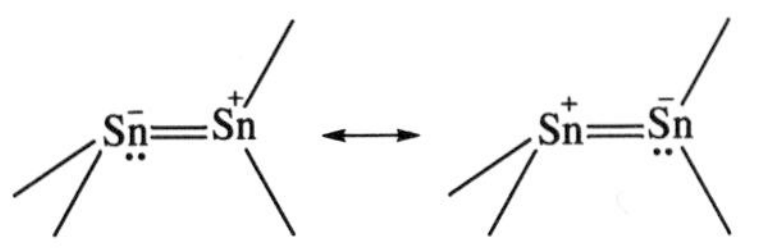

Only a few unsymmetrical compounds are known, e.g., (mes)$_2$Ge=Si(mes)$_2$ and (Is)$_2$Sn=Ge(mes)$_2$,[122] Is = 2,4,6-triisopropylphenyl.

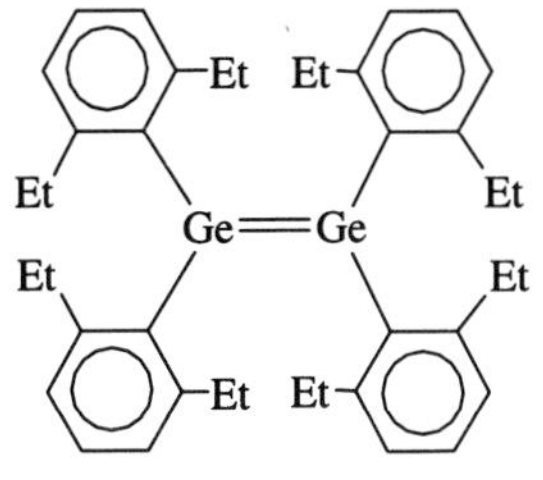

(8-XV)

There are also many compounds with formulas R$_2$E=CR$_2$. For example, silene compounds with Si=C bonds[123] can be made by elimination reactions of the type

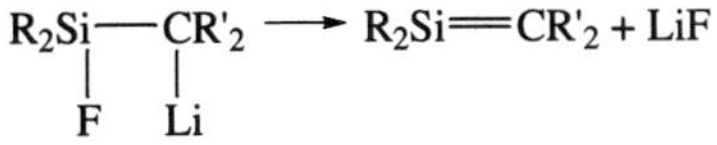

The first stable compound in this class was (Me$_3$Si)$_2$Si=C[(OSiMe$_3$)(1-adamantyl)]. In general, the Si=C bonds are *ca.* 10% shorter than typical Si—C single bonds.[124] The core around the Si=C unit is nearly planar. The germanium analogues[125] are fewer. They also show a significant shortening of the Ge=C bond relative to the Ge—C bond, but there are relatively large twist angles in some compounds. There

[120]S. Masamune *et al.*, *J. Am. Chem. Soc.* **1990**, *112*, 9394.
[121]K. W. Klinkhammer and W. Schwarz, *Angew. Chem. Int. Ed. Engl.* **1995**, *34*, 1334.
[122]J. Escundié *et al.*, *Chem. Commun.* **1996**, 2621.
[123]A. G. Brook and M. A. Brook, *Adv. Organomet. Chem.* **1996**, *39*, 71.
[124]N. C. Norman, *Polyhedron* **1993**, *12*, 2431.
[125]J. Barrau *et al.*, *Chem. Rev.* **1990**, *90*, 283.

are also tin analogues, some with relatively short Sn—C bonds, but the majority have significantly longer bonds.[126]

A large variety of compounds with multiple bonding character between Si—N,[127] Si—P,[128] Si—S, Si—As, Ge—N, Ge—P, Ge—S, Ge—O, Sn—N, and Sn—P units have been prepared. Examples are $(2,4,6\text{-}Pr^i_3C_6H_2)Si{=}ESiR_3$, E = P and As,[129] and $Bu^t_2Si{=}NSiBu^t_3$. Double bonding character to transition metals is also known (Section 8-19); an example is $[Cp^*(PMe_3)_2Ru{=}SiMe_2][B(C_6F_5)_4]$.[130]

A few examples of triple bonds are also known. A silane nitrile has been detected by matrix isolation techniques:[131]

$$:Si{=}N{-}H \xrightarrow[h\nu\ (254\text{nm})]{h\nu\ (193\text{nm})} H{-}Si{\equiv}N$$

while a relatively stable complex with a two-coordinate germanium atom and a formal Mo≡Ge bond[132] is made according to

$$Na[MoCp(CO)_3] + 2,6\text{-mes}_2C_6H_3GeCl \xrightarrow[-CO]{THF,\ 50^\circ C} Cp(CO)_2MoGeC_6H_3\text{-}2,6\text{-mes}_2 + NaCl$$

Another species which can *formally* be described as having an M≡Si triple bond is $[Cp^*(PMe_3)_2RuSi(STol\text{-}p)(phen)]^{2+}$ [133] (8-XVI); it has a Ru—Si bond distance of 2.269(5) Å.

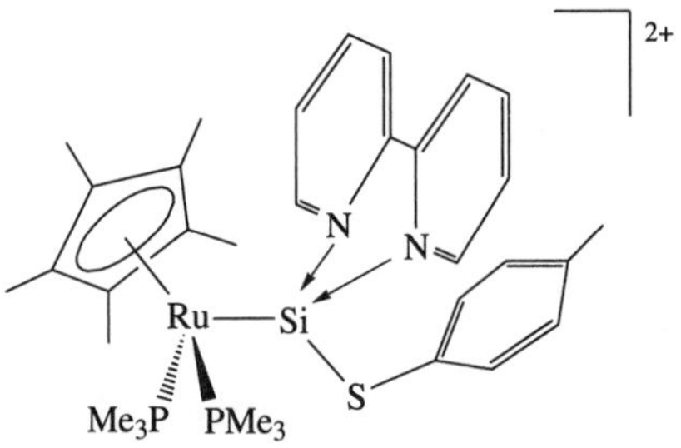

(8-XVI)

THE DIVALENT STATE

8-14   Divalent Organo Compounds

The *carbene analogues*, ER_2, have been intensively investigated. Their stability increases with an increase in the atomic weight of the element. Silenes (also referred

[126]H. Grützmacher *et al., Angew. Chem. Int. Ed. Engl.* **1994,** *33,* 456; M. Weidenbruch *et al., J. Chem. Soc., Chem. Commun.* **1995,** 1157.
[127]M. Driess, *Adv. Organomet. Chem.* **1996,** *39,* 193 (also Si═As)
[128]I. Hemme and U. Klingebil, *Adv. Organomet. Chem.* **1996,** *39,* 159.
[129]M. Driess *et al., J. Chem. Soc., Chem. Commun.* **1995,** 253.
[130]T. D. Tilley and S. K. Grumbine, *J. Am. Chem. Soc.* **1994,** *116,* 5495.
[131]G. Maier and J. Glatthaar, *Angew. Chem. Int. Ed. Engl.* **1994,** *33,* 473.
[132]R. S. Simons and P. P. Power, *J. Am. Chem. Soc.* **1996,** *118,* 11966.
[133]T. D. Tilley *et al., J. Am. Chem. Soc.* **1992,** *114,* 1513.

to as silanylenes), SiR_2, are short-lived species;[134] no stable mononuclear compound, even one containing bulky substituents, has been isolated so far.

The simplest silene (Me_2Si) can be obtained only in the gas phase or in solution by the reaction

$$c\text{-}(Me_2Si)_6 \xrightarrow{\Delta} c\text{-}(Me_2Si)_5 + Me_2Si$$

Similarly, early compounds of stoichiometry GeR_2 or SnR_2 proved to be cyclogermanes or stannanes such as $(Me_2Sn)_6$ or various other polymers with $M^{II}-M^{II}$ entities. However, germylenes (and stannylenes) which have been stabilized by electronic or steric effects are now well known.[135] Many transient free germylenes can be prepared from 7,7-disubstituted-7-germanobenzonorbornadienes according to:

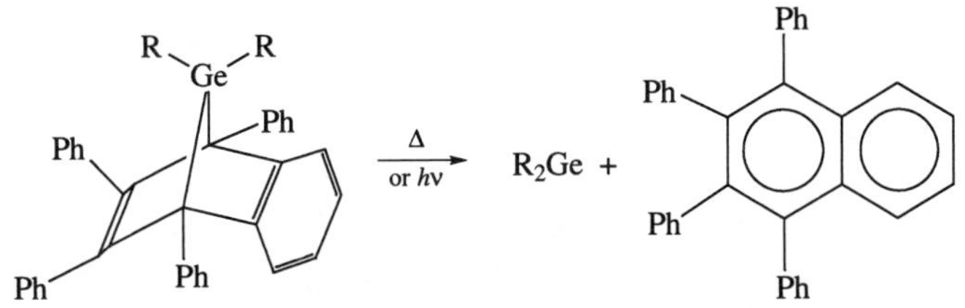

The organic group R on R_2Ge is often Me, but can also be Et, Bu, Ph, and 4-MePh (this is sometimes convenient, because the 4-Me group can be used by ^1H nmr as a tracer to follow what happens). Stable mononuclear $:GeR_2$ compounds, R = $CH(SiMe_3)_2$ or $2,6\text{-}C_6H_3mes_2$,[136] for which there are also Sn and Pb analogues, can be made according to:

$$GeCl_2 \cdot dioxane + 2LiR \xrightarrow[0°C]{Et_2O} :GeR_2 + 2LiCl$$

The C—Ge—C bond angles of 114° for $Ge(mes^*)_2$ (Fig. 8-10) gives a clear indication that the electron pair is stereochemically active.

A stable 2-coordinate diaryllead(II),[137] and its Sn analogue,[138] have been prepared in a similar way:

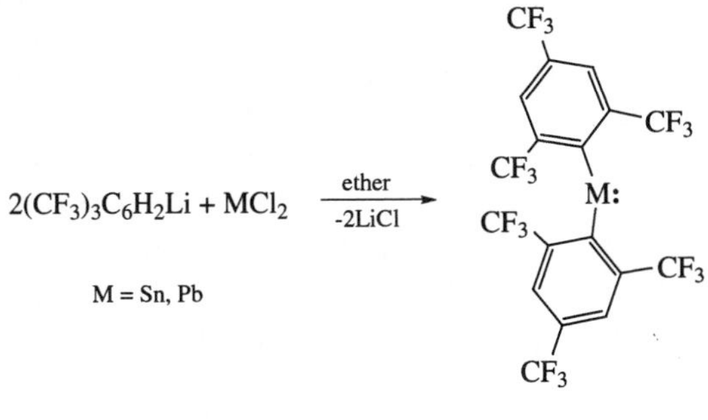

$$2(CF_3)_3C_6H_2Li + MCl_2 \xrightarrow[-2LiCl]{ether}$$

M = Sn, Pb

[134]P. P. Gaspar *et al.*, *Acc. Chem. Res.* **1987**, *20*, 329.

[135]W. P. Neumann, *Chem. Rev.* **1991**, *91*, 311; W. Ando *et al.*, *Organometallics*, **1989**, *8*, 2759.

[136]P. P. Power *et al.*, *Organometallics* **1997**, *16*, 1920.

[137]F. T. Edelmann *et al.*, *Organometallics* **1991**, *10*, 25.

[138]H. Grützmacher *et al.*, *Organometallics* **1991**, *10*, 23.

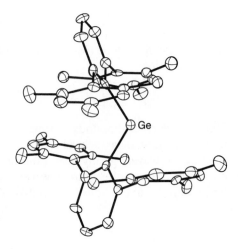

Figure 8-10 A drawing of the molecular structure of Ge{C$_6$H$_3$-2,6-mes$_2$}$_2$ (mes = 2,4,6-Me$_3$C$_6$H$_2$) (P. P. Power *et al., Organometallics* **1997,** *16,* 1920).

An analogous complex can be made with the bulky anionic ligand Si(SiMe$_3$)$_3^-$.[139] The monoorganolead complex {Pb[C(SiMe$_2$Ph)$_3$](μ-Cl)}$_2$[140] (8-XVII) is yellow-orange and highly air-sensitive. It decomposes upon exposure to light. The geometry around the metal can be regarded as distorted tetrahedral with a stereochemically active lone pair occupying one of the coordination sites.

$$\text{(PhMe}_2\text{Si)}_3\text{C} \diagdown \underset{\diagup \text{Cl}}{\overset{\text{Cl}}{\text{Pb}\cdots\cdots\text{Pb}}} \diagdown \text{C(SiMe}_2\text{Ph)}_3$$

(8-XVII)

Of the *cyclopentadienyl* complexes, the air-sensitive (Me$_5$C$_5$)$_2$Si has two forms in the solid state, one with parallel rings, the other bent like Cp$_2$Sn. With the exception of the decaphenyl, all other (η-C$_5$R$_5$)$_2$M, M = Ge, Sn, and Pb, are angular ($\sim$30–50°) in the gas phase and in solution with stereochemically active lone pairs. The decaphenyl (η^5-Ph$_5$C$_5$)$_2$Sn has a nonangular, ferrocene-like structure; the bright yellow color, high stability, and low basicity suggest that the lone-pair electrons are delocalized over the aromatic, electron-withdrawing system. Nucleophilic addition of MgCp$_2$ to SnCp$_2$ in THF gives [Mg(THF)$_6$][Sn(η^3-Cp)$_3$]$_2$.[141] Crystals of (C$_5$H$_5$)$_2$Pb, grown by sublimation, have a zigzag chain with alternating bridging and nonbridging C$_5$H$_5$ groups; however, crystallization from toluene solutions results in modification of the conformation of the polymer chain[142] as shown in (8-XVIII) and cyclization into hexamers (8-XIX). Reaction of PbCp$_2$ with LiCp in the presence of a crown ether gives {Li(12-crown-4)$_2^+$}$_2$\{[(Cp)$_9$Pb$_4^-$][(Cp)$_5$Pb$_2^-$]\}.[143]

[139]K. W. Klinkhammer and W. Schwarz, *Angew. Chem. Int. Ed. Engl.* **1995,** *34,* 1334.
[140]C. Eaborn *et al., J. Chem. Soc., Chem. Commun.* **1995,** 1829.
[141]D. Stalke *et al., J. Chem. Soc. Dalton Trans.* **1993,** 1465.
[142]M. V. Paver *et al., Chem. Commun.* **1997,** 109.
[143]D. S. Wright *et al., J. Chem. Soc., Chem. Commun.* **1995,** 1141.

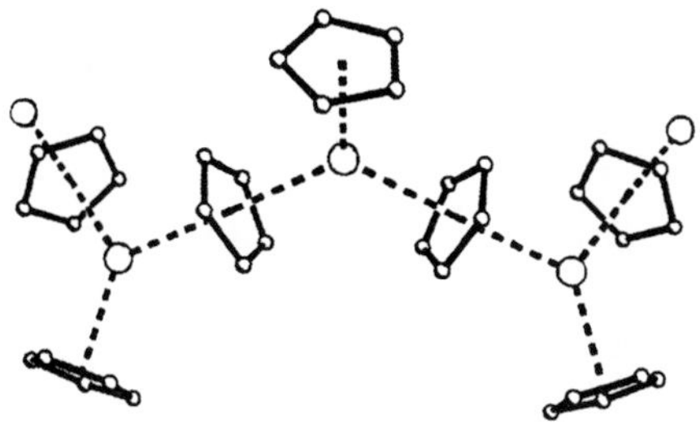

(8-XVIII) 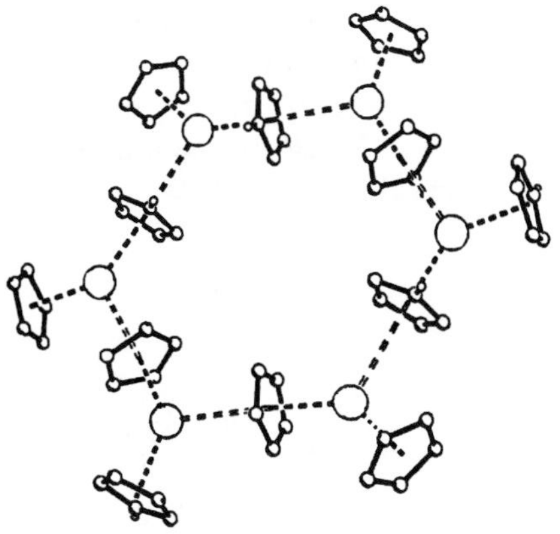 (8-XIX)

8-15 Halides, Oxides, Salts, and Complexes

Silicon

Most divalent Si^{II} species are thermodynamically unstable under normal conditions, but the SiX_2 species occur in high temperature reactions and have been trapped by rapid chilling to liquid nitrogen temperature.

At $\sim 1150°C$ and low pressures SiF_4 and Si react to give SiF_2 in $\sim 50\%$ yield:

$$SiF_4 + Si \rightleftharpoons 2SiF_2$$

The compound is stable for a few minutes at *ca.* 10^{-6} atm pressure, whereas CF_2 has $t_{1/2} \approx 1$ s and GeF_2 is a stable solid at room temperature. It is diamagnetic and the molecule is angular, with a bond angle of $101°$ both in the vapor and in the condensed phase. The reddish-brown solid gives an esr spectrum and presumably also contains $\cdot SiF_2(SiF_2)_n SiF_2\cdot$ radicals. When warmed it becomes white, cracking to give fluorosilanes up to $Si_{16}F_{34}$.

In the gas phase SiF_2 reacts with oxygen but is otherwise not very reactive. Allowing the solid to warm in the presence of various compounds (e.g., CF_3I, H_2S, GeH_4, or H_2O) yields insertion products such as $SiF_2H(OH)$ and H_3GeSiF_2H.

Silicon dichloride has been made by the reaction

$$Si_2Cl_6 \xrightarrow[1450°C]{KSi} 3SiCl_2$$

and in various other ways.

Some polymeric, formally divalent iodides, have been made by treating $(SiPh_2)_n$ with HI and $AlCl_3$; these are Si_4I_8, Si_5I_{10}, and Si_6I_{12}.

Among the few stable divalent silicon compounds that have been isolated, there are the tetracoordinate (ψ-tbp) $Si[(PMe_2)_2C(SiMe_3)]_2$[144] (8-XX) and 2-coordinate

[144]H. H. Karsch *et al.*, *Angew. Chem. Int. Ed. Engl.* **1990**, *29*, 295.

silynes of the type $Si(RNCHCHNR)^{145}$ (8-XXI). The latter do not decompose in sealed nmr tubes over several months and can be distilled without decomposition; they are, however, quite reactive and undergo oxidative addition reactions with ethanol and MeI.[146] In the presence of N_3CPh_3, they react to form compounds with Si=N bonds.[147]

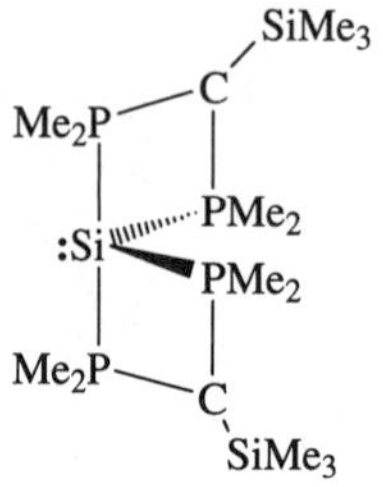

(8-XX) 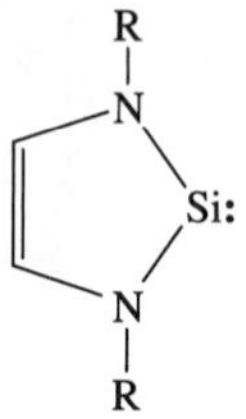   (8-XXI)

Germanium

The germanium dihalides are quite stable. Germanium(II) fluoride (GeF_2), a white crystalline solid (mp 111°C), is formed by the action of anhydrous HF on Ge in a bomb at 200°C or by the reaction of Ge and GeF_4 above 100°C. It is a fluorine-bridged polymer, the Ge atom having a distorted trigonal bipyramid arrangement of four atoms and an equatorial lone pair. The compound reacts exothermically with solutions of alkali metal fluorides to give the hydrolytically stable ion GeF_3^-; in fluoride solutions the ion is oxidized by air, and in strong acid solutions by H^+, to give GeF_6^{2-}. The vapors of GeF_2 contain oligomers $(GeF_2)_n$ where $n = 1$ to 3.

The other dihalides are less stable than GeF_2 and similar to each other. They can be prepared by the reactions

$$Ge + GeX_4 \longrightarrow 2GeX_2$$

In the gas phase or isolated in noble gas matrices, they are bent with angles of 90 to 100°. The solids react to complete their octets (e.g., with donors, to produce pyramidal $LGeX_2$ molecules), or with butadiene:

$$GeI_2 + R_3P \longrightarrow R_3PGeI_2$$

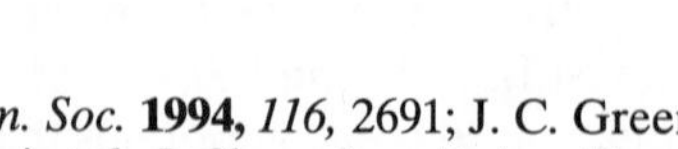

[145]M. Denk *et al., J. Am. Chem. Soc.* **1994,** *116,* 2691; J. C. Green *et al., J. Chem. Soc. Dalton Trans.* **1994,** 2405; T. Veszprémi *et al., J. Chem. Soc., Dalton Trans.* **1996,** 1475; A. J. Arduengo, III *et al., J. Am. Chem. Soc.* **1994,** *116,* 6641.
[146]B. Gehrhus *et al., J. Chem. Soc., Chem. Commun.* **1995,** 1931.
[147]M. Denk *et al., J. Am. Chem. Soc.* **1994,** *116,* 10813.

Other divalent germanium compounds include salts of $GeCl_3^-$, such as the substituted pyrazolyl complexes $[Ge_2(pz^*)_3][GeCl_3]$ (8-XXII) which is obtained in the reaction of $Ba(pz^*)_2$ with $GeCl_2 \cdot$ dioxane.[148] There are also a sulfide, GeS, several thiolato complexes,[149] and a white to yellow hydroxide of no definite stoichiometry that is converted by NaOH to a brown material that has Ge—H bonds.

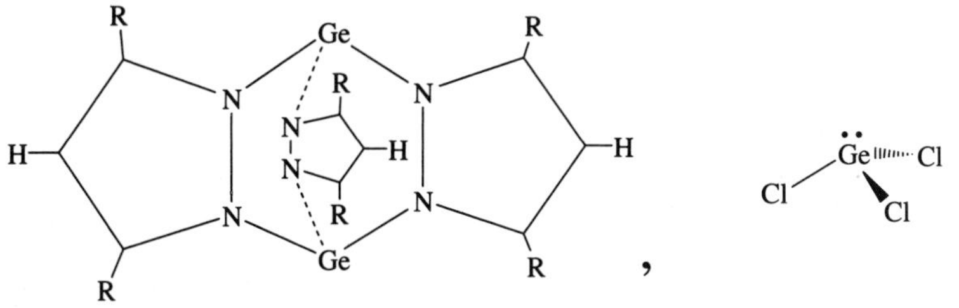

(8-XXII)

Germanium and tin form stable, diamagnetic β-diketonato complexes such as $M(acac)_2$ and $M(acac)X$, where M is Ge, Sn, and X is Cl or I. The $M(dike)_2$ types can be distilled or sublimed; they are soluble and mononuclear in benzene and other hydrocarbon solvents.

A series of thermally stable, homoleptic, and bulky phosphanyl-substituted carbene analogues of germanium, tin, and lead have been prepared[150] by the following method:

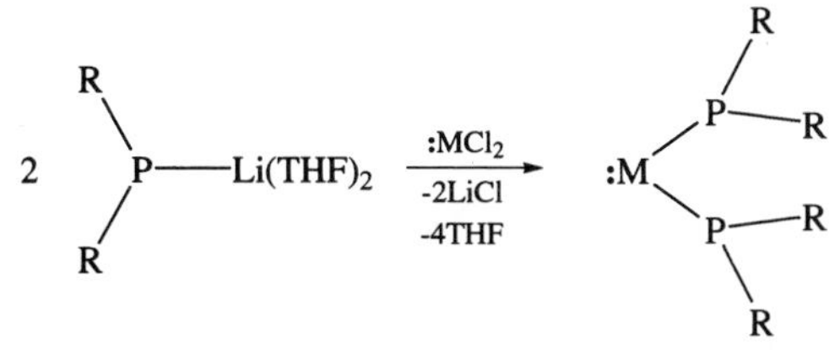

Tin

The fluoride and chloride are obtained by reaction of Sn with gaseous HF or HCl. Tin(II) bromide ($SnBr_2$) is obtained by dissolving tin in aqueous HBr, distilling off constant boiling HBr/H_2O, and cooling. The tin atoms are 9- and 8-coordinate in $SnCl_2$ and $SnBr_2$, respectively. Tin(II) fluoride (SnF_2) has a unique structure with an eight-membered ring of alternating Sn and F atoms and one terminal F on each trigonal pyramidal Sn atom. Water hydrolyzes the halides, but they dissolve in solutions containing excess halide ion to give SnX_3^- ions.[151] With the fluoride, SnF^+ and $Sn_2F_5^-$ can also be detected, and the latter is known in crystalline salts. The SnF_2 forms $[SnF]^+[SbF_6]^-$ and similar compounds with F^- acceptors. Tin(II) fluoride is used in toothpastes, as a source of fluoride to harden dental enamel.

[148]A. Steiner and D. Stalke, *Inorg. Chem.* **1995**, *34*, 4846.
[149]B. Kersting and B. Krebs, *Inorg. Chem.* **1994**, *33*, 3886.
[150]M. Driess *et al., Angew. Chem. Int. Ed. Engl.* **1995**, *34*, 1614.
[151]A. Vogler *et al., Inorg. Chem.* **1992**, *31*, 3277.

Many salts of the ion $SnCl_3^-$ are known,[152] but few for the ψ-*tbp* $SnCl_4^{2-}$ ion and various solid phases (for example, Cs_4SnCl_6) have been characterized.

The halides dissolve in donor solvents such as acetone, pyridine, or DMSO to give pyramidal adducts SnX_2L.

The air-sensitive Sn^{2+} ion occurs in acid perchlorate solutions, which may be obtained by the reaction

$$Cu(ClO_4)_2 + Sn/Hg \rightleftharpoons Cu/Hg + Sn^{2+} + 2ClO_4^-$$

A salt of bare Sn^{2+} ion appears to be $Sn^{2+}(SbF_6^-)_2(AsF_3)_2$ which is formed when SnF_2 and SbF_5 are dissolved in AsF_3. The ion is 9-coordinate with six F atoms from SbF_6^- and three F atoms from AsF_3 but there is distortion due to effects of the lone pair.

The hydrolysis of Sn^{II} in aqueous solution can produce $[Sn(OH)]^+$, $[Sn_2(OH)_2]^{2+}$, $[Sn_3(OH)_4]^{2+}$, and finally the hydrous Sn^{II} oxide $Sn_3O_2(OH)_2$; a crystaline example is $[Sn_3(OH)_4](NO_3)_2$[153] (8-XXIII). A true tin(II) hydroxide has been made by the anhydrous reaction

$$2R_3SnOH + SnCl_2 \longrightarrow Sn(OH)_2 + 2R_3SnCl$$

as a white, amorphous solid. The hydrolysis of SnF_2 produces a polymeric oxide, Sn_4OF_6.[154]

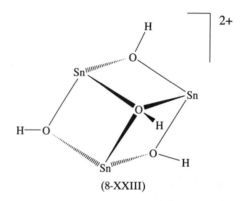

(8-XXIII)

All Sn^{II} solutions are readily oxidized by oxygen and, unless stringently protected from air, normally contain some Sn^{IV}. The chloride solutions are often used as mild reducing agents:

$$SnCl_6^{2-} + 2e = SnCl_3^- + 3Cl^- \qquad E^0 = {\sim}0.0 \text{ V } (1 \ M \text{ HCl, } 4 \ M \text{ Cl}^-)$$

The addition of aqueous ammonia to Sn^{II} solutions gives the white hydrous oxide that is dehydrated to black SnO when heated in suspension at 60 to 70°C in $2 \ M$ NH_4OH. It disproportionates above ~250°C to give Sn and Sn_3O_4, and over 500°C, Sn and SnO_2 are formed.

[152]S. M. Godfrey *et al.*, *Chem. Soc. Dalton. Trans.* **1995,** 1593.
[153]J. D. Donaldson *et al.*, *J. Chem. Soc. Dalton Trans.* **1995,** 2273.
[154]I. Abrahams *et al.*, *J. Chem. Soc. Dalton Trans.* **1994,** 2581.

The anhydrous oxide is amphoteric and dissolves in alkali hydroxide to give solutions of *stannites*, which may contain the ion $[Sn(OH)_6]^{4-}$. These solutions are quite strong reducing agents; on storage they deposit SnO, and at 70 to 100°C they disproportionate slowly to β-tin and Sn^{IV}. A characterized oxostannite is the deep yellow $K_2Sn_2O_3$ made by direct interaction at 550°C; it has a perovskite structure with half the anions missing.

Other Sn^{II} compounds include carboxylates and carboxylato anion complexes such as $Sn(O_2CCF_3)_2$ and $[Sn(O_2CMe)_3]^-$. There are also some mixed valence carboxylates made by the interaction of Sn_2Ph_6 with carboxylic acids. Examples are $[Sn^{II}Sn^{IV}O(O_2CCF_3)_4]_2$ and $[Sn^{II}Sn^{IV}O(O_2CR)_4\{O(OCR)_2\}]_2$; the latter has μ_3-O and μ-O_2CR groups.

The thiolate ion $[Sn(SPh)_3]^-$ is pyramidal like the Pb analogue, while a phosphine complex $Sn[C(PMe_2)_3]_2$ has two P atoms of the anion $C(PMe_2)_3^-$ coordinated in a ψ-*tbp* structure; a similar type of coordination is found in several acylpyrazolone derivatives.[155]

Some Sn^{II} (and also Pb^{II}) chalcogenide, halide, and mixed compounds (e.g., $CsSnBr_3$ and Pb_4Cl_6S) are intensely colored, even black, the color becoming more intense as the atomic number of the chalcogen or halogen increases. These appear to have unusual structures and bonding. For example, $CsSnBr_3$ has an ideal perovskite structure (no stereochemically active lone pair) and is a semiconductor; the halide and chalcogenide ions may use outer d orbitals to form energy bands that are partially populated by the $5s$ or $6s$ electrons of Sn^{II} or Pb^{II}.

Lead.[156] Lead has a well-defined cationic chemistry. There are several crystalline salts, most of which are insoluble ($PbSO_4$, $PbCrO_4$, $2PbCO_3 \cdot Pb(OH)_2$[157]) or sparingly soluble (PbF_2, $PbCl_2$) in water. The anhydrous PbF_2 can be made by reaction between lead carbonate and a solution of 40% HF; the colorless precipitate is then dried at 500°C under an argon stream.[158] The soluble ones, $Pb(ClO_4)_2 \cdot 3H_2O$, $Pb(NO_3)_2$, and $Pb(O_2CMe)_2 \cdot 3H_2O$, give the hydrated ions Pb^{2+}, $PbNO_3^+$, and $Pb(O_2CMe)^+$, respectively, in solution. The hydrated lead(II) dimethylacetate is soluble in many common organic solvents such as ethanol and benzene. It forms hexanuclear units of $[Pb_6(O_2CCHMe_2)_{12}] \cdot 4H_2O$[159] which have pairs of Pb atoms with 5-, 6-, and 7-fold coordination in the molecular unit. The halides, unlike those of Sn, are always anhydrous and have distorted close-packed halogen lattices. The solubility of PbI_2 is low in polar organic solvents but increases upon the addition of alkali metal iodide because of complexation. Various anions such as $[PbI_3]^-$, $[PbI_5]^{3-}$,[160] octahedral $[PbI_6]^{4-}$, and polynuclear $[Pb_{18}I_{44}]^{8-}$ are known.[161]

The *oxide* has two forms: *litharge*, red with a layer structure, and *massicott*, yellow with a chain structure. The oxide Pb_3O_4 (red lead), which is made by heating PbO or PbO_2 in air, behaves chemically as a mixture of PbO and PbO_2, but the

[155]C. Pettinari *et al.*, *Polyhedron* **1994**, *13*, 939.

[156]J. Parr, *Polyhedron* **1997**, *16*, 551.

[157]D. A. Ciomartan *et al.*, *J. Chem. Soc., Dalton Trans.* **1996**, 3639.

[158]A. de Kozak *et al.*, *Z. anorg. allg. Chem.* **1996**, *622*, 1200.

[159]B. O. West *et al.*, *Polyhedron* **1997**, *16*, 19.

[160]D. B. Mitzi *et al.*, *J. Am. Chem. Soc.* **1995**, *117*, 5297.

[161]H. Krautscheid and F. Vielsack, *Angew. Chem. Int. Ed. Engl.* **1995**, *34*, 2035; A. B. Corradi *et al.*, *Inorg. Chim. Acta* **1997**, *254*, 137.

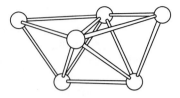

Figure 8-11 The three face-sharing tetrahedra of Pb atoms in the $Pb_6O(OH)_6^{4+}$ cluster. The middle tetrahedron has an O atom at its center. The end tetrahedra have μ_2-OH groups.

crystal contains $Pb^{IV}O_6$ octahedra linked in chains by sharing of opposite edges, the chains being linked by Pb^{II} atoms each bound to three oxygen atoms. In all Pb^{II} oxides and mixed oxides the stereochemical influence of the lone pair is apparent.

The lead(II) ion is partially hydrolyzed in water.[162] In perchlorate solutions the first equilibrium appears to be

$$Pb^{2+} + H_2O = PbOH^+ + H^+ \qquad \log K \approx -7.9$$

but the main aqueous species is $Pb_4(OH)_4^{4+}$, which has a cubane structure with μ_3-OH bridges in crystalline salts.[163]

Another polymeric ion obtained by dissolving PbO in $HClO_4$ and adding an appropriate amount of base has a structure shown in Fig. 8-11. A μ_4-O unit in the center of a Pb_4O_4 cube also occurs in $Pb_4O(OSiPh_3)_6$ which is formed by the action of Ph_3SiOH on Cp_2Pb, while $Pb_9O_4Br_{10}$ has a Pb_4O_4 cube with each O atom bound to an additional Pb giving a $Pb_4(OPb)_4$ unit.

Addition of base to lead(II) solutions gives the hydrous oxide, which dissolves in an excess of base to give the plumbate ion. If aqueous ammonia is added to a lead acetate solution, a white precipitate of a basic acetate is formed. On suspension in warm aqueous ammonia and subsequent drying, it yields a very pure lead oxide, which is most stable in the tetragonal red form.

Lead(II) also forms numerous complexes that are mostly octahedral, although a phosphorodithioate, $Pb(S_2PPr^i_2)_2$, is polymeric, with six Pb—S bonds and a stereochemically active lone pair. Interaction of the nitrate and NaSPh allows isolation of $[Ph_4As][Pb(SPh)_3]$, which provides an example of trigonal pyramidal Pb^{II}. A crystalline sulfide, PbS, can be prepared by reaction of elemental Pb and sulfur in liquid ammonia.[164]

8-16 Dialkylamides and Alkoxides[165]

These are somewhat similar to the alkyl or aryl compounds, being mononuclear with bulky ligands and dinuclear with less bulky ones. Examples are (8-XXIV) and (8-XXV). The orange $Ge(NBu^t_2)_2$ is stable but the maroon tin analogue is thermo- and photosensitive. Metathetical exchange[166] between carbon dioxide and bisamides

[162]V. Romano *et al., J. Chem. Soc. Dalton Trans.* **1996,** 4597.
[163]S. M. Grimes *et al., J. Chem. Soc., Dalton Trans.* **1995,** 2081.
[164]I. P. Parkin *et al., Chem. Commun.* **1996,** 1095.
[165]D. S. Wright *et al., Angew. Chem. Int. Ed. Engl.* **1995,** *34,* 1545; J. Barrau *et al., Inorg. Chim. Acta* **1996,** *241,* 9.
[166]L. R. Sita *et al., J. Am. Chem. Soc.* **1996,** *118,* 10912.

of silicon and germanium proceeds according to

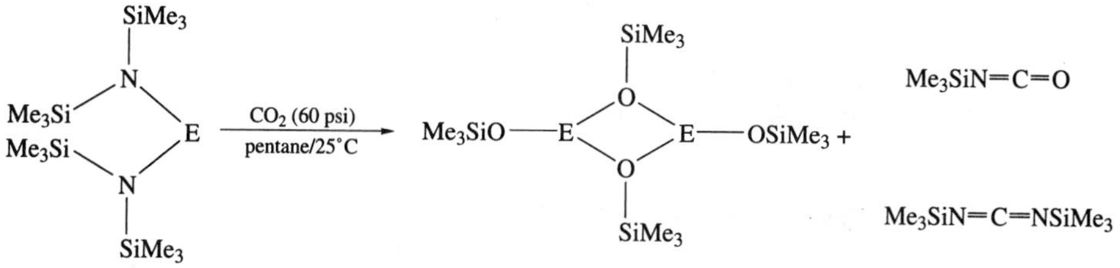
SiMe3
Me3Si
Me3Si
N
E
N
SiMe3
CO2 (60 psi)
pentane/25°C
Me3SiO
E
O
O
E
OSiMe3 +
SiMe3
SiMe3
Me3SiN=C=O
Me3SiN=C=NSiMe3

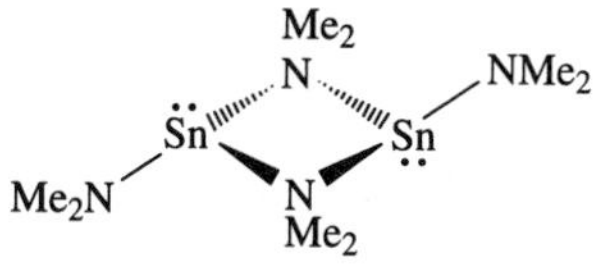
Me2
N
NMe2
Sn
Sn
Me2N
N
Me2

(8-XXIV)

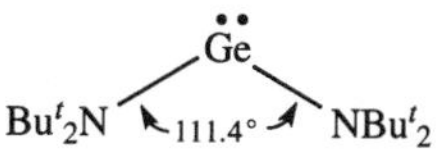
Ge
Bu'2N
111.4°
NBu'2

(8-XXV)

The *t*-butoxide $[Ge(OCMe_3)_2]_2$ made by the action of *t*-butanol on Cp_2Ge is trans dimeric but the bulkier $Ge(OCBu^t_3)_2$ is mononuclear. There is a similar compound $[Sn(OBu^t)_2]_2$ and salts $M[Sn(OBu^t)_3]$, bulky thiolates $M(SR)_2$, and more complicated species like the cubane $Sn_4(\mu\text{-}NR)_4$.

A very large alkoxide is made by reacting $Pb(NO_3)_2$ with the sodium salts of polyols such as cyclodextrin. The crystals of Pb_{16}(cyclodextrinate) have an arrangement of atoms in the form of an eight-point star with each of the alkoxide O atoms coordinated in a bridging fashion to two lead atoms.[167] Mixed metal alkoxides such as $Pb_4Zr_2(OPr^i)_{16}$ and $Pb_2Zr_4(OPr^i)_{20}$ can be made from the reaction of $Pb(OPr^i)_2$ and $Zr(OPr^i)_4$.[168] Some alkoxides and dialkylamides have been used as precursors for the preparation of SnO_2 and GeO_2.[169]

THE ELEMENTS AND THEIR COMPOUNDS AS LIGANDS

8-17 The Elements

There is a variety of compounds that have Si bound to transition metals in relatively simple molecules such as $Si[Re(CO)_5]_4$ or in clusters such as $[Co_9(\mu_8\text{-}Si)(CO)_{21}]^{2-}$. Germanium, tin, and lead also form clusters such as $[Co_5Ge(CO)_{16}]$, $Co_4Pb(CO)_{16}(PR_3)_4$, and $[Fe_4Pb(CO)_{16}]^{2-}$; metallated Zintl ions are found in anions such as $[Sn_6\{Cr(CO)_5\}_6]^{2-}$ and $[Ge_9(\mu_{10}\text{-}Ge)Ni(PPh_3)]^{2-}$.[170]

In addition, there are compounds such as (8-XXVI), (8-XXVII), and (8-XXVIII).

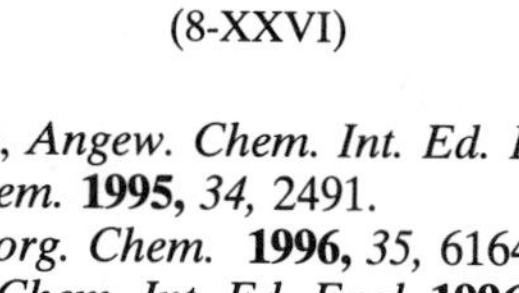
(CO)5W
Sn
W(CO)5
(CO)5W

(8-XXVI)

[167]P. Klüfers and J. Schuhmacher, *Angew. Chem. Int. Ed. Engl.* **1994,** *33,* 1863.
[168]K. G. Caulton *et al., Inorg. Chem.* **1995,** *34,* 2491.
[169]S. Suh and D. M. Hoffman, *Inorg. Chem.* **1996,** *35,* 6164.
[170]B. W. Eichhorn *et al., Angew. Chem. Int. Ed. Engl.* **1996,** *35,* 2852.

$$Ge[W(CO)_5]_4 \qquad Cp(CO)_2Mn{=}Pb{=}Mn(CO)_2Cp$$

$$(8\text{-}XXVII) \qquad\qquad (8\text{-}XXVIII)$$

The latter is made by photolyzing $CpMn(CO)_3$ and $PbCl_2$ in THF.
In (8-XXVI) the W_3Sn group is planar.

8-18　Silicon, Germanium, and Tin(IV)

The major types of complex containing Si,[171] Ge, or Sn ligands are those with
SiR_3, GeR_3, and SnR_3 groups that are similar in stoichiometry to those of carbon
compounds. They can be obtained by reactions such as

$$(\eta\text{-}C_5H_5)(CO)_3WNa + ClSiMe_3 \longrightarrow (\eta\text{-}C_5H_5)(CO)_3WSiMe_3 + NaCl$$

$$Cp(CO)_2FeNa + ClSiMe_2(SiMe_2)_nMe \longrightarrow Cp(CO)_2FeSiMe_2(SiMe_2)_nMe + NaCl$$

or by oxidative-addition reactions, for example,

$$Me_3SnCl + Pt(PR_3)_2 \longrightarrow Me_3SnPtCl(PR_3)_2$$

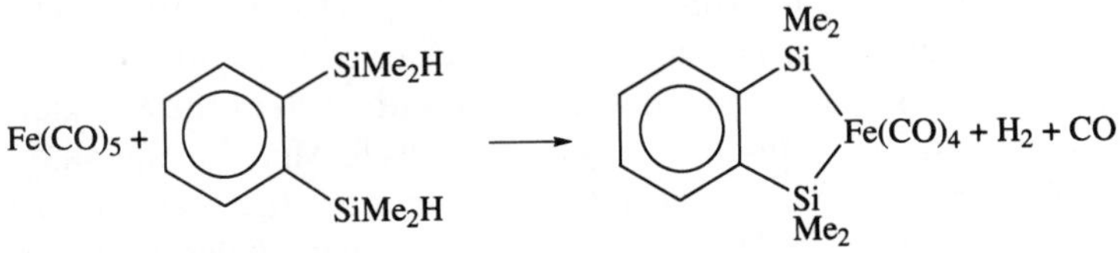

or activated tin addition, as in

$$Sn + 2Co_2(CO)_8 \longrightarrow Sn[Co(CO)_4]_4$$

The SiR_3 groups have a very high trans effect. There are also compounds with
bridging groups such as the following:

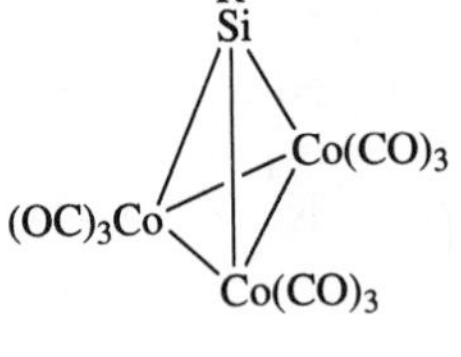

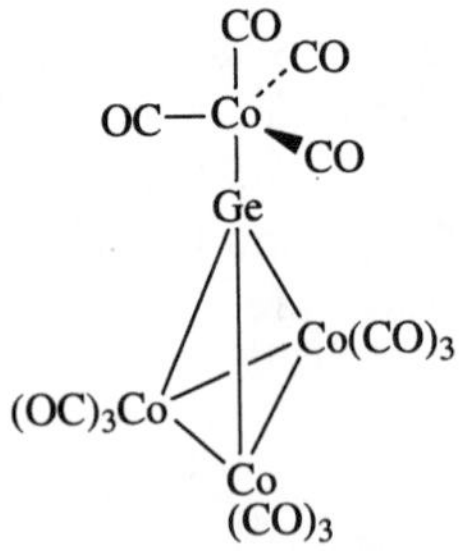

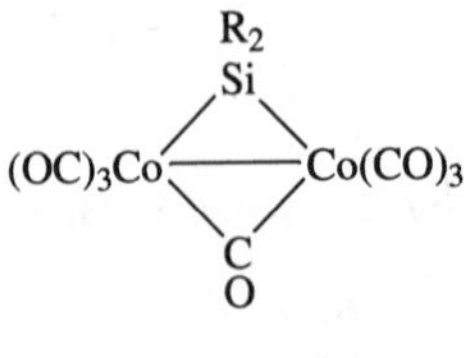

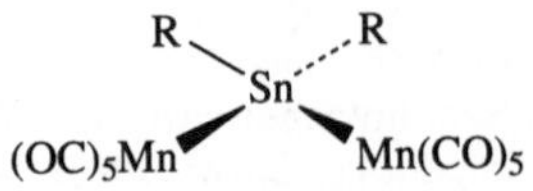

[171]See for example: H. K. Sharma and K. H. Pannell, *Chem. Rev.* **1995,** *95,* 1351.

Non-transition metal compounds are Li or Mg ones such as $[Mg(SiMe_3)_2]_2(tmen)$, which can be used to synthesize other compounds.

8-19 Divalent Compounds[172]

Most of the divalent compounds of Si, Ge, Sn, and Pb have a lone pair and hence can act as donors. Examples are neutral compounds like Cp_2Sn, $Sn(acac)_2$, $Sn[CH(SiMe_3)_2]_2$, or $Sn(NR_2)_2$. Some anionic species are, for example, $GeCl_3^-$ and $SnCl_3^-$. As well as acting as donors, the MX_2 compounds may behave as reducing agents or insert into metal halogen bonds, for example,

$$(CO)_5MnBr + Sn(NR_2)_2 \longrightarrow (CO)_5MnSnBr(NR_2)_2$$

$$[(BuNC)_6MoCl]^+ + SnCl_2 \longrightarrow [(BuNC)_6MoSnCl_3]^+$$

Tin compounds have been most studied,[173] especially those of the $SnCl_3^-$ ion, by ^{119}Sn nmr and Mössbauer spectra. The $SnCl_3^-$ ligand has a high trans effect and trans influence but has a rather low nucleophilicity. Thus in contrast to N, P, and As donors, $SnCl_2$ will not cleave halide bridges in compounds with, for example, $Pt(\mu\text{-}Cl)_2Pt$ bridges.

The $SnCl_3^-$ ion can generally displace Cl^-, CO, and more weakly bound ligands, for example,

$$[PtCl_4]^{2-} \xrightarrow[-2Cl^-]{2SnCl_3^-} [PtCl_2(SnCl_3)_2]^{2-} \xrightarrow{3SnCl_3^-} [Pt(SnCl_3)_5]^{3-}$$

Typical complexes are $[RuCl_2(SnCl_3)_4]^{4-}$, $[Os(SnCl_3)_6]^{4-}$, $[PtH(SnCl_3)_2(PEt_3)_2]^-$, and *cis*- or *trans*-$[Pt(CO)Cl_2(SnCl_3)]^-$.

The $SnCl_3^-$ ligand is a good leaving group and can thus provide a site for substrate coordination in catalytic reactions; platinum metal systems will catalyze hydrogenation or hydroformylation of alkenes, water-gas shift, and other reactions in solutions or in melts of quaternary salts ($R_4N^+SnCl_3^-$).

Some metal-metal bonded complexes react with $SnCl_2$ in a type of insertion reaction where $SnCl_2$ could be considered to be showing carbene-like behavior, but the resulting compounds are clearly compounds of tin(IV):

$$(\eta\text{-}C_5H_5)_2Fe_2(CO)_4 \xrightarrow{SnCl_2} (\eta\text{-}C_5H_5)(CO)_2Fe\!-\!\!\underset{\underset{Cl}{|}}{\overset{\overset{Cl}{|}}{Sn}}\!-\!\!Fe(CO)_2(\eta\text{-}C_5H_5)$$

A truly divalent germanium complex with a Cr—Ge bond is found in the macrocycle-containing compound $(tmtaa)Ge$—$Cr(CO)_5$[174] in which the local geome-

[172]See for example: U. Siemeling, *Angew. Chem. Int. Ed. Engl.* **1997**, *36*, 831 and references therein.
[173]J. H. Nelson *et al., Chem. Rev.* **1989**, *89*, 11.
[174]A. H. Cowley *et al., Inorg. Chem.* **1993**, *32*, 4671.

tries around the chromium and germanium atoms are octahedral and square pyramidal, respectively. There is a series of compounds with very short M—SiR$_3$ distances which suggest silylene character;[175] an example is $[(\eta^5\text{-}C_5Me_5)(PMe_3)_2Ru=SiR_2]BF_4$ where R = S-*p*-tolyl or SEt. Other compounds with M=E multiple bonds have been noted in Section 8-13.

Additional References

N. Auner and J. Weis, Eds., *Organosilicon Chemistry From Molecules to Materials,* VCH Publishers, New York, 1994.
Chemical Reviews **1995,** *95,* 1135–1673 (several reviews on silicon chemistry).
H. Chon *et al.,* Eds., *Recent Advances and New Horizons in Zeolite Science and Technology,* Studies in Surface Science and Catalysis, vol. 102, Elsevier Science, Amsterdam, 1996.
R. G. Jones, Ed., *Silicone-Containing Polymers,* Royal Society of Chemistry, London, 1995.
B. Marciniec and J. Chojnowski, Eds., *Progress in Organosilicon Chemistry,* Gordon and Breach Science Publishers, Basil, Switzerland, 1995.
S. Patai, Ed., *The Chemistry of Organic Germanium, Tin and Lead Compounds,* John Wiley & Sons, New York, 1995.
P. P. Power, *J. Chem. Soc., Dalton Trans.* **1998,** 2939 (a review of heavier main group multiple bonds, where it is argued that only for Si and some Ge compounds of type R$_2$EER$_2$ should be called metallenes).

---

[175]P. D. Lickiss, *Chem. Soc. Rev.* **1992,** 271; P. Braunstein *et al., Angew. Chem. Int. Ed. Engl.* **1994,** *33,* 2440.

Chapter 9

NITROGEN

GENERAL REMARKS

9-1 Introduction

The electronic configuration of the nitrogen atom in its ground state (^4S) is $1s^2 2s^2 2p^3$, with the three $2p$ electrons distributed among the p_x, p_y, and p_z orbitals with spins parallel. Nitrogen forms an exceedingly large number of compounds, most of which are to be considered organic rather than inorganic. It is one of the most electronegative elements, only oxygen and fluorine exceeding it in this respect.

The nitrogen atom may complete its octet in several ways:

1. *Electron Gain to Form the Nitride Ion,* N^{3-}. This ion occurs only in the saltlike nitrides of the most electropositive elements, for example, Li_3N. Many nonionic nitrides exist and are discussed later in this chapter.

2. *Formation of Electron-Pair Bonds.* The octet can be completed either by the formation of three single bonds, as in NH_3 or NF_3, or by multiple-bond formation, as in nitrogen itself ($:N\equiv N:$), azo compounds ($-N=N-$), nitro compounds (RNO_2), and so on.

3. *Formation of Electron-Pair Bonds with Electron Gain.* The completed octet is achieved in this way in ions such as the amide ion NH_2^- and the imide ion NH^{2-}.

4. *Formation of Electron-Pair Bonds with Electron Loss.* Nitrogen can form four bonds, provided an electron is lost, to give positively charged ions R_4N^+ such as NH_4^+, $N_2H_5^+$, and $(C_2H_5)_4N^+$. The ions may sometimes be regarded as being formed by protonation of the lone pair:

$$H_3N: + H^+ \longrightarrow [NH_4]^+$$

or generally

$$R_3N: + RX \longrightarrow R_4N^+ + X^-$$

Failure to Complete the Octet

There are a few relatively stable species in which, *formally*, the octet of nitrogen is incomplete. The classic examples are NO and NO_2 together with nitroxides R_2NO and the ion $(O_3S)_2NO^{2-}$; all have one unpaired electron. Nitroxide radicals are used as *spin labels*, since they can be attached to proteins, membranes, and so on, and inferences about their environment drawn from the characteristics of the observed esr signals.

Expansion of the Octet

Normally, no expansion of the octet is permissible, but there are reactions that may indicate pentacoordinate nitrogen intermediates, or more likely, transition states.

Thus when molten ammonium trifluoroacetate (mp 130°C) is treated with LiH, deuteration studies show isotopic scrambling. Although it could be that D^- impinges only on the hydrogen of NH_4^+, another possibility is

$$D^- + NH_4^+ \rightleftharpoons \quad D-\underset{\underset{H \quad H}{|}}{\overset{\overset{H}{|}}{N}}-H \quad \rightleftharpoons \quad NH_3D^+ + H^-$$

Also, the synthesis of NF_4^+ from NF_3, F_2, and AsF_5 could involve an NF_5 species, though this seems unlikely (see Section 9-8).

Formal Oxidation Numbers

Classically, formal oxidation numbers ranging from -3 (e.g., in NH_3) to $+5$ (e.g., in HNO_3) have been assigned to nitrogen; though useful in balancing redox equations, they have no physical significance.

9-2   Types of Covalence in Nitrogen; Stereochemistry

In common with other first-row elements, nitrogen has only four orbitals available for bond formation, and a maximum of four *2c-2e* bonds may be formed. However, since formation of three electron-pair bonds completes the octet $:N(:R)_3$, and the nitrogen atom then possesses a lone pair of electrons, four *2c-2e* bonds can only be formed either (*a*) by coordination, as in donor-acceptor complexes, for example, $F_3B^- -N^+(CH_3)_3$, or in amine oxides, for example, $(CH_3)_3N^+ -O^-$, or (*b*) by loss of an electron, as in ammonium ions (NH_4^+ and NR_4^+). This loss of an electron gives a valence state configuration for nitrogen (as N^+) with four unpaired electrons in sp^3 hybrid orbitals analogous to that of neutral carbon, while, as noted previously, gain of an electron (as in NH_2^-) leaves only two electrons for bond formation. In this case the nitrogen atom (as N^-) is isoelectronic with the neutral oxygen atom, and angular bonds are formed. We can thus compare, sterically, the following isoelectronic species:

Tetrahedral	*Angular*
NH_4^+ and CH_4 (also BH_4^-)	$H\overset{\ddot{O}}{\diagdown}{}_{\diagup H}\quad H{}_{\diagup}\overset{\ddot{N}\cdot}{\diagdown}{}_{H}$

It may be noted that the ions NH_2^-, OH^-, and F^- are isoelectronic and have comparable sizes. The amide, imide, and nitride ions, which can be considered to be members of the isoelectronic series NH_4^+, NH_3, NH_2^-, NH^{2-}, and N^{3-}, occur as discrete ions only in salts of highly electropositive elements.

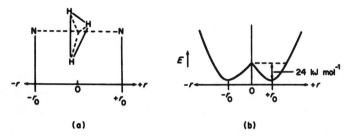

Figure 9-1 Diagrams illustrating the inversion of NH_3 (see text).

In all nitrogen compounds where the atom forms two or three bonds, there remain, respectively, two pairs or one pair of nonbonding or lone-pair electrons. The lone pairs have a profound effect on stereochemistry and are also responsible for the donor properties of the atom possessing them. To illustrate the important chemical consequences of nonbonding electron pairs, we consider one of the most important types of molecule, namely NR_3, as exemplified by NH_3 and amines.

Three-Covalent Nitrogen

Molecules of this type are invariably pyramidal, except in special cases to be discussed later. The bond angles vary according to the groups attached to the nitrogen atom. In principle, pyramidal molecules of the kind $NRR'R''$ should be chiral. However, molecules of this type execute a motion known as inversion, in which the nitrogen atom oscillates through the plane of the three R groups much as an umbrella can turn inside out. As the nitrogen atom crosses from one side of the plane to the other [from one equilibrium position, say $+r_0$, to the other, $-r_0$, in Fig. 9-1(a)], the molecule goes through a state of higher potential energy, as shown in the potential energy curve [Fig. 9-1(b)].

The inversion frequency of NH_3, 2.38×10^{10} s^{-1}, corresponds to a tunnelling barrier of 12.6 kJ mol^{-1}, about half of the thermodynamic barrier, $\Delta H_{inv}^{\ddagger} = 24.5$ kJ mol^{-1}. On substitution of H by other groups, e.g., F, CH_3, or CF_3, the gas phase structure and other molecular properties like $\Delta H_{inv}^{\ddagger}$ are dramatically changed. With some ligands considerable flattening, even to planarity, is observed. For example, NPr_3^i has an NC_3 skeleton with a trigonal planar equilibrium structure, despite the C—N distance of 1.46 Å being comparable to that in pyramidal NMe_3.[1]

Since barriers in simple amines are in the range 16–40 kJ mol^{-1}, isolation of optical isomers is not possible. However, substitution and incorporation of N into strained rings raises the barrier and in some cases invertomers have been separated. For the heavier elements (P, As, Sb, Bi), the inversion barriers of EH_3 and ER_3 molecules are much higher.

Multiple Bonding in Nitrogen and Its Compounds

Like its neighbors carbon and oxygen, nitrogen readily forms multiple bonds, and many compounds for which there are no analogues among the heavier elements. Thus, whereas P, As, and Sb form tetrahedral molecules P_4, As_4, and Sb_4, nitrogen

[1]H. Bock *et al., Angew. Chem. Int. Ed. Engl.* **1991**, *30*, 187.

forms the multiply-bonded diatomic molecule :N≡N:, with an extremely short internuclear distance (1.094 Å) and very high bond strength. Nitrogen also forms triple bonds to other elements including carbon ($H_3CC≡N$), sulfur ($F_3S≡N$), and some transition metals ($O_3Os≡N^-$).

In compounds where nitrogen forms one single and one double bond, the grouping X—N=Y is nonlinear. This can be explained by assuming that nitrogen uses a set of sp^2 orbitals, two of which form σ bonds to X and Y, while the third houses the lone pair. A π bond to Y is then formed using the nitrogen p_z orbital. In certain cases stereoisomers result from the nonlinearity, for example, *cis*-, and *trans*-azobenzenes (9-Ia and b) and the oximes (9-IIa and b). These are interconverted more easily than are *cis*- and *trans*-olefins, but not readily.

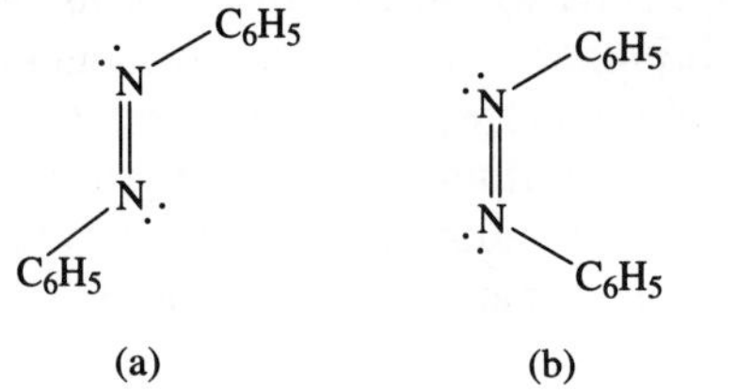

(a)   (b)

(9-I)

(a)   (b)

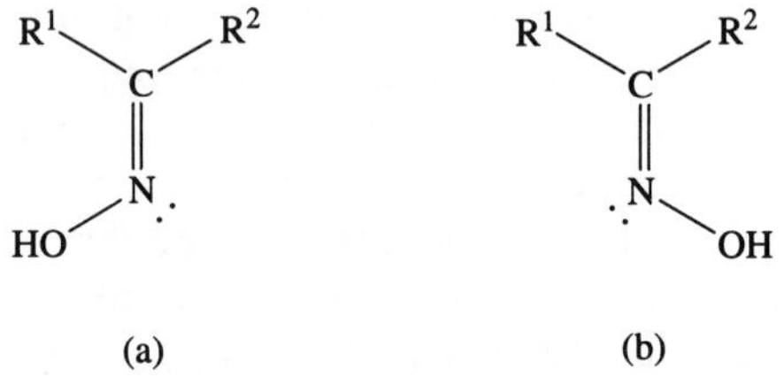

(9-II)

Multiple bonding occurs also in oxo compounds. For example, NO_2^- (9-III) and NO_3^- (9-IV) can be regarded as resonance hybrids in the valence bond approach. From the MO viewpoint, one considers the existence of a π MO extending symmetrically over the entire ion and containing the two π electrons.

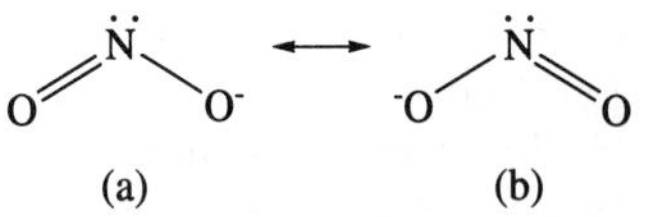

(a)   (b)

(9-III)

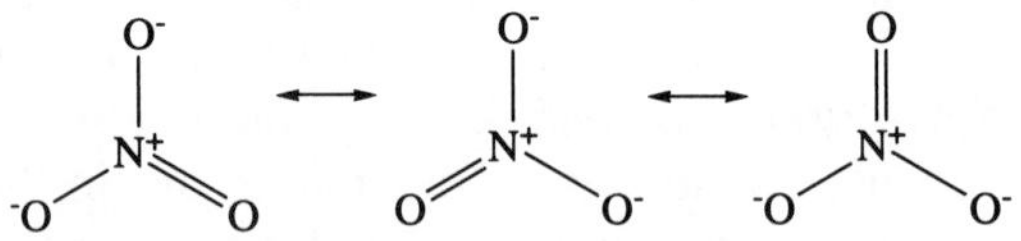

(9-IV)

Planar Three-Coordinate Nitrogen

As noted above, planar N atoms exist. Some early ones were in silylamines such as $N(SiH_3)_3$ and $NMe_2(SiMe_3)$; the planarity was commonly ascribed to π-bonding involving $3d$ orbitals of Si. This does not always appear to be the case and indeed some silylamines and hydroxylamines are pyramidal[2] while a hydrazine de-

[2]H. Schmidbauer *et al., Organometallics* **1994,** *13,* 1762; J. F. Harrod *et al., Organometallics* **1994,** *13,* 2496.

rivative[3] contains both planar and pyramidal N atoms. Since the p orbital with the lone pair is perpendicular to the plane, the lone pairs can interact with other groups in the molecule such as an aromatic π-system.[4] The N-centered complex $[Ir_3N(SO_4)_6(H_2O)_3]^{6-}$ (Section 9-17) has a planar NIr_3 core.

Donor Properties of Three-Covalent Nitrogen; Four-Covalent Nitrogen

As noted previously, the formation of approximately tetrahedral bonds to nitrogen occurs principally in ammonium cations (R_4N^+), amine oxides ($R_3N^+\!-\!O^-$), and Lewis acid-base adducts (e.g., $R_3N^+\!-\!B^-X_3$). In the amine oxides and these adducts, the bonds must have considerable polarity; in the amine oxides, for instance, $N \rightarrow O$ donation cannot be effectively counterbalanced by any back-donation to N. In accord with this, the stability of amine oxides decreases as the R_3N basicity decreases, since the ability of N to donate to O is the major factor. Similarly, $R_3N \rightarrow BX_3$ complexes have stabilities that are roughly parallel to R_3N basicity for given BX_3. When R is fluorine, basicity is minimal and $F_3N \rightarrow BX_3$ compounds are unknown. It is, therefore, curious that F_3NO is an isolable compound. Evidently the extreme electronegativity of fluorine coupled with the availability of $p\pi$ electrons on oxygen allows the structures in (9-V) to contribute to stability.

$$\underset{F^-}{\overset{F}{\underset{\Big|}{\overset{\diagdown}{F-N^+\!=\!O}}}} \longleftrightarrow \overset{F}{\underset{F}{\overset{\diagdown}{\diagup}}}N^+\!=\!O^- \longleftrightarrow \overset{F^-}{\underset{F}{\overset{}{\diagup}}}N^+\!=\!O$$

(9-V)

Catenation and N—N Single-Bond Energies

Unlike carbon and a few other elements, nitrogen has little tendency for catenation, primarily owing to the weakness of the N—N single bond. If we compare the approximate single-bond energies in $H_3C\!-\!CH_3$, $H_2N\!-\!NH_2$, $HO\!-\!OH$, and F—F ($\sim$350, 160, 140, and 150 kJ mol^{-1}, respectively), it is clear that there is a profound drop between C and N. This difference is most probably attributable to the effects of repulsion between nonbonding lone-pair electrons. The strength of the N—N bond, and also of the O—O bond, decreases with increasing electronegativity of the attached groups; increasing electronegativity would perhaps have been expected to reduce repulsion between lone pairs, but it obviously will also weaken any homonuclear σ bond.

There are a few types of compound containing chains of three or more nitrogen atoms with some multiple bonds such as $R_2N\!-\!N\!=\!NR$, $RN\!=\!N\!-\!NR\!-\!NR_2$, $RN\!=\!N\!-\!NR\!-\!N\!=\!NR$, and $RN\!=\!N\!-\!NR\!-\!N\!=\!N\!-\!NR\!-\!N\!=\!NR$, where R represents an organic radical (some R groups may be H but known compounds contain only a few H atoms). There are also cyclic compounds containing rings with up to five consecutive nitrogen atoms. Many of these compounds are not particularly stable, and all are traditionally in the realm of organic chemistry.

[3]H. Schmidbauer *et al., Organometallics* **1994,** *13*, 1762.
[4]G. Wilkinson *et al., J. Chem. Soc., Dalton Trans.* **1994,** 2223 and references therein.

THE ELEMENT

9-3 Occurrence and Properties

Nitrogen occurs in Nature mainly as dinitrogen (N_2), an inert diatomic gas (mp 63.1 K, bp 77.3 K) that comprises 78% by volume of the earth's atmosphere. No other allotrope except N_2 is known or likely to be, although the possible stability of many others, such as N_3, N_4, and many forms of N_5 has been examined theoretically.[5] Naturally occurring nitrogen consists of ^{14}N and ^{15}N with an absolute ratio $^{14}N/^{15}N$ = 272.0 ± 0.3. The isotope ^{15}N is useful as a tracer and especially for nmr studies. There is a wide range of chemical shifts (~900 ppm) compared to ^1H and ^{13}C; the abundance (0.375%) is low and the sensitivity is only ~10^{-3} that of ^1H but spectra can be quite readily obtained. Nitrogen-14 has a quadrupole moment but sharp lines can be observed in certain cases such as NH_4^+, NF_4^+, MeCN, NO_3^-, and NO_2^+ since high local symmetry enables sufficiently slow relaxation of the electric quadrupole and hence of nuclear spin.

Nitric acid containing up to 99.8% ^{15}N can be obtained by efficient fractionation of the system

$$^{15}NO(g) + H^{14}NO_3(aq) = {}^{14}NO(g) + H^{15}NO_3(aq) \qquad K = 1.055$$

The $H^{15}NO_3$ produced can be used to prepare other ^{15}N-labeled nitrogen compounds.

The N—N bond distance in N_2 is 1.097 Å in the gas and 1.15 Å in the solid. The heat of dissociation is extremely large:

$$N_2(g) = 2N(g) \qquad \Delta H = 944.7 \text{ kJ mol}^{-1} \qquad K_{25°C} = 10^{-120}$$

Because the reaction is endothermic, the equilibrium constant increases with increasing temperature, but still, even at 3000°C and ordinary pressures, there is no appreciable dissociation. The great strength of the N≡N bond is principally responsible for the chemical inertness of N_2 and for the endothermicity of most simple nitrogen compounds, even though they may contain strong bonds. Thus $E(N≡N) \approx 6E(N—N)$, whereas $E(C≡C) \approx 2.5E(C—C)$. Dinitrogen is notably unreactive in comparison with isoelectronic, triply bonded systems such as X—C≡C—X, :C≡O:, X—C≡N:, and X—N≡C:. Both —C≡C— and —C≡N groups are known to serve as donors by using their π electrons. The lower ability of N_2 to form stable linkages in this way may be attributed to its electron configuration, which is . . . $(\pi)^4(\sigma)^2$; that is, the π-bonding electrons are even more tightly bound than the σ-bonding electrons, and the latter are themselves tightly bound (ΔH_{ion} = 1496 kJ mol^{-1}). In acetylene, on the other hand, the electron configuration is . . . $(\sigma_g)^2(\pi_u)^4$, and the ΔH_{ion} of the π_u electrons is only 1100 kJ mol^{-1}.

Dinitrogen is obtained commercially by liquefaction and fractionation of air; it usually contains some argon and, depending on the quality, upwards of ~30 ppm of oxygen. The oxygen may be removed by passing the gas over Cr^{2+} on silica gel, through sodium benzophenone ketyl in THF, aqueous V^{2+}, and so on. Spectroscopically pure N_2 is made by heating NaN_3 or $Ba(N_3)_2$.

Although generally nonreactive except at elevated temperatures, N_2 will react at room temperature with Li to give Li_3N and with a variety of transition metal

[5]N. T. Nguyen and T.-K. Ha, *Chem. Ber., Inorg. Organometal. Chem.* **1996,** *129,* 1157.

complexes to give the $M(N_2)$ species. It is utilized directly by nitrogen fixing bacteria, either free living or symbiotic, on root nodules of clover, peas, beans, and so on. Despite extensive study on mild conversions of N_2 to NH_3, N_2H_4, and other compounds, no economic catalytic way has yet been discovered. However, *coordinated* dinitrogen can be attacked under mild conditions (Section 9-9).

At elevated temperatures N_2 becomes more reactive, especially when catalyzed, typical reactions being

$$N_2(g) + 3H_2(g) = 2NH_3(g) \qquad K_{25°C} = 10^3 \text{ atm}^{-2}$$

$$N_2(g) + O_2(g) = 2NO(g) \qquad K_{25°C} = 5 \times 10^{-31}$$

$$N_2(g) + 3Mg(s) = Mg_3N_2(s)$$

$$N_2(g) + CaC_2(s) = C(s) + CaNCN(s)$$

The $N-N$ bonds of N_2 can be cleaved at room temperature on interaction with some 3-coordinate transition metal complexes[6] having bulky ligands. The planar molybdenum complex, $Mo(NR'R'')_3$, $R' = C(CD_3)_2Me$, $R'' = 3,5\text{-}C_6H_3Me_2$, designated MoL_3, appears to react as follows to give a nitrido compound.

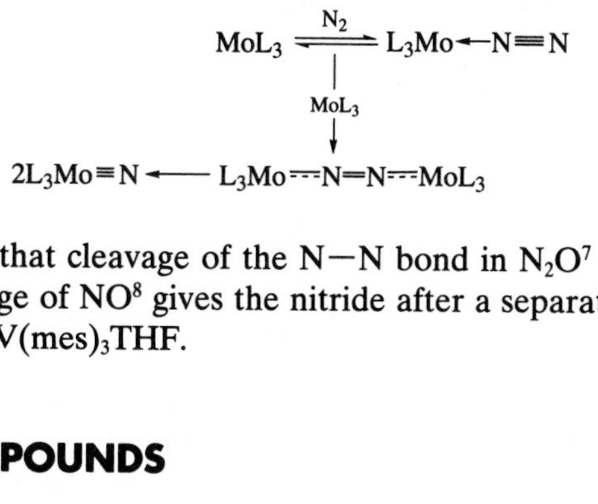

$$MoL_3 \xrightarrow{\quad N_2 \quad} L_3Mo \leftarrow N \equiv N$$

$$\downarrow MoL_3$$

$$2L_3Mo \equiv N \longleftarrow L_3Mo \text{---} N \equiv N \text{---} MoL_3$$

It can be noted also that cleavage of the $N-N$ bond in N_2O[7] gives L_3MoNO and L_3MoN, while cleavage of NO[8] gives the nitride after a separate deoxygenation of the NO complex by $V(mes)_3THF$.

NITROGEN COMPOUNDS

9-4 Nitrides[9]

As with carbides, there are three general classes. Ionic nitrides are formed by Li, Mg, Ca, Sr, Ba, Zn, and some others. Their formulas correspond to what would result from combination of the normal metal ions with N^{3-} ions. They are all essentially ionic and are properly written as $(Ca^{2+})_3(N^{3-})_2$, $(Li^+)_3N^{3-}$, and so on. Nitrides of the M_3N_2 type are often anti-isomorphous with oxides of the M_2O_3 type. This does not in itself mean that, like the oxides, they are ionic. However, their

[6]C. C. Cummins *et al.*, *Science* **1995**, *268*, 861; K. Morokuma *et al.*, *J. Am. Chem. Soc.* **1995**, *117*, 12366.
[7]C. C. Cummins *et al.*, *J. Am. Chem. Soc.* **1995**, *117*, 4999.
[8]C. C. Cummins *et al.*, *J. Am. Chem. Soc.* **1995**, *117*, 6613.
[9]N. R. Brese and M. O'Keefe, *Structure and Bonding* **1992**, *79*, 307 (nitrides, azides, amides, etc.); W. Schnick, *Angew. Chem. Int. Ed. Engl.* **1993**, *32*, 806 (non-metal nitrides).

ready hydrolysis to NH_3 and the metal hydroxides makes this seem likely. The ionic nitrides are prepared by direct union of the elements or by loss of ammonia from amides on heating, for example,

$$3Ba(NH_2)_2 \longrightarrow Ba_3N_2 + 4NH_3$$

There are various covalent *nitrides* (BN, S_4N_4, P_3N_5, etc.), and their properties vary greatly depending on the element with which it is combined. Such substances are therefore discussed under the appropriate element.

Some transition metal nitrides,[10] MN, of Ti, Zr, and Hf have cubic (NaCl type) structures. Others which are often not exactly stoichiometric (being N deficient), are chemically very inert and extremely hard with high melting points. The electronic band structure of the metal persists, the appearance is metallic and the compounds are electrically conducting. As an example, VN has mp 2570°C and hardness 9-10.

These metal compounds can be made by high temperature (>100°C) reactions such as that of NH_3 with the metal or metal oxide, or by lower temperature routes where metal halides are heated with Li_3N or Na_3N[11]; the latter method has been used for Ti, Cr, Mn, Hf, Y, and Sm nitrides. Thin films of nitrides can be made by chemical vapor deposition[12] using amido compounds such as $Ti(NMe_2)_4$ and NH_3 at 150–450°C at 1 atm.

There are large numbers of tertiary nitrides such as Ca_3CrN_3, Sr_3CrN_3, and $LaSi_6N_{11}$ which usually have complicated structures.[13] In Ca_3CrN_3 there are trigonal planar $[CrN_3]^{2-}$ ions; another anion is $[FeN_2]^{4-}$.

9-5 Nitrogen Hydrides

Ammonia

Anhydrous NH_3 is made on a large scale[14] by the Haber process, in which the reaction

$$N_2(g) + 3H_2(g) = 2NH_3(g) \qquad \Delta H = \text{-}46 \text{ kJ mol}^{-1} \qquad K_{25°C} = 10^3 \text{ atm}^{-2}$$

is carried out in the presence of a catalyst at pressures of 10^2 to 10^3 atm and temperatures of 400 to 550°C. Although the equilibrium is most favorable at low temperature, elevated temperatures are invariably required to obtain satisfactory rates. The original catalyst is α iron containing some oxide to widen the lattice and enlarge the active interface, but Ru on activated carbon promoted by alkali metals or Ru clusters on basic zeolites are replacing iron.[15]

Ammonia is a colorless pungent gas with a normal boiling point of $-33.35°C$ and a freezing point of $-77.7°C$. The liquid has a large heat of evaporation

[10]H.-C. zur Loye *et al., Inorg. Chem.* **1996,** *35,* 581.
[11]E. G. Gillan and R. B. Kanar, *Inorg. Chem.* **1994,** *33,* 5693; A. L. Hector and I. P. Parkin, *Polyhedron* **1995,** *14,* 913; *Chem. Mater.* **1995,** *7,* 1728.
[12]D. M. Hoffman, *Polyhedron* **1994,** *13,* 1169.
[13]M. G. Barker *et al., J. Chem. Soc. Dalton Trans.* **1996,** 1; M. Woike and W. Jeitschko, *Inorg. Chem.* **1995,** *34,* 5105; M. T. Green and T. Hughbanks, *Inorg. Chem.* **1993,** *32,* 5611.
[14]J. R. Jennings *ed., Catalytic Ammonia Synthesis; Fundamentals and Practice,* Plenum, New York, 1991.
[15]R. J. Davis *et al., Chem. Commun.* **1996,** 649.

(1.37 kJ g⁻¹ at the boiling point) and is therefore fairly easily handled in ordinary laboratory equipment. Liquid ammonia resembles water in its physical behavior, being highly associated because of the polarity of the molecules and strong hydrogen bonding. Its dielectric constant ($\sim$22 at $-34°C$; *cf.* 81 for H_2O at 25°C) is sufficiently high to make it a fair ionizing solvent and electrochemical studies can be made in the liquid near the critical pressure (133°C, 112.5 atm):

$$2NH_3 = NH_4^+ + NH_2^-$$

$$K_{-50°C} = [NH_4^+][NH_2^-] = \sim10^{-30}$$

Liquid ammonia has lower reactivity than H_2O toward electropositive metals, which may dissolve, giving blue solutions (Section 3-3).

Because $NH_3(1)$ has a much lower dielectric constant than water, it is a better solvent for organic compounds but generally a poorer one for ionic inorganic compounds. Exceptions occur when complexing by NH_3 is superior to that by water. Thus AgI is exceedingly insoluble in water but $NH_3(1)$ at 25°C dissolves 207 g/100 cm³. Primary solvation numbers of cations in $NH_3(1)$ appear similar to those in H_2O (e.g., 5.0 ± 0.2 and 6.0 ± 0.5 for Mg^{2+} and Al^{3+}, respectively), but there may be some exceptions. Thus Ag^+ appears to be primarily linearly 2-coordinate in H_2O but tetrahedrally coordinated as $[Ag(NH_3)_4]^+$ in $NH_3(1)$. It has also been suggested that $[Zn(NH_3)_4]^{2+}$ may be the principal species in $NH_3(1)$ as compared to $[Zn(H_2O)_6]^{2+}$ in H_2O.

Reactions of Ammonia

Ammonia reacts with oxygen as in Eq. 9-1.

$$4NH_3(g) + 3O_2(g) = 2N_2(g) + 6H_2O(g) \qquad K_{25°C} = 10^{228} \tag{1}$$

However, ammonia can be made to react as in Eq. 9-2.

$$4NH_3 + 5O_2 = 4NO + 6H_2O \qquad K_{25°C} = 10^{168} \tag{2}$$

even though the process of eq. 9-1 is thermodynamically much more favorable, by carrying out the reaction at 750 to 900°C in the presence of a platinum or platinum-rhodium catalyst. This can easily be demonstrated in the laboratory by introducing a piece of glowing platinum foil into a jar containing gaseous NH_3 and O_2; the foil will continue to glow because of the heat of reaction 9-2, which occurs only on the surface of the metal, and brown fumes will appear, owing to the reaction of NO with the excess of O_2 to produce NO_2. Industrially the mixed oxides of nitrogen are then absorbed in water to form nitric acid:

$$2NO + O_2 \longrightarrow 2NO_2$$

$$3NO_2 + H_2O \longrightarrow 2HNO_3 + NO \text{ and so on.}$$

Thus the sequence in industrial utilization of atmospheric nitrogen is

$$N_2 \xrightarrow{\ H_2\ } NH_3 \xrightarrow{\ O_2\ } NO \xrightarrow{\ O_2 + H_2O\ } HNO_3(aq)$$

Haber process   Ostwald process

Ammonia is extremely soluble in water and gives *hydrates*. The hydrates $NH_3 \cdot H_2O$, $NH_3 \cdot 2H_2O$, and $2NH_3 \cdot H_2O$ have NH_3 and H_2O molecules linked by H bonds; no NH_4^+ or OH^- ions or NH_4OH molecules exist in the hydrates. The hydrate $NH_3 \cdot H_2O$ has chains of H_2O molecules linked by H bonds; these chains are cross linked by NH_3 into a three-dimensional lattice by $O-H \cdots N$ and $O \cdots H-N$ bonds. In aqueous solution ammonia is probably hydrated in a similar manner. Although aqueous solutions are commonly referred to as solutions of the weak base NH_4OH, called "ammonium hydroxide," there is reason to believe that it probably does not exist. Solutions of ammonia are best described as $NH_3(aq)$, with the equilibrium written as

$$NH_3(aq) + H_2O = NH_4^+ + OH^-$$

$$K_{25°C} = \frac{[NH_4^+][OH^-]}{[NH_3]} = 1.77 \times 10^{-5} \quad (pK_b = 4.75)$$

In an odd sense NH_4OH might be considered a strong base, since it is completely dissociated in water. A 1 M solution of NH_3 is only 0.0042 M in NH_4^+ and OH^-.

Nuclear magnetic resonance measurements show that the hydrogen atoms of NH_3 rapidly exchange with those of water by the process

$$H_2O + NH_3 = OH^- + NH_4^+$$

but there is only slow exchange between NH_3 molecules in the vapor phase.

Ammonium Salts[16]

There are many rather stable crystalline salts of the tetrahedral NH_4^+ ion; most of them are water soluble. Salts of strong acids are fully ionized, and the solutions are slightly acidic:

$$NH_4Cl = NH_4^+ + Cl^- \qquad K \sim \infty$$

$$NH_4^+ + H_2O = NH_3 + H_3O^+ \qquad K_{25°C} = 5.5 \times 10^{-10}$$

Thus a 1 M solution will have a pH of ~4.7. The constant for the second reaction is sometimes called the hydrolysis constant; however, it may also be considered the acidity constant of the cationic acid NH_4^+, and the system regarded as an acid-base system in the following sense:

$$NH_4^+ + H_2O = H_3O^+ + NH_3(aq)$$

Acid   Base   Acid   Base

[16]G. Meyer, *Adv. Synth. React. Solids* **1994**, *2*, 1.

Ammonium salts generally resemble those of K^+ and Rb^+ in solubility and, except where hydrogen-bonding effects are important for the structure, since the three ions are of comparable (Pauling) radii: $NH_4^+ = 1.48$ Å, $K^+ = 1.33$ Å, $Rb^+ = 1.48$ Å. Many ammonium salts volatilize with dissociation around 300°C, for example,

$$NH_4Cl(s) = NH_3(g) + HCl(g) \qquad \Delta H = 177 \text{ kJ mol}^{-1} \qquad K_{25°C} = 10^{-16}$$

$$NH_4NO_3(s) = NH_3(g) + HNO_3(g) \qquad \Delta H = 171 \text{ kJ mol}^{-1}$$

Some salts that contain oxidizing anions decompose when heated, with oxidation of the ammonia to N_2O, N_2, or both. For example,

$$(NH_4)_2Cr_2O_7(s) = N_2(g) + 4H_2O(g) + Cr_2O_3(s) \qquad \Delta H = -315 \text{ kJ mol}^{-1}$$

$$NH_4NO_3(s) = N_2O(g) + 2H_2O(g) \qquad \Delta H = -23 \text{ kJ mol}^{-1}$$

Ammonium halides are very useful for conversion of oxides to halides, oxohalides, or ammonium salts of halogeno anions at elevated temperatures.

Ammonium nitrate volatilizes reversibly at moderate temperatures; at higher temperatures, irreversible decomposition occurs exothermically, giving mainly N_2O. This is the reaction by which N_2O is prepared commercially. At still higher temperatures, the N_2O itself decomposes into nitrogen and oxygen. The decomposition of liquid ammonium nitrate can also become explosively rapid, particularly when catalyzed by traces of acid and chloride; there are instances of disastrous explosions of ammonium nitrate in bulk following fires. However, NH_4NO_3 with diesel oil, alone or mixed with kaolin and detonated by dynamite, is now the standard explosive for blasting and quarrying.

Tetraalkylammonium ions, R_4N^+, especially those with bulky R groups, are selective and robust phase transfer catalysts and also of use where large cations are required to allow precipitation of crystalline solids. Radicals R_4N in the form of amalgams ($\sim$12Hg/NR_4) can be obtained electrolytically or by reduction of R_4NX with Na/Hg in media where the resulting NaX is insoluble.

Hydrazine[17]

This compound is a base somewhat weaker than ammonia and two series of hydrazinium salts are obtainable.

$$N_2H_4(aq) + H_2O = N_2H_5^+ + OH^- \qquad K_{25°C} = 8.5 \times 10^{-7}$$

$$N_2H_5^+ + H_2O = N_2H_6^{2+} + OH^- \qquad K_{25°C} = 8.9 \times 10^{-16}$$

Those of $N_2H_5^+$ are stable in water, and those of $N_2H_6^{2+}$ are, as expected from the foregoing equilibrium constant, extensively hydrolyzed. From the reaction of $[N_2H_5]NO_3$ and aqueous H_2SO_4 crystals of $[N_2H_6]SO_4$ can be isolated.[18]

[17]For many references see Y. Zhong and P. K. Lim, *J. Am. Chem. Soc.* **1989**, *111*, 8398.
[18]T. M. Klapötke *et al.*, *Polyhedron* **1996**, *15*, 2579.

As another consequence of its basicity, hydrazine, like NH_3, can form coordination complexes with both Lewis acids and metal ions. Just as with the proton, electrostatic considerations (and, in these cases, steric considerations) militate against bifunctional behavior.

Aqueous solutions of N_2H_4 have been made by the Raschig synthesis.[19] The overall reaction, carried out in aqueous solution, is

$$2NH_3 + NaOCl \longrightarrow N_2H_4 + NaCl + H_2O$$

The reaction proceeds in two steps, the first being nearly quantitative at 0°C with a 3:1 NH_3 to OCl^- ratio:

$$NH_3 + NaOCl \longrightarrow NaOH + NH_2Cl \quad \text{(fast)}$$

$$NH_3 + NH_2Cl + NaOH \longrightarrow N_2H_4 + NaCl + H_2O$$

However, there is a competing and parasitic reaction that is rather fast once some N_2H_4 has been formed:

$$2NH_2Cl + N_2H_4 \longrightarrow 2NH_4Cl + N_2$$

Addition of gelatine is said to suppress this reaction by scavenging heavy metal ions. An improved process[20] depends on the following cycle, where a phosphate catalyst is used to speed up the reaction:

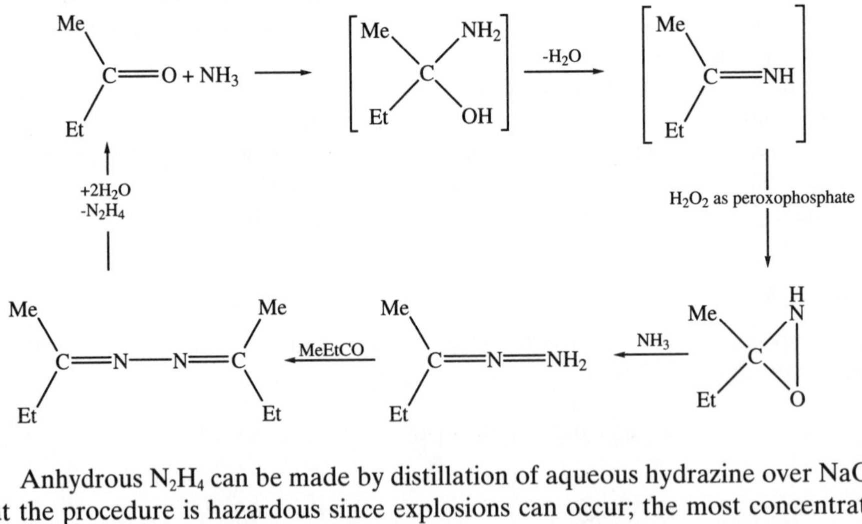

Anhydrous N_2H_4 can be made by distillation of aqueous hydrazine over NaOH but the procedure is hazardous since explosions can occur; the most concentrated product that is made is $N_2H_4 \cdot H_2O$. Pure N_2H_4 can be obtained from aqueous solutions in the form of inclusion compounds with hydroquinone or *p*-methoxyphenol and these solids can be used in synthetic reactions.[21]

[19]M. Ferriol *et al., Inorg. Chem.* **1989,** *28,* 3808.
[20]W. Büchner *et al., Industrial Inorganic Chemistry,* VCH, New York, 1989, Sect. 14.2.1.
[21]F. Toda *et al., J. Chem. Soc., Chem. Commun.* **1995,** 1531.

The pure compound, mp 2°C, bp 111.5°C, is endothermic ($\Delta H_f^0 = 50$ kJ mol^{-1}). At 25°C it is wholly in the *gauche* form (9-VI).

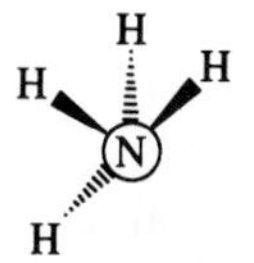

$$N\text{—}N = 1.47 \text{ Å}$$
$$\angle N\text{—}N\text{—}H = 112°$$
(9-VI)

Aqueous hydrazine is a powerful reducing agent in basic solution; in many such reactions diimine (see below) is an intermediate. One reaction, which is quantitative with some oxidants (e.g., I_2), is:

$$N_2 + 4H_2O + 4e = 4OH^- + N_2H_4(aq) \qquad E^0 = -1.16 \text{ V}$$

However, NH_3 and HN_3 are also obtained under various conditions. The reaction of hydrazine with air or O_2, especially when catalyzed by multivalent metal ions in basic solution, produces H_2O_2:

$$2O_2 + N_2H_4(aq) \longrightarrow 2H_2O_2(aq) + N_2$$

but further reaction occurs in the presence of some metal ions:

$$N_2H_4 + 2H_2O_2 \longrightarrow N_2 + 4H_2O$$

Hydrazine reacts with HNO_3 or HNO_2,[22]

$$N_2H_5^+ + HNO_2 \longrightarrow HN_3 + H^+ + 2H_2O$$

It can also be reduced to NH_3 by $MoFe_3S_4$ polycarbonate clusters,[23] a process that may be relevant in the natural reduction of N_2 to NH_3 by nitrogenase.

The main uses of hydrazine are for synthesis of herbicides and pesticides, blowing agents for foam rubber and plastics, rocket fuel, and removal of O_2 in boilers.

Diazene (Diimide, N_2H_2) and Other Nitrogen Hydrides

Diazene can be obtained with NH_3, by microwave discharge in gaseous N_2H_4, and in the pure state by the reaction:

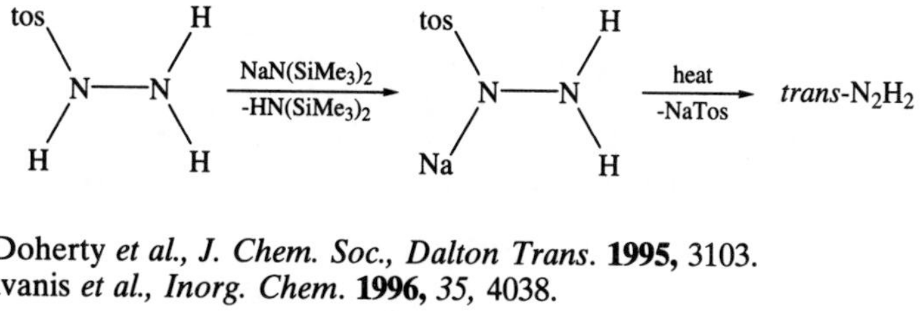

[22]A. M. M. Doherty *et al.*, *J. Chem. Soc., Dalton Trans.* **1995**, 3103.
[23]D. Coucouvanis *et al.*, *Inorg. Chem.* **1996**, *35*, 4038.

and in aqueous solution from the azodiformate anion[24]:

$$(NCO_2)_2^{2-} + 2H^+ \longrightarrow N_2H_2(aq) + 2CO_2$$

The pure yellow product is very unstable, decomposing above *ca.* $-130°C$ to N_2 and N_2H_4.

Diazene can be detected by chemical reactions such as the stereospecific, highly selective *cis*-hydrogenation of C=C bonds by hydrazine and an oxidant, and by mass spectrometry in the gas phase decomposition of both NH_3 and N_2H_4 on rhodium surfaces.[25]

Other unstable or transient hydrides detected are $N=NH_2$, $HN=N-NH_2$, $H_2N-N=N-NH_2$, and $(H_2NNH)_2$.

Hydrazoic Acid and Azides[26]

Although HN_3 is a hydride of nitrogen in a formal sense, it has no essential relationship to NH_3 and N_2H_4. The sodium salt is prepared by the reactions

$$3NaNH_2 + NaNO_3 \xrightarrow{175°C} NaN_3 + 3NaOH + NH_3$$

$$2NaNH_2 + N_2O(g) \xrightarrow{195°C} NaN_3 + NaOH + NH_3$$

$$3N_2O + 4Na + NH_3 \xrightarrow{NH_3(l)} NaN_3 + 3NaOH + 2N_2$$

and the free acid can be obtained in aqueous solution by the reaction

$$N_2H_5^+ + HNO_2 \longrightarrow HN_3 + H^+ + 2H_2O$$

The pure acid (bp 37°C) is dangerously explosive but can be obtained on heating NaN_3 with stearic acid at 110–130°C; it is toxic even in low concentration in air.

Alkali metal azides are not explosive, sodium azide for example decomposing at *ca.* 300°C to give Na and N_2. Tetramethylammonium azide[27] can be made in high yield and purity:

$$Me_4N^+F^- + Me_3SiN_3 \xrightarrow{MeCN} Me_4N^+N_3^- + Me_3SiF$$

This crystalline solid is also safe to handle, but when this or NaN_3 are used in syntheses, these should be carried out in solution, particularly when NaN_3 is added to acid solutions as in the synthesis of $[Ru(N)Cl_4]^-$.

Azides of heavy metals are generally explosive; those of Hg and Pb explode on being struck sharply and are used in detonation caps.

[24]D. M. Stanbury *et al., J. Am. Chem. Soc.* **1995,** *117,* 8967.
[25]J. Prasad and J. L. Gland, *J. Am. Chem. Soc.* **1991,** *113,* 1577.
[26]For references see: T. M. Klapötke *et al., Inorg. Chem.* **1995,** *34,* 4343; I. C. Tornieporth-Oetting and T. M. Klapötke, *Angew. Chem. Int. Ed. Engl.* **1995,** *34,* 511 (covalent azides; 78 references and much data).
[27]K. O. Christe *et al., J. Am. Chem. Soc.* **1992,** *114,* 341.

The N_3^- ion is symmetrical and linear (N—N = 1.16 Å) while azides of the type RN_3, R = H, CH_3, C_6H_5, etc., have angular R—N—N groups.[28]

Organic azides[29] are reasonably stable, but care is always required in their synthesis and use.

Interactions of HN_3 and Lewis acids such as BF_3, SbF_5, or AsF_5 in liquid HF form the aminodiazonium salts, e.g., $H_2N_3^+AsF_6^-$ as stable solids. The cation $(H)_2NNN^+$ has a pyramidal NH_2 group with N—N distances corresponding to single and triple bonds,[30] M_2N—N≡N.

Halogen azides,[26] XN_3, X = F, Cl, Br, and I are explosive. The first three have bent structures, but the iodide has chains of the type —(N_3)—I—(N_3)—I—. The cation $[I(N_3)_2]^+$ can be made[31]:

$$[ICl_2]^+AsF_6^- + 2AgN_3 \longrightarrow [I(N_3)_2]^+AsF_6^- + 2AgCl$$

Hydroxylamine

Replacement of one H in NH_3 by OH gives NH_2OH. The amine is a weaker base than NH_3 and has a *trans*-structure with an N—O distance of 1.46 Å.[32]

Hydroxylamine can be made by reduction of NO in dilute solution by H_2 using platinized charcoal as catalyst or by reduction of nitrates electrolytically or with SO_2. The solid (mp 33°C) decomposes above *ca.* 0°C and the amine is normally encountered as stable, water soluble salts of NH_3OH^+.

Although NH_2OH can act as an oxidizing agent, it is usually used as a reductant. Its main use is in synthesis of caprolactam and oxime derivatives, but it is also used as an O_2 scavenger.

Reactions such as that with Fe^{3+} are quantitative

$$4Fe^{3+} + 2NH_3OH^+ = N_2O + 4Fe^{2+} + 6H^+ + H_2O$$

and can be used for standardization. Other reactions such as the oxidation by halogens,[33]

$$2I_2 + 2NH_2OH = N_2O + 4I^- + 4H^+ + H_2O$$

or by $[IrCl_6]^{-34}$ are mechanistically very complex.

9-6 Oxides of Nitrogen

The oxides of nitrogen are listed in Table 9-1 and their structures are shown in Fig. 9-2.

[28]T. M. Klapötke *et al., Polyhedron* **1996**, *15*, 1405.
[29]E. F. V. Scriven, *Chem. Rev.* **1988**, *88*, 297.
[30]K. O. Christe *et al., J. Am. Chem. Soc.* **1993**, *115*, 1836.
[31]T. M. Klapötke *et al., Inorg. Chem.* **1993**, *32*, 5640.
[32]D. M. Gange and E. A. Kallel, *J. Chem. Soc., Chem. Commun.* **1992**, 824.
[33]D. W. Margerum *et al., Inorg. Chem.* **1995**, *34*, 6093.
[34]D. M. Stanbury *et al., Inorg. Chem.* **1994**, *33*, 5108.

Table 9-1   Oxides of Nitrogen

Formula	Name	Color	Temperature (°C)	Remarks
N_2O	Nitrous oxide	Colorless	mp −90.8; bp −88.5	Rather unreactive
NO	Nitric oxide	Colorless	mp −163.6; bp −151.8	Moderately reactive
N_2O_3	Dinitrogen trioxide	Dark blue	fp −100.6; dec. 3.5	Extensively dissociated as gas
NO_2	Nitrogen dioxide	Brown		Rather reactive
N_2O_4	Dinitrogen tetroxide	Colorless	fp −11.2; bp 21.15	Extensively dissociated to NO_2 as gas and partly as liquid
N_2O_5	Dinitrogen pentoxide	Colorless	mp 30; dec. 47	Unstable as gas; ionic solid
NO_3; N_2O_6				Not well characterized and quite unstable

Dinitrogen Oxide (Nitrous Oxide)[35]

This oxide can be made on a laboratory scale by heating NH_4NO_3 in the melt at 250–260°C when water is lost; some NO as contaminant can be removed by passage through $FeSO_4$ solution. It is available on a large scale as a by-product in the manufacture of nylon and is available in cylinders.

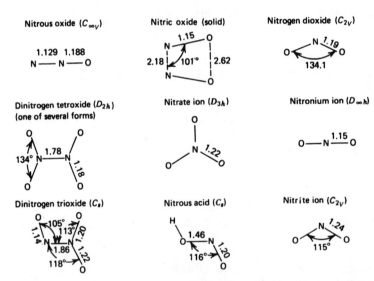

Figure 9-2   The structures and point group symmetries of some nitrogen oxides and anions (angles in degrees, bond lengths in Å). Note that the N—NO_2 bond length in N_2O_4 and N_2O_3 is long compared to the single bond distance in, for example, H_2N—NH_2 of 1.47 Å.

[35]For many references see J. T. Groves and J. S. Roman, *J. Am. Chem. Soc.* **1995**, *117*, 5594.

Although N_2O has, in the past, been regarded as an inert molecule, this is no longer strictly true, although N_2O does not react with halogens, alkali metals, and ozone at room temperature. It is kinetically, though not thermodynamically (free energy of formation, $\Delta G_f^0 = +104.18$ kJ mol^{-1}), inert in the absence of a transition metal center. It has been known for some time[36] as a potent oxygen transfer reagent:

$$N_2O = N_2 + O \qquad \Delta G_f^0 = 104.6 \text{ kJ mol}^{-1}$$

and examples are the following:

$$W_2(OBu^t)_6 + N_2O \xrightarrow[\text{pentane}]{-15°C} W_3O_2(OBu^t)_8$$

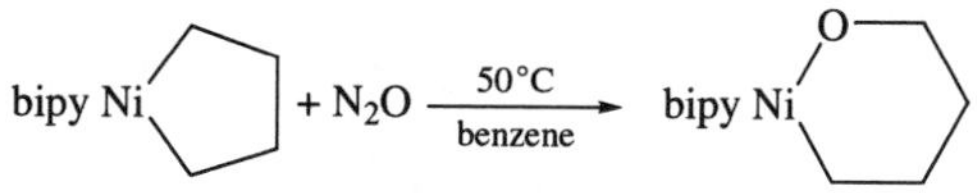

$$Ru^{II}(porph)(THF)_2 + N_2O \xrightarrow{\text{toluene}} Ru^{VI}(porph)(O)_2$$

Many such reactions are now known, some being very rapid. The cleavage reaction of N_2O on a 3-coordinate molybdenum amido species has been noted earlier (Section 1-3). Complexes of N_2O will be considered later.

Nitric Oxide

This is formed in many reactions involving reduction of nitric acid and solutions of nitrates and nitrites. For example, with 8 M nitric acid,

$$8HNO_3 + 3Cu \longrightarrow 3Cu(NO_3)_2 + 4H_2O + 2NO$$

Reasonably pure NO is obtained by the aqueous reactions

$$2NaNO_2 + 2NaI + 4H_2SO_4 \longrightarrow I_2 + 4NaHSO_4 + 2H_2O + 2NO$$

$$2NaNO_2 + 2FeSO_4 + 3H_2SO_4 \longrightarrow Fe_2(SO_4)_3 + 2NaHSO_4 + 2H_2O + 2NO$$

or dry,

$$3KNO_2(l) + KNO_3(l) + Cr_2O_3(s) \longrightarrow 2K_2CrO_4(s) + 4NO$$

Commercially it is obtained by catalytic oxidation of ammonia as already noted. Direct combination of the elements occurs only at very high temperatures, and to isolate the small amounts so formed (a few volume percent at 3000°C) the equilib-

[36]G. L. Hillhouse *et al., Organometallics* **1995,** *14,* 456; S. Gambarotta *et al., Inorg. Chem.* **1996,** *35,* 1874; R. H. Holm, *Chem. Rev.* **1987,** *87,* 1401; R. H. Holm and J. P. Donahue, *Polyhedron* **1993,** *12,* 571.

rium mixture must be rapidly chilled. Though much studied, this reaction has not been developed into a practical commercial synthesis.

In the gas phase NO reacts with O_2, probably *via* an intermediate, possibly ONOONO, to give NO_2, but since the reaction is second order in NO, the rate of reaction is insignificant at concentrations in air of *ca.* 500 ppb.

In aqueous solution the reaction of NO and O_2 is very complex[37] but the overall reaction appears to be

$$4NO + O_2 + 2H_2O \longrightarrow 4HNO_2$$

Nitric oxide also reacts with F_2, Cl_2, and Br_2 to form the nitrosyl halides XNO (Section 9-8) and with CF_3I to give CF_3NO and I_2. It is oxidized to nitric acid by several strong oxidizing agents; the reaction with permanganate is quantitative and provides a method of analysis. It is reduced to N_2O by SO_2 and to NH_2OH by chromium(II) ion, in acid solution in both cases.

Nitric oxide is thermodynamically unstable at 25°C and 1 atm and under pressure readily decomposes in the range 30 to 50°C and reacts chemically as NO_2 or N_2O_3:

$$3NO \longrightarrow N_2O + NO_2$$

Over Cu^{2+}-ZSM-5 zeolite disproportionation to N_2 and O_2 occurs.

The NO molecule has the electron configuration $(\sigma_g)^2(\sigma_u)^2(\sigma_g, \pi_u)^6(\pi_g)^1$. The unpaired π antibonding electron renders the molecule paramagnetic and partly cancels the effect of the π-bonding electrons. Thus the bond order is 2.5, consistent with an interatomic distance of 1.15 Å, which is intermediate between the triple-bond distance in NO^+ (see later) of 1.06 Å and representative double-bond distances of ~1.20 Å.

Nitric oxide dimerizes in the solid state (Fig. 9-2). Dimers persist in the vapor at the boiling point where the structure is similar but the N—N distance is longer (2.237 Å) and the ONN angle is 99.6°. The binding energy of the dimer, ~16 kJ mol^{-1}, is consistent with the long N—N bond. The dimer has no unpaired spins but feeble intrinsic temperature-independent paramagnetism. Unstable forms can be isolated in matrices.

The electron in the π^* orbital is relatively easily lost ($\Delta H_{ion} = 891$ kJ mol^{-1}), to give the *nitrosonium ion* (NO^+), which has an extensive and important chemistry. Because the electron removed comes out of an antibonding orbital, the bond is stronger in NO^+ than in NO: the bond length decreases by 0.09 Å and the vibration frequency rises from 1840 cm^{-1} in NO to 2150 to 2400 cm^{-1} (depending on environment) in NO^+.

The redox equilibria[38] show a marked dependence on solvent, but for $NO^+BF_4^-$ in MeCN we have

$$NO^+ + e \rightleftharpoons NO \qquad E^0 = 1.28 \text{ V } \textit{vs} \text{ S.C.E.}$$

[37]S. Goldstein and G. Czapski, *J. Am. Chem. Soc.* **1996,** *118,* 3419; **1995,** *117,* 12078.
[38]J. K. Kochi *et al., Inorg. Chem.* **1990,** *29,* 4196.

The NO^+ ion has been isolated in a variety of salts obtained in reactions such as

$$NO + MoF_6 \longrightarrow NO^+MoF_6^-$$

$$ClNO + SbCl_5 \longrightarrow NO^+SbCl_6^-$$

$$N_2O_3 + 3H_2SO_4 \longrightarrow 2NO^+ + 3HSO_4^- + H_3O^+$$

All are readily hydrolyzed by water. The ion is isoelectronic with CO and its complexes are discussed in Section 9-10. The compound responsible for the brown ring test for nitrates is $[Fe(H_2O)_5NO]^{2+}$.

Although it might seem unlikely in view of its reactivity and toxicity, NO synthesized *in vivo* is of major importance as a messenger in physiological processes in humans and other animals, e.g., in blood clotting and blood pressure control, neurotransmissions, and so on; over 3500 papers referring to NO were published in 1995.[39]

These discoveries initiated other studies such as trapping agents to monitor NO production in cell cultures and living organisms.[40] It was shown that NO arises from the guanidine N atom of *L*-arginine. Since NO is now so important *in vivo*, an electrochemical sensor involving graphite-epoxy electrodes and an iron(III) Schiffs base complex has been devised.[41]

Reduction of NO to N_2O or N_2 is important in biological denitrification where anaerobic organisms are involved and model studies using copper catalysts have been studied.[42] Much study on other catalytic reductions of NO (and of NO_2) has been made in connection with atmospheric pollution.[43]

Oxygen transfer reactions of NO are quite common and several reactions of the type

$$(Cp_2TiCl)_2 + 2NO \longrightarrow (Cp_2TiCl)_2\mu\text{-}O + N_2O$$

have been studied[44]; the reaction sequence

$$Cp_2NbMe_2 \xrightarrow{NO} Cp_2(Me)Nb\underset{NMe}{\overset{O}{<}} \xrightarrow{\Delta} Cp_2Nb\overset{O}{<}_{Me} + \tfrac{1}{2}MeN=NMe$$

involves an insertion reaction initially.

[39]R. Rawls, *C&EN* **1996,** *May 6,* 38; J. S. Stamler *et al., Nature* **1996,** *380,* 221; S. Goldstein *et al., J. Am. Chem. Soc.* **1996,** *118,* 3419; Y. Wang and B. A. Averill, *J. Am. Chem. Soc.* **1996,** *118,* 3972; D. L. H. Williams, *Chem. Commun.* **1996,** 1085; R. J. P. Williams, *Chem. Soc. Rev.* **1996,** *25,* 77.

[40]H. G. Korth *et al., J. Am. Chem. Soc.* **1994,** *116,* 2767; K. J. Reska *et al., J. Am. Chem. Soc.* **1994,** *116,* 4119; M. R. C. Symons *et al., J. Chem. Soc., Chem. Commun.* **1993,** 1099.

[41]D. O'Hare *et al., Chem. Commun.* **1996,** 23.

[42]W. B. Tolman *et al., Angew. Chem. Int. Ed. Engl.* **1994,** *33,* 895; *J. Chem. Soc., Chem. Commun.* **1994,** 1625.

[43]H. Bosch and F. J. G. Janssen, *Catalytic Reduction of Nitrogen Oxides, Catalysis Today,* 1988, Vol 2, No. 4, Elsevier, New York; Y. Iwasawa *et al., J. Chem. Soc., Chem. Commun.* **1993,** 184 (and references therein).

[44]F. Bottomley *et al., Organometallics* **1995,** *14,* 691.

Reductions of NO leading to oxo compounds and N_2O appear to proceed *via* unstable hyponitrite complexes and thus resemble the reduction of NO by sodium to give hyponitrites (Section 9-7) by 1e electron transfers:

$$M^{\cdot} + {}^{\cdot}NO \longrightarrow M^+ + NO^-$$

$$NO^- + {}^{\cdot}NO \longrightarrow N_2O_2^{\cdot\,-}$$

$$M^+N_2O_2^{\cdot\,-} + M^{\cdot} \longrightarrow 2M^+N_2O_2^{2-}$$

The hyponitrites decompose quite readily:

$$N_2O_2^{2-} \longrightarrow N_2O + O^{2-}$$

Infrared data from reductions of NO by tetramesityliridium and by Cp_2Co,[45] both of which have a single unpaired electron, indicate that unstable hyponitrites are intermediates. For Cp_2Co there is hence the sequence:

$$2Cp_2Co^{\cdot} + 2NO^{\cdot} \longrightarrow [Cp_2Co]_2(N_2O_2) \longrightarrow N_2O + CpCo[C_5H_4(\mu-O)C_5H_4]CoCp$$

This results in O transfer to give a remarkably reactive ether.

Dinitrogen Trioxide

The molecule N_2O_3 is obtained by interaction of stoichiometric amounts of NO and O_2 or NO and N_2O_4. It exists pure only in the pale blue solid (mp *ca.* $-100°C$) because of the ready dissociation to give NO and NO_2. There is also some dissociation in the liquid to NO^+ and NO_2^-. The oxide, formally the anhydride of HNO_2, is blue in water and in non-polar solvents. It is a powerful nitrosating agent for organic compounds. The molecule is planar in the gas phase, Fig. 9-2; in the solid state[46] there are two phases (-107 to $-170°C$) and the N—N bond is somewhat longer, 1.89 Å, than in the gas phase, 1.86 Å. The molecule is polar in the sense $ON^{\delta+}-NO_2^{\delta-}$. In addition to the planarity, the NO_2 group is tilted about 3.8° within the molecular plane towards the NO group.

Nitrogen Dioxide and Dinitrogen Tetroxide

These two oxides exist in a strongly temperature-dependent equilibrium:

$$N_2O_4 \rightleftharpoons 2NO_2 \qquad K = 1.4 \times 10^{-5} \text{ mol dm}^{-3} \text{ (303 K)}$$

The dissociation energy of N_2O_4 in the gas phase is 57 kJ mol^{-1}. In the solid, the oxide is entirely the colorless diamagnetic molecule N_2O_4. Partial dissociation occurs in the liquid; it is pale yellow at the freezing point and contains 0.01% of NO_2, which increases to 0.1% in the deep red-brown liquid at the boiling point, 21.15°C. In the vapor at 100°C the composition is NO_2 90%, N_2O_4 10%, and dissociation is

[45]G. Wilkinson *et al., J. Chem. Soc., Dalton Trans.* **1994**, 2223.
[46]A. Simon *et al., Chem. Eur. J.* **1995**, *1*, 389.

complete above 140°C. Molecular beam mass spectrometric studies indicate that trimers and tetramers occur, but there is no evidence for such species under normal conditions.

The monomer NO_2 has an unpaired electron and its properties, red-brown color, and ready dimerization to colorless and diamagnetic N_2O_4, are not unexpected for such a radical. Nitrogen dioxide can also lose its odd electron fairly readily ($\Delta H_{ion} = 928$ kJ mol^{-1}) to give NO_2^+, the *nitronium ion*, discussed later.

Although other forms can exist in inert matrices, the most stable form of N_2O_4 is that shown in Fig. 9-2. This molecule has unusual features, namely, the planarity and the long N—N bond. Molecular orbital calculations suggest that although the bond is of a σ type, it is long because of delocalization of the electron pair over the whole molecule, with large repulsion between doubly occupied MO's of NO_2. The coplanarity results from a delicate balance of forces operating in the skew and planar forms. The barrier to rotation about the N—N bond is estimated to be ~9.6 kJ mol^{-1}.

The mixed oxides are obtained by heating metal nitrates, by oxidation of NO in air, and by reduction of nitric acid and nitrates by metals and other reducing agents. The gases are highly toxic and reactive.

The hydrolysis reaction also involves electron self exchange[47]:

$$2NO_2 + H_2O = NO_2^- + NO_3^- + 2H^+$$

$$NO_2 + {}^{15}NO_2^- = NO_2^- + {}^{15}NO_2$$

The thermal decomposition

$$2NO_2 \rightleftharpoons 2NO + O_2$$

begins at 150°C and is complete at 600°C.

The oxides are fairly strong oxidizing agents in aqueous solution, comparable in strength to bromine:

$$N_2O_4 + 2H^+ + 2e = 2HNO_2 \qquad E^0 = +1.07 \text{ V}$$

The mixed oxides, "nitrous fumes," or NO_x are used in organic chemistry as selective oxidizing agents; the first step is hydrogen abstraction,

$$RH + NO_2 = R^{\cdot} + HONO$$

and the strength of the C—H bond generally determines the nature of the reaction.

Dinitrogen tetroxide undergoes self-ionization endothermically:

$$N_2O_4 \rightleftharpoons NO^+ + NO_3^- \qquad K = 7.1 \times 10^{-8} \text{ mol dm}^{-3} \text{ (303 K)}$$

$$N_2O_4 \rightleftharpoons NO_2^+ + NO_2^- \qquad K = 10^{-22} \text{ mol dm}^{-3} \text{ (303 K)}$$

It has been used as a solvent and forms molecular addition compounds with a great variety of nitrogen, oxygen, and aromatic donor compounds, e.g., $N_2O_4 \cdot$THF.

[47]D. M. Stanbury *et al., J. Am. Chem. Soc.* **1989,** *111,* 5497.

Systems involving liquid N_2O_4 mixed with an organic solvent are often very reactive; for example, they dissolve relatively noble metals to form nitrates, often solvated with N_2O_4. Thus Cu reacts vigorously with N_2O_4 in ethyl acetate to give crystalline $Cu(NO_3)_2 \cdot N_2O_4$, from which anhydrous, volatile (at 150–200°C) Cu^{II} nitrate is obtained. Some of the compounds obtained in this way are nitrosonium salts, for example, $Zn(NO_3)_2 \cdot 2N_2O_4$ is $(NO^+)_2[Zn(NO_3)_4]^{2-}$, and $Cu(NO_3)_2 \cdot N_2O_4$ consists of NO^+ ions and polymeric nitrato anions. The complex $Fe(NO_3)_3 \cdot 1.5N_2O_4$ also appears to have a nitrato anion $[Fe(NO_3)_4]^-$, but with a cation $N_4O_6^+$ that can be regarded as NO_3^- bound to three NO^+ groups.

In anhydrous acids N_2O_4 dissociates ionically and in anhydrous HNO_3 dissociation is almost complete:

$$N_2O_4 = NO^+ + NO_3^-$$

The dissociation in H_2SO_4 is complete in dilute solution; at higher concentrations undissociated N_2O_4 is present, and at very high concentrations HNO_3 is formed:

$$N_2O_4 + 3H_2SO_4 = NO^+HSO_4^- + HNO_3 + HSO_4^- + SO_3 + H_3O^+$$

The $NOHSO_4$ actually crystallizes out. The detailed mechanism and intermediates are undoubtedly complex.

The involvement of NO_2 and NO (NO_x) is of great importance in atmospheric pollution, arising mainly from automobile exhausts, but this topic will not be dealt with here.

Dinitrogen Pentoxide

The oxide N_2O_5 has the structure (9-VII) in the gas phase at $-11°C$ but in the solid, the stable form is $NO_2^+NO_3^-$.

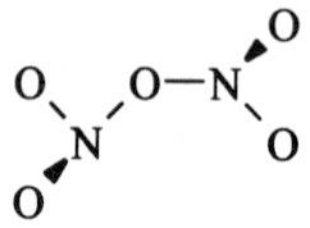

N—O(bridge) = 1.498 Å
N—O(terminal) = 1.188 Å
∠NON = 11.8°
∠ONO = 133.2°

(9-VII)

The oxide is obtained by dehydration of HNO_3 with P_2O_5 or by interaction of FNO_2 and $LiNO_3$; it is not very stable (sometimes exploding) and is distilled in a current of ozonized O_2:

$$2HNO_3 + P_2O_5 \longrightarrow 2HPO_3 + N_2O_5$$

It is, conversely, the anhydride of nitric acid:

$$N_2O_5 + H_2O = 2HNO_3$$

It is deliquescent, readily producing nitric acid by this reaction.

As with N_2O_4, ionic dissociation occurs in anhydrous H_2SO_4, HNO_3, or H_3PO_4 to produce NO_2^+, for instance,

$$N_2O_5 + 3H_2SO_4 \rightleftharpoons 2NO_2^+ + 3HSO_4^- + H_3O^+$$

Many gas phase reactions of N_2O_5 depend on dissociation to NO_2 and NO_3, with the latter then reacting further as an oxidizing agent. These reactions are among the better understood complex inorganic reactions.

In the N_2O_5-catalyzed decomposition of O_3, the steady state concentration of NO_3 can be high enough to allow its absorption spectrum to be recorded.

Other Oxides

Nitrosylazide, N_4O, is a pale yellow highly unstable solid made by the low temperature reaction

$$NaN_3(\text{excess}) + NOCl \longrightarrow NaCl + N_4O$$

It will sublime in vacuum but decomposes exothermically at *ca.* $-50°C$ to N_2O and N_2. Raman spectra and *ab initio* calculations indicate a planar, chair-like structure (9-VIII).[48]

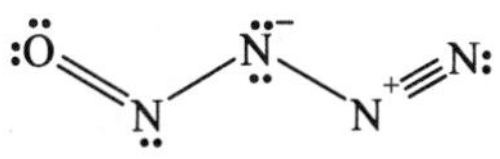

(9-VIII)

There is also some evidence for an unstable "oxide" N_4O_2 or $NO_2^+N_3^-$ as well as the radical NO_3^- found in equilibrium with NO_2 in gaseous N_2O_5.[49]

9-7 Oxo Acids and Anions of Nitrogen

Hyponitrites[50]

The sodium salt, $Na_2N_2O_2$, is made by Na reduction of NO in 1,2-dimethoxyethane containing benzophenone or quantitatively by absorption of N_2O into Na_2O[51]:

$$Na_2O + N_2O \xrightarrow{300\text{-}400°C} Na_2(ONNO)$$

It is soluble in water and recrystallizable from ethanol.

The silver salt is precipitated from aqueous solutions by $AgNO_3$. In the solid state the hyponitrite ion has been found in the *cis* conformation (9-IX); the *trans* ion has been reported but its existence has been questioned. The anion is stable in water but is decomposed by dilute acid. Protonation on only one oxygen atom

[48]T. M. Klapötke *et al.*, *Angew. Chem. Int. Ed. Engl.* **1993**, *32*, 1610.
[49]C. A. Cantrell *et al.*, *J. Chem. Phys.* **1988**, *88*, 4997.
[50]F. T. Bonner and M. N. Hughes, *Comments Inorg. Chem.* **1988**, *7*, 215.
[51]C. Feldmann and M. Jansen, *Angew. Chem. Int. Ed. Engl.* **1996**, *35*, 1728.

leads to decomposition:

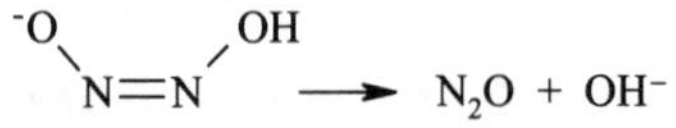

$$\underset{\displaystyle (9\text{-})}{\overset{\displaystyle {}^-O \diagdown \qquad \diagup OH}{N{=}N}} \longrightarrow N_2O + OH^-$$

However, the free acid $(NOH)_2$ can be obtained in more strongly acid solution, being more stable than ON_2OH^-. An isomer of the acid is nitramide, H_2NNO_2.

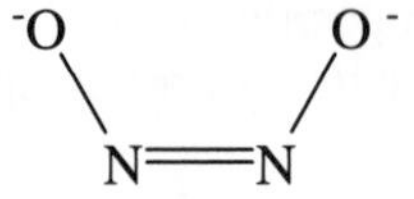

$$\overset{\displaystyle {}^-O \diagdown \qquad\qquad O^-\diagup}{\underset{\displaystyle (9\text{-IX})}{N{=\!=}N}}$$

As noted above, some metal complexes are known, another example being $(Ph_3P)_2Pt(\eta^2\text{-ONNO})$.

The Trioxodinitrate Ion

The interaction of NH_2OH and an alkyl nitrate in methanol containing NaOMe at $0°C$ gives the salt $Na_2N_2O_3$, which is stable but decomposes in neutral or alkaline media:

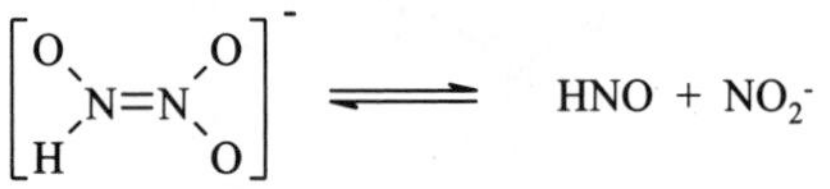

$$\left[\begin{array}{c} O\diagdown \qquad \diagup O \\ N{=}N \\ H\diagup \qquad \diagdown O \end{array}\right]^- \rightleftharpoons HNO + NO_2^-$$

$$2HNO \longrightarrow N_2O + H_2O$$

The planar anion forms transition metal complexes such as 5-coordinate $Zn(bipy)(H_2O)(N_2O_3)$ and 6-coordinate $Co(bipy)_2(N_2O_3)$, where the $N_2O_3^{2-}$ anion forms a five-membered ring with the metal ion.[52]

Nitrous Acid, Nitrites

Nitrites of electropositive metals can be made by heating nitrates, usually with C, Pb, or Fe as reducing agent. Aqueous solutions of the weak acid $(pK_a = 5.22)$ are formed on interaction of $Ba(NO_2)_2$ and H_2SO_4. The *trans*-form (Fig. 9-2) is more stable than the *cis*-form by *ca.* 2.1 kJ mol^{-1}. Aqueous solutions of the acid are unstable and normally contain HNO_2, NO_2^-, NO^+, NO, and NO_3^-; they decompose rapidly on heating to give NO, H^+, NO_3^-, and H_2O.

Nitrous acid can act either as an oxidant or a reductant:

$$HNO_2 + H^+ + e = NO + H_2O \qquad E^0 = 1.0 \text{ V}$$

$$NO_3^- + 3H^+ + 2e = HNO_2 + H_2O \qquad E^0 = 0.94 \text{ V}$$

[52]E. Libby *et al., Inorg. Chem.* **1997,** *36,* 2004.

Organic nitrosation reactions at low acidity and in the absence of nucleophiles are second order in HONO, consistent with the presence of N_2O_3 as the active nitrosating agent. Protonation of NO_2^- by strong acids in the presence of NO suggests some sort of acidium ion[53]:

$$H^+ + HNO_2 \rightleftharpoons H_2NO_2^+ \rightleftharpoons NO^+ + H_2O$$

Finally we note that enzymatic reduction of NO_2^- and NO_3^- is important in the natural nitrogen cycle. Copper-containing reductases are known and, as models, copper(I) nitrite complexes with triazacyclononane ligands have been shown to evolve NO.[54]

Peroxonitrites[55]

The yellow $ONOO^-$ ion can be made in pure form by passing NO through a solution of tetramethylammonium superoxide in ammonia[56]; $ONOO^-$ from other syntheses can be contaminated by reactants and nitrate. ONOOH is a weak acid with a pK_a near 6.8 and isomerizes to NO_3 and H^+ in seconds.[57] At pH 12 the anion is stable for several hours at 0°C. It occurs solely as the cis isomer, due to a partial double bond between N and the first peroxide O.[58] Peroxynitrite is toxic; *in vivo* it is formed through the diffusion-controlled reaction of $O_2^{\cdot-}$ with $NO^{\cdot}$.[57] ONOOH oxidizes *via* direct and indirect pathways.[59] Under physiological conditions the main reaction partners for $ONOO^-$ and ONOOH appear to be carbon dioxide[60] and thiols,[61] respectively. The $ONOO-CO_2-$adduct nitrates tyrosines.

It is convenient to mention here that *peroxonitric* acid can be made by interaction of HNO_2 with an excess of H_2O_2 at or below 0°C. In acid solution at room temperature its half life is *ca.* 30 min; decomposition is rapid in alkaline solution giving NO_2^- and O_2.[62]

Nitric Acid and Nitrates

The acid has already been discussed (Section 2-12). Nitrates of almost all metallic elements are known. They are frequently hydrated and most are soluble in water. Many metal nitrates can be obtained anhydrous, and a number of these, for example, $Cu(NO_3)_2$, sublime without decomposition.

Anhydrous nitrates often have high solubility in organic solvents and are best regarded as covalent. Alkali metal nitrates sublime in vacuum at 350 to 500°C, but

[53]F. T. Bonner *et al.*, *Inorg. Chem.* **1993**, *32*, 3316; G. K. S. Prakash *et al.*, *Inorg. Chem.* **1990**, *29*, 4965.
[54]W. B. Tolman *et al.*, *J. Am. Chem. Soc.* **1996**, *118*, 763.
[55]J. O. Edwards and R. C. Plumb, *Prog. Inorg. Chem.* **1993**, *432*, 599.
[56]D. S. Bohle *et al.*, *J. Am. Chem. Soc.* **1994**, *116*, 7423.
[57]W. H. Koppenol *et al.*, *Chem. Res. Toxicol.* **1997**, *10*, 1285.
[58]W. H. Koppenol *et al.*, *J. Phys. Chem.* **1996**, *100*, 15087.
[59]E. Goldstein and G. Czapski, *Inorg. Chem.* **1995**, *34*, 4041.
[60]S. V. Lymar and J. K. Hurst, *J. Am. Chem. Soc.* **1995**, *117*, 8867.
[61]R. Radi *et al.*, *J. Biol. Chem.* **1991**, *266*, 4244.
[62]E. H. Appelman and D. J. Gosztola, *Inorg. Chem.* **1995**, *34*, 787.

decomposition occurs at higher temperatures to yield nitrites and oxygen or above 700°C to yield oxides or peroxides.

The NO_3^- ion is reduced only with difficulty, e.g., by Al in NaOH, which gives NH_3. In view of the problem of nitrate (and nitrite) pollution[63] through the use of nitrogen fertilizers, many studies of their removal as NH_3, NH_4^+, or N_2 have been made, but no very practical method yet exists. A recent example of homogeneous reduction is that using ethanol, ethers, or carboxylic acids in sulfuric acid to give ammonium sulfate.[64]

The Nitronium Ion

This ion (NO_2^+) is directly involved, not only in the dissociation of HNO_3 itself, but also in nitration reactions and in solutions of nitrogen oxides in HNO_3 and other strong acids.

Detailed kinetic studies on the nitration of aromatic compounds first led to the idea that the attacking species was the NO_2^+ ion generated by ionizations of the following types:

$$2HNO_3 = NO_2^+ + NO_3^- + H_2O$$

$$HNO_3 + H_2SO_4 = NO_2^+ + HSO_4^- + H_2O$$

The importance of the first type is reflected in the fact that addition of ionized nitrate salts to the reaction mixture will retard the reaction. The actual nitration process can then be formulated as

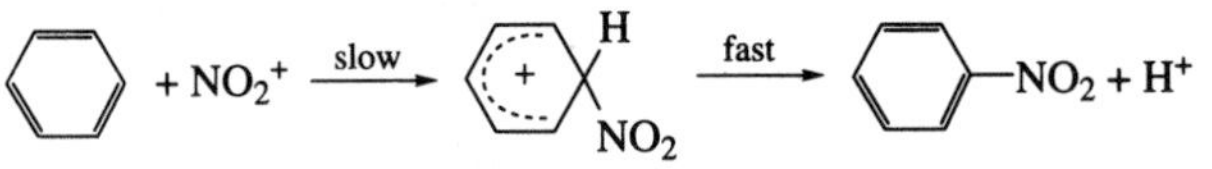

The dissociation of HNO_3 in various media is shown by cryoscopic and spectroscopic studies, and nitrogen oxides have also been found to dissociate to produce nitronium ions as noted previously.

Crystalline *nitronium salts* can be made by reactions such as

$$N_2O_5 + HClO_4 \longrightarrow [NO_2^+ClO_4^-] + HNO_3$$

$$N_2O_5 + FSO_3H \longrightarrow [NO_2^+FSO_3^-] + HNO_3$$

$$HNO_3 + 2SO_3 \longrightarrow [NO_2^+HS_2O_7^-]$$

The first two of these reactions are really just metatheses, since N_2O_5 in the solid and in anhydrous acid solution is $NO_2^+NO_3^-$. The other reaction is one between an acid anhydride, SO_3, and a base (!), namely, $NO_2^+OH^-$.

[63]C. Glidewell, *Chem. Brit.* **1990**, *Feb.*, p. 137.
[64]A. C. Hutson and A. Sen, *J. Am. Chem. Soc.* **1994**, *116*, 4527.

Nitronium salts are thermodynamically stable, but very reactive chemically. They are rapidly hydrolyzed by moisture; in addition, $NO_2^+ClO_4^-$, for example, reacts violently with organic matter, but it can actually be used to carry out nitrations in nitrobenzene solution.

The Hydroxylamine-N,N-Disulfonate and Nitroso Disulfonate Ions

The interaction of nitrite and bisulfite ions gives the hydroxylamine-N,N-disulfonate ion, $HON(SO_3)_2^{2-}$, or, in base, $ON(SO_3)_2^{3-}$:

$$2HSO_3^- + HNO_2 \longrightarrow HON(SO_3)_2^{2-} \xrightarrow{\ OH^-\ } ON(SO_3)_2^{3-}$$

In $Na_3[ON(SO_3)_2] \cdot H_2O$ the N atom is pyramidal and bound to O and to two S atoms. The ion can undergo $1e$ oxidation, for example, by air, to give the *nitrosodisulfonate* ion $\cdot ON(SO_3)_2^{2-}$. The ion in solution and in salts of large cations is paramagnetic but the potassium salt (Fremy's salt) is dimorphic with yellow and orange-brown forms. The yellow monoclinic version, though nearly diamagnetic, has a thermally accessible triplet state. The triclinic form is paramagnetic with magnetic interaction between neighboring ions, which are arranged:

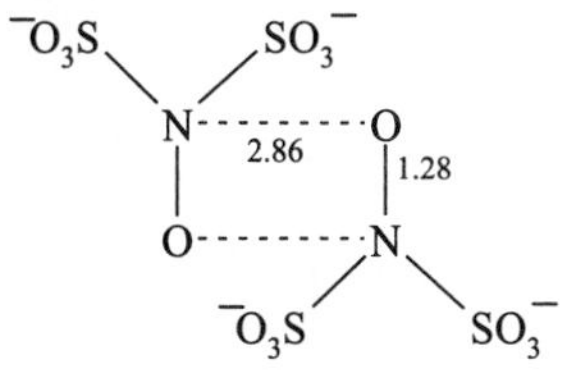

9-8 Halogen Compounds of Nitrogen[65]

Binary Compounds

Apart from the halogen azides discussed above, there are the halides NX_3, X = F, Cl, Br, I, N_2F_2, N_2F_4, NF_2Cl, and $NFCl_2$. With the exception of NF_3 all are reactive and potentially hazardous substances. Only the fluorides are important.

Nitrogen Trifluoride

The electrolysis of NH_4F in anhydrous HF yields NF_3 plus small amounts of N_2F_2, whereas electrolysis of molten NH_4F constitutes a preferred preparative method for N_2F_2. Other synthetic methods are

$$2NF_2H \xrightarrow[\text{pH 1-2}]{FeCl_3(aq)} N_2F_4 \qquad (\sim 100\%)$$

$$2NF_2H + 2KF \longrightarrow 2KHF_2 + N_2F_2$$

[65]J. M. Schreeve, *Adv. Inorg. Chem.* **1989**, *33*, 139 (N$-$F and related compounds—309 references).

$$\text{NH}_3 + \text{F}_2 \text{ (diluted by N}_2) \xrightarrow[\text{reactor}]{\text{copper packed}} \begin{cases} \text{NF}_3 \\ \text{N}_2\text{F}_4 \\ \text{N}_2\text{F}_2 \\ \text{NHF}_2 \end{cases}$$

The predominant product depends on conditions, especially the F_2/NH_3 ratio. Dinitrogen difluoride may also be prepared by photolysis of N_2F_4 in the presence of Br_2.

Nitrogen trifluoride (bp $-129°\text{C}$) is a very stable gas that normally is reactive only at *ca.* $250–300°\text{C}$ with powders of Ti, Si, or Sn[66] or with AlCl_3 at *ca.* $70°\text{C}$ when N_2, Cl_2, and AlF_3 are formed.

It is unaffected by water or most other reagents at room temperature and thermally stable in the absence of reducing metals; when heated in the presence of fluorine acceptors such as Cu, the metal is fluorinated and N_2F_4 is obtained. The NF_3 molecule has a pyramidal structure but a very low dipole moment, and it appears to be devoid of donor properties. The interaction of NF_3, F_2, and a strong Lewis acid (such as BF_3, AsF_5, or SbF_5) under pressure, uv irradiation at low temperatures, or glow discharge in a sapphire apparatus gives salts of the *tetrafluoroammonium* ion (NF_4^+), for example,

$$\text{NF}_3 + \text{F}_2 + \text{BF}_3 \xrightarrow{h\nu} \text{NF}_4\text{BF}_4$$

$$\text{NF}_3 + \text{F}_2 + \text{AsF}_5 \longrightarrow \text{NF}_4\text{AsF}_6$$

The NF_4^+ salts are hydrolyzed giving NF_3, H_2O_2, O_2, and HF.

In view of the stability of NF_4^+, the potential stability of NF_5 has been considered, but the compound is unlikely to be stable due to severe overcrowding. Attempts to isolate NF_4^+F^- showed that even at $-142°\text{C}$, exothermic decomposition to NF_3 and F_2 occurred.[67]

The unstable difluoroammonium ion can be made from the explosive difluoroamine, NHF_2 (see below), by the reaction

$$\text{NHF}_2 + \text{HF} + \text{AsF}_5 \xrightarrow{-78°\text{C}} \text{NH}_2\text{F}_2\text{AsF}_6$$

Tetrafluorohydrazine

This is a gas (bp $-73°\text{C}$) best prepared by the reaction of NF_3 with copper at high temperature. Its structure is similar to that of hydrazine, but differs in having comparable amounts of gauche and anti forms, the latter being slightly more stable, by $\sim 2 \text{ kJ mol}^{-1}$. Tetrafluorohydrazine dissociates readily in the gas and liquid phases according to the equation

$$\text{N}_2\text{F}_4 = 2\text{NF}_2 \qquad \Delta H_{298°\text{C}} = 78 \pm 6 \text{ kJ mol}^{-1}$$

which accounts for its high reactivity. The esr and electronic spectra of the difluoroamino radical NF_2 indicate that it is bent (*cf.* OF_2, O_3, SO_2, ClO_2) with the odd electron in a relatively pure π MO.

[66]E. Vileno *et al.*, *Chem. Mater.* **1995**, *7*, 683; **1996**, *8*, 1217.
[67]K. O. Christe and W. W. Wilson, *J. Am. Chem. Soc.* **1992**, *114*, 9934.

Since N_2F_4 dissociates so readily, it shows reactions typical of free radicals; thus it abstracts hydrogen from thiols,

$$2NF_2 + 2RSH \longrightarrow 2HNF_2 + RSSR$$

and undergoes other reactions such as

$$N_2F_4 + Cl_2 \xrightarrow{h\nu} 2NF_2Cl \qquad K_{25°C} = 1 \times 10^{-3}$$

$$2RI + N_2F_4 \xrightarrow{h\nu} 2RNF_2 + I_2$$

$$RCHO + N_2F_4 \longrightarrow RCONF_2 + NHF_2$$

It reacts explosively with H_2 in a radical chain reaction. It also reacts at 300°C with NO and rapid chilling in liquid nitrogen gives the unstable purple nitrosodifluoroamine ($ONNF_2$). Tetrafluorohydrazine is hydrolyzed by water, but only after an induction period. In HF(l) it reacts with SbF_5 to give $N_2F_3^+SbF_6^-$.

Difluorodiazene (Dinitrogen Difluoride)

This gas (N_2F_2) consists of two isomers (9-X) and (9-XI):

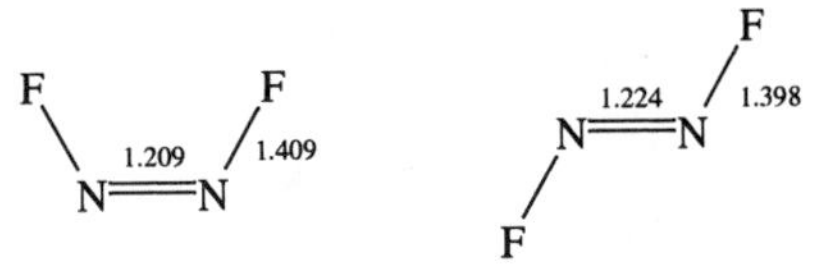

bp -105.7°   bp -111.4°

(9-X)   (9-XI)

The cis isomer predominates (~90%) at 25°C and is the more reactive. Isomerization to the equilibrium mixture is catalyzed by stainless steel. The pure transform can be obtained in ~45% yield by the reaction

$$2N_2F_4 + 2AlCl_3 \longrightarrow N_2F_2 + 3Cl_2 + 2AlF_3 + N_2$$

Interaction of *trans*-N_2F_2 with AsF_5 at 70°C gives $N_2F^+AsF_6^-$; the cation can oxidize Xe to XeF^+ and ClF to ClF^+.[68] The salts $N_2F^+AsF_6^-$ and $NF_4^+AsF_6^-$ act as fluorinating agents and will fluorinate even CH_4 in liquid HF.[69]

Other Nitrogen Trihalides[70]

The chloride is formed by the action of Cl_2 on slightly acid (pH 1–6) solutions of NH_4Cl and may be extracted into CCl_4 continuously; the reaction system has been studied in detail.[71] When pure it is a pale yellow oil (bp ~71°C) that is endothermic ($\Delta H_f^0 = 232$ kJ mol^{-1}), dangerously explosive, photosensitive, and very reactive. The molecule is pyramidal.

[68]K. O. Christe et al., J. Am. Chem. Soc. **1991**, 113, 3795.
[69]G. A. Olah et al., J. Am. Chem. Soc. **1994**, 116, 5671.
[70]For references and reactions see D. W. Margerum et al., Inorg. Chem. **1995**, 34, 3536
[71]M. Knothe and W. Hasenpusch, Inorg. Chem. **1996**, 35, 4529.

The cation $[NCl_4]^+$ is formed by the reaction in liquid SO_2:

$$2NCl_3 + Cl_2 + 3AsF_5 \xrightarrow[-78°C]{SO_2(l)} 2NCl_4^+AsF_6^- + AsF_3$$

It is stable for days at $-40°C$ and is not explosive.[72] Nitrogen bromide is similar to NCl_3.

Interaction of I_2 with concentrated NH_3 solution at room temperature gives black crystals of $(NI_3 \cdot NH_3)_n$, $n = 1, 3, 5$, that are explosive when dry. The crystals have zigzag chains of NI_4 tetrahedra sharing corners with NH_3 molecules linking the chains together. Pure NI_3 has been made as red-black explosive, volatile crystals by interaction of boron nitride with IF in $CFCl_3$.[73]

Haloamines

These are compounds of the type H_2NX and HNX_2, where H may be replaced by an alkyl radical. Only H_2NCl (chloramine), HNF_2, and H_2NF have been *isolated*; $HNCl_2$, H_2NBr, and $HNBr_2$ probably exist but are quite unstable.

Chloramine, fp $-66°C$, is formed in the gas phase reaction

$$2NH_3 + Cl_2 \longrightarrow NH_2Cl(g) + NH_4Cl(s)$$

It is stable as a gas or in solutions but the liquid and solid are explosive. The chlorination of aqueous NH_3 has been studied in detail and NH_2Cl is readily obtained by interaction of NH_3 and OCl^- at pH >8. At pH 3 to 5 $NHCl_2$, and at pH <3, NCl_3 is formed according to the reactions

$$2NH_2Cl + H^+ \longrightarrow NHCl_2 + NH_4^+$$

$$3NHCl_2 + H^+ \longrightarrow 2NCl_3 + NH_4^+$$

Difluoroamine, a colorless, explosive liquid (mp $-117°C$, bp $-23°C$), can be obtained as just described or by H_2SO_4 acidification of fluorinated aqueous solutions of urea; the first product H_2NCONF_2, gives HNF_2 on hydrolysis. It can be converted into chlorodifluoroamine ($ClNF_2$) by the action of Cl_2 and KF. Unlike NF_3, NHF_2 is a weak donor:

$$NHF_2(g) + BF_3(g) = HF_2NBF_3(s) \qquad \Delta H = -88 \text{ kJ mol}^{-1}$$

The X-ray crystal structure has been determined.[74]

Oxohalides

The *nitrosyl halides* XNO are gases at ambient temperature with boiling points ranging from $-60°C$ for FNO to $0°C$ for BrNO. They can be made by direct

[72]R. Minkewitz *et al.*, *Angew. Chem. Int. Ed. Engl.* **1990,** *29,* 181.
[73]I. C. Tornieporth-Oetting and T. M. Klapötke, *Angew. Chem. Int. Ed. Engl.* **1990,** *29,* 677.
[74]H. Willner *et al.*, *Angew. Chem. Int. Ed. Engl.* **1996,** *35,* 320.

interaction or in other ways, e.g., NO_2/N_2O_4 and KCl give red crystals on cooling (mp −61°C). Nitrosyl chloride is a major component of *aqua regia* and the most reactive. All are powerful oxidants and attack many metals, and must be handled with care. In the gas phase the structures, determined by microwave spectroscopy, are bent, but the structure of NOCl in the solid state is different from that in the vapor in that while the angle is almost the same, the distances are N—O 1.089 Å, N—Cl 2.173 Å (solid) *vs.* N—O 1.139 Å, N—Cl 1.975 Å.[75]

The *nitryl halides* are FNO_2 and $ClNO_2$ which are colorless gases (bp −72°C and −15°C, respectively) with planar structures in the gas phase and for $ClNO_2$ in the solid state by X-ray diffraction.

They can be made by the reactions

$$N_2O_4 + 2CoF_3(s) \xrightarrow{300°C} 2FNO_2 + 2CoF_2(s)$$

$$ClSO_3H + HNO_3(\text{anhydrous}) \xrightarrow{0°C} ClNO_2 + H_2SO_4$$

Both compounds are reactive and hydrolyzed to HNO_3 + HX.

The molecular compounds $ClONO_2$ (bp 22.3°C) and $FONO_2$ (bp −46°C) are again, as the above halides, highly reactive and readily hydrolyzed. They can be made by the reactions with anhydrous HNO_3:

$$HNO_3 + ClF \longrightarrow ClONO_2 + HF$$

$$HNO_3 + F_2 \longrightarrow FONO_2 + HF$$

In the gas phase both molecules are planar and the structure of $ClONO_2$ in the solid is similar.

Finally there are other compounds such as

(a) ONOF, made by interaction of NO_2 and F_2 at −30°C,
(b) $[ONCl_2]^+SbCl_6^-$, made by interaction of NCl_3, $SOCl_2$, and $SbCl_5$ in CCl_4,
(c) AsF_6^- salts of $NClF^+$, $ON(CF_3)F^+$, and $HON(CH_2)CF_3^+$.[76]

Nitrogen Trifluoride Oxide[77]

F_3NO, which is isoelectronic with NF_4^+, is a remarkably stable molecule though toxic, reactive, and oxidizing; it is resistant to hydrolysis. It reacts with AsF_5 or SbF_5 to give salts of $[ONF_2]^+$, with NO to give FNO, and with organic compounds such as amines, tertiary phosphines, etc. Although it has been made by electric discharge through NF_3 and O_2 or oxidation of FNO by IrF_6, it can now be made efficiently by vacuum pyrolysis of $NF_2O^+Sb_2F_{11}^-$ at 190–230°C in the presence of excess NaF:

$$ONF_2^+Sb_2F_{11}^- + 2NaF \longrightarrow 2Na^+SbF_6^- + ONF_3$$

[75] A. Simon *et al., J. Am. Chem. Soc.* **1995**, *117*, 7887 (structures of ClNO, ClNO₂, and ClNO₃ with discussion of bonding).
[76] R. Minkwitz *et al., Inorg. Chem.* **1991**, *30*, 2157.
[77] K. O. Christe, *J. Am. Chem. Soc.* **1995**, *117*, 6136 and references therein.

The NF_2O^+ salt is made by oxidation of NF_3 with N_2O at 150°C:

$$NF_3 + N_2O + 2SbF_5 \longrightarrow ONF_2^+Sb_2F_{11}^- + N_2$$

Above 250°C, ONF_3 undergoes the reactions

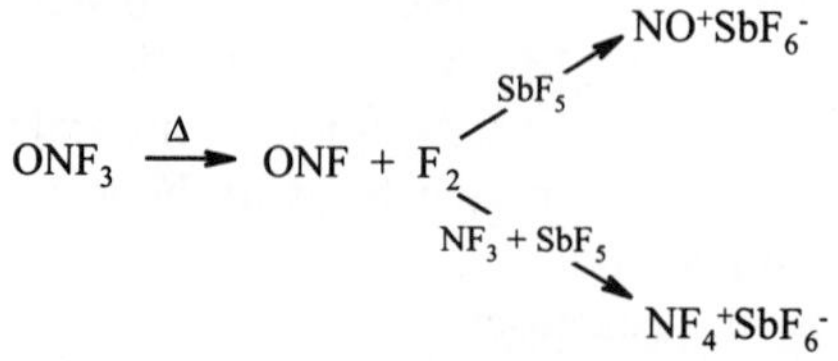

The compound is not a typical amine oxide, R_3NO, since the short N—O bond of 1.159 Å indicates a high degree of double bond character. It is best described as a resonance hybrid of $ONF_2^+F^-$ structures; this is consistent with the rather long polar N—F bonds, 1.432 Å.

DINITROGEN AND NITROGEN COMPOUNDS AS LIGANDS

Virtually all nitrogenous compounds can act as N ligands. The bonding may be solely through N but vast numbers of compounds can have N/C, N/O, N/P, etc., bonding. We deal here only briefly with some of the more important or interesting types.

9-9 Dinitrogen[78]

The first N_2 complex, $[Ru(N_2)(NH_3)_5]^{2+}$, was discovered by Allan and Senoff in 1965 by the reaction of $RuCl_3(aq)$ and hydrazine. Since then, many compounds have been made, much of the interest being to find improved, milder catalytic routes to NH_3 and other compounds; so far such hopes have not been realized.

The bonding types are as follows:

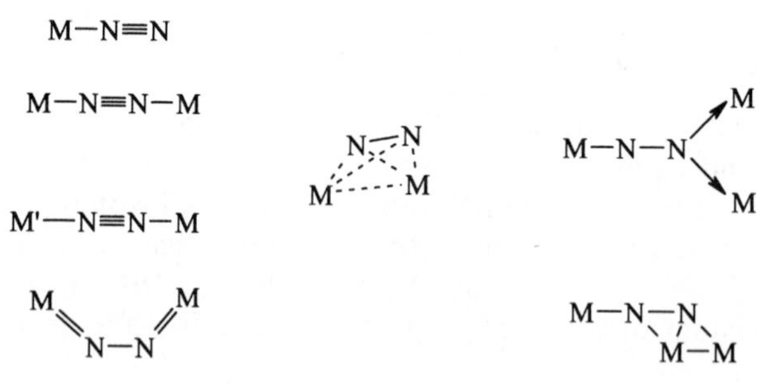

The end-on MNN η^1 type is most common. In such complexes the N—N distances, 1.10–1.16 Å, are not much larger than in N_2 itself, 1.097 Å. For bridging comp-

[78]M. Hidai and Y. Mizobe, *Chem. Rev.* **1995**, *95*, 1115 (159 refs); M. R. Blomberg and P. E. M. Siegbahn, *J. Am. Chem. Soc.* **1993**, *115*, 6908 (*ab initio* calculations).

lexes,[79] of which the linear mode MNNM is most common, the distances are in the range 1.12–1.36 Å. In the μ-η^2: η^2-N$_2$ type complex, [Cp$^*_2$Sm]$_2$(N$_2$), the N—N distance, 1.088 Å, is again close to that for N$_2$, indicating weak bonding to the metal; in a similar Zr complex the bond distance was 1.548 Å. A bridge of this type has also been reported for lithium[80] in the cation (THF)$_3$Li(N$_2$)Li(THF)$_3^{2+}$.

Syntheses

Nitrogen complexes are usually made by reduction of halides or halide complexes with other ligands using sodium amalgam[81] or other chemical reductants in an atmosphere of N$_2$. They can also be made by displacement of η^2-H$_2$ in dihydrogen compounds (Section 2-16) by N$_2$[82]; representative syntheses are as follows:

$$\text{FeCl}_2(\text{depe})_2 + \text{Na(naphthalene)} \xrightarrow[\text{N}_2]{\text{THF}} \text{Fe(N}_2)(\text{depe})_2$$

$$[\text{FeH}(\eta^2\text{-H}_2)(\text{dmpe})_2]^+ \xrightarrow[\text{N}_2]{\text{MeOH}} [\text{FeH(N}_2)(\text{dmpe})_2]^+$$

Reactivity of Bound N$_2$[83]

Although the mechanism of reduction of N$_2$ in biological systems is by no means clear, the reactions of coordinated N$_2$ in simple systems are reasonably well understood. The principal types of reaction sequence involve *hydrazido*, *imido*, and *nitrido* species (see Section 9-19).

Electrophilic attack by H$^+$ or other electrophiles can lead to N—H or N—C bond formation and in some cases to NH$_3$ and N$_2$H$_4$. Examples are

$$(\text{PhMe}_2\text{P})_4\text{Mo(N}_2)_2 + 2\text{HX} \longrightarrow [(\text{PhMe}_2\text{P})_4\text{XMoNNH}_2]\text{X} + \text{N}_2$$

$$(\text{diphos})_2\text{W(N}_2)_2 + \text{PhCOCl} \longrightarrow (\text{diphos})_2\text{ClWNNC(O)Ph} + \text{N}_2$$

For some compounds, and depending on the nature of the acid and of the solvent, there can be formation of (*a*) NH bonds or MH bonds, (*b*) the complex is oxidized with the loss of N$_2$ and of H$_2$, and (*c*) oxidative addition of HX occurs,

[79]M. D. Fryzuk *et al.*, *J. Am. Chem. Soc.* **1994,** *116,* 9529; **1993,** *115,* 2782; N. M. Doherty *et al.*, *Organometallics* **1993,** *12,* 2420.
[80]D. W. Stephan *et al.*, *J. Am. Chem. Soc.* **1993,** *115,* 3792.
[81]S. Komiya *et al.*, *J. Chem. Soc., Chem. Commun.* **1993,** 787.
[82]G. J. Leigh *et al.*, *J. Chem. Soc., Dalton Trans.* **1993,** 3041.
[83]M. Hidai and Y. Mizobe in *Reactions of Coordinated Ligands,* P. S. Braterman ed., Plenum, New York, Vol. 2, 1989 (202 references); J. P. Collman *et al.*, *Angew. Chem. Int. Ed. Engl.* **1994,** *33,* 1537 (Electrochemical conversions of N$_2$ and NH compounds in ruthenium porphyrin systems); H. Deng and R. Hoffmann, *Angew. Chem. Int. Ed. Engl.* **1993,** *32,* 1062; D. Sellman, *Angew. Chem. Int. Ed. Engl.* **1993,** *32,* 64; A. Cusanelli and D. Sutton, *Organometallics* **1996,** *15,* 1457.

for example,

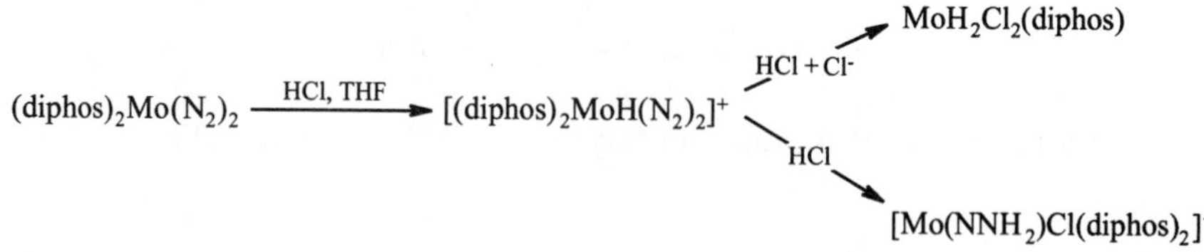

$$(diphos)_2Mo(N_2)_2 \xrightarrow{\text{HCl, THF}} [(diphos)_2MoH(N_2)_2]^+$$

with products:
- $MoH_2Cl_2(diphos)$ (via $HCl + Cl^-$)
- $[Mo(NNH_2)Cl(diphos)_2]^+$ (via HCl)

Addition of alkyl halides can lead to $C-X$ bond cleavage and *radical reactions*:

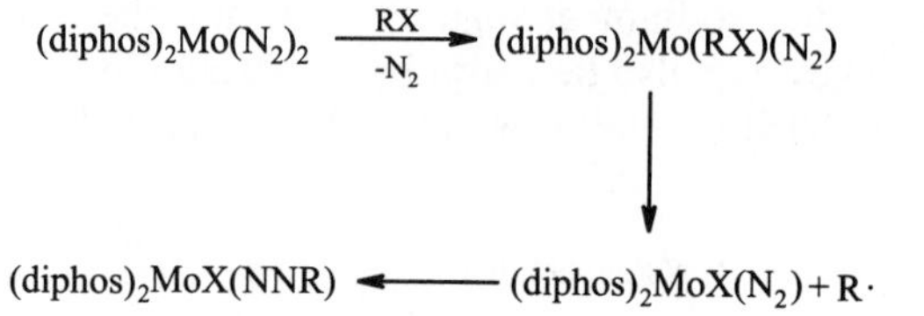

$$(diphos)_2Mo(N_2)_2 \xrightarrow[-N_2]{RX} (diphos)_2Mo(RX)(N_2)$$

$$(diphos)_2MoX(NNR) \longleftarrow (diphos)_2MoX(N_2) + R\cdot$$

Nucleophilic attack is also possible as in the reaction

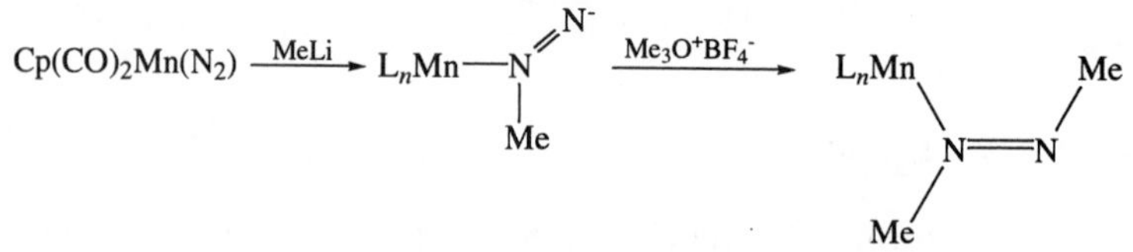

$$Cp(CO)_2Mn(N_2) \xrightarrow{MeLi} L_nMn-N\underset{Me}{\overset{N^-}{\diagdown}} \xrightarrow{Me_3O^+BF_4^-} L_nMn\diagdown N=N\diagdown Me$$

Alkyl magnesiums can react

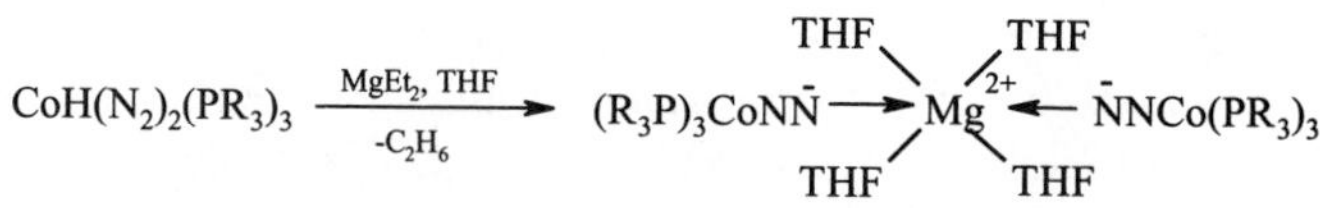

$$CoH(N_2)_2(PR_3)_3 \xrightarrow[-C_2H_6]{MgEt_2, THF} (R_3P)_3CoN\bar{N} \xrightarrow{} \underset{THF}{\overset{THF}{Mg^{2+}}} \xleftarrow{} \bar{N}NCo(PR_3)_3$$

MN_2 species can act as simple donors to Lewis acids such as $AlCl_3$ or $AlMe_3$.

Finally, in freshly synthesized $CpRe(CO)_2(^{15}N^{14}N)$, the ^{15}N is bound to Re but at 291 K in acetone isomerization occurs,[84] a possible pathway being:

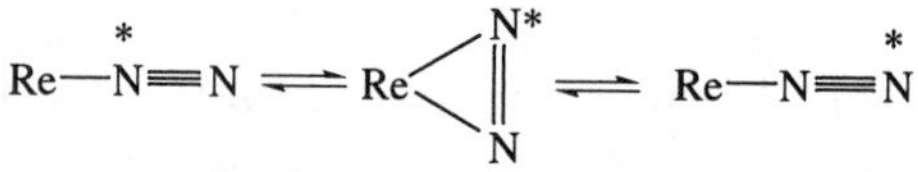

$$Re-\overset{*}{N}\equiv N \rightleftharpoons Re\diagdown\overset{N^*}{\underset{N}{\|}} \rightleftharpoons Re-N\equiv\overset{*}{N}$$

9-10   Nitric Oxide, Nitrous Oxide, Thionitrosyls

Nitric oxide compounds[85] have long been known, as have reactions such as

$$[Fe(CN)_5NO]^{2-} + 2OH^- \rightleftharpoons [Fe(CN)_5NO_2]^{4-} + H_2O$$

The compounds usually have only one NO, but di- and trinitrosyls such as Fe $(CO)_2(NO)_2$ and $Mn(CO)(NO)_3$ exist. The complexes can be made from carbonyls

[84]A. Cusanelli and D. Sutton, *Organometallics* **1996,** *15,* 1457.
[85]G. B. Richter-Addo and P. Legzdins, *Chem. Rev.* **1988,** *88,* 991 (Organometallic compounds); A. R. Butler *et al., Adv. Inorg. Chem.* **1988,** *32,* 335 (FeSNO clusters); D. M. P. Mingos and D. J. Sherman, *Adv. Inorg. Chem.* **1989,** *34,* 293 (279 references).

by NO substitution, from $NOPF_6$, by reductive processes from NO_2^- or NO_3^-, or by interaction of oxo species with hydroxylamine. Some representative syntheses are

$$Co_2(CO)_8 + 2NO \longrightarrow 2Co(NO)(CO)_3 + 2CO$$

$$Cp_2Ni + NO \longrightarrow CpNiNO + ?$$

$$Fe(CO)_5 + PPN^+NO_2^- \xrightarrow{MeCN} PPN^+[Fe(CO)_3NO]^- + CO_2 + CO$$

$$OsO_4 \xrightarrow[NCS^-]{NH_2OH \cdot HCl} [Os(NO)(NCS)_5]^{2-}$$

$$RhCl(CO)(PPr_3)_2 \xrightarrow{NOBF_4} [RhCl(NO)(CO)(PPr_3)_2]BF_4$$

The reaction with OsO_4 is similar to a common reaction of oxo anions such as VO_4^{3-} or ReO_4^- in alkaline or acidic solution containing coordinating anions with hydroxylamine[86]; intermediates with partially deprotonated ligands H_2NO and HNO may be isolated. Nitric acid can also act as a source of NO, especially for ruthenium, any of whose solutions treated with HNO_3 can be assumed to contain $Ru(NO)$ species.

Bonding Modes

There are numerous bonding modes, the most important being the terminal linear M—NO in which the NO is formally regarded as binding as a $2e$ donor NO^+ with transfer of the odd electron to the metal, thus lowering its formal oxidation state. Overall, NO is then a $3e$ donor. In many cases the MNO angle is below 180°, especially where the group is not in an axially symmetrical environment and "slightly bent" MNO groups may have angles between ~160 and 180°. They are still termed "linear" MNO groups.

Bent MNO groups have angles between ~120 and 140°. The NO can be considered as a $1e$ donor and compared to compounds like ClNO or Bu^tNO with a M—N single bond.

Bridging NO groups bound through N may be μ_2-symmetric or unsymmetric[87] and μ_3 and μ_4 in clusters such as $Cp_3Mn_3(NO)_4$ ($3\mu_2, 1\mu_3$). A recent new form in a molybdenum compound is $\mu\text{-}\eta^1 : \eta^2\text{-} NO$.[88]

There are also bridges where lone pairs on the O atom bind to Na^+ (9-XIIa) or there is an interaction of the type (9-XIIb).[89]

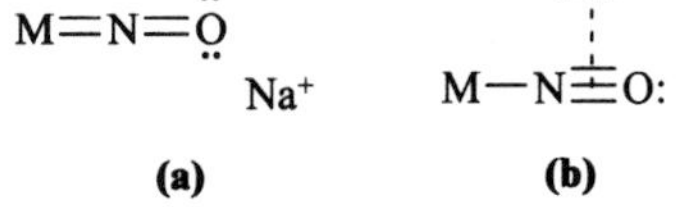

(9-XII)

[86]P. Gouzerh et al., *Inorg. Chem.* **1993**, *32*, 5291.
[87]J. L. Hubbard et al., *J. Am. Chem. Soc.* **1991**, *113*, 9176.
[88]P. Legzdins et al., *Organometallics* **1993**, *12*, 3575.
[89]K. Wieghardt et al., *Angew. Chem. Int. Ed. Engl.* **1994**, *33*, 1473.

It may be noted in this respect that Lewis acids such as BCl_3 or $AlMe_3$ can give bent bonds, e.g., $MNOBCl_3$.[90]

For mononitrosyls, the stoichiometries and properties can be accounted for by considering the $\{MNO\}^n$ unit as a functional group, n being the number of d electrons on the metal M plus the number of NO electrons in excess of those on NO^+. Examples are

$$\{MNO\}^5 \text{ in } [Cr^I(NO)(CN)_5]^{3-}$$

$$\{MNO\}^8 \text{ in } Mn^{-I}(NO)(CO)_4, Ir^I(NO)Cl_2(PR_3)_2$$

Whether the MNO group is linear or bent depends on

1. the coordination geometry and number,
2. the value of n,
3. the nature of the one-electron MOs.

For $n \leq 6$ all $\{MNO\}^n$ groups are linear or nearly so in octahedral coordination, but bent if $n \geq 7$.

For 5-coordination MNO is linear for $n \leq 6$; $\{MNO\}^8$ is linear for *tbp* but bent in *sp*. For 4-coordination, $n = 10$, tetrahedral has linear but planar has bent MNO.

Infrared spectra allow a distinction between "linear" and bent groups, the former generally having $\nu(NO) > 1550$ cm^{-1}; bridged species have $\nu(NO)$ for μ_2 1455 to 1510 cm^{-1} and for μ_3 *ca.* 1320 cm^{-1}. There is some overlap in the ir regions and a better spectroscopic criterion is the ^{15}N nmr spectrum.[91]

Linear MNO groups have shifts (relative to $MeNO_2$, high frequency positive) near to the value for NO^+ in the region -75 to $+180$ ppm and slightly bent nitrosyls (angles 160–180°) are slightly deshielded relative to linear ones. For bent species, there is a very substantial deshielding, increasing with the degree of bending of ~500 to 800 ppm (*cf.* Bu^{t}NO, δ 594). Magic-angle nmr spectra of solids also show a large chemical shift anisotropy for bent MNO. Specific examples are $[Ru(NO)(NH_3)_5]Cl_3$, $\angle RuNO = 172.8°$, $\delta -29$ and $Co(NO)salen$, $\angle CoNO = 127.0°$, δ 725.4.

Compounds that have both bent and linear groups can be obtained, one example being $[RuCl(NO)_2(PPh_3)_2]^+$. The nmr studies show that the ion is nonrigid with a linear-bent exchange giving isomers in solution:

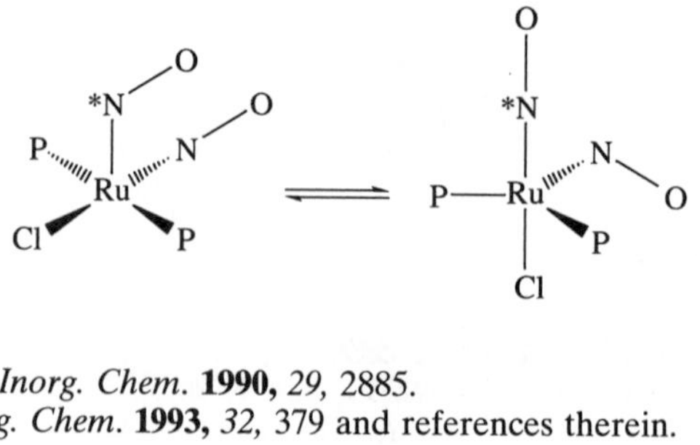

[90]J. A. Gladysz *et al., Inorg. Chem.* **1990,** *29,* 2885.
[91]J. Mason *et al., Inorg. Chem.* **1993,** *32,* 379 and references therein.

In the solid, the square pyramidal ion has bent apical and linear basal NO groups.

The linear-bent transition probably requires little energy and may well be involved in reactions of dinitrosyl intermediates in reactions such as that noted previously, leading to a hyponitrite complex, or in reactions of NO and CO, catalyzed by nitrosyl complexes, to give N_2O and CO_2.

Increasing the electron density on the metal may also lead to a linear $\rightarrow$ bent transition. This can be done by addition of a donor ligand or by reduction as in the conversion of $CpM(CO)_2NO$ to $CpM(CO)_2NO^-$.

Reactions of Coordinated NO

In view of the differences between the electronic structures of linear, bent, and bridging NO groups, considerable differences in their chemical reactivity can be expected. Generally, linear MNO groups are attacked by nucleophiles while bent ones with, in effect, a lone pair on N are susceptible to electrophilic attack, but this is by no means a rule and there are many exceptions. A few of the more common reactions of nitrosyls are now described.

Nucleophilic Attack

Interaction of linear NO complexes with nucleophiles such as OH^-, OR^-, SR^-, and NH_2R usually occurs at the N atom but attacks on O may also occur. A classic example is that of nitroprusside with hydroxide ion to give a nitro complex,[92]

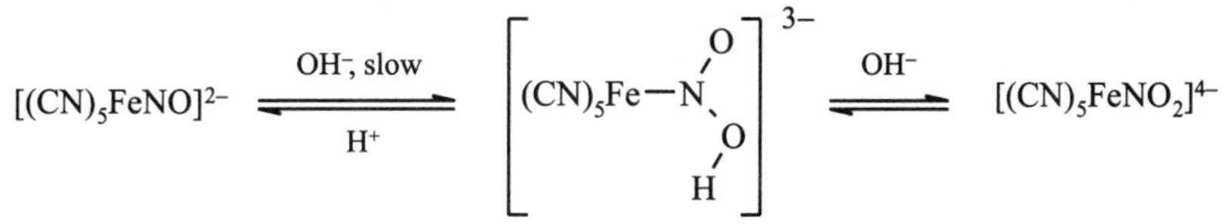

$$[(CN)_5FeNO]^{2-} \underset{H^+}{\overset{OH^-, slow}{\rightleftharpoons}} \left[(CN)_5Fe-N\overset{O}{\underset{O-H}{}} \right]^{3-} \overset{OH^-}{\rightleftharpoons} [(CN)_5FeNO_2]^{4-}$$

and with thiols to give a red coloration, a reaction long used for detection of RS^-,

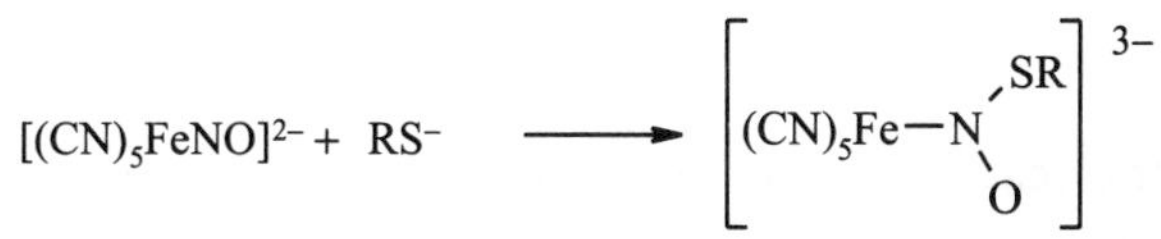

$$[(CN)_5FeNO]^{2-} + RS^- \longrightarrow \left[(CN)_5Fe-N\overset{SR}{\underset{O}{}} \right]^{3-}$$

The reaction of nitroprusside with sulfite may indeed be the first example of the reaction of a coordinated ligand (Boedeker, 1861). The type of red unstable species is probably indicated by the stable ruthenium complex:

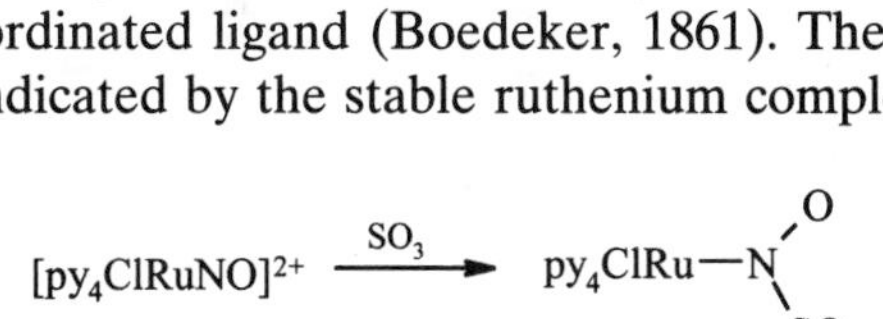

$$[py_4ClRuNO]^{2+} \overset{SO_3}{\longrightarrow} py_4ClRu-N\overset{O}{\underset{SO_3}{}}$$

where the unusual nitrosylsulfito ligand has a long (1.82 Å) weak N—S bond.

This type of ruthenium complex also illustrates a further type of reaction probably proceeding *via* nucleophilic attack, namely, condensation and elimination

[92]For Ru analogues see H. Tanaka *et al.*, *Inorg. Chem.* **1992**, *31*, 1971.

of water:

$$[bipy_2ClRuNO]^{2+} + H_2NPh \longrightarrow [bipy_2ClRuN=NPh]^{2+} + H_2O$$

Electrophilic Attacks

Attacks by H^+ or Me^+ may be at either N or O, for example,

$$(Ph_3P)_2(CO)ClOsNO \xrightarrow{\text{HCl}} (Ph_3P)_2(CO)Cl_2Os\!-\!N\underset{O}{\overset{H}{\diagdown}}$$

Bridged NO species can also be attacked, for example,

$$[Ru_3(CO)_{10}(\mu\text{-NO})]^- \xrightarrow{CF_3SO_3Me} [Ru_3(CO)_9(\mu_3\text{-CO})(\mu\text{-NOMe})]$$

In the reaction of a manganese cluster $[M_3(\mu_3\text{-NO})]$, where $M = (\eta^5\text{-}C_5H_4Me\text{-}MnNO)$, the hydroximido product can undergo further reaction with HBF_4 to give μ_3-NH.

$$M_3(\mu_3\text{-NO}) \underset{Et_3N}{\overset{H^+}{\rightleftharpoons}} M_3(\mu_3\text{-NOH})^+ \xrightarrow[2e]{2H^+} M_3(\mu_3\text{-NH})^+$$

Oxygenation

Some NO complexes react with O_2:

$$L_nMNO + {}^1\!/_2O_2 = L_nMNO_2$$

Since the nitro complexes can be reduced to MNO, for example, by alkenes, there is the possibility of oxidation reactions that are catalytic.

Alkene Reactions

The stoichiometric reactions of μ-NO species with alkenes can provide a route to amines:

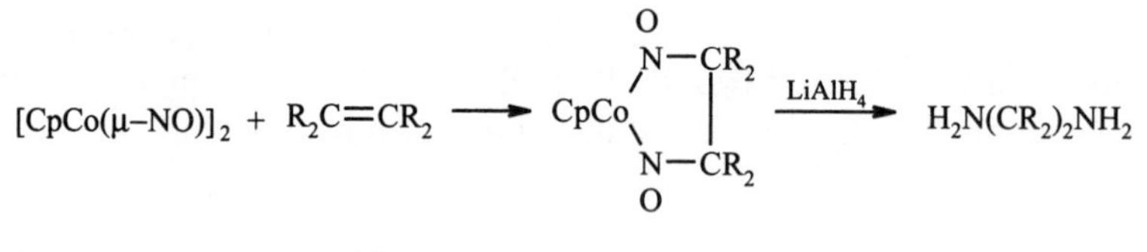

$$[CpCo(\mu\text{-NO})]_2 + R_2C\!=\!CR_2 \longrightarrow CpCo\underset{\substack{\diagdown \\ N-CR_2 \\ \,\,O}}{\overset{\substack{O \\ N-CR_2 \\ \diagup}}{\big|}} \xrightarrow{LiAlH_4} H_2N(CR_2)_2NH_2$$

Reductions

Reduction of L_nMNO by boro- or aluminohydrides or other reducing agents can give species with H—M—NO units that can be further reduced, for example,

$$FeCl(NO)_3 \xrightarrow{Al} Fe_2(NO)_4(\mu\text{-NH}_2)_2$$

$$[RuClpy_4NO]^{2+} \xrightarrow{Zn, H^+} [RuClpy_4NH_3]^+ + H_2O$$

Many NO species can, of course undergo reversible one-electron reductions such as

$$(Et_2NCS_2)_2Mo(NO)_2 \underset{-e}{\overset{e}{\rightleftarrows}} (Et_2NCS_2)_2Mo(NO)_2]^-$$

Reduction of a molybdenum tetrathiolate "S_4"$Mo(NO)_2$ by N_2H_4 can also give a hydroxylamino complex[93]:

$$"S_4"Mo(NO)_2 + \tfrac{1}{2}N_2H_4 \xrightarrow{\text{THF}} "S_4"Mo(\eta^2\text{-}N,O\text{-}H_2NO)NO + \tfrac{1}{2}N_2$$

Thionitrosyls[94]

Relatively few compounds with MNS groups are known. They have been made by reactions involving attack on $M\equiv N$ by S_8 or SCl_2, or by reactions involving KCNS, $(NSCl)_3$, or NS^+. Some examples are

$$(Et_2NCS_2)_2Mo\equiv N + \tfrac{1}{8}S_8 \longrightarrow (Et_2NCS_2)_2MoNS$$

$$Na[CpCr(CO)_3] + \tfrac{1}{3}S_3N_3Cl_3 \longrightarrow CpCr(CO)_2NS + CO + NaCl$$

$$[CpFe(CO)_2SO_2]AsF_6 + NSAsF_6 \longrightarrow [CpFe(CO)_2NS]^{2+} + 2AsF_6^- + SO_2$$

Although ir spectra suggest that, like NO, NS can give linear, bent, and bridged species, the only X-ray diffraction studies have been on compounds with linear or almost linear groups.

Nitrous Oxide

Only a few complexes are known, the best example being $[Ru(NH_3)_5(N_2O)]^{2+}$.

9-11 Ammonia and Amines

Ammonia and amine complexes are among the longest known and most intensively studied. As well as classical Werner-type complexes such as $[Co(NH_3)_6]^{3+}$, metal halides commonly form adducts with NH_3 and amines. The complexes are discussed in the appropriate chapters for the elements.

Macrocyclic Amines[95]

There is a vast array of macrocycles, not only with solely N donors, but also with mixed N, O; N, S; N, O, S; N, O, P; and so on, donors.

[93]D. Sellman and B. Seubert, *Angew. Chem. Int. Ed. Engl.* **1992**, *31*, 205.
[94]K. K. Pandey, *Prog. Inorg. Chem.* **1992**, *40*, 2145 (also NSO$^-$ and N$_3$S$^-$ ligands); K. K. Pandey *et al.*, *Polyhedron* **1995**, *13–14*, 1987.
[95]L. F. Lindoy, *The Chemistry of Macrocyclic Ligand Complexes*, C.U.P. Cambridge, 1992; A. Bencini *et al.*, *Coord. Chem. Rev.* **1992**, *120*, 51; R. Bhular *et al.*, *Coord. Chem. Rev.* **1988**, *91*, 89; I. Bernal, ed., *Stereochemistry of Organometallic and Inorganic Compounds*, Vol. 2, Elsevier, New York, 1987.

Depending on the donor atoms, these can be designated N_4, N_2O_2, N_2S_2, and so on. The heterocycles can be broadly classed into those without and those with conjugated π systems; the latter are discussed later.

Macrocyclic complexes have the following characteristics:

1. A marked kinetic inertness both to the formation of the complexes from the ligand and metal ion, and to the reverse, the extrusion of the metal ion from the ligand.

2. They can stabilize high oxidation states that are not normally readily attainable, such as Cu^{III} or Ni^{III}.

3. They have high thermodynamic stability—the formation constants for N_4 macrocycles may be orders of magnitude greater than the formation constants for nonmacrocyclic N_4 ligands. Thus for Ni^{2+} the formation constant for the macrocycle cyclam (9-XIII) is about five orders of magnitude greater than that for the nonmacrocycle tetradentate (9-XIV):

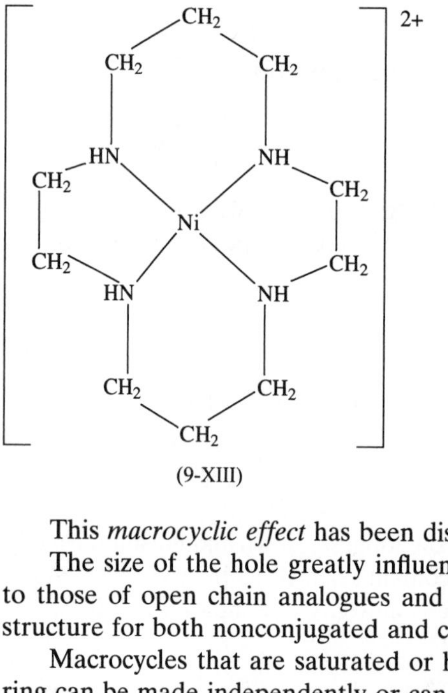

(9-XIII) 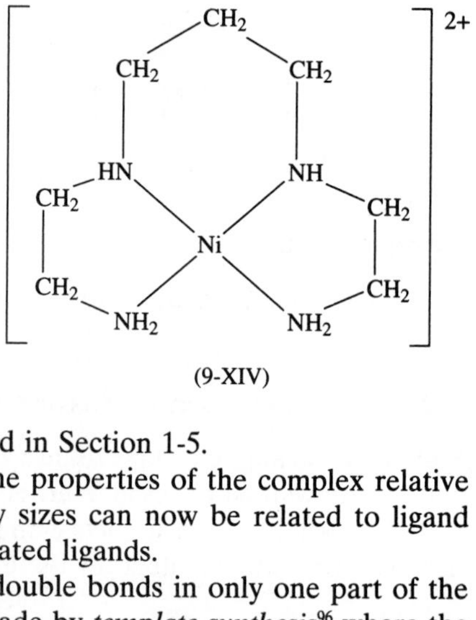   (9-XIV)

This *macrocyclic effect* has been discussed in Section 1-5.

The size of the hole greatly influences the properties of the complex relative to those of open chain analogues and cavity sizes can now be related to ligand structure for both nonconjugated and conjugated ligands.

Macrocycles that are saturated or have double bonds in only one part of the ring can be made independently or can be made by *template synthesis*[96] where the presence of a metal ion controls the ligand synthesis. An important route is the Schiff base condensation reaction (eq. 9-3),

$$\begin{array}{c} R \\ R' \end{array}\!\!C{=}O \; + \; H_2NR'' \; \longrightarrow \; \begin{array}{c} R \\ R' \end{array}\!\!C{=}N{-}R'' \; + \; H_2O \qquad\qquad (3)$$

often (but not necessarily) with a metal ion as template, and with subsequent hydrogenation to obtain a saturated system not subject to hydrolytic degradation

[96]R. Hoss and F. Vögtle, *Angew. Chem. Int. Ed. Engl.* **1994,** *33,* 375 (50 references).

by reversal of reaction 9-3. Some representative preparative reactions are eqs. 9-4 and 9-5.

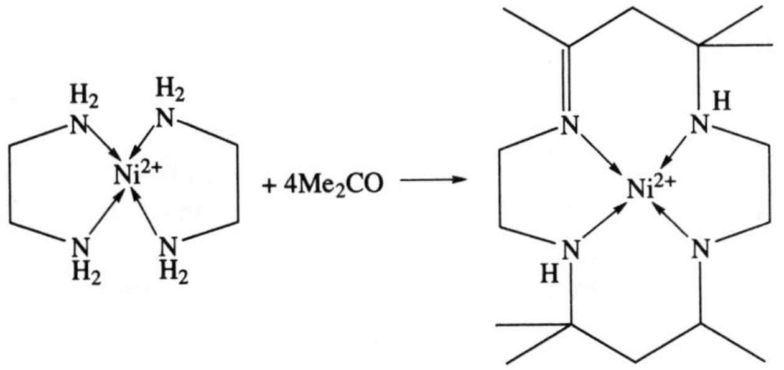

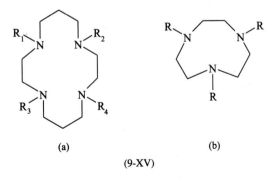

$$+ 4Me_2CO \longrightarrow \qquad (4)$$

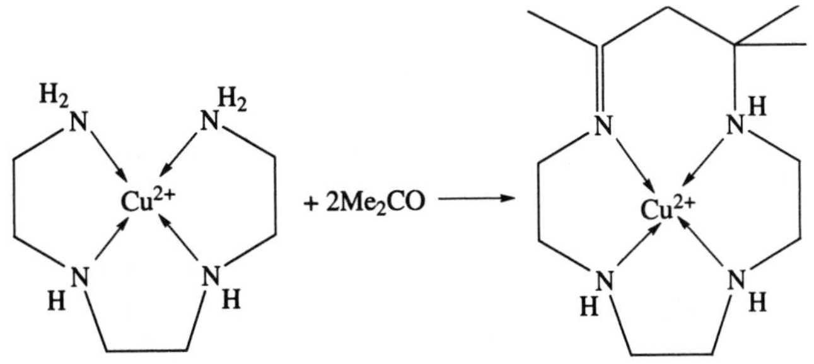

$$+ 2Me_2CO \longrightarrow \qquad (5)$$

Other ligands that involve cages with various sized cavities are called *sepulchrates*,[97] one example of which is

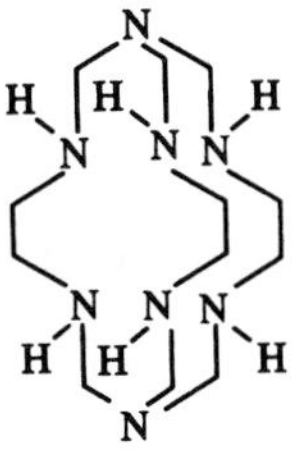

A great variety of other types of molecules with large cavities is known.[98]

Two important macrocycles are (9-XVa,b)

(a)   (b)

(9-XV)

[97]A. M. Sargeson *et al., J. Chem. Soc. Chem. Commun.* **1994,** 667, 1513.
[98]C. Seel and F. Vögtle, *Angew. Chem. Int. Ed. Engl.* **1992,** *31,* 529 (159 references).

The cyclononane types[99] are particularly useful ligands, as are the similar triazacyclo-hexanes.[100] Giant size azamacrocycles[101] having 30 to 60 atoms have very large cavities.

Pyridines

These aromatic ligands differ quite substantially from aliphatic amines, generally giving complexes that are more stable than those of aliphatic amines. Pyridines can also bind as arenes,[102] i.e., η^6, and also as η^2-ligands, an example[103] being the chelate siloxo ligand (silox) complex (9-XVI) with η^2-NC$_5$H$_5$-*N,C* bonding.

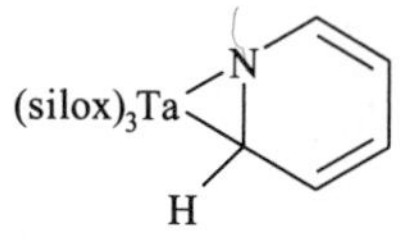

(9-XVI)

The η^2-*N,C* bonding can also occur in deprotonated pyridines,[104] i.e., *pyridyl complexes* made by reactions such as (9-6).

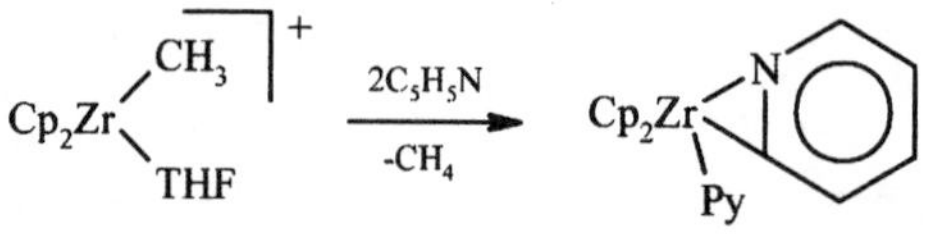

(6)

Such η^2-*N,C* pyridines can also undergo bond cleavages and the reactions provide models for hydrodenitrogenation catalysis.[105]

Polyimines[106]

Heterocycles such as 2,2′-bipyridine, 1,10-phenanthroline, terpyridine (9-XVII a–c) are best considered as *α-diimines* having the group (9-XVIId).

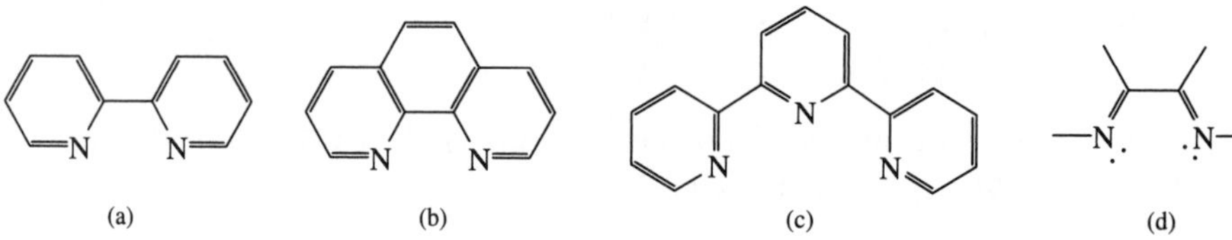

(a)   (b)   (c)   (d)

(9-XVII)

[99]K. Wieghardt *et al., Inorg. Chem.* **1995,** *34,* 6440; R. D. Peacock *et al., J. Chem. Soc., Dalton Trans.* **1993,** 2759.
[100]R. D. Köhn and G. Kociok-Köhn, *Angew. Chem. Int. Ed. Engl.* **1994,** *33,* 1871.
[101]A. Bencini *et al., J. Chem. Soc., Chem. Commun.* **1994,** 1119.
[102]See P. T. Wolczanski *et al., Inorg. Chem.* **1991,** *30,* 2494 for references and extensive discussion of η^1 *vs* η^2-*N,C* bonding of pyridines.
[103]R. F. Jordan *et al., Organometallics* **1990,** *9,* 1546; J. D. Scoll *et al., Organometallics* **1995,** *14,* 5478.
[104]K. I. Hardcastle *et al., J. Chem. Soc., Dalton Trans.* **1992,** 1607.
[105]D. E. Wigley *et al., Organometallics* **1995,** *14,* 5588; *J. Am. Chem. Soc.* **1995,** *117,* 10678.
[106]E. C. Constable, *Adv. Inorg. Chem. Radiochem.* **1989,** *34,* 1 (994 references); E. C. Constable and P. J. Steel, *Coord. Chem. Rev.* **1989,** *93,* 205 (*N,N* chelates); P. J. Steel, *Coord. Chem. Rev.* **1990,** *106,* 227 (bridging ligands).

A characteristic feature is that such ligands can form metal complexes in a wide range of oxidation states, as in the redox series

$$[Cr(bipy)_3]^{3+} \underset{-e}{\overset{+e}{\rightleftarrows}} [Cr(bipy)_3]^{2+} \underset{-e}{\overset{+e}{\rightleftarrows}} [Cr(bipy)_3]^{+} \underset{-e}{\overset{+e}{\rightleftarrows}} [Cr(bipy)_3]^{0} \underset{-e}{\overset{+e}{\rightleftarrows}} [Cr(bipy)_3]^{-}$$

For metal ions in "normal" oxidation states, the interaction of metal $d\pi$ orbitals with the ligand π^* orbitals is significant, but not exceptional. However, these ligands can stabilize metal atoms in very low formal oxidation states and in such complexes there is extensive occupation of the ligand π^* orbitals, so that the compounds can often be best formulated as having radical anion ligands $L^{\cdot-}$.

Although 2,2'-bipyridine and 1,10-phenanthroline usually give chelate complexes, *unidentate complexes* can be formed as in $[PtCl(PEt_3)_2(\eta^1\text{-phen})]^+$; in solution they are fluxional and the platinum atom moves from one nitrogen atom to the other.

Aqueous solutions of aromatic amine complexes such as $[Fe(phen)_3]^{3+}$ or $[PtCl_2(py)_4]^{2+}$ often show unexpected kinetic, equilibrium, and spectral behavior in the presence of nucleophiles like OH^-, CN^-, or OR^-. Since certain heterocyclic amines—*not* py, bipy, or phen, however—can undergo attack at C by OH^- or H_2O, and so on, it was suggested that when phen or similar ligands were *activated* by complexation, they too could be attacked at carbon, eq. 9-7.

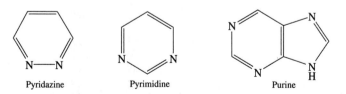

Of particular importance are *oligopyridine ligands* such as 2,2':6',2",2''':6''',2''''-pentapyridine. These molecules allow synthesis of helical, inter-linked and knotted molecular arrays.[107] Similar species with interlocking rings are the *catenands* and *torands*.[108] Some examples of oligopyridines are (9-XVIII) and (9-XIX). Other nitrogen heterocycles are the following:

Pyridazine   Pyrimidine   Purine

[107]E. C. Constable *et al., J. Chem. Soc., Dalton Trans.* **1993,** 194, *Angew. Chem. Int. Ed. Engl.* **1993,** *32,* 1465; D. J. Cram *et al., J. Chem. Soc., Chem. Commun.* **1993,** 1323.
[108]J. P. Sauvage *et al., Angew. Chem. Int. Ed. Engl.* **1993,** *32,* 1435; V. Balzani *et al., J. Chem. Soc., Dalton Trans.* **1993,** 3241.

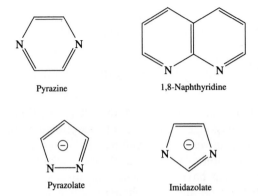

Pyrazine	1,8-Naphthyridine
Pyrazolate	Imidazolate

Pyrazolates can give N-bound η^1 complexes as well as η^2 and several types of bridged species (9-XX).[109]

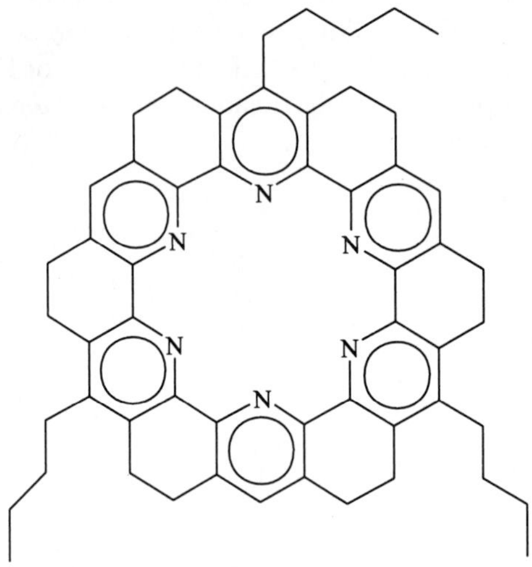

Torand

(9-XVIII)

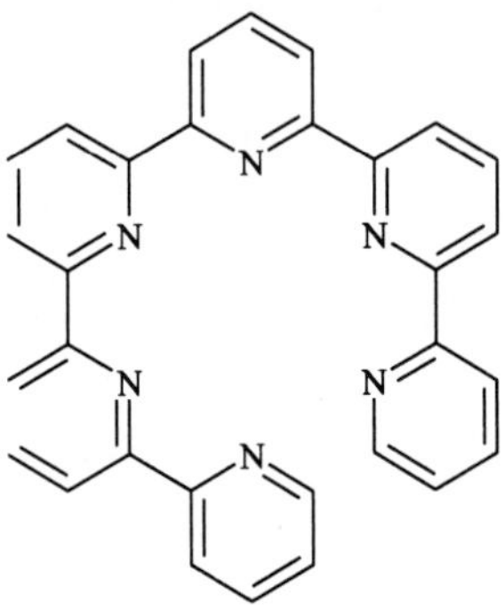

2,2':6',2":6",2"':6"',2"":6"",2"""

hexapyridine

(9-XIX)

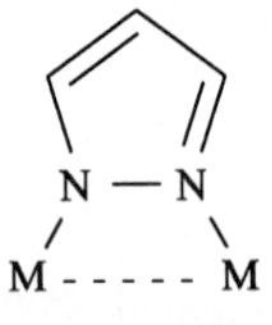

(9-XX)

1,8-Naphthyridine complexes are mostly chelates but there are some bridged complexes. The 4,4'-bipyrimidine (9-XXI) has high π-acceptor ability and can chelate *N,N'*, or η^1 through the end N atoms.

[109]See, e.g., D. A. Johnson *et al., J. Coord. Chem.* **1994**, *32*, 1; G. Erker *et al., Organometallics* **1994**, *13*, 3897.

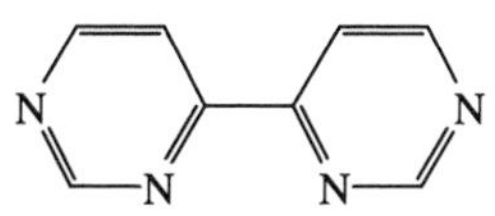

(9-XXI)

Nucleotides, purines, and *pyrimidines* have been studied as ligands largely on account of their presence in nucleic acids. The action of certain metal complexes, notably *cis*-PtCl$_2$(NH$_3$)$_2$, as anticancer agents, is believed to arise through binding to nucleic acids. Other aspects of the binding of metals to nucleic acid include the attachment of lanthanide ions as shift reagents and fluorescent probes and the use of heavy metals to assist in X-ray structural determinations.

For unsubstituted *purines,* the most probable site for coordination is the imidazole nitrogen (N-9), which is protonated in the free neutral ligand. An example is the cobalt complex of adenine, [Co(ad)$_2$(H$_2$O)$_4$]$^+$ (9-XXII).

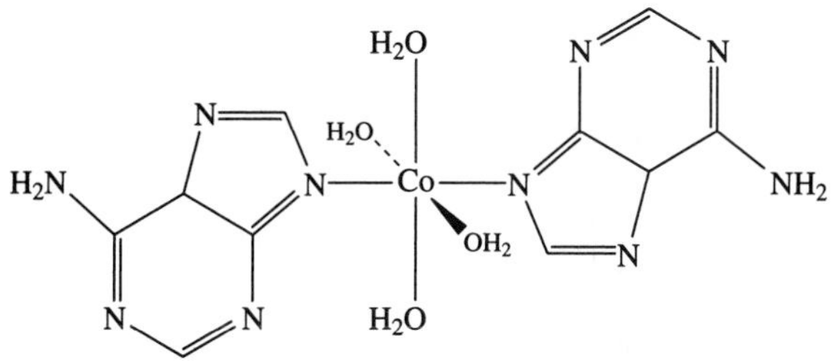

(9-XXII)

If the 9-position is blocked, the other imidazole nitrogen, N-7, is coordinated. Binding appears somewhat less likely through N-1 than through N-7; there is also binding with both N-1 and N-7.

Imidazoles have been widely studied. Although the binding is usually through the N atom (9-XXIIIa), in some Ru^{II}, Ru^{III}, Fe^0, and Cr^0 complexes it is possible to have C-bonded groups (9-XXIIIb). The C-bonded entity can be regarded as a carbene (9-XXIIIc) or as a C-bound amidine (9-XXIIId). An example of a C-bonded species is the ruthenium(II) complex obtained as follows:

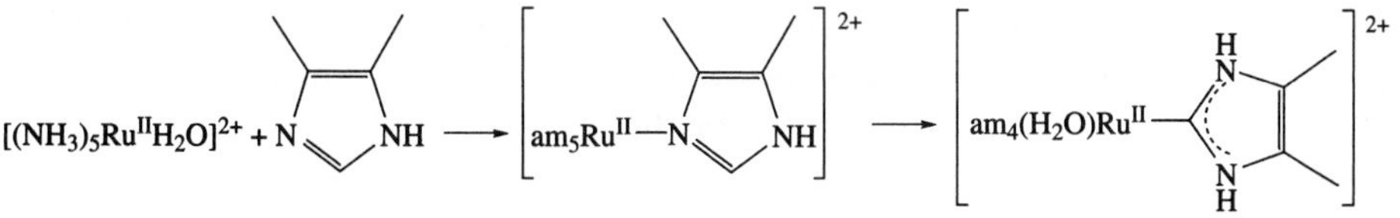

The N-bonded imidazolate monoanion forms bridges. Biimidazoles can act as mono- or dianion, for example, in the complexes

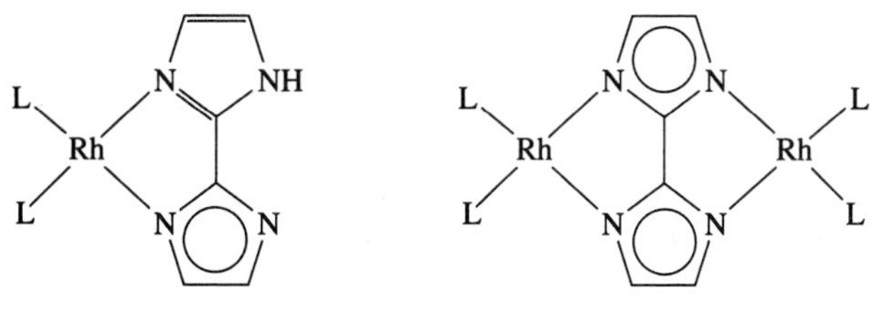

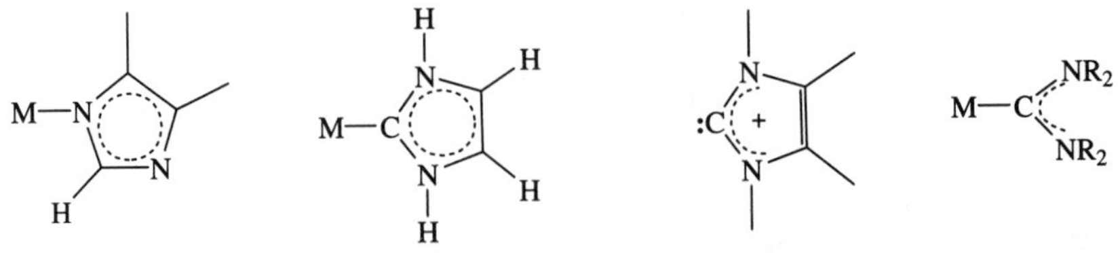

(a)   (b)   (c)   (d)

(9-XXIII)

Bridging Bidentate Anions. In the chemistry of binuclear, metal—metal bonded compounds, ligands of the generic type 9-XXIV play a major role because of their relatively high basicity and their ability to hold the metal atoms in close proximity. Actual examples are:

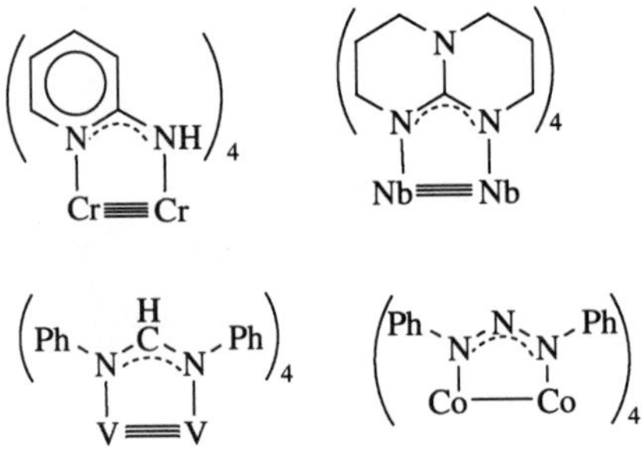

(9-XXIV)

9-12 Macrocyclic Ligands with Conjugated π Systems[110]

There is often a very substantial difference between the conjugated and saturated macrocycles with regard to the rates of substitution reactions, which may be up to 10^{12} times greater for the conjugated systems. The enormous rate enhancement can be correlated with the kinetic lability of the leaving ligand in the inner coordination sphere of the metal complex. This profound difference presumably explains why biologically active metal systems invariably have highly unsaturated macrocyclic ligands.

Phthalocyanines[111]

These constitute one of the earliest classes of synthetic N_4 macrocycles to be prepared. They are obtained by interaction of phthalonitrile with metal halides in

[110]L. F. Lindoy, *Chemistry of Macrocyclic Ligand Complexes*, Cambridge University Press, 1992.
[111]F. H. Moser and A. L. Thomas, eds., *The Phthalocyanines*, Vols. I and II, CRC Press, Boca Raton, Florida, 1983; N. Kobayashi *et al., J. Am. Chem. Soc.* **1994,** *116,* 879.

which the metal ion plays an essential role as a template. Complexes such as 9-XXV characteristically have exceptional thermal stabilities, subliming in vacuum around 500°C. They are also intensely colored and are an important class of commercial pigments. The solubility is usually very low, but sulfonated derivatives are soluble in polar solvents. Polymers of the type (9-XXVI) can be made where metal phthalocyanines are linked in chains by CN^- or by ligands such as pyrazines.

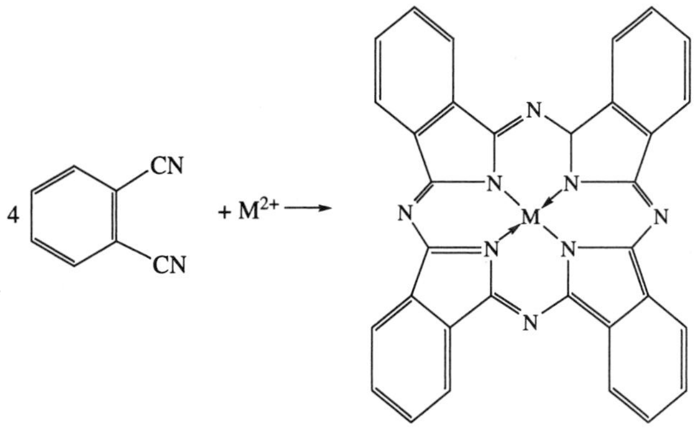

(9-XXV)

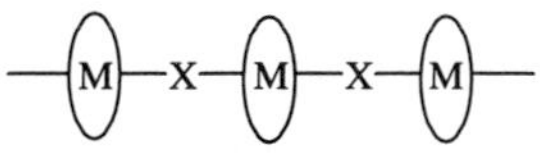

(9-XXVI)

Porphyrins[112]

These ligands are especially important because many naturally occurring, metal-containing compounds such as chlorophylls, heme, cytochromes, etc., contain macrocycles. Some examples of these ligands are shown in Fig. 9-3.

Many substituted species are known, common ones being octaethylporphyrin (H_2OEP) and tetraphenylporphyrin (H_2TPP); examples of more complicated ones are given below.

Almost every metal in the Periodic Table has been coordinated to a porphyrin of one sort or another. The complexes are usually made by interaction of the ligand with a metal salt in a solvent such as DMF; other ways include interaction of the H_2porph with metal carbonyls, alkyls, hydrides, or acetylacetonates. The radius of the central hole is fixed, but puckering of the rings can alter it; normally it lies between 1.929 and 2.098 Å. It is also common for larger metal ions to sit above

[112]C. Floriani, *Chem. Commun.* **1996,** 1257 (artificial porphyrins 21 references); D. Dolphin, ed., *The Porphyrins,* Academic Press, New York (reference work in 7 volumes); B. Frank and A. Nonn, *Angew. Chem. Int. Ed. Engl.* **1995,** *34,* 1795 (biomimetic syntheses of porphyrinoids); R. Sheldon and Y. Naruta, *Metalloporphyrins in Catalytic Oxidations,* Dekker, New York, 1994; F. Montanari and L. Casella, eds., *Metalloporphyrin Catalysed Oxidations,* Kluwer, Dordrecht, 1994.

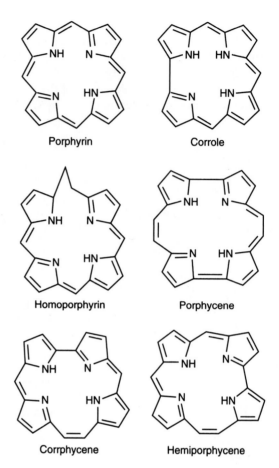

Porphyrin

Corrole

Homoporphyrin

Porphycene

Corrphycene

Hemiporphycene

Figure 9-3 Porphynoid rings with eighteen π-electrons.

the hole. Some examples are

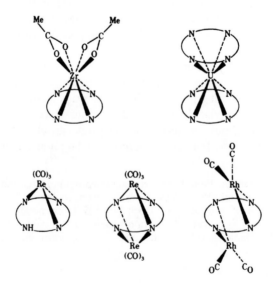

Porphyrin complexes commonly undergo reversible redox reactions, but in certain systems transfer of a group from the metal to a N atom can occur, and this again leads to tridentate binding, namely,

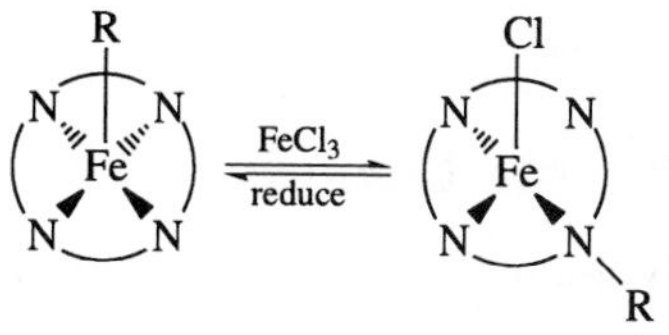
R
Fe
FeCl3
reduce
Cl
Fe
R

Alkylidene complexes, for example, (porph)Fe$=$$CCl_2$ may be formed in reactions of metalloporphyrin with halogenated solvents and such species may be responsible for the toxicity of halogenated insecticides and solvents like CCl_4.

Quite complicated porphyrin systems have been constructed with the object of modeling natural systems with biological functions. Some have holes, pockets, and protection for the metal from oxidation by O_2.

Some, shown diagramatically, are the following, where the coiled lines designate a chain such as

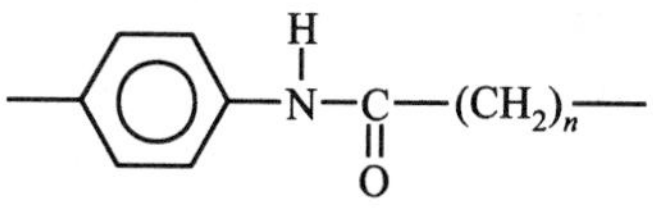

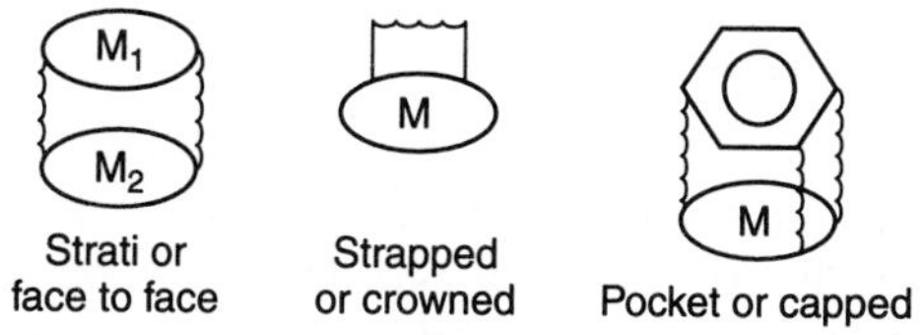
M1
M2
Strati or
face to face
M
Strapped
or crowned
M
Pocket or capped

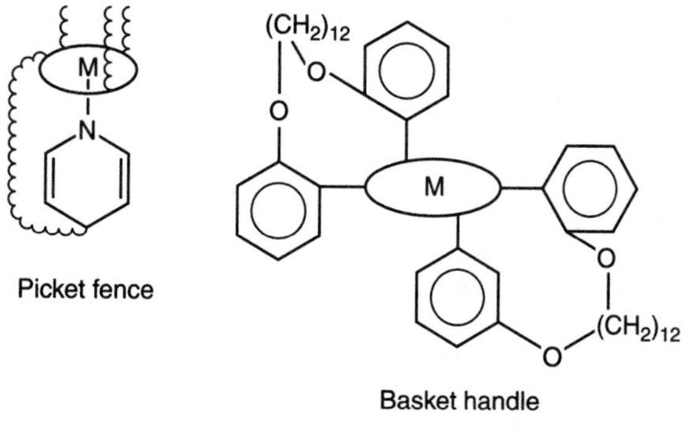
M
N
Picket fence
(CH2)12
O
O
M
O
(CH2)12
O
Basket handle

Other important macrocycles are *corrins*, which are involved in vitamin B_{12}, and *corroles*, which are intermediate between porphyrins and corrins. The corrins are more flexible than porphyrins, while corroles have a smaller cavity.[113] The macrocycle tetramethyldibenzo[b,i][1,4,8,11]tetraazacyclotetradecine (H_2TMTAA) with its porphyrin-like dianion forms both mono and binuclear complexes such as LMo^{II}-$Mo^{II}L$.[114]

Finally, supramolecular macrocycles involving porphyrins can be made.[115]

9-13   Polypyrazolylborate Ligands[116]

These ligands are extremely versatile since the pyrazole rings can be replaced by triazole or tetrazoles.[117]

The simple ligands are of the type (9-XXVIIa,b) designated $R_2Bpz_2^-$ and $RBpz_3^-$ respectively. They are made by interaction of a pyrazole and $NaBH_4$ or substituted borohydrides. A commonly used ligand is hydrido *tris*-(3,5-dimethylpyrazolyl)borate (Tp*).

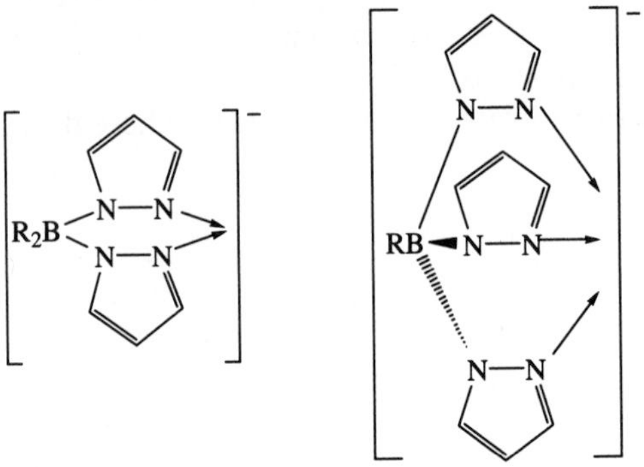

(a)   (b)

(9-XXVII)

The dipyrazolyl anions $R_2Bpz_2^-$ have a formal analogy to the β-diketonate ions and, like them, form complexes of the type $(R_2Bpz_2)_2M$. However, because of the much greater steric requirements of the $R_2Bpz_2^-$ ligand, such compounds are always monomeric. For steric reasons it appears to be difficult to make tris complexes; one example is the anion $[V(pz_2BH_2)_3]^-$.

The $RBpz_3^-$ ligands give many important complexes. These ligands themselves are unusual in being trigonally tridentate, uninegative ligands. They form trigonally distorted octahedral complexes, $(RBpz_3)_2M^{0,+}$, with di- and trivalent metal ions, most of which are exceptionally stable. At least to a degree, an analogy can be

[113]S. Licoccia *et al., Inorg. Chem.* **1994,** *33,* 1171; E. Vogel *et al., Angew. Chem. Int. Ed. Engl.* **1994,** *33,* 731.

[114]F. A. Cotton and J. Czuchajowska, *Polyhedron* **1990,** *9,* 2553; J. E. Guerchais *et al., Inorg. Chem.* **1993,** *32,* 713.

[115]J.-M. Lehn *et al., Chem. Commun.* **1996,** 337.

[116]N. Kitajima and W. B. Tolman, *Progr. Inorg. Chem.* **1995,** *43,* 419 (278 references); S. Trofimenko *et al., Chem. Rev.* **1993,** *93,* 943 (421 references).

[117]C. Janiak and H. Hemling, *J. Chem. Soc., Dalton Trans.* **1994,** 2947; D. L. Reger *et al., Inorg. Chem.* **1994,** *33,* 1803.

made between RBpz$_3^-$ and the cyclopentadienyl anion C$_5$H$_5^-$; both are six-electron, uninegative ligands. There are some mono-RBpz$_3^-$ complexes that bear considerable resemblance to half-sandwich complexes (CpML$_x$); thus Mo(CO)$_6$ reacts with Na(RBpz$_3$) and NaCp to give (RBpz$_3$)Mo(CO)$_3^-$ and CpMo(CO)$_3^-$, respectively; another example is Rh(C$_2$H$_4$)$_2$(HBpz$_3$), which can be compared to CpRh(CO)$_2$. The compound Yb(HBpz$_3$)$_3$ has two tri- and one bidentate ligand.

Recent developments include synthesis of optically active trispyrazolylborates[118] and both *bis* and *tris* ligands with CF$_3$ groups,[119] e.g., [HB(3,5(CF$_3$)$_2$)pz$_3$]$^-$. It may be noted that main group elements such as tin can form pyrazolylborates.[120]

9-14 Nitriles

Nitriles are excellent donors and the bonding is usually η^1 with the M $\leftarrow$ N$\equiv$CR moiety linear or nearly so. Bisnitriles can act as chelates; cyano complexes such as bipy$_2$ Ru(CN)$_2$ can also act as ligands though the result is merely a cyanide bridge, M$-$CN$-$M or M$-$CN$-$M' (Section 7-11).

There are a number of η^2 *C,N* bonded side-on compounds[121] such as Cp$_2$Mo(η^2-MeCN), [WCl(PMe$_3$)$_2$(bipy)(η^2-RCN)]$^+$, and WCl$_2$(PMe$_3$)(η^2-MeCN). In the first, MeCN acts as a 2e donor, but in W^{II} complexes the bonding is considered to be 4e, similar to the 4e alkyne, RC$\equiv$CR complexes.[122] Bridging nitriles, μ,η^2 are also known.[123] Bridging nitriles of several other types are known[124]: μ-η^1:η^2, μ_3-η^2, and μ_4-η^2.

Nitriles may reduce halides in higher oxidation states [e.g., interaction of ReCl$_5$ or WCl$_6$ gives ReCl$_4$(MeCN)$_2$ and WCl$_4$(MeCN)$_2$, respectively] together with chlorinated organic N compounds.

Reactions of Nitriles

Most metals form complexes with nitriles. Coordinated nitriles are electrophilic at the α-carbon atom and many *nucleophilic attacks* have been demonstrated, some of the more important being the following[125]:

alcohols → iminoethers

amines → amidines

carbanions → imines

base catalyzed hydrolysis → amidates and amides

[118]W. B. Tolman *et al.*, *Inorg. Chem.* **1994**, *33*, 636.

[119]H. V. R. Dias *et al.*, *Inorg. Chem.* **1996**, *35*, 2317; *Organometallics* **1996**, *15*, 2994.

[120]G. G. Lobbia *et al.*, *J. Chem. Soc., Dalton Trans.* **1996**, 2475.

[121]C. G. Young *et al.*, *Organometallics* **1996**, *15*, 2428; T. G. Richmond, *Chem. Commun.* **1996**, 1691; W. D. Harman, *et al.*, *Organometallics* **1993**, *12*, 428; F. A. Cotton and F. E. Kühn, *J. Am. Chem. Soc.* **1996**, *118*, 5826.

[122]C. G. Young *et al.*, *Organometallics* **1996**, *15*, 2428; T. G. Richmond, *Chem. Commun.* **1996**, 1691; W. D. Harman, *et al.*, *Organometallics* **1993**, *12*, 438.

[123]J. L. Eglin *et al.*, *Inorg. Chim. Acta* **1995**, *229*, 113; F. A. Cotton and F. E. Kühn, *J. Am. Chem. Soc.* **1996**, *118*, 5826.

[124]A. Tiripicchio *et al.*, *Organometallics* **1992**, *11*, 801.

[125]See G. Natile *et al.*, *Inorg. Chem.* **1995**, *34*, 1130 for references.

Specific examples are:

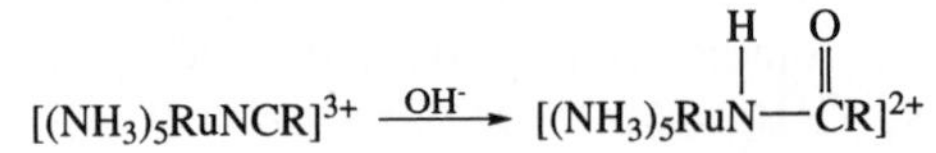

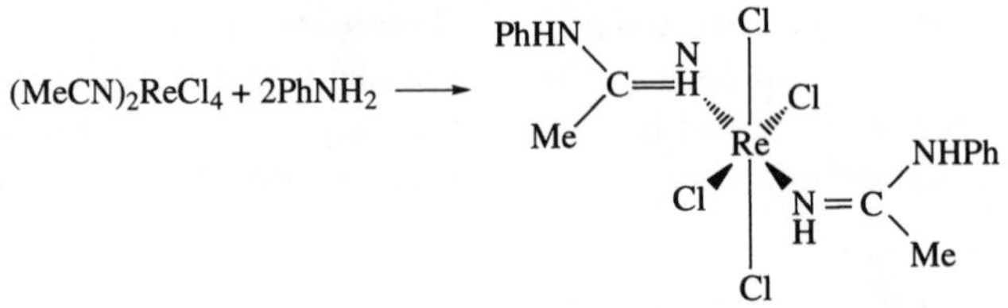

$$[(NH_3)_5RuNCR]^{3+} \xrightarrow{\text{OH}^-} [(NH_3)_5RuN\overset{\overset{\displaystyle H}{|}}{-}\overset{\overset{\displaystyle O}{\|}}{C}R]^{2+}$$

$$(MeCN)_2ReCl_4 + 2PhNH_2 \longrightarrow$$

Reduction can occur in several ways. Hydride transfer giving an alkylideneamido group occurs in the reaction

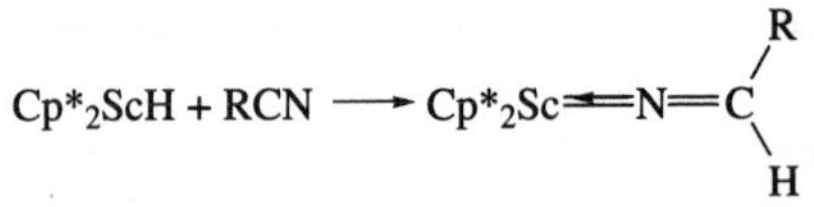

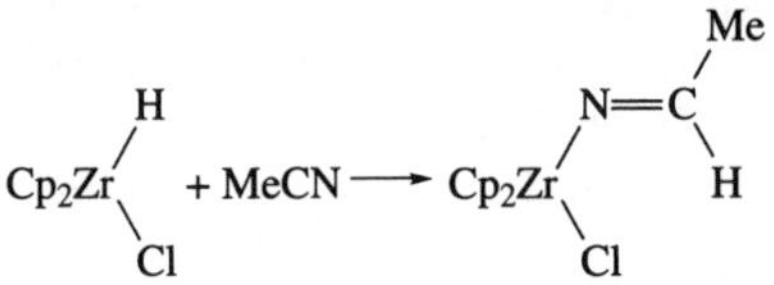

$$Cp^*_2ScH + RCN \longrightarrow Cp^*_2Sc\!=\!\!=\!N\!=\!C\overset{\displaystyle R}{\underset{\displaystyle H}{<}}$$

$$Cp_2Zr\overset{\displaystyle H}{\underset{\displaystyle Cl}{<}} + MeCN \longrightarrow Cp_2Zr\overset{\displaystyle N=C<^{Me}_{H}}{\underset{\displaystyle Cl}{<}}$$

Catalytic reductions, for example, of CH_3CN to $CH_3CH_2NH_2$ can also be achieved by use of phosphine complexes such as $[diphos_2Ru(NCMe)_2]^{2+}$ or $RhH(PPr)_3$ and by Cp^*_2ScR under hydrogen pressure. Stoichiometric reductions with $NaBH_4$ in MeOH can convert MNCMe to MNH_2Et and this can also be achieved by sequential H^- and H^+ addition.[126]

A different form of reduction, involving coupling by C—C bond formation, can occur when metal halides or halide complexes are reduced. These reactions give species with groupings such as

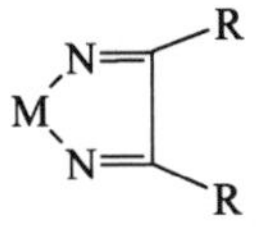

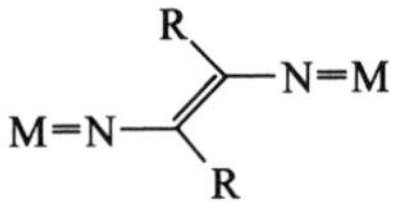

Both 2e and 4e reductive coupling are known.[127]

9-15   Amido and Related Ligands[128]

This important class of compounds is generally of the type NHR^- or NR_2^- where R may be alkyl or aryl groups, but there are, of course, many compounds with

[126]W.-Y. Yeh *et al., Organometallics* **1995,** *14,* 1417 and references therein.
[127]C. G. Young *et al., Inorg. Chem.* **1995,** *34,* 6412; J. L. Templeton *et al., Organometallics* **1994,** *13,* 1214.
[128]H. E. Bryndza and W. Tam, *Chem. Rev.* **1988,** *88,* 1163; M. D. Fryzuk and C. D. Montgomery, *Coord. Chem. Rev.* **1989,** *95,* 1.

NH_2^- groups. The complexes are commonly made by interaction of the halides with the corresponding lithium reagents. They are closely related to alkoxides and to alkyls, often having similar stoichiometries, structures, and volatilities, for example, $Cr(NEt_2)_4$, $Cr(OBu^t)_4$, and $Cr(CH_2SiMe_3)_4$. Compounds with NHR^- groups are less well studied but are important in syntheses of imido species discussed later; a typical example is $W(NBu^t)_4(NHBu^t)_2$.

As in the case of alkyls and alkoxides, the use of very bulky ligands can prevent dimerization or polymerization and stabilize compounds with low coordination numbers, such as two in $Fe[N(SiMe_3)_2]_2$. The amido group is usually planar due to double bonding of the type

$$L_nM \rightleftharpoons \ddot{N}R_2$$

but nevertheless a few cases of pyramidal amido ligands are known.[129] There can be restricted rotation about the M—N axis leading to inequivalence of the R groups, as shown by nmr spectra.

Another oddity is that the NHR group can act as an iminyl radical:

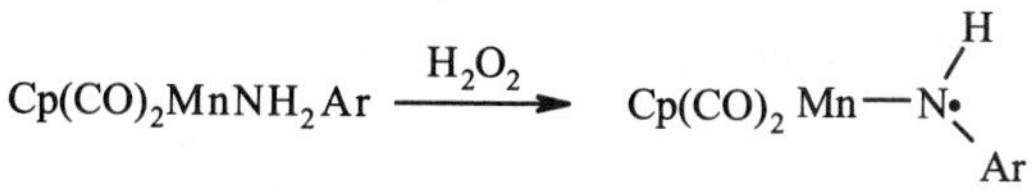

$$Cp(CO)_2MnNH_2Ar \xrightarrow{H_2O_2} Cp(CO)_2\,Mn-N\cdot\begin{matrix}H\\ \diagup\\ \diagdown\\ Ar\end{matrix}$$

Both dialkylamides and alkylamides can undergo insertion reactions with CO_2, CS_2, and other small molecules, such as isocyanates[130]:

$$Me_3Ta(NMe_2)_2 + 2CS_2 \longrightarrow Me_3Ta(S_2CNMe_2)_2$$

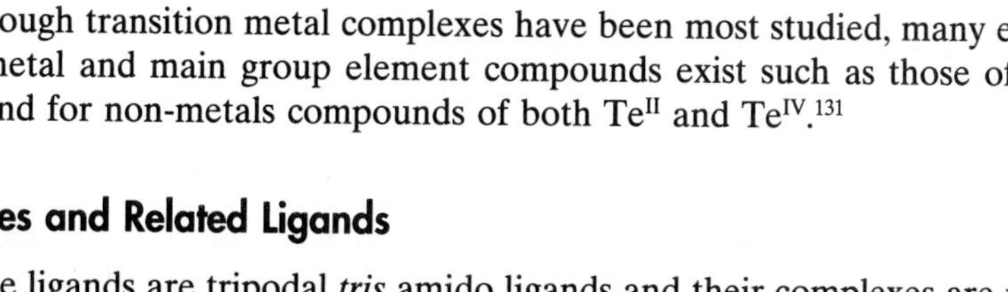

$$Mo(NBu^t)_2(NHBu^t)_2 + 2Bu^tNCO \longrightarrow Bu^tHNC \diagdown\!\!\!\begin{matrix}Bu^t\\ N\\ \|\\ N-Mo-N\\ \diagdown\ \|\ \diagup\\ O\ N\ O\\ \|\\ N\\ Bu^t\end{matrix}\!\!\!\diagup CNHBu^t$$

Although transition metal complexes have been most studied, many examples of non-metal and main group element compounds exist such as those of Sn, Sb, and Bi and for non-metals compounds of both Te^{II} and Te^{IV}.[131]

Azatranes and Related Ligands

These ligands are tripodal *tris* amido ligands and their complexes are made by interaction of the *tris* lithium salts such as $Li_3[(Me_3SiNCH_2CH_2)_3N]$ with metal

[129]J. A. Gladysz et al., J. Chem. Soc., Chem. Commun. 1991, 712.
[130]G. Wilkinson et al., J. Chem. Soc., Dalton Trans. 1993, 781.
[131]D. S. Wright et al., J. Chem. Soc., Dalton Trans. 1996, 1727.

halides.[132] Compounds are known for main group and transition elements, actinides, and lanthanides.

Complexes of the type (9-XXVIII) where the geometry is trigonal pyramidal are typical.

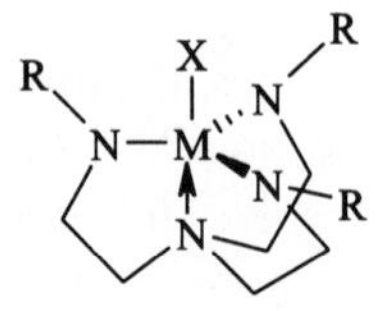

(9-XXVIII)

The oxidation states for M are usually 3 or 4, but in $LMo\equiv N$, 6. The variations can be extensive: (*a*) the central M atoms can be *d* group transition metals, actinides, lanthanides, or main group elements such as Si, Sn, P, etc.; (*b*) the X groups can be uninegative groups like halides, alkyls or aryls, alkoxides, neutral donor molecules, $\equiv P$, or $\equiv N$ groups; (*c*) the R groups can be varied but are usually alkyls or silyls like $SiMe_3$. Compounds with a particular central atom are referred to as, for example, azavanadatranes (V), azagermanatranes (Ge), and so on. The use of very bulky groups attached to nitrogen atoms can alter the ligand properties substantially.

The N atom of the center can be replaced by CR' to give lithium compounds of the type $Li_3[(RNCH_2CH_2)_3CR']$ and corresponding complexes. A tridentate ligand L, $[Me_3SiN\{CH_2CH_2N(SiMe_3)\}_2]^{2-}$ gives complexes such as $LZr(BH_4)_2$.[133]

Chelate Amido and Related Complexes

Many chelate amido complexes are known, recent examples having sterically demanding ligands such as $ArN(CH_2)_3NAr^{2-}$ where $Ar = 2,6-Pr_2^iC_6H_3$.[134] These can give complexes like $LZr^{IV}(NMe_2)_2$. Another ligand is $[mes_2BNCH_2CH_2NBmes_2]^{2-}$ which gives Ti and Zr compounds.[135]

It is convenient to note here that the oxidative dehydrogenation of diamines such as ethylenediamine and cage type amines[136] can lead to *α-diimine complexes*.

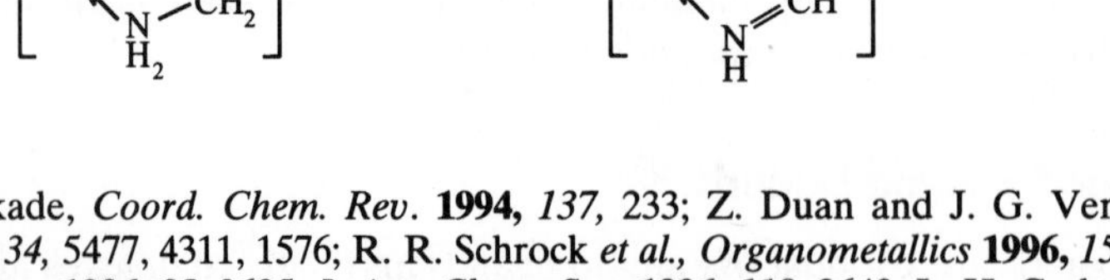

[132]J. G. Verkade, *Coord. Chem. Rev.* **1994**, *137*, 233; Z. Duan and J. G. Verkade, *Inorg. Chem.* **1995**, *34*, 5477, 4311, 1576; R. R. Schrock *et al.*, *Organometallics* **1996**, *15*, 2777, 1470, 5; *Inorg. Chem.* **1996**, *35*, 3695; *J. Am. Chem. Soc.* **1996**, *118*, 3643; L. H. Gade *et al.*, *Inorg. Chem.* **1995**, *34*, 4062; *J. Chem. Soc., Dalton Trans.* **1996**, 125; H. C. Aspinall and M. R. Tillotsen, *Inorg. Chem.* **1996**, *35*, 2163 (lanthanides).
[133]F. G. N. Cloke *et al.*, *J. Chem. Soc., Dalton Trans.* **1995**, 25.
[134]D. H. McConville *et al.*, *Organometallics* **1995**, *14*, 5478.
[135]R. R. Schrock *et al.*, *Organometallics* **1996**, *15*, 562.
[136]P. Bernard and A. M. Sargeson, *J. Am. Chem. Soc.* **1989**, *111*, 597.

The reactions of aromatic 1,2-diamines such as *o*-phenylenediamine or 9,10-phenanthraquinonediamine can give species of the types:

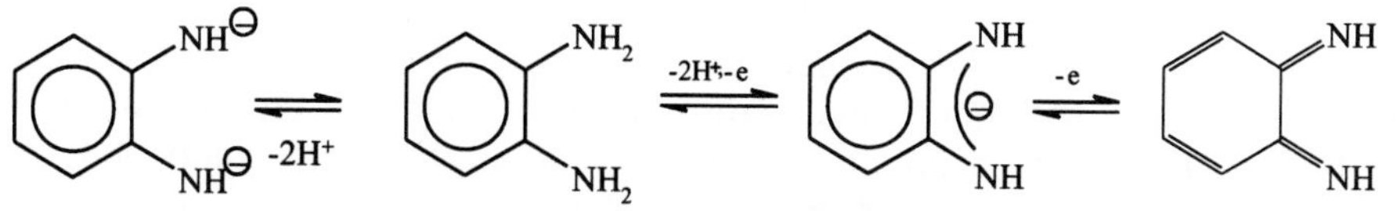

All these species can act as ligands.[137]

Finally there are many types of uninegative *N,N'* chelate ligands,[138] examples being (9-XXIXa) and (9-XXIXb); the former is similar to acetylacetonates while the latter is analogous to salicylaldiminato ligands.

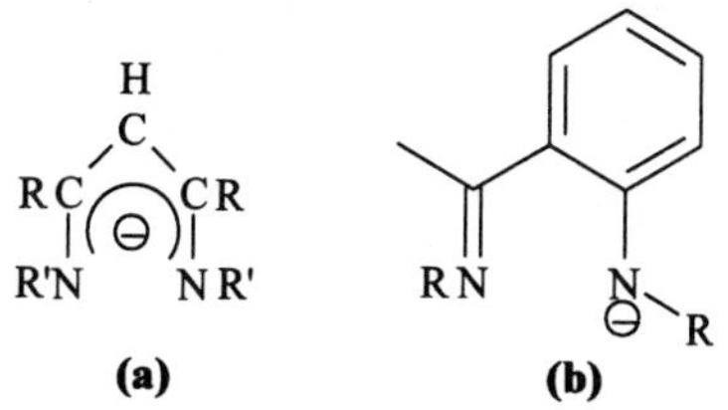

(a)   (b)

(9-XXIX)

9-16 Imido Complexes[139]

Just as deprotonation of NH_3 gives the sequence $NH_3 \rightarrow NH_2^- \rightarrow NH^{2-} \rightarrow N^{3-}$, alkyl or aryl amines give the NHR^- and NR_2^- species of the last section and finally NR^{2-} which is isoelectronic with O^{2-}. There is, however, a difference in that imido ligands can be either linear or bent (see below). In imido compounds the R groups can be H, alkyl, aryl, tosylate, $CR_2'OH$, $CH_2CH=CH_2$, 1-pyrolylimido, etc.

Such complexes have been referred to as nitrene complexes but, strictly, nitrenes are RN: species and while zero valent NR compounds have been reported, none has been isolated. Imido compounds are commonly found in transition metal complexes with oxidation states 3 and above. The high capacity for electron donation by the imido group does of course, like that of O^{2-}, act to stabilize high oxidation states; prime examples of this are $Os(NBu^t)_4$ *vs.* OsO_4 and $(Bu^tN)_3MnCl$ *vs.* O_3MnCl.

The principal bonding modes are:

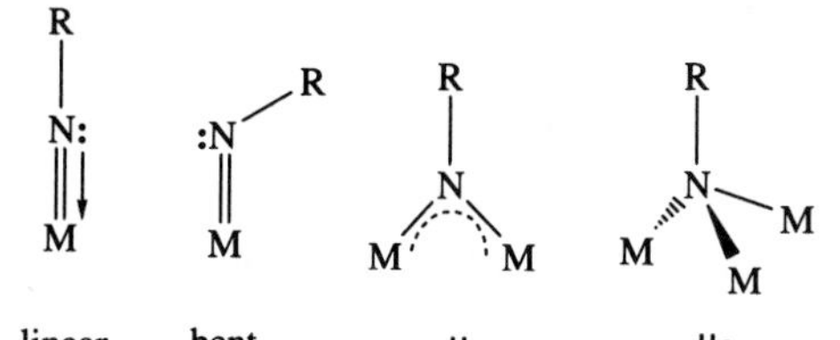

linear   bent μ_2 μ_3

[137]S.-M. Peng *et al., J. Chem. Soc., Chem. Commun.* **1994,** 1645; G. Wilkinson *et al., J. Chem. Soc., Dalton Trans.* **1990,** 315.
[138]For references see G. Wilkinson *et al., J. Chem. Soc., Dalton Trans.* **1995,** 205, *Polyhedron* **1996,** *15,* 3605; M. F. Lappert *et al., J. Chem. Soc., Chem. Commun.* **1994,** 1699.
[139]D. E. Wigley, *Prog. Inorg. Chem.* **1994,** *42,* 239; B. F. Sullivan *et al., Coord. Chem. Rev.* **1991,** *111,* 27; W. E. Nugent and J. M. Mayer, *Metal-Ligand Multiple Bonds,* Wiley, New York, 1988; K. Dehnicke and J. Strähle, *Chem. Rev.* **1993,** *93,* 981 (MNCl complexes); T. R. Cundari *et al., Inorg. Chem.* **1995,** *34,* 2348 (bonding).

In addition to many structure determinations by X-ray diffraction, nmr studies using ^{14}N and ^{15}N have allowed the comparison of NR with other N-bonded ligands like NO, NNR, etc.[140]

The terminal MNR ligands may have angles between 170–180° but there is almost a continuum of bond angles from linear to truly bent groups with angles *ca.* 140°. In tending to maintain an 18e count in electron rich compounds, and depending upon the nature of the ligands, bending may occur as required. In $Os(NBu^t)_4$ for example,[141] all four OsNC bonds are bent with an angle of 156.4° according to electron diffraction studies. Another case is that of $(dtc)_2Mo(NPh)_2$ where there is one severely bent imido since the linear groups would lead to a 20e count for Mo.

Imido compounds can be obtained by many different methods.[142] Much used routes are RN transfer reactions using organic thiocyanates, RNCO, phosphineimines, $R_3P{=}NR'$, hydrazines, RNHNHR', silylamines, or RNH_2 in presence of Me_3SiCl in THF, dme, or similar solvents. Examples are

$$O{=}WCl_4 + RNCO \longrightarrow (RN)WCl_4 + CO_2$$

$$(Me_3SiO)ReO_3 + 3Bu^t(Me_3Si)NH \longrightarrow (Me_3SiO)Re(NBu^t)_3 + 3Me_3SiOH$$

$$O{=}ReCl_3(PPh_3)_2 + ArNH_2 \longrightarrow (ArN)ReCl_3(PPh_3)_2 + H_2O$$

Other methods involve oxidation by azides[143] and cleavage of $PhN{=}NPh$[144] or $RN{=}CHPh$:

$$Cp^*_2V + PhN_3 \longrightarrow Cp^*_2V(NPh) + N_2$$

$$Bu^tC(H){=}TaCl_3(THF)_2 \xrightarrow{\ RN{=}CHPh\ } (RN)TaCl_3(THF)_2 + Bu^tCH{=}CHPh$$

Reactions of lithium amides can also be used:

$$\tfrac{1}{2}[Cp^*IrCl(\mu{-}Cl)]_2 \xrightarrow{\ LiNHBu^t\ } Cp^*Ir(NHBu^t)_2 \xrightarrow{\ -Bu^tNH_2\ } Cp^*Ir(NBu^t)$$

$$L_nM{\equiv}N \xrightarrow{\ LiMe\ } Li[L_nM{=}NMe]$$

Although imido compounds of transition metals have long been known, they had other ligands present. Homoleptic compounds[145] have been made by the reac-

[140]D. C. Bradley *et al., J. Chem. Soc., Dalton Trans.* **1992**, 1663.
[141]D. W. H. Rankin *et al., J. Chem. Soc., Dalton Trans.* **1994**, 1563.
[142]For references see J. R. Dilworth *et al., Polyhedron* **1996**, *15*, 2041.
[143]R. G. Bergman *et al., Angew. Chem. Int. Ed. Engl.* **1996**, *35*, 653.
[144]O. Eisenstein *et al., J. Am. Chem. Soc.* **1996**, *118*, 2762.
[145]A. A. Danopoulos *et al., Polyhedron* **1989**, *8*, 2657; *J. Chem. Soc., Dalton Trans.* **1989**, 2753; **1991**, 269, 1858.

tions:

$$(Bu^tN)_3ReCl \xrightarrow{Na/Hg} Re_2(NBu^t)_6$$

$$(Bu^tN)_2MCl_2 \xrightarrow{LiNHBu} (Bu^tN)_2M(NHBu^t)_2 \xrightarrow{LiMe} Li_2M(NBu^t)_4$$

$$M = Cr, Mo, W$$

$$OsO_4 + NH(SiMe_3)Bu^t \longrightarrow Os(NBu^t)_4$$

By contrast to the last reaction, interactions of the aryl isocyanate, $2,6\text{-}Pr_2^i(C_6H_3NCO)$ with OsO_4 produces the reduced, planar $Os^{VI}(NAr)_3$.[146]

Exchange reactions can also be effected by use of isocyanates[147] or amines,[148] e.g.,

$$CrCl_2(NBu^t)_2 + 2RNCO \longrightarrow CrCl_2(NR)_2 + 2Bu^tNCO$$

In addition to compounds of d group transition metals and actinides, numerous imido compounds of main group elements have been made. A useful synthetic reagent is $[Mg(NPh)THF]_6$, obtained by the reaction of $PhNH_2$ and Bu_2Mg in THF/heptane. The 1-naphthyl $[MgN(1\text{-}nap)THF]_6 \cdot 2.25THF$ was also made. The Mg compounds have a core of alternating Mg—N atoms in a hexagonal prismatic cage; each N is bound to Ph and Mg to the O atom of solvating molecules such as THF or $(Me_2N)_3PO$. The $[Mg(NPh)THF]_6$ can be used to make, *inter alia*, $(GeNPh)_4$ and $(MNPh)_4 \cdot 0.5PhMe$, M = Sn and Pb.[149] Other tin compounds include $[Sn^{II}NBu^t]_4$.[150]

Non-metal imides can also be obtained. Thus the metaphosphate ion, PO_3^- has its analogue in $[Li(THF)_4]^+[P(NAr)_3]^-$, where the three NAr groups are propeller-like with the PN_3 core planar.[151]

Sulfur compounds[152] $S(NR)_2$ and $S(NSiMe_3)_3$ are analogues of SO_2 and SO_3, respectively; treatment of $S(NBu^t)_2$ with $LiNHBu^t$ leads to $\{Li_2[S(NBu^t)_3]\}_2$, which formally is derived from $SO_3^{2-}-E(NR)_3^-$; compounds are known for E = Ge, Sn, and Pb. Related imido compounds are known for Se and Te,[153] examples being $Li_2Te(NBu^t)_3$ and $Se(NBu^t)_2$.

The imido groups of transition metal complexes undergo a wide range of reactions including electrophilic attacks by protons, NR transfer, and many other reactions.[154]

[146]R. R. Schrock et al., J. Am. Chem. Soc. **1990**, 112, 1642.

[147]G. Wilkinson et al., J. Chem. Soc., Dalton Trans. **1995**, 2111.

[148]P. Mountford et al., J. Chem. Soc., Dalton Trans. **1995**, 374.

[149]P. P. Power et al., Inorg. Chem. **1996**, 35, 3254.

[150]D. S. Wright et al., Chem. Commun. **1996**, 150.

[151]E. Niecke et al., Angew. Chem. Int. Ed. Engl. **1994**, 33, 2111.

[152]D. Stalke et al., Angew. Chem. Int. Ed. Engl. **1996**, 35, 204.

[153]T. Chivers et al., Inorg. Chem. **1996**, 35, 553, 4094, 4336.

[154]See for example, J. L. Bennett and P. T. Wolczanski, J. Am. Chem. Soc. **1994**, 116, 2179; T. R. Cundari, Organometallics **1993**, 12, 4971; J. Am. Chem. Soc. **1992**, 114, 7879; S. Y. Lee and R. G. Bergman, J. Am. Chem. Soc. **1996**, 118, 6396.

Insertion reactions can proceed with isocyanides, isocyanates, organic azides, and so on,[155] e.g.,

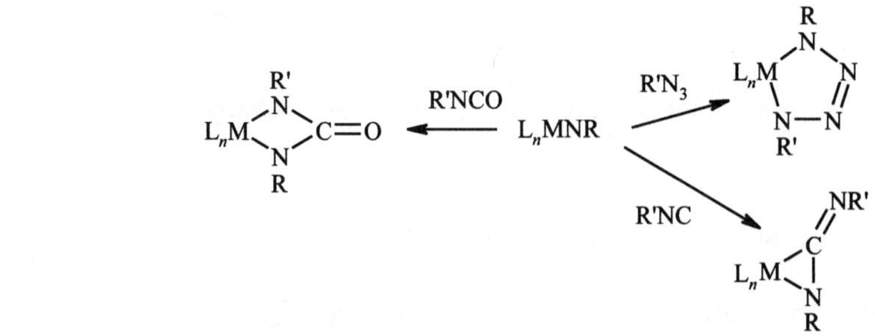

Other reactions are noted elsewhere in the text.

9-17   Nitrido Complexes[156,157]

The N^{3-} ion is one of the strongest of π-donors; $M\equiv N$ distances are in the region 1.6–1.8 Å. Nitrido complexes are obtained by a variety of routes as shown in the following examples:

$$W_2(OBu^t)_6 + RCN \longrightarrow (Bu^tO)_3W\equiv N + (Bu^tO)_3W\equiv CR$$

$$OsO_4 + OH^- + NH_3 \longrightarrow Os(N)O_3^- + 2H_2O$$

$$4Cp^*Ti(NMe_2)_3 + 4NH_3 \xrightarrow[\text{toluene}]{90°C} (Cp^*Ti)_4(\mu_3\text{-}N)_4 + 12Me_2NH$$

$$TPPMnCl + PhIO + 2NH_3 \longrightarrow TPPMn\equiv N + NH_4Cl + H_2O + PhI$$

$$porphCr(N_3) \xrightarrow{h\nu} porphCrN + N_2$$

The splitting of N_2 to give nitrido species was noted in Section 9-3.

Unusual syntheses involve elimination of *iso*-butylene from *tert*-butylamido complexes,[158] e.g.,

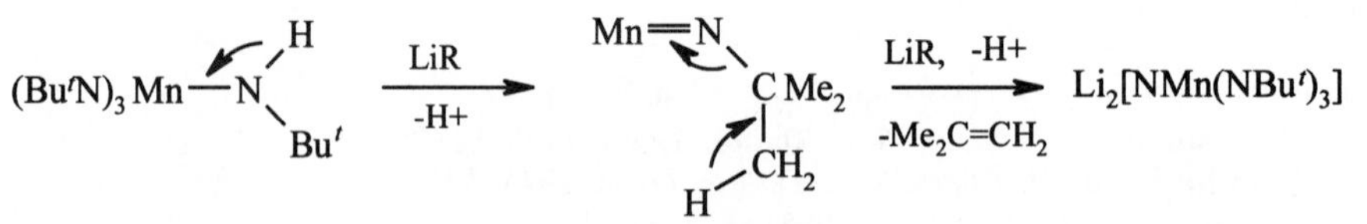

[155]G. Wilkinson *et al., J. Chem. Soc., Dalton Trans.,* **1996,** 271; G. Proulx and R. G. Bergman, *Organometallics* **1996,** *15,* 684.
[156]K. Dehnicke and J. Strahle, *Angew. Chem. Int. Ed. Engl.* **1992,** *31,* 955 (172 references); G. Frenking *et al., Inorg. Chem.* **1994,** *33,* 5278 (structural data and theory—Mo, W, Re, Os).
[157]H.-T. Chiu *et al., Chem. Commun.* **1996,** 139.
[158]G. Wilkinson *et al., J. Chem. Soc., Dalton Trans.* **1995,** 205; **1991,** 2791.

Bridged Nitrido Species[159]

The linear species can be symmetrical or asymmetrical. There are several other bridged types:

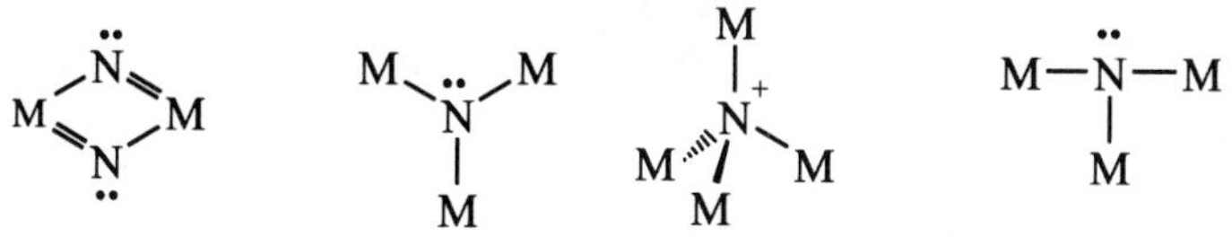

Although usually only transition metals give mixed species such as $(Me_3SiO)_3V\equiv N-Rh(PPh_3)_2$,[160] main group complexes are known, e.g., $Cl_3Ga-NReCl_2(PMe_2Ph)_3$.[161] The metals can also have mixed valences as in $V_2(\mu\text{-}N)Cl_5(tmeda)_2$[162], which has $V^V\equiv N: \rightarrow V^{III}$; another example involves $Fe^{III}-N\equiv Fe^{IV}$, which can be oxidized by Br_2 to $Fe^{IV}=N=Fe^{IV}$.[163]

The triangular, planar μ_3 species are less common than M_3O oxo-centered ones, but $[Ir_3N(SO_4)_6(H_2O)_3]^{6-}$ is an example. The μ_4 tetrahedra are found in $(MeHg)_4N^+$ and $(Ph_3PAu)_4N^+$ while N atoms can be found at the center of clusters such as $[Ru_6N(CO)_{16}]^-$.

There are now many examples of polymeric species with bridging μ_2-N as in the following:

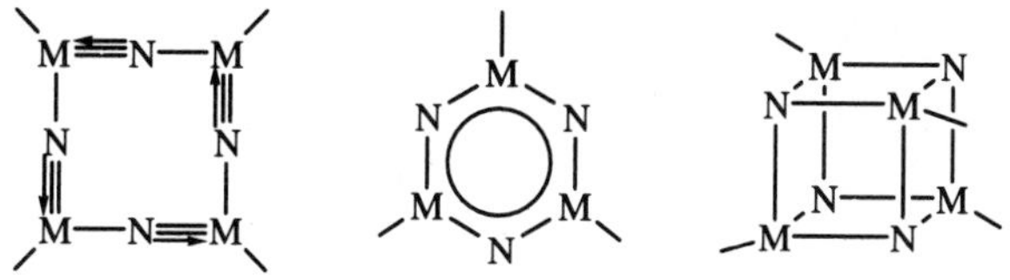

Tetrameric species have alternating long and short M—N distances, e.g., in $\{MoNCl_3[O(C_4H_9)_2]\}_4$, of 1.65 and 2.16 Å while in a Tc complex they range between 1.681 to 1.977 Å[164]; a titanium complex $(Cp^*Ti)_4(\mu_3\text{-}N)_4$ has a cubane structure.[165]

In trimers such as $[Cp^*TaCl(\mu\text{-}N)]_3$ and $\{[(CF_3)Me_2CO]W\mu\text{-}N\}_3$ again bond distances alternate but the ring is planar.[166] The compound $[(Bu^tO)_3W\equiv N]_\infty$ however is polymeric with short triple and long dative bonds.

Finally it may be noted that N transfer from nitrido species such as salen or porphyrin complexes of manganese, e.g., porph $Mn^V\equiv N$, to organic substrates can occur. Examples are aziridation of cyclooctene or amination of silylenol ethers.[167]

[159]F. Bottomley et al., Organometallics **1996**, 15, 1750; A. L. Odom and C. C. Cummins, Organometallics **1996**, 15, 898; M. Hidai et al., Inorg. Chem. **1994**, 33, 3619; J. Baldas et al., J. Chem. Soc. Chem. Commun. **1994**, 2153.
[160]C. M. Jones and N. M. Doherty, Polyhedron **1995**, 14, 81.
[161]U. Abram et al., J. Chem. Soc., Chem. Commun. **1995**, 2047.
[162]N. M. Doherty et al., Inorg. Chem. **1992**, 31, 2679.
[163]K. Wieghardt et al., Angew. Chem. Int. Ed. Engl. **1995**, 34, 669.
[164]J. Baldas et al., J. Chem. Soc., Chem. Commun. **1994**, 2153.
[165]M. Mena et al., J. Chem. Soc., Chem. Commun. **1995**, 2185.
[166]See M. H. Chisholm et al., Angew. Chem. Int. Ed. Engl. **1995**, 34, 110.
[167]See E. M. Carreira et al., J. Am. Chem. Soc. **1996**, 118, 915.

9-18   1,2-Diazenes, Azadienes, and Related Compounds

1,2-Diazenes[168]

These compounds, RN=NR', are commonly called azo compounds; azobenzene, PhN=NPh is one of the best known. As ligands they can act in η^1-, η^2-, and bridging modes; η^1 ligands are *trans* but the bridging modes can have *cis* or *trans* groups.

Diazene, HN=NH, and HN=NR compounds are known and N_2H_2 complexes may be involved in dinitrogen reducing systems.[169]

The diazene compounds can be obtained by oxidation of hydrazine complexes and in other ways. Typical examples are $Cp*Me_3W(\eta^2\text{-MeNNMe})$ and the bridged species $[Cp(THF)Yb]_2(\mu\text{-Ph}_2N_2)\cdot C_7H_8$.

Azadienes

The most important class are the *diazabutadienes*.[170] These ligands, usually designated *dab* or *dad* are of the type RN=CH—CH=NR and can give complexes of the following types:

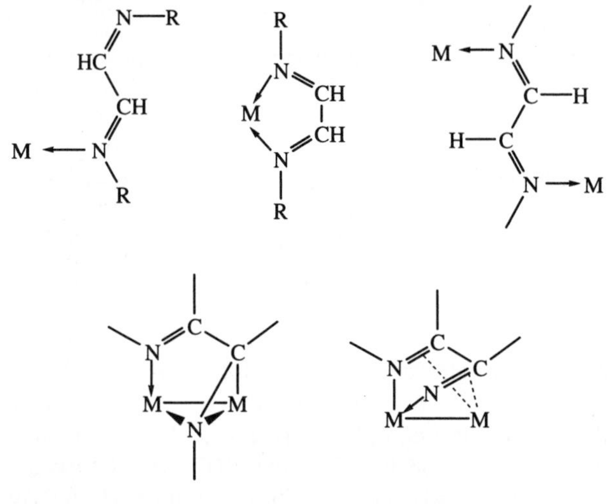

Most transition metals[171] as well as some main group elements form complexes, often in low oxidation states. For Al, Si, and Ge,[172] 2-coordination is stabilized as in (9-XXX).

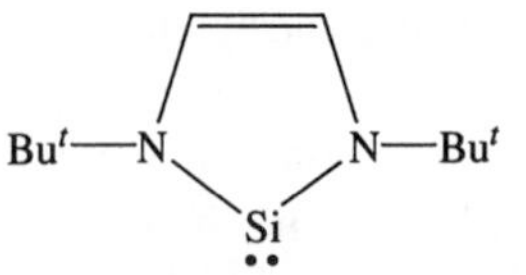

(9-XXX)

[168]D. Sutton, *Chem. Rev.* **1993**, *93*, 995; M. Retbøll and K. A. Jørgensen, *Inorg. Chem.* **1994**, *33*, 6403 (structures and bonding).

[169]G. L. Hillhouse *et al., J. Am. Chem. Soc.* **1994**, *116*, 204.

[170]N. Kaltsoyannis, *J. Chem. Soc., Dalton Trans.* **1996**, 1583 (computations).

[171]For references see L. D. Field *et al., Inorg. Chem.* **1994**, *33*, 1539; K. H. Thiele *et al., Angew. Chem. Int. Ed. Engl.* **1995**, *34*, 2649.

[172]For references see A. H. Cowley *et al., Inorg. Chem.* **1995**, *34*, 615.

The compounds can be made by substitution of CO, C_2H_4, or other labile ligands by the dienes, or by interaction of halides alone or in the presence of reducing agents. The variety of binding types arises from the flexibility of the dab skeleton and strong σ-donor and π-acceptor behavior, coupled with different substituents on N or C. The ligands can act as 2, 4, 6, or 8 electron donors: in the 8e case, $\pi C = N$ orbitals are used for η^2-C,N bonding.

Lithium substitution of H atoms in $Bu^tN=CH-CH=NBu^t$ to give (9-XXXI) allows the synthesis of $[(dab)Li]^-$ complexes of lanthanides[173] by reaction with the trichlorides $LnX_3(THF)_3$.

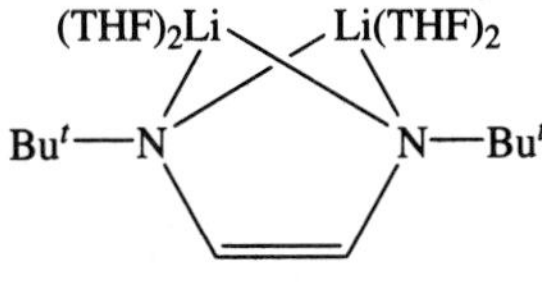

(9-XXXI)

Tetraazabutadienes[174] as such do not exist but complexes can be formed from reaction of organic azides, RN_3, with compounds like $Ni(COD)_3$, by interaction of halides with $Li_2[PhNN=NNPh]$, or by 1,3-dipolar cycloadditions with imido complexes.[175] The compounds have been variously termed as tetraazadiene, tetraazene, or tetraazenido. They normally have short and long bonds as in (9-XXXII) but there is doubtless electron delocalization over the ring system.

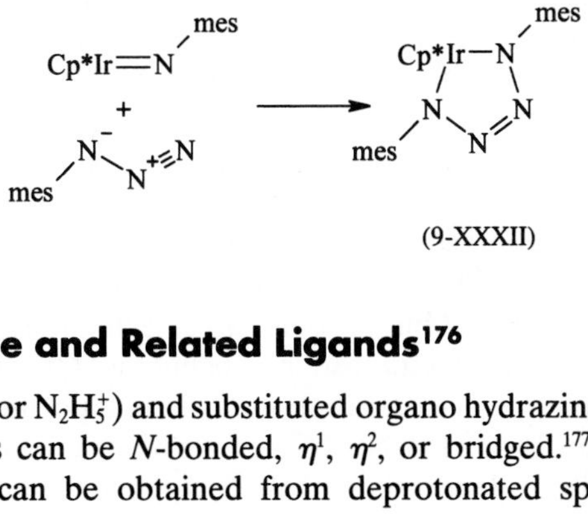

(9-XXXII)

9-19 Hydrazine and Related Ligands[176]

Complexes of N_2H_4 (or $N_2H_5^+$) and substituted organo hydrazines are well characterized and the ligands can be N-bonded, η^1, η^2, or bridged.[177] Similar terminal or bridged complexes can be obtained from deprotonated species, H_2NNH^- and $HNNH^{2-}$.[178]

An example of a substituted hydrazine (1-) complex is $Ge[N(Me)NMe_2]_4$ which has a distorted tetrahedral structure; complexes of other main group elements (as well as transition metal complexes) are known.[179] An example of a $NHNH_2^-$ species is $Cp^*Me_4W(\eta^2\text{-}NHNH_2)$.[180]

[173]J. Scholz et al., Angew. Chem. Int. Ed. Engl. 1995, 34, 673.
[174]W. C. Trogler, Acc. Chem. Res. 1990, 23, 426; S. W. Lee and W. C. Trogler, Inorg. Chem. 1990, 29, 1659.
[175]G. Wilkinson et al., J. Chem. Soc., Dalton Trans. 1996, 3771; R. G. Bergman et al., Organometallics 1993, 12, 2741.
[176]D. Sutton, Chem. Rev. 1993, 93, 995 (266 references).
[177]R. R. Schrock et al., J. Am. Chem. Soc. 1993, 115, 1760; Organometallics 1993, 12, 1140.
[178]W. J. Evans et al., Inorg. Chem. 1992, 31, 3592.
[179]D. W. H. Rankin et al., J. Chem. Soc., Dalton Trans. 1996, 2095.
[180]R. R. Schrock et al., Inorg. Chem. 1988, 27, 3574.

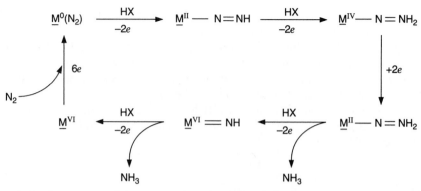

Figure 9-4 A summary of the protonation chemistry of dinitrogen complexes of molybdenum and tungsten based on a wide range of chemical, electrochemical, and kinetic researches. Species characteristic of all these stages except M^{VI} have been isolated and characterized. In a typical case $\underline{M} = M(Ph_2PCH_2CH_2PPh_2)_2$, $X = Cl$, and $M = Mo$ or W (redrawn from R. R. Eady and G. J. Leigh, *J. Chem. Soc., Dalton Trans.* **1994,** 2739).

The compounds can be obtained in a variety of ways[181] from the hydrazines or their Li derivatives, from $Me_3SiNHNR_2$, and so on. Ligands of the type NNH_2 or NNR_2 as the (1-) and (2-) anions have long been known, the (1-) species being less common. They have been much studied because they have been proposed as intermediates in the metal catalyzed reductions of N_2 to NH_3 and N_2H_4.

The current state of the reactions and of the biological and biochemical studies has been reviewed[182] and a summary of the reactions is shown in Fig. 9-4.

There has been some concern about the nature of NNR_2^{2-} species since in some compounds the oxidation state of the metal is either too high, or impossible and the electron count may also exceed 18e. In such cases it appears that the ligand is the neutral *isodiazene*, $N^-=N^+R_2$. It is not easy from structural data to identify the oxidation state of the metal but in some cases, techniques such as EXAFS, epr, or particularly Mössbauer spectra can be definitive. Of course, if considering NNR_2^{2-} would give an impossible oxidation state, there is no alternative. An example is that of $WCl_3(OSiMe_3)(N_2C_9H_{18})$, $NC_9H_{18} = 2,2,6,6$-tetramethylpiperid-1-yl, where the (2-) ligand gives an oxidation state of W^{VIII}; the true oxidation state is W^{IV} with a neutral isodiazene.[183] A particularly clear example is the porphyrinato complex, $Fe(porph)(N_2C_9H_{18})$,[184] which was conclusively shown by several techniques, in particular by Mössbauer spectra, to have the oxidation state Fe^{II}. The Fe—N—N group is linear with Fe—N = 1.809 Å and N—N = 1.232 Å, both distances being short and indicating π-bonding and multiple bond character, characteristic of isodiazene (4e) species.

It had been noted in 1982 by J. Chatt *et al.* for $WCl_3(NNH_2)(PMe_2Ph)_2$ that "Structurally this compound appears closer to isodiazene than hydrazide (2-) but we retain the latter name because of its common usage and the IUPAC recommendation."

[181]P. Gouzerh *et al., Inorg. Chem.* **1994,** *33,* 1427, 4937 ($NNR_2^{1-,2-}$).
[182]R. R. Eady and G. J. Leigh, *J. Chem. Soc., Dalton Trans.* **1994,** 2739.
[183]G. Wilkinson *et al., J. Chem. Soc., Dalton Trans.* **1994,** 907 and references therein.
[184]J.-P. Mahy *et al., J. Am. Chem. Soc.* **1984,** *106,* 1699.

This practice should now be abandoned and it is clear that many NNR_2 compounds that could and have been designated as $NNR_2(2-)$ ones are isodiazenes(0)—a situation comparable to $M(H_2)$ vs. $M(H)_2$ in Chapter 2. Compounds in high oxidation states such as Mo^{VI} and W^{VI}, e.g., $Mo(NNPh_2)_2ClPPh_3$ formulated as Mo^{VI} with $NNPh_2(2-)$,[185] could well be diazenes especially when MeCN, a good reducing agent, is used as solvent.

9-20   Diazenido Complexes[186,187]

The ligand NNR has been referred to as alkyl- or arylazo, diazo, or diazenato, but diazenido is preferred. The aryl compounds are commonest; they can be obtained from diazonium compounds (ArN_2^+) and from hydrazines such as $PhNNH_2$ or $ArC(O)NNH$. They are also formed by electrophilic or nucleophilic attacks on dinitrogen compounds, for example,

$$(diphos)_2Mo(N_2) + RCOCl + HCl \xrightarrow{NEt_3} (diphos)_2ClMo(N_2COR)$$

$$Cp(CO)_2MnN_2 \xrightarrow{LiPh} Li[Cp(CO)_2MnN_2Ph]$$

The ligand can be regarded as RN_2^+ (the analogue of NO^+), as RN_2^-, (the analogue of NO^-), or as neutral RN_2 (the analogue of CO), and there has been the same type of discussion concerning the extent of metal-ligand π bonding. As with NO compounds, the $N=N$ stretches vary widely from ~ 2095 cm^{-1} (indicating $M-N$ π bonding) down to ~ 1440 cm^{-1} in bridging species. The main distinction is between the following types of complex:

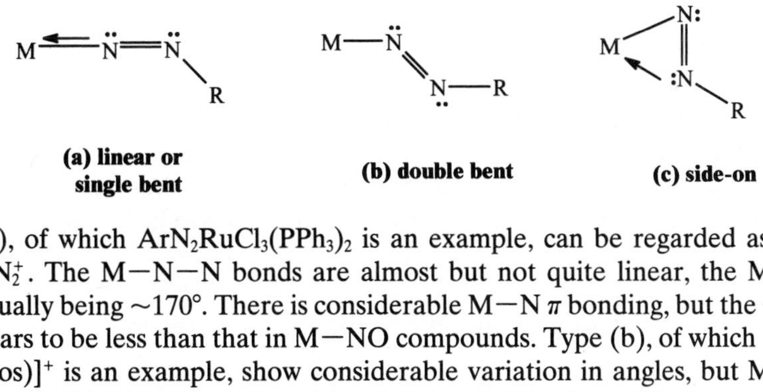

(a) linear or single bent   **(b) double bent**   **(c) side-on**

Type (a), of which $ArN_2RuCl_3(PPh_3)_2$ is an example, can be regarded as derived from RN_2^+. The $M-N-N$ bonds are almost but not quite linear, the $M-N-N$ angle usually being $\sim 170°$. There is considerable $M-N$ π bonding, but the π character appears to be less than that in $M-NO$ compounds. Type (b), of which $[ArN_2Rh Cl(triphos)]^+$ is an example, show considerable variation in angles, but $M-N-N$ is usually $\sim 120°$. They can be considered to be derived from RN_2^-.

There are some iridium complexes of the *neutral* group $C_5Cl_4N_2$ acting as a two-electron donor [e.g., $IrCl(N_2C_5Cl_4)(PPh_3)_2$], which can be compared with $IrCl(CO)(PPh_3)_2$; these also have a type (a) bent structure.

The η^2-bonding is found in $CpTiCl_2(\eta^2\text{-NNPh})$. The RN_2 compounds undergo a number of reactions, e.g., hydrogenation to a hydrazino complex:

$$[(Ph_3P)_2PtN_2Ar]^+ \xrightarrow{H_2} [(Ph_3P)_2Pt(H)NH_2NHAr]^+$$

[185]P. Gouzerh et al., Inorg. Chem. **1994**, 33, 1427, 4937 ($NNR_2^{1-,2-}$).
[186]D. Sutton, Chem. Rev. **1993**, 93, 995 (266 references).
[187]D. Sutton et al., Inorg. Chem. **1995**, 34, 6163.

More important, however, is attack by electrophiles, especially H^+, which shows that diazenido complexes are related to diazene complexes:

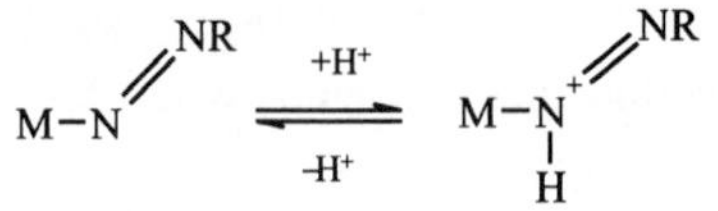

$$M-N \overset{NR}{\diagup} \quad \underset{-H^+}{\overset{+H^+}{\rightleftharpoons}} \quad M-\overset{+}{\underset{H}{N}} \overset{NR}{\diagup}$$

diazenido   **diazene**

The NNR group of ligands also includes *aryldiazonium complexes*[188] derived from diazonium salts, ArN_2^+, that have the following coordination modes:

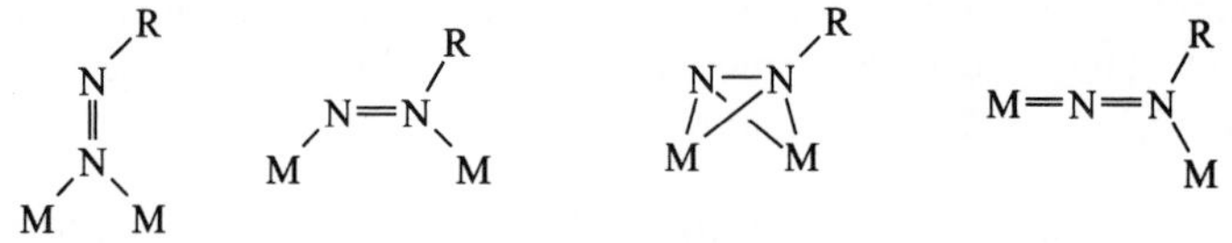

$$\underset{M \quad M}{\overset{R}{\underset{N}{\overset{N}{\|}}}} \qquad \underset{M \qquad M}{\overset{R}{N=N}} \qquad \underset{M \quad M}{\overset{R}{N-N}} \qquad \underset{M}{M=N=N}^{R}$$

Diazoalkanes,[189] R_2CN_2, form complexes with η^1, η^2, and bridging modes, and η^1 has linear or bent groups.

9-21 Alkylidineamido and Related Complexes

The species can arise by base attack on bent imido complexes (Section 9-16), for example,[190]

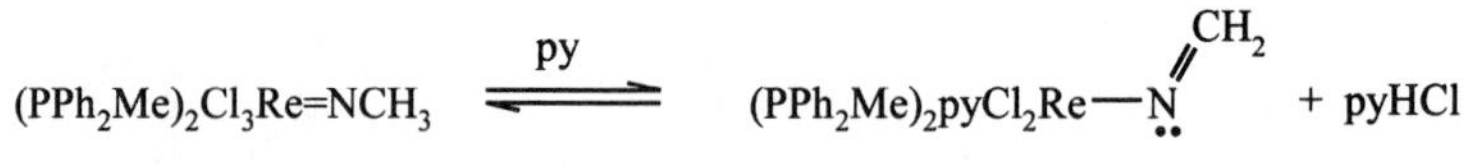

$$(PPh_2Me)_2Cl_3Re{=}NCH_3 \quad \overset{py}{\rightleftharpoons} \quad (PPh_2Me)_2pyCl_2Re{-}\overset{CH_2}{\underset{\cdot\cdot}{N}} + pyHCl$$

A more general synthesis is the reaction of halides with $LiN{=}CR_2$ that can be obtained by the reaction of $HN{=}CR_2$ with *n*-butyllithium.

The ligands can be considered to resemble NO in that they can act as 1e or 3e ligands giving bent or linear π-bonded groups respectively. The bonding modes are:

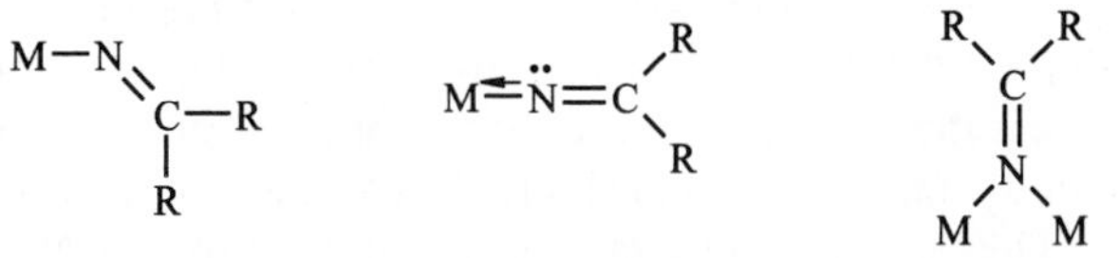

$$M-N \overset{}{\underset{R}{\diagdown}} \underset{R}{\overset{}{C-R}} \qquad M{\doteq}\overset{\cdot\cdot}{N}{=}C \overset{R}{\underset{R}{\diagdown}} \qquad \underset{M \quad M}{\overset{R \quad R}{\underset{N}{\overset{C}{\|}}}}$$

bent   **linear**   **bridging**

α-Imine complexes[191] with the related groups $M-N^+(H){=}CR_2$ or $M-N^+$ $(R'){=}CR_2$ can be made from the free imines or in other ways. Imine ligands can

[188]F. Neve *et al., Inorg. Chem.* **1992,** *31,* 2979.
[189]M. Cowie *et al., Organometallics* **1986,** *5,* 860; M. Hidai and Y. Ishi, *Bull. Soc. Chem. Japan* **1996,** *69,* 819 (review with 49 references).
[190]J. A. Gladysz *et al., Organometallics* **1993,** *12,* 4523.
[191]R. J. Whitby *et al., Organometallics* **1994,** *13,* 190.

bond in η^2 modes as follows:

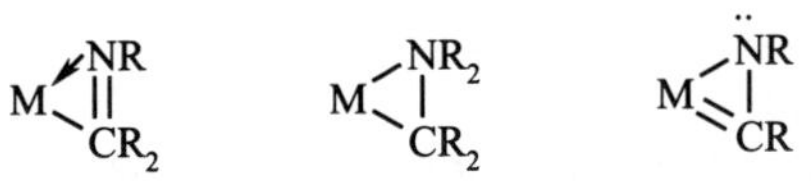

$$M \overset{NR}{\underset{CR_2}{\langle\!\!\parallel}} \qquad M \overset{NR_2}{\underset{CR_2}{\langle\!\!\mid}} \qquad M \overset{\overset{\cdot\cdot}{NR}}{\underset{CR}{\langle\!\!\mid}}$$

Azavinylidenes are of the type $M=N=CRR'$ and can be obtained from nitrile cations.[192]

$$L_n MNCR^+ + {}^-OR' \longrightarrow M{\overset{\longleftarrow}{=}}NCR(OR')$$

9-22 Hydroxylamido(1-, 2-), C-Nitroso, and Oxime Complexes

Hydroxylamine (H_2NOH) or substituted hydroxylamines can give neutral unidentate complexes with Co^{III}, Pt^{II}, or Ni^{II}, but the deprotonated species, *hydroxylamido* (1-) (H_2NO^-), *hydroxylamido* (2-) (HNO), and their alkyl or aryl substituted analogues are more important. Fully dehydrogenated H_2NOH is, of course NO, and indeed NO complexes can often be obtained by reactions of H_2NOH in basic solution in addition to deprotonated species, for example,

$$MoO_4^{2-} + 4SCN^- + 2NH_2OH \longrightarrow [MoNO(\eta^2\text{-}H_2NO)(NCS)_4]^{2-} + 4OH^-$$

Hydroxylamido complexes[193] are commonly obtained by the reactions:

$$L_n MCl_2 + 2LiONR_2 \longrightarrow L_n M(ONR_2)_2 + 2LiCl$$

$$L_n M(OR)_2 + 2R'_2NOH \longrightarrow L_n M(ONR'_2)_2 + 2ROH$$

$$M(NMe_2)_4 + 4Me_2NOH \longrightarrow M(ONMe_2)_4 + 4NHMe_2$$

The dialkylhydroxylaminato (R_2NO^-) complexes are generally side-on η^2 with N,O bonding.[194] However, η^1-O bonded and bridged species are known for Me_2NO^-.[195] As NH_2OH is isoelectronic with H_2O_2 there is some formal similarity between H_2NO^- and HOO^-.

Hydroxylamido(2-) complexes[196] can be made by reactions such as:

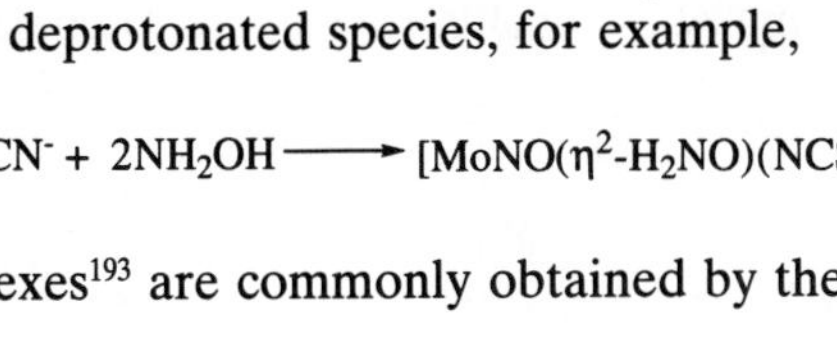

$$L_n M \overset{O}{\underset{O}{\langle\!\!\mid}} + RNH_2 \xrightarrow{-H_2O} L_n M \overset{O}{\underset{NR}{\langle\!\!\mid}}$$

$$L_n M + RNO \nearrow$$

[192]J. L. Templeton *et al.*, *J. Am. Chem. Soc.* **1994**, *116*, 8613.
[193]J. A. McCleverty, *Transition Met. Chem.* **1987**, *12*, 282.
[194]D. W. H. Rankin *et al.*, *J. Chem. Soc., Dalton Trans.* **1996**, 2089.
[195]G. Wilkinson *et al.*, *J. Chem. Soc., Dalton Trans.* **1992**, 555.
[196]E. R. Moeller and K. A. Jorgensen, *J. Am. Chem. Soc.* **1993**, *115*, 11814 and references therein.

where the reaction with the nitroso compound is an oxidative addition. Such compounds are isoelectronic with peroxo, O^{2-}, species. The η^2-NO complexes can also be called *metallaoxaziridines*. The neutral molecules RNO are, of course, bent C-*nitroso* complexes and numerous complexes usually of $(Bu^tNO)_2$, which is in equilibrium with the monomer in solution are known.[197] These can be made in various ways, e.g.,

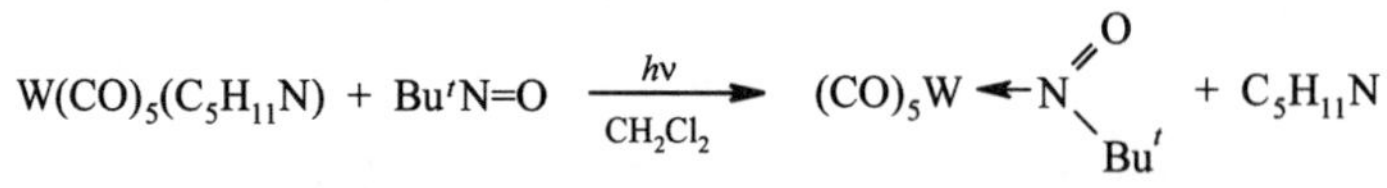

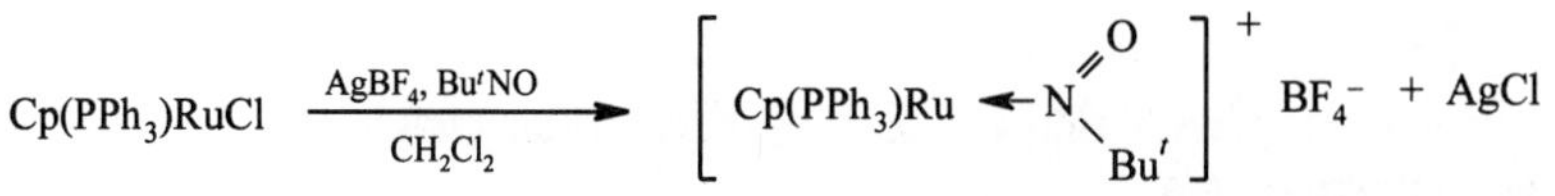

$$W(CO)_5(C_5H_{11}N) + Bu'N{=}O \xrightarrow[CH_2Cl_2]{h\nu} (CO)_5W{\leftarrow}N\overset{O}{\underset{Bu'}{\diagup}} + C_5H_{11}N$$

$$Cp(PPh_3)RuCl \xrightarrow[CH_2Cl_2]{AgBF_4,\ Bu'NO} \left[Cp(PPh_3)Ru{\leftarrow}N\overset{O}{\underset{Bu'}{\diagup}} \right]^+ BF_4^- + AgCl$$

Other N-bonded complexes are $(COD)ClRh^I(ONC_6H_3\text{-}2,4\text{-}Me_2)$ and Pd^I $Cl_2(ONPh)_2$; compounds with O-bonding and with *N,O*-bridged bonding are known.[198]

The insertion of NO into metal alkyls proceeds *via* RNO intermediates, but these can react further depending on whether the initial product is paramagnetic, in which case an *N*-methyl-*N*-nitrosohydroxylaminato chelate is formed by N—N bond formation,

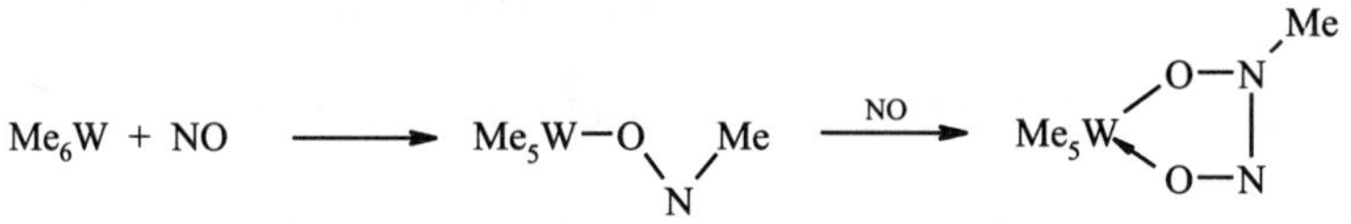

$$Me_6W + NO \longrightarrow Me_5W{-}O\underset{N}{\diagdown}Me \xrightarrow{NO} Me_5W\overset{O-N\diagup Me}{\underset{O-N}{\diagdown}}$$

or is diamagnetic, in which case MeN is transferred to NMe and azomethane eliminated:

$$2ReOMe_4 + 2NO \longrightarrow 2ReO(ONMe)Me_3 \longrightarrow 2Me_3Re(O)_2 + MeN{=}NMe$$

Other C—NO Complexes

Apart from those of RNO discussed previously, others include those of nitroxide radicals such as Bu_2^tNO and 2,2,6,6-tetramethylpiperidinyl-1-oxy (tempo) (9-XXXIII) used as probes in biological systems. The bonding can be η^1-O or η^2-N,O where the ligand is bound as R_2NO^-.

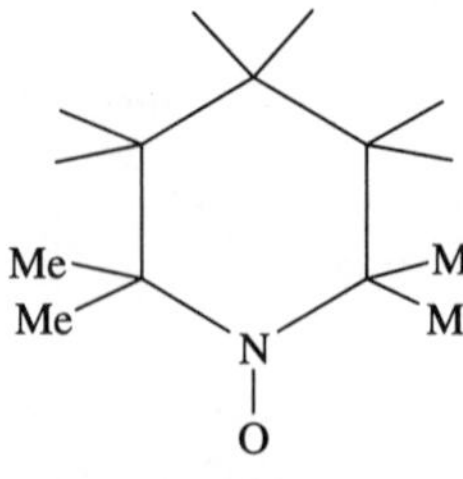

(9-XXXIII)

[197]W. L. Gladfelter *et al., Organometallics* **1994,** *13,* 4137; G. L. Geoffroy *et al., Organometallics* **1990,** *9,* 312.
[198]G. Matsubayashi and T. Tanaka, *J. Chem. Soc., Dalton Trans.* **1990,** 437.

Oxime complexes are made by condensation of hydroxylamines with aldehydes or ketones and can be of the types

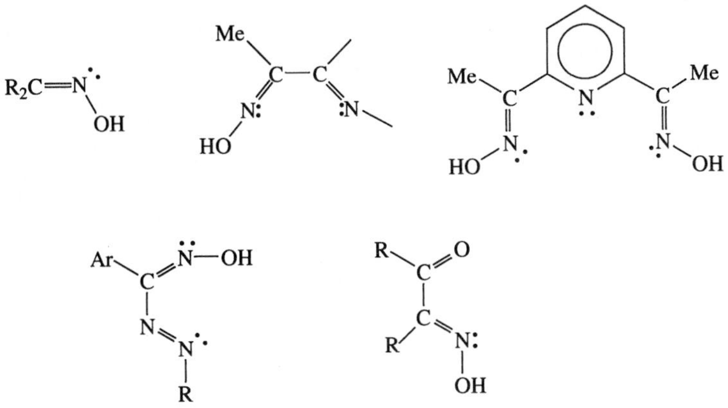

There is an extensive chemistry of *trans*-bisdioximes such as dimethylglyoxime. They provide model systems for vitamin B_{12} and can be oxygen carriers and catalysts.[199]

The $O-H\cdots O$ hydrogen bridge of dimethylglyoxime complexes can also be replaced by $O-BF_2-O$ or $O-M-O$ as in the reaction

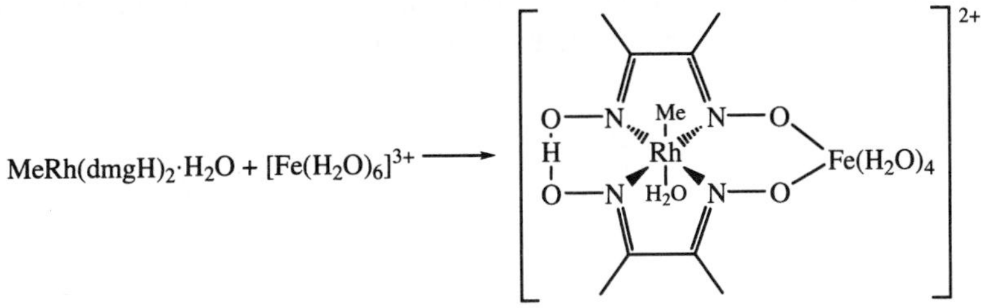

$$MeRh(dmgH)_2 \cdot H_2O + [Fe(H_2O)_6]^{3+} \longrightarrow$$

Oxime complexes are used commercially for extraction of metals such as copper by complexing and solvent extraction.

9-23 Schiff Base Ligands[200]

These diverse ligands usually contain both N and O donor atoms although purely N as well as N, S donors exist.

One of the best known Schiff base ligands is bis(salicylaldehyde)ethylenediimine (salen):

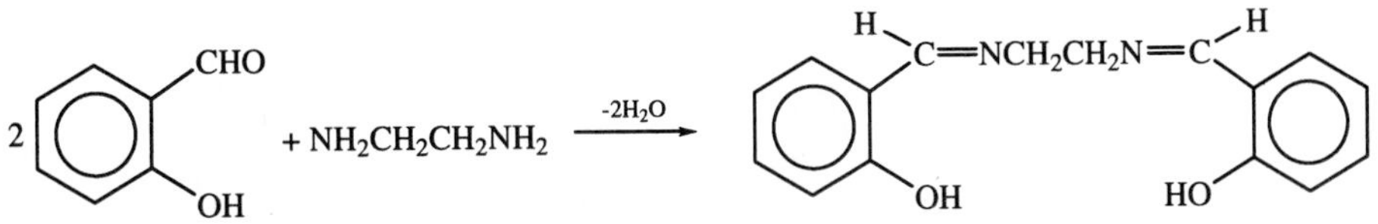

This is an acidic (two OH groups), tetradentate (2N, 2O) ligand. Other Schiff bases can be mono-, di-, or tetrafunctional and can have denticities of six or more with

---

[199]M. R. Churchill *et al.*, *Inorg. Chem.* **1992**, *31*, 859 and references therein.
[200]J. Costamagna, *Coord. Chem. Rev.* **1992**, *119*, 67.

various donor atom combinations (e.g., for pentadentate, N_3O_2; N_2O_3; N_2O_2P; N_2O_2S). Complexes of un-ionized or partly ionized Schiff bases are also known (e.g., $LaCl_3salenH_2 \cdot aq$).

Some representative types of complex that illustrate the formation of not only mononuclear but of binuclear and polymeric species are (9-XXXIV).

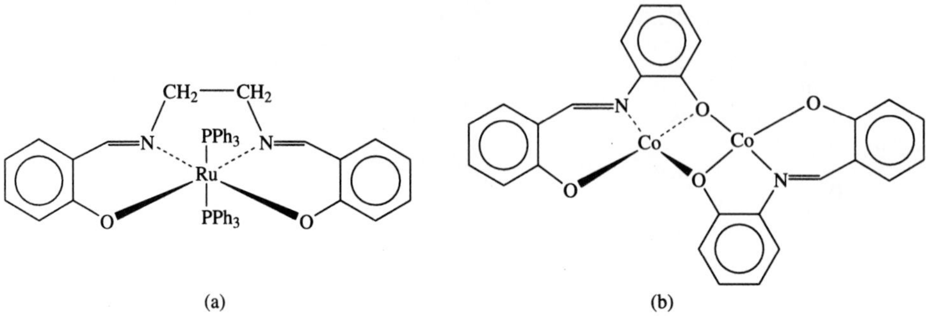

(a)　　　　　　　　　　(b)

(9-XXXIV)

Like the porphyrin ligands noted earlier, face-to-face Schiff base ligands can be constructed to hold two metals such as Mn^{II}, Mn^{II} or Mn^{II}, Mn^{III}.

A type of ligand that can hold metals side by side is called *compartmental*. They can be made by condensing 1,3,5 triketones with α, ω alkanediamines. An example is (9-XXXVa) which can be used to bind two metals as in (9-XXXVb).

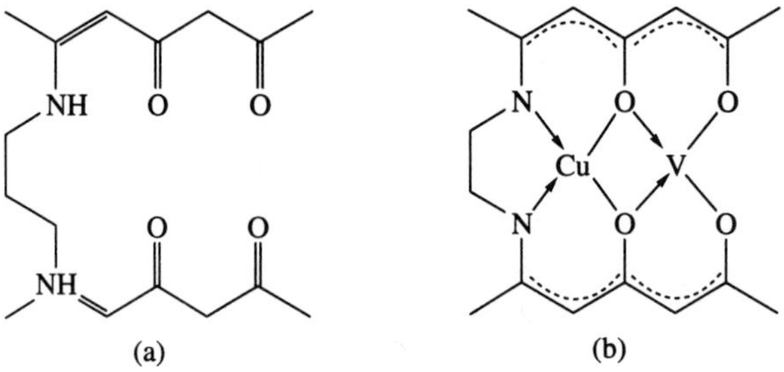

(a)　　　　　　　　　　(b)

(9-XXXV)

9-24　Azides and Other Anionic Ligands

Azides[201]

The N_3^- ion can give terminal η^1 and several types of bridges:

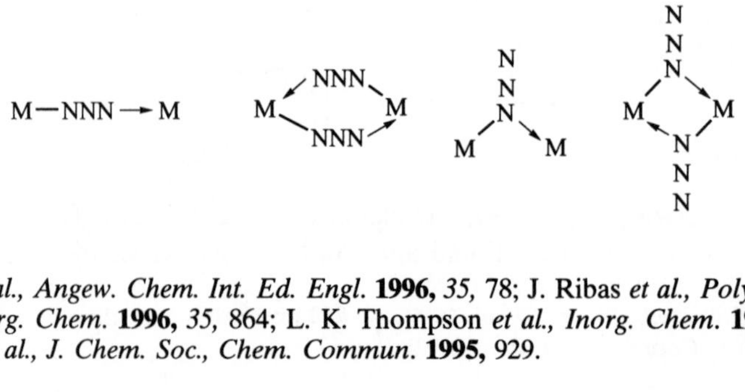

[201]T. Rojo *et al.*, *Angew. Chem. Int. Ed. Engl.* **1996**, *35*, 78; J. Ribas *et al.*, *Polyhedron* **1996**, *15*, 1091; *Inorg. Chem.* **1996**, *35*, 864; L. K. Thompson *et al.*, *Inorg. Chem.* **1995**, *34*, 2356; P. D. Beer *et al.*, *J. Chem. Soc., Chem. Commun.* **1995**, 929.

The complexes are usually made by interaction of NaN_3 or AgN_3 with metal halide or halide complexes. Azides of halogens and other non-metals are well known in addition to metal complexes; examples are IN_3 and $Sb(N_3)_3$.[202]

Organic azides, RN_3, usually react losing N_2 to give imido compounds (Section 9-16), but an azide intermediate has been isolated and characterized in the reaction[203]:

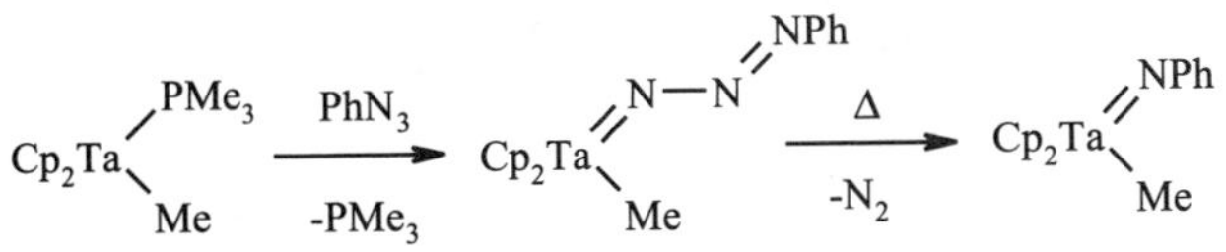

A similar sequence[204] for a vanadium azido complex is

$$L_nV(N_3)\text{mes} \xrightarrow[-N_2]{\Delta} L_nV N\text{mes}$$

Thiocyanate and Cyanate[205]

The NCS^- ion is ambidentate, binding through N or S or as a bridge $M-NCS-M$. The distinction between N and S bonding can be made by ^{14}N nuclear quadrupole studies and by MAS nmr ^{13}C spectra.

The bonding depends on the metal, steric factors, and the nature of other ligands present. Heavier metals are commonly S-bonded and in such cases $M-SCN$ is usually bent with an angle $\sim 110°$ but N-bonded species may also have nonlinear $M-N-C$ groups.

Pseudo Allyl Ligands

These ligands are similar to carboxylates, $RC(O)O^-$, but of the form $RNXNR^-$:

X = N, 1,3-triazenido[206]

X = CR', amidinato or amidino[207]

X = SR', iminoaminosulfinato[208]

X = PR$_2$, iminoaminophosphoranato[209]

[202]T. M. Klapötke et al., J. Chem. Soc., Dalton Trans. **1996**, 2895.
[203]G. Proulx and R. G. Bergman, Organometallics **1996**, 15, 684; R. G. Bergman et al., Angew. Chem. Int. Ed. Engl. **1996**, 35, 653.
[204]C. C. Cummings, J. Am. Chem. Soc. **1995**, 117, 6384.
[205]A. M. Golub et al., The Chemistry of Pseudohalides, Elsevier, Amsterdam, 1986; R. J. H. Clark and D. G. Humphrey, Inorg. Chem. **1996**, 35, 2053.
[206]H. G. Ang et al., J. Chem. Soc. Dalton Trans. **1996**, 1573; D. S. Moore and S. D. Robinson, Adv. Inorg. Chem. **1986**, 30, 1; N. G. Connelly et al., J. Chem. Soc., Dalton Trans. **1994**, 2025.
[207]J. R. Hagadorn and J. Arnold, J. Am. Chem. Soc. **1996**, 118, 893; Organometallics **1996**, 15, 984; F. A. Cotton et al., Inorg. Chem. **1996**, 35, 498; J. Am. Chem. Soc. **1996**, 118, 4830.
[208]D. Stalke et al., J. Chem. Soc., Dalton Trans. **1993**, 3479.
[209]A. Steiner and D. Stalke, Inorg. Chem. **1993**, 32, 1977.

2-Substituted Pyridines

These ligands arise from compounds such as 2-methylaminopyridine,[210] 2-hydroxy-pyridine,[211] and 2-thiolates.[212] The thiolates have been much studied and the bonds of the following types are typical:

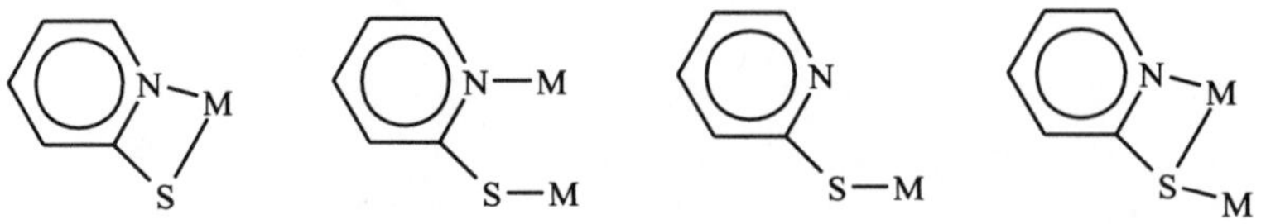

9-25 Miscellaneous Ligands

Organic Isocyanates, RNCO. These are normally bound η^2-N,C.[213]

Nitrogen-Sulfur Ligands. Many types of these are known for transition metals. Two examples are the heterocyclic dianion[214] $Ph_4P_2N_4S_2^{2-}$ and the sulfenamido anion $RNSR'^-$; the latter forms compounds such as $Zr(\eta^2\text{-}Bu^tNSPh)_4$ by interaction of $ZrCl_4$ with $Bu^tN(Li)SPh$.[215]

Carbodiimides[216]. The $RN{=}C{=}NR$ groups can bind to metals in several ways, but normally the binding is η^1-N or η^2-NC.

Iminoacyls[217]. These are formed by isocyanide, RNC, insertions into MH, metal alkyl or -aryl bonds. Although η^2-C,N bonding is most common, bridging groups are known.

Ureato and Carbamato Complexes. These can be made in a variety of ways such as the interaction of metal carbonyls with RNCO or RN_3. A more general method involves the sequence[218]:

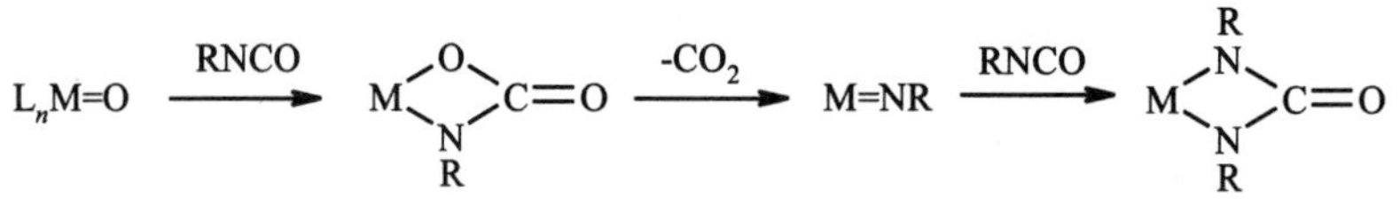

[210]R. Kempe *et al., Inorg. Chem.* **1996**, *35*, 2644, *Organometallics* **1996**, *15*, 1071.
[211]J. M. Rawson and R. E. P. Winpenny, *Coord. Chem. Rev.* **1995**, *139*, 313.
[212]J. Zubieta *et al., Inorg. Chem.* **1996**, *35*, 3548; A. R. Barron *et al., Polyhedron* **1996**, *15*, 391; J. G. Reynolds *et al., Inorg. Chem.* **1995**, *34*, 5745.
[213]P. Braunstein and D. Nobel, *Chem. Rev.* **1989**, *89*, 1927; A. Otero *et al., J. Chem. Soc., Dalton Trans.* **1995**, 1007.
[214]T. Chivers *et al., Inorg. Chem.* **1995**, *34*, 1681; *Coord. Chem. Rev.* **1994**, *137*, 201 (N,S ligand review).
[215]G. Wilkinson *et al., J. Chem. Soc., Dalton Trans.* **1996**, 1309.
[216]H. Werner *et al., Organometallics* **1993**, *12*, 1775.
[217]E. Carmona *et al., J. Am. Chem. Soc.* **1995**, *117*, 1759.
[218]G. Wilkinson *et al., J. Chem. Soc., Dalton Trans.* **1993**, 781 and references therein.

Additional References

C. W. Allen, *Organophosphorus Chem.* **1992,** *23,* 313 (298 references); phosphazine complexes of transition metals.

M. R. A. Blomberg, P. E. M. Siegbahn, and M. Svennson, *Inorg. Chem.* **1993,** *32,* 4218; theoretical study of activation of N—H bonds in NH_3 by metals.

N. M. Doherty *et al., Inorg. Chem.* **1994,** *33,* 4360; phosphoraminato ligands, NPR_3^-.

A. J. L. Pombeiro, *New J. Chem.* **1994,** *18,* 163; coordination chemistry of unsaturated molecules, RCN, NO, $NCNH_2$, N=CHR (review).

A. Togni, and L. M. Venanzi, *Nitrogen Donors in Organometallic Chemistry and Homogeneous Catalysis, Angew. Chem. Int. Ed. Engl.* **1994,** *33,* 497 (170 references).

J. D. Woollins, *J. Chem. Soc., Dalton Trans.* **1996,** 2893; metallocycles with P, N, S/Se compounds *N*-bonded to transition metals.

H. Zollinger, *Diazo Chemistry II,* VCH, New York, 1995. Aliphatic inorganic and organometallic compounds.

Chapter 10

THE GROUP 15 ELEMENTS: P, As, Sb, Bi

GENERAL REMARKS

10-1 Group Trends and Stereochemistry

The electronic structures and some other properties of the elements in Group 15 are listed in Table 10-1. The valence shells have a structure formally similar to that of nitrogen, but beyond the stoichiometries of some of the simpler compounds—NH_3, PH_3, NCl_3, $BiCl_3$, for example—there is little resemblance between the characteristics of compounds of these elements and those of nitrogen.

The only naturally occurring isotope of phosphorus (^{31}P) has a nuclear spin of ½ and a large magnetic moment. Nuclear magnetic resonance spectroscopy has accordingly played an extremely important role in the study of phosphorus compounds.[1] For antimony, the isotope ^{121}Sb is suitable for Mössbauer spectroscopy.

The elements P, As, Sb, and Bi show a considerable range in chemical behavior. There are fairly continuous variations in certain properties and characteristics, although in several instances, for example, in the ability of the pentoxides to act as oxidizing agents, there is no regular trend. Phosphorus, like nitrogen, is essentially covalent in all its chemistry, whereas As, Sb, and Bi show increasing tendencies to cationic behavior. Although the electronic structure of the next noble gas could be achieved by electron gain, considerable energies are involved (e.g., ~1450 kJ mol^{-1} to form P^{3-} from P); thus significantly ionic compounds such as Na_3P are few. The loss of valence electrons is similarly difficult to achieve because of the high ionization enthalpies. The 5+ ions do not exist, but for trivalent antimony and bismuth cationic behavior does occur. Bismuth trifluoride seems predominantly ionic, and salts such as $Sb_2(SO_4)_3$ and $Bi(NO_3)_3 \cdot 5H_2O$, as well as salts of the oxo ions SbO^+ and BiO^+, exist.

Some of the more important trends are shown by the oxides, which change from acidic for phosphorus to basic for bismuth, and by the halides, which have increasingly ionic character: PCl_3 is instantly hydrolyzed by water to $HPO(OH)_2$, and the other trihalides give initially clear solutions that hydrolyze to As_2O_3, SbOCl, and BiOCl, respectively. There is also an increase in the stability of the lower

---

[1]L. D. Quin and J. G. Verkade, Eds., *Phosphorus-31 NMR Spectral Properties in Compound Characterization and Structural Analysis,* VCH Publishers, 1994.

Table 10-1 Some Properties of P, As, Sb, and Bi

Property	P	As	Sb	Bi
Electronic structure	$[Ne]3s^23p^3$	$[Ar]3d^{10}4s^24p^3$	$[Kr]4d^{10}5s^25p^3$	$[Xe]4f^{14}5d^{10}6s^26p^3$
Sum of 1st three ionization enthalpies $[(kJ\ mol^{-1})/10^3]$	5.83	5.60	5.05	5.02
Electronegativity[a]	2.06	2.20	1.82	1.67
Radii (Å)				
Ionic	$2.12(P^{3-})$		$0.92(Sb^{3+})$	$1.08(Bi^{3+})$
Covalent[b]	1.10	1.21	1.41	1.52
Melting point (°C)	44.1 (α-form)	814 (36 atm)	630.5	271.3

[a]Allred-Rochow type.
[b]For trivalent state.

oxidation state with increasing atomic number; thus Bi_2O_5 is the most difficult to prepare and the least stable pentoxide.

Oxidation states or oxidation numbers of III ($+3$) or V ($+5$) can be, and often are, assigned to these elements in their compounds and are useful for purposes of classification. However, they are purely formal. What matters is the number of covalent bonds formed and the stereochemistries. The general types of compound and stereochemical possibilities are given in Table 10-2.

There are few reported examples of an oxidation state other than III or V. The only simple ionic one concerns univalent bismuth, obtained by dissolving the metal in a solution of $BiCl_3$ in HCl. There is evidence that Bi^+ ions are present[2] but little else is known.

Table 10-2 Major Stereochemical Patterns for Compounds of Group 15

Formal Valence	Number of Bonds Formed	Geometry[a]	Examples
1	3	Trigonal plane	$PhP[Mn(CO)_2Cp]_2$
3	2	Angular	PH_2^-, $(Me_2N)_2P^+$, $R_2N-P=NR$, $R-P=CR_2$
	3	Pyramidal	PH_3, $AsCl_3$, $SbPh_3$
	4	Tetrahedral	PH_4^+, $P(CH_2OH)_4^+$, $AsPh_4^+$
		$\psi\ tbp$	KSb_2F_7, $SbCl_3(PhNH_2)$, $K_2[Sb_2(tart)_2]\cdot3H_2O$, $SbOCl$
	5	ψ octahedral	SbF_5^{2-}, $[Sb_4F_{16}]^{4-}$, Sb_2S_3, $SbCl_3(PhNH_2)_2$
	6	Octahedral	$[Bi_6O_6(OH)_3]^{3+}$, $[SbBr_6]^{3-}$
5	3	Planar	$P[N(SiMe_3)_2](NSiMe_3)_2$
	4	Tetrahedral	PCl_4^+, $(RO)_3PO$, $RPO(OH)_2$
	5	tbp	PF_5, AsF_5, $SbCl_5$, $AsPh_5$
		sp	$SbPh_5$
	6	Octahedral	PF_6^-, $Sb(OH)_6^-$, $SbBr_6^-$

[a]ψ means with one position occupied by a lone pair of electrons.

[2]S. Ulvenlund and L. A. Bengtsson, *Acta Chem. Scand.* **1994**, *48*, 635.

The differences between the chemistries of N and P, which are due to the same factors that are responsible for the C—Si and O—S differences, can be summarized as follows:

Nitrogen	*Phosphorus*
(a) Very strong $p\pi$-$p\pi$ bonds	Weak $p\pi$-$p\pi$ bonds
(b) $p\pi$-$d\pi$ bonding is rare	Weak to moderate but important $d\pi$-$p\pi$ and $d\pi$-$d\pi$ bonding
(c) No valence expansion	Valence expansion

Point (a) leads to facts such as the existence of $P(OR)_3$ but not of $N(OR)_3$, nitrogen giving instead $O{=}N(OR)$, and the marked differences between the oxides and oxo acids of nitrogen on the one hand and those of phosphorus on the other.

Point (b) is associated with rearrangements such as

$$\begin{array}{c}\diagdown\\ \diagup\end{array}\!\!P{-}OH \;\rightleftharpoons\; H{-}\!\!\begin{array}{c}\diagdown\\ \diagup\end{array}\!\!P{=}O$$

and with the existence of phosphonitrilic compounds $(PNCl_2)_n$. Furthermore, although PX_3, AsX_3, and SbX_3 (X = halogen, alkyl or aryl), like NR_3 compounds, behave as donors owing to the presence of lone pairs, there is one major difference. The nitrogen atom can have no function other than simple donation, because no other orbital is accessible, whereas P, As, and Sb have empty d orbitals of fairly low energy. Thus when the atom to which the P, As, or Sb donates has electrons in orbitals of the same symmetry as the empty d orbitals, back-donation resulting in overall multiple-bond character may result. This factor is especially important for the stability of complexes with transition elements where $d\pi$-$d\pi$ overlap contributes substantially to the bonding.

The consequences of vacant d orbitals are also evident on comparing the amine oxides, R_3NO, with R_3PO or R_3AsO. In the *N*-oxide the electronic structure can be represented by the single canonical structure $R_3N^+{-}O^-$, whereas for the others the bonds to oxygen have multiple character and are represented as resonance hybrids:

$$R_3P^+{-}O^- \longleftrightarrow R_3P{=}O \longleftrightarrow R_3P^-{\equiv}O^+$$

These views are substantiated by the shortness of the P—O bonds ($\sim$1.45 Å as compared with $\sim$1.6 Å for the sum of the single-bond radii) and by the normal bond lengths and high polarities of N—O bonds. The amine oxides are also more chemically reactive, the P—O bonds being very stable indeed, as would be expected from their strength, $\sim$500 kJ mol^{-1}.

Point (c) is responsible for phenomena such as the Wittig reaction (Section 10-14) and for the existence of compounds such as $P(C_6H_5)_5$, $P(OR)_5$, $[P(OR)_6]^-$, and $[PCl_4]^+[PCl_6]^-$ in which coordination numbers 5 or 6 occur. The extent to which the $3d$ orbitals are involved is somewhat uncertain, since the d levels are rather high in energy and the higher coordination numbers may depend considerably on electrostatic forces; it is significant that the higher coordination numbers for P^V are most readily obtained with more electronegative groups such as halogens, OR, or phenol.

Stereochemistry with Formal Valence Three

The overwhelming majority of compounds having these elements in formal valence state three have three pyramidally directed single bonds, with a lone pair occupying the fourth tetrahedral position. The tetrahedron can be completed by "quaternization" (formal addition of R^+ or H^+) to give cations, for example, PR_4^+.

An inversion process is possible for all EX_3 molecules,[3,4] but the way in which it occurs is not necessarily by so-called vertex inversion, as in the case of amines. For the hydrides, it is believed that vertex inversion is the preferred pathway and the barrier (in kJ mol^{-1}) increases down the group: NH_3, 24; PH_3, 155; AsH_3, 170; SbH_3, 176. On the other hand, with the EF_3 molecules, the preferred pathway (except for NF_3) is so-called edge inversion where the transition state is a T-shaped structure, and the barriers are ordered in the opposite sense: PF_3, 221; AsF_3, 191; SbF_3, 160. In general, inversion barriers for all EX_3 molecules are above about 125 kJ mol^{-1} so that inversion should be effectively stopped at room temperature. Thus, optical isomers of $EXX'X''$ species might be isolable, but as yet, with X, X', and X'' being halogens, no such compounds have been made. However, for organo compounds of the types $ERR'R''$ and $EXRR'$ many examples of resolution have been reported.[5]

The only stable EH_4^+ ion is PH_4^+, but only in the iodide, formed as colorless crystals on mixing gaseous PH_3 and HI. The chloride and bromide exist only at very low temperatures. The PH_4^+ ion is completely hydrolyzed by the much stronger base, water:

$$PH_4I(s) + H_2O \longrightarrow H_3O^+ + I^- + PH_3(g)$$

There are, however, numerous and important R_4E^+ compounds, mostly made by a straightforward reaction of the type

$$R_3E + RX \longrightarrow [R_4E^+]X^-$$

All R_4E^+ ions are tetrahedral. With Bi only the Ph_4Bi^+ and Me_4Bi^+ ions are known.

There are a few examples of two-coordination with phosphorus, as in PH_2^-, $(R_2N)_2P^+$, and others listed in Table 10-2.

With higher coordination numbers the question of the stereochemical activity of the lone pair arises. In many cases ψ-tbp and ψ-octahedral species are observed, but in some cases the lone pair seems to occupy an s orbital and not compete with the bonding pairs for a position in the coordination sphere (see Section 10-6).

Stereochemistry with Formal Valence Five

An interesting feature in the stereochemistry of the five-valent compounds is that the tbp and sp configurations for the five-connected species differ very little in

[3]D. A. Dixon and A. J. Arduengo, III, *J. Am. Chem. Soc.* **1987,** *109,* 338.
[4]P. Schwerdtfeger *et al., J. Am. Chem. Soc.* **1994,** *116,* 9620.
[5]S. B. Wild *et al., Inorg. Chem.* **1996,** *35,* 1244.

energy. The *tbp* is nearly always the more stable, but not by a great deal, for EX_5 molecules in which all X are separate groups (i.e., not connected to give rings). A series of ER_5 (E = As, Sb, Bi) molecules furnishes examples.[6] As a general rule *sp* structures are favored by the presence of two unsaturated five-membered rings (10-I). The similar energies of the *tbp* and *sp* configurations provide a pathway for stereochemical nonrigidity, *via* the Berry pseudorotation path (Section 1-3).

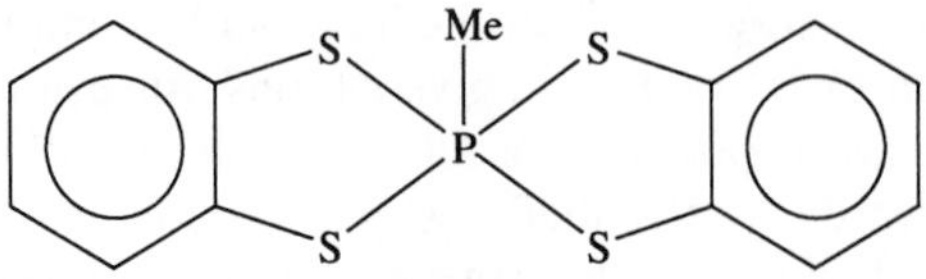

(10-I)

One general rule for mixed species (EX_nY_{5-n}) with *tbp* structures is that the more electronegative groups prefer the axial sites (Section 10-5).

Planar P, As, and Sb Compounds

There are unusual compounds in which the elements, formally in the I oxidation state, form trigonal planar compounds of type (10-II); $d\pi$-$p\pi$ interaction is doubtless involved in the bonding. In another (10-III) the phosphorus is formally pentavalent.

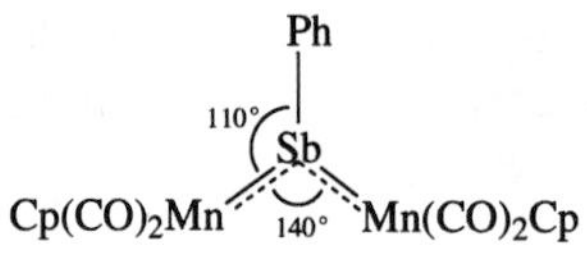

(10-II) 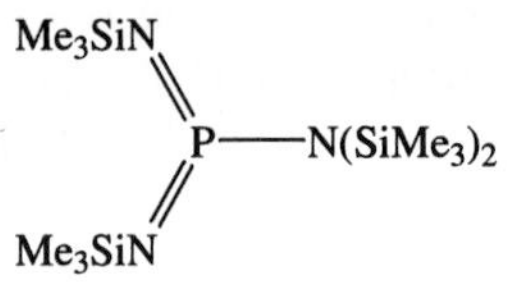   (10-III)

Phosphorus Radicals

The action of light on an R_2PX molecule in the presence of an activated alkene as halogen acceptor can lead to R_2P radicals, namely,

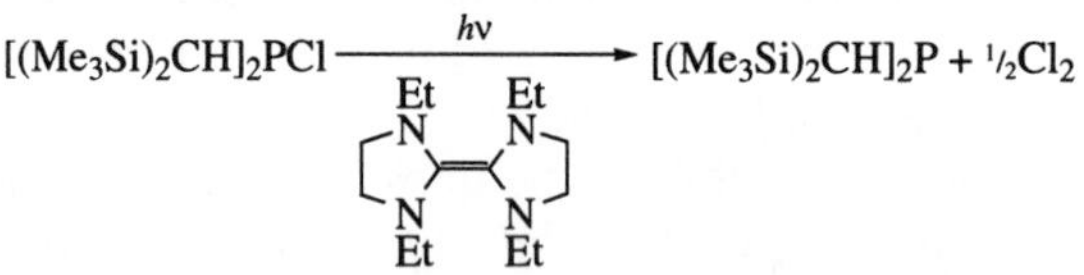

The stability of the radical is connected with the bulk of the bis(trimethylsilyl)-methyl group.

10-2 The Elements

All are found in Nature as compounds except for minute amounts of elemental arsenic and bismuth. Their abundances decrease in order of increasing atomic

[6]S. Wallenhauer and K. Seppelt, *Inorg. Chem.* **1995**, *34*, 116.

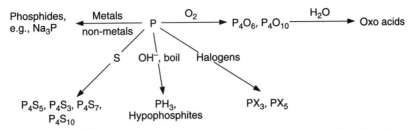

Figure 10-1 Some typical and important reactions of red and white phosphorus.

number with phosphorus being considerably more abundant than all of the others combined. Though more than 200 phosphate minerals are known, only those of the apatite type, for example, *fluoroapatite*, $[3Ca_3(PO_4)_2 \cdot Ca(F, Cl)_2]$, are commercial sources. Arsenic and antimony are commonly associated with sulfide minerals, particularly those of copper, lead, and silver. The main bismuth ores are Bi_2O_3 and various sulfides.

Phosphorus

Commercial production is by reduction of phosphate rock with coke and silica in an electric furnace. The element volatilizes and is condensed under water as white phosphorus, a soft, white, waxy solid (mp 44.1°C, bp 280°C):

$$2Ca_3(PO_4)_2 + 6SiO_2 + 10C \longrightarrow P_4 + 6CaSiO_3 + 10CO$$

Most of the phosphorus produced is converted to phosphoric acid for use in making fertilizer, or in other commercial processes. Since the conversion of the element to P_2O_5 or H_3PO_4 is very exothermic, a process in which the heat from this second step is used in place of electricity to heat the kiln in which the reduction is carried out has been developed.

There are three main forms of phosphorus—white, black, and red. Numerous other allotropes, some of dubious validity, have been claimed.

White phosphorus in the liquid and solid forms[7] consists of tetrahedral P_4 molecules ($P-P$ = 2.21 Å), which persist in the vapor phase up to 800°C where measurable dissociation to P_2 molecules ($P \equiv P$ = 1.89 Å) begins. The process $P_4(g) \rightarrow 2P_2(g)$ is endothermic (217 kJ/mol), and proceeds cleanly in a unimolecular fashion without intermediates or other fragments.

Although white phosphorus is the allotrope to which all others convert when melted, at 1 atm pressure, it is highly reactive and toxic, and is commonly stored under water to protect it from air. It is soluble in organic solvents such as CS_2 and benzene and by virtue of the lone pairs projecting from each P atom, the P_4 molecule can serve as a kind of phosphine ligand (Section 10-17). Some reactions of white (and red, *vide infra*) phosphorus are shown in Fig. 10-1.

The stability of the P_4 molecule, which appears to be very strained relative to a less strained cubic P_8 molecule, has provoked theoretical studies. These show, counter-intuitively, that a $2P_4 \rightarrow P_8$ conversion is disfavored enthalpically as well

[7]A. Simon *et al., Chem. Ber.* **1997,** *130,* 1235.

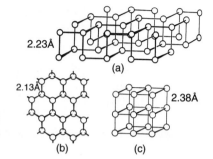

Figure 10-2 Three forms of black phosphorus. (a) The orthorhombic form (one double layer). (b) The rhombohedral form. (c) The cubic form.

as entropically. The ready formation of white phosphorus, despite its being thermodynamically unstable relative to the red and black allotropes, has also been the subject of theoretical studies.[8]

Orthorhombic black phosphorus, the most thermodynamically stable and least reactive form, is obtained by heating white phosphorus under pressure. It has a graphitic appearance and consists of double layers as shown in Fig. 10-2(a). When subjected to pressures above 12 kbar, the orthorhombic form transforms successively to the rhombohedral and cubic forms, Fig. 10-2(b) and (c).

Red phosphorus is of intermediate reactivity, relatively non-toxic, and is used commercially. It is easily obtained by heating white phosphorus in a sealed vessel at ~400°C. X-ray structural characterization of "red phosphorus" is not possible because it is amorphous. It also appears that there are several structural motifs, but all of them consist of P_8, P_9, or P_{10} polyhedra linked into long chains by P_2 units.[9]

Arsenic, Antimony, and Bismuth

Extractive methods are varied depending on the ore. For arsenic, industrial scale recovery is by thermolysis of the ore FeAsS to FeS and As. For antimony, a typical process is reaction of Sb_2S_3 with Fe to give FeS + Sb.

These elements have fewer allotropic forms than phosphorus. For As and Sb, unstable yellow allotropes comparable to white phosphorus are obtainable by rapid condensation of vapors. They readily transform to the bright, "metallic" α-rhombohedral forms similar to rhombohedral black phosphorus. This is also the commonest form for bismuth. Other reported allotropes are not well characterized.

The elements, like phosphorus, all combine readily with O_2, and they react readily with halogens and some other nonmetals. They are unaffected by dilute nonoxidizing acids, but with nitric acid As gives arsenic acid, Sb gives the trioxide and Bi dissolves to form the nitrate.

BINARY COMPOUNDS

10-3　Phosphides, Arsenides, Antimonides, and Bismuthides

The Group 15 elements (E) form binary compounds with virtually every metallic element in the Periodic Table. The compounds vary enormously in their chemical

[8]S. Böcher and M. Häser, *Z. anorg. allg. Chem.* **1995,** *621,* 258.
[9]H. Hartl, *Angew. Chem. Int. Ed. Engl.* **1995,** *34,* 2637.

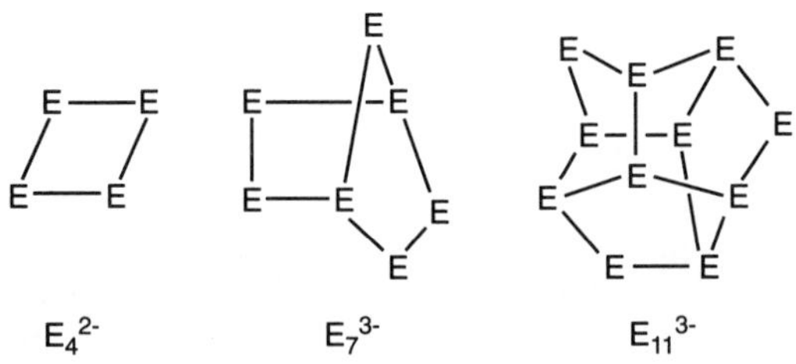

E_4^{2-} E_7^{3-} E_{11}^{3-}

Figure 10-3 Three types of Zintl ions formed by Group 15 elements.

and physical properties, and their structures range from simple ones (e.g., many ME compounds have the zinc blende structure) to very elaborate ones containing chains, sheets, spirals, and polyhedra formed by the nonmetallic element. There are some compounds that have formulas consistent with the presence of E^{3-} ions (e.g., Sr_3P_2 and Na_3As) but they are metallic rather than ionic in character. Most are prepared simply by reaction of stoichiometric quantities of the elements.

The transition metals generally form compounds that have compositions ME_n, where (n = 1, 2, or 3). Most of them are hard, insoluble, relatively unreactive, and semiconducting. Typical are $CoAs_3$ (*skutterudite*), which has a cubic structure in which square As_4 rings are found, and NiAs, in which each As is in a trigonal prism of Ni atoms while each Ni atom is in the center of a hexagonal bipyramid of six As and two Ni atoms. This type of structure has a pronounced tendency to be non-stoichiometric and phases with compositions approaching ME_2 can be obtained.

For As, Sb, and Bi, there is an important series of compounds formed with the Group 13 elements. These are commonly called III-V compounds (based on the traditional group numbering in the Periodic Table) and they have important applications as semiconductors. To illustrate, GaAs is isoelectronic to silicon, but a greater range of properties can be achieved because composition can be more easily varied. The disadvantage of III-V compounds relative to silicon based semiconductors is that they are reactive towards moisture and must be encapsulated.

Zintl Anions

The Group 15 elements all react with alkali and alkaline earth elements to form a considerable number of these which contain E_n^{m-} anions. Characterization of the E_n^{m-} ions has often been accomplished by dissolving the solids in liquid NH_3 or $H_2NCH_2CH_2NH_2$ and adding 2,2,2-crypt to complex the cations. The resulting crystalline materials, such as $[K(2,2,2\text{-crypt})]_2Bi_4$, can then be examined by X-ray crystallography. In this way, it has been shown that the following principal types of Group 15 Zintl anions exist.

1. Square E_4^{2-} for As, Sb, and Bi (Fig. 10-3).
2. The E_7^{3-} ions (Fig. 10-3). Note that P_4S_3 is isostructural and isoelectronic with P_7^{3-}.
3. As_5^{3-}, As_6^{4-}, P_{11}^{3-}, and As_{11}^{3-} are all known.[10]

It has recently been shown that structural information can be obtained directly on solutions of the Zintl phases by EXAFS. Zintl anions need not be discrete, as

[10]N. Korber and J. Daniels, *Polyhedron* **1996,** *15*, 2681; N. Korber and H. G. von Schnering, *Chem. Ber.* **1996,** *129*, 155.

shown by KSb_2, in which there are ribbons of fused Sb_6 chair conformers with K^+ ions in between.[11]

10-4   Hydrides

The gases EH_3 can be obtained by treating phosphides or arsenides of electropositive metals with acids or by reduction of sulfuric acid solutions of arsenic, antimony, or bismuth with an electropositive metal or electrolytically. The stability falls rapidly down the group, so the SbH_3 and BiH_3 are very unstable thermally, the latter having been obtained only in traces. The average bond energies are in accord with this trend in stabilities: E_{N-H}, 391; E_{P-H}, 322; E_{As-H}, 247; and E_{Sb-H}, 255 kJ mol^{-1}.

Phosphine (PH_3) is readily obtained by the action of dilute acid on calcium or aluminum phosphide, by pyrolysis of H_3PO_3 or, in a purer state, by the action of KOH in PH_4I. On an industrial scale, PH_3 is made by the action of NaOH or KOH on white phosphorus, a reaction which also forms alkali hypophosphite under other conditions (Section 10-12). One of the remarkable reactions of PH_3 is its reaction with formaldehyde in HCl solution:

$$PH_3 + 4CH_2O + HCl \longrightarrow [P(CH_2OH)_4^+]Cl^-$$

The PH_3 molecule is pyramidal with an HPH angle of 93.7°. Phosphine, when pure, is not spontaneously flammable, but often inflames owing to traces of P_2H_4 or P_4 vapor. It is readily oxidized by air when ignited, and explosive mixtures may be formed. It is also exceedingly poisonous. Unlike NH_3, it is not associated in the liquid state and it is only sparingly soluble in water; pH measurements show that the solutions are neither basic nor acidic—the acid constant is $\sim 10^{-29}$ and the base constant $\sim 10^{-26}$.

The proton affinities of PH_3 and NH_3 (eq. 10-1) differ considerably.

$$EH_3(g) + H^+(g) = EH_4^+(g)$$

$$\Delta H^0 = -770 \text{ kJ mol}^{-1} \text{ for E = P} \tag{1}$$

$$\Delta H^0 = -866 \text{ kJ mol}^{-1} \text{ for E = N}$$

Also, the barrier to inversion for PH_3 is 155 kJ mol^{-1} as compared with only 24 kJ mol^{-1} for NH_3. Quite generally the barriers for R_3P and R_3N compounds differ by about this much. Like other PX_3 compounds, PH_3 (and also AsH_3) forms complexes with transition metals, e.g., *cis*-$Cr(CO)_3(PH_3)_3$.

Arsine (AsH_3) is extremely poisonous. Its ready thermal decomposition to arsenic, which is deposited on hot surfaces as a mirror, is utilized in tests for arsenic, for example, the well-known Marsh test, where arsenic compounds are first reduced by zinc in HCl solution.

Stibine is very similar to arsine but even less stable. *Bismuthine* is too unstable to be of any importance.

[11]S. M. Kauzlarich *et al., Inorg. Chem.* **1995,** *34,* 6218.

All these hydrides are strong reducing agents and react with solutions of many metal ions, such as Ag^I and Cu^{II}, to give the phosphides, arsenides, or stibnides, or a mixture of these with the metals.

Phosphorus alone forms other hydrides. *Diphosphine* (P_2H_4) is obtained along with phosphine by hydrolysis of calcium phosphide and can be condensed as a yellow liquid (mp $-99°C$). It is spontaneously flammable and decomposes on storage to form polymeric, amorphous yellow solids, insoluble in common solvents and of stoichiometry approximating to, but varying around, P_2H. Unlike N_2H_4, diphosphine has no basic properties. On photolysis it gives P_3H_5. It exists mainly in the gauche conformation.

10-5 Halides

The major types are the trihalides (EX_3), the pentahalides (EX_5), and the dielement tetrahalides (E_2X_4). We shall discuss them in that order. All trihalides except PF_3 are best obtained by direct halogenation, keeping the element in excess, whereas the pentahalides may all be prepared by reaction of the elements with excess halogen. Special methods are used for E_2X_4 compounds.

Trihalides

These are listed in Table 10-3; all 16 are known, but only a few merit discussion.

All the trihalides are rapidly hydrolyzed by water and are rather volatile. The gaseous molecules have pyramidal structures, and some form molecular lattices. Mixed trihalides can be detected in mixtures, but it appears unlikely, despite some older claims, that any can be isolated in pure form since equilibria, such as

$$PCl_3 + PBr_3 \rightleftharpoons PCl_2Br + PClBr_2$$

are labile.

Phosphorus trifluoride is a colorless gas, best made by fluorination of PCl_3 with AsF_3 or ZnF_2. It forms complexes with transition metals similar to those formed by carbon monoxide. Like CO, it is highly poisonous because of the formation of a hemoglobin complex. Unlike the other trihalides, PF_3 is hydrolyzed only slowly by water, but it is attacked by alkalis.

Phosphorus trichloride has an extensive chemistry (see Fig. 10-4) and is made on an industrial scale. In addition to the reactions shown in Fig. 10-4, PCl_3 reacts with the AgCN to give $P(CN)_3$.

Table 10-3 The Trihalides of the Group 15 Elements (mp and bp in °C)

Fluorides			Chlorides			Bromides			Iodides		
PF_3	bp	-101.8	PCl_3	bp	76.1	PBr_3	bp	173.2	PI_3	mp	61.2
AsF_3	bp	62.8	$AsCl_3$	bp	103.2	$AsBr_3$	mp	31.2	AsI_3	mp	140.0
							bp	221.0			
SbF_3	mp	292.0	$SbCl_3$	mp	73.17	$SbBr_3$	mp	97.0	SbI_3	mp	171.0
BiF_3	mp	725.0	$BiCl_3$	mp	233.5	$BiBr_3$	mp	219.0	BiI_3	mp	408.6

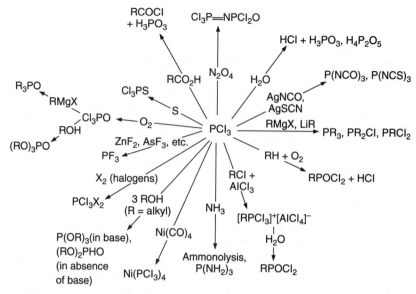

Figure 10-4 Some important reactions of PCl$_3$. Many of these are typical for other MX$_3$ compounds as well as for MOX$_3$ compounds.

Phosphorus triiodide has been shown to be a useful oxygen atom abstracting agent converting R$_2$SO to R$_2$S and RCH$_2$NO$_2$ to RCN, for example.

Arsenic, antimony, and bismuth trihalides comprise 12 compounds that exhibit diversity in their physical and chemical properties, as well as considerable variation in their structures. Bismuth trifluoride has an ionic lattice (like that of YF$_3$) with 9-coordination of the bismuth. SbF$_3$ has an intermediate structure in which SbF$_3$ molecules (Sb—F = 1.92 Å) are linked through F bridges (Sb···F, 2.61 Å) to give each Sb^{III} a very distorted octahedral environment. BiCl$_3$ also has a quasi-molecular structure. Others, such as AsF$_3$, AsCl$_3$, AsBr$_3$, SbCl$_3$, and SbBr$_3$ are essentially molecular, and they give pyramidal EX$_3$ molecules readily in the vapor phase. For the iodides, the solids have close-packed arrays of I atoms with E atoms in octahedral interstices but located off-center so that incipient EI$_3$ molecules can be considered to exist.

Bismuth trichloride differs from other EX$_3$ compounds in being readily hydrolyzed by water to BiOCl, but this redissolves in concentrated aqueous HCl, and BiCl$_3$ is recovered on evaporation. Antimony trichloride has somewhat similar behavior.

Arsenic trifluoride and especially SbF$_3$ (called the Swarts reagent) are very useful reagents for fluorination of various nonmetallic substrates, as indicated by the following conversions:

$$RCl \longrightarrow RF$$

$$SiCl_4 \longrightarrow SiCl_3F, SiCl_2F_2, SiClF_3$$

$$R_3PS \longrightarrow R_3PF_2$$

Table 10-4 Pentahalides of the Elements P, As, Sb, and Bi

Fluorides	Chlorides	Bromide	Iodide
	A. Binary		
PF_5	PCl_5	PBr_5	PI_5
AsF_5	$AsCl_5$		
SbF_5	$SbCl_5$		
BiF_5			
	B. Mixed		
PCl_4F	PCl_3F_2	PBr_2F_3	
	$[AsCl_4][AsF_6]$		
$SbCl_4F$	$SbCl_3F_2, SbCl_2F_3$		

Arsenic trichloride and $SbCl_3$ also have some utility as nonaqueous solvents. It is doubtful that they undergo significant self-ionization (although this has been proposed, namely, to $ECl_2^+ + ECl_4^-$) but they have low viscosities, high dielectric constants, liquid ranges of ~150°C, and are good media for Cl^- transfer reactions.

The binary azides, $As(N_3)_3$ and $[As(N_3)_4]^+$, have recently been reported.[12] Both are highly explosive.

Pentahalides

As indicated in Table 10-4, the four pentafluorides are all well known. Phosphorus pentafluoride, which may be prepared by fluorinating PCl_5 with AsF_3 or CaF_2 is molecular, with a *tbp* structure in the solid state[13] and under all other conditions. It is a colorless gas (bp -102°C), which reacts readily with Lewis bases such as amines and ethers, as well as F^-, to form six-coordinate complexes. The fluxional character of the molecule, whereby the axial and equatorial F atoms change places rapidly on the nmr time scale is generally explained as a Berry pseudorotation.

Arsenic pentafluoride is generally similar to PF_5, but SbF_5 is markedly different. It is associated through F bridges even in the gas phase. In the liquid state it is extremely viscous due to the formation of linear polymers believed to be as shown in 10-IV and in the solid there are cyclic tetramers (10-V). Bismuth pentafluoride forms white needle crystals built of infinite linear chains of BiF_6 octahedra linked by trans bridges. The three heavier pentafluorides, especially BiF_5, are powerful fluorinating and oxidizing agents.

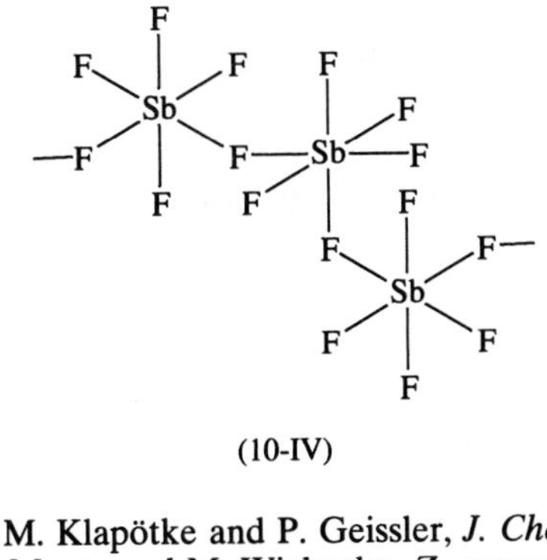

(10-IV) 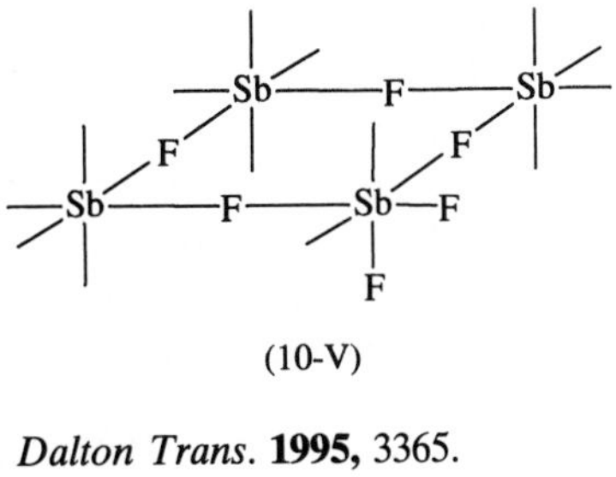   (10-V)

[12]T. M. Klapötke and P. Geissler, *J. Chem. Soc. Dalton Trans.* **1995**, 3365.
[13]D. Mootz and M. Wiebecke, *Z. anorg. allg. Chem.* **1987**, *545*, 39.

Arsenic pentafluoride and SbF_5, and, to a lesser extent PF_5, are potent fluoride ion acceptors, forming MF_6^- ions or more complex species. The PF_6^- ion is a common and convenient noncomplexing anion, which has even less coordinating ability than ClO_4^- or BF_4^-.

In liquid HF, PF_5 is a nonelectrolyte, but AsF_5 and SbF_5 give conducting solutions, presumably because of the reactions

$$2HF + 2AsF_5 \longrightarrow H_2F^+ + As_2F_{11}^-\ \text{(can be isolated as } Et_4N^+ \text{ salt)}$$

$$2HF + SbF_5 \longrightarrow H_2F^+ + SbF_6^-$$

$$SbF_6^- + SbF_5 \longrightarrow Sb_2F_{11}^-$$

The strong X^- acceptor capacity of the SbX_5 compounds leads to their use as Friedel-Crafts catalysts and to the ability of SbF_5 to enhance the acidities of HF(l) and HSO_3F(l).

The other pentahalides all have complex structural behavior centering around the existence of fairly stable EX_4^+ ions. Not all of the possible binary pentahalides are known, but there are quite a few mixed pentahalides. All of the binary ones and the better known mixed ones are listed in Table 10-4. Some of the important structural features will now be discussed.

Phosphorus pentachloride is molecular in the gas and liquid phases; when dissolved in nonpolar solvents, it may dimerize to a certain extent in some of them. However, it crystallizes as $[PCl_4^+][PCl_6^-]$, and when dissolved in polar solvents (e.g., MeCN and $PhNO_2$) there are two forms of ionization whose relative importance depends in an obvious way on concentration:

$$2PCl_5 = [PCl_4^+] + [PCl_6^-]$$

$$PCl_5 = [PCl_4^+] + Cl^-$$

There is also a metastable solid form consisting of $[PCl_4^+]_2[PCl_6^-]Cl^-$, which has an analogue with Br^- in place of Cl^-.[14] Compounds containing the PCl_4^+ and PCl_6^- ions form under other circumstances as well. For example, in the PCl_5-TiCl_4 system one can obtain $[PCl_4]_3[TiCl_6][PCl_6]$.

Phosphorus pentabromide appears to be wholly dissociated to $PBr_3 + Br_2$ in the gas phase and no liquid is known. The solid normally consists of $[PBr_4^+]Br^-$, but a form consisting of $[PBr_4^+][Br_3^-]PBr_3$ has been made by rapidly cooling the vapor at 15 K. The pentaiodide is not well known; it appears to be $[PI_4^+]I^-$ in solution.

Arsenic pentachloride can be prepared at low temperature but decomposes above $\sim-50°C$. Antimony pentachloride is made by reaction of Cl_2 with $SbCl_3$ and is stable up to $\sim140°C$. The solid (mp 4°C) contains $Cl_4Sb(\mu\text{-}Cl)_2SbCl_4$ dimers. The curious instability of $AsCl_5$ relative to PCl_5 and $SbCl_5$ has been attributed to a stabilization of the $4s^2$ electron pair in all the elements immediately following the

[14]P. N. Gates *et al., J. Chem. Soc., Dalton Trans.* **1995,** 2719.

first transition series. Other consequences of this stabilization are found in the chemistry of Ga, Ge, Se, and Br, generally having to do with lower stability of the highest oxidation states.

The nonexistence of $BiCl_5$ and of EBr_5 and EI_5 compounds other than those of phosphorus can be attributed to the inability of the formally uninegative halogens to coexist with the oxidation state +5 of the Group 15 elements (*cf.* the nonexistence of CuI_2 and TlI_3). There are many mixed halides, which are often ionic in nature. The halide PCl_4F has both a molecular and an ionic form, that is, $[PCl_4^+][PCl_4F_2^-]$. In the PCl_xF_{5-x} molecules, the chlorine atoms always prefer the equatorial and the fluorine atoms the axial positions, in line with the general rule that the more electronegative groups prefer the axial positions in PX_xY_{5-x} molecules.

In strong acid media mixed halo cations $[PX_nY_{4-n}]^+$ can be generated by reactions of PX_3 with Y_2.

Lower Halides

For P, As, and Sb, these are of the stoichiometry E_2X_4, and only some are known. The halides P_2Cl_4 and P_2Br_4 are least well characterized, while P_2F_4, P_2I_4, As_2I_4, and Sb_2I_4 are well defined. They all have an X_2E-EX_2 type of structure, and all are reactive. Diarsenic tetraiodide and Sb_2I_4 decompose on standing to $EX_3 + E$.

No Bi_2X_4 compound is known, but it has long been known that when metallic bismuth is dissolved in molten $BiCl_3$ a black solid of approximate composition BiCl can be obtained. This solid is $Bi_{24}Cl_{28}$, and it has an elaborate constitution, consisting of four $BiCl_5^{2-}$, one $Bi_2Cl_8^{2-}$, and two Bi_9^{5+} ions, the structures of which are depicted in Fig. 10-5. The electronic structure of the Bi_9^{5+} ion, a metal atom cluster, is best understood in terms of delocalized molecular orbitals. Other low-valent species present in various molten salt solutions are Bi^+, Bi^{3+}, Bi_3^{3+}, and Bi_8^{2+}. The last, in $Bi_8(AlCl_4)_2$, has a square antiprismatic structure.

Several intermediate antimony halides, such as $(SbF_3)_x(SbF_5)_y$, with $x = y = 1$; $x = 6$ and $y = 5$; $x = 2$ and $y = 1$; $x = 3$ and $y = 1$ are known. Their existence, and their structures, depend on the high fluoride ion affinity of SbF_5, whereby SbF_6^- and $Sb_2F_{11}^-$ ions and complex polymeric cations, such as $(Sb_3F_8^+)_\infty$ are formed.

10-6   Complexes in Oxidation State III

Halide Complexes

Phosphorus trihalides have virtually no Lewis acidity, though a PBr_4^- ion has been observed.

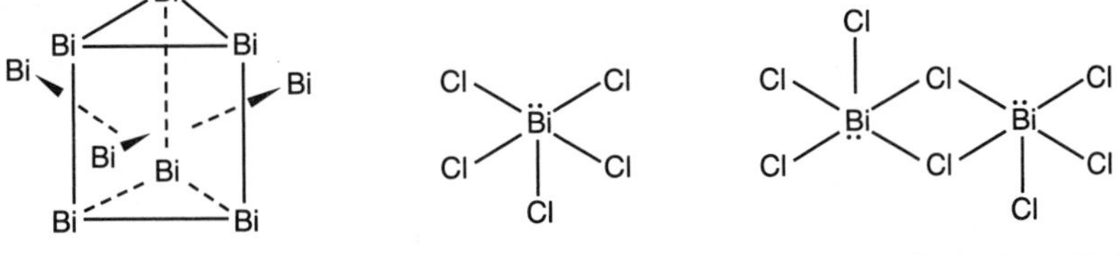

Figure 10-5   The structures of the species present in BiCl, which is, in fact, $Bi_{24}Cl_{28}$.

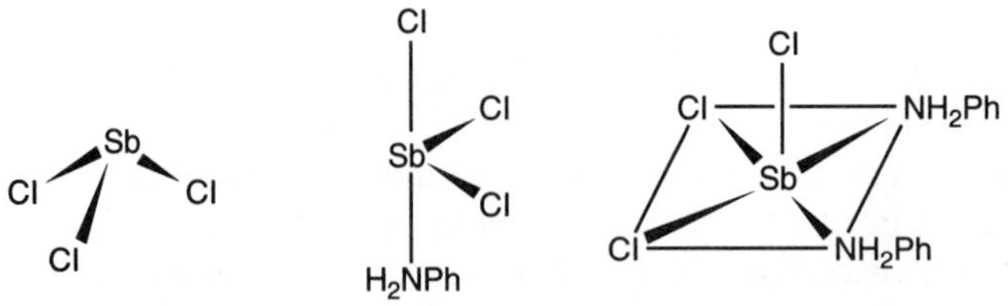

Figure 10-6   The structures of $SbCl_3$ and two of its complexes.

Arsenic(III) halides display limited but significant Lewis acidity and form the $As_2Cl_8^{2-}$, $As_2Br_8^{2-}$, and $(AsBr_4^-)_n$, anions. The first two consist of AsX_5 square pyramids sharing an edge (similar to $Bi_2Cl_8^{2-}$ as shown in Fig. 10-5), and the last is an infinite chain similar to that in $pyHSbCl_4$ to be discussed later.

It is with Sb^{III} and Bi^{III} that really extensive complex formation occurs. Antimony trichloride forms complexes with neutral donors; those with $PhNH_2$ have structures showing that the N—Sb bonds are relatively weak and that the lone pair is stereo-chemically active, as illustrated in Fig. 10-6. Antimony trichloride also forms 2:1 and 1:1 complexes with aromatic hydrocarbons such as naphthalene or *p*-xylene, where weak interaction between $SbCl_3$ and the π cloud occurs.

There are some EX_4^- containing compounds, where the EX_4^- ions have the same type of ψ-*tbp* structure as isoelectronic molecules such as SeF_4. On the other hand, it is very common for compounds whose formulas (e.g., $KSbF_4$) might suggest the presence of discrete EX_4^- ions to have much more complex anions resulting from association *via* μ-X groups. Similarly, while a few discrete MX_5^{2-} ions are found (with ψ-octahedral structures, like IF_5), association into larger, X-bridged aggregates is very common. Some representative examples are shown in Figs. 10-7 and 10-8.

There are also EX_6^{3-} complexes, and many of them have been shown to have regular octahedral structures, despite the fact that they have an "extra" pair of valence shell electrons, that is, a total of 14 rather than 12. Examples are $(NH_4)_2SbBr_6$, which is not an Sb^{IV} compound, but rather consists of $(NH_4)_4(Sb^{III}Br_6)(Sb^VBr_6)$ and various compounds containing discrete octahedral $BiBr_6^{3-}$ ions.[15] The probable explanation for this lack of a stereochemical effect due to the extra pair of electrons is that they occupy the *ns* orbital, which has been relativistically lowered in energy, and do not play a direct role in bonding. There are also $E_2X_9^{3-}$ species with confacial bioctahedral structures and even one, $Bi_3I_{12}^{3-}$, that has a linear chain of three face-sharing BiI_6 octahedra.[16]

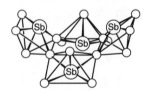

Figure 10-7   The structure of the $Sb_4F_{16}^{4-}$ ion in $KSbF_4$.

[15]F. Lazarini, *Acta Cryst.* **1985**, *C41*, 1617.
[16]N. C. Norman *et al., Z. anorg. allg. Chem.* **1995**, *621*, 47.

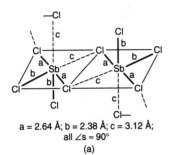
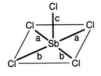

a = 2.64 Å; b = 2.38 Å; c = 3.12 Å;
all ∠s ≈ 90°

(a)

a = 2.58 Å; b = 2.69 Å;
c = 2.36 Å

(b)

Figure 10-8 (a) A portion of the anion chain in (pyH)SbCl$_4$. (b) The SbCl$_5^{2-}$ ion found in (NH$_4$)$_2$SbCl$_5$.

It is not easy to forecast when there will be a stereochemically active lone pair in these E^{III} complexes, but the following general trends can be noted. Stereochemical activity of the lone pair decreases with (1) increasing coordination numbers, (2) increasing atomic number of the halogen, and especially (3) increasing atomic number of E (i.e., As > Sb > Bi).

In support of these generalizations we note that in the [Sb$_8$I$_{28}$]$^{4-}$ ion,[15] six of the Sb atoms are octahedral 6-coordinate while two are ψ-octahedral. On the other hand, in two cases where Bi and I are the components, namely the infinite chain [Bi$_2$I$_7$]$_n^{n-}$ ion and the [Bi$_4$I$_{16}$]$^{4-}$ ion, all Bi^{3+} have regular octahedral coordination.[17] Note that the [Bi$_4$I$_{16}$]$^{4-}$ ion has the same stoichiometry as the [Sb$_4$F$_{16}$]$^{4-}$ ion cited earlier where all Sb^{3+} are in ψ-octahedral coordination.

Other Complexes

In keeping with its being the most electropositive member of Group 15, bismuth forms complexes that have no analogues with P, As, or Sb. The most striking is the aqua ion [Bi(H$_2$O)$_9$]$^{3+}$, which has been found in [Bi(H$_2$O)$_9$](SO$_3$CF$_3$)$_3$, where the coordination polyhedron is a tricapped trigonal prism.[18] Also recently reported are Bi^{3+} complexes with the ligands triethylenetetraaminehexaacetic acid and *trans*-cyclohexane-1,2-diaminetetraacetic acid with coordination numbers of 10 and 8, respectively.[19]

Other, rather unusual complexes are the cation {Bi[OP(NMe$_2$)$_3$]$_2$-[Fe(CO)$_2$Cp]$_2$}$^{+20}$ and the Sb^{III} [Sb(S$_2$C$_6$H$_4$)$_2$]$^-$ species[21] which has ψ-*tbp* coordination. This Sb^{III} complex is similar to [Sb(S$_2$CNBu$_2$)$_2$]$^+$, but it can react with O$_2$ and another mole of S$_2$C$_6$H$_4^-$ to give a nearly octahedral [Sb(S$_2$C$_6$H$_4$)$_3$]$^-$ ion containing Sb^V.

Apart from the [Bi(H$_2$O)$_9$]$^{3+}$ ion mentioned earlier in a crystalline compound, there is no other evidence for a simple [Bi(H$_2$O)$_n$]$^{3+}$ ion. In neutral perchlorate solutions the main species is [Bi$_6$O$_6$]$^{6+}$ or its hydrated form, [Bi$_6$(OH)$_{12}$]$^{6+}$, and at higher pH [Bi$_6$O$_6$(OH)$_3$]$^{3+}$ is formed. The [Bi$_6$(OH)$_{12}$]$^{6+}$ species contains an octahedron of Bi^{III} ions with an OH$^-$ bridging each edge.

[17]N. C. Norman *et al., Z. Naturforsch.* **1995,** *50b,* 1591.
[18]W. Frank *et al., Angew. Chem. Int. Ed. Engl.* **1995,** *34,* 2416.
[19]M. Devillers *et al., J. Chem. Soc. Dalton Trans.* **1996,** 2023.
[20]N. C. Norman *et al., J. Chem. Soc. Dalton Trans.* **1996,** 455.
[21]D. M. Giolando *et al., J. Chem. Soc. Dalton Trans.* **1994,** 1213.

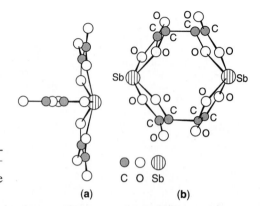

Figure 10-9 The structures of two antimony(III) complexes. (a) The $Sb(C_2O_4)_3^{3-}$ ion. (b) The $Sb_2(d\text{-}C_4H_2O_6)_2^{2-}$ anion in the tartrate complex called tartar emetic.

There is considerable evidence for the association of the Bi^{3+} ion with nitrate ions in aqueous solution. The nitrate ions appear to be mainly bidentate, and all members of the set $Bi(NO_3)(H_2O)_n^{2+}$---$Bi(NO_3)_4^-$ appear to occur. From acid solution various hydrated crystalline salts such as $Bi(NO_3)_3 \cdot 5H_2O$, $Bi_2(SO_4)_3$, and double nitrates of the type $M^{II}_3[Bi(NO_3)_6]_2 \cdot 24H_2O$ can be obtained. Treatment of Bi_2O_3 with HNO_3 gives basic salts such as $BiO(NO_3)$ and $Bi_2O_2(OH)(NO_3)$. Similar salts, generally insoluble in water, are precipitated on dilution of strongly acid solutions of Bi compounds. The nitrate $Bi_6O_4(OH)_4(NO_3)_6 \cdot 6H_2O$ has a Bi_6 octahedron with face-bridging μ_3-oxo groups, and the $[Bi_6O_4(OH)_4]^{6+}$ unit is known to persist in solution.

There is little if any evidence for the existence of P or As cationic species under aqueous conditions, although the following reactions may occur to some slight extent.

$$H_2O + OH^- + AsO^+ \longleftarrow As(OH)_3 \longrightarrow As^{3+} + 3OH^-$$

Antimony has some definite cationic chemistry, but only in the trivalent state, the basic character of Sb_2O_5 being negligible. Cationic compounds of Sb^{III} are mostly of the "antimonyl" ion, SbO^+, although some of the "Sb^{3+}" ion, such as $Sb_2(SO_4)_3$, are known. Antimony salts readily form complexes with various acids in which the antimony forms the nucleus of an anion.

For Sb^{III} in sulfuric acid, the species present vary markedly with the acid concentration, namely,

$$SbO^+ \text{ and/or } Sb(OH)_2^+ \quad < 1.5\ M\ H_2SO_4$$

$$SbOSO_4^-, Sb(SO_4)_2^- \quad \sim 1.0 \text{ to } 18\ M\ H_2SO_4$$

In salts the $Sb(C_2O_4)_3^{3-}$ ion has a ψ-pentagonal bipyramidal structure [Fig. 10-9(a)] with a lone pair at one axial position.

The tartrate complex of antimony(III), *tartar emetic*, $K_2[Sb_2(d\text{-}C_4O_6H_2)_2] \cdot 3H_2O$, has been known in medicine for over 300 years and is used for treatment of schistosomiasis and leishmaniasis; the toxic side effects can be mediated by penicillamine. In the salts the ion has a binuclear structure [Fig. 10-9(b)] and the Sb atom has ψ-tbp geometry. This coordination is also found in the complex $K[As(C_6H_4O_2)_2]$ derived from catechol (o-dihydroxybenzene).

Phosphorus forms a series of two-coordinate cations, called phosphenium ions. The general method of preparation is the following:

$$RR'PCl + \tfrac{1}{2}Al_2Cl_6 \longrightarrow [RR'P][AlCl_4]$$

R and R' groups may be R_2N, $(Me_3Si)_2N$, t-Bu, Cl, and Cp*. For arsenic and antimony there are only a few analogous species, for example, the $[Cp*_2E]BF_4$ compounds. A compound $[Ph_3PPPPh_3][AlCl_4]$ has also been reported that formally contains a complexed P^+ ion.

10-7 Complexes in Oxidation State V

Halide Complexes

The overwhelming tendency is for the EX_5 molecules to interact with an electron donor (L) to give octahedral or distorted octahedral complexes, $EX_5L^{0,-}$. When L = X, which is perhaps the most important case, the strictly octahedral hexahalo anions (EX_6^-) are obtained.

The PF_6^- ion is important in preparing compounds where an anion of very low coordinating power is required; along with BF_4^- and $CF_3SO_3^-$ it affords a safe alternative to ClO_4^-, which should be avoided because of its tendency to give explosive compounds. The PCl_6^- and PBr_6^- ions are also known.

Complexes of Sb^V with fluorine and chlorine are numerous. The simple SbF_6^- and $SbCl_6^-$ ions are well characterized and occur with a great variety of cations, including some very exotic ones such as O_2^+, S_2N^+, and polyhalogeno cations such as ICl_2^+ or IBr_2^+. More elaborate anions can be formed by association through F bridges. Thus, the $Sb_2F_{11}^-$ and $Sb_3F_{16}^-$ ions (10-VI and 10-VII) are known. The angle at the μ-F atom of the $Sb_2F_{11}^-$ ion varies somewhat but is generally 150 to 160°.

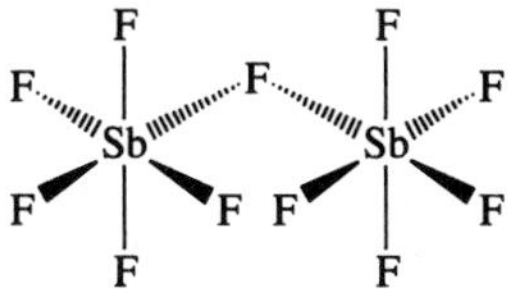

(10-VI) 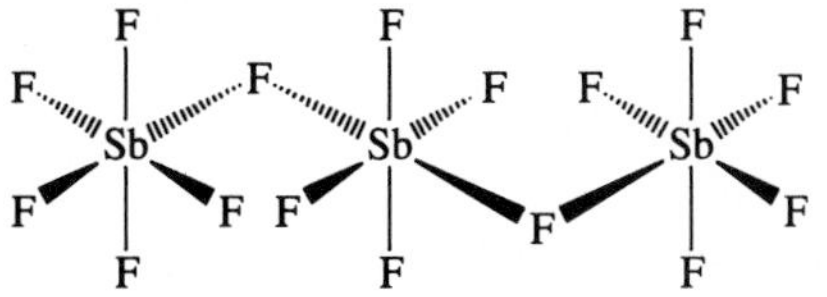   (10-VII)

Other Complexes

The existence of numerous ER_4^+ cations has been noted (Section 10-1), as has the occurrence of EX_4^+ cations, such as PCl_4^+ in crystalline PCl_5. While there is no known example of a BiX_4^+ ion, numerous other EX_4^+ cations are well established with a variety of counterions. The entire series of PX_4^+ and AsX_4^+ ions, with X = Cl, Br, and I, are known, as well as the $SbCl_4^+$ and $SbBr_4^+$ ions.[22]

[22]G. J. Schrobilgen et al., Inorg. Chem. **1996,** 35, 929.

A very unusual type of P^V complex is formed with tetraphenylporphyrin, 10-VIII, where X = OH or OR.[23]

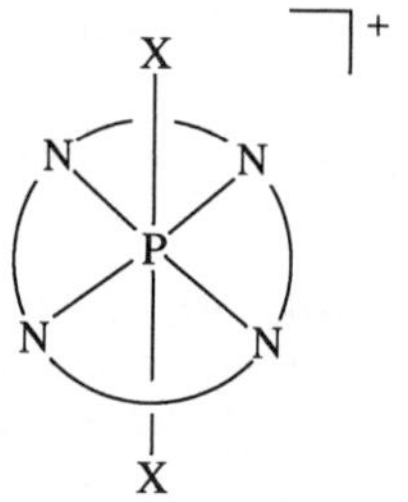

(10-VIII)

The $[Sb(S_2C_6H_4)_3]^-$ ion was noted earlier in Section 10-6. It has also been reported that in 0.5 to 12 M H_2SO_4, Sb^V appears to be present as the $[Sb_3O_9]^{3-}$ ion.

10-8   Oxides

The following are the established oxides; those in [ ] are poorly characterized:

P_4O_6	$As_4O_6(As_2O_3)$	Sb_4O_6 (Sb_2O_3)	Bi_2O_3
P_4O_7			
P_4O_8		Sb_2O_4	
P_4O_9			
P_4O_{10}	As_2O_5	Sb_2O_5	$[Bi_2O_5]$
$[P_2O_6]$			

Phosphorus Oxides

The five well-characterized molecular ones form a structurally homologous series. The structures of the end members, P_4O_6 and P_4O_{10}, are shown in Fig. 10-10. The intermediate ones have 1, 2, or 3 exo oxygen atoms added to the P_4O_6 structure. Preparative methods for the intermediate members are varied and not simple and will not be given here.

The *trioxide*, P_4O_6, so named for historical reasons, has an adamantane-like structure of tetrahedral symmetry [Fig. 10-10]. It is a colorless, volatile (mp 23.8°C, bp 175°C) compound formed in ~50% yield when white phosphorus is burned in an oxygen-deficient atmosphere. It is difficult to separate by distillation from traces of unchanged phosphorus, but irradiation with uv light changes the white phosphorus into red, from which the P_4O_6 can be separated by dissolution in organic solvents. The chemistry of P_4O_6 is complex and not fully understood.

When heated above 210°C, P_4O_6 decomposes into red P and other oxides (PO_x). It reacts vigorously with chlorine and bromide to give the oxo halides and with iodine in a sealed tube to give P_2I_4, but is stable to oxygen at room temperature. When it is shaken vigorously with an excess of cold water, it is hydrated exclusively to phosphorous acid (H_3PO_3), of which it is formally the anhydride; P_4O_6 apparently

[23]J.-H. Chen *et al.*, *Polyhedron* **1994**, *13*, 2887.

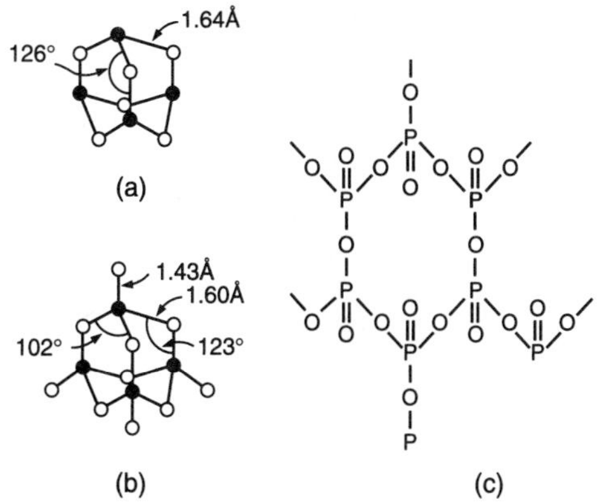

Figure 10-10 The structures of (a) the P_4O_6 molecule (T_d); (b) the P_4O_{10} molecule (T_d); (c) P_4O_{10} sheets.

cannot be obtained by dehydration of phosphorous acid. The reaction of phosphorus trioxide with *hot* water is very complicated, producing among other products PH_3, phosphoric acid, and elemental P; it may be noted in partial explanation that phosphorous acid itself, and all trivalent phosphorus acids generally, are thermally unstable, for example,

$$4H_3PO_3 \longrightarrow 3H_3PO_4 + PH_3$$

The *pentoxide*, P_4O_{10}, also so-called for historical reasons, is usually the main product of burning phosphorus; conditions can be optimized to make it the sole product. It is a white, crystalline solid that sublimes at 360°C (1 atm) and this affords a good method of purifying it from its commonest impurities, which are nonvolatile products of incipient hydrolysis. The hexagonal crystalline form (H-form), obtained on sublimation contains P_4O_{10} molecules [Fig. 10-10(b)] in which the P_4O_6 cage is augmented by four external oxygen atoms, one on each P atom. These molecules, which persist in the gas phase, have full tetrahedral (T_d) symmetry, and the $P-O(br)$ and $P-O(t)$ distances are 1.604 and 1.429 Å, respectively. Other crystalline and glassy forms, obtained by heating the H form, have infinite sheets[24] [Fig. 10-10(c)]; note that the local environment of each phosphorus atom is the same as that in the P_4O_{10} molecule.

The most important chemical property of P_4O_{10} is its avidity for water. It is one of the most effective drying agents known at temperatures below 100°C. It reacts with water to form a mixture of phosphoric acids whose composition depends on the quantity of water and other conditions. It will even extract the elements of water from many other substances themselves considered good dehydrating agents; for example, it converts pure HNO_3 into N_2O_5 and H_2SO_4 into SO_3. It also dehydrates many organic compounds (e.g., converting amides into nitriles). With alcohols it gives esters of simple and polymeric phosphoric acids depending on reaction conditions.

The breakdown of P_4O_{10} with various reagents (alcohols, water, phenols, ethers, alkyl phosphates, etc.) is a very general one and is illustrative also of the general reaction schemes for the breakdown of P_4S_{10} and for the reaction of P_4 with alkali

[24]D. Stachel, *Acta Cryst.* **1995,** *C51,* 1049.

to give PH_3, hypophosphite, and so on. The reaction initially involves breaking a P—O—P bridge. Thus an alcohol reacts with P_4O_{10},

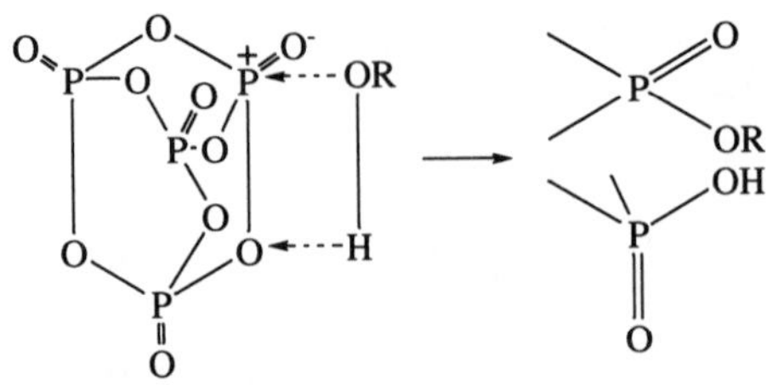

followed by further reaction at the next most anhydride-like linkage, until eventually products containing only one P atom are produced:

$$P_4O_{10} + 6\,ROH \longrightarrow 2\,(RO)_2P(O)OH + 2\,ROPO(OH)_2$$

The fusion of P_4O_{10} with basic oxides gives solid phosphates of various types, their nature depending on experimental conditions.

The oxide, PO, though known to occur in interstellar space, is known on earth only as a ligand where it caps a triangle of metal atoms.[25] This sort of arrangement can also be thought of as a trimetallic analogue of the familiar X_3PO and R_3PO compounds.

Arsenic Oxides

Arsenic trioxide (As_4O_6), formed on burning the metal in air, has in the ordinary form the same structure as P_4O_6. In another crystalline form there are AsO_3 pyramids joined through oxygen atoms to form layers. The ordinary form is soluble in various organic solvents as As_4O_6 molecules and in water to give solutions of "arsenious acid," which is probably $As(OH)_3$ though it has never been firmly characterized. Arsenic trioxide dissolves in aqueous bases to give arsenite ions such as $[AsO(OH)_2]^-$, $[AsO_2(OH)]^{2-}$, and $[AsO_3]^{3-}$.

Arsenic pentoxide cannot be obtained by direct reaction of arsenic with oxygen. It is prepared by oxidation of As with nitric acid followed by dehydration of the arsenic acid hydrates so obtained. It readily loses oxygen when heated to give the trioxide. It is very soluble in water, giving solutions of arsenic acid. Its structure consists of AsO_6 octahedra and AsO_4 tetrahedra completely linked by corner sharing to give a very complex overall pattern.

Antimony Oxides

Antimony trioxide is obtained by direct reaction of the element with oxygen. In the vapor and in the solid below 570°C it consists of P_4O_6 type molecules; the high-temperature solid form is polymeric. It is insoluble in water or dilute nitric and sulfuric acids, but soluble in hydrochloric and certain organic acids. It dissolves in

[25]A. J. Carty *et al., Organometallics* **1996,** 15, 2770.

bases to give solutions of antimonates(III). The yellow *pentoxide* is made mainly by the action of oxygen at high pressure and temperature on Sb_2O_3; it has a structure similar to that of Nb_2O_5 with octahedral SbO_6 groups. Treatment of elemental Sb with concentrated HNO_3 gives a mixture of Sb_2O_3 and $Sb_4O_4(OH)_2(NO_3)_2$; the latter decomposes on heating at 135°C in air to Sb_2O_3.

When either the tri- or the pentoxide is heated in air at about 900°C, a white insoluble powder of composition SbO_2 is obtained. Both α- and β-forms are recognized. Both consist of sheets of linked Sb^VO_6 octahedra with $Sb^{III}O_4$ ψ-tetragonal pyramids joining the layers.[26]

Bismuth Oxides

The only well-established oxide of bismuth is Bi_2O_3, a yellow powder soluble in acids to give bismuth salts. It lacks acidic character and is insoluble in alkalis. From solutions of bismuth salts, alkali, or ammonium hydroxide precipitates a *hydroxide*, $Bi(OH)_3$. Like the oxide, this compound is completely basic. *Bismuth(V) oxide* is extremely unstable and has never been obtained in pure form. The action of extremely powerful oxidizing agents on Bi_2O_3 gives a red-brown powder that rapidly loses oxygen at 100°C.

The oxides of the Group 15 elements clearly exemplify two important trends that are manifest to some extent in all main groups of the Periodic Table: (1) the stability of the higher oxidation state decreases with increasing atomic number, and (2) in a given oxidation state the metallic character of the elements, therefore the basicity of the oxides, increases with increasing atomic number. Thus P^{III} and As^{III} oxides are acidic, Sb^{III} oxide is amphoteric, and Bi^{III} oxide is strictly basic.

10-9 Sulfides and Other Chalcogenides

Like the oxides, sulfides, and other chalcogenides are mostly based structurally on a tetrahedral array of P or As atoms with bridging S or Se atoms and, less commonly, terminal S or Se, but the details differ a good deal. We shall discuss in detail only the phosphorus sulfides. Some of the known structures are shown in Fig. 10-11. Some are known quantitatively from X-ray crystallography, but for others only the qualitative structure is available from ^{31}P nmr. At least four more, including two forms of P_4S_8, are known.

The molecular sulfides P_4S_3, P_4S_7, P_4S_9, and P_4S_{10} are prepared by heating together red phosphorus and sulfur in the corresponding formula ratios. Melts of composition intermediate between P/S = 4:3 and P/S = 4:7 contain at least five other molecular phosphorus sulfides. These cannot be efficiently separated, but may be prepared separately in other ways.[27]

α- and β-P_4S_4

Iodine adds across a P—P bond in P_4S_3 to give β-$P_4S_3I_2$, which slowly isomerizes to α-$P_4S_3I_2$. The latter may also be obtained directly by heating the elements together. The following reaction shows the quantitative conversion of α- and β-$P_4S_3I_2$

[26]I. Rasines *et al.*, *Inorg. Chem.* **1988**, *27*, 1367.
[27]B. Blachnik *et al.*, *Z. anorg. allg. Chem.* **1995**, *621*, 1637.

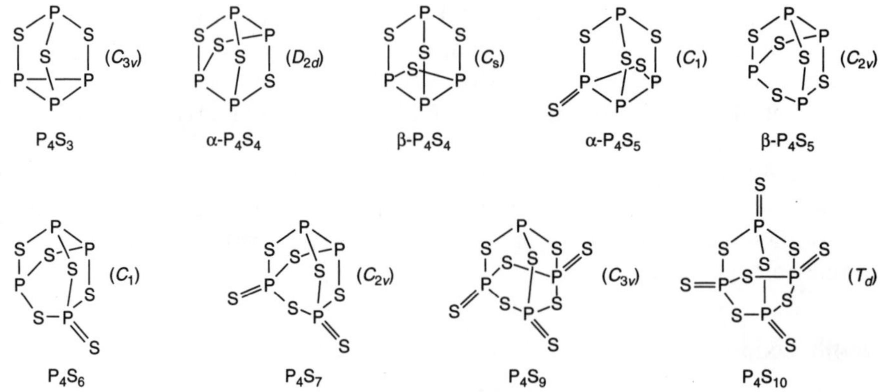

Figure 10-11   Structures of some P_4S_n molecules.

to α- and β-P_4S_4 by bis(trimethytin)sulfide in CS_2:

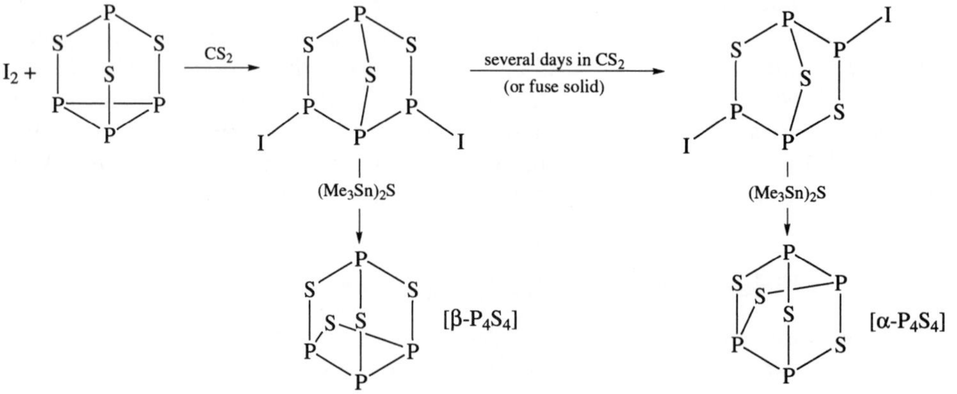

α- and β-P_4S_5 are obtained by the following reactions:

$$5P_4S_3 + 12Br_2 \longrightarrow 3\alpha\text{-}P_4S_5 + 8PBr_3$$

$$P_4S_7 + 2Ph_3P \longrightarrow \beta\text{-}P_4S_5 + 2Ph_3PS$$

In the second reaction P_4S_6 is an intermediate, but has not been isolated pure as it forms mixed crystals with P_4S_7. Its structure is inferred from those of P_4S_7 and β-P_4S_5. Triphenylphosphine is a general reagent for removal of terminal S and can also be used to prepare P_4S_9 from P_4S_{10} and β-P_4S_4 from α-P_4S_5.

All the P_4S_n compounds are stable in CS_2 solution except for P_4S_4, which slowly disproportionates into α-P_4S_5 and P_4S_3. All structures derive from a P_4 tetrahedron by replacement of P—P units by P—S—P and by addition of terminal S atoms. The sulfide P_4S_3 is industrially used in making "strike-anywhere" matches, while P_4S_{10} is used in organic chemistry to convert organic CO, CO_2H, CONH, and OH groups to their sulfur analogues, and in the preparation of industrial lubricant

Table 10-5 Structurally Analogous Group 15 Molecular Chalcogenides

$P_4S_3{}^a$	$\alpha\text{-}P_4S_4$	$\beta\text{-}P_4S_4$	$\alpha\text{-}P_4S_5$	P_4S_7	P_4S_{10}
P_4Se_3	P_4Se_4		P_4Se_5	P_4Se_7	P_4Se_{10}
As_4S_3	$\alpha\text{-},\ \beta\text{-}As_4S_4{}^b$	$\gamma\text{-}As_4S_4$	As_4S_5		As_4S_{10}
As_4Se_3	As_4Se_4				

^{a}P_7^{3-} and As_7^{3-} have the same type of structure, as do P_7H_3 and P_7R_3 compounds, where PH or PR replaces S.
^{b}$\alpha\text{-}$ and $\beta\text{-}As_4S_4$ have the same molecular structure but different crystal packing.

additives. All of these sulfides are more or less readily hydrolyzed to H_3PO_4 or other oxo acids.

The pentaselenide, P_2Se_5, is structurally quite different from the pentoxide and pentasulfide.[28] It has the norbornane-like structure 10-IX. In P_4Se_4 there are P_4Se_3 norbornane units connected by Se atoms, 10-X.[29]

(10-IX)   (10-X)

A mixed oxoselenide, P_4O_6Se, is formed when P_4O_6 and Se dissolved in CS_2 react under irradiation; the Se atom occupies one outer position on the P_4O_6 cage.[30]

There are other Group 15 chalcogenides that conform to the structural pattern of the phosphorus sulfides. Some of these are displayed in Table 10-5. In addition to these, arsenic forms As_2S_3 and As_2Se_3, which have layer structures, and As_2S_5 (of unknown structure). The sulfides As_2S_3 and As_2S_5 can be precipitated from aqueous solutions of As^{III} and As^{V} by H_2S. They are insoluble in water but acidic enough to dissolve in alkali sulfide solutions to form thio anions.

Antimony forms Sb_2S_3 either by direct combination of the elements or by precipitation with H_2S from Sb^{III} solutions; it dissolves in an excess of sulfide to give anionic thio complexes, probably mainly SbS_3^{3-}. Antimony(III) sulfide (Sb_2S_3), as well as Sb_2Se_3 and Bi_2S_3, have a ribbonlike polymeric structure in which each Sb atom and each S atom is bound to three atoms of the opposite kind, forming interlocking SbS_3 and SSb_3 pyramids (see Section 12-7). So-called antimony(V) sulfide (Sb_2S_5) is not a stoichiometric substance and according to Mössbauer spectroscopy contains only Sb^{III}.

Bismuth gives dark brown Bi_2S_3 on precipitation of Bi^{III} solutions by H_2S; it is not acidic. A sulfide BiS_2 is obtained as gray needles by direct interaction at 1250°C and 50 kbar; its structure is unknown but may be $Bi^{3+}[BiS_4]^{3-}$.

In addition to the neutral, binary chalcogen compounds, there are ionic and ternary ones. Examples of the former are the $P_7S_3^{3-}$ ion[31] and the $As_{10}Te_3^{2-}$ ion.[32]

[28]B. Blachnik *et al., Acta Cryst.* **1994,** *C50,* 659.
[29]M. Ruck, *Z. anorg. allg. Chem.* **1994,** *620,* 1832.
[30]M. Jansen *et al., Z. anorg. allg. Chem.* **1995,** *621,* 2065.
[31]M. Baudler and A. Floruss, *Z. anorg. allg. Chem.* **1994,** *620,* 2070.
[32]R. C. Haushalter, *J. Chem. Soc., Chem. Commun.* **1987,** 196.

There is also a family of compounds containing the $P_2Se_6^{4-}$ unit, combined with either M^{II} cations, e.g., $Pb_2P_2Se_6$, or with Sb^{III} or Bi^{III} to give $Sb_4(P_2Se_6)_3$ and $Bi_4(P_2Se_6)_3$.[33]

OTHER COMPOUNDS

10-10   Oxo Halides

The most important oxo halides are the phosphoryl halides (X_3PO), in which X may be F, Cl, Br, or I. The commonest (Cl_3PO) is obtained by the reactions

$$2PCl_3 + O_2 \longrightarrow 2Cl_3PO$$

$$P_4O_{10} + 6PCl_5 \longrightarrow 10Cl_3PO$$

The reactions of Cl_3PO are much like those of PCl_3 (Fig. 10-4). The halogens can be replaced by alkyl or aryl groups by means of Grignard reagents, and by alkoxo groups by means of alcohols; hydrolysis by water yields phosphoric acid. Phosphoryl chloride (Cl_3PO) also has donor properties toward metal ions, and many complexes are known. Distillation of the Cl_3PO complexes of $ZrCl_4$ and $HfCl_4$ can be used to separate zirconium and hafnium, and the very strong $Cl_3PO-AlCl_3$ complex has been utilized to remove Al_2Cl_6 from adducts with Friedel-Crafts reaction products.

All X_3PO molecules have a pyramidal PX_3 group, with the oxygen atom occupying the fourth position to complete a distorted tetrahedron. The P—O bond lengths are about 1.55 Å, consistent with the existence of double bonds. Several mixed phosphoryl halides, phosphoryl pseudohalides, as well as X_3PS and X_3PSe compounds are also known. All of these compounds are prone to ready hydrolysis.

Some more complex oxohalides are known, of which the *pyrophosphoryl halides*, $X_2P(O)-O-P(O)X_2$, X = Cl, F are best known.

There are no definite, analogous X_3EO compounds except Cl_3AsO, which is unstable above $-25°C$. There are, however, some polymeric As and Sb compounds with μ-O atoms, such as $As_2O_3Cl_4$ (not isostructural with its P analogue), $Sb_4O_5Cl_2$, $Sb_8O_{11}Cl_2$, and Sb_8OCl_{22}. There are also some oxohalo anionic species, such as $[F_5EOEF_5]^{2-}$, E = As, Sb, and $[F_4As(\mu\text{-}O)_2AsF_4]^{2-}$.

Antimony and *bismuth* form the important oxo halides SbOCl and BiOCl, which are insoluble in water. They are precipitated when solutions of Sb^{III} and Bi^{III} in concentrated HCl are diluted. They have quite different but complicated, layer structures.

10-11   Phosphorus-Nitrogen Compounds

There is a very extensive chemistry of compounds with P—N and P=N bonds; for the most part the molecules are oligomeric or polymeric. Among the few relatively simple ones are $(Me_2N)_3P$ and $(Me_2N)_3PO$ (hexamethylphosphoramide), several X_2P-NMe_2 (X = Cl, F, and CF_3), and $PF_3(NH_2)_2$ (which has equatorial NH_2 groups). The so-called PNP^+ (sometimes PPN^+) ion ($Ph_3PNPPh_3^+$) is widely used to isolate

[33]M. Ruck, *Z. anorg. allg. Chem.* **1995,** *621*, 1344.

large anions in crystalline form or to give organic-soluble reagents for use in synthesis.

Actually, the PNP cation is but one special example of a group of compounds called *phosphazenes*. Monophosphazenes ($R_3P=NR'$), diphosphazenes ($R_3P=N-PR_2'$), and polyphosphazenes ($R_3P=N-(PR_2=N)_n\cdots PR_2$) are all known. Hundreds of monophosphazenes have been characterized, and a general synthetic route is

$$R_3PCl_2 + R'NH_2 \longrightarrow R_3P=NR' + 2HCl$$

However, it is the cyclophosphazenes and long-chain polyphosphazenes that currently attract the greatest attention.

Cyclophosphazenes[34]

The important ones are those with 6- and 8-membered rings, 10-XI and 10-XII. The most common ones are $(Cl_2P-N)_3$ and $(Cl_2P-N)_4$, which are commercially available. A mixture of these, together with some polymer, results from the following reaction:

$$nPCl_5 + nNH_4Cl \xrightarrow{\text{C}_2\text{H}_2\text{Cl}_4 \text{ or } \text{C}_6\text{H}_5\text{Cl}} (NPCl_2)_n + 4nHCl$$

Conditions can be controlled to maximize the yield of the trimer, which can easily be separated and purified. It is a white, crystalline solid (mp 113°C) that can be sublimed. It readily undergoes reactions whereby the Cl atoms are replaced by other groups:

$$(NPCl_2)_3 + 6NaF \xrightarrow{\text{MeCN}} (NPF_2)_3 + 6NaCl$$

$$(NPCl_2)_3 + 6NaOR \longrightarrow [NP(OR)_2]_3 + 6NaCl$$

$$[NP(OR)_2]_3 + xRNH^- \longrightarrow N_3P_3(OR)_{6-x}(NHR)_x + xOR^-$$

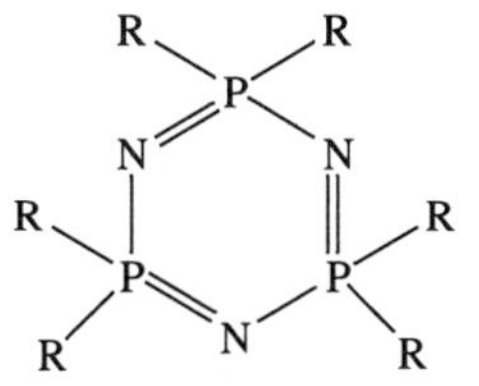

(10-XI) 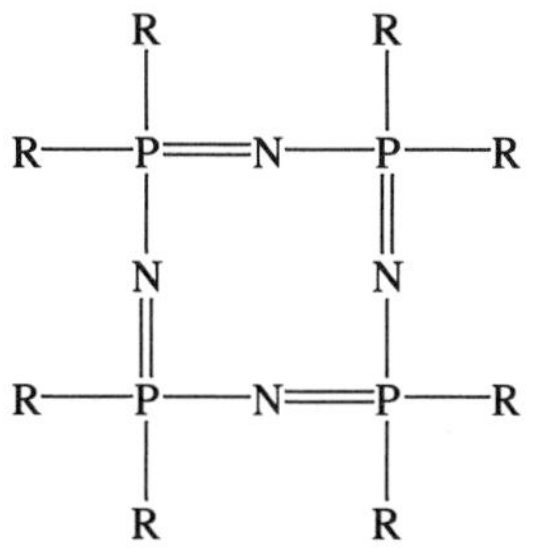   (10-XII)

The rings in $(NPF_2)_x$ where $x = 3$ or 4 are planar but larger rings are not planar. For other $(NPX_2)_n$ compounds the six rings are planar or nearly so, but larger rings

[34]C. W. Allen, *Coord. Chem. Rev.* **1994**, *130*, 137.

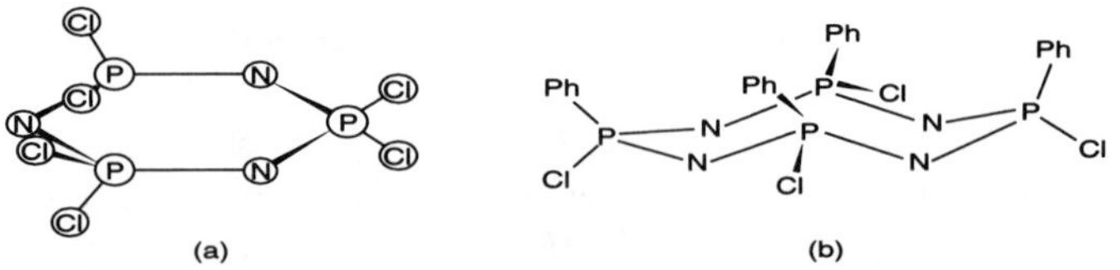

cotton/advances 6e 10FG12 s/s

Figure 10-12   The structures of two representative cyclic phosphazenes: (a) $(NPCl_2)_3$; (b) all-*cis*-$(NPClPh)_4$.

are generally nonplanar with NPN angles of ~120° and PNP angles of ~132°. Figure 10-12 shows the structures of $(NPCl_2)_3$ and $(NPClPh)_4$. The P—N distances, which are generally equal or very nearly so in these ring systems, lie in the range 1.55 to 1.61 Å; they are thus shorter than the expected single-bond length of ~1.75 to 1.80 Å. Considerable attention has been paid to the nature of the P—N π bonding, which the P—N distances indicate is appreciable, but the matter is still subject to controversy. The main question concerns the extent of delocalization, that is, whether there is complete delocalization all around the rings to give them a kind of aromatic character, or whether there are more localized "islands" within the NPN segments. Of course there may be considerable differences between the essentially planar rings and those that are puckered. The problem is a complicated one owing to the large number of orbitals potentially involved and to the general lack of ring planarity, which means that rigorous assignment of σ and π character to individual orbitals is impossible.

These rings are conformationally flexible, and π bonding is only one of many factors that influence the conformations. However, satisfactory force fields for use in molecular mechanics type calculations have not yet been developed.

Linear Polyphosphazenes[35]

A great variety of these are known, mainly from the work of H. R. Allcock, and they have considerable practical interest because their properties can be varied by the choice of R groups in the $(R_2P—N)_n$ chain, 10-XIII. More than 700 different examples are known.

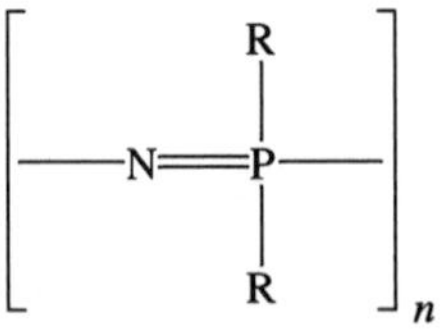

(10-XIII)

The most widely used synthetic access to these materials begins with very pure $(NPCl_2)_3$ from which all PCl_5 has been eliminated. Failure to eliminate all PCl_5 leads to a cross-linked, insoluble product not suitable for further chemistry. When $(NPCl_3)_3$ is heated above 230°C it gives the polymer, where the degree of polymeriza-

[35]J. E. Mark *et al., Inorganic Polymers,* Prentice-Hall, Englewood Cliffs, NJ, 1992, Chap. 3.

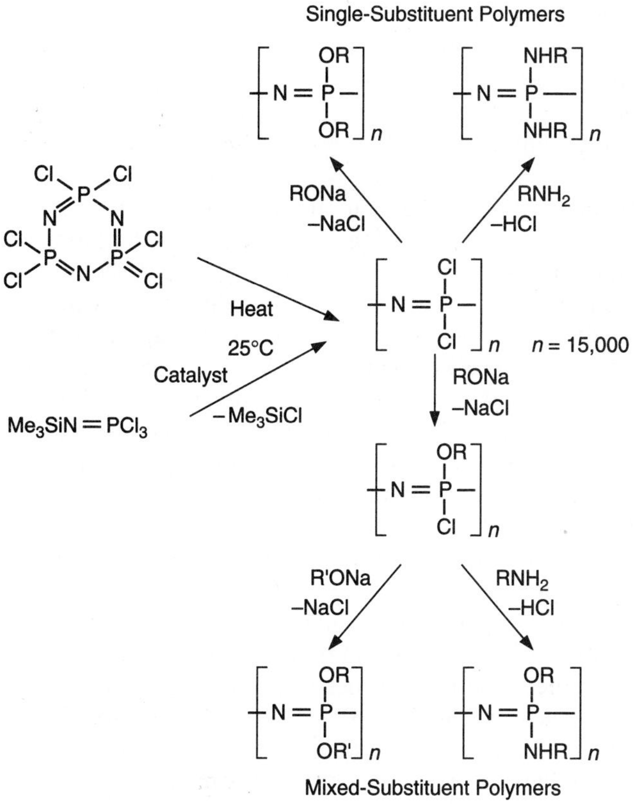

Single-Substituent Polymers

Figure 10-13 Some synthetic routes for polyphosphazenes.

Mixed-Substituent Polymers

tion, n, can be $\geq 15{,}000$. This translates to molecular weights in the region of 2 million. This polymer is exceptional in its high reactivity to nucleophiles such as alkoxides, aryloxides, amines, or organometallic reagents, since essentially all 30,000 chlorine atoms per molecule can be replaced by organic or organometallic groups under moderate reaction conditions, and usually without cleavage of the inorganic backbone. Because two or more different nucleophiles can participate in this reaction either simultaneously or sequentially, the scope for preparation of mixed-substituent polymers is almost infinite. For some examples see Fig. 10-13.

The properties of these polymers depend on both the inorganic backbone and the organic or organometallic side groups. The backbone confers flexibility, fire resistance, stability to oxidation and high energy radiation and, in some cases, biocompatibility. The side groups control solubility, hydrolytic stability, surface characteristics, secondary reactivity, thermal stability, and a wide range of optical, electrical, and biological properties. The size and polarity of the side groups also influence material properties such as crystallinity, liquid crystallinity, and glass transition temperature. Thus, the opportunities for designing and fine-tuning polymers for different uses are broad.

Commercial development of polyphosphazenes began with the production of high performance elastomers with fluoroalkoxy or aryloxy side groups. Other polyphosphazenes are used in medical devices.[36] Another development is the use of

[36]R. Langer *et al., J. Am. Chem. Soc.* **1990,** *112,* 7832.

alkyl ether side chain polyphosphazenes as solid solvents for salts such as lithium or silver triflate.[37] These ion-conducting materials can be fabricated into lightweight, thin-film rechargeable lithium batteries. Additional areas where polyphosphazenes are under detailed investigation are as fibers, membranes, hydrogels, bone-replacement materials, surface coatings, liquid crystalline and non-linear optical devices, and composite materials.

Advances in synthetic chemistry include new methods for the preparation of poly(dichlorophosphazene). For example, the high temperature condensation of $O=PCl_2-N=PCl_3$ gives moderate molecular weight $(NPCl_2)_n$. Processes for the BCl_3 catalyzed polymerization of $(NPCl_2)_3$ have been developed, and polymerizations in solution are now possible. Narrow molecular weight distribution $(NPCl_2)_n$ can be prepared *via* the PCl_5 catalyzed condensation of $Me_3Si-N=PCl_3$ at room temperature.[38] This "living" cationic polymerization allows precise control of the polymer chain length and provides access to block copolymers with other monomers. Methods are also available for the direct synthesis of organophosphazene polymers without macromolecular substitution. For example, the condensation polymerization of $Me_3Si-N=PR_2(OCH_2CF_3)$ at 200°C gives $(NPR_2)_n$ directly.[39] Some partly and fully organic-substituted phosphazene cyclic trimers undergo ring-opening polymerization to the corresponding polymer.[40]

In addition, a variety of hybrid systems have been synthesized which combine the phosphazene skeleton with other inorganic units. For example, it has been demonstrated that cyclocarbo- and thiophosphazenes undergo ring-opening polymerization to macromolecules such as 10-XIV, 10-XV, and 10-XVI, which can undergo macromolecular substitution in the same way as $(NPCl_2)_n$.[41] Hybrid phosphazene-organosilicon polymers have been produced which combine some of the properties of phosphazenes with those of silicones. Phosphazene polymers with carborane or borazine side groups are known, and a range of polyphosphazenes with transition metal side group units, such as ferrocene groups, have been prepared and studied.

(10-XIV)

(10-XV)

(10-XVI)

[37]H. R. Allcock *et al.*, *Macromolecules* **1996,** *29,* 1951.

[38]I. Manners *et al.*, *J. Am. Chem. Soc.* **1995,** *117,* 7035.

[39]R. H. Neilson and P. Wisian-Neilson, *Chem. Rev.* **1988,** *88,* 541.

[40]H. R. Allcock, *ACS Symp. Ser.* **1992,** *496,* 236.

[41]H. R. Allcock *et al.*, *J. Am. Chem. Soc.* **1989,** *111,* 5478; *Macromolecules* **1993,** *26,* 11; M. Liang and I. Manners, *J. Am. Chem. Soc.* **1991,** *113,* 4044.

A material known as *phospham*, produced by the reaction

$$4NH_3(g) + 2P(red) \xrightleftharpoons{500\,°C} 2PN_2H(s) + 5H_2$$

is a highly cross-linked polyphosphazene (10-XVII):

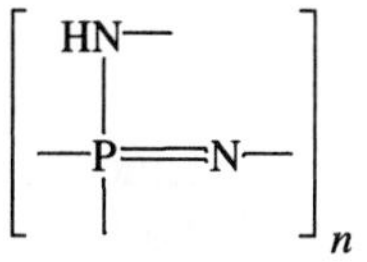

$$\left[\begin{array}{c} HN— \\ | \\ —P{=}N— \\ | \end{array}\right]_n$$

(10-XVII)

Other P—N Compounds

In the large and important class of compounds called phosphazenes we are formally dealing with P^V and N^{III}. There are other P—N compounds that are formally P^{III}—N^{III} types. In these the P—N bonds are all nominally single, but, of course, partial double-bond character arises by $N p\pi$—$P d\pi$ donor bonding, indicated as follows in resonance terms:

$$P—\ddot{N} \longleftrightarrow \bar{P}{=}\overset{+}{N}$$

The *aminophosphanes*, $(R_2N)_nPX_{3-n}$, are monomeric, but the aminoiminophosphanes $(R_2N—P{=}NR')$ may be either monomers when the R and R' groups are bulky, or dimers (10-XVIII) with less bulky substituents.

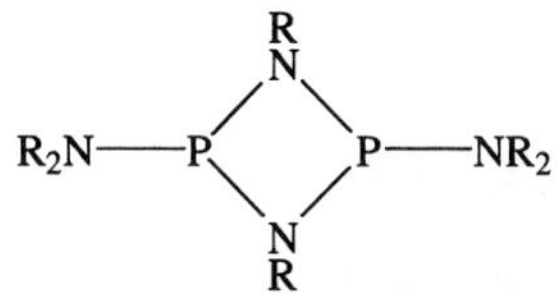

$$R_2N—P\underset{\underset{R}{N}}{\overset{\overset{R}{N}}{\diamond}}P—NR_2$$

(10-XVIII)

There are also compounds with larger rings, such as (10-XIX). The cyclic dimers have essentially planar rings, but the substituents on the phosphorus atoms (which have overall pyramidal sets of bonds) can adopt either cis or trans relationships relative to the ring plane.

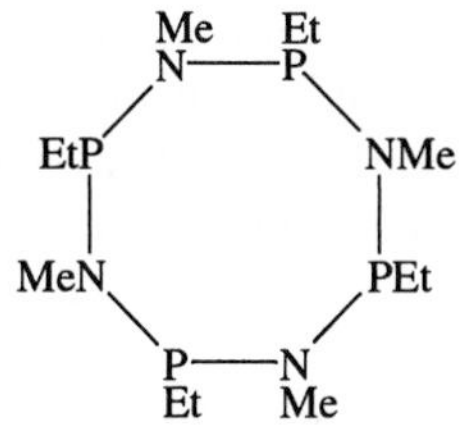

(10-XIX)

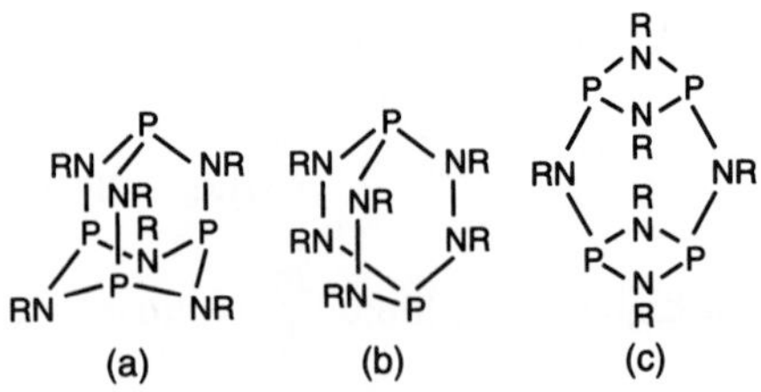
(a)
(b)
(c)

Figure 10-14 Three polycyclic types of P—N compounds.

A variety of rings with still other elements included can be made, as in the following examples (where, again, we have formally P^V):

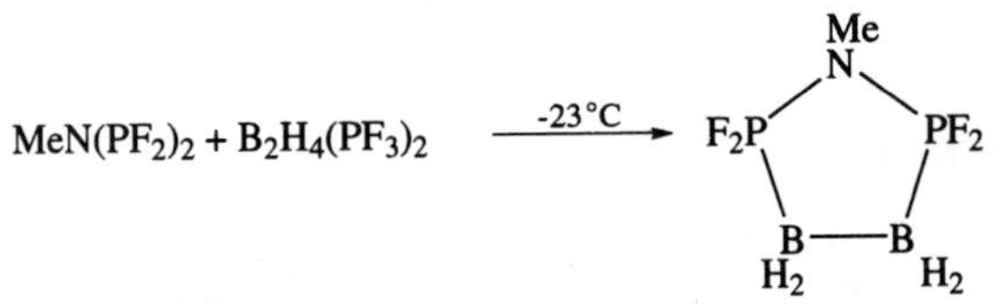
MeN(PF2)2 + B2H4(PF3)2
-23°C
Me
N
F2P
PF2
B
B
H2
H2

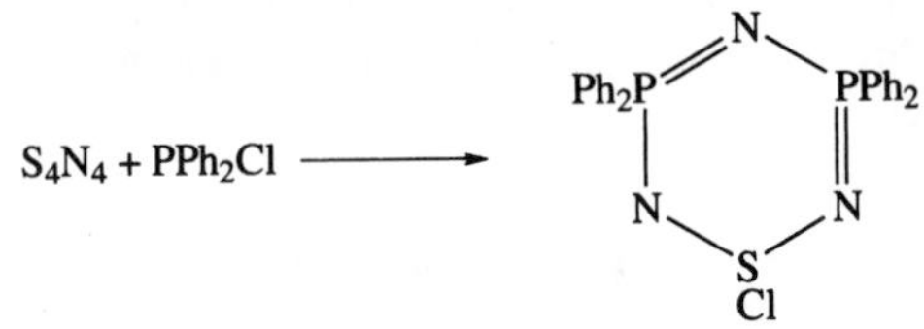
S4N4 + PPh2Cl
N
Ph2P
PPh2
N
N
S
Cl

Finally, there are polycyclic P—N compounds such as those shown in Fig. 10-14. Note that (a) which has an adamantane-type structure and (c) are isomers; when R = Me, the adamantane structure is thermally stable, but when R = Me_2HC, the (a) structure converts to the (c) structure on heating for 12 days at 157°C.

THE OXO ACIDS AND ANIONS

The oxo acids and anions in both lower and higher states are a very important part of the chemistry of phosphorus and arsenic and comprise the only real aqueous chemistry of these elements. For the more metallic antimony and bismuth, oxo anion formation is less pronounced, and for bismuth only ill-defined "bismuthates" exist.

10-12 Oxo Acids and Anions of Phosphorus

All phosphorus oxo acids have POH groups in which the hydrogen atom is ionizable; hydrogen atoms of the P—H type are not ionized. There is a vast number of oxo acids or ions, some of them of great technical importance, but we can deal only with some structural principles, and some of the more important individual compounds. The *oxo anions* are of main importance, since in many cases the free acid cannot be isolated, though its salts are stable. Both lower (P^{III}) and higher (P^V) acids are known.

The principal higher acid is orthophosphoric acid (10-XXa) and its various anions (10-XXb)-(10-XXd), all of which are tetrahedral. The phosphorus(III) acid, which might naively have been considered to be P(OH)$_3$, has in fact the four-connected tetrahedral structure (10-XXIa); it is only difunctional, and its anions are (10-XXIb) and (10-XXIc). Only in the triesters of phosphorous acid [P(OR)$_3$] do we encounter three-connected phosphorus atoms, and even these, as will be seen later, have a tendency to rearrange to four-connected species.

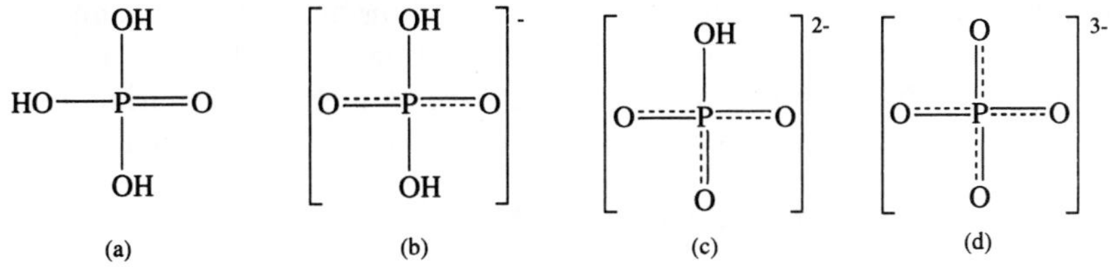

(10-XX)

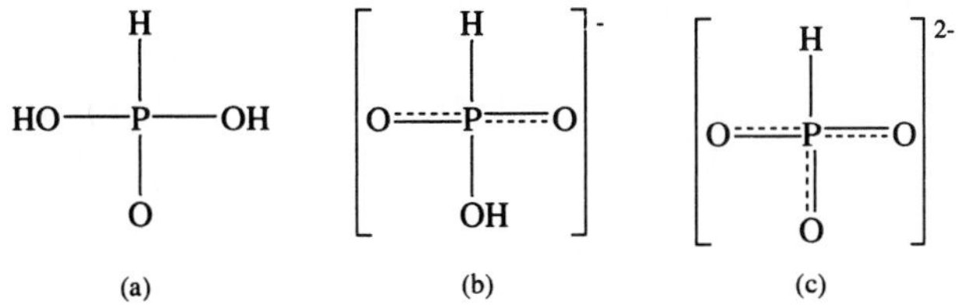

(10-XXI)

Similarly, the acid of formula H$_3$PO$_2$ (hypophosphorous acid) also has a four-connected structure (10-XXIIa), as does its anion (10-XXIIb).

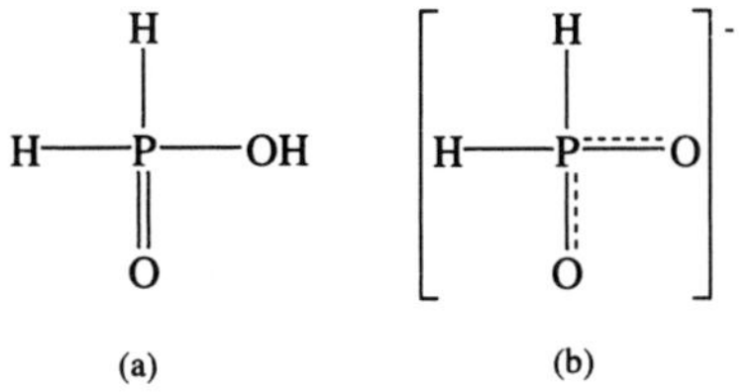

(10-XXII)

Lower Acids

Hypophosphorous Acid

The salts of PH$_2$(O)(OH) are usually prepared by boiling white phosphorus with alkali or alkaline earth hydroxide. The main reaction appears to be

$$P_4 + 4OH^- + 4H_2O \longrightarrow 4H_2PO_2^- + 2H_2$$

The calcium salt is soluble in water, unlike that of phosphite or phosphate; the free acid can be made from it or obtained by oxidation of phosphine with iodine in water. Both the acid and its salts are powerful reducing agents, being oxidized to orthophosphate. The pure white crystalline solid is a monobasic acid (pK = 1.2). The more correct name of this acid, phosphinic acid, is rarely used, but when the

H—P bonds are replaced by R—P bonds, to give $R_2P(O)OH$, these are regularly called phosphinic acids and their salts phosphinates.

Phosphorous Acid

This acid, $PH(O)(OH)_2$, is obtained by treating PCl_3 or P_4O_6 with water; when pure, it is a deliquescent colorless solid (mp 70.1°C, $pK = 1.3$). It can be oxidized to orthophosphate by various agents, but the reactions are slow and complex. The mono-, di-, and triesters can be obtained from reactions of alcohols or phenols with PCl_3 alone or in the presence of an organic base as hydrogen chloride acceptor. They can also be obtained directly from white phosphorus by the reaction

$$P_4 + 6OR^- + 6CCl_4 + 6ROH = 4P(OR)_3 + 6CHCl_3 + 6Cl^-$$

The systematic but little used name for this acid is phosphonic acid. When the P—H bond is replaced by a P—R bond, to give $RP(O)(OH)_2$, these acids are regularly called phosphonic acids and their salts phosphonates.

Although compounds of composition $P(OR)_3$ are well known and are useful as ligands (Section 10-17) and in other ways, they are subject to the Arbuzov reaction with alkyl halides, which converts them to diesters of phosphonic acids:

$$P(OR)_3 + R'X \longrightarrow [(RO)_3PR'X] \longrightarrow RO-\overset{\displaystyle O}{\underset{\displaystyle OR}{\overset{\|}{P}}}-R' + RX$$

phosphonium intermediate

The methyl ester easily undergoes spontaneous isomerization to the dimethyl ester of methylphosphonic acid:

$$P(OCH_3)_3 \longrightarrow CH_3PO(OCH_3)_2$$

When a $P(OR)_3$ molecule serves as a ligand to an organometallic cation, the same sort of reaction can occur,[42] for example:

$$CpFe(CO)_2[P(OMe)_3]^+Cl^- \longrightarrow CpFe(CO)_2[PO(OMe)_2] + MeCl$$

Higher Acids

Orthophosphoric Acid

The acid H_3PO_4 or $PO(OH)_3$, commonly called phosphoric acid, is one of the oldest known and most important phosphorus compounds. It is made in vast quantities, usually as 85% syrupy acid, by the direct reaction of ground phosphate rock with sulfuric acid and also by the direct burning of phosphorus and subsequent hydration of the oxide P_4O_{10}. The pure acid is a colorless crystalline solid (mp 42.35°C), consisting of a three-dimensional strongly hydrogen-bonded array.[43] It is very stable and has essentially no oxidizing properties below 350 to 400°C. At elevated temperatures it is fairly reactive toward metals and is reduced; it will also then attack quartz. Fresh molten H_3PO_4 has appreciable ionic conductivity,

[42]I. Kuksis and M. C. Baird, *J. Organomet. Chem.* **1996,** *512,* 253.
[43]R. H. Blessing, *Acta Cryst.* **1988,** *B44,* 334.

suggesting autoprotolysis:

$$2H_3PO_4 \rightleftharpoons H_4PO_4^+ + H_2PO_4^-$$

Pyrophosphoric acid is also produced:

$$2H_3PO_4 \longrightarrow H_2O + H_4P_2O_7$$

but this conversion is temperature dependent and is slow at room temperature.

The acid is tribasic: at 25°C, $pK_1 = 2.15$, $pK_2 = 7.1$, $pK_3 \approx 12.4$. The pure acid and its crystalline hydrates have tetrahedral PO_4 groups connected by hydrogen bonds. These persist in the concentrated solutions and are responsible for the syrupy nature. For solutions of concentration $< \sim 50\%$, the phosphate anions are hydrogen bonded to the liquid water rather than to other phosphate anions.

Orthophosphates of most metals are known, but by far the most important are those of calcium, particularly those of the apatite type,[44] for which the general formula is $Ca_5(PO_4)_3X$. *Fluoroapatite,* where X = F, is the major constituent of phosphate rock which occurs in vast deposits in many places in the world and is the chief ore of phosphorus. Other important apatites are those in which X = Cl or OH. Biologically, apatites are indispensable since they are key components of both teeth and bone. Synthetic apatites that permit bone grafts are now available.[45]

Orthophosphoric acid and phosphates form complexes with many transition metal ions. The precipitation of insoluble phosphates from fairly concentrated acid solution (3-6 M HNO_3) is characteristic of 4+ cations such as those of Ce, Th, Zr, U, and Pu. Phosphates of B, Al, Zr, and so on, are used industrially as catalysts for a variety of reactions. The phosphate ions, $H_2PO_4^-$, HPO_4^{2-}, and PO_4^{3-} are well known as ligands, of monodentate, chelating, or bridging types; examples of the last two are

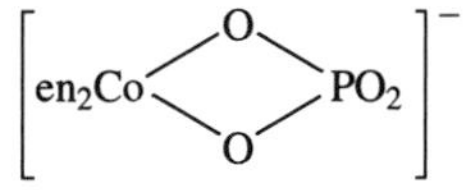

$$\left[en_2Co \overset{O}{\underset{O}{\diagup\diagdown}} PO_2 \right]^-$$

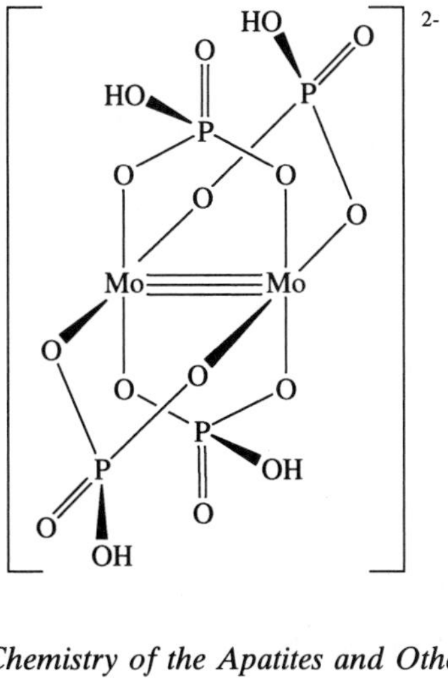

[44]J. C. Elliot, *Structure and Chemistry of the Apatites and Other Calcium Orthophosphates,* Elsevier, Amsterdam, 1994.
[45]R. L. Lagow *et al., J. Oral Maxillofac. Surg.* **1993,** *51,* 1363.

Although the majority of phosphate esters have four-connected phosphorus, there are a few in which five or even six oxygen atoms surround phosphorus, for example, $[P(OMe)_6]^-$ and $[P(o\text{-}C_6H_4O_2)_3]^-$.

Thio Acids and Esters

There are sulfur analogues to some of the types of oxo-acids and esters, many of which have been extensively studied,[46] and some of which have important industrial uses. The most important of these are the dithiophosphinates (or phosphinodithioates) and the dithiophosphates which are the anions of $R_2P(S)SH$ and $(RO)_2PS(SH)$, respectively. The dithiophosphates are essential additives to motor oils. They are easily made:

$$P_2S_5 + 3NaOR + ROH \longrightarrow 2(RO)_2P(S)S^-Na^+ + NaSH$$

Condensed Phosphates

Condensed phosphates are those containing more than one P atom and having P—O—P bonds. We may note that the *lower* acids can also give condensed species. We shall deal here only with a few examples of condensed phosphates. There are also numerous polyphosphates, and/or the corresponding acids, that contain one or more direct P—P bonds (10-XXIIIa) and (10-XXIIIb) but these will not be discussed further.

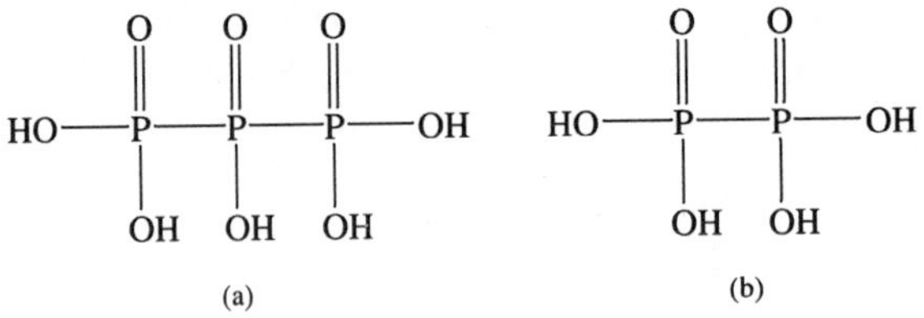

(a)   (b)

(10-XXIII)

There are three main building units in condensed phosphates: the end unit (10-XXIVa), the middle unit (10-XXIVb), and the branching unit (10-XXIVc).

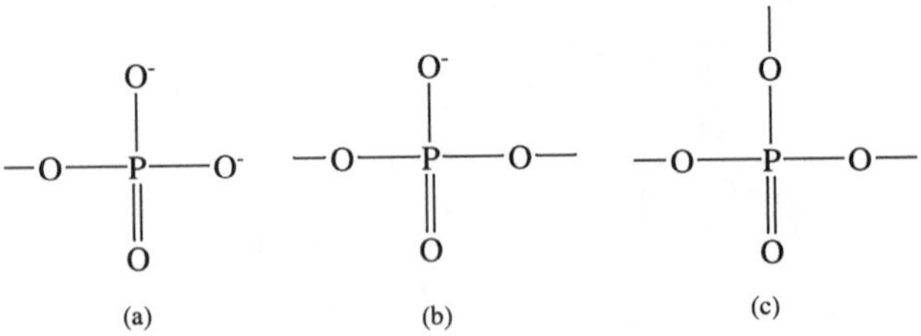

(a)   (b)   (c)

(10-XXIV)

[46]I. Haiduc *et al.*, *Polyhedron* **1995**, *14*, 3389; I. Haiduc and D. B. Sowerby, *Polyhedron* **1996**, *15*, 2469.

These units can be distinguished by chemical reactivity differences (branch points are rapidly hydrolyzed) and by ^{31}P nmr spectra. Combinations of these units can give rise to four main types of polyphosphates:

1. Polyphosphates (10-XXV), which have been isolated pure with $n = 1$ to 16 and general formula $[P_nO_{3n+1}]^{(n+2)-}$.

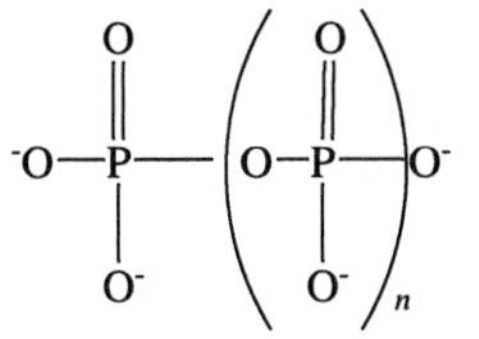

$$\text{-O}-\overset{\overset{\displaystyle O}{\|}}{\underset{\underset{\displaystyle O^-}{|}}{P}}-\left(\text{O}-\overset{\overset{\displaystyle O}{\|}}{\underset{\underset{\displaystyle O^-}{|}}{P}}-\text{O}^-\right)_n$$

(10-XXV)

2. Infinite chain *metaphosphates*, $[(PO_3)_n]^{n-}$.
3. Cyclic metaphosphates, $[(PO_3)_n]^{n-}$, with $n = 3$ or more.
4. Ultraphosphates, which contain branching units. The ultimate ultraphosphate is P_4O_{10}, consisting entirely of branch units.

Some representative linear polyphosphates are the alkali salts $M^I_4P_2O_7$ (pyrophosphates) and $M^I_5P_3O_{10}$ (tripolyphosphates). Many detergents contain $Na_5P_3O_{10}$, and it has other industrial uses. In detergents the triphosphate serves to sequester Ca^{2+} and Mg^{2+} ions. Using Ln^{3+} ions as models, the mode of complexation was shown by nmr to involve attachment of two tetradentate $P_3O_{10}^{5-}$ ions.

An infinite chain metaphosphate is $Li_2(NH_4)P_3O_9$. The most common type of metaphosphates are the cyclotriphosphates ($M_3P_3O_9$) and, to a lesser extent, cyclic tetraphosphates, which contain the anions (10-XXVIa) and (10-XXVIb). The $P_3O_9^{3-}$ rings have chair conformations;[47] the $P_4O_{12}^{4-}$ rings are puckered but conformations vary from compound to compound. An efficient synthesis of a cyclotriphosphate is by the following reaction:

$$2(MeO)_3PO + (Me_2N)_3PO \longrightarrow [Me_4N]_3(P_3O_9)$$

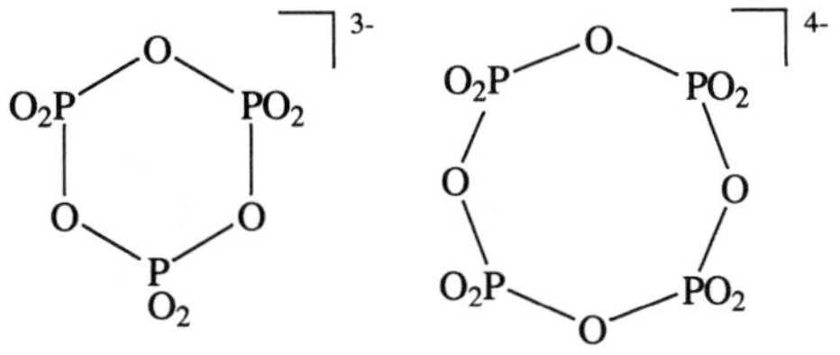

(a)   (b)

(10-XXVI)

Condensed phosphates are usually prepared by dehydration of orthophosphates under various conditions of temperature (300–1200°C) and also by appropriate

[47]M. T. Averbuch-Pouchot *et al.*, *Acta Cryst.* **1988**, *C44*, 1907, 1909.

hydration of dehydrated species, as, for example,

$$(n-2)NaH_2PO_4 + 2Na_2HPO_4 \xrightarrow{\text{heat}} Na_{n+2}P_nO_{3n+1} + (n-1)H_2O$$

Polyphosphate

$$nNaH_2PO_4 \xrightarrow{\text{heat}} (NaPO_3)_n + nH_2O$$

Metaphosphate

They can also be prepared by controlled addition of water or other reagents to P_4O_{10}, by treating chlorophosphates with silver phosphates, and so on. The complex mixtures of anions that can be obtained are separated by using ion-exchange or chromatographic procedures.

Tripolyphosphate, $P_3O_{10}^{5-}$, can be dehydrated to cyclotrimetaphosphate by sulphamide:[48]

$$H_{n+2}P_3O_{10}^{-3+n} + (NH_2)_2SO_2 \longrightarrow H_nP_3O_9^{-3+n} + (NH_4)(H_2NSO_3)$$

Carbodiimides in solvents such as DMSO or tetramethylurea can dehydrate H_3PO_4, polyphosphoric acids, or ring acids to the bicyclic ultraphosphonic acid $(H_2P_4O_{11})$, as illustrated by the following reaction:

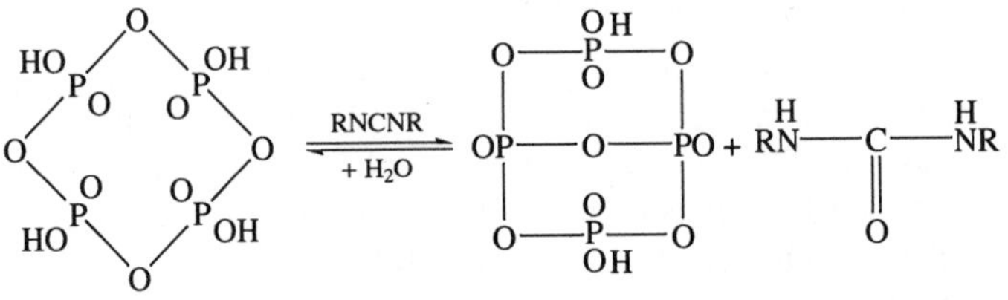

Polyphosphoric acids are obtained by reaction of P_4O_{10} with water or by heating H_3PO_4 to remove water. A P_4O_{10}/H_2O mixture containing 72.42% P_4O_{10} corresponds to pure H_3PO_4, but the usual commercial grades of the acid contain more water. As the P_4O_{10} content increases, *pyrophosphoric acid*, $(HO)_2OPOPO(OH)_2$, forms along with the P_3 through P_8 polyphosphoric acids. Equilibria are achieved only slowly, and half-lives for the formation or hydrolysis of P—O—P linkages when catalysts are excluded can be of the order of years. That is why enzymes called phosphorylases or nucleases are so important in biochemistry, to achieve needed interconversions of mono-, di-, tri-, and polyphosphate esters.

Phosphate Esters in Biology

Many of the essential chemicals in life processes are phosphate esters. These include the genetic substances DNA and RNA [representative fragments of the chains appear as (10-XXVII) and (10-XXVIII), respectively] as well as cyclic AMP (adenosine monophosphate), (10-XXIX). In addition, the transfer of phosphate groups between ATP and ADP

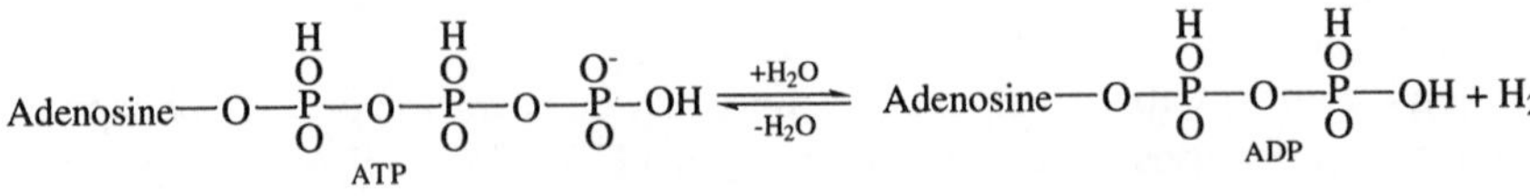

[48]D. R. Gard and E. J. Griffith, *J. Chem. Soc., Chem. Commun.* **1990**, 881.

is of fundamental importance in the energetics of biological systems. All the biological reactions involving formation and hydrolysis of these and other phosphate esters and polyphosphates are effected by enzyme catalysts, many of which contain metal ions as parts of their structure, or require them as coenzymes.

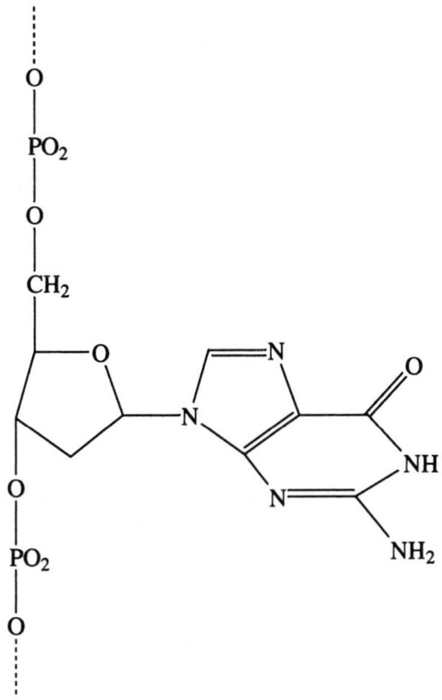

(10-XXVII)

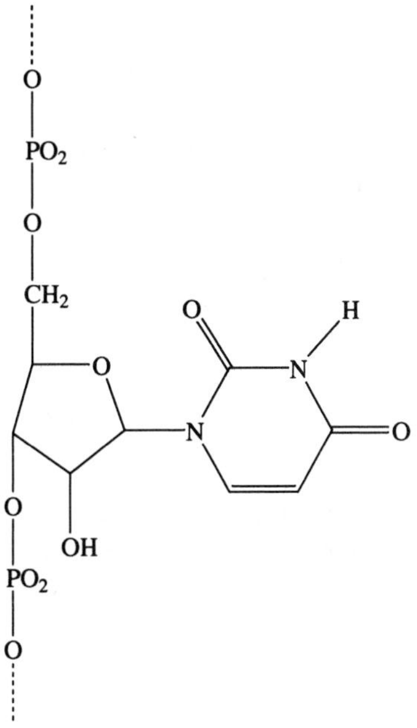

(10-XXVIII)

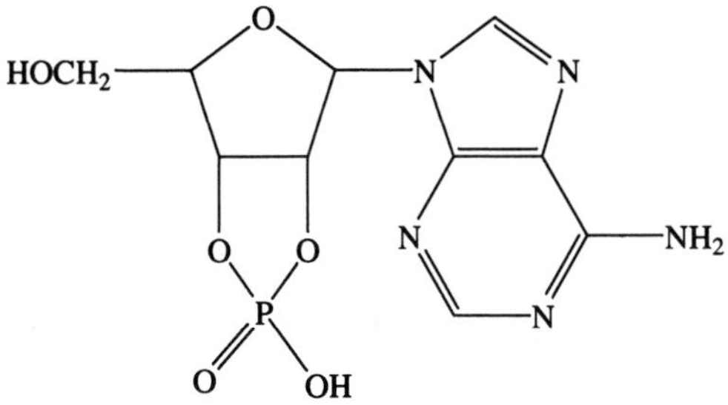

(10-XXIX)

In no small measure because of the importance of such substances and processes as those just mentioned, the hydrolysis of phosphate esters has received much fundamental study. Triesters are attacked by OH^- at phosphorus and by H_2O at carbon:

$$OP(OR)_3 \xrightarrow{{}^{18}OH^-} OP(OR)_2({}^{18}OH) + RO^-$$
$$OP(OR)_3 \xrightarrow{H_2{}^{18}O} OP(OR)_2(OH) + R^{18}OH$$

The strongly acidic diesters are completely in the anionic form at normal (and physiological) pH's:

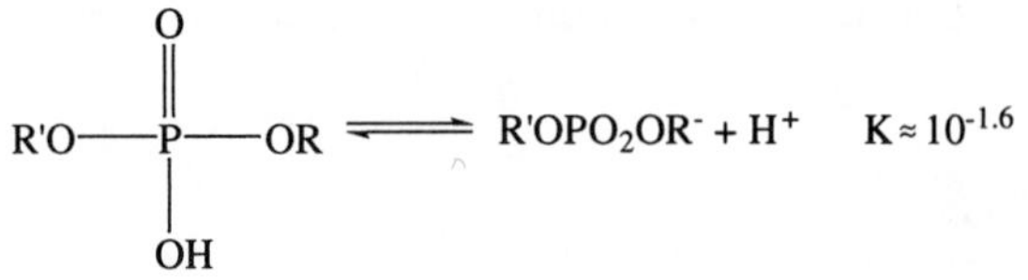

$$R'O-\overset{\overset{\displaystyle O}{\|}}{\underset{\underset{\displaystyle OH}{|}}{P}}-OR \rightleftharpoons R'OPO_2OR^- + H^+ \quad K \approx 10^{-1.6}$$

They are thus relatively resistant to nucleophilic attack by either OH^- or H_2O, and this is why enzymatic catalysis is indispensable to achieve useful rates of reaction. Three major pathways have been discussed for phosphate ester hydrolyses:

1. One-step nucleophilic displacement (S_N2) with inversion:

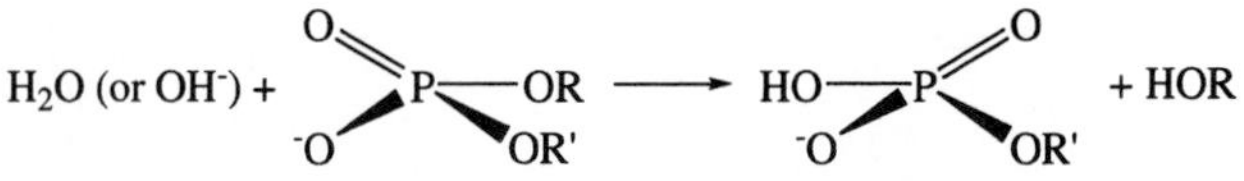

2. Nucleophilic attack in which a cyclic 5-coordinate intermediate is formed, which then pseudorotates:

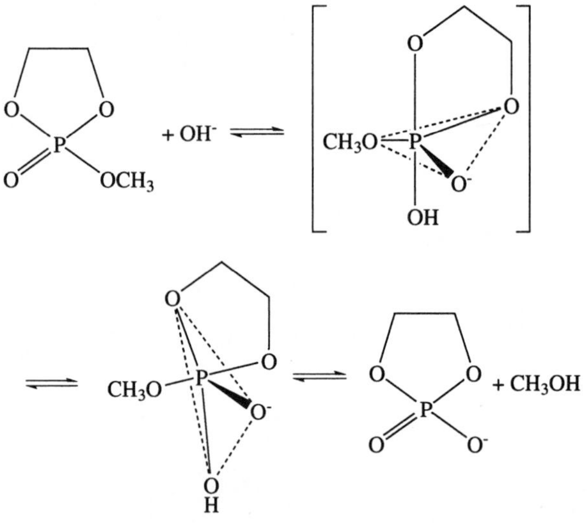

3. The somewhat controversial "metaphosphate" path, which entails release of a short-lived PO_3^- ion, which is rapidly converted by H_2O to $H_2PO_4^-$. The PO_3^- ion has been shown to be very stable and unreactive in the gas phase, and it appears likely that this pathway may occur in aprotic solvents (e.g., CH_3CN), but its occurrence in aqueous media remains uncertain.[49]

[49]J. R. Knowles *et al., J. Am. Chem. Soc.* **1988,** *110,* 1268.

10-13 Oxo Acids and Anions of As, Sb, and Bi

Arsenic

Arsenic acid (H_3AsO_4) is obtained by treating arsenic with concentrated nitric acid to give white crystals, $H_3AsO_4 \cdot \frac{1}{2}H_2O$, or by catalyzed treatment of As_2O_3 with air and H_2O under pressure. Unlike phosphoric acid, it is a moderately strong oxidizing agent in acid solution, the potentials being

$$H_3AsO_4 + 2H^+ + 2e = HAsO_2 + 2H_2O \qquad E^0 = 0.559 \text{ V}$$

$$H_3PO_4 + 2H^+ + 2e = H_3PO_3 + H_2O \qquad E^0 = -0.276 \text{ V}$$

Arsenic acid is tribasic but somewhat weaker ($pK_1 = 2.3$) than phosphoric acid. The arsenates generally resemble orthophosphates and often form isomorphous crystals.

Condensed arsenic anions are much less stable than the condensed phosphates and, owing to rapid hydrolysis, do not exist in aqueous solution. Dehydration of KH_2AsO_4 gives three forms, stable at different temperatures, of metaarsenate; one form is known to contain an infinite chain polyanion, and another contains the cyclotriarsenate ion, $[As_3O_9]^{3-}$.

Raman spectra show that in acid solutions of As_4O_6 the only detectable species is the pyramidal $As(OH)_3$. In basic solutions ($[OH^-]/[As^{III}]$ ratios of 3.5-15) the four pyramidal species $As(OH)_3$, $As(OH)_2O^-$, $As(OH)O_2^{2-}$, and AsO_3^{3-} appear to be present. In solid salts the arsenite ion is known in the AsO_3^{3-} form, as well as in more complex ones. Alkali metal arsenites are very soluble in water while those of heavy metals are more or less insoluble.

Antimony

No acid derived from Sb^{III} is known, although $Sb_2O_3 \cdot nH_2O$, of unknown structure has been reported. A few well-defined antimonites exist but are of little importance.

There is no well-defined antimony(V) acid either, nor has an SbO_4^{3-} ion ever been observed. All known antimonates are based on SbO_6 octahedra, and there are three classes. The simplest, obtained by addition of dilute alkali hydroxides to $SbCl_5$, contain the $Sb(OH)_6^-$ ion. Several alkali metal salts of these are known. Other antimonates that consist of linked SbO_6 octahedra are obtained by heating mixed oxides and have formulas such as $M^I SbO_3$, $M^{II}_2 Sb_2O_7$, and $M^{III} SbO_4$. The $Sb(OH)_6^-$ ion is known to oligomerize in aqueous solution. Knowledge of what is present in the complex mixture of oligomers in solution is sketchy but a $[Sb_8O_{12}(OH)_{20}]^{4-}$ ion has recently been obtained.[50] It consists of four edge-sharing bioctahedra associated via μ_3-O atoms as shown in Fig. 10-15.

Bismuth

When $Bi(OH)_3$ in strongly alkaline solution is treated with chlorine or other strong oxidizing agents, "bismuthates" are obtained, but never in a state of high purity.

[50]Y. Ozawa et al., J. Am. Chem. Soc. 1995, 117, 12007.

Figure 10-15   The structure of the $[Sb_8O_{12}(OH)_{20}]^{4-}$ from Y. Ozawa *et al., J. Am. Chem. Soc.* **1995,** *117,* 12007.

They can be made, for example, by heating Na_2O_2 and Bi_2O_3, which gives $NaBi^VO_3$. This yellow-brown solid dissolves in 0.5 *M* $HClO_4$ to give a solution that is stable in absence of light for several days. The Bi^V/Bi^{III} potential is estimated to be + 2.03 V, which suggests that Bi^V is one of the most powerful oxidants in aqueous solution, being comparable to peroxodisulfate ($S_2O_8^{2-}/2SO_4^{2-}$, E^0 = 2.01 V) or ozone (O_3, $2H^+/O_2$, H_2O, E^0 = 2.07 V). The precise nature of the bismuth (V) species is unknown, but it could be $[Bi(OH)_6]^-$.

10-14 Organic Compounds

There is a vast chemistry of organophosphorus compounds, and even for arsenic, antimony, and bismuth, the literature is voluminous. Consequently only a few topics can be discussed here. It must also be noted that we discuss only the compounds that have P—C bonds. Many compounds sometimes referred to as organophosphorus compounds that are widely used as insecticides, nerve poisons, and so on, as a result of their anticholinesterase activity, do not, in general, contain P—C bonds. They are usually organic esters of phosphates or thiophosphates; examples are the well-known malathion and parathion, $(EtO)_2P^V(S)(OC_6H_4NO_2)$. Compounds with P—C bonds are almost entirely synthetic, though a few rare examples occur in Nature.

Most of the organo derivatives are compounds with only three or four bonds to the central atom. They may be prepared in a great variety of ways, the simplest being by treatment of halides or oxo halides with Grignard reagents:

$$(O)MX_3 + 3RMgX \longrightarrow (O)MR_3 + 3MgX_2$$

Trimethylphosphine is spontaneously flammable in air, but the higher trialkyls are oxidized more slowly. The phosphine oxides (R_3PO), which may be obtained from the oxo halides as shown above or by oxidation of the corresponding R_3P compounds by H_2O_2 or air, are all very stable. The P—O bonds are very short (e.g., 1.483 Å in Ph_3PO), suggesting a bond order >2. In R_3PS compounds the P—S distances are close to the double-bond values.

In general there are not only R_3E molecules but also REX_2 and R_2EX types for all four elements. The R_2EX (or Ar_2EX) types are the more important, in part because they can be used to introduce R_2E groups into other compounds in two ways, either directly:[51]

$$Ph_2BiCl + TePh^- \longrightarrow Ph_2BiTePh + Cl^-$$

[51]F. Calderazzo *et al., Inorg. Chem.* **1988,** *27,* 3730.

or after being converted to anions accompanied by Li^+ [52] or Mg^{2+}:[53]

$$Ar_2PLi + RCl \longrightarrow Ar_2PR + LiCl$$

This is the way that, for example, the widely used ligand Ph_2MeP is made.

In view of the fact that Bi has the greatest tendency to cationic behavior, it is not surprising that Ar_2BiX compounds give $Ar_2BiL_2^+$ ions:[54]

$$Ph_2BiBr + AgBF_4 + 2py \longrightarrow [Ph_2Bi \, py_2]BF_4 + AgBr$$

The Ar_2BiX and $ArBiX_2$ molecules also have extensive reactivity as Lewis acids.[55]

There are good methods for preparing optically pure dissymmetric phosphine oxides, abcPO, for example, $(CH_3)(C_3H_7)(C_6H_5)PO$. It is then possible to reduce these to optically pure phosphines. The reductant $HSiCl_3$ accomplishes this with either retention or inversion, depending on the base used in conjunction with it. Hexachlorodisilane reduces with inversion, and to account for this the following mechanism has been proposed:

$$abcPO + Si_2Cl_6 \longrightarrow [abcPOSiCl_3]^+ + SiCl_3^-$$

$$SiCl_3^- + [abcPOSiCl_3]^+ \longrightarrow [Cl_3SiPabc]^+ + SiCl_3O^- \text{ (inversion)}$$

$$[Cl_3SiPabc]^+ + SiCl_3O^- \longrightarrow Cl_3SiOSiCl_3 + Pabc$$
$$\text{(attack of } Cl_3SiO^- \text{ on } SiCl_3)$$

Interestingly, the same reagent removes S from abcPS with retention; it is presumed that the first step is similar, but that $SiCl_3^-$ then attacks sulfur, rather than phosphorus:

$$abcPS + Si_2Cl_6 \longrightarrow [abcPSSiCl_3]^+ + SiCl_3^-$$

$$[abcPSSiCl_3]^+ + SiCl_3^- \longrightarrow abcP + Cl_3SiSSiCl_3$$

Trialkyl- and triarylphosphines, -arsines, and -stibines, and chelating di- and triphosphines and -arsines are widely used as π-acid ligands (Section 10-17).

Toward trivalent boron compounds, gas phase calorimetric studies of the reactions (E = P, As, and Sb)

$$Me_3E(g) + BX_3(g) = Me_3EBX_3(s)$$

show that the order of base strength is P > As > Sb.

Triphenylphosphine, a white crystalline solid (mp 80°C), is a particularly important ligand for transition metal complexes and is used industrially in the rhodium-

[52]H. Aspinall and M. R. Tillotson, *Inorg. Chem.* **1996**, *35*, 5.
[53]M. Weaterhauser and A. Pfitzner, *J. Organometal. Chem.* **1995**, *487*, 187.
[54]N. C. Norman *et al.*, *J. Chem. Soc. Dalton Trans.* **1996**, 443.
[55]N. C. Norman *et al.*, *J. Organometal. Chem.* **1995**, *496*, 59.

catalyzed hydroformylation process. It is also widely used in the *Wittig reaction* for olefin synthesis. This reaction involves the formation of alkylidenetriphenylphosphoranes from the action of butyllithium or other base on the quaternary halide, for example,

$$[(C_6H_5)_3PCH_3]^+Br^- \xrightarrow{\text{Bu}^n\text{Li}} (C_6H_5)_3P{=}CH_2$$

This intermediate reacts very rapidly with aldehydes and ketones to give zwitterionic compounds (10-XXXa), which eliminate triphenylphosphine oxide under mild conditions to give olefins (10-XXXb):

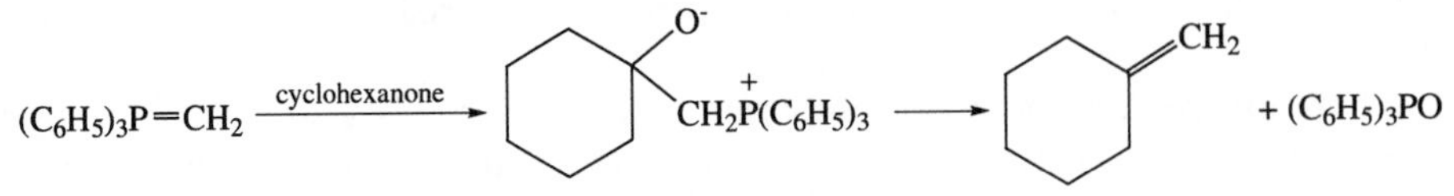

$$(C_6H_5)_3P{=}CH_2 \xrightarrow{\text{cyclohexanone}}$$

(a)   (b)

(10-XXX)

Alkylidene Phosphoranes

In addition to the important Wittig reagent ($Ph_3P{=}CH_2$) there are many other $R_3P{=}CR'R''$ compounds in which (usually) R' or both R' and R'' are hydrogen. Examples are $Me_3P{=}CH_2$, $Et_3P{=}CH_2$, $Me_2EtP{=}CH_2$, and $Et_3P{=}CHMe$, all of which are colorless liquids, stable for long periods in an inert atmosphere. There are arsenic analogues of these compounds.

The P=C distances range from 1.66 to 1.74 Å, clearly indicative of considerable, if not full double-bond character, and in general their electronic structures may be represented in resonance terms as follows:

$$R_3P{=}CR'_2 \longleftrightarrow R_3\overset{+}{P}{-}\overset{-}{C}R'_2$$

These compounds, which are often called phosphine ylides, are prepared by deprotonation of a quaternary phosphonium ion, namely,

$$Me_4PBr + NaNH_2 \xrightarrow{\text{reflux}} Me_3P{=}CH_2 + NaBr + NH_3$$

$$Ph_2Me_2PBr + KH \xrightarrow{\text{THF}} Ph_2MeP{=}CH_2 + KBr + H_2$$

These molecules can be further deprotonated to give the ylide anions, $Me_2P(CH_2)_2^-$ and $Ph_2P(CH_2)_2^-$, which can serve as bridging ligands, as in (10-XXXI) and (10-XXXII).

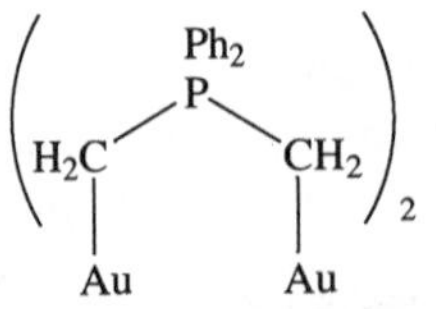

(10-XXXI) 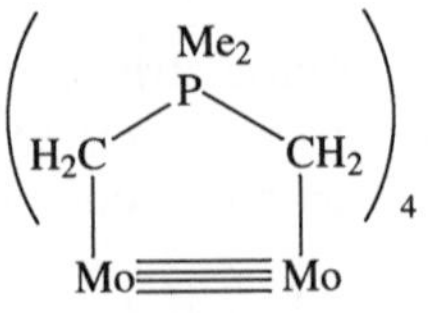   (10-XXXII)

An unusual, cumulenelike phosphine ylide is $Ph_3P{=}C{=}PPh_3$, prepared by dehydrobromination of $[Ph_3PCH_2PPh_3]Br_2$ with potassium in diglyme. Somewhat surprisingly, the P—C—P angle is 137°, suggesting that again there are polar contributions to the electronic structure:

Phosphaalkenes

Not to be confused with the phosphine ylides, these are compounds of the general formula $R_2C=P-R'$. They are accessible by the following general route:

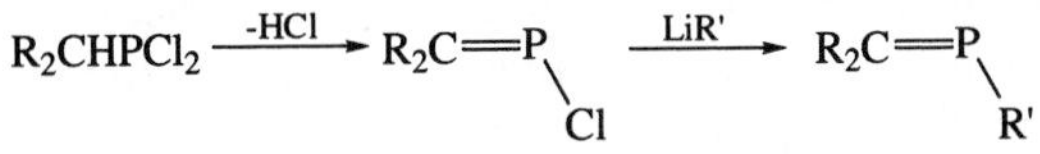

$$R,R' = Me_3Si, 2,4,6\text{-}Bu^t_3C_6H_2, \text{ mesityl, and so on}$$

The molecules are bent ($\sim110°$), with $P=C$ distances of ~1.68 Å. They can form complexes similar to those formed by olefins in which the bonding arises by donation of the π electrons, as in (10-XXXIII).

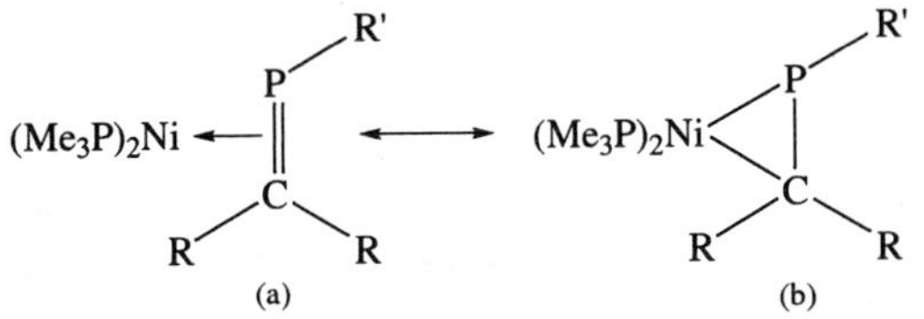

(10-XXXIII)

There is also a P^{III} cumulene type molecule, $(2,4,6\text{-}Bu^t_3C_6H_2)P=C=P(2,4,6\text{-}Bu^t_3 C_6H_2)$, in which the central P—C—P unit is nearly linear (*ca.* 173°).

Phospha- and Arsaalkynes

There are several compounds of the type $RC\equiv P$[56] as well as a few arsaalkynes.[57] The $RC\equiv P$ and $RC\equiv As$ molecules are prone to oligomerize unless the R groups are very large. For example, while $(2,4,6\text{-}Bu^t_3C_6H_2)C\equiv As$ is isolable and its structure has been determined ($C\equiv As$ bond length is 1.657 Å), $Me_3CC\equiv As$ has only a transient existence and in the presence of $CoCl_2$ it rapidly forms the cuboidal tetramer 10-XXXIV.

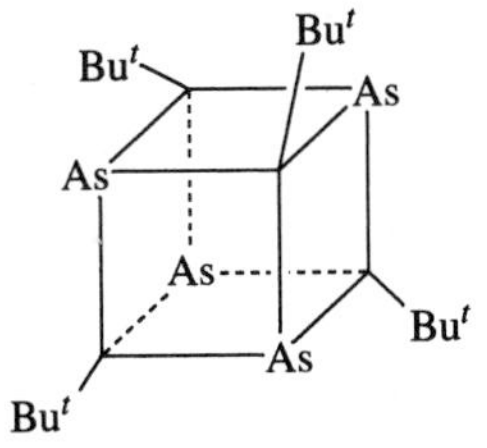

(10-XXXIV)

[56]J. F. Nixon, *Coord. Chem. Rev.* **1995**, *145*, 201; R. Streubel, *Angew. Chem. Int. Ed. Engl.* **1995**, *34*, 436; L. Weber, *Chem. Ber.* **1996**, *129*, 367.
[57]J. F. Nixon *et al.*, *J. Chem. Soc., Chem. Commun.* **1994**, 2061.

Formation of polycyclic oligomers, not necessarily the cuboidal tetramer, is characteristic of the phosphaalkynes.[58] The P≡C and As≡C bonds can serve as π-donor ligands in much the same way as nitriles and acetylenes. The fact that the carbon atom is more nucleophilic than the P atom has been explained on the basis of electron density mapping for Bu^{t}C≡P.[59]

Pyridine Analogues

These molecules (10-XXXV) are known for all four elements. Older trivial names for these are phosphabenzene, arsabenzene, etc., but the current literature[60] favors the names phosphinine, arsinine, stibinine, and bismuthinine. The bismuth compound, C_5H_5Sb, is extremely unstable and polymerizes at room temperature. Phosphinine and arsinine are thermally stable (distillable) and can be prepared by the reactions:

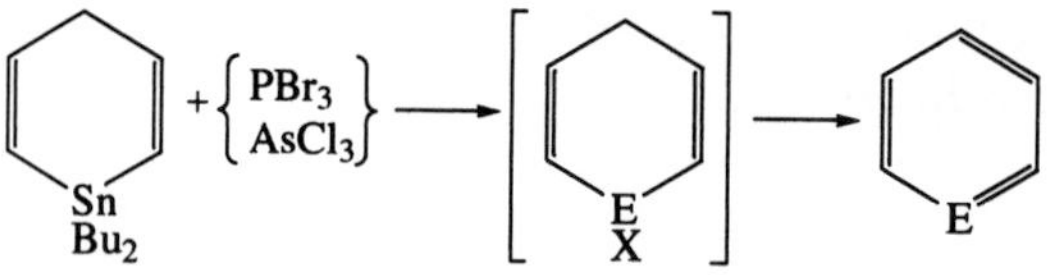

Not only are derivatives, such as the 2,4,6-triphenyl ones, known for both C_5H_5P and C_5H_5As, there are polycyclic compounds such as 1-arsanaphthalene, 9-arsaanthracene, and 2,2'-biphosphines.[61]

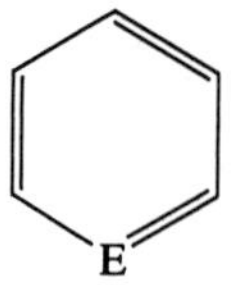

E = P, As, Sb, and Bi

(10-XXXV)

Both C_5H_5P and C_5H_5As have planar structures with C—C and C—E bond lengths indicative of aromatic delocalization of the π electrons. However, the scope of their aromatic organic chemistry is not nearly as extensive as that of pyridine. Phosphinine and its derivatives form both σ and η^6 type complexes, for example, 10-XXXVI and 10-XXXVII.

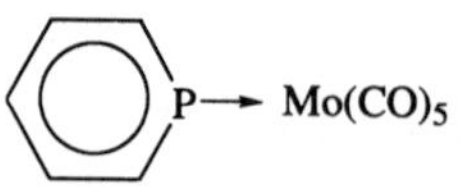

(10-XXXVI) 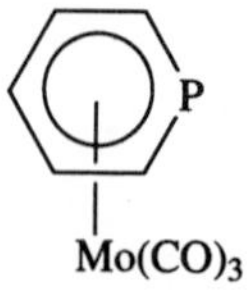   (10-XXXVII)

[58]J. F. Nixon *et al., Angew. Chem. Int. Ed. Engl.* **1994,** *33,* 2202; **1995,** *34,* 484.
[59]M. Yu. Antipin, *J. Chem. Soc., Chem. Commun.* **1995,** 505.
[60]H. T. Teunissen and F. Bickelhaupt, *Organometallics* **1996,** *15,* 802.
[61]F. Mathey *et al., Inorg. Chem.* **1995,** *34,* 5070; F. Mathey and P. LeFloch, *Chem. Ber.* **1996,** *129,* 263.

Other Heteroaromatics

There are a number of 5-membered ring compounds. Pyrrole analogues (10-XXXVIII, E = N) are known with the heavier elements P, As, and Sb, but not Bi.

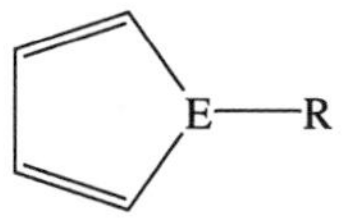

(10-XXXVIII)

An interesting feature of their chemistry is that they can all form cyclopentadienyl-like anions. There are also multihetero ring anions[62,63] such as $1,3\text{-}(CH)_3PAs^-$ which has been shown to form a ferrocene analogue:

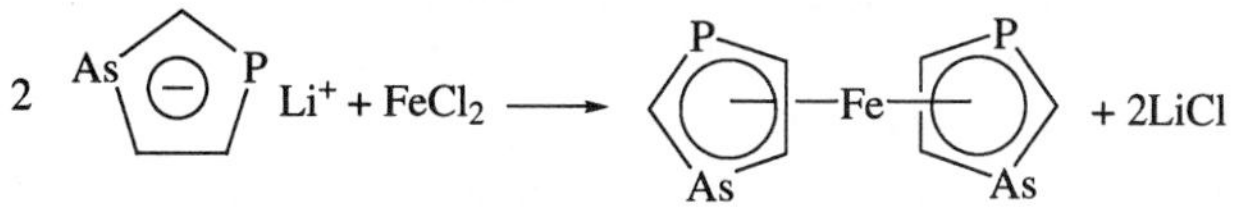

The entire series of $(CH)_nP_{5-n}^-$ ions are also known. In a very bizarre reaction[64] it was found that cocondensation of V atoms and an excess of $Bu^tC\equiv P$ formed the sandwich compound 10-XXXIX. With iron, there are compounds containing $Cp*Fe(C_2Bu_2^tP_3)$ and $Cp*Fe(C_3Bu_3^tP_2)$ either as such or in the former case having one P atom forming a donor bond to the Mo or W atom of an $M(CO)_5$ moiety.[65]

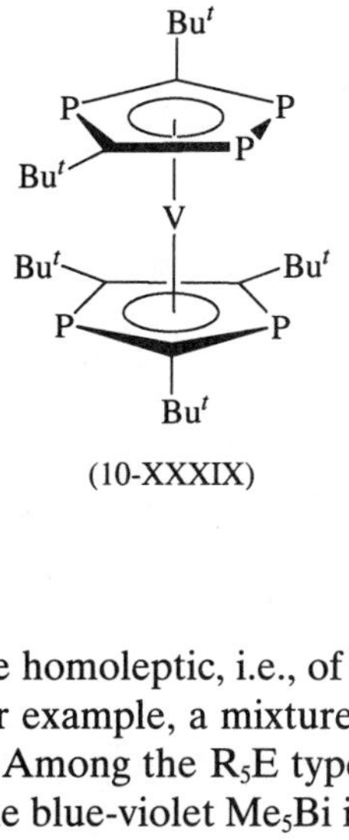

(10-XXXIX)

Pentavalent Compounds

Compounds of this kind may be homoleptic, i.e., of the R_5E or Ar_5E types, or they may have mixed ligand sets, for example, a mixture of Ar and R groups, or also a mixture of R/Ar and halogens. Among the R_5E types the Me_5E species are known only for E = As, Sb, and Bi. The blue-violet Me_5Bi is very unstable. It is prepared[66]

[62]F. Mathey, *J. Organomet. Chem.* **1994**, *475*, 25.
[63]S. L. Buchwald *et al.*, *Organometallics* **1994**, *13*, 5160.
[64]J. F. Nixon *et al.*, *J. Chem. Soc., Chem. Commun.* **1995**, 1659.
[65]R. Schmutzler *et al.*, *J. Organomet. Chem.* **1996**, *512*, 141.
[66]S. Wallenhauer and K. Seppelt, *Angew. Chem. Int. Ed. Engl.* **1994**, *33*, 976; B. Neumüller and K. Dehnicke, *Angew. Chem. Int. Ed. Engl.* **1994**, *33*, 1726.

by the reaction of a stoichiometric amount of LiMe with $BiMe_3Cl_2$, while the use of an additional equivalent of LiMe gives the $BiMe_6^-$ ion:

$$BiMe_3Cl_2 + 2LiMe \xrightarrow[-95°C]{Et_2O} BiMe_5 + 2LiCl$$

$$BiMe_3Cl_2 + 3LiMe \xrightarrow{THF} [Li(THF)_4][BiMe_6]$$

The only R_4Bi^+ ion known is prepared by the reaction of "magic methyl" with $BiMe_3$:

$$BiMe_3 + MeOSO_2CF_3 \longrightarrow [BiMe_4][CF_3SO_3]$$

Many Ar_5E, Ar_3R_2E, and Ar_3EX_2 compounds are known. Those of phosphorus and arsenic favor the *tbp* structure, those of Bi mostly have the *sp* structure (although the distinction is often blurred) and for those of Sb, both types as well as those in between are common. There is also fluxional behavior in solution such that all R and Ar groups are nmr equivalent.

The structures of the Ar_nEX_{5-n} and R_nEX_{5-n} compounds in the solid state have been problematic. For example, even though all the Me_3EBr_2 compounds of As and Sb appear to be molecular in solution, with the Br atoms in axial positions of a trigonal bipyramid, they differ in the solid state.[67] Until recently the difference was not clear because of crystallographic ambiguities. While Me_3AsCl_2, Me_3SbCl_2, Me_3SbBr_2, and Me_3SbI_2 are all *tbp* molecules in the crystalline state, Me_3AsBr_2 consists of infinite, quasi-ionic chains (10-XL).

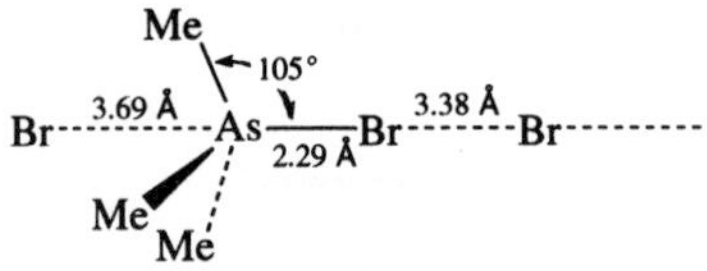

(10-XL)

It should also be noted that still more complex structures can be found among compounds formed in reactions that superficially have been expected to give R_3EX_2 or Ar_3EX_2, especially for E = As[68] or P.[69] The products have structures that change with changes in R, Ar, and X. A few representative examples are: (1) Ph_3AsBr_2 has a *tbp* structure but Ph_3AsI_2 has a $Ph_3As-I-I$ structure; (2) $(p-FC_6H_4)_3As$ reacts with I_2 to give Ar_3AsI_4, whose actual composition is $[Ar_3AsI]I_3$; (3) reaction of I_2 with Ph_3P proceeds as follows:

$$Ph_3P \xrightarrow{I_2} [Ph_3PI]I \xrightarrow{I_2} [Ph_3PI]I_3$$

$$2[Ph_3PI]^+ + I_3^- \longrightarrow [(Ph_3PI)_2I_3]^+ \xrightarrow{I_3^-} [(Ph_3PI)_2I_3]I_3$$

[67]A. L. Rheingold *et al.*, *J. Organomet. Chem.* **1996**, *512*, 21.
[68]S. M. Godfrey *et al.*, *J. Chem. Soc. Dalton Trans.* **1995**, 3873.
[69]F. A. Cotton and P. Kibala, *J. Am. Chem. Soc.* **1987**, *109*, 3308.

The mixed chloroalkyl phosphoranes, R_nPCl_{5-n} behave as Lewis acids, for example,

$$PMeCl_4 + R_4NCl \longrightarrow [R_4N][PMeCl_5]$$

but can also react as Cl^- donors, namely,

$$PMeCl_4 \xrightarrow{SbCl_5} [PMeCl_3][SbCl_6] \xrightarrow{bipy} [PMeCl_3bipy][SbCl_6]$$

Attempts to make pentamethylphosphorane by the same method as succeeds for Ph_5P gives the ylid, namely,

$$Ph_4PBr + PhLi \longrightarrow Ph_5P + LiBr$$

but

$$Me_4PI + MeLi \longrightarrow Me_3P{=}CH_2 + CH_4 + LiI$$

and, similarly,

$$Ph_4PBr + MeLi \longrightarrow Ph_3P{=}CH_2 + C_6H_6 + LiBr$$

However, the formation of ylids can be prevented by incorporation of P into heterocycles as in the following examples:

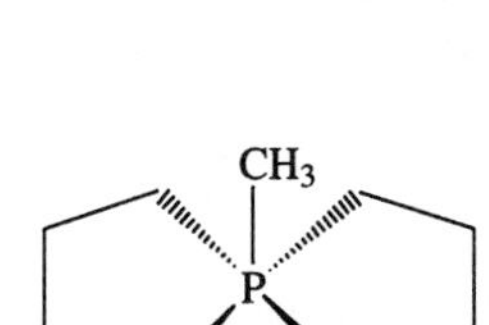

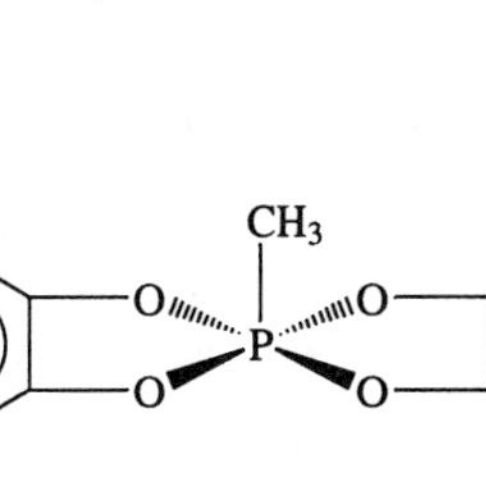

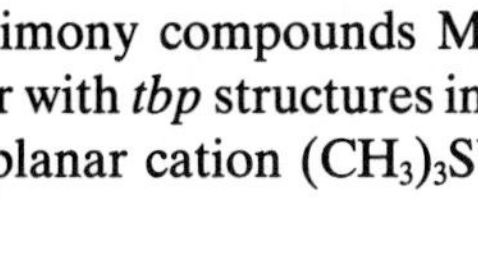

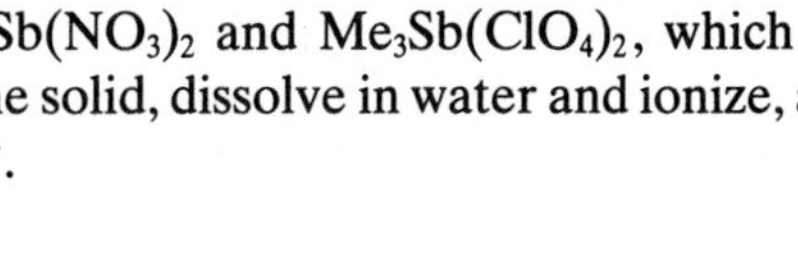

The spirophosphoranes and phosphorus(V) esters are approximately square pyramidal.

The antimony compounds $Me_3Sb(NO_3)_2$ and $Me_3Sb(ClO_4)_2$, which appear to be molecular with *tbp* structures in the solid, dissolve in water and ionize, apparently to give the planar cation $(CH_3)_3Sb^{2+}$.

10-15   Compounds with Element-Element Single Bonds

The elements themselves as well as the halides (E_2X_4) and the hydrogen compounds (H_2PPH_2 and P_3H_5) have already been mentioned. Also previously noted (Section 10-3) are the many Zintl anions, which contain E—E bonds that are approximately single bonds, although often better discussed in terms of a delocalized picture. There is, finally, a very large number of molecular compounds containing organic groups together with one or more E—E bonds, and it is to these that we now turn.

The simplest are the R_2E—ER_2 molecules, formed by all four elements. A general method of preparation is by coupling identical R_2EX units, for example,

$$2Ph_2BiCl + 2Na \xrightarrow{NH_3(l)} Ph_2Bi\text{——}BiPh_2 + 2NaCl$$

The *diphosphines* (R_2PPR_2) readily react with O_2 or S_n to give the $R_2P(O)P(O)R_2$ and $R_2P(S)P(S)R_2$ molecules, but the latter, also available in other ways, can be desulfurized as a way of preparing R_2PPR_2 compounds.

Among the *diarsines* (R_2As—AsR_2) the methyl compound (Me_4As_2) was called in older literature dicacodyl, and its Me_2AsX derivatives are called cacodyls, for example, cacodyl chloride (Me_2AsCl).

Distibines (R_2Sb—SbR_2) *and dibismuthines* (R_2Bi—BiR_2) with a wide variety of R groups are known. All of them have a yellow or orange color when melted, but those with small R groups give solids that are orange, red, blue, or violet. It is believed that these changes in color are due to intermolecular association into $\cdots$E—E$\cdots$E—E$\cdots$ chains in the solid. The Ph_2E—EPh_2 compounds, which show no such chains remain yellow as solids, whereas the solid Me_2E—EMe_2 compounds are blue.

Other classes of compounds containing one P—P bond are the $(RO)_2PP(OR)_2$ and $(RO)_2P(O)P(O)(OR)_2$ types, as well as RClPPClR and $(R_2N)ClPPCl(NR_2)$. On mixing Me_2EEMe_2 and $Me_2E'E'Me_2$, the $Me_2EE'Me_2$ compounds are formed in equilibrium with the homonuclear ones.

Organocyclophosphanes (PR)$_n$, *and -arsanes* (AsR)$_n$, with n = 3 to 6, are well known. Ring size preference is largely determined by the size or electronegativity of the R groups. Small alkyl substituents as well as phenyl favor (PR)$_5$, whereas bulky groups favor (PR)$_4$ and even (PR)$_3$. The compounds are thermally stable though reactive and typical preparative reactions are:

$$5CH_3PCl_2 + 10LiH \longrightarrow (PCH_3)_5 + 10LiCl + 5H_2$$

$$3Me_3CPCl_2 + 3Mg \longrightarrow (PCMe_3)_3 + 3MgCl_2$$

$$nRAsO_3Na_2 + (\text{excess}) H_3PO_2 \longrightarrow (RAs)_n$$

The larger rings are, as expected, puckered (Fig. 10-16), while in (PR)$_3$ compounds only two of the R groups lie on one side of the ring plane.

There are also linear triphosphines containing CF_3 groups, such as $(CF_3)P[P(CF_3)_2]_2$, and $CH_3P[P(CF_3)_2]_2$, and other partly cyclic species such as

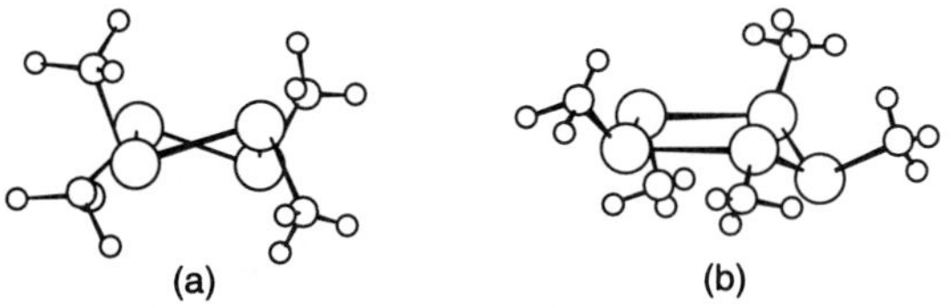

Figure 10-16 The structures of (a) $(CF_3P)_4$ and (b) $(CF_3P)_5$. Large, medium, and small circles represent P, C, and F atoms, respectively.

(10-XLI), (10-XLII), and (10-XLIII) (where R = Me$_3$C):

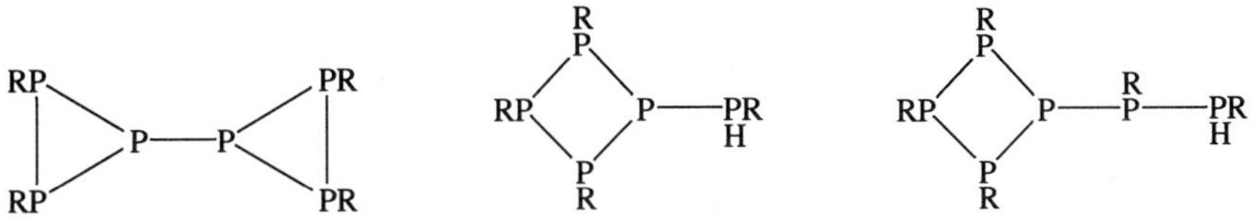

(10-XLI)   (10-XLII)   (10-XLIII)

Even larger molecules are known, for which polycyclic structures such as 10-XLIV–10-XLVI have been proposed.[70]

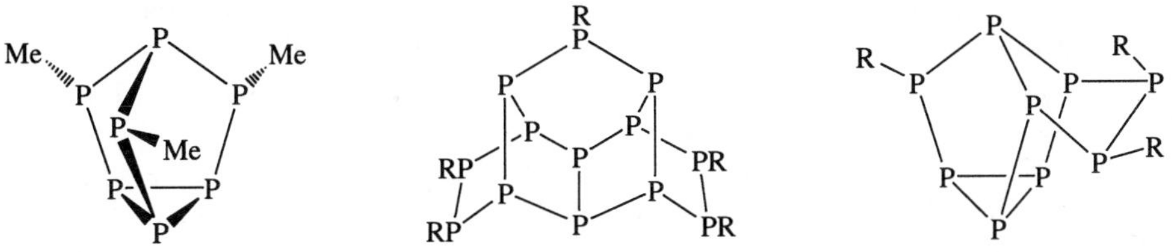

(10-XLIV)   (10-XLV)   (10-XLVI)

10-16 Compounds with Element-Element Double Bonds[71]

While N=N bonds abound, until recently other Group 15 E=E bonds were unknown. Today, however, there are stable compounds that contain P=P, P=As, P=Sb, As=As, and Bi=Bi[72] bonds. Compounds containing Sb=Sb bonds are so far known only as ligands (10-XLVII). The best calculations show that the π bond strengths of HN=NH and HP=PH are 256 and 150 kJ mol^{-1}, respectively. Thus, while much weaker than the N—N π bond, the P—P π bond has considerable strength.

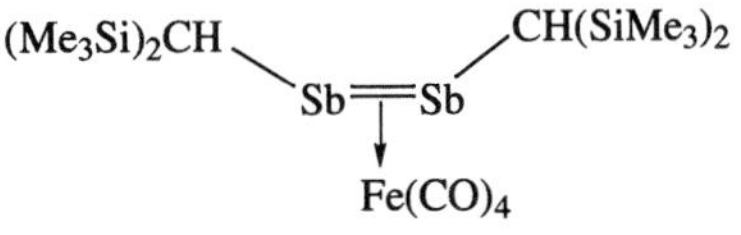

(10-XLVII)

[70]M. Baudler *et al.*, *Z. anorg. allg. Chem.* **1994,** *620,* 2021.
[71]L. Weber, *Chem. Rev.* **1992,** *92,* 1839; N. C. Norman, *Polyhedron* **1993,** *12,* 2431.
[72]N. Tokitoh *et al.*, *Science* **1997,** *277,* 78.

Table 10-6   Structures of Some RE=E′R′ Compounds[a]

E	E′	R	R′	E=E′ distance (Å)	Sum of Double Bond Radii (Å)
P	P	Ar[b]	Ar[b]	2.034(2)	2.00
P	P	$(Me_3Si)_3C$	$(Me_3Si)_3C$	2.014(6)	2.00
				2.004(6)	2.00
P	P	$(Me_3Si)BuN$	$(Me_3Si)BuN$	2.034	2.00
P	As	Ar	$(Me_3Si)_2CH$	2.124(2)	2.11
As	As	Ar	$(Me_3Si)_2CH$	2.224(2)	2.22
As	As	$(Me_3Si)_3C$	$(Me_3Si)_3C$	2.244	2.22
Bi	Bi	Ar	Ar	2.82	—

[a]See ref. 71 for others.
[b]Ar = 2,4,6-$(Me_3C)C_6H_2$ (called supermesityl).

The basic difficulty in obtaining stable compounds that contain E=E′ bonds is that such bonds are unstable relative to cyclic oligomers such as $(RP)_n$ or $(RAs)_n$, but oligomerization can be thwarted by employing large R groups. This strategy is mainly a kinetic one (the compounds are relatively unreactive in other ways as well) but may have a thermodynamic element as well in that cyclic $(RE)_n$ molecules are doubtless destabilized when R is very large.

Some known RE=E′R′ molecules are listed in Table 10-6 along with their E=E′ bond lengths. The bond lengths are all ~0.20 Å shorter than the corresponding E—E′ lengths and in good accord with estimates from Pauling's double-bond radii. The molecules are all planar in their central X—E=E—X portions. The structure of $(Me_3Si)_3CAs=AsC(SiMe_3)_3$ is shown in Fig. 10-17.

Methods of preparation are mainly the following, although others are known:

$$2RPCl_2 + 2Mg \longrightarrow RP{=}PR + 2MgCl_2$$

$$RECl_2 + H_2E′R′ \xrightarrow{\text{base}} RE{\equiv}E′R′ + 2HCl$$

$$R_2NP\overset{\overset{\displaystyle Cl}{|}}{\underset{\underset{\displaystyle H}{|}}{-}}PNR_2 \xrightarrow{(Bu)(Me_3Si)NLi} RNP{=}PNR$$

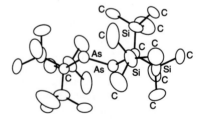

Figure 10-17   The structure of $(Me_3Si)_3CAs{=}AsC(SiMe_3)_3$.

Although the large R groups inhibit oligomerization, as well as other forms of reactivity, some of the reactions that have been observed are

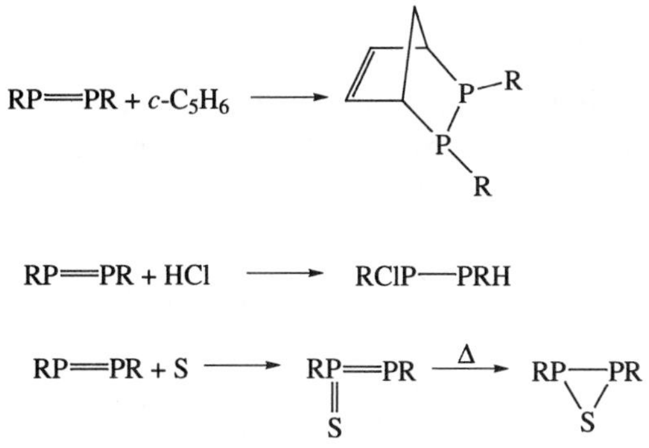

$$RP{=}PR + c\text{-}C_5H_6 \longrightarrow$$

$$RP{=}PR + HCl \longrightarrow RClP{-}PRH$$

$$RP{=}PR + S \longrightarrow RP{=}PR \overset{\Delta}{\longrightarrow} RP{-}PR$$

There are also reactions with metal-centered Lewis acids, whereby a lone pair on P or As is employed to form a donor bond; compounds such as (10-XLVIII)-(10-L) are thus known [Ar = 2,4,6-$(Me_3C)_3C_6H_2$].

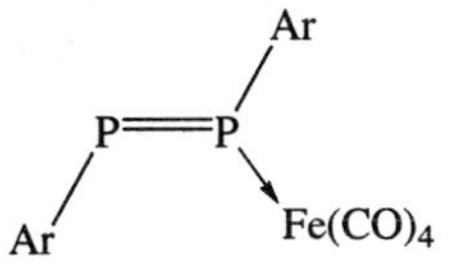

(10-XLVIII) 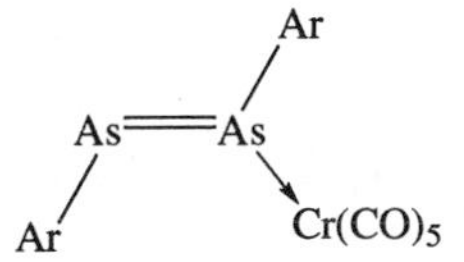   (10-XLIX) 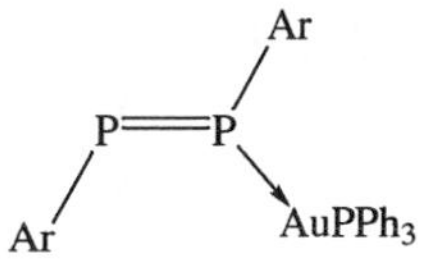   (10-L)

There are a few examples of hetero-double bonds where elements from Groups 15 and 16 are coupled. An example[73] is (10-LI).

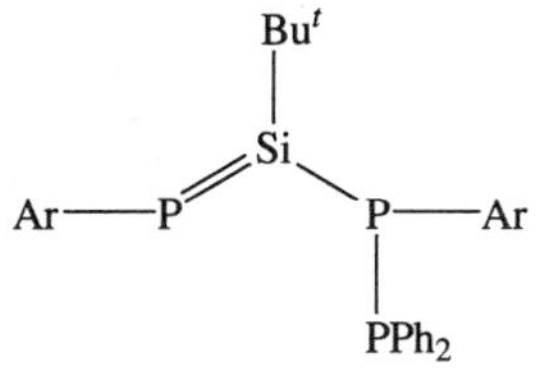

(10-LI)

A molecule containing both P—P and P=P bonds is $Bu_2^tP{-}P{=}PBu_2^tBr$, where the bond lengths are 2.20 and 2.08 Å, respectively.[74] Be it noted however that the formal valence states are III, III, V.

10-17 Ligands Formed by the Group 15 Elements

The huge role played by nitrogen-based ligands in aqueous coordination chemistry is matched by the importance of phosphorus-based ligands in the organometallic

[73]E. Niecke et al., J. Am. Chem. Soc. 1993, 115, 3314.
[74]G. Fritz et al., Z. anorg. allg. Chem. 1994, 620, 1818.

chemistry of the transition elements. Only a few of the salient features of this enormous subject will be mentioned here.

The Elements and Their Ions

Single atoms, which can be formally regarded as E^{3-} are found as capping ligands (10-LII), as centering ligands in planar EM_3 groups, or occupying the center of a metal atom cluster[75] such as $Zr_6I_{12}(\mu_6\text{-}P)$.

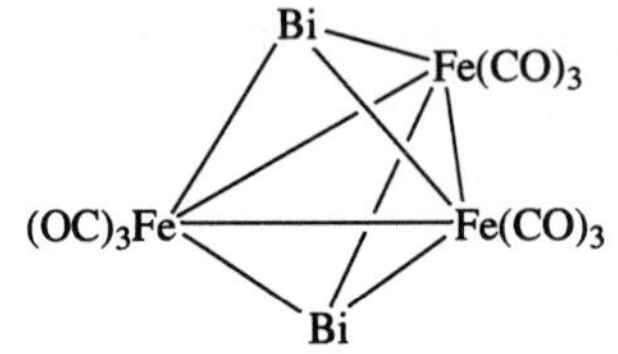

(10-LII)

Although $M\equiv N$ bonds are well known, the first two examples of $M\equiv P$ bonds were only recently reported, with $M = Mo$,[76] W.[77] There are a number of ligands that are made up of several Group 15 atoms, such as the P_3 unit found, for example, in complexes such as (10-LIII) and (10-LIV) where L represents a tripod ligand such as $MeC(CH_2PPh_2)_3$ or $N(C_2H_4PPh_2)_3$. The P_4 molecule itself is known to act as a uni- or bidentate ligand.

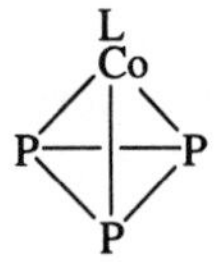

(10-LIII) 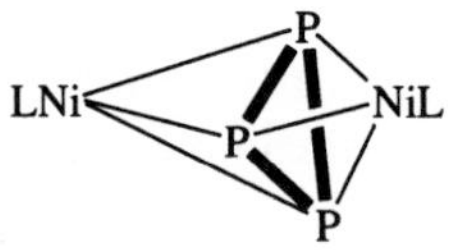   (10-LIV)

The P_2 and As_2 molecules can bind in a variety of ways, formally as four-, six-, or eight-electron donors:

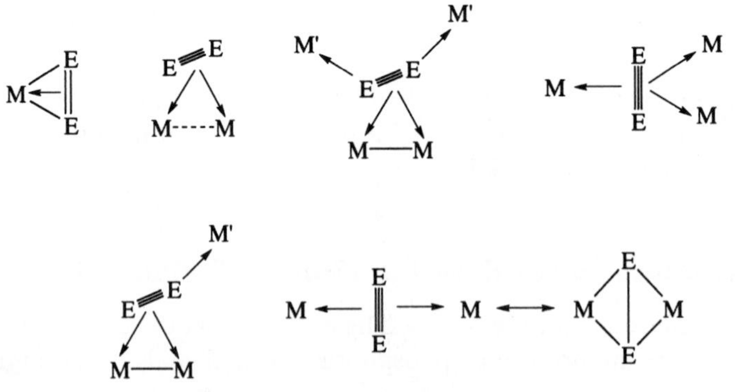

[75]G. Rosenthal and J. D. Corbett, *Inorg. Chem.* **1988**, *27*, 53.
[76]R. R. Schrock *et al., Angew. Chem. Int. Ed. Engl.* **1995**, *34*, 2044.
[77]C. C. Cummins *et al., Angew. Chem. Int. Ed. Engl.* **1995**, *34*, 2042.

There are resemblances in several cases to the binding of acetylene. Examples are $[Cp(CO)_2Mo]_2As_2$, $[(OC)_5W]_3As_2$, and $(diphos)_2NiP_2$.

There are five isomers of $(CH)_6$, of which benzene is the most stable. Each of these, in principle, has an E_6 analogue. The monocyclic E_6 analogues of benzene, P_6 and As_6, have both been found in compounds of the type[78] 10-LV. In addition there are compounds containing As_6 units that are topologically analogous to several other $(CH)_6$ isomers,[79] for example, 10-LVI as well as a compound with Cp^*Co moieties on both faces of an As_4 ring. There are also a few compounds having metal atoms on both sides of P_5 and As_5 rings which are somewhat distorted from 5-fold symmetry.

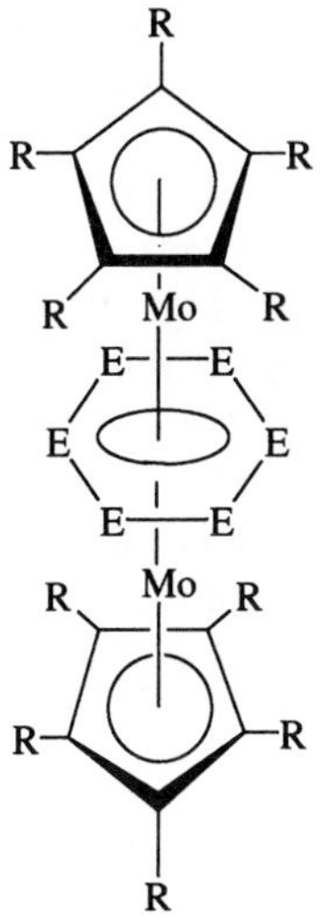

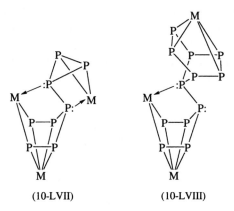

(10-LV)

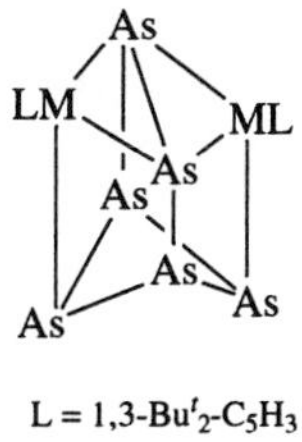

(10-LVI)

$L = 1,3-Bu^t_2-C_5H_3$

Recently, compounds containing P_8 and P_{12} ligands have been reported.[80] These large skeins of phosphorus atoms are found coordinated to $(1,2,4-Bu^i_3C_5H_2)Co$ moieties as shown in 10-LVII and 10-LVIII, where M represents the metal moiety.

(10-LVII)

(10-LVIII)

[78]O. J. Sherer et al., Angew. Chem. Int. Ed. Engl. **1988**, 27, 212.
[79]O. J. Sherer et al., J. Organomet. Chem. **1992**, 425, 141; **1994**, 484, C5.
[80]O. J. Sherer et al., Chem. Ber. **1996**, 129, 53.

Phosphorus Oxides and Sulfides

Cage molecules such as P_4O_6 and P_4S_3 can employ a lone pair on a phosphorus atom to form complexes such as $Fe(CO)_4(\eta^1\text{-}P_4O_6)$ and $(triphos)Ni(\eta^1\text{-}P_4S_3)$; bridge formation may also occur, as in $[Pt(PPh_3)(\mu\text{-}P_4S_3)]_3$. Removal of $P{=}O$ from P_4O_{10} to give $P_3O_9^{3-}$ allows binding of three oxygen atoms to a metal.

Phosphines as Ligands

One of the most important classes of ligands in transition metal chemistry are the PR_3, PR_2Ar, $PRAr_2$, and PAr_3 molecules. Closely related are the $P(OR)_3$ and $P(OAr)_3$ molecules as well as mixed haloalkyl and haloaryl ligands. The key properties of these ligands, all of which are subject to deliberate control, are their steric demands, their σ-donor ability and their π-acidity.

Analogous compounds are formed by As, Sb, and to a small extent Bi. The donor ability decreases P > As > Sb > Bi, other things being equal. The As compounds are generally similar to their P analogues, but the phosphorus compounds are so much more important than those of As and Sb that we shall restrict discussion to them alone.

The steric demand of a monophosphine ligand is difficult to define precisely, especially for those with more than one kind of group attached to the P atom. A rough but useful estimate of the solid angle subtended at the metal atom to which the ligand is bound is the so-called *cone angle*, θ, defined as follows:

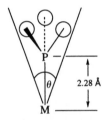

As for the σ-donor and π-acceptor properties, a useful generalization is that as the average electronegativity of the three groups attached to P increases, the ligand becomes a poorer σ-donor but a better π-acceptor. With polyphosphines the σ-donor/π-acceptor properties of each phosphorus atom will be governed by the same rules that apply in a monophosphine, but additional steric considerations come into play based on how the two or more phosphorus atoms are connected.

Secondary phosphines, R_2PH, usually serve simply as monodentate ligands, but they react with M—Cl bonds to eliminate HX and form phosphido, R_2P—M, or μ-phosphido, M—$P(R'_2)$—M, ligands. Other routes to phosphido ligands are discussed later.

Monophosphines

Two of the most commonly encountered are triphenylphosphine and trimethylphosphine, but 40 to 50 other monophosphines are commercially available, and many more are reported in the literature. Many of them have longer-chain R groups and others have mixed R groups, e.g., PMe$_2$Ph, PHEt$_2$, etc. An interesting type recently reported are those containing *N*-pyrrolyl groups, which lower the basicity of the P atom almost as much as C$_n$F$_{2n+1}$ groups.[81]

Monophosphines can be combined with other ligands to afford many interesting mixed-function ligands (10-LIX) or polydentate ligands such as (10-LX) and (10-LXI).

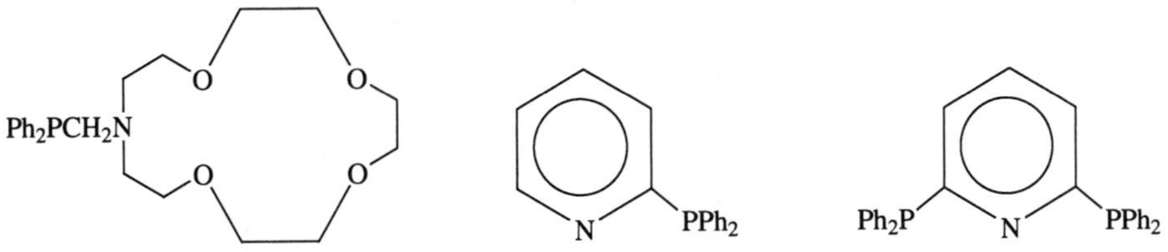

(10-LIX)   (10-LX)   (10-LXI)

Examples of phosphine ligands that are used to make water-soluble complexes are sulphonated triphenylphosphines, e.g., Ph$_2$P(*m*-SO$_3$C$_6$H$_4$)$^-$ and 1,3,5-triaza-7-phosphaadamantane, TPA[82] (10-LXII).

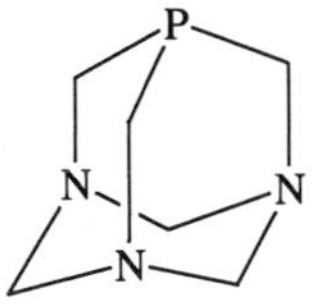

(10-LXII)

Arylphosphines may also be bonded to metal atoms as substituted η^6-arenes. The dangling PR$_2$ may then act as a donor to another metal atom, thus bridging across M—M bonded units:

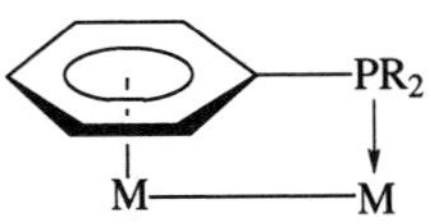

An interesting application[83] of this is a reaction that probably proceeds through such an intermediate to afford a designed synthesis for a Mo$\overset{4}{-}$W compound:

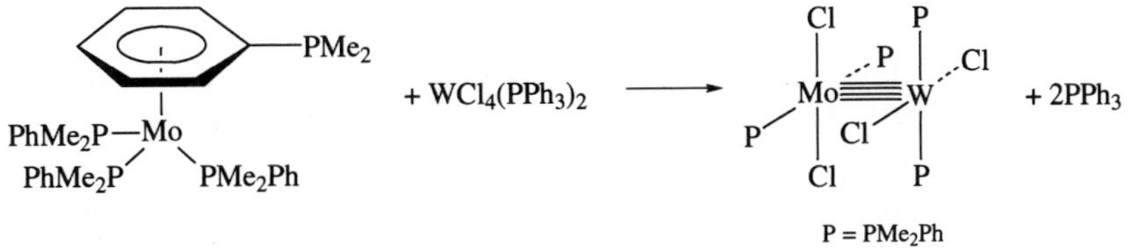

P = PMe$_2$Ph

[81]K. G. Moloy and J. L. Petersen, *J. Am. Chem. Soc.* **1995,** *117,* 7696.
[82]D. J. Darensbourg *et al., Inorg. Chem.* **1994,** *33,* 175.
[83]R. L. Luck *et al., Inorg. Chem.* **1987,** *26,* 2422.

Diphosphine Ligands

These ligands, of which the most widely used are $Me_2PCH_2CH_2PMe_2$ [bis(dimethyl-phosphinoethane) (dmpe)], $Ph_2PCH_2CH_2PPh_2$ [bis(diphenylphosphinoethane) (dppe)], $Me_2PCH_2PMe_2$ [bis(dimethylphosphinomethane) (dmpm)], and $Ph_2PCH_2PPh_2$ [bis(diphenylphosphinomethane) (dppm)], can form either chelated or bridged complexes, as shown by the examples (10-LXIII) to (10-LXVI).

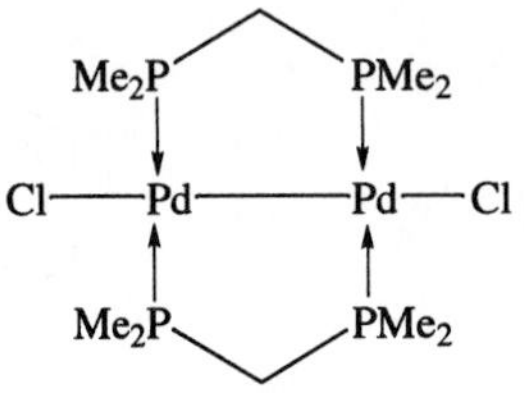

(10-LXIII) 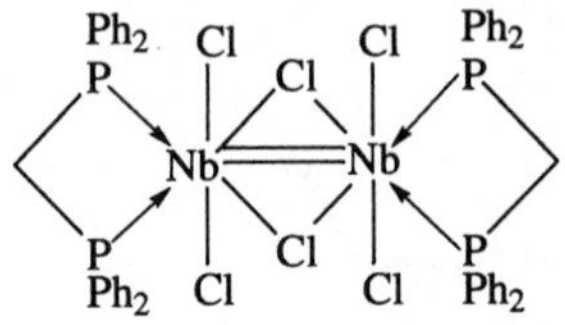   (10-LXIV)

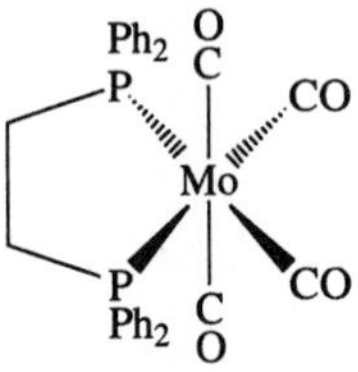

(10-LXV) 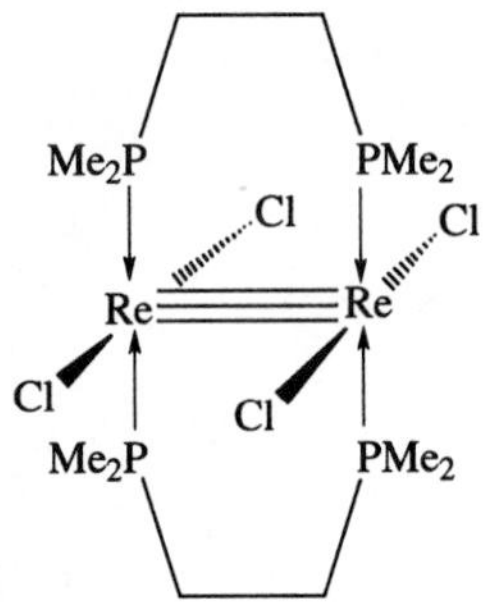   (10-LXVI)

On the other hand, ligands can be designed to do only one thing. For example, the ligand 10-LXVII is capable of chelation only.[84]

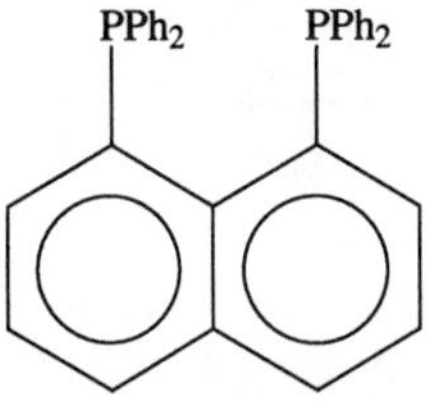

(10-LXVII)

Also, in dmpm or dppm the CH_2 group can be readily deprotonated to give the anionic ligand $[R_2PCHPR_2]^-$, which can act as a P donor and/or form a metal to carbon bond. An example is $La[(Ph_2P)_2CH]_3$.

The diphosphinoamines, $Ph_2PN(R)PPh_2$, have assumed increasing importance recently.[85]

[84]P. G. Pringle *et al.*, *J. Organomet. Chem.* **1993**, *458*, C3.
[85]F. A. Cotton *et al.*, *Inorg. Chim. Acta* **1996**, *252*, 251.

There are important chiral diphosphinoethanes, such as (10-LXVIII)-(10-LXX), which can be used to impose chirality on complexes or on products formed when the chiral complexes serve as catalysts.

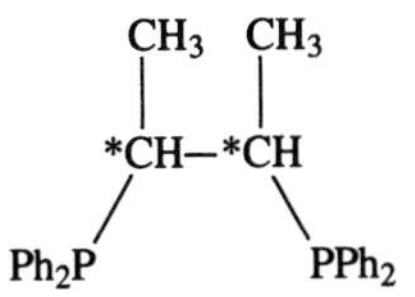

(10-LXVIII) 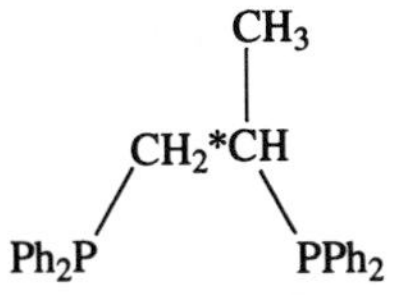   (10-LXIX) 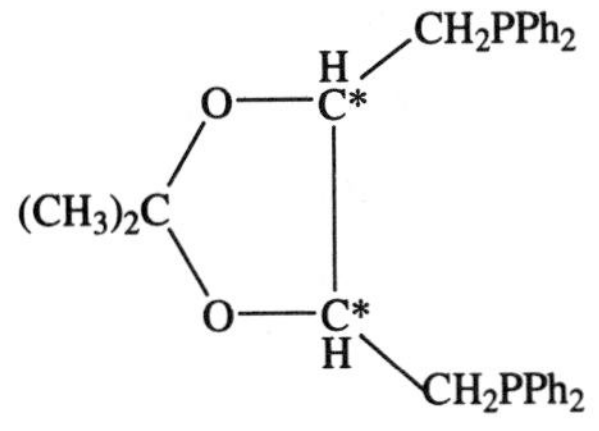   (10-LXX)

There is often a considerable difference in the products given by dmpm versus dppm or dmpe versus dppe. Such differences depend in part on the different steric requirements of the Me_2P and Ph_2P groups, but also on their different basicities. For example, dmpe allows the formation of phosphine complexes of metals that otherwise form phosphine complexes only with difficulty, such as U^{IV}, Tc^{IV}, and Cr^{II}.

Diphosphines with long chains between the P atoms, for example, $Bu_2^t P(CH_2)_{5-8}PBu_2^t$, can span trans positions in square complexes or give large ring systems, (10-LXXI) and (10-LXXII), respectively. A similar situation has also been found in an octahedral complex,[86] where the trans isomer has been shown to be more stable than the cis.

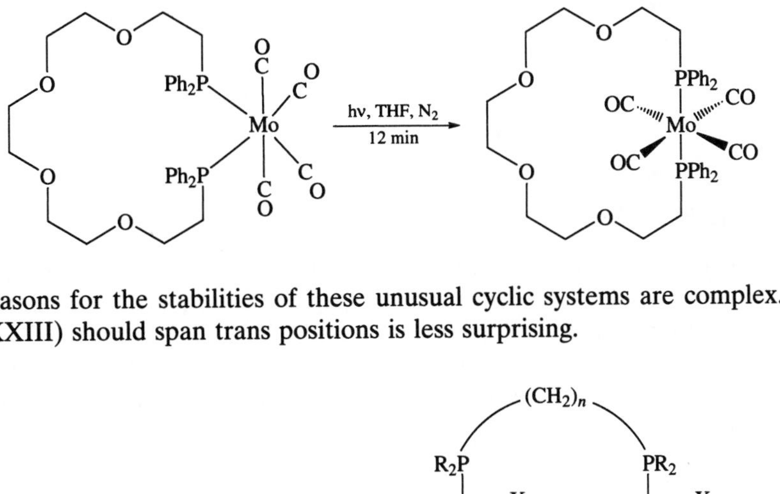

The reasons for the stabilities of these unusual cyclic systems are complex. That (10-LXXIII) should span trans positions is less surprising.

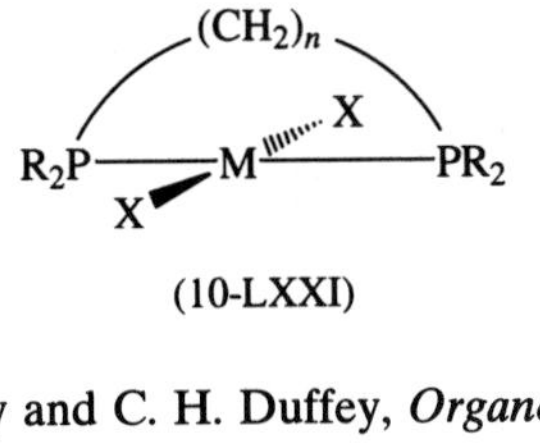

(10-LXXI) 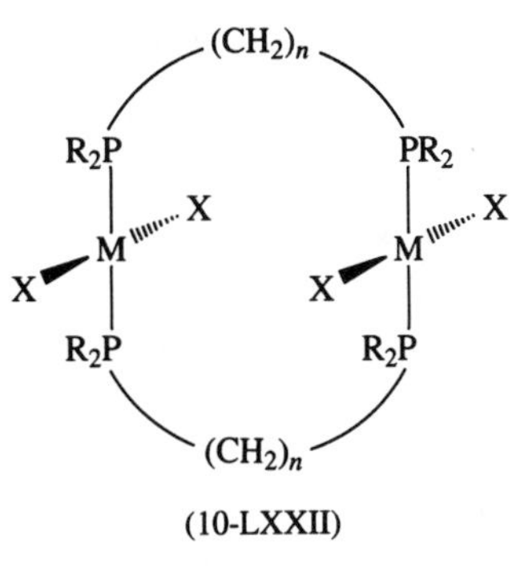   (10-LXXII)

[86]G. M. Gray and C. H. Duffey, *Organometallics* **1994**, *13*, 1542.

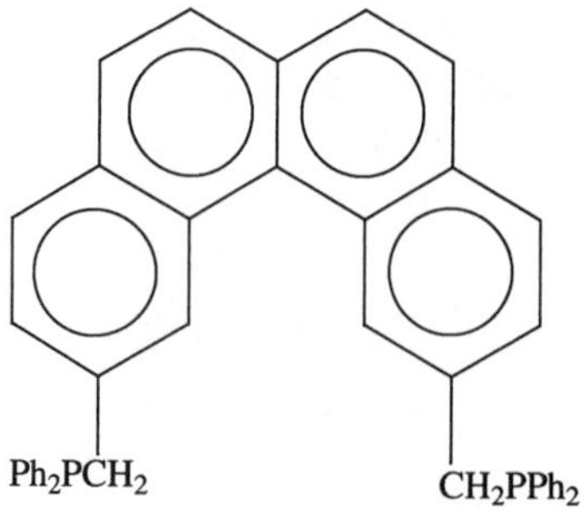

Ph₂PCH₂   CH₂PPh₂

(10-LXXIII)

Other important biphosphines are the asymmetric binaphthyls (10-LXXIV) and others as well as 1,1′-diphosphinoferrocenes (10-LXXV) that can act as either chelating or bridging ligands.

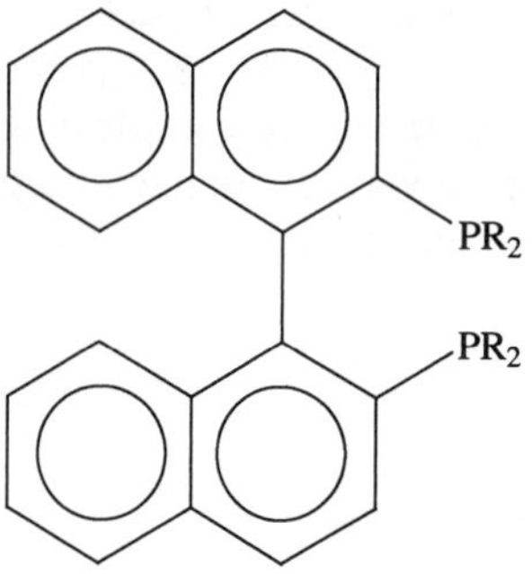

(10-LXXIV) 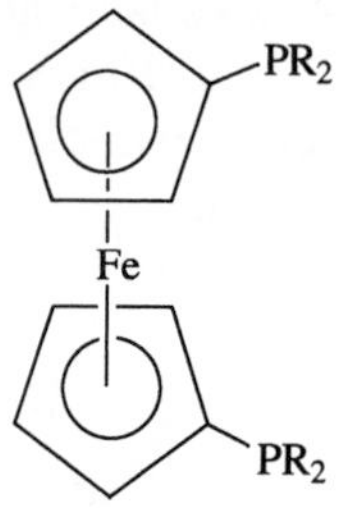   (10-LXXV)

Polydentate Phosphines[87]

Perhaps the most important of these are the tripods, such as $HC(CH_2PR_2)_3$, as well as ligands such as $N(CH_2CH_2PPh_2)_3$ or $P(CH_2CH_2PMe_2)_3$, which are potentially tetradentate. Linear tridentate phosphines can behave in a number of ways, as shown in (10-LXXVIa) through (10-LXXVId). Also, there is the situation where a diphosphinoarsine first forms a binuclear macrocycle, which can then add another metal atom:

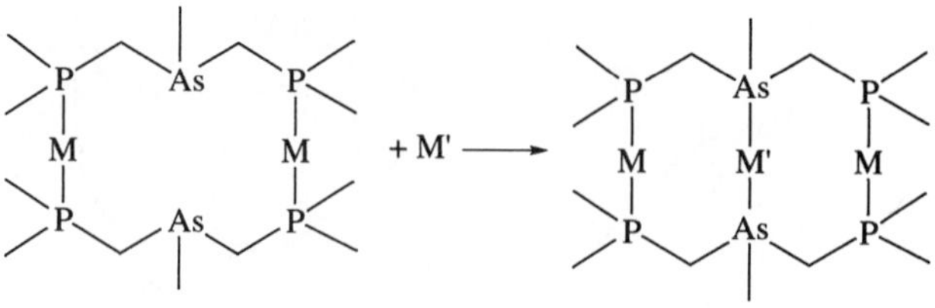

[87]F. A. Cotton and B. Hong, *Prog. Inorg. Chem.* **1992**, *40*, 179.

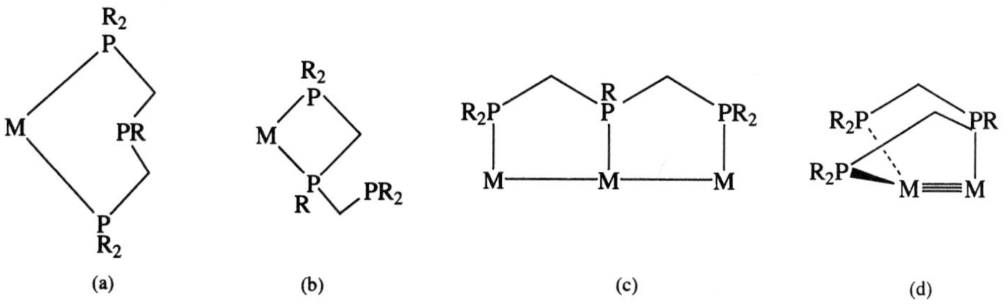

(10-LXXVI)

The hydroformylation reaction

$$RCH = CH_2 + CO + H_2 \longrightarrow RCH_2CH_2CHO \qquad \textbf{(10-1a)}$$

$$RCH = CH_2 + CO + H_2 \longrightarrow RCH(CHO)-CH_3 \qquad \textbf{(10-1b)}$$

is the largest-scale industrial process based on homogeneous catalysis, and the catalysts are all phosphine complexes of low-valent rhodium. These catalysts play three roles: to increase conversion, to favor reaction 10-1a over 10-1b, and to minimize undesired side reactions such as hydrogenation or isomerization of the olefin. It has recently been shown[88] that a catalyst superior to those now in use may result from the use of a linear tetraphosphine, 10-LXXVII.

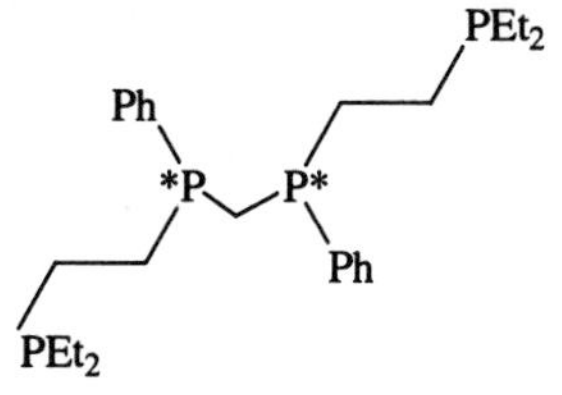

racemic

(10-LXXVII)

This ligand reacts with $[Rh(nbd)_2]BF_4$ to form 10-LXXVIII, which under reaction conditions reacts with the CO/H_2 mixture to form the actual catalyst. The latter is thought to be 10-LXXIX.

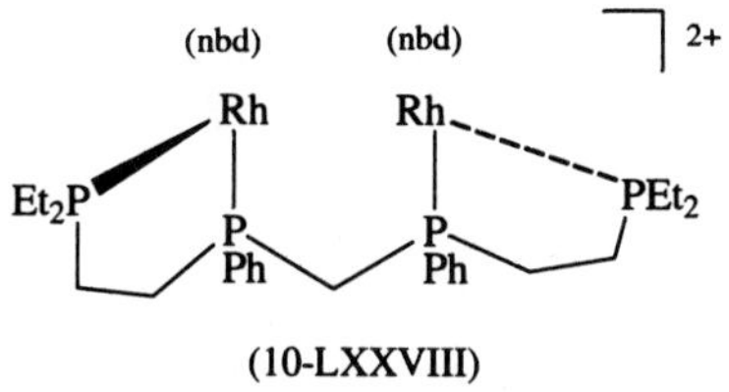

(10-LXXVIII) 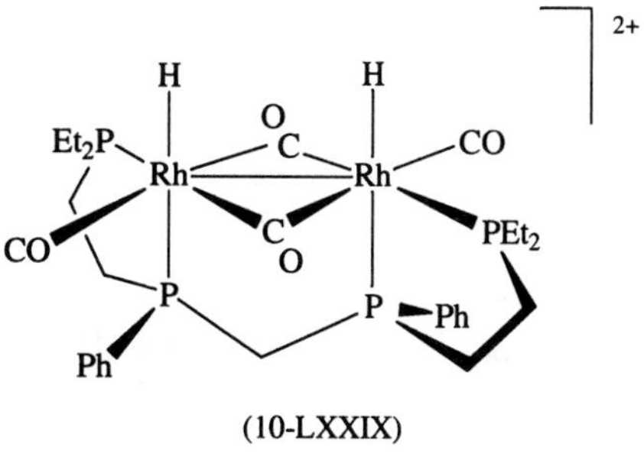   (10-LXXIX)

[88]G. G. Stanley et al., Science **1993**, 260, 1784; Angew. Chem. Int. Ed. Engl. **1996**, 35, 2253.

Macrocyclic Phosphines

There are a number of macrocycles with phosphorus donors only, or with mixed sets of P and N or P and S donors, for example, (10-LXXXa and b).

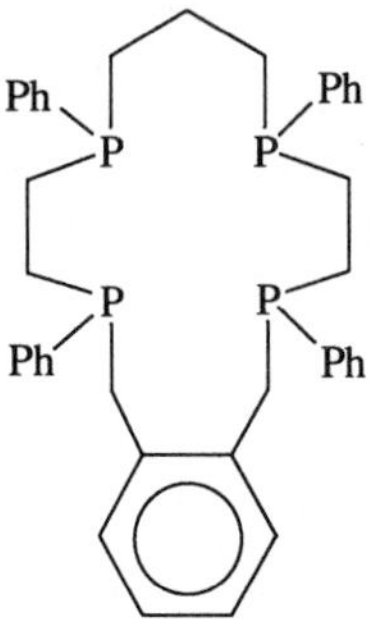

(a)

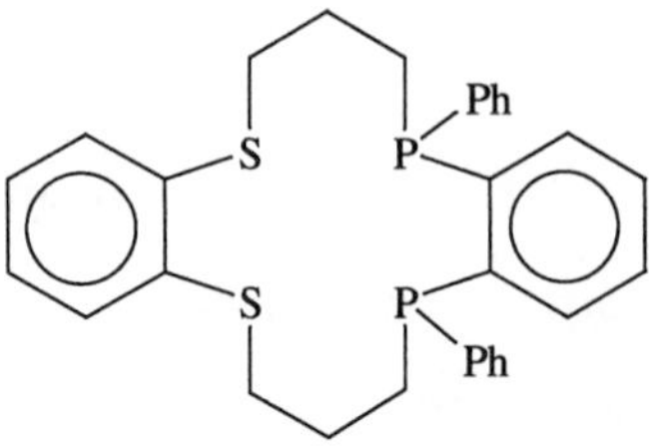

(b)

(10-LXXX)

Phosphite Ligands

There are many ligands similar to the phosphine ligands with OR instead of R groups, such as $P(OMe)_3$ or $(RO)_2POP(OR)_2$. Their complexes are often similar to those of phosphines but since the phosphites tend to be more basic and less sterically hindered [e.g., $P(OPh)_3$ and, *a fortiori*, $P(OMe)_3$ compared with PPh_3], there are important differences. Thus, there is no phosphine analogue of $Re_2(CO)_{10}$, but $Re_2[P(OMe)_3]_{10}$ can be made. In some cases where a substitution-labile ligand is also present on the metal, the reaction with an alkyl phosphite gives a coordinated phosphonate. Somewhat comparable methyl migration reactions are the following:

$$L_nMX + P(OMe)_3 \longrightarrow L_nM\overset{\displaystyle O}{\overset{\|}{-}}P(OMe)_2 + CH_3X$$

$$Ru[P(OMe)_3]_5 \xrightarrow{150°C} [(MeO)_3P]_4Ru\overset{\displaystyle CH_3}{\underset{\underset{O}{\|}}{-}}P(OMe)_2$$

It may be noted that while $(RO)_2POP(OR)_2$ ligands can be unidentate or bridging, because of the wide POP angle, they do not chelate.

Many phosphites and their amido derivatives play a major role in industrial hydrometallurgy as extractants, because of their ability to complex selectively certain metal ions.

Heterocyclic Phosphorus Ligands

The ligand capacity of phosphabenzene has already been noted (Section 10-14). The phospholes (10-LXXXI) can also behave simply as η^1-donors, or also engage their π-electron density, whence they generally become bridging ligands, as in

(10-LXXXII). There is also a P,N ligand analogous to 2,2'-bipyridyl, namely, (10-LXXXIII). Phosphazenes can also act as ligands.

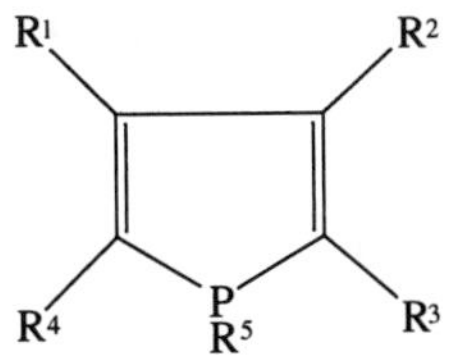

(10-LXXXI) 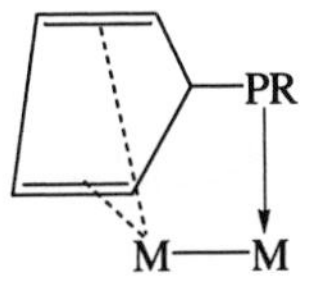 (10-LXXXII) 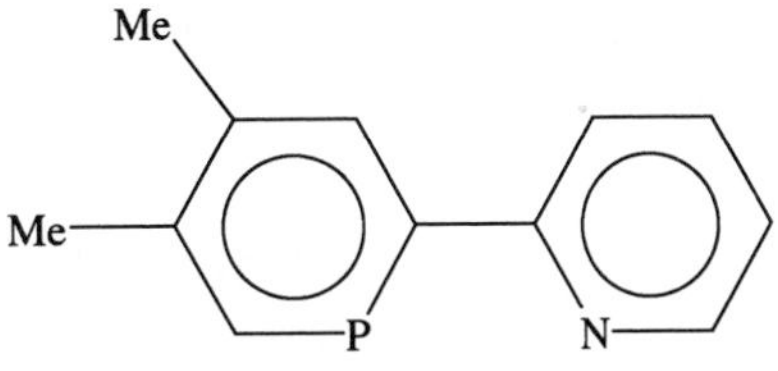 (10-LXXXIII)

Cyclometallated Phosphine Complexes

Phosphine ligands often engage in H-transfer reactions to give M—C bonds. Simple examples are provided by reactions of PMe_3:

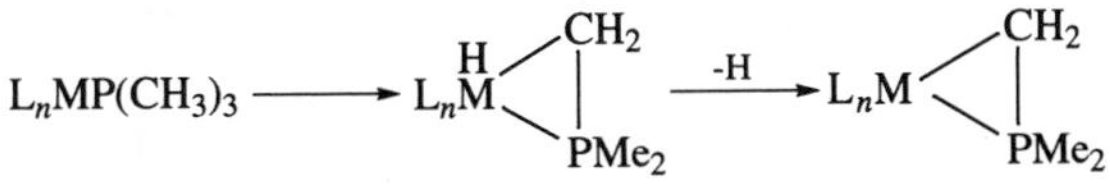

$$L_nMP(CH_3)_3 \longrightarrow L_nM \overset{H}{\underset{PMe_2}{\diagdown}} CH_2 \xrightarrow{-H} L_nM \overset{CH_2}{\underset{PMe_2}{\diagdown}}$$

With longer-chain alkyl phosphines, β-, γ-, or δ-hydrogen atoms may be lost leading to 4-, 5-, or 6-membered heterocyclic rings. Hydrogen transfer from phenyl rings has also been observed to give products with the units (10-LXXXIV) or (10-LXXXV). It is also possible for CH_2PMe_2 or CH_2PPh_2 groups to be bound only through carbon and for such compounds the "dangling" phosphorus atom can coordinate further, for example, (10-LXXXVI).

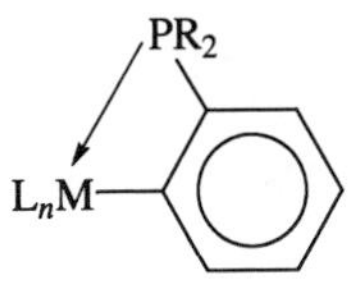

(10-LXXXIV) 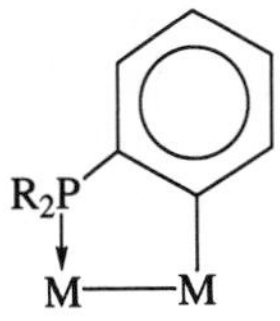 (10-LXXXV) 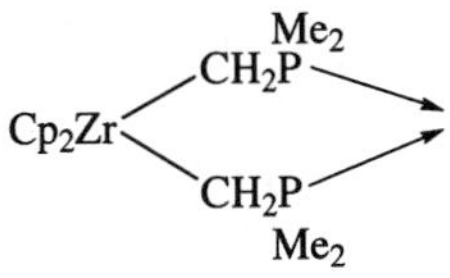 (10-LXXXVI)

Dialkyl-, Diarylphosphido Ligands

Major methods of preparing complexes of R_2P ligands, which may be either terminal or bridging, are illustrated by the following reactions:

$$Cp_2ZrCl_2 + 2LiPPh_2 \longrightarrow Cp_2Zr(PPh_2)_2 + 2LiCl$$

$$PtCl_2(PPh_2Cl)_2 + 2NaMn(CO)_5 \longrightarrow Cl_2Pt[(\mu\text{-}PPh_2)Mn(CO)_5]_2 + 2NaCl$$

$$L_nMCH_3 + HPPh_2 \longrightarrow L_nMPPh_2 + CH_4$$

Cleavage of R_2PPR_2 or P—C bonds can also be employed. A variety of R_2As compounds as well as R_2Sb and R_2Bi compounds are also known.

Although bridging R_2P groups are the most common, there are a number of compounds analogous to dialkylamido compounds, such as $Mo[P(c\text{-}C_6H_{11})_2]_4$ and $Zr[P(c\text{-}C_6H_{11})_2]_5^-$. However, homoleptic complexes are relatively rare and usually other ligands such as PR_3 or CO are present. The nonbridging species can react to form bridges, as in the following example:

$$Cp*_2Th(PR_2)_2 + Ni(CO)_4 \longrightarrow Cp*_2Th(\mu\text{-}PR_2)_2Ni(CO)_2 + 2CO$$

Phosphine-phosphido ligands can be made, for example, from $R_2P(CH_2)_nPRH$, and give complexes such as the following:

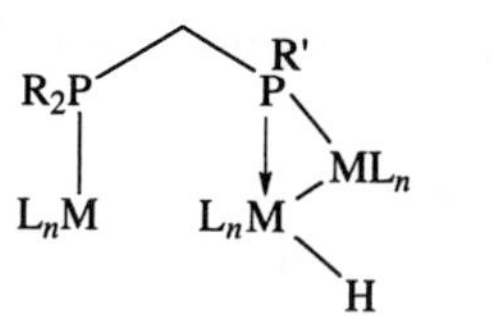

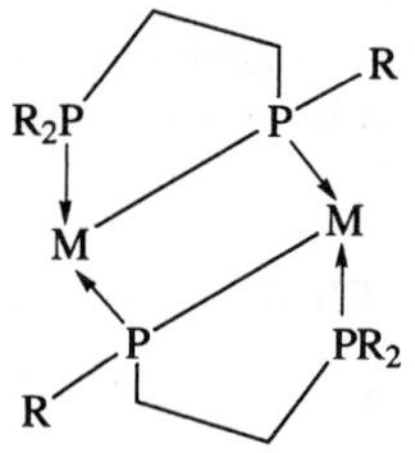

Phosphinidine Ligands

Unlike $M=NR$ compounds which are often quite stable, compounds with $M=PR$ units are usually transient intermediates. However, with bulky R groups, compounds such as $Cp_2MoP(C_6H_2Bu_3^t\text{-}2,4,6)$ can be isolated; the $Mo-P-C$ angle, 115.8°, is consistent with a $Mo=P$ bond and a lone pair on P. Usually ER (E = P, As, Sb) groups serve as bridges, of μ_2, μ_3, or μ_4 types. For the μ_2 type, the structures may be either open (i.e., no $M-M$ bond) or closed, depending on the overall electronic requirements of the metal atoms, as illustrated by the following examples (where Ar = 2,4,6-$Bu_3^tC_6H_2$):

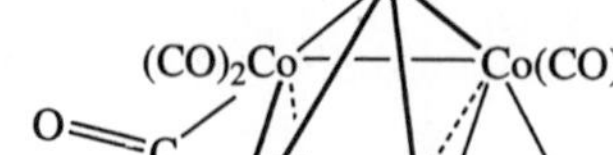

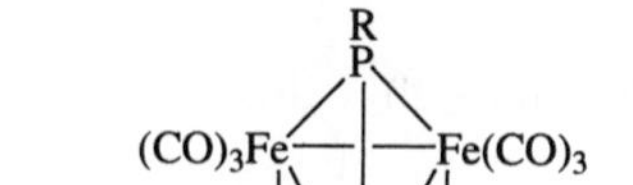

Examples of μ_3- and μ_4-RP complexes are

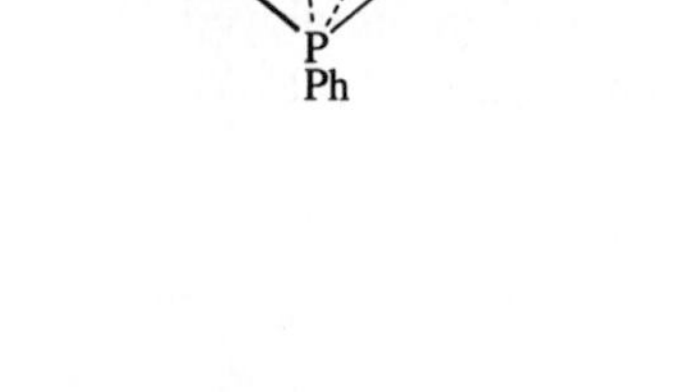

Ligands with P=P, P=C, and P≡C Bonds

Numerous P and As compounds that contain multiple bonds can form a variety of complexes, the following being important types:

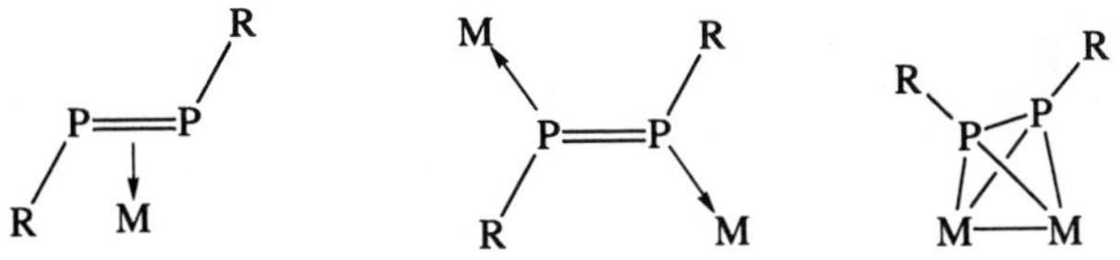

Donation can occur from the multiple bond, from one or both lone pairs, or even all three. Some real examples (Ar = 2,4,6-Bu$_3^t$C$_6$H$_2$) are

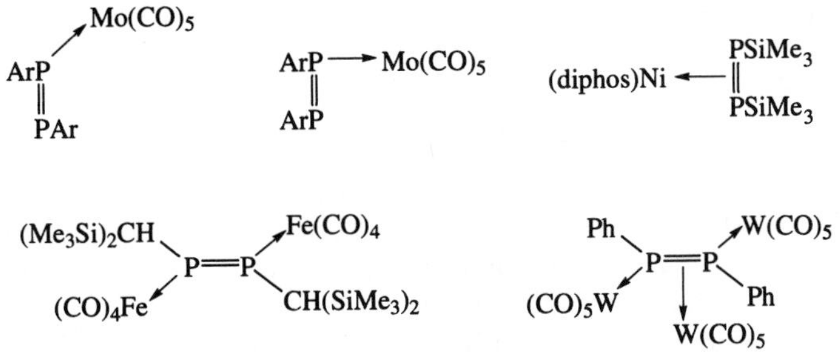

Phosphaalkenes (RP=CR$_2'$) can act as either η^1 or η^2 ligands

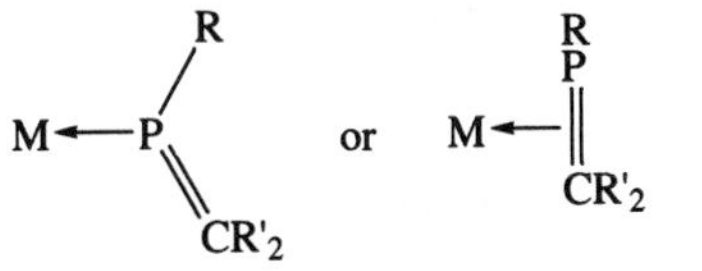

while phosphaalkynes have so far been seen only as η^2 ligands.

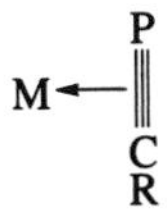

Additional References

M. Baudler, *Angew. Chem. Int. Ed. Engl.* **1987,** *26,* 419 (polyphosphorus compounds).

Chemical Reviews **1994,** *94,* (No. 5) 1161-1456. A collection of reviews on phosphorus chemistry.

H. C. Clark and M. J. Hampden-Smith, *Coord. Chem. Rev.* **1987,** *79,* 229 (bulky P ligand interactions).

D. E. C. Corbridge, *The Structural Chemistry of Phosphorus,* Elsevier, Amsterdam, 1974.

D. E. C. Corbridge, *Phosphorus, An Outline of its Chemistry, Biochemistry and Technology,* 5th ed., Elsevier, Amsterdam, 1995.

A. Durif, *Crystal Chemistry of Condensed Phosphates,* Plenum, New York, 1995.

R. S. Edmunsen, Ed., *Dictionary of Organophosphorus Compounds,* Chapman Hall, London, 1988.

Chapter 11

OXYGEN

GENERAL REMARKS

11-1 Types of Oxides

The oxygen atom has the electronic structure $1s^2 2s^2 2p^4$. Oxygen forms compounds with all the elements except He, Ne, and possibly Ar, and it combines directly with all the other elements except the halogens, a few noble metals, and the noble gases, either at room or at elevated temperatures. The earth's crust contains ~50% by weight of oxygen. Most inorganic chemistry is concerned with its compounds, if only in the sense that so much chemistry involves water.

Oxygen follows the octet rule, and the closed-shell configuration can be achieved in ways that are similar to those for nitrogen, namely, by (*a*) electron gain to form O^{2-}, (*b*) formation of two single covalent bonds (e.g., R—O—R) or a double bond (e.g., O=C=O), (*c*) gain of one electron and formation of one single bond (e.g., in OH^-), and (*d*) formation of three or four covalent bonds (e.g., R_2OH^+).

There is a variety of disparate binary oxide compounds. The change of physical properties is attributable to the range of bond types from essentially ionic to essentially covalent.

The formation of the O^{2-} ion from O_2 requires the expenditure of considerable energy, ~1000 kJ mol^{-1}:

$$\tfrac{1}{2}O_2(g) = O(g) \qquad \Delta H = 248 \text{ kJ mol}^{-1}$$

$$O(g) + 2e = O^{2-}(g) \qquad \Delta H = 752 \text{ kJ mol}^{-1}$$

Moreover, in the formation of an ionic oxide, energy must be expended in vaporizing and ionizing the metal atoms. Nevertheless, many essentially ionic oxides exist (e.g., CaO) and are very stable because the energies of lattices containing the relatively small (1.40 Å) O^{2-} ion are quite high. In fact, the lattice energies are often sufficiently high to allow the ionization of metal atoms to unusually high oxidation states. Many metals form oxides in oxidation states not encountered in their other compounds, except perhaps in fluorides or some complexes. Examples of such higher oxides are MnO_2, AgO, and PrO_2, many of which are nonstoichiometric.

In some cases the lattice energy is still insufficient to permit complete ionization, and oxides having substantial covalent character, such as BeO or B_2O_3, are formed. Finally, at the other extreme there are gases or volatile solids or liquids, such as

CO_2, the N and P oxides, SO_2, and SO_3, that are essentially covalent molecular compounds. Even in "covalent" oxides, unusually high *formal* oxidation states are often found, as in OsO_4, CrO_3, XeO_4, and so on.

In some oxides containing transition metals in very low oxidation states, metal "*d* electrons" enter delocalized conduction bands and the materials have metallic properties; an example is NbO.

In terms of chemical behavior, it is convenient to classify oxides according to their acid or base character in the aqueous system.

Basic Oxides

Although X-ray studies show the existence of discrete oxide ions (O^{2-}) [and also peroxide, (O_2^{2-}), and superoxide (O_2^-) ions, discussed later], these ions cannot exist in any appreciable concentration in aqueous solution owing to the hydrolytic reaction

$$O^{2-} + H_2O = 2OH^- \quad K > 10^{22}$$

For the per- and superoxide ions we also have:

$$O_2^{2-} + H_2O = HO_2^- + OH^-$$

$$2O_2^- + H_2O = O_2 + HO_2^- + OH^-$$

Thus only those ionic oxides that are insoluble in water are inert to it. Ionic oxides function, therefore, as *basic anhydrides*. When insoluble in water, they usually dissolve in dilute acids, for example,

$$MgO(s) + 2H^+(aq) \longrightarrow Mg^{2+}(aq) + H_2O$$

although in some cases, MgO being one, high-temperature ignition produces a very inert material, quite resistant to acid attack.

Acidic Oxides

The covalent oxides of the nonmetals are usually acidic, dissolving in water to produce solutions of acids. They are termed *acid anhydrides*. Insoluble oxides of some less electropositive metals of this class generally dissolve in bases. Thus,

$$N_2O_5(s) + H_2O \longrightarrow 2H^+(aq) + 2NO_3^-(aq)$$

$$Sb_2O_5(s) + 2OH^- + 5H_2O \longrightarrow 2Sb(OH)_6^-$$

Certain oxides, including SiO_2, Al_2O_3, TiO_2, and Ta_2O_5 possess surface acid sites which often lead to interesting catalytic properties.[1]

[1]G. Guiu and P. Grange, *J. Catal.* **1995,** *156,* 132; K. Dyrek and M. Che, *Chem. Rev.* **1997,** *97,* 305.

Basic and acidic oxides will often combine directly to produce salts, such as

$$Na_2O + SiO_2 \xrightarrow{\text{fusion}} Na_2SiO_3$$

Amphoteric Oxides

These oxides behave acidically toward strong bases and as bases toward strong acids:

$$ZnO + 2H^+(aq) \longrightarrow Zn^{2+} + H_2O$$

$$ZnO + 2OH^- + H_2O \longrightarrow Zn(OH)_4^{2-}$$

Other Oxides

There are various other oxides, some of which are relatively inert, dissolving in neither acids nor bases, for instance, N_2O, CO, and MnO_2; when MnO_2 (or PbO_2) does react with acids (e.g., concentrated HCl), it is a redox, not an acid-base, reaction.

There are also many oxides that are nonstoichiometric. These commonly consist of arrays of close-packed oxide ions with some of the interstices filled by metal ions. However, if there is variability in the oxidation state of the metal, nonstoichiometric materials result. Thus iron(II) oxide generally has a composition in the range $FeO_{0.90}$ to $FeO_{0.95}$, depending on the manner of preparation. There is an extensive chemistry of mixed metal oxides.

It may be noted further that when a given element forms several oxides, the oxide with the element in the highest formal oxidation state (usually meaning more covalent) is more acidic. Thus for Cr we have: CrO, basic; Cr_2O_3, amphoteric; and CrO_3, fully acidic.

Amorphous metal oxides with basic and amphoteric surfaces are active catalysts for a variety of important reactions including olefin isomerization, hydrogenation, and functional group elimination.[2]

The Hydroxide Ion

Discrete OH^- ions exist only in the hydroxides of the more electropositive elements such as the alkali metals and alkaline earths. For such an ionic material, dissolution in water results in formation of aquated metal ions and aquated hydroxide ions:

$$M^+OH^-(s) + nH_2O \longrightarrow M^+(aq) + OH^-(aq)$$

and the substance is a strong base. In the limit of an extremely covalent M—O bond, dissociation will occur to varying degrees as follows:

$$MOH + nH_2O \rightleftharpoons MO^-(aq) + H_3O^+(aq)$$

[2]K. Tanabe *et al., New Solid Acids and Bases*, Elsevier, Amsterdam, 1989 ; J. Haw *et al., J. Am. Chem. Soc.* **1994,** *116,* 10839.

and the substance must be considered an acid. Amphoteric hydroxides are those in which there is the possibility of either kind of dissociation, the one being favored by the presence of a strong acid:

$$M—O—H + H^+ = M^+ + H_2O$$

the other by strong base:

$$M—O—H + OH^- = MO^- + H_2O$$

because the formation of water is so highly favored, that is,

$$H^+ + OH^- = H_2O \quad K_{25°C} = 10^{14}$$

Hydrolytic reactions of metal ions can be written

$$M(H_2O)_x^{n+} = [M(H_2O)_{x-1}(OH)]^{(n-1)+} + H^+$$

Thus we may consider that the more covalent the M—O bond tends to be, the more acidic are the hydrogen atoms in the aquated ion, but at present there is no extensive correlation of the acidities of aqua ions with properties of the metal.

The formation of hydroxo bridges occurs at an early stage in the precipitation of hydroxides or, in some cases more accurately, hydrous oxides. In the case of Fe^{3+}, precipitation of $Fe_2O_3 \cdot xH_2O$ [commonly, but incorrectly, written $Fe(OH)_3$] proceeds through the stages

$$[Fe(H_2O)_6]^{3+} \xrightarrow{-H^+} [Fe(H_2O)_5OH]^{2+} \longrightarrow [(H_2O)_4Fe(OH)_2Fe(H_2O)_4]^{4+} \xrightarrow{-xH^+}$$
$$\text{pH} < 0 \qquad\qquad 0 < \text{pH} < 2 \qquad\qquad \sim 2 < \text{pH} < \sim 3$$

$$\text{colloidal } Fe_2O_3 \cdot xH_2O \xrightarrow{-yH_2O} Fe_2O_3 \cdot zH_2O(\text{ppt})$$
$$\sim 3 < \text{pH} < \sim 5 \qquad\qquad \text{pH} \sim 5$$

A few compounds contain the hydrated hydroxide anion $H_3O_2^-$.[3]

11-2   Covalent Compounds; Stereochemistry of Oxygen

Two-Coordinate Oxygen

The majority of oxygen compounds contain 2-coordinate oxygen, in which the O atom forms two single bonds to other atoms and has two unshared pairs of electrons in its valence shell. Such compounds include water, alcohols, ethers, and a variety of other covalent oxides. In the simple systems without significant π bonding the X—O—X group is bent; typical angles are 104.5° in H_2O and 111° in $(CH_3)_2O$.

In many cases, where the X atoms of the X—O—X group have orbitals (usually d orbitals) capable of interacting with the lone-pair orbitals of the O atom, the X—O bonds acquire some π character. Such interaction causes shortening of the

[3]M. Julve et al., Inorg. Chem. **1994**, 33, 1585.

X—O bonds and generally widens the X—O—X angle. The former effect is not easy to document, since an unambiguous standard of reference for the pure single bond is generally lacking. However, the increases in the angle are self-evident, for example, $(C_6H_5)_2O$ (124°) and the Si—O—Si angle in quartz (142°). In the case of H_3Si—O—SiH_3 the angle is apparently > 150°.

The limiting case of π interaction in X—O—X systems occurs when the σ bonds are formed by two linear *sp* hybrid orbitals on oxygen, thus leaving two pairs of π electrons in pure *p* orbitals; these can then interact with empty $d\pi$ orbitals on the X atoms so as to stabilize the linear arrangement. Many examples of this are known, e.g., $[Cl_5Ru$—O—$RuCl_5]^{4-}$ and $[(OAc)(py)(Cl)_2Os$—O—$Os(Cl)_3(py)_2]$.[4]

Three-Coordinate Oxygen

Pyramidal and planar geometries occur, the former being represented by *oxonium ions*[5] [e.g., H_3O^+, R_3O^+ and $(R_3Si)_3O^+$] and by donor-acceptor complexes such as $(C_2H_5)_2OBF_3$. There are examples of both geometric types among trinuclear μ_3-oxo complexes (Section 11-9). The formation of oxonium ions is analogous to the formation of ammonium ions such as NH_4^+, RNH_3^+, ..., R_4N^+, except that oxygen is less basic and the oxonium ions are therefore less stable. Water, alcohol, and ether molecules serving as ligands to metal ions presumably also have pyramidal structures, at least for the most part. Like NR_3 compounds (Section 9-2), OR_3^+ species undergo rapid inversion.

Four-Coordinate Oxygen

Tetrahedral geometry is found in certain ionic or partly ionic oxides (e.g., PbO), and in polynuclear complexes such as $Mg_4OBr_6\cdot4THF$, $Cu_4OCl_6(Ph_3PO)_4$, $M_4O(O_2CR)_6$, where M = Be, Co, and Zn, $M_4O(formamidinate)_6$ compounds, where M = Fe, Co, Mn and Zn,[6] and $Fe_{16}Mn(\mu_4\text{-}O)_6(\mu_3\text{-}O)_4(\mu_3\text{-}OH)_8(\mu\text{-}OH)_2$.[7] Other geometries are also known, for example, in the tetranuclear anion $\{Nb_4OCl_8[(PhC)_4]_2\}^{2-}$, there is an oxygen atom at the center of a rectangle of niobium atoms.[8]

Higher Coordinate Oxygen

Five and six-coordination is less common but examples are found in polynuclear complexes with cores such as $Mn^{IV}Mn^{III}_6Mn^{II}_6(\mu_5\text{-}O)_6(\mu_3\text{-}O)_2(\mu_3\text{-}OEt)_6$[9] and $Fe_6(\mu_6\text{-}O)$.[10]

Unicoordinate, Multiply Bonded Oxygen

There are many examples of XO groups, where the order of the XO bond may vary from essentially unity, as in amine oxides, $\longrightarrow \overset{+}{N} \overset{\cdot\cdot}{\underset{\cdot\cdot}{O}}\colon{}^-$, though varying degrees

[4]K. Umakoshi *et al.*, *Inorg. Chem.* **1995**, *34*, 813.
[5]G. A. Olah *et al.*, *J. Am. Chem. Soc.* **1993**, *115*, 1277; *J. Am. Chem. Soc.* **1995**, *117*, 8962.
[6]F. A. Cotton *et al.*, *Inorg. Chim. Acta,* **1997**, *266*, 91.
[7]W. Micklitz and S. J. Lippard, *J. Am. Chem. Soc.* **1989**, *111*, 6856.
[8]F. A. Cotton and M. Shang, *J. Am. Chem. Soc.* **1990**, *112*, 1584.
[9]D. N. Hendrickson *et al.*, *Inorg. Chem.* **1996**, *35*, 6640.
[10]K. Hegetschweiler *et al.*, *Inorg. Chem.* **1992**, *31*, 1299.

of π bonding up to a total bond order of 2, or a little more. The simplest π- bonding situation occurs in ketones, where one well-defined π bond occurs perpendicular to the molecular plane. In most inorganic situations, such as R_3PO or R_3AsO compounds, tetrahedral ions such as PO_3^{4-}, ClO_4^-, MnO_4^-, OsO_4, or species such as $OsCl_4O_2^{2-}$, the opportunity exists for two π interactions between X and O, in mutually perpendicular planes that intersect along the X—O line. Indeed, the symmetry of the molecule or ion is such that the two π interactions *must* be of equal extent. Thus, in principle the extreme limiting structure (11-Ib) must be considered to be mixed with (11-Ia). In general, available evidence suggests that a partially polar bond of order approaching 2 results; it is important, however, to note the distinction from the situation in a ketone, since a π bond order of 1 in this context does not mean one full π interaction but rather two mutually perpendicular π interactions of order 0.5.

$$\overset{+}{X} \overset{-}{:\ddot{O}:} \longleftrightarrow \overset{-}{X} ::: \overset{+}{O} :$$

(a)   (b)

(11-I)

Catenation

As with nitrogen, catenation occurs only to a very limited extent. In peroxides and superoxides there are two consecutive oxygen atoms. Only in O_3, O_3^-, and the few $R_FO_3R_F$ molecules are there well-established chains of three oxygen atoms. There is a four-atom chain in O_4F_2.

The weakness of O—O single bonds in H_2O_2 and O_2 and O_2^{2-} is doubtless due to repulsive effects of the electron pairs in these small atoms.

THE ELEMENT

11-3 Occurrence, Properties, and Allotropes

Oxygen occurs in Nature in three isotopic species: ^{16}O (99.759%), ^{17}O (0.0374%), and ^{18}O (0.2039%). The rare isotopes, particularly ^{18}O, can be concentrated by fractional distillation of water, and concentrates containing up to 99 at. % ^{18}O or up to 90 at. % ^{17}O as well as other labeled compounds are commercially available. Oxygen-18 has been widely used as a tracer in studying reaction mechanisms of oxygen compounds. Oxygen-17 has a nuclear spin $5/2$, but because of the low abundance of this isotope and appreciable quadrupole moment, enriched materials and Fourier transform nmr techniques are required.

Dioxygen occurs in two allotropic forms; O_2 and ozone (O_3). The O_2 form is paramagnetic in the gaseous, liquid, and solid states and has the rather high dissociation energy of 496 kJ mol^{-1}. Molecular orbital theory, even in first approximation, correctly accounts for the triplet ground state ($^3\Sigma_g^-$) having a double bond. There are several low-lying singlet states that are important in photochemical oxidations; these are discussed shortly. Like NO, which has one unpaired electron in an antibonding (π^*) MO, oxygen molecules associate only weakly, and true electron pairing to form a symmetrical O_4 species apparently does not occur even in the solid. Both liquid and solid O_2 are pale blue.

Table 11-1   Various Bond Values for Dioxygen Species

Species	O—O Distance (Å)	Number of π^* electrons	$\nu_{oo}(cm^{-1})$
O_2^+	1.12	1	1905
O_2	1.21	2	1580
O_2^-	1.33	3	1097
O_2^{2-}	1.49	4	802

Oxygen species $O_2^{n\pm}$ from O_2^+ to O_2^{2-} illustrate the effect of varying the number of antibonding electrons on the length and stretching frequency of the O—O bond (Table 11-1).

Chemical Properties of Dioxygen

Oxygen combines with almost all other elements, usually, however, only on heating. Dioxygen will sometimes react reversibly[11] with certain transition metal complexes and the ligand behavior of O_2 and its ions is discussed in Section 11-10. The potentials in aqueous solution are

$$O_2 + 4H^+ + 4e = 2H_2O \qquad E^0 = +1.229 \text{ V}$$

$$O_2 + 2H_2O + 4e = 4OH^- \qquad E^0 = +0.401 \text{ V}$$

$$O_2 + 4H^+(10^{-7} M) + 4e = 2H_2O \qquad E^0 = +0.815 \text{ V}$$

Thus neutral water saturated with O_2 is a fairly good oxidizing agent. For example, although Cr^{2+} is just stable toward oxidation in water, in air-saturated water it is rapidly oxidized; Fe^{2+} is oxidized (only slowly in acid, but rapidly in base) to Fe^{3+} in the presence of air, although in air-free water Fe^{2+} is quite stable ($Fe^{3+} + e = Fe^{2+}$, $E^0 = +0.77$ V).

The rate of oxidation of various substances (e.g., ascorbic acid) may be vastly increased by catalytic amounts of transition metal ions, especially Cu^{2+}, where a Cu^I—Cu^{II} redox cycle is involved.

Oxygen is readily soluble in organic solvents, and merely pouring such liquids in air serves to saturate them with oxygen. This should be kept in mind when determining the reactivity of air-sensitive materials in solution in organic solvents. Note that many organic substances such as ethers readily form peroxides or hydroperoxides in air.

Measurements of the electronic spectra of alcohols, ethers, benzene, and even saturated hydrocarbons show that there is interaction of the charge-transfer type with the oxygen molecule. However, there is no true complex formation, since the heats of formation are negligible and the spectral changes are due to contact between the molecules at van der Waals distances. The classic example is that of

[11]K. D. Karlin *et al.*, *Inorg. Chem.* **1994**, *33*, 1953.

N,N-dimethylaniline, which becomes yellow in air or oxygen but colorless again when the oxygen is removed. Such weak charge-transfer complexes make certain electronic transitions in molecules more intense; they are also a plausible first stage in photooxidations.

Many studies have been made on the mechanism[12] of the reduction of O_2. There is no evidence for four- or two-electron reduction steps, as would be suggested by the overall reactions noted previously, or the following:

$$O_2 + 2H^+ + 2e = H_2O_2 \qquad E^0 = +0.682 \text{ V}$$

$$O_2 + H_2O + 2e = OH^- + HO_2^- \quad E^0 = -0.076 \text{ V}$$

The first step is a one-electron reduction to the *superoxide radical ion* O_2^- (Section 11-10). The potential for the reduction

$$O_2 + e = O_2^-$$

ranges from ~ -0.2 to -0.5 V depending on the medium. The O_2^- ion is a moderate reducing agent, comparable to dithionite, and a *very* weak oxidizing agent. Thus most of the oxidation by O_2 is due to *peroxide* ions HO_2^- and O_2^{2-}, formed by reactions such as

$$H^+ + O_2^- + e = HO_2^-$$

$$2O_2^- + H_2O = O_2 + HO_2^- + OH^-$$

The oxidation of water to O_2 in photosynthetic systems in plants has been greatly studied[13] as have synthetic models such as those of manganese complexes; the oxidation seems most likely to proceed *via* H_2O_2 rather than OH· radicals and probably two or four manganese centers are involved.[14]

Cleavage of oxygen can occur in certain reactions of transition metal compounds to give M=O bonds, for example,

$$W(CO)_6 + O_2 \xrightarrow[\text{uv}]{\text{matrix}} [W(\eta^2\text{-}O_2)(CO)_4] \longrightarrow (CO)_3W(O)_2 \longrightarrow \text{and so on}$$

$$(Me_3SiCH_2)_6Ru_2 + O_2 \xrightarrow{-50\,°C} [(Me_3SiCH_2)_3(O)Ru]_2$$

Singlet Oxygen[15]

The lowest energy electron configuration of O_2, which contains two electrons in π^* orbitals, gives rise to three states, Table 11-2.

[12]R. van Eldik *et al.*, *Inorg. Chem.* **1994**, *33*, 687.
[13]A. J. Bard and M. A. Fox, *Acc. Chem. Res.* **1995**, *28*, 141; J. P. Collman *et al.*, *Angew. Chem. Int. Ed. Engl.* **1994**, *33*, 1537.
[14]See for example: W. H. Armstrong *et al.*, *J. Am. Chem. Soc.* **1996**, *118*, 10910 and references therein.
[15]E. A. Lissi *et al.*, *Chem. Rev.* **1993**, *93*, 699.

Table 11-2   States of the O_2 Molecule

State	π_a^*	π_b^*	Energy
$^1\Sigma_g^+$	↑	↓	155 kJ ($\sim$13,000 cm^{-1})
$^1\Delta_g$	↑↓	—	92 kJ ($\sim$8000 cm^{-1})
$^3\Sigma_g^-$	↑	↑	0 (ground state)

The $^1\Delta_g$ state has a much longer lifetime than the $^1\Sigma_g^+$ state.[16] The lifetime is solvent dependent, varying by orders of magnitude in organic solvents.

There are two main ways of generating the singlet oxygen molecules: (1) photochemically by irradiation in presence of a sensitizer and (2) chemically. The photochemical route proceeds as follows, where "sens" represents the photosensitizer (typically a fluorescein derivative, methylene blue, certain porphyrins, or certain polycyclic aromatic hydrocarbons):

$$^1\text{sens} \xrightarrow{h\nu} {}^1\text{sens*}$$

$$^1\text{sens*} \longrightarrow {}^3\text{sens*}$$

$$^3\text{sens*} + {}^3O_2 \longrightarrow {}^1\text{sens} + {}^1O_2$$

$$^1O_2 + \text{substrate} \longrightarrow \text{products}$$

Energy transfer from triplet excited ^{3}sens* to $^3O_2$ to give $^1O_2$ is a spin allowed process. A representative example is the use of excited azoalkanes with O_2.[17]

Singlet oxygen can be generated by a variety of chemical reactions including decomposition of H_2O_2, peroxyacids, and thermal decomposition of ozonides (Section 11-3) such as $(PhO)_3PO_3$ or Na_2MO_8, M=Mo or W, in basic aqueous systems.[18]

In the reactions

$$H_2O_2 + Cl_2 = 2Cl^- + 2H^+ + O_2$$

$$H_2O_2 + ClO^- = Cl^- + H_2O + O_2$$

the accompanying red chemiluminescent glow is due to the excited oxygen molecules trapped in the bubbles.

Singlet oxygen molecules are ubiquitous, being implicated in biological oxidations, photoconversions of air pollutants, and degradation of synthetic polymers and may well be generated in living cells as side products of enzyme reactions. Large scale photochemical synthesis in the fine chemicals industry is for selective

[16]P.-T. Chou *et al., J. Am. Chem. Soc.* **1996,** *118,* 3031.
[17]W. M. Nau *et al., J. Am. Chem. Soc.* **1996,** *118,* 2742.
[18]Q. J. Niu and C. S. Foote, *Inorg. Chem.* **1992,** *31,* 3472.

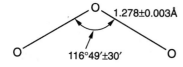

Figure 11-1 The structure of ozone (O_3).

oxidations, since singlet O_2 reacts electrophilically rather than in a free-radical fashion. An example is a Diels-Alder-like 1,4-addition to a 1,3-diene.

Ozone

The diamagnetic triatomic molecule has the structure shown in Fig. 11-1. Since the O—O distance in H_2O_2 is 1.49 Å and in O_2 is 1.21 Å, the bonds in O_3 must have considerable double-bond character.

Ozone is commonly prepared by silent electric discharge in oxygen, which gives up to ~ 10% O_3. The gas is blue. Pure ozone can be obtained by fractional liquefaction of O_2—O_3 mixtures. There is a two-phase liquid system; the one with 25% of O_3 is stable, but a deep purple phase with 70% of O_3 is explosive, as is the deep blue pure liquid (bp $-112°C$). The solid (mp $-193°C$) is black violet. Small quantities of O_3 are formed in electrolysis of dilute sulfuric acid, in some chemical reactions producing O_2, and by the action of uv light on O_2; the combination of certain peroxo radicals and some chemical reactions may also result in O_3.[19]

Ozone is very endothermic,

$$O_3 = \tfrac{3}{2}O_2 \quad \Delta H = \text{-142 kJ mol}^{-1}$$

but it decomposes only slowly at 250°C in the absence of catalysts and uv light; many studies have been done on the photodissociation process.[20]

Ozone is an important natural constituent of the atmosphere,[21] being principally concentrated (up to ~27% by weight) between altitudes of 15 and 25 km. It is formed by solar uv radiation in the range 240 to 300 nm *via* the reactions

$$O_2 \xrightarrow{h\nu} 2O \qquad (11\text{-}1)$$

$$O + O_2 \longrightarrow O_3 \qquad (11\text{-}2)$$

Ozone absorbs uv radiation from 200 to 360 nm. This leads partly to a reversal of reaction 11-2 and thus a steady state concentration is established. The net result of all these processes is absorption and conversion to heat of considerable solar uv radiation that would otherwise strike the earth's surface. Destruction of any signifi-

[19]C. E. Castro, *J. Am. Chem. Soc.* **1996**, *118*, 3984.
[20]D. J. Tannor, *J. Am. Chem. Soc.* **1989**, *111*, 2772.
[21]P. J. Crutzen, *Angew. Chem. Int. Ed. Engl.* **1996**, *35*, 1758.

cant percentage of this ozone could have serious effects (e.g., increased surface temperature, high incidence of skin cancer), and some human activities are capable of destroying stratospheric ozone. Supersonic aircraft, which fly in the ozone layer, discharge NO and NO_2, and these can catalyze the decomposition of ozone *via* the following reactions:

$$O_3 + NO \longrightarrow O_2 + NO_2$$

$$NO_2 + O \longrightarrow O_2 + NO$$

$$\text{or} \quad NO_2 + O_3 \longrightarrow O_2 + NO_3$$

$$NO_3 \xrightarrow{h\nu} O_2 + NO$$

Chlorofluorocarbons such as $CFCl_3$ and CF_2Cl_2, used as foam-blowing agents, aerosol propellants, and refrigerants, are photochemically decomposed to give Cl atoms, and these catalyze ozone decomposition[22] *via* the mechanism

$$O_3 + Cl \longrightarrow O_2 + ClO$$

$$ClO + O \longrightarrow O_2 + Cl$$

There has been much study of the very complicated pattern of O_3 decomposition in the upper atmosphere and also of other effects, for example, the increasing CO_2 concentration. It is not yet known if serious damage has already been done. Near the ground, ozone is a serious pollutant.

Chemical Properties of Ozone

The O_3 molecule is a much more powerful oxidant than is O_2 and reacts with most substances at 25°C. It is often used in organic chemistry.[23] The oxidations doubtless involve free-radical chains as well as peroxo intermediates.

Some compounds form adducts, for example,

$$(PhO)_3P + O_3 \xrightarrow{-78°C} (PhO)_3P(O_3) \xrightarrow{-15°C} (PhO)_3PO + O_2(^1\Delta_g)$$

Complexes are also formed with aromatic compounds and other π systems. The reaction

$$O_3 + 2KI + H_2O = I_2 + 2KOH + O_2$$

[22]L. H. Allen, Jr. in *Stratospheric Ozone Depletion/UV-B Radiation in the Biosphere*, R. H. Biggs and M. E. B. Joyner, Eds., Springer-Verlag, Berlin, 1994.
[23]R. L. Kuczkowski, *Chem. Soc. Rev.* **1992**, 79; O. Reiser *et al., J. Org. Chem.* **1993**, *58*, 3169; G. Zvilichovsky and V. Gurvich, *Tetrahedron* **1995**, *51*, 5479.

is quantitative and can be used to determine O_3. The overall potentials in aqueous solution are

$$O_3 + 2H^+ + 2e = O_2 + H_2O \qquad E^0 = +2.07 \text{ V}$$

$$O_3 + 2H_2O + 2e = O_2 + 2OH^- \qquad E^0 = +1.24 \text{ V}$$

$$O_3 + 2H^+ (10^{-7}M) + 2e = O_2 + H_2O \qquad E^0 = +1.65 \text{ V}$$

In acid solution O_3 is exceeded in oxidizing power only by fluorine, the perxenate ion, atomic oxygen, OH radicals, and a few other such species. The rate of decomposition of ozone drops sharply in alkaline solutions, the half-life being ~2 min in 1 M NaOH at 25°C, 40 min at 5 M, and 83 h at 20 M:

$$O_3 + OH^- \longrightarrow HO_2^- + O_2$$

OXYGEN COMPOUNDS

Most oxygen compounds are described in this book during treatment of the chemistry of other elements. Water and the hydroxonium ion have been discussed in Chapter 2.

11-4 Oxygen Fluorides

Since fluorine is more electronegative than oxygen, it is logical to call binary compounds oxygen fluorides rather than fluorine oxides, although the latter names are sometimes seen. Oxygen fluorides have been intensively studied as potential rocket fuel oxidizers.

Oxygen Difluoride (OF$_2$)

This is prepared by passing fluorine rapidly through 2% sodium hydroxide solution, by electrolysis of aqueous HF-KF solutions, or by the action of F_2 on moist KF. It is a pale yellow poisonous gas (bp 145°C). It is relatively unreactive and can be mixed with H_2, CH_4, or CO without reaction, although sparking causes violent explosion. Mixtures of OF_2 with Cl_2, Br_2, or I_2 explode at room temperature. It is fairly readily hydrolyzed by base:

$$OF_2 + 2OH^- \longrightarrow O_2 + 2F^- + H_2O$$

It reacts more slowly with water, but explodes with steam:

$$OF_2 + H_2O \longrightarrow O_2 + 2HF$$

and it liberates other halogens from their acids or salts:

$$OF_2 + 4HX(aq) \longrightarrow 2X_2 + 2HF + H_2O$$

Metals and nonmetals are oxidized and/or fluorinated; in an electric discharge even Xe reacts to give a mixture of fluoride and oxide fluoride.

Dioxygen Difluoride (O_2F_2)[24]

This is a yellow-orange solid (mp 109.7 K), obtained by high-voltage electric discharges[25] on mixtures of O_2 and F_2 at 10 to 20 mm pressure and temperatures of 77 to 90 K. It decomposes into O_2 and F_2 in the gas at $-50°C$ with a half-life of ~3 h. It is an extremely potent fluorinating and oxidizing agent, and under controlled conditions OOF groups may be transferred to a substrate. Many substances explode on exposure to O_2F_2 at low temperatures, and even C_2F_4 is converted into COF_2, CF_4, CF_3OOCF_3, and so on. In the presence of F^- acceptors it forms dioxygenyl salts:

$$O_2F_2 + BF_3 \longrightarrow O_2^+BF_4^- + \tfrac{1}{2}F_2$$

Dioxygen difluoride has been used for oxidizing primary aliphatic amines to the corresponding nitroso compounds.

The structure of O_2F_2 (11-II) is notable for the shortness of the O—O bond (1.217 Å, cf. 1.48 Å in H_2O_2 and 1.49 in O_2^{2-}) and the relatively long O—F bonds (1.575 Å) compared with those in OF_2 (11-III) (1.409 Å).

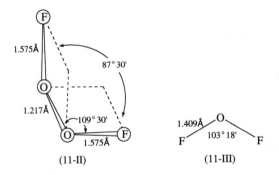

(11-II)   (11-III)

Dioxygen Monofluoride[26]

This is also known and can be generated by reaction of fluorine atoms and oxygen

$$O_2 + F \longrightarrow O_2F$$

Other, very unstable oxygen fluorides have been reported; for example, O_4F_2 decomposes slowly even at $-183°C$. Some oxygen fluoride chloride compounds are also known.

[24]J. L. Lyman, *J. Phys. Chem. Ref. Data* **1989,** *18,* 799.
[25]S. A. Kinkead *et al., Inorg. Chem.* **1990,** *29,* 1779.
[26]J. L. Lyman and R. Holland, *J. Phys. Chem.* **1988,** *92,* 7232.

11-5 Hydrogen Peroxide, Peroxides, and Peroxo Compounds

Hydrogen Peroxide[27]

The main process for the synthesis of H_2O_2 is the autooxidation in an organic solvent such as alkylbenzenes of an alkyl anthraquinol:

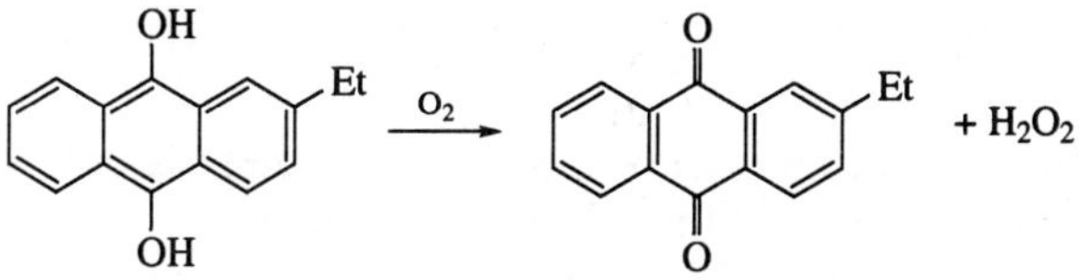

The H_2O_2 is extracted with water and the 20 to 40% solution so obtained purified by solvent extraction. The anthraquinone solution has to be purified by removal of degradation products before reduction back to the quinol on a supported platinum or nickel catalyst:

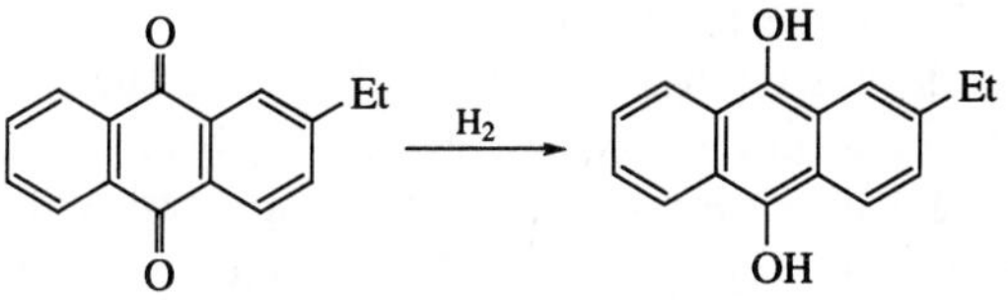

A second process involves oxidation of isopropanol to acetone and H_2O_2 in either vapor or liquid phases at 15 to 20 atm and ~100°C.

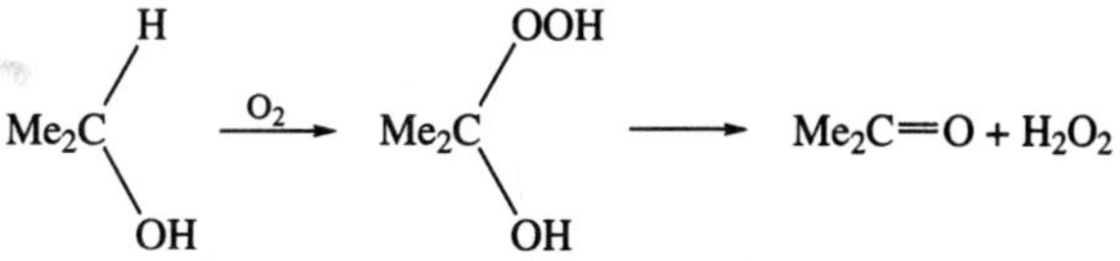

Some secondary reactions also occur giving peroxides of aldehydes and acids. The water-peroxide-acetone-isopropanol mixture is fractionated by distillation.

An electrolytic cell permitting gas-liquid reaction has promise for direct conversion of O_2 into ~2% H_2O_2 in 1 M NaOH, a solution used for bleaching of wood pulp. The cell reactions are

$$2OH^- = \tfrac{1}{2}O_2 + H_2O + 2e$$

$$O_2 + H_2O + 2e = HO_2^- + OH^-$$

$$\overline{\hspace{4cm}}$$

$$OH^- + \tfrac{1}{2}O_2 = HO_2^-$$

Pure H_2O_2, obtained by the concentration of dilute solutions tion, is a colorless liquid (bp 150.2°C, mp −0.43°C) that resemb

[27]V. N. Kislenko and A. A. Berlin, *Russ. Chem. Rev. (Engl. Transl.)* **19**

of its physical properties, although it is denser (1.44 at 25°C). The pure liquid has a dielectric constant at 20°C of 73.2, and a 65% solution in water has a dielectric constant of 120. Thus both the pure liquid and its aqueous solutions are potentially excellent ionizing solvents, but its utility in this respect is limited by its strongly oxidizing nature, and its ready decomposition (eq. 11-3) in the presence of even traces of many heavy metals (see later):

$$2H_2O_2 = 2H_2O + O_2 \quad \Delta H = \text{-99 kJ mol}^{-1} \quad (11\text{-}3)$$

The structure of the molecule is shown in Fig. 11-2; there is only a low barrier to internal rotation about the O—O bond in the gas phase.

In dilute aqueous solution H_2O_2 is more acidic than water:

$$H_2O_2 = H^+ + HO_2^- \quad K_{20°C} = 1.5 \times 10^{-12}$$

There is ordinarily no exchange of oxygen isotopes between H_2O_2 and H_2O in the liquid phase, even in the presence of strong acids. However, HSO_3F catalyzes the exchange, possibly *via* the intermediate $H_3O_2^+$.

Hydrogen peroxide has been estimated to be more than 106 times less *basic* than H_2O. However, on addition of concentrated H_2O_2 to tetrafluoroboric acid in tetrahydrothiophene 1,1-dioxide (sulfolane) the conjugate cation ($H_3O_2^+$) can be obtained. The solutions are very powerful, but unselective, oxidants for benzene, cyclohexane, and other organic materials.

The chemistry in aqueous solution is summarized by the following potentials:

$$H_2O_2 + 2H^+ + 2e = 2H_2O \quad E^0 = 1.77 \text{ V}$$

$$O_2 + 2H^+ + 2e = H_2O_2 \quad E^0 = 0.68 \text{ V}$$

$$HO_2^- + H_2O + 2e = 3OH^- \quad E^0 = 0.87 \text{ V}$$

from which it appears that H_2O_2 is a strong oxidizing agent in either acid or basic solution; only toward very strong oxidizing agents such as MnO_4^- will it behave as a reducing agent.

Dilute or 30% H_2O_2 solutions are widely used as oxidants. In acid solution oxidations with H_2O_2 are most often slow, whereas in basic solution they are usually fast. Decomposition of H_2O_2 according to reaction 11-3, which may be considered a self-oxidation, occurs most rapidly in basic solution; hence an excess of H_2O_2 may best be destroyed by heating in basic solution.

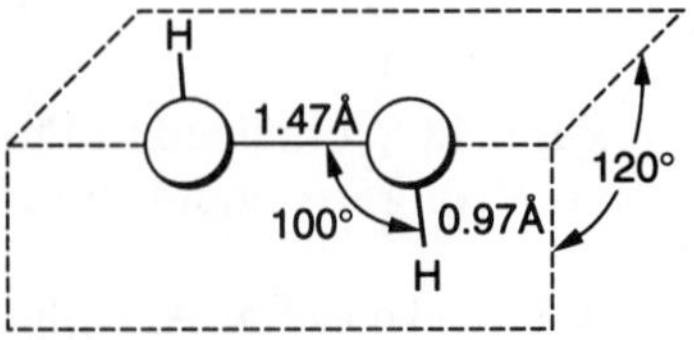

Figure 11-2 The structure of the free H_2O_2 molecule. In crystalline solids such as $H_2O_2(s)$, $Na_2C_2O_4 \cdot H_2O_2$, $(NH_2)_2 \cdot H_2O_2$, and so on, the parameters may vary. In $H_2O_2(s)$ the O—O distance is 1.453 Å with hydrogen bonding O—H···O, 2.8 Å versus 2.76 Å in ice.

The oxidation of H_2O_2 in aqueous solution by Cl_2, MnO_4^-, Ce^{4+}, and so on, and the catalytic decomposition caused by Fe^{3+}, I_2, MnO_2, and so on, have been studied using labeled H_2O_2. In both cases the oxygen produced is derived entirely from H_2O_2, not from water. This suggests that oxidizing agents do not break the $O—O$ bond but simply remove electrons. In the case of oxidation by Cl_2, a mechanism of the following kind is consistent with the lack of exchange of ^{18}O between H_2O_2 and H_2O:

$$Cl_2 + H_2{}^{18}O_2 \longrightarrow H^+ + Cl^- + H^{18}O^{18}OCl$$

$$H^{18}O^{18}OCl \longrightarrow H^+ + Cl^- + {}^{18}O_2$$

As just noted, traces of transition metal ions catalyze the decomposition of H_2O_2,[28] the Fe^{2+}-H_2O_2 system, called Fenton's reagent,[29] can oxidize or dehydrogenate organic substances:

$$Fe^{2+}{}_{aq} + H_2O_2 \longrightarrow Fe^{III}(OH)^{2+}{}_{aq} + OH^\bullet$$

Studies in MeCN suggest that an iron(IV) "ferryl" ion, FeO^{2+}, is initially formed; that is,

$$Fe^{2+} + H_2O_2 \xrightarrow{\text{MeCN}} FeO^{2+} + H_2O$$

followed by

$$FeO^{2+} + H_2O_2 = Fe^{2+} + O_2({}^1\Delta_g) + H_2O$$

In dry MeCN, Fenton-type reactions, which involve $OH^\bullet$ radicals, do not occur.

Ionic Peroxides

Peroxides that contain O_2^{2-} ions are known for the alkali metals, Ca, Sr, and Ba. Sodium peroxide is made commercially by air oxidation of Na, first to Na_2O, then to Na_2O_2; it is a yellowish powder, very hygroscopic though thermally stable to 500°C, which also contains, according to esr studies, ~10% NaO_2. Barium peroxide, which was originally used for making dilute solutions of H_2O_2 by treatment with dilute H_2SO_4, is made by the action of air or O_2 on BaO; the reaction is slow below 500°C and BaO_2 decomposes above 600°C.

The ionic peroxides with water or dilute acids give H_2O_2, and all are powerful oxidizing agents. They convert organic materials into carbonate even at moderate temperatures. Sodium peroxide vigorously oxidizes some metals; for example, Fe violently gives FeO_4^{2-}, and Na_2O_2 can be generally employed for oxidizing fusions. The alkali peroxides react with CO_2:

$$2CO_2(g) + 2M_2O_2 \longrightarrow 2M_2CO_3 + O_2$$

[28]J. Kim and J. O. Edwards, *Inorg. Chim. Acta* **1995**, *235*, 9.
[29]D. T. Sawyer *et al.*, *Acc. Chem. Res.* **1996**, *29*, 409.

Peroxides can also serve as reducing agents for such strongly oxidizing substances as MnO_4^-.

Other electropositive metals such as magnesium, the lanthanides, or uranyl ion also give peroxides that are intermediate in character between the ionic ones and the essentially covalent peroxides of metals such as Zn, Cd, and Hg.

A characteristic feature of the ionic peroxides is the formation of well-crystallized hydrates. Thus $Na_2O_2 \cdot 8H_2O$ can be obtained by adding ethanol to 30% H_2O_2 in concentrated NaOH at 15°C, or by rapid crystallization of Na_2O_2 from ice water. The alkaline earths all form the octahydrates ($M^{II}O_2 \cdot 8H_2O$). They are isostructural, containing discrete O_2^{2-} ions to which the water molecules are hydrogen bonded, giving chains of the type $\cdots O_2^{2-} \cdots (H_2O)_8 \cdots O_2^{2-} \cdots (H_2O)_8 \cdots$. Some salts crystallize with interstitial H_2O_2 (cf. H_2O), which can be H bonded to the anions. This class of peroxohydrates has to be distinguished from true peroxo compounds. Some examples are $M_2CO_3 \cdot nH_2O_2$, $n = 1$ (NH_4), 1.5 (Na), and 3 (K, Rb, Cs) and $KF \cdot H_2O_2$.

Peroxo Compounds

There are numerous species where $-O-$ is replaced by $-O-O-$. The peroxo carbonate ion O_2COOH^- occurs in $NaHCO_4 \cdot H_2O$ and alkali metal salts of stoichiometry $M_2C_2O_6$ exist in the solid state. In solution in aqueous H_2O_2 there appear to be the reactions:

$$C_2O_6^{2-} + H_2O \xrightarrow{\text{fast}} HCO_4^- + HCO_3^-$$

$$HCO_4^- + H_2O \rightleftharpoons HCO_3^- + H_2O_2$$

Many *organic peroxides* and *hydroperoxides* are known.[30] Peroxo carboxylic acids (e.g., peroxoacetic acid, $CH_3CO \cdot OOH$) can be obtained by the action of H_2O_2 on acid anhydrides. Peroxoacetic acid is made as 10 to 55% aqueous solutions containing some acetic acid by interaction of 50% H_2O_2 and acetic acid, with H_2SO_4 as catalyst at 45 to 60°C; the dilute acid is distilled under reduced pressure. It is also made by air oxidation of acetaldehyde. The peroxo acids are useful oxidants and sources of free radicals [e.g., by treatment with Fe^{2+}(aq)]. Dibenzoyl peroxide, di-*t*-butyl peroxide, and cumyl hydroperoxide are moderately stable and widely used as polymerization initiators and for other purposes where free-radical initiation is required.

Organic peroxo compounds are also obtained by *autooxidation* of ethers, unsaturated hydrocarbons, and other organic materials on exposure to air. A free-radical chain reaction is initiated almost certainly by radicals generated by the interaction of oxygen and traces of metals such as copper, cobalt, or iron. The attack on specific reactive C—H bonds by a radical $X^{\cdot}$ gives first $R^{\cdot}$, then hydroperoxides, which can react further:

$$RH + X^{\cdot} \longrightarrow R^{\cdot} + HX$$

$$R^{\cdot} + O_2 \longrightarrow RO_2^{\cdot}$$

$$RO_2^{\cdot} + RH \longrightarrow RO_2H + R^{\cdot}$$

[30]D. M. Davies and N. D. Gillitt, *J. Chem. Soc., Dalton Trans.* **1995,** 3323.

Peroxide formation can lead to explosions if oxidized solvents are distilled. Peroxides are best removed by washing with acidified $FeSO_4$ solution or, for ethers and hydrocarbons, by passage through a column of activated alumina. Peroxides are absent when the $Fe^{2+} + SCN^-$ reagent gives no red color.

There are a number of sulfur-containing peroxo compounds such as Caro's acid, $HOS(O)_2(OOH)$, prepared and crystallized by addition of H_2O_2 to oleum[31] (Section 12-13). Others that contain fluorine groups are reasonably stable.[32] Some examples are

$(FSO_2)OO(SO_2F)$	Peroxodisulfonyldifluoride
SF_5OOSF_5	Bis(pentafluorosulfur)peroxide
CF_3OOCF_3	Bis(perfluoromethyl)peroxide

The compounds are usually prepared by fluorination of oxygen compounds, for example,

$$SO_3 + F_2 \xrightarrow[160°C]{AgF_2} (FSO_2)OO(SO_2F)$$

$$COF_2 + ClF_3 \xrightarrow[250°C]{KF} CF_3OOCF_3$$

11-6 Superoxides, the Superoxide Ion, and Ozonides

The formation of the orange crystalline *superoxides* (MO_2) by the action of O_2 on K, Rb, and Cs has been noted (Section 3-5). Sodium superoxide is obtained only at 500°C and 300 atm while LiO_2 has not been isolated. For Mg, Ca, Sr, and Ba low concentrations exist in the peroxides. The K, Rb, and Cs superoxides have the CaC_2 structure (Section 7-3) while NaO_2 is cubic owing to O_2^- ion disorder. The main source for O_2^- is KO_2, which is normally only ~96% pure having K_2O_2 and KOH impurities. In KO_2, the O—O distance is 1.28 Å. The electronic structure of O_2^- is similar to that of NO and the unpaired electron is in an antibonding (π^*) MO between the two oxygen atoms and consequently its chemical reactions show little radical character.

The superoxides react with water

$$2O_2^- + H_2O = O_2 + HO_2^- + OH^-$$

$$2HO_2^- = 2OH^- + O_2 \quad \text{(slow)}$$

and with CO_2, a reaction of use in regeneration of oxygen in closed systems:

$$4MO_2(s) + 2CO_2(g) = 2M_2CO_3(s) + 3O_2(g)$$

Although solid superoxides have long been known, the reaction chemistry of the O_2^- ion *in solutions* was initiated by the discovery of O_2^- from its esr spectrum

[31]W. Frank and B. Bertsch-Frank, *Angew. Chem. Int. Ed. Engl.* **1992**, *31*, 436.
[32]H. Sawada, *Chem. Rev.* **1996**, *96*, 1779.

during an enzyme reaction involving O_2^-. Metallaproteins known as superoxide dismutase (SOD) appear to protect living cells against the toxic effects of O_2^-. The prime action of superoxide dismutase is to catalyze the reaction[33]

$$2O_2^- + 2H^+ = O_2 + H_2O_2$$

which proceeds by two, one-electron transfers.

Solutions of O_2^- can be readily obtained. Thus KO_2 can be solubilized by 18-C-6 in DMSO although it slowly reacts with the solvent. Potassium superoxide can be metathesized with $(Me_4N)_2CO_3$ to $(Me_4N)O_2$, ~93% pure, which is soluble in MeCN and other aprotic solvents.

Transient quantities of O_2^- in aqueous solution can be generated by pulse radiolysis of O_2 and by photolysis of H_2O_2 in aqueous media. In aprotic media stable solutions of O_2^- can be prepared by electrochemical reduction of molecular oxygen and by the base-induced decomposition of H_2O_2. Superoxide species can also be made from basic dioxygen-saturated solutions of aniline in dimethyl sulfoxide:[34]

$$2PhNH_2 + 2OH^- + 3O_2 \longrightarrow 2O_2^- + PhN=NPh + H_2O_2 + 2H_2O$$

In aprotic solvents O_2^- acts as a nucleophile, for example, to alkyl halides, and as a one-electron oxidant for organic compounds and metal ions such as Cu^{2+}. Transition metal complexes are discussed in Section 11-10.

Ozonides

Interaction of K or Rb superoxide with $O_3 + O_2$ followed by extraction and crystallization from liquid ammonia gives the deep red MO_3. Ozonides of other alkali ions and of NR_4^+ are known;[35] all are colored owing to an absorption band in the 400- to 600-nm region. The O_3^- ion has C_{2v} symmetry with bond angles between 113.5 and 119.5°; the bond lengths between 1.285 and 1.345 Å are somewhat longer than in O_3 itself (1.278 Å). Thermal stabilities decrease from Cs to Li. The O_3^- ion may be formed in the decomposition of alkaline H_2O_2 and in radiolytic reactions. A number of organic ozonides have been characterized with alkenes; rapid addition of O_3 occurs even at 100°C to give a 1,2,3 trioxide, which isomerizes at higher temperature to a 1,2,4 trioxolane:

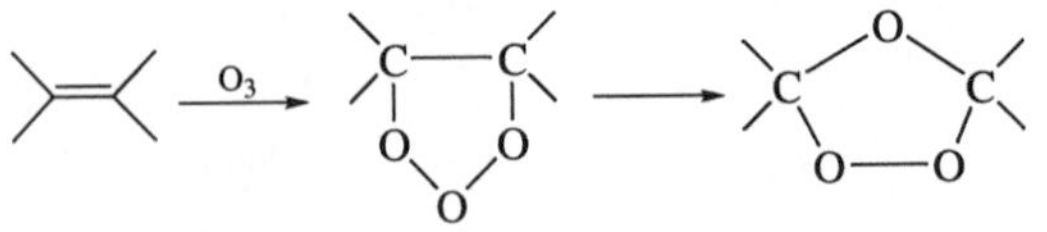

Ozonation of saturated compounds appears to proceed by hydride transfer to give the ion pair $[R^+HO_3^-]$, which decomposes by a radical path giving high yields of singlet oxygen; other paths give H_2O_2.

[33]J.-L. Pierre *et al., J. Am. Chem. Soc.* **1995,** *117,* 1965.
[34]S. Jeon and D. T. Sawyer, *Inorg. Chem.* **1990,** *29,* 4612.
[35]D. Reinen *et al., Inorg. Chem.* **1991,** *30,* 1923; M. Jansen and W. Assenmacher, *Z. Kristallog.* **1991,** *194,* 315.

11-7 The Dioxygenyl Cation

The O_2^+ ion was first obtained by the interaction of oxygen with PtF_6, which gives the orange solid O_2PtF_6, isomorphous with $KPtF_6$. Other salts can be made by reactions such as

$$2O_2 + 2BF_3 + F_2 \xrightarrow[-78\,°C]{hv} 2O_2BF_4$$

$$2O_2 + 2GeF_4 + F_2 \xrightarrow{hv} 2O_2GeF_5$$

$$O_2 + Pt + 3F_2 \xrightarrow{280\,°C} O_2PtF_6$$

Clearly, large, inoxidizable anions are required to stabilize O_2^+. Some of the salts are quite volatile (e.g., O_2RhF_6 will sublime at room temperature), but are readily hydrolyzed by water. The O_2^+ ion is paramagnetic and the $O-O$ stretching frequency is 1905 cm^{-1}. Spectroscopic study of gaseous O_2^+ gives an $O-O$ distance of 1.12 Å (cf. 1.09 Å in isoelectronic NO).

OXYGEN AND OXYGEN COMPOUNDS AS LIGANDS

11-8 Water and the Hydroxide Ion

In aqueous solution metal ions are surrounded by water molecules.[36] In some cases, such as the alkali ions, they are weakly bound, whereas in others, such as $[Cr(H_2O)_6]^{3+}$ or $[Rh(H_2O)_6]^{3+}$, they may be firmly bound and exchange with solvent water molecules only very slowly; for the lanthanides water exchange decreases with decreasing ionic radii.[37] Coordination numbers vary extensively, depending on the size of the metal ion. For example, coordination number four is common for lithium, six is most frequently found for transition metal ions;[38] higher coordination numbers are not unusual for larger ions, e.g., Bi^{3+} can form $Bi(H_2O)_9^{3+}$.[39]

The bound water molecules may be acidic, giving rise to *hydroxo* species and the acidities of aqua ions can vary by orders of magnitude, for example,

$$[Pt(NH_3)_4(H_2O)_2]^{4+} = [Pt(NH_3)_4(H_2O)(OH)]^{3+} + H^+ \qquad K \approx 10^{-2}$$

$$[Co(NH_3)_5(H_2O)]^{3+} = [Co(NH_3)_5(OH)]^{2+} + H^+ \qquad K \approx 10^{-5.7}$$

Some molten hydrates (e.g., $ZnCl_2 \cdot 4H_2O$) can act as extremely strong acids, even though the aqua ion $[Zn(H_2O)_4]^{2+}$ in aqueous solution is a very weak acid.

Of particular interest are structural changes, octahedral → tetrahedral, on going from certain aqua ions to higher hydroxo species, for example, $[M(H_2O)_6]^{3+} \rightarrow [M(OH)_4]^{2-}$ for Al^{III} and Co^{II}. For $[Hg(H_2O)_6]^{2+}$ there is a change to linear

[36]H. Ohtaki and T. Radnai, *Chem. Rev.* **1993**, *93*, 1157.
[37]A. E. Merbach *et al.*, *Inorg. Chim. Acta* **1995**, *235*, 311.
[38]F. A. Cotton *et al.*, *Inorg. Chem.* **1993**, *32*, 4861.
[39]W. Frank *et al.*, *Angew. Chem. Int. Ed. Engl.* **1995**, *34*, 2416.

$(H_2O)Hg(OH)_{aq}^+$ and $(HO)Hg(OH)_{aq}$ species where two Hg—O bonds are shortened and the other four lengthened. Protonation of transition metal oxo complexes can affect the geometry and magnetic behavior of the compounds.[40]

A common feature of hydroxo complexes is the formation of hydroxo bridges of the following types:

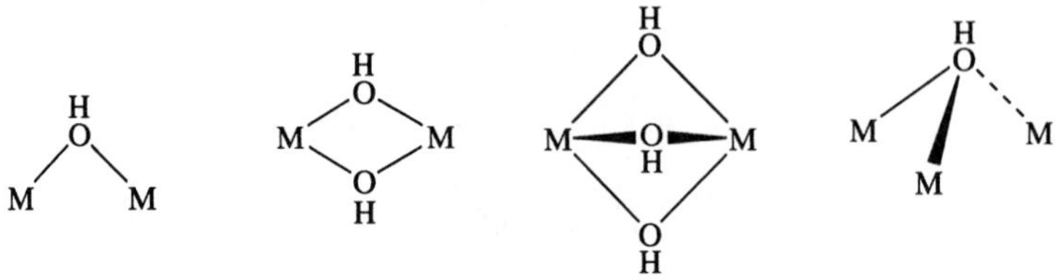

Single bridges are found in complexes such as $(porph)Fe(\mu\text{-}OH)Fe(porph)$,[41] but double bridges are more common.[42] Three μ_2-bridges are found in the π-arene complex $[ArRu(OH)_3RuAr]^+$, whereas μ_3-hydroxo groups occur in the cubanelike ions $[(ArRuOH)_4]^{4+}$ (11-IV), $[Pt(OH)Me_3]_4$, $[Pb_4(OH)_4]^{4+}$ [43] and some others.

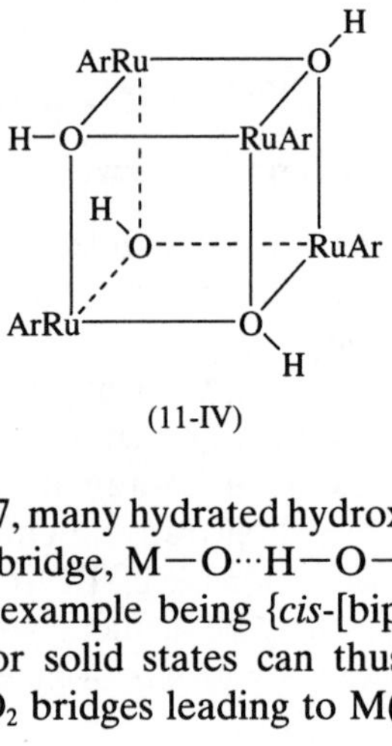

(11-IV)

As discussed in Section 2-7, many hydrated hydroxo complexes have a hydrogen bonded $H_3O_2^-$ ion acting as a bridge, M—O···H—O—M, and numerous complexes have been established,[44] one example being $\{cis\text{-}[bipy_2Cr^{III}(H_3O_2)]\}^{4+}$. The *olation reactions* in either solution or solid states can thus be considered as involving elimination of H_2O from H_3O_2 bridges leading to $M(\mu\text{-}OH)M$ bridges:

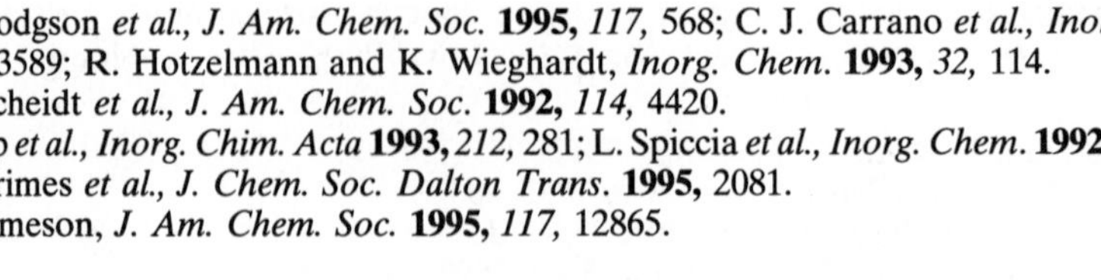

[40]K. O. Hodgson *et al., J. Am. Chem. Soc.* **1995,** *117,* 568; C. J. Carrano *et al., Inorg. Chem.* **1993,** *32,* 3589; R. Hotzelmann and K. Wieghardt, *Inorg. Chem.* **1993,** *32,* 114.
[41]W. R. Scheidt *et al., J. Am. Chem. Soc.* **1992,** *114,* 4420.
[42]J. Glerup *et al., Inorg. Chim. Acta* **1993,** *212,* 281; L. Spiccia *et al., Inorg. Chem.* **1992,** *31,* 4894.
[43]S. M. Grimes *et al., J. Chem. Soc. Dalton Trans.* **1995,** 2081.
[44]G. B. Jameson, *J. Am. Chem. Soc.* **1995,** *117,* 12865.

11-9 Oxo Compounds[45]

The loss of a second proton from coordinated water can lead to the formation of *oxo compounds*, which can be of several types, some of which have been noted in Section 11-2.

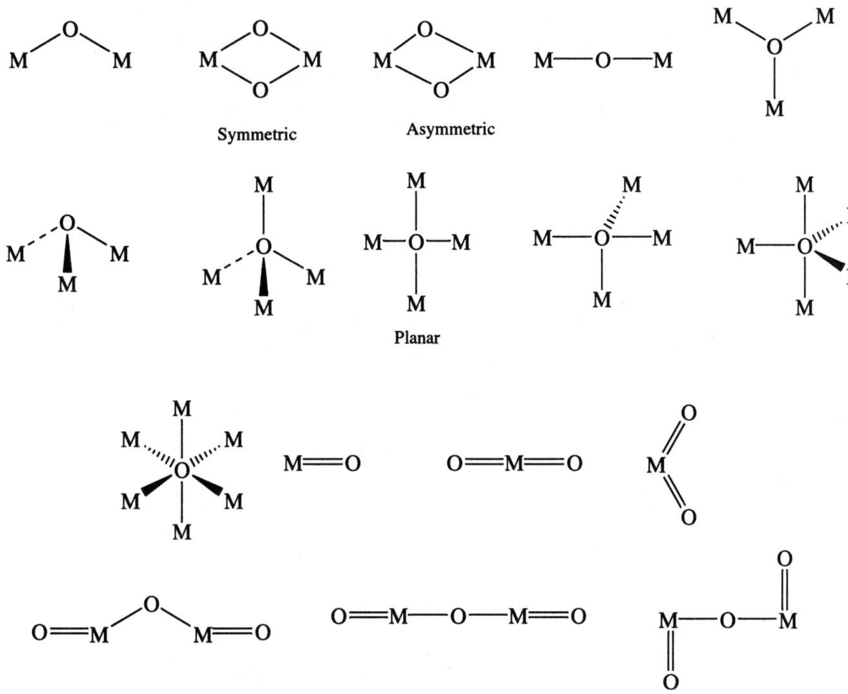

Symmetric

Asymmetric

Planar

The multiply-bonded oxo group (M=O) is found not only in oxo compounds and oxo anions of non-transition elements such as SO_4^{2-}, $O=PCl_3$, and PO_4^{3-}, but also in transition metal compounds such as vanadyl ($O=V^{2+}$), uranyl ($O=U=O^{2+}$), permanganate (MnO_4^-), and osmium tetroxide (OsO_4). In all these cases the bond distances ($\sim$1.59-1.66 Å) correspond to a double bond, and the M—O ir stretching frequencies usually lie in the 800 to 1000 cm^{-1} region for transition metal species. In the latter the π component is best regarded as arising from $Op\pi \rightarrow Md\pi$ electron flow. Since this is the opposite of electron flow in π-bonding ligands of the CO type, it is not surprising that the latter are most stable in low oxidation states whereas M=O bonds are most likely in high oxidation states.

Protonation of M=O to M—OH occurs on interaction with strong acids (Section 11-8); in some cases, condensation reactions with amines can give M=NR compounds (Section 9-16):

$$L_nM=O + RNH_2 \longrightarrow L_nM=NR + H_2O$$

The M=O bonding is commonly affected by the nature of groups trans to oxygen—and oxygen has a strong trans influence. Donors that increase electron

[45]D. M. Kurtz, Jr., *Chem. Rev.* **1990,** *90,* 585.

density on the metal tend to reduce its acceptor properties, thus *lowering* the M—O multiple bond character, hence the M—O stretching frequency. Because of the strong trans influence, ligands trans to oxygen may be labile.

Dioxo compounds may be linear[46] (trans) as in $O{=}U{=}O^{2+}$ or angular (cis) as in some molybdenum complexes[47] and $ReO_2(bipy)(py)_2^+$.[48]

Several cases of bridging groups of the type (11-V) are known. In $(Me_3SiCH_2)_6(O)_2Ru_2$, Ru=O is 1.733 Å and O → Ru is 2.208 Å. In [OsO_2(cyclohexane-1,2-diolate)quinuclidene]_2 there is dimerization only *via* oxygen bridging (1.78, 2.22 Å), but other compounds have metal-metal bonds also.

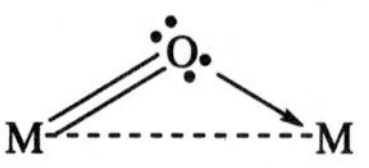

(11-V)

Exchange of bridge and terminal oxo groups is known to occur in solution.[49]

Single MOM[50] *bridges* may be *bent* or *linear* with the angle varying from ~140 to 180° and being determined to a large extent by the steric requirements of other ligands attached to the metal. Bent bridges are found in many dimeric species such as $Cr_2O_7^{2-}$, $Mo_2O_7^{2-}$, $P_2O_7^{4-}$, $[Fe_2OCl_6]^{2-}$, and in polymeric species such as $Cp_6Ti_6O_8$ and the ions (11-VI) and (11-VII),

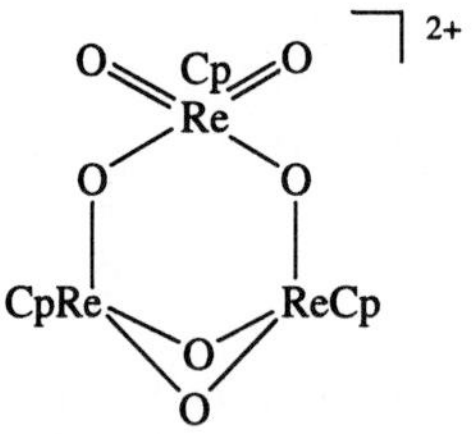

(11-VI) 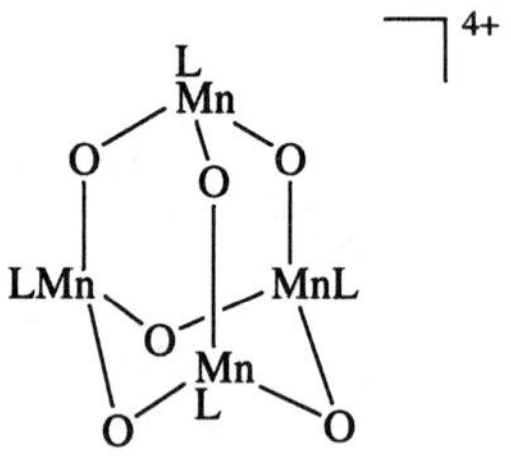   (11-VII)

where L = 1,4,7-triazacyclononane.

Other linear M—O—M bridges[51] are found in some complexes and hetero binuclear porphyrin and cyclopentadienyl species.[52] In the Ru and Os ions, $[M_2OX_{10}]^{4-}$, the M—O—M unit forms an electronically unique independent chromophore. The linearity results from $d\pi-p\pi$ bonding through overlap of the p_x and p_y orbitals on O with d_{xz} and d_{yz} orbitals on the metal atoms. Linear M—O—M groups have ir vibrations lower than those in bent bridges. Units of the type $Mo_2O_3^{4+}$ and $Re_2O_3^{6+}$ are common for Mo^V and Re^{VI} complexes; there are linear,

[46]V. Staemmler *et al., Inorg. Chem.* **1994,** *33,* 6219; K. Umakoshi *et al., Inorg. Chem.* **1995,** *34,* 813.
[47]C. Floriani *et al., J. Chem. Soc. Dalton Trans.* **1989,** 145.
[48]J. T. Hupp *et al., Inorg. Chem.* **1990,** *29,* 1791.
[49]N. J. Cooper *et al., Inorg. Chem.* **1993,** *32,* 6067.
[50]S. M. Gorun and S. J. Lippard, *Inorg. Chem.* **1991,** *30,* 1625.
[51]I. Vernik and D. V. Stynes, *Inorg. Chem.* **1996,** *35,* 2006.
[52]F. Bottomley, *Polyhedron* **1992,** *11,* 1707.

O=M—O—M=O, and cis or trans types:

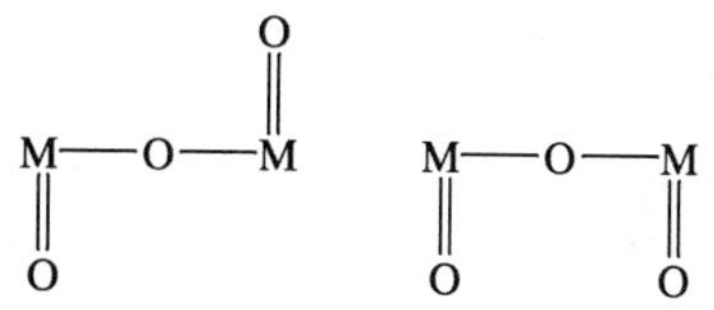

The *pyramidal* $(\mu_3$-O)M$_3$ unit occurs widely in species such as OHg$_3^+$, [Nb$_3(\mu_3$-O)$_2$(O$_2$CR)$_6$(THF)$_3$]$^+$, and Ru$_3(\mu_3$-O)$(\mu_3$-CO)(CO)$_5(\mu$-η^2dppm)$_2$; several species have multiple μ_3-O caps as in [CpCr$(\mu_3$-O)]$_4$ and many large polyoxometalates;[53] the capping oxygen acts as a 4e ligand.

Oxo-centered complexes can have μ_4-tetrahedral oxygen (Section 11-2) in the center of a tetrahedron of divalent metal atoms as in M$_4$O(O$_2$CMe)$_6$ or in clusters. The best-known complex is Be$_4$O(O$_2$CMe)$_6$, but Zn^{II} and Co^{II} analogues are also known. Iron and Mn complexes[54], for example, Fe$^{III}_4$O(O$_2$CMe)$_{10}$, also have μ_4-O, and higher oxo iron polymers may have trigonal bipyramidal coordination, as in [Fe$_5$O(O$_2$CMe)$_{12}$]$^+$, or a square-planar oxide bridge as in [Fe$_8(\mu_4$-O)$(\mu_3$-O)$_4$(OAc)$_8$-(tren)$_4$]$^{6+}$.[55] Compounds with higher coordination numbers have been noted (Section 11-2).

Oxygen-centered triangles[56] are an important unit found widely in so-called *"basic"* carboxylates (and some other bridging anions) of many metals, for example, Ga[57], V, Cr, Mn, Fe,[58] Co, Ru,[59] Rh, Ir, and Pt. For the +3 state they have the general formula [M$_3$O(O$_2$CR)$_6$L$_3$]$^+$ where L is a ligand such as H$_2$O or pyridine and the structure (11-VIII) is found.

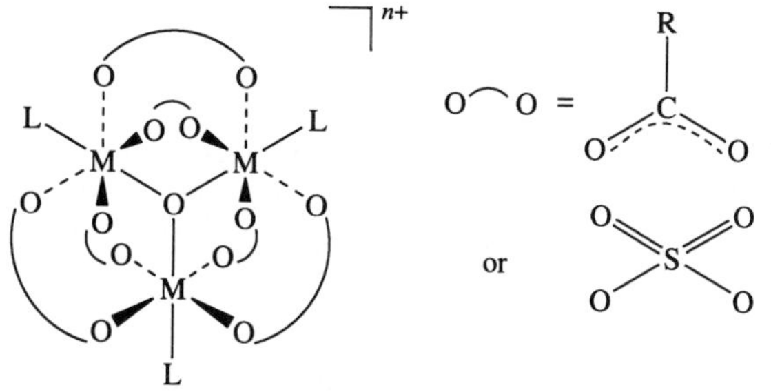

(11-VIII)

The M$_3$O group is usually planar on account of π-bonding, though in a few cases, as in [Fe$_3$O(O$_2$CCMe$_3$)$_6$(MeOH)$_3$]$^+$, the O atom may be slightly out of plane.

[53]J. Evans *et al.*, *J. Chem. Soc., Dalton Trans.* **1996**, 2951; M. I. Khan and J. Zubieta, *Prog. Inorg. Chem.* **1995**, *43*, 1.
[54]D. N. Hendrickson *et al.*, *J. Chem. Soc., Chem. Commun.* **1994**, 1031.
[55]V. S. Nair and K. S. Hagen, *Inorg. Chem.* **1994**, *33*, 185.
[56]R. D. Cannon and R. P. White, *Prog. Inorg. Chem.* **1988**, *36*, 195.
[57]S. A. Duraj *et al.*, *J. Am. Chem. Soc.* **1992**, *114*, 786.
[58]S. J. Lippard *et al.*, *J. Am. Chem. Soc.* **1991**, *113*, 4645.
[59]D. T. Richens *et al.*, *Inorg. Chim. Acta* **1995**, *232*, 167.

11-10   Dioxygen, Superoxo, and Peroxo Ligands

The O_2 molecule and its two reduced species O_2^- and O_2^{2-} can act as ligands to transition metals. Molecular oxygen reacts reversibly with some metal complexes[60] and such reversible reactions are involved in the oxygenation of hemoglobin and myoglobin (Section 17-E-10).

Generally, electron density is transferred from metal to O_2 with formal oxidation of the metal and reduction of O_2:

$$L_xM^{n+} + O_2 \rightleftharpoons L_xM^{(n+1)+}O_2^{\bullet-}$$

initially giving a *superoxo* complex. Specific examples are

$$Co^{II}(acacen) + O_2 + Me_2NCHO \rightleftharpoons Co^{III}(O_2)(acacen)(Me_2NCHO)$$

$$SmTp^*_2 \xrightarrow[-78°C, \text{ toluene}]{O_2(g), 1 \text{ atm}} \xrightarrow{\text{room temperature}} Tp^*_2Sm\overset{O}{\underset{O}{\diagdown\mid}}$$

However, in certain cases we can have a two-electron transfer giving a peroxo complex, as in oxidative-addition reactions:

$$trans\text{-}Ir^ICl(CO)(PPh_3)_3 + O_2 \rightleftharpoons Ir^{III}(O_2)Cl(CO)(PPh_3)_2 + PPh_3$$

The reaction of a superoxo complex with a second metal can lead to a bridging μ-peroxo species

$$L_xM^{n+1}\text{-}O_2^- + L_xM^{n+} \rightleftharpoons L_xM^{n+1}\text{-}(\mu\text{-}O_2^{2-})\text{-}ML_x^{n+1}$$

which in turn can be oxidized to a μ-superoxo species:

$$L_xM^{n+1}\text{-}(O_2^{2-})\text{-}ML_x^{n+1} \underset{+e}{\overset{-e}{\rightleftharpoons}} L_xM^{n+1}\text{-}(\mu\text{-}O_2^-)\text{-}M^{n+1}L_x$$

Peroxo complexes are often obtainable in aqueous solution from H_2O_2 especially for the early transition metals like Ti, V,[61] Nb, Cr, Mo, W[62] and Mn.[63] Sometimes the same complex can be obtained from either O_2 or H_2O_2, for example,

$$[Co^I(diars)_2]^+ \xrightarrow{O_2} [O_2Co^{III}(diars)_2]^+ \xrightarrow[\substack{-2H_2O \\ -2H^+}]{H_2O_2} cis\text{-}[Co^{III}(H_2O)_2(diars)_2]^{3+}$$

but more typically,

$$HCrO_4^- + 2H_2O_2 + H^+ \longrightarrow CrO(O_2)_2(H_2O) + 2H_2O$$

For copper,[64] the compounds are typically prepared by reaction with O_2.

[60]K. D. Karlin *et al., J. Am. Chem. Soc.* **1995,** *117,* 12498; **1994,** *116,* 1324; N. Kitajima *et al., J. Am. Chem. Soc.* **1994,** *116,* 9071; R. Guilard *et al., J. Am. Chem. Soc.* **1994,** *116,* 10202; T. Imamura *et al., J. Chem. Soc., Chem. Commun.* **1993,** 1436.

[61]A. Butler *et al., Chem. Rev.* **1994,** *94,* 625.

[62]M. H. Dickman and M. T. Pope, *Chem. Rev.* **1994,** *94,* 569.

[63]V. L. Pecoraro *et al., Chem. Rev.* **1994,** *94,* 807.

[64]N. Kitajima and Y. Moro-oka, *Chem. Rev.* **1994,** *94,* 737; W. Tollman, *Acc. Chem. Res.* **1997,** *30,* 227.

The $M(O_2)$ complexes can broadly be classed in two groups:

1. Those containing *peroxo groups* (O_2^{2-}) that are (a) part of a three-membered ring, (b) bridging staggered,[65] (c) bridging symmetrical, or (d) bridging through a single oxygen atom.[66]

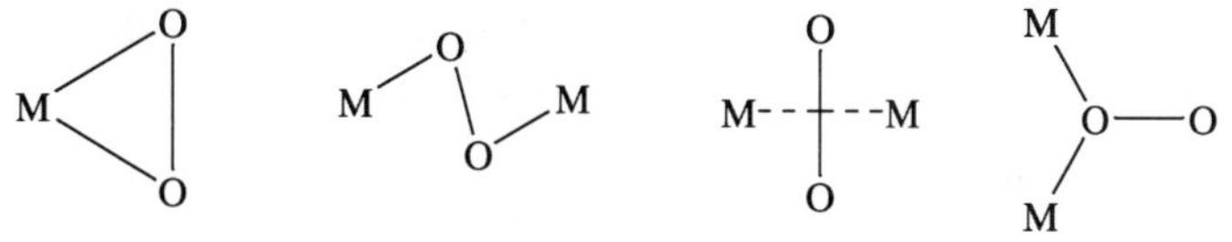

In peroxo compounds the O—O bond distances are *fairly constant* in the range 1.40 to 1.50 Å (O_2^{2-} = 1.49 Å in Na_2O_2) and *do not depend on the nature of the metal and its ligands*. The O—O stretching frequencies are in the 740 to 930 cm^{-1} region (for Na_2O_2, ν = 738 cm^{-1}). For the triangular η^2-O_2 species these frequencies are around 850 cm^{-1}. It makes no difference whether compounds are made from O_2 or H_2O_2, and there is no correlation between reversible oxidation by O_2 and any bond parameters.

Although the bonding in the three-membered ring is most easily described by localized bonding, it can also be described by an MO treatment similar to that for the bonding of olefins or acetylenes. Crudely, a σ bond is formed by filled $Op\pi \rightarrow Md\sigma$ bonding and back-bonding is due to $Md\pi \rightarrow O\pi^*$. Representative examples of three-membered ring compounds are oxygen adducts of planar d^8 metal complexes such as $Ir(O_2)Cl(CO)(PPh_3)_2$ (11-IX), Cr, Mo, and W peroxo complexes such as the dodecahedral $[Cr(O_2)_4]^{3-}$ ion (11-X), and organometallic rhenium complexes such as $(HMPA)(CH_3)ORe(O_2)_2$ (11-XI).[67]

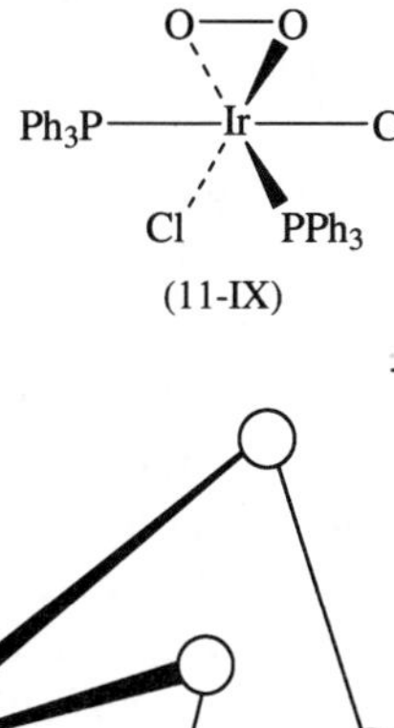

(11-IX)

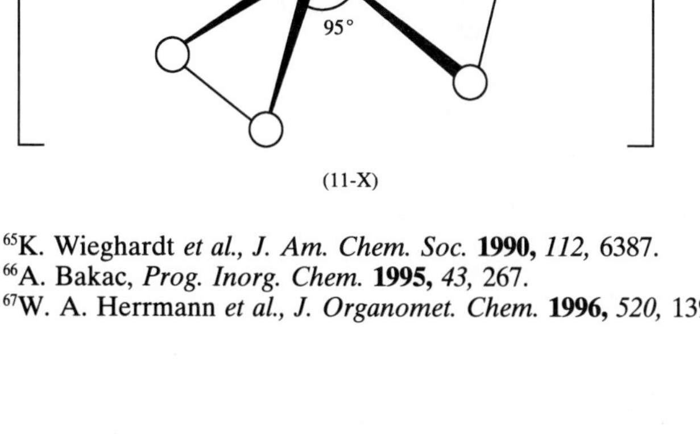

(11-X)

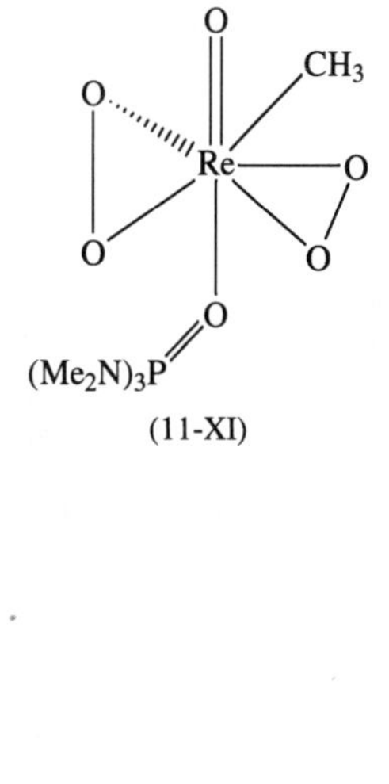

(11-XI)

[65]K. Wieghardt *et al.*, *J. Am. Chem. Soc.* **1990**, *112*, 6387.
[66]A. Bakac, *Prog. Inorg. Chem.* **1995**, *43*, 267.
[67]W. A. Herrmann *et al.*, *J. Organomet. Chem.* **1996**, *520*, 139.

An example of a staggered peroxo-bridged complex which also has a metal-metal bond is $[Ir_2I_2(CO)_2(\mu\text{-}O_2)(dppm)_2]$.[68] Symmetrical bridges are known in peroxo-uranates as well as in the species $[\{(TMPA)Cu\}_2(O_2)]^{2+}$ in which the peroxo group is found to be in equilibrium between two configurations, bent side-on and symmetrical side-on.[69] This dicopper species is also remarkable because oxygen is reversibly exchanged. This type of symmetrical side-on bonding is considered biologically significant because it is known to occur in some hemocyanins.[70]

2. Those containing the *superoxo* O_2^- ion which can be bound in the following ways:

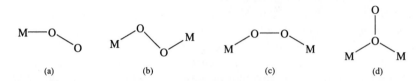

(a)   (b)   (c)   (d)

Unidentate, end-on and bent superoxo groups occur mainly in complexes of cobalt(III) such as $[Co(O_2)(CN)_5]^{3-}$ and rhodium(III) such as $[Rh(O_2)en_2(H_2O)]^{2+}$.

The superoxo *bridged species* of types (b) and (c) are again found mostly in Co^{III} and Rh^{III} species by oxidation of peroxo complexes, for example:

$$\left[am_4Co\overset{O_2}{\underset{NH_2}{<\,>}}Co\,am_4 \right]^{3+} \underset{+e}{\overset{-e}{\rightleftharpoons}} \left[am_4Co\overset{O_2}{\underset{NH_2}{<\,>}}Co\,am_4 \right]^{4+}$$

The bridged superoxo complexes have $O-O$ distances in the range 1.10 to 1.32 Å, for example, 1.319(5), 1.262(8), and 1.23(3) Å in $MTp^*(\eta^2\text{-}O_2)$ complexes of Sm,[71] Co,[72] and Cu,[73] respectively, which can be compared to that in O_2^- (1.33 Å in KO_2). The $O-O$ stretching frequencies for superoxo species generally fall in the 1070 to 1200 cm^{-1} region (KO_2, 1145 cm^{-1}).

The oxygen adducts of metal complexes of macrocyclic N ligands have been intensively studied because of the relation to natural oxygen transport molecules containing Fe and Cu.[74] Considerable ingenuity has gone into trying to make truly reversible synthetic models for heme (Chapter 17-E). The problem is to prevent irreversible oxidation of the iron atom in the macrocycle from Fe^{II} to Fe^{III}.

One approach has been to construct what are termed "picket-fence" or basket handle porphyrins, whereby the way the oxygen molecule can approach and leave the iron atom axially is sterically restricted by bulky groups.

Although most O_2 "carriers" have organic ligands, a purely inorganic one, $[GeWMnO_{39}]^{6-}$ acts this way at low temperatures.

[68]M. Cowie *et al., J. Am. Chem. Soc.* **1990,** *112,* 9425.
[69]K. D. Karlin *et al., J. Am. Chem. Soc.* **1996,** *118,* 3763.
[70]J. Sanders-Loehr *et al., J. Am. Chem. Soc.* **1994,** *116,* 7682.
[71]J. Takats *et al., J. Am. Chem. Soc.* **1995,** *117,* 7828.
[72]A. L. Rheingold *et al., J. Am. Chem. Soc.* **1990,** *112,* 2445.
[73]N. Kitajima *et al., J. Am. Chem. Soc.* **1994,** *116,* 12079.
[74]D. H. Busch and N. W. Alcock, *Chem. Rev.,* **1994,** *94,* 585.

Hydroperoxo complexes are prepared[75] by protonation of peroxo complexes, by insertion of dioxygen into metal–hydrogen bonds, by hydrogen abstraction by metal dioxygen complexes, by reduction of superoxo complexes or by reaction of the metal ion with hydrogen peroxide. Well-defined stable species have been characterized for Cu,[76] Ir, Pt, and other metals, for example, by syntheses of the type:

$$\text{diphos Pt(OH)(CF}_3\text{)} \underset{\text{ROOH}}{\overset{\text{H}_2\text{O}_2}{\rightleftarrows}} \begin{array}{l} \text{diphos Pt(OOH)(CF}_3\text{)} + \text{H}_2\text{O} \\ \text{diphos Pt(OOR)(CF}_3\text{)} + \text{H}_2\text{O} \end{array}$$

$$\text{IrCl(CO)(PR}_3\text{)}_2 \xrightarrow{\text{Bu}^t\text{OOH}} \textit{cis}\text{-IrCl(O}_2\text{Bu}^t\text{)}_2\text{(CO)(PR}_3\text{)}_2$$

From aquation reactions of $[\text{Co(CN)}_5\text{OOH}]^{3-}$ it has been determined that the relative affinity for Co^{III} is $\text{H}_2\text{O}_2 < \text{H}_2\text{O} < \text{HO}_2^-$.

The selective oxidation of organic compounds is an area of great interest in the chemical industry.[77] Metal activation of dioxygen is frequently used. For example, in the reaction

$$\text{C}_4\text{H}_{10} + 3.5\text{O}_2 \longrightarrow \begin{array}{c} \\ \end{array} + 4\text{H}_2\text{O}$$

a vanadium phosphorus oxide is used as catalyst. Bismuth—molybdenum oxides are used in the ammoxidation of propene to acrylonitrile:

$$2\text{CH}_2\text{=CHCH}_3 + 3\text{O}_2 + 2\text{NH}_3 \longrightarrow 2\text{CH}_2\text{=CHCN} + 6\text{H}_2\text{O}$$

The selective oxidation of ethylene to ethylene oxide is carried out over an Ag/Al$_2$O$_3$ mixture and requires an oxygen atom transfer to an olefin. Some possible intermediates of this process have been pepared from polyoxoanion-supported iridium-olefin complexes.[78] The Wacker process producing acetaldehyde from ethylene in water[79] is catalyzed by PdCl_2, CuCl_2, and O_2 (Chapter 21). Many other homogeneous metal-catalyzed oxidations by O_2 are known.[80] In some, the complex binds O_2 which then reacts with the substrate; other reactions might involve peroxides. Most of the oxidations are radical in nature but some do proceed by nonradical paths. In Nature, enzymes called monooxygenases (Section 17-E-10) catalyze oxidations such as hydroxylation of non-activated C—H bonds:[81]

$$\text{\textgreater}\text{C—H} + \text{O}_2 + 2e^- + 2\text{H}^+ \xrightarrow{\text{monoxygenase}} \text{\textgreater}\text{C—OH} + \text{H}_2\text{O}$$

[75]A. F. Williams *et al.*, *Inorg. Chem.* **1996,** *35,* 1332; A. Bakac and W.-D. Wang, *J. Am. Chem. Soc.* **1996,** *118,* 10325.

[76]K. D. Karlin *et al.*, *Inorg. Chem.* **1992,** *31,* 3001.

[77]D. Riley *et al.*, *The Activation of Dioxygen and Homogeneous Catalytic Oxidation,* D. H. R. Barton *et al.*, eds., Plenum Press, New York, 1993, p. 31.

[78]V. W. Day *et al.*, *J. Am. Chem. Soc.* **1990,** *112,* 2031.

[79]T. Hosokawa and S.-I. Murahashi, *Acc. Chem. Res.* **1990,** *23,* 49.

[80]R. S. Drago, *Coord. Chem. Rev.* **1992,** *117,* 185.

[81]A. M. Khenkin and A. E. Shilov, *New J. Chem.* **1989,** *13,* 659; R. C. Mehrotra and A. Singh, *Prog. Inorg. Chem.* **1997,** *46,* 239. (Syntheses, properties, structures, chemical reactivity, and uses; 590 references.)

11-11   Alkoxides and Aryloxides

In solution in alcohols, particularly methanol, metal ions may be solvated just as in water, but the solvent molecules are usually readily displaced by stronger donor ligands such as water itself.

Just as coordinated water can lose a proton (to give hydroxo complexes), so can alcohols:

$$M-O{<}^{R}_{H} \rightleftharpoons [MOR]^- + H^+$$

Alkoxide ligands,[82] RO^-, and related oxygen donor ligands such as aryloxides and trialkyl- or triarylsiloxides are hard σ-donors and may also act as π-donors as a result of their filled oxygen $p\pi$-orbitals. They may therefore stabilize metal atoms in relatively high (unusually high) oxidation states, e.g., $Mo(OMe)_6$, and by the selection of appropriate steric properties, they may stabilize metal ions in unusual coordination environments as in $Cr(OBu^t)_4$. As ancillary ligands, the RO σ/π-donation may be adjusted by selection of R. The degree of π-donation is responsive to substrate uptake and release in an analogous manner to say, the slippage of indenyl ligands ($\eta^5 \rightarrow \eta^3 \rightarrow \eta^1$) or the bending of the $M-N-O$ angle of a nitrosyl ligand.

Although $M-OR$ groups are usually bent, in $[ZrCl_4(OMe)(MeOH)]^-$, the ZrOC angle is 171.4°, compatible with a triple bond $Zr \equiv O-Me$.

Alkoxides are known for practically every element of the $s, p, d,$[83] and f[84] blocks of the Periodic Table and form many heterometallic complexes.[85] They are usually made by reaction of metal halides and alcohols (or phenols) in the presence of a HX acceptor, for example,

$$TiCl_4 + 4EtOH + 4Et_3N \longrightarrow Ti(OEt)_4 + 4Et_3NH^+Cl^-$$

or by use of alkali metal or Tl^I alkoxides. They are normally readily hydrolyzed but thermally stable, distillable liquids or volatile solids. These properties have been used for the preparation of sol-gels for the production of ceramic materials[86] and high purity oxides of the refractory metals by pyrolysis of metal alkoxides. They undergo a wide variety of other reactions such as insertions with small molecules, for example,

$$L_nMOR + CO_2 \longrightarrow L_nM{<}^{O}_{O}{>}COR$$

[82]M. H. Chisholm, *Chem. Soc. Rev.* **1995,** 79.
[83]W. A. Herrmann *et al., Angew. Chem. Int. Ed. Engl.* **1995,** *34,* 2187.
[84]W. G. Van Der Sluys and A. P. Sattelberger, *Chem. Rev.* **1990,** *90,* 1027.
[85]K. G. Caulton and L. G. Hubert-Pfalzgraf, *Chem. Rev.* **1990,** *90,* 969.
[86]R. C. Mehrotra and A. Singh, *Chem. Soc. Rev.* **1996,** 1.

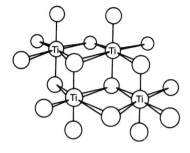

Figure 11-3 The tetranuclear structure of crystalline [Ti(OC$_2$H$_5$)$_4$]$_4$ showing both μ_2 and μ_3 bridges. Only Ti and O atoms are shown. Note that the more bulky trifluoromethyl compound is mononuclear.

and hydrogenolysis, for example,

$$(CuOBu^t)_4 \xrightarrow[PR_3]{H_2} (HCuPR_3)_6 + Bu^tOH$$

Nitriles, such as CH$_3$CN, can produce metathesis reactions with dinuclear alkoxides such W$_2$(OBu^t)$_6$ to yield (Bu^{t}O)$_3$W≡N and (Bu^{t}O)$_3$W≡CMe. In contrast W$_2$[OCMe$_2$(CF$_3$)]$_6$ reacts reversibly to give an adduct, W$_2$[OCMe$_2$(CF$_3$)]$_6$(NCMe)$_2$.[87]

Most alkoxides with simple groups are polymeric to the extent that maximum coordination of the metal is achieved. Some common types are shown in (11-XII) to (11-XIV) and the structure of another is illustrated in Fig. 11-3.

(NaOBu^t)$_4$
(11-XII)

(CuOBu^t)$_4$
(11-XIII)

[Nb(OMe)$_5$]$_2$
(11-XIV)

The use of very bulky alkoxide or aryloxide groups such as Bu$_3^t$CO$^-$, Bu$_3^t$SiO$^-$, 2,6-di-*t*-butylphenoxide and adamantoxides can, however, give simple mononuclear species with low coordination numbers.

Alkoxo groups not engaged in bridging may, of course act as donors to other metal species. Thus U(OPr^i)$_6$ gives adducts with Li, Mg, and Al alkyls such as

[87]M. H. Chisholm, *J. Chem. Soc., Dalton Trans.* **1996**, 1781.

(11-XV), while in the lithium salt of the $W(OPh)_6^-$ ion, two phenoxides are bound to Li^+ (11-XVI).

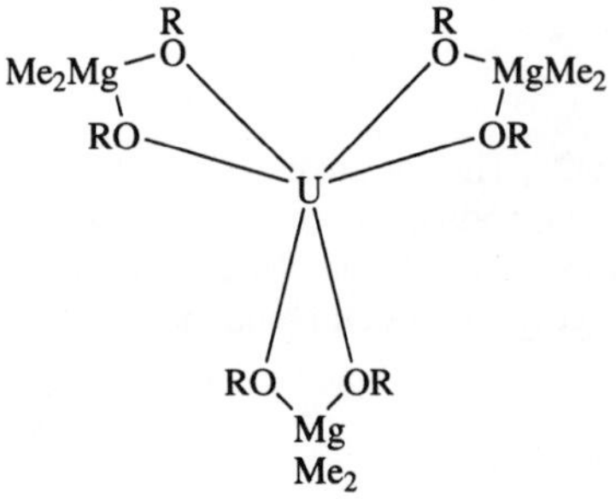

(11-XV)

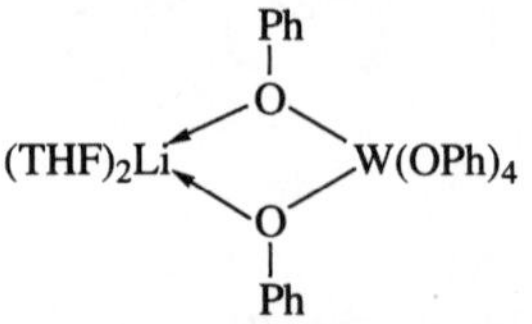

(11-XVI)

Although aryloxides can form unidentate or bridge groups as in $W(OPh)_6$ and $(PhO)Cl_2Ti(\mu\text{-OPh})_2TiCl_2(OPh)$, respectively, for Ru, Rh and Ir, the phenoxide ion can be bound as a η-1-5-oxocyclohexadienyl (11-XVII) where the C—O group becomes more keto-like and the bonding is delocalized.

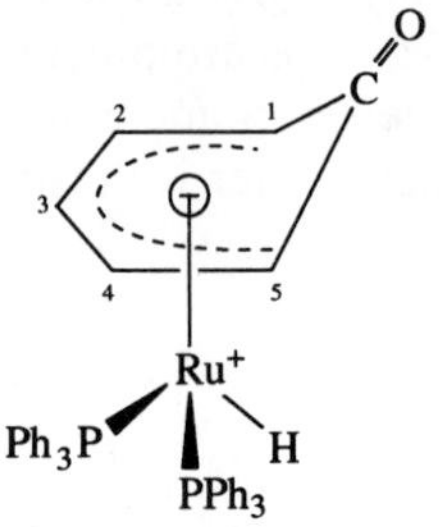

(11-XVII)

Chiral alkoxides[88] can be made using sugars, as in MoO_2L_2 and $MoO_2L_2(phen)$, where L is 11-XVIII.

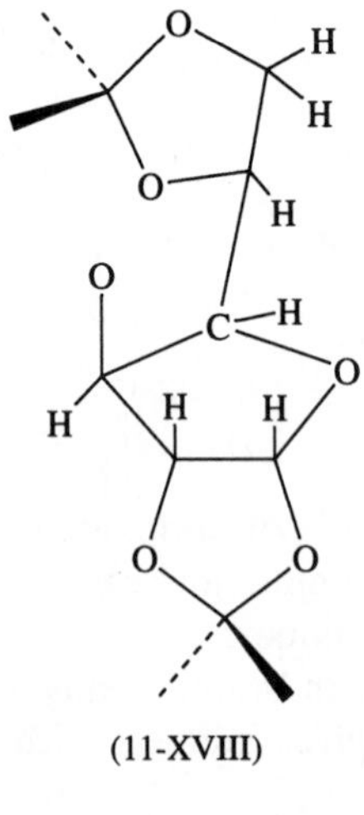

(11-XVIII)

[88]C. Floriani *et al., J. Chem. Soc., Dalton Trans.* **1995**, 3329; *Prog. Inorg. Chem.* **1997**, *45*, 293; *Chem. Commun.* **1997**, 183.

An interesting class of aryloxides are the *calixarenes*[89] (calix, Greek for chalice), which are molecules having a torus shape similar to that of cyclodextrins with potential to mimic enzymes.[90] The number of phenolic residues in the macrocycle is designated by the value of n (4, 6, or 8) in the general term calix[n]arene. The cyclic polymers, such as calix[4]arene (11-XIX), can have a closed (11-XX) or open cavity (11-XXI). They can form complexes with small molecules, behaving as hosts, but can also transport metal ions through hydrophobic liquid membranes, as well as giving phenoxides by loss of protons.

(11-XIX)   (11-XX)   (11-XXI)

Some complexes can be made, for example, by reaction of calixarenes with $M(NMe_2)_n$, M = Ti, Fe, Co, or $TiCl_4$. They can form mononuclear[91] (11-XXII) or polynuclear[92] compounds including those with metal-metal bonds.[93]

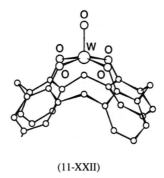

(11-XXII)

[89]D. M. Roundhill, *Prog. Inorg. Chem.* **1995**, *43*, 533; A. Ikeda and S. Shinkai, *Chem. Rev.* **1997**, *97*, 1713.
[90]L. Echegoyen et al., *J. Am. Chem. Soc.* **1994**, *116*, 3580.
[91]C. Floriani et al., *Inorg. Chem.* **1991**, *30*, 4465.
[92]C. L. Raston et al., *Chem. Commun.* **1996**, 2491; J. L. Atwood et al., *Chem. Commun.* **1996**, 2487.
[93]S. J. Lippard et al., *Inorg. Chim. Acta* **1995**, *229*, 5.

11-12 Catecholates[94] and *o*-Quinone Complexes[95]

Phenols with ortho hydroxy groups such as pyrocatechol can give chelates by bonding the 2^- anion as in (11-XXIII) producing complex anions such as $[Cr(O_2C_6H_4)_3]^{3-}$ and $[Cr(CO)_3(O_2C_6H_4)]^{2-}$.[96]

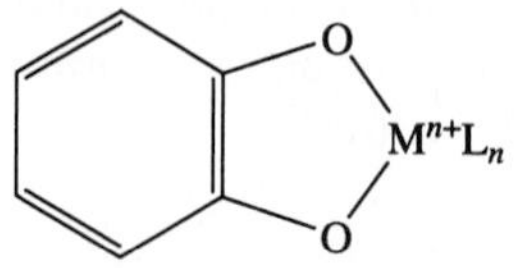

(11-XXIII)

Formally similar complexes (11-XXIV) can be given by *orthoquinones*. Thus interaction of $Cr(CO)_6$ with tetrachlorobenzoquinone (L) gives CrL_3; the oxygen atoms can also bind to separate metals so that a bridged quinone results. The tris species tend to form distorted trigonal prismatic complexes.

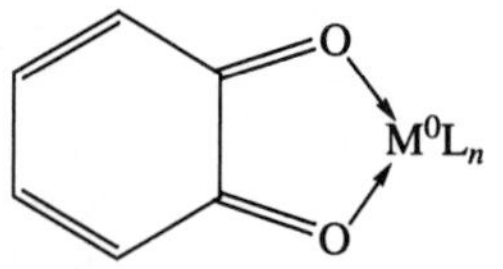

(11-XXIV)

The quinone groups can be considered similar to 1,2-dithiolenes and *o*-quinoneimines and undergo similar types of oxidation-reduction sequences:[97]

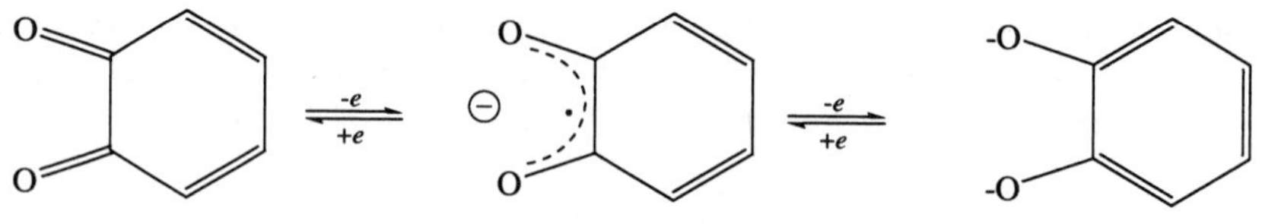

Quinone (Q)   Semiquinone (SQ)   Catecholate (Cat)

In their complexes, distinction between the three forms, Q, SQ, and Cat, is not always straightforward because oxidation-reduction processes can be metal-based or ligand-based (11-XXV) but spectroscopic techniques and X-ray crystallography can be helpful; for example, the C—O lengths are around 1.23 Å for benzoquinones, 1.29 Å for semiquinones, and 1.35 Å for catecholates.[98] In the paramagnetic semiquinone complex the electron is located on the ligand. The relative importance of the quinone versus dihydroxo species will depend on the basicity of the metal and the oxidizing ability of the quinone.

[94]C. G. Pierpont and C. W. Lange, *Prog. Inorg. Chem.* **1994,** *41,* 331; L. Que, Jr. and R. Y. N. Ho, *Chem. Rev.* **1996,** *96,* 2607.
[95]R. Usón *et al., J. Am. Chem. Soc.* **1994,** *116,* 7160.
[96]D. J. Darensbourg *et al., Inorg. Chem.* **1995,** *34,* 4676.
[97]C. G. Pierpont *et al., Inorg. Chem.* **1995,** *34,* 4427; **1996,** *35,* 1033.
[98]See for example: O. Carugo *et al., J. Chem. Soc. Dalton Trans.* **1992,** 837.

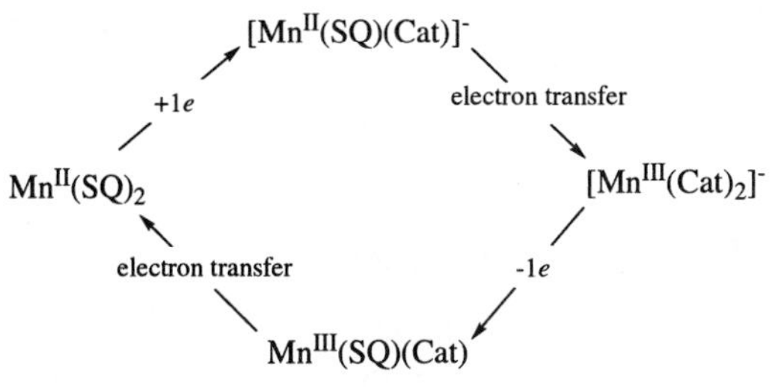

$$[Mn^{II}(SQ)(Cat)]^-$$

(11-XXV)

Paraquinones and their mono- and dianions also form complexes: the di-anions can form *bridges* between two metal atoms.

The quinone, semiquinone, and catecholate species particularly of iron are important in electron-transfer processes in photosynthesis and respiration.

11-13 Ketones and Esters

Ketones and esters are generally rather weak bases except toward strong Lewis acids such as BF_3. However, most covalent metal halides will form adducts. They are prepared by substitution reactions of ligands such as CH_2Cl_2[99] or H_2O,[100] for example:

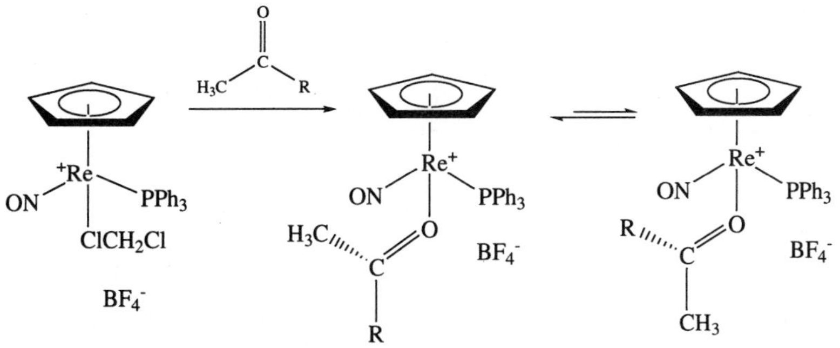

$R = CH_3, CH_2CH_3, CH(CH_3)_2, C(CH_3)_3,$ and C_6H_5

$$cis\text{-}[Ru(NH_3)_4(H_2O)_2](PF_6)_2 + CH_3COCH_3 \longrightarrow cis\text{-}[Ru(NH_3)_4(acetone)_2](PF_6)_2$$

The latter complex is a useful precursor for *cis*-tetraammineruthenium(II) chemistry (Section 18-F-3).

Although ketone bonding is normally through oxygen, η^2-complexes with C, O bonding can be formed. With certain Pt compounds, acetone can form C-bonded acetonate species.[101] Coupling of ketones can also occur under some conditions as with diethylketomalonate:

[99]J. A. Gladysz *et al.*, *J. Am. Chem. Soc.* **1990**, *112*, 9198.
[100]T. Sugaya and M. Sano, *Inorg. Chem.* **1993**, *32*, 5878.
[101]K. Matsumoto *et al.*, *J. Am. Chem. Soc.* **1996**, *118*, 8959.

$$\text{Cp}_2\text{Ti(CO)}_2 + 2(\text{EtCO}_2)_2\text{C}=\text{O} \xrightarrow{-2\text{CO}} \text{Cp}_2\text{Ti} \Big\langle {}^{\text{O}-\text{C(CO}_2\text{Et)}_2}_{\text{O}-\text{C(CO}_2\text{Et)}_2}$$

11-14 Ethers: Crown Ethers and Cryptates[102]

Metal halides are commonly soluble in, and can form solvates with, THF. Tetrahydrofuran and related ethers, such as dimethoxyethane and ethylene glycol dimethyl ether, are often used as solvents in reactions of transition metal halides with lithium or magnesium alkyls, reductions using sodium, and so on, and the complexing of the alkali metal cation by the ether is of undoubted importance. Indeed some complexes such as $[\text{K(diglyme)}_3]^+$, $[\text{K(THF)}_6]^+$, and $[\text{Li(THF)}_4]^+$ [103] have been characterized. Vanadium complexes with dimethoxyethane such as *fac*-$\text{VCl}_3(\text{DME})(\text{THF})$[104] are also known. Other complexes have ligands such as polyethylene glycol[105] and tetraglyme.[106]

Of more importance, however, are the heterocyclic ether ligands we now discuss.

Crown Ethers[107] and Cryptates[108]

The macrocyclic polyethers, termed "crown ethers" from their structural resemblance to crowns, were first synthesized by C. J. Pedersen in 1967 by reactions such as the following:

Dibenzo-18-crown-6

Ethers with from 3 to 20 oxygen atoms have been synthesized. The hydrogenated derivative of dibenzo-18-crown-6 is formally 2,5,8,15,18,21-hexaoxatricyclo[20.4.0.0]-hexacosane but usually is called cyclohexyl-18-crown-6. A general abbreviation is *n*-C-*m* where *n* is the ring size and *m* is the number of O atoms, for example, 18-C-6.

Related macropolycycles are the *cryptates*[109] or cryptands, which are N,O compounds such as $\text{N[CH}_2\text{CH}_2\text{OCH}_2\text{CH}_2\text{OCH}_2\text{CH}_2]_3\text{N}$, the structure of whose complex with Rb^+ is shown in Fig. 3-2.

[102]G. W. Gokel, *Crown Ethers and Cryptands* in *Monographs in Supramolecular Chemistry*, J. F. Stoddart, Ed., Royal Society of Chemistry, Cambridge, 1991.
[103]T. C. W. Mak *et al.*, *Polyhedron* **1997**, *16*, 345.
[104]G. Pampaloni and U. Englert, *Inorg. Chim. Acta* **1995**, *231*, 167.
[105]R. D. Rogers *et al.*, *Inorg. Chem.* **1996**, *35*, 6964.
[106]M. J. Hampden-Smith *et al.*, *Inorg. Chem.* **1996**, *35*, 6995.
[107]L. Echegoyen *et al.*, *J. Am. Chem. Soc.* **1994**, *116*, 3087; W. Y. Feng and C. Lifshitz, *J. Am. Chem. Soc.* **1995**, *117*, 11548; R. D. Rogers and A. H. Bond, *Inorg. Chim. Acta* **1996**, *250*, 105.
[108]R. J. Motekaitis and A. E. Martell, *J. Am. Chem. Soc.* **1988**, *110*, 7715.
[109]L. Echegoyen *et al.*, *Inorg. Chem.* **1993**, *32*, 572.

Crown ethers have particularly large complexing constants for alkali metals; equilibrium constants for cyclohexyl-crown-6, for example, are in the order $K^+ >$ $Rb^+ > Cs^+ > Na^+ > Li^+$. The cryptates have high complexing ability especially for M^{2+} ions and will render even $BaSO_4$ soluble. They also have good complexing ability for transition metal ions (e.g., for lanthanides).

Crown ethers find many uses.[110] They will render salts such as $KMnO_4$ or KOH soluble in benzene or other aromatic hydrocarbons, thus increasing the facility for oxidation or base reactions. Species such as Sn_9^{4+} or Pb_5^{2-} can be isolated as salts of crown-solvated alkali ions. The ethers are widely used as solvents in a variety of organic and organometallic reactions where solvation of alkali ions can effect improvements in rates.

Natural macrocycles concerned with complexing of Na^+ and K^+ and their transport through the hydrophobic lipid bilayer of cell membranes have been noted (see Fig. 3-3).

There are many other heterocycles containing not only oxygen but N and/or S. Crown-type groupings can also be attached to other ligand systems to give different types of binding sites, and in "lariat" ethers, which are crowns with side chains.

An important property of crown ethers (also cyclodextrins) is that they can act as *second coordination sphere ligands*. Thus $[Pt(bipy)(NH_3)_2]^{2+}$ gives crystalline $\{Pt(bipy)(NH_3)_2[18\text{-}C\text{-}6]\}^{2+}$, where there is $N-H\cdots O$ bonding; ammonium ions can be similarly coordinated. A cheap ether for catalysis of solid-liquid phase reactions is $N(CH_2CH_2OCH_2CH_2OMe)_3$ known as TDA1.

11-15 β-Ketoenolato and Related Ligands[111]

β-Diketones can form anions as a result of enolization and ionization:

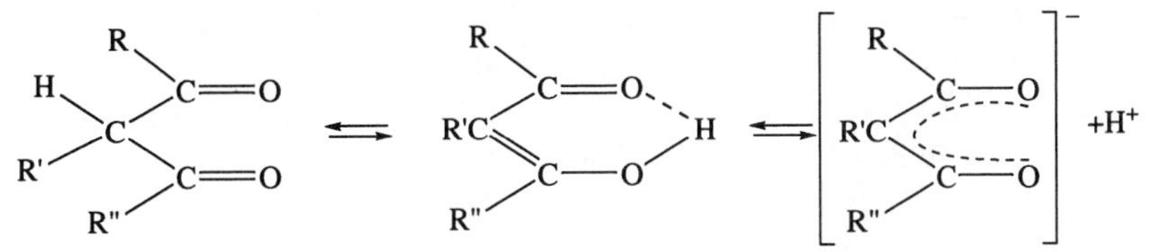

These β-ketoenolate ions form very stable chelate complexes with most metal ions. The commonest ligand is the acetylacetonate ion $(acac)^-$, in which $R = R'' = CH_3$ and $R' = H$. A general abbreviation for β-ketoenolate ions in general is dike.

Among the commonest types of diketo complex are those with the stoichiometries $M(dike)_3$ and $M(dike)_2$. The former all have structures based on an octahedral disposition of the six oxygen atoms. The tris(chelate) molecules then actually have D_3 symmetry and exist as enantiomers. When there are unsymmetrical diketo ligands (i.e., those with $R \neq R''$), geometrical isomers also exist, as indicated in (11-XXVI). Such compounds have been of value in investigations of the mechanism of racemization of tris(chelate) complexes. Chiral diketonate complexes have found many

[110]Y. Li and L. Echegoyen, *J. Am. Chem. Soc.* **1994,** *116,* 6832; M. J. Hampden-Smith *et al.,* *Inorg. Chem.* **1996,** *35,* 6638.
[111]D. A. Thornton, *Coord. Chem. Rev.* **1990,** *104,* 173.

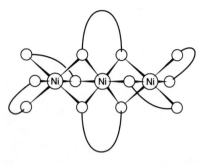

Figure 11-4   The trimeric structure of nickel(II) acetylacetonate. The unlabeled circles represent O atoms and the curved lines connecting them in pairs represent the remaining portions of the acetylacetonate rings.

applications in fields such as catalysis and gas-chromatographic separation of enantiomers.[112] Tetradiketo complexes M(β-dike)$_4$ are usually nonrigid.

$$\textit{cis} \qquad \textit{trans}$$

(11-XXVI)

Substances of composition M(dike)$_2$ are very often oligomeric, thereby allowing coordinative saturation of the metal.

Thus acetylacetonates of Zn, Ni, and Mn[II] are trinuclear, Fig. 11-4, while Co(acac)$_2$ is tetranuclear; all have bridging β-diketonate groups. The presence of bulky substituents on the β-diketones such as Me$_3$C sterically impedes oligomerization and monomers are formed. However, these are commonly solvated by H$_2$O, ROH, or py to give 5- or 6-coordinate complexes, *trans*-M(dike)$_2$L$_{1,2}$.

The linking of β-diketonates by bridges allows the formation of "face-to-face" complexes (11-XXVII) similar to those of face-to-face porphyrins (Section 9-12); small molecules may occupy the central hole. With trivalent ions such as Ti^{3+}, V^{3+}, Mn^{3+}, or Fe^{3+}, triple-helical structures consisting of two 6-coordinate M[III] ions chelated by three bis(dike) ligands can be obtained.[113]

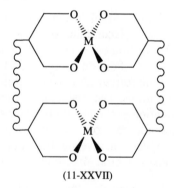

(11-XXVII)

[112]A. Salzer *et al., Inorg. Chem.* **1995,** *34,* 6231 and references therein.
[113]G. Christou *et al., Chem. Commun.* **1997,** 1561.

Dinuclear tetradiketonate complexes with metal-metal bonds have been prepared for $Mo^{II\ 114}$ and Pt^{III}.[115] Also familiar are complexes with mixed sets of ligands such as $VO(acac)_2$, $V(acac)Cl_2(THF)_2$, and $V(acac)Cl(THF)_3$.[116] Many of the diketonate compounds are volatile and have been extensively investigated as precursors for metal organic chemical vapor deposition (MOCVD).[117]

For neutral compounds, especially of acetylacetone, the methine CH group of the ring can undergo a wide range of substitution reactions similar to those of aromatic substances, even though the rings have little or no aromatic character. However, certain hexafluoroacetylacetonates may be attacked at carbon by nucleophiles:

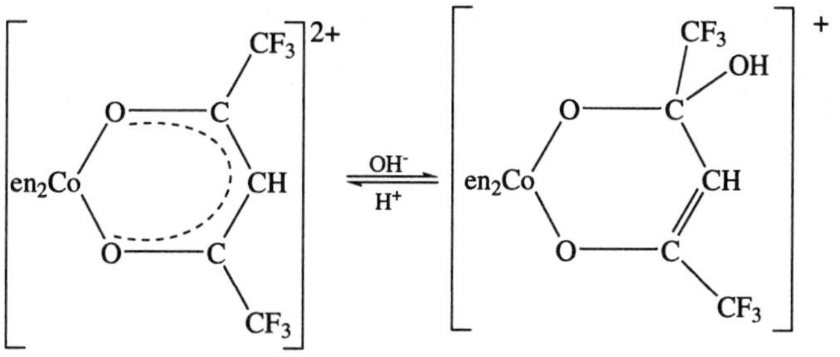

Another unusual reaction is the substitution of two hydrogen atoms from a methyl group of an acac ligand[118] in the reaction of $VO(acac)_2$ and a formamidine:

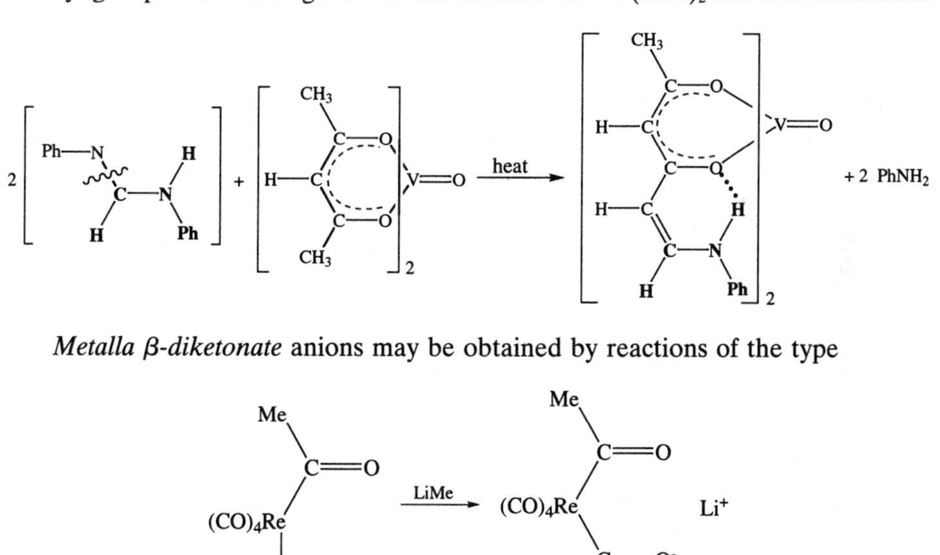

Metalla β-diketonate anions may be obtained by reactions of the type

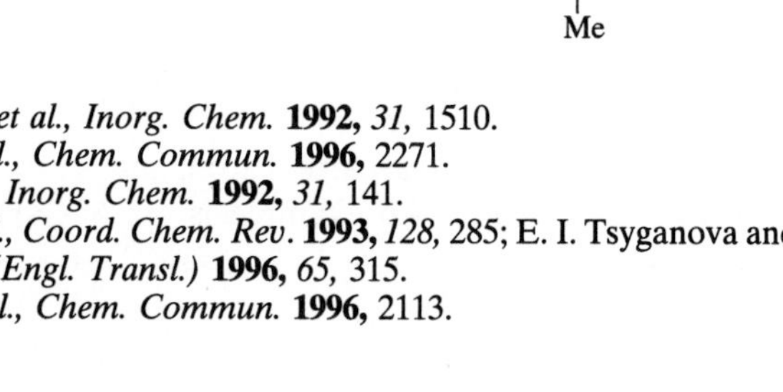

[114]M. H. Chisholm *et al.*, *Inorg. Chem.* **1992**, *31*, 1510.
[115]G. A. Heath *et al.*, *Chem. Commun.* **1996**, 2271.
[116]C. Floriani *et al.*, *Inorg. Chem.* **1992**, *31*, 141.
[117]R. E. Sievers *et al.*, *Coord. Chem. Rev.* **1993**, *128*, 285; E. I. Tsyganova and L. M. Dyagileva, *Russ. Chem. Rev. (Engl. Transl.)* **1996**, *65*, 315.
[118]F. A. Cotton *et al.*, *Chem. Commun.* **1996**, 2113.

Related species can also be formed when the metal has R_2PO or R_2POH groups in cis positions (11-XXVIII). They all give 2- anions that form complexes with other metals.

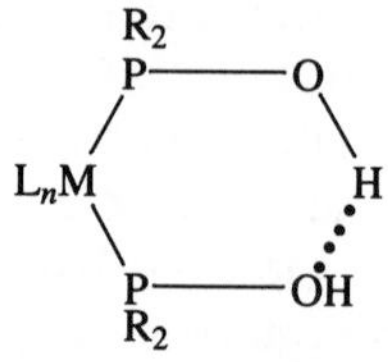

(11-XXVIII)

The hydrogen form of the rhenioacac (11-XXIX) has a very short symmetrical H-bond, 2.4 Å, with an $O-H-O$ angle of 172°; nevertheless, it is best considered as having a localized π system as shown with an asymmetric H-bond rather than a delocalized one.

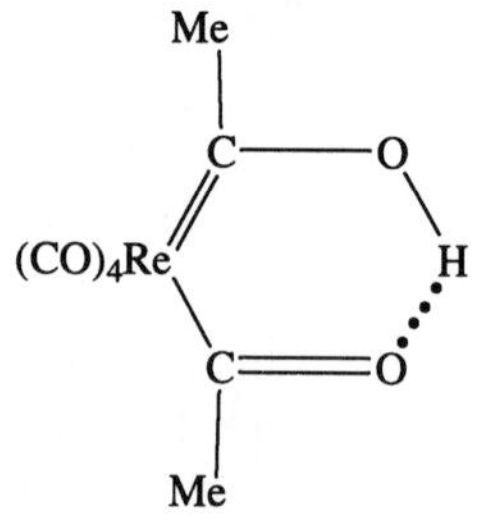

(11-XXIX)

By use of 2 moles of MeLi a tridentate ligand can be obtained; this is an analogue of $(MeCO)_3C^-$ and also resembles the tridentate $RBpz_3^-$ ligand (Section 9-13):

$$MeCORe(CO)_5 + 2LiMe \longrightarrow [\mathit{fac}\text{-}(CO)_3Re(COMe)_3]^{2-}$$

The keto groups can also undergo base condensations with amines to give ketoimines (11-XXXa) and be capped by BF_2 groups to give species of type (11-XXXb)

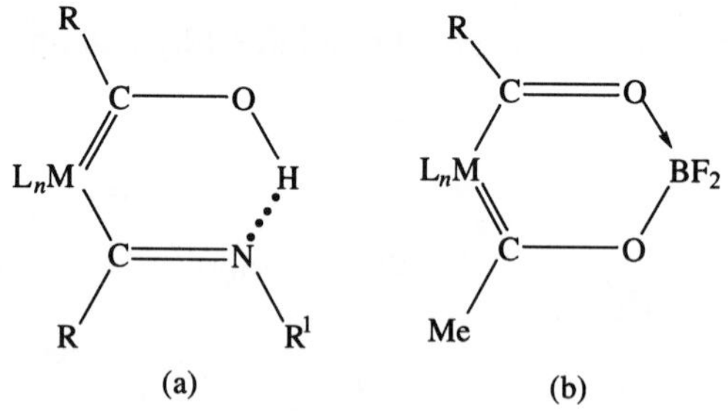

(a)   (b)

(11-XXX)

C-Bonded β-Diketonate Complexes

These have the metal bound to the central carbon atom and are known for a number of metals especially of the Pt group. An example is (11-XXXI), which has both normal and C-bonded groups.

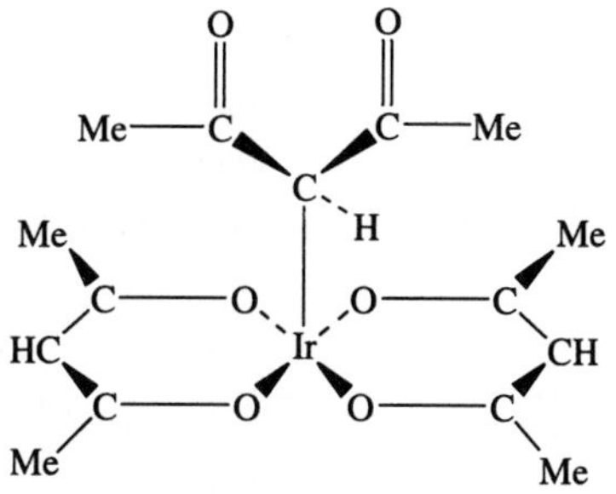

(11-XXXI)

Note that the C-bonded groups have two free $\overset{\diagdown}{\underset{\diagup}{C}}{=}O$ groups that can allow chelation to other metals[119] as in (11-XXXII).

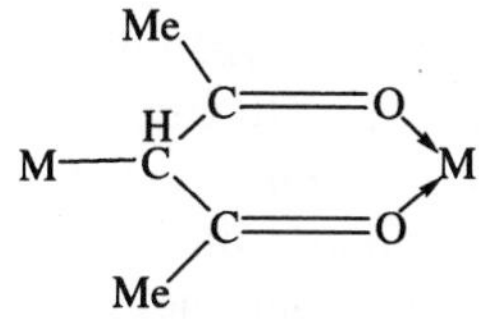

(11-XXXII)

1,3,5-Triketonates[120]

These and similar polyketones can give rise to *compartmental ligands* and link two (11-XXXIII) or more (11-XXXIV) metal centers that can then undergo redox reactions.

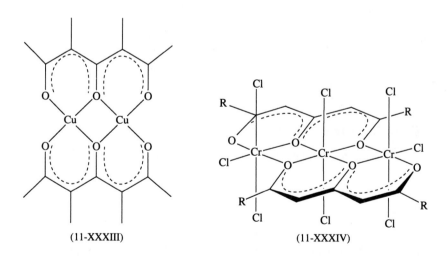

(11-XXXIII)   (11-XXXIV)

[119]Y. Sasaki *et al., J. Am. Chem. Soc.* **1990,** *112,* 4038; J. Forniés *et al., Organometallics* **1993,** *12,* 940; **1996,** *15,* 1813.

[120]L. L. Borer and E. Sinn, *Inorg. Chem.* **1990,** *29,* 2514.

Tropolonates[121]

A system similar to the β-diketonates is provided by *tropolone* and its anion (11-XXXV). The tropolonato ion gives many complexes that are broadly similar to analogous β-diketonate complexes, although there are often very significant differences.

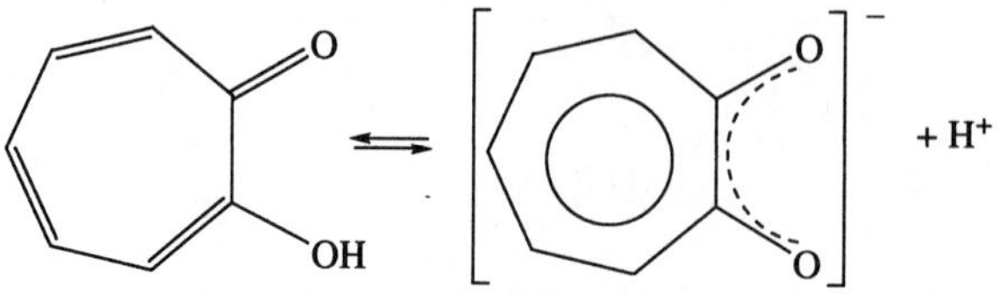

(11-XXXV)

It should be noted that the tropolonato ion forms a five-membered chelate ring and that the "bite," that is, the oxygen-to-oxygen distance, is smaller than in the β-diketonato ions. This leads the tristropolonato complexes to considerable distortion from an octahedral set of oxygen atoms. Thus in $Fe(O_2C_7H_5)_3$ the O—Fe—O ring angles are only 78° and the entire configuration is twisted about the threefold axis toward a more prismatic structure. The upper set of Fe—O bonds is twisted only 40° instead of 60° relative to the lower set.

Finally, note that β-diketones can occasionally act as *neutral* ligands, being bound either through oxygen as in (11-XXXVI) or for certain metals that can form olefin complexes as shown in (11-XXXVII).

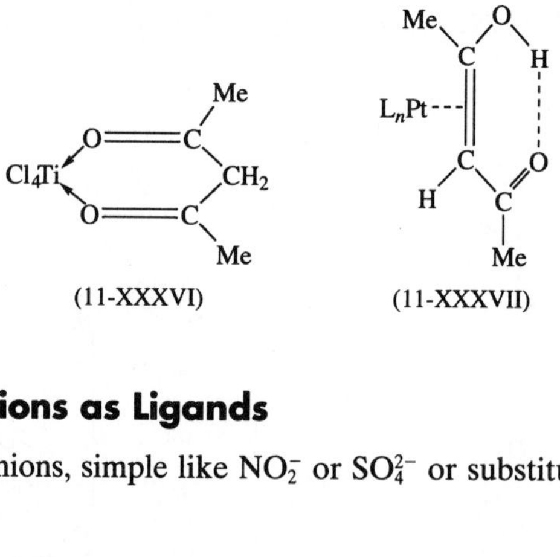

(11-XXXVI)   (11-XXXVII)

11-16 Oxo Anions as Ligands

Essentially all oxo anions, simple like NO_2^- or SO_4^{2-} or substituted like RCO_2^-, act as ligands.

Carbon Oxo Anions

The most important are *carbonate* and *bicarbonate* for which a great variety of bonding modes has been established;[122] these can be designated by the link and

[121]M. A. J. Moss and C. J. Jones, *J. Chem. Soc., Dalton Trans.* **1990,** 581.
[122]R. Vincente *et al., J. Chem. Soc. Dalton Trans.* **1997,** 2315; A. G. Blackman *et al., Inorg. Chem.* **1995,** *34,* 2795; P. B. Stein *et al., Inorg. Chem.* **1993,** *32,* 4976; L. Spiccia *et al., Inorg. Chem.* **1992,** *31,* 1066; L. Que, Jr. *et al., Inorg. Chem.* **1990,** *29,* 4629.

metal number (L, M):

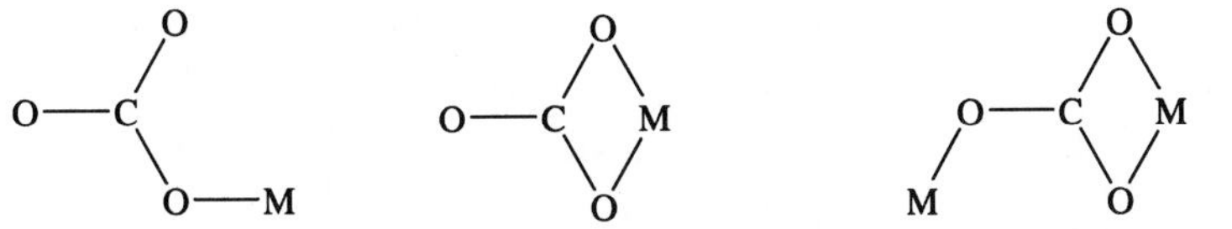

1L1M, η^1

2L1M, η^2

3L2M

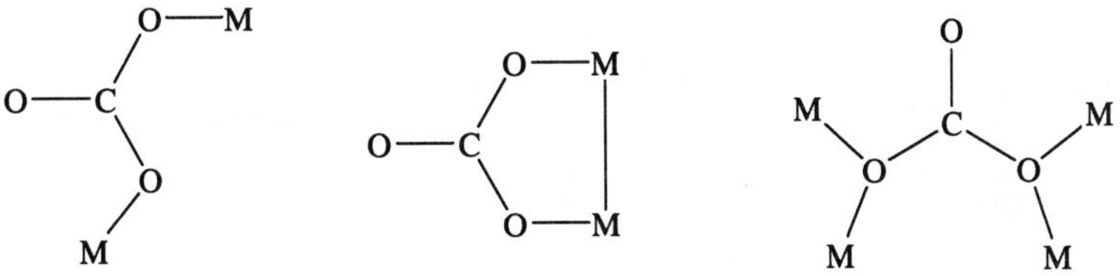

syn, anti; 2L2M

syn, syn; 2L2M

2L4M

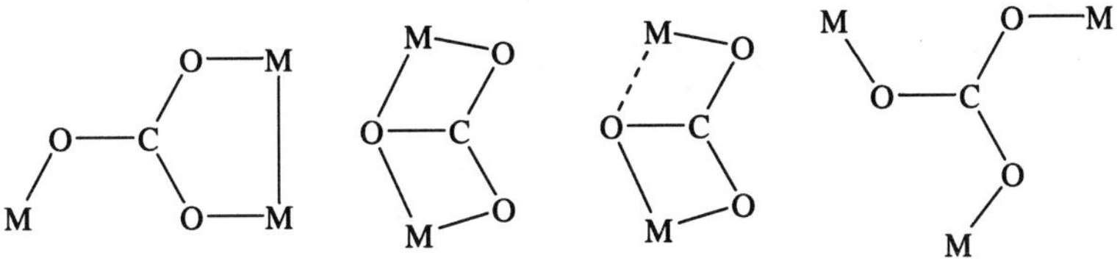

3L3M

3L2M

3L2M

3L3M

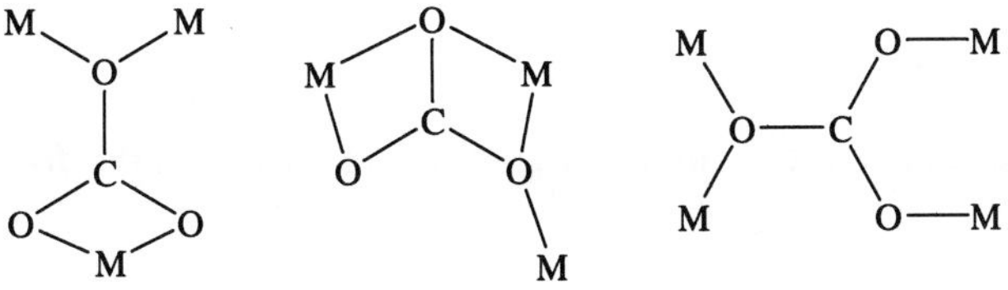

3L3M

3L3M

3L4M

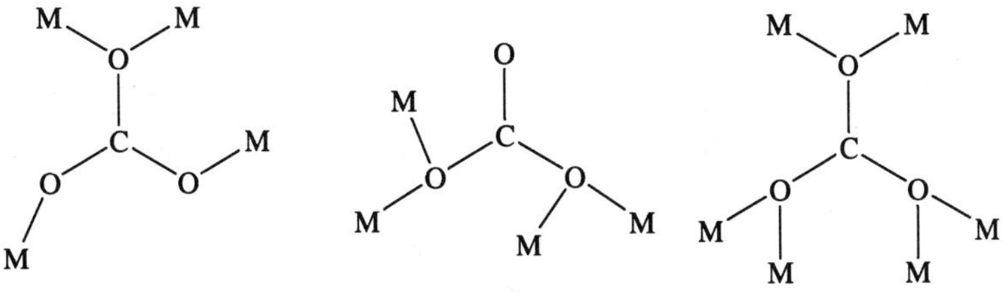

3L4M

3L4M

3L6M

Although carbonate complexes are usually made from CO_3^{2-}, HCO_3^-, or CO_2, they can be formed on oxidation, for example,

$$[CpMo(CO)_3]^- \xrightarrow{\ O_2\ } [CpMo(CO)_2(\eta^2\text{-}O_2CO)]^-$$

A few compounds are formed by spontaneous fixation of CO_2 in air such as $[Ni_3(Me_2en)_6(CO_3)(H_2O)_4](ClO_4)_4$.[123]

Oxalato complexes[124] are also common, the main linkage types being

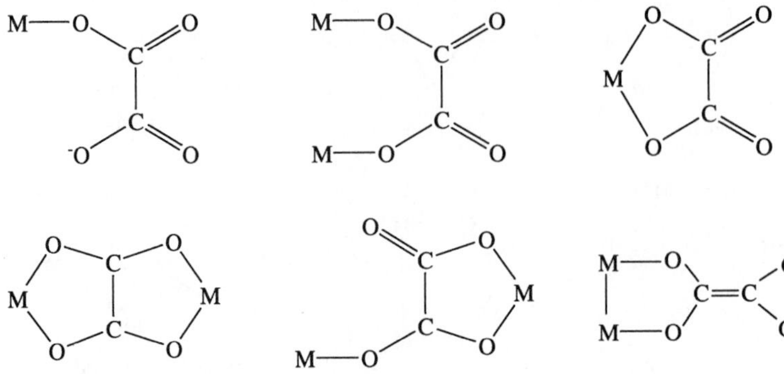

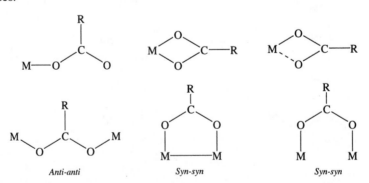

Typical chelate complexes are trisoxalato ions, for example, $[Co\ ox_3]^{3-}$. Many mono-, di-, and trithio oxalates are also known.[125]

Squarates[126] may act as chelates or bridges or form chains:

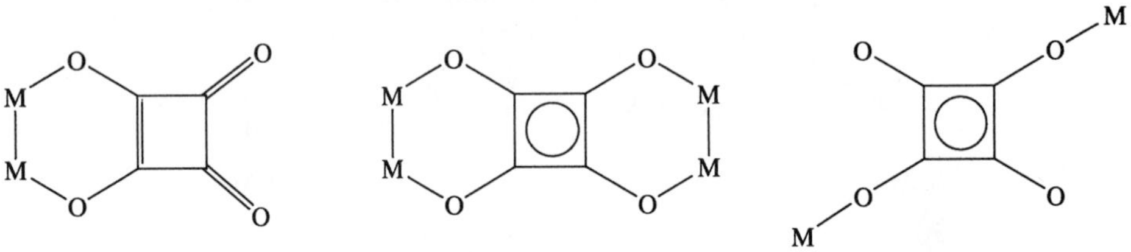

Carboxylates[127]

The carboxylates are a very important class of ligands with the following bonding modes.

[123]T. Tanase, *Inorg. Chem.* **1992**, *31*, 1058.
[124]J. Glerup *et al., Inorg. Chem.* **1995**, *34*, 6255; A. Bianchi *et al., Inorg. Chem.* **1990**, *29*, 963.
[125]W. Dietzsch *et al., Coord. Chem. Rev.* **1992**, *121*, 43.
[126]J. Sletten *et al., Inorg. Chem.* **1995**, *34*, 4903; J.-C. Trombe *et al., Inorg. Chim. Acta* **1995**, *230*, 1; **1992**, *195*, 193; **1990**, *167*, 69; J. M. Williams *et al., Inorg. Chim. Acta* **1992**, *192*, 195; M. J. Sisley and R. B. Jordan, *Inorg. Chem.* **1991**, *30*, 2190.
[127]T. G. Appleton *et al., Inorg. Chem.* **1995**, *34*, 5646; T. C. W. Mak *et al., Polyhedron* **1997**, *16*, 897 and references therein; M. R. Sundberg *et al., Inorg. Chim. Acta* **1995**, *232*, 1; X.-M. Chen and T. C. W. Mak, *J. Chem. Soc. Dalton Trans.* **1992**, 1585.

Syn-skew
non-planar

Skew-skew non-planar
metals on same side

Skew-skew non-planar
metals on different sides

Anti-syn

Monoatomic

The most common forms are unidentate, symmetrical chelate, and symmetrical syn-syn bridging. Acetates and trifluoroacetates of weak Lewis acids are usually symmetric, those of very strong Lewis acids are often asymmetric. The other forms are not common, but anti-anti single bridging occurs in [Mn salenCO$_2$Me]$_n$, and anti-syn in [(PhCH$_2$)$_3$SnO$_2$CMe]$_n$. It is not uncommon to have more than one coordination type in the same molecule.[128]

The main types of bonding can often be distinguished by ir and nmr spectra. The syn-syn bridging RCO$_2^-$ ligand is extremely common and important in compounds with M—M quadruple bonds.

In "ionic" acetates or in aqueous solution, the "free" CH$_3$CO$_2$ ion has symmetric and antisymmetric C—O stretching modes at ~1415 and 1570 cm^{-1}. These frequencies can vary by ±20 cm^{-1}. Since the symmetry of even the free ion is low and it gives two ir-active bands, evidence for the mode of coordination must be derived from the positions rather than the number of bands. When the carboxyl group is unidentate, one of the C—O bonds should have enhanced double-bond character and should give rise to a high-frequency band. Such bands are observed in the 1590 to 1650 cm^{-1} region and are considered to be diagnostic of unidentate coordination.

Symmetrical bidentate coordination, as in Zn(CH$_3$CO$_2$)$_2$·2H$_2$O and Na[U-O$_2$(CH$_3$CO$_2$)$_3$], and symmetrical bridging, as in the M$_2$(O$_2$CCH$_3$)$_4$L$_2$ and M$_3$O(O$_2$CCH$_3$)$_6$L$_3$ types of molecules, leaves the C—O bonds still equivalent, and the effect on the frequencies is not easily predictable. In fact, no criteria for distinguishing these cases have been found. In general, multiple bands appear between 1400 and 1550 cm^{-1}, the multiplicity being attributable to coupling between CH$_3$CO$_2$ groups bonded to the same metal atom(s).

While acetates and other lower carboxylate complexes are prepared from the acids or alkali metal salts, for *formates* a different synthesis is the insertion reaction of CO$_2$ into M—H bonds, for example,

$$(Me_3P)_5MoH_2 + CO_2 \longrightarrow (Me_3P)_5MoH(\eta^2\text{-}O_2CH)$$

[128]J. D. Martin and R. F. Hess, *Chem. Commun.* **1996**, 2419; H. W. Roesky *et al., Inorg. Chem.* **1996**, *35*, 7181; P. Lahuerta *et al., Inorg. Chim. Acta* **1995**, *229*, 203; F. A. Cotton and J. Su, *J. Cluster Sci.* **1995**, *6*, 39.

Probably the only significant reaction of the carboxylate ligand is decarboxylation, which is catalyzed by transition metal species.[129]

Straight-chain alkyl carboxylic acids derived from petroleum that also have a terminal cyclohexyl or cyclopentyl group are known as naphthenic acids. They form complexes, presumably polymeric, with many transition metals, and these compounds are freely soluble in petroleum. Copper naphthenates are used as fungicides, aluminum naphthenate was used as a gelling agent in "napalm," and cobalt naphthenates are used in paints.

There are, of course, more complicated carboxylic acids such as ethylenediaminetetraacetic acid that can function as multidentate ligands with both N and O bound to the metal. Also, *hydroxo carboxylic acids* such as citrate[130] readily form complexes in which both carboxylate and hydroxo groups are involved. Of such acids, probably the best studied are *tartrato complexes.*[131] A fairly common type of structure is one with bridges linking two metal atoms. A particular example is the antimony complex "tartar emetic." Because of the chirality and multiplicity of bonding possibilities, many isomers of tartrate complexes are possible, and the relative stabilities of these can be explained in terms of steric constraints of the binuclear structure and conformation of the tartrato groups, and depend strongly on the coordination geometry about the metal.

Carbamates[132]

Complexes of $R_2NCO_2^-$ are not as extensive or as useful as their sulfur analogues the dithiocarbamates (Sections 7-8 and 12-16).

Carbamates are often obtained by insertion reactions of CO_2:

$$Ti(NMe_2)_4 + 4CO_2 = Ti(O_2CNMe_2)_4$$

Such insertions may be carried out *in situ* as in reactions of cobalt or lanthanide halides with R_2NH and CO_2 in hexane to give, for example, $Co_6(O_2CNEt_2)_{12}$ or $[Yb(O_2CNPr^i_2)_4]$:

$$MCl_n + 2nNHR_2 + nCO_2 \longrightarrow M(O_2CNR_2)_n + nNH_2R_2Cl$$

Although carbamates are mostly chelate, they can be unidentate and have bridging modes similar to CO_3^{2-} and RCO_2^{-}[133]

[129]D. J. Darensbourg *et al., J. Am. Chem. Soc.* **1993,** *115,* 8839; M. Yashiro *et al., Inorg. Chem.* **1994,** *33,* 1003.
[130]D. W. Wright et al., *Inorg. Chem.* **1995,** *34,* 4194.
[131]J. J. Cruywagen *et al., J. Chem. Soc. Dalton Trans.* **1990,** 1951.
[132]W. L. Gladfelter *et al., J. Am. Chem. Soc.* **1992,** *114,* 8933.
[133]M. H. Chisholm *et al., J. Am. Chem. Soc.* **1989,** *111,* 8149.

The Nitrate Ion

This ion has several structural roles,[134] as follows:

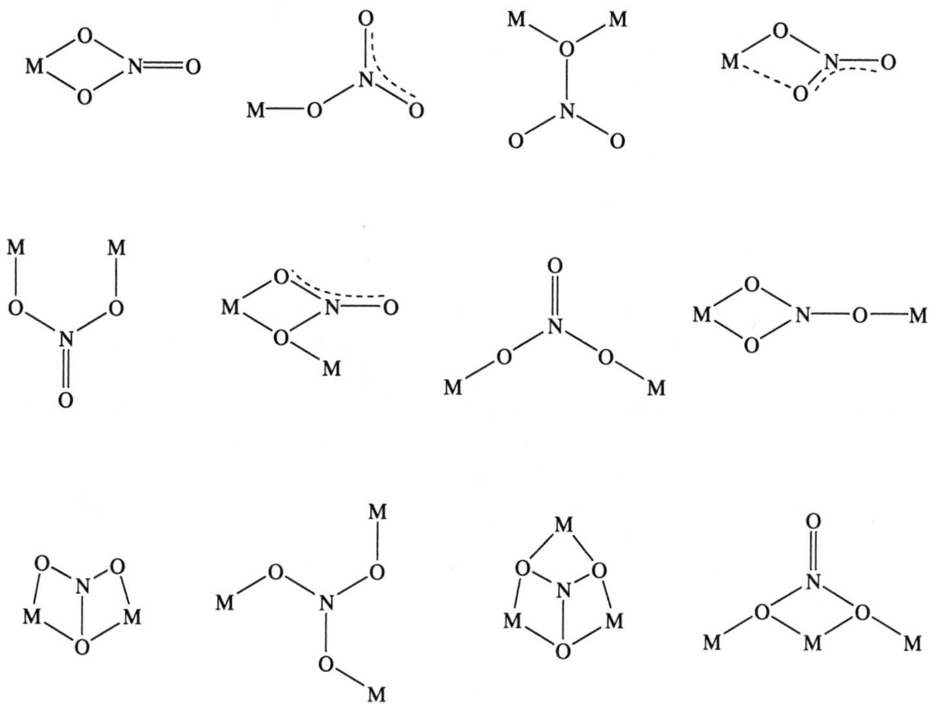

The most common forms of the nitrate ion are symmetrical bidentate followed by η^1. The free NO_3^- ion has relatively high symmetry (D_{3h}); thus its ir spectrum is fairly simple. The totally symmetric N—O stretching mode is not ir active, but the doubly degenerate N—O stretching mode gives rise to a strong band at ~1390 cm^{-1}. There are also two ir-active deformation modes, one of which is doubly degenerate, at 830 and 720 cm^{-1}. When NO_3^- is coordinated, its effective symmetry is reduced, causing the degeneracies to split and all modes (six) to be ir active. Hence, it is possible to distinguish between ionic and coordinated NO$_3$ groups.

Because the two commonest forms of coordinated NO_3^- have the same effective symmetry, hence the same number of ir-active vibrational modes, criteria for distinguishing between them must be based on the positions of the bands rather than their number. In practice, the situation is quite complex and there are no entirely straightforward criteria. This is because the array of frequencies depends on both the geometry and strength of coordination.

[134]C. Orvig et al., Inorg. Chem. 1995, 34, 4921; P. N. V. P. Kumar and D. S. Marynick, Inorg. Chem. 1993, 32, 1857; R. Han and G. Parkin, J. Am. Chem. Soc. 1991, 113, 9707; O. Yamauchi et al., Inorg. Chem. 1996, 35, 7148.

Nitrite Ion[135]

The NO_2^- ion can act as an N-ligand to form *nitro* compounds, and as oxygen-bonded *nitrito:*

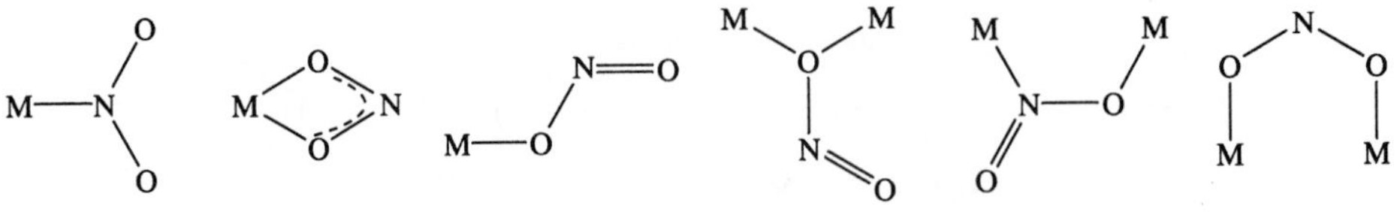

The NO_2^- ion has low symmetry (C_{2v}) and its three vibrational modes, symmetric N—O stretching, ν_s; antisymmetric N—O stretching, ν_{as}; and bending, δ; are all ir active to begin with. Thus the number of bands cannot change on coordination, and the use of ir spectra to infer structure must depend on interpretation of shifts in the frequencies. The δ vibration is rather insensitive to coordination geometry, but there are characteristic shifts of ν_s and ν_{as} that often can distinguish reliably between the nitro and the nitrito structures. Thus for *nitro* complexes the two frequencies are similar, typical values being 1300 to 1340 cm^{-1} for ν_{as} and 1360 to 1430 cm^{-1} for ν_s. This is in keeping with the equivalence of the N—O bond orders in the nitro case. For *nitrito* bonding, the two N—O bonds have very different strengths and the two N—O stretching frequencies are typically in the ranges 1400 to 1500 cm^{-1} for N=O and 1000 to 1100 cm^{-1} for N—O.

Related to the nitrite ion are the ions derived from the *aci* form of *nitroalkanes,*[136] for example,

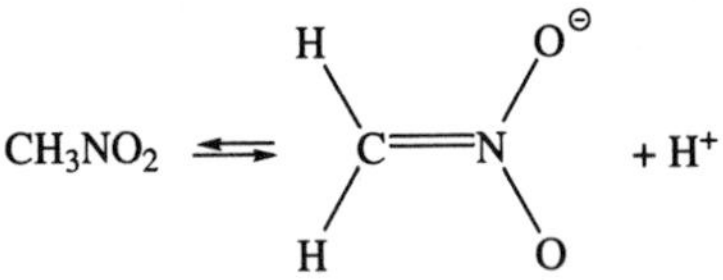

Alkane nitronates such as $Zr(O_2N-CR_2)_4$ are known. In $Ni(O_2N=CHPh)_2(tmed)$ the ligand is chelate through both oxygen atoms as it is apparently also in $Ru(O_2N=CH_2)H(PPh_3)_3$ and some copper compounds such as $[Cu(phen)O_2N=CH_2]^+$. Some tin, lead, and mercury compounds have unidentate groups $MON(O)CR'R''$. However, C bonding, as $M-CH_2NO_2$, is also possible, as in the platinum compounds made by the reaction

$$PtCl_2(PEt_3)_2 + CH_3NO_2 \xrightarrow{Ag_2O} cis\text{-}PtCl(CH_2NO_2)(PEt_3)_2$$

This reaction implies initial oxidative addition as H and CH_2NO_2, of nitromethane, and indeed such additions are known. The anion *nitrosodicyanomethanide,*

[135]D. V. Fomitchev and P. Coppens, *Inorg. Chem.* **1996,** *35,* 7021; B. H. Huynh *et al., J. Am. Chem. Soc.* **1991,** *113,* 717; W. B. Tolman, *Inorg. Chem.* **1991,** *30,* 4877; J. P. Costes *et al., Inorg. Chim. Acta* **1995,** *239,* 53.

[136]I. Erden *et al., J. Am. Chem. Soc.* **1993,** *115,* 9834.

[ONC(CN)$_2$]$^-$, is a pseudo halide found O-bonded in coordination compounds such as Fe(porph){η'-ONC(CN)$_2$} or N-bonded in Ir{η'-N(O)C(CN)$_2$}(CO)(PPh$_3$)$_2$.[137]

Perchlorate Ion[138]

The ClO$_4^-$ ion is a hard base with low tendency to coordinate,[139] though cases have been established with unidentate, chelating, and bridges[140] such as:

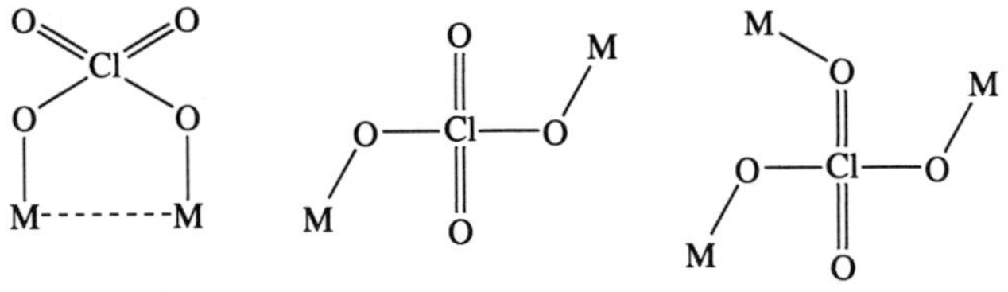

There is evidence for complexing by ClO$_4^-$ in solutions and the reduction of ClO$_4^-$ by metal ions such as Ru^{2+} doubtless involves initial complexing.

Perchlorates should *always* be avoided because of potential explosion hazards and CF$_3$SO$_3^-$, PF$_6^-$, BF$_4^-$, and so on, used instead.

Phosphorus Oxo Acids[141]

The many types of phosphorus oxo acids, which may contain either P^{III} or P^V, are discussed in Chapter 10 along with organic-substituted ions such as dialkylphosphinates (R$_2$PO$_2^-$) and acids such as R$_2$POH and (RO)$_2$POH.

Phosphato complexes may have PO$_4^{3-}$, HPO$_4^{2-}$, or H$_2$PO$_4^-$ coordinated.[142] In Co^{III}en$_2$(PO$_4$) there is a bidentate chelate, whereas the pyrophosphate Co^{III}en$_2$(HP$_2$O$_7$) has a six-membered ring:

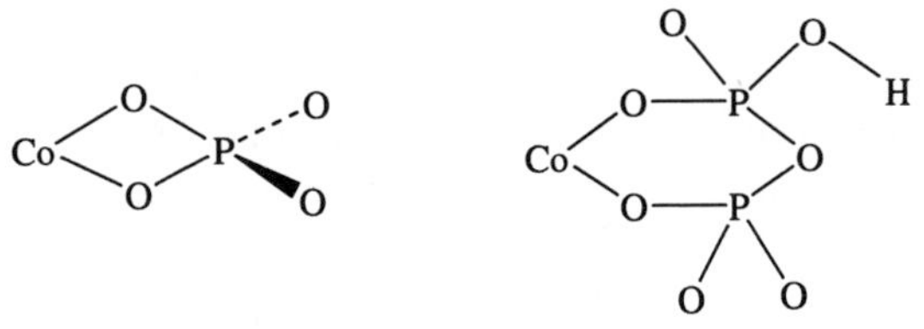

Extended structures such as those in Zr(HPO$_4$)$_2$·2H$_2$O and Zr(PO$_4$)(H$_2$PO$_4$)·2H$_2$O are also known. The phosphite ligand (HO$_2$P)$_2$O$_2^-$ is noted in Section 18-H-5.

[137]D. S. Bohle et al., Inorg. Chem. **1995**, 34, 2569.
[138]C. Kimblin and G. Parkin, Inorg. Chem. **1996**, 35, 6912; G. A. Olah et al., J. Am. Chem. Soc. **1990**, 112, 5991.
[139]G. Johansson and H. Yokoyama, Inorg. Chem. **1990**, 29, 2460.
[140]F. Favier et al., J. Chem. Soc. Dalton Trans. **1994**, 3119.
[141]G. Cao et al., Acc. Chem. Res. **1992**, 25, 420.
[142]A. Clearfield et al., J. Chem. Soc. Dalton Trans. **1995**, 111; Chem. Rev. **1988**, 88, 125; Z. Wang et al., Inorg. Chim. Acta **1995**, 232, 83.

The various polyphosphoric anions such as $[P_3O_{10}]^{5-}$,[143] and diphosphonates, $[O_3PC(R)R'PO_3]^{4-}$, also give complexes.

Sulfur Oxo Acids

The sulfate ions HSO_4^- and SO_4^{2-} form numerous complexes. *Sulfate* can have the following bonding modes:[144]

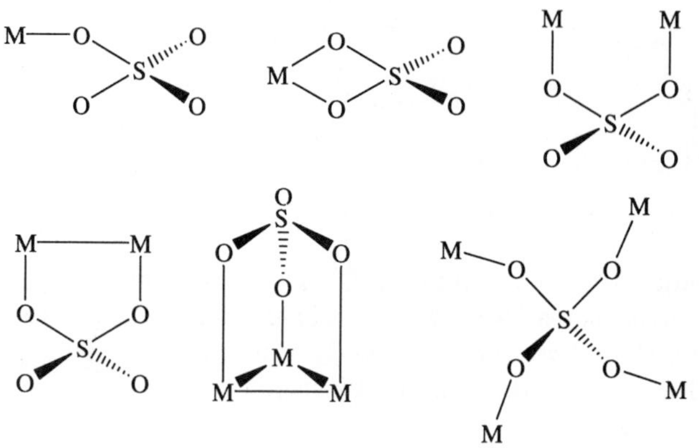

The μ_3 mode is found only in $H_2Os_3(CO)_9(\mu_3\text{-}O_3SO)$, and a μ_4-mode is found in species such as $\{[(SO_4)Mo(O)(\mu\text{-}S)_2Mo(O)(SO_4)]_2(SO_4)\}^{6+}$.[145]

The free sulfate ion is tetrahedral (T_d), but when it functions as a unidentate ligand, the coordinated oxygen atom is no longer equivalent to the other three and the effective symmetry is lowered to C_{3v}. Since the M—O—S chain is normally bent, the actual symmetry is even lower, but this perturbation of C_{3v} symmetry does not measurably affect the ir spectra. When two oxygen atoms become coordinated, either to the same metal ion or to different ones, the symmetry is lowered still further to C_{2v}.

The distinction between uncoordinated, unidentate, and bidentate SO_4^{2-} by ir spectra is very straightforward. Table 11-3 summarizes the selection rules for the S—O stretching modes in the three cases. It can be seen that uncoordinated SO_4^{2-} should have one, unidentate SO_4^{2-} three, and bidentate SO_4^{2-} four S—O stretching bands in the infrared. Note that the appearance of four bands for the bidentate ion is expected regardless of whether it is chelating or bridging.

Observed spectra are in accord with these predictions, except that ν_1 does appear weakly in the spectrum of the uncoordinated SO_4^{2-} ion. This is due to nonbonded interactions of SO_4^{2-} with its neighbors in the crystal, which perturb the T_d symmetry; the same environmental effects also cause the ν_3 band to be very broad.

Even though bridging and chelating sulfates cannot be distinguished on the basis of the number of bands they give, it appears that the former have bands at

[143]E. Kimura *et al., J. Am. Chem. Soc.* **1996,** *118,* 3091.
[144]C. A. Murillo *et al., Inorg. Chim. Acta* **1995,** *229,* 27; K. Nag *et al., J. Chem. Soc., Dalton Trans.* **1993,** 2241; T. Hori *et al., J. Chem. Soc., Dalton Trans.* **1992,** 275; F. A. Cotton *et al., Eur. J. Solid State Inorg. Chem.* **1994,** *31,* 535.
[145]C. G. Kim and D. Coucouvanis, *Inorg. Chem.* **1993,** *32,* 2232.

Table 11-3 Correlation of the Types and Activities of S—O Stretching Modes of SO_4^{2-}

State of SO_4^{2-}	Effective symmetry		Types and activities of modes[a] (R = Raman; I = ir)
Uncoordinated	T_d	$\nu_1(A_1, R)$	$\nu_3(T_2,I,R)$
		$\downarrow$	
Unidentate	C_{3v}	$\nu_1(A_1,I,R)$	$\nu_{3a}(A_1,I,R)\ \nu_{3b}(E,I,R)$
		$\downarrow$	
Bidentate	C_{2v}	$\nu_1(A_1,I,R)$	$\nu_{3a}(A_1,I,R)\ \nu_{3b}(B_1,I,R)\ \nu_{3c}(B_2,I,R)$

[a] ν_2 and ν_4 are not listed because they are O—S—O bending modes. Note also that the arrows drawn have only rough qualitative significance, since all modes of the same symmetry will be mixed in the higher symmetry.

different frequencies than the latter, typical values in cm^{-1} being as follows:

Bridging	Chelating
1160-1200	1210-1240
~1110	1090-1176
~1120	995-1075
960-1000	930-1000

Many *thiosulfate* complexes such as those containing the $[OsO_2(S_2O_3)_2]^{2-}$ anion have also been prepared and their vibrational data have been reported.[146]

Sulfites, SO_3^{2-}; *sulfinates*, RSO_2^-; and *sulfenates*,[147] RSO^-, may be bound η^1 through O or through S. In cobalt complexes such as $[(NH_3)_5CoSO_3]^+$, the S-bonded ligand has a high trans effect; S-bonded sulfites are more stable than O-bonded ones. There is evidence for bridging SO_3^{2-} and the chelating anion has been clearly established in the product of the reaction:[148]

$$Et_4N[CpMo(O)(\mu\text{-}S)_2Mo(O)(S)] \xrightarrow[CH_3CN,\,0°C]{SO_2} Et_4N[CpMo(O)(\mu\text{-}S)_2Mo(O)(\eta^2\text{-}O_2SO)]$$

If the reaction is done at room temperature, the corresponding chelating thiosulfate derivative is obtained; the sulfate is isolated when the reaction is carried out at reflux temperature.

Selenites[149] of type $MSeO_3{\cdot}H_2O$ (M = Mn, Co, Ni, Zn, Cd) have also been prepared and many are isotypic with the corresponding sulfite analogues.

Trifluoromethanesulfonate (triflate), $CF_3SO_3^-$, is generally a good leaving group in organic chemistry and in inorganic complexes is readily ionized or displaced, being like ClO_4^- a relatively poor ligand, although under certain conditions it can be a better ligand than water, i.e., *trans*-$[Rh(CO)(PPh_3)_2(OSO_2CF_3)]$ does not react

[146]C. F. Edwards *et al.*, *J. Chem. Soc. Dalton Trans.* **1992**, 145.
[147]M. Y. Darensbourg *et al.*, *J. Am. Chem. Soc.* **1995**, *117*, 963.
[148]C. G. Kim and D. Coucouvanis, *Inorg. Chem.* **1993**, *32*, 1881.
[149]B. Engelen *et al.*, *Z. anorg. allg. Chem.* **1996**, *622*, 1886.

with H_2O in CH_2Cl_2.[150] Uni-, bi-, and tridentate and also bridged complexes are known, examples being $(\eta^5\text{-}Cp^*)(CO)_2Fe(OSO_2CF_3)$, $W_2(NMe_2)_4(OSO_2CF_3)_2$,[151] and $TiCl_2(OSO_2CF_3)_2$. Fluorosulfates (FSO_3^-) and other sulfonic acids (RSO_3^-) are similar; examples are $Pt(FSO_3)_4$, $[Cr(pySO_3)_2(H_2O)_2]_n$,[152] and $Ca(H_3C(H_2N)C_6H_3SO_3)_2 \cdot 7H_2O$.[153]

A vibrational study of the triflate anion has been published.[154]

Hydroxamates

Hydroxamates and their monothio and dithio analogues form chelate complexes usually. They have been much studied because hydroxamates are present in sidero-phores, the microbial iron transport systems (Section 17-E-10) and also because of their use in sequestering agents for actinide ions and in analytical studies.

The general system is

where loss of the acidic NH proton gives a complex hydroxamate(2-) anion.[155]

11-17   Other η^1-O Donors
Dialkyl Sulfoxides[156]

The most common ligand is Me_2SO which, though a relatively poor nucleophile, forms many complexes that can be η^1-O or η^1-S. These types are usually readily distinguished by ir or nmr spectra. It is also found in complexes containing the entities $\mu_2\text{-}O,S$ and $\mu_3\text{-}O,O,S$.[157] Palladium(II) and Pt^{II} are frequently S-bonded, as in $[Pd(en)(Me_2SO)Cl]\,ClO_4$, but it is O-bonded in $[Pd(bipy)(Me_2SO)Cl]BF_4$. The latter equilibrate in solution to give a mixture of O- and S-bonded isomers.[158]

Pyridine and Dipyridyl *N*-Oxides; Tertiary Phosphine Oxides

These ligands bind only η^1-O. Long chain alkyl phosphine oxides such as *n*-octyl are used for solvent extraction of actinide and lanthanide ions. The ^{31}P chemical shifts for R_3PO provide correlations with the basicity of the corresponding R_3P and other data.

[150] A. Svetlanova-Larsen and J. L. Hubbard, *Inorg. Chem.* **1996,** *35,* 3073.
[151] M. H. Chisholm *et al., Inorg. Chem.* **1992,** *31,* 4469.
[152] F. A. Cotton *et al., Polyhedron* **1992,** *11,* 2475.
[153] B. J. Gunderman and P. J. Squattrito, *Inorg. Chem.* **1995,** *34,* 2399.
[154] D. H. Johnston and D. F. Shriver, *Inorg. Chem.* **1993,** *32,* 1045.
[155] D. Goldfarb *et al., J. Am. Chem. Soc.* **1995,** *117,* 383, 12771.
[156] M. Calligaris and O. Carugo, *Coord. Chem. Rev.* **1996,** *153,* 83.
[157] B. R. James *et al., Inorg. Chim. Acta* **1993,** *207,* 97.
[158] G. Annibale *et al., J. Chem. Soc. Dalton Trans.* **1989,** 1265.

Pentafluorooxoanions of S, Se, and Te[159]

The F_5XO^- ions are both bulky and electronegative. They behave to some extent like halide anions, especially fluorine for its ability to stabilize high oxidation states and give species such as $(CO)_5Mn(OTeF_5)$, cis-$ReO_2(OTeF_5)_4^-$,[160] $O_2Xe(OTeF_5)_2$, and $Xe(OSeF_5)_2$; they can also bridge.

Phenyloxoiodine or iodosylbenzene (PhIO) is much used as an oxygen transfer species but appears to form η^1-O complexes.

Additional References

Z. B. Alfassi, Ed., *Peroxyl Radicals*, Wiley, New York, 1997.
D. H. R. Barton, A. E. Martell, and D. T. Sawyer, Eds., *The Activation of Dioxygen and Homogeneous Catalytic Oxidation*, Plenum Press, New York, 1993.
Catalysis Today, **1997,** *33,* 1-118 (a special issue on recent developments in catalysis and photocatalysis on metal oxides).
Chemical Reviews. **1994,** *94,* 567-856 (a special issue on *Metal-Dioxygen Complexes*).
M. H. Chisholm, Ed., *Early Transition Metal Clusters with π-Donor Ligands*, VCH Publishers, New York, 1995 (polyoxo and alkoxide clusters).
Coordination Chemistry Reviews, **1996,** *148,* (several reviews on crown ethers).
A. E. Martell and D. T. Sawyer, Eds., *Oxygen Complexes and Oxygen Activation by Transition Metals*, Plenum Press, New York, 1988.

[159]L. Turowsky and K. Seppelt, *Z. anorg. allg. Chem.* **1992,** *609,* 153; S. H. Strauss, *Chem. Rev.* **1993,** *93,* 927.
[160]G. J. Schrobilgen *et al., Inorg. Chem.* **1996,** *35,* 7279.

Chapter 12

THE GROUP 16 ELEMENTS: S, Se, Te, Po

12-1 Group Characteristics and Trends

Some properties of the elements in Group 16 are given in Table 12-1. They are commonly called the *chalcogens*.

The atoms are two electrons short of the configuration of the next noble gas, and the elements show essentially nonmetallic covalent chemistry except for polonium and to a very slight extent tellurium. They may complete the noble gas configuration by forming (a) the *chalcogenide* ions S^{2-}, Se^{2-}, and Te^{2-}, although these ions exist only in the salts of the most electropositive elements, (b) two electron-pair bonds [e.g., $(CH_3)_2S$, H_2S, SCl_2], (c) ionic species with one bond and one negative charge (e.g., HS^- and RS^-), or (d) three bonds and one positive charge (e.g., R_3S^+).

In addition to such divalent species, the elements form compounds in *formal* oxidation states IV and VI with four, five, or six bonds; tellurium may give an 8-coordinate ion TeF_8^{2-}. Some examples of compounds of Group 16 elements and their stereochemistries are listed in Table 12-2. They also form cyclic and polycyclic cations (e.g., S_4^{2+}, Te_6^{4+}) under special conditions.

There are great differences between the chemistry of oxygen and that of sulfur, with more gradual variations through the sequence S, Se, Te, and Po. Differences from oxygen are attributable, among other things, to the following:

1. The lower electronegativities of the S—Po elements lessen the ionic character of compounds that are formally analogous to those of oxygen, alter the relative stabilities of various kinds of bonds, and drastically lessen the importance of hydrogen bonding, although weak X—H···S and S—H···X bonds do exist.

2. The maximum coordination number is not limited to 4, nor is the valence limited to 2, as in the case of oxygen, since *d* orbitals may be utilized in bonding. Thus sulfur forms several hexacoordinate compounds (e.g., SF_6), and for tellurium 6 is the characteristic coordination number.

3. Sulfur has a strong tendency to catenation, which manifests itself not only in the many forms of the element that all contain S_n rings of various sizes, but in polysulfide ions S_n^{2-} which may be discrete in highly ionic salts or serve as chelating ligands towards transition metals, sulfanes (XS_nX) (where X

Table 12-1   Some Properties of the Group 16 Elements

Element	Electronic structure	Melting point (°C)	Boiling point (°C)	Radius X^{2-} (Å)	Covalent radius —X— (Å)	Electronegativity
S	$[Ne]3s^23p^4$	119[a]	444.6	1.70	1.03	2.44
Se	$[Ar]3d^{10}4s^24p^4$	221[b]	684.8	1.84	1.17	2.48
Te	$[Kr]4d^{10}5s^25p^4$	450	990	2.07	1.37	2.01
Po	$[Xe]4f^{14}5d^{10}6s^26p^4$	254	962			1.76

[a]For monoclinic S_β see text.
[b]For grey Se.

may be H, halogen, CN, or NR_2), the polythionic acids ($HO_3SS_nSO_3H$) and their salts, to mention only the longer chains. In addition, the S—S unit occurs in many contexts, especially in organic and biological systems. In the structures of most protein molecules one or more —S—S— bridges are found to play a crucial role.

Although selenium and tellurium have a smaller tendency to catenation, they form rings (Se only) and long chains in their elemental forms. None of these chains is branched, because the valence of the element is only 2.

Table 12-2   Compounds of Group 16 Elements and Their Stereochemistries

Valence	Number of bonds	Geometry	Examples
II	2	Angular	Me_2S, H_2Te, S_n
	3	Pyramidal	Me_3S^+
	4	Square	$Te[SC(NH_2)_2]_2Cl_2$
IV	2	Angular	SO_2
	3	Pyramidal	SF_3^+, OSF_2, SO_3^{2-}, Me_3TeBPh_4
		Trigonal planar	$(SeO_2)_n$
	4	ψ-Trigonal bipyramidal	SF_4, RSF_3, Me_2TeCl_2
		Tetrahedral	Me_3SO^+
	5	ψ-Octahedral (square pyramidal)	$SeOCl_2py_2$, SF_5^-, TeF_5^-
	6	Octahedral[a]	$SeBr_6^{2-}$, PoI_6^{2-}, $TeBr_6^{2-}$
	7	Distorted pentagonal bipyramidal	$TePh(S_2CNEt_2)_3$
	8	Distorted dodecahedral	$Te(S_2CNEt_2)_4$
VI	3	Trigonal planar	$SO_3(g)$, $S(NCR)_3$
	4	Tetrahedral	SeO_4^{2-}, $SO_3(s)$, SeO_2Cl_2
	5	Trigonal bipyramidal	SOF_4
	6	Octahedral	RSF_5, SeF_6, $Te(OH)_6$, $TeMe_6$[b]

[a]These octahedra are frequently distorted because the electron pair remaining in the valence shell is not always completely sterically inactive. The degree of distortion for a given EX_6^{2-} ion varies with the size and shape of the accompanying cation. See, for example, A. du Bois and W. Abriel, *Z. Naturforsch.* **1990**, *45b*, 573; M. J. Taylor *et al.*, *Polyhedron* **1995**, *14*, 401.
[b]A. Haaland *et al.*, *J. Am. Chem. Soc.* **1995**, *117*, 7554; L. Ahmed and J. A. Morrison, *J. Am. Chem. Soc.* **1990**, *112*, 7411. No S or Se analogues known.

Gradual changes of properties are evident with increasing size, decreasing electronegativity, and so on, such as:

1. Decreasing thermal stability of the H_2E compounds. Thus H_2Te is considerably endothermic.
2. Increasing metallic character of the elements.
3. Increasing tendency to form anionic complexes such as $SeBr_6^{2-}$, $TeBr_6^{2-}$, and PoI_6^{2-}.
4. Decreasing stability of compounds in high formal positive oxidation states.
5. Emergence of cationic properties for Po and, very marginally, for Te. Thus TeO_2 and PoO_2 appear to have ionic lattices and they react with hydrohalic acids to give Te^{IV} and Po^{IV} halides, and PoO_2 forms a hydroxide $Po(OH)_4$. There are also some ill defined "salts" of Te and Po, such as $Po(SO_4)_2$ and $TeO_2 \cdot SO_3$.

Use of d Orbitals. The S—Po elements employ *d* orbitals together with their *s* and *p* orbitals to form more than four σ bonds (as in SF_6), and they also make frequent use of $d\pi$ orbitals to form multiple bonds. Thus, for example, in the sulfate ion, where the *s* and *p* orbitals are used in σ bonding, the shortness of the S—O bonds suggests that there must be considerable multiple-bond character. The usual explanation for this is that empty $d\pi$ orbitals of sulfur accept electrons from filled $p\pi$ orbitals of oxygen. Similar $d\pi$—$p\pi$ bonding occurs in some phosphorus compounds, but it seems to be more prominent with sulfur.

Other Remarks. In general, selenium and tellurium compounds are toxic. On the other hand selenium, in *trace* quantities, is biologically essential and may have anticancer properties.

THE ELEMENTS

12-2 Occurrence and Uses

Sulfur occurs widely in Nature as the element, as H_2S and SO_2, in numerous sulfide minerals, and in sulfates such as the many forms of *gypsum* ($CaSO_4 \cdot 2H_2O$). It occurs in crude oils and in coal and as H_2S in natural gas, from which it is recovered in large quantities *via* the reaction

$$2H_2S + SO_2 = 3S + 2H_2O$$

Sulfur is used mainly to manufacture sulfuric acid, but also in vulcanizing rubber and to make CS_2, P_2S_5, and a few other sulfides.

Selenium and tellurium are much less abundant than sulfur and no ores are rich in these elements. They are recovered from the anode slime deposited in the electrolytic purification of copper (having been present as impurities in the copper sulfide ores), as by-products in other sulfide ore processing, and in sulfuric acid manufacture.

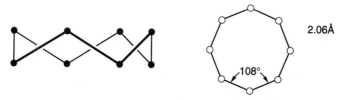

2.06Å

108°

Figure 12-1   The crown S_8 ring, as found in α sulfur.

There is no stable isotope of polonium. The isotope $^{210}Po(\alpha, 138.4d)$ occurs in U and Th minerals as an intermediate in the radioactive decay series, and was discovered by M. S. Curie in 1898. The only practical source of polonium (in gram quantities) is from nuclear reactors by the process

$$^{209}Bi(n,\gamma)^{210}Bi \longrightarrow {}^{210}Po + \beta^-$$

The study of polonium chemistry is difficult owing to the intense α radiation, which causes damage to solutions and solids, evolves much heat, and makes special handling techniques necessary for protection of the chemist.

Selenium and tellurium are mainly used as additives in metallurgy, though selenium is also used in glass and as a photoconductor (especially in Xerox machines). Polonium is used as a heat or power source in satellites.

12-3 Elemental Sulfur

Because of its ability for catenation, sulfur forms open and cyclic S_n species from $n = 2$ to $n = 20$ for cycles and higher for chains. This leads to enormous complexity in the physical and chemical behavior of the element.

Solid Sulfur

Solid forms of sulfur containing S_n rings with $n = 6, 7, 9$-$15, 18$, and 20 are also known.[1] These rings are all similar in their S—S distances (~ 2.05 Å), S—S—S angles ($\sim 106°$), and S—S bond energies (~ 265 kJ mol^{-1}), although the S—S—S—S torsion angles vary considerably. All other rings (and chains) are thermodynamically unstable (at 25°C) relative to the S_8 ring with a crown conformation (Fig. 12-1). Methods of obtaining the metastable rings are quite varied, but a reasonably systematic approach is to react linear components with the correct total number of sulfur atoms, for example,

$$H_2S_4 + S_2Cl_2 \longrightarrow cyclo\text{-}S_6 + 2HCl$$

$$H_2S_8 + S_4Cl_2 \longrightarrow cyclo\text{-}S_{12} + 2HCl$$

$$Cp_2TiS_5 + S_2Cl_2 \longrightarrow cyclo\text{-}S_7 + Cp_2TiCl_2$$

[1]R. Strauss and R. Steudel, *Z. Naturforsch.* **1988,** *43b,* 1151; R. Steudel *et al. Angew. Chem. Int. Ed.,* **1998,** *37,* 2377.

The common, stable form of sulfur at 25°C is orthorhombic α sulfur, containing *cyclo*-S_8 molecules. At 368.46 K (95.5°C) sulfur transforms to the high-temperature β form, monoclinic sulfur. The enthalpy of the transition is small (0.4 kJ g-atom^{-1} at 95.5°C) and the process is slow, so that it is possible by rapid heating of S_α to attain the melting point of S_α (112.8°C); S_β melts at 119°C. Monoclinic S_β crystallizes from sulfur melts, and although slow conversion to S_α occurs, the crystals can be preserved for weeks. Its structure contains S_8 rings as in S_α but differently packed. Another form of sulfur (S_γ, mp 106.8°C) is obtained by decomposition of copper(I) ethylxanthate in pyridine. It transforms slowly into S_β and/or S_α but is stable in the 95 to 115°C region. It too contains crown S_8 rings.

In addition to homoatomic S_n rings there are now numerous heteroatomic ones incorporating selenium and a few with Te.[2] An example is the 1,2-Se_2S_6 ring which can be photorearranged to the 1,3- and other isomers.[3]

$$(\text{tmen})\text{ZnS}_6 + \text{Se}_2\text{Cl}_2 \quad \xrightarrow{-\text{ZnCl}_2(\text{tmen})}$$

When α sulfur is dissolved in polar solvents (e.g., MeOH or MeCN) an equilibrium is set up in which ~1% of the sulfur is present as S_6 and S_7 rings. Since these rings are more reactive than S_8 rings, they may provide the pathway for reactions of elemental sulfur in polar media.

Catenasulfur

When molten sulfur is poured into ice water, the so-called plastic sulfur is obtained; although normally this has S_8 inclusions, it can be obtained as long fibers by heating S_α in nitrogen at 300°C for 5 min and quenching a thin stream in ice water. These fibers can be stretched under water and appear to contain helical chains of sulfur atoms with ~3.5 atoms per turn. Unlike the other sulfur allotropes, catenasulfur is insoluble in CS_2; it transforms slowly to S_α.

Liquid Sulfur

Precisely what happens when S_8 melts is still not fully understood; doubtless much depends on the level of impurities.

On melting, S_8 first gives a yellow, transparent, mobile liquid, which becomes brown and increasingly viscous above ~160°C. The viscosity reaches a maximum at ~200°C and thereafter falls until at the boiling point (444.60°C) the sulfur is again a rather mobile, dark red liquid. Figure 12-2(a) shows the viscosity and specific heat as a function of temperature. Although S_8 rings persist in the liquid up to ~193°C, the changes in viscosity are due to ring cleavage and the formation of chains, as well as other sulfur ring species with $n = 6$ to 20, and even >20 in equilibrium. The average degree of polymerization is shown in Fig. 12-2(b). The sulfur chains must have radical ends; they reach their greatest average length, 5 to 8×10^5 atoms, at about 200°C, where the viscosity is highest. The quantitative

[2]R. S. Laitinen *et al., Coord. Chem. Rev.* **1994,** *130,* 1.
[3]A. K. Verma and T. B. Rauchfuss, *Inorg. Chem.* **1995,** *34,* 6199.

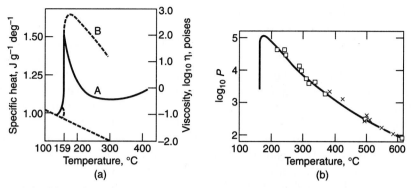

Figure 12-2 (a) Specific heat (A) and viscosity (B) of liquid sulfur. (b) Chain length (P) as a function of temperature, × from magnetic susceptibility measurements and □ from esr measurements.

behavior of the system is sensitive to certain impurities, such as iodine, which can stabilize chain ends by formation of S—I bonds. In the formation of polymers, almost every broken S—S bond of an S_8 ring is replaced by an S—S bond in a linear polymer, and the overall heat of the polymerization is thus expected to be close to zero. An enthalpy of 13.4 kJ mol^{-1} of S_8 converted into polymer has been found at the critical polymerization temperature (159°C).

The color changes on melting are due to an increase in the intensity and a shift of an absorption band to the red. This is associated with the formation above ~250°C of the red species S_3 and S_4, which comprise 1 to 3% of sulfur at its boiling point.

Sulfur Vapor[4]

The vapor contains S_n species with n from 2 to 8, with S_7 predominating at ~600°C while S_2 becomes dominant above ~1000°C. The color of the vapor changes as its composition changes. Only at very high temperatures (>2200°C) and low pressures (<10^{-4} mm Hg) do sulfur atoms become predominant. The S_2 molecule has a strong S=S bond (422 kJ mol^{-1}) based on a triplet ground state analogous to that of O_2. The S=S distance, 1.89 Å, is appreciably shorter than the S—S single bond distance of ~2.05 Å.

12-4 Elemental Selenium, Tellurium, and Polonium

Grey selenium (mp 494 K, metallic) is the stable form. It may be obtained crystalline from hot solutions of Se in aniline or from the melt. The structure, which has no sulfur analogue, contains infinite, spiral chains of selenium atoms. Although there are fairly strong single bonds between adjacent atoms in each chain, there is evidently weak metallic interaction between the neighboring atoms of different chains. Selenium is not comparable with most true metals in its electrical conductivity in

[4]J. Corset *et al., Ber. Bunsenges. Phys. Chem.,* **1988,** *92,* 859.

the dark, but it is markedly photoconductive and is widely used in photoelectric devices, most notably in xerography.

In the gas above grey or liquid selenium there is a temperature-dependent mixture of Se_n, $n = 2$ to 10, species, and in solution cyclic Se_6, Se_7, and Se_8 are present. Metastable crystalline solids containing Se_6 and Se_8 (three polymorphs) are known, but any external force (heat, pressure) initiates conversion to grey Se.

The one form of tellurium is silvery-white, semimetallic, and isomorphous with grey Se. Like the latter it is virtually insoluble in all liquids except those with which it reacts. Even though Te_8 rings are not found in elemental tellurium, they do occur in the compound Cs_3Te_{22},[5] which also contains unusual planar nets of Te atoms.

Whereas sulfur is a true insulator (specific resistivity, in $\mu\Omega$-cm $= 2 \times 10^{23}$), selenium (2×10^{11}) and tellurium (2×10^5) are intermediate in their electrical conductivities, and the temperature coefficient of resistivity in all three cases is negative, which is usually considered characteristic of nonmetals.

The trend toward greater metallic character of the elements is complete at polonium, which has two allotropes, both with typically metallic structures: α cubic, converting at 36°C to β rhombohedral (mp 254°C). Each of these has resistivity typical of a true metal with a positive temperature coefficient.

12-5 Reactions of the Elements

The allotropes of S and Se containing cyclo species are soluble in CS_2 and other nonpolar solvents such as benzene and cyclohexane. The solutions are light sensitive, becoming cloudy on exposure, and may also be reactive toward air unless very special precautions are taken, sulfur contains traces of H_2S and other impurities that can have substantial effects on rates of reactions. The nature of the sulfur produced on photolysis is not well established, but such material reverts to S_8 slowly in the dark or rapidly in the presence of triethylamine. From the solvent CHI_3 sulfur crystallizes as a charge-transfer compound $CHI_3 \cdot 3S_8$, with I$\cdots$S interactions; isomorphous compounds with PI_3, AsI_3, and SbI_3 are also known. It is probable that similar charge-transfer complexes wherein the S_8 ring is retained are first formed in reactions of sulfur with, for example, bromine. When heated, S, Se, Te, and Po burn in air to give the dioxides EO_2, and the elements react when heated with halogens, most metals, and nonmetals. Sulfur, Se, and Te are not affected by nonoxidizing acids, but the more metallic polonium will dissolve in concentrated HCl as well as in H_2SO_4 and HNO_3, giving solutions of Po^{II} and then Po^{IV}.

Although Po is distinctly metallic in some ways, [e.g., it forms saltlike Po^{II} and Po^{IV} compounds such as $PoCl_2$, $PoBr_2$, $Po(SO_4)_2$, and $Po(CrO_4)_2$] it also shows nonmetal characteristics by forming numerous polonides (MPo), which are often isostructural with tellurides and appear to be fairly ionic.

Sulfur dissolves in liquid ammonia to give S_4N^-, S_6^{2-} and S_3^- as primary products:

$$10S + 4NH_3 \longrightarrow S_4N^- + S_6^{2-} + 3NH_4^+$$

$$\Updownarrow$$

$$2S_3^-$$

On addition of alkali metal amides S_3N^- and S_4^{2-} are believed to form.[6]

[5]W. S. Sheldrick and M. Wachhold, *Angew. Chem. Int. Ed. Engl.* **1995**, *34*, 450.
[6]J. P. Lelieur *et al., Inorg. Chem.* **1988**, *27*, 73, 1883, 3032.

Sulfur also dissolves, with reaction, in organic amines such as piperidine, to give colored solutions containing N,N'-polythiobisamines in which there are free radicals ($\sim$ 1 per 10^4 S atoms):

$$2RR'NH + S_n \longrightarrow (RR'N)_2S_{n-1} + H_2S$$

Many sulfur reactions are catalyzed by amines, and in such S—S bond-breaking reactions free radicals may be involved.

Sulfur and selenium react with many organic molecules. For example, saturated hydrocarbons are dehydrogenated. The reaction of sulfur with alkenes and other unsaturated hydrocarbons is of enormous technical importance: hot sulfurization results in the vulcanization (formation of S bridges between carbon chains) of natural and synthetic rubbers.

It is clear that all reactions of S_8 or other cyclo species require that the initial attack open the ring to give sulfur chains or chain compounds. Many common reactions can be rationalized by considering a nucleophilic attack on S—S bonds. Some typical reactions are

$$S_8 + 8CN^- \longrightarrow 8SCN^-$$

$$S_8 + 8Na_2SO_3 \longrightarrow 8Na_2S_2O_3$$

$$S_8 + 8Ph_3P \longrightarrow 8Ph_3PS$$

It appears that the rate-determining step is the initial attack on the S_8 ring and that subsequent steps proceed very rapidly, so that the reactions can be assumed to proceed as follows:

$$S_8 + CN^- \longrightarrow {}^{\bullet}SSSSSSSSCN^-$$

$${}^{\bullet}S_6-S-SCN^- + CN^- \longrightarrow {}^{\bullet}S_6SCN^- + SCN^- \text{ and so on}$$

Cationic Compounds

It has long been known that sulfur, selenium, and tellurium will dissolve in oleums to give blue, green, and red solutions, respectively, which are unstable and change in color when kept or warmed. The colored species are cyclic polycations in which the element is formally in a fractional oxidation state. It is difficult to isolate solids from oleum solutions, and crystalline salts can be much more easily obtained by selective oxidations of the elements with SbF_5 or AsF_5 in liquid HF or SO_2,[7] or with $S_2O_6F_2$ in HSO_3F.

Some representative preparative reactions are given here and the major cations are listed in Table 12-3.

$$S_8 + 3SbF_5 \longrightarrow S_8^{2+} + 2SbF_6^- + SbF_3$$

$$4Se + S_2O_6F_2 \longrightarrow Se_4^{2+} + 2SO_3F^-$$

[7]M. P. Murchie *et al., Inorg. Synth.* **1997,** *31,* 102.

Table 12-3 Polyatomic Cations of Sulfur, Selenium, and Tellurium

Cation Type	S$^-$	Se	Te	Mixed
E$_{19}^{2+}$	S$_{19}^{2+}$			
E$_{10}^{2+}$		Se$_{10}^{2+}$		Te$_2$S$_8^{2+}$
E$_8^{2+}$	S$_8^{2+}$	Se$_8^{2+}$		Te$_2$Se$_6^{2+}$
E$_6^{2+}$				Te$_3$Se$_3^{2+}$, Te$_2$Se$_4^{2+}$
E$_4^{2+}$	S$_4^{2+}$	Se$_4^{2+}$	Te$_4^{2+}$	Te$_2$Se$_2^{2+}$
E$_6^{4+}$			Te$_6^{4+}$	

$$4Se + 5SbF_5 \xrightarrow{\text{SO}_2} Se_4{}^{2+} + 2Sb_2F_{11}{}^- + SbF_3$$

$$7Te + TeCl_4 + 4AlCl_3 \longrightarrow 2Te_4{}^{2+} + 4AlCl_4{}^-$$

$$8Te + 2WCl_6 \longrightarrow Te_8{}^{2+} + 2WCl_6{}^-$$

Many of the structures have been established by X-ray crystallography and by ^{77}Se and ^{125}Te nmr. In the solids, the E$_n^{2+}$ cations are closely surrounded by fluoroanions (e.g., AsF$_6^-$, Sb$_2$F$_{11}^-$, Sb$_3$F$_{16}^-$, and SO$_3$F$^-$) and the finer details of the structures are undoubtedly affected by interactions with these anions.

The E$_4^{2+}$ species are thoroughly characterized. They have square structures with bond lengths slightly (~3%) shorter than the conventional E—E single bond lengths. Their electronic structures may be formulated as in (12-I) (with the three analogous structures in resonance with the one shown) or as in (12-II), where a net π bond order of one may be envisioned.

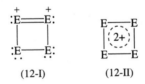

(12-I)   (12-II)

A number of other structures have been established, especially with tellurium[8,9] and a few are illustrated schematically in (12-III) through (12-V). These structures are not easy to account for in a fully satisfactory way. The dashed lines in (12-III) and (12-IV) indicate partial bonding, but in none of these three structures is there a clear understanding of how electron density and charge are distributed. The S$_{19}^{2+}$ ion has two S$_7$ rings linked by an S$_5$ chain and the Te$_6^{4+}$ ion is a trigonal prism with two Te$_3$ triangles (Te—Te, ~2.68 Å) joined by three longer bonds (Te—Te, ~3.10 Å).

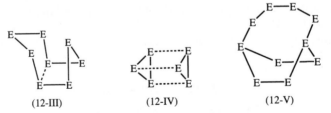

(12-III)   (12-IV)   (12-V)

[8]J. Beck, *Angew. Chem. Int. Ed. Engl.* **1994**, *33*, 163; *Chem. Ber.*, **1995**, *128*, 23.
[9]J. Beck and G. Bock, *Z. anorg. allg. Chem.* **1996**, *622*, 823.

Other complex cations are known, such as Se_{17}^{2+} which consists of two seven-membered rings linked by an Se_3 unit[10] and an infinite chain $[Te_7^{2+}]_\infty$ cation. There are also many examples of mixed chalcogenide cations, such as $Se_4Te_2^{2+}$ obtained by reacting a Se/Te mixture with AsF_5 or SbF_5.[11]

BINARY COMPOUNDS

12-6 Hydrides

The dihydrides, H_2S, H_2Se, and H_2Te, are extremely poisonous gases with revolting odors; the toxicity of H_2S far exceeds that of HCN. They are readily obtained by the action of acids on metal chalcogenides. Polonium hydride has been prepared only in trace quantities, by dissolving magnesium foil plated with Po in 0.2 M HCl. The thermal stability and bond strengths decrease from H_2S to H_2Po. Although pure H_2Se is thermally stable to 280°C, H_2Te and H_2Po appear to be thermodynamically unstable with respect to their constituent elements. All behave as very weak acids in aqueous solution, and the general reactivity increases with increasing atomic number. *Hydrogen sulfide* dissolves in water to give a solution ~0.1 M under 1 atm pressure. The dissociation constants are

$$H_2S + H_2O = H_3O^+ + HS^- \qquad pK = 6.88 \pm 0.02$$

$$HS^- + H_2O = H_3O^+ + S^{2-} \qquad pK = 14.15 \pm 0.05$$

In acid solution H_2S is also a mild reducing agent.

Sulfanes

The compounds H_2S through H_2S_8 have been isolated in pure states, whereas higher members, H_2S_n, are so far known only in mixtures. All are reactive yellow liquids whose viscosities increase with chain length. They may be prepared in large quantities by reactions such as

$$Na_2S_n(aq) + 2HCl(aq) \longrightarrow 2NaCl(aq) + H_2S_n(l) \qquad (n = 4\text{-}6)$$

$$S_nCl_2(l) + 2H_2S(l) \longrightarrow 2HCl(g) + H_2S_{n+2}(l)$$

$$S_nCl_2(l) + 2H_2S_2(l) \longrightarrow 2HCl(g) + H_2S_{n+4}(l)$$

The oils from the first reaction can be cracked and fractionated to give pure H_2S_2 through H_2S_5, whereas the higher sulfanes are obtained from the other reactions. Although the sulfanes are all thermodynamically unstable with respect to the reactions

$$H_2S_n(l) = H_2S(g) + (n\text{-}1)S(s)$$

[10]J. Beck and J. Wetterau, *Inorg. Chem.* **1995**, *34*, 6202.
[11]R. J. Gillespie *et al.*, *Inorg. Chem.* **1988**, *27*, 1807.

These are believed to be free radical in nature, are sufficiently slow for the compounds to be stable for considerable periods.

12-7   Simple Chalcogenides and Polychalcogenides

Most metallic elements react directly with S, Se, Te, and, so far as is known, Po. Often they react very readily, mercury and sulfur, for example, at room temperature. Binary compounds of great variety and complexity of structure can be obtained. The nature of the products usually also depends on the ratios of reactants, the temperature of reaction, and other conditions. Many elements form several compounds and sometimes long series of compounds with a given chalcogenide. We give here only the briefest account, with emphasis on the more important sulfur compounds. Many selenides and tellurides are similar. More complex chalcogenides will be discussed in the next section.

Ionic Sulfides; Sulfide Ions

Only the alkalis and alkaline earths form sulfides that appear to be mainly ionic. They are the only sulfides that dissolve in water and they crystallize in simple ionic lattices, for example, an anti-fluorite lattice for the alkali sulfides and a rock salt lattice for the alkaline earth sulfides. Essentially only SH^- ions are present in aqueous solution, owing to the low second dissociation constant of H_2S. Although S^{2-} is present in concentrated alkali solutions, it cannot be detected below ~ 8 M NaOH owing to the reaction

$$S^{2-} + H_2O = SH^- + OH^-   K = \sim 1$$

The alkali and alkaline earth hydrosulfides can be made by the action of H_2S on the metal in liquid ammonia.

When aqueous sulfide solutions are heated with sulfur, solutions containing largely S_3^{2-} and S_4^{2-} are obtained. These polysulfide ions are the only ones stable in solution but a number of crystalline compounds with S_n^{2-} ions from $n = 3$ to $n = 8$ can be prepared, especially by using large cations (e.g., Cs^+, NH_4^+ and enH_2^{2+} and R_3NH^+). In all those with $n = 4$ to 8 the S—S distances run from 2.00 to 2.11 Å and the S—S—S angles are *ca.* 110°. Structures of some S_n^{2-} ions are shown in Fig. 12-3.

High-density power sources can be obtained from lithium– and sodium–sulfur batteries. The sulfides present in these systems are M_2S, M_2S_2, M_2S_4, and M_2S_5.

Although polyselenide and polytelluride ions are less common, the Se_3^{2-}, Se_4^{2-}, Se_5^{2-}, Te_3^{2-}, Se_6^{2-}, Te_4^{2-}, and Te_5^{2-} ions are known.[12,13,14]

When alkali polysulfides are dissolved in polar solvents such as acetone, DMF, or DMSO, deep blue solutions are formed. Certain blue sulfur-containing minerals, notably *lapis lazuli* and *ultramarine,* owe their color primarily to the 610-nm absorption associated with the S_3^- radical anion,[15] though in green ultramarine some S_2^- ($\lambda_{max} = 400$ nm) is also present.

[12]J. A. Ibers *et al., Inorg. Chem.* **1988,** *27,* 940.
[13]J. C. Huffman and R. C. Haushalter, *Z. anorg. allg. Chem.* **1984,** *518,* 203.
[14]R. C. Haushalter *et al., Inorg. Chem.* **1983,** *22,* 1809.
[15]H. Johansen, *Acta Chem. Scand.* **1995,** *49,* 79.

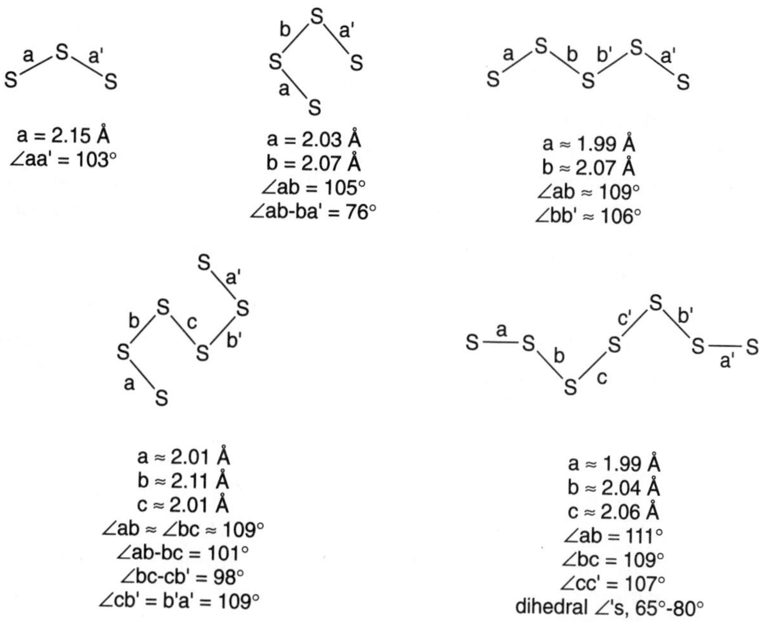

a = 2.15 Å
∠aa' = 103°

a = 2.03 Å
b = 2.07 Å
∠ab = 105°
∠ab-ba' = 76°

a ≈ 1.99 Å
b ≈ 2.07 Å
∠ab ≈ 109°
∠bb' ≈ 106°

a ≈ 2.01 Å
b ≈ 2.11 Å
c ≈ 2.01 Å
∠ab ≈ ∠bc ≈ 109°
∠ab-bc = 101°
∠bc-cb' = 98°
∠cb' = b'a' = 109°

a ≈ 1.99 Å
b ≈ 2.04 Å
c ≈ 2.06 Å
∠ab = 111°
∠bc = 109°
∠cc' = 107°
dihedral ∠'s, 65°-80°

Figure 12-3 Structures of representative polysulfide anions, S_n^{2-}.

Transition Metal Sulfides

Metal sulfides frequently have peculiar stoichiometries, and are often nonstoichiometric phases rather than compounds in a classical sense. They are often polymorphic, and many of them are alloylike or semimetallic in behavior. Sulfides tend to be much more covalent than the corresponding oxides, with the result that quite often there is only limited and occasionally no stoichiometric analogy between the oxides and the sulfides of a given metal. Very often, indeed possibly in most cases, when there is a sulfide and an oxide of identical empirical formula they have different structures.

Several transition metal sulfides (e.g., FeS, CoS, and NiS) adopt a structure called the *nickel arsenide structure*, illustrated in Fig. 12-4. In this structure each metal atom is surrounded octahedrally by six sulfur atoms, but also is approached

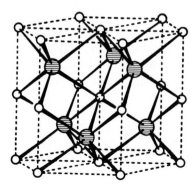

Figure 12-4 Structure of NiAs (As atoms shaded). The Ni atom in the center of the diagram is surrounded octahedrally by six As atoms and has also two near Ni neighbors, which are coplanar with four of the As atoms.

fairly closely by two other metal atoms. These metal-metal distances are 2.60 to 2.68 Å, in FeS, CoS, and NiS, and at such distances there must be a considerable amount of metal-metal bonding, thus accounting for their alloylike or semimetallic character. Note that such a structure is not in the least likely for a predominantly ionic salt, requiring as it would the close approach of dipositive ions.

Another class of sulfides of considerable importance are the *disulfides*, represented by FeS_2, CoS_2, and others. All these contain discrete S_2 units with an S—S distance almost exactly equal to that to be expected for an S—S single bond. These assume one of two closely related structures. First there is the *pyrite structure* named after the polymorph of FeS_2 that exhibits it. This structure may be visualized as a distorted NaCl structure. The Fe atoms occupy Na positions and the S_2 groups are placed with their centers at the Cl positions but turned in such a way that they are not parallel to any of the cube axes. The *marcasite structure* is very similar but somewhat less regular.

Iron sulfide is a good example of a well-characterized nonstoichiometric sulfide. It has long been known that a sample with an Fe/S ratio precisely unity is rarely encountered, and in the older literature such formulas as Fe_6S_7 and $Fe_{11}S_{12}$ have been assigned to it. The iron–sulfur system assumes the nickel arsenide structure over the composition range 50 to 55.5 at. % of sulfur and, when the S/Fe ratio exceeds unity, some of the iron positions in the lattice are vacant in a random way. Thus the very attempt to assign stoichiometric formulas such as Fe_6S_7 is meaningless. We are dealing not with *one* compound, in the classical sense, but with a *phase* that may be (but rarely is) perfect (FeS), or may be deficient in iron. The particular specimen that happens to have the composition Fe_6S_7 is better described as $Fe_{0.86}\square_{0.14}S$.

An even more extreme example of nonstoichiometry is provided by the Co—Te (and the analogous Ni—Te) system. Here, a phase with the nickel arsenide structure is stable over the entire composition range CoTe to $CoTe_2$. It is possible to pass continuously from the former to the latter by progressive loss of Co atoms from alternate planes (see Fig. 12-4) until, at $CoTe_2$, every other plane of Co atoms present in CoTe has completely vanished.

Typical of a system in which many different phases occur (each with a small range of existence so that each may be encountered in nonstoichiometric form) is the Cr—S system, where six phases occur in the composition range $CrS_{0.95}$ to $CrS_{1.5}$.

Although there are differences, the structural chemistry of the simpler selenides and tellurides is generally similar to that of sulfides.

Nonmetallic Binary Sulfides

Most nonmetallic (or metalloid) elements form sulfides that if not molecular, have polymeric structures involving sulfide bridges. Thus silicon disulfide (12-VI) consists of infinite chains of SiS_4 tetrahedra sharing edges, whereas the isomorphous Sb_2S_3 and Bi_2S_3 (12-VII) form infinite bands that are then held in parallel strips in the crystal by weak secondary bonds.

(12-VI)

(12-VII)

12-8 More Complex Metal Chalcogenides

In addition to the relatively simple chalcogenides already mentioned, there are others, formed mainly by selenium and tellurium, that have much more complicated structures. Space limitations do not permit us to do more than cite a few examples: (1) S_4 or Te_4 rectangular units connecting sets of four metal atoms;[16] gold and mercury telluride anions[17] such as $[Au_4Te_4]^{2-}$, $[Au_2Te_4]^{2-}$, and $[Hg_4Te_{12}]^{4-}$; (3) the $Sn_2Te_6^{2-}$ ion;[18] (4) complex chain, ring, and sheet structures found in the Cs_nTe_m ($n = 2$, m $= 13$; $n = 4$, $m = 28$; $n = 3$, $m = 22$) compounds.[19] Many of them may be considered as Zintl anions (see Section 1-9).

12-9 Cyclic Chalcogeno-Nitrogen Compounds

These are mostly sulfur-nitrogen compounds, although some cogeneric compounds are known. Sulfur and nitrogen have similar electronegativities and strong tendencies to form single and multiple covalent bonds. Hence, it is not surprising that there is a very extensive chemistry of sulfur-nitrogen compounds. Most of them are cyclic (or polymeric) and entail S—N π bonding as well as σ bonding. While NS (unlike NO) is not stable under normal conditions, it, along with SNS and NSS, has been observed[20] in an argon matrix at 12 K and it is also known as a ligand.

Binary Neutral S—N Compounds

Tetrasulfur Tetranitride. Aside from its intrinsic interest, S_4N_4 is a source compound for the preparation of other important S—N compounds. It is best made by the action of NH_3 on S_2Cl_2 or SCl_2; the reaction is complex and not well understood, although NSCl and $S_4N_3^+Cl^-$ are probable intermediates. Tetrasulfur tetranitride forms thermochromic crystals (mp 185°C) that are orange yellow at 25°C, red above 100°C, and almost colorless at −190°C. The compound must be handled with care, since grinding, percussion, friction, or rapid heating can cause it to explode.

The structure of S_4N_4 is a cage with a square set of N atoms and a bisphenoid of S atoms (Fig. 12-5), which is in interesting contrast to the structure of As_4N_4

[16]K. Isobe *et al., Angew. Chem. Int. Ed. Engl.* **1994,** *33,* 1882; B. K. Das and M. G. Kanatzidis, *Inorg. Chem.* **1995,** *34,* 1011.
[17]R. C. Haushalter, *Angew. Chem. Int. Ed. Engl.* **1985,** *24,* 432; R. C. Haushalter, *Inorg. Chem.* **1985,** *102,* L37.
[18]R. C. Haushalter *et al., Inorg. Chem.* **1984,** *23,* 2312.
[19]W. S. Sheldrick and M. Weinhold, *Chem. Commun.* **1996,** 607.
[20]P. Hassanzadeh and L. Andrews, *J. Am. Chem. Soc.* **1992,** *114,* 83.

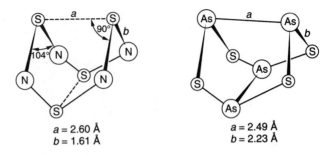

Figure 12-5 Structures of N_4S_4 and As_4S_4. Both have D_{2d} symmetry.

$a = 2.60$ Å
$b = 1.61$ Å

$a = 2.49$ Å
$b = 2.23$ Å

(*realgar*), also shown in Fig. 12-5. The S···S distance (2.60 Å) is longer than the normal S—S single-bond distance (~2.08 Å) but short enough to indicate significant interaction; even the S-to-S (linked by N) distance (2.71 Å) is indicative of direct S···S interaction. A low-temperature X-ray study indicates electron density along the shorter S···S line. The N—S distances and the angles in N_4S_4 (as well as $S_4N_5^-$) suggest the presence of lone pairs on the N and S atoms. There is an Se_4N_4 that has a similar molecular structure.[21]

Tetrasulfur tetranitride is prototypal for other cyclic S—N compounds in the sense that its electronic structure is not accounted for by any single classical bonding diagram. In terms of a valence bond/resonance approach, the following canonical structures all merit consideration:

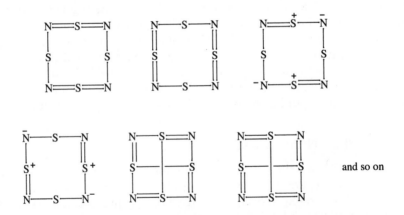

and so on

Molecular orbital calculations at various levels have given reasonably satisfactory descriptions of the structure and bonding in S_4N_4 and other related species.

Some reactions of S_4N_4 are given in Fig. 12-6. In addition to its conversion to S_2N_2 and $(SN)_x$, discussed later, it undergoes two main types of reaction: (a) reactions in which the S—N ring is preserved as in adducts of BF_3 or $SbCl_5$ or in the reduction to $S_4N_4H_2$; and (b) reactions in which ring cleavage occurs with reorganization to form other S—N rings. These products may be cationic as when S_4N_4 reacts with Lewis acids like SbF_5; examples are the ions $S_3N_2^+$ and $S_4N_4^{2+}$. They may also be anionic, as in the reaction of S_4N_4 and N_3^- (see below).

[21]K. Dehnicke *et al.*, *Z. anorg. allg. Chem.* **1994**, *620*, 1011.

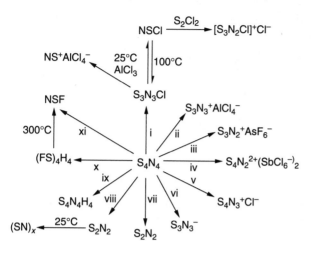

Figure 12-6 Some reactions of S_4N_4: (i) Cl_2; (ii) $AlCl_3$; (iii) AsF_5; (iv) $SbCl_5$; (v) HCl or S_2Cl_2; (vi) N_3^- or electrolytic reduction; (vii) S_8, 100°C; (viii) Ag, 250°C/1 mm; (ix) $SnCl_2$ in EtOH; (x) AgF in CCl_4; (xi) HgF_2 reflux in CCl_4.

Disulfur Dinitride. Colorless, crystalline S_2N_2 is made by pumping S_4N_4 vapor through silver wool or gauze, or in other ways. It has an approximately square structure (12-VIII). Like S_4N_4 it is potentially explosive and can be polymerized to $(SN)_x$, probably by a ring-opening radical mechanism.

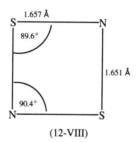

(12-VIII)

Polythiazyl. This polymer, $(SN)_x$, known since 1910, can now be obtained in a pure state. It is golden bronze in color and displays metallic type electrical conductance; more remarkable still is the fact that at 0.26 K it becomes a superconductor. In the crystal the kinked, nearly planar, chains (12-IX) lie parallel and conductance takes place along the chains, in which π electrons are extensively delocalized according to molecular quantum mechanical calculations. A partially brominated substance, $(SNBr_{0.4})_n$, is an even better conductor.

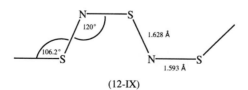

(12-IX)

The reduction of S_4N_4 by $SnCl_2$ to $S_4N_4H_4$ gives another cyclic molecule with the structure 12-X in which the S—N bond order is about 1 or slightly greater.[22]

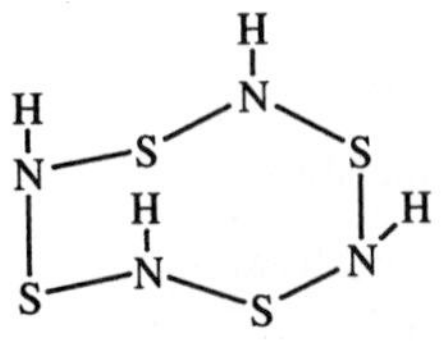

(12-X)

S_xN_y Compounds. A number of such compounds are known, for example, S_4N_2, $S_{11}N_2$, $(S_7N)_2S_x$, and S_5N_6, and have been shown to have cyclic or polycyclic structures, as shown in 12-XI to 12-XIII:

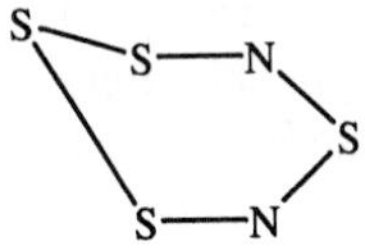

(12-XI) 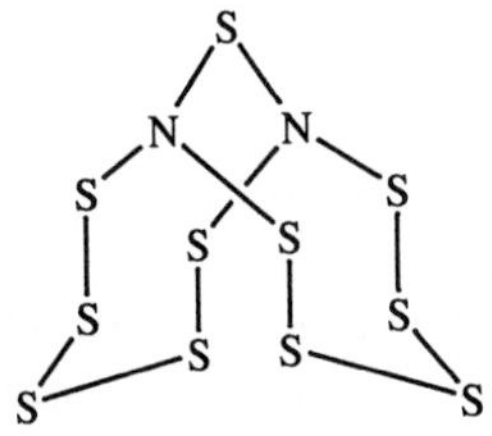   (12-XII) 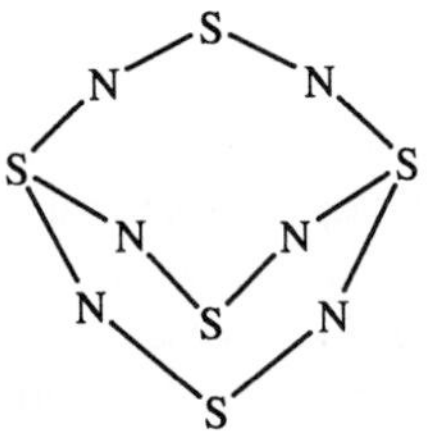   (12-XIII)

Sulfur-Nitrogen Cations and Anions. The simple cation, NS_2^+, is the analogue of the nitronium (NO_2^+) ion; it is made by the reaction:

$$S_7NH \xrightarrow[\text{in SO}_2]{SbCl_5} [S_2N^+][SbCl_6^-]$$

Three other well-established cations are prepared by the following reactions:

$$3S_3N_2Cl_2 + S_2Cl_2 \longrightarrow 2[S_4N_3^+]Cl^- + 3SCl_2$$

$$S_3N_3Cl_3 + (Me_3SiN)_2S \longrightarrow [S_4N_5^+]Cl^-$$

$$S_4N_4 + 4SbF_5 \longrightarrow [S_4N_4^{2+}][SbF_6^-][Sb_3F_{14}^-]$$

The structures of the first two are shown in (12 XIV) and (12-XV). The $S_4N_4^{2+}$ cation has an analogue $S_2Te_2N_4^{2+}$ in which there is a direct Te—Te bond (2.88 Å).[23]

[22]H. Fuess *et al., J. Am. Chem. Soc.* **1988**, *110*, 8488.
[23]A. Haas and M. Pyrka, *Chem. Ber.* **1995**, *128*, 11.

(12-XIV)   (12-XV)

Well-characterized S—N anions include S_4N^-, $S_3N_3^-$, and $S_4N_5^-$, whose structures are shown as (12-XVI) to (12-XVIII).

(12-XVI)   (12-XVII)

The $S_4N_5^-$ ion can be obtained from S_4N_4 in various ways, and its structure (12-XVIII) is clearly derived from that of S_4N_4 by addition of a bridging N^- unit.

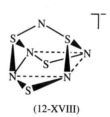

(12-XVIII)

The $S_3N_3^-$ ion, with its planar, partially π-bonded structure (12-XVII) is also derived from S_4N_4 by the action of azide ion or metallic potassium. The S_4N^- ion has been obtained by thermal decomposition of $S_4N_5^-$.

Other S-N Compounds. In addition to compounds containing only S and N, there are numerous others in which S—N bonds occur along with other bonds. There is, for example, a series of cyclic $S_x(NH)_y$ compounds with $x + y = 8$. These include S_7NH, $S_6(NH)_2$ (3 isomers), $S_5(NH)_3$ (2 isomers), and $S_4(NH)_4$ (the 1,3,5,7 isomer). These rings contain S—S and S—N bonds but no N—N bonds. There are also many compounds containing rings with additional elements such as C, Si, P, As, Sn, and O.

The S—N compounds containing sulfur in higher oxidation states are also important. Thus, there are cyclic compounds of the types shown in Fig. 12-7, and there are N—S bonds in sulfamic acid ($H_3N^+SO_3^-$), the sulfamate anion ($H_2NSO_3^-$), and sulfamides, $(R_2N)_2SO_2$. There are also some Se-N and S-Se-N compounds.[24]

[24]A. Haas *et al.*, *Chem. Ber.* **1991**, *124*, 1895.

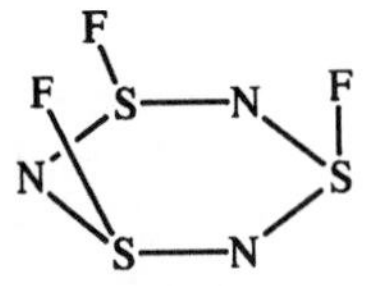

$S_3N_3F_3$

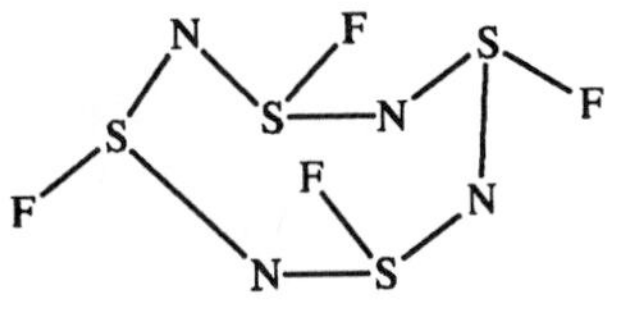

$S_4N_4F_4$

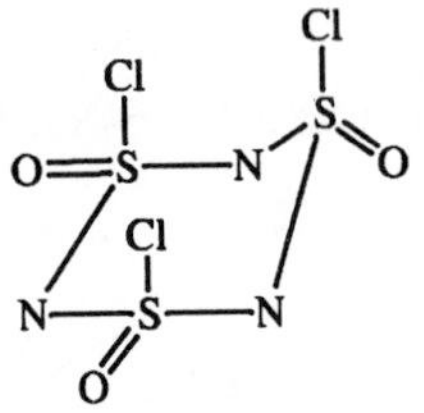

$S_3N_3O_3Cl(cis)$

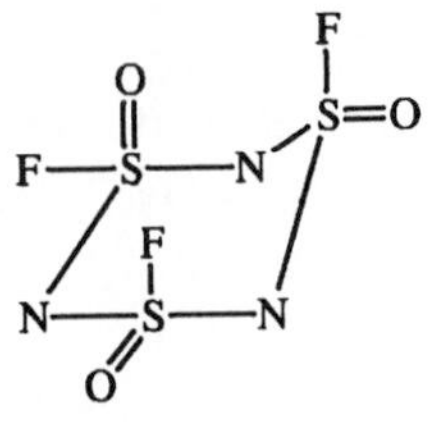

$S_3N_3O_3F_3(trans)$

Figure 12-7 Some cyclic S—N compounds with sulfur also bonded to F, Cl, and O.

12-10 Halides

Sulfur Fluorides

These are rather different from the other halides and are appropriately dealt with separately. There are seven sulfur fluorides, six of which are depicted, along with their properties, in Table 12-4. The isomeric *disulfur difluorides* are actually thermodynamically unstable. The compound FSSF prepared by fluorination of sulfur with AgF, readily isomerizes to SSF_2, which can also be obtained in other ways. However, SSF_2 itself in the presence of acid catalysts (e.g., HF or BF_3) rapidly disproportionates:

$$2SSF_2 \longrightarrow \tfrac{3}{8}S_8 + SF_4$$

Sulfur difluoride is difficult to make and is known only as a dilute gas. It dimerizes to form *disulfur tetrafluoride* (S_2F_4), which has a curious structure (12-XIX). One sulfur atom (S^1) can be considered to occupy an incomplete trigonal bipyramid with two axial F atoms and with a lone pair, an F atom and an S—F group in the equatorial plane.

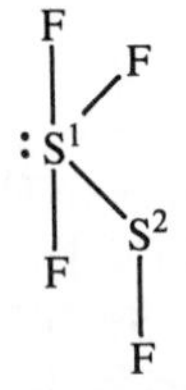

(12-XIX)

Sulfur tetrafluoride (SF_4) is made by reaction of SCl_2 with a 70/30 mixture of HF/pyridine. Sulfur tetrafluoride is an extremely reactive substance (as indicated in Fig. 12-8), instantly hydrolyzed by water to SO_2 and HF, but its fluorinating action is remarkably selective. It will convert C=O and P=O groups smoothly into CF_2 and PF_2, and CO_2H and P(O)OH groups into CF_3 and PF_3 groups, without attack on most other functional or reactive groups that may be present. Compounds of the type $ROSF_3$, which may be intermediates in the reaction with keto groups,

Table 12-4 Binary Sulfur Fluorides[a]

Compound	(°C)	Structure	Symmetry
FSSF	mp −133 bp +15	108° F, 1.64, S—S 1.89, F. Torsion angle: 88°	C_2
SSF_2	mp −165 bp −10.6	S—1.86—S···1.60 F, 108°, F	C_s
SF_2	Known only in gas phase	S, 1.59, F 98° F	C_{2v}
SF_4	mp −121 bp −38	F—1.65—S—F, 1.54 F, F	C_{2v}
SF_6	mp −50.5 subl −63.8	F 1.56, F···S···F, F, F, F	O_h
S_2F_{10}	mp −52.7 bp +30	F, F, F F, F—S—2.21—S—F, F, F F F, 1.56	D_{4d}

[a]Distances in Å; angles in deg.; mp = melting point and bp = boiling point.

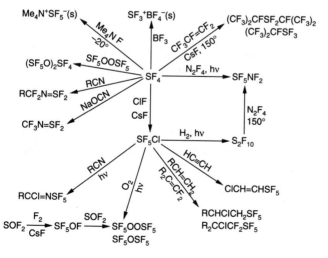

Figure 12-8 Some reactions of sulfur–fluorine compounds.

have been prepared. Sulfur tetrafluoride is also quite useful for converting metal oxides into fluorides, which are (usually) in the same oxidation state.

Aryl-substituted fluorides can be readily obtained by the reaction

$$(C_6H_5)_2S_2 + 6AgF_2 \longrightarrow 2C_6H_5SF_3 + 6AgF$$

which is carried out in trichloro- or trifluoromethane. The arylsulfur trifluorides are more convenient laboratory fluorinating agents than SF_4 in that they do not require pressure above atmospheric. The structure of SF_4 and of substituted derivatives RSF_3 is that of a trigonal bipyramid with an equatorial position occupied by the lone pair; SF_3CN retains the tbp structure with the CN in an equatorial position.[25]

Sulfur hexafluoride is chemically very inert, reacting only with such reagents as red hot metals and sodium in liquid ammonia. Molten KOH and steam at 500°C are without effect. Not surprisingly it is nontoxic and finds use as an insulating gas in high voltage electrical equipment. The low reactivity, particularly toward hydrolysis, which contrasts with the very high reactivity of SF_4 is mainly due to kinetic factors, not to thermodynamic stability, since the reaction of SF_6 with H_2O to give SO_3 and HF would be decidedly favorable ($\Delta G = -460$ kJ mol^{-1}), and the average bond energy in SF_4 (326 kJ mol^{-1}) is slightly higher than that of SF_6. The possibility of electrophilic attack on SF_6 has been confirmed by its reactions with certain Lewis acids. Thus Al_2Cl_6 at 180 to 200°C gives AlF_3, Cl_2, and sulfur chlorides, and the thermodynamically allowed reaction

$$SF_6 + 2SO_3 \longrightarrow 3SO_2F_2$$

proceeds slowly at 250°C. Sulfur hexafluoride also reacts rapidly and quantitatively with sodium in ethyleneglycol dimethyl ether containing biphenyl at room temperature:

$$8Na + SF_6 \longrightarrow Na_2S + 6NaF$$

Electron transfer from a biphenyl radical ion to an SF_6 molecule to give an unstable SF_6^- ion is probably involved.

Disulfur decafluoride is best obtained by the photochemical reaction

$$2SF_5Cl + H_2 \xrightarrow{h\nu} S_2F_{10} + 2HCl$$

In S_2F_{10} each S atom is octahedral and the S—S bond is unusually long (2.21 Å versus ~2.08 Å expected for a single bond), whereas the S—F bonds are, as in SF_6, ~0.2 Å shorter than expected for an S—F single bond. It is extremely poisonous (the reason for which is not clear), being similar to phosgene in its physiological action. It is not dissolved or hydrolyzed by water or alkalis and at room temperature it shows scarcely any chemical reactivity, though it oxidizes the iodide in an acetone solution of KI. At elevated temperatures, however, it is a powerful oxidizing agent, generally causing destructive oxidation and fluorination, presumably owing to initial breakdown to free radicals:

$$S_2F_{10} \longrightarrow 2SF_5^{\cdot}$$

$$SF_5^{\cdot} \longrightarrow SF_4 + F^{\cdot}$$

---

[25]H. Oberhammer, *Inorg. Chem.* **1996**, *35*, 806.

Substituted Sulfur Fluorides

There is an extensive chemistry of substituted sulfur fluorides of the types RSF_3 and RSF_5; examples of the former were mentioned previously. The SF_5 derivatives bear considerable resemblance to CF_3 derivatives, with the principal difference that in reactions with organometallic compounds the SF_5 group is fairly readily reduced, whereas the CF_3 group is not.

The mixed halide SF_5Cl is an important intermediate (Fig. 12-8). Although it can be made by interaction of S_2F_{10} with Cl_2 at 200 to 250°C (as can SF_5Br using S_2F_{10} and Br_2), it is best made by the CsF-catalyzed reaction

$$SF_4 + ClF \xrightarrow{\text{25°C, 1 h, CsF}} SF_5Cl$$

A probable intermediate is the salt $CsSF_5$, which dissociates significantly above 150°C:

$$CsF + SF_4 \xrightleftharpoons[150°C]{100°C} CsSF_5$$

SF_5Cl is a colorless gas (bp −15.1°C, mp −64°C), which is more reactive than SF_6, being readily attacked by OH^- and other nucleophiles, though it is inert to acids. Its hydrolysis and its powerful oxidizing action toward many organic substances are consistent with the charge distribution $F_5S^{\delta-}—Cl^{\delta+}$. Its radical reactions with olefins and fluoroolefins resemble those of CF_3I. Pentafluorobromosulfur is a pale yellow liquid boiling at 3°C, and also readily hydrolyzed.

The very reactive yellow *pentafluorosulfur hypofluorite* is one of the few known hypofluorites; it is obtained by the catalytic reaction

$$SOF_2 + 2F_2 \xrightarrow{\text{CsF, 25°C}} SF_5OF$$

In the absence of CsF, SOF_4 is obtained and this reacts separately with CsF to give SF_3OF.

The hydroxide SF_5OH and the hydroperoxide SF_5OOH are also known; the former gives the anion SF_5O^-.

Finally, the carbon compounds F_5SCH_3 and $F_4S{=}CH_2$ can be made by the sequence

$$SF_5Cl \xrightarrow{H_2C=C=O} SF_5CH_2COCl \xrightarrow{H_2O} SF_5CH_2CO_2H$$

$$SF_5CH_2CO_2Ag \xrightarrow[-CO_2]{Br_2} SF_5CH_2Br \xrightarrow{Zn/HCl} SF_5CH_3$$

$$\xrightarrow{BuLi}$$

$$SF_5CH_2Li \xrightarrow[-LiF]{-70°C} F_4S{=}CH_2$$

Both are quite stable gases. The compound F_4SCH_2 contains a C=S double bond, 1.554(4) Å.

Selenium and Tellurium Fluorides

While SeF_2, $FSeSeF$, and $SeSeF_2$ have been prepared only in minute amounts and their chemistry is little known, SeF_4 and TeF_4 can be prepared by the reaction

$$EO_2 + 2SF_4 \longrightarrow EF_4 + 2SOF_2$$

The compound SeF$_4$ (mp $\sim$ $-39°$C, bp 106°C) resembles SF$_4$, but being a liquid and somewhat easier to handle, it has some advantages as a fluorinating agent. Tellurium tetrafluoride is a colorless crystalline solid (mp 130°C) with a polymeric structure as shown in (12-XX).

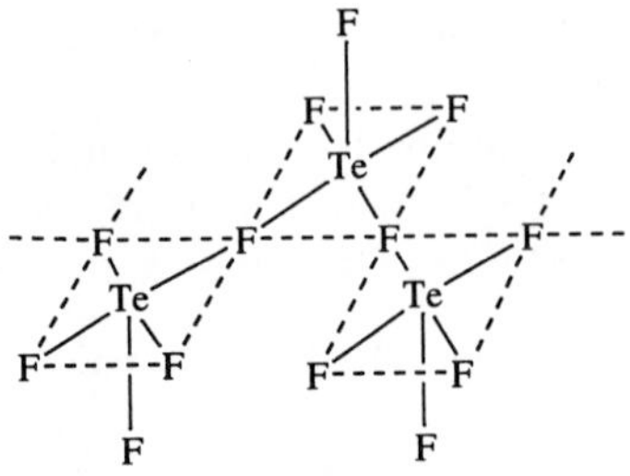

(12-XX)

Selenium hexafluoride and TeF$_6$ are colorless gases, somewhat more reactive than SF$_6$. Selenium hexafluoride though unreactive toward water is very toxic, while TeF$_6$ slowly hydrolyzes through Te(OH)$_n$F$_{6-n}$ intermediates to Te(OH)$_6$. The compounds Te$_2$F$_{10}$, TeF$_5$Cl, and TeF$_5$Br are also known.

Other Halides of Group 16

The other halides of S, Se, and Te are listed in Table 12-5.

Sulfur Chlorides. The chlorination of molten sulfur gives S$_2$Cl$_2$, an orange liquid of revolting smell. By using an excess of chlorine and traces of FeCl$_3$, SnCl$_4$, I$_2$, and so on, as catalyst at room temperature, an equilibrium mixture containing

Table 12-5 The Group 16 Binary Halides

Chlorides[a]	Bromides[a]	Iodides[a]
	Sulfur	
S$_2$Cl$_2$,[b] mp -80, bp 138	S$_2$Br$_2$,[b] mp -46, d 90	S$_2$I$_2$,[b] dec -30
SCl$_2$, mp -78, bp 59		
SCl$_4$, dec -31		
	Selenium	
Se$_2$Cl$_2$	Se$_2$Br$_2$, dec in vapor	
SeCl$_2$, dec in vapor	SeBr$_2$, dec in vapor	
SeCl$_4$, subl 191	SeBr$_4$	
	Tellurium[c]	
Te$_4$Cl$_{16}$, mp 223, bp 390	Te$_4$Br$_{16}$, mp 388, bp 414	Te$_4$I$_{16}$, mp 280
	dec in vapor	dec 100

[a]mp = melting point; bp = boiling point; dec = decomposes; subl = sublimes; all in °C.
[b]There are also dichloro- and dibromosulfanes, XS$_n$X with 2 < n < 100.
[c]Subhalides such as Te$_2$Br and TeI have also been reported; see R. Kniep and A. Rabenau, *Topics in Current Chemistry*, F. L. Boschke, Ed., Springer-Verlag, New York, 1983, p. 145.

~85% of SCl_2 is obtained. The dichloride readily dissociates within a few hours:

$$2SCl_2 \rightleftharpoons S_2Cl_2 + Cl_2$$

but it can be obtained pure as a dark red liquid by fractional distillation in the presence of some PCl_5, small amounts of which will stabilize SCl_2 for some weeks.

Sulfur chlorides are used as a solvent for sulfur (giving dichlorosulfanes up to about $S_{100}Cl_2$), in the vulcanization of rubber, as chlorinating agents, and as intermediates. Specific higher chlorosulfanes can be obtained by reactions such as

$$2SCl_2 + H_2S_4 \xrightarrow{-80°C} S_6Cl_2 + 2HCl$$

All sulfur chlorides are readily hydrolyzed by water. In the vapor S_2Cl_2 has a Cl—S—S—Cl chain with C_2 symmetry (S—S = 1.95 Å, S—Cl = 2.05 Å), and a similar structure occurs in the crystal. Sulfur dichloride has a bent structure with S—Cl = 2.014(5) Å and an angle of 102.8(2)°. Both of these chlorides have important industrial uses and are produced in ton quantities. Yellow crystalline SCl_4 (which may be, but there is no firm evidence, $SCl_3^+Cl^-$) is produced by the action of Cl_2 on S_2Cl_2 or SCl_2 at −80°C; it dissociates to $SCl_2 + Cl_2$ above −31°C.

Sulfur Bromide. This compound (S_2Br_2) is isostructural with S_2Cl_2, but the existence of SBr_2 is uncertain.

The existence of *sulfur iodides* is controversial, although S_2I_2 may be genuine. The best evidence for S—I bonds is in $[S_7I^+][SbF_6^-]$ and $[(S_7I)_2I^{3+}][SbF_6^-]_3$, prepared from reaction of iodine with sulfur in SbF_5 solution. Comparable Se_6I^+ cations are also known.

Selenium and Tellurium Halides. For selenium, these halides are mostly of marginal stability and the characterization is incomplete; the most stable ones are Se_2Cl_2 and Se_2Br_2, whose structures are known. In acetonitrile there are the following equilibria:

$$3SeCl_2 = Se_2Cl_2 + SeCl_4 \qquad K \approx \leq 2 \times 10^2$$

$$2SeBr_2 = Se_2Br_2 + Br_2 \qquad K \approx 1 \times 10^{-2}$$

In the vapor state $TeCl_4$ has a ψ-*tbp* structure but forms Te_4Cl_{16} molecules with a cubane-type structure (12-XXI) in the crystal.

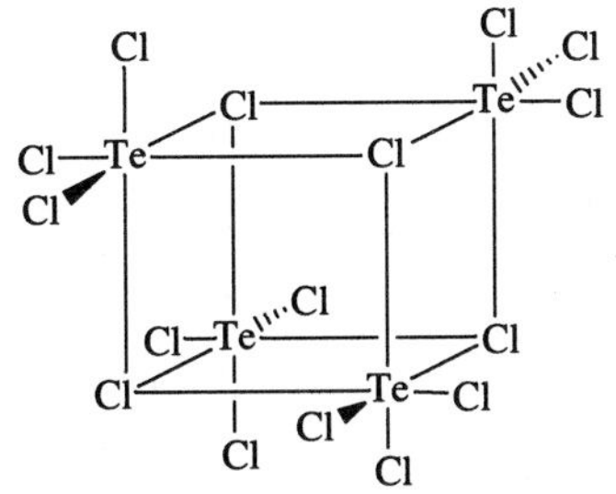

(12-XXI)

12-11   Oxides

The principal oxides are listed in Table 12-6. The oxides SO, SeO, and TeO have only transient existences, but PoO is a stable though easily oxidized solid. The oxide S_2O is an unstable colorless gas. The cyclosulfur oxides, S_nO, $n = 5$ to 8, which are formed on treating cyclosulfurs with CF_3CO_3H, have ring structures with one S=O unit.

Dioxides

The dioxides are obtained by burning the elements in air, though small amounts of SO_3 also form in the burning of sulfur. Sulfur dioxide is also produced when many sulfides are roasted in air and when sulfur-containing fuels such as oils and coals are burned. It presents a major pollution and ecological problem. For example, in Norway, the pH of lakes is decreasing because of the sulfuric acid formed from SO_2 pollution originating in the British Isles and Europe. Sulfur dioxide can be removed from flue gases by solid slurries of, say, calcium hydroxide, but there is then still an enormous sludge problem.

Selenium and tellurium dioxides are also obtained by treating the elements with hot nitric acid to form H_2SeO_3 and $2TeO_2 \cdot HNO_3$, respectively, and then heating these to drive off water or nitric acid.

The dioxides differ considerably in structure. Sulfur dioxide is a gas, SeO_2 is a white volatile solid, while TeO_2 and PoO_2 are nonvolatile solids. The chain polymer structure of SeO_2 is shown in (12-XXII).

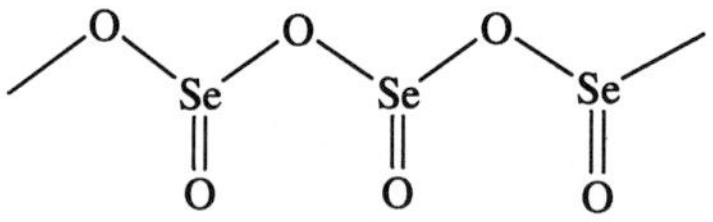

(12-XXII)

Tellurium dioxide has a white (α) and a yellow (β) modification, in both of which Te atoms have a coordination number of 4 and the structure is three dimensional. Polonium dioxide has the fluorite structure.

Sulfur dioxide is by far the most important of the dioxides. It is a weak reductant in aqueous acid but stronger in base where the sulfite ion (Section 12-13) is formed.

Table 12-6   Oxides of S, Se, Te, and Po

$\begin{cases} S_2O \\ SO \end{cases}$			PoO
SO_2	SeO_2	TeO_2	PoO_2 [$PoO(OH)_2$]
bp $-10.07°C$	subl $315°C$	mp $733°C$	
mp $-75.5°C$			
SO_3	SeO_3	TeO_3	
mp $16.8°$ (γ)	mp $120°C$	dec $400°C$	
bp $44.8°$		Te_2O_5	
		dec $> 400°C$	
S_nO, $n = 5$–8			
S_7O_2			

The pure liquid (bp $-10°C$) is a useful nonaqueous solvent despite its low dielectric constant (~15), and lack of any self-ionization. It is particularly useful as a solvent for superacid systems.

Sulfur dioxide has lone pairs and can act as a Lewis base; it can also act as a Lewis acid. With certain amines, crystalline 1:1 charge-transfer complexes are formed in which electrons from nitrogen are presumably transferred to antibonding acceptor orbitals localized on sulfur. One of the most stable is $Me_3N \cdot SO_2$ (12-XXIII) where the dimensions of the SO_2 molecule appear to be unchanged by complex formation.

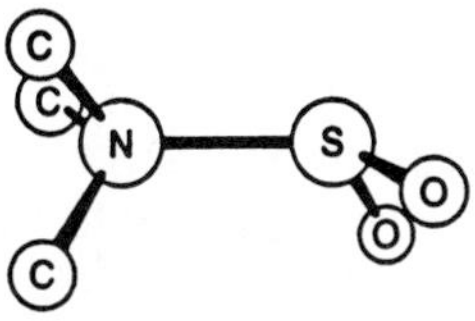

(12-XXIII)

Although crystals are formed with quinol and some other hydrogen-bonding compounds, these are clathrates or, in the case of $SO_2 \cdot 7H_2O$, a clathrate hydrate.

Sulfur dioxide also forms complexes with a number of transition metal species (Section 12-16).

Trioxides

The only important trioxide in this group, SO_3, is obtained by reaction of sulfur dioxide with molecular oxygen, a reaction that is thermodynamically very favorable but extremely slow in the absence of a catalyst. Platinum sponge, V_2O_5, and NO serve as catalysts under various conditions. Sulfur trioxide reacts vigorously with water to form sulfuric acid. Commercially, for practical reasons, SO_3 is absorbed in concentrated sulfuric acid, to give oleum, which is then diluted. Sulfur trioxide is used as such for preparing sulfonated oils and alkyl arenesulfonate detergents. It is also a powerful but generally indiscriminate oxidizing agent; however, it will selectively oxidize pentachlorotoluene and similar compounds to the alcohol.

The free molecule, in the gas phase, has a planar, triangular structure that may be considered to be a resonance hybrid involving $p\pi-p\pi$ S—O bonding, as in (12-XXIV), with additional π bonding *via* overlap of filled oxygen $p\pi$ orbitals with empty sulfur $d\pi$ orbitals, to account for the very short S—O distance of 1.41 Å.

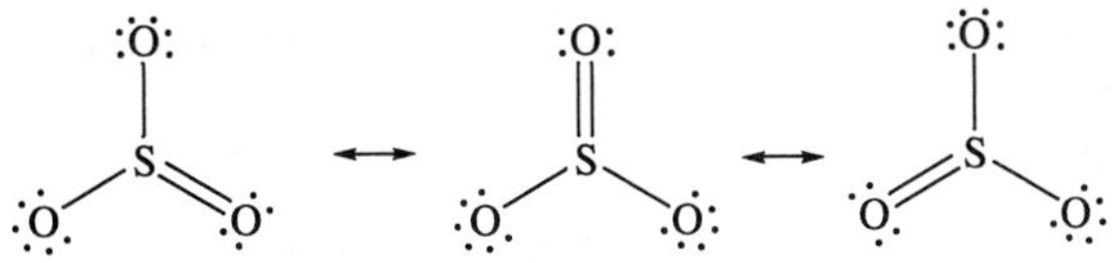

(12-XXIV)

In view of this affinity of S in SO_3 for electrons, it is not surprising that SO_3 functions as a fairly strong Lewis acid toward the bases that it does not preferentially oxidize. Thus the trioxide gives crystalline complexes with pyridine, trimethylamine, or dioxane, which can be used, like SO_3 itself, as sulfonating agents for organic compounds.

The structure of solid SO_3 is complex. At least three well-defined phases are known. γ-Sulfur trioxide, formed by condensation of vapors at $-80°C$ or below, is an icelike solid containing cyclic trimers with structure (12-XXV).

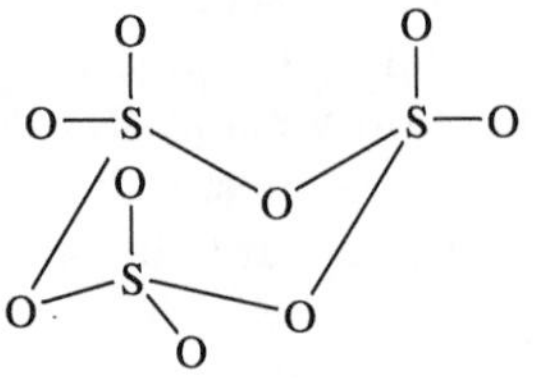

(12-XXV)

A more stable, asbestos-like phase (β-SO_3) has infinite helical chains of linked SO_4 tetrahedra (12-XXVI), and the most stable form, α-SO_3, which also has an asbestos-like appearance, presumably has similar chains crosslinked into layers.

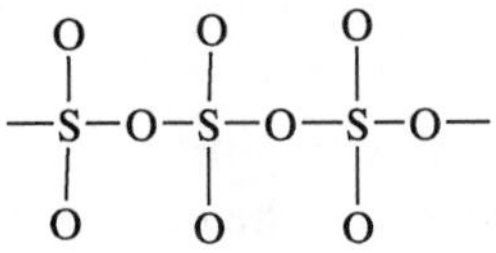

(12-XXVI)

Liquid γ-SO_3, which is a monomer-trimer mixture, can be stabilized by the addition of boric acid. In the pure state it is readily polymerized by traces of water. It reacts smoothly with R^FSiMe_3 compounds to form $R^FS(O)_2OSiMe_3$ esters that can be hydrolyzed to afford the perfluorosulfonic acids.[26]

Selenium trioxide is made by dehydration of H_2SeO_4 by P_2O_5 at 150 to 160°C; it is a strong oxidant and is rapidly rehydrated by water. Selenium trioxide dissolves in liquid HF to give *fluoroselenic acid*, $FSeO_3H$, a viscous fuming liquid.

Tellurium trioxide is made by dehydration of $Te(OH)_6$. This orange compound reacts only slowly with water but dissolves rapidly in bases to give tellurates.

12-12 Oxohalides

These are principally formed by sulfur and selenium and the main ones are the EOX_2, $EOXY$, EO_2X_2, and EO_2XY (X, Y = F, Cl, Br) molecular compounds, having the structures shown in Fig. 12-9. The E=O bonds are short[27] (1.39-1.44 Å for S=O) and the multiple bond character depends on oxygen to sulfur (or selenium) $p\pi \rightarrow d\pi$ bonding. This is enhanced by the electronegativity of the halogens, and the E=O bond lengths decrease in both series of compounds as the halogens change from Br to Cl to F.

The thionyl and selenyl halides are

SOF_4	SOF_2	$SOCl_2$	$SOBr_2$	$SOFCl$	$SOF(CN)$
$SeOF_4$	$SeOF_2$	$SeOCl_2$	$SeOBr_2$		

With the exception of SOF_2, which reacts only slowly with water, the compounds are rapidly, sometimes violently hydrolyzed; $SOF(CN)$ has only recently been made.[28]

[26]R. L. Kirchmeier *et al., Inorg. Chem.* **1994,** *33,* 6369.
[27]D. Mootz and A. Merschenz-Quack, *Acta Cryst.* **1988,** *C44,* 924, 926.
[28]J. Jacobs and H. Wildner, *Z. anorg. allg. Chem.* **1993,** *619,* 1221.

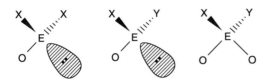

Figure 12-9 The structures of EOX$_2$, EOXY and EO$_2$XY.

The most common compound is *thionyl chloride*, which is made by the reaction

$$SO_2 + PCl_5 = SOCl_2 + POCl_3$$

It is used to prepare anhydrous metal halides from hydrated chlorides or hydroxides, since the only products of hydrolysis are gases:

$$SOCl_2 + H_2O = 2HCl + SO_2$$

Other representative syntheses are

$$SOCl_2 + 2HF(1) = SOF_2 + 2HCl$$

$$SOCl_2 + 2HBr \xrightarrow{0°C} SOBr_2 + 2HCl$$

$$SeO_2 + SeCl_4 \xrightarrow{CCl_4} 2SeOCl_2$$

The compounds are stable at ordinary temperatures but decompose on heating. They can act both as weak Lewis bases using lone pairs on oxygen, and as weak Lewis acids using vacant *d* orbitals. For example, SeOCl$_2$(pyrazole)$_2$ has a square pyramidal structure with apical O and trans Cl atoms in the base.

Sulfuryl halides are SO$_2$F$_2$, SO$_2$Cl$_2$, SO$_2$FCl, and SO$_2$FBr. *Sulfuryl chloride* is formed by the action of Cl$_2$ on SO$_2$ in the presence of a little FeCl$_3$ catalyst. It is stable to almost 300°C. It fumes in moist air and is hydrolyzed fairly rapidly with liquid water. It can be used as a chlorinating agent. Sulfuryl difluoride, a rather inert gas, is made by fluorination of SO$_2$Cl$_2$ or by heating barium fluorosulfate

$$Ba(SO_3F)_2 \xrightarrow{500°C} SO_2F_2 + BaSO_4$$

SO$_2$F$_2$ is soluble in water without hydrolysis but reacts in base:

$$SO_2F_2 + OH^- \longrightarrow \left[HO-\overset{\displaystyle O}{\underset{\displaystyle O}{S}}\overset{F}{\underset{F}{}} \right]^- \longrightarrow HO-\overset{\displaystyle O}{\underset{\displaystyle O}{S}}F + F^-$$

It also reacts with other nucleophiles in aqueous solution, probably again by nucleophilic attack on S and displacement of F$^-$. Thus NH$_3$ gives SO$_2$(NH$_2$)$_2$ and C$_6$H$_5$O$^-$ gives C$_6$H$_5$OSO$_2$F.

For Se only SeO$_2$F$_2$ is known. For Te there are no TeO$_2$X$_2$ compounds, but recently TeOF$_2$ and Te$_2$O$_3$F$_2$ have been reported. The structure of TeOF$_2$ is unknown but Te$_2$O$_3$F$_2$ has a complex polymeric structure. Both selenium and tellurium, as well as sulfur, form a number of more complex oxofluorides, a few of which are shown, with their structures, where known, in Table 12-7.

Table 12-7   Some Group 16 Oxofluorides

SO_3F_2	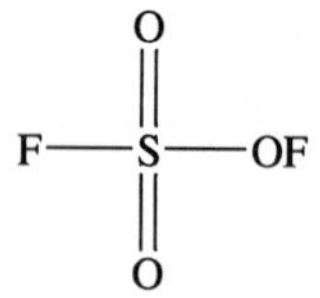
$S_2O_5F_2$	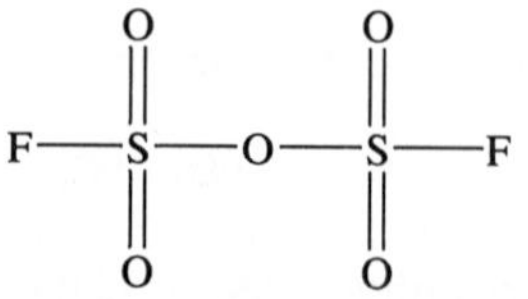
$S_3O_8F_2$	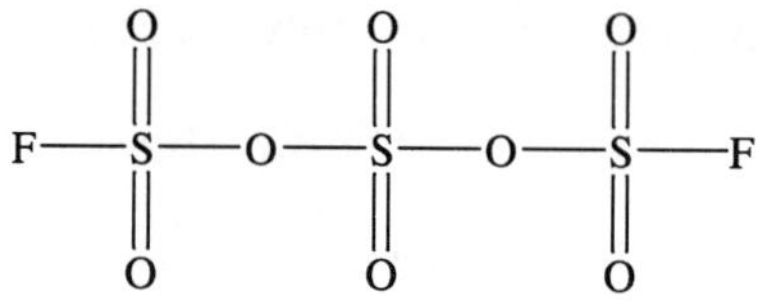
$E_2O_2F_8$ (Se, Te)	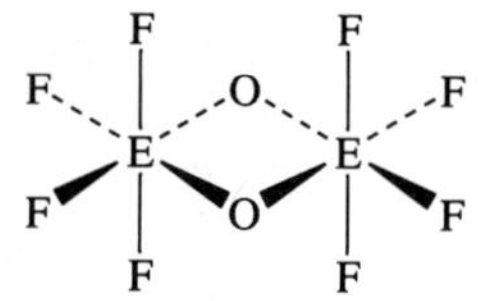
E_2OF_{10} (S, Se)	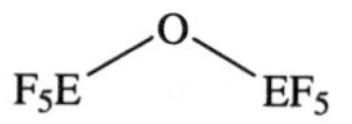
$Te_5O_4F_{22}$	$F_2Te(OTeF_5)_4$ (cis and trans isomers)

12-13   Oxo Acids of Sulfur

The oxo acids of sulfur are numerous and many are of importance. In some cases the acid is not known as such but the anion and its salts are known. Table 12-8 lists the major types of sulfur oxo acids according to structural type. This classification is to some extent arbitrary, but it corresponds approximately with the order in which we discuss these acids. Sulfuric acid has already been discussed (Section 2-11). None of the oxo acids in which there are S—S bonds has any known Se or Te analogue.

Table 12-8   Principal Oxo Acids of Sulfur

Name	Formula	Structure[a]
Acids Containing One Sulfur Atom		
Sulfurous	H_2SO_3[b]	SO_3^{2-} (in sulfites)
Sulfuric	H_2SO_4	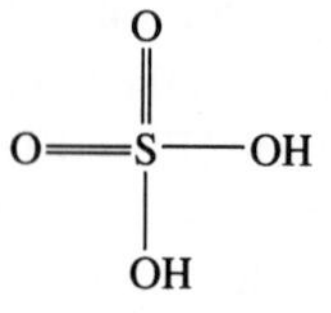

Table 12-8 (*Continued*)

Name	Formula	Structure[a]
Acids Containing Two Sulfur Atoms		
Thiosulfuric	$H_2S_2O_3$	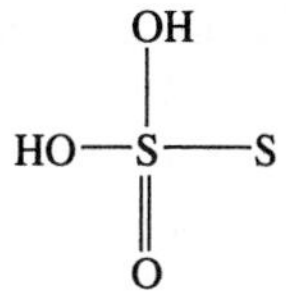 $HO-S(OH)(=O)-S$
Dithionous	$H_2S_2O_4$[b]	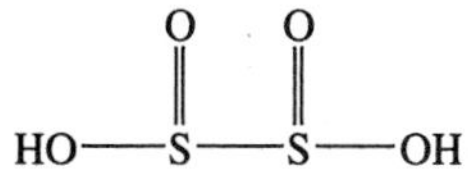 $HO-S(=O)-S(=O)-OH$
Disulfurous	$H_2S_2O_5$[b]	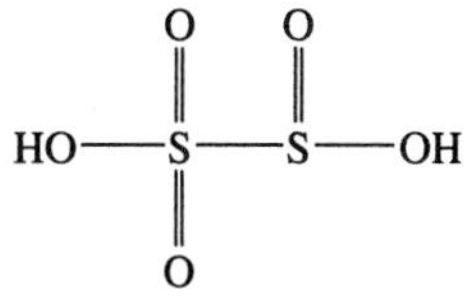 $HO-S(=O)-S(=O)(=O)-OH$
Dithionic	$H_2S_2O_6$	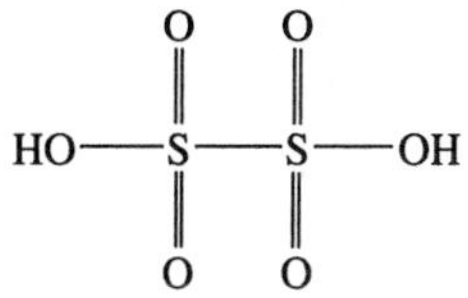 $HO-S(=O)(=O)-S(=O)(=O)-OH$
Disulfuric	$H_2S_2O_7$	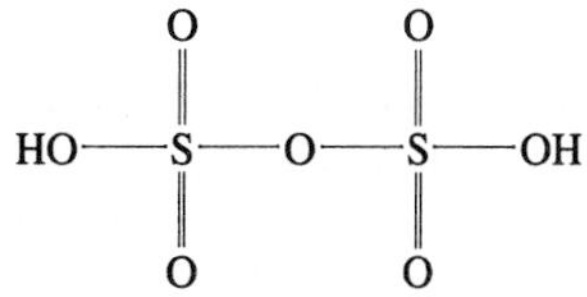 $HO-S(=O)(=O)-O-S(=O)(=O)-OH$
Acids Containing Three or More Sulfur Atoms		
Polythionic	$H_2S_{n+2}O_6$	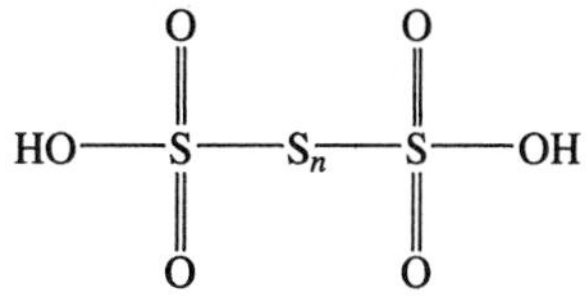 $HO-S(=O)(=O)-S_n-S(=O)(=O)-OH$
Peroxo Acids		
Peroxomonosulfuric	H_2SO_5	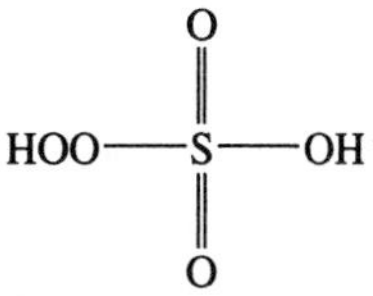 $HOO-S(=O)(=O)-OH$
Peroxodisulfuric	$H_2S_2O_8$	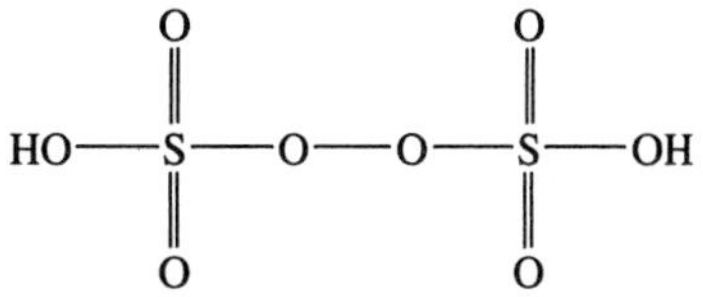 $HO-S(=O)(=O)-O-O-S(=O)(=O)-OH$

[a]In most cases the structure given is inferred from the structure of anions in salts of the acids.
[b]Free acid unknown.

Sulfurous and Disulfurous Acids

Neither of these, H_2SO_3 and $H_2S_2O_5$ exists as such in normal chemistry, although there is evidence for $(HO)_2SO$ in the gas phase. The HSO_3^-, SO_3^{2-}, $HS_2O_5^-$, and $S_2O_5^{2-}$ ions, however, are well known.

Sulfur dioxide is quite soluble in water; such solutions, which possess acidic properties, have long been referred to as solutions of sulfurous acid (H_2SO_3). However, H_2SO_3 either is not present or is present only in infinitesimal quantities in such solutions. The so-called hydrate $H_2SO_3 \cdot \sim 6H_2O$ is really a gas hydrate $SO_2 \cdot \sim 7H_2O$. The equilibria in aqueous solutions of SO_2 are best represented as

$$SO_2 + xH_2O = SO_2 \cdot xH_2O (\text{hydrated } SO_2)$$

$$SO_2 \cdot xH_2O = H_2SO_3 \qquad K \lll 1$$

$$SO_2 \cdot xH_2O = HSO_3^-(aq) + H_3O^+ + (x-2)H_2O$$

and the first acid dissociation constant for "sulfurous acid" is properly defined as follows:

$$K_1 = \frac{[HSO_3^-][H^+]}{[\text{total dissolved } SO_2] - [HSO_3^-] - [SO_3^{2-}]} = 1.3 \times 10^{-2}$$

The lighter alkali ions give sulfites or bisulfites (see later), and it appears that larger ions such as Rb^+, Cs^+, or R_4N^+ are necessary to stabilize HSO_3^-. In the solid state both ions are pyramidal and the H atom in HSO_3^- appears to be on the sulfur atom. In solution ^{17}O nmr shows that the $SO_2(OH)$ form also exists and the $[SO_2(OH)^-]/[HSO_3^-]$ ratio is ~ 5 at 20°C.

Heating solid bisulfites or passing SO_2 into their aqueous solutions affords disulfites:

$$2MHSO_3 \xrightleftharpoons{\text{heat}} M_2S_2O_5 + H_2O$$

$$HSO_3^-(aq) + SO_2 = HS_2O_5^-(aq)$$

Although condensation processes of this type usually result in an oxobridged diacid or anion, as in disulfuric acid ($H_2S_2O_7$), the $S_2O_5^{2-}$ ion is unsymmetrical and has an S—S bond, as shown in Table 12-8. Some important reactions of sulfites are shown in Fig. 12-10. The reducing character of SO_2 and sulfites is expressed quantitatively by the following potentials:

$$SO_4^{2-} + 4H^+ + (x-2)H_2O + 2e = SO_2 \cdot xH_2O \qquad E^0 = 0.17 \text{ V}$$

$$SO_4^{2-} + H_2O + 2e = SO_3^{2-} + 2OH^- \qquad E^0 = -0.93 \text{ V}$$

The diesters of sulfurous acid exist in tautomeric forms, $(RO)_2SO$ and RSO_2OR.

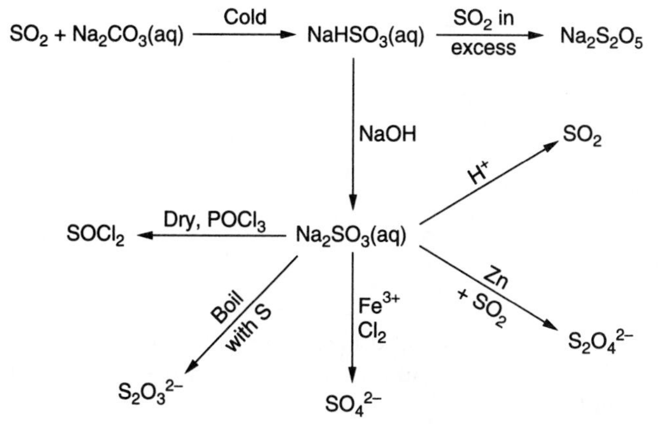

$$SO_2 + Na_2CO_3(aq) \xrightarrow{\text{Cold}} NaHSO_3(aq) \xrightarrow[\text{excess}]{SO_2 \text{ in}} Na_2S_2O_5$$

with $NaOH$ leading to $Na_2SO_3(aq)$, and branches:

- $\xrightarrow{H^+} SO_2$
- $\xrightarrow{\text{Dry, POCl}_3} SOCl_2$
- $\xrightarrow[\text{with S}]{\text{Boil}} S_2O_3^{2-}$
- $\xrightarrow[Cl_2]{Fe^{3+}} SO_4^{2-}$
- $\xrightarrow[+ SO_2]{Zn} S_2O_4^{2-}$

Figure 12-10 Some reactions of sulfites.

Thiosulfuric Acid ($H_2S_2O_3$)

The thiosulfate ion ($S_2O_3^{2-}$) has a structure comparable to that of the SO_4^{2-} ion with one oxygen atom replaced by a sulfur atom; the S—S and S—O distances [2.013(3) and 1.468(4) Å, respectively] imply that little S—S π bonding and much S—O π bonding are present. Solutions of hydrated sodium thiosulfate, $Na_2S_2O_3 \cdot 5H_2O$ (hypo), are used as a "fixer" in photography because it can dissolve unphotolyzed AgBr from the emulsion by complexing the Ag^+ ion:

$$AgBr(s) + 3S_2O_3^{2-}(aq) = Ag(S_2O_3)_3^{5-} + Br^-$$

The moderate reducing character of $S_2O_3^{2-}$ provides a basis for analytical determination of iodine:

$$2S_2O_3^{2-} + I_2 \longrightarrow S_4O_6^{2-} + 2I^-$$

Thiosulfuric acid cannot be generated in aqueous solution by acidifying solutions of thiosulfates because it rapidly decomposes to H_2SO_4 and a mixture of S, H_2S, H_2S_n, and SO_2. It can, however, be prepared by nonaqueous reactions such as

$$Na_2S_2O_3 + 2HCl \xrightarrow[-78°C]{Et_2O} H_2S_2O_3 \cdot 2Et_2O + 2NaCl$$

$$HSO_3Cl + H_2S \xrightarrow[<10°C]{\text{no solvent}} H_2S_2O_3 + HCl$$

Dithionite Ion ($S_2O_4^{2-}$)

The reduction of sulfites in aqueous solutions containing an excess of SO_2, usually by zinc dust, or of SO_3 by formate in aqueous methanol, gives the dithionite ion $S_2O_4^{2-}$. Solutions of this ion are not very stable and decompose according to the stoichiometry:

$$2S_2O_4^{2-} + H_2O \longrightarrow S_2O_3^{2-} + 2HSO_3^-$$

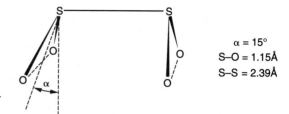

$\alpha = 15°$
S–O = 1.15Å
S–S = 2.39Å

Figure 12-11   Structure of the dithionite ion ($S_2O_4^{2-}$) in $Na_2S_2O_4$.

In acid solution decomposition is extremely rapid and $H_2S_2O_4$ is unknown under any conditions. The Zn and Na salts are commonly used as powerful and rapid reducing agents in alkaline solution:

$$2SO_3^{2-} + 2H_2O + 2e = 4OH^- + S_2O_4^{2-} \qquad E^0 = -1.12 \text{ V}$$

Because of a reversible dissociation to afford the SO_2^- ion radical,[29]

$$S_2O_4^{2-} \rightleftharpoons 2SO_2^-$$

dithionite solutions show strong esr signals and give kinetic rate laws that are half-order in $S_2O_4^{2-}$. When aqueous $Na_2S_2O_4$ is mixed with 2-anthraquinonesulfonate as a catalyst (Fieser's solution), the solution efficiently removes dioxygen from gases.

The structure of the dithionite ion (Fig. 12-11) has several remarkable features. The oxygen atoms, which must bear considerable negative charge, are closely juxtaposed by the eclipsed (C_{2v}) configuration and by the small value of the angle α, which would be 35° for sp^3 tetrahedral hybridization at the sulfur atom. Second, the S—S distance is much longer than S—S bonds in disulfides, polysulfides, and so on, which are in the range 2.00 to 2.05 Å. The long and weak bond is consistent with dissociation to SO_2^-.

Dithionic Acid ($H_2S_2O_6$)

This cannot be regarded as one of the polythionic acids since it contains no sulfur atom bound only to other sulfur atoms, as in $H_2S_3O_6$ and its $H_2S_nO_6$ homologues. The dithionate ion has a D_{3d} O_3SSO_3 structure with approximately tetrahedral bond angles about each sulfur atom, and the S—O bond length (1.45 Å; cf. 1.44 Å in SO_4^{2-}) again suggests considerable double-bond character.

Dithionate is usually obtained by oxidation of sulfite or SO_2 solutions with manganese(IV) oxide:

$$MnO_2 + 2SO_3^{2-} + 4H^+ = Mn^{2+} + S_2O_6^{2-} + 2H_2O$$

Other oxo acids of sulfur that are formed as by-products are precipitated with barium hydroxide, and $BaS_2O_6\cdot2H_2O$ is then crystallized. Treatment of aqueous

[29]S. P. Mishra and M. R. C. Symons, *J. Chem. Soc. Dalton Trans.* **1994,** 1271.

solutions of this with sulfuric acid gives solutions of the free acid, which may be used to prepare other salts by neutralization of the appropriate bases. Dithionic acid is a moderately stable strong acid that decomposes slowly in concentrated solutions and when warmed. The ion itself is stable, and solutions of its salts may be boiled without decomposition. Although it contains sulfur in an intermediate oxidation state, it resists most oxidizing and reducing agents, presumably for kinetic reasons.

Polythionates

These anions, $S_nO_6^{2-}$, with n up to 22 or higher, can be separated chromatographically; only lower members are well characterized. They are normally named according to the total number of sulfur atoms, namely, trithionate, $S_3O_6^{2-}$, and so on. They are occasionally called *sulfane-disulfonates*, for example, disulfanedisulfonate for $S_4O_6^{2-}$. The free acids are unstable as are also the $HS_nO_6^-$ ions.

Polythionates are obtained by reduction of thiosulfate solutions with SO_2 in the presence of As_2O_3 and by the reaction of H_2S with an aqueous solution of SO_2, which produces a solution called Wackenroder's liquid. A general reaction reported to produce polythionates up to very great chain lengths is

$$6S_2O_3{}^{2-} + (2n-9)H_2S + (n-3)SO_2 \longrightarrow 3S_nO_6{}^{2-} + (2n-12)H_2O + 6OH^-$$

Many polythionates are best made by selective preparations, for example, by the action of H_2O_2 on cold saturated sodium thiosulfate:

$$2S_2O_3{}^{2-} + 4H_2O_2 \longrightarrow S_2O_6{}^{2-} + SO_4{}^{2-} + 4H_2O$$

Tetrathionates are obtained by treatment of thiosulfates with iodine as in the reaction used in the volumetric determination of iodine mentioned earlier.

Various anions containing Se and Te are also known, such as $Se_nS_2O_6^{2-}$ ($2 \leq n \leq 6$), $O_3S_2SeS_2O_3^{2-}$, and $O_3S_2TeS_2O_3^{2-}$.

There are also some $O_2S(R)S_nS(R)O_2$ and $O_2S(R)Se_nS(R)O_2$ molecules, with $n = 1, 2$ and R = alkyl or aryl that are structurally related.

Peroxo Acids

None are known for Se or Te, but those of sulfur, $H_2S_2O_8$, and H_2SO_5, are well known and of importance.

Peroxodisulfuric acid can be obtained from its NH_4^+ or Na^+ salts, which can be crystallized from solutions after electrolysis of the corresponding sulfates at low temperatures and high current densities. The $S_2O_8^{2-}$ ion has the structure $O_3S-O-O-SO_3$, with approximately tetrahedral angles about each S atom.

The peroxodisulfate ion is one of the most powerful and useful of oxidizing agents:

$$S_2O_8{}^{2-} + 2e = 2SO_4{}^{2-} \qquad E^0 = 2.01 \text{ V}$$

However, the reactions may be complicated mechanistically and in some of them there is evidence for the formation of the radical ion SO_4^- by one-electron reduction:

$$S_2O_8^{2-} + e = SO_4^{2-} + SO_4^-$$

Oxidations by $S_2O_8^{2-}$ often proceed slowly but become more rapid in the presence of the silver ion. The mechanism is not entirely clear, but it appears that a weak 1:1 complex is first formed between Ag^+ and $S_2O_8^{2-}$, from which the rapidly reacting oxidizing species, Ag^{II}, arises.

Peroxomonosulfuric acid (Caro's acid) is obtained by hydrolysis of peroxodisulfuric acid:

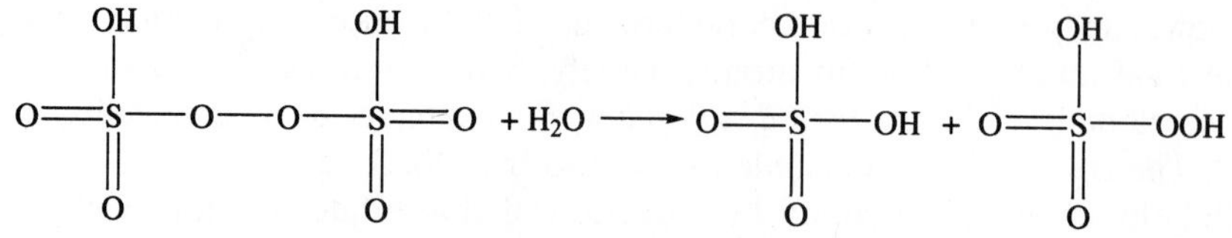

and also by the action of concentrated hydrogen peroxide on sulfuric acid or chlorosulfuric acid:

$$H_2O_2 + H_2SO_4 \longrightarrow HOOSO_2OH + H_2O$$

$$H_2O_2 + HSO_3Cl \longrightarrow HOOSO_2OH + HCl$$

Salts such as $KHSO_5$ and $K_5(HSO_5)_2(HSO_4)(SO_4)$ can be prepared and crystallographic study has shown that the HSO_5^- ion has the structure (12-XXVII), with three S—O distances of ~1.44 Å and one (to O_2H) of 1.63 Å.

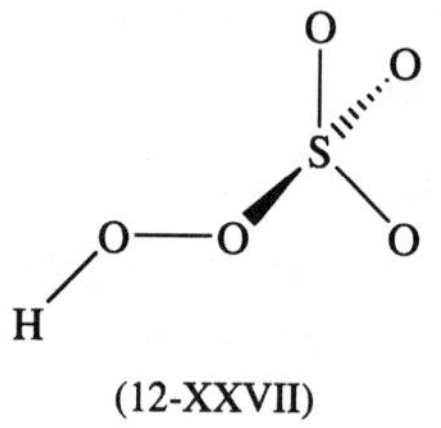

(12-XXVII)

Polysulfates

By thermal dehydration of bisulfates it is possible to make compounds containing the disulfate ion ($S_2O_7^{2-}$) in which two tetrahedra share an oxygen atom. The longer chain $S_3O_{10}^{2-}$, $S_4O_{13}^{2-}$ and $S_5O_{16}^{2-}$ ions are also known.

Halooxo Acids

The ones of principal importance are derived from sulfur.

Fluorosulfurous acid exists only as salts, which are formed by the action of SO_2 on alkali fluorides, for example,

$$KF + SO_2 \rightleftharpoons KSO_2F$$

The salts have a measurable dissociation pressure at normal temperatures but are useful and convenient mild fluorinating agents, for example,

$$(PNCl_2)_3 + 6KSO_2F \longrightarrow (PNF_2)_3 + 6KCl + 6SO_2$$

$$C_6H_5COCl + KSO_2F \longrightarrow C_6H_5COF + KCl + SO_2$$

Halogenosulfuric acids, FSO_3H (Section 3-12), $ClSO_3H$ and $BrSO_3H$, can be regarded as derived from SO_2X_2 by replacement of one halogen by OH.

Chlorosulfuric acid, a colorless fuming liquid, explosively hydrolyzed by water, forms no salts. It is made by treating SO_3 with dry HCl. Its main use is for the sulfonation of organic compounds.

Bromosulfuric acid, prepared from HBr and SO_3 in liquid SO_2 at $-35°C$, decomposes at its melting point ($8°C$) into Br_2, SO_2, and H_2SO_4.

12-14 Oxo Acids of Selenium and Tellurium

Selenous and Tellurous Acids and Salts

Selenium dioxide dissolves in water to give solutions of selenous acid, $(HO)_2SeO$, while H_2TeO_3 (structure uncertain for both solution and solid) is best obtained by hydrolysis of a tetrahalide. Both are weak acids ($K_1 \approx 10^{-3}$ and $K_2 \approx 10^{-8}$). Solid selenous acid has layers of pyramidal SeO_3 groups linked by hydrogen bonds. It is a moderately strong oxidizing agent:

$$H_2SeO_3 + 4H^+ + 4e^- = Se + 3H_2O \qquad E^0 = 0.74 \text{ V}$$

Salts of the $HSeO_3^-$, $HTeO_3^-$, SeO_3^{2-}, and TeO_3^{2-} ions can be obtained by neutralization of the acids, but the SeO_3^{2-} and TeO_3^{2-} salts are often prepared by heating a mixture of the metal oxide with SeO_2 or TeO_2.

Selenic Acid (H_2SeO_4)

Vigorous oxidation of selenites or fusion of selenium with potassium nitrate gives selenic acid (or its salts). The free acid forms colorless crystals (mp 57°C). It is very similar to sulfuric acid in its formation of hydrates, in acid strength, and in the properties of its salts, most of which are isomorphous with the corresponding sulfates and hydrogen sulfates. It differs mainly in being less stable, and it evolves oxygen when heated above $\sim$200°C. It is a strong, though usually not kinetically fast, oxidizing agent:

$$SeO_4^{2-} + 4H^+ + 2e = H_2SeO_3 + H_2O \qquad E^0 = 1.15 \text{ V}$$

Telluric Acid $Te(OH)_6$

This is very different from sulfuric and selenic acids. It consists of hydrogen-bonded octahedral molecules $Te(OH)_6$ in both its cubic and monoclinic crystalline forms. The acid or its salts may be prepared by oxidation of tellurium or TeO_2 by H_2O_2,

Na_2O_2, CrO_3, or other powerful oxidizing agents. It is a moderately strong, but, like selenic acid, kinetically slow oxidizing agent ($E^0 = 1.02$ V). It is a very weak dibasic acid with $K_1 = 10^{-7}$. Tellurates of various stoichiometries are known and most, if not all, contain TeO_6 octahedra. Examples are $K[TeO(OH)_5] \cdot H_2O$, $Ag_2[TeO_2(OH)_4]$, and Hg_3TeO_6. Tellurates such as $BaTeO_4$ that can be made by heating TeO_2 and metal oxides are not isostructural with sulfates. Magnesium tellurate ($MgTeO_4$) is isostructural with $MgWO_4$ and contains TeO_6 octahedra.

SeF$_5$OH, TeF$_5$OH, and Their Anions

These acids are obtainable by the reactions:

$$SeO_2F_2 + 3HF \longrightarrow SeF_5OH + H_2O$$

$$Te(OH)_6 + 5HSO_3F \longrightarrow TeF_5OH + 5H_2SO_4$$

They can be used to prepare $Xe(OEF_5)_2$ compounds. The tellurium compound has been used to prepare large, weakly coordinating anions[30] such as $[M(OTeF_5)_4]^{2-}$, M = Co, Ni, Cu, Zn, Pd.

12-15 Other Se and Te Compounds

In keeping with the usual group trend, selenium and even more so tellurium show cationic behavior, which manifests itself in the formation of complexes and organo derivatives. We can deal here only with a few leading examples.

Organotellurium compounds are formed in the II, IV, and VI oxidation states. Alkylhalide compounds add to TeR_2 to give Te(IV) species, for example,

$$Et_2Te + EtBr \longrightarrow Et_3TeBr$$

These R_3TeX compounds may have either tetrameric or dimeric structures, as shown in Fig. 12-12. The $Te \cdots X \cdots Te$ distances are quite long and are to be regarded as secondary, essentially ionic interactions. In compounds like $Me_3Te^+BF_4^-$ the ionic limit is approached even more closely.

Other Te^{IV} organo compounds are of the R_2TeX_2[31] and $RTeX_3$ types. These may associate through secondary μ-X bonds to dimers. By reaction of Ph_3TeF with XeF_2 it is oxidized to Ph_3TeF_3, which has a meridional octahedral structure.

Sulfur as well as Se and Te form many compounds formally containing EX_3^+ ions, with X = F, Cl, Br, or I. In all of these, however, there are strong secondary interactions between these "cations" and the accompanying anions so that the coordination numbers of the element E, counting both primary and secondary bonds, reaches 7, 8, and even 9.

Tellurium forms other, more complex halogeno cations such as $[Te_6I_2]^{2+}$ and the infinite polymeric[32] one, $[Te_{15}X_2^{2+}]_\infty$.

[30]S. H. Strauss, *Chem. Rev.* **1993**, *93*, 927.
[31]J. E. Drake *et al., Inorg. Chem.* **1994**, *33*, 6154.
[32]J. Beck *et al., Z. anorg. allg. Chem.* **1996**, *622*, 473.

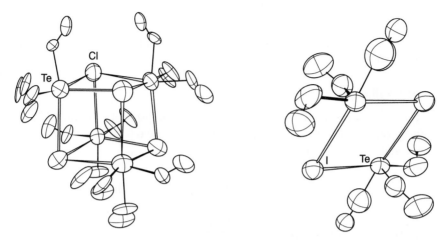

Figure 12-12 The structure of [Et$_3$TeCl]$_4$ (left) and [Et$_3$TeI]$_2$ (right). The solid Te—C bonds are normal covalent bonds while the open Te—Cl and Te—I bonds represent longer and weaker interactions.

Octahedral EX$_6^{2-}$ ions of Se^{IV} and Te^{IV} with Cl$^-$ and Br$^-$ have long been known and have been of interest because of their octahedral structures despite the presence of a valence shell electron pair. The stereochemical inactivity of this pair of electrons indicates that they occupy the valence shell ns orbital. These EX$_6^{2-}$ complexes are generally obtained by saturating aqueous solutions of EO$_2$ in KX with HX, and adding appropriate large cations. However, the equilibria in these solutions are very complex and species such as SeOBr$_3^-$ and TeCl$_4$(OH)$^-$ also form. An interesting case where the lone pair is stereochemically active is the ψ-octahedral TeF$_5^-$ ion in BaTe$_2$F$_{10}$.[33]

Tellurium forms a number of complex halides in the II as well as the IV state, some of the best known being those with thiourea (tu) or substituted thioureas as ligands. The red Te^{IV} compounds are made by treating TeO$_2$ in concentrated hydrochloric acid solution with, for example, tetramethylthiourea (Me$_4$tu):

$$TeO_2 + 4HCl + 2Me_4tu \longrightarrow trans\text{-}Te(Me_4tu)_2Cl_4 + 2H_2O$$

The structures of these compounds are octahedral with trans sulfur atoms showing no evidence (cf. TeX$_6^{2-}$) of stereochemical influence of the lone pair.

The Me$_4$tu ligand can further act as a reducing agent in methanolic 4 M HCl.

$$Te(Me_4tu)_2Cl_4 \xrightarrow{\text{heat}} Te^{II}(Me_4tu)Cl_2 + (Me_4tu)^{2+} + 2Cl^-$$

$$\underset{\text{Red}}{Te^{II}(Me_4tu)Cl_2} + Me_4tu \xrightarrow[\text{heat}]{\text{MeOH}} \underset{\text{Yellow}}{Te^{II}(Me_4tu)_2Cl_2}$$

In TeCl$_2$(ethylenethiourea)$_4$·2H$_2$O there is a [TeS$_4$]$^{2+}$ square.

[33]B. Frit *et al.*, *J. Fluorine Chem.* **1996**, *77*, 15.

12-16 Chalcogens and Chalcogen Compounds as Ligands[34]

We deal primarily with sulfur ligands, but there are many Se or Te compounds that behave similarly.

Hydrogen Sulfide

The action of H_2S on metal species commonly gives insoluble sulfides. However, some complexes of H_2S are known, although these may be readily oxidized to sulfur or deprotonated to SH^- complexes.

Thus, in the presence of Eu^{2+}, to keep the system reduced, we have

$$[am_5Ru(OH_2)]^{2+} + H_2S = [am_5Ru(SH_2)]^{2+}   K = 1.5 \times 10^{-3} M^{-1}$$

Other examples are $(CO)_5WSH_2$ and $[(CO)_5WSH]^-$, $Mo(SH)(S_2CNEt_2)_3$ and the bridged SH species $(Cp^*_2Mo)_2(\mu\text{-}S)(\mu\text{-}SH)$.

Sulfide (S_n^{2-}) Ions

There are large numbers of compounds with the simple sulfide ion (S^{2-}) as a ligand comparable to O^{2-} that are mononuclear like $S{=}WCl_3$ or bi- or polynuclear with sulfur bridges. These and related species derived from polysulfide ions (S_n^{2-}) have been intensively studied in part because sulfur bridged species occur in Nature in ferredoxins and related compounds.

Although sulfide species are commonly made from sulfur, alkali sulfides, or polysulfides, a useful procedure for converting oxoanions or alkoxides to sulfides is by use of trimethylsilyl sulfide, for example,

$$Mo_2O_7^{2-} \xrightarrow{(Me_3Si)_2S} [MoO_{(3-x)}S_x(OSiMe_3)]^- + [MoS_3(OSiMe_3)]^-$$

$$Nb(OEt)_5 \xrightarrow{(Me_3Si)_2S} [Nb_6S_{17}]^{4-}$$

There are numerous compounds in which one or more terminal S ligands (less commonly Se or Te ligands) are present. Examples are the ME_4^{n-} ions (e.g., MoS_4^{2-}), the $WOTe_3^{2-}$ ion,[35] the $NbSe_3(SCMe_3)^{2-}$ ion[36] and many halosulfides, such as $NbSBr_3$, $MoSBr_3$, and $ReSCl_4$. Monosulfide bridges of the following types are known:

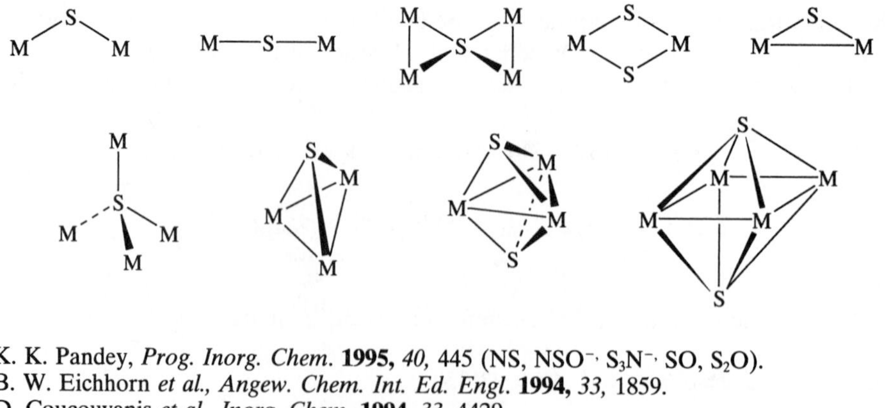

[34]K. K. Pandey, *Prog. Inorg. Chem.* **1995**, *40*, 445 (NS, NSO$^-$· S$_3$N$^-$· SO, S$_2$O).
[35]B. W. Eichhorn *et al., Angew. Chem. Int. Ed. Engl.* **1994**, *33*, 1859.
[36]D. Coucouvanis *et al., Inorg. Chem.* **1994**, *33*, 4429.

The linear bridges like linear MOM bridges have π-bonding interactions between metal $d\pi$ and sulfur $d\pi + p\pi$ orbitals. Planar units of the type M_3O do not appear to exist for sulfur.

Of special importance in ferredoxins and model compounds for ferredoxins are M_4S_4 units, which have a cubane or distorted cubane structure (12-XXVIII).

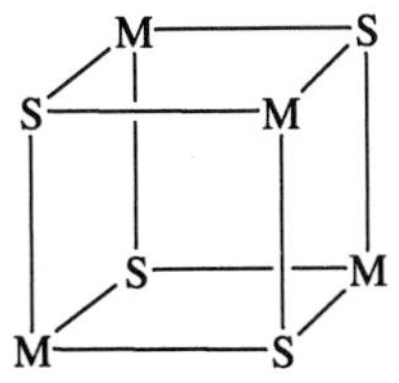

(12-XXVIII)

Examples of such compounds are $[(\eta\text{-Cp})MoS]_4$ and $[(RS)_4Fe_4S_4]^{2-}$. There are also compounds of S, Se, and Te, formally in high oxidation states, that have cumulated multiple bonds.

Examples are $Cp(CO)_2Cr\equiv S\equiv Cr(CO)_2Cp$ made by interaction of $[CpCr(CO)_3]^-$ with $N_3S_3Cl_3$, and $(triphos)Co=S=Co(triphos)$.

Polysulfide Ions

The disulfide ion (S_2^{2-}) can, like O_2^{2-}, coordinate either side-on or as a bridge:

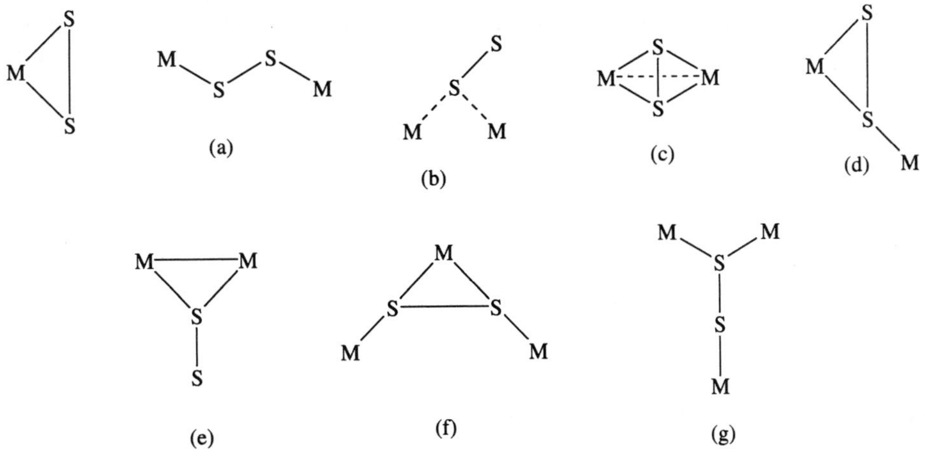

There are relatively few compounds with $\eta^2\text{-}S_2^{2-}$ but two examples are $Os(S_2)(CO)_2(PPh_3)_2$ and $[(C_2O_4)(O)Mo(S_2)_2]^{2-}$.

Bridges of type (a) are very common.[37] The compound $(Cp^*)_2Cr_2S_5$ is remarkable in having three types of bridge: bent μ-S, as well as types (c) and (e). The higher sulfane anions S_n^{2-} commonly give puckered MS_n rings, $n = 4$ to 6 or higher.[38]

[37]M. Millar *et al.*, *Inorg. Chem.* **1995**, *34*, 1981.
[38]T. B. Rauchfuss *et al.*, *J. Am. Chem. Soc.* **1990**, *112*, 6385.

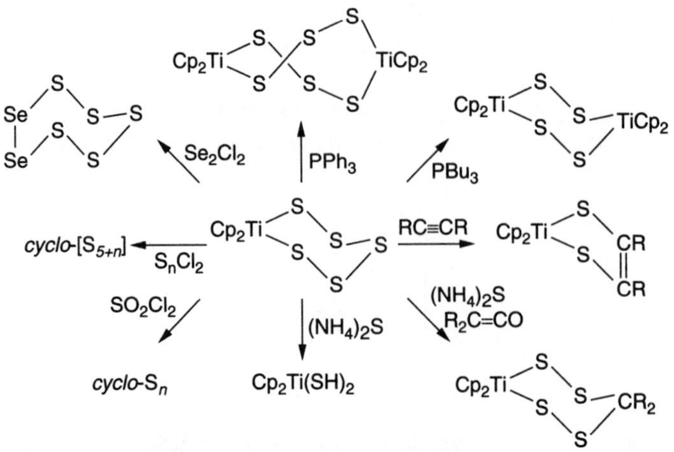

Figure 12-13   Reactions of the sulfur ring of $(\eta^5\text{-}C_5H_5)_2TiS_5$.

Normally made by interaction of ions or halides with alkali polysulfides, they can be formed from sulfur, as in the reaction:

$$Cp_2WH_2 + 5S = Cp_2WS_4 + H_2S$$

Large numbers of metal complex anions containing S_2^{2-}, S_3^{2-}, and S_n^{2-} have been characterized. Examples are $[Pt(S_5)_3]^{2-}$, which provides an unusual example of a chiral, purely inorganic, molecule, $[S=Mo(S_4)_2]^{2-}$, and $[AuS_9]^-$, which has a 10-membered ring. Sulfur moieties such as S_4^{2-} or S_5^{2-} can also bridge two metals as in (12-XXIX) and can bind in a η^3 mode *via* donation from a central S atom (12-XXX). There exists a tellurium analogue to this structure, with M = Hg.[39] Sulfur ring systems can undergo various reactions as shown in Fig. 12-13.

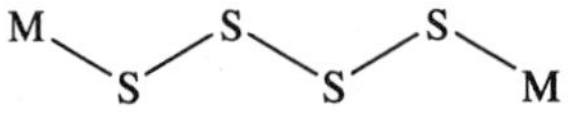

(12-XXIX) 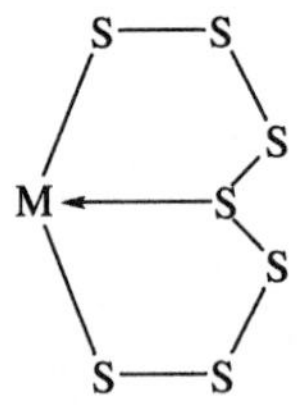   (12-XXX)

Polyselenide and Polytelluride Ions

Although these were neglected until recently, the last decade has seen a rapid development of chemistry having these E_n^{2-} species as ligands.[40] The polyselenides are more tolerant of higher metal oxidation states than the S_n^{2-} ions and hence capable of an even more extensive chemistry. The Se_4^{2-} ligand is predominant as a chelating ligand and gives, for example, the $[Pt(Se_4)_3]^{2-}$ ion, comparable to the

[39]J. A. Ibers *et al., Inorg. Chem.* **1993,** *32,* 3201; K. Dehnicke *et al., Z. anorg. allg. Chem.* **1993,** *619,* 500.
[40]M. G. Kanatzidis and S. Huang, *Coord. Chem. Rev.* **1994,** *130,* 509.

$[Pt(S_5)_3]^{2-}$ ion. With tin there is a complex having a mixture of ligands, *viz.*, $[Sn(S_4)_2(Se_6)]^{2-}$.[41]

A very interesting recent observation is that a $W(Te)_2$ may be converted to a $W(\eta^2\text{-}Te_2)$ by changing the other ligand about the W atom.[42]

Thiolates

Many studies have been made on thiolates (RS^-) because they have provided models for ferredoxin and related natural sulfur compounds. Although unidentate groups are well characterized, bridge groups are common and for μ_2 both syn and anti isomers can be obtained; syn-anti isomerization proceeds *via* a bridge opening mechanism. The main types are

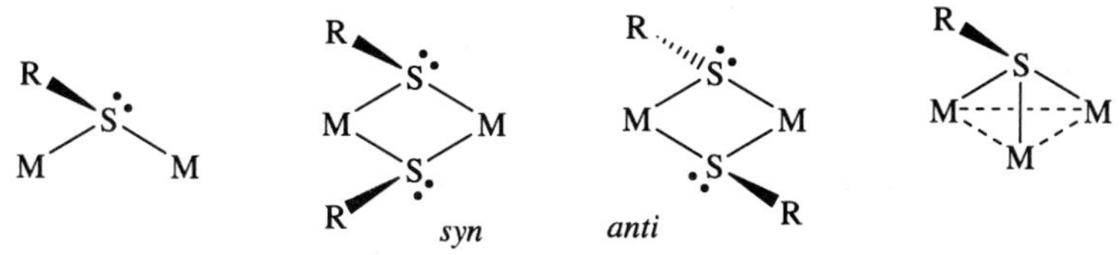

and μ_2 species with metal bonds are also known. Mononuclear thiolates, $[M(SPh)_4]^{2-}$, are formed by ions such as Zn^{II}, Cd^{II}, Mn^{II}, Fe^{II}, Co^{II}, Ni^{II}, and Hg^{II}. Often the use of highly hindered ArS^- groups enhances their chemical stability.[43]

Polynuclear compounds commonly have an $M_4(\mu\text{-}SR)_6$ adamantane-like cage structure. It should be noted that selenolate, RSe^-, and tellurolate, RTe^-, ligands are also well known.[44]

Thiolates are usually made by direct reactions such as that leading to 12-XXXI

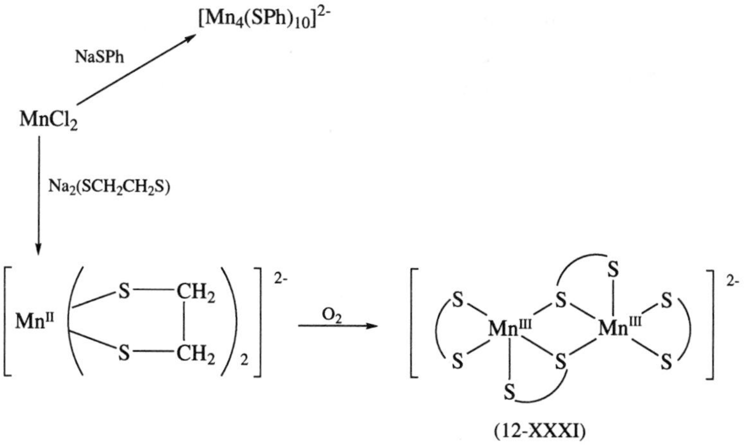

(12-XXXI)

Some species, however, can be made in aqueous solution from Fe^{2+}, S, $^-SCH_2CH_2OH$, and so on.

The *ethane-l,2-dithiolates* have a variety of structures ranging from chelated octahedra as in $[Ti(S_2C_2H_4)_3]^{2-}$, tetrahedra in $[Co(S_2C_2H_4)_2]^{2-}$, or planar in

[41]M. G. Kanatzidis et al., Inorg. Chem. **1993**, 32, 2453.
[42]D. Rabinovich and G. Parkin, J. Am. Chem. Soc. **1993**, 115, 9822.
[43]S. A. Koch and M. Millar et al., Inorg. Chim. Acta **1993**, 205, 9.
[44]J. Arnold, Prog. Inorg. Chem. **1995**, 43, 353.

$[Cr(S_2C_2H_4)_2]^{2-}$, to various types of bridges, one example being the manganese(III) complex (12-XXXI). Thiols such as $HSCH_2CO_2Et$ can give more complicated species such as $Ni_8(SCH_2CO_2Et)_{16}$.

Finally, thiolates (RSS−) can give species with the groupings:

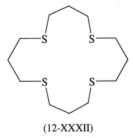

Disulfides (RSSR)

These are capable of serving, intact, as ligands[45] but they frequently undergo cleavage and give oxidative addition products, *viz.*,

$$Mo\!\equiv\!Mo + EtSSEt \longrightarrow Mo\!\equiv\!Mo$$

Thioethers

These can be simple such as Me_2S or tetrahydrothiophene (THT), or more complex such as the macrocyclic analogues of the crown ethers. There are a few selenium analogues.[46] Like their oxygen and nitrogen analogues, they are commonly abbreviated as illustrated by either 16S4 or [16]aneS$_4$ for 1,5,9,13-tetrathiacyclohexadecane 12-XXXII

(12-XXXII)

They are useful ligands for transition metals in a variety of circumstances.[47] Recently an elegant catalytic method of synthesizing the 12S3, 16S4, and 24S6 compounds by oligomerization of thietane $SCH_2CH_2CH_2$, has been described.[48]

[45]D. Carrillo, *Coord. Chem. Rev.* **1992**, *119*, 137.
[46]G. Reid *et al.*, *Inorg. Chem.* **1995**, *34*, 651.
[47]S. R. Cooper, Ed. *Crown Compounds: Toward Future Applications*, VCH Publishers, New York, 1992, Chap. 15.
[48]R. D. Adams and S. B. Falloon, *Organometallics* **1995**, *14*, 1748.

Sulfur Oxides

Lower Oxides. Despite the instability of SO, S_2O, and S_2O_2, all can be trapped as ligands. Sulfur monoxide gives bent MSO groups and in addition can have μ_2, μ,η^2, and μ_3 bridging modes.

Some synthetic procedures for these complexes are the following

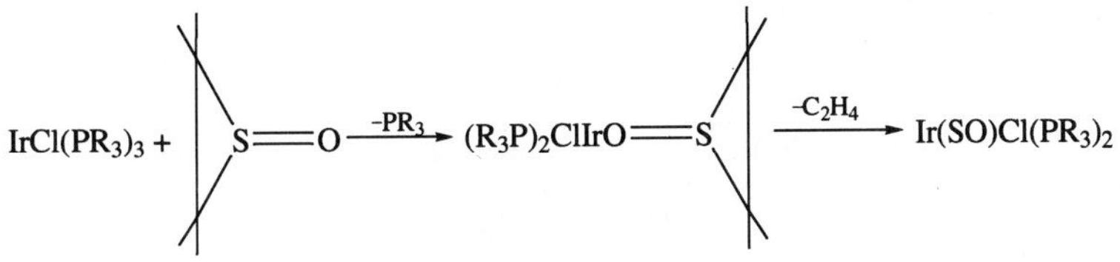

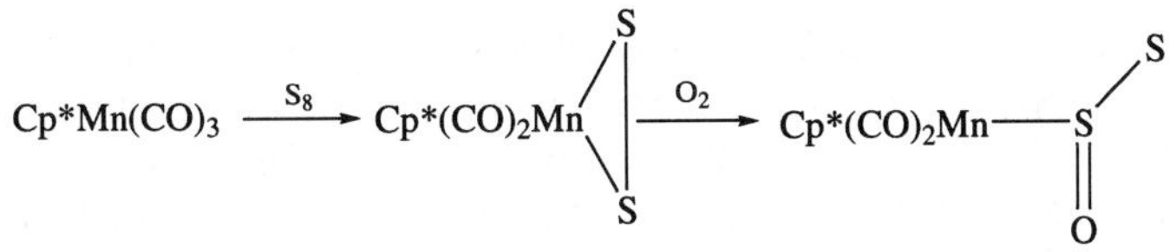

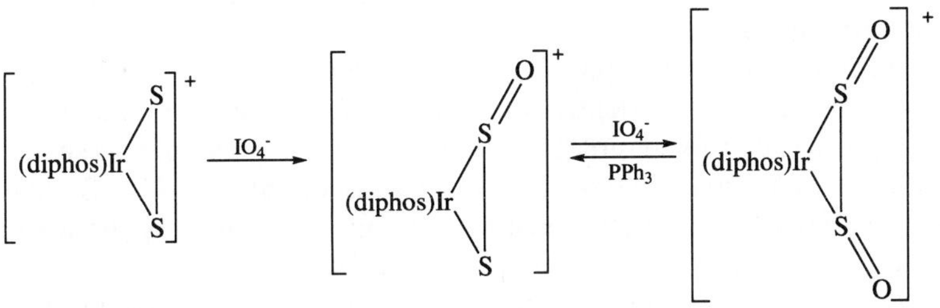

Sulfur Dioxide[49]. The dioxide can be bound in several ways:

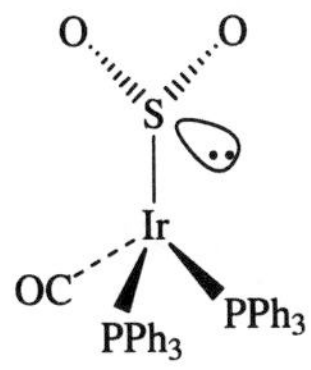

| Pyramidal M—SO₂ | Planar M—SO₂ | η², side bound | 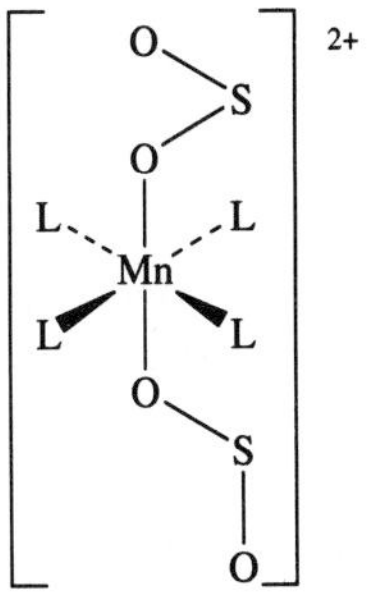 O-bound (L = OPPh₃) |

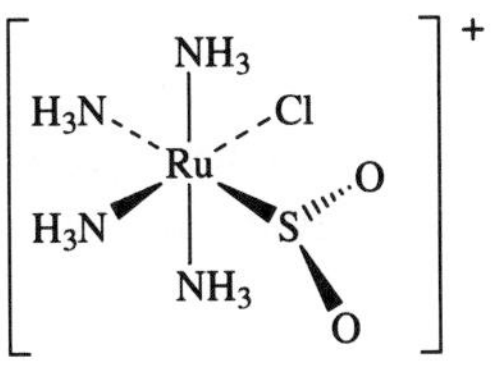

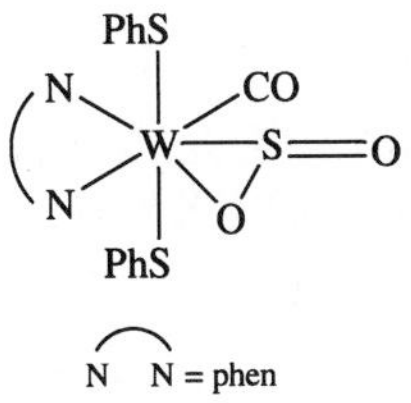

Examples of bridged species are

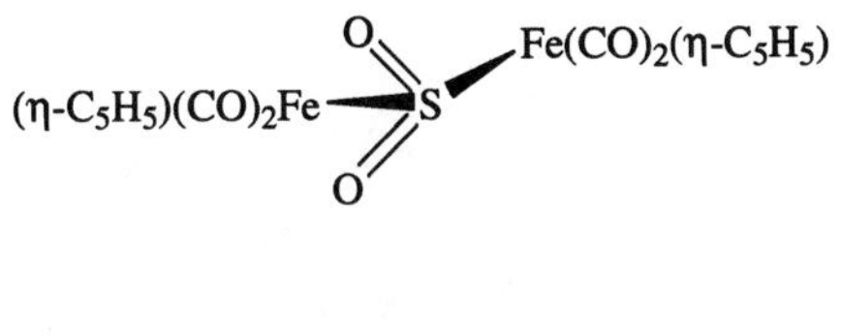

[49]G. J. Kubas *et al., Acc. Chem. Res.* **1994,** *27,* 183; C. D. Hoff *et al., J. Am. Chem. Soc.* **1994,** *116,* 9747.

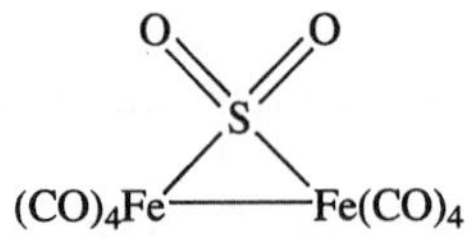

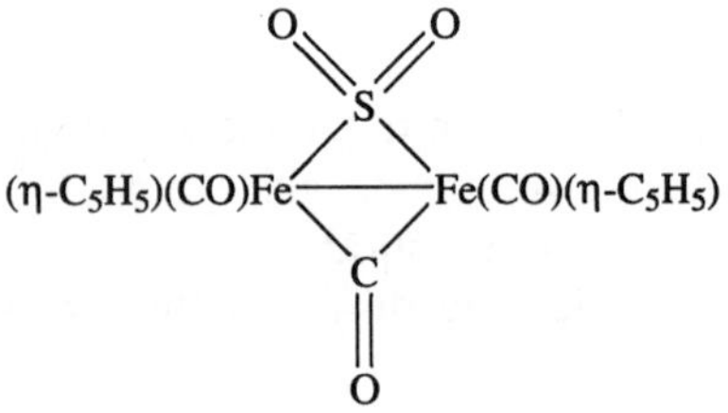

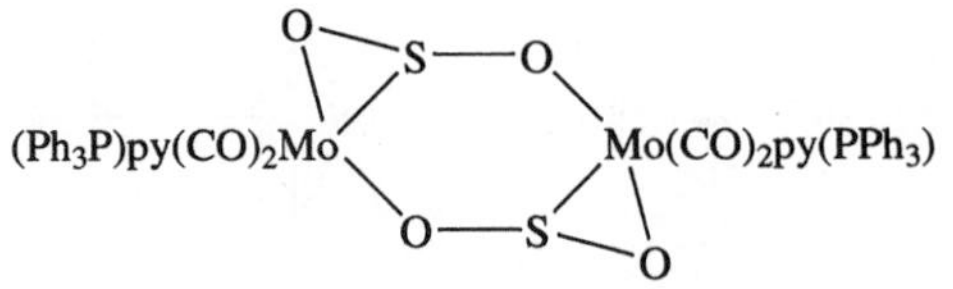

The η^1 pyramidal geometry is found only when the ML_n unit acts as a σ base while η^2 bonding is favored by π donation from ML_n. An unusual η^2 complex is the above tungsten compound, which has the interesting property of going on to lose CO and sulfur thus leaving $phen(PhS)_2WO_2$. A rare example of O-bonded SO_2 is *trans*-$[Mn(OPPh_3)_4(SO_2)_2]I_2$, but O-bonded complexes of main group metals, for example, F_5SbOSO are known (see later).

In the compounds such as $[am_4ClRuSO_2]^+$ and $CpRh(C_2H_4)SO_2$ that have planar $M-SO_2$ groups, there is evidently some $M-S$ π bonding with S acting as σ donor and π acceptor. The bridged molecules with $M-M$ bonds can of course be regarded as derivatives of sulfuryl dichloride (SO_2Cl_2) rather than of SO_2. Similarly, what can be considered to be a bound sulfinate anion is obtained by the action of SO_2 on $CpFe(CO)_2K$ to give the anion $CpFe(CO)_2SO_2^-$.

Finally, cations "solvated" by SO_2 can be obtained for Mn, Fe, Co, Ni, Cu, Zn, and Mg by oxidation of the metal by AsF_5 in liquid SO_2 of the type $[Fe(SO_2)_2(AsF_6)_2]_2$. The magnesium compound has bent *trans*$-MgOSO$ groups with the octahedral coordination completed by F atoms in $Mg-F-As$ bridges.

Sulfur Oxoanions

As noted in Section 12-13, *sulfites* can be S or O bonded, or sometimes both, as in the bridged species[50] (12-XXXIII). Related species in which R may be alkyl or aryl are

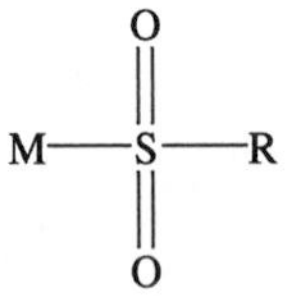

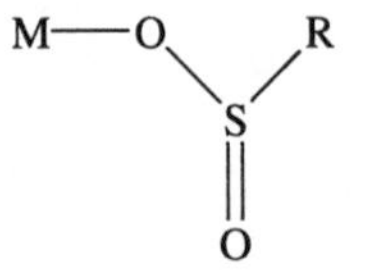

Sulfinate

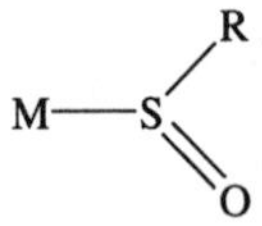

Sulfenate

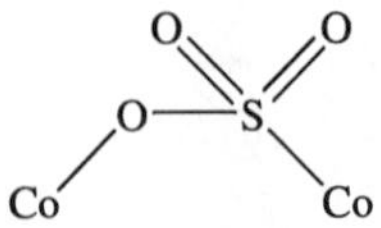

(12-XXXIII)

[50]A. Bino *et al., Inorg. Chim. Acta* **1991,** *188,* 91.

The unsubstituted sulfinate (L_nMSO_2H) can be obtained in certain cases:

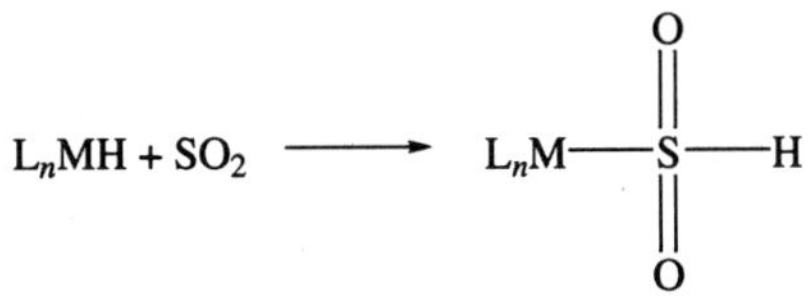

$$L_nMH + SO_2 \longrightarrow L_nM\overset{\displaystyle O}{\underset{\displaystyle O}{\overset{\|}{\underset{\|}{S}}}}\text{—H}$$

but the reaction can go further to give L_nMS and H_2O. Thus $CpW(CO)_3H$ reacts with excess SO_2 in MeCN to give $[CpW(CO)_3]_2$ and $[CpW(CO)_3]_2S$. However, the corresponding pentamethylcyclopentadienyl gives a unique SO_2 adduct of a bridged sulfide with the group:

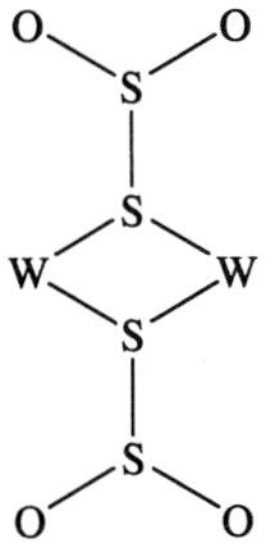

Sulfinates are also made by insertion of SO_2 into M—C bonds.
Thiosulfite as a bridge has been made by oxidation of a μ-S_2 species.

Dithiocarbamates and Related Anions (1,2-Dithiolates)

There is a wide variety of compounds of the following anions:

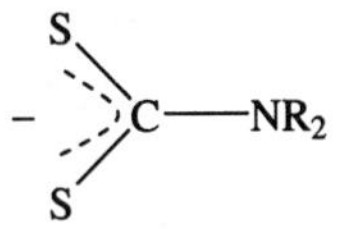

Dithiocarbamate

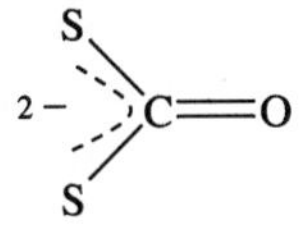

Dithiocarbonate

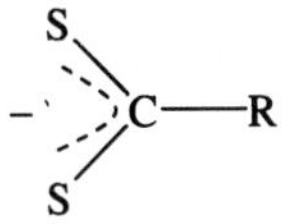

Dithiocarboxylate

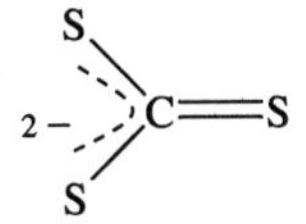

Trithiocarbonate

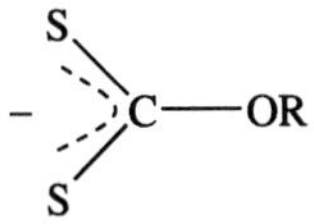

Xanthate

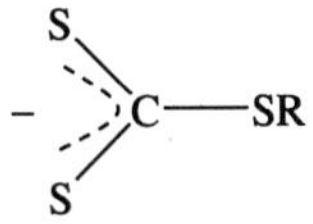

Thioxanthate

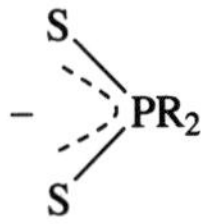

Dithiophosphinate

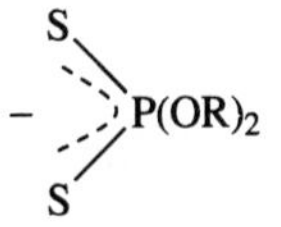

Dialkyldithiophosphate

Each of these ions may have a corresponding monothio analogue that is S,O bound, for example,

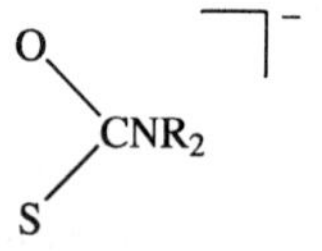

Thiocarbamate 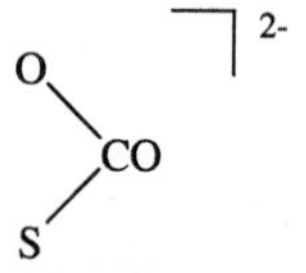   Thiocarbonate

Related species are

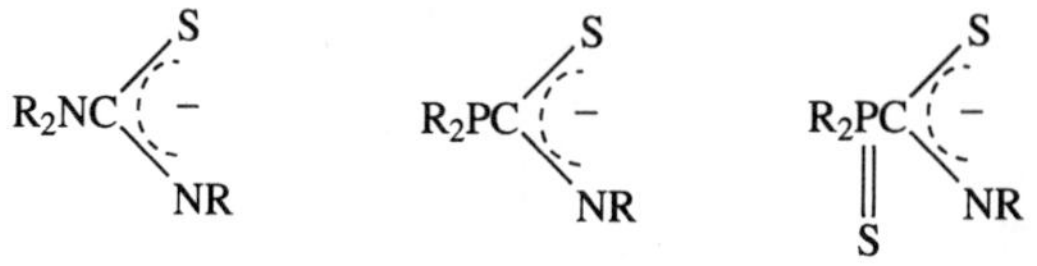

where the phosphorus compounds can also donate *via* the P atom. There are also linear and macrocyclic dithiocarbamates that have high affinity for UO_2^{2+}; one example is

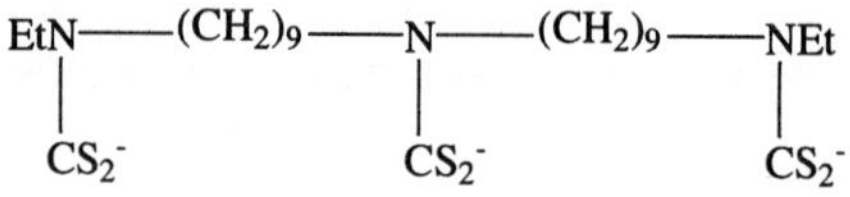

Dithiocarbamate and dithiophosphinate complexes have important uses. The former are used as fungicides and for solvent extraction and the latter as high pressure lubricants. Dithiocarbamates stabilize high oxidation states as in $[Fe^{IV}(dtc)_3]^+$ or $[Ni^{IV}(dtc)_3]^+$. Although dithiocarbamates are usually made from sodium salts such as NaS_2CNMe_2 or by oxidations using thiuram disulfides, they can also be made by insertion reactions of CS_2 with dialkylamides, for example,

$$Ti(NR_2)_4 + 4CS_2 \longrightarrow Ti(S_2CNR_2)_4$$

The types of bonding are

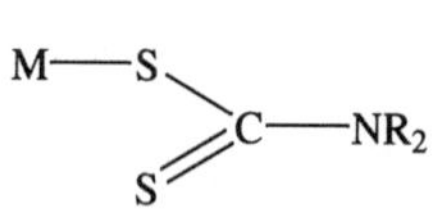

unidentate 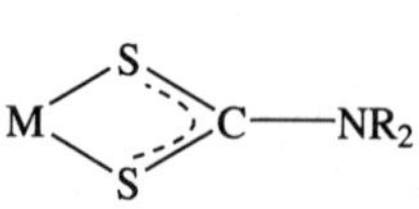   sym. chelate 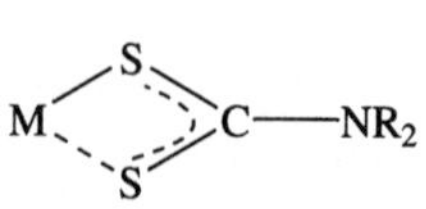   unsym. chelate

The bridging $^-S_2CNR_2$ groups (shown diagrammatically) can have several forms

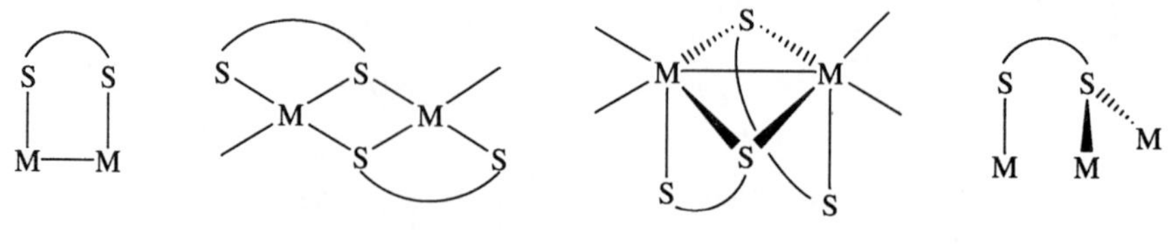

The compound $[Ru_2(S_2CNEt_2)_5]^+$, for example, has two isomers with chelate and different types of bridging dithiocarbamate:

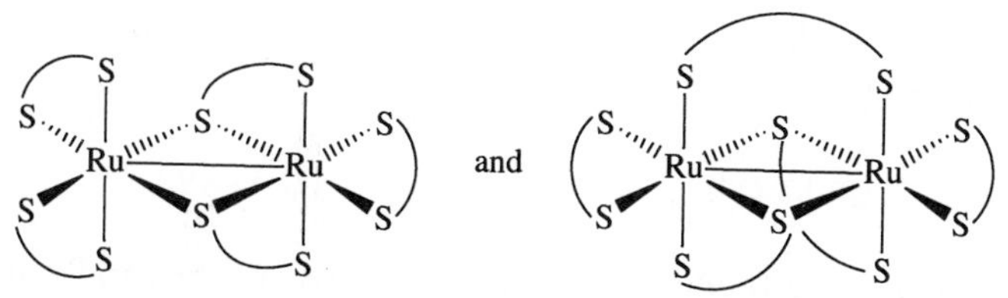

Thiocarbamates are similar, with bonding modes such as:

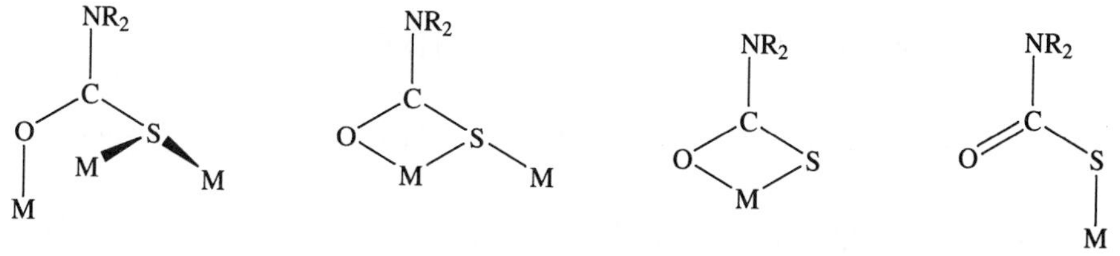

The *dithiophosphinate* or diorganophosphinodithioate ligands generally prefer to form four-membered chelate rings as in square $Ni(S_2PMe_2)_4$, but some bridged species, such as $[Zn(S_2PEt_2)_2]_2$ are known.

Tetrathiometallate and Related Anions[51]

There is an extensive chemistry of complexes in which ions such as MoS_4^{2-}, $WO_2S_2^{2-}$, WOS_3^{2-}, or ReS_4^-, serve as ligands. Two typical examples are

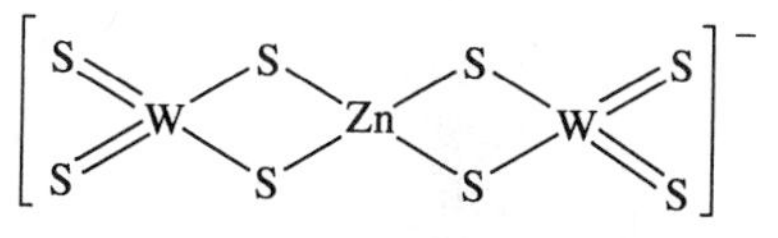

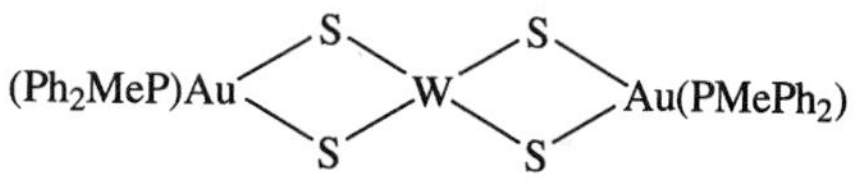

1,2-Dithiolenes

The 1,2-dithiolenes are a class of ligands that form a wide variety of compounds with metals in apparently many different oxidation states. This is more apparent than real because with ligands having extended π systems, delocalization of electrons onto the ligands occurs. It is very characteristic of dithiolene-type ligands (12-XXXIV) and related classes (12-XXXV), (12-XXXVI), and (12-XXXVII)

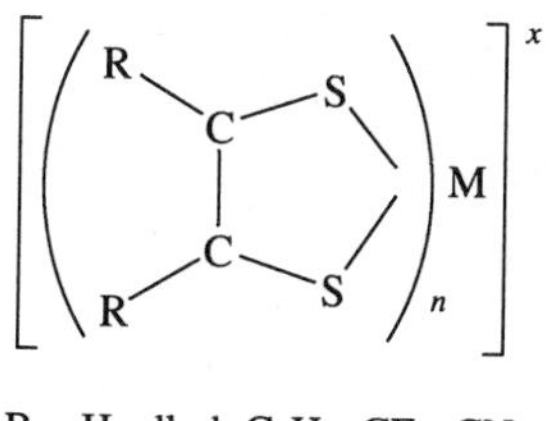

R = H, alkyl, C_6H_5, CF_3, CN
$n = 2; x = 0, -1, -2$
$n = 3; x = 0, -1, -2, -3$

(12-XXXIV)

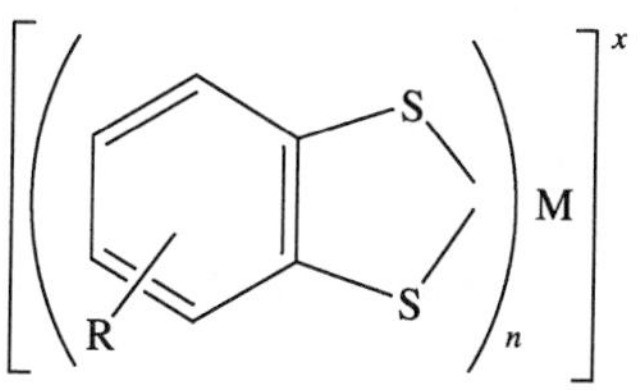

R = alkyl
$n = 2; x = 0, -1, -2$

(12-XXXV)

[51]R. Jostes and A. Müller, *J. Mol. Struct.* **1988,** *164,* 211.

(12-XXXVI)   (12-XXXVII)

that reversible oxidation-reduction sequences between structurally similar molecules differing only in their electron populations can occur. Examples are

$$\{Ni[S_2C_2(CN)_2]_2\} \underset{-e}{\overset{+e}{\rightleftharpoons}} \{Ni[S_2C_2(CN)_2]_2\}^{1-} \underset{-e}{\overset{+e}{\rightleftharpoons}} \{Ni[S_2C_2(CN)_2]_2\}^{2-}$$

$$[CoL_2]_2^0 \underset{-e}{\overset{+e}{\rightleftharpoons}} [CoL_2]_2^{1-} \underset{-e}{\overset{+e}{\rightleftharpoons}} [CoL_2]_2^{2-} \underset{-2e}{\overset{+2e}{\rightleftharpoons}} 2[CoL_2]^{2-} \qquad [L = S_2C_2(CF_3)_2]$$

$$[CrL_3]^0 \underset{-e}{\overset{+e}{\rightleftharpoons}} [CrL_3]^{1-} \underset{-e}{\overset{+e}{\rightleftharpoons}} [CrL_3]^{2-} \underset{-e}{\overset{+e}{\rightleftharpoons}} [CrL_3]^{3-} \qquad [L = S_2C_2(CN)_2]$$

A few representative syntheses of dithiolene complexes are the following:

$$NiCl_2 + Na_2S_2C_2(CN)_2 \xrightarrow{Et_4N^+} (Et_4N)_2\{Ni[S_2C_2(CN)_2]_2\}$$

$$Ni(CO)_4 + 2PhC\equiv CPh + 4S \longrightarrow Ni(S_2C_2Ph_2)_2 + 4CO$$

$$Ni(CO)_4 + 2 \begin{array}{c} S\text{---}C\text{---}CF_3 \\ | \quad\quad || \\ S\text{---}C\text{---}CF_3 \end{array} \longrightarrow Ni[S_2C_2(CF_3)_2]_2 + 4CO$$

$$Fe_2S_{12}^{2-} + MeO_2CC\equiv CCO_2Me \longrightarrow Fe_2[S_2C_2(CO_2Me)_2]_4^{2-}$$

For complexes containing only dithiolene ligands, four types of structure have been observed (Fig. 12-14). The planar D_{2h} structure is found for a majority of the structurally characterized *bis* complexes. The second structure type, observed in

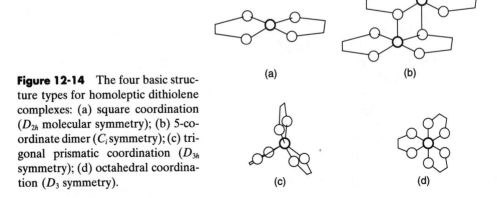

Figure 12-14 The four basic structure types for homoleptic dithiolene complexes: (a) square coordination (D_{2h} molecular symmetry); (b) 5-coordinate dimer (C_i symmetry); (c) trigonal prismatic coordination (D_{3h} symmetry); (d) octahedral coordination (D_3 symmetry).

(a)   (b)

(c)   (d)

the remaining *bis* complexes, is a centrosymmetric dimer, each metal atom being 5-coordinate. The metal atoms are significantly displaced from the planes of the dithiolene ligands (by 0.2-0.4 Å), but the bridging linkages are relatively weak. The third type of structure is one having trigonal prismatic D_{3h} coordination geometry. The interligand S···S distances in this structure are rather short (3.0-3.1 Å), indicative of weak interactions directly between the sulfur atoms; this structure is found only in a few of the more highly oxidized or neutral tris complexes, one example being the seleno analogue, $Mo[Se_2C_2(CF_3)_2]_3$.

In addition to homoleptic dithiolene complexes, compounds are known with additional ligands such as CO, Cp, and olefins.

The electronic structures of the 1,2-dithiolene complexes can be written in two extreme forms:

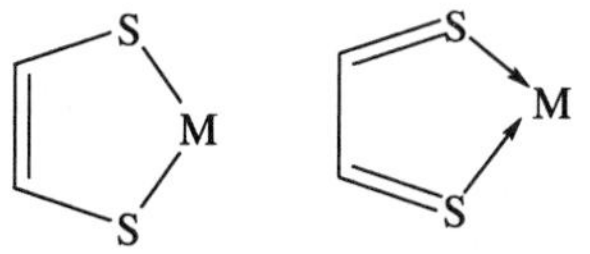

The formal oxidation number of the metal differs by two in these two cases. In molecular orbital terms the problem is one of the extent to which electrons are in metal *d* orbitals or delocalized over the ligand. Undoubtedly, in general, considerable delocalization occurs, which accounts for the ability of these complexes to exist with such a range of electron populations. The exact specification of orbital populations in any given case is a difficult and subtle question. The same problem arises for quinone and diimine complexes.

Dithioacetylacetonate. The sulfur analogue of acetylacetone does not exist in the free state but occurs as the inert dimer tetrathiaadamantane. However, complexes of the anion (sacsac) can be made by template or ligand trapping reactions of diketones and metal ions in the presence of H_2S.

Other Sulfur Ligands

Sulfur nitrides give a variety of complexes with NS groups by interaction of N_4S_4, S_7NH, and similar compounds with metal carbonyls or halides. Where hydroxylic solvents are used, hydrogen abstraction occurs to give complexes with N—H bonds. Some examples of the varied types of compounds are the following:

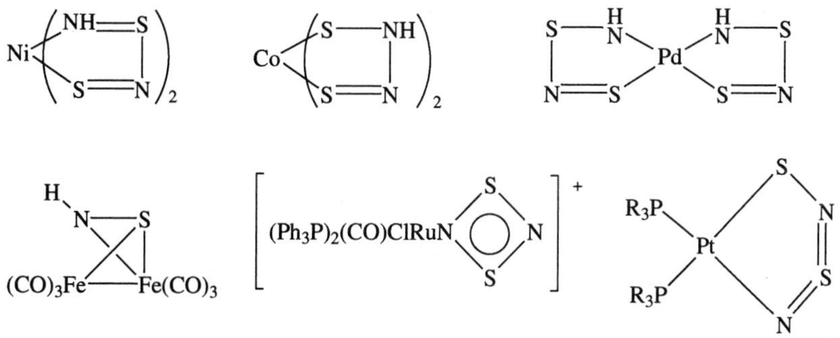

The *thiocyanate* ion is ambidentate[52] but commonly S-bonded with heavier metals. The *sulfides* of P and As such as P_4S_3 and As_2S_3 can act as ligands.

There are numerous compounds with S=C and S=P bonds, as well as others, that bond through sulfur. Most important perhaps are thiourea and substituted thioureas. These can be bound either as terminal or bridging ligands:

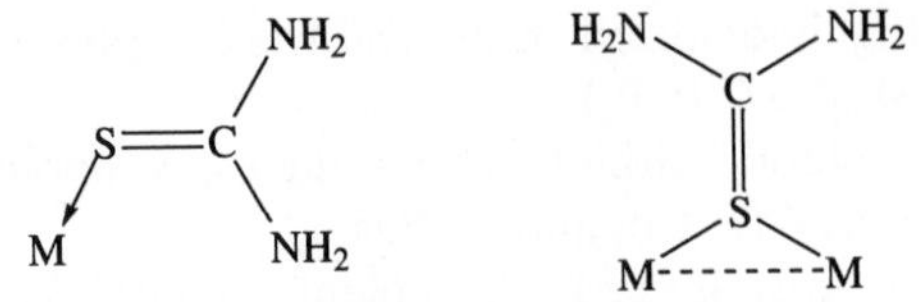

Additional References

Phosphorus, Sulfur and Related Elements (formerly *Int. J. Sulfur Chem.*), Gordon & Breach, New York. London.

E. Block, Ed., *Advances in Sulfur Chemistry,* Vol. 1, 1994.

S. Patai, Z. Rappoport, and G. Sterling, Eds., *Chemistry of Sulfoxides and Sulfanes,* John Wiley & Sons, 1988.

[52]M. Semrau and W. Preetz, *Z. anorg. allg. Chem.* **1996,** *622,* 1953.

Chapter 13

THE GROUP 17 ELEMENTS: F, Cl, Br, I, At

GENERAL REMARKS

13-1 Electronic Structures and Valences

Some important properties of the Group 17 elements (halogens) are given in Table 13-1. Since the atoms are only one electron short of the noble gas configuration, the elements readily form the anion X^- or a single covalent bond. Their chemistries are almost completely nonmetallic, and in general, the properties of the elements and their compounds change progressively with increasing size. As in other groups there is a much greater change between the first-row element fluorine and the second-row element chlorine than between other pairs, but with the exception of the Li—Cs group there are closer similarities within the group than in any other in the Periodic Table.

The high reactivity of fluorine results from a combination of the low F—F bond energy and the high strength of bonds from fluorine to other atoms. The small size and high electronegativity of the F atom account for many of the other differences between fluorine and the other halogens.

The coordination number of the halogens in the -1 state is normally only 1, but exceptions are found in HXH^+ cations, in polyhalide ions such as $FClF^-$ and I_3^-, and when X^- ions occur as bridging ligands where the coordination number is 2. There are also triply bridging X^- ions in some metal atom cluster compounds.

Positive formal oxidation numbers and higher coordination numbers may be assigned to the central halogen atoms in several classes of compounds such as the halogen fluorides (e.g., CF_3, ClF_5, BrF_5, and IF_7), oxo compounds (e.g., Cl_2O_7 or I_2O_5), oxofluorides (e.g., F_3BrO, or FIO_3), and a few other cases such as CF_3IF_2. It should be remembered that such formal oxidation numbers, though pragmatically useful in certain ways, bear no relation to actual charges.

For fluorine there is little evidence for positive behavior even in the formal sense. In oxygen fluorides the F atom is probably somewhat negative with respect to oxygen; whereas in ClF evidence from chlorine nuclear quadrupole coupling shows that the actual charge distribution involves partial positive charge on Cl.

Bond polarities in other halogen compounds indicate the importance of forms such as I^+Cl^- in ICl or I^+CN^- in ICN. In general, when a halogen atom forms a bond to another atom or group more electronegative than itself, the bond will be

547

Table 13-1   Some Properties of the Halogen Atoms and Molecules

	F	Cl	Br	I	At
At. No.	9	17	35	53	85
At. Wt.	19.00	35.45	79.90	126.90	209.99[a]
Stable isotopes	^{19}F	^{35}Cl, ^{37}Cl	^{79}Br, ^{81}Br	^{127}I	None
Valence shell	$2s^2 2p^5$	$3s^2 3p^5$	$4s^2 4p^5$	$5s^2 5p^5$	$6s^2 6p^5$
mp (°C)	−218.6	−101.0	−7.3	113.6[b]	
bp (°C)	−188.1	−34.9	59.5	185.2[b]	
X—X distance (Å)	1.43	1.99	2.28	2.66	
Covalent radius (Å)	0.64	0.99	1.14	1.33	
van der Waals radius (Å)	1.35	1.80	1.95	2.15	
Radius of X⁻ (Å)	1.19	1.67	1.82	2.06	
ΔH_{diss} of X_2 (kJ mol⁻¹)	158	242	193	151	
Electron attachment enthalpy (kJ mol⁻¹)	−328	−349	−325	−295	−270 ± 20
Atomic ionization enthalpy (kJ mol⁻¹)	1681	1255	1140	1008	

[a] Longest lived isotope.
[b] Vapor pressure is 90.5 Torr at 113.6°C.

polar with a partial positive charge on the halogen. Examples are the Cl atoms in SF_5OCl, CF_3OCl, FSO_2OCl, and O_3ClOCl, the bromine atoms in O_3ClOBr and FSO_2OBr, and the iodine atom in $I(NO_3)_3$.

Even for the heaviest member of the group, astatine, there is little evidence for any unambiguously "metallic" behavior.

THE ELEMENTS

Because of their reactivity, none of the halogens occurs in the elemental state in Nature. The elements all consist of nonpolar diatomic molecules between which the forces are of the same character as those between noble gas atoms. The trends in melting and boiling points are therefore qualitatively the same for the two groups of elements. In each case the dominant factor is the increasing magnitude of van der Waals forces as the size and polarizability of the atoms or molecules increase. The tendency to be colored of both the elements and their compounds increases with increasing atomic number as absorption bands move to longer wavelengths.

13-2   Fluorine

Fluorine is more abundant in the earth's crust (0.065%) than chlorine (0.055%) and forms concentrated deposits in such minerals as *fluorite* (or fluorspar), CaF_2, *cryolite*, Na_3AlF_6, and *fluorapatite*, $3Ca_3(PO_4)_2 \cdot Ca(F,Cl)_2$.

Only the isotope ^{19}F is found in Nature but ^{18}F with a half-life of 109.7 m is available and can be used, albeit with difficulty, as a tracer.

The estimated standard potential of fluorine ($E^0 = +2.85$ V) clearly indicates why early attempts to prepare the element by electrolytic methods in aqueous solution suitable for chlorine ($E^0 = +1.36$ V) failed. The element was first isolated in 1886 by Moissan, who pioneered the chemistry of fluorine and its compounds. The yellow gas is obtained by electrolysis of HF. Although anhydrous HF is a poor conductor, the addition of anhydrous KF gives conducting solutions. The most commonly used electrolytes are KF·2-3HF, which is molten at 70 to 100°C, and KF−HF, which is molten at 150 to 270°C. When the melting point begins to be too high because of HF consumption, the electrolyte can be regenerated by resaturation with HF from a storage tank. There have been many designs for fluorine cells; these are constructed of steel, copper, or Monel metal, which become coated with an unreactive layer of fluoride.

Steel or copper cathodes with ungraphitized carbon anodes are used. Although fluorine is often handled in metal apparatus, it can be handled in the laboratory in glass apparatus provided traces of HF, which attack glass rapidly, are removed by passing the gas through sodium or potassium fluoride with which HF forms the bifluorides MHF_2.

The only chemical synthesis of F_2 (which is not useful) is by the reaction

$$K_2MnF_6 + 2SbF_5 \longrightarrow 2KSbF_6 + MnF_3 + \tfrac{1}{2}F_2$$

The underlying principle is that the stronger Lewis acid, SbF_5, can displace the weaker one, MnF_4, from its salt; the MnF_4 is unstable and decomposes rapidly.

Fluorine is the chemically most reactive of all the elements and combines directly at ordinary or elevated temperatures with all the elements other than N_2, O_2, and the lighter noble gases. It also attacks organic material violently. However, selective fluorinations of organic compounds are feasible using inorganic agents such as CoF_3 and the unstable K_2NiF_6 and NiF_4 in HF(l),[1] SbF_5, UF_6,[2] and xenon fluorides.

Other fluorinating agents are CF_3OF, N_2F^+, or $NF_4^+SbF_6^-$ in $py(HF)_n$ solution[3] which can convert CH_4 to CH_3F and Me_3SnF.[4] R. Lagow showed that fluorine itself, when diluted in N_2 or Ar and using solvents like $CFCl_3$, CH_2Cl_2, or liquid HF, can be used for some selective fluorinations and for non-destructive perfluorination of many organic molecules.

Many fluorinating agents also act as oxidants, e.g., PtF_6 converts O_2 to $O_2^+PtF_6^-$. A quantitative scale, based on thermodynamic data, for oxidizing strength of 36 oxidizers, e.g., KrF^+, ClF_2^+, and NF_2O^+, has been proposed.[5] Some trends were noted such that for a given central atom, the oxidizing power increases with an increase in the formal oxidation state, e.g., $ClF_2^+ > ClF_4 > ClF_6^+$. The presence of one or more free valence electron pairs on the central atom, *decreases* the oxidizing strength, while the strengths of ψ-octahedral and ψ-tetrahedral species are low relative to strengths of ψ-pentagonal or ψ-tbp and ψ-trigonal planar anions.

The low bond energy, which is so important in the high reactivity of F_2 (in both the kinetic and thermodynamic senses), is best explained by the small size and high nuclear charge of the fluorine atom, which causes decreased overlap of the bonding orbitals and increased repulsion between the nonbonding orbitals on the two fluorine

[1]N. Bartlett *et al., Chem. Commun.* **1996,** 1049.
[2]J. H. Holloway *et al., J. Chem. Soc., Chem. Commun.* **1993,** 1429.
[3]G. Olah *et al., J. Am. Chem. Soc.* **1994,** *116,* 5671
[4]H. W. Roesky *et al., Organometallics* **1994,** *13,* 1251.
[5]K. O. Christe and D. A. Nixon, *J. Am. Chem. Soc.* **1992,** *114,* 2978.

atoms. It should be noted that the O—O bond in peroxides and the N—N bond in hydrazines are also relatively weak for similar reasons. The importance of repulsion between nonbonding electrons on the two F atoms in F_2 is evident when we note (*cf.* Table 13-1) that for the other halogens the X—X distance in X_2 molecules is essentially twice the covalent radius, whereas for F_2 the internuclear distance 1.43 Å is considerably larger than that (1.28 Å). This is because in a molecule such as X_3CF there are no nonbonding electrons on the atom to which F is bonded as compared to the situation in F_2. For the other halogens the X—X distances are large enough that lone-pair to lone-pair repulsions are much less important.

13-3   Chlorine

Chlorine occurs in Nature mainly as sodium chloride in seawater and in various inland salt lakes, and as solid deposits originating presumably from the prehistoric evaporation of salt lakes. Chlorine is prepared industrially mainly by electrolysis of brine:

$$Na^+ + Cl^- + H_2O \longrightarrow Na^+ + OH^- + \tfrac{1}{2}Cl_2 + \tfrac{1}{2}H_2$$

The traditional mercury cells, which pose an environmental hazard because of mercury loss, are being replaced by newer membrane cells.

Although chlorine can be recovered from HCl and its solutions by electrolytic and oxidative methods, these are inefficient. The old Deacon process was based on the copper catalyzed reaction

$$2HCl + \tfrac{1}{2}O_2 = Cl_2 + H_2O$$

This has been made more efficient by passing HCl over CuO at 200°C to give $CuCl_2$ and H_2O after which $CuCl_2$ at 300°C is treated with O_2.[6]

Chlorine is a greenish gas and moderately soluble in, but reactive with, water (see Section 13-12); it can form a gas hydrate $Cl_2 \cdot 7.3H_2O$ from Cl_2 action on aqueous $CaCl_2$ at 0°C.

13-4   Bromine

Bromine occurs principally as bromide salts of the alkali and alkaline earth elements in much smaller amounts than, but along with, chlorides. Bromine is obtained from brines and seawater by chlorination at a pH of ~3.5 and is swept out in a current of air.

Bromine is a dense, mobile, dark red liquid at room temperature. It is moderately soluble in water (33.6 g L^{-1} at 25°C) and miscible with nonpolar solvents such as CS_2 and CCl_4. Like Cl_2 it gives a crystalline hydrate, which appears to have a unique, noncubic structure, with a formula approximating to $Br_2 \cdot 7.9H_2O$.

[6]*C&EN* **1995,** *Sept 11*; *Ind. Eng. Chem. Res.* **1994,** *33,* 2996.

Table 13-2   Characteristics of Iodine Solutions

Solvent	Color	Absorption maximum (nm)
C_nH_{2n+2}, CCl_4	Violet	520–540
Aromatic hydrocarbons	Pink-red	490–510
Alcohols, amines	Brown	450–480

13-5 Iodine

Iodine occurs as iodide in brines and in the form of sodium and calcium iodates. Also, various forms of marine life concentrate iodine. Production of iodine involves either oxidizing I^- or reducing iodates to I^- followed by oxidation to the elemental state. Exact methods vary considerably depending on the raw materials. A commonly used oxidation reaction, and one suited to laboratory use when necessary, is oxidation of I^- in acid solution with MnO_2 (also used for preparation of Cl_2 and Br_2 from X^-).

Iodine is a black solid with a slight metallic luster. At atmospheric pressure it sublimes giving a violet vapor. Its solubility in water is slight (0.33 g L^{-1} at 25°C). It is readily soluble in nonpolar solvents such as CS_2 and CCl_4 to give violet solutions; spectroscopic studies indicate that "dimerization" occurs in solutions to some extent:

$$2I_2 \rightleftharpoons I_4$$

Iodine solutions are brown in solvents such as unsaturated hydrocarbons, liquid SO_2, alcohols, and ketones, and pinkish brown in benzene (see Table 13-2).

Iodine forms a well-known blue complex with the amylose form of starch. From resonance Raman and ^{129}I Mossbauer spectroscopy it has been shown that the color is caused by a linear array of I_5^- ($I_2I^-I_2$) repeating units held inside the amylose helix.

13-6 Astatine[7]

Astatine isotopes have been identified as short-lived products in the natural decay series of uranium and thorium. The element was first obtained in quantities sufficient to afford a knowledge of some of its properties by the cyclotron reaction $^{209}Bi(\alpha,2n)^{211}At$, and was named astatine from the Greek for "unstable." About 20 isotopes are known, the longest lived and their half-lives being ^{210}At (8.3h) and ^{211}At (7.2h). Consequently macroscopic quantities cannot be accumulated, although some compounds, HAt, CH_3At, AtI, AtBr, and AtCl have been detected mass spectrometrically. Our knowledge of the chemistry of At is based mainly on tracer studies, which show that it behaves about as one might expect by extrapolation from the other halogens. The element is rather volatile, and somewhat soluble in water, from which it may, like iodine, be extracted into benzene or carbon tetrachloride. Unlike iodine, however, it cannot be extracted from basic solutions.

The At^- ion is produced by reduction with SO_2 or zinc and can be carried down in AgI or TlI precipitates. Astatine is also carried by $[Ipy_2]^+$ salts, which is evidence

[7]G. W. M. Visser, *Radiochem. Acta* **1989,** *47,* 97; I. Brown, *Adv. Inorg. Chem.* **1987,** *31,* 43.

for oxidation to the At^I ion; powerful oxidants give AtO_3^- (carried by $AgIO_3$), and there is inconclusive evidence for an oxoastatine(VII) species. The aqueous potentials (0.1 M H^+) are estimated to be as follows:

$$At^- \xrightarrow{0.3\ V} At(0) \xrightarrow{1.0\ V} HAtO(?) \xrightarrow{1.5\ V} AtO_3^- \xrightarrow{>1.6\ V} At(VII)$$

Complex ions with halogens, N, O, and S ligands have been characterized for At^I as well as a few organic compounds; examples are $AtBr_2^-$ and $PhAt$.[8]

13-7 Charge-Transfer or Electron Donor-Acceptor Complexes of Halogens[9]

It is well known that iodine dissolves in a variety of organic solvents to give a whole range of colors. The general trend is summarized in Table 13-2. The reason for this is well understood. The iodine molecule has the electronic structure $1\sigma_g^2 1\sigma_u^2 \pi_u^4 2\sigma_g^2 \pi_g^4$ with an empty $2\sigma_u$ orbital, which is antibonding. The normal violet color of gaseous iodine is attributable to absorption caused by an allowed $\pi_g \rightarrow 2\sigma_u$ transition. If the I_2 molecule is dissolved in a solvent, S, that is an electron-pair donor, the situation depicted in Fig. 13-1 will occur. The more strongly the solvent molecule S engages in forming an $S \rightarrow I_2$ donor-acceptor or charge-transfer complex with I_2, the more the energy separation of the π_g and $2\sigma_u$ orbitals will be increased and the higher the energy of the $\pi_g \rightarrow 2\sigma_u$ transition will become. With poor donors such as saturated hydrocarbons the $S \rightarrow I_2$ interaction is extremely weak and hardly perturbs the I_2 molecule. Aromatic solvents donate their π electrons more effectively and regular electron-pair donors ROH, R_2O, RNH_2, and so on are quite effective donors. Of course with extremely good donors real reactions may occur, such as

$$2I_2 + 2py \longrightarrow 2pyI_2 \longrightarrow [I\ py_2^+][I_3^-]$$

Good evidence for the formation of charge-transfer complexes comes from the isolation and structure determination of solids. Chlorine and Br_2, as well as the interhalogens IBr and ICl, have been found to behave similarly to iodine. Some examples of crystal structures are:

1. $Me_3N \rightarrow I_2$, linear, $N-I = 2.27$ Å, $I-I = 2.83$ Å. Note that the $I-I$ bond has been weakened, as expected, when there is partial donation to the σ antibonding orbital.

2. Many infinite chain structures such as $Br_2 \cdot C_6H_6$, $Br_2 \cdot C_4H_8O_2$, and $Br_2 \cdot 2CH_3OH$ in which each Br_2 molecule is engaged by donors at both ends; the $S \cdots Br - Br \cdots S$ units are essentially linear, as expected for a σ-type acceptor orbital.

Note the significance of the term "charge-transfer complex." When the $S \rightarrow X_2$ complex forms there is actual mixing of the lone pair orbital on S with the $2\sigma_u$

[8]R. Dreyer *et al., Polyhedron* **1991**, *10*, 11 and references therein.
[9]For theory, data and references see T. Dahl and I. Reggen, *J. Am. Chem. Soc.* **1996**, *118*, 4152; H. Bock *et al., Chem. Commun.* **1996**, 1529.

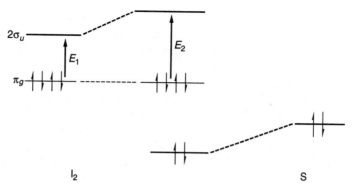

Figure 13-1 Diagram showing how interaction of I_2 with a donor solvent molecule S causes an increase in the separation of the π_g and $2\sigma_u$ orbitals from E_1 to E_2.

orbital of X_2; the molecule is polar and in resonance terms can be described as a hybrid $S \cdots X_2 \leftrightarrow S^+ X_2^-$, with the nonpolar form predominating. In fact, there is another electronic transition, well out into the uv (250-300 nm) in these complexes that can be described as a transition from a ground state where $S \cdots X_2$ predominates to an excited state where $S^+ X_2^-$ predominates. This is called a *charge-transfer transition* and such transitions are also important in other areas of chemistry.

Although charge transfers generally involve organics, iodine adducts of metal complexes are known. An example is $[CpMo(NBu^t)(\mu\text{-}S)I_2]_2$[10] where there is an I—I—S bond similar to those found for organic sulfur compounds such as 1,4,7-trithiacyclononane.[11]

Bromine π-complexes of alkenes can be detected spectroscopically; with bulky alkenes these intermediates can be stopped at the π-complex stage,[12] but normally the sequence is:

HALIDES

13-8 General Remarks

Except for helium, neon, and argon, all the elements in the Periodic Table form halides, often in several oxidation states, and halides generally are among the most important and common compounds. The ionic and covalent radii of the halogens are shown in Table 13-1.

[10]M. R. DuBois *et al.*, *Inorg. Chem.* **1994,** *33,* 2505.
[11]M. Schroder *et al.*, *J. Chem. Soc., Chem. Commun.* **1993,** 491.
[12]G. Bellucci *et al.*, *J. Am. Chem. Soc.* **1995,** *117,* 12001.

There are many ways of classifying halides. There are not only binary halides that can range from simple molecules with molecular lattices to complicated polymers and ionic arrays, but also oxohalides, hydroxo halides, and other complex halides of various structural types.

The syntheses for anhydrous metal and non-metal halides are given throughout the text. The simplest and most common is direct interaction of the elements but for halides other than fluorides elevated temperatures are usually required. Anhydrous halides can sometimes be obtained by dehydration of hydrated halides using reagents such as $SOCl_2$ or Me_3SiCl, while for metal fluorides and chlorides halogenation of oxides by heating with hexachlorobutadiene, hexachloropropane, NH_4Cl, or CCl_4 may be used. For fluorides special methods are available using fluorinating agents such as CoF_3, BrF_3, or XeF_2.

Widely used starting materials, particularly in organometallic chemistry, are the solvated halide (Cl, Br, I) complexes such as $CrCl_3(THF)_3$ or $RhCl_3(THT)_3$ or similar acetonitrile complexes; these are usually more reactive then anhydrous metal halides in organic solvents such as tetrahydrofuran, 1,2-dimethoxyethane, and other ethers unless the halides are soluble.

13-9   Ionic Halides

Most metal halides are substances of predominantly ionic character, although partial covalence is important in some. Actually, of course, there is a uniform gradation from halides that are for all practical purposes purely ionic, through those of intermediate character, to those that are essentially covalent. As a rough guide, we consider as basically ionic the halides in which the lattice consists of discrete ions rather than definite molecular units, although there may still be considerable covalence in the metal-halogen interaction, and the description "ionic" should never be taken entirely literally. As a borderline case *par excellence,* which clearly indicates the danger of taking such rough classifications as "ionic" and "covalent" or even "ionic" and "molecular" too seriously, we have $AlCl_3$; this has an extended structure in which aluminum atoms occupy octahedral interstices in a close-packed array of chlorine atoms; this kind of nonmolecular structure could accommodate an appreciably ionic substance. Yet $AlCl_3$ melts at a low temperature (193°C) to a molecular liquid containing Al_2Cl_6, these molecules being much like the Al_2Br_6 molecules that occur in both solid and liquid states of the bromide. Thus although $AlCl_3$ cannot be called simply a molecular halide, it is an oversimplification to call it an ionic one.

The relatively small radius of F^- (1.19 Å) is close to that of the oxide O^{2-} ion (1.26 Å); consequently many fluorides of monovalent metals and oxides of bivalent metals are ionic with similar formulas and crystal structures—for example, CaO and NaF. The compounds of the other halogens with the same formula usually form quite different structures due to the change in the anion/cation radius ratios and may even give molecular lattices. Thus chlorides and other halides often resemble sulfides, just as the fluorides often resemble oxides. In several cases the fluorides are completely ionic, whereas the other halides are covalent; for example, CdF_2 and SrF_2 have the CaF_2 lattice (nearly all difluorides have the fluorite or rutile structure), but $CdCl_2$ and $MgCl_2$ have layer lattices with the metal atoms octahedrally surrounded by chlorine atoms.

Many metals show their highest oxidation state in the fluorides. Let us consider the following Born-Haber cycle:

$$M(s) \xrightarrow{S} M(g) \xrightarrow{\Delta H^{(4)}_{ion}} M^{4+}(g) \searrow$$
$$MX_4(s)$$
$$2X_2(g) \xrightarrow{2D} 4X(g) \xrightarrow{4A} 4X^-(g) \nearrow$$

The value of $(A - D/2)$, the energy change in forming 1 g-ion of X^- from 2 mol of X_2, is ~250 kJ for all the halogens, and S is small compared to $\Delta H^{(4)}_{ion}$ in all cases. Although the structure of MX_4, hence the lattice energy, may not be known to allow us to say whether $4(A - D/2)$ plus the lattice energy will compensate for $(\Delta H^{(4)}_{ion} + S)$, we can say that the lattice energy, hence the potential for forming an ionic halide in a high oxidation state, will be greatest for fluoride. Generally, for a given cation size, the greatest lattice energy will be available for the smallest anion, that is, F^-.

However, for very high oxidation states, which are formed notably with transition metals, for example, WF_6 or OsF_6, the energy available is quite insufficient to allow ionic crystals with, say, W^{6+} or Os^{6+} ions; consequently such fluorides are gases, volatile liquids, or solids resembling closely the covalent fluorides of the nonmetals. It cannot be reliably predicted whether a metal fluoride will be ionic or molecular, and the distinction between the types is not always sharp.

In addition to the tendency of high cation charge to militate against ionicity, as just noted, coordination number plays an important role in determining the character of a halide. For a halide of formula MX_n where M is relatively large and has a high coordination number and n is small, the coordination number of M can be satisfied only by having a packing arrangement whereby each X atom is shared so that more than n of them may surround each M atom. Usually such structures are in fact essentially ionic, but the nonmolecular, hence nonvolatile, character of the substance is a consequence of the packing regardless of the degree of ionicity of the bonds. At the other extreme, if n is large and M has a coordination number of n, the halide will be molecular and probably volatile. The sequence KCl, $CaCl_2$, $ScCl_3$, and $TiCl_4$ shows these effects. The first three are nonmolecular solids and are not volatile, whereas $TiCl_4$ is a molecular solid, relatively easy to vaporize. It should not, of course, be thought that there is any sudden discontinuity in the nature of the M—Cl bonds between $ScCl_3$ and $TiCl_4$ even though the physical properties show a qualitative change. Similarly, for a given metal with various oxidation numbers, the lower halide(s) tend to be nonvolatile (and ionic), and the higher one(s) are more covalent and molecular. This is illustrated by $PbCl_2$ versus $PbCl_4$ and UF_4 (a solid of low volatility) versus UF_6, which is a gas at 25°C.

Most ionic halides dissolve in water to give hydrated metal ions and halide ions. However, the lanthanide and actinide elements in the +3 and +4 oxidation states form fluorides insoluble in water. Fluorides of Li, Ca, Sr, and Ba also are sparingly soluble, the lithium compound being precipitated by ammonium fluoride. Lead gives a sparingly soluble salt PbClF, which can be used for gravimetric determination of F^-. The chlorides, bromides, and iodides of Ag^I, Cu^I, Hg^I, and Pb^{II} are also quite insoluble. The solubility through a series of mainly ionic halides of a given element, $MF_n \sim MI_n$, may vary in either order. When all four halides are essentially ionic, the solubility order is iodide > bromide > chloride > fluoride, since the governing factor is the lattice energies, which increase as the ionic radii decrease. This order is found among the alkali, alkaline earth, and lanthanide halides. On the other hand, if covalence is fairly important, it can reverse the trend, making the fluoride most and the iodide least soluble, as in the familiar cases of silver and mercury(I) halides.

"Naked" Fluoride Ions. While F^- ions in solution and in many solid compounds are hydrated, e.g., as $F(OH_2)_4^-$ the question of "naked" fluoride ions has been discussed for compounds with large cations. Anhydrous compounds such as $Me_4N^+F^-$,[13] Cp_2CoF,[14] and others[15] have been made. These anhydrous compounds are sources of highly nucleophilic F^- and $Me_4N^+F^-$ for example is much more reactive than CsF. They are used in syntheses of organic and inorganic fluorides, such as XeF_5^-, BrF_6^-, PF_4^-, and so on.[16]

Halide Ions as Ligands. All of the halide ions have the ability to function as ligands and form complexes of enormous diversity. Some are homoleptic, MX_x^{n-} (e.g., SiF_6^{2-}, $MoCl_6^{3-}$, $FeBr_4^-$, HgI_4^{2-}) and many have both halide ions and other ligands in the coordination sphere (e.g., $Mo(PEt_3)_3Cl_3$, $[Co(NH_3)_4Cl_2]^{2+}$, $Pt(NH_3)_2Cl_2$). The fluoride ion can form complexes with metal ions in all possible oxidation states including the highest ones[16] whereas I^- can be associated only with the metal ions in lower (and hence relatively non-oxidizing) oxidation states.

Halide Bridges. These are formed in both simple molecular compounds and in complexes; the main types are shown in (13-I). Triple bridges are not common[17] while even rarer are μ_4[18] and μ_6[19] bridges.

(13-I)

The single bridges can be either linear or bent with angles ranging from *ca.* 115° to 180°. Single bridges are most common for F^-. The nearer the bridge is to

[13]K. O. Christe *et al.*, *J. Am. Chem. Soc.* **1990,** *112,* 7619.
[14]T. G. Richmond *et al.*, *J. Am. Chem. Soc.* **1994,** *116,* 11165.
[15]K. Seppelt *et al.*, *Chem. Eur. J.* **1995,** *1,* 261.
[16]N. Doherty and N. W. Hoffman, *Chem. Rev.* **1991,** *91,* 553.
[17]See e.g., H. W. Roesky *et al.*, *Chem. Commun.* **1996,** 29.
[18]R. J. Puddephatt *et al.*, *J. Am. Chem. Soc.* **1993,** *115,* 6546.
[19]W. Clegg *et al.*, *J. Chem. Soc., Chem. Commun.* **1992,** 1010.

being linear, the stronger is the metal-fluorine π-bonding. A representative example is (13-II).

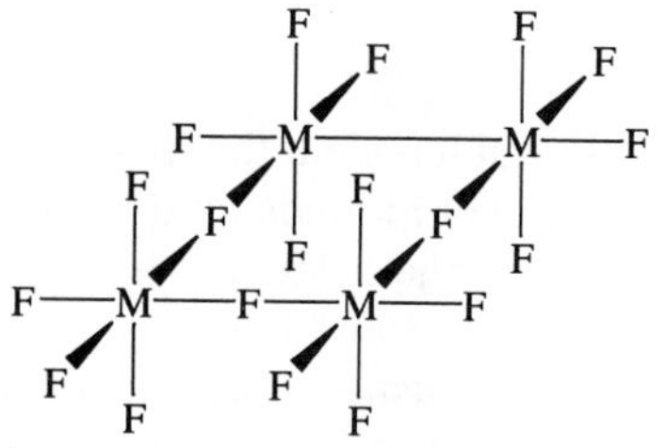

(13-II)

For Cl, Br, and I, single bridges are much less common, the usual type being double bridges; such bridges are commonly cleavable by Lewis bases such as pyridine and other amines, tertiary phosphines, and so on. In double bridges the M—X—M angles vary widely—for Cl and Br these are in the range 70-100°. There are three types of structures shown in (13-III): these types are adopted also by other bridging groups such as NR^{2-}, O^{2-}, OR^-, SCN^-, S^{2-}, SR^-, and so on.

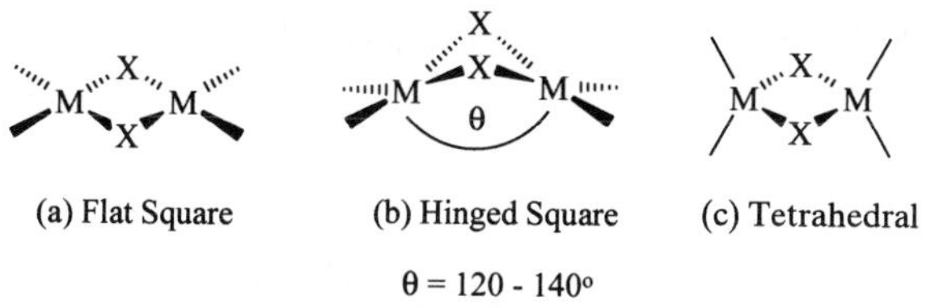

(a) Flat Square   (b) Hinged Square   (c) Tetrahedral

$\theta = 120 - 140°$

(13-III)

The type (b) is often said to have a "butterfly" geometry. Type (c) is often found for PR_3 groups while the M—M distance is shorter than in (a) or (b). Subtle electronic packing effects in crystals can influence the M—M distance, while the adoption of a particular structure is related to the π-donor capacity of the bridge ligand.

13-10 Molecular Halides

Molecular halides are usually volatile, though this will not be so if they are polymeric, as, for example, Teflon $(-CF_2-)_n$. There are also many cases (e.g., $AlCl_3$ and SnF_4) of substances that can exist as molecules in the gas phase, but because of the tendency of the metal atom to have a higher coordination number (4 and 6, respectively, in these cases) the solids have extended array structures. Most of the electronegative elements and the metals in high oxidation states (V, VI) form molecular halides. A unique but very important group of molecular halides are the hydrogen halides. These form molecular crystals and are volatile, although they readily and extensively dissociate in polar media such as H_2O. The H—X bond energies and the thermal stabilities decrease markedly in the order HF > HCl > HBr > HI, that is, with increasing atomic number of the halogen. The same trend is found, in varying degrees, among the halides of all elements giving a set of molecular halides, such as those of B, C, Si, and P. Interhalogen compounds are discussed in the following sections.

Molecular Fluorides

Many molecular fluorides exist, but it is clear that because of the high electronegativity of fluorine, the bonds in such compounds tend to be very polar. Because of the low dissociation energy of F_2 and the relatively high energy of many bonds to F (e.g., C—F, 486; N—F, 272; P—F, 490 kJ mol^{-1}), molecular fluorides are often formed very exothermically; this is just the opposite of the situation with nitrogen, where the great strength of the bond in N_2 makes nitrogen compounds mostly endothermic. Interestingly, in what might be considered a direct confrontation between these two effects, the tendency of fluorine to form exothermic compounds wins. Thus for NF_3 we have

$$\tfrac{1}{2}N_2(g) = N(g) \qquad\qquad \Delta H \sim 475 \text{ kJ mol}^{-1}$$

$$\tfrac{3}{2}F_2(g) = 3F(g) \qquad\qquad \Delta H \sim 232 \text{ kJ mol}^{-1}$$

$$N(g) + 3F(g) = NF_3(g) \qquad -3E_{N\text{-}F} \approx -3(272) = -816 \text{ kJ mol}^{-1}$$

Therefore,

$$\tfrac{1}{2}N_2(g) + \tfrac{3}{2}F_2(g) = NF_3(g) \qquad \Delta H \approx -109 \text{ kJ mol}^{-1}$$

The high electronegativity of fluorine often has a profound effect on the properties of molecules in which several F atoms occur. Representative are facts such as (a) CF_3CO_2H is a strong acid, (b) $(CF_3)_3N$ and NF_3 have no basicity, and (c) CF_3 derivatives in general are attacked much less readily by electrophilic reagents in anionic substitutions than are CH_3 compounds. The CF_3 group may be considered as a kind of large pseudohalogen with an electronegativity about comparable to that of Cl.

Reactivity

The detailed properties of a given molecular halide depend on the particular elements involved, and these are discussed where appropriate in other chapters. However, a fairly general property of molecular halides is their easy hydrolysis to produce the hydrohalic acid and an acid of the other element. Typical examples are

$$BCl_3 + 3H_2O \longrightarrow B(OH)_3 + 3H^+ + 3Cl^-$$

$$PBr_3 + 3H_2O \longrightarrow HPO(OH)_2 + 3H^+ + 3Br^-$$

$$SiCl_4 + 4H_2O \longrightarrow Si(OH)_4 + 4H^+ + 4Cl^-$$

When the central atom of a molecular halide has its maximum stable coordination number, as in CCl_4 or SF_6, the substance is usually quite unreactive toward water or even OH^-. This does not mean, however, that reaction is thermodynamically unfavorable, but only that it is kinetically inhibited, since there is no room for nucleophilic attack. Thus for CF_4 the equilibrium constant for the reaction

$$CF_4(g) + 2H_2O(l) = CO_2(g) + 4HF(g)$$

is $\sim 10^{23}$. The necessity for a means of attack is well illustrated by the failure of SF_6 to be hydrolyzed, whereas SeF_6 and TeF_6 are hydrolyzed at 25°C through expansion of the coordination sphere, which is not possible for sulfur.

Organic Halides as Ligands

There is an extensive chemistry of organic halides as ligands[20] that includes now R_3SiX.[21] The compounds can be obtained by reactions such as

$$L_nMCl \ + \ AgPF_6 \ + \ RI \ \xrightarrow[-AgCl]{CH_2Cl_2} \ [L_nM\!\leftarrow\!IR]^+PF_6^-$$

$$L_nRuI \ + \ CF_3SO_3Me \ \longrightarrow [L_nRu\!\leftarrow\!IMe]^+CF_3SO_3^-$$

$$Cp(NO)(Ph_3P)ReCH_3 \ \xrightarrow[\substack{+CH_2Cl_2 \\ -CH_4}]{HBF_4\cdot OEt_2} \ [Cp(NO)ReClCH_2Cl]^+BF_4^-$$

$$\downarrow RI$$

$$[Cp(NO)ReIR]^+BF_4^-$$

Although the compounds are mainly of the angular type $M\cdots XR$, there are numerous examples of compounds with bridging or chelate groups arising from halides such as CH_2I_2, $Cl(CH_2)_2Cl$, $I(CH_2)_3I$, and so on. An example is $[Ir^{III}H_2(PPh_3)_2(\eta^2\text{-}1,2\text{-}C_6H_4I_2)]^+SbF_6^-$. Silver complexes have been particularly well studied.[22] A different type of complex has an $M\cdots X-CH_2-M$ unit with an $M-C$ bond;[23] aryl fluorides have also been shown to coordinate to $RuCl_2(PPh_3)_3$.[24] Compounds with organic halide ligands can be useful starting materials as the halides can often be easily replaced by other ligands.

[20]R. J. Kulawiec and R. H. Crabtree, *Coord. Chem. Rev.* **1990**, *99*, 89.
[21]R. U. Kirss, *Inorg. Chem.* **1992**, *31*, 3451.
[22]J. Powell *et al.*, *J. Chem. Soc., Dalton Trans.* **1996**, 1669; R. D. Gillard *et al.*, *Polyhedron* **1996**, *15*, 2409.
[23]Y. Zhou and J. A. Gladysz, *Organometallics* **1993**, *12*, 1073.
[24]S. D. Perera and B. L. Shaw, *Inorg. Chim. Acta* **1995**, *228*, 127.

Table 13-3 Binary Oxides of the Halogens[a]

Fluorine	bp (°C)	mp (°C)	Chlorine	bp (°C)	mp (°C)	Bromine[b]	bp (°C)	Iodine
F_2O	−145	−224	Cl_2O	~4	−116	Br_2O	−18	I_2O_4
			$Cl_2O_3{}^c$			Br_2O_3		
F_2O_2	−57	−163	ClO_2	~10	−5.9			I_4O_9
			Cl_2O_4	44.5	−117			
			Cl_2O_6		3.5	Br_2O_5		I_2O_5
			Cl_2O_7	82	−91.5			

[a]ClO may be involved in ozone depletion in the stratosphere, possibly as a radical complex ClO·HOH (J. S. Francisco and S. P. Spander, *J. Am. Chem. Soc.* **1995**, *117*, 9917). The unstable chloryl chloride ClClO₂ was characterized spectroscopically in the gas and matrices (H. S. P. Müller and H. Willner, *Inorg. Chem.* **1992**, *31*, 2527).
[b]Oxides structurally characterized but highly unstable.
[c]Explodes below 0°C.
Note. The short lived radical ClO_4' is formed in thermolysis of Cl_2O_6 or Cl_2O_7: vibrational spectra indicate a tetrahedral structure (H. Grolke and H. Willner, *Angew. Chem. Int. Ed. Engl.* **1996**, *35*, 768).

HALOGEN OXIDES AND OXO COMPOUNDS

13-11 Binary Oxides

These are listed in Table 13-3; oxygen fluorides were discussed in Chapter 11.

Chlorine Oxides

These are all highly endothermic and because of this they cannot be obtained by the reaction of Cl_2 and O_2, and they must be handled with great caution because of their tendency to explode.

Dichlorine monoxide is a brown-yellow gas stable at room temperature. It has two isomers, OClCl and ClOCl; although the latter is thermodynamically more stable by *ca.* 12.5 kJ mol⁻¹, it is exceedingly reactive. The kinetically stable form ClOCl is bent (110.9°) with Cl—O = 1.696 Å. It explodes rather easily when heated or sparked. It dissolves readily in water to give an orange solution in which the equilibrium $Cl_2O + H_2O = 2HOCl$ exists. It can be used to chlorinate organic compounds. The photochemistry of Cl_2O has been studied (along with that of ClO_2 and bromine oxides) because of its role in the depletion of polar ozone through production of Cl atoms.[25]

The oxide is made both commercially and on a laboratory scale by chlorination of HgO dry or in suspension in CCl₄:

$$2Cl_2 + 2HgO \longrightarrow Cl_2O + HgCl_2 + HgO$$

Chlorine dioxide is similarly explosive and reactive but can be handled at temperatures below −40°C and pressures below 50 mbar. At room temperature it

[25]R. C. Dunn and J. D. Simon, *J. Am. Chem. Soc.* **1992**, *114*, 4856. For ClO₂ see C. S. Foote *et al.*, *J. Am. Chem. Soc.* **1993**, *115*, 5307.

is a yellowish gas and the molecule, like Cl_2O, is angular (117.4°, 1.470 Å). As a 19e AB_2 type molecule with an unpaired electron it was thought that ClO_2^- should not dimerize because of repulsion between non-bonding pairs on Cl and the location of the electron in the $2b_1(\pi^*)$ MO. However, an X-ray study at −150°C indicates a weakly bonded dimer, $OCl(\mu\text{-}O)_2ClO$, in the *trans* form which is diamagnetic below −84°C.[26]

Chlorine dioxide can be made on a commercial scale for immediate use in bleaching by the exothermic reaction of $NaClO_3$ in 4-4.5 M sulfuric acid containing 0.05-0.25 M Cl^- with SO_2:

$$2NaClO_3 + SO_2 + H_2SO_4 \longrightarrow 2ClO_2 + 2NaHSO_4$$

On a small scale $KClO_3$ can be reduced by moist oxalic acid at 90°C when CO_2 liberated serves as a diluent.

$$2KClO_3 + H_2C_2O_4 \longrightarrow 2ClO_2 + 2CO_2 + 2KOH$$

The oxide is soluble in water where it is a mild but rapid oxidant (ClO_2/ClO_2^- E^0 = 0.936 V) and electron transfer agent.[27] The solutions are photosensitive and decompose slowly to HCl and $HClO_4$; they are stable in the dark. In alkaline solution ClO_2 reacts fairly rapidly:

$$2ClO_2 + 2OH^- \longrightarrow ClO_2^- + ClO_3^- + H_2O$$

The oxidation of ClO_2 should give ClO_2^+ and this cation occurs in several salts in reactions such as

$$ClO_2F + RuF_5 \longrightarrow ClO_2^+RuF_6^-$$

The ion is angular[28] and similar to ClF_2^+ discussed below.

Dichlorine Tetraoxide. This extremely unstable compound is actually $Cl^+ClO_4^-$ made by the reaction

$$CsClO_4 + ClSO_3F \longrightarrow Cl^+ClO_4^- + CsSO_3F$$

Pyrolysis and matrix isolation allows the radical ClO_3^- to be detected.[29]

Dichlorine Hexaoxide. This is also a perchlorate, $ClO_2^+ClO_4^-$, obtained on inter-action of ClO_2 with O_3/O_2 at −10°C. The impure red liquid on distillation +10° to

[26]A. Rehr and M. Jansen, *Inorg. Chem.* **1992**, *31*, 4740.
[27]H. H. Awad and D. M. Stanbury, *J. Am. Chem. Soc.* **1993**, *115*, 3636.
[28]R. Bougon *et al.*, *Inorg. Chem.* **1991**, *30*, 102.
[29]H. Grothe and H. Willner, *Angew. Chem. Int. Ed. Engl.* **1994**, *33*, 1482.
[30]F. Favier *et al.*, *J. Chem. Soc., Dalton Trans.* **1992**, 1977; **1994**, 3119.

−20°C gives ruby crystals.[30] It has been used in synthesis of unsolvated perchlorates but can explode on contact with organic matter.

Dichlorine Heptaoxide. This is the most stable oxide, obtained by dehydration of $HClO_4$ by P_2O_5 at −10°C followed by careful vacuum distillation. The structure in the solid state is $O_3Cl(\mu\text{-}O)ClO_3$[31] and the same in the gas phase by electron diffraction. It resembles Mn_2O_7 structurally but there are differences in bond lengths and angles. As the anhydride of perchloric acid it reacts with H_2O and OH^- to form ClO_4^-.

Chlorine Tetraoxide. This radical, $ClO_4^{\cdot}$, was made by thermolysis at 230°C of a dilute 1:500 stream of Cl_2O_6 or Cl_2O_7 in Ar or Ne.[32] It has a lifetime of only a few milliseconds but could be obtained in a matrix at low temperature and characterized spectroscopically along with ClO_2 and ClO_3 from the reactions

$$Cl_2O_6 \longrightarrow ClO_4^{\cdot} + ClO_2^{\cdot}$$

$$Cl_2O_7 \longrightarrow ClO_4^{\cdot} + ClO_3^{\cdot}$$

Bromine Oxides[33]

These are the least well known halogen oxides due to instability and intractability. All are explosive.

Dibromine monooxide, Br_2O, which decomposes above 250 K, was first made by interaction of Br_2 and HgO (*cf.* Cl_2O) at low temperature but it can be obtained by controlled decomposition of "BrO_2." The latter is obtained as a yellow solid by ozonization of Br_2 at −78°C or by high voltage discharge through Br_2-O_2 mixtures at low pressure and temperature. Infrared spectra of Br_2O in a N_2 matrix at 12 K indicates a C_{2v} symmetry like Cl_2O and K-edge EXAFS gives $\angle BrOBr = 112°$ and $Br-O = 1.85\text{Å}$.[34]

The "BrO_2" from ozonization of Br_2 in $CFCl_3$ is partly soluble in CH_2Cl_2 leaving a white solid. From CH_2Cl_2, orange needles crystallize at −90°C but these decompose above −40°C and detonate when rapidly warmed. The X-ray structure confirms this to be bromine bromate $BrOBrO_2$.[35] Ozonization of Br_2 in CH_2Cl_2 at −78°C leads to the colorless product noted above (which can detonate above −40°C) *via* the sequence

$$Br_2 \xrightarrow{O_3} BrOBrO_2 \xrightarrow{O_3} O_2BrOBrO_2$$

[31]A. Simon and H. Borrmann, *Angew. Chem. Int. Ed. Engl.* **1988,** *27,* 1339.
[32]H. Grothe and H. Willner, *Angew. Chem. Int. Ed. Engl.* **1996,** *35,* 768.
[33]K. Seppelt, *Acc. Chem. Res.* **1997,** *30,* 111.
[34]J. S. Ogden *et al., J. Am. Chem. Soc.* **1990,** *112,* 1019.
[35]R. Kuschel and K. Seppelt, *Angew. Chem. Int. Ed. Engl.* **1993,** *32,* 1632.

This can be crystallized from propionitrile as $Br_2O_5 \cdot (EtCN)_3$ and the crystal structure shows $O_2BrOBrO_2$ with the Br^V atoms pyramidally surrounded by three O atoms ($\angle BrOBr = 121.2°$) and the terminal O atoms eclipsed, unlike I_2O_5 where they are staggered.[36]

Iodine Oxides[37]

Of these, white crystalline iodine pentoxide is the most important and is made by the reaction

$$2HIO_3 \xrightleftharpoons{240°C} I_2O_5 + H_2O$$

It has IO_3 pyramids sharing one oxygen to give O_2IOIO_2 units, but quite strong intermolecular $I\cdots O$ interactions lead to a three-dimensional network. This compound is stable up to ~300°C, where it melts with decomposition to iodine and oxygen. It is the anhydride of iodic acid and reacts immediately with water. It reacts as an oxidizing agent with various substances such as H_2S, HCl, and CO. One of its important uses is as a reagent for the determination of CO, the iodine that is produced quantitatively being then determined by standard iodometric procedures:

$$5CO + I_2O_5 \longrightarrow I_2 + 5CO_2$$

The other oxides of iodine I_2O_4 and I_4O_9 are of less certain nature. They decompose when heated at ~100°C to I_2O_5 and iodine, or to iodine and oxygen. The yellow solid I_2O_4, which is obtained by partial hydrolysis of $(IO)_2SO_4$ (discussed later), appears to have a network built up of polymeric I—O chains that are cross linked by IO_3 groups. I_4O_9, which can be made by treating I_2 with ozonized oxygen, can be regarded as $I(IO_3)_3$, similarly cross linked.

Iodine trioxide is made by the interaction of H_5IO_6 with concentrated H_2SO_4 at 70°C. It precipitates as a yellow solid which has a polymeric layer structure made of I_4O_{12} units.

13-12 Oxo Acids and Anions

The known oxo acids of the halogens are listed in Table 13-4. The chemistry of these acids and their salts is very complicated. Solutions of all the acids and of several of the anions can be obtained by reaction of the free halogens with water or aqueous bases. We discuss these reactions first; the term "halogen" refers to chlorine, bromine, and iodine, only.

Although one often finds hypochlorous and chlorous acids written HClO and $HClO_2$, this is unfortunate and should be discontinued since they are actually HOCl and HOClO. This is not a purely pedantic matter since in studies of gas phase kinetics in the earth's ozone layer, the distinction between the unstable isomer, HClO, and HOCl is significant. Hypochlorous acid HOCl is ~280 kJ mol^{-1} more stable than HClO.

[36]D. Leopold and K. Seppelt, *Angew. Chem. Int. Ed. Engl.* **1994,** *33,* 975.
[37]T. Kraft and M. Jansen, *J. Am. Chem. Soc.* **1995,** *117,* 6795; H. Fjellvåg and A. Kjekshus, *Acta Chem. Scand.* **1994,** *48,* 815.

Table 13-4   Oxo Acids of the Halogens

Fluorine	Chlorine	Bromine	Iodine		
HOF	$HOCl^a$	$HOBr^a$	HOI^a		
	$HOClO^a$	$HOBrO^a$			
	$HOClO_2^a$	$HOBrO_2^a$	$HOIO_2$		
	$HOClO_3$	$HOBrO_3^a$	$(HOIO_3)_n$, $(HO)_5IO$, $H_4I_2O_9$		

^{a}Stable only in solution.

Reaction of Halogens with H_2O and OH^-

A considerable degree of order can be found in this area if full and proper use is made of thermodynamic data in the form of oxidation potentials and equilibrium constants and if the relative rates of competing reactions are also considered. We shall consider first the basic thermodynamic data which are given in Table 13-5. From these, all necessary potentials and equilibrium constants can be derived by use of basic thermodynamic relationships.

The halogens are all to some extent soluble in water. However, in all such solutions there are species other than solvated halogen molecules, since a disproportionation reaction occurs *rapidly*. Two equilibria serve to define the nature of the solution:

$$X_2(g,l,s) = X_2(aq) \qquad K_1$$

$$X_2(aq) = H^+ + X^- + HOX \qquad K_2$$

The values of K_1 for the various halogens are Cl_2, 0.062; Br_2, 0.21; I, 0.0013. The values of K_2 can be computed from the potentials in Table 14-4 to be 4.2×10^{-4} for Cl_2, 7.2×10^{-9} for Br_2, and 2.0×10^{-13} for I_2. We can also estimate from

$$\tfrac{1}{2}X_2 + e = X^-$$

and

$$O_2 + 4H^+ + 4e = 2H_2O \qquad E^0 = 1.23 \text{ V}$$

that the potentials for the reactions

$$2H^+ + 2X^- + \tfrac{1}{2}O_2 = X_2 + H_2O$$

are -1.62 V for fluorine, -0.13 V for chlorine, 0.16 V for bromine, and 0.69 V for iodine.

Thus for saturated solutions of the halogens in water at 25°C we have the results shown in Table 13-6. There is an appreciable concentration of hypochlorous acid in a saturated aqueous solution of chlorine, a smaller concentration of HOBr

Table 13-5 Standard Potentials (V) for Reactions of the Halogens

Reaction	Cl	Br	I
$H^+ + HOX + e = \frac{1}{2}X_2(g,l,s) + H_2O$	1.63	1.59	1.45
$3H^+ + HOXO + 3e = \frac{1}{2}X_2(g,l,s) + 2H_2O$	1.64		
$6H^+ + XO_3^- + 5e = \frac{1}{2}X_2(g,l,s) + 3H_2O$	1.47	1.52	1.20
$8H^+ + XO_4^- + 7e = \frac{1}{2}X_2(g,l,s) + 4H_2O$	1.42	1.59	1.34
$\frac{1}{2}X_2(g,l,s) + e = X^-$	1.36	1.07	0.54[a]
$XO^- + H_2O + 2e = X^- + 2OH^-$	0.89	0.76	0.49
$XO_2^- + 2H_2O + 4e = X^- + 4OH^-$	0.78		
$XO_3^- + 3H_2O + 6e = X^- + 6OH^-$	0.63	0.61	0.26
$XO_4^- + 4H_2O + 8e = X^- + 8OH^-$	0.56	0.69	0.39

[a]Indicates that I^- can be oxidized by oxygen in aqueous solution.

in a saturated solution of Br_2, but only a negligible concentration of HOI in a saturated solution of iodine.

While the equilibrium situation is well covered by the foregoing thermodynamic data, less has been known about the kinetics. There are now fairly detailed kinetic studies leading to mechanistic understanding for the hydrolyses of Cl_2,[38] Br_2,[39] and I_2.[40]

Hypohalous Acids

The compound HOF, a colorless solid, melts at $-117°C$ to a pale yellow liquid. It is a gas at ambient temperature, highly reactive towards water, and has a half-life for spontaneous decomposition to HF and O_2 of ~30 min at 25°C. It is prepared, with difficulty, by reaction of F_2 with H_2O at low temperature. The crystalline compound has HOF molecules ($\angle HOF = 101°$, $O-F = 1.44$ Å) linked by hydrogen bonds into planar, zig-zag chains.[41] Reaction of F_2 diluted in N_2 with H_2O (10%) in MeCN gives a solution of HOF in MeCN that is useful as an oxygen atom transfer agent.[42] A number of compounds containing covalently bound OF groups, and called hypofluorites, are known, examples being CF_3OF, SF_5OF, O_3ClOF, and FSO_2OF.

Table 13-6 Equilibrium Concentrations in Aqueous Solutions of the Halogens at 25°C (mol L^{-1})

	Cl_2	Br_2	I_2
Total solubility	0.091	0.21	0.0013
Concentration X_2(aq)	0.061	0.21	0.0013
$[H^+] = [X^-] = [HOX]$	0.030	1.15×10^{-3}	6.4×10^{-6}

[38]T. X. Wang and D. W. Margerum, *Inorg. Chem.* **1994**, *33*, 1050.
[39]D. W. Margerum *et al.*, *Inorg. Chem.* **1996**, *35*, 995.
[40]K. Kustin *et al.*, *Inorg. Chem.* **1993**, *32*, 5880.
[41]D. Mootz *et al.*, *Angew. Chem. Int. Ed. Engl.* **1988**, *27*, 392.
[42]E. H. Appelman *et al.*, *J. Fluorine Chem.* **1992**, *56*, 199; S. Rozen and Y. Bareket, *J. Chem. Soc. Chem. Commun.* **1994**, 1959.

The decomposition of aqueous HOCl has been studied in some detail.[43] The overall stoichiometry is described by

$$2HOCl + OCl^- \longrightarrow ClO_3^- + 2H^+ + 2Cl^-$$

The other HOX compounds are also unstable. In water their dissociation constants are HOCl, 3.4×10^{-8}; HOBr, 2×10^{-9}; HOI, 2×10^{-11}. As can be readily seen, reaction of halogens with water does not constitute a suitable method for preparing aqueous solutions of the hypohalous acids owing to the unfavorable equilibria. A useful general method is interaction of the halogen and a well-agitated suspension of mercuric oxide:

$$2X_2 + 2HgO + H_2O \longrightarrow HgO \cdot HgX_2 + 2HOX$$

The hypohalous acids are good oxidizing agents, especially in acid solution (see Table 13-5) and this accounts for the chief industrial use of hypochlorites, on a tonnage scale, for bleaching and sterilizing. Hypohalites and the acids are also used to halogenate both aromatic and aliphatic organic compounds.

The hypohalite ions can all be produced in principle by dissolving the halogens in cold base according to the general reaction

$$X_2 + 2OH^- \longrightarrow X^- + XO^- + H_2O$$

and for these reactions the equilibrium constants are all favorable—7.5×10^{15} for Cl_2, 2×10^8 for Br_2, and 30 for I_2—and the reactions are rapid.

The situation, however, is complicated by the tendency of the hypohalite ions to disproportionate further in basic solution to produce the halate ions:

$$3XO^- = 2X^- + XO_3^-$$

For this reaction the equilibrium constant is in each case very favorable, that is, 10^{27} for ClO^-, 10^{15} for BrO^-, and 10^{20} for IO^-. Thus the actual products obtained on dissolving the halogens in base depend on the rates at which the hypohalite ions initially produced undergo disproportionation, and these rates vary from one to the other and with temperature.

The disproportionation of ClO^- is slow at and below room temperature. Thus when chlorine reacts with base "in the cold," reasonably pure solutions of Cl^- and ClO^- are obtained. In hot solutions ($\sim75°C$) the rate of disproportionation is fairly rapid and under proper conditions, good yields of ClO_3^- can be secured.

The disproportionation of BrO^- is moderately fast even at room temperature. Consequently solutions of BrO^- can only be made and/or kept at around 0°C. At temperatures of 50 to 80°C quantitative yields of BrO_3^- are obtained:

$$3Br_2 + 6OH^- \longrightarrow 5Br^- + BrO_3^- + 3H_2O$$

[43]G. Gordon *et al., Inorg. Chem.* **1992,** *31,* 3534.

The rate of disproportionation of IO^- is very fast at all temperatures, so that it is unknown in solution. Reaction of iodine with base gives IO_3^- quantitatively according to an equation analogous to that for Br_2.

Halous Acids and Halites

These do not arise in the hydrolysis of the halogens. The compound HOIO apparently does not exist, HOBrO is doubtful, and HOClO is not formed by disproportionation of HOCl if for no other reason than that the equilibrium constant is quite unfavorable:

$$2HOCl = Cl^- + H^+ + HOClO \qquad K \sim 10^{-5}$$

The reaction

$$2ClO^- = Cl^- + ClO_2^- \qquad K \sim 10^7$$

is favorable, but the disproportionation of ClO^- to ClO_3^- and Cl^- (cited previously) is so much more favorable that the first reaction is not observed.

Finally, we must consider the possibility of production of perhalate ions by disproportionation of the halate ions. Since the acids $HOXO_2$ and $HOXO_3$ are all strong, these equilibria are independent of pH. The reaction

$$4ClO_3^- = Cl^- + 3ClO_4^-$$

has an equilibrium constant of 10^{29}, but it takes place only very slowly in solution even near 100°C; hence perchlorates are not readily produced. Neither perbromate nor periodate can be obtained in comparable disproportionation reactions because the equilibrium constants are 10^{-33} and 10^{-53}, respectively.

The only definitely known halous acid, chlorous acid, is obtained in aqueous solution by treating a suspension of barium chlorite with sulfuric acid and filtering off the precipitate of barium sulfate. It is a relatively weak acid ($K_A \approx 10^{-2}$) and cannot be isolated in the free state. Chlorites ($MClO_2$) themselves are obtained by reaction of ClO_2 with solutions of bases:

$$2ClO_2 + 2OH^- \longrightarrow ClO_2^- + ClO_3^- + H_2O$$

Sodium chlorite is manufactured on a ton scale for use as a bleaching agent. In alkaline solution the ion is stable to prolonged boiling and up to a year at 25°C in the absence of light. In acid solutions, however, the decomposition is rapid and is catalyzed by Cl^-:

$$5HOClO \longrightarrow 4ClO_2 + Cl^- + H^+ + 2H_2O$$

but the reaction sequence is complicated.

The ClO_2^- and BrO_2^- ions are both angular.[44] The yellow salt $NaBrO_2 \cdot 3H_2O$ is a mild, selective oxidant in organic chemistry. The ClO_2^- ion forms a few weak complexes in solution,[45] such as $CuClO_2^+$.

Halic Acids and Halates

Only iodic acid (HIO_3) is known out of aqueous solution. The white solid consists of pyramidal $IO_2(OH)$ molecules connected by hydrogen bonds. The IO_3^- ion as well as the BrO_3^- and ClO_3^- ions are pyramidal. Iodic acid is easily made by oxidizing I_2 with concentrated nitric acid or other strong oxidizing agents. It can be dehydrated to I_2O_5 and yields salts of the types $M^IH(IO_3)_2$ and $M^IH_2(IO_3)_3$ as well as M^IIO_3. The acids, HXO_3, are all strong.

While all the halates can be obtained by reactions of X_2 with hot aqueous alkali,

$$3X_2 + 6OH^- \longrightarrow XO_3^- + 5X^- + 3H_2O$$

this is unsuitable for the large-scale manufacture of $NaClO_3$ since only $1/6$ of the Cl_2 is converted. Instead an electrolytic oxidation of brine is used, which can give conversions of up to 90% efficiency based on current used.

The halic acids are all good oxidizing agents, but their chemical behavior is extremely complex and varied depending on temperature, acidity, and other factors. Both iodate and bromate participate in reactions with complex time-dependent behavior. There is the long-known Landolt chemical clock reaction in which HIO_3 and Na_2SO_3 react in acid solution in the presence of starch so as to generate the blue color indicative of free iodine only after predictable time intervals, and when sulfite is in excess in a repetitive, periodic manner. Other periodic reactions include that of HIO_3 with H_2O_2 (Bray reaction), the BrO_3^-/I^- reaction, and finally, the celebrated Belousov-Zhabotinskii oscillating reaction, which occurs in solutions of $KBrO_3$, Ce^{IV}, and malonic acid in stirred sulfuric acid solution.

Perchlorates

Although disproportionation of ClO_3^- to ClO_4^- and Cl^- is thermodynamically very favorable, the reaction occurs only very slowly in solution and does not constitute a useful preparative procedure. Perchlorates are commonly prepared by electrolytic oxidation of chlorates. The properties of perchloric acid were discussed in Section 2-12.

Perchlorates of almost all electropositive metals are known. Except for a few with large cations of low charge, such as $CsClO_4$, $RbClO_4$, and $KClO_4$, they are readily soluble in water. Solid perchlorates containing the tetrahedral ClO_4^- ion are often isomorphous with salts of other tetrahedral anions (e.g., MnO_4^-, SO_4^{2-}, and BF_4^-). A particularly important property of the perchlorate ion is its slight tendency to serve as a ligand in complexes. Thus perchlorates are widely used in studies of complex ion formation, the *assumption* being made that no appreciable correction for the concentration of perchlorate complexes need be considered. This

---

[44]W. Levason *et al.*, *J. Chem. Soc. Dalton Trans.* **1990**, 349. This reference contains extensive structural data on compounds with Br—O bonds.
[45]I. Fábián and G. Gordon, *Inorg. Chem.* **1991**, *30*, 3785.

may often be true for aqueous solutions, but it is well known that when no other donor is present to compete, perchlorate ion exercises a donor capacity and can be monodentate, bridging bidentate, or chelating bidentate (see also Section 11-16). This is illustrated by structures of compounds such as $(CH_3)_3SnClO_4$, $Co(MeSC_2H_4SMe)_2(ClO_4)_2$, $(Ph_3BiOBiPh_3)(ClO_4)_2$, $Sb_2Cl_6(OH)(O)(ClO_4)$, and $Ti(ClO_4)_4$.

The use of perchlorate as an ion for the isolation of crystalline salts of cations containing organic ligands should be avoided where possible as they can detonate dangerously. Although replacement ions such as $CF_3SO_3^-$, PF_6^-, BF_4^-, and $[B(C_6H_5)_4]^-$ can be used, some of these can also act as ligands.

Although ClO_4^- is potentially a good oxidant

$$ClO_4^- + 2H^+ + 2e = ClO_3^- + H_2O \qquad E^0 = 1.23 \text{ V}$$

in aqueous solution it is reduced only by Ru^{II} (to ClO_3^-), and by V^{II}, V^{III}, Mo^{III}, Mo^{III}_2, and Ti^{III} to Cl^-. Despite the more favorable potential for reduction by Eu^{2+} or Cr^{2+}, no reaction occurs, possibly because intermediates with M=O bonds are required in the reductions.

Perbromic Acid and Perbromates

Perbromates were prepared only in 1968; previously there were many papers theoretically justifying their nonexistence. This provides an excellent example of the folly of concluding the nonexistence of certain compounds until all conceivable preparative methods have been exhausted.

The potential

$$BrO_4^- + 2H^+ + 2e = BrO_3^- + H_2O \qquad E^0 = +1.76 \text{ V}$$

shows that only the strongest oxidants can form perbromate. Kinetic factors must be responsible for the failure of ozone $(E^\circ = +2.07 \text{ V})$ and $S_2O_8^{2-}$ $(E^\circ = +2.01 \text{ V})$ to cause oxidation.

Small amounts of perbromic acid or perbromates can be obtained by oxidation of BrO_3^- electrolytically or by the action of XeF_2. The best preparation involves oxidation of BrO_3^- by fluorine in 5 M NaOH solution; by a rather complicated procedure, pure solutions can be obtained:

$$BrO_3^- + F_2 + 2OH^- \longrightarrow BrO_4^- + 2F^- + H_2O$$

Solutions of $HBrO_4$ can be concentrated up to 6 M (55%) without decomposition and are stable indefinitely even at 100°C. More concentrated solutions, up to 83%, can be obtained but these are unstable; the hydrate $HBrO_4 \cdot 2H_2O$ can be crystallized.

Salts such as $KBrO_4$ (dec $\sim$ 275°C) and NH_4BrO_4 (dec $\sim$ 170°C) are fairly stable thermally and are isomorphous with their perchlorate analogues.

Perbromate is an even stronger oxidant than ClO_4^- (1.23 V) or IO_4^- (1.64 V) but it is very sluggish at 25°C. The dilute acid is only slowly reduced by I^-, Br^- (not at all by Cl^-), and by other reagents. However, the 3 M acid readily oxidizes stainless steel and the 12 M acid rapidly oxidizes Cl⁻ and explodes on contact with cellulose.

The high oxidizing power of BrO_4^- can be related to its being thermodynamically less stable than either ClO_4^- or IO_4^-. The ΔG_f^0 values (kJ mol⁻¹ at 298°C) for the three potassium salts are as follows: $KClO_4$, −302; $KBrO_4$, −174; KIO_4, −349. This anomaly may be compared with some similar behavior of selenates and arsenates, and is part of the general tendency of these three elements to have properties that cannot be quantitatively interpolated from those of their lighter and heavier congeners.

Periodic Acid and Periodates

This chemistry is more complex than for the other perhalates because there are at least four structural types. In aqueous solution the acid HIO_4, which is strong, and the series based on H_5IO_6 are predominant. The following rapidly established equilibria are of major importance in acid solutions:

$$I(OH)_6^+ = IO(OH)_5 + H^+ \qquad\qquad K \sim 6$$

$$IO(OH)_5 = IO_2(OH)_4^- + H^+ \qquad\qquad K \sim 5 \times 10^{-4}$$

$$IO_2(OH)_4^- = IO_4^- + 2H_2O \qquad\qquad K \sim 29$$

$$IO_2(OH)_4^- = IO_3(OH)_3^{2-} + H^+ \qquad\qquad K \sim 5 \times 10^{-9}$$

$$2IO_3(OH)_3^{2-} = (HO)_2I_2O_8^{4-} + 2H_2O \qquad K \sim 820$$

Microcrystalline HIO_4 (called periodic acid) can be made by careful dehydration of H_5IO_6 (see below) with oleum in H_2SO_4 at 50°C; it consists of one-dimensional infinite chains built up of distorted, edge-sharing IO_6 octahedra.[46] The IO_4^- ion (called periodate or, sometimes, metaperiodate) is tetrahedral. The other species, are derived from the acid H_5IO_6, called orthoperiodic acid, all are based on IO_6 octahedra; they are oligomeric and have the formula $H_{6+n}I_{2+n}O_{10+4n}$. The $(HO)_2I_2O_8^{4-}$ ion consists of two octahedra sharing a pair of oxygen atoms and when LiH_4IO_6 is dehydrated (ultimately to $LiIO_4$) it gives an intermediate compound, $Li_2H_4I_2O_{10}$, which also has two octahedra bridged by O atoms.[47]

Periodates are best obtained by oxidizing iodate (or even I⁻ or I_2) electrolytically or with Cl_2 in strongly basic solution to give $Na_3H_2IO_6$, or by thermal disproportionation of an iodate:

$$5Ba(IO_3)_2 \longrightarrow Ba_5(IO_6)_2 + 4I_2 + 9O_2$$

The white crystalline acid (H_5IO_6) can then be obtained from strongly acid solutions, and this can be dehydrated at 100°C in vacuum to HIO_4. Under other conditions

[46]T. Kraft and M. Jansen, *Angew. Chem. Int. Ed. Engl.* **1997,** *36,* 1753.
[47]M. Jansen and R. Müller, *Z. anorg. allg. Chem.* **1996,** *622,* 1901.

Table 13-7 Oxohalogen Fluorides, F_nXO_m

Formal oxidation No. of X	Chlorine	Bromine	Iodine
III	FClO		
V	$FClO_2$	$FBrO_2$	FIO_2
	F_3ClO	F_3BrO	F_3IO
VII	$FClO_3$	$FBrO_3$	FIO_3
	F_3ClO_2		F_3IO_2
			F_5IO

$H_7I_3O_{14}$ (structure unknown) is obtained. There is also a square pyramidal IO_5^{3-} ion that occurs in K_3IO_5. With H^+ in concentrated acid solutions the compounds $[I(OH)_6]HSO_4$ and $[I(OH)_6]_2SO_4$ can be obtained. Colorless $I(OH)_6^+$ is a member of the isostructural series $Sn(OH)_6^{2-}$, $Sb(OH)_6^-$, and $Te(OH)_6$.

Periodate anions have an extensive chemistry as ligands, being able to stabilize transition metals in high oxidation states. Examples[48] are $CeHIO_6 \cdot 4H_2O$, $Na_6\{OsO_2[IO_5(OH)_2]_2\} \cdot 18H_2O$ and $RuO_2(bipy)[IO_3(OH)_3] \cdot 1.5H_2O$. The latter is an epoxidation catalyst for alkenes using $NaIO_4$ or $(Bu_4^nN)IO_4$ as co-oxidants.

Periodic acids themselves are strong and usually rapid oxidants that are often selective.

13-13 Oxohalogen Fluorides

This class comprises those compounds containing F, O, and X that have fluorine to halogen bonds. They are often called halogen oxofluorides but this term is ambiguous. Compounds containing $F-O-X$ groups are discussed in the next section. The known compounds are listed in Table 13-7. No one halogen has yet been shown to form all six types. The structures of the molecular species, some of which are shown in Fig. 13-2, are all predictable from the VSEPR formalism. Thus FXO is bent (two lone pairs), FXO_2 is pyramidal (one lone pair), FXO_3 has C_{3v} symmetry. F_3XO is an incomplete *tbp* with F, O, and a lone pair in the equatorial plane and F_3XO_2 is a complete *tbp* with F, O, O equatorial; F_5IO is octahedral with C_{4v} symmetry.

While most of the molecules are monomers, F_3IO_2 both dimerizes and trimerizes *via* oxygen bridges. In some cases solid adducts with other molecules are formed. Thus, there is $(F_3IO_2SbF_5)_2$, which is cyclic with alternating SbF_4 and IF_5 units linked by oxygen atoms, and $(F_3IO_2 \cdot F_3IO)_2$, with alternating F_2IO and IF_4 units linked by oxygen atoms.

In the 6-coordinate F_5IO (C_{4v}) *all* $I-F$ bonds are the same length, indicating that there is no *trans* effect for the axial fluoride *trans* to the O atom which is above the plane of the four equatorial F atoms.[49]

[48]W. Levason *et al.*, *Polyhedron* **1996**, *15*, 409; *J. Chem. Soc., Dalton Trans.* **1994**, 1483; W. P. Griffith *et al.*, *J. Chem. Soc., Chem. Commun.* **1994,** 1833.
[49]K. O. Christe *et al.*, *J. Am. Chem. Soc.* **1993**, *115*, 9655.

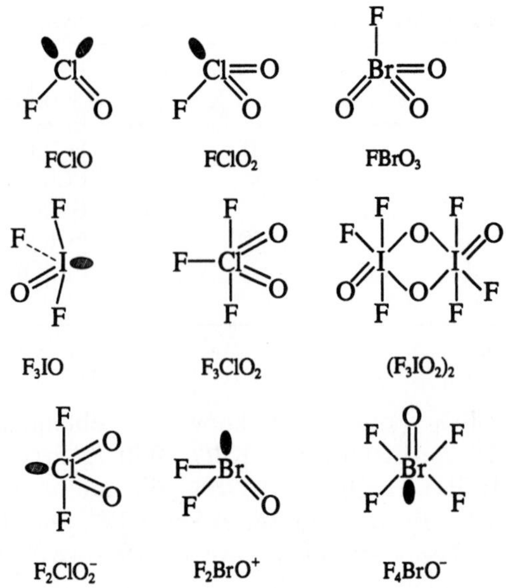

Figure 13-2 Some representative structures of oxohalogen fluoride molecules and ions.

A variety of *cations* and *anions* are derived formally (and as a general rule also in practice) from some of the neutral molecules by removal or addition of F⁻. These are the types shown in Table 13-8. Their structures are again in accord with VSEPR formalism.

Examples of the synthetic methods are

$$IF_7 + OPF_3 \longrightarrow OIF_5 + PF_5$$

$$F_3BrO + BF_3 \longrightarrow [F_2BrO]^+[BF_4]^-$$

$$F_5IO + Me_4NF \xrightarrow[-31\,°C]{MeCN} [Me_4N]^+[F_6IO]^-$$

Table 13-8 Cations and Anions of General Formula $F_nXO_m^{+,-}$

"Parent" molecule	Cations	Anions
FXO_2	ClO_2^+	$F_2ClO_2^-$
	BrO_2^+	$F_2BrO_2^-$
		$F_2IO_2^-$
F_3XO	F_2ClO^+	F_4ClO^-
	F_2BrO^+	F_4BrO^-
F_3XO_2	$F_2ClO_2^+$	F_6IO^{-a}

[a]Rigid in solution by nmr, see Ref. 50.

Most of the compounds are highly reactive oxidizing and fluorinating agents. The structure of F_6IO^- is pentagonal bipyramidal with five F atoms in the equatorial plane.[50]

Trioxohalofluorides, FXO₃

The most important compound, $FClO_3$, is best made by the reaction

$$KClO_4 + 2HF + SbF_5 \xrightarrow{40\text{-}50°C} FClO_3 + KSbF_6 + H_2O$$

The toxic gas (mp $-147.8°C$, bp $-46.7°C$) is thermally stable to 500°C and resists hydrolysis. At elevated temperatures it is a powerful oxidizing agent and has selective fluorinating properties, especially for replacement of H by F in CH_2 groups. It can also be used to introduce ClO_3 groups into organic compounds (e.g., C_6H_5Li gives $C_6H_5ClO_3$). In these reactions it appears that the nucleophile attacks the chlorine atom and the F^- ion is expelled:

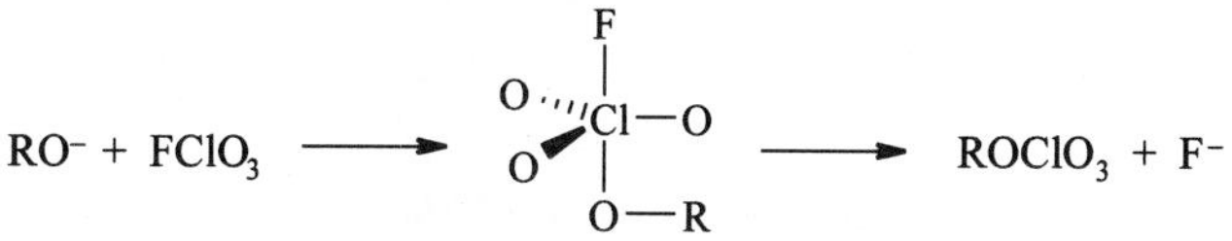

$$RO^- + FClO_3 \longrightarrow \underset{\underset{O-R}{\overset{F}{|}}}{\overset{O}{\underset{}{O}}\text{Cl}-O} \longrightarrow ROClO_3 + F^-$$

The bromide $FBrO_3$ (mp $-110°C$) is made similarly, but it is more reactive than $FClO_3$ and is hydrolyzed by base:

$$FBrO_3 + 2OH^- = BrO_4^- + H_2O + F^-$$

presumably by initial attack of OH^- on Br.

13-14 Other Oxo Compounds

The compounds Cl_2O_4 and Cl_2O_6 have been discussed in Section 13-11. *Fluorine perchlorate*, $FOClO_3$ is a colorless gas (bp 16°C) made by thermal decomposition of NF_4ClO_4 together with NF_3. It has been used in the synthesis of anhydrous metal perchlorates. There are four halogen fluorosulfates, $XOSO_2F$, but all are reactive and thermally unstable. The fluoride is the best known and is made by interaction of F_2 with SO_3; the others are made by reacting X_2 with $S_2O_6F_2$. There are similarly four nitrates, $XONO_2$, but only the compounds with X = F, Cl, and Br are well established. All are thermally unstable and highly reactive. The fluoride, made by interaction of F_2 and KNO_3 has a planar structure, $\angle NOF$ angle = 106°.[51] Other compounds involving carbon[52] are $FC(O)OF$, $CF_3C(O)OX$, X = F, Cl, and $F_2C(OCl)(OF)$.

[50]K. O. Christe *et al., J. Am. Chem. Soc.* **1993,** *115,* 2696.
[51]H. Oberhammer *et al., J. Am. Chem. Soc.* **1994,** *116,* 8317.
[52]H. Willner *et al., Inorg. Chem.* **1995,** *34,* 2089; A. Russo and D. D. Desmarteau, *Inorg. Chem.* **1995,** *34,* 6221.

13-15   Other Compounds with Formally Positive Halogens

The +1 Oxidation State

Although Br^+ has been postulated to occur in reactions of Br_2, the evidence is usually only kinetic. However, the solvated linear ions L_2Br^+ and L_2I^+, where L may be pyridine, MeCN, quinuclidine, etc.,[53] are known.

These mild oxidants can be made in reactions like

$$UF_6 + \tfrac{1}{2}I_2 \xrightarrow{\text{MeCN}} [I(NCMe)_2]^+UF_6^-$$

but Br^+ is too electrophilic to exist in acetonitrile. The reaction in liquid SO_2

$$I_3^+AsF_6^- + MeCN \longrightarrow [MeCNI]^+AsF_6^- + I_2$$

gives a 95% yield of the salt,[54] while in $CHCl_3$ there is the reaction

$$I_2 + [2,2,2\text{-cryptate}] \longrightarrow [2,2,2\text{-C}]I^+ + I^- \qquad K > 10^7$$

which is reminiscent of the disproportionation $2Na = Na^+ + Na^-$.

The +3 Oxidation State

Many I^{III} compounds are known.[55] The T-shaped $PhICl_2$, made by direct reaction of PhI with Cl_2, is a very useful synthetic source of Cl_2, and iodosylbenzene (PhIO) finds wide use as an oxygen-transfer agent, as, for example, in the conversion of Fe^{III} porphyrins to Fe^VO porphyrins. The polymeric solid, $-I(Ph)-O-I(Ph)-O-$, is not generally very soluble except in alcohols, especially methanol, where $PhI(OMe)_2$ is formed. Among the compounds in which I^{III} is combined with oxoanions are $I(OSO_2F)_3$, $I(NO_3)_3$, $I(OCOCH_3)_3$, IPO_4, and $I(OClO_3)_3$. These compounds contain essentially covalent I—O bonds. Preparative methods include the following:

$$2I_2 + 3AgClO_4 \xrightarrow{\text{ether}} I(OClO_3)_3 + 3AgI$$

$$I_2 + 6ClOClO_3 \xrightarrow{-85\,^\circ C} 2I(OClO_3)_3 + 3Cl_2$$

$$CsI + 4ClOClO_3 \longrightarrow Cs[I(OClO_3)_4] + 2Cl_2$$

$$I_2 + HNO_3\,(\text{fuming}) + (CH_3CO)_2O \longrightarrow I(OCOCH_3)_3$$

$$I_2 + HNO_3\,(\text{conc}) + H_3PO_4 \longrightarrow IPO_4$$

[53]J. M. Winfield *et al., J. Chem. Soc., Dalton Trans.* **1995**, 3837.
[54]P. K. Gowik and T. M. Klapötke, *J. Chem. Soc., Chem. Commun.* **1990**, 1433.
[55]P. P. Power *et al., Inorg. Chem.* **1995**, *34*, 3210; J. Protasiewicz, *J. Chem. Soc., Chem. Commun.* **1995**, 1115.

The compounds are sensitive to moisture and are not stable much above room temperature. They are hydrolyzed with disproportionation of the I^{III}, as illustrated for IPO_4 thus:

$$5IPO_4 + 9H_2O \longrightarrow I_2 + 3HIO_3 + 5H_3PO_4$$

Covalent I^{III} is known also in the compound triphenyliodine $(C_6H_5)_3I$ and a large number of diaryliodonium salts, such as $(C_6H_5)_2I^+X^-$, where X may be one of a number of common anions.

The I^{III} compounds (13-IV) and (13-V) are also well characterized, with the expected T geometry at the iodine atom.

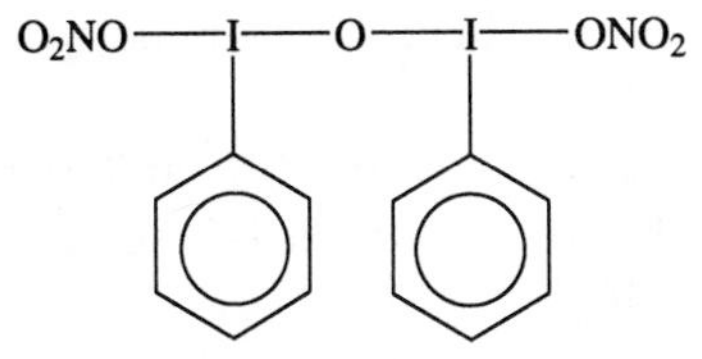

(13-IV) 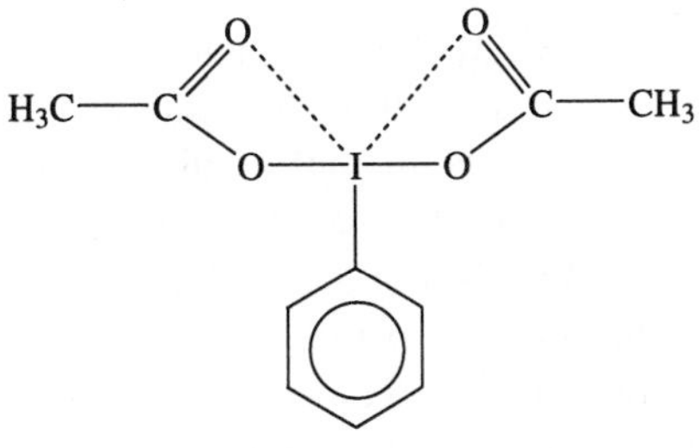   (13-V)

The structure of $I(OCOMe)_3$ is T-shaped as expected with three primary I—O—C bonds and longer I—O interactions. In the Bu_4N^+ salt the anion $[I(OTeF_5)_4]^-$ has a square IO_4 unit.[56] Iodine(III) species with alkynyl phenyl groups such as $[RC\equiv C-I-Ph]^+BF_4^-$ are useful synthons for organic compounds.[57]

The +5 Oxidation State

A number of ψ-tbp organo compounds are known, two examples being (13-VI) and (13-VII).[58]

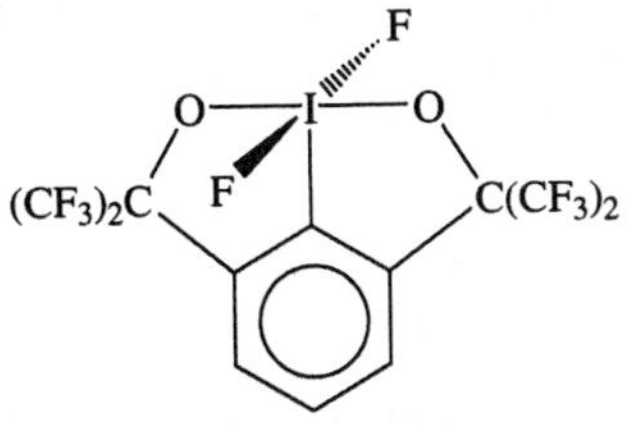

(13-VI) 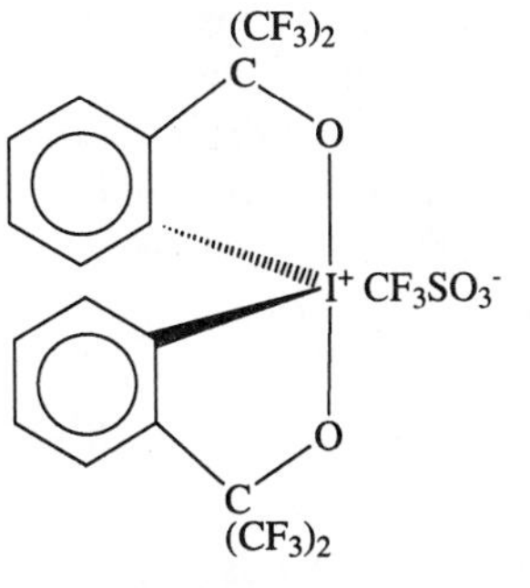   (13-VII)

[56]L. Turowsky and K. Seppelt, Z. anorg. allg. Chem. 1991, 602, 79.
[57]P. J. Stang, et al., Angew. Chem. Int. Ed. Engl. 1992, 31, 274 (Review).
[58]See J. C. Martin et al., J. Am. Chem. Soc. 1993, 115, 2488.

Less stable cations of I^V such as $PhIF_3^+$ are known only in solution. Entirely inorganic compounds are $I(OTeF_5)_5$, $O{=}I(OTeF_5)_3$ and the square pyramidal anion in $(Bu_4N)[O{=}I(OTeF_5)_4]$.

INTERHALOGEN AND POLYHALOGEN MOLECULES, CATIONS, AND ANIONS

13-16   General Survey

The halogens form many compounds that are (or contain local units that are) binary or even ternary combinations of halogen atoms. The majority of these (the only important exceptions being BrCl, ICl, IBr, and ICl_3 among neutral molecules) contain fluorine. There are four basic types:

1. *Neutral Molecules.* These are all binary compounds such as those just mentioned as well as, for example, BrF_3 and IF_7. They are all of the type XX'_n where n is an odd number. They are all closed shell, diamagnetic species, mostly rather volatile.

2. *Cations.* These are both homonuclear, for example, I_2^+ and Cl_3^+, and heteronuclear, for example, $I_3Cl_2^+$, and may be either paramagnetic (I_2^+) or diamagnetic (ICl_2^+). There is one ternary cation, $IBrCl^+$.

3. *Anions.* A great many of these are known, in addition to the classical I_3^- ion. It is in this group that ternary combinations, such as $IBrCl^-$ and $IBrCl_3^-$, are mostly found. They generally occur in combination with large cations as crystalline solids.

4. *Covalent Organo Derivatives.* These are limited to a relatively few aryl species, mostly of the types $ArXF_n$, where X = I or Br and n = 2 or 4, although RXF_2 and Ar_2XF types are also known. A typical preparative route involves low-temperature fluorination of RX or ArX (X = Br and I) compounds, but attack on R or Ar occurs under more strenuous conditions. With C_6F_5Br and C_6F_5I, however, the $C_6F_5XF_4$ as well as the $C_6F_5XF_2$ products are readily accessible. With iodine only it is also possible to obtain compounds such as $ArICl_2$, which has just been discussed.

For interhalogen compounds of types (1)-(3) the following sections will provide more detailed information. Chemically, the molecular interhalogens are all rather reactive. They are corrosive oxidizing substances and attack most other elements, producing mixtures of the halides. They are all more or less readily hydrolyzed (some, e.g., BrF_3, being dangerously explosive in this respect), in some cases according to the equation

$$XX' + H_2O = H^+ + X'^- + HOX$$

The diatomic compounds often add to ethylenic double bonds and may react with the heavier alkali and alkaline earth metals to give polyhalide salts.

The *diatomic compounds* are ClF, BrF, BrCl, IBr, and ICl. In their physical properties they are usually intermediate between the constituent halogens. They are of course polar, whereas the halogen molecules are not. Chlorine monofluoride

(ClF) is a colorless gas, whereas BrF, BrCl (bp 20°C), and ICl (mp 27°C, dec 97°C) are red or red brown, and IBr is dark gray (mp 41°C, subl 50°C). The hydrolysis of interhalogens is very rapid compared to the rates for the halogens themselves.[59]

Iodine monofluoride (IF) is unknown except in minute amounts observed spectroscopically. It is apparently too unstable with respect to disproportionation to IF_5 and I_2 to permit its isolation. The other isolable diatomic compounds have varying degrees of stability with respect to disproportionation and fall in the following stability order, where the numbers in parentheses represent the disproportionation constants for the gaseous compounds and the elements in their standard states at 25°C: ClF (2.9×10^{-11}) > ICl (1.8×10^{-3}) > BrF (8×10^{-3}) > IBr (5×10^{-2}) > BrCl (0.34).

Bromine monofluoride also disproportionates according to

$$3BrF = BrF_3 + Br_2$$

Chlorine monofluoride may be prepared by direct interaction at 220 to 250°C and it is readily freed from ClF_3 by distillation, but it is best prepared by interaction of Cl_2 and ClF_3 at 250 to 350°C. Bromine monofluoride also results on direct reaction of Br_2 with F_2, but it has never been obtained in high purity because of its ready disproportionation. Iodine monochloride is obtained as brownish-red tablets (β form) by treating liquid chlorine with solid iodine in stoichiometric amount, and cooling to solidify the liquid product. It readily transforms to the α form, ruby-red needles. Bromine monochloride is prone to dissociation:

$$2BrCl \rightleftharpoons Br_2 + Cl_2 \qquad K = 0.145 \ (25°C \ in \ CCl_4)$$

Iodine monobromide, a solid resulting from direct combination of the elements, is endothermic and extensively dissociated in the vapor. It is used instead of Br_2 in some industrial processes. Despite the general instability of the BrX compounds, the fluorosulfate ($BrOSO_2F$) obtained by treating Br_2 with $S_2O_6F_2$, is stable to 150°C. Iodine trichloride (ICl_3) is also formed (like ICl) by treatment of liquid chlorine with the stoichiometric quantity of iodine, or with a deficiency of iodine followed by evaporation of the excess of chlorine. It is a fluffy orange powder, unstable much above room temperature, with a planar dimeric structure (13-VIII).

a = 2.70 Å, b = 2.39 Å

(13-VIII)

The most important class of interhalogen molecules are the fluorides, XF_n (n = 1, 3, 5, and 7), to which we now turn.

[59]D. W. Margerum et al., *Inorg. Chem.* **1991**, *30*, 4838.

Table 13-9   Some Physical Properties of Halogen Fluorides

	mp (°C)	bp (°C)	Structure
ClF	−156.6	−100.1	
ClF$_3$	−76.3	11.75	Planar; distorted T
ClF$_5$	−103	−14	Square pyramidal
BrF	−33	20	
BrF$_3$	9	126	Planar; distorted T
BrF$_5$	−60	41	Square pyramidal
IF$_3$			
IF$_5$	10	101	Square pyramidal
IF$_7$	6.45		Pentagonal bipyramidal[a]

[a] nmr spectra show high fluxionality. K. O. Christe *et al., J. Am. Chem. Soc.* **1993,** *115,* 2696.

13-17   Halogen Fluoride Molecules

These compounds and some of their important physical properties are listed in Table 13-9. The preparations of ClF and BrF have already been mentioned. Chlorine trifluoride may be prepared by direct combination of the elements at 200 to 300°C and is available commercially. It is purified by converting it into KClF$_4$ by the action of KF and thermally decomposing the salt at 130 to 150°C. Chlorine pentafluoride (ClF$_5$) can be made efficiently either by interaction of F$_2$ with ClF$_3$ or by the reaction:

$$KCl + 3F_2 \xrightarrow[\text{bomb}]{200\,°C} KF + ClF_5$$

ClF$_5$ is a colorless gas; it is less stable thermally but also less reactive than ClF$_3$, and above 165°C there is the equilibrium

$$ClF_5 \rightleftharpoons ClF_3 + F_2$$

The other halides are best prepared by the reactions

$$Br_2 + 3F_2 \xrightarrow{200\,°C} 2BrF_3$$

$$BrF_3 + F_2 \longrightarrow BrF_5$$

$$I_2 + 5F_2 \xrightarrow{25\,°C} 2IF_5$$

$$KI + 4F_2 \longrightarrow KF + IF_7$$

Bromine trifluoride and IF$_5$ are formed by the reaction of X$_2$ with AgF in HF to give AgX plus XF, followed by disproportionation of XF, but the products are not easily purified. Iodine trifluoride is a yellow powder obtained by fluorination of I$_2$ in Freon at −78°C; it decomposes to I$_2$ and IF$_5$ above −35°C.

The halogen fluorides are very reactive, and with water or organic substances they react vigorously or explosively. They are powerful fluorinating agents for inorganic compounds or, when diluted with nitrogen, for organic compounds. The most useful compounds are ClF, ClF_3, and BrF_3. The order of reactivity is approximately $ClF_3 > BrF_3 > BrF_5 > IF_7 > ClF > IF_5 > BrF$.

Certain compounds, notably ClF, BrF_3, and IF_5, have high entropies of vaporization, and BrF_3 has appreciable electrical conductance. To account for these observations, association by fluorine bridging, in addition to self-dissociation,

$$2BrF_3 \longrightarrow BrF_2^+ + BrF_4^-$$

have been postulated. An analogy with other solvent systems can be made. In liquid BrF_3, for example, the "acid" would be BrF_2^+ and the "base" BrF_4^-. Indeed, suitable compounds such as $BrF_3 \cdot SbF_5$, (or $BrF_2^+ \cdot SbF_6^-$) and $KBrF_4$ dissolve in BrF_3 to give highly conducting solutions.

A characteristic property of most halogen fluorides is their amphoteric character; that is, with strong bases, such as alkali metal fluorides, they can form anions, and with strong Lewis acids, such as SbF_5, they can form cations:

$$XF_{n+1}^- \xleftarrow{\ +F^-\ } XF_n \xrightarrow{\ -F^-\ } XF_{n-1}^+$$

These anions and cations are discussed later. Some representative reactions are:

$$2ClF + AsF_5 \longrightarrow FCl_2^+AsF_6^-$$

$$ClF + CsF \longrightarrow Cs^+ClF_2^-$$

$$ClF_3 + CsF \longrightarrow Cs^+ClF_4^-$$

$$ClF_5 + SbF_5 \longrightarrow ClF_4^+SbF_6^-$$

$$ClF_3 + AsF_5 \longrightarrow ClF_2^+AsF_6^-$$

$$IF_5 + CsF \longrightarrow Cs^+IF_6^-$$

13-18 The Homoatomic X_n^+ Ions[60]

Iodine has long been known to dissolve in highly acidic, oxidizing media, for example, oleum, to give bright blue paramagnetic solutions; these solutions contain the I_2^+ ion and salts with anions of extremely low basicity can be isolated. The Br_2^+ ion may be similarly generated, but Cl_2^+ is still unknown except in the gas phase. These

[60]W. H. E. Schwartz *et al.*, *Inorg. Chem.* **1996,** *35,* 100 (Electronic structures and physical properties of X_3^+ and XY_2^+ ions.)

cations arise by removal of a π^* electron and, accordingly, the internuclear distances (r) decrease and the stretching frequencies (ν) increase from those in X_2, as shown here:

	Br_2	Br_2^+	I_2	I_2^+
r(Å)	2.28	2.13	2.67	2.56
ν(cm^{-1})	319	360	215	238

Salts of the X_2^+ ions are best made in HSO_3F as solvent using $S_2O_6F_2$ as oxidant followed by addition of SbF_5:

$$X_2 + S_2O_6F_2 \xrightarrow{\text{in } HSO_3F} \xrightarrow{SbF_5} X_2^+[Sb_3F_{16}^-]$$

The stability of these ions depends on the nature of the superacid solvent. Thus I_2^+ is stable in HF solutions of the highest acidity (HF, 0.2 M SbF_5, $-H_0 = 20.65$) but disproportionates to I_3^+, I_5^+, and I_2 as the acidity is decreased by adding different amounts of NbF_5 or TaF_5 or NaF,[61] e.g.,

$$14I_2^+ + 5F^- \longrightarrow 9I_3^+ + IF_5$$

These salts are thermodynamically quite stable: bright red $[Br_2^+][Sb_3F_{16}^-]$ melts at 85°C to a red liquid and $[I_2^+][Sb_2F_{11}^-]$ melts sharply at 127°C to a blue liquid.

For the I_2^+ ion, dimerization occurs to give a brown diamagnetic I_4^{2+} ion when solutions are cooled. Salts containing this cation, e.g., $[I_4^{2+}][AsF_6^-]_2$, have been isolated and the cations shown to have structure (13-IX), where the closest approaches of neighboring fluorine atoms are also indicated.

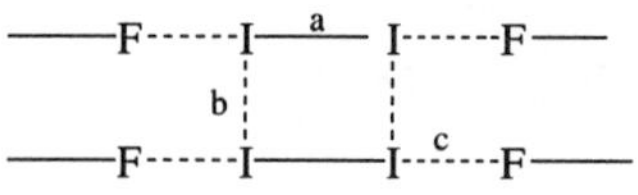

(13-IX)

This weak dimerization can be attributed to overlap of half-filled π^* orbitals on the two I_2^+ units.

The X_3^+ (X = Cl, Br, and I), Br_5^+, and I_5^+ ions are well established, with the I_7^+ ion being possible but uncertain. Salts of these ions can be prepared in a variety of ways, for example,

$$Cl_2 + ClF + AsF_5 \xrightarrow{-78°C} [Cl_3^+][AsF_6^-]$$

$$Au(s) + 4BrSO_3F \xrightarrow{\text{in } BrSO_3F} [Br_3^+][Au(SO_3F)_4^-] + \tfrac{1}{2}Br_2$$

$$[Br_3^+][Au(SO_3F)_4^-] + Br_2 \longrightarrow [Br_5^+][Au(SO_3F)_4^-]$$

[61]J. Besida and T. A. O'Donnell, *Inorg. Chem.* **1989**, *28*, 1669.

$$8Br_2 + 3XeF^+AsF_6^- \longrightarrow 3[Br_5^+][AsF_6^-] + 3Xe + BrF_3$$

$$3I_2 + 3AsF_5 \xrightarrow{\text{in } SO_2} 2[I_3^+][AsF_6^-] + AsF_3$$

$$2I_2 + ICl + AlCl_3 \longrightarrow [I_5^+][AlCl_4^-]$$

The X_3^+ are angular while X_5^+ ions have planar, *trans* Z-type structures;[62] the latter are also found for $I_3Br_2^+$ and $I_3Cl_2^+$.

13-19 Interhalogen Cations

These may be tri-, penta-, or heptaatomic (Table 13-10). The fluorocations are colorless or pale yellow and are explosively reactive oxidizers. Other cations are orange, red, or deep purple.

The structures normally are those predicted by VSEPR theory. Triatomics are angular, BrF_4^+ has a structure like SF_4, while IF_6^+ is octahedral; Cl_2F is bent and asymmetric, $Cl-Cl-F^+$.[63]

The syntheses involve interaction with an X^- acceptor that can also be oxidized. Some examples[64] are

$$2ClF + AsF_5 \longrightarrow [FCl_2^+][AsF_6^-]$$

$$I_2Cl_6 + 2SbCl_5 \longrightarrow 2[ICl_2^+][SbCl_6^-]$$

$$I_2 + 3Cl_2 + 2SbCl_5 \longrightarrow 2[ICl_2^+][SbCl_6^-]$$

$$ClF_3 + AsF_5 \longrightarrow [ClF_2^+][AsF_6^-]$$

$$BrF_5 + 2SbF_5 \longrightarrow [BrF_4^+][Sb_2F_{11}^-]$$

$$IF_7 + BF_3 \longrightarrow [IF_6^+][BF_4^-]$$

Table 13-10 Interhalogen Cations

Triatomic		Pentaatomic	Heptaatomic
ClF_2^+	I_2Cl^+	ClF_4^+	ClF_6^+
Cl_2F^+	IBr_2^+	BrF_4^+	BrF_6^+
BrF_2^+	I_2Br^+	IF_4^+	IF_6^+
IF_2^+	$IBrCl^+$	$I_3Cl_2^+$	
ICl_2^+			

[62]H. Hartl *et al.*, *Angew. Chem. Int. Ed. Engl.* **1991**, *30*, 328; K. O. Christe *et al.*, *Z. anorg. allg. Chem.* **1991**, *593*, 46.
[63]G. Frenkin and W. Koch, *Inorg. Chem.* **1990**, *29*, 4513.
[64]R. Bougon *et al.*, *Inorg. Chem.* **1991**, *30*, 102.

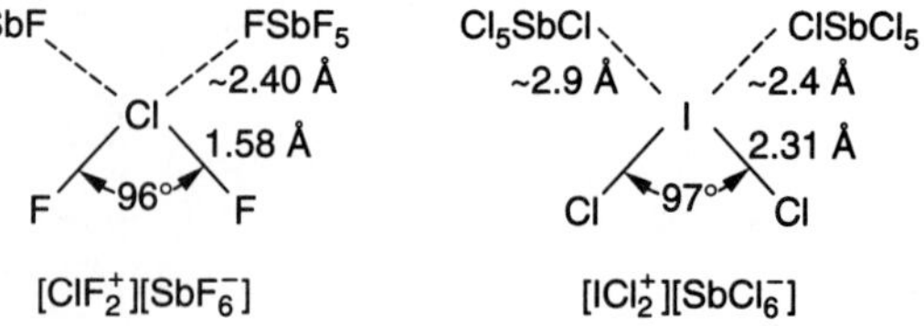

Figure 13-3 Two structures showing the close anion–cation interactions typical of polyhalogen cations.

Since ClF_7 and BrF_7 do not exist, the cations are made using strong oxidants that provide the necessary anions:

$$2ClF_5 + 2PtF_6 \longrightarrow [ClF_6^+][PtF_6^-] + [ClF_4^+][PtF_6^-]$$

$$BrF_5 + [KrF^+][AsF_6^-] \longrightarrow [BrF_6^+][AsF_6^-] + Kr$$

It is important to note that in all known structures, the anions make very close contacts, *via* halogen bridges, with the cations. Two representative structures illustrating this point are shown in Fig. 13-3.

The $I_3Cl_2^+$ ion is known only in the $SbCl_6^-$ salt. Its structure differs from the others being more like I_5^+.

13-20 Homoatomic Polyanions

These are formed mainly by iodine. Iodide ions in either aqueous or non-aqueous solutions commonly interact with I_2 to give polyiodides. In water at 25°C,

$$I^- + I_2 = I_3^- \qquad K \ ca \ 700$$

In solution I_3^- ions appear to be linear and symmetrical but in crystals, depending on the nature of the cation the anion may be unsymmetrical.

In addition to I_3^- there are numerous other polyiodine anions, especially those shown in Fig. 13-4. These are all formed by the relatively loose association of several I_2 molecules with one or two I^- ions.[65,66] For I_5^- there may be either discrete bent ions or infinite chains. The I_7^- ion can be described as an I_3^- ion to which two I_2 molecules are attached, while I_9^- seems to be a very loose aggregation. The I_4^{2-} ion has two I^- ions associated with an I_2 molecule and I_8^{2-} can be described as I_4^{2-} to which two more I_2 molecules are attached. The well-known blue starch-iodine compound has linear I_5^- ions in the amylose helix.

There are even larger arrays, I_n^- reported, such as I_{16}^-.[67]

The Cl_3^- and Br_3^- ions are similar to I_3^-. The Cl_3^- ion is known only in an unsymmetrical form with distances of 2.23 and 2.31 Å. The Br_3^- ion does not exist in aqueous solution but is formed on addition of Br_2 to Bu_4^nNBr in 1,2-dichloroethane

[65]K.-F. Tebbe and T. Gilles, *Z. anorg. allg. Chem.* **1996,** *622,* 1587.
[66]P. B. Hitchcock *et al., J. Chem. Soc., Dalton Trans.* **1994,** 3683.
[67]T. B. Rauchfuss *et al., J. Am. Chem. Soc.* **1990,** *112,* 1860.

Formula	Composition	Structure

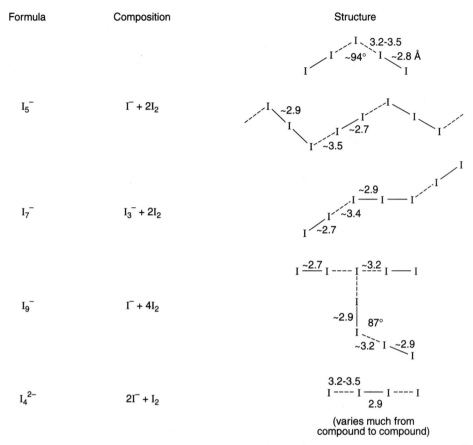

Figure 13-4 Some higher polyiodide ions (distances in Å). Distances > 3.0 and < 3.0 Å are indicated by $\cdots$ and $-$, respectively. other anions include I_{12}^{2-}, I_{14}^{2-}, I_{15}^{3-}, I_{16}^{2-}, and I_{16}^{4-} (see M. Schroder *et al., Angew. Chem. Int. Ed. Engl.* **1995**, *34*, 2374).

or SO_2. The Br_5^- ion, formed on interaction of Bu_4^nNBr with excess Br_2 in solutions, has been characterized spectroscopically.[68] There are also Br_4^- and Br_{10}^{2-} ions.[69]

13-21 Interhalogen Anions

Anions of the type $X_lY_mZ_n^-$ ($l + m + n$ odd) are listed in Table 13-11. The central atom is always the one with the highest atomic number, e.g., $[Cl-I-Br]^-$. The triatomic ions are linear and the pentaatomic ones are square-planar. Salts of the important fluoroanions with Na^+, Cs^+, and $[Me_4N]^+$ can be obtained[70] by reactions

[68]G. Bellucci *et al., J. Am. Chem. Soc.* **1989**, *111*, 199.
[69]C. W. Cunningham *et al., Inorg. Chim. Acta* **1990**, *169*, 135.
[70]W. W. Wilson and K. O. Christe, *Inorg. Chem.* **1989**, *28*, 4172; K. O. Christe *et al., Inorg. Chem.* **1990**, *29*, 3506.

Table 13-11   Interhalogen Anions

	Tri-	Penta-	Hepta-	Nonaatomic
	Br_2Cl^-	ICl_3F^-	ClF_6^-	IF_8^-
	$BrCl_2^-$	ICl_4^-	BrF_6^-	
	I_2Cl^-	$IBrCl_3^-$	IF_6^-	
	$BrIBr^-$	$I_2Cl_3^-$		
	$ClICl^-$	$I_2BrCl_2^-$		
	$ClIBr^-$	$I_2Br_2Cl^-$		
	$FIBr^-$	$I_2Br_3^-$		
	$FClF^-$	I_4Cl^-		
		ClF_4^-		
		BrF_4^-		
		IF_4^-		

in MeCN such as

$$ClF + CsF \longrightarrow Cs^+ClF_2^-$$

$$ClF_3 + CsF \longrightarrow Cs^+ClF_4^-$$

$$Me_4NF + CsClF_4 \longrightarrow CsF + [Me_4N^+]ClF_4^-$$

$$Me_4NF + CsBrF_6 \longrightarrow CsF + [Me_4N^+]BrF_6^-$$

The anions are weaker oxidants than the parent molecules and less reactive towards organic compounds. Some are white crystalline solids of quite high thermal stability: $(Me_4N)ClF_4$ is the least stable, decomposing at *ca.* 100°C. However, many are of very limited thermal stability. The hexafluorochlorate(V) anion, ClF_6^-, obtained from Me_4NF or CsF and ClF_5 in MeCN forms salts that are thermally unstable even at −31°C and can explode. Raman spectra at low temperatures confirm that ClF_6^- is similar to BrF_6^- and IF_6^-. These AX_6E anions have six ligands and a lone pair of electrons that, depending on the size of the central atom A can be sterically active or inactive. In $[Me_4N^+][IF_6^-]$ the anion is distorted octahedral with a lone pair whereas ClF_6^- and BrF_6^- are octahedral.[71] The IF_2^- and BrF_2^- ions are also very unstable.[72]

The reaction chemistry of the fluorides shows that the maximum coordination of Cl and Br by F is 6 whereas that of I is 8. The unique IF_8^- anion[73] can be made as its Me_4N^+ salt by interaction of IF_7 with Me_4NF in MeCN, but other salts, e.g., of Cs^+ and NO^+ are also known. The IF_8^- ion has a square-antiprismatic structure.

[71]K. Seppelt *et al., Chem. Eur. J.* **1995,** *1,* 261.
[72]X. Zhang and K. Seppelt, *Z. anorg. allg. Chem.* **1997,** *623,* 491.
[73]K. Seppelt *et al., Chem. Eur. J.* **1996,** *2,* 371.

Additional References

R. E. Banks, B. Smart, and J. C. Tatlow, *Organofluorine Chemistry: Principles and Commercial Applications*, Plenum, New York, 1994.

N. Bartlett *et al.*, *Am. Chem. Soc.*, Symposium Series 1994, No. 555, *Inorganic Fluorine Chemistry*, A.C.S., Washington, D.C.

Chemical Reviews **1996,** *96* (July/Aug): A collection of reviews on fluorine chemistry.

N. M. Doherty and N. W. Hoffman, *Chem. Rev.* **1991,** *91,* 553. Transition metal compounds of fluorine with CO, PR₃, As, and Sb ligands.

R. K. Harris and P. Jackson, *Chem. Rev.* **1991,** *91,* 1427. Solid state ¹⁹F nmr.

J. L. Kiplinger and T. G. Richmond, *Chem. Rev.* **1994,** *94,* 373. (Activation of C—F bonds by transition metal complexes, 400 refs.)

M. Meyer and D. O'Hagen, *Chem. Brit.* **1992,** *Sept.,* p. 785. (Fluorine compounds in plants and bacteria.)

G. A. Olah, R. D. Chambers, and C. K. S. Prakash, Eds., *Synthetic Fluorine Chemistry*, Wiley, New York, 1992.

S. P. Sander *et al.*, *Adv. Phys. Chem.* **1995,** *3,* 876. Experimental and theoretical studies on atmospheric chlorine.

J. Schiers, Ed., *Modern Fluoropolymers*, John Wiley & Sons, New York, 1997.

S. H. Strauss and J. S. Thrasher, Eds., *Inorganic Fluorine Chemistry: Towards the 21st Century*, Am. Chem. Soc. Symposium Series No. 555, 1994, A.C.S., Washington, D.C.

Chapter 14

THE GROUP 18 ELEMENTS:
He, Ne, Ar, Kr, Xe, Rn

THE ELEMENTS

14-1 Group Trends

The closed-shell electronic structures of the noble gas atoms are extremely stable, as shown by the high ionization enthalpies, especially of the lighter members (Table 14-1). The elements are all low-boiling gases whose physical properties vary systematically with atomic number. The boiling point of helium is the lowest of any known substance. The boiling points and heats of vaporization increase monotonically with increasing atomic number.

The heats of vaporization are measures of the work that must be done to overcome interatomic attractive forces. Since there are no ordinary electron-pair interactions between noble gas atoms, these weak forces (of the van der Waals or London type) are proportional to the polarizability and inversely proportional to the ionization enthalpies of the atoms; they increase therefore as the size and diffuseness of the electron clouds increase.

The ability of the gases to combine with other atoms is relatively limited. Only Kr, Xe, and Rn have so far been induced to do so. This ability would be expected to increase with decreasing ionization enthalpy and decreasing energy of promotion to states with unpaired electrons. The data in Table 14-1 for ionization enthalpies and for the lowest-energy promotion process show that chemical activity should increase down the group. Apparently the threshold of actual chemical activity is reached only at Kr; that of Xe is markedly greater while that of Rn is presumably still greater, but it is difficult to assess because the half-life of the longest-lived isotope, ^{222}Rn, is only 3.825 days.

14-2 Occurrence, Isolation, and Applications

The noble gases occur as minor constituents of the atmosphere (Table 14-1). Helium is also found as a component (up to *ca.* 7%) in certain natural hydrocarbon gases in the United States. This helium undoubtedly originated from decay of radioactive

586

Table 14-1   Some Properties of the Noble Gases

Element	Outer Shell Configuration	Atomic Number	First Ionization Enthalpy (kJ mol^{-1})	Normal bp (K)	ΔH_{vap} (kJ mol^{-1})	% by Volume in the Atmosphere	Promotion Energy (kJ mol^{-1}) $ns^2np^6 \rightarrow$ $ns^2np^5(n+1)s$
He	$1s^2$	2	2372	4.18	0.09	5.24×10^{-4}	
Ne	$2s^2 2p^6$	10	2080	27.13	18.0	1.8×10^{-3}	1601
Ar	$3s^2 3p^6$	18	1520	87.29	6.3	0.934	1110
Kr	$4s^2 4p^6$	36	1351	120.26	9.7	1.14×10^{-3}	955
Xe	$5s^2 5p^6$	54	1169	166.06	13.7	8.7×10^{-6}	801
Rn	$6s^2 6p^6$	86	1037	208.16	18.0		656

elements in rocks and certain radioactive minerals contain occluded He that can be released on heating. All isotopes of Rn are radioactive and are occasionally given specific names (e.g., actinon, thoron) derived from their source in the radioactive decay series; ^{222}Rn is normally obtained by pumping off the gas from radium chloride solutions. Neon, Ar, Kr, and Xe are obtainable as products of fractionation of liquid air. The main uses are in providing inert atmospheres, for example, Ar in welding and gas-filled electric light bulbs, Ne in discharge tubes. Liquid He is used extensively in cryoscopy.

Liquid Ar, Kr, and Xe are commonly used as solvents for spectroscopic and synthetic studies;[1] supercritical Xe is transparent over the range vacuum uv to far ir and is also useful for nmr studies. Examples of spectroscopic studies are those on N,O species[2] such as ONNO, ONNO$_2$, N$_2$O$_4$, and NO and photochemical reactions of compounds such as Cp*Rh(CO)$_2$, Cp*Mn(CO)$_3$, or Cp*V(CO)$_4$;[3] ir spectra of ZrH$_4$ and HfH$_4$ trapped in solid argon have been obtained.[4]

The noble gases can be trapped in clathrates such as Xe$\cdot$3C$_6$H$_4$(OH)$_2$ but the gas hydrates are the most important. For Xe, the formation and occupancy of both the smaller and larger cages can be followed by ^{129}Xe nmr spectra[5]; such spectra can also indicate dynamic behavior of Xe trapped in zeolites.[6] Small water clusters such as Ar$_2$H$_2$O, Ar$_3$H$_2$O, Ar(H$_2$O)$_2$, and Ar(H$_2$O)$_3$ have been detected by rotational spectra.[7]

Radon constitutes a major health hazard since radioactive decay of uranium in minerals, especially in granite areas, generates the gas that can diffuse into homes and other buildings. The decay of Rn produces Po, Bi, and Pb isotopes that can be absorbed on dust particles and breathed into the lungs causing cancer.[8]

[1]For vibrational spectra in supercritical fluids including noble gases see M. Poliakoff, *et al.*, *Angew. Chem. Int. Ed. Engl.* **1995**, *34*, 1275; *Chem. Brit.* **1995**, 118.
[2]E. J. Sluyts and B. J. Van der Viken, *J. Mol. Struct.* **1994**, *320*, 249.
[3]C. B. Moore *et al.*, *J. Am. Chem. Soc.* **1994**, *116*, 9585; S. G. Kazerian *et al.*, *Organometallics* **1994**, *13*, 1767.
[4]G. V. Chertihin and L. Andrews, *J. Am. Chem. Soc.* **1995**, *117*, 6402.
[5]T. Pietrass *et al.*, *J. Am. Chem. Soc.* **1995**, *117*, 7520.
[6]C. I. Ratcliffe and J. A. Ripmeester, *J. Am. Chem. Soc.* **1995**, *117*, 1445.
[7]H. S. Gutowsky *et al.*, *J. Am. Chem. Soc.* **1994**, *116*, 8418.
[8]A. F. Gardner *et al.*, *Chem. Brit.* **1992**, 344; C. F. Cochern, *Revs. Environ. Contam. Toxicol.*, Vol. III, Springer, New York, 1990.

Table 14-2 Principal Xenon Compounds

Oxidation State	Compound	Form	mp (°C)	Structure	Remarks
II	XeF_2	Colorless crystals	129	Linear	Hydrolyzed to Xe + O_2; very soluble in HF(l)
IV	XeF_4	Colorless crystals	117	Square	Stable, $\Delta H_f^{298K} = -284$ kJ mol^{-1}
VI	XeF_6	Colorless crystals	49.6	Complex	Stable, $\Delta H_f^{298K} = -402$ kJ mol^{-1}
	$CsXeF_7$	Colorless solid		See text	dec. > 50°C
	Cs_2XeF_8	Yellow solid		Archimedian antiprism^a	Stable to 400°C
	$XeOF_4$	Colorless liquid	−46	ψ Octahedral^b	Stable
	XeO_2F_2	Colorless crystals	31	ψ tbp^b	Metastable
	XeO_3	Colorless crystals		ψ Tetrahedral^b	Explosive, $\Delta H_f^{298K} = +402$ kJ mol^{-1}; hygroscopic; stable in solution
VIII	$K_n^+[XeO_3F^-]_n$	Colorless crystals		sp (F bridges)	Very stable
	XeO_4	Colorless gas	−35.9	Tetrahedral	Highly explosive
	XeO_3F_2	Colorless gas	−54.1		
	XeO_6^{4-}	Colorless salts		Octahedral	Anions $HXeO_6^{3-}$, $H_2XeO_6^{2-}$, $H_3XeO_6^-$, also exist

^{a} In the salt $(NO^+)_2[XeF_8]^{2-}$ from XeF_6 and NOF.
^{b} Lone pair present.

THE CHEMISTRY OF THE NOBLE GASES

After his observation that O_2 reacts with PtF_6 to give the compound $[O_2^+][PtF_6^-]$, N. Bartlett in 1962 recognized that since the ionization enthalpy of Xe is almost identical with that of O_2, an analogous reaction should occur with Xe. By interaction of Xe and PtF_6 he isolated a red solid, said to be $XePtF_6$. Present knowledge indicates the oxidation sequence:

$$Xe + 2PtF_6 \xrightarrow{25°C} [XeF^+][PtF_6^-] + PtF_5 \xrightarrow{60°C} [XeF^+][Pt_2F_{11}^-]$$

as in similar oxidations described below. It is known that although Bartlett published very quickly, others (R. Hoppe and a group at Argonne National Laboratory) had independently made the discovery of either the Bartlett type material or even XeF_2.[9]

14-3 The Chemistry of Xenon

Xenon reacts directly only with F_2, but compounds in oxidation states from II to VIII are known, some of which are exceedingly stable and can be obtained in large quantities. The more important compounds and some of their properties are given in Table 14-2.

[9]P. Lazlo and G. J. Schrobilgen, *Angew. Chem. Int. Ed. Engl.* **1988**, *27*, 479.

Fluorides

The equilibrium constants for the reactions

$$Xe + F_2 = XeF_2$$

$$XeF_2 + F_2 = XeF_4$$

$$XeF_4 + F_2 = XeF_6$$

for the range 25 to 500°C show unequivocally that only these three binary fluorides exist. The equilibria are established rapidly only above 250°C, which is the lower limit for thermal synthesis. All three fluorides are volatile, readily subliming at room temperature. They can be stored indefinitely in nickel or Monel metal containers, but XeF_4 and XeF_6 are particularly susceptible to hydrolysis and traces of water must be rigorously excluded, since the explosive XeO_3 is formed (see below).

Xenon Difluoride

This is made by interaction of F_2 and excess Xe under pressure or by oxidation of Xe by the dark blue solutions of AgF_2 in HF(l) in the presence of BF_3:[10]

$$2AgF_2 + 2BF_3 + Xe = XeF_2 + 2AgBF_4$$

It is soluble in water, giving solutions 0.15 M at 0°C that evidently contain XeF_2 molecules. The hydrolysis is slow in dilute acid but rapid in basic solution:

$$XeF_2 + 2OH^- \longrightarrow Xe + \frac{1}{2}O_2 + 2F^- + H_2O$$

The solutions, which have a pungent odor due to XeF_2, are powerful oxidizing agents (e.g., HCl gives Cl_2, Ce^{III} gives Ce^{IV}), and the estimated potential is

$$XeF_2(aq) + 2H^+ + 2e = Xe + 2HF(aq) \qquad E^0 = +2.64 \text{ V}$$

XeF_2 also acts as a mild fluorinating agent for organic compounds; for example, in solution or in the vapor phase benzene is converted into C_6H_5F. The dissociation energy of XeF_2 to $XeF + F$ is *ca.* 252 kJ mol^{-1}.[11]

Xenon Tetrafluoride

This can also be made by direct interaction like XeF_2 but with heating under pressure, by oxidation of Xe with O_2F_2,[12] or by photolysis of $Xe + F_2$. Hydrolysis of XeF_4 produces the dangerously explosive XeO_3. The fluoride has been used for specific fluorination of the ring in substituted arenes like toluene.

[10]N. Bartlett *et al.*, *J. Am. Chem. Soc.* **1990**, *112*, 4846.
[11]G. Bucher and J. C. Scaiano, *J. Am. Chem. Soc.* **1994**, *116*, 10076.
[12]S. A. Kinkead *et al.*, *Inorg. Chem.* **1990**, *29*, 1779.

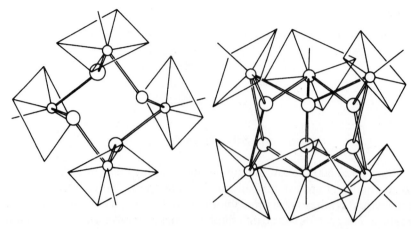

Figure 14-1 The tetrameric and hexameric units in one of the crystal forms of XeF_6.

Xenon Hexafluoride

This can be made by interaction of XeF_4 with F_2 under pressure but use of a special "hot wire" reactor allows milder conditions to be used.[12] Solid XeF_6 is colorless but it gives a yellow liquid and vapor. In the melt and in solution it exists as a monomer and tetramer in equilibrium; in solution at low temperature only the tetramer is observed. It is an extremely powerful fluorinating agent. It reacts rapidly with quartz to give $XeOF_4$ and SiF_4 and is extremely readily hydrolyzed to the explosive XeO_3. The free XeF_6 molecule has a stereoactive lone pair on Xe which leads to a distorted pseudo 7-coordinate structure and highly fluxional behavior. The solid has various crystalline forms, of which three are tetrameric and a fourth has both hexamers and tetramers. These oligomers, see Fig. 14-1, have square pyramidal XeF_5^+ cations and bridging μ-F anions.

Fluorocations

Cationic fluoro species can be made by interaction of the binary xenon fluorides with compounds that are strong F^- ion acceptors such as TaF_5 or PtF_5. They are known for oxidation states II–VI and are of the stoichiometry $Xe_nF_m^+$. Such compounds may not be fully ionic, however, since the fluoroanions can form F-bridges to the cations. An example of this is shown in Fig. 14-2 for $[XeF][RuF_6]$. The structure of $Xe_2F_3^+$ is shown in (14-I).[13]

$$\overset{\displaystyle F}{\underset{\displaystyle F}{Xe}}\ \overset{2.14\,\text{Å}}{\underset{151°}{\bigcup}}\ \overset{\displaystyle Xe}{\underset{\displaystyle F}{}}\ {1.90\,\text{Å}}$$

(14-I)

[13]For discussion of bonding see D. A. Dixon *et al.*, *Inorg. Chem.* **1989**, *28*, 4589.

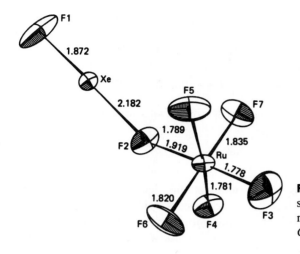

Figure 14-2 The $[XeF^+][RuF_6^-]$ structural unit (reproduced by permission from N. Bartlett *et al., Inorg. Chem.* **1973**, *12*, 1717).

The reaction of $[XeF][Sb_2F_{11}]$ in a HF/SbF$_5$ medium affords (among other products) $[Xe_2^+][Sb_4F_{21}^-]$.

$$XeF^+Sb_2F_{11}^- + 3Xe + 2SbF_5 \rightleftharpoons 2Xe_2^+Sb_4F_{21}^-$$

The Xe_2^+ radical cation[14] has an Xe—Xe distance of 3.09 Å.

For the +4 oxidation state, interaction of XeF_4 and BiF_5 gives $[XeF_3]^+[BiF_6]^-$ while the compound $XeF_5^+AuF_4^-$ is formed in the reaction

$$BrF_3 \cdot AuF_3 + XeF_6(excess) \longrightarrow XeF_5^+AuF_4^- + BrF_3$$

Other similar compounds, e.g., $XeF_5^+RuF_6^-$ are known and, as noted above, the crystalline forms of XeF_6 have F-bridged XeF_5^+ units. Such F-bridges are also found in $Xe_2F_{11}^+AuF_6^-$ that is made[10] by oxidation of $XeF_5^+AuF_4^-$ with the powerful oxidant KrF_2. Although calculations[15] suggest that XeF_7^+ is potentially stable this ion is not yet known; it would be in the isoelectronic series TeF_7^-, IF_7, XeF_7^+.

Fluoroanions

Both XeF_4 and XeF_6 can act as F^- acceptors as well as F^- donors and thus form anionic species in reactions such as

$$XeF_4 + NaF \longrightarrow Na^+XeF_5^-$$

$$XeF_6 + CsF \longrightarrow Cs^+XeF_7^-$$

$$2XeF_6 + NO_2F \longrightarrow NO_2^+Xe_2F_{13}^-$$

[14]T. Drews and K. Seppelt, *Angew. Chem. Int. Ed. Engl.* **1997**, *36*, 273.
[15]K. O. Christe *et al., J. Am. Chem. Soc.* **1993**, *115*, 9461.

The XeF_5^- anion has an unusual pentagonal planar structure that is much more rigid than the planar equatorial fluorines in the fluxional IF_7 molecule. The rigidity is attributed to increased repulsions from the Xe electron pairs.[16]

$CsXeF_7$ crystallizes from BrF_5 as yellow crystals at 4°C. The XeF_7^- ion has a capped octahedral structure[17] with the cap having a long Xe—F bond of 2.1 Å due to interaction with 3 Cs^+ ions.

Octafluoroxenates are formed on heating Rb or Cs salts of XeF_7^-:

$$2MXeF_7 \longrightarrow XeF_6 + M_2XeF_8$$

The Rb and Cs octafluoroxenates are thermally stable, decomposing only above 400°C; the NaF adduct of XeF_6, however, decomposes below 100°C and can be used to purify XeF_6. The XeF_8^{2-} ions are square antiprismatic similar to WF_8^{2-} and ReF_8^{2-}.[18] The salt $NO_2^+Xe_2F_{13}^-$ has XeF_7^- and XeF_6 units where 2 F atoms of the former interact with the Xe atom of the latter.[17]

Compounds with Xe—O Bonds

Both XeF_4 and XeF_6 are violently hydrolyzed by water to give stable aqueous solutions up to 11 *M* of *xenon trioxide* (XeO_3). The oxide is also obtained by interaction of XeF_6 with $HOPOF_2$. Xenon trioxide is a white deliquescent solid and is dangerously explosive; its formation is why great care must be taken to avoid water in studies of XeF_4 and XeF_6. The molecule is pyramidal (C_{3v}). It can be quantitatively reduced by iodide:

$$XeO_3 + 6H^+ + 9I^- \longrightarrow Xe + 3H_2O + 3I_3^-$$

Xenate esters may be formed in violent reactions with alcohols. In water XeO_3 appears to be present as XeO_3 molecules, but in basic solutions we have the main equilibrium

$$XeO_3 + OH^- \longrightarrow HXeO_4^- \qquad K = 1.5 \times 10^{-3}$$

where $HXeO_4^-$ slowly disproportionates to produce Xe^{VIII} and Xe:

$$2\,HXeO_4^- + 2OH^- \longrightarrow XeO_6^{4-} + Xe + O_2 + 2H_2O$$

Aqueous Xe^{VIII} arises also when O_3 is passed through a dilute solution of Xe^{VI} in base. These yellow *perxenate* solutions are powerful and rapid oxidizing agents. The stable salts $Na_4XeO_6 \cdot 8H_2O$ and $Na_4XeO_6 \cdot 6H_2O$ contain XeO_6^{4-} octahedra. The solutions of sodium perxenate are alkaline owing to hydrolysis, and the following

[16]K. O. Christe *et al., J. Am. Chem. Soc.* **1991,** *113,* 3351.
[17]K. Seppelt *et al., Angew. Chem. Int. Ed. Engl.* **1996,** *35,* 1123.
[18]K. Seppelt *et al., Chem. Eur. J.* **1996,** *2,* 371.

equilibrium constants have been estimated:

$$HXeO_6^{3-} + OH^- = XeO_6^{4-} + H_2O \qquad K < 3$$

$$H_2XeO_6^{2-} + OH^- = HXeO_6^{3-} + H_2O \qquad K \sim 4 \times 10^3$$

so that even at pH 11 to 13 the main species is $HXeO_6^{3-}$. From the equilibria it follows that for H_4XeO_6, pK_3 and pK_4 are $\sim 4 \times 10^{-11}$ and $< 10^{-14}$, respectively. Hence, by comparison with H_6TeO_6 and H_5IO_6, H_4XeO_6 appears to be an anomalously weak acid.

Solutions of perxenates are reduced by water at pH 11.5 at a rate of $\sim 1\%/h$, but in acid solutions almost instantaneously:

$$H_2XeO_6^{2-} + H^+ \longrightarrow HXeO_4^- + \tfrac{1}{2}O_2 + H_2O$$

This reduction appears to proceed almost entirely by a radical process involving hydroxyl radicals and H_2O_2.

Xenon tetroxide is a highly unstable and explosive gas formed by the action of concentrated H_2SO_4 on barium perxenate. It is tetrahedral (Xe$-$O $= 1.736$ Å), according to electron diffraction studies.

The aqueous chemistry of xenon is briefly summarized by the potentials

$$H_4XeO_6 \xrightarrow{\text{2.36 V}} XeO_3 \xrightarrow{\text{2.12 V}} Xe$$

Acid solution

$$XeF_2 \xrightarrow{\text{2.64 V}} Xe$$

Alkaline solution $\quad HXeO_6^{3-} \xrightarrow{\text{0.94 V}} HXeO_4^- \xrightarrow{\text{1.26 V}} Xe$

Oxofluorides

Several of these are known. Representative syntheses are

$$XeF_6 + NaNO_3 \longrightarrow XeOF_4 + FNO_2 + NaF$$

$$XeF_6 + OPF_3 \longrightarrow XeOF_4 + PF_5$$

$$XeF_6 + 2XeO_3 \longrightarrow 3XeO_2F_2$$

Xenon oxotetrafluoride has a square pyramidal structure, whereas XeO_2F_2 is like SF_4 with F atoms in axial positions (C_{2v}). Xenon oxodifluoride ($XeOF_2$) and XeO_3F_2 are unstable.

There are also some derivatives of the oxofluorides with the ligand $OTeF_5^-$; these are $XeO(OTeF_5)_4$ and $XeO_2(OTeF_5)_2$. The only compound of Xe^{IV} bound exclusively to oxygen is $Xe(OTeF_5)_4$ made by the reaction

$$3XeF_4 + 4B(OTeF_5)_3 \xrightarrow{0°C} 3Xe(OTeF_5)_4 + 4BF_3$$

Just as the xenon fluorides react with F^- acceptors to give cations, so do oxofluorides. The structure of the cation in $XeOF_3^+SbF_6^-$ is similar to that of $ClOF_3$.[19] There are several cations of Xe^{II} such as $Xe(OTeF_5)^+$, $Xe(OSO_2F)^+$, as well as two other examples of cations in higher oxidation states, XeO_2F^+ and $O_2Xe(OTeF_5)^+$.[20]

The $[XeOF_5]^-$ ion can be obtained by the following general reaction, where $M^+ = Cs^+$, Me_4N^+, NO^+:

$$XeOF_4 + MF \longrightarrow M[(XeOF_4)_3F] \xrightarrow[25°C]{pump} M[XeOF_5]$$

Vibrational spectra for the explosive $[Me_4N][XeOF_5]$[21] and X-ray crystallography for $[NO][XeOF_5]$[22] show that the anion has a ψ-pentagonal bipyramid structure with O at one apex and a lone pair at the other.

Xenon trifluoroacetates and trifluoromethane sulfonates can be made *via* the sequence:[23]

$$XeF_2 + 2CF_3CO_2H \xrightarrow[-40°C]{CCl_3F} Xe(OCOCF_3)_2 + 2HF$$

$$\Big\downarrow CF_3SO_3H$$

$$[XeAr]^+OSO_2CF_3^- \xleftarrow{ArH} Xe(OCOCF_3)(OSO_2CF_3)$$

Other species with bonds to oxygen are formed when XeF_2 reacts with strong oxo acids (e.g., HSO_3F) or their anhydrides [e.g., $(F_2OP)_2O$] to give products in which one or both F atoms are replaced. Examples are

$$XeF_2 + HSO_3F \longrightarrow FXeOSO_2F, \; Xe(OSO_2F)_2$$

$$XeF_2 + P_2O_3F_4 \longrightarrow FXe(OPO_2F), \; Xe(OPO_2F)_2$$

These products are unstable at about room temperature and several are highly explosive. Xenon tetrafluoride and XeF_6 react with HSO_3F as follows:

$$XeF_4 + HSO_3F \longrightarrow FXe(OSO_2F), \; Xe(OSO_2F)_2 + S_2O_6F_2$$

$$XeF_6 + HSO_3F \longrightarrow F_5Xe(OSO_2F) + HF$$

$F_5Xe(OSO_2F)$ is a white solid, stable at 22°C but decomposing above 73°C.

[19]G. J. Schrobilgen *et al., Inorg. Chem.* **1993,** *32,* 386.
[20]G. J. Schrobilgen *et al., Inorg. Chem.* **1992,** *31,* 3381.
[21]K. O. Christe *et al., Inorg. Chem.* **1995,** *34,* 1868.
[22]A. Ellern and K. Seppelt, *Angew. Chem. Int. Engl.* **1995,** *34,* 1586.
[23]D. Naumann *et al., J. Chem. Soc., Chem. Commun.* **1994,** 2651.

Xenon Compounds with Bonds to Other Elements

Carbon. The earliest compound, $Xe(CF_3)_2$ was made by action of $CF_3{}^{\cdot}$ radicals on XeF_2 but it was stable for only *ca.* 30 minutes at room temperature. Stable species, mostly cationic in nature, can be made. Thus the reaction of fluorinated aryls, e.g., $B(C_6F_5)_3$, with XeF_2 in the presence of $BF_3(OMe_2)$, gives $[ArXe]^+[BF_4]^-$. The stability of such compounds depends on the nature of the aryl group and decomposition temperatures range from about $-14°C$ for $4\text{-}FC_6H_4$ to $130°C$ for $2,6\text{-}F_2C_6H_3$.[24]

The AsF_6^- salts of alkynyls, $[XeC{\equiv}CR]^+$, and alkenyls[25] have been made, an example of the latter *via* the reaction

$$[C_6F_5Xe][AsF_6] \xrightarrow{\text{XeF}_2/\text{HF(1)}} [XeC_6F_9][AsF_6] \longrightarrow [XeC_6F_{11}][AsF_6]$$

where the XeF_2 in HF has transformed C_6F_5 into the radicals 14-II and 14-III.

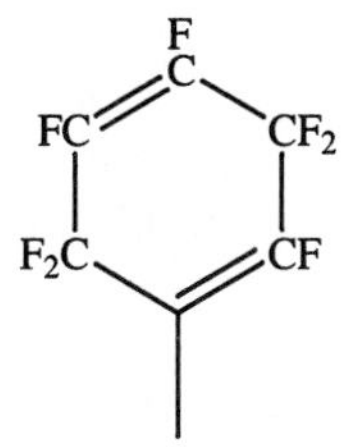

(14-II) 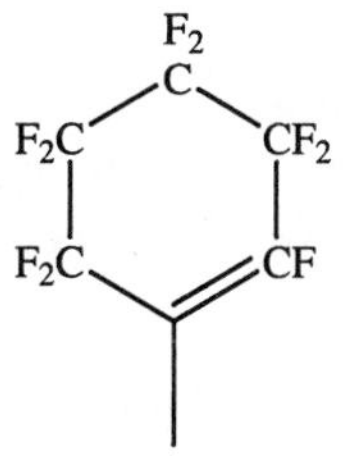 (14-III)

The reaction of $[C_6F_5Xe][AsF_6]$ with $CsO_2CC_6F_5$ affords $C_6F_5XeOCOC_6F_5$, which is thermally stable,[26] whereas $FXeOCOCF_3$ is prone to detonate.

Silicon. The F_3SiXe^+ ion can be obtained under mass spectrometric conditions by the reaction[27]

$$F_3Si\text{-}FH^+ + Xe \longrightarrow F_3SiXe^+ + HF$$

Nitrogen. The first fully characterized compound was $FXeN(SO_2F)_2$ which is a readily hydrolyzed white solid that is stable to 70°C. It was made by the reaction

$$XeF_2 + HN(SO_2F)_2 \longrightarrow FXeN(SO_2F)_2 + HF$$

[24]D. Naumann *et al., Inorg. Chem.* **1993**, *32*, 861.
[25]H. J. Frohn and V. V. Bardin, *J. Chem. Soc., Chem. Commun.* **1993**, 1072.
[26]H. J. Frohn *et al., Angew. Chem. Int. Ed. Engl.* **1993**, *32*, 99.
[27]R. Cipollini and F. Grandinetti, *J. Chem. Soc., Chem. Commun.* **1995**, 773.

and it is converted by more of the amide to $Xe[N(SO_2F)_2]_2$ and HF. Other similar compounds include $F[XeN(SO_2F)_2]_2^+$ and $[XeN(SO_2F)_2]AsF_6$.

The hydrogen cyanide complex,[28] $[HC\equiv NXeF][AsF_6]$, is made by interaction of XeF^+ or $Xe_2F_3^+$ cations with HCN in anhydrous HF. The cation is linear with a weak Xe—N bond and solutions in HF are extensively solvolyzed to give XeF_2 and $HC\equiv NH^+$. $[C_6F_5Xe][AsF_6]$ has been shown to form compounds with pyridine and substituted pyridines.[29]

Chlorine. Xenon trioxide reacts with RbCl and CsCl to give compounds with the composition $M_9(XeO_3Cl_2)_4Cl$, which consist of M^+ cations, Cl^- anions, and infinite chain anions in which each Xe atom is surrounded by a very distorted octahedron of three oxygen atoms (Xe—O ≈ 1.77 Å) and three Cl atoms (Xe—Cl ≈ 2.96 Å). These are the only examples of compounds stable at room temperature that contain Xe—Cl bonds.

Although "$XeCl_2$" has been trapped in solid Xe after photolysis of included Cl_2, it appears to be a Xe—Cl_2 complex with weak van der Waals bonding.[30]

Metals. Laser photolysis of $M(CO)_6$, M = Cr and W in the presence of Xe(l) (and also Kr) gives unstable carbonyls[31]; $W(CO)_5Xe$ has a lifetime of only *ca.* 1.5 min at 170 K in liquid Xe. There is a dissociative equilibrium,

$$W(CO)_5Xe \rightleftharpoons W(CO)_5 + Xe$$

and CO readily replaces the Xe or Kr to reform $M(CO)_6$. For $W(CO)_5Xe$ a W—Xe bond energy of 35.1 kJ mol^{-1} was determined. There is also evidence[32] that compounds such as $Cp*Rh(CO)Xe$ can be obtained after photolysis of $Cp*Rh(CO)_2$ in liquid Kr or Xe. The interaction energies are in the region 20-40 kJ mol^{-1}.

Theoretical studies have also been made of the interaction of noble gases with Au^+ and Au—Xe bond energies of *ca.* 87 kJ mol^{-1} predicted.[33]

14-4   The Chemistry of Krypton and Radon

The chemistry of these elements as far as it is known is similar to that of Xe.

Krypton difluoride[34] is obtained on photolysis of Kr—F_2 mixtures at $-196°C$ when it forms a white solid only slightly soluble in liquid fluorine. It slowly decom-

---

[28]A. A. A. Emara and G. J. Schrobilgen, *et al.*, *Inorg. Chem.* **1992**, *31*, 1323.

[29]H. J. Frohn *et al.*, *Z. Naturforsch.* **1995**, *50*, 1799.

[30]R. Hoffmann *et al.*, *J. Am. Chem. Soc.* **1991**, *113*, 7184.

[31]B. H. Weiller, *et al.*, *J. Am. Chem. Soc.* **1992**, *114*, 10910.

[32]R. G. Bergman *et al.*, *J. Am. Chem. Soc.* **1995**, *117*, 3879.

[33]P. Pyykkö, *J. Am. Chem. Soc.* **1995**, *117*, 2067.

[34]J. H. Holloway *et al.*, *J. Chem. Soc., Dalton Trans.* **1991**, 2381.

poses at room temperature. It is a powerful fluorinating and oxidizing agent; one example is the conversion of RuO_4 to $RuOF_4$ in liquid HF.[35]

The instability of KrF_2 compared to XeF_2 can be seen from the enthalpies:

$$KrF_2(g) = Kr(g) + F_2(g) \qquad \Delta H^0 = \text{-63 kJ mol}^{-1}$$

$$XeF_2(g) = Xe(g) + F_2(g) \qquad \Delta H^0 = \text{105 kJ mol}^{-1}$$

These energetic relationships are understandable on the basis of rigorous quantum mechanical calculations, which justify the view that in both difluorides there is considerable ionic character. The bonding can be more simply represented by the resonance picture:

$$F—Xe^+F^- \longleftrightarrow F^-Xe^+—F$$

Since the ionization enthalpies (Table 14-1) of Kr and Xe differ by ~182 kJ mol^{-1}, the experimental difference in ΔH_f^0 values, namely, $105 - (-63) = 168$ kJ mol^{-1} is well explained by this picture.

There are cationic species similar to those of Xe but less stable; they include KrF^+, $Kr_2F_3^+$, and $[CF_3CNKrF]^+$.

Since *radon* has only a short half-life, study is difficult, but tracer studies allow some properties to be deduced, for example, the formation of RnF_2, $RnF^+TaF_6^-$, and possibly RnO_3. Oxidation of Rn by ClF_3 and study on a fluorinated ion-exchange material (Nafion) suggests that Rn^+ can displace Na^+ or K^+.

Additional References

G. Schrobilgen, in *Synthetic Fluorine Chemistry*, G. A. Olah *et al.*, Eds., Wiley, New York, 1992, Chap. 1 (noble gas fluorocations).

J. Wilks, and D. S. Betts, *An Introduction to Liquid Helium,* Oxford University Press, 2nd ed., 1988.

B. Žemva, in *Encyclopedia of Inorganic Chemistry*, R. B. King, Ed., Vol. 5, Wiley, Chichester, UK, 1994, p. 2660.

---

[35]L. Meublat *et al., Can. J. Chem.* **1989,** *67,* 1729.

Chapter 15

THE GROUP 12 ELEMENTS: Zn, Cd, Hg

15-1 General Remarks

The elements Zn, Cd, and Hg follow Cu, Ag, and Au, respectively. Each has a filled $(n-1)d$ shell plus two ns electrons. While Cu, Ag, and Au all give rise to ions or complexes in which one or even two d electrons are lost, that is to compounds in oxidation states II and III, no such compounds have ever been isolated for the Group 12 metals. Thus, while Cu, Ag, and Au are classified as transition elements, Zn, Cd, and Hg are not. For mercury it has been claimed that at $-78°C$ [Hg cyclam]$(BF_4)_2$ can be oxidized electrochemically in acetonitrile to give the [Hg cyclam]$^{3+}$ ion, with a half-life of $ca.$ 5s, and there have also been theoretical arguments that Hg^{IV} might exist, for example, in HgF_4.[1]

The only deviations from strict divalence are (a) the M_2^{2+} ions, of which only Hg_2^{2+} is stable under normal conditions, and (b) compounds containing Hg_3^{2+} and Hg_4^{2+} ions.

Table 15-1 summarizes some fundamental properties of the Group 12 elements. These three metals are all low melting (and soft) and Zn and Cd are much more electropositive than their Group 11 neighbors. Although the formation of complexes with ammonia, amines, halide, and pseudohalide ions is reminiscent of transition metal ion behavior, the ability of the Group 12 ions to serve as $d\pi$ donors is so low that they form none of the other typical sorts of transition metal complexes, such as carbonyls, nitrosyls, or π-complexes with olefins. The only exceptions to this are [Hg(CO)$_2$][Sb$_2$F$_{11}$]$_2$ and [Hg$_2$(CO)$_2$][Sb$_2$F$_{11}$]$_2$, which are relatively stable but show no evidence for Hg—CO π-bonding. Indeed, the first of these has the highest reported value of ν_{co}, $ca.$ 2280 cm^{-1}, compared to 2146 cm^{-1} for CO itself.[2]

There is close homology in the chemistry of zinc and cadmium, but mercury diverges markedly. For example, ZnCl$_2$ and CdCl$_2$ have typical ionic structures, but HgCl$_2$ crystals consist of linear molecules. While all three M^{2+} ions form numerous complexes, the formation constants are orders of magnitude higher for those of Hg^{2+}. Zinc and Cd are very electropositive, but mercury is "noble"; Zn(OH)$_2$ is basic but also amphoteric, giving, e.g., [Zn(OH)$_3$]$^-$, and Cd(OH)$_2$ is more basic and not amphoteric, but Hg(OH)$_2$ is an extremely weak base.

---

[1]M. Kaupp *et al., Inorg. Chem.* **1994**, *33*, 2122.
[2]H. Willner *et al., J. Chem. Soc., Chem. Commun.* **1994**, 1189.

Table 15-1 Some Properties of the Group 12 Elements

Property	Zn	Cd	Hg
Outer configuration	$3d^{10}4s^2$	$4d^{10}5s^2$	$5d^{10}6s^2$
Ionization enthalpies (kJ mol^{-1})			
1st	906	876	1007
2nd	1734	1630	1809
3rd	3831	3615	3300
Melting point (°C)	419	321	−38.87
Boiling point (°C)	907	767	357
Heat of evaporation (kJ mol^{-1})	130.8	112	61.3
E^0 for $M^{2+} + 2e = M$ (V)	−0.762	−0.402	0.854
Ionic radii (C.N.6) of M^{2+} (Å)	0.88	1.09	1.16

The Group 12 elements differ markedly from those in Group 2 in nearly all respects except having II as their only important oxidation state. Thus, while the Zn^{2+} and Mg^{2+} ions are very similar in their 6-coordinate radii (0.88 Å and 0.86 Å, respectively), Zn^{2+} has a relatively polarizable $3d^{10}$ shell whereas the neon core of Mg^{2+} is very "hard." This special combination of "softness" and a high charge-to-radius ratio appears to be responsible for the unique role played by zinc in biochemistry (see Section 15-17).

While only a little useful nmr work can be done on compounds of either Zn or Hg, ^{113}Cd has proven to be very practical for nmr studies.[3]

15-2 Stereochemistry

The M^{2+} ions with their d^{10} configurations show no stereochemical preferences arising from ligand field stabilization effects. Therefore, they display a variety of coordination numbers and geometries based on the interplay of electrostatic forces, covalence, and the size factor. Because of size, Cd^{2+} is more often found with a coordination number of 6 than is Zn^{2+}. This is exemplified by the fact that both $ZnCl_2$ and $CdCl_2$ have structures based on close-packed arrays of Cl^- ions, but Zn^{2+} occupies tetrahedral interstices while Cd^{2+} occupies octahedral ones. Similarly CdO has the rock salt structure while ZnO forms structures in which ZnO_4 tetrahedra are linked.

As shown in Table 15-2, coordination numbers 4, 5, and 6 are the common ones for all three elements, although linear 2-coordination is often seen for Hg^{2+} and has been attributed to relativistic effects on the $6s$ orbital.[4]

15-3 The Elements: Sources and Properties

Both zinc and mercury have been known for hundreds of years, despite having relatively low abundance, because they are easily extracted from the minerals in which they occur.

[3]P. D. Ellis *et al.*, *J. Am. Chem. Soc.* **1993**, *115*, 755.
[4]M. Kaupp and H. G. von Schnering, *Inorg. Chem.* **1994**, *33*, 2555.

Table 15-2 Stereochemistry[a] of Divalent Zinc, Cadmium, and Mercury

Coordination number	Geometry	Examples
2	Linear	$Zn(CH_3)_2$, HgO, $Hg(CN)_2$, $Cd[N(SiMe_3)_2]_2$
3	Planar	$[Me_3S]^+HgX_3^-$, $[MeHg\ bipy]^+$, $Hg(SiMe_3)_3^-$, $Zn_2(\mu\text{-}OH)_2[C(SiMe_2Ph)_3]_2$
4	Tetrahedral	$[Zn(CN)_4]^{2-}$, $ZnCl_2(s)$, ZnO, $[Cd(NH_3)_4]^{2+}$, $HgCl_2(OAsPh_3)_2$
5	*tbp*	$(terpy)ZnCl_2$, $[Zn(SCN)\ tren]^+$, $[Co(NH_3)_6][CdCl_5]$
	sp	$Zn(acac)_2 \cdot H_2O$
6	Octahedral	$[Zn(NH_3)_6]^{2+}$ (solid only), CdO, $CdCl_2$, $[Hg(en)_3]^{2+}$, $[Hg(C_5H_4NO)_6]^{2+}$
7	Pentagonal bipyramid	$[Zn(H_2dapp)(H_2O)_2]^{2+b}$
	Distorted pentagonal bipyramid	$Cd(quin)(H_2O)(NO_3)_2$
8	Distorted square antiprism	$[Hg(NO_2)_4]^{2-}$
	Dodecahedral	$(Ph_4As)_2[Zn(NO_3)_4]$

[a] According to H. Sigel and R. B. Martin, *Chem. Soc. Rev.*, **1994**, *28*, 83, Cd^{II} adopts coordination numbers 4 (20%), 5 (8%), and 6 (56%) of the time.
[b] H_2dapp = 2,6-diacetylpyridine(2′-pyridylhydrazone).

The major ore of zinc (though it occurs in numerous other minerals) is *sphalerite*, a form of ZnS usually found to also contain iron and commonly associated with *galena* (PbS). Cadmium is mostly found isomorphously replacing zinc in zinc minerals. To obtain metallic zinc from its ore, the ore is first calcined to ZnO and this is then roasted at *ca.* 1000°C with charcoal in the absence of air, whence the metal distills out. Cadmium can be separated from sulfate solutions of both zinc and cadmium by addition of zinc dust:

$$Zn + Cd^{2+} = Zn^{2+} + Cd \qquad E^0 = +0.36\ V$$

Mercury is obtained from its principal ore, *cinnabar* (HgS), by roasting to form HgO and then decomposing this at *ca.* 500°C.

Zinc is widely used as a protective coating (galvanizing) and forms many useful alloys, such as brass. Zinc is not known to be toxic in any form, but cadmium and mercury are exceedingly toxic and must be handled accordingly.

Zinc and cadmium have distorted hexagonal close-packed structures, whereas mercury is a shiny liquid down to *ca.* −39°C. The very low melting point of mercury (−39°C *vs.* 1064°C for Au and 303°C for Tl) is attributable to the relativistic contraction of its $6s^2$ shell.[5] All three elements are unusually volatile for metals, especially mercury. The vapor of mercury (1.3×10^{-3} mm at 20°C) is monotomic and mercury also dissolves appreciably in many liquids, such as water, where at room temperature the solubility is about 6×10^{-5} g/L. Mercury should always be kept in tightly closed

[5] L. J. Norrby, *J. Chem. Educ.* **1991**, *68*, 110.

containers and handled only in well-ventilated areas. It may be generated in, and escape from, solutions of its salts due to adventitious reduction and/or disproportion-ation of Hg_2^{2+}.

While mercury is inert to non-oxidizing acids and bases, both zinc and cadmium readily dissolve in simple acids to give the M^{2+} ions. Zinc will dissolve in concentrated bases to form zincate ions:

$$Zn + 2OH^- + 2H_2O \longrightarrow [Zn(OH)_4]^{2-} + H_2$$

Cadmiate ions are much less stable and cadmium therefore does not dissolve in bases.

While zinc and cadmium react with oxygen on heating to form ZnO and CdO, the behavior of mercury is more complex. Mercury combines with oxygen, but exceedingly slowly until temperatures of 300–350°C are reached. However, above 400°C the oxide again decomposes:

$$HgO(s) \longrightarrow Hg(s) + \tfrac{1}{2}O_2 \qquad \Delta H = 160 \text{ kJ mol}^{-1}$$

This behavior was known to Priestley and Lavoisier who employed it in their studies of elemental oxygen.

Finally, we note that mercury forms alloys, known as *amalgams*, with many other metals. Some transition metals do not so react (Fe, for example) and can be used as containers for mercury, but others, such as Na and K react vigorously and form compounds of definite composition, such as $NaHg_2$. Sodium and zinc amalgams are often used as powerful reducing agents in preparative chemistry.

15-4   The Univalent State for Zinc and Cadmium

While this state, in the form of M_2^{2+} ions, is not of great importance for Zn and Cd, it does exist in a few stable compounds. Zinc metal dissolves in molten $ZnCl_2$ (500–700°C) to afford, on cooling, a yellow, diamagnetic glass. Similarly, Cd dis-solves in molten $CdCl_2$. Reaction of cadmium metal with molten $Cd(AlCl_4)_2$ affords crystalline $Cd_2(AlCl_4)_2$, in which Cd_2^{2+} ions (Cd—Cd distance, 2.576 Å) are found.

The compound $HB(2,4-Me_2pz)_3Cd-Cd(2,4-Me_2pz)_3BH$, while not structurally characterized has been shown to have a $^{111}Cd-^{113}Cd$ coupling that is indicative of a Cd—Cd bond.[6] No structural data are available for the Zn_2^{2+} ion, but force con-stants, obtained from Raman spectroscopy, indicate that the M—M bond strengths are in the (expected) order: $Zn_2^{2+} < Cd_2^{2+} < Hg_2^{2+}$.

15-5   The Univalent State for Mercury

For mercury, the Hg_2^{2+} ion is a very important species. Its much greater stability relative to the Cd_2^{2+} and Zn_2^{2+} ions is due to the large ionization enthalpy of the Hg

[6]D. L. Reger *et al., J. Am. Chem. Soc.* **1993**, *115*, 10406.

atom (135 kJ mol^{-1} greater than that for Cd), which is, in turn, due to the relativistic stabilization of the $6s$ orbital. This provides an energetic advantage when two Hg^+ ions share a pair of $6s$ electrons.

In aqueous chemistry the Hg_2^{2+} ion is readily obtained by reduction of the Hg^{2+} ion, and numerous crystalline compounds containing it can be obtained. The Hg—Hg distance varies from about 2.51 Å when the anions are weakly coordinating (e.g., SO_4^{2-}, BrO_3^-, F^-, NO_3^-) to 2.69 Å in Hg_2I_2.

To understand the aqueous chemistry of the mercurous ion, Hg_2^{2+}, it is useful to begin with the potentials of the following half reactions:

$$Hg_2^{2+} + 2e = 2Hg(l) \qquad E^0 = 0.7960 \text{ V}$$

$$2Hg^{2+} + 2e = Hg_2^{2+} \qquad E^0 = 0.9110 \text{ V}$$

$$Hg^{2+} + 2e = Hg(l) \qquad E^0 = 0.8535 \text{ V}$$

From these the potential of the following rapid, reversible reaction is then calculated:

$$Hg_2^{2+} = Hg(l) + Hg^{2+} \qquad E^0 = -0.115 \text{ V}$$

From this, the following equilibrium constant is obtained:

$$K = \frac{[Hg^{2+}]}{[Hg_2^{2+}]} = 1.14 \times 10^{-2}$$

It follows, then, that when a solution of Hg^{2+} ion is treated with an equimolar (or greater) quantity of Hg, a solution of Hg_2^{2+} is formed. This conclusion, as well as the accuracy of all the quantitative information in all of the above equations assume that we are dealing with uncomplexed aqua ions. This, of course, is seldom the case.

Since many anions tend to complex more strongly with Hg^{2+} than with Hg_2^{2+}, the marginal stability of the latter ion against disproportionation is easily altered and thus there are relatively few stable Hg_2^{2+} compounds. All anions or ligands such as NH_3, amines, OH^-, CN^-, SCN^-, S^{2-}, and $acac^-$ that complex or precipitate Hg^{2+} promote the disproportionation of Hg_2^{2+}. Examples are

$$Hg_2^{2+} + 2OH^- \longrightarrow Hg(l) + HgO(s) + H_2O$$

$$Hg_2^{2+} + S^{2-} \longrightarrow Hg(l) + HgS(s)$$

$$Hg_2^{2+} + 2CN^- \longrightarrow Hg(l) + Hg(CN)_2(aq)$$

The rate-determining step in these disproportionations has been shown to be cleavage of the Hg—Hg bond. The reactions with ammonia and strongly basic amines are more complex as will be explained in Section 15-13.

The Hg—Hg bond is believed to be a simple σ bond, which can be described by an electron configuration σ^2. Irradiation of an Hg_2^{2+} solution at *ca.* 240 nm is believed to excite it into a non-bonded state, $\sigma\sigma^*$, which dissociates to $2Hg^+$. In presence of O_2, the following reaction occurs.[7]

$$2Hg^+ + O_2 + 2H^+ \longrightarrow 2Hg^{2+} + H_2O_2$$

Mercurous (Hg_2^{2+}) Compounds

The four halides, Hg_2X_2, are all known, the chloride, bromide, and iodide all being very insoluble in water. Hg_2F_2 is rapidly hydrolyzed to HF, Hg(l), and HgO. $Hg_2(NO_3)_2 \cdot 2H_2O$ and $Hg_2(ClO_4)_2 \cdot 4H_2O$ are very soluble in water to give stable solutions from which the insoluble halides can easily be precipitated. Other compounds that contain weakly coordinating anions, e.g., sulfate, chlorate, bromate, iodate, and acetate, are known. In general, oxygen-donor ligands such as oxalate, succinate, $P_2O_7^{4-}$, and $P_3O_{10}^{5-}$, which do not form strong complexes with Hg^{2+}, give stable complexes, such as $[Hg_2(P_2O_7)_2]^{6-}$, with mercurous ion. Also, the less basic nitrogen ligands, such as $PhNH_2$ and 1,10-phenanthroline, for example, will form stable complexes such as $[Hg_2(PhNH_2)_2]^{2+}$.

15-6 Lower Oxidation States for Mercury

Compounds containing Hg_3^{2+} and Hg_4^{2+} as well as Hg_2^{2+} are known. AsF_5 in liquid SO_2 (also used to oxidize S, Se, and Te) allows the following reactions to be achieved:

$$2Hg + 3AsF_5 = Hg_2(AsF_6)_2 + AsF_3 \qquad \text{colorless solution}$$

$$3Hg + 3AsF_5 = Hg_3(AsF_6)_2 + AsF_3 \qquad \text{yellow solution}$$

$$4Hg + 3AsF_5 = Hg_4(AsF_6)_2 + AsF_3 \qquad \text{red solution}$$

It is also possible to dissolve Hg in molten HgX_2 compounds to obtain Hg_2X_2 and Hg_3X_2 molecules, identified by their Raman spectra, although not isolated.[8] Dissolution of mercury in FSO_3H also gives a yellow solution containing the Hg_3^{2+} ion.

The Hg_3^{2+} and Hg_4^{2+} units are essentially linear and can be characterized in solution by ^{199}Hg nmr spectra as well as Raman spectra.

If reactions of mercury with AsF_5 are carried out using even lower ratios of oxidant to mercury than those shown in the equations above, golden metal-like solids, $Hg_{2.82}AsF_6$ or $Hg_{2.9}SbF_6$ are obtained. These have non-intersecting, mutually perpendicular chains of Hg atoms, separated by EF_6^- ions. Silvery Hg_3EF_6 com-

[7]A. Vogler and H. Kunkely, *Inorg. Chim. Acta* **1989**, *162*, 169.
[8]G. A. Vogiatzis and G. N. Papatheodorou, *Inorg. Chem.* **1992**, *31*, 1945.

pounds containing sheets of close-packed Hg atoms separated by layers of EF_6^- ions have also been made.

DIVALENT ZINC AND CADMIUM COMPOUNDS

15-7 Aqua Ions, Oxides, and Hydroxides

Zinc and cadmium form numerous water soluble salts, for example, the nitrates, sulfates, sulfites, perchlorates, and acetates, that contain the aqua ions, $[M(H_2O)_6]^{2+}$. These aqua ions are also present in aqueous solutions.[9] Some salts of Zn^{2+} and Cd^{2+} are isomorphous with those of Mg^{2+}. The $[M(H_2O)_6]^{2+}$ ions are slightly hydrolyzed in water:

$$[M(H_2O)_6]^{2+} = [M(H_2O)_5(OH)]^+ + H^+ \qquad K \approx 10^{-9}(Zn),\ 10^{-10}(Cd)$$

In more concentrated solutions, polynuclear species, beginning with $M_2(OH)^{3+}(aq)$, are formed.

Addition of alkali metal hydroxides causes precipitation of $Zn(OH)_2$ or $Cd(OH)_2$, for which the solubility products are $\sim 10^{-11}$ and 10^{-14}, respectively. However, $Zn(OH)_2$ dissolves as such to the extent of ca. 10^{-6} molar. It also dissolves completely in excess aqueous base to give principally $[Zn(OH)_3(H_2O)]^-$. $Cd(OH)_2$ will also dissolve in very concentrated base. Both $Zn(OH)_2$ and $Cd(OH)_2$ dissolve in ammonium hydroxide but for a different reason: amine complexes, $[Zn(NH_3)_4]^{2+}$ and $[Cd(NH_3)_6]^{2+}$, are formed.

The anhydrous oxides, ZnO and CdO, are formed on pyrolysis of the nitrates or carbonates, or by roasting the metals in air. Both oxides can be sublimed at very high temperatures. Both ZnO and CdO tend to have lattice defects leading to coloration. For ZnO the pure white compound turns yellow on heating (deficit of oxygen), while CdO may be brown or even black, depending on how much it has been heated.

Zinc oxide has various uses, of which the most important is as a co-catalyst, with CuO (supported on Al_2O_3) for low pressure synthesis of methanol by oxidation of methane.[10]

On heating $Zn(O_2CCH_3)_2$ in vacuum the basic acetate, $Zn_4O(O_2CCH_3)_6$, plus acetic anhydride, are formed. The molecule has an oxo-centered Zn_4 tetrahedron with a CH_3CO_2 bridge on each edge. Unlike the analogous $Be_4O(O_2CCH_3)_6$, it is easily hydrolyzed. Cadmium, on the other hand forms dinuclear carboxylates of the type $Cd_2(\mu\text{-}O_2CR)_4L_2$. There are a number of other $Zn_4O(O_2CR)_6$ compounds, as well as the structurally similar $Zn_4O(O_2CNEt_2)_6$,[11] $ZnO(7\text{-azaindolate})_6$,[12] and $Zn_4O(DPhF)_6$.[13]

There are also basic zinc carboxylates having other types of structures, such as $Zn_5(OH)_2(O_2CCHCHCH_3)_8$, a chain polymer.[14]

[9]J. Lindgren et al., Inorg. Chem. 1992, 31, 150.
[10]K. Klier et al., I&EC Research 1991, 29, 61.
[11]F. Calderazzo et al., Inorg. Chem. 1991, 30, 3778.
[12]S.-M. Peng et al., J. Chem. Soc., Dalton Trans. 1993, 467.
[13]F. A. Cotton et al., Inorg. Chim. Acta 1997, 266, 91.
[14]B. P. Straughn et al., Inorg. Chim. Acta 1991, 186, 51; O. Berkesi et al., ibid. 1992, 195, 169.

Table 15-3 Structures[a] of Zn and Cd Oxides and Chalcogenides[b]

Metal	O	S	Se	Te
Zn	W, Z	Z, W	Z	Z, W
Cd	NaCl	W, Z	W, Z	Z

[a]W = wurtzite structure; Z = zinc blende structure; NaCl = rock salt structure.
[b]Where two polymorphs occur, the one stable at lower temperatures is listed first.

15-8 Sulfides, Selenides, and Tellurides

In addition to the oxides, the other six chalcogenides are also known. Table 15-3 shows the structures of the eight compounds. Clearly, with the sole exception of CdO, the chalcogenides of zinc and cadmium prefer tetrahedral coordination, though preference for the cubic zinc blende structure or the hexagonal wurtzite structure varies irregularly.

15-9 Halides and Halo Complexes

Both elements form all four halides and some basic properties are listed in Table 15-4. The Madelung energies of the fluorides are much higher than those of the other halides, giving rise to very high-melting points and boiling points as well as much lower solubilities. In each case, the metal ion in the fluoride has a higher coordination number than it does in the other halides of the same element.

All the halides are fairly or even highly soluble in water, as well as other donor solvents such as ethanol, acetone, and THF. The solubility of $ZnCl_2$ in water is extraordinarily great and the highly concentrated solutions correspond to liquid hydrates, i.e., $ZnCl_2 \cdot nH_2O$, n = 1-4, with all Cl^- ions bridging the Zn^{2+} ions.

In less concentrated solutions of $ZnCl_2$ discrete complexes such as $[ZnCl_4]^{2-}$, $[ZnCl_4(H_2O)_2]^{2-}$, $[ZnCl_2(H_2O)_4]$, and eventually $[Zn(H_2O)_6]^{2+}$ are present. Roughly the same behavior is displayed by the other halides of both zinc and cadmium

Table 15-4 Some properties of the Zinc and Cadmium Halides

Halide	Solubility in water (mol L^{-1})	(°C)	mp (°C)	bp (°C)	Structure
ZnF_2	1.57	20	872	1502	Rutile
$ZnCl_2$[a]	31.8	25	275	756	*ccp* anions with Zn in tetrahedral interstices
$ZnBr_2$	20.9	25	394	697	
ZnI_2	13	25	446	(sublimes)	
CdF_2	0.29	25	1110	1747	Fluorite
$CdCl_2$	7.7	20	868	980	Close-packed anions with Cd in octahedral interstices
$CdBr_2$	4.2	20	568	1136	
CdI_2	2.3	20	387	(sublimes)	

[a]Very pure $ZnCl_2$ has but one form; other reported polymorphs contain OH groups.

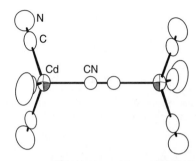

Figure 15-1   The $[Cd_2(CN)_7]^{3-}$ ion in $[PPh_4]_3[Cd_2(CN)_7]$. The bridging CN^- ion is disordered so that each atom appears to be 0.5C/0.5N. See T. Kitazawa and M. Takeda, *J. Chem. Soc., Chem. Commun.,* **1993,** 309.

except for the fluorides. For example, in a 0.5 M solution of $CdBr_2$ at 25°C we have the following molar concentrations: Cd^{2+} (0.01), $CdBr^+$ (0.26), $CdBr_2$ (0.16), $CdBr_3^-$ (0.04), $CdBr_4^{2-}$ (0.02), and Br^- (0.20). For both Zn^{2+} and Cd^{2+} complexing by F^- is very weak and no species beyond MF^+ has any importance.[15]

Tetrahedral $[ZnX_4]^{2-}$ and $[CdX_4]^{2-}$ ions occur in crystalline compounds with medium to large organic cations, e.g., $[xanthane^+]_2[ZnCl_4^{2-}]^{[16]}$ and $[H_2pip][CdBr_4]$.[17] In acetonitrile, complexes up to $[MX_4]^{2-}$ as well as some polynuclear species such as $Cd_2I_5^-$ form readily.[18] They, as well as other halo-complexes,[19] appear to involve mainly tetrahedral coordination.

15-10   Zinc and Cadmium Cyanides and Cyanide Complexes

With Zn^{2+} and Cd^{2+} the cyanide ion has a marked tendency to form linear bridges, $M-C\equiv N-M$. The simplest instance of this tendency is exemplified by the $[Cd_2(CN)_7]^{3-}$ ion, shown in Fig. 15-1. In general, bridging CN ions tend to be disordered so that C and N are not distinguishable. This makes possible the formation of a great variety of unusual framework structures. In fact, isomorphous $Zn(CN)_2$ and $Cd(CN)_2$ themselves have a very unusual structure that consists of two interpenetrating diamond-like networks, as shown in Fig. 15-2. Because of the length of the $M-C\equiv N-M$ units (5.11 Å for Zn and 5.46 Å for Cd), each network has distinct cavities, each of which is occupied by an atom of metal and its connected CN groups of the other one. The cavities, in this case, have an "adamantane" shape.

When $Cd(CN)_2$ is crystallized in the presence of other molecules that can "stuff" cavities or tunnels, many different structures are formed depending on the size and shape of the guests that stuff the cavities. Similar behavior is, of course, found elsewhere, e.g., in gas hydrates and hydrothermal synthesis of zeolites. These cadmium cyanide structures may be considered as a new class of clathrates.

The simplest ones are those obtained merely by leaving an aqueous solution of $CdCl_2 + K_2[Cd(CN)_4]$ to crystallize in contact with neopentane, CCl_4, or any

[15]G. T. Hefter *et al., Polyhedron* **1990,** *9,* 901.
[16]E. Dubler *et al., Inorg. Chim. Acta* **1992,** *197,* 135.
[17]A. B. Corradi *et al., Inorg. Chim. Acta* **1991,** *187,* 141.
[18]D. P. Graddon and C. S. Khoo, *Polyhedron* **1988,** *7,* 2129.
[19]C. A. McAuliffe *et al., J. Chem. Soc. Commun.* **1992,** 944.

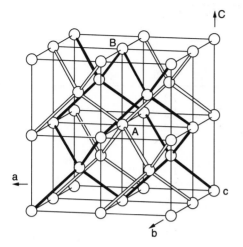

Figure 15-2   Eight unit cells of the Zn(CN)$_2$/Cd(CN)$_2$ structure, showing how two diamond-like networks, each with large adamantane-like cavities interpenetrate each other. See B. F. Hoskins and R. Robson, *J. Am. Chem. Soc.*, **1990**, *112*, 1546.

of the intermediate CMe$_n$Cl$_{4-n}$ species. In this way cubic crystals containing an arrangement of Cd/2CN analogous to that in one sublattice of Cd(CN)$_2$ is obtained with each adamantane-like cavity occupied by one CMe$_n$Cl$_{4-n}$ molecule. Clearly related to these structurally are mixed metal clathrates such as CdHg(CN)$_4$·2CCl$_4$, CdZn(CN)$_4$·2CCl$_4$, (NMe$_4$)(CCl$_4$)CdCu(CN)$_4$, and others.[20]

More commonly, clathrates are formed by linking both octahedrally and tetrahedrally coordinated cadmium ions. For example, simply by crystallization of Cd(CN)$_2$ from aqueous alcohols[21] (or in similar ways[22]) one obtains clathrates having infinite channels, as shown in Fig. 15-3 for the products obtained with Pr^{n}OH and Pr^{i}OH. Similar structures with similar (or even wider) channels are obtained with DMF and DMSO,[23] and in the compound Cd(CN)$_2$·⅔H$_2$O·Bu^{i}OH.[24] There are other clathrates that include a combination of water and ether molecules,[25] and still

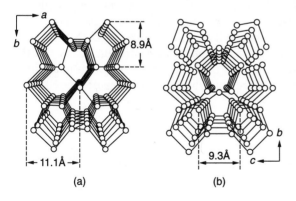

Figure 15-3   Perspective view of [Cd(CN)$_2$]$_n$ frameworks: (a) in Cd(CN)$_2$·²/₅H$_2$O·⁶/₅Pr$_n$OH and (b) in Cd(CN)$_2$·²/₃H$_2$O·²/₃Pr^{i}OH. See K. Kim et al., *J. Chem. Soc., Chem. Commun.*, **1993**, 1400.

[20]T. Iwamoto *et al.*, *J. Chem. Soc., Dalton Trans.* **1994**, 1029.
[21]K. Kim *et al.*, *J. Chem. Soc., Chem. Commun.* **1993**, 1400.
[22]C. I. Ratcliff *et al.*, *J. Am. Chem. Soc.* **1992**, *114*, 8590.
[23]K. Kim *et al.*, *J. Chem. Soc., Chem. Commun.* **1994**, 637.
[24]R. Robson *et al.*, *J. Chem. Soc., Chem. Commun.* **1990**, 60; *J. Am. Chem. Soc.* **1992**, *114*, 10641; J. A. Ripmeester *et al.*, *J. Chem. Soc., Chem. Commun.* **1991**, 735.
[25]T. Kitazawa *et al.*, *J. Chem. Soc., Dalton Trans.* **1994**, 2933.

Table 15-5 Equilibrium Constants for Some Complexes of Zn, Cd, and Hg ($M^{2+} + 4X = [MX_4]$; $K = [MX_4]/[M^{2+}][X]^4$)

	M Ion		
X	Zn^{2+}	Cd^{2+}	Hg^{2+}
NO_2^-[a]	<1	<1	7.3×10^{11}
Cl^-	1	10^3	1.3×10^{15}
Br^-	10^{-1}	10^4	9.2×10^{20}
I^-	10^{-2}	10^6	5.6×10^{29}
NH_3	10^9	10^7	2.0×10^{19}
CN^-	10^{21}	10^{19}	1.9×10^{41}

[a]F. B. Erim and E. Avsar, *Polyhedron,* **1988,** *7,* 213.

others that have hexamethylene tetramine[26] or ionic guests,[27] e.g., $(NMe_4 \cdot 2EtCN)^+$ in the cavities.

15-11 Other Compounds and Complexes of Zinc and Cadmium

A good comparison of stabilities of Zn^{2+} and Cd^{2+} complexes with various ligands in solution is provided by Table 15-5. In the solid state there is a wealth of data showing that an enormous range of coordination numbers and geometries occur, depending on ligand type and steric bulk.

Chalcogen Ligands

Zinc and cadmium (as well as mercury, which will be dealt with later) have a strong affinity for chalcogen ligands, especially RS^- and ArS^- groups. They generally prefer S or Se to P as a ligating atom[28] although species containing both, e.g., $(R_3P)(PhSe)Cd(\mu\text{-}PhSe)_2Cd(SePh)(PR_3)$, can be made.[29] Numerous compounds and complexes containing only the metal ions and the RE^- ligands are known and display great structural diversity. Some such as $Zn(SPh)_2$ and $[Zn(SPh)_4]^{2-}$ are simple, but there are, for example, adamantane-like structures, as in the $[Cd_4(SPh)_{10}]^{2-}$ ions. There are also complexes that contain other ligands in addition to RE^-, such as $[Cd_{10}S_4(SPh)_{16}]^{4-}$ and $[Zn_4Cl_2(SPh)_8]^{2-}$. Structures of types 15-I to 15-V are commonly encountered.

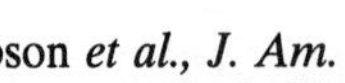

(15-I)

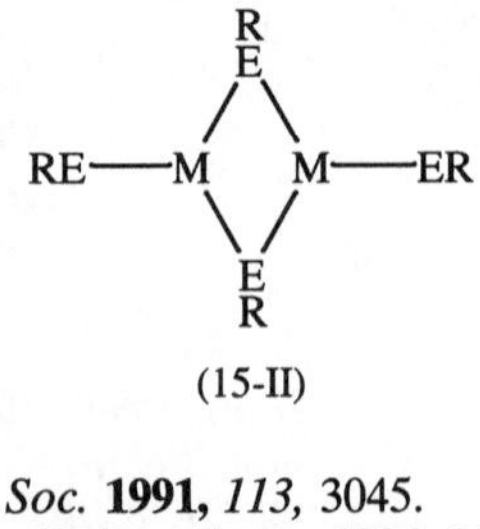

(15-II)

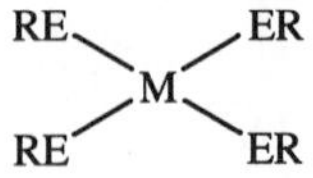

(15-III)

[26]R. Robson *et al., J. Am. Chem. Soc.* **1991,** *113,* 3045.
[27]T. Iwamoto *et al., J. Chem. Soc., Dalton Trans.* **1994,** 3695.
[28]A. M. Bond *et al., Inorg. Chem.* **1989,** *28,* 4509.
[29]T. Siegrist *J. Am. Chem. Soc.* **1989,** *111,* 4141.

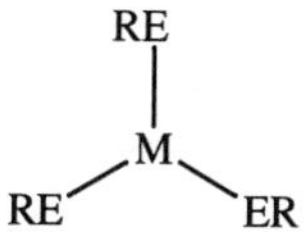

(15-IV) 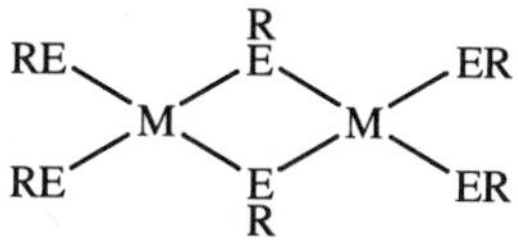   (15-V)

The "simple" $M(ER)_2$ and $M(EAr)_2$ compounds have structures that strongly depend on the bulkiness of the R or Ar groups. For the most part they are polymers. The $Cd(SC_6H_4X)_2$ (X = H, Me, Cl, etc.) compounds have solid state structures in which adamantane-like $Cd_4(SAr)_6$ cages are linked by the additional SAr groups into infinite three-dimensional arrays.[30] The adamantane motif persists even when the compounds are dissolved in DMF,[31] and a discrete adamantane cage has been characterized in $[Cd_4(\mu\text{-}SPr^i)_6](PPh_3)_2(ClO_4)_2$.[32] Even with an Ar group as large as 15-VI the $M(EAr)_2$ molecule dimerizes (15-II), although it is partly dissociated in solution.[33] With the R group 15-VII, a nearly linear monomer is formed.[34] For the compound $Zn[TeSi(SiMe_3)_3]_2$ the bulk is also not great enough to prevent dimerization as in 15-II.[35]

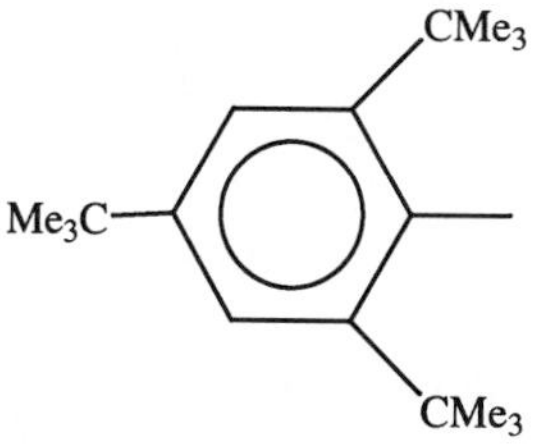

(15-VI) 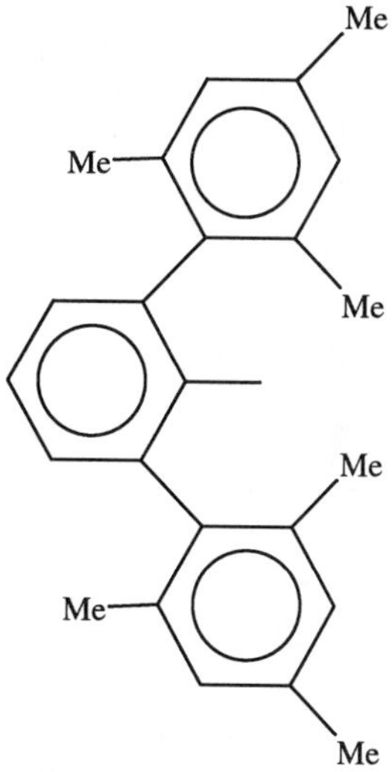   (15-VII)

The polymeric $M(SMe)_2$ compounds undergo efficient pyrolysis to form the sulfides.[36]

$$M(SMe)_2 \longrightarrow MS + Me_2S$$

[30] I. G. Dance *et al.*, *Inorg. Chem.* **1987**, *26*, 4057.
[31] I. G. Dance *et al.*, *Inorg. Chem.* **1990**, *29*, 603.
[32] P. A. W. Dean *et al.*, *Inorg. Chem.* **1993**, *32*, 4632.
[33] M. Bochmann *et al.*, *J. Chem. Soc. Dalton Trans.* **1991**, 2317; *Inorg. Chem.* **1993**, *32*, 532.
[34] J. J. Ellison and P. P. Power, *Inorg. Chem.* **1994**, *33*, 4231.
[35] P. J. Bonasia and J. Arnold, *Inorg. Chem.* **1992**, *31*, 2508.
[36] K. Osakada and T. Yamamoto, *Inorg. Chem.* **1991**, *30*, 2328.

Turning now to anionic species, trigonal $[M(SR)_3]^-$ species, 15-IV, are formed with moderately bulky ligands,[37] whereas with $[Zn_2(SPh)_6]^{2-}$ the structure is of type 15-V.[38] There are numerous examples of tetrahedral or distorted tetrahedral (15-III) type $[M(SAr)_4]^{2-}$ anions.[39,40] An unusual $[Cd_3(\mu\text{-}SAr)_3(\mu_3\text{-}SAr)(SAr)_3]^-$ anion (a defective Cd_4S_4 cube) has also been reported.[41]

When both ER^- and E^{2-} ligands are present, quite elaborate structures with large clusters, e.g., $[Cd_{10}S_4(SPh)_{16}]^{4-}$, $[TeCd_8(SePh)_{16}]^{2-}$, and others are formed by cadmium.[42]

Numerous other types of complexes with chalcogenide ligands are also known. Such ligands include polysulfides, as in $[Cd(S_n)(S_m)]^{2-}$ ($n, m = 5, 6$),[43] macrocycles such as tetrathiacyclononane[44] and tetrathiacyclohexadecane,[45] and $Me_2NCH_2\text{-}CH_2S^-$, which gives tetrahedral monomers.[46,47] Further, the main classes of disulfur ligands, i.e., $^-S_2CNR_2$[48] and their selenium analogues,[49] 1,2-phenylene dithiols, and $^-S_2CC(CN)_2$[50] form Zn and Cd complexes.

Dithiocarbamates, especially those of zinc, are well known and have important industrial applications as antioxidants/antiabrasives in motor oils and as vulcanization accelerators in rubber. They are normally dimerized as in 15-VIII, thereby giving each zinc atom a distorted trigonal bipyramidal coordination.[51] $Zn(Me_2dtc)_2$ is dimerized in a different way, 15-IX, but vaporizes as monomers.[52] Both zinc and cadmium dithiocarbamates are cleaved by monoamines to give 5-coordinate

(15-VIII)   (15-IX)

[37]E. S. Gruff and S. A. Koch, *J. Am. Chem. Soc.* **1989**, *111*, 8762; **1990**, *112*, 1245.

[38]M. A. Walters *et al.*, *Inorg. Chem.* **1991**, *30*, 4280.

[39]E. Block *et al.*, *Inorg. Chem.* **1989**, *28*, 1263.

[40]I. G. Dance *et al.*, *Inorg. Chem.* **1989**, *28*, 1862.

[41]K. Tang *et al.*, *J. Chem. Soc., Chem. Commun.* **1991**, 1590.

[42]I. G. Dance *et al.*, *J. Am. Chem. Soc.* **1990**, *112*, 6435; *Inorg. Chem.* **1993**, *32*, 66.

[43]I. G. Dance *et al.*, *Inorg. Chem.* **1989**, *28*, 1862.

[44]G. S. Wilson *et al.*, *Inorg. Chim. Acta* **1993**, *207*, 241.

[45]W. N. Setzer *et al.*, *Inorg. Chem.* **1991**, *30*, 3652.

[46]P. González-Duarte, *Polyhedron* **1990**, *9*, 763.

[47]V. I. Minkin *et al.*, *Polyhedron* **1991**, *10*, 771.

[48]L. A. Glinskaya *et al.*, *Polyhedron* **1992**, *11*, 2951.

[49]M. B. Hursthouse *et al.*, *Polyhedron* **1992**, *11*, 45.

[50]H. Y. Li and E. L. Amma, *Inorg. Chim. Acta* **1990**, *177*, 5.

[51]P. M. Maitlis *et al.*, *Polyhedron* **1988**, *7*, 1861; P. T. Manoharan *et al.*, *Inorg. Chem.* **1990**, *29*, 4011.

[52]D. A. Rice *et al.*, *Inorg. Chem.* **1989**, *28*, 3239.

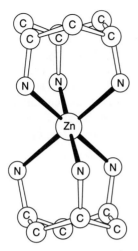

Figure 15-4   The $Zn[(C_6H_9)(NH_2)_3]_2^{2+}$ cation. See U. Brand and H. Vahrenkamp, *Inorg. Chim. Acta*, **1992**, *198–200*, 663.

monomers or by bipy or phen to give monomers with distorted octahedral coordination.[53]

Reaction with additional R_2dtc^- ions gives $[M(R_2dtc)_3]^-$ anions. Zinc dithiophosphate esters, $Zn[S_2P(OR)_2]_2$, show many similarities to the dithiocarbamates.

Amine Ligands

Both zinc and cadmium form numerous complexes with amine ligands, ranging from NH_3 through diamines such as ethylene diamine and substituted propylene diamines,[54] to a variety of macrocycles.[55,56] Pyridines and pyrazoles are also good ligands.[57,58] In the case of 1,3,5-triaminocyclo-hexane, a regular octahedral complex cation is formed, as shown in Fig. 15-4. With 3-Me-pyrazine, tetrahedral $[Zn(C_4N_2H_6)_4]^{2+}$ is formed.[59]

Amides. The diamides $M(NR_2)_2$, are useful starting materials to make, for example, alkoxides[60]

$$Cd[N(SiMe_3)_2]_2 + 2ROH \longrightarrow Cd(OR)_2 + 2(Me_3Si)_2NH$$

Several of them, with $N(SiMe_3)_2$, $N(CMe_3)(SiMe_3)$,[61] and $N(SiMePh_2)_2$[62] have linear $N-Zn-N$ units.

[53]N. A. Bell *et al.*, *Inorg. Chim. Acta* **1989,** *156*, 205; J. R. Lechat *et al.*, *Inorg. Chim. Acta* **1990,** *174*, 103.
[54]A. Seroni *et al.*, *Inorg. Chim. Acta* **1989,** *159*, 173; *Inorg. Chem.* **1992,** *31*, 1401.
[55]P. Moore *et al.*, *J. Chem. Soc., Chem. Commun.* **1991,** 706.
[56]J. L. Sessler *et al.*, *Inorg. Chem.* **1989,** *28*, 1333.
[57]B. Kamenar *et al.*, *Inorg. Chim. Acta* **1991,** *188*, 151.
[58]M. Fujita *et al.*, *J. Chem. Soc., Chem. Commun.* **1994,** 1977.
[59]X.-M. Chen *et al.*, *Acta Cryst.* **1996,** *52C*, 2482.
[60]L. G. Hubert-Pfalzgraf *et al.*, *Polyhedron* **1992,** *11*, 1331.
[61]W. S. Rees, Jr. *et al.*, *Polyhedron* **1992,** *11*, 1697.
[62]P. P. Power *et al.*, *Inorg. Chem.* **1991,** *30*, 5013.

Tripyrazolylborate Complexes

With the $HB(pyr)_3^-$ type ligand, where pyr is a pyrazolyl or substituted pyrazolyl group, a number of interesting complexes can be made. With the 3-butylpyrazolyl ligand, a large number of tetrahedral species of the type 15-X, with X = H, Cl, Br, I, CN, N_3, NCS, SH, $OSiMe_3$, OAc, C≡CPh have been made.[63] It has been shown that a similar cadmium compound (X = Cl) also exists.[64] Both zinc and cadmium can form octahedral $[HB(pyr)_3]_2M$ complexes.[65] The zinc compound with the bis-pyrazolyl ligand, $[H_2B(3-Bu^tpyr)_2]ZnCMe_3$ is a mononuclear 3-coordinate complex.[66]

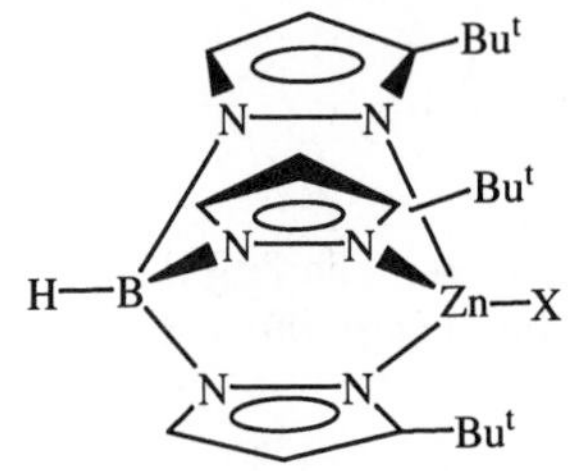

(15-X)

Oxygen Ligands

Ligands having oxygen as donor atom are also able to form many complexes with Zn and Cd. These include phosphine oxide complexes such as tetrahedral $[Zn(OPPh_3)_4](ClO_4)_2$,[67] pyridine oxide complexes, the simplest of which is $[Cd(pyO)_6][CdCl_4]$,[68] alkoxides, $M(OR)_2$, carbamates,[69] and phosphates.[70]

Hydrides

In addition to the pyrazolylborate hydride mentioned earlier, there is a tetrameric $[Me_3COZnH]_4$ that can be made from ZnH_2.[71]

DIVALENT MERCURY

15-12   Binary Compounds and Simple Salts

The addition of OH^- to aqueous Hg^{2+} precipitates HgO, as a yellow solid of fine particles. When prepared in other ways, e.g., by gentle thermolysis of $Hg_2(NO_3)_2$ or $Hg(NO_3)_2$, or by direct combination of Hg and O_2, it is red. The stable form of the oxide has zigzag $-O-Hg-O-Hg-$ chains with the $O-Hg-O$ units linear

[63]G. Parkin et al., J. Chem. Soc., Chem. Commun. **1990**, 220; Inorg. Chem. **1992**, 31, 1656.
[64]D. L. Reger et al., Inorg. Chem. **1993**, 32, 4345; G. Parkin et al., Inorg. Chem. **1994**, 33, 1158.
[65]D. L. Reger et al., Inorg. Chem. **1993**, 32, 5216; C. Janiak and H. Hemling, J. Chem. Soc., Chem. Commun. **1994**, 2947.
[66]G. Parkin et al., J. Am. Chem. Soc. **1990**, 112, 4068.
[67]P. Rubini et al., Polyhedron **1992**, 11, 1795.
[68]C. J. Wilkins et al., Polyhedron **1991**, 10, 2111.
[69]I. Abrahams et al., J. Chem. Soc., Dalton Trans. **1995**, 1043.
[70]A. Clearfield et al., Inorg. Chem. **1989**, 28, 2608.
[71]T. L. Neils and J. M. Burlitch, Inorg. Chem. **1989**, 28, 1607.

Table 15-6 Hg—X Distances in Mercuric Halides (Å)

Compound	Two at	Two at	Two at	Vapor
		Solid		
HgF$_2$		Eight at 2.40		
HgCl$_2$	2.25	3.34	3.63	2.28 ± 0.04
HgBr$_2$	2.48	3.23	3.23	2.40 ± 0.04
HgI$_2$ (red)		Four at 2.78		2.57 ± 0.04

and Hg—O = 2.03 Å. While no solid hydroxide is known, the following hydrolysis reactions occur in perchlorate solution.

$$Hg^{2+} + H_2O = Hg(OH)^+ + H^+ \quad K = 2.6 \times 10^{-4}$$

$$Hg(OH)^+ + H_2O = Hg(OH)_2 + H^+ \quad K = 2.6 \times 10^{-3}$$

The addition of H_2S or alkali metal sulfides to aqueous Hg^{2+} precipitates the highly insoluble, black mercuric sulfide, HgS. This black solid when heated or treated in other ways is changed into a red form that is identical to the mineral *cinnabar*. In this red form, HgS has a distorted NaCl structure in which $(Hg—S)_\infty$ chains can be recognized. Red cinnabar on irradiation in aqueous KI is converted to black cinnabar, which has the ZnS structure and also occurs in Nature.

All four halides, HgX_2, are known; their structures are summarized in Table 15-6 and some properties are given in Table 15-7. Mercury(II) fluoride has the fluorite structure and is not volatile. It is decomposed by water, as might be expected since HgO and HF are both weakly dissociated. No fluoro complex of Hg^{2+} is known.

The other three halides can all be vaporized as XHgX molecules and such molecules also occur in solution, although dimerization to $XHg(\mu\text{-}X)_2HgX$ species also occurs in some organic solvents. Solid mercuric chloride has an essentially molecular (linear ClHgCl) structure, whereas in solid $HgBr_2$ linear BrHgBr units are approached more closely by Br atoms of adjacent molecules. The room-temperature red form of HgI_2 has layers of linked HgI_4 tetrahedra, but above 126°C it changes to a yellow, molecular form. All three heavier halides dissolve in water to a small extent, apparently mainly in molecular form, although some hydrolysis of $HgCl_2$ to Hg(OH)Cl occurs.

Table 15-7 Some Properties of Mercuric Halides

Halide	mp (°C)	bp (°C)	H$_2$O	C$_2$H$_5$OH	C$_2$H$_5$OCOCH$_3$	C$_6$H$_6$
			Solubility (mol/100 mol at 25°C)			
HgF$_2$	645 dec		Hydrolyzes	Insoluble	Insoluble	Insoluble
HgCl$_2$	280	303	0.48	8.14	9.42	0.152
HgBr$_2$	238	318	0.031	3.83		
HgI$_2$	257	351	0.00023	0.396	0.566	0.067

Salts of oxo anions, such as the nitrate, perchlorate, and sulfate are appreciably dissociated in aqueous solution, but, because of the weakness of mercuric oxide as a base, the solutions must be acidified to be stable. An aqua ion, $[Hg(H_2O)_6]^{2+}$, apparently exists, but this readily hydrolyzes to $Hg(OH)_{aq}^+$ and then to $Hg(OH)_2(aq)$ in which there is a linear $HO-Hg-OH$ unit. The dissolved nitrate is mainly present as $Hg(NO_3)_2$, $Hg(NO_3)^+$ and Hg^{2+}, but in the presence of excess NO_3^- the complexes $Hg(NO_3)_3^-$ and $Hg(NO_3)_4^{2-}$ are formed.

The carboxylates can be made by dissolving HgO in hot acid, from which they then separate on cooling. The acetate and trifluoroacetate react with alkenes and alkynes (see Section 15-15) and are soluble in a number of organic solvents (especially $Hg(O_2CCF_3)_2$) as well as water.

Some salts such as the oxalate and phosphate are only sparingly soluble.

15-13 Mercury(II) Complexes

Mercury(II) is a distinctly "soft" cation, showing a strong preference for Cl, Br, I, P, S, Se, and certain N-type ligands. It displays coordination numbers of 2 through 6, with a preference for the lower ones. Its marked preference for linear 2-coordination is a distinctive feature. An immense number of crystal structures have been reported in recent years and only a small fraction of these can be explicitly cited here.

Halogen and Pseudohalogen Complexes

In aqueous solution mercury forms complexes HgX_n^{-n+2} for $n = 1-4$. At 10^{-1} M Cl^-, for example, approximately equal amounts of $HgCl_2$, $HgCl_3^-$, and $HgCl_4^{2-}$ are present, but at 1 M Cl^- essentially only $HgCl_4^{2-}$ is present. In other solvents, such as tributyl phosphate or methanol,[72] HgX_3^- species are favored, which may have trigonal bipyramidal $[HgX_3S_2]^-$ or tetrahedral $[HgX_3S]^-$ structures (S = solvent).

In the solid state the tetrahedral $HgCl_4^{2-}$ ion (sometimes distorted) is fairly common, and $Hg_2X_6^{2-}$ ions consisting of two tetrahedra sharing an edge are also well known.[73] With large protonated amine cations a variety of polynuclear species can be stabilized.[74] These include $Hg_2Cl_8^{4-}$, 15-XI, $(HgCl_3^-)_\infty$, 15-XII, and $Hg_3Cl_{12}^{6-}$, 15-XIII.

(15-XI)   (15-XII)

[72]T. R. Griffiths and R. A. Anderson, *Inorg. Chem.* **1991**, *30*, 1912.
[73]M. G. B. Drew *et al.*, *Inorg. Chim. Acta* **1989**, *155*, 39; G. Marangoni *et al.*, *J. Chem. Soc., Dalton Trans.* **1990**, 915.
[74]D. A. House *et al.*, *Inorg. Chim. Acta* **1992**, *193*, 77; O. A. Dyachenko *et al.*, *Russ. Chem. Bull.* **1996**, *45*, 370.

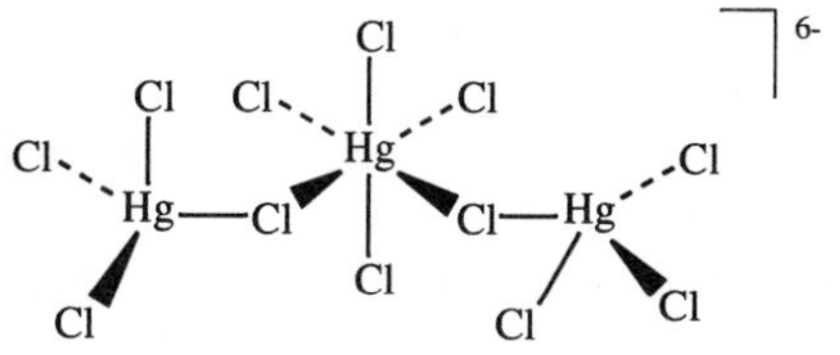

(15-XIII)

Mercury(II) cyanide, which consists of linear molecules, NC—Hg—CN, is soluble in CN^- solutions to give $Hg(CN)_3^-$ and $Hg(CN)_4^{2-}$ ions. Broadly similar behavior is observed in the thiocyanate system.

Ammonia

This ligand reacts in a unique way with mercury(II) compounds. With $HgCl_2$ it forms dark products, the exact nature of which depends on conditions. As shown in the following equations, two of these result from displacement of H atoms by Hg:

$$HgCl_2 + 2NH_3 \rightleftharpoons Hg(NH_3)_2Cl_2(s)$$

$$HgCl_2 + 2NH_3 \rightleftharpoons HgNH_2Cl(s) + NH_4^+ + Cl^-$$

$$2HgCl_2 + 4NH_3 + H_2O \rightleftharpoons Hg_2NCl + H_2O + 3NH_4^+ + 3Cl^-$$

By direct action of NH_3 on HgO one obtains $Hg_2NOH \cdot 2H_2O$ (Millon's base), which is a clathrate. The framework is made of $(Hg_2N)_\infty$ and the cavities and channels contain the OH^- ions and water molecules. In addition to the chloride mentioned above, many other salts of Millon's base, $Hg_2NX \cdot nH_2O$, with X = NO_3^-, ClO_4^-, Br^-, I^-, and $n = 0-2$, are known. These compounds are a type of ion exchanger.

Chalcogenide Complexes

As with Zn and Cd, but to an even greater extent, mercury(II) has a great affinity for ligands with sulfur and the other chalcogenides as the ligating atom, and forms more complexes with such ligands than with any other type. Indeed, the name mercaptan for thiols arose from their affinity for mercury. In biological systems Hg^{II} invariably binds to cysteine thiolate groups. In at least one case, however, a ligand capable of coordinating through S or N is found coordinated through N. This complex[75] is shown as 15-XIV.

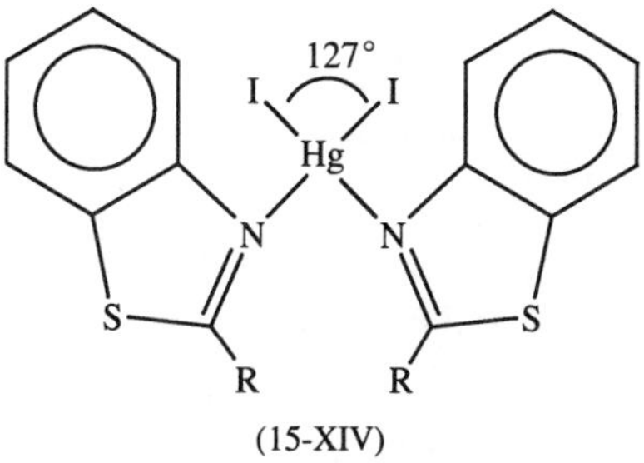

(15-XIV)

[75]I. M. Vezzosi *et al.*, *Polyhedron* **1993**, *12*, 2235.

It may first be noted that the $Hg(SR)_2$ compounds can be easily made by reaction of HgO with 2RSH in CH_2Cl_2 or ethanol.[76] Many are soluble in nonpolar solvents and presumably contain linear molecules. In some cases this has been proven by X-ray crystallography.[77]

Mercury(II) thiolate complexes[78] may be homoleptic, i.e., of the planar $[Hg(SR)_3]^-$ or tetrahedral $[Hg(SR)_4]^{2-}$ types,[79] or more complicated, as in the polymeric[80] $HgCl_2(SC_3H_6NHMe_2)$, (15-XV), and $[Hg(SC_3H_6NMe_3)_2](PF_6)_2$, (15-XVI), or the adamantane-like compounds, $Hg_4(ER)_6X_4$ where E = S, Se, Te, and X = Cl, Br, I (15-XVII).[81]

(15-XV)   (15-XVI)   (15-XVII)

Besides thiolate, RSe^- and RTe^- complexes, there are also $HgX_2(ER_2)$ complexes, which generally dimerize to give approximately tetrahedral Hg.[82] There are also numerous complexes of Hg^{II} with macrocyclic ligands of sulfur[83] as well as a number of polysulfide complexes, such as $[Hg(S_6)_2]^{2-}$ formed on addition of S_n^{2-} to Hg^{II} acetate in methanol. Polytelluride complexes have recently been characterized.[84]

Mercury(II) dithiocarbamates, ranging from the simple, e.g., $Hg(Et_2dtc)_2$, to the more complex,[85] e.g., $[Hg_3(Et_2dtc)_4]^{2+}$ and $[Hg_5(Et_2dtc)_8]^{2+}$, are also known.

Other Ligands

Mercury(II) forms complexes with most other types of ligand. With *oxo anions* it forms, for example, $[Hg(SO_3)_2]^{2-}$, $[Hg(ox)_2]^{2-}$, and $[Hg(NO_2)_4]^{2-}$. *Phosphine* complexes are numerous, many being of the types $HgX_2(PR_3)_n$, readily formed on mixing HgX_2 and PR_3 in a nonaqueous solvent.[86] For n = 2 these are tetrahedral monomers, while for n = 1 they are usually $(R_3P)XHg(\mu-X)_2HgX(PR_3)$ dimers or perhaps polymers. The existence of the linear molecule $Bu_2^tPHgPBu_2^t$ might also be noted here.[87]

[76]L. Carlton and D. White, *Polyhedron* **1990**, *9*, 2717.
[77]E. Block *et al.*, *Inorg. Chem.* **1990**, *29*, 3172.
[78]J. G. Wright *et al.*, *Prog. Inorg. Chem.* **1990**, *38*, 323.
[79]S. Koch *et al.*, *Inorg. Chem.* **1992**, *31*, 5343.
[80]P. González-Duarte *et al.*, *Polyhedron* **1988**, *7*, 2509; *Inorg. Chim. Acta* **1991**, *184*, 167.
[81]P. A. W. Dean *et al.*, *Inorg. Chem.* **1990**, *29*, 2997; **1994**, *33*, 2180.
[82]B. L. Khandelival *et al.*, *Polyhedron* **1990**, *9*, 2041.
[83]M. Schröder *et al.*, *Polyhedron* **1990**, *9*, 2931.
[84]J. A. Ibers *et al.*, *Inorg. Chem.* **1993**, *32*, 3201.
[85]A. M. Bond *et al.*, *Inorg. Chim. Acta* **1990**, *168*, 233.
[86]G. A. Bowmaker *et al.*, *Inorg. Chim. Acta* **1993**, *210*, 107.
[87]A. H. Cowley *et al.*, *J. Am. Chem. Soc.* **1989**, *111*, 4986.

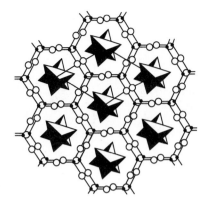

Figure 15-5 The structure of $Hg_5P_2Br_4$. The mercury atoms (open circles) in the chains are disordered. See A. V. Shevelkov *et al., J. Chem. Soc. Dalton Trans.,* **1996,** 147.

Amine complexes of Hg^{II} are easily formed and adopt a variety of structures. With HgI_2 discrete tetrahedral structures are found, as in HgI_2py_2 and $HgI_2(LL)$, where LL is a 2,2'-bipy type ligand.[88] With $HgCl_2$ higher coordination numbers are found, as in $(LL)ClHg(\mu\text{-}Cl)_2HgCl(LL)$, which has trigonal bipyramidal units sharing an edge[86] and $HgCl_2py_2$ which forms chains of octahedra, although the presumably tetrahedral monomer is present in solution. Octahedral HgN_6 units are found in $[Hg\ en_3]^{2+}$, $[Hg\ py_6]^{2+}$,[89] and (with distortion) in $[Hg([18]aneN_6)]^{2+}$.[90]

Recently a remarkable phosphorus analogue of Millon's base has been reported.[91] It has a framework structure, Fig. 15-5, formed by Hg_2P units, in which infinite chains of $HgBr_5$ trigonal bipyramids occupy channels.

15-14 Organozinc and Organocadmium Compounds[92,93]

The discovery of organozinc compounds by Frankland in 1849 is of great historical importance because these were the first organometallic compounds and, as such, they played a significant role in the development of chemical bonding concepts.

For zinc there are two major classes of organometallic compounds, R_2Zn and RZnX, whereas only the R_2Cd type are important. In the solid state the R_2M compounds are probably all linear molecules, whereas the RZnX compounds are oligomers or polymers with μ_2—X or μ_3—X atoms. Both zinc and cadmium form

[88]N. Hadjiliadis *et al., Inorg. Chim. Acta* **1994,** *227,* 129.
[89]R. Akesson *et al., Acta Chem. Scand.* **1991,** *45,* 1165.
[90]M. A. Santos *et al., Polyhedron* **1993,** *12,* 931.
[91]A. V. Shevelkov *et al., J. Chem. Soc. Dalton Trans.* **1996,** 147.
[92]P. O'Brien in *Comprehensive Organometallic Chemistry,* 2nd ed., Vol. 3, Chap. 4, Pergamon, New York, 1995, pp. 175.
[93]J. L. Wardell, Ed., *Organometallic Compounds of Zinc, Cadmium and Mercury,* Chapman and Hall, London, 1985.

cyclotrimeric RMX compounds in which R or X or both are very bulky, such as 15-XVIII[94] and 15-XIX.[95]

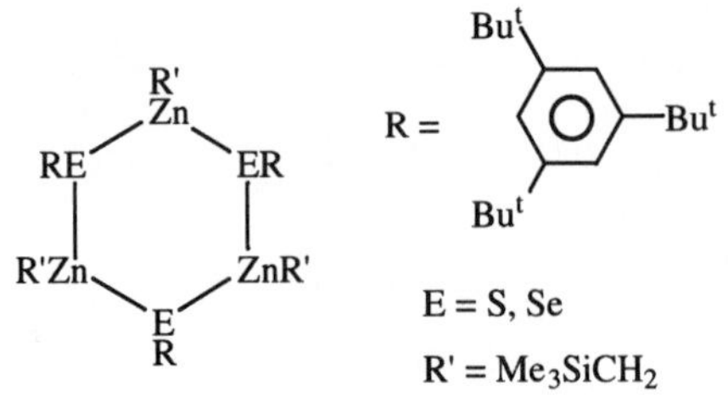

(15-XVIII) 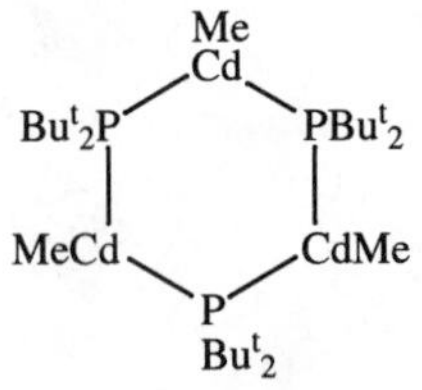   (15-XIX)

The R_2Zn compounds may be obtained in several ways: (1) by reactions of $ZnCl_2$ with RLi, R_2Mg, or R_3Al compounds; (2) for the aryls especially, the reaction of the metal with an organomercury compound:

$$R_2Hg + Zn \longrightarrow R_2Zn + Hg$$

(3) zinc alkyls can be obtained through the following two-step process:

$$RI + Zn(Cu) \longrightarrow RZnI \xrightarrow{\text{heat}} R_2Zn + ZnI_2$$

A general route to R_2Cd compounds is by treatment of an anhydrous cadmium halide with RLi or RMgX. A halide-free route[96] is the following:

$$2RLi + Cd[N(SiMe_3)_2]_2 \longrightarrow R_2Cd + 2LiN(SiMe_3)_2$$

Compounds with perfluoroalkyl or aryl groups[97] are useful for introducing CF_3 and other C_nF_{2n+1} groups in place of halogens.[98] $(CF_3)_2Cd$, best made by the exchange reaction

$$(CF_3)_2Hg + Me_2Cd \xrightarrow{\text{THF}} (CF_3)_2Cd(THF) + Me_2Hg$$

is more reactive than $(CF_3)_2Hg$ and often more useful as a source of CF_3. The zinc analogue, $(CF_3)_2Zn$, as well as several $(CF_3)MX$ compounds of both metals are also known.

The R_2M compounds are low-melting solids or liquids, soluble in organic media. $(Me_3CCH_2)_2Cd$, mp 40°, is typical.[99] The R_2M compounds are all reactive to O_2 or H_2O, the lower zinc alkyls being spontaneously flammable. They undergo many reactions similar to those of RLi or RMgX, but their lower reactivity is often

[94]P. P. Power et al., J. Am. Chem. Soc. 1991, 113, 3379; Inorg. Chem. 1993, 32, 4505.
[95]A. H. Cowley et al., J. Am. Chem. Soc. 1989, 111, 4986.
[96]D. S. Wright et al., J. Chem. Soc., Chem. Commun. 1994, 1627.
[97]I. Büsching and H. Strasdeit, J. Chem. Soc., Chem. Commun. 1994, 2789.
[98]W. Dukat and D. Naumann, J. Chem. Soc., Dalton Trans. 1989, 739.
[99]P. O'Brien et al., Polyhedron 1990, 9, 1483.

advantageous in obtaining selectivity. An example is the use of cadmium alkyls to synthesize ketones from acyl chlorides, where RLi or RMgX would be too reactive:

$$2RCOCl + R'_2Cd \longrightarrow 2RC(O)R' + CdCl_2$$

Both metals form bis-cyclopentadienyl compounds and numerous substituted derivatives. None of these have anything approaching a ferrocene-like structure since the M^{2+} ions have closed d-shells. Cp_2Zn is an infinite polymer while Cp_2Cd is a monomer with η^1-C_5H_5 groups.

Perhaps the most important reaction of organozinc reagents is the Simmons-Smith reaction:[100]

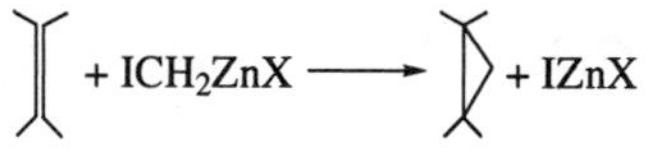

In the well-known Reformatsky reaction,

$$BrCH_2CO_2R + Zn \xrightarrow[\text{(2) } H_2O]{\text{(1) } R'_2CO} R'_2C(OH)CH_2CO_2R$$

there are cyclic organozinc intermediates such as 15-XX.

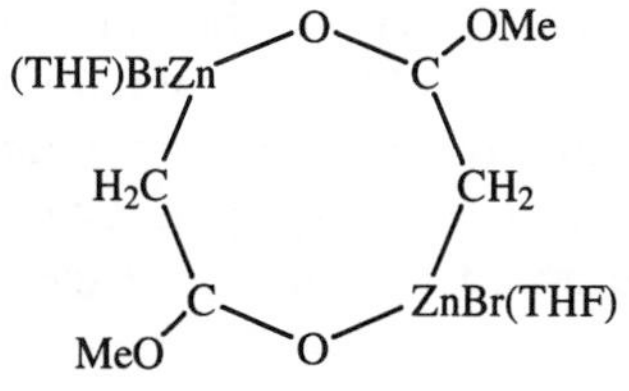

(15-XX)

For RMX compounds in solution there are equilibria of the type

$$R_2M + MX_2 = 2RMX$$

that are quite solvent dependent.

15-15 Organomercury Compounds

The vast majority of organomercury compounds are of the RHgX or R_2Hg types, in which the C—Hg—X or C—Hg—C units are linear. The most common way of making them is by reaction of $HgCl_2$ with RMgX followed by changing X. The case of R = CH_3 is of special interest and will be discussed in detail later.

[100]S. E. Denmark *et al., J. Am. Chem. Soc.* **1992,** *114,* 2592.

The RHgX compounds with X = Cl, Br, I, CN, SCN, OH or any other group that covalently bonds to the mercury atom are crystalline solids, soluble in organic solvents. With I^- and SCN^- complexes such as $RHgX_2^-$ and $RHgX_3^{2-}$ can be formed. With X = SO_4^{2-}, NO_3^-, and RCO_2^- the compounds are more salt-like. RHgH compounds exist in solution but decompose to Hg and RH when isolation is attempted.[101]

The dialkyls and diaryls are colorless liquids or low-melting solids. They react less readily than their zinc and cadmium analogues with air and water, but they are thermally and photochemically unstable, with homolytic Hg—C bond cleavage being the key in their decomposition. They may be kept for several months in the dark at room temperature. All are toxic.

The dialkyl- and diarylmercury compounds are very useful for the preparation of other organo compounds by reactions of the general type:

$$n\text{R}_2\text{Hg} + 2\text{M} \longrightarrow 2\text{R}_n\text{M} + n\text{Hg}$$

In most cases the reaction goes effectively to completion (e.g., with the alkali and alkaline earth metals, Zn, Al, Ga, Sn, Pb, Bi, Se, and Te) but in a few cases (In, Tl, and Cd) reversible equilibria are set up.

Another useful synthetic application is partial alkylation of other halides, viz.,

$$\text{AsCl}_3 + \text{Pr}_2\text{Hg} \longrightarrow \text{PrHgCl} + \text{PrAsCl}_2$$

Cyclopentadienyl compounds of mercury are limited to $(C_5H_5)_2Hg$, which is a molecular compound with linear C—Hg—C bonding and C_5Me_5HgCl, also having linear (C—Hg—Cl) bonds to Hg. $(C_5Me_5)_2Hg$ apparently cannot be made.[102]

The *mechanisms* of reaction of organomercury compounds have been extensively investigated, but only a couple of basic points can be covered here. Reactions of the types shown below proceed via cyclic intermediates (or transition states), as shown.

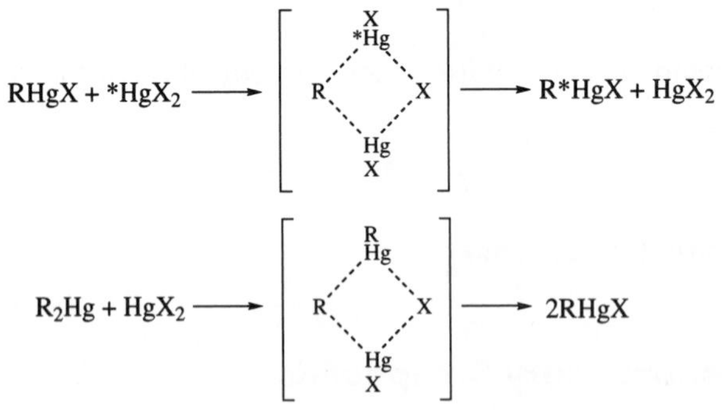

$$\text{RHgX} + {}^*\text{HgX}_2 \longrightarrow \left[\begin{array}{c} \text{X} \\ {}^*\text{Hg} \\ \text{R} \diamond \text{X} \\ \text{Hg} \\ \text{X} \end{array} \right] \longrightarrow \text{R}{}^*\text{HgX} + \text{HgX}_2$$

$$\text{R}_2\text{Hg} + \text{HgX}_2 \longrightarrow \left[\begin{array}{c} \text{R} \\ \text{Hg} \\ \text{R} \diamond \text{X} \\ \text{Hg} \\ \text{X} \end{array} \right] \longrightarrow 2\text{RHgX}$$

Optically active R groups (e.g., $MeEtCH^-$) are transferred with retention of configuration.

[101]K. Kwetkat and W. Kitching, *J. Chem. Soc., Chem. Commun.* **1994,** 345.
[102]B. Fischer, Ph.D. Thesis, Utrecht, 1989.

The β-diketonates of Hg^{II}, like those of Pt^{II}, have bonds to C rather than only to oxygen.

Perhalogeno alkyls are useful reagents for transferring CX_3, CX_2, or CX groups to other elements. A fairly general preparative method is

$$PhHgCl + HCX_3 + Bu^tOK \longrightarrow PhHgCX_3 + KCl + Bu^tOH$$

For $Hg(CF_3)_2$, which is an excellent reagent for making other CF_3 derivatives[103] such as $Se(CF_3)_2$, $P(CF_3)_3$, etc., the most convenient method is

$$Hg(CF_3CO_2)_2 \xrightarrow{\Delta} Hg(CF_3)_2 + 2CO_2$$

Tetramercurimethanes, $C(HgX)_4$, compounds with $X = CO_2R$ and I are also known.

Methylmercury(II) Compounds

The environmental toxicity of mercury is well known and severe episodes (e.g., in Minimata, Japan) have occurred. Generally, if not always, the culprit is the CH_3Hg^+ ion, or a compound thereof, which has irreversible negative effects on the central nervous system. When elemental mercury is discharged into the environment or when mercury compounds used in agricultural fungicides are released, CH_3Hg^+ is formed by biological methylation carried out by microorganisms. Vitamin B_{12} evidently plays a role in this.

The CH_3Hg^+ ion exists in aqueous solution in one or more of the following forms, depending on pH: $CH_3Hg(H_2O)^+$, CH_3HgOH, $(CH_3Hg)_2O$, $(CH_3Hg)_3O^+$. The CH_3Hg^+ ion is very persistent and can form a great variety of CH_3HgX compounds,[104] where X may be a unidentate group[105] or a polydentate one such as a tripod ligand, whereby a tetrahedral complex is formed.[106] Many complexes with purine and pyrimidine bases have also been characterized.[107] Reactions of CH_3Hg^+ with proteins, peptides, nucleotides and other biological molecules is presumably the reason for its toxicity. It may be noted that Hg^{2+} is also toxic, probably by reaction with SH groups in cysteine.

Mercuration and Oxomercuration

Mercury to carbon bonds can be formed by the reaction of certain mercury(II) compounds, especially the acetate, trifluoroacetate, or nitrate, with olefinic or aromatic organic molecules.

[103]J. A. Morrison *et al.*, *Inorg. Chem.* **1988**, *27*, 4535.
[104]G. Geier and H. Gross, *Inorg. Chim. Acta* **1989**, *156*, 91.
[105]P. Stoppioni *et al.*, *Polyhedron* **1990**, *9*, 2477; D. M. Roundhill *et al.*, *Polyhedron* **1990**, *9*, 597; M. Bockmann *et al.*, *Polyhedron* **1992**, *11*, 507.
[106]C. A. Ghilardi *et al.*, *J. Chem. Soc., Chem. Commun.* **1992**, 1691; S. Midollini *et al.*, *Inorg. Chem.* **1994**, *33*, 6163.
[107]W. S. Sheldrick *et al.*, *Inorg. Chim. Acta* **1989**, *156*, 139; **1989**, *160*, 265; **1989**, *163*, 181; **1992**, *194*, 67.

Reactions with olefins are usually carried out in alcohols and result in oxomercuration products as in the following equation:

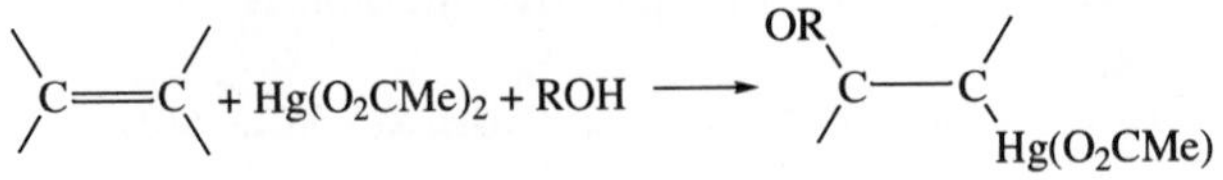

$$\text{C=C} + Hg(O_2CMe)_2 + ROH \longrightarrow \underset{Hg(O_2CMe)}{\overset{OR}{\text{C}-\text{C}}}$$

These reactions are believed to proceed through a mercurinium ion intermediate (15-XXI) which in turn reacts with the alcohol.

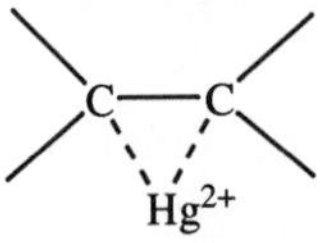

(15-XXI)

Evidence for the existence of mercurinium ions has been obtained by study of solutions in $FSO_3H-SbF_5-SO_2$ at $-70°$ where they are long-lived, employing reactions such as

$$CH_3OCH_2CH_2HgCl \xrightarrow{H^+} CH_3OH_2^+ + HCl + CH_2CH_2Hg^{2+}$$

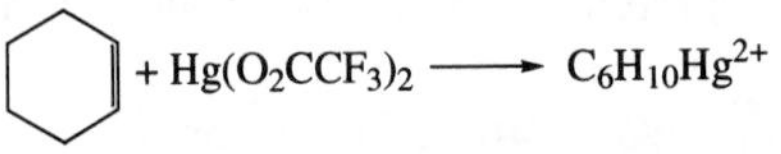

$$\bigcirc + Hg(O_2CCF_3)_2 \longrightarrow C_6H_{10}Hg^{2+}$$

Mercury(II) salts also catalyze the hydration of alkynes through similar mercurinium intermediates, with the overall process being

$$CR{\equiv}CR' + H_2O \longrightarrow \underset{OH\;\;H}{RC{=}CR'} \longrightarrow \underset{O}{RC{-}CH_2R'}$$

With acetylene itself, acetaldehyde is obtained.

Mercuration of aromatic compounds, as in the following example,

$$C_6H_6 + Hg(O_2CMe)_2 \longrightarrow C_6H_5Hg(O_2CMe) + MeCO_2H$$

is a general reaction. It may proceed through arene—Hg complexes, some of which have been isolated:

$$Hg(SbF_6)_2 + 2C_6H_6 \xrightarrow{SO_2(l)} (C_6H_6)_2Hg(SbF_6)_2$$

It is known that Hg^{2+} forms a weak, reversible π-complex with benzene:

$$Hg^{2+} + C_6H_6 \rightleftharpoons Hg(C_6H_6)^{2+} \qquad K \approx 0.5 \text{ L mol}^{-1}$$

Finally, we note that reversible CO insertion into HgO bonds can occur, for example:

$$MeCO_2Hg-OCH_3 + CO \underset{\Delta}{\overset{25°}{\rightleftharpoons}} MeCO_2HgC(O)OCH_3$$

The reaction is reversible on heating. At 25°C reactions such as the following one can be carried out:

$$RHgNO_3 + CO + CH_3OH \longrightarrow RCOOCH_3 + Hg + HNO_3$$

15-16 Intermetallic Compounds

All three metals have marked ability to form covalent bonds to the transition metals, although for mercury there are many more examples.

For many of these compounds the method of preparation is to react an MX_2 compound with a hydrido complex or a transition metal carbonylate anion, as in the following reactions:

$$2CpFe(CO)_2Na + ZnCl_2 \longrightarrow [CpFe(CO)_2]_2Zn + 2NaCl$$

$$2CpMo(CO)_3Na + HgCl_2 \longrightarrow [CpMo(CO)_3]_2Hg + 2NaCl$$

$$(Ph_2MeAs)_3RhCl_2H + HgCl_2 \longrightarrow (Ph_2MeAs)_3Rh(HgCl)Cl_2 + HCl$$

Apart from relatively simple linear trinuclear structures such as that of the above $[CpFe(CO)_2]_2Zn$, $Zn[Co(CO)_4]_2$, $[Fe(SiR_3)(CO)_3(PR_3)CdBr]_2$, and related compounds with $FeCd(\mu\text{-}X)_2CdFe$ and $Fe-Cd-Fe$ groups,[108] there are cyclic systems such as 15-XXII, 15-XXIII, and 15-XXIV, and a compound containing a Zn_4O_4 cubane core, $[MeOZnFe(CO)_2Cp]_4$.

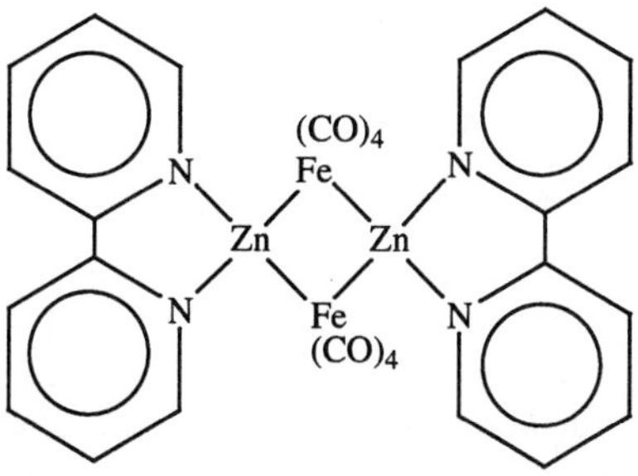

(15-XXII)

[108]P. Braunstein *et al., Inorg. Chem.* **1993**, *32*, 1656.

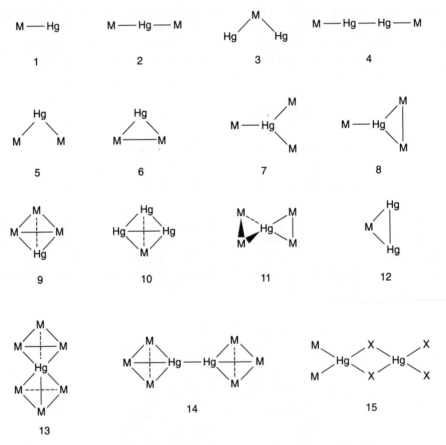

(CO)$_4$Fe—Cd—Fe(CO)$_4$ ...

(15-XXIII)   (15-XXIV)

Cp$_2$Zn can also be used to make intermetallic compounds. For example, it reacts with Ni(COD)$_2$ to give Cp$_6$Ni$_2$Zn$_4$ containing a central octahedron of metal atoms, and with HCo(N$_2$)$_2$(PPh$_3$)$_2$ to give CpCo(ZnCp)$_2$(PPh$_3$) in which there is a bent Zn—Co—Zn chain.

The structural variety and range of compounds with Hg—M bonds is enormous. All of the arrangements shown in Fig. 15-6 have been reported.[109] Space limitations do not permit the inclusion of more than a few examples that illustrate some of these (referring to the numbers in Fig. 15-6).

Figure 15-6   Some of the structural arrangements found in compounds with mercury atoms bonded to transition metal atoms.

[109]P. Braustein *et al.*, *J. Chem. Soc., Dalton Trans.* **1992**, 911.

The simplest arrangement (**1**) is found in $CpMo(CO)_3HgPPh_3$[110] as well as many other compounds. More complicated examples of **1** are found in large molecules with $Hg-Pt$[111] and $Hg-Os$[112] bonds as well as in a series of octahedral ruthenium complexes of the type $Ru(CO)(NO)(PPh_3)(Cl)(X)(HgX)$ where $X = Cl$, I, CN, SCN.[113] Often the formation of these and other types of compounds entail the action of HgX_2 or $RHgX$ as an oxidative addition reagent, for example,

$$trans\text{-}IrCl(CO)(PPh_3)_2 + HgCl_2 \longrightarrow IrCl_2(HgCl)(CO)(PPh_3)_2$$

In species such as $Fe(CO)_4Hg_2X_2$ there are $Hg\cdots Hg$ distances in the $3.10 - 3.22$ Å range[114] and it is therefore debatable whether these should be assigned to type **1** or **12**.

The linear arrangement **2** is found, for example, in a $[(MeO)_3Si(PR_3)(CO)_3$-$Fe]_2Hg$ molecule.[115]

Arrangement **6** is extremely common, and recently reported examples involve binuclear platinum species,[116,117] trinuclear osmium species,[118] as well as some Ru_2 species to which further HgX_2 may be added[119] to give an arrangement of type **15**:

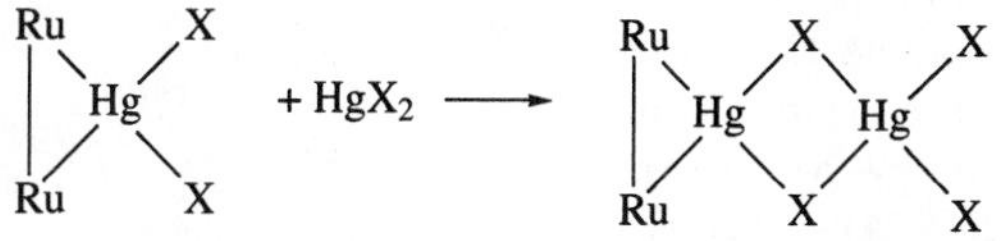

Two recent examples of **9** are provided by $HgPt_3(dppm)_3CO$ and $HgPt_3(dppm)_3Hg$ in which there are tetrahedral and trigonal bipyramidal clusters of metal atoms.[120]

Examples of arrangement **11** also continue to multiply, as exemplified by $\{[SeFe_3(CO)_9]_2Hg\}^{2-}$ [121] and $\{[Ru_6C(CO)_{16}]_2Hg\}^{2-}$.[122]

15-17 Bioinorganic Chemistry

We shall be concerned here only with zinc, since cadmium and mercury have, so far as is known, toxicity as their only biochemical function. Zinc, however, is one of the preeminently important metals in life processes, along with iron and copper.

[110]D. F. Mullica et al., Polyhedron **1992**, 11, 2265.
[111]F. Ruffo et al., J. Chem. Soc., Dalton Trans. **1993**, 3421.
[112]J. Lewis et al., J. Chem. Soc., Dalton Trans. **1992**, 921.
[113]L. Ballester-Reventós et al., Polyhedron **1991**, 10, 1013.
[114]A. E. Mauro et al., Polyhedron **1992**, 11, 799.
[115]P. Braunstein et al., Inorg. Chem. **1992**, 31, 3685.
[116]K. Vrieze et al., Inorg. Chim. Acta **1992**, 195, 203.
[117]D. V. Toronto and A. L. Balch, Inorg. Chem. **1994**, 33, 6132.
[118]M. J. Rosales et al., Polyhedron **1988**, 7, 2159.
[119]J. A. Cabeza et al., J. Chem. Soc., Chem. Commun. **1991**, 168.
[120]R. J. Puddephatt et al., J. Am. Chem. Soc. **1990**, 112, 6400.
[121]M. Shieh and Y.-C. Tsai, Inorg. Chem. **1994**, 33, 2303.
[122]J. Lewis et al., J. Chem. Soc., Dalton Trans. **1991**, 1037.

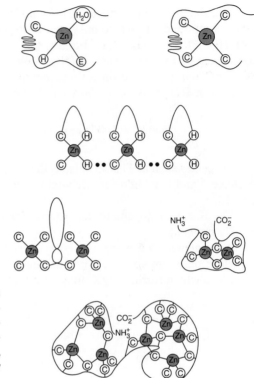

Figure 15-7 Schematic representations of the binding of Zn^{2+} in its various biochemical roles. H = histadine, C = cysteine, and E = glutamate are the most common ligands. Adapted, by permission, from B. L. Vallee and D. S. Auld, *Faraday Discuss.*, **1992**, *93*, 47.

Zinc is recognized to be essential to all forms of life,[123] and a large number of diseases and congenital disorders have been traced to zinc deficiency. In the adult human body there are 2 to 3 g of zinc, as compared to 4 to 6 g of iron and only ~250 mg of copper. Biochemists were somewhat slow to appreciate the presence and importance of zinc because it is colorless, non-magnetic, and generally not as easily noticed as iron and copper. However, there are today effective methods for measuring zinc at levels as low as 10^{-14} g.[124]

In 1940 carbonic anhydrase was shown to be a zinc enzyme, and in 1955 carboxypeptidase became the second recognized zinc enzyme. Since then more than 300 other zinc enzymes have been reported, and functionally they are of many kinds, including alcohol dehydrogenases, aldolases, peptidases, carboxypeptidases, proteases, phosphatases, transphosphorylases, a transcarbamylase, and DNA- and RNA-polymerases.

In addition to its role at the active sites of hundreds of enzymes, there are cases where zinc serves a purely structural role. The most notable of these are the "zinc fingers" and some related "twists" and "clusters." These occur in DNA binding proteins where they stabilize the correct binding sites. Figure 15-7 shows

[123]C. F. Mills, ed., *Zinc in Human Biology*, Springer-Verlag, New York, 1989; A. S. Prasad, *Biochemistry of Zinc*, Plenum, New York, 1993; B. L. Vallee and D. S. Auld, *Biochemistry* **1990**, *29*, 5647; I. Bertini *et al.*, *Bioinorganic Chemistry*, University Science Books, Mill Valley, CA, 1994.
[124]B. L. Vallee and K. H. Falchuk, *Physiol. Rev.* **1993**, *73*, 79.

some of the important structures[125] for zinc *in vivo*. It may be seen that the sulfur atoms of cysteine and the nitrogen atoms of histidine are the most frequent ligating atoms in zinc peptides. Finally, there are the metallothionines, which store zinc. Each of these types of zinc-containing proteins will now be briefly described.

Zinc Enzymes

Of the more than 300 known zinc enzymes, about 20 have known structures based on X-ray study in the crystalline state and/or nmr data for solutions.[126] With structural data it is possible to draw inferences as to the mechanisms. This has been done in some detail for several, of which *carbonic acid anhydrase* and *carboxypeptidase* are important examples.

In the presence of carbonic anhydrase the reaction

$$CO_2(aq) + H_2O \rightleftharpoons H_2CO_3$$

which would occur very slowly at about pH = 7, is replaced by the fast reaction

$$CO_2(aq) + OH^- \rightleftharpoons HCO_3^-$$

that, absent the enzyme, would require a pH of 10 or greater. The enzyme apparently does this because a water molecule coordinated to zinc is converted to a bound hydroxide, Zn—OH, that is then positioned by its surroundings at the active site of the enzyme to form a complex, 15-XXV, that can go on to produce a bicarbonate ion.

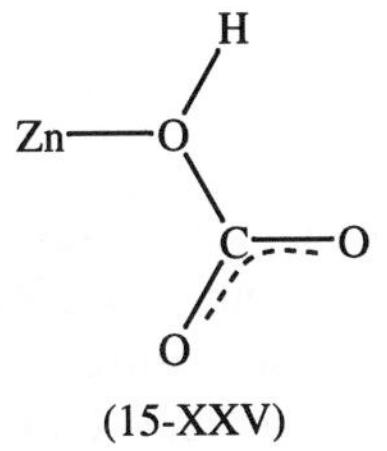

(15-XXV)

In carboxypeptidase, which cleaves peptide linkages (and also esters), there is first a complex in which the peptide carbonyl oxygen atom coordinates to the zinc ion and then things proceed (at least approximately) as shown in Fig. 15-8. These two mechanisms involve, in somewhat different ways, the electrophilicity of the Zn^{2+} ion, and it is known or believed that it is this property of the Zn^{2+} ion, exercised in one way or another, that is critical in all other enzymes having Zn^{2+} at the active site.

In many cases the Zn^{2+} ion is known to be tetrahedrally coordinated in the resting enzyme, with one ligand being H_2O, and that it can be replaced by Co^{2+}, which provides a spectroscopic probe of the active site. The Co^{2+} ion does not significantly depress the enzymatic activity and sometimes even enhances it.

[125]B. L. Vallee and D. S. Auld, *Faraday Discuss.* **1992,** *93,* 47.
[126]B. L. Vallee and D. S. Auld, *Proc. Natl. Acad. Sci. USA* **1990,** *87,* 220.

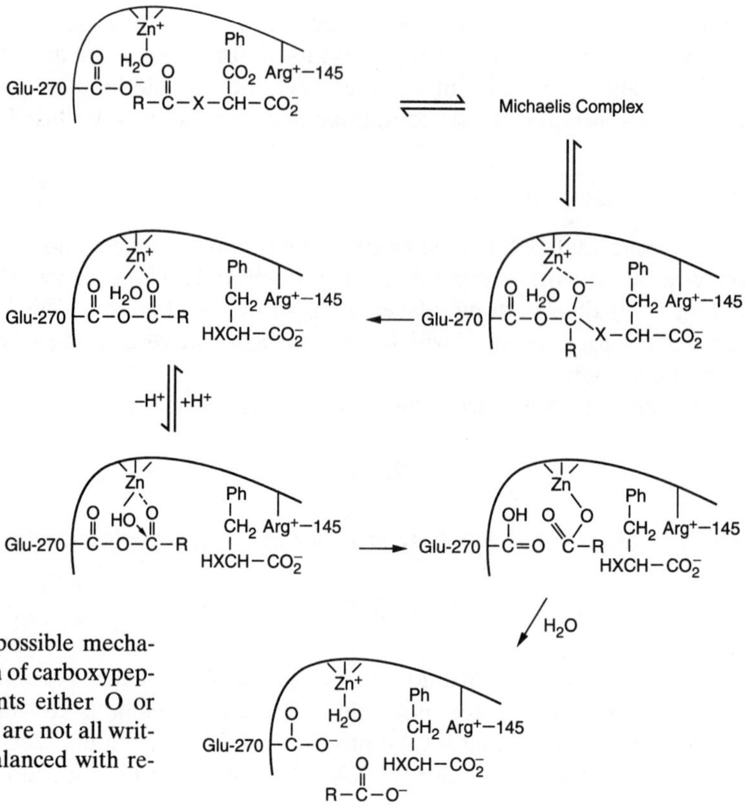

Figure 15-8 A possible mechanism for the action of carboxypeptidase. X represents either O or NH, and the steps are not all written so as to be balanced with respect to protons.

Structural Zinc Ions[127]

There are a number of cases in which Zn^{2+} ions are present to provide structural rigidity to a certain region of the protein. Such Zn^{2+} ions are always tetrahedrally coordinated, and usually by four cysteine sulfur atoms, as shown by the zinc twists and zinc clusters. In the zinc fingers, which are DNA binding domains, the ligands are usually, if not always, two cysteine sulfur atoms and two histidine nitrogen atoms.[128] There is direct crystallographic evidence for this coordination and how it stabilizes the region of the peptide so that it presents the necessary appearance to bind to a DNA molecule.[129]

Metallothioneins (MTs)[130]

Discovered by B. L. Vallee in 1957, these zinc containing proteins remain unique in character but are encountered both in animals and microorganisms. Remarkably,

[127]B. L. Vallee and D. S. Auld, *Proc. Natl. Acad. Sci. USA* **1991,** *88,* 999.
[128]J. M. Berg *et al., Inorg. Chem.* **1993,** *32,* 937.
[129]N. P. Pavletich and C. O. Pabo, *Science* **1991,** *252,* 809.
[130]B. L. Vallee and W. Maret, in *Metallothionem III,* K. T. Suzuki, N. Imura, and M. Kimura, eds., Bierkhäuser Verlag, Basel, Switzerland, 1993, 1-27; B. L. Vallee, *Neurochem. Int.* **1995,** *27,* 23.

their function is still not well established. Most likely they are involved in both storage and transfer of Zn^{2+} (which is essentially absent in the free state in cells). MTs also strongly bind cadmium, mercury, and copper and another role may be to tie up the toxic elements Cd and Hg, but this is unproven. Since zinc is so essential to so many biological processes, the need for a storage and transport agent seems reasonable. Despite the high binding constant (*ca.* 10^{12} mol^{-1}) of MT for zinc, the zinc ions are kinetically very labile.

The structure of metallothioneins (Fig. 15-7) is unique. It consists of a peptide chain of about 60 amino acids, about a third of which are cysteines. The cysteines are highly conserved from species to species. There are two regions which together bind seven metal atoms. A group of these metal atoms is held together by three bridging and six terminal sulfur atoms and a second group of four metal atoms is supported by five bridging and six terminal sulfur atoms. The shortest Zn···Zn distances are *ca.* 4.1 Å.

Part 3

THE CHEMISTRY OF THE TRANSITION ELEMENTS

Chapter 16

SURVEY OF TRANSITION METAL CHEMISTRY

16-1 Definition and General Characteristics of Transition Metals

The purpose of this chapter is to deal with various topics that concern the transition elements, especially those in the d-block, as a class. The transition elements have general forms of chemical behavior that are entirely, or largely, restricted to themselves and not shared by the main-group elements. The most important of these topics are presented here in a general way before taking up each of the transition elements in detail.

The transition elements may be defined strictly as those that, as elements, have partly filled d or f shells. Here we shall adopt a slightly broader definition and include also elements that have partly filled d or f shells in any of their commonly occurring oxidation states. This means that we treat the coinage metals copper, silver, and gold as transition metals, since Cu^{II} has a $3d^9$ configuration, Ag^{II} a $4d^9$ configuration, and Au^{III} a $5d^8$ configuration. From a purely chemical point of view it is also appropriate to consider these elements as transition elements because their chemical behavior is, on the whole, quite similar to that of other transition elements.

With our broad definition in mind, we find that there are now some 56 transition elements, counting the heaviest elements through the one of atomic number 104. Clearly the majority of all known elements are transition elements. All these transition elements have certain general properties in common:

1. They are all metals.
2. They are almost all hard, strong, high-melting, high-boiling metals that conduct heat and electricity well. In short, they are "typical" metals of the sort we meet in ordinary circumstances.
3. They form alloys with one another and with other metallic elements.
4. Many of them are sufficiently electropositive to dissolve in mineral acids, although a few are "noble" — that is, they have such positive electrode potentials that they are unaffected by simple acids.
5. With very few exceptions they exhibit variable valence, and their ions and compounds are colored in one if not all oxidation states.
6. Because of partially filled shells, they form at least some paramagnetic compounds.

This large number of transition elements is subdivided into three main groups: (a) the main transition elements or *d*-block elements, (b) the lanthanide elements, and (c) the actinide elements.

The main transition group or *d* block includes the elements that have partially filled *d* shells only. Thus the element scandium, with the outer electron configuration $4s^2 3d$, is the lightest member. The eight succeeding elements, Ti, V, Cr, Mn, Fe, Co, Ni, and Cu, all have partly filled 3*d* shells either in the ground state of the free atom (all except Cu) or in one or more of their chemically important ions (all except Sc). This group of elements is called the *first transition series*. At zinc the configuration is $3d^{10} 4s^2$, and this element forms no compound in which the 3*d* shell is ionized, nor does this ionization occur in any of the next nine elements. It is not until we come to yttrium, with ground state outer electron configuration $5s^2 4d$, that we meet the next transition element. The following eight elements, Zr, Nb, Mo, Tc, Ru, Rh, Pd, and Ag, all have partially filled 4*d* shells either in the free element (all but Ag) or in one or more of the chemically important ions (all but Y). This group of nine elements constitutes the *second transition series*.

Again there follows a sequence of elements in which there are never *d*-shell vacancies under chemically significant conditions until we reach the element lanthanum, with an outer electron configuration in the ground state of $6s^2 5d$. Now, if the pattern we have observed twice before were to be repeated, there would follow 8 elements with enlarged but not complete sets of 5*d* electrons. This does not happen, however. The 4*f* shell now becomes slightly more stable than the 5*d* shell, and through the next 14 elements, electrons enter the 4*f* shell until at lutetium it becomes filled. Lutetium thus has the outer electron configuration $4f^{14} 5d 6s^2$. Since both La and Lu have partially filled *d* shells and no other partially filled shells, it might be argued that both these should be considered as *d*-block elements. However, for chemical reasons, it would be unwise to classify them in this way, since all the 15 elements La (Z = 57) through Lu (Z = 71) have very similar chemical and physical properties, those of lanthanum being in a sense prototypal; hence these elements are called the *lanthanides*, and their chemistry is considered separately in Chapter 19. Since the properties of Y are extremely similar to, and those of Sc mainly like, those of the lanthanide elements proper, and quite different from those of the regular *d*-block elements, we treat them also in Chapter 19.

For practical purposes, then, the *third transition series* begins with hafnium, having the ground state outer electron configuration $6s^2 5d^2$, and embraces the elements Ta, W, Re, Os, Ir, Pt, and Au, all of which have partially filled 5*d* shells in one or more chemically important oxidation states as well as (except Au) in the neutral atom.

Continuing on from mercury, which follows gold, we come *via* the noble gas radon and the radioelements Fr and Ra to actinium, with the outer electron configuration $7s^2 6d$. Here we might expect, by analogy to what happened at lanthanum, that in the following elements electrons would enter the 5*f* orbitals, producing a lanthanide-like series of 15 elements. What actually occurs is not so simple. Although immediately following lanthanum the 4*f* orbitals become decisively more favorable than the 5*d* orbitals for the electrons entering in the succeeding elements, there is apparently not so great a difference between 5*f* and 6*d* orbitals until later. Hence for the elements immediately following Ac, and their ions, there may be electrons in the 5*f* or 6*d* orbitals or both. Since it appears that later on, after four or five more electrons have been added to the Ac configuration, the $5f$ orbitals do become definitely the more stable, and since the elements from about americium on do show moderately homologous chemical behavior, it has become accepted practice to call the 15 elements beginning with Ac the *actinide elements*.

There is an important distinction, based on electronic structures, between the three classes of transition elements. For the d-block elements the partially filled shells are d shells, $3d$, $4d$, or $5d$. These d orbitals project well out to the periphery of the atoms and ions so that the electrons occupying them are strongly influenced by the surroundings of the ion and, in turn, are able to influence the environments very significantly. Thus many of the properties of an ion with a partly filled d shell are quite sensitive to the number and arrangement of the d electrons present. In marked contrast to this, the $4f$ orbitals in the lanthanide elements are rather deeply buried in the atoms and ions. The electrons that occupy them are largely screened from the surroundings by the overlying shells ($5s$, $5p$) of electrons; therefore, reciprocal interactions of the $4f$ electrons and the surroundings of the atom or the ion are of relatively little chemical significance. This is why the chemistry of all the lanthanides is so homologous, whereas there are seemingly erratic and irregular variations in chemical properties as one passes through a series of d-block elements. The behavior of the actinide element lies between those of the two types just described because the $5f$ orbitals are not so well shielded as are the $4f$ orbitals, although not so exposed as are the d orbitals in the d-block elements.

16-2 Oxidation State

The concept of oxidation state (or number) plays a more prominent part in describing the chemistry of the transition than it does for the main group elements. This is mainly because in the more ionic bonding that is characteristic of these metallic elements, there is more physical reality to the assignment of such numbers than is the case with the largely covalent chemistry of the main group elements. Of course, even for metallic elements, as oxidation states of $+4$ or greater are attained, the numbers cannot be given physical significance. Thus while a Ti^{3+} ion has physical reality, there is no real M^{4+} ion and certainly the "Mn^{7+}" in MnO_4^- is purely a formality. Because of this it is best to use, for example, Mn^{VII} rather than Mn^{7+}, that is, a Roman superscript, in designating all oxidation states, even the lower ones.

Generally speaking, and especially for the elements of the first transition series, the highest oxidation state attained in chemically significant circumstances may not correspond to the total number of valence shell electrons. For example, while Sc^{III}, Ti^{IV}, V^V, Cr^{VI}, and Mn^{VII} all occur in stable compounds, beyond this the highest oxidation states are lower, namely, Fe^{VI}, Co^V, Ni^{IV}, and Cu^{III}.

A graph showing the highest oxidation states of the transition elements is shown in Fig. 16-1. It can be seen that the heavier transition metals can, from Group 7 on, be induced to form compounds in higher states than their first series congeners, but this is almost always confined to their fluorides. In general on descending any group, from 3 to 11, the higher oxidation states become more stable. Thus, Ti^{III} compounds are numerous and important whereas Hf^{III} compounds are relatively rare and Hf^{IV} is predominant. As another example, MnO_4^- is a powerful oxidizing agent, whereas ReO_4^- is not easily reduced.

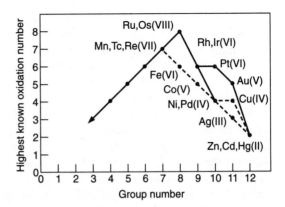

Figure 16-1 Highest known oxidation states for the transition metals. Adapted from M. Kwapp *et al., Inorg. Chem.* **1994,** *33*, 2122.

16-3 Metal Carbonyls and Kindred Compounds

Although a number of relatively unstable compounds having a CO bonded to metal or metalloid atoms not belonging to the *d*-block have been reported,[1] the formulation of complexes with one or several (up to 6) CO groups bound through carbon atoms to metal atoms is characteristically a property of the transition metals. We shall discuss the compounds in detail in connection with the individual metallic elements, but here some generalizations, especially concerning bonding, will be introduced.

Before discussing bonding as such, we note that the composition of the homoleptic carbonyls is generally given by the so-called 18-electron rule, which states that if each CO group is regarded as a two-electron donor, the sum of the number of electrons in the valence orbitals of the metal atom plus two for each CO group should be equal to 18. In this way formulas such as $M(CO)_6$ ($M = Cr$, Mo, W, Mn^+, Re^+) and $Ni(CO)_4$ are accounted for. When M has an odd number of electrons, e.g., Mn, Co, dinuclear species in which there are M—M bonds are formed and the shared electron pair counts as two electrons for each metal atom. Thus we have $(OC)_5M—M(CO)_5$ ($M = $ Mn, Tc, Re) and $Co_2(CO)_8$. As will be noted specifically later, CO groups can also occupy bridging positions. A notable exception to the 18-electron rule is $V(CO)_6$ which has only 17, but such exceptions are rare and the rule extends to mixed organo/carbonyl species, e.g., $C_6H_6Cr(CO)_3$, $C_5H_5Mn(CO)_3$, and $C_5H_5V(CO)_4$.

Let us turn now to the nature of the bond that is normally found in M—CO groups. In most cases, it is adequate for practical, everyday purposes to regard a ligand simply as an electron pair donor and to think of the bond to the central atom simply as L→M. However, there are important classes of compounds for which this simple concept is seriously inadequate. The most prominent examples are the metal carbonyls, metal nitrosyls, and compounds of low-valent metals containing phosphines or isonitriles as ligands.

The explanation for this is believed to be that, as shown in Fig. 16-2, the C→M σ dative bond is reinforced by a donation of electrons in the π-type *d* orbitals of M to empty π* orbitals of the CO. Indeed, some calculations ascribe more importance to the π-type "back-bonding" than to the σ donation.

[1]J. E. Ellis and W. Beck, *Angew. Chem. Int. Ed. Engl.* **1995,** *34*, 2489.

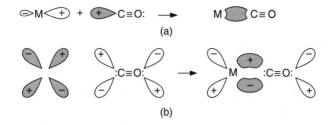

Figure 16-2 (a) The formation of the metal ← carbon σ bond using an unshared pair of the C atom. (b) The formation of the metal → carbon π bond. The outer orbitals on the CO are omitted for clarity.

This bonding mechanism is *synergic*, since the drift of metal electrons, referred to as back-bonding, into CO orbitals, will tend to make the CO as a whole negative, hence to increase its basicity *via* the σ orbital of carbon; at the same time the drift of electrons to the metal in the σ bond tends to make the CO positive, thus enhancing the acceptor strength of the π orbitals. Thus up to a point the effects of σ-bond formation strengthen the π bonding, and vice versa.

Two important and broadly useful lines of physical evidence supporting the multiple nature of M—CO bonds are bond lengths and vibrational spectra. According to the preceding description of the bonding, as the extent of back-donation from M to CO increases, the M—C bond becomes stronger and the C≡O bond becomes weaker. Thus, the multiple bonding should be evidenced by shorter M—C and longer C—O bonds as compared to M—C single bonds and C≡O triple bonds, respectively. Actually very little information can be obtained from the CO bond lengths, because in the range of bond orders (2-3) concerned, CO bond length is relatively insensitive to bond order. The bond length in CO itself is 1.128 Å, while the bond lengths in metal carbonyl molecules are ~1.15 Å, a shift in the proper direction but of little quantitative significance owing to its small magnitude and the uncertainties (~0.01 Å) in the individual distances. For M—C distances, the sensitivity to bond order in the range concerned (1-2) is relatively high, probably ~0.3 to 0.4 Å per unit of bond order, and good evidence for multiple bonding can therefore be expected from such data. To do this we measure the lengths of M—CO bonds in the same molecule in which some other bond, M—X exists, such that this bond must be single. Then, using the known covalent radius for X, estimating the single bond covalent radius of C to be 0.70 Å when an *sp* hybrid orbital is used (the greater *s* character makes this ~0.07 Å shorter than that for *sp*^{3} carbon), the length for a single M—CO bond *in this molecule* can be estimated and compared with the observed value.

Figure 16-3 shows some substitution products of Cr(CO)$_6$ in which three or four CO groups have been replaced by ligands such as aliphatic amine nitrogen,

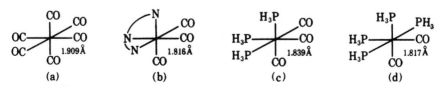

Figure 16-3 Cr—C bond distances in (a) Cr(CO)$_6$, (b) *fac*-[H$_2$N(CH$_2$)$_2$NH(CH$_2$)$_2$NH$_2$]-Cr(CO)$_3$, (c) *fac*-(PH$_3$)$_3$Cr(CO)$_3$, and (d) *cis*-(PH$_3$)$_4$Cr(CO)$_2$.

which has no capacity to compete with CO trans to it for π bonding or PH_3, which has very little capacity to do so. We see that in such cases the remaining CO groups have even shorter Cr—C bonds because of even more extensive development of Cr—C π-back-bonding. The shortening is greater in (b) than in (c), since there is slight π bonding to the phosphorus atoms of PH_3. The shortening is also greater in (d) than in (c) because there are only two CO groups in (d) to compete for the available $d\pi$ electrons of the chromium atom.

From the vibrational spectra of metal carbonyls, it is also possible to infer the existence and extent of M—C multiple bonding. This is most easily done by studying the CO stretching frequencies rather than the MC stretching frequencies, since the former give rise to strong sharp bands, well separated from all other vibrational modes of the molecules. The inferring of M—C bond orders from the behavior of C—O vibrations depends on the assumption that the valence of C is constant, so that a given increase in the M—C bond order must cause an equal decrease in the C—O bond order; this, in turn, will cause a drop in the CO vibrational frequency.

From the direct comparison of CO stretching frequencies in carbonyl molecules with the stretching frequency of CO itself, qualitative conclusions can be drawn. The CO molecule has a stretching frequency of 2143 cm^{-1}. Terminal CO groups in most neutral metal carbonyl molecules are found in the range 2125 to 1850 cm^{-1}, showing the reduction in CO bond orders. Moreover, when changes are made that should increase the extent of M—C back-bonding, the CO frequencies are shifted to even lower values. Thus if some CO groups are replaced by ligands with low or negligible back-accepting ability, those CO groups that remain must accept $d\pi$ electrons from the metal to a greater extent to prevent the accumulation of negative charge on the metal atom. Thus the frequencies for $Cr(CO)_6$ are ~2100, 2000, and 1985 cm^{-1} (exact values vary with phase and solvent) whereas, when three CO's are replaced by amine groups that have essentially no ability to back-accept, as Cr dien(CO)$_3$ (Fig. 16-3b), there are two CO stretching modes with frequencies of ~1900 and 1760 cm^{-1}. Similarly, when we go from $Cr(CO)_6$ to the isoelectronic $V(CO)_6^-$, when more negative charge must be taken from the metal atom, a band is found at ~1860 cm^{-1} corresponding to the one found at ~2000 cm^{-1} in $Cr(CO)_6$. A series of these isoelectronic species illustrating this trend, with their infrared-active CO stretching frequencies (cm^{-1}) is $Ni(CO)_4$ (~2060); $Co(CO)_4^-$ (~1890); $Fe(CO)_4^{2-}$ (~1790). Conversely, a change that would tend to inhibit the shift of electrons from metal to CO π orbitals, such as placing a positive charge on the metal, should cause the CO frequencies to *rise*, and this effect has been observed in several cases, the following being representative:

Mn(CO)$_6^+$, ~2090   Mn dien(CO)$_3^+$, ~2020, ~1900
Cr(CO)$_6$, ~2000   Cr dien(CO)$_3$, ~1900, ~1760
V(CO)$_6^-$, ~1860

More direct evidence for M—CO π bonding can be obtained by photoelectron spectroscopy. For $Cr(CO)_6$ and $W(CO)_6$, the photoionization of one of those electrons that must be responsible for whatever M—CO π bonding exists was carried out and the effect of removing such an electron on the frequency of the totally symmetric M—C stretching vibration in the resulting $M(CO)_6^+$ ion measured. The frequency of this vibration was, in each case, found to be significantly lower than that of the corresponding vibration in the neutral $M(CO)_6$ molecule, that is, by 10%

for $W(CO)_6^+$ and 15% for $Cr(CO)_6^+$. This unequivocally shows that the orbitals from which the electrons have been removed contribute importantly to the M—C bond strengths. Further analysis of these same data shows that the loss of one such electron increases the Cr—C and W—C distances by ~0.14 and 0.10 Å, respectively.

The general picture just presented of the bonding in metal carbonyls is very unlikely to be wrong, but recently several examples of metal carbonyl complexes whose existence and properties are not explained by this formulation have been discovered.[2] All of these are cationic species, namely, $[Pd(CO)_4]^{2+}$, $[Pt(CO)_4]^{2+}$, $[AgCO]^+$, $[Ag(CO)_2]^{2+}$, $[Hg(CO)_2]^{2+}$, $[Au(CO)_2]^{2+}$, $[Os(CO)_6]^{2+}$, and $[Ir(CO)_6]^{3+}$. They have been called "non-classical" carbonyls.[3] In every case, however, the accompanying anions are the same, namely, $[Sb_2F_{11}]^-$, and there is evidence that these interact more strongly than might normally have been expected with the carbonyl cations. The reason why these cationic carbonyl complexes do not come within the scope of the traditional back-bonding model is that they have CO stretching frequencies that are *higher* than that of CO itself (2143 cm^{-1}). In the extreme case of $[Ir(CO)_6]^{3+}$ the three normal modes of CO stretching are at 2295, 2276, and 2254 cm^{-1} whereas in the analogous "classical" carbonyl, $W(CO)_6$, which is isoelectronic and isostructural, the values are 2115, 1998, and 1977 cm^{-1}. Clearly, these cationic complexes pose bonding questions that have not yet been answered. They have been described as having "σ-only" M—CO bonds,[4] but this may be too simple an explanation.

Other π-Acid Ligands

In addition to CO, other important π-acid ligands are RNC, N_2, CS, and NO. To a lesser extent phosphines may participate in the same sort of synergic interaction, but because there are also steric factors of importance for phosphines they will be discussed separately in the next section.

Isocyanide Ligands

Isocyanides are capable of displaying as great π-acceptor capacity as CO when linked to a metal center capable of strong π donation, and this may be described using the same pattern of orbital overlaps as previously discussed for M—CO bonding. Both spectroscopic evidence (lowering of C—N stretching frequencies) and structural evidence [short M—C distances, such as 1.94 Å in $Cr(CNPh)_6$] demonstrate this. The important difference between CO and RNC is that isocyanides are good Lewis bases and can form bonds to metal ions where there is essentially only σ donation involved.

It may be noted here that the cyanide ion (CN^-) is also isoelectronic with CO and RNC, but its capacity to serve as a π acid is much lower and from the point of view of bonding it is perhaps best regarded as a pseudohalide rather than as a π-acid ligand.

[2]F. Aubke and H. Willner *et al.*, *J. Chem. Soc., Chem. Commun.* **1994**, 1189; **1995**, 2072; *Inorg. Chem.* **1996**, *35*, 82; *Angew. Chem. Int. Ed. Engl.* **1996**, *35*, 1974.
[3]S. H. Strauss *et al.*, *J. Am. Chem. Soc.* **1994**, *116*, 10003.
[4]F. Aubke and C. Wang, *Coord. Chem. Rev.* **1994**, *137*, 483.

Dinitrogen

Although N_2 is isoelectronic with CO and RNC, and isosteric with the former, it is far more inert and the first dinitrogen complex, $[Ru(NH_3)_5(N_2)]^{2+}$, was discovered only in 1965. Dinitrogen can be bound to metal atoms only in the presence of other ligands; there are no homoleptic complexes analogous to those formed by CO or RNC.

The bonding in *linear M−N−N groups* is qualitatively similar to that in terminal M−CO groups; the same two basic components, $M \leftarrow N_2$ σ donation and $M \rightarrow N_2$ π acceptance, are involved. The major quantitative differences, which account for the lower stability of N_2 complexes, appear to arise from small differences in the energies of the MO's of CO and N_2. For CO the σ-donor orbital is weakly antibonding, whereas the corresponding orbital for N_2 is of bonding character. Thus N_2 is a significantly poorer σ donor than is CO. It is observed that in pairs of N_2 and CO complexes where the metal and other ligands are identical, the fractional lowerings of N_2 and CO frequencies are nearly identical. For the CO complexes, weakening of the CO bond, insofar as electronic factors are concerned, is due entirely to back-donation from metal $d\pi$ orbitals to CO π^* orbitals, with the σ donation slightly canceling some of this effect. For N_2 complexes, on the other hand, N−N bond weakening results from both σ donation and π back-acceptance. The very similar changes in stretching frequencies for these two ligands suggest then that N_2 is weaker than CO in both its σ-donor and π-acceptor functions. This in turn would account for the poor stability of N_2 complexes in general. Terminal dinitrogen compounds have N−N stretching frequencies in the region 1930 to 2230 cm^{-1} (N_2 has $\nu = 2331$ cm^{-1}).

The CS Ligand

Although the CS molecule cannot be isolated, complexes containing the M−CS unit are known. The synergic σ/π interactions in M−CS units are similar to those in M−CO units, but perhaps somewhat stronger.

The NO Ligand[5]

The NO molecule is closely akin to the CO molecule except that it contains one more electron, which resides in a π^* orbital. Loss of this electron gives NO^+ which is isoelectronic with CO.

Just as the CO group reacts with a metal atom that presents an empty σ orbital and a pair of filled $d\pi$ orbitals to give a linear MCO grouping with a $C \rightarrow M$ σ bond and a significant degree of $M \rightarrow C$ π bonding, so the NO group engages in a structurally and electronically analogous reaction with a metal atom that may be considered, at least formally, to present an empty σ-orbital and a pair of $d\pi$ orbitals containing only three electrons. The full set of four electrons for the $Md\pi \rightarrow \pi^*(NO)$ interactions is thus made up of three electrons from M and one from NO. In effect, NO contributes three electrons to the total bonding configuration under circumstances where CO contributes only two. Thus for purposes of formal electron

[5]G. B. Richter-Addo and P. Legzdins, *Metal Nitrosyls,* Oxford University Press, London 1992.

"bookkeeping," the ligand NO can be regarded as a three-electron donor in the same sense as the ligand CO is considered a two-electron donor. This leads to the following very useful general rules concerning stoichiometry, which may be applied without specifically allocating the difference in the number of electrons to any particular (i.e., σ or π) orbitals:

1. Compounds isoelectronic with one containing an $M(CO)_n$ grouping are those containing $M'(CO)_{n-1}(NO)$, $M''(CO)_{n-2}(NO)_2$, and so on, where M', M'', and so on, have atomic numbers that are $1, 2, \ldots$, and so on, $<M$. Some examples are $(\eta\text{-}C_5H_5)CuCO$, $(\eta\text{-}C_5H_5)NiNO$; $Ni(CO)_4$, $Co(CO)_3NO$, $Fe(CO)_2(NO)_2$, $Mn(CO)(NO)_3$; $Fe(CO)_5$, $Mn(CO)_4NO$.

2. Three CO groups can be replaced by two NO groups. Examples of pairs of compounds so related are

$Fe(CO)_5$	$Fe(CO)_2(NO)_2$
$Mn(CO)_4NO$	$Mn(CO)(NO)_3$
$Cr(CO)_6$	$Cr(NO)_4$

It may be noted that $Cr(NO)_4$ is the only known homoleptic metal nitrosyl complex. It is isoelectronic with $Ni(CO)_4$ and similarly has a tetrahedral structure, with a very short Cr—N distance (1.763 Å). Claims for $Co(NO)_3$, $Fe(NO)_4$, and $Ru(NO)_4$ have never been confirmed.

It should be noted that the designation "linear MNO group" does not disallow a small amount of bending in cases where the group is not in an axially symmetric environment, just as with terminal MCO groups. Thus MNO angles of 161 to 175° may be found in "linear" MNO groups. Truly "bent MNO groups" have angles of 120 to 140°.

In compounds containing both MCO and linear MNO groups, the M—C and M—N bond lengths differ by a fairly constant amount, ~0.07 Å, approximately equal to the expected difference in the C and N radii, and suggest that under comparable circumstances M—CO and M—NO bonds are typically about equally strong. In a chemical sense the M—N bonds appear to be stronger, since substitution reactions on mixed carbonyl nitrosyl compounds typically result in displacement of CO in preference to NO. For example, $Co(CO)_3NO$ reacts with a variety of R_3P, X_3P, amine, and RNC compounds, invariably to yield the $Co(CO)_2(NO)L$ product.

The NO vibration frequencies for linear MNO groups substantiate the idea of extensive M—N π bonding, leading to appreciable population of NO π^* orbitals. Both the NO and O_2^+ species contain one π^* electron and their stretching frequencies are 1860 and 1876 cm^{-1}, respectively. Thus the observed frequencies in the range 1800–1900 cm^{-1}, which are typical of linear MNO groups in molecules with small or zero charge, indicate the presence of approximately one electron pair shared between metal $d\pi$ and NO π^* orbitals.

Since nitrosyl halides and nitroso alkanes are bent, similar bent metal complexes can be anticipated:

$$\overset{\cdot\cdot}{N}=\overset{\cdot\cdot}{\underset{\cdot\cdot}{O}} \quad \text{and} \quad \overset{\cdot\cdot}{N}=\overset{\cdot\cdot}{\underset{\cdot\cdot}{O}}$$
$$\diagup \qquad\qquad\qquad \diagup$$
$$X(R) \qquad\qquad\qquad L_nM$$

A typical example is $IrCl_2(NO)(PPh_3)_2$ where the MNO angle is 123°. The NO here is best regarded as $1e$ donor or as NO^-; π bonding is not involved. Bridging NO groups, where NO can be regarded as a $3e$ donor, are also known.

16-4 Compounds with PX₃ Ligands

Although they might appear very different from CO, NO, etc., PX_3 compounds (also AsX_3, SbX_3, and even SX_2 and SeX_2 species) can be π-acceptor ligands when X is fairly electronegative (Ph, OR) or very electronegative (Cl, F). Phosphorus trifluoride forms many compounds comparable to those of CO, for example, $Ni(PF_3)_4$ and $Cr(PF_3)_6$. Tertiary phosphines and phosphites, however, are also much better Lewis bases than CO and can form many complexes where π acidity plays little or no role. This is observed with the phosphine complexes of the early transition metals and with metal atoms of any kind in their higher oxidation states where the M$-$P distances show no evidence of significant π bonding.

In almost any CO-containing molecule, one or more CO groups can be replaced by a PX_3 or similar ligand. The ability of PR_3 ligands (R = Me, Et, or Bu) to compete with CO groups for metal $d\pi$ electrons can be studied in detail by photoelectron spectroscopy and vibrational spectroscopy in $Mo(CO)_n(PR_3)_{6-n}$, species and similar ones. The significant σ-donor ability and steric requirements of such ligands is also important in regard to the structures of $M(CO)_x(PR_3)_y$ type molecules.

While the occurrence of M $\rightarrow$ P dative π bonding is a generally acknowledged fact, the explanation for it entails controversy. The classical, and still widely credited, picture is that shown in Fig. 16-4, in which phosphorus specifically employs a pair of its d orbitals to accept metal electrons. However, it has been proposed, on the basis of quantum mechanical calculations that phosphorus p orbitals and the P$-$X σ^* orbitals may play a major role in accepting metal $d\pi$ electrons, even to the complete exclusion of the phosphorus $d\pi$ orbitals. Experimental evidence for or against such ideas is lacking.

Of course, the Lewis basicities of the PX_3 ligands vary considerably, and not entirely predictably with X. This must be taken into consideration because, as with other classes of ligands, the synergistic relationship between σ donation and π acceptance (regardless of the exact mechanism for the latter) is also a source of ambiguity in understanding this class of ligands.

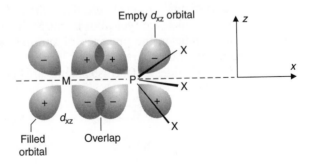

Figure 16-4 The back-bonding from a filled metal d orbital to an empty phosphorus $3d$ orbital in the PX_3 ligand taking the internuclear axis as the x axis. An exactly similar overlap occurs in the xy plane using the d_{xy} orbitals.

The extent of both donation from the lone pair on the P atom and back-donation depends on the nature of the groups attached to P. For PH_3 and $P(alkyl)_3$, π-acceptor ability is very low, but it becomes important with more electronegative groups. Analogous PX_3, AsX_3, and SbX_3 compounds differ very little, but the ligands having a nitrogen atom, which lacks π orbitals, cause significantly lower frequencies for the CO vibrations, as indicated by the CO stretching frequencies (cm^{-1}) in the following series of compounds:

$(PCl_3)_3Mo(CO)_3$	2040, 1991
$(AsCl_3)_3Mo(CO)_3$	2031, 1992
$(SbCl_3)_3Mo(CO)_3$	2045, 1991
dien $Mo(CO)_3$	1898, 1758

The pronounced effect of the electronegativity of the groups X is shown by the following CO stretching frequencies

$[(C_2H_5)_3P]_3Mo(CO)_3$	1937, 1841
$[(C_6H_5O)_3P]_3Mo(CO)_3$	1994, 1922
$[Cl_2(C_2H_5O)P]_3Mo(CO)_3$	2027, 1969
$(Cl_3P)_3Mo(CO)_3$	2040, 1991
$(F_3P)_3Mo(CO)_3$	2090, 2055

The most electronegative substituent, F in PF_3, will reduce very substantially the σ-donor character so that there will be less $P \rightarrow M$ electron transfer, and $Md\pi \rightarrow Pd\pi$ transfer should be aided. The result is that PF_3 and CO are quite comparable in their π-bonding capacity.

Attempts to order a large number of PX_3, AsX_3, and SbX_3 ligands according to their net π-accepting capacity have been made using $LNi(CO)_3$ compounds, and to a lesser extent others such as $LCr(CO)_5$. Both the CO stretching frequencies and the ^{13}C chemical shifts of the CO groups have been used as the indicative experimental parameter and the two orderings are in general agreement. Based on these data some of the more common phosphorus ligands come out in the following order (As and Sb ligands come very close to the corresponding P ligands):

$$PF_3 > PCl_3 > P(OAr)_3 > P(OR)_3 > P(Ar)_3 > PR(Ar)_2 > PR_2Ar > PR_3$$

There is also structural evidence to support this general order. For example, in $(PhO)_3PCr(CO)_5$ the $P-Cr$ bond is 0.11 Å shorter than that in $Ph_3PCr(CO)_5$.

Of at least as great importance to the chemistry of PX_3 compounds as the electronic factors are *steric factors*.[6] Indeed these may be more important or even dominant in determining the stereochemistries and structures of compounds. Steric factors also affect rates and equilibria of dissociation reactions and the stereochemistry of phosphine ligands is the prime factor in many highly selective catalytic reactions of phosphine complexes, such as hydroformylation and asymmetric hydrogenation.

The steric requirement of a PX_3 ligand is usually expressed by the cone angle, θ, as defined earlier (Section 10-17).

[6]T. L. Brown and K. J. Lee, *Coord. Chem. Rev.* **1993**, *128*, 89.

16-5 Metal to Ligand Multiple Bonds[7]

We have already discussed the case of π-acceptor ligands (CO, NO, etc. in Section 16-3) in which there is at least fractional π-character based on metal to ligand back donation. In this section we are concerned with bonds in which there are full double and triple bonds. Typical of these are the following:

$$M{=}O \quad M{=}S \quad M{=}CR_2 \quad M{\equiv}CR \quad M{\equiv}N$$

as well as some M—X—M systems where multiple bond character is present in the M—X—M bonds. In all of these cases, overlap between ligand atom $p\pi$ orbitals and metal atom $d\pi$ orbitals must be invoked to account for bonding in addition to the M—X σ bonds.

M=O and M⋯O⋯M Bonds

The M—O bonds are extremely numerous. In general, the distances are very short, namely, 1.5 to 1.8 Å depending on oxidation numbers, other ligands, and so on. In many cases the bond order is probably significantly >2 because there are two filled $p\pi$ orbitals on the oxygen atom, both of which can engage in bonding to appropriately directed $d\pi$ orbitals on the metal atom, as shown in (16-I). There is no reason why, in many cases, both of these interactions cannot proceed to the point where the sum of the two partial dative π bonds exceeds the bonding that would be provided by just one such interaction at its full extent. Thus, when a double bond, M=O, is written, this should be regarded as a formalism and not an accurate specification of bond order.

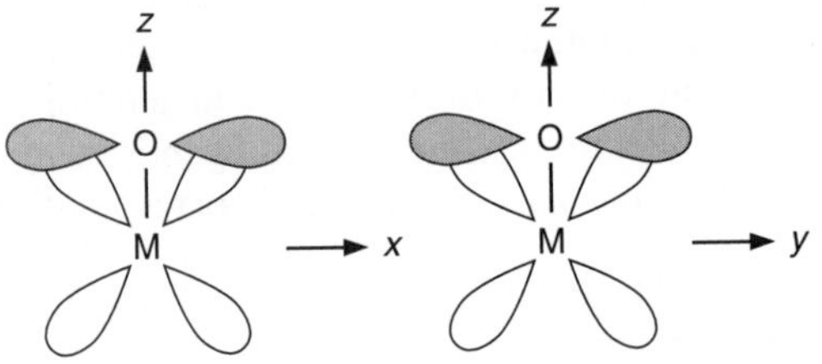

(16-I)

There are also many examples of linear M—O—M units, and in these more delocalized π bonding occurs. The $[Cl_5MOMCl_5]^{n-}$ ions provide straightforward examples. It may be assumed that by using a pair of sp hybrid orbitals on the oxygen atom a linear pair of M—O—M σ bonds is formed. There are then two orthogonal, occupied $p\pi$ orbitals on the oxygen atom that can interact with suitable metal $d\pi$ orbitals. Now, however, there is a total of four such $d\pi$ orbitals and two sets of π interactions (16-II) (one in the xz plane and the other just like it in the yz plane)

[7]W. A. Nugent and J. M. Mayer, *Metal-Ligand Multiple Bonds,* John Wiley & Sons, New York, 1988.

will be established. With both of the two components of the E_u orbital occupied, there is a total of four σ electrons and four π electrons, and thus each M—O bond may be assigned (formally) a bond order of 2. This is consistent with the occurrence of M—O distances in the range of 1.75 to 1.90 Å for various M—O—M systems.

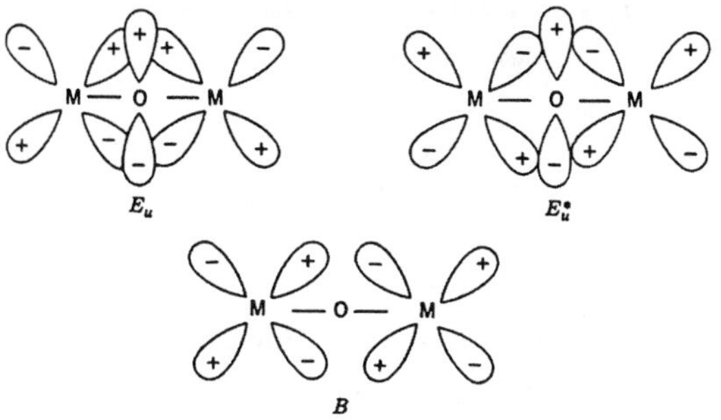

(16-II)

Metal-Nitrogen Multiple Bonds

As with oxygen there are two types, M≡N and M=N=M, but the former show a pronounced tendency towards association, either into linear chains (16-III) or cycles (16-IV) as found in MoNCl$_3$.

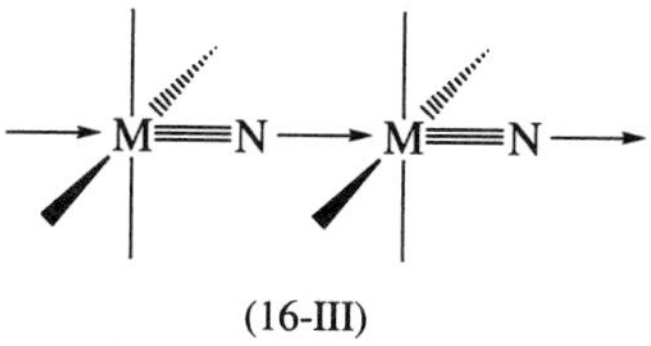

(16-III)    (16-IV)

The types of orbital overlaps entailed in the formation of M≡N and M=N=M bonds are entirely analogous to those already discussed for the oxo cases. As full triple bonds, the M≡N linkages are generally very short (e.g., 1.66 Å in MoCl$_4$N$^-$, and 1.58 Å in ReCl$_4$N), but the bonds in M=N=M systems are longer (~1.85 Å).

Formal metal to nitrogen double bonds may occur in M=NR units, but the situation is more complex. Compounds containing these linkages are often better represented by $\overline{M}{\equiv}\overset{+}{N}R$ and in most cases by something in between. There are actually few cases where the presence of a long M—N bond ($\geq$ 1.8 Å) and a distinctly non-linear arrangement ($\angle$ M—N—R $\leq$ 160°) are found. In most cases the distances are *ca.* 1.7 Å and the angles > 170°, especially when the arrangement of the other ligands about the metal atom is symmetrical enough to allow the occurrence of two M to N π interactions of equal importance. A representative example is shown in 16-V.

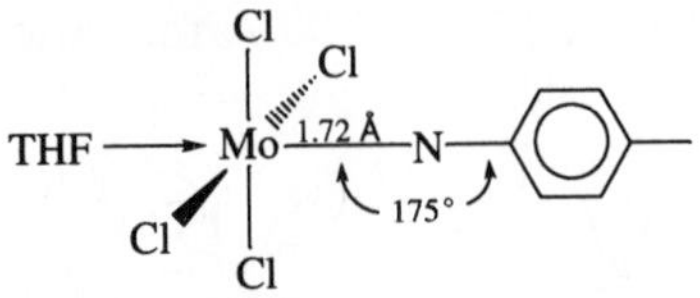

(16-V)

M=CR₂ and M≡CR Bonds

These play an enormous role in organometallic chemistry and will be mentioned frequently. The nature of the M to C bonding is very much dependent on the nature of M and R. At one extreme there are compounds in which M is in a high valence state and the R group(s) not of π-donor character. In these cases, the bonds are comparable to those just discussed for M=NR and M≡N. For these types of compounds, the terms *alkylidene* (M=CR₂) and *alkylidyne* (M≡CR) have been favored. On the other hand, when the metal is in a low valence state and the substituents on carbon are π donors, the M—C bonds are not full double or triple bonds and the systems are rendered stable by the migration of charge from the substituents (such as OR or NR) onto the carbon atom, as shown in resonance terms in (16-VI). For these compounds it is customary to use the terms *carbene complex* and *carbyne complex*. The chemistry is qualitatively different for the two classes.

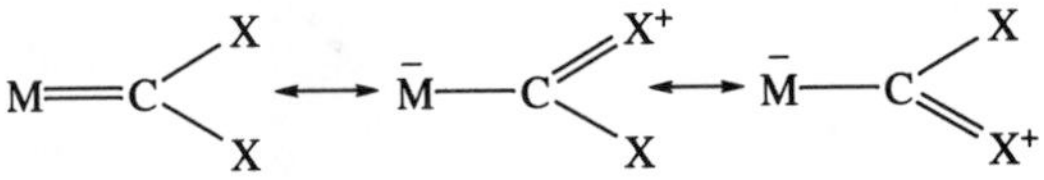

(16-VI)

Heavier Ligand Atoms

We have just discussed the formation of M—X multiple bonds for the principal cases where X is an element from the first short period, namely, C, N, or O. It is to be noted that recently analogues for some of these, in which the heavier elements, Si, Ge, P, As, S, and Se occur, have been prepared, and doubtless this development will continue. As a first approximation the M—X bonding to the heavier X atoms can be described in the same ways as for the light X atoms, but it is possible that outer d orbitals may make some contribution for the heavier X atoms. The greater size of the heavier elements may also be important in some cases. Whatever the reasons, compounds such as linear (C₅Me₅)(CO)₂Mn=Ge=Mn(CO)₂(C₅Me₅) and bent (C₅Me₅)(CO)₂Mn=Te=Mn(CO)₂(C₅Me₅) have no C or O analogues; M=C=M bonds are found in some porphyrin complexes, however.

Another type of linkage formed only by the heavier elements is displayed by the compounds Cp(CO)₂CrXCr(CO)₂Cp, where X = S, Se but not O. The very short bonds (e.g., 2.20 Å for Cr—Se), together with other considerations, have been taken to imply that Cr≡X triple bonds exist.

16-6 Multiple Metal—Metal Bonds

There are many compounds now known in which transition metal atoms are united by bonds of order greater than 1, including those of order 4.[8] *Quadruple bonds* are based on d orbital overlaps and they are to be expected only among the transition elements. General recognition of direct M—M bonds, including multiple ones, has occurred only since the 1960s, and compounds containing such bonds are not embraced by classical (Werner) coordination theory. However, thousands are now known and they form an essential part of transition metal chemistry. We shall now survey the four principal types. Details concerning compounds of specific elements will be found in appropriate places later.

1. *Edge-Sharing Bioctahedra.* These are numerous and may in principle exhibit M—M bond orders of 1 to 3, by overlap of pairs of metal d orbitals, as shown in Fig. 16-5. Examples of single (σ^2) and double ($\sigma^2\pi^2$) bonds are provided by molecules such as (16-VII) and (16-VIII), respectively. In the case where two d^4 metal atoms interact, there will also be a net double bond because both the δ and δ' orbitals will be filled. The existence of a triple bond in this context, however, is problematical. The energy level diagram in Fig. 16-5 shows the δ orbital to be more stable than δ', but when the different interactions of these two orbitals with ligand orbitals are taken into account, the order may be, and in many cases probably is, reversed. Thus, in a d^3—d^3 system the bond order may be 1 rather than 3. Whatever the order of the δ and δ' orbitals, they are generally close in energy; in some cases they are so close that a singlet state (δ^2 or δ^{*2}) and a triplet state ($\delta\delta^*$) are only a few hundred cm^{-1} apart, and spin-state equilibria (singlet-triplet) are observed.

(16-VII)   (16-VIII)

2. *Face-Sharing Bioctahedra.* What we can now recognize as the first example of a M—M triple bond was discovered within this structural context, namely, the W≡W bond in the $[W_2Cl_9]^{3-}$ ion (16-IX). There are very important Nb^{III} and Ta^{III} complexes of general type (16-X) which have M—M double bonds.

[8] F. A. Cotton and R. A. Walton, *Multiple Bonds between Metal Atoms,* 2nd ed., Oxford University Press, 1993.

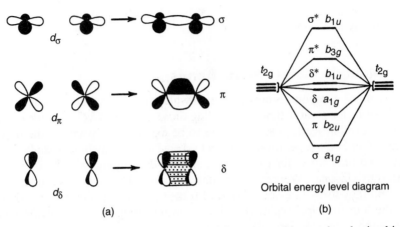

Figure 16-5 (a) The three $d-d$ overlaps that can be expected in an edge-sharing bioctahedral structure. (b) The pattern of energy levels expected when only the direct overlaps are considered.

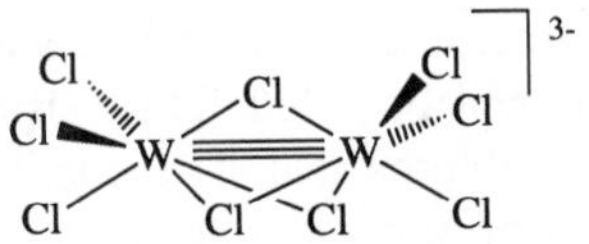

(16-IX)

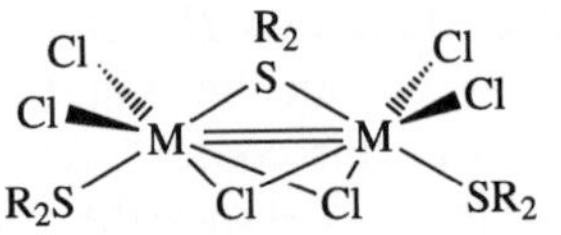

(16-X)

3. *Tetragonal Prismatic Structures.* These provide a large and thoroughly investigated group of compounds with M—M bonds up to quadruple. The prototypal examples are the $[Re_2Cl_8]^{2-}$ and $[Mo_2Cl_8]^{4-}$ ions, whose general, idealized structure is shown in (16-XI). To understand the M—M bonding possibilities within this structural context we consider the five possible overlaps of the d orbitals on two metal atoms, as shown in Fig. 16-6. The relative values of the three types of overlap shown, σ, π, and δ, decrease in that

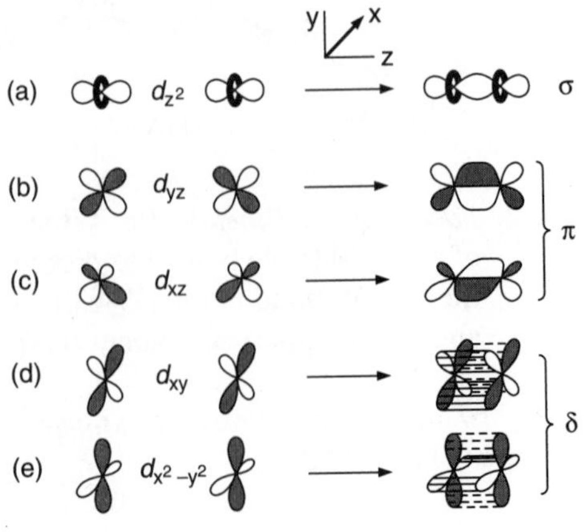

(a) d_{z^2} σ

(b) d_{yz}

(c) d_{xz} π

(d) d_{xy}

(e) $d_{x^2-y^2}$ δ

Figure 16-6 The σ, π, and δ overlaps between two sets of d orbitals on adjacent metal atoms with the internuclear axis as the z axis.

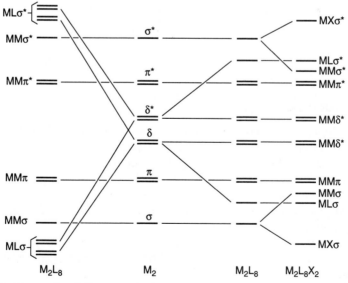

(16-XI)

order and thus the ordering of bonding and antibonding orbitals might be expected to appear as in the column labeled M_2 of Fig. 16-7.

When four ligand atoms are brought up to each metal atom to give the arrangement shown in (16-XI), the order of the MO's is modified. If we define the M—M axis as the z direction and the M—L bond axes as $\pm x$ and $\pm y$, the eight lobes of the two $d_{x^2-y^2}$ orbitals will become engaged in the formation of metal-to-ligand σ bonds. Therefore one member of the δ set of M_2 drops to lower energy and becomes an MLσ orbital; at the same time one member of the δ' pair rises in energy and becomes an MLσ^* orbital. The exact extent to which these MLσ and MLσ^* orbitals move relative to the other M—M bonding and antibonding orbitals will vary from case to case, and the arrangement shown in the M_2L_8 column of Fig. 16-7 is only one possibility. In any case, however, the MLσ orbital will be filled by electrons that contribute to M—L bonding and will play no further role in M—M bonding.

The remaining orbitals that result primarily from M—M overlaps are the MMσ, MMπ (a degenerate pair), and MMδ orbitals. In a case where the metal atoms each have four d electrons to contribute, these four orbitals

Figure 16-7 Energy level diagrams showing schematically how d-orbital overlaps between two metal atoms (M_2) can be modified by bonding ligands to give triple bonds in M_2L_6, strong quadruple bonds in M_2L_8, and weaker quadruple bonds in $M_2L_8X_2$.

are filled to give a metal–metal quadruple bond, the electron configuration of which is $\sigma^2\pi^4\delta^2$. Such a bond has two characteristic properties: (1) it is very strong, therefore very short, and (2) because of the angular properties of the d_{xy} orbitals that overlap to form the δ bond, it has an inherent dependence on the angle of internal rotation. The δ bond is strongest ($d_{xy} - d_{xy}$ overlap maximizes) when the two ML_4 halves have an eclipsed relationship. However, L···L nonbonded repulsions are also maximized in this conformation. Therefore, the rotational conformation about a quadruple bond might in some cases be expected to be twisted somewhat away from the exactly eclipsed one. Indeed, the $d_{xy} - d_{xy}$ overlap decreases only slightly through the first few degrees of rotation, so that little δ-bond energy is lost by small rotations. Examples of rotations of up to 20° have been observed; the majority of quadruple bonds are essentially eclipsed, however.

4. *Trigonal Antiprismatic Structure.* The fourth major structural pattern among compounds with M—M multiple bonds is that shown in (16-XII). The bonding possibilities here may be examined by referring again to Fig. 16-7. The approach of six ligands in the manner shown in (16-XII) (an ethane-like pattern) causes all the δ and δ^* orbitals to shift their energies, as shown, and leaves three M—M bonding type orbitals, σ and (doubly degenerate) π. Thus, this particular structural context is especially favorable for the formation of triple bonds with $\sigma^2\pi^4$ configurations. Compounds belonging to this class are formed mainly by molybdenum(III) and tungsten(III), and they will be discussed in detail in Chapter 18-C.

(16-XII)

More on Quadruple Bonds

Since the existence of a quadruple bond was first recognized and explained in 1964 in the case of the $[Re_2Cl_8]^{2-}$ ion, hundreds of compounds containing such bonds have been prepared. They are formed by the elements Cr, Mo, W, Tc, and Re, and important examples will be discussed under the chemistry of each of these elements.

Theoretical and spectroscopic studies have provided a more thorough and sophisticated description of the bonding, but the simple orbital overlap picture presented above remains qualitatively valid, except possibly with respect to the role of the σ component of the bond, which may be less important than previously thought. For the crucial δ and π components, however, all theoretical and experimental results are in accord.

The δ interaction in the quadruple bond is undoubtedly its most interesting aspect, especially when both the δ (bonding) and δ^* (antibonding) orbitals are considered. Because the δ orbital is only weakly bonding and the δ^* orbital only weakly antibonding (and hence the separation between them is very small) a number of interesting chemical and spectroscopic consequences ensue.

The most characteristic spectroscopic feature of all quadruply bonded species is an absorption band in the visible region due to the excitation of an electron from a singlet $\sigma^2\pi^4\delta^2$ ground state to give a singlet $\sigma^2\pi^4\delta\delta^*$ excited state. Although these transitions are relatively weak considering that they are quantum mechanically allowed (the weakness is traceable to the small orbital overlap in the δ bond), they are intense enough to confer strong, characteristic colors on the species in which they occur. The characteristic royal blue of $[Re_2Cl_8]^{2-}$, intense red of $[Mo_2Cl_8]^{4-}$, and yellow of $Mo_2(O_2CCH_3)_4$ are due to $\delta \rightarrow \delta^*$ transitions at 7000, 5250, and 4350 Å, respectively. An important fact about these transitions, however, is that their energies are determined more by interelectronic repulsive forces (correlation energies) than by orbital energy differences. In Mo_2^{4+} species, the total transition energy of ~18,000 cm^{-1} is, crudely speaking, the sum of about 6000 cm^{-1} orbital energy difference and 12,000 cm^{-1} electron correlation energy.

The behavior of the δ bonding as a function of the angle of internal rotation, from 0° to 45°, has been very throughly investigated, both experimentally and theoretically, because it provides a test bed for the theory of 2-electron covalent bonds in general. The basis of experimental studies lies in the fact that the twist angle can be varied from 0° (eclipsed) to *ca.* 40° in a series of $Mo_2Cl_4(diphos)_2$ molecules. The experimental[9] and theoretical[10] results agree very well and fully confirm the validity of the LCAO–MO description of a 2-electron bond.

Because of the weakness of the δ bond, the gain or loss of electrons from δ or δ^* orbitals has only a slight effect on the strength of the M–M bond. One interesting consequence of this is that the M–M bond length may be influenced as much or more by the change in effective atomic charge when an electron is removed from a δ or δ^* orbital. This is evident in the following series of structures.

$Re_2Cl_4(PMe_2Ph)_4$	$\sigma^2\pi^4\delta^2\delta^{*2}$	2.241(1) Å
$[Re_2Cl_4(PMe_2Ph)_4]^+$	$\sigma^2\pi^4\delta^2\delta^{*1}$	2.218(1) Å
$[Re_2Cl_4(PMe_2Ph)_4]^{2+}$	$\sigma^2\pi^4\delta^2\delta^{*0}$	2.215(1) Å

As the δ^* electrons are removed the formal bond order changes from 3.0 through 3.5 to 4.0, but the changes in Re–Re distance are very small. This is because at the same time the oxidation state of the metal atoms is also increasing, which causes the d orbitals to contract and thus the overlap in the π bonds becomes poorer. This bond weakening effect approximately offsets the strengthening that would be expected when antibonding electrons are lost. In the case of $[Tc_2Cl_8]^{3-}$ *vs.* $[Tc_2Cl_8]^{2-}$ there is actually a lengthening of the Tc–Tc bond by about 0.03 Å when the δ^* electron is lost.

Other Bond Orders in the Tetragonal Context

Because of the nearly nonbonding character of both δ and δ^* orbitals, both deletion and addition of either one or two electrons to the $\sigma^2\pi^4\delta^2$ configuration might be

[9]F. A. Cotton, *Pure Appl. Chem.* **1992**, *64*, 1383 and references therein; D. G. Nocera, *Acc. Chem. Res.* **1995**, *28*, 209.
[10]D. C. Smith and W. A. Goddard, *J. Am. Chem. Soc.* **1987**, *109*, 5580; F. A. Cotton and X. Feng, *J. Am. Chem. Soc.* **1993**, *115*, 1074.

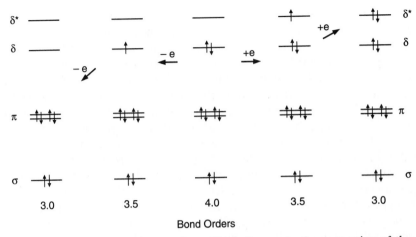

Figure 16-8   A schematic representation of how changes in the occupation of the δ and $\delta*$ orbitals change the M—M bond order.

expected to occur easily. On this basis, species that are structurally within the tetragonal prismatic framework having bond orders of 3 and 3.5 as well as 4 may be expected, as indicated schematically in Fig. 16-8. Species of all the types represented there are known. Examples range from those with V_2^{4+}, Nb_2^{4+}, and Mo_2^{6+} cores that have triple bonds ($\sigma^2\pi^4$) to those with Tc_2^{4+}, Re_2^{4+}, and Os_2^{6+} cores that also have triple bonds ($\sigma^2\pi^4\delta^2\delta*^2$). These two types of triple bonds are often called electron-poor and electron-rich, respectively.

Addition of further electrons, which enter the $\pi*$ and finally the $\sigma*$ orbitals, causes decreases in the formal bond orders, by half-integer units from 2.5 eventually to 0. In the vast number of dirhodium compounds there is a Rh_2^{4+} core which has a single bond based on the $\sigma^2\pi^4\delta^2\delta*^2\pi*^4$ configuration. Recently a set of diplatinum compounds shown in Table 16-1 has provided examples of the progression from bond order 1 to 0.

16-7   Metal Atom Cluster Compounds[11]

Although a few compounds containing metal atom clusters were known much earlier than the advent of X-ray crystallography, their true nature was not understood. It was not until the 1930s and 1940s that the existence of a few metal–metal bonds became known, e.g., in $W_2Cl_9^{3-}$ and $Mo_6Cl_8^{4+}$. However, no sense of the generality of M—M bonding or metal atom cluster formation developed until the early 1960s. Key steps were the discovery of the $[Re_3Cl_{12}]^{3-}$ ion, since this awakened an interest in the entire class, and the structural elucidation of metal carbonyl cluster compounds such as the tetrahedral $M_4(CO)_{12}$ (M = Co, Rh, Ir) molecules. Since then the growth of this field has been exponential.

Although the term *metal atom cluster* is sometimes used more broadly, it is best restricted to cases where two or more metal atoms form a group in which there are direct bonds between metal atoms. This excludes more classical polynuclear

[11]D. F. Shriver *et al.*, Eds., *The Chemistry of Metal Cluster Complexes,* VCH, New York, 1990; G. Gonzales-Moraga, *Cluster Chemistry,* Springer-Verlag, Berlin, 1994.

Table 16-1 Three Homologous Compounds That Show a Progression Through the Pt—Pt Bond Orders 1, ½, 0[a]

$$\left[\left(Ar-\underset{\underset{X-Pt-}{|}}{N}\overset{\overset{H}{C}}{\diagup}\underset{\underset{-Pt-X}{|}}{N}-Ar \right)_4 \right]^{n+}$$

Bond order	X	n	Pt—Pt, Å	Configuration
1	Cl	0	2.517	$\sigma^2\pi^4\delta^2\delta^{*2}\pi^{*4}$
½	—	1	2.530	$\sigma^2\pi^4\delta^2\sigma^{*2}\pi^{*4}\sigma^*$
0	—	0	2.649	$\sigma^2\pi^4\delta^2\delta^{*2}\pi^{*4}\sigma^{*2}$

[a] F. A. Cotton et al., Inorg. Chim. Acta **1997**, 264, 61.

complexes in which the metal atoms are held together solely by M—X—M bridge bonds. There are, of course, borderline cases, such as the Fe_4S_4 aggregates in ferredoxins (Section 17-E-10), where some degree of direct M—M bonding probably exists. There are also many copper sulfide, selenide, and telluride aggregates that are commonly called "clusters" (Section 17-H-2), but there is little or no direct metal–metal bonding and thus they are not metal atom clusters according to the definition we are using.

Broadly speaking, there are two main classes of metal atom cluster compounds:

(1) Those with the metal atoms in very low formal oxidation states, where the ligands are mostly CO groups. These also tend to occur mostly with the later transition elements, groups 7-10.

(2) Those in which the metal atoms are in somewhat higher oxidation states (+2 to +4) and the ligands are typically halide, sulfide, or oxide ions and some others of the same ilk as those in mononuclear Werner complexes. Clusters of this type are most common among the early transition elements, groups 5-7.

There is not a great deal of chemistry interconnecting these two classes and it is therefore convenient to discuss them separately.

Metal Carbonyl Type Clusters

These range in size from dinuclear ($Mn_2(CO)_{10}$, $Co_2(CO)_8$) to some of very high nuclearity, with twenty or more metal atoms. It is also notable that there are many mixed metal clusters, such as $H_2FeRu_2Os(CO)_{13}$. The presence of intimately bound—or even encapsulated—non-metal atoms, such as C, O, N, S, is also a common occurrence. Hydrogen atoms are often present as bridges and, occasionally, encapsulated.

Trinuclear clusters are nearly always triangular, often heteronuclear, and very numerous. The $M_3(CO)_{12}$ clusters of Fe, Ru, and Os have been much studied and can be used to illustrate some typical cluster chemistry. The $Os_3(CO)_{12}$ cluster with the structure (16-XIII), where CO groups are denoted simply by lines, are electronically precise, namely, they have exactly the right number of electrons to provide each metal atom with an 18-electron, closed shell configuration. In such systems there is a total of 48 electrons, and each M—M bond is of order 1. As noted later, the $H_2Os_3(CO)_{10}$ cluster can be derived from $Os_3(CO)_{12}$. This is an unsaturated but still electron precise system; it possesses a total of only 46 electrons. From the structure (16-XIV) in which one edge of the Os_3 triangle is shorter (2.68 Å) than the others (2.88 Å) and bridged by hydrogen atoms, it is inferred that there is one Os=Os double bond, thus allowing all Os atoms to achieve formal 18-electron configurations.

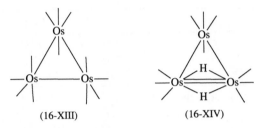

While the $Ru_3(CO)_{12}$ and $Os_3(CO)_{12}$ clusters are somewhat unreactive, they can be converted into more reactive derivatives, from which a large amount of interesting chemistry can then be developed. Three important points of departure are the following:

1. By replacement of one or two CO ligands with CH_3CN; these substituted species are then relatively reactive, by loss of CH_3CN, toward various ligands, *inter alia*, olefins. Some of the chemistry for the osmium system is summarized in Fig. 16-9. Similar reactions occur for $Ru_3(CO)_{12}$.

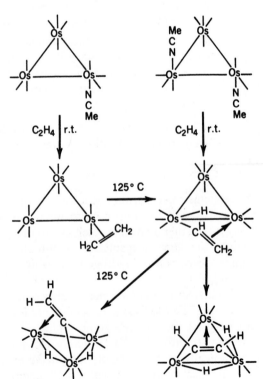

Figure 16-9 Some reactions of Os_3 $(CO)_{11}(MeCN)$ and $Os_3(CO)_{10}(MeCN)_2$.

2. Reaction of $Os_3(CO)_{12}$ with H_2 at 125°C displaces two CO groups, giving $H_2Os(CO)_{10}$, which, as already mentioned, is unsaturated and reactive towards a variety of reagents, such as Lewis bases:

$$H_2Os_3(CO)_{10} \xrightarrow{\text{L}}$$

3. Treatment of $Os_3(CO)_{12}$ with KOH in MeOH gives the $[HOs_3(CO)_{11}]^-$ ion, which can be converted *via* reactions such as those shown in Fig. 16-10 into a methyl derivative, $Os_3(CO)_{10}(H)CH_3$. This methyl compound is electronically unsaturated (46 electrons) and responds to this by developing a strong interaction between one C—H bond and an adjacent osmium atom (see Section 2-16 for "agostic" hydrogen atoms). This peculiar methyl compound

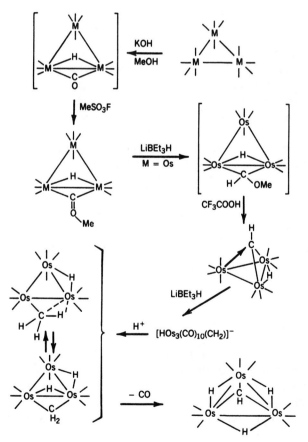

Figure 16-10 Some reactions of $Os_3(CO)_{12}$ and, in part, of its Fe and Ru analogues.

is actually in equilibrium with the methylene bridged isomer (which is a 48 electron species) and the latter has been isolated from solution and crystallographically characterized. On loss of CO it forms a stable CH-capped cluster.

Four-atom clusters are also very numerous, particularly in heteronuclear form. The majority of them display tetrahedral structures and the best known homonuclear prototypes are the $M_4(CO)_{12}$ (M = Co, Rh, and Ir) molecules. In these the 18-electron rule is satisfied for each metal atom through the formation of six M—M single bonds and the total electron count is 60. Some representative heteronuclear analogues are shown as (16-XV) and (16-XVI); these are also 60 electron systems.

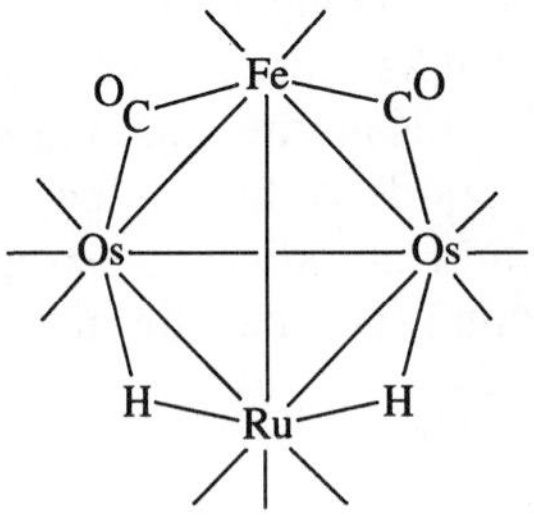

(16-XV) 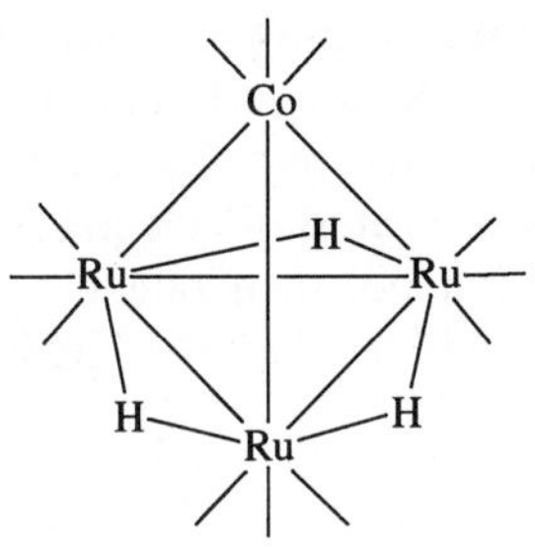   (16-XVI)

The compound $H_4Re_4(CO)_{12}$ has only 56 electrons but nevertheless has a regular tetrahedral Re_4 core. Since this structure precludes the existence of discrete, localized Re≡Re double bonds [comparable to the Os≡Os bond in $H_2Os_3(CO)_{10}$], both resonating double bonds and the formation of three-center (face) bonding have been proposed.

Besides the tetrahedral geometry, four-atom clusters are known with butterfly (16-XVII) and planar (16-XVIII) structures as well as still others.

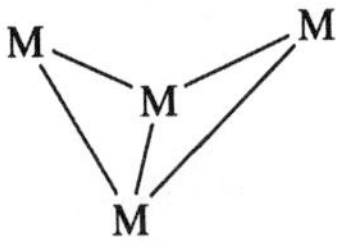

(16-XVII) 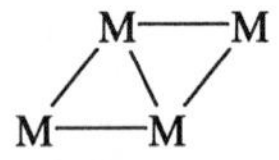   (16-XVIII)

Anionic and Hydrido Clusters

Many of the carbonyl cluster species are anions (cations being virtually unknown), hydrido species, or both. The relationships of these to each other and to neutral clusters in terms of electron count are the same as in simpler metal carbonyls, namely, one CO can be replaced by two hydrogen atoms, one H and one negative charge, or two negative charges. Protonation or deprotonation reactions are usually facile, and cause no change in the cluster structure, although there are exceptions; an example[12] is

$$[Os_6(CO)_{18}]^{2-} \xrightarrow{+H^+} [Os_6(CO)_{18}H]^- \xrightarrow{+H^+} Os_6(CO)_{18}H_2$$

Os$_6$ octahedra

The fact that many large clusters are relatively strong polyacids has made their isolation and characterization more difficult since several species at different stages of ionization may coexist. The locations of the hydrogen atoms in the hydrido clusters (and in some cases even the total number present) are often difficult to ascertain. There are, however, cases where clear results have been obtained, either directly by neutron diffraction, or less directly by X-ray diffraction coupled with steric considerations. In $[FeRu_3(CO)_{13}H]^-$ a neutron diffraction study has shown that one Ru—Ru edge of the metal atom tetrahedron is symmetrically bridged by the hydrogen, and in $Os_4H_4(CO)_{11}P(OMe)_3$ four edges are bridged, in the pattern shown in (16-XIX). There are many cases where the H atoms reside within an octahedral cluster, e.g., $[Co_6(CO)_{15}H]^-$, $[Ru_6(CO)_{18}H]^-$, $[Ni_{12}(CO)_{21}H_2]^{2-}$.

(16-XIX)

Larger Carbonyl Clusters

The first six-atom cluster, $Rh_6(CO)_{16}$, was structurally elucidated by Corey, Dahl, and Beck in 1963; today there is an enormous number of them and we can sketch only a few of the main facts. In addition to their intrinsic interest, large clusters, especially the very largest ones, are of importance because they may be expected to show properties verging on those of bulk metals. They provide one approach to answering the question: How large does a particle of metal have to be before bulk metal properties begin to appear?

Structural Patterns

There is great structural diversity, but four major types can be recognized:
1. Closed polyhedra; the octahedron is particularly common, examples being $[Os_6(CO)_{18}]^{2-}$, $Rh_6(CO)_{16}$, and $Ru_6(CO)_{18}H_2$. In some of these a hetero atom such as C, N, or H may be encapsulated, for example, $[Co_6(CO)_{15}H]^-$. Other

[12]R. Bau et al., J. Am. Chem. Soc. **1997**, 119, 11992.

polyhedra found less commonly are the trigonal bipyramid, found in $[Ni_5(CO)_{12}]^{2-}$, $Ni_3Mo_2(CO)_{13}$, and $Os_5(CO)_{16}$, the trigonally distorted octahedron (a trigonal antiprism) as in $[Ni_6(CO)_{12}]^{2-}$, and occasionally, the cube, as in $Ni_8Se_6(PR_3)_6$.[13]

2. Close-packed arrays, in which the metal atoms are grouped nearly or even exactly as in bulk metal. Rhodium and Pt provide many examples of these.

3. Stacked triangular arrays, as in the $[Ni_3(CO)_6]_n^{2-}$ $(n = 2\text{-}5)$ species, and similar platinum ones.

4. Rafts, that is, large, triangulated, planar nets. Thus far this sort of structure has been found only for some osmium species.

All of these types of structure possess exposed triangular faces and a very common mode of cluster enlargement is by capping one or more of these faces by another metal atom [e.g., by an $M(CO)_3$ group], thus generating a tetrahedron fused to the main or central unit. An example is $Os_7(CO)_{21}$. An additional metal atom (plus some ligands) may also associate itself with one of the edges of a polyhedral or close-packed central array.

Methods of Synthesis

These are largely of empirical origin and mechanistic insight is, in general, but with exceptions, rather limited and superficial. Syntheses generally require the generation of reactive fragments that can combine to form the clusters, and these fragments may be produced thermally, photolytically or by reduction processes.

Pyrolysis reactions provide syntheses that are perhaps the least rational, but often surprisingly efficient. The entire field of large osmium clusters was opened up by vacuum pyrolysis of $Os_3(CO)_{12}$, and by optimizing conditions high yields of selected products [e.g., 60% of $Os_6(CO)_{18}$] can be obtained. A useful modification of pyrolytic syntheses is often to first replace one or two CO groups, under mild conditions, by pyridine or CH_3CN (much more labile ligands) and then conduct a pyrolysis reaction. For example, the following preparation proceeds in 65% yield:

$$Os_3(CO)_{11}py \xrightarrow[64\ h]{250°C} \xrightarrow[\text{extract}]{\text{acetone}} \xrightarrow{+ PPN^+} (PPN)_2[Os_{10}(CO)_{24}C]$$

This same example also illustrates the fact that pyrolytic methods of synthesis are quite prone to give carbido clusters.

Thermal reactions of a milder nature may also be employed. The use of partially substituted starting materials provides a means to enlarge medium size clusters. Thus, $Os(CO)_4H_2$ reacts to displace CH_3CN ligands from $Os_6(CO)_{18-n}(MeCN)_n$ species to give heptanuclear clusters, $Os_7(CO)_nH_2$ $(n = 20, 21, \text{and } 22)$.

Another mild type of thermal pathway for synthesis of larger clusters utilizes ligands that have unshared electron pairs. Among a number of examples that might be discussed we can cite reactions between a cluster containing one or more labile ligands and a cluster with a capping sulfur atom whose lone pair can come into play:

[13]D. Fenske *et al., Angew. Chem. Int. Ed. Engl.* **1992,** *31,* 321.

$$Os_3(CO)_{10}(MeCN)_2 + Os_3(CO)_{10}S \xrightarrow{\text{-CO, 2MeCN}}$$

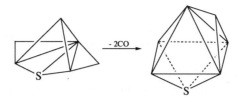

Large clusters of the cobalt and nickel subgroups are commonly prepared by reductively generating reactive anionic intermediates that condense. For example, $Co_4(CO)_{12}$ is converted by treatment with alkali metals in THF to the $[Co_6(CO)_{15}]^{2-}$ ion. Similarly, the action of strong, basic reductants (Na/THF, NaBH$_4$/THF) on $Ni(CO)_4$ leads to cluster anions such as $[Ni_5(CO)_{12}]^{2-}$, $[Ni_6(CO)_{12}]^{2-}$, $[Ni_9(CO)_{18}]^{2-}$, or $[Ni_{12}(CO)_{21}]^{4-}$ depending on the exact conditions.

These reductive pathways are particularly suitable for making clusters that encapsulate hetero atoms, simply by introducing a suitable source of these atoms, such as CCl_4 for C, C_2Cl_6 for C_2, H_2S for S, or Ph_3As for As.

Carbonyl metallates may also react with neutral carbonyls containing a different metal to produce heteronuclear products, namely:

$$[Rh_6(CO)_{15}]^{2-} + Ni(CO)_4 \xrightarrow{\text{-3CO}} [Rh_6Ni(CO)_{16}]^{2-}$$

$$[Pt_3(CO)_6]^{2-} + 3Fe(CO)_5 \xrightarrow{\text{-6CO}} [Fe_3Pt_3(CO)_{15}]^{2-}$$

Even simple metal halides may react with carbonyl anions. Examples are provided by the following reactions:

$$[Ni_6(CO)_{12}]^{2-} + PtCl_2 \longrightarrow [Ni_{38}Pt_6(CO)_{48}H_2]^{4-}$$

$$[Ni_6(CO)_{12}]^{2-} + Ph_3PAuCl \longrightarrow [Au_6Ni_{12}(CO)_{24}]^{2-}$$

The structure[14] of the gold/nickel cluster has tetrahedral symmetry with the metal core shown in Fig. 16-11.

Superlarge Clusters

In recent years there have been several reports of clusters so large that they have internal structures that are very clearly metal-like. A striking example is the recently reported[15] anion, $[Pd_{33}Ni_9(CO)_{41}(PPh_3)_6]^{4-}$, which contains five central layers of metal atoms, stacked in the same manner as that found in a hexagonally close-packed metal. The outer metal atoms have the 41 CO groups and 6 Ph$_3$P ligands attached. Numerous other large clusters, such as $Pd_{23}(CO)_{22}(PEt_3)_{10}$ and $Pd_{34}(CO)_{24}(PEt_3)_{12}$[16] have been reported.

[14]A. J. Whoolery and L. F. Dahl, *J. Am. Chem. Soc.* **1991,** *113,* 6683.
[15]L. F. Dahl *et al., J. Am. Chem. Soc.* **1996,** *118,* 7869.
[16]E. G. Mednikov and N. I. Kanteeva, *Russ. Chem. Bull.* **1995,** *44,* 163.

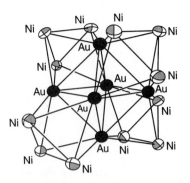

Figure 16-11 Au_6Ni_{12} core of the 236-electron $[Au_6Ni_{12}(CO)_{24}]^{2-}$ dianion. This 18-vertex polyhedron of cubic T_d (43m) symmetry may be viewed as a face-to-face condensation of four octahedral Au_3Ni_3 fragments at alternate faces of a central Au_6 octahedron (A. J. Whoolery and L. F. Dahl, *J. Am. Chem. Soc.* **1991,** *113,* 6683.)

Heteroatoms in Clusters

The presence of heteroatoms (i.e., nonmetal atoms) within or intimately associated with metal atom clusters has already been briefly mentioned. Apart from H and C atoms, N, P, As, and S are fairly common. Carbon and nitrogen atoms are most common, and for carbon some representative examples are shown in Fig. 16-12. Others such as $[Co_{11}C_2(CO)_{22}]^{3-}$, $Rh_{12}C_2(CO)_{25}$, and $Co_3Ni_7C_2(CO)_{16}]^{2-}$ contain C_2

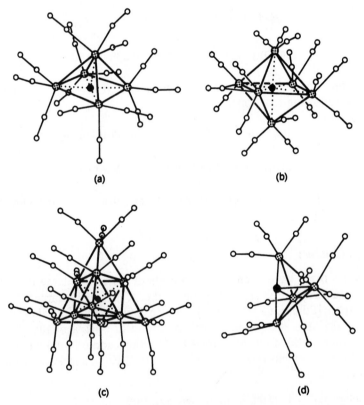

Figure 16-12 The structures of some representative carbido carbonyl cluster species. (a) $FeC(CO)_{15}$, (b) $Ru_6C(CO)_{17}$, (c) $[Os_{10}C(CO)_{24}]^{2-}$, and (d) $Fe_4C(CO)_{13}$.

units with $C-C$ distances on the order of 1.5-1.6 Å (Section 7-9). The C atoms or C_2 units often originate by the process $2CO \rightarrow$ "C" $+ CO_2$ or from $CHCl_3$ or CCl_4.

Nitrogen-containing clusters are often formed with NO, N_3^-, or NCO^- as the nitrogen source, as in the reactions.

$$[H_3Ru_4(CO)_{12}]^- + NO^+ \longrightarrow HRu_4N(CO)_{12} + H_3Ru_4N(CO)_{11}$$

$$2Ru_3(CO)_{12} + N_3^- \longrightarrow [Ru_6N(CO)_{16}]^- + 8CO + N_2$$

Electron Counting in Medium Size Clusters[17]

For these clusters, many of which contain heavier metal atoms, conventional molecular quantum mechanical treatments have not been feasible. Instead much effort has been devoted to seeking relationships between their structures and the number of electrons available for cluster bonding. All of these large clusters are regarded as electron deficient in the sense that there are insufficient electrons to permit the assignment of an electron-pair bond between each adjacent pair of metal atoms.

For polyhedral clusters (sometimes called deltahedral, because the faces are all triangles resembling the Greek letter delta) the ancestor of all electron counting schemes is the correlation proposed by Wade between borane (or carborane) cages and metal carbonyl cages. Wade first drew attention to the similarity of a $M(CO)_3$ unit and a BH (or CH) unit, a relationship that we would now call isolobality (Section 1-6). He then proposed that the $2n + 2$ rule for *closo* boranes (Chapter 5) would also apply to *closo* metal cluster species such as $[Os_6(CO)_{18}]^{2-}$, and that $2n + 4$ and $2n + 6$ electron counts would, similarly, be appropriate for stable M_n clusters with *nido* and *arachno* structures. Hydrogen atoms are assumed to contribute one electron each, an interstitial carbon atom four electrons, and so on.

More elaborate or differently derived treatments of the relationships between structure and electron count for large clusters have appeared, but we shall restrict our discussion to the basic ideas proposed by Wade, plus a few added rules that will be useful in subsequent sections.

While the number of skeletal electron pairs, denoted S, is at the root of the correlation procedures, discussions found in the literature are often couched in terms of other, related parameters, particularly the *total electron count* (TEC). The TEC is obtained from the formula by adding the following contributions:

1. The number of valence electrons for each metal atom. For example, in an Os_6 cluster, $6 \times 8 = 48$.
2. Two electrons for each CO group, regardless of its structural type (terminal, μ_2, or μ_3).
3. One electron for each negative charge.
4. The number of valence electrons for each hetero and/or interstitial atom. For example, 1 for H, 4 for C, 5 for P. Column 2 of Table 16-2 presents some examples of total electron counts.

[17]D. M. P. Mingos and D. J. Wales, *Introduction to Cluster Chemistry*, Prentice Hall, Englewood Cliffs, NJ, 1990.

Table 16-2   Examples of Correlating Structures with Total Electron Count
on a Wade's Rules Basis

Cluster	Total electron count (TEC)	No. of skeletal electron pairs (S)	Vertices of parent polyhedron	Structural conclusion
$Rh_6(CO)_{16}$	$(6 \times 9) + (2 \times 16) = 86$	$\frac{1}{2}[86 - (6 \times 12)] = 7$	6	*Closo*
$Os_5(CO)_{16}$	$(5 \times 8) + (2 \times 16) = 72$	$\frac{1}{2}[72 - (5 \times 12)] = 6$	5	*Closo*
$Os_5C(CO)_{15}$	$(5 \times 8) + (2 \times 15) + 4 = 74$	$\frac{1}{2}[74 - (5 \times 12)] = 7$	6	*Nido*
$[Fe_4C(CO)_{12}]^{2-}$	$(4 \times 8) + (2 \times 12) + 4 + 2 = 62$	$\frac{1}{2}[62 - (4 \times 12)] = 7$	6	*Arachno*
$[H_3Ru_4(CO)_{12}]^-$	$(4 \times 8) + (2 \times 12) + 3 + 1 = 60$	$\frac{1}{2}[60 - (4 \times 12)] = 6$	5	*Nido*

Let us now ask how we could predict the correct total electron count, as just
defined, for a stable cluster of known structure (i.e., *closo*, *nido*, or *arachno*). To
do this for metal carbonyl clusters, it is postulated that in addition to the electrons
necessary for skeletal bonding each metal atom will also have 12 nonskeletal electrons. The basis for this assumption is that in the pyramidal $M(CO)_3$ unit each
M$-$CO bond will comprise two formally "carbon σ" electrons that are donated
to the metal atom and two formally "metal" π electrons that backbond, at least
partially, to the CO ligand. Thus, in predicting the *total electron count* for a closo
polyhedral cluster of n vertices, the result would be $12n + 2(n + 1)$. Similarly, for
nido and *arachno* clusters that are derived from an n-vertex polyhedron (their
parent polyhedron) by removal of one or two vertices, respectively, there will be
12 and 24 fewer total electrons, respectively.

The predictions for TEC can therefore be stated in the following equations
(where n is the number of vertices in the parent polyhedron for the *nido* and
arachno cases—not the actual number of metal atoms in the cluster itself):

closo	$12n + 2(n + 1)$
nido	$12(n - 1) + 2(n + 1)$
arachno	$12(n - 2) + 2(n + 1)$

In practice, these relations are often employed "in reverse" so to speak. From the
actual TEC (easily derived from the formula as already explained) $12e$ are subtracted
for each metal atom and the number of skeletal pairs (S) thus determined. On the
basis that $(S + 1)$ pairs are required for a polyhedron of S vertices, one may
select the most likely structure or structures. Some illustrations will make this
procedure clear.

For $Rh_6(CO)_{16}$ the TEC is $(6 \times 9) + (2 \times 16) = 86$. The number of skeletal
pairs (S) will be obtained by subtracting 6×12 from 86 and dividing by 2; the
result is 7. Since the number of vertices of the parent polyhedron is $S - 1$, we
conclude that this will be a six-vertex polyhedron (e.g., an octahedron). Since there
are six metal atoms, we conclude that a *closo* structure, probably an octahedron,
is appropriate. This is correct. In general 86-electron clusters consisting of six metal
atoms are octahedral.

To present more concisely some further examples, we employ Table 16-2. The
$Rh_6(CO)_{16}$ example just discussed is included in the table to help illustrate the way

it works. The $Os_5(CO)_{16}$ cluster affords another example of how a *closo* structure (a trigonal bipyramid) is correctly predicted.

The next two examples in Table 16-2, $Os_5C(CO)_{15}$ and $[Fe_4C(CO)_{12}]^{2-}$ show how *nido* and *arachno* structures are predicted. The *nido* structure already shown in Fig. 16-12(a) for the iron analogue, $Fe_5C(CO)_{15}$, is derived by removal of one vertex from the parent polyhedron, which is an octahedron. The $[Fe_4C(CO)_{12}]^{2-}$ cluster should be isostructural with the $Fe_4C(CO)_{13}$ cluster, whose structure has been shown in Fig. 16-12(d). It is derived from a parent octahedron by removal of two vertices.

The final example in Table 16-2 illustrates the point that a tetrahedron is treated as a trigonal bipyramid that is missing one vertex, that is, as a *nido* structure. While this may seem artificial, it is not illogical and it is necessary in order to maintain consistency with the general rules.

The Capping Rule

Many carbonyl-type clusters consist of a central polyhedron to which one or more additional metal atoms are appended by placement over a triangular face. These appended metal atoms are called capping atoms. The electron counting rules we have discussed can be extended to cover such structures by the rule that the addition of a capping atom does not change the skeletal electron requirement for the central cluster. Thus, the addition of a capping atom simply increases the total electron count by 12 electrons.

The $Os_7(CO)_{21}$ cluster provides a simple illustration. The TEC is given by $(7 \times 8) + (21 \times 2)$, namely, 98. Subtracting $7 \times 12 = 84$ and dividing by 2, we obtain $S = 7$, which implicates a six-vertex central polyhedron. This is consistent with there being a *closo* six-atom deltahedron (namely, an octahedron) capped on one face—which is, indeed, the actual structure of $Os_7(CO)_{21}$. Correct results are similarly obtained for $[Os_8(CO)_{22}]^{2-}$, which is a bicapped octahedron (16-XX) and $[Os_{10}C(CO)_{24}]^{2-}$, which is a tetracapped octahedron. For the latter, a capsule summary of the analysis is as follows:

$$\text{TEC:} \quad (10 \times 8) + 4 + (2 \times 24) + 2 = 134$$

$$S: \quad \tfrac{1}{2}[134 - (10 \times 12)] = 7$$

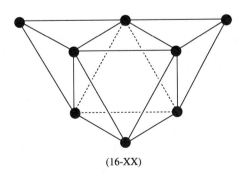

(16-XX)

A final word about the capping rule concerns a limitation. The $[Os_8(CO)_{22}]^{2-}$ cluster has the bicapped structure shown in (16-XX). However, two other isomeric bicapped M_8 structures, (16-XXI) and (16-XXII), are possible and the capping rule *per se* cannot predict which one to expect, nor can any other rule presently known. The *trans*-capped structure (16-XXI) is known for $[Re_8C(CO)_{24}]^{2-}$, but no example of (16-XXII) has yet been reported. It may be that when the steric requirements of the CO groups are considered, this *cis*-capped arrangement is, for that reason, not possible.

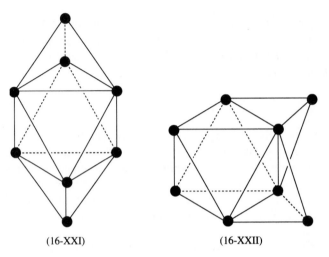

(16-XXI)   (16-XXII)

Clusters of the Fe, Ru, Os Group

These metals are prolific formers of clusters, most of which obey the electron counting rules just presented. Osmium forms the greatest number, and is the only one to form clusters with more than six metal atoms. Indeed, osmium cluster compounds are so numerous (including many heteronuclear ones) that they afford examples of many subtle structural variations.

There is, for example $H_2Os_6(CO)_{18}$, which has the same electron count as $[Os_6(CO)_{18}]^{2-}$ but, as shown in Fig. 16-13, has a different skeletal arrangement, one that can be regarded as a capped square pyramid. It is, however, also consistent with the electron counting rule.

By moving one $Os(CO)_3$ out of the cluster we delete $12e$, but the remaining $Os_5(CO)_{15}$ unit still has $S = 7$ and should therefore be a *nido* octahedron. By now using the deleted $Os(CO)_3$ unit to cap this, we require no change in the central square pyramid. The general implication of this example is that in addition to a set of Os_{6+n} clusters obtainable by adding n caps to an octahedron, there will be another set of structures obtainable by adding $(n + 1)$ caps to the square pyramids and equally in accord with the rules.

The principle can be extended further. A set of structures can be derived by capping an *arachno* parent with $S = 7$, as shown for the structure of $H_2Os_5(CO)_{16}$, also in Fig. 16-13.

Osmium cluster chemistry also includes a type called *rafts*, for example $Os_6(CO)_{17}[P(OMe)_3]_4$, which is formally a substitution product of $Os_6(CO)_{21}$. The

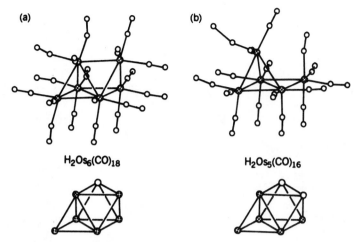

(a) H₂Os₆(CO)₁₈ **(b)** H₂Os₅(CO)₁₆

Figure 16-13 Two osmium clusters whose structures illustrate the capping of incomplete polyhedra: (a) capping of a *nido* polyhedron; (b) capping of an *arachno* polyhedron.

structure, shown schematically in (16-XXIII), is easily understood electronically, with every Os—Os bond being a *2c-2e* bond and each metal atom having an 18-electron configuration.

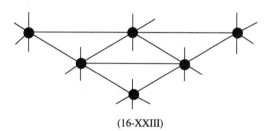

(16-XXIII)

Clusters of the Cobalt, Rhodium, Iridium Group

Cobalt forms clusters with 5 to at least 16 metal atoms and rhodium forms them with 5 to at least 22 metal atoms. The number of iridium clusters is more limited.

Rhodium clusters are prone to adopt structures having an interstitial rhodium atom within a polyhedron of rhodium atoms. The most prominent structure of this type is that for the $[Rh_{13}(CO)_{24}H_n]^{(5-n)-}$ species, which is shown in Fig. 16-14a. This metal atom arrangement is precisely a fragment of a hexagonal close-packed array. By additions to it larger clusters are built up, as also shown in Fig. 16-14.

Rhodium also forms an unusually large number of clusters containing encapsulated hetero atoms as, for example, the $[Rh_{17}(S)_2(CO)_{32}]^{3-}$ species whose skeletal structure is shown in Fig. 16-15, and a number that contain carbon atoms. It also forms both smaller, for example, $Rh_6(CO)_{16}$, and larger, for example, $Rh_{12}H_2(CO)_{25}$, clusters that have no encapsulated atoms. The Rh_{12} species has a very novel structure in which two Rh_6 octahedra are linked together to form an Rh_6 octahedron between them. Finally, rhodium also forms many mixed metal clusters, especially with plati-

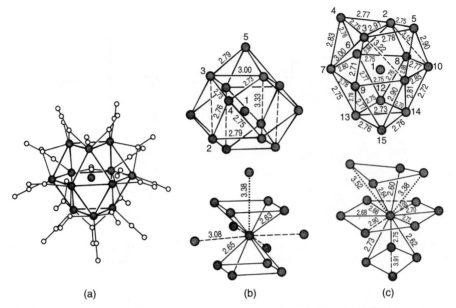

Figure 16-14 The configurations of the metal atom cores in three high-nuclearity rhodium clusters that have internal metal atoms: (a) $[Rh_{13}(CO)_{24}H_{2,3}]^{3,2-}$, (b) $[Rh_{14}(CO)_{25}]^{4+}$, and (c) $[Rh_{15}(CO)_{27}]^{3-}$.

num. The $[PtRh_8(CO)_{19}]^{2-}$ ion consists of two octahedra fused together, with the Pt atom at one of the shared vertices.

Clusters of the Nickel, Palladium, Platinum Group

These metals are prolific formers of clusters. The earliest studies dealt mainly with the clusters of Ni and Pt in which $M_3(\mu\text{-}CO)_3(CO)_3$ units are stacked to form columns; there are no Pd analogues of these.

The first nickel clusters to be recognized were $[Ni_5(CO)_{12}]^{2-}$ and $[Ni_6(CO)_{12}]^{2-}$, whose structures are shown in Fig. 16-16. Both are generated by reduction of $Ni(CO)_4$ with alkali metals in THF or with alkali hydroxides in methanol. Note that $[Ni_5(CO)_{12}]^{2-}$ does not obey the electron counting rules previously explained; unlike $[Os_5(CO)_{15}]^{2-}$ (which has 72 electrons and does) it has 76 electrons and

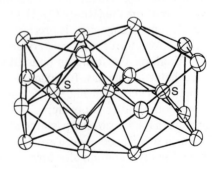

Figure 16-15 Diagram of $[Rh_{17}(S)_2(CO)_{32}]^{3-}$, omitting the carbon monoxide ligands.

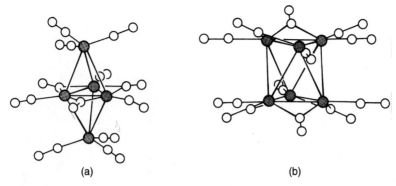

Figure 16-16 The structures of the nickel carbonylate ions (a) $[Ni_5(CO)_{12}]^{2-}$ and (b) $[Ni_6(CO)_{12}]^{2-}$.

would be predicted to have an *arachno* structure based on a seven-vertex parent polyhedron. By oxidation of $[Ni_6(CO)_{12}]^{2-}$ with $FeCl_3$, the $[Ni_9(CO)_{18}]^{2-}$ ion is obtained almost quantitatively. It has a structure in which one more $Ni_3(CO)_6$ unit is added to the $[Ni_6(CO)_{12}]^{2-}$ structure, but eclipsed to one face; the result is thus an octahedron and a trigonal prism with a common triangular face.

Larger nickel clusters with different structures are stabilized by introduction of interstitial carbon atoms. Reaction of $[Ni_6(CO)_{12}]^{2-}$ with CCl_4 produces the $[Ni_8(CO)_{16}C]^{2-}$, $[Ni_9(CO)_{17}C]^{2-}$, and $[Ni_{10}(CO)_{18}C]^{2-}$ species, the first two of which have been structurally characterized. The Ni_8 species has a square antiprism of metal atoms. The Ni_9 cluster has a cap on one square face and presumably the Ni_{10} species has a cap on the other square face as well. The reaction of $[Ni_6(CO)_{12}]^{2-}$ with C_2Cl_6 affords $[Ni_{10}(CO)_{16}C_2]^{2-}$ in which there is a C_2 unit with C—C = 1.40 Å. The 10 nickel atoms are arranged as in (16-XXIV). When this cluster anion is treated with Ph_3P it is quantitatively converted to the $[Ni_{16}(CO)_{23}(C_2)_2]^{4-}$ ion, which consists of a large polyhedral metal atom cage within which are two separate C_2 units (C—C = 1.38 Å). The structure is shown in Fig. 16-17. The cubic $Ni_8Se_6(PR_3)_6$ cluster has already been mentioned.

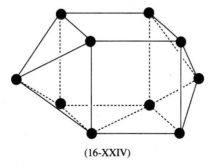

(16-XXIV)

Palladium clusters with a mixture of CO and phosphine ligands have a variety of structures. The species $Pd_7(CO)_7(PMe_3)_7$ and $Pd_{10}(CO)_{18-n}(PMe_3)_n$ contain a central Pd_6 octahedron with one or four capping atoms, while $Pd_{23}(CO)_{22}(PEt_3)_{10}$ has

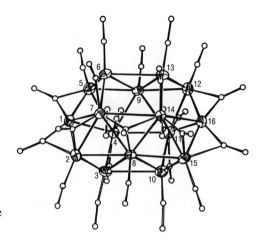

Figure 16-17 The structure of the $[Ni_{16}(CO)_{23}(C_2)_2]^{2-}$ ion.

an extraordinary structure based on a cubooctahedron with one encapsulated Pd atom, a cap on each square face and four edge-bridging Pd atoms.

Platinum clusters are of two broad classes:

1. Columnar species, $[Pt_3(CO)_6]_n^{2-}$ (n = 2-5), which are built up by stacking planar $Pt_3(CO)_6$ units. Structures of three of these are shown in Fig. 16-18. The details of these structures are curious and not easily explained. The stacking is essentially prismatic rather than antiprismatic, but in addition there are offsets and slight twists from one layer to the next. It has been speculated that these seemingly irregular structures result from a compromise between the competing effects of Pt—Pt bonding and nonbonded repulsions, although the avoidance of antiprismatic stacking where a larger number of interlayer Pt—Pt bonds could be formed is puzzling. The intratriangle Pt—Pt distances (~2.65 Å) are much shorter than the interlayer ones (~3.05 Å).

2. Large platinum carbonyl anions, such as $[Pt_{19}(CO)_{22}]^{4-}$, $[Pt_{26}(CO)_{32}]^{3-}$, and $[Pt_{38}(CO)_{44}H_x]^{2-}$ (where the value of x is uncertain). The last two consist of approximately cubic close-packed arrays of platinum atoms, and can be thought of almost as small chunks of the metal solubilized by a coating of CO groups together with the negative charge and/or hydrogen atoms. These two platinum cluster anions come closest of all known metal atom clusters

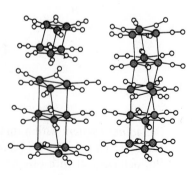

Figure 16-18 The structures of $[Pt_3(CO)_6]_n^{2-}$ ions with n = 2, 3, and 5.

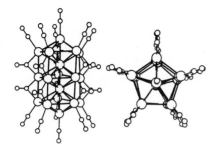

Figure 16-19 Two views of the $[Pt_{19}(CO)_{22}]^{4-}$ ion, which has idealized D_{5h} symmetry.

to approaching the borderline between molecular and bulk metal properties. The $[Pt_{19}(CO)_{22}]^{4-}$ ion has the remarkable structure shown in Fig. 16-19; this fivefold (D_{5h}) symmetry is not related to close-packed structures found in bulk metals, but it is similar to that reported for certain microcrystalline materials such as whiskers and dendrites.

We may note, finally, that nickel and platinum form various mixed clusters, some very large, as in the case of the $[Ni_{38}Pt_6(CO)_{48}H_{6-n}]^{n-}$ ions. A medium size cluster, $[Ni_9Pt_3(CO)_{21}H]^{3-}$, has been structurally characterized. The metal framework is as shown in (16-XXV).

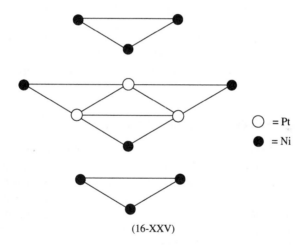

$\bigcirc$ = Pt
$\bullet$ = Ni

(16-XXV)

All three metals also form smaller clusters with isocyanides, namely, Ni_4L_6, Ni_4L_7, Pd_3L_6, and Pt_3L_6.

Catalysis by Clusters

Metal cluster compounds of the carbonyl type have often served as "catalyst precursors," that is, as the source of fragments that are the actual catalysts,[18] although there have been relatively few fully detailed descriptions of catalytic cycles based on clusters.[19] Triruthenium cluster compounds have been found to catalyze hydrogen

[18]G. Süss-Fink and G. Meister, *Adv. Organomet. Chem.* **1993,** *35,* 41.
[19]*Catalysis by Di- and Polynuclear Metal Cluster Complexes,* R. D. Adams and F. A. Cotton, Eds., Wiley-VCH, New York, 1998.

and hydrosilation of alkanes and recently it was shown that the addition of platinum to the triruthenium clusters improves catalytic activity.[20] For example, the complex $Pt_3Ru_6(CO)_{20}(\mu$-H$)(\mu_3$-H$)(\mu_3$-PhC$_2$Ph$)$ (16-XXVI) is an effective catalyst for conversion of diphenylacetylene to *cis*-stilbene. To date, however, there are no molecular cluster catalysts that are as active as the best mononuclear homogeneous catalysts (Chapter 22).

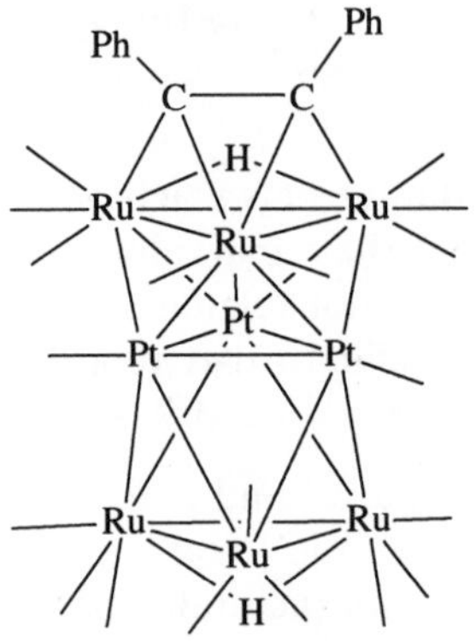

(16-XXVI)

There has been interest in using metal cluster compounds as precursors to heterogeneous catalysts on oxide supports.[21] It is already known that metal alloys dispersed on supports are heterogeneous catalysts and they are widely used in petroleum refining.

Non-Carbonyl (Higher Oxidation State) Clusters

While there is great variety in this class of compound,[22] the majority have the metal atoms in oxidation states of II to IV and have octahedral or triangular structures.

Octahedral clusters display two prototypical structures as shown in Fig. 16-20. In one there are μ_3-X atoms capping the faces and in the other there are twelve μ_2-X atoms spanning the edges. In addition, in each case, there are six additional ligands in the external positions of the metal atoms. In some cases the clusters are electron-precise. For example, in the $[Mo_6(\mu_3$-Cl$)_8]^{4+}$ cluster, there are four electrons per Mo atom and each octahedron edge may be considered as a single Mo—Mo bond, and in the triangular $[Re_3Cl_{12}]^{3-}$ ion, each Re—Re edge may be considered as a double bond. In general, however, it is not possible to assign distinct localized bonds of integral order and a more flexible molecular orbital approach is required.

There are also clusters that achieve stability only by utilizing additional electrons that are provided by other atoms that are encapsulated or attached.[23] This is well

[20]R. D. Adams *et al., J. Am. Chem. Soc.* **1994,** *116,* 9103.
[21]P. Braustein and J. Rose, in *Catalysis by Di- and Polynuclear Metal Cluster Complexes,* R. D. Adams and F. A. Cotton, Eds., Wiley-VCH, New York, 1998.
[22]J. D. Corbett, *J. Chem. Soc., Dalton Trans.* **1996,** 575.
[23]J. D. Corbett, *J. Alloys Compounds.* **1995,** *229,* 10.

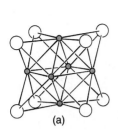 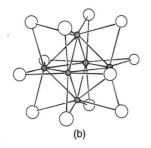

Figure 16-20 Prototypical structures for octahedral cluster species: (a) the M_6X_8 type; (b) the M_6X_{12} type.

illustrated by octahedral zirconium clusters,[24] which have a preferred number of cluster-based electrons equal to 14. This is obviously insufficient to allow the formation of twelve Zr—Zr bonds, even of order 1. Even 14 electrons are not available from the six Zr atoms alone, but in cases such as $Zr_6Cl_{15}N$ and $[Zr_6Cl_{15}B]^{2-}$ the additional electrons contributed by the encapsulated atoms (N, B) make up the deficit. In cases like $[Zr_6Cl_{18}H_5]^{3-}$, hydrogen atoms on the octahedral faces[25] provide the electrons.

Chevrel Phases. These are ternary molybdenum chalcogenides, named for Roger Chevrel who was the first (1971) to recognize and characterize them structurally. Their general formula is $M_xMo_6X_8$, where M represents one or several of the following: a vacancy, a rare earth element, a transition or main group metallic element (e.g., Pb, Sn, Cu, Co, Fe), and X may be S, Se, or Te. The idealized structure of $PbMo_6S_8$ is shown in Fig. 16-21. It is easily seen that one component of the structure, packed together with the Pb atoms in a CsCl pattern, is an Mo_6S_8 unit that has the type of structure shown in Fig. 16-20(a). These Mo_6S_8 units are linked together because some of the S atoms are bonded to the external coordination sites of their neighbors.

The binary phase (Mo_6S_8) contains quite distorted Mo_6 octahedra, elongated on a threefold axis to become trigonal antiprisms. This can be attributed to the fact that there are only 20 cluster electrons. This number is insufficient to form a full set of Mo—Mo bonds, that is, to populate all the 12 bonding orbitals corresponding to an electron-precise structure, resulting in a distorted octahedron with somewhat longer Mo—Mo bonds compared to the 24-electron clusters. In the compounds containing additional metal atoms, some electrons are added to the Mo_6 core and it becomes more regular in structure with shorter Mo—Mo bonds.

The Chevrel phases have aroused great interest because they display remarkable superconductivity, as well as some other notable physical properties. Their superconductivity is exceptional in its persistence in the presence of magnetic fields, a property that is obviously important if superconductivity is to be exploited in making powerful magnets. In recent years a great many chemical variations on the basic $M_xMo_6X_8$ composition have been devised, for example, compounds such as $Mo_6Br_6S_3$, which contains $Mo_6(\mu_3\text{-Br})_4(\mu_3\text{-S})_4$ units.

Triangular clusters come in a great variety of structural types. For rhenium in oxidation state III there is a type of cluster not known for any other element

[24]F. A. Cotton *et al.*, in *Early Transition Metal Clusters with π-Donor Ligands*, M. H. Chisholm, Ed., VCH Publishers, Inc., New York, 1995, pp 1-26.
[25]F. A. Cotton *et al.*, *Inorg. Chem.* **1997**, *36*, 4047.

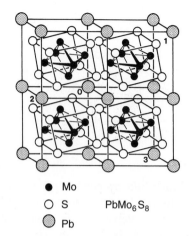

Figure 16-21   The idealized structure of a representative Chevrel compound.

● Mo
○ S $PbMo_6S_8$
◐ Pb

(Section 18-D-5), in which there are $Re{=}Re$ double bonds. These were the first M—M double bonds to be recognized and are still the only ones known in metal clusters of this class with nuclearity three or higher.

Numerous trinuclear cluster compounds are formed by molybdenum, and to a lesser but significant extent by tungsten and niobium. Four principal types are known, and their structures are shown schematically in Fig. 16-22. In all of these the M—M bond orders are in the range ⅔ to 1. For molybdenum particularly, there are numerous mixed oxides in which the type of structure shown in Fig. 16-22(b) figures prominently.

As a kind of extension of this particular type of trinuclear structure, there are numerous compounds that contain two such units fused together on a common edge to give either a discrete molecule as in $W_4(OEt)_{16}$ (Fig. 16-23), and $Mo_4O_8(OPr^i)_4py_4$ or extended arrays of such a unit, joined by shared nonmetal atoms, as in MNb_4Cl_{11} (Fig. 18-B-13) or certain mixed oxides of molybdenum such as $Ba_{1.14}Mo_8O_{16}$.

Solid state extended arrays. In traditional solid state chemistry M—M bonding was not observed because the compounds studied were mainly oxides and halides containing the metal atoms in their higher (or highest) oxidation states, where they

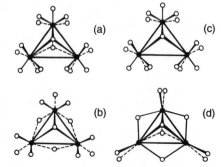

Figure 16-22   Four of the most important types of trinuclear cluster structures. Solid circles are metal atoms and open circles nonmetal (ligand) atoms.

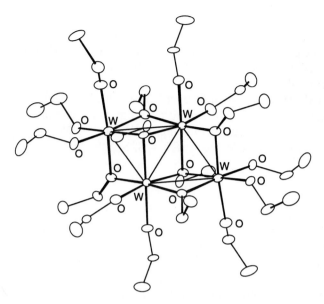

Figure 16-23   The $W_4(OEt)_{16}$ molecule.

do not form M—M bonds. In recent years investigations of chalcogenides, oxides, and halides in which the oxidation states of the metal atoms are kept low have revealed the existence of a new realm of solid state materials in which M—M bonding plays a widespread and essential role. We have already referred to some solid state systems such as the Chevrel phases, in which metal atom clusters occur. We now draw attention to substances containing infinite arrays of M—M bonds.

Among the most interesting systems are some ternary oxides of lower-valent molybdenum that contain infinite chains of octahedral clusters fused on opposite edges. This type of structure is similar to those found in the lower-halides of

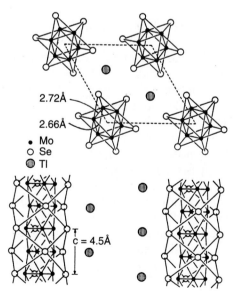

2.72Å

2.66Å

• Mo
○ Se
◉ Tl

c = 4.5Å

Figure 16-24   Two views of the structure of $Tl_2Mo_6Se_6$.

scandium (Section 19-6) and the lanthanides (Section 19-8). Some compounds that display such structures are $NaMo_4O_6$, $Ba_5(Mo_4O_6)_8$, and $Sc_3Zn_5(Mo_4O_7)_4$. At present a thorough understanding of the electronic structures of these chains of edge-sharing octahedra has not been achieved. It is evident, however, from the Mo—Mo distances that bond orders in the vicinity of 1 are present.

Another type of extended array is also built up of octahedra, but with the sharing of opposite faces. Compounds in this class range from those containing only two or three fused octahedra to infinite chain compounds (Fig. 16-24), which have deceptively simple formulas such as $M_2Mo_6X_6$ (X = S, Se, Te; M = In, Tl, Na, K).

16-8   Organometallic Chemistry

The multifarious ways in which transition metal atoms can form bonds to carbon atoms goes far beyond what can happen with the main group metals, where the chemistry is largely (though not entirely) confined to M—C single bonds as in $Al(C_2H_5)_3$, $Pb(C_2H_5)_4$, or $Sb(C_2H_5)_5$. For the transition metals, organometallic compounds of this type are known and often important, but are outnumbered by those involving several other types of metal—carbon bonding, that require the participation of *d* orbitals.

In this section the main types of transition metal organometallic compounds will be briefly introduced. Specific examples are cited in Chapters 17 and 18 under the individual elements. The many sorts of reactions that organotransition metal compounds undergo, with particular emphasis on their relationship to catalysis, are reviewed in detail in Chapter 21. Catalysis *via* organometallic compounds is presented in Chapter 22.

BONDING TO SINGLE CARBON ATOMS
Metal-to-Carbon Single Bonds

While the failure to isolate simple transition metal alkyl compounds was long believed due to intrinsic weakness of the M—C bonds, this is now known not to be so. The bond energies, 120 to 160 kJ mol^{-1}, are similar to those with non-transition metals and the critical factors are kinetic rather than thermodynamic. This generalization is well illustrated by the comparison between the thermally stable $PbMe_4$ (bp 200°C) and $TiMe_4$ which is unstable above about −50°C, despite the fact that the Ti—C bond energy (*ca.* 260 kJ mol^{-1}) is higher than the Pb—C bond energy (*ca.* 170 kJ mol^{-1}). Vacant valence orbitals on Ti are available to allow decomposition, but when octahedral coordination is completed, as in $TiMe_4$bipy, thermal stability is greatly increased; $TiMe_4$bipy is stable up to *ca.* 15°C.

Two major thermal decomposition paths are recognized: (a) homolytic M—C bond cleavage, giving R radicals, and (b) hydrogen transfer reactions, of which β-transfers are the most common. An interesting illustration is provided by the decomposition of two similar alkyls, only one of which has a readily transferable β-hydrogen atom: $Bu_3PCuCH_2CH_2CH_2CH_3$ and $Bu_3PCuCH_2CMe_2Ph$. The former decomposes to give Cu, H_2, but-1-ene, and butane; the latter gives as products $PhCMe_3$, $PhCHMe_2$, $PhCH_2C(Me)=CH_2$, and $(PhMe_2CCH_2.)_2$, which can come only from homolytic fission of the Cu—C bond and subsequent radical reactions. These and other decomposition mechanisms for alkyls and aryls of transition metals are discussed in detail in Section 21-3.

The main classes of transition metal alkyls and aryls are the following:

1. *Mononuclear.* Homoleptic examples are $Ti(CH_2Ph)_4$, $Mn(1\text{-norbornyl})_4$, WMe_6, and $OsPh_4$. Oxo compounds, often with the metal atoms in very high oxidation states are also well known, an important example being CH_3ReO_3 (Section 18-D-12).

   There are hundreds of alkyl compounds in which one (or more) metal—alkyl bond exists in the presence of arene ligands. Examples are $CpFe(CO)_2R$ (where R may be another C_5H_5 that is σ-bonded through only one carbon atom), Cp_2ZrMe_2, and $CpMo(CO)_3R$.

   It may also be noted that octahedral compounds of Cr^{IV}, Co^{III}, and Rh^{III}, such as $CrPh_3(THF)_3$, $[CrR(H_2O)_5]^{2+}$, and $[RhC_2H_5(NH_3)_5]^{2+}$ are stable because these metal ions are coordinately saturated and inert to substitution reactions.

2. *Binuclear with Metal—Metal Bonds.* Examples are $Li_2Re_2(CH_3)_8$, $Mo_2(CH_2SiMe_3)_6$, and $Ru_2(CH_2CMe_3)_6$ in which there are quadruple or triple M to M bonds.

3. *Bi- or Polynuclear with Alkyl Bridges.* Two examples are the following:

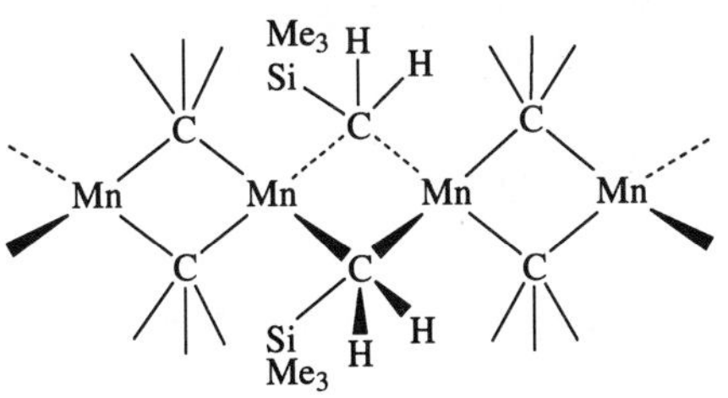

(a) 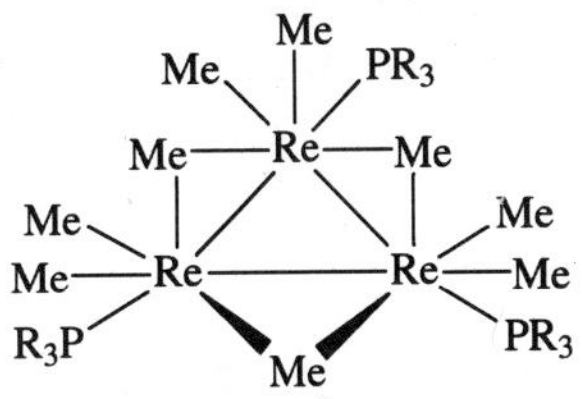   (b)

4. *Alkylate Anions.* Alkyls or aryls that are coordinately unsaturated can combine with additional alkylate anions to give anions such as $[Co(C_6F_5)_4]^{2-}$, $[ThMe_7]^{3-}$, and $[CpLuMe_3]^{2-}$. These are often obtained by using lithium alkyls in the presence of tmen and crystallize as $[Li\ tmen]^+$ salts. The alkylate anions of copper are extensively used in organic syntheses.

5. *Metallacyles.* These may have either saturated or unsaturated ring systems.

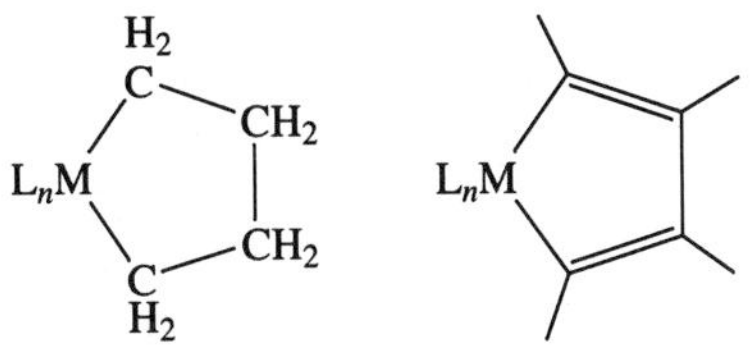

The saturated metallacycles can be made using dimagnesium reagents $XMg(CH_2)_nMgX$ but also by coupling of two ethylene molecules:

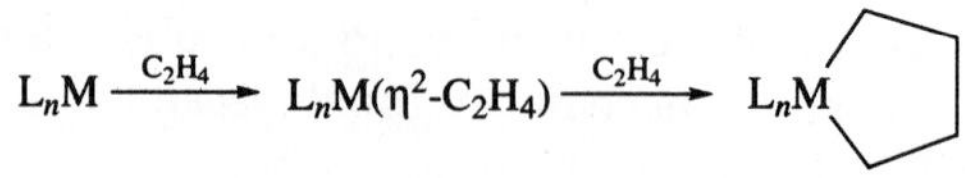

$$L_nM \xrightarrow{C_2H_4} L_nM(\eta^2\text{-}C_2H_4) \xrightarrow{C_2H_4} L_nM \overset{\Large\bigcirc}{}$$

Similarly, coupling of two alkynes can form the diene metallacycles.

Another important class are those obtained by *orthometallation* of phenyl rings, as in the following general reaction:

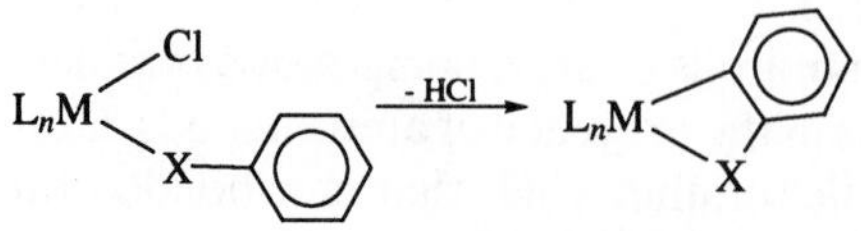

6. *Acetylides.* These are numerous and often quite stable, and are easily formed from the lithium reagents such as $PhC{\equiv}CLi$. Examples are $PhCC{-}Ru_2(2\text{-}$anilinopyridine), $CpCr(CO)_3CCPh$, and $[Ni(CCR)_4]^{2-}$.

7. *Fluoroalkyls and Aryls.* Many of these are known with C_6F_5, CF_3, C_2F_4H, and so on. They are often more stable than their hydrocarbon analogues, not so much because the $M{-}C$ bonds are stronger (though this is probably so) but due to the lack of decomposition pathways entailing hydrogen transfers.

8. *Acyls.* Compounds with an acyl or formyl group are especially important as they are formed by reactions of CO such as

$$(CO)_5MnCH_3 + CO \longrightarrow (CO)_5MnC(O)CH_3$$

that are discussed in Chapter 21. They can also sometimes be made by oxidative-addition of aldehydes:

$$RhCl(PMe_3)_3 + RCHO \longrightarrow (Me_3P)_3ClRh(H)(RCO)$$

Acylate anions can be made by nucleophilic attacks of lithium alkyls on metal carbonyls, for example, $W(CO)_6 + LiR \rightarrow Li[(CO)_5WC(O)R]$. They are precursors of alkylidenes which will be discussed shortly.

Acyl groups have several bonding modes in addition to η^1 as follows:

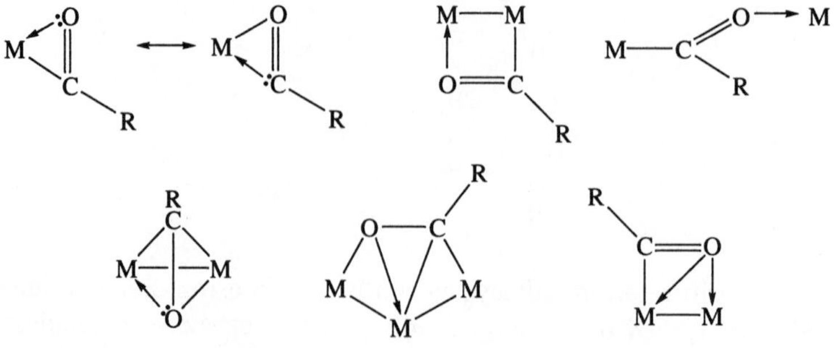

Metal-to-Carbon Double Bonds

Such bonds were first prepared by E. O. Fischer in compounds with the metal atoms in low oxidation states and 18-electron configurations, as in the following reaction:

$$Cr(CO)_6 + LiR \longrightarrow \left[(OC)_5Cr\!\!<\!\!\begin{array}{c}O^- \\ R\end{array}\right] \xrightarrow{R'_3O^+} (OC)_5Cr\!=\!\!C\!\!<\!\!\begin{array}{c}OR' \\ R\end{array}$$

They may be viewed as having a RR'C: ligand replacing a CO group, and they are therefore called *carbene complexes*, or, more simply *(metal) carbenes*.

Later Schrock showed that compounds in high oxidation states, for example, $Cl_2H(PR_3)_3Ta^V(CHCMe_3)$, with fewer than 18 electrons could be made by α-hydrogen loss from an alkyl group. These have been termed *alkylidene complexes*. Informally, the two types are often called *Fischer carbenes* and *Schrock carbenes*. The IUPAC recommends calling them all *alkylidene complexes*, with the term carbene being restricted to a free CR_2 species.

The low-valent alkylidenes are sometimes represented simply as having $M=C$ bonds, but other canonical forms evidently are involved:

$$M\!=\!\!C\!\!<\!\!\begin{array}{c}R' \\ R\end{array} \longleftrightarrow \overset{+}{M}\!:\!\overset{-}{C}\!\!<\!\!\begin{array}{c}R' \\ R\end{array} \longleftrightarrow \overset{-}{M}\!:\!C\!\!<\!\!\begin{array}{c}\overset{+}{R}' \\ R\end{array} \longleftrightarrow \overset{-}{M}\!:\!C\!\!<\!\!\begin{array}{c}R' \\ R^+\end{array}$$

Many of the known complexes have OR or NR_2 groups attached to carbon, and electron flow from lone pairs on O or N leads to a strong contribution in which there is O—C or N—C $p\pi \to p\pi$ bonding. However, several complexes without such groups are known.

The structures of many of the low-valent alkylidenes are known and they support the above resonance picture. The M—CRR' group is always planar and the M—C distances as well as C—O and C—N distances are shortened in accord with the π-bonding shown:

$$cis\text{-}(OC)_4(Ph_3P)Cr \xrightarrow{2.00(2)} C\!\!<\!\!\begin{array}{c}\overset{1.32(2)}{OCH_3} \\ \underset{1.53(3)}{CH_3}\end{array} \qquad (OC)_5Cr \xrightarrow{2.16(1)} C\!\!<\!\!\begin{array}{c}\overset{1.31(1)}{N(CH_3)_2} \\ \underset{1.50(1)}{CH_3}\end{array}$$

The high-valent alkylidenes are formed mostly by the elements Nb, Ta, Mo, and W. Some typical preparative reactions are:

$$(Me_3CCH_2)_3TaCl_2 \xrightarrow{Me_3CCH_2Li} (Me_3CCH_2)_3Ta\!=\!\!C\!\!<\!\!\begin{array}{c}H \\ CMe_3\end{array}$$

$$(\eta\text{-}C_5H_5)_2TaMe_3 \xrightarrow{Ph_3C^+BF_4^-} (\eta\text{-}C_5H_5)_2TaMe_2^+ \xrightarrow{NaOMe} (\eta\text{-}C_5H_5)_2Ta\!\!<\!\!\begin{array}{c}CH_3 \\ CH_2\end{array}$$

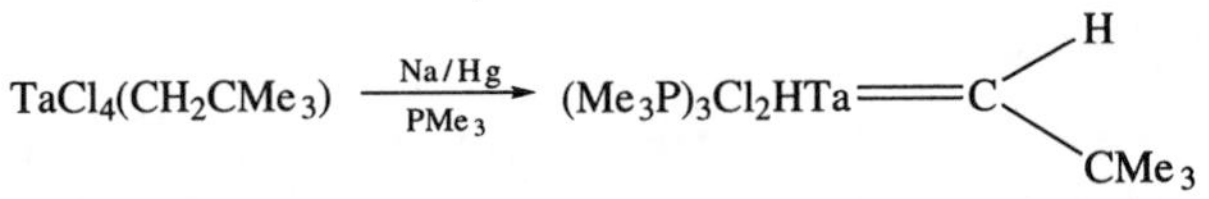

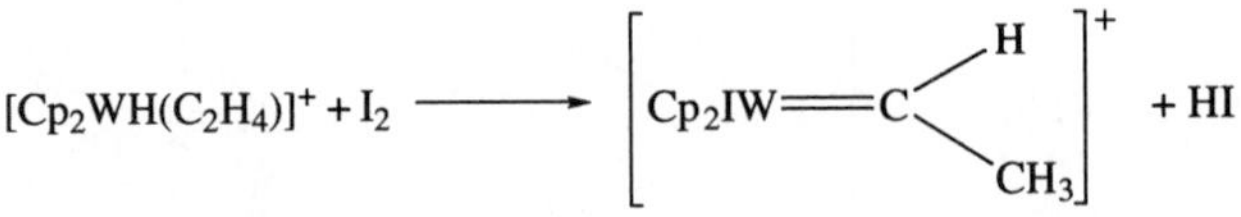

$$TaCl_4(CH_2CMe_3) \xrightarrow[PMe_3]{Na/Hg} (Me_3P)_3Cl_2HTa=\!\!=\!\!C\!\!\begin{array}{c} H \\ \diagup \\ \diagdown \\ CMe_3 \end{array}$$

$$[Cp_2WH(C_2H_4)]^+ + I_2 \longrightarrow \left[Cp_2IW=\!\!=\!\!C\!\!\begin{array}{c} H \\ \diagup \\ \diagdown \\ CH_3 \end{array} \right]^+ + HI$$

As an extension of the alkylidene complexes, there are *metallocumulenes*, examples of which are shown as 16-XXVII, 16-XXVIII, and 16-XXIX.

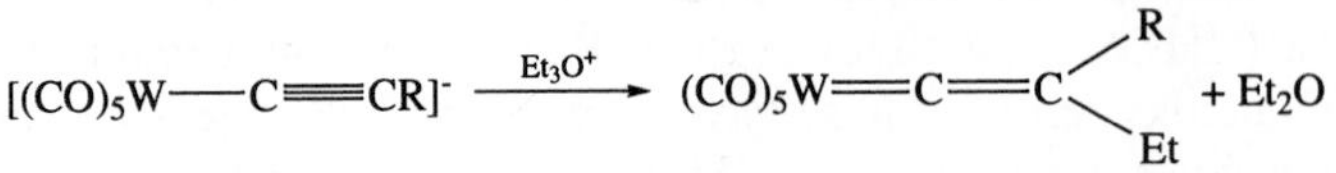

$$[(CO)_5W\!\!-\!\!C\!\!\equiv\!\!CR]^- \xrightarrow{Et_3O^+} (CO)_5W=\!\!=\!\!C=\!\!=\!\!C\!\!\begin{array}{c} R \\ \diagup \\ \diagdown \\ Et \end{array} + Et_2O$$

(16-XXVII)

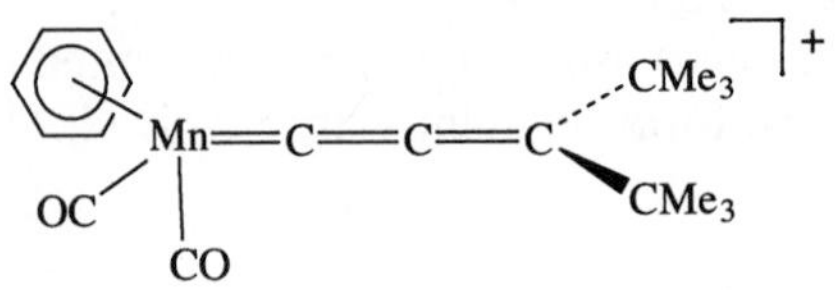

(16-XXVIII)

$$(CO)_5W=\!\!=\!\!C=\!\!=\!\!C=\!\!=\!\!CR_2$$

(16-XXIX)

Metal-to-Carbon Triple Bonds

Collectively called *alkylidynes*, these again comprise both low- and high-valent types. The former, often called carbynes, are obtained from the "carbenes" by reactions such as:

$$cis\text{-}(OC)_4(Me_3P)Cr=\!\!CMe(OMe) \xrightarrow[pentane]{BX_3} (OC)_3(Me_3P)XCr\equiv CMe + CO + (BX_2OMe)$$

Structural studies of these complexes show that in some cases the $M-C-C$ group is not linear [e.g., *trans*-$I(CO)_4W\equiv CPh$ has an angle of 162°], whereas in other compounds [e.g., *trans*-$I(CO)_4Cr\equiv CCH_3$] the $Cr-C-C$ group is linear. The force constants for $M\equiv C$ bonds are comparable to those for $M\equiv N$ compounds.

The high-valent alkylidynes are obtained in deprotonation of alkylidenes having a $C-H$ group as in the reaction (dmp = N,N'-dimethylpiperazine)

$$(Me_3CCH_2)_3Ta=\!\!C(H)CMe_3 \xrightarrow[dmp]{BuLi} (Li\,dmp)[(Me_3CCH_2)_3Ta\equiv CCMe_3]$$

or by direct α-elimination *via* a methyl compound:

$$WCl_2(PMe_3)_4 \xrightarrow{AlMe_3} trans\text{-}W(CH)Cl(PMe_3)_4$$

Other syntheses include electrophilic attacks on isocyanides:

$$(dppe)_2ClRe(CNR) \xrightarrow{\ H^+\ } [(dppe)_2ClRe \equiv CN(H)R]^+$$

Many different types of compounds are now known, examples being $(RO)_3WC$-Bu^t, $[Cl_4WCBu^t]^-$, $(Bu^tCH_2)_3WCBu^t$, and *trans*-$(Me_3P)_2W(CH_2CMe_3)(CHCMe_3)$-$(CCMe_3)$, which has $W-C$, $W=C$, and $W\equiv C$ bonds.

BONDING TO OLEFINS, POLYOLEFINS, AND ALKYNES

Alkene Complexes

Although the first such complex, Zeise's salt, $K[PtCl_3(C_2H_4)]$, was made in 1827, no more appeared until the 1950s. Ethylene itself forms many more such complexes, usually of the same simple η^2 type, but in a few cases compounds such as $[(R_3P)_2Cl_3Zr]_2(\mu,\eta^2\text{-}C_2H_4)$, $Cp_2Y(\mu,\eta^2\text{-}C_2H_4)Pt(PPh_3)_2$, and others[26] where it forms a symmetrical bridge with the C_2H_4 plane perpendicular to the $M\cdots M'$ line. The structures of Zeises salt and two other simple η^2 complexes are shown in Fig. 16-25.

The most generally useful description of the $(\eta^2\text{-}C_2H_4)-M$ bonding was developed first for copper—alkene complexes by M. J. S. Dewar and later extended to other transition metals. Figure 16-26 illustrates the assumption that as with other π-bonding ligands like CO (Section 16-3), there are *two* components to the total bonding: (a) overlap of the π-electron density of the olefin with a σ-type acceptor orbital on the metal atom and (b) a "back-bond" resulting from flow of electron density from filled metal d_{xz} or other $d\pi-p\pi$ hybrid orbitals into *antibonding* orbitals of the carbon atoms. This view is thus similar to that discussed for the bonding of carbon monoxide and similar weakly basic ligands and implies the retention of appreciable "double-bond" character in the olefin. Of course, the donation of π-bonding electrons to the metal σ orbital and the introduction of electrons into the π-antibonding orbital both weaken the π bonding in the olefin, and in every

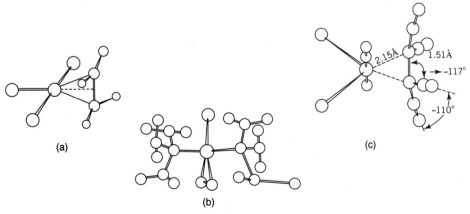

(a)

(b)

(c)

Figure 16-25 (a) The structure of the ion in Zeise's salt, (b) the structure of *trans*-$RhCl(C_2H_4)(PPr^i_3)_2$, (c) the structure of the tetracyanoethylene complex $IrBr(CO)$-$[(CN)_2C=C(CN)_2](PPh_3)_2$.

[26]P. Royo *et al.*, Organometallics **1997,** *16,* 1553; F. A. Cotton and P. A. Kibala, *Inorg. Chem.,* **1990,** *29,* 3192.

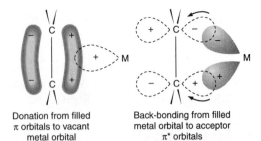

Figure 16-26 The molecular orbital view of alkene—metal bonding according to Dewar.

Donation from filled π orbitals to vacant metal orbital

Back-bonding from filled metal orbital to acceptor π* orbitals

case except the anion of Zeise's salt there is significant lengthening of the olefin C—C bond.

In complexes containing tetracyanoethylene (Fig. 16-25c) or $F_2C=CF_2$, the C—C bond is about as long as a normal single bond and the angles within the $C_2(CN)_4$ or C_2F_4 ligand suggest that the carbon atoms bound to the metal approach tetrahedral hybridization. Indeed, it is possible to formulate the bonding as involving two normal 2c-2e metal—carbon bonds in a metallacycle with approximately sp^3 hybridized carbon.

Actually, the metallacycle view and the π-donor view are neither incompatible nor mutually exclusive but are complimentary, with a smooth graduation of one description into the other. The one to be preferred in any given case depends on the extent to which the double bond of the ligand has been reduced to a single bond. From a formal point of view, however, the metallacycle view entails a problem with oxidation state. For example, a compound such as $Ni(C_2F_4)(CO)_3$ could be regarded as a nickel(II) rather than a nickel(0) complex. Clearly, in a compound such as $Pt(C_2H_4)_3$, it would be absurd to propose Pt^{VI}. It is, in general, best to regard molecules bound sideways as neutral ligands that do not alter the formal oxidation state.

Complexes formed by *non-conjugated polyenes* can be regarded as having two or more non-interacting $(\eta^2\text{-}C_2H_4)$—M type bonds. A representative case is $(COD)PtCl_2$ (16-XXX).

(16-XXX)

With *conjugated polyenes*, the situation is more complex, although qualitatively the two types of basic, synergic components are involved. The case of the 1,3-butadiene unit is an important one and shows why it would be a drastic oversimplification to treat such cases as simply collections of separate monoolefin—metal interactions.

Two extreme formal representations of the bonding of 1,3-butadiene to a metal atom are possible (Fig. 16-27). The structure (b) would imply that bonds 1-2 and 3-4 should be longer than bond 2-3. In $C_4H_6Fe(CO)_3$ the bond lengths are approximately the same and ^{13}C—H coupling constants in the nmr spectra indicate that the hybridization at carbon still approximates to sp^2. However, in some other compounds of conjugated cyclic alkenes, the pattern is of the long—short—long type, indicating some contribution from this extreme structure.

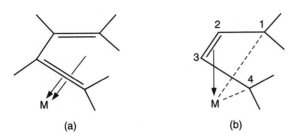

Figure 16-27 Two extreme formal representations of the bonding of a 1,3-butadiene group to a metal atom: (a) implies that there are two more or less independent monoolefin metal interactions; (b) depicts σ bonds to C-1 and C-4 coupled with a monoolefin metal interaction to C-2 and C-3.

In addition to the bonding mode just discussed (which is much the most common), butadiene sometimes bonds in other ways, e.g., (16-XXXI).

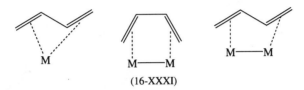

(16-XXXI)

It is also possible for one or two metal atoms to bond to a subset of a chain of conjugated double bonds, as illustrated in (16-XXXII), all involving cyclooctatetraene.

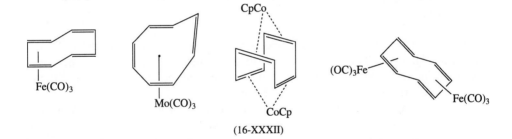

(16-XXXII)

Alkyne Complexes

The presence of two mutually perpendicular π bonds in an alkyne makes for considerably more varied and complicated ligand behavior than that shown by alkenes. There is, of course, the simple use of one π bond and the associated π^* orbital in exactly the same way as found in olefin complexes. This results in a lengthening of the C—C distance and a marked deviation from linearity. Typical results are shown schematically in Fig. 16-28.

An alkyne can also behave as a formal four-electron donor to one metal atom. In addition to the type of interaction just discussed, the other pair of π electrons may be partially donated to a metal $d\pi$ orbital lying perpendicular to the plane of

the three-membered $M \begin{smallmatrix} \diagup C \\ \mid \\ \diagdown C \end{smallmatrix}$ ring. Again, a synergistic back-donation may also occur

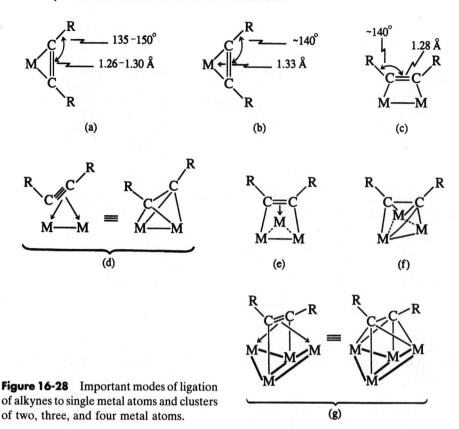

(a)   (b)   (c)

(d)   (e)   (f)

(g)

Figure 16-28   Important modes of ligation of alkynes to single metal atoms and clusters of two, three, and four metal atoms.

if there is a filled *d* orbital perpendicular to *both* of those already used. This is shown in Fig. 16-29. This typically leads to even greater lengthening of the C—C distance. Figure 16-28(b) shows typical dimensions and indicates how an additional arrow is used to represent this four-electron bonding.

When acting as a ligand toward two or more metal atoms, alkynes have a large repertoire of roles. With respect to a dimetal unit, there are two major types of structures, both describable as η^2, μ_2; they are shown in Fig. 16-28(c) and (d). In the planar structure of Fig. 16-28(c), the alkyne, is a two-electron donor. In structure

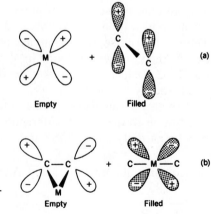

Figure 16-29   The orbitals employed in the second synergic alkyne-to-metal bond.

(d), which is a distorted tetrahedral shape, the alkyne is a four-electron donor. The bonding in Fig. 16-28(d) can be envisioned as donation of one pair of π electrons to each metal atom (with appropriate back-donation) or as the formation of two single bonds to each metal atom. The latter view is certainly too extreme since the C—C distances in these cases are in the range 1.30 to 1.35 Å.

With trinuclear sets of metal atoms there are again two principal geometries, Fig. 16-28(e) and (f), in each of which the alkyne is a formal four-electron donor. Finally, we note that towards a tetranuclear set of metal atoms, the geometry shown in Fig. 16-28(g), where the metal atoms have the nonplanar "butterfly" arrangement is common. Here, again, the alkyne is a formal four-electron donor.

Open Enyl Ligands

Anions of the type (16-XXXIII) can interact with a metal atom (or atoms) to serve as $(2n+1)$-electron donors. The two cases of most importance are allyl and pentadienyl groups. The allyl ligand can, of course, be monodentate, (16-XXXIV(a)) but more interestingly can be bonded in the η^3 fashion, (16-XXXIV(b)).

$$H(CH{=\!=}CH)_nCH_2^-$$

(16-XXIII)   (a) 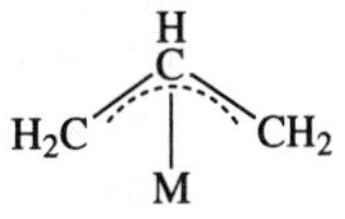   (b)

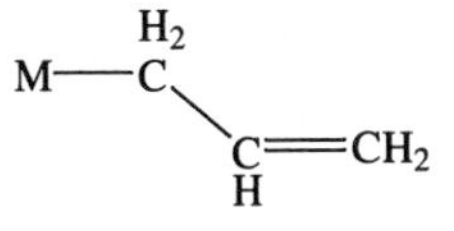

(16-XXXIV)

The η^2-pentadienyl ligand, as a five-electron donor in its semicircular, delocalized form represented schematically in 16-XXXV is capable of forming a variety of metal complexes, including a so-called "open" ferrocene, $(C_5H_7)_2Fe$.

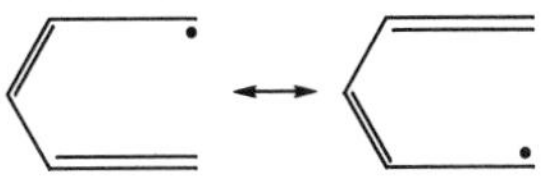

(16-XXXV)

It may also be noted that η^3-allyl and η^5-pentadienyl units are often found within rings, especially six-membered rings, namely,

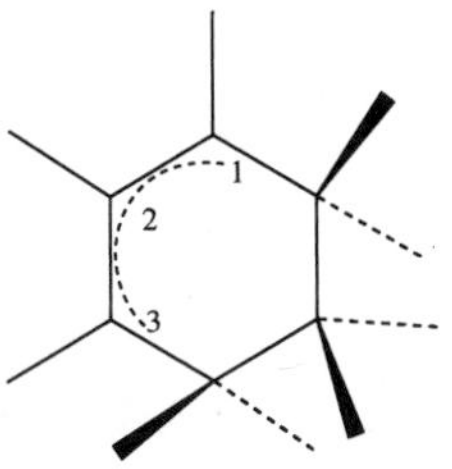

η^3-allyl 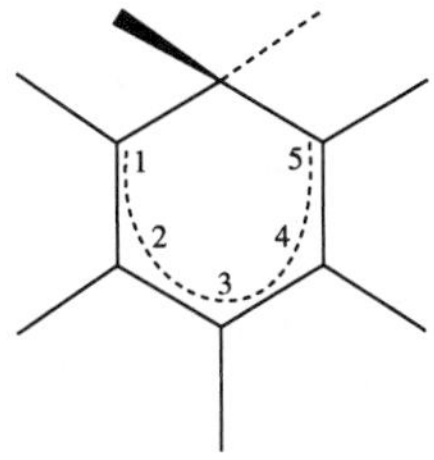 η^5-dienyl

Arene Complexes

Just as the π electrons of alkenes can interact with metal d orbitals, so can certain of the delocalized π-electron ring systems of aromatic molecules overlap with d_{xz} and d_{yz} metal orbitals.

The first example of this type of complex was the molecule $Fe(C_5H_5)_2$, now known as *ferrocene*, in which the 6π-electron system of the ion $C_5H_5^-$ is bound to the metal. Other aromatic systems with the "magic numbers" of 2, 6, and 10 for the aromatic electronic configuration are the carbocycles:

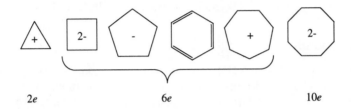

The C_5H_5, C_6H_6, and C_8H_8 rings are the most common in complexes, but the C_7H_7 and C_4H_4 systems also occur frequently. It should also be noted that for purposes of electron counting the ring system and the metal atom may be considered as neutral. For example, the total of 18 electrons in ferrocene can be regarded as 5 per C_5H_5 ring plus 8 from Fe.

Compounds are known that have only π-bonded rings such as ferrocene (16-XXXVI), dibenzenechromium (16-XXXVII), and $(C_8H_8)_2U$ (16-XXXVIII), but there are many compounds with one ring and other ligands such as halogens, CO, RNC, and R_3P. Examples are η-$C_5H_5Mn(CO)_3$ and η-$C_5H_5Fe(CO)_2Cl$. The symbol η is used to signify that all carbon atoms of the ring are bonded to the metal atom. There are also molecules in which two different types of arene ring are present, such that the total number of π electrons they provide, plus those possessed by the metal atom itself, add to 18. For example, in (16-XXXIX), there are five π electrons from C_5H_5, four from C_4R_4, and nine from Co. Similarly, we have $(\eta$-$C_5H_5)$-$(\eta$-$C_6H_6)$Mn.

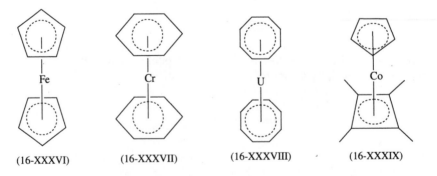

(16-XXXVI)   (16-XXXVII)   (16-XXXVIII)   (16-XXXIX)

It is also possible for heterocyclic arene rings to form complexes, examples being $(\eta$-$C_4H_4N)Mn(CO)_3$, $(\eta$-$C_4H_4S)Cr(CO)_3$, $(\eta$-$C_5H_5)(\eta$-$C_4H_4N)Fe$, $(\eta$-$C_5H_5)$-$(\eta$-$C_4H_4P)Fe$, and (16-XL).

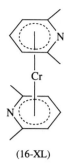

(16-XL)

The basic qualitative features of the bonding in ferrocene are well understood, and will serve to illustrate the basic principles for all (η-C_nH_n)M bonding. The discussion of bonding does not depend critically on whether the preferred rotational orientation of the rings (see Fig. 16-30) in an (η-C_5H_5)$_2$M compound is staggered (D_{5d}) or eclipsed (D_{5h}); in any event, the barriers to ring rotation in all types of arene-metal complex are very low, *ca.* 10-20 kJ mol^{-1}.

The bonding is best treated in the linear combination of atomic orbitals (LCAO—MO) approximation. A semiquantitative energy level diagram is given in Fig. 16-31. Each C_5H_5 ring, taken as a regular pentagon, has five π MO's, one strongly bonding (a), a degenerate pair that are weakly bonding (e_1), and a degenerate pair that are markedly antibonding (e_2), as shown in Fig. 16-32. The pair of rings taken together then has ten π orbitals and, if D_{5d} symmetry is assumed, so that there is a center of symmetry in the (η-C_5H_5)$_2$M molecule, there will be symmetric (g) and antisymmetric (u) combinations. This is the origin of the set of orbitals shown on the left of Fig. 16-31. On the right are the valence shell ($3d$, $4s$, $4p$) orbitals of the iron atom. In the center are the MO's formed when the ring π orbitals and the valence orbitals of the iron atom interact.

For (η-C_5H_5)$_2$Fe, there are 18 valence electrons to be accommodated: 5 π-electrons from each C_5H_5 ring and 8 valence shell electrons from the iron atom. It will be seen that the pattern of MO's is such that there are exactly 9 bonding or nonbonding MO's and 10 antibonding ones. Hence the 18 electrons can just fill the bonding and nonbonding MO's, giving a closed configuration. Since the occupied orbitals are either of a type (which are each symmetric around the 5-fold molecular axis) or they are *pairs* of e_1 or e_2 type, which are also, in pairs, symmetrical about the axis, no intrinsic barrier to internal rotation is predicted. The very low barriers observed may be attributed to van der Waals forces directly between the rings.

Figure 16-31 indicates that among the principal bonding interactions is that giving rise to the strongly bonding e_{1g} and strongly antibonding e_{1g}^* orbitals. To give one concrete example of how ring and metal orbitals overlap, the nature of this

Figure 16-30 Staggered and eclipsed configurations of an (η-C_5H_5)$_2$M compound. In crystalline ferrocene there are molecules of different orientations randomly distributed throughout the crystal.

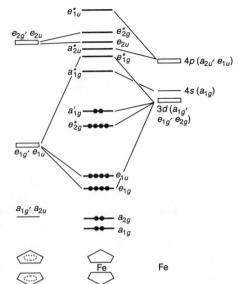

Figure 16-31 An approximate MO diagram for ferrocene. Different workers often disagree about the exact order of the MO's; the order shown here, especially for the antibonding MO's, may be incorrect in detail, but the general pattern is widely accepted.

particular important interaction is illustrated in Fig. 16-33. This particular interaction is in general the most important single one because the directional properties of the e_1-type d orbitals (d_{xz}, d_{yz}) give an excellent overlap with the e_1-type ring π orbitals, as Fig. 16-33 shows.

Systems containing only one η-C_5H_5 ring include (η-C_5H_5)Mn(CO)$_3$, (η-C_5H_5)Co(CO)$_2$, (η-C_5H_5)NiNO, and (η-C_5H_5)CuPR$_3$. The ring-to-metal bonding in these cases can be accounted for by a conceptually simple modification of the picture given previously for (η-C_5H_5)$_2$M systems. In each case a principal axis of

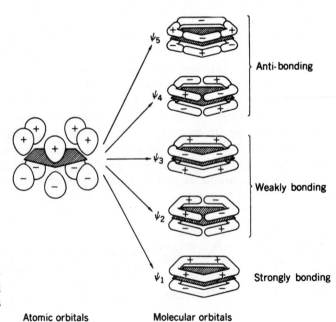

Figure 16-32 The π molecular orbitals formed from the set of $p\pi$ orbitals of the C_5H_5 ring.

Atomic orbitals   Molecular orbitals

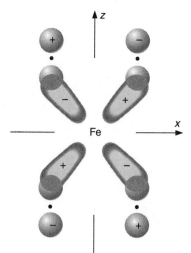

Figure 16-33 Overlapping of one of the e_1-type d orbitals d_{xz} with an e_1-type π orbital to give a delocalized metal-ring bond: cross-sectional view taken in the xz plane.

symmetry can be chosen so as to pass through the metal atom and intersect the ring plane perpendicularly at the ring center; in other words, the C_5H_5M group is a pentagonal pyramid, symmetry C_{5v}. The single ring may then be considered to interact with the various metal orbitals in about the same way as do each of the rings in the sandwich system. The only difference is that opposite to this single ring is a different set of ligands which interact with the opposite lobes of, for example, the de_1 orbitals, to form their own appropriate bonds to the metal atom.

Turning now to a few specific examples of arene-metal compounds, we note that those containing $R_3C_3^+$ are very rare, while those of R_4C_4 are more numerous, though still not common. Butadiene itself is antiaromatic and unstable in the free state but can be stabilized by bonding to a metal atom with a suitable electron configuration. The first such compound was the Ni compound obtained as shown in Fig. 16-34. This chloride can then be reduced in the presence of bipyridine to give the Ni^0 complex $(Me_4C_4)Nibipy$. A true sandwich compound has been obtained as a blue solid resistant to air, water, and carbon monoxide:

$$\xrightarrow{\text{Li, NiBr}_2} Ni(C_4Ph_4)_2$$

Unsubstituted compounds can be made by reactions such as

$$\xrightarrow{\text{Na}_2\text{Mo(CO)}_5} C_4H_4Mo(CO)_4$$

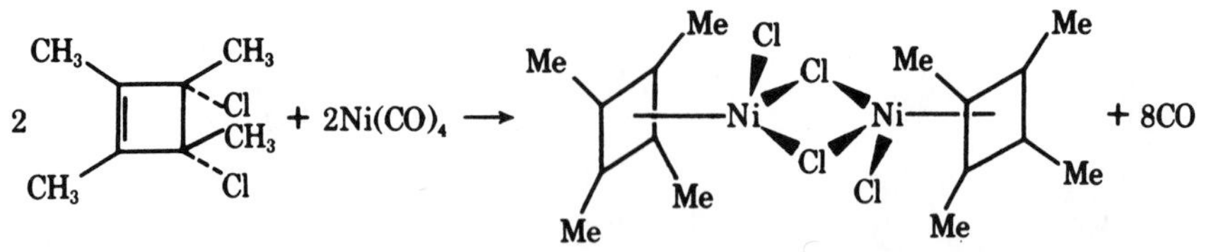

Figure 16-34   The preparation and structure of $[(CH_3C)_4NiCl_2]_2$.

Like the C_5H_5 ring in ferrocene and $CpMn(CO)_3$ (see later), the C_4H_4 ring undergoes typical aromatic electrophilic substitution reactions. There is quite an extensive organic chemistry of the derivatives. Finally, it is possible to obtain free cyclobutadiene for use in *in situ* organic reactions by oxidative decomposition:

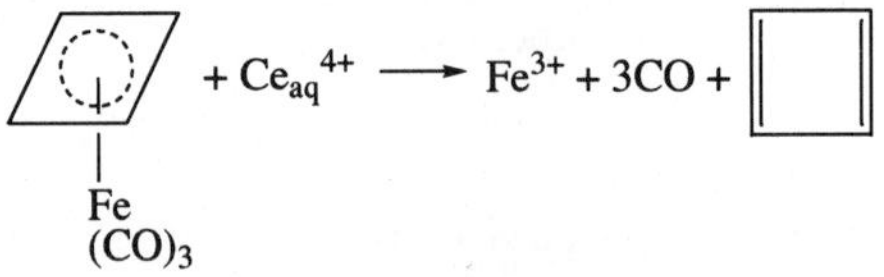

Cyclopentadienyl compounds are by far the most important of all carbocyclic π complexes, and the C_5H_5 ligand or substituted derivatives such as MeC_5H_4 or Me_5C_5 are widely used. The C_5H_5 group is commonly abbreviated as Cp, Me_5C_5 as Cp*, and other substituted species as Cp' or Cp".

The synthetic methods for C_5H_5 compounds depend on the fact that cyclopentadiene is a weak acid ($pK_a \sim 20$) and with strong bases gives salts of the symmetrical cyclopentadienide ion, $C_5H_5^-$.

The most general method is the formation of the sodium salt by action of Na or NaH on C_5H_6 in THF and the subsequent reaction of this solution with metal halides, carbonyls, and so on,

$$2C_5H_6 + 2Na = 2C_5H_5^- + 2Na^+ + H_2 \qquad \text{(main reaction)}$$

$$3C_5H_6 + 2Na = 2C_5H_5^- + 2Na^+ + C_5H_8$$

$$FeCl_2 + 2C_5H_5Na = (\eta\text{-}C_5H_5)_2Fe + 2NaCl$$

$$W(CO)_6 + C_5H_5Na = Na^+[(\eta\text{-}C_5H_5)W(CO)_3]^- + 3CO$$

Excess of a strong organic base can also be used as an acceptor for HCl, for example,

$$2C_5H_6 + CoCl_2 + 2Et_2NH \xrightarrow{THF} Cp_2Co + 2Et_2NH_2Cl$$

Some direct interactions of C_5H_6 or dicyclopentadiene with metals or metal carbonyls gives the complex, for example,

$$2C_5H_6(g) + Mg(s) \xrightarrow{\Delta} Cp_2Mg + H_2$$

$$Fe(CO)_5 + C_{10}H_{12} \longrightarrow [CpFe(CO)_2]_2$$

Di-η^5-cyclopentadienyls

Since the anion functions as a uninegative ligand, the compounds are of the type $[Cp_2M]X_{n-2}$, where the oxidation state of the metal M is n and X is a uninegative ion. Hence in the II oxidation state we obtain neutral, sublimable, and organic solvent-soluble molecules like Cp_2Fe and Cp_2Cr, and in III, IV, and V oxidation states species such as Cp_2Co^+, Cp_2TiCl_2, and Cp_2NbBr_3, respectively. Some representative compounds are given in Table 16-3.

Neutral compounds are known for V, Cr, Mn, Fe, Co, and Ni. The Mn and Ti compounds are anomalous. Although Cp_2Mn does have the sandwich structure, it behaves as an ionic cyclopentadienide and is, for example, miscible in all proportions with Cp_2Mg, whereas Cp_2Fe and the others are not.

The neutral species can undergo electron transfer reactions to give species such as Cp_2Cr^- or Cp_2Cr^+ and ferrocene is often used as a standard in electrochemical studies.

Except for Cp_2Ru and Cp_2Os the second and third row elements do not form stable isolable Cp_2M compounds and conventional syntheses may lead to hydrides such as Cp_2ReH or Cp_2WH_2 by H abstraction. A "rhenocene" is a dimer with a

Table 16-3 Some Di-η^5-cyclopentadienyl Metal Compounds

Compound	Appearance mp (°C)	Unpaired electrons	Other properties
$(\eta^5\text{-}C_5H_5)_2Fe$	Orange crystals; 174	0	Oxidized by Ag^+(aq), dil. HNO_3; $Cp_2Fe = Cp_2Fe^+$ $E^0 = 0.3$ V (vs SCE); stable thermally to >500°C
$(\eta^5\text{-}C_5H_5)_2Cr$	Scarlet crystals; 173	2	Very air sensitive; soluble in HCl giving C_5H_6 and blue cation, probably $[\eta^5\text{-}C_5H_5CrCl(H_2O)_n]^+$
$(\eta^5\text{-}C_5H_5)_2Ni$	Bright green; dec at 173	2	Fairly air stable as solid; oxidized to Cp_2Ni^+; NO gives CpNiNO; Na/Hg in C_2H_5OH gives $CpNiC_5H_7$
$(\eta^5\text{-}C_5H_5)_2Co^+$	Yellow ion in aqueous solution	0	Forms numerous salts and a stable, strong base (absorbs CO_2 from air); thermally stable to ~400°C
$(\eta^5\text{-}C_5H_5)_2TiCl_2$	Bright red crystals; 230	0	Slightly soluble in H_2O giving Cp_2TiOH^+; C_6H_5Li gives $Cp_2Ti(C_6H_5)_2$; reducible to Cp_2TiCl; Al alkyls give polymerization catalyst
$(\eta^5\text{-}C_5H_5)_2 WH_2$	Yellow crystals; 163	0	Moderately stable in air, soluble benzene, etc.; soluble in acids giving $Cp_2WH_3^+$ ion

Re—Re bond. Some very reactive species like "Cp$_2$W" may exist as intermediates that have carbenelike behavior and have been trapped in matrices at low temperature.

Many studies on pentamethylcyclopentadienyls have been made, partly because there are no hydrogen atoms on the ring that can be displaced, but more importantly, because the greater steric requirement of the Cp* ring (as Me$_5$C$_5$ is designated) leads to different sorts of chemistry, often enhanced stability. The inductive effect of the methyl groups also changes the redox behavior.

In addition to the di-η^5-cyclopentadienyl compounds, there are many mono-η^5-cyclopentadienyls, such as CpMn(CO)$_3$, CpCo(CO)$_2$, and [CpFe(CO)$_2$]$_2$. Substituted Cp rings, such as *indenyl* and coupled rings, as in *fulvalenes* are also known to form metal complexes, examples being (16-XLI) and (16-XLII).

Mn(CO)$_3$ (OC)$_2$Ru————Ru(CO)$_2$

(16-XLI)   (16-XLII)

Six-membered arene rings also form a very large number of complexes.

The prototype neutral compound dibenzenechromium (C$_6$H$_6$)$_2$Cr (16-XLIII) can be obtained from the Grignard reaction of CrCl$_3$, but a more effective method of wider applicability to other metals is the direct interaction of an aromatic hydrocarbon and metal halide in the presence of Al powder as a reducing agent and halogen acceptor and AlCl$_3$ as a Friedel-Crafts-type activator. Although the neutral species are formed directly in the case of Cr, the usual procedure is to hydrolyze the reaction mixture with dilute acid, which gives the cations, such as (C$_6$H$_6$)$_2$Cr$^+$ and (mesitylene)$_2$Ru$_2^+$, which can be reduced to the neutral molecules by reducing agents such as hypophosphorous acid.

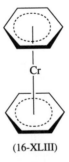

(16-XLIII)

Dibenzenechromium, which forms dark brown crystals, is much more sensitive to air than is ferrocene, with which it is isoelectronic; it does not survive the reaction conditions of aromatic substitution. Structural studies on arene complexes show that the C—C bond lengths are usually equivalent, or nearly so. Many other transition metals, including some lanthanides and uranium form arene complexes, and there are mono complexes such as $C_6H_6Cr(CO)_3$, as well as "sandwiches." Arene to metal bonding can be understood similarly to Cp—M bonding.

As a historical note, the very first arene metal compounds were made as long ago as 1919 by Franz Hein, but were then believed to be "polyphenyl" compounds of chromium, e.g., $Cr(C_6H_5)_3^+$. This is now known to be (16-XLIV).

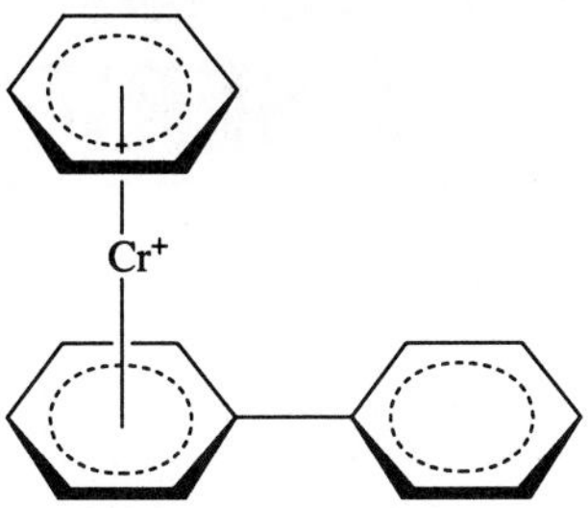

(16-XLIV)

Aromatic heterocycles such as thiophene, pyridine, and pyrrole are also able to form arene complexes, for example $(\eta^5\text{-}C_4H_4N)(\eta^5\text{-}C_5H_5)Fe$ (azaferrocene), $(\eta^5\text{-}C_4H_4S)Cr(CO)_3$, and $(\eta^6\text{-}C_5H_5N)W(CO)_3$.

Arene bridges are also known, as in (16-XLV) and (16-XLVI).

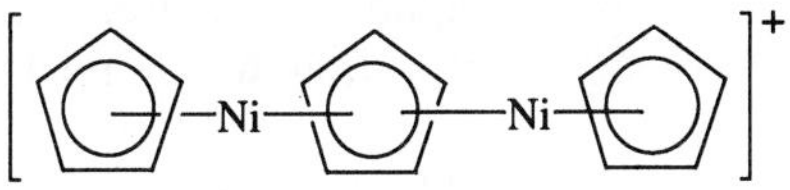

(16-XLV)

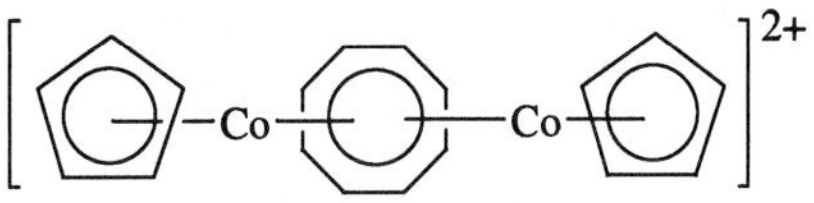

(16-XLVI)

Chapter 17

THE ELEMENTS OF THE FIRST TRANSITION SERIES

GENERAL REMARKS

We discuss in this chapter the elements of the first transition series, titanium through copper. There are two main reasons for considering these elements apart from their heavier congeners of the second and third transition series: (1) in each group (e.g., V, Nb, and Ta) the first-series element always differs appreciably from the heavier elements, and comparisons are of limited use, and (2) the aqueous chemistry of the first-series elements is much simpler, and the use of ligand field theory in explaining both the spectra and magnetic properties of compounds has been far more extensive.

In each of the Sections 17-A through 17-H there is a table to summarize the oxidation states and stereochemistries for each element. There are many Jahn-Teller distortions from perfect geometries, namely, those in octahedral d^1 and d^2 (slight), high-spin octahedral d^4 (two long coaxial bonds), low-spin octahedral d^4 (slight), high-spin octahedral d^6, d^7 (slight), or low-spin octahedral d^7 molecules (two long coaxial bonds), and octahedral d^9 molecules (two long coaxial bonds). These distortions are not specified in the tables.

The energies of the $3d$ and $4s$ orbitals in the neutral atoms are quite similar so that while most configurations are of the $3d^n4s^1$ type, the exchange-energy stabilization of filled and half-filled shells gives $3d^54s^1$ for Cr and $3d^{10}4s^1$ for Cu. When the atoms are ionized, the $3d$ orbitals become appreciably more stable than the $4s$ orbitals and the ions all have $3d^n$ configurations. The high values of third ionization enthalpies (see Appendix 2) account for the difficulty in obtaining Ni^{III} and Cu^{III} compounds, although a few do occur. Copper is the only element in the series to have a M^I state that exists regularly in the absence of π-acceptor ligands.

It should be recognized that while ionization enthalpies give some guidance concerning the relative stabilities of oxidation states, this problem is a very complex one and not amenable to ready generalization. Indeed it is often futile to discuss relative stabilities of oxidation states because some oxidation states may be perfectly stable under certain conditions (e.g., in solid compounds, in fused melts, in the vapor at high temperatures, in absence of air) but nonexistent in aqueous solutions or in air. Thus there is no aqueous chemistry of Ti^{2+} yet crystalline $TiCl_2$ is stable up to ~400°C in the absence of air; also, in fused potassium chloride, titanium and titanium trichloride give Ti^{II} as the main species and Ti^{IV} is in vanishingly small

Table 17-1 Standard Potentials (strongly acid solution) for the +1, +2, and +3
Oxidation States (V)

Couple	Ti	V	Cr	Mn	Fe	Co	Ni	Cu
$M^+ + e = M$								+0.520
$M^{2+} + 2e = M$	a	−1.13	−0.90	−1.18	−0.44	−0.277	−0.257	+0.340
$M^{3+} + 3e = M$	−1.21		−0.74		−0.04			
$M^{3+} + e = M^{2+}$	a	−0.255	−0.424	+1.54	+0.771	+1.92		

^{a}Values have been reported for these but are dubious because Ti^{2+}(aq) has little (if any) stability.

concentrations; on the other hand, in aqueous solutions in air only Ti^{IV} species
are stable.

However, it is sometimes profitable to compare the relative stabilities of ions
differing by unit charge when surrounded by similar ligands with similar stereochemistry, as in the case of the $Fe^{3+} - Fe^{2+}$ potentials (Table 17-1), or with different
anions. In these cases, as elsewhere, many factors are usually involved; some of
these have already been discussed, but they include (*a*) ionization enthalpies of the
metal atoms, (*b*) ionic radii of the metal ions, (*c*) electronic structure of the metal
ions, (*d*) the nature of the anions or ligands involved with respect to their polarizability, donor $p\pi$- or acceptor $d\pi$-bonding capacities, (*e*) the stereochemistry either
in a complex ion or a crystalline lattice, and (*f*) nature of solvents or other media.
In spite of the complexities there are a few trends to be found, namely:

1. From Ti to Mn the highest valence, which is usually found only in oxo
compounds or fluorides or chlorides, corresponds to the total number of *d* and *s*
electrons in the atom. The stability of the highest state decreases from Ti^{IV} to Mn^{VII}.
After Mn (i.e., for Fe, Co, and Ni) the higher oxidation states are difficult to obtain.

2. In the characteristic oxo anions of the valence states IV to VII, the metal
atom is tetrahedrally surrounded by oxygen atoms, whereas in the oxides of valences
up to IV the atoms are usually octahedrally coordinated.

3. The oxides of a given element become more acidic with increasing oxidation
state and the halides more covalent and susceptible to hydrolysis by water.

4. In the II and III states, complexes in aqueous solution or in crystals are
usually in either four- or six-coordination and, across the first series, generally
similar in respect to stoichiometry and chemical properties.

5. The oxidation states <II, except for Cu^I, are found only with π-acid-type
ligands or in organometallic compounds.

Finally, we reemphasize that the occurrence of a given oxidation state as well
as its stereochemistry depend very much on the experimental conditions and that
species that cannot have independent existence under ordinary conditions of temperature and pressure in air may be the dominant species under other conditions.
In this connection we note that transition metal ions may be obtained in a configuration that is difficult to produce by other means through incorporation by isomorphous substitution in a crystalline host lattice, for example, tetrahedral Co^{3+} in
other oxides, tetrahedral V^{3+} in the $NaAlCl_4$ lattice, as well as by using rigid ligands
such as phthalocyanins.

Although some discussion of the relationships between the first, second, and third transition series is useful, we defer this until the next chapter.

In the discussion of individual elements we have kept to the traditional order; that is, elemental chemistries are considered separately, with reference to their oxidation states. However, it is possible to organize the subject matter from the standpoint of the d^n electronic configuration of the metal. This can bring out useful similarities in spectra and magnetic properties in certain cases and has a basis in theory; however the differences in chemical properties of d^n species due to differences in the nature of the metal, its energy levels, and especially the charge on the ion, often exceed the similarities. Nonetheless, such cross considerations (e.g., in the d^6 series V^{-I}, Cr^0, Mn^I, Fe^{II}, Co^{III}, and Ni^{IV}) can provide a useful exercise for students.

Before we consider the chemistries, a few general remarks on the various oxidation states of the first-row elements are pertinent.

The Oxidation States Below II. All the elements form at least a few compounds in the oxidation states I, 0, and -I, but only with ligands of the π-acid type or π-complexing ligands. An exception, of course, is the Cu^I state for copper, where some insoluble binary Cu^I compounds such as $CuCl$ are known, as well as complex compounds. In the absence of complexing ligands the Cu^+ ion has only a transitory existence in water, although it is quite stable in acetonitrile as $[Cu(MeCN)_4]^+$.

Lower formal oxidation numbers are known for a few of the elements. Examples of these "super-reduced" species include $[K(15\text{-}C\text{-}5)_2]_2[Ti(CO)_6]$, $Na_3[Co(CO)_3]$, and $Na_4[Cr(CO)_4]$.

The II State. All the elements Ti to Cu inclusive form well-defined binary compounds in the divalent state, such as oxides and halides, which are essentially ionic. Except for Ti, they form well-defined aqua ions $[M(H_2O)_6]^{2+}$; the potentials are summarized in Table 17-1.

In addition, all the elements form a wide range of complex compounds, which may be cationic, neutral, or anionic depending on the nature of the ligands.

The III State. All the elements form at least some compounds in this state, which is the highest known for copper, and then only in certain complex compounds. The fluorides and oxides are again generally ionic, although the chlorides may have considerable covalent character, as in $FeCl_3$.

The elements Ti to Co form aqua ions, although the Co^{III} and Mn^{III} ones are readily reduced. In aqueous solution certain anions readily form complex species and for Fe^{3+}, for example, one can be sure of obtaining the $Fe(H_2O)^{3+}$ ion only at high acidity (to prevent hydrolysis) and when non-complexing anions such as ClO_4 or CF_3SO_3, are present. There is an especially extensive aqueous complex chemistry of the substitution-inert octahedral complexes of Cr^{III} and Co^{III}.

The trivalent halides, and indeed also the halides of other oxidation states, generally act readily as Lewis acids and form neutral compounds with donor ligands [e.g., $TiCl_3(NMe_3)_2$] and anionic species with corresponding halide ions (e.g., VCl_4^- and FeF_6^{3-}).

The IV State. This is the most important oxidation state of Ti, and the main chemistry is that of TiO_2 and $TICl_4$ and its derivatives. It is also an important state for vanadium, which forms the vanadyl ion VO^2 and many derivatives, cationic, anionic, and neutral, containing the VO group. For the remaining elements Cr to Ni, the IV state is found mainly in fluorides, fluoro complex anions, and cationic complexes; however, an important class of compounds are the salts of the oxo ions and other oxo species.

The V, VI, and VII States. These occur only as $Cr^{V,VI}$, $Mn^{V,VI,VII}$, $Fe^{V,VI}$, and Co^V, and apart from the fluorides CrF_5, CrF_6, and oxofluorides, MnO_3F, the main chemistry is that of the oxo anions $M^nO_4^{(8-n)-}$. All the compounds in these oxidation states are powerful oxidizing agents.

17-A   TITANIUM: GROUP 4

Titanium has four valence electrons $3d^24s^2$. The most stable and common oxidation state is Ti^{IV}; compounds in lower oxidation states, $-I$, 0, II, and III, are quite readily oxidized to Ti^{IV} by air, water, or other reagents. The energy for removal of four electrons is high, so that the Ti^{4+} ion may not exist and Ti^{IV} compounds are generally covalent. In this IV state there are some resemblances to the elements Si, Ge, Sn, and Pb, especially Sn. The estimated ionic radii ($Sn^{4+} = 0.71$, $Ti^{4+} = 0.68$ Å) and the octahedral covalent radii ($Sn^{IV} = 1.45$, $Ti^{IV} = 1.36$ Å) are similar; thus TiO_2 (*rutile*) is isomorphous with SnO_2 (*cassiterite*) and is similarly yellow when hot. Titanium tetrachloride, like $SnCl_4$, is a distillable liquid readily hydrolyzed by water, behaving as a Lewis acid, and giving adducts with donor molecules; $SiCl_4$ and $GeCl_4$ do not give stable, solid, molecular addition compounds with ethers, although $TiCl_4$ and $SnCl_4$ do so—a difference that may be attributed to the ability of the halogen atoms to fill the coordination sphere of the smaller Si and Ge atoms. There are also similar halogeno anions such as TiF_6^{2-}, GeF_6^{2-}, $TiCl_6^{2-}$, $SnCl_6^{2-}$, and $PbCl_6^{2-}$, some of whose salts are isomorphous; the Sn and Ti nitrates, $M(NO_3)_4$, are also isomorphous. There are other similarities such as the behavior of the tetrachlorides on ammonolysis to give amido species. It is a characteristic of Ti^{IV} compounds that they undergo hydrolysis to species with $Ti-O$ bonds, in many of which there is octahedral coordination by oxygen; compounds with $Ti-O-Ti$, $Ti-O-C$, $Ti-O-Si$, and $Ti-O-Sn$ bonds are known.

The stereochemistry of Ti compounds is summarized in Table 17-A-1.

17-A-1   The Element[1]

Titanium is relatively abundant in the earth's crust (0.6%). The main ores are *ilmenite* ($FeTiO_3$) and *rutile*, one of the several crystalline varieties of TiO_2. It is not possible to obtain the metal by the common method of reduction with carbon because a very stable carbide is produced; moreover, the metal is rather reactive toward oxygen and nitrogen at elevated temperatures. Because the metal has uniquely useful properties, however, expensive methods for its purification are justified. In addition to a proprietary electrolytic method, there is the older Kroll

---

[1]D. J. Jones, *Chem. Brit.* **1998,** 1135.

Table 17-A-1   Oxidation States and Stereochemistry of Titanium

Oxidation state	Coordination number	Geometry	Examples
Ti^0, d^4		π-complex	$CpTi(CO)_4^-$
	7		$Ti(CO)_2(PF_3)(dmpe)_2$, $Ti(CO)_4(PMe_3)_3$
Ti^{II}, d^2		π-complex	$Cp_2Ti(CO)_2$, $Ti(\eta^6\text{-}2,6\text{-}Me_2C_5H_3N)_2$
	6^a	Octahedral	$TiCl_2$, *trans*-$TiCl_2(py)_4$
Ti^{III}, d^1	3	Planar	$Ti[N(SiMe_3)_2]_3$
	5	*tbp*	$TiBr_3(NMe_3)_2$
	6^a	Octahedral	TiF_6^{3-}, $Ti(H_2O)_6^{3+}$, $TiCl_3(THF)_3$
	7	Distorted	$TiN(SiMe_3)_2(\eta^2\text{-}BH_4)_2(py)_2$
	9	Distorted	$Ti(BH_4)_3$
Ti^{IV}, d^0	4^a	Tetrahedral	$TiCl_4$
		Distorted tetrahedral	Cp_2TiCl_2
	5	*tbp*	$TiCl_5^-$
		sp	$TiO(porphyrin)$
	6	Octahedral	TiF_6^{2-}, $Ti(acac)_2Cl_2$
			TiO_2, $[Cl_3POTiCl_4]_2$
	7	ZrF_7^{3-} type	$[Ti(O_2)F_5]^{3-}$
	8	Distorted dodecahedral	$TiCl_4(diars)_2$
		Dodecahedral	$Ti(ClO_4)_4$

[a]Most common state.

process in which *ilmenite* or *rutile* is treated at red heat with carbon and chlorine to give $TiCl_4$, which is fractionated to free it from impurities such as $FeCl_3$. The $TiCl_4$ is then reduced with Na or Mg at ~900°C in an atmosphere of argon. The latter gives Ti metal as a spongy mass from which $MgCl_2$ is tapped off and removed periodically; Mg and Cl_2 are then recovered by electrolysis—an important economic feature of the process. The sponge may then be fused in an atmosphere of Ar in an electric arc and cast into ingots.

Extremely pure Ti can be made on a small scale by the van Arkel–de Boer method (also used for other metals) in which pure TiI_4 vapor is decomposed on a hot wire at low pressure.

Below 882.5°C the metal has a hexagonal close-packed lattice. It resembles other transition metals such as Fe and Ni in being hard, refractory (mp 1668°C, bp 3260°C), and a good conductor of heat and electricity. Its density is quite low in comparison to other metals of similar mechanical and thermal properties and it is unusually resistant to corrosion due to a self-healing thin protective oxide coating which reforms almost instantaneously after mechanical damage; therefore, it has come into demand for special applications[2] in turbine engines and industrial, chemical, aircraft, and marine equipment. It is also used in medicine in hip and knee replacements.

Although rather unreactive at ordinary temperatures, Ti combines directly with most nonmetals, for example, H_2, the halogens, O_2, N_2, C, B, Si, and S, at elevated

[2]P. A. Blenkinsop, *J. Phys. IV.* **1993**, *C7*, 161.

temperatures. The nitride (TiN), and borides (TiB and TiB_2) are interstitial compounds that are very stable, hard, and refractory. Highly reflective, gold-colored TiN films can be made by chemical vapor deposition of $TiCl_4(NH_3)_2$.[3]

The metal is not attacked by mineral acids at room temperature or even by hot aqueous alkali. It dissolves in hot HCl, giving Ti^{III} species, whereas hot HNO_3 converts it into a hydrous oxide that is rather insoluble in acid or base. The best solvents are HF or acids to which fluoride ions have been added. Such media dissolve Ti and hold it in solution as fluoro complexes.

TITANIUM COMPOUNDS

17-A-2 The Chemistry of Titanium(IV), d^0

Binary Compounds

Halides. The tetrachloride ($TiCl_4$), the usual starting point for the preparation of most other Ti compounds, is a colorless liquid (mp $-23°C$, bp $136°C$) with a pungent odor. The Ti—Cl bond is short, 2.17 Å, possibly due to π donation $Ti \Leftarrow Cl$. Titanium tetrachloride fumes strongly in moist air and is vigorously, though not violently, hydrolyzed by water:

$$TiCl_4 + 2H_2O \longrightarrow TiO_2 + 4HCl$$

With some HCl present or a deficit of H_2O, partial hydrolysis occurs, giving oxo chlorides (see later).

Titanium tetrabromide[4] and a metastable form of TiI_4 are crystalline at room temperature and are isomorphous with SiI_4, GeI_4, and SnI_4, having molecular lattices. The fluoride is obtained as a white powder by the action of F_2 on Ti at 200°C or by solvothermal decomposition of $(O_2)_4Ti_9F_{30}$; it sublimes readily and is hygroscopic; its structure is dominated by isolated columns of corner-linked TiF_6—octahedra.[5] All the halides behave as Lewis acids; with neutral donors such as ethers they give adducts (see later).

Titanium Oxide;[6] Complex Oxides; Sulfide. The naturally occurring dioxide TiO_2 has three crystal modifications, *rutile*, *anatase*, and *brookite*. In *rutile*, the commonest, the Ti is octahedral and this structure is a common one for MX_2 compounds. In *anatase* and *brookite* there are very distorted octahedra of oxygen atoms about each titanium, two being relatively close. Although *rutile* has been assumed to be the most stable form because of its common occurrence, *anatase* is 8 to 12 kJ mol^{-1} more stable than *rutile*.

The dioxide is used as a white pigment.[7] Naturally occurring forms are usually colored, sometimes even black, owing to the presence of impurities such as iron.

[3]C. H. Winter *et al., Inorg. Chem.* **1994,** *33*, 1227.
[4]S. I. Troyanov *et al., Russ. J. Inorg. Chem.* **1990,** *35*, 494.
[5]B. G. Müller *et al., Z. anorg. allg. Chem.* **1995,** *621*, 1227.
[6]K. I. Hadjiivanov and G. Klissurski, *Chem. Soc. Rev.* **1996,** 61.
[7]A. Baidins *et al., Progr. Org. Coatings* **1992,** *20*, 105.

Pigment-grade material is generally made by hydrolysis of titanium(IV) sulfate solution or vapor-phase reaction of $TiCl_4$ with oxygen. The solubility of TiO_2 depends considerably on its chemical and thermal history. Strongly roasted specimens are chemically inert but under hydrothermal conditions they can be used to prepare zeolites.[8] It reacts also with glycol in the presence of alkali metal hydroxides to yield soluble titanium glycolates.[9] The dioxide impregnated with some metal complexes has been much studied as a catalyst for photodecomposition of water.[10] Ultraviolet irradiation of a gas/solid interface of microcrystalline TiO_2 in the presence of H_2O and CO_2 leads to the formation of CO, H_2, and CH_4.[11] Many other uses are known; examples are found in catalysis[12] and preparation of devices such as film electrodes.[13]

Hydrous titanium dioxide is made by adding base to Ti^{IV} solutions and by action of acids on alkali titanates. The hydrous material dissolves in bases giving hydrous "titanates."

Many materials called "titanates" are known; nearly all have one of the three major mixed metal oxide structures. Indeed the names of two of the structures are those of the Ti compounds that were the first found to possess them, namely, *ilmenite* ($FeTiO_3$) and *perovskite* ($CaTiO_3$). Other titanites with the ilmenite structure are $MgTiO_3$, $MnTiO_3$, $CoTiO_3$, and $NiTiO_3$, and others with the perovskite structure are $SrTiO_3$ and $BaTiO_3$. There are also titanates with the spinel structure such as Mg_2TiO_4, Zn_2TiO_4, and Co_2TiO_4. The Li_4TiO_4 is isostructural with Li_4GeO_4 and contains tetrahedrally coordinated Ti^{IV} ions and tetragonally packed oxide ions;[14] Rb_3NaTiO_4 is isotypic with Rb_3NaPbO_4.[15]

Barium oxide and TiO_2 react to form an extensive series of phases from simple ones such as $BaTiO_3$ (commonly called barium titanate) and Ba_2TiO_4 to $Ba_4Ti_{13}O_{30}$ and $Ba_6Ti_{17}O_{40}$, the general formula being $Ba_xTi_yO_{x+2y}$. All are of technical interest because of their ferroelectric properties, which may be qualitatively understood as follows. The Ba^{2+} ion is so large relative to the small ion Ti^{4+} that the latter can literally "rattle around" in its octahedral hole. When an electric field is applied to a crystal of this material, it can be highly polarized because each of the Ti^{4+} ions is drawn over to one side of its octahedron, thus causing an enormous electrical polarization of the crystal as a whole.

Titanium disulfide, like the disulfides of Zr, Hf, V, Nb, and Ta, has a layer structure; two adjacent close-packed layers of S atoms have Ti atoms in octahedral interstices. These "sandwiches" are then stacked so that there are adjacent layers of S atoms. Lewis bases such as aliphatic amines can be intercalated between these adjacent sulfur layers; similar intercalation compounds can be made with MS_2 and MSe_2 compounds for M = Ti, Zr, Hf, V, Nb, and Ta. Many of these have potentially useful electrical properties, including use as cathode material for lithium batteries, and superconductivity, and may be compared with the intercalation compounds of

[8]X. Liu and J. K. Thomas, *J. Chem. Soc., Chem. Commun.* **1996,** 1435.
[9]L. Lensink *et al., Inorg. Chem.* **1995,** *34,* 746.
[10]See for example: J. T. Yates, Jr. *et al., Chem. Rev.* **1995,** *95,* 735.
[11]I. Kamber *et al., J. Chem. Soc., Chem. Commun.* **1995,** 533.
[12]K. I. Hadjiivanov and D. G. Klissurski, *Chem. Soc. Rev.* **1996,** 61.
[13]T. Gerfin *et al., Prog. Inorg. Chem.* **1997,** *44,* 345.
[14]R. P. Gunawardane *et al., J. Solid State Chem.* **1994,** *112,* 70.
[15]C. Weiss and R. Hoppe, *Z. anorg. allg. Chem.* **1996,** *622,* 603.

graphite (Section 7-2). Thin films of TiS_2 can be prepared by chemical vapor deposition methods from $TiCl_4(HSR)_2$, R = C_6H_{11}, C_5H_9.[16] There are also alkali metal intercalates M^IMY_2 (where M^I is any alkali metal, M = Ti, Zr, Hf, V, Nb, Ta and Y = S, Se, Te). The lithium intercalates can be made very smoothly by treating MY_2 with butyllithium.

Titanium(IV) Complexes

Aqueous Chemistry: Oxo Salts. No aquated Ti^{4+} salts can be isolated, but Ti^{4+} possibly exists in solution though the main species are hydrolyzed. Many studies in different media show that species such as $Ti(OH)^{3+}$, $Ti(OH)_2^{2+}$, or possibly $Ti_2(OH)_6^{2+}$ may be present. In strong acid the TiO^{2+} ion is in equilibrium with $Ti(OH)_2^{2+}$ and $Ti_{(aq)}^{4+}$. At higher pH there are species such as $Ti_3O_4^{4+}$, leading eventually to colloidal or precipitated $TiO_2 \cdot nH_2O$.

There are a number of basic salts that would appear to have the Ti=O group, for example, $TiOSO_4 \cdot H_2O$. However, the latter has infinite zigzag Ti—O—Ti—O chains with the octahedral coordination about Ti completed by H_2O and SO_4^{2-}. Similarly $TiO(acac)_2$ is a dinuclear compound with a $Ti(\mu\text{-}O)_2Ti$ ring, while $(NH_4)_2TiO(C_2O_4)_2H_2O$ contains cyclic tetrameric anions with a central eight-membered $(-Ti-O-)_4$ ring.

However, there are a few true titanyl examples, mostly porphyrins and related complexes that do have Ti=O bonds. The Ti=O bond appears to be short ~1.62 Å while ir bands are in the region 890 to 972 cm^{-1} (Raman, v = 975 in 2 M HCl) consistent with a Ti=O double bond. Octahedral titanyl complexes, such as $(Me_3tacn)Cl_2TiO$, are readily available from Ti^{III} precursors by oxidation with dioxygen in an aprotic solvent.[17] A cationic titanyl complex which is stable in aqueous media[18] can be made according to

$$TiCl_4 + H_2O \xrightarrow[-2\ HCl]{DMSO}$$

The Ti=O group is more basic and more susceptible to electrophilic attack by protons than the corresponding vanadyl complexes leading to Ti—O—Ti[19] and $[Ti=O-Ti-O-Ti-O=Ti]_n$[20] species. Another interesting complex with a Ti—O—Ti motif, $[(NH_3)_5Ti-O-Ti(NH_3)_5]I_4$,[21] is prepared by reaction of TiI_3 with ammonia in the presence of traces of water or oxygen.

An important class of oxo complexes are the titanosilicates which have properties analogous to those of the aluminosilicate clays (Section 8-7). These materials are very good catalysts for hydrocarbon oxidations in the presence of hydrogen

[16]C. H. Winter *et al.*, *Inorg. Chem.* **1993**, *32*, 3807.
[17]K. Wieghardt *et al.*, *Inorg. Chem.* **1994**, *33*, 2462.
[18]M. Enders *et al.*, *Z. Natursforsch.* **1997**, *52b* 496.
[19]P. Comba and A. Merbach, *Inorg. Chem.* **1987**, *26*, 1315.
[20]C. Floriani *et al.*, *Chem. Eur. J.* **1996**, *2*, 1466.
[21]H. Jacobs *et al.*, *Z. anorg. allg. Chem.* **1996**, *622*, 2074.

peroxide, *tert*-butyl peroxyhydroxides and even—in some circumstances—in the presence of oxygen or air. With the exception of *fresnoite* ($Ba_2TiSi_2O_8$), which contains square-pyramidal TiO_5 polyhedra, all other minerals contain 6-coordinate titanium atoms. A recently prepared non-centrosymmetric layered solid, $Na_4Ti_2Si_8O_{22} \cdot 4H_2O$, also contains TiO_5 square pyramids in which each of the vertices of the base is linked to SiO_4 tetrahedral to form a continuous sheet.[22]

Anionic Complexes. The solutions obtained by dissolving Ti, TiF_4, or hydrous oxides in aqueous HF contain various fluoro complex ions but predominantly the very stable TiF_6^{2-} ion, which can be isolated as cystalline salts, in which it is distorted octahedral. The hydrolyzed species $(NH_4)_2TiF_4O$ is formed by heating an aqueous solution of H_2TiF_6 and urea. The structure consists of kinked parallel chains of TiF_4O_2 octahedra linked through the two oxygen atoms in trans positions.[23] In moderately polar solvents (e.g., SO_2 and CH_3CN) a variety of mono- and polynuclear fluoro complexes, such as $[TiF_5(solv)]^-$ and $Ti_2F_9^-$ are formed; more complex species such as $CsTi_8F_{33}$[24] have also been made. Various chloro and bromo anions can be obtained by interaction of TiX_4 with R_4N^+ or other cations in CH_2Cl_2, $SOCl_2$, and so on. Large cations give $[TiX_5]^-$ salts as in $(MgL_6)(TiCl_5L_2)$, L = THF or CH_3CN,[25] while smaller cations may form $[Ti_2Cl_{10}]^{2-}$, $[Ti_2Cl_9]^-$,[26] and $[TiCl_6]^{2-}$. The latter ion is formed in >12 *M* HCl solutions, probably together with *trans*-$[Ti(OH)_2Cl_4]^{2-}$ and $[Ti(OH)(H_2O)Cl_4]^-$.

Cationic Species. These are rare but examples are $[TiCl_3(MeCN)_3]SbCl_6$ and $[(DMSO)_5TiO]Cl_2$; a cationic amido complex[27] can be prepared according to:

$$Ti(NMe_2)_4 \xrightarrow[\text{py}]{\text{[HNEt}_3\text{]BPh}_4} [Ti(NMe_2)_3(py)_2]BPh_4$$

Adducts of TiX₄

The halides readily give adducts TiX_4L and TiX_4L_2 with donor molecules; many are crystalline solids soluble in organic solvents. Some representative examples are $TiCl_4L_2$, L = OMe_2, THF, $OPCl_3$, CH_3CN, mesCHO, and NH_3. Most adducts contain octahedrally coordinated Ti, the monoadducts being dimerized through halogen bridges, for example, $[TiCl_4(OPCl_3)]_2$, $[TiCl_4(Me_2CO)]_2$,[28] and $[TiCl_4(MeCO_2Et)]_2$. With certain chelating ligands such as diars, both 6-coordinate 17-A-I and 8-coordinate 17-A-II complexes are obtained.

[22]M. A. Roberts *et al., Nature* **1996,** *381,* 401.
[23]J. Patarin *et al., Eur. J. Solid State Inorg. Chem.* **1994,** *31,* 501.
[24]H. Bialowons and B. G. Müller, *Z. anorg. allg. Chem.* **1995,** *621,* 1223.
[25]G. R. Willey *et al., J. Chem. Soc., Dalton Trans.* **1994,** 1799.
[26]T. Gloger *et al., Z. Kristallogr.* **1996,** *211,* 821; B. Bajab and H.-J. Meyer, *Z. Kristallogr.* **1996,** *211,* 817.
[27]M. Ephritikhine *et al., J. Orgamomet. Chem.* **1997,** *531,* 115.
[28]C. Floriani *et al., Chem. Ber.* **1996,** *129,* 1361; Y. Dang, *Coord. Chem. Rev.* **1994,** *135/136,* 93; P. L. W. Tregenna-Piggott and S. P. Best, *Inorg. Chem.* **1996,** *35,* 5730.

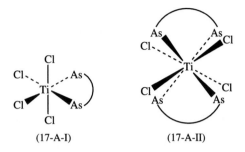

(17-A-I)   (17-A-II)

β-Diketonate and Other Chelate Complexes. Titanium tetrahalides react with many compounds containing active hydrogen atoms such as those of OH groups, undergoing solvolysis with loss of HX, but reactions with alkali salts are commonly used. Examples are

$$TiCl_4 + (salen)H_2 \longrightarrow (salen)TiCl_2 + 2HCl$$

$$TiCl_4 + 2Na(acac) \longrightarrow cis\text{-}TiCl_2(acac)_2 + 2NaCl$$

$$TiCl_4 + K[HB(pz)_3] \longrightarrow [HB(pz)_3]TiCl_3 + KCl$$

$$TiCl_4 + nNaS_2CNR_2 \longrightarrow TiCl_{4-n}(S_2CNR_2)_n + nNaCl$$

Hydrous TiO_2 will also dissolve in catechol to give $[Ti(cat)_3]^{2-}$. The diketonates may also be of the type $Ti(\beta\text{-dik})X_3$ that are usually halogen-bridged dimers.

The 6-, 7-, and 8-coordinate complexes are almost invariably octahedral, pentagonal bipyramidal, and dodecahedral, respectively.

A curiosity is the volatile dodecahedral $Ti(NO_3)_4$, which resembles other anhydrous nitrates of Sn^{IV}, Co^{III}, and Cu^{II}.

The perchlorate, $Ti(ClO_4)_4$, which has $\eta^2\text{-}ClO_4$ groups, is made by interaction of $TiCl_4$ with Cl_2O_6. A sulfate is made by the reaction

$$TiCl_4 + 6SO_3 \longrightarrow Ti(SO_4)_2 + 2S_2O_5Cl_2$$

while the phosphate acts as an ion-exchange material. The hydrogen phosphate, $Ti(HPO_4)_2 \cdot H_2O$,[29] has a layered structure in which the hydration water molecule is retained in the interlayer region by H-bonds. It is isostructural with the Si, Ge, Sn, Pb, Zr, and Hf analogues.

The titanium "teflate" anion $[Ti(OTeF_5)_6]^{2-}$ is non-basic and weakly coordinating, allowing the isolation of chlorocarbon complexes such as $[Ag(Cl_2CH_2)_2]_2$-$[Ti(OTeF_5)_6]$.[30]

[29]S. Bruque *et al.*, *Inorg. Chem.* **1995**, *34*, 893.
[30]S. Strauss *et al.*, *J. Am. Chem. Soc.* **1992**, *114*, 10995.

Alkoxides. These are among the most useful complexes; one application is the production of ceramic thin films.[31] They are made by reactions such as

$$TiCl_4 + 4ROH + 4NH_3 \longrightarrow Ti(OR)_4 + 4NH_4Cl$$

$$TiCl_4 + 3EtOH \longrightarrow 2HCl + TiCl_2(OEt)_2 \cdot EtOH$$

They are liquids or solids that can be distilled or sublimed and are soluble in organic solvents such as benzene. They are readily hydrolyzed by even traces of water, the ease decreasing with increasing chain length of the alkyl group and the degree of condensation. Under controlled conditions such reactions give polymeric species with OR or OH and oxygen bridges as in $[Ti_xO_y](OR)_{4x-2y}$.[32] Another example of a hydrolysis product is $[Ti_{18}O_{28}H][OBu^t]_{17}$, in which the metal-oxygen framework has a pentacapped Keggin structure.[33]

It is characteristic of alkoxides that unless the OR groups are extremely bulky or the compounds are in extremely dilute solutions, they exist as oligomers with bridging OR groups; the structure of $[Ti(OEt)_4]_4$ shown in Fig. 11-3, Section 11-11, is representative. In solutions, there may be various degrees and types of polymerization depending on solvent and OR group. In benzene, primary OR groups give trimers, whereas secondary and tertiary OR give only mononuclear species. Some alkoxides react with Lewis bases to form complexes such as $Ti[OCH(CF_3)_2]_4L_2$, L = CH_3CN and THF.[34]

Partially substituted compounds like $[TiCl_2(OPh)_2]_2$ and $[(EtO)_3Ti(glycinate)]_2$[35] have bridging phenoxide groups but with bulky aryloxides a variety of different types may be formed.[36]

Nitrogen Compounds

Titanium tetrachloride interacts with ammonia in dichloromethane to give an insoluble yellow solid, $TiCl_4(NH_3)_4$, from which upon sublimation $TiCl_4(NH_3)_2$ is obtained. The ammonia molecules in the latter are easily displaced by tripiperidinephosphine oxide (TPPO) to yield *trans*-$TiCl_4(TPPO)_2$. The reaction of Cp*TiMe_3 and ammonia gives the nitrido cluster $[Cp*Ti(\mu\text{-}NH)]_3(\mu_3\text{-}N)$.[37] A relatively air-stable complex is the phosphorane iminato $Ti(NPPh_3)_4$,[38] prepared according to

$$TiCl_2(NPPh_3)_2 \xrightarrow[R = Me, Cp]{LiR} Ti(NPPh_3)_4$$

Dialkylamides. The action of $LiNR_2$ on $TiCl_4$ leads to liquid or solid compounds of the type $Ti(NEt_2)_4$, which, like the alkoxides, are readily hydrolyzed by

[31]T. J. Boyle *et al.*, *Inorg. Chem.* **1995**, *34*, 1110.
[32]V. W. Day *et al.*, *Inorg. Chim. Acta* **1995**, *229*, 391.
[33]V. W. Day *et al.*, *J. Chem. Soc., Dalton Trans.* **1996**, 691.
[34]W. G. Van Der Sluys *et al.*, *Inorg. Chem.* **1994**, *33*, 4950.
[35]U. Schubert *et al.*, *Inorg. Chem.* **1995**, *34*, 995.
[36]See for example: M. D. Rausch *et al.*, *Organometallics* **1995**, *14*, 1827.
[37]H. W. Roesky *et al.*, *Angew. Chem. Int. Ed. Engl.* **1989**, *28*, 754.
[38]J. Li *et al.*, *Z. anorg. allg. Chem.* **1997**, *623*, 1035.

water with liberation of amine. Similarly, dialkylamides are known for both Ti^{III} and Ti^{II}. The Ti(NR$_2$)$_4$ compounds undergo insertion reactions; with CS$_2$, for example, the dithiocarbamates Ti(S$_2$CNR$_2$)$_4$ are formed. A good way to prepare TiCl$_{4-n}$(NR$_2$)$_n$ is by reacting stoichiometric amounts of TiCl$_4$ and Ti(NR$_2$)$_4$. The complex TiCl$_2$(NMe$_2$)$_2$ serves as starting material for mixed dialkylamides such as (Me$_2$N)$_2$Ti(NRAr)$_2$, which further react with methyl iodide to give titanium iodide complexes, e.g., I$_2$Ti(NRAr)$_2$.[39]

In some cases NR$_2$ groups function as bridges in cyclic oligomers; thus TiF$_4$ reacts with Ti(NMe$_2$)$_4$ to give [TiF$_2$(NMe$_2$)$_2$]$_4$ in which distorted TiF$_3$N$_3$ octahedra are linked by μ-F and μ-NMe$_2$ groups. Reaction of TiF$_4$ and Me$_3$SiNPPh$_3$ in boiling dichloromethane gives an *imido*-containing centrosymmetric dinuclear complex (17-A-III) which has two μ—F bridges.[40]

(17-A-III)

Whereas Groups 5-7 metals form numerous complexes with M=E double bonds, this feature has been less prominent in Group 4. However, an increasing number of titanium imido compounds have recently been prepared:[41]

The Ti—N bond in this type of complex is short, *ca.* 1.7 Å. The reduction of TiCl$_4$(NH$_3$)$_2$ with NaH in the presence of L = Ph$_3$PO gives the imide complex Ti(=NH)Cl$_2$L$_2$.[42]

Sulfur Compounds[43]

Treatment of TiCl$_4$ with 2 equivalents of thiol at ambient temperature gives yellow addition products of the type *cis*-TiCl$_4$(HSR)$_2$, while the reaction with potassium salts in the THF affords molecular Ti(SR)$_4$[44] complexes with a distorted tetrahedral TiS$_4$ core. When R = Me$_3$C$_6$H$_4$, the Ti—S bonds are considerably shorter than

[39]C. C. Cummins *et al.*, *Organometallics* **1994**, *13*, 2907.
[40]K. Dehnicke *et al.*, *Z. anorg. allg. Chem.* **1996**, *622*, 1091.
[41]H. W. Roesky *et al.*, *Angew. Chem. Int. Ed. Engl.* **1990**, *29*, 669; See also I. P. Rothwell *et al.*, *Angew. Chem. Int. Ed. Engl.* **1990**, *29*, 664; R. Duchateau *et al.*, *Inorg. Chem.* **1991**, *30*, 4863; D. L. Thorn *et al.*, *Inorg. Chem.* **1992**, *31*, 3917; P. Mountford *et al.*, *J. Chem. Soc., Chem. Commun.* **1994**, 2007.
[42]C. H. Winter *et al.*, *Inorg. Chem.* **1996**, *35*, 5968.
[43]See for example: D. W. Stephan and T. T. Nadasdi, *Coord. Chem. Rev.* **1996**, *147*, 147; J. A. Davies and S. G. Dutremez, *Inorg. Synth.* **1997**, *31*, 11.
[44]S. A. Koch *et al.*, *Inorg. Chim. Acta* **1995**, *229*, 335.

the corresponding bonds in $[Ti(SCH_2CH_2S)_3]^{2-}$ in accord with the difference in tetrahedral versus octahedral coordination. The Ti—S bond distance is considerably longer than the corresponding Ti—O in related alkoxides, indicating a lesser capability to form π-interactions. Other known titanium thiolate complexes include $Ti(diars)Cl_2(SC(CH_3)_3)$ and $Cp_2Ti(SR)_2$.

The toluene-3,4-dithiolate complex $[Ti(S_2C_6H_3CH_3)_3]^{2-}$ is octahedral, although many similar dithiolene complexes are trigonal prismatic, which is consistent with the view that the higher-lying d orbitals of Ti do not interact effectively with the lone pairs on the sulfur atoms. In $[NH_2(CH_3)_2]_2[Ti(S_2C_6H_4)_3]$ the titanium center contains an unusual TiS_6 coordination in which the S_6 polyhedron can be described as a pentagonal bipyramid with a missing equatorial vertex.[45] Polynuclear compounds such as $[NMe_2H_2][Ti_2(SMe)_9]$ and $Ti_3(SMe)_{12}$ are prepared by reacting $Ti(NMe_2)_4$ with different amounts of HSMe.[46]

Peroxo Complexes

It has long been known that aqueous solutions of Ti^{IV} and its complexes develop an intense orange color with H_2O_2, which can be used for analytical determinations. These colors are due to formation of stable peroxo complexes such as $[Ti(O_2)OH\ aq]^+$, $[Ti(O_2)\ edta]^{2-}$, and so on.

A number of crystalline salts have been isolated; for example, interaction of TiO_2, MF, 40% HF, and 30% H_2O_2 at pH 6 gives $M_3[Ti(O_2)F_5]$ and at pH 9, $M_2[Ti(O_2)_2F_2]$. Peroxo complexes sometimes cocrystallize with titanyl complexes (see above). An example is found when the reaction mixture of the dinuclear $[\{(Me_3tacn)_2Ti_2^{III}(NCO)_4\}(\mu\text{-}O)]$ is dissolved in acetonitrile and reacts with O_2; red-orange crystals of $(Me_3tacn)Ti(O_2)(NCO)_2$ form which also contain the colorless $(Me_3tacn)Ti(O)(NCO)_2$ molecules as shown by the ir bands at 907 cm^{-1} [$\nu(O-O)$] and 933 cm^{-1} (Ti=O).[17] Another example is shown below:

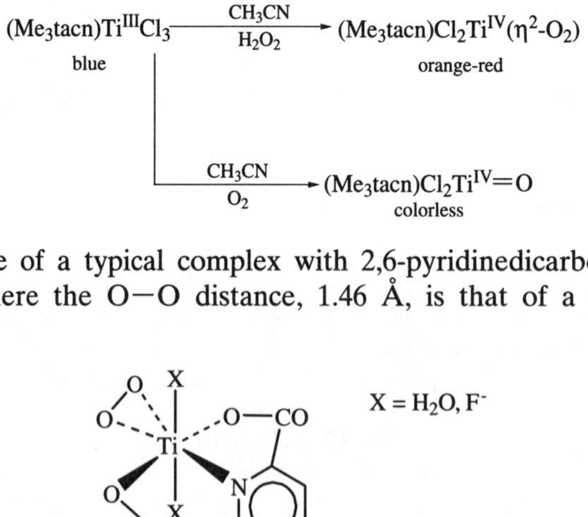

$$(Me_3tacn)Ti^{III}Cl_3 \xrightarrow[H_2O_2]{CH_3CN} (Me_3tacn)Cl_2Ti^{IV}(\eta^2\text{-}O_2)$$

blue   orange-red

$$\xrightarrow[O_2]{CH_3CN} (Me_3tacn)Cl_2Ti^{IV}{=}O$$

colorless

The structure of a typical complex with 2,6-pyridinedicarboxylate as ligand is (17-A-IV), where the O—O distance, 1.46 Å, is that of a η^2-peroxo ligand (Section 11-10).

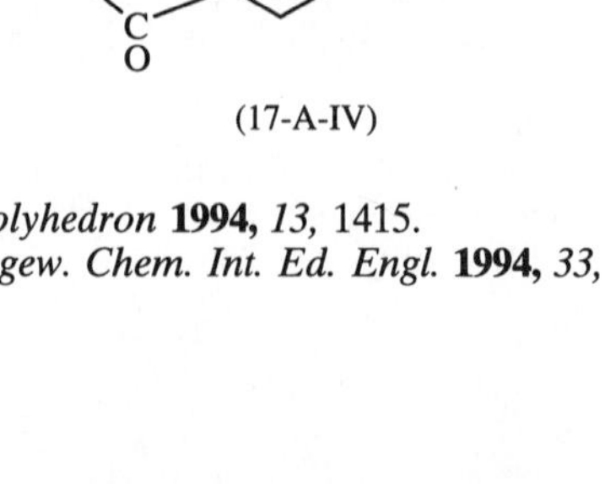

X = H_2O, F^-

(17-A-IV)

[45]D. M. Giorlando *et al.*, *Polyhedron* **1994**, *13*, 1415.
[46]D. M. Giolando *et al.*, *Angew. Chem. Int. Ed. Engl.* **1994**, *33*, 1981.

Heterobimetallic Complexes. Monohalide Ti^{IV} compounds containing amido,[47] alkoxides,[48] or tripodal amido[49] groups react with many metal carbonyl cationic species to give unsupported heterodinuclear complexes such as 17-A-V.

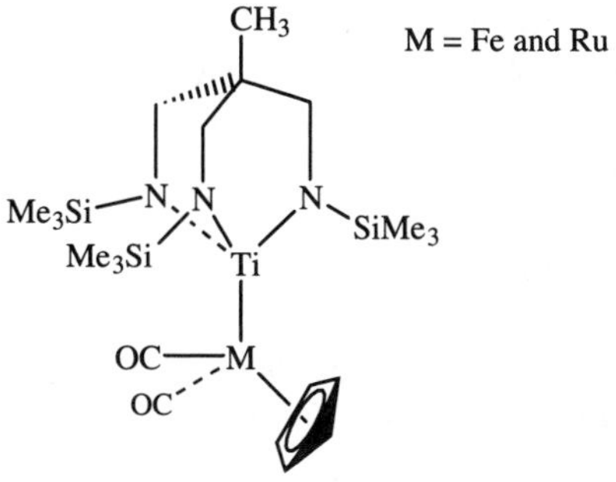

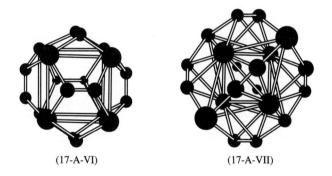

M = Fe and Ru

(17-A-V)

Metallocarbohedrene and Other Cluster Compounds. The compound Ti_8C_{12}[50] is the prototype in the M_8C_{12} family (Section 7-9); it shows a peak in the mass spectrum corresponding to that mass. It is proposed to have either the structure (17-A-VI), in which six C_2 groups are bonded over the faces of a Ti_8 cube, parallel to the edges, with symmetry T_h or the structure (17-A-VII) with T_d symmetry. Other species have also been observed, such as the closely related $Ti_7C_{12}^+$, $Ti_8C_{11}^+$, $Ti_7C_{13}^+$, and $Ti_8C_{13}^+$ clusters.[51] None of those species have yet been isolated but another cluster, $[Ti_6C]Cl_{14}$, was obtained by reduction of $TiCl_3$ with Na in the presence of a carbon source. The latter is the only known interstitially stabilized Ti_6 cluster;[52] it is isostructural with Nb_6Cl_{14} (Section 18-B-7).

(17-A-VI)   (17-A-VII)

17-A-3 The Chemistry of Titanium(III), d^1

There is an extensive chemistry of solid compounds and Ti^{III} species in solution.

[47]W. S. Sartain and J. P. Selegue, *Organometallics* **1989**, *8*, 2153.
[48]D. Selent *et al.*, *Organometallics* **1993**, *12*, 2857.
[49]L. H. Gade *et al.*, *Inorg. Chem.* **1996**, *35*, 2433; L. H. Gade *et al.*, *Angew. Chem. Int. Ed. Engl.* **1994**, *33*, 676.
[50]A. W. Castleman *et al.*, *Science* **1992**, *256*, 515; I. Dance, *J. Am. Chem. Soc.* **1996**, *118*, 6309.
[51]M. T. Bowers, *Acc. Chem. Res.* **1994**, *27*, 324.
[52]D. J. Hinz and G. Meyer, *J. Chem. Soc., Chem. Commun.* **1994**, 125.

Binary Compounds. The *chloride*, $TiCl_3$, has several crystalline forms. It can be made by H_2 reduction of $TiCl_4$ vapor at 500 to 1200°C; this and other high-temperature methods give the violet α form. The reduction of $TiCl_4$ by aluminum alkyls in inert solvents gives a brown β form, which is converted into the α form at 250 to 300°C. Two other forms are known, but these and the α form have layer lattices containing $TiCl_6$ groups, whereas β-$TiCl_3$ is fibrous with single chains of $TiCl_6$ octahedra sharing edges. The latter is of particular importance because the stereospecific polymerization of propene depends critically on the structure of the β form.

The trichloride is oxidized by air and reacts with donor molecules to give adducts of the general formula $TiCl_3L_n$ (n = 1-6), for example, $TiCl_3(THF)_3$. When heated above 500°C, $TiCl_3$ disproportionates (see later). The fluoride, bromide, and iodide of Ti^{III} are also known. At low temperature $TiBr_3$ contains titanium ions with a distorted octahedron of bromine atoms.[53]

The *oxide* Ti_2O_3 (corundum structure) is obtained by reducing TiO_2 at 1000°C in a stream of H_2. It is rather inert and is attacked only by oxidizing acids. The addition of OH^- ions to aqueous Ti^{III} solutions gives a purple precipitate of the hydrous oxide.

Solution Chemistry and Complexes of Titanium(III)

Aqueous solutions of the $[Ti(H_2O)_6]^{3+}$ ion can be readily obtained by reducing aqueous Ti^{IV} either electrolytically or with zinc. The violet solutions reduce oxygen and hence must be handled under N_2 or H_2. The potential in the acidity range $0.3\ M \leq [H^+] \leq 12\ M$ can be taken as

$$Ti^{4+} + e = Ti^{3+} \qquad E^0 = 0.2\ V$$

and the following potential has been estimated:

$$Ti(OH)_2^{2+} + 2H^+ + e \rightleftharpoons Ti^{3+} + 2H_2O \quad E^0 = 7.7 \times 10^{-3}\ V$$

The Ti^{3+} ion is extensively used as a reducing agent.

In all its complexes both neutral and ionic Ti^{III} is normally octahedral. In dilute acids the main species is $[Ti(H_2O)_6]^{3+}$, which hydrolyzes ($pK \approx 1$) to give $[Ti(OH)(H_2O)_5]^{2+}$. The hexaaqua ion occurs in alums such as $CsTi(SO_4)_2 \cdot 12H_2O$ and in simple salts such as $[Ti(H_2O)_6](p\text{-}MeC_6H_4SO_3)_3 \cdot 3H_2O$.[54]

In more concentrated HCl or HBr solutions there are species such as $[TiX(H_2O)_5]^{2+}$, $[TiX_5(H_2O)]^{2-}$, and $[TiX_6]^{3-}$. Crystallization of HCl solutions gives *trans*-$[TiCl_2(H_2O)_4]Cl \cdot 2H_2O$; a comparable ion is *trans*-$[TiCl_2(THF)_4]^+$ obtained by interaction of $ZnCl_2$ and $TiCl_3$ in THF. The high acidity of $Ti^{III}(aq)$ prevents the formation of certain derivatives of weak acids such as Ti^{III} cyanide complexes but they can be prepared in non-aqueous media[55] as follows:

$$TiCl_3 + 3HO_3SCF_3 \xrightarrow{\text{MeCN}} Ti(O_3SCF_3)_3(MeCN)_3 + 3HCl$$

$$Ti(O_3SCF_3)_3(MeCN)_3 + 6(NEt_4)CN \xrightarrow{\text{MeCN}} (NEt_4)_3[Ti(CN)_6] \cdot 4MeCN + 3(NEt_4)(O_3SCF_3)$$

[53]S. I. Troyanov *et al.*, *Russ. J. Inorg. Chem.* **1994**, *39*, 374.
[54]W. Clegg *et al.*, *Acta Cryst.* **1995**, *C51*, 560.
[55]G. S. Girolami *et al.*, *J. Am. Chem. Soc.* **1997**, *119*, 6251.

The Ti(CN)$_6^{3-}$ anions adopt nearly ideal octahedral geometry, not the distorted geometry expected from a d^1 Jahn-Teller active ion. This behavior is common to many other Ti^{III} compounds.

There are various adducts of TiX$_3$ with donors, examples being TiCl$_3$(MeCN)$_3$, TiCl$_3$(NMe$_3$)$_3$, and [TiCl$_4$(THF)$_2$]$^-$.[56] The reaction of Ti and molten InBr$_3$ gives In$_3$Ti$_2$Br$_9$ which contains antiferromagnetically coupled Ti^{III} atoms in face-sharing bioctahedral Ti$_2$Br$_9^{3-}$ units;[57] exchange interactions are also found in the Ti$_2$Cl$_9^{3-}$ species.[58] Other species include the 7-coordinate [Ti(H-edta)(H$_2$O)]·H$_2$O and TiN(SiMe$_3$)$_2$(η^2-BH$_4$)$_2$(py)$_2$,[59] 5-coordinate TiN(SiMe$_3$)$_2$Cl$_2$(THF)$_2$, 4-coordinate {Ti[N(SiMe$_3$)$_2$]Cl$_2$}$^-$ anion,[60] 3-coordinate bulky Ti[N(SiMe$_3$)$_2$]$_3$, and the edge-sharing Ti$_2$Cl$_6$(PMe$_3$)$_4$[61] which can be converted to the face-sharing [Ti$_2$(μ-Cl)$_3$Cl$_4$(PR$_3$)$_2$]$^-$ anions.[62]

Other titanium(III) complexes with nitrogen containing ligands are common. Reaction of TiCl$_3$(THF)$_3$ and LiL (L = trimethylsilylbenzamidinate) in TMEDA yields L$_2$Ti(μ-Cl)$_2$LiTMEDA which reacts further with (allyl)MgBr or LiBH$_4$ to give 6-coordinate L$_2$Ti(allyl) and L$_2$Ti(μ-H)$_2$BH$_2$ (17-A-VIII), respectively.[63]

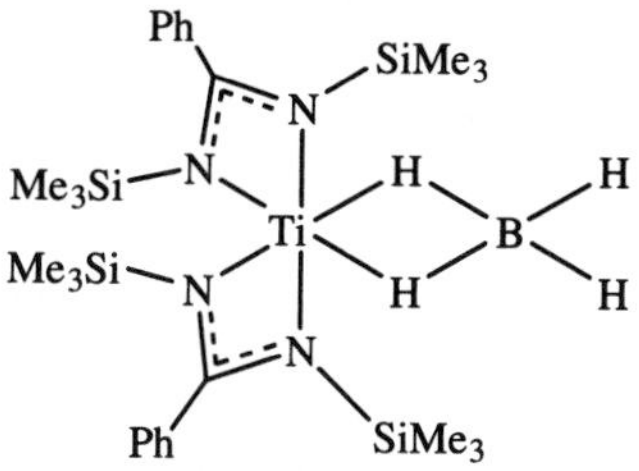

(17-A-VIII)

In the presence of Na/Hg, TiCl$_4$ reacts with Li(PhNC(H)NPh), to give mononuclear Ti(DPhF)$_3$, but the use of HSn(Bu^n)$_3$ as reducing agents yields the dinuclear compound (Ti$_2$(μ-DPhF)$_2$(DPhF)(μ-NPh)$_2$ (17-A-IX).

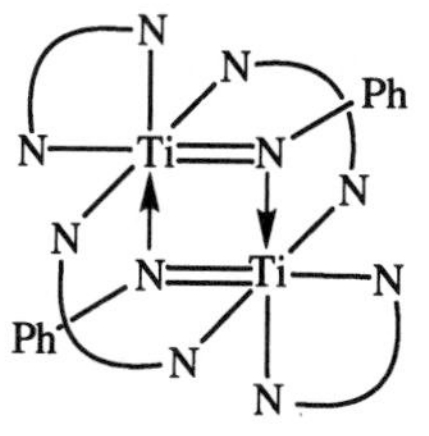

(17-A-IX)$^-$

Compounds of the type Ti$_2$(RNC(H)NR)$_2$)(μ-RNC(H)NR)$_2$(μ-Cl)$_2$, with R = phenyl and biphenyl, are diamagnetic, a manifestation of Ti—Ti bonding.

[56]P. Sobota *et al.*, *Polyhedron* **1989**, *8*, 2013.
[57]R. Dronskowski, *Chem. Eur. J.* **1995**, *1*, 118.
[58]A. Ceulemans *et al.*, *Inorg. Chim. Acta* **1996**, *251*, 15; G. Christou *et al.*, *Chem. Commun.* **1996**, 2177.
[59]L. Scoles and S. Gambarotta, *Inorg. Chim. Acta* **1995**, *235*, 375.
[60]K. Dehnicke *et al.*, *Chem. Ber.* **1996**, *129*, 1401.
[61]F. A. Cotton and W. A. Wojtczak, *Gazz. Chim. Ital.* **1993**, *123*, 499.
[62]F. A. Cotton *et al.*, *Inorg. Chem.* **1996**, *35*, 7358.
[63]S. Gambarotta *et al.*, *Inorg. Chem.* **1993**, *32*, 1959.

A few μ-oxobridged compounds have been structurally characterized; examples are [{(py)$_3$TiBr$_2$}$_2$(μ-O)], [(TPP)Ti]$_2$(μ-O),[64] and [{LTi(NCO)$_2$}$_2$(μ-O)][65] (L = 1,4,7-trimethyl-1,4,7-triazacyclononane). The latter has only a weak antiferromagnetic coupling contrary to the usually strong coupling found in other M(III) species,[66] M = V, Cr, Mn, and Fe. The orthophosphate TiPO$_4$ is isostructural with the vanadium analogue and CrVO$_4$;[67] the magnetic ions form chains which run along the c-axis with antiferromagnetic intra-chain and inter-chain ordering.

Electronic Structure

The Ti^{III} ion is a d^1 system, and in an octahedral ligand field the configuration must be t_{2g}. One absorption band is expected ($t_{2g} \to e_g$ transition), and has been observed in several compounds. The [Ti(H$_2$O)$_6$]$^{3+}$ ion has a single absorption band ($t_{2g} \to e_g$) which permits some blue and most red light to be transmitted.

A d^1 ion in an electrostatic field of perfect O_h symmetry should show a highly temperature-dependent magnetic moment as a result of spin-orbit coupling, but the combined effects of distortion and covalence cause a leveling out of μ_{eff}, which does in general vary from ~1.5 BM at 80 K to ~1.8 BM at about 300 K. Values of μ_{eff} are generally close to 1.7 BM at 293 K.

17-A-4 The Chemistry of Titanium(II), d^{2} [68]

Other than those discussed in the next section, compounds of Ti^{II} are few. No aqueous chemistry exists because of oxidation by water, although ice-cold solutions of TiO in dilute HCl are said to contain Ti^{II} ions, which persist for some time. The *halides* are obtained by reduction of the tetrahalides with titanium,

$$TiX_4 + Ti \longrightarrow 2TiX_2$$

or by disproportionation of the trihalides,

$$2TiX_3 \xrightarrow{\Delta} TiX_2 + TiX_4$$

In the presence of AlCl$_3$, Ti^{II} and Ti^{III} become more volatile, presumably because of the formation of species such as TiAl$_2$Cl$_8$, TiAlCl$_6$, and TiAl$_3$Cl$_{11}$ (Section 6-7). Halides such as Ti$_7$Cl$_{16}$ contain Ti—Ti bonds. The compound Ti(AlBr$_4$)$_2$ is prepared by prolonged heating of TiBr$_4$ and aluminum powder.[69] Other inorganic compounds are Na$_2$Ti$_3$Cl$_8$[70] and Na$_2$TiCl$_4$.[71] At room temperature the latter is isotypic with the Mn analogue; the Ti atoms are essentially isolated from each other, but at low

[64]L. K. Woo, *Inorg. Chem.* **1996**, *35*, 7601.
[65]K. Wieghardt *et al., Inorg. Chem.* **1994**, *33*, 47.
[66]See for example: V. Staemmler *et al., Inorg. Chem.* **1994**, *33*, 6219.
[67]R. Glaum *et al., J. Solid State Chem.* **1996**, *126*, 15.
[68]K. Dehnicke *et al., Z. Naturforsch.* **1996**, *51b*, 602.
[69]S. I. Troyanov *et al., Russian J. Inorg. Chem.* **1990**, *35*, 494.
[70]G. Meyer *et al., Angew. Chem. Int. Ed. Engl.* **1995**, *34*, 71.
[71]W. Urland *et al., Z. anorg. allg. Chem.* **1994**, *620*, 801.

temperature they form $(Ti_3)^{6+}$ clusters. Quaternary salts of the type $NaM^{II}Zr_2F_{11}$ can be made for M = Ti, V, and Cu.[72]

The *oxide*, which is made by heating Ti and TiO_2, has the NaCl structure but is normally nonstoichiometric.

The Ti^{2+} ion isolated in place of Na^+ in an NaCl crystal shows the expected $d-d$ transitions, namely, $^3T_{1g} \rightarrow {}^3T_{2g}$ and $^3T_{1g} \rightarrow {}^3T_{1g}(P)$ from which $\Delta_0 = 8520$ cm^{-1} and $\beta = 572$ cm^{-1} are calculated.

There are a few complex halides such as $[TiCl_5]^{3-}$ and $[TiCl_4]^{2-}$ and adducts of $TiCl_2$ with MeCN and the complexes *trans*-$[TiX_2dmpe_2]$, X = Cl, CH$_3$, and BH$_4$.[73] The reactions of $TiCl_4$ and Na or $TiCl_3(THF)_3$ and KC_8 in THF produces blue solution of $TiCl_2(THF)_n$; the THF can be substituted by other ligands such as DME[74] and pyridine from which *trans*-$TiCl_2(py)_4$ can be crystallized.[75]

Complexes with bulky aryloxides, such as *cis*-$Ti(OAr)_2(bipy)_2$ and *trans*-$Ti(OPh)_2(dmpe)_2$,[76] have been synthesized; tetratolylporphyrinato compounds with other ligands, e.g., pyridine and picoline, are also known.[77]

Titanium(II) compounds are very strong reducing agents; many react with various small molecules. In the presence of N_2, *trans*-$[TiCl_2(TMEDA)_2]$ reacts with the sterically demanding lithium amide $LiN(SiMe_3)_2$, in pyridine, to give 17-A-X and similar complexes;[78] these N_2 complexes contain very short (*ca.* 1.7 Å) Ti=N bonds and are best considered as Ti^{IV} compounds.

(17-A-X)

Side-on coordination of N_2 is observed in the formally mixed-valent Ti(I)/Ti(II) amido complex $[Li(TMEDA)][\{((Me_3Si)_2N)_2Ti\}_2(\mu\text{-}\eta^2: \eta^2\text{-}N_2)_2]$[79] (17-A-XI).

(17-A-XI)

Other Low Valent Species

A series of compounds prepared by reaction of $TiCl_3$ with reducing agents such as $LiAlH_4$, KC_8, Mg, and Zn/Cu are widely used in organic synthesis for coupling of

[72]H. Bialowons and B. G. Müller, *Z. anorg. allg. Chem.* **1996**, *622*, 1187.
[73]G. S. Girolami *et al.*, *J. Am. Chem. Soc.* **1987**, *109*, 8094.
[74]G. Pampaloni *et al.*, *Z. Naturforsch.* **1996**, *51b*, 506.
[75]F. A. Cotton *et al.*, *Inorg. Chem.* **1995**, *34*, 5424; M. Klinga *et al.*, *Z. Kristallog.* **1996**, *211*, 506.
[76]R. J. Morris and G. S. Girolami, *Inorg. Chem.* **1990**, *29*, 4169.
[77]L. K. Woo *et al.*, *Inorg. Chem.* **1993**, *32*, 4186.
[78]S. Gambarotta *et al.*, *J. Chem. Soc., Chem. Commun.* **1992**, 244; J. R. Hagadorn and J. Arnold, *J. Am. Chem. Soc.* **1996**, *118*, 893.
[79]S. Gambarotta *et al.*, *J. Am. Chem. Soc.* **1991**, *113*, 8986.

carbonyl compounds in the McMurry reaction:[80]

(17-A-XII)

The nature of the reduced species is generally unclear, but EXAFS has shown the structure of some of the active species such as $[HTiCl(THF)_{\sim 0.5}]_x$,[81] and the inorganic Grignard complex $Ti(MgCl)_2 \cdot x THF$[82] which has a short Ti—Mg separation of 2.72(1) Å. Another known compound is $TiMgCl_2 \cdot x THF$.

The reduction of $TiBr_4 \cdot 2THF$ with $K[BEt_3H]$ yields a hydrogen-free organosol $Ti \cdot 0.5THF$ which seems to consist of very small Ti particles in the zerovalent state stabilized by THF molecules.[83] This nanoparticulate metal colloid is extremely oxophilic, very soluble in THF but insoluble in hydrocarbons.

17-A-5   Organo Compounds: Titanium(IV), (III), (II), (0), (-I), and (-II)

Organotitanium compounds have been intensively studied, initially mainly because of the discovery by Ziegler and Natta that ethylene and propylene can be polymerized by $TiCl_3$-aluminum alkyl mixtures in hydrocarbons at 25°C and 1 atm pressure[84] (Section 22-9). Organic compounds have been found to react with N_2 and to act as catalysts in a number of other reactions.

Molecular alkyls of both Ti^{IV} and Ti^{III} can be made using bulky, elimination-stabilized groups. Examples are $Ti(CH_2Ph)_4$, $Ti(CH_2SiMe_3)_4$, and $Ti[CH(SiMe_3)_2]_3$. Although CH_3TiCl_3 is stable at 25°C, the yellow $TiMe_4$ is unstable above ~ -40°C. However both compounds form thermally stable adducts with donor ligands, for example, Me_4Ti bipy, although even these are sensitive to air and water. The adduct $TiMe_4 \cdot THF$ has a *tbp* structure.[85] Comproportionation of $TiMe_4$ with TiX_4 gives a series of complexes $TiMe_nX_{4-n}$, the thermal stability of which increases for X = Cl < Br < I and $n = 3 < 2 < 1$; thus Me_3TiCl decomposes at ~ -40°C but $MeTiI_3$ is stable to 126°C.

Cyclopentadienyls. The η^5-C_5H_5 (Cp), η^5-C_5Me_5 (Cp*), and other substituted species are among the most widely studied titanium compounds.[86] The red crystalline

[80]A. Fürstner and B. Bogdanović, *Angew. Chem. Int. Ed. Engl.,* **1996,** *35,* 2443; J. E. McMurry, *Chem. Rev.* **1989,** *89,* 1513.
[81]L. E. Aleandri *et al., J. Organomet. Chem.* **1994,** *472,* 97.
[82]L. E. Aleandri *et al., J. Organomet. Chem.* **1993,** *459,* 87.
[83]J. Hormes *et al., J. Am. Chem. Soc.* **1996,** *118,* 12090.
[84]See for example: M. Bochmann, *J. Chem. Soc., Dalton Trans.* **1996,** 255; C. Pellecchia *et al., J. Am. Chem. Soc.* **1995,** *117,* 6593; J. R. Stille *et al., J. Am. Chem. Soc.* **1994,** *116,* 8912.
[85]K. H. Thiele *et al., Z. anorg. allg. Chem.* **1995,** *621,* 861.
[86]See for example: Y. Dang, *Coord. Chem. Rev.* **1994,** *135/136,* 93; M. C. Baird *et al., J. Organomet. Chem.* **1997,** *527,* 7.

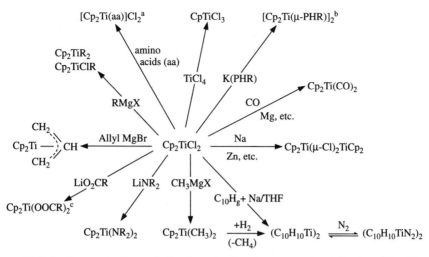

Figure 17-A-1 Some reactions of dicyclopentadienyl compounds of titanium(II), -III, and -IV. (a) J. Ho and D. W. Stephan, *Inorg. Chem.* **1994,** *33,* 4595; (b) I. C. Tornieporth-Oetting *et al., Angew. Chem. Int. Ed. Engl.* **1994,** *33,* 1518; (c) C. Floriani *et al., Chem. Ber.* **1996,** *129,* 1361; Y. Dang, *Coord. Chem. Rev.* **1994,** *135/136,* 93.

Cp_2TiCl_2, which has a quasi-tetrahedral structure, is the principal starting material for much of the chemistry and some of its reactions are shown in Fig. 17-A-1.

The *alkyl and aryl derivatives* may be quite stable and chiral compounds (17-A-XIII) have been resolved.[87] The complex Cp_4Ti, which has the dialkyl type structure 17-A-XIV, is nonrigid; the σ-bonded rings undergo rapid 1,2 shifts while the two types of rings interchange their roles rapidly so that at 25°C, all 20 ring protons give only a single broad ^1H nmr line.

R = CHMePh; R' = Ph
R = CMe$_2$Ph; R' = OPh

(17-A-XIII)   (17-A-XIV)

Interaction of Cp_2TiCl_2 with $AlMe_3$ gives a bridged complex

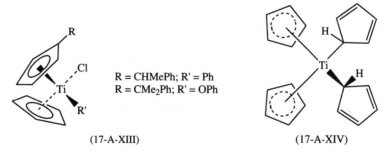

which acts as a CH_2 transfer agent, homologating olefins and converting, for example, R_2CO to $R_2C=CH_2$; it has also been used as a model system in alkene metathesis

[87]R. O. Duthaler and A. Hafner, *Chem. Rev.* **1992,** *92,* 807.

studies. Reaction of Cp_2TiL_2 (L = PMe_3) and 3,3-dimethylcyclopropene gives the mononuclear titanium alkylidene complex $Cp_2Ti(=CH-CH=CMe_2)(L)$.[88] The thermolysis of Cp^*TiMe_3 proceeds cleanly to the cubane $[Cp^*Ti(\mu_3\text{-}CH)]_4$.[89]

The dimethyl compound reacts with silanes, for example,

$$Cp_2TiMe_2 \xrightarrow{R_3SiH} (Cp_2TiH)_2\mu\text{-}H + R_3SiMe + R_6Si_2 + CH_4$$

to give a blue hydride.

A green compound produced by interaction of $TiCl_2$ and CpNa, hydrogenolysis of Cp_2TiMe_2, or reductions of Cp_2TiCl_2 has been much studied and is believed to have the structure 17-A-XV, while a more reactive form obtained by reduction of Cp_2TiCl_2 with potassium naphthalenide has been shown to be 17-A-XVI.

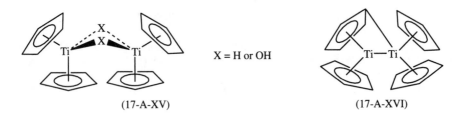

X = H or OH

(17-A-XV)   (17-A-XVI)

An important application of Cp_2TiMe_2 is found in catalytic processes such as dehydrocoupling of ammonia and silanes.[90]

The *pentamethylcyclopentadienyls* often differ from those of the $\eta^5\text{-}C_5H_5$ analogues since loss of a hydrogen atom from the ring (as in 17-A-XVI) cannot occur. Thus the true monomer $(\eta^5\text{-}C_5Me_5)_2Ti$ can be obtained by dissociation of $[Cp^*_2Ti]_2\mu\text{-}N_2$. It has two unpaired electrons and reacts with hydrogen to give $Cp^*_2TiH_2$. In part because of interest in metathesis there has been much study of titanium cyclobutanes and similar compounds, again containing the Cp_2Ti or Cp^*_2Ti unit. There is also a variety of titanium(III) compounds of the type Cp^*_2TiX, where X = H, Cl, BH_4, and $OCMe_3$; in contrast to the Cp derivatives these are mononuclear (*vide infra*).

There are a number of oxo species that are obtained by oxidation of $Cp_2M(CO)_2$ (or for other metal species Cp_2M) by N_2O of general formula $Cp_nM_nO_m$, the titanium compound being $Cp_6Ti_6O_8$. These have cage structures with $\mu_3\text{-}O$ groups; others are $Cp_5V_5O_6$ and $Cp_4Cr_4O_4$. An oxo species made by interaction of $(Cp_2TiCl)_2$ and N_2O is $(Cp_2TiCl)_2(\mu\text{-}O)$, while $Cp_2Ti(CO)_2$ and RNO gives $[Cp_2TiO]_n$. The reaction of Cp^*TiF_3 with $O(SnBu_3)_2$ affords the eight-membered ring compound $[Cp^*TiF(\mu\text{-}O)]_4$;[91] the Ti_4O_4 ring is almost planar contrary to that found in the Br analogue.[92,93] Some organotitanium oxides are used as supports of metal carbonyl species as in $[Cp^*_3Ti_3(\mu\text{-}O)_3Me][(\mu\text{-}OC)M(CO)_2Cp]_2$, M = Mo, W.[94]

Several Ti complexes containing terminal or bridging hydroxo ligands have

[88]P. Binger *et al.*, *Angew. Chem. Int. Ed. Engl.* **1989**, *28*, 610.
[89]M. Mena *et al.*, *Angew. Chem. Int. Ed. Engl.* **1997**, *36*, 115.
[90]H. Q. Liu and J. F. Harrod, *Organometallics* **1992**, *11*, 822.
[91]H. W. Roesky *et al.*, *Inorg. Chem.* **1993**, *32*, 5102.
[92]F. Palacios *et al.*, *J. Organomet. Chem.* **1989**, *375*, 5.
[93]S. I. Troyanov *et al.*, *J. Organomet. Chem.* **1991**, *402*, 201.
[94]M. Mena *et al.*, *J. Chem. Soc., Chem. Commun.* **1995**, 551.

been structurally characterized;[95] in [Cp*$_2$Ti(OH)(HNCPh$_2$)]BPh$_4$ the coordination of the titanium atom is pseudotetrahedral and the Ti—OH distance of 1.853(7) Å is short, which indicates there is substantial Ti—O double-bond character. The compounds are normally prepared by controlled hydrolysis but sometimes aqua complexes such as [Cp*$_2$Ti(H$_2$O)$_2$](CF$_3$SO$_3$)$_2$[96] can be isolated.

Sulfur species such as Cp$_5$Ti$_5$S$_6$ and Cp*$_2$Ti(SH)$_2$ have been made from the dicarbonyls and H$_2$S.

Compounds of the type Cp$_2$TiR (R = alkyl or aryl) have been shown to be unstable unless complexation by a donor molecule containing nitrogen or oxygen occurs. However, the stability is enhanced significantly when the active site is blocked sterically or by intramolecular coordination.

A general method for the synthesis of Cp$_2$TiR (R = alkyl, aryl) compounds consists of the reaction between [Cp$_2$TiCl]$_2$ and a Grignard or organolithium reagent; most of the Ti^{III} compounds have been synthesized in this manner.[97]

Arenes. These were originally made by metal vapor syntheses from mesitylene, naphthalene, biphenyl (17-A-XVII), and so on but other methods have been developed,[98] for example,

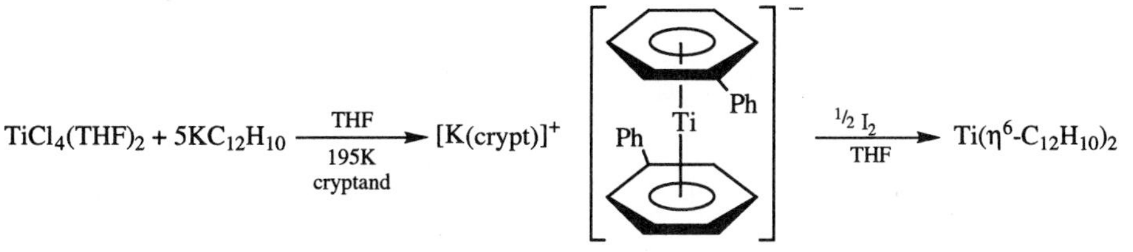

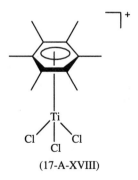

(17-A-XVII)

In chlorinated solvents, TiCl$_4$ reacts with hexamethyl benzene to give [(η^6-C$_6$Me$_6$)-TiCl$_3$](Ti$_2$Cl$_9$).[99] When AlX$_3$ is present, arene solutions of TiX$_4$ readily give [TiX$_3$-(η^6-arene)]AlX$_4$ (17-A-XVIII):[100]

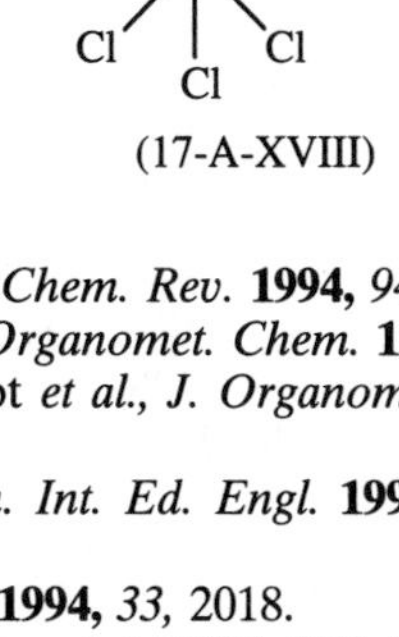

(17-A-XVIII)

[95]J. W. Gilje and H. W. Roesky, *Chem. Rev.* **1994,** *94,* 895.
[96]U. Thewalt and B. Honald, *J. Organomet. Chem.* **1988,** *348,* 291.
[97]See for example: F. Bickelhaupt *et al., J. Organomet. Chem.* **1997,** *527,* 1 and references therein.
[98]J. E. Ellis *et al., Angew. Chem. Int. Ed. Engl.* **1992,** *31,* 1495; *J. Am. Chem. Soc.* **1993,** *115,* 11616.
[99]C. Floriani *et al., Inorg. Chem.* **1994,** *33,* 2018.
[100]G. Pampaloni *et al., Inorg. Chim. Acta* **1995,** *229,* 179.

Alkene, Alkyne, Alkylidene,[101] ***and Carbonyl Complexes.*** While titanium alkene complexes are unquestionably involved in polymerizations, relatively few have been isolated. Interactions of $TiCl_4(dmpe)_2$ and butadiene under reducing conditions give $Ti(\eta-C_4H_6)_2(dmpe)$, which can be converted by CO and PF_3 to $Ti(CO)_2(PF_3)$-$(dmpe)_2$ and $Ti(CO)_3(dmpe)_2$. The latter reacts with K in the presence of biphenyl or naphthalene and then with CO to give $Ti(CO)_6^{2-}$ species which are isolable as $K(cryptate)^+$ salts:[102]

$$Ti(CO)_3(dmpe)_2 \xrightarrow[\text{2.2.2 crypt}]{KC_{10}H_8,\ THF} \xrightarrow{CO} [K(crypt)_2]_2[Ti(CO)_6]$$

Binary carbonyls like the 16-electron species $Ti(CO)_6$ are unstable but can be isolated if an $[R_3Sn]^-$ unit is present as in $[R_3SnTi(CO)_6]^-$;[103] the $[Ti(CO)_6AuPEt_3]^{-}$ [104] anion is also known.

Most of the compounds, however, have Cp or Cp* groups, examples being $Cp_2Ti(CO)_2$, $CpTi(CO)_4$, $Cp*_2Ti(\eta^2-C_2H_4)$, and $Cp_2Ti(\eta^2PhC\equiv CPh)PMe_3$. There are similar species like $Cp_2Ti(PMe_3)_2$ or $Cp_2Ti(dmpe)$; these Ti^{II} compounds can undergo oxidative addition reactions, for example,

and can provide precursors for hydrogenation catalysts.[105]

They are usually made by reducing Cp_2TiCl_2 in the presence of the corresponding phosphine ligand,[106] for example:

$$Cp_2TiCl_2 + 2PMe_3 + Mg \longrightarrow Cp_2Ti(PMe_3)_2 + MgCl_2$$

17-B VANADIUM: GROUP 5

The maximum oxidation state of vanadium is V, but apart from this there is little similarity, other than in some of the stoichiometry, to the chemistry of elements of the P group. The chemistry of V^{IV} is dominated by the formation of oxo species, and a wide range of compounds with VO^{2+} groups is known. There are four well-defined cationic species, $[V^{II}(H_2O)_6]^{2+}$, $[V^{III}(H_2O)_6]^{3+}$, $V^{IV}O^{2+}(aq)$, and $V^VO_2^+(aq)$, and none of these disproportionates because the ions become better oxidants as the oxidation state increases; both V^{II} and V^{III} ions are oxidized by air. As with Ti, and in common with other transition elements, the vanadium halides and oxohalides

[101]J. Feldman and R. R. Schrock, *Prog. Inorg. Chem.* **1991**, *39*, 1.
[102]J. E. Ellis *et al., J. Am. Chem. Soc.* **1988**, *110*, 383.
[103]J. E. Ellis and P. Yuen, *Inorg. Chem.* **1993**, *32*, 4998.
[104]J. E. Ellis *et al., Chem. Commun.* **1997**, 1249.
[105]G. S. Girolami *et al., Organometallics* **1994**, *13*, 4655; M. C. Baird *et al., J. Am. Chem. Soc.* **1994**, *116*, 6435; E. M. Carreira *et al., J. Am. Chem. Soc.* **1994**, *116*, 8837.
[106]K. Mach *et al., Organometallics* **1995**, *14*, 1410.

behave as Lewis acids, forming adducts with neutral ligands and complex ions with halide ions.

The oxidation states and stereochemistries for vanadium are summarized in Table 17-B-1. We shall discuss the oxidation states V, IV, III, and II individually, and some compounds with mixed oxidation states, metal-metal bonds, organometallic complexes, and bioinorganic chemistry.

17-B-1 The Element

Vanadium has an abundance in Nature of ~0.014%. It is widely spread, but there are a few concentrated deposits. Important minerals are *patronite* (a complex sulfide),

Table 17-B-1 Oxidation States and Stereochemistry of Vanadium

Oxidation state	Coordination number	Geometry	Examples
V^{-1}, d^6	6	Octahedral	$V(CO)_6^-$, $Li[V(bipy)_3]\cdot4C_4H_8O$
V^0, d^5	6	Octahedral	$V(CO)_6$, $V(bipy)_3$, $V[C_2H_4(PMe_2)_2]_3$
V^I, d^4	6	Octahedral	$[V(bipy)_3]^+$
		Tetragonal pyramidal	η^5-$C_5H_5V(CO)_4$
	7	Monocapped octahedral	$[V(CO)_3(PMe_3)_4]^+$, $V(CO)_2(dmpe)_2Cl$
V^{II}, d^3	4	Almost planar	$[Li(THF)]_2V(2,6\text{-diisopropylphenolate})_4$
	5	*sp*	$(2,6\text{-}Ph_2C_6H_3O)_2V(py)_3$
	6^a	Octahedral	$[V(H_2O)_6]^{2+}$, $[V(CN)_6]^{4-}$
V^{III}, d^2	3	Planar	$V[N(SiMe_3)_2]_3$, $V[CH(SiMe_3)_3]_3$
	4	Tetrahedral	$[VCl_4]^-$
	5	*tbp*	$trans$-$VCl_3(SMe_2)_2$, $VCl_3(NMe_3)_2$
	6^a	Octahedral	$[V(NH_3)_6]^{3+}$, $[V(C_2O_4)_3]^{3-}$, VF_3
	7	Pentagonal bipyramidal	$K_4[V(CN)_7]\cdot2H_2O$
V^{IV}, d^1	4	Tetrahedral	VCl_4, $V(NEt_2)_4$, $V(CH_2SiMe_3)_4$
	5	Tetragonal pyramidal	$VO(acac)_2$, $PCl_4^+VCl_5^-$
		?	$[VO(SCN)_4]^{2-}$
		sp	$[V_4O_4(OH)_2(PhPO_3)_4]^{2+}$, see text
		tbp	$VOCl_2$ $trans$-$(NMe_3)_2$
	6^a	Octahedral	VO_2 (rutile), K_2VCl_6, $VO(acac)_2(py)$, $V(acac)_2Cl_2$
	8	Dodecahedral	$VCl_4(diars)_2$, $V(S_2CMe)_4$
V^V, d^0	4	Tetrahedral (C_{3v})	$VOCl_3$
	5	*tbp*	$VF_5(g)$, $VNCl_2(quinuclidine)_2$
		sp	$CsVOF_4$
	6^a	Octahedral	$VF_5(s)$, VF_6^-, V_2O_5 (very distorted, almost *tbp* with one distant O); $[VO_2(ox)_2]^{3-}$, V_2S_5
	7	Pentagonal bipyramidal	$VO(NO_3)_3\cdot CH_3CN$, $VO(Et_2NCS_2)_3$

^{a}Most important states.

vanadinite [$Pb_5(VO_4)_3Cl$], and *carnotite* [$K(UO_2)VO_4 \cdot \frac{3}{2}H_2O$]. The last of these is more important as a uranium ore, but the vanadium is usually recovered as well. Vanadium also occurs widely in certain petroleums, notably those from Venezuela. Vanadium pentoxide (V_2O_5) is recovered from flue dusts after combustion.

Very pure vanadium is rare because, like Ti, it is quite reactive toward oxygen, nitrogen, and carbon at the elevated temperatures used in conventional thermomet-allurgical processes. Since its chief commercial use is in alloy steels and cast iron, to which it lends ductility and shock resistance, commercial production is mainly as an iron alloy, ferrovanadium. The very pure metal can be prepared by the de Boer-van Arkel process (Section 17-A-1). It is reported to melt at ~1700°C, but addition of carbon (interstitially) raises the melting point markedly: vanadium containing 10% of carbon melts at ~2700°C. The pure, or nearly pure, metal resembles Ti in being corrosion resistant, hard, and steel gray. In the massive state it is not attacked by air, water, alkalis, or non-oxidizing acids other than HF at room temperature. It dissolves in nitric acid and aqua regia.

At elevated temperatures it combines with most nonmetals. With oxygen it gives V_2O_5 contaminated with lower oxides, and with nitrogen the interstitial nitride VN. Arsenides, silicides, carbides, and other such compounds, many of which are interstitial and nonstoichiometric, are also obtained by direct reaction of the elements.

VANADIUM COMPOUNDS

17-B-2 Vanadium Halides

The halides of vanadium are listed in Table 17-B-2 together with some of their reactions.

The tetrachloride is obtained not only from V + Cl_2 but also by the action of CCl_4 on red-hot V_2O_5 and by chlorination of ferrovanadium (followed by distillation to separate VCl_4 from Fe_2Cl_6). It is an oil that is violently hydrolyzed by water to give solutions of oxovanadium(IV) chloride; its magnetic and spectral properties confirm its nonassociated tetrahedral nature. It has a high dissociation pressure and loses chlorine slowly when kept, but rapidly on boiling, leaving VCl_3. The latter may be decomposed to VCl_2, which is then stable (mp 1350°C):

$$2VCl_3(s) \longrightarrow VCl_2(s) + VCl_4(g)$$

$$VCl_3(s) \longrightarrow VCl_2(s) + \frac{1}{2}Cl_2(g)$$

The bromide system is similar but there is only indirect evidence for VI_4 in the vapor phase. VF_4 is polymeric with layers of corner-sharing tetragonally compressed VF_6 moieties.[1] The trihalides have the BiI_3 structure in which each metal atom is at the center of a nearly perfect octahedron of halogen atoms.

The pentafluoride appears to be a trigonal bipyramid in the vapor, but with some unusual vibrational amplitudes or distortion. In the crystal it gives an infinite

[1]S. Becker and B. G. Müller, *Angew. Chem. Int. Ed. Engl.* **1990**, 29, 406.

Table 17-B-2 The Halides of Vanadium

$$VF_5{}^a \xrightarrow{\text{PCl}_3} VF_4{}^b \xrightarrow{\sim 150°C^c} VF_3 \xrightarrow[115°C]{H_2 + HF} VF_2$$

$VF_5{}^a$	$VF_4{}^b$	VF_3	VF_2
Colorless mp. 19.5°C bp. 48°C	Lime green subl > 150°C	Yellow green	Blue

25°C | HF in CClF$_3$ 600°C | HF(g)   600°C | HF(g)

$$VCl_4{}^a \xrightleftharpoons[\text{Cl}_2]{\text{reflux}} VCl_3 \xrightarrow{>450°C} VCl_2$$

$VCl_4{}^a$ Red brown, bp. 154°C	VCl_3 Violet	VCl_2 Pale green

$$[VBr_4{}^d] \xrightleftharpoons[\text{Br}_2]{>-23°C} VBr_3{}^a \xrightarrow{>280°C} VBr_2$$

[$VBr_4{}^d$] Magenta	$VBr_3{}^a$ Black	VBr_2 Red brown

$$[VI_4(g)] \longleftarrow VI_3{}^e \xrightarrow{>280°C} VI_2$$

[$VI_4(g)$]	$VI_3{}^e$ Brown	VI_2 Dark violet

[a] Made by direct interaction at elevated temperatures; F$_2$, 300°C; Cl$_2$, 500°C; Br$_2$, 150°C.
[b] Also made by direct interaction of the metal powder with a very slight excess of F$_2$, in a closed system at 280°C (see reference 1).
[c] Disproportionation reaction (e.g., 2VCl$_3$ = VCl$_2$ + VCl$_4$).
[d] Isolated from vapor at $\sim$550°C by rapid cooling; decomposes above -23°C.
[e] Made in a temperature gradient with V at >400°C, I$_2$ at 250 to 300°C.

polymer (17-B-I), fragments of which persist in the liquid, thus explaining the high viscosity.

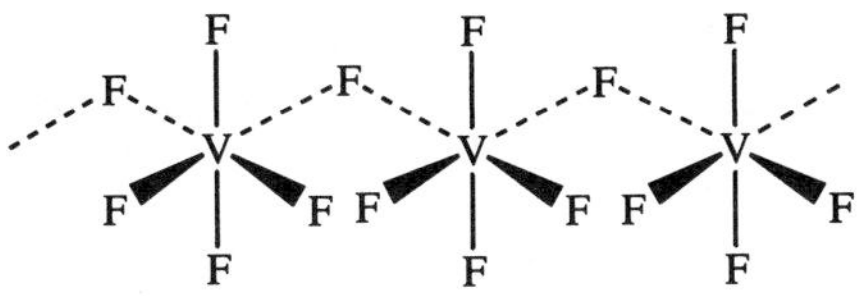

(17-B-I)

All the halides are Lewis acids and form complexes, which are discussed later under the pertinent oxidation states. The oxohalides are also discussed later.

17-B-3 The Chemistry of Vanadium(V)

Vanadium(V) Oxide

Vanadium(V) oxide is obtained on burning the finely divided metal in an excess of oxygen, although some quantities of lower oxides are also formed. The usual method of preparation is by heating ammonium metavanadate:

$$2NH_4VO_3 \longrightarrow V_2O_5 + 2NH_3 + H_2O$$

It is thus obtained as an orange powder that melts at ~650°C and solidifies on cooling to orange, rhombic needle crystals. Addition of dilute H_2SO_4 to solutions of NH_4VO_3 gives a brick red precipitate of V_2O_5. This has slight solubility in water (~0.007 g L^{-1}) to give pale yellow acidic solutions. Although mainly acidic, hence readily soluble in bases, V_2O_5 also dissolves in acids. That the V^V species so formed are moderately strong oxidizing agents is indicated by the evolution of chlorine when V_2O_5 is dissolved in hydrochloric acid; V^{IV} is produced. This oxide is also reduced by warm sulfuric acid. The following standard potential has been estimated:

$$VO_2^+ + 2H^+ + e = VO^{2+} + H_2O \qquad E^0 = 1.0 \text{ V}$$

Vanadates[2]. Vanadium pentoxide dissolves in sodium hydroxide to give colorless solutions and in the highly alkaline region, pH > 13, the main ion is VO_4^{3-}. As the basicity is reduced, a series of complicated reactions occurs. A protonated species is first formed:

$$VO_4^{3-} + H_2O = VO_3(OH)^{2-} + OH^- \qquad pK = 1.0$$

and this then aggregates into binuclear and subsequently more complex species depending on the concentration and pH.

In the pH range 2 to 6 the main species is the orange *decavanadate ion*, which can exist in several protonated forms.

$$V_{10}O_{28}^{6-} + H^+ \rightleftharpoons V_{10}O_{27}(OH)^{5-}$$

$$V_{10}O_{27}(OH)^{5-} + H^+ \rightleftharpoons V_{10}O_{26}(OH)_2^{4-}$$

$$V_{10}O_{26}(OH)_2^{4-} + H^+ \rightleftharpoons V_{10}O_{25}(OH)_3^{3-}$$

$$V_{10}O_{25}(OH)_3^{3-} + H^+ \rightleftharpoons V_{10}O_{24}(OH)_4^{2-} \xrightarrow[OH^-]{H^+} VO_2^+$$

The $V_{10}O_{24}(OH)_4^{2-}$ ion is very unstable and with further acid rapidly gives the dioxovanadium(V) ion (VO_2^+). In alkaline solution breakup of the $V_{10}O_{28}$ unit is much slower.

Many salts of the decavanadate ion, for example, $Ca_3V_{10}O_{28} \cdot 18H_2O$ and $(EtC_5H_4\text{-}NH)_4[V_{10}O_{28}H_2]$, can be crystallized and the structure, shown in Fig. 17-B-1(a), is well established. It consists of ten VO_6 octahedra fused together. Spectroscopic studies (Raman, ^{17}O and ^{51}V nmr) indicate that this structure persists in solution. Silicate anions combine with aqueous $[HVO_4]^{2-}$ in aqueous alkaline solutions to form $[H_2VSiO_7]^{3-}$, $[H_3VSiO_7]^{2-}$, and various related monovanado-oligosilicate species.[3]

[2]Q. Chen and J. Zubieta, *Coord. Chem. Rev.* **1992**, *114,* 107.
[3]O. W. Howarth and J. J. Hastings, *J. Chem. Soc., Dalton Trans.* **1996,** 4189.

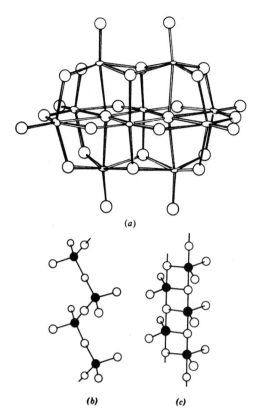

Figure 17-B-1 The structures in the crystalline state of (a) the decavanadate ion $V_{10}O_{28}^{6-}$, (b) the anion in KVO_3, (c) the anion in $KVO_3 \cdot H_2O$.

Many other vanadates are known.[4] The metavanadates, MVO_3 (M = NH_4, K, Rb, and Cs) are isomorphous, and have infinite chains of corner-sharing VO_4 tetrahedra [Fig. 17-B-1(b)], while the hydrate $KVO_3 \cdot H_2O$ [Fig.17-B-1(c)] has chains of linked VO_5 polyhedra. Most of the dinegative metavanadates, MV_2O_6 (M = Mg, Ca, Mn, Co, Cu, Zn, and Cd) have structures consisting of zigzag chains of VO_5 square pyramids sharing edges but with larger metal cations such as Ba, the structure adopted contains VO_4 tetrahedral units.[5] The latter structural motif is typically found in orthovanadates such as $LiMnVO_4$[6] and $LnVO_4$.[7] More complex structures are also known, like that of $KBaCd_2(VO_4)(V_2O_7)$, which contains both tetrahedral units of VO_4 and V_2O_7.[8] Other compounds contain the dinuclear motif $H_xV_2O_7^{(4-1x)}$ (x = 0, 1, 2). The ion $V_4O_{12}^{4-}$ has a cyclic tetrametaphosphate-like structure[9] while $V_5O_{15}^{5-}$ is pentanuclear.[10]

Oxovanadium—organophosphate species are legion.[11] An interesting dodecavanadate clathrate is formed as dark-red crystals of $(Bu_4^nN)_4[CH_3CN \subset (V_{12}O_{32})]$ by

[4]W. G. Klemperer *et al.*, *Angew. Chem. Int. Ed. Engl.* **1992**, *31*, 49.
[5]Y. Oka *et al.*, *Inorg. Chim. Acta* **1995**, *238*, 165.
[6]M. Sato *et al.*, *J. Mater. Chem.* **1996**, *6*, 1191.
[7]D. Mullica *et al.*, *Inorg. Chim. Acta* **1996**, *248*, 85.
[8]S. Münchau and Hk. Müller-Buschbaum, *Z. anorg. allg. Chem.* **1996**, *622*, 622.
[9]P. Román *et. al.*, *Inorg. Chem.* **1993**, *32*, 775.
[10]D. C. Crans, *Comments Inorg. Chem.* **1994**, *16*, 1.
[11]See for example: A. Müller *et al.*, *Angew. Chem. Int. Ed. Engl.* **1995**, *34*, 779; J. Salta and J. Zubieta, *J. Cluster Sci.* **1996**, *7*, 531.

refluxing an acetonitrile solution of $(Bu_4^nN)_4V_{10}O_{28}H_2^-$ for 1-2 min. The acetonitrile molecule is suspended in the center of a basket-like $V_{12}O_{32}^{4-}$ cage.[12]

Oxohalides and Their Complexes. The oxohalides are VOX_3 (X = F, Cl, and Br), VO_2F, and VO_2Cl. They form adducts with donors to give compounds such as the 5-coordinate $VO_2Cl(py)_2$.[13] Hydrothermal synthesis in aqueous HF and a mixture of V_2O_5, BaF_2 and NaF produces $NaVO_2F_2$,[14] the structure of which is related to that of $NaMoO_3F$. The VO_4F_2 polyhedra are linked by O corners but $[VO_2F_2]^-$ can also exist as a mononuclear species.[15] Vanadium oxotrichloride ($VOCl_3$), made by the action of Cl_2 on V_2O_5 + C at ~300°C is the most important oxohalide species. It is a yellow liquid, readily hydrolyzed, and consists of $VOCl_3$ molecules having C_{3v} symmetry. It readily interacts with Cl^- to give $VOCl_4^-$ and with many neutral ligands to give $VOCl_3L$ or $VOCl_3L_2$ adducts. The VO unit in the $VOCl_4^-$ ion and the $VOCl_3L$ compounds is similar in bond length (~1.55–1.60 Å) and stretching frequency (950–1000 cm^{-1}) to those in other VO^{2+} complexes.

Other Vanadium(V) Compounds. When vanadates are strongly acidified, the *cis*-$[VO_2(H_2O)_4]^+$ ion is formed, and it can give complexes by displacement of water molecules, for example, $[VO_2Cl_4]^{3-}$, $[VO_2(EDTA)]^{3-}$, and $[VO_2(ox)_2]^{3-}$. The *cis* VO_2 arrangement is favored (as for other d^0 systems) because this allows better $Op\pi \rightarrow Md\pi$ bonding than a linear arrangement would allow.

Formally similar to the oxohalides are oxo species in which some or all of the halogens are replaced by OR groups, namely, $VO(OR)_nCl_{3-n}$. Most of the V^V alkoxo-halide compounds are 6-coordinate but a few are 5- and 4-coordinate,[16] e.g., $[VOCl(OCMe_2CMe_2O)]_2$ and $[VOCl(OCH_2CH_2O)]_2$, respectively. Compounds of the type $TpVO(OR)Cl$ contain the shortest V—OR bonds yet reported (1.70–1.75 Å).[17] Related species are of the type $VO(NR_2)_3$ and $[VO(SCH_2CH_2)_3N]$.[18] There are some compounds which contain V—N multiple bonding, for example 6-coordinate $V(NPMe_2Ph)Cl(PMe_2Ph)$,[19] 5-coordinate $VCl_2(NPR_3)_3$, and 4-coordinate $[V(NPR_3)_4]Cl$ compounds.[20] Other non-oxo V^V complexes include the green $Tp^*V(NBu^t)Cl_2$[21] which are made by simple substitution reactions according to:

$$VCl_3(NBu^t) + KTp^* \longrightarrow Tp^*V(NBu^t)Cl_2 + KCl$$

Like oxo groups, imido, nitrido, and phosphoraniminato ligands can stabilize electron-poor metals such as V^V by π-donation. The latter has been regarded as

[12]V. W. Day *et al., J. Am. Chem. Soc.* **1989,** *111,* 5959.
[13]A. C. Sullivan *et al., Polyhedron* **1996,** *15,* 2387.
[14]M.-P. Crosnier-Lopez *et al., Eur. J. Solid State Inorg. Chem.* **1994,** *31,* 957.
[15]G. J. Leigh *et al., J. Chem. Soc., Dalton Trans.* **1994,** 2591.
[16]D. C. Crans *et al., Inorg. Chem.* **1993,** *32,* 247 and references therein.
[17]C. J. Carrano *et al., Inorg. Chem.* **1994,** *33,* 646.
[18]A. W. Addison *et al., Inorg. Chem.* **1996,** *35,* 1.
[19]G. J. Leigh *et al., J. Chem. Soc., Dalton Trans.* **1993,** 3609.
[20]N. M. Doherty *et al., Inorg. Chem.* **1994,** *33,* 4360.
[21]J. Sundermeyer *et al., Chem. Ber.* **1994,** *127,* 1201; M. Etienne, *Coord. Chem. Rev.* **1996,** *156,* 201 (this review has many references to other hydridotris(pyrazolylborato)vanadium complexes in oxidation states II–V).

bonding to the metal in limiting forms such as $\overline{M}{\equiv}N{-}\overset{+}{P}R_3$ and $M{-}\ddot{N}{=}PR_3$. However, one should be cautious with these simple representations. For example, it has been found that the $V{-}N$ stretching frequency in the phosphoraniminato complexes is more than $100\ cm^{-1}$ higher than that of the nitrido compounds[22] ($V{\equiv}N$). They have been prepared by reaction of imido complexes such as $Cl_3VNSiMe_3$ and various types of pyridines, for example:

$$Cl_3VNSiMe_3 + 2py \xrightarrow[\text{benzene}]{} [VNCl_2py_2]_n + ClSiMe_3$$

The nitride nitrogen seems to be very basic and polymerization of the type $V{\equiv}N{\rightarrow}V{\equiv}N\cdots$ is common, as observed in $[VNCl_2(py)_2]_n$. If a bulkier amine like quinuclidine or 4-Bu^{t}py is used instead of py, mononuclear complexes of the type $Cl_2L_2V{\equiv}N$ are formed. The VN stretching frequency for the mononuclear complex $VNCl_2(PMe_2Ph)_2$ is at $1125\ cm^{-1}$ while the polymeric compound $[VNCl_2(py)_2]_n$ shows $\nu(V{-}N)$ at $ca.$ $970\ cm^{-1}$. A similar decrease is observed in polymeric vanadyl containing complexes (Section 17-B-4).

Many vanadium(V) complexes with Schiff bases also are known.[23]

Peroxo Complexes[24]

The dissolution of V_2O_5 in 30% H_2O_2, or the addition of H_2O_2 to acidic V^V solutions, gives red peroxo complexes in which oxygen atoms on the vanadate ions are replaced by one or more O_2^{2-} groups. There are oxoperoxo complexes such as $VO(O_2)$-$(pic)(py)_2$, pentagonal pyramidal oxodiperoxo complexes like $H[VO(O_2)_2bipy]$, and pentagonal bipyramidal oxodiperoxo complexes such as $Na_3[VO(O_2)_2(ox)]$.[25] Triperoxo, e.g., $K[V(O_2)_3]\cdot3H_2O$[26] and tetraperoxo compounds like $K_3[VO(O_2)_4]$[27] have also been made.

The following equilibria, written in terms of predominant species at pH 7, have been studied by means of ^{51}V nmr:[28]

$$VO_2(OH)_2^- + 2H_2O_2 = VO(O_2)_2^- + 3H_2O \qquad K_1 = 4 \times 10^8 \qquad (1)$$

$$VO(O_2)_2^- + H_2O = VO(OH)(O_2)_2^{2-} + H^+ \qquad K_2 = 6 \times 10^{-8} \qquad (2)$$

$$VO(O_2)_2^- + H_2O_2 = V(OH)(O_2)_3^{2-} + H^+ \qquad K_3 = 2.7 \times 10^{-6} \qquad (3)$$

$$V(OH)(O_2)_3^{2-} + HO_2^- = V(O_2)_4^{3-} + H_2O \qquad K_4 = 2.8 \qquad (4)$$

[22]N. M. Doherty et al., J. Am. Chem. Soc. **1988**, 110, 8071.
[23]D. Rehder et al., Inorg. Chem. **1993**, 32, 1844.
[24]A. Butler et al., Chem. Rev. **1994**, 94, 625 and references therein.
[25]P. Schwendt and M. Pišarčik, Spectrochim. Acta **1990**, 46A, 397.
[26]M. C. Chakravorty et al., Polyhedron **1995**, 13, 695.
[27]W. P. Griffith et al., Polyhedron **1989**, 8, 1379.
[28]J. S. Jaswal and A. S. Tracey., Inorg. Chem. **1991**, 30, 3718; R. C. Thompson et al., Inorg. Chem. **1995**, 34, 4499.

$$V(O_2)_4^{3-} + OH^- = VO(O_2)_3^{3-} + HO_2^- \qquad K_{5a} = 40 \qquad (5a)$$

$$V(OH)(O_2)_3^{2-} = VO(O_2)_3^{3-} + H^+ \qquad K_{5b} = 10^{-12} \qquad (5b)$$

The peroxide group typically binds in a bidentate (η^2) fashion, which creates a local C_{2v} environment, giving rise to three ir and Raman active modes. These vibrations occur at approximately 880, 600, and 500 cm^{-1}.

Photooxidation of $V(CO)_6$ in low-temperature matrices containing oxygen produces a superoxo species, namely, $[VO_2(\eta^1\text{-}O_2)]$.[29]

Binuclear vanadium(V) peroxo complexes such as $(NH_4)_5[V_2O_2(O_2)_4PO_4]\cdot H_2O$ are also known.[30] They have been implicated as the reactive species in the bromide oxidation by hydrogen peroxide,[31] a process which is important in some biological systems (Section 17-B-8). Some stable peroxo compounds display insulin-mimetic activity both *in vivo* and *in vitro*.[32]

A number of V complexes of various nuclearities are now known that contain persulfide groups in either terminal or bridging modes, mostly in the III and IV oxidation states, but at least two pentavalent vanadium compounds have been characterized,[33] namely $(Me_3NCH_2Ph)_2[VS_2(S_2)(SPh)]$ and $(NEt_4)[VO(S_2)_2(bipy)]$.

Thiovanadates. The red tetrathiovanadate ion (VS_4^{3-}) can be prepared according to

$$VO(OMe)_3 + 4(Me_3Si)_2S + 3LiOMe \xrightarrow[\text{5h}]{\text{MeCN}} Li_3VS_4 + (Me_3Si)_2O + 6Me_3SiOMe$$

The $Li_3VS_4\cdot 2tmeda$ solvate can be obtained highly pure upon recrystallization from tmeda; its structure shows slightly distorted tetrahedral, D_{2d}, $[MS_4]^{3-}$ ions, similar to that of the Nb and Ta analogues.[34] The V^V—S distance of 2.157(1) Å is longer than the typical V^{IV}=S distance of 2.07–2.12 Å found in 5-coordinate species such as $[VS(SPh_4)_4]^{2-}$, $[VS(ethanedithiolate)_2]^{2-}$, and VS(acetylacetonylideniminate); a similar pattern is found for the Nb and Ta analogues. This can be explained by symmetry considerations, as the π-bond order in a tetrahedral d^0 complex cannot exceed the 1.5–2 range whereas, in the pseudotetragonal arrangements usually encountered in the other complexes, a higher bond order is formally possible. The reactivity of the VS_4^{3-} anion has been studied extensively,[35] mainly because of the possible relevance of V—S bonds in V—Nitrogenase (see later); some reactions are shown in Fig. 17-B-2.

17-B-4   The Chemistry of Vanadium(IV), *d*¹

The halides (VX_4), have been noted (Section 17-B-2), and there are compounds derived from them such as $V(CH_2SiMe_3)_4$, $V(NR_2)_4$,[36] $[V(OR)_4]_x$, and dithiocarba-

[29]M. J. Almond and R. W. Atkins, *J. Chem. Soc., Dalton Trans.* **1994,** 835.
[30]P. Schwendt *et al., Inorg. Chem.* **1995,** *34,* 1964.
[31]M. J. Clague and A. Butler, *J. Am. Chem. Soc.* **1995,** *117,* 3475.
[32]See for example: A. Shaver *et al., Inorg. Chem.* **1993,** *32,* 3109.
[33]G. Christou *et al., Inorg. Chem.* **1993,** *32,* 204, 2978.
[34]R. H. Holm *et al., Inorg. Chem.* **1992,** *31,* 4333.
[35]Q. Liu *et al., Inorg. Chem.* **1997,** *36,* 214 and references therein.
[36]F. Maury *et al., J. Mater. Chem.* **1996,** *6,* 1501.

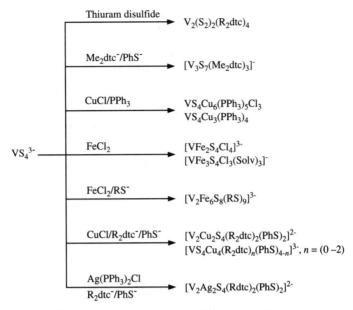

Figure 17-B-2 Some reactions of VS_4^{3-}, dtc = dithiocarbamate.

mates, $V(S_2CNR_2)_4$, but the chemistry of vanadium(IV) is dominated by oxygen compounds.

Vanadium(IV) Oxide and Oxo Anions

The dark blue oxide VO_2 is obtained by mild reduction of V_2O_5, a classic method being by fusion with oxalic acid; it is amphoteric, being about equally readily soluble in both noncomplexing acids, to give the blue ion $[VO(H_2O)_5]^{2+}$, and in base. It has a distorted rutile structure; one bond (the V=O) is much shorter than the others in the VO_6 unit (note that the Ti—O distances in TiO_2 are essentially equal).

When strong base is added to a solution of $[VO(H_2O)_5]^{2+}$, a gray hydrous oxide $VO_2 \cdot nH_2O$, is formed at ~pH 4. This dissolves to give brown solutions from which brown-black salts, e.g., $Na_{12}V_{18}O_{42} \cdot 24H_2O$, can be crystallized. These contain the $V_{18}O_{42}^{12-}$ ion (Fig. 17-B-3), which is somewhat unstable in dilute solution. The more stable anions, $[H_4V_{18}O_{42}X]^{9-}$, X = Cl, Br, I, might contain various encapsulated species.[37] For example neutral molecules such as acetonitrile or others such as halide ions,[38] in $(Ph_4P)_4\{V_{18}O_{25}(H_2O)_2PhPO_3)_{20}\}Cl_4$ chloride ions are found in four cavities.[39] Many other examples are known in which arsenic or phosphorus are part of the cluster, as in the anion $[V_{15}As_6O_{42}(H_2O)]^{6-}$,[40] but perhaps the largest number of compounds described so far have a mixture of V^V, V^{IV}, and V^{III} ions.[41] In $[V_{19}O_{41}(OH)_9]^{8-}$ all V^{IV} sites are octahedral and antiferromagnetically coupled[42]

[37]A. Müller et al., Angew. Chem. Int. Ed. Engl. **1995,** 34, 2328.
[38]A. Müller et al., Angew. Chem. Int. Ed. Engl. **1990,** 29, 926.
[39]J. Zubieta et al., Angew. Chem. Int. Ed. Engl. **1994,** 33, 757.
[40]A. Müller and J. Döring, Angew. Chem. Int. Ed. Engl. **1988,** 27, 1721.
[41]M. T. Pope and A. Müller, Angew. Chem. Int. Ed. Engl. **1991,** 30, 34.
[42]A. Müller et al., Angew. Chem. Int. Ed. Engl. **1988,** 27, 1719.

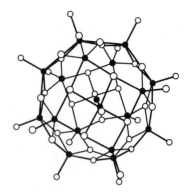

Figure 17-B-3   The structure of the vanadium(IV) iso-
polyanion $V_{18}O_{42}^{12-}$.

while all V^V are in a tetrahedral environment. A large number of oxovanadium-
organophosphonate compounds, containing anions such as $[V_5O_7(OCH_3)_2$-
$(PhPO_3)_5]^-$ and $[V_7O_{12}(PhPO_3)_6Cl]^{2-}$, are known.[43] Most of these compounds are
formed by hydrothermal synthesis. An interesting trilayer compound with polar and
non-polar domains, $[Et_2NH_2][Me_2NH_2][V_4O_4(OH)_2(PhPO_3)_4]$,[44] is made by reacting
aqueous solutions of $RbVO_3$, $PhPO(OH)_2$, and R_2NH_2Cl at 160°C for 4 days. The
structure shows a layer of $Et_2NH_2^+$ cations sandwiched between inorganic V/P/O
layers which are in turn bounded by organic bilayers consisting of phenyl groups
from adjacent $V/O/C_6H_5PO_3$ slabs as shown in Fig. 17-B-4. The structure consists
of corner-sharing $\{VO_5\}$ square pyramids and $\{CPO_3\}$ tetrahedra, with phenyl groups
directed exclusively to one face of the layer while the $\{V=O\}$ groups are directed
to the other. Since the $\{V=O\}$ groups of the adjacent layer are directed toward

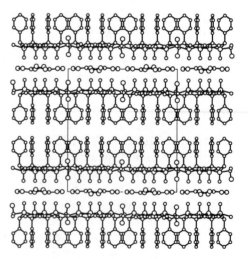

Figure 17-B-4   View of the *ac* plane of
$[Et_2NH_2][(Me_2)_2NH_2][V_4O_4(OH)_2(C_6H_5PO_3)_4]$
showing the layer of $Et_2NH_2^+$ cations intercalated between inorganic V/P/O layers, which
are in turn separated by bilayers of phenyl
groups. The $(Me_2)_2NH_2^+$ cations also penetrate the V/P/O layers.

[43]Q. Chen and J. Zubieta, *Angew. Chem. Int. Ed. Engl.* **1993**, *32*, 261; J. Zubieta *et al.*, *Angew. Chem. Int. Ed. Engl.* **1994**, *33*, 325; *Inorg. Chem.* **1996**, *35*, 5603.
[44]J. Zubieta *et al.*, *J. Am. Chem. Soc.* **1994**, *116*, 4525.

the vanadyl face of the neighboring layer, a highly hydrophilic region is produced which accommodates the layer of $Et_2NH_2^+$ cationic templates. Another interesting type of mixed-valence materials are the zeolite-like compounds based on pentavanadate phosphate building blocks. Examples are $[HN(CH_2CH_2)_3NH]K_{1.35}[V_5O_9(PO_4)_2]\cdot xH_2O$ and $Cs_3[V_5O_9(PO_4)_2]\cdot xH_2O$. These materials are composed of VO_5 square pyramids and PO_4 tetrahedra arranged in such a way as to produce very large cavities, thus having very low densities.[45]

The Oxovanadium(IV) Ion (VO²⁺) and Its Complexes. This ion dominates vanadium(IV) chemistry. It is obtained by mild reduction of the VO_2^+ ion or by oxidation by air of V^{3+} solutions:

$$VO^{2+} + 2H^+ + e = V^{3+} + H_2O \qquad E^0 = +0.34 \text{ V}$$

$$VO_2^+ + 2H^+ + e = VO^{2+} + H_2O \qquad E^0 = +1.0 \text{ V}$$

The interaction of V_2O_5 with ethanolic HCl gives a solution containing $VOCl_5^{3-}$ that can conveniently be used as a source of V^{IV} oxo complexes. Also, $VOCl_3$ can be reduced by hydrogen to the deliquescent solid $VOCl_2$; anhydrous $VOCl_2(g)$ has been detected in the vapor over a mixture of V_2O_3 and Cl_2 at 550–620 K.[46]

Many compounds containing the VO^{2+} unit are blue and display two other characteristic physical properties: (1) an epr spectrum with characteristic *g* values and ^{51}V hyperfine coupling (8 lines) and (2) a strong V=O stretching band in the ir in the range 950 to 1035 cm^{-1} (a little lower, *ca.* 880 cm^{-1}, for compounds which contain V—O—V chains in the solid state). Because, as noted later, VO^{2+} complexes may be five- or 6-coordinate with the additional ligand (trans to the V=O bond) showing various degrees of interaction, all of the spectral features reflect the exact nature of the ligand set. The V=O bond is very strong, possibly having partial triple-bond character and the V=O distances are thus very short (1.55–1.68 Å) with a dependence on the nature of the ligand set, especially the strength of coordination in the position trans to the V=O bond. In some solid compounds the V=O units are stacked to give V=O···V=O chains.

Representative solid compounds of VO^{2+} include $VOSO_4\cdot nH_2O$ (n = 5 or 6), which contain discrete $VO(H_2O)_4^{2+}$ and $VO(H_2O)_5^{2+}$ ions, and many compounds with oxo anions in which three-dimensional networks arise by coordination of bridging anions, for example, $VOSO_4\cdot 3H_2O$, $VOMoO_4$, $(VO)_2P_2O_7$, $(VO)_2H_4P_2O_9$, and some $VO(RPO_3)$ or $VO(HOPO_3)$ compounds that exhibit a rich coordination chemistry.[47]

Discrete complexes of VO^{2+} are legion. For the most part they have square pyramidal structures, akin to that shown in (17-B-II) for the acetylacetonate; to these a sixth ligand may be attached trans to V=O. An interesting structure which contains an almost linear chain of the type $[V^{IV}{=}O{\rightarrow}V^{IV}{-}O{-}V^{IV}{\leftarrow}O{=}V^{IV}]^{2+}$ has been found in $[V(salen)OV(salen)OV(salen)OV(salen)][BF_4]_2$.[48] In a few cases, for

[45]R. C. Haushalter *et al.*, *Chem. Mater.* **1996**, *8*, 43.

[46]R. Gruehn *et al.*, *Z. anorg. allg. Chem.* **1996**, *622*, 1651.

[47]J. Zubieta *et al.*, *Inorg. Chem.* **1994**, *33*, 3855; R. C. Haushalter *et al.*, *Angew. Chem. Int. Ed. Engl.* **1995**, *34*, 223.

[48]D. L. Hughes *et al.*, *J. Chem. Soc., Dalton Trans.* **1994**, 2457.

example, certain Schiff base complexes, the structure is distorted trigonal bipyramidal [e.g., (17-B-III)]; these are often not blue, but instead yellow or maroon. In a majority of cases the ligands are bi- or polydentate but simple $[VOX_4]^{2-}$ and $[VOX_5]^{3-}$ species (X = F, Cl, CN, SCN, etc.) have been well characterized. There are also bridged binuclear species[49] such as those shown in (17-B-IV), (17-B-V), and (17-B-VI). Complexes of VS^{2+} can be obtained by treating VO^{2+} complexes with B_2S_3. *Trans*-$VOCl_2L_2(H_2O)$, L = THF and CH_3CN, are convenient starting materials.[50]

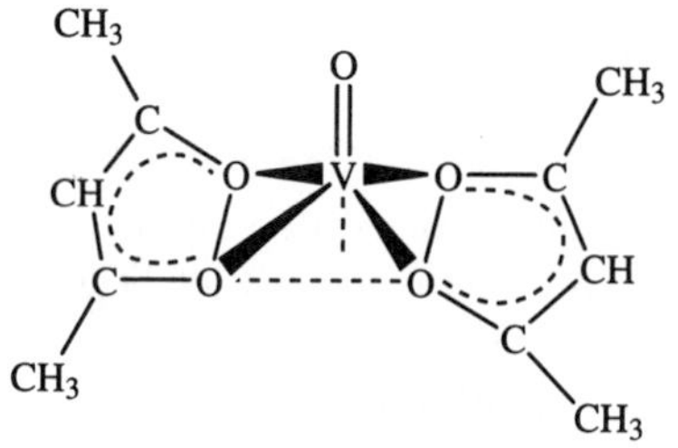

(17-B-II)

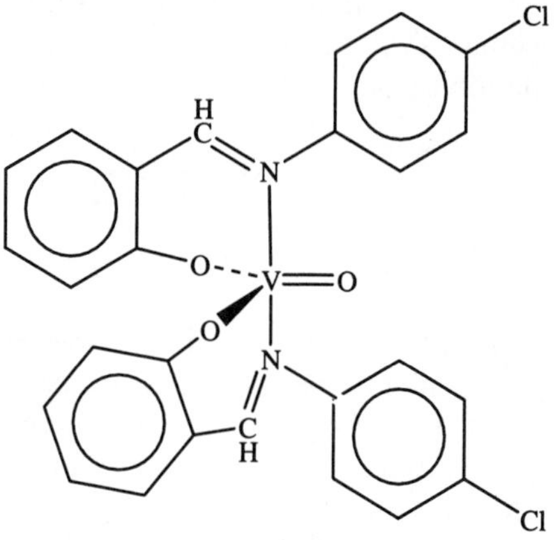

(17-B-III)

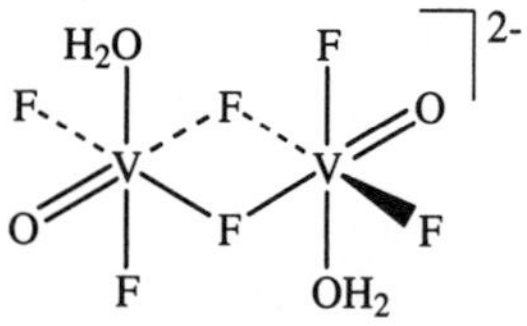

(17-B-IV)

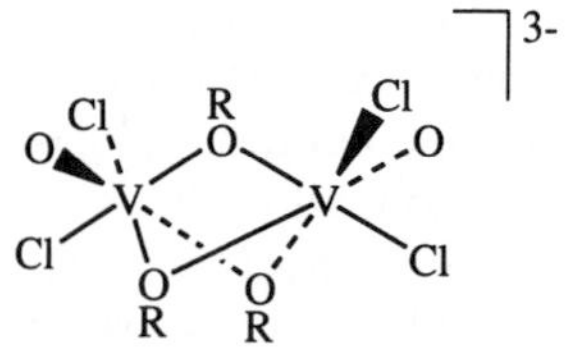

(17-B-V)

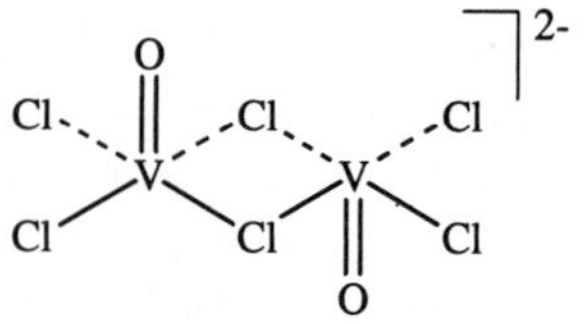

(17-B-VI)

The occurrence of a VO^{2+} porphyrin complex in petroleum or shale oil is well known. Many other VO(porph) complexes can be made and have the expected square pyramidal structures; they can be reduced electrochemically.[51]

Other Complexes. While the overwhelming majority of V^{IV} complexes contain the VO^{2+} ion, *(vide supra)* there are a few which do not, examples being a *tris*-catecholate, $[V(cat)_3]^{2-}$, and a series of *cis*- and *trans*-$VX_2(LL)_2$ (X = Cl and Br), in which LL is a bidentate ligand employing (O, O), (N, O), or (N, N) donors;[52] these are obtained by deoxygenation of the corresponding VO(LL)_2 complexes

[49]R. E. McCarley *et al., J. Am. Chem. Soc.* **1996**, *118*, 5302; J. Zubieta *et al., Inorg. Chem.* **1994**, *33*, 6340; W. Plass, *Angew. Chem. Int. Ed. Engl.* **1996**, *35*, 627; J. C. Pessoa *et al., J. Chem. Soc., Dalton Trans.* **1996**, 1989; C. J. Carrano *et al., Inorg. Chem.* **1996**, *35*, 7643.
[50]D. Collins *et al., Inorg. Chem.* **1993**, *32*, 664; W. Priensch and D. Rehder, *Inorg. Chem.* **1990**, *29*, 3013.
[51]K. M. Kadish *et al., Inorg. Chem.* **1989**, *28*, 2528.
[52]E. Ludwig *et al., Z. anorg. allg. Chem,* **1995**, *621*, 23.

with $SOCl_2$. Other "bare" complexes have been synthesized by electrochemical methods[53] from V^{III} precursors.

17-B-5   The Chemistry of Vanadium(III), d^2

Vanadium(III) Oxide

This black, refractory substance is made by reduction of V_2O_5 with hydrogen or carbon monoxide. It has the corundum structure but is difficult to obtain pure, since it has a marked tendency to become oxygen deficient without change in structure. Compositions as low in oxygen as $VO_{1.35}$ are reported to retain the corundum structure.

Divanadium trioxide is entirely basic and dissolves in acids to give solutions of the V^{III} aqua ion or its complexes. From these solutions addition of OH^- gives the hydrous oxide, which is very easily oxidized in air.

The Aqua Ion and Complexes

The greenish blue aqua ion $[V(H_2O)_6]^{3+}$ can be obtained as above or by electrolytic or chemical reduction of V^{IV} or V^V solutions. Such solutions, and also others, of V^{III} are subject to aerial oxidation in view of the potential

$$VO^{2+} + 2H^+ + e = V^{3+} + H_2O \qquad E^0 = 0.34 \text{ V}$$

The ion hydrolyzes partially to VO^+ and $V(OH)^{2+}$.

When solutions of V^{2+} and VO^{2+} are mixed, V^{3+} is formed, but VOV^{4+}, a brown intermediate species that has an oxo bridge, occurs. Studies of simple μ-oxo divanadium(III) complexes have identified two general classes of compounds: (1) unsupported, monobridged μ-oxo species in which the V—O—V bridging angle falls in the range 165–180° and (2) tribridged species in which the oxo-bridge is supported by two co-bridging carboxylate or phosphate groups resulting in a bridging V—O—V angle that falls in the range 130–145°.[54] Strong ferromagnetic exchange interactions are found in dinuclear compounds of the type $[L_2V_2(\mu\text{-}O)(\mu\text{-}RCOO)_2]$, L = tacn or Tp which switch to antiferromagnetic interactions upon protonation to give $[L_2V_2(\mu\text{-}OH)(\mu\text{-}RCOO)_2]$;[55] the V—O—V angle decreases from 133 to 123°.

In the thiolate $V_2O(SCH_2CH_2NMe_2)_4$ there is also a linear VOV^{4+} unit. In $V_2O(SPh)_4(Me_2bipy)_2$ which has a $V_2(\mu\text{-}O)(\mu\text{-}SPh)_2^{2+}$ core with a terminal PhS^- and a chelating Me_2bipy group at each metal atom, the V—V separation of 2.579(3) Å is suggestive of a V^{III}—V^{III} single bond.[56] A more complex species, $[V_4O_2(O_2CEt)_7\text{-}(bipy)_2]$,[57] contains μ_3-O groups and has a butterfly-shaped core (17-B-VII).

[53]A. Neves et al., J. Chem. Soc., Chem. Commun. **1992**, 652.

[54]R. S. Czernuszewicz et al., Inorg. Chem. **1994**, 33, 6116; C. J. Carrano et al., Chem. Commun. **1996**, 37.

[55]P. Knopp and K. Wieghardt, Inorg. Chem. **1991**, 30, 4061; C. J. Carrano et al., Inorg. Chem. **1993**, 32, 3589.

[56]G. Christou et al., Inorg. Chem. **1994**, 34, 1608.

[57]D. N. Hendrickson et al., J. Chem. Soc., Chem. Commun. **1995**, 2517.

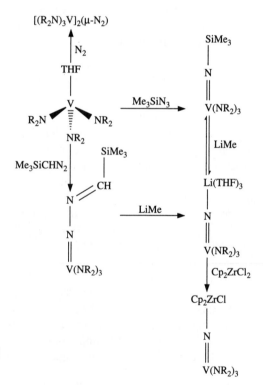

(17-B-VII)

Vanadium(III) forms a number of complex ions, mostly octahedral, for example, $[V(H_2O)_6]^{3+}$, $[VCl_2(MeOH)_4]^+$, $[V(ox)_3]^{3-}$, $[V(NCS)_6]^{3-}$, and $[VCl_2(THF)_2(H_2O)_2]^+$. A 7-coordinate cyanide complex $K_4[V(CN)_7]\cdot 2H_2O$, can be obtained as red crystals by the action of KCN on VCl_3 in dilute HCl. Others like $V(NR_2)_3(THF)$, R = Ph, Pr^i, Cy, are tetrahedral and very reactive[58] (Fig 17-B-5); with dinitrogen they yield diamagnetic $[(R_2N)_3V]_2(\mu-N_2)$.[59] Pyramidal complexes are obtained using the triami-

Figure 17-B-5 Some reactions of $V(NR_2)_3(THF)$ compounds.

[58]See for example: J.-I. Song and S. Gambarotta, *Chem. Eur. J.* **1996,** *2*, 1258; S. Gambarotta *et al., Chem. Commun.* **1996,** 779.

[59]S. Gambarotta *et al., J. Am. Chem. Soc.* **1994,** *116*, 6927.

doamine $(N_3N_F)^{3-}$ anion,[60] where $N_3N_F^{3-} = [(C_6F_5NCH_2CH_2)_3N]^{3-}$:

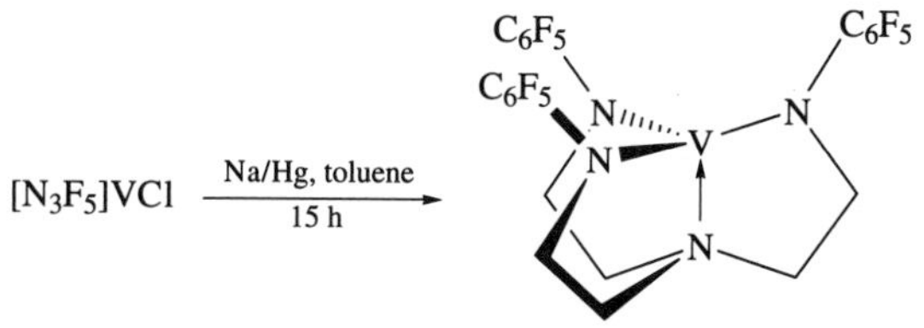

The $[V(H_2O)_6]^{3+}$ ion occurs in *alums* $M^IV(SO_4)_2 \cdot 12H_2O$.[61] The ammonium alum is obtained as air-stable, blue-violet crystals by electrolytic reduction of NH_4VO_3 in H_2SO_4. The hydrated halides $VX_3 \cdot 6H_2O$, have the structure *trans*-$[VCl_2(H_2O)_4]Cl \cdot 2H_2O$ as found in similar hydrates of Fe^{III} and Cr^{III}. The bromide and some bromo complexes can be made by heating V_2O_5 with ethanolic HBr (as noted previously, HCl gives the VO^{2+} species).

The V^{III} halides form numerous 6-coordinate adducts such as *mer*-$VCl_3(CNCMe_3)_3$, $VCl_3(THF)_2(H_2O)$,[62] and *mer*-$VCl_3(THF)_3$; the latter is an excellent starting material for synthesis of other V^{III} complexes. In addition, with bulkier ligands, 5-coordinate, trigonal bipyramidal complexes such as $VCl_3(NMe_3)_2$ are formed. Several types of vanadium(III) phosphine complexes, such as $[HPEt_3]VCl_4$-$(PEt_3)_2$, *mer*-$VCl_3(OPEt_3)_3$, and $V_2Cl_6(PMe_3)_4$ are obtained from the reaction of VCl_4 and the corresponding phosphine.[63] Hydrothermal reactions of V_2O_5, V, H_3PO_3, water, and M^I cation sources produce $M[V(HPO_4)_2]$ compounds which contain octahedral VO_6 units.[64]

17-B-6   The Chemistry of Vanadium(II)[65]

Binary compounds are few, those of importance being the halides (Section 17-B-2) and the black oxide (VO), which has an NaCl lattice but is prone to nonstoichiometry (obtainable with 45–55 at. % oxygen). It has somewhat metallic physical characteristics. Chemically it is basic, dissolving in mineral acids to give V^{II} solutions.

The Aqua Ion. Electrolytic or zinc reduction of acidic solutions of V^V, V^{IV}, or V^{III} or dissolution of the metal in acid produces violet air-sensitive solutions containing the $[V(H_2O)_6]^{2+}$ ion. These are strongly reducing (Table 17-1) and are oxidized by water with evolution of hydrogen even though the standard potential V^{3+}/V^{2+} would indicate otherwise. The oxidation of V^{2+} by air is complicated and appears to proceed in part by direct oxidation to VO^{2+} and in part by way of an intermediate species of type VOV^{4+}.

[60]R. R. Schrock *et al.*, *Inorg. Chem.* **1997,** *36,* 123.
[61]J. K. Beattie *et al.*, *J. Chem. Soc., Dalton Trans.* **1996,** 1481.
[62]P. Sobota *et al.*, *Polyhedron* **1996,** *15,* 381.
[63]F. A. Cotton *et al.*, *Inorg. Chim. Acta* **1994,** *215,* 47.
[64]R. C. Haushalter *et al.*, *Inorg. Chim. Acta* **1995,** *232,* 83.
[65]G. J. Leigh and J. S. de Souza, *Coord. Chem. Rev.* **1996,** *154,* 71.

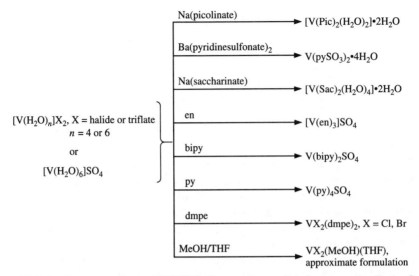

Figure 17-B-6 Some reactions of $V(H_2O)_nX_2$ complexes. See for example, F. A. Cotton *et al., Polyhedron* **1992,** *21,* 2767; C. A. Murillo *et al., Inorg. Chim. Acta* **1995,** *229,* 27; D. G. L. Holt *et al., Inorg. Chim. Acta* **1993,** *207,* 11.

Several crystalline salts contain the $[V(H_2O)_6]^{2+}$ ion, although the hydrate $VCl_2 \cdot 4H_2O$ is actually *trans*-$VCl_2(H_2O)_4$. The most important are the sulfates $VSO_4 \cdot nH_2O$, $n = 6, 7,$[66] which are formed as violet crystals on addition of ethanol to reduced sulfate solutions, and the double sulfates (Tutton salts) $M_2[V(H_2O)_6](SO_4)_2,$[67] where M = NH_4^+, K^+, Rb^+, or Cs^+. The electronic absorption spectra are consistent with octahedral aqua ions both in crystals and in solution, and the energy level diagram is analogous to that for Cr^{III}. The magnetic moments of the sulfates lie close to the spin-only value.

In spite of the d^3 configuration, which would be expected to be kinetically inert, most substitution reactions of aqueous solutions of $[V(H_2O)_6]^{2+}$ are not slow. Some representative reactions are shown in Fig. 17-B-6.

Complexes. The characteristic coordination is octahedral, and examples are *trans*-$VCl_2(py)_4$, *trans*-$[VX_2(dmpe)_2]$ (X = Cl, Br, I),[68] *trans*-VI_2(isonitrile)$_4$,[69] $[V(MeCN)_6][I_4]$,[70] $[V(NH_3)_6]I_2$,[71] $[V(NCS)_6]^{4-}$ salts, $[VCl_4(CH_3CO_2H)_2]^{2-}$ salts,[72] the porphyrin complexes[73] prepared by reduction of $V^{IV}(porph)X_2$ with zinc amalgam

[66]F. A. Cotton *et al., Inorg. Chem.* **1994,** *33,* 5391.
[67]F. A. Cotton *et al., Inorg. Chem.* **1993,** *32,* 4861.
[68]F. Süssmilch *et al., J. Organomet. Chem.* **1994,** *472,* 119.
[69]D. Rehder *et al., J. Organomet. Chem.* **1995,** *496,* 43.
[70]P. B. Hitchcock *et al., J. Chem. Soc., Dalton Trans.* **1994,** 3683.
[71]H. Jacobs *et al., Z. anorg. allg. Chem.* **1996,** *622,* 1161.
[72]L. F. Larkworthy *et al., Polyhedron* **1995,** *14,* 1453.
[73]See for example: H. Brand and J. Arnold, *Coord. Chem. Rev.* **1995,** *140,* 137 (this review also contains many references to porphyrin complexes of vanadium in higher oxidation states).

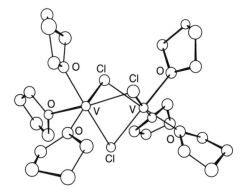

Figure 17-B-7 The structure of the $[V_2(\mu\text{-Cl})_3(THF)_6]^+$ ion.

to give *trans*-V(porph)L_2, L = PPhMe$_2$, or THF and the crown-thioether complex $[VI_2(THF)([9]aneS_3)]$.[74]

Alkoxide compounds show various coordination numbers and geometries which depend on the bulkiness of the ligand. Examples are the nearly planar 4-coordinate $[Li(THF)]_2V(2,6\text{-diisopropylphenolate})_4$,[75] square pyramidal $(2,6\text{-Ph}_2PhO)_2V(py)_3$,[76] and octahedral guaicolato $[Na(THF)VG_3]_2$.[77]

Dinuclear compounds of the type $[V_2(\mu\text{-X})_3(THF)_6]_2[M_2Cl_6]$, X = Cl, Br[78] (Fig. 17-B-7) are prepared by reaction of VX_3 and metals such as Zn, Mn, and Fe in THF; their solubility is relatively limited in organic solvents and they are used for cross coupling of aldehydes.[79] Some or all of the THF ligands in the $[V_2Cl_3(THF)_6]^+$ ion can be replaced by other ligands such as PMe$_3$ or PPh$_3$. The cation has V^{II}—V^{II} electronic coupling that makes normally spin-forbidden excitations of the V^{II} ions appear strongly in the spectrum. A similar coupling is detected for V^{II} doped into CsMgCl$_3$.

The largest class of vanadium(II) compounds known contains nitrogen donor atoms such as amines and formamidines. Some complexes are shown in Fig 17-B-8. Other examples are the square pyramidal $[2,5\text{-}(CH_3)_2C_4H_2N]_2V(py)_3$,[80] octahedral $Vpy_2(DTolF)_2$,[81] and $VCl_2(tmeda)_2$. The latter rearranges in solution to give a *tri-angulo*-$\{V_3(\mu\text{-Cl})_3\}$ unit, according to a facile equilibrium:[82]

$$3[VCl_2(tmeda)_2] \rightleftharpoons [V_3(\mu\text{-Cl})_3(\mu_3\text{-Cl})_2(tmeda)_3]^+ + Cl^- + 3tmeda$$

The ability of several V^{II} containing systems, namely, V(OH)$_2$/Mg(OH)$_2$ slurry, V(OH)$_2$/ZrO$_2$·H$_2$O, and certain catechol complexes of V^{II} to reduce N$_2$ to NH$_3$

[74]R. L. Richards *et al.*, *Inorg. Chim. Acta* **1996**, *251*, 13.
[75]W. H. Armstrong *et al.*, *J. Am. Chem. Soc.* **1990**, *112*, 2429.
[76]S. Gambarotta *et al.*, *J. Am. Chem. Soc.* **1993**, *115*, 6710.
[77]C. Floriani *et al.*, *Angew. Chem. Int. Ed. Engl.* **1988**, *27*, 576.
[78]See for example: F. A. Cotton *et al.*, *Polyhedron* **1988**, *7*, 737; G. Pampaloni and U. Englert, *Inorg. Chim. Acta* **1995**, *231*, 167.
[79]S. F. Pederson *et al.*, *J. Am. Chem. Soc.* **1994**, *116*, 1316.
[80]S. Gambarotta *et al.*, *Inorg. Chem.* **1991**, *30*, 2062.
[81]F. A. Cotton and R. Poli, *Inorg. Chim. Acta* **1988**, *141*, 91.
[82]G. J. Leigh *et al.*, *J. Chem. Soc., Dalton Trans.* **1997**, 1127.

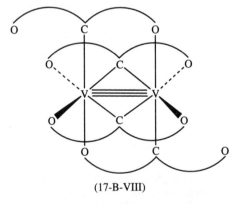

Figure 17-B-8   Some reactions of (tmeda)$_2$VCl$_2$ with nitrogen-containing ligands.

and/or N$_2$H$_4$ has been demonstrated and the energetics of the process have been studied by theoretical calculations.[83] A few well-characterized dinitrogen complexes have been isolated[84]

V—V Bonding. Since its congeners, Nb and Ta, as well as its neighbor Cr form numerous compounds with strong M—M bonds, similar behavior by vanadium might have been expected, but so far only a few are known. A triple bond (V—V = 2.20 Å) occurs in V$_2$(2,6-dimethoxyphenyl)$_4$, which has an edge-sharing bioctahedral structure shown schematically in 17-B-VIII.

(17-B-VIII)

Two compounds have been reported that appear to contain double bonds; both are rather complicated and it is not clear that either one provides a general approach.

[83]A. Sgamellotti *et al., Inorg. Chem.* **1994,** *33,* 4390; **1995,** *34,* 3410.
[84]C. Floriani *et al., Angew. Chem. Int. Ed. Engl.* **1993,** *32,* 396; S. Gambarotta *et al., J. Am. Chem. Soc.* **1994,** *116,* 7417; D. Rehder *et al., J. Chem. Soc., Dalton Trans.* **1994,** 3471 (lower valent V-dinitrogen complexes).

Their formulas are $V_2(PMePh_2)_4(H_2ZnH_2BH_2)_2$ and $V_2(salophen)_2Na_2(THF)_6$. *Ab initio* SCF/CI calculations predict only antiferromagnetic V—V coupling in $(CpV)_2C_8H_4$.[85] No $V_2(O_2CR)_4$ compounds have been reported; most attempts to make them have led to oxo-centered trinuclear species of the type $[V_3O(O_2CR)_6]^{0,+}$. In $(\eta^5\text{-}RC_5H_4)_2V_2S_4$ molecules, the V—V distance of 2.610 Å implies the existence of a single bond, but it has been argued that a triple bond exists in $[(\eta^6\text{-}C_6H_5CH_3)V(CO)_2]_2$ (17-B-IX) which has a V—V distance of 2.388(2) Å.

There is no doubt that a triple bond exists in $V_2(formamidinate)_4$ compounds; in the paddlewheel $V_2(DTolF)_4$[86] complex (17-B-X) the V—V bond distance is only 1.978(2) Å. The compound is diamagnetic with a large magnetic anisotropy. A key in its preparation was the use of $NaHBEt_3$ to produce a soluble $VCl_2(THF)_n$ species.[87]

$$VCl_3(THF)_3 + NaHBEt_3 \xrightarrow{\text{THF}} VCl_2(THF)_n + NaCl + {}^1\!/_2H_2$$

$$2VCl_2(THF)_n + 4LiDTolF \xrightarrow{\text{THF}} V_2(DTolF)_4 + 4LiCl$$

$V_2(DTolF)_4$ reacts with pyridine to give $V(py)_2(DTolF)_2$:

$$V_2(DTolF)_4 \xrightarrow[\Delta]{\text{pyridine}} 2 \text{ } trans\text{-}V(DTolF)_2(py)_2$$

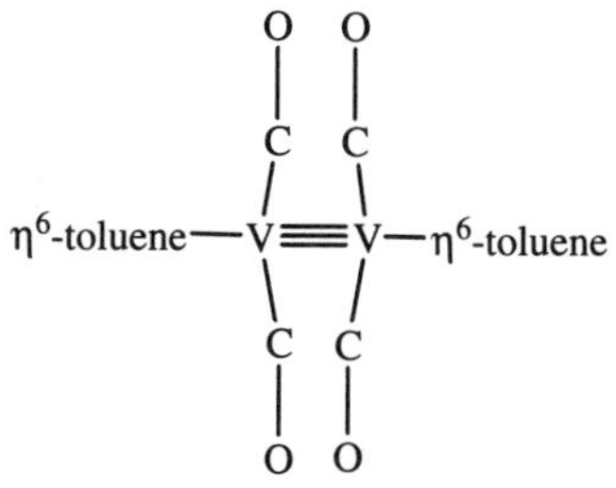

(17-B-IX) 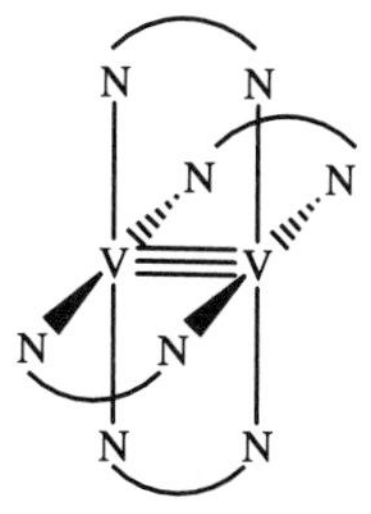   (17-B-X)

17-B-7 Carbonyl and Organometallic Compounds

Hexacarbonylvanadium is a 17-electron mononuclear compound which is readily reduced to the $V(CO)_6^-$ anion. It is prepared by high-temperature reductive carbonylation of VCl_3 under CO.

$$VCl_3(THF)_3 \xrightarrow[\text{CO}, \Delta, \text{ pressure}]{\text{Mg, Zn, I}_2} NaV(CO)_6 \xrightarrow{H^+} V(CO)_6$$

In liquid ammonia further reduction of $[Na(diglyme)_2][V(CO)_6]$ can be accomplished to produce $Na_3V(CO)_6$.

[85]M. Bénard *et al., Inorg. Chem.* **1990,** *29,* 2387.
[86]F. A. Cotton *et al., Inorg. Chem.* **1993,** *32,* 2881.
[87]F. A. Cotton *et al., Angew. Chem. Int. Ed. Engl.* **1992,** *31,* 737.

Substitution of the carbonyl groups gives a large variety of compounds such as phosphine-substituted $V(CO)_{6-n}P_n$ and $HV(CO)_{6-n}P_n$[88] complexes.

Metal complexes of arenes and heteroarenes[89] can be made by means of metal-atom ligand-vapor cocondensation or by conventional chemistry such as the reduction of $(\eta^6\text{-}1,3,5\text{-}Me_3C_6H_3)_2VI$:

$$V(at) + 2[2,4,6\text{-}Bu^t_3\text{-}\eta^6\text{-}C_5H_2P] \longrightarrow [2,4,6\text{-}Bu^t_3\text{-}\eta^6\text{-}C_5H_2P]_2V$$

$$(\eta^6\text{-}1,3,5\text{-}Me_3C_6H_3)_2VI + Cp_2Co \xrightarrow{\text{THF}} (\eta^6\text{-}1,3,5\text{-}Me_3C_6H_3)_2V + Cp_2CoI$$

Complexes containing the $V(mes)_2^+$ cation[90] can be made in high yields by reaction of VCl_3 with $Al/AlCl_3$ in refluxing mesitylene according to

$$VCl_3 + 2\,mes \xrightarrow[\Delta]{Al/AlCl_3} V(mes)_2AlCl_4$$

The discrete $V(mes)_2^+$ species contains symmetrically bonded, and almost parallel rings similar to the neutral species. Carbonyl and phosphine adducts of 17-electron open vanadocenes such as those derived from pentadienyls[91] are also known, e.g., $(1\text{-}C_6H_9)_2V(CO)$ (17-B-XI).

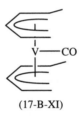

(17-B-XI)

Cyclopentadienyl derivatives are legion. Reduction of VCl_3 in THF with zinc dust and then addition of LiCp produces paramagnetic Cp_2V whose centrosymmetric molecules in the crystal have an ideal staggered conformation of the Cp-rings.[92] Vanadocene can be oxidized to the important synthetic precursors Cp_2VX and Cp_2VX_2. But the most common derivatives are the mono(η^5-cyclopentadienyl) complexes[93] which are known in various oxidation states such as $CpV(CO)_4$, $CpVCl(THF)$, $CpVCl_2(PR_3)_2$, $CpVX_3$, and $CpV(O)X_2$. Vanadium 51 has been used successfully to study the solution behavior of various compounds.[94] Cyclopendienyl complexes have both the highest ^{51}V shielding, for the $[CpV(CO)_3SnPPh_3]^-$ anion (-2059 ppm), and the lowest ^{51}V shielding, for $Cp^*_2V_2Te_2Se_2$ ($+2375$ ppm).

The chemistry of σ-alkyls and aryls is less well developed than for some other elements, but $V[CH(SiMe_3)_2]_3$, $V(CH_2SiMe_3)_4$, $Li_4(Et_2O)_4[VPh_6]$,[95] $VR_2(dmpe)_2$

[88] See for example: D. Rehder *et al., J. Organomet. Chem.,* **1991,** *411,* 357.

[89] E. Elschenbroich *et al., Organometallics* **1993,** *12,* 3373.

[90] F. Calderazzo *et al., J. Organomet. Chem.* **1991,** *417,* C16.

[91] A. L. Rheingold *et al., Organometallics* **1992,** *11,* 1693.

[92] M. Yu. Antipin *et al., J. Organomet. Chem.* **1996,** *508,* 259.

[93] See for example: N. J. Coville *et al., Coord. Chem. Rev.* **1992,** *116,* 1 and references therein.

[94] D. Rehder, *Coord. Chem. Rev.* **1991,** *110,* 161 and references therein.

[95] P. P. Power *et al., Organometallics* **1988,** *7,* 1380.

(R = CH$_3$, CH$_2$SiMe$_3$),[96] VO(CH$_2$SiMe$_3$)$_3$, [(Me$_3$CCH$_2$)$_3$V]$_2$(μ-N$_2$),[97] and [V(mes)$_3$(THF)] are all isolable. The latter is a convenient starting material because it is easily prepared and reacts readily with protic sources such as α-amino acids.[98] Unstable alkyls are present in the solutions of vanadium oxides or halides and Al alkyls,[99] which are used in the Ziegler-Natta type reaction for the copolymerization of styrene, butadiene, and dicyclopentadiene to give synthetic rubbers.

17-B-8 Bioinorganic Chemistry[100]

It is known that vanadium in extremely small amounts is a nutritional requirement for many types of organism, including higher animals. Among marine organisms which accumulate vanadium are members of an order of tunicates, the Ascidiacea (sea squirts). Some lichens and a toadstool are also known to contain vanadium in the active sites of some enzymes. The high abundance of vanadium in fossil matter of animal and plant origin (for example, the crude oil from Venezuela) seems to indicate that this element was perhaps more prevalent in early life.

The two most important vanadium enzymes described to date are the vanadium nitrogenase (V$-$Nase) and haloperoxidase.

Vanadium nitrogenase is produced by certain bacteria grown in molybdenum-deficient environments. It is effective in the reduction of N$_2$ and other nitrogenase substrates, although with less activity than the Mo$-$Nase. The enzyme resembles the Mo analogue (see Sections 17-E-10 and 18-C-13) in the construction and structure of the prosthetic groups, as well as in its functions.[101] It consists of a FeV protein, FeVco, and an iron protein (a 4Fe$-$4S ferredoxin).

The V content of FeVco is variable. The original values reported based on a $\alpha_2\beta_2$ structure vary from 0.7 to 2 V atoms, 9 to 23 Fe atoms, and 20 acid-labile sulfur atoms. More recently the V:Fe:S ratios for different species of *A. vinelandii* have been given as 1:19:19 for $\alpha\beta_2$ and 2:30:34 for $\alpha_2\beta_2$ forms.

EXAFS studies have provided important clues on the coordination of the metal atoms. At the V K-edge the features of the absorption edge are consistent with V^{II} or V^{IV} in a distorted octahedral environment with Fe, S, and O or N atoms as the nearest neighbors to the V atom. The Fe K-edge is dominated by Fe$-$S and Fe$-$Fe interactions consistent with the structure proposed for FeMoco. These results emphasize the structural similarity between the cofactor centers of MoFe and VFe proteins.

Many complexes have been prepared as models for V-Nase.[102] An interesting group contains the [VFe$_3$S$_4$]$^{2-}$ cuboidal core[103] (17-B-XII), e.g. (Me$_4$N)[(DMF)$_3$-

[96]G. S. Girolami *et al.*, *J. Organomet. Chem.*, **1994**, *480*, 1.
[97]J. H. Teuben *et al.*, *Organometallics* **1993**, *12*, 2004.
[98]C. Floriani *et al.*, *Inorg. Chem.* **1993**, *32*, 2729; *Organometallics* **1993**, *12*, 1802.
[99]See for example: F. J. Feher and R. L. Blanski, *J. Am. Chem. Soc.* **1992**, *114*, 5886.
[100]N. D. Chasteen (Ed.): *Vanadium in Biological Systems*, Kluwer, Dordrecht, 1990; H. Sigel and A. Sigel, eds., *Metal Ions in Biological Systems*, Vol. 31, Marcel Dekker, Inc. New York, NY (1995); D. Rehder, *Angew. Chem. Int. Ed. Engl.* **1991**, *30*, 148; A. Butler and C. J. Carrano, *Coord. Chem. Rev.* **1991**, *109*, 61.
[101]R. E. Eady, *Chem. Rev.* **1996**, *96*, 3013.
[102]R. H. Holm *et al.*, *J. Am. Chem. Soc.* **1993**, *115*, 9515.
[103]D. Coucouvanis *et al.*, *J. Am. Chem. Soc.* **1995**, *117*, 3126.

VFe$_3$S$_4$X$_3$] (X = Cl, Br, and I), (Me$_4$N)$_2$[TpVFeS$_4$Cl$_3$], and (Me$_4$N)[(NH$_3$)-(bipy)Fe$_3$S$_4$Cl$_3$]. A study of the catalytic reduction of hydrazine (a nitrogenase substrate) to ammonia in the presence of an external source of electrons and protons shows that the rate of reduction decreases as the number of labile solvent molecules coordinated to the V atom decreases but does not depend on the nature of the atom attached to the Fe atoms.

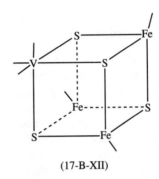

(17-B-XII)

In another vanadium-dependent enzyme, a haloperoxidase from the marine brown alga *Ascophyllum nodosum*, the active site contains two V^V per molecule. EXAFS[104] studies suggest a coordination environment for the active V^V site consisting of a V=O bond (1.61 Å), some short V—O bonds (1.72 Å) which have been attributed to esterlike bonds, and some longer bonds (2.11 Å) which could be V—O or V—N bonds. Vanadium haloperoxidase functions by coordinating hydrogen peroxide and then oxidizing halides; the nature of the oxidized halide species as enzyme-retained or enzyme-released seems to depend on the nature of the organic substrate.[105]

17-C CHROMIUM: GROUP 6

For chromium, as for Ti and V, the highest oxidation state is that corresponding to the total number of 3*d* and 4*s* electrons. Although Ti^{IV} is the most stable state for titanium and V^V is only mildly oxidizing, chromium(VI), which exists typically in the form of oxo compounds such as CrO$_3$, CrO$_4^{2-}$, and CrO$_2$F$_2$, is strongly oxidizing. Apart from stoichiometric similarities, chromium resembles the Group 16 elements of the sulfur group only in the acidity of the trioxide and the covalent nature and ready hydrolysis of CrO$_2$Cl$_2$.

The intermediate states Cr^V and Cr^{IV} have a somewhat restricted chemistry. The very low formal oxidation states are found largely in carbonyl- and organometallic-type compounds. With ligands such as aryl isocyanides, bipy, terpy, and phen, which are better donors and somewhat less obligatory back-acceptors than CO, it is possible to generate stable Cr^I compounds. With the isocyanides Cr(CNR)$_6$, electrochemical or Ag$^+$ oxidation may be used to obtain the +1 and +2 ions [Cr(CNR)$_6$]$^{n+}$, with one and two unpaired electrons, respectively. With the chelating

[104]S. S. Hasnain *et al., Biochemistry* **1989**, *28*, 7968.
[105]R. A. Tschirret-Guth and A. Butler, *J. Am. Chem. Soc.* **1994**, *116*, 411.

amine ligands the entire series of $[Cr(LL)_3]^n$ with $n = -1, 0, +1, +2,$ and $+3$ is obtained and the compounds are electrochemically interconvertible. In the more electron-rich ones electrons enter orbitals with appreciable (for -1, predominant) ligand character. Complexes containing the $[CrNO]^{2+}$ unit, which contain Cr^I, the Cr^I species $[Cr(CO)_2(dmpe)_2]^+$, and arene complexes such as $[Cr^I(C_6H_6)_2]^+$, are also known.

The most stable and generally important states are Cr^{II} and Cr^{III}. This dominance of the II and III states that begins here persists through the following transition elements. We shall discuss these states first.

The oxidation states and stereochemistry are summarized in Table 17-C-1.

Table 17-C-1   Oxidation States and Stereochemistry of Chromium

Oxidation state	Coordination number	Geometry	Examples
Cr^{-IV}	4	Tetrahedral	$Na_4Cr(CO)_4$
Cr^{-II}	5	*tbp*	$Na_2[Cr(CO)_5]$
Cr^{-I}	6	Octahedral	$Na_2[Cr_2(CO)_{10}]$
Cr^0, d^6	6	Octahedral	$Cr(CO)_6$, $[Cr(CO)_5I]^-$, $Cr(bipy)_3$
Cr^I, d^5	6	Octahedral	$[Cr(bipy)_3]^+$, $[Cr(CNR)_6]^+$
Cr^{II}, d^4	2	Bent (111°)	$Cr[N(Ph)B(mes)_2]_2^c$
	3	T shape	$Cr(OCBu^t_3)_2LiCl(THF)$
	3	Trigonal planar	$[Cr(NPr^i_2)_2]_2^d$
	4^b	Square	$CrCl_2(Me_3py)_2$, $Cr(acac)_2$
	4	Distorted tetrahedral	$CrCl_2(MeCN)_2$, $CrI_2(OPPh_3)_2$
	5	*tbp*	$[Cr(Me_6tren)Br]^+$
	6	Distorted^a octahedral	CrF_2, $CrCl_2$, CrS
	5 or 6	Cr—Cr quadruple bond	$Cr_2(O_2CR)_4L_2$, $Cr_2[(CH_2)_2P(CH_3)_2]_4$ $[Cr(CO)_2(diars)_2X]X$
	7	?	
Cr^{III}, d^3	3	Planar	$Cr[N(SiMe_3)_2]_3$
	4	Distorted tetrahedral	$[PCl_4]^+[CrCl_4]^-$, $[Cr(CH_2SiMe_3)_4]^-$
	5	*tbp*	$CrCl_3(NMe_3)_2$
	5	*sp*	$CrCl(tmtaa)^e$ (see text)
	6^b	Octahedral	$[Cr(NH_3)_6]^{3+}$, $Cr(acac)_3$, $K_3[Cr(CN)_6]$
$Cr^{IV} d^2$	4	Tetrahedral	$Cr(OC_4H_9)_4$, Ba_2CrO_4, $Cr(CH_2SiMe_3)_4$
	6	Octahedral	K_2CrF_6, $[Cr(O_2)_2(en)]H_2O$, *trans-*$[Cr(NCHMe)_2(dmpe)_2]^{2+}$
	8	Dodecahedral	$CrH_4(dmpe)_2$
Cr^V, d^1	4	Tetrahedral	CrO_4^{3-}
	5	Distorted *tbp*	$CrF_5(g)$
	5	*sp*	$CrOCl_4^-$
	6	Octahedral	$K_2[CrOCl_5]$, $(CrF_4)_n$
	8	Quasi-dodecahedral	$K_3Cr(O_2)_4$
Cr^{VI}, d^0	4	Tetrahedral	CrO_4^{2-}, CrO_2Cl_2, CrO_3

^{a} Four short and two long bonds.
^{b} Most stable states.
^{c} P. P. Power *et al., J. Am. Chem. Soc.* **1990,** *112,* 1048.
^{d} S. Gambarotta *et al., Inorg. Chem.* **1989,** *28,* 812.
^{e} F. A. Cotton *et al., Inorg. Chim. Acta* **1990,** *172,* 135.

17-C-1   The Element

The chief ore[1] is *chromite* ($FeCr_2O_4$), which is a spinel with Cr^{III} on octahedral sites and Fe^{II} on the tetrahedral ones. If pure chromium is not required—as for use in ferrous alloys such as the chromium additive used to make stainless steel—the chromite is reduced with carbon in a furnace, affording the carbon-containing alloy ferrochromium:

$$FeCr_2O_4 + 4C \longrightarrow Fe + 2Cr + 4CO$$

When pure chromium is required, the chromite is first treated with molten alkali and oxygen to convert the Cr^{III} to chromate(VI), which is dissolved in water and eventually precipitated as sodium dichromate. This is then reduced with carbon to Cr^{III} oxide:

$$Na_2Cr_2O_7 + 2C \longrightarrow Cr_2O_3 + Na_2CO_3 + CO$$

This oxide is then reduced with aluminum:

$$Cr_2O_3 + 2Al \longrightarrow Al_2O_3 + 2Cr$$

Chromium is a white, hard, lustrous, and brittle metal (mp $1903 \pm 10°C$). It is extremely resistant to ordinary corrosive agents, which accounts for its extensive use as an electroplated protective coating. The metal dissolves fairly readily in nonoxidizing mineral acids, for example, hydrochloric and sulfuric acids, but not in cold aqua regia or nitric acid, either concentrated or dilute. The last two reagents passivate the metal in a manner that is not well understood. The electrode potentials of the metal are

$$Cr^{2+} + 2e = Cr \qquad E^0 = -0.91 \text{ V}$$

$$Cr^{3+} + 3e = Cr \qquad E^0 = -0.74 \text{ V}$$

Thus it is rather easily oxidized when not passivated, and it readily displaces copper, tin, and nickel from aqueous solutions of their salts.

At elevated temperatures chromium unites directly with the halogens, sulfur, silicon, boron, nitrogen, carbon, and oxygen.

CHROMIUM COMPOUNDS

17-C-2   Binary Compounds

Halides

These are listed in Table 17-C-2. The anhydrous Cr^{II} halides are obtained by action of HF, HCl, HBr, or I_2 on the metal at 600 to 700°C or by reduction of the trihalides with H_2 at 500 to 600°C. Chromium dichloride is the most common and most important of these halides, dissolving in oxygen-free water to give a blue solution of Cr^{2+} ion.

---

[1]C. W. Stowe, Ed., *Evolution of Chromium Ore Fields,* Van Nostrand Reinhold, N.Y., 1987.

Table 17-C-2 Halides of Chromium

Halogen	Cr^{II}	Cr^{III}	Higher and mixed oxidation states	
F	CrF$_2$ Green, mp 894°C	CrF$_3$^{a} Green, mp 1404°C	CrF$_4$^{b} Green, subl 100°C Cr$_2$F$_5$^{c}	CrF$_5$ Red, mp 30°C^e
Cl	CrCl$_2$ White, mp 820°C	CrCl$_3$ Violet, mp 1150°C	CrCl$_4$^{d}	
Br	CrBr$_2$ White, mp 842°C	CrBr$_3$ Very dark green, mp 113°C	CrBr$_4$^{d}	
I	CrI$_2$ Red-Brown, mp 868°C	CrI$_3$ Green-Black, dec		

^{a} Melts only in a closed system; in an open system disproportionates above 600°C to give CrF$_5$.
^{b} Colorless in vapor phase. Becomes brown on slightest contact with moisture.
^{c} Often nonstoichiometric; contains Cr^{III}F$_6$ and highly distorted Cr^{II}F$_6$ octahedra sharing corners and edges.
^{d} Not known as solids; appear to exist in vapors formed when the trihalides are heated in an excess of the halogen. The tetrachloride is tetrahedral and CrCl$_3$ planar in the vapor.
^{e} Yellow in vapor phase.

Of the Cr^{III} halides, the red-violet anhydrous chloride CrCl$_3$, which can be prepared in a variety of ways (e.g., by the action of SOCl$_2$ on the hydrated salt CrCl$_3$·6H$_2$O) is singularly important. It can be sublimed in a stream of chlorine at ~600°C, but if heated to such a temperature in the absence of chlorine it decomposes to Cr^{II} chloride and chlorine. The flaky or leaflet form of CrCl$_3$ is a consequence of its crystal structure, which is of an unusual type. It consists of a cubic close-packed array of chlorine atoms in which two thirds of the octahedral holes between *every other* pair of Cl planes are occupied by metal atoms. The alternate layers of chlorine atoms with no metal atoms between them are held together only by van der Waals' forces; thus the crystal has pronounced cleavage parallel to the layers. Chromium trichloride is the only substance known to have this exact structure, but CrBr$_3$, as well as FeCl$_3$ and triiodides of As, Sb, and Bi, have a structure that differs only in that the halogen atoms are in hexagonal rather than cubic close packing.

Chromium(III) chloride does not dissolve at a significant rate in pure water, but it dissolves readily in the presence of Cr^{II} ion or reducing agents such as SnCl$_2$ that can generate some Cr^{II} from the CrCl$_3$. This is because the process of solution can then take place by electron transfer from Cr^{II} in solution *via* a Cl bridge to the Cr^{III} in the crystal. This Cr^{II} can then leave the crystal and act on a Cr^{III} ion elsewhere on the crystal surface. The "solubilizing" effect of reducing agents is probably related in this or some similar way to the mechanism by which chromium(II) ions cause decomposition of otherwise inert Cr^{III} complexes in solution.

Chromium (III) chloride forms adducts with a variety of donor ligands. The tetrahydrofuran complex CrCl$_3$·3THF, which is obtained as violet crystals by action of a little zinc on CrCl$_3$ in THF, is a particularly useful material for the preparation of other chromium compounds such as carbonyls or organo compounds, as it is soluble in organic solvents.

CrCl$_4$ is unstable but may be trapped in an argon matrix; it is tetrahedral.

Chromium(IV) fluoride is made by fluorination of the metal at 350°C. The colorless monomer has T_d symmetry.[2] The highest fluoride of chromium is CrF_5, which may be obtained by fluorination of CrO_3 under relatively mild conditions or by more vigorous fluorination of the metal. It is a powerful fluorinating agent. CrF_5 is a blood-red low melting solid (mp 29–30°C) and is yellow in the vapor phase. It has a Jahn-Teller distorted *tbp* structure.[2] The formation of CrF_6 has been reported but not confirmed.[3]

A number of oxohalides are known, including the volatile complexes $CrOF_4$ and CrO_2X_2 (X = F, Cl, Br); they are red to purple. Of these only chromyl chloride CrO_2Cl_2 is of practical importance; the red liquid is made by treating CrO_3 with HCl. It is monomeric with a tetrahedral structure and two short (1.58 Å) Cr=O bonds.

Oxides

Only Cr_2O_3, CrO_2, and CrO_3 are of importance. The green oxide α-Cr_2O_3, which has the corundum structure, is formed on burning the metal in oxygen, on thermal decomposition of Cr^{VI} oxide or ammonium dichromate, or on roasting the hydrous oxide $Cr_2O_3 \cdot nH_2O$. The latter is normally obtained by adding hydroxide to aqueous Cr^{III} at room temperature and has variable water content. It is often called chromic hydroxide, but there is in fact a true, crystalline hydroxide, $Cr(OH)_3(H_2O)_3$, that can be prepared by slow addition of base to a cold solution of $[Cr(H_2O)_6]^{3+}$. The crystalline material quickly becomes amorphous at higher temperatures.

If ignited too strongly, Cr_2O_3 becomes inert toward both acid and base, but otherwise it and its hydrous form are amphoteric, dissolving readily in acid to give aqua ions $[Cr(H_2O)_6]^{3+}$, and in concentrated alkali to form "chromites."

Chromium oxide and chromium supported on other oxides such as Al_2O_3 are important catalysts for a wide variety of reactions.

The black-brown chromium(IV) oxide (CrO_2) is normally synthesized by hydrothermal reduction of CrO_3. It has an undistorted rutile structure (i.e., no M—M bonds as in MoO_2). It is ferromagnetic and has metallic conductance, presumably because of delocalization of electrons into energy bands formed by overlap of metal d and oxygen $p\pi$ orbitals. Because of its magnetic properties it is used in magnetic recording tape. Above 250°C CrO_2 decomposes to Cr_2O_3.

Chromium(VI) oxide (CrO_3) can be obtained as an orange-red precipitate on adding sulfuric acid to solutions of Na or K dichromate. The red solid, which consists of infinite chains of CrO_4 tetrahedra sharing vertices, is unstable above its melting point 197°C, losing oxygen to give Cr_2O_3 after various intermediate stages. It is readily soluble in water and is highly poisonous (see Section 17-C-10).

Interaction of CrO_3 and organic substances is vigorous and may be explosive. However, CrO_3 is widely used in organic chemistry as an oxidant, commonly in acetic acid as solvent. The mechanism has been greatly studied and is believed to proceed initially by the formation of chromate esters (when pure, they are highly explosive) that undergo C—H bond cleavage as the rate-determining step to give

[2]E. Jacobs and H. Willner, *Chem. Ber.* **1990**, *123*, 1319; L. Hedberg *et al.*, *Acta Chem. Scand.*, **1988**, *A42*, 318.
[3]H. Willner *et al.*, *Inorg. Chem.* **1992**, *31*, 5357.

Cr^{IV} as the first product; the general scheme appears to be:

$$H_2A + Cr^{VI} \rightleftharpoons Cr^{IV} + A \quad \text{(slow)}$$

$$Cr^{IV} + Cr^{VI} \rightleftharpoons 2Cr^{V}$$

$$Cr^{V} + H_2A \rightleftharpoons Cr^{III} + A$$

There are various mixed metal oxides; those containing the higher oxidation states are discussed later. Chromium(III) oxide can be fused with a number of M^{II} oxides to give crystalline $M^{II}O \cdot Cr_2O_3$ compounds having the spinel structure with Cr^{III} ions in the octahedral holes. Sodium metal reacts with each of the oxides Cr_2O_3, CrO_2, CrO_3, as well as with Na_2CrO_4, to give the "chromite" $NaCrO_2$, in which both cations have octahedral coordination.

Other Binary Compounds

The chromium sulfide system is very complex, with two forms of Cr_2S_3 and several intermediate phases between these and CrS. Rhombohedral Cr_2S_3 has complex electrical and magnetic properties. The mixed-valent telluride Cr_3Te_4 is ferromagnetic.

17-C-3   The Chemistry of Chromium(II)

Mononuclear Compounds; The Aqua Ion

This ion is bright blue and is best obtained in solution by dissolving the very pure metal in deoxygenated, dilute mineral acids or by reducing Cr^{III} solutions electrolytically or with Zn/Hg. The ion is readily oxidized:

$$Cr^{3+} + e = Cr^{2+} \quad E^0 = -0.41 \text{ V}$$

and the solutions must be protected from air—even then, they decompose at rates varying with acidity and the anions present, by reducing water with liberation of hydrogen.

The Cr^{2+} aqua ion undergoes hydrolysis

$$Cr^{2+}(aq) = CrOH^+(aq) + H^+(aq) \quad pK = -5.30$$

and is not strongly complexed in aqueous solution.

The aqua ion has been extensively used as a reductant in studies on the mechanism of electron transfer reactions, best exemplified by the classical example of the reaction of $Cr^{2+}(aq)$ with $[Co^{III}(NH_3)_5X]^{2+}$. This reaction proceeds *via* an inner-sphere (ligand-bridged) mechanism, as in the general reaction sequence

$$MX^{m+} + N^{n+} \longrightarrow \underset{\text{precursor complex}}{[M^{m+}\text{—}X\text{—}N^{n+}]} \longrightarrow \underset{\text{successor complex}}{[M^{(m-1)+}\text{—}X\text{—}N^{(n+1)+}]} \longrightarrow \underset{\text{products}}{M^{(m-1)+} + XN^{(n+1)+}}$$

In the reaction between Co^{III} and Cr^{2+}, the kinetically inert complex $[Co(NH_3)_5X]^{2+}$ reacts with the labile Cr^{II} aqua ion to give labile $Co^{2+}(aq)$ and a substitution-inert Cr^{III} chloro complex which must have been formed *via* a bridged intermediate.

$$Cr^{II}(H_2O)_6^{2+} + Co^{III}(NH_3)_5Cl^{2+} \longrightarrow [(H_2O)_5Cr^{II}ClCo^{III}(NH_3)_5]^{4+}$$

$$\updownarrow \text{electron transfer}$$

$$Cr(H_2O)_5Cl^{2+} + Co(NH_3)_5(H_2O)^{2+} \longleftarrow [(H_2O)_5Cr^{III}ClCo^{II}(NH_3)_5]^{4+}$$

$$\downarrow \text{H}^+, \text{H}_2\text{O}$$

$$Co(H_2O)_6^{2+} + 5NH_4^+$$

Similiar electron transfer reactions involve other anions, e.g., F^-, Br^-, I^-, SO_4^{2-}, NCS^-, N_3^-, PO_4^{3-}, $CH_3CO_2^-$, etc. In the reaction of Cr^{2+} with $[Co(NH_3)_5X]^{2+}$, and of Cr^{2+} with CrX^{2+}, the rates decrease as X is varied, in the order $X = I^- > Br^- > Cl^- > F^-$. Note, however, that the opposite order is found in electron transfer reactions of $[Co(NH_3)_5X]^{2+}$ with Fe^{2+} or Eu^{2+}. It should be kept in mind, however, that in general ligand transfer is not an essential or diagnostic feature of an inner-sphere electron transfer mechanism.

It has been proposed that oxidation of the Cr^{2+} aqua ion by O_2 gives first the CrO_2Cr group, which undergoes protonation to $[(H_2O)_4Cr(\mu\text{-OH})_2Cr(H_2O)_4]^{4+}$; this in turn splits to give $[Cr(H_2O)_6]^{3+}$, which according to labeling studies, contains all atoms originally in the O_2.

The $Cr^{2+}(aq)$ ion reacts readily with alkyl halides, apparently by a radical mechanism, to generate $[R-Cr(H_2O)_5]^{2+}$ ions; the $R-Cr$ bonds can then be cleaved by H_3O^+. On the basis of these fundamental processes, $Cr^{2+}(aq)$ finds important use as a reductant for organic compounds, especially in aqueous DMF as solvent, with ethylenediamine present to complex the Cr^{III} produced.

Mononuclear Complexes. Chromium(II) forms three major types of discrete complexes: (1) High-spin ($S = 2$) "octahedral" complexes, which show marked tetragonal distortion owing to the Jahn–Teller effect; (2) low-spin ($S = 1$) octahedral complexes; (3) square complexes. The low-spin octahedral complexes, which require strong-field ligands, are represented by $[Cr(CN)_6]^{4-}$, $[Cr(bipy)_3]^{2+}$, $[Cr(phen)_3]^{2+}$, $[Cr(phen)_2(NCS)_2]$, and $[CrX_2(L-L)_2]$ ($L-L$ = diphos,[4] diars; X = Cl, Br, and I). The high-spin "octahedral" complexes are relatively unstable towards hydrolysis, though the macrocycle complex *trans*-$[Cr([14]aneN_4)(H_2O)_2]^{2+}$ is stable to acids.[5]

The square complexes, which may be thought of as the asymptotic limit of tetragonal Jahn–Teller distortion of high-spin 6-coordinate species, are numerous and stable. Representative complexes are $Cr(acac)_2$, the bis(pyrazolatoborates) $Cr(R_2Bpz_2)_2$ (R = H or Et), a series of Schiff base complexes such as (17-C-I) and (17-C-II), and the tetraphenylporphyrin complex, which takes up to two pyridine ligands to become low-spin $[Cr(porph)py_2]$. A *trans*-$CrBr_2(H_2O)_2$ unit is found in the compound $CrBr_2(H_2O)_2\cdot(pyH)Br$.

[4]G. Wilkinson *et al.*, *Polyhedron* **1993**, *12*, 363.
[5]A. Bakac and J. H. Espenson, *Inorg. Chem.* **1992**, *31*, 1108.

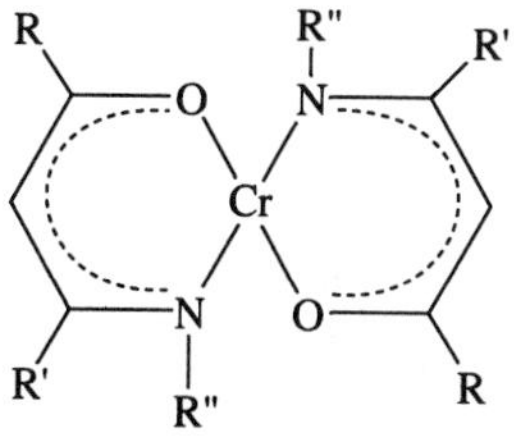

(17-C-I) 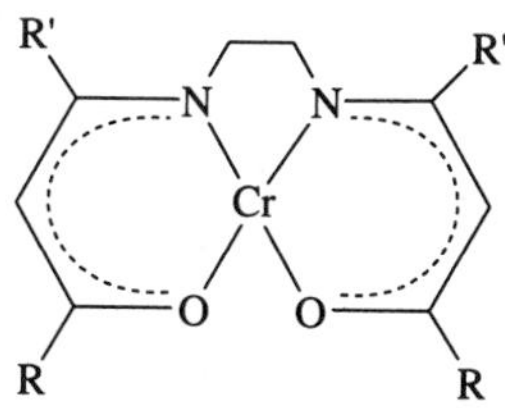   (17-C-II)

Alkoxides of Cr^{II} tend to be polymeric or, as in $(Pr^iO)_8Cr_2Na_4(THF)_4$, OR-bridged dimers. The very bulky compound $Cr(OCBu_3^t)_2 \cdot LiCl(THF)_2$ is T-shaped, while $Cr[OSi(OBu^t)_3]_2(NHEt_2)_2$ is square.[6] Such species are relevant in the context of Cr-based ethylene polymerization catalysts. The diamagnetic Cr^{II} alkyl $(Pr_2^iN)(NO)Cr(CH_2SiMe_3)_2$ is a rare example of a low-spin tetrahedral d^4 complex.[7]

Complex halides of the types $M_2^ICrX_4$ and M^ICrX_3 contain infinite polymeric anions. Many of the former have the K_2NiF_4 structure or a similar one and they display complex and unusual magnetic properties (e.g., ferromagnetism) at very low temperatures. The M^ICrCl_3 compounds have high-spin Cr^{II} ions in distorted octahedra and show antiferromagnetism. The anion $[Cr_3Cl_{12}]^{6-}$, which is $[Cl_3Cr(\mu\text{-}Cl)_3Cr(\mu\text{-}Cl)_3CrCl_3]^{6-}$ is also known, e.g., as the $NH_2Me_2^+$ salt.

Binuclear Compounds: Quadruple Bonds. One of the earliest Cr^{II} compounds discovered (1844) was the acetate hydrate $Cr_2(O_2CCH_3)_4(H_2O)_2$. It was long recognized as anomalous because it is red and diamagnetic, whereas the mononuclear Cr^{II} compounds are blue-violet and strongly paramagnetic. This compound, as well as numerous others of the general formula $Cr_2(O_2CR)_4L_2$, have the structure (17-C-III), whereas carboxylates free of donor ligands L tend to associate, as in (17-C-IV).[8]

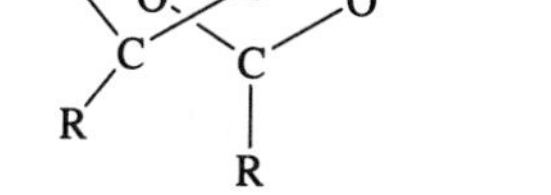

(17-C-III) 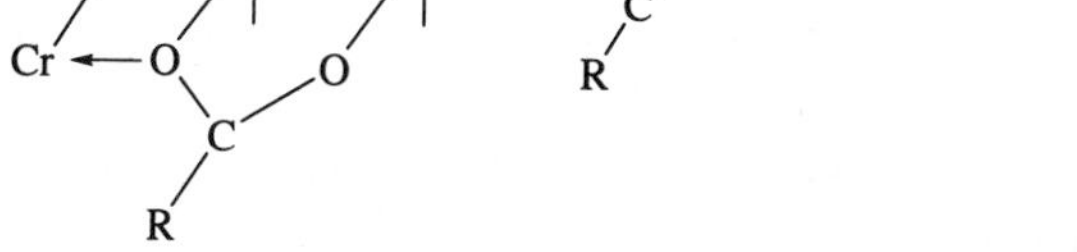   (17-C-IV)

The chromium atoms are united by a quadruple bond (Section 16-6) but the strength of this interaction is a sensitive inverse function of the strength of the axial

[6]T. D. Tilley *et al., Inorg. Chem.* **1993**, *32*, 5402.
[7]P. Legzdins *et al., Organometallics* **1997**, *16*, 3569.
[8]F. A. Cotton *et al., Inorg. Chem.* **1992**, *31*, 1865.

Table 17-C-3 Some Bridging Ligands That Occur in Cr^{II} Compounds with Strong Quadruple Bonds

Ligand	d_{Cr-Cr} (Å)	Ligand	d_{Cr-Cr} (Å)
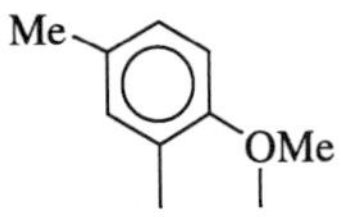 Me—(ring)—OMe	1.828(2)	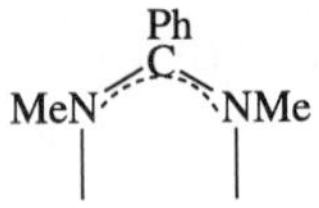 Ph–C(–NMe)(MeN–)	1.873(2)
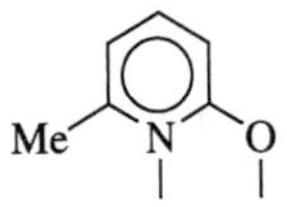 Me–(pyridine ring)–N–O	1.889(1)	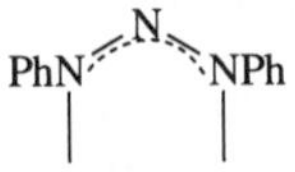 N(PhN–)(–NPh)	1.858(1)
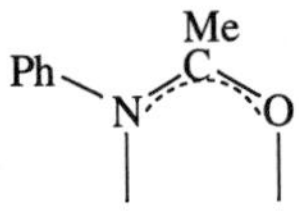 Ph–N–C(Me)–O	1.873(4)	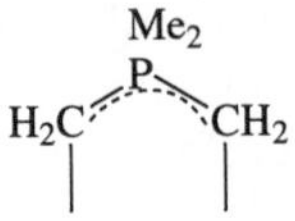 Me_2 P(H_2C–)(–CH_2)	1.895(3)

ligand bonding and is also (to a much lesser extent) sensitive to the basicity of the RCO_2^- groups. The Cr—Cr distances vary from 2.28 to ~2.54 Å. In the $Cr_2(O_2CCH_3)_4$ molecule itself, which was studied in the gas phase by electron diffraction because axial interactions of some kind always occur in condensed phases, the Cr—Cr distance is 1.96 Å, which is similar to those in numerous other Cr_2^{4+} compounds where there are no axial ligands and hence very strong Cr—Cr quadruple bonds. These other compounds are nearly all of the type in which the $[Cr—Cr]^{4+}$ unit is bridged by a ligand with two donor atoms separated by one other atom, and some representative ones, together with the Cr—Cr distances they span, are listed in Table 17-C-3.

17-C-4 The Chemistry of Chromium(III), d^3

Chromium(III) Complexes

There are literally thousands of chromium(III) complexes that, with a few exceptions, are hexacoordinate and "octahedral." An important characteristic of these complexes in aqueous solutions is their relative kinetic inertness. Ligand displacement reactions of Cr^{III} complexes are only ~10 times faster than those of Co^{III}, with half-times in the range of several hours. It is largely because of this kinetic inertness that so many complex species can be isolated as solids and that they persist for relatively long periods of time in solution, even under conditions of marked thermodynamic instability.

The hexaaqua ion $[Cr(H_2O)_6]^{3+}$, which is regular octahedral, occurs in aqueous solution and in numerous salts such as the violet hydrate $[Cr(H_2O)_6]Cl_3$ and in an extensive series of alums $M^ICr(SO_4)_2 \cdot 12H_2O$. The chloride has three isomers, the others being the dark green *trans*-$[CrCl_2(H_2O)_4]Cl \cdot 2H_2O$, which is the normal commercially available salt, and pale green $[CrCl(H_2O)_5]Cl_2 \cdot H_2O$. The aqua ion is acidic ($pK = 4$), and the hydroxo ion condenses to give a dimeric hydroxo bridged species:

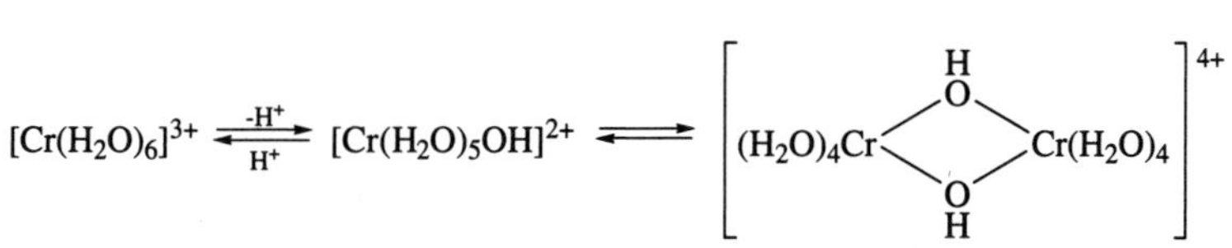

$$[Cr(H_2O)_6]^{3+} \underset{H^+}{\overset{-H^+}{\rightleftharpoons}} [Cr(H_2O)_5OH]^{2+} \rightleftharpoons \left[(H_2O)_4Cr \underset{\underset{H}{O}}{\overset{\overset{H}{O}}{<}} Cr(H_2O)_4 \right]^{4+}$$

Further polymerization proceeds stepwise *via* OH-bridges to give trimers $[Cr_3(\mu\text{-}OH)_4(OH)_{(n+1)}]^{(4-n)+}$ ($n = 0, 1, 2$), tetramers, hexamers, etc.[9] Anion association may be important in these condensation reactions. Larger aggregates such as $[Cr_{12}(OH)_{28}(H_2O)_{12}]^{8+}$, reminiscent of the Al_{13} species $[AlO_4Al_{12}(OH)_{24}(H_2O)_{12}]^{7+}$, have also been postulated.[10]

On addition of further base, a precipitate is formed that consists of H-bonded layers of $Cr(OH)_3(H_2O)_3$ and readily redissolves in acid. Within a minute or less, however, this precipitate begins "aging" to an oligomeric or polymeric structure that is much less soluble. The hydroxide is amphoteric and in strongly alkaline solution forms green $[Cr(OH)_6]^{3-}$.

The reaction of O_2 with the aqueous Cr^{2+} ion gives a number of Cr^{III} complexes including the hydroperoxo complex $[(H_2O)_5Cr(OOH)]^{2+}$; it is reduced by Fe^{2+} to Cr^{3+} *via* $[(H_2O)_5CrO]^{2+}$ as an intermediate.[11] The structure of the nitrosyl complex $[(H_2O)_5CrNO]^{2+}$, previously formulated as a Cr^I compound, suggests that it is more properly described as a Cr^{III} complex of NO^-.[12]

The *ammonia and amine complexes* are the most numerous chromium derivatives and the most extensively studied. They include the pure ammine $[CrAm_6]^{3+}$, the mixed ammine-aqua types, that is, $[CrAm_{6-n}(H_2O)_n]^{3+}$ ($n = 0$–4, 6), the mixed ammine-acido types, that is, $[CrAm_{6-n}X_n]^{(3-n)+}$ ($n = 1$–4, 6), and mixed ammine-aqua-acido types, for example, $[CrAm_{6-n-m}(H_2O)_nX_m]^{(3-m)+}$ (here Am represents the monodentate ligand NH_3 or half of a polydentate amine such as ethylenediamine, and X an acido ligand such as halide, nitrite, or sulfate ion). These complexes provide examples of virtually all kinds of isomerism possible in octahedral complexes.

The preparation of polyammine complexes sometimes presents difficulties, partly because in neutral or basic solution hydroxo- or oxo-bridged polynuclear complexes are often formed. Such polyammines are often conveniently prepared from the Cr^{IV} peroxo species, noted below; thus the action of HCl on $[Cr^{IV}en(H_2O)(O_2)_2]\cdot H_2O$ forms the blue salt $[Cr^{III}en(H_2O)_2Cl_2]Cl$.

The majority of polynuclear complexes are of one of the types (17-C-V) or (17-C-VI). In the former there is a single bridging group, which is usually O or OH. Some representative reactions involving such compounds are shown below.

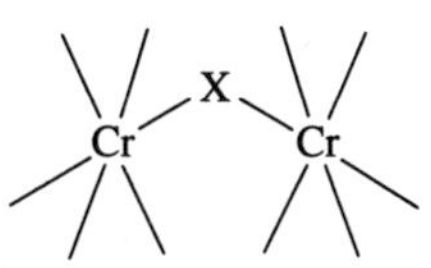

(17-C-V) 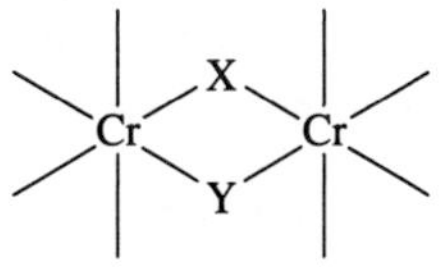   (17-C-VI)

[9]Drljaca and L. Spiccia, *Polyhedron* **1996**, *15*, 2875.
[10]R. A. Kydd *et al.*, *J. Chem. Soc., Dalton Trans.* **1993**, 2415.
[11]J. H. Espenson *et al.*, *Inorg. Chem.* **1993**, *32*, 2005, 5034.
[12]M. Ardon and S. Cohen, *Inorg. Chem.* **1993**, *32*, 3241.

$$[(NH_3)_5Cr(OH)Cr(NH_3)_5]^{5+} \underset{H^+}{\overset{OH^-}{\rightleftharpoons}} [(NH_3)_5CrOCr(NH_3)_5]^{4+}$$

$\downarrow$ H$_2$O, 1 day, 100°C | $\downarrow$ OH$^-$, several days, 25°C

$$[(NH_3)_5Cr(OH)Cr(NH_3)_4(H_2O)]^{5+} \xleftarrow{\;H^+\;} [(NH_3)_5Cr(OH)Cr(NH_3)_4(OH)]^{4+}$$

The oxo-bridged complex has a linear Cr—O—Cr group, indicating $d\pi$–$p\pi$ bonding as in other cases of M—O—M groups. Even in the "acid rhodo" complex $[(NH_3)_5Cr(OH)Cr(NH_3)_5]^{5+}$ the bridge is nearly linear ($\angle$ Cr—O—Cr = 166°) and there is considerable magnetic coupling. A large number of the (17-C-VI) type of complex are known, especially with X=Y=OH. All show a repulsion between the Cr^{III} atoms (distances >3.0 Å) and weak but significant magnetic interactions that are nearly always antiferromagnetic.

Dinuclear triply-bridged complexes $[L_2Cr_2(\mu\text{-}O)(\mu\text{-}OH)_2]^{2+}$ and $[L_2Cr_2(OH)_3]^{3+}$ (L = 1,5,9-triazacyclononane) possess a face-sharing bioctahedral structure. The complexes show strong antiferromagnetic coupling, the magnitude of which correlates with the Cr—Cr distance (2.5–2.7 Å).[13]

Anionic complexes are also common and are of the type $[CrX_6]^{3-}$, where X may be F$^-$, Cl$^-$, NCS$^-$, or CN$^-$, but they may also have lower charges if neutral ligands are present as in the ion $[Cr(NCS)_4(NH_3)_2]^-$. Complexes of bi- or polydentate anions are also known, one example being $[Cr(oxalate)_3]^{3-}$.

A different type of anionic complex is represented by the $Cr_2X_9^{3-}$ ions, which have a face-sharing bioctahedral structure similar to $W_2Cl_9^{3-}$ except that the Cr^{3+} ions repel each other from the centers of their octahedra and the magnetic moments are normal, indicating that there is no Cr—Cr bond. Compounds of formula M^{I}CrX$_4$ such as KCrF$_4$ and [PCl$_4$][CrCl$_4$] contain CrX$_6$ octahedra with some sharing of the X atoms.

As expected, Cr^{III} can also form complexes of other types, including neutral complexes with β-diketonates and similar ligands [e.g., Cr(acac)$_3$ and Cr(OCOCF$_3$)$_3$]. It also forms oxo-centered trinuclear carboxylates (Section 11-9) such as $[Cr_3O(O_2CMe)_6(H_2O)_3]^+$, which has the structure (11-VIII).

Mixed-metal oxo-centered complexes $[Cr_2M(\mu_3\text{-}O)(OAc)_6py_3]$ are made by heating Cr$_2$(OAc)$_4$·2H$_2$O with M^{II}(OAc)$_2$ in pyridine (M = Mn, Co).[14] Heating $[Cr_3(O)(OAc)_6(H_2O)_3]$Cl with 1,10-phenanthroline gives tetranuclear $[Cr_4O_2(OAc)_7$-(phen)$_2$]Cl which has a $[Cr_4(\mu_3\text{-}O)_2]^{8+}$ core with butterfly structure.[15] Reduction of [PPh$_4$]$_2$[Cr$_2$O$_7$] with sodium amalgam in the presence of SOCl$_2$ gives [PPh$_4$]$_3$[Cr$_4$O-(SO$_4$)$_2$Cl$_9$], with a Cr$_4$(μ_4-O) core.[16]

Non-Octahedral Cr^{III} Complexes. Distorted tetrahedral coordination, presumably due to extreme steric factors, is found in the compound LiCr[OCH-(CMe$_3$)$_2$]$_4$·THF.

The coordination number 3 occurs in dialkylamides [e.g., Cr(NPr$_2^i$)$_3$]. A combination of steric factors and multiple Cr—N bonding has been proposed to explain the stability of such monomers. Another of the rare nonoctahedral Cr^{III} complexes is that shown in (17-C-VII), where there is a distorted pentagonal bipyramid.

[13]K. Wieghardt *et al., Agnew. Chem. Int. Ed. Engl.* **1992,** *31,* 311.
[14]R. D. Cannon *et al., J. Chem. Soc., Dalton Trans.* **1993,** 2005.
[15]J. B. Vincent *et al., Inorg. Chem.* **1994,** *33,* 5522.
[16]R. J. Errington *et al., J. Chem. Soc., Chem. Commun.* **1990,** 1565.

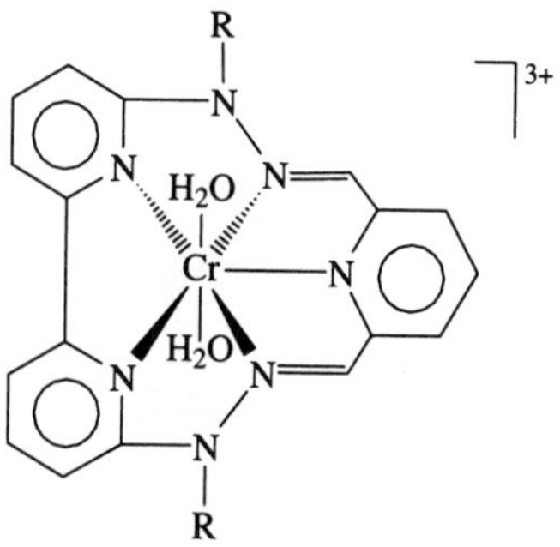

(17-C-VII)

Five-coordination is often observed when $CrCl_3$ forms numerous adducts with ethers, nitriles, amines, and phosphines, which have formulas $CrCl_3 \cdot 2L$ or $CrCl_3 \cdot 3L$. An example is $CrCl_3 \cdot 2NMe_3$ where X-ray studies confirm the trigonal bipyramidal structure with axial amine groups. The only known square pyramidal complex (17-C-VIIIa) is formed with the tetradentate macrocycle, tmtaa, (17-C-VIIIb).[17]

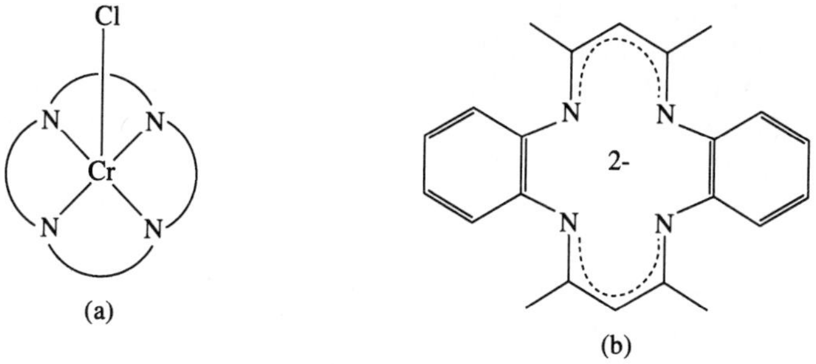

(a)

(b)

(17-C-VIII)

Electronic Structures of Octahedral Chromium(III) Complexes

The magnetic properties of the octahedral Cr^{III} complexes are uncomplicated. All such complexes must have three unpaired electrons, irrespective of the strength of the ligand field, and this has been confirmed for all known mononuclear complexes. More sophisticated theory further predicts that the magnetic moments should be very close to, but slightly below, the spin-only value of 3.88 BM; this, too, is observed experimentally.

The spectra of Cr^{III} complexes are also well understood in their main features. A partial energy level diagram (Fig. 17-C-1) indicates that three spin-allowed transitions are expected, and these have been observed in many complexes. Indeed, the spectrochemical series was originally established by Tsuchida using data for Cr^{III} and Co^{III} complexes. In the aqua ion the bands are found at 17,400, 24,700, and 37,000 cm^{-1}.

Ruby, natural or synthetic, is α-Al_2O_3 containing occasional Cr^{III} ions in place of Al^{III} ions. The environment of the Cr^{III} in ruby is thus a slightly distorted (D_{3d}) octahedron of oxide ions. The frequencies of the spin-allowed bands of Cr^{III} in ruby

[17]F. A. Cotton *et al.*, *Inorg. Chim. Acta*, **1990**, *172*, 135.

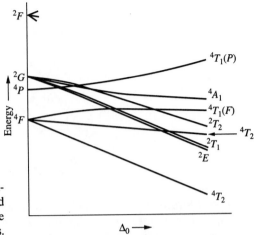

Figure 17-C-1   Partial energy level diagram for a d^3 ion in an octahedral field (also for a d^7 ion in a tetrahedral field). The quartet states are drawn with heavier lines.

indicate that the Cr^{III} ions are under considerable compression, since the value of Δ_0 calculated is significantly higher than in the $[Cr(H_2O)_6]^{3+}$ ion or in other oxide lattices and glasses. Also, in ruby, spin-forbidden transitions from the $^4A_2$ ground state to the doublet states arising from the ^2G state of the free ion are observed. The transitions to the ^2E and $^2T_1$ states give rise to extremely sharp lines because the slopes of the energy lines for these states are the same as that for the ground state (except in extremely weak fields).

The same doublet states play a key role in the operation of the ruby laser. In this device a large single crystal of ruby is irradiated with light of the proper frequency to cause excitation to the $^4T_2(F)$ state. The exact magnitudes of certain energy differences and relaxation times in the ruby are such that the system rapidly makes a radiationless transition (i.e., by loss of energy to the crystal lattice in the form of vibrations) to the ^2E and $^2T_1$ states, instead of decaying directly back to the ground state. The systems then return from these doublet states to the ground state by stimulated emission of very sharp lines that are in phase with the stimulating radiation. Bursts of extremely intense, monochromatic, and coherent (all emitters in phase) radiation are thus obtained, which are of use in communication and as sources of energy.

Organochromium(III) Compounds

The reaction of Cr^{2+} with alkyl halides to generate $[R-Cr^{III}(H_2O)_5]^{2+}$ has already been mentioned. These alkyls decompose in aqueous acid by either a heterolytic (protonation) or a homolytic (radical, R·) pathway, depending on R.[18] In addition, there are a number of other alkyls and aryls of formula CrR_3L_n, where L is often THF. These are obtained by the action of lithium alkyls or Grignard reagents on $CrCl_3 \cdot 3THF$. The alkyls are rather unstable and their modes of decomposition have been greatly studied, but the aryls such as $Cr(C_6H_5)_3 \cdot 3THF$ are considerably more stable. Some anionic methyl complex anions, for example, $Li_3[Cr(CH_3)_6]$ are also reasonably stable, as is the 6-coordinate phosphine ylid complex $[Cr\{(CH_2)_2PMe_2\}_3]$.

[18] Z. Zhang and R. B. Jordan, *Inorg. Chem.* **1994,** *33,* 680.

Chromocene ($CrCp_2$) supported on silica is used to generate certain chromium-based catalysts for the polymerization of ethylene (e.g., Phillips and Union Carbide catalysts). The nature of the organometallic species responsible for the catalysis is not known with certainty, though it is noteworthy that some Cr^{III} alkyls such as $[Cp^*Cr(CH_2Ph)(THF)_2]^+BPh_4^-$ catalyze the polymerization of ethylene.[19]

17-C-5 The Chemistry of Chromium(IV), d^2

The chemistry of Cr^{IV} is limited but growing.[20] The oxide CrO_2 and the fluorides CrF_4 and $CrOF_2$ have been mentioned before. The fluorination of $CrCl_3/MCl$ mixtures gives alkali or alkaline earth salts of the CrF_6^{2-} or CrF_5^- ions; these are also present in $XeF_2 \cdot CrF_4$ and $[(XeF_5^+CrF_5^-)_4 \cdot XeF_4]$.[21] Mixed oxides of Cr^{IV} are obtained by high temperature reactions (1000°C):

$$NaCrO_2 + 2Na_2O \xrightarrow{1000°C} Na_4CrO_4 + Na$$

$$SrCrO_4 + Cr_2O_3 + 5Sr(OH)_2 \xrightarrow{1000°C} 3Sr_2CrO_4 + 5H_2O$$

The compounds $M_3^{II}CrO_5$ and $M_4^{II}CrO_6$ are also known. Peroxo species are described in Section 17-C-8. The chromyl ion $[O{=}Cr(H_2O)_5]^{2+}$ exists in acidic aqueous solution where it is a strong oxidizing agent,[22] and is also known in a porphyrin complex where the Cr=O distance is 1.62 Å.

Chromium(IV) complexes sometimes arise by disproportionation on heating Cr^{III} complexes. Thus the reaction of $CrCl_3$ with $LiNEt_2$ followed by distillation gives the volatile Cr^{IV} amide $Cr(NEt_2)_4$. The green amide reacts with some alcohols to give the blue to green Cr^{IV} alkoxides $Cr(OR)_4$ (R = Bu^t, $CHBu_2^i$, etc.). The alkylation of $CrCl_3 \cdot 3THF$ or of $Cr(OR)_4$ with lithium alkyls or Grignard reagents leads to Cr^{IV} alkyls CrR_4 (R = Me, Pr^i, Bu^t, CH_2CMe_3, etc.). All these monomeric complexes are tetrahedral.

Other examples of Cr^{IV} complexes include the dodecahedral hydride $CrH_4(dmpe)_2$, and octahedral species such as $[Cr(O_3SCF_3)_4\{o\text{-}(H_2N)_2C_6H_4\}]$[23] and the ketimido complex $[Cr(N{=}CHMe)_2(dmpe)_2]^{2+}$.

17-C-6 The Chemisty of Chromium(V), d^1

In addition to the stable compounds mentioned below, the d^1 Cr^V ion is readily detected by esr spectroscopy as a transient species in the reduction of Cr^{VI}, as well as in certain oxide lattices and catalysts. The reduction of Cr^{VI} by oxalic acid, citric acid, isopropanol, and various other organic reductants gives esr-detectable Cr^V

[19]K. H. Theopold et al., Organometallics 1995, 14, 738; 1996, 15, 5473.
[20]E. S. Gould, Coord. Chem. Rev. 1994, 135/136, 651.
[21]K. Lutar et al., Eur. J. Solid State Inorg. Chem. 1992, 29, 713.
[22]J. H. Espenson et al., J. Am. Chem. Soc. 1992, 114, 4205; Inorg. Chem. 1993, 32, 5792; 1994, 33, 1011.
[23]G. Wilkinson et al., J. Chem. Soc., Dalton Trans. 1992, 1803.

intermediates of varying lifetimes, and on dissolving chromates(VI) in 65% oleum, blue solutions containing Cr^V species are formed.

The stable compounds of Cr^V nearly all involve oxygen and/or the halogens. Dark green, hygroscopic chromates(V), such as Li_3CrO_4, Na_3CrO_4, and $Ca_3[CrO_4]_2$, contain discrete tetrahedral CrO_4^{3-} ions, and one-electron reduction of $HCrO_4^-$ in acid solution apparently gives H_3CrO_4. It should be noted that the series of $M^ICr_3O_8$ compounds do *not* contain Cr^V, but rather are built of $Cr^{VI}O_4$ tetrahedra and $Cr^{III}O_6$ octahedra.

The pentafluoride has been mentioned (Section 17-C-2). The oxofluoride can be prepared in several ways (e.g., by action of ClF_3 or BrF_3 on CrO_3) but the pure crystalline solid is obtained by the reaction:

$$XeF_2 + 2CrO_2F_2 \longrightarrow 2CrOF_3 + Xe + O_2$$

The crystal structure consists of an infinite three-dimensional array of corner-shared $CrOF_5$ octahedra. Chromium oxotrichloride ($CrOCl_3$), which can be made by reaction of $SOCl_2$ with CrO_3 or BCl_3 with CrO_2Cl_2, is volatile. Although its crystal structure is unknown, it has been studied spectroscopically in Kr and Ar matrices where it has C_{3v} molecular symmetry. Heating a mixture of Li_3N, Ba_3N_2 and $CrN/Cr_2N(1:1)$ to 700°C gives black $Ba_5[CrN_4]N$ which contains the tetrahedral $[Cr^VN_4]^{7-}$ anion.[24]

Among the more stable Cr^V compounds are complexes of the $[Cr=O]^{3+}$ ion. Many compounds containing the $[CrOX_4]^-$ (X = F, Cl, and Br) ions have been prepared and well characterized. These species are square pyramidal, with very short $Cr-O$ distances (~ 1.52 Å), magnetic moments of ~ 1.7 BM, and ν_{Cr-O} values of ~ 1000 cm^{-1}. There are a number of complexes with O- and N-donor ligands of similar structure, for example (17-C-IX) and (17-C-X) and related diolato and oxalato species,[25] as well as the cation (17-C-XI). Compound (17-C-X) is reduced to Cr^{III} by iodide *via* a Cr^{IV} intermediate.[26]

Imido analogues of these oxo complexes are also known, e.g., the anion $[Cr^V(NBu^t)Cl_4(H_2O)]^-$;[27] further examples are shown later in Figure 17-C-4. The heteropolytungstates $[X^{n+}W_{11}O_{39}Cr^VO]^{(9-n)-}$ ($X^{n+} = P^{5+}$ and Si^{4+}) catalyze the oxidation of alkanes, alkenes, etc. with OCl^-, H_2O_2, or PhIO.[28]

(17-C-IX)   (17-C-X)

[24]A. Tennstedt *et al.*, *Z. anorg. allg. Chem.* **1995**, *621*, 511.
[25]R. Bramley *et al.*, *Inorg. Chem.* **1990**, *29*, 3089; **1991**, *30*, 1557; H. Nishino and J. K. Kochi, *Inorg. Chim. Acta* **1990**, *174*, 93; M. Branca *et al.*, *Inorg. Chem.* **1992**, *31*, 2404.
[26]E. S. Gould *et al.*, *Inorg. Chem.* **1987**, *26*, 899.
[27]W. H. Leung *et al.*, *J. Chem. Soc., Dalton Trans.* **1994**, 1659.
[28]A. M. Khenkin and C. L. Hill *et al.*, *J. Am. Chem. Soc.* **1993**, *115*, 8178.

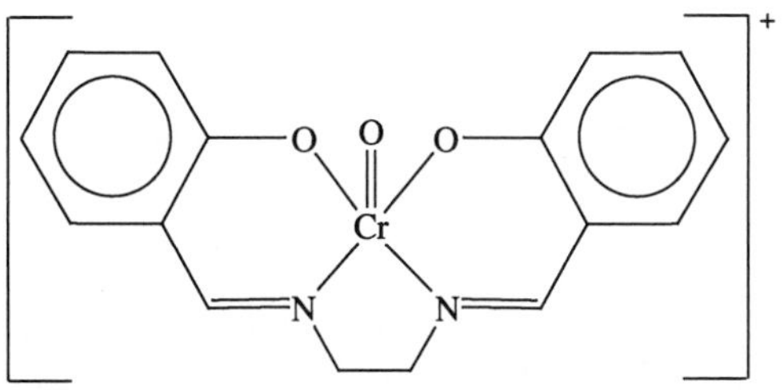

(17-C-XI)

17-C-7   The Chemistry of Chromium(VI), d^0

Chromate and Dichromate Ions. In basic solutions above pH 6, CrO_3 forms the tetrahedral yellow *chromate* ion CrO_4^{2-}; between pH 2 and 6, $HCrO_4^-$ and the orange-red *dichromate* ion $Cr_2O_7^{2-}$ are in equilibrium; and at pH values <1 the main species is H_2CrO_4. The equilibria are the following:

$$HCrO_4^- \rightleftharpoons CrO_4^{2-} + H^+ \qquad K = 10^{-5.9}$$

$$H_2CrO_4 \rightleftharpoons HCrO_4^- + H^+ \qquad K = 4.1$$

$$Cr_2O_7^{2-} + H_2O \rightleftharpoons 2HCrO_4^- \qquad K = 10^{-2.2}$$

which have been studied kinetically for a variety of bases.

The pH-dependent equilibria are quite labile, and on addition of cations that form insoluble chromates (e.g., Ba^{2+}, Pb^{2+}, and Ag^+) the chromates and not the dichromates are precipitated. Furthermore, the species present depend on the acid used, and only for HNO_3 and $HClO_4$ are the equilibria as given. When hydrochloric acid is used, there is essentially quantitative conversion into the chlorochromate ion; with sulfuric acid a sulfato complex results:

$$[CrO_3(OH)]^- + H^+ + Cl^- \longrightarrow CrO_3Cl^- + H_2O$$

$$[CrO_3(OH)]^- + HSO_4^- \longrightarrow [CrO_3(OSO_3)]^{2-} + H_2O$$

Orange potassium chlorochromate can be prepared simply by dissolving $K_2Cr_2O_7$ in hot 6 *M* HCl and crystallizing, or by chlorination of $Cr_2O_7^{2-}$ with oxalylchloride. It can be recrystallized from HCl but is hydrolyzed by water:

$$CrO_3Cl^- + H_2O \longrightarrow [CrO_3(OH)]^- + H^+ + Cl^-$$

Reaction of CrO_3, HF and an organic base afford a salt of CrO_3F^-, the structure of which was determined.[29] The potassium salts of CrO_3Br^- and CrO_3I^- are also

[29]P. Gili *et al. Eur. J. Solid State Inorg. Chem.* **1995**, *32*, 353.

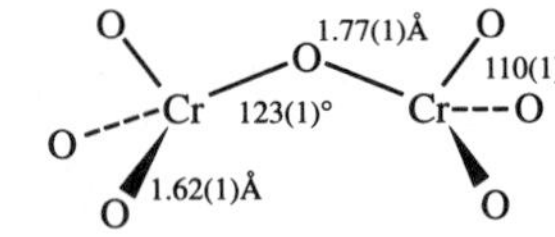

Figure 17-C-2   The structure of the dichromate ion as found in $Rb_2Cr_2O_7$.

known. They owe their existence to the fact that dichromate, though a powerful oxidizing agent, is kinetically slow in its oxidizing action toward halide ions. The pyridinium and related salts of CrO_3F^- find use as oxidants in organic chemistry.[30]

Acid solutions of dichromate are strong oxidants:

$$Cr_2O_7^{2-} + 14H^+ + 6e = 2Cr^{3+} + 7H_2O \qquad E^0 = 1.33 \text{ V}$$

The mechanism of oxidation of Fe^{2+} and other common ions by Cr^{VI} has been studied in detail; with one- and two-electron reductants, respectively, Cr^V and Cr^{IV} are initially formed. The reaction with H_2O_2 in acid solution has a very complex and imperfectly understood mechanism.

The chromate ion in basic solution, however, is much less oxidizing:

$$CrO_4^{2-} + 4H_2O + 3e = Cr(OH)_3(s) + 5OH^- \qquad E^0 = -0.13 \text{ V}$$

Chromium(VI) does not give rise to the extensive and complex series of polyacids and polyanions characteristic of the somewhat less acidic oxides of V^V, Mo^{VI}, and W^{VI}. The reason for this is perhaps the greater extent of multiple bonding ($Cr=O$) for the smaller chromium ion. However, with suitably large cations salts containing the deep red $Cr_3O_{10}^{2-}$ and $Cr_4O_{13}^{2-}$ ions are formed;[30,31] the structures continue the pattern set by the $Cr_2O_7^{2-}$ ion (Fig. 17-C-2) in having chains of CrO_4 tetrahedra sharing corners. In the limit of course we have CrO_3, which consists of infinite chains of corner-sharing tetrahedra $-O-CrO_2-O-$.

Oxohalides.   Five of these are known. The only oxohalide with the $CrOX_4$ stoichiometry, $CrOF_4$ (mp 55°C, bp 95°C), can be obtained by fluorination of CrO_3[32] or by reaction of CrO_2F_2 with KrF_2 in liquid HF. In the gaseous state as well as in various solutions and inert gas matrices it consists of square pyramidal molecules; the solid has a fluorine-bridged polymeric structure. It shows little activity as either a Lewis acid or base, but does form some salts, for example, $[NO][CrOF_5]$ and $Cs[CrOF_5]$.

The most important oxohalide is chromyl chloride CrO_2Cl_2 (bp 117°C); it is formed by the action of HCl on chromium(VI) oxide:

$$CrO_3 + 2HCl \longrightarrow CrO_2Cl_2 + H_2O$$

or by warming dichromate with an alkali metal chloride in concentrated sulfuric acid:

$$K_2Cr_2O_7 + 4KCl + 3H_2SO_4 \longrightarrow 2CrO_2Cl_2 + 3K_2SO_4 + 3H_2O$$

and in other ways. It is photosensitive but otherwise rather stable, although it vigorously oxidizes organic matter, sometimes selectively. Even saturated hydrocar-

[30]R. D. Willett *et al.*, *Inorg. Chem.* **1990**, *29*, 2872; M. K. Chaudhuri *et al.*, *Bull. Chem. Soc. Jpn.* **1994**, *67*, 1894.
[31]R. D. Willett *et al.*, *Inorg. Chem.* **1988**, *27*, 260.
[32]J. M. Mayer *et al.*, *J. Am. Chem. Soc.* **1994**, *116*, 1855; **1995**, *117*, 7139.

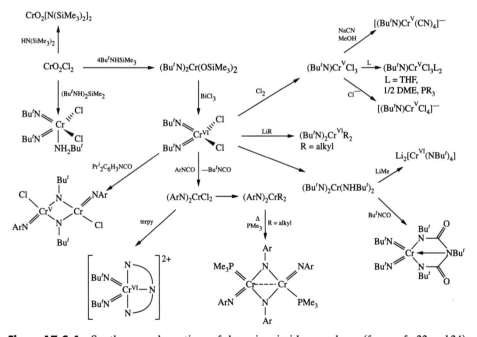

Figure 17-C-3 The molecular structures of CrO_2F_2 and CrO_2Cl_2 (K. Hedberg *et al., Inorg. Chem.* **1982,** *21,* 1115; **1983,** *22,* 892).

bons are oxidized, e.g., cyclohexane or isobutane to chlorocyclohexane and *t*-butylchloride, respectively.[32] CrO_2Cl_2 is hydrolyzed to CrO_4^{2-} and HCl. There are analogous CrO_2F_2, CrO_2Br_2, and CrO_2ClBr molecules (although the last is incompletely characterized) as well as CrO_2X_2 (X = O_2CCF_3, OR, ONO_2); the nitrate contains asymmetrically coordinated NO_3 ligands in the gas phase, with a tendency towards an octahedral structure.[33] The structures of CrO_2F_2 and CrO_2Cl_2 are shown in Fig. 17-C-3.

Imido Complexes. These complexes of general composition $(RN)_2CrX_2$ are the nitrogen analogues of the oxo species described above. For example, the reaction of CrO_2Cl_2 with $Bu^tNHSiMe_3$ gives $Cr(NBu^t)_2Cl_2$. Numerous imido complexes of chromium(IV, V, VI) are now known; some of these are shown in Fig. 17-C-4.[34] They are considerably less oxidizing than the oxo complexes. Mixed complexes

Figure 17-C-4 Syntheses and reactions of chromium imido complexes (from refs. 33 and 34).

[33] K. Hedberg *et al., Inorg. Chem.* **1991,** *30,* 4761.
[34] G. Wilkinson *et al., J. Chem. Soc., Chem. Commun.* **1990,** 1678; *J. Chem. Soc., Dalton Trans.* **1988,** 53; **1991,** 2051; **1993,** 781 and 1477; **1995,** 2111; *Polyhedron* **1990,** *9,* 2625; **1996,** *15,* 873; V. C. Gibson *et al., Polyhedron* **1995,** *14,* 2455.

such as $Cr(O)(NR)Cl_2$ also exist. The alkyls $(Bu^tN)_2CrR_2$ and $[(Bu^tN)_2CrR]^+$ (e.g., $R=CH_2Ph$) are relevant as ethylene polymerization catalysts or catalyst precursors.[35,36]

Chromium(VI) nitrido complexes $N\equiv CrX_3$ ($X=NR_2$, OBu^t) can be obtained in the former case by deoxygenation of $Cr-NO$ complexes[37] by $Vmes_3$ or in the latter case from mixtures of $(NH_4)_2Cr_2O_7$, Me_3SiCl, $HN(SiMe_3)_2$, NEt_3, and Bu^tOH. The $Cr-N$ bonds are short, 1.544 Å and 1.538 Å, respectively.[38] The reduction of $N\equiv CrX_3$ with sodium amalgam gives $[Cr^V(\mu-N)X_2]_2$ ($X = NPr^i_2$).[39]

17-C-8   Peroxo Complexes of Chromium(IV), (V), and (VI)

Like other transition metals, notably Ti, V, Nb, Ta, Mo, and W, chromium forms peroxo compounds in the higher oxidation states. Most of them are unstable, and in the solid state some of them are dangerously explosive or flammable in air. Bis(peroxo)chromium(IV) complexes with nitrogen ligands are remarkably stable, e.g., (17-C-XII).[40]

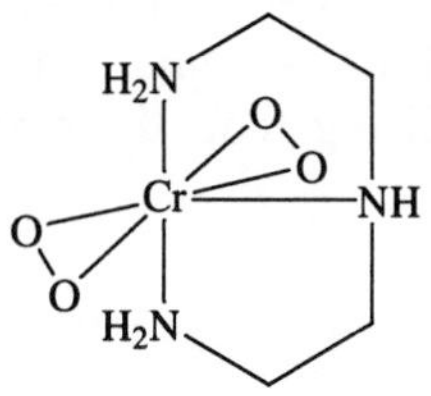

(17-C-XII)

When acidic dichromate solutions are treated with H_2O_2, a deep blue color rapidly appears but does not persist long. The overall reaction is:

$$2HCrO_4^- + 3H_2O_2 + 8H^+ \longrightarrow 2Cr^{3+} + 3O_2 + 8H_2O$$

but depending on the conditions, the intermediate species may be characterized. At temperatures below 0°C, green cationic species are formed:

$$2HCrO_4^- + 4H_2O_2 + 6H^+ \longrightarrow [Cr_2(O_2)]^{4+} + 3O_2 + 8H_2O$$

$$6HCrO_4^- + 13H_2O_2 + 16H^+ \longrightarrow 2[Cr_3(O_2)_2]^{5+} + 9O_2 + 24H_2O$$

The blue species, which is one of the products at room temperature,

$$HCrO_4^- + 2H_2O_2 + H^+ \longrightarrow CrO(O_2)_2 + 3H_2O$$

decomposes fairly readily, giving Cr^{3+}, but it may be extracted into ether where it is more stable and, on addition of pyridine to the ether solution, the compound

[35] C. J. Schaverien *et al.*, *Organometallics* **1990**, *9*, 774.
[36] V. C. Gibson *et al.*, *J. Chem. Soc., Chem. Commun.* **1995**, 1709.
[37] A. L. Odom and C. C. Cummins, *J. Am. Chem. Soc.* **1995**, *117*, 6613.
[38] H. T. Chiu *et al.*, *Chem. Commun.* **1996**, 139.
[39] A. L. Odom and C. C. Cummins, *Organometallics* **1996**, *15*, 898.
[40] S. K. Ghosh and E. S. Gould, *Inorg. Chem.* **1989**, *28*, 3651.

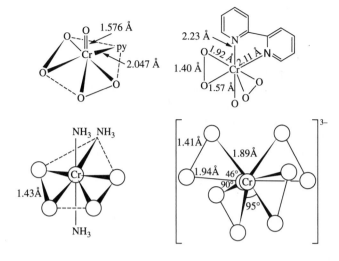

Figure 17-C-5 The structures of some chromium peroxo complexes mentioned in the text.

$CrO_5(py)$, a monomer in benzene and essentially diamagnetic, is obtained. The structures of two typical adducts, $CrO_5(py)$ and $CrO_5(bipy)$, are shown in Fig. 17-C-5. These compounds can be used to hydroxylate hydrocarbons.

The action of H_2O_2 on neutral or slightly acidic solutions of K, NH_4, or Tl dichromate leads to diamagnetic, blue-violet, violently explosive salts, believed to contain the ion $[Cr^{VI}O(O_2)_2OH]^-$.

On treatment of alkaline chromate solutions with 30% hydrogen peroxide, and after further manipulations, the red-brown peroxochromates $M_3^ICrO_8$ can be isolated. They are paramagnetic with one unpaired electron per formula unit, and K_3CrO_8 forms mixed crystals with K_3NbO_8 and K_3TaO_8; the heavy metals are pentavalent in both. The $[Cr^V(O_2)_4]^{3-}$ ion has the dodecahedral structure shown in Fig. 17-C-5. This ion can be converted to the previously mentioned $[Cr(O_2)_2O(OH)]^-$ ion, and there is evidence that the following equilibrium is attained:

$$[Cr(O_2)_4]^{3-} + 2H^+ + H_2O = [Cr(O_2)_2(O)(OH)]^- + \frac{3}{2}H_2O_2$$

Red brown   Blue

When the reaction mixture used in preparing $(NH_4)_4CrO_8$ is heated to 50°C and then cooled to 0°C, brown crystals of $(NH_3)_3CrO_4$ are obtained (Fig. 17-C-5).

17-C-9 Chromium in Low Oxidation States

In organometallic compounds and in complexes with good π-acceptor ligands, chromium can be present in oxidation states I, 0, -II, or even -IV. The first organometallic chromium complexes reported by F. Hein in 1918 and originally formulated as $Cr(C_6H_5)_nX$ (n = 3, 4, and 5) are actually ionic benzene and biphenyl π-arene complexes of Cr^I.[41] The classical examples of this type, the sandwich complexes $[Cr(\eta^6-C_6H_6)_2]^{n+}$ (n = 0, 1) are prepared by reduction of $CrCl_3$ with aluminum and $AlCl_3$ in benzene.

[41]E. Uhlig, *Organometallics* **1993**, *12*, 4751.

$$CrCl_3 + {}^2/_3Al \xrightarrow[C_6H_6]{AlCl_3} [Cr(C_6H_6)_2]AlCl_4 \xrightarrow{S_2O_4 + AlCl_3} Cr(\eta\text{-}C_6H_6)_2$$

Other *bis*(π-arene) complexes are made by cocondensation of metal vapor with arene ligands. Heating $Cr(CO)_6$ or $Cr(CO)_3(NCMe)_3$ with arenes gives the synthetically important, air- and moisture-stable 18-electron complexes (π-arene)$Cr(CO)_3$. Reduction of $CrCl_2$ with sodium amalgam in the presence of $Bu^tN{\equiv}C$ gives the isocyanide complex $Cr(CNBu^t)_6$ which reacts with HI to give C—C-coupling.[42] Isocyanide complexes can be oxidized to $[Cr^I(CNR)_6]^+$ cations. $Cr(bipy)_3$ is an example of a non-organometallic Cr^0 complex.

The reduction of $Cr(CO)_6$ leads to the carbonylates $Na_2[Cr_2(CO)_{10}]$, $Na_2[Cr(CO)_5]$, and the highly reduced $Na_4[Cr(CO)_4]$.[43]

17-C-10   Biological Aspects of Chromium Chemistry

There is no evidence for any toxic effects of chromium(III), which is an essential trace element in mammals (required daily intake 50–200 μg) and participates in glucose and lipid metabolism. In the "low-molecular-weight Cr binding substance" (LMWCr), an oligopeptide, a tetranuclear Cr^{III} carboxylate complex may be present.[44]

On the other hand, chromium(VI) is toxic and possesses mutagenic and carcinogenic activity.[45] Chromate CrO_4^{2-} is taken up by the cell where it is reduced *via* Cr^V and Cr^{IV} intermediates to Cr^{III}, most probably by reacting with the SH function of peptides, such as glutathione (GSH, 17-C-XIII) to give $GS{-}CrO_3^-$ and, eventually, disulfides RSSR.[46] Related isolable chromium(VI) thioesters $[RSCrO_3]^-$ can be obtained from CrO_4^{2-} and RSH (R = aryl, alkyl).[47] The intermediate Cr^V species in the Cr^{VI} reduction sequence has been shown to coordinate to phosphates, e.g., phosphate units of DNA and may thus be involved in DNA damage including DNA strand cleavage and DNA-protein crosslinks.[48]

$$^-O_2C{-}\underset{\underset{NH_3^+}{|}}{CH}{-}CH_2CH_2{-}\overset{\overset{O}{\|}}{C}{-}NH{-}\underset{\underset{CH_2SH}{|}}{CH}{-}\overset{\overset{O}{\|}}{C}{-}NH{-}CH_2CO_2^-$$

(17-C-XIII)

[42]S. J. Lippard *et al., Organometallics* **1994,** *13,* 1294.
[43]Review; J. E. Ellis, *Adv. Organomet. Chem.* **1990,** *31,* 1.
[44]J. B. Vincent *et al., Polyhedron* **1995,** *14,* 971; *Inorg. Chem.* **1994,** *33,* 5522; D. M. Stearns and W. H. Armstrong, *Inorg. Chem.* **1992,** *31,* 5178.
[45]Review: M. Cieslak-Golonka, *Polyhedron* **1996,** *15,* 3667.
[46]R. N. Bose *Inorg. Chem.* **1992,** *31,* 1987; K. E. Wetterhahn *et al., Inorg. Chem.* **1996,** *35,* 373.
[47]W. Mazurek *et al., Polyhedron* **1990,** *9,* 777; **1991,** *10,* 753.
[48]K. D. Sugden and K. E. Wetterhahn, *Inorg. Chem.* **1996,** *35,* 3727; N. R. Gordon *et al., Inorg. Chim. Acta* **1991,** *188,* 85.

17-D MANGANESE: GROUP 7

Manganese resembles Ti, V, and Cr in that the highest oxidation state, Mn^{VII}, corresponds to the total number of $3d$ and $4s$ electrons. In common with V and Cr the higher oxidation states form $Mn=O$ and $Mn=NBu^t$ multiple bonds. The oxo species are strong oxidants and some are very unstable and explosive.

The oxidation states and stereochemistries are given in Table 17-D-1.

Table 17-D-1 Oxidation States and Stereochemistry of Manganese

Oxidation state	Coordination number	Geometry	Examples
Mn^{-III}	4	Tetrahedral	$Mn(NO)_3CO$
Mn^{-II}	4 or 6	Square	$[Mn(phthalocyanine)]^{2-}$
Mn^{-I}	5	*tbp*	$Mn(CO)_5^-$, $[Mn(CO)_4PR_3]^-$
	4 or 6	Square	$[Mn(phthalocyanine)]^-$
Mn^0	6	Octahedral	$Mn_2(CO)_{10}$
Mn^I, d^6	6	Octahedral	$Mn(CO)_5Cl$, $K_5[Mn(CN)_6]$, $[Mn(CNR)_6]^+$
Mn^{II}, d^5	2	Linear	$Mn[C(SiMe_3)_3]_2$
	3	Trig. Planar	$[Li(thf)_4][Mn(mes)_3]$
	4	Tetrahedral	$[MnCl_4]^{2-}$, $[Mn(CH_2SiMe_3)_2]_n$, $MnBr_2(OPR_3)_2$
	4	Square	$[Mn(H_2O)_4]SO_4\cdot H_2O$, $Mn(S_2CNEt_2)_2$
	5	Distorted *tbp*	$MnBr_2[(MeHN)_2CO]_3$
		tbp	$[Mn(trenMe_6)Br]Br$
	6	Octahedral	$[Mn(H_2O)_6]^{2+}$, $[Mn(NCS)_6]^{4-}$
	7	NbF_7^{2-} structure	$[Mn(EDTA)H_2O]^{2-}$
		Pentagonal bipyramidal	$MnX_2(N_5$ macrocycle$)$
	8^a	Dodecahedral	$(Ph_4As)_2Mn(NO_3)_4$
Mn^{III}, d^4	3	Trig. Planar	$Mn[N(SiMe_3)_2]_3$
	4	Square	$[Mn(S_2C_6H_3Me)_2]^-$
	5	*sp*	$MnXsal_2en$, $[bipyH_2]MnCl_5$
	5	*tbp*	$MnI_3(PMe_3)_2$
	6	Octahedral	$Mn(acac)_3$, $[Mn(ox)_3]^{3-}$, MnF_3(distorted), $Mn(S_2CNR_2)_3$
	7		$[Mn(EDTA)H_2O]^-$, $MnH_3(dmpe)_2$
Mn^{IV}, d^3	4	Tetrahedral	$Mn(1$-norbornyl$)_4$
	6	Octahedral	MnO_2, $MnMe_4(dmpe)$, $MnCl_6^{2-}$, $Mn(S_2CNR_2)_3^+$
Mn^V, d^2	4	Tetrahedral	MnO_4^{3-}, $[Mn(NBu^t)_2(\mu\text{-}NBu^t)]_2^{2-}$
Mn^{VI}, d^1	4	Tetrahedral	MnO_4^{2-}, $[Mn(NBu^t)_2(\mu\text{-}NBu^t)]_2$
Mn^{VII}, d^0	3	Planar	MnO_3^+
	4	Tetrahedral	MnO_4^-, MnO_3F, $MnCl(NBu^t)_3$

^{a} For an 8-coordinate N_4, O_4 cryptate, see K. S. Hagen, *Angew. Chem. Int. Ed. Engl.* **1992**, *31*, 765.

17-D-1   The Element

Manganese is relatively abundant, constituting about 0.085% of the earth's crust. Among the heavy metals, only Fe is more abundant. Although widely distributed, it occurs in a number of substantial deposits, mainly oxides, hydrous oxides, or carbonate. It also occurs in nodules on the Pacific seabed together with Ni, Cu, and Co.

The metal is obtained from the oxides by reduction with Al. A large use of Mn is in ferromanganese for steels.

Manganese is roughly similar to Fe in its physical and chemical properties, the chief difference being that it is harder and more brittle but less refractory (mp 1247°C). It is quite electropositive and readily dissolves in dilute, non-oxidizing acids, e.g., trifluoroacetic acid gives $Mn(CF_3CO_2)_2$.[1]

Manganese is not particularly reactive towards non-metals at room temperatures, but at elevated temperatures it reacts vigorously with many. Thus it burns in Cl_2 to give $MnCl_2$, reacts with F_2 to give MnF_2 and MnF_3, burns in N_2 above 1200°C to give Mn_3N_2, and combines with O_2, giving Mn_3O_4, at high temperatures. The powder also reacts with R_3PX_2, X = I and Br, to give, e.g., $MnI_2(PR_3)_2$.[2]

MANGANESE COMPOUNDS

17-D-2   The Chemistry of Manganese(II), d^5

The divalent state is the common and most stable oxidation state. In neutral or acid aqueous solution there is the very pale pink hexaaqua ion $[Mn(H_2O)_6]^{2+}$, which is resistant to oxidation as shown by the potentials

$$MnO_4^- \qquad Mn^{3+} \xrightarrow{1.6\ V} Mn^{2+} \xrightarrow{-1.18\ V} Mn$$
$$\underset{1.5\ V}{\rule{5cm}{0.4pt}}$$

In basic media the hydroxide $Mn(OH)_2$ is formed; this is very easily oxidized by air, as shown by the potentials:

$$MnO_2 \cdot yH_2O \xrightarrow{-0.1\ V} Mn_2O_3 \cdot xH_2O \xrightarrow{-0.2\ V} Mn(OH)_2$$

Binary Compounds

Manganese(II) *oxide* is a gray-green to dark green powder made by roasting the carbonate in H_2 or N_2 or by the action of steam on $MnCl_2$ at 600°C. It has the rocksalt structure and is insoluble in water. Manganese(II) *hydroxide* is precipitated from Mn^{2+} solutions by alkali metal hydroxides as a gelatinous white solid that rapidly darkens because of oxidation by atmospheric O_2. The $Mn(OH)_2$ has the same crystal structure as $Mg(OH)_2$. It is only very slightly amphoteric:

$$Mn(OH)_2 + OH^- = Mn(OH)_3^- \qquad K \sim 10^{-5}$$

[1]H. W. Roesky *et al., Chem. Ber.,* 1991, **124,** 515; B. A. Moyer *et al., Inorg. Chem.* **1995,** *34,* 209.
[2]C. A. McAuliffe *et al., J. Chem. Soc., Dalton Trans.* **1993,** *371,* 2229.

Hydrous manganese(II) *sulfide* is a salmon-colored substance precipitated by alkaline sulfide solutions. It has a relatively high K_{sp} (10^{-14}) and redissolves easily in dilute acids. It becomes brown when left in air owing to oxidation; if air is excluded, the material changes on long storage, or more rapidly on boiling, into green, crystalline, anhydrous MnS.

Manganese sulfide, MnSe, and MnTe have the NaCl structure. They are all strongly antiferromagnetic, as are also the anhydrous halides. A superexchange mechanism is believed responsible for their antiferromagnetism.

Manganese(II) Salts

Manganese(II) forms an extensive series of salts with all common anions. Most are soluble in water, although the phosphate and carbonate are only slightly so. Most of the salts crystallize from water as hydrates; those with weakly coordinating anions, such as $Mn(ClO_4)_2 \cdot 6H_2O$ and $MnSO_4 \cdot 7H_2O$, contain $[Mn(H_2O)_6]^{2+}$. However, $MnCl_2 \cdot 4H_2O$ contains *cis*-$MnCl_2(H_2O)_4$ units, and $MnCl_2 \cdot 2H_2O$ has polymeric chains with *trans*-$Mn(H_2O)_2Cl_4$ octahedra sharing edges. The sulfate, $MnSO_4$, obtained on fuming down H_2SO_4 solutions, is very stable and may be used for Mn analysis, provided no other cations giving nonvolatile sulfates are present.

Manganese(II) Complexes

The equilibrium constants for complex formation in aqueous solution are not high compared to those for $Fe^{2+}-Cu^{2+}$ because the Mn^{II} ion is the largest of these and it has no ligand field stabilization energy in its complexes (except in the few of low spin).

Halide Ligands. The formation constants in aqueous solution are very low, for example,

$$Mn^{2+}(aq) + Cl^-(aq) \rightleftharpoons MnCl^+(aq) \qquad K \sim 3.85$$

but when ethanol or acetic acid is used as solvent, salts of complex anions of varying types may be isolated, such as:

MnX_3^-	Octahedral with perovskite structure
MnX_4^{2-}	Tetrahedral (green yellow) or polymeric octahedral with halide bridges (pink)
$MnCl_6^{4-}$	Only Na and K salts known; octahedral
$Mn_2Cl_7^{3-}$	With Me_3NH^+, linear chain of face-sharing $MnCl_6$ octahedra; also discrete $MnCl_4^{2-}$, tetrahedral

The precise nature of the product obtained depends on the cation used and also on the halide and the solvent. There is a variety of halide complexes with other ligands present such as $MnI_2(thf)_3$, *trans*-$[MnCl_4(H_2O)_2]^{2-}$, and amine and phosphine species. Homoleptic pseudohalide complexes are represented by $[Mn(NCS)_6]^{4-}$ and $[Mn(CN)_6]^{4-}$, as well as Prussian blue-like analogues, e.g., $K_2Mn^{II}[Mn^{II}(CN)_6]$, that are ferrimagnetic.[3]

[3]W. R. Entley and G. S. Girolami, *Inorg. Chem.* **1994**, *33*, 5165.

Nitrogen Ligands. Amine complexes are not particularly stable but there are cations like $[Mn(bipy)_3]^{2+}$ and tertiary amine complexes, e.g., $[MnI_3(NMe_3)]^-$ and $[MnI_2(NEt_3)]_2$.[4]

Amides[5] can be 3-coordinate with bridges as in $Mn_2(\mu\text{-}NPr_2^i)_2(NPr_2^i)_2$. Lithium compounds, e.g., $[Li\ thf][Mn\{N(SiMe_3)_2\}_3]$, also have 3-coordinate Mn but adducts like $Mn[N(SiMe_3)_2]_2py_2$ are tetrahedral. The *azide*, $(Me_4N)[Mn(N_3)_3]$, has end to end $\mu\text{-}N_3$ bridges in a 3-dimensional network.[6]

Tetrahedral diimine complexes such as

$$Mn[N(Bu^t)=C(H)C_6H_3(Me)NH]_2$$

are similar to *N,N'* analogues of diketonates and Schiff bases.[7]

Square planar geometry is found only in the *porphyrin* and *phthalocyanin* complexes. Few examples of 7- and 8-coordination are known (Table 17-D-1). A recent one $[MnL](PF_6)_2$, L = the tripodal heptadentate ligand *tris*{2-[*N*-(2-pyridyl-methyl)amino]ethyl}, acts as a superoxide scavenger.[8] The octahedral acetonitrile complex $[Mn(MeCN)_6][MnI_4]$ with a tetrahedral anion is obtained as yellow crystals from MeCN solutions of MnI_2.[9]

Phosphine Ligands. Although complexes of stoichiometry $MnX_2(PR_3)$ are polymeric with halide bridges, there are octahedral chelates such as *trans-*$MnX_2(dmpe)_2$ and a distorted tetrahedral monomer $MnI_2(PEt_3)_2$.

The unidentate phosphine complexes $[MnX_2(PR_3)]_n$ in THF at low temperatures react with a number of small molecules, O_2, CO, SO_2, and C_2H_4, evidently reversibly. The O_2 species are deep blue or purple and probably peroxo ones; irreversible oxidation of R_3P to R_3PO also occurs slowly to give, e.g., $MnI_2(OPMePh_2)(PMePh_2)$.[10]

Phosphine complexes with alkyl ligands are discussed in Section 17-D-9.

Oxygen Ligands. The *acetylacetonate*, like other M^{II} β-diketonates, polymerizes to increase the coordination number and is trimeric, $[cf.\ \{Ni(acac)_2\}_3]$, but the two end Mn atoms have trigonal prismatic rather than the octahedral coordination found for nickel.

Alkoxides. These are usually made by interaction of MnX_2 with LiOR in the alcohol. Use of bulky OR groups[11] can give monomeric species or bridged dimers such as $[Mn(OC_6Cl_3H_2)(\mu\text{-}OC_6Cl_3H_2)bipy]_2$.

[4]M. Godfrey *et al., J. Chem. Soc., Dalton Trans.* **1995,** 701.

[5]P. P. Power *et al., Chem. Commun.* **1996,** 1573; *Inorg. Chem.* **1991,** *30,* 1783, 2487; H. W. Roesky *et al., Z. Naturforsch.* **1992,** *47B,* 9.

[6]T. Rojo *et al., Angew. Chem. Int. Ed. Engl.* **1996,** *35,* 78.

[7]G. Wilkinson *et al., Polyhedron* **1996,** *15,* 3605.

[8]I. Morgenstern-Baderau *et al., J. Am. Chem. Soc.* **1996,** *118,* 4567.

[9]K. Dehnicke *et al., Z. Naturforsch.* **1996,** *51B,* 298.

[10]C. A. McAuliffe *et al., J. Chem. Soc., Dalton Trans.* **1993,** 2053, 3373.

[11]J. A. Osborn *et al., Polyhedron* **1990,** *18,* 1311; *Angew, Chem. Int. Ed. Engl.* **1994,** *33,* 1592.

Several tetrameric alkoxides are known;[12] these have Mn_4O_4 cores similar to those found for Fe^{II} and other Mn species that have cubes as building blocks in higher nuclearity structures to be discussed in Section 17-D-6. A typical tetramer is $[Mn(OMe)(MeOH)DPM]_4$ (HDPM = dipyrazolylmethane). This has four μ_3-OMe groups and four Mn^{II} atoms at alternating vertices of the cube with the chelate diketonate and MeOH bound to the Mn^{II} atoms which are thus octahedrally coordinated.

Other oxo compounds include the oxalate $[Mn\ ox_3]^{4-}$, and the catecholate, $[Mn\ cat_3]^{4-}$; these can be readily oxidized to the Mn^{III} state. Although oxocentered Mn^{III}_3 carboxylates (see later) have long been known, only recently has a Mn^{II} complex with μ_3-OH and acetate groups been made.[13] This is $py_5Mn_3(\mu_2\text{-}OAc)_3(\mu_3\text{-}OH)(\text{catechol})$ obtained by interaction of manganese acetate, 1,2-catechol, pyridine, and ethanolic Et_4NOH. The Fe^{II}, Co^{II}, and Ni^{II} analogues were also made.

Sulfur Ligands. The *thiolates*[14] may be of the type $[Mn(SR)_4]^{2-}$, $[Mn_2(SR)_4(\mu\text{-}SR)_2]^{2-}$, and $[Mn_4(SPh)_{10}]^{2-}$. Thiolates with bulky aryl groups[15] can stabilize low coordination numbers as in the 3-coordinate $[(ArS)Mn^{II}(\mu\text{-}SAr)]_2$, Ar = $2,4,6\text{-}Bu^t_3C_6H_2S$.

As well as the sulfido anion,[16] $[Mn(S_6)_2]^{2-}$, there are dithiocarbamates that are infinite polymers with bridging S atoms; Se[14,17] and Te[17] analogues of some of the sulfur compounds are known.

Electronic Spectra of Manganese(II) Compounds

The high-spin d^5 configuration has certain unique properties and manganese(II) is the most prominent example of this configuration. The majority of Mn^{II} complexes are high-spin. In *octahedral fields* this configuration gives spin-forbidden as well as parity-forbidden transitions, thus accounting for the extremely pale color of such compounds. In *tetrahedral environments*, the transitions are still spin-forbidden but no longer parity-forbidden; these transitions are therefore ~100 times stronger and the compounds have a noticeable pale yellow-green color. Tetrahedral Mn^{2+} can also show intense yellow-green fluorescence and commercial phosphors are Mn^{II} activated Zn compounds like Zn_2SiO_4. The high-spin d^5 configuration gives an essentially spin-only, temperature-independent magnetic moment of ~5.9 BM.

At sufficiently high values of Δ_0, a t_{2g}^5 configuration gives rise to a doublet ground state; for Mn^{II} the pairing energy is high and only a few of the strongest ligand sets, for example, those in $[Mn(CN)_6]^{4-}$, $[Mn(CN)_5NO]^{3-}$, and $[Mn(CNR)_6]^{2+}$, can accomplish this.

[12]S. J. Lippard *et al.*, *J. Am. Chem. Soc.* **1996**, *118*, 3069.
[13]D. Coucouvanis *et al.*, *Inorg. Chem.* **1996**, *35*, 2724.
[14]H.-O. Stephan and G. Henkel, *Polyhedron* **1996**, *15*, 501.
[15]P. P. Power and S. C. Shoner, *Angew. Chem. Int. Ed. Engl.* **1991**, *30*, 330.
[16]S.-B. Yu and R. H. Holm, *Polyhedron* **1993**, *12*, 263.
[17]P. P. Power *et al.*, *Inorg. Chem.* **1995**, *34*, 49; M. Bochmann *et al.*, *J. Chem. Soc., Dalton Trans.* **1995**, 1645.

17-D-3   The Chemistry of Manganese(III), d^4

Binary Compounds

Oxides. The first product of oxidation by O_2 of Mn or MnO at 470–600°C is Mn_2O_3; at 1000°C this gives black Mn_3O_4 which is found as the mineral *hausmannite*, a spinel, $Mn^{II}Mn_2^{III}O_4$. Oxidation of $Mn(OH)_2$ gives a brown hydrous oxide MnO(OH). There are also mixed oxide systems that contain Mn^{III} such as $LiMnO_2$, Na_5MnO_4, and $K_6Mn_2O_6$.

Halides. The fluoride is formed on fluorination of $MnCl_2$ as a red-purple solid instantly hydrolyzed by water. The black chloride has been obtained by chlorination of MnO_2 with HCl gas in CCl_4 but it is poorly characterized as it decomposes *ca.* −40°C. Solutions containing $MnCl_3L_3$, L = Et_2O, MeCN, etc., can be made.

The Manganese(III) Ion

The aqua ion can be obtained by electrolytic or peroxosulfate oxidation of Mn^{2+} solutions, or by reduction of MnO_4^-. The ion plays a central role in the complex redox reactions of the higher oxidation states of manganese in aqueous solutions. It is most stable in acid solutions, since it is very readily hydrolyzed:

$$Mn^{3+} + H_2O = Mn(OH)^{2+} + H^+ \quad K = 0.93$$

the initial monomer slowly polymerizing.

Under suitable conditions the Mn^{3+}–Mn^{2+} couple is reversible ($E^0 = 1.559$ V in 3 M $LiClO_4$). The Mn^{3+} ion is slowly reduced by water:

$$2Mn^{3+} + H_2O = 2Mn^{2+} + 2H^+ + \tfrac{1}{2}O_2$$

Manganese(III) Complexes

Halides. For fluorine both anhydrous and hydrated salts of the anions MnF_4^-, MnF_5^{2-}, and MnF_6^{3-} are known.[18]

The salt $Tl_2MnF_5 \cdot H_2O$ has chains of $(MnF_5)_n^{2n-}$ units.[19] Some other salts are $[MnF_3(urea)_3] \cdot 3H_2O$ and $K_3[MnF_2(C_2O_4)_2] \cdot 3H_2O$.

Chlorides. Although several methods of making $MnCl_3$ in solution are known,[20] by far the best way is the interaction of the carboxylate, $[Mn_{12}^{III}O_{12}(O_2CMe)_{16}(H_2O)_4] \cdot MeCO_2H \cdot 3H_2O$,[21] in acetonitrile or some other solvents, with Me_3SiCl. The purple solutions (which also contain silicon species of various sorts) are useful starting materials for synthesis of the chloroanion, $(Et_4N)_2MnCl_5$, amine complexes such

[18] W. Massa and D. Bakel, *Chem. Rev.* **1988**, *88*, 275.
[19] A. Tressaud *et al.*, *Inorg. Chem.* **1992**, *31*, 770.
[20] See e.g., R. Uson *et al.*, *Transition Met. Chem.* **1976**, *1*, 122.
[21] G. Christou *et al.*, *Inorg. Chem.* **1991**, *30*, 1665; *J. Am. Chem. Soc.* **1993**, *115*, 1804.

as $[MnCl_3(bipy)]_n$, and the imido species $MnCl(NBu^t)_3$ (Section 17-D-8). Other important compounds are the Schiff base species such as MnCl(salen), porphyrins, e.g., ClMn(porph), and related macrocyclic complexes.

A dark green 5-coordinate complex, *mer*-$MnI_3(PMe_3)_2$, is formed on oxidation of MnI_2 in excess PMe_3. The anion $[MnI_4]^-$ was noted earlier.

Oxygen Ligands. The well known octahedral, brown $Mn(acac)_3$, which is soluble in organic solvents and a useful starting material, is obtained by interaction of acetylacetone with a concentrated aqueous solution of $KMnO_4$. It can be oxidized to the unstable cation $[Mn^{IV}(acac)_3]^+$ by electrochemical oxidation in MeCN. In aqueous solution the Mn^{III} state can be stabilized by complexing anions such as $C_2O_4^{2-}$, SO_4^{2-}, and $EDTA^{4-}$, but even these species undergo slow decomposition. A representative cationic example is $[Mn(urea)_6]^{3+}$. Although "manganic acetate" has long been known to have the oxo-centered $[Mn^{III}O(\mu\text{-}O_2CMe)_6]^+$ core it is still often referred to as $Mn(OAc)_3 \cdot 2H_2O$.[22] The carboxylates can be extremely complicated and often have mixed valencies, e.g., Mn^{II}, Mn^{III} (see Section 17-D-6). Other multinuclear oxo species are known, one example being $[Mn_{10}O_4(biphen)_4Cl_{12}]^{4-}$, that has mixed valencies, here $4Mn^{III}$ + $6Mn^{II}$ with $4\mu_4$-oxo ligands.[23]

Nitrogen Ligands. There are numerous porphyrin[24] and Schiff base complexes,[25] generally LMnX. They have been much studied because of their involvement in catalytic oxidation reactions in presence of H_2O_2, PhIO, or O_2 (see later discussion).

Amido species include the 3-coordinate planar $Mn[N(SiMe_3)_2]_3$ that was obtained by oxidation of $Mn[N(SiMe_3)_2]_2$ with $BrN(SiMe_3)_2$[26] and a trigonal pyramidal complex[27] $Mn(N_3N)$, where N_3N = $[(Bu^tMe_2Si)NCH_2CH_2]_3N$; four N atoms are bound to Mn and the empty axial coordination site is protected by the bulkiness of the trialkylsilyl substituent on the triamido ligand.

The interaction of alkaline earth nitrides and manganese nitride at high temperatures leads to compounds such as Ca_3MnN_3 or Ca_6MnN_5 that have sheet structures involving the trigonal planar $[Mn(N)_3]^{6-}$ anion (*cf.* CO_3^{2-}).[28]

Other well-established Mn^{III} compounds are salts of $[Mn(CN)_6]^{3-}$ made by air oxidation of Mn^{II} solutions in presence of an excess of CN^- and the phosphine complex $MnH_3(dmpe)_2$. Both compounds are low spin.

Although manganese dioxygen complexes are probably involved in many enzymatic reactions, few such complexes have been structurally characterized— manganese(III) species include side-on (η^2-O_2)Mn(porph), a μ_3-oxo,μ-peroxo-

[22]See e.g., I. Ryu and H. Alper, *J. Am. Chem. Soc.* **1993**, *115*, 7543.

[23]R. Sessoli *et al.*, *J. Am. Chem. Soc.* **1995**, *117*, 5789.

[24]R. S. Armstrong *et al.*, *Inorg. Chem.* **1992**, *31*, 1463.

[25]C.-M. Che *et al.*, *J. Chem. Soc., Dalton Trans.* **1996**, 405; P. A. Grieco *et al.*, *J. Am. Chem. Soc.* **1993**, *115*, 11648; T. Katsuki, *Coord. Chem. Rev.* **1995**, *140*, 189.

[26]P. P. Power *et al.*, *J. Am. Chem. Soc.* **1989**, *111*, 8045.

[27]R. R. Schrock *et al.*, *Angew. Chem. Int. Ed. Engl.* **1992**, *31*, 1501.

[28]D. H. Gregory *et al.*, *Inorg. Chem.* **1995**, *14*, 5195.

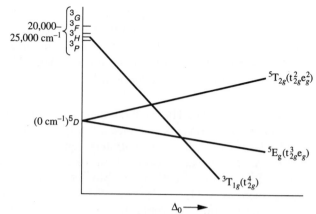

Figure   17-D-1 Simplified energy level diagram for the d^4 system Mn^{III} in octahedral surroundings.

Mn(III,III,III)   trimer,   and   a   pyrazolate-pyrazolylborate, $Mn(\eta^2\text{-}O_2)(3,5\text{-}Pr^i pzH)(HB(3,5\text{-}Pr^i_2 pz)_3)$, with side-on O_2^{2-}.[29]

Electronic Structure of Mn^{III} Compounds

The $^5E_g(t^3_{2g}e_g)$ state for octahedral Mn^{III} is subject to a Jahn-Teller distortion. Because of the odd number of e_g electrons, this distortion should be appreciable and resemble the distortion in Cr^{II} and Cu^{II} compounds. Indeed, a considerable elongation of two trans bonds with little difference in the lengths of the other four has been observed in many Mn^{III} compounds. For example, MnF_3 has the same basic structure as VF_3 where each V^{3+} ion is surrounded by a regular octahedron of F^- ions, except that two Mn—F distances are 1.79 Å, two more are 1.91 Å, and the remaining two are 2.09 Å. There is also distortion of the spinel structure of Mn_3O_4 where Mn^{2+} ions are in tetrahedral interstices and Mn^{3+} in octahedral interstices: Each of the latter tends to distort its own octahedron, and the cumulative effect is that the entire lattice is distorted from cubic to elongated tetragonal.

The complex $Mn(acac)_3$ has two forms, one of which shows a substantial tetragonal elongation (two Mn—O = 2.12 Å; four Mn—O = 1.93 Å) as do other high-spin complexes such as porphyrin adducts; the other form shows a moderate tetragonal compression (two Mn—O = 1.95 Å; four Mn—O = 2.00 Å).

In $[Mn(urea)_6]^{3+}$ all six Mn—O distances are equal (1.986 Å) but analysis of displacements of the O atoms and electronic spectra are in agreement with a dynamic Jahn-Teller effect.

The very strong acidity of the aqua ion has been attributed to a strong ligand field stabilizing a distorted ion, possibly $[Mn(OH)(H_2O)_5]^{2+}$.

A simplified energy level diagram for d^4 systems is shown in Fig. 17-D-1. It is consistent with the existence of both high-spin and low-spin octahedral complexes. Because the next quintet state (^5F, derived from the d^3s configuration) lying ~110,000 cm^{-1} above the ^5D ground state of the free ion is of such high energy, only one spin-allowed absorption band ($^5E_g \rightarrow {}^5T_{2g}$) is to be expected in the visible region. For $[Mn(H_2O)_6]^{3+}$ and tris(oxalato)- and tris(acetylacetonato)manga-

[29] N. Kitajima, *J. Am. Chem. Soc.* **1994,** *116,* 11596.

nese(III) a rather broad band appears around 20,000 cm^{-1} and the red or red-brown colors of high-spin Mn^{III} compounds may be attributed to such absorption bands. However, the spectra of some 6-coordinate Mn^{III} complexes are not so simple, and they are difficult to interpret in all their details, presumably because both static and dynamic Jahn-Teller effects perturb the simple picture based on O$_h$ symmetry.

Low-spin manganese(III) compounds are the diamagnetic MnH$_3$(dmpe)$_2$ and salts of the [Mn(CN)$_6$]$^{3-}$ ion. Manganese(II) in the presence of an excess of CN$^-$ is readily oxidized, even by a current of air, with the production of this ion that is first isolated from the solution as the Mn^{II} salt Mn$_3^{II}$[Mn(CN)$_6$]$_2$, from which other salts are obtained.

17-D-4 The Chemistry of Manganese(IV), d^3

Binary Compounds

Manganese dioxide, a grey to black solid, occurs in ores such as *pyrolusite*, where it is usually nonstoichiometric. When made by the action of O$_2$ on Mn at a high temperature, it has the rutile structure found for many other oxides MO$_2$ (e.g., those of Ru, Mo, W, Re, Os, Ir, and Rh). However, as normally made by heating Mn(NO$_3$)$_2$·6H$_2$O in air (~530°C), it is nonstoichiometric. A hydrated form is obtained by reduction of aqueous KMnO$_4$ in basic solution. Manganese(IV) occurs in a number of mixed oxides.

Manganese dioxide is inert to most acids except when heated, but it does not dissolve to give Mn^{IV} in solution; instead it functions as an oxidizing agent, the exact manner of this depending on the acid. With HCl, chlorine is evolved,

$$MnO_2 + 4HCl \longrightarrow MnCl_2 + Cl_2 + 2H_2O$$

and this reaction is often used for small-scale generation of Cl$_2$ in the laboratory. With H$_2$SO$_4$ at 110°C, O$_2$ is evolved and an Mn^{III} acid sulfate is formed.

The oxide is used in dry cell batteries, in ferrite production and as a catalyst for oxidation of alcohols and other organic compounds.[30]

The *tetrafluoride* is an unstable blue solid that decomposes to MnF$_3$ + F$_2$. Both this fluoride and MnF$_3$ form adducts with SbF$_5$ that are non-ionic with fluoride bridges.[31]

Non-oxo Complexes

These are relatively few. The dithiocarbamate cations [Mn(S$_2$CNR$_2$)$_3$]$^+$ are readily obtained by an oxidation of the Mn^{III} compounds in CH$_2$Cl$_2$ in presence of BF$_3$; they are isolated as purple BF$_4^-$ salts. The salt K$_2$MnF$_6$ is formed by interaction of KMnO$_4$ in 40% HF and is stable, unlike the [MnCl$_6$]$^{2-}$ ion that can only be trapped in a matrix by interaction of KMnO$_4$ and concentrated HCl in the presence of K$_2$SnCl$_6$.

The pyrazolylborate {Mn[HB(3,5-Me$_2$pz)$_3$]$_2$}$^{2+}$ has been made by oxidation of Mn^{2+} perchlorate in presence of KTp* by NaMnO$_4$.[32]

[30]H. Cao and S. L. Suib, *J. Am. Chem. Soc.* **1994**, *116*, 5334.
[31]W. Sawodny and K. M. Rau, *J. Fluorine Chem.* **1993**, *61*, 111.
[32]M. K. Chan and W. H. Armstrong, *Inorg. Chem.* **1989**, *28*, 3777.

The addition of $(Bu_4^nN)MnO_4$ to an ethanol solution of Mn^{II} acetate and biguanide leads to oxidation to a red octahedral cation,[33] $[MnL_3]^+$ where L is the anion, (17-D-I),

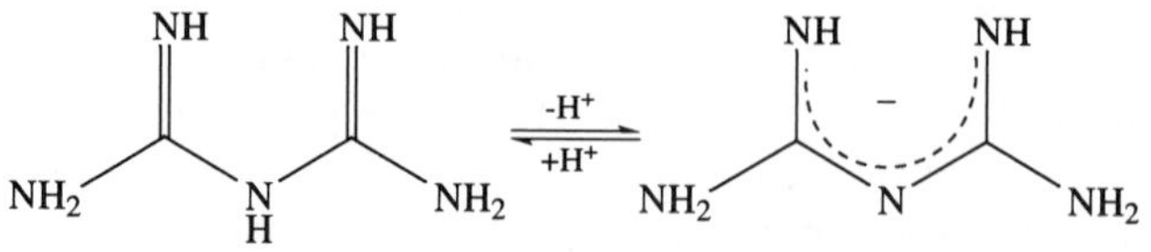

(17-D-I)

Surprisingly the NH_2 groups are not oxidized.

Another surprising compound is the tetraazide $Mn(N_3)_4$bipy where the two azido groups *trans* to the N atoms of *bipy* show stronger Mn–N bonds than those in the *cis* positions.[34]

Oxo Complexes

These are of great variety and importance. Catecholates,[35] such as $Mn^{IV}py_2(3,5-DBcat)_2$ (3,5-DBcat = 3,5-di-*tert*-butyl-1,2-benzoquinone), show dramatic thermochromic effects on heating in toluene due to a 2e transfer to give the Mn^{II} semiquinone tautomer:

$$Mn^{IV}py_2(3,5-DBcat)_2 \rightleftharpoons Mn^{II}py_2(3,5-DBSQ)_2$$
$$\text{purple (240 K)} \qquad \text{pale green-brown (298 K)}$$

The relatively simple O,N, Schiff base $salpn^{2-}$ complex, dichloro[*N,N'*(salicylidene)-1,3- propanediamino]manganese(IV), $Mn^{IV}(salpn)Cl_2$, made by action of HCl on $Mn^{III}(salpn)Cl$ in MeCN, is unusual in that it can halogenate alkenes.[36]

Compounds with oxo bridges are very characteristic of Mn^{IV} and a common feature of such species is their reduction in 1e steps to the corresponding III-IV and III-III complexes. A representative example is $[(phen)_2Mn^{IV}(\mu\text{-}O)_2Mn^{IV}(phen)_2]^{4+}$. Schiff base (L) complexes[37] also may have μ-O groups, for example $LMn(\mu\text{-}O)_2MnL$, while triazacyclononane ligands (L) form $(\mu\text{-}O)_3$ species as in $[Mn_2(\mu\text{-}O)_3L_2]^{2+}$.[38]

Many porphyrin complexes have Mn=O bonds;[39] as noted later these have been much studied as oxygen transfer agents, for example, in the epoxidation of C=C bonds.[40]

[33] S. R. Cooper *et al., J. Chem. Soc., Chem. Commun.* **1992**, 894.
[34] B. C. Dave and R. S. Czernuswicz, *J. Coord. Chem.* **1994**, *33*, 257.
[35] A. S. Attia and C. G. Pierpoint, *Inorg. Chem.* **1995**, *34*, 1172.
[36] V. L. Pecoraro *et al., J. Chem. Soc., Chem. Commun.* **1995**, 2015.
[37] C. P. Horwitz *et al., Inorg. Chem.* **1993**, *32*, 82, 5951.
[38] J. H. Koek *et al., J. Chem. Soc., Dalton Trans.* **1996**, 353.
[39] See e.g., T. C. Bruice *et al., J. Am. Chem. Soc.* **1993**, *115*, 7985.
[40] C. D. Garner *et al., Angew. Chem. Int. Ed. Engl.* **1995**, *34*, 343.

17-D-5 The Chemistry of Manganese(V), d^2

Most Mn^V chemistry involves oxo compounds and these are discussed in the next section. *Nitrido compounds*[41] with $Mn\equiv N$ bonds are commonly made by photolysis of azide precursors:

$$L_nMn^{III}(N_3) \xrightarrow{\ hv\ } L_nMn^V\equiv N + N_2$$

The ligands, L, can be porphyrins, phthalocyanins, triazacyclononanes[42] and so on.

The reduction of the Mn^{VII} complex, $MnCl(NBu^t)_3$ (Section 17-D-8), with lithium in 1,2-dimethoxyethane gives the Mn_2^V dimer $[Li(dme)_2]_2[Mn(NBu^t)_2(\mu\text{-}NBu^t)]_2$ while a similar reduction by Na/Hg in THF gives the mixed valence V,VI anion $(NaTHF)[Mn(NBu^t)_2(\mu\text{-}NBu^t)]_2$. These species are readily oxidized by O_2 to give the Mn^{VI} dimer $[(Bu^tN)_2Mn(\mu\text{-}NBu^t)]_2$. This compound is partly reduced by I_2 to give a linear mixed valence, VI, V, VI, trimer, $\{[Mn(NBu^t)_2(\mu\text{-}NBu^t)_2]_2Mn\}^+I_3^-$.[43]

17-D-6 Mixed Oxidation State Complexes

Since about 1983 there has been an increasing deluge of publications on manganese compounds that might provide models for—or at least provide some insight into—the role of Mn in photosynthesis and several enzyme systems. While we cannot attempt to discuss these complicated systems,[44] for photosynthesis it appears that tetranuclear Mn complexes are involved that undergo oxidation/reduction cycles in photosystem II where the overall reduction involved is

$$2H_2O \longrightarrow O_2 + 4H^+ + 4e$$

Oxygen formation occurs after four successive light flashes and a catalytic cycle (Kok cycle) has been devised to explain the sequence for the oxidation states where electrons are released to photosystem I. Models for the Mn_4 core and reaction mechanisms have been published.[45] However, until recently no model systems had produced oxygen from water. Use of the Schiff base, 3,5-dichlorosalicylidine-1,2-diaminoethane, gives the dimeric cation $[Mn(salen)(H_2O)_2]_2^{2+}$ that has the planar ligand N,O bound to each Mn^{III} and the Mn atoms are bridged by H-bonds between phenoxy O atoms of the ligand and H_2O molecules bound to Mn. This produces a cleft that allows the approach of a *p*-benzoquinone molecule when photolysis with visible light gives O_2.[46]

A similar Mn^{III} cation having porphyrins, rigidly linked *via* a 1,2-C_6H_4 ring, also gives O_2 in a 4e oxidation of water on electrolysis in MeCN, without going through H_2O_2 as an intermediate.[47]

[41]K. Dehnicke and J. Strahle, *Angew. Chem. Int. Ed. Engl.* **1992**, *31*, 95 (extensive review).
[42]K. Wieghardt *et al.*, *Inorg. Chem.* **1996**, *35*, 906.
[43]G. Wilkinson *et al.*, *J. Chem. Soc., Dalton Trans.* **1995**, 937.
[44]See K. Wieghardt, *Angew. Chem. Int. Ed. Engl.* **1994**, *33*, 725; M. P. Klein *et al.*, *J. Am. Chem. Soc.* **1994**, *116*, 5239; C. F. Yocum, *J. Am. Chem. Soc.* **1996**, *118*, 2400.
[45]D. N. Henderson *et al.*, *J. Am. Chem. Soc.* **1994**, *116*, 8376.
[46]C. A. McAuliffe *et al.*, *J. Chem. Soc., Chem. Commun.* **1994**, *1153*, 2141.
[47]Y. Naruta *et al.*, *Angew. Chem. Int. Ed. Engl.* **1994, 33**, 1839.

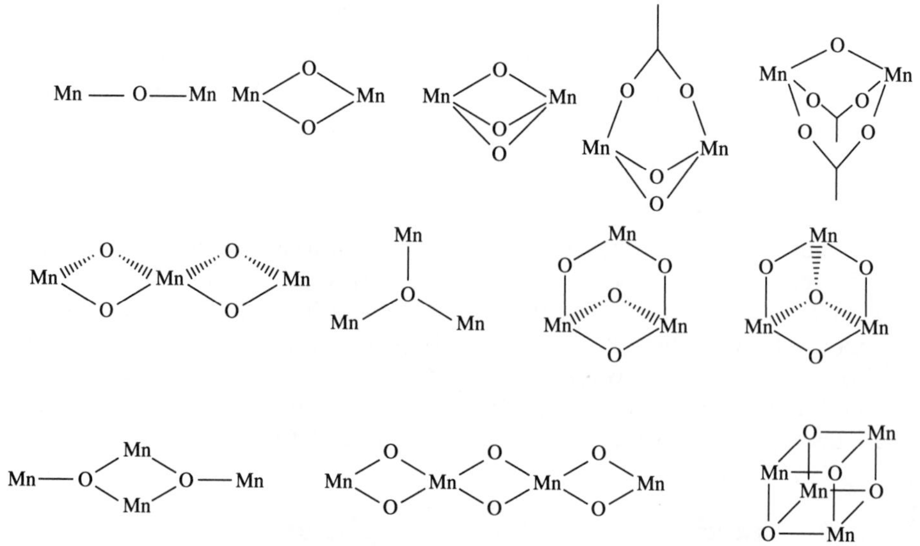

Figure 17-D-2 Various types of manganese oxo, alkoxo, or carboxylato core structures capable of existing in differing oxidation states and as part of very complex molecules or anions.

In the search for model systems, hundreds of new compounds have been made, those with carboxylate ligands being the most common. However, other ligands employed have been bipyridine, pyrazolylborates, triazacyclononanes, and Schiff bases of various sorts. These have been obtained in oxidation states II-V with both unitary and mixed valencies. The structures vary from quite simple to extremely complicated. Some of the core types are shown in Fig. 17-D-2.

The more complex species can have structures built up of different core units. The Mn_4O_4 cubes can have different charges, e.g., 4+(Mn^{III}), 8+(Mn^{IV}); there are also rings of various sorts, clusters such as $[Mn_4(\mu_3\text{-}O)_2]^{8+}$ and $[Mn_{18}O_{16}]^{22+}$ in which 18 Mn^{III} ions are linked by 16 $\mu_3\text{-}O^{2-}$ ions. Some specific examples[48] are the following:

$[Mn^{IV}(salpn)(\mu\text{-}O)]_2$

$[Mn^{III}Mn^{IV}(\mu\text{-}O)_2(bipy)_4]^{3+}$

$[Mn_{12}O_{12}(O_2CEt)_{16}(H_2O)_4]^-$, Mn^{II}, $7Mn^{III}$, $4Mn^{IV}$

$Mn_6^{II,III}O_2(O_2CBu^t)_{10}py_4$

Other naturally occurring manganese compounds other than chlorophylls in Nature are those found in various enzymes. *Catalases*[49] catalyze the cell reaction

$$2H_2O_2 \longrightarrow 2H_2O + O_2$$

[48]G. Christou, *Acc. Chem. Res.* **1989,** *22,* 328; *J. Am. Chem. Soc.* **1995,** *117,* 6463, 7275; R. D. Britt *et al., J. Am. Chem. Soc.* **1995,** *117,* 11780; W. H. Armstrong *et al., Inorg. Chem.* **1995,** *34,* 4708; M. V. Rajasekhavan *et al., Inorg. Chem.* **1996,** *35,* 2283; A. Boussac *et al., J. Am. Chem. Soc.* **1996,** *118,* 2669.
[49]W. H. Armstrong *et al., J. Am. Chem. Soc.* **1994,** *116,* 2392; A. Haddy *et al., Inorg. Chem.* **1994,** *33,* 2677.

Superoxide dismutases[50] catalyze the dismutation of superoxide ions to give non-radical products:

$$O_2^{\cdot-} + HO_2^{\cdot} + H^+ \longrightarrow O_2 + H_2O_2$$

Models for these and other enzymes have been much studied.

Oxidations[51]

Porphyrin complexes such as (porph)MnCl or (porph)Mn=O have been much studied for the oxidation of alkanes, alkenes, and other organic compounds using O-transfer agents such as PhIO, H_2O_2, or O_2. Schiff base complexes have also been used since chiral complexes (e.g., of the salen type) can be readily made on a ton scale and the enantiomers separated in pure forms. Thus the epoxidation

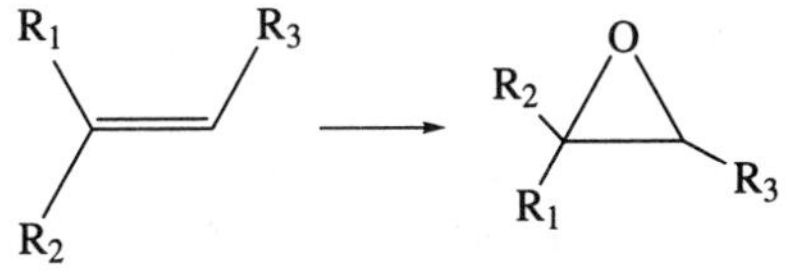

can be achieved with >98% yield using catalyst precursors of the type (17-D-II). Aqueous NaOCl is the oxidant with the Mn catalyst and the substrate in an organic solvent such as toluene or chlorobenzene. The intermediates are doubtless species with M=O or M–O bonds, possibly of the type $LMn^{IV}(\mu\text{-O})_2Mn^{IV}L$, since bridging O atoms are known to participate in O-transfer reactions.

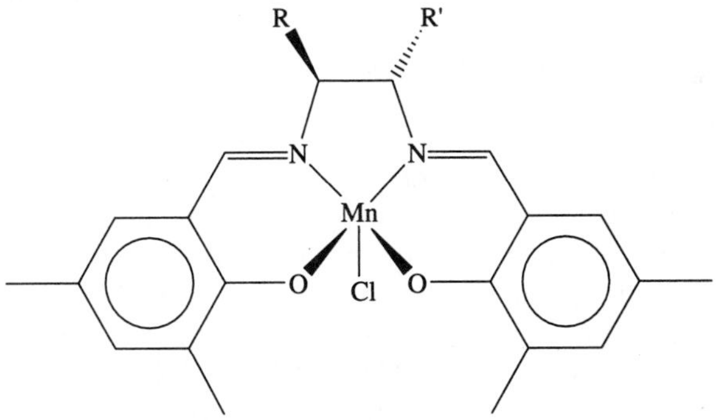

(17-D-II)

17-D-7   Oxo Chemistry of Manganese(V)-(VII)

A few species of manganese(IV) compounds with Mn=O bonds have been noted earlier. However, the main chemistry of Mn=O compounds is in the higher oxidation states.

[50]D. P. Riley and R. H. Weiss, *J. Am. Chem. Soc.* **1994,** *116,* 387.
[51]N. Jacobson, *Comprehensive Organometallic Chemistry* II, Pergamon, Oxford, 1955, Vol. 12, Chap. 11.1; W. Nam and J. S. Valentine, *J. Am. Chem. Soc.* **1993,** *115,* 1772.

Tetraoxomanganates

The tetrahedral oxo anions have been intensively studied by pulse radiolysis and spectroscopic methods; unstable protonated forms, $O_3Mn^V(OH)^{2-}$ and $O_3Mn^{VI}(OH)^-$, also exist.[52]

The three oxo anions, blue $Mn^VO_4^{3-}$, green $Mn^{VI}O_4^{2-}$, and purple $Mn^{VII}O_4^-$, have been known for over 150 years. They have been isolated as crystalline salts and shown to be tetrahedral.[53]

A mixed valence compound, $K_3[Mn^{VII}O_4][Mn^{VI}O_4]$, has been studied by neutron diffraction,[54] while more complex structures have been found in salts like $Na_{10}Li_2[MnO_4]_4$.[55]

The *hypomanganate(3−) ion* can be obtained in melt reactions:

$$2MnO_2 + 6KOH + \tfrac{1}{2}O_2 \xrightarrow{\;ca\;400\,°C\;} 2K_3MnO_4 + 3H_2O$$

It can be made[56,57] in solution by reduction of MnO_4^{2-} by H_2O_2 or sodium formate in 10 M KOH, where it exists mainly as $MnO_3(OH)^-$ in equilibrium with MnO_4^{3-} and $MnO_2(OH)_2^-$.

The *manganate(2−) ion* is obtained by oxidizing the fused K_3MnO_4 with O_2 or by a KNO_3 fusion of MnO_2. It can also be made in solution by adding KOH (5–10 M) to $KMnO_4$. The best known ion, *permanganate*, MnO_4^-, is made on a large scale by electrolysis or air-oxidation of alkaline solutions of MnO_4^{2-} usually as the Na^+ or K^+ salts. The potentially explosive tetraalkylammonium salts[58] are soluble in organic solvents; $KMnO_4$ can be solubilized in benzene by crown ethers or cryptands.

While MnO_4^{3-} is unstable in solution when the base concentration is below *ca.* 8 M, the MnO_4^{2-} ion disproportionates only when the base strength is below 0.1 M, forming MnO_4^- and MnO_4^{3-}, which decomposes, so that the overall reaction is

$$3MnO_4^{2-} + 4H^+ = 2MnO_4^- + MnO_2(s) + 2H_2O$$

Interaction of $Ba(MnO_4)_2$ and H_2SO_4 allowed isolation of crystals of $HMnO_4$ and $H_5O_2^+MnO_4^-$; both are violent oxidants.

Permanganates are common oxidants in inorganic and organic chemistry. Reduction with SO_2 in the presence of F^- provides evidence for Mn^{III} intermediates, i.e., $K_2[MnF_3SO_4]$ and $K_2[MnF_5]·H_2O$, before the final Mn^{II} product.[59] Reductive nitrosylation of MnO_4^- in aqueous alkaline solution with hydroxylamine in presence of CN^- ion gives cyano species with $Mn(NO)^{2+}$, $Mn(NO)_2^+$, or $Mn(NO)_3^+$ groups. This reaction is similar to those with other oxoanions, MoO_4^{2-}, CrO_4^{2-}, and VO_4^{3-}.[60]

[52]J. D. Rush and B. H. J. Bielski, *Inorg. Chem.* **1995**, *34*, 5832; E. Záhonyi-Budó and L. I. Simándi, *Inorg. Chim. Acta* **1995**, *237*, 173.
[53]M. B. Hursthouse *et al., J. Chem. Soc., Faraday Trans.* **1992**, *88*, 3071.
[54]R. D. Cannon *et al., J. Chem. Soc., Chem. Commun.* **1992**, 1445.
[55]D. Fischer and R. Hoppe, *Angew, Chem. Int. Ed. Engl.* **1990**, *29*, 800.
[56]J. D. Rush and B. H. J. Bielski, *Inorg. Chem.* **1995**, *34*, 5832.
[57]D. G. Lee and T. Chen, *J. Am. Chem. Soc.* **1989**, *111*, 7535.
[58]G. Christou *et al., Inorg. Chem.* **1992**, *31*, 5185.
[59]C. Bhattacharyya *et al., J. Chem. Soc., Dalton Trans.* **1993**, 3497.
[60]C. Bhattacharyya *et al., J. Chem. Soc., Dalton Trans.* **1989**, 1963.

The oxidation of organic compounds by MnO_4^- and MnO_4^{2-} has often been studied. Although lower-valent manganese esters have been involved as intermediates, recent studies[61] suggest a 2+2 addition of C–H bonds across Mn=O.

Oxide. The highly explosive oxide, Mn_2O_7, that has the structure[62] $O_3MnOMnO_3$ (MnOMn angle = 120.7°) is an oil (mp 5–9°C). It is obtained by interaction of $KMnO_4$ and concentrated H_2SO_4; with low $KMnO_4$ concentrations a green solution of the MnO_3^+ ion is obtained:

$$KMnO_4 + 3H_2SO_4 \rightleftharpoons K^+ + MnO_3^+ + H_3O^+ + 3HSO_4^-$$

Extraction of Mn_2O_7 into CCl_4 or chlorofluorocarbons gives reasonably stable and safe solutions.[63]

Oxohalides. Although MnO_3Cl was made by Wöhler in 1828, all the oxohalides, green Mn^VOCl_3, brown $Mn^{VI}O_2Cl_2$, and green $Mn^{VII}O_3Cl$, are highly unstable and explosive liquids. Spectroscopic studies are reported for MnO_3F, made by interaction of $KMnO_4$ and FSO_3H or IF_5.[64]

Other Oxo Species. There are a few Mn^V=O species such as the macrocyclic tetraamide (17-D-III); the bond distance, 1.555 Å, suggests Mn≡O.[65]

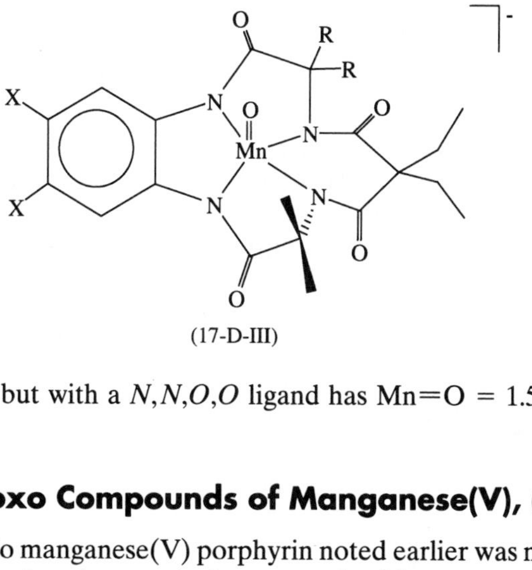

(17-D-III)

A similar complex but with a N,N,O,O ligand has Mn=O = 1.558 Å.[66]

17-D-8 Non-oxo Compounds of Manganese(V), (VI), and (VII)

Although the nitrido manganese(V) porphyrin noted earlier was made in 1983, only recently has a comprehensive range of compounds without oxo groups been made.[67]

[61]D. G. Lee and T. Chen, *J. Am. Chem. Soc.*, **1993**, *115*, 11231.
[62]B. Krebs *et al.*, *Z. anorg. allg. Chem.* **1988**, *588*, 7.
[63]M. Trömel and M. Russ, *Angew. Chem. Int. Ed. Engl.* **1987**, *26*, 1007.
[64]A. K. Brisdon *et al.*, *J. Chem. Soc. Dalton Trans.* **1991**, 3127.
[65]T. J. Collins *et al.*, *Inorg. Chem.* **1992**, *31*, 1548.
[66]T. V. O. Halloran *et al.*, *J. Am. Chem. Soc.* **1994**, *116*, 7431.
[67]G. Wilkinson *et al.*, *J. Chem. Soc., Dalton Trans.* **1994**, 1037; **1995**, 205.

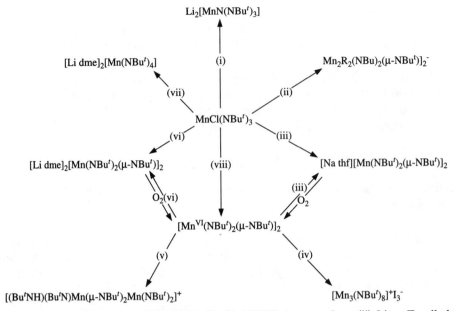

Figure 17-D-3   Reactions of MnCl(NBu^t)$_3$. (i) LiNHBu^t excess dme; (ii) Li or Zn alkyls, R = Me, Et, CH$_2$Ph, etc.; (iii) Na/Hg, thf; (iv) I$_2$, CH$_2$Cl$_2$; (v) CF$_3$SO$_3$H; (vi) Li, dme; (vii) LiNHBu^t dme; (viii) LiHNBu^t, Et$_2$O, then O$_2$.

The *tert*-butylimide analogue of MnO$_3$Cl, i.e., Mn(NBu^t)$_3$Cl, is a green crystalline solid, mp 94°C, sublimable in vacuum. It is stable in air and is unaffected by water. The compound was obtained by interaction of the purple solution of MnCl$_3$ in MeCN with NH(Bu^t)(SiMe$_3$). There is a remarkable reaction involving a 4e oxidation, i.e., +3 to +7, in a solution where the only obvious oxidant is solvated MnCl$_3$. It was proposed that oxidation occurs by repetition of the following process:

$$2Mn^{III}Cl_3 \xrightarrow[-Me_3SiCl]{NHBu^t(SiMe_3)} Cl_2Mn^{III}\!-\!N \xrightarrow{NHBu^t(SiMe_3)} \begin{array}{l} Cl_2Mn^{IV}\!=\!NBu^t \\ + MnCl_2 \\ + Me_3SiCl \\ + Bu^tNH_2 \end{array}$$

These lead to Mn^{VI}(NBu^t)$_3$, which is then chlorinated by MnCl$_3$, giving the overall stoichiometry

$$5MnCl_3 + 6NHBu^t(SiMe_3) \longrightarrow MnCl(NBu^t)_3 + 4MnCl_2 + 6Me_3SiCl + 3Bu^tNH_2$$

The yield of MnCl(NBu^t)$_3$ corresponds to this stoichiometry and Mn^{II} is found in the residual products as the MnCl$_4^{2-}$ salt.

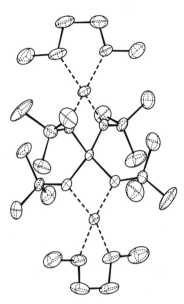

Figure 17-D-4   The structure of [Li dme]$_2$[Mn^{VI}(NBu^t)$_4$]. Reproduced by permission from A. A. Danopoulos *et al., J. Chem. Soc., Dalton Trans.* **1994,** 1044. The Mn center is tetrahedral.

The Cl atom in MnCl(NBu^t)$_3$ can be substituted by Br, O$_2$CR, OCH(CF$_3$)$_2$, SC$_6$F$_5$, C$_6$F$_5$, and NHBu^t; all of these compounds are green. As noted earlier (Section 17-D-5), reduction of MnCl(NBu^t)$_3$ can give Mn^V and Mn^{VI} binuclear complexes. The principal reactions are shown in Fig. 17-D-3. Over 40 imido compounds are now known.

Of particular interest are reactions of MnCl(NBu^t)$_3$ with LiNHBu^t from which, depending on the conditions—solvent, temperature and ratio of the reagent to Mn$^-$, different products are obtained. The anion of [Li dme]$_2$[Mn^{VI}(NBu^t)$_4$], which is the analogue of the MnO$_4^{2-}$ anion, is one of them; its structure is shown in Fig. 17-D-4.

Another unusual anion is [Mn^{VII}($\equiv$N)(NBu^t)$_3$]$^{2-}$, which is formed by a reaction involving elimination of isobutylene:

In addition to the complexes in Fig. 17-D-3 there are compounds such as 17-D-IV.

$$
\begin{array}{ccc}
& Bu^t \quad\quad Bu^t & \\
& N \quad\quad\quad N & \\
X & \diagdown\;\diagup \quad Mn \quad \diagdown\;\diagup & X \\
& N \quad\quad\quad N & \\
& Bu^t \quad\quad Bu^t &
\end{array}
$$

X = AlMe$_2$ or ZnMe

(17-D-IV)

17-D-9 Organometallic Compounds

There is a vast chemistry of compounds with Mn—C bonds. This includes ligands such as CO, alkyls and aryls, η^5-C$_5$H$_5$, η^5-C$_5$Me$_5$, η^6-arenes, alkenes, acetylenes, and allyls. Only some simple compounds can be noted here.

The *low oxidation state* chemistry of manganese −1, 0, 1 mostly involves carbon monoxide compounds of which the first known was Mn$_2$(CO)$_{10}$.[68] It has an extensive derivative chemistry[69] with compounds such as [Mn(CO)$_5$]$^-$, Mn(CO)$_5$Cl, Mn$_2$(μ-Cl)$_2$(CO)$_8$, *cis*-[Mn(CO)$_4$(NO$_2$)$_2$]$^-$, and [Mn$_2$(μ-Cl)$_3$(CO)$_6$]$^-$.[70] Nitric oxide compounds are also known, one example being [Mn(CN)$_5$NO]$^+$.[71] Other low-valent species include HMn(CO)$_5$, *trans*-MnH(C$_2$H$_4$)(dmpe)$_2$, MnH(CO)$_2$(η^6-C$_6$Me$_6$), and the (η^6-arene)Mn(CO)$_3^+$ ion.

Manganese(II) alkyls are well known (see Fig. 17-D-4). The trimeric mesityl, Mn$_3^{II}$mes$_6$,[72] has terminal and bridging aryl groups while the fluoroaryl, Mn(C$_6$F$_5$)$_2$(THF)$_2$, was obtained by reaction of Mn with Hg(C$_6$F$_5$)$_2$ in tetrahydrofuran.[73] Both monomeric and dimeric, bridged phosphine complexes are known, e.g., Mn(CH$_2$Ph)$_2$(dmpe) and Mn$_2$(CH$_2$SiMe$_3$)$_4$(PEt$_3$)$_2$.

The use of bulky ligands has allowed the synthesis[74] of monomeric quasi-linear 2-coordinate compounds such as Mn(2,4,6-Bu$_3^t$C$_6$H$_2$)$_2$ and Mn[C(SiMe$_3$)$_3$]$_2$; similar Mg and Fe compounds were made.

Ato complexes include [MnMe$_4$]$^{2-}$ [75] and the planar mesityl [Mn mes$_3$]$^-$.[76] An amido compound with a CH$_3$ bridge is {Mn(μ-Me)[N(SiMe$_3$)$_2$AlMe$_3$]$_2$}$_2$.[77]

Manganese(III) and (IV) Compounds. The earliest example of a high oxidation state manganese compound was the green Mn(1-norbornyl)$_4$ which was obtained by interaction of Li(1-norbornyl) and MnX$_2$; some type of disproportionation or, more likely, oxidation of an ato complex, may account for the formation of Mn^{IV}. Other Mn^{IV} compounds include Me$_4$Mn (dmpe) and the anion[78] [MnMe$_6$]$^{2-}$.

[68]B. J. Baerends *et al., Inorg. Chem.* **1996,** *35,* 2886.
[69]B. F. G. Johnson *et al., J. Chem. Soc., Dalton Trans.* **1996,** 1419.
[70]B. F. G. Johnson *et al., J. Chem. Soc., Dalton Trans.* **1996,** 1419.
[71]D. F. Mullica *et al., Inorg. Chim. Acta* **1995,** *237,* 111.
[72]C. Floriani *et al., Organometallics* **1995,** *14,* 2265.
[73]G. B. Deacon *et al., Polyhedron* **1993,** *12,* 497.
[74]R. J. Wehmschulte and P. P. Power, *Organometallics* **1995,** *14,* 3264.
[75]R. J. Morris and G. S. Girolami, *Organometallics* **1991,** *10,* 792.
[76]P. P. Power *et al., Organometallics* **1988,** *7,* 180.
[77]P. P. Power *et al., Chem. Commun.* **1996,** 1573.
[78]R. J. Morris and G. S. Girolami, *Organometallics* **1991,** *10,* 792.

There are large numbers of cyclopentadienyl and substituted cyclopentadienyl and arene compounds. The first known, Cp_2Mn, is brown, polymeric, and ferromagnetic but undergoes a phase change on heating to a pink form with the same mp as Cp_2Fe that has 5 unpaired electrons as expected for Mn^{2+}. The reactivity of Cp_2Mn suggests a largely ionic structure. By contrast, Cp^*_2Mn is low spin (1e). Biscyclopentadienylmanganese reacts with phosphines to give bent species such as Cp_2MnPMe_3. The only commercial use of manganese organometallics involves $(\eta^5\text{-}C_5H_4Me)Mn(CO)_3$, long known as a combustion catalyst and used originally in Concorde engines. It was developed as an anti-knock compound for automobiles by the Ethyl Corp.[79]

Additional Reference

B. B. Snider, *Chem. Rev.* **1996,** *96,* 339. Manganese(III) based oxidative free-radical cyclizations of organic compounds.

17-E   IRON: GROUP 8

With iron the trends already noted in the relative stabilities of oxidation states continue, except that there is now no compound or chemically important circumstance in which the oxidation state is equal to the total number of valence shell electrons, which in this case is eight. The highest oxidation state known is VI, and it is rare. The only oxidation states of importance in the ordinary aqueous and related chemistry of iron are II and III. The oxidation states and stereochemistries are given in Table 17-E-1.

17-E-1   The Element

Iron is the second most abundant metal after Al and the fourth most abundant element in the earth's crust. The earth's core is believed to consist mainly of iron and nickel, and the occurrence of iron meteorites suggests that it is abundant throughout the solar system. The major iron ores are *hematite* (Fe_2O_3), *magnetite* (Fe_3O_4), *limonite* [FeO(OH)], and *siderite* ($FeCO_3$).

Because of its high abundance, iron is often found as an impurity in other materials. For example, *corundum* ($\gamma\text{-}Al_2O_3$) of gem quality is sapphire, and its colors are caused by small amounts of Fe^{IV}.

Chemically pure Fe can be prepared by reduction of pure iron oxide (which is obtained by thermal decomposition of iron(II) oxalate, carbonate, or nitrate) with H_2, by electrodeposition from aqueous solutions of Fe salts, or by thermal decomposition of $Fe(CO)_5$.

Iron is a white, lustrous metal (mp 1528°C). It is not particularly hard, and it is quite reactive. In moist air it is rather rapidly oxidized to give a hydrous Fe^{III}

[79]*C&EN* **1995,** *October 30,* p. 8.

Table 17-E-1 Oxidation States and Stereochemistry of Iron

Oxidation state	Coordination number	Geometry	Examples
Fe^{-II}	4	Tetrahedral	$Fe(CO)_4^{2-}$, $Fe(CO)_2(NO)_2$
Fe^0	5	*tbp*	$Fe(CO)_5$, $(Ph_3P)_2Fe(CO)_3$, $Fe(PF_3)_5$
	6	Octahedral(?)	$Fe(CO)_5H^+$, $Fe(CO)_4PPh_3H^+$
Fe^I, d^7	6	Octahedral	$[Fe(H_2O)_5NO]^{2+}$
Fe^{II}, d^6	2	Bent	$[Fe\{(2,4,6-Bu_3^tC_6H_2)\}_2]$
	3	Planar	FeO_3^{4-}, $[Fe(mes)(\mu\text{-mes})]_2$ (two 3-coordinated Fe atoms)
	4	Tetrahedral	$FeCl_4^{2-}$, $FeCl_2(PPh_3)_2$
	4	Square	$Fe(tpp)$
	5	*tbp*	$[FeBr(Me_6tren)]Br$
	5	*sp*	$[Fe(ClO_4)(OAsMe_3)_4]ClO_4$, $Fe(\eta^2\text{-OAc})(HB(3,5\text{-}Pr_2^ipz)_3)$
	6	Octahedral	$[Fe(H_2O)_6]^{2+}$, $[Fe(CN)_6]^{4-}$
	8	Dodecahedral (D_{2h})	$[Fe(1,8\text{-naphthyridine})_4](ClO_4)_2$
Fe^{III}, d^5	3	Trigonal	$Fe[N(SiMe_3)_2]_3$
	4	Tetrahedral	$FeCl_4^-$, Fe^{III} in Fe_3O_4
	5	*sp*	$FeCl(dtc)_2$, $Fe(acac)_2Cl$
	5	*tbp*	$Fe(N_3)_5^{2-}$, $FeCl_5^{2-}$
	6	Octahedral	Fe_2O_3, $[Fe(C_2O_4)_3]^{3-}$, $Fe(acac)_3$, $FeCl_3$
	7	Approx. pentagonal bipyramidal	$[FeEDTA(H_2O)]^-$
	8	Dodecahedral	$[Fe(NO_3)_4]^-$
Fe^{IV}, d^4	4	Tetrahedral	$Fe(1\text{-norbornyl})_4$
	6	Octahedral	$[Fe(diars)_2Cl_2]^{2+}$
	?	Organometallic	$(\eta^6\text{-arene})Fe(H)_2(SiX_3)_2$ [a]
Fe^{VI}, d^2	4	Tetrahedral	FeO_4^{2-}

[a] Z. Yao and K. J. Klabunde, *Inorg. Chem.* **1997,** *36,* 2119.

oxide, commonly known as rust; it affords no protection because it flakes off, exposing fresh metal surfaces. Finely divided Fe is pyrophoric. Iron combines vigorously with Cl_2 on mild heating and also with a variety of other nonmetals including the other halogens, S, P, B, C, and Si. The carbide and silicide phases play a major role in the technical metallurgy of iron.

The metal dissolves readily in dilute mineral acids, in the absence of air and with nonoxidizing acids, to give Fe^{II}. With air present or when warm dilute HNO_3 is used, some of the iron goes to Fe^{III}. Very strongly oxidizing media such as concentrated HNO_3 or acids containing $Cr_2O_7^{2-}$ passivate iron. Air-free water and dilute air-free hydroxides have little effect on the metal, but hot concentrated NaOH attacks it.

IRON COMPOUNDS

17-E-2 The Hydroxides and Oxides of Iron

Because of the interrelationships between them it is convenient to discuss the hydroxides and oxides together.

Iron(II) Oxides

Thermal decomposition of iron(II) oxalate in a vacuum gives a pyrophoric black powder that becomes less reactive if heated to higher temperatures. The crystalline substance can be obtained only by establishing equilibrium conditions at high temperature, then rapidly quenching the system, since at lower temperatures FeO is unstable with respect to Fe and Fe_3O_4; slow cooling allows disproportionation. Iron(II) oxide has the rock salt structure. The FeO referred to thus far is iron defective (see later), having a typical composition of $Fe_{0.95}O$. Essentially stoichiometric FeO has been prepared from $Fe_{0.95}O$ and Fe at 1050 K and 50 katm; it is ~0.4% less dense.

The white hydroxide $Fe(OH)_2$ is precipitated from Fe^{2+} solutions by base; it can be obtained crystalline with the $Mg(OH)_2$ structure. The hydroxide is rapidly oxidized in air turning red brown and eventually giving $Fe_2O_3 \cdot nH_2O$. It is soluble in acids and in strong base; on boiling Fe powder with 50% KOH blue-green crystals of $K_4[Fe(OH)_6]$ can be obtained. Interaction of FeO and Na_2O give Na_4FeO_3, which has a planar FeO_3^{4-} ion.

Iron(III) Oxides

Hydrous iron(III) oxides, FeO(OH), have different types (α, β, γ) of chain structures with FeO_6 octahedra sharing edges and they resemble those of AlO(OH). They are generally red-brown gels and are a major constituent of soils. Oxidation of Fe^{2+} in basic solution [or $Fe(OH)_2$] by air is complicated but can lead to γ-FeO(OH) and also Fe_3O_4. Cobalt catalyzed oxidation of $FeSO_4$ with H_2O_2 in acid solution gives crystals of what appears to be $Fe(OH)_3$.

The hydrous oxide is soluble in acids and to some extent in bases. On boiling $[Fe(H_2O)_6]^{3+}$ with $Ba(OH)_2$, crystalline $Ba_3[Fe(OH)_6]_2$ can be obtained and moderate concentrations of $[Fe(OH)_6]^{3-}$ can be maintained in strong base solution.

The oxide obtained on heating the hydrous oxide at 200°C is α-Fe_2O_3, which occurs as the mineral *hematite*. This has the corundum structure where the oxide ions form a *hexagonally* close-packed array with Fe^{III} ions occupying octahedral interstices. By careful oxidation of Fe_3O_4 or by heating one of the modifications of FeO(OH) (*lepidocrocite*) one obtains α-Fe_2O_3, which may be regarded as a *cubic* close-packed array of oxide ions with the Fe^{III} ions distributed randomly over both the octahedral and the tetrahedral interstices. There is also a rare form designated β-Fe_2O_3.

The oxide Fe_3O_4 occurs as the mineral *magnetite*. It is a Fe^{II}-Fe^{III} mixed oxide with the inverse spinel structure with Fe^{II} in octahedral interstices and Fe^{III} ions half in tetrahedral and half in octahedral interstices of a cubic close-packed array of oxide ions. The electrical conductivity ($\sim 10^6$ that of Fe_2O_3) is probably due to rapid valence oscillation between the Fe sites. It can be made by oxidation of Fe^{2+} with alkaline KNO_3 in the presence of phosphite or by reaction of acidified jarosites, $M^I Fe_3(SO_4)_2(OH)_6$, and $FeSO_4$ in ammonia:[1]

$$2(H_3O)Fe_3(SO_4)_2(OH)_6 + 3FeSO_4 + 14NH_4OH \longrightarrow 3Fe_3O_4 + 7(NH_4)_2SO_4 + 16H_2O$$

After controlled thermal treatment, this magnetite can be used as a black pigment.

[1] J. Boháček *et al., J. Mat. Sci.* **1993,** *28,* 2827.

There are also large oxo-hydroxo iron species which are produced by controlled hydrolysis of Fe^{II} or Fe^{III} precursors. They are discussed in more detail in Section 17-E-10.

Ferrites

These are important mixed oxide materials with octahedral Fe^{3+} ions used in a variety of electronic appliances ranging from magnetic recording devices to cores of power transformers. Examples are the spinels $M^{II}Fe_2O_4$, $BaFe_{12}O_{19}$, and $Ba_2Mn_2^{II}Fe_{12}O_{22}$. However, some discrete ions are known.

Ferrites can be made by fusing Fe_2O_3 with carbonates or by various hydrothermal methods from $Fe(OH)_2$ or $FeO(OH)$ suspensions in the presence of metal ions.

Perovskite-type oxides with Fe^{IV}, for example, $SrFeO_3$, can be made under high O_2 pressures and spinels such as Ba_2FeO_4 can be made by the reaction

$$M^{II}{}_3[Fe(OH)_6]_2 + M^{II}(OH)_2 + \tfrac{1}{2}O_2 \xrightarrow{\;800\text{-}900\,°C\;} 2M^{II}{}_2FeO_4 + 7H_2O$$

Mixed metal ferrites such as $Ba_3Fe_{24}Ti_7O_{53}$ can be made by heating stoichiometric mixtures of $BaCO_3$, Fe_3O_4, and TiO_2 at high temperatures (1375°C) for long periods of time (2 d).[2]

17-E-3 Halides and Sulfides

Iron(II) fluoride, chloride, and bromide are made by interaction of Fe with HX; FeI_2 can be made using the halogen, which does not oxidize Fe^{II} to Fe^{III}. Iron dichloride ($FeCl_2$) is best made by reduction of $FeCl_3$ with Fe in THF; FeF_2 is prepared quantitatively from $(NH_4)_3FeF_6$ at 850°C.[3]

Iron(III) halides are obtained by direct halogenation of Fe; but $FeBr_3$ and FeI_3 are best prepared by a photochemical reaction:[4]

$$(OC)_4FeX_2 + \tfrac{1}{2}X_2 \xrightarrow[\text{hexane}]{h\nu} FeX_3 + 4CO$$

The fluoride is white, having only spin-forbidden electronic transitions in the visible spectrum (*cf.* Mn^{II}); it has no low-energy charge-transfer band and has FeF_6 octahedra. Iron trichloride and $FeBr_3$ are red brown because of charge-transfer transitions and have nonmolecular crystal structures with Fe^{III} ions occupying two thirds of the octahedral holes in alternate layers. The molecules are dinuclear in the gas phase and Fe_2Cl_6 has a puckered four-membered Fe_2Cl_2 ring. Iron triiodide is black and looses I_2 easily. The electronic spectra of the series of trihaloiron(III) compounds, $FeCl_3$, $FeBr_3$, and FeI_3, show a progressive red shift together with an increasing number of absorption bands.

Iron forms many binary compounds with the Group 15 and Group 16 elements. Many are nonstoichiometric and/or interstitial. The sulfides are the most common.

[2]R. D. Adams *et al.*, *Chem. Ber.* **1996,** *129,* 1441.
[3]N. I. Kuznetsova *et al.*, *Zh. Neorg. Khim.* **1991,** *36,* 603.
[4]K. B. Yoon and J. K. Kochi, *Inorg. Chem.* **1990,** *29,* 869.

Iron(II) sulfide (FeS) and FeS_2 have been discussed (Chapter 12). Iron(III) sulfide is unstable and can be prepared and stored only with difficulty. By quantitative treatment of aqueous Fe^{III} with Na_2S at 0°C or below it is obtained as a black air-sensitive solid.

17-E-4 Aqueous and Coordination Chemistry of Iron(II), d^6

Aqueous solutions of Fe^{II} in the absence of complexing anions contain the pale blue-green ion $[Fe(H_2O)_6]^{2+}$, which is oxidized in acid solution:

$$2Fe^{2+} + \tfrac{1}{2}O_2 + 2H^+ = 2Fe^{3+} + H_2O \qquad E^0 = 0.46 \text{ V}$$

As noted earlier, oxidation is easier in basic solution, but neutral and acid solutions of Fe^{2+} oxidize *less* rapidly with increasing acidity (even though the potential of the oxidation reaction becomes more positive). This is because Fe^{III} is actually present in the form of hydroxo complexes, except in extremely acid solutions, and there may also be kinetic reasons.

The Fe^{2+} ion is also oxidized by other common oxidants. The action of NO_3^- or NO_2^- involves the transient formation of a brown nitrosyl $[FeNO(H_2O)_5]^{2+}$ in the overall reaction:

$$3Fe^{2+} + 4H^+ + NO_3^- = 3Fe^{3+} + 2H_2O + NO$$

This nitrosyl chemistry may have relevance for the reduction of NO_2^- and NO_3^- by microorganisms in soils.

The mixture of Fe^{2+} and H_2O_2 or $S_2O_8^{2-}$ is called Fenton's reagent; it oxidizes due to the presence of $OH\cdot$ and SO_4^- radicals although Fe^{IV} oxo species may be present in the complicated solutions. Iron(II) forms salts with virtually every stable anion, generally as green, hydrated, crystalline substances isolated by evaporation of aqueous solutions. The sulfate and perchlorate contain octahedral $[Fe(H_2O)_6]^{2+}$ ions. Mohr's salt, $(NH_4)_2[Fe(H_2O)_6](SO_4)_2$, is fairly stable toward both air oxidation and loss of water because of the extensive hydrogen bonding network which protects the metal center.[5] It is commonly used in volumetric analysis to prepare standard solutions of Fe^{II} and as a calibration substance in magnetic measurements. The hydrated halides (F, Cl, and Br) can be made by dissolving Fe in the aqueous acids and crystallizing. The compound $FeCl_2\cdot6H_2O$ contains *trans*-$[FeCl_2(H_2O)_4]$; all the species from $[Fe(H_2O)_6]^{2+}$ to $[FeCl_4]^{2-}$ are known in solution.

Hydrides which contain the $[FeH_6]^{4-}$ anion[6] can be made by interaction of Grignard reagents with the iron halides or by reaction of metallic iron with metal hydrides at high temperature. In $FeH_6(MgX)_4(THF)_8$ there is an octahedral $[FeH_6]^{4-}$ ion, which is isoelectronic with ReH_6^{3-}, and is surrounded by a tetrahedral array of solvated $MgX(THF)_2^+$ units lying over four of the octahedral faces as shown in Fig. 17-E-1. The latter, a green unsolvated Mg_2FeH_6 compound, has octahedral low-spin $[FeH_6]^{4-}$ ions. Other hydrides such as FeH, FeH_2, and FeH_3 have been observed in argon matrices after condensation of iron atoms with hydrogen molecules.[7]

[5]F. A. Cotton *et al., Inorg. Chem.* **1993**, *32*, 4861.
[6]D. E. Linn, Jr. and S. G. Gibbins, *Inorg. Chem.* **1997**, *36*, 3461.
[7]G. V. Chertihin and L. Andrews, *J. Phys. Chem.* **1995**, *99*, 12131.

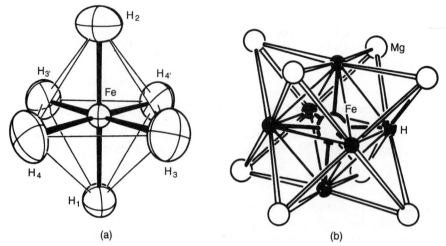

Figure 17-E-1   (a) The structure of the $[FeH_6]^{4-}$ ion in $FeH_6(MgX)_4(THF)_8$. (b) The structural unit of Mg_2FeH_6 which has the K_2PtCl_6 type structure.

Complexes

Most Fe^{II} complexes are octahedral. There are a few *tetrahedral* complexes which include the $[FeCl_4]^{2-}$ ion in salts with large cations, $[Fe(OPPh_3)_4]^{2+}$, $FeBr_2(PEt_3)_2$, and a few others. *Pentacoordinate* species, usually distorted *tbp*, include those of tripod ligands such as $[Fe(np^3)X]^+$. *Dodecahedral* complexes include $[Fe(naphthyridine)_4]^{2+}$ and the crown ether complex noted below. There are also 3-coordinate organometallic complexes such as $[\{Fe(mes)(\mu\text{-mes})\}_2]$ and some unusual 2-coordinate complexes (17-E-I) prepared with bulkier[8] groups such as $2,4,6\text{-}Bu_3^tC_6H_2$, mes*, according to

$$2FeBr_2 \cdot 2THF + 4(mes^*)MgBr \xrightarrow{\text{THF}} Fe_2Mg_3Br_6(mes^*)_4(THF)_2 + MgBr_2$$

$$Fe_2Mg_3Br_6(mes^*)_4(THF)_2 \xrightarrow{\text{THF/dioxane}} \xrightarrow{\text{n-pentane}} 2[Fe(mes^*)_2] + 3MgBr_2(dioxane)_2$$

(17-E-I)

[8]W. Seidel *et al., Angew. Chem. Int. Ed. Engl.* **1995,** *34,* 325 and references therein.

The aqueous system provides a good example of the effect of complexing ligands on the redox potentials

$$[Fe(CN)_6]^{3-} + e = [Fe(CN)_6]^{4-} \qquad E^0 = 0.36 \text{ V}$$

$$[Fe(H_2O)_6]^{3+} + e = [Fe(H_2O)_6]^{2+} \qquad E^0 = 0.77 \text{ V}$$

$$[Fe(phen)_3]^{3+} + e = [Fe(phen)_3]^{2+} \qquad E^0 = 1.12 \text{ V}$$

Halides. The best known species are the tetrahedral $[FeX_4]^{2-}$ but discrete homoleptic dinuclear iron(II) halide anions, $[Fe_2X_6]^{2-}$, are also known.[9] More common are complexes with Fe_2X_2 cores.[10]

Nitrogen Ligands. Iron(II) halides and other salts absorb NH_3 in excess giving the $[Fe(NH_3)_6]^{2+}$ anion but ammonia complexes are stable only in saturated excess ammonia. Stable complexes with chelating amines are formed, for example, for ethylenediamine:

$$[Fe(H_2O)_6]^{2+} + en = [Fe(en)(H_2O)_4]^{2+} + 2H_2O \qquad K = 10^{4.3}$$

$$[Fe(en)(H_2O)_4]^{2+} + en = [Fe(en)_2(H_2O)_2]^{2+} + 2H_2O \qquad K = 10^{3.3}$$

$$[Fe(en)_2(H_2O)_2]^{2+} + en = [Fe(en)_3]^{2+} + 2H_2O \qquad K = 10^2$$

Ligands like phen, bipy, and others supplying imine nitrogen donor atoms give stable low-spin (i.e., diamagnetic), octahedral, or distorted octahedral complexes. In $[Fe(2,6-diacetylpyridine)bis(2-thenoylhydrazone)]Cl_2$ the iron atom exhibits a pentagonal bipyramidal coordination geometry with the equatorial plane defined by the N_3O_2 donor set of the hydrazone ligand; the two chlorine atoms occupy the axial positions.[11] The formamidine HDTolF (and others) reacts with $FeCl_2$ to give tetrahedral $FeCl_2(HDTolF)_2$[12] which is soluble in non-polar solvents such as benzene (Section 17-E-9).

The most important ligands, however, are porphyrins as many enzyme systems are iron porphyrins (Section 17-E-10).

Oxygen Ligands. With simple oxygen donor ligands, iron(II) complexes are normally high spin. Reaction of $FeCl_3$ and Fe in THF gives a nearly colorless crystalline solid, $Fe_4Cl_8(THF)_6$, which consists of centrosymmetric tetranuclear molecules with two different types of Fe^{II} atoms, two of them are in trigonal bipyramidal environments and the others in octahedral environments.[13] In the presence of stoi-

[9]See, for example, K. R. Dunbar and A. Quillevéré, *Angew. Chem. Int. Ed. Engl.* **1993,** *32,* 293 and references therein.

[10]L. Que *et al., Inorg. Chim. Acta* **1993,** *213,* 41 and references therein; G. A. Olah *et al., J. Am. Chem. Soc.* **1995,** *117,* 8962.

[11]C. Pelizzi *et al., J. Chem. Soc., Dalton Trans.* **1990,** 2771.

[12]F. A. Cotton *et al., Polyhedron* **1994,** *13,* 815.

[13]F. A. Cotton *et al., Inorg. Chim. Acta* **1991,** *179,* 11; B. M. Bulychev *et al., Inorg. Chim. Acta,* **1985,** *96,* 123.

chiometric amounts of $OPMe_3$, $Fe_4Cl_8(THF)_6$ reacts to give polymeric [$FeCl_2$(OP-Me$_3$)]$_\infty$, which consists of linear chains of $FeCl_2(OPMe)$ units where each Fe atom is linked by four bridging chlorine atoms and one $OPMe_3$ molecule; use of an excess $OPMe_3$ in the reaction gives the mononuclear tetrahedral $FeCl_2(OPMe_3)_2$.[14] The latter is a member of the family of FeX_2L_2 compounds where L = R_3AsO, R_3PO, or quinoline and X = Cl or Br. The β-diketonates, like those of other M^{II} ions, are polymeric and [$Fe(acac)_2]_4$ is tetranuclear in the solid with 6-coordinate Fe^{II} as a result of both oxygen bridges and weak Fe—C bonds.

A crown (12-C-4) ether complex (17-E-II), which has a "sandwich" structure, can be made by irradiation of a [(n^5-C_5H_5)Fe(n^6-C_6H_5Me)]$^+$ sandwich cation in CH_2Cl_2 in the presence of ethylene oxide. The Fe^{2+} is 8-coordinate in a slightly twisted cube.

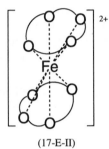

(17-E-II)

There are also carboxylates[15] (Section 17-E-10), phosphates, catecholates, and oxalates. Most of the divalent iron oxalato compounds have a chain structure but the product of photoreduction of mononuclear [bipyH][$Fe^{III}(ox)_2(H_2O)_2$] gives a three-dimensional anionic polymeric network of [$Fe_2^{II}(ox)_3]_n^{2n-}$ and [$Fe^{II}(bipy)_3]^{2+}$ ions.[16] Magnetic exchange typically occurs through the oxalate bridges.[17] Six-coordinate iron(II) α-amino acid complexes are readily obtained from the reaction of the 4-coordinate $Fe(mes)_2(phen)$ and the protic amino acid, AH, in THF:[18]

$$[Fe(mes)_2phen)] + 2AH \longrightarrow [Fe(A)_2(phen)] + 2mesH$$

A simple oxo-anion is [$O_2FeOFeO_2]^{6-}$;[19] $Fe_4O(DPhF)_6$ is a tetranuclear basic oxo-compound in which there is a "tetrahedron" of iron atoms with an oxygen atom at its center; two opposite Fe⋯Fe edges are doubly bridged by formamidinate ligands, another two opposite edges are singly bridged and the remaining two edges are unbridged.[20] This type of structure (Fig. 17-E-2) is somewhat related to that of the basic beryllium acetate.

Sulfur and Selenium Ligands. Many studies have been made on thiolates because of their relevance to natural Fe—S proteins (Section 17-E-10). Some are

[14]F. A. Cotton *et al., Inorg. Chim. Acta* **1991,** *184,* 177.
[15]See for example: L. Que, Jr. *et al., Angew. Chem. Int. Ed. Engl.* **1994,** *33,* 1660 and references therein.
[16]S. Decurtins *et al., Inorg. Chem.* **1993,** 32, 1888.
[17]J. Glerup *et al., Inorg. Chem.* **1995,** *34,* 6255.
[18]C. Floriani *et al., Inorg. Chem.* **1994,** *33,* 1928.
[19]H.-P. Müller and R. Hoppe, *Z. anorg. allg. Chem.* **1993,** *619,* 193.
[20]F. A. Cotton *et al., Inorg. Chim. Acta* **1997,** *266,* 91.

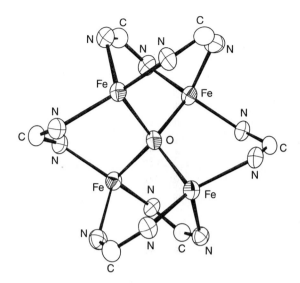

Figure 17-E-2 Core of the tetra-nuclear $Fe_4O(DPhF)_6$ complex. The phenyl groups bonded to the N atoms are not shown.

mononuclear and tetrahedral like $[Fe(SPh)_4]^{2-}$, but others can form adamantanelike cages, $[Fe_4(SPh)_4]^{2-}$, or six-membered rings, for example, $[FeCl_2SR]_3^{3-}$. Tetrahedral compounds such as (17-E-III) provide models for Fe in rubredoxin. There are very few reports of structurally characterized neutral 4-coordinate thiolato iron(II) complexes, such as $Fe[SC(NMe_2)_2]_2(S\text{-}2,4,6\text{-}Pr_3^iC_6H_2)_2$.[21] In the 2-coordinate $[Fe\{S(2,6\text{-}mes_2C_6H_3)\}_2]$ the sulfur atoms bind strongly to the iron(II) center; there are however weak interactions with the π-electron density of the *ortho*-mesityl group of the thiolato ligand.[22] Low-coordinate Fe^{II} thiolate complexes are rare but important[23] because several iron atoms in the cofactor of nitrogenase have been described as 3-coordinate with a distorted trigonal geometry. Other $Fe{-}S$ species are noted later (Section 17-E-10).

A few neutral selenolato compounds like $Fe(Se\text{-}2,6\text{-}Pr_2C_6H_3)_2(PR_3)_2$[24] are also known. In $Fe_{12}(SePh)_{24}$[25] each iron atom is surrounded by four Se atoms of the μ_2-SePh ligands in a distorted tetrahedral fashion. As a result, edge-sharing Se_4 tetrahedra connect to form a ring.

(17-E-III)

[21]S. Pohl *et al.*, *Z. Naturforsch. B* **1991**, *46*, 1629.
[22]P. P. Powers *et al.*, *Angew. Chem. Int. Ed. Engl.* **1994**, *33*, 1178.
[23]D. J. Evans *et al.*, *Inorg. Chem.* **1997**, *36*, 747.
[24]R. H. Morris *et al.*, *Inorg. Chem.* **1994**, *33*, 5647.
[25]D. Fenske and A. Fischer, *Angew. Chem. Int. Ed. Engl.* **1995**, *34*, 307.

Phosphorus Ligands. These ligands give a variety of complexes such as *cis*- and *trans*-FeX$_2$(diphos)$_2$ where X can be halogen, H, alkyl group, and so on.[26] There are also tetrahedral FeX$_2$(PR$_3$)$_2$ compounds that react with (Me$_3$Si)$_2$S or LiS in THF to give complexes of the type Fe$_6$S$_6$(PR$_3$)$_4$.[27]

Carbon Ligands. Probably the most important compounds are *ferrocene*, (n^5-C$_5$H$_5$)$_2$Fe, its derivatives and the carbonyl compounds (see later).

The hexacyanoferrate(II) ion, commonly called ferrocyanide, is very stable and non-labile because of its low-spin d^6 configuration (*cf.* Co^{3+}). The free acid H$_4$[Fe(CN)$_6$] can be precipitated as an ether addition compound (probably containing oxonium ions R$_2$OH$^+$) by adding ether to a solution of the ion in strongly acidic solution; the ether can then be removed to leave the acid as a white powder. It is a strong tetrabasic acid when dissolved in water; in the solid the protons are bound to the nitrogen atoms of the CN groups and there is intermolecular hydrogen bonding. Hexacyanoferrate dissolves without decomposition in liquid HF and is protonated to give [Fe(CNH)$_6$]$^{2+}$.

Electronic Structures of Iron(II) Complexes

The ground state ^5D of a d^6 configuration is split by octahedral and tetrahedral ligand fields into $^5T_2$ and ^5E states; there are no other quintet states; hence only one spin-allowed $d-d$ transition occurs if one of these is the ground state. All tetrahedral complexes are high spin, and the $^5E \to {}^5T_2$ band typically occurs at ~4000 cm^{-1}. The magnetic moments are normally 5.0 to 5.5 BM, owing to the spins of the four unpaired electrons and a small, second-order orbital contribution. For high-spin octahedral complexes, for example, Fe(H$_2$O)$_6^{2+}$, the $^5T_{2g} \to {}^5E_g$ transition occurs in the visible or near-ir region (~10,000 cm^{-1} for the aqua ion) and is broad or even resolvably split owing to a Jahn-Teller effect in the excited state, which derives from a $t_{2g}^3 e_g^3$ configuration.

For Fe^{II} quite strong ligand fields are required to cause spin pairing, but a number of low-spin complexes such as [Fe(CN)$_6$]$^{4-}$, [Fe(CNR)$_6$]$^{2+}$, and [Fe(phen)$_3$]$^{2+}$ are known. These essentially diamagnetic complexes are often intensely colored due to charge transfer transitions which frequently obscure the two predicted spin allowed transitions $^1A_{1g} \to {}^1T_{1g}$ and $^1A_{1g} \to {}^1T_{2g}$.

Although for strict octahedral symmetry no d^6 ion can have a ground state with two unpaired electrons (only 4 or 0), this might be possible in 6-coordinate complexes in which there are significant departures from O_h symmetry in the ligand field. Perhaps the best documented examples are complexes of the type [Fe(LL)$_2$ox] and [Fe(LL)$_2$mal], where LL represents a bidentate diamine ligand such as o-phen or bipy, and ox and mal represent oxalato and maleato ions. These complexes have magnetic susceptibilities that follow the Curie-Weiss law over a broad temperature range, with $\mu \approx 3.90$ BM (part of which is due to a temperature-independent paramagnetism).

[26]A. R. Hermes and G. S. Girolami, *Inorg. Chem.* **1988,** 27, 1775; R. A. Henderson, *J. Chem. Soc., Dalton Trans.* **1988,** 509.
[27]B. S. Snyder and R. H. Holm, *Inorg. Chem.* **1988,** 27, 2339.

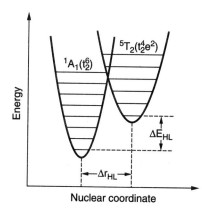

Figure 17-E-3 Schematic representation of the potential wells for the LS ($^1A_1$) and the HS ($^5T_2$) states of an iron(II) spin crossover complex. The nuclear coordinate is the bond length. Thermal spin crossover is observed when $\Delta E_{HL} \simeq kT$.

For a strictly square, 4-coordinate complex such as phthalocyanine iron(II), the extreme tetragonality of the ligand field apparently places one d orbital ($d_{x^2-y^2}$) at high energy, and the six electrons adopt a high-spin distribution among the remaining four, thus giving a triplet ground state, independent of temperature.

Spin-State Crossovers in Iron(II) Complexes[28]

There are many examples of complexes where a high-spin (HS) and a low-spin (LS) are separated by only about the thermal energy prevailing at or below room temperature. The magnetic properties change anomously (that is, differently from ordinary Curie, or Curie-Weiss behavior) as a function of temperature; they can change from LS to HS under special conditions (see below) giving rise to so-called "spin-crossover behavior."

By far the majority of spin-crossover complexes contain nitrogen atoms bound to iron in 6-coordinate $Fe^{II}-N_6$ species. Although the phenomenon is also observable for other coordination geometries and other metal ions with d^4, d^5, d^6 and d^7 electronic configurations; we shall discuss only the iron(II) case. Due to the absence of electrons in the e_g antibonding orbitals of the LS octahedral complexes, the Fe—N distances are typically 0.2 Å shorter than those found in the HS complexes. This difference in bond lengths and bond strengths leads, in part, to significant differences in the frequency for the Fe—N stretching vibration for the two spin states.

Fig. 17-E-3 has a representation of the potential wells for the HS and LS configuration relative to the Fe—donor atom distance. Crossover occurs when the vertical separation of the zero-point energies of the wells (ΔE_{HL}) is small (of the order of kT); that is, when the spin-pairing energy and the magnitude of the orbital splitting (Δ_0) are about equal.

When a thermally induced spin transition occurs, a strong temperature dependence of the properties of the system is observed, ranging from those associated with the purely singlet state to those associated with the purely quintet state. From

[28]P. Gütlich *et al., NATO-ASI-Series E: Applied Sciences,* E. Coronado, ed., **1996,** *321,* 327 (169 references).

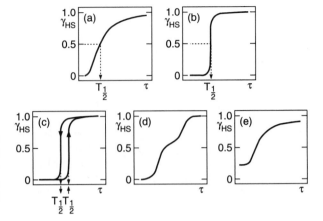

Figure 17-E-4 Classification of spin transition curves: (a) gradual, (b) abrupt, (c) discontinuous with hysteresis, (d) two-step, (e) incomplete spin transition.

the changes observed it is possible to evaluate the fraction of molecules in each spin state, γ_{HS} or γ_{LS} ($\gamma_{HS} + \gamma_{LS} = 1$), at a particular temperature. The course of the spin transition is then generally indicated by a plot of γ_{HS} *vs.* temperature, generally known as a *spin transition curve.* Such curves can be very diagnostic of the nature of the spin crossover. For solid systems they reveal a range of different behaviors as illustrated in Fig 17-E-4. In (a) a gradual, continuous change in γ_{HS} is indicated over an extended temperature range. The transition may be more abrupt and occur with in a very narrow temperature range as in (b). The transition may be associated with a thermal hysteresis loop (c) or it may be a two-step process (d). Finally, transitions may be incomplete at either or both extremes of the spin transition curve $\gamma_{HS}(T)$, as indicated in (e). The occurrence of hysteresis is often indicative of a structural phase change accompanying the spin change. In this instance the transition is described as discontinuous, in contrast to those which do not show hysteresis and are classified as continuous.

The spin crossover is normally induced by a change in temperature, but it can be brought about by a change in pressure (because of the difference in the Fe—L bond lengths). Switching between the HS and LS states can also be achieved by irradiation with light of different wavelengths; this phenomenon has been termed "Light-Induced Excited Spin State Trapping (LIESST)."

The experimental methods used to study Fe^{II} spin crossovers include measurement of bulk magnetic susceptibility, vibrational spectroscopy (because M—L bond strengths differ appreciably between the HS and LS states), crystallography, and Mössbauer spectroscopy.

Some complexes show a strong interdependence between crystal structure and spin-transition features. In the series of compounds $[Fe(Rtz)_6](BF_4)_2$ (Rtz = 1-alkyltetrazole) the spin crossover behavior varies with the substituent R and is strongly influenced by cooperative effects. For example, the propyl derivative shows a quantitative spin transition, which is accompanied by a first-order crystallographic phase transition; in the methyl and ethyl derivatives the Fe^{II} complexes occupy two nonequivalent lattice sites, only one of which shows a thermal spin transition.[29]

---

[29]See for example, A. Hausser *et al., Inorg. Chem.* **1994,** *33,* 567 and references therein; P. Gütlich *et al., Chem. Eur. J.* **1996,** *2,* 1427.

17-E-5   Aqueous and Coordination Chemistry of Iron(III), d^5

Iron(III) occurs in salts with most anions, except those with reducing capacity. Examples obtained as pale pink to nearly white hydrates from aqueous solutions are $Fe(ClO_4)_3 \cdot 10H_2O$, $Fe(NO_3)_3 \cdot 9(\text{or } 6)H_2O$, and $Fe_2(SO_4)_3 \cdot 10H_2O$.

Aqueous Chemistry

One of the most conspicuous features of iron(III) in aqueous solution is its tendency to hydrolyze and/or to form complexes.

The hydrolysis of the pale purple ion $[Fe(H_2O)_6]^{3+}$ in noncomplexing media is complicated and condition dependent.

At pH < 1 the sole species is the aqua ion but above pH 1 stepwise hydrolysis occurs. At the lower concentrations the main equilibrium is

$$[Fe(H_2O)_6]^{3+} = [Fe(OH)(H_2O)_5]^{2+} + H^+ \qquad K = 1.84 \times 10^{-3}$$

A small amount of $Fe(OH)_2^+$ may be formed but the second main species is believed to be the diamagnetic μ-oxo dimer:

$$[Fe(OH)(H_2O)_5]^{2+} \rightleftharpoons [(H_2O)_5FeOFe(H_2O)_5]^{4+} + H_2O$$

In the 1 to 2 pH range still other types of oxo species may be formed. At pH > 2 more condensed species and colloidal gels are formed leading to precipitation of the red brown gelatinous hydrous oxide.

The hydroxo species are yellow because of charge-transfer bands in the ultraviolet region that have tails in the visible.

In the presence of complexing anions such as Cl^-, the hydrolysis of Fe^{3+} or of $FeCl_3$ is more complicated giving chloro, aqua, and hydroxy species as well as $[FeCl_4]^-$ at high Cl^- concentration.

Iron(III) has a high affinity for F^- as shown by the constants

$$Fe^{3+} + F^- = FeF^{2+} \qquad K_1 \approx 10^5$$

$$FeF^{2+} + F^- = FeF_2^+ \qquad K_2 \approx 10^5$$

$$FeF_2^+ + F^- = FeF_3 \qquad K_3 \approx 10^3$$

which are $\sim 10^4$ greater than those for Cl. Thiocyanate forms intense red complexes and $[Fe(NCS)_4]^-$ can be extracted into ether combined with $(H_3O)^+$.

Complex Compounds

Halides. The halogen complex anions have just been noted and there are crystalline salts such as K_3FeF_6 and $CsFeF_4$, which has octahedral FeF_6 sharing F

---

[30]U. Bentrup and W. Massa, *Z. Naturforsch.* **1991**, *B46*, 395.

ions; $[Fe_2F_8(H_2O)_2]^{2-}$ [30] has two bridging F ions. There are also tetrahedral $FeCl_4^-$, trigonal bipyramidal $FeCl_5^{2-}$,[31] octahedral $FeCl_6^{3-}$, and face-sharing bioctahedral $Fe_2Cl_9^{3-}$ which can be obtained with large cations. However, a compound formulated as $(pyH)_3Fe_2Br_9$ is better described by the formula $2\{(pyH)[FeBr_4]\}\cdot(pyH)Br$.[32]

Iron(III) chloride forms adducts with donors, giving compounds with tetrahedral geometry such as $FeCl_3L$, L = ether,[33] large phosphine;[34] if the phosphine is smaller, mononuclear trigonal-bipyramidal structures of formula $FeCl_3L_2$ are found.[35] Very few FeI_3 complexes have been fully characterized. The best known are complexes of the type FeI_4^- and FeI_3(tetramethylthiourea).[36] The former can be made by using solid state reactions,[37] solution chemistry from carbonyl complexes,[38] or FeI_3 according to

$$Fe + MI + {}^3/_2 I_2 \xrightarrow[\text{sealed quartz ampule}]{300\,°C} MFeI_4, \ M = K, Rb, Cs$$

$$Fe + NR_4I + {}^3/_2 I_2 \xrightarrow[\text{nitromethane}]{} NR_4FeI_4$$

$$4\,Ph_3SbI_2 + Fe_2(CO)_9 \xrightarrow[\text{Et}_2\text{O}]{} 2\{[Ph_4Sb][FeI_4]\cdot Ph_3SbI_2\} + 9\,CO$$

$$FeI_3 + I^- \longrightarrow FeI_4^-$$

Nitrogen Ligands. The affinity of Fe^{III} for amines is low. No simple ammine complex exists in aqueous solution; addition of aqueous ammonia only precipitates the hydrous oxide. For the reaction

$$M-NH_3 + H_2O \longrightarrow M-OH^- + NH_4^+$$

$\log K = 3.2$.[39] Chelating amines, for example, EDTA, form complexes, among which is the 7-coordinate $[Fe(EDTA)H_2O]^-$ ion. Also, ligands such as bipy and phen that produce ligand fields strong enough to cause spin pairing form fairly stable complexes, isolable in crystalline form with large anions.

Oxygen Ligands. These have high affinity for Fe^{III} and complexes are formed by phosphate,[40] organic phosphates[41] and oxalate ions, glycerol, sugars, and so on, while β-diketones give octahedral neutral $Fe(dike)_3$.

[31]B. D. James *et al., Inorg. Chem.* **1995,** *34,* 2054.
[32]R. L. Carlin *et al., Inorg. Chem.* **1994,** *33,* 3051.
[33]F. A. Cotton *et al., Acta Cryst.* **1990,** *C46,* 1424.
[34]J. D. Walker and R. Poli, *Inorg. Chem.* **1990,** *29,* 756.
[35]J. D. Walker amd R. Poli, *Inorg. Chem.* **1989,** *28,* 1793.
[36]S. Pohl *et al., Angew. Chem. Int. Ed. Engl.* **1989,** *28,* 776.
[37]G. Thiele *et al., Z. anorg. allg. Chem.* **1996,** *622,* 795.
[38]S. M. Godfrey *et al., J. Chem. Soc., Dalton Trans.* **1994,** 3249.
[39]K. Hegetschweiler *et al., Inorg. Chem.* **1995,** *34,* 1950.
[40]N. Anisimova *et al., Z. anorg. allg. Chem.* **1996,** *622,* 1920; K.-H. Lii and Y.-F. Huang, *J. Chem. Soc., Dalton Trans.* **1997,** 2221.
[41]A. Warshawsky *et al., J. Chem. Soc., Dalton Trans.* **1990,** 2081.

Largely because of interest in iron transport and storage molecules (sidero-phores and ferritin) in living systems, there has been much study of model com-pounds especially with hydroxamate, hydroxypyridinone, catecholate, catechol-amide, and related ligands. There are relatively simple compounds like the alkoxides, $[Fe(OR)_4]^-$ and $Fe_2(\mu\text{-}OSiMe_3)_2(OSiMe_3)_4$.

A characteristic feature is the formation of *oxo* and/or *hydroxo bridges*.[42] Binu-clear systems may be (a) doubly bridged and (b) singly bridged with a linear or bent FeOFe. The bent ones are known with 4-, 5-, and 6-coordinate iron(III); an example is $[(salen)Fe]_2(\mu\text{-}O)$. Porphyrin and phthalocyanine complexes are linear bridged species of the type $(porph\ Fe)_2\mu\text{-}X$ where X = O, N, and C.

In the simple ion $[Fe_2(\mu\text{-}O)Cl_6]^{2-}$ the Fe—O—Fe angle can vary from 147 to almost linear. The geometry of the Cl_3FeO unit is approximately tetrahedral.[43] The Fe—O—Fe angles are smaller (120–135°) in many oxo-bridged diiron complexes used as models for *met*hemerythin which have two iron atoms bridged simulta-neously by an oxo and one or two carboxylato groups.[44]

In complexes with linear FeOFe links there is π bonding across the bridge and even bent species have large bond angles; this can lead to antiferromagnetic coupling of the electron spins on each Fe^{III}.

It may be noted that there are fewer examples of a sulfur bridge. An example is $[Fe(salen)]_2S$; it has an angle much smaller (121.8°) than that of $[Fe(salen)]_2O$ (145°).

Like OH, OR can also bridge and the alkoxide $[(acac)_2Fe(\mu\text{-}OEt)]_2$ has a planar Fe_2O_4 ring.

(17-E-IV)

Oxo bridges can also be formed from OH groups on multidentate ligands as in (17-E-IV).

Iron(III) forms extensive series of *basic carboxylates* of the type $[Fe_3O\text{-}(O_2CR)_6L_3]^+$, discussed in Section 11-9. For some, reduction gives mixed valence species; thus $Fe_3O(O_2CMe)_6(py)_3$ has a localized Fe_2^{III}, Fe^{II} electronic structure in the solid at 100 K according to Mössbauer spectra. Many more complicated poly-meric oxo species are also known, such as tetranuclear $Li_2(HDPhF)_2Fe_4O_4(DPhF)_6$ which has a centrosymmetric eight-membered ring of alternating iron and oxygen atoms.[45] Controlled polymerization of iron complexes in nonaqueous media has yielded structurally characterized oxo-compounds with nuclearities up to 19.[46] A

[42]D. M. Kurtz, Jr., *Chem. Rev.* **1990**, *90*, 585.
[43]K. Wieghardt *et al., Inorg. Chem.* **1993**, *32*, 520.
[44]See for example Table III in: L Que, Jr. and A. E. True, *Prog. Inorg. Chem.* **1990**, *38*, 97.
[45]F. A. Cotton *et al., Inorg. Chim. Acta* **1996**, *252*, 293.
[46]See for example: S. J. Lippard *et al., J. Am. Chem. Soc.* **1994**, *116*, 8061.

well-known example is that of $Fe_{11}O_6(PhCO_2)_{15}(OH)_6$; such species may provide models for the core of ferritin (see later).

Cyanide. In contrast to $[Fe(CN)_6]^{4-}$, the $[Fe(CN)_6]^{3-}$ ion is quite poisonous; for kinetic reasons the latter dissociates and reacts rapidly, whereas the former is not labile. There is a variety of substituted ions $[Fe(CN)_5X]^{n-}$ (X= H_2O, NO_2, etc.), of which the best known is the nitroprusside ion $[Fe(CN)_5NO]^{2-}$. Crystalline sodium nitroprusside dihydrate exhibits extremely long-lived electronic excited states.[47] A metastable state I and a metastable state II have lifetimes greater than 10^4 s at temperatures below 185 and 140 K, respectively. The nitroprusside anion is attacked by OH^- to give $[Fe(CN)_5NO_2]^{2-}$.

The reaction of $[Fe(CN)_6]^{3-}$ and Schiff base manganese complexes give a polynuclear compound with [Fe—C≡N—Mn] units.[48] A dinuclear $[Fe_2(CN)_{10}]^{4-}$ species is also known and has been studied extensively.[49]

Electronic Structures of Iron(III) Compounds

Iron(III) is isoelectronic with Mn^{II}, but less is known of the details of Fe^{III} spectra because of the much greater tendency of the trivalent ion to have charge-transfer bands in the near-uv region with strong low-energy shoulders in the visible that obscure the very weak, spin-forbidden $d-d$ bands. Insofar as they are known, however, the spectral features of Fe^{III} ions in octahedral surroundings are in accord with theoretical expectations.

Iron(III), like manganese(II), is high spin in nearly all its complexes, except those with the strongest ligands, exemplified by $[Fe(CN)_6]^{3-}$, $[Fe(bipy)_3]^{3+}$, $[Fe(phen)_3]^{3+}$, and other tris complexes with imine nitrogen atoms as donors. In the high-spin complexes the magnetic moments are always very close to the spin-only value of 5.9 BM because the ground state (derived from the ^6S state of the free ion) has no orbital angular momentum and there is no effective mechanism for introducing any coupling with excited states. The low-spin complexes, with t_{2g}^5 configurations, usually have considerable orbital contributions to their moments at about room temperature, values of ~2.3 BM being obtained. The moments are, however, intrinsically temperature dependent, and at liquid nitrogen temperature (77 K) they decrease to ~1.9 BM. There is evidence of very high covalence and electron delocalization in low-spin complexes such as $[Fe(phen)_3]^{3+}$ and $[Fe(bipy)_3]^{3+}$.

Five-coordinate complexes may be high- or low-spin depending on the ligands; some of the important high-spin complexes have already been encountered in oxo-bridged dinuclear complexes. The $Fe(S_2CNR_2)_2X$ (X= Cl, Br, and I) complexes that form readily on treating $Fe(S_2CNR_2)_3$ with halogens have three unpaired electrons. These molecules have a very distorted *sp* configuration with X axial (actual symmetry C_{2v}), and with coordinate axes as defined in (17-E-V) the electron configuration is, $d_{x^2-y^2}^2$, d_{xz}^1, d_{yz}^1, $d_{z^2}^1$, d_{xy}^0, according to magnetic anisotropy measurements.

[47]See for example, M. R. Pressprich *et al.*, *J. Am. Chem. Soc.* **1994**, *116*, 5233; J. Schefer *et al.*, *Z. Kristallog.* **1997**, *212*, 29.
[48]N. Matsumoto *et al.*, *J. Am. Chem. Soc.* **1996**, *118*, 981.
[49]T. P. Dasgupta *et al.*, *J. Chem. Soc., Dalton Trans.* **1993**, 3605.

(17-E-V)

Dithiocarbamates and Schiff base complexes provide many good examples of spin crossovers and low-spin—high-spin equilibria.

17-E-6 Mixed Valence Compounds of Iron

Prussian Blues[50]

The blue precipitates obtained on mixing Fe^{3+} with $[Fe^{II}(CN)_6]^{4-}$ or Fe^{2+} and $[Fe^{III}(CN)_6]^{3-}$ have long been known; both products are $Fe_4^{III}[Fe^{II}(CN)_6]_3 \cdot 15H_2O$. This compound presents an intense absorption band at 700 nm due to transition from the ground state to an excited state in which an electron is transferred from an Fe^{II} to an Fe^{III} site. Prussian blue may be considered as the archetype of mixed valence compounds; it contains two identical metals in different oxidation states. These compounds have played, and continue to play, a crucial role in the study of electron transfer phenomena.[51]

The structure of Prussian blue, which has been used as a pigment, and other similar materials such as $Cu_2^{II}[Fe^{II}(CN)_6](aq)$ or $Mn_3^{II}[Co^{III}(CN)_6]_2(aq)$ are based on a three-dimensional cubic framework with M^A and M^B atoms at the corners of a cube and with M^A—N—C—M^B links. There can be empty metal and CN sites depending on the stoichiometry, that is, on the valence of M^A and M^B. Water molecules can also be bound to Fe^{III} in Prussian blue as well as being interstitial as in zeolites. Reduction of Prussian blue gives Everitt's salt $K_2[Fe^{II}Fe^{II}(CN)_6]$.

Iron-Sulfur Complexes[52]

Iron-sulfur systems have been intensively studied because of their relationship to nonheme iron-sulfur proteins.

The compounds, many of which have thiolato groups, can have iron in II, III, and mixed oxidation states; they are of types such as $[Fe(SR)_4]^{1-,2-}$, $[Fe_2(\mu\text{-}S)_2(SR)_4]^{2-}$, $[Fe_4(\mu_3\text{-}S)_4(SR)_4]^{1-,2-,3-,4-}$, and $[Fe_3S_4(SR)_4]^{3-}$ where there are $Fe(\mu\text{-}S)_2Fe(\mu\text{-}S)_2Fe$ groups, and $[Fe_6S_9(SR)_2]^{4-}$, which has both μ-S and μ_3-S groups (Section 17-E-10). Two simpler cases, where charges are omitted, are shown in (17-E-VI) and

[50]O. Kahn, *Nature* **1995**, *378*, 667; M. Verdaguer, *Science* **1996**, *272*, 697.
[51]See for example: H. Takagi and W. Swaddle, *Inorg. Chem.* **1992**, *31*, 4669 and references therein; D. C. Arnett *et al.*, *J. Am. Chem. Soc.* **1995**, *117*, 12262.
[52]R. Holm, *Adv. Inorg. Chem.* **1992**, *38*, 1.; *J. Am. Chem. Soc.* **1996**, *118*, 11844; E. Münck *et al.*, *J. Am. Chem. Soc.* **1996**, *118*, 1966.

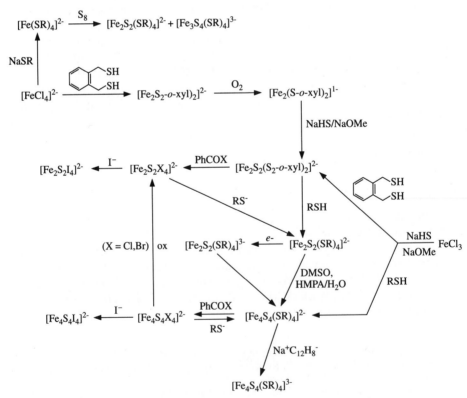

(17-E-VI)   (17-E-VII)

(17-E-VII); the latter is one of the most pervasive electron transfer centers in biology (Section 17-E-10).

In most cases the group X is an RS— or half of an —S—S— ligand, but compounds in which X = Cl, Br, I, or OR are known.

A prime feature of many of these systems and others with both Fe and Mo, are reversible electron transfer redox reactions and these give a clue to the importance of such polynuclear species in Nature. The compounds are made quite readily from $FeCl_2$, $FeCl_3$, $[FeCl_4]^{1-,2-}$, or $[Fe(SR)_4]^{1-,2-}$ as shown in Fig. 17-E-5.

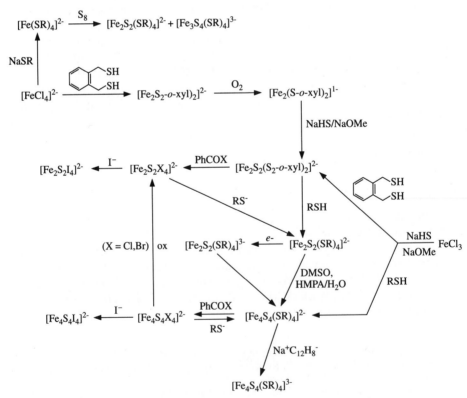

Figure 17-E-5 Some important reactions of iron-sulfur clusters.

17-E-7 Porphyrins

Complexes of conjugated macrocycles (Section 9-12) are of prime importance in living systems and for similar reasons other N_4 macrocyclic complexes have been intensively studied (Section 17-E-10).

The square Fe^{II} porphyrins initially add O_2 reversibly but, unlike heme and ligand protected systems, undergo autoxidation even at low temperatures to give μ-peroxo species, [porph $Fe^{III}]_2\mu$-(O_2), which can react with bases to give (base)(porph)Fe^{IV}=O.[53] The Fe^{IV} species are unstable, decomposing to (porph Fe^{III})$_2\mu$-O[54] which can be protonated.[55] The oxygenated species can oxidize organic substrates. Most (porph Fe^{III})$_2(\mu$-$O)$ species have porphyrin rings essentially parallel with nearly linear Fe—O—Fe angles. In general, they can undergo a series of one-electron oxidations and reductions. For example, when porph is 2,3,12,13-tetrabromo-5,10,15,20-tetraphenylporphyrinate, seven steps are observed (four oxidations and three reductions).[56]

There are also other porphyrin compounds such as (porph)FePh, bridged species (porph Fe)$_2$X, X = O, N, or C and carbenes, (porph)Fe=CCl$_2$, that are of the type implicated in CCl_4 and DDT toxicity. There are also other Fe^{IV} species such as (porph)Fe(X)Ph, (porph)Fe=NR, and so on. Reduced Fe(oep)$^{n-}$ species with Fe^0 and Fe^I are also known.

17-E-8 The Higher Oxidation States

Iron(IV)

Although relatively few iron(IV) compounds have been characterized, transient $Fe^{IV}O$ species appear to be involved in many natural oxidation systems, especially in porphyrin and related complexes[57] (see above). Several low-spin μ-nitridodiiron complexes, [Fe—N—Fe]$^{n+}$, are known for $n = 5$, 6[58] and they are prepared as shown in Fig. 17-E-6. Mixed oxides containing Fe^{IV} have been noted. In an octahedral environment, high-spin iron(IV) oxides show the typical distortion of the FeO$_6$ moiety due to the Jahn-Teller effect.[59]

Various cationic complexes can be made by chemical or electrochemical oxidation of Fe^{III} analogues; examples are [Fe(S$_2$CNR$_2$)$_3$]$^+$ and [Fe(bipy)$_3$]$^{4+}$ that are octahedral with two unpaired electrons; phosphine complexes of the type trans-[diphos$_2$FeCl$_2$]$^{2+}$ have a t_{2g}^4 configuration.

The stability of the tetrahedral purple alkyl Fe(1-norbornyl)$_4$ is doubtless due to the steric bulk of the ligand. The (Me$_5$C$_5$)$_2$Fe^{2+} ion is also reasonably stable in AlCl$_3$-1-butylpyridinium chloride melts; the yellow compound (η^6-toluene)Fe(H)$_2$-(SiCl$_3$)$_2$ can tolerate short air exposures.[60]

[53]See for example: T. Kitagawa and Y. Mizutani, *Coord. Chem. Rev.* **1995**, *135/136*, 685.
[54]See for example: E. Vogel *et al.*, *Angew. Chem. Int. Ed. Engl.* **1994**, *33*, 736.
[55]W. R. Scheidt *et al.*, *J. Am. Chem. Soc.* **1992**, *114*, 4420.
[56]K. M. Kadish *et al.*, *Inorg. Chem.* **1997**, *36*, 204.
[57]E. Vogel *et al.*, *Angew. Chem. Int. Ed. Engl.* **1994**, *33*, 731.
[58]See for example: K. Wieghardt *et al.*, *Angew. Chem. Int. Ed. Engl.* **1995**, *34*, 669 and references therein.
[59]P. Hagenmuller *et al.*, *C. R. Acad. Sci. Paris* **1987**, *304*, 633.
[60]K. J. Klabunde *et al.*, *J. Am. Chem. Soc.* **1994**, *116*, 5493.

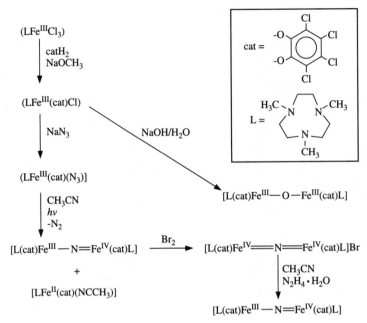

Figure 17-E-6   Preparation of high-oxidation nitrido complexes.

Iron(V)

In aqueous solution iron(V) decays quickly. It can be prepared from aqueous K_2FeO_4 using pulse radiolytically generated free radicals.[61] It exists in at least three protonated forms:

$$H_3FeO_4 \rightleftharpoons H_2FeO_4^- + H^+, \quad 5.5 \le pK_1 \le 6.5$$

$$H_2FeO_4^- \rightleftharpoons HFeO_4^{2-} + H^+, \quad pK_2 \approx 7.2$$

$$HFeO_4^{2-} \rightleftharpoons FeO_4^{3-} + H^+, \quad pK_3 = 10.1$$

Various porphyrin compounds such as NFe^VOEP are known.[62] The $\nu(Fe\equiv N)$ stretching frequency of 853 cm^{-1} is lower than that of the Cr and Mn analogues; the nitrido Fe^V complex is stable only at very low temperature ($\sim$30 K). It is widely accepted that one of the intermediate species in the catalytic cycle of cytochrome P-450 (see later) contains pentavalent iron. The reduction of an R_3CFe^V moiety is an accepted part of the mechanism of dismutation of hydrogen peroxide to water and oxygen catalyzed by catalase.

Selective oxidation of saturated hydrocarbons with oxygen in the presence of Fe^{II} in the so-called Gif reactions[63] is also postulated to proceed through an $Fe^V=O$ intermediate species.

[61]J. D. Rush and B. H. J. Bielski, *Inorg. Chem.* **1994**, *33*, 5499.
[62]W.-D. Wagner and K. Nakamoto, *J. Am. Chem. Soc.* **1989**, *111*, 1590.
[63]D. H. R. Barton and D. Doller, *Acc. Chem. Res.* **1992**, *25*, 504; *Chem. Soc. Rev.* **1996**, *25*, 237; M. J. Perkins, *Chem. Soc. Rev.* **1996**, *25*, 229.

Iron(VI)

Alkali metal salts can be obtained by hypochlorite oxidation of iron(III) nitrate in strong alkali solution. The blue K_2FeO_4 is isomorphous with K_2CrO_4 and K_2MnO_4 having the discrete FeO_4^{2-} ion;[64] it is a stronger oxidant than MnO_4^-. Its visible and near-ir absorption spectrum is characterized by two bands at ~ 500 and 800 nm. Crystals of K_2CrO_4 doped with FeO_4^{2-} exhibit a sharp-line luminesce around 1.6 um.[65]

17-E-9 Compounds with Short Iron—Iron Distances

Contrary to the well-accepted notion that metal to metal bonds are present in many compounds with metals such as Cr, Mo, W, Ru, Os and many others,[66] iron-iron bonding is still controversial but there are compounds with short intermetallic distances which indicate some degree of bonding. The overwhelming majority of such compounds are iron carbonyls. A typical example is the face-sharing bioctahedral complex $Fe_2(CO)_9$ in which each metal center has a formal electron count of 17 and consequently the two iron atoms share a single bond. The iron-iron separation of 2.523 Å and the diamagnetism provide additional support for this description; there are however some theoretical calculations which indicate that only a small direct Fe—Fe attractive interaction exists.[67] A group of compounds in which double metal-metal bonds result from coordinatively unsaturation are of the type $[Fe_2(CO)_5(\mu\text{-}PR_2)]$.[68] Iron-iron distances in these compounds are *ca.* 2.46–2.48 Å which are considerably shorter than those in the hexacarbonyls $Fe_2(CO)_6(\mu\text{-}PR_2)$ (2.62–2.82 Å).

Another type of organometallic compound with relatively short iron-iron separations is obtained from the reaction of Fe_2mes_4 and Bu^tNC. The iron-iron distance drops from 2.61 Å in Fe_2mes_4 to 2.371 Å in $[\{\eta^2\text{-}C(mes)=NBu^t\}_2Fe_2\{\mu\text{-}C\text{-}(mes)=NBu^t\}_2]$;[69] this is even shorter than the one found in metallic iron (2.48 Å). However, recent theoretical calculations seem to indicate that direct metal to metal bonding[70] is questionable in this type of molecule.

A very short diiron separation is found in a new type of dinuclear metal-metal bonded compound formed by reacting $FeCl_2(HDPhF)_2$, $NaHBEt_3$, and butyllithium in THF. The isolated product, $Fe_2(DPhF)_3$,[71] contains the mixed-valent dinuclear unit bridged by only three formamidinato groups (17-E-VIII); the iron-iron distance is 2.232 Å. Its epr spectrum shows two signals at g values of 1.99 and 7.94 which are consistent with the presence of seven unpaired electrons. If the above reaction is carried out using methyllithium instead of butyllithium, a distorted lantern-type dinuclear compound, $Fe_2(DPhF)_4$, is obtained; the iron-iron distance is 2.462 Å.[72]

[64]D. Reinen *et al.*, *Inorg. Chem.* **1995**, *34*, 1934.
[65]M. Herren and H. U. Güdel, *Inorg. Chem.* **1992**, *31*, 3683.
[66]F. A. Cotton and R. A. Walton, *Multiple Bonds between Metal Atoms*, 2nd ed., Oxford University Press, New York, NY, 1993.
[67]J. Reinhold *et al.*, *New. J. Chem.* **1994**, *18*, 465.
[68]See for example: B. Walther *et al.*, *Organometallics* **1992**, *11*, 1542; A. Wojcicki *et al.*, *Inorg. Chem.* **1992**, *31*, 2.
[69]C. Floriani *et al.*, *J. Am. Chem. Soc.*, **1994**, *116*, 9123.
[70]A. Sgamellotti *et al.*, *Inorg. Chem.* **1996**, *35*, 7776.
[71]F. A. Cotton *et al.*, *Inorg. Chim. Acta* **1997**, *256*, 269, 303.
[72]F. A. Cotton *et al.*, *Inorg. Chim. Acta* **1997**, *256*, 277.

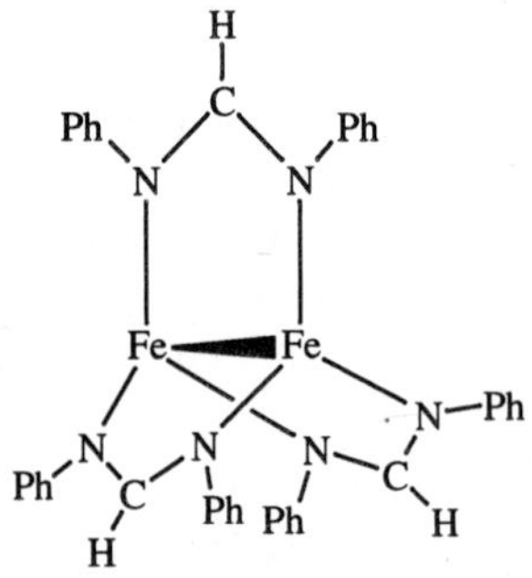

(17-E-VIII)

17-E-10   Bioinorganic Chemistry of Iron

Iron is truly ubiquitous in living systems. Its versatility is unique. It is at the active center of molecules responsible for oxygen transport and electron transport and it is found in, or with, such diverse metalloenzymes as various oxidases, hydrogenases, reductases, dehydrogenases, deoxygenases, and dehydrases.

Not only is iron involved in an enormous range of functions, it is also found in the whole gamut of life forms from bacteria to man. Iron is extremely abundant in the earth's crust and it has two readily interconverted oxidation states; doubtless these properties have led to evolutionary selection for use in many life processes. Table 17-E-2 shows the principal forms in which iron is found in the human body.

In biological systems there are three well-characterized iron systems: proteins that contain one or more iron-porphyrin units such as hemoglobin, myoglobin, and cytochrome P_{450}; a diverse group of proteins that contain non-heme iron, in particular the iron-sulfur clusters like nitrogenase, rubredoxin, and ferredoxins; and the non-heme diiron oxo-bridged species, most of which have carboxylates such as hemerythrin, methane monooxygenase, and ribonucleotide reductase.

Thanks to extensive studies of electronic spectra, Mössbauer, infrared, resonance Raman, ^{1}H and ^{17}O nmr spectroscopies, magnetism, EXAFS and X-ray crystallography, among others, on both the natural proteins or on synthetic model compounds, a lot is known about the active sites.

The very large number of known iron enzymes makes an exhaustive treatment of their chemistry unrealistic for a book such as this one. We arbitrarily decided to include only a few examples starting with heme proteins, followed by non-heme oxo-bridged proteins, iron storage proteins, and finally iron sulfur proteins. In all cases a very small segment is dedicated to the discussion of model compounds; the reader is advised to review the references given in the text for further details.

Iron Porphyrin Proteins[73]

They play very diverse and important roles such as O_2 transport and storage (hemoglobin and myoglobin), catalytic dehydrogenation or oxidation of organic molecules

[73]P. G. Debrunner. *Iron Prophyrins Part 3,* A. B. P. Lever and H. B. Gray, Eds.; VCH: New York, **1989**, pp. 137-234.

Table 17-E-2 Distribution of Iron-Containing Proteins in a Normal Adult

Protein	Molecular weight of protein	Amount of protein (g)	Amount of iron (g)	% of total body iron	Nature of iron heme (H) or nonheme (N)	Number of iron atoms bound per molecule	Valence state	Function
Hemoglobin	64,500	750	2.60	65	H	4	Fe^{2+}	Oxygen transport in plasma
Myoglobin	17,000	40	0.13	6	H	1	Fe^{2+}	Oxygen storage in muscle
Transferrin	76,000	20	0.007	0.2	N	2	Fe^{3+}	Iron transport *via* plasma
Ferritin	444,000	2.4	0.52	13	N	0–4,500	Fe^{3+}	Iron storage in cells
Hemosiderin		1.6	0.48	12	N	5,000	Fe^{3+}	Iron storage in cells
Catalase	280,000	5.0	0.004	0.1	H		Fe^{2+}	Metabolism of H_2O_2
Cytochrome *c*	12,500	0.8	0.004	0.1	H		Fe^{2+}/Fe^{3+}	Terminal oxidation
Peroxidase	44,100				H	1		Metabolism of H_2O_2
Cytochromes and oxidase			0.02	<0.5	H		Fe^{2+}/Fe^{3+}	Terminal oxidation
Flavoprotein dehydrogenases, oxidases and oxygenases					N		Fe^{2+}	Oxidation reactions, incorporation of molecular oxygen

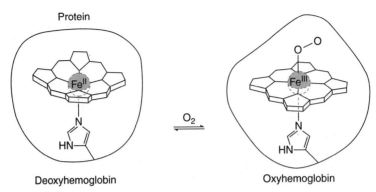

Figure 17-E-7   Reversible binding of dioxygen in Hb.

(peroxidase, cytochrome P_{450}), O_2 reduction (cytochrome *c* oxidase), and electron transport (cytochrome *c*).

In all cases, the active site is defined by an iron center coordinated by four nitrogen atoms provided by one heme group. At least another base is found in another coordination site. The type of function is determined by the protein tertiary structure.

Hemoglobin and Myoglobin[74,75]

Both are important in the transport and storage of oxygen in vertebrates. Hemoglobin, Hb, an $\alpha_2\beta_2$ heterotetramer that binds O_2 cooperatively, is found in red blood cells. Myoglobin, Mb, an α-helical mononuclear globular protein, stores oxygen in the muscle tissue until it is required for oxidative phosphorylation.

Our present understanding of how they work has arisen from the interplay of studies on both the proteins and on synthetic chemical models.

In the native deoxy form the active site consists of an iron(II) protoporphyrin IX encapsulated in a water resistant pocket and bound to the protein through the imidazole group of the histidine residue F8. The ferrous ion is 5-coordinate which allows the reversible binding of molecular oxygen in the sixth coordination site as shown in Fig. 17-E-7.

In the deoxy form, the iron(II) atom is in the high-spin (S = 2) state and lies about 0.5 Å out of the heme plane in the direction of the histidine group. In the oxygenated form the iron(III) ion is in the low-spin state and is nearly centered in the porphyrin plane. The dioxygen unit is bound in a bent fashion similar to the η^1-superoxo type found in many model compounds. There is strong evidence for hydrogen bonding between an imidazol N—H of a distal histidine and the bound O_2 group.

Using the structural changes that occur in the heme, a "trigger" mechanism for the cooperativity in the process of oxygenation in Hb has been proposed. According to this mechanism, the out-of-plane iron atom in the deoxy Hb is held in a "tense" (T) 5-coordinate conformation. The successive binding of O_2 relieves this tension as the oxygenated protein adopts a "relaxed" (R) conformation. An

[74]M. F. Perutz, *Q. Rev. Biophys.* **1989**, 22, 139; *Annu. Rev. Physiol.* **1990**, 52, 1.
[75]B. A. Springer *et al.*, *Chem. Rev.* **1994**, 94, 699.

important parameter that affects the oxygen binding ability is the acidity of the media; under acid conditions the equilibrium between deoxy- and oxyHb is shifted in favor of the deoxygenation process.

Hb and Mb Model Compounds[76]

Much has been learned from synthetic complexes (Section 17-E-7). The requirements to mimic oxyhemoglobin are the formation of a 5-coordinate heme precursor having a proximal base (imidazol, pyridine, or other) and hindering pathways that would lead to irreversible formation of μ-peroxo dimers. The lifetime of the working models is increased by exclusion of acidic protons and nucleophiles from the O_2 binding site and working at low temperatures.

To limit the formation of 6-coordinate iron(II) porphyrins some kind of steric hinderance is deliberately introduced, such as in the "picket fences" approach[77] (17-E-IX) or the "strapped" (17-E-X) and "roofed models" (17-E-XI).

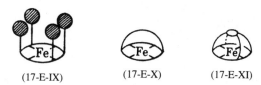

| (17-E-IX) | (17-E-X) | (17-E-XI) |

A few model complexes are known to bind dioxygen reversibly at room temperature with high affinities.[78]

Cytochromes[79]

They are heme proteins that act as electron carriers, linking the oxidation of substrates to the reduction of O_2. They operate by shuttling the iron atom between various oxidation states (Fe^{II}—Fe^V). One of the most studied is cytochrome P_{450},[80] a monooxygenase which converts C—H groups to C—OH groups according to the reaction:

$$RH + O_2 + 2e + 2H^+ \longrightarrow ROH + H_2O$$

In the divalent state, the fifth coordination site is occupied by a thiolate sulfur from a cysteine residue.

The key features of the presently accepted catalytic cycle for a cytochrome P_{450} enzyme are summarized in Fig. 17-E-8. The multistep process includes four stable

[76]M. Momenteau and C. A. Reed, *Chem. Rev.* **1994,** *94,* 659; D. H. Busch *et al., J. Am. Chem. Soc.* **1993,** *115,* 3623.

[77]See for example: E. Tsuchida *et al., J. Chem. Soc., Dalton Trans.* **1990,** 2713; E. Oldfield *et al., J. Am. Chem. Soc.* **1991,** *113,* 8680.

[78]J. P. Collman *et al., J. Am. Chem. Soc.* **1994,** *116,* 6245 and references therein; *Chem. Commun.* **1997,** 193.

[79]J. Everse, K. E. Everse and M. B. Grisham, Eds., *Peroxidases in Chemistry and Biology*; CRC Press: Boca Raton, FL, **1991,** Vols 1 and 2; B. Meunier, *Chem. Rev.* **1992,** *92,* 1411; D. Ostovic and T. Bruice, *Acc. Chem. Res.,* **1992,** *25,* 314; M. Momenteau and C. A. Reed, *Chem. Rev.* **1994,** *94,* 659.

[80]M. Sono *et al., Chem. Rev.* **1996,** *96,* 2841.

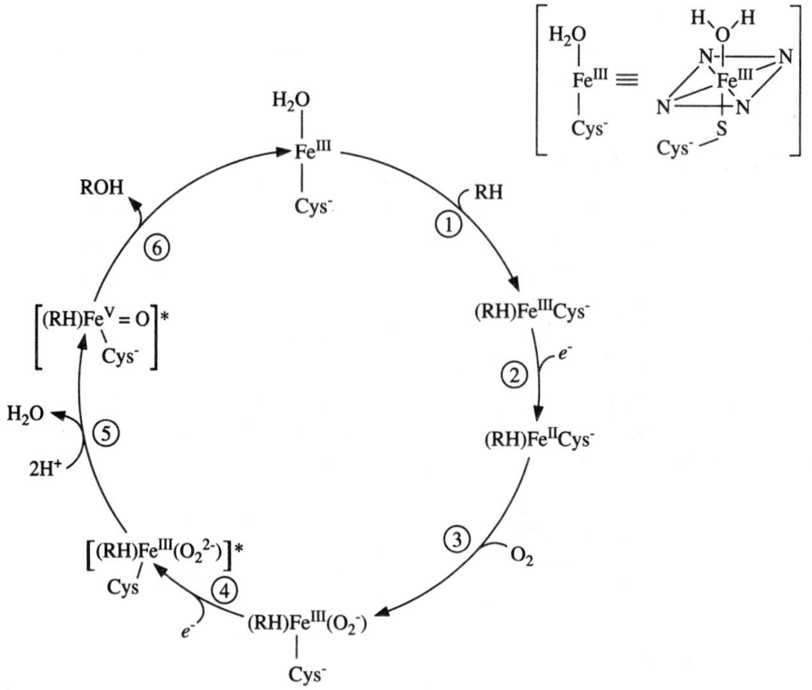

Figure 17-E-8 A schematic catalytic cycle for cytochrome P-450. The porphyrin group is not shown except in the view at the top. An asterisk indicates a proposed intermediate.

and isolable states and two proposed intermediates. The low-spin ferric resting state binds substrate at or near the iron atom (step 1) to give a ferric complex which is reduced by another enzymatic system (step 2) to a high-spin ferrous state. Then oxygen is incorporated, much as Hb or Mb would do, to give a low-spin dioxygen complex (step 3). Another one electron reduction (step 4) leads to the active oxygenating species after loss of water (step 5). The active species is believed to be an oxyferryl complex (Fe^V=O) although it has been formulated also as an iron(IV) porphyrin cation radical.

Many model compounds have been prepared[81] by reacting hemes with alkyl or aryl thiolate in organic solution. Several of them bind CO reversibly and exhibit a strong spectral resemblance to carbonylated cytochromes P_{450}.

Non-heme Oxo-bridged Iron Proteins[82]

Because of the ubiquitous presence of bridging carboxylato groups, they are also known as carboxylate-bridged non-heme iron proteins. They are known for a large number of nuclearities which range from one iron center to many iron centers; the most common contain the diiron unit. Several of these metalloenzymes are involved

[81]M. J. Gunter and P. Turner, *Coord. Chem. Rev.* **1991**, *108*, 115; N. Ueyama *et al.*, *J. Am. Chem. Soc.* **1996**, *118*, 12826.
[82]L. Que, Jr., In *Bioinorganic Catalysis*; J. Reedijk, Ed.; Marcel Dekker; New York, 1993; pp. 347–393. J. Sanders-Loehr., In *Iron Carriers and Iron Proteins*; T. M. Loehr, Ed.; VCH Publishers: New York, 1989; Vol. 5; pp. 373–466.

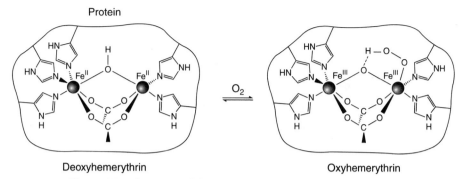

Figure 17-E-9 Reversible binding of dioxygen to hemerythrin.

in the biological utilization of dioxygen such as the binding and transport of O_2 (hemerythrin, Hr), various redox reactions (purple phosphatase,[83] PAP, and methane monooxygenase, MMO), and other non-redox processes (ribonucleotide reductase, RR).

Hemerythrin, Hr[84]

It is the oxygen-carrier protein found in marine invertebrates; it binds oxygen reversibly in a manner similar to Hb and Mb. Several oligomers are known, but a diiron subunit is common to all of them.

The active site consists of two iron atoms, roughly 3.25–3.5 Å apart, bound to the protein ligand by five histidine side-chain residues. In the reduced deoxy form the core is asymmetric, having one 5-coordinate and one 6-coordinate iron atom. Two protein carboxylate units and one hydroxo group derived from water are bridged across the iron atoms as shown in Fig. 17-E-9.

Dioxygen binds at the vacant site of the dioxy form and the proton from the hydroxo bridge shifts to the bound O_2 as two electrons from the Fe_2^{4+} core are transferred to O_2, thus reducing it. Resonance raman spectroscopy revealed the presence of a $\mu(O-O)$ stretching band at 844 cm^{-1}, characteristic of bound peroxide ions. The minimal structural changes associated with the diiron core upon oxygenation and deoxygenation are consistent with its reversible dioxygen binding function, which requires that the deoxy and oxy forms be in thermodynamic equilibrium.

Methane Monooxygenase, MMO[85]

It is a mixed-function oxidase in which one atom of O_2 is transferred to a substrate and the other forms water. The system converts methane to methanol in a process that is coupled to the oxidation of NADH, according to

$$CH_4 + O_2 + H^+ + NADH \longrightarrow CH_3OH + H_2O + NAD^+$$

[83]W. Haase *et al., Inorg. Chim. Acta* **1996,** *252,* 13 and references therein.
[84]R. E. Stenkamp, *Chem. Rev.* **1994,** *94,* 715; M. A. Holmes *et al., J. Mol. Biol.* **1991,** *218,* 583.
[85]S. J. Lippard *et al., Nature* **1993,** *366,* 537; B. Krebs and N. Sträter, *Angew. Chem. Int. Ed. Engl.* **1994,** *33,* 841; A. L. Feig and S. J. Lippard, *Chem. Rev.* **1994,** *94,* 759.

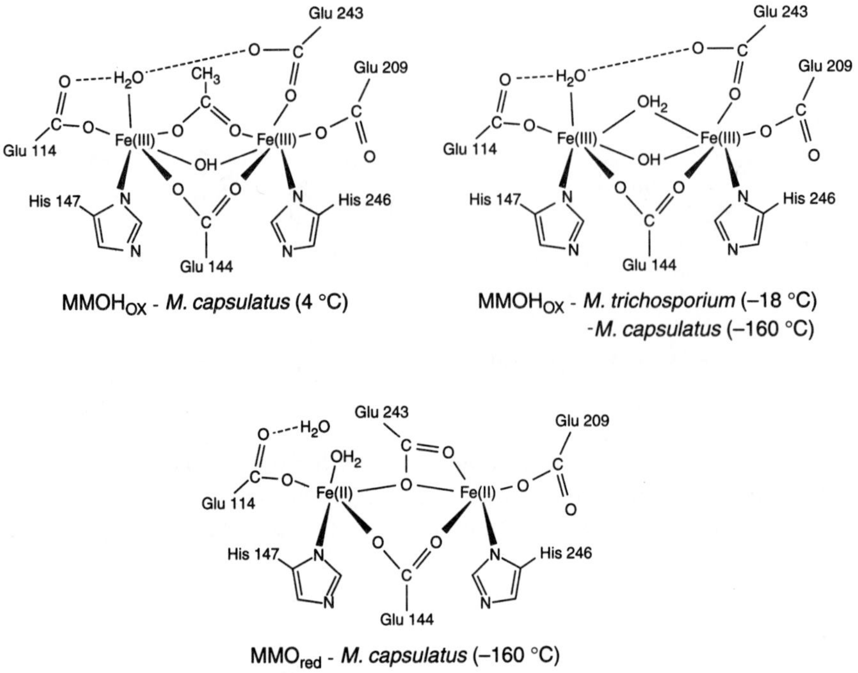

Figure 17-E-10   Structural representation of the binuclear iron in various forms of MMO.

The soluble MMO in several methanotrophs consists of three proteins: a hydroxylase, a reductive, and a coupling protein. All three are required for enzymatic activity. The hydroxylase contains dinuclear iron centers responsible for methane hydroxylation. The crystal structures of the oxidized M. *trichosporium* and M. *capsulatus* enzymes and also the reduced form of the latter are known. With that information and other spectroscopic evidence the picture shown in Fig. 17-E-10 has emerged for the active site. In the reduced enzymes, each Fe^{II} ion is terminally coordinated by one histidine and one carboxylate group and bridged to the other Fe^{II} ion by two carboxylate moieties. Thus both Fe^{II} ions have available sites for O_2 to bind. Upon oxidation, one of the bridging carboxylate groups shifts to a terminal position.

The fully oxidized enzyme is inactive; the reduced form is the one that reacts with dioxygen. A mixed-valent hydroxylase is also inactive. A bridging exogeneous acetate observed in the oxidized form has its origin in the buffer solution used to crystallize the enzyme. It might be the site at which dioxygen, substrate, or methoxide product interacts with the core.

Model Compounds[86]. Very few ferrous complexes having non-porphyrin ligands react with dioxygen to form stable adducts. Unlike the pseudo-heme model complexes, all of the adducts prepared so far appear to be peroxo rather than superoxo species. To mimic the protein fragment on the non-heme proteins multidentate

[86]B. J. Waller and J. D. Lipscomb, *Chem. Rev.* **1996,** *96,* 2625; D. M. Kurtz, Jr., *Chem. Rev.* **1990,** *90,* 585; B. A. Averill *et al., Chem. Rev.* **1990,** *90,* 1447; L. Que, Jr. and Y. Dong, *Acc. Chem. Res.* **1996,** *29,* 190; A. Uehara *et al., J. Am. Chem. Soc.* **1996,** *118,* 701.

ligands are commonly used,[87] for example substituted trispyrazolylborate anions, Tp*. One synthetic procedure employs mononuclear Fe^{II} complexes as illustrated below:

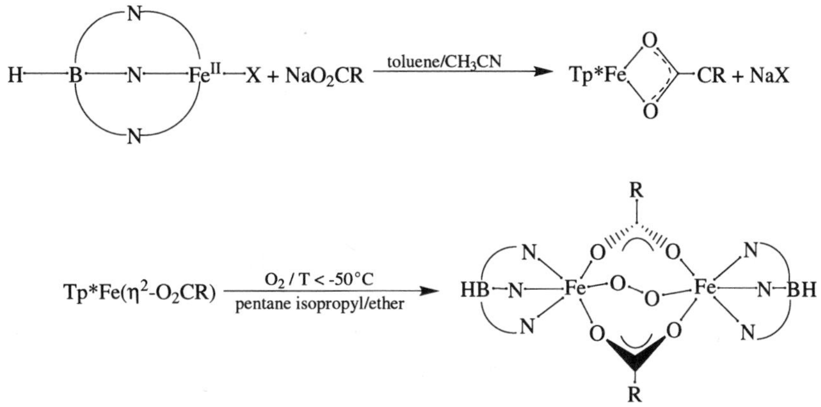

Ferritin, Ft[88]

It is one of the two principal iron storage compounds in the human body (the other is hemosiderin). Ferritin is important in iron homeostasis. Its twenty-four chains of two types, H and L, assemble as a hollow shell forming a 65–70 Å in diameter iron-storage cavity capable of holding up to 4500 iron atoms in microcrystalline particles. The current model of Ft formation requires a site for the initial binding and oxidation to Fe^{III}, then migration into the cavity. Ferritins are composed of a mineral iron(III) oxide/hydroxide core, similar to *ferrihydrite* [$FeO(OH) \cdot H_2O$] (also referred to as $5Fe_2O_3 \cdot 9H_2O$). The detailed structure of the inorganic core remains ill-defined but studies indicate that the Fe^{III} atoms have octahedral geometry; they are surrounded by oxygen donor atoms.

Iron is oxidized for incorporation into the mineralized core by either a protein enzymatic mechanism involving a putative dinuclear Fe ferroxidase site on the H chain subunit or a mineral surface mechanism. The net stoichiometric reactions for the two kinetic pathways are given by the following equations:

$$2Fe^{2+} + O_2 + 4H_2O \longrightarrow 2FeO(OH)_{core} + H_2O_2 + 4H^+$$

$$4Fe^{2+} + O_2 + 6H_2O \longrightarrow 4FeO(OH)_{core} + 8H^+$$

The L chain subunit appears to be involved mostly in mineralization of the core but modulates the ferroxidase activity of the H chain subunit as well.

[87]See for example: K. Kim and S. J. Lippard, *J. Am. Chem. Soc.* **1996**, *118*, 4914; S. J. Lippard *et al., Inorg. Chem.* **1994**, *33*, 636; N. Kitajima *et al., J. Am. Chem. Soc.* **1994**, *116*, 9071; N. Kitajima *et al., Inorg. Chem.* **1992**, *32*, 3342.

[88]P. J. Artymiuk *et al., Nature* **1991**, *349*, 541; P. M. Proulx-Curry and N. D. Chasteen, *Coord. Chem. Rev.* **1995**, *144*, 347; R. R. Crichton, *Inorganic Biochemistry of Iron Metabolism*; Horwood: New York, 1991; P. J. Artymiuk *et al.*, in *Iron Biominerals*; R. B. Frankel and R. P. Blakemore, Eds.; Plenum: New York, 1991; pp. 269–294.

Synthetic Models[89.] Other than the iron minerals mentioned earlier, no iron compound containing only oxo and hydroxo groups has been crystallized and structurally characterized yet. However, there are several discrete, relatively high-nuclearity iron oxo complexes with carboxylato or alkoxy groups which are normally produced by hydrolysis of iron(III) reagents in the presence of the organic ligands to mitigate the aggregation process which otherwise would lead to the formation of ferrihydrite in aqueous solution. The iron/oxo/hydroxo complexes which have been prepared vary extensively in their structure and the number of atoms. The O and OH groups can be found occupying terminal or bridging positions; the oxo groups can have a coordination number as high as six (Section 11-2) as in $[Fe_6(\mu_6\text{-}O)(\mu_2\text{-}OMe)_{12}(OMe)_6]^{2-}$.[90] Compounds have been prepared in which the number of iron atoms varies from 1 to 19.[91] They can be as simple as the well-known $Fe_3O(O_2CR)_6L_3$ complexes (Section 11-9) (17-E-XII) or more complex species such as $[(tacn)_6Fe_8(\mu_3\text{-}O)_2(\mu\text{-}OH)_{12}Br_7(H_2O)]^+$ (17-E-XIII), $[Fe_{19}(\mu_3\text{-}O)\text{-}$

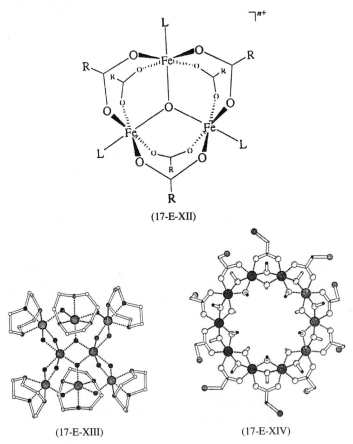

(17-E-XII)

(17-E-XIII)   (17-E-XIV)

[89]A. K. Powell *et al., Chem. Eur. J.* **1996,** *2,* 634; D. Gatteschi *et al., Chem. Soc. Rev.* **1996,** 101.
[90]K. Hegetschweiler *et al., Inorg. Chem.* **1992,** *31,* 1299.
[91]A. K. Powell *et al., J. Am. Chem. Soc.* **1995,** *117,* 2491; D. Gatteschi *et al., Angew. Chem. Int. Ed. Engl.* **1995,** *34,* 2716; S. J. Lippard and W. Micklitz, *J. Am. Chem. Soc.* **1989,** *111,* 6856; D. Gatteschi *et al., Chem. Eur. J.* **1996,** *2,* 1379; R. W. Saalfrank *et al., Angew. Chem. Int. Ed. Engl.* **1996** *35,* 2206.

$(\mu_2\text{-OH})_8(\text{heidi})_{10}(H_2O)_{12}]^+$ or the peculiar decanuclear *ferric wheel* [Fe(OMe)$_2$-(O$_2$CH$_2$Cl)]$_{10}$ (17-E-XIV).

Iron Sulfur Proteins[92]

These are the most widely encountered iron proteins. They display important biological roles as redox centers (ferredoxins, rubredoxins)[93] and chemical catalysts[94] (aconitase,[95] several (de)hydratases, sulfite reductase,[96] hydrogenase,[97] CO dehydrogenase,[98] and nitrogenase).

Ferredoxins, Fd[91,99]

They are relatively small proteins whose main function is to act as electron transfer proteins. The biological [Fe—S] centers which have been structurally characterized fall into three major categories: [2Fe—2S], [3Fe—4S,] and [4Fe—4S]; the last is the most common. In each of these clusters, the iron atoms are linked together by μ_2- or μ_3-S bridging atoms.[100] Schematic structures for some of them are provided in Fig. 17-E-11.

The Fe$_4$S$_4$ clusters formally contain two Fe^{III} and two Fe^{II} ions in the oxidized state and three Fe^{II} and one Fe^{III} ion in the reduced state. The active site of the oxidized form has a ground state S = 0 and is epr silent. The Mössbauer data are consistent with four equivalent iron atoms with an average oxidation number 2.5+. When fully reduced, the proteins have a ground state S = ½ and give typical epr spectra.

The [2Fe—2S] ferrodoxins are very acidic proteins. The active site consists of two di-μ-sulfido bridged high-spin tetrahedral Fe^{III} atoms (S = 5/2), which are antiferromagnetically coupled and therefore epr silent. The core is coordinated by four cysteine residues (RS$^-$) to give [Fe$_2$S$_2$(SR)$_4$]$^{2-}$.

Synthetic Analogues. An extremely large number of synthetic [Fe$_m$S$_x$(SR)$_y$]$^{n-}$ species is known. The family of homoleptic iron thiolates is currently composed of the types [Fe^{II}(SR)$_2$], [Fe^{II}(SR)$_3$]$^-$, [Fe^{II}(SR)$_4$]$^{2-}$, [Fe^{III}(SR)$_4$]$^-$, [Fe$_2^{II}$(SR)$_4$], [Fe$_2^{III}$(SR)$_8$]$^{2-}$, [Fe$_2^{II}$(SR)$_6$]$^{2-}$, and [Fe$_4^{II}$(SR)$_{10}$]$^{2-}$.[101] Many other types are found in which there are sulfido, thiolato, and other types or ligands attached to the iron center(s)[102] (Sections 17-E-5 and 17-E-6).

[92]R. Cammack. In *Advances in Inorganic Chemistry, Vol. 38: Iron Sulfur Proteins*, R. Cammack and A. G. Sykes, Eds., Academic Press: San Diego, CA, 1992, pp. 281–322.
[93]See for example: A. G. Webb *et al.*, *Inorg. Chem.* **1996** *35*, 5902.
[94]D. H. Flint and R. M. Allen, *Chem. Rev.* **1996**, *96*, 2315.
[95]H. Beinet *et al.*, *Chem. Rev.* **1996**, *96*, 2335.
[96]R. H. Holm *et al.*, *Inorg. Chem.* **1996**, *35*, 2767.
[97]M. Frey *et al.*, *Nature* **1995**, *373*, 580; *J. Am. Chem. Soc.* **1996**, *118*, 12989.
[98]J. Xia and P. A. Lindahl, *J. Am. Chem. Soc.* **1996**, *118*, 483.
[99]I. Bertini *et al.*, *J. Am. Chem. Soc.* **1994**, *116*, 651; R. H. Holm *et al.*, *Inorg. Chem.* **1994**, *33*, 4861 and many other papers from his group; A. G. Sykes *et al.*, *J. Am. Chem. Soc.* **1995**, *117*, 3635; D. TranOui and J. C. Jesior, *Acta Crystallogr.* **1995**, *D51*, 155.
[100]R. H. Holm *et al.*, *Chem. Rev.* **1996**, *96*, 2239; P. J. Stephens *et al.*, *Chem. Rev.* **1996**, *96*, 2491.
[101]See for example: R. H. Holm *et al.*, *Inorg. Chem.* **1995**, *34*, 1815 and references therein.
[102]See for example: L. Cai and R. H. Holm, *J. Am. Chem. Soc.* **1994**, *116*, 7177.

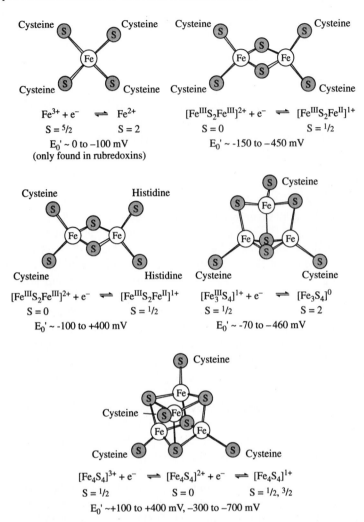

Figure 17-E-11   Schematic structure of Fe–S redox centers of nuclearities 1–4, and their electron-transfer reactions and approximate ranges of potentials. Individual iron atom spin states are indicated.

Nitrogenases[103]

These enzymes catalyze the activation and reduction of dinitrogen to ammonia according to:

$$N_2 + 8e + 8H^+ + 16MgATP \longrightarrow 2NH_3 + H_2 + 16MgADP$$

[103]R. R. Eady and G. J. Leigh, *J. Chem. Soc. Dalton* **1994**, 2739; D. Coucouvanis *et al., Molybdenum Enzymes, Cofactors and Model Systems*, ACS Symposium Series 535, American Chemical Society, Washington, DC, 1993, Chapters 10–23; M. J. Dilworth and A. P. Glenn, in *Biology and Biochemistry of Nitrogen Fixation*, Elsevier: Amsterdam, 1991; G. Stacey *et al., Biological Nitrogen Fixation*, Chapman and Hall, New York, 1992; J. Kim and D. C. Rees, *Biochemistry* **1994**, *33*, 389; J. T. Bolin *et al.,* in *New Horizons in Nitrogen Fixation*, P. Palacios *et al.* Eds., Kluwer, Dordrecht, The Netherlands, 1993; K. O. Hodgson *et al., J. Am. Chem. Soc.* **1994,** *116*, 2418.

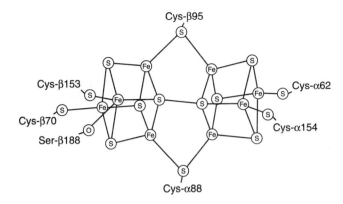

Figure 17-E-12 Current model for the structure of the P cluster in nitrogenase.

They are composed of two metalloproteins which are commonly referred to as the Fe protein and the MFe^{104} (M = Mo, V, and Fe) cofactor. The Fe protein, common to all three nitrogenases, is a dimer with a single Fe_4S_4 cluster bound between two equivalent subunits as shown in Fig. 17-E-12. It is frequently known as the "P cluster." There is some recent evidence that indicates that the P cluster probably is of the type 8Fe—7S instead of the 8Fe—8S type.[105] The function of the Fe protein is believed to involve MgATP-activated electron transfer to the MFe protein. The 4Fe—4S cluster of the Fe protein is generally considered to undergo a one-electron redox cycle between the $[4Fe-4S]^{2+}$ state and the $[4Fe-4S]^{+}$ state. The MoFe protein has been the most studied; it contains the "P cluster" and the FeMo cofactor (FeMoco). The diamagnetic P cluster is likely involved in electron transfer between the Fe protein and the FeMoco cofactor. The latter is believed to be the site of substrate reduction. The structure of the FeMoco cluster isolated from *A. vinelandii* and *C. pasteurianum* is shown in Fig. 17-E-13

The two entities, $MoFe_3S_3$ (left) and Fe_4S_3 (right) are bridged by three sulfur atoms. The Fe—Fe distances between bridged iron sites average *ca.* 2.5 Å, close enough to be considered metal—metal bonded. Only two protein ligands, Cys275 and His442, coordinate the cofactor to the protein, resulting in the unusual situation in which six Fe atoms bridged by nonprotein ligands are 3-coordinate. This type of coordination is relatively rare but not unprecedented in the chemistry of iron with sulfur ligands.[106]

EXAFS studies indicate very little change in the coordination of the Mo atom in Mo-nitrogenase during enzyme turnover; an indication that N_2 probably does not bind to molybdenum but instead to iron. However, this issue has not been completely resolved.

Model Compounds.[107] Before the structure of FeMoco was known, large numbers of the Fe/Mo/S compounds were prepared. Most of them do not resemble the

[104]B. K. Burgess and D. J. Lowe, *Chem. Rev.* **1996**, *96*, 2983; R. R. Eady, *Chem. Rev.* **1996**, *96*, 3013.
[105]See note added in proof in J. B. Howard and D. C. Rees, *Chem. Rev.* **1996**, *96*, 2965.
[106]See for example: P. P. Power and S. C. Shoner, *Angew. Chem. Int. Ed. Engl.* **1991**, *30*, 330.
[107]D. Coucouvanis, *Acc. Chem. Res.* **1991**, *24*, 1; L. J. Laughlin and D. Coucouvanis, *J. Am. Chem. Soc.* **1995**, *117*, 3118.

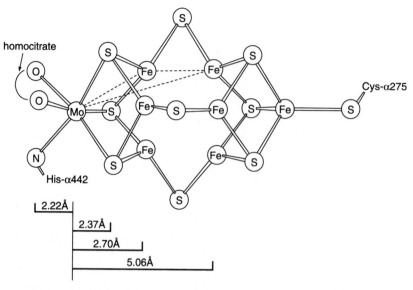

Figure 17-E-13   Schematic representation of the core of FeMoco.

actual structure but compounds with $MoFe_3S_4$ cores are now known as are those with MFe_4S_6 cores, M = V, Mo. For example in $MFe_4S_4(PEt_3)_4Cl$ (17-E-XV) the $Fe_4(\mu_3$-$S)_3(\mu_2$-$S)_3$ portion structurally resembles the core of FeMoco.[108]

(17-E-XV)

17-E-11   Organometallic Chemistry

The organometallic chemistry of iron is very extensive and is dominated by reactions with π-acid ligands, especially CO and Cp derivatives.

Carbonyls

The simplest homoleptic carbonyl species is $Fe(CO)_5$. The 18 electron complex is easily made by direct combination of highly dispersed metal and CO at high tempera-

[108]R. H. Holm *et al., Inorg. Chem.* **1994,** *33,* 5809 and references therein.

ture and pressure. The yellow liquid (mp $-20°C$, bp $103°C$) has a slightly distorted *tbp* structure[109] (17-E-XVI) in which all Fe—C distances are very similar (*ca.* 1.80–1.81 Å at 198 K). Irradiation with uv light promotes CO elimination and the formation of the very insoluble and non-volatile golden solid $Fe_2(CO)_9$ which has the structure of a face-sharing bioctahedron (17-E-XVII) with three μ-CO groups. It is frequently used as a source of reactive $Fe(CO)_4$ fragments. Another well-known polynuclear carbonyl compound is the green-black $Fe_3(CO)_{12}$ (mp $\sim$ 140–150°C with decomposition). Structure 17-E-XVIII shows a triangle of Fe atoms, two of which have unsymmetrical carbonyl bridges. The molecule is fluxional even in the solid state, as shown by solid-state magic angle spinning ^{13}C nmr spectra. The chemistry of the carbonyl compounds, particularly $Fe(CO)_5$, is dominated by reactions in which there is dissociative loss of one or more CO ligands followed by complexation and/or oxidative addition to a new substrate or nucleophilic attack at the carbon atom of CO, as indicated in Fig. 17-E-14.

(17E-XVI)   (17E-XVII)   (17E-XVIII)

Substitution reactions such as those carried out using PR_3 groups normally give mixtures of mono- and disubstituted complexes. However, selective synthesis can be accomplished. For example, refluxing $Fe(CO)_5$, PR_3 and $NaBH_4$, or NaOH in *n*-butanol gives ax,ax-$Fe(CO)_3(PR_3)_2$ exclusively.[110] Many other electron donors can be used to substitute the CO groups, for example, GeH_3, SnR_2,[111] N_2, NO^+, AsR_3, thiolates, and others.

The reaction of $Fe(CO)_5$ and aqueous alkali solutions gives water-soluble complexes which contain an equilibrium of $HFe(CO)_4^-$ and $Fe(CO)_4^{2-}$ ions:

$$HFe(CO)_4^- \rightleftharpoons H^+ + Fe(CO)_4^{2-} \qquad pK_A = 12.68\ (20°C)$$

The $HFe(CO)_4^-$ anion can be protonated by strong acids:

$$H_2Fe(CO)_4 \rightleftharpoons H^+ + HFe(CO)_4^- \qquad pK_A = 4.0\ (20°C)$$

The first step in the formation of $HFe(CO)_4^-$, from the reaction of $Fe(CO)_5$ and aqueous hydroxide ions, involves an attack of the OH^- ion on the CO group to give a metallocarboxylic acid.[112]

$$Fe(CO)_5 + OH^- \longrightarrow [(CO)_4FeCO_2H^-]$$

[109]D. Braga *et al.*, *Organometallics* **1993**, *12*, 1481.
[110]See for example: R. Glaser *et al.*, *Inorg. Chem.* **1996**, *35*, 1758 and references therein.
[111]M. Weidenbruch *et al.*, *Chem. Ber.* **1996**, *129*, 1565.
[112]See for example: H. des Abbayes *et al.*, *J. Organomet. Chem.* **1989**, *359*, 205.

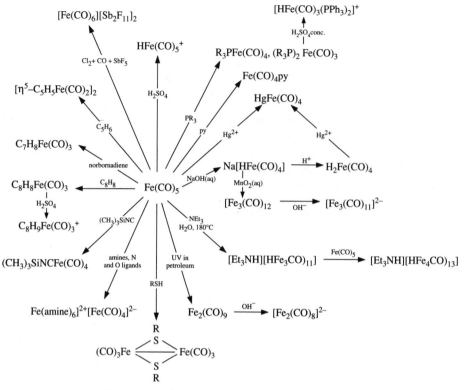

Figure 17-E-14   Some reactions of iron carbonyls.

The decarboxylation of this intermediate is believed to proceed *via* the dianion $[(CO)_4FeCO_2]^{2-}$ as follows:

$$[(CO)_4FeCO_2H^-] \xrightarrow[-H_2O]{OH^-} [(CO)_4FeCO_2]^{2-} \xrightarrow{-CO_2} [Fe(CO)_4]^{2-} \xrightarrow[-OH^-]{H_2O} [HFe(CO)_4]^-$$

The hydride iron carbonyl anion, $[HFe(CO)_4]^-$, has a structure that is best described as a distorted *tbp* with the hydride ligand occupying an axial site. The anion is fluxional both in solution and in the solid state.[113] The main use is found in organic synthesis and catalysis.[114] A typical reaction is the reduction of olefins

$$RCH{=}CH_2 \xrightarrow[\text{EtOH, CO, 60°C, 25h}]{KHFe(CO)_4} RCH_2CH_3$$

or α, β unsaturated carbonyl compounds,

$$RCH{=}CH{-}\underset{\underset{O}{\|}}{C}{-}R' \xrightarrow[\text{95\% MeOH, 0-60°C}]{KHFe(CO)_4} RCH_2CH_2{-}\underset{\underset{O}{\|}}{C}{-}R'$$

[113]B. E. Hanson and K. H. Whitmire, *J. Am. Chem. Soc.* **1990**, *112*, 974.
[114]J.-J. Brunet, *Chem. Rev.* **1990**, *90*, 1041 (an extensive review with preparation methods, stoichiometric and catalytic reactions).

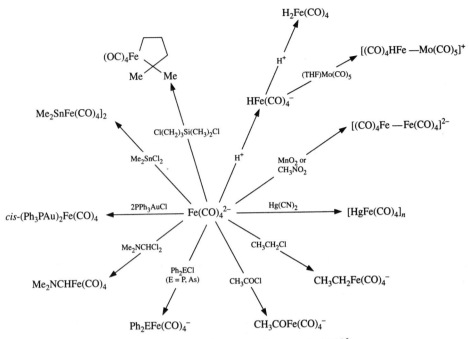

Figure 17-E-15 Some reactions of $Fe(CO)_4^{2-}$.

Salts of $Na_2Fe(CO)_4$ are also used in organic synthesis as illustrated by the following reactions:

$$Na_2Fe(CO)_4 + RX \longrightarrow [RFe(CO)_4]^-$$

```
                    R'X
          RCR'  ◄──────┐   CO
           ‖         [RCFe(CO)4]⁻
           O            O
                  H⁺ │ O₂,  │ I₂, R'OH
                     │ H₂O  │
          RCHO     RCO₂H    RCOOR'
```

Other reactions are shown in Fig. 17-E-15.

Other hydrido carbonyl ferrates and dianionic carbonyl ferrates are $H_2Fe_2(CO)_8^-$, $HFe_3(CO)_{11}^-$ (17-E-XIX), $HFe_4(CO)_{13}^-$, $Fe_2(CO)_8^{2-}$, and $Fe_4(CO)_{13}^{2-}$[115] (17-E-XX).

(17-E-XIX)

[115]See for example: H. Vahrenkamp *et al.*, *J. Organomet. Chem.* **1991**, *411*, 431.

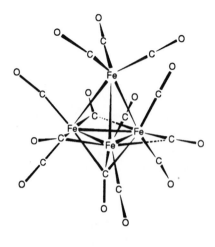

(17-E-XX)

A unique reaction of $Fe(CO)_5$ is the oxidative carbonylation in liquid SbF_5 in a CO atmosphere with AsF_5 or Cl_2 as oxidizing agents[116] according to

$$Fe(CO)_5 + 5SbF_5 + AsF_5 + CO \xrightarrow[\substack{60-90°C}]{\substack{1 \text{ atm CO, } SbF_5}} [Fe(CO)_6][Sb_2F_{11}]_2 + AsF_3 \cdot SbF_5$$

The homoleptic carbonyl cation $[Fe(CO)_6]^{2+}$, the first of the type formed by a first transition series metal, is isostructural with those of Ru and Os. The ir stretching force constant f_{CO} of 19.82×10^2 Nm^{-1} suggest only insignificant metal to CO π-backbonding. The ^{57}Fe Mössbauer spectrum indicates that the complex is octahedral around the Fe atom and that the ground state of the Fe^{II} atom corresponds to a $3d^6$ low-spin configuration. There is an indication that when excess Cl_2 is used in the synthesis an $[Fe(CO)_6]^{3+}$ ion also forms.

Cyclopentadienyl Compounds

There is a very large number of complexes which contain one or two Cp groups per iron atom; the best known is the unusually stable ferrocene for which a vast organic chemistry is known.[117] It undergoes Friedel-Crafts acylation, sulfonation, and metallation by alkyllithiums, and so on. An example is the reaction

$$Cp_2Fe + CO_2 \xrightarrow{AlCl_3} CpFeC_5H_4CO_2H$$

Some of the substituted ferrocenes can be chiral[118] and find uses in asymmetric catalysis, enantioselective synthesis, and in diverse materials science areas.

Very strong acids give the cation Cp_2FeH^+ which has bent rings. It has been suggested that electrophilic attacks on Cp_2Fe proceed *via* initial attack on the metal

[116]H. Willner *et al., Inorg. Chem.* **1997**, *36*, 158.
[117]Annual Surveys of Ferrocene Chemistry: B. W. Rockett and G. Marr, *J. Organomet. Chem.* **1990**, *392*, 93; **1991**, *416*, 327.
[118]V. Snieckus *et al., J. Am. Chem. Soc.* **1996**, *118*, 685.

followed by rearrangements as in

However, it seems more likely that the electrophile attacks the ring in the *exo* position with the *endo* H then moving to the metal and being subsequently lost, as ferrocene is such a weak base.

Ferrocene can be oxidized by various oxidizing agents such as *p*-benzoquinone in $HBF_4 \cdot OEt_2$ to yield the blue salt $[Cp_2Fe][BF_4]$. The ferrocenium salts are mild one-electron oxidants usually regarded as outer-sphere reagents. Because they are readily available and easily handled, they are widely used in both stoichiometric reactions and redox catalysis.[119] Another advantage is that the oxidation potential of the ferrocenium ion can be systematically altered by ring substitution providing a range of oxidants with E^0 (*vs.* Cp_2Fe) values varying from -0.63 V for $Fe(\eta\text{-}C_5H_4NMe_2)_2$ to $+0.64$ V for $Fe(\eta\text{-}C_5H_4CF_3)_2$.

Other Complexes with Fe—C Bonds

A few compounds are known to contain η^2-alkenes or alkynes but they are frequently unstable and have to be made at low temperature, but there are large numbers of complexes with η^3-allyl groups,[120] η^4-ligands, η^6-ligands,[121] and others.

Simple alkyls or aryls have also been made; many such as $Fe(CO)_2$-$(PMe_3)_2MeX$[122] (17-E-XXI) have been widely studied because of the importance of the CO insertion reaction which gives $Fe(CO)_2(PMe_3)_2(COMe)X$ and is used in catalysis. Another important subgroup is that of compounds of the type (porph)FeR[123] which have been cited before.

The best characterized homoleptic aryls are $[Li(OEt_2)]_4[FePh_4]$[124] and $[Li(OEt_2)]_2[Fe(naphthalenide)_4]$ as well as $(mes)Fe(\mu\text{-}mes)_2Fe(mes)$[125] and many of its derivatives (Section 17-E-9). The Fe^0 complex $[Li(OEt_2)]_4[FePh_4]$ has remarkable activity toward dinitrogen and is synthesized by reaction of $FeCl_3$ and 10 equivalents of phenyl lithium in diethyl ether:

$$FeCl_3 + \text{excess PhLi} \xrightarrow{\text{Et}_2\text{O}} [Li(Et_2O)_4][FePh_4]$$

[119]N. G. Connelly and W. E. Geiger, *Chem. Rev.* **1996**, *96*, 877.
[120]G. M. Williams and D. E. Rudisill, *Inorg. Chem.* **1989**, *28*, 797.
[121]R. M. G. Roberts *et al.*, *J. Organomet. Chem.* **1989**, *359*, 331.
[122]G. Cardaci *et al.*, *Inorg. Chem.* **1992**, *31*, 3018.
[123]A. L. Bach *et al.*, *Inorg. Chem.* **1994**, *33*, 2815.
[124]S. U. Koschmieder and G. Wilkinson, *Polyhedron* **1991**, *10*, 135.
[125]H. Müller *et al.*, *Z. anorg. allg. Chem.* **1996**, *622*, 1968.

$$\underset{\text{(17-E-XXI)}}{\overset{\displaystyle PMe_3}{\underset{\displaystyle PMe_3}{\overset{\displaystyle \underset{OC}{\overset{OC}{\diagdown}}\,\,Fe\,\,\overset{Me}{\underset{X}{\diagup}}}{|}}}}$$

(17-E-XXI)

Derivatives of the type $Fe(dppe)R_2$, R $=Bz$, $p\text{-}CH_2C_6H_4Me$, CH_2CMe_2Ph, and CH_2SiMe_3 can be prepared by reaction of the halides and the corresponding Grignard reagents:

$$FeBr_2(dppe) + 2RMgX \longrightarrow Fe(dppe)R_2 + 2MgXBr$$

There are also carbene, carbyne,[126] and alkylidyne[127] complexes. By far the most common route to carbenoid complexes $R_2C{=}Fe(CO)_4$ is the classic Fisher synthesis, involving attack by a strong nucleophile (usually a carbanion) upon an iron carbonyl group to form an acyl anion followed by quenching at the oxygen atom with an alkylating agent to form the corresponding carbenoid, for example:

$$Fe(CO)_5 + CO + LiMe \longrightarrow Li[Fe(CO)_4(COMe)] \xrightarrow{\text{LiR}} (CO)_4Fe{=}C(OMe)R$$

Additional References

I. Bertini, S. Ciurli, C. Luchinat, and W. J. A. Maaskant, *Structure and Bonding 83: Iron-Sulphur Proteins/Perovskites*, Springer, New York, 1995.

Chemical Reviews, **1996,** *96* (contains several reviews on bioinorganic chemistry of iron and other metals).

T. Funabiki, Ed., *Oxygenases and Model Systems* in *Catalysis by Metal Complexes, Vol. 19*, Kluwer Academic Publishers, Dordrecht, 1997.

O. Hayaishi, *Molecular Mechanism of Oxygen Activation*, Academic Press, New York, 1994.

S. J. Lippard and J. M. Berg, *Principles of Bioinorganic Chemistry*, University Science Books, Mill Valley, CA, 1994.

P. R. Ortiz de Montellano, Ed., *Cytochrome P450. Structure, Mechanism and Biochemistry*, 2nd ed., Plenum, New York, 1995.

D. Sellmann *et al.*, *Angew. Chem. Int. Ed. Engl.* **1997,** *36*, 1201 (a recent paper describing a high yield synthesis of an Fe^{IV} complex; it gives several references on this topic. See also Section 17-E-8).

17-F   COBALT: GROUP 9

The trends toward decreased stability of the very high oxidation states and the increased stability of the II state relative to the III state, which have been noted through the series Ti, V, Cr, Mn, and Fe, persist with cobalt. The oxidation states IV and V are represented by only a few compounds. The III state is relatively unstable in simple compounds, but the low-spin complexes are exceedingly numerous and stable, especially where the donor atoms (usually N) make strong contributions to the ligand field. There are also some important complexes of Co^I; this

---

[126]D. Seyferth *et al., Organometallics* **1989,** *8,* 836.
[127]D. Lentz, *Coord. Chem. Rev.* **1995,** *143,* 383.

oxidation state is better known for cobalt than for any other element of the first transition series except copper.

The oxidation states and stereochemistry are summarized in Table 17-F-1.

17-F-1 The Element

Cobalt always occurs in Nature in association with Ni and usually also with arsenic. The most important Co minerals are *smaltite* ($CoAs_2$) and *cobaltite* (CoAsS), but the chief technical sources of Co are residues called "speisses," which are obtained in the smelting of arsenical ores of Ni, Cu, and Pb.

Cobalt is a hard, bluish-white metal (mp 1493°C, bp 3100°C). It is ferromagnetic with a Curie temperature of 1121°C. It dissolves slowly in dilute mineral acids, the Co^{2+}/Co potential being -0.227 V, but it is relatively unreactive. While it does not

Table 17-F-1 Oxidation States and Stereochemistry of Cobalt

Oxidation state	Coordination number	Geometry	Examples
Co^{-1}, d^{10}	4	Tetrahedral	$[Co(CO)_4]^-$, $Co(CO)_3NO$
Co^0, d^9	4	Tetrahedral	$K_4[Co(CN)_4]$, $Co(PMe_3)_4$
Co^I, d^8	3	Planar	$(tempo)Co(CO)_2$
	4	Tetrahedral	$CoBr(PR_3)_3$
	5[a]	*tbp*	$[Co(CO)_3(PR_3)_2]^+$, $HCo(PR_3)_4$, $[Co(NCMe)_5]^+$
	5	*sp*	$[Co(NCPh)_5]ClO_4$
	6	Octahedral	$[Co(bipy)_3]^+$
Co^{II}, d^7	3	Trigonal	$\{Co(OCBu_3)_2[N(SiMe_3)_2]\}^-$, $Co_2(NPh_2)_4$
	4[a]	Tetrahedral	$[CoCl_4]^{2-}$, $CoBr_2(PR_3)_2$, Co^{II} in Co_3O_4
	4	Square	$[(Ph_3P)_2N]_2[Co(CN)_4]$, $[Co(py)_4](Cl)(PF_6)$[b]
	5	*tbp*	$[Co(Me_6tren)Br]^+$, $CoH(BH_4)(PCy_3)_2$
	5	*sp*	$[Co(ClO_4)(MePh_2AsO)_4]^+$, $[Co(CN)_5]^{3-}$, $[Co(CNPh)_5]^{2+}$
	6[a]	Octahedral	$[Co(NH_3)_6]^{2+}$
	8	Dodecahedral	$(Ph_4As)_2[Co(NO_3)_4]$
Co^{III}, d^6	4	Tetrahedral	In a 12-heteropolytungstate; in garnets
	4	Square	$[Co(SR)_4]^-$
	5	*sp*	$RCo(saloph)$
	5	*tbp*	$CoCl(TC-4,4)$[c]
	6[a]	Octahedral	$[Co(en)_2Cl_2]^+$, $[Cr(CN)_6]^{3-}$, $ZnCo_2O_4$, CoF_3, $[CoF_6]^{3-}$
Co^{IV}, d^5	4	Tetrahedral	$Co(1\text{-norbornyl})_4$
	6	Octahedral	$[CoF_6]^{2-}$, $[Co(dtc)_3]^+$
Co^V, d^4	4	Tetrahedral	$[Co(1\text{-norbornyl})_4]^+$

[a]Most common states.
[b]J. K. Beattie *et al.*, *Polyhedron* **1996**, *15*, 473.

[c]S. J. Lippard *et al.*, *J. Am. Chem. Soc.* **1992**, *114*, 9670, TC-4,4 =

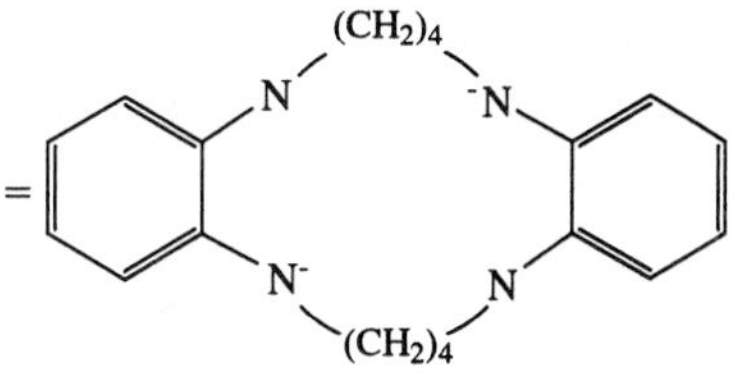

combine directly with C, P, and S on heating, it is attacked by atmospheric O_2 and by water vapor at elevated temperatures, giving CoO. Very reactive finely divided metal particles can be made by reduction of $CoCl_2$ with Li naphthalenide in glyme.

COBALT COMPOUNDS

17-F-2   Binary Cobalt Compounds and Simple Salts

Oxides

On heating Co^{II} carbonate or nitrate olive-green cobalt(II) oxide is obtained. It normally has a slight excess of oxygen and is a *p*-type semiconductor. It has the NaCl structure.

At 400 to 500°C in air the oxide Co_3O_4 is obtained. This is a normal spinel with Co^{2+} ions in tetrahedral interstices and Co^{3+} ions in octahedral interstices. Other oxides Co_2O_3 and CoO_2 and a red oxocobaltate(II) $M_{10}[Co_4O_9]$ (M = Na, K) are known; these contain trigonal-planar CoO_3 building blocks.[1]

The hydrous oxide CoO(OH) occurs as the mineral *heterogenite*. It has a layer lattice with $Co^{III}O_6$ octahedra and gives Co_3O_4 on heating.

Halides

The anhydrous halides CoX_2 may be made from the hydrated halides by heating or treatment with $SOCl_2$. Cobalt difluoride is obtained by reacting $CoCl_2$ with HF; it has the rutile structure. Like the dichlorides of Mg, Mn, and Fe, blue $CoCl_2$ adopts the $CdCl_2$ structure.

The action of fluorine or other fluorinating agents on cobalt halides at 300 to 400°C gives dark brown CoF_3, commonly used as a fluorinating agent. It is reduced by water.

Sulfide

From Co^{2+} solutions, a black solid CoS is precipitated by the action of H_2S.

Salts

Cobalt(II) forms an extensive group of simple and hydrated salts. The latter are red or pink and contain the $[Co(H_2O)_6]^{2+}$ ion or other octahedrally coordinated ions. Addition of OH^- to Co^{2+} solutions gives $Co(OH)_2$, which may be pink or blue depending on the conditions; only the pink form is stable. It is amphoteric, dissolving in concentrated hydroxide to give a deep blue solution containing $[Co(OH)_4]^{2-}$ ions, from which crystalline salts can be obtained. The color change from anhydrous blue to hydrated octahedral pink Co(II) is used as a moisture indicator in silica gel drying agents.

Cobalt(III) forms few simple salts, but the green hydrated fluoride $CoF_3 \cdot 3.5H_2O$ and the blue hydrated sulfate $Co_2(SO_4)_3 \cdot 18H_2O$ separate on electrolytic oxidation

[1]R. Hoppe *et al.*, *Z. anorg. allg. Chem.* **1993**, *619*, 1807.

of Co^{2+} in 40% HF and 8 M H$_2$SO$_4$, respectively. Alums, M^{I}Co(SO$_4$)$_2$·12H$_2$O, are dark blue; they are reduced by water.

In aqueous solutions containing no complexing agents, oxidation of [Co(H$_2$O)$_6$]$^{2+}$ to Co^{III} is very unfavorable:

$$[Co(H_2O)_6]^{3+} + e = [Co(H_2O)_6]^{2+} \qquad E^0 = 1.84 \text{ V}$$

However, electrolytic or O$_3$ oxidation of cold acidic perchlorate solutions of Co^{2+} gives [Co(H$_2$O$_6$]$^{3+}$, which is in equilibrium with [Co(OH)(H$_2$O)$_5$]$^{2+}$. At 0°C, the half-life of these diamagnetic ions is about a month. In the presence of complexing agents such as NH$_3$, which form stable complexes with Co^{III}, the stability of Co^{III} is greatly improved.

$$[Co(NH_3)_6]^{3+} + e = [Co(NH_3)_6]^{2+} \qquad E^0 = 0.1 \text{ V}$$

In basic media we have

$$CoO(OH)(s) + H_2O + e = Co(OH)_2(s) + OH^- \qquad E^0 = 0.17 \text{ V}$$

Water rapidly reduces uncomplexed Co^{3+} at room temperature. This relative instability is evidenced by the rarity of simple salts and binary compounds, whereas Co^{II} forms such compounds in abundance.

17-F-3 Complexes of Cobalt(II), d^7

Cobalt(II) forms numerous complexes, mostly either octahedral or tetrahedral but five-coordinate and square species are also known. There are more *tetrahedral complexes* of Co^{II} than for other transition metal ions. This is in accord with the fact that for a d^7 ion, ligand field stabilization energies disfavor the tetrahedral configuration relative to the octahedral one to a smaller extent than for any other $d^n (1 \leq n \leq 9)$ configuration, although it should be noted that this argument is valid only in comparing the behavior of one metal ion to another, not for assessing the absolute stabilities of the configurations for any particular ion. The only d^7 ion of common occurrence is Co^{2+}.

Because of the small stability difference between octahedral and tetrahedral Co^{II} complexes, there are several cases in which the two types with the same ligand are both known and may be in equilibrium. An example is that of the thiocyanate complexes in methanol. There is always some [Co(H$_2$O)$_4$]$^{2+}$ in equilibrium with [Co(H$_2$O)$_6$]$^{2+}$ in aqueous solution.

Tetrahedral complexes, [CoX$_4$]$^{2-}$, are generally formed with monodentate anionic ligands such as Cl$^-$, Br$^-$, I$^-$, SCN$^-$, N$_3^-$, and OH$^-$; with a combination of two such ligands and two neutral ones, tetrahedral complexes of the type CoX$_2$L$_2$ are formed. With ligands that are bidentate monoanions, tetrahedral complexes are formed in some cases (e.g., with *N*-alkylsalicylaldiminato and bulky β-diketonate anions). With the less hindered ligands of this type, association to give a higher coordination number often occurs. Thus in bis(*N*-methylsalicylaldiminato)cobalt(II) a dimer with five-coordinate Co atoms is formed [Fig. 17-F-1(a)], whereas Co(acac)$_2$ is a tetramer in which each Co atom is six-coordinate [Fig. 17-F-1(b)]. The cobalt

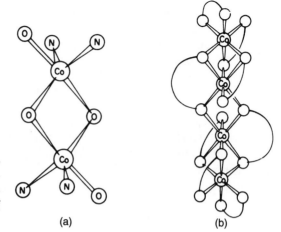

Figure 17-F-1 Schematic representations of the structures of (a) the dimer of bis(*N*-methylsalicylaldimin-
ato)cobalt(II) and (b) the tetramer of
bis(acetylacetonato)cobalt(II).

(a)   (b)

phosphate $[H_3NCH_2CH_2NH_3]^{2+}[(CoPO_4)^-]_2$ forms a zeolite-like open framework with alternating CoO_4 and PO_4 tetrahedra.[2]

Planar complexes are formed with several bidentate monoanions such as dimethylglyoximate, *o*-aminophenoxide, dithioacetylacetonate, and dithiolate-type ligands. Several neutral bidentate ligands also give planar complexes, although it is either known or reasonable to presume that the accompanying anions are coordinated *to some degree*, so that these complexes could also be considered as very distorted octahedral ones. Examples are $[Co(en)_2](AgI_2)_2$ and $[Co(CH_3SC_2H_4SCH_3)_2](ClO_4)_2$. With tetradentate ligands such as salen and porphyrins, planar complexes are also obtained.

Octahedral complexes are common with halides, pseudohalides, and O-donors. The dark-blue THF complex $CoCl_2 \cdot 1.5THF$ has the structure $Co_4(\mu_3\text{-}Cl)_2(\mu_2\text{-}Cl)_4Cl_2(THF)_6$ with octahedral $CoCl_4O_2$ and *tbp* $CoCl_4O$ units.[3] $CoCl_3 \cdot 3DMSO$ is ionic, $[Co(DMSO)_6]^{2+}[CoCl_4]^{2-}$, with O-bound DMSO.[4]

The addition of KCN to aqueous Co^{2+} gives a green solution and a purple solid. The anion in the diamagnetic purple salt $Ba_3[Co_2(CN)_{10}] \cdot 13H_2O$ has the $Mn_2(CO)_{10}$ type structure with a Co—Co bond. In solution the primary species is probably paramagnetic $[Co(CN)_5(H_2O)]^{3-}$, but there are also complexes with less CN^-, and the solution is unstable, reacting with water to give $[Co(CN)_5H]^{3-}$, $[Co(CN)_5OH]^{3-}$, and other products. In solvents like MeCN or DMF salts such as $Li_3[Co(CN)_5] \cdot 3DMF$ or $(PPN)_2[Co(CN)_4] \cdot 4DMF$ can be obtained. The green aqueous solution reacts with hydrogen

$$2[Co(CN)_5]^{3-} + H_2 = 2[Co^{III}(CN)_5H]^{3-}$$

and is a catalyst for homogeneous hydrogenation of conjugated alkenes, especially under phase transfer conditions. The ion also reacts with C_2F_4, C_2H_2, SO_2, or $SnCl_2$ to give cobalt(III) complexes in which the small molecule is "inserted" between

[2]J. M. Thomas *et al.*, *Angew. Chem. Int. Ed. Engl.* **1994**, *33*, 639.
[3]P. Sobota *et al.*, *Polyhedron* **1993**, *12*, 613.
[4]A. Ciccarese *et al.*, *J. Crystallogr. Spectrosc. Res.* **1993**, *23*, 223.

two cobalt atoms as in $[(NC)_5Co-CF_2CF_2-Co(CN)_5]^{6-}$ and $K_6[(CN)_5CoCH=CH-Co(CN)_5]$, where the configuration is trans about the double bond. The red-brown $[Co(CN)_5O_2]^{3-}$ is best regarded as a superoxo complex of Co^{III} (see Section 17-F-5).

There are a number of dinuclear carboxylate complexes of Co^{II} which show interesting magnetic behavior and contain the $[Co_2(\mu\text{-}OH_2)(RCO_2)_2]^{2+}$ or $[Co_2(\mu\text{-}X)(\mu\text{-}RCO_2)_2]^+$ core (X = OH, Cl, Br), stabilized by chelating nitrogen ligands to complete the octahedral coordination. The latter can be oxidized to mixed-valence $Co^{II}Co^{III}$ compounds in which the valences are localized.[5]

With suitable ligands large clusters or infinite structures with pores and channels can be constructed, all based on Co^{II} octahedra. For example, the hydrothermal reaction of $CoCl_2$ with squaric acid and KOH at 200° gives $\{[Co_3(\mu_3\text{-}OH)_2(C_4O_4)_2]\cdot 3H_2O\}_n$ which forms a lattice with 7.5 Å channels.[6] The reaction of $Co(NO_3)_2$ with 1,2-bis(4-pyridyl)ethane (bpe) gives $[Co(NO_3)_2(bpe)_{1.5}]_n$, and a coordination polymer which can adopt a number of conformational isomers leading to different types of network structures which include voids of up to 10×10 Å size.[7] On the other hand, the potentially chelating ligand 2-methyl-6-hydroxypyridine (Hmhp) allows the isolation of the cluster $Co_{24}(OH)_{18}$-$(OMe)_2Cl_6(mhp)_{22}$ which, like the squarate complex mentioned above, is based on incomplete Co_3O_4 cubes as building units.[8]

Five coordinate species that are *tbp* or *sp* are phosphine adducts such as $CoBr_2(PMe_3)_3$ or $Co(CN)_2(PMe_2Ph)_3$ that bind O_2. Other types are mainly with polydentate ligands. The geometry varies, some approaching the *tbp* and others the *sp* limiting cases, and many have an intermediate (C_{2v}) arrangement. Interest in these complexes has centered mainly on correlating their electronic structures with molecular symmetry and the atoms constituting the ligand set; these points are mentioned later in connection with electronic structures.

Bridged metal—metal bonded species are becoming more numerous. Although the long known benzoate, $Co_2(O_2CPh)_4L_2$ (L = quinoline), has a very long Co to Co distance and no Co—Co bond, $Co_2(amidinato)_4$ and $Co_2(triazenato)_4$ compounds are well known and do have Co—Co bonds.[9] The complex with PhNC(Ph)NPh bridges can be both oxidized and reduced to stable +1 and −1 ions.[10] There are also $Co_2(amidinato)_3$ species[11] that have a short Co—Co distance (*ca.* 2.39 Å) but a bond order of only 0.5. Oxidation of this species by one electron severs this bond.

In a number of cases with sterically hindered ligands trigonal-planar coordination is observed, for example, in dialkylamides, $Co_2(\mu\text{-}NR_2)_2(NR_2)_2$, with large R. The magnetic moments of these compounds vary considerably from μ_B = 4.83 (R = SiMe_3) to 1.72 (R = Ph) BM. The bulky thiolate $Co_2(\mu\text{-}SAr'')_2(SAr'')_2$ is also three-coordinate,[12] while other thiolates, e.g. $[Co(SPh)_4]^{2-}$ and $[Co_2(\mu\text{-}SPr^i)_3(SPr^i)_2]^-$, are tetrahedral.[13]

[5]K. Wieghardt *et al.*, *J. Chem. Soc., Dalton Trans.* **1990**, 271.
[6]P. T. Wood *et al.*, *Angew. Chem. Int. Ed. Engl.* **1997**, *36*, 991.
[7]M. J. Zaworotko *et al.*, *Angew. Chem. Int. Ed. Engl.* **1997**, 36, 972.
[8]R. E. P. Winpenny *et al.*, *Chem. Commun.* **1997**, 653.
[9]F. A. Cotton *et al.*, *Inorg. Chim. Acta* **1997**, *256*, 291 and references therein.
[10]J. L. Bear *et al.*, *Inorg. Chem.* **1992**, *31*, 620.
[11]F. A. Cotton *et al.*, *Inorg. Chim. Acta* **1997**, *256*, 283, 303.
[12]P. P. Power and S. C. Shoner, *Angew. Chem. Int. Ed. Engl.* **1991**, *30*, 330.
[13]G. Henkel and S. Weissgräber, *Angew. Chem. Int. Ed. Engl.* **1992**, *31*, 1368.

Electronic Structures of Cobalt(II) Compounds

As already noted, cobalt(II) occurs in a great variety of structural environments; because of this the electronic structures, hence the spectral and magnetic properties of the ion, are extremely varied.

High-Spin Octahedral and Tetrahedral Complexes. For qualitative purposes the partial energy level diagram in Fig. 17-F-2 is useful. In each case there is a quartet ground state and three spin-allowed electronic transitions to the excited quartet states. Quantitatively the two cases differ considerably, as might be inferred from the simple observation that octahedral complexes are typically pale red or purple, whereas many common tetrahedral ones are an intense blue. In each case the visible spectrum is dominated by the highest energy transition, $^4A_2 \rightarrow {}^4T_1(P)$ for tetrahedral and $^4T_{1g}(F) \rightarrow {}^4T_{1g}(P)$ for octahedral complexes. However, in the octahedral systems the $^4A_{2g}$ level is usually close to the $^4T_{1g}(P)$ level and the transitions to these two levels are close together. Since the $^4A_{2g}$ state is derived from a $t_{2g}^3 e_g^4$ electron configuration, and the $^4T_{1g}(F)$ ground state is derived mainly from a $t_{2g}^5 e_g^2$ configuration, the $^4T_{1g}(F) \rightarrow {}^4A_{2g}$ transition is essentially a two-electron process; it is therefore weaker by about a factor of 10^{-2} than the other transitions. In the tetrahedral systems, as illustrated in Fig. 17-F-3, the visible transition is generally about an order of magnitude more intense and displaced to lower energies, in accord with the observed colors mentioned previously. For octahedral complexes, there is one more spin-allowed transition $[^4T_{1g}(F) \rightarrow {}^4T_{2g}]$ which generally occurs in the near-ir region. For tetrahedral complexes there is also a transition in the near-ir region $[^4A_2 \rightarrow {}^4T_1(F)]$, as well as one of quite low energy ($^4A_2 \rightarrow {}^4T_2$), which is seldom observed because it is in an inconvenient region of the spectrum (1000–2000 nm) and it is orbitally forbidden. The visible transitions in both cases, but particularly in the tetrahedral case, generally have complex envelopes because a number of transitions to doublet excited states occur in the same region, and these acquire some intensity by means of spin-orbit coupling.

The octahedral and tetrahedral complexes also differ in their magnetic properties. Because of the intrinsic orbital angular momentum in the octahedral ground state, there is consistently a considerable orbital contribution, and effective magnetic

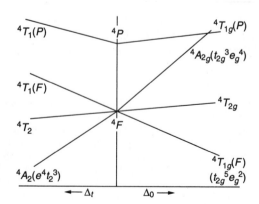

Figure 17-F-2 Schematic energy level diagram for quartet states of a d^7 ion in tetrahedral and octahedral ligand fields.

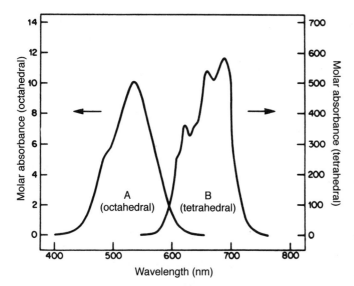

Figure 17-F-3 The visible spectra of $[Co(H_2O)_6]^{2+}$ (curve A) and $[CoCl_4]^{2-}$ (curve B).

moments for such compounds around room temperature are between 4.7 and 5.2 BM. For tetrahedral complexes the ground state acquires angular momentum only indirectly through mixing with the $^4T_2$ state by a spin-orbit coupling perturbation. First-order perturbation theory leads to the expression

$$\mu = 3.89 - 15.59\lambda'/\Delta_t$$

where 3.89 is the spin-only moment for three unpaired electrons and λ' is the effective value of the spin-orbit coupling constant (which is inherently negative). Since λ' varies little from one complex to another, orbital contributions vary inversely with the strength of the ligand field. For example, among the tetrahalo complexes we have: $[CoCl_4]^{2-}$, 4.59 BM; $[CoBr_4]^{2-}$, 4.69 BM; $[CoI_4]^{2-}$, 4.77 BM.

Low-Spin Octahedral Complexes. A sufficiently strong ligand field ($\Delta_o \geq$ 15,000 cm^{-1}) can cause a ^2E state originating in the ^2G state of the free ion to become the ground state. The electron configuration here is mainly $t_{2g}^6e_g$, and a Jahn-Teller distortion would be expected. Consequently octahedral low-spin Co^{II} complexes are rare, tending to lose ligands and form low-spin 4- or 5-coordinate species. An authentic example, however, is a complex of an S$_6$ crown, $[Co(18S6)]^{2+}$, but unlike a comparable Ni^{II} complex there is a tetragonal distortion (equatorial Co—S 2.25 to 2.29 Å, axial Co—S 2.48 Å) and there is a strong resemblance to Jahn-Teller distorted Cu^{2+} complexes structurally and also magnetically. Another example is the $[Co(CNPh)_6]^{2+}$ ion which is axially distorted.

Square Complexes. All these are low-spin with magnetic moments of 2.2 to 2.7 BM at 300 K. Their spectra are complex, and neither magnetic nor spectral properties of such compounds have been treated in detail. There are some data to suggest that the unpaired electron occupies the d_{z^2} orbital, as might be expected.

Five-Coordinate Complexes. Both high-spin (three unpaired electrons) and low-spin (one unpaired electron) configurations are found for both trigonal-bipyramidal and square pyramidal as well as for intermediate configurations. The following configurations are for the four spin-structure combinations:

Spin	D_{3h}	C_{4v}
High	$(e'')^4(e')^2(a'_1)$	$b_2^2 e^3 a_1 b_1$
Low	$(e'')^4(e')^3$	$b_2^2 e^4 a_1$

It appears that the relationships between spin state, geometry, and nature of the ligand atoms are closely interlocked, so that no simple relationship between any two of these factors has been found. However, it does appear that spin state and the nature of the donor atom set are roughly correlated independently of geometry in such a way that the more heavy atom (e.g., P, As, Br, or S) donors (as compared with O and N) are present, the greater is the tendency to spin pairing; this is hardly surprising. For high-spin complexes with fairly regular geometry, for example, $[Co(Me_6 tren)Br]^+$ (C_{3v}) and $[Co(Ph_2 MeAsO)_4(ClO_4)]^+$ (C_{4v}), detailed and reasonably convincing spectral assignments have been made. For irregular geometries and for low-spin complexes there are more uncertainties.

17-F-4 Complexes of Cobalt(III), d^6

The complexes of Co^{III} are exceedingly numerous. Because they generally undergo ligand-exchange reactions relatively slowly, they have, from the days of Werner and Jørgensen, been extensively studied and a large fraction of our knowledge of the isomerism, modes of reaction, and general properties of octahedral complexes as a class is based on studies of Co^{III} complexes. Almost all discrete Co^{III} complexes are octahedral, though a few are tetrahedral, planar, and square antiprismatic.

Cobalt(III) shows a particular affinity for nitrogen donors, and the majority of its complexes contain NH_3, ethylenediamine, NO_2 groups, or N-bonded SCN groups, as well halide ions and water molecules. In general, these complexes are synthesized in several steps beginning with one in which the aqua Co^{II} ion is oxidized in solution, typically by O_2 (see later discussion) or H_2O_2 and often a surface-active catalyst such as activated charcoal, in the presence of the ligands. For example, when a vigorous stream of air is drawn for several hours through a solution of a Co^{II} salt, CoX_2 (X = Cl, Br, or NO_3), containing NH_3, the corresponding ammonium salt and some activated charcoal, good yields of hexammine salts are obtained:

$$4CoX_2 + 4NH_4X + 20NH_3 + O_2 \longrightarrow 4[Co(NH_3)_6]X_3 + 2H_2O$$

In the absence of charcoal, ligand replacement usually occurs to give, for example, $[Co(am)_5Cl]^{2+}$ and $[Co(am)_4(CO_3)]^+$. There is a similar oxidation:

$$4CoCl_2 + 8en + 4en \cdot HCl + O_2 = 4[Co(en)_3]Cl_3 + 2H_2O$$

However, a similar reaction in acid solution with the hydrochloride gives the green salt *trans*-$[Co(en)_2Cl_2][H_5O_2]Cl_2$, which loses HCl on heating. This *trans* isomer may be isomerized to the red racemic *cis* isomer on evaporation of a neutral aqueous solution at 90 to 100°C. Both the *cis* and the *trans* isomers are aquated when heated in water:

$$[Co(en)_2Cl_2]^+ + H_2O \longrightarrow [Co(en)_2Cl(H_2O)]^{2+} + Cl^-$$

$$[Co(en)_2Cl(H_2O)]^{2+} + H_2O \longrightarrow [Co(en)_2(H_2O)_2]^{3+} + Cl^-$$

and on treatment with solutions of other anions are converted into other $[Co(en)_2X_2]^+$ species, for example,

$$[Co(en)_2Cl_2]^+ + 2NCS^- \longrightarrow [Co(en)_2(NCS)_2]^+ + 2Cl^-$$

The stability of $[Co(am)_6]^{3+}$ complexes allows the template construction of tricyclic ligands around the metal ion. An example is the "sarcophagine" complex (17-F-I) in which the metal center is irreversibly encapsulated but retains its redox activity. With suitable substituents, such complexes can act as intercalators in DNA or as photosensitizers.[14]

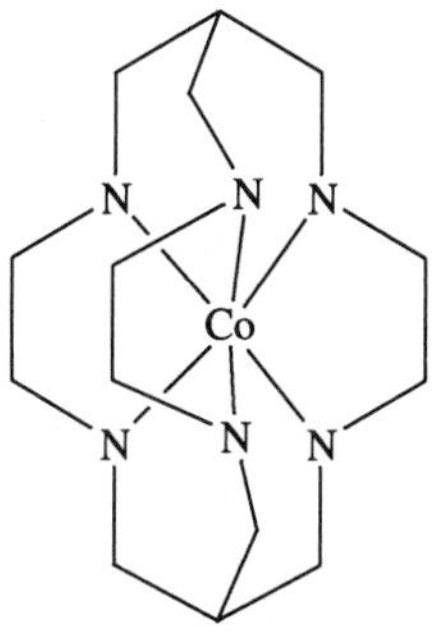

(17-F-I)

These few reactions are illustrative of the very extensive chemistry of Co^{III} complexes with nitrogen-coordinating ligands.

In addition to the numerous mononuclear ammine complexes of Co^{III}, there are a number of polynuclear complexes in which hydroxo (OH$^-$), peroxo (O$_2^{2-}$), amido (NH$_2^-$), and imido (NH^{2-}) groups function as bridges. Some typical complexes of this class are: $[(am)_3Co(\mu\text{-OH})_3Co(am)_3]^{3+}$, $[(am)_3Co(OH)_3Co(OH)_3Co(am)_3]^{3+}$, $[(am)_5Co-O-O-Co(am)_5]^{4+}$, and $[(am)_3Co(NH_2)_3Co(am)_3]^{3+}$ (am = NH$_3$, amine).[15]

Some other Co^{III} complexes of significance are the hexacyano complex $[Co(CN)_6]^{3-}$, the oxygen-coordinated complexes such as carbonates, for example, the green ion $[Co(CO_3)_3]^{3-}$ and *cis*-$[Co(CO_3)_2(py_2)]^-$, Co(acac)$_3$, and salts of the trisoxalatocobalt(III) anion. The oxidation of Co(OAc)$_2$·4H$_2$O can lead to a mixture of species including Co$_2(\mu$-OH)$_2$(OAc)$_4$ and the dark green oxo-centered cluster $[Co_3(\mu_3\text{-O})(OAc)_6(HOAc)_3]^+$,[16] in which each Co ion is octahedral. The pyridine

[14]A. M. Sargeson *et al.*, *Chem. Eur. J.* **1997**, *3*, 1283.
[15]W. Frank *et al.*, *Angew. Chem. Int. Ed. Engl.* **1990**, *29*, 1158.
[16]A. B. Blake *et al.*, *J. Chem. Soc., Dalton Trans.* **1990**, 3719.

adduct $[Co_3(\mu_3\text{-}O)(OAc)_6(py)_3]ClO_4$ contains a planar Co_3ON_3 core.[17] Condensation reactions of Co^{III} hydroxo complexes may also lead to compounds with Co_4O_4 cubane structures, e.g., $[Co_4O_4(OAc)_2(bipy)_4]^{2+}$.[18] Cobalt(III) trifluoroacetate $Co(O_2CCF_3)_3$ oxidizes methane stoichiometrically to $CH_3O_2CCF_3$ (150–180°C, 10–40 bar); the reaction becomes catalytic in the presence of O_2.[19] Cobalt(III) complexes of chelating amines, e.g., *cis*-$[(tren)Co(OH_2)_2]^{3+}$ are highly effective in hydrolyzing carboxylate esters, amides, nitriles, and phosphate esters.[20]

The reaction of cobalt powder with Ph_3PI_2 gives $[PPh_3I][Co^{II}I_3(PPh_3)]$ but Me_3PI_2 leads to $CoI_3(PMe_3)_2$ with a *tbp* structure. Ph_3SbI_2 reacts similarly to give dark green $CoI_3(SbPh_3)_2$, a rare example of a paramagnetic Co^{III} complex.[21]

Electronic Structures of Cobalt(III) Complexes

The free Co^{III} ion (d^6) has qualitatively the same energy level diagram as does Fe^{II}. However, with Co^{III} the $^1A_{1g}$ state originating in one of the high-energy singlet states of the free ion drops very rapidly and crosses the $^5T_{2g}$ state at a very low value of Δ. Thus all known octahedral Co^{III} complexes, including even $[Co(H_2O)_6]^{3+}$ and $[Co(NH_3)_6]^{3+}$, have diamagnetic ground states, except for $[Co(H_2O)_3F_3]$ and $[CoF_6]^{3-}$, which are paramagnetic with four unpaired electrons.

The visible absorption spectra of Co^{III} complexes may be expected to consist of transitions from the $^1A_{1g}$ ground state to other singlet states. The two absorption bands found in the visible spectra of regular octahedral Co^{III} complexes represent transitions to the upper states $^1T_{1g}$ and $^1T_{2g}$. In complexes of the type CoA_4B_2, which can exist in both cis and trans configurations, certain spectral features are diagnostic of the cis or trans configuration (Fig. 17-F-4).

The origin of these features lies in the splitting of the $^1T_{1g}$ state by the environments of lower that O_h symmetry, as also shown diagrammatically in Fig. 17-F-4. Theory shows that splitting of the $^1T_{2g}$ state will always be slight, whereas the $^1T_{1g}$ state will be split markedly in the trans isomer whenever there is a substantial difference in the positions of the ligands, A and B, in the spectrochemical series. Moreouver, because the cis isomer lacks a center of symmetry it may be expected to have a somewhat more intense spectrum than the trans isomer. These predictions are nicely borne out by the spectra of *cis*- and *trans*-$[Co(en)_2F_2]^+$.

17-F-5   The Oxidation of Cobalt(II) Complexes by Molecular Oxygen; Peroxo and Superoxo Species; Oxygen Carriers[22]

The interaction of Co^{II} complexes in solution with O_2 has been the subject of intensive study. Under certain conditions oxidation ultimately to Co^{III} complexes

[17]J. K. Beattie *et al.*, *Polyhedron* **1996**, *15*, 2141.
[18]G. Christou *et al.*, *J. Am. Chem. Soc.* **1993**, *115*, 6432.
[19]I. I. Moiseev *et al.*, *J. Chem. Soc., Chem. Commun.* **1990**, 1049.
[20]J. Chin *et al.*, *Inorg. Chem.* **1994**, *33*, 665; **1995**, *34*, 1094.
[21]C. A. McAuliffe *et al.*, *Angew. Chem. Int. Ed. Engl.* **1992**, *31*, 919; *J. Chem. Soc., Dalton Trans.* **1993**, 1599; *J. Chem. Soc., Chem. Commun.* **1994**, 45.
[22]H. R. Mäcke and A. F. Williams, in: *Photoinduced Electron Transfer* (Eds. M. A. Fox and M. Chanon) Elsevier, Amsterdam, **1988**, Part D, p. 28; T. Gajda *et al.*, *Inorg. Chem.* **1997**, *36*, 1850 and references therein.

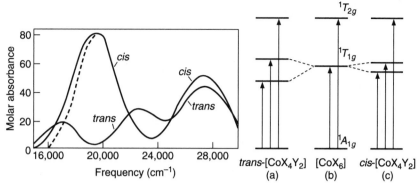

Figure 17-F-4   The visible spectra of *cis*- and *trans*-[Co(en)$_2$F$_2$]$^+$. The broken line shows where the low-frequency side of the $A_{1g} \rightarrow T_{1g}$ band of the cis isomer would be if the band were completely symmetrical. The asymmetry is caused by slight spitting of the $^1T_{1g}$ state. *Right:* diagrammatic representation (not to scale) of the energy levels involved in the transitions responsible for the observed bands of octahedral Co^{II} complexes; (b) the levels for a regular octahedral complex [CoX$_6$]; (a) and (c) the splittings caused by the replacement of two ligands X by two ligands Y.

occurs. However, in the absence of charcoal or other catalysts, or by the choice of suitable ligands, the intermediate peroxo (O$_2^{2-}$) and superoxo (O$_2^-$) species can be isolated. Some of these complexes behave as reversible carriers of O$_2$ and have been much studied because of their potential utility and as models for natural oxygen transport systems.

We first consider the situation for ligands like NH$_3$ or CN$^-$. The first in the sequence of steps may involve the addition of O$_2$ to give a transient mononuclear (superoxo) complex, which then reacts with another Co^{II} species to give a binuclear peroxo-bridged species such as [(am)$_5$Co(μ-O$_2$)(Co(am)$_5$]$^{4+}$ or [(NC)$_5$Co(μ-O$_2$)-Co(CN)$_5$]$^{6-}$. These are isolable as moderately stable solid salts but decompose fairly easily in water or acids. The open-chain complex [(am)$_5$Co(μ-O$_2$)Co(am)$_5$]$^{4+}$ can be cyclized in the presence of base to give (17-F-IIa). It seems safe to assume that all such species, open-chain or cyclic, contain low-spin Co^{III} and bridging peroxide ions; in [(am)$_5$Co(μ-O$_2$)Co(am)$_5$]$^{4+}$ the O—O distance (1.47 Å) is the same as that in H$_2$O$_2$.

$$\left[\begin{array}{c} \text{O}\!-\!\text{O} \\ \text{(am)}_4\text{Co} \qquad \text{Co(am)}_4 \\ \text{N} \\ \text{H}_2 \end{array} \right]^{3+} \underset{+e^-}{\overset{-e^-}{\rightleftharpoons}} \left[\begin{array}{c} \text{O}\!-\!\text{O} \\ \text{(am)}_4\text{Co} \qquad \text{Co(am)}_4 \\ \text{N} \\ \text{H}_2 \end{array} \right]^{4+}$$

(a)   (b)

(17-F-II)

These μ-O$_2$ complexes can often be oxidized in a one-electron step to species such as [(am)$_5$CoO$_2$Co(am)$_5$]$^{5+}$ and (17-F-IIb). These ions were first prepared by A. Werner, who formulated them as peroxo-bridged complexes of Co^{III} and Co^{IV}. The esr data show that the single unpaired electron is distributed equally

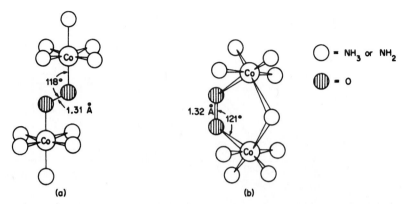

Figure 17-F-5   The structures of (a) $[(NH_3)_5Co(\mu\text{-}O_2)Co(NH_3)_3]^{5+}$ and (b) $[Co_2(NH_3)_4\text{-}(\mu\text{-}O_2)(\mu\text{-}NH_2)]^{4+}$, showing the octahedral coordination about each cobalt ion and the angles and distances at the bridging superoxo groups. The five-membered ring in (b) is essentially planar. Both peroxo and superoxo complexes have these two types of bridged structure.

over both cobalt ions, thus ruling out that description. The problem of formulation has been settled by X-ray structural determination as in Fig. 17-F-5. The O—O distances may be compared with that (1.28 Å) characteristic of the superoxide ion, (O_2^-). The unpaired electron formally belonging to O_2^- resides in a molecular orbital of π symmetry relative to the planar $Co(\mu\text{-}O_2)Co$ groupings and is delocalized over these four atoms. The cobalt atoms are formally described as Co^{III} ions.

In the case of cyanide we have the system

$$2[Co(CN)_5]^{3-} \xrightarrow{\ O_2\ } \{[(CN)_5Co^{III}]_2\mu\text{-}O_2\}^{6-}$$
$$\text{Peroxo, diamagnetic}$$
$$\downarrow {\scriptstyle -e}$$
$$\{[(am)_2Co]_2\mu\text{-}O_2\}^{5+} \xrightarrow{\ CN^-\ } \{[(CN)_5Co]_2\mu\text{-}O_2\}^{5-}$$
$$\text{Superoxo} \qquad\qquad\qquad \text{Superoxo}$$

The best studied oxygen complexes are those of Schiff bases such as $Co^{II}(acacen)$ (17-F-III), which in solution in pyridine, DMF, or similar solvents will pick up O_2:

$$+ DMF + O_2 \longrightarrow Co(acacen)(DMF)O_2$$
$$K = 21 \text{ at } -10°C$$

(17-F-III)

The oxygenation reactions are usually reversible only at low temperatures, since otherwise the complex is either irreversibly oxidized or bridged peroxo or superoxo species are formed:

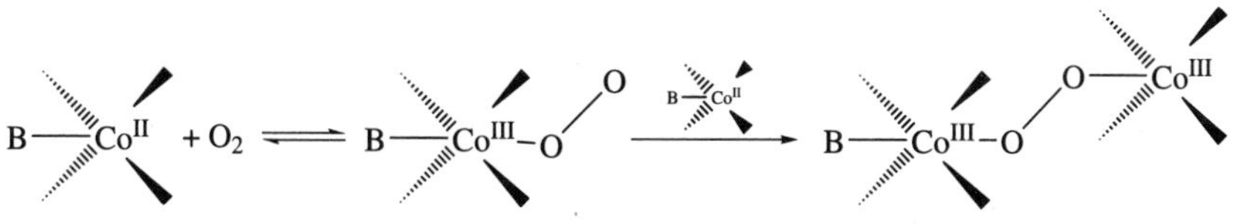

Dimerization can be avoided at low temperatures in dilute solutions or by choice of appropriate ligands and axial bases B, and some monomeric species stable at room temperature are known. X-ray structures confirm that O_2 bound to Co^{III} is bent and best regarded as the superoxide ion, O_2^-.

The $O-O$ distances so far determined seem to indicate that they are dependent on all electronic factors present in the complexes. Some are shorter than the $O-O$ distance in O_2^- (1.28 Å) and some longer. The esr spectra show that in the O_2 adducts the unpaired electron is largely located on the oxygen atoms.

The brown paramagnetic tris(pyrazolato)borate complex $(Tp^*)Co(O_2)$ is formed on exposure of $(Tp^*)Co^I(CO)$ to O_2 and exists in equilibrium with an unusual green, diamagnetic bis-O_2 dimer $(O-O\ 1.35\ Å)$.[23]

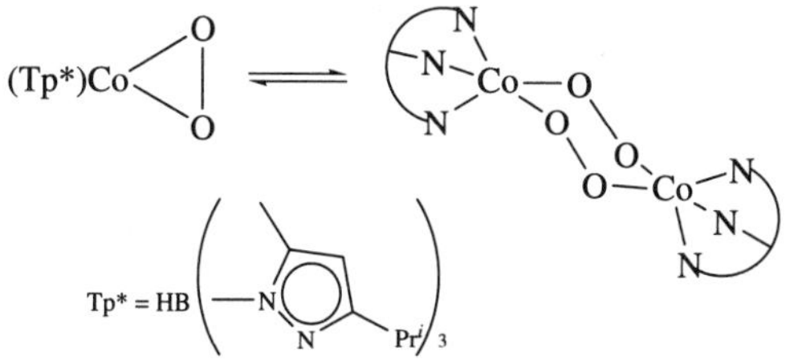

Several Co complexes with chelating nitrogen ligands catalyze oxidation reactions, e.g. $[Co^{II}(bipy)_2]^{2+}$ activates O_2 and oxidizes N-methyl-anilines, benzyl alcohols, and aldehydes. In the absence of organic substrates O_2 is reduced to H_2O_2.[24] Cobalt(II) complexes of tetraaza macrocyclic ligands (N_4) reversibly form O_2 adducts $[(N_4)CoOO]^{2+}$ which are rapidly reduced to $[(N_4)CoOOH]^{2+}$; these species are involved in the electro-reduction of O_2 to H_2O_2.[25]

Porphyrin cobalt complexes, too, activate O_2. The confacial diporphyrin (17-F-IV) is unusual since it does not bind O_2 in the $[PCo^{II}Co^{II}P]$ state but gives remarkably strong O_2 adducts on oxidation to $[PCo^{II}Co^{III}P]^+$ and $[PCo^{III}Co^{III}P]^{2+}$.[26]

[23]K. H. Theopold *et al.*, *Angew. Chem. Int. Ed. Engl.* **1995**, *34*, 2051.
[24]A. Sobkowiak and D. T. Sawyer, *J. Am. Chem. Soc.* **1991**, *113*, 9520.
[25]C. Kang and F. C. Anson, *Inorg. Chem.* **1995**, *34*, 2771.
[26]Y. Le Mest and M. L'Her, *J. Chem. Soc., Chem. Commun.* **1995**, 1441.

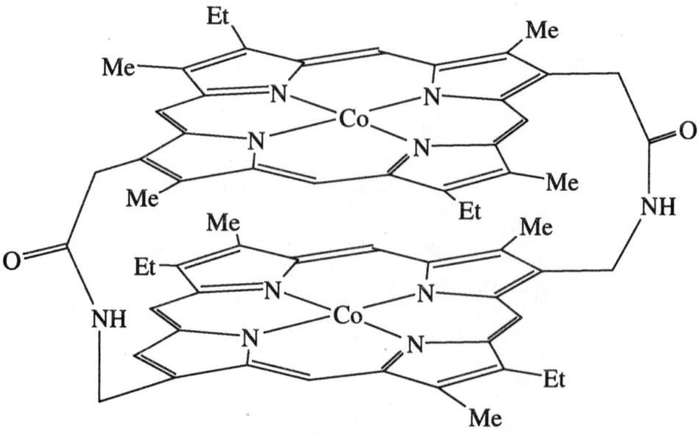

(17-F-IV)

Polydentate Co^{II} complexes with ligands capable of intercalation into DNA strands (17-F-V) are capable of inducing DNA cleavage under photochemical conditions.[27]

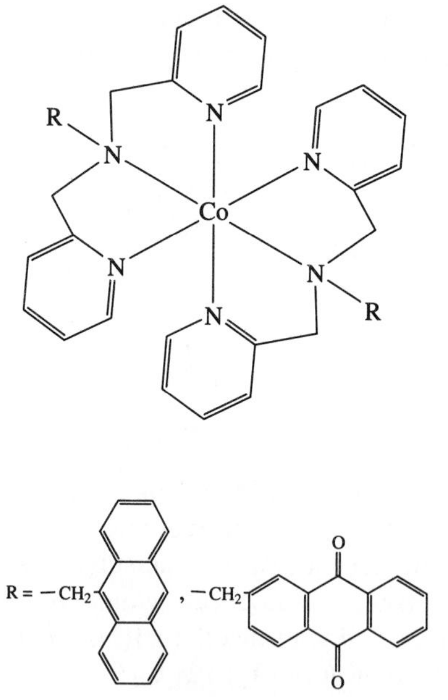

(17-F-V)

The binuclear Cu/Co compound (17-F-VI) acts as a functional model for cytochrome *c* oxidase. The natural enzyme contains a copper atom coordinated to

[27]S. Bhattacharya and S. S. Mandal, *Chem. Commun.* **1996**, 1515.

histidine residues and heme-Fe, in place of the Co in (17-F-VI); it catalyzes the four-electron reduction of O_2 to water, an exergonic process that drives adenosine triphosphate (ATP) formation. Whereas the Cu-free cobalt complex forms an O_2 adduct reversibly, the Cu/Co compound gives a very stable $1:1$ oxygen adduct, probably a peroxide bridging both metal centers. Attached to a graphite electrode, (17-F-VI) catalyzes the reduction of O_2 to H_2O.[28]

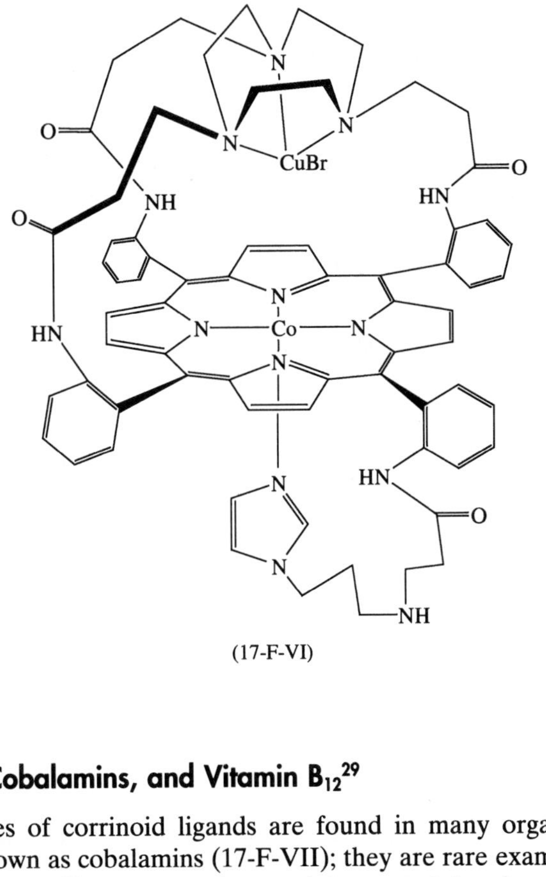

(17-F-VI)

Cobaloximes, Cobalamins, and Vitamin B$_{12}$[29]

Cobalt complexes of corrinoid ligands are found in many organisms, including man, and are known as cobalamins (17-F-VII); they are rare examples of naturally occurring organometallic compounds containing a metal-carbon σ-bond. Several cobalamins are biologically active; R = adenosyl (AdoCbl), R = OH (OH-Cbl), and R = CH$_3$ (MeCbl). The cyano derivative (R = CN) is known as vitamin B$_{12}$, AdoCbl as co-enzyme B$_{12}$. CN—Cbl is converted into AdoCbl and MeCbl under physiological conditions. The human body contains *ca.* 5 mg of cobalamins; their lack causes the disease pernicious anaemia.

[28]J. P. Collman *et al., Science* **1997**, *275*, 949.
[29]B. Kräutler and C. Kratky, *Angew. Chem. Int. Ed. Engl.* **1996**, *35*, 167.

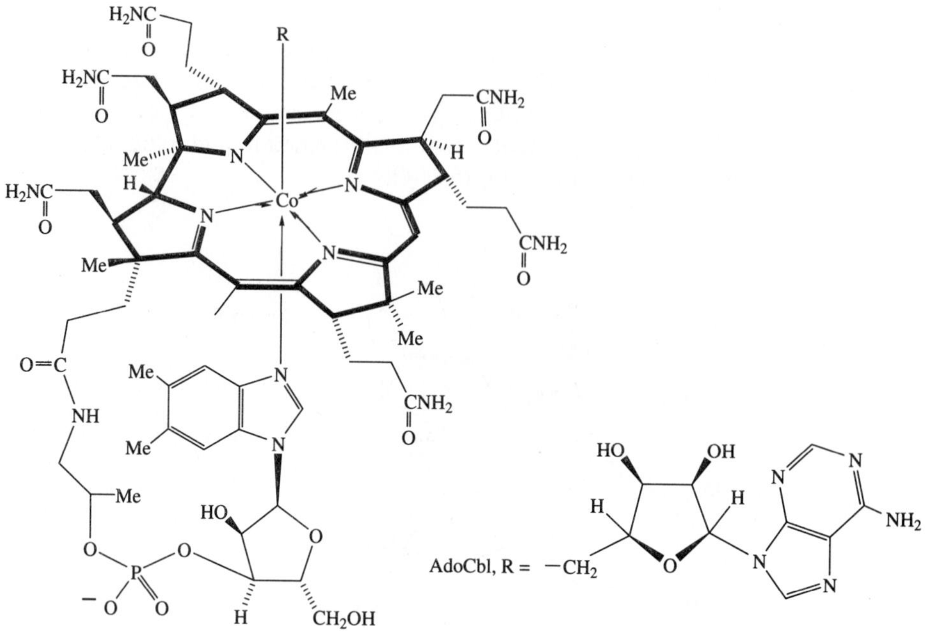

(17-F-VII)

Vitamin B_{12} is required in humans for several transformations, such as the AdoCbl-dependent conversion of (*R*)-methylmalonyl co-enzyme A (CoA) into succinyl-CoA:

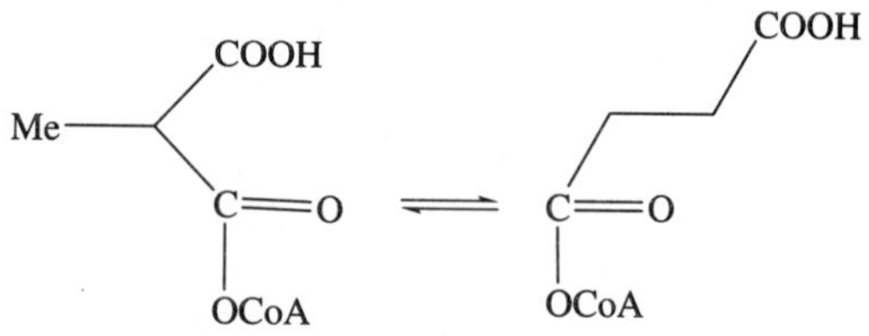

and the MeCbl-dependent conversion of (*S*)-homocysteine into methionine:

$$HSCH_2CH_2CH(NH_3^+)COO^- \longrightarrow MeSCH_2CH_2CH(NH_3^+)COO^-$$

There are a number of related reactions in which a substrate >CH—CX< is rearranged to >CX—CH<. The mechanistic details are obscure but may involve radicals derived from the homolysis of the Co—R bond.[30]

As cobalt(III) complexes (d^6), cobalamins strictly adhere to an octahedral geometry, with an axial benzimidazole nitrogen ligand. It appears that in protein-bound B_{12} this axial ligand is replaced by a histidine provided by the peptide chain, with consequent configurational changes that influence the reactivity of the

[30]B. T. Golding, *Chem. Brit.* **1990**, 950.

cobalamin unit.[27] Cobalamins have a pronounced redox chemistry:

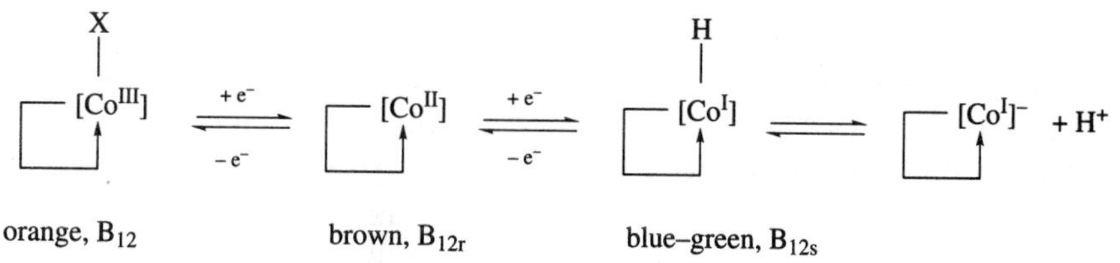

orange, B_{12} brown, B_{12r} blue–green, B_{12s}

Cobaloximes and Related Compounds

Since the recognition of Vitamin B_{12} and the existence in Nature of Co–C bonds, there has been much study of "model" systems. These have been mainly complexes of dimethyl and other glyoximes (the so-called cobaloximes), and Schiff base and macrocyclic ligand complexes. The methyl complex $[Co(CH_3)(NH_3)_5]^{2+}$ may be regarded as the simplest model; it has electronic characteristics very similar to those of $[Co(NH_3)_6]^{3+}$.[31] The main interest has centered on the reduction of Co^I species and on the formation of Co–C bonds and their reactions.

Reduction of cobaloximes, like the reduction of B_{12}, gives blue or green Co^I species that are very powerful reducing agents and nucleophiles. Thus they may react with water to give hydridocobalt(III) species, especially if tertiary phosphine ligands are also present.

Reduced Vitamin B_{12} itself will reduce ClO_3^- to Cl^- and, at pH 1.5 to 2.5, NO_3^- to NH_4^+. The latter is an overall $8e$ reduction and probably initially proceeds by a hydride transfer

$$Co^I-H + NO_3^- \longrightarrow NO_2^- + Co^{III} + OH^-$$

and overall

$$8Co^I + NO_3^- + 10H^+ = 8Co^{II} + NH_4^+ + 3H_2O$$

In the Schiff base complexes such as (17-F-VIII) the strength of the bonding of the trans ligand is determined mainly by the inductive effects of the R groups, although there is some evidence that the nature of the cis ligand atoms is important. The reactivity of the Co–C bond and its ease of photolysis also are powerfully affected by the nature of the trans ligand.

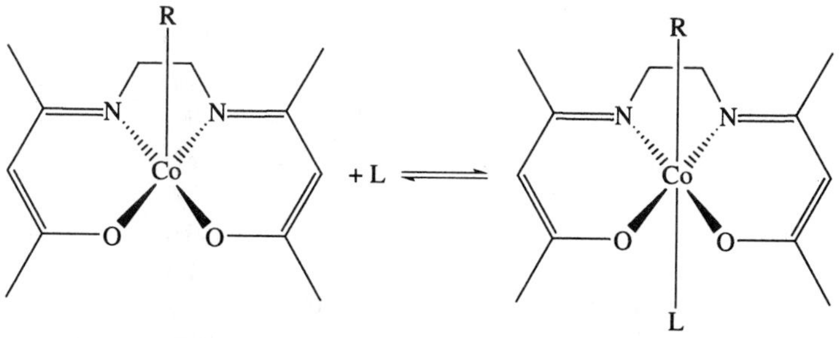

(17-F-VIII)

[31]P. Kofod, *Inorg. Chem.* **1995**, *34*, 2768.

Alkylcobalt(III) cobaloximes can also undergo a one-electron oxidation to give organocobalt(IV) radical ions, stable at low temperatures ($< -50°$). The C atom bound to the metal is very susceptible to nucleophilic attack.

Reduction of Co(dmg)$_2$(R)(L) gives the "supernucleophilic" cobaloxime(I) which undergoes oxidative addition with alkyl halides to give Co^{III} alkyls. The [Co(dmgBF$_2$)$_2$(py)]$^-$anion is square-pyramidal, with very short Co—N bonds.[32]

17-F-6 Cobalt(IV), d^5 and Cobalt(V), d^4

Fusing NaO$_2$ and Co$_3$O$_4$ gives Na$_4$CoO$_4$, which has tetrahedral Co^{IV}O$_4$ anions with Na$^+$ coordinated to the oxygen atoms. The compound Ba$_2$CoO$_4$, a red-brown substance obtained by oxidation of 2Ba(OH)$_2$ and 2Co(OH)$_2$ at 1050°C, and a hetero-polymolybdate of Co^{IV}, namely, 3K$_2$O·CoO$_2$·9MoO$_3$·6H$_2$O, have been reported. The action of oxidizing agents (e.g., Cl$_2$, O$_2$, or O$_3$) on strongly alkaline Co^{II} solutions produces a black material believed to be hydrous CoO$_2$, at least in part, but it is ill characterized. The black mixed-valence compound Sr$_6$Co$_5$O$_{15}$ contains infinite chains of face-sharing CoO$_6$ octahedra.[33]

Fluorination of Cs$_2$CoCl$_4$ gives Cs$_2$CoF$_6$ which is isomorphous with Cs$_2$SiF$_6$; its magnetic moment rises from 2.46 BM at 90 K to 2.97 BM at 294 K. The reflectance spectrum has been assigned to an octahedrally coordinated t_{2g}^5 ion. The high magnetic moments could be due to a large orbital contribution in the $^2T_{2g}$ ground state or to partial population of a $^6A_{1g}$ ($t_{2g}^3e_g^2$) state.

Oxidation of Co^{III} dithiocarbamates gives [Co(dtc)$_3$]$^+$ complexes which are low-spin d^5 with magnetic moments between 2.2 and 2.7 BM. For the macrocyclic complex (17-F-IX) (formally Co^{IV}) square geometry was confirmed by X-ray diffraction; the compound has one unpaired electron and slowly oxidizes water, with reduction to H[Co^{III}(L)].[34]

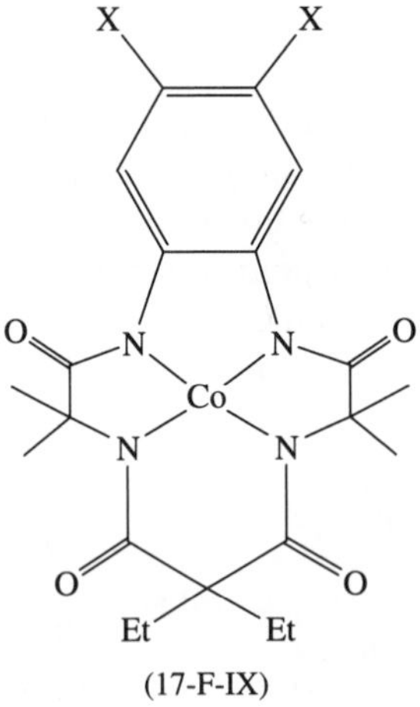

(17-F-IX)

A well-characterized Co^{IV} compound is the remarkable alkyl complex tetrakis-(1-norbornyl)cobalt, which is made from CoCl$_2$ and lithium norbornyl (nor). The

[32]J. H. Espenson *et al., Inorg. Chem.* **1991**, *30*, 3407.
[33]W. T. A. Harrison *et al., J. Chem. Soc., Chem. Commun.* **1995**, 1953.
[34]T. J. Collins *et al., J. Am. Chem. Soc.* **1991**, *113*, 8419.

Table 17-F-2 Phosphine or Phospite Complexes of Cobalt

Co^{-I}	Co^0	Co^I	Co^{II}	Co^{III}
K(N$_2$)Co(PMe$_3$)$_3$^{a}	Co(PMe$_3$)$_4$	CoCl(PR$_3$)$_3$	CoCl$_2$(PR$_3$)$_3$	CoH$_3$(PPh$_3$)$_3$
M[Co(PMe$_3$)$_4$]	Co$_2$(P(OMe)$_3$)$_8$	CoCl(CO)$_2$(PR$_3$)$_2$	[CoH{P(OPh)$_3$}$_4$]$^+$	CoMe$_3$(PMe$_3$)$_3$
(M = Li, Na, K)		CoH(N$_2$)(PPh$_3$)$_3$		[CoH$_2${P(OR)$_3$}$_4$]$^+$
		CoH(PMe$_3$)$_3$		
		CoH(CO)(PPh$_3$)$_3$		
		[Co(CO)$_2$(PR$_3$)$_3$]$^+$		
		[Co{P(OEt)$_3$}$_5$]$^+$		
		[Co(C$_2$H$_4$)(MeCN)(PMe$_3$)$_2$]$^+$		

^{a}H.-F. Klein et al., Organometallics **1987**, 6, 1341.

brown compound is paramagnetic, low spin tetrahedral and reasonably stable to air and heat. It is oxidized by AgBF$_4$ to the diamagnetic tetrahedral Co^V ion [Co(nor)$_4$]$^+$ and can be reduced to [Co^{III}(nor)$_4$]$^-$.

17-F-7 Complexes of Cobalt(-I), (I), (II), and (III) with Phosphorus and Related Ligands

There is an extensive chemistry of cobalt in which phosphines or phosphites are coordinated alone or with other ligands. Examples are given in Table 17-F-2. The Co^I compounds are typically made by reductions such as

$$CoCl_2 + 4PMe_3 \xrightarrow{Na/Hg} Co(PMe_3)_4 + 2NaCl$$

$$CoCl_2 + 3PR_3 \xrightarrow{Zn/MeCN} CoCl(PR_3)_3 + ZnCl_2$$

$$Co(acac)_3 + N_2 \xrightarrow[PR_3]{AlEt_3} CoH(N_2)(PR_3)_3$$

Disproportionation of Co$^{II} \to$ Co^I + Co^{III} is possible with certain ligands, especially phosphites, for example,

$$2Co^{2+} + 11PR_3 \longrightarrow [Co(PR_3)_5]^+ + [Co(PR_3)_6]^{3+}$$

Hydrido species such as HCo[P(OPh)$_3$]$_3$(MeCN) catalyze the homogeneous hydrogenation of unsaturated substances. In the crystal HCo(CO)$_2$(PPh$_3$)$_2$ can adopt both *sp* and *tbp* structures.[35] The ion [Co(bipy)$_3$]$^+$ reduces water to H$_2$ and CO$_2$ to CO; the complex can be generated photochemically from [Co(bipy)$_3$]$^{n+}$ precursors.[36]

Phosphine derivatives obtained from Co$_2$(CO)$_8$ invariably have CO groups present also, but these are either dimeric Co^0 species or Co^I and Co^{-I} species formed by disproportionation:

$$Co_2(CO)_8 + 2PPh_3 \xrightarrow{0°, benzene} Co_2(CO)_6(PPh_3)_2 + 2CO$$

$$Co_2(CO)_8 + 2PR_3 \xrightarrow{polar solvent} [Co(CO)_3(PR_3)_2][Co(CO)_4] + CO$$

[35]D. Zhao and L. Brammer, *Inorg. Chem.* **1994**, 33, 5897.
[36]D. A. Reitsma and F. R. Keene, *Organometallics* **1994**, 13, 1351.

Examples of other paramagnetic Co^0 species are the maleic anhydride (L) complex $CoL[P(OMe)_3]_3$, $[(triphos)Co]_2(\mu\text{-}N_2)$, and $[(Me_3P)_3Co]_2(\eta\text{-norborna-}$ diene), where both double bonds of norbornadiene are coordinated and there is strong antiferromagnetic coupling of the d^9 centers, and $Co[P(OPr^i)_3]_4$. For the phosphites, $P(OMe)_3$ and $P(OEt)_3$, there are both paramagnetic monomers and diamagnetic dimers, presumably because steric hindrance is less than for $P(OPr^i)_3$. The isopropyl phosphite also gives the compound $Na[Co[P(OPr^i)_3]_5]$, which has strong ion pairs with Na^+ bound to the O atom of the phosphite. Co-condensation of Co atoms with $Bu^tCH{=}CH{-}CH{=}CHBu^t$ gives the paramagnetic Co^0 butadiene complex $Co(C_4H_4Bu_2^i)_2$ which can be reduced to $[Co^{-1}(C_4H_4Bu_2^i)_2]^{-}$.[37]

Several Co^{III} complexes are known with H and CH_3 ligands. The former can be made by oxidative-addition reactions:

$$CoH(N_2)(PPh_3)_3 + H_2 = CoH_3(PPh_3)_3 + N_2$$

$$CoH[P(OR)_3]_4 + HX = CoH_2[P(OR)_3]_4^+ + X^-$$

The Co^{III} methyl complex $[CoMe(NH_3)_5]^{2+}$ is obtained by reaction of $[Co(NH_3)_6]^{2+}$ with methylhydrazine. There is a series of stable dialkyls $cis\text{-}[R_2Co(bipy)_2]^+$ (R = Me, Et, CH_2Ph, etc.).

Isocyanide complexes may be formed either by reduction of $CoCl_2(CNR)_4$ with an active metal, N_2H_4 or $S_2O_4^{2-}$, or by interaction of RNC with $Co_2(CO)_8$, which leads to disproportionation:

$$Co_2(CO)_8 + 5RNC = [Co(CNR)_5]^+[Co(CO)_4]^- + 4CO$$

In $[Co(NCMe)_5]ClO_4$ the cation has *tbp* geometry, but in $[Co(NCPh)_5]ClO_4$ it is *sp*. Anionic isocyanide complexes are accessible by reduction of cobaltocene:[38]

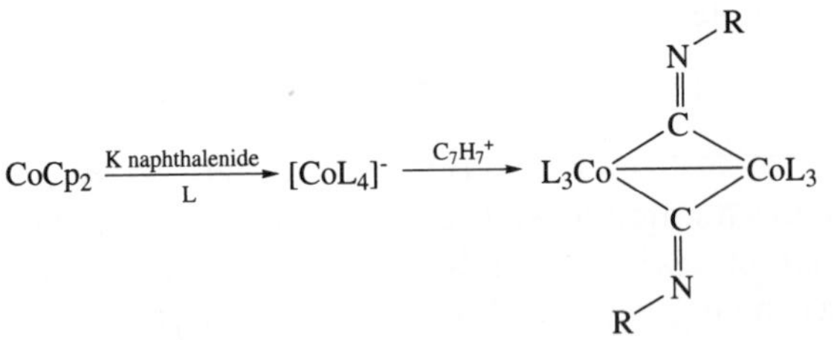

There is an extensive Co^I chemisty of *monocyclopentadienyl* compounds, of which $CpCo(CO)_2$ is one example. The CO may be replaced by phosphines, alkenes, isocyanides, and so on. The dicarbonyl itself can be reduced to a radical ion that contains cobalt in the formal oxidation state 0.5, and can be isolated as its $(Ph_3P)_2N^+$ salt:

[37]F. G. N. Cloke *et al., J. Chem. Soc. Chem. Commun.* **1993**, 248.
[38]N. J. Cooper *et al., J. Am. Chem. Soc.* **1994**, *116*, 8566.

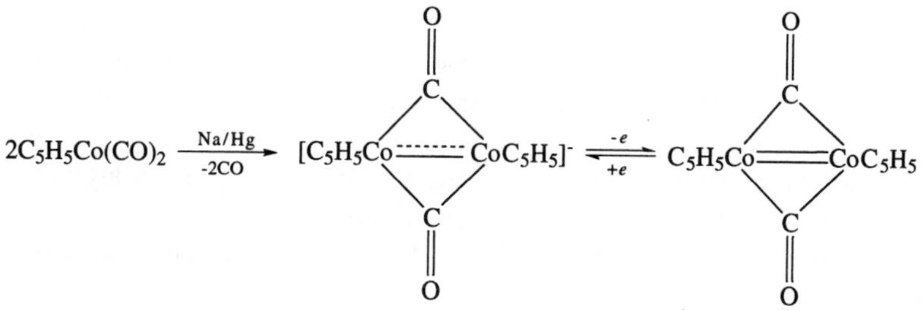

$$2C_5H_5Co(CO)_2 \xrightarrow[-2CO]{Na/Hg} [C_5H_5Co\text{------}CoC_5H_5]^- \underset{+e}{\overset{-e}{\rightleftharpoons}} C_5H_5Co\text{====}CoC_5H_5$$

The Co^I arene complex $[Co(C_6Me_6)_2]^+$ has a sandwich structure, with two η^6-bound arene ligands. The complex is paramagnetic, with two unpaired electrons.[39]

Alkyl cobalt(I) carbonyls, generally made by the reactions

$$NaCo(CO)_4 + RI = RCo(CO)_4 + NaI$$

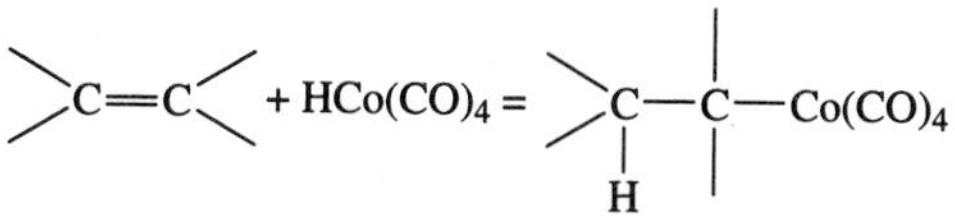

$$\underset{}{>}C{=}C{<} + HCo(CO)_4 = {>}\underset{\underset{H}{|}}{C}{-}\underset{|}{C}{-}Co(CO)_4$$

are important intermediates in catalytic reactions (Chapter 21).

The hydride $HCo(CO)_4$ formed by treatment of $Co_2(CO)_8$ with H_2 under pressure, or by protonation of $[Co(CO)_4]^-$, is a key intermediate in many catalytic reactions.

In addition to the numerous complexes with π-donor ligands, there are some homoleptic Co^{II} alkyls and aryls which show electron-deficient alkyl bridge bonding reminiscent of alkyls of magnesium or aluminum. Examples are the tetrahedral $Co[(\mu\text{-}R)_2Li(TMEDA)]_2$ $(R = CH_2SiMe_3)$[40] and the trigonal planar $[Co(Mes)_2]^-$ and $Co_2(\mu\text{-Mes})_2(Mes)_2$.[41] If $R = 2,4,6\text{-}C_6H_2(CF_3)_3$ a blue, volatile and apparently two-coordinate complex CoR_2 is obtained. There is also a similar nickel complex.[42]

Additional Reference

Oxygen Complexes and Oxygen Activation by Transition Metals, Eds. A. E. Martell and D. T. Sawyer, Plenum Press, New York, 1988.

17-G   NICKEL: GROUP 10

The trend toward decreased stability of higher oxidation states continues with nickel, so that only Ni^{II} occurs in the ordinary chemistry of the element. However, there is a complex array of stereochemistries associated with this species. The higher

[39]D. O'Hare *et al.*, *J. Chem. Soc., Dalton Trans.* **1993**, 3071.
[40]G. Wilkinson *et al.*, *Polyhedron* **1990**, *9*, 931.
[41]K. H. Theopold *et al.*, *Organometallics* **1989**, *8*, 2001.
[42]M. Belay and F. T. Edelmann, *J. Organomet. Chem.* **1994**, *479*, C21.

oxidation states are relatively rare but are gaining in importance. Both Ni^I and Ni^{II} species are being studied increasingly because of the possible involvement of these oxidation states in nickel-containing metalloenzymes. Zerovalent nickel compounds, in particular phosphine complexes, are important catalyst precursors.

Nickel has an enormous, and important, organometallic and catalytic chemistry, and some of the these aspects are noted in Chapter 21.

The oxidation states and stereochemistry of nickel are summarized in Table 17-G-1.

17-G-1   The Element

Nickel occurs in Nature mainly in combination with arsenic, antimony, and sulfur, for example, as *millerite* (NiS), as a red nickel ore that is mainly NiAs, and in deposits consisting chiefly of NiSb, $NiAs_2$, NiAsS, or NiSbS. The most important deposits commercially are *garnierite*, a magnesium-nickel silicate of variable composition, and certain varieties of the iron mineral *pyrrhotite* (Fe_nS_{n+1}), which contain 3 to 5% Ni. Elemental nickel is also found alloyed with iron in many meteors, and the central regions of the earth are believed to contain considerable quantities. The metallurgy of nickel is complicated in its details, many of which vary a good deal with the particular ore being processed. In general, the ore is transformed to Ni_2S_3, which is roasted in air to give NiO, and this is reduced with carbon to give the metal. Some high-purity nickel is made by the *carbonyl process*: carbon monoxide reacts with impure nickel at 50°C and ordinary pressure or with nickel—copper matte under more strenuous conditions, giving volatile $Ni(CO)_4$, from which metal of 99.9 to 99.99% purity is obtained on thermal decomposition at 200°C.

Table 17-G-1   Oxidation States and Stereochemistry of Nickel

Oxidation state	Coordination number	Geometry	Examples
Ni^0	2	Linear	Ni(1,3-diarylimidazolylidene)$_2$
	3	?	Ni[P(OC$_6$H$_4$-O-Me)$_3$]$_3$
	4	Tetrahedral	Ni(PF$_3$)$_4$, [Ni(CN)$_4$]$^{4-}$, Ni(CO)$_4$
	5	?	NiH[P(OEt)$_3$]$_4^+$
Ni^I, d^9	4	Tetrahedral	Ni(PPh$_3$)$_3$Br
Ni^{II}, d^8	3	Trigonal planar	[Ni(NPh$_2$)$_3$]$^-$, Ni$_2$(NR$_2$)$_4$
	4^a	Square	NiBr$_2$(PEt$_3$)$_2$, [Ni(CN)$_4$]$^{2-}$, Ni(DMGH)$_2$
	4^a	Tetrahedral	[NiCl$_4$]$^{2-}$, NiCl$_2$(PPh$_3$)$_2$
	5	*sp*	[Ni(CN)$_5$]$^{3-}$, BaNiS, [Ni$_2$Cl$_8$]$^{4-}$
	5	*tbp*	[NiX(trident]$^+$, [Ni(CN)$_5$]$^{3-}$, Ni(SiCl$_3$)$_2$(CO)$_3$
	6^a	Octahedral	NiO, [Ni(NCS)$_6$]$^{4-}$, KNiF$_3$, [Ni(NH$_3$)$_6$]$^{2+}$, [Ni(bipy)$_3$]$^{2+}$
	6	Trigonal prism	NiAs
Ni^{III}, d^7	5	*tbp*	NiBr$_3$(PR$_3$)$_2$
	6	Octahedral (distorted)	[Ni(diars)$_2$Cl$_2$]$^+$, [NiF$_6$]$^{3-}$
Ni^{IV}, d^6	6	Octahedral (distorted)	K$_2$NiF$_6$, {Ni[Bu$_2$(dtc)]$_3$}$^+$, {Ni[Se$_2$C$_2$(CN)$_2$]$_3$}$^{2-}$

^{a}Most common states.

Nickel is silver-white, with high electrical and thermal conductivities (both ~15% of those of silver) and mp 1452°C, and it can be drawn, rolled, forged, and polished. It is quite resistant to attack by air or water at ordinary temperatures when compact and is therefore often electroplated as a protective coating. Because nickel reacts but slowly with fluorine, the metal and certain alloys (Monel) are used to handle F_2 and other corrosive fluorides. It is also ferromagnetic, but not as much as iron.

The metal is moderately electropositive:

$$Ni^{2+} + 2e = Ni \qquad E^0 = -0.24 \text{ V}$$

and dissolves readily in dilute mineral acids. Like iron, it does not dissolve in concentrated nitric acid because it is rendered passive by this reagent. Nickel-based hydrogenation catalysts are obtained by treating a nickel-aluminum alloy with aqueous NaOH to leach out the aluminum under an inert gas atmosphere. The finely divided nickel particles generated by this process are known as *Raney nickel*; they are pyrophoric in air. Raney nickel is used on an industrial scale, mainly as a hydrogenation catalyst.

NICKEL COMPOUNDS

17-G-2 The Chemistry of Nickel(II), d^8

Binary Compounds

Nickel(II) oxide, a green solid with the rock salt structure, is formed when the hydroxide, carbonate, oxalate, or nitrate of nickel(II) is heated. It is insoluble in water but dissolves readily in acids.

The *hydroxide Ni(OH)$_2$* may be precipitated from aqueous solutions of Ni^{II} salts on addition of alkali metal hydroxides, forming a voluminous green gel that crystallizes [Mg(OH)$_2$ structure] on prolonged storage. It is readily soluble in acid and also in aqueous ammonia owing to the formation of ammine complexes. Ni(OH)$_2$ is not amphoteric.

Addition of sulfide ions to aqueous solutions of Ni^{2+} precipitates black NiS. This is initially freely soluble in acid, but like CoS, on exposure to air it soon becomes insoluble owing to conversion to Ni(OH)S. Fusion of Ni, S, and BaS gives BaNiS$_2$, which forms black plates; this product is metallic and has Ni in square pyramidal coordination.

All four nickel *halides* are known in the anhydrous state. Except for the fluoride, which is best made indirectly, they can be prepared by direct reaction of the elements. All the halides are soluble in water (the fluoride only moderately so), and from aqueous solutions they can be crystallized as the hexahydrates, except for the fluoride, which gives NiF$_2$·3H$_2$O. Solid NiCl$_2$·6H$_2$O has the structure *trans*-[NiCl$_2$(H$_2$O)$_4$]·2H$_2$O but gives the hexaquo ion in solution. On heating in vacuum, anhydrous NiCl$_2$ is formed as a yellow solid. Addition of excess Cl$^-$ to nickel chloride solutions gives the NiCl$_4^{2-}$ ion which, like the tetrahalo ions of other divalent metals, is tetrahedral.

On addition of CN ions to aqueous Ni^{II} the *cyanide* is precipitated in a green hydrated form. When heated at 180 to 200°C the hydrate is converted into the yellow-brown, anhydrous Ni(CN)$_2$. The green precipitate readily redissolves in an excess of cyanide to form the yellow $[Ni(CN)_4]^{2-}$ ion, which is both thermodynamically very stable (log $\text{ß}_4 \approx 30.5$) and kinetically slow to release CN^- ion. Many hydrated salts of this ion, for example, $Na_2[Ni(CN)_4]\cdot3H_2O$, may be crystallized from such solutions. In strong cyanide solutions a further CN^- is taken up to give the red $[Ni(CN)_5]^{3-}$ ion which is known in both *tbp* and *sp* geometry; the latter is preferred. Nickel(II) thiocyanate is also known, as a yellow-brown hydrated solid that reacts with an excess of SCN^- to form the complex ions $[Ni(NCS)_4]^{2-}$ and $[Ni(NCS)_6]^{4-}$.

Other binary nickel compounds, probably all containing Ni^{II} but not all stoichiometric, may be obtained by the direct reaction of nickel with various nonmetals such as P, As, Sb, S, Se, Te, C, and B. Nickel appears to form a nitride Ni_3N. The oxide K_2NiO_2 contains linear $[O-Ni^{II}-O]^{2-}$ units.[1]

Salts of Oxo Acids

A large number of these are known. They occur most commonly as hydrates, for example, $Ni(NO_3)_2\cdot6H_2O$, $NiSO_4\cdot7H_2O$, and most of them are soluble in water. Exceptions are the carbonate $NiCO_3\cdot6H_2O$, which is precipitated on addition of alkali hydrogen carbonates to solutions of Ni^{II}, and the phosphate $Ni_3(PO_4)_2\cdot nH_2O$. Aqueous solutions of Ni^{II} not containing strong complexing agents contain the green hexaaquanickel(II) ion, $[Ni(H_2O)_6]^{2+}$.

17-G-3   Stereochemistry and Electronic Structures of Nickel(II) Complexes

Nickel(II) forms a large number of complexes with coordination numbers 3 to 6 (Table 17-G-1). Softer ligands, such as P and S ligands, generally give 4-coordinate species, with a strong preference for square-planar. Nickel does, however, have a tendency to add a further ligand to give 5-coordinate compounds; it differs in this respect from its homologues Pd and Pt. Consequently, ligand exchange processes for Ni^{II} tend to be associative, while with Pd and Pt dissociative pathways predominate.

A particular characteristic is the existence of complicated equilibria, commonly temperature and concentration dependent, involving different structural types.

Octahedral Complexes

The maximum coordination number of nickel(II) is 6. A considerable number of neutral ligands, especially amines, displace some or all of the water molecules in the octahedral $[Ni(H_2O)_6]^{2+}$ ion to form complexes such as *trans*-$[Ni(H_2O)_2$-$(NH_3)_4](NO_3)_2$, $[Ni(NH_3)_6](ClO_4)_2$, and $[Ni(en)_3]SO_4$. These complexes are characteristically blue or purple in contrast to the bright green of the hexaaquanickel ion. This is because of shifts in the absorption bands when H_2O ligands are replaced by others lying toward the stronger end of the spectrochemical series. This can be seen in Fig. 17-G-1, where the spectra of $[Ni(H_2O)_6]^{2+}$ and $[Ni(en)_3]^{2+}$ are shown. Three spin-allowed transitions are expected from the energy level diagram for d^8 ions and the three observed bands in each spectrum may thus be assigned as shown in Table 17-G-2. It is a characteristic feature of the spectra of octahedral nickel(II)

[1]R. Hoppe *et al., Inorg. Chem.* **1988,** *27,* 2506.

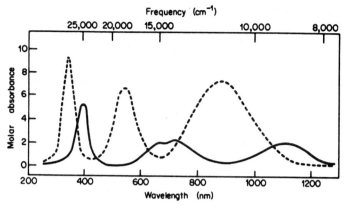

Figure 17-G-1 Absorption spectra of $[Ni(H_2O)_6]^{2+}$ (solid curve) and $[Ni(en)_3]^{2+}$ (dashed curve).

complexes, exemplified by those of $[Ni(H_2O)_6]^{2+}$ and $[Ni(en)_3]^{2+}$, that molar absorbances of the bands are at the low end of the range (1-100) for octahedral complexes of the first transition series in general, namely, between 1 and 10. The splitting of the middle band in the $[Ni(H_2O)_6]^{2+}$ spectrum is due to spin-orbit coupling that mixes the $^3T_{1g}(F)$ and ^1E_g states, which are very close in energy at the Δ_0 value given by 6H$_2$O, whereas in the stronger field of the 3en they are so far apart that no significant mixing occurs. An analysis of the spectrum at the level of the previous discussion is adequate for almost all chemical purposes, but it is worth noting that at higher resolution the spectra show much greater complexity and require much more sophisticated analysis.

Magnetically, octahedral nickel(II) complexes have relatively simple behavior. From both d-orbital splitting and energy level diagrams it follows that they all should have two unpaired electrons, and this is found always to be the case, the magnetic moments ranging from 2.9 to 3.4 BM, depending on the magnitude of the orbital contribution.

The type of structure found in CsNiCl$_3$ is often considered to be the generic structure for a large number of $M^IM^{II}X_3$ compounds, where M^I is a large univalent cation and M^{II} is a transition metal ion. In this structure there are infinite parallel chains of NiX$_6$ octahedra sharing opposite faces, and the M^I ions lie between these chains in an ordered pattern. The M^{II} ions generally show significant antiferromagnetic coupling that is believed to occur by a superexchange mechanism *via* the bridging X ions.

Table 17-G-2 Spectra of Octahedral Nickel(II) Complexes

Transitions	Approximate band positions (cm^{-1})	
	$[Ni(H_2O)_6]^{2+}$	$[Ni(en)_3]^{2+}$
$^3A_{2g} \rightarrow {}^3T_{2g}$	9,000	11,000
$^3A_{2g} \rightarrow {}^3T_{1g}(F)$	14,000	18,500
$^3A_{2g} \rightarrow {}^3T_{1g}(P)$	25,000	30,000

Five-Coordinate Nickel(II) Complexes

A considerable number of both trigonal bipyramidal and square pyramidal complexes occur and high- ($S = 1$) and low-spin ($S = 0$) examples of each geometry are known. Many of the trigonal bipyramidal complexes contain one of the tripod ligands such as *pp₃* and *np₃* and have the type structure shown in (17-G-I).

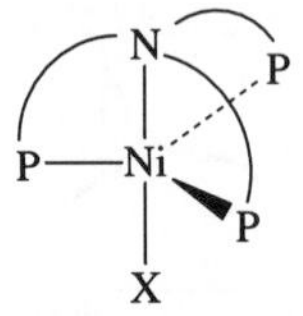

(17-G-I)

The fifth ligand is typically a halide ion, though hydride, thiolate, alkyls, and other anionic ligands are also known; the complexes are therefore +1 cations. The complexes are usually diamagnetic and possess approximately the *tbp* structure. Other examples of low-spin, five-coordinate Ni^{II} complexes include [NiL$_5$]$^{2+}$ and [NiL$_3$X$_2$], as well as [NiL$_4$X]$^+$ types (L = typically a phosphine or arsine ligand). In many cases the ligands L are unidentate, as in [Ni(SbMe$_3$)$_5$]$^{2+}$, [NiBr(PMe$_3$)$_4$]$^+$, [NiMe(PMe$_3$)$_4$]$^+$, and NiX$_2$(SbMe$_3$)$_3$.

The [Ni(CN)$_5$]$^{3-}$ ion forms salts with suitable large cations and, as mentioned above, usually adopts *sp* geometry. In [Cr(en)$_3$][Ni(CN)$_5$]·1.5H$_2$O there are two crystallographically independent [Ni(CN)$_5$]$^{3-}$ ions, one with *sp* and the other with *tbp* geometry. However, when this compound is dehydrated or subjected to pressure the crystal structure changes and the *tbp* one becomes *sp*. There are a number of five-coordinate complexes with weak field oxygen ligands, for example, [Ni(Me$_3$AsO)$_5$](ClO$_4$)$_2$; these are high-spin.

Tetrahedral Complexes

These are mainly of the following stoichiometric types: NiX$_4^{2-}$, NiX$_3$L$^-$, NiL$_2$X$_2$, and Ni(L—L)$_2$, where X represents a halogen or SPh, L a neutral ligand such as phosphine, phosphine oxide, or arsine, and L—L is one of several types of bidentate ligand [e.g., (17-G-II) to (17-G-IV)].

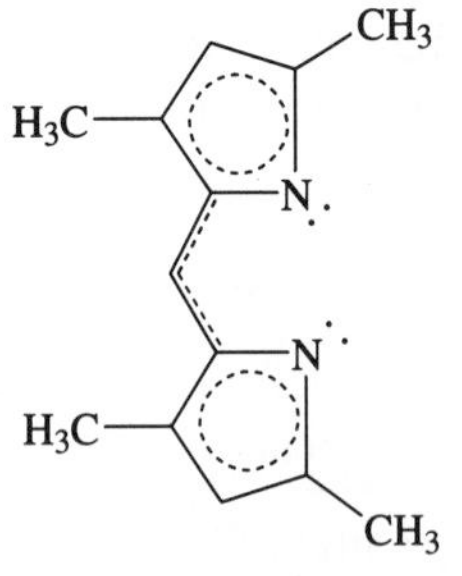

(17-G-II) 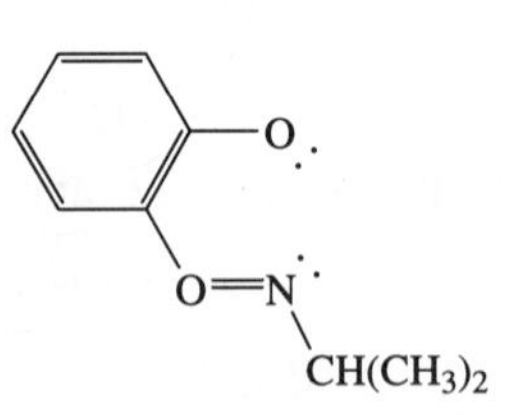   (17-G-III) 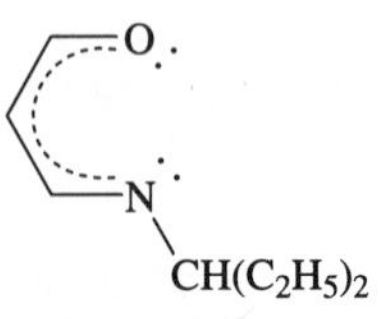   (17-G-IV)

These three bidentate anions all contain sufficiently bulky substituents on, or adjacent to, the nitrogen atoms to render planarity of the Ni(L—L)$_2$ molecule sterically impossible. When small substituents are present, planar, or nearly planar complexes are formed. It must be stressed that except for the NiX_4^{2-} species, a rigorously tetrahedral configuration cannot be expected. However, in some cases there are marked distortions even from the highest symmetry possible, given the inherent shapes of the ligands. Thus in Ni(L—L)$_2$ molecules the most symmetrical configuration possible would have the planes of the two L—L ligands perpendicular. Most often, however, this dihedral angle differs considerably from 90°; for example, when L—L is (17-G-II) it is only 76°. Thus the term "tetrahedral" is sometimes used very loosely (i.e., does not imply a regular tetrahedron); since all the so-called tetrahedral species are paramagnetic with two unpaired electrons, it would perhaps be better to simply call them paramagnetic rather than tetrahedral. Indeed, the most meaningful way to distinguish between tetrahedral and "planar" four-coordinate nickel(II) complexes is to consider that for a given ligand set ABCD, there is a critical value of the dihedral angle between two planes, such as A—Ni—B and C—Ni—D. When the angle exceeds this value the molecule will be paramagnetic; it may be called tetrahedral even though the dihedral angle is appreciably <90°. Conversely, when the angle is below the critical value the complex will be diamagnetic; it may be called planar even if the limit of strict planarity is not actually attained.

For regular or nearly regular tetrahedral complexes there are characteristic spectral and magnetic properties. Naturally the more irregular the geometry of a paramagnetic nickel(II) complex the less likely it is to conform to these specifications. In T_d symmetry the d^8 configuration gives rise to a $^3T_1(F)$ ground state. The transition from this to the $^3T(P)$ state occurs in the visible region ($\sim15,000$ cm^{-1}) and is relatively strong ($\varepsilon \approx 10^2$) compared to the corresponding $^3A_{2g} \rightarrow {}^3T_{1g}$ transition in octahedral complexes. Thus tetrahedral complexes are generally strongly colored and tend to be blue or green unless the ligands also have absorption bands in the visible region. Because the ground state $^3T_1(F)$ has much inherent orbital angular momentum, the magnetic moment of truly tetrahedral Ni^{II} should be ~4.2 BM at room temperature. However, even slight distortions reduce this markedly (by splitting the orbital degeneracy). Thus fairly regular tetrahedral complexes have moments of 3.5 to 4.0 BM; for the more distorted ones the moments are 3.0 to 3.5 BM (i.e., in the same range as for six-coordinate complexes).

The $NiCl_4^{2-}$ ion, a representative tetrahedral complex, has been studied in great detail spectroscopically at 2.2 K. The observations can all be accounted for by a parameterized crystal field model, with $\Delta_t \approx 3500$ cm^{-1}.

Planar Complexes

For the vast majority of four-coordinate nickel(II) complexes, planar geometry is preferred. This is a natural consequence of the d^8 configuration, since the planar ligand set causes one of the d orbitals ($d_{x^2-y^2}$) to be uniquely high in energy and the eight electrons can occupy the other four d orbitals but leave this strongly antibonding one vacant. In tetrahedral coordination, on the other hand, occupation of antibonding orbitals is unavoidable. With the congeneric d^8 systems Pd^{II} and Pt^{II} this factor becomes so important that no tetrahedral complex is formed.

Almost all planar complexes of Ni^{II} are thus diamagnetic (there are only one or two exceptions, see below). They are frequently red, yellow, or brown owing to the presence of an absorption band of medium intensity ($\varepsilon \sim 60$) in the range 450 to 600 nm, but other colors do occur when additional absorption bands are present.

As important examples of square complexes, we may mention yellow $[Ni(CN)_4]^{2-}$, red bis(dimethylglyoximato)nickel(II) (17-G-V), which gives a stacked polymer, and the yellow to orange-brown compounds $NiX_2(PR_3)_2$ (R = alkyl or aryl).

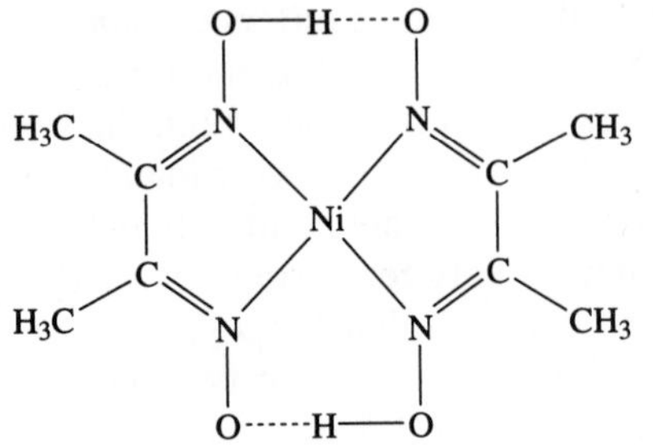

(17-G-V)

Compound (17-G-V) is used in analysis for the gravimetric determination of nickel. Phosphine complexes of the type *trans*-NiX_2L_2 and (L—L)NiX_2 are used as precursors for numerous catalytic reactions; some of these have also been discussed as potential anti-tumor agents.[2] An unusual class of compounds are the metalladithiolenes, for example, (17-G-VI) and (17-G-VII), which are readily oxidized to give, formally, Ni^{III} and Ni^{IV} species. These compounds show metallic conductivity and may show low-temperature superconductivity.[3]

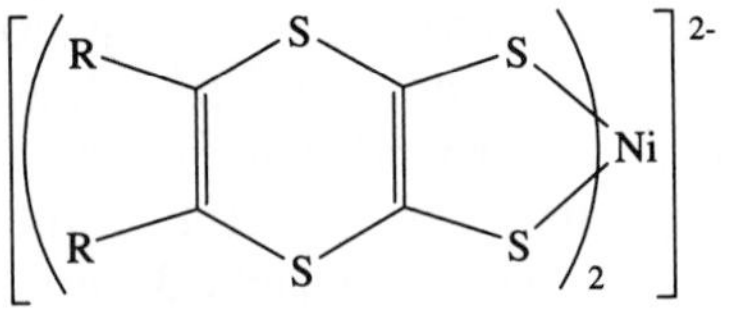

(17-G-VI)

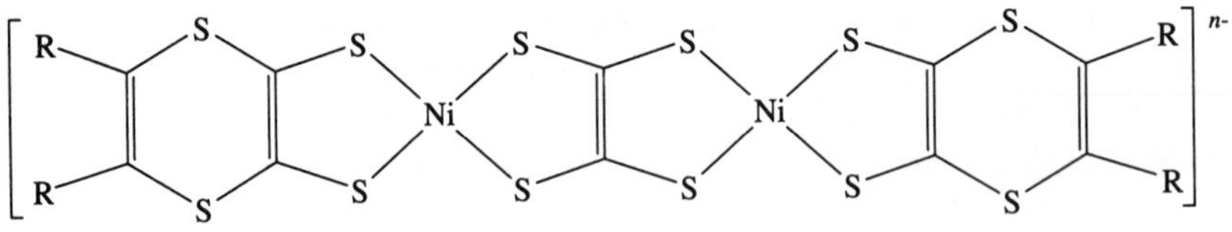

$n = 0, 2; R = COOMe$

(17-G-VII)

Trigonal Complexes

Whereas there are numerous such examples for Ni^0, they are rare for Ni^{II}. Examples are the dialkylamides $[Ni(NR_2)_3]^-$ and $Ni_2(\mu$-$NR_2)_2(NR_2)_2$, as well as the blue mesityl complex $[Ni(mes)_3]^-$ which has an approximately T-shaped geometry.[4]

[2]P. S. Jarrett and P. J. Sadler, *Inorg. Chem.* **1991,** *30,* 2098.
[3]T. B. Rauchfuss *et al., J. Chem. Soc., Chem. Commun.* **1994,** 821.
[4]G. Wilkinson *et al., Polyhedron* **1996,** *15,* 3163.

17-G-4 "Anomalous" Properties of Nickel(II) Complexes; Conformational Changes

A considerable number of nickel(II) complexes do not behave consistently in accord with expectations for any discrete structural types, and they have in the past been termed "anomalous." All the anomalies can be satisfactorily explained in terms of several types of conformational or other structural change and, ironically, so many examples are now known that the term anomalous is no longer appropriate. The three main structural and conformational changes that nickel(II) complexes undergo are described and illustrated next.

1. *Formation of 5- and 6-Coordinate Complexes by Addition of Ligands to Square Complexes.* For any square NiL_4, the following equilibria with additional ligands L' must in principle exist:

$$ML_4 + L' = ML_4L'$$

$$ML_4L' + L' = ML_4L_2'$$

In the case where $L = L' = CN$, only the 5-coordinate species is formed, but in most cases equilibria strongly favor the 6-coordinate species that have *trans* structures and two unpaired electrons.

An interesting group of compounds is provided by the series based on $[NiL]^{2+}$, shown as (17-G-VIII). In the perchlorate $[NiL](ClO_4)_2$ the nickel atom is 4-coordinate and the compound is red and diamagnetic. The compounds $[NiLX]X$ (X = Cl, Br, and I) are blue or green and have two unpaired electrons; the cationic complex is 5-coordinate *sp*. The complex $[NiL(NCS)_2]$ is octahedral, violet, and has two unpaired electrons.

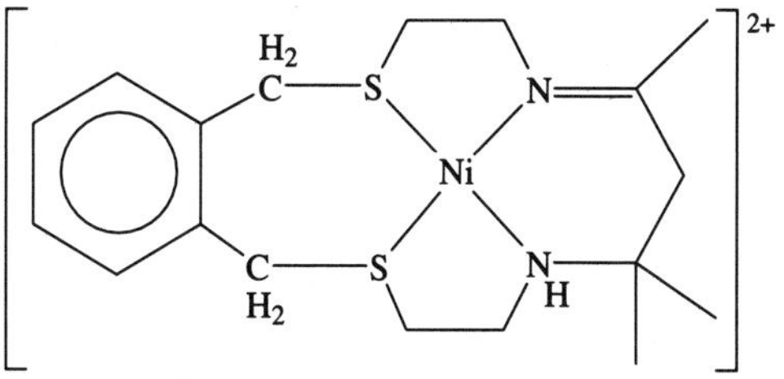

(17-G-VIII)

Well-known examples of the square-octahedral ambivalence are the Lifschitz salts, complexes of nickel(II) with substituted ethylenediamines, especially the stilbenediamines, one of which is illustrated in (17-G-IX). Many years ago Lifschitz and others observed that such complexes were sometimes blue and paramagnetic and other times yellow and diamagnetic, depending on factors such as temperature, identity of the anions present, the solvent in which they are dissolved or from which they were crystallized, exposure to atmospheric water vapor, and the particular diamine involved. The bare experimental facts bewildered chemists for several decades. It is now recognized that the yellow species are square complexes, as typified by (17-G-IX), and the blue ones are octahedral complexes, derived from the square complexes by coordination of two additional ligands—solvent molecules, water molecules, or anions—above and below the plane of the square complex.

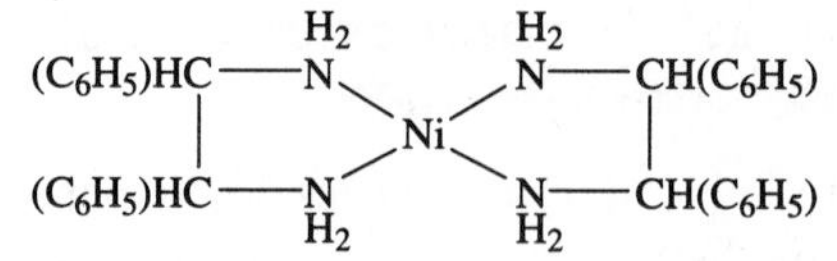

$$(C_6H_5)HC \overset{\displaystyle H_2}{\underset{\displaystyle H_2}{\overset{N}{\underset{N}{}}}} Ni \overset{\displaystyle H_2}{\underset{\displaystyle H_2}{\overset{N}{\underset{N}{}}}} CH(C_6H_5)$$

(17-G-IX)

The structures of many Ni^{II} complexes are very sensitive to steric effects; these influences are also reflected in the redox potential. Complexes of the "cyclam" ligand (17-G-X) are particularly stable and may exist as 4-, 5-, or 6-coordinate compounds in the solid state, depending on R. If R = Me both square and *sp* structures exist in the same unit cell.[5]

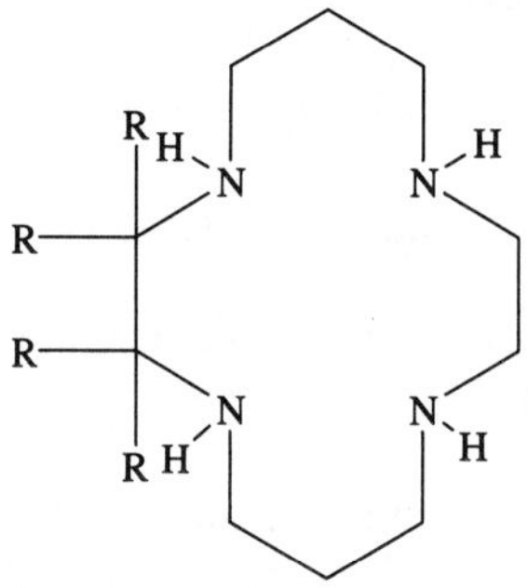

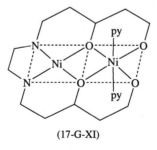

(17-G-X)

The complex (17-G-XI) is interesting in that it contains, side by side, low-spin, square-coordinated and high-spin, octahedrally coordinated nickel atoms.

(17-G-XI)

2. *Monomer-Polymer Equilibria.* In many cases 4-coordinate complexes associate or polymerize, to give species in which the nickel ions become 5- or 6-coordinate. In some cases the association is very strong and the 4-coordinate monomers are observed only at high temperatures; in others the position of the equilibrium is such that both red, diamagnetic monomers and green or blue, paramagnetic polymers are present in a temperature- and concentration-dependent equilibrium around room temperature. A clear example of this situation is provided by various β-ketoenolate complexes, (17-G-XII). If R is bulky (for example CMe_3), a red, diamagnetic square

[5]K. Kobiro *et al., Inorg. Chem.* **1992**, *31*, 676.

mononuclear complex results, whereas if R = Me the complex exists as a green paramagnetic hydrate which, on removal of the coordinated water, gives the trimer shown in Fig. 11-4, with NiO_6 octahedra. This trimer is very stable and only at temperatures around 200°C (in a non-coordinating solvent) do detectable quantities of monomer appear. It is, however, readily cleaved by donors such as H_2O or pyridine, to give 6-coordinate monomers. If the R groups are sterically intermediate between CH_3 and $C(CH_3)_3$, temperature- and concentration-dependent purple monomer–green trimer equilibria are observed in non-coordinating solvents.

(17-G-XII)

3. *Planar-Tetrahedral Equilibria and Isomerism.* We have already indicated that nickel(II) complexes of certain stoichiometric types, namely, the bishalo bisphosphine and bis(salicylaldiminato) types may have either square or tetrahedral structures, depending on the identity of the ligands. For example, in the NiX_2L_2 cases, when L is triphenylphosphine, tetrahedral structures are found, whereas the complexes with trialkylphosphines are generally square. Perhaps it is then not very surprising that a number of NiX_2L_2 complexes, in which L represents a mixed alkylarylphosphine, exist in solution in an equilibrium distribution between the tetrahedral and square forms. The influence of the varying R groups in the phosphines is almost entirely electronic rather than steric. For example, in CH_2Cl_2 solution, for PPh_2Bu and Pcy_3, the mole fractions of the tetrahedral form are 1.00 and 0.00, respectively, even though Pcy_3 is much bulkier (cone angles *ca.* 140° and 170°, respectively). At 25°C the rate constants for conversion of tetrahedral into planar isomers are in the range of 10^5 to 10^6 s^{-1} with enthalpies of activation of around 45 kJ mol^{-1}.

In some cases it is possible to isolate two crystalline forms of the compound, one yellow to red and diamagnetic, the other green or blue with two unpaired electrons. There is even $NiBr_2[PPh_2(CH_2Ph)]_2$, in which both tetrahedral and square complexes are found together in the same crystalline substance.

The anomalous magnetic behavior of chelate complexes of type (17-G-XIII) is also due to square-tetrahedral isomerism. Thus (17-G-XIII, R = Et) is diamagnetic in the solid state but tetrahedral and paramagnetic in solution [μ_{eff} = 1.52 (353K) to 0.98 B.M. (273K)]; these changes are accompanied by drastic changes in the electronic spectrum. An interesting exception to this explanation of magnetic behavior is, however, the closely related O-compound (17-G-XIV): both the planar olive-green and the dark blue (distorted) tetrahedral isomers of this complex are paramagnetic.[6]

[6]W. Kuchen *et al., Angew. Chem. Int. Ed. Engl.* **1992,** *31,* 612; **1993,** *32,* 907; *Chem. Phys. Lett.* **1994,** *231,* 235.

(17-G-XIII)   (17-G-XIV)

Further examples of the steric influence on coordination geometry are thiolate complexes, e.g., $[Ni(SPh)_4]^{2-}$ (tetrahedral), $[Ni_2(\mu\text{-}SR)_2(SR)_4]^{2-}$ (R = $p\text{-}C_6H_4Cl$, square-planar), and $[Ni_2(\mu\text{-}SR)_3(SR)_2]^-$ (R = $C_6H_2Pr^i_3$, face-sharing bitetrahedral).[7]

Thermochromism

This phenomenon is frequently encountered among Ni^{II} complexes. It comes in general from temperature-dependent variability of structure, which causes variation in the *d-d* absorption bands. For example, in $(NR_xH_{4-x})_2[NiCl_4]$ (x = 1, 2, and 3), reversible thermochromism, from yellow brown or green at low temperature to blue at high temperature, appears to be due to changes from octahedral (with bridging Cl atoms) to tetrahedral coordination.

17-G-5   The Chemistry of Nickel(III), $d^{7,8}$

Binary Compounds

Nickel(III) fluoride has been prepared as a nearly black crystalline solid by the thermal decomposition of NiF_4 at 0°C. There are various modifications. NiF_3 is stable at 20°C but loses F_2 on warming to give NiF_2. It is a very strong oxidizing agent.[9]

There is no good evidence for Ni_2O_3, but there are two proved crystalline forms of black NiO(OH). The more common β-NiO(OH) is obtained by the oxidation of nickel(II) nitrate solutions with bromine in aqueous potassium hydroxide below 25°C. It is readily soluble in acids; on aging, or by oxidation in hot solutions, a Ni^{II}—Ni^{III} hydroxide of stoichiometry $Ni_3O_2(OH)_4$ is obtained. The oxidation of alkaline nickel sulfate solutions by NaOCl gives a black "peroxide" $NiO_2 \cdot nH_2O$. This is unstable, being readily reduced by water, but it is a useful oxidizing agent for organic compounds.

The compound $NaNiO_2$ and several related ones also seem to be genuine. They can be made by bubbling oxygen through molten alkali metal hydroxides contained in nickel vessels at about 800°C. Other oxides and oxide phases can be made by heating NiO with alkali or alkaline earth oxides in oxygen. These mixed oxides evolve oxygen on treatment with water or acid.

The Edison or nickel-iron battery, which uses KOH as the electrolyte, is based on the reaction

$$Fe + 2NiO(OH) + 2H_2O \underset{\text{charge}}{\overset{\text{discharge}}{\rightleftharpoons}} Fe(OH)_2 + 2Ni(OH)_2 \quad E^0 \sim 1.3 \text{ V}$$

but the mechanism and the true nature of the oxidized species are not fully understood.

---

[7]A. Silver and M. Millar, *J. Chem. Soc., Chem. Commun.* **1992,** 948.
[8]A. G. Lappin and A. McAuley, *Adv. Inorg. Chem.* **1988,** *32,* 241.
[9]N. Bartlett *et al., J. Am. Chem. Soc.* **1995,** *117,* 10025.

Complexes

For Ni^{III} as a d^7 ion, with a $t_{2g}^6 e_g^1$ configuration, Jahn-Teller distortion is expected. This is indeed found, e.g., in K_3NiF_6 and $[Ni(bipy)_3]^{3+}$, which show tetragonally distorted octahedral geometries with short axial bonds.

Interest in Ni^{III} complexes has increased significantly due to the suspected involvement of this oxidation state in nickel-containing metalloenzymes (see below). Like other trivalent ions (V^{3+}, Cr^{3+}, Fe^{3+}, Mn^{3+}, Ru^{3+}, Rh^{3+}, Ir^{3+}) Ni^{III} forms a "basic acetate," $[Ni_3(\mu_3\text{-O})(OOCR)_6L_3]^+$. Oxidation of the brown chelate complexes $Ni^{II}I_2(L-L)_2$ with concentrated HNO_3 affords green $[NiI_2(L-L)_2]^+$ ($L-L$ = diphosphine or diarsine).[10] The compounds NiX_3(diphosphine) are square-pyramidal, while the mono-phosphine compounds NiX_3L_2 have the *tbp* structure (X = halide, L = trialkylphosphine). The reaction of Me_3PI_2 with granular nickel metal gives black $NiI_3(PMe_3)_2$ quantitatively.[11]

The open chain ligands such as $EDTA^{4-}$ and deprotonated peptides give stable complexes in aqueous solution.

There are a number of Ni^{III} complexes with macrocyclic nitrogen ligands. They are usually made by one-electron oxidation of the Ni^{II} species; the ring size and size of cavity in the macrocycles has an influence on the Ni^{II}/Ni^{III} redox potential. Some Ni^{III} macrocycles can also be oxidized to Ni^{IV}. A structurally characterized example of a Ni^{III} macrocycle is the deep purple complex (17-G-XV) which contains a square-planar NiN_4 core.[12]

(17-G-XV)

17-G-6 The Chemistry of Nickel(IV), d^6

Fluorination of alkali metal Ni^{II} fluorides gives the red or purple alkali salts M_2NiF_6. Ni^{IV} also occurs in $BaNiO_3$, $[NiNb_{12}O_{38}]^{12-}$, and in the periodate salts $MNiIO_6$ (M = alkali cation, NH_4) which are accessible by oxidation of nickel(II) salts in hot aqueous solution in the presence of periodate ions.[13] The addition of F^- acceptors (BF_3, AsF_5) to NiF_6^{2-} salts in anhydrous HF at $-65°$ leads to NiF_4 as a tan solid which is thermally unstable and decomposes to NiF_3.[9]

A well-established complex of Ni^{IV} is the diacetylpyridine dioximato complex

[10]W. Levason et al., Inorg. Chem. **1991**, 30, 331; **1995**, 34, 1626.
[11]C. A. McAuliffe, J. Chem. Soc., Dalton Trans. **1993**, 2875.
[12]T. J. Collins et al., J. Am. Chem. Soc. **1991**, 113, 4708.
[13]M. T. Weller et al., J. Chem. Soc., Dalton Trans. **1994**, 1483.

(17-G-XVI); it is fairly stable in aqueous solution at low pH values. The compound is readily reduced to Ni^{II}.[14]

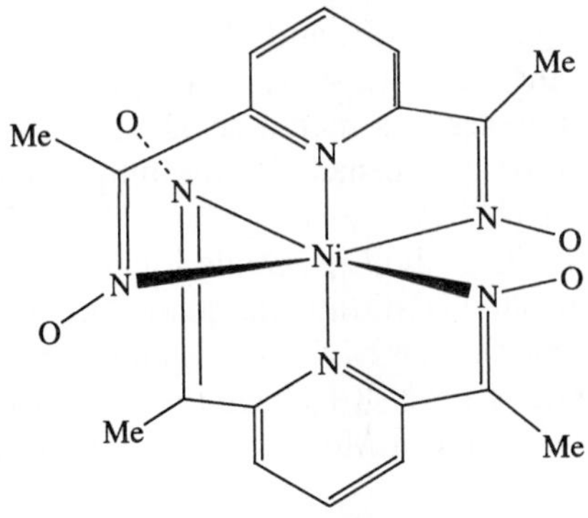

(17-G-XVI)

There are a number of similar redox-active nickel chelate complexes. However, some of these, such as 1,2-dithiolates, contain the oxidized form of the ligand and are more properly considered as Ni^{II} compounds.

Other examples of Ni^{IV} species include the "dicarbollide" complex $Ni(closo$-1,2-$C_2B_9H_{11})_2$ which reacts with electron donors to give electrically conducting charge-transfer salts of the $[Ni(C_2B_9H_{11})_2]^-$ radical anion,[15] and the polyselenide anion $[Ni_4Se_4(Se_3)_5(Se_4)]^{4-}$ with a cubic Ni_4Se_4 core.[16]

17-G-7 Nickel in Lower Oxidation States (−I), (O), (+I), and Mixed-Valence Compounds

The low-valent oxides $K_3[NiO_2]$ and $KNa_2[NiO_2]$ contain linear $[O-Ni^I-O]^{3-}$ units, similar to analogous Fe and Co compounds.[17] There are a number of Ni^I complexes with metal-metal bonds of varying lengths, for example, (17-G-XVII) and the mixed-valence complex (17-G-XVIII).

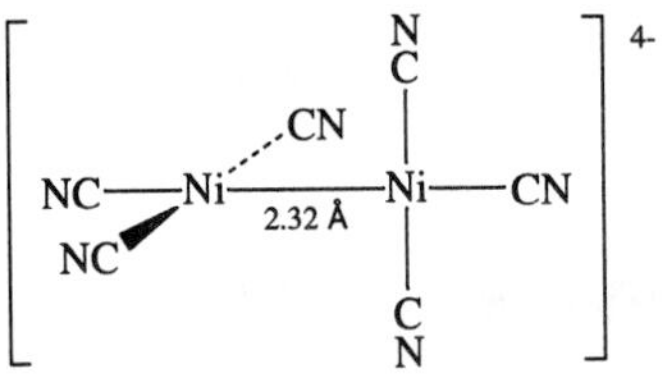

(17-G-XVII) 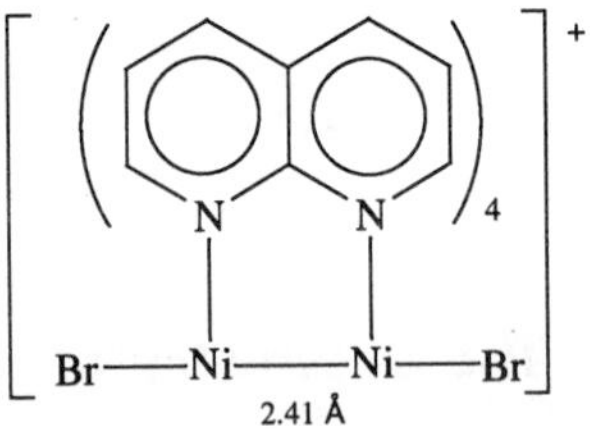   (17-G-XVIII)

The majority of *nickel(I) complexes* contain phosphine ligands, or closely related ones, and have tetrahedral or *tbp* structures. They are paramagnetic as expected for d^9 configurations. The tetrahedral compounds $NiX(PPh_3)_3$ (X = Cl, Br, and I) were among the first of this sort to be isolated; they decompose only slowly in air and are stable for long periods of time under nitrogen. Electrolytic reductions of

[14]S. Mandal and E. S. Gould, *Inorg. Chem.* **1995**, *34*, 3993.
[15]P. A. Chetcuti *et al., Organometallics* **1995**, *14*, 666.
[16]J. A. Ibers *et al., J. Am. Chem. Soc.* **1991**, *113*, 7078.
[17]A. Möller *et al., Inorg. Chem.* **1995**, *34*, 2684.

Ni^{2+} in MeCN in the presence of PR_3 can stabilize Ni^I but phosphites stabilize Ni^0 or Ni^{II}, although due to kinetic effects $\{Ni[P(OEt)_3]_4\}^+$ can be isolated.

With the tripod ligand np_3 an entire series of compounds with *tbp* structures $Ni(np_3)X$ (X = Cl, Br, I, CN, CO, and H) can be prepared by reaction of nickel halides with the np_3 ligand in the presence of borohydride, followed, if necessary, by metathesis with other X groups. The comproportionation of the Ni^0 complex $Ni_2(CO)_2(\mu\text{-}CO)(\mu\text{-}dppm)_2$ with $NiCl_2(dppm)_2$ (dppm = $Ph_2PCH_2PPh_2$) gives the fluxional A-frame compound (17-G-XIX).[18]

$$
\begin{array}{c}
Ph_2P \overset{\displaystyle O}{\underset{\displaystyle C}{\diagup \quad \diagdown}} PPh_2 \\[2pt]
Cl\!-\!Ni \Longleftarrow \quad \Longrightarrow Ni \\[2pt]
Ph_2P \diagdown \quad Cl \diagup PPh_2
\end{array}
$$

(17-G-XIX)

Nickel(I) complexes of N_4 macrocycles can be prepared by reduction of the corresponding Ni^{II} complexes with sodium amalgam. They possess more or less distorted square-planar structures.[19] By contrast, the one-electron reduction of Ni^{II} porphyrin complexes may result in Ni^I porphyrins or Ni^{II} π-anion radicals, depending on the reaction conditions.[20] Complexes of this kind are useful models for the Ni sites in certain metalloenzymes (see below).

Thiolato and chalcogenide ligands stabilize a wide range of mixed-valent Ni complexes. Examples are $Ni_4(SPr^i)_8X$ (X = Br, I),[21] $[Ni_8S(SC_4H_9)_9]^-$, and $[Ni_5S\text{-}(SBu^t)_5]^-$; the latter possesses an unusual star-shaped structure with a μ_5-S ligand (17-G-XX).[22] Examples for a range of chalcogenide clusters with the metal in varying oxidation states are $[Ni_{20}Se_{12}(SeMe)_{10}]^{2-}$, $Ni_{34}Se_{22}(PPh_3)_{10}$, and $Ni_9Te_6(PEt_3)_8$.[23] By contrast to the polyhedral structures of these compounds, the Ni^{II} sulfide $Ni_{15}(\mu_3\text{-}S)_6(\mu\text{-}S)_9(PPh_3)_3$ consists of a stack of five planar Ni_3S_3 layers.[24]

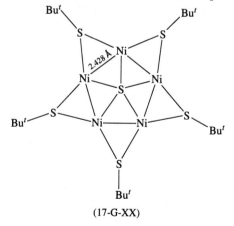

(17-G-XX)

[18]L. Manojlovic-Muir *et al., Inorg. Chem.* **1992,** *31,* 904.
[19]M. P. Suh *et al., Inorg. Chem.* **1992,** *31,* 3620; *J. Chem. Soc., Dalton Trans.* **1995,** 1577.
[20]K. M. Kadish *et al., J. Am. Chem. Soc.* **1991,** *113,* 512.
[21]B. Krebs *et al., Angew. Chem. Int. Ed. Engl.* **1992,** *31,* 54.
[22]A. Müller and G. Henkel, *Chem. Commun.* **1996,** 1005.
[23]M. L. Steigerwald *et al., J. Am. Chem. Soc.* **1989,** *111,* 9240.
[24]H. Liu *et al., J. Chem. Soc., Chem. Commun.* **1990,** 1210.

There are numerous complexes of zerovalent nickel, usually involving N- or P-donor ligands or organometallic complexes with olefin or CO ligands. The phosphine adducts $Ni(PR_3)_4$ are colorless to orange solids of tetrahedral structure. There is a tendency to form 16-electron trigonal-planar complexes; for example, white $Ni(PEt_3)_4$ dissolves in hexane to give purple solutions of highly reactive $Ni(PEt_3)_3$. The reactivity of Ni^0 species is strongly ligand dependent; for example, NiL_4 is pyrophoric in air if L = strong electron donor, such as PMe_3, but stable to air and water if L = $P(OPh)_3$. Nickel(0) is also stabilized by π-acceptor ligands such as 2,2'-bipyridyl (bipy) and 1,4-diazadienes.[25] With tridentate phosphines np_3 = $N(CH_2CH_2PPh_2)_3$, Ni^0 also forms complexes with yellow phosphorus in which one P atom acts as a simple donor, $(np_3)Ni(\eta^1\text{-}P_4)$.

Although anionic species with nickel in the oxidation state -I have been postulated, such as $[Ni_2(CO)_6]^{2-}$, these have not been confirmed. Nickel does, however, form a number of anionic carbonyl clusters with higher nuclearity, e.g., $[Ni_5(CO)_{12}]^{2-}$, in which the metal carries a partial negative charge.

17-G-8   Nickel in Biological Systems

It is now known that nickel is an essential component in at least four types of enzymes: urease, carbon monoxide dehydrogenase (CODH, or acetyl coenzyme A synthase), hydrogenase, and methyl-*S*-coenzyme M reductase. Nickel can exist under physiological conditions in the oxidation states I, II, and III. Urease catalyzes the hydrolysis of urea to ammonium and carbamate ($H_2NCO_2^-$) ions; it appears to contain a redox-inactive Ni_2^{III} unit in which the nickel atoms are octahedrally coordinated and probably act as Lewis acid for substrate binding. Dinuclear urea complexes such as (17-G-XXI) have been suggested as models.[26] CO-Dehydrogenase (CODH) catalyzes the reaction

$$CO + H_2O \rightleftharpoons [C_1] \rightleftharpoons CO_2 + 2H^+ + 2e$$

via an intermediate $[C_1]$. Nickel containing enzymes of certain methanogenic organisms also catalyze the synthesis of acetyl-coenzyme A:

$$Me{-}X + CoASH + CO \longrightarrow CoAS{-}COMe + HX \qquad (X = \text{tetrahydrofolate})$$

It is likely therefore that organometallic methyl and carbonyl species are involved; the *tbp* NS_3 complexes $[(NS_3)Ni^{II}X]^+$ (X = H, Me, COMe) (17-G-XXII) and $[(NS_3)Ni^ICO]^+$ are thought to be models for such intermediates.[27]

There is spectroscopic evidence for nickel-methyl complexes in CODH of anaerobic bacteria. Here the nickel center is linked to an Fe_4S_4 cubane cluster; it is this, not Ni, which appears to act as the binding site for CO. In a subsequent step the methyl ligand is then transferred from Ni to the coordinated CO to generate the acetyl function of acetyl coenzyme A.[28]

[25]K. R. Pörschke *et al., J. Organomet. Chem.* **1990,** *397,* 255.
[26]S. J. Lippard *et al., Inorg. Chem.* **1993,** *32,* 4985.
[27]R. H. Holm *et al., J. Am. Chem. Soc.* **1991,** *113,* 8485.
[28]S. W. Ragsdale *et al., Science* **1995,** *270,* 628; **1994,** *264,* 817.

The redox-active Ni center in hydrogenases probably has a N_3S_2 coordination sphere; in some types of hydrogenases Ni is coordinated to a Se-donor of a selenocysteine residue. These metal centers are characterized by unusually low Ni^{III}/Ni^{II} redox potentials. Various model complexes have been used to mimic the spectroscopies and electrochemical properties of the enzymes, for example, the distorted-octahedral anion $[Ni^{II}(pdtc)_2]^{2-}$ which is readily oxidized by iodine to $[Ni^{III}(pdtc)_2]^-$ $[(d_{z^2})^1$ ground state] (17-G-XXIII) [pdtc = pyridine-2,6-bis(monothiocarboxylate)],[29] as well as complexes with distorted *tbp* geometries, such as (17-G-XXIV), which can readily be reduced to Ni^I or oxidized to Ni^{III} species given the right N-donor ligands. Only the Ni^I species is able to bind CO.[30] The protonation of such Ni^I complexes leads to the Ni-mediated formation of H_2; several pathways are possible:[31]

$$Ni^I + H^+ \longrightarrow Ni^{III}-H^- \xrightarrow{e} Ni^{II}-H^- \xrightarrow{H^+} Ni^{II} + H_2$$

$$2Ni^{III}-H^- \longrightarrow 2Ni^{II} + H_2$$

Methyl-coenzyme M reductase contains a prosthetic group consisting of a redox-active NiN_4 macrocycle (factor F_{430}) (17-G-XXV).[32]

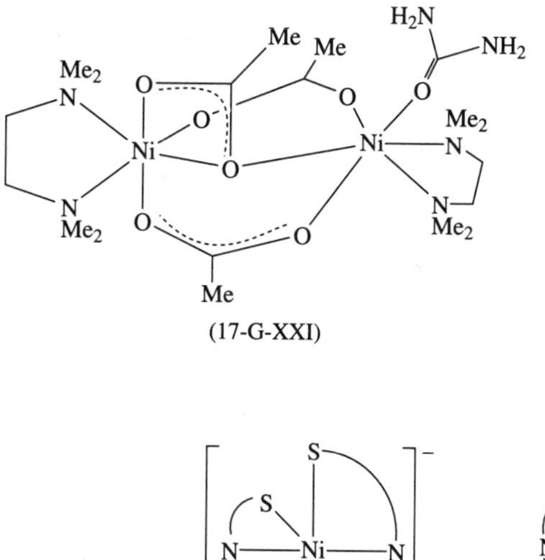

(17-G-XXI)

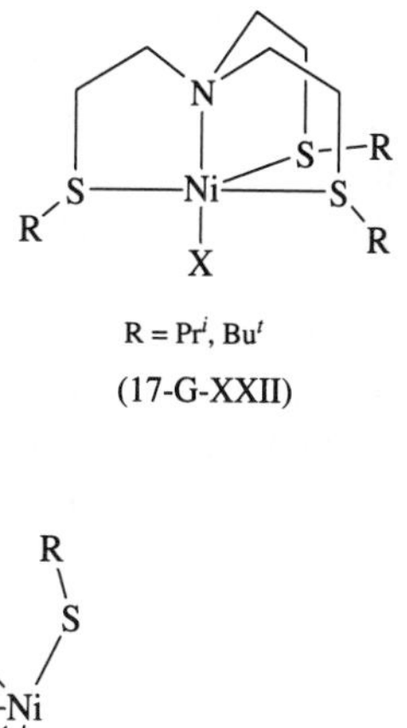

R = Pr^i, Bu^t

(17-G-XXII)

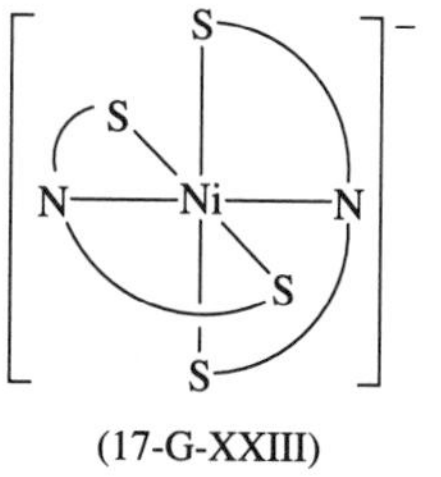

(17-G-XXIII)

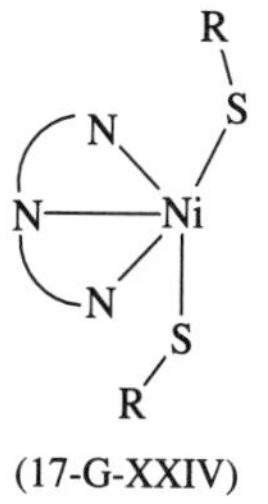

(17-G-XXIV)

[29]H. J. Krüger and R. H. Holm, *J. Am. Chem. Soc.* **1990**, *112*, 2955.
[30]P. K. Mascharak *et al.*, *J. Am. Chem. Soc.* **1995**, *117*, 1584.
[31]R. H. Holm *et al.*, *Inorg. Chem.* **1996**, *35*, 4148.
[32]L. Latos-Grazynski *et al.*, *Inorg. Chem.* **1989**, *28*, 4065.

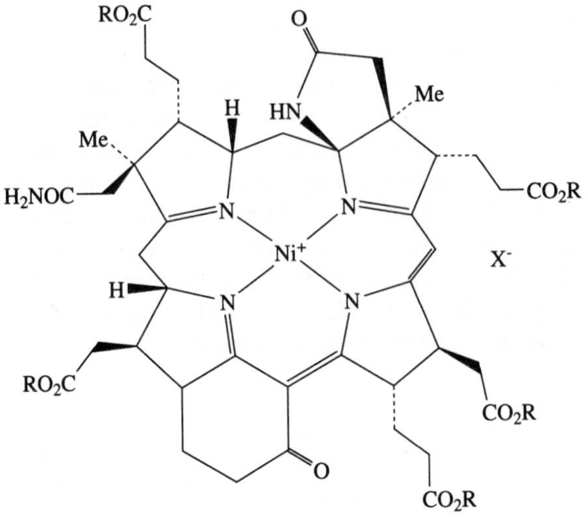

(17-G-XXV)

17-G-9 Organometallic Nickel Complexes

Nickel forms organometallic complexes in the oxidation states 0, I, and II; there are also a few examples of Ni^{III} and Ni^{IV}. Nickel tetracarbonyl, the first carbonyl of a zerovalent metal to be made (L. Mond, 1888), is readily accessible from metallic nickel and CO as an almost colorless volatile liquid; it is tetrahedral (d^{10}). The reduction of Ni^{II} salts in the presence of 1,5-cyclooctadiene (COD) affords the yellow crystalline $Ni(COD)_2$. The COD ligands are labile, and the tetrahedral compound is an excellent starting material. Tris(ethylene)nickel, by contrast, is trigonal-planar; the in-plane conformation of the C_2H_4 ligands maximizes back-donation. There are numerous analogous complexes stabilized by phosphine ligands, such as tetrahedral $Ni(CO)_{4-n}L_n$ and trigonal $L_2Ni(alkene)$ and $L_2Ni(alkyne)$. The bis(alkyne) complex $Ni(RC\equiv CR)_2$ (R = CMe_2OH) contains two perpendicular $C\equiv C$ ligands.[33] Stable carbenes of the imidazol type react with $Ni(COD)_2$ to give two-coordinate 14-electron complexes (17-G-XXVI).[34] Related stable silylenes react with $Ni(CO)_4$ to give mono- and bis-silylene complexes with Ni=Si distances of 2.21 Å.[35]

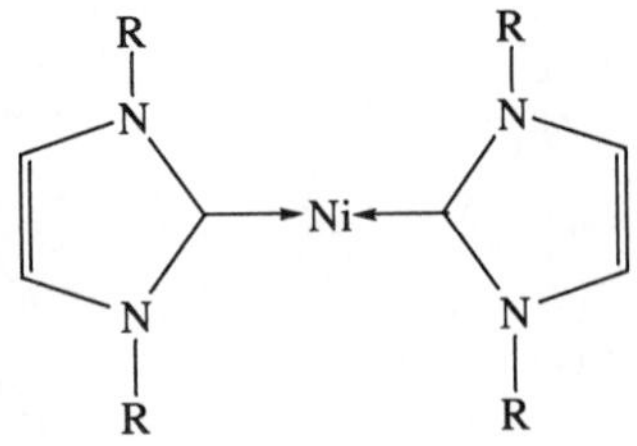

(17-G-XXVI)

[33]D. Walther *et al., Angew. Chem. Int. Ed. Engl.* **1994,** *33,* 1373.
[34]A. J. Arduengo *et al., J. Am. Chem. Soc.* **1994,** *116,* 4391; see also W. A. Herrmann *et al., Organometallics* **1997,** *16,* 2209.
[35]R. West *et al., J. Chem. Soc., Chem. Commun.* **1994,** 33.

The alkylation of Ni^{II} halides NiX$_2$L$_2$ (L = phosphine etc.) with lithium or Grignard reagents leads to alkyl, aryl, or benzyl complexes NiRXL$_2$ and NiR$_2$L$_2$. Similar products are obtained by the oxidative addition of R—X to Ni^0 complexes; Ni complexes with basic phosphines even react with usually inert aryl—Cl bonds. While neutral Ni^{II} alkyls are square, both square 4- and *tbp* 5-coordinate cationic species [L$_{4,5}$Ni-R]$^+$ are known. There are only very few examples for CO complexes of Ni^{II}; they exist only in the presence of strong electron donor ligands. NiCl$_2$(PMe$_3$)$_2$ takes up CO reversibly to give NiX$_2$(PMe$_3$)$_2$(CO); while in (C$_6$H$_6$)Ni(SiCl$_3$)$_2$ the benzene ligand is displaced to give Ni(SiCl$_3$)$_2$(CO)$_3$.[36]

There are numerous Ni^{II} π-allyl complexes, such as [(η^3-C$_3$H$_5$)Ni(μ-Br)]$_2$ and Ni(η-C$_3$H$_5$)$_2$. The latter has a sandwich structure. In the majority of these complexes 16-electron species are preferred.

Nickelocene NiCp$_2$ is readily obtained from the reaction of Cp$^-$ with Ni^{II} salts. It is a green paramagnetic 20-electron sandwich complex (2 unpaired electrons) with a tendency to form 18-electron products in most reactions; for example, treatment with allylMgBr gives the orange diamagnetic complex CpNi(η^3-allyl), and comproportionation of NiCp$_2$ with Ni(CO)$_4$ affords the dimeric Ni^I complex (17-G-XXVII). The analogous PEt$_3$ compound contains an unsupported Ni—Ni bond (17-G-XXVIII).[37] NiCp$_2$ can be oxidized electrochemically to [NiCp$_2$]$^+$ and [NiCp$_2$]$^{2+}$, though the latter is very susceptible to nucleophilic attack and decomposes readily. The more electron-rich decamethylnickelocene Ni(C$_5$Me$_5$)$_2$ undergoes two reversible one-electron oxidation steps (−0.73 and +0.37V in MeCN, vs. standard calomel electrode) and forms stable radical cation-radical anion salts with electron acceptors, [Ni(C$_5$Me$_5$)$_2$]$^{\cdot+}$ X$^{\cdot-}$ (X = e.g., tetracyanobenzoquinone, fullerene C$_{60}$) (17-G-XXIX); their structure consists of linear donor-acceptor chains.[38] Compounds of this kind are of interest as magnetic materials.

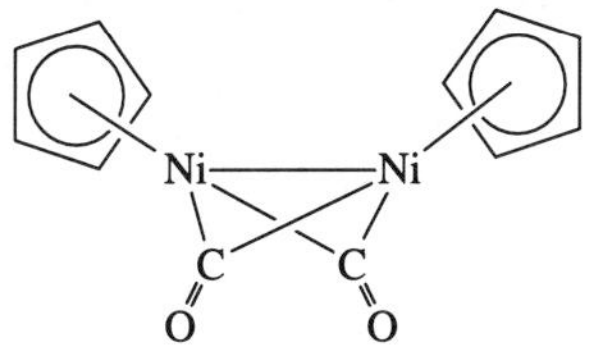

(17-G-XXVII)

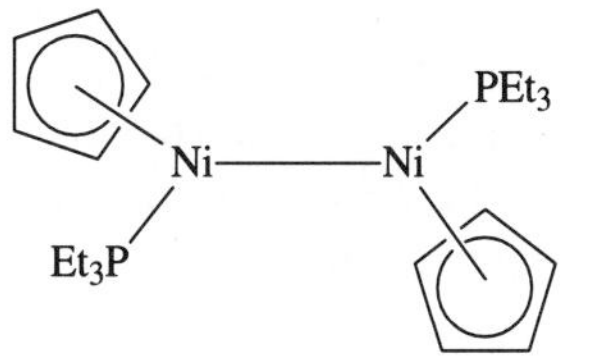

(17-G-XXVIII)

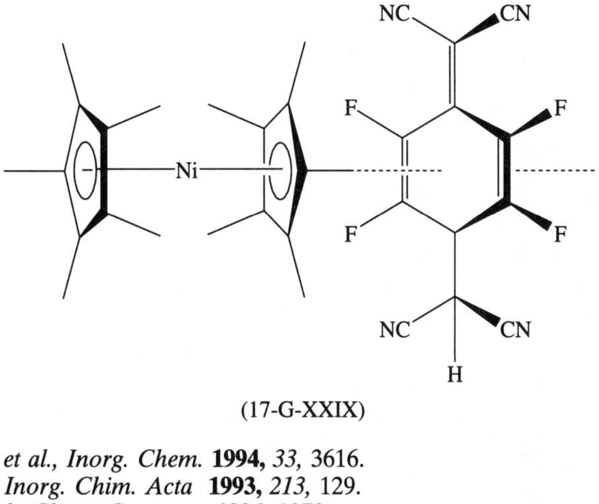

(17-G-XXIX)

[36] R. H. Crabtree *et al., Inorg. Chem.* **1994,** *33,* 3616.

[37] G. Wilke *et al., Inorg. Chim. Acta* **1993,** *213,* 129.

[38] J. S. Miller *et al., Chem. Commun.* **1996,** 1979.

Additional Reference

J. R. Lancaster, *The Bioinorganic Chemistry of Nickel,* VCH Publishers, New York, 1988.

17-H   COPPER: GROUP 11

Copper has a single *s* electron outside the filled 3*d* shell but essentially nothing in common with the alkalis except formal stoichiometries in the +1 oxidation state. The filled *d* shell is much less effective than a noble gas shell in shielding the *s* electron from the nuclear charge, so that the first ionization enthalpy of Cu is higher than those of the alkalis. Since the electrons of the *d* shell are also involved in metallic bonding, the heat of sublimation and the melting point of Cu are also much higher than those of the alkalis. These factors are responsible for the more noble character of copper, and the effect is to make the compounds more covalent and to give them higher lattice energies, which are not offset by the somewhat smaller radius of the unipositive ion compared to the nearest alkali ions—Cu^+, 0.93; Na^+, 0.95; and K^+, 1.33 Å.

The second and third ionization enthalpies of Cu are very much lower than those of the alkalis and account in part for the transition metal character shown by the existence of colored paramagnetic ions and complexes in the II and III oxidation states. Even in the I oxidation state numerous transition-metal-like complexes are formed (e.g., those with olefins).

There is only moderate similarity between Cu and the heavier elements, Ag and Au, but some points are noted in the later discussions of these elements (Chapter 18).

In recent years a great deal of study of copper complexes has been driven by the hope of modeling biological molecules that contain copper (see Section 17-H-5).

The oxidation states and stereochemistry of Cu are summarized in Table 17-H-1. Stable copper(0) compounds are not confirmed, but reactive intermediates appear to occur in some reactions.

17-H-1   The Element

Copper is widely distributed in Nature as metal, in sulfides, arsenides, chlorides, carbonates, and so on. It is extracted from ores, usually by wet processes, for example, by leaching with dilute sulfuric acid, or by solvent extraction using salicylaldoximes and similar ligands. Copper is refined by electrolysis.

Copper is a tough, soft, and ductile reddish metal, second only to Ag in its high thermal and electrical conductivities. It is used in alloys such as brasses and is completely miscible with Au. It is only superficially oxidized in air, sometimes acquiring a green coating of hydroxo carbonate and hydroxo sulfate.

Copper reacts at red heat with O_2 to give CuO and, at higher temperatures, Cu_2O; with sulfur it gives Cu_2S or a nonstoichiometric form of this compound. It is attacked by halogens but is unaffected by non-oxidizing or noncomplexing dilute acids in absence of air. Copper readily dissolves in HNO_3 and H_2SO_4 in the presence of O_2. It is soluble in NH_4OH, ammonium carbonate, or KCN solutions in the presence of O_2, as indicated by the potentials

$$Cu + 2NH_3 \xrightarrow{\text{-0.12 V}} [Cu(NH_3)_2]^+ \xrightarrow{\text{-0.01 V}} [Cu(NH_3)_4]^{2+}$$

Table 17-H-1   Oxidation States and Stereochemistry of Copper

Oxidation state	Coordination number	Geometry	Examples
Cu^I, d^{10}	2	Linear	Cu_2O, $KCuO$, $CuCl_2^-$, $CuBr_2^-$
	3	Planar	$K[Cu(CN)_2]$, $[Cu(SPMe_3)_3]ClO_4$
	4^a	Tetrahedral	CuI, $[Cu(CN)_4]^{3-}$, $[Cu(MeCN)_4]^+$
	4	Distorted planar	CuL^c
	5	sp	$[CuLCO]^c$
Cu^{II}, d^9	3	Trigonal planar	$Cu_2(\mu\text{-}Br)_2Br_2$
	$4^{a,b}$	Tetrahedral (distorted)	(N-Isopropylsalicylaldiminato)$_2$Cu, $Cs_2[CuCl_4]$
	$4^{a,b}$	Square	CuO, $[Cu(py)_4]^{2+}$, $(NH_4)_2[CuCl_4]$
	$6^{a,b}$	Distorted octahedral	K_2CuF_4, $K_2[CuEDTA]$, $CuCl_2$, $Cu(ClO_4)_2{}^d$
	5	tbp	$[Cu(bipy)_2I]^+$, $[CuCl_5]^{2-}$, $[Cu_2Cl_8]^{4-}$
	5	sp	$[Cu(dmg)_2]_2(s)$, $[Cu(NH_3)_5]^{2+e}$
	6	Octahedral	$K_2Pb[Cu(NO_2)_6]$
	7	Pentagonal bipyramidal	$[Cu(H_2O)_2(dps)]^{2+f}$
	8	Distorted dodecahedron	$Ca[Cu(CO_2Me)_4]\cdot 6H_2O$
Cu^{III}, d^8	4	Square	$KCuO_2$, $CuBr_2(S_2CNBu_2)$
	6	Octahedral	K_3CuF_6

[a]Most common state.
[b]These three cases are often not sharply distinguished; see text.
[c]L = a macrocyclic N_4 anionic ligand.
[d]F. Favier et al., J. Chem. Soc. Dalton Trans. **1994**, 3119.
[e]In $K[Cu(NH_3)_5](PF_6)_3$.
[f]dps = 2,6-diacetylpyridine bis(semicarbazone).

It is also soluble in acid solutions containing thiourea, which stabilizes Cu^I as a complex; acid thiourea solutions are also used to dissolve copper deposits in boilers.

COPPER COMPOUNDS

17-H-2   The Chemistry of Copper(I), d^{10}

Copper(I) compounds are diamagnetic and, except where color results from the anion or charge-transfer bands, colorless.

The relative stabilities of the Cu^I and Cu^{II} states in normal aqueous solution are indicated by the following potential data:

$$Cu^+ + e = Cu \qquad E^0 = 0.52 \text{ V}$$

$$Cu^{2+} + e = Cu^+ \qquad E^0 = 0.153 \text{ V}$$

whence

$$Cu + Cu^{2+} = 2Cu^+ \qquad E^0 = -0.37 \text{ V} \qquad K = [Cu^{2+}]/[Cu^+]^2 = \sim 10^6$$

The relative stabilities of Cu^I and Cu^{II} in aqueous solution depend very strongly on the nature of anions or other ligands present and vary considerably with solvent or the nature of neighboring atoms in a crystal.

In aqueous solution only low equilibrium concentrations of Cu^+ ($<10^{-2}$ M) can exist (see later) and the only simple compounds that are stable to water are the highly insoluble ones such as CuCl or CuCN. This instability toward water is due partly to the greater lattice and solvation energies and higher formation constants for complexes of the Cu^{II} ion, so that ionic Cu^I derivatives are unstable. Of course numerous Cu^I cationic or anionic complexes are stable in aqueous solution.

The equilibrium $2Cu^I \rightleftharpoons Cu + Cu^{II}$ can readily be displaced in either direction. Thus with CN^-, I^-, and Me_2S, Cu^{II} reacts to give the Cu^I compound; with anions that cannot give covalent bonds or bridging groups (e.g., ClO_4^- and SO_4^{2-}) or with complexing agents that have their greater affinity for Cu^{II}, the Cu^{II} state is favored. For example, ethylenediamine reacts with CuCl in aqueous potassium chloride solution:

$$2CuCl + 2en = [Cu\ en_2]^{2+} + 2Cl^- + Cu^0$$

That the latter reaction also depends on the geometry of the ligand, that is, on its chelate nature, is shown by differences in the $[Cu^{2+}]/[Cu^+]^2$ equilibrium with chelating and nonchelating amines. Thus for ethylenediamine, K is $\sim 10^5$, for pentamethylenediamine (which does not chelate) 3×10^{-2}, and for ammonia 2×10^{-2}. Hence in the last case the favored reaction is

$$[Cu(NH_3)_4]^{2+} + Cu^0 = 2[Cu(NH_3)_2]^+$$

In ClO_4^- solution from 6.4 to 14.2 M in NH_3 only $[Cu(NH_3)_2]^+$ and $[Cu(NH_3)_3]^+$ occur.

The lifetime of the Cu^+ aqua ion in water depends strongly on conditions. Usually disproportionation is very fast (<1 s), but ~ 0.01 M solutions prepared in 0.1 M $HClO_4$ at 0°C by the reaction

or by reduction of Cu^{2+} with V^{2+} or Cr^{2+}, may last for several hours if air is excluded.

An excellent illustration of how the stability of Cu^I relative to Cu^{II} may be affected by solvent is provided by acetonitrile. The Cu^+ ion is very effectively solvated by MeCN, and the copper(I) halides have relatively high solubilities (e.g., CuI, 35 g/kg MeCN), versus negligible solubilities in H_2O. Copper(I) is more stable than Cu^{II} in MeCN and the latter is, in fact, a comparatively powerful oxidizing agent. The tetrahedral ion $[Cu(MeCN)_4]^+$ can be isolated in salts with large anions (e.g., ClO_4^- and PF_6^-).

Copper(I) Binary Compounds

The oxide and the sulfide are more stable than the corresponding Cu^{II} compounds at high temperatures. Copper(I) oxide (Cu_2O) is made as a yellow powder by controlled reduction of an alkaline solution of a Cu^{2+} salt with hydrazine or, as red crystals, by thermal decomposition of CuO. A yellow "hydroxide" is precipitated from the metastable Cu^+ solution mentioned previously. Copper(I) sulfide (Cu_2S) is a black crystalline solid prepared by heating copper and sulfur in the absence of air.

Alkali metal oxocuprates are made by heating Cu_2O with Na_2O, K_2O, and so on. The compounds $M_4^I Cu_4O_4$ have $[Cu_4O_4]^{4-}$ rings with the four Cu atoms coplanar. There are similar Ag and Au compounds.

Copper(I) chloride and bromide are made by boiling an acidic solution of the Cu^{II} salt with an excess of Cu; on dilution, white CuCl or pale yellow CuBr is precipitated. Addition of I^- to a solution of Cu^{2+} forms a precipitate that rapidly and quantitatively decomposes to CuI and iodine. Copper(I) fluoride is unknown. The halides have the zinc blende structure (tetrahedrally coordinated Cu^+). Copper(I) chloride and CuBr are polymeric in the vapor state, and for CuCl the principal species appears to be a six ring of alternating Cu and Cl atoms with Cu—Cl, ~2.16 Å. White CuCl becomes deep blue at 178°C and melts to a deep green liquid. The halides are very insoluble in water but are solubilized by complex formation

$$CuX(s) + nX^- \rightleftharpoons [CuX_{n+1}]^{n-} \quad n = 2\text{-}4$$

and also by complexing with CN^- or NH_3.

Other relatively common Cu^I compounds are the cyanide, conveniently prepared by the reaction

$$2Cu^{2+}(aq) + 4CN^-(aq) \longrightarrow 2CuCN(s) + C_2N_2$$

and soluble in an excess of cyanide to give the ions $Cu(CN)_2^-$, $Cu(CN)_3^{2-}$, and $Cu(CN)_4^{3-}$. Although compounds such as $KCu(CN)_2$ and $NaCu(CN)_2 \cdot 2H_2O$ are known, they do not contain a simple ion analogous to $[Ag(CN)_2]^-$, but instead have infinite chains. However, in $Na_2[Cu(CN)_3] \cdot 3H_2O$ there is a discrete 3-coordinate ion.

Copper(I) Carboxylates, Triflate, Alkoxides, and Dialkylamides. The carboxylates have varied structures. The acetate that is obtained as white air-sensitive crystals by reduction of Cu^{II} acetate by Cu in pyridine or MeCN has a planar chain structure (17-H-I). By contrast the trifluoroacetate $[CuO_2CCF_3]_4 \cdot 2C_6H_6$, and benzoate $[CuO_2CPh]_4$ complexes are tetramers with bridging carboxylates as in (17-H-II). This is only one type of Cu_4 polynuclear structure (see later). There are also bridged pyrazole and pyrazolylborate compounds.

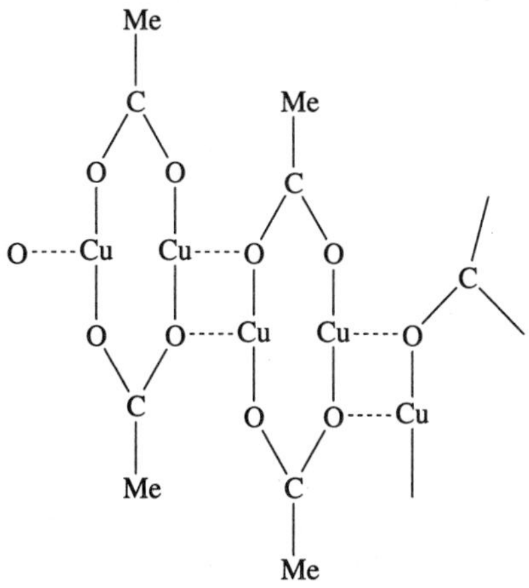

(17-H-I) 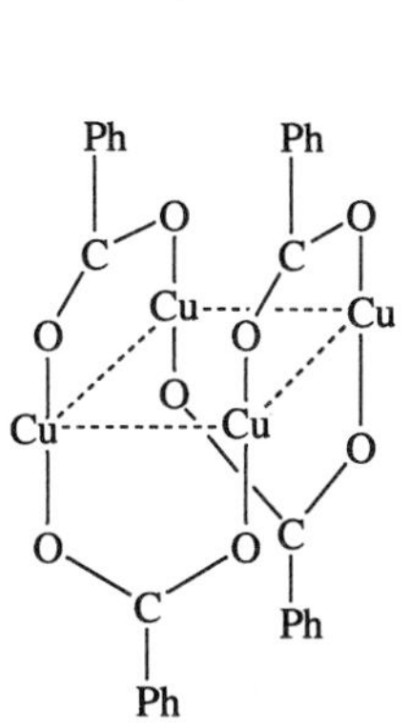   (17-H-II)

Copper(I) trifluoromethanesulfonate can be isolated as a white crystalline, but air-sensitive complex $[Cu(O_3SCF_3)]_2 \cdot C_6H_6$ by interaction of Cu_2O and trifluoromethanesulfonic anhydride in benzene. The benzene is readily displaced by a variety of olefins to give cationic olefin complexes (see later). The complex also catalyzes the cyclopropanation of olefins by use of diazoalkanes $RCHN_2$.

Copper(I) alkoxides (CuOR) *and aryloxides* (CuOAr) are yellow substances that can be made, for example, by the reactions

$$CuCl + LiOR = LiCl + CuOR$$

$$(CuMe)_n + nC_6H_5OH = nCuOC_6H_5 + nCH_4$$

The methoxide is insoluble, but other alkoxides can be sublimed and are soluble in ethers. The *t*-butoxide is a tetramer $[CuOCMe_3]_4$ with alkoxo bridges.

Alkoxides react with organic halides R'I to give ethers ROR'. The butoxide will metallate acidic hydrocarbons such as C_5H_6 or $PhC\equiv CH$ and reacts with CO_2 and NHR_2 to give the carbamate.

Copper(I) Dialkylamides. Only a few are known so far; colorless $[Cu(NEt_2)]_4$ is a tetramer.

Copper(I) Complexes

Since Cu^I has no $d-d$ transitions, its complexes are usually colorless, but there are some that are red or orange because of charge transfer transitions (both LMCT and MLCT).[1] Copper(I) halide and other complexes are usually obtained by (a) direct interaction of ligands (which includes additional halide ions) with copper(I) halides or the triflate, (b) reduction of corresponding copper(II) compounds, or (c) reduction of Cu^{2+} in the presence of, or by, the ligand.

The stoichiometries of the compounds give little clue to their structures, which can be very complicated, being mononuclear, binuclear with halide bridges, polynuclear and the copper atom 2-, 3-, or 4-coordinate, or infinite chains.

Mononuclear species can be of many structural types such as the following, (L = neutral ligands):

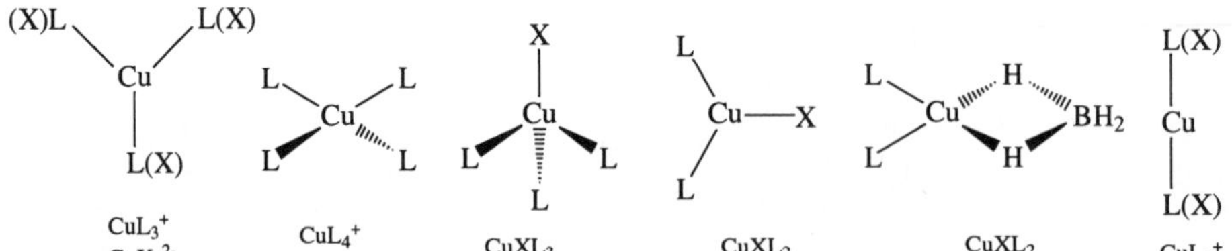

Compounds of a particular stoichiometry, for example, CuXL or $CuXL_2$ may have more than one structure depending on the nature of L. Discrete halogeno anions such as $CuCl_2^-$, $CuBr_2^-$, and $CuBr_3^{2-}$ are known. Many tetrahedral L_4Cu^I and trigonal

[1]M. Melnik *et al., Coord. Chem. Rev.* **1993**, *126*, 71.

L_3Cu^I complexes can be prepared by employing the tetrahedral $[Cu(NCCH_3)_4]PF_6$ as a starting material.[2]

Binuclear Species. Typical formulas Cu_2X_2L, $Cu_2X_2L_4$, $Cu_2X_2L_3$,[3] and $Cu_2X_4^{2-}$ [4] have structures of the following types:

Chain Structures. Halido anions can have the following types:

1. Infinite chains of $CuCl_4$ tetrahedral units sharing edges, e.g., $[Cu(NH_3)_4]$-Cu_2Cl_4.
2. Infinite chains of $CuCl_4$ tetrahedral units sharing corners, e.g., K_2CuCl_3.
3. Infinite double chains of $CuCl_4$ tetrahedra sharing corners, e.g., $CsCu_2Cl_3$.
4. Iodocuprates $[Cu_xI_y]^{(y-x)-}$ show tremendous structural diversity. One of the most remarkable structures is that of $\{[Ph_4P][Cu_3I_4]\}_\infty$ which contains helical chains of face-sharing CuI_4 tetrahedra.[5] Another polymeric structure is formed by tetrahedra sharing edges.[6] In addition to the $Cu_2I_4^{2-}$ ion, there is a $Cu_3I_6^{3-}$ ion formed by two tetrahedra CuI_4 and one trigonal CuI_3 unit sharing I atoms.[7] Even larger anions such as $Cu_8I_{13}^{5-}$ and $Cu_{36}I_{56}^{20-}$ have been characterized.

Spiral chains with almost planar trigonal Cu^I are formed in $[Cu(CN)_2]^-$ ions, $-Cu(CN)(\mu\text{-}CN)Cu(CN)(\mu\text{-}CN)Cu(CN)-$.

Tetranuclear Structures. Copper benzoate and other compounds discussed previously have four Cu atoms that although part of a ring, in themselves form a parallelogram. Tetrameric Cu_4^I complexes may have structures in which the four Cu atoms are in (a) a parallelogram, rectangle, or a square; (b) most commonly, at the vertices of a tetrahedron, regular or slightly distorted, and (c) in a halogen-bridged step structure.

Another class[8] are those with monatomic ligands $(CuXL)_4$ where L is usually R_3P or py and X is usually a halogen but can be SR. These have two main limiting structures, each of which may be distorted. The first is the *cubane structure* (17-H-III) in which there is a Cu_4 tetrahedron with a triply bridging halide on each face and a ligand on each 4-coordinate Cu atom. Many of these compounds are luminescent.[9]

[2]J. R. Black and W. Levason, *J. Chem. Soc., Dalton Trans.* **1994**, 3225.
[3]T. Kräuter and B. Neumüller, *Polyhedron* **1996**, *15*, 2851.
[4]S. Ramaprabhu *et al., Inorg. Chim. Acta* **1994**, *227*, 153.
[5]H. Hartl and F. Mahdjour-Hassan-Abadi, *Angew. Chem. Int. Ed. Engl.* **1994**, *33*, 1841.
[6]G. Hu and E. M. Holt, *Acta Cryst.* **1994**, *C50*, 1578.
[7]G. Hu and E. M. Holt, *Acta Cryst.* **1994**, *C50*, 1576.
[8]M. Munakata *et al., J. Chem. Soc., Dalton Trans.* **1994**, 2771.
[9]P. C. Ford *et al., Inorg. Chem.* **1994**, *33*, 561.

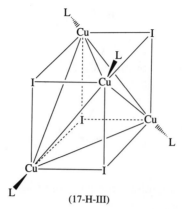

(17-H-III)

The second has the *step form* (17-H-IV) with double and triple halide bridges and two 4-coordinate, tetrahedral and two 3-coordinate, trigonal copper atoms.

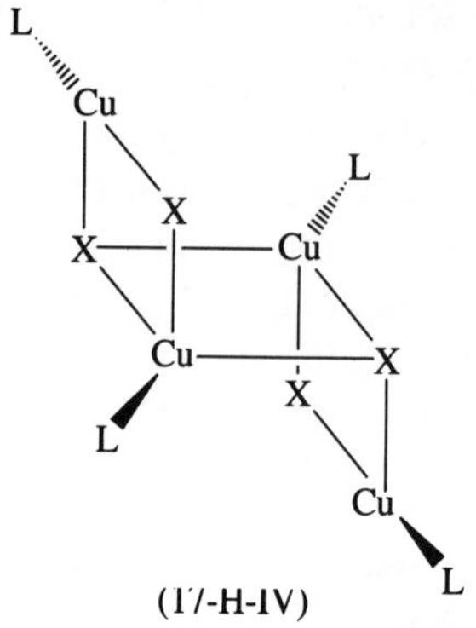

(17-H-IV)

The silver analogues (Section 18-I-2) are similar. Which structure a complex has depends on the sizes of the metal and halide atoms, on the steric bulk of the ligand, and on the solvent from which it crystallized. Some, for example, $[Ph_3PCuX]_4$, X = Br and I, can have both forms. Certain other compounds may have the cubane form but are linked by corner Cu—X bonds into a polymer, for example, $[Cu_4I_4-(NEt_3)_3]_\infty$, while the step form can also give an infinite chain stair structure (17-H-V). Compounds with planar cores $(CuR)_4$ have squares with 2-coordinate copper and bridge groups giving 8-membered rings (17-H-VI). For R = CH_2SiMe_3 and OBu^t the ring is almost planar but for both amido (NR_2)[10] and imido (N—R)[11] bridged species, the N atoms lie alternately above and below the Cu_4 plane.

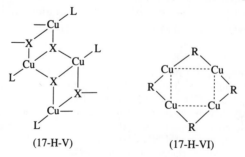

(17-H-V)   (17-H-VI)

[10]P. P. Power *et al., J. Chem. Soc., Dalton Trans.* **1992,** 451.
[11]D. S. Wright *et al., J. Chem. Soc., Dalton Trans.* **1995,** 2707.

Pentanuclear Complexes. In $[Cu_5(\mu\text{-}SBu^i)_6]^-$ there is *tbp* Cu^I with μ-SR and this structure is also found in the aryl species (see later), $MgCu_4Ph_6$, $[Cu_5Ph_6]^-$, and $[Li_2Cu_3Ph_6]^-$. The phenyl groups bridge metal atoms that may be Cu only, Cu + Li, or Cu + Mg.

Other Polynuclear Complexes. $H_8Cu_8(dppp)_4$, dppp $= Ph_2P(CH_2)_3PPh_2$, has a distorted dodecahedron of Cu atoms. Other complexes mainly have sulfur ligands with Cu_8S_{12} and $Cu_{10}S_6$ cores or, for $[CuSR]_8$ a 16-membered ring with alternate Cu and S atoms.

Cu—Cu Interaction in Cu^I Polynuclear Compounds. In many of these structures there are relatively short Cu—Cu distances 2.38 to 3.2 Å (Cu—Cu in metal 2.40 Å). However, there are always bridges present and since Cu^I is d^{10}, metal-metal bonding would appear to be weak if existing at all. The compounds are hence best referred to as aggregates or cages and not metal atom clusters whose defining property is the existence of M—M bonds.

Copper(I) Sulfido Anions. There is a remarkable array of polynuclear anions containing sulfur rings or chains that can be isolated as salts of large cations. They are formed by dissolution of copper sulfides in polysulfide solutions or by the action of sulfur and H_2S on ethanolic solutions of Cu^{II} acetate. Examples are ions such as $[Cu_2S_{20}]^{4-}$, $[Cu_3S_{12}]^{3-}$, and $[Cu_4S_{18}]^{4-}$. Many of the structures are quite complex, for example, (17-H-VII) and (17-H-VIII).

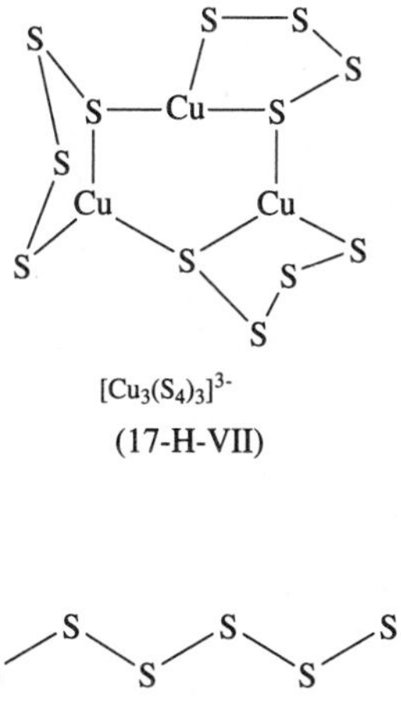

$[Cu_3(S_4)_3]^{3-}$
(17-H-VII)

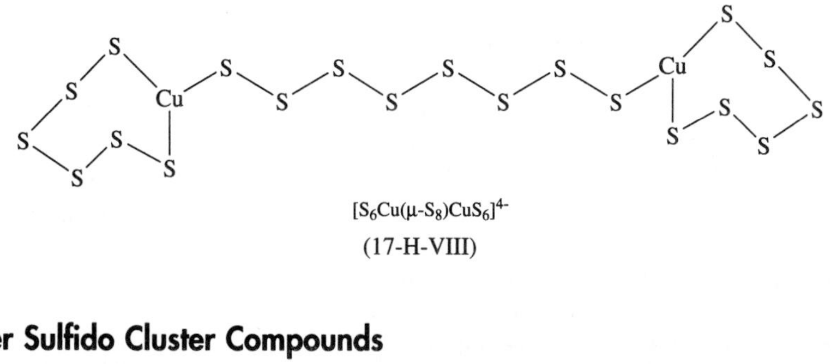

$[S_6Cu(\mu\text{-}S_8)CuS_6]^{4-}$
(17-H-VIII)

Copper Sulfido Cluster Compounds

From the time the first one was discovered in 1968, this class of compounds has grown rapidly in number. The principal types have tetrahedral, octahedral, or cubic

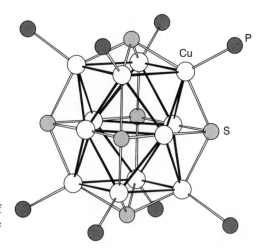

Figure 17-H-1   The cluster framework of $Cu_{12}S_6(PEtPh_2)_8$, adapted from D. Fenske *et al.*, *J. Cluster Sci.* **1966,** *7,* 351.

sets of Cu atoms, as well as larger groups, and in all cases there are sulfur ligands and/or sulfur atoms present. Some of the very large ones incorporate Se or Te atoms. The valence of the copper is exactly or approximately one.

The first one recognized was $Cu_4[S_2P(OPr^i)_2]_4$. It has a distorted tetrahedron of Cu atoms with each face bridged by a pair of S atoms. With the $(RO)_2PS_2^-$ type of ligand there are also distorted octahedral, $Cu_6[S_2P(OR)_2]_6$, and cubic, $Cu_8[S_2P(OR)_2]_8(\mu_8\text{-}S)$, clusters, the latter having an S atom at the center of the cube.[12] There are also cubic clusters such as $Cu_8[S_2CC(CN)_2]_6^{4-}$ that do not have any atom inside the cube. This can be converted to octahedral and tetrahedral species, the latter, $Cu_4[S_2CC(CN)_2]_4$ having the same type structure as the dithiophosphate species cited above. There are still other Cu_4 and Cu_5 clusters.[13]

Recently, new and larger clusters, $Cu_{12}S_6(PR_3)_8$ and $Cu_{20}S_{10}(PR_3)_8$, have been reported,[14] and even bigger ones are known with Se and Te atoms. The largest so far is $Cu_{146}Se_{73}(PPh_3)_3$. In general the oxidation state of copper is again +1. In at least two cases, clusters with the same stoichiometry may differ in structure when they have different PR_3 ligands. The cluster framework of $Cu_{12}S_6(PEtPh_2)_8$ is shown in Fig. 17-H-1.

Copper(I) Organo Compounds

Copper(I), but not copper(II), forms a variety of compounds with Cu—C bonds. Many are useful in organic synthesis.[15]

The *alkyls* and *aryls* may be obtained by interaction of copper(I) halides with lithium or Grignard reagents. The alkyls usually decompose readily but methyl copper, a bright yellow polymer insoluble in organic solvents, is reasonably stable; it can be used in certain organic syntheses, but the use of lithium alkyl cuprates

[12]J. P. Fackler, Jr. *et al., J. Am. Chem. Soc.* **1995,** *117,* 9778.
[13]D. Coucouvanis *et al., J. Am. Chem. Soc.* **1993,** *115,* 11271.
[14]D. Fenske *et al., J. Cluster Sci.* **1996,** *7,* 351.
[15]R. J. K. Taylor, ed., *Organocopper Reagents: A Practical Approach,* Oxford University Press, 1995.

(see later) is more common. Alkyls can be stabilized by phosphine ligands as in, for example, $Bu_3PCu(CH_2)_3CH_3$.

Like the $[CuCH_2SiMe_3]_4$ noted earlier, the aryls are aggregates of Cu_4, Cu_5, Cu_6, or Cu_8 with organic groups acting as bridges, for example, the mesityl $(Cu\ mes)_5$ is a puckered 10-membered ring, which undergoes ring contraction on treatment with thiophene to give $Cu_4\ mes_4(SC_4H_8)_2$ that now has a $Cu_4C_4S_4$ ring.

The fluoroaryls, for example, $[CuC_6F_5]_4$, and perfluoroalkyls are more stable than hydrocarbon analogues; the aryl group $(2-Me_2N)C_6H_4$ has been much used, since the N atom can also occupy a coordination site and stabilize the molecules.

Mixed aryl compounds such as $Ar_6Cu_6(C{\equiv}CAr')_2$ and $Ar_4Cu_4Ag_2(CF_3SO_3)_2$ are also known.

The decarboxylation of copper perfluorobenzoate in quinoline gives $(CuC_6F_5)_n$. The catalytic action of copper or copper salts on decarboxylation reactions of carboxylic acids presumably involves organo intermediates.

Lithium Alkyl Cuprates. These important species are commonly used in ether or a similar solvent for a wide variety of organic syntheses. They are especially useful for C—C bond formation by interaction with organic halides;

$$LiCuR_2 + R'X \longrightarrow R{-}R' + LiX + CuR$$

The species in solution depend on the solvent and the ratio of Cu to LiR; in THF there are several species in equilibrium

$$2LiMe + 2CuMe \longrightarrow LiMe + LiCu_2Me_3 \rightleftharpoons Li_2Cu_2Me_4$$

In the presence of lithium salts these different species can give different reactions depending on the Cu to LiR ratio.

The structures of several of the isolated compounds have been determined. The simpler species are the linear ions $[Cu\ mes_2]^-$ and $\{Cu[C(SiMe_3)_3]_2\}^-$ and the neutral complex $Li_2Cu_2(C_6H_4CH_2NMe_2)_4$.

Alkene Complexes. Of all the metals involved in biological systems only Cu reacts with ethylene, and a Cu^I complex is involved in binding C_2H_4 which acts as a hormone in plants. Formation constants can be determined in solutions and several complexes have been isolated, for example, $Cu(C_2H_4)(bipy)^+$. The reaction

$$Cu + Cu^{2+} + 2C_2H_4 \underset{}{\overset{H_2O}{\rightleftharpoons}} 2[Cu(C_2H_4)(H_2O)_2]^+$$

occurs with perchlorate anions. Other alkene complexes can be made by direct interaction of CuCl or $CuCF_3SO_3$ with olefins or by reduction of Cu^{II} salts by trialkylphosphites in ethanol in the presence of alkene. Crystalline compounds are obtained when using chelating alkenes such as norbornadiene or cyclic polyenes.

Cationic complexes are readily obtained by displacement of benzene from $(CuSO_3CF_3)_2 \cdot C_6H_6$. These are thermally stable and often soluble in organic solvents. They are of the types LCu^+ and L_2Cu^+, where L is a chelating diolefin such as cycloocta-l,5-diene.

Acetylene Compounds. Copper(I) chloride in concentrated hydrochloric acid absorbs acetylene to give colorless species such as $CuCl(C_2H_2)$ and $[CuCl_2(C_2H_2)]^-$. These halide solutions can also catalyze the conversion of acetylene into vinylacetylene (in concentrated alkali chloride solution) or to vinyl chloride (at high HCl concentration), and the reaction of acetylene with hydrogen cyanide to give acrylonitrile is also catalyzed.

Copper(I) ammine solutions react with acetylenes containing the $HC{\equiv}C$ group to give yellow or red precipitates, which are believed to have the $RC{\equiv}C{-}Cu$ unit π bonded to another Cu atom. The trimer $[Et_3PCuC{\equiv}CMe]_3$ is of this type.

Copper(I) acetylides provide a useful route to the synthesis of a variety of organic acetylenic compounds and heterocycles, by reaction with aryl and other halides. A particularly important indirect use, where acetylides are probable intermediates, is the oxidative dimerization of acetylenes. A common procedure is to use the N,N,N',N'-tetramethylethylenediamine complex of CuCl in a solvent, or CuCl in pyridine-methanol, and oxygen as a reoxidant for Cu^+:

$$2RC{\equiv}CH + 2Cu^{2+} + 2py \longrightarrow RC{\equiv}C{-}C{\equiv}CR + 2Cu^+ + 2pyH^+$$

Compounds with μ_2-η^1 bridging acetylides can also be prepared,[16] an example being $(PMePh_2)_2Cu(\mu_2\text{-}\eta^1\text{-}C{\equiv}CPh)_2Cu(PMePh_2)_2$ in which the Cu atoms are in distorted tetrahedral coordination.

Carbonyls. Simple carbonyl complexes are not stable. Thus $CuCl(CO)$ exists only under CO pressure. At a CO pressure of 1 atm the following reaction has been observed:

$$Cu + Cu^{2+} + 2CO \underset{}{\overset{H_2O}{\rightleftharpoons}} 2[Cu(CO)(H_2O)_2]^+$$

It is also reported that $Cu(CO)_3^+$ and $Cu(CO)_4^+$ are formed in concentrated H_2SO_4 under pressure. When other ligands are present stable species such as $CpCu(CO)$, $TpCu(CO)$, and $[(dien)Cu(CO)]^+$ are formed.

Hydrides. A series of $Li_nCu_mH_{m+n}$ have been obtained by interaction of lithium methylcuprates with $LiAlH_4$ in ether, for example,

$$LiCuMe_2 + LiAlH_4 \longrightarrow LiCuH_2 + LiAlH_2Me_2$$

More complex compounds include $Cu_2(\mu\text{-}H)_2(triphos)_2$ and the $[HCuPR_3]_6$ molecules that have an octahedron of Cu atoms with the hydrogen atoms (not seen by X-ray) probably bridging.

17-H-3 The Chemistry of Copper(II), d^9

Most Cu^I compounds are fairly readily oxidized to Cu^{II} compounds, but further oxidation to Cu^{III} is more difficult. There is a well-defined aqueous chemistry of

[16]V. W. Yam *et al.*, *J. Chem. Soc., Dalton Trans.* **1996**, 2889.

Table 17-H-2   Interatomic Distances in Some Copper(II) Coordination Polyhedra

Compound	Distances (Å)
$CuCl_2$	4 Cl at 2.30, 2 Cl at 2.95
$CsCuCl_3$	4 Cl at 2.30, 2 Cl at 2.65
$CuCl_2 \cdot 2H_2O$	2 O at 2.01, 2 Cl at 2.31, 2 Cl at 2.98
$CuBr_2$	4 Br at 2.40, 2 Br at 3.18
CuF_2	4 F at 1.93, 2 F at 2.27
$[Cu(H_2O)_2(NH_3)_4]$ in $CuSO_4 \cdot 4NH_3 \cdot H_2O$	4 N at 2.05, 1 O at 2.59, 1 O at 3.37
$Cu(ClO_4)_2$	4 O at 1.96, 2 O at 2.46

Cu^{2+}, and a large number of salts of various anions, many of which are water soluble, exist in addition to a wealth of complexes. The aqua ion $[Cu(H_2O)_6]^{2+}$ is tetragonally distorted both in crystals and in solution.

Stereochemistry

The d^9 configuration makes Cu^{II} subject to Jahn-Teller distortion if placed in an environment of cubic (i.e., regular octahedral or tetrahedral) symmetry, and this has a profound effect on all its stereochemistry. With but a few exceptions, mentioned below, it is never observed in these regular environments. When 6-coordinate, the "octahedron" is usually severely distorted, as indicated by the data in Table 17-H-2. The typical distortion is an elongation along one fourfold axis, so that there is a planar array of four short Cu—L bonds and two *trans* long ones. In the limit, of course, the elongation leads to a situation indistinguishable from square coordination as found in CuO and many discrete complexes of Cu^{II}. Thus the cases of tetragonally distorted octahedral coordination and square coordination cannot be sharply differentiated.

There are numerous cases in which apparently octahedral Cu^{II} complexes execute a dynamic (pulsating) Jahn-Teller behavior, where the direction of the elongation varies rapidly.[17] Other unusual effects are a switching of the direction of elongation with pressure change.[18] The only cases in which Cu^{II} seems to be in a truly undistorted octahedral environment are $K_2Pb[Cu(NO_2)_6]$ and $[Cu(H_2O)_6]SiF_6$. In the latter[19] three-fourths of the $[Cu(H_2O)_6]^{2+}$ octahedra are distorted in the usual way, while one-fourth are regular.

Chloro Complexes

Chloro complexes are formed in aqueous solutions and many salts have been isolated. Salts of stoichiometry M^ICuCl_3 usually contain $[Cu_2Cl_6]^{2-}$ ions and large cations (e.g., Ph_4P^+) to keep the anions well separated; the $[Cu_2Cl_6]^{2-}$ ions are formed by two tetrahedra sharing an edge. For smaller cations, the dimers become linked by long Cu—Cl bonds giving infinite chains with 5- or 6-coordinate Cu^{II}. The

[17]J. S. Wood *et al., Inorg. Chem.* **1996,** *35,* 1214.
[18]C. J. Simmons *et al., J. Am. Chem. Soc.* **1993,** *115,* 11304.
[19]F. A. Cotton *et al., Inorg. Chem.* **1993,** *32,* 4861.

chloro complex $CsCuCl_3$ is unique in forming infinite chains of distorted octahedra sharing opposite triangular faces.

For $[CuCl_4]^{2-}$, theory predicts the flattened tetrahedral (D_{2d}) geometry in $M_2^I[CuCl_4]$ when the cation is very large, thus isolating the anions (Fig. 17-H-2). With smaller cations, linking of $CuCl_4$ units occurs as for the $[Cu_2Cl_6]^{2-}$ units just mentioned and a linked two-dimensional layer structure results in which Cu^{II} is in a tetragonally elongated octahedron. The compound $(NH_4)_2CuCl_4$ contains square $[CuCl_4]^{2-}$ ions.

Some compounds with formulas corresponding to $[CuCl_5]^{3-}$ do, in fact, contain such discrete *tbp* anions if the cations are large and tripositive, for example, $[M(NH_3)_6]^{3+}$. However, with smaller cations (e.g., $M_3^ICuCl_5$ of the alkalis) there are discrete $[CuCl_4]^{2-}$ ions of D_{2d} structure together with isolated Cl^- ions.

Other Complexes

Distorted tetrahedral species occur and $[Cu(C_6H_{11}NH_2)_4](NO_3)_2$, for example, is intermediate between tetrahedral and square. Distorted tetrahedral structures with imidazole, cysteine, methionine, and similar ligands have been greatly studied as models for Cu^{II} in azurin and plastacyanin.

Similar distorted Cu^{II} is found in some Schiff base complexes with bulky substituents on N and in some dipyrromethane species. With these few exceptions neutral 4-coordinate Cu^{II} complexes with chelate ligands have planar coordination.

Variants on this include some cases where additional ligands complete a very elongated octahedron and many where there is dimerization of the type shown schematically in (17-H-IX) for the β form of bis(8-quinolinolato)copper(II), in which each metal atom becomes 5-coordinate.

(17-H-IX)

$Cu(dike)_2$ complexes, which are planar (as well as Cu^I derivatives such as $Cu(hfac)L$) have been extensively studied as agents for chemical vapor deposition

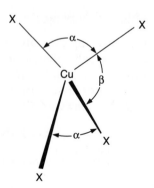

Figure 17-H-2   Squashed tetrahedral structures of $[CuX_4]^{2-}$ ions in Cs_2CuX_4 salts: $\alpha > \beta$.

(CVD) of metallic copper films.[20] The search for precursors for (potentially super-conducting) $LnM_2^{II}Cu_3O_{7-x}$ phases has led to synthesis of heteronuclear complexes containing Cu^{II} combined with lanthanide and/or alkaline earth (M^{II}) ions.[21]

Trigonal bipyramidal coordination is found in other cases, mostly with distortion. Examples are $Cu(terpy)Cl_2$, $[Cu(bipy)_2I]^+$, and $[Cu(phen)_2(H_2O)]^{2+}$.

Spectral and Magnetic Properties

Because of the relatively low symmetry (i.e., less than cubic) of the environments in which the Cu^{2+} ion is characteristically found, detailed interpretations of the spectra and magnetic properties are somewhat complicated, even though one is dealing with the equivalent of a 1-electron case. Virtually all complexes and compounds are blue or green. Exceptions are generally caused by strong uv bands—charge-transfer bands—tailing off into the blue end of the visible spectrum, thus causing the substances to appear red or brown. The blue or green colors are due to the presence of an absorption band in the 600 to 900 nm region of the spectrum. The envelopes of these bands are generally unsymmetrical, seeming to encompass several overlapping transitions, but definitive resolution into the proper number of subbands with correct locations is difficult. Only when polarized spectra of single crystals have been measured has this resolution been achieved unambiguously.

The magnetic moments of mononuclear Cu^{II} complexes are generally in the range 1.75 to 2.20 BM, regardless of stereochemistry and independently of temperature except at extremely low temperatures (<5 K). Magnetic exchange between Cu^{2+} ions in $Cu(\mu-X)_2Cu$ situations (X = Cl, Br, etc.) remain a subject of study, with the object of developing generalizations about the relationship between structure parameters such as the CuXCu angle and the $Cu\cdots Cu$ distance. Such exchanges are generally antiferromagnetic and take place through the X bridges rather than by direct $Cu\cdots Cu$ interaction.[22] Magnetic coupling which may be either ferromagnetic or antiferromagnetic, depending on the compositional and structural details, has been studied in many other dinuclear Cu^{II} complexes with a wide variety of structures.[23]

Binary Copper(II) Compounds

Black crystalline CuO is obtained by pyrolysis of the nitrate or other oxo salts; above 800°C it decomposes to Cu_2O. The hydroxide is obtained as a blue bulky precipitate on addition of alkali hydroxide to cupric solutions; warming an aqueous slurry dehydrates this to the oxide. The hydroxide is readily soluble in strong acids and also in concentrated alkali hydroxides, to give deep blue anions, for example, $[Cu(OH)_4]^{2-}$ and $[Cu(OH)_6]^{4-}$. In aqueous ammonia deep blue ammine complexes are formed, as described later.

Copper sulfide, CuS, is more complex structurally than the formula suggests. It contains S_2^{2-} as well as S^{2-} ions and is probably $Cu_2^+Cu^{2+}(S^{2-})(S_2^{2-})$. There are a

[20]M. J. Hampden-Smith and T. T. Kodas, *Polyhedron* **1995**, *14*, 699.
[21]S. R. Breeze and S. Wang, *Inorg. Chem.* **1994**, *33*, 5113.
[22]M. A. Romero *et al.*, *Inorg. Chem.* **1994**, *33*, 5477.
[23]L. K. Thompson *et al.*, *Inorg. Chem.* **1994**, *33*, 5555.

number of ternary sulfides such as $NaCuS_4$ that contain only Cu^I together with S_4^{2-} ions.[24]

Copper(II), and also copper(III), form oxometallate ions of various structural types, but mainly containing linked CuO_4 planar units. Examples are Ba_2CuO_3, Sr_2CuO_3, and $BaCuO_2$. Mixed oxides of stoichiometries of the type $La_{2-x}M_x^{II}CuO_{4-y}$, with $x \le 0.2$, $y \simeq 0$, have been under intense study as high temperature superconducting materials.

The *halides* are the colorless CuF_2, with a distorted rutile structure, the yellow chloride and the almost black bromide, the last two having structures with infinite parallel bands of square CuX_4 units sharing edges. The bands are arranged so that a tetragonally elongated octahedron is completed about each copper atom by bromine atoms of neighboring chains. Copper dichloride and $CuBr_2$ are readily soluble in water, from which hydrates may be crystallized, and also in donor solvents such as acetone, alcohol, and pyridine.

Salts of Oxo Acids. The most familiar compound is the blue hydrated sulfate $CuSO_4 \cdot 5H_2O$, which contains four water molecules in the plane with O atoms of SO_4^{2-} groups occupying the axial positions, and the fifth water molecule hydrogen bonded in the lattice. It may be dehydrated to the virtually white anhydrous substance. The hydrated nitrate cannot be fully dehydrated without decomposition. The anhydrous nitrate is prepared by dissolving the metal in a solution of N_2O_4 in ethyl acetate and crystallizing the salt $Cu(NO_3)_2 \cdot N_2O_4$, which probably has the constitution $[NO^+][Cu(NO_3)_3^-]$. When heated at 90°C this solvate gives the blue $Cu(NO_3)_2$, which can be sublimed without decomposition in a vacuum at 150 to 200°C. There are two forms of the solid, both possessing complex structures in which Cu^{II} ions are linked together by nitrate ions in an infinite array. The vapor has planar molecules with chelated NO_3^-.

The triflate, $Cu(CF_3SO_3)_2$, is soluble in MeOH, MeCN, and DMF.

Aqueous Chemistry and Complexes of Copper(II)

Most Cu^{II} salts dissolve readily in water and give the aqua ion. Addition of ligands to such aqueous solutions leads to the formation of complexes by successive displacement of water molecules. With NH_3, for example, the species $[Cu(NH_3)(H_2O)_5]^{2+}$, . . . , $[Cu(NH_3)_4(H_2O)_2]^{2+}$ are formed in the normal way. Addition of the fifth NH_3 can occur in aqueous solution, but the sixth occurs only in liquid ammonia. The reason for this unusual behavior is connected with the Jahn-Teller effect. Because of it, the Cu^{II} ion does not bind the fifth and sixth ligands strongly (even the H_2O). When this intrinsic weak binding of the fifth and sixth ligands is added to the normally expected decrease in the stepwise formation constants, the formation constants K_5 and K_6 are very small indeed. Similarly, it is found with ethylenediamine that $[Cu\ en(H_2O)_4]^{2+}$ and $[Cu\ en_2(H_2O)_2]^{2+}$ form readily, but $[Cu\ en_3]^{2+}$ is formed only at extremely high concentrations of en. Many other amine complexes of Cu^{II} are known, and all are much more intensely blue than the aqua ion. This is because the amines produce a stronger ligand field, which causes the absorption band to move from the far red to the middle of the red region of the spectrum.

---

[24]M. Kanatzidis *et al.*, *J. Am. Chem. Soc.* **1996,** *118,* 693 and earlier references therein.

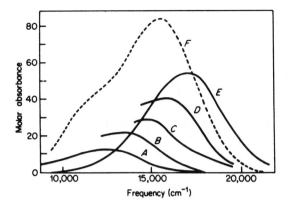

Figure 17-H-3 Absorption spectra of (A) $[Cu(H_2O)_6]^{2+}$ and of the ammines in 2 M ammonium nitrate at 25°C: (B) $[Cu(NH_3)(H_2O)_5]^{2+}$; (C) $[Cu(NH_3)_2(H_2O)_4]^{2+}$; (D) $[Cu(NH_3)_3(H_2O)_3]^{2+}$; (E) $[Cu(NH_3)_4(H_2O)_2]^{2+}$; and (F) $[Cu(NH_3)_5H_2O]^{2+}$.

For example, in the aqua ion the absorption maximum is at ~800 nm, whereas in $[Cu(NH_3)_4(H_2O)_2]^{2+}$ it is at ~600 nm, as shown in Fig. 17-H-3. The reversal of the shifts with increasing takeup of ammonia for the fifth ammonia is to be noted, indicating again the weaker bonding of the fifth ammonia molecule.

In halide solutions the equilibrium concentrations of the various possible species depend on the conditions; although $CuCl_5^{3-}$ has only a low formation constant, it is precipitated from solutions by large cations of similar charge (see above).

Many other Cu^{II} complexes may be isolated by treating aqueous solutions with ligands. When the ligands are such as to form neutral, water-insoluble complexes, as in the following equation, the complexes are precipitated and can be purified by recrystallization from organic solvents.

$$Cu^{2+}(aq) + 2 \quad \text{[structure]} \longrightarrow \text{[structure]}$$

The bis(acetylacetonato)copper(II) complex is another example of this type.

Although as noted previously, addition of CN^- normally leads to reduction to CuCN, in the presence of nitrogen donors like 1,10-phenanthroline reduction is inhibited and 5-coordinate complexes like $[Cu\ phen_2(CN)]^+$ and $[Cu\ phen_2(CN)_2]$ are obtained.

Multidentate ligands that coordinate through oxygen or nitrogen, such as amino acids, form Cu^{II} complexes, often of considerable complexity.

The well-known blue solutions formed by addition of tartrate to Cu^{2+} solutions (known as Fehling's solution when basic and when meso-tartrate is used) may contain monomeric, dimeric, or polymeric species at different pH values. One of the dimers, $Na_2[Cu\{(\pm)C_4O_6H_2\}]\cdot 5H_2O$, has square coordinate Cu^{II}, two tartrate bridges, and a $Cu-Cu$ distance of 2.99 Å.

Halide complexes have been noted earlier. The main characteristic is the large number of salts of differing stoichiometries and cation-dependent structures that may be crystallized from solutions.

Cage Polynuclear Copper(II) Complexes

There are a number of compounds whose structures are based on a central core, Cu_4O, with six edge-bridging ligand atoms and four more that occupy external positions. The prototypal species $[Cu_4(\mu_4\text{-}O)(\mu_2\text{-}Cl)_6Cl_4]^{4-}$ is shown in 17-H-X. Ideally this structure has full T_d symmetry and each Cu^{II} is in a *tbp* environment. In many derivatives some or all of the Cl ligands are replaced by others.[25] In all such compounds, the magnetic properties show marked antiferromagnetism.

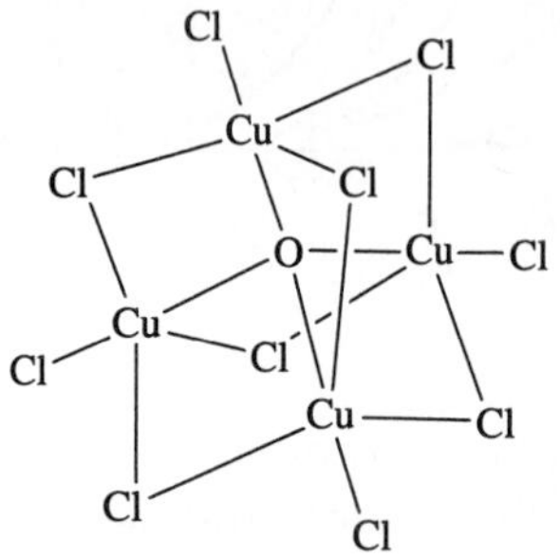

(17-H-X)

A different type of compound, $Cu_4Cl_4(OR)_4$, where the alkoxide is 2-diethyl-aminoethanolate, has a cubane-type structure with a Cu_4O_4 cube and tetrahedral Cu^{II}.

Dimeric or polymeric compounds are also known for alkoxides or aryloxides, for example, $[Cu(OPh)_2en]_2 \cdot 2PhOH$, which has distorted *sp* copper.

Copper(II) Carboxylates. These are readily made by interaction of the acid with $CuCO_3$. They are binuclear with four carboxylate bridges and may have end groups as in (17-H-XI). The 1,3-triazenate (17-H-XII) is similar. In these compounds the Cu—Cu distance varies from ~2.44 to 2.81 Å. There is weak coupling of the unpaired electrons, one on each Cu^{II} ion, giving rise to a singlet ground state with a triplet state lying only a few kilojoules per mole above it; the latter state is thus appreciably populated at normal temperatures and the compounds are paramagnetic. At 25°C, μ_{eff} is typically ~1.4 BM/Cu atom and the temperature dependence is very pronounced.

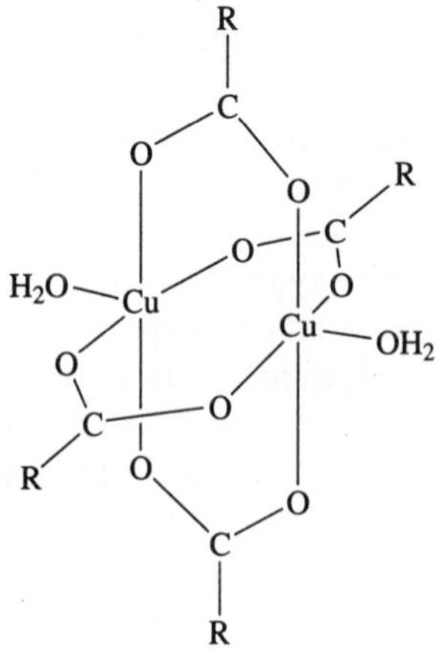

(17-H-XI)

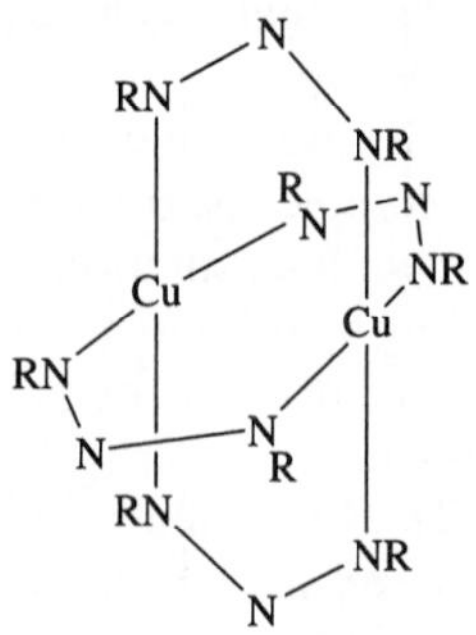

(17-H-XII)

[25]B. Krebs *et al.*, *J. Chem. Soc., Dalton Trans.* **1995**, 2649.

Phenomenologically the interaction in the dinuclear acetate and many other compounds with similar temperature dependences of the magnetic moment can be described as antiferromagnetic couplings of the unpaired spins on the adjacent Cu^{II} atoms, without invoking any $Cu-Cu$ bonding.

Attempts to specify this interaction in detail have been plagued by controversy, and there are still differences of opinion. The interaction appears basically to be one between orbitals of δ symmetry, but whether this is primarily a direct interaction between $d_{x^2-y^2}$ orbitals of the two metal atoms or one that is substantially transmitted through the π orbitals of the bridging RCO_2^- groups is not clear.

Mixed Valence Complexes. A number of compounds containing both Cu^I and Cu^{II} are well established. The sulfides, for example, $K_3Cu_8S_6$ made by heating K_2CO_3, Cu, and S, may have metal-like electrical conductivities and magnetic properties.

Oxidation Catalysis by Copper; Peroxo and Superoxo Complexes. Copper ions and compounds participate in or catalyze a variety of oxidation reactions that consume O_2. This is one of the several key biochemical roles of copper and much of the recent work on the subject has been done in efforts to model the biological systems. In some (non-biological) cases, e.g., the Wacker process, copper(II) itself may be the actual oxidant, but usually it serves as a carrier of oxygen.

There are compounds in which two Cu^+ ions, held in proximity by a binucleating bridging ligand, react with O_2 to give a violet Cu^{2+} species with a peroxo bridge:

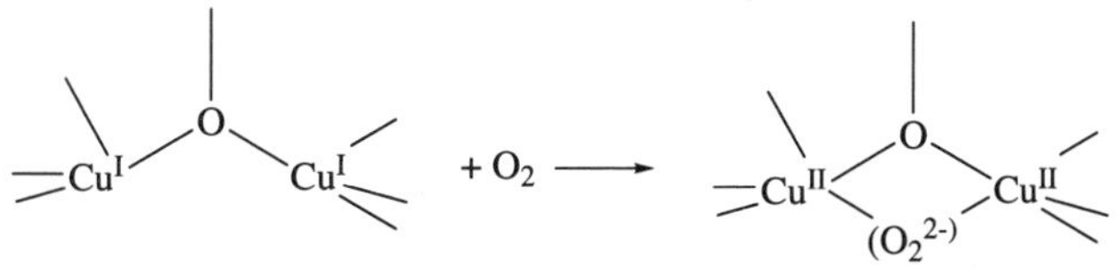

The manner in which the peroxide is held is not certain but there are other cases in which a peroxo complex can be isolated and characterized. In one case[26] the arrangement is as in 17-H-XIII(a), whereas in another[27] it is as in 17-H-XIII(b).

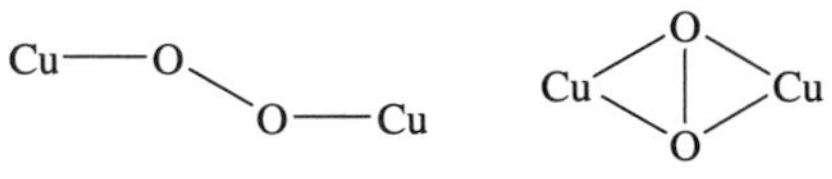

(a) (b)

(17-H-XIII)

There is also evidence[28] that a peroxo species of the second type can be in equilibrium with a $(\mu\text{-}O)_2$ species, which then contains Cu^{III}:

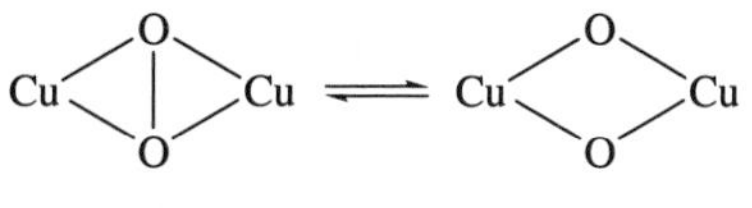

[26]K. D. Karlin *et al., J. Am. Chem. Soc.* **1993,** *115,* 2677.
[27]N. Kitajima *et al., J. Am. Chem. Soc.* **1992,** *114,* 1277.
[28]W. B. Tolman *et al., Science* **1996,** *271,* 1397.

It is believed that $Cu^{II}(O_2^-)$ complexes play a key role in numerous monoxygenases and a model compound, Tp^*CuO_2, has been prepared.[29] The superoxide ion is bound to the copper atom in a symmetrical, side-on fashion, with $Cu-O$ distances of 1.84(1) Å and an $O-O$ distance at 1.22(3) Å, consistent with O_2^- rather than O_2^{2-}.

17-H-4   The Chemistry of Copper(III)

This oxidation state is uncommon but far from unknown and Cu^{III} appears to have a biochemical role (see Section 17-H-5).

Among the simpler compounds of Cu^{III} are some alkali and alkaline earth mixed oxides, e.g., $NaCuO_2$, obtained by heating the mixed oxides in O_2. Fluorination of a mixture of 3KCl and CuCl affords K_3CuF_6 as a pale green paramagnetic solid, the only high-spin Cu^{III} compound known.

Some other examples of Cu^{III} compounds are:

1. $CuBr_2(S_2CNBu_2^i)$, violet needles obtained by treatment of $Cu(S_2CNBu_2^i)$ with Br_2 in CS_2.
2. $KCu\,bi_2$ (where $biH_2 = H_2NCONHCONH_2$) which has square planar Cu^{III}.
3. The products of alkaline ClO_2^- oxidation of Cu^{2+} in presence of iodate or tellurate, viz., $[Cu(IO_5OH)_2]^{5-}$ and $\{Cu[TeO_4(OH)_2]_2\}^{5-}$.
4. A stable, intensely violet anion (shown in 17-H-XIV) which is obtained by oxidation of Cu^{2+} in the presence oxalodihydrazide and acetaldehyde. This 5-coordinate structure was established by X-ray crystallography.

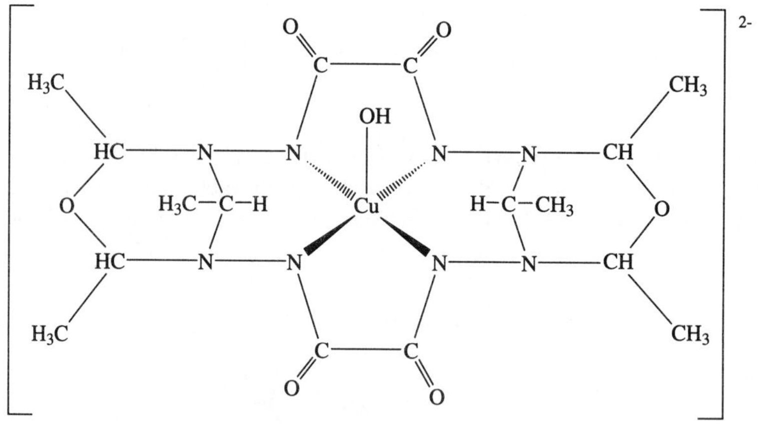

(17-H-XIV)

A variety of deprotonated peptide complexes of Cu^{III} are reasonably stable in alkaline solution. The $Cu^{III}-Cu^{II}$ potentials are very sensitive to the nature of the ligand and vary from 0.45 to 1.02 V; for example, for the tetraglycine (GH_4) complex (17-H-XV).

$$[Cu^{III}H_{-3}G]^- + e = [Cu^{II}H_{-3}G]^{2-} \qquad E^0 = +0.631 \text{ V}$$

[29]N. Kitajima *et al.*, *J. Am. Chem. Soc.* **1994**, *116*, 12079.

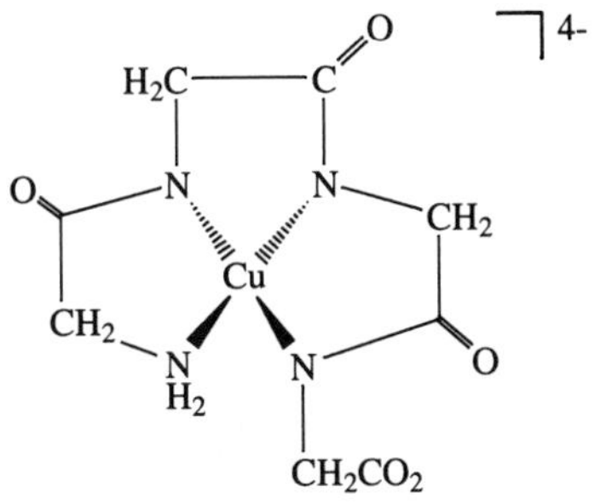

(17-H-XV)

A cationic Cu^III complex that is stable in acid solution has been made by electrolysis; it has a deprotonated diglycylethylenediamine ($NH_2CH_2CONHCH_2$-$CH_2NHCOCH_2NH_2$) as ligand. Another deprotonated peptide complex is the triglycylglycine complex (17-H-XV).

Copper(III) also occurs in N_4-macrocyclic complexes where the Cu^{II}/Cu^{III} potential depends on the ring size, and in tetradentate Schiff base complexes.

17-H-5 The Biochemistry of Copper[30]

Copper is the third most abundant metallic element in the human body, following iron and zinc. It also occurs in all other forms of life and it plays a role in the action of a multitude of enzymes that catalyze a great variety of reactions. There are two cross-cutting ways to classify the copper-containing enzymes: (1) According to the structural and spectroscopic characteristics of the copper complex at the active site. (2) According to the function of the enzyme. We shall base discussion on the first of these, with allusions to function as we go along.

There are many kinds of copper sites, which are the enzyme active sites. Most common are those containing copper(II) and they have been the most studied because of their accessibility by epr and uv-vis spectroscopies. There are also Cu^I sites. The three principal types of Cu^{II} sites are:

1. Normal Cu^{II} sites where the spectroscopic characteristics are those of the familiar tetragonally coordinated Cu^{2+} ion.

2. "Blue" copper sites,[31] where Cu^{II} is in an unusual geometry, as for example, that shown in Fig. 17-H-4. The blue color of these sites is caused by very strong absorption in the red part of the visible spectrum that can be assigned for a S→Cu LMCT transition. These Cu^{II} sites also have anomalous epr spectra.

3. Coupled two-copper centers, where the antiferromagnetic coupling leads to diamagnetism and an abnormal visible spectra. There are still other less common but more complex sites.[32]

[30]K. D. Karlin and Z. Tyeklar, Eds., *Bioinorganic Chemistry of Copper,* Chapman and Hall, 1993; W. Kaim and J. Rall, *Angew. Chem. Int. Ed. Engl.* **1996,** *35,* 43; N. Katajima and Y. Moro-oka, *Chem. Rev.* **1994,** *94,* 737.
[31]S. K. Chapman, *Perspectives on Bioinorganic Chemistry,* Vol. 1, R. W. Hay, J. R. Dilworth, and K. B. Nolan, Eds., JAI Press, 1991.
[32]E. I. Solomon *et al., Chem. Rev.* **1996,** *96,* 2563.

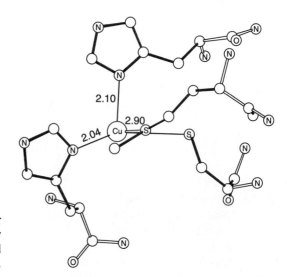

Figure 17-H-4 The "blue" copper site in plastocyanin as determined by X-ray crystallography; the Cu—ligand bond lengths are given in Angstroms.

Superoxide dismutase (CuZn SOD) provides an example of a "normal" copper site. Its function is to catalyze the following reaction:

$$2O_2^- + 2H^+ \longrightarrow H_2O_2 + O_2$$

This rather small (MW $\approx$ 16,000) enzyme has both a Cu^{2+} and a Zn^{2+} ion at its active site, which is shown schematically in Fig. 17-H-5. It is clear that the copper atom is the functional one while the zinc plays a supportive, structural role, perhaps that of holding the bridging imidazolate group in place.[33] The zinc can be replaced by Co or Cd with 70 to 100% retention of activity, but without copper there is no activity. The catalytic cycle is probably as indicated in the following equations:

$$(HisH)_3Cu^{II}(His^-)Zn + O_2^- + H^+ \longrightarrow (HisH)_3Cu^I + (HisH)Zn + O_2$$

$$(HisH)_3Cu^I + O_2^- + H^+ \longrightarrow (HisH)_3Cu^{II}(His^-)Zn + H_2O_2$$

Figure 17-H-5 Schematic drawing of the active site of bovine superoxide dismutase.

[33] J. S. Valentine *et al.*, *J. Am. Chem. Soc.* **1994,** *116,* 9743.

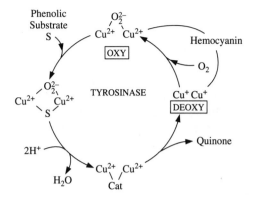

Figure 17-H-6 A proposed catalytic cycle for the role of tyrosinase in oxidizing phenol to *o*-quinone. Cat = catecholate dianion.

The "blue" copper sites seem generally to be involved in electron transfer, with the copper cycling between the +1 and +2 states. The mechanistic details of the reversible redox cycle can, however, display unexpected complexity.[34]

Many of the coupled two-copper (and some multicopper) sites are to be found in oxidases. Their function is to assist in the four-electron reduction of O_2 to water with concomitant oxidation of a substrate. To do this they must first receive and bind the O_2 molecule and the peroxo complexes mentioned earlier appear to be models for the way they do this. A cycle of activity for tyrosinase illustrates this, Fig. 17-H-6.

It should be noted that besides the multicopper site oxygenases, there are also oxygenases having a single-copper site.[35] These interact with the O_2 molecule by forming a $Cu^{II}-O_2^-$ complex rather than a $Cu^{II}(O_2^{2-})Cu^{II}$ complex.

Another function of two-copper sites is seen in the *hemocyanins*.[36] These are the respiratory enzymes found in most molluscs and arthropods. They are proteins of high molecular weight but they consist of subunits. Each subunit has one pair of Cu atoms and can bind one O_2 molecule. It appears that the subunits are able to act cooperatively (as in hemoglobin) but the structural details in this case are unknown. The deoxy form is colorless as it contains Cu^I, while the blue oxy form contains Cu^{II} but gives no epr signal because of the strong pairwise antiferromagnetic coupling.

Cytochrome c Oxidase. This is a ubiquitous enzyme that catalyzes the terminal step in the four-electron reduction of O_2 to $2H_2O$. It contains both heme iron and copper and there are two kinds of copper sites, Cu_A and Cu_B, in each of which there is an association of copper with heme iron. The Cu_B site seems to have one copper atom linked to a heme unit, but details are still obscure.[37] From modeling studies and EXAFS data, opinion presently favors some sort of Fe—X—Cu bridge

[34]A. G. Sykes *et al., J. Chem. Soc., Dalton Trans.* **1994,** 3017.
[35]N. Kitajima and Y. Moro-oka, *Chem. Rev.* **1994,** *94,* 737.
[36]K. A. Magnus *et al., Chem. Rev.* **1994, 94,** 727.
[37]M. J. Scott and R. H. Holm, *J. Am. Chem. Soc.* **1994,** *116,* 11357.

which can be converted to an $Fe-O-O-Cu$ bridge when oxygen is taken up. The identity of the bridging group, X, is unknown.[38]

The Cu_A site may be of exceptional significance in the context of inorganic chemistry, since there is EXAFS evidence that it contains two copper atoms only about 2.50 Å apart.[39] Although this has been questioned,[40] it is also supported by X-ray evidence.[41] It has been suggested that at such a distance, one may postulate a $Cu-Cu$ bond, the first metal-metal bond to be found in a biological context. That concept, however, is controversial. The site has two tetrahedra sharing an edge with μ-S atoms from cysteine residues, and it cycles between Cu^{+1}/Cu^{+1} and $Cu^{+1.5}/Cu^{+1.5}$ states.

[38]R. H. Holm *et al., Inorg. Chem.* **1994,** *33,* 4651; G. Henkel *et al., Angew. Chem. Int. Ed. Engl.* **1995,** *34,* 1488.
[39]N. J. Blackburn *et al., Biochem.* **1994,** *33,* 10401.
[40]H. Bertagnolli and W. Kaim, *Angew. Chem. Int. Ed. Engl.* **1995,** *34,* 771.
[41]E. Sauer-Eriksson *et al., Proc. Natl. Acad. Sci. USA* **1995,** *92,* 11955.

Chapter 18

THE ELEMENTS OF THE SECOND AND THIRD TRANSITION SERIES

GENERAL COMPARISONS WITH THE FIRST TRANSITION SERIES

In general, the second and the third transition series elements of a given group have similar chemical properties but both show pronounced differences from their light congener. A few examples will illustrate this generalization. Although Co^{II} forms a considerable number of tetrahedral and octahedral complexes and is an important species in ordinary aqueous chemistry, Rh^{II} is virtually unknown in mononuclear complexes (but occurs in many Rh_2^{4+} compounds) and Ir^{II} is very rare. Similarly, the Mn^{2+} ion is very stable, but for Tc and Re the II oxidation state is known in relatively few complexes, mostly those with metal-metal bonding or strongly π-acidic ligands. Chromium III forms an enormous number of cationic amine complexes, whereas Mo^{III} and W^{III} form only a few such complexes, none of which is especially important. Again, Cr^{VI} species are powerful oxidizing agents, but there are no large polynuclear oxo anions, whereas Mo^{VI}, W^{VI}, Tc^{VII}, and Re^{VII} are quite stable and give rise to an extensive series of polynuclear oxo anions.

This is not to say that there is no valid analogy between the chemistry of the three series of transition elements. For example, the chemistry of Rh^{III} complexes is in general similar to that of Co^{III} complexes, and here, as elsewhere, the ligand field bands in the spectra of complexes in corresponding oxidation states are similar. On the whole, however, there are certain consistent differences of which the above-mentioned comparisons are only a few among many obvious manifestations.

Some important features of the elements and comparison of these with the corresponding features of the first series are the following:

1. *Radii*. The filling of the $4f$ orbitals (as well as relativistic effects) through the lanthanide elements cause a steady contraction, called the *lanthanide contraction* (Section 19-1), in atomic and ionic sizes. Thus the expected size increases of elements of the third transition series relative to those of the second transition series, due to an increased number of electrons and the higher principal quantum numbers of the outer ones, are almost exactly offset, and there is in general little difference in atomic and ionic sizes between the two heavy atoms of a group, whereas the corresponding atoms

877

and ions of the first transition series are significantly smaller. Relativistic effects in the third transition series also have a significant effect on radii, ionization energies, and other properties (see page 38).

2. *Oxidation States.* For the heavier transition elements, higher oxidation states are in general much more stable than for the elements of the first series. Thus the elements Mo, W, Tc, and Re form oxo anions in high valence states which are not easily reduced, whereas the analogous compounds of the first transition series elements, when they exist, are strong oxidizing agents. Indeed, the heavier elements form many compounds such as RuO_4, WCl_6, and PtF_6 that have no analogues among the lighter ones. At the same time, the chemistry of complexes and aquo ions of the lower valence states, especially II and III, which plays such a large part for the lighter elements, is of relatively little importance for most of the heavier ones.

3. *Aqueous Chemistry.* Aqua ions of low and medium valence states are not in general well defined or important for any of the heavier transition elements, and some, such as Zr, Hf, Nb, and Ta, do not seem to form simple cationic complexes. For most of them anionic oxo and halo complexes play a major role in their aqueous chemistry although some, such as Ru, Rh, Pd, and Pt, do form important cationic complexes as well.

4. *Metal-Metal Bonding.* In general, although not invariably, the heavier transition elements are far more prone to form strong M—M bonds than are their congeners in the first transition series. The main exceptions to this are the polynuclear metal carbonyl compounds and some related ones, where analogous or similar structures are found for all three elements of a given family. Aside from these, however, it is common to find that the first series metal will form few or no M—M bonded species, whereas the heavier congeners form an extensive series. Examples are the $M_6X_{12}^{n+}$ species formed by Nb and Ta, with no V analogues at all, the Mo_3^{IV} and W_3^{IV} oxo clusters for which no chromium analogues exist, and the quadruply-bonded $Tc_2Cl_8^{2-}$ and $Re_2Cl_8^{2-}$ ions, which have no manganese analogue.

5. *Magnetic Properties.* In general, the heavier elements have magnetic properties that are less useful to the chemist than was the case in the first transition series. For one thing there is a much greater tendency to form low-spin complexes, which means that those with an even number of electrons are usually diamagnetic and therefore lack informative magnetic characteristics. The paramagnetic complexes usually have complicated behavior in which magnetic moments differ considerably from spin-only values and often vary markedly with temperature.

18-A ZIRCONIUM AND HAFNIUM: GROUP 4

The chemistries of zirconium and hafnium are more nearly identical than for any other two congeneric elements. This is due in considerable measure the result of the lanthanide contraction having made both the atomic and ionic radii (1.45 and 0.86 Å for Zr and Zr^{4+}; 1.44 and 0.85 Å for Hf and Hf^{4+}) essentially identical.

Table 18-A-1 Oxidation States and Stereochemistry of Zirconium and Hafnium

Oxidation state	Coordination number	Geometry	Examples
Zr^{-II}, Hf^{-II}	6	Octahedral	$[M(CO)_6]^{2-}$
Zr^0, Hf^0	6	Octahedral	$[Zr(bipy)_3]$, $Zr(C_4H_6)_2dmpe$
		π complex	$(arene)_2HfPMe_3$
	7	Pentagonal bipyramidal	$[Zr(CO)_5(SnMe_3)_2]^{2-}$
Zr^I, Hf^I, d^3			
		Complex sheet and cluster structures; see text	
Zr^{II}, d^2			
Zr^{II}, Hf^{II}	8		$Cp_2M(CO)_2$, $CpZrCl(dmpe)_2$,
			$[Zr(CO)_4(SnMe_3)_4]^{2-}$
Zr^{III}, Hf^{III}, d^1	6	Octahedral	$ZrCl_3$, $ZrBr_3$, ZrI_3, HfI_3; see text
Zr^{IV}, Hf^{IV}, d^0	4	Tetrahedral	$ZrCl_4(g)$, $Zr(CH_2C_6H_5)_4$, $Hf(NPh_2)_4$[a]
	5	Trigonal bipyramidal	$K_2Zr_2(OBu^t)_{10}$ (dinuclear compound composed of two five-coordinate Zr centers)
		?	$Zr(NMe_2)_3(diaminoborate)$[b]
	6	Octahedral	Li_2ZrF_6, $Zr(acac)_2Cl_2$, $ZrCl_6^{2-}$, $ZrCl_4(s)$
		Trigonal prismatic	$ZrMe_6^{2-}$
	7	Pentagonal bipyramidal	Na_3ZrF_7, Na_3HfF_7, $K_2CuZr_2F_{12}\cdot 6H_2O$
		Capped trigonal prism	$(NH_4)_3ZrF_7$
		See text, (Fig. 18-A-2)	ZrO_2, HfO_2(monoclinic)
	8	Square antiprism	$Zr(acac)_4$, $Zr(SO_4)_2\cdot 4H_2O$
		Dodecahedral	$[Zr(C_2O_4)_4]^{4-}$, $[ZrX_4(diars)_2]$, $[Zr_4(OH)_8(H_2O)_{16}]^{8+}$
	10	Complex prism	$[Zr(NO_3)_5]^-$ (mononuclear anion in which the Zr atom is surrounded by five chelating ONO_2^- species)[c]

[a]M. Polamo et al., Acta Cryst. **1996,** C52, 1348.
[b]M. Kol et al., J. Chem. Soc., Chem. Commun. **1997,** 229.
[c]E. Stumpp et al., Z. anorg. allg. Chem. **1997,** 623, 449.

The oxidation states and stereochemistries of zirconium and hafnium are summarized in Table 18-A-1. These elements, because of the larger atoms and ions, differ from Ti in having more basic oxides, having somewhat more extensive aqueous chemistry, and more commonly attaining higher coordination numbers, 7 and 8. They have a more limited chemistry of the III oxidation state.

18-A-1 The Elements

Zirconium occurs widely over the earth's crust but not in very concentrated deposits. The major minerals are *baddeleyite,* a form of ZrO_2, and *zircon* ($ZrSiO_4$). The chemical similarity of Zr and Hf is well exemplified in their geochemistry, for Hf

Figure 18-A-1   The zigzag $ZrCl_6$ chains in $ZrCl_4$.

is found in Nature in all zirconium minerals in the range of fractions of a percentage of the Zr content. Separation of the two elements is difficult, even more so than for adjacent lanthanides, but it can be accomplished satisfactorily by ion-exchange, solvent extraction, or electrochemical methods.[1]

Zirconium metal (mp 1855°C ± 15°C), like titanium, is hard and corrosion resistant, resembling stainless steel in appearance. It is made by the Kroll process (Section 17-A-1). Hafnium metal (mp 2222°C ± 30°C) is similar. Like titanium, these metals are fairly resistant to acids, and they are best dissolved in HF where the formation of anionic fluoro complexes is important in the stabilization of the solutions. Zirconium will burn in air at high temperatures, reacting more rapidly with nitrogen than with oxygen, to give a mixture of nitride, oxide, and oxide nitride (Zr_2ON_2).

18-A-2   Compounds of Zirconium(IV) and Hafnium(IV)

Halides

The tetrahalides MCl_4, MBr_4, and MI_4 are all tetrahedral and mononuclear in the gas phase, but the solids are polymers with halide bridging. Zirconium tetrachloride, a white solid subliming at 331°C, has the structure shown in Fig. 18-A-1 with zigzag chains of $ZrCl_6$ octahedra; $ZrBr_4$, $HfCl_4$,[2] and $HfBr_4$ are known to be isotypic. Zirconium tetrachloride resembles $TiCl_4$ in its chemical properties. It may be prepared by chlorination of heated zirconium, zirconium carbide, or a mixture of ZrO_2 and charcoal; it fumes in moist air, and it is hydrolyzed vigorously by water. Hydrolysis proceeds only part way at room temperature, affording the stable oxide chloride

$$ZrCl_4 + 9H_2O \longrightarrow ZrOCl_2 \cdot 8H_2O + 2HCl$$

Zirconium tetrachloride is extremely insoluble in noncoordinating solvents but can be easily solubilized in CH_2Cl_2 in the presence of certain aromatic hydrocarbons such as mesitylene or $1,2,4,5\text{-}Me_4C_6H_2$;[3] $ZrBr_4$ and ZrI_4 are similar to $ZrCl_4$. Zirconium tetrafluoride is a white crystalline solid subliming at 903°C; unlike the other halides, it is insoluble in donor solvents; it has an 8-coordinate structure with square antiprisms joined by shared fluorine atoms. Hydrated fluorides $ZrF_4 \cdot 1$ or $3H_2O$, can be crystallized from HF-HNO$_3$ solutions. The trihydrate has an 8-coordinate structure with two bridging fluorides, $(H_2O)_3F_3ZrF_2ZrF_3(H_2O)_3$. The hafnium trihydrate has the same stoichiometry but a different structure with chains of $HfF_4(H_2O)$

[1]V. Y. Korovin *et al., Russ. J. Appl. Chem.* **1996,** *67,* 678.
[2]R. Niewa and H. Jacobs, *Z. Kristallogr.* **1995,** *210,* 687.
[3]C. Floriani *et al., Angew. Chem. Int. Ed. Engl.* **1995,** *34,* 1510.

units linked through four bridging fluorine atoms. The hafnium monohydrate can also be prepared by reacting metallic hafnium and 40% aqueous HF:

$$Hf + 4HF(aq) \longrightarrow HfF_4 \cdot H_2O + 2H_2$$

and can be dehydrated by heating to 350°C for seven days under a flow of F_2 and N_2, which avoids the formation of $HfOF_2$. The anhydrous HfF_4 obtained by this method is isotypic with β-ZrF_4 and ThF_4.[4]

Hexahalozirconates(IV) and hafnates(IV) are known for F^-, Cl^-, Br^-, and I^-. None of these MX_6^{2-} ions is very resistant to hydrolysis and the iodide species are best obtained by reaction of CsI and MI_4 in a sealed tube.

The tetrachlorides, MCl_4, and some of the other halides also combine with neutral donors to form adducts of varied stereochemistry. In many cases the structures appear to be, or are known to be, octahedral, typically *trans*- or *cis*-MX_4L_2 such as in $HfCl_4(THF)_2$,[5] *cis*-$ZrCl_4(pinacol)_2$[6] and *trans*-$ZrI_4(PMe_2Ph)_2$[7] or *cis*-$MX_4(LL)$ such as in $HfCl_4(dppe)$.[8] However, $ZrCl_4$ also combines with 2 moles of certain chelating bidentate ligands to form $ZrCl_4(LL)_2$ complexes that are usually dodecahedral [e.g., for LL = diars or $MeSCH_2CH_2SMe$] but which can be strongly distorted toward square antiprismatic with ligands that are not compatible with the dodecahedral arrangement. The halides also react with anionic polydentate ligands to give products in which there is partial or even complete replacement of halide ions. For example, the ligands $[N(SiMe_2CH_2PR_2)_2]^-$ react with $ZrCl_4$ or $HfCl_4$ to give, depending on preparative procedures, complexes of the types (18-A-I) or (18-A-II). Ammonolysis of β-ZrF_4 or $ZrF_4 \cdot NH_3$ yields ZrNF, which is isotypic to *baddeleyite* (ZrO_2).[9] Colorless $ZrF_4 \cdot NH_3$ and its Hf analogue can be made by oxidation of the corresponding metal powder with NH_4HF_2; their structures consist of bicapped trigonal prisms $[M(NH_3)F_7]$ connected via edges and corners to form corrugated layers that are held together by hydrogen bonding.[10]

(18-A-I)   (18-A-II)

[4]G. Benner and B. G. Müller, *Z. anorg. allg. Chem.* **1990**, *588*, 33.
[5]S. A. Duraj *et al., Acta Cryst.* **1990**, *C46*, 890 and references therein.
[6]J. D. Wuest *et al., Inorg. Chem.* **1990**, *29*, 951.
[7]F. A. Cotton and W. A. Wojtczak, *Acta Cryst.* **1994**, *C50*, 1662.
[8]F. A. Cotton *et al., Acta Cryst.* **1991**, *C47*, 89.
[9]E. Schweda *et al., Z. anorg. allg. Chem.* **1993**, *619*, 367.
[10]G. Meyer *et al., Z. anorg. allg. Chem.* **1997**, *623*, 79.

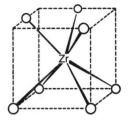

Figure 18-A-2   Coordination geometry in the baddeleyite form of ZrO_2.

Zirconium Oxide and Mixed Oxides

Addition of hydroxide to zirconium(IV) solutions causes the precipitation of white gelatinous $ZrO_2 \cdot nH_2O$, where the water content is variable. On strong heating, this hydrous oxide gives hard, white, insoluble ZrO_2.[11] This has an extremely high melting point (2700°C), exceptional resistance to attack by both acids and alkalis, and good mechanical properties; it is used for crucibles and furnace cores. Zirconium dioxide in its monoclinic (baddeleyite) form and one form of HfO_2 are isomorphous and have a structure in which the metal atoms are 7-coordinate, as shown in Fig. 18-A-2. Three other forms of ZrO_2 have been described, but none has the rutile structure so often found among MO_2 compounds.

A number of compounds called "zirconates" may be made by combining oxides, hydroxides, nitrates, and so on, of other metals with similar zirconium compounds and firing the mixtures at 1000 to 2500°C. These, like their titanium analogues, are mixed metal oxides; there are no discrete zirconate ions known. The zirconate $CaZrO_3$ is isomorphous with perovskite. By dissolving ZrO_2 in molten KOH and evaporating off the excess of solvent at 1050°C, the crystalline compounds $K_2Zr_2O_5$ and K_2ZrO_3 may be obtained. The former contains ZrO_6 octahedra sharing faces to form chains that in turn share edges and corners with other chains. The latter contains infinite chains of ZrO_5 square pyramids (18-A-III).

$$
\begin{array}{c}
\text{(structure 18-A-III: chain of Zr atoms bridged by O)}
\end{array}
$$

(18-A-III)

The other chalcogenides, ZrE_2 and HfE_2 (E = S, Se, and Te) have layered structures and are intrinsic semiconductors. They can also form intercalation compounds which generally have layered structures.

Aqueous Chemistry

This is not very extensive because a +4 ion, even a large one, tends to be extensively hydrolyzed. Only at very low concentration ($\sim 10^{-4}$ M) and high acidity ($[H^+]$ of 1–2 M) does the Zr^{4+}(aq) ion appear to exist. Zirconium dioxide is more basic than TiO_2 and is virtually insoluble in excess base. No ZrO^{2+} ion has been detected

[11]See for example: E. Kato *et al.*, *J. Mater. Sci.* **1997**, *32*, 1789.

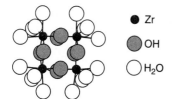

Figure 18-A-3 The structure (schematic) of the $[Zr_4(OH)_8(H_2O)_{16}]^{8+}$ ion.

convincingly. Instead, there seems to be a more or less direct conversion of Zr^{4+}(aq) to tetranuclear $[Zr_4(OH)_8(H_2O)_{16}]^{8+}$ ion and octanuclear $[Zr_8(OH)_{20}(H_2O)_{24}]^{12+}$ species which exist in equilibrium, such that, for 0.05 M Zr^{IV} in highly acidic solutions ($[H^+] \geq 0.6$ M), the tetranuclear ion is the dominant species; and at conditions with $[H^+] \leq 0.5$ M, the octanuclear species becomes predominant.[12] For the chloride, the equilibrium can be described as follows:

$$2[Zr_4(OH)_8(H_2O)_{16}Cl_6]^{2+} \rightleftharpoons [Zr_8(OH)_{20}(H_2O)_{24}Cl_{12}] + 4H^+ + 4H_2O$$

On refluxing or on increasing the pH of the solutions, the tetranuclear units aggregate as two-dimensional sheets which eventually form gels or precipitates.[13] From dilute HX solutions compounds of composition $ZrOX_2 \cdot 8H_2O$ for X = Cl, Br, and I can be crystallized and they contain the tetranuclear ion, whose structure is shown in Fig. 18-A-3. The latter can be used to prepare other zirconium products, i.e., the reaction of $ZrOCl_2 \cdot 8H_2O$ with 2,2'-oxydiacetic acid in aqueous media gives a mononuclear 8-coordinate complex $[Zr\{O(CH_2CO_2)_2\}_2(H_2O)_2] \cdot 4H_2O$[14] or the reaction with a mixture of KF and H_2O_2 which yields the peroxo complex $K_6[Zr_3F_{12}(O_2)_3(H_2O)_2] \cdot 2H_2O$[15] with bidentate bridging peroxo groups.

Both zirconium and hafnium form many other basic salts, such as sulfates and chromates, in which there are infinite chains of composition $[M(\mu\text{-}OH)_2]_n^{2n+}$. In addition to the bridging OH groups, the metal ions are coordinated by oxygen atoms of the anions and achieve coordination numbers of 7 or 8, with pentagonal bipyramidal and square antiprismatic geometries, respectively. For example, the structure of the basic zirconium sulfate, $Zr(OH)_2SO_4$, prepared by hydrothermal synthesis from solutions of $Zr(SO_4)_2 \cdot 4H_2O$, shows infinite chains of 8-coordinate $[ZrO_4(OH)_2]_n$ moieties joined by SO_4^{2-} anions.[16]

Other Complexes and Compounds

The Zr^{4+} ion is relatively large, highly charged, and spherical, with no partly filled shell to give it stereochemical preferences. Thus it is not surprising that zirconium(IV) compounds exhibit high coordination numbers and a great variety of coordination polyhedra. This is well illustrated by the fluoro complexes. Compounds containing (a) octahedral ZrF_6^{2-} ions, (b) ZrF_7^{3-} ions with both pentagonal bipyramidal and capped trigonal prismatic structures, (c) $Zr_2F_{10}^{2-}$ [17] and $Zr_2F_{12}^{4-}$ ions

[12]L. M. Toth et al., J. Am. Chem. Soc. **1996**, 118, 11529.
[13]A. Clearfield, J. Mater. Res. **1990**, 5, 161.
[14]M. Perec et al., Inorg. Chem. **1995**, 34, 1961.
[15]B. N. Chernyshov et al., Zh. Neorg. Khim. **1989**, 34, 2786.
[16]M. E. Brahimi et al., Eur. J. Solid State Inorg. Chem. **1988**, 25, 185.
[17]B. Frit et al., J. Solid State Chem. **1988**, 72, 181.

formed by edge-sharing pentagonal bipyramids, (d) square antiprismatic ZrF_8^{4-} ions, (e) $Zr_2F_{14}^{6-}$ ions formed by edge-sharing square antiprisms, and (f) still other structures, are known.[18] Similar behavior is observed for the hafnium fluorides.[19]

Mononuclear and dinuclear chlorides have been made by reacting CPh_3Cl and MCl_4,[20] M = Zr and Hf, according to

$$MCl_4 + Cl^- \longrightarrow [MCl_5^-]$$

$$[MCl_5^-] \xrightarrow[\text{crystallization}]{\text{THF}} [MCl_5(THF)]^-$$

$$[MCl_5^-] \xrightarrow[\text{crystallization}]{CH_2Cl_2} [M_2Cl_{10}]^{2-}$$

$$[MCl_5^-] \xrightarrow[\text{crystallization}]{MCl_4, CH_2Cl_2} [M_2Cl_9]^-$$

The face-sharing bioctahedral monoanion $Hf_2Cl_9^-$ can also be isolated by reacting $HfCl_4$ with hexamethylbenzene:

$$3HfCl_4 + Me_6C_6 \longrightarrow [(\eta^6\text{-}Me_6C_6)HfCl_3]Hf_2Cl_9$$

Curiously the analogous $Zr_2Cl_9^-$ anion does not form from the same type of reaction which yields only $(\eta^6\text{-}Me_6C_6)Cl_2Zr(\mu\text{-}Cl)_3ZrCl_3$, although the $Zr_2Cl_9^-$ species can be isolated from other reactions such as that of $[Zr_6Cl_{18}H_5]^{2-}$ and $TiCl_4$ (Section 18-A-4).

While the coordination chemistry is not highly developed it shows considerable variety. There are acetylacetonates of the 6-coordinate $M(acac)_2X_2$ and 7-coordinate $M(acac)_3X$ types, as well as 8-coordinate $M(dik)_4$ and the corresponding thio-complexes $M(Sacac)_4$. Six-coordinate pyrazolylborate complexes, $RB(pz)_3ZrCl_3$,[21] and anionic 6-coordinate complexes such as $[Zr(LL)_3]^{2-}$, LL = $Ph_2Si(O)OSi(O)Ph_2^{2-}$ have been structurally characterized.

Some seemingly simple zirconium salts are best regarded as essentially covalent molecules or as complexes; examples are the carboxylates $Zr(OCOR)_4$, the tetraacetylacetonate, the oxalate, and the nitrate. Like its Ti analogue, the last of these is made by heating the initial solid adduct of N_2O_5 and N_2O_4 obtained in the reaction

$$ZrCl_4 + (4+x+y)N_2O_5 \xrightarrow{30°C} Zr(NO_3)_4 \cdot xN_2O_5 \cdot yN_2O_4 + 4NO_2Cl$$

It forms colorless sublimable crystals, and ir and Raman spectra suggest that the molecule is isostructural with $Ti(NO_3)_4$ and $Sn(NO_3)_4$; it is soluble in water but insoluble in toluene. Hafnium gives $Hf(NO_3)_4 \cdot N_2O_5$. Nitrato anions such as $M(NO_3)_6^{2-}$ are also known.

[18]See for example: J. P. Laval and A. Abaouz, *J. Solid State Chem.* **1992**, *100*, 90; Y. Gao *et al.*, *Eur. J. Solid State Inorg. Chem.* **1992**, *29*, 1243; V. Gaumet *et al.*, *Eur. J. Solid State Inorg. Chem.* **1997**, *34*, 283.
[19]See for example: P. P. Fedorov *et al.*, *Russian J. Inorg. Chem.* **1993**, *38*, 1503.
[20]G. Pampaloni *et al.*, *J. Organomet. Chem.* **1996**, *518*, 189.
[21]For other examples see: S. Trofimenko, *Chem. Rev.* **1993**, *93*, 943.

Some other complexes include "tetrahedral" hydroborates such as $Zr(H_3BH)_4$ and $Zr(H_3BR)_4$, molecular species in which 12 hydrogen atoms are in contact with the metal atom,[22] thiolate complexes,[23] mono- and dithiocarbamate complexes, $Zr(OSCNR_2)_4$ and $Zr(S_2CNR_2)_4$, which have dodecahedral structures, tetrahedral $MCl[N(SiMe_3)_2]_3$ compounds, distorted octahedral complexes of the type $M(R_2N)_2Cl_2(THF)_2$ which are prepared by comproportionation of the corresponding $M(NR_2)_4$ and MCl_4 in THF,[24] porphyrin complexes of the type $Zr(porph)Cl_2$, and the sandwich $Zr(porph)_2$.[25] Alkoxo molecules such as $Zr(OR)_4$ and $ZrCl(OR)_3$ can be mononuclear, dinuclear, or polynuclear depending on the size of the R group or the reaction conditions[26] and are easily hydrolyzed. The hydrolysis of the bimetallic compound $BaZr_4(OPr^i)_{18}$ with 1 equivalent of water proceeds in THF at room temperature with replacement of a μ_3-isopropoxide with a μ_3-hydroxide in a simple acid-base reaction[27] according to

$$BaZr_4(OPr^i)_{18} + H_2O \longrightarrow BaZr_4(OH)(OPr^i)_{17} + HOPr^i$$

The structure of the mixed hydroxo-alkoxide species consists of two face-sharing bioctahedral $Zr_2X_9^-$ units each bound to Ba^{2+} by four X groups. Recent studies[28] have shown that for the system $Al(OPr^i)_3-M(OPr^i)_4$, M = Zr and Hf, in isopropanol there is only one type of complex formed, namely, $[(OPr^i)_2Al(\mu\text{-}OPr^i)_2]_2M(OPr^i)_2$. A rare heterotermetallic alkoxide which contains Cd, Ba, and Zr has been prepared[29] according to

$$CdI_2 + KZr_2(OPr^i)_9 \longrightarrow ICd\{Zr_2(OPr^i)_9\} + KI$$

$$ICd\{Zr_2(OPr^i)_9\} + KBa(OPr^i)_3 \longrightarrow \tfrac{1}{2}[\{Cd(OPr^i)_3\}Ba\{Zr_2(OPr^i)_9\}]_2 + KI$$

Homoleptic hydroxylamides of Zr and Hf as well as Ti are prepared as follows:[30]

$$M(NR_2)_4 + 4R_2NOH \longrightarrow M(ONR_2)_4 + 4NR_2H$$

An interesting reaction takes place when $ZrCl_3[N(SiMe_2CH_2PPr^i_2)_2]$ and Na/Hg are in contact with N_2: blue crystals of $\{[Pr^i_2PCH_2SiMe_2)_2]N]ZrCl\}_2(N_2)$ (18-A-IV) are isolated in which the dinitrogen unit is found in a planar side-on bridging mode.[31]

[22] U. Englert et al., Inorg. Chim. Acta **1995**, 231, 175.
[23] D. Coucouvanis et al., Inorg. Chem. **1994**, 33, 3645; K. Tatsumi et al., Inorg. Chem. **1996**, 35, 4391.
[24] R. Kempe et al., Z. anorg. allg. Chem. **1995**, 621, 2021.
[25] See for example: G. S. Girolami et al., Inorg. Chem. **1994**, 33, 626.
[26] N. Ya. Turova et al., Russian Chem. Bull. **1995**, 44, 735; K. G. Caulton et al., Inorg. Chem. **1994**, 33, 6289.
[27] K. G. Caulton et al., Inorg. Chem. **1993**, 32, 1272.
[28] N. Ya. Turova et al., Polyhedron **1997**, 16, 663.
[29] M. Veith et al., J. Am. Chem. Soc. **1996**, 118, 903.
[30] D. W. H. Rankin et al., J. Chem. Soc., Dalton Trans. **1996**, 2089.
[31] M. D. Fryzuk et al., J. Am. Chem. Soc. **1990**, 112, 8185.

Figure 18-A-4   Structure of some polar unsupported heterobimetallic compounds, M = Ti, Zr, and Hf and M′ = Fe and Ru.

(18-A-IV)

Compounds containing unsupported metal-to-metal bonds can be made from MCl(tripodal amido ligand), M = Ti, Zr and Hf, and K[M′Cp(CO)$_2$] salts,[32] M′ = Fe or Ru, as shown in Fig. 18-A-4.

α-Zirconium phosphate,[33] Zr(HPO$_4$)$_2$·H$_2$O, has a layer structure shown in Fig. 18-A-5. The structure consists of octahedral metal atoms situated nearly in a plane which are bridged by phosphate oxygen atoms. Three oxygen atoms of each phosphate are bonded to three different metal atoms while the fourth atom bonds to a proton and points away from the layer. The spaces between the sheets, where the

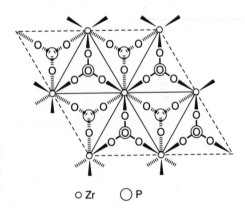

Figure 18-A-5   A schematic representation of the layer structure of α-Zr(HPO$_4$)$_2$·H$_2$O. The water molecules are held in the interlayer cavities and are hydrogen bonded to the phosphate OH groups.

o Zr   O P

[32] L. H. Gade *et al.*, *J. Chem. Soc., Chem. Commun.* **1996,** 219.
[33] A. Clearfield, *Comments Inorg. Chem.* **1990,** *10,* 89.

Figure 18-A-6 Schematic structure of ZrI_3 showing the two independent Zr—Zr distances (Å).

H_2O molecules and P—OH groups are, allow for ion-exchange activity, whereby some or all of the hydrogen ions as well as the H_2O molecules can be displaced by metal cations. α-Zirconium phosphate crystals exchange Na^+ ions to form phases such as $Zr(NaPO_4)(HPO_4)\cdot 5H_2O$, $Zr(NaPO_4)_2\cdot 3H_2O$ and compounds such as $Zr_2(NaPO_4)_4\cdot 6H_2O$.[34] β-Zirconium phosphate, $Zr(HPO_4)_2\cdot 2H_2O$, has a similar structure and properties.[35] When certain transition metal cations are introduced, catalytic properties can develop. It is also possible to modify these zirconium phosphates by replacing some HPO_4^{2-} groups by $ROPO_3^{2-}$ [36] or $HAsO_4^{2-}$ ions. A compound with a cavity structure can be prepared by hydrothermal synthesis[37] according to

$$ZrOCl_2\cdot 8H_2O \xrightarrow[180°C, 6\,days]{en,\ HF,\ H_3PO_4} [(en\ H_2)_{0.5}][Zr_2(PO_4)(HPO_4)F]\cdot H_2O$$

The zirconium fluoride phosphate has a microporous 3D structure similar to that found in some aluminophosphate molecular sieves.

18-A-3 Oxidation State III

There is no aqueous and very little conventional solution chemistry known. The trihalides MX_3, X = Cl, Br, and I, are all well-established compounds. The structure is readily described as hexagonally close-packed halide ions with metal atoms in one-third of the octahedral interstices, not in layers as in β-$TiCl_3$ but centered in one-third of the semiinfinite strings of octahedral sites that run normal to the halogen layers. The structure thus contains chains of confacial trigonal antiprisms ("octahedra"), $\frac{1}{\infty}[MX_{6/2}]$, that are held together only by weak forces, in accord with the very fibrous character usually noted for these crystals. There has been some dispute, however, whether this model is correct in detail regarding the regular spacing of the metal atoms along the chains. However, a careful crystallographic study of ZrI_3[38] indicates that this compound has unsymmetrical Zr—Zr distances (Fig. 18-A-6). The short distance of 3.172 Å, similar to that in $Cs_3Zr_2I_9$, is an indication of the presence of a metal-metal bond. There are various synthetic methods, such as reduction of ZrX_4 with Al or Zr in molten AlX_3. In this process the HfX_4 compounds are reduced more slowly (10 times for the chloride), thus allowing separation of Zr from contaminating Hf.

The trihalides can also be synthesized by reactions of ZrX_4 with ZrX in sealed systems at temperatures of 435 to 600°C. It is found that they are nonstoichiometric phases rather than distinct compounds. Their ranges of phase stability (i.e., values of x in ZrX_x) are for Cl, 2.94 to 3.03; for Br, 2.87 to 3.23; for I, 2.83 to 3.43. The

[34] D. M. Poojary and A. Clearfield, *Inorg. Chem.* **1994**, *33*, 3685.
[35] A. N. Christensen *et al.*, *Acta Chem. Scand.* **1990**, *44*, 865.
[36] A. Clearfield *et al.*, *Angew. Chem. Int. Ed. Engl.* **1994**, *33*, 2324.
[37] E. Kemnitz *et al.*, *Angew. Chem. Int. Ed. Engl.* **1996**, *35*, 2677.
[38] J. D. Corbett *et al.*, *Inorg. Chem.* **1990**, *29*, 2242 and references therein.

color of each phase varies markedly over its composition range (e.g., from olive green to bluish black for $ZrBr_x$). For the HfI_3 phase a composition as high as $HfI_{3.5}$ can be obtained. Whether these "off-stoichiometric" compositions involve replacements of the type $4M^{III} \rightarrow 3M^{IV} + [\ ]$ or some other variation is not known.

The coordination chemistry is very limited. There are a few well-characterized compounds such as *cis-cis*-$Zr_2Cl_6(PBu_3)_4$ (which has PMe_3,[39] PEt_3, and PPr_3 homologues); it has the structure (18-A-V) and since it is diamagnetic with a Zr—Zr distance of 3.182 Å, the existence of a Zr—Zr single bond may be assumed.

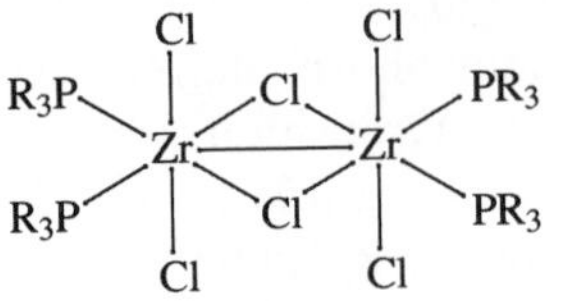

(18-A-V)

The phosphine ligands can be substituted by pyridine-like groups[40] as in

$$Zr_2Cl_6(PBu_3)_4 + 4py\text{-}R \xrightarrow[1\ h]{hexanes} Zr_2Cl_6(py\text{-}R)_4 + 4PBu_3$$

The anionic species $[1,4\text{-}Zr_2I_7(PEt_3)_2]^{-}$ [41] can be prepared according to

$$ZrI_4 + HSn(Bu)_3 \xrightarrow[PEt_3]{NBu_4I} [NBu_4][1,4\text{-}Zr_2I_7(PEt_3)_2]$$

Since the reduction potentials of hafnium are the most negative of the transition elements, very few Hf^{III} coordination complexes are known; examples are the dinuclear complexes of the type $Hf_2X_6(phosphine)_4$ for X = Cl or I,[42] in which the first Hf—Hf bonds to be recognized exist.

18-A-4   Oxidation States Below III

Our initial knowledge of this chemistry was due to the work of J. D. Corbett and his co-workers,[43] but new developments have changed some of the original views.

Zirconium(0)

Apart from organometallic compounds there are a few other compounds of zerovalent zirconium. The reduction of $ZrCl_4$ with lithium in the presence of bipyridine in THF gives violet $Zr(bipy)_4$ where, doubtless, there is considerable delocalization of electrons over the ligands. The reaction of KCN or RbCN with $ZrCl_3$ in liquid ammonia is reported to give $M^I_5Zr(CN)_5$.

[39]F. A. Cotton and W. A. Wojtczak, *Gazz. Chim. Ital.* **1993**, *123*, 499.
[40]D. M. Hoffman and S. Lee, *Inorg. Chem.* **1992**, *31*, 2675.
[41]F. A. Cotton and W. A. Wojtczak, *Inorg. Chim. Acta* **1994**, *216*, 9.
[42]G. S. Girolami *et al.*, *Inorg. Chem.* **1990**, *29*, 3200; F. A. Cotton *et al.*, *Inorg. Chim. Acta* **1990**, *177*, 1; F. A. Cotton *et al.*, *Inorg. Chem.* **1991**, *30*, 3670.
[43]R. P. Ziebart and J. D. Corbett, *Acc. Chem. Res.* **1989**, *22*, 256.

Zirconium(I)[44]

The compounds ZrCl and ZrBr (which have hafnium analogues) are obtained by reactions of MX_4 with M at 800 to 850°C. The structures consist of stacked, hexagonally packed layers of either all metal atoms or all X atoms, with a stacking sequence ---XMMX---XMMX---. There are two slightly different stacking patterns. The X---X interlayer distances (e.g., ~3.60 Å for chlorides) are normal van der Waals contacts, and the M—X distances (e.g., 2.63 Å in ZrCl) are appropriate for single bonds. Within the adjacent metal atom layers there are two sets of distances. Within one layer Zr—Zr distances are 3.42 Å, and between layers they are 3.09 Å; these may be compared to an average distance of 3.20 Å in α-Zr. The monohalides have great thermal stability (mp > 1100°C) and metallic character; they cleave like graphite. A compound of composition Hf_2S also appears to have a structure with double layers of metal atoms.

Both ZrCl and ZrBr react with hydrogen, oxygen, or zirconium dioxide to form phases in which H or O atoms have been introduced within the double layer of metal atoms. With hydrogen, compositions of $ZrXH_{0.5}$ and ZrXH are well defined, while with oxygen there is a greater tendency to nonstoichiometry, $ZrXO_{0.43}$ and $ZrBr_{0.23}$ being typical compositions that have been studied. It may be noted that nonstoichiometric ZrH_x ($x < 2$) and ZrO_y ($y < 1$) are also known. A binary mixed-valent chalcogenide, Hf_3Te_2, has a layered structure.[45]

Zirconium(II)

There are no true coordination compounds known in this oxidation state. All ZrX_2, X = Cl, Br, and I, are now known to consist of M_6X_{12} units which also contain other atoms (see below). There are some binary compounds such as Hf_3P_2 which crystallizes in the anti-Sb_2S_3 structure type consisting of singly capped and bicapped trigonal Hf_6P prisms.[46] A series of compounds of the type $M_2X_4(\eta^6\text{-}C_6H_5PMe_2)_2$-$(PMe_2Ph)_2$, M = Zr and Hf, X = Cl, Br, and I,[47] have a phosphine phenyl group coordinated to the metal atom. They are therefore best described as organometallic species, as are those obtained by reaction of MX_4 and Al/AlX_3 in an aromatic hydrocarbon[48]:

$$3MX_4 + 4AlX_3 + 2Al + 3 \text{ arene} \longrightarrow 3[(\text{arene})M(AlX_4)_2]$$

Zirconium Clusters[49]

High temperature reactions of ZrX_4, X = Cl, Br, and I, and Zr powder give compounds which were assigned formulas ZrX_2 but were later formulated as

[44]R. L. Daake and J. D. Corbett, Inorg. Synth. 1995, 30, 6.
[45]R. L. Abdon and T. Hughbanks, Angew. Chem. Int. Ed. Engl. 1994, 33, 2328.
[46]H. Kleinke and H. F. Franzen, Acta Cryst. 1996, C52, 2127.
[47]F. A. Cotton et al., Organometallics 1991, 10, 2626; F. A. Cotton and W. A. Wojtczak, Inorg. Chim. Acta 1994, 217, 187.
[48]F. Calderazzo et al., J. Chem. Soc., Dalton Trans. 1990, 1813.
[49]F. A. Cotton et al., Early Transition Metal Clusters with π-donor Ligands, M. H. Chisholm, Ed., VCH Publishers, New York, NY (1995), pp. 1–26; J. D. Corbett, In Modern Perspectives in Inorganic Crystal Chemistry, E. Parthé, Ed., NATO ASI Series C; Kluwer Academic Publishers, The Netherlands, (1992), pp. 27–56.

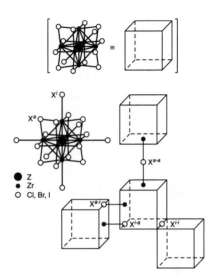

Figure 18-A-7   Diagram illustrating the general notation for intercluster linkage. Superscripts on the linking atoms are appropriate for the central (circled) atoms. (Adapted from ref. 49.)

$[Zr_6Z]X_{12}$-containing units. The building block consists of an octahedron of Zr atoms with Zr—Zr distances of ~3.2 Å. The interstitial central atom, Z, can be any of a large variety such as Be, B, C, N, Al, Si, Ge, and P or transition elements such as Cr, Mn, Fe, Co, and Ni. Small variations in the relative amounts of the starting materials gives products which contain metal atoms in other oxidation states but the common formula is $[Zr_6(Z)X_{12}^{n+}]X_n$. Addition of alkali or alkaline earth metal cations, A, to the high temperature reactions give compounds of the type $A_y^I[Zr_6(Z)X_{12}]X_m$. In the solid state the octahedral units are associated *via* Zr—X—Zr bridges to various degrees. The structure dimensions and the connectivity between individual cluster units in the phases are primarily determined by the values of y and m in the general formula and the type of interstitial atom (Z). Each of the cluster units in any given phase requires 12 edge bridging halogen atoms and 6 exo bonds. The halide ligands can be placed into one of five categories, labeled as X^a, X^i, X^{a-a}, and X^{a-i} (X^{i-a}). A halide labeled as X^{a-a} is one of six outer (ausser) halides shared by *two* clusters that it serves to link together (thus the double superscript). Halides designated as X^a or X^i are respectively ausser or inner(i) with respect to a given cluster, but not involved in connecting different clusters together. The two remaining types of halides (X^{i-i}, and X^{a-i}) also serve to link the separate Zr_6 clusters together to form the extended solid phase. Fig. 18-A-7 illustrates a cluster (circled) and the labeling of halides that link it to clusters in its environment.

Clusters in which intercluster linkages are minimized can be excised. For example, by mixing 1-ethyl-3-methylimidazolium chloride (ImCl) with $AlCl_3$ and $(Zr_6Fe)Br_{14}$ in a melt reaction, the compound $(Im)_4[(Zr_6Fe)Cl_{18}]$ can be isolated.[50]

Molecular orbital calculations[51] have indicated that the optimum number of cluster-based electrons (i.e., those that occupy the molecular orbitals responsible for bonding) is 14 for main-group-centered clusters and 18 for transition-metal-

[50] C. E. Runyan, Jr. and T. Hughbanks, *J. Am. Chem. Soc.* **1994,** *116,* 7909.
[51] M. R. Bond and T. Hughbanks, *Inorg. Chem.* **1992,** *31,* 5015 and references therein.

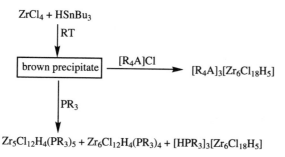

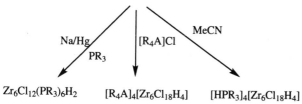

Figure 18-A-8 Formation and reactivity of some clusters. The number of hydrogen atoms shown was obtained by ^{1}H nmr spectroscopic measurements, X-ray or neutron diffraction.

centered clusters. For example, for an anionic species such as $[(Zr_6C)Cl_{15}]^-$ the electron counting is done as follows:

$$(6 \times 4) + 4 - 15 + 1 = 14$$

and for $[(Zr_6Co)Cl_{15}]$:

$$(6 \times 4) + 9 - 15 = 18$$

Examples of diamagnetic 14-electron systems are $[NEt_4]_4[(Zr_6C)Cl_{18}]$, $K(Zr_6C)Cl_5$, $(Zr_6Be)Cl_{12}(PEt_3)_6$, and $[(Zr_6N)Cl_{15}]$. Representative examples of 18-electron clusters are $[PEt_4]_4[(Zr_6Fe)Cl_{18}]$, $K[(Zr_6Fe)Cl_{15}]$, $Cs[(Zr_6Mn)I_{14}]$, and $(Zr_6Co)Cl_{15}$. The vast majority of the known clusters follow the above pattern but a few exceptions have been found. Clusters such as $[NEt_4]_4[(Zr_6Be)Cl_{18}]$, $[PPh_4]_4[(Zr_6B)Cl_{18}]\cdot$-$[PPh_4]_2ZrCl_6$, and $[(Zr_6B)I_{12}]$ have electron counts of 12, 13, and 15, respectively.

Hydride-containing clusters can be prepared under very mild conditions by reacting $ZrCl_4$ and various amounts of $HSnBu_3$ followed by addition of phosphines or halide salts[52] (phosphonium or ammonium) as shown in Fig. 18-A-8.

Neutron diffraction studies[53] have shown that the hydrogen atoms occupy, in a random manner, the triangular faces of the octahedron of Zr atoms. Some of those cluster compounds can be quite reactive,[54] as shown in Fig. 18-A-9, and most are diamagnetic with 14 electron cores but a few are paramagnetic, e.g., $[PPh_4]_2[Zr_6Cl_{18}H_5]$, with only 13 cluster-based electrons. Pentanuclear compounds, $H_4Zr_5Cl_{12}(PR_3)$, have also been prepared.[55]

A few hafnium clusters of the type $Hf_6(Z)Cl_{12}Cl_n$ are also known.[56] However, the Hf—HfCl$_4$ reactions are very slow and so far only B and C interstitials (Z) have given cluster phases. Some of the known compounds are of the type

[52]F. A. Cotton et al., Angew. Chem. Int. Ed. Engl. **1995,** 34, 1877.
[53]F. A. Cotton et al., C. R. Acad. Sci. Paris IIb **1996,** 323, 539.
[54]L. Chen and F. A. Cotton, Inorg. Chem. **1996,** 35, 7364.
[55]F. A. Cotton et al., J. Am. Chem. Soc. **1994,** 116, 4364.
[56]R.-Y. Qi and J. D. Corbett, Inorg. Chem. **1994,** 33, 5727.

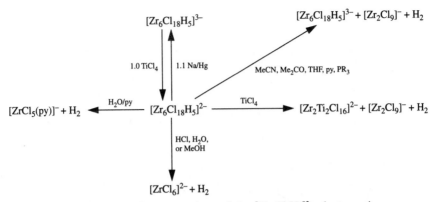

Figure 18-A-9 Some reactions of the $[Zr_6Cl_8H_5]^{2-}$ cluster anion.

$A(Hf_6Z)Cl_{14}$ with A = Li, Na, K, or nothing and Z = B or C, and $A(Hf_6B)Cl_{15}$ with A = K or Cs.

There is also a series of ternary zirconium chalcogenides such as Zr_6MTe_2,[57] M = Mn, Fe, Co, Ni, Ru, or Pt, which exhibit metal-centered tricapped trigonal prismatic clusters as the basic structural unit.

18-A-5   Organometallic Compounds

Organozirconium and organohafnium chemistries resemble organotitanium chemistry (Section 17-A-5). There are some stable MR_4 compounds with β-elimination stabilized alkyls such as $PhCH_2$ or the trigonal prismatic $ZrMe_6^{2-}$ anion,[58] but the most extensive and important chemistry in both the IV and II states involves $(C_5H_5)_2M$ and $(Me_5C_5)_2M$ groups. The M^{IV} compounds are of the Cp_2MXY type,[59] where X and Y may be R, Ar, Si-, Ge-, Sn-, N-, P-, As-, Sb-, O-, S-, Se-, Te-, Cl, Br, H, and others such as $Cp_2^*Zr(E)py$, E = O, S, Se, Te.[60] An important set of compounds have cationic species such as Cp_2MMe^+; many of which form ion-pairs and are used as homogeneous polymerization catalysts.[61] Zirconocenes can be synthesized in a general way by reaction of MCl_4 with the lithium salts of the corresponding ligands in THF under an inert atmosphere. The unsubstituted Cp_2MCl_2 have a bent sandwich structure (18-A-VI).

(18-A-VI)

[57] C. Wang and T. Hughbanks, *Inorg. Chem.* **1996**, *35*, 6987.

[58] P. M. Morse and G. S. Girolami, *J. Am. Chem. Soc.* **1989**, *111*, 4114.

[59] See for example: E. Hey-Hawkins, *Chem. Rev.* **1994**, *94*, 1661; J. E. Bercaw *et al.*, *Polyhedron* **1988**, *7*, 1409.

[60] W. A. Howard and G. Parkin, *J. Am. Chem. Soc.* **1994**, *116*, 606.

[61] See for example: P. A. Deck and T. J. Marks, *J. Am. Chem. Soc.* **1995**, *117*, 6128; M. Bochmann, *J. Chem. Soc., Dalton Trans.* **1996**, 255; C. Pellecchia *et al.*, *J. Am. Chem. Soc.* **1993**, *115*, 1160; T. J. Marks *et al.*, *J. Am. Chem. Soc.* **1996**, *118*, 7900.

By substitution of the Cp ligands, especially using more stereorigid ligands with a bridge to form strapped zirconocenes, also called *ansa*-zirconocenes, chiral molecules with C_{2v} symmetry are obtained. They can polymerize α-olefins to give highly isotactic polymers[62] such as polypropylene (see Chapter 21). There are relatively few $CpMX_3$[63] or Cp_3MX compounds and some $[Cp_3ZrL]^+$ cationic species,[64] e.g., L = RCN. The Cp_4M molecules are also known; one of the Cp rings is monohapto.[65] In all mononuclear cases the four ligands provide an approximately tetrahedral array about the metal atom. There are also tetranuclear compounds such as $[Cp^*MF_3]_4$, M = Zr, Hf, and Ti;[66] they can be prepared according to

$$4CpMCl_3 + 12Me_3SnF \longrightarrow [CpMF_3]_4 + 12Me_3SnCl$$

An important use of Cp_2ZrHCl is in the so-called hydrozirconation reaction with alkenes and alkynes which generates a Zr-alkyl group which can then be converted to alcohols or other products.[67] In the reactions of CO with Zr and Hf compounds the oxophilicity of these elements is important, because it leads to the formation of a variety of compounds that have M—O bonds.

Reduced M^{III} complexes can be made from M^{IV} precursors using reducing agents such as Na/Hg and $M^I(C_{10}H_8)$, M = alkali metal. Most M^{III} compounds are extremely air and moisture sensitive except for those with large substituents on the cyclopentadienyl groups. Complexes such as $[C_5H_3(Bu^t)_2]_2MCl$[68] are mononuclear and paramagnetic while those with smaller rings such as $[Cp_2MX]_2$ are typically dinuclear and diamagnetic, as in $[CpM(PR_2)]_2$.[69] The geometry of the $M(\mu\text{-}X)_2M$ core is determined by a balance of the repulsive forces of the Cp and the X groups and the Zr$\cdots$Zr interactions. With the smaller groups, the Zr_2X_2 four-membered rings are planar.

Metallocene complexes of Zr^{II} and Hf^{II} are generally prepared by the reduction of metallocene(IV) compounds in the presence of stabilizing ligands[70] such as CO, PR_3, or alkenes as in

$$Cp_2MCl_2 + Mg + 2CO \longrightarrow Cp_2M(CO)_2 + MgCl_2$$

Other examples of M^{II} compounds are $CpZrCl(dmpe)_2$, $Cp_2Zr(alkyne)$, and $Li_2[Zr(CH_2Ph)_4]$.

Lower-valent compounds are scarcer but their existence is well established.

[62]See for example: J. E. Bercaw *et al.*, *J. Am. Chem. Soc.* **1996,** *118,* 1045, 11988; H. H. Brintzinger *et al.*, *Angew. Chem. Int. Ed. Engl.* **1995,** *34,* 1143; A. M. Thayer, *C&EN*, Sept. 11, **1995,** 15.
[63]See for example: M. L. H. Green *et al.*, *J. Organomet. Chem.* **1995,** *491,* 153.
[64]G. Erker *et al.*, *Chem. Ber./Recueil* **1997,** *130,* 899.
[65]See for example: R. J. Strittmatter and B. E. Bursten, *J. Am. Chem. Soc.* **1991,** *113,* 552 and references therein.
[66]H. W. Roesky *et al.*, *Inorg. Chem.* **1996,** *35,* 23 and references therein; *Angew. Chem. Int. Ed. Engl.* **1994,** *33,* 967.
[67]B. H. Lipshutz and E. L. Ellsworth, *J. Am. Chem. Soc.* **1990,** *112,* 7440.
[68]D. A. Lemenovskii *et al.*, *J. Organomet. Chem.,* **1989,** *368,* 287.
[69]D. W. Stephan *et al.*, *Organometallics* **1993,** *12,* 3145.
[70]D. J. Sikora *et al.*, *Inorg. Synth.* **1986,** *24,* 147.

Sixteen-electron bis(arene) complexes of formally zero-valent zirconium, hafnium, and titanium have been made by cocondensation of the metal atoms with bulky ligands such as 1,3,5-tri-*t*-butylbenzene[71]:

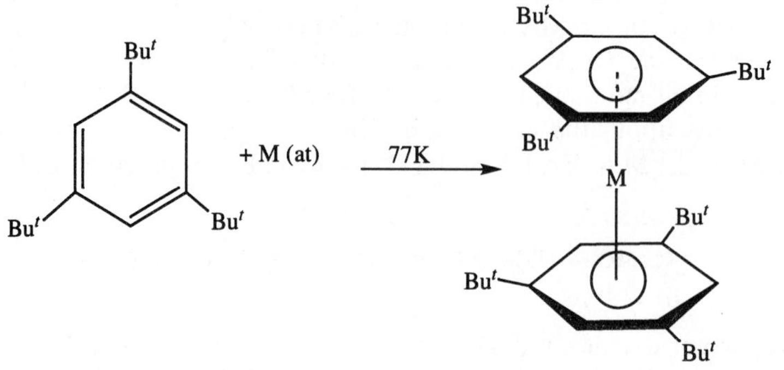

M = Ti, Zr, Hf

Hf(η^6-Bu$^t_3$C$_6$H$_3$)$_2$ reacts irreversibly with CO to give the stable 18-electron adduct (18-A-VII) while the Zr analogues react reversibly.

Reduction of ZrCl$_4$ with Na/Hg at < -10 °C in the presence of cycloheptatriene gives Zr(η^6-C$_7$H$_8$)$_2$; it exhibits a non-parallel arrangement of the cycloheptatrienyl ligands (18-A-VIII) and can be used to prepare a series of zero- and monovalent 16-*e* zirconium complexes such as Zr(η^7-C$_7$H$_7$)(η^5-C$_7$H$_9$)(PMe$_3$), Zr(η^7-C$_7$H$_7$)-(tmeda)Cl, and others.[72]

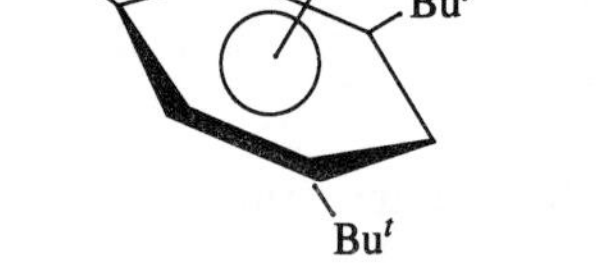

(18-A-VII) 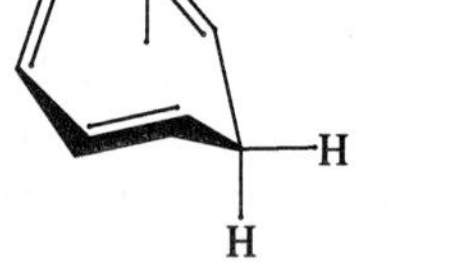   (18-A-VIII)

Mixed Cp/CO species such as [CpHf(CO)$_4$]$^-$ can be synthesized from CpHfCl$_3$ and a stoichiometric amount of K(naphthalenide) under a CO atmosphere[73] in DME:

$$\text{CpHfCl}_3 \xrightarrow[\text{-70 to 20°C}]{\text{K(naphthalenide)}} \xrightarrow[\text{15-crown-5}]{\text{CO}} [\text{K}(15\text{-C-5})_2][\text{CpHf(CO)}_4]$$

[71]F. N. Cloke *et al.*, *Polyhedron* **1989**, *8*, 1641.
[72]M. L. H. Green *et al.*, *J. Chem. Soc., Dalton Trans.* **1991**, 173; **1992**, 2259.
[73]S. R. Frerichs and J. E. Ellis, *J. Organomet. Chem.* **1989**, *359*, C41.

Even lower formal oxidation states can be found in compounds prepared by reacting $ZrCl_4(THF)_2$ and six equivalents of K(naphthalenide):[74]

$$ZrCl_4(THF)_2 + 6KC_{10}H_8 + 2(15\text{-}C\text{-}5) \xrightarrow[\text{-60 to 0}\,^\circ\text{C}]{\text{DME}} [K(15\text{-}C\text{-}5)_2]_2[Zr(C_{10}H_8)_3] + 3C_{10}H_8 + 4KCl + 2THF$$

which reacts further to give the binary carbonylzirconium anion $[Zr(CO)_6]^{2-}$ upon addition of CO.[75]

Following similar reaction strategies, compounds containing anions such as 7-coordinate $[Zr(CO)_5(SnMe_3)_2]^{2-}$ and $[Zr(CO)_4(dppe)SnMe_3]^-$ or 8-coordinate $[Zr(CO)_4(SnMe_3)_4]^{2-}$ can be synthesized.[76] The formal oxidation number of the Zr atom is 2 for the latter compound and 0 for the others.

Additional References

Catalysis Today, **1994**, *20*, 185–320. (An issue of the journal dedicated to "Zirconium in Catalysis").
Ziegler Catalysts, G. Fink *et al.*, Eds., Springer-Verlag, Berlin, 1995 (several chapters discuss various aspects of olefin polymerization catalysis by zirconium and hafnium complexes).

18-B   NIOBIUM AND TANTALUM: GROUP 5

These two elements have very similar chemistries, though not so nearly identical as in the case of zirconium and hafnium. They have very little cationic behavior, but they form many complexes in oxidation states II, III, IV, and V. In oxidation states II and III M—M bonds are fairly common and in addition there are numerous compounds in lower oxidation states where metal atom clusters exist. An overview of oxidation states and stereochemistry (excluding the cluster compounds) is presented in Table 18-B-1. In discussing these elements it will be convenient to discuss some aspects (e.g., oxygen compounds, halides, and clusters) as classes without regard to oxidation state, while the complexes are more conveniently treated according to oxidation state.

The elements themselves require little comment. Niobium is 10 to 12 times more abundant in the earth's crust than tantalum. The main commercial sources of both are the *columbite-tantalite* series of minerals, which have the general composition $(Fe/Mn)(Nb/Ta)_2O_6$, with the ratios Fe/Mn and Nb/Ta continuously variable. Niobium is also obtained from *pyrochlore*, a mixed calcium-sodium niobate. Separation and production of the metals is complex. Both metals are bright, high melting (Nb, 2468°C; Ta, 2996°C), and very resistant to acids. They can be dissolved with vigor in an HNO_3—HF mixture, and very slowly in fused alkalis.

[74]M. Jang and J. E. Ellis, *Angew. Chem. Int. Ed. Engl.* **1994,** *33*, 1973.
[75]J. E. Ellis *et al.*, *Angew. Chem. Int. Ed. Engl.* **1987,** *26*, 1190.
[76]J. E. Ellis *et al.*, *J. Organomet. Chem.* **1996,** *507*, 283.

Table 18-B-1   Oxidation States and Stereochemistries of Niobium and Tantalum

Oxidation state	Coordination number	Geometry	Examples
Nb^{-III}, Ta^{-III}	5	*tbp*	$[M(CO)_5]^{3-}$
Nb^{-I}, Ta^{-I}	6	Octahedral	$[M(CO)_6]^-$
	7	?	$HTa(PF_3)_6$
Nb^0, Ta^0, d^5	6	Octahedral	$Ta(CO)_4$(diphos), $17e^-$ compound
		π Complex	$M(\eta^6\text{-mes})_2$
Nb^I, Ta^I, d^4	7	π Complex	$(C_5H_5)M(CO)_4$
	7	Distorted capped octahedron (nonrigid)	$TaH(CO)_2$(diphos)$_2$
Nb^{II}, Ta^{II}, d^3		See text	NbO
	6	Octahedral	$TaCl_2$(dmpe)$_2$, $NbCl_2$(py)$_4$
Nb^{III}, Ta^{III}, d^2	6	Trigonal prism	$LiNbO_2$
		Octahedral	$Nb_2Cl_9^{3-}$, $M_2Cl_6(SMe_2)_3$
	7	Complex	$TaCl_3(CO)(PMe_2Ph)_3 \cdot EtOH$
	8	Dodecahedral	$K_5[Nb(CN)_8]$
Nb^{IV}, Ta^{IV}, d^1	6	Octahedral	$(NbCl_4)_x$, $TaCl_4py_2$, MCl_6^{2-}
	7	Distorted pentagonal bipyramidal	K_3NbF_7
	7	Capped octahedron	$MCl_4(PMe_3)_3$
	8	Nonrigid in solution	TaH_4(diphos)$_2$
		Square antiprism	$Nb(\beta\text{-dike})_4$, $M_2Cl_8(PMe_3)_4$, $Nb(SCN)_4$(dipy)$_2$,
		Dodecahedral	$K_4Nb(CN)_8 \cdot 2H_2O$
		π Complex	Cp_2NbMe_2
Nb^V, Ta^V, d^0	4	Tetrahedral	$ScNbO_4$
	5	*tbp*	MCl_5(vapor), $TaMe_5$, $Nb(NR_2)_5$
		Distorted tetragonal pyramid	$Nb(NMe_2)_5$
	6	Octahedral	$NaMO_3$ (perovskite), $NbCl_5 \cdot OPCl_3$, $TaCl_5 \cdot S(CH_3)_2$, TaF_6^-, $NbOCl_3$, M_2Cl_{10}, MCl_6^-
	6	Trigonal prism	$[M(S_2C_6H_4)_3]^-$
	7	Distorted pentagonal bipyramidal	$NbO(S_2CNEt_2)_3$
		Pentagonal bipyramidal, fluxional	$STa(S_2CNEt_2)_3$, $Ta(NMe_2)(S_2CNMe_2)_3$, $(S_2CNR_2)_2TaMe_3$
	8	Bicapped trigonal prism	$[Nb(trop)_4]^+$
		Square antiprism	Na_3TaF_8
		Dodecahedral	$Ta(S_2CNMe_2)_4^+$
	9	π Complex	$(\eta^5\text{-}C_5H_5)_2TaH_3$

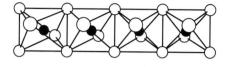

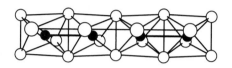

Figure 18-B-1 Chains of *trans*-edge-sharing NbO_6 octahedra in high-temperature NbO_2 (top) and low-temperature NbO_2 (bottom).

18-B-1 Oxygen Compounds

The oxides are numerous and structurally elaborate. Both Nb_2O_5 and Ta_2O_5 are relatively inert white solids. Concentrated HF is the only acid that attacks them, but they can be dissolved by fusion with an alkali hydrogen sulfate, carbonate, or hydroxide. They are obtained by roasting the hydrous oxides (so-called niobic and tantalic acids) or other compounds in excess oxygen. The hydrous oxides themselves are gelatinous white precipitates obtained on neutralizing acid solutions of Nb^V and Ta^V halides. Their composition is variable owing to the inconsistency of the water content which varies depending on the method of preparation and drying.[1] At the surface of certain metal oxide supports, niobic acid ($Nb_2O_5 \cdot nH_2O$) exhibits a high acidity, equivalent to the acidity of 70% sulfuric acid; it is widely used as a catalyst[2] in reactions with participation or elimination of water such as esterification, hydration, dehydration, hydrolysis, and others. On heating, the hydrated niobium pentoxide slowly loses water; then the Nb_2O_5 goes through a series of structural changes. Finally it loses a certain amount of oxygen with formation of oxygen-deficient non-stoichiometric niobium oxides.

Both of the pentoxides, but especially Nb_2O_5, have complex structural relationships.[3] They are built of MO_6 octahedra sharing edges and corners, but this can be (and is) done in an almost unlimited number of ways. The Ta_2O_5 phase exists with an excess of Ta atoms from $TaO_{2.5}$ to nearly TaO_2.

Tantalum dioxide has the rutile structure; so also does NbO_2 at higher temperatures, but either of two similar phases with alternating bonding (2.74 Å) and non-bonding (3.30 Å) Nb—Nb bond distances may be obtained at low temperatures as shown in Fig. 18-B-1.

Niobate and tantalate isopolyanions can be obtained by fusing the oxides in an excess of alkali hydroxide or carbonate and dissolving the melts in water. The solutions are stable only at higher pH: precipitation occurs below pH ~7 for niobates and ~10 for tantalates. At pH > 14.5, mononuclear $NbO_2(OH)_4^{3-}$ anions are likely to be present in dilute aqueous solution. At lower pH the only species that appear to be present in solution are the $[H_xM_6O_{19}]^{(8-x)-}$ ions ($x = 0, 1, 2,$ or 3),[4] despite frequent claims for others. The structure of the $M_6O_{19}^{8-}$ ions, found in crystals and believed to persist in solutions, is shown in Fig. 18-B-2.

[1] F. A. Chernyshkova, *Russian Chem. Rev.* **1993**, *62*, 743.
[2] K. Taube, *Catal. Today* **1990**, *8*, 1.
[3] E. I. Ko and J. G. Weissman, *Catal. Today* **1990**, *8*, 27.
[4] N. Etxebarria *et al.*, *J. Chem. Soc., Dalton Trans.* **1994**, 3055.

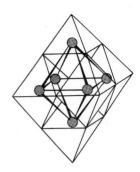

Figure 18-B-2   The structure of $M_6O_{19}^{8-}$ ions (M = Nb or Ta).

Heteropolyniobates and -tantalates are not well known, but a few of the former have been prepared and characterized.

With the exception of a few insoluble lanthanide niobates and tantalates (e.g., ScNbO$_4$) that contain discrete, tetrahedral MO_4^{3-} ions, the coordination number of Nb^V and Ta^V with oxygen is essentially always 6 (e.g., the ferroelectric compounds LiMO$_3$,[5] M = Nb and Ta). These compounds are built mainly from NbO$_6$ octahedra without direct M–M bonds even when they are metallic. In contrast, the more reduced nonmetallic oxides ANbO$_2$ (A = Li, Na) have layers of trigonal prismatic NbO$_2$; the A atoms occupy octahedral interstices. This type of atom arrangement is not unusual for a d^2 configuration. Short (*ca.* 2.90 Å) Nb–Nb distances give evidence of metal-metal bonding.

In the lower oxidation states, oxides and oxometalates typically have structures in which the metal-metal bonding is extensive (see Section 18-B-7). The oxometalates are commonly synthesized by solid state reactions from appropriate mixtures of binary or ternary starting materials with carbon or NbO and Nb as reducing agents, at temperatures above 1500 K, using sealed Au, Pt, or Nb ampoules or quartz ampoules protected by corundum crucibles. Mineralizers or fluxes such as alkali or alkaline earth halides or borates have been used successfully to reduce reaction temperatures and to obtain single-crystal specimens. However, some fluxes can lead to composition changes; for example, phosphates give NbP instead.

18-B-2   Halides and Oxohalides

Simple binary halides are known for oxidation states IV and V, those in lower states being cluster compounds (see Section 18-B-7).

Fluorides

The pentafluorides are made by direct fluorination of the metals or the pentachlorides. Both are volatile white solids (Nb: mp 80°C, bp 235°C; Ta: mp 95°C, bp 229°C), giving colorless liquids and vapors. They have the tetranuclear structures shown in Fig. 18-B-3. Niobium(IV) fluoride is a black, nonvolatile, paramagnetic solid; TaF$_4$ is unknown.

[5]R. Hsu *et al., Acta Cryst.* **1997**, *B53*, 420.

Figure 18-B-3 The tetranuclear structures of NbF_5 and TaF_5 (also MoF_5 and, with slight distortion, RuF_5 and OsF_5) Nb–F bond lengths: 2.06 Å (bridging), 1.77 Å (nonbridging).

Other Halides of Niobium(V) and Tantalum(V)

These are yellow to brown or purple-red solids best prepared by direct reaction of the metals with excess of the halogen. The halides are soluble in various organic liquids such as ethers and CCl_4. They are quickly hydrolyzed by water to the hydrous pentoxides and the hydrohalic acid. The chlorides give clear solutions in concentrated hydrochloric acid, forming oxochloro complexes. All the pentahalides, which melt and boil between 200 and 300°C, can be sublimed without decomposition in an atmosphere of the appropriate halogen; in the vapor they are mononuclear and trigonal bipyramidal like WCl_5 and $ReCl_5$.[6] Several polymorphs of the dinuclear $NbCl_5$ species are known.[7] The crystalline structure of the α polymorph is shown in Fig. 18-B-4(a); $NbBr_5$, $TaCl_5$, $TaBr_5$, and TaI_5 are isostructural. In CCl_4 and $MeNO_2$, both $NbCl_5$ and $TaCl_5$ are dinuclear but in coordinating solvents adducts are formed. It appears probable that NbI_5 has a hexagonal close-packed array of iodine atoms with niobium atoms in octahedral interstices; mononuclear trigonal-bipyramidal NbI_5 molecules have been observed occupying interstices formed by the host molecule $Nb_3S_2I_7$ but attempts to remove such molecules have been unsuccessful.[8]

Other Halides of Niobium(IV) and Tantalum(IV)

All six are known and resemble one another in their structures and in being diamagnetic. The *tetrachlorides* and *tetrabromides* are all brown-black or black isomorphous solids,[9] obtained by reduction of the pentahalides with H_2, Al, Nb, or Ta at elevated

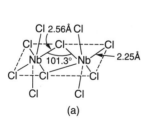

(a)

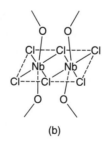

(b)

Figure 18-B-4 (a) The dinuclear structure of $NbCl_5$ in the solid; the octahedra are distorted as shown. The Nb–Nb distance, 3.951 Å indicates no metal–metal bonding. (b) The structure of $NbOCl_3$ in the crystal. The oxygen atoms form bridges between infinite chains of the planar Nb_2Cl_6 groups.

[6]B. Bajan and H.-J. Meyer, *Z. Kristallogr.* **1996**, *211*, 818.
[7]F. A. Cotton *et al.*, *Acta Cryst.* **1991**, *C47*, 2435.
[8]G. J. Miller and J. Lin, *Angew. Chem. Int. Ed. Engl.* **1994**, *33*, 335.
[9]A. Haaland *et al.*, *J. Chem. Soc., Dalton Trans.* **1997**, 1013.

temperatures. Crystalline $NbCl_4$ has linear chains with chlorine bridges (18-B-I) and alternating Nb—Nb bonds.

(18-B-I)

Niobium(IV) iodide is obtained on heating NbI_5 to 300°C. It is diamagnetic and trimorphic. One form contains infinite chains of octahedra with the Nb atoms off center so as to form pairs with Nb—Nb distances of 3.31 Å. Tantalum tetraiodide appears to be similar. The latter can be made most easily by allowing TaI_5 to react with an excess of pyridine to give $TaI_4(py)_2$ which, on heating, loses pyridine to give TaI_4. The structures of both $NbBr_4$ and NbI_4 have been studied by electron diffraction techniques; in the gas phase the molecules are mononuclear and tetrahedral like VCl_4 and VBr_4.[10]

Oxohalides

A large number are known, although only a few are of common occurrence. They are of three main types. For the M^VOX_3 and M^VO_2X compounds all possible types (16 in all) are known. For the $M^{IV}OX_2$ type, the six with M = Nb or Ta and X = Cl, Br, and I are known. The M_3O_7Cl and $Nb_5O_{11}Cl_3$ compounds have also been reported.

The MOX_3 compounds with X = Cl, Br, and I can generally be obtained by controlled reactions between MX_5 and wet O_2, and in some special ways, for example, by pyrolysis of $TaCl_5OEt_2$ for $TaOCl_3$. All are readily hydrolyzed, but from solutions in concentrated HX larger cations will precipitate salts of ions such as $[NbOCl_5]^{2-}$, $[NbOF_5]^{2-}$, or $[NbOCl_4]^-$. Neutral adducts such as $NbOCl_3(PR_3)_2$ can also be formed.[11] The white chlorides and yellow bromides are volatile, giving mononuclear vapors with molecules having C_{3v} symmetry[12] although the solids have polymeric structures, as shown in Fig. 18-B-4(b) for $NbOCl_3$.

Mixed halide complexes with other ligands are also known, for example, mononuclear complexes of the type Tp^*MOCl_2,[13] the dinuclear μ-oxo complex $[\{NbCl_3(O_2CPh)\}_2O]$,[14] trinuclear complexes of type $[NbOCl_3(OCR_2)]_3$,[15] and tetranuclear $[\{TaOCl_2(O_2CR)\}_4]$.[16] There are thio analogues (MSX_3) that are less well

[10]I. N. Belova *et al.*, *J. Struct. Chem. (Engl. Transl.)* **1996**, *37*, 232.

[11]V. C. Gibson and T. P. Kee, *J. Chem. Soc., Dalton Trans.* **1993**, 1657.

[12]I. N. Belova *et al.*, *J. Struct. Chem. (Engl. Transl.)* **1996**, *37*, 224.

[13]J. Sundermeyer *et al.*, *Chem. Ber.* **1994**, *127*, 1201.

[14]D. A. Brown *et al.*, *J. Chem. Soc., Dalton Trans.* **1993**, 2037.

[15]S. A. Duraj *et al.*, *Inorg. Chim. Acta* **1989**, *156*, 41.

[16]D. A. Brown *et al.*, *J. Chem. Soc., Dalton Trans.* **1993**, 1163.

known. Like the MOX_3 compounds, they form 5- and 6-coordinate adducts with neutral donors.

18-B-3 Oxidation State V

Fluoride Complexes

The metals and the pentoxides dissolve in aqueous HF to give fluoro complexes, whose composition depends markedly on the conditions. Addition of CsF to niobium in 50% HF precipitates $CsNbF_6$, while in weakly acidic solutions hydrolyzed species $[NbO_xF_y \cdot 3H_2O]^{5-2x-y}$ occur, and components such as $K_2NbOF_5 \cdot H_2O$ may be isolated. Raman and ^{19}F nmr spectra show that $[NbOF_5]^{2-}$ is present in aqueous solution up to ~35% in HF, and $[NbF_6]^-$ becomes detectable beginning at ~25% HF. The ion $[NbF_6]^-$ is normally the highest fluoro complex of Nb^V formed in solution, although in 95 to 100% HF, NbF_7^{2-} ions may possibly be present, but in the molten system $LiF-NaF-KF-K_2NbN_7$, they are undoubtedly present, according to Raman studies.[17] Salts containing the NbF_7^{2-} ion can be crystallized from solutions with very high F^- concentrations. From solutions of low acidity and high F^- concentration, salts of the $[NbOF_6]^{3-}$ ion can be isolated.

The TaF_6^- anion can also be prepared by reaction of Ta_3N_5 with 50% HF. The NH_4TaF_6 salt produced readily yields TaO_2F via $(NH_4)_2Ta_2O_3F_6$ as an intermediate.[18] Other crystalline tantalum fluoro compounds are $KTaF_6$, K_2TaF_7, and K_3TaF_8. The TaF_8^{3-} ion, like NbF_7^{2-}, may be stabilized by crystal forces, since in aqueous HF or NH_4F solutions Raman spectra show the presence of TaF_6^- and TaF_7^{2-} ions only. In anhydrous HF solutions of $KTaF_6$ and K_2TaF_7 the only species identified by ^{19}F nmr spectra is TaF_6^-. In HF solutions tantalum can be separated from niobium by selective extraction into isobutyl methyl ketone.

The hexafluoro anions can also be made by the dry reaction:

$$M_2O_5 + 2KCl \xrightarrow{\text{BrF}_3} 2KMF_6$$

and the $[MOF_6]^{3-}$ salts can be prepared by bromination of the metals in methanol followed by addition of NH_4F or KF.

The $M_2F_{11}^-$ ions, obtained by the reactions

$$MF_5 + Bu^n_4NBF_4 \xrightarrow{\text{CH}_2\text{Cl}_2} (Bu^n_4N)[MF_6]$$

$$(Bu^n_4N)[MF_6] + MF_5 \longrightarrow (Bu^n_4N)[M_2F_{11}]$$

have two octahedra linked by an $M-F-M$ bridge. When H_2O_2 is added to solutions of the pentoxides in HF, mixed fluoro-peroxo complexes, $[MF_4(O_2)_2]^{3-}$ or $[MF_5O_2]^{2-}$,[19] can be precipitated. The latter can also be made by reaction of an

[17] J. H. von Barner et al., Inorg. Chem. 1991, 30, 561.
[18] C. Grimberg et al., Eur. J. Solid State Inorg. Chem. 1994, 31, 449.
[19] S. Subramanian et al., Inorg. Chem. 1991, 30, 1630.

aqueous solution of $HNbF_6$ and H_2O_2:

$$HNbF_6 + H_2O_2 \xrightarrow{\quad\quad} \xrightarrow{\text{KOH}} K_2Nb(\eta^2\text{-}O_2)F_5 \cdot H_2O$$

Pentahalides as Lewis Acids

All of the pentahalides can combine with halide ions to form MX_6^- ions or with neutral Lewis bases (N, O, P, or S donors, commonly) to form MX_5L species such as $NbCl_5(MeCN)$.[20] With amines, $NbCl_5$ often undergoes reduction leading to Nb^{IV} complexes such as $NbCl_4py_2$.

Because of their Lewis acidity, the MX_5 species, usually the fluorides or the chlorides, can act as catalysts in cyclotrimerizing or linearly polymerizing acetylenes (although in the latter case metal alkylidene intermediates may be important) and in Friedel-Crafts and related alkylation reactions. They also polymerize THF and can catalyze the hydrolysis of acetonitrile.[21]

The chlorides and bromides can also abstract oxygen from certain donor solvents (cf. VCl_4 and $MoCl_5$) to give the oxohalides:

$$NbCl_5 + 3Me_2SO \xrightarrow{\quad\quad} NbOCl_3 \cdot 2Me_2SO + Me_2SCl_2$$

$$Me_2SCl_2 \xrightarrow{\quad\quad} ClCH_2SMe + HCl$$

Other Complexes and Compounds

In addition to the anionic or neutral complexes formed by Lewis acid behavior of MX_5 (or MOX_3), there is an extensive chemistry of compounds in which halogen is replaced by alkoxide (OR), dialkylamide (NR_2), and alkyl (CR_3) groups. These compounds may be coordinately unsaturated, and they form neutral adducts or anionic complexes like the slightly distorted trigonal prismatic $[Ta(C_6H_5)_6]^-$.[22]

Some reactions of $TaCl_5$ that are also, in general, typical for $NbCl_5$ are given in Fig. 18-B-5.

Oxygen Ligands

The dinuclear *alkoxides*, $M_2(OR)_{10}$, obtained by the action of alcohols and an amine on the pentachlorides, have two alkoxo bridges.[23] These may be cleaved by donors to give mononuclear compounds like $Nb(OMe)_5py$. Partially substituted species such as $MCl_x(OR)_{5-x}$ and $Nb_2Cl_4(OEt)_4(O_2CR)_2$[24] can be made. With extremely bulky OR groups (e.g., R = 1- or 2-adamantyl) mononuclear species can be isolated. Numerous alkoxoniobiates of general composition $M[Nb(OR)_6]$[25] and a related

[20]G. R. Willey *et al.*, *Polyhedron* **1997**, *16*, 351.
[21]G. R. Lee and J. A. Crayston, *Polyhedron* **1996**, *15*, 1817.
[22]K. Seppelt *et al.*, *Chem. Ber./Recueil* **1997**, *130*, 903.
[23]N. Ya. Turova *et al.*, *Polyhedron* **1996**, *15*, 3869.
[24]D. A. Brown *et al.*, *Inorg. Chim. Acta* **1994**, *227*, 99.
[25]N. Ya. Turova *et al.*, *Polyhedron* **1995**, *14*, 1531.

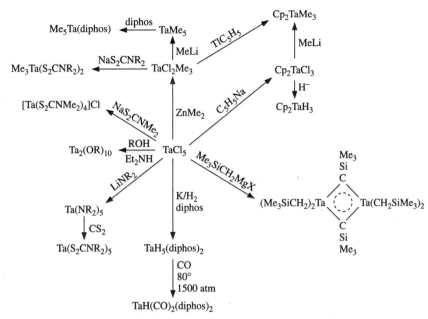

Figure 18-B-5 Some reactions of tantalum pentachloride.

species $[Nb(OTeF_5)_6]^-$ [26] have been described. Oxoalkoxides of different nuclearity are common.[27] Hydrido aryloxides of tantalum can catalyze the regio- and stereo-selective hydrogenation of arene rings and alkenes.[28] Others, such as $(Bu_3^tSiO)_3TaH_2$, are active carbonylation agents.[29]

Cationic 8-coordinate complexes with tropolonato and β-diketonato anions, for example, $[M(\beta\text{-dike})_4]^+$, are also well known.

Nitrogen Ligands

While they form few complexes with simple amines, the pentavalent metals form important $M(NR_2)_xX_{5-x}$ compounds. The $M(NMe_2)_5$ compounds, obtained by reaction of MCl_5 with $LiNMe_2$, have been greatly studied and used for synthesis of other compounds. They are mononuclear in the gas phase with apparently square pyramidal geometry[30] but in the solid $Ta(NMe_2)_5$ is *tbp* while $Nb(NMe_2)_5$ approaches an *sp* structure. Mixed species can be obtained by reactions such as

$$Ta(NMe_2)_5 + 2Me_3SiCl \longrightarrow TaCl_2(NMe_2)_3 + 2Me_3SiNMe_2$$

$$TaCl_5 + 5HNMe_2 \longrightarrow TaCl_3(NMe_2)_2(HNMe_2) + 2Me_2NH_2Cl$$

[26]S. H. Strauss *et al.*, *J. Am. Chem. Soc.* **1992,** *114,* 10995.
[27]L. G. Hubert-Pfalzgraf *et al.*, *J. Chem. Soc., Chem. Commun.* **1994,** 601; W. A. Herrmann *et al.*, *Angew. Chem. Int. Ed. Engl.* **1995,** *34,* 2187.
[28]I. P. Rothwell *et al.*, *J. Am. Chem. Soc.* **1992,** *114,* 1927; *J. Chem. Soc., Chem. Commun.* **1995,** 553.
[29]P. T. Wolczanski *et al.*, *J. Am. Chem. Soc.* **1993,** *115,* 5570.
[30]D. A. Rice *et al.*, *Inorg. Chem.* **1992,** *31,* 4733.

There are a number of compounds containing M=NR units, especially with tantalum, examples being $Ta(NMe_2)_3(NBu^t)$, $TaCl_3(THF)_2(NR)$ and $Ta(S_2CNMe_2)_3$-$(NSiMe_3)$, and $Ta(DPhF)_3(NPh)$.[31] A particularly interesting compound of this type is the binuclear dinitrogen complex,

$$(benzyl)_3P(THF)Cl_3Ta{=}N{-}N{=}TaCl_3(THF)P(benzyl)_3$$

which could be formulated as consisting of an N_2 molecule donating to two Ta^{III} atoms, but, on the basis of the observed $N{-}N$ (1.28 Å) and $N{-}Ta$ (1.80 Å) distances is probably better represented as a diimine complex involving two Ta^V atoms.

Reaction of $M(NR_2)_5$ compounds with alcohols provides an excellent route to the alkoxides. Moreover, the $M(NR_2)_5$ compounds undergo insertion reactions with CS_2 and CO_2:

$$Nb(NMe_2)_5 + 5CO_2 \longrightarrow Nb(O_2CNMe_2)_5$$

$$Ta(NMe_2)_5 + 5CS_2 \longrightarrow Ta(S_2CNMe_2)_5$$

The metal atoms in the resulting carbamates and dithiocarbamates are 8-coordinate, with two monodentate and three bidentate ligands. The CO_2 insertion is partially reversible so that the carbamates can exchange with labeled CO_2:

$$Nb(O_2CNMe_2)_5 \rightleftharpoons Nb(O_2CNMe_2)_4(NMe_2) + CO_2$$

$$\downarrow {}^*CO_2$$

$$Nb(O_2CNMe)_4(O_2{}^*CNMe_2)$$

Some $(porph)M(\mu\text{-}O)_3M(porph)$ complexes in which the metal atom lies well out of the porphyrin plane and is bonded to three oxygen atoms are known.[32] Cationic sandwich compounds of the type $[Ta(porph)_2][TaCl_6]$ can be prepared from $TaCl_5$ and $Li_2(porph)$ while neutral "half-sandwich" compounds such as $Ta(porph)Me_3$ can be made from $TaMe_3Cl_2$ and $Li_2(porph)$.[33]

Sulfur Ligands[34]

Besides the $M(S_2CNMe_2)_5$ compounds just mentioned, many other dithiocarbamates are known, of which the cationic species $[M(S_2CNR_2)_4]^+$ are most important. They generally have dodecahedral structures in which the S_2CNR_2 ligands span the four m edges (see Fig. 1-4) of the dodecahedron, but many are fluxional. Compounds of this type can be made by the reaction:

$$TaCl_5 + 4NaS_2CNR_2 \xrightarrow[\text{reflux}]{CH_2Cl_2} [Ta(S_2CNR_2)_4]Cl + 4NaCl$$

[31]F. A. Cotton et al., Bull. Soc. Chim. Fr. **1996**, *133*, 711.
[32]H. Brand and J. Arnold, Coord. Chem. Rev. **1995**, *140*, 137.
[33]J. Arnold et al., J. Am. Chem. Soc. **1994**, *116*, 9797.
[34]D. W. Stephan and T. T. Nadasdi, Coord. Chem. Rev. **1996**, *147*, 147 (references to early transition metal thiolates in various oxidation states).

There are some trigonal prismatically coordinated complexes formed by reaction of $M(NR_2)_5$ with dithiolate dianions, for example, $[M(1,2\text{-}S_2C_6H_4)_3]^-$. Several mixed species such as $NbE(S_2CNEt_2)_3$, E = O, S, or S_2 are also known.[35] Simpler chalcogenometalate species of the type $[M(E)_3(Bu^tS)]^{2-}$ can be prepared by reaction of MCl_5 with $NaSBu^t$ in the presence of sulfur or selenium in CH_3CN solutions.[36] The tetrathiometalates $[MS_4]^{3-}$ can be made in the reaction system $M(OEt)_5/(Me_3Si)_2S/$ LiOMe in acetonitrile; the anions are discrete and essentially tetrahedral.[37] Polysulfides such as $K_4Nb_2S_{11}$, $K_4Nb_2S_4$, and $K_6Nb_4S_{25}$ are also known.[38]

Other Compounds

These include nitrides, silicides, selenides, and phosphides,[39] as well as many alloys. Definite hydride phases also appear to exist.

Freshly precipitated $Ta_2O_5 \cdot nH_2O$ reacts with basic H_2O_2 to yield colorless crystals of the peroxo compound $K_3[Ta(O_2)_4]$ which is isotypic with the chromium analogue.[40] There are no simple salts such as sulfates and nitrates. Sulfates such as $Nb_2O_2(SO_4)_3$ probably have oxo bridges and coordinated sulfato groups. In HNO_3, H_2SO_4, or HCl solutions, Nb^V can exist as cationic, neutral, and anionic species, hydrolyzed, polymeric, and colloidal forms in equilibrium, depending on the conditions.

Various halo-alkyl carboxylates have been reported, examples are $TaCl(OAc)_4$ and $[MCl_4(O_2CR)]$. The latter contain dinuclear molecules with $syn-syn$ bridging carboxylate groups (18-B-II) and are prepared from the pentahalides and the corresponding acid[41] as in:

$$2TaCl_5 + 2RCOOH \xrightarrow[\text{R.T. or higher}]{CH_2Cl_2 \text{ or } CCl_4} [TaCl_4(O_2CR)]_2 + 2HCl$$

Some contamination with oxo compounds is frequently found, especially for the niobium compounds.

(18-B-II)

[35]C. G. Young *et al.*, *J. Chem. Soc., Dalton Trans.* **1994**, 1765.

[36]D. Coucouvanis *et al.*, *Inorg. Chem.* **1994**, *33*, 4429; **1995**, *34*, 2267.

[37]R. H. Holm *et al.*, *Inorg. Chem.* **1992**, *31*, 4333.

[38]W. Bensch and P. Dürichen, *Eur. J. Solid State Inorg. Chem.* **1996**, *33*, 527; W. Bensch and P. Dürichen, *Inorg. Chim. Acta* **1997**, *261*, 103.

[39]M. Greenblatt *et al.*, *Inorg. Chem.* **1996**, *35*, 845.

[40]G. Wehrum and R. Hoppe, *Z. anorg. allg. Chem.* **1993**, *619*, 1315.

[41]M. G. Wallbridge *et al.*, *Polyhedron* **1994**, *13*, 2265.

18-B-4 Oxidation State IV

The oxides, halides, and oxohalides have been mentioned. The disulfides (MS_2) have layer structures with adjacent layers of sulfur atoms. Intercalation compounds can be formed by insertion between these sheets. Both simple (e.g., $NbS_2 \cdot \frac{1}{2}py$) and ionic [e.g., $TaS_2^{x-}(NH_4^+)_x(NH_3)_{1-x}$] compounds can be prepared.

There are several Nb^{IV} alkoxides such as $Nb(OEt)_4$, $[NbCl(OEt)_3py]_2$, $NbCl_3(OR)bipy$, and salts of the ion $[NbCl_5OEt]^{2-}$.

Halide and Pseudohalide Complexes

They are usually prepared by reacting the ligand with the tetrahalide or by reduction of the pentahalides, either electrolytically or with metals like Na, Zn, Al, Mg, and Nb. Substitution of weakly coordinating THF in $NbCl_4(THF)_2$ is also a convenient route to $NbCl_4$ adducts.[42] Typical homoleptic complexes are the octahedral $[MX_6]^{2-}$ ions (X = Cl, Br, and SCN), pentagonal bipyramidal $[NbF_7]^{3-}$, and dodecahedral $[Nb(CN)_8]^{4-}$.

Adducts of the types MX_4L_n ($n = 1–4$) are numerous, with the octahedral (MX_4L_2) complexes being most common. Some are cis [e.g., $NbX_4(MeCN)_2$ and $TaCl_4(PMe_2Ph)_2$] and others are trans [e.g., $TaCl_4py_2$ [43]]. It is possible that the red and green forms of $NbBr_4py_2$ may be cis and trans isomers. All mononuclear species display paramagnetism corresponding to one d electron and many give sharp esr signals. Electronic structure calculations of $trans$-$NbCl_4(PR_3)_2$ correctly predict the observed Jahn-Teller distortion found in the solid state.[44]

For the smallest phosphines, PMe_3 and PMe_2Ph, higher coordination numbers are found. Thus, there are $MX_4(PR_3)_3$ species, which have capped octahedral structures (18-B-III) and there are several examples of diamagnetic $M_2X_8(PR_3)_4$ complexes having the type of structure shown in Fig. 18-B-6, with an M—M single bond. Other isolable 8-coordinate species, $MX_4(LL)_2$, are obtained with diphosphines, diarsines, and bipyridyl. For $TaCl_4(dmpe)_2$ the structure is approximately square antiprismatic (18-B-IV), which contrasts with the dodecahedral structure of the corresponding $[TaCl_4(dmpe)_2]^+$ ion.

(18-B-III) 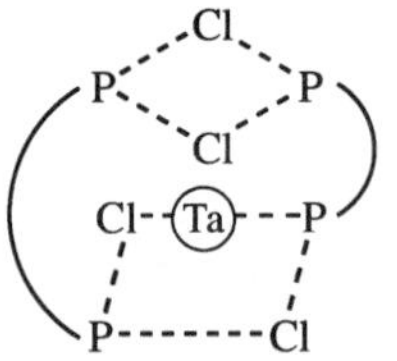   (18-B-IV)

Other Complexes

Mixed amido complexes can be prepared by reducing Ta^V precursors, for example:

$$Ta(NR_2)_2Cl_3 \xrightarrow{Na/Hg} Ta(NR_2)_2Cl_2; \quad R = SiMe_3, Et$$

[42]S. F. Pedersen et al., Inorg. Synth. 1992, 29, 119.
[43]G. J. Miller et al., Acta Cryst. 1993, C49, 1770.
[44]M. B. Hall et al., Inorg. Chem. 1994, 33, 1473.

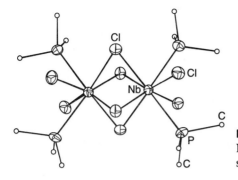

Figure 18-B-6 The structure of $Nb_2Cl_8(PMe_3)_4$, consisting of two square antiprisms sharing a Cl_4 face.

They have distorted tetrahedral geometries,[45] while complexes such as $NbCl_3\{N(CH_2CH_2XR_2)_2\}$, X = N, P, are 6-coordinate.[46] Other complexes with nitrogen-containing ligands are those derived from the tridentate trispyrazolylborate anion.[47] The reaction of $TaCl_5$ and KTp in neat MeCN gives a mixture of Nb^V and Nb^{IV} complexes but in a mixture of MeCN and CH_2Cl_2 a dinuclear Nb^{IV} species is the only product:

$$NbCl_5 + KTp \xrightarrow[-30°C]{MeCN} K[TpNbCl_5] + Tp_2Nb_2Cl_6$$

$$\Big\downarrow \xrightarrow[reflux]{MeCN/CH_2Cl_2} Tp_2Nb_2Cl_6$$

If $NbCl_4(MeCN)_2$ is used as starting material only an ionic product is isolated, according to

$$NbCl_4(MeCN)_2 + KTp \xrightarrow{MeCN} K[TpNbCl_4]$$

There are also numerous mixed ligand complexes, such as $NbCl_3(\beta\text{-dike})_2$, $[NbCl_5(OR)]^{2-}$, $[NbCl_3(OR)bipy]$, $NbCl_2(phthalocyanine)$, and so on. Frequently, Ta^{IV} reacts by abstracting oxygen and thus some Ta analogues of Nb complexes with oxygen-containing ligands have not been isolated.

Various dinuclear compounds have the structure of an edge-sharing bioctahedron, in which there is a metal-metal bond of order 1. This results in diamagnetism and M—M distances of 2.80 to 2.90 Å. Some examples are $[(THT)_2Cl_2Nb]_2(\mu\text{-S})_2$, $[(MeCN)_2Cl_2Nb]_2(\mu\text{-S})_2$, $[(MeOH)(MeO)Cl_2Nb]_2(\mu\text{-OMe})_2$, and complexes of the type $M_2(\mu\text{-S})_2Cl_4L_4$.[48] A similar compound with one S^{2-} and one S_2^{2-} bridge is $[(THT)_2Br_2Nb]_2(\mu\text{-S})(\mu\text{-S}_2)$. In the anion $Nb_2(\mu\text{-S}_2)_2(NCS)_8^{4-}$ there are two S_2^{2-} bridges.[49]

Several hydrido compounds of Nb^{IV} and Ta^{IV} have been obtained from low-

[45]S. Suh and D. M. Hoffman, *Inorg. Chem.* **1996,** 35, 5015.
[46]P. G. Edwards *et al., J. Chem. Soc., Dalton Trans.* **1995,** 355.
[47]M. Etienne, *Coord. Chem. Rev.* **1996,** 156, 201.
[48]F. A. Cotton *et al., Inorg. Chem.* **1990,** 29, 4002; E. Babaian-Kibala and F. A. Cotton, *Inorg. Chim. Acta* **1991,** 182, 77.
[49]M. Sokolov *et al., Inorg. Chem.* **1994,** 33, 3503.

valent phosphine complexes. The reaction of M^{II} adducts with H_2 gives stable paramagnetic $MH_2Cl_2L_4$, where L = PMe_3 or ½ dmpe. The paramagnetic dihydride and also the tetrahydride $TaH_4(diphos)_2$ are obtained from the M^V hydrides by H-abstraction using di-*tert*-butylperoxide, namely,

$$\left.\begin{array}{l} Cp_2NbH_3 \\ (diphos)_2TaH_5 \end{array}\right\} \xrightarrow{(Bu^tO)_2} \left\{\begin{array}{l} Cp_2NbH_2 \\ (diphos)_2TaH_4 \end{array}\right.$$

A related compound, $[TaH_5(dmpe)_2][LiHBEt_3]$, is obtained from the reaction of $TaCl_4(dmpe)_2$ and $Li(HBEt_3)$.[50]

Dinuclear species of the general formula $[TaCl_3(PMe_3)]_2(\mu\text{-}Cl)_n(\mu\text{-}H)_{4-n}$ where n = 0, 1, and 2 are formed upon reacting $[TaCl_2(PMe_3)_2]_2(\mu\text{-}H)_2$ with H_2, HCl, and Cl_2, respectively. An amide-containing complex $[(Cy_2N)_2ClTa(\mu\text{-}H)]_2$[51] has a $Ta(\mu\text{-}H)_2Ta$ core with no metal–metal bond.

18-B-5 Oxidation State III

This oxidation state is readily accessible *via* reduction of pentahalides with two equivalents of sodium amalgam in the presence of neutral ligands like SC_4H_8 (tetrahydrothiophene, THT), SMe_2, or PMe_3. The sulfur donors afford confacial bioctahedral $[M_2Cl_{6+x}L_{3-x}]^{x-}$ species[52] while with PMe_3 the edge-sharing bioctahedral molecule, $M_2Cl_6(PMe_2)_4$ is obtained.

The THT and SMe_2 adducts have structures of the type (18-B-V). Their chemistry has been extensively studied and it is summarized in Fig. 18-B-7. The diverse, and in some cases unique, reactivity of these compounds includes: substitution with preservation of the geometry or with conversion to $(MX_4)_2(\mu\text{-}X)_2$ species, oxidative-addition,[53] cluster formation, splitting of C–N bonds,[54] and above all coupling of the molecules with triply bonded carbon atom.[55] They catalytically trimerize and polymerize terminal acetylenes, and dimerize nitriles and isonitriles with incorporation of the new ligand into the complex. Another remarkable reaction of $M_2Cl_6L_3$ is the metathesis of M≡M and N≡N bonds into two M≡N bonds upon reaction with azobenzene.

(18-B-V)

The reactivity of $Ta_2Cl_6(PMe_3)_4$ (18-B-VI) is also extraordinary. It provided the first example of oxidative-addition of molecular hydrogen to a multiple M–M bond, by forming $[TaCl_2(PMe_3)_2]_2(\mu\text{-}Cl)_2(\mu\text{-}H)_2$ (18-B-VII). The latter, a Ta^{IV} complex, can

[50]J. H. Teuben *et al., Inorg. Chim. Acta* **1997,** *259,* 237.
[51]F. A. Cotton *et al., J. Am. Chem. Soc.* **1996,** *118,* 12449.
[52]F. A. Cotton *et al., Inorg. Chem.* **1994,** *33,* 3055.
[53]F. A. Cotton *et al., Inorg. Chem.* **1997,** *36,* 896.
[54]D. E. Wigley *et al., J. Am. Chem. Soc.* **1992,** *114,* 5462.
[55]F. A. Cotton and M. Shang, *Organometallics* **1990,** *9,* 2131.

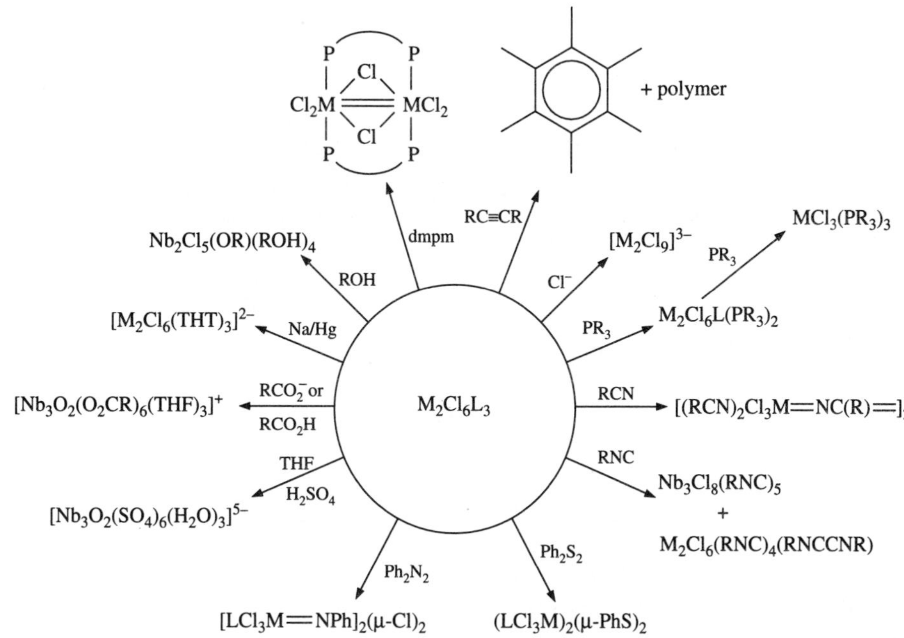

Figure 18-B-7   Some reactions of the $M_2Cl_6L_3$ species (M = Nb and Ta; L = Me_2S and THT).

be reduced to the dinuclear, $Ta_2Cl_4(PMe_3)_4H_2$, compound which contains bridging-hydrides and the TaX_4 squares are eclipsed. The Ta—Ta bond distance in this compound, 2.545 Å, is the shortest metal-metal bond so far found for this element. The compound $Ta_2Cl_6(PMe_3)_4$ also reacts with ethylene, yielding a mononuclear adduct of composition $TaCl_3(PMe_3)_2(C_2H_4)$.

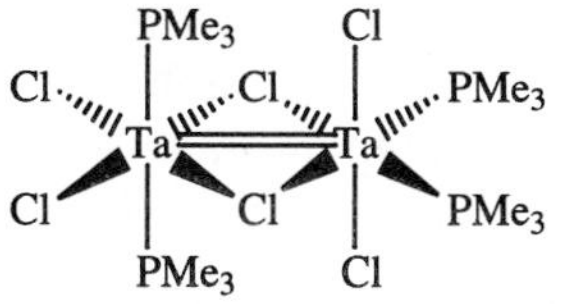

(18-B-VI) 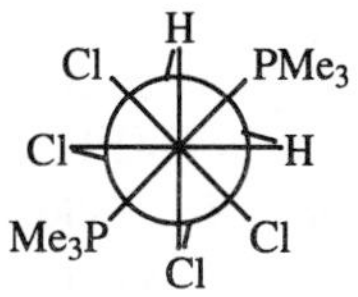   (18-B-VII)

In general the trivalent oxidation state is dominated by highly reactive dinuclear complexes, usually possessing an M—M double bond with a typical length around 2.7Å, although M—M distances >2.8 Å are also encountered. Due to this tendency to form dinuclear species with a strong interaction between metal atoms the stabilization of mononuclear compounds requires the presence of excess ligand or coordinative saturation. The latter is exemplified by $Nb(CN)_8^{5-}$, $NbBr_3(PMe_2Ph)_3$, $MH_2X(PMe_3)_4$, and $CpNb(Bu^tNC)_4Cl^+$.

The three different Ta^{III} species obtained by reduction of $TaCl_5$ in the presence of PMe_3 provide an excellent illustration of this point. In toluene either $Ta_2Cl_6(PMe_3)_4$ or $TaCl_3(PMe_3)_3$ is obtained. The preparation of the latter requires excess ligand, and the mononuclear complex dimerizes with loss of PMe_3 upon dissolution. A reduction in neat phosphine leads to deprotonation of methyl groups producing $Ta(PMe_3)_3(\eta^2\text{-CHPMe}_2)(\eta^2\text{-CH}_2PMe_2)$. Similarly, there are two types of

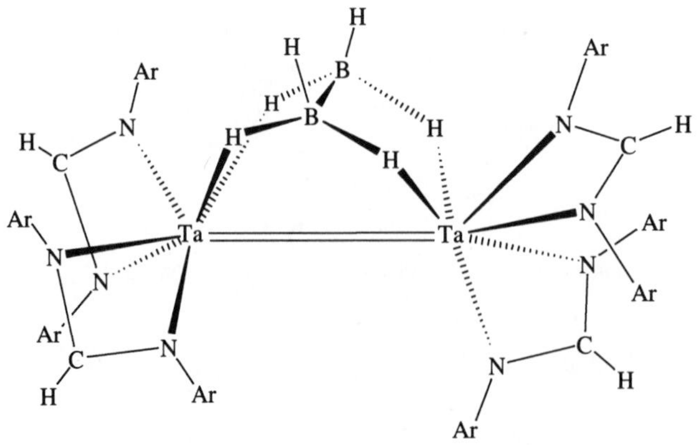

(18-B-VIII)

TaCl$_3$/py compounds, one is mononuclear *mer*-TaCl$_3$(py)$_3$[56] and the other one is dinuclear Ta$_2$Cl$_6$(py)$_4$.[57]

Reduction of TaCl$_5$ with LiBH$_4$ followed by addition of LiDArF, Ar = phenyl or tolyl, gives dinuclear complexes of trivalent tantalum with B$_2$H$_6^{2-}$ species that bridge across the Ta=Ta bond[58] (18-B-VIII).

18-B-6 Oxidation State II

There are various types of compounds: mononuclear, binuclear, cluster (see Section 18-B-7), and organometallic and relatively few of each type. Paramagnetic, octahedral MX$_2$L$_4$ species with either metal and X = Cl, L = PMe$_3$, or ½dmpe, and with M = Nb, X = OAr, and L = ½dmpe,[59] can be prepared by reduction of higher-valent chlorides or aryl oxides with Na/Hg. The MCl$_2$(PMe$_3$)$_4$ compounds undergo phosphine exchange with dmpe and all the chloro complexes react with H$_2$ to give M^{IV} hydrides of composition M(H)$_2$Cl$_2$L$_4$. Potassium graphite, KC$_8$, works best for the reduction of NbCl$_4$(THF)$_2$ in pyridine to prepare *trans*-NbCl$_2$(py)$_4$.[60]

Other triply bonded dinuclear M^{II} species are [Cl$_3$M(μ-THT)$_3$MCl$_3$]$^{2-}$. For tantalum, the bright red species is obtained by reduction of Ta$_2$Cl$_6$(THT)$_3$ with Na/Hg; it is stable towards water and, in general, chemically inert (Ta—Ta = 2.626 Å). Another anionic species is [Nb$_2$Cl$_5$(THT)(py)$_5$]$^-$.[61]

Very recently a new type of divalent Nb compound containing a triple bond in a tetragonal lantern environment (Fig. 18-B-8) was prepared by the following reaction:

$$[NbCl_3(DME)]_n + nKC_8 + 2nLihpp \longrightarrow {}^n/_2Nb_2(hpp)_4 + nKCl + 2nLiCl + 8nC$$

[56]F. A. Cotton *et al.*, *Inorg. Chim. Acta* **1996**, *245*, 115.
[57]E. Babaian-Kibala and F. A. Cotton, *Inorg. Chim. Acta* **1990**, *171*, 71.
[58]F. A. Cotton *et al.*, *J. Am. Chem. Soc.* **1996**, *118*, 4830.
[59]I. P. Rothwell, *J. Am. Chem. Soc.* **1989**, *111*, 4742.
[60]F. A. Cotton *et al.*, *Inorg. Chem.* **1995**, *34*, 5424.
[61]F. A. Cotton and M. Shang, *Inorg. Chim. Acta* **1994**, *227*, 191.

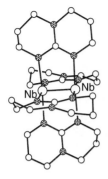

Figure 18-B-8 The structure of $Nb_2(hpp)_4$. Dotted and open circles correspond to N and C atoms, respectively; hydrogen atoms are not shown.

The diamagnetic complex has a short Nb—Nb distance of 2.2035(9) Å[62] which is by far the shortest metal—metal bond found for either niobium or tantalum.

18-B-7 Niobium and Tantalum Cluster Complexes

Both of these elements show a marked tendency to form metal atom cluster compounds in their lower oxidation states. The best known are oxo and halide cluster complexes.

Structures of Reduced Oxides and Oxoniobates[63]

Niobium oxide has a unique and intriguing structure as the coordination of both Nb and O atoms appears to be square planar, an unusual coordination geometry for oxygen and for a d^3 metal, as shown in Fig. 18-B-9. The arrangement of atoms is that of a defect rock salt structure in which all the atoms at the center and vertices of the unit cells are removed. It has been shown that the formation of the ordered defect structure augments Nb—Nb bonding and stabilizes the square planar oxygen atom by Nb—O π-bonding. Thus the compound is properly formulated as $Nb_{0.75}O_{0.75}$[64] and it is actually an aggregate of Nb_6O_{12} clusters. Tantalum oxide, TaO, has the regular NaCl structure. There is also a distinct Ta_4O phase.

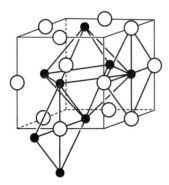

Figure 18-B-9 Schematic drawing of the NbO structure emphasizing square planar coordination around Nb and O atoms. Open circles represent O; closed circles repesent Nb.

[62]F. A. Cotton et al., J. Am. Chem. Soc. **1997,** 119, 7889.
[63]J. Köhler et al., Angew. Chem. Int. Ed. Engl. **1992, 31,** 1437.
[64]J. F. Mitchell, Inorg. Chem. **1993,** 32, 5004.

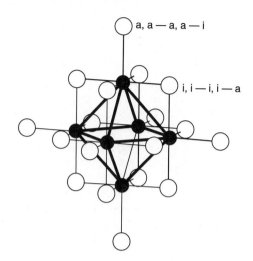

Figure 18-B-10 $Nb_6O^i_{12}O^a_6$ cluster.

A large number of low-valent oxoniobate compounds are known (Section 18-B-1). Most of the structures consist of close-packing of units that have comparable sizes. These units are O atoms, Nb_6 octahedra, and A atoms which serve as counter cations. Large electropositive metals A (A = Na, K, Rb, Sr, La, Eu) form a conventional close-packing; smaller M atoms fill tetrahedral (M = Al, Si, Mg, Mn), octahedral (M = Nb, V, Mg, Mn), and *tbp* (M = V, Nb) interstices of the crystals. In the reduced oxoniobates the Nb_6 octahedron is normally surrounded by twelve oxygen atoms located above the edges and belong to the inner sphere and form a cuboctahedron. They are designated O^i if they belong to one cluster, or O^{i-i} if they connect adjacent clusters. As many as six O atoms above the apices of the octahedron constitute the outer ligand sphere and may belong to one cluster only (O^a), or link clusters with identical (O^{a-a}), as well as different functionalities (O^{a-i}, O^{i-a}). A $Nb_6O^i_{12}O^a_6$ cluster is shown in Fig. 18-B-10.

Most of the Nb—Nb distances in clusters containing discrete Nb_6O_{12} units fall within a narrow range of 2.79–2.89 Å.

Halide Containing Clusters[65]

They are obtained by reduction of the pentahalides by the metal alone, or in the presence of sodium chloride. Their key structural unit, an $[M_6X_{12}]^{n+}$ group, is shown in Fig. 18-B-11. A typical reaction is

$$14NbCl_5 + 16Nb + 20NaCl \xrightarrow{\text{12h, 850°C}} 5Na_4Nb_6Cl_{18}$$

The extraction of the melt with very dilute HCl gives a green solution from which, by addition of concentrated HCl, the black solid $Nb_6Cl_{14} \cdot 8H_2O$ is obtained. Corresponding tantalum compounds are obtained similarly, and mixed metal species, for example, $[(Ta_5MoCl_{12})Cl_6]^{2-}$, can be obtained by the reduction of $TaCl_5$ + $MoCl_5$ in $NaAlCl_4$—$AlCl_3$ melts by Al.

The M_6X_{12}, X = F, Cl, Br, units are usually found with a charge of 2+ and one additional ligand (X$^-$, H_2O, MeOH, OH$^-$, etc.) is coordinated to the external site

[65]C. Perrin *et al., New J. Chem.* **1988,** *12,* 321.

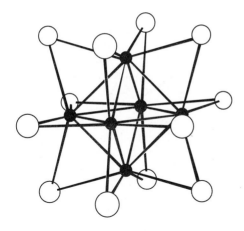

Figure 18-B-11 The structure of the $[M_6X_{12}]^{n+}$ unit found in many halogen compounds of lower-valent niobium and tantalum.

on each metal atom.[66] Within the M_6 octahedron there are M—M bonds of fractional order; in $[Nb_6Cl_{12}]^{2+}$ for example, there are 16 electrons so that a bond order of $16/24 = 2/3$ can be assigned, consistent with the Nb—Nb bond distances of 2.80 Å. One- or two-electron oxidation by various oxidizing agents, e.g., $NOBF_4$, give hexanuclear species with charges of 3+ or 4+. Electrochemical reductions in a basic melt can give monocharged cations.[67] Contrary to the zirconium and hafnium clusters (Section 18-A-4), in the Nb and Ta clusters there is no interstitial atom.[68]

Iodide compounds with formulas such as Nb_6I_{11}, $CsNb_6I_{11}$, $Nb_6I_{11}H$, and $CsNb_6I_{11}H$ can be made. These also contain octahedral Nb_6 clusters, but of a different type in which there are eight face-bridging (i.e., μ_3) iodine atoms (see Section 18-C-11 for similar Mo and W clusters). In the hydride species the hydrogen atom is inside the Nb_6 octahedron.

The existence of trigonal-prismatic niobium clusters has been suggested for a few compounds but this type of structure was not proven until recently; in $Rb_3[Nb_6SBr_{17}]$, 18 Br atoms are coordinated to a trigonal-prismatic, sulfur-centered Nb_6S unit. The long Nb—Nb distances of 3.28 Å along the edges of the rectangles contrast with the shorter distances (2.97 Å) along the edges of the triangles of the prism.[69]

Triangular Clusters

There are also many compounds in which triangular M_3 clusters are found. Among the earliest examples were the halides Nb_3X_8 in which Nb atoms occupy octahedral interstices in a close-packed array of halide ions so as to form Nb_3 clusters. In Nb_3Cl_8, for example, the Nb—Nb bond distances are ~2.80 Å. Each cluster is surrounded by nine more Cl atoms, three of which bridge to two other clusters and six of which bridge to another cluster as shown in Fig. 18-B-12. Thus, the formula of Nb_3Cl_8 can be written in the notation introduced some years ago by H. Schäfer and others as $Nb_3Cl_4Cl_{3/3}Cl_{6/2}$. Discrete trinuclear complexes of the type $Nb_3Cl_7(PR_3)_6$ or anionic species $M_3Cl_{10}(PR_3)_3^-$, M = Nb, Ta can be made by adding phosphines

[66]N. Brničević *et al.*, *Inorg. Chem.* **1992,** *31*, 3924; *J. Chem. Soc., Dalton Trans.* **1995,** 1441.
[67]C. L. Hussey *et al.*, *Inorg. Chem.* **1992,** *31*, 1255; **1995,** *34*, 370.
[68]A. Simon *et al.*, *Z. anorg. allg. Chem.* **1997,** *623*, 59.
[69]H. Womelsdorf and J.-J. Meyer, *Angew. Chem. Int. Ed. Engl.* **1994,** *33*, 1943.

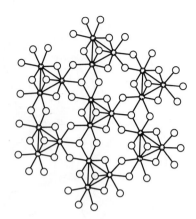

Figure 18-B-12   Portion of the infinite array of atoms in Nb_3Cl_8.

to higher valent halide compounds and then reducing with Na/Hg.[70] There are also $M^INb_4X_{11}$ compounds that contain an Nb_6 cluster, which is effectively two triangular clusters fused together, as shown in Fig. 18-B-13.

Another example, long known (1912) but only recently structurally defined is the red-brown compound obtained by electrochemical reduction of Nb_2O_5 in H_2SO_4. It contains a trinuclear, bioxocapped anion of composition $[Nb_3O_2(SO_4)_6(H_2O)_3]^{5-}$. It was the first example of such a trinuclear species outside the Group 6 transition elements. A more convenient method for its preparation involves reaction of $Nb_2Cl_6(THT)_3$ in THF with H_2SO_4. Carboxylato analogues of the sulfate can also be obtained. All of them contain an equilateral triangle of Nb atoms <2.9 Å apart. The reaction between $Nb_2Cl_6(THT)_3$ and concentrated HCl affords an as yet unidentified green aquo complex, which, upon reaction with SCN^-, gives another type of trinuclear cluster containing a capping sulfur atom and three bridging oxygen atoms, namely, $Nb_3(S)(O)_3(NCS)_9^{6-}$; the corresponding aqua ion has also been reported.[71]

Two trinuclear carbonyl and isonitrile compounds, $Cp_3Nb_3(CO)_7$ and

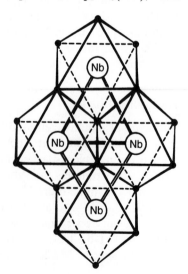

Figure 18-B-13   The key structural unit in $M^INb_4X_{11}$ compounds. Small solid circles are halogen atoms and double lines represent M–M bonds. Many of the peripheral halogen atoms are shared with other units.

[70]F. A. Cotton *et al.*, *Inorg. Chem.* **1988,** *27,* 3413; F. A. Cotton and M. Shang, *Inorg. Chem.* **1993,** *32,* 969.
[71]S. Minhas and D. T. Richens, *J. Chem. Soc., Dalton Trans.* **1996,** 703.

$Nb_3Cl_8(Bu^tNC)_5$, contain CO and Bu^tNC acting as six-electron donors with μ_3-C, μ_2-O and μ_3-C, μ_2-N bonding, respectively, over the isosceles triangles of metal atoms.

ORGANOMETALLIC COMPOUNDS

18-B-8 High Oxidation States (IV and V)

There is an extensive chemistry of Nb^V and Ta^V with M—C σ bonds, η-C_5H_5, cycloheptatriene,[72] and cyclooctatetraene groups.

Niobium and tantalum methyls, Me_3MCl_2, were among the first known stable transition metal methyls. Other partially substituted alkyls, as well as the pentamethyl, are also known. The structure of $TaMe_5$ is square-pyramidal[73] but Me_3TaF_2 is trigonal-bipyramidal.[74] These compounds are made by the reactions:

$$2TaCl_5 + 3ZnMe_2 \xrightarrow{\text{ether}} 2TaCl_2Me_3 + 3ZnCl_2$$

$$TaCl_2Me_3 + 2LiMe \xrightarrow{\text{ether}} TaMe_5 + 2LiCl$$

The niobium compounds are less stable than the tantalum ones. For example, $NbMe_5$ and $TaMe_5$ undergo decomposition readily at $-30°C$ and room temperature, respectively; the reaction is autocatalytic. The more sterically crowded $Ta(CH_2Ph)_5$ is more stable. The formation of adducts such as $TaMe_5(diphos)$ also increases the stability. All these compounds (a) act as Lewis bases, adding neutral ligands or forming anions, and (b) can undergo insertion reactions into the M—CH_3 group (e.g., with NO or CS_2). Furthermore, the halide atoms can be replaced by other groups such as acac, Schiff bases, NR_2, or OR, to give compounds such as $R_3Ta(OR')_2$, which are mononuclear if R and R' are very bulky. The latter serve as precursors for the homogeneous hydrogenation of arene rings[75] as shown by the following reaction:

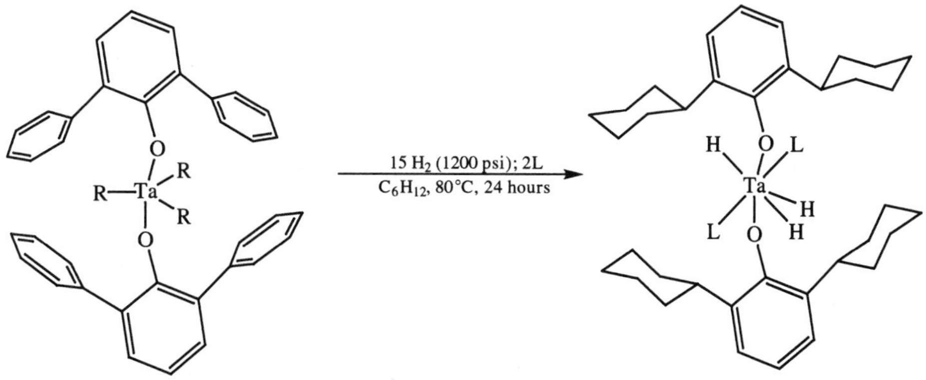

$$\xrightarrow[\text{C}_6\text{H}_{12},\, 80°\text{C},\, 24\text{ hours}]{15\ \text{H}_2\ (1200\ \text{psi});\ 2\text{L}}$$

[72]M. L. H. Green and D. K. P. Ng, *Chem. Rev.* **1995**, *95*, 439.

[73]A. Haaland *et al.*, *Angew. Chem. Int. Ed. Engl.* **1992**, *31*, 1464.

[74]J. Kadel and H. Oberhammer, *Inorg. Chem.* **1994**, *33*, 3197.

[75]See for example: I. P. Rothwell *et al.*, *J. Am. Chem. Soc.* **1991**, *113*, 4710; J. S. Yu and I. P. Rothwell, *J. Chem. Soc., Chem. Commun.* **1992**, 632.

Other important types of compounds contain $Ta=CH_2$, $Ta=CR_2$, and $Ta\equiv CR$ species. Some of those compounds such as $(Me_3SiCH_2)_3Ta=CHSiMe_3$ have been observed as part of the decomposition products of $Ta(CH_2SiMe_3)_5$.[76] Metal-to-ligand double bonds are also obtained by reaction of Me_3TaCl_2 and $LiNR_2$,[77] for example:

$$Me_3TaCl_2 + 2LiNHSiBu^t_3 \xrightarrow[25°C, -2LiCl]{hexanes} (Bu^t_3SiNH)Me_2Ta=NSiBu^t_3 + CH_4$$

Compounds of the type Cp_2MX_3 are also very important and the hydrides are particularly interesting;[78] they catalyze the exchange of D_2 with aromatic compounds at 1 to 2 atm and 80 to 100°C. The pentahydride, $TaH_5(diphos)_2$, will also do this. The mechanism involves dissociation to form the mono-hydrido species and oxidative addition of benzene, namely,

$$Cp_2MH_3 \underset{-H_2}{\rightleftharpoons} Cp_2MH \underset{+C_6H_6}{\rightleftharpoons} Cp_2M\begin{matrix} C_6H_5 \\ -H \\ H \end{matrix}$$

In the presence of stoichiometric amounts of tertiary phosphines, monocyclopentadienyl compounds such as $CpNbCl_4PR_3$ are formed but in excess phosphine reduction to Nb^{IV} species is observed.[79] The relative rate of reduction depends on the nature of the phosphine, increasing in the order $PMePh_2 < PMe_2Ph < PMe_3$.

Monoalkyls of the type Cp^*TaRCl_3, $R = CH_2C_6H_4$-p-Me, CH_2CMe_3, react with CO to give isolable tantalum enolate$-O$ and η^2-acyl complexes,[80] according to

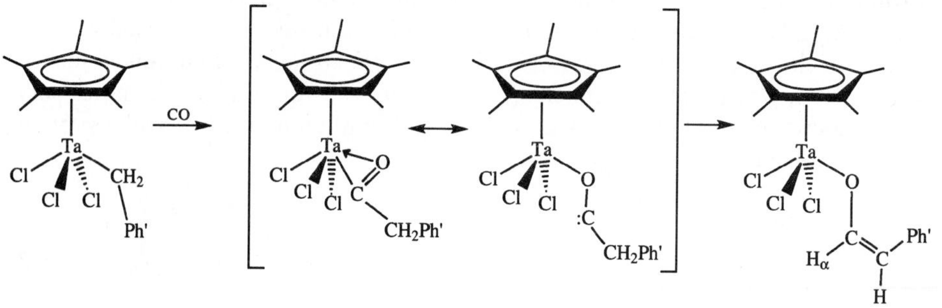

The cyclooctatetraene complexes, for example, $[Nb(C_8H_8)_3]^-$ and $CH_3Ta(C_8H_8)_2$, have η^4, butadienelike, and η^3-allyllike bonding of the rings to the metal; they are nonrigid.

There are also tetravalent cyclopentadienyl complexes of stoichiometry Cp_2MX_2 where X is a monovalent group like H^-, BH_4^-, halide, pseudohalide, alkyl, aryl,

[76]Y.-D. Wu et al., J. Am. Chem. Soc. 1995, 117, 9259 and references therein.
[77]C. P. Schaller and P. T. Wolczanski, Inorg. Chem. 1993, 32, 131.
[78]W. D. Jones et al., Inorg. Chem. 1990, 29, 686; D. M. Heinekey, J. Am. Chem. Soc. 1991, 113, 6074.
[79]R. Poli et al., Inorg. Chem. 1995, 34, 2343.
[80]L. Messerle et al., Inorg. Chem. 1990, 29, 4045.

alkoxide, carboxylate, thiolate, or dithiolene.[81] The compound Cp_2NbCl_2 can be prepared in good yields from $NbCl_5$ and $NaCp$[82] according to

$$NbCl_5 + 6NaCp \longrightarrow 5NaCl + Cp_4Nb + \text{organic products}$$

$$2Cp_4Nb + 4HCl + [O] \xrightarrow[O_2]{Br_2} [\{Cp_2NbCl\}_2O]Cl_2 + 2CpH$$

$$5Cl^- + [\{Cp_2NbCl\}_2O]^{2+} + SnCl_3^- + 2H^+ \longrightarrow 2Cp_2NbCl_2 + SnCl_6^{2-} + H_2O$$

It can be oxidized by Cp_2FeX to form cationic species such as $[Cp_2NbCl_2]X$:[83]

$$Cp_2NbCl_2 + Cp_2FeX \xrightarrow[0°C]{CH_2Cl_2 \text{ or } THF} [Cp_2NbCl_2]X + Cp_2Fe$$

$$X = BF_4 \text{ or } PF_6$$

18-B-9 Intermediate Oxidation States (II and III)

A number of M^{III} compounds containing one or two cyclopentadienyl anions is known. The former are of stoichiometry $CpMX_5$ with X_5 being an appropriate combination of monovalent (halide, H) and neutral ligands (CO, phosphine), for example, $CpMCl_2(CO)_3$, $CpMX_2(CO)_2(PR_3)$,[84] $CpMX_2(CO)(diphos)$, and $[CpMH(CO)_2(diphos)]^+$. Bis(cyclopentadienyl) derivatives are usually of composition $Cp_2MX(L)$, with the most common X and L being an alkyl and CO, respectively.

An interesting type of trivalent compound is $CpM(\eta^4\text{-diene})X_2$ (18-B-IX) X = Cl, Me, and triflate which can be prepared from $CpMCl_4$ and two equivalents of methylated allyl Grignard reagents in THF. In the presence of methylaluminoxane (MAO), these compounds polymerize ethylene[85] similarly to the Ti analogues (Section 17-A-5).

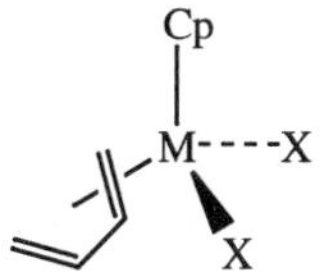

(18-B-IX)

Dinuclear M^{II} complexes with η^5-Cp or substituted Cp ligands possess only single M—M bonds of length around 3 Å. They are of composition $(CpNbL_2)_2$-$(\mu\text{-Cl})_2$, where L = CO or ½PhCCPh, and are obtained by a sodium amalgam reduction of Nb^{III} complexes of general formula $(CpNbCl_2L_2)_n$, with the carbonyl species being dinuclear while the acetylene one is mononuclear. The profound structural difference between the Nb^{II}-carbonyl and Nb^{II}-acetylene compounds,

[81]F. Guyon et al., J. Chem. Soc., Dalton Trans. **1996**, 4093.
[82]C. R. Lucas, Inorg. Synth. **1990**, 28, 267.
[83]K.-H. Thiele et al., Z. anorg. allg. Chem. **1990**, 587, 80.
[84]V. C. Gibson et al., J. Chem. Soc., Dalton Trans. **1990**, 3199.
[85]K. Mashima et al., Organometallics **1995**, 14, 2633.

namely butterfly versus planar Nb_2Cl_2 moiety, and much shorter Nb—Cl(terminal) distances in the latter, introduces ambiguity as to the oxidation state assignment. The usually observed strong interaction within the MC_2 unit may require its formal treatment as $M^{2+}-C_2^{2-}$, rather than as a π complex. In the above case this would account for the difference between seemingly analogous compounds with the acetylene derivative being formally a Nb^{IV} complex.

Compounds of the type $M_2(\eta^6\text{-arene})_2Cl_4$[86] can be prepared in good yields as follows:

$$MX_5 + Al + AlX_3 + \text{arene} \xrightarrow[\text{reflux}]{\text{benzene}} M(\eta^6\text{-arene})(AlX_4)_2 \xrightarrow{\text{THF}} {}^1\!/_2 M_2(\eta^6\text{-arene})_2X_4 + 2\,AlX_3(THF)_3$$

They are good precursors to other divalent compounds. In the presence of CO_2 and an alkylamine they produce diamagnetic carbamato-containing complexes.

18-B-10 Low and Very Low Formal Oxidation States (I or less)

The majority of these contain CO ligands. The neutral homoleptic carbonyls have not been isolated, but anionic $M(CO)_6^-$ as well as highly reduced $M(CO)_5^{3-}$ species are known. The original synthesis of the -1 species required elevated temperatures and high pressure but recently two simple, atmospheric pressure methods have been developed. They involve reduction of pentahalides in pyridine with Zn/Mg or in dimethoxyethane with sodium naphthalenide under an atmosphere of CO. These yellow salts contain discrete $M(CO)_6^-$ anions. The facile syntheses of the octahedral hexacarbonyl anions allowed systematic exploration of the previously difficult to access area of low-valent complexes of Nb and Ta. Since $M(CO)_6^-$ are rather inert towards displacement of CO the substitution products of general formula $M(CO)_{6-n}L_n$ have to be obtained by other routes, for example, by reduction of $MX(CO)_{6-n}L_n$ compounds. The monosubstituted derivatives are conveniently prepared *via* the following method[87]:

$$M(CO)_6^- \xrightarrow[-78°C, -NaCl]{Na, NH_3(l)} HM(CO)_5^{2-} \xrightarrow[NH_3, -70°C, -NaCl, H_2]{NH_4^+} M(CO)_6NH_3^- \xrightarrow[-60 \text{ to } 0°C]{L} [M^{-I}(CO)_5L]^{z-}$$

where L can be PMe_3, PPh_3, $P(OMe)_3$, $AsPh_3$, $SbPh_3$, $CNBu^t$, and CN^- and $z = -1$ for the neutral ligands or -2 for CN^-.

The highly reduced $M(CO)_5^{3-}$ is stable only at low temperature and explodes upon warming especially when dry. It can undergo monoprotonation and stannylation to form $[HTa(CO)_5]^{2-}$ and $[Ph_3SnTa(CO)_5]^{2-}$ anions, respectively. The hydride is a useful precursor to $[(Ph_3PAu)_3Ta(CO)_5]$, which is the only known gold cluster of tantalum.

Reduced solutions of tantalum naphthalenide react with PF_3 to give CO analogues,[88] which can be protonated in concentrated sulfuric acid to give a yellow

[86]G. Pampaloni *et al.*, *J. Chem. Soc., Dalton Trans.* **1996**, 311.

[87]J. E. Ellis *et al.*, *Inorg. Chim. Acta* **1995**, *240*, 379.

[88]J. E. Ellis *et al.*, *Chem. Eur. J.* **1995**, *1*, 521; *Inorg. Chem.* **1998**, *37*, 6518.

volatile hydride compound according to:

$$\text{TaCl}_5 + 6\text{NaC}_{10}\text{H}_8 \xrightarrow[-60\,°\text{C, -NaCl}]{\text{DME}} \text{Na[Ta(C}_{10}\text{H}_8)] \xrightarrow[-60\text{ to }20\,°\text{C}]{\text{PF}_3 \,(\leq 1\text{ atm})} \xrightarrow{\text{Et}_4\text{NBr}} [\text{Et}_4\text{N}][\text{Ta(PF}_3)_6]$$

$$\xrightarrow[20\text{-}25\,°\text{C}]{\text{H}_2\text{SO}_4} [\text{HTa(PF}_3)]\uparrow + \text{Et}_4\text{N(HSO}_4)$$

Oxidation of $M(CO)_6^-$ with halogens affords $M_2(CO)_8X_3^-$ anions, but in the presence of PMe_3, neutral mononuclear species $MX(CO)_3(PMe_3)_3$ are formed. The former contain three bridging halide atoms, which can be substituted by MeO^- or $CH_3CO_2^-$ by reaction with methanol or acetic acid. The metal-metal distance of over 3.5 Å precludes direct interaction. The dinuclear chloro anions are converted to $CpM(CO)_4$ by reaction with LiCp and to $(\eta^6\text{-arene})M(CO)_4^+$ in the presence of arene and $AlBr_3$.

There are also face-sharing bioctahedral M^I compounds of the type $[M(\mu\text{-X})(CO)_2(RCCR)_2]_2$[89] and 7-coordinate mononuclear complexes of stoichiometries $MX(CO)_4(diphos)$ and $MX(CO)_2(diphos)_2$ for $X = H$, Cl, and CH_3.[90] In one such compound, $TaCl(CO)(dmpe)_2$, there is, under appropriate conditions, reductive coupling of the two ligating carbonyls as shown by the isolation of $Ta(Me_3SiO\text{-}C\equiv COSiMe_3)(dmpe)_2Cl$.[91] Other reactions reported for low-valent group V complexes include reductive coupling of nitriles, isocyanides, the activation of alkynes, the synthesis and characterization of polyhydrides, and formation of dinitrogen complexes.[92]

A unique 17-electron compound can be prepared by reaction of $TaH(CO)_4(dppe)$ and $(Bu^tC_6H_4)_3C^·$ radicals generated from the corresponding bromide compound according to

$$(Bu^tC_6H_4)_3C^· + TaH(CO)_4(dppe) \longrightarrow (Bu^tC_6H_4)_3CH + Ta(CO)_4(dppe)$$

The compound exists in solution as an equilibrium mixture of the paramagnetic monomer (the dominant species) and the diamagnetic carbonyl-bridged dimer $[Ta(CO)_4(dppe)]_2$.[93]

By metal vapor synthesis, zerovalent Nb and Ta compounds such as $M(arene)_2$[94] and $M(dmpe)_3$ can be prepared in substantial quantities. The mesitylene compound $Nb(mes)_2$ can also be made by treating a mixture of $NbCl_5/Al/AlCl_3/mesitylene$ in DME at low temperature. It reacts with CO to give $[Nb(mes)_2CO][Nb(CO)_6]$.[95] In the presence of $[FeCp_2]BPh_4$ it forms $[Nb(\eta^6\text{-mes})_2]BPh_4$. Upon heating the later compound in the presence of an alkene, red-pyrophoric crystals are obtained which contain two arene groups coordinating the metal center[96] (18-B-X).

[89]D. Rehder et al., Inorg. Chim. Acta **1992**, 202, 121.
[90]M. C. Baird et al., Inorg. Chem. **1996**, 35, 6937.
[91]S. J. Lippard et al., J. Am. Chem. Soc. **1993**, 115, 808.
[92]See for example: E. M. Carnahan and S. J. Lippard, J. Am. Chem. Soc. **1992**, 114, 4166; S. J. Lippard et al., Inorg. Chem. **1992**, 31, 4134 and references therein.
[93]M. C. Baird et al., Organometallics **1996**, 15, 3289.
[94]D. L. Clark et al., Inorg. Chim. Acta **1996**, 244, 269.
[95]F. Calderazzo et al., Angew. Chem. Int. Ed. Engl. **1991**, 30, 102.
[96]G. Pampaloni et al., Angew. Chem. Int. Ed. Engl. **1992**, 31, 1235.

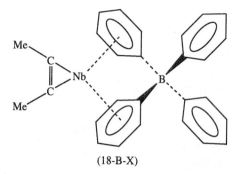

(18-B-X)

An anion containing two 7-coordinate Nb atoms that share a triangular face of Cl atoms can be synthesized from the reaction of Nb(mes)$_2$X and CO[97] according to

$$3Nb(mes)_2X + 9CO \longrightarrow [Nb(mes)_2(CO)][Nb_2(\mu\text{-}X)_3(CO)_8] + 4mes$$

Additional Reference

K. Tanabe, Ed., *Catalysis Today,* **1966**, *28,* 1-205 (a special issue on "catalytic properties of niobium materials").

18-C   MOLYBDENUM AND TUNGSTEN: GROUP 6

Molybdenum and tungsten are similar chemically, although there are differences between them in various types of compounds that are not easy to explain. Thus some compounds of the same type differ noticeably in their reactivities toward various reagents: for example, Mo(CO)$_6$ but not W(CO)$_6$ reacts with acetic acid to give the quadruply-bonded dimetal tetraacetate.

Except for compounds with π-acid ligands, there is not a great deal of similarity to chromium. The divalent state, well defined for Cr, is not well known for Mo and W except in strongly M—M bonded compounds; and the abundance of highly stable Cr^{III} complexes has no counterpart in Mo or W chemistry. For the heavier elements, the higher oxidation states are more common and more stable against reduction.

Uranium has sometimes been classed with Mo and W in Group 6, and indeed there are some valid, though often rather superficial, similarities; the three elements form volatile hexafluorides, oxide halides, and oxo anions that are similar in certain respects. There is little resemblance to the sulfur group except in regard to stoichiometric similarities (e.g., SeF$_6$, WF$_6$, SO$_4^{2-}$, and MoO$_4^{2-}$), and such comparisons are not profitable.

Both Mo and W have a wide variety of stereochemistries in addition to the

[97]G. Pampaloni *et al., Chem. Ber.* **1992,** *125,* 1005.

Table 18-C-1 Oxidation States and Stereochemistry of Molybdenum and Tungsten

Oxidation state	Coordination number	Geometry	Examples
Mo^{-II}, W^{-II}	5	?	$[Mo(CO)_5]^{2-}$
Mo^0, W^0, d^6	6	Octahedral	$W(CO)_6$, $(py)_3Mo(CO)_3$, $[Mo(CO)_5I]^-$, $[Mo(CN)_5NO]^{4-}$, $Mo(N_2)_2(diphos)_2$
Mo^I, W^I, d^5	6^a	π Complex	$(C_6H_6)_2Mo^+$, η^5-$C_5H_5MoC_6H_6$
	7^a		$[\eta^5$-$C_5H_5Mo(CO)_3]_2$
	6	?	$MoCl(N_2)(diphos)_2$
Mo^{II}, W^{II}, d^4	5	M—M quadruple bond	$Mo_2(O_2CR)_4$, $[Mo_2Cl_8]^{4-}$, $W_2Cl_4(dppe)_2$
	6	Octahedral	$Mo(diars)_2X_2$, $trans$-$Me_2W(PMe_3)_4$
	7^a	π Complex	η^5-$C_5H_5W(CO)_3Cl$
		Capped trigonal prism	$[Mo(CNR)_7]^{2+}$
		Pentagonal bipyramidal	$[Mo(CN)_7]^{5-}$
		4:3	$[WBr(CO)_2(PR_3)_2]_2C_2O_4$
	9	Cluster compounds	Mo_6Cl_{12}, W_6Cl_{12}
Mo^{III}, W^{III}, d^3	4	M—M triple bond	$Mo_2(OR)_6$, $W_2(NR_2)_6$, $[Mo_2(HPO_4)_4]^{2-}$
	6	Octahedral	$[Mo(NCS)_6]^{3-}$, $[MoCl_6]^{3-}$, $[W_2Cl_9]^{3-}$
	7	?	$[W(diars)(CO)_3Br_2]^+$
	8	Dodecahedral (?)	$[Mo(CN)_7(H_2O)]^{4-}$
Mo^{IV}, W^{IV}, d^2	8^a	π Complex	$(\eta^5$-$C_5H_5)_2WH_2$, $(\eta^5$-$C_5H_5)_2MoCl_2$
	9^a	π Complex	$(\eta^5$-$C_5H_5)_2WH_3^+$
	4	Tetrahedral	$Mo(NMe_2)_4$, $Mo(SBu^t)_4$, $Mo[TeSi(SiMe_3)_3]_4^b$
	6	Octahedral	$[Mo(NCS)_6]^{2-}$, $[Mo(diars)_2Br_2]^{2+}$, $WBr_4(MeCN)_2$, $MoOCl_2(PR_3)_3$
	6	Trigonal prism	MoS_2
	8	Dodecahedral or square antiprism	$[M(CN)_8]^{4-}$, $Mo(S_2CNMe_2)_4$, $M(picolinate)_4$
Mo^V, W^V, d^1	5	tbp	$MoCl_5(g)$
	6	Octahedral	$Mo_2Cl_{10}(s)$, $[MoOCl_5]^{2-}$, WF_6^-
	8	Dodecahedral or square antiprism	$[M(CN)_8]^{3-}$
Mo^{VI}, W^{VI}, d^0	4	Tetrahedral	MO_4^{2-}, MoO_2Cl_2, WO_2Cl_2
	5?	?	$WOCl_4$, $MoOF_4$
	6	Octahedral	MoO_6, WO_6 in polyacids, WCl_6, $Mo(OMe)_6$, $WOCl_4(s)$, MoF_6, $[MoO_2F_4]^{2-}$, MoO_3(distorted), WO_3(distorted)
		Distorted trigonal prism	$W(CH_3)_6$
	7	Distorted pentagonal bipyramid	$WOCl_4(diars)$, $K_2[MoO(O_2)_2(ox)]$
	8	?	MoF_8^{2-}, WF_8^{2-}, $[WMe_8]^{2-}$
	9	?	$WH_6(Me_2PhP)_3$

[a] Assuming η^6-C_6H_6 or η^5-C_5H_5 occupy three coordination sites.
[b] J. Arnold *et al.*, *Chem. Commun.* **1996**, 2565.

variety of oxidation states, and their chemistry is among the most complex of the transition elements. The oxidation states and stereochemistry are summarized in Table 18-C-1.

Molybdenum has long been known as one of the biologically active transition elements. It is intimately involved in the functioning of enzymes called nitrogenases, which cause atmospheric N_2 to be reduced to NH_3 or its derivatives, in enzymes concerned with reduction of nitrate, and in still other biological processes. Several aspects of molybdenum chemistry have been vigorously studied in the past decade principally because of their possible relation to the biological processes. There are now well-established tungsten enzymes as well.

18-C-1   The Elements

Both molybdenum and tungsten have an abundance of *ca.* $10^{-4}\%$ in the earth's crust. Molybdenum occurs chiefly as *molybdenite* (MoS_2) but also as molybdates such as $PbMoO_4$ (*wulfenite*) and others. Tungsten is found almost exclusively in the form of tungstates, the chief ones being *wolframite* (a solid solution and/or mixture of the isomorphous substances $FeWO_4$ and $MnWO_4$), *scheelite* ($CaWO_4$), and *stolzite* ($PbWO_4$).

The MoS_2 in ores is concentrated by the foam flotation process; the concentrate is then converted by roasting into MoO_3 which, after purification, is reduced with hydrogen to the metal. Reduction with carbon must be avoided because this yields carbides rather than the metal.

Tungsten ores are concentrated by mechanical and magnetic processes and the concentrates attacked by fusion with NaOH. The cooled melts are leached with water, giving solutions of sodium tungstate from which hydrous WO_3 is precipitated on acidification. The hydrous oxide is dried and reduced to metal by hydrogen.

In the powder form in which they are first obtained both metals are dull gray, but when converted into the massive state by fusion are lustrous, silver-white substances of typically metallic appearance and properties. They have electrical conductances about 30% that of silver. They are extremely refractory; Mo melts at 1620°C and W at 3380°C.

Neither metal is readily attacked by acids. Concentrated nitric acid initially attacks molybdenum, but the metal surface is soon passivated. Both metals can be dissolved—tungsten only slowly, however—by a mixture of concentrated nitric and hydrofluoric acids. Oxidizing alkaline melts such as fused KNO_3-NaOH or Na_2O_2 attack them rapidly, but aqueous alkalis are without effect.

Both metals are inert to oxygen at ordinary temperatures, but at red heat they combine with it readily to give the trioxides. They both combine with chlorine when heated, but even at room temperature they are attacked by fluorine, yielding the hexafluorides. The metals also react on heating with B, N, and Si. Molybdenum disilicide ($MoSi_2$) is used in resistance heating elements, and WC is used to tip cutting tools.

Molybdenum is used in a variety of catalysts, especially combined with cobalt in desulfurization of petroleum. Another major use for both metals is in alloy steels to which they impart hardness and strength. Tungsten is also used in lamp filaments.

18-C-2 Oxides, Sulfides, Simple Oxo and Sulfido Anions

Oxides

Many molybdenum and tungsten oxides are known. The simple ones are MoO_3, WO_3, MoO_2, and WO_2. Other, nonstoichiometric, oxides have been characterized and have complicated structures.

The ultimate products of heating the metals or other compounds such as the sulfides in oxygen are the *trioxides*. They are not attacked by acids but dissolve in bases to form molybdate and tungstate solutions, which are discussed later.

Molybdenum trioxide is a white solid at room temperature but becomes yellow when hot and melts at 795°C to a deep yellow liquid. It is the anhydride of molybdic acid, but it does not form hydrates directly, although these are known (see later). One of its two polymorphs, the stable α-form, has a rare type of layer structure in which each molybdenum atom is surrounded by a distorted octahedron of oxygen atoms.

Tungsten trioxide is a lemon yellow solid (mp 1200°C); it has a slightly distorted form of the cubic rhenium trioxide structure (Fig. 18-D-1).

Molybdenum(IV) oxide (MoO_2), is obtained by reducing MoO_3 with hydrogen or NH_3 below 470°C (above this temperature reduction proceeds to the metal) or by reaction of molybdenum with steam at 800°C. It is a brown-violet solid with a coppery luster, insoluble in nonoxidizing mineral acids but soluble in concentrated nitric acid with oxidation of the molybdenum to Mo^{VI}. The structure is derived from that of rutile but distorted so that strong $Mo-Mo$ bonds are formed. Tungsten dioxide is similar. $Mo-Mo$ and $W-W$ distances are 2.51 and 2.49 Å, respectively.

Although older literature describes simple oxides in intermediate oxidation states [e.g., Mo_2O_5 and $MoO(OH)_3$], these are apparently not genuine. There is, however, a large number of oxides of composition MO_x ($2 < x < 3$) obtainable by simply heating MoO_3 with Mo at 700°C or WO_3 with W at 1000°C or by heating the trioxides in a vacuum. The structures of these intensely blue or purple solids are varied. One important structural class are the *shear structures,* an example of which is shown in Fig. 18-C-1. The principle on which these are built is that beginning with a structure of composition MoO_3 in which every MoO_6 octahedron shares every corner (but no edges) with neighboring octahedra, edge sharing is systematically introduced. In this way oxygen is lost and phases of compositions M_nO_{3n-1} (e.g., M_8O_{23} in the example shown), M_nO_{3n-2} (e.g., $W_{20}O_{58}$), and so on, are obtained.

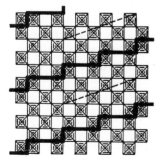

Figure 18-C-1   One plane of the idealized Mo_8O_{23} structure. The shear planes that disturb the otherwise perfect checkerboard arrangement of MoO_6 octahedra sharing only corners are shown by heavy lines.

There are also lower oxides in which some Mo atoms are in pentagonal bipyramidal coordination (e.g., Mo_5O_{14} and $W_{18}O_{49}$) or tetrahedral coordination (e.g., Mo_4O_{11}).

Mixed oxide-hydroxide materials, called *blue oxides,* are obtained by reduction of acidified solutions of molybdates or tungstates (or suspensions of MO_3) with Sn^{II}, SO_2, and so on. Because they are noncrystalline, the structures are not known and the cause of the color is uncertain, but speculation abounds.[1]

On fusion of MoO_3 or WO_3 with alkali or alkaline earth oxides, mixed oxide systems are obtained. These consist, so far as is known, of infinite chains or rings of MO_6 octahedra, and are unrelated to the aqueous molybdates and tungstates.

Tungsten Bronzes

The reduction of sodium tungstate with hydrogen at red heat gives a chemically inert substance with a bronze-like appearance. Similar materials are obtained by vapor phase reduction of WO_3 with alkali metals.

The tungsten bronzes are nonstoichiometric substances of general formula $M_n^IWO_3$ ($0 < n < 1$). The colors vary greatly with composition from golden yellow for $n \approx 0.9$ to blue violet for $n \approx 0.3$. Tungsten bronzes with $n > 0.3$ are extremely inert and have semimetallic properties, especially metallic luster, and good electrical conductivity in which the charge carriers are electrons. Those with $n < 0.3$ are semiconductors. They are insoluble in water and resistant to all acids except hydrofluoric, and they can be oxidized by oxygen in the presence of base to give tungstates(VI):

$$4NaWO_3 + 4NaOH + O_2 \longrightarrow 4Na_2WO_4 + 2H_2O$$

Structurally the sodium tungsten bronzes may be regarded as defective M^IWO_3 phases having the perovskite structure. In the defective phase $M_n^IWO_3$ there are $(1 - n)W^{VI}$ atoms, and $(1 - n)$ of the Na sites of the pure $NaWO_3$ phase are unoccupied. It appears that completely pure $NaWO_3$ has not been prepared, although phases with sodium enrichment up to perhaps $n \approx 0.95$ are known. The cubic structure collapses to rhombic and then triclinic for $n < 0.3$. In the limit of $n = 0$ we have, of course, WO_3, which is known to have a triclinically distorted ReO_3 structure (Fig. 18-D-1). The cubic ReO_3 structure is the same as the perovskite structure with all the large cations removed. Thus the actual range of composition of the tungsten bronzes is approximately $Na_{0.3}WO_3-Na_{0.95}WO_3$.

The semimetallic properties of the tungsten bronzes are associated with the fact that no distinction can be made between W^V and W^{VI} atoms in the lattice, all W atoms appearing equivalent. Thus the n "extra" electrons per mole (over the number for WO_3) are distributed throughout the lattice, delocalized in energy bands somewhat similar to those of metals.

[1] A. Müller *et al., Angew. Chem. Int. Ed. Engl.* **1996,** *35,* 1206.

Sulfides

The MS_2 and MS_3 compounds are those of importance, although others, such as hydrated Mo_2S_5 (precipitated from Mo^V solutions), anhydrous Mo_2S_5, and a few others (Mo_2S_3, MoS_4) are known.

Molybdenum disulfide (MoS_2) occurs in nature as molybdenite, the most important ore of molybdenum. It can be prepared by direct combination of the elements, but a better method,[2] which *mutatis mutandis*, is equally good for WS_2, $MoSe_2$, and WSe_2 is the following:

$$2MoCl_5 + 5Na_2S \longrightarrow 2MoS_2 + 10NaCl + S$$

It is the most stable sulfide at higher temperatures, and the others that are richer in sulfur revert to it when heated in a vacuum. It dissolves only in strongly oxidizing acids such as aqua regia and boiling concentrated sulfuric acid. Chlorine and oxygen attack it at elevated temperatures giving $MoCl_5$ and MoO_3, respectively.

Molybdenum disulfide has a structure built of close-packed layers of sulfur atoms stacked to create trigonal prismatic interstices that are occupied by Mo atoms. Tungsten disulfide is similar. The stacking is such as to permit easy slippage of alternate layers; thus MoS_2 has mechanical properties (lubricity) similar to those of graphite. It is possible to make many intercalation compounds.[3] Among these Li_xMoS_2 is particularly interesting. The action of Bu^nLi can be controlled to give $LiMoS_2$. This reacts with water, under sonication, to give a suspension of single layers of MoS_2.[4]

Brown hydrous MoS_3, obtained on passing H_2S into slightly acidified solutions of molybdates, dissolves on digestion with alkali sulfide solution to give brown-red thiomolybdates.

Simple Molybdates and Tungstates

Defined as those containing only the simple MO_4^{2-} ions, they can be obtained from solutions of MO_3 in aqueous alkali. The MO_4^{2-} ions persist as such in basic solution. Although both molybdates and tungstates can be reduced in solution (see later), they lack the powerful oxidizing property so characteristic of chromates(VI). The normal tungstates and molybdates of many other metals can be prepared by metathetical reactions. The alkali metal, ammonium, magnesium, and thallous salts are soluble in water, whereas those of other metals are nearly all insoluble.

When solutions of molybdates and tungstates are made weakly acidic, polymeric anions are formed, but from more strongly acid solutions substances often called molybdic or tungstic acid are obtained. At room temperature the yellow $MoO_3 \cdot 2H_2O$ and the isomorphous $WO_3 \cdot 2H_2O$ crystallize, the former very slowly. From hot solutions, monohydrates are obtained rapidly. These compounds are oxide hydrates. $MoO_3 \cdot 2H_2O$ contains sheets of MoO_6 octahedra sharing corners

[2]P. R. Bonneau *et al.*, *Inorg. Synth.* **1995**, *30*, 33; R. B. Kayer *et al.*, *Inorg. Chem.* **1992**, *31*, 2127.
[3]M. G. Kanatzidis *et al.*, *Materials* **1993**, *5*, 595.
[4]M. G. Kanatzidis *et al.*, *J. Chem. Soc., Chem. Commun.* **1993**, 1582.

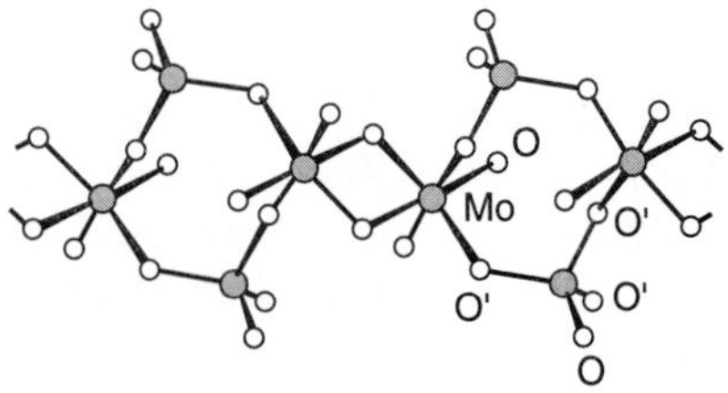

Figure 18-C-2 The polymeric anion present in $(NH_4)_2Mo_2O_7$.

and is best formulated as $[MoO_{4/2}O(H_2O)] \cdot H_2O$ with one H_2O bound to Mo, the other one hydrogen bonded in the lattice.

The discrete $[Mo_2O_7]^{2-}$ ion, analogous to the dichromate ion, has been obtained by addition of $[Bu_4^nN]OH$ to a solution of $[Bu_4^nN]_4Mo_8O_{26}$ in CH_3CN. It retains its structure in organic solvents, but on addition of small cations it is converted to $[Mo_7O_{24}]^{6-}$. Other compounds with the composition $M_2^IMo_2O_7$ contain polymeric anions. An example is the commercially important $(NH_4)_2Mo_2O_7$, in which the anion is an infinite polymer of linked MoO_6 octahedra and MoO_4 tetrahedra, as shown in Fig. 18-C-2.

Thiomolybdates and Thiotungstates

The perthio ions (MS_4^{2-}) have been long known and well characterized. They show marked ability to serve as ligands, most commonly bidentate (Section 12-16), and are important starting materials for preparation of many of the more elaborate thio complexes to be discussed later. In addition, the mixed thio/oxo ions MOS_3^{2-} and $MO_2S_2^{2-}$ are known for both metals as well as the MoO_3S^{2-} ion. The MSe_4^{2-} ions are also known. While the MTe_4^{2-} ions are not known, $WOTe_3^{2-}$ has recently been reported.[5]

18-C-3 Isopoly and Heteropoly Acids and Salts[6]

A prominent feature of the chemistry of molybdenum and tungsten is the formation of numerous polymolybdate(VI) and polytungstate(VI) acids and their salts. Vanadium(V), Nb^V, Ta^V, and U^{VI} show comparable behavior, but to a more limited extent.

Molybdenum, in particular, is known to form a very large number of structurally diverse polyanions in association with other oxo ligands such as squarate $(C_4O_4^{2-})$, methoxide, malate, phosphate, and arsenate anions.[7] Space does not permit discussion of this plethora of interesting compounds, nor of still others. We shall focus mainly on the so-called *classic* polymolybdates and polytungstates.

The classic poly acids of molybdenum and tungsten are of two types: (a) the *isopoly acids* and their related anions, which contain only molybdenum or tungsten along with oxygen and hydrogen, and (b) the *heteropoly acids* and anions, which

[5]B. W. Eichorn *et al., Angew. Chem. Int. Ed. Engl.* **1994,** *33,* 1859.
[6]M. T. Pope and A. Müller, *Angew. Chem. Int. Ed. Engl.* **1991,** *30,* 34; *Polyoxometallates: From Platonic Solids to Anti-Retroviral Activity,* Kluwer Academic Publishers, Dorderecht, The Netherlands, 1994; M. T. Pope, *Prog. Inorg. Chem.* **1991,** *39,* 81.
[7]M. I. Khan and J. Zubieta, *Prog. Inorg. Chem.* **1995,** *43,* 1.

contain one or two atoms of another element in addition to molybdenum or tungsten, oxygen, and hydrogen. The relationship of both these classes to metal oxide type catalysts, and even the actual use of some polyoxoanions as catalysts, broadens the base of interest in them, and ^{17}O nmr as well as X-ray crystallography have made them attractive research topics.

It is to be noted that while many structures that have been determined crystallographically apparently persist in solution, this is not always true and it can never be taken for granted. There are also many problems concerning the formation and relative stabilities of these substances. The polyanions are built primarily of MO_6 octahedra, but they are prepared by starting with the tetrahedral MO_4^{2-} ions. Moreover, in Group 6 only the metals Mo and W but not Cr form polyanions.

Let us consider the initial steps leading to isopolymolybdates. In strongly basic solution Mo^{VI} is present only as MoO_4^{2-}. On addition of acid, the following protonation equilibria are established:

	K	ΔH kJ mol^{-1}	ΔS J mol^{-1} K^{-1}	
$MoO_4^{2-} + H^+ \underset{}{\overset{K_1}{\rightleftharpoons}} HMoO_4^-$	$10^{3.7}$	20	140	(18-C-1)
$HMoO_4^- + H^+ + 2H_2O \underset{}{\overset{K_2}{\rightleftharpoons}} Mo(OH)_6$	$10^{3.7}$	-49	-92	(18-C-2)

It might be considered surprising that K_2 is as large as K_1, especially since the incorporation of two water molecules should cause a very unfavorable entropy change, as indeed it does. The ΔH and ΔS values for the first protonation are perfectly normal for a reaction of its type. The high value of K_2 is due to the large negative enthalpy change, and this may be attributed to the formation of two new Mo—O bonds while the number of O—H bonds is maintained (remember that H^+ belongs to H_3O^+). From these data we can see why the reaction

$$2HMoO_4^- \rightleftharpoons [Mo_2O_7]^{2-} + H_2O \qquad (18\text{-}C\text{-}3)$$

is not observed, even though the analogous reaction is very important for Cr^{VI}.

It is also remarkable that no polynuclear Mo^{VI} species containing fewer than seven molybdenum atoms is observed in solution. In other words, in addition to the equilibria 18-C-1 and 18-C-2, the system is described by

$$7MoO_4^{2-} + 8H^+ \rightleftharpoons Mo_7O_{24}^{6-} + 4H_2O \qquad (18\text{-}C\text{-}4)$$

together with, at more acid pH's,

$$Mo_7O_{24}^{6-} + 3H^+ + HMoO_4^- \rightleftharpoons Mo_8O_{26}^{4-} + 2H_2O \qquad (18\text{-}C\text{-}5)$$

The absence of any detectable amounts of polynuclear species between the mononuclear ones and $Mo_7O_{24}^{6-}$ may be understood as follows. In view of the large value for K_2, $Mo(OH)_6$ is present in concentrations less than that of $HMoO_4^-$ by only a factor of $K_2[H^+]$. Thus reactions 18-C-6

$$Mo(OH)_6 + HMoO_4^- = [(HO)_5MoOMoO_3]^- + H_2O$$

$$[(HO)_5MoOMoO_3]^- + HMoO_4^- = [(HO)_4Mo(OMoO_3)_2]^{2-} + H_2O$$

$$\vdots$$

$$[(HO)Mo(OMoO_3)_5]^{5-} + HMoO_4^- = [Mo(OMoO_3)_6]^{6-} + H_2O$$

$$[Mo(OMoO_3)_6]^{6-} \equiv [Mo_7O_{24}]^{6-} \qquad (18\text{-}C\text{-}6)$$

can compete effectively with reaction 18-C-3. The species $[Mo(OMoO_3)_6]^{6-}$, with six MoO_4 tetrahedra, each attached through an oxygen atom to the central Mo atom, can rearrange internally to afford the final $Mo_7O_{24}^{6-}$ structure, which consists entirely of octahedra. Of course such a rearrangement may take place partially at earlier stages, and the extent of protonation of any species can vary, so the foregoing should not be taken literally. For tungsten in the pH range 5 to 7.8, the equilibria involve WO_4^{2-}, $[W_6O_{20}(OH)_2]^{6-}$, $[W_7O_{24}]^{6-}$, $[HW_7O_{24}]^{5-}$, and $[H_2W_{12}O_{42}]^{10-}$.

In addition to the species consisting of only molybdenum or only tungsten, there are mixed anions, of both iso- and heteropoly types in which the metal sites can be occupied by both molybdenum and tungsten atoms.[8] Partial substitution by vanadium atoms or other Group 5 (e.g., Nb) atoms can also occur.[9]

Isopolymolybdates and -tungstates

These are broadly similar in modes of preparation, general properties, and in being built up almost entirely by the condensation of MO_6 octahedra through sharing of vertices or edges, but not faces. However, they differ as far as detailed compositions and structures go. Figure 18-C-3 shows some important structures in a conventional, stylized way. One important species, $[M_6O_{19}]^{2-}$, is common to both metals. This very neat, compact structure is also found in mixed (i.e., Mo_5V and Mo_4V_2) species and in the $[M_6O_{19}]^{8-}$ ions where M = Nb, Ta. It should be noted that the octahedra are actually not at all regular. For oxygen atoms bonded to only one metal atom the M—O bond lengths are ~1.70 Å, while the M—O—M bond lengths are longer, ~1.90 Å, and any deeply buried oxygen atoms, such as the central one in $[M_6O_{19}]^{2-}$, form very long bonds, ~2.30 Å.

An important structure that does not consist entirely of Mo_6 octahedra is the α-$[Mo_8O_{26}]^{4-}$ structure shown in Fig. 18-C-4. This structure consists of a central ring, crown-shaped, of six edge-sharing distorted octahedra capped by two tetrahedral MoO_4 units. The α and β structures (Figs. 18-C-3 and 18-C-4) can coexist in solution, where they interconvert rapidly by an intramolecular pathway, and the equilibrium distribution and the products precipitated depend strongly on the accompanying cations. Crystalline compounds containing pure α or pure β isomers can be obtained. The dynamic properties of the α-$[Mo_8O_{26}]^{4-}$ structure have been partly elucidated by ^{17}O nmr studies of the Mo_8 ion itself and derivatives thereof

[8]O. W. Howarth *et al., J. Chem. Soc., Dalton Trans.* **1994**, 1061.
[9]R. G. Finke *et al., Inorg. Chem.* **1996**, 35, 7905.

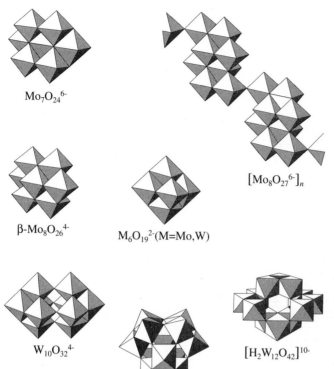

$Mo_7O_{24}^{6-}$

$[Mo_8O_{27}^{6-}]_n$

β-$Mo_8O_{26}^{4-}$

$M_6O_{19}^{2-}(M=Mo,W)$

$W_{10}O_{32}^{4-}$

$[H_2W_{12}O_{40}]^{6-}$

$[H_2W_{12}O_{42}]^{10-}$

Figure 18-C-3 Some of the important isopoly anion structures. Structures known only for molybdates or tungstates are shown at top and bottom. The $[M_6O_{19}]^{2-}$ structure known for both is in the center.

in which one or both of the capping MoO_4^{2-} tetrahedra are replaced by isostructural units such as $PhAsO_3^-$, as indicated in Fig. 18-C-4(b).

The sort of mixed species just discussed is but one type in this relatively new branch of polyoxoanion chemistry. This new area is of interest because of its potential relationship to catalysis by metal oxide surfaces. Besides the derivatives of the α-$[Mo_8O_{26}]^{4-}$ ion just mentioned, there are others such as $[(Me_2As)Mo_4O_{14}OH]^{2-}$, easily prepared by acidification of a mixture of MoO_4^{2-} and $Me_2AsO_2^-$ ions. More complex systems can be made by reaction of organometal and organometalloidal halides such as $RSnCl_3$ or $RAsCl_2$, with fragments of the Keggin ion

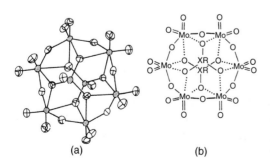

(a)

(b)

Figure 18-C-4 Two representations of the α-$[Mo_8O_{26}]^{4-}$ structure. (a) The crystallographically determined structure. (b) A schematic drawing emphasizing the variable nature of the capping XR groups.

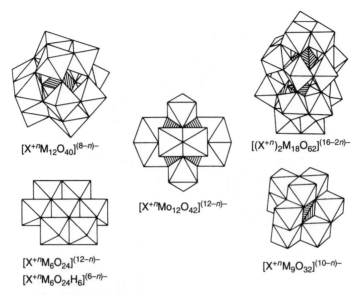

$[X^{+n}M_{12}O_{40}]^{(8-n)-}$

$[(X^{+n})_2M_{18}O_{62}]^{(16-2n)-}$

$[X^{+n}Mo_{12}O_{42}]^{(12-n)-}$

$[X^{+n}M_6O_{24}]^{(12-n)-}$
$[X^{+n}M_6O_{24}H_6]^{(6-n)-}$

$[X^{+n}M_9O_{32}]^{(10-n)-}$

Figure 18-C-5 Structures of important types of heteropolymolybdate and -tungstate ions with "enshrouded" heteroatoms. The general formula of each is shown below it.

(see Fig. 18-C-5), the latter being, for example, $[PW_{11}O_{39}]^{7-}$ or $[SiMo_{11}O_{39}]^{8-}$. The entering RSn or RAs unit binds to three oxygen atoms replacing the missing MoO or WO groups. If a $CpTiCl_3$ group is used the CpTi unit is introduced. Some formylated species with $Mo-O-CHO$ and $Mo-OCH_2O-$ groups have also been made.

Heteropoly Anions

These can be formed either by acidification of solutions containing the requisite simple anions, or by introduction of the heteroelement after first acidifying the molybdate or tungstate:

$$HPO_4^{2-} + MoO_4^{2-} \xrightarrow{H^+,\ 25\,°C} [PMo_{12}O_{40}]^{3-}$$

$$WO_4^{2-} \xrightarrow{H^+ \text{ to } pH = 6} \xrightarrow{Co^{2+},\ 100\,°C} [Co_2W_{11}O_{40}H_2]^{8-}$$

Transformations are also possible, namely,

$$[P_2W_{18}O_{62}]^{6-} \xrightarrow{HCO_3^-,\ 25\,°C} [P_2W_{17}O_{61}]^{10-}$$

Many of the heteropoly anions are quite robust toward excess acid and may be protonated to give the *heteropoly acids* both in solution and as crystalline hydrates.

This is a truly enormous class of compounds, and we note here only the main

Table 18-C-2 Some Heteropoly Anions with Buried Heteroatoms

Formula type	Central group	M = Mo	M = W
$X^{n+}M_{12}O_{40}^{(8-n)-}$	XO_4	Si^{IV}, Ge^{IV}, P^{V}, As^{V}, Ti^{IV}, Zr^{IV}	B^{III}, Si^{IV}, Ge^{IV}, P^{V}, As^{V}, Al^{III}, Fe^{III}, Co^{II}, Co^{III}, Cu^{I}, Cu^{II}, Zn^{II}, Cr^{III}, Mn^{IV}, Te^{IVa}, Ga^{IIIb}
$X_2^{n1}M_{18}O_{62}^{(16-2n)-}$	XO_4	P^{V}, As^{V}	P^{V}, As^{V}
$X_2^{n+}Z_4^{+m}M_{18}O_{70}H_4^{(28-2n-4m)-}$	XO_4		X = P^{V}, As^{V} Z = Mn^{II}, Co^{II}, Ni^{II}, Cu^{II}, Zn^{II}
$X^{n+}M_9O_{32}^{(10-n)-}$	XO_6	Mn^{IV}, Ni^{IV}	
$X^{n+}M_6O_{24}^{(12-n)-}$	XO_6	Te^{VI}, I^{VII}	Ni^{IV}, Te^{VI}, I^{VII}
$X^{n+}M_6O_{24}H_6^{(6-n)-}$	XO_6	Al^{III}, Cr^{III}, Co^{III}, Fe^{III}, Ga^{III}, Rh^{III}, Mn^{II}, Co^{II}, Ni^{II}, Cu^{II}, Zn^{II}	Ni^{II}
$X_2^{n+}M_{10}O_{38}H_4^{(12-2n)-}$	XO_6	Co^{III}	
$X^{n+}M_{12}O_{42}^{(12-n)-}$	XO_{12}	Ce^{IV}, Th^{IV}, U^{IV}	

^{a}Existence of anion, or membership of series, requires confirmation.
^{b}Closely related 11-tungstate.

types. The largest and best known group is composed of those with the hetero atom(s) enshrouded by a cage of MO_6 octahedra. Table 18-C-2 lists most of those that have known structures. Six of the important structures that predominate in this group are shown schematically in Fig. 18-C-5. Two of these are often referred to eponymously.

The $[X^{n+}M_{12}O_{40}]^{(8-n)}$ structure is often called the *Keggin structure* after its discoverer. It has full tetrahedral (T_d) symmetry and although very compact, it accommodates a variety of heterocations that differ considerably in size, as Table 18-C-2 shows. There are many heteropolyanions with structures that can be regarded as modifications of the Keggin structure. One such modification is obtained by rotating one of the four sets of three octahedra in the Keggin structure by 60° and reattaching all the same corners. This spoils the T_d symmetry, leaving only one three-fold axis, that around which the 60° rotation was made. This "isomeric" Keggin structure differs little in stability from the true Keggin structure, and the anions $[XW_{12}O_{40}]^{4-}$ with X = Si and Ge, for example, can be isolated in either form by varying the conditions of acidification in the preparation.

The structure of the $[(X^{n+})_2M_{18}O_{62}]^{(16-2n)-}$ ion (often called the *Dawson structure*) is larger but closely related to that of the Keggin ion. If three adjacent corner-linked Mo_6 octahedra are removed from the Keggin structure, to leave a fragment with a set of three octahedra over a ring of six, we have one-half of this M_{18} anion. These two halves are then linked by corner sharing as shown in Fig. 18-C-5. Actually there are two such ways to link the halves. The way shown gives a structure of D_{3h} symmetry, but by rotating one half 60° relative to the other an isomer with D_{3d} symmetry is obtained; both occur. A new variation of the Dawson structure is found in the $[P_2Mo_{18}O_{61}]^{4-}$ anion,[10] which has a P_2O_7 unit (with an enforced linear P—O—P unit) instead of two PO_4 units as in $[P_2Mo_{18}O_{62}]^{6-}$.

The $[X^{n+}M_6O_{24}]^{(12-n)-}$ ion, its protonated form $[X^{n+}M_6O_{24}H_6]^{(6-n)-}$, and the $[X^{n+}M_9O_{32}]^{(10-n)-}$ ion are among those in which the heteroatom finds itself in an octahedron of oxygen atoms. The latter is known for only two cases, namely, those in which $X^{n+} = Mn^{4+}$ or Ni^{4+} and M = Mo, but these are of interest because the hetero ions are examples of unusual oxidation states stabilized by the unusual "ligand."

The $[X^{n+}M_{12}O_{42}]^{(12-n)-}$ ion provides an icosahedral set of 12 oxygen ligands for the hetero cation and it seems to be especially hospitable to +4 ions, such as Ce^{4+}, Th^{4+}, and U^{4+}.

Just as the heteropoly (and isopoly) ions are built up by acidification of solutions of the mononuclear oxo ions, the action of strong base on the heteropoly ions (and on the isopoly ions) will eventually degrade them entirely to mononuclear anions. There are well-defined steps in both processes, and intermediates can be observed and even isolated in crystalline compounds. Examples of species that have been well established are fragments of Keggin anions, especially those in which formal loss of an "MO" unit has occurred, for example, $[PW_{11}O_{39}]^{7-}$ or $[SiMo_{11}O_{39}]^{8-}$. There are also the "enneatungsto" compounds, such as $Na_{10}SiW_9O_{34}\cdot18H_2O$, that are thought to contain half-Dawson structures, but this lacks proof. The $[P_2Mo_5O_{23}]^{6-}$ ion is an example of one of the intermediates that has been characterized structurally. It consists of a cyclic assemblage of MoO_6 octahedra with four edge junctions and one corner junction, and there is a PO_4 tetrahedron fused over the hole on each side. Another example of this type of heteropoly anion in which the heteroatoms are exposed rather than buried is the $[As_2Mo_6O_{26}]^{6-}$ ion, whose structure has been determined by ^{17}O nmr. This structure is much like the $[X^{n+}M_6O_{24}]^{(12-n)-}$ structure except that the central, octahedrally coordinated X^{n+} ion is replaced by two AsO^{3+} groups, one above and the other below the central cavity in the six-membered ring of MoO_6 octahedra.

Heteropoly Blues

These are dark blue species obtained by reduction of both isopoly- and heteropoly molybdates and tungstates, and thus the class is broader than the name implies. It appears that in general, these reductions are reversible and major structural features are retained. In the case of the $[M_6O_{19}]^{2-}$ ions, reduction by one electron gives a 3-ion in which the added electron is weakly trapped at one metal atom but is thermally mobile.

When the Keggin anion, $[PMo_{12}O_{40}]^{3-}$, is reduced, from one to four electrons may be added, but the behavior is markedly solvent dependent. It has been found that the tungsten compound $NaH_6[PW_{12}O_{40}]$ to which four electrons have been added can be isolated.[11] It appears that in these cases, each electron, up to a total of four, goes to a different M_3O_{13} subunit. However, in the related $[H_2W_{12}O_{40}]^{12-}$

[10]U. Kortz and M. T. Pope, *Inorg. Chem.* **1994**, *33*, 5643.
[11]W. Enbo *et al.*, *Polyhedron* **1996**, *15*, 1383.

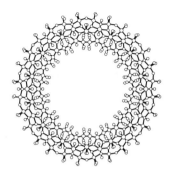

Figure 18-C-6 The structure of the giant anion $[Mo_{154}(NO)_{14}O_{420}(OH)_{28}(H_2O)_{70}]^{(25\pm5)-}$ (taken from Ref. 12).

ion, obtained by adding six electrons to the metatungstate ion (Fig. 18-C-3), which is also a Keggin type structure, it appears that all six added electrons go to one W_3O_{13} subunit, converting it to a metal atom cluster of the $W_3O_4^{4+}$ type (see Section 18-C-6).

The Dawson $[X_2Mo_{18}O_{62}]^{6-}$ ions can also be reversibly reduced to blue species. In general, the question of how delocalized the added electrons are in all these blue species remains open to further investigation.

Other Polyoxo Ions

In addition to the traditional iso- and heteropoly anions just discussed, recent work has revealed that others, many of them very large, with crown-type structures can be obtained. Examples of these are the $[NaP_5W_{30}O_{110}]^{14-}$ ion, which has a double crown structure with five-fold symmetry, $[H_7P_8W_{48}O_{184}]^{33-}$, which has a double crown structure with four-fold symmetry and two structurally similar crown species of molybdenum, $[Mo_{36}O_{112}]^{8-}$, and $Mo_{36}O_{110}(NO)_4(H_2O)_{14}$.

Undoubtedly, the most spectacular giant polymolybdate structure that is still a discrete anion is $[Mo_{154}(NO)_{14}O_{420}(OH)_{28}(H_2O)_{70}]^{(25\pm5)-}$. The structure[12] of this is shown in Fig. 18-C-6.

An interesting general reaction of polyoxo anions that has been recently explored is with RNCO and ArNCO reagents. Thus, with the $[M_6O_{19}]^{2-}$ ions, products such as $[W_6O_{18}NAr]^{2-}$, $[Mo_6O_{18}(NPh)_2]^{2-}$, and $Mo_6O_{15}(NAr)_4]^{2-}$ have been obtained.[13] Of course this type of reaction,

$$M{=}O + RNCO \longrightarrow M{=}NR + CO_2$$

is more general and has been applied to simpler species as well (see Section 18-C-9).

[12] A. Müller *et al.*, *Angew. Chem. Int. Ed. Engl.* **1995**, *34*, 2122.
[13] E. A. Maatta *et al.*, *J. Am. Chem. Soc.* **1994**, *116*, 3601; A. Proust *et al.*, *Inorg. Chim. Acta* **1994**, *224*, 81; E. A. Maatta *et al.*, *Inorg. Chem.* **1995**, *34*, 9.

Table 18-C-3   Halides of Molybdenum and Tungsten[a,b]

III	IV	V[b]	VI[b]
MoF_3	MoF_4	$(MoF_5)_4$	MoF_6
Yellow, nonvolatile	Light green, nonvolatile	Yellow, mp 67°C, disprop. 165°C	Colorless, mp 17.5°C, bp 35.0°C
	WF_4	$(WF_5)_4$	WF_6
	Red brown, nonvolatile	Yellow, disprop. 25°C	Colorless, mp 2.3°C, bp 17.0°C
$MoCl_3$	$MoCl_4{}^c$	$(MoCl_5)_2$	
Dark red	Black	Black, mp 194°C, bp 268°C	
WCl_3	WCl_4	$(WCl_5)_2$	WCl_6
Red	Black	Green black, mp 242°C, bp 286°C	Blue black, mp 275°C, bp 346°C
$MoBr_3$	$MoBr_4$		
Green, d 977°C	Black		
WBr_3	WBr_4	WBr_5	
Black	Black	Black	
MoI_3			
Black, d 927°C			

[a]For halides of Mo^{II} and W^{II} see Sections 18-C-9 and 18-C-10.
[b]Mixed halides, for example, MoF_3Cl_2, WF_5Cl, WCl_5F, and WCl_4F_2 are also known.
[c]$MoCl_4$ exists in three forms.

18-C-4 Halides

The halides are listed in Table 18-C-3. Others containing metal-metal bonds or metal atom clusters are discussed in Sections 18-C-10 and 18-C-11. Those that might be called "classical" in metal oxidation states III to VI, are discussed here.

Hexahalides

The MF_6 compounds are volatile, colorless liquids, readily hydrolyzed. Molybdenum hexafluoride is more reactive, less stable, and a considerably stronger oxidizing agent. The existence of $MoCl_6$ is very doubtful, but WCl_6 and WBr_6 are both obtained by direct halogenation of the metal. Tungsten hexachloride cocrystallizes with S_8 and is found there with a regular octahedral structure.[14] It can be volatilized to a monomeric vapor and is soluble in liquids such as CS_2, CCl_4, EtOH, and Et_2O, whereas WBr_6, a dark blue solid, gives WBr_5 on moderate heating. Both are hydrolyzed to tungstic acid.

Pentahalides

Treatment of molybdenum carbonyl with fluorine diluted in nitrogen at −75°C gives a product of composition Mo_2F_9. The structure of this substance has not been

[14]F. A. Cotton *et al., Acta Cryst.* **1989,** *C45,* 1287.

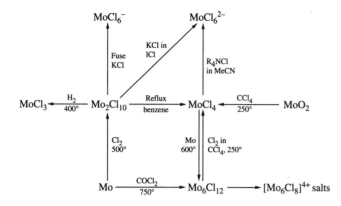

Figure 18-C-7 Preparation of molybdenum chlorides and chloro complexes.

investigated, but when it is heated to 150°C it yields the nonvolatile MoF_4 as a residue and the volatile MoF_5 condenses in cooler regions of the apparatus. Molybdenum pentafluoride is also obtained by the reactions

$$5MoF_6 + Mo(CO)_6 \longrightarrow 6MoF_5 + 6CO$$

$$Mo + 5MoF_6 \longrightarrow 6MoF_5$$

$$2Mo + 5F_2(\text{dilute}) \xrightarrow{400°C} 2MoF_5$$

WF_5 is obtained by quenching the products of reaction of W with WF_6 at 800 to 1000 K. It disproportionates above 320 K into WF_4 and WF_6. Crystalline MoF_5 and WF_5 (and also WOF_4) have the tetrameric structure common to many pentafluorides (Section 13-11).

Molybdenum(V) and W^V chlorides are obtained by direct chlorination of the metals under proper conditions, and the solids consist of edge-sharing bioctahedra (M_2Cl_{10}) in which there are no M—M bonds (Mo—Mo = 3.84 Å; W—W = 3.81 Å); they are paramagnetic. A study of mononuclear matrix-isolated WCl_5 and WBr_5 showed that they have *tbp* structures.[15] In the vapor state Mo_2Cl_{10} is entirely or at least mainly monomeric but the *tbp* structure is uncertain. It is soluble, and monomeric, in many organic media. It is very reactive, being hydrolyzed by water, reduced by amines to give amido complexes, and it abstracts oxygen from some oxygenated solvents to give Mo=O species. Further reactions are shown in Figs. 18-C-7 and 18-C-8. Little is known about W^V bromide.

Tetrahalides

These include MoF_4 and WF_4, the former arising on disproportionation of Mo_2F_9 as noted previously, and both by reduction of the hexahalides with hydrocarbons

[15] J. S. Ogden *et al.*, *J. Chem. Soc. Dalton Trans.* **1989,** 313.

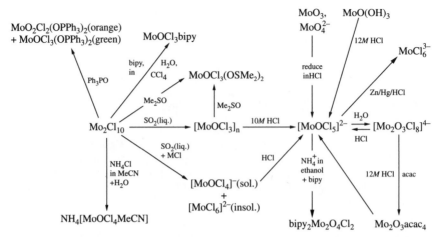

Figure 18-C-8 Some preparations and reactions of molybdenum pentachloride and of oxymolybdenum compounds.

(e.g., benzene at ~110°C). Both are nonvolatile. Molybdenum tetrachloride, which is very sensitive to oxidation and hydrolysis, exists in three forms. By the reaction

$$MoCl_5 + C_2Cl_4 \longrightarrow \alpha\text{-}MoCl_4$$

the α form isomorphous with NbCl$_4$ (Section 18-B-2) is obtained. The same form, contaminated with carbon, is also obtained when MoCl$_5$ is reduced with hydrocarbons. A β form can be obtained by reduction of MoCl$_5$ with metallic Mo. The α-MoCl$_4$ has partial spin pairing through Mo-Mo interactions, whereas the β form has molybdenum atoms occupying selected octahedral interstices in an *hcp* array of Cl atoms, giving hexagonal arrays but with no Mo—Mo bonding (Mo···Mo = 3.67 Å). The third form, obtained by treatment of MoO$_2$ with carbon in an N$_2$-borne stream of CCl$_4$ vapor, has a magnetic moment of ~1.9 BM and its structure is unknown. MoBr$_4$ is also known; it can be made[16] by the action of HBr on MoCl$_5$.

Tungsten tetrachloride is a very useful starting material for preparing many lower-valent tungsten compounds. The best methods of preparation are:

$$2WCl_6 + W(CO)_6 \xrightarrow[\text{reflux}]{C_6H_5Cl} 3WCl_4$$

$$3WCl_6 + \tfrac{1}{2}P_4 \xrightarrow[\text{tube}]{\text{sealed}} 3WCl_4 + 2PCl_3$$

The method employing P$_4$ has the advantage of eliminating all contamination by oxohalides of tungsten, since POCl$_3$ forms preferentially with any oxygen present. Tungsten tetrachloride disproportionates at 500°C to WCl$_2$ + 2WCl$_5$. The tetrabromide WBr$_4$ is not well characterized.

[16]G. Pampaloni *et al.*, *J. Chem. Soc., Dalton Trans.* **1993,** 655.

Trihalides

These include MoF_3, $MoCl_3$, $MoBr_3$, MoI_3, WCl_3, WBr_3, and WI_3. Molybdenum trifluoride, a nonvolatile yellow solid in which Mo atoms are found in octahedra of F atoms, is obtained by reaction of Mo with MoF_6 at ~400°C. Molybdenum trichloride has two polymorphs, with *hcp* and *ccp* arrays of Cl atoms and, in each case, Mo atoms in pairs of adjacent octahedral holes at a distance of 2.76 Å across a common edge. The magnetic properties confirm the expected M—M interaction. WCl_3 has an especially interesting cluster structure (see Section 18-C-11). The $MoX_3(THF)_3$ compounds are very useful synthetic starting materials.

18-C-5   Oxide Halides

These are formed by both metals mainly in oxidation states V and VI. Some of them are notorious as unwanted by-products in the preparation of anhydrous binary halides, since the formation of M=O bonds from any O_2, H_2O, or other sources of oxygen is strongly favored thermodynamically. The established oxide halides are listed in Table 18-C-4.

Methods of preparation are varied. Some of the important preparative reactions are as follows (where M represents both Mo and W):

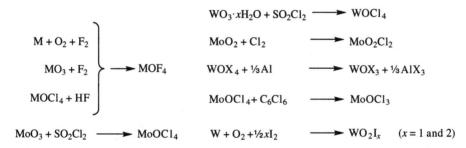

$$WO_3 \cdot xH_2O + SO_2Cl_2 \longrightarrow WOCl_4$$

$$MoO_2 + Cl_2 \longrightarrow MoO_2Cl_2$$

$$\left.\begin{array}{l} M + O_2 + F_2 \\[4pt] MO_3 + F_2 \\[4pt] MoCl_4 + HF \end{array}\right\} \longrightarrow MOF_4$$

$$WOX_4 + \tfrac{1}{3}Al \longrightarrow WOX_3 + \tfrac{1}{3}AlX_3$$

$$MoOCl_4 + C_6Cl_6 \longrightarrow MoOCl_3$$

$$MoO_3 + SO_2Cl_2 \longrightarrow MoOCl_4$$

$$W + O_2 + \tfrac{1}{2}xI_2 \longrightarrow WO_2I_x \quad (x = 1 \text{ and } 2)$$

Table 18-C-4   Oxide Halides of Molybdenum and Tungsten[a]

V	VI	VI
	MoOF$_4$, colorless, mp 97°C, bp 186°C	MoO$_2$F$_2$, colorless, subl 270°C
	WOF$_4$, colorless, mp 107°C, bp 186°C	WO$_2$F$_2$?
MoOCl$_3$, black, d > 215°C	MoOCl$_4$, green, mp 102°C, bp 159°C	MoO$_2$Cl$_2$, yellow, mp 170°C, bp 250°C
WOCl$_3$	WOCl$_4$, red, mp 208°C	WO$_2$Cl$_2$
MoOBr$_3$, black subl 270°C	MoOBr$_4$	MoO$_2$Br$_2$, brown
WOBr$_3$	WOBr$_4$, purple black	WO$_2$Br$_2$ MoO$_2$I$_2$
WO$_2$I, blue black		WO$_2$I$_2$, dark green

[a]Also WOCl$_2$Br and WOCl$_2$Br$_2$

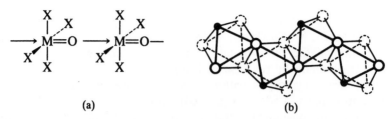

Figure 18-C-9   (a) The mode of association of MOX_4 molecules in the solid state. (b) The infinite chain structure of $MoOCl_3$, consisting of $MoOCl_5$ octahedra sharing cis pairs of Cl atoms.

The oxide halides are all rather reactive, though less so than the binary halides. While they normally hydrolyze completely in contact with water, $MoO_2Cl_2(H_2O)_2$ stabilized by association with 2,5,8-trioxononane can be crystallized from a solution of MoO_3 in aqueous HCl.[17] The MOX_4 species readily form adducts with neutral donors, for example, $MoOCl_5(MeCN)$. In the vapor phase the MOX_4 molecules are tetragonal pyramids, but they associate into infinite chains with M—O—M units, Fig. 18-C-9(a), in the crystal, which lengthens the M=O bonds by ~0.08 Å. All of the MOX_3 compound are crystallographically isotypic with $NbOCl_3$, but $MoOCl_3$ has a second form, shown in Fig. 18-C-9(b).

A few sulfide halides are known but little studied. The best known are $MoSCl_3$, $MoSCl_2$, WSF_4,[18] and $WSCl_4$.[19]

18-C-6   Halogen Containing Complexes

Homoleptic Complexes

Tungsten(V) gives the octahedral WX_6^- (X = F, Cl, and Br) species and also WF_8^{3-}; the only molybdenum(V) analogues are the fluoro complexes MoF_6^- and MoF_8^{3-}. Molybdenum(V) is prone to form oxo complexes instead (Section 18-C-7). The MX_6^{2-} complexes of M^{IV} are known, including the recently reported WF_6^{2-} ion,[20] and a very unusual $[W_4F_{18}]^{2-}$ ion with a tetrahedral array of $W^{IV}F_3$ units bridged by six more F atoms.[21]

For oxidation state III, molybdenum forms MoX_6^{3-} ions for X = F, Cl, and Br, but no such complex (nor many other octahedral ones) has been obtained for tungsten(III). Instead, there are the face-sharing bioctahedral species $W_2Cl_9^{3-}$ and $W_2Br_9^{3-}$ in which strong W≡W triple bonds are formed. The $Mo_2X_9^{3-}$ analogues exist, but they have variable Mo—Mo distances with only partial bonding and are in varying degrees paramagnetic. In the $Mo_2Cl_9^{3-}$ ion, the distances vary from 2.52 Å to 2.82 Å and theory shows that the best formulation of the Mo—Mo bonding is in terms of a σ bond plus an antiferromagnetic coupling of the remaining pairs of

[17]F. J. Arnaiz et al., Polyhedron **1994**, 13, 2745.
[18]A. K. Brisdon et al., J. Chem. Soc., Dalton Trans. **1996**, 2975.
[19]F. A. Cotton et al., Inorg. Chem. **1989**, 28, 2485.
[20]J. H. Reibenspies et al., Z. Krist. **1995**, 210, 882.
[21]T. S. Cameron et al., J. Chem. Soc., Dalton Trans. **1993**, 659.

electrons.[22] The $W_2X_9^{3-}$ species can be oxidized to $W_2X_9^{2-}$, which have one un-paired electron.

Mixed Ligand Complexes

Complexes of the type $[MX_nL_{6-n}]$ are fairly numerous. The pentahalides can form adducts (MX_5L) with many donors, although with amines or RCN they more commonly react to give M^{IV} complexes:

$$MCl_{5\ or\ 6} + \begin{Bmatrix} py \\ bipy \\ RCN \end{Bmatrix} \longrightarrow \begin{Bmatrix} MCl_4py_2 \\ MCl_4bipy \\ MCl_4(RCN)_2 \end{Bmatrix}$$

From these many other MX_4L_2 complexes may be obtained by ligand replacement reactions. In some cases coordination number 7 is reached, as in the $MCl_4(PMe_2Ph)_3$ complexes.

The Mo^{III} complexes $(NH_4)_2[MoX_5(H_2O)]$, X = Cl or Br, are known and the bromo compound reacts with pyridine to give $(NH_4)[MoBr_4py_2]$ as the cis isomer exclusively.[23]

As already noted, few simple octahedral W^{III} complexes are known, but mer-$WCl_3(PMe_2Ph)_3$ can be made[24] by zinc reduction of $WCl_4(PMe_2Ph)_3$.

The pentachlorides react with ROH and RO^- to yield various types of alkoxo complex. Tungsten forms an extensive series of complexes that includes the paramagnetic $[M(OR)Cl_5]^-$ and $[M(OR)_2Cl_4]^-$ ions as well as the diamagnetic $W_2Cl_2(OR)_8$ and $W_2Cl_4(OR)_6$ molecules, which presumably contain octahedrally coordinated metal atoms, bridging Cl atoms, and W—W bonds. Molybdenum pentachloride reacts with alcohols and amines to give products of the types $MoCl_3(OR)_2$ and $MoCl_3(NRR')_2$, which appear generally to be dinuclear with bridging chlorine atoms. The chief products of the reaction with phenols are of the type $[MoCl_2(OAr)_3]_3$. Reactions of halides with $LiNR_2$ are mentioned later in connection with M—M triple bonds.

For molybdenum(III), the $MoCl_3L_3$ complexes with L = THF or RCN are useful reactants, as is also $MoI_3(THF)_3$. These molecules have meridional structures and readily undergo ligand replacement reactions with phosphines.

Another route to some MX_nL_{6-n} complexes is by oxidation of cis-$M(CO)_4L_2$ with Cl_2 or Br_2, for example,

$$cis\text{-}W(CO)_4(PMe_2Ph)_2 + 2Br_2 \longrightarrow WBr_4(PMe_2Ph)_2 + 4CO$$

There are also some monomeric Mo^{II} complexes, such as $MoCl_2(dmpe)_2$ which can be both oxidized and reduced:

$$[MoCl_2(dmpe)_2]PF_6 \xleftarrow{\ AgPF_6\ } MoCl_2(dmpe)_2 \xrightarrow[N_2]{Na/Hg} Mo(dmpe)_2(N_2)_2$$

[22]F. A. Cotton and X. Feng, *Intl. J. Quantum Chem.* **1996**, *58*, 671.
[23]J. V. Brenčič *et al.*, *Z. anorg. allg. Chem.* **1994**, *620*, 950.
[24]G. J. Leigh *et al.*, *J. Chem. Soc., Dalton Trans.* **1991**, 1515.

and $MoH(SR)(dppe)_2$, as well as tungsten(II) phosphine complexes such as $WCl_2(PR_3)_4$.

18-C-7   Aqua and Oxo Complexes

Aqua Ions

Genuine aqua ions (containing only Mo and H_2O) are known only for oxidation states II and III. For Mo^{II} the red aqua ion is $Mo_2^{4+}(aq)$, formed when a solution of $K_4[Mo_2(SO_4)_4]$ is treated with $Ba(CF_3SO_3)_2$ in dilute CF_3SO_3H solution. This quadruply-bonded species is a powerful reducing agent but thermally stable at 25°C. The yellow $[Mo(H_2O)_6]^{3+}$ ion can be obtained by dissolving an $[MoCl_6]^{3-}$ or $[MoCl_5(H_2O)]^{2-}$ salt in aqueous CF_3SO_3H and separating the $Mo^{3+}(aq)$ ion from Cl^- ions on a cation-exchange column. Its formulation as $[Mo(H_2O)_6]^{3+}$ is supported by its visible spectrum and magnetic moment (3.69 BM). The ion has been characterized by X-ray study of an alum. No tungsten analogue is known.

The "aqua" ions for oxidation states IV and V are actually di- or trinuclear oxo species, and Mo^{III} also forms a binuclear ion that is either $[(H_2O)_4Mo(OH)_2-Mo(H_2O)_4]^{4+}$ or possibly $[(H_2O)_5MoOMo(H_2O)_5]^{4+}$. The Mo^{IV} aqua ion has now been well established to be $[Mo_3O_4(H_2O)_9]^{4+}$, from which (*vide infra*) numerous $Mo_3O_4^{4+}$ complexes can be derived. There is a similar $W_3O_4^{4+}(aq)$ ion. The Mo^V aqua ion, $Mo_2O_4^{2+}$, is diamagnetic and probably has the structure (18-C-I) in which a $Mo-Mo$ single bond exists. Numerous complexes containing this $Mo_2O_4^{2+}$ core, and the related trans one, are known and will be discussed later. The $W_2O_4^{2+}(aq)$ ion is also known. At pH < 1 Mo^{VI} is believed to exist as the $[MoO_2(OH)(H_2O)_3]^+$ ion.

(18-C-I)

Trinuclear M^{IV} Oxo Complexes

We have already mentioned the $Mo_3O_4^{4+}(aq)$ ion. The essential structure of this, shown in Fig. 18-C-10(a), can be described in four complementary ways: (1) As an M_3 triangle capped on one side, with three edge bridges on the other side and three more nonbridging ligands on each metal atom. (2) As three MO_6 octahedra sharing a common vertex and also having one edge common to each pair. (3) As two adjacent close-packed layers of ligand atoms with metal atoms occupying octahedral interstices—but moved off the centers of these interstices toward each other. (4) As an incomplete cuboid M_4O_4 structure that lacks one M atom. Description (2) emphasizes the relationship of these $M-M$ bonded species to the M_3O_{13} subunits in various iso- and heteropoly anion structures, such as the Keggin structure (Fig. 18-C-3), while description (3) relates this unit to the structures of mixed oxide

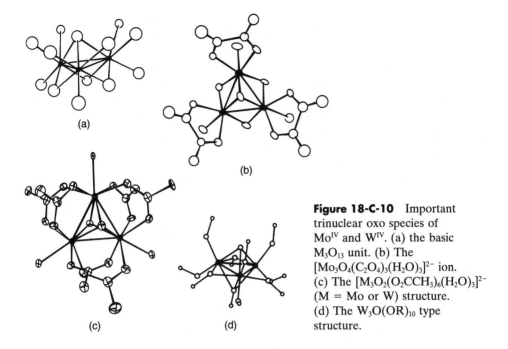

(a)

(b)

(c)   (d)

Figure 18-C-10 Important trinuclear oxo species of Mo^{IV} and W^{IV}. (a) the basic M_3O_{13} unit. (b) The $[Mo_3O_4(C_2O_4)_3(H_2O)_3]^{2-}$ ion. (c) The $[M_3O_2(O_2CCH_3)_6(H_2O)_3]^{2-}$ (M = Mo or W) structure. (d) The $W_3O(OR)_{10}$ type structure.

systems such as $Zn_2Mo_3O_8$.[25] Description (4) relates it to the very large class of cuboidal species M_4X_4 and $M_3M'X_4$, whose idealized general structure is 18-C-II.

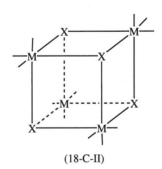

(18-C-II)

From the $Mo_3O_4^{4+}$ aqua ion many complexes can be obtained by replacing some, or all of the nine outer water molecules by other ligands. Examples are the $[Mo_3O_4$-$(C_2O_4)_3(H_2O)_3]^{2-}$ ion, Fig. 18-C-10(b), $[Mo_3O_4F_9]^{5-}$, and $\{Mo_3O_4[(O_2CCH_2)_2NCH_3]_3\}^{2-}$. There are some analogous tungsten complexes, for example, $[W_3O_4F_9]^{5-}$ and $[W_3O_4L_3]^{4+}$ where L = 1,4,7-triazacyclononane.

A second major type of oxo trinuclear cluster is the type with two μ^3-O caps, formed equally by molybdenum and tungsten. They are usually obtained by treating $M(CO)_6$ with a mixture of a carboxylic acid and its anhydride, $RCO_2H/(RCO)_2O$, pouring the reaction mixture, after dilution with water, on a cation exchange column

[25]S. J. Hibble and I. D. Fawcett, *Inorg. Chem.* **1995**, *34*, 500.

and eluting with an acid such as CF_3SO_3H or HBF_4. The products contain trinuclear cations of the general type $[M_3O_2(RCO_2)_6(H_2O)_3]^{2+}$, one of which is illustrated in Fig. 18-C-10(c). There are some remarkable variations on the basic M_3O_2 structural theme. It is possible to obtain an essentially identical type of complex except that one capping oxygen atom is absent.[26] Even more remarkable is that under certain preparative conditions products are obtained in which one or both μ^3-O ligands are replaced by μ^3-CR (alkylidyne) caps. The persistence of these systems in acid, aqueous media, and in the presence of oxygen is surprising. The probable reason is their kinetic inertness. Counting the M—M bonds, each metal atom is 9-coordinate and employs all nine valence orbitals in bonding, thus making associative reaction pathways inaccessible. A striking manifestation of this kinetic inertness is provided by the fact that the $[M_3O_2(O_2CC_6H_5)_6(H_2O)_3]^{2+}$ ions can be fully nitrated in all 12 meta positions, employing a mixture of concentrated sulfuric and nitric acids at 40°C, without detectable core degradation.

Another important type of trinuclear oxo structure is that of $Mo_3O(OR)_{10}$, Fig. 18-C-10(d). This has two different capping groups, μ_3-O and μ_3-OR, plus the μ_2-OR groups and six terminal ones.

As a general rule, though not in all cases, oxo triangular species formed by Mo_3 or W_3 are also obtainable for the Mo_2W and MoW_2 mixtures.[27]

Mixed Oxo-halo Trinuclear Clusters

While the $Mo_3O_4^{4+}$ core is electron-precise, in the sense that it has just the six electrons needed to establish three Mo—Mo bonds, this is not true for the structurally similar $M_3(\mu_3\text{-O})(\mu_2\text{-Cl})_3^{6+}$ cores. It has been shown theoretically[28] that they prefer to have eight electrons (two of which are essentially non-bonding) and in fact quite a few 8- and even 9-electron compounds with a $MOCl_3^{6+}$ core are known.[29]

Mononuclear M^{IV} Oxo Complexes

These contain the $[M=O]^{2+}$ unit and have quasi-octahedral structures with a weaker than normal bond trans to the oxo group. Examples are $[WOBr(dmpe)_2]Br$,[30] the $WOBr_2(PR_3)_3$ compounds,[31] and the 12[ane](NMe)$_3$ (=L) compounds, $LMoOX_2$ (X = Br, I, OMe) and $LMo(O)X—O—Mo(O)XL$.[32]

M^{V} Oxo Species

An extensive range of Mo^V compounds can be obtained by reduction of molybdates or MoO_3 in acid solution either chemically (e.g., by shaking with mercury) or electrolytically. They contain an MoO^{3+} core with a strong, short (1.6–1.7 Å) bond,

[26] A. Bino et al., Inorg. Chem. **1988**, 27, 3592.

[27] D. T. Richens et al., J. Chem. Soc., Dalton Trans. **1993**, 767.

[28] F. A. Cotton and X. Feng, Inorg. Chem. **1991**, 30, 3666.

[29] F. A. Cotton et al., J. Am. Chem. Soc. **1991**, 113, 3007, 6917; J. Cluster Sci. **1992**, 3, 123; Inorg. Chem. **1994**, 33, 3195.

[30] F. A. Cotton et al., Acta Cryst. **1989**, C45, 1126.

[31] F. A. Cotton and S. K. Mandal, Inorg. Chim. Acta **1992**, 194, 179.

[32] K. Wieghardt et al., J. Chem. Soc., Dalton Trans. **1993**, 1987.

four other ligands forming the base of a square pyramid and a sixth weakly bound ligand (sometimes missing) trans to the MoO bond, as in 18-C-III. An important species, often a source material for preparation of other Mo^V compounds, is the emerald green ion $[MoOCl_5]^{2-}$ or the closely related $[MoOCl_4]^-$ and $[MoOCl_4(H_2O)]^-$ ions, all of which are readily interconverted by varying (or removing) the ligand trans to the Mo=O bond. There are also bromo and iodo analogues. These ions can be obtained by reduction of Mo^{VI} in aqueous HX, by oxidation of $Mo_2(O_2CCH_3)_4$ in aqueous HX, or by dissolving $MoCl_5$ in aqueous acid. Molybdenum pentachloride will also react with many organic compounds or solvents (e.g., Me_2SO or Ph_3PO) to abstract oxygen and form oxo-molybdenum(V) complexes such as $MoOCl_3(OSMe_2)_2$.

(18-C-III)

Because of their possible presence in biological molecules and because they also exhibit remarkable fluorescence, the $[X_4MoO]^-$ ions have been much studied, spectroscopically and theoretically. The electron occupies the d_{xy} orbital and the remaining orbitals are ordered[33] in increasing energy (d_{xz}, d_{yz}), $d_{x^2-y^2}$, d_{z^2}. The fluorescence is due to the $(xz,yz) \rightarrow xy$ transition.[34]

Many more complicated MoO^{3+} complexes are known, such as $(HBpz_3)$(catechol)MoO[35] and a peroxo species $[MoO(O_2)(ox)]^{2-}$ with a roughly pentagonal bipyramidal structure.[36]

The only compound containing a $[cis\text{-}Mo^VO_2]^+$ ion was recently reported.[37]

When WO_4^{2-} is reduced in 12 M HCl, the blue $[WOCl_5]^{2-}$ ion is obtained. There are also various neutral complexes of the types $WOCl_3L_2$ and $WOCl_3(LL)$.

Although the mononuclear complexes just mentioned are important, the oxo chemistry of Mo^V is dominated by dinuclear complexes. Similar dinuclear oxo complexes of W^V are rare. The dinuclear oxo complexes of Mo^V are of two main types; singly bridged ones that exist in cis and trans rotamers (18-C-IVa) and (18-C-IVb), and doubly bridged ones (18-C-IVc), which are cis. In most cases some or all of the ligands (not shown in these sketches) are chelating. Examples of type (18-C-IVa) are $Mo_2O_3(S_2COEt)_4$ and $Mo_2O_3(S_2CNPr_2)_4$; complexes of the type (18-C-IVb) are $Mo_2O_3[S_2P(OEt)_2]_4$ and $Mo_2O_3(LL)_4$ in which LL represents o-thiopyridine. The doubly bridged structure (18-C-IVc) is found in $[Mo_2O_4(C_2O_4)_2(H_2O)_2]^{2-}$, for example, where the water molecules are only weakly bound trans to the Mo=O bonds. In all mono- and dinuclear oxo complexes of Mo^V, ligands trans to Mo=O bonds are rather weakly bonded and are often entirely absent.

[33] E. I. Solomon *et al.*, *J. Am. Chem. Soc.* **1994**, *116*, 11, 865.
[34] A. W. Maverick *et al.*, *Inorg. Chem.* **1992**, *31*, 4441; *Inorg. Chim. Acta* **1994**, *226*, 25.
[35] J. H. Enemark *et al.*, *Inorg. Chem.* **1995**, *34*, 405.
[36] M. S. Reynolds *et al.*, *Inorg. Chem.* **1994**, *33*, 4977.
[37] Z. Xiao *et al.*, *J. Am. Chem. Soc.* **1996**, *118*, 2912.

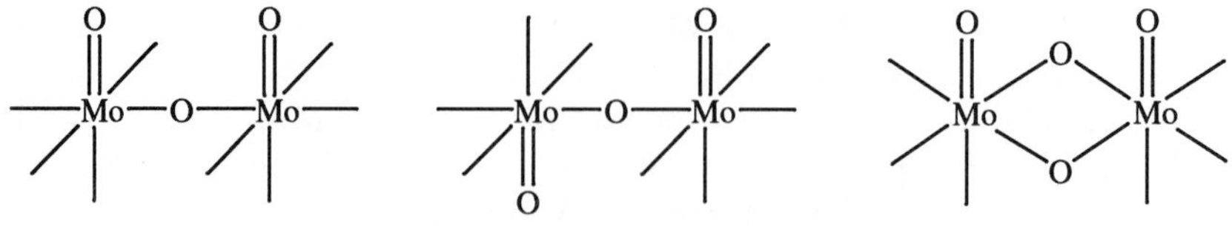

(a)   (b)   (c)

(18-C-IV)

M^{VI} Oxo Species

Tungsten(VI) forms only a few of these, most of which resemble their molybdenum analogues, but those of molybdenum are numerous and important. With but few exceptions such as (triamine)MoO_3 the mononuclear ones are MoO_2^{2+} complexes, all of which have distorted octahedral structures with *cis* oxygen atoms, Mo—O distances of 1.70 ± 0.02 Å, and O—Mo—O angles of $106 \pm 3°$. They are off-white to orange in color and have strong Mo=O stretching bands in the ir at ~880 and 920 cm^{-1}. Examples are the $[MoO_2Cl_4]^{2-}$ and $[MoO_2Cl_2(H_2O)_2]$ complexes obtained by dissolving MoO_3 in aqueous HCl, the former predominating in 12 M and the latter in 6 M acid. Adducts such as $MoO_2Cl_2(OPPh_3)_2$ and species with chelating ligands, for example, $[MoO_2(acac)_2]$ and $[MoO_2(S_2CNEt_2)_2]$, are also well known. Tungsten(VI) forms a few oxo complexes such as WOF_5^-, $WO_2F_4^{2-}$, $WO_2Cl_4^{2-}$, and $WO_3F_3^{3-}$.

With Mo^{VI} there are also a few binuclear species having both Mo=O and Mo—O—Mo groups. The anion in $K_2[Mo_2O_5(C_2O_4)_2(H_2O)_2]$ has the centrosymmetric structure in Fig. 18-C-11; the bridge Mo—O—Mo group is linear and symmetrical.

$$L_4MoO_2 + X \longrightarrow L_4MoO + OX \tag{18-C-7}$$

$$L_4MoO_2 + L_4MoO = L_4\overset{O}{\underset{\|}{Mo}}—O—\overset{O}{\underset{\|}{Mo}}L_4 \tag{18-C-8}$$

Dioxomolybdenum(VI) species can serve as oxygen atom transfer agents,[38] eq. 18-C-7. The acceptor may be a thiol, a tertiary phosphine, or any one of various organic molecules. There is, however, a secondary reaction, eq. 18-C-8, that often makes this chemistry more complicated. The presence of unreacted dioxomolybdenum(VI) complex together with the 5-coordinated oxomolybdenum(IV) product can give rise to the oxo-bridged dimolybdenum(V) compound. This is an equilibrium reaction, and the position of the equilibrium is strongly dependent on the nature of the ligands (L) especially their steric properties. With sufficiently bulky polydentate ligands reaction 18-C-8 can be blocked. In cases where it cannot be entirely suppressed it is necessary to include it in carrying out a correct kinetic analysis of the coupled reactions. There are two reasons for being interested in the nature of

[38]R. H. Holm *et al.*, *Coord. Chem. Rev.* **1990**, *100*, 183.

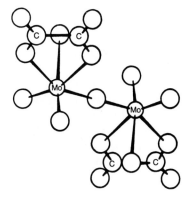

Figure 18-C-11 The structure of the anion in $K_2[Mo_2O_5(C_2O_4)_2(H_2O)_2]$.

reaction 18-C-7 uncomplicated by reaction 18-C-8. If the L_4MoO_2 is to be used fully and efficiently as an oxidant, one does not wish to tie up half of it in the dimolybdenum(V) species. In addition, there is much evidence that a number of molybdenum-containing enzymes (*cf.* Section 18-C-12) actually employ a reaction like eq. 18-C-7, but in the enzymes massive steric hindrance entirely suppresses the dimerization; good models for the enzymatic activity must therefore also be designed to prevent reaction 18-C-8. Considerable success in designing good enzyme models has been reported in which appropriately bulky tridentate ligands, as shown in (18-C-V), are present.

(18-C-V)

Some unusual oxo tungsten(VI) species have been reported, such as $[W_2Cl_6O_4]^{2-}$ with the structure shown as (18-C-VI) and $W_2O_3(CH_2CMe_3)_6$, the only organo oxo W^{VI} species, (18-C-VII).

(18-C-VI)   (18-C-VII)

18-C-8 Thia and Selena Compounds

Over the past decade there has arisen a large chemistry of sulfur (and to a lesser extent selenium) compounds of molybdenum and tungsten.[39] The simple tetrahedral M^{VI} ions have already been mentioned (Section 18-C-2).

There are many S and some Se analogs to the complexes of $Mo_2O_4^{2+}$ (18-C-IVc). Since S prefers the bridging positions, the thia species are of the types 18-C-VIII only. Tungsten analogues are known[40] too. These can be used for the systematic preparation of all the mixed O/S compounds of the $Mo_3S_nO_{4-n}^{4+}$ core type (of which there are six, 18-C-IX). Representative reactions are shown in eq. 18-C-9. The initial cuboidal species can be relatively easily degraded by excision of one Mo atom.

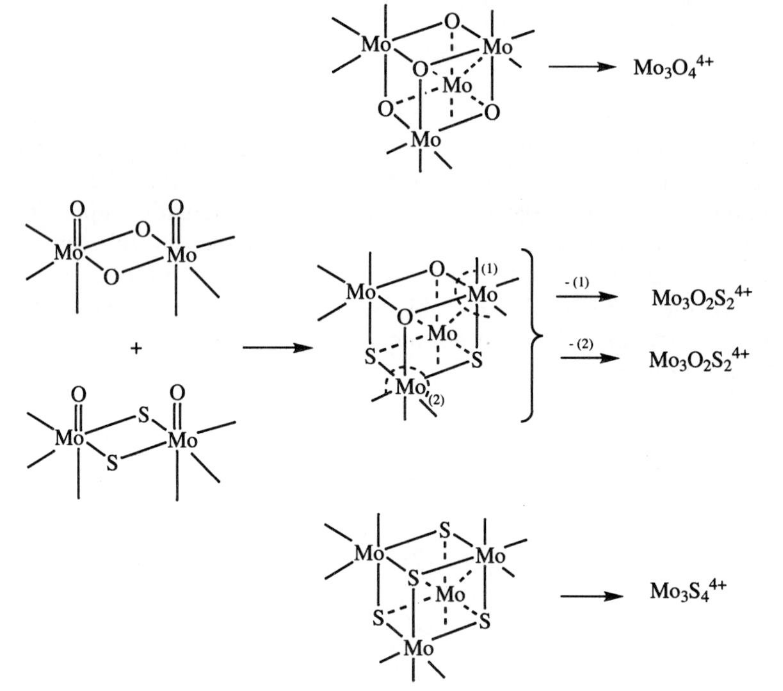

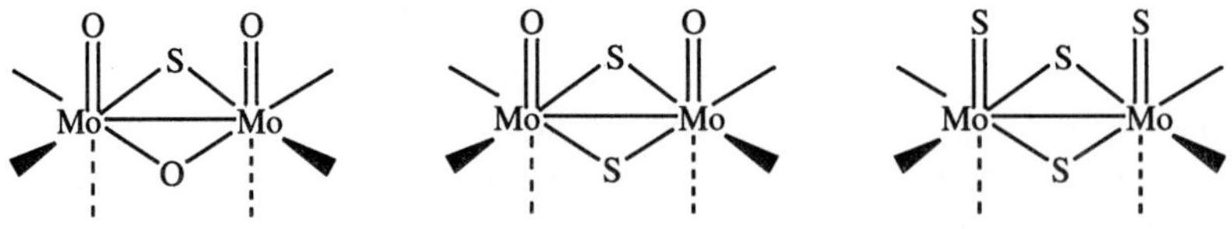

(18-C-VIII)

M_2S_4, M_3S_4, and $M_3S(S_2)_3$ Cores

This is now an extensive field of chemistry. The cuboidal $[Mo_4S_4]^{n+}$, with $n = 4$, 5, or, more rarely, 6, have no oxo analogues other than in one solid state compound,

[39]T. S. Shibahara, *Coord. Chem. Rev.* **1993**, *123*, 73.
[40]F. Sécheresse *et al.*, *J. Chem. Soc., Dalton Trans.* **1994**, 1311.

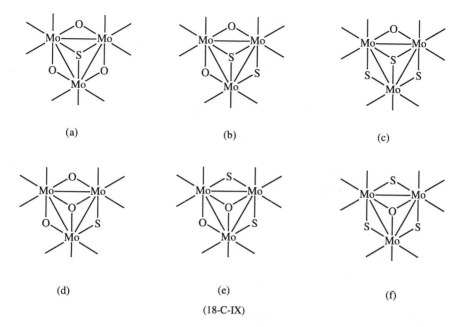

(a)

(b)

(c)

(d)

(e)

(f)

(18-C-IX)

$Cs_3Mo_4P_3O_{16}$.[41] All are known as aqua ions, $[Mo_4S_4(H_2O)_{12}]^{n+}$, where the water molecules complete an octahedron of ligands about each metal atom, as shown in 18-C-X, and also in the form of anionic complexes, such as $[Mo_4S_4(SCN)_{12}]^{6-}$. Their redox and other properties have been much studied[42] and they may be prepared in various ways.[43]

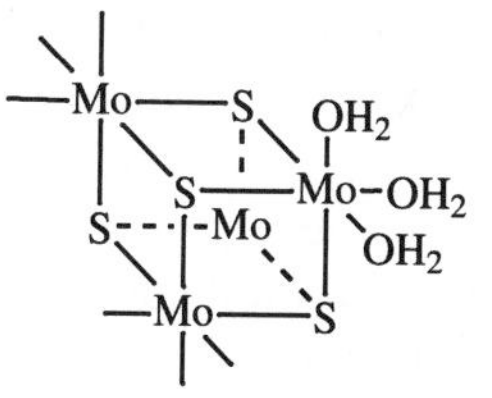

(18-C-X)

The $[M_3(\mu_3\text{-}E)(E_2)_3]^{n+}$ core has the structure shown in 18-C-XI for $Mo_3(\mu_3\text{-}S)(S_2)_3$. This species and its seleno analogue are both well known as the aquo ions,[44] $[Mo_3(\mu_3\text{-}E)(E_2)_3(H_2O)_6]^{4+}$. Many anionic derivatives are known such as the $[Mo_3(\mu_3\text{-}S)(S_2)_3(S_2)_3]^{2-}$ ion[45] where the additional S_2^{2-} ions occupy the six outer positions, its seleno analogue,[46] and the $[M_3(\mu_3\text{-}S)(S_2)_3X_6]^{2-}$ ions (X = Cl, Br) that are easily made from the solid state compounds $M_3(\mu_3\text{-}S)(S_2)_3X_4$.[47]

[41]R. C. Haushalter, *J. Chem. Soc., Chem. Commun.* **1987,** 1566.
[42]A. G. Sykes *et al., J. Chem. Soc., Dalton Trans.* **1993,** 2613; M. C. Ghosh and E. S. Gould, *Inorg. Chim. Acta* **1994,** *225,* 297.
[43]M. Brorson *et al., Inorg. Chim. Acta* **1995,** *232,* 171.
[44]A. G. Sykes *et al., J. Chem. Soc., Chem. Commun.* **1994,** 2685.
[45]A. Müller and E. Krickemeyer, *Inorg. Synth.* **1990,** *27,* 47.
[46]M. G. Kanatzidis *et al., Inorg. Chem.* **1995,** *34,* 2658.
[47]V. P. Fedin *et al., Inorg. Chim. Acta* **1990,** *175,* 217.

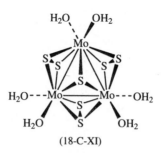

(18-C-XI)

The $[Mo_3(\mu_3\text{-}S)(S_2)_5(S_4)]^{2-}$ ion is found along with the $[Mo_3(\mu_3\text{-}S)(S_2)_6]^{2-}$ ion in one compound[48] and there is no reason why other such replacements of external S_2^{2-} ligands by other S_n^{2-} ligands should not occur.

Among the many non-aqueous routes to $M(\mu_3\text{-}S)(S)_3^{4+}$ compounds one (originally discovered by Federov) begins with $Mo_3(\mu_3\text{-}S)(S_2)_3X_4$, a solid state compound with intermolecular μ-X bridges (X = Cl, Br).[49] A representative reaction[50] is

$$Mo_3S_7Cl_4 + 6Et_3P + 2H_2O \longrightarrow Mo_3S_4Cl_4(H_2O)_2(PEt_3)_3 + 3Et_3PS$$

Another[51] is to react $MoCl_3(THF)_3$, WCl_4, or WBr_4 with NaSH and diphosphines, PP, whereby $[M_3S_4Cl_3(PP)_3]^+$ cations are obtained. On treatment of the tungsten compounds with $LiBH_4$ the hydrido analogues, e.g., $[W_3S_4H_3(dmpe)_3]^+$ can be obtained.[52]

It is typical of the $M_3(\mu_3\text{-}E)(\mu\text{-}E_2)_3$ type cores to be attacked by S acceptors (e.g., CN^-, R_3P) to form $Mo_3(\mu_3\text{-}E)(\mu\text{-}E)_3$ plus ECN^- or R_3PE.[53] It has recently been shown that when E = Te this does not happen and the CN complex of $[Mo_3(\mu_3\text{-}Te)(\mu\text{-}Te_2)_3]^{n+}$ can be isolated.[54] There are also ways of completely desulfurizing $Mo_3(\mu_3\text{-}S)(S_2)_3^{8+}$ to give $Mo_3O_4^{4+}$. A very unusual reaction of $[Mo_3(\mu_3\text{-}S)(S_2)_6]^{2-}$ with PMe_3 leads to the molecule $Mo_3(\mu_3\text{-}S)_2(\mu\text{-}S)_3(PMe_3)_6$; some sulfur is removed as Me_3PS but a rearrangement also occurs to give the $Mo_3(\mu_3\text{-}S)_2(S)_3^0$ core, in which the Mo has been reduced to the oxidation state 10/3 from 4.[55]

It may be noted that there are numerous other cases in which S_2^{2-} units serve as bridges, lying above and in a plane perpendicular to an M—M bond.[56]

$[M_3S_4]^{4+}$ Cores

These are well known, being obtainable by extrusion of one metal atom from M_4S_4 units,[57] by desulfurization of $[M_3(\mu_3\text{-}S)(S_2)_6]^{2-}$, and in other ways.[58] There are also

[48]P. K. Dorhout et al., Inorg. Chem. 1994, 33, 2703.
[49]F. A. Cotton et al., Inorg. Chem. 1989, 28, 2623.
[50]F. A. Cotton et al., Inorg. Chem. 1991, 30, 548.
[51]F. A. Cotton and S. K. Mandal, Inorg. Chim. Acta 1992, 192, 71.
[52]F. A. Cotton et al., J. Am. Chem. Soc. 1989, 111, 4332.
[53]K. Hegetschweiler et al., Inorg. Chim. Acta 1993, 213, 157.
[54]T. Saito et al., Inorg. Chem. 1995, 34, 5097.
[55]T. Saito et al., Inorg. Chem. 1995, 34, 3404.
[56]C. G. Young et al., Inorg. Chem. 1994, 33, 6252.
[57]A. G. Sykes et al., Inorg. Chem. 1992, 31, 3011.
[58]T. Shibahara et al., Inorg. Chem. 1992, 31, 640.

$[Mo_3S(S_2)_6]^{2-}$

$[Mo_2S_4(S_2)_2]^{2-}$

$[Mo_2(S_2)_6]^{2-}$

$[Mo_2S_4(S_4)_2]^{2-}$

Figure 18-C-12 The structures of four representative Mo_2 or Mo_3 anions containing S^{2-}, S_2^{2-}, or S_4^{2-} ions.

various compounds of the M_3E_4 type known with the mixed metal sets, Mo_2W and MoW_2.[59,60] All of these have the six-electron $[M_3(\mu_3\text{-}E)(E)_3]^{4+}$ core. As noted earlier, all possible mixed $[Mo_3O_nS_{4-n}]^{4+}$ cores are also known.[61]

A very characteristic reaction of the M_3S_4 species is to incorporate other metal ions, M′, into a hetero cuboidal $M_3M'S_4$ unit. This can also be viewed as the M_3S_4 unit acting as a *fac*-tridentate ligand toward M′, which may be Cr, Fe, Co, Ni, Cu, Sn, Pb, Hg, In, and probably others.[62,63,64]

In conclusion, a few of the many more S and Se complexes of other types may be noted. Both the $[Mo_2S_4(S_2)_2]^{2-}$ ion and its seleno analogue[65] are known, each having the structure shown in Fig. 18-C-12. Similarly, the $M_3S_8^{2-}$, $M_3S_9^{2-}$, and $M_3S_{10}^{2-}$ ions are known, particularly for W.[66] Selenium analogues have been made and characterized.[67] The mixing of M=E, μ_2-E, μ_2-η^2-E, etc. linkages seems to be possible to an almost unlimited extent.

18-C-9 Other Complexes

These elements form an enormous variety of complexes besides the halide, oxide, and sulfide complexes already considered. Some of these others will now be mentioned. Most of those with metal-to-metal bonds are covered in sections 18-C-10 and 18-C-11.

[59]D. T. Richens *et al., J. Chem. Soc., Dalton Trans.* **1993,** 767.
[60]T. Shibahara *et al., Inorg. Chem.* **1994,** *33,* 292.
[61]T. Shibahara *et al., Inorg. Chem.* **1995,** *34,* 42.
[62]G. Sakane and T. Shibahara, *Inorg. Chem.* **1993,** *32,* 777.
[63]A. Müller *et al., Inorg. Chem.* **1994,** *33,* 2243.
[64]A. G. Sykes *et al., J. Chem. Soc., Dalton Trans.* **1994,** 1275, 2809; *Inorg. Chim. Acta* **1994,** *225,* 157; *J. Cluster Sci.* **1995,** *6,* 449; *J. Chem. Soc., Dalton Trans.* **1996,** 2623.
[65]B. W. Eichhorn *et al., Inorg. Chem.* **1993,** *32,* 5412.
[66]F. Sécheresse *et al., J. Chem. Soc., Dalton Trans.* **1994,** 1311.
[67]J. A. Ibers *et al., Inorg. Chem.* **1988,** *27,* 1747.

Complexes of NO and N₂

The *nitrosyl* chlorides, $M(NO)_2Cl_2$, are polymeric solids that react with a variety of Lewis bases to give $M(NO)_2Cl_2L_2$ adducts that have *cis*-L_2 for small (e.g., CH_3CN) and *trans*-L_2 for large L (e.g., R_3P). The reaction of Mo_2Cl_{10} with NO, followed by the addition of other ligands gives rise to mononitrosyl complexes of the types $[MoCl_4(NO)L]^-$, $MoCl_3(NO)L_2$, $[MoCl_3(NO)L_2]^-$, and $MoCl(NO)L_4$. Other derivatives include $WH(CO)_2(NO)(PPh_3)_2$, $W(CO)_3NO(PR_3)_2^+$, and several pyrazolylborate species.

Reactions of $M_2(OR)_6$ compounds also give nitrosyl complexes such as (18-C-XII). These are unusual among nitrosyl complexes in having only 14-electron configurations, very short M—N distances, and very low (1560-1640 cm^{-1}) N—O stretching frequencies.

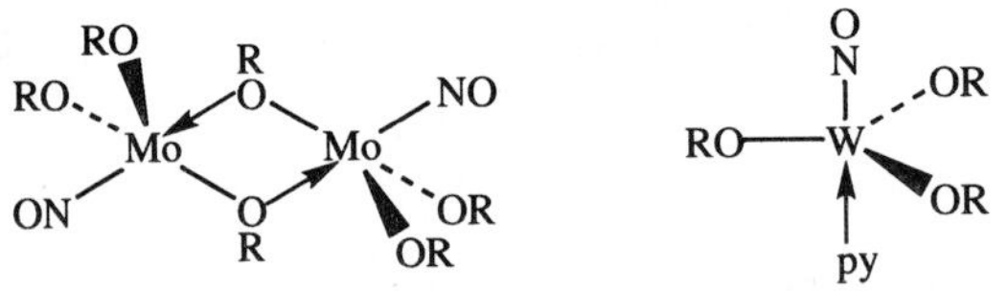

(a)   (b)

(18-C-XII)

It is also possible to make a variety of polyoxomolybdate species containing NO ligands.[68] The NO source is generally $Me_2C{=}NOH$.

Dinitrogen complexes have been studied extensively[69] with a view to understanding natural nitrogen-fixing systems. The bis(dinitrogen) complexes are generally prepared by reactions of higher-valent halo complexes already containing the phosphine ligands with strong reducing agents (e.g., Na/Hg) in the presence of N_2 gas. For example,

$$WCl_4(PMe_2Ph)_2 + 2N_2 \xrightarrow[\text{Na/Hg}]{\text{excess PMe}_2\text{Ph}} W(N_2)_2(PMe_2Ph)_4$$

$$MoCl_5 + 4PR_3 + 2N_2 \xrightarrow[\text{Na/Hg}]{\text{THF}} \textit{trans-}Mo(N_2)_2(PR_3)_4$$

On treatment with anhydrous HCl or HBr in organic solvents the N_2 complexes afford both hydrazine and ammonia in varying yields. The pathways to both N_2H_4 and NH_3 are believed to involve successive protonation of the N_2 ligand, with subtle electronic and/or steric factors controlling the product distribution.

It has been shown that the coordinated N_2 may be converted into various organonitrogen complexes such as organohydrazido,[70] organodiazinido, and diazoal-

[68]P. Gouzerh *et al.*, *J. Chem. Soc., Chem. Commun.* **1993,** 836; *Inorg. Chem.* **1993,** *32,* 5291, 5299; A. Proust *et al.*, *J. Chem. Soc., Dalton Trans.* **1994,** 825.
[69]G. J. Leigh, *Acc. Chem. Res.* **1992,** *25,* 177.
[70]M. Hidai *et al.*, *J. Organomet. Chem.* **1992,** *423,* 39.

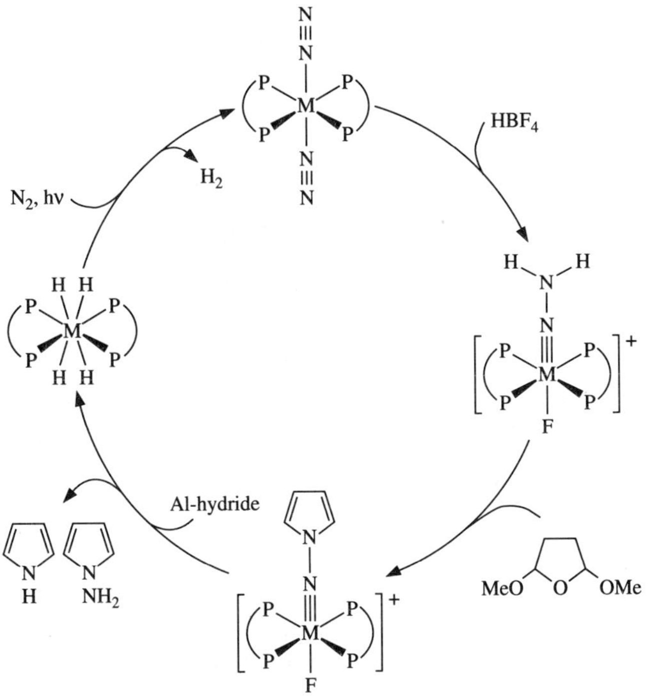

Figure 18-C-13 A cycle whereby N_2 is converted to useful organic products (adapted from ref. 71).

kane complexes. However the subsequent release of the organic moieties has been limited to alkylamines and azines. Efforts to extend this kind of chemistry continue and it has recently been found that the sequence of reactions shown in Fig. 18-C-13 can be carried out to produce pyrrole and *N*-aminopyrrole.[71]

In addition to the bis-dinitrogen complexes some mono-N_2 complexes, such as $Mo(PMe_3)_5(N_2)$, can also be made.[72] Complexes containing the N_2 ligand are frequently useful starting materials[73,74] because the N_2 ligand is easily displaced by others, e.g.,

$$trans\text{-}Mo(dppe)_2(N_2)_2 \xrightarrow{BzS_3Bz} trans\text{-}Mo(dppe)_2(S)_2 + N_2$$

$$Mo(N_2)_2(dppe)_2 + 2C_2H_4 \longrightarrow Mo(C_2H_4)_2(dppe)_2$$

The reaction need not always be so straightforward, however, as in the case where $W(N_2)_2(PMe_2Ph)_4$ reacts with $(Me_3Si)_2S$ to give a tetranuclear cluster, $W_4S_6(SH)_2(PMe_2Ph)_6$,[75] which can be reduced to $W_4S_6(PMe_2Ph)_4$, whose structure

[71]M. Hidai *et al., J. Am. Chem. Soc.* **1995**, *117*, 12181.
[72]E. Carmona *et al., J. Chem. Soc., Dalton Trans.* **1995**, 3801.
[73]G. J. Kubas *et al., Inorg. Chem.* **1994**, *33*, 5219.
[74]F. A. Cotton and G. Schmid, *Inorg. Chem.* **1997**, *36*, 2267.
[75]M. Hidai *et al., J. Chem. Soc., Chem. Commun.* **1995**, 1057.

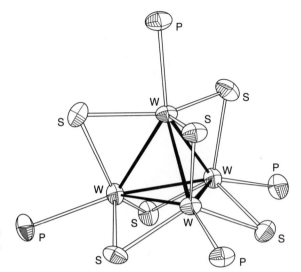

Figure 18-C-14   The $W_4S_6P_4$ core of the $W_4S_6(PMe_2Ph)_4$ molecule (adapted from ref. 75).

is shown in Fig. 18-C-14. The W_4 tetrahedron has simple W—W single bonds (2.634 Å) and is the only such M—M bonded metal tetrahedron known.

Cyano Complexes

The best known and most thoroughly studied cyano complexes are the octacyano ions $M(CN)_8^{3-}$ and $M(CN)_8^{4-}$. Their structures appear to vary with environment. Because of the similar energies of dodecahedral (D_{2d}) and square antiprismatic (D_{4d}) structures, fluxional character is to be expected. In some solid compounds the structures found for the $M(CN)_8^{n-}$ ions are as follows (where M indicates that both the Mo and W compounds have the stated structure):

$$\left.\begin{array}{ll} \text{Na}_3\text{M(CN)}_8 \cdot 4\text{H}_2\text{O} & D_{4d} \\[4pt] [(n\text{-C}_4\text{H}_9)_4\text{N}]_3[\text{Mo(CN)}_8] & D_{2d} \end{array}\right\} \text{M}^{\text{V}} \text{ species}$$

$$\left.\begin{array}{ll} \text{K}_4\text{M(CN)}_8 \cdot 2\text{H}_2\text{O} & D_{2d} \\[4pt] \text{K}_4\text{M(CN)}_8 \cdot 6\text{H}_2\text{O} & D_{4d} \end{array}\right\} \text{M}^{\text{IV}} \text{ species}$$

The D_{4d} structure has recently been found in several more compounds.[76] The surroundings clearly play a decisive role in stabilizing one or the other geometry. Most Raman and ir studies of solutions have been inconclusive, though Raman and ^{99}Mo nmr spectra unequivocally favor the D_{2d} structure for $Mo(CN)_8^{4-}$ in aqueous solution.

The $M(CN)_8^{4-}$ ions in aqueous solution are photochemically converted, through several intermediates, into isolable species long believed to be $[M(CN)_4(OH)_4]^{4-}$ but more recently shown to be 6-coordinate, *trans*-dioxo complexes $[MO_2(CN)_4]^{4-}$.

A cyano complex reported for Mo^{III} is $K_4Mo(CN)_7 \cdot 2H_2O$, which readily oxidizes

[76]W. Meske and D. Babel, *Z. Naturforsch.* **1988,** 43b, 1167.

to $K_4Mo(CN)_8 \cdot 2H_2O$. It has $\mu_{eff} = 1.75$ BM at room temperature. The apparent presence of only one unpaired electron has been attributed to d-orbital splitting in the necessarily low symmetry of either an $[Mo(CN)_7]^{4-}$ or an $[Mo(CN)_7(H_2O)]^{4-}$ ion. The molybdenum(II) ion, $[Mo(CN)_7]^{5-}$, has a pentagonal bipyramidal structure.

Thiocyanate complexes are formed by molybdenum in the III, IV, and V oxidation states, the last being of the oxo type, for example, $[MoO(NCS)_5]^{2-}$. The $[Mo(NCS)_6]^{3-}$ ion has been shown conclusively to have N-bonded thiocyanate ions, and this appears likely to be the case also in all other molybdenum thiocyanato species.

We may also mention that there are numerous *isocyanide complexes*. Isoleptic species are obtained smoothly by the reaction of RNC with $Mo_2(O_2CCH_3)_4$ and isolated as $[Mo(RNC)_7](PF_6)_2$, and in other ways, some of which allow preparation of analogous W^{II} complexes. Substitution of RCN ligands by phosphines, phen, bipy, and so on, lead to many other complexes, particularly the types $[M(RCN)_nL_{7-n}]^{2+}$. The $[MoX_2(CO)_4]_2$ compounds react with RNC to give $Mo(RNC)_5X_2$. The $[Mo(CNR)_6I]^+$ and $[Mo(CNR)_7]^{2+}$ ions can be made by reaction of $Mo(CO)_6$ with RNC and I_2 or by alkylation and reduction of $[Mo(CN)_8]^{4-}$. Both types have capped trigonal prismatic geometry.

Hydrido and η^2-H_2 Complexes

Both elements form many of these. Perhaps the most notable is shown in 18-C-XIII. This is the compound in which G. J. Kubas first recognized that H_2 could serve as a ligand while retaining a significant fraction of the H—H bond. There are now many more H_2 complexes of both Mo and W, and both elements also form classical hydrido complexes, for example, $W(H)_2I(PMe_3)_3(SiMe_3)$, shown systematically in 18-C-XIV and $Mo(H)_2(PMe_3)_5$, which has a similar pentagonal bipyramidal structure in which Me_3Si and I are replaced by PMe_3.

(18-C-XIII)   (18-C-XIV)

Nitrido Complexes

The largest known class are species such as $[MNCl_4]^{n-}$ which contain M^V when $n = 2$ or M^{VI} when $n = 1$. They have very short $M \equiv N$ bonds (*ca.* 1.70 Å). The $[MNCl_4]^-$ complexes are derived from the $MNCl_3$ compounds, which are both known as cyclic tetramers.[77] Similar to these are the $(RO)_3MN$ molecules. There are also ions with bridging nitrido atoms, such as $[Cl_5W-N-WCl_5]^{n-}$ ($n = 1, 2$), $[Cl_5W-N-W-(Cl_4)-N-WCl_5]^{2-}$, and $[Cl_3py_2Mo-N-Mopy_2Cl_3]^-$.[78]

[77]M. R. Close and R. E. McCarley, *Inorg. Chem.* **1994**, *33*, 4198.
[78]E. A. Maatta *et al.*, *J. Chem. Soc., Chem. Commun.* **1994**, 2163.

An entirely new class which has been discovered and investigated in the last decade are those based on MN_4^{6-} tetrahedra. In some the tetrahedra are discrete[79] while in others they are linked by shared N atoms.[80] The tetrahedra are isoelectronic with the MO_4^{2-} ions, but these nitrido compounds are highly sensitive to water. They are made by high-temperature ($> 600°C$) bomb reactions of the metals with alkali or alkaline earth amides.

Finally, for Mo, new compounds of the type $(ArRN)_3MoN$ have been made[81] by using very bulky Ar and R groups, whereby $(ArRN)_3Mo$ is isolable (whereas, as discussed in Section 18-C-10, with the less bulky amides, Me_2N or Et_2N, dinuclear, triply-bonded $M_2(NR_2)_6$ are formed). Direct reactions to split $N\equiv N$ bonds occur, as in the following reactions:

$$(ArRN)_3Mo + N_2O \longrightarrow (ArRN)_3MoN + NO$$

$$2(ArRN)_3Mo + N_2 \longrightarrow 2(ArRN)_3MoN$$

Not only does the bulk of the three NArR groups protect the $(ArRN)_3Mo$ molecule from reacting with other than "skinny" reactants (N_2, N_2O), but it keeps the resulting nitride from oligomerizing or polymerizing, whereas the $(RO)_3MoN$ molecules form either cyclic trimers[82] ($R = (CH_3)_2CF_3C$) or infinite chains[83] ($R = C(CH_3)_3$).

It may be noted here that the same $(ArRN)_3Mo$ compound mentioned above also reacts with P_4 to give $(ArRN)_3MoP$.[84]

Imido (Nitrene) and Hydrazido Complexes

There are many of these and they bear a formal analogy to the M=O oxo species:

$$M{=}O \qquad M{=}N{-}R \qquad M{=}N{-}NR_2$$

The $M{-}N{-}R$ and $M{-}N{-}N$ chains are very nearly linear and the M to N bonds are very short ($1.70{-}1.80$ Å) and thus a contribution from a polarized triple bond structure, $\bar{M}{\equiv}\overset{+}{N}{-}$ has to be considered. Some of the M=RN bonds can be obtained directly from M=O bonds by the reactions[85]

$$M{=}O + PPh_3{=}NR \longrightarrow M{=}NR + Ph_3PO$$

$$M{=}O + RNCO \longrightarrow M{=}NR + CO_2$$

but there are numerous other synthetic methods.[86]

[79]C. Wachsmann and H. Jacobs, Z. anorg. allg. Chem. 1996, 622, 885.
[80]R. Niewa and H. Jacobs, Z. anorg. allg. Chem. 1996, 622, 881.
[81]C. C. Cummins et al., J. Am. Chem. Soc. 1996, 118, 8623.
[82]M. H. Chisholm et al., Angew. Chem. Int. Ed. Engl. 1995, 34, 110.
[83]M. H. Chisholm et al., Inorg. Chem. 1983, 22, 2903.
[84]C. C. Cummins et al., Angew. Chem. Int. Ed. Engl. 1995, 34, 2042.
[85]E. A. Maata et al., J. Am. Chem. Soc. 1992, 114, 345; A. Proust et al., Inorg. Chim. Acta 1994, 224, 81.
[86]E. A. Maata et al., Inorg. Chem. 1994, 33, 6415; D. Carrillo et al., Inorg. Chem. 1994, 33, 4937; V. C. Gibson et al., J. Chem. Soc., Chem. Commun. 1994, 2247.

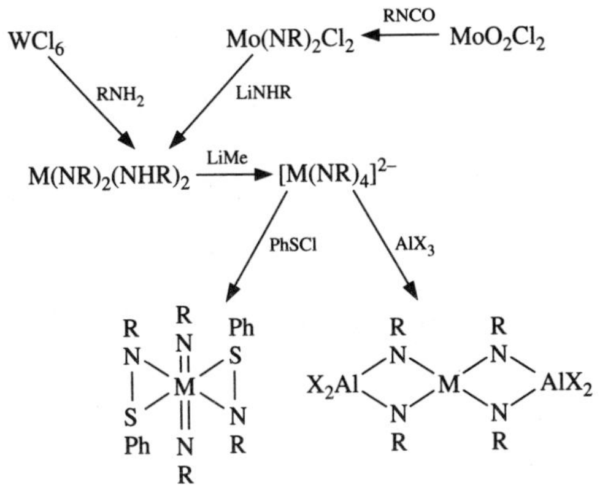

Figure 18-C-15 Some chemistry of the $[M(NR)_4]^{2-}$ ions, M = Mo or W, R = *t*-butyl.

The $[M(NBu^t)_4]^{2-}$ ions are unusually interesting because they are analogues of the MO_4^{2-} ions. The chemistry leading to and then derived from them[87] is summarized in Fig. 18-C-15.

The M(PMe₃)₆ Molecules

These rather surprising molecules are known for both Mo and W.[88] Both have a strong tendency to undergo the following reversible transformation:

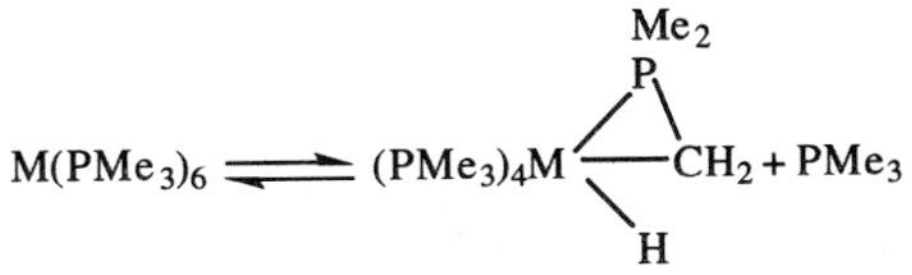

For Mo the equilibrium lies to the left while for W it lies to the right. Both of these compounds are highly reactive and can be used to prepare many other complexes in which four or five PMe₃ are retained but new ligands are introduced, as in the following reactions:[89]

$$Mo(PMe_3)_6 + 2E \longrightarrow Mo(PMe_3)_4(E)_2 + 2PMe_3$$

(E = Se, Te)

Peroxo Complexes of Mⱽᴵ

There are many of these, and one important structural motif is that in which two peroxo groups and another O atom form a distorted pentagon about the M atom, to which a doubly bonded oxygen atom and another ligand atom are attached above and below the median $M(O_2)_2O$ plane.[90] Certain of these compounds are

[87]G. Wilkinson *et al.*, *Polyhedron* **1989**, *8*, 2947; *J. Chem. Soc., Dalton Trans.* **1990**, 2753; *J. Chem. Soc., Dalton Trans.* **1995**, 1059.
[88]G. Parkin *et al.*, *J. Am. Chem. Soc.* **1992**, *114*, 4611; *Inorg. Chem.* **1995**, *34*, 6341.
[89]V. J. Murphy and G. Parkin, *J. Am. Chem. Soc.* **1995**, *117*, 3522.
[90]W. P. Griffith *et al.*, *J. Chem. Soc., Dalton Trans.* **1995**, *1833*, 3131.

stoichiometric reagents for epoxidation of olefins while others are effective catalysts for oxidation of organic compounds by H_2O_2. Two examples are shown as 18-C-XV and 18-C-XVI. The simple species $[M(O_2)_4]^{2-}$ and $[MO(O_2)_3]^{2-}$ are believed to exist in solution.[91]

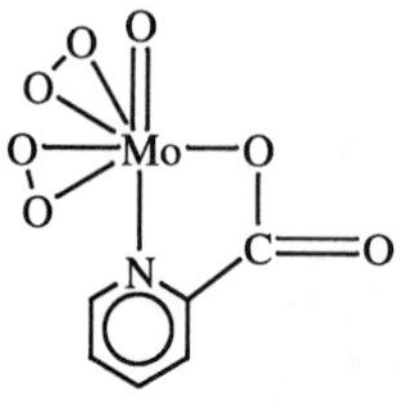

(18-C-XV) 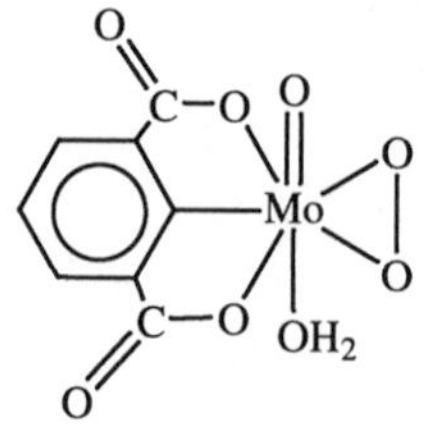   (18-C-XVI)

Face-Sharing and Edge-Sharing Bioctahedral Complexes, FSBOs and ESBOs

Octahedral complexes MX_3L_3, can condense to give ESBOs[92] and then FSBOs by loss of ligands, L, as shown in eq. 18-C-10. There are nine chemically reasonable isomers for the $M_2X_6L_4$ ESBO molecules (the commonest type) many of which have been observed[93] but some not. Not all such systems participate in equilibria of the type shown in eq. 18-C-10 but when they do the equilibria can be monitored by nmr, especially for Mo,[94,95] and factors controlling such equilibria have been discussed.[96]

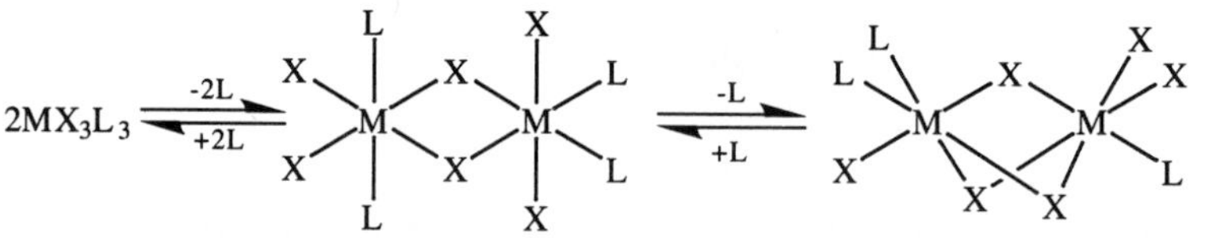

(18-C-10)

The FSBO structure occurs not only for neutral molecules such as $Mo_2Cl_6(PEt_3)_3$[97] but for anionic complexes of the types $[MX_7L_2]^-$ and $[M_2X_8L]^-$, where L is a phosphine.[98] The $[M_2X_9]^{3-}$ ions have been noted in Section 18-C-6. In both ESBO and FSBO structures a greater variety of ligands may also be present, as in the ESBO 18-C-XVII,[99] and in the FSBO $[Cl_3Mo(\mu\text{-}Cl)_2(\mu\text{-}H)MoCl_3]^{3-}$. It is a curious fact that while many FSBO $W_2X_6L_3$ and ESBO $W_2X_6L_4$ compounds are known,[100] it is not possible to convert $WCl_3(PMe_2Ph)_3$, (neither the *mer* nor the *fac*

[91] W. P. Griffith *et al., J. Chem. Soc., Dalton Trans.* **1989,** 1203.
[92] H.-B. Kraatz and P. M. Boorman, *Coord. Chem. Rev.* **1995,** *143,* 35.
[93] F. A. Cotton *et al., Inorg. Chem.* **1993,** *32,* 687.
[94] R. Poli *et al., J. Am. Chem. Soc.* **1990,** *112,* 2446; *Polyhedron* **1991,** *10,* 1667.
[95] F. A. Cotton and J. Su, *Inorg. Chim. Acta* **1996,** *251,* 101.
[96] R. Poli and R. C. Torralba, *Inorg. Chim. Acta* **1993,** *212,* 123.
[97] F. A. Cotton *et al., Inorg. Chim. Acta* **1990,** *173,* 131.
[98] K. Vidyasagar, *Inorg. Chim. Acta* **1995,** *229,* 473.
[99] K. A. Hall and J. M. Mayer, *Inorg. Chem.* **1995,** *34,* 1145.
[100] M. H. Chisholm *et al., Inorg. Chem.* **1993,** *32,* 2322.

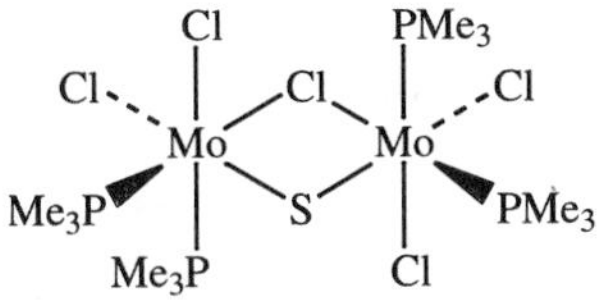

(18-C-XVII)

isomer) thermally to the FSBO or ESBO type compound,[101] since decomposition occurs more easily.

The electronic structures and spectra of the Mo_2 FSBOs are problematic,[102,103] while for tungsten there is no doubt that strong $W\equiv W$ triple bonds exist.

18-C-10 Multiple M—M Bonds[104]

Molybdenum is the most prolific former of such bonds among all metals. Tungsten forms many analogous compounds but they are sometimes more reactive, and/or less easily prepared. Table 18-C-5 lists representative examples of quadruple and triple bonds. It will be noted that the Mo—Mo distances are 0.10 ± 0.04 Å shorter than corresponding W—W distances. This reflects the fact that at the distances involved the much denser cores of tungsten atoms create a significantly greater core-core repulsion than occurs for a pair of molybdenum atoms, thus lengthening the W—W bonds. This, in turn, leads to a weakening of the δ component of the W—W quadruple bond. This is the principal reason why in general quadruply bonded W_2 species are less easy to prepare and more reactive than their molybdenum analogues. However, recent work has afforded better access to many of the quadruply bonded ditungsten compounds.[105] The effect of the greater W—W distance for the triple bonds is negligible and there is great similarity between the two sets of triply bonded compounds. The general discussion of M—M multiple bonds in Section 16-6 should be consulted in this connection.

Quadruple Bonds

Molybdenum forms hundreds of compounds containing Mo—Mo quadruple bonds. The most generally useful entries to this chemistry are shown in the following equations:

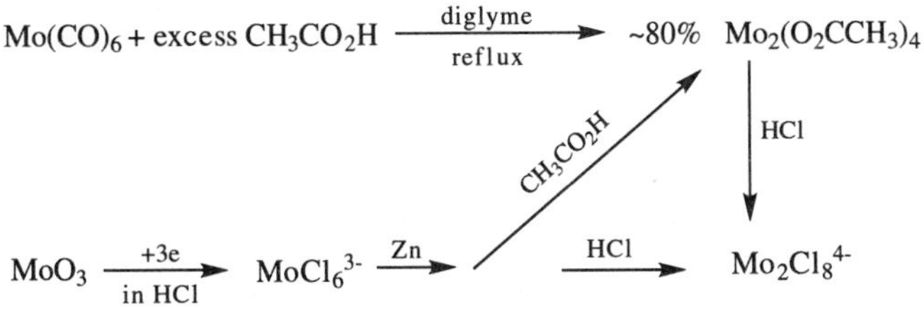

[101]K. G. Caulton et al., Inorg. Chem. **1993**, 32, 4573.
[102]R. Stranger et al., Inorg. Chem. **1993**, 32, 4555; Inorg. Chem. **1994**, 33, 3976.
[103]F. A. Cotton and X. Feng, Intl. J. Quantum Chem., **1996**, 58, 671.
[104]F. A. Cotton and R. A. Walton, Multiple Bonds between Metal Atoms, 2nd Ed., Oxford University Press, Oxford, 1993, especially Chap. 3.
[105]F. A. Cotton et al., Inorg. Chem. **1993**, 32, 681.

Table 18-C-5 Representative Triple and Quadruple Mo—Mo and W—W Bonds and Their Lengths (Å)

	Quadruple Bonds		
$Mo_2(O_2CCH_3)_4$	2.09	$W_2(O_2CC_2H_5)_4$	2.19
$[Mo_2Cl_8]^{4-}$	2.14	$[W_2Cl_8]^{4-}$	2.25
$Mo_2Cl_4(PMe_3)_4$	2.13	$W_2Cl_4(PMe_3)_4$	2.26
$Mo_2(mhp)_4{}^a$	2.07	$W_2(mhp)_4$	2.16
	Triple Bonds		
$Mo_2(CH_2SiMe_3)_6$	2.17	$W_2(CH_2SiMe_3)_6$	2.26
$Mo_2(NMe_2)_6$	2.21	$W_2(NMe_2)_6$	2.29
$Mo_2Cl_2(NMe_2)_4$	2.20	$W_2Cl_2(NMe_2)_4$	2.29
$Mo_2(OCH_2CMe_3)_6$	2.22		
$[Mo_2(HPO_4)_4]^{2-}$	2.23		
		$W_2(O_2CNMe_2)_6$	2.28

^{a}mph$^-$ = anion of 2-hydrido-6-methylpyridine.

Yellow $Mo_2(O_2CCH_3)_4$ (18-C-XVIII) is thermally stable but very slowly decomposes in air. The purple $[Mo_2Cl_8]^{4-}$ ion (18-C-XIX) can be isolated in a variety of air-stable salts. Some important reactions of these key complexes are shown in Fig. 18-C-16.

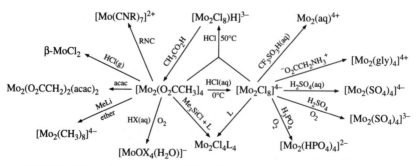

Figure 18-C-16 Some important reactions of species with Mo—Mo quadruple bonds.

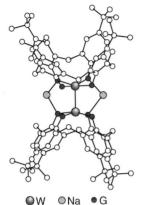

Figure 18-C-17   A calix[4]arene complex of the $(W\equiv W)^{6+}$ unit (see C. Floriani *et al., Angew. Chem. Int. Ed. Engl.* **1997,** *36,* 753).

⬤W   ◯Na   ●G

Among the many reactions of $Mo_2(O_2CCH_3)_4$ and $[Mo_2Cl_8]^{4-}$ are many simple (i.e., nonredox) ligand-exchange reactions in which the quadruply bonded Mo_2^{4+} unit remains intact. The new ligands may be simple or exotic, e.g., calixarines,[106,107] as shown in Fig. 18-C-17. There are also reactions in which the products are mononuclear without oxidation, such as $[Mo(CNR)_7]^{2+}$ and $[MoOX_4(H_2O)]^-$, where oxidation has occurred. The reaction of $[Mo_2Cl_8]^{4-}$ with sulfuric acid in the presence of O_2 gives $[Mo_2(SO_4)_4]^{3-}$, in which the loss of one electron reduces the Mo—Mo bond order to 3.5, and reaction with phosphoric acid in O_2 gives $[Mo_2(HPO_4)_4]^{2-}$, which contains only a triple bond. The decrease in bond order from 4.0 to 3.5 to 3.0 in the series $[Mo_2(SO_4)_4]^{4-}$, $[Mo_2(SO_4)_4]^{3-}$, $[Mo_2(HPO_4)_4]^{2-}$ is accompanied by a steady increase in bond lengths (2.11, 2.16, and 2.23 Å, respectively).

The $[Mo_2(CH_3CN)_{10}]^{4+}$ ion can be prepared[108] and provides a very useful synthon.[109]

The reaction of $Mo_2(O_2CCH_3)_4$ with gaseous HCl, HBr, or HI at ~300°C gives dihalides, β-MoX_2, which are different from the long known "$MoCl_2$," which (see Section 18-C-10) is a Mo_6 cluster compound. Though these powders have not been characterized by crystallography, their chemistry indicates that the quadruply bonded Mo_2 unit is present. The reaction of $Mo_2(O_2CCH_3)_4$ or $Mo_2Cl_8^{4-}$ with aqueous HX at 50°C gives the hydrido bridged $[Mo_2X_9H]^{3-}$ ions, (18-C-XX), which retain a strong Mo—Mo bond, and can be reconverted to $Mo_2(O_2CCH_3)_4$ by CH_3CO_2H. These reactions are examples of oxidative-addition and reductive-elimination reactions involving dinuclear rather than the usual mononuclear complexes.

$$\left[\begin{array}{c} \text{Cl} \\ \text{Cl} \diagdown \quad \diagup \text{H} \diagdown \quad \text{Cl} \\ \quad \text{Mo} \text{——} \text{Mo} \quad \\ \text{Cl} \diagup \quad \diagdown \quad \diagup \diagdown \text{Cl} \\ \text{Cl} \quad \text{Cl} \end{array} \right]^{3-}$$

(18-C-XX)

[106] J. A. Acho and S. J. Lippard, *Inorg. Chim. Acta* **1995,** *229,* 5.
[107] C. Floriani *et al., Angew. Chem. Int. Ed. Engl.* **1997,** *36,* 753.
[108] F. A. Cotton and K. J. Wiesinger, *Inorg. Synth.* **1992,** *29,* 134.
[109] F. A. Cotton and F. E. Kühn, *J. Am. Chem. Soc.* **1996,** *118,* 5826.

Treatment of M—M quadruply bonded (and also triply bonded) compounds with strongly π-accepting ligands (e.g., CO, RNC, and NO) causes fission of the M—M bond and affords mononuclear products. Such reactions may often constitute a good way (sometimes the best way, and even the only way) to prepare such products. Examples are provided by the following reactions:

$$Mo_2(O_2CCH_3)_4 + 14CH_3NC \longrightarrow 2[Mo(CNCH_3)_7](CH_3CO_2)_2$$

$$Mo_2(OR)_6 + 2NO + 2py \longrightarrow 2(RO)_3Mo(NO)py$$

$$W_2Cl_4(PEt_3)_4 \xrightarrow[\text{-Cl}^-]{\text{CO}} W(CO)_3(PEt_3)_2Cl_2 + W(CO)_4(PEt_3)_2 + W(CO)_5(PEt_3)$$

An interesting reaction of the quadruple bonds is self-addition to produce cyclobutadiyne-like products,

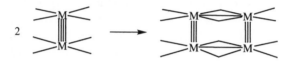

It is possible to prepare compounds of the type $Mo_2X_4(LL)_2$, where LL represents a diphosphine such as dppm or dppe, by several routes, and depending on preparative details, either or both of two isomers (18-C-XXI) and (18-C-XXII), α and β, respectively, may be obtained. In the β(bridged) isomers with dmpe or dppe, the steric requirements of the bridging diphosphines result in the rotational conformations about the Mo—Mo bonds being twisted to varying degrees away from the eclipsed conformation that prevails in species such as $Mo_2(O_2CR)_4$, $[Mo_2Cl_8]^{4-}$, and $Mo_2Cl_4(PR_3)_4$. Thus the purely schematic representation shown in (18-C-XXII) should not be taken literally. Figure 18-C-18 shows a real example of the twisted conformation of such a molecule. By studying the dependence of the Mo—Mo distance and the position of the $\delta \to \delta^*$ absorption band in the visible spectrum on the magnitude of the twist angle, considerable insight into the properties of the quadruple bond has been obtained. It is also notable that such a twisted molecule is chiral and displays pronounced ORD and CD spectra. It is indicative of how well we understand the nature of quadruple bonds (or at least the δ components thereof) that we can predict correctly from first principles how the sign of the $\delta \to \delta^*$ CD band correlates with the absolute rotational conformation of the molecule. The compound shown in Fig. 18-C-18 is one in which only one helical isomer is locked in by use of a chiral ligand.

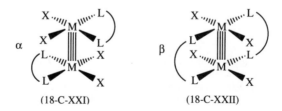

α (18-C-XXI)   β (18-C-XXII)

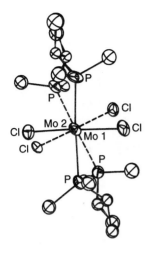

Figure 18-C-18 A view down the Mo(1)—Mo(2) axis of the β-Mo$_2$Cl$_4$[Ph$_2$PCH(CH$_3$)CH(CH$_3$)PPh$_2$]$_2$ molecule, showing the internal rotation. The ligand has the S configuration at each asymmetric carbon atom thus giving rise to only one of the two helical conformations about the Mo—Mo bond.

A final, interesting point concerning these Mo$_2$Cl$_4$(LL)$_2$ molecules is that in general the β isomers appear to be more stable thermodynamically than the α isomers and on heating the α isomers, either in the solid state or in solutions, they are transformed to the β isomers. It is believed that these isomerizations take place by a simple flip of the Mo$_2$ unit within the cage formed by the ligands.

There are tungsten analogues to most of the quadruply bonded compounds of molybdenum, but because of the weakness of the W—W δ bond they are more easily oxidized, and hence often more difficult to prepare. They generally cannot be exposed to protonic acids because protonation of the δ bond occurs:

$$\text{W} \equiv\!\equiv\!\equiv \text{W} \quad + 2HX \longrightarrow \quad \text{W} = \!=\!= \text{W}$$

There is now a very elegant method to prepare certain MoW^{4+} compounds, Eq. 18-C-11, discovered by Luck and Morris,[110] and since heavily utilized,[111] with some modifications. In the case of octaethylporphyrin (OEP) compounds, it is even possible to make heteronuclear species that combine Mo with metal atoms in other groups, e.g., (OEP)MoRu(OEP) and (OEP)MoOs(OEP).[112]

$$\text{(structure)} + WCl_4(PPh_3)_2 \longrightarrow \text{Mo} \equiv\!\equiv\!\equiv \text{W} + 2PPh_3 \qquad \text{(18-C-12)}$$

[110]R. L. Luck and R. H. Morris, *J. Am. Chem. Soc.* **1984**, *106*, 7978.
[111]F. A. Cotton and C. A. James, *Inorg. Chem.* **1992**, *31*, 5298.
[112]J. P. Collman *et al.*, *J. Am. Chem. Soc.* **1994**, *116*, 9761.

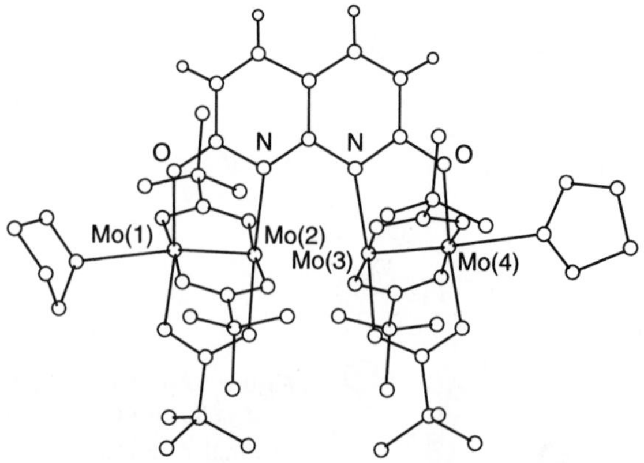

Figure 18-C-19 Two conjoined units each containing a quadruple M—M bond.

One-dimensional polymers can be made incorporating the quadruple bonds between molybdenum and tungsten atoms by reactions involving either the $M_2(O_2CR)_4$ and a dicarboxylic $HOOC-R-COOH$ or by the metathetic reaction between $M_2(O_2CR)_2(NCMe)_4^{2+}(PF_6^-)_2$ and the sodium salt of a bridging ligand such as the anion derived from the deprotonation of 2,7-dihydroxynaphthyridine. By the judicious choice of the bridge, the M—M axis can be aligned parallel or perpendicular to the polymer axis. Discrete dinuclear compounds such as $[Mo_2(O_2CBu^t)_3]_2$-$(\mu\text{-}O_2N_2C_8H_4)$ shown in Fig. 18-C-19 provide models for the subunits in such polymers.[113]

The n-alkanoates of molybdenum $Mo_2(O_2C(CH_2)_nCH_3)_4$ where $n = 3$ to 9 have been shown to exhibit hexagonal columnar liquid crystalline phases in the temperature range 100–150°C. Varying the substituents in the alkyl chain can alter the temperature range of the meso phase from room temperature to 200°C. The M—M axis was shown to align perpendicular to an applied magnetic field.[114]

Triple Bonds

These are formed mainly between metal atoms in d^3-d^3 dinuclear complexes for which perhaps the most important group comprises the ethane-like $X_3M\equiv MX_3$ (D_{3d}) compounds, where X = alkyl, amide, alkoxide, thiolate, selenate, and mixtures thereof.[115] The synthetic entry into these compounds involves principally the reactions of $M_2(NMe_2)_6$, as shown in Fig. 18-C-20, and these in turn are made by metathetic reactions involving either $MoCl_3$ or WCl_4 and $LiNMe_2$.

There are many other geometries accessible to d^3-d^3 dinuclear complexes involving coordination numbers of metal atoms ranging from 3 to 6 and their interconversions have been extensively studied by Chisholm and co-workers.[116]

Within this class of compounds the reactivity of the $M_2(OR)_6$ compounds is particularly noteworthy. These have been shown to be templates for organometallic chemistry by the addition of unsaturated molecules such as CO, alkynes, ethene, allene, 1,3-dienes, α,β-unsaturated ketones, etc.[117] Some reactivity unique to di-

[113] M. H. Chisholm *et al.*, *J. Am. Chem. Soc.* **1991**, *113*, 8709.

[114] M. H. Chisholm *et al.*, *J. Am. Chem. Soc.* **1994**, *116*, 4551.

[115] M. H. Chisholm, *Acc. Chem. Res.* **1990**, *23*, 419.

[116] M. H. Chisholm, *J. Organomet. Chem.* **1990**, *400*, 235; *Pure Appl. Chem.* **1990**, *63*, 665.

[117] M. H. Chisholm, *J. Chem. Soc., Dalton Trans.* **1996**, 1781.

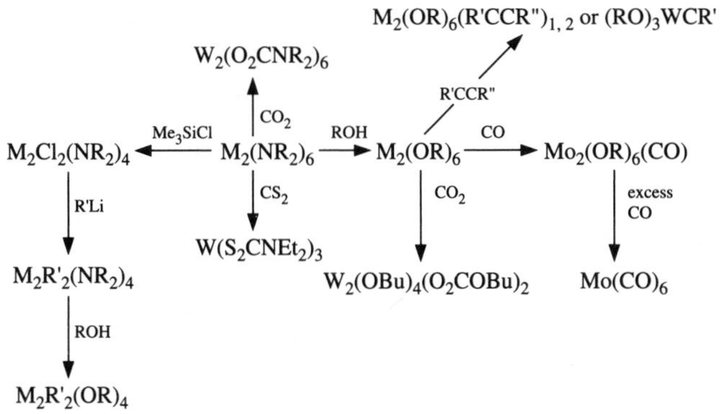

Figure 18-C-20 Some reactions of Mo and W compounds with metal-to-metal triple bonds.

nuclear chemistry has been observed, including unusual catalysis such as the selective hydrogenation of 1,3-dienes to 1-enes.

A combined carboxylation and thermolysis reaction, shown below, leads to smooth conversion of a triple bond to a quadruple bond:

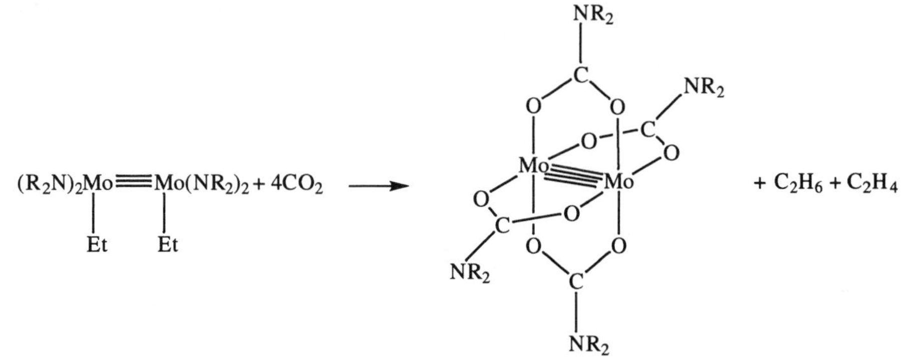

Particularly important advances have been made recently in the chemistry of $M_2(OR)_6$ compounds. Many examples of simple crosswise addition of a $R'C\equiv CR''$ molecule to $M_2(OR)_6$, in the presence of pyridine, have been observed. The products differ in the details of structure and composition depending on the nature of R, R', and R'', and on whether M is Mo or W, but in each case there is a quasitetrahedral M_2C_2 cluster in which an $M=M$ bond (2.55 to 2.66 Å in length) is present. These simple adducts will react further with alkynes to give products in which two or more alkyne molecules have been linked, as in (18-C-XXIII). Further reaction affords alkyne cyclotrimerization products (substituted benzenes).

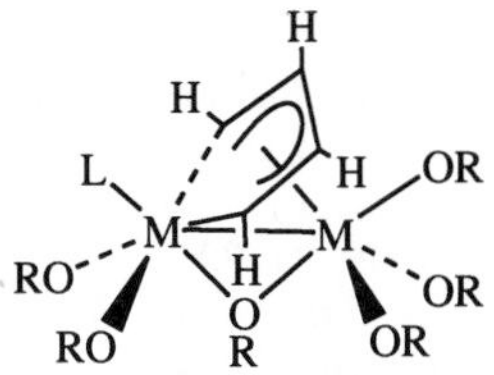

(18-C-XXIII)

The $M_2(OR)_6$ compounds are also templates for cluster synthesis and the reversible coupling of two $M\equiv M$ bonds to give a metallacyclobutadiene, where $M = W$, $R = Pr^i$, is a particularly noteworthy reaction.

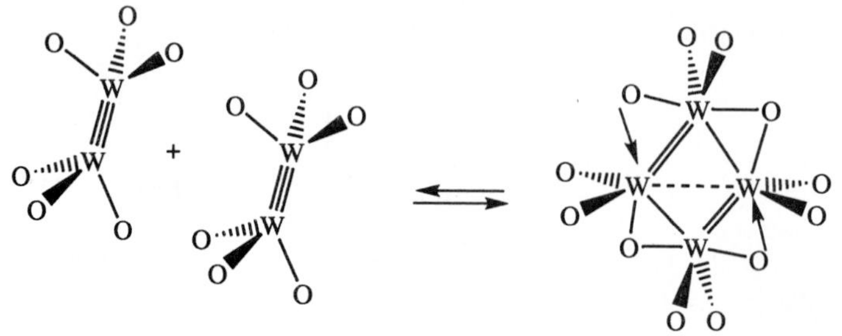

These compounds have also provided an entry point into cluster hydrido alkoxides of general formula $H_xM_y(OR)_z$.[118]

Finally, of note are some multiply bonded compounds formed between metal atoms in different oxidation states. Thus, while $Mo_2(OPr^i)_4(PMe_3)_4$ has a d^4-d^4 quadruple bond of the type normally seen for $M_2X_4L_4$ compounds of molybdenum, the compound $Mo_2(OPr^i)_4(dmpe)_2$ has a staggered geometry wherein one Mo atom is bonded to four OPr^i ligands and the other to two η^2-dmpe ligands. This type of mixed valence $M-M$ bonding is quite common when both hard and soft ligands are present at the dinuclear center.[119]

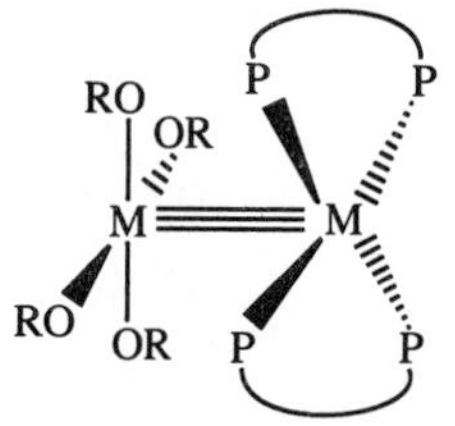

M—M Double Bonds

These are relatively rare and are always accompanied by bridging ligands. Their reactions are less developed compared to those of triple and quadruple bonds, but a recent report on the reactions of $W_2(OCH_2Bu^t)_8$ suggests that this could become a rich area of chemistry. Some of the reactions are shown in Fig. 18-C-21.[120]

18-C-11   Metal Atom Cluster Compounds

Molybdenum and tungsten in their lower oxidation states have a profound tendency to form $M-M$ bonds, and therefore cluster compounds, as well as the dinuclear species just discussed in Section 18-C-10. One class of these clusters has already been presented in Sections 18-C-7 and 18-C-8, namely, the trinuclear species such as the $M_3O_4^{4+}$ and $M_3S_4^{4+}$ cores and the cuboidal $M_4S_4^{n+}$ ones. This section will now deal with a number of other clusters of these elements.

[118]M. H. Chisholm, *Chem. Soc. Rev.* **1995,** *24,* 79.
[119]M. H. Chisholm *et al., Angew. Chem. Int. Ed. Engl.* **1995,** *34,* 891.
[120]M. H. Chisholm *et al., Angew. Chem. Int. Ed. Engl.* **1997,** *36,* 52.

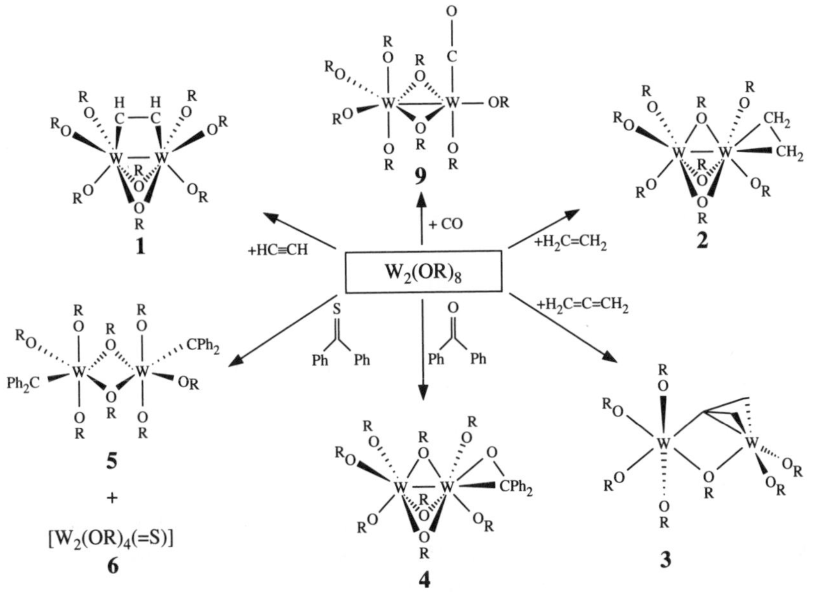

Figure 18-C-21 Some reactions of the doubly-bonded $W_2(OR)_8$ type molecule (taken from Ref. 120).

The existence of Mo^{II} and W^{II} halides has been known for more than 130 years, but only in the last 40 years have their structures and properties been fully appreciated. The key structural unit is an octahedron of metal atoms surrounded by a cubic array of halogen atoms, as shown in Fig. 18-C-22. In addition, each metal atom can coordinate another ligand along a four-fold axis of the octahedron and, so far as is known, they always do so. In the "dihalides" themselves, the M_6X_8 units are linked together by four additional X atoms that are shared between two units and there are also two unshared X atoms per cluster, thus satisfying the net composition $M_6X_{12} \equiv MX_2$. Numerous compounds can be isolated in which there is a complete set of six external X atoms, thereby giving $[M_6X_{14}]^{2-}$ ions, which form nicely crystalline salts with a variety of large cations. Examples of structurally characterized compounds are the $[(C_4H_9)_4N]_2[W_6X_{14}]$ (X = Cl, Br, and I) series. Numerous other related compounds of both Mo and W are well characterized structurally. The outer six ligands are kinetically labile and it is possible to obtain a wide variety of complexes such as $[Mo_6Cl_8(OR)_6]^{2-}$ [121] and others of the general formula $[(Mo_6Cl_8)Cl_nL_{6-n}]^{(4-n)+}$. In aqueous solution the $[M_6X_8]^{4+}$ units are unstable toward strongly nucleophilic groups such as OH^-, SH^-, and CN^-.

Replacement reactions for the μ_3-X atoms forming the X_8 cube are very slow, but mixed species can be obtained, examples being $Mo_6S_6X_2$ (X = Br and I). The M_6S_8 and M_6Se_8 type clusters will be discussed below under the heading of *Chevrel Phases*.

The usual synthetic routes to the halo type cluster compounds begin with preparation of the anhydrous halides. The preparation of Mo_6Cl_{12}, a yellow, nonvola-

[121]N. Perchenek and A. Simon, *Z. anorg. allg. Chem.* **1993**, *619*, 98, 103.

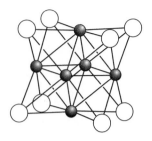

Figure 18-C-22   The key structural unit $M_6X_8^{4+}$ found in metal atom cluster compounds of Mo^{II}.

tile solid is shown in Fig. 18-C-7. The tungsten halides W_6X_{12} (X = Cl or Br) are obtained on disproportionation of WX_4 at 450 to 500°C or by reduction of WCl_5 or WBr_5 with aluminum or WCl_6 with W in an appropriate temperature gradient. The compound W_6I_{12} is obtained by fusing W_6Cl_{12} with a 10-fold excess of a KI-LiI mixture at 540°C.

There is little evidence that oxidation (or reduction) is possible for the $(Mo_6X_8)^{4+}$ species, but W_6Cl_{12} and W_6Br_{12} are oxidized by free halogen at elevated temperatures. The reaction of Cl_2 with W_6Cl_{12} at 100°C results in an intriguing structural change. The product is of stoichiometry WCl_3 and has been shown to contain the $(W_6Cl_{12})^{6+}$ unit, isostructural with the $(M_6X_{12})^{n+}$ units found characteristically in the cluster compounds of Nb and Ta (Section 18-B-8); the complete formulation of WCl_3 is $(W_6Cl_{12})Cl_6$. In the case of W_6Br_{12} the products are W_6Br_{14}, W_6Br_{16}, and W_6Br_{18} if the temperature is kept below 150°C (above which WBr_6 is obtained). In all these there has been a two-electron oxidation of the W_6Br_8 group. The compound W_6Br_{14} may be formulated as $(W_6Br_8)Br_6$; the others contain bridging Br_4^{2-} units and are formulated as $(W_6Br_8)Br_4(Br_4)_{2/2}$ and $(W_6Br_8)Br_2(Br_4)_{4/2}$. A clean one-electron oxidation of $[W_6Br_{14}]^{2-}$ to $[W_6Br_{14}]^-$ can be achieved by use of $NOPF_6$. In aqueous solution the Mo species are not particularly redox active but the W compounds are fairly good reducing agents.

The photochemical and photophysical properties of the $[M_6X_{14}]^{2-}$ ions have been extensively studied because they provide a new class of photoreceptors for light-induced chemical reactions that are chemically very stable and undergo facile electron-transfer reactions in their ground and excited states.

Other Cluster Species

There is, in addition to the M_6X_8 based clusters, a plethora of other cluster compounds of Mo and W in which the number of metal atoms ranges from 3 to 5. Space does not permit the discussion of all of these and we will specifically mention a few such as the group of tungsten iodide clusters recently reported.[122] The reactions of $W(CO)_6$ with I_2, which are a function of temperature, together with subsequent treatment of the products, leads to the following set of cluster species: $[W_3I_9]^{1-}$, $[W_4I_{11}]^{1-}$, $[W_5I_{13}]^{1-}$, and $[W_6Cl_{14}]^{2-}$. In all of these there are W—W distances (2.48–2.82 Å) that indicate direct W—W bonding.

A class of Mo cluster compounds that have no known W analogues is repre-

[122] R. H. Holm *et al.*, *J. Am. Chem. Soc.* **1995**, *117*, 8139.

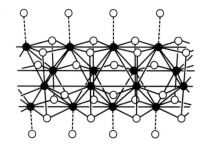

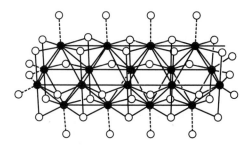

Figure 18-C-23 The infinite chains in the $NaMo_4O_6$ and the $Mo_{18}O_{28}$ oligomers (adapted from ref. 124).

sented by compounds of general formula $M_{n-x}Mo_{4n+2}O_{6n+4}$, in which there are infinite oligomers of the same type.[123] The infinite chain and the oligomer with $n = 4$ are shown in Fig. 18-C-23. The Mo—Mo distances are around 2.8 Å, indicating metal—metal bonding with bond orders between 0.5 and 1.0. The structure/bonding relationships are quite complex.[124]

The existence of *trans* edge-sharing chains seems to occur only when the Mo oxidation state is < 3. With higher oxidation states (+3 to +5) other structures are found, such as the bicapped octahedron, Mo_8, in $LaMo_{7.70}O_{14}$[125] and the edge-bridged octahedral W_6O_{12} units in $Sn_{10}W_{16}O_{44}$.[126]

Chevrel Phases

These are ternary molybdenum chalcogenides, named for R. Chevrel, who with M. Sergent was the first (1971) to recognize and characterize them structurally. Their general formula is $M_xMo_6X_8$, where M represents one or several of the following: a vacancy, a rare earth element, or a transition or main group metallic element (e.g., Pb, Sn, Cu, Co, Fe), and X may be S, Se, or Te. The idealized structure of $PbMo_6S_8$ is shown in Fig. 18-C-24. It is easily seen that one component of the structure, packed together with the Pb atoms in a CsCl pattern, is an Mo_6S_8 unit that has the type of structure shown in Fig. 18-C-22. These Mo_6S_8 units are linked together because some of the S atoms are bonded to the external coordination sites of their neighbors.

[123]R. E. McCarley *et al.*, *Inorg. Chem.* **1995**, *34*, 6130; **1994**, *33*, 1259.
[124]R. A. Wheeler and R. Hoffmann, *J. Am. Chem. Soc.* **1988**, *110*, 7315.
[125]K. Leligny *et al.*, *J. Solid State Chem.* **1990**, *87*, 35.
[126]S. J. Hibble and S. A. Grellis, *J. Chem. Soc., Dalton Trans.* **1995**, 1947.

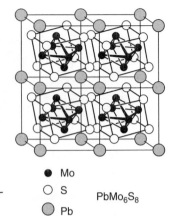

● Mo
○ S
◉ Pb

$PbMo_6S_8$

Figure 18-C-24 The idealized structure of a representative Chevrel compound.

It is curious that despite efforts to do so, no tungsten analogues have yet been made. Recently, both $W_6S_8L_6$ and $W_6Se_8L_6$ compounds[127] have been reported so that perhaps Chevrel analogues will soon emerge.

The binary phase (Mo_6S_8) contains quite distorted Mo_6 octahedra, elongated on a three-fold axis to become trigonal antiprisms. This can be attributed to the fact that there are only 20 cluster electrons. This number is insufficient to form a full set of Mo—Mo bonds, that is, to populate all the 12 bonding orbitals with the result that there is a distorted octahedron with somewhat longer Mo—Mo bonds compared to the 24-electron clusters. In the Chevrel phases the additional metal atoms add electrons to the Mo_6 core and it becomes more regular in structure with shorter Mo—Mo bonds.

The Chevrel phases have aroused great interest because they display remarkable superconductivity, as well as some other notable physical properties. Their superconductivity is exceptional in its persistence in the presence of magnetic fields, a property that is obviously important if superconductivity is to be exploited in making powerful magnets. In recent years a great many chemical variations on the basic MMo_6X_8 composition have been devised.

18-C-12 Organometallic and Carbonyl Chemistry

The Metal Carbonyls

These metals form only one type of binary carbonyl, $M(CO)_6$, in which there are molecules of rigorous O_h symmetry.[128] These molecules undergo a variety of reactions entailing partial or total replacement of CO ligands. Fig. 18-C-25 shows some of these reactions for $Mo(CO)_6$; the chemistry of $W(CO)_6$ is very similar. As can be seen, reactions in which CO groups are directly replaced by organic moieties provide routes into much of the organometallic chemistry of Mo and W.

[127]X. Zhang and R. E. McCarley, *Inorg. Chem.* **1995**, *34*, 2678, 6124; T. Saito *et al., Inorg. Chem.* **1989**, *28*, 3588.
[128]F. Heinemann *et al., Z. Krist.* **1992**, *198*, 123.

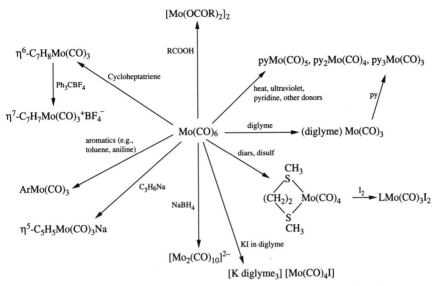

Figure 18-C-25 Some reactions of molybdenum hexacarbonyl.

There are carbonylate anions, $M(CO)_5^{2-}$, and carbonyl hydride anions, $M(CO)_5H^-$, $[M(CO)_5]_2(\mu\text{-}H)^-$, as well as carbonyl halide anions, $M(CO)_5X^-$ and similar species[129] (e.g., $M(CO)_5(NO_2)^-$). An attempt to isolate $W(CO)_5F^-$ failed because it readily loses CO to give $[(OC)_3W(\mu\text{-}F)_3W(CO)_3]^{3-}$.[130]

Only tungsten has been shown to form the remarkable dianion 18-C-XXIV, in which all $W(CO)_4$ fragments are eclipsed and the $W-W$ distances are 2.79 Å.

$$\left[\begin{array}{ccc} \overset{\displaystyle O}{\underset{\displaystyle |}{\overset{\displaystyle C}{}}} & \overset{\displaystyle O}{\underset{\displaystyle |}{\overset{\displaystyle C}{}}} & \overset{\displaystyle O}{\underset{\displaystyle |}{\overset{\displaystyle C}{}}} \\ & CO & CO & CO \\ OC-W-W-W-CO \\ OC & OC & OC \\ \overset{\displaystyle C}{\underset{\displaystyle O}{}} & \overset{\displaystyle C}{\underset{\displaystyle O}{}} & \overset{\displaystyle C}{\underset{\displaystyle O}{}} \end{array}\right]^{2-}$$

(18-C-XXIV)

An extremely important chemical reaction of $W(CO)_6$ is the following one which led to the discovery of the entire class of carbene complexes by E. O. Fischer (1964):

$$W(CO)_6 + PhLi \longrightarrow (OC)_5W=C\overset{\displaystyle OLi}{\underset{\displaystyle Ph}{\big<}}$$

[129]B. F. G. Johnson et al., J. Chem. Soc., Dalton Trans. **1995**, 1391.
[130]D. J. Darensbourg et al., Inorg. Chem. **1995**, 34, 4933.

When this was followed by reaction with a source of "cationic" CH_3, the first known carbene complex was formed:

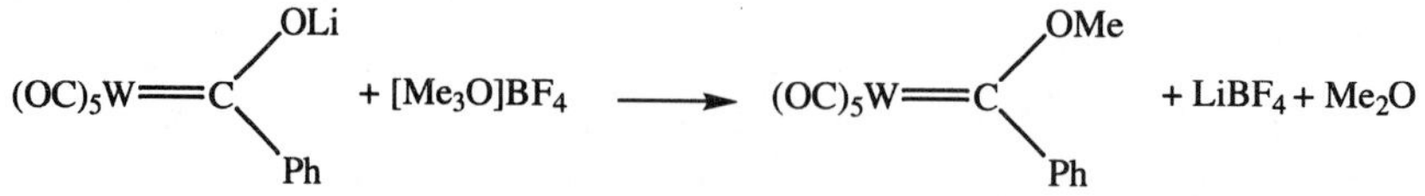

$$(OC)_5W{=}C\overset{OLi}{\underset{Ph}{\big<}} + [Me_3O]BF_4 \longrightarrow (OC)_5W{=}C\overset{OMe}{\underset{Ph}{\big<}} + LiBF_4 + Me_2O$$

An important reaction of the carbenes (or alkylidenes as they can also be called) of Mo and W (also Re) is to trigger a catalytic process of *olefin metathesis*.[131] The essence of this is indicated in the following reaction sequence:

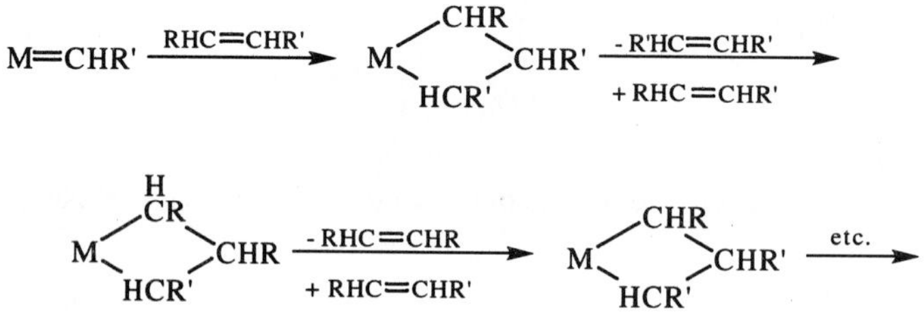

$$M{=}CHR' \xrightarrow{RHC=CHR'} M\overset{CHR}{\underset{HCR'}{\big<}}CHR' \xrightarrow[+RHC=CHR']{-R'HC=CHR'}$$

$$M\overset{\overset{H}{CR}}{\underset{HCR'}{\big<}}CHR \xrightarrow[+RHC=CHR]{-RHC=CHR} M\overset{CHR}{\underset{HCR'}{\big<}}CHR' \xrightarrow{etc.}$$

The overall net process is

$$2\,RHC{=}CHR' \rightleftharpoons RHC{=}CHR + R'HC{=}CHR'$$

This type of process has been extended to allow the catalytic polymerization of cyclic olefins.

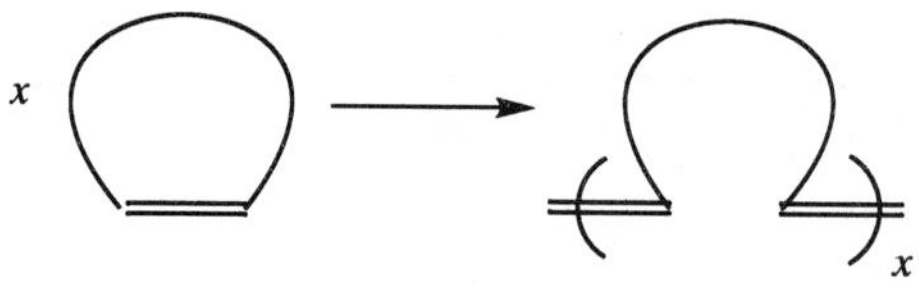

It was from such carbene complexes that, again, E. O. Fischer (1973) prepared the first carbyne complexes, as illustrated by the following reactions:[132]

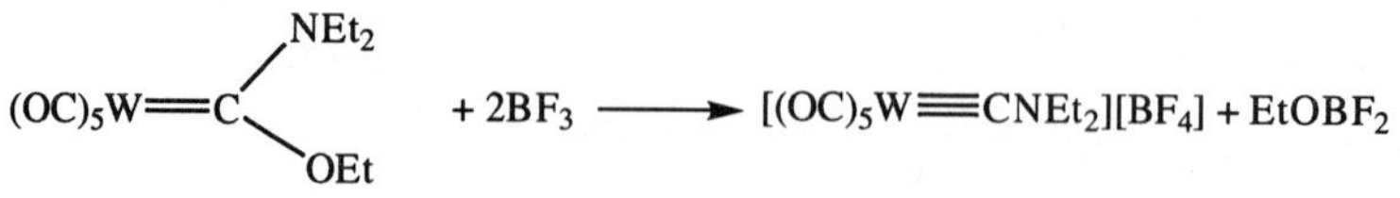

$$(OC)_5W{=}C\overset{NEt_2}{\underset{OEt}{\big<}} + 2BF_3 \longrightarrow [(OC)_5W{\equiv}CNEt_2][BF_4] + EtOBF_2$$

[131]R. R. Schrock *et al., Inorg. Chem.* **1992**, *31*, 2287.
[132]E. O. Fischer and D. Wittmann, *Inorg. Synth.* **1989**, *26*, 40.

Treatment of these carbynes with CpNa or Cp*Na, as in the following reaction sequence,[133] yields an important class of carbyne complexes:

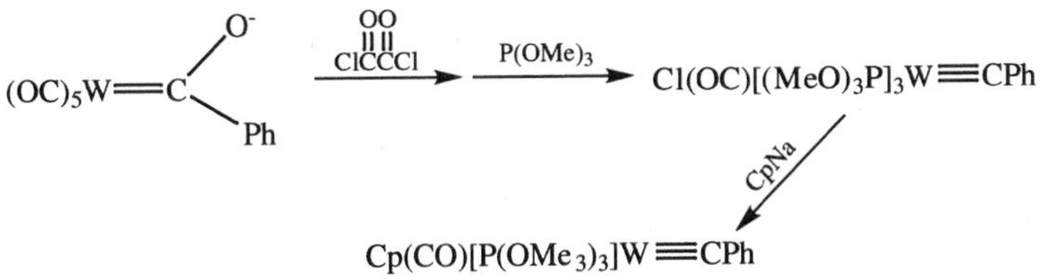

In connection with carbene and carbyne complexes, an especially interesting tungsten compound is 18-C-XXV, which contains $W \equiv C$, $W = C$, and $W - C$ bonds, whose lengths are 1.79, 1.94, and 2.26 Å, respectively.

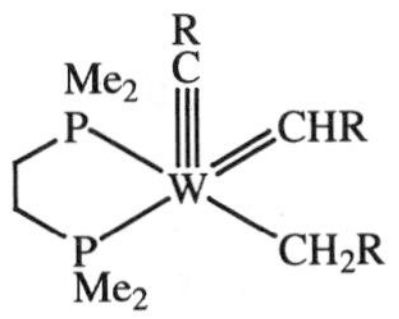

R = CMe$_3$

(18-C-XXV)

A remarkable reaction[134] leading to amido carbyne complexes is the thermal elimination of H$_2$ from an alkyl compound, eq. 18-C-13. When the alkyl is CH$_2$R, the intermediate alkyl cannot be isolated.

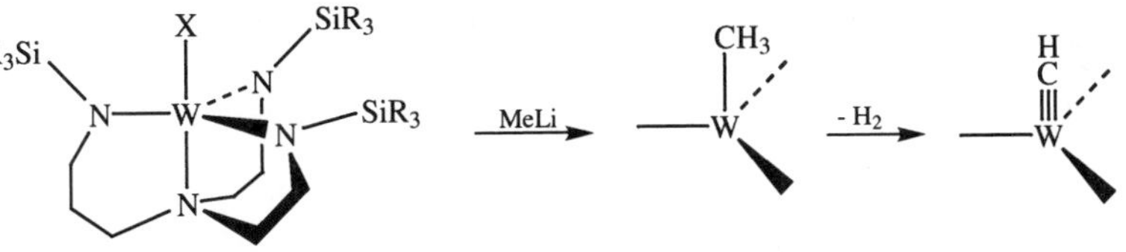

(18-C-13)

Cyclopentadienyl Compounds

Simple homoleptic ones such as Cp$_2$M or Cp$_2^*$M are unknown (although Cp$_2$Cr is), but mono-Cp or -Cp* compounds are abundant and important.

The first such compounds [CpM(CO)$_3$]$_2$ were obtained by direct reaction of cyclopentadiene with the M(CO)$_6$. They can better be made by reaction of M(CO)$_6$ with LiCp or NaCp followed by oxidation. These have a direct M—M bond, 18-C-XXVI, which is long (ca. 3.25 Å) and easily cleaved to give an almost endless variety of CpM(CO)$_3$X species, and related ones may be obtained with substituted

[133]L. McElwee-White et al., J. Am. Chem. Soc. **1991**, 113, 2947.
[134]R. R. Schrock et al., J. Am. Chem. Soc. **1994**, 116, 12103.

Cp rings and/or other ligands.[135] It is also possible to replace X and one CO by a simple three-electron donor such as NO or allyl. Still other variations are possible.[136] In addition to these 18-electron species, there are moderately stable, paramagnetic 16-electron species, Cp*MoClL$_2$ (L = a phosphine), that react readily to add a great variety of ligands (PMe$_3$, N$_2$, CO, H$_2$).[137]

(18-C-XXVI)

Reaction of Cp*W(CO)$_3$Me with PF$_5$ gives dinuclear (Cp*WCl$_4$)$_2$ (structure unknown) from which a great variety of other Cp*W compounds, such as Cp*W(CH$_3$)$_4$ can be obtained.[138]

The [CpM(CO)$_2$]$_2$ compounds (also similar ones with substituted Cp rings) can be made by thermolyzing or photolyzing [CpMo(CO)$_3$]$_2$. These molecules have partially multiple M—M bonds (ca. 2.50 Å). The CpM≡MCp units (Cp means ring centroid) is nearly linear and the CO groups are semi-bridging. Such molecules are very reactive toward nucleophiles, re-adding CO or other ligands, to give products such as (OC)$_2$PR$_3$M—M(CO)$_2$PR$_3$ or one with a crosswise alkyne bridging an (OC)$_2$M—M(CO)$_2$ unit. This sort of chemistry is very extensive, especially for molybdenum.

Arene Compounds

The bis arenes, of which (C$_6$H$_6$)$_2$M are the prototypes, and even the M(C$_6$H$_3$F$_3$)$_2$ and M(C$_6$F$_6$)$_2$ derivatives,[139] are known. However, they are difficult to synthesize and it is the (arene)M(CO)$_3$ compounds, easily obtainable by reactions of arenes with M(CO)$_6$ or M(CO)$_3$(THF)$_3$ that are important. These can undergo three major classes of reactions.

Thermal reaction with ligands such as phosphines usually leads to displacement of the arene, thus affording fac-M(CO)$_3$L$_3$ products. Photochemical conditions, on the other hand, allow replacement of a CO group.

Attacks on the arene ring by Li, nucleophiles, and electrophiles can occur, but this sort of chemistry has been studied in far more detail for the (arene)Cr(CO)$_3$ compounds.

W(CH$_3$)$_6$

This unique homoleptic alkyl in Group 6 (it has [M(CH$_3$)$_6$]$^{n-}$ (M = Zr, Nb, Ta, and Re(CH$_3$)$_6$ analogues) was first made in 1973, but only recently has its structure

[135]P. Leoni et al., J. Chem. Soc., Dalton Trans. **1989**, *155*, 959.
[136]M. L. H. Green et al., J. Chem. Soc., Dalton Trans. **1994**, 2851, 2975.
[137]R. Poli et al., Inorg. Chem. **1992**, *31*, 662; J. Chem. Soc., Chem. Commun. **1994**, 2317.
[138]R. R. Schrock et al., J. Am. Chem. Soc. **1987**, *109*, 4282; Inorg. Chem. **1992**, *31*, 1112.
[139]P. L. Timms et al., J. Chem. Soc., Dalton Trans. **1993**, 3097.

been elucidated.[140] Like the others it is *not* octahedral but trigonal prismatic.[141,142] It has also been shown that the prismatic structure is significantly distorted in a way that is very difficult to account for.[142]

18-C-13 Bioinorganic Chemistry[143]

Molybdenum is very important in the biochemistry of animals, plants, and microorganisms. It is the only element in the second transition series known to have natural biological functions. It occurs in more than 30 enzymes, in some of which it may be replaced by tungsten or vanadium. Tungsten is the only element in the third transition series known to have natural biological functions. Not only does it sometimes occur in enzymes that usually contain molybdenum, but there are some enzymes that are known only with tungsten.

Although neither element is very abundant in the earth's crust, molybdenum in particular has a concentration in natural waters that is greater than those of other biologically important elements such as iron, copper, and zinc.

Enzymes containing molybdenum are of two types: (1) Nitrogenases, which are required for converting atmospheric nitrogen to nitrogen compounds (NH_3, for example); nitrogenases contain a characteristic polymetal atom cluster species called the iron-molybdenum cofactor, FeMoco (Section 17-E-10). (2) The other Mo enzymes, all of which have some variant of a characteristic molybdenum cofactor, Moco.

A few representative Mo enzymes are listed in Table 18-C-6. Note that most have very high molecular weights, and this has delayed structure determinations. In the case of DMSO-reductase, however, the lower molecular weight has made possible a structure determination. The Mo site is found to be as shown in 18-C-XXVII. In a tungsten-containing enzyme, ferredoxin aldehyde oxidoreductase, the metal site is as shown in 18-C-XXVIII.

O (of serine)

(18-C-XXVII)   (18-C-XXVIII)

In addition to X-ray studies aimed at defining directly the structures of the Mo enzymes and especially the active sites, efforts to study them by indirect methods such as epr and endor continue. These methods are useful only for elucidating the nature of Mo^V species (Mo^{IV} and Mo^{VI} being esr silent) which may, or may not, be

[140]K. Seppelt *et al.*, *Chem. Eur. J.* **1998**, *4*, 1687.
[141]A. Haaland *et al.*, *J. Am. Chem. Soc.* **1990**, *112*, 4547.
[142]M. Kraup, *J. Am. Chem. Soc.* **1996**, *118*, 3018.
[143]See articles in *Chem. Rev.*, **1996**, *96*, No. 7, by R. Hille, by M. Johnson, D. C. Rees and M. W. W. Adams, by J. B. Howard and D. C. Rees, and by B. K. Burgess and D. J. Lowe. The coverage in these articles is so thorough and up to date that few further references will be given here.

Table 18-C-6   Some Enzymes that Contain Molybdenum

Name	Function	Remarks
Nitrogenases	N_2 to NH_3	Found in bacteria and blue-green algae
		$MW \approx 220,000$
Aldehyde oxidases	RCHO to RCOOH	$MW \approx 270,000$
Nitrate reductase	NO_3^- to NO_2^-	MW $200,000 - 360,000$
Sulfite oxidase	SO_3^{2-} to SO_4^{2-}	
Xanthine oxidase	Xanthine to uric acid	$MW \approx 300,000$
Formate dehydrogenase	HCO_2H^- to $CO_2 + H_2O$	
DMSO reductase	Me_2SO to Me_2S	$MW \approx 85,000$

directly involved in the catalytic process. The study of model systems, for both the Mo and W[144,145] enzymes continues.

It seems clear that, broadly speaking though not in detail, the role of the Moco sites in molybdenum enzymes (and *mutatis mutandi* in tungsten enzymes) is to facilitate the following type of process:[146]

$$Mo^{VI}O + Sub \rightleftharpoons Mo^{IV} + OSub$$

where Sub or OSub represents the substrate, such as Me_2SO, SO_3^{2-} or NO_2^-. The Mo (or W) atom cycles back and forth from M^{IV} to M^{VI} via oxygen atom transfer.

As already noted, the nitrogenases have a more complex active center FeMoco. The structure of this center is now known in fair detail (see Fig. 17-E-8) as a result of an X-ray diffraction study of one of the bacterial nitrogenases. In all prior efforts to devise models for this site it was taken for granted that an Mo atom served to bind an N_2 molecule. It was an enormous surprise therefore that the molybdenum atom is almost certainly *not* the bonding site for the N_2 molecule. Its presence is probably not mandatory since nitrogenases having no Mo have been discovered.

18-D   TECHNETIUM AND RHENIUM: GROUP 7

Although these elements form some compounds that are stoichiometrically analogous to those of manganese, e.g., MO_4^-, $M_2(CO)_{10}$, and $CpM(CO)_3$, there is very little genuine chemical similarity. Many of the stoichiometrically analogous species are quite different in their chemistry (*cf.* MnO_4^- *vs.* TcO_4^- and ReO_4^-) and there are many types of compounds known only for Tc and Re and not for Mn, and vice versa. Metal—metal bonding is very important for Re^{II} and Re^{III} and to a lesser extent for Tc, but virtually unknown for Mn. Again, the chemistry of Mn is largely that of aqueous Mn^{II} (with some for Mn^{III} and Mn^{IV}), whereas Tc and Re have extensive chemistry in oxidation states IV and V, very little of it aqueous.

The oxidation states and stereochemistries of compounds of the elements are summarized in Table 18-D-1.

[144]C. G. Young *et al.*, *J. Am. Chem. Soc.* **1994**, *116*, 9749.
[145]S. Sarkar *et al.*, *J. Am. Chem. Soc.* **1996**, *118*, 1387.
[146]Z. Xiao *et al.*, *Inorg. Chem.* **1996**, *35*, 7508.

Table 18-D-1 Oxidation States and Stereochemistry of Technetium^a and Rhenium^{b,c}

Oxidation state	Coordination number	Geometry	Examples
Tc^{-I}, Re^{-I}	5	?	[M(CO)$_5$]$^-$
Tc^0, Re^0, d^7	6	Octahedral	M$_2$(CO)$_{10}$
Tc^I, Re^I, d^6	5	tbp	ReCl(dppe)$_2$, ReCl(CO)$_2$(PPh$_3$)$_2$
		π Complex	η^5-C$_5$H$_5$Re(CO)$_2$C$_5$H$_8$, η^5-C$_5$H$_5$Re(CO)$_3$
	6	Octahedral	Re(CO)$_5$Cl, K$_5$[Re(CN)$_6$], Re(CO)$_3$(py)$_2$Cl, [(CH$_3$C$_6$H$_4$NC)$_6$Re]$^+$, ReCl(N$_2$)(PR$_3$)$_4$
Tc^{II}, Re^{II}, d^5	6	Octahedral	ReCl(N$_2$)(PR$_3$)$_3^+$, ReH$_2$(NO)(PPh$_3$)$_3$, ReCl$_2$(NO)(PPh$_3$)$_2$L, M(diars)$_2$Cl$_2$, [Re(bipy)$_3$]$^{2+}$, [Tc(MeCN)$_6$]$^{2+}$
	4	Dinuclear (3° bond)	M$_2$Cl$_4$(PR$_3$)$_4$
Tc^{III}, Re^{III}, d^4		Distorted tetrahedral	Re[P(C$_6$H$_{11}$)$_2$]$_4^-$
		π Complex	(η^5-C$_5$H$_5$)$_2$ReH, (η^5-C$_5$H$_5$)$_2$ReH$_2^+$
	5	tbp	Ph$_3$Re(PPhEt$_2$)$_2$
	6	Octahedral	[Tc(diars)$_2$Cl$_2$]$^+$, ReCl$_2$(acac)(PPh$_3$)$_2$, $trans$-TcCl(acac)$_2$PPh$_3$, mer-MCl$_3$(PR$_3$)$_3$
		Trig. prism	Re(S$_2$C$_2$Ph$_2$)$_3$
	7	Pentagonal bipyramidal	ReH$_3$(diphos)$_2$, K$_4$[M(CN)$_7$]·2H$_2$O, Tc(CO)(Etdtc)$_3$
		Dinuclear (4° bond)	M$_2$X$_8^{2-}$
		Metal cluster	Re$_3$X$_9$L$_3$, Re$_3$Cl$_3$(CH$_2$SiMe$_3$)$_6$, Re$_3$Me$_9$
Tc^{IV}, Re^{IV}, d^3	4	Tetrahedral	Re(o-MeC$_6$H$_4$)$_4$
		sp	Re(O-2,6-C$_6$H$_3$Pr$_2^i$)$_4^d$
	5	?	(Me$_3$SiCH$_2$)$_4$Re(N$_2$)Re(CH$_2$SiMe$_4$)$_4$
	6	Octahedral	MX$_6^{2-}$, ReI$_4$(py)$_2$, MCl$_4$, Tc(C$_2$O$_4$)$_3^{2-}$, [Re$_2$OCl$_{10}$]$^{4-}$, ReCl$_4$(diars), MO$_2$
	8	?	Tc(S$_2$CNC$_4$H$_8$O)$_4$
	7	?	[ReCO(diars)$_2$I$_2$](ClO$_4$)$_2$
		Metal cluster	Re$_3$(CH$_2$SiMe$_3$)$_{12}$
Tc^V, Re^V, d^2	5	tbp or distorted tbp	ReCl$_5$(g), ReF$_5$, NReCl$_2$(PPh$_3$)$_2$, ReIO$_2$(PPh$_3$)$_2$
		sp	[MOX$_4$]$^-$
	6	Octahedral	ReOCl$_3$(PPh$_3$)$_2$, [MOCl$_5$]$^{2-}$, Re$_2$Cl$_{10}$, Tc(NCS)$_6^-$
	7	?	ReOCl$_3$TAS
	8	Dodecahedral(?)	[M(diars)$_2$Cl$_4$]$^+$
		Distorted (C_s)	ReH$_5$(PEtPh$_2$)$_3$
Tc^{VI}, Re^{VI}, d^1	4	Distorted tetrahedral	ReO$_2$(mesityl)$_2$, ReO$_4^{2-}$
	5	sp	ReOMe$_4$, ReOCl$_4$
	6	Octahedral	MF$_6$, Re(OMe)$_6$
		Trig. prism	ReMe$_6$
	7	?	ReOCl$_6^{2-}$
	8	Square antiprism	ReF$_8^{2-}$
		Dodecahedral	[ReMe$_8$]$^{2-}$

Table 18-D-1   (*Continued*)

Oxidation state	Coordination number	Geometry	Examples
Tc^{VII}, Re^{VII}, d^0	4	Tetrahedral	MO_4^-, MO_3Cl, $Me_3SiOReO_3$, $MeReO_3$
	5	*tbp*	cis-$Re(O)_2Me_3$, $ReCl_3(NBu^t)_2$
	6	Octahedral	$ReO_3Cl_3^{2-}$
	7	Pentagonal bipyramidal	ReF_7
	9	Tricapped trigonal prism	ReH_9^{2-}

[a] G. Bandoli *et al.*, *Coord. Chem. Rev.*, **1994**, *135*, 325.
[b] Some compounds in nonintegral oxidation states are also known (see text).
[c] Use of M implies that both the Tc and the Re species are well known.
[d] D. E. Wigley *et al.*, *Inorg. Chem.*, **1989**, *28*, 1769.

Applications in Nuclear Medicine[1]

The use of a radioactive compound to localize in an organ in the body, either to examine that organ by a scanning device and thus provide a non-invasive method of examination or to utilize the radioactivity to treat a pathology, are the two major forms of nuclear medicine. No element has been more important than technetium in the first of these. The metastable isotope ^{99m}Tc has properties that are practically ideal for the purpose. It decays to ^{99}Tc with $t_{1/2}$ of 6 hr by emitting a gamma ray that is sufficiently intense to emerge from inside the body and be focused to provide an image, but neither this radiation nor the soft β from ^{99}Tc are dangerous to the patient. It is estimated that about seven million doses of ^{99m}Tc radiopharmaceuticals are given annually in the United States.

The radiopharmaceuticals are generated by using an alumina column loaded with $^{99}MoO_4^{2-}$ which decays to $^{99m}TcO_4^-$ ($t_{1/2} = 66$ h). The $^{99m}TcO_4^-$ is eluted selectively from the column (along with $^{99}TcO_4^-$ already formed). The eluate is then reduced in the presence of appropriate ligands, whereby a radiopharmaceutical with desired physiological properties (particularly specific organ or tissue preferences) is formed *in situ* and then injected intravenously.

The isotopes ^{186}Re and ^{188}Re are also being employed, using similar chemistry, but with a view to having their high-energy β emissions serve a therapeutic function, particularly with regard to bone cancer.

Nmr spectroscopy of Tc and Re nuclides is not ordinarily useful because of the high nuclear quadrupole moments, although in diamagnetic and cubic complexes such as MO_4^- or $[M(CNR)_6]^+$ useful data may be obtained.[2]

[1] M. Nicolini *et al.*, Eds. *Technetium and Rhenium in Chemistry and Nuclear Medicine*, Raven Press, New York, 1990; A. J. Fishman *et al.*, *J. Nucl. Med.* **1993**, *34*, 2253; B. Johannsen *et al.*, *Radiochem. Acta* **1993**, *63*, 133; K. Schwochau, *Angew. Chem. Int. Ed. Engl.* **1994**, *33*, 2258; B. Johannsen and S. Jurisson, *Top. Current Chem.* **1996**, *176*, 77; W. A. Volkert and S. Jurisson, *Top. Current Chem.* **1996**, *176*, 123; K. Haskimoto and K. Yoshihara, *Top. Current Chem.* **1996**, *176*, 275; S. Jurisson *et al.*, *Chem. Rev.* **1993**, *93*, 1137.
[2] A. Davison *et al.*, *Inorg. Chem.* **1988**, *27*, 3245.

18-D-1   The Elements

Although the Periodic Table predicted the existence of these elements at an early date, the first detection of rhenium *via* its X-ray spectrum did not occur until 1925. Not long after, it began to be isolated in gram quantities. It occurs in Nature with *molybdenite* (MoS_2) and some copper ores. It is usually isolated as the perrhenate ion, ReO_4^-, and can be precipitated as the slightly soluble $KReO_4$. Naturally occurring rhenium consist of two stable isotopes, ^{185}Re (37.4%) and ^{187}Re (62.6%).

Technetium was the first of the artificially produced elements (1937) when it was obtained, as the isotopes ^{95}Tc and ^{97}Tc, by bombarding Mo with deuterons. Today twenty-one isotopes, all radioactive, are known with mass numbers 90–111. The longest lived isotope is ^{98}Tc ($t_{1/2} = 4 \times 10^6$ y), but the commonest is ^{99}Tc ($t_{1/2} = 2 \times 10^5$ y). It is isolated in fairly large quantities from spent nuclear fuel, where it constitutes *ca.* 6% of the fission products. The total amount of ^{99}Tc is about 78 tons, which exceeds the known amount of rhenium in the earth's crust. ^{99}Tc emits a soft (293.6 Kev) β particle and can be handled with only very modest precautions.

In much of its chemistry Tc resembles Re, especially in its organometallic chemistry. However, important differences occur, such as the much greater reducibility of the higher oxidation states of Tc. Both metals dissolve in neutral or alkaline solutions of H_2O_2 to give solutions of TcO_4^- and ReO_4^- ions. Rhenium is distinguished by its extremely high melting point (3180°C), which is exceeded only by that of W (3400°C). Both metals burn in air above 400°C to give the heptoxides, M_2O_7.

18-D-2   Chalcogenides

Oxides

The M_2O_7 oxides of both metals are well known as the products of burning the metals in an excess of oxygen. Both are volatile and the vapors consist of molecules in which MO_4 tetrahedra share a vertex to give a linear M—O—M chain. Tc_2O_7 remains molecular in the solid state but Re_2O_7 has a layer structure in which half of the Re atoms are tetrahedrally coordinated while the others are octahedrally coordinated. This difference in structure accounts for the difference in their melting points (119°C *vs.* 300°C). It is not known if liquid Re_2O_7 (bp 360° *vs.* 310° for Tc_2O_7) is molecular but this seems likely.

The heptaoxides can also be obtained from solutions of the MO_4^- ions and conversely the oxides dissolve readily in water, giving acid solutions of HMO_4.

Both elements form dioxides, MO_2, that have distorted rutile structures in which there are M—M bonds. ReO_3 is an important oxide but TcO_3 is not known to exist. The structure of red ReO_3, shown in Fig. 18-D-1, is also adopted by CrO_3, WO_3, and others and is very similar to the perovskite structure. ReO_3 has metallic type electrical conductivity due to delocalization of the Re^{VI} electrons in a conduction band of the solid.

A few other oxides of rhenium have been reported, $Re_2O_3 \cdot xH_2O$ and Re_2O_5, but are not well characterized or important.

Sulfides, Selenides, and Tellurides

The black heptasulfides, M_2S_7, are precipitated when HCl solutions of the MO_4^- ions are saturated with H_2S, or when neutral solutions of the MO_4^- ions are treated

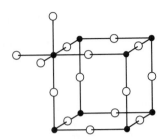

Figure 18-D-1 The ReO_3 structure. The perovskite structure is obtained from this one by insertion of a large cation at the center of the cube.

with thioacetamide or $Na_2S_2O_3$. They decompose to $MS_2 + S$ on heating. Neither compound has been obtained crystalline, but an EXAFS study[3] of Re_2S_7 shows it to contain a rhomboidal Re_4 unit with Re—Re distances of *ca.* 2.74 Å. Re_2S_7 reacts with aqueous KCN to produce cluster species such as cuboidal $[Re_4S_4(CN)_{12}]^{2-}$ and $[Re_4S_4(SO_2)_4(CN)_{10}]^{8-}$, which is consistent with its structure.

It is possible chemically or electrochemically to oxidize $Na_4Re_6S_{12}$, with concomitant loss of the sodium to give the binary sulfide Re_6S_{12}, which retains the essential structure of $Na_4Re_6S_{12}$ (i.e., Re_6S_8 clusters linked by S_2 bridges). On heating to 600° Re_6S_{12} loses S_2 to form virtually amorphous Re_6S_{10}.[4]

The MS_2, MSe_2, and MTe_2 compounds are all known, but there is (or may be) still some confusion about their structures. Recently it was shown[5] that ReS_2 is not, as previously supposed, isomorphous with $ReSe_2$, although both have layer structures.

18-D-3 Halides and Oxohalides

As shown in Table 18-D-2, quite a number of halides are known, although the list includes only three halides of technetium. Either TcF_6 or TcF_5 can be obtained by direct fluorination of the metal, depending on conditions. $TcCl_4$ is obtained by direct chlorination of the metal or by reaction of CCl_4 with Tc_2O_7 in a bomb. Structurally $TcCl_4$ is similar to $ZrCl_4$, with infinite chains of edge-sharing octahedra. The Tc···Tc distances are equal at 3.62 Å and neither this nor the paramagnetism suggests metal—metal bonding.

Rhenium heptafluoride, the only heptafluoride known other than IF_7, is obtained by fluorination of the metal in a bomb at 400°C. At atmospheric pressure and 120°C, ReF_6 is formed. ReF_6 is readily hydrolyzed, first to $ReOF_4$ and then to ReO_2, $HReO_4$, and HF. It can be reduced to give ReF_5 and ReF_4.

$ReCl_6$ is of little known importance but rhenium pentachloride, Re_2Cl_{10}, is very important. It can be prepared by reaction of Cl_2 with the metal at about 400°C or by the action of CCl_4 at about 400°C on Re_2O_7 (a method also applicable to the chlorides of other metals). The edge-sharing bioctahedral molecules of Re_2Cl_{10} (Re···Re = 3.74 Å) show a magnetic interaction but no actual Re—Re bond. Re_2Cl_{10} is rapidly hydrolyzed with disproportionation:

$$3Re_2Cl_{10} + 24H_2O \longrightarrow 4ReO_2 \cdot 2H_2O + 2HReO_4 + 30Cl^- + 30H^+$$

[3]S. J. Hubble and R. I. Walton, *J. Chem. Soc., Chem. Commun.* **1996**, 2135.
[4]A. Nemundry and R. Schöllhorn, *J. Chem. Soc., Chem. Commun.* **1994**, 2617.
[5]H. H. Murray *et al.*, *Inorg. Chem.* **1994**, *33*, 4418.

Table 18-D-2   Binary Halides of Technetium and Rhenium[a]

Oxidation state	Technetium		Rhenium		
+7		ReF_7 pale yellow mp 48°			
+6	TcF_6 yellow mp 37°	ReF_6 yellow mp 18.7°	$ReCl_6$ dark green mp 29°		
+5	TcF_5 yellow mp 50°	ReF_5 green-yellow mp 48°	Re_2Cl_{10} red-brown mp 261°	$ReBr_5$ brown	
+4	$TcCl_4$ red-brown subl >300°	ReF_4 blue subl >300°	$ReCl_4$ black	$ReBr_4$ red	ReI_4 black
+3			Re_3Cl_9 dark red	Re_3Br_9 brown	Re_3I_9 black

[a]The reported existence of ReI_2 is uncertain.

It is also readily reduced by potential ligands (or solvents) such as CH_3CN and CH_3CO_2H to afford Re^{IV} or Re^{III} compounds, and when heated cautiously in a nitrogen stream decomposes to rhenium(III) chloride (vide infra).

Rhenium(IV) chloride can be prepared by the following reactions:

$$Re_3Cl_9 + 3ReCl_5 \longrightarrow 6ReCl_4$$

$$2ReCl_5 + SbCl_3 \longrightarrow 2ReCl_4 + SbCl_5$$

$$2ReCl_5 + Cl_2C{=}CCl_2 \longrightarrow 2ReCl_4 + C_2Cl_6$$

It was discovered only in 1967 when the first of these reactions was serendipitously carried out. It has a structure not found in any other tetrahalide, consisting of zigzag chains of Re_2Cl_9 confacial bioctahedra (Re—Re = 2.73 Å) with one Cl atom on each end shared with the neighboring Re_2Cl_9 units. The tetrabromide and tetraiodide, whose structures are not known, are formed by evaporation of solutions of $HReO_4$ in aqueous HBr or HI. Neither has been much studied.

The *rhenium(III) halides*, with Cl, Br, and I are all well established. All three can be prepared by a variety of methods and all have been shown to have structures built of Re_3X_9 units of D_{3h} symmetry with Re≡Re double bonds. The structure of the crystalline bromide was only recently reported.[6] Although these halides are often designated as ReX_3, especially in older literature, they should be written as Re_3X_9, since only on this basis does their chemistry make sense.

A diagram showing some relationships between the rhenium(III) and rhenium(V) chlorides and complexes derived from them is given in Fig. 18-D-2.

The structure of the Re_3X_9 unit, with the three external sites for ligation is shown in Fig. 18-D-3, along with the structure of Re_3Cl_9 in which these sites are

[6]L. A. Aslanov *et al., Ukrain. Chem. J.* **1992,** *58,* 279.

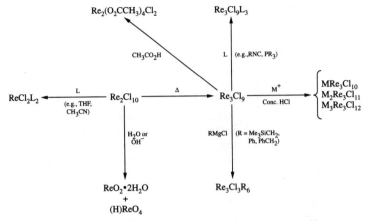

Figure 18-D-2 Some reactions of Re_2Cl_{10} and Re_3Cl_9.

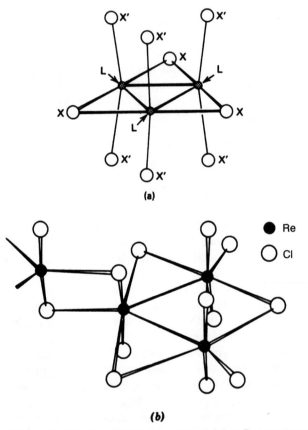

(a)

(b)

Figure 18-D-3 (a) Sketch of the isolated Re_3X_9 unit which has D_{3h} symmetry. The positions on each Re where additional ligands may be coordinated are marked by ←L. When L is a halide ion, we obtain the anions $[(Re_3X_3)X'_{6+n}]^{n-}$ (n = 1, 2, or 3 depending on the nature of the accompanying cation). (b) The structure of Re_3Cl_9 showing the linking of Re_3 units by chloride bridges.

occupied by terminal chlorine atoms from adjacent molecules. Freshly prepared Re_3Cl_9 is difficult to dissolve, but after exposure to a solvent and then removal of solvent, by pumping or heating, a powder much more readily soluble in many solvents is obtained. This is due to initial disruption of the tight Re_3Cl_9 structure, which remains disrupted after removal of the first solvent.

In the $Re_3(\mu\text{-}X)_3X_6L_3$ type structure the lability of the ligands varies greatly in the order $L > X >> \mu\text{-}X$. By virtue of this a great variety of species with mixed sets of ligands are obtained, such as $Re_3Br_9(PR_3)_3$, $Re_3Cl_9Br_3^{3-}$, or $[Re_3Cl_3Br_7(H_2O)_2]^-$.

Although simple reactions of $Re_3X_9L_3$ species under mild conditions leave the $Re_3(\mu\text{-}X)_3X_6'$ core structure intact, under more forcing conditions much more complex reactions occur. Thus, for example, with phosphines, dinuclear species such as $Re_2Cl_5(PR_3)_3$ and $Re_2Cl_4(PR_3)_4$ arise, although these are more conveniently available from $[Re_2Cl_8]^{2-}$, as noted in Section 18-D-5.

The *oxohalides* are fairly numerous, as indicated by Table 18-D-3, and occur for the oxidation states +5, +6, and +7. In general the solid MOF_4 compounds consist of infinite chains or rings of MOF_5 octahedra in which there are $\mu\text{-}F$ atoms. MOF_4 molecules of C_{4v} symmetry are observed in the vapor phase. The most important oxohalide is $ReOCl_4$, prepared by treatment of Re_2Cl_{10} with O_2 or Re with SO_2Cl_2 at elevated temperatures. The solid is built of weakly interacting molecules.

Besides those in Table 18-D-3, $Re_2O_4Cl_5$, an adduct of $ReOCl_4$ and ReO_3Cl with an Re—O—Re unit, is structurally characterized. ReO_2F and $ReOF_2$ have been claimed and the thiofluorides $ReSF_3$, $ReSF_4$, and $ReSF_5$ are known.

Like the halides, the oxohalides are hydrolyzed readily; those of M^{VII} go to MO_4^-, while those in lower oxidation states disproportionate to give MO_4^- and lower oxides.

Table 18-D-3 Oxohalides of Technetium and Rhenium

Oxidation state	Technetium			Rhenium	
+7	TcO_3F yellow mp 18.3°	TcO_3Cl colorless bp ~25°	ReO_3F yellow mp 147°	ReO_3Cl colorless bp 130°	ReO_3Br colorless mp 39.5°
	$TcO_2F_3^a$ yellow mp 200°		ReO_2F_3 yellow mp 90°		
	$TcOF_5^b$ orange mp 57°		$ReOF_5$ cream mp 44°		
+6	$TcOF_4$ blue mp 134°	$TcOCl_4$ purple mp ~35°	$ReOF_4$ blue mp 108°	$ReOCl_4$ brown mp 30°	$ReOBr_4$ blue dec. >80°
+5	$[TcOCl_3]_n$ brown subl ~500°	$[TcOBr_3]_n$ black	$[ReOF_3]_n$ black	$ReOCl_3$	$ReOBr_3$

[a] H. P. A. Mercier and G. J. Schrobilgen, *Inorg. Chem.* **1993**, *32*, 145.
[b] N. LeBland and G. J. Schrobilgen, *J. Chem. Soc., Chem. Commun.* **1996**, 2479.

18-D-4   The Pertechnetate and Perrhenate Ions

The aqueous chemistry of both elements is dominated by the MO_4^- ions. They are the ultimate products of oxidizing other Tc and Re compounds. Solutions of the pure acids, HMO_4, can be obtained by dissolving the heptoxides in water. Pure perrhenic acid has not been isolated, but dark red, hygroscopic crystals of $HTcO_4$ have been described. However, in the absence of structural data, the exact nature of this substance is uncertain.

Both anions are tetrahedral and numerous salts are known. Neither of the MO_4^- ions shows any tendency to form polyanions, in striking contrast to the case of the MO_4^{2-} (M = Mo, W) ions. The $AsPh_4^+$ ion and other large cations give highly insoluble precipitates suitable for analytical purposes. Neither anion is a good oxidizing agent and only ReO_4^- shows a significant tendency to serve as a ligand. It is remarkable that TcO_4^- (but not ReO_4^-) shows catalytic properties (i.e., in the oxidation of N_2H_4 by NO_3^- or ClO_4^-) and is a remarkable corrosion inhibitor for iron and steel even at extremely low concentrations.

Solutions of the $(Me_4N)MO_4$ compounds in acetonitrile can be cathodically reduced to give the MO_4^{2-} ions, which are both very sensitive to oxygen.

The ReS_4^- ion has been well-established and there is also evidence for the ReO_3S^- ion. In neither case is the chemistry very extensive, but ReS_4^- reacts with alkenes and alkynes to give $S=Re(S_2C_2R_{2\,or\,4})_2$ square pyramidal products.[7]

18-D-5   M—M Multiple Bonds[8] and M_n Clusters

Both elements form a great many compounds with M—M multiple bonds, but there are several major differences between the two chemistries:

(i) For Re trinuclear clusters are very important, whereas none are known for Tc.

(ii) Re forms many Re_2^{6+} complexes with Re—Re quadruple bonds, whereas Tc prefers Tc_2^{5+} or Tc_4^{4+} complexes with lower bond orders.

(iii) Tc forms Tc_6 and Tc_8 prismatic clusters within which there are three or four Tc≡Tc units; Re forms Re_4 rectangular clusters with two Re≡Re bonds.

Trinuclear Rhenium Compounds

All of the binary halides of Re^{III} (with Cl, Br, I) have, as already noted, an Re_3X_9 core with double bonds between the Re atoms. An extensive chemistry is derived from these halides in which the Re_3 core persists. The simplest derivatives are those in which one, two, or three of the external coordination sites is occupied by a ligand, to give $Re_3X_9L_{1-3}$ species such as $[Re_3Cl_{12}]^{3-}$, $[Re_3Br_{11}]^{2-}$, and $Re_3Cl_9(H_2O)_3$. It is also possible to convert Re_3^{9+} species into Re_2^{4+} compounds, e.g.,

$$2Re_3I_9 + 12PR_3 \longrightarrow 3Re_2I_4(PR_3)_4 + ?$$

[7]T. B. Rauchfuss et al., J. Am. Chem. Soc. 1996, 118, 674.
[8]F. A. Cotton and R. A. Walton, Multiple Bonds between Metal Atoms, Oxford University Press, Oxford, 1993, pp. 28–138, 553–563.

The reactions of Re$_3$Cl$_9$ or Re$_3$Br$_9$ with pyridine or other amines lead to polymeric materials, [ReX$_2$am]$_n$, believed to retain the Re$_3$X$_3$ core, but not further characterized structurally. A stacking of Re$_3$X$_3$am$_3$ units with μ-X bridges has been proposed.

Even more remarkable are the reactions of Re$_3$Cl$_9$ or Re$_3$Cl$_9$(THF)$_3$ with Grignard reagents to give Re$_3$Cl$_3$R$_6$ (R = Me$_3$SiCH$_2$, PhCH$_2$, Ph and others). With R = Me$_3$SiCH$_2$, two isomers are known, Re$_3$(μ-Cl)$_3$(CH$_2$SiMe$_3$)$_6$ and Re$_3$(μ-CH$_2$SiMe$_3$)$_3$Cl$_3$(CH$_2$SiMe$_3$)$_3$. With R = Me, the permethyl species, Re$_3$Me$_9$, can be obtained.

The Tc$_2^{n+}$ and Re$_2^{n+}$ Compounds

The [Re$_2$Cl$_8$]$^{2-}$ ion was the first stable chemical entity shown to possess a quadruple bond. This deep blue, air-stable anion can be made in several ways, of which the best is by reaction of (Bu$_4$N)$_2$ReO$_4$ with PhCOCl, whereby the synthetically useful compound (Bu$_4$N)$_2$[Re$_2$Cl$_8$] is obtained. [Re$_2$X$_8$]$^{2-}$ ions are known with X = F, Cl, Br, I, NCS, CH$_3$. All of them have the eclipsed (D_{4h}) structure shown in 18-D-I, with Re—Re distances of about 2.24 Å.

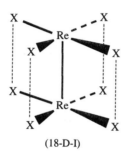

(18-D-I)

The [Re$_2$Cl$_8$]$^{2-}$ ion reacts in three principal ways:

(*i*) Simple ligand replacement, of which the following are typical. The reactions with RCO$_2$H are reversible.

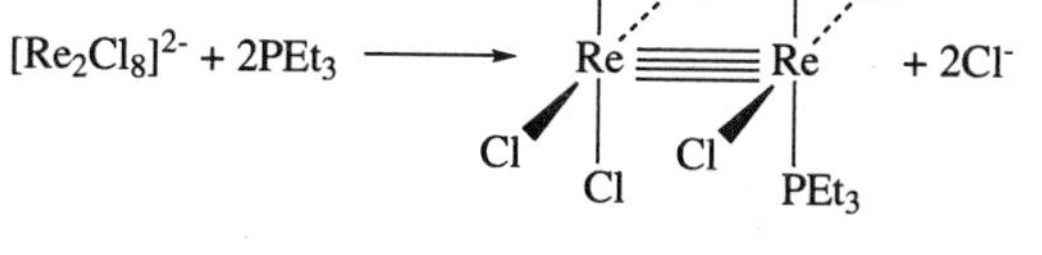

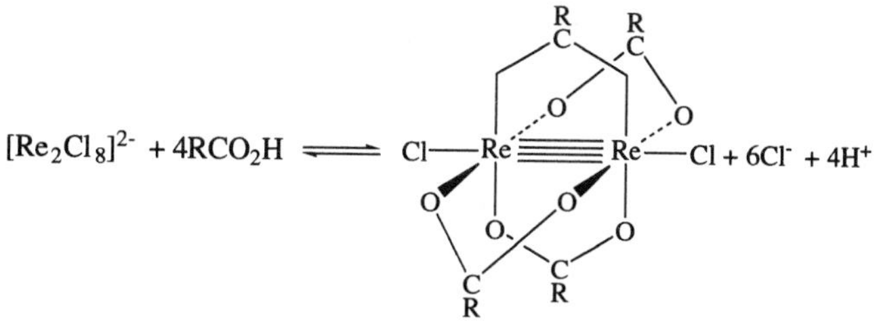

$$[Re_2Cl_8]^{2-} + 2PEt_3 \longrightarrow \quad + 2Cl^-$$

$$[Re_2Cl_8]^{2-} + 4RCO_2H \rightleftharpoons \quad + 6Cl^- + 4H^+$$

(*ii*) Ligand replacement with reduction:

$$[Re_2Cl_8]^{2+} + 5PMe_3 \xrightarrow[PMe_3]{ProH}$$

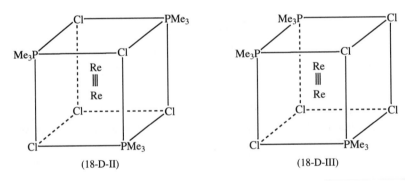

The $Re_2Cl_4(PMe_3)_4$ and related compounds contain a so-called "electron rich" triple bond, based on the $\sigma^2\pi^4\delta^2\delta^{*2}$ configuration, where the delta bond is annulled. This is equivalent to simply retaining a d_{xy}^2 configuration on each metal atom. These $Re_2Cl_4(PR_3)_4$ species can be reoxidized, stepwise and reversibly:

$$Re_2Cl_4(PR_3)_4 \underset{+e}{\overset{-e}{\rightleftharpoons}} [Re_2Cl_4(PR_3)_4]^+ \underset{+e}{\overset{-e}{\rightleftharpoons}} [Re_2Cl_4(PR_3)_4]^{2+}$$

Another way of oxidizing them is by use of $PhICl_2$ (a convenient chemical equivalent of Cl_2):

$$Re_2Cl_4(PMe_3)_4 \xrightarrow{PhICl_2} Re_2Cl_5(PMe_3)_3$$

The product of this reaction exists in isomeric forms, 18-D-II and 18-D-III.

(18-D-II)   (18-D-III)

(*iii*) Re—Re bond splitting occurs when strongly π-accepting ligands (RNC, CO, NO) are used:

$$Re_2(O_2CR)_4Cl_2 \xrightarrow{RNC} [Re(CNR)_6]^+$$

$$[Re_2Cl_8]^{2-} \xrightarrow{RNC} [Re(CNR)_4Cl_2]^+$$

$$Re_2Cl_4(PR_3)_4 \xrightarrow{CO} ReCl_2(CO)_2(PR_3)_2$$

(*iv*) A few other novel reactions of the dirhenium compounds are the following:

$$[Re_2Cl_8]^{2-} \xrightarrow{NaBH_4, PR_3} Re_2(\mu\text{-}H)_4H_4(PR_3)_4$$

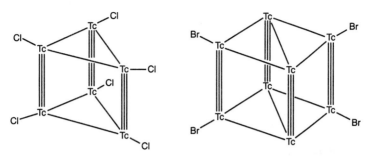

Figure 18-D-4 The $[Tc_6(\mu\text{-Cl})_6Cl_6]^-$ and $[Tc_8(\mu\text{-Br}_8)Br_4]^{0,+}$ clusters. All external triangular edges have μ-Cl or μ-Br atoms, omitted here for clarity.

The first Tc$_2$ compound reported contained the $[Tc_2Cl_8]^{3-}$ ion, reflecting the easier reduction of the second row transition metal. The $[Tc_2Cl_8]^{2-}$ ion has been obtained by reduction of $[TcCl_6]^{2-}$ in HCl with zinc, and halogen exchange with HBr gives $[Tc_2Br_8]^{2-}$. The chemistry of $[Tc_2Cl_8]^{2-}$ is difficult because of its ready reduction, but it can be converted to $[Tc_2(O_2CR)_4]^{2+}$ by reaction with RCO$_2$H. Access to the Tc$_2$Cl$_4$(PR$_3$)$_4$ compounds is available by the reaction[9]

$$TcCl_4(PR_3)_2 \xrightarrow[PR_3]{Zn} Tc_2Cl_4(PR_3)_4$$

This can be converted to the $[Tc_2(NCCH_3)_{10}]^{4+}$ by the reaction[10]

$$Tc_2Cl_4(PR_3)_4 \xrightarrow[\text{in CH}_3\text{CN}]{HBF_4\cdot Et_2O} [Tc_2(NCCH_3)_{10}](BF_4)_4$$

Tc$_6$ and Tc$_8$ Clusters

These are obtained by reduction of TcO$_4^-$ or $[TcCl_6]^{2-}$ by hydrogen under pressure. The chemistry is complex, highly dependent on conditions, and polymeric compounds, such as K$_{2x}[Tc_2Cl_6]_x$, are also obtained. The Tc$_6$ and Tc$_8$ clusters are shown schematically in Fig. 18-D-4. The Tc—Tc distances in the triangles are in the range 2.5–2.7 Å, while the vertical ones are about 2.15 Å. On this basis, these structures may be regarded as consisting of Tc≡Tc units linked by Tc—Tc single bonds. There is only one Re analogue, $[Re_6Br_{12}]$,[11] but rhenium also does something similar in forming Re$_4$ clusters with two Re≡Re units linked by Re—Re single bonds,[12] as in 18-D-IV, which shows the $[Re_4(\mu\text{-Cl})_2(\mu\text{-O})_2Cl_8]^{2-}$ ion.

[9]F. A. Cotton et al., Inorg. Chem. **1994**, 33, 2257.
[10]F. A. Cotton et al., Inorg. Chem. **1995**, 34, 1875.
[11]P. A. Koz'min et al., Dokl. Akad. Nauk SSSR **1987**, 295, 647.
[12]F. A. Cotton and E. V. Dikarev, J. Cluster Sci., **1995**, 6, 411.

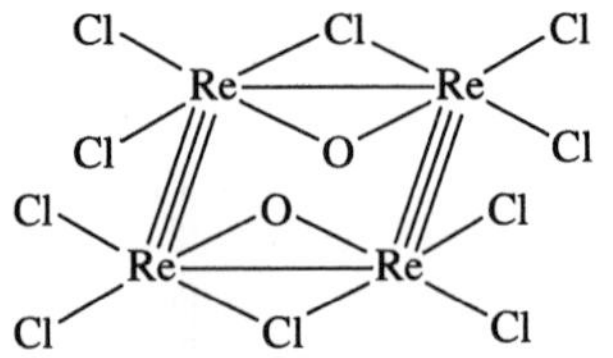

(18-D-IV)

Octahedral Clusters

There is an extensive and complex chemistry of $[Re_6(\mu\text{-}X)_n(\mu\text{-}Y)_{8-n}]$ species, where X may be exclusively S or Se, or there may be a mixture of chalcogenide and halide (Y) bridges.[13] There are also technetium clusters with Tc_6 octahedra and allegedly only 5 face bridges, e.g., $[Tc_6(\mu_3\text{-}Br_5)Br_6]^{-1,-2}$, but they require better characterization.

TECHNETIUM AND RHENIUM COMPLEXES

For both elements, complexes are formed in oxidation states from -I to VII. There are several broad classes of complex that cut across oxidation states and these, namely, oxo complexes (Section 18-D-6) and nitrido (and related) complexes (Section 18-D-7) will be discussed first. After a tour through the oxidation states (Sections 18-D-8 and 18-D-9) other classes of compounds (that could as well be called complexes) will be taken up.

18-D-6   Oxo Complexes

Although the majority of these are in oxidation state V, they are formed by both elements in oxidation state IV–VII. Thus they are best discussed as a class.

Broadly, for both elements there are three major types:

(*i*) 5- or 6- coordinate species with a MO^{3+} core.

(*ii*) 6- coordinate species with a linear $O{=}M{=}O^+$ core.

(*iii*) Species based on the linear $O{=}M{-}O{-}M{=}O^{4+}$ core to which ligands are attached to form an octahedron around each metal atom.

The chemistry is qualitatively the same for both Tc and Re, but differences in detail arise because Tc is more easily reduced than Re. It is also true that some types of compounds are either better known or little known for Tc because research on Tc has been so heavily driven by the search for imaging agents based on the TcO^{3+} core.

For both elements, the usual starting material is the MO_4^- ion, which, in concentrated HCl, gives rise to the *fac*-$[MO_3Cl_3]^{2-}$ ion. For Tc, merely heating leads to reduction:

$$[TcO_3Cl_3]^{2-} + 3HCl + H_3O^+ \longrightarrow [TcOCl_4]^- + 3H_2O + Cl_2$$

[13]A. Perrin *et al.*, *Eur. J. Solid State Inorg. Chem.* **1991**, *28*, 919; A. Perrin and M. Sergent, *New J. Chem.* **1988**, *12*, 337; V. P. Fedin *et al.*, *Polyhedron* **1996**, *15*, 1229.

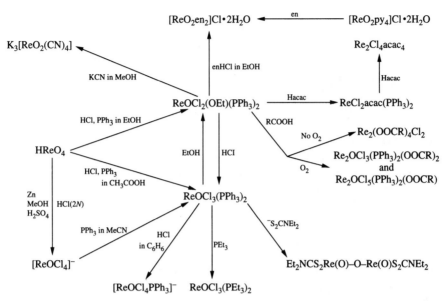

Figure 18-D-5 Some reactions of $ReOCl_3(PPh_3)_2$. Related compounds (i.e., with various PR_3, AsR_3, or SbR_3 groups in place of PPh_3) are known and have similar though not always identical reactions.

whereas for Re it is generally necessary to add a reducing agent. Thus, in the preparation of the important compound $ReOCl_3(PPh_3)_2$, (Fig. 18-D-5), PPh_3 serves as a reducing agent:

$$ReO_4^- + 4HCl + 3R_3P \longrightarrow ReOCl_3(PR_3)_2 + R_3PO + 2H_2O + Cl^-$$

With Tc, careful attention to maintaining a low temperature is necessary to avoid further reduction:

$$[TcOCl_4]^- + 3HCl \xrightarrow{H_2O} [TcCl_6]^{2-} + H_3O^+ + \tfrac{1}{2}Cl_2$$

$TcOX_3(PR_3)_2$ compounds cannot be obtained as above for $ReOX_3(PR_3)_2$ because of easy further reduction to give $TcX_4(PR_3)_2$.

The $[MOX_4]^-$ and $[MOX_5]^{2-}$ ions can be precipitated with larger cations favoring the latter one. Reactions of the $[MOX_4]^-$ compounds with a great variety of ligands, especially bi-, tri-, and tetradentate ones allows the preparation of many species having square pyramidal structures,[14] e.g., 18-D-V and 18-D-VI, or, less often, distorted octahedral,[15] 18-D-VII.

The $M{=}O^{3+}$ units characteristically show strong bands in the infrared in the 950–1000 cm^{-1} range. They also have very short bond lengths, 1.60–1.70 Å, and

[14]A. Davison *et al., Inorg. Chem.* **1988,** *27,* 2154; R. M. Baldwin *et al., Inorg. Chem.* **1994,** *33,* 5579.

[15]E. Deutsch *et al., Inorg. Chem.* **1988,** *27,* 4121; E. Duatti *et al., Inorg. Chem.* **1988,** *27,* 4208.

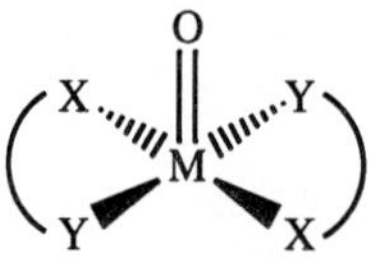

(18-D-V) 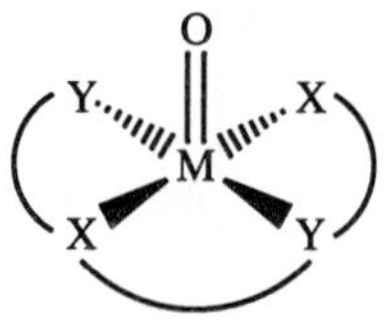   (18-D-VI) 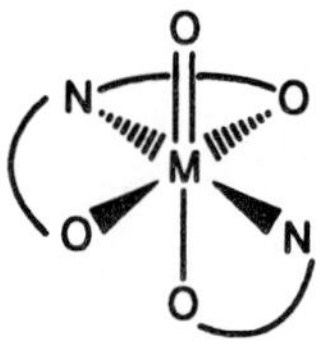   (18-D-VII)

for both these reasons the electronic structure is best represented as a resonance hybrid, $M{=}O \leftrightarrow \bar{M}{\overset{\leftarrow}{\overset{+}{=}}}O$ (*cf.* the metal nitrenes, Section 18-D-7).

Less common than the MO^{3+} species are the $O{=}M{=}O^{+}$ complexes. These are made in a variety of ways, as shown in Fig. 18-D-5 and by the following recent examples.[16,17]

$$ReO_4^- + dppe \xrightarrow{\ Na\ } [ReO_2(dppe)_2]^-$$

$$[TcOCl_4]^- + 6py + H_2O \longrightarrow [TcO_2py_4]^+ + 2Cl^- + 2pyHCl$$

Reactions of the latter type, with moist pyridine, also convert $ReOCl_3(PPh_3)_2$, $ReOCl_4^-$, and $ReO(OEt)Cl_2(PPh_3)_2$ to $[ReO_2py_4]Cl$. $ReOCl_3(PPh_3)_2$ will also react easily with tetraamine ligands of the type $H_2N(CH_2)_xNH(CH_2)_yNH(CH_2)_zNH_2$ ($x, y, z = 2$ or 3) to form very stable complexes.[18]

$[O{=}M{-}O{-}M{=}O]^{4+}$ cores are also easily formed, for instance, in the reaction[19]

$$2ReOCl_5^{2-} + 6OH^- + 2H_2L \longrightarrow Re_2O_3L_2 + 5H_2O + 10Cl^-$$

where H_2L represents a tetradentate ligand of the type 18-D-VIII. Technetium complexes of this type as well as many other Tc and Re complexes having the $O{=}M{-}O{-}M{=}O$ core, e.g., 18-D-IX, are known.[20] The $M{=}O$ and $M{-}O$ distances in these compounds are in the ranges 1.65–1.70 Å and 1.90–1.92 Å, respectively.

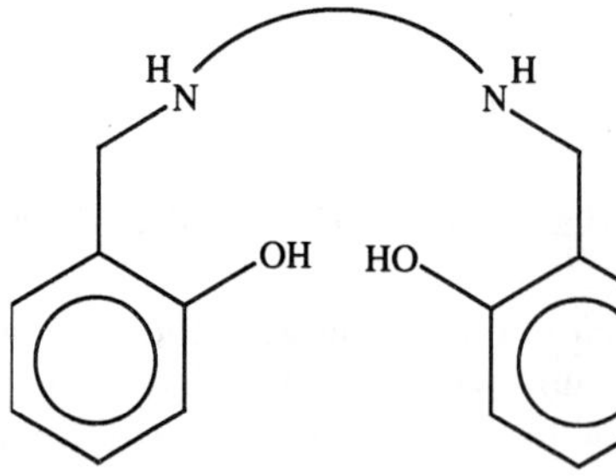

(18-D-VIII) 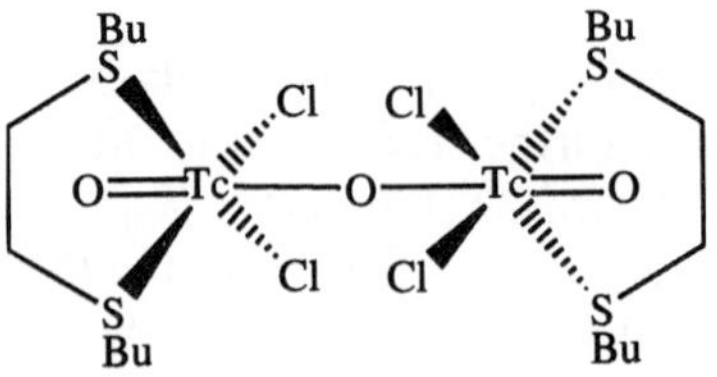   (18-D-IX)

[16]R. L. Luck *et al.*, *Inorg. Chem.* **1994**, *33*, 879.
[17]A. Davison *et al.*, *Inorg. Chem.* **1988**, *27*, 2409.
[18]D. Parker and P. S. Roy, *Inorg. Chem.* **1988**, *27*, 4127.
[19]M. R. A. Pillai *et al.*, *Polyhedron* **1994**, *13*, 701.
[20]H. Spies *et al.*, *Polyhedron* **1993**, *12*, 187.

Other Oxo Complexes

Besides the *fac*-[MO_3X_3]$^{2-}$ species, there are other M^{VII} oxo complexes, especially for Re. Some are of the $MO_3X(LL)$ type, e.g., $TcO_3Cl(bipy)$ and the rhenium analogues with X = F, Br, CN, NCS prepared[21] by the reactions

$$Re_2O_7 \xrightarrow[\text{or } ZnX_2]{[Bu_4N]X} ReO_3X \xrightarrow{\text{bipy}} ReO_3X(bipy)$$

Another one of interest is the $LReO_3^+$ (L = 1,4,7-triazacyclononane) ion, which is similar to the corresponding Mo^{VI} and W^{VI} compounds, which are neutral.

18-D-7 Nitrido and Nitrene (imido) Complexes

Nitrido Species

Both metals, M, form strong, triple bonds to nitrogen in oxidation states V and VI, i.e., MN^{2+} and MN^{3+}, and the complexes of these units are among the most important for both metals. Although for Tc the first nitrido complex was reported only in 1981, there are now over sixty whose crystal structures are known.[22] The M≡N bonds are of the order 3 ($\sigma+2\pi$) and have the following common characteristics:

1. M≡N distances in the range 1.60–1.70 Å.
2. M≡N stretching vibrations in the range 1000–1100 cm^{-1}.
3. Formation of complexes of the type 18-D-X, in which the NMX_4 portion is square pyramidal, with M well above the basal (X_4) plane and the ligand Y weakly bonded or even absent.

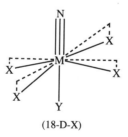

(18-D-X)

4. Chemical stability such that the MN unit persists through many reactions in which the other ligands change.[23] For example:

$$ReNBr_2(PPh_3)_2 \xrightarrow{\text{excess KCN(aq)}} K_2[ReN(CN)_4]$$

$$[TcNCl_4]^- \xrightarrow{\text{excess NCS}^-} [TcN(NCS)_4]^-$$

$$TcNCl_2(PPh_3)_2 + \left\{ \begin{array}{c} \text{tetradentate} \\ \text{or} \\ \text{two bidentate} \\ \text{ligands} \end{array} \right\} \longrightarrow TcN(LL)_2 \text{ or } TcNL^4$$

[21]W. A. Herrmann *et al.*, *Chem. Ber.* **1994**, *127*, 47.
[22]J. Baldas, *Topics in Current Chem.* **1996**, *176*, 37.
[23]G. Cros *et al.*, *Inorg. Chim. Acta* **1994**, *227*, 25.

Such exchange reactions have been used to prepare scores of TcN^{2+} complexes with complex, usually polydentate ligands for trial as possible ^{99m}Tc imaging agents.

For rhenium the most important nitrido complexes are those of Re^V with general formula $ReNX_2(PR_3)_x$ where the number of PR_3 ligands depends on the bulk; with PPh_3, $x = 2$, but with PMe_2Ph, $x = 3$. Similar $TcNX_2(PR_3)_2$ compounds are also known.

In the $M\equiv N$ unit there is a lone pair on the nitrogen atom that can be donated to a suitable acceptor, such as BX_3 or GaX_3,[24] or to another $L_4M\equiv N$ moiety to give either linear chains or a square tetramer,[25] 18-D-XI.

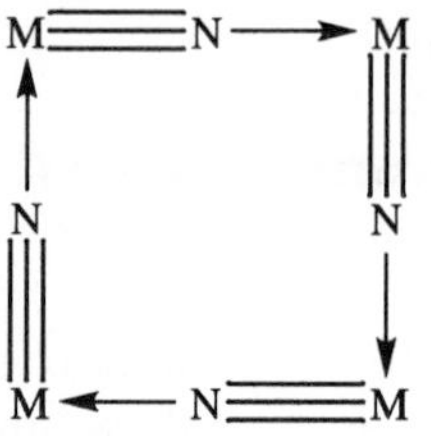

(18-D-XI)

It is both characteristic of and peculiar to the TcN^{3+} unit to form two types of oxo bridged dimers, $[NTc-O-TcN]^{4+}$ and $[NTc(\mu-O)_2TcN]^{2+}$. These cores are then surrounded by the additional ligands needed to complete distorted octahedra or square pyramids about each Tc^{VI}. The following reactions are representative:

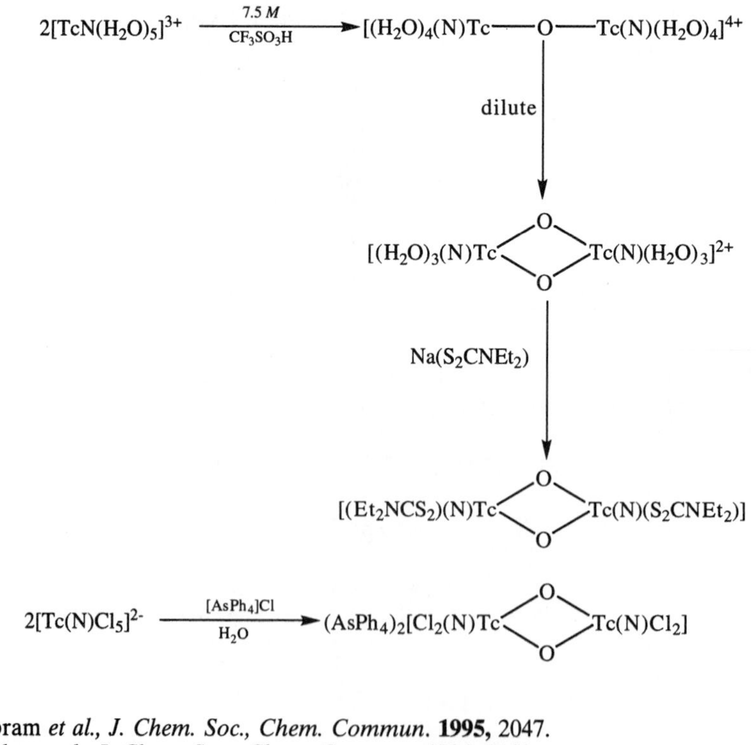

[24]U. Abram et al., J. Chem. Soc., Chem. Commun. 1995, 2047.
[25]J. Baldas et al., J. Chem. Soc., Chem. Commun. 1994, 2153.

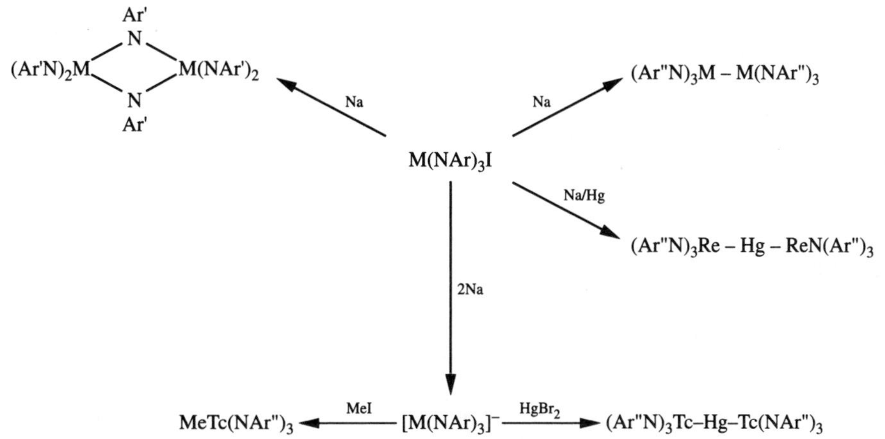

Figure 18-D-6 Some chemistry of Tc^{VII} and Re^{VII} polynitrenes. M = either metal; Ar = either Ar' or Ar" as shown in 18-D-XII.

In many ways this chemistry of $Tc^{II}N$ is parallel to that of the isoelectronic $Mo^{V}O$ unit, where $[OMo^V-O-Mo^VO]^{4+}$ and $[OMo^V(\mu\text{-}O)_2Mo^VO]^{2+}$ species are well known.

Nitrene Species

These contain MNR^{n+} cores and are known for both elements. Although the presence of only a M=NR double bond might be expected, the M—N—C units are virtually linear with M—N distances at the upper end of the range for M≡N bonds. Thus they are better represented by $M\overset{\leftarrow}{=}N-R$.

There is some fascinating chemistry involving the $M(=NAr)_3$ units,[26] where Ar represents either of the arenes shown in 18-D-XII and Fig. 18-D-6 summarizes some of the chemistry.

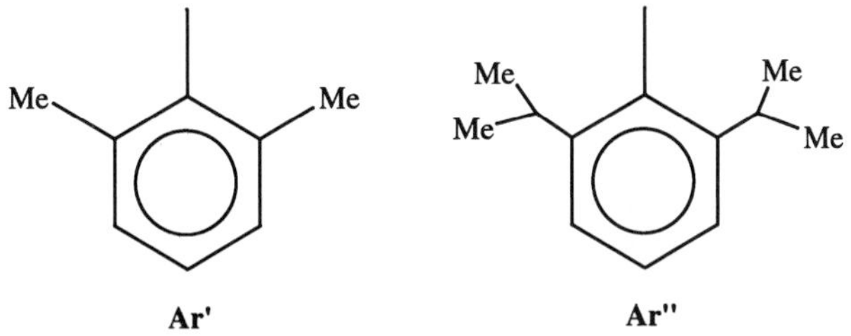

(18-D-XII)

Hydrazido complexes, $L_nM\overset{\leftarrow}{=}N-NR_2$, are also well known;[27] in these the M—N—N chain is linear with M—N distances of *ca.* 1.76 Å.

[26]J. C. Bryan *et al.*, *Polyhedron* **1993**, *12*, 1769; *J. Am. Chem. Soc.* **1994**, *116*, 3813.
[27]J. R. Dilworth *et al.*, *J. Chem. Soc., Dalton Trans.* **1994**, 1251; J. Zubieta *et al.*, *Inorg. Chem.* **1994**, *33*, 5864.

18-D-8   Complexes in Oxidation States I–IV

Oxidation States I, II

Classical complexes in these oxidation states are generally octahedral with low-spin d^6 and d^5 electron configurations, and they can often be interconverted electrochemically. Apart from carbonyl-containing species (e.g., $M(CO)_5X$) the most stable M^I complexes are the $[M(CN)_6]^-$ and $[M(CNR)_6]^+$ ions. Salts of the $[Tc(CNR)_6]^+$ cations, which are among the best ^{99m}Tc heart-imaging agents, can be synthesized by the reaction

$$TcO_4^- \xrightarrow[\text{excess RNC}]{S_2O_4^{2-}} [Tc(CNR)_6]^+$$

The $MCl(dppe)_2$ compounds,[28] which have *tbp* structures with Cl equatorial, are notable for their ability to react directly with RCN, N_2, and even H_2 to form octahedral complexes. Reaction of H_2 with $TcCl(dppe)_2$ gives a side-on η^2–H_2 adduct (see Section 18-D-10). The octahedral *trans*-$[Tc(H_2O)_2(dppe)_2]^+$ and *trans*-$[TcCl(NO)(diars)_2]^+$ ions are also known.[29]

N₂ Complexes

These are formed mainly by Re^I but some can be oxidized with retention of the N_2 ligand. The best known types are $Re(N_2)X(PR_3)_4$, $Re(N_2)X(CO)_2(PR_3)_2$, and $Re(N_2)X(diphos)_2$, but there are also some with linear bridging N_2 groups such as $(PhMe_2P)_4ClReNNMoCl_4$.

Technetium(II) and rhenium(II) complexes are somewhat more numerous; they resemble each other closely, though the Re^{II} complexes are less reactive[30] and are nearly always octahedral. They include the $[M(bipy)_3]^{2+}$ ions,[31] and numerous octahedral complexes containing two halide ions and some neutral ligands. Examples include *trans*-$MX_2(LL)_2$, where LL may be dppe or diars, and $TcCl_2(PMe_2Ph)_2$-(bipy).[32] $[Tc(9S3)_2]^{2+}$, where 9S3 is cyclonona-1,4,7-thiane, and its Re analogue[33] are approximately octahedral.[34] The $[Tc(NCCH_3)_6]^{2+}$ ion is available[35] *via* the following photochemical reaction:

$$[Tc_2(NCCH_3)_{10}]^{4+} \xrightarrow[CH_3CN]{h\nu} 2[Tc(NCCH_3)_6]^{2+}$$

The report[36] of the tetrahedral $TcBr_4^{2-}$ ion is open to suspicion.

[28]A. J. L. Pombeiro *et al., J. Chem. Soc., Dalton Trans.* **1994**, 3299.
[29]U. Abram *et al., Acta Cryst. C* **1994**, *50*, 188; H. J. Banbery and T. A. Hamor, *Acta Cryst. C* **1994**, *50*, 44.
[30]J. Barrera *et al., Inorg. Chem.* **1996**, *35*, 335.
[31]J. R. Dilworth *et al., J. Chem. Soc., Dalton Trans.* **1993**, 461.
[32]E. Deutsch *et al., Inorg. Chem.* **1989**, *28*, 3917.
[33]S. R. Cooper *et al., J. Chem. Soc., Chem. Commun.* **1995**, 161.
[34]S. R. Cooper *et al., Inorg. Chem.* **1992**, *31*, 5351.
[35]F. A. Cotton *et. al., J. Am. Chem. Soc.* **1996**, *118*, 5486.
[36]C. M. Archer *et al., J. Chem. Soc., Dalton Trans.* **1992**, 183.

NO and NS Complexes

Very low, formal oxidation states can be assigned to certain complexes following the convention of treating the NO in a linear $M-NO$ unit as NO^+. For example, there are the Re^0 complexes, $Re(NO)_2X_2L_2$,[37] and the Tc^I complexes of types $[Tc(NO)(NH_3)_4H_2O]Cl_2$ and $Tc(NO)Cl(PPh_3)_2(S_2COR)$.[38] Nitrosyl complexes of M^{II} include the $[M(NO)X_4]^-$ ions (X = Cl, Br, I, NCS) and derivatives $M(NO)X_3L_2$.

Some Tc^INS complexes and their methods of preparation[39] are:

$$[TcNCl_4]^- \xrightarrow[\text{pic}]{S_2O_4{}^{2-}} Tc(NS)Cl_2(pic)_3$$

$$TcNCl_2(PR_3)_3 \xrightarrow{S_2Cl_2} Tc(NS)Cl_2(PR_3)_3$$

For Tc^{II} compounds of the type $[Tc(NS)X_4]^-$ preparation may be accomplished by the reaction

$$[TcX_6]^{2-} \xrightarrow{(NSCl)_3} [Tc(NS)X_4]^-$$

but such compounds readily lose the sulfur atom:[40]

$$[Tc(NS)X_4]^- \longrightarrow [TcNX_4]^-$$

Oxidation State III

It is in this oxidation state that the known chemistry of the two elements is most different. For rhenium di- and trinuclear cluster species predominate (Section 18-D-5), whereas for Tc there are only a few Tc_2^{6+} and no Tc_3^{9+} complexes. Mononuclear Tc^{3+} species do occur. The only simple $[MX_6]^{3-}$ complex known for either element is the $[Tc(NCS)_6]^{3-}$ ion, but the *mer*-$MCl_3(PR_3)_3$, MCl_3py_3, and $[MX_2(LL)_2]^+$ complexes with X = halide, NCS^-, or MeS^- and LL a diphosphine or diarsine are important.[41] Some of these complexes can be reduced with retention of structure in the II and even I oxidation states.[42]

Other simple, 6-coordinate M^{III} compounds include the $Tc(dike)_3$ and analogous monothio dike molecules. With very bulky ArS ligands (e.g., Ar = 2,4,6-trisisopropylphenyl), 5-coordinate $Tc(SAr)_3L_2$ complexes have been obtained.[43]

There are also numerous 7-coordinate complexes, such as the $[Re(terpy)_2X]^{2+}$ ions (X = OH, Cl, NCS),[44] in which the coordination is a distorted monocapped

[37]J. O. Dziegielewski *et al.*, *Polyhedron* **1995**, *14*, 555.
[38]T. Nicholson *et al.*, *Inorg. Chim. Acta* **1996**, *241*, 95.
[39]J. Lu and M. J. Clark, *Inorg. Chem.* **1990**, *29*, 4123; U. Abram *et al.*, *Inorg. Chim. Acta* **1991**, *181*, 161.
[40]U. Abram *et al.*, *Inorg. Chim. Acta* **1993**, *206*, 9.
[41]E. Deutsch *et al.*, *Inorg. Chem.* **1988**, *27*, 4113.
[42]E. Deutsch *et al.*, *Inorg. Chem.* **1988**, *27*, 3608, 3614.
[43]A. Davison *et al.*, *Inorg. Chem.* **1988**, *27*, 1574.
[44]E. Deutsch *et al.*, *Inorg. Chem.* **1994**, *33*, 3442.

trigonal prism. The simplest are the $[M(CN)_7]^{4-}$ ions which have a pentagonal bipyramidal shape.

M^{IV} Complexes

Both elements form a number of straightforward MX_6^{2-} complexes with X = F, Cl, Br, I, NCS, etc. There are also many *trans*-$MX_4(PR_3)_2$ compounds,[45] and another class of octahedral complexes are the tris-chelated species,[46] $[Tc(LL)_3]^+$, where LL may be a ligand such as 18-D-XIII.

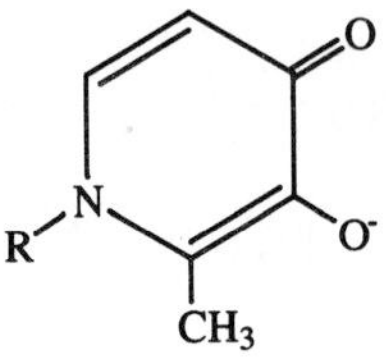

(18-D-XIII)

Eight-coordinate complexes such as $MCl_4(diars)_2$ are also known.

18-D-9   Complexes in Oxidation States V–VII

For both elements the dominant types of complexes in these oxidation states, especially V, are those containing MO^{3+}, MO_2^+, $M_2O_3^{4+}$, and MN^{2+} cores, and these have been covered in Sections 18-D-6 and 18-D-7. Thus there is only a little more to add here.

For oxidation state V we may note the rhenium complexes $[PCl_4][ReCl_6]$, made by reaction of the metal with PCl_5 at 500°C, and the $[TcCl_4(diars)_2]^+$ ion, as well as $[Re(CN)_8]^{3-}$.

For oxidation state VI there is a square prismatic anion in $K_2[ReF_8]$. In the compound $Re(S_2C_2Ph_2)_3$, one may also choose to assign the oxidation state as Re^0:

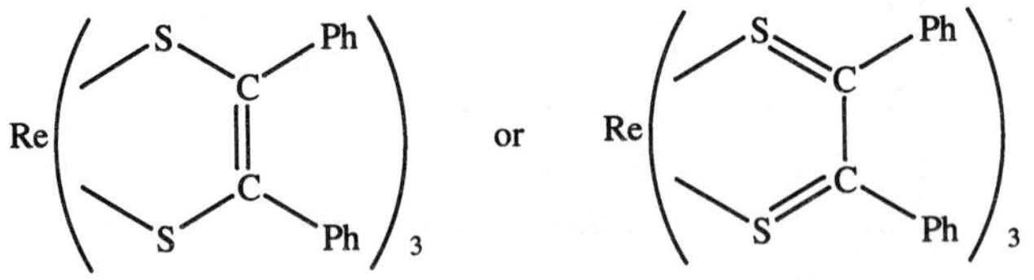

(18-D-XXIII)

For oxidation state VII there are no complexes other than oxo complexes or the organometallics (e.g., RMO_3) which are covered in Section 18-D-12.

[45] F. A. Cotton *et al.*, *Acta Cryst.* **1990**, *C46*, 1623; H. Kraudelt *et al.*, *Z. anorg. allg. Chem.* **1995**, *621*, 1797.
[46] C. Orwig *et al.*, *Inorg. Chem.* **1994**, *33*, 5607.

18-D-10 Hydrido and Dihydrogen Complexes

Both TcO_4^- and ReO_4^- can be reduced by Na or K in ethanol to give the only transition metal homoleptic hydrido complexes, $[MH_9]^{2-}$, which have been isolated as the Na, K, and (for ReH_9^{2-} only) Ba salts.[47] In the solid state the H atoms adopt a tricapped trigonal prismatic arrangement around M, but ^1H and ^{99}Tc nmr show that in solution the TcH_9^{2-} ion is fluxional with the two types of hydrogen atoms undergoing rapid exchange.

The only substituted derivative of $[TcH_9]^{2-}$ is $TcH_7(PEt_2Ph)_2$, but for rhenium there are a number of mixed phosphino/hydrido species. Thus, treatment of $ReOCl_3(PR_3)_2$ with hydridic reducing agents ($LiAlH_4$, $LiBH_4$[48]) gives $ReH_5(PR_3)_3$ as well as $ReH_7(PR_3)_2$. The arsine/hydrido compound, $ReH_7(AsPh_3)_2$ is also known.[49]

The structure[50] of $ReH_5(PMePh_2)_3$ shows a distorted dodecahedral arrangement. For the $ReH_7(PR_3)_2$ type compound, there is one case, $ReH_7(Ph_2PC_2H_4PPh_2)$, where the tricapped trigonal prism, as in $[ReH_9]^{2-}$, is preserved, with the dppe occupying two positions that define one vertical edge of the trigonal prism.[51] In $ReH_7(Ptol_3)_2$, however, this sort of structure is partly collapsed along one H···H edge to give an H···H distance of only 1.36 Å; this implies that the proper formulation of this compound may be intermediate between $ReH_7(PR_3)_2$ and $ReH_5(H_2)(PR_3)_2$.[52]

$ReH_7(PR_3)_2$ compounds undergo two important reactions:

$$Re_2H_8(PR_3)_4 \longleftarrow ReH_7(PR_3)_2 \xrightarrow{+H^+} [ReH_8(PR_3)_2]^+$$

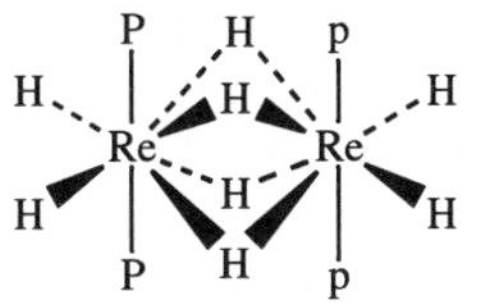

(18-D-XIV) 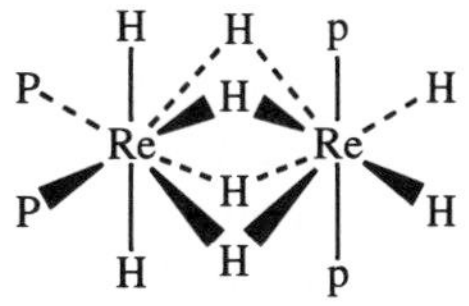   (18-D-XV)

Turning first to the binuclear octahydrido compounds, $[(H)_2(PR_3)_2Re]_2(\mu\text{-}H)_4$, we note that they are now most efficiently made by the action of $NaBH_4$ on $Re_2Cl_4(PR_3)_4$.[53] It is found that both eclipsed (18-D-XIV) and staggered structures (18-D-XV) are adopted, depending on the phosphine and the crystallization conditions.[54] There are also many other less symmetrical dirhenium polyhydrides.[55] For example, the following reaction sequence occurs:

$$Re_2H_8(PR_3)_4 \xrightarrow[-H_2]{+PR_3} Re_2H_6(PR_3)_5 \xrightarrow[-H_2]{+PR_3} Re_2H_4(PR_3)_6$$

[47]K. Yvon et al., Inorg. Chem. 1994, 33, 4598.
[48]B. L. Shaw et al., J. Chem. Soc., Dalton Trans. 1994, 917.
[49]M. L. Loza and R. H. Crabtree, Inorg. Chim. Acta 1995, 236, 63.
[50]T. F. Koetzle et al., Inorg. Chem. 1984, 23, 4012.
[51]J. A. K. Howard et al., J. Chem. Soc., Chem. Commun. 1988, 1502.
[52]L. Brammer et al., J. Chem. Soc., Chem. Commun. 1991, 241.
[53]R. A. Walton et al., Inorg. Chem. 1989, 28, 395, 3203; 1990, 29, 43.
[54]F. A. Cotton and R. L. Luck, Inorg. Chem. 1989, 28, 4522.
[55]See F. A. Cotton and R. A. Walton, Multiple Bonds between Metal Atoms, 2nd Ed., Oxford University Press, Oxford, UK, pp. 116–118.

For the $Re_2H_6(PR_3)_5$ molecules the structure[56] is that shown in 18-D-XVI.

(18-D-XVI)

Turning now to H_2 complexes, several compounds that are likely to be H_2 complexes have been reported. For technetium it would seem to be nearly certain that $TcCl(dppe)_2(H_2)$ is an example,[57] although neutron diffraction evidence is still lacking. However, it appears that there may be a related complex, $TcH_3(dppe)_2$, containing three H^- ligands, but obtained by a different synthetic route.[58]

With rhenium there are several possible or probable H_2 complexes as well as some cases that are very equivocal. It is believed that the $[Re(PR_3)_2H_8]^+$ ion, obtained by protonation of $ReH_7(PR_3)_2$ must contain at least one H_2 ligand, since a classical structure containing eight H^- ligands exceeds the possible valence of rhenium. Other protonated species, e.g., $[Re(PR_3)_3H_6]^+$, may well be classical polyhydrides. It is to be noted that protonation occurs only at low temperature; otherwise there is a reaction such as the following in which H_2 is evolved:

$$Re(PPh_3)_2H_7 + HBF_4 \xrightarrow{CH_3CN} [Re(PPh_3)_2(CH_3CN)_4H]^+$$

The compound $ReCl(H_2)(PMePh_2)_4$ seems to be an H_2 complex rather than a dihydride,[59] but the structure is not entirely certain and might contain an η^1-H_2 (i.e., end-on) or something intermediate between the η^1 and η^2 types.

For the $ReH_5(PR_3)_3$ compounds (PR_3 = PPh_3, PMe_2Ph),[60] there is some uncertainty as to whether the $Re(H_2)H_3(PR_3)_3$ or $ReH_5(PR_3)_3$ formulation is correct, although for $ReH_5(PMePh_2)_3$ there is a neutron diffraction study[61] to show that in the crystalline state all hydrogen atoms are hydride ions. There are also $ReH_3(PR_3)_4$ compounds where the $Re(H_2)H(PR_3)_4$ formulation has not been definitely ruled out in solution[62] but seems unlikely in the solid. The formulation $ReCl(H_2)(dppe)_2$ is preferred over $ReClH_2(dppe)_2$ but lacks direct proof.[63]

18-D-11   Tc and Re Carbonyl Compounds[64]

Both elements form the white $M_2(CO)_{10}$ compounds which are isomorphous and sublimable, like $Mn_2(CO)_{10}$. They are made by treatment of the oxides or MO_4^- salts with CO at high pressure (100–400 atm) and temperature (100–250°C) in

[56]R. A. Walton et al., Inorg. Chem. 1990, 29, 4437.
[57]G. J. Kubas et al., J. Am. Chem. Soc. 1994, 116, 1575.
[58]L. Kaden et al., Z. Chem. 1981, 21, 232.
[59]F. A. Cotton and R. L. Luck, Inorg. Chem. 1991, 30, 767.
[60] F. A. Cotton and R. L. Luck, Inorg. Chem. 1989, 28, 6; J. Am. Chem. Soc. 1989, 111, 5757.
[61]T. F. Koetzle et al., Inorg. Chem. 1984, 23, 4012.
[62]F. A. Cotton and R. L. Luck, Inorg. Chem. 1989, 28, 2181.
[63]R. L. Luck et al., Inorg. Chem. 1994, 33, 879.
[64]Comprehensive Organometallic Chemistry, 2nd ed., E. W. Abel, F. G. A. Stone, and G. Wilkinson, Eds., 1995, Pergamon Press, New York, Vol. 6.

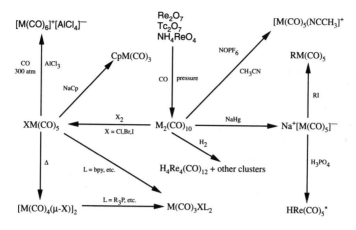

*No hydridotechnetium carbonyl compounds

Figure 18-D-7 Some important reactions of the $M_2(CO)_{10}$ compounds, M = Tc, Re.

yields of 70–95%.[65] Recently a 62% yield of $Re_2(CO)_{10}$ has been obtained by reaction of CO at atmospheric pressure with NH_4ReO_4 reduced with $(Bu^i)_2AlH$.[66] They have the structure shown as 18-D-XVII with long weak M—M bonds (3.04 Å for Tc and 3.02 Å for Re). These weak bonds are readily cleaved to give a variety of products,[67] some of which are shown in Fig. 18-D-7. There are also a number of triangular and tetrahedral hydrido carbonyl clusters, such as $[Re_3H_2(CO)_{12}]^-$, $Re_4H_4(CO)_{12}$, $[Re_4(CO)_{16}]^{2-}$, and substitution products thereof, such as $Re_3H_3(CO)_{11}(PPh_3)$.

$$\begin{array}{c} \text{OC—M} \underset{}{\overset{}{\longrightarrow}} \text{M—CO} \end{array}$$

(18-D-XVII)

Among the important cleavage products of the $M_2(CO)_{10}$ molecules are the halopentacarbonyls, $XM(CO)_5$ and the $M(\mu\text{-}X)_2(CO)_8$ molecules that are formed by thermal loss of CO. From either of these halocarbonyls more CO can be displaced by neutral ligands to give tricarbonyls (which are usually the facial isomer) such as the bpy compound,[68] 18-D-XVIII.

[65]W. A. Herrmann *et al.*, *Chem. Ber.* **1991**, *124*, 1107; F. Calderazzo *et al.*, *Gazz. Chim. Ital.* **1989**, *119*, 241; D. M. Heinekey *et al.*, *J. Organomet. Chem.* **1988**, *342*, 243.
[66]G. Jaouen *et al.*, *J. Chem. Soc., Chem. Commun.* **1996**, 3611.
[67]C. C. Ramão in *Encyclopedia of Inorganic Chemistry*, R. B. King, Ed., John Wiley and Sons, New York, **1994**, p. 3437; R. J. Clark, *loc. cit.*, p. 4099.
[68]A. Juris *et al.*, *Inorg. Chem.* **1988**, *27*, 4007.

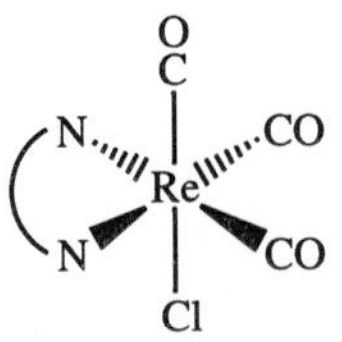

(18-D-XVIII)

18-D-12   Organometallic Chemistry[63]

Both elements have a significant organometallic chemistry, although that of Re is far more extensive. Indeed, it is so extensive that only a few key compounds of rhenium can be mentioned here, followed by brief remarks on technetium compounds.

Alkyl and Aryl Compounds

A few homoleptic compounds are known. These include $ReMe_6$ (explosive) which is formed by the action of $AlMe_3$ on $ReOMe_4$, and reacts with LiMe to generate the $[ReMe_8]^{2-}$ ion. There is also the $[Re_2Me_8]^{2-}$ ion, derived from $[Re_2Cl_8]^{2-}$ and $[Re_3Me_9]_n$ derived from Re_3Cl_9. $Re_2(allyl)_4$ and a number of volatile Re_3R_{12} (R = Me, Me_3SiCH_2, and Me_3CCH_2) compounds are also known. The aryl $Re(o\text{-}MeC_6H_4)_4$ has a slightly flattened tetrahedral structure.

Among the most important alkyl and aryl compounds are those in which the ligands CO, NO, NR, and O are also present. The carbonyl compounds $CpRe(CO)_3$ and $Cp^*Re(CO)_3$ are very important starting materials for much other chemistry and are made by the following reactions:

$$BrRe(CO)_5 + CpSnMe_3 \text{ or } NaCp \longrightarrow CpRe(CO)_3$$

$$Re_2(CO)_{10} + Cp^*H \longrightarrow Cp^*Re(CO)_3$$

Among the many reactions of these species is the conversion to chiral species such as 18-D-XIX. This compound can be obtained in enantiomerically pure form and can be converted to the $CpRe(NO)PPh_3^+$ ion.[69] This ion is a chiral Lewis base that binds a variety of prochiral molecules (olefins, ketones, aldehydes, amines). With these adducts one may conduct numerous reactions where enantiomeric excesses >98% are obtained. As an example, a prochiral methyl ketone will bind selectively, as in 18-D-XX and is then subject to attack by R'X to produce only one enantiomer of the $RR'MeCO^-$ product.

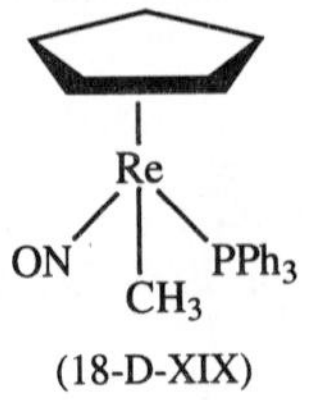

(18-D-XIX) 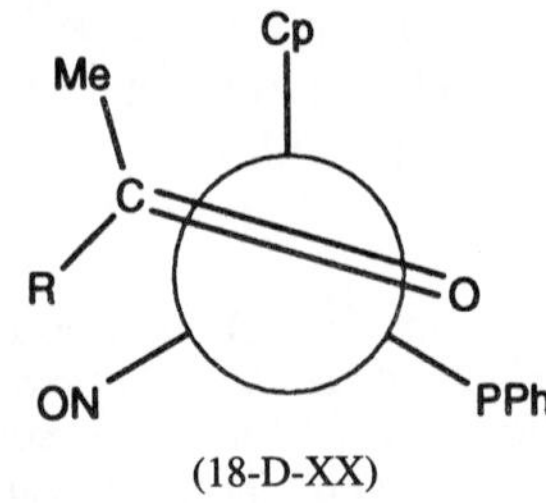   (18-D-XX)

[69]J. A. Gladysz et al., J. Organomet. Chem. **1995**, 503, 235, and earlier references therein.

RReO₃ Compounds

This very important class of compounds has recently been the subject of intensive study. Two of the important R groups are CH_3 and C_5R_5 (R = Me, Et, etc.). Turning first to CH_3ReO_3 (MTO), it was discovered as a product of long exposure of Me_4ReO or Me_3ReO_2 to dry air. The development of efficient syntheses[70] and the discovery of its many catalytic capabilities,[71] e.g., for olefin epoxidation, olefin metathesis, oxidation of organic sulfides, arenes, phenols, aldehydes, and others,[72] has led to MTO being one of the most thoroughly studied organometallic compounds. The most efficient synthesis (virtually quantitative) employs the following reaction:

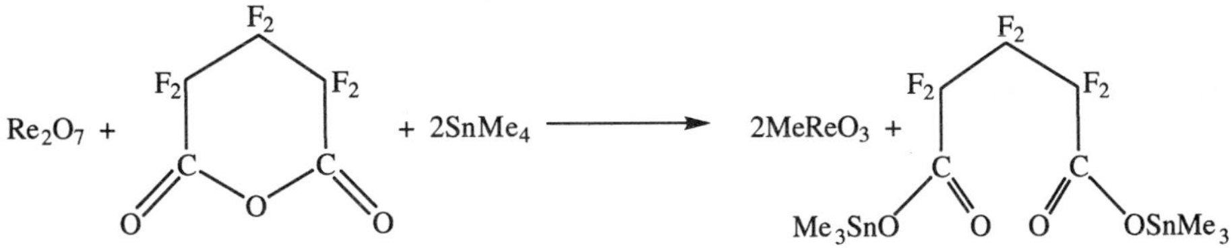

Because of the numerous and diverse oxidation reactions catalyzed by MTO, in all of which aqueous H_2O_2 is the oxidant, the interaction of MTO and H_2O_2 has been much studied.[73] The key intermediates are MTO/H_2O_2 adducts, formed by the following reactions:

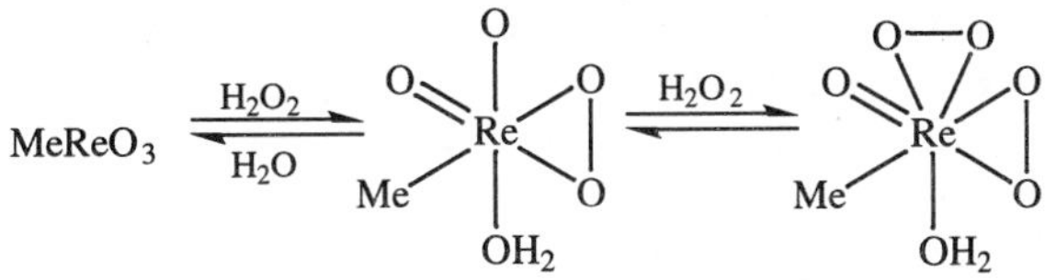

One or both of the adducts may be active in various oxidations, and it is remarkable that the MTO/H_2O_2 system can effect both nucleophilic attack (Bayer-Villiger reaction) and electrophilic attack (olefin epoxidation).

$(C_5Me_5)ReO_3$ (Cp*ReO₃) is the other prototypical RReO₃ compound. It has the expected pianostool structure, 18-D-XXI. For unknown reasons it has proven impossible to prepare the C_5H_5 analogue. While Cp*ReO₃ has not displayed catalytic activity, it has an extensive chemistry,[74] as shown in Fig. 18-D-8.

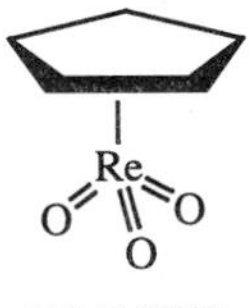

(18-D-XXI)

[70]W. A. Herrmann et al., Inorg. Chem. **1992**, 31, 4431.
[71]W. A. Herrmann et al., Angew. Chem. Int. Ed. Engl. **1991**, 30, 1636, 1638, 1641.
[72]W. A. Herrmann et al., Organometallics **1994**, 13, 4531.
[73]J. H. Espenson et al., J. Chem. Soc., Dalton Trans. **1995**, 133; J. Am. Chem. Soc. **1996**, 118, 4966.
[74]W. A. Herrmann, Angew. Chem. Int. Ed. Engl. **1988**, 27, 1297.

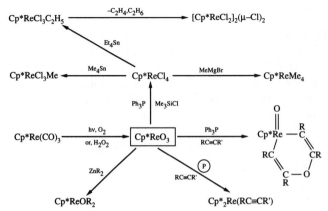

Figure 18-D-8 Some representative reactions of Cp*ReO$_3$ and Cp*ReCl$_4$. (P) represents polymer-bound Ph$_3$P.

Technetium Compounds

Organometallic compounds of Tc have been only lightly investigated since they have neither pharmaceutical nor industrial applications. What little is known suggests that they will be generally similar to those of rhenium. Thus there are numerous Cp and Cp* compounds such as the carbonyls Cp*Tc(μ-CO)$_3$TcCp*, arene compounds such as Ar$_2$Tc$^+$ and ArTc(CO)$_3^+$, and extremely volatile MeTcO$_3$, which seems to have catalytic properties similar to those of MeReO$_3$.

Preparation of MeTcO$_3$ by treatment of Tc$_2$O$_7$ with SnMe$_4$ must be done at low temperature else a mixture of Me$_3$SnOTcO$_3$ and the dinuclear complex,[75] 18-D-XXII, results. There is, however, no Tc analogue of Cp*ReO$_3$, nor does the alleged Cp*Tc$_2$O$_3$ exist.[76]

$$\text{(18-D-XXII)}$$

Additional References

J. Baldas, *Adv. Inorg. Chem.* **1994,** *41,* 1. General overview of Tc chemistry, *Radiochimica Acta* **1993,** *63:* Special Issue on Behavior and Utilization of Technetium.

K. Yoshihara and T. Omosi, Eds., *Topics in Current Chemistry,* Vol. 176, Springer Verlag, Berlin, 1996. Specialized reviews on technetium and a chapter on rhenium in nuclear medicine.

[75]W. A. Herrmann *et al., Angew. Chem. Int. Ed. Engl.* **1990,** *29,* 189.
[76]F. A. Cotton *et al., Inorg. Chem.* **1995,** *34,* 4253.

18-E THE PLATINUM GROUP METALS[1]

In this section we consider only the occurrence, the metals, and their most important binary compounds. Other chemistry is discussed under the sections for the individual elements.

18-E-1 Occurrence

Ruthenium, osmium, rhodium, iridium, palladium, and platinum are the six heaviest members of Groups 8-10. They are rare elements; Pt is the most common with an abundance of $\sim 10^{-6}\%$, whereas the others have abundances $\sim 10^{-7}\%$ of the earth's crust. They occur as metals, often as alloys such as "osmiridium," and in sulfide, arsenide, and other ores. They are commonly associated with Cu, Ag, and Au. Abnormally high Ir in some rocks has been assumed to come from a meteorite 65 million years ago. The main sources of the metals are South Africa, the CIS, and Canada; in all the ores the "values" of the platinum metals are in grams per ton, and concentration by gravitation and flotation is required. Extraction methods depend on the ore, but the concentrate is smelted with coke, lime, and sand and bessemerized in a converter. The resulting Ni-Cu sulfide "matte" is cast into anodes. On electrolysis in sulfuric acid solution, Cu is deposited at the cathode, and Ni remains in solution, from which it is subsequently recovered by electrodeposition. The platinum metals, Ag, and Au collect in the anode slimes. The subsequent procedures for separation of the elements are complicated. Although most of the separations used to involve classical precipitations, with fractional crystallization of salts and so on, current procedures involve ion exchange and particularly solvent extraction, for example, by 2-nonylpyridine-1-oxide. Ruthenium and Os are recovered first, however, as the volatile oxides, MO_4, from oxidizing solutions.

18-E-2 The Metals

Some properties of the platinum metals are given in Table 18-E-1.

The metals are obtained initially as sponge or powder by heating ammonium salts of the hexachloro anions. Almost all complex and binary compounds of the elements give the metal when heated above 200°C in air or oxygen; Os is oxidized to the volatile OsO_4, and at dull red heat Ru gives RuO_2, so that reduction in hydrogen is necessary. The finely divided metals are also obtained by reduction of acidic solutions of salts or complexes by Mg, Zn, H_2, or other reducing agents such as citrate, oxalic, or formic acid or by electrolysis under proper conditions.

The metals, as gauze or foil, and especially on supports such as charcoal or alumina onto which the metal salts are absorbed and reduced *in situ* under specified conditions, are widely used as catalysts for an extremely large range of reactions in the gas phase or in solution. One of the larger uses of Pt is for the reforming of hydrocarbons. Commercial uses in homogeneous reactions are fewer, but palladium is used in the Smidt process and Rh in hydroformylation and in acetic acid synthesis.

[1]F. R. Hartley, Ed., *Chemistry of the Platinum Group Metals: Recent Developments,* Elsevier, Amsterdam, 1991.

Table 18-E-1   Some Properties of the Platinum Metals

Element	mp (°C)	Form	Best solvent
Ru	2546	Gray-white, brittle, fairly hard	Alkaline oxidizing fusion
Os	~3050	Gray-white, brittle, fairly hard	Alkaline oxidizing fusion
Rh	1960	Silver-white, soft, ductile	Hot concentrated H_2SO_4; concentrated HCl + $NaClO_3$ at 125–150°C
Ir	2443	Silver-white, hard, brittle	Concentrated HCl + $NaClO_3$ at 125–150°C
Pd	1552	Gray-white, lustrous, malleable, ductile	Concentrated HNO_3, HCl + Cl_2
Pt	1769	Gray-white, lustrous, malleable, ductile	Aqua regia

Industrially, as well as in the laboratory, catalytic reductions are especially important.

Platinum or its alloys are used for electrical contacts, for printed circuitry, for plating and on a large scale, together with Rh, for automobile exhaust convertors.

Ruthenium and Os are unaffected by mineral acids below ~100°C and are best dissolved by alkaline oxidizing fusion (e.g., NaOH + Na_2O_2, $KClO_3$). Rhodium and Ir are extremely resistant to attack by acids, neither metal dissolving even in aqua regia when in the massive state. Finely divided Rh can be dissolved in aqua regia or hot concentrated H_2SO_4. Both metals also dissolve in concentrated HCl under pressure of oxygen or in presence of sodium chlorate in a sealed tube at 125–150°C. At red heat interaction with Cl_2 leads to the trichlorides.

Palladium and Pt are more reactive than the other metals. Palladium is dissolved by nitric acid, giving $Pd^{IV}(NO_3)_2(OH)_2$; in the massive state the attack is slow, but it is accelerated by oxygen and oxides of nitrogen. As a sponge, Pd also dissolves slowly in HCl in the presence of chlorine or oxygen. Platinum is considerably more resistant to acids and is not attacked by any single mineral acid, although it readily dissolves in aqua regia and even slowly in HCl in the presence of air since

$$PtCl_4^{2-} + 2e = Pt + 4Cl^- \qquad E^0 = 0.75 \text{ V}$$

$$PtCl_6^{2-} + 2e = PtCl_4^{2-} + 2Cl^- \qquad E^0 = 0.77 \text{ V}$$

Platinum is not the inert material that it is often considered to be. There are a great many oxidation-reduction and decomposition reactions in solution that are catalyzed by metallic platinum. Examples are the Ce^{IV}—Br^- reaction and the decomposition of N_2H_4 to N_2 and NH_3. It is possible to predict whether catalysis can occur from a knowledge of the electrochemical properties of the reacting couples.

Both Pd and Pt are rapidly attacked by fused alkali oxides, and especially by their peroxides, and by F_2 and Cl_2 at red heat. In using platinum vessels or other equipment it should be noted that the metal reacts with elemental P, Si, Pb, As,

Sb, S, and Se and also with their compounds when these are heated under reducing conditions.

Both Pd and Pt are capable of absorbing large volumes of molecular hydrogen, and Pd is used for the purification of H_2 by diffusion.

BINARY COMPOUNDS

18-E-3 Oxides, Sulfides, Phosphides, and Similar Compounds

Oxides

The best known anhydrous oxides are listed in Table 18-E-2; the tetraoxides of Ru and Os are discussed later (Section 18-F-1). The oxides, generally rather inert to aqueous acids, are reduced to the metal by hydrogen, and dissociate on heating. There are *mixed metal oxides*, e.g., $BaRuO_3$, and platinum and palladium "bronzes" of formula $M_x^I Pt_3O_4$ ($x = 0–1$). Some oxides like $M^{II}Pt_3O_6$ have Pt—Pt bonds. Mixed oxides are used for electrodes in H_2—O_2 fuel cells and in the chloralkali process.

Table 18-E-2 Anhydrous Oxides of Platinum Metals[a]

Oxide	Color/form	Structure	Comment
RuO_2	Blue-black	Rutile	From O_2 on Ru at 1250°C or $RuCl_3$ at 500–700°C; usually O-defective
RuO_4	Orange-yellow crystals, mp 25°C, bp 129.6°C	Tetrahedral molecules	See Section 18-F-1
OsO_2	Coppery	Rutile	Heat Os in NO or OsO_4 or dry $OsO_2 \cdot nH_2O$
OsO_4	Colorless crystals, mp 40°C, bp 129.7°C	Tetrahedral molecules	Normal product of heating Os in air; see Section 18-F-1
Rh_2O_3	Brown	Corundum	Heat Rh^{III} nitrate or Rh_2O_3(aq)
RhO_2	Black	Rutile	Heat Rh_2O_3(aq) at 700–800°C in high-pressure O_2
Ir_2O_3	Brown		Impure by heating $K_2IrCl_6 + Na_2CO_3$
IrO_2	Black	Rutile	Normal product of Ir + O_2; dissociates >1100°C
PdO	Black		From Pd + O_2; dissociates at 875°C; insoluble in all acids
PtO_2	Brown		Dehydrate PtO_2(aq); dec 650°C

[a]In oxygen at 800–1500°C, gaseous oxides exist: RuO_3, OsO_3, RhO_2, IrO_3, and PtO_2.

Hydrous oxides are commonly precipitated when NaOH is added to aqueous metal solutions, but they are difficult to free from alkali ions and sometimes readily become colloidal. When freshly precipitated, they may be soluble in acids, but only with great difficulty or not at all after aging.

Representative are various forms of $RuO_2 \cdot nH_2O$[2] and yellow $Rh_2O_3 \cdot nH_2O$. In base solution, powerful oxidants convert the latter into $RhO_2 \cdot nH_2O$ which loses oxygen on dehydration.

The hydrous oxide, $Ir_2O_3 \cdot nH_2O$, can be obtained only in moist atmospheres; it is at least partially oxidized by air to $IrO_2 \cdot nH_2O$, which is formed either by the action of mild oxidants on $Ir_2O_3 \cdot nH_2O$ or by addition of OH^- to $IrCl_6^{2-}$ in the presence of H_2O_2. The precipitation of the oxides of Rh and Ir from buffered $NaHCO_3$ solutions by the action of ClO_2^- or BrO_3^- provides a rather selective separation of these elements.

The hydrous oxide, $PdO \cdot nH_2O$, is a yellow gelatinous precipitate that dries in air to a brown, less hydrated form and at 100°C loses more water, eventually becoming black; it cannot be dehydrated completely without loss of oxygen.

When $PtCl_6^{2-}$ is boiled with Na_2CO_3, red-brown $PtO_2 \cdot nH_2O$ is obtained. It dissolves in acids and also in strong alkalis to give solutions of hexahydroxoplatinate(IV), $[Pt(OH)_6]^{2-}$. The hydrous oxide becomes insoluble on heating to ~200°C. The brown material formed by fusion of $NaNO_3$ and chloroplatinic acid at ~550°C followed by extraction of soluble salts with water is known as Adams's catalyst and is widely used in organic chemistry for catalytic hydrogenations.

A very unstable Pt^{II} hydrous oxide is obtained by addition of OH^- to $PtCl_4^{2-}$, after drying in CO_2 at 120–150°C it approximates to $Pt(OH)_2$, but at higher temperatures gives PtO_2 and Pt.

Sulfides, Phosphides, and Similar Compounds

Direct interaction of the metal and other elements such as S, Se, Te, P, As, Sn, or Pb under selected conditions produces dark, often semi-metallic solids that are resistant to acids other than HNO_3. These products may be stoichiometric compounds and/or nonstoichiometric phases depending on the conditions of preparation.

The chalcogenides and phosphides are generally rather similar to those of other transition metals; indeed many of the phosphides, for example, are isostructural with those of the iron group, namely, Ru_2P with Co_2P; RuP with FeP and CoP; RhP_3, PdP_3 with CoP_3 and NiP_3.

Sulfides can also be obtained by passing H_2S into platinum metal salt solutions. Thus from $PtCl_4^{2-}$ and $PtCl_6^{2-}$ are obtained PtS and PtS_2 respectively; from Pd^{II} solutions PdS, which when heated with S gives PdS_2; the Rh^{III} and Ir^{III} sulfides are assumed to be $M_2S_3 \cdot nH_2O$, but exact compositions are uncertain.

The disulfides have pyrite (Ru[3] and Os), distorted pyrite (Pd) and CdI_2 (Pt) structures; all are diamagnetic. There are various mixed sulfides such as Ta_2PdS_6.

[2]A. Mills and H. L. Davies, *Inorg. Chim. Acta* **1991,** *189,* 149.
[3]Y.-S. Huang and S.-S. Lin, *Mat. Res. Bull.* **1988,** *23,* 277.

Table 18-E-3 Fluorides of the Platinum Metals

II	III	IV	V	VI
	RuF_3	RuF_4	Ru_4F_{20}	RuF_6
	Brown	Pink	Dark green[a]	Dark brown
			mp 86.5°C;	mp 54°C
			bp 227°C	
		OsF_4	Os_4F_{20}	OsF_6[b]
		Yellow-orange	Blue[a]	Pale yellow[a]
			mp 70°C;	mp 33.2°C
	RhF_3	RhF_4	Rh_4F_{20}	RhF_6
	Red	Purple-red	Dark red	Black
	IrF_3	IrF_4	Ir_4F_{20}	IrF_6
	Dark brown	Red-brown	Yellow	Yellow
			mp 105°C	mp 43.8°C; bp 53.6°C
PdF_2		PdF_4		
Violet		Brick-red		
		PtF_4	PtF_5	PtF_6[b]
		Yellow-brown	Deep red	Dark red
			mp 80°C	mp 61.3°C; bp 69.1°C

[a]Colorless vapor.
[b]Slightly distorted octahedral (K. Seppelt *et al.*, *J. Chem. Phys.* **1996**, *104*, 7658).

18-E-4 Halides of the Platinum Metals

This section describes binary halides; halide complexes are discussed later. The fluorides are listed in Table 18-E-3.

Hexafluorides

The hexafluorides[4] can be obtained by interaction of the metal and F_2 under pressure at elevated temperatures, but for RuF_6 and RhF_6, which are the least stable, though storable in "passivated" form, a preferred synthesis is fluorination of $(MF_5)_4$ at *ca.* 50–60 atm.

The hexafluorides are extraordinarily reactive, volatile, and corrosive substances and normally must be handled in Ni or Monel apparatus, although quartz can be used. Only PtF_6 and RhF_6 actually react with glass (even when rigorously dry) at room temperature. In addition to thermal dissociation, uv radiation causes decomposition to lower fluorides, even OsF_6 giving OsF_5. The vapors hydrolyze with water vapor, and liquid water reacts violently; for example, IrF_6 gives HF, O_2, O_3, and $IrO_2(aq)$; OsF_6 gives OsF_4, HF, and OsF_6^-.

Platinum hexafluoride is one of the most powerful oxidizing agents, reacting with O_2 to give $O_2^+PtF_6^-$ and with Xe to give a solid whose composition approximates $Xe(PtF_6)$. The molecules are octahedral, monomeric in the vapor phase with M—F bond lengths in the range 1.81–1.84 Å; spectra and magnetic properties have been determined.

[4]W. Levason *et al.*, *J. Chem. Soc., Dalton Trans.* **1992**, *139*, 447.

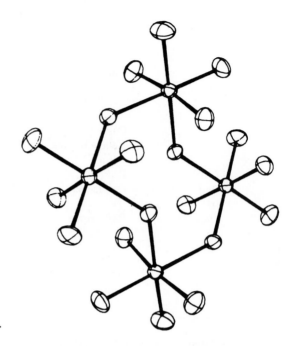

Figure 18-E-1   Structure of Rh_4F_{20}.

Pentafluorides

These can be obtained by fluorination at *ca.* 300–400°C. However, $(OsF_5)_4$ is best made by reduction of OsF_6, e.g., by I_2 in IF_5.[5] All are very reactive and readily hydrolyzed. In the solid state the structures are tetramers with rings as shown in the representative structure of Rh_4F_{20}, Fig. 18-E-1; there are bent M—F—M bridges and differences in the non-bridging M—F bond distances.[6]

Where structures in the vapor phase are available, it appears that the aggregation depends on the temperature; for Ru and Os there are trimers, $M_3(\mu\text{-}F)_3F_{12}$, and dimers, the former having rings with bent M–F–M bonds, but boat and chair forms cannot be distinguished.[5] It may be noted here that all the above fluorides, and indeed transition metal binary fluorides in general, have MF_6 octahedra whether they are monomeric or polymeric; for M^V there are two bridging F atoms (not always bent; for Nb they are linear), for M^{IV} four and for M^{III}, all six are bridging (for references see ref. 6).

Tetrafluorides

These may be obtained by reactions such as

$$10RuF_5 + I_2 = 10RuF_4 + 2IF_5$$

$$2IrF_5 + H_2 = 2IrF_4 + 2HF$$

[5]D. A. Rice *et al., Inorg. Chem.* **1993,** *32,* 4311 (Ru, Os).
[6]N. Bartlett *et al., Inorg. Chem.* **1992,** *31,* 3124 (RuF_n, n = 3, 4, 5).

$$RhCl_3 \xrightarrow{BrF_3(l)} RhF_4 \cdot 2BrF_3 \xrightarrow{heat} RhF_4$$

$$Pt \xrightarrow{BrF_3(l)} PtF_4 \cdot 2BrF_3 \xrightarrow{heat} PtF_4 \xleftarrow{F_2} PtBr_4$$

$$PdBr_2 \xrightarrow{BrF_3(l)} Pd^{II}Pd^{IV}F_6 \xrightarrow{F_2,\ 100\ psi,\ 150\,°C} PdF_4$$

The deep pink RuF_4 can be made by interaction of RuF_6^{2-} with AsF_5 in liquid HF.[6] As noted above the isostructural compounds are octahedral with bridging fluorides in a 3-dimensional puckered sheet structure.

Trifluorides

These are obtained by reactions of the type

$$5RuF_5 + 2Ir \xrightarrow{250\,°C} 5RuF_3 + 2IrF_5 \uparrow$$

$$RhCl_3 + 3F_2 \xrightarrow{500\,°C} RhF_3 + 3ClF$$

$$RuF_5 + SF_4 \longrightarrow SF_3^+RuF_6^- \xrightarrow{heat} SF_6 + RuF_3$$

In the structure[6] of RuF_3 as an example there is an infinite hexagonal array of nearly close-packed F atoms and each Ru atom is octahedral with bent Ru–F–Ru bridges.

The only *difluoride* is the paramagnetic PdF_2 that is obtained by the route

$$PdBr_2 \xrightarrow{BrF_3} Pd^{2+}[PdF_6]^{2-} \xrightarrow[reflux]{SeF_4} PdF_2$$

Chlorides, Bromides, and Iodides

The anhydrous halides (other than fluorides—see Table 18-E-3) are listed in Table 18-E-4; they are normally obtained by direct interaction under selected conditions. We discuss only some of the more important chlorides.

Ruthenium trichloride[7] has two forms. Reaction of Cl_2/N_2 with $Ru_3(CO)_{12}$ at 360°C gives the brown, powdery β-form which is converted to the black crystalline α-form on heating in Cl_2. The α-form is antiferromagnetic at low temperature but the β-form is paramagnetic; the difference is ascribed to differences in the lattice structure; layer *vs.* linear chain.[7]

Osmium tetrachloride is formed when an excess of Cl_2 is used at temperatures above 650°C; otherwise a mixture with the trichloride is formed. The latter is obtained when $OsCl_4$ is decomposed at 470°C in a flow system with a low pressure of chlorine. Osmium tetrachloride in its orthorhombic high-temperature form has infinite chains of octahedra sharing opposite edges with no structural indication of metal-metal bonding.

[7]H.-J. Cantow *et al.*, *Angew. Chem. Int. Ed. Engl.* **1990**, *29*, 537; Y. Kobayashi *et al.*, *Inorg. Chem.* **1992**, *31*, 4570.

Table 18-E-4   Anhydrous Chlorides, Bromides, and Iodides of Platinum Metals

Oxidation state	Ru	Os	Rh	Ir	Pd	Pt
II					$PdCl_2$ Dark red $PdBr_2$ Dark red PdI_2 Black	$PtCl_2$ Red-black $PtBr_2$ Brown PtI_2 Black
III	$RuCl_3$ $RuBr_3$ Dark brown RuI_3 Black	$OsCl_3$ Dark gray $OsBr_3$ Black OsI_3 Black	$RhCl_3$ Red $RhBr_3$ Dark red $RhI_3(?)$ Black	$IrCl_3$ Brown-red $IrBr_3$ Yellow IrI_3 Black		$PtCl_3$ Green-black $PtBr_3$ Green-black $PtI_3(?)$ Black
IV		$OsCl_4$ Black $OsBr_4$ Black		$IrCl_4(?)$ $IrBr_4(?)$ $IrI_4(?)$		$PtCl_4$ Red-brown mp 370°C $PtBr_4$ Dark red PtI_4 Brown-black
V		$(OsCl_5)_2$ Black				

The *pentachloride*, made by interaction of OsO_4 and SCl_2, is isomorphous with Re_2Cl_{10}, but the Os–Os distance is too long for metal-metal bonding and the antiferromagnetic coupling is through the two μ-Cl atoms.

Rhodium trichloride has a layer lattice isostructural with $AlCl_3$ and is exceedingly inert. However, when $RhCl_3 \cdot 3H_2O$ is dehydrated in dry HCl at 180°C, the red product is much more reactive and dissolves in water or THF; this property is lost on heating at 300°C.

Iridium trichloride is also very inert and useless as a starting material (see Section 18-G-3).

Palladium(II) chloride. Two, or possibly more, forms are known, but the α and β forms are well characterized. They can be obtained from chlorination of the metal

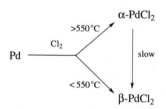

but the β-form, Pd_6Cl_{12} is best made otherwise. Solutions of $PdCl_2(PhCN)_2$ in benzene/$CHCl_3$ on standing give dark red prisms.[8] Crystals are also obtained by allowing

[8] M. M. Olmstead *et al., J. Am. Chem. Soc.* **1996**, *118*, 7737.

solutions of $Pd_3(O_2CMe)_6$ in glacial acetic acid to react with HCl. Similarly, any polycrystalline form of $PdCl_2$ suspended in $SOCl_2$ under CO (1 atm) gives solutions in which a carbonyl is formed:

$$2PdCl_2 + 2CO \xrightleftharpoons[]{SOCl_2} [Pd_2Cl_4(CO)_2]_{solv}$$

Over several days rhombohedral crystals of Pt_6Cl_{12} are obtained at room temperature.[9] Unlike the Ni halides or PdF_2, which are ionic and paramagnetic, these chlorides are molecular or polymeric and diamagnetic, and Born-Haber calculations indicate that ionic lattices would be endothermic.

The structures of the α- and β-forms are shown in diagrams 18-E-I and 18-E-II respectively. The α-form has linear chains with square coordination of Pd and bridging chlorine atoms while the β-form has 6 metal atoms with square planar coordination in an octahedron whose tetragonal planes have four Cl bridges. The Pd_6Cl_{12} is insoluble in CH_2Cl_2 or $CHCl_3$ but is readily soluble in aromatic solvents. Arenes such as mesitylene can form 1:1 adducts and a fullerene compound $Pd_6Cl_{12} \cdot 0.5C_{60} \cdot 1.5C_6H_6$ has also been characterized by X-ray diffraction. Interaction of a benzene solution with pyridine, PPh_3, RCN, and similar ligands can lead to complexes of the types L_2PdCl_2 and $[LPdCl\mu\text{-}Cl]_2$. With concentrated HCl, Pd_6Cl_{12} gives $[PdCl_4]^{2-}$.

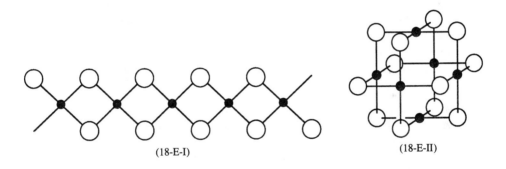

(18-E-I)   (18-E-II)

Platinum(II) chloride. This also has α and β-forms. The β-form is isomorphous with Pd_6Cl_{12}, but the structure of the α-form is still uncertain. The compounds are obtained by the reactions:

$$H_2PtCl_6 \cdot 6H_2O \xrightarrow{Cl_2 \sim 500\,°C} PtCl_4 \xrightarrow{>350\,°C} \beta\text{-}PtCl_2$$

$$\xrightarrow[\text{1-2 days}]{500\,°C}$$

$$Pt \xrightarrow[\text{gradient}]{Cl_2 \sim 650°C \longrightarrow 500\,°C} \alpha\text{-}PtCl_2$$

[9]D. B. Dell Amico *et al.*, *Angew. Chem. Int. Ed. Engl.* **1996**, *35*, 1331.

As for Pd_6Cl_{12}, adducts with benzene, CS_2, $CHCl_3$, and Br are known. On heating with $AlCl_3$, Pt_6Cl_{12} gives a purple vapor of an aluminum chloride complex.

The greenish-black "*trichloride*" $PtCl_3$ contains both Pt^{II} and Pt^{IV} with units of $[Pt_6Cl_{12}]$ and an infinite chain, $[PtCl_2Cl_{4/2}]_\infty$, containing distorted $PtCl_6$ octahedra linked by common edges, similar to the chain of PtI_4. The chloride and the similar $PtBr_3$ and PtI_3 are made by thermal gradient reactions of Pt and halogen.

Platinum(IV) chloride is commonly made by heating $(H_3O)_2PtCl_6$ to 300°C in Cl_2, but is best made by interaction of Pt and SO_2Cl_2. The reddish-brown crystals are readily soluble in water, presumably giving ions such as $[PtCl_4(OH)_2]^{2-}$, and in polar solvents. The structure, isomorphous to those of $PtBr_4$ and α-PtI_4, consists of infinite chains of $PtCl_6$ octahedra sharing edges, with the unshared Cl atoms in *cis* positions.[10]

Oxohalides

The best characterized are the fluorides.[11] The Os^{VII} compound *cis*-OsO_2F_4[12] is a purple solid, mp 90°C, made by interaction of OsO_4 with KrF_2 in HF(l). The structure was determined in the vapor phase by electron diffraction. The halide reacts with strong F^- acceptors, AsF_5 and SbF_5 in HF(l), to form orange salts of the cation $[(cis\text{-}OsO_2F_3)_2(\mu\text{-}F)]^+$ with AsF_6^- and $Sb_2F_{11}^-$.[13]

The other Os^{VIII} compound is the yellow OsO_3F_2 that has a chain structure in the solid with symmetrical, nonlinear F bridges. It is made by interaction of OsO_4 with ClF_3; $OsOF_5$ made simultaneously is sublimed away.[14] The compound, $Os^{VII}OF_5$ can also be obtained by interaction of OsO_4 with OsF_6; it is paramagnetic with a distorted octahedral structure.[15] Other fluorides are $OsOF_4$ and $RuOF_4$, the latter decomposes to RuF_4 at *ca.* 70°C. Interaction of OsO_4 and BCl_3 leads to the red-brown $OsOCl_4$, mp 32°C, while interaction of RuS_2 and S_2Cl_2 gives $RuSCl_4$.

18-F   RUTHENIUM AND OSMIUM: GROUP 8

The chemistry of Ru and Os[1] bears little resemblance to that of Fe except in some solids such as sulfides or phosphides and in some complexes with π-bonding ligands. The higher oxidation states are much more easily obtained than for iron. There is an extensive chemistry of M=O species for both elements.

The oxidation states and stereochemistries are summarized in Table 18-F-1.

[10]M. F. Pilbrow, *Chem. Commun.* **1972,** 270.
[11]See R. Bougon *et al., J. Fluorine Chem.* **1994,** *67,* 271 for preparations and references.
[12]K. O. Christe *et al., J. Am. Chem. Soc.* **1993,** *115,* 11279.
[13]G. J. Schrobilgen *et al., Inorg. Chem.* **1996,** *35,* 4310.
[14]K. Seppelt *et al., Chem. Ber.* **1993,** *126,* 1331; A. Veldkamp and G. Frenkling, *Chem. Ber.* **1993,** *126,* 1375 (theoretical).
[15]J. B. Raynor *et al., J. Chem. Soc., Dalton Trans.* **1992,** 1131.
[1]B. K. Gosh and A. Chakravorty, *Coord. Chem. Rev.* **1989,** *95,* 239 (electrochemistry, potentials etc., 176 refs); P. A. Lay and D. W. Harmon, *Adv. Inorg. Chem.* **1991,** *37,* 219 (osmium chemistry, 646 refs); C. M. Che and V. W. W. Lam, *Adv. Inorg. Chem.* **1992,** *39,* 233 (chemistry of Ru and Os, IV-VIII, 372 refs).

Table 18-F-1 Oxidation States and Stereochemistry of Ruthenium and Osmium

Oxidation state	Coordination number	Geometry	Examples
Ru^{-II}	4	Tetrahedral(?)	Ru(CO)$_4^{2-}$(?), [Ru(diphos)$_2$]$^{2-}$
Ru^0, Os^0, d^8	5	*tbp*	Ru(CO)$_5$, Os(CO)$_5$, Ru(CO)$_3$(PPh$_3$)$_2$
Ru^I, Os^I, d^7	6^a		[η^5-C$_5$H$_5$Ru(CO)$_2$]$_2$, [Os(CO)$_4$X]$_2$
Ru^{II}, Os^{II}, d^6	5	See text	RuCl$_2$(PPh$_3$)$_3$
	5	*tbp*	RuHCl(PPh$_3$)$_3$
	5	Distorted *sp*	OsCl$_2$(PPh$_3$)$_3$
	6^b	Octahedral	[RuNOCl$_4$]$^{2-}$, [Ru(bipy)$_3$]$^{2+}$, [Ru(NH$_3$)$_6$]$^{2+}$, [Os(CN)$_6$]$^{4-}$, RuCl$_2$CO(PEtPh$_2$)$_3$, OsHCl(diphos)$_2$
Ru^{III}, Os^{III}, d^5	4	Distorted tetrahedral	Ru$_2$(CH$_2$SiMe$_3$)$_6$^{c}
	6^{b,d}	Octahedral	[Ru(NH$_3$)$_5$Cl]$^{2+}$, [RuCl$_5$H$_2$O]$^{2-}$, [Os(dipy)$_3$]$^{3+}$, K$_3$RuF$_6$, [OsCl$_6$]$^{3-}$
Ru^{IV}, Os^{IV}, d^4	4	Square	Ru[N(2,6-Pr^{i}C$_6$H$_3$)]$_2$(PMe$_3$)$_2$
	4	Tetrahedral	Ru(c-C$_6$H$_{11}$)$_4$, OsPh$_4$
	5	*tbp*	Ru(SR)$_4$(MeCN)
	6^{b,d}	Octahedral	K$_2$OsCl$_6$, K$_2$RuCl$_6$, [Os(diars)$_2$X$_2$]$^{2+}$, RuO$_2$^{c}
	7	Distorted Pentagonal bipyramidal	OsH$_4$(PMe$_2$Ph)$_3$, OsH$_3$(PPh$_3$)$^{4+}$ [RuCl(S$_2$CNMe$_2$)$_3$]
Ru^V, Os^V, d^3	5		Ru$_2$(O)$_2$(CH$_2$SiMe$_3$)$_6$^{c}
	6	Octahedral	KRuF$_6$, NaOsF$_6$, (RuF$_5$)$_4$
Ru^{VI}, Os^{VI}, d^2	3	Trigonal planar	Os[N(2,6-Pr^{i}C$_6$H$_3$)]$_3$
	4	Tetrahedral	RuO$_4^{2-}$, OsO$_2$(mesityl)$_2$
	5	*sp*	Os(NMe)(CH$_2$SiMe$_3$)$_4$, OsOCl$_4$, [RuNCl$_4$]$^-$
		tbp	[RuO$_3$(OH)$_2$]$^{2-}$
	6^d	Octahedral	RuF$_6$, OsF$_6$, [OsO$_2$Cl$_4$]$^{2-}$, [OsO$_2$(OH)$_4$]$^{2-}$, [OsNCl$_5$]$^{2-}$
	8	Dodecahedral	OsH$_6$[PPh(Pr$^i_2$)]$_2$
Ru^{VII}, Os^{VII}, d^1	4	Tetrahedral	RuO$_4^-$, OsO$_4^-$
	6	Octahedral	OsOF$_5$
Ru^{VIII}, Os^{VIII}, d^0	4	Tetrahedral	RuO$_4$, OsO$_4$, [OsO$_3$N]$^-$, Os(NBu^t)$_4$
	5	Distorted *tbp*	[(OsO$_4$)$_2$(μ-OH)]$^-$
		tbp	OsO$_4$py
	6	Octahedral	[OsO$_3$F$_3$]$^-$, [OsO$_4$(OH)$_2$]$^{2-}$

^{a}If η^5-C$_5$H$_5$ is assumed to occupy three coordination sites.
^{b}Most common states for Ru.
^{c}Metal–metal bond present.
^{d}Most common states for Os.

CHEMISTRY OF RUTHENIUM(II), (III), AND (IV)

Although lower oxidation states are known (Table 18-F-1), these are mainly with π-bonding ligands and are discussed later (Section 18-F-13). For the oxidation states II-IV, there is great complexity since many compounds can undergo reversible oxidation and reduction reactions to give species with the same structure but a different charge. It is therefore more useful to consider the chemistry according to the ligands present.

18-F-1 Aqua Ions

There are aqua ions in the II, III, and IV states of which the divalent, diamagnetic, pink $[Ru(H_2O)_6]^{2+}$ is the most important, being a useful starting material for the synthesis of numerous ruthenium complexes. It is best made[2] by the reduction of an aqueous solution of RuO_4 (made from RuO_2 by interaction with $NaIO_4$) with lead. The Pb^{2+} is removed with H_2SO_4 and the $[Ru(H_2O)_6]^{2+}$ isolated as the toluene sulfonate after ion exchange treatment. The tosylate is readily converted to the triflate by ion exchange.

The +2 ion is readily oxidized by air to the yellow $[Ru(H_2O)_6]^{3+}$ for which the reduction potential is 0.23 V. The water in $[Ru(H_2O)_6]^{2+}$ can be readily substituted by Cl^- and other anions and by neutral ligands such as MeCN and DMSO; rate constants for monosubstitution have been obtained.[3]

The 2+ aqua ion catalyzes[4] several reactions in solution such as the dimerization of ethylene *via* the intermediates $[Ru(C_2H_4)(H_2O)_5]^{2+}$ and $[Ru(C_2H_4)_2(H_2O)_4]^{2+}$ that were isolated as tosylates.[5] It also catalyzes polymerization of norbornenes.[6]

Ruthenium(IV) aqua ions, $H_n[Ru_4O_6(H_2O)_{12}]^{(4+n)+}$, are obtained on electrochemical oxidation of $[Ru(H_2O)_6]^{2+}$ in aqueous HBF_4. Although the structures are not fully characterized the $Ru_4O_6^{4+}$ core with two single and two double oxo-bridges appears to be established.[7]

18-F-2 Halide Complexes

The most important of these are the chlorides. A very common starting material is the commercial "$RuCl_3 \cdot nH_2O$" made by evaporation of concentrated HCl solutions of RuO_4. This dark red, deliquescent product is a mixture mainly of Ru^{IV} complexes. It is best dissolved in HCl(aq) and the solution boiled and evaporated a few times to form Ru^{III} solutions which, with NaCl, give $Na_2[RuCl_5(H_2O)]$.

Ruthenium(II) chloro complexes, whose precise nature is still uncertain, are obtained when "$RuCl_3 \cdot nH_2O$" in HCl solution is reduced electrochemically or chemically using ethanol or H_2/Pt black. Deep blue, very air-sensitive solutions can

[2]P. Bernhard *et al.*, *Polyhedron* **1990**, *9*, 1095.
[3]A. E. Merbach *et al.*, *Inorg. Chem.* **1993**, *32*, 2810.
[4]See A. C. Benyei, *NATO-ASI Ser. 3* **1995**, *5*, 159 (review, 24 refs).
[5]A. E. Merbach, *et al.*, *J. Chem. Soc., Chem. Commun.* **1993**, 187; T. Karlen and A. Ludi, *J. Am. Chem. Soc.* **1994**, *116*, 11375.
[6]R. H. Grubbs *et al.*, *J. Am. Chem. Soc.* **1996**, *118*, 784.
[7]A. Patel and D. J. Richens, *Inorg. Chem.* **1991**, *30*, 3789.

also be obtained in the reversible reaction:

$$Ru_2(CO_2Me)_4 \underset{NaOAc}{\overset{HCl}{\rightleftharpoons}} \text{blue solution}$$

but no crystalline product has so far been obtained from these "ruthenium blue" solutions. A green salt from blue HCl solutions of $[Ru_2(O_2CMe)_4]Cl$ turned out to be $[Cl_3Ru^{III}(\mu\text{-}Cl)_3Ru^{II}(\mu\text{-}Cl)_3Ru^{III}Cl_3]^{4-}$.

The blue solutions made by reduction have been much used as starting materials for syntheses of Cp_2Ru, carboxylates with Ru_2^{n+} groups (Section 18-F-5), phosphine complexes, bipyridyl complexes,[8] $[Ru(bipy)_3]^{2+}$, cis-$[RuCl_2(bipy)]^+$, $[RuCl_4(bipy)]^-$, $[RuCl_6]^{3-}$, a pyridylpeptide complex,[9] and $Ru(acac)_3$.[10] Examples of other simple Ru^{II} complexes are $RuCl_2(bipy)_2$,[11] $RuCl_2(tht)_4$, and $RuCl_2(NCMe)_4$[12] that have $trans$ Cl atoms.

Ruthenium(III) Chloro Species

These range from $[RuCl_6]^{3-}$ to $[RuCl(H_2O)_5]^{2+}$. The aquation rates from the hexachloride vary enormously:[13]

$$[RuCl_6]^{3-} + H_2O \longrightarrow [RuCl_5(H_2O)]^{2-} + Cl^- \qquad \text{seconds}$$

$$[RuCl(H_2O)_5]^{2+} + H_2O \longrightarrow [Ru(H_2O)_6]^{3+} + Cl^- \qquad t_{1/2}\,ca\,1\,\text{year}$$

The most stable complexes are the cis and $trans$ isomers of $[RuCl_2(H_2O)_4]^+$. The $trans$ ion and also $trans$-$[RuCl_4(H_2O)_2]^-$ are formed when "ruthenium blue" oxidizes in water in absence of air when H_2 is evolved. Some Ru^{III} complexes of the type $trans$-$[RuCl_4L_2]^-$, where L = imidazole, have antitumor activity.[14] There are, of course, many other chloro complexes having additional ligands. Some typical examples are mer-$[RuCl_3(Ph_2SO)_3]$[15] and $trans$-$[(MeCN)_2RuCl_4]^-$.[16]

Ruthenium(IV) chloro complexes with OH and H_2O ligands can be obtained, but the $[RuCl_6]^{2-}$ ion is formed only in high concentrations of Cl^- by chlorine oxidation of Ru^{III} chloro complexes. The purple or brown salts are isomorphous with other $[MCl_6]^{2-}$ species of Os, Ir, Pd, and Pt.

Dinuclear complexes are of the type $[Ru_2X_9]^{n-}$, X = Cl, Br, n = 1–4; all have triple halide bridges and a confacial, bioctahedral geometry. When n = 4, there is mixed valence, II-III, and reduced species have been observed in electrochemical studies.[17]

[8]M. Mukaida *et al., Inorg. Chim. Acta* **1992**, *195*, 221.
[9]M. R. Ghadiri, *J. Am. Chem. Soc.* **1992**, *114*, 4000.
[10]T. S. Knowles *et al., Polyhedron* **1994**, *13*, 2197.
[11]M. D. Ward *et al., J. Chem. Soc., Dalton Trans.* **1996**, 2527.
[12]M. Brown and D. C. R. Hockless, *Acta Cryst.* **1996**, *C52*, 1105.
[13]G. Ramachandraiah *et al., Polyhedron* **1992**, *11*, 3075.
[14]B. K. Keppler *et al., Chem. Commun.* **1996**, 1741.
[15]M. Calligaris, *J. Chem. Soc., Dalton Trans.* **1995**, 1653.
[16]G. A. Heath *et al., Acta Cryst.* **1995**, *C51*, 1805.
[17]B. J. Kennedy *et al., Inorg. Chim. Acta* **1991**, *190*, 265.

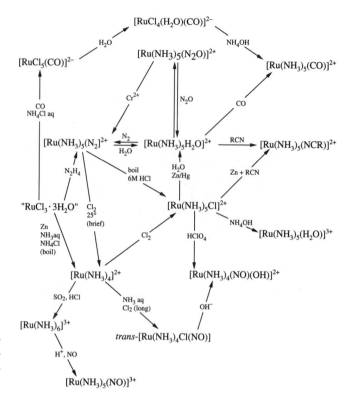

Figure 18-F-1 Some reactions of ruthenium ammines in aqueous solutions.

Fluoroanions, $[RuF_6]^{n-}$, $n = 1–3$ are well established and can be obtained by reactions such as

$$RuCl_3 + M^ICl + F_2 \xrightarrow{300°C} M^IRu^VF_6$$

$$Ru + M^{II}Cl_2 + BrF_3 \longrightarrow M^{II}Ru^{IV}F_6$$

The $[RuF_6]^-$ salts[18] are reduced by water to $[RuF_6]^{2-}$ and O_2 is evolved.

18-F-3 Nitrogen Ligand Complexes

Ruthenium complexes with N ligands are extremely numerous, forming perhaps the most extensive and important areas of ruthenium chemistry, especially in oxidation states II and III.

Ammonia Complexes

Some reactions involving ammonia complexes[19] are shown in Figs. 18-F-1 and 18-F-2.

[18]N. Bartlett *et al.*, *Inorg. Chem.* **1995**, *34*, 2692; *J. Fluorine Chem.* **1995**, *73*, 157.
[19]D. W. Franco, *Coord. Chem. Rev.* **1992**, *119*, 199.

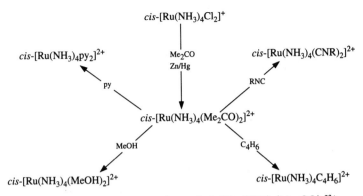

Figure 18-F-2 Some reactions of *cis*-[Ru(NH$_3$)$_4$(Me$_2$CO)$_2$]$^{2+}$.

The orange hexammine, [Ru^{II}(NH$_3$)$_6$]Cl$_2$, is formed when ammoniacal solutions of "RuCl$_3$(aq)" containing excess NH$_4$Cl are reduced by zinc. The hexammine is a reductant:

$$[Ru^{III}(NH_3)_6]^{3+} + e = [Ru^{II}(NH_3)_6]^{2+} \qquad E^0 = 0.24 \text{ V}$$

Although the rate of aquation is pH dependent the product, [Ru(NH$_3$)$_5$H$_2$O]$^{2+}$, is a very versatile complex since the H$_2$O can be substituted by a wide range of neutral and anionic ligands, including CO, N$_2$, N$_2$O, alkenes, and alkynes. This occurs because Ru(NH$_3$)$_5^{2+}$ (low spin t_{2g}^6) is a remarkably good π-donor and the π-donation extends also to η^2-bonding of C—C multiple bonds in alkenes,[20] alkynes, and arenes, as well as in η^2-acetone complexes. More highly substituted ammines can be made from [Ru(NH$_3$)$_5$(H$_2$O)]$^{2+}$. The acetone complex,[21] *cis*-[Ru(NH$_3$)$_4$(Me$_2$CO)$_2$]$^{2+}$, (Fig. 18-F-2) is a versatile precursor for *cis*-tetrammine compounds containing the ion [Ru(NH$_3$)$_5$(diene)$_2$]$^{2+}$. It is readily made by interaction of acetone on *cis*-[Ru(NH$_3$)$_4$(H$_2$O)$_2$]$^{2+}$, which in turn is obtained by reduction of *cis*-[Ru(NH$_3$)$_4$Cl$_2$]$^+$ by Zn/Hg in aqueous solution at pH 2. Other useful synthons are the complexes [Ru(NH$_3$)$_5$(Me$_2$CO)]$^{2+,3+}$ which have linkage isomers η^1-OCMe$_2$ and η^2-C,O.[22]

A historically important complex, that also depends on the π-donor capacity as above, is [Ru(NH$_3$)$_5$(N$_2$)]$^{2+}$. This was the first dinitrogen compound to be recognized and it opened up an enormous development in N$_2$ complex chemistry (see Section 9-9). Although originally made *via* hydrazine reduction of "RuCl$_3 \cdot n$H$_2$O," it is best prepared by the reactions

$$[Ru(NH_3)_5(H_2O)]^{2+} + N_2 \rightleftharpoons [Ru(NH_3)_5(N_2)]^{2+} + H_2O$$

$$[Ru(NH_3)_5(N_2)]^{2+} + [Ru(NH_3)_5(H_2O)]^{2+} \rightleftharpoons [(NH_3)_5RuN_2Ru(NH_3)_5]^{4+}$$

[20]H. Chen and W. D. Harman, *J. Am. Chem. Soc.* **1996**, *118*, 5672.
[21]T. Sugaya *et al.*, *Inorg. Chem.* **1996**, *35*, 2692.
[22]D. W. Powell and P. A. Lay, *Inorg. Chem.* **1992**, *31*, 3542.

The second reaction, forming the bridged ion, exemplifies the donor properties of coordinated N_2. The $Ru-N-N-Ru$ group is nearly linear and the $N-N$ distance (1.124 Å) is only slightly longer than in N_2 itself (1.0976 Å). The bonding can be described by MO theory similar to that for $Ru-O-Ru$.

The N_2O complex is very rapidly reduced by Cr^{2+} to form the N_2 complex, the overall reaction being

$$[Ru(NH_3)_5N_2O]^{2+} + 2Cr^{2+} + 2H^+ \longrightarrow [Ru(NH_3)_5N_2]^{2+} + 2Cr^{3+} + H_2O$$

It is evident that coordination of N_2O lowers the $N-O$ bond strength, but the precise course of the reduction is still uncertain. Azido Ru^{III} ammines are unstable:

$$[Ru(NH_3)_5N_3]^{2+} \longrightarrow [Ru(NH_3)_5N_2]^{2+} + \tfrac{1}{2}N_2$$

but in the presence of acid, bridged dimers are also formed, and a nitrene complex $Ru=NH$ may be involved.

Still another reaction leading to $[Ru(NH_3)_5N_2]^{2+}$ is the following:

$$[Ru(NH_3)_6]^{3+} + NO + OH^- \longrightarrow [Ru(NH_3)_5N_2]^{2+} + 2H_2O$$

It is interesting that in acid solution only the simple replacement of NH_3 by NO occurs:

$$[Ru(NH_3)_6]^{3+} + NO + H^+ \longrightarrow [Ru(NH_3)_5NO]^{3+} + NH_4^+$$

Action of concentrated HCl on $[Ru(NH_3)_6]^{3+}$ gives blue solutions (*cf.*, blue chlorides above) and the product is the mixed valence (II, III) ion $[(NH_3)_3Ru(\mu\text{-}Cl)_3Ru(NH_3)_3]^{2+}$ (*cf.*, also Creutz-Taube complexes below).

Ruthenium(III), with a t_{2g}^5 configuration, is, in contrast to Ru^{II}, a very good π-acceptor. This is dramatically demonstrated by the rates of hydrolysis of free and coordinated nitriles:[23]

$$[Ru(NH_3)_5NCR]^{3+} + H_2O \longrightarrow [Ru(NH_3)_5NH_2C(O)R]^{3+}$$

$$NCR + H_2O \longrightarrow H_2NC(O)R$$

The reaction of the coordinated RCN is 10^8 to 10^9 times faster because the π-acceptor character of the $Ru(NH_3)_5^{3+}$ moiety stabilizes the transition state or intermediate resulting from the attack of OH^- on the carbon atom by taking electron density from the $C\equiv N$ bond.

The reactions of ammines of the type $[Ru(NH_3)_5L]^{3+}$ with Cr^{2+} and other reducing agents are similar to those of Co^{III} complexes except that the electron enters

[23]For refs. see J. J. F. Alves and D. W. Franco, *Polyhedron* **1996**, *15*, 3299.

the t_{2g} rather than the e_g level and the resulting Ru^{II} complexes are diamagnetic. Bridged species are probably intermediates.

In some reactions, the rate-determining step appears to be attack on the d-electron density of the metal atom. Thus the hexaammine $[Ru(NH_3)_6]^{3+}$ undergoes aquation only very slowly at room temperature but reacts rapidly with NO, as noted previously. Substitution in reactions such as:

$$\textit{trans-}[Ru(NH_3)_4(PR_3)(H_2O)]^{2+} + L \rightleftharpoons \textit{trans-}[Ru(NH_3)_4(PR_3)L]^{2+} + H_2O$$

where L = imidazole, has allowed estimation of the following *trans* influence series: $P(OBu)_3 < P(OMe)_3 < PPh_3 < PBu_3$.[24]

Photolabilization studies of *trans*-$[Ru(NH_3)_4\{P(OR)_3\}_2]^{2+}$ have also been made, but no dependence on the cone angle of PR_3 was observed.[25]

The Creutz-Taube and Related Complexes

There is a series of complexes whose common feature is the presence of two (or more) Ru atoms bridged by bidentate ligands through which a potentially adjustable degree of electron transfer can take place. Such complexes have been extensively studied because of the information they may give about the processes of electron exchange in bimolecular redox reactions and about the general character of electron transmission through chemical systems, including the spectroscopic phenomena associated with such processes.

The prototype, intentionally synthesized, ion (18-F-I) has been the most intensively studied and also has been named after the discoverers, Carol Creutz and Henry Taube.

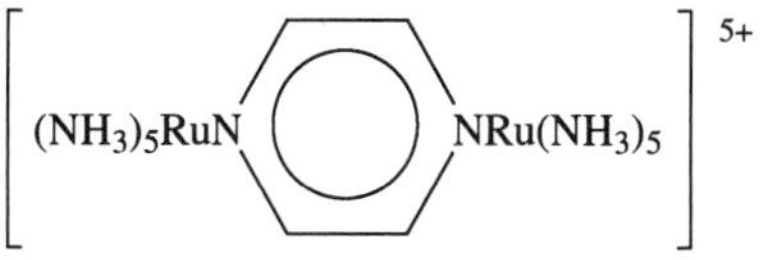

(18-F-I)

The ion has mixed valencies, II-III (5+), but II-II (4+) and III-III (6+) salts have been isolated. Crystal structures of these species are consistent with symmetrical ions even in the "II-III" case. Although there has been much discussion and experimentation,[26] only recently has resonance Raman spectroscopy in the near infrared provided verification of 3-site, valence delocalization.[27]

In some related ions localization is found, e.g., in 18-F-II, while in 18-F-III, which is unsymmetrical, the localization is such that Ru^{III} is at the pentammine end.

[24]B. S. Lima-Neto *et al., Polyhedron* **1996,** *15,* 1965.
[25]D. W. Franco *et al., Inorg. Chem.* **1996,** *35,* 3509.
[26]W. Kaim *et al., Inorg. Chem.* **1993,** *32,* 2640.
[27]J. T. Hupp *et al., J. Am. Chem. Soc.* **1994,** *116,* 2171.

This ion provides a model for study of intervalence transfer absorption bands characteristic of "trapped" valence species that correspond to the process

$$[Ru^{II}-Ru^{III}]^{3+} \xrightarrow{h\nu} [Ru^{III}-Ru^{II}]^{3+}$$

$$\left[(bipy)_2ClRuN \bigcirc NRuCl(bipy)_2 \right]^{3+}$$

(18-F-II)

$$\left[(NH_3)_5RuN \bigcirc NRuCl(bipy)_2 \right]^{4+}$$

(18-F-III)

Aromatic Amine Complexes

Complexes of pyridines, bipyridyls, terpyridyls, and other polypyridines differ considerably from those of NH_3 and aliphatic amines. They have been intensively studied because many of them have unusual properties for photoinduced energy migrations and charge separation, luminescence, photocatalytic reactions, and water activation. They are also involved in the construction of double-helical complexes, complexes that bind to DNA, and molecular rods and wires[28] for fast electron transfers.[29]

The simple pyridine complexes are exemplified by [Ru py$_6$]$^{2+}$ and *trans*-RuCl$_2$py$_4$.[29]

The bipyridyls of Ru and Os, [M(bipy)$_3$]$^{2+}$, are quite similar and can occur in singly, doubly, and triply reduced forms: [Ru bipy(bipy$^-$)]$^+$, [Ru bipy(bipy$^-$)$_2$]^0, [Ru(bipy$^-$)$_3$]$^-$; optical isomers such as Δ-[Ru(bipy)$_3$]$^{2+}$ have been resolved.[30]

The compound [Ru(bipy)$_3$]$^{2+}$, commonly called "rubipy," is extensively used as a sensitizer in photodriven chemical and physical processes such as photolysis of water. The electronic structure of the excited ion *[Ru(bipy)$_3$]$^{2+}$ appears to result from transfer of an electron from a metal t_{2g} orbital to a ligand π^* orbital. This excited state can be described as Ru^{III}—L$^-$, and it reverts to the ground state by photoemission without chemical reaction unless suitable reactants called quenchers are present, as discussed shortly. This is in distinct contrast to the photochemical behavior of the simple ammines, which respond to photoexcitation by promptly aquating or oxidizing:

$$[Ru(NH_3)_5L]^{2+} + H_2O \xrightarrow{h\nu} [Ru(NH_3)_5H_2O]^{2+} \text{ or } [Ru(NH_3)_4L(H_2O)]^{2+}$$

$$[Ru(NH_3)_5L]^{2+} + H^+ \xrightarrow{h\nu} [Ru(NH_3)_5L]^{3+} + \tfrac{1}{2}H_2$$

[28]A. Harriman and R. Ziessel, *Chem. Commun.* **1996**, 1707; B. L. Iverson *et al., J. Am. Chem. Soc.* **1996**, *118*, 3656; V. Grosshenny *et al., Platinum Metals Rev.* **1996**, *40*, 72; J-M. Lehn *et al., Chem. Commun.* **1996**, 701; J. J. Meyer *et al., Inorg. Chem.* **1996**, *35*, 4575; C.-M. Che *et al., Chem. Commun.* **1996**, 1197; J. Kelley *et al., Chem. Commun.* **1996**, 1013; S. Campagna *et al., J. Am. Chem. Soc.* **1995**, *117*, 1754.
[29]B. J. Coe *et al., Inorg. Chem.* **1995**, *34*, 593.
[30]B. Noble and R. D. Peacock, *Inorg. Chem.* **1996**, *35*, 1616.

Examples of more complicated ligands whose complexes show luminescent behavior are as follows:

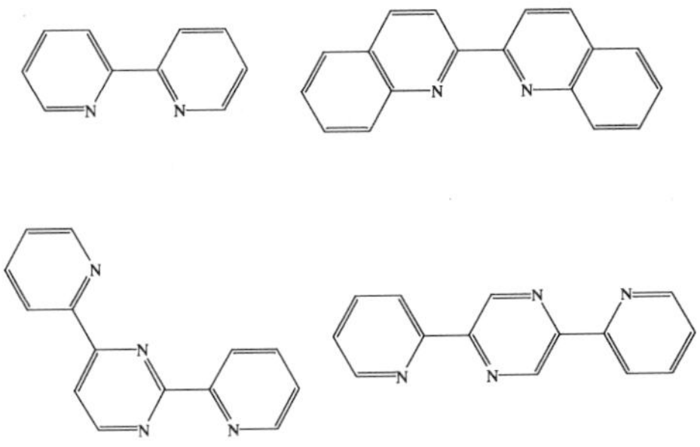

Other N-Ligands

These include *pyrazolylborates* such as Tp*RuH(H$_2$)$_2$,[31] and *diazadiene* compounds of RN=CR'—CR'=NR ligands.[32] Nitrido and imido compounds are discussed in Section 18-F-11; there are a few *amido* complexes known.[33]

Nitrile complexes can be obtained by substitutions as noted earlier; thus the reaction with [RuCl$_6$]$^-$ can give all the stages to [Ru(NCR)$_6$]$^{2+}$. These series allow all the potentials for complexes involving the couples Ru$^{II/III}$, Ru$^{III/IV}$, and Ru$^{IV/V}$ to be determined.[34]

Nitriles also form dimers, e.g., [Ru$_2$(NCMe)$_{10}$]$^{4+}$; other N-ligand species are *amidinate* and *triazenido* compounds such as Ru$_2$(PhNCHNPh)$_2$Cl$_2$[35] and [Ru(1,3-diaryltriazenide)$_2$(PPh$_3$)$_2$]$^{0,+}$.[36]

Porphyrins[37] and Phthalocyanins[38]

Complexes of both types have been intensively studied. As well as monomeric species of classical type such as Ru(TTP)NO(OMe), Ru(oep)X$_2$, Ru(pc)L$_2$, L = NH$_3$, Me$_2$SO, etc., there are dimeric species, Ru$_2$(porph)$_2$, that have Ru—Ru bonds; the dimers are cleaved by K in THF to give anions [Ru(porph)]$^-$.

Nitric Oxide

Any ruthenium solution treated with HNO$_3$ can be suspected of having coordinated NO groups. Such complexes have proved very troublesome in the processing and

[31]B. Chaudret *et al., J. Am. Chem. Soc.* **1995**, *117*, 7441.
[32]H. Tom Dieck *et al., J. Organomet. Chem.* **1991**, *411*, 445.
[33]C.-M. Che *et al., Inorg. Chem.* **1996**, *35*, 3369.
[34]C. M. Duff and G. A. Heath, *Inorg. Chem.* **1991**, *20*, 2528.
[35]S. D. Robinson *et al., J. Chem. Soc., Dalton Trans.* **1992**, 3199.
[36]A. Chakravorty *et al., Inorg. Chem.* **1995**, *34*, 1361.
[37]D. S. Bohle *et al., Polyhedron* **1996**, *15*, 3147; K. M. Kadish *et al., Inorg. Chem.* **1996**, *35*, 1343.
[38]R. C. Brooks *et al., Inorg. Chem.* **1995**, *34*, 1524; G. Rossi *et al., Inorg. Chem.* **1996**, *35*, 4643.

recovery of U and Pu from irradiated uranium fuel in nuclear reactors. Ruthenium isotopes are among the fission products in the nitrate solutions from which the neutral species $Ru(NO)(NO_3)_3(H_2O)_2$ can be extracted by Bu_3PO.

Almost any ligand can be present with the RuNO unit, examples being $[Ru(NO)(CN)_5]^{2-}$ and $Ru(NO)(S_2CNMe_2)_3$, which has two bi- and one unidentate dithiocarbamate. The source of NO can be NO_3^-, NO_2^-, or NO.

The NO group is electrophilic and susceptible to nucleophilic attack by OH^-, SH^-, N_3^-, RNH_2, and so on, for example,

$$[RuX\ bipy_2NO]^{2+} + 2OH^- \longrightarrow [RuX\ bipy_2NO_2] + H_2O$$

$RuNO^{3+}$ can also be electrolytically reduced to $RuNH_3^{2+}$.

The vast majority of RuNO complexes are of the general type $Ru(NO)L_5$, in which the metal atom is *formally* in the divalent state and may be designated $\{RuNO\}^6$ complexes. There is a significant number of $\{RuNO\}^8$ and $\{RuNO\}^{10}$ complexes, but these all contain π-acid ligands such as phosphines. In all $\{RuNO\}^6$ complexes the Ru—N—O chains are essentially linear.

18-F-4 Phosphine Complexes

In common with other platinum metals, an important area of ruthenium chemistry involves trialkyl- and triarylphosphines, and the corresponding phosphites. An extremely wide range of complexes is known, mainly of the II state, although compounds in the 0, III, and less commonly, IV state are known; other ligands commonly associated with the PR_3 group are halogens, alkyl and aryl groups, CO, NO, and alkenes as well as H and H_2. Similar chemistry is found for osmium.

The main preparative routes are as follows:

1. Interaction of "$RuCl_3 \cdot nH_2O$" or halide species with PR_3 in an alcohol or other solvent. In many of these reactions, either hydride or CO may be abstracted from the solvent molecule, leading to hydrido or carbonyl species. Sodium borohydride is also used as a reducing agent.
2. Complexes in the 0 oxidation state may be obtained either by reduction of halides such as $RuCl_2(PPh_3)_3$ with Na or Zn in the presence of CO or other ligands such as RNC, or by reaction of metal carbonyls with phosphines. Reactions of polynuclear carbonyls such as $Ru_3(CO)_{12}$ (Section 18-F-13) with phosphines tend to preserve the cluster structure.
3. Carbonyl-containing complex ions such as $Ru(CO)Cl_5^{3-}$, *cis*-$Ru(CO)_2Cl_4^{2-}$, and $Ru(CO)_3Cl_3^-$ are formed by the action of CO or formic acid on Ru chloro-complexes; addition of phosphines to such solutions gives replacement products.

Some typical reactions for ruthenium (which are generally representative also for osmium) are given in Fig. 18-F-3. Phosphines with differing steric and basic properties often give different types of product. The reaction conditions are also

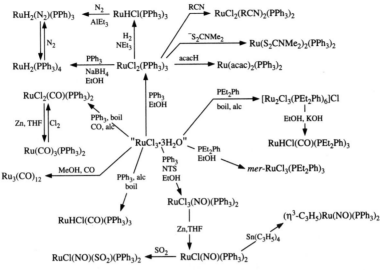

Figure 18-F-3 Some reactions of tertiary phosphine complexes of ruthenium (alc = 2-methoxyethanol).

critical: for example, for trialkylphosphines and $RuCl_3$ in ethanol, short reaction times give $RuCl_3(PR_3)_3$, whereas prolonged reactions give $[Ru_2Cl_3(PR_3)_6]Cl$.

Although Fig. 18-F-3 notes only reactions of PPh_3, hundreds of compounds with tertiary phosphines of all types, including bi- and tridentate ones have been made. Representative examples are $RuH_2(PMe_3)_4$, $Ru(C_2H_4)(PMe_3)_4$,[39] and $RuCl[P(CH_2CH_2PPh_2)_3]$. Complexes of water-soluble phosphines like $P(C_6H_4SO_3)_3^{3-}$ have been used as catalysts for hydroformylation and other reactions in two-phase systems.[40]

The compound $RuHCl(PPh_3)_3$ made by the reaction

$$RuCl_2(PPh_3)_3 + H_2 + Et_3N \longrightarrow RuHCl(PPh_3)_3 + Et_3NHCl$$

and $RuH_2(PPh_3)_3$ are very active catalysts for hydrogenation of C=C bonds, as is also $RuH(CO)(PPh_3)_3$. These complexes in solution dissociate PPh_3 leaving vacant sites for substrate coordination and H-transfers.

Ruthenium complexes of *chiral phosphines* notably BINAP, 2,2′-bis(diphenyl-phosphino)-1,1′-binaphthyl are very useful for industrial hydrogenations and H-transfer hydrogenation of alkenes and ketones. This is due to high turnover numbers and enantiomeric excesses (ee) of the products. A specific example[41] of an

[39]R. G. Bergman *et al.*, *Organometallics* **1995**, *14*, 137.
[40]J. M. Basset *et al.*, *J. Mol. Catal.* **1992**, *72*, 331.

(R)-BINAP compound of Ru^{II} is

which will catalyze hydrosilations of alkenes and ketones.

The CO stretching frequency in the 5-coordinate RuHX(CO)(PMeBu$^t_2$)$_2$ has been used to determine the donating ability of X when X = halogens, alkoxides, OH, NHPh, SPh, and C≡CPh.[42]

18-F-5 Oxygen and Sulfur Ligand Complexes

There are many complexes such as [Ru ox$_3$]$^{3-}$, [Ru EDTA(H$_2$O)]$^{2-}$, and *cis* and *trans*-RuX$_2$(Me$_2$SO)$_4$, the latter having either O and/or S bonding. The *β-diketonate* complexes have been much studied, the simplest being Ru(acac)$_3$;[43] these can be reduced to [Ru^{II}(β-dike)$_3$]$^-$.[44]

There are many *carboxylates*; some having Ru=O or μ-O groups are discussed in Section 18-F-11. The most important are carboxylate bridged compounds with Ru$_2^{n+}$, n = 4, 5, 6,[45] that can exist in different oxidation states with no change in the bridged structure: an example is Ru$_2$(μ-O$_2$CMe)$_4$(THF)$_2$. Mixed valence compounds such as [Ru^{II}Ru^{III}(O$_2$CR)$_4$]$^+$ have long been known. The II,III acetate can be converted to the carbonate, Na$_3$[Ru$_2$(μ-O$_2$CO)$_4$]·6H$_2$O, on the interaction with Na$_2$CO$_3$ and similar sulfato compounds can be made.

Dimethyl and other sulfoxide complexes have cytotoxic activity.[46] Other complexes include *catecholate* and *quinone* compounds[47] such as Ru^{III}(3,5-DBQ)$_3$; the tridentate P$_3$O$_9^{3-}$ ligand gives anions such as [(C$_6$H$_6$)RuO$_9$P$_3$]$^-$.[48]

There are relatively few ruthenium complexes of *sulfur ligands*. Although *dithiocarbonates*, such as Ru(S$_2$CNR$_2$)$_3$ and [Ru(S$_2$CNEt$_2$)$_2$(PPh$_3$)$_2$]$^{0,+}$, are known,[49] *thiolates* are not common. Examples include LRu(SR)$_4$, L = CO, MeCN, and

[41]For extensive reference see S. H. Bergens *et al., Organometallics* **1996,** *15,* 3782,

[42]O. Eisenstein *et al., Inorg. Chem.* **1994,** *33,* 1476.

[43]T. S. Knowles *et al., Polyhedron* **1994,** *13,* 2197.

[44]J. Tamura *et al., Inorg. Chim. Acta* **1995,** *236,* 37.

[45]For refs. see F. A. Cotton and R. A. Walton, *Multiple Bonds between Metal Atoms,* 2nd ed., Oxford University Press, Oxford, UK, 1993, Chap. 6.

[46]E. Alessio *et al., Inorg. Chim. Acta* **1993,** *203,* 205.

[47]S. Bhattacharya and C. G. Pierpont, *Inorg. Chem.* **1994,** *33,* 6038.

[48]V. W. Day *et al., Inorg. Chem.* **1993,** *32,* 1629.

[49]A. Chakravorty *et al., J. Chem. Soc., Dalton Trans.* **1995,** 1543.

$[Ru(SR)_3(MeCN)_2]^+$ where R is the bulky group 2,4,6-$Pr^i_3C_6H_2$.[50] Interaction of H_2S with $Ru(CO)_2(PPh_3)_3$ gives $Ru(SH)(CO)_2(PPh_3)_3$.[51] There are cyclic thio-ether compounds such as $\{[9]aneS_3Ru(MeCN)_3\}^{2+}$ [52] and one with a polysulfide anion,[53] $[Ru(NO)(NH_3)(S_4)_2]^-$. Compounds with $Ru^{II}SSRu^{III}$ cores such as $\{RuCl[P(OMe)_3]_2\}_2(\mu\text{-}Cl)_2(\mu\text{-}S_2)$ are also known as well as similar species with $\mu\text{-}MoS_4$ bridges.[54]

CHEMISTRY OF OSMIUM IN LOWER OXIDATION STATES

18-F-6 Halide Complexes

The interaction of the osmate ion $[OsO_2(OH)_4]^{2-}$, discussed in Section 18-F-11, with aqueous HCl leads *via* complex transformations to species in oxidation states III to VI some with Os=O or OH groups, e.g.,

$$[Os^{VI}O_2Cl_3H_2O]^- \underset{H_2O}{\overset{Cl^-}{\rightleftharpoons}} [OsO_2Cl_4]^{2-} \overset{3e}{\rightleftharpoons} [OsCl_4(H_2O)_2]^-$$

$$\Big\updownarrow H_2O \,\| \, Cl^-$$

$$[OsCl_6]^{2-} \underset{-e}{\overset{+e}{\rightleftharpoons}} [OsCl_6]^{3-} \underset{Cl^-}{\overset{H_2O}{\rightleftharpoons}} [OsCl_5(H_2O)]^{2-}$$

The most common and useful salts for synthesis are those of the *hexachloroosmate(IV) ion*, $[OsCl_6]^{2-}$, which is readily obtained by reducing an HCl solution of OsO_4 by Fe^{2+}; the salts are brown or orange. The ion can be further reduced only to $[OsCl_6]^{3-}$.

The Os^V ion, $[OsCl_6]^-$, is obtained by the reaction

$$OsO_4 + SCl_2 + Cl_2 \xrightarrow{25\,°C} [SCl_3]^+[OsCl_6]^- \xrightarrow{70\,°C} Os_2Cl_{10} + SCl_2 + Cl_2$$

or by PbO_2 oxidation of $[OsCl_6]^{2-}$. The ion is a strong oxidant ($OsCl_6^-/OsCl_6^{2-}$, $E^0 = 1.44$ V) undergoing one-electron transfers.[55] The OsX_6^{n-} anions, $n = 1, 2, 3$, are all octahedral; the halides can be substituted by π-bonding and other ligands to give complexes such as $[OsX_4(CNR)_2]^-$, $[OsX_4(CO)L]^-$, L = MeCN, py, PEt_3, and others.[56]

Interaction of $OsCl_5$ (Section 18-E-4) and MeCN gives the red *cis*-$OsCl_4(MeCN)_2$ compound.[57]

The $[OsF_6]^-$ ion can be made by the reaction:

$$OsCl_4 + M^ICl + BrF_3 \longrightarrow M^I[OsF_6]$$

[50]S. A. Koch *et al.*, *Inorg. Chem.* **1992**, *31*, 5160.
[51]B. R. James *et al.*, *Inorg. Chem.* **1992**, *31*, 4601.
[52]C. Landgrafe and W. S. Sheldrick, *J. Chem. Soc., Dalton Trans.* **1996**, 989.
[53]A. Müller *et al.*, *Inorg. Chem.* **1991**, *30*, 2040.
[54]K. Matsumoto *et al.*, *J. Am. Chem. Soc.* **1996**, *118*, 3597.
[55]L. Eberson and M. Nilsen, *J. Chem. Soc., Chem. Commun.* **1992**, 1041.
[56]R. A. Walton *et al.*, *Inorg. Chem.* **1991**, *30*, 4146.
[57]K. Dehnicke *et al.*, *Z. Naturforsch.* **1990**, *45B*, 1210.

Unlike its Ru analogue, it is not affected by H_2O, but on addition of base, oxygen evolution and reduction to $[OsF_6]^{2-}$ occur.

18-F-7 Osmium Complexes with N-Ligands

Ammonia and Amines[58]

Much of the ammonia chemistry resembles that of Ru shown in Fig. 18-F-1 and -2 with complexes in oxidation states II-IV. However, there are differences in that the $Os(NH_3)_5^{2+}$ and $Os(NH_3)_4^{2+}$ moieties have exceptional capacities for complexing with π-bonding ligands. In forming these, the triflates, $[Os(NH_3)_5CF_3SO_3]$-$(CF_3SO_3)_2$ and $[Os(NH_3)_4(CF_3SO_3)_2](CF_3SO_3)$ are useful starting materials.[59] Thus reduction of the pentaammine complex in aqueous solution with Zn/Hg leads to the η^2-acetone complex $[Os(NH_3)_5(\eta^2\text{-OCMe}_2)]^{2+}$.[60] While CO and N_2 give end-on complexes, most of the other complexes from alkenes like ethylene, 1,5-cycloocta-diene or acrylonitrile, acetylenes, arenes, and anilines[61] give η^2-compounds. A particularly unusual example is the alkene-alkyne complex (18-F-IV) formed with ethylenediamine.[62]

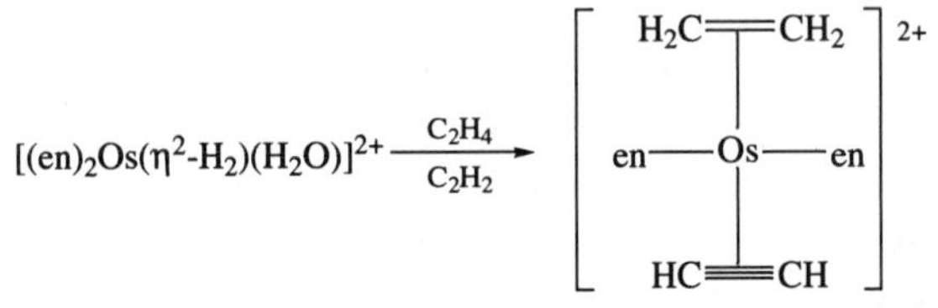

$$[(en)_2Os(\eta^2\text{-}H_2)(H_2O)]^{2+} \xrightarrow{\ C_2H_4\ }_{C_2H_2} \left[\begin{array}{c} H_2C{=}CH_2 \\ en{-}Os{-}en \\ HC{\equiv}CH \end{array}\right]^{2+}$$

(18-F-IV)

Another unusual species is the carbene complex formed on reduction of $[Os(NH_3)_4(CF_3SO_3)_2]CF_3SO_3$ by Na/Hg in presence of a cyclic ether as reduction medium that also has a $\eta^2\text{-}H_2$ group and gives *anti* and *syn* forms[63] (18-F-V):

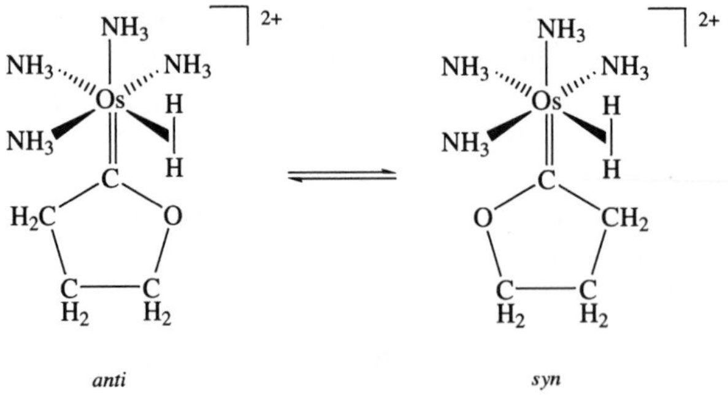

anti   *syn*

(18-F-V)

[58]P. A. Lay and W. D. Harman, *Adv. Inorg. Chem.* **1991**, *37*, 219.
[59]H. Taube *et al.*, *Inorg. Chem.* **1994**, *33*, 3635, **1996**, *35*, 4622.
[60]M. G. Finn *et al.*, *Inorg. Chem.* **1993**, *32*, 2123.
[61]W. D. Harman *et al.*, *Organometallics* **1996**, *15*, 245.
[62]H. Taube *et al.*, *J. Am. Chem. Soc.* **1993**, *115*, 2345.
[63]Z.-W. Li and H. Taube, *J. Am. Chem. Soc.* **1994**, *116*, 11584.

The use of ethylenediamine and NH_3 complexes in connection with J_{HD} of dihydrogen complexes has been noted in Section 2-16 and the mechanism of solvent exchange with *trans*-$[Os(en)_2(\eta^2\text{-}H_2)S]^{2+}$, $S = H_2O$ and MeCN, has been studied.[64]

Finally, it may be noted that no direct analogues of ruthenium reds (Section 18-F-11) are known; the air oxidation of aqueous ammonia solutions of $[OsCl_6]^{2-}$ gives nitrido species as in $[(H_2O)(NH_3)_4Os(\mu\text{-}N)Os(H_2O)_4(\mu\text{-}N)Os(NH_3)_4(H_2O)]^{6+}$ and $[Os_3(N)_2(CN)_{10}(H_2O)_4]^{4-}$.

Aromatic amine complexes[65] of osmium, and also complexes of N macrocycles, are generally similar to those of Ru, but $[Os(bipy)_3]^{2+}$ and $[Os(terpy)_2]^{2+}$ are more labile and reactive than their Ru analogues. As for Ru, some Os^{II} complexes are luminescent and potentially of use in photochemical molecular devices and for detection of DNA.[66]

An unusual sequence of multielectron chemistry unique to Os (and Ru) is the reversible interconversion of NH_3 and NO in polypyridyl complexes:[67]

$$[Os(bipy)_2pyNO]^{3+} \xrightleftharpoons[-6e - 5H^+ + H_2O]{+6e + 5H^+ - H_2O} [Os(bipy)_2py(NH_3)]^{2+}$$

Similar electrolytic oxidation of NH_3 in $[(bipy)_2NH_3Ru(\mu\text{-}O)]_2^{4+}$ gives N_2.[68] There are 1*e* intermediate steps beginning with $[Os^{II}(bipy)_2 \, py \, NO]^{2+}$ on reduction and $[Os^{III}(bipy)_2 \, py(NH_3)]^{2+}$ on oxidation. A similar 4*e* electron change occurs in the reaction

$$[Os^{VI}(terpy)Cl_2(N)]^+ \xrightleftharpoons[-4e - 3H^+]{+4e + 3H^+} [Os^{II}(terpy)Cl_2(NH_3)]$$

where there are Os^{III}, Os^{IV}, and Os^V intermediates.

Porphyrin and phthalocyanine complexes of Os in several oxidation states are generally similar to those of Ru. Examples are $Os(oep)^-$, $[Os(oep)_2]_2^{2+}$, $Os(porph)X_2$, $Os^{VI}(porph)(O)_2$,[69] and $[Os \, Pc(CN)_2]^{2-}$.[70]

Schiff base complexes of $Os^{III,IV,V}$ are known with N,O or N,N′ ligands of the salen type,[71] an example being $Os^{IV}(salen)(OPr^i)_2$.

Nitrile complexes such as $[Os(MeCN)_6]^{2+}$ are known.[72]

Imido compounds are discussed in Section 18-F-11.

[64]U. Frey *et al.*, *Inorg. Chem.* **1996**, *35*, 981.
[65]W. Sliwa, *Transition Met. Chem.* **1995**, *20*, 1.
[66]P.-W. Wang and M. A. Fox, *Inorg. Chem.* **1995**, *34*, 36; R. E. Holmlin and J. K. Barton, *Inorg. Chem.* **1995**, *34*, 7.
[67]T. J. Meyer *et al.*, *J. Am. Chem. Soc.* **1995**, *117*, 823.
[68]T. J. Meyer *et al.*, *Inorg. Chem.* **1996**, *35*, 2167.
[69]J. P. Collman *et al.*, *J. Chem. Soc., Chem. Commun.* **1994**, 11; J. A. Smiieja *et al.*, *Polyhedron* **1994**, *13*, 339; W. R. Scheidt and H. Nasri, *Inorg. Chem.* **1995**, *34*, 2190.
[70]M. Sekota and T. Nyokong, *Polyhedron* **1996**, *15*, 2901.
[71]C.-M. Che *et al.*, *J. Chem. Soc., Dalton Trans.* **1992**, 91.
[72]K. R. Mann *et al.*, *Inorg. Chem.* **1995**, *34*, 2617.

18-F-8 Oxygen, Sulfur, and Phosphorus Ligand Complexes

There are numerous *catecholate* and *quinone* complexes such as $Os^{VI}(O_2C_6H_4)_3$, $Os^{IV}(cat)_2bipy$,[73] and also the triply-bonded Os_2^{6+} species[74], e.g., $Os_2^{III}(O_2CMe)_4Cl_2$, to be mentioned in Section 18-F-13.

Thiolate complexes[75] can be made by the reaction

$$OsO_4 \xrightarrow[\text{PR}'_3]{\text{RSH}} Os^{IV}(SR)_4(PR'_3)_3$$

Paramagnetic octahedral compounds such as $[Os(SC_6F_5)_2(O_2CR)(PMe_2Ph)_2]$ are obtained by interaction of $[Os(SC_6F_5)_2(PMe_2Ph)_2]$ with fluorocarboxylic acids.[76] Other thiolates are $OsCl(SR)(PR_3)_2$ and $Os(SR)_3(PR_3)_2$ as well as acetonitrile complexes that have *tbp* structures, $Os(SR)_4(MeCN)$ and $[Os(SR)_3(MeCN)_2]^+$.[77]

There are also *dithiocarbamates*, $[Os(S_2CNEt_2)_2(PPh_3)_2]^{0,+}$, similar *xanthates*, and *pyridine 2-thiolates*.[78]

A range of compounds with only S atoms as ligands are those of cyclic thioethers such as the octahedral cation $[Os([9]ane\ S_3)_2]^{2+}$.[79] Finally, Os^{VI} complexes of WS_4^{2-} (as well as ReO_4^{2-}, CrO_4^{2-}, WO_4^{2-}) are known, e.g., $[Os(N)(CH_2S_2Me_3)_2-(\mu-S_2)WS_4]^-$.[80]

Compounds with phosphorus ligands are again very similar to those of Ru, typical being *trans*-$[OsX_2(PR_3)_4]^{0,+,2+}$, $Os(H)_2Cl_2(PPr_3)_2$, and $OsH_6(PPr^i_3)_2$[81] and the dodecahedral cation, $[OsH_5(PMe_2Ph)_3]^+$.[82] The latter acts as a catalyst in the hydrogenation and hydroformylation of alkenes.

OXO, IMIDO, AND NITRIDO COMPOUNDS OF Ru AND Os IN HIGH OXIDATION STATES

The higher oxidation states of both Ru and Os have a very extensive and important chemistry. The oxo species are widely used as oxidizing agents for organic compounds. Since $M{=}O$ and $M{=}NR$ are isoelectronic, it is convenient to discuss them together and to include nitrido chemistry here also.

18-F-9 Ruthenium Oxo Compounds

Ruthenium Tetraoxide

This volatile, toxic solid, mp 25.5°C, has a characteristic ozone-like odor. It is obtained when aqueous solutions of "$RuCl_3(aq)$" are oxidized, e.g., by KIO_4, and

[73] S. Battacharya and C. G. Pierpont, *Inorg. Chem.* **1992**, *31*, 35.
[74] F. A. Cotton *et al.*, *Inorg. Chem.* **1993**, *32*, 965.
[75] R. L. Richards *et al.*, *J. Chem. Soc., Dalton Trans.* **1994**, 1819.
[76] R. L. Richards *et al.*, *Polyhedron* **1996**, *15*, 3623.
[77] S. A. Koch *et al.*, *Inorg. Chem.* **1992**, *31*, 5160.
[78] A. Chakravorty *et al.*, *J. Chem. Soc., Dalton Trans.* **1993**, 237.
[79] M. Schröder *et al.*, *J. Chem. Soc., Dalton Trans.* **1992**, 2977.
[80] P. A. Shapley *et al.*, *Inorg. Chem.* **1993**, *32*, 5646.
[81] K. G. Caulton *et al.*, *J. Am. Chem. Soc.* **1995**, *117*, 9473.
[82] O. Eisenstein *et al.*, *J. Am. Chem. Soc.* **1995**, *117*, 281; *Inorg. Chem.* **1994**, *33*, 4966.

the RuO$_4$ swept out by a gas stream or extracted into CCl$_4$. It is decomposed by light and when pure can explode above *ca.* 180°C, giving RuO$_2$ and O$_2$. The oxide is very soluble in CCl$_4$ and moderately soluble in dilute H$_2$SO$_4$, giving golden yellow solutions. It is a very vigorous, but generally non-selective, oxidant for organic substances. However, a two-phase solvent/water system[83] with "RuCl$_3$(aq)" and NaIO$_4$ at 0°C can give catalytic hydroxylation of alkenes, a reaction best known for OsO$_4$ as discussed below.

Ruthenates(VII) and (VI)[84]

Although RuO$_4$ can be reduced by hydroxide ion with liberation of O$_2$, the ruthenate ions are best made in other ways.

The green *perruthenate ion*, RuO$_4^-$, is obtained as the quaternary ammonium salt when RuO$_4$ in a gas stream is passed into a solution of (Pr$_4^n$N)OH. The salt is soluble in organic solvents. A more convenient way to obtain solutions of the ion is by oxidation of aqueous solutions of "RuCl$_3$(aq)" containing Na$_2$CO$_3$ with sodium bromate; the solutions are stable in the pH range 8–12. The ion can also be obtained by chlorine oxidation of ruthenates in solution. In salts such as KRuO$_4$, the anion is tetrahedral; (Pr$_4^n$N)RuO$_4$ is a useful mild oxidant and in presence of a co-oxidant, *N*-methylmorpholine-*N*-oxide, the reactions can be made catalytic.

The orange, paramagnetic *ruthenate* ion can be made by fusion of RuO$_2$ in molten KOH + KNO$_3$ followed by crystallization from water, but solutions of the anion are more easily made by oxidation of "RuCl$_3$(aq)" in KOH solutions of K$_2$S$_2$O$_8$. The anion was long considered to be [RuO$_4$]$^{2-}$ and so it is in Cs$_2$RuO$_4$, but crystal structures of the K and Ba salts showed it to be the *tbp* [Ru(OH)$_2$O$_3$]$^-$ with axial OH groups.[84] Compounds such as SrRuO$_4$ that are made by high temperature methods are again different, having RuO$_6$ octahedra.[85]

The kinetics of electron transfer and self exchange for the perruthenate-ruthenate couple have been studied.[86]

Other Oxo Compounds

There is a vast array of oxoruthenium compounds in oxidation states from III to VII. These may have Ru=O, *cis* and *trans* Ru(O)$_2$, and μ-O bridge groups (which may be linear or bent) and there are also numerous complexes having triangular O-centered Ru$_3$O units noted below. Some representative examples are *cis*-[Ru(O)(bipy)$_2$py]$^{2+}$ [87] that has optically active forms, [Ru(O)(O$_2$CR)Cl$_2$]$^-$,[88] porphyrins such as (porph)Ru^{VI}(O)$_2$, 1,4,7-triazacyclononane (tacn or Me$_3$tacn) complexes, [Ru(O)(bipy)tacn]$^{2+}$, and *cis*-[Ru(O)$_2$(Me$_3$tacn)(CF$_3$CO$_2$)].[89]

[83]T. K. M. Shing *et al., Angew. Chem. Int. Ed. Engl.* **1994,** *33,* 2312.
[84]W. P. Griffith *et al., Inorg. Chem.* **1993,** *32,* 268; W. P. Griffith, *Chem. Soc. Rev.* **1992,** 179 (Ru oxo compounds as oxidants).
[85]Q. Huang *et al., J. Solid State Chem.* **1994,** *112,* 355.
[86]T.-C. Lau and S. L. L. Kong, *J. Chem. Soc., Dalton Trans.* **1995,** 2221.
[87]X. Hua and A. G. Lappin, *Inorg. Chem.* **1995,** *34,* 992.
[88]W. P. Griffith and J. M. Jollife, *J. Chem. Soc., Dalton Trans.* **1992,** 3483.
[89]K. Wieghardt *et al., J. Chem. Soc., Dalton Trans.* **1994,** 48.

Many of these oxo ruthenium (and similar osmium) species can be involved in reversible redox reactions in which OH or H_2O participate, for example,

$$[py(bipy)_2Ru^{IV}O]^{2+} \rightleftharpoons [py(bipy)_2Ru^{III}(OH)]^{2+} \rightleftharpoons [py(bipy)Ru^{II}(OH_2)]^{2+}$$

For the bridged complexes, the best known case of a *linear* Ru—O—Ru group is the ion $[Ru_2^{IV}OCl_{10}]^{4-}$ that is formed on reduction of RuO_2 by HCl. The Ru—O—Ru π-bonding leads to electron pairing and diamagnetism.[90] Other complexes with linear groups are $[Cl(porph)Ru]_2(\mu\text{-}O)$ and species with the $Ru_2^{III}(\mu\text{-}O)$-$(\mu\text{-}O_2CR)_2^{2+}$ cores.[91]

Nonlinear groups occur in $\{[(bipy)_2(H_2O)Ru]_2(\mu\text{-}O)\}^{n+}$ species that have oxidation states 3,3 and 5,5; they are involved in oxidation of H_2O to O_2[92] and in binding to DNA.[93]

Ruthenium red is the name given to highly colored complexes formed when ammoniacal solutions of "$RuCl_3(aq)$" are allowed to oxidize in air. A crystalline compound, $[Ru_3(O)_2(NH_3)_{14}]Cl_6\cdot4H_2O$, has a cation with the linear structure $[(NH_3)_5Ru^{III}ORu^{IV}(NH_3)_4ORu^{III}(NH_3)_5]^{6+}$; the complex $[(NH_3)_5RuORu(en)_2ORu$-$(NH_3)_5]^{6+}$ is similar. The diamagnetism of the species can be ascribed to RuORu π-bonding as above. Oxidation of the red ion in acid solution by Ce^{4+} gives a brown paramagnetic ion of the same stoichiometry but with a charge of +7.

Ruthenium red has been used as a cytological stain and it inhibits mitochondrial uptake of Ca^{2+}. This effect appears to be due to an impurity characterized as $\{[Ru(NH_3)_4(HCO_2)]_2(\mu\text{-}O)\}Cl_3$. The active species in solutions appears to be $\{[X(NH_3)_4Ru]_2(\mu\text{-}O)\}^{3+}$, X = Cl or OH.[94]

The *oxo-centered carboxylates*, $[Ru_3^{III}(\mu_3\text{-}O)(O_2CR)_6L_3]^+$, may be formed when "$RuCl_3(aq)$" is refluxed with, e.g., sodium acetate in acetic acid. The ligand L can be H_2O, py, etc.[95] The normal III,III,III complexes can undergo reversible redox steps. In the presence of phosphines they can act as hydrogenation catalysts and they also act as mild oxidants. There are also dinuclear compounds[96] such as $[\{Ru(bipy)(imidazole)\}_2(\mu\text{-}O)(\mu\text{-}O_2CMe)_2]^{2+}$. These μ-O compounds can also undergo redox reactions.

A variety of oxo species have been used for oxidation of organic substances,[97] commonly in the presence of co-oxidants such as IO_4^-, H_2O_2 or O_2. Some systems will oxidize even alkenes;[98] $[Ru(O)(terpy)(bipy)]^{2+}$ oxidizes sugars and nucleotides.[99]

[90]For theory see Z. Lin and M. B. Hall, *Inorg. Chem.* **1991**, *30*, 3817.

[91]R. J. H. Clark *et al., J. Chem. Soc., Dalton Trans.* **1995**, 2417; A. Syamala and A. R. Chakravarty, *Polyhedron* **1995**, *14*, 231.

[92]Y. Lei and J. K. Hunt, *Inorg. Chem.* **1994**, *33*, 4660.

[93]H. H. Thorp *et al., Inorg. Chem.* **1994**, *33*, 3544.

[94]M. J. Clarke *et al., J. Am. Chem. Soc.* **1993**, *115*, 11799.

[95]K. Uosaki *et al., Inorg. Chem.* **1995**, *34*, 4527; M. Abe *et al., Inorg. Chem.* **1995**, *34*, 4490; A. Bino *et al., Inorg. Chim. Acta* **1995**, *232*, 167.

[96]C. Suda and A. R. Chakravarty, *J. Chem. Soc., Dalton Trans.* **1996**, 3289.

[97]W. P. Griffith, *Chem. Soc. Rev.* **1992**, *179;* W. P. Griffith *et al., J. Chem. Soc., Dalton Trans.* **1995**, 3537.

[98]R. S. Drago *et al., J. Am. Chem. Soc.* **1994**, *116*, 2424.

[99]H. H. Thorpe *et al., J. Am. Chem. Soc.* **1995**, *117*, 1463.

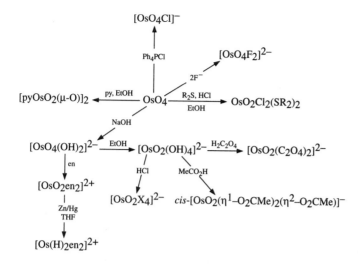

Figure 18-F-4 Some reactions of OsO$_4$ and [OsO$_2$(OH)$_4$]$^{2-}$; all species have *trans*-O=Os=O groups, except as noted.

Finally some oxoalkyls, [Ru(O)(CH$_2$SiMe$_3$)$_3$]$_2$ and Ru(O)(CH$_2$SiMe$_3$)$_4$, are known.[100]

18-F-10 Osmium Oxo Compounds

Osmium Tetroxide

This yellow volatile solid, mp 40°C, is obtained by burning Os in O$_2$ or more usually by oxidation of osmium halide solutions with NaOCl. It is more stable thermally than RuO$_4$ but is equally toxic, being particularly hazardous to the eyes on account of facile reduction to a black oxide. This feature has long been used in biological staining employing dilute aqueous solutions of OsO$_4$. Some reactions of OsO$_4$ are given in Fig. 18-F-4.

The oxide gives very stable *tbp* adducts with a variety of amines like pyridine or quinuclidine. It also forms highly colored but weak complexes with arenes.[101] This has been exploited[102] to convert benzene to polycyclohexane derivatives.

A distorted octahedral, deep red complex of 1,2-diamine (*R,R'*)-*trans*-1,2-bis(*N*-pyrrolidino)cyclohexane (L), L·OsO$_4$ has been made at low temperatures.[103] It is very reactive and is formally a 20*e* outer valence shell species. The isolation of this and relevant 1,2-diamine complexes throws light upon the mechanism of enantioselective dihydroxylation of alkenes noted below. The tetroxide has important commercial use in the stoichiometric and catalytic *cis*-hydroxylation of alkenes. There has been much discussion of the mechanism, particularly for the enantiospecificity achieved in the presence of cinchona alkaloids or other chiral agents.[104]

[100]P. A. Shapley *et al.*, *Organometallics* **1990**, *9*, 1341.
[101]J. M. Wallis and J. K. Kochi, *J. Am. Chem. Soc.* **1988**, *110*, 8207.
[102]W. B. Motherwell and A. S. Williams, *Angew. Chem. Int. Ed. Engl.* **1995**, *34*, 2031.
[103]E. J. Corey *et al.*, *J. Am. Chem. Soc.* **1996**, *118*, 7851.
[104]K. B. Sharpless *et al.*, *Chem. Rev.* **1994**, *94*, 2483; *Organometallics* **1994**, *13*, 344; *J. Am. Chem. Soc.* **1994**, *116*, 1278; A. Veldkamp and G. Frinking, *J. Am. Chem. Soc.* **1994**, *116*, 4937.

At its simplest, the initial step of *cis*-hydroxylation involves a [3+2] or [2+2] cycloaddition (18-F-VI):

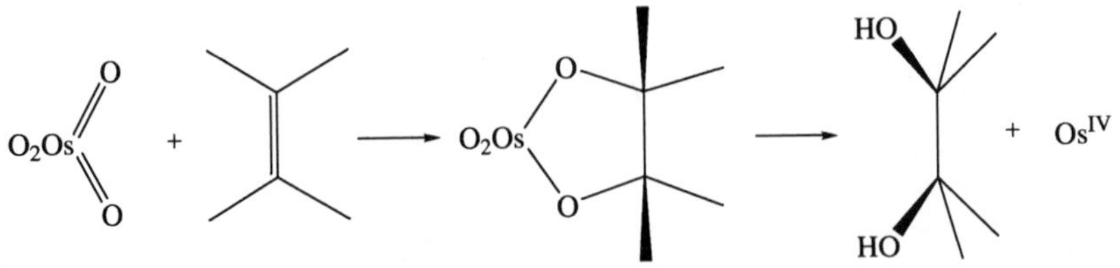

(18-F-VI)

The initial complexes then form more complicated ones such as those shown in (18-F-VII) by dimerization or reaction with additional alkene. The *cis*-diol product is then isolated by reduction with Na_2SO_3 or in other ways. Methods for amino hydroxylation, i.e., addition of amino and OH groups to C=C bonds have also been developed.[105]

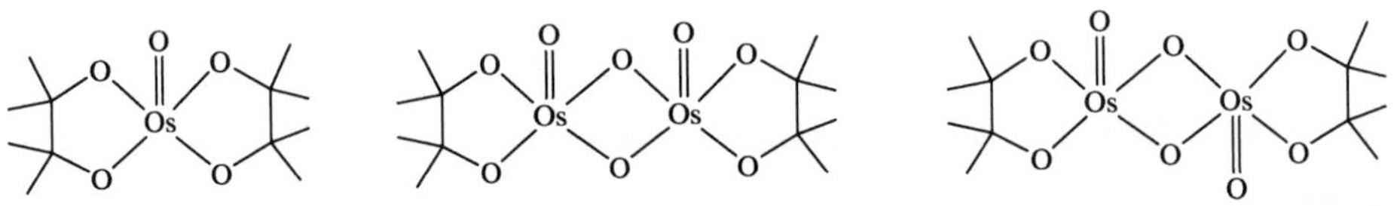

(18-F-VII)

Osmates

Salts of the tetrahedral ion $[OsO_4]^-$, which is similar to $[RuO_4]^-$, are obtained when OsO_4 in CH_2Cl_2 is treated with Ph_4AsI, KI, etc. Interaction of OsO_4 with RbOH or CsOH in appropriate ratios give the species $[OsO_4(OH)]^-$, $[O_4Os(\mu\text{-}OH)OsO_4]^-$, and *trans*-$[OsO_4(OH)_2]^{2-}$. However, KOH and NaOH with OsO_4 give only *trans*-$[OsO_4(OH)_2]^{2-}$. This ion is reduced by ethanol to the dark purple $[Os^{VI}(O)_2(OH)_4]^{2-}$. The latter and its substituted derivatives such as $[Os(O)_2Cl_4]^{2-}$ are diamagnetic. This happens because the oxide ligands form Os=O bonds by π overlap mainly with d_{xz} and d_{yz}, thus destabilizing those orbitals, leaving a low-lying d_{xy} orbital that will be occupied by the two electrons, leading to diamagnetism.

The $[Os(O)_2(OH)_4]^-$ salts are useful synthons for other oxo compounds (Fig. 18-F-4). There are many other oxo species, cationic, anionic, and neutral, some examples being $[OsO_2(en)_2]^{2+}$, $[OsO_2(OAc)_3]^-$, $OsO_3(NBu^t)$, and $OsO_2(mes)$ described in Section 18-F-14.

Only two *peroxo* complexes appear to be known, namely *trans*-$[Os(\eta^2\text{-}O_2)X(\overparen{PP})_2]BPh_4$, X = H or Cl, $\overparen{PP}$ = 1,2-bis(dicyclohexylphosphino)ethane, which are formed by interaction of O_2 and $[OsX(\overparen{PP})_2]^+$.[106]

[105]O. Reiser, *Angew. Chem. Int. Ed. Engl.* **1996,** *35,* 1308 (review and refs.); K. B. Sharpless *et al., Angew. Chem. Int. Ed. Engl.* **1996,** *35,* 451.
[106]A. Mazzetti *et al., J. Chem. Soc., Chem. Commun.* **1994,** 1597.

18-F-11 Imido and Nitrido Compounds of Osmium and Ruthenium

Osmium

The partial substitution of NR^{2-} for O^{2-} in OsO_4 has long been known in the compounds $OsO_3(NR)$ and $OsO(NR)_3$ that are formed on reaction of OsO_4 with amines. Complete substitution has been achieved using neat $Bu^tNH(SiMe_3)$ which gives a mixture of the orange-red $Os(NBu^t)_4$, mp $ca.$ 30°C, and a tetranuclear osmium(VI) complex $[(Bu^tN)_2Os(\mu\text{-}NBu^t)_2Os(NBu^t)\mu\text{-}O]_2$.[107] The $Os(NBu^t)_4$ molecule has a distorted tetrahedral structure with bent OsNC bonds according to electron diffraction data.[108] The bent bonding can be interpreted a compromise of $Os\equiv N-$ and $Os=N-$ bonding.

Reduction of the tetraimido complex by Na/Hg gives the Os^{VI} dimer $[(Bu^tN)_2\text{-}Os^{VI}(\mu\text{-}NBu^t)]_2$ which with $Me_3O^+BF_4^-$ gives $[(Bu^tN)_2Os^{VII}\mu\text{-}NBu^t]_2(BF_4)_2$. A number of other derivatives have been characterized. The interaction of OsO_4 with $NHAr(SiMe_3)$[109,110] by contrast gives the reduced Os^{VI} arylamido complexes $Os(NAr)_3$. The $2,6\text{-}Pr^i C_6H_3$ compound was shown to have a trigonal planar structure. Interaction of $Os(NAr)_3$ with PMe_2Ph gives the phosphineimine by NAr transfer and the square $trans\text{-}Os^{IV}(NAr)_2(PR_3)_2$. Osmium(II) arene compounds such as $(\eta^6\text{-}Ar)Os(NBu^t)$, $Ar = p\text{-}cymene$ and C_6Me_6, are also known.[111] So far, despite many efforts, it has not proved possible to make the corresponding homoleptic ruthenium compounds. Some bridging and terminal imido compounds have been made, but the only stucturally characterized terminal ones are $trans\text{-}(Me_3P)_2Ru^{IV}[N(2,6\text{-}Pr^i_2C_6H_3)]_2$[112] and $(p\text{-}cymene)Ru[N(2,4,6\text{-}Bu^t_3C_6H_2)]$.[113]

18-F-12 Nitrido Compounds of Osmium

The interaction of OsO_4 with ammonia in aqueous KOH has long been known to produce the "osmiamate" $K[N\equiv OsO_3]$, as orange-red crystals. Interaction of $[Bu^n_4N][NOsO_3]$ with $Au^I(PPh_3)CF_3SO_3$ gives $(PPh_3)AuNOsO_3$; $cis\text{-}[(Me_3P)_2Pt(NOsO_3)_2]$ is also known.[114]

The anion $[OsNCl_4]^-$ can be obtained as its $(Bu^t_4N)^+$ salt by interaction of sodium azide solution with an HCl solution of potassium osmate. It reacts with maleonitrile dithiolate to give $(Bu^t_4N)_2[OsN(mnt)_2]$.[115]

Osmium nitrido species may be luminescent and have other photophysical and chemical properties and Os^{VI} compounds such as $[Os(N)(bipy)Cl_2(H_2O)]^+$ and $[Os(N)(NH_3)_4]^{3+}$ have been studied in some detail.[116] Interaction of $[Os(N)Cl_4]^-$ and

[107]G. Wilkinson $et\ al.$, $J.\ Chem.\ Soc.,\ Dalton\ Trans.$ **1991,** 269, 1855.
[108]D. W. H. Rankin $et\ al.$, $J.\ Chem.\ Soc.,\ Dalton\ Trans.$ **1994,** 1563.
[109]G. Wilkinson $et\ al.$, $J.\ Chem.\ Soc.,\ Dalton\ Trans.$ **1991,** 269, 1855.
[110]R. R. Schrock $et\ al.$, $Inorg.\ Chem.$ **1991,** 30, 3595.
[111]R. A. Andersen $et\ al.$, $Organometallics$ **1993,** 12, 2741.
[112]G. Wilkinson $et\ al.$, $Polyhedron$ **1992,** 11, 2961.
[113]A. K. Burrell, and A. J. Steedman $J.\ Chem.\ Soc.,\ Chem.\ Commun.$ **1995,** 2109.
[114]W.-H. Leung $et\ al.$, $J.\ Chem.\ Soc.,\ Dalton\ Trans.$ **1996,** 3153.
[115]W.-H. Leung $et\ al.$, $J.\ Chem.\ Soc.,\ Dalton\ Trans.$ **1994,** 2519.
[116]C.-M. Che $et\ al.$, $J.\ Chem.\ Soc.,\ Dalton\ Trans.$ **1995,** 657.

terpyridine give *trans*-[Os(N)Cl$_2$(terpy)]Cl, which slowly isomerizes to the *cis* cation in MeOH.[117] An organometallic nitride[118] is the thiolate anion [Ru^{VI}N(Me)$_3$-(SSiMe$_3$)]$^-$ made from the corresponding bromide and NaSSiMe$_3$.

Relatively few nitrido complexes of Ru are known; examples are [Ru(N)X$_4$]$^-$ [119] and some organometallic ones mentioned in the next section. Examples of osmium hydrides are OsH$_3$(BH$_4$)(PR$_3$)$_2$ that has an Osμ-H$_2$BH$_2$ bridge, OsH$_6$(PR$_3$)$_2$, [OsH$_5$(PR$_3$)$_3$]$^+$, and OsH$_4$(PR$_3$)$_3$.

18-F-13   Compounds with M—M Multiple Bonds[120]

Both Ru and Os form many of these. Both elements form many compounds in which the multiply-bonded M$_2^{n+}$ cores are bridged by four ligands such as carboxyl anions, or those shown below:

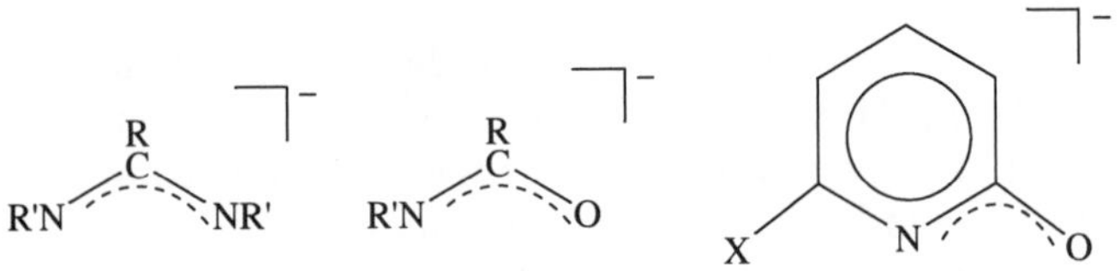

For ruthenium the charge in Ru$_2^{n+}$ may be 4+, 5+, or 6+ and the Ru—Ru distances range from 2.24 to 2.56 Å depending on the oxidation state and the nature of the ligands present.[121] The Ru—Ru bond configurations range over $\sigma^2\pi^4\delta^2\pi^{*4}$ (Ru$_2^{4+}$, low-spin), $\sigma^2\pi^4\delta\delta^*\pi^{*2}$ (Ru$_2^{6+}$, high-spin), and $\sigma^2\pi^4\delta^2\delta^*\pi^{*2}$ (Ru$_2^{5+}$, high-spin), the last one occurring in the very common Ru$_2$(O$_2$CR)$_4$X compounds.

Osmium compounds are mostly of the types Os$_2$(O$_2$CR)$_4$X$_2$ ($\sigma^2\pi^4\delta^2\delta^{*2}$ and [Os$_2$X$_8$]$^{2-}$, neither of which has any Ru analogues.

Both Ru and Os form dinuclear porphyrin complexes such as M$_2$(oep)$_2$, and some related compounds with other macrocycles.

18-F-14   Compounds with M—C Bonds

The chemistry of Ru and Os compounds with metal-carbon bonds is extremely extensive and complicated. Much of this chemistry involves organic groups such as C$_5$H$_5^-$ and its substituted derivatives and arenes, effectively as "spectator" ligands, other groups being present as ligands. There is also an extensive chemistry of carbonyls like Ru$_3$(CO)$_{12}$ and other Ru and Os clusters. We can deal here only with some of the simpler compounds, especially those with M—C, M=C, MCp, and M(arene) bonds.

[117]T. J. Meyer *et al.*, *Inorg. Chem.* **1995**, *34*, 586.
[118]H.-C. Liang and P. A. Shapley, *Organometallics* **1996**, *15*, 1331.
[119]C.-M. Che *et al.*, *Inorg. Chem.* **1996**, *35*, 540.
[120]F. A. Cotton and R. A. Walton, *Multiple Bonds between Metal Atoms*, 2nd ed., Oxford University Press, Oxford, UK 1996, Chap. 6.
[121]F. A. Cotton and A. Yokochi, *Inorg. Chem.* **1997, 36**, 567.

Homoleptic Alkyls and Aryls

The first homoleptic alkyl compounds to be made were Ru_2R_6, R = CH_2SiMe_3 and CH_2CMe_3 where there is a metal-metal triple bond. The compounds were obtained by an unusual disproportionation reaction:

$$[Ru^{II,III}_2(\mu\text{-}O_2CMe)_4]Cl \xrightarrow{RMgX} Ru^{III}_2R_6 + [Ru^{II}_2(\mu\text{-}O_2CMe)_4](THF)_2$$

The alkylation of $Os_2(\mu\text{-}O_2CMe)_4Cl_2$ gave only the partially substituted $Os^{III}_2(\mu\text{-}O_2CMe_2)(CH_2SiMe_3)_4$. The air-sensitive thermally unstable *ato* complex [Li tmed]$_3$[RuMe$_6$][122] made by treating $RuCl_3(tht)_3$ with MeLi in Et_2O and adding tmed is another Ru^{III} complex. For both elements M^{IV} compounds are the cyclohexyls, $M(C_6H_{11})_4$, and aryls, MAr_4, Ar = *o*-tolyl, mesityl; these are air stable tetrahedral compounds.[123] Some of the aryls undergo reversible redox reactions, e.g.,

$$[Ru^{III}(o\text{-tol})_4]^- \underset{+e}{\overset{-e}{\rightleftharpoons}} Ru^{IV}(o\text{-tol})_4 \underset{+e}{\overset{-e}{\rightleftharpoons}} [Ru^V(o\text{-tol})_4]^+$$

Although the anions have not been isolated, M^V compounds can be obtained by oxidation of the M^{IV} compound with AgO_3SCF_3, $AgBF_4$, or $NOPF_6$ in CH_2Cl_2, e.g.,

$$Ru^{IV}(mes)_4 \longrightarrow [Ru^V(mes)_4]PF_6$$

| green-black | blue-black |
| diamagnetic | paramagnetic |

The osmium *o*-tolyl undergoes coupling of aryl groups of the type proposed earlier by H. Zeiss to explain the formation of chromium π-aryls (Hein's "polyphenyl" chromium compounds):

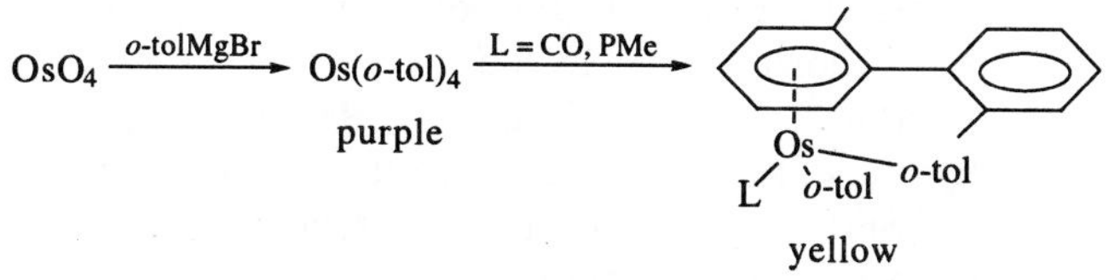

$$OsO_4 \xrightarrow{o\text{-tolMgBr}} \underset{\text{purple}}{Os(o\text{-tol})_4} \xrightarrow{L = CO, PMe}$$

yellow

Kinetic studies[124] suggest an initial associative mechanism. With an excess of PMe_3 the biaryl is eliminated to give $Os(o\text{-tol})_2(PMe_3)_4$.

Oxo and Imido Compounds

Several alkyls and aryls with M=O (or M=NR) groups have been characterized. A simple methyl, $OsOMe_4$ is obtainable from OsO_4. The dioxo alkyl, $OsO_2(CH_2Bu^t)_2$,[125] can be converted into alkylidene and alkylidyne compounds,

[122]G. Wilkinson *et al.*, *Polyhedron* **1990**, *9*, 2071.
[123]G. Wilkinson *et al.*, *J. Chem. Soc., Dalton Trans.* **1987**, 557, **1988**, 669, **1989**, 2149, **1992**, 3477.
[124]G. B. Young *et al.*, *Polyhedron* **1996**, *15*, 1363.
[125]P. P. Schrock *et al.*, *J. Am. Chem. Soc.* **1995**, *117*, 482.

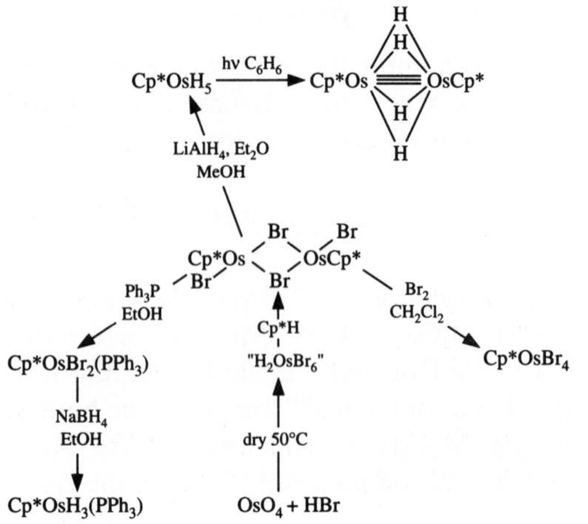

Figure 18-F-5 Some reactions of Cp*Os compounds (adapted from G. S. Girolami *et al., J. Am. Chem. Soc.* **1994,** *116,* 10294).

e.g., $Os(CHBu^t)_2(CH_2Bu^t)_2$. The dioxo mesityl[126] $OsO_2(mes)_2$ undergoes an unusual reaction with NO_2 to give $[OsO_2(\eta^1\text{-}ONO_2)_2mes]^-$ together with NO and nitro-mesitylene. Being coordinatively unsaturated, $OsO_2(mes)_2$ also reacts with pyridines to give adducts, and with N-heterocycles;[127] pyrazine gives the bridged dimer $[OsO_2(mes)_2]_2(\mu\text{-pyz})$.

Cyclopentadienyl and Arene Compounds

The first organometallic compound known for ruthenium, $(\eta^5\text{-}C_5H_5)_2Ru$, and the second "sandwich" compound to be made was isolated shortly after the recognition of the sandwich structure of $Fe(C_5H_5)_2$ by interaction of $Ru(acac)_3$ with CpMgBr.

The cyclopentadienide anion, $C_5H_5^-$, and benzene and a wide variety of their substituted compounds, notably $C_5Me_5^-$ or Cp*, form major groups of compounds, commonly of the type $CpML_n$ or $ArML_n$ for both Ru and Os. Some representative reactions for Cp*Os are shown in Fig. 18-F-5.

Representative[128] arene compounds are $Cp*Ru(NO)Ph_2$, $CpRu(\eta^6\text{-}C_6H_6)^+$, $[(\eta^6\text{-}C_6H_6)Os(MeCN)_3]^{2+}$, which gives $[Os(MeCN)_6]^{2+}$ on photolysis in MeCN, and $[CpOs(NMe)(CH_2SiMe_3)_2][SO_3CF_3]$. The η^6-arene $(p\text{-}MeC_6H_4Pr^i)RuN(2,4,6\text{-}Bu^t_3C_6H_2)$ undergoes an insertion reaction of NAr' with mesityl azide to give a tetrazene complex:[129]

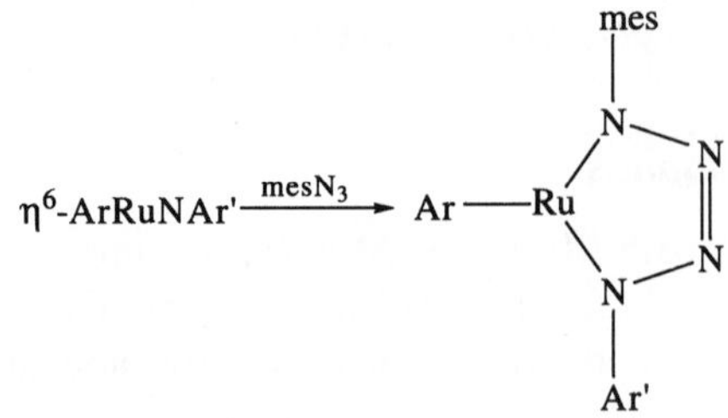

[126]G. Wilkinson *et al., J. Chem. Soc., Dalton Trans.* **1990,** 2465.
[127]W.-H. Leung *et al., Organometallics* **1996,** 15, 1497.
[128]C. D. Tagge and B. G. Berman, *J. Am. Chem. Soc.* **1996,** *118,* 6908; K. R. Mann *et al., Inorg. Chem.* **1995,** *34,* 2617; P. A. Shapley *et al., Organometallics* **1996,** *15,* 1622.
[129]G. Wilkinson *et al., J. Chem. Soc., Dalton Trans.* **1996,** 3771.

For Os^{II} complexes such as (p-cymene)OsX_2PMe_3, extensive ^{187}Os nmr studies have been made.[130]

Carbene Complexes

The formation of alkylidene and alkylidyne compounds was noted above. Quite a number of these are now known. The use of electron rich alkenes to make carbene complexes was pioneered by Lappert[131] as in the reaction

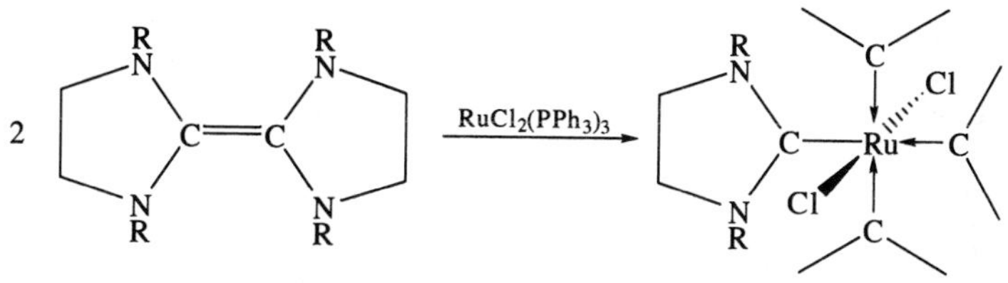

A recent application of this synthesis[132] is the following:

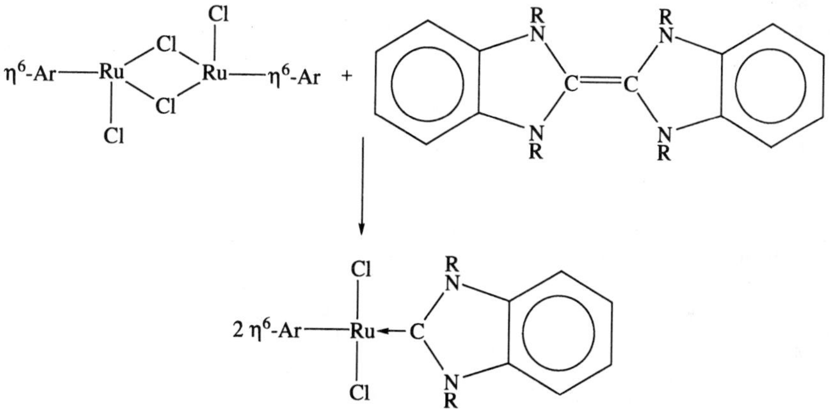

However, following the work of Arduengo on stable carbenes[133] of the type 18-F-VIII,

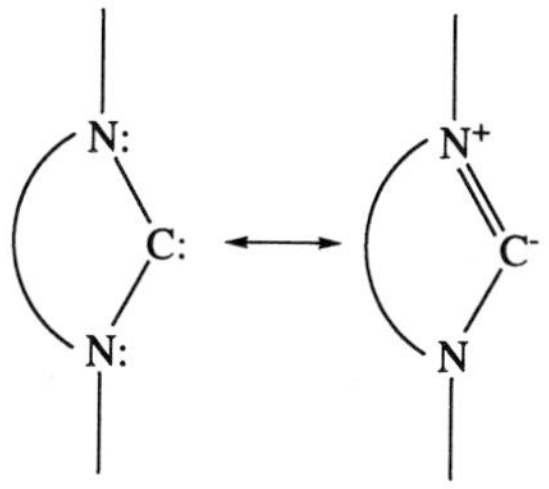

(18-F-VIII)

[130]W. von Philipsborn et al., Organometallics **1996,** 15, 3124.
[131]M. F. Lappert et al., J. Chem. Soc., Dalton Trans. **1994,** 2355.
[132]B. Cetinkaya et al., Organometallics **1996,** 15, 2434.
[133]A. J. Arduengo III et al., J. Am. Chem. Soc. **1995,** 117, 572, 11027.

many metal complexes can be made using these carbenes or salts of type 18-F-IX. Ruthenium(II) and osmium(II) compounds have been made by the reactions[134]

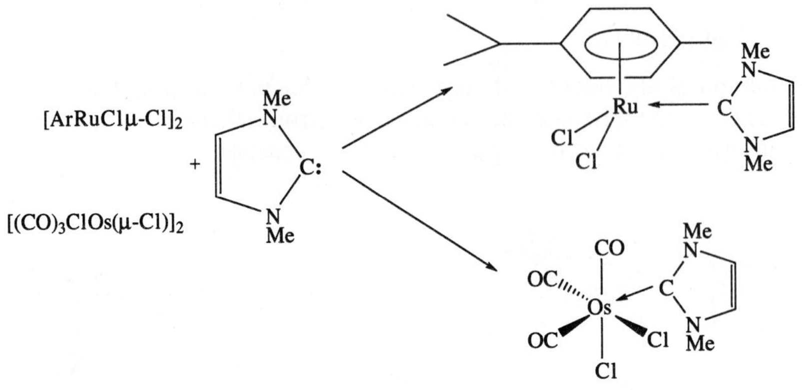

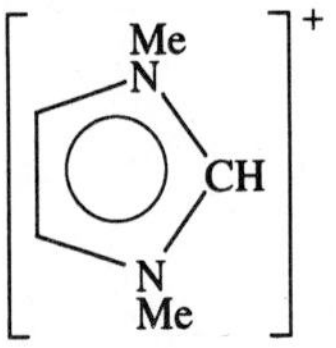

(18-F-IX)

Compounds with Ru=C bonds can be obtained by the reaction[135]

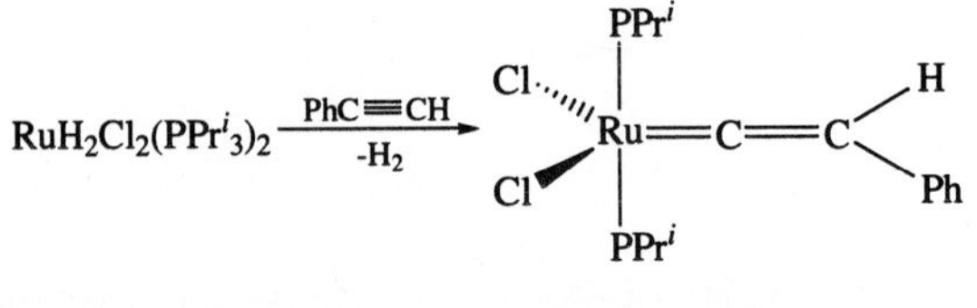

Other methods developed by Grubbs[136] for ruthenium complexes which can act as catalysts for ring opening metathesis polymerization (ROMP) and metathesis of conjugated and cumulated alkenes include the following:

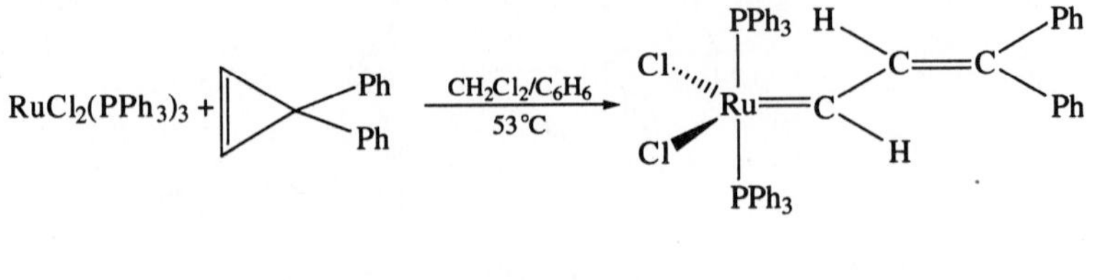

[134]W. A. Herrmann et al., Chem. Eur. J. **1996**, 2, 773.
[135]H. Werner et al., Organometallics **1996**, 15, 1960.
[136]R. H. Grubbs et al., J. Am. Chem. Soc. **1996**, 118, 100.

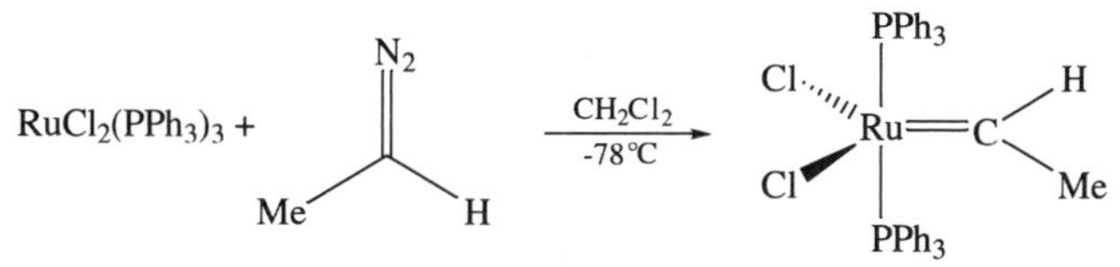

We can only note that alkene, alkyne, allyl, and related compounds of many types are known. Important ruthenium starting materials are $RuCl_2(COD)$ and $Ru(COD)(2$-methylallyl$)_2$, both of which are commercially available.

Carbonyls

The carbonyls of Ru and Os have generally similar chemistry. The most important Ru carbonyl is the white crystalline solid $Ru_3(CO)_{12}$ that is obtained by carbonylation of "$RuCl_3(aq)$" in MeOH at 125°C under CO pressure. The pentacarbonyl, $Ru(CO)_5$, made by carbonylation of $Ru(acac)_3$, can be obtained as a liquid, mp 17°C. It rapidly loses CO in solutions and the equilibrium with $Ru_3(CO)_{12}$ has been studied in detail:

$$Ru_3(CO)_{12} + 3CO \rightleftharpoons 3Ru(CO)_5$$

The dinuclear $Ru_2(CO)_9$ compound has been identified spectroscopically in hexane at low temperatures. Its structure and that of $Os_2(CO)_9$ differ from that of $Fe_2(CO)_9$; they have two $M(CO)_4$ groups united by an M—M bond and a μ-CO.

Ruthenium and, particularly, osmium form a great many high nuclearity carbonyl clusters. Only a few illustrative examples can be mentioned here. Figure 18-F-6 shows some of the topological relationships among the osmium compounds. The scope of the ruthenium chemistry has been extended recently with the preparation of the $[Ru_{10}H_2(CO)_{25}]^{2-}$ and $[Ru_{11}H(CO)_{27}]^{3-}$ ions.[137]

A Ru^0 substituted carbonyl[138] has been made by the reaction

$$RuCl_2(CO)_2(PMeBu^t{}_2)_2 \xrightarrow[25°C]{Mg, THF} Ru(CO)_2(PMeBu^t{}_2)_2$$

It has a distorted structure similar to that obtained if an equatorial CO were removed from tbp-$Ru(CO)_3L_2$.

Stable, homoleptic cations, $[M(CO)_6][Sb_2F_{11}]_2$, have been made by interaction of the fluorosulfates $M(SO_3F)_3$ in liquid SbF_5 under CO at 60–90°C.[139] There is also a wide range of halide complexes, examples being $[Ru(CO)_2Cl_2]_n$, $Ru(CO)_3Cl_2$, $[Ru(CO)Cl_5]_n^{2-}$, and $[Ru(CO)_3Cl_3]^-$ as well as Os analogues. Oxidation of $Os_3(CO)_{12}$ by XeF_2 in HF(l) gives cis-$OsF_2(CO)_4$ and other products.[140]

[137]J. Lewis *et al.*, *J. Chem. Soc., Dalton Trans.* **1996**, 3515.
[138]O. Eisenstein *et al.*, *J. Am. Chem. Soc.* **1995**, *117*, 8869.
[139]F. Aubke *et al.*, *J. Chem. Soc., Chem. Commun.* **1995**, 2071.
[140]E. G. Hope *et al.*, *J. Chem. Soc., Dalton Trans.* **1994**, 1062.

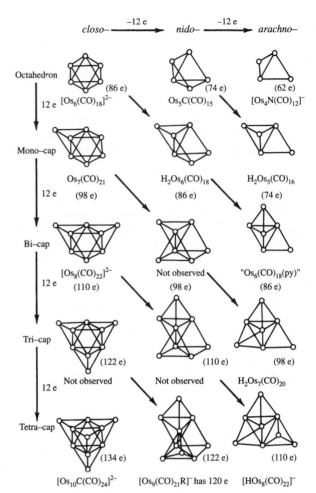

Figure 18-F-6 Structures and total electron counts for osmium clusters with seven skeletal electron pairs. Structures related by the diagonal arrows are alternatives for the same total electron count.

Cyanides and Isocyanides

The Ru and Os compounds are generally similar to those of the other transition metals. Typical cyanides[141] are $[M(CN)_6]^{4-}$ and $Na_2[Os(CN)_5NO] \cdot 2H_2O$. *Isocyanides* of several types are known, a homoleptic example[142] being $[Ru(CNBu^t)_4]^{2-}$, which is the analogue of $[Fe(CO)_4]^{2-}$. This anion is made by interaction of potassium naphthalenide with $Ru(CNBu^t)_4Cl_2$ and can undergo a variety of reactions to give derivatives. Thus Ph_3SnCl gives *trans*-$Ru(CNBu^t)_4SnPh_3$.

18-F-15 Hydride and Hydrogen Compounds

Ruthenium compounds with phosphine ligands, such as $RuHCl(CO)(PR_3)_3$ and $RuH_2(CO)(PR_3)_3$, are useful starting materials and also are active catalysts for the

[141]J. A. Olabe *et al.*, *Inorg. Chem.* **1994**, *33*, 5890; H. Taube *et al.*, *Inorg. Chem.* **1994**, *33*, 2874.
[142]N. J. Cooper *et al.*, *Angew. Chem. Int. Ed. Engl.* **1992**, *31*, 83.

hydrogenation of alkenes and other reactions.[143] The similar complex, $OsHCl(CO)$-$(PPr_3^i)_3$, reacts with H_2 to give the *trans*-hydrido dihydrogen compound $OsHCl$-$(\eta^2\text{-}H_2)(CO)(PPr_3^i)_2$.[144] The catalytic hydrogenation cycle for conversion of benzylidine- acetone to 4-phenylbutan-2-one involves both of these Os species. The H—H distance in the $\eta^2\text{-}H_2$ compound was determined by variable temperature ^1H T_1 measurements to be *ca.* 0.8 Å. Some other new osmium dihydrogen complexes, *trans*-$[Os(H\cdots H)X(Ph_2PCH_2CH_2PPh_2)_2]^+$, X = Cl, Br by contrast have stretched H$\cdots$H distances of 1.11(6) Å and 1.22(3) Å by X-ray and neutron diffraction studies.[145] The Os^{IV} complex ion $[OsH_3(Cy_2PCH_2CH_2PCy_2)_2]^+$ has also been made as well as $OsH_3(H_2BH_2)(PPr_3^i)_2$,[146] which has a $\mu\text{-}BH_4$ ligand and a pentagonal bipyramidal structure. Protonation[147] of $Os(H)_2Cl(PPr_3^i)_2$ not only removes Cl but induces aggregation to give $\{[(Pr_3^iP)_2OsH_2]_2(\mu\text{-}Cl)_3\}^+$. It is interesting that no non-octahedral complex has ever been protonated at hydride in preference to Cl, but the general rules to establish selectivity are not yet clear.

Ruthenium complexes[148] with stretched H$\cdots$H bonds have been made by the reaction of $RuH_2(\eta^2\text{-}H_2)_2(PCy_3)_2$ with pyridines having OH, and NH_2 substituents (LXH) give complexes $RuH(H\cdots H)(LX)(PCy_3)_2$. The stretching is attributed to good σ-donor ligands in the *trans* position. As noted in Chapter 2, a number of polyhydrides that were originally thought to be in high oxidation states have had to be reformulated. An example is $RuH_6(PCy_3)_2$, which is actually $RuH_2(\eta^2\text{-}H_2)_2(PCy_3)_2$.

Additional References

M. I. Bruce *et al.*, *Polyhedron* **1991**, *10*, 277 (Ru clusters with NR_3, RCN, NO, and other N ligands).

R. Krämer, *Angew. Chem. Int. Ed. Engl.* **1996**, *35*, 1197 (π-arene Ru complexes in peptide synthesis and labeling).

R. Sánchez-Delgado *et al.*, *J. Mol. Catal. A: Chem.* **1995**, *96*, 231(Homogeneous catalyses by osmium compounds >53 refs).

E. A. Seddon and K. R. Seddon, *The Chemistry of Ruthenium*, Elsevier, Amsterdam, 1984.

18-G   RHODIUM AND IRIDIUM: GROUP 9

18-G-1   General Remarks

The chemistry of Rh centers around oxidation states I-III while for Ir, the IV and V states are also significant. The oxidation states and stereochemistries are summarized in Table 18-G-1.

Since their chemistry is very important for the d^8 species Rh^I, Ir^I, Pd^{II}, and Pt^{II}, it is convenient to discuss here three classes of binuclear compounds (18-G-I),

---

[143]For examples and references see: H. Werner *et al.*, *Organometallics* **1996**, *15*, 1960; M. A. Esternelas *et al.*, *Organometallics* **1996**, *15*, 3423; V. I. Bakhmutov *et al.*, *Chem. Eur. J.* **1996**, *2*, 815.

[144]V. I. Bakhmutov *et al.*, *Chem. Eur. J.* **1996**, *2*, 815.

[145]R. H. Morris *et al.*, *J. Am. Chem. Soc.* **1996**, *118*, 5396.

[146]A. Liedós *et al.*, *J. Am. Chem. Soc.* **1996**, *118*, 8338.

[147]K. G. Caulton *et al.*, *J. Am. Chem. Soc.* **1996**, *118*, 6934.

[148]B. Chaudret *et al.*, *Organometallics* **1996**, *15*, 3472.

Table 18-G-1   Oxidation States of Rhodium and Iridium

Oxidation state	Coordination number	Geometry	Examples
Rh^{-I}, Ir^{-I},	4	Tetrahedral	$Rh(CO)_4^-$, $Ir(CO)_3PPh_3^-$
Rh^0, Ir^0, d^9	4	Distorted tetrahedral	$Rh[P(OPr^i)_3]_4$
Rh^I, Ir^I, d^8	3	Planar	$RhCl(PCy_3)_2$, $Rh[N(SiMe_3)_2](PPh_3)_2$, $Rh(PPh_3)_3^+ClO_4^-$
	4	Planar[a]	$RhCl(PMe_3)_3$, $[RhCl(CO)_2]_2$, $IrCl(CO)(PR_3)_2$
		Tetrahedral	$[Rh(PMe_3)_4]^+$
	5	*tbp*	$HRh(diphos)_2$, $HIrCO(PPh_3)_3$, $HRh(PF_3)_4$, $[Rh(SnCl_3)_5]^{4-}$
Rh^{II}, Ir^{II}, d^7	4	Square	$RhCl_2[P(o\text{-}MeC_6H_4)_3]_2$, $Rh(C_6Cl_5)_2(COD)$, $Ir(mes)_2(SEt_2)_2$
	5	*sp*	$Ir_2(oep)_2$, $Rh_2(O_2CR)_4$ (M—M bonds)
	6	Octahedral	$[(Ph_3P)Rh(\mu\text{-}O_2CR)_2]_2$
Rh^{III}, Ir^{III}, d^6	3	See text	$Rh(mes)_3$
	5	*tbp*	$IrH_3(PR_3)_2$
	5	*sp*	$RhI_2(CH_3)(PPh_3)_2$, $[Rh(C_6F_5)_5]^{2-}$
	6	Octahedral	$[Rh(H_2O)_6]^{3+}$, $RhCl_6^{3-}$, $IrH_3(PPh_3)_3$, $RhCl_3(PEt_3)_3$, $IrCl_6^{3-}$, RhF_3, $[Rh(diars)_2Cl_2]^+$, IrF_3 (FeO_3 type)
Rh^{IV}, Ir^{IV}, d^5	4	Tetrahedral	$Ir(mes)_4$
	6	Octahedral	$IrCl_6^{2-}$, $Ir(C_2O_4)_3^{2-}$, RhF_6^{2-}
Rh^V, Ir^V, d^4	4	Tetrahedral	$[Ir(mes)_4]^+$, $Ir(mes)_3O$
	6	Octahedral	$CsMF_6$
	7	Pentagonal bipyramid	$IrH_5(PPr^i_3)_2$ (classical M—H)
Rh^{VI}, Ir^{VI}, d^3	6	Octahedral	RhF_6, IrF_6

[a]With bulky phosphines some 4-coordinate complexes such as $RhCl(CO)(PBu^t_3)_2$ have flattened tetrahedral geometry; see R. L. Harlow *et al.*, *Inorg. Chem.* **1992**, *31*, 323.

(18-G-II), and (18-G-III) that are formed by bifunctional phosphines such as $Ph_2PCH_2PPh_2$ or $Ph_2P(1\text{-pyridyl})$.[1]

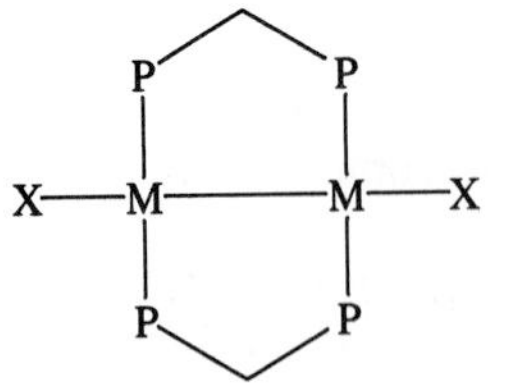

Side by side   (18-G-I)   A-frame   (18-G-II)   Face to face   (18-G-III)

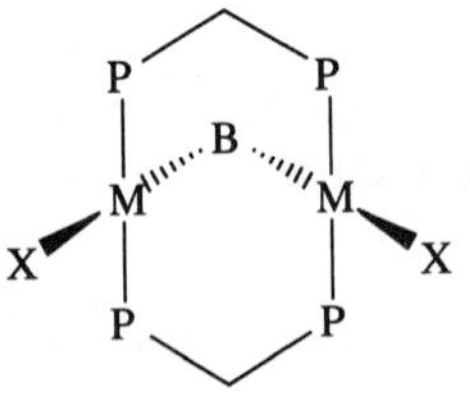

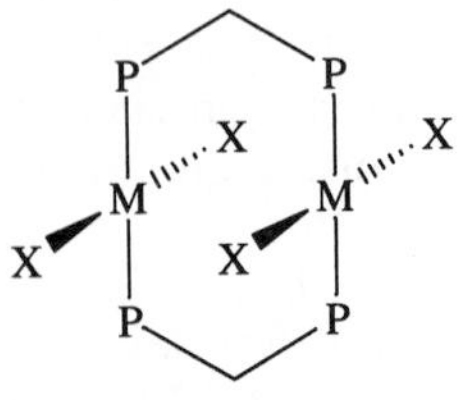

[1]For references and examples see R. Eisenberg *et al.*, *Inorg. Chem.* **1996**, *35*, 2688.

These complexes show unique reactions due to the proximity of the metals including (a) oxidation of one or both metal atoms with formation of an M—M bond, (b) addition of neutral molecules with formation of M—M bonds, and (c) formation of bridge groups such as μ-CO, μ-CH$_2$, and η^2-CO or η^2-RNC.

The *A-frames* are especially important, having square M^I, a nonphosphine bridge (μ-B) that may be neutral or anionic, and a vacant site to which small molecules or ions (L) can be bound. Thus we have reactions such as those in eqs. 18-G-1 and 18-G-2 ($\frown$ = phosphine bridge) where B can be neutral, anionic, or cationic:

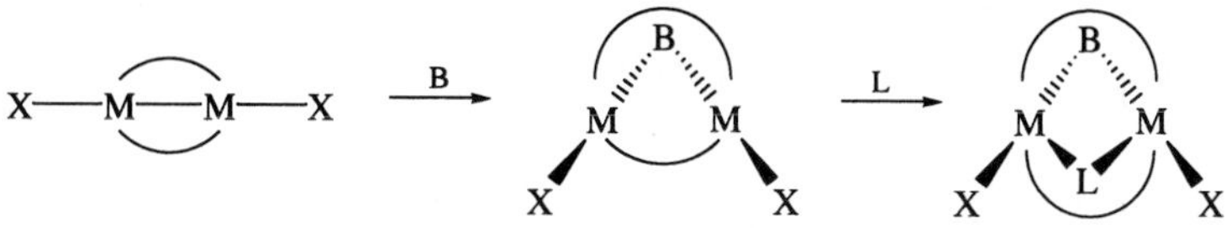

$$\text{(18-C-1)}$$

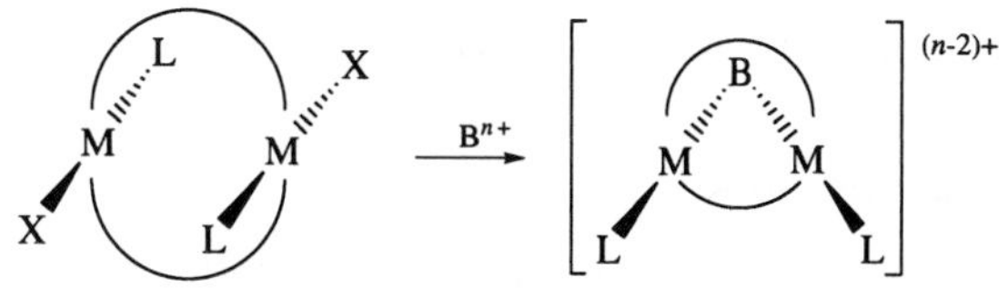

$$\text{(18-C-2)}$$

In some cases, the *A*-frame structure is not obtained and oxidation giving octahedral coordination occurs as in eq. 18-G-3.

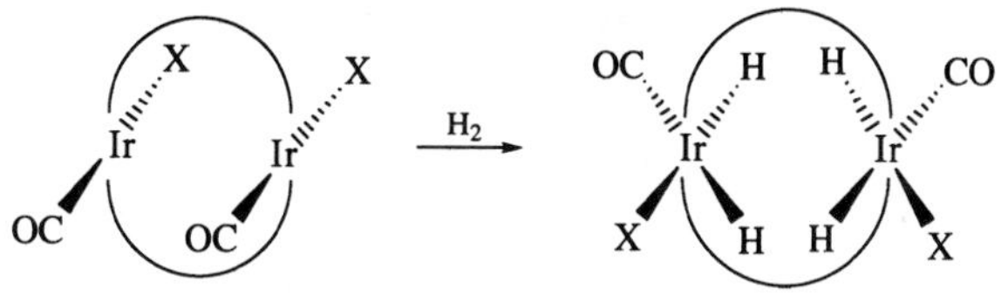

$$\text{(18-C-3)}$$

Some representative A frames are [Rh$_2$(μ-Cl)(CO)$_2$dppm$_2$]$^+$, [Rh$_2$(μ-σ,π, C$\equiv$CBu)-(CO)$_2$dppe$_2$]$^+$, Rh$_2$(μ-S)(CO)$_2$dppm$_2$, and [Ir$_2$(μ-CO)(CO)$_2$(dppm)$_2$]$^+$.

18-G-2 Complexes of Rhodium(I) and Iridium(I), d^8

The chemistry is almost entirely one involving π-bonding ligands such as CO, PR$_3$, RNC, alkenes, cyclopentadienyls, and arenes. It is important because a wide variety of catalytic reactions involve Rh^I and Ir^I species; there are indeed thousands of references in this area.

Square, tetrahedral, and five-coordinate species are formed. The latter are commonly produced by addition of neutral ligands to the other two, for example,

$$\textit{trans-}\text{IrCl(CO)(PPh}_3)_2 + \text{CO} \rightleftharpoons \text{IrCl(CO)}_2(\text{PPh}_3)_2$$

The criteria for relative stability of five- and four-coordinate species are by no means fully established. Substitution reactions of square species, which are often rapid, proceed by an associative pathway involving five-coordinate intermediates, for example,

$$RhCl(C_8H_{12})SbR_3 + amine \rightleftharpoons RhCl(C_8H_{12})amine + SbR_3$$

Most of the square complexes undergo oxidative addition reactions and this leads to octahedral M^{III} complexes with π-bonding ligands; the $M^I - M^{III}$ oxidation changes are important in catalytic cycles.

The Rh^I and Ir^I complexes are usually prepared by reduction of similar M^{III} complexes or of halide species such as $RhCl_3 \cdot 3H_2O$ or Na_3IrCl_6 in the presence of the complexing ligand. Alcohols, aldehydes or formic acid may furnish CO and/or H and the ligand itself may act as a reducing agent.

Rhodium

Hundreds of complexes are known and only a few of the more important ones can be discussed. Some preparations and reactions are shown in Fig. 18-G-1.

Tetracarbonyldichlorodirhodium

This compound is readily made by passing CO saturated with EtOH over powdered $RhCl_3 \cdot 3H_2O$ at 100°C when it sublimes as red needles. The molecule has square planar Rh with chlorine bridges, $[(CO)_2Rh(\mu\text{-}Cl)]_2$; there is a dihedral angle of 56° along the Cl—Cl line.

The carbonyl chloride is an excellent source of other rhodium(I) species, and the halogen bridges are readily cleaved by a wide variety of donor ligands to give cis-dicarbonyl complexes:

$$[Rh(CO)_2Cl]_2 + 2L \longrightarrow 2RhCl(CO)_2L$$

$$[Rh(CO)_2Cl]_2 + 2Cl^- \longrightarrow 2[Rh(CO)_2Cl_2]^-$$

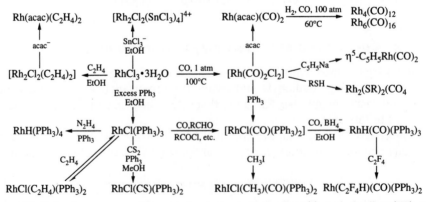

Figure 18-G-1 Some preparations and reactions of rhodium(I) and rhodium(III) compounds.

Some of the complexes produced thus may be made directly from rhodium trichloride, however. Interaction of the carbonyl chloride with a Schiff base (salen type) leads to metallomesogens (liquid crystals) such as (18-G-IV).[2]

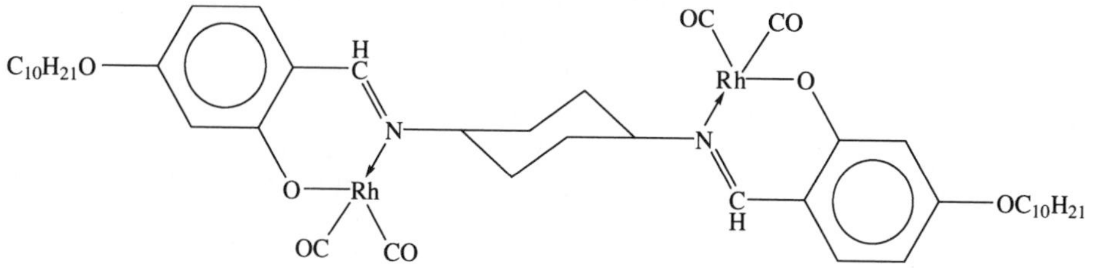

(18-G-IV)

The carbonyl iodides of Rh^{I} and Rh^{III} play an important role in the carbonylation of CH_3OH to CH_3CO_2H (see Chapter 21) in which LiI and the rhodium species are involved in a rather complex catalytic cycle.[3]

The oxidative addition of methyl iodide is one of the key steps:

$$[Rh^{I}(CO)_2(I)_2]^- + CH_3I \longrightarrow [Rh^{III}(CO)_2CH_3(I)_3]^-$$

This is followed by CO insertion to give the acyl $[Rh(CO)(COMe)(I)_3]^-$. In the carbonylation of other alcohols and ethylene to give carboxylic acids similar reactions occur:[4]

$$[Rh(CO)_2(I)_2]^- \xrightarrow{HI} [HRh(CO)_2(I)_3]^-$$

$$[Rh(CO)_2(I)_2]^- \xrightarrow{HI + C_2H_4} [Rh(CO)(I)_3(EtCO)]_2^{2-}$$

It may be noted that the iridium(III) anion *cis*-$[Ir(CO)_2(I)_4]^-$ has been isolated[5] from the products of carbonylation of alcohols using $IrCl_3$(aq) and HI as catalyst precursors. The ethylene analogue of the carbonyl, i.e., $[RhCl(C_2H_4)_2]_2$, is obtained by bubbling C_2H_4 through $RhCl_3$(aq) in aqueous methanol. This compound and its cyclooctene analogue are very useful synthons for other complexes. It is interesting that rotation of the C_2H_4 molecule in the derivative $Rh(acac)(C_2H_4)_2$ occurs even

[2]P. M. Maitlis *et al., J. Chem. Soc., Chem. Commun.* **1994,** 1313.
[3]B. L. Smith *et al., J. Mol. Catal.* **1987,** *39,* 115; P. M. Maitlis *et al., J. Chem. Soc., Dalton Trans.* **1996,** 2187; D. B. Cook *et al., J. Am. Chem. Soc.* **1996,** *118,* 3029.
[4]E. E. Bunel *et al., J. Am. Chem. Soc.* **1994,** *116,* 1163.
[5]S. B. Padye *et al., Polyhedron* **1996,** *15,* 194.

in the solid state.[6] This complex also provides a classic example of differences in substitution rates between coordinatively unsaturated (16e) and saturated (18e) species since the rate of exchange with ethylene gas is 10^{14} times the rate for the 18e complex $CpRh(C_2H_4)_2$.

A major aspect of the chemistry of both Rh^I and Ir^I is the formation of tertiary phosphine complexes. Much used are the phosphines PPh_3, PPr^i_3, PMe_3 and mixed phosphines such as PMe_2Ph. Triphenylphosphine is used in catalytic hydroformylations of alkenes, while water soluble phosphines, notably $P(C_6H_4SO_3H)_3$, are used in two-phase systems. The important *trans*-chlorocarbonylbis(triphenylphosphine)rhodium(I) is obtained as yellow crystals on reduction of $RhCl_3(aq)$ in EtOH with formaldehyde. It can also be made by the reactions

$$[Rh(CO)_2Cl]_2 \xrightarrow{\text{PPh}_3} RhCl(CO)(PPh_3)_2$$
$$RhCl(PPh_3)_3 \xrightarrow{\text{CO}}$$

It is less basic than $IrCl(CO)(PPh_3)_2$, discussed below, and does not react with triplet oxygen although it does give a metastable adduct, $RhCl(CO)(PPh_3)_2(\eta^2\text{-}O_2)$, with singlet oxygen. By contrast Vaska's compound, $IrCl(CO)(PPh_3)_2$, reacts 10^9 times faster with $^1O_2$ than with triplet oxygen.[7] The lower basicity of the Rh^I compound is also shown by the fact that in oxidative addition reactions such as

$$\textit{trans-}Rh^ICl(CO)(PPh_3)_2 + HCl \rightleftharpoons Rh^{III}HCl_2(CO)(PPh_3)_2$$

the equilibria generally lie well to the Rh^I side. The interaction of $Rh(Cl)(CO)(PPh_3)_2$ with $AgSO_3CF_3$ and similar salts leads to complexes with anions more weakly coordinating than Cl^- and hence useful in substitutions[8]; examples are *trans*-$[Rh(CO)(PPh_3)_2(H_2O)]SO_3CF_3$ and $Rh(OPOF_2)(CO)(PPh_3)_2$. A key intermediate in hydroformylation is $RhH(CO)(PPh_3)_3$ made by the reaction

$$RhCl(CO)(PPh_3)_2 + PPh_3 \xrightarrow[\text{EtOH}]{\text{NaBH}_4} RhH(CO)(PPh_3)_3$$

The yellow crystalline compound is also formed when many Rh compounds, e.g., $Rh(CO)_2(acac)$, are treated with $CO + H_2$ under mild conditions in presence of excess PPh_3. It dissociates in solution

$$RhH(CO)(PPh_3)_3 \xrightleftharpoons{-\text{PPh}_3} RhH(CO)(PPh_3)_2 \xrightleftharpoons{-\text{PPh}_3} RhH(CO)(PPh_3)$$

and sites so created can coordinate alkenes. With alk-1-enes this leads to catalytic isomerization to alk-2-enes and under $CO + H_2$ to formation of aldehydes (see Chapter 21).

[6]C. E. Barnes *et al.*, *J. Am. Chem. Soc.* **1994**, *116*, 7445.
[7]C. S. Foote *et al.*, *Inorg. Chem.* **1995**, *34*, 5715.
[8]A. Svetlanova-Larson and J. L. Hubbard, *Inorg. Chem.* **1996**, *35*, 3076.

Chlorotris(triphenylphosphine)rhodium (Wilkinson's Catalyst)[9]

This remarkable compound is perhaps the most studied of all Rh^I phosphine species because of the wide range of its stoichiometric and catalytic reactions. It is formed when ethanolic solutions of $RhCl_3$(aq) are refluxed with excess PPh_3. It exists in two forms, the normal red-violet and orange. Both have structures that are square with a distortion to tetrahedral and in both there are close contacts with *ortho* hydrogen atoms on a phenyl ring. The complex was the first compound to be discovered that allowed the catalytic hydrogenation of alkenes and other unsaturated substances in homogeneous solutions at room temperature and pressure, and its discovery stimulated an enormous development in the synthesis of related complexes of rhodium (and other metals) with tertiary phosphine ligands. Not only monophosphine but chelating phosphine complexes behave similarly. There are a number of species involved in the catalytic cycle (Chapter 21) some only recently observed by using parahydrogen-induced polarization,[10] e.g., the complex $H_2Rh(PPh_3)_2(\mu\text{-}Cl)_2Rh(PPh_3)$(olefin).

Some reactions of $RhCl(PPh_3)_3$ are shown in Fig. 18-G-1. The compound readily reacts with CO and will abstract CO from other molecules stoichiometrically to give $RhCl(CO)(PPh_3)_2$. Interaction of $RhCl(PPh_3)_3$ with aqueous KOH in a 2-phase system with benzene produces $[(Ph_3P)_2Rh\mu\text{-}OH]_2$ which is a useful starting material for other syntheses.[11]

The hydrides, $HRh(PR_3)_{3,4}$, can be made by reduction of $RhCl_3$(aq) and PR_3 under H_2 by Na/Hg while under N_2, rather unstable $RhH(N_2)(PR_3)_2$ species are formed.

There are many trimethylphosphine complexes, the simplest being $RhCl(PMe_3)_3$ and $[Rh(PMe_3)_4]^+$; these also have extensive chemistry, reacting for example with NaSR to give the thiolates, $Rh(SR)(PMe_3)_3$.[12]

The *isopropyl* compound, $RhCl(PPr^i_3)_3$,[13] is also a highly reactive and useful starting material; it is made by interaction of $[RhCl(cyclooctene)_2]_2$ and excess of the phosphine.

Complexes of a bidentate phosphine[14] can be obtained by the reaction in the equation:

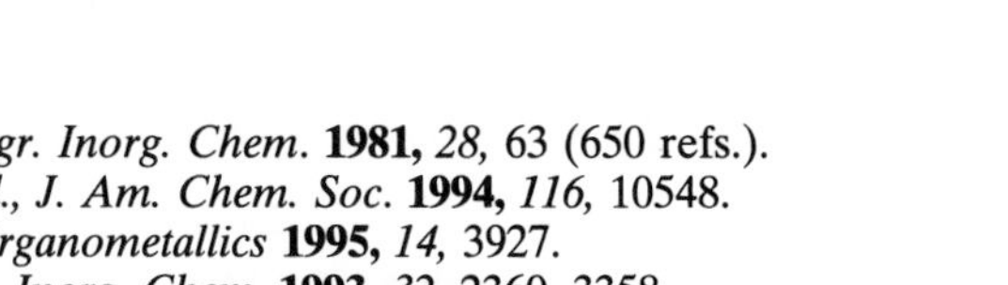

[9]F. H. Jardine, *Progr. Inorg. Chem.* **1981,** *28,* 63 (650 refs.).
[10]R. Eisenberg *et al., J. Am. Chem. Soc.* **1994,** *116,* 10548.
[11]H. Alper *et al., Organometallics* **1995,** *14,* 3927.
[12]K. Osakada *et al., Inorg. Chem.* **1993,** *32,* 2360, 3358.
[13]See papers by H. Werner *et al.,* for other PPr^i_3 complexes, e.g., *Chem. Ber.* **1996,** *129,* 29; *Organometallics* **1996,** *15,* 2806; $[Rh(\mu\text{-}OH)(PPr^i_3)_2]$ and reactions.
[14]D. Milstein *et al., Organometallics* **1996,** *15,* 1839.

The N_2 molecule can be substituted by CO_2, C_2H_4 and η^2-H_2.

Isocyanides can give cationic species:

$$RhCl(CO)(PPh_3)_3 \xrightarrow{\text{RNC}} [Rh(CNR)_4]^+Cl^- + CO + 3PPh_3$$

In polar solvents such as MeCN the solutions may be yellow when dilute, blue when concentrated. This is attributable to polymerization:

$$n[Rh(CNR)_4]^+ \rightleftharpoons [Rh(CNR)_4]_2^{2+}, [Rh(CNR)_4]_3^{3+}, \text{etc.}$$

For the dimer, there are two planar face to face $Rh(CNR)_4$ units with a weak Rh—Rh bond.

Bipyridyls give quite complicated air-sensitive systems whose nature, for Rh, depends on pH and concentration. Like the isocyanides there are mono- and dimeric species, for example, $[Rh(bipy)_2]^+$,[15] $[Rh(bipy)_2]_2^{2+}$, and $[Rh^{II}(bipy)_2]^{2+}$, as well as oxidized species formed on acidification, for example, $[Rh(bipy)_2H(H_2O)]^{2+}$. Some of these may be involved in photochemical reactions leading to water cleavage and other reactions such as the water gas shift reaction.

Iridium

The most important of Ir^+ complexes is *trans*-$IrCl(CO)(PPh_3)_2$ (Vaska's compound). It has many derivatives and analogues with other phosphines. A useful synthesis[16] starts with carbonylation of $IrCl_3(aq)$ in the presence of *p*-toluidine (L) to give $IrCl(CO)_2L$ which can react as follows:

$$IrCl(CO)_2L \xrightarrow[\text{-CO}]{2PR_3} IrCl(CO)(PR_3)_2L \xrightarrow[\text{vacuum}]{75\,°C} IrCl(CO)(PR_3)_2 + L$$

Many studies have been made on these types of compounds[17] in addition to simple substitutions like the conversion by action of 50% NaOH, in the presence of $[(PhCH_2)Et_3N]Cl$ as a phase transfer catalyst, to give $Ir(OH)(CO)(PPh_3)_2$.[18]

Major study has been made of oxidative-addition reactions (Chapter 21) since the equilibria

$$\textit{trans-}Ir^IX(CO)(PR_3)_2 + AB \rightleftharpoons Ir^{III}XAB(CO)(PR_3)_2$$

lie well to the oxidized side; the oxidized compounds are usually stable octahedral complexes, unlike many of the less stable Rh^{III} analogues.

The carbonyl is readily converted into the five-coordinate hydride:

$$\textit{trans-}IrCl(CO)(PPh_3)_2 \xrightarrow{CO} IrCl(CO)_2(PPh_3)_2 \xrightarrow[\text{EtOH}]{NaBH_4} IrH(CO)_2(PPh_3)_2$$

[15]K. Shinozaki and N. Takahashi, *Inorg. Chem.* **1996,** *35,* 3917.
[16]M. Rahim and K. J. Ahmed, *Inorg. Chem.* **1994,** *33,* 3003.
[17]See e.g., M. R. Churchill *et al., Organometallics* **1994,** *13,* 5080.
[18]S. A. Al-Jibori, *J. Organomet. Chem.* **1996,** *506,* 119.

and this is of interest in that it is much more stable than its Rh analogue, and hence allows isolation of many prototypes for intermediates in the hydroformylation sequence.

Interaction of $IrCl(CO)(PPh_3)_2$ with $LiNHAr^{19}$ gives rise to monoamido, *trans*-$Ir(NHAr)(CO)(PPh_3)_2$ as well as bridged binuclear species $[(CO)_2Ir(\mu\text{-}NHAr)]_2$. The latter can undergo oxidative addition reactions to give Ir^I—Ir^{III} dimers with Cl_2 or Ir^I—Ir^{II} dimers with I_2. The Ir^I dimers can undergo deprotonation with $LiBu^n$ to give μ-imido complexes $Li_2[(CO)_2Ir(\mu\text{-}NAr)]_2$.

Trimethylphosphine complexes have also been much studied, typical ones being $IrCl(PMe_3)_3$, $[Ir(PMe_3)_4]^+X^-$, and $[Ir(PMe_3)_3(COD)]^+Cl^-$. There are also mono- and bidentate *isocyanide* complexes similar to those of rhodium.[20]

Finally it may be noted that the Ir analogue of $[Rh(CO)_2(\mu\text{-}Cl)]_2$ has been obtained only in solution by carbonylation of $[IrClcyclooctene]_2$ in acetone or MeCN. Evaporation of the blue solutions gives an amorphous polymer $[Ir(CO)_2Cl]_n$ that dissolves only in donor solvents to re-form the dimer. It is a convenient source for making $[Ir(CO)_2Cl_2]^-$ and $Ir(CO)_2(acac)$.[21]

18-G-3 Complexes of Rhodium(III) and Iridium(III), d^6

Both Rh and Ir form many octahedral complexes, cationic, neutral, and anionic; in contrast to Co^{III} complexes, reduction of Rh^{III} or Ir^{III} does not normally give rise to divalent complexes. Thus depending on the nature of the ligands and on the conditions, reduction may lead to the metal—usually with halogens, water, or amine ligands present—or to hydridic species of M^{III} or to M^I when π-bonding ligands are involved.

Though similar to Co^{III} in giving complex anions with CN^- and NO_2^-, Rh and Ir differ in readily giving octahedral complexes with halides, for example, $[RhCl_5H_2O]^{2-}$ and $[IrCl_6]^{3-}$, and with oxygen ligands such as oxalate and EDTA.

The cationic and neutral complexes of all three elements are generally kinetically inert, but the anionic complexes of Rh^{III} are usually labile. By contrast, anionic Ir^{III} complexes are inert, and the preparation of such complexes is significantly harder than for the corresponding Rh species.

Rhodium complex cations have proved particularly suitable for studying *trans* effects in octahedral complexes.

In their magnetic and spectral properties the Rh^{III} complexes are fairly simple. All the complexes, and indeed all compounds of rhodium(III), are diamagnetic. This includes even the $[RhF_6]^{3-}$ ion, of which the cobalt analogue constitutes the only example of a high-spin Co^{III}, Rh^{III}, or Ir^{III} ion in octahedral coordination. Thus the inherent tendency of the octahedral d^6 configuration to adopt the low-spin t_{2g}^6 arrangement, together with the relatively high ligand field strengths prevailing in these complexes of tripositive higher transition series ions, as well as the fact that all $4d^n$ and $5d^n$ configurations are more prone to spin pairing than their $3d^n$ analogues, provide a combination of factors that evidently leaves no possibility of there being any high spin octahedral complex of Rh^{III} or Ir^{III}.

[19]M. K. Kolel-Veetil and K. J. Ahmed, *Inorg. Chem.* **1994,** *33,* 4945.
[20]K. R. Mann et al., *Inorg. Chem.* **1993,** *32,* 783.
[21]D. Roberto et al., *Organometallics* **1994,** *13,* 4227.

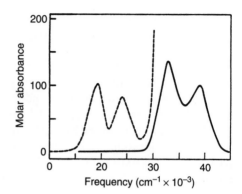

Figure 18-G-2 The visible spectra of $[RhCl_6]^{3-}$ (dashed curve) and the $[Rh(NH_3)_6]^{3+}$ (solid curve) ions.

The visible spectra of Rh^{III} complexes have the same explanation as do those of Co^{III} complexes. As illustrated in Fig. 18-G-2 for the $[Rh(NH_3)_6]^{3+}$ and $[RhCl_6]^{3-}$ ions, there are in general two bands toward the blue end of the visible region, which together with any additional absorption in the blue due to charge-transfer transitions, are responsible for the characteristic orange, red, yellow, or brown colors of rhodium(III) compounds. These bands are assigned as transitions from the $^1A_{1g}$ ground state to the $^1T_{1g}$ and $^1T_{2g}$ upper states just as in the energy level diagram for Fe^{II} and Co^{III}. The spectra of Ir^{III} complexes have a similar interpretation.

The Rhodium and Iridium Aqua Ions

Unlike cobalt, rhodium gives a stable yellow aqua ion $[Rh(H_2O)_6]^{3+}$. It is obtained by dissolution of Rh_2O_3(aq) in cold mineral acids, or, as the perchlorate by repeated evaporation of $HClO_4$ solutions of $RhCl_3$(aq). Exchange studies with $H_2^{18}O$ confirm the hydration number as 5.9 ± 0.4, as well as showing that exchange between the first sphere H_2O protons and bulk water is highly pH dependent. The acid dissociation constant for the aqua ion is $pK_{a1} = 3.6$ and for $[Rh(H_2O)_5(OH)]^{2+}$, $pK_{a2} = 4.7$.[22]

The addition of NaOH to solutions of the perchlorate gives the dimeric ion $[(H_2O)_4Rh(\mu\text{-}OH)]_2^{4+}$ which is similar to an aquahydroxochromium species. A linear OH-bridged complex is also formed,[23] the heterobinuclear ion $[(H_2O)_4Rh^{III}(\mu\text{-}OH)_2\text{-}Cr^{III}(H_2O)_4]^{4+}$ is also known.[24]

The $[Ir(H_2O)_6]^{3+}$ ion is made by dissolution of the hydrous oxide in $HClO_4$ and crystallization as the perchlorate. Unlike $[Rh(H_2O)_6]^{3+}$ it is readily oxidized by air to give Ir complexes in higher oxidation states (Section 18-G-6). The water exchange is the slowest yet known for a homoleptic mononuclear metal (residence time *ca.* 300 years).[25] Alums for both Rh and Ir, $CsM(SO_4)_2 \cdot 12H_2O$, and a few other salts contain the aqua ions. At near neutral pH, $[Ir(H_2O)_6]^{3+}$ precipitates the hydrated oxide, a major component of this probably being $Ir(H_2O)_3(OH)_3$. It dissolves in 1 *M* base to give mainly $[Ir(OH)_6]^{3-}$ plus some oligomers, and is also readily oxidized by air.[26]

[22]I. Bányai *et al.*, *Inorg. Chem.* **1995**, *34*, 2423.
[23]J. Glaser *et al.*, *Inorg. Chem.* **1992**, *31*, 4155; L. Spiccia *et al.*, *Inorg. Chem.* **1996**, *35*, 985.
[24]S. J. Crimp and L. Spiccia, *J. Chem. Soc., Dalton Trans.*, 1996, 1051.
[25]A. E. Merbach *et al.*, *J. Am. Chem. Soc.* **1996**, *118*, 5265.
[26]H. Gamsjäger *et al.*, *Inorg. Chem.* **1989**, *28*, 379.

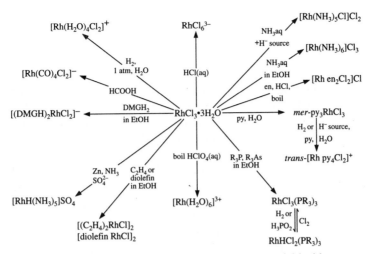

Figure 18-G-3 Some reactions of rhodium trichloride.

Rhodium(III) Chloride and Related Complexes

One of the most important of Rh^{III} compounds and the usual starting material for the preparation of rhodium complexes (see Figs. 18-G-1 and 18-G-3) is the dark red, crystalline, deliquescent trichloride, $RhCl_3(H_2O)_3$. This is obtained by dissolving hydrous Rh_2O_3 in aqueous hydrochloric acid and evaporating the hot solution. It is very soluble in water and alcohols, giving red-brown solutions. Water solutions are extensively hydrolyzed,

$$RhCl_3(H_2O)_3 \xrightleftharpoons{\quad H_2O \quad} RhCl_2(OH)(H_2O)_3 + H^+ + Cl^-$$

and ^{1}H nmr of the solutions show that the *fac* and *mer* isomers of $RhCl_2(OH)(H_2O)_3$ are present.[27] The *fac* and *mer* isomers of $[RhCl_3(MeCN)_3]$ are both obtained by interaction of the hydrate with MeCN.[28]

Some reactions of $RhCl_3 \cdot 3H_2O$ are shown in Fig. 18-G-3. On boiling aqueous solutions of the trichloride, $[Rh(H_2O)_6]^{3+}$ is formed and with excess HCl, the rose-pink hexachlororhodate ion, $[RhCl_6]^{3-}$ is obtained. Between these two species there are clearly several intermediates. On aquation $[RhCl_6]^{3-}$ produces $[RhCl_5(H_2O)]^{2-}$, *cis*-$[RhCl_4(H_2O)_2]^-$, and *fac*-$RhCl_3(H_2O)_3$. Use of ^{103}Rh nmr spectra assisted the identification of bromide species.[29]

With large cations halogen-bridged dimers such as $[Rh_2Cl_9]^{3-}$, $Rh_2Cl_6(PEt_3)_3$, and $[Rh_2Cl_7(PR_3)_2]^-$ can be obtained.[30]

In addition to $[MX_6]^{3-}$ the *pseudohalide* anions $[Rh(CN)_6]^{3-}$ and $[Rh(SCN)_6]^{3-}$ are known.[31]

[27]H. Patin *et al.*, *Inorg. Chem.* **1987**, *26*, 2922.
[28]L. M. Venanzi *et al.*, *Inorg. Chim. Acta* **1995**, *240*, 575.
[29]J. Glaser *et al.*, *J. Chem. Soc., Dalton Trans.* **1992**, 233.
[30]F. A. Cotton *et al.*, *Inorg. Chim. Acta* **1993**, *206*, 29; *Inorg. Chem.* **1993**, *32*, 2336.
[31]M. C. Read *et al.*, *J. Chem. Soc., Dalton Trans.* **1994**, 3243.

Cationic Complexes

For Rh (and also Ir) cationic species with NH_3 and other amine ligands are known, generally of the types $[ML_6]^{3+}$, $[ML_5X]^{2+}$, and $[ML_4X_2]^+$; they are usually made from aqueous solutions of $RhCl_3$.

The formation of complex ions from $RhCl_3(aq)$, $[Rh(H_2O)Cl_5]^{2-}$, or $[RhCl_6]^{3-}$ is often catalyzed by the addition of reducing agents that can furnish hydride ions; ligands such as ethylenediamine may also themselves act in this way. The effect of ethanol was discovered by Delépine long before the general nature of such catalysis was recognized. It now appears that many rhodium complexes have been made only because ethanol was used as a solvent. One example of the catalysis is the action of pyridine, which with $RhCl_3(aq)$ gives mainly Rh py_3Cl_3 and with aqueous $[Rh(H_2O)Cl_5]^{2-}$ gives $[Rh\, py_2Cl_4]^-$. On addition of alcohol, hydrazine, BH_4^-, or other reducing substances—even molecular hydrogen at 25°C and ≤ 1 atm—conversion into trans-$[Rh\, py_4Cl_2]^+$ rapidly occurs. Kinetic studies of this reaction suggest that Rh^I complexes, rather than hydridic ones, are involved in the catalysis, since $[Rh(CO)_2Cl]_2$ and $Rh(acac)(CO)_2$ are more effective than hydride-producing substances. Further evidence comes from the reaction of $[Rh(H_2O)_6]^{3+}$ in ethanolic ClO_4^- solution with 2,2'-bipyridine, when an air-sensitive brown complex $[Rh^I bipy_2]ClO_4 \cdot 3H_2O$ can be isolated.

The reduction of $[Rh\, en_2Cl_2]^+$ at a mercury electrode does *not* give a Rh^{II} species as first thought, but a mercury complex $\{[en_2Rh]_2Hg\}^+$; the formation of bonds to mercury is a common phenomenon for Rh, Ru, and some other elements, and the formation of mercury complexes can always be suspected whenever mercury or its compounds are used.

Rhodium and Iridium Peroxo Complexes

Numerous claims have been made that electrochemical or chemical oxidation of basic solutions of $[RhCl_5(H_2O)]^{2-}$ and similar complexes gives blue or purple solutions containing Rh^{IV} or Rh^V species. All these colored species appear to be superoxo complexes.[32]

A wide range of bridged superoxo species having a variety of other ligands exists–one typical example is trans-$[py_4ClRh(\mu\text{-}O_2)RhClpy_4]^{3+}$. The presence of O_2^- is confirmed by epr and Raman spectra and in some cases by X-ray crystallography. The complex called "Claus' blue," originally formulated as $[Rh^{VI}O_4]^{2-}$ and made by chlorination of alkaline solutions of $RhCl_3(aq)$ is now formulated $[Rh_2^{III}(OH)_2(H_2O)_n(\mu\text{-}O_2)]^{3+}$ where coordinated OH is oxidized first to O_2^{2-} and then to O_2^-.

Similar oxidation can occur in acid solution when green $Rh_2^{4+}(aq)$ (see below) in 3 M $HClO_4$ is oxygenated to give violet $\{[(H_2O)_4(OH)Rh]_2(\mu\text{-}O_2)\}^{3+}$; this cation can undergo a 1e reduction to a complex with a $\mu\text{-}O_2^{2-}$ group.[33]

The compound $RhCl(\eta^2\text{-}O_2)(PEt_2Ph)_3$ reacts with CO_2 to give the peroxocarbonate,

$$Cl(PhEt_2P)_3RhOOC(O)O$$

[32]R. D. Gillard et al., J. Chem. Soc., Dalton Trans. **1994**, 2531; Polyhedron **1994**, 13, 1351.
[33]E. S. Gould et al., Inorg. Chem. **1993**, 32, 4780.

via cleavage of the O—O bond and CO$_2$ insertion.[34] Rhodium and iridium cationic complexes [M^I(CO)(MeCN)(PPh$_3$)$_2$]$^+$ClO$_4^-$ undergo photooxidation with singlet oxygen[35] to give the corresponding η^2-O$_2$ complexes [M^{III}(CO)(MeCN)(PPh$_3$)$_2$O$_2$]$^+$. Although the Ir^{III} complex is stable at room temperature the peroxorhodium species, which decomposes to give Ph$_3$PO, is stable only below 0°C. There are also a few rhodium complexes with hydroperoxo groups[36] such as K$_2$[Rh(OOH)(CN)$_4$H$_2$O] and [Cp*Ir(μ-pz)Rh(OOH)dppe]$^+$ that have been structurally characterized.

Iridium Chloride Complexes

The nature of commercial black "iridium trichloride hydrate" is uncertain; it is quite different from RhCl$_3$(H$_2$O)$_3$,[37] and can vary considerably depending on how it is made and dried. At *ca.* 120°C for short periods, lattice water is converted at least in part to coordinated water, as shown by ir spectra. The reactivity of the substance depends on the thermal treatment and overheating gives unreactive products. Removal of water by SOCl$_2$ treatment also changes the reactivity.

Dissolution of the hydrated chloride or hydrous oxide in HCl or aquation of [IrCl$_6$]$^{3-}$ gives green aqua complexes, [IrCl$_5$(H$_2$O)]$^{2-}$, [IrCl$_4$(H$_2$O)$_4$]$^-$, and possibly IrCl$_3$(H$_2$O)$_3$ (in solution only).

The formation of iridium complexes is normally very slow but, as for rhodium, can be catalyzed. Thus to convert Na$_2$Ir^{IV}Cl$_6$ into py$_3$IrCl$_3$[38] and *trans*-[Ir py$_4$Cl$_2$]Cl, a bomb reaction was formerly used. Quite rapid conversions are obtained as follows:

$$\text{Na}_2\text{IrCl}_6 \xrightarrow[\text{boil 30 min}]{\text{NaH}_2\text{PO}_2\text{(aq)} + \text{py}} \textit{fac}\text{-py}_3\text{IrCl}_3 \xrightarrow{\text{6 h}} [\text{Ir py}_4\text{Cl}_2]\text{Cl}$$

$$\text{H}_2\text{IrCl}_6 \xrightarrow[\substack{\text{2-methoxyethanol,}\\\text{boil 10 min}}]{\text{HCl in}} \text{IrCl}_6^{3-} \xrightarrow{+ \text{ py, boil 1 h}} \textit{mer}\text{-py}_3\text{IrCl}_3$$

trans-[Ir en$_2$Cl$_2$]$^+$ can be similarly obtained by using hypophosphorous acid as catalyst.

Hydrido Complexes of Rhodium(III) and Iridium(III)

With NH$_3$ or amines, quite stable octahedral hydrido complexes can be obtained for rhodium. Thus the reduction of RhCl$_3$·3H$_2$O in NH$_4$OH by Zn in the presence of SO$_4^{2-}$ leads to the white, air-stable, crystalline salt [RhH am$_5$]SO$_4$. In aqueous solution the ion hydrolyzes:

$$[\text{RhH am}_5]^{2+} + \text{H}_2\text{O} = [\text{RhH am}_4\text{H}_2\text{O}]^{2+} + \text{NH}_3 \qquad K \sim 2 \times 10^{-4}$$

Various substitution reactions with other amines can be carried out, and with alkenes remarkably stable alkyl derivatives, for example, [RhC$_2$H$_5$ am$_5$]SO$_4$, can be obtained. The structure of [RhH am$_5$](ClO$_4$)$_2$ shows a distinct hydridic trans weaken-

[34]M. Aresta *et al., Inorg. Chem.* **1996**, *35*, 4254.
[35]C. S. Foote *et al., Inorg. Chem.* **1996**, *35*, 4519.
[36]D. Carmona *et al., J. Chem. Soc., Chem. Commun.* **1994**, 575.
[37]G. Wilkinson *et al., J. Chem. Soc., Dalton Trans.* **1993**, 3219.
[38]R. D. Gillard and S. H. Mitchell, *Polyhedron* **1989**, *8*, 2245.

ing effect (0.165 Å) on the trans Rh—N bond. The cyanide $[RhH(CN)_5]^{3-}$ is also known.

The use of a triazacyclononane ligand (L) has led to isolation of other N-ligand species such as *anti*-$[L_2Rh_2(H)_2(\mu\text{-}H)](PF_6)_2$ while HBPz$_3^*$ as ligand gave $(HBPz_3^*)Rh(H_2)(H)_2$.[39] An unusually stable cation, $[LRh(PMe_3)(H)(Me)]^+$, is also known.[40]

Most hydrido or dihydrogen complexes have tertiary phosphine, CO, or other π-bonding ligands. Examples of cationic compounds are $[IrH(OH)(PMe_3)_4][PF_6]$[41] and $[mer\text{-}(Me_3P)_3IrH(\eta^2\text{-}O_2CPh)][PF_6]$[42] while neutral species are *fac*-$RhH_3(PMe_3)_3$ and *mer*-$IrH_3(PPh_3)_3$.

Iridium hydrido compounds are particularly stable. Many of these are obtained from square M^I compounds by oxidative addition of H_2 or HX.

Simple Rhodium(III) and Iridium(III) Compounds with M—C Bonds

Although organometallic compounds are noted in Section 18-G-8, it is convenient to discuss simple aryls and alkyls here.

The interaction of mesityl magnesium bromide with *fac*-$RhCl_3(tht)_3$ produces the air and thermally stable mesityl, $Rh(mes)_3$, which has a quasi-octahedral structure in the crystal with three σ-bonded aryl groups in a *fac* configuration with the other positions occupied by agostic H-atoms of the three o-CH_3 groups in the mesityl ligand. In solution, however, nmr data indicate that, in common with other $M(mes)_3$ species, the molecule is planar with propeller-like rotation of the mesityl groups about the Rh—C bonds.[43] The Ir^{III} analogue[44] is similar except that it is exceedingly air-sensitive, being oxidized by O_2 or by pyNO to the tetrahedral $(mes)_3Ir^V=O$.

The pentafluorophenyls $[Rh(C_6F_5)_5]^{2-}$ and $[Rh(C_6F_5)_5(PEt_3)]^{2-}$ have square pyramidal geometries;[45] many derivatives with the $Rh(C_6F_5)_3$ moiety are known, e.g., $[Rh(C_6F_5)_3acac]^-$ and $[Rh(C_6H_5)_3(\mu\text{-}OH)]_2^{2-}$. Although alkyls of the type $MMe_3(PR_3)_2$ or $MMe_2(PR_3)_2(\eta^1\text{-}FBF_3)$[46] and so on are known, the only homoleptic alkyl compounds are the *ato* Rh and Ir complexes $(Li\ tmed)_3(MMe_6)$.[47]

Other Complexes of Rhodium(III) and Iridium(III)

An array of halides with CO, PR_3, aromatic amines, sulfur ligands, etc., can be obtained from $RhCl_3(aq)$ or Na_3IrCl_6 or by oxidative additions to M^I complexes. Some representative examples are *fac* and *mer*-$RhCl_3(MeCN)_3$, $RhCl_3(tht)_3$, and $IrCl_3py_3$. Compounds without halogens include $Ir(SC_6H_4PPh_2)_3$[48] and $[Ir(Se_4)_3]^{3-}$ as its $[K(18C6)]^+$ salt.[49]

[39]K. Wieghardt *et al.*, *Inorg. Chem.* **1993**, *32*, 4300 and references therein.
[40]J. W. Ziller *et al.*, *J. Am. Chem. Soc.* **1995**, *117*, 1647.
[41]R. Bau *et al.*, *J. Chem. Soc., Dalton Trans.* **1990**, 1429.
[42]F. T. Ladipo and J. S. Merola, *Inorg. Chem.* **1993**, *32*, 5201.
[43]G. Wilkinson *et al.*, *J. Chem. Soc., Dalton Trans.* **1991**, 2821.
[44]G. Wilkinson *et al.*, *J. Chem. Soc., Dalton Trans.* **1992**, 3477.
[45]L. A. Oro *et al.*, *J. Organomet. Chem.* **1992**, *438*, 229.
[46]K. G. Caulton *et al.*, *Organometallics* **1990**, *9*, 2254.
[47]G. Wilkinson *et al.*, *Polyhedron* **1990**, *9*, 2071.
[48]J. R. Dilworth *et al.*, *J. Chem. Soc., Dalton Trans.* **1995**, 1957.
[49]J. A. Ibers *et al.*, *Inorg. Chem.* **1995**, *34*, 5101.

Dimethylglyoximates, e.g., (dmgH)$_2$RhClpy,[50] and Schiff base complexes[51] are similar to those of Co^{III}.

The oxidation of Ir$_4$(CO)$_{12}$ by XeF$_2$ in HF(l) gives the *fac* and *mer* isomers of IrF$_3$(CO)$_3$,[52] while oxidation of Ir metal with S$_2$O$_6$F$_2$ in HSO$_3$F at 120–140°C gives Ir(SO$_3$F)$_3$ that reacts with CO in HSO$_3$F to give the similar carbonyl *fac* and *mer* Ir(CO)$_3$(SO$_3$F)$_3$.[53]

18-G-4 Complexes of Rhodium(II),[54] d^7

Paramagnetic Rh^{II} species have often been detected by epr spectra on reduction or photolysis of Rh^{III} compounds, e.g., RhCl(PPh$_3$)$_3$ usually has traces of Rh^{II} species, but stable compounds are known and many have been structurally characterized.

Mononuclear Complexes

Compounds with PR$_3$ ligands have been known for a long time, an example being RhCl$_2$(PPr$^i_3$)$_2$;[55] all are *trans*-square planar.

A formally Rh^{II} complex, Rh(NO)Cl$_2$(PPh$_3$)$_2$, which is 5-coordinate, reacts with NaS$_2$CNR$_2$ to give either the square Rh(S$_2$CNR$_2$)$_2$ or Rh(S$_2$CNR$_2$)$_2$(PPh$_3$), which is square pyramidal. Another sulfur ligand complex has been made in solution by reduction of the 1,4,7-trithianonane complex [Rh(9S3)$_2$]$^{3+}$.[56]

A very stable complex, [Rh(η^3-TMPP)$_2$](BF$_4$)$_2$, TMPP = *tris*-(2,4,6-trimethoxy-phenylphosphine), is octahedral with four weak Rh—O and two strong Rh—P bonds.[57] It is made from [Rh$_2$(MeCN)$_{10}$]$^{4+}$ and can be converted to the isocyanide [Rh(η^3-TMPP)$_2$(CNBu^t)$_2$]$^{2+}$, which has a distorted planar geometry with *trans*-P ligands.

The aryl complex, *trans*-Rh(2,4,6-Pr$^i_3$C$_6$H$_2$)$_2$(tht)$_2$, is obtained when *fac*-RhCl$_3$(tht)$_3$ is reacted with the lithium aryl.[43] Various square planar pentachlorophe-nyls are obtained by chlorination of the Rh^I species, Li[Rh(C$_6$Cl$_5$)$_2$L$_2$], made by reacting LiC$_6$Cl$_5$ with [Rh^I(μ-Cl)L$_2$]$_2$; examples are Rh(C$_6$Cl$_5$)$_2$(COD) and Rh(C$_6$H$_5$)$_2$py$_2$.[58]

Binuclear Complexes

These diamagnetic compounds can be considered as being derived from Rh$_2^{4+}$, which has a Rh—Rh single bond. The ligands can be unidentate as in [Rh$_2$(MeCN)$_{10}$]$^{4+}$ or bridging as in the carboxylates,[59] Rh$_2$(μ-O$_2$CR)$_4$; the latter type have the same structure as similar carboxylates of Mo, W, Re, Ru, Cu,

[50]B. Giese *et al.*, *Chem. Ber.* **1993**, *126*, 1193.
[51]D. A. Anderson and R. Eisenberg, *Inorg. Chem.* **1994**, *33*, 5378.
[52]E. G. Hope *et al.*, *J. Chem. Soc., Dalton Trans.* **1995**, 2945.
[53]F. Aubke *et al.*, *Inorg. Chem.* **1996**, *35*, 1279.
[54]D. G. DeWit, *Coord. Chem. Rev.* **1996**, *147*, 209. Reactions of mononuclear Rh^{II} compounds (161 refs.).
[55]R. L. Harlow *et al.*, *Inorg. Chem.* **1992**, *31*, 993.
[56]S. R. Cooper *et al.*, *J. Am. Chem. Soc.* **1991**, *113*, 1600.
[57]K. R. Dunbar *et al.*, *Organometallics* **1992**, *11*, 1431; *J. Am. Chem. Soc.* **1991**, *113*, 9540.
[58]M. P. García *et al.*, *Organometallics* **1993**, *12*, 3257.
[59]F. A. Cotton *et al.*, *Inorg. Chim. Acta* **1995**, *237*, 19.

etc. Although carboxylates are best known, bridged compounds can be made with anions such as $CF_3C(O)NH^-$, $CH_3C(O)S^-$ and N,N' ligands, $RNC(R')NR^-$, $RNNNR^-$, and pyrazolates.[60] The *acetate* is readily obtained on heating methanol solutions of $RhCl_3$(aq) with sodium acetate. Although the dimers can be solvent-free, usually the ends have donor ligands. With O-ligands the carboxylates are usually green or blue, with π-bonding ligands like PPh_3, red. Long chain carboxylate complexes are mesogenic.[61]

The bridged compounds can undergo redox reactions to give mixed valence, I,II or II,III species:[62]

$$Rh_2^{3+} \underset{+e}{\overset{-e}{\rightleftharpoons}} Rh_2^{4+} \underset{+e}{\overset{-e}{\rightleftharpoons}} Rh_2^{5+}$$

The rhodium(I) carbonyl, $[Rh(CO)_2(\mu\text{-}RN_3R)]_2$ is a useful precursor for synthesis of compounds with Rh_2^{n+}, $n = 2 - 4$ cores.[63] The acetate $Rh_2(O_2CMe)_4(MeOH)_2$ and PPh_3 on boiling in acetic acid give an *o*-metallated complex[64] Rh_2-$(O_2CMe)_2(PPh_2C_2H_4)_2(MeCO_2H)_2$. Other examples of *o*-metallation with PPh_3 ligands are known. Carboxylates with nitrogen ligands such as bipy have been studied in some detail because of the possible connection with the mode of bonding of $Rh_2(O_2CR)_4$ compounds to templates in DNA synthesis, believed to be responsible for the carcinostatic activity of the Rh^{II} species against tumors in mice.[65] Rhodium(II) acetate has also been widely studied for its catalysis of organic reactions, notably those involving diazo compounds leading to metal carbene transformations of various types.[66] The use of optically active carboxylates allows enantioselective catalysis.[67] Finally the O_2CR bridges can be completely or partially replaced by acetonitrile, polypyridyl, and other ligands[68] as in $[Rh_2(O_2CMe)_2(MeCN)_6]^{2+}$ and $[Rh_2(MeCN)_{10}]^{4+}$.

The green aqua ion Rh_2^{4+}(aq) made by reduction of $[Rh(H_2O)_6]^{3+}$ by Cr^{2+} in water can be used to make bridged species such as $[Rh_2(\mu\text{-}SO_4)_4]^{4-}$ and $[Rh_2(\mu\text{-}CO_3)_4]^{4-}$. The aqua ion is unstable and decomposes (*ca.* 1 h) to Rh metal and $[Rh(H_2O)_6]^{3+}$. Structural data on over 100 bridged Rh_2^{II} species are available[69] as are data for bridgeless dimers.[70] ^1H nmr data for $[Rh_2(MeCN)_{10}]^{4+}$ show a 4:1 equato-

[60]P. Piraino *et al., Inorg. Chem.* **1996,** *35,* 1377; see M. P. Doyle *et al., J. Am. Chem. Soc.* **1993,** *115,* 9968; R. J. H. Clark *et al., Inorg. Chem.* **1992,** *31,* 456; M. B. Hursthouse *et al., Polyhedron* **1993,** *12,* 563.

[61]L. A. Oro *et al., Inorg. Chem.* **1992,** *31,* 732; J-C. Marchon *et al., Inorg. Chem.* **1990,** *29,* 4851.

[62]N. G. Connelly *et al., Inorg. Chem.* **1994,** *33,* 960.

[63]N. G. Connelly *et al., J. Chem. Soc., Dalton Trans.* **1996,** 2511.

[64]P. Lahuerta *et al., J. Chem. Soc., Dalton Trans.* **1994,** 539, 545.

[65]K. R. Dunbar *et al., Inorg. Chem.* **1993,** *32,* 3125; D. Waysbort *et al., Inorg. Chem.* **1993,** *32,* 4774.

[66]See e.g., A. Padwa and D. J. Austin, *Angew. Chem. Int. Ed. Engl.* **1994,** *33,* 1797; M. C. Pirrung and A. J. Moorhead, Jr., *J. Am. Chem. Soc.* **1994,** *116,* 8991.

[67]H. Brunner, *Angew. Chem. Int. Ed. Engl.* **1992,** *31,* 1183.

[68]See F. P. Pruchnik *et al., Inorg. Chem.* **1996,** *35,* 4261; K. R. Dunbar *et al., Inorg. Chem.* **1994,** *33,* 25; F. A. Cotton and S. J. Kang, *Inorg. Chim. Acta* **1993,** *209,* 23; P. Lahuerta *et al., Inorg. Chim. Acta* **1993,** *209,* 91.

[69]G. Aullón and S. Alvarez, *Inorg. Chem.* **1993,** *32,* 3712.

[70]M. T. Chen *et al., Organometallics* **1996,** *15,* 2338.

rial to axial ratio and the axial MeCN is very labile. For the above green aqua anion the ^{17}O nmr spectrum is consistent with $[Rh_2(H_2O_{equat.})_8(H_2O_{ax})_2]^{4+}$.[71]

Porphyrins give complexes generally $Rh_2(porph)_2$;[72] they can be made by reactions such as that for the octaethyltetrazo (OETAP) compound:

$$[Rh(CO)_2Cl]_2 \xrightarrow[\text{CHCl}_3 + I_2]{\text{OETAPH}_2} (OETAP)RhI \xrightarrow[\text{MeI}]{\text{NaBH}_4} (OETAP)RhMe \xrightarrow{hv} [(OETAP)Rh]_2$$

There is an extensive chemistry of monoporphyrinates of the types LRhX, LRh⁻, LRhCH₂CH₂RhL, and a variety of reactions such as CO reductive coupling[73] and alkane activation.[74] The reaction of (OEP)RhI with AgBF₄ gives the cation $[OEPRh]_2^{2+}$ which has a Rh—Rh bond and also interporphyrin, cation-radical π-π interactions. Confacial dimers of the type 18-G-V can also be made[75]; these may undergo Rh—Rh bond cleavage reactions such as that shown, which proceed *via* carbonyl and radical intermediates.

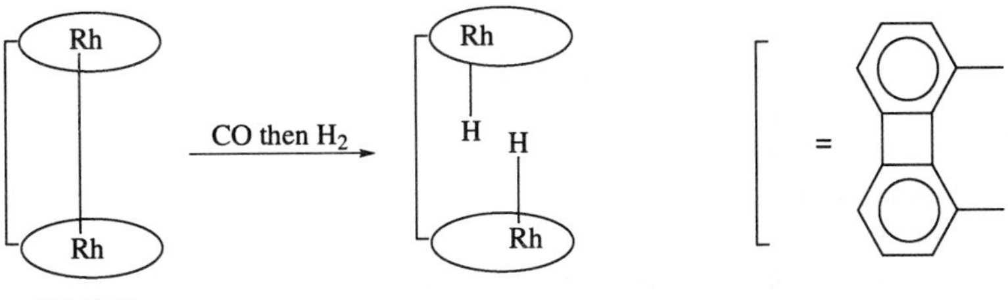

(18-G-V)

Other dimers are dimethylglyoximates, $Rh_2(dmgH)_4(PPh_3)_2$, and isocyanides[76] made by the reaction

$$Rh^I(CNR)_4^+ + [Rh^{III}X_2(CNR)_4]^+ \longrightarrow [Rh^{II}_2X_2(CNR)_8]^{2+}$$

Schiff base, salen (L) type dimers have weak Rh—Rh bonds and dissociate on heating in toluene to give paramagnetic monomers.[77] Some reactions are:

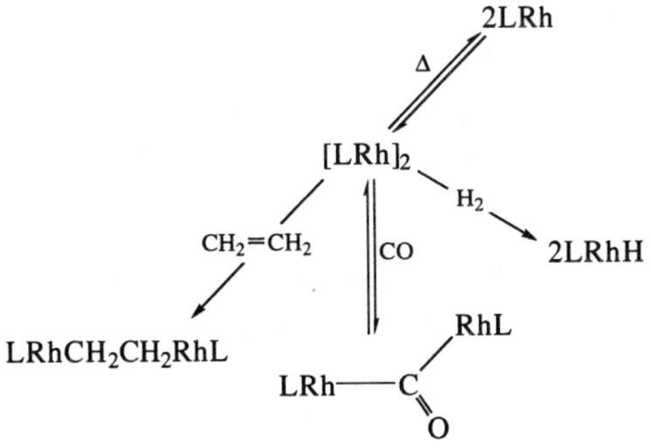

[71]A. E. Merbach *et al.*, *Inorg. Chim. Acta* **1993**, *206*, 135.
[72]B. B. Wayland *et al.*, *Inorg. Chem.* **1994**, *33*, 2029.
[73]B. B. Wayland *et al.*, *J. Am. Chem. Soc.* **1992**, *114*, 1673.
[74]X-X. Zhang and B. B. Wayland, *J. Am. Chem. Soc.* **1994**, *116*, 7897.
[75]J. P. Collman *et al.*, *Acc. Chem. Res.* **1993**, *26*, 586; *Inorg. Chem.* **1993**, *32*, 1788.
[76]H. B. Gray *et al.*, *J. Am. Chem. Soc.* **1990**, *112*, 3754.
[77]B. B. Wayland *et al.*, *Organometallics* **1994**, *13*, 3390.

18-G-5   Complexes of Iridium(II), d^7

Monomeric complexes provided the earliest examples of Ir^{II} species in the red, square oxoaryl phosphines (18-G-VI) although $IrBr_3(NO)(PPh_3)_2$ and polymeric carbonyl halides $[Ir(CO)_2X_2]_n$ were known previously.

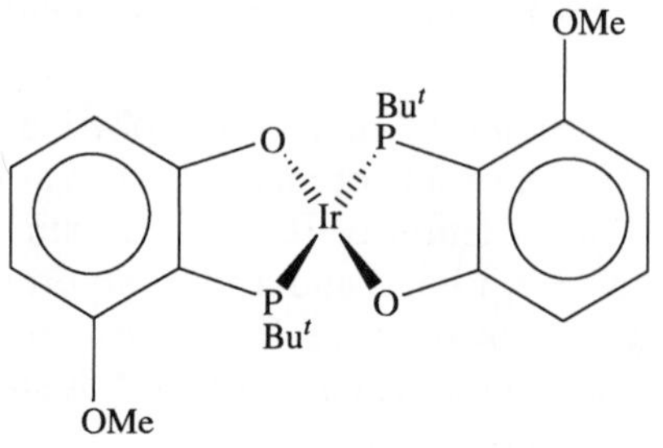

(18-G-VI)

Paramagnetic square mesityls, *trans*-$Ir(mes)_2L_2$, L = SEt_2, PMe_3, similar to the rhodium(II) mesityls, have been characterized[78] as well as the pentachlorophenyls[79] $[Ir(C_6Cl_5)_4]^{2-}$ and $Ir(C_6Cl_5)_2(COD)$.

Binuclear compounds similar to those of rhodium(II) are unknown with but one exception.[80]

A more unusual bridged species is the amido complex[81] (18-G-VII) obtained by addition of I_2 to the Ir^I dimer $Ir_2(CO)_4(\mu\text{-}NHp\text{-}tol)_2$.

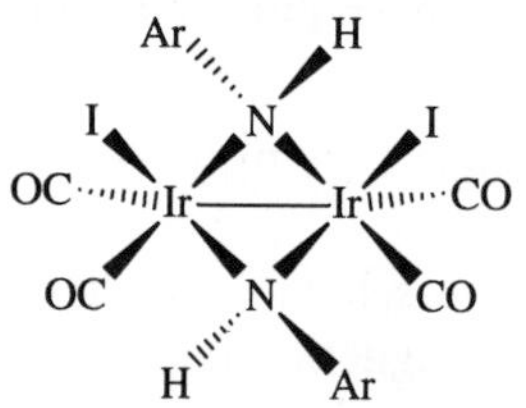

(18-G-VII)

Porphyrinates are similar to those of rhodium(II); there are also compounds such as $Ir_2(\mu\text{-}I)_2(I)_2(COD)$, $[Ir_2(\text{diisocyanide})_4I_2]^{2+}$ and $[IrMe(CO)(PR_3)_3]^+$.

18-G-6   Complexes of Rhodium(IV) and Iridium(IV), d^5

Rhodium

As noted earlier, many species long considered to be in higher oxidation states have been shown to be superoxo complexes. Well-defined compounds are salts of $[RhX_6]^{2-}$, X = F^{82} and $Cl,^{83}$ that are paramagnetic (*ca.* 1.8 BM), consistent with a

[78]G. Wilkinson *et al., J. Chem. Soc., Dalton Trans.* **1992**, 3165.
[79]M. P. García *et al., Organometallics* **1993**, *12*, 4660.
[80]F. A. Cotton and R. Poli, *Polyhedron* **1987**, *6*, 1625.
[81]K. J. Ahmed *et al., Polyhedron* **1994**, *13*, 919.
[82]W. Levason *et al., J. Chem. Soc., Dalton Trans.* **1992**, 447.
[83]I. J. Ellison and R. D. Gillard, *Polyhedron* **1996**, *15*, 339.

t_{2g}^5 configuration. The fluoride is made by treating $RhCl_3$ and an alkali metal chloride with F_2 or BrF_5. The green $Cs_2[RhCl_6]$ compound is obtained by treating $Cs_3[RhCl_6]$ with a chlorine saturated solution of $(NH_4)_2[Ce(NO_3)_6]$.[84] Electrochemical oxidation of M^{III} dithiocarbamates of both Ir^{III} and Rh^{III} gives the cations $[M(dtc)_3]^+$ in solution.

There are several Rh^{IV} compounds with $Rh-C$ bonds. First are several types of stable complexes $[Cp*Rh(\mu\text{-}CH_2)]_2X_2$, a specific example being $[\{Cp_2^*Rh_2\text{-}(\mu\text{-}CH_2)_2\}Me(\eta^2\text{-}C_2H_4)]^+$.[85] The second type is obtained when $RhCl_3(tht)_3$ is reacted with neopentyllithium in presence of Me_3NO. The diamagnetic $[(Me_3CCH_2)_3\text{-}Rh]_2\mu\text{-}O$ has a linear $Rh-O-Rh$ group[86] similar to well known groups such as $Ru-O-Ru$ where the electrons are spin-paired.[87]

Iridium

Probably the most important halogen compounds are salts of the *hexachloroiridate(IV) ion* that can be made by chlorinating the metal plus an alkali metal chloride or by adding MCl to a suspension of hydrous IrO_2 in aqueous HCl. Commercial $(NH_4)_2IrCl_6$ can be converted to $Na_2IrCl_6 \cdot 6H_2O$ by boiling with HCl, evaporating with excess NaCl in a Cl_2 atmosphere, extraction with acetone, and precipitation with $CHCl_3$. The black crystalline solid is the usual starting material for many syntheses; some commercial "$IrCl_4$" is hydrated $IrCl_3 \cdot nH_2O$.[88] The $[IrCl_6]^{2-}$ ion has been much studied as a convenient one-electron oxidant. Octahedral Ir^{IV}, t_{2g}^5, has one unpaired electron; μ_{eff} values are low (1.6 − 1.7 BM) owing to antiferrimagnetic interactions which disappear on dilution with isomorphous $[PtCl_6]^{2-}$ salts.

Action of *aqua regia* on the ammonium salt gives "chloroiridic acid," which is soluble in ether and hydroxylated solvents and is probably $(H_3O)_2IrCl_6 \cdot 4H_2O$. In neutral or even weakly acidic solution the dark red-brown $IrCl_6^{2-}$ undergoes spontaneous reduction within minutes to pale yellow-green $IrCl_6^{3-}$:

$$2IrCl_6^{2-} + 2OH^- \rightleftharpoons 2IrCl_6^{3-} + \tfrac{1}{2}O_2 + H_2O$$

From known potentials the *acid* reaction can be written:

$$2IrCl_6^{2-} + H_2O \rightleftharpoons 2IrCl_6^{3-} + \tfrac{1}{2}O_2 + 2H^+$$

$$K = 7 \times 10^{-8}\,atm^{1/2}mol^2\,l^{-2}\,(25\,°C)$$

Thus in strong acid, say 12 M HCl, $IrCl_6^{3-}$ is partially oxidized to $IrCl_6^{2-}$ in the cold and completely on heating, whereas in strong base (pH > 11), $IrCl_6^{2-}$ is rapidly and quantitatively reduced to $IrCl_6^{3-}$ with evolution of O_2. The ion $IrCl_6^{2-}$ is readily and quantitatively reduced to $IrCl_6^{3-}$ by KI or sodium oxalate and photochemically (254 nm) to $[IrCl_5(H_2O)]^{2-}$.

[84]I. J. Ellison and R. D. Gillard, *Polyhedron* **1996**, *15*, 339.
[85]P. M. Maitlis *et al.*, *Polyhedron* **1995**, *14*, 2767.
[86]G. Wilkinson *et al.*, *Polyhedron* **1990**, *9*, 2071.
[87]Z. Lin and M. B. Hall, *Inorg. Chem.* **1991**, *30*, 3817.
[88]For comments and references see G. Wilkinson *et al.*, *J. Chem. Soc., Dalton Trans.* **1993**, 3219.

Heating $(Et_4N)_2IrCl_6$ in CF_3CO_2H produces the edge-sharing bioctahedral anion $[Ir_2Cl_{10}]^{2-}$ that is similar to its Re, Os, and Pt analogues.[89] Salts of IrF_6^{2-} are well characterized.[90] There is also a variety of aquated species such as $[IrCl_3(H_2O)_3]^+$, $[IrCl_5(H_2O)]^-$, $IrCl_4(H_2O)_2$ and so on, most of which are readily reduced to Ir^{III}.

Oxygen Ligands

Extensive studies have been made of the oxidation of $[Ir(H_2O)_6]^{3+}$ in solution electrochemically or by use of Ce^{4+} in non-complexing acids.[91] Initial oxidation gives a brown-green unstable ($t_{1/2}$, *ca.* 5 h, 25°C) Ir^V species that can be reduced to blue-purple Ir^{IV} and then to Ir^{III}, the later as two dimers separable by ion exchange and assigned as $[(H_2O)_4Ir(OH)_2Ir(H_2O)_4]^{4+}$ and $[(H_2O)_5Ir(OH)Ir(H_2O)_5]^{5+}$. Electrochemical oxidation of these gives blue and purple Ir^{IV} products probably $[(H_2O)_4Ir(OH)_2Ir(H_2O)_4]^{6+}$ and $[(H_2O)_5IrOIr(H_2O)_5]^{6+}$ respectively. No crystalline products were isolated and structures were proposed on spectroscopic and other grounds. Oxidation by O_3 of the oxo-centered triangular complexes $[Ir_3^{III}(\mu_3-O)-(O_2CMe)_6(H_2O)_3]^+$ gives blue, mixed valence, III,III,IV, complexes with the same structure.[92]

Oxidation also occurs on interaction of $[IrBr_6]^{2-}$ with 100% HNO_3 to give the purple ion $[Ir_3(\mu_3-O)(NO_3)_9]^{4+}$; if N_2O_5 is used, the ion $[Ir(NO_3)_6]^{2-}$ is obtained. A related N-centered complex is $[Ir_3(\mu_3-N)(SO_4)_6(H_2O)_3]^{6-}$ obtained on heating $IrCl_3 \cdot nH_2O$ with H_2SO_4 and $(NH_4)_2SO_4$. The use of the tridentate oxygen ligand $C(Ph_2P{=}O)_3^-$ (triso) allows the synthesis[93] of (triso)$Ir^I(C_2H_4)_2$, which with silanes such as Ph_3SiH undergoes oxidative addition to give not only (triso)$Ir^{IV}H(SiPh_3)(C_2H_4)$ but also (triso)$Ir^V(H)_2(SiMePh_2)_2$.

Interaction of alkali metal and iridium oxides at 740°C gives crystals of e.g., $Cs_4[Ir^{IV}O_4]$, which has a planar anion.[94]

Phosphine complexes such as $IrCl_4(PMe_2Ph)_2$ are soluble in organic solvents;[95] the hydrides,[96] $Ir(H)_2Cl_2(PR_3)_4$, are photochemically reduced to Ir^{II} species. Another unusual complex,[97] $[(Me_3P)_3Ir(biphBF)Cl]^+$, has a BF moiety inserted into the $Ir-C$ bond of the biphenyl-2,2'-diyl group.

The only tetrahedral Ir^{IV} compound so far known is the tetramesityl,[98] $Ir(mes)_4$, made by the interaction of Li mes with partially dehydrated $IrCl_3 \cdot nH_2O$ in Et_2O. It can be oxidized by $AgPF_6$ to $[Ir^V(mes)_4]PF_6$. This cation is also tetrahedral. Polarography shows the reversible system $Ir^{III}(mes)_4^- \rightleftharpoons Ir^{IV}(mes)_4 \rightleftharpoons Ir^V(mes)_4^+$ but the anion could not be isolated.

[89]P. Hollmann and W. Preetz, *Z. Naturforsch.* **1992**, *47b*, 1115.
[90]W. Levason *et al., J. Chem. Soc., Dalton Trans.* **1992**, 139.
[91]D. T. Richens *et al., Inorg. Chem.* **1989**, *28*, 954.
[92]A. Bino *et al., Inorg. Chim. Acta* **1993**, *213*, 99; R. D. Cannon and R. P. White, *Progr. Inorg. Chem.* **1988**, *36*, 195.
[93]R. S. Tanke and R. H. Crabtree, *Organometallics* **1991**, *10*, 415.
[94]K. Mader and R. Hoppe, *Z. anorg. allg. Chem.* **1992**, *604*, 30.
[95]W. A. Levason *et al., J. Chem. Soc., Dalton Trans.* **1987**, 199.
[96]D. Attanasio *et al., New J. Chem.* **1992**, *16*, 347.
[97]R. H. Crabtree *et al., J. Chem. Soc., Chem. Commun.* **1993**, 1877.
[98]G. Wilkinson *et al., J. Chem. Soc., Dalton Trans.* **1992**, 1477.

18-G-7   Complexes of Rhodium and Iridium(V) and (VI), d^4, d^3

Salts of the hexafluoroanions,[99] $M^V F_6^-$, can be made by reactions such as

$$RhF_5 + CsF \xrightarrow{\text{IF}_5} CsRhF_6$$

$$IrBr_3 + CsCl \xrightarrow{\text{BrF}_3} CsIrF_6$$

The IrF_6^- ion can be further oxidized to IrF_6 (Table 18-E-3) by Ag^{2+} in liquid HF.[100] Fluorination of $AgIrF_6$ by F_2[101] gives a salt $AgF^+ IrF_6^-$ that has a chain structure with μ-F bridges.

There are several non-fluoro compounds of Ir^V. The Ir^V tetramesityl cation and the *triso* complex were noted above as was the green oxo compound $(mes)_3Ir{=}O$ made by action of O_2 on $Ir(mes)_3$.

The species, probably in oxidation state V, made by air oxidation of $[Ir(H_2O)_6]^{3+}$ solutions, was also noted earlier.

There are numerous *polyhydrides*. Phosphine compounds such as $IrH_5(PR_3)_2$ can be made by interaction of $LiAlH_4$ on chloro complexes of PR_3 followed by hydrolysis with methanol. Other species include the 7-coordinate IrH_5-$(Pr^i_2SiOSiPr^i_2)$ that has Ir—Si bonds[102] and a variety of Cp* compounds such as Cp^*IrH_4, $[Cp^*IrH_3(PMe_3)]^+$, and $Cp^*IrH_3(SnPh_3)$. Iridium(V) can also be present in metalloborane clusters and there are several other Cp* complexes[103] formed in reactions such as

$$(Cp^*Ir)_2Cl_4 \xrightarrow{\text{Al}_2\text{Me}_6} [Cp^*IrMe_3]_2AlMe \xrightarrow{\text{O}_2} Cp^*IrMe_4 + [Cp^*IrMe_3OAlO]_n$$

$$(Cp^*Rh)_2Cl_4 \underset{\text{HCl}}{\overset{\text{Et}_3\text{SiH}}{\rightleftarrows}} Cp^*Rh(H)_2(SiEt_3)_2 \xrightarrow{\text{HBF}_4} [(Cp^*RhH)_4]^{2+}$$

Finally there are some perovskite type mixed oxides such as Ba_2SrIrO_6 that have Ir^{VI} according to Mössbauer spectra.[104]

18-G-8   Organometallic Compounds[105]

In previous sections some organometallic compounds of Rh and Ir in high oxidation states, III-V as well as some CO compounds have been described. There is a vast chemistry for both elements involving cyclopentadienyl or substituted cyclopenta-

[99]W. A. Levason *et al., J. Chem. Soc., Dalton Trans.* **1992,** 139, 447.

[100]N. Bartlett *et al., J. Fluorine Chem.* **1995,** 72, 157.

[101]N. Bartlett *et al., Inorg. Chem.* **1995,** 34, 2692.

[102]J. W. Faller *et al., Inorg. Chem.* **1995,** 34, 2937.

[103]P. M. Maitlis *et al., J. Chem. Soc., Dalton Trans.* **1987,** 2709.

[104]G. Demazeau and D-Y. Jung, *Eur. J. Solid State Inorg. Chem.* **1995,** 32, 383.

[105]F. H. Jardine in *Chemistry of Platinum Group Metals,* F. R. Hartley ed., Elsevier, New York, 1991; J. C. W. Lohrenz and H. Jacobson, *Angew. Chem. Int. Ed. Engl.* **1996,** 35, 1305 (cationic Ir^{III} complexes; C—H activators).

dienyl ligands, arene, alkene, alkyne, and many other types of organic ligand; both elements form *tris* allyls $M(CH_2CH=CH_2)_3$. Only a few examples can be given here.

Carbonyls

Some compounds and their reactions have been noted in Section 18-G-2. The homoleptic carbonyls are clusters such as $M_4(CO)_{12}$ and $M_6(CO)_{18}$. The hydride, $HRh(CO)_4$, is very much less stable than $HCo(CO)_4$ and has been made only under *ca.* 1400 atm pressure since it readily loses H_2 to give clusters. Both Rh and Ir give anions $[M(CO)_4]^-$ and $[M(CO)_3]_4^{3-}$ as R_3NH^+ salts as well as various cluster anions such as $[Ir_8(CO)_{22}]^{2-}$ and $[Rh_5(CO)_{15}]^-$. Hydrido and other substituted polynuclear carbonyls are known.

Cyclopentadienyls

There are many of these compounds, the first of which was Cp_2Rh^+. The chemistry of Cp* compounds has been extensively studied. A simple example is $Cp*Ir^I(C_2H_4)_2$. The trivalent compounds $[Cp*M^{III}Cl(\mu\text{-}Cl)]_2$ are very useful starting materials, a recent reaction[106] being:

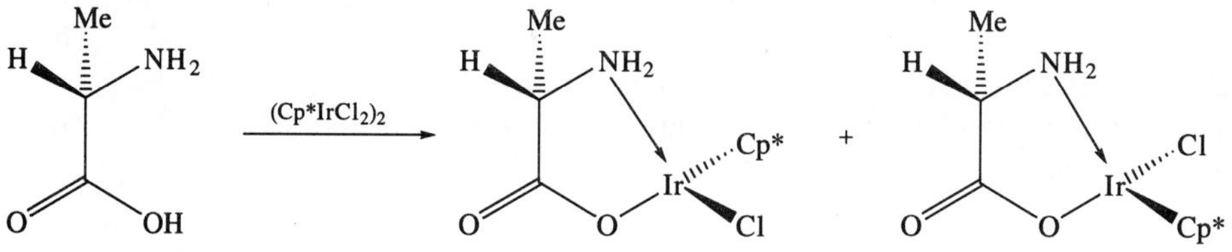

Interaction of the corresponding rhodium chloride with water and silver triflate gives the *tris* hydrate:

$$2[Cp*RhCl(\mu\text{-}Cl)]_2 \xrightarrow[CH_2Cl_2]{H_2O/AgOTf} [Cp*Rh(H_2O)_3](OTf)_2$$

The *hydrate* is but one example of several aqua organometallic compounds now known.[107] It undergoes reversible conversion to a bridged hydroxy complex:

$$2[Cp*Rh(H_2O)_3]^{2+} \underset{3H^+}{\overset{3OH^-}{\rightleftharpoons}} [Cp*Rh(\mu\text{-}OH)_3RhCp*]^+ + H_2O$$

The *imido* complexes $Cp*Ir=NR$ have many insertion reactions[108]; for example, $Cp*Ir=NBu^t$ reacts with Bu^tNC to give the η^2-*N,C*-carbodiimide complex $Cp*Ir(\eta^2\text{-}Bu^tNCNBu^t)(CNBu^t)$. However, other isocyanides may react differently.[109]

[106]J. L. Hubbard *et al., Organometallics* **1996**, *15*, 1230.
[107]R. H. Fish *et al., Organometallics* **1996**, *15*, 2009; **1995**, *14*, 2806.
[108]See R. G. Bergman *et al., J. Am. Chem. Soc.* **1991**, *113*, 2041; *Organometallics* **1994**, *13*, 4594.
[109]G. Wilkinson *et al., J. Chem. Soc., Dalton Trans.* **1996**, 3771.

It may be noted that the proposed carbene intermediate has precedents in the stable carbenes[110] with groups of the type 18-G-VIII.

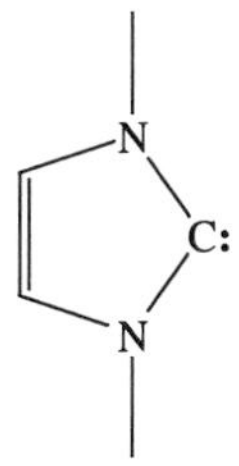

(18-G-VIII)

Although Cp*Rh=NR compounds have not been isolated from reactions of [Cp*RhCl$_2$]$_2$ with LiNHR an indication that Cp*Rh=N(2,6-Pr$_2^i$C$_6$H$_3$) exists for a short time at 0°C is the trapping reaction with mesityl azide, mesN$_3$, which gives the N,N'-diaryl tetrazine complex by a 1,3-dipolar cycloaddition:[111]

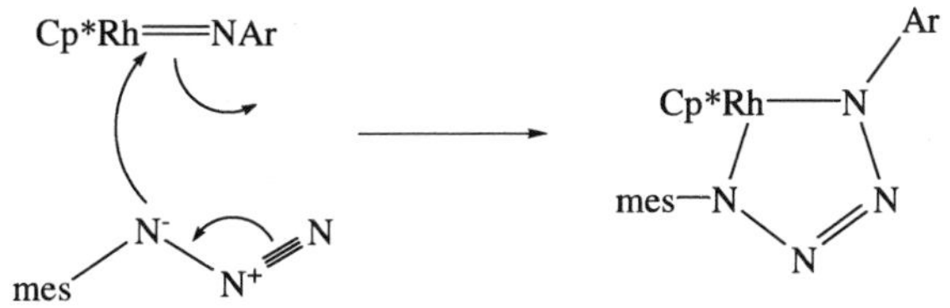

A different type of complex is a metallabenzene,[112] 18-G-IX.

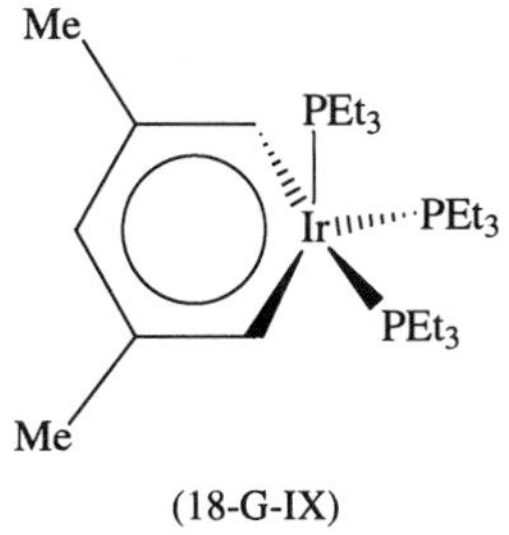

(18-G-IX)

It reacts with small molecules such as N$_2$O, CO$_2$, and SO$_2$.

Although carbene complexes of platinum group metals are fairly rare, iridium forms several. One is made by the following sequence:[113]

---

[110]See papers by A. J. Arduengo, III *et al., J. Am. Chem. Soc.* **1995,** *117,* 11027; R. R. Savers, *Tetrahedron Lett.* **1996,** *37,* 149.

[111]G. Wilkinson *et al., J. Chem. Soc., Dalton Trans.* **1996,** 3771.

[112]J. R. Bleeke *et al., J. Am. Chem. Soc.* **1994,** *116,* 4093.

[113]R. G. Bergman *et al., J. Am. Chem. Soc.* **1996,** *118,* 2517.

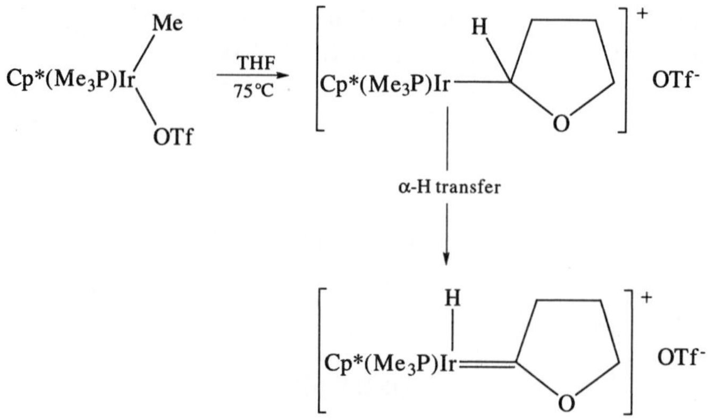

There are other compounds that have double bonds to carbon, mainly for rhodium. These are vinylidine and allenylidine complexes:[114]

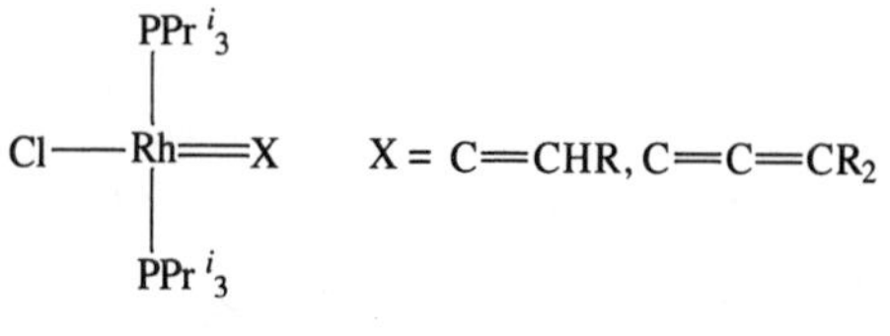

$$X = C{=}CHR, C{=}C{=}CR_2$$

These can undergo various substitutions of the Cl atoms, e.g., by OH, PhO, MeC(O)O, and so on. The vinylidines can be converted for example to acetylides with *trans* groups OC—Rh—C≡CR or to cumulene complexes by the reaction

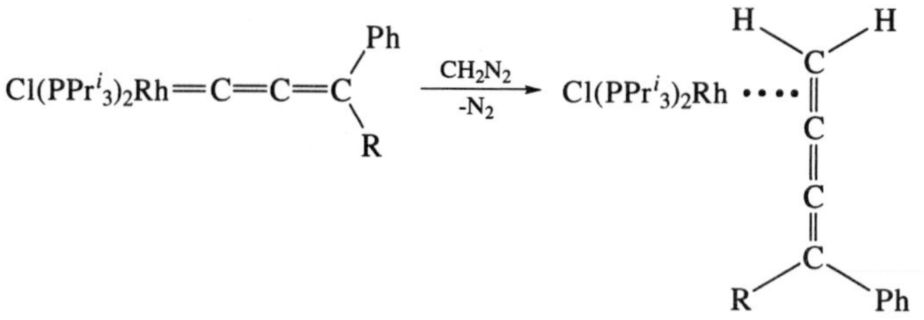

Complexes often similar to these with Cp* as ligand are *pyrazolylborates* such as hydrotris(3,5-Me$_2$pyrazolyl)borate. An example[115] is Tp*Ir^{III}H(C$_6$H$_5$)P(OMe)$_3$; another[116] is the reaction

$$Tp*Ir(C_2H_4)_2 \xrightarrow{MeCN} Tp*Ir(CH{=}CH_2)Et(MeCN)$$

---

[114]H. Werner *et al., Angew. Chem. Int. Ed. Engl.* **1996,** *35,* 1237; *J. Chem. Soc., Chem. Commun.* **1996,** 1413.
[115]S. Sostero *et al., Inorg. Chem.* **1996,** *35,* 1602.
[116]E. Carmona *et al., Organometallics* **1996,** *15,* 2192.

Complexes with 1,4-diazabutadienes (dab) are also known in [Cp*Ir^{III}Cl(dab)]$^+$ from which reaction with NaBH$_3$CN gives Cp*Ir(dab).[117]

Finally, it may be noted that in addition to homogeneous catalysis by Rh and Ir complexes, a number of compounds have been shown to activate C—H bonds in alkanes and alkenes. An example is the activation of CH$_4$ by CpM(PMe$_3$) species generated photochemically in matrices at low temperatures.[118] Mixtures of CH$_4$, CO, and O$_2$ give acetic acid when passed over RhCl$_3$ in the presence of I$^-$ or Pd on charcoal.[119] Catalytic dehydrogenation of alkanes to alkenes by IrCl(H)$_2$(PPr$_3^i$)$_2$ at 150°C in the presence of *tert*-butylethylene as H acceptor appears to proceed *via* an alkane association mechanism and H migration to the *tert*-butylethylene.[120]

18-H PALLADIUM AND PLATINUM: GROUP 10

18-H-1 General Remarks; Stereochemistry[1]

The principal oxidation states of Pd and Pt are II and IV, but there is extensive chemistry in the I and III states where M—M bonds are involved, and in the 0 state, where PR$_3$, CO, or other π-acid ligands are present and metal clusters are also found. Formal negative oxidation states occur in certain carbonyl anions of general formula [Pt$_3$(CO)$_3$(μ-CO)$_3$]$_n^{2-}$. The higher states V and VI occur only in a few fluoro compounds (Section 18-E-4).

There are also compounds with mixed oxidation states, usually II and IV but some with III.

The II State, d^8

Pd(II) and Pt(II) form almost invariably square-planar complexes. They are most frequently of the type MX$_2$L$_2$ (X = monodentate anion; L = donor ligand) which may exist as *cis* and *trans* isomers. Other possible types are also found, such as ML$_4^{2+}$, ML$_3$X$^+$, MLX$_3^-$, and MX$_4^{2-}$. An exception is PdF$_2$ in which the metal ion is octahedrally coordinated and paramagnetic although it gives aqueous solutions presumably containing diamagnetic [Pd(H$_2$O)$_4$]$^{2+}$.

Palladium and platinum(II) generally show low affinity for "hard" (F$^-$ and O) ligands and a preference for heavier halogens and ligands that can π-bond, such as R$_3$P, R$_2$S, CN$^-$, NO$_2^-$, alkenes, and alkynes.

The formation of cationic species even with non-π-bonding ligands and anionic species with halide ions contrasts with the chemistry of the isoelectronic Rh^I and Ir^I where most of the complexes involve π-bonding. The difference is presumably a reflection of the higher charge. Furthermore, although Pd^{II} and Pt^{II} species add neutral molecules to give 5- and 6-coordinate species, they do so with much less

[117]W. Klein *et al., Inorg. Chem.* **1996**, *35*, 3998.

[118]B. A. Arndtsen and R. G. Bergman, *Science* **1995**, *270*, 1970 for [Cp*(PMe$_3$)IrMe(ClCH$_2$Cl)]$^+$.

[119]M. Poliakoff *et al., Organometallics* **1996**, *15*, 1804 for Cp*Ir(CO)$_2$.

[120]J. Belli and C. M. Jensen, *Organometallics* **1996**, *15*, 1532.

[1]For a recent summary see *Studies in Inorganic Chemistry, Vol. 11: Chemistry of the Platinum Group Metals, Recent Developments*, F. R. Hartley (Ed.), Elsevier, Amsterdam, 1991.

ease; also the oxidative addition reactions characteristic of square d^8 complexes tend to be reversible except with strong oxidants, presumably owing to the greater promotional energy for $M^{II}-M^{IV}$ than for M^I-M^{III}.

Palladium(II) complexes are somewhat less stable in both the thermodynamic and the kinetic sense than their Pt^{II} analogues, but otherwise the complexes are usually similar. The kinetic inertness of the Pt^{II} (and also Pt^{IV}) complexes has allowed them to play an important role in the development of coordination chemistry such as studies of geometrical isomerism and reaction mechanisms.

The IV State, d^6

Although Pd^{IV} compounds exist, they are generally less stable than those of Pt^{IV}. The coordination number is invariably 6. The substitution reactions of Pt^{IV} complexes are greatly accelerated by the presence of Pt^{II} species. Solutions also readily undergo photochemical reactions in light.

The oxidation states and stereochemistries are summarized in Table 18-H-1.

Platinum nmr

The naturally occurring isotope ^{195}Pt has an abundance of 33.8 percent and a spin $I = 1/2$. It has a very wide chemical shift range (*ca.* 15,000 ppm) which makes ^{195}Pt nmr suitable for the identification of species in solution; $[PtCl_6]^{2-}$ in water is

Table 18-H-1 Oxidation States and Stereochemistry of Palladium and Platinum

Oxidation state	Coordination number	Geometry	Examples
Pd^0, Pt^0, d^{10}	2	Linear	$Pd(PBu_2^tPh)_2$
	3	Planar	$Pd(PPh_3)_3$, $Pt(SO_2)(PCy_3)_2$, $Pt[Sn(NR_2)]_3$
	4	Distorted	$Pt(CO)(PPh_3)_3$
	4	Tetrahedral	$Pt(Ph_2PCH_2CH_2PPh_2)_2$, $Pd(PF_3)_4$
Pd^I, d^9	4	Square	$[Pd(N_4macrocycle)]^+$, $[Pd_2(PMe_3)_6]^{2+}$
Pd^{II}, Pt^{II}, d^8	$4^{a,b}$	Planar	$[PdCl_2]_n$, $[Pd(NH_3)_4]Cl_2$, $PtCl_4^{2-}$, $PtHBr(PEt_3)_2$, $[Pd(CN)_4]^{2-}$
	5	*tbp*	$[Pd(diars)_2Cl]^+$, $[Pt(SnCl_3)_5]^{3-}$
		sp	$[PtI(PMe_3)_4]^+$
	6	Octahedral	PdF_2 (rutile type), $[Pt(NO)Cl_5]^{2-}$, $Pd(diars)_2I_2$, $Pd(dmgH)_2^c$
Pd^{III}, Pt^{III}, d^7	4	Square	$[Pt(C_6Cl_5)_4]^-$
	6	Octahedral	$[Pt_2(\mu-SO_4)_4(H_2O)_2]^{2-}$, $Pd_2(hpp)_4Cl_2^d$
Pd^{IV}, Pt^{IV}, d^6		Octahedral	$[Pt(en)_2Cl_2]^{2+}$, $PdCl_6^{2-}$, $[Pt(NH_3)_6]^{4+}$ $[Me_3PtCl]_4$, *fac*-Me$_3$Pd(I)(bipy)
Pt^V, d^5		Octahedral	$[PtF_5]_4$, PtF_6^-
Pt^{VI}, d^4	6	Octahedral	PtF_6

^{a}Most common states for Pd.
^{b}Most common states for Pt.
^{c}Has planar set of N atoms with weak Pd—Pd bonds completing a distorted octahedron.
^{d}F. A. Cotton *et al.*, *J. Amer. Chem. Soc.*, **1998**, *120*, 13280. (Pd—Pd bond).

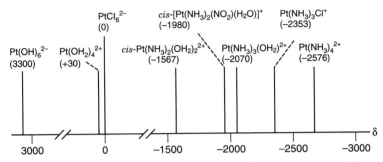

Figure 18-H-1 Typical platinum complexes and their ^{195}Pt nmr chemical shifts, relative to [PtCl$_6$]$^{2-}$ (modified, from W. Levason and D. Pletcher, *Platinum Metals Review* **1993**, *37*, 17).

commonly used as the reference ($\delta = 0$ ppm). Chemical shifts of some typical complexes are shown in Figure 18-H-1.

18-H-2 Palladium and Platinum Complexes in the Oxidation State 0, d^{10}

There are numerous zerovalent complexes of palladium and platinum. They are usually stabilized by phosphine ligands and, like their nickel analogues, tend to have a tetrahedral structure. Examples include Pd(PBu$_3^t$)$_2$, Pd(CO)(PPh$_3$)$_3$, Pt(PPh$_3$)$_3$, and Pt[P(OR)$_3$]$_4$. Analogues of Ni(CO)$_4$ do not exist, but Pt(PF$_3$)$_4$ and Pt[PF(CF$_3$)$_2$]$_4$ are stable, although Pd(PF$_3$)$_4$ decomposes above $-20°$C. The complexes can be made by a variety of reactions including reductions or eliminations, for example,

$$K_2MCl_4 + 4PPh_3 \xrightarrow[\text{EtOH}]{N_2H_4} M(PPh_3)_4$$

$$PtEt_2(PEt_3)_2 \xrightarrow{\Delta} Pt(C_2H_4)(PEt_3)_2 + C_2H_6$$

The tendency of M(PR$_3$)$_4$ molecules to dissociate, giving 3-coordinate M(PR$_3$)$_3$ and 2-coordinate M(PR$_3$)$_2$ in solution, depends mainly on the cone angle and the basicity of the phosphine. For triaryl and alkyldiaryl phosphines dissociation is extensive, whereas for trialkyl or dialkylaryl phosphines it is not; PMe$_3$ gives only Pt(PMe$_3$)$_4$, Pt(PEt$_3$)$_4$ on heating will give Pt(PEt$_3$)$_2$, while with very bulky phosphines such as PCy$_3$, only Pt(PR$_3$)$_2$ can be isolated.

The complexes undergo displacement reactions with CO, C$_2$H$_4$, dienes, alkynes and other donors to give complexes such as L$_2$Pt(RC≡CR). The alkene and alkyne ligands in such compounds lie in the ML$_2$ plane to allow for maximum back-bonding. With the chelating phosphine Ph$_2$PCH$_2$PPh$_2$ (dppm) dimers may be formed, Pt$_2$ (μ-dppm)$_3$. However, the most important aspect of the chemistry of Pd^0 and Pt^0 is the ease with which they undergo oxidative addition reactions (Fig. 18-H-2); this is the reason why they are extensively used in catalysis, notably in C—C coupling reactions.

The oxygen adducts formally having Pt^{II}, Pt(η^2-O$_2$)(PR$_3$)$_2$, have an extensive

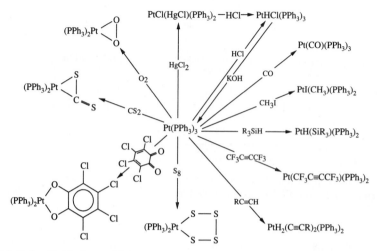

Figure 18-H-2 Oxidative addition and related reactions of Pt(PPh$_3$)$_3$. Reactions of Pd(PPh$_3$)$_4$ and Pt(PPh$_3$)$_4$ are similar.

chemistry and will react with CO_2, SO_2, and organic molecules initially to give peroxo compounds, for example,

$$(Ph_3P)_2Pt\overset{O}{\underset{O}{\diagdown}} + CO_2 \longrightarrow (Ph_3P)_2Pt\overset{O-O}{\underset{O}{\diagdown}}C=O$$

Palladium(0) and triphenylphosphine form PdL$_4$, which dissociates in solution to PdL$_3$. Before oxidative addition (for example, of an aryl halide) can take place, further dissociation to "PdL$_2$" must occur. When "PdL$_2$" is generated *in situ* in the presence of halide ions, anionic complexes PdL$_2$X$_2^{2-}$, [PdL$_2$X]$_2^{2-}$, and PdL$_2$X$^-$ are formed, the stability of which may influence the rate of oxidative addition.[2]

With strongly basic trialkylphosphines water may be oxidatively added to give cationic hydrido complexes, as in the case of the water-soluble ligand P(CH$_2$CH$_2$OH)$_3$:[3]

$$PtL_4 + H_2O \rightleftharpoons [HPtL_4]^+OH^-$$

The thermolysis of Pt(PPh$_3$)$_3$ and related compounds gives di- and trinuclear phosphido complexes, evidently *via* oxidative P—C bond cleavage.

While the carbonyls M(CO)$_4$ (M = Pd, Pt) are thermally unstable and observed only in noble gas or CO matrices at low temperature, phosphine carbonyl complexes are stable, e.g., M(CO)(PPh$_3$)$_3$ and Pt(CO)$_2$(PR$_3$)$_2$; they are tetrahedral.

[2]C. Amatore *et al.*, *J. Am. Chem. Soc.* **1991,** *113,* 8375.
[3]P. G. Pringle *et al.*, *J. Chem. Soc., Dalton Trans.* **1993,** 269.

Cluster Compounds

Low-valent palladium and platinum form numerous clusters, usually based on the M_3 triangle, as in the carbonyl phosphine complexes 18-H-I. Higher nuclearity clusters can be built up from edge-sharing triangles, e.g., the "butterfly" structure 18-H-II.

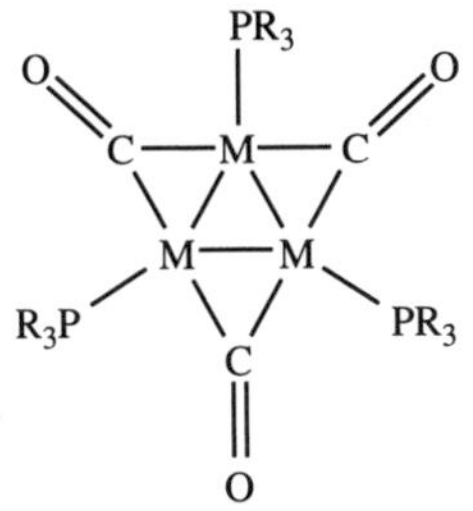

(18-H-I) 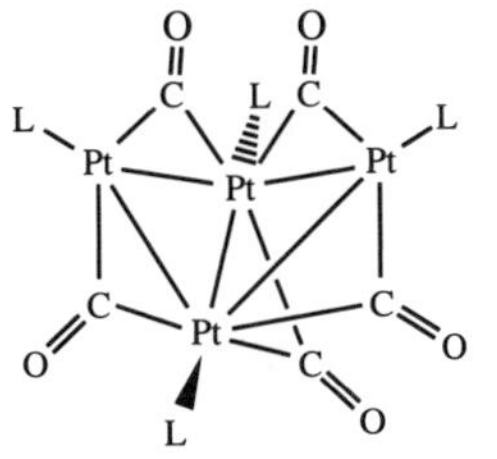   (18-H-II)

Other electron-acceptor ligands can take the place of CO, for example SO_2. There is a series of higher clusters $Pd_n(CO)_x(PR_3)_y$ (n = 3–8, 10, 16, and 23); they form on reduction of Pd salts in the presence of CO and PR_3. Mixed-valence clusters may also arise in a "self-assembly" fashion, for example, (18-H-III) is formed from palladium acetate, CO, H^+, and $Ph_2PCH_2PPh_2$ (dppm).[4]

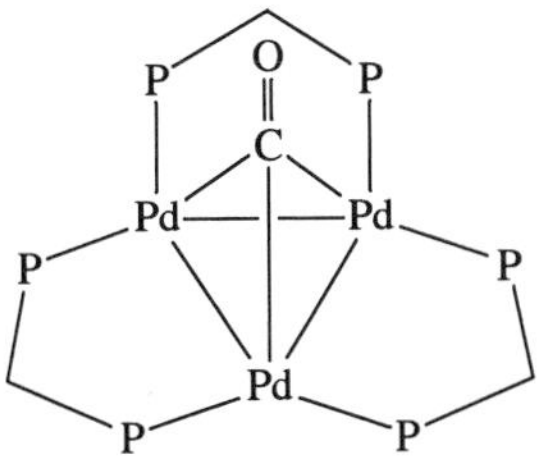

(18-H-III)

Polynuclear carbonyl anions of the general formula $[Pt_3(CO)_6]_n^{2-}$, where n depends on the alkali metal-to-Pt ratio, are obtained on reduction of $Na_2PtCl_6 \cdot 6H_2O$ or $PtCl_2(CO)_2$ under CO (1 atm, 25°C). They are usually crystallized with large cations such as $(C_4H_9)_4N^+$ or Ph_4P^+. The structures of three representative ions have been discussed in Section 16-6 and indicate $Pt_3(CO)_3(\mu\text{-}CO)_3$ units arranged to form skewed and twisted stacks.

All the species $[Pt_3(CO)_6]_n^{2-}$ are quite reactive, the specific character of their reactivity depending on n, since the greater the value of n, the less "reduced" is the metal. The larger the value of n, the greater is the reactivity toward nucleophilic and reducing agents; with small n it is greatest toward electrophiles and oxidizing agents. Thus the lower members react very avidly:

$$(n + 1)\{[(Pt_3(CO)_6]_n\}^{2-} + \tfrac{1}{2}O_2 + H_2O \longrightarrow n\{[Pt_3(CO)_6]_{n+1}\}^{2-} + 2OH^-$$

[4]R. J. Puddephatt *et al.*, *Organometallics* **1993**, *12*, 1231.

The base so formed favors reductive carbonylation, tending to reverse this change from n to $n + 1$ if CO is present. Moreover, the various species tend to react with each other, namely,

$$\{[Pt_3(CO)_6]_2\}^{2-} + \{[Pt_3(CO)_6]_4\}^{2-} = 2\{[Pt_3(CO)_6]_3\}^{2-}$$

The cluster $[Pt_{15}(CO)_{30}]^{2-}$ catalyzes the hydrogenation of MeCN, PhCHO, and other organic substances.

Reduction of, for example, $PtCl_2(RNC)_2$ leads to isocyanide polynuclear species such as $Pt_3(RNC)_6$ or $Pt_7(RNC)_{12}$. The Pt_3 and Pd_3 compounds have three terminal and three μ-RNC groups (*cf.* structure 18-H-I).

Alkene and Alkyne Complexes

Reduction of (COD)PtCl$_2$ (COD = 1,5-cyclooctadiene) in the presence of COD gives the labile alkene complex Pt(COD)$_2$, which reacts with ethene to give Pt(C$_2$H$_4$)$_3$ and with diphenylacetylene to give tetrahedral Pt(C$_2$Ph$_2$)$_2$. The trigonal-planar structure of Pt(C$_2$H$_4$)$_3$ maximizes metal-ligand back-bonding; the phosphine complexes (R$_3$P)$_2$ML (L = alkene, alkyne), too, are planar.

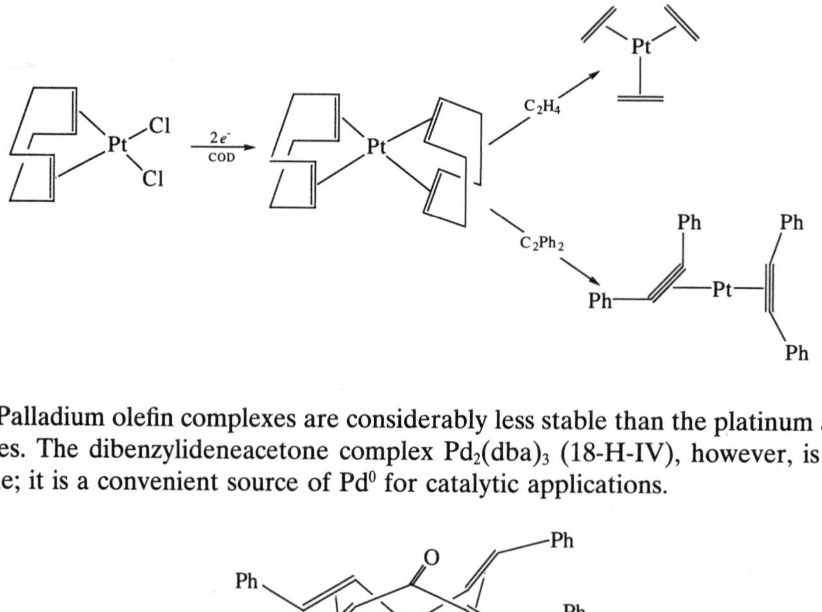

Palladium olefin complexes are considerably less stable than the platinum analogues. The dibenzylideneacetone complex Pd$_2$(dba)$_3$ (18-H-IV), however, is air-stable; it is a convenient source of Pd^0 for catalytic applications.

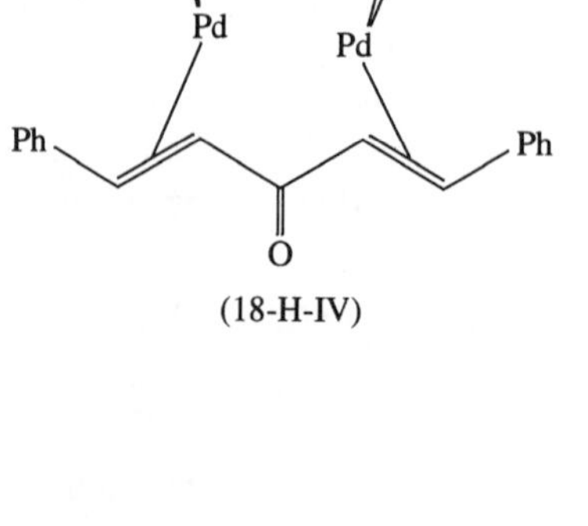

(18-H-IV)

The reaction of Pt(COD)$_2$ with stable carbenes gives 2-coordinate carbene complexes (18-H-V):[5]

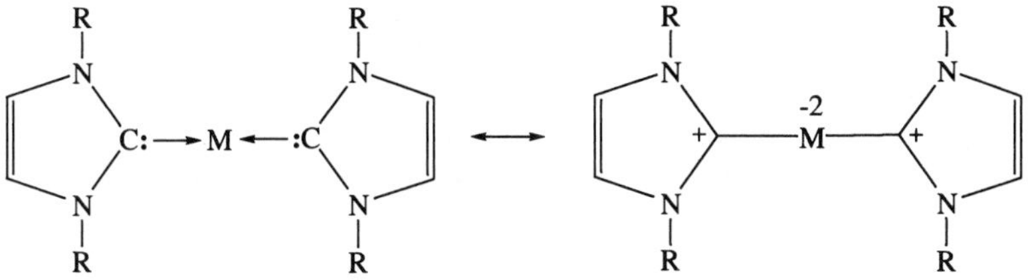

(18-H-V)

18-H-3 Complexes of Palladium(I) and Platinum(I), d^9

Complexes of palladium and platinum are accessible by the comproportionation of M^0 with M^{II} compounds,[6] by reduction of M^{II} precursors,[7] or by the judicious oxidation of M^0 complexes. The M^I products form metal-metal bonded dimers:

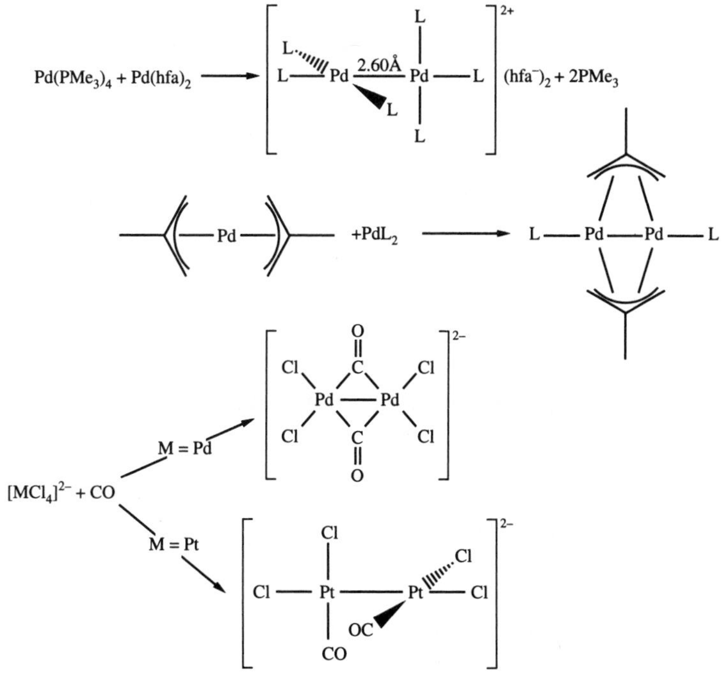

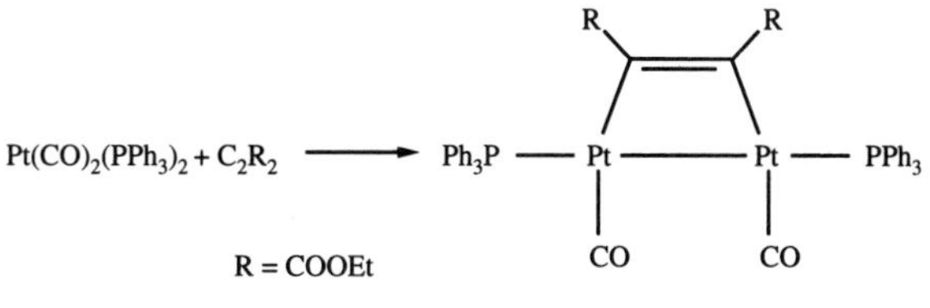

[5]A. J. Arduengo, III *et al., J. Am. Chem. Soc.* **1994,** *116,* 4391.
[6]G. S. Girolami *et al., Inorg. Chem.* **1994,** *33,* 2265.
[7]P. L. Goggin *et al., J. Chem. Soc., Dalton Trans.* **1974,** 534; **1983,** 1101.

The CO ligands in $[Pd_2(CO)_2Cl_4]^{2-}$ are bridging but in the Pt analogue are not. Isocyanide complexes, e.g., $[Pd_2(CNR)_6]^{2+}$ (accessible from $Pd_2(dba)_3$ and $[Pd(CNR)_4]^{2+}$) are non-bridged dimers, with each metal center being in a square-planar environment; there are also linear trinuclear examples $[M_3(RNC)_8]^{2+}$.[8] The chelating phosphine $Ph_2PCH_2PPh_2$ (dppm) gives bridged systems with reactive M—M bonds which readily add CO, SO_2, CH_2N_2, S, or HCl to give M^I or M^{II} "A-frame" complexes, while bipy and phen act as non-bridging chelating ligands.[9]

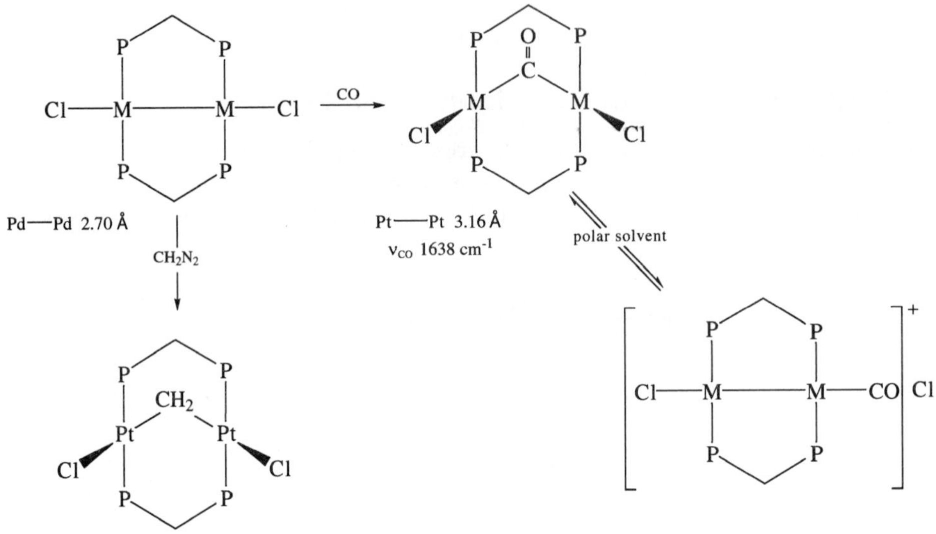

18-H-4 Complexes of Palladium(II) and Platinum(II), d^8

Aqua Ions and Simple Cations

The $[Pd(H_2O)_4]^{2+}$ ion is obtained on dissolution of PdO in $HClO_4$ and in ~ 4 M $HClO_4$, the Pd/Pd^{2+} potential is 0.98 V.

The $[Pt(H_2O)_4]^{2+}$ ion is obtained by interaction of K_2PtCl_4 and $AgClO_4$. An nmr study using ^{195}Pt shows that it reacts with various anions to give substituted species. Thus OH^- gives $[Pt(OH)_4]^{2-}$, although in concentrated solution disproportionation to Pt + $[Pt(OH)_6]^{2-}$ occurs. The $[Pt(H_2O)_4]^{2+}$ ion also reacts about 10^6 times less rapidly in water exchange than does $[Pd(H_2O)_4]^{2+}$.

Of interest in view of the antitumor properties of cis-$PtCl_2(NH_3)_2$ (see below) are the cations such as cis-$[Pt(NH_3)_2(H_2O)_2]^{2+}$. Exposure of phosphate containing solutions to air gives blue species that may be related to the platinum blues discussed later. There are also well-established amine complexes such as $[M(am)_4]^{2+}$ and $[M(en)_2]^{2+}$.

[8]Y. Yamamoto et al., J. Chem. Soc., Dalton Trans. **1996**, 1815, 3059.
[9]G. S. Girolami et al., Inorg. Chem. **1994**, 33, 2265.

Halogen and Related Anions

Salts of the ions $[MCl_4]^{2-}$ are common source materials for the preparation of other complexes in the II and the 0 oxidation states. The yellowish $[PdCl_4]^{2-}$ ion is formed when $PdCl_2$ is dissolved in aqueous HCl or when $[PdCl_6]^{2-}$ is reduced with Pd sponge. The red $[PtCl_4]^{2-}$ ion is normally made by reduction of $[PtCl_6]^{2-}$ with a stoichiometric amount of N_2H_4—HCl, oxalic acid, or other reducing agent. The sodium salt cannot be obtained pure. In water, solvolysis of $[PtCl_4]^{2-}$ is extensive at 25°C, but the rate is slow:

$$[PtCl_4]^{2-} + H_2O = [PtCl_3(H_2O)]^- + Cl^- \qquad K = 1.34 \times 10^{-2}\,M$$

$$[PtCl_3(H_2O)]^- + H_2O = PtCl_2(H_2O)_2 + Cl^- \qquad K = 1.1 \times 10^{-3}\,M$$

so that a $10^{-3}\,M$ solution of K_2PtCl_4 at equilibrium contains only 5% of $[PtCl_4]^{2-}$ with 53% of mono- and 42% of bisaqua species.

For both metals, bromo and iodo complex anions occur, and if large cations such as Et_4N^+ are used, salts of halogen-bridged ions $[M_2X_6]^{2-}$ may be obtained. Both $[MX_4]^{2-}$ and $[M_2X_6]^{2-}$ are square, but in crystals the ions in K_2MCl_4 are stacked one above the other. However, unlike other stacks containing $[MX_4]^{2-}$ ions discussed in Section 18-H-7, the M—M distances (Pd, 4.10 Å; Pt, 4.13 Å) are too large for any chemical bonding; similarly in the dimeric ions there is no evidence for metal-metal interaction.

The $[PdCl_4]^{2-}$ and $[Pd(phen)_2]^{2+}$ ions associate with Cl^- in solution to give 5-coordinate species but the Pt analogues appear not to do so. An important feature of chloro species of both Pd and Pt is the reaction with $SnCl_2$ and $SnCl_3^-$ (or $GeCl_3^-$), whereby $SnCl_3^-$ (or $GeCl_3^-$) complexes are obtained. The nature of the species depends on the conditions, but those such as cis-$[PtCl_2(SnCl_3)_2]^{2-}$ and $[Pt(SnCl_3)_5]^{3-}$ can be obtained in crystalline salts and are well characterized. The $SnCl_3$ ligand exerts a strong trans effect and labilizes the Cl ligand opposite; Pt—Sn complexes are therefore active catalysts, e.g., for the hydrogenation of alkenes.

The nitrato anion $[Pt(NO_3)_4]^{2-}$ can be obtained by treating $[Pt(H_2O)_4]^{2+}$ with KNO_3. The cyano anions $[M(CN)_4]^{2-}$ of both Pd and Pt are extremely stable with high formation constants. Other square anions are the σ alkyls and aryls, of which the perhalogenated anions such as $[Pd(C_6F_5)_4]^{2-}$ and $[Pt(C_6Cl_5)_4]^{2-}$ are most stable.

Neutral Complexes

There are many Pd and Pt complexes of the general formula $MXYL^1L^2$ where X and Y are anionic and L^1 and L^2 are neutral donors. X and Y can be halides, oxalate, SO_4^{2-}, H, alkyl, or aryl, while L can be NR_3, PR_3, CO, and so on.

The complexes are almost invariably planar, but with bulky ligands some distortion can occur. Cis and trans isomers of MX_2L_2 can interconvert, either by a dissociative pathway via a T-shaped 14-electron intermediate MX_2L (as, for example, in the case of PtR_2L_2; R = alkyl or aryl, L = S-ligand)[10] or by an associative pathway involving 5-coordinate intermediates, where the process is catalyzed by an excess of phosphine or a coordinating solvent (e.g., MeCN).

[10]D. Minetti, *Inorg. Chem.* **1994**, *33*, 2631.

The *cis-trans* isomerization of $PtCl_2(Bu_3^nP)_2$ and similar Pd complexes, where the isomerization is immeasurably slow in the absence of an excess of phosphine, is very fast when free phosphine is present. The isomerization doubtless proceeds by pseudorotation of the 5-coordinate state. In this case an ionic mechanism is unlikely, since polar solvents actually slow the reaction. Similar palladium complexes establish *cis/trans* equilibrium mixtures rapidly. Halide ligand substitution reactions usually follow an associative mechanism with *tbp* intermediates. Photochemical isomerizations, on the other hand, appear to proceed through tetrahedral intermediates.

Although square-planar four-coordination is the rule, 5-coordinate *tbp* structures may become the ground state. In $Pt(N-N)(L)X_2$, for example, this is the case if L = strong π-acceptor (N—N = chelating N-donor); if not, even a typical chelating ligand such as phenanthroline acts as a monodentate ligand.[11] Stable complexes of tetradentate phosphines $[Pd(pp_3)X]^+$ are also trigonal-bipyramidal; kinetic and spectroscopic studies have shown that here the σ- and π-donor abilities of the axial monodentate ligands X decrease in the order $Cl^- > Br^- > I^- >> P(OMe)_3$. An associative pathway is suggested for ligand exchange.[12] Strong σ-donors favor 5-coordination, as in $PtI_2(PMe_3)_3$, $[Pt(SnCl_3)_5]^{3-}$, and the tripod complex (18-H-VI).

(18-H-VI)

Bridged Complexes

There are three types of bridged complexes (18-H-VIIa, b, and c). The structure of types (a) and (b) are commonly found when μ-X = halogen, S^{2-}, and SR, whereas type (c) is confined to PR_2 and where the M—M distance is shorter than in (a) and (b). Subtle electronic and packing effects in crystals can influence the distance while the adoption of a particular geometry is related to the π-donor capability of the *bridge* ligand. Representative examples of type (a) are $[Pt_2Br_6]^{2-}$ and of type (b) $Pt_2(\mu-S)_2(PMe_2Ph)_4$, where $\theta = 121°$. Note that other d^8 metal complexes can adopt similar structures, for example, $Rh_2Cl_2(CO)_4$ (Section 18-G-2) is (a) type and $Ir_2(\mu-PPh_2)_2(CO)_2(PMe_3)_2$ is (c) type.

(a) Flat Square   (b) Hinged Square   (c) Tetrahedral
                          $\theta = 120 - 140°$

(18-H-VII)

[11]G. Natile *et al., J. Chem. Soc., Chem. Commun.* **1992**, 333.
[12]S. Funahashi *et al., Inorg. Chem.* **1996**, 35, 5163.

Bridges like SCN may form *linkage isomers*. For Pt^{II} the bridge tendencies are in the order $SnCl_3^- < RSO_2^- < Cl^- < Br^- < I^- < R_2PO^- < SR^- < PR_2^-$. Bridged species are generally subject to cleavage by donors giving mononuclear species, for example,

$$\left[\begin{array}{c} Bu_3P \diagdown \diagup Cl \diagdown \diagup Cl \\ Pd \quad Pd \\ Cl \diagup \diagdown Cl \diagdown PBu_3 \end{array}\right] + 2PhNH_2 \longrightarrow 2 \left[\begin{array}{c} Bu_3P \diagdown \diagup Cl \\ Pd \\ Cl \diagup \diagdown NH_2Ph \end{array}\right]$$

Oxygen and Sulfur Ligands

Although Pd^{II} and Pt^{II} are generally viewed as having low affinity for oxygen donors, in the later case there are some notable exceptions. There are various stable μ-OH dimers and trimers. For example, removal of halide anions from L_2PtCl_2 by Ag^+ in the presence of water gives bridged hydroxo complexes $[L_2Pt(\mu\text{-}OH)]_2^{2+}$ (L = PPh_3, 1/2 dppm). Deprotonation with a strong base such as $LiN(SiMe_3)_2$ leads to the oxo complex $[L_2Pt(\mu\text{-}O)]_2 \cdot LiBF_4$ which is of the hinged-bridge type (18-H-VIIb); if L = PMe_2Ph, it decomposes to the oxo cluster $[Pt_3(\mu_3\text{-}O)_2L_6](BF_4)_2$.[13]

Palladium(II) acetate, one of the most important palladium compounds, is used as a source for other Pd compounds and has been greatly studied in a wide variety of Pd catalyzed reactions of organic compounds. It is readily obtained by dissolving Pd in acetic acid containing some concentrated HNO_3 and forms brown crystals. It acts somewhat like $Hg(O_2CMe)_2$ and $Pb(O_2CMe)_4$ in attacking aromatic hydrocarbons electrophilically in acidic media, although in acetic acid it will attack the CH_3 group of toluene.

In the crystal the acetate is trimeric (18-H-VIII) but dissociates in hot benzene. Platinum(II) acetate, which is much more difficult to make, has a quite different structure without Pt—Pt single bonds (18-H-IX). A similar compound with a tetrabridged structure is $Pt_2(\mu\text{-}form)_4$, where *form* is N, N-bis(p-tolylformamidate).[14]

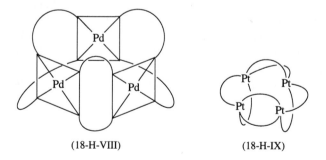

(18-H-VIII)   (18-H-IX)

The acetate readily undergoes cleavage with donors to give yellow *trans* compounds, $Pd(O_2CMe)_2L_2$ and is readily reduced to Pd^0, hence its use in catalytic reactions.

[13]P. R. Sharp *et al., Inorg. Chem.* **1996,** *35,* 604.
[14]F. A. Cotton *et al., Inorg. Chem.* **1996,** *35,* 498.

In $[Pt(OAc)_2]_4$ the acetate ligands in the plane of the Pt_4 cluster are labile while those perpendicular to it are inert to substitution. Thus, reaction with ethylenediamine gives $[Pt_4(\mu\text{-}OAc)_4(en)_4]^{4+}$ with chelating en ligands (18-H-X):[15]

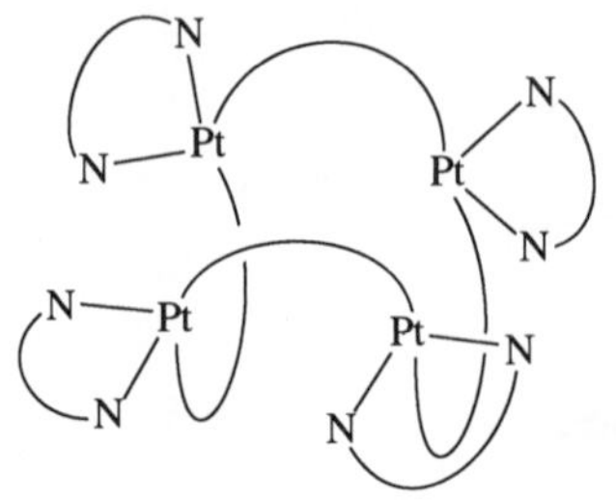

(18-H-X)

β-Diketonates form conventional square compounds with O,O′ binding but *C*-bonded complexes can be obtained as in the reaction:

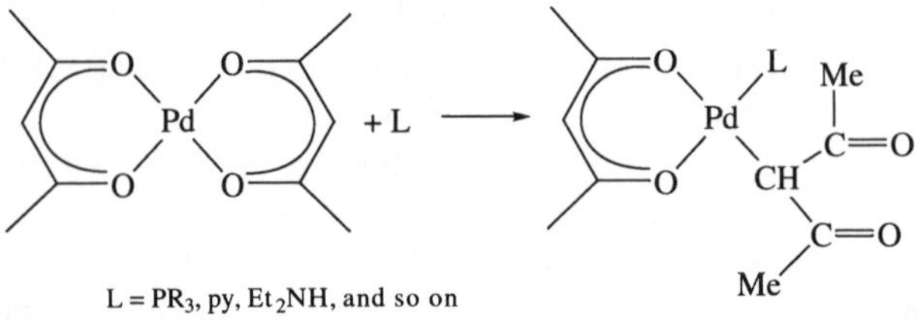

L = PR_3, py, Et_2NH, and so on

The hexafluoroacetylacetonates are strong Lewis acids and give 5-coordinate adducts, usually of a distorted structure.

The platinum(II) phosphite complex $[Pt_2(P_2O_5H_2)_4]^{4-}$ has a tetrabridged structure. It is readily oxidized to Pt^{III} dimers $[Pt_2(P_2O_5H_2)_4L_2]^{2-}$. On photolysis they oxidize organic substrates and are capable of cleaving DNA.[16]

Other complexes with oxygen ligands include the heteropolytungstate $[Pd_2W_{10}O_{36}]^{8-}$, in which two $[W_5O_{18}]^{6-}$ anions act as bidentate ligands to two square-planar Pd^{2+} ions[17] and the phenoxides $M(OPh)_2L_2$ (M = Pd, Pt, L = nitrogen ligand). The phenoxy hydride $Pt(H)(OPh)L_2$ is obtained by the oxidative addition of PhOH to $L_2Pt(C_2H_4)$. The phenoxy ligand is highly nucleophilic and is O-alkylated by iodomethane.[18] Amido complexes, common for early transition metals but rare for noble metals, have now become accessible; for example, *trans*-$PdPh(NHPh)L_2$ is made from $PdPh(I)L_2$ and KNHPh (L = PMe_3).[19]

Sulfur Ligands

Palladium and platinum have a high affinity for sulfur donors; e.g., SO_2 and SO_3^{2-} are coordinated *via* S rather than O. Polysulfide dianions S_n^{2-} form complexes with

[15]T. Ito *et al., Inorg. Chem.* **1993**, *32*, 4996.
[16]H. H. Thorp *et al., Inorg. Chem.* **1994**, *33*, 3313.
[17]G. A. Lawrence *et al., J. Chem. Soc., Chem. Commun.* **1994**, 523.
[18]W. C. Trogler *et al., Inorg. Chem.* **1991**, *30*, 3371.
[19]J. M. Boncella, *Organometallics* **1994**, *13*, 3921.

puckered MS_n rings, e.g., $[Pt(S_5)_2]^{2-}$ and $[Pd_2(\mu\text{-}S_7)_4]^{2-}$; there are also selenium and tellurium analogues $[M(E_4)_2]^{2-}$ (M = Pd, Pt; E = Se, Te). K_2PtCl_4 and K_2S_4 afford $[Pt_4S_{22}]^{4-}$ which consists of a $[Pt_4S_4]^{4+}$ cubane core surrounded by six bridging S_3^{2-} ligands.[20] PdL_4 reacts with $Et_3P=Te$ under mild conditions to give $[L_2Pd(\mu\text{-}Te)]_2$ and the cluster $Pd_6Te_6L_8$, while heating leads to $PdTe$.[21]

The oxidative addition of H_2S to (triphos)Pt(PPh$_3$) gives a rare example of a stable M-SH complex, (triphos)Pt(H)(SH); the compound is square and fluxional.[22] Thiolato complexes, on the other hand, are numerous, examples being the polynuclear compounds $[Pd(SPr^n)_2]_n$ (n = 6, 8)[23] and $[Pd_6(SCH_2CH_2SR)_8]^{4+},$[24] as well as anions such as $[M(SR)_4]^{2-}$ and $[M_2(\mu\text{-}SR)_2(SR)_4]^{2-}$.

Dithiocarbamates and related sulfur ligands form stable complexes with both Pd and Pt. They usually act as bidentate ligands. The dithioacetates $M_2(\mu\text{-}S_2CMe)_4$ have a tetrabridged dimer structure. The complexes of $[R_2C_2E_2]^{2-}$ ligands (E = S, Se) (18-H-XI) tend to form stacks and show an unusual electrochemistry; they are of interest as highly conducting molecular solids ("molecular metals").[25]

$n = 0, -1, -2$

(18-H-XI)

(18-H-XII)

Thioethers form stable complexes, particularly if they are polydentate. Whereas only two sulfur atoms are coordinated in $Ph_2Pt(9S3)$,[26] larger macrocycles, as in $[Pt(12S4)]^{2+}$, can encapsulate the metal atom.[27] The L-methionine complex (18-H-XII) is a metabolite of the anti-cancer drug cisplatin (see below).[28]

Ammine, Nucleotide, and Related Compounds

The simple ammonia adducts $[Pt(NH_3)_4]Cl_2$ and *cis*- and *trans*-$PtCl_2(NH_3)_2$ are among the oldest known platinum(II) complexes. $[Pt(NH_3)_4][PtCl_4]$ is known as Magnus's green salt; it consists of stacks of planar cations and anions forming chains of metal atoms (*cf.* Section 18-H-7). Treatment of $[PtCl_4]^{2-}$ with NH_3 gives *cis*-$PtCl_2(NH_3)_2$, whereas the reaction of HCl with $[Pt(NH_3)_4]^{2+}$ leads to *trans*-$PtCl_2(NH_3)_2$, a consequence of the *trans*-effect of Cl^-.

[20]K. W. Kim and M. G. Kanatzidis, *Inorg. Chem.* **1993,** *32,* 4161.
[21]M. L. Steigerwald *et al., J. Am. Chem. Soc.* **1990,** *112,* 9233.
[22]S. Midollini *et al., J. Chem. Soc., Dalton Trans.* **1991,** 1129.
[23]J. D. Higgins and J. W. Suggs, *Inorg. Chim. Acta* **1988,** *145,* 247.
[24]D. W. Stephan *et al., Inorg. Chem.* **1993,** *32,* 3022.
[25]B. M. Hoffman *et al., Inorg. Chem.* **1988,** *27,* 1474.
[26]M. A. Bennett *et al., Inorg. Chem.* **1993,** *32,* 1951.
[27]M. A. Watzky *et al., Inorg. Chem.* **1993,** *32,* 4882.
[28]P. J. Sadler *et al., Inorg. Chem.* **1993,** *32,* 2249.

The discovery, *ca.* 1968, by B. Rosenberg, that the *cis* isomer has anti-tumor activity stimulated the synthesis and screening of over 2000 different types of complexes with different amines and anionic ligands. Platinum compounds are among the most effective medications for the treatment of advanced cancer; $PtCl_2(NH_3)_2$ ("Cisplatin") is mainly used for the treatment of testicular and ovarian cancer. One of its drawbacks are the severe toxic side-effects which may be related to the facile hydrolysis of Cl ligands under physiological conditions:

$$cis\text{-}PtCl_2(NH_3)_2 \underset{}{\overset{H_2O}{\rightleftharpoons}} cis\text{-}[PtCl(H_2O)(NH_3)_2]^+ \underset{}{\overset{H_2O}{\rightleftharpoons}} cis\text{-}[Pt(H_2O)_2(NH_3)_2]^{2+}$$

Toxicity is much reduced with the less labile platinum carboxylates, for example, (18-H-XIII) ("Carboplatin") and (18-H-XIV); Pt^{IV} compounds are also being tested, e.g., (18-H-XV) ("Tetraplatin").[29,30]

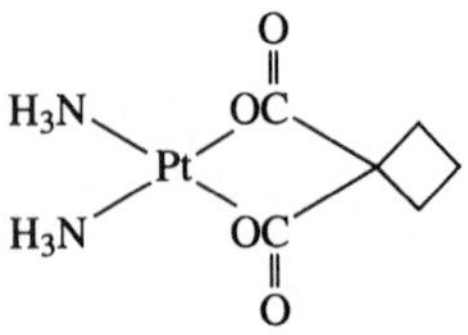

(18-H-XIII) 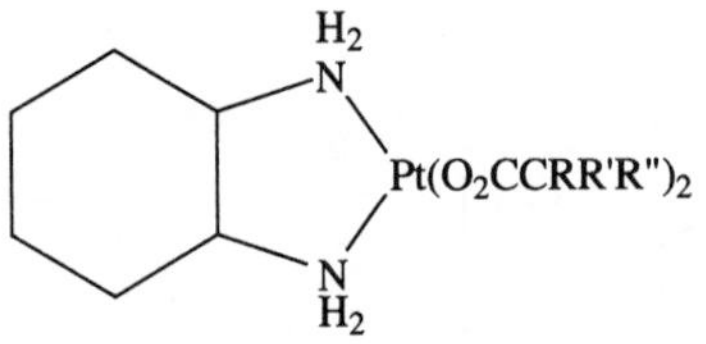   (18-H-XIV) 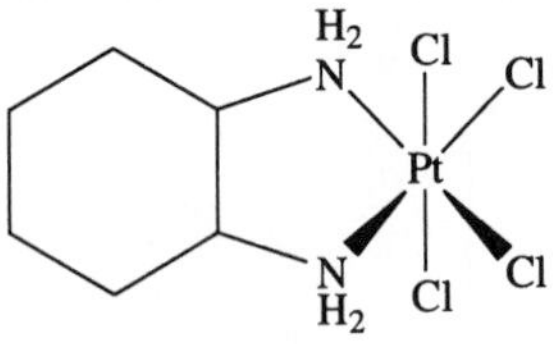   (18-H-XV)

These drugs deliver platinum(II) to the cell where it binds to DNA and interferes with gene transcription (and hence the anomalously fast cell growth in cancers). The Pt^{2+} ion coordinates preferentially to N-7 of a DNA guanine base:

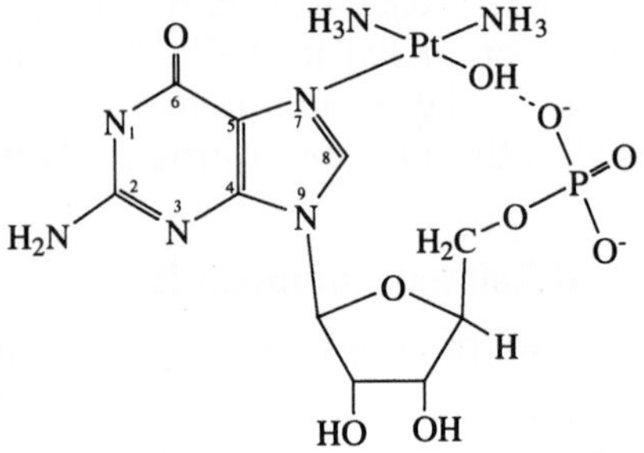

The binding of Pt^{2+} to DNA bases is strong and appears to be essentially irreversible. It has been shown that in trinucleotides such binding leads to significant conformational changes and would alter the structure of the DNA backbone.[31] Suitably tailored Pd^{II} macrocyclic complexes are capable of acting as receptors for aromatic amines and DNA nucleobases, which are bound by a combination of coordination to the metal, hydrogen bonding, and π-stacking.[32]

[29]L. R. Kelland *et al., Platinum Metals Rev.* **1992,** *36,* 178.
[30]J. Reedijk, *Chem. Commun.* **1996,** 801.
[31]J. Reedijk *et al., J. Am. Chem. Soc.* **1992,** *114,* 930.
[32]S. J. Loeb *et al., Chem. Eur. J.* **1997,** *3,* 1203.

Hydride Complexes

Hydride complexes of palladium and platinum are almost invariably stabilized by phosphine ligands and play an important role in catalytic processes such as hydrogenation. Examples are $Pt(H)ClL_2$ and $Pt(H)_2L_2$, as well as hydrido alkyls and aryls, $trans$-$Pt(H)(R)L_2$. There are cis and $trans$ isomers. A typical reaction is the insertion of alkenes and alkynes into the Pt—H bond:[33]

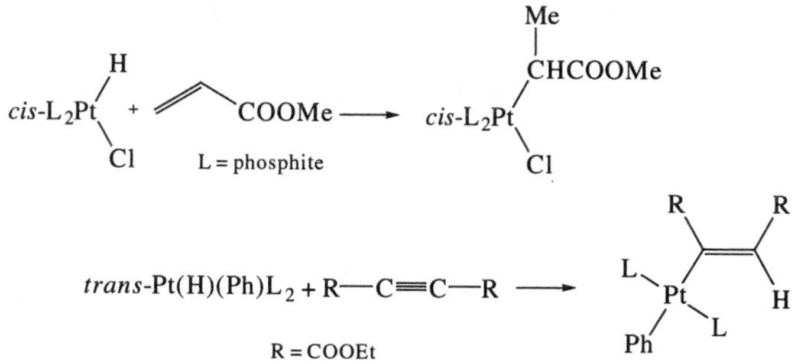

$$trans\text{-}Pt(H)(Ph)L_2 + R\text{—}C\equiv C\text{—}R \longrightarrow$$

R = COOEt

The protonation of phosphine complexes ML_3 with strong acids leads to cationic hydrides $[L_3MH]^+$.

The metal hydride solid-state phases A_2PdH_2 (A = Li, Na), are metallic, with linear PdH_2 moieties, whereas the non-conducting tetrahydrides K_2PdH_4 and Na_4PtH_4 contain square-planar MH_4^{2-} ions.[34]

Complexes of Pd^{II} and Pt^{II} with Bonds to Carbon

Both metals form comparatively stable π-complexes with olefins. The best known example is Zeise's salt, $K[PtCl_3(\eta^2\text{-}C_2H_4)]$ (see p. 679). Its palladium analogue is an intermediate in the oxidation of ethylene to acetaldehyde (Wacker process). Although olefin coordination to transition metals is usually strengthened by back-bonding, this contribution is quite weak in complexes of Pd^{II} and Pt^{II}. The carbonyl complexes $[M(CO)Cl(\mu\text{-}Cl)]_2$ and $MCl_2(CO)_2$ were the first metal carbonyls made (P. Schützenberger, 1868-1870). The monomeric and dimeric species are related by the sequence

$$Pt_2(CO)_2Cl_2(\mu\text{-}Cl)_2 \xrightleftharpoons{CO} trans\text{-}PtCl_2(CO)_2 \longrightarrow cis\text{-}PtCl_2(CO)_2$$

In these carbonyl complexes π-back-bonding to CO is absent. In consequence, the CO stretching frequency is significantly higher than that of the free CO (2143 cm^{-1}); for example, for $Pt(CO)_2Cl_2$ $\nu_{co} = 2181$ cm^{-1}. Square planar cationic tetracarbonyls, formed as salts of the very weakly coordinating anion $Sb_2F_{11}^-$, have even higher CO stretching frequencies, $[Pd(CO)_4]^{2+}$ (2248 cm^{-1}), and $[Pt(CO)_4]^{2+}$ (2244 cm^{-1}).[35]

[33]W. R. Meyer and L. M. Venanzi, *Angew. Chem. Int. Ed. Engl.* **1984,** *23,* 529; D. P. Arnold and M. A. Bennett, *Inorg. Chem.* **1984,** *23,* 2110.
[34]K. Kadir *et al., J. Less-Common Met.* **1991,** *172,* 36.
[35]L. Weber, *Angew. Chem. Int. Ed. Engl.* **1994,** *33,* 1077.

Palladium and platinum form numerous alkyl complexes of the types $M(R)(X)L_2$ and MR_2L_2, usually stabilized by phosphine ligands. The palladium compounds are prone to reductive elimination; this is important in palladium catalyzed cross-coupling reactions (Section 21-14):

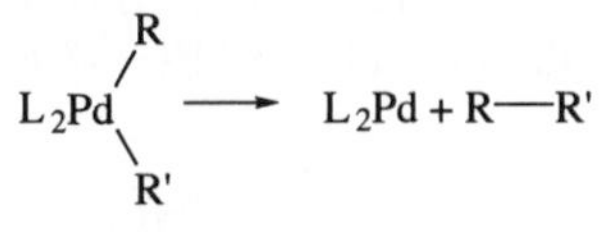

$$L_2Pd \overset{R}{\underset{R'}{\diagup\diagdown}} \longrightarrow L_2Pd + R\text{—}R'$$

The metallation of aromatic and benzylic hydrogen atoms is often a facile route to metallacycles:[36]

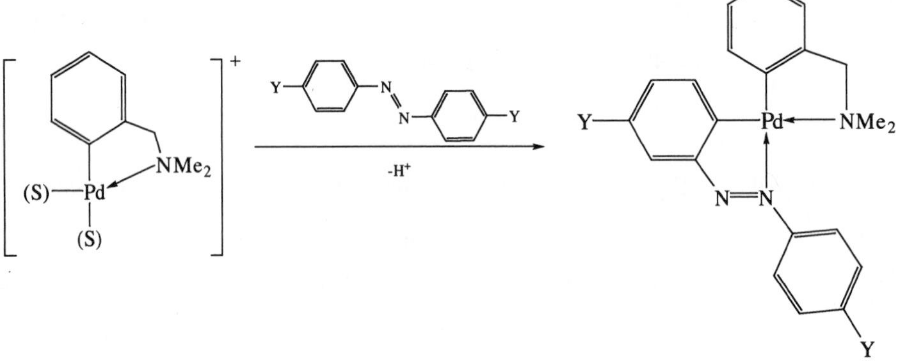

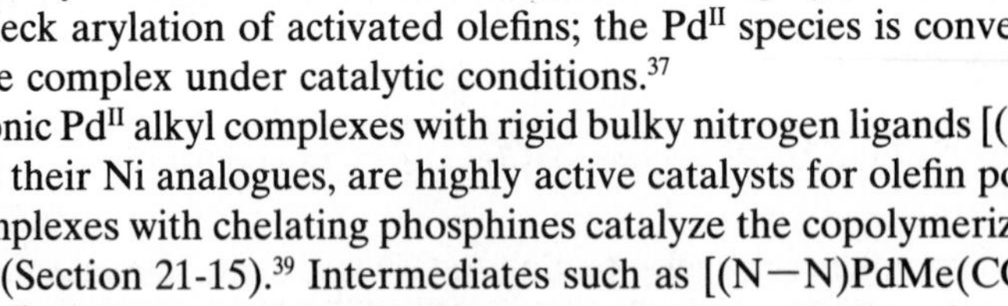

The palladacyle formed in the latter reaction is a highly active catalyst precursor for the Heck arylation of activated olefins; the Pd^{II} species is converted into a Pd^0 phosphine complex under catalytic conditions.[37]

Cationic Pd^{II} alkyl complexes with rigid bulky nitrogen ligands $[(N\text{—}N)PdMe]^+$, as well as their Ni analogues, are highly active catalysts for olefin polymerization,[38] while complexes with chelating phosphines catalyze the copolymerization of olefins with CO (Section 21-15).[39] Intermediates such as $[(N\text{—}N)PdMe(CO)]^+$ have been identified.[40] Platinum hydridoalkyl complexes of very bulky phosphines give on

[36]A. D. Ryabov and R. van Eldik, *Angew. Chem. Int. Ed. Engl.* **1994**, *33*, 783.
[37]W. A. Herrmann *et al.*, *Angew. Chem. Int. Ed. Engl.* **1995**, *34*, 1844; J. Louie and J. F. Hartwig, *Angew. Chem. Int. Ed. Engl.* **1996**, *35*, 2359.
[38]M. Brookhart *et al.*, *J. Am. Chem. Soc.* **1995**, *117*, 6414.
[39]E. Drent and P. H. M. Budzelaar, *Chem. Rev.* **1996**, *96*, 663.
[40]R. J. Puddephatt *et al.*, *J. Chem. Soc., Dalton Trans.* **1996**, 1809.

reductive elimination highly reactive LPt^0 ($L = Bu_2^tPCH_2PBu_2^t$), which is capable of reacting even with the $Si-C$ bond of $SiMe_4$ to give $LPt(Me)SiMe_3$.[41]

There are numerous Pd and Pt carbene complexes. A silicon analogue of a Fischer-type platinum carbene is also known, $[H(PR_3)_2Pt=Si(SEt)_2]^+$.[42]

1-Alkene complexes of $PdCl_2$ in the presence of base give palladium π-allyl complexes:

$$2PdCl_2 + 2H_2C=CHCH_3 \xrightarrow[-2HCl]{+2OH^-} \left[\begin{array}{c} \diagup \\ \left\langle \hspace{-0.3em} \begin{array}{c} \\ \end{array} \right. \hspace{-0.3em} PdCl \\ \diagdown \end{array} \right]_2$$

The same products can be obtained by oxidative addition of allylchloride to Pd^0. On addition of phosphines, cationic complexes $[(allyl)PdL_2]^+$ result. Such complexes have considerable synthetic applications.

18-H-5   Complexes of Palladium(III) and Platinum(III), d^7

There are numerous complexes, mostly binuclear with metal—metal bonds, that can *formally* be described as having Pt^{III}. Some compounds have a mean oxidation number of III while others contain Pt^{II} and Pt^{IV}, and it is often difficult to decide whether there is $Pt^{II}-Pt^{IV}$ or $Pt^{III}-Pt^{III}$. In other cases, the unpaired electrons may reside on the ligand rather than on the metal as in some bipyridyl and dithiolene complexes.

Mononuclear compounds are so far rare. A blue aryl anion $[Pt^{III}(C_6Cl_5)_4]^-$ is obtained by Cl_2 oxidation of yellow $[Pt(C_6Cl_5)_4]^{2-}$; it is paramagnetic ($\mu_{eff} = 2.4–2.5$ BM) and square. Certain sulfur macrocycles such as 1,4,7-trithiacyclononane (9S3) can also stabilize Pd^{III} and Pt^{III} as shown by oxidation, for example of $[Pt(9S3)_2]^{2+}$, which is square pyramidal with a non-bonded S atom, to a paramagnetic Pt^{III} species with a characteristic esr spectrum. The palladium analogue is one of the few confirmed monomeric Pd^{III} complexes.

Although one of the simplest *binuclear compounds* is $[Pt_2(\mu-SO_4)_4(H_2O)_2]^{2-}$, the most extensively studied compounds are those of the pyrophosphite $(HO_2P)_2O^{2-}$, abbreviated POP. The Pt_2^{III} species are made by oxidation of $K_4[Pt_2^{II}(POP)_4]$ with halogens or CH_3I and have the tetrabridged type structure (18-H-XVI).

$$\left[\begin{array}{c} X \diagdown \quad O \diagdown \quad O \diagup \\ HO-P \diagdown \quad P-OH \\ \vert \quad \vert \\ X-Pt-Pt-X \\ \underset{\smile\smile}{} \end{array} \right]^{4-} \qquad = HO_2POPO_2H^{2-}$$

(18-H-XVI)

Heating a mixture of $K_2[Pt(NO_2)_4]$, glacial acetic acid, and perchloric acid in air gives the tetrabridged acetate $[Pt_2(\mu-OAc)_4(H_2O)_2]^{2+}$, which contains a

[41]P. Hoffmann *et al.*, *Angew. Chem. Int. Ed. Engl.* **1990**, *29*, 880.
[42]T. D. Tilley *et al.*, *J. Am. Chem. Soc.* **1993**, *115*, 7884.

metal—metal single bond,[43] as does $Pt_2(\mu\text{-form})_4Cl_2$.[44] An organometallic Pt^{III} species is the methyl complex $Pt_2Me_4(OAc)_2(py)_2$. There are also the intermediate (II)-(III) mixed oxidation state species such as $[Pt_2(POP)_4Br]^{4-}$, which has infinite chains with Pt—Pt—Br—Pt—Pt—Br linkages [*cf.* $Ru_2^{II,III}(O_2CMe)_4Cl$].

18-H-6   Complexes of Palladium(IV) and Platinum(IV), d^6

Palladium

Apart from $Pd(NO_3)_2(OH)_2$ formed on dissolution of Pd in concentrated nitric acid, the complexes are mainly the octahedral halide anions, MX_6^{2-}. The fluoro complexes of Ni, Pd, and Pt are all very similar and are rapidly hydrolyzed by water. The chloro and bromo ions are stable to hydrolysis but are decomposed by hot water to give the Pd^{III} complex and halogen. The red $PdCl_6^{2-}$ ion is formed when Pd is dissolved in aqua regia or when $PdCl_4^{2-}$ solutions are treated with chlorine.

Oxidation of K_2PdCl_4 by persulfate in the presence of KCN gives the yellow salt $K_2[Pd(CN)_6]$.

Pyridine and bipyridine complexes such as $[Pd(py)X_5]^-$ and *cis*-[(bipy)PdCl$_4$] and similar complexes with PR_3 and AsR_3 donors can be obtained by halogen oxidation of the Pd^{II} complexes. Other stable complexes include $Pd(am)_2(NO_2)_2Cl_2$, $[Pd(S_2CNR_2)_3]^+$, and $Pd(S_2CNR_2)_2X_2$. Alkyl complexes of Pd^{IV} of the type $PdR_3(X)(L)$ (L = bipy, phen) have recently been found to be surprisingly stable (see below).

Platinum

In marked contrast to Pd^{IV}, Pt^{IV} forms many thermally stable and kinetically inert complexes that are invariably octahedral.

Apart from $PtCl_4$ and other halides (Section 18-E-4), the only simple compound is $Pt(SO_3F)_4$, made by interaction of Pt with $S_2O_6F_2$—HSO_3F at 120°C as orange crystals. Since spectra show both uni- and bidentate (μ-) SO_3F groups, a polymeric structure is most likely. The anion $[Pt(SO_3F)_6]^{2-}$ can also be obtained.

The most important Pt^{IV} compounds are salts of the red *hexachloroplatinate* ion, $PtCl_6^{2-}$. The "acid," commonly referred to as chloroplatinic acid, has the composition $H_2[PtCl_6]\cdot nH_2O$ (n = 2, 4, and 6) and contains H_3O^+, $H_5O_2^+$, and $H_7O_3^+$ ions. On thermal decomposition it eventually gives mainly the metal, but Pt_6Cl_{12} is one of the intermediates. The acid or its Na or K salts are the normal starting materials for the preparation of many Pt compounds. The ion is formed on dissolving Pt in aqua regia or HCl saturated with Cl_2. In aqueous alcohol solution the ion is reduced to $PtCl_4^{2-}$ and eventually to Pt in the presence of visible light. Heating $PtCl_6^{2-}$ in CF_3COOH gives $Pt_2Cl_{10}^{2-}$, with an edge-shared bioctahedral structure.[45]

Other simple anionic species are the azide $[Pt(N_3)_6]^{2-}$ and the polysulfide

[43]T. G. Appleton *et al., J. Am. Chem. Soc.* **1992**, *114*, 7305.
[44]F. A. Cotton *et al., Inorg. Chim. Acta* **1997**, *264*, 61.
[45]P. Hollmann and W. Preetz, *Z. anorg. allg. Chem.* **1991**, *601*, 47.

$[Pt(S_5)_3]^{2-}$ that has PtS_5 rings; the anion is chiral and is a rare example of a completely inorganic structure that can be resolved into d and l isomers.

Amine Ligands

The ammonia complexes span the entire range from $[Pt(NH_3)_6]X_4$, including all intermediates such as $[Pt(NH_3)_4X_2]X_2$ to $M^I[Pt(NH_3)X_5]$. Some are notable as examples of the classical evidence that led A. Werner to assign the coordination number 6 to Pt^{IV}. The amine ligands include ammonia, ethylenediamine, hydrazine, and hydroxylamine and the acido groups include halide, SCN^-, NO_2^-, and so on. Although not all these groups are known to occur in all possible combinations in all types of compounds, it can be said that with a few exceptions, they are generally interchangeable.

Interaction of $[Pt(NH_3)_6]Cl_4$ with β-diketones in basic solution gives iminato complexes such as (18-H-XVII) and (18-H-XVIII).

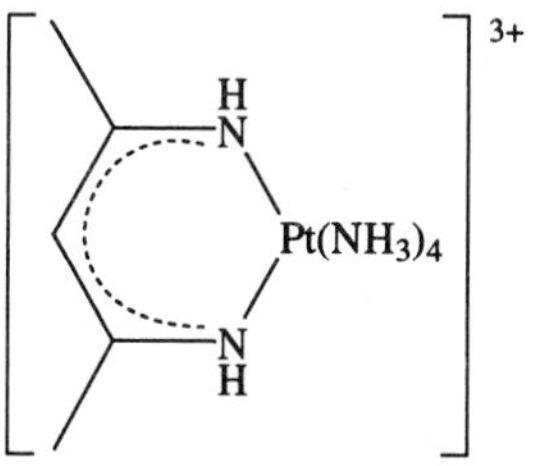

(18-H-XVII) 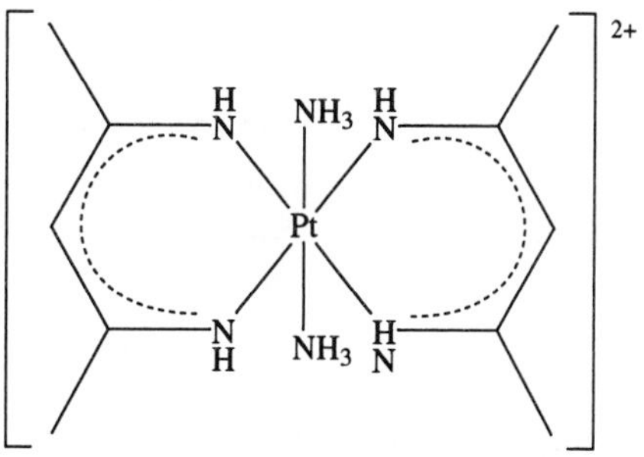   (18-H-XVIII)

Hydrido complexes such as $Pt^{IV}H_2Cl_2(PR_3)_2$ are usually made by oxidative addition to the Pt^{II} complexes. They tend rather easily to eliminate H_2 and reform Pt^{II}.

Complexes of Pd^{IV} and Pt^{IV} with Bonds to Carbon

The first isolable alkyl complex of a transition metal was $[PtMe_3(\mu\text{-}I)]_4$ (Pope, 1907). It has a cubane structure with triply-bridging iodide and octahedral metal centers. A compound first thought to be $PtMe_4$ is actually $[PtMe_3(\mu\text{-}OH)]_4$. In aqueous solutions the stable ion $[PtMe_3(H_2O)_3]^+$ is formed with weakly coordinating anions such as BF_4^- or ClO_4^-. The aryl complex $Pt(C_6Cl_5)_4$, the only confirmed homoleptic Pt^{IV} alkyl, possesses a monomeric structure and achieves octahedral geometry by coordination to two ortho-Cl atoms (18-H-XIX).[46] In the cyclopentadienyl complex $CpPtMe_3$ the Cp ligand occupies effectively three coordination sites of an octahedron. The compound is volatile and has been used for the deposition of platinum metal films.[47] Platinum(IV) alkyl complexes are readily formed by oxidative addition; for example, *cis*-$PtR_2(dppm)$ reacts with MeI to give $Pt(Me)(I)R_2(dppm)$ in which the Me and I ligands are *trans*.[48]

[46]J. Fourniés *et al.*, *J. Am. Chem. Soc.* **1995**, *117*, 4295.
[47]H. D. Kaesz *et al.*, *J. Am. Chem. Soc.* **1989**, *111*, 8779.
[48]B. L. Shaw *et al.*, *J. Chem. Soc., Dalton Trans.* **1985**, 1501.

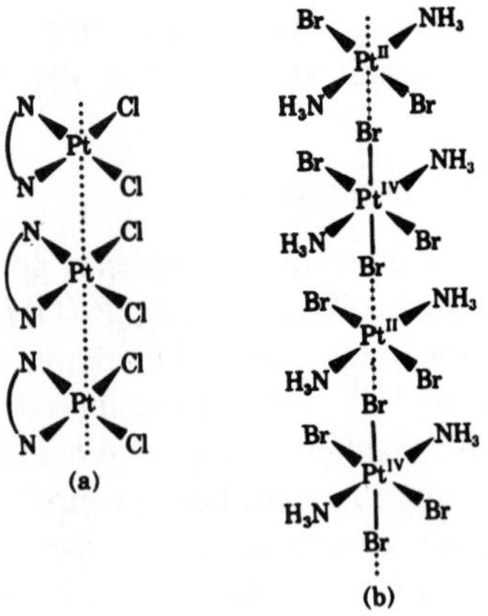

Figure 18-H-3 (a) Linear stacks of planar Pt(en)Cl$_2$ molecules. (b) Chains of alternating Pt^{II} and Pt^{IV} atoms with bridging bromide ions in Pt(NH$_3$)$_2$Br$_3$.

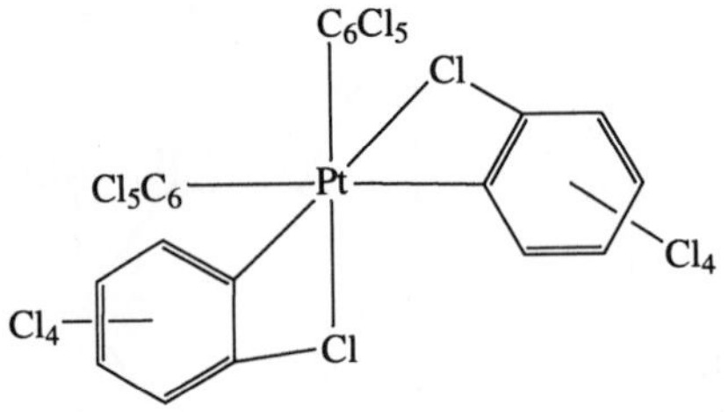

(18-H-XIX)

Organometallic complexes of Pd^{IV} were almost unknown until a few years ago. They are however accessible by the oxidative addition of MeI or PhCH$_2$Br to palladium dialkyl complexes stabilized by nitrogen ligands such as bipy. Most are thermally sensitive and reductively eliminate R—Me at room temperature.[49] The first Pd^{IV} silyl complex was obtained in a clean methyl exchange reaction between (dmpe)PdMe$_2$ and 1,2-C$_6$H$_4$(SiH$_3$)$_2$; the compound is thermally remarkably stable.[50]

18-H-7   Mixed Valence (II, IV) Linear Chain Compounds

Many square complexes of Pd^{II} or Pt^{II} are packed in crystals to form infinite chains of metal atoms (Fig. 18-H-3a) close enough to interact electronically with one another, giving rise to marked spectral dichroism and electrical conductivity.

A related class of compounds with chainlike structures contains both M^{II} and M^{IV} but differs from the above in that the metal units are linked by halide bridges. Both show high electrical conductivity along the direction of the —Cl—M^{II}—Cl—M^{IV} chains, for example, in [Pd^{II}(NH$_3$)$_2$Cl$_2$][Pd^{IV}(NH$_3$)$_2$Cl$_4$]. One of the best known, *Wolfram's red salt*, has octahedral [Pt(EtNH$_2$)$_4$Cl$_2$]$^{2+}$ and planar [Pt(EtNH$_2$)$_4$]$^{2+}$ ions linked in chains, the other four Cl$^-$ ions being within the lattice. The structure in Fig. 18-H-3b is typical. One of the best known examples is K$_2$[Pt(CN)$_4$Br$_{0.30}$]·3H$_2$O, a stacked chain with Pt—Pt distances of 2.89 Å. Like Rb$_2$[Pt(CN)$_4$Cl$_{0.3}$]·3H$_2$O it is obtained by partial oxidation of [Pt(CN)$_4$]$^{2-}$ salts. These compounds behave as electrically conducting one-dimensional metals.

[49]A. J. Canty, *Platinum Metals Rev.* **1993,** 37, 2; *Acc. Chem. Res.* **1995,** 28, 406.
[50]M. Tanaka *et al., Angew. Chem. Int. Ed. Engl.* **1996,** 35, 1856.

Other examples are the 1,2-diaminopropane complex $[Pt^{II}(pn)_2][Pt^{IV}(pn)_2X_2]^{4+}$ and $[Pt(NH_3)_4][Pt(NH_3)_4Br_2]^{4+}$. These compounds commonly show dichroic behavior, intense absorption polarized parallel to the chain, exceptionally strong resonance Raman spectra, and electrical conductance. All these can be associated with intervalence electron transfer. Finally, as well as Pt^{II} and Pt^{IV}, it is possible to have Pt_2^{III} anions in the same compound, as in the copper colored needles, $\{[Pt^{II}(en)_2][Pt^{IV}(en)_2X_2]\}^{4+}[Pt_2^{III}(H_2P_2O_5)_4X_2]^{4-}$.

Platinum Blues

Although not directly related to the previously mentioned chain species, a remarkable class of compounds generally called platinum blues contain limited chains with Pt—Pt interaction and variable oxidation states, $Pt^{II,III,IV}$. They have long been known from hydrolysis of cis-$PtCl_2(MeCN)_2$ in the presence of Ag^+ salts, but such materials could never be crystallized.

Blue compounds are formed generally when cis-$Pt(NH_3)_2Cl_2$ is treated in aqueous solution with uracils and uridines, and with thymine and other related pyrimidines. The pyrimidine blues have anti-tumor activity [cf. cis-$PtCl_2(NH_3)_2$].

Much progress has been made by studying the products of oxidation of Pt^{II} ammines in the presence of potentially bridging ligands such as

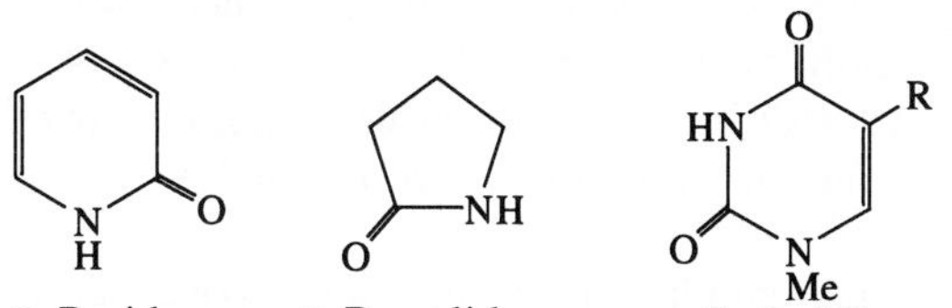

α-Pyridone   α-Pyrrolidone   Pyrimidine

The structure of $[Pt_4(en)_4(pyr)_4]^{5+}$, a typical pyrimidine blue (18-H-XX), has been determined by X-ray diffraction (pyr = α-pyridinone anion).

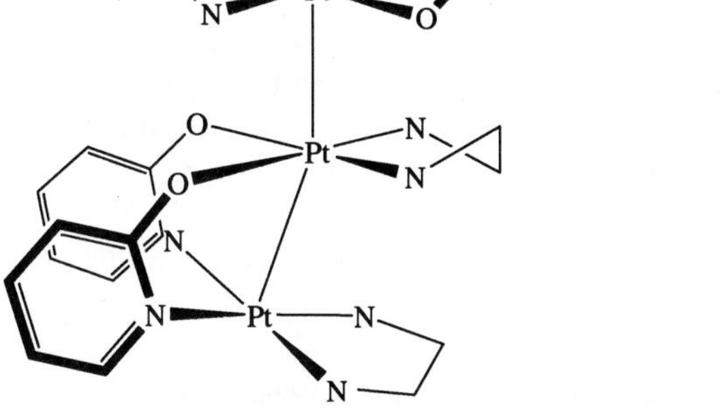

(18-H-XX)

Such di- and tetranuclear complexes can have an extensive redox chemistry. For example, the Pt^{II} compound $[Pt_2Cl(NH_3)_3(thym)_2]^+$ (thym = methylthymine anion) is oxidized to Pt^{III} *via* intensely colored mixed-valence intermediates.[51]

$$2[Pt(2.0)]_2 \xrightarrow{-e} [Pt(2.25)]_4 \xrightarrow{-e} [Pt(2.5)]_4 \xrightarrow{-e} [Pt(2.75)]_4 \xrightarrow{-e} 2[Pt(3.0)]_2$$

Other partially oxidized, nonstoichiometric complexes can be green, blue, or tan, depending on the oxidation state and without a change in the basic structure of the Pt_4 chain; examples are the green α-pyrrolidinone complex $[Pt_4(NH_3)_8\text{-}(C_4H_6NO)_4]^{5+} \cdot 3H_2O$ and the yellow $[Pt_4^{III}(NH_3)_8(C_4H_6NO)_4]^{6+}$.

Materials similar to the original Pt blue can be made by treating aqueous acidic K_2PtCl_4 with amides.

18-I   SILVER AND GOLD: GROUP 11

Like copper, silver and gold have a single *s* electron outside the completed *d* shell, but in spite of the similarity in electronic structures and ionization potential, the chemistries of Ag, Au, and Cu differ more than might be expected. There are no simple explanations for many of the differences although some of the differences between Ag and Au may be traced to relativistic effects on the $6s$ electrons of the latter. The covalent radii of the triad follow the trend Cu < Ag ≥ Au, i.e., gold has about the same or a slightly smaller covalent radius than silver in comparable compounds, a phenomenon frequently referred to as "relativistic contraction" (*cf.* lanthanide contraction).

Apart from obviously similar stoichiometries of compounds in the same oxidation state (which do not always have the same structure), there are some similarities within the group—or at least between two of the three elements:

1. The metals all crystallize with the same face centered cubic (*ccp*) lattice.
2. Cu_2O and Ag_2O have the same body centered cubic structure where the metal atom has two close oxygen neighbors and every oxygen atom is tetrahedrally surrounded by four metal atoms.
3. Although the stability constant sequence for halide complexes of many metals is F > Cl > Br > I, Cu^I and Ag^I belong to the group of ions of the more noble metals, for which it is the reverse.
4. Cu^I and Ag^I (and to a lesser extent Au^I) form very much the same types of complex ions and compounds, such as $[MCl_2]^-$, $[Et_3AsMI]_4$, and K_2MCl_3.

The only aqua ions are $Ag^+(aq)$ and $Ag^{2+}(aq)$. However, the former binds H_2O molecules only weakly, and the crystalline salts of the Ag^+ ion are rarely, if ever, hydrated. By contrast the Au^+ ion does not exist except in complexes; it is exceedingly unstable with respect to the disproportionation

$$3Au^+(aq) = Au^{3+}(aq) + 2Au(s) \qquad K \approx 10^{10}$$

[51]B. Lippert *et al., Angew. Chem. Int. Ed. Engl.* **1990**, *29*, 84.

Table 18-I-1 Oxidation States and Stereochemistry of Silver and Gold

Oxidation state	Coordination number	Geometry	Examples
Ag^I, d^{10}	2[a]	Linear	[Ag(CN)$_2$]$^-$, [Ag(NH$_3$)$_2$]$^+$, AgSCN
	3	Trigonal	[Ag(PCy$_2$Ph)$_3$]BF$_4$, [Au(PCyPh$_2$)$_3$]CO$_4$
	4[a]	Tetrahedral	[Ag(SCN)$_4$]$^{3-}$, [AgI(PR$_3$)]$_4$, [Ag(py)$_4$]$^+$, [Ag(PPh$_3$)$_4$]ClO$_4$
	5	Distorted pentagonal plane	[Ag(L)]$^{+\,b}$
	5	Pentagonal pyramidal	[Ag(L)]$_2^{2+\,b}$
	6	Octahedral	AgF, AgCl, AgBr (NaCl structure), [Ag(18S6)]CF$_3$SO$_3^c$
Ag^{II}, d^9	4	Planar	[Ag(py)$_4$]$^{2+}$
	6	Distorted octahedral	Ag(2,6-pyridinedicarboxylate)$_2\cdot$H$_2$O
Ag^{III}, d^8	4	Planar	AgF$_4^-$, half Ag atoms in AgO, [Ag(ebbg)$_2$]$^{3+\,d}$
	6	Octahedral	[Ag(IO$_6$)$_2$]$^{7-}$, Cs$_2$KAgF$_6$
Au^I, d^{10}	2[a]	Linear	[Au(CN)$_2$]$^-$, Et$_3$PAuC≡CPh; (AuI)$_n$
	3	Trigonal	AuCl(PPh$_3$)$_2$
	4	Tetrahedral	[Au(diars)$_2$]$^+$
Au^{II}, d^9	4	Square	[Au(mnt)$_2$]$^{2-}$
Au^{III}, d^8	4[a]	Planar	AuBr$_4^-$, Au$_2$Cl$_6$, [(C$_2$H$_5$)$_2$AuBr]$_2$, R$_3$PAuX$_3$, [AuPh$_4$]$^-$
	5	*tbp*	[Au(diars)$_2$I]$^{2+}$
		sp	AuCl$_3$(2,9-Me$_2$-1,10-phen), AuCl(TPP)
	6	Octahedral	AuBr$_6^{3-}$, *trans*-[Au(diars)$_2$I$_2$]$^+$
Au^V, d^6	6	Octahedral	[Xe$_2$F$_{11}$]$^+$[AuF$_6$]$^-$, [AuF$_5$]$_{2,3}$ (g)

[a]Most common states.
[b]L is an N$_5$ macrocycle:

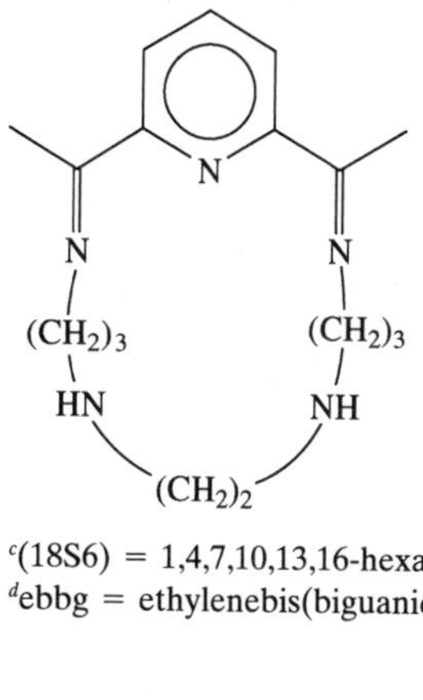

[c](18S6) = 1,4,7,10,13,16-hexathiacyclo-octadecane (M. Schröder *et al.*, *Polyhedron* **1989**, *8*, 513).
[d]ebbg = ethylenebis(biguanide).

Gold(III) also is invariably complexed in all solutions, usually as anionic species such as [AuCl$_3$OH]$^-$. Generally speaking, oxidation states, Ag^{II}, Ag^{III}, and Au^{III}, are either unstable to water or exist only in insoluble compounds or complexed species. Intercomparisons of the standard potentials are of limited utility, particularly since these strongly depend on the nature of the anion; some useful ones are

$$Ag^{2+} \xrightarrow{\ 1.98\ } Ag^+ \xrightarrow{\ 0.799\ } Ag$$

$$Ag(CN)_2^- \xrightarrow{\ -0.31\ } Ag + 2CN^-$$

$$AuCl_4^- \xrightarrow{\ 1.00\ } Au + 4Cl^-$$

$$Au(CN)_2^- \xrightarrow{\ -0.6\ } Au + 2CN^-$$

The chemistry of gold is more diversified than that of silver. Six oxidation states, from -I to III and V, occur in its chemistry. Gold(-I) and Au^V have no counterparts in the chemistry of silver. Solvated electrons in liquid ammonia can reduce gold to give the Au^- ion which is stable in liquid ammonia ($E^0 = -2.15$ V). In the series of binary compounds MAu (M = Na, K, Rb, Cs), the metallic character decreases from Na to Cs. CsAu is a semiconductor with the CsCl structure and is best described as an ionic compound, Cs^+Au^-. The electron affinity of gold (-222.7 kJ mol^{-1}) is comparable to that of iodine (-295.3 kJ mol^{-1}). Gold in the oxidation state -I is also found in the oxides $(M^+)_3Au^-O^{2-}$ (M = Rb, Cs); these, too, have semiconducting properties.[1]

The evidence for the existence of oxidation state +IV is not yet convincing. Gold(V) is found in the fluorides AuF_5 and AuF_6^-, in which it has a low-spin $[Xe]4f^{14}5d^6$ configuration. It is a powerful oxidizing agent, the strongest of the MF_6^- species. AuF_7 has been claimed but not yet confirmed.

The oxidation states and stereochemistries are summarized in Table 18-I-1.

18-I-1   The Elements

Silver and gold are widely distributed in Nature. They occur as metals and also as numerous sulfide ores, usually accompanied by sulfides of Fe, Cu, Ni, and so on. Gold, with its high density (19.3 g cm^{-3}), erodes through weathering and mechanical erosion and accumulates in placer deposits. The main deposits are in South Africa and Russia. Silver also occurs as *horn silver* (AgCl).

After flotation or other concentration processes, the crucial chemical steps are cyanide leaching and zinc precipitation, for example,

$$4Au + 8KCN + O_2 + 2H_2O = 4KAu(CN)_2 + 4KOH$$

$$2KAu(CN)_2 + Zn \longrightarrow K_2Zn(CN)_4 + 2Au$$

Silver and gold are normally purified by electrolysis.

---

[1]M. Jansen *et al., J. Am. Chem. Soc.* **1995,** *117,* 11749.

Silver is a white, lustrous, soft, and malleable metal (mp 961°C) with the highest known electrical and thermal conductivities. It is chemically less reactive than copper, except toward sulfur and hydrogen sulfide, which rapidly blacken silver surfaces. The metal dissolves in oxidizing acids (concentrated HNO_3) and in cyanide solutions in the presence of oxygen or hydrogen peroxide.

Gold is a soft, yellow metal (mp 1063°C) with the highest ductility and malleability of any element. It is chemically unreactive and is not attacked by oxygen or sulfur, but reacts readily with halogens or with solutions containing or generating chlorine such as aqua regia; it also dissolves in cyanide solutions in the presence of air or hydrogen peroxide to form $[Au(CN)_2]^-$. A solution of Cl_2 in acetonitrile in the presence of $[NMe_3H]Cl$ dissolves both silver and gold at 30°C at a faster rate than aqua regia, whereas I_2/KI in refluxing methanol dissolves only gold. The system I_2/NaI/acetone allows the extraction of Ag and Au from low-grade ores.[2] Gold extraction from metal sulfide mineral veins can also be facilitated by sulfur-consuming bacteria.

The reduction of solutions of $AuCl_4^-$ by various agents may, under suitable conditions, give highly colored solutions containing colloidal gold. Colloids of silver and gold are also obtained by ultrasound irradiation of $AgClO_4$, $AgNO_3$, or $HAuCl_4$ solutions. The reduction to the metal presumably involves H and OH radicals produced in the sonication process.[3] The "purple of Cassius" obtained using Sn^{II} reduction of gold salts is used for coloring ceramics.

18-I-2 Compounds of Silver(I)

This is the common oxidation state. The salts $AgNO_3$, $AgClO_3$, and $AgClO_4$ are water soluble, while Ag_2SO_4 and AgO_2CCH_3 are but sparingly so. The salts of oxo anions are primarily ionic. The water insoluble halides AgCl and AgBr have the NaCl structure, with appreciable covalent character in the Ag···X interactions. The pseudohalides AgCN, AgNCO, and AgSCN have chain structures:

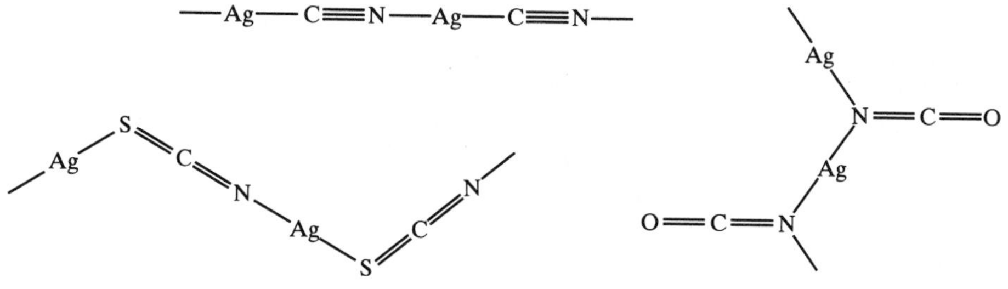

As illustrated in Table 18-I-1, Ag^I and Au^I show a pronounced tendency to exhibit linear twofold coordination.

[2]Y. Nakao, *J. Chem. Soc., Chem. Commun.* **1992**, 426.
[3]Y. Nagata *et al., J. Chem. Soc., Chem. Commun.* **1992**, 1620; F. Grieser *et al., ibid.* **1993**, 378.

Binary Compounds

Silver(I) Oxide. The addition of alkali hydroxide to Ag^+ produces a dark brown precipitate that is difficult to free from alkali ions. It is strongly basic, and its aqueous suspensions are alkaline:

$$\tfrac{1}{2}Ag_2O(s) + \tfrac{1}{2}H_2O = Ag^+ + OH^- \qquad \log K = -7.42 \; (25°C, 3M \; NaClO_4)$$

$$\tfrac{1}{2}Ag_2O(s) + \tfrac{1}{2}H_2O = AgOH \qquad \log K = -5.75$$

They absorb carbon dioxide from the air to give Ag_2CO_3. The oxide decomposes above ~160°C and is readily reduced to the metal by hydrogen. Silver oxide is more soluble in strongly alkaline solution than in water, and AgOH and $[Ag(OH)_2]^-$ are formed. The treatment of water-soluble halides with a suspension of silver oxide is a useful way of preparing hydroxides, since the silver halides are insoluble. Analogously to copper and gold, alkali metal silver oxides contain $Ag_4O_4^{4-}$ units.

Silver(I) Sulfide. The action of hydrogen sulfide on a silver(I) solution gives black Ag_2S, which is the least soluble in water of all silver compounds ($K_{sp} \approx 10^{-50}$). The black coating often found on silver articles is the sulfide; this can be readily reduced by contact with aluminum in dilute sodium carbonate solution.

Silver(I) Halides. The *fluoride* is unique in forming hydrates such as Ag-F·4H₂O, which are obtained by crystallizing solutions of Ag_2O in aqueous HF. The other well-known halides are precipitated by the addition of X^- to Ag^+ solutions; the color and insolubility in water increase Cl < Br < I.

Both AgCl and AgBr have a rock salt structure, though they are covalent insulators. AgI has both zinc blende and wurtzite structures with tetrahedral coordination about Ag.

Silver halides are light sensitive; this property forms the basis of *photography*. Photographic film contains a coating of a silver halide "emulsion" in gelatine. During the exposure process, some Ag^+ ions in an AgX crystal are photochemically reduced to Ag^0 (ligand-to-metal charge transfer), usually at crystal defects. On subsequent exposure to a reducing agent, such as hydroquinone ($HO-C_6H_4-OH$) ("developer"), those crystals containing seed silver atoms ("latent image") are reduced to silver particles, while reduction of crystals that have not been exposed to light is significantly slower; these are dissolved in a complexing agent, e.g., thiosulfate, ("fixer"):

$$AgX + 2S_2O_3^{2-} = [Ag(S_2O_3)_2]^{3-} + X^-$$

The result is a negative consisting of silver particles of varying density. Size and morphology of the AgX crystallites determine the sensitivity of the film, with larger crystallites giving highly sensitive (and more grainy) film; hence methods of crystal engineering have become a subject of intense research. AgBr is most commonly used, together with an organic sensitizer which allows photochemical excitation over a wide range of the visible spectrum. Highly sensitive films contain AgI.

Complexes of Ag^I

Nitrogen Ligands. Many of these form readily in aqueous solution, NH$_3$ being the most important of the nitrogen ligands. In general, large formation constants are seen only for the AgL$^+$ and AgL$_2^+$ complexes. This is due to the preference for linear coordination; the linear [H$_3$N—Ag—NH$_3$]$^+$ ion has been observed crystallographically in several compounds.

Since the formation of this ion is dependent on [Ag$^+$] and [NH$_3$], it can be used for the separation of silver halides. The solubility products of AgX at 25°C are:

$$X = Cl \qquad 1.77 \times 10^{-10}$$

$$X = Br \qquad 5.35 \times 10^{-13}$$

$$X = I \qquad 8.51 \times 10^{-17}$$

The solubility of AgCl is therefore sufficient to give [Ag(NH$_3$)$_2$]Cl on treatment with aqueous ammonium carbonate, while AgBr dissolves only in aqueous ammonia, in which AgI is poorly soluble. Aqueous pyridine and substituted pyridines form [Ag(py)]$^+$ and [Ag(py)$_2$]$^+$ ions, but in non-aqueous conditions tetrahedral complexes, such as [Ag(py)$_4$]ClO$_4$, may be obtained. The tetrahedral acetonitrile adduct [Ag(NCMe)$_4$]$^+$ is also known and is quite stable. There are a number of argentate(I) complexes, such as [Ag(NCO)$_2$]$^-$ and [Ag(ONO$_2$)$_2$]$^-$; they are linear, with 2-coordinate Ag$^+$.[4] Linear coordination is also found in the tetrameric silver amide [Ag{N(SiMe$_3$)$_2$}]$_4$.[5]

Halogeno Complexes. The halides AgX react with halide anions X$^-$ to give a series of anionic complexes [Ag$_k$X$_l$]$^{(l-k)-}$, with relative stabilities I$^-$ > Br$^-$ > Cl$^-$. The formation constants of [AgX$_2$]$^-$ ions vary considerably with the solvent and are quite low in water. Nonetheless, AgCl is about 100 times more soluble in 1 *M* HCl than in pure water.

The type of complex anion formed is strongly influenced by the countercation and the nature of the halide, and there is considerable structural diversity. The ions [X—Ag—X]$^-$ (X = Cl, Br) are linear,[6] whereas [Ag$_2$X$_4$]$^{2-}$ (X = Cl, Br, I) and [AgI$_3$]$^-$ contain 3-coordinate silver.[7] Examples of anionic iodo complexes are shown in Figure 18-I-1.[8]

[4]R. D. Gillard *et al.*, *Polyhedron* **1990**, *9*, 2127.
[5]M. F. Lappert *et al.*, *Chem. Commun.* **1996**, 1189.
[6]G. Helgesson and S. Jagner, *Inorg. Chem.* **1991**, *30*, 2574.
[7]G. A. Bowmaker *et al.*, *J. Chem. Soc., Dalton Trans.* **1990**, 727.
[8]S. Jagner *et al.*, *Inorg. Chim. Acta* **1994**, *217*, 15.

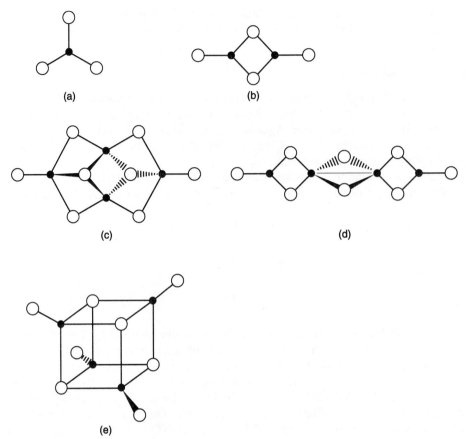

Figure 18-I-1   Structures of the anionic iodoargentate complexes (a) $[AgI_3]^{2-}$; (b) $[Ag_2I_4]^{2-}$; (c)–(e) isomers of $[Ag_4I_8]^{4-}$.

Cationic species such as Ag_2X^+ and $Ag_3X_2^+$ are formed when the silver halides are dissolved in aqueous solutions of $AgNO_3$ or $AgClO_4$.

Other Ligands. Silver(I) has a relatively low affinity for oxygen donors, although compounds and complexes containing carboxylate ions, DMSO, DMF, and crown ethers are known. However, it forms numerous complexes with the donor atoms S, Se, P, and As. With sulfur the thiosulfate complexes, $[Ag(S_2O_3)]^-$ and $[Ag(S_2O_3)_2]^{3-}$, are quite stable and silver chloride and bromide will dissolve in aqueous thiosulfate, thus providing a means of "fixing" photographic images (see above). Other important sulfur ligands for Ag^I are thiolate anions, which give oligomers, $(AgSR)_n$, dithiocarbamate ions, SCN^-, thioureas, and thioethers. Silver(I) binds to peptides and proteins, with a preference for –SR and imidazole nitrogen functionalities.

The dithiocarbamate ion $Pr_2NCS_2^-$ forms a hexanuclear complex with a trigonal-antiprismatic array of silver ions. Similar anionic clusters are also formed with the related $(NC)_2C=CS_2^{2-}$ ligand, e.g., $[Ag_6L_6]^{6-}$ and $[Ag_8L_6]^{4-}$. Four-, five-, and six-coordination is found in Ag^I complexes with sulfur macrocycles. For example, in [Ag(16S6)]ClO$_4$ only four of the six S atoms coordinate to give a tetrahedral geometry (16S6 = hexathiacyclohexadecane),[9] while [Ag(18S6)]$^+$ and [Ag(9S3)$_2$]$^+$ are octahedral. The latter complex can be oxidized to the silver(II) compound.[10] The reaction of Ag^I with tetrathiometallates MS_4^{2-} gives polymers [Ag(MS$_4$)]$_n^{n-}$ (M = Mo, W). In [NH$_4$]$_n$[Ag(WS$_4$)]$_n$ this polyanion consists of a chain of S-bridged eight-membered rings (18-I-I).[11]

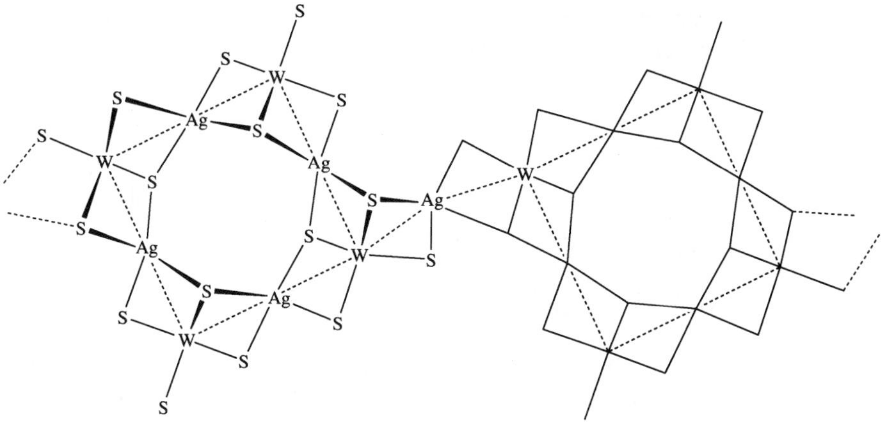

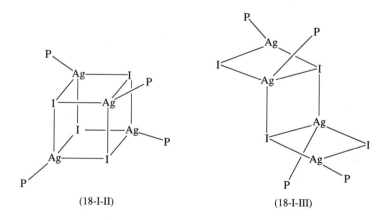

(18-I-I)

Numerous *phosphine complexes* (and some arsine and stibine analogues) are known. With monophosphines these are mainly of the type (R$_3$P)$_n$AgX, with n = 1–4. The 1:1 complexes are tetrameric, with either cubane (18-I-II) or chair (18-I-III) structures depending on the steric requirements of both X and R$_3$P; Ag$_4$I$_4$(PPh$_3$)$_4$ undergoes cube-chair isomerization, and the two structural types may be obtained by crystallization from different solvents.

(18-I-II)   (18-I-III)

[9]B. de Groot and S. J. Loeb, *Inorg. Chem.* **1991**, *30*, 3103.
[10]M. Schröder *et al.*, *Polyhedron* **1989**, *8*, 513; S. R. Cooper *et al.*, *Inorg. Chem.* **1989**, *28*, 4040.
[11]Q. Huang *et al.*, *Angew. Chem. Int. Ed. Engl.* **1996**, *35*, 868.

The complexes $AgX(PR_3)_2$ are generally dimeric with bridging X. Triphenylphosphine forms 4-coordinate $[Ag(PPh_3)_4]^+$. Slightly larger phosphines, such as $PCyPh_2$ and PCy_2Ph give 3-coordinate $[AgL_3]^+$,[12] while bulky phosphines such as PBu_3^t or $P(mes)_3$ form linear cations $[L-Ag-L]^+$. A comparison of the latter with the isostructural gold(I) complex shows that the covalent radius for Ag^I is ca. 6% larger than that for Au^I, 1.33 vs. 1.25 Å., in line with theoretical calculations which indicate the contraction of gold due to relativistic effects.[13]

Complexes with bridging bidentate phosphines tend to be dimers or tetramers with bridging phosphines, as in (18-I-IV). Phosphines with a large bite angle, as in (18-I-V), may however form mononuclear chelates.

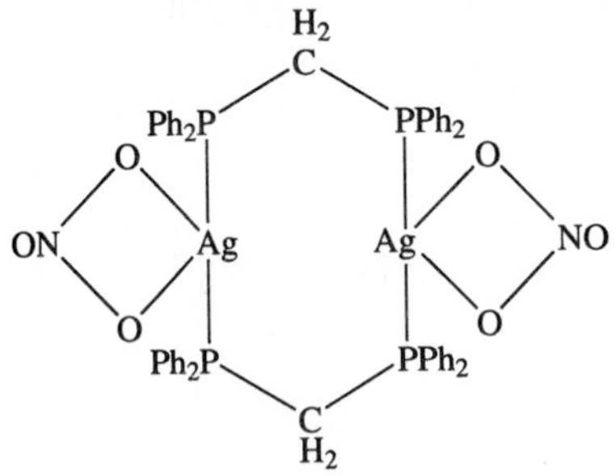

(18-I-IV)

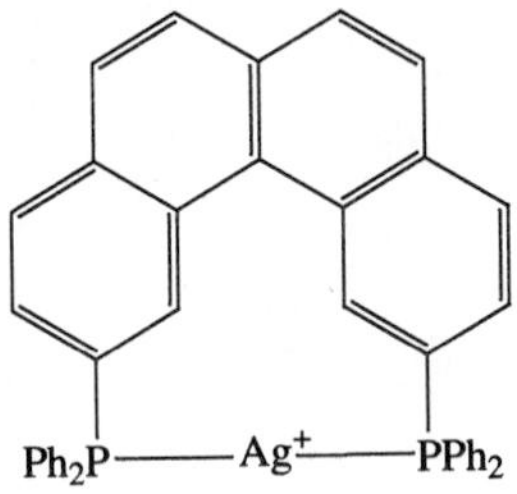

(18-I-V)

Halocarbons do not usually form stable complexes. However, in the presence of very weakly coordinating anions $[M(OTeF_5)_6]^{n-}$ ($n = 1$, M = Nb; $n = 2$, M = Ti), Ag^I becomes sufficiently electrophilic to form crystalline complexes with dihalomethanes, $[Ag(CH_2X_2)_3]^+$ (X = Cl, Br), in which the halocarbons act as chelating ligands. The Ag-Pd compound (18-I-VI) is another example.[14]

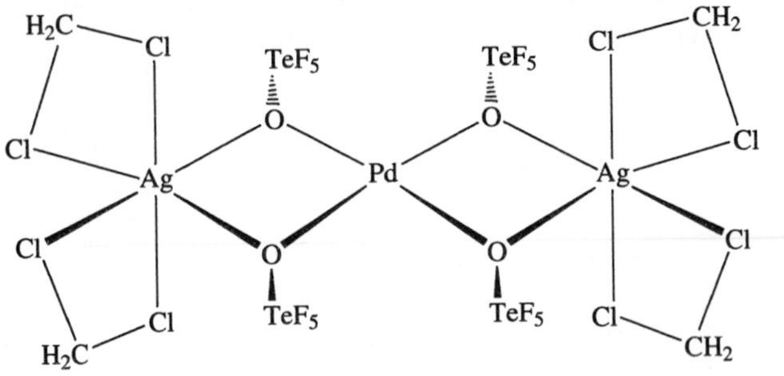

(18-I-VI)

Organosilver Compounds

There are strong similarities between the organometallic compounds of copper and silver. The simple alkyl complexes AgR are thermally unstable. The stability

[12]R. D. Willett et al., Inorg. Chem. 1990, 29, 4805.
[13]H. Schmidbauer et al., J. Am. Chem. Soc. 1996, 118, 7006.
[14]S. H. Strauss, Inorg. Chem. 1995, 34, 3453; J. Am. Chem. Soc. 1989, 111, 3762.

decreases with increasing chain length of R; thermolysis gives a mixture of alkenes and alkanes. Vinyl and aryl silver compounds are more readily accessible, usually by the reaction of a mild agent (ZnR_2, PbR_4) with silver nitrate:

$$AgNO_3 + ZnAr_2 \longrightarrow AgAr + ArZn(NO)_3$$

They are sensitive to light, air, and moisture. Alkyl compounds stabilized by donor ligands are significantly more stable, for example (18-I-VII)[15] and (18-I-VIII).[16] The linear L—Ag—C arrangement is typical.

L = 2-py, Ph_2P=S

(18-I-VII)   (18-I-VIII)

Silver perfluoroalkyl compounds are thermally much more stable. For example, $(CF_3)_2CFAg(NCMe)$ decomposes above 60°C. The adducts CF_3AgPR_3 are photosensitive but are oxidized by air to the stable Ag^{III} anion $[Ag(CF_3)_4]^-$.[17] Complexes with the strongly electron-withdrawing pentafluoro- and pentachlorophenyl ligands are stable to water and oxygen, e.g., the ylide adduct $(C_6F_5)Ag(CH_2PPh_3)$.[18]

Alkynylsilver compounds are readily accessible from $AgNO_3$ and $HC{\equiv}CR$ in methanol or water in the presence of base. They are thermally stable. Like many heavy metal acetylides, silver compounds of unsubstituted acetylene are shock sensitive and explosive.

With an excess of alkylating agents linear argentate(I) anions are formed, $[R—Ag—R]^-$. These may associate to clusters, such that as in $[Li_2Ag_3Ph_6]^-$, where three $[Ph—Ag—Ph]^-$ units are held together by Ph$\cdots$Li interactions.[19]

Complexes with alkenes and arenes are formed when the hydrocarbons are shaken with aqueous solutions of silver(I) salts. Di- or polyalkenes often give crystalline compounds with Ag^+ bound to one to three double bonds. The formation of alkene complexes of varying stability may be used for the purification of alkenes, or for the separation of isomeric mixtures (e.g., 1,3-, 1,4-, and 1,5-cyclooctadienes), or of the optical isomers of α- and β-pinene. There is very little back-bonding contribution in the formation of Ag^I π-complexes. For example, the planar complex $(hfa)Ag(Ph–C{\equiv}C–Ph)$ contains an almost linear acetylene ligand with a $C{\equiv}C$

[15]J. P. Fackler, Jr. *et al.*, *Organometallics* **1990**, *9*, 1973.
[16]J. Vicente *et al.*, *Organometallics* **1989**, *8*, 767.
[17]H. K. Nair and J. A. Morrison, *J. Organomet. Chem.* **1989**, *376*, 149.
[18]R. Usón *et al.*, *J. Chem. Soc., Dalton Trans.* **1988**, 701.
[19]R. Bau *et al.*, *J. Am. Chem. Soc.* **1985**, *107*, 1679.

bond that is only marginally longer than in free diphenylacetylene, while the Ag—C bonds are long (2.26 Å).[20] Arene ligands are η^2- and not η^6-bonded, in contrast to most transition metal arene complexes.

Until recently, isolable CO complexes of silver were unknown. However, if the counteranion possesses only very weak basicity, fairly stable CO adducts are obtained. Carbon monoxide acts in these compounds as a σ-donor ligand only. Back-bonding is virtually absent, and the C—O stretching frequencies are substantially higher than that of free CO (2143 cm^{-1}). The compounds [Ag(CO)]$^+$X$^-$ (ν_{CO} = 2204 cm^{-1}) and [OC—Ag—CO]$^+$X$^-$ (ν_{CO} = 2198 cm^{-1}) [X = B(OTeF$_5$)$_4$] were characterized by X-ray diffraction.[21] The [Ag(CO)$_3$]$^+$ ion is formed under CO pressure.

18-I-3 Compounds of Silver(II), d^9

There are only a few binary compounds. The black oxide AgO, obtained by ozonization of Ag$_2$O in water or by electrolysis of 2 M AgNO$_3$ solutions, is actually Ag^{I}Ag^{III}O$_2$. Silver(II) fluoride is obtained as a dark brown solid by fluorination of AgF or other Ag compounds at elevated temperatures, or from the reaction of AgF$_3$ with xenon:

$$2AgF_3 + Xe \longrightarrow 2AgF_2 + XeF_2$$

AgF$_2$ (d^9) is antiferromagnetic, with a magnetic moment well below that expected for one unpaired electron. It is a potent fluorinating agent and is rapidly hydrolyzed by water. Solutions of AgF$_2$ in liquid HF in the presence of Lewis acids such as SbF$_5$ or AsF$_5$ give deep blue salts [AgF][MF$_6$] which contain linear to kinked chains, [AgF]$_n^{n+}$, in which Ag^{II} is linearly coordinated to two F ligands. In Ag[SbF$_6$]$_2$, by contrast, Ag^{II} has a distorted octahedral geometry with four short and two long Ag—F contacts, as expected for a Jahn–Teller distorted d^9 system.[22] In liquid HF at −78°C the Ag^{II} ion oxidizes O$_2$ to O$_2^+$.

A mixed-valent Ag^{II,III} compound, Ag[AgF$_4$]$_2$, also exists; it is paramagnetic.[23]

Silver(II) fluorosulfate is made by heating Ag with S$_2$O$_6$F$_2$ at 70°C and is stable to 210°C.

The aqua ion, [Ag(H$_2$O)$_4$]$^{2+}$, which is paramagnetic with one unpaired electron, is obtained in HClO$_4$ or HNO$_3$ solution by oxidation of Ag$^+$ with ozone or by dissolution of AgO in acid.

The potentials for the Ag^{2+}/Ag$^+$ couple, + 2.00 V in 4 M HClO$_4$ and + 1.93 V in 4 M HNO$_3$, show that Ag^{2+} is a powerful oxidizing agent. There is evidence for complexing by NO$_3^-$, SO$_4^{2-}$, and ClO$_4^-$ in solution, and the electronic spectra in HClO$_4$ solutions are dependent on acid concentration. The ion is reduced by water, even in strongly acid solution, but the mechanism is complicated.

[20]K. M. Chi *et al., Organometallics* **1996,** *15,* 2660.
[21]S. H. Strauss *et al., Inorg. Chem.* **1993,** *32,* 373; *J. Am. Chem. Soc.* **1994,** *116,* 10003.
[22]N. Bartlett *et al., Inorg. Chem.* **1995,** *34,* 2692; *J. Fluorine Chem.* **1995,** *72,* 157.
[23]N. Bartlett *et al., J. Solid State Chem.* **1992,** *96,* 84.

Although the disproportionations

$$2Cu^+ = Cu + Cu^{2+} \text{ and } 3Au^+ = 2Au + Au^{3+}$$

have been long known, only recently has a ligand-induced disproportionation of Ag^+ been discovered. The low heat of hydration of Ag^{2+} makes the equilibrium unfavorable in water.

$$2Ag^+ = Ag + Ag^{2+} \qquad K \approx 10^{-20}$$

However, using a tetraazamacrocyclic ligand (L), this has been achieved in both H_2O and CN^- solutions, and a paramagnetic complex AgL^{2+} has been isolated. Normally, however, Ag^{II} species can be obtained only by oxidation.

Many oxidations (e.g., of oxalate) by the peroxodisulfate ion are catalyzed by Ag^+ ion, and the kinetics are best interpreted by assuming initial oxidation to Ag^{2+}, which is then reduced by the substrate. Decarboxylation of carboxylic acids are also promoted by Ag^{II} complexes, such as (18-I-IX) and others.

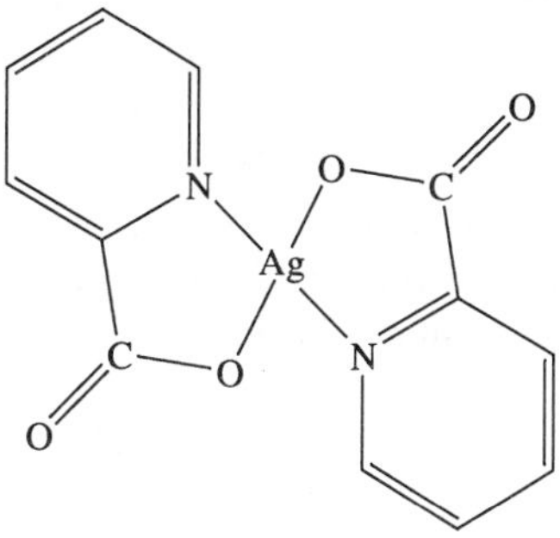

(18-I-IX)

With neutral ligands, cationic species such as $[Ag(py)_4]^{2+}$, $[Ag(bipy)_2]^{2+}$, and $[Ag(phen)_2]^{2+}$ form crystalline salts, whereas with uninegative chelating ligands such as 2-pyridinecarboxylate, neutral species such as (18-I-IX) are obtained. In acid solution the $[Ag(bipy)_2]^{2+}$ ion is in equilibrium with the $[Ag(bipy)]^{2+}$ ion, and it is the latter that oxidizes H_2O_2 in acid solution.

The Ag^{II} complexes have $\mu_{eff} = 1.75$ to 2.2 BM, consistent with the d^9 configuration, and their electronic spectra accord with square coordination. The salt $[Ag(py)_4]S_2O_8$ and bispicolinate (18-I-IX) are isomorphous with the planar copper(II) analogues.

Some neutral Ag^{II} porphyrin complexes, which tend to dimerize in solution, are also known. Silver(II) phthalocyanins are very stable and can be oxidized electrochemically to Ag^{III} phthalocyanine; further oxidation gives a Ag^{III} phthalocyanine cation radical.[24] The oxidation of Ag^I complexes of sulfur macrocycles also leads reversibly to Ag^{II} compounds; for example, $[Ag(15S5)]^+$ in acetonitrile shows a redox couple at $E_{1/2} = 0.76$ V $vs.$ Cp_2Fe/Cp_2Fe^+.[25]

[24]A. B. P. Lever et al., Inorg. Chem. 1990, 29, 4090.
[25]M. Schröder et al., J. Chem. Soc., Dalton Trans. 1993, 521.

18-I-4   Compounds of Silver(III), d^8

The anodic oxidation of neutral aqueous solutions of $AgClO_4$ or $AgBF_4$ yields Ag_2O_3 as black, metallic shining crystals which contain square-planar Ag^{III}. Oxidation at lower potentials leads to the silver(II, III) oxide Ag_3O_4.[26] As mentioned previously, Ag^{III} is present in AgO, where it has square coordination while the Ag^I coordination is linear. When AgO is dissolved in acid a comproportionation occurs,

$$AgO^+ + Ag^+ + 2H^+ \longrightarrow 2Ag^{2+} + H_2O$$

but, in the presence of complexing agents in alkaline solution, Ag^{III} complexes (see later) are obtained. The separation of Ag^I and Ag^{III} can be effected by the reaction

$$4AgO + 6KOH + 4KIO_4 \longrightarrow 2K_5H_2[Ag^{III}(IO_6)_2] + Ag_2O + H_2O$$

Unusual salts of stoichiometry $Ag_7O_8^+HF_2^-$, which are obtained as black needles by electrolysis of aqueous solutions of AgF, contain a polyhedral ion, $[Ag_6O_8]^+$, that acts as a clathrate to enclose Ag^+ and HF_2^-. The salts thus contain Ag^I and Ag^{III} with an average oxidation state of $2^3/_7$. The oxidation of $AgNO_3$ solutions gives $Ag_7O_8(NO_3)$.[26] The fluorination of $KNO_3/AgNO_3$ mixtures gives $K[AgF_4]$, which contains the diamagnetic square-planar d^8 ion AgF_4^-. Analogous AuF_4^- salts are isomorphous. By contrast, Cs_2KAgF_6 contains the octahedral high-spin ion AgF_6^{3-}.[27]

The addition of BF_3, PF_5, or AsF_5 to AgF_4^- salts gives AgF_3 as a red, diamagnetic solid:

$$AgF_4^- + BF_3 \longrightarrow AgF_3 + BF_4^-$$

The compound is thermodynamically unstable and in HF solution at 20°C slowly releases F_2 to give Ag_3F_8. The coordination geometry of Ag in AgF_3 is an elongated octahedron, with Ag—F distances of *ca.* 1.9 Å (equatorial) and 2.54 Å (axial).[29]

Complexes of Silver(III)

The anodic oxidation of silver metal in strong KOH solution gives the yellow tetrahydroxoargentate(III) ion $[Ag(OH)_4]^-$, which has a half-life of ~100 min in 1.2 *M* NaOH, (<30 min in 0.1 *M* NaOH), decomposing to AgO and O_2. The anion reacts rapidly with tellurate or periodate to give very stable complexes, which are usually obtained by direct oxidation of Ag^+ alkaline solutions by $K_2S_2O_8$ in the presence of periodate or tellurate ions. Representative are the ions $\{Ag[TeO_4(OH)_2]_2\}^{5-}$ and $\{Ag[IO_5(OH)]_2\}^{5-}$; copper(III) complexes (Section

[26]B. Standke and M. Jansen, *Angew. Chem. Int. Ed. Engl.* **1985**, *24*, 118; **1986**, *25*, 77.
[27]N. Bartlett *et al.*, *J. Am. Chem. Soc.* **1991**, *113*, 4192.

(18-I-X)   (18-I-XI)

17-H-4) are analogous, and all the ions appear to have Te—OH or I—OH groups as in (18-I-X).

In acid solution, the AgO^+ ion is extensively hydrolyzed even in 1.5 to 6 M acid. An Ag^{III} complex of remarkable stability is formed with ethylenebis(biguanide) (18-I-XI), which is obtained as the red sulfate when Ag_2SO_4 is treated with aqueous potassium peroxodisulfate in the presence of ethylenebis(biguanidinium) sulfate. The complex is diamagnetic and a good oxidizing agent.[28] Other comparatively stable complexes are Ag^{III} porphyrins and phthalocyanines (see Ag^{II}).

18-I-5   Subvalent Silver Compounds and Silver Clusters

Although Ag^+ contains a filled d^{10} shell, the structures of silver compounds can in many instances be understood only if one assumes the existence of weak attractive Ag^+—Ag^+ interactions. Similar attractive forces have been shown to be responsible for the formation of gold cluster compounds. In Ag^+ solid state compounds the unoccupied $5s$ and $5p$ orbitals in such assemblies give rise to empty bands which may be partially occupied, thus resulting in subvalent silver compounds; examples are Ag_3O and Ag_2F. The compound Ag_5SiO_4, prepared from Ag and SiO_2 at 350°C under a high pressure of oxygen, contains $[Ag_6]^{4+}$ clusters (average oxidation state 2/3) and should be formulated at $[Ag_6]^{4+}(Ag^+)_4(SiO_4^{4-})_2$.[29]

Silver and gold form numerous cluster compounds in which the metals are zero valent or subvalent. Examples are $Ag_6Fe_3(CO)_{12}\{HC(PPh_2)_3\}$ and the paramagnetic cluster anion $[Ag_{13}Fe_8(CO)_{32}]^{4-}$.[30] The latter contains the centered icosahedral Ag_{13} unit which has also been found as the basic building block of a series of "superclusters," for example, $Ag_{20}Au_{18}Cl_{14}(PPh_3)_{12}$, $[Ag_{19}Au_{18}Br_{11}\{P(p\text{-}tolyl)_3\}_{12}]^{2+}$, and $[Ag_{12}Au_{13}Br_8(PPh_3)_{10}]^+$;[31] the structure of the $Ag_{12}Au_{13}$ skeleton is shown in (18-I-XII).

[28]E. S. Gould et al., Inorg. Chim. Acta. 1994, 225, 83; S. Mukhopadhyay et al., J. Chem. Soc., Dalton Trans. 1994, 1349.
[29]C. Linke and M. Jansen, Inorg. Chem. 1994, 33, 2614.
[30]V. G. Albano et al., J. Am. Chem. Soc. 1992, 114, 5708.
[31]B. K. Teo et al., J. Am. Chem. Soc. 1987, 109, 3494; Angew. Chem. Int. Ed. Engl. 1987, 26, 897; J. Am. Chem. Soc. 1991, 113, 4329; Angew. Chem. Int. Ed. Engl. 1992, 31, 445.

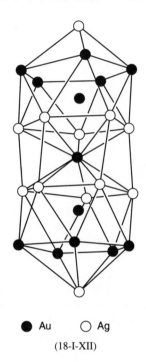

● Au   ○ Ag

(18-I-XII)

18-I-6   Compounds of Gold(I), d^{10}

Simple Compounds

No Au^I oxide is known although AuO contains Au^I along with Au^{III}. The gold sulfide (Au_2S) is a very insoluble, luminescent solid. The three gold(I) halides, AuCl, AuBr, and AuI, are all well characterized. They form chain structures with $X-Au-X$ groups linked by angular (72–90°) $Au-X-Au$ bridges.

Complexes

Most Au^I complexes are linear and of the type $L-Au-X$, where L is a 2-electron donor ligand (PR_3, R_2S, CO) and X is a halogen or pseudohalogen. The phosphine complexes are particularly stable, whereas R_2S is more readily displaced; hence R_2S complexes are useful preparative intermediates.

With higher concentrations of R_3P ligands higher coordination numbers are achieved, as in $(Ph_3P)_3AuCl$, $(Ph_3P)_3AuSCN$, and $[Au(PMePh_2)_4][PF_6]$, where tetrahedral coordination is found. The chelate complex $[Au(dppe)_2]Cl$ is active against certain types of cancer cells, possibly because of gold coordination to thiolate groups of proteins involved in gene transcription.[32]

Dinuclear phosphine complexes, such as $[Au_2(dppm)_2]^{2+}$ as well as tri- and tetranuclear analogues, contain linear $P-Au-P$ moieties and exhibit pronounced photoluminescence in solution.[33] A particularly striking example was recently found

[32]S. P. Frickler, Gold Bull. **1996**, *29*, 53.
[33]C. M. Che *et al., J. Chem. Soc., Chem. Commun.* **1989,** 885; *J. Chem. Soc., Dalton Trans.* **1993,** 189, 195.

for crystals of (18-I-XIII) which consist of infinite trigonal-prismatic stacks of Au_3 units. Following irradiation with near-UV light, contact of the crystals with organic solvents triggers intense yellow luminescence. The effect is probably due to the unique columnar solid-state structure and is not shown by closely related Au_3 derivatives which give a different crystal packing.[34]

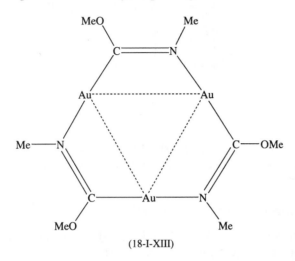

(18-I-XIII)

The aquo complex $[Au(OH_2)(PEt_3)]^+$ is generated by treating $AuCl(PEt_3)$ with $AgNO_3$ in aqueous solution. The water ligand is readily displaced by X^- to give $AuX(PEt_3)$ (X = Br, CN, SCN).[35] Thiocyanate complexes of Au^I are S-bound. Nitriles form labile complexes; e.g., $[Au(NCPh)_2]^+$ reacts with ammonia gas in acetonitrile solution to give $[H_3N—Au—NH_3]^+$.[36]

The complex anions $[AuX_2]^-$ (X = Cl, Br, I, N_3, SCN) are known from numerous salts; $[AuI_2]^-$ forms when gold is oxidized by I_3^-/I^- solutions:[37]

$$2Au + I_3^- + I^- = 2[AuI_2]^-$$

Unless excess halide is present $[AuCl_2]^-$ and $[AuBr_2]^-$ disproportionate in aqueous solution:

$$3AuX_2^- = AuX_4^- + 2Au + 2X^-$$

Gold(I) compounds of sulfur ligands are well known as drugs for the treatment of rheumatoid arthritis, for example, (triethylphosphine)(thioglucose)gold(I)[34] (18-I-XIV) ("Auranofin") and gold(I)thiomalate $[NaO_2CCH_2CH(CO_2Na)SAu]_n$ ("Myocrisin"). For the latter, a polymeric structure with linear S—Au—S units has been suggested[38] although it may in fact be tetrameric.[39] It appears that the S— ligands in these drugs are readily exchanged *in vivo* and Au^I binds to —SH

[34]L. H. Gade, *Angew. Chem. Int. Ed. Engl.* **1997**, *36*, 1171; A. L. Balch *et al., ibid.,* **1997**, *36*, 1179.
[35]M. M. El-Etri and W. M. Scovell, *Inorg. Chem.* **1990**, *29*, 480.
[36]D. M. P. Mingos *et al., J. Chem. Soc., Dalton Trans.* **1995**, 319.
[37]Y. Nakao and K. Sone, *Chem. Commun.* **1996**, 897.
[38]W. E. Smith *et al., Inorg. Chem.* **1990**, *29*, 5190.
[39]C. J. L. Lock *et al., Chem. Commun.* **1996**, 1391.

and $-S-S-$ units of proteins such as blood serum albumin which transports gold compounds throughout the body.

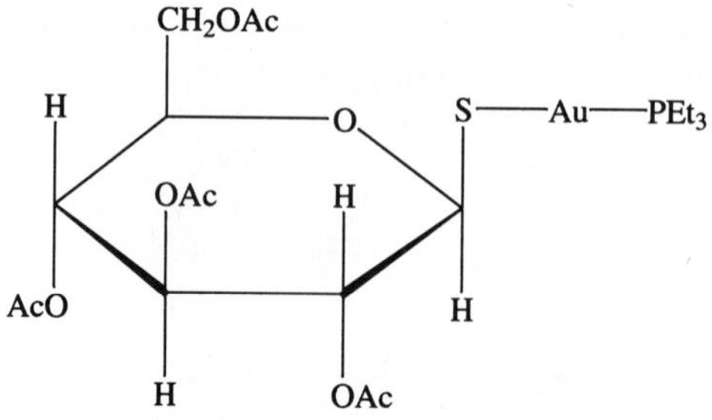

(18-I-XIV)

Gold thiolates formed from sulfurized terpene resins are used for decorating glass and china; they decompose to gold films on heating.

18-I-7 Compounds of Gold(II), d^9

A comparison of ionization potentials indicates that it is more difficult to obtain Au^{2+} than either Cu^{2+} or Ag^{2+}, whereas Au^{3+} forms more readily than the $+3$ ions of its lighter homologues. Genuine gold(II) compounds are therefore rare, and many formally divalent gold complexes are in fact mixed-valence Au^I/Au^{III} compounds, such as "$AuCl_2$" $= [Au^I(Au^{III}Cl_4)]_2$ (18-I-XV), $Au_2(SO_4)_2$, and the black complex halides $CsAuX_3$ ($= Cs_2Au^IAu^{III}X_6$, $X = Cl$, Br, I).[40] These mixed-valence compounds are diamagnetic. On the other hand, the nonstoichiometric compound $CsAu_{0.6}Br_{2.6}$, prepared by partial reduction of $AuBr_4^-$ in the presence of $CsBr$, is paramagnetic and has been shown by esr spectroscopy to contain octahedral Au^{II}.[41]

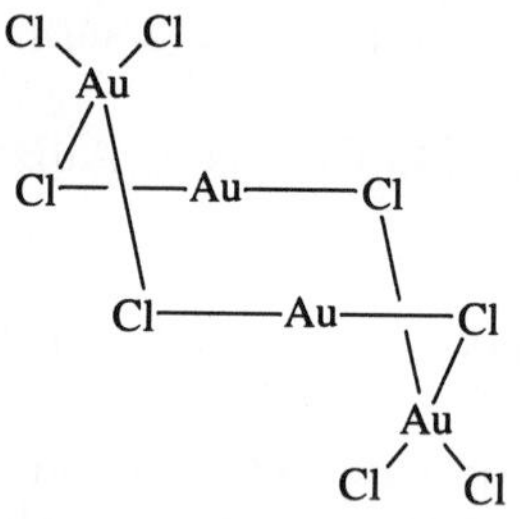

(18-I-XV)

Solvated Au^{2+} is obtained by reducing $Au(SO_3F)_3$ with gold powder in HSO_3F solution. Chelating ligands are able to stabilize Au^{II}, for example, in the monomeric sulfur macrocycle complex $[Au(9S3)_2](BF_4)_2$ ($9S3 = 1,4,7$-trithiacyclononane) which contains tetragonally Jahn-Teller distorted octahedral Au^{II}. The compound is air-stable and readily reduced to Au^I or oxidized to Au^{III}.[42]

[40]H. Kitagawa et al., J. Chem. Soc., Dalton Trans. **1991**, 3211.
[41]H. Kitagawa et al., J. Chem. Soc., Dalton Trans. **1991**, 3115.
[42]M. Schröder et al., Angew. Chem. Int. Ed. Engl. **1990**, 29, 197.

Monomeric Au^{II} has also been established for unsaturated S-donor ligands such as $R_2NCS_2^-$ and $^-S-C(CN)=(CN)-S^-$, and in the dicarbollyl complex $[Au(B_9C_2H_{11})_2]^{2-}$. Most Au^{II} compounds, however, contain metal—metal bonds.[43] Good examples are the phosphorus ylide complexes shown in Fig. 18-I-3 (see Section 18-I-10). The cation $[\{AuCl(dppn)\}_2]^{2+}$ contains a $Au^{II}—Au^{II}$ bond of 2.61 Å, unsupported by bridging ligands [dppn = 1, 8-bis-(diphenylphosphino)naphthalene].[44]

18-I-8 Compounds of Gold(III), d^8

Simple Compounds

These, except the halides, play a minor role in gold(III) chemistry. Gold trioxide (Au_2O_3) is obtained in hydrated form as an amorphous brown precipitate on addition of base to $AuCl_4^-$ solutions. It decomposes on heating to Au, O_2, and H_2O. It is weakly acidic and dissolves in excess strong base, probably as $[Au(OH)_4]^-$. Crystalline Au_2O_3, made by hydrothermal methods, contains square AuO_4 groups sharing O atoms. Au_2E_3 (E = S, Se, and Te) exist but their structures are unknown.

Halides

Gold(III) fluoride is best made by fluorination of Au_2Cl_6 at 300°C and forms orange crystals, which decompose to the metal at 500°C. It has a unique structure with square AuF_4 units linked into a chain by cis fluoride bridges; the F atoms of adjacent chains interact weakly with the axial sites.

The *chloride* and *bromide*, which form red crystals, are made by direct interaction, but Au_2Cl_6 is best made by the reaction

$$2H_3O^+AuCl_4^- + 2SOCl_2 = 2SO_2 + 6HCl + Au_2Cl_6$$

Both are dimers in the solid and in the vapor. They dissolve in water, undergoing hydrolysis, but in HX the ions AuX_4^- are formed. Gold(III) chloride is a powerful oxidant, being reduced to Au; it will, for example, oxidize ruthenium complexes in aqueous solution to RuO_4.

Complexes

Gold(III), like Pt^{II}, displays predominantly square coordination. Although there is little evidence for any persisting 5-coordinate complexes, there is no doubt that, as with Pt^{II} species, substitution reactions proceed *via* 5-coordinate intermediates. There is no evidence for a simple aqua ion, $[Au(H_2O)_4]^{3+}$, but mixed chloro-aqua and chloro-hydroxo complexes are formed by hydrolysis of $[AuCl_4]^-$. Distorted octahedral coordination occurs rarely, $[Au(en)_2Cl_2]^+$ being an example.

There are numerous complexes containing the $[AuX_4]^-$ ion (X = halogen or pseudohalogen). Treating a mixture of gold and an alkali metal chloride with BrF_3 gives M^IAuF_4.

[43]S. Ghosh *et al.*, *Inorg. Chem.* **1992**, *31*, 305.
[44]V. W. W. Yam *et al.*, *Chem. Commun.* **1996**, 1173.

When gold is dissolved in aqua regia or Au_2Cl_6 is dissolved in HCl and the solution of $AuCl_4^-$ is evaporated, *chloroauric acid* can be obtained as yellow crystals $[H_3O]^+[AuCl_4]^- \cdot 3H_2O$. Other water-soluble salts such as $KAuCl_4$ and $NaAuCl_4 \cdot 2H_2O$ are readily obtained. In water, hydrolysis to $[AuCl_3(OH)]^-$ occurs. From the dilute hydrochloric acid solutions the gold can be solvent-extracted with a very high partition coefficient into ethyl acetate or diethyl ether; the species in the organic solvent appears to be $[AuCl_3(OH)]^-$, which is presumably associated in an ion pair with an oxonium ion. Gold is readily recovered from such solutions (e.g., by precipitation with SO_2).

The $[AuBr_4]^-$ ion is similar to $[AuCl_4]^-$. Both are formed by the disproportionation of $[AuX_2]^-$ in aqueous solution in the absence of excess halide. The tetraiodoaurate ion, $[AuI_4]^-$, is obtained by oxidation of $[AuI_2]^-$ with iodine. It is unstable, and in acetonitrile the temperature-dependent reaction is reversible:

$$3[AuI_2]^- + 2I_2 \underset{82°C}{\overset{20°C}{\rightleftharpoons}} 2Au + [AuI_4]^- + 2I_3^-$$

Many neutral complexes can be obtained as simple adducts of R_3E (E = P, As, and Sb) with AuX_3, for example, $(Et_3P)AuCl_3$ and $(Et_3P)Au(CN)_3$. With bidentate chelating anionic ligands, species such as (18-I-XVI) can be obtained.

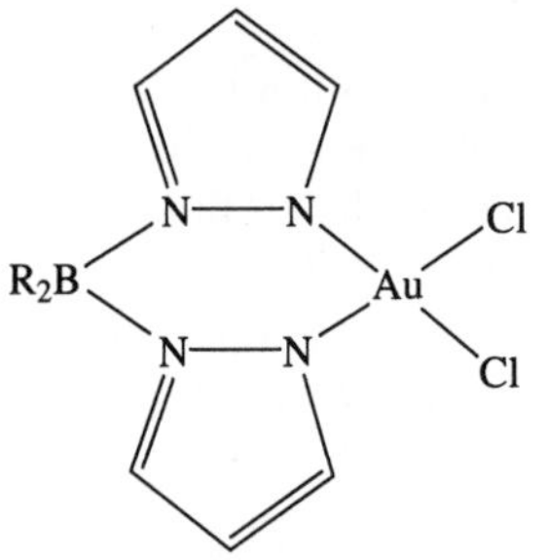

(18-I-XVI)

The cationic complex $[Au(NH_3)_4]^{3+}$ is well characterized and numerous others exist, for example, $[Au(py)_2Cl_2]^+$ and $[Au(dien)Cl]^{2+}$.

The mixed-valence compounds $CsAuX_3$ have been mentioned before. A different type of Au^I/Au^{III} complex is obtained if the trinuclear complexes $[Au^I(\mu\text{-}R_2pz)]_3$ are oxidized to the $Au_2^IAu^{III}$ complex $[Au(R_2pz)]_3Cl_2$ (R_2pz = 3,5-disubstituted pyrazolato anion) (18-I-XVII).[45]

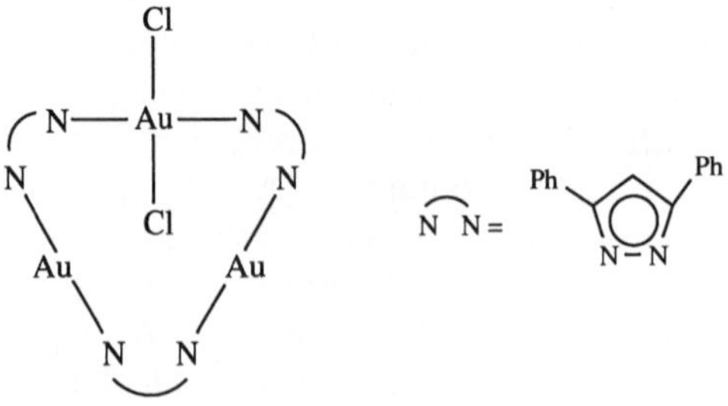

(18-I-XVII)

[45]R. G. Raptis and J. P. Fackler Jr., *Inorg. Chem.* **1990**, *29*, 5003.

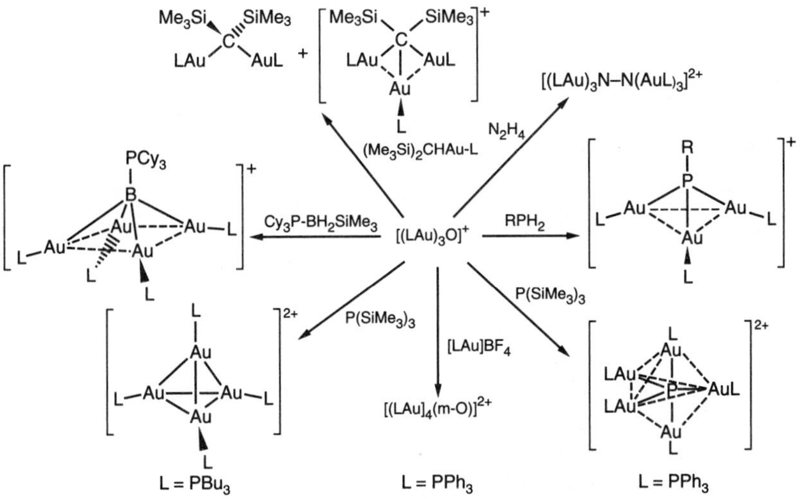

Figure 18-I-2 Formation of gold clusters from $[(L\text{-}Au)_3(\mu_3\text{-}O)]^+$.

18-I-9 Gold Clusters

Gold(I) forms numerous cluster compounds containing main group heteroatoms in which the gold units are held together by weak attractive forces. A good example is the oxide cluster $[(AuL)_3(\mu_3\text{-}O)]^+$ which has a trigonal-pyramidal structure. In the solid state two triangular units form a loosely associated dimer in which two vertices are connected by long Au⋯Au contacts (3.1 Å). The cluster is a useful starting material for a range of tri-, tetra-, and pentanuclear compounds, usually incorporating a heteroatom such as B, C, N, P, etc. (Fig. 18-I-2).[46,47]

A hexanuclear cluster in which a carbon atom is octahedrally surrounded by six Au—L units can be made by the reaction[48]

$$C[B(OMe)_2]_4 + 6LAuCl + 6CsF \longrightarrow [C(AuL)_6]^{2+}[MeOBF_3^-]_2 + 6CsCl + 2B(OMe)_3$$
$$(L = PPh_3)$$

Again, the cluster is held together by weak Au⋯Au interactions.

There are also numerous gold clusters without incorporated heteroatoms, of the general formula $[Au_{n+1}(PR_3)_{n-y}X_y] = [Au(AuL)_{n-y}(AuX)_y]$. They are usually made by reduction of L—AuX (X = Cl, Br, NO_3, etc.). A well-known example is $[Au_9(PPh_3)_8]^{3+}$ ($n = 8$; $y = 0$) (18-I-XVIII). Depending on the nature of the phosphine ligands, a green and a brown isomer may be isolated. If L = PPh_3, the green form (18-I-XVIII) is converted reversibly by high pressure (60 kbar) into the brown isomer (18-I-XIX).[49]

[46]H. Schmidbaur *et al.*, *J. Chem. Soc., Chem. Commun.* **1993**, 1005; *Organometallics* **1993**, *12*, 2408; *Angew. Chem. Int. Ed. Engl.* **1991**, *30*, 433; *Inorg. Chem.* **1993**, *32*, 3203; *Nature*, **1995**, *377*, 503.
[47]P. R. Sharp, *J. Am. Chem. Soc.* **1992**, *114*, 1526.
[48]H. Schmidbaur *et al.*, *Angew. Chem. Int. Ed. Engl.* **1988**, *27*, 1544; **1991**, *30*, 1488.
[49]J. R. Shapley *et al.*, *Inorg. Chem.* **1990**, *29*, 3900.

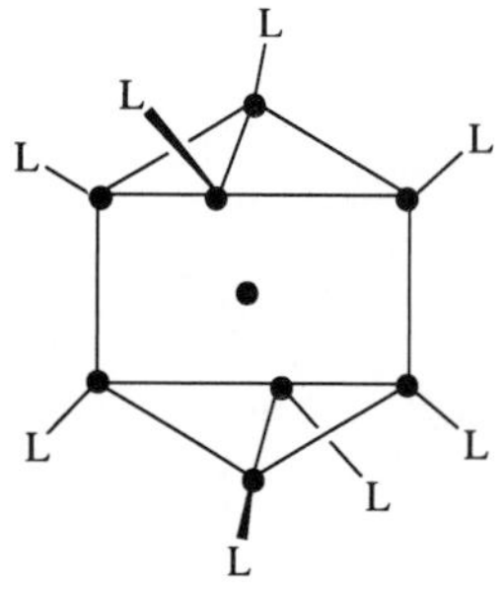

(18-I-XVIII) 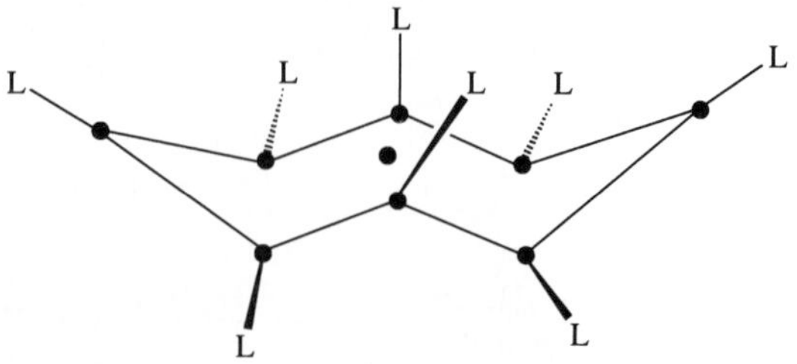   (18-I-XIX)

The cluster $[Au_{13}Cl_2(PMe_2Ph)_{10}](PF_6)_3$ has a centered icosahedral Au_{13} core (18-I-XX). Such an arrangement of 13 metal atoms is particularly stable and is also present in heteronuclear clusters, such as the $Ag_{12}Au_{13}$ cluster (18-I-XII) and in $[Ag_{20}Au_{18}Cl_{14}\{P(p\text{-}tolyl)_3\}_{12}]$ in which three M_{13} icosahedra share three vertices in a triangular array.[50] "Superclusters" of associated Au_{13} units have been identified by mass spectrometric techniques which confirm a cubic close-packed arrangement of Au_{13} units in $Au_{55}Cl_6(PPh_3)_{12}$ and have identified clusters of very high nuclearity in the gas phase, e.g. $(Au_{13})_n$, $n = 31, 39, 47,$ and $55.$[51] On the other hand, the compound $[Au_{39}Cl_6(PPh_3)_{14}]Cl_2$ consists of a $1:9:9:1:9:9:1$ layer structure of gold atoms.[52]

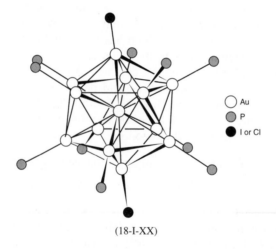

○ Au
● P
● I or Cl

(18-I-XX)

18-I-10 Organogold Compounds

Both gold(I) and gold(III) form σ-bonds to carbon. Olefin, carbene, and carbonyl complexes are also known.

Gold(I) alkyl complexes of the types R—Au—L (L = donor ligand such as Me_2S, PR_3, or isocyanide), $[R—Au—R]^-$, and $(RAu)_n$ are usually made by the

[50]B. K. Teo et al., J. Am. Chem. Soc. 1990, 112, 8552.
[51]A. Benninghoven et al., J. Am. Chem. Soc. 1990, 112, 8166.
[52]B. K. Teo et al., J. Am. Chem. Soc. 1992, 114, 2743.

reaction of gold halides with lithium or Grignard reagents. The linear anion $[AuMe_2]^-$ is fairly sensitive, whereas the C-bonded acac complex $[Au(acac)_2]^-$ is air-stable; it reacts with a variety of C-H acidic reagents HX [X = $CH(CN)_2$, $CH(COOMe)_2$, $CH(PPh_2)_2$, C≡CR, etc.] to give $[AuX_2]^-$.[53] The hexanuclear acetylide $[AuC≡CBu^t]_6$ has a cyclic structure containing $[Au(C_2R)_2]^-$ units, $[Au^+\{Au(C≡CR)_2^-\}]_3$.[54] Gold(I) aryl complexes are quite stable; $[Au(mes)]_5$ is a pentamer of structure (18-I-XXI), as is the analogous silver complex.[55] It reacts with Cl⁻ to give the mononuclear $[ClAu(mes)]^-$ anion which is oxidized by Cl_2 to the gold(III) complex $[Au(mes)Cl_3]^-$, without cleavage of the Au—C bond.[56]

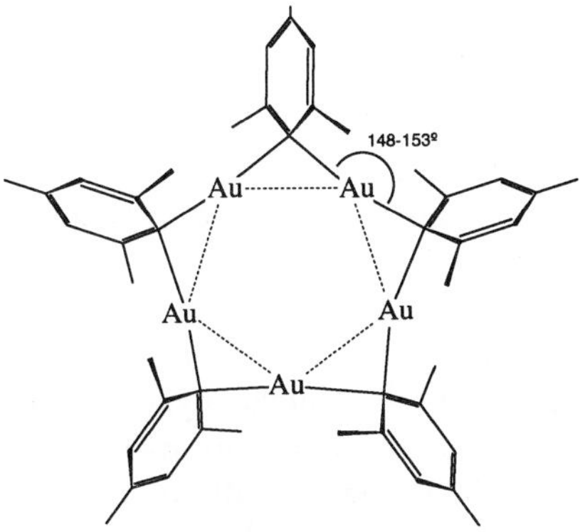

(18-I-XXI)

Gold(I) carbene complexes are readily obtained from heterocyclic aryl ligands:[57]

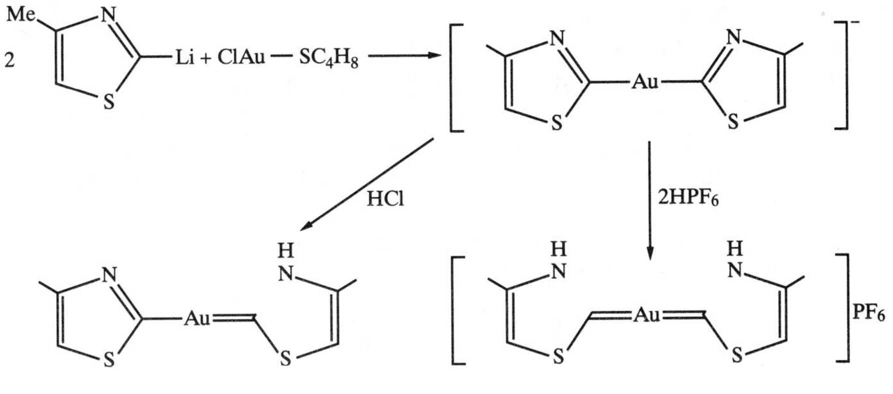

[53]J. Vicente *et al.*, *J. Chem. Soc., Chem. Commun.* **1992**, 915.
[54]D. M. P. Mingos *et al.*, *Angew. Chem. Int. Ed. Engl.* **1995**, *34*, 1894.
[55]C. Floriani *et al.*, *Organometallics* **1989**, *8*, 1067.
[56]A. Laguna *et al.*, *J. Chem. Soc., Dalton Trans.* **1994**, 2515.
[57]H. G. Raubenheimer *et al.*, *J. Chem. Soc., Chem. Commun.* **1990**, 1723.

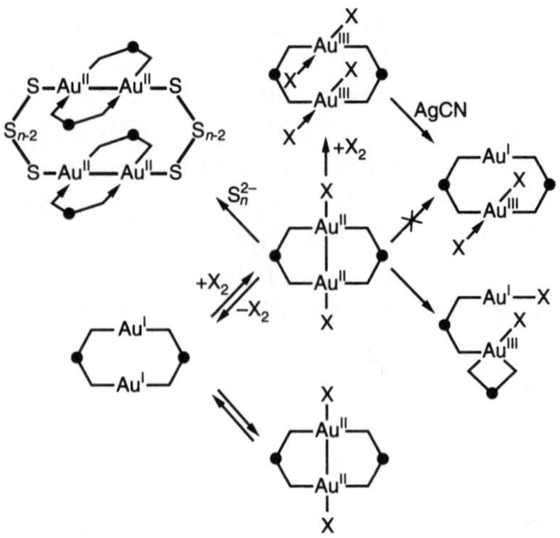

Figure 18-I-3 Chemistry of binuclear ylid complexes of Au^I, Au^{II}, and Au^{III}.   R = Me, CCl$_3$; X = Cl, Br, I; $\wedge$ = H$_2$CPR$_2$CH$_2$(R = Me, Ph)

A particularly interesting class are the binuclear ylid complexes $[Au_2\{\mu\text{-}(CH_2)_2PR_2\}_2]$ which react with Cl$_2$, Br$_2$, or I$_2$ reversibly to give Au—Au bonded Au^{II} alkyls; further oxidation leads to Au^{III} complexes without metal-metal bonds (Fig. 18-I-3).

The Au^I carbonyl complex AuCl(CO) has been known since 1925. The linear dicarbonyl cation $[Au(CO)_2]^+$ has recently been obtained as thermally stable UF$_6^{2-}$ and Sb$_2$F$_{11}^{2-}$ salts.[58] In these compounds back-bonding to CO is almost entirely absent, as shown by the high IR ν_{co} frequencies of 2152 cm^{-1} and 2217 cm^{-1} for AuCl(CO) and $[Au(CO)_2]^+$, respectively.

Gold(III) compounds are mainly square-planar complexes of the types R$_3$AuL, $[R_2AuL_2]^+$, and R$_2$AuXL. The reaction of Me$_3$Au(PPh$_3$) with methyllithium gives the very stable $[AuMe_4]^-$ anion. Treatment of $[Me_2Au(\mu\text{-}Br)]_2$ with aqueous AgNO$_3$ affords the stable aquo ion $[Me_2Au(H_2O)_2]^+$. Aryl gold(III) compounds are also well known; those containing the C$_6$F$_5^-$ group, e.g., $[Au(C_6F_5)_4]^-$, are particularly numerous and stable. Another notable group of Au^{III} organometallic compounds are those containing ylides, for example, $(CH_3)_3AuCH_2S(O)(CH_3)_2$ and $(CH_3)_3AuCH_2PPh_3$.

Although the formation of Au^{III} from Au^I compounds could be expected, oxidative-addition reactions such as

$$MeAu^IPR_3 + MeI = Me_2Au^{III}(I)PR_3$$

are slow and are complicated mechanistically. Additions to anionic Au^I species proceed more rapidly, for example,

$$RAuPEt_3 + R'Li(= Li[AuRR'PEt_3]) \xrightarrow{R''X} RR'R''AuPEt_3 + LiX$$

[58]M. Adelhelm *et al., Chem. Ber.* **1991,** *124,* 1559; H. Willner *et al., J. Am. Chem. Soc.* **1992,** *114,* 8972.

Additional References

R. C. Elder, *Chem. Rev.* **1987,** *87,* 1027. (Gold drugs).

C. E. Housecroft, *Coord. Chem. Rev.* **1995,** *142,* 101 (*Gold*); **1995,** *146,* 211 (*Silver*).

Clusters: B. K. Teo and H. Zhang, *Coord. Chem. Rev.* **1995,** *143,* 611.

Gold, Progress in Chemistry, Biochemistry and Technology, H. Schmidbauer, Ed. John Wiley, New York, 1999.

Chapter 19

THE GROUP 3 ELEMENTS
AND THE LANTHANIDES

19-1 Overview

The Group 3 elements are scandium, yttrium, and lanthanum. Strictly speaking actinium should also be included, but it is the general practice to associate it with the elements that follow it (the actinides) and treat them all separately, as we do in this book in Chapter 20.

There are fourteen elements that follow lanthanum and these are called the lanthanides. They are listed along with lanthanum in Table 19-1, where some of their principal characteristics are also listed. In passing from La to the last of the lanthanides, lutecium, Lu, the electron configuration goes from $[Xe]5d6s^2$ to $[Xe]4f^{14}5d6s^2$. That is, the fourteen $4f$ electrons are added, although as shown in Table 19-1, not without slight irregularities at several places. Among the Ln^{3+} ions (where Ln is a generic symbol for any one of the lanthanides) the progression is completely regular, from $4f^0$ (La), $4f^1$ (Ce), to $4f^{14}$ (Lu). The lanthanide metals, and La, are all relatively electropositive metals that strongly, although not exclusively, favor the tripositive oxidation state. Because of their very similar chemistries, they generally occur together in Nature and separating them is a non-trivial problem (see Section 19-3).

The lanthanide contraction is the name given to the phenomenon that is evident in the last column of Table 19-1, namely, that there is a steady decrease in the radii of the Ln^{3+} ions, from La to Lu, amounting overall to 0.17 Å. A similar contraction occurs for the metallic radii and is reflected in many smooth and systematic changes in the properties of the elements (Section 19-3). The lanthanide contraction also has consequences for all the elements following Lu. Thus, while the Zr atom is larger than the Ti atom with major consequences in differentiating their chemistries, Hf as well as Hf^{4+} are virtually identical in size to Zr and Zr^{4+} and their chemistries are very, very similar, although not identical. This sort of situation continues (i.e., for the pairs Nb/Ta, Mo/W, etc.) although the divergence in chemical properties progressively increases.

The lanthanide contraction used to be ascribed simply to the fact that while the $4f$ electrons have a considerable part of their wave function within the outer parts of the $4d$, $5s$, and $5p$ orbitals (i.e., the Xe core) thus making them unreactive, they still provide only incomplete shielding of the outer electrons from the steadily

Table 19-1 Some Properties of Lanthanide Atoms and Ions

Atomic number	Name	Symbol	Electronic configuration		$E^0(V)^a$	Radius M^{3+} (Å)
			Atom	M^{3+}		
57	Lanthanum	La	$5d6s^2$	[Xe]	−2.38	1.17
58	Cerium	Ce	$4f^15d^16s^2$	$4f^1$	−2.34	1.15
59	Praseodymium	Pr	$4f^36s^2$	$4f^2$	−2.35	1.13
60	Neodymium	Nd	$4f^46s^2$	$4f^3$	−2.32	1.12
61	Promethium	Pm	$4f^56s^2$	$4f^4$	−2.29	1.11
62	Samarium	Sm	$4f^66s^2$	$4f^5$	−2.30	1.10
63	Europium	Eu	$4f^76s^2$	$4f^6$	−1.99	1.09
64	Gadolinium	Gd	$4f^75d6s^2$	$4f^7$	−2.28	1.08
65	Terbium	Tb	$4f^96s^2$	$4f^8$	−2.31	1.06
66	Dysprosium	Dy	$4f^{10}6s^2$	$4f^9$	−2.29	1.05
67	Holmium	Ho	$4f^{11}6s^2$	$4f^{10}$	−2.33	1.04
68	Erbium	Er	$4f^{12}6s^2$	$4f^{11}$	−2.32	1.03
69	Thulium	Tm	$4f^{13}6s^2$	$4f^{12}$	−2.32	1.02
70	Ytterbium	Yb	$4f^{14}6s^2$	$4f^{13}$	−2.22	1.01
71	Lutetium	Lu	$4f^{14}5d6s^2$	$4f^{14}$	−2.30	1.00

^{a}Ln^{3+} + 3e = Ln.

increasing nuclear charge. Therefore, the electron cloud as a whole steadily shrinks as the $4f$ shell is filled. In recent years there has been theoretical work suggesting that relativistic effects may also play a role.

The term *rare earth* elements is sometimes applied to the elements La–Lu plus yttrium. The convenience of including La, which, strictly speaking, is not a lanthanide, is obvious. The reason for including Y is that Y has radii (atomic, metallic, ionic) that fall close to those of erbium and holmium and all of its chemistry is in the trivalent state. Hence it resembles the later lanthanides very closely in its chemistry and occurs with them in Nature.

Scandium, on the other hand, is far smaller than any of the lanthanides (Sc^{3+} radius, 0.89 Å) and its chemical behavior deviates in many ways from that of the lanthanides, being in significant ways similar to that of aluminum or gallium.

Although the +3 oxidation state is by far the most common one for the rare earth elements, for some of them others (+2, +4) are of importance. Cerium, and to a much lesser extent Pr and Tb, can form Ln^{4+} ions (formally speaking) but these are strongly oxidizing. Sm, Eu, and to a lesser extent Yb form Ln^{2+} ions. These deviations from "normal" behavior (i.e., formation of only Ln^{3+}) are sometimes attributed to the special stability of empty, half-filled or filled shells: Ce^{4+} ($4f^0$), Eu^{2+} ($4f^7$), Yb^{2+} ($4f^{14}$), but Pr^{4+} ($4f^1$) and Sm^{2+} ($4f^6$) do not fully satisfy this criterion. This idea is better considered as a mnenonic than as an explanation.

Because ligand field stabilization energies are very small for the Ln^{3+} ions, the thermodynamic properties of their compounds as well as their electrode potentials can be fairly accurately correlated by equations based solely on the electrostatic consequences of charge and size.

19-2   Coordination Numbers and Stereochemistry

Examples of the various coordination numbers and stereochemistries are presented in Table 19-2. The most common coordination numbers are 8 and 9. Many seeming examples of coordination number 6 are invalid because coordinated solvent molecules are present and raise the actual coordination number to 7, 8, or 9.

Table 19-2 Oxidation States, Coordination Numbers, and Stereochemistry of Lanthanide Ions

Oxidation state	Coordination number	Geometry	Examples
+2	6	NaCl type	EuTe, SmO, YbSe
	6	CdI$_2$ type	YbI$_2$
	6	Octahedral	Yb(PPh$_2$)$_2$(THF)$_4$
	7	Pent. bipyramid	SmI$_2$(THF)$_5$
	8	CaF$_2$ type	SmF$_2$
+3	3	Pyramidal	Ln[N(SiMe$_3$)$_2$]$_3$
	4	Distorted tetrahedral	La[N(SiMe$_3$)$_2$]$_3$OPPh$_3$
		Tetrahedral	[Lu(mes)$_4$]$^-$, [Y(CH$_2$SiMe$_3$)$_4$]$^-$
	6	Octahedral	[Er(NCS)$_6$]$^{3-}$, [Ln$_2$Cl$_9$]$^{3-}$, LnX$_6^{3-}$
	6	AlCl$_3$ type	LnCl$_3$ (Tb−Lu)
	6	Distorted trigonal prism	Pr[S$_2$P(C$_6$H$_{11}$)$_2$]$_3$
	7	Monocapped trigonal prism	Gd$_2$S$_3$, Y(acac)$_3$·H$_2$O
	7	ZrO$_2$ type	ScOF
	8	Distorted square antiprism	Y(acac)$_3$·3H$_2$O, La(acac)$_3$(H$_2$O)$_2$
	8	Dodecahedral	Cs[Y(CF$_3$COCHCOCF$_3$)$_4$]
	8	Distorted dodecahedral	Na[Lu(S$_2$CNEt$_2$)$_4$]
	8	Cubic	La[(bipyO$_2$)$_4$]$^{3+}$
	8	Bicapped trigonal prism	Gd$_2$S$_3$, LnX$_3$(PuBr$_3$ type)
	9	Tricapped distorted trigonal prism	[Nd(H$_2$O)$_9$]$^{3+}$, Y(OH)$_3$, K[La(EDTA)]·8H$_2$O, La$_2$(SO$_4$)$_3$·9H$_2$O
	9	Capped square antiprism	Ln(NO$_3$)$_3$(H$_2$O)$_3$[a]
	9	UCl$_3$ type	LaF$_3$, LnCl$_3$ (La−Gd)
	10	Complex	La$_2$(CO$_3$)$_3$·8H$_2$O
	10	Bicapped dodecahedron	Ce(NO$_3$)$_5^{2-}$, La(NO$_3$)$_3$(DMSO)$_4$
	11	Complex	La(NO$_3$)$_3$(H$_2$O)$_5$·H$_2$O
	12	Distorted icosahedron	[Pr(1,8-naphthyridine)$_6$]$^{3+}$, Ce(NO$_3$)$_6^{3-}$
+4	6	Octahedral	Cs$_2$CeCl$_6$
	8	Square antiprism	Ce(acac)$_4$
	8	Distorted square antiprism (chains)	(NH$_4$)$_2$CeF$_6$
	8	CaF$_2$ type	CeO$_2$
	10	Complex	Ce(NO$_3$)$_4$(OPPh$_3$)$_2$
	12	Distorted icosahedron	(NH$_4$)$_2$[Ce(NO$_3$)$_6$]

[a]X. Gan *et al., Polyhedron* **1996,** *15,* 2607.

The considerable change in size of the Ln^{3+} ions from La^{3+} (1.17 Å) to Lu^{3+} (1.00 Å) means that homologous compounds of lanthanides with appreciably different radii may differ in structure. Examples are the following:

(a) The anhydrous halides MCl_3 of La to Gd are all nine-coordinate with a UCl_3 type lattice; $TbCl_3$ and one form of $DyCl_3$ are of the $PuBr_3$ type, and the Tb—Lu chlorides have the octahedral $AlCl_3$ type structure.

(b) In the M_2S_3 series there are three main structure types: La to Dy (ortho-rhombic, 8- or 7-coordinate); Ho_2S_3 actually has half its atoms 6- and half 7-coordinate); Yb and Lu (corundum type, 6-coordinate).

(c) The enthalpies of sublimation of certain volatile chelates show irregular decreases from Pr to Lu, owing to structural differences.

(d) Solvated $LnCl_3$ compounds display great structural diversity.[1]

Aqua ions found in crystalline compounds are generally 9-coordinate with the tricapped trigonal prism being the favored structure.[2] The $[Ln(H_2O)_n]^{3+}$ ions in aqueous solution are either 8- or 9-coordinate but this may vary with ionic strength and concentration.[3] The number, in 1 M perchlorate solution, appears to change from 9 for the earliest (largest) Ln^{3+} ions to 8 for the latest (smallest), with a middle one such as Sm^{3+} having some of each.[4]

There are many lanthanide complexes with polydentate macrocyclic ligands (e.g., crown ethers) plus anions, H_2O, etc. with high coordination numbers.[5]

19-3 Sources, Extraction, Applications

The rare earths (RE) are actually far from rare.[6] The known reserves are more than 84×10^6 tons, a 2300 year supply at the present rates of consumption. Even the scarcest one, thulium, is more common than Bi, As, Cd, Hg, or Se. The largest reserves (51%) are in mainland China; other major deposits are in the USA (15%), Australia (6%), and India (3%).

Typical ores are *monazite* and *xenotime* which are lanthanide orthophosphates that also contain much thorium and *bastnasite*, which is, approximately, $LnF(CO_3)$. The relative amounts of the elements vary widely from ore to ore. The Chinese ores and concentrates, relatively new to the market, tend to be richer in some elements (Dy, Sm, Nd, Pr) than those previously in commerce.

Promethium is not available from RE ores. It occurs in Nature only in traces in uranium ores where it forms by spontaneous fission of ^{238}U. It was first isolated as ^{147}Pm by ion exchange methods from products of nuclear fission reactions. Isotopes from ^{140}Pm to ^{156}Pm are known, with ^{145}Pm being the longest lived ($t_{1/2} = 18$ y). Kilograms of ^{147}Pm ($t_{1/2} = 2.64$ y) have been isolated, but only about 100 mg can be safely handled in a glove box in the laboratory. Several dozen compounds have been structurally characterized.

[1]W. J. Evans *et al., Inorg. Chem.* **1995,** *34,* 576.
[2]T. Kurisaki *et al., J. Alloys. Comp.* **1993,** *192,* 293.
[3]H. Kanno and H. Yokoyama, *Polyhedron* **1996,** *15,* 1437.
[4]A. E. Merbach *et al., J. Am. Chem. Soc.* **1995,** *117,* 3790; *New J. Chem.* **1995,** *19,* 27.
[5]Y. Hirashima and G. Adachi, in *Cation Binding by Macrocycles,* Y. Inoue and G. W. Gokel, eds., Marcel Dekker, Inc., 1991.
[6]P. Falconnet, *J. Alloys. Comp.* **1993,** *192,* 114.

Some uses of the lanthanides, such as in certain steel alloys or as cigarette lighter "flints," require no separation of the metals as obtained from certain ores. In other applications, such as phosphors for television screens, medical immunoassays, X-ray photography, and, of course, the study of their chemistry, the individual elements must be separated.[7]

The first general separation procedures, introduced in the 1950s, were based on complexation-enhanced ion exchange processes. A cation exchange resin alone can provide separations since the affinities of the lanthanide ions for the resin vary inversely with their hydrated radii and these in turn vary inversely with the crystallographic radii. Thus, the order of elution is Lu···La. For a practical, large-scale process, this trend is enhanced by the use of complexing agents at the appropriate pH. The complexing agents prefer the smaller bare ions and thus enhance the tendency of the ions to elute in the order Lu···La. Among the useful ligands are α-hydroxyisobutyric acid, EDTA, and 2-hydroxy-EDTA. The eluates are then treated with oxalate ion and the precipitated RE oxalates ignited to the oxides.

In the mid-1960s liquid—liquid extraction processes were introduced and today all large-scale commercial production is done in this way. An aqueous solution of the Ln^{3+} ions is extracted in a continuous countercurrent process into a nonpolar organic liquid containing tri-n-butylphosphine oxide or bis(2-ethylhexyl)phosphinic acid (DEHPA). Typical separation factors for adjacent rare earths using DEHPA are 2.5 per extraction step so that under automatic multistep or countercurrent conditions purities of 99 to 99.9% are routinely achieved.

In general, lanthanides can be separated from mixtures with other elements by precipitation as oxalates or fluorides. Cerium and europium can conveniently be removed from the others, the former by oxidation to Ce^{IV} and precipitation as the iodate, and the latter by reduction to Eu^{2+}, which can be precipitated as $EuSO_4$.

19-4   The Metals

The lighter metals (La—Gd) are obtained by reduction of the trichlorides with Ca at 1000°C or more, whereas for others (Tb, Dy, Ho, Er, Tm, and also Y) the trifluorides are used because the chlorides are too volatile. Promethium metal is made by reduction of PmF_3 with Li. Trichlorides of Eu, Sm, and Yb are reduced only to the dihalides by Ca, but the metals can be prepared by reduction of the oxides M_2O_3 with La at high temperatures.

The metals are silvery-white and very reactive. They all react directly with water, slowly in the cold, rapidly on heating, to liberate hydrogen. Their high potentials (Table 19-1) show their electropositive character. They tarnish readily in air and all burn easily to give the sesquioxides, except cerium, which gives CeO_2. Yttrium is remarkably resistant to air even up to 1000°C owing to formation of a protective oxide coating. The metals react exothermically with hydrogen, though heating to 300 to 400°C is often required to initiate the reaction. The resulting phases MH_2 and MH_3, which are usually in a defect state, have remarkable thermal stability, in some cases up to 900°C. The metals also react readily with C, N_2, Si, P, S, halogens, and other nonmetals at elevated temperatures.

[7]G. Blasse, *J. Alloys. Comp.* **1993,** *192,* 17.

Metallic Eu and Yb dissolve in liquid ammonia at $-78°C$ to give blue solutions, golden when concentrated. The spectra of the blue solutions, which decolorize slowly, are those expected for M^{2+} and solvated electrons.

Unlike most of the Ln^{III} compounds, the metals themselves show some marked irregularities in their properties, especially those elements having the greatest tendency to have chemistry in the divalent state. An example is provided by the enthalpies of sublimation (Fig. 19-1), and another (closely related, of course) occurs in the range of vapor pressures, which vary irregularly over 9 orders of magnitude. The electric conductivities also show marked deviations from monotonic behavior.

Such irregularities arise from the changing tendency, as a function of atomic number, to delocalize valence shell electrons into the conduction band of the metal. The most deviant of the metals are Sm, Eu, and Yb, which are just the elements that most readily form Ln^{II} compounds and, correspondingly, in the metallic state donate only two electrons to the conduction band.

19-5 Magnetism and Spectra

In their magnetic and spectroscopic properties the lanthanides show important differences from the d-block elements. This happens because the $4f$ electrons are pretty well (although not totally) shielded from external fields by the overlying $5s^2$ and $5p^6$ shells. The states arising from the various $4f^n$ configurations therefore tend to remain nearly invariant for a given ion.

The states of the $4f^n$ configurations are all given, to a useful approximation, by the Russell-Saunders coupling scheme. In addition, the spin-orbit coupling constants are quite large (order of 1000 cm^{-1}). The result of all this is that with only a few exceptions, the lanthanide ions have ground states with a single well-defined value of the total angular momentum, J, with the next lowest J state at energies many times kT (at ordinary temperatures equal to $\sim$200 cm^{-1}) above, hence virtually unpopulated.

Thus the susceptibilities and magnetic moments should be given straightforwardly by formulas considering only this one well-defined J state, and indeed such

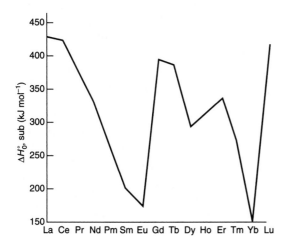

Figure 19-1 Enthalpies of sublimation of the lanthanide metals (data from E. D. Cater, *J. Chem. Educ.*, **1978,** *55,* 697).

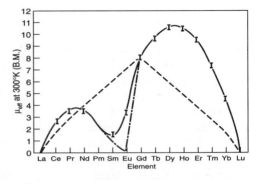

Figure 19-2 Measured and calculated effective magnetic moments of lanthanide M^{3+} ions at 300 K: I's are ranges of experimental values; solid curve gives values calculated for appropriate J ground states with allowance for the Sm and Eu anomalies; dash-dot curve gives values calculated without allowance for the Sm and Eu anomalies; dashed curve gives calculated spin-only values.

cotton/advances 6e 19FG02 s/s

calculations give results that are, with only two exceptions, in excellent agreement with experimental values (Fig. 19-2). For Sm^{3+} and Eu^{3+}, it turns out that the first excited J-state is sufficiently close to the ground state for this state (and in the case of Eu^{3+} even the second and third excited states) to be appreciably populated at ordinary temperatures. Since these excited states have higher J values than the ground state, the actual magnetic moments are higher than those calculated by considering the ground states only. Calculations taking into account the population of excited states afford results in excellent agreement with experiment (Fig. 19-2).

It should be emphasized that magnetic behavior depending on J values is qualitatively different from that depending on S values—that is, the "spin-only" behavior—which gives a fair approximation for many of the d-block transition elements. Only for the f^0, f^7, and f^{14} cases, where there is no orbital angular momentum ($J = S$), do the two treatments give the same answer. For the lanthanides the external fields do not either appreciably split the free-ion terms nor quench the orbital angular momentum.

Because the f orbitals are so well shielded from the surroundings of the ions, the various states arising from the f^n configurations are split by external fields only to the extent of ~100 cm^{-1}. Thus when electronic transitions, called f—f transitions, occur from one J state of an f^n configuration to another J state of this configuration the absorption bands are *extremely sharp*. They are similar to those for free atoms and are quite unlike the broad bands observed for the d—d transitions. Virtually all the absorption bands found in the visible and near-uv spectra of the lanthanide $+3$ ions have this linelike character. The intensities of the f—f bands show measurable sensitivity to the nature of the coordination sphere but the relationship is complex and not quantitatively understood.

The colors and electronic ground states of the M^{3+} ions are given in Table 19-3; the color sequence in the La to Gd series is accidentally repeated in the series Lu to Gd. As implied by the earlier discussion, insofar as the colors are due to the f—f transitions, they are virtually independent of the environment of the ions.

An important feature in the spectroscopic behavior is that of fluorescence or luminescence of certain lanthanide ions, notably Tb, Ho, and Eu. They are used commercially in oxide phosphors for television tubes and related devices. The luminescence of the Eu^{3+} ion can be used as a probe of its environment, giving

Table 19-3 Colors and Electronic Ground States of the M^{3+} Ions

Ion	Ground state	Color	Ion	Ground state
La	$^1S_0$	Colorless	Lu	$^1S_0$
Ce	$^2F_{5/2}$	Colorless	Yb	$^2F_{7/2}$
Pr	$^3H_4$	Green	Tm	$^3H_6$
Nd	$^4I_{9/2}$	Lilac	Er	$^4I_{15/2}$
Pm	$^5I_4$	Pink; yellow	Ho	$^5I_8$
Sm	$^6H_{5/2}$	Yellow	Dy	$^6H_{15/2}$
Eu	$^7F_0$	Pale pink	Tb	$^7F_6$
Gd	$^8S_{7/2}$	Colorless	Gd	$^8S_{7/2}$

information on ligand charges, binding constants, site symmetry, number of OH or NH bonds in the first coordination sphere, and ligand exchange rates.

Finally, it may be noted that several of the paramagnetic lanthanide ions, especially Pr^{3+}, Eu^{3+}, and Yb^{3+} are useful as nmr shift reagents. When an organic molecule with a complex nmr spectrum is coordinated to one of these ions, the large magnetic moment of the ion causes displacements and a spreading out of the spectrum and this often helps in assigning and interpreting the peaks.

19-6 Binary and Ternary Compounds

The trivalent state is the dominant one for all the lanthanides. They form oxides, M_2O_3, whose properties are like those of CaO and BaO; they absorb CO_2 and H_2O from the air to form carbonates and hydroxides, respectively.

The *hydroxides* $M(OH)_3$ are definite compounds, not merely hydrous oxides, and may be obtained crystalline by aging M_2O_3 in strong alkali at high temperature and pressure. They have hexagonal structures with nine-coordinate tricapped trigonal prismatic coordination. The basicities of the hydroxides decrease with increasing atomic number, as would be expected from the decrease in ionic radius. The hydroxides are precipitated from aqueous solutions by ammonia or dilute alkalis as gelatinous precipitates and are not amphoteric.

A range of mixed oxides, many with commercial uses, are known. Perhaps the most interesting are those containing La, Ba or Sr, and Cu, such as $La_{2-x}Sr_xCuO_4$, which are superconducting at and even above liquid nitrogen temperature.

Among the *halides*, the fluorides are of particular importance because of their insolubility. Addition of hydrofluoric acid or fluoride ions precipitates the fluorides from Ln^{3+} solutions even 3 M in nitric acid and is a characteristic test for lanthanide ions. The fluorides, particularly of the heavier lanthanides, are slightly soluble in an excess of HF owing to complex formation. They may be redissolved in 3 M nitric acid saturated with boric acid, which removes F^- as BF_4^-. The chlorides are soluble in water, from which they crystallize as hydrates, the La to Nd group often with $7H_2O$, and the Nd to Lu group (including Y) with $6H_2O$; other hydrates may also be obtained.

The anhydrous *chlorides* cannot generally be obtained simply by heating the hydrates because these lose hydrochloric acid (to give the oxochlorides, LnOCl) more readily than they lose water. The chlorides may be made by heating oxides with an excess of ammonium chloride,

$$Ln_2O_3 + 6NH_4Cl \xrightarrow{\sim 300°C} 2LnCl_3 + 3H_2O + 6NH_3$$

or as methanolates, $LnCl_3 \cdot 4CH_3OH$, by treating the hydrates with 2,2-dimethoxy-propane. They may be made as etherates, $LnCl_3(ether)_n$, by reaction of the oxides or carbonates with HCl produced *in situ* from $SOCl_2$ and H_2O in the presence of glyme ($MeOCH_2CH_2OMe$) under mild conditions.[8] The halides can also be prepared from the metals by action of HCl(g), Br_2, or I_2. At high temperatures they react with glass:

$$2LnX_3 + SiO_2 \longrightarrow 2LnOX + SiX_4$$

The oxohalides (LnOX) form readily when the halides are heated in the presence of water vapor. There are several structural types depending on mode of preparation and the Ln element, as well as a series with the composition M_3O_4X.[9]

Numerous other binary compounds are obtained by direct interaction at elevated temperatures; examples are the semiconducting sulfides Ln_2S_3, which can also be made by reaction of $LnCl_3$ with H_2S at 1100°C, Group 15 compounds LnX (X = N, P, As, Sb, or Bi) that have the NaCl structure, a variety of borides, carbides (LnC_2) and (Ln_2C_3), and hydrides that tend strongly to be nonstoichiometric, $LnH_{1.8}-LnH_{3.0}$.

19-7 Oxo Salts

Hydrated salts of common acids, which contain the ions $[Ln(H_2O)_n]^{3+}$, are readily obtained by dissolving the oxide in acid and crystallizing.

Double salts are very common, the most important being the double nitrates and double sulfates such as $2Ln(NO_3)_3 \cdot 3Mg(NO_3)_2 \cdot 24H_2O$, $Ln(NO_3)_3 \cdot 2NH_4NO_3 \cdot 4H_2O$, and $Ln_2(SO_4)_3 \cdot 3Na_2SO_4 \cdot 12H_2O$. The solubilities of double sulfates of this type fall roughly into two classes: the cerium group La to Eu, and the yttrium group Gd to Lu and Y. Those of the Ce group are only sparingly soluble in sodium sulfate, whereas those of the Y group are appreciably soluble. Thus a fairly rapid separation of the entire group of lanthanides into two subgroups is possible. Various of the double nitrates were used in the past for further separations by fractional crystallization procedures.

The precipitation of the *oxalates* from dilute nitric acid solution is a quantitative and fairly specific separation procedure for the lanthanides, which can be determined gravimetrically in this way, with subsequent ignition to the oxides. The actual nature of the oxalate precipitate depends on conditions. In nitric acid solutions, where the main ion is Hox^-, ammonium ion gives double salts $NH_4Ln(ox)_2 \cdot yH_2O$ ($y = 1$ or 3). In neutral solution, ammonium oxalate gives the normal oxalate with lighter, but mixtures with heavier, lanthanides. Washing the double salts with 0.01 M HNO_3 gives, with some ions, the normal oxalates. The phosphates are sparingly soluble

[8] D. B. Dell'Amico *et al., Inorg. Chim. Acta* **1995,** 240, 1.
[9] P. Hagenmuller *et al., J. Electrochem. Soc.* **1988,** 135, 2099.

in dilute acid solution. Although carbonates exist, many are basic; the normal carbonates are best made by hydrolysis of the chloroacetates:

$$2Ln(Cl_3CCO_2)_3 + (x + 3)H_2O \longrightarrow Ln_2(CO_3)_3 \cdot xH_2O + 3CO_2 + 6CHCl_3$$

19-8 Other Compounds and Complexes

The aqua ions hydrolyze, increasingly so from La to Lu as they become smaller.

$$[Ln(H_2O)_n]^{3+} + H_2O = [Ln(H_2O)_{n-1}(OH)]^{2+} + H_3O^+$$

Polymeric species such as $Ln_2(OH)_3^{3+}$ and $Ln_3(OH)_3^{6+}$ may also be present for La.

Halide complexes are of relatively little importance in aqueous media, but several types can be isolated in the solid state, for example, $(NH_4)_3LnCl_6$, Ph_4PLnBr_4, $Cs_3Ln_2Cl_9$, and so on.

Oxygen Ligands

By far the most stable and common of lanthanide complexes are those with chelating oxygen ligands. The use of EDTA-type anions and hydroxo acids such as tartaric or citric, for the formation of water-soluble complexes, is of great importance in ion-exchange separations, as noted previously. All of these can be assumed to have coordination numbers exceeding 6, as in $[La(OH_2)_4EDTAH] \cdot 3H_2O$.

The *β-diketonates* have been extensively studied, particularly since some of the fluorinated derivatives give complexes that are volatile and suitable for gas-chromatographic separation. The preparation of β-diketonates by conventional methods *invariably* gives hydrated or solvated species such as $[Ln(acac)_3] \cdot C_2H_5OH \cdot 3H_2O$. Anhydrous species obtained by vacuum dehydration appear to be polymeric, not octahedral. The neutral β-diketonates can complex further to give anionic species such as the eight-coordinate thenoyltrifluoroacetate $[Nd(TTA)_4]^-$ The alkali metal salts of $[Ln(\beta\text{-dike})_4]^-$ are sometimes appreciably volatile and can be sublimed.

Under highly basic conditions anionic catecholate complexes, for example, $[Gd(cat)_4]^{5-}$ and $[Gd_2(cat)_6]^{6-}$ in which the coordination numbers are 8 and 7, respectively, can be obtained as crystalline salts.

In aqueous solution many other types of polyoxo ligands effectively complex the Ln^{3+} ions, including polyphosphates, mixed phosphoryl-carbonyl species, and crown ethers.

Complexes of monodentate oxygen ligands are less stable than those of chelates and tend to dissociate in aqueous solution, but many crystalline compounds or salts have been obtained from the lanthanide salts in ethanolic solutions with hexamethylphosphoramide, which gives six-coordinate species, for example, $Pr(HMPA)_3Cl_3$ and $[Ln(HMPA)_6](ClO_4)_3$, with triphenylphosphine oxide or triphenyl arsine oxide, and pyridine-*N*-oxides, for example, $Ln(NO_3)_3(OAsPh_3)_4$ and $[Ln(pyO)_8](ClO_4)_3$, and with DMSO, for example, $(DMSO)_nM(NO_3)_3$.

Several aryloxo complexes are known such as $Sm(OAr)_3(THF)$ where Ar is very hindered[10] and $K[Ln(OAr)_4]$ (Ln = La, Nd, Er) where again Ar is very hindered.[11]

[10]Q. Shen *et al.*, *Polyhedron* **1995**, *14*, 413.
[11]D. L. Clark *et al.*, *Inorg. Chem.* **1994**, *33*, 5903.

A mixed siloxo/amido compound, $Y[OSiBu^tAr_2][N(Me_3Si)_2]_2$, has also been characterized.[12]

There are, of course, numerous complexes in which polydentate ligands with both oxygen and nitrogen donor atoms are present, generally giving high coordination numbers.[13] The ligand diethylenetriaminepentaacetic acid forms La^{3+} and Eu^{3+} complexes with a coordination number of 11.[14]

A lanthanide superoxide complex, $[HB(3,5-Me_2pz)_3]_2Sm(\eta^2-O_2)$, has recently been reported.[15] It is simply made by the reaction of the pyrazolylborato compound of Sm^{II} with O_2.

Nitrogen Ligands

Conventional complexes of *amines* such as en, dien, and dipy are known, together with *N*-bonded *thiocyanato* complexes. With few exceptions they have high coordination numbers. Examples of such complexes are $Ln\ en_3Cl_3$, $[La\ en_4(CF_3SO_3)]^{2+}$, $LaL(CF_3SO_3)_2]^+$, and $[YbL(CF_3SO_3)]^{2+}$ (where L represents a macrocyclic octaamine), in which coordination numbers of 9, 9, 10, and 9, respectively, are displayed. In aqueous solution thiocyanato complexes have appreciable formation constants, and SCN^- can be used as elutant for ion-exchange separations. The $[Ln(NCS)_6]^{3-}$ ions are octahedral.

Amido ligands with bulky substituents such as $(Me_3Si)_2N^-$ or $(Me_3Si)ArN^-$ form stable compounds with Sc^{III}, Nd^{III}, Eu^{III}, and Gd^{III} such as $[(Me_3Si)_2N]_3Ln(THF)$ and $[(Me_3Si)(C_6H_5)N]_3Nd(THF)$.[16]

Complexes with porphyrins, porphyrinogens,[17] and phthalocyanine[18] are known, with the latter being sandwich-like, $Li[(Pc)Ln(Pc)]$.

Sulfur and Phosphorus Ligands

There are relatively few Ln^{3+} complexes of sulfur ligands. The best known ones contain dithiocarbamates and dithiophosphinates and are prepared by reaction of $LnCl_3$ with $Na[S_2CNR_2]$ or $Na[S_2PMe_2]_4]^-$ in ethanol. Examples include the eight-coordinate $[Ln(S_2CNR_2)_4]^-$ and $[Ln(S_2PMe_2)_4]^-$ ions and six-coordinate trigonal prismatic $Pr[S_2P(C_6H_{11})_2]_3$. Mixed sulfur/nitrogen compounds, $\{Ln[N(SiMe_3)_2]-[\mu-SBu^t]\}_2$, Ln = Eu, Gd, and Y are also known.

Only a few phosphido compounds[19] have been reported, these being $Ln[N(SiMe_3)_2]_2(PPh_2)$ (Ln = La, Eu) and $Tm[P(SiMe_3)_2]_3(THF)_2$.

19-9   Cyclopentadienyl Compounds

The organometallic chemistry of the lanthanides is predominantly, though not exclusively, that of cyclopentadienyl and substituted cyclopentadienyl compounds.

[12]D. J. Berg *et al.*, *Inorg. Chem.* **1994,** *33,* 6334.
[13]F. Uggeri *et al.*, *Inorg. Chem.* **1995,** *34,* 633.
[14]S. J. Franklin and K. N. Raymond, *Inorg. Chem.* **1994,** *33,* 5794.
[15]J. Takats *et al.*, *J. Am. Chem. Soc.* **1995,** *117,* 7828.
[16]H. Schumann *et al.*, *Z. anorg. allg. Chem.* **1995,** *621,* 122; W.-K. Wong, *Polyhedron* **1997,** *16,* 345.
[17]S. Gambarotta *et al.*, *J. Chem. Soc., Chem. Commun.* **1994,** 2641.
[18]A. Iwase *et al.*, *J. Alloys. Comp.* **1993,** *192,* 280.
[19]G. W. Rabe *et al.*, *J. Chem. Soc., Chem. Commun.* **1995,** 577.

The properties of these cyclopentadienyl compounds are influenced markedly by the relationship between the size of the lanthanide atom and the steric demand of the cyclopentadienyl group. The former varies from La to Lu according to the lanthanide contraction, while the latter varies from C_5H_5 (Cp), which makes the least steric demand, to heavily substituted cyclopentadienyl rings, such as C_5Me_5 (Cp*), which is appreciably larger. The organometallic chemistry of Sc and Y is very similar to that of the lanthanides with proper allowance for the relative sizes of these atoms.

It should be noted that all organolanthanide compounds are markedly sensitive to oxygen and moisture.

The first cyclopentadienyl compounds of the lanthanides (and indeed the first organolanthanides of any kind) were the Cp_3Ln compounds (Ln = Sc, Y, La, Ce, Pr, Nd, Sm, Gd, Dy, Er, Yb), prepared by the reaction

$$LnCl_3 + 3NaCp \longrightarrow Cp_3Ln + 3NaCl$$

These react with various ligands, such as R_3P, RNC, THF, $(EtO)_3PO$[20] to give quasi-tetrahedral molecules, Cp_3LnL. The size of the Cp* ligand was thought to rule out the existence of Cp_3^*Ln compounds, but recently one of these, Cp_3^*Sm, has been reported.[21] The quasi-tetrahedral $Cp_3CeOCMe_3$ molecule is also known.[22] Previous reports of Cp_4Ce and Cp_3CeCl have never been confirmed.

The most common and generally important compounds are those with two Cp or Cp* rings attached to a Ln^{III} or, for Sm, Eu and Yb, a Ln^{II} atom. A great deal of this chemistry has entailed the reactions of $Cp_2^*Sm^{II}$ or an adduct[23] to form $Cp_2^*Sm^{III}X$ or $(Cp_2^*Sm^{III})_2Y$ products, as illustrated by Fig. 19-3, where reactions with HCCH,[24] alkenes,[25] dinitrogen,[26] and polycyclic aromatics[27] are shown.

The Cp_3Ln compounds frequently react with loss of one Cp ring and the formation of $Cp_2Ln(\mu\text{-}X)_2LnCp_2$ products;[28] for example, with alcohols:

$$2Cp_3Dy + 2ROH \longrightarrow Cp_2Dy(\mu\text{-}OR)_2DyCp_2 + 2C_5H_6$$

There is a great deal of interesting chemistry of the Cp_2Ln^{III} or $Cp_2^*Ln^{III}$ groups. For example, Cp_2Ln^{III} can serve as the center for catalytic polymerization of olefins. In support of the idea that these catalytic properties are the result of a homogeneous Ziegler-Natta type mechanism, it has been shown[29] that both a metal-carbon σ-bond and a metal-olefin bond can be formed simultaneously, as would be required, and that insertion of unactivated alkenes and alkynes into Ln—R bonds proceeds easily.[30]

[20]B. Kanellakopoulos et al., *J. Organomet. Chem.* **1994**, *483*, 193.
[21]W. J. Evans et al., *Organometallics* **1996**, *15*, 527.
[22]W. J. Evans et al., *Organometallics* **1989**, *8*, 1581.
[23]W. J. Evans et al., *J. Organomet. Chem.* **1994**, *483*, 39.
[24]W. J. Evans et al., *J. Organomet. Chem.* **1994**, *483*, 21.
[25]W. J. Evans et al., *J. Am. Chem. Soc.* **1990**, *112*, 219, 2314.
[26]W. J. Evans et al., *J. Am. Chem. Soc.* **1988**, *110*, 6877.
[27]W. J. Evans et al., *J. Am. Chem. Soc.* **1994**, *116*, 2600.
[28]Z. Huang et al., *Polyhedron* **1996**, *15*, 13, 127.
[29]C. P. Casey et al., *J. Am. Chem. Soc.* **1995**, *117*, 9770.
[30]Y. Li and T. J. Marks, *J. Am. Chem. Soc.* **1996**, *118*, 707 and numerous references therein.

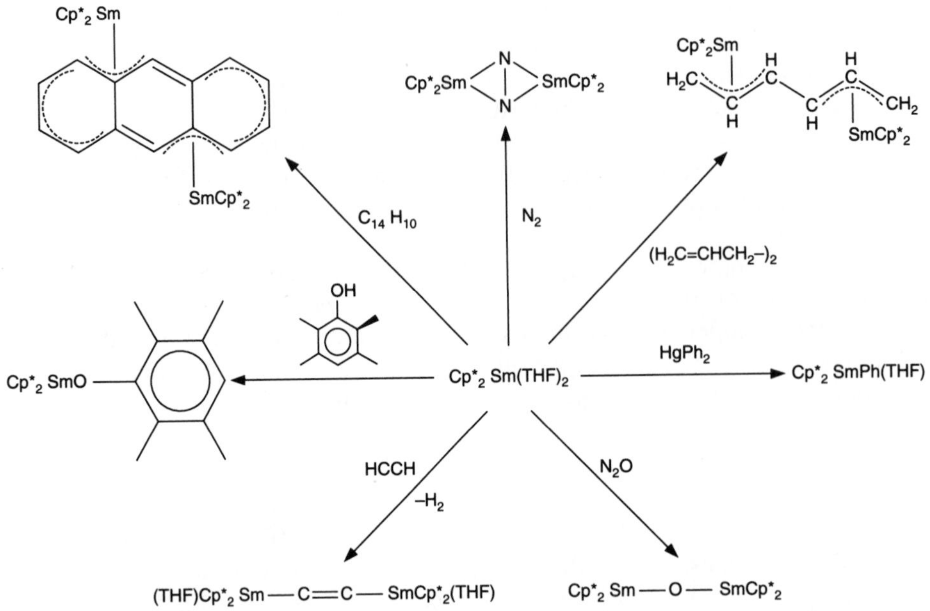

Figure 19-3   Some reactions of Cp$_2^*$Sm.

With Cp$_2^*$LuZ (Z = H, CH$_3$) where the large Cp* rings are on the smallest Ln, the following remarkable reactions occur:

$$Cp*_2LuZ + SiMe_4 \longrightarrow Cp*_2LuCH_2SiMe_3 + ZH$$

$$Cp*_2LuZ + C_6H_6 \longrightarrow Cp*_2LuC_6H_5 + ZH$$

$$Cp*_2LuC_6H_5 + Cp*_2LuZ \longrightarrow Cp*_2Lu\text{—}\boxed{\bigcirc}\text{—}Lu(Cp*)_2 + ZH$$

$$Cp*_2LuH + D_2 \longrightarrow Cp*_2LuD + HD$$

$$Cp*_2LuZ + *CH_4 \longrightarrow Cp*_2Lu*CH_3 + ZH$$

The last reaction was the first unequivocal example of the reaction of methane homogeneously to form an M—C bond, and it proceeds under relatively mild conditions.

There are many structurally and chemically interesting hydrido derivatives of the Cp and Cp* lanthanide compounds. In many of these there are double hydrido bridges (19-I), but a number of more complex structures (e.g., 19-II) are also known. The hydrido compounds are generally quite reactive and often show activity as hydrogenation catalysts. They are most commonly obtained by hydrogenolysis of Cp$_2$LnR compounds. Ln—R bonds undergo many other reactions, of which those with CO are often especially interesting, for example:

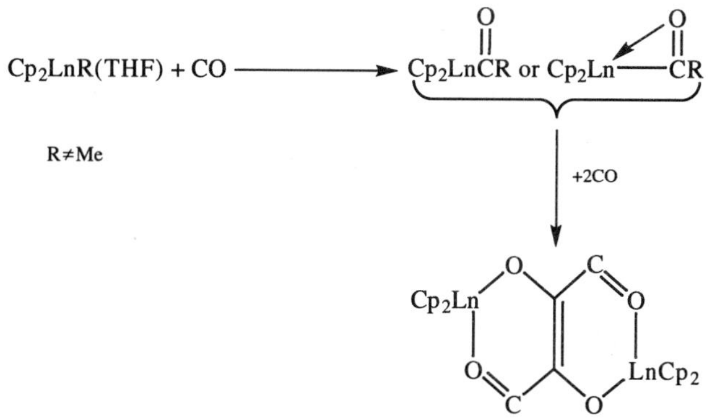

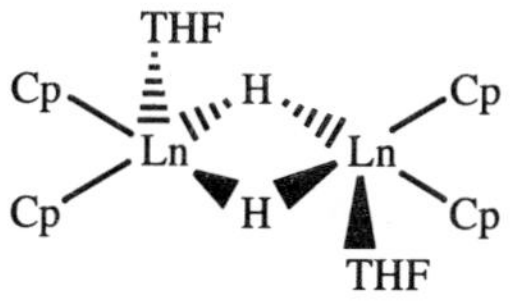

$$Cp_2LnR(THF) + CO \longrightarrow Cp_2LnCR \text{ or } Cp_2Ln{-}CR$$

R≠Me

(19-I) 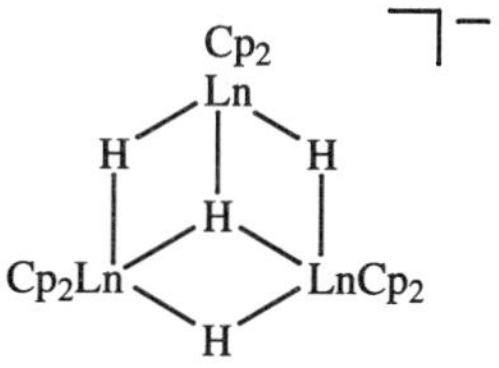 (19-II)

Generally speaking, the chemistry of the $CpLnX_2$ type compounds, though known, is limited and less important than the Cp_3Ln and Cp_2LnX chemistries.

19-10 Other Organometallic Compounds[31,32]

Because of the limited radial extension of the $4f$ orbitals, coupled with the preference of the rather large Ln^{III} atoms for high coordination numbers, simple LnR_n or $LnAr_n$ compounds are rare and difficult to make. Those known in unsolvated form include the $Ln(C_6F_5)_2$ (Ln = Eu, Yb) compounds and $Yb[C(SiMe_3)_3]_3$. Some $LnR_2(Solv)_n$ compounds are formed with smaller R groups.

All LnR_3 or $LnAr_3$ compounds are solvated, as exemplified by $Ph_3Er(THF)_3$. There are also anionic species such as that in $[Li(tmeda)]_3HoMe_6$, where interaction with the methyl groups probably enhances the stability, and the $[Ln(allyl)_4]^-$ ions. The latter are highly active catalysts for the polymerization of butadiene. Numerous other compounds containing one or more Ln—C sigma bonds are known.[33]

The three divalent lanthanides, Sm, Eu, and Yb, form Grignard-like compounds that are useful in synthesis and which, like the Grignard reagents, are structurally more complex than the usual formulas suggest. From direct reactions of RI with the metals the RLnI compounds are prepared in THF solution and used, like Grignards, as alkylating agents.

[31]F. T. Edelmann, *Angew. Chem. Int. Ed. Engl.* **1995**, *34*, 2466. An extensive review.
[32]G. B. Deacon and Q. Shen, *J. Organomet. Chem.* **1996**, *506*, 1.
[33]M. F. Lappert *et al., J. Chem. Soc., Chem. Commun.* **1994**, 2691.

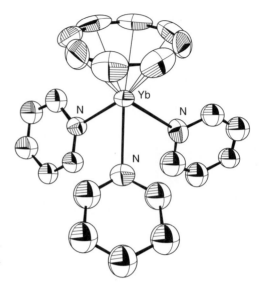

Figure 19-4 The structure of (η^8-C_8H_8)Ybpy$_3$ (Adapted from A. L. Wayda *et al., Organometallics,* **1987,** *6,* 1238.)

SmI$_2$ (and Cp$_2$Sm also) finds use as a coupling agent[34] as in the so-called samarium Barbier reaction:

$$^1R{\diagdown}{}_{}^{}C{=}O + RI \xrightarrow[\text{THF}]{SmI_2} \xrightarrow{H_3O^+} R^2{\diagup}{}^{R^1}_{R}C{-}OH$$

Presumably these and other coupling reactions promoted by Sm^{II} compounds proceed through intermediates containing Sm$-$R bonds.

Cyclooctatetraene Compounds

In all of these the C$_8$H$_8$ ring is a planar octagon that can be considered formally as a dianion. Ln(C$_8$H$_8$) compounds are rare, insoluble, and assumed to be polymeric. On treatment with pyridine, soluble monomers such as that shown in Fig. 19-4 are formed. Sandwich anions, [Ln(C$_8$H$_8$)]$^{2-}$, are also known.

The most extensive chemistry is with the trivalent lanthanides. Reactions of LnCl$_3$ with K$_2$C$_8$H$_8$ lead to [Ln(C$_8$H$_8$)$_2$]$^-$ anions as well as dimeric [(C$_8$H$_8$)-Ln(THF)$_2$]$_2$(μ-Cl)$_2$ compounds.

There are also Ln$_2$(C$_8$H$_8$)$_3$ compounds of unknown structure made by atom vapor reactions. In solvated form they have a [Ln(C$_8$H$_8$)$_2$]$^-$ bound through two carbon atoms to a [Ln(C$_8$H$_8$)(THF)$_2$]$^+$ unit.

The only C$_8$H$_8$ complexes of a tetravalent lanthanide are Ce(C$_8$H$_8$)$_2$ and some substituted derivatives.[35]

Other unsaturated organic molecules form bonds to the lanthanides. The dimethyl pentadienyl ligand (19-III), C$_7$H$_{11}$, forms both solvated and unsolvated

[34]H. B. Kagan *et al., J. Alloys. Comp.* **1993,** *192,* 191.
[35]R. D. Fischer, *Angew. Chem. Int. Ed. Engl.* **1994,** *33,* 2165.

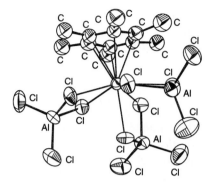

Figure 19-5   The structure of $(C_6Me_6)Sm(AlCl_4)_3$.

$Ln(C_7H_{11})_3$ compounds.[36] Aromatics such as 1,3,5-tri-*t*-butylbenzene and hexamethylbenzene act effectively as neutral donors, as in $(\eta^2\text{-}C_6Me_6)Sm(\eta^2\text{-}AlCl_4)_3$, whose structure is shown in Fig. 19-5, or in the $Ln^0(arene)_2$species.[37] This is in contrast to the formally anionic (i.e., $C_8H_8^{2-}$) character of cyclooctatetraene.

(19-III)

19-11   Scandium

Scandium is more like an element of the first transition series (or like aluminum) than like the rare earths. For example, it characteristically has a coordination number of 6 (6-coordinate radius, 0.89 Å compared to 1.00–1.17 Å for the lanthanides), although coordination numbers 8 and 9 are known. Examples of CN 8 are $Na_5[Sc(CO_3)_4]\cdot6H_2O$ and the tropolonate, $[SeT_4]^-$. There is but one example of CN 9, namely the tricapped trigonal prismatic $[Sc(H_2O)_9]^{3+}$ ion found in $Sc(CF_3SO_3)_3\cdot9H_2O$.[38] The aqua ion in solution, $[Sc(H_2O)_6]^{3+}$, is appreciably hydrolyzed and OH-bridged di- and trinuclear species are formed.

Binary compounds include the oxide, Sc_2O_3, which is more similar to Al_2O_3 than the Ln_2O_3 compounds. It is precipitated from aqueous solution in hydrous form and $ScO(OH)$, isostructural with $AlO(OH)$, is known. The hydrous oxide will dissolve in concentrated NaOH and $Na_3[Sc(OH)_6]\cdot2H_2O$ can be crystallized from solutions with >8 *M* hydroxide ion. "Scandates" such as $LiScO_2$ are made by fusing the two oxides together.

The fluoride is insoluble in water but dissolves readily in an excess of HF or in NH_4F to give fluoro complexes such as ScF_6^{3-}, and the similarity to Al is confirmed by the existence of a cryolite phase Na_3ScF_6, as well as $NaScF_4$ in the NaF—ScF_3 system. The chloride $ScCl_3$, and $ScBr_3$, can be obtained by P_2O_5 dehydration of

[36]D. Baudry *et al.*, *J. Organomet. Chem.* **1994**, *482*, 125.
[37]F. G. N. Cloke *et al.*, *J. Chem. Soc., Chem. Commun.* **1989**, 53.
[38]C. B. Castellani *et al.*, *Eur. J. Solid State Inorg. Chem.* **1995**, *32*, 1089.

hydrated halides; the anhydrous chloride sublimes at a much lower temperature than the lanthanide halides, and this can be associated with the different structure of the solid, which is isomorphous with $FeCl_3$. Unlike $AlCl_3$ it does not act as a Friedel-Crafts catalyst.

Oxo Salts and Complexes

Simple hydrated oxo salts are known, as well as some double salts such as $K_2SO_4 \cdot Sc_2(SO_4)_3 \cdot nH_2O$, which is very insoluble in K_2SO_4 solution. Ammonium double salts such as the tartrate, phosphate, and oxalate are also insoluble in water.

The β-diketonates resemble those of Al rather than those of the lanthanides. Thus the acetylacetonate is normally anhydrous and may be sublimed at around 200°C; it has a trigonally distorted octahedral structure.

The TTA complex can be extracted from aqueous solutions at pH 1.5 to 2 by benzene, and the 8-quinolinolate (*cf.* Al) can be quantitatively extracted by $CHCl_3$; Sc^{3+} can also be extracted from aqueous sulfate solutions by a quaternary ammonium salt.

Lower Oxidation States

Apart from some Sc^{II} compounds of the type $M^ISc X_3$ as well as a series of nonstoichiometric $M^ISc_xCl_3$ phases, the most important compounds are those highly air-sensitive substances obtained by prolonged reaction of Sc and ScX_3 at high temperatures. In addition to the monochloride, which is similar to the LnCl compounds, scandium forms Sc_5Cl_8, Sc_7Cl_{10}, and Sc_7Cl_{12}. All of these have in common the presence of Sc_6 octahedra.

In Sc_5Cl_8 there are chains of these sharing opposite edges (Fig. 19-6) linked together by Sc^{3+} ions that form chains of edge-sharing $ScCl_6$ octahedra. In Sc_7Cl_{10},

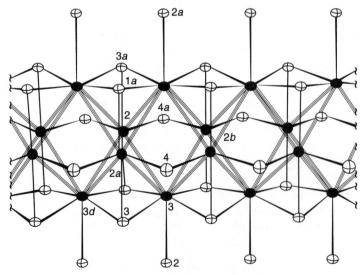

Figure 19-6   The anionic chain of Sc_6 octahedra $(Sc_4Cl_8)_x$ in Sc_5Cl_8. These are tied together by Sc^{3+} ions.

which has been shown to have metallic conduction and puzzling magnetic properties, there are double Sc_6 octahedra sharing edges to form infinite chains and, again, Sc^{3+} ions tying these chains together. In Sc_7Cl_{12} there are discrete $[Sc_6Cl_{12}]^{3-}$ groups (with Sc—Sc distances of 3.20–3.23 Å) similar to the M_6Cl_{12} clusters formed by Nb and Ta, packed with Sc^{3+} ions that tie them together.

19-12   The Oxidation State IV

This oxidation state is very important for cerium and of minor, though not negligible, importance for praseodymium and terbium. With the possible exception of a few fluorides, for example, Cs_3LnF_7 for Nd and Dy, no other Ln^{IV} compounds are known.

Oxides[39]

For Ce, Pr, and Tb, there are higher oxide phases spanning the range Ln_2O_3—LnO_2, but the relative stabilities of the highest oxides, LnO_2 can be judged by the equilibrium O_2 pressures: 10^{-40} atm over CeO_2, 0.2 atm over PrO_2, and 10^3 atm over TbO_2. The details of the intermediate phases are complicated, but some of them have been described by formulae such as Pr_6O_{11}, Pr_9O_{16}, $Pr_{10}O_{18}$, Tb_4O_7, $Tb_{11}O_{20}$, and $Tb_{12}O_{22}$.

Cerium(IV)

The terms cerous and ceric are commonly used for Ce^{III} and Ce^{IV}. Solid ceric compounds include CeO_2, the hydrous oxide $CeO_2 \cdot nH_2O$ and CeF_4. The dioxide CeO_2, white when pure, is obtained by heating cerium metal, $Ce(OH)_3$, or any of several Ce^{III} salts of oxo acids such as the oxalate, carbonate, or nitrate, in air or oxygen; it is a rather inert substance, not attacked by either strong acids or alkalis; it can, however, be dissolved by acids in the presence of reducing agents (H_2O_2, Sn^{II}, etc.), giving Ce^{III} solutions. Hydrous ceric oxide $CeO_2 \cdot nH_2O$ is a yellow, gelatinous precipitate obtained on treating Ce^{IV} solutions with bases; it redissolves fairly easily in acids. CeF_4 is prepared by treating anhydrous $CeCl_3$ or CeF_3 with fluorine at room temperature; it is relatively inert to cold water and is reduced to CeF_3 by hydrogen at 200 to 300°C.

Cerium (IV) in solution is obtained by treatment of Ce^{III} solutions with very powerful oxidizing agents, for example, peroxodisulfate or bismuthate in nitric acid. The aqueous chemistry of Ce^{IV} is similar to that of Zr, Hf, and, particularly, tetravalent actinides. Thus Ce^{IV} gives a phosphate insoluble in 4 M HNO_3 and an iodate insoluble in 6 M HNO_3, as well as an insoluble oxalate. The phosphate and iodate precipitations can be used to separate Ce^{IV} from the trivalent lanthanides. Ce^{IV} is also much more readily extracted into organic solvents by tributyl phosphate and similar extractants than are the Ln^{III} lanthanide ions.

The hydrated ion $[Ce(H_2O)_n]^{4+}$ is a fairly strong acid, and except at very low pH, hydrolysis and polymerization occur. It is probable that the $[Ce(H_2O)_n]^{4+}$ ion exists only in concentrated perchloric acid solution. In other acid media there is coordination of anions, which accounts for the dependence of the potential of the

[39]J. Zhang *et al.*, *J. Alloys. Comp.* **1993**, *192*, 57; L. Eyring *et al.*, *Z. anorg. allg. Chem.* **1996**, *622*, 465.

Ce^{IV}/Ce^{III} couple on the nature of the acid medium:

$$Ce^{IV} + e = Ce^{III} \qquad E^0 = +1.44 \text{ V } (1\ M\ H_2SO_4),\ +1.61 \text{ V } (1\ M\ HNO_3),\ +1.70 \text{ V } (1\ M\ HClO_4)$$

Comparison of the potential in sulfuric acid, where at high SO_4^{2-} concentrations the major species is $[Ce(SO_4)_3]^{2-}$, with that for the oxidation of water,

$$O_2 + 4H^+ + 4e = 2H_2O \qquad E^0 = +1.229 \text{ V}$$

shows that the acid Ce^{IV} solutions commonly used in analysis are metastable. The oxidation of water is kinetically controlled but can be temporarily catalyzed by fresh glass surfaces.

Cerium(IV) is used as an oxidant both in analysis and in preparative chemistry. Dissolved in acetic acid it oxidizes aldehydes and ketones at the α-carbon atom; a solution of ammonium hexanitratocerate(IV) will oxidize toluene to benzaldehyde. Aqueous Ce^{IV} oxidizes concentrated Cl^- to Cl_2, but the rate depends strongly on the pH and the presence of a catalyst.[40]

Many complex ions are formed, such as the $[Ce(NO_3)_6]^{2-}$ ion, where bidentate NO_3^- ions afford 12-coordination. Fluoro complexes have been isolated as $(NH_4)_4[CeF_8]$, which on heating gives $(NH_4)_2[CeF_6]$ and $(NH_4)_3CeF_7 \cdot H_2O$, which comes out of 28% NH_4F solution. Reaction of CeO_2 with HCl in dioxane gives orange needles of the oxonium salt of $[CeCl_6]^{2-}$, and various other salts such as the pyridinium salt can be made.

Ceric alkoxides are obtainable by several reactions:[41]

$$(pyH)_2CeCl_6 + 4ROH + 6NH_3 \longrightarrow Ce(OR)_4 + 2py + 6NH_4Cl$$

$$(NH_4)_2Ce(NO_3)_6 + 4ROH + 6NaOMe \longrightarrow Ce(OR)_4 + 6NaNO_3 + 6MeOH + 2NH_3$$

$$(NH_4)_2Ce(NO_3)_6 + 5NaOCMe_3 \longrightarrow Ce(OCMe_3)_3(NO_3)(THF)_2 + 2NH_3 + 5NaNO_3 + 2Me_3COH$$

Some of the $Ce(OR)_4$ compounds (e.g., R = $CHMe_2$) are crystalline and sublime rather easily while others are nonvolatile and presumably oligomers with $Ce-O(R)-Ce$ bridges.

Praseodymium(IV)

Only a few solid compounds are known, the commonest being the black nonstoichiometric oxides, often formulated as Pr_6O_{11}, formed on heating Pr^{III} salts or Pr_2O_3 in air. "Pr_6O_{11}" dissolves in acids to give aqueous Pr^{III} and liberate oxygen, chlorine, and so on, depending on the acid used.

When alkali fluorides mixed in the correct stoichiometric ratio with Pr salts are heated in F_2 at 300 to 500°C, compounds such as $NaPrF_5$ or Na_2PrF_6 are obtained. The action of dry HF on the latter gives PrF_4, although this cannot be obtained by direct fluorination of PrF_3. There is some evidence that $Pr(NO_3)_4$ is partially formed by action of N_2O_5 and O_3 on PrO_2.

[40]A. Mills and G. Meadows, *J. Chem. Soc., Chem. Commun.* **1994,** 2059.
[41]W. J. Evans *et al., Inorg. Chem.* **1989,** *28,* 4027.

Table 19-4 Properties of the Lanthanide M^{2+} Ions

Ion	Color	$E^0(V)^a$	Crystal radius (Å)
Sm^{2+}	Blood red	-1.40	1.36
Eu^{2+}	Colorless	-0.34	1.31
Yb^{2+}	Yellow	-1.04	1.16

$^a Ln^{3+} + e = Ln^{2+}$.

Praseodymium(IV) is a very powerful oxidizing agent, the Pr^{IV}/Pr^{III} couple being estimated as $+2.9$ V. This potential is such that Pr^{IV} would oxidize water itself, so that its nonexistence in solution is not surprising.

Terbium(IV)

The chemistry resembles that of Pr^{IV}. When oxo salts are ignited under ordinary conditions, oxides with compositions ranging from $TbO_{1.71}$ to $TbO_{1.81}$ are obtained, depending on the preparative details. Terbium dioxide, with a fluorite structure, can be obtained by oxidation of Tb_2O_3 with atomic oxygen at 450°C. Colorless TbF_4, isostructural with CeF_4 and ThF_4, is obtained by treating TbF_3 with gaseous fluorine at 300 to 400°C, and compounds of the type $M_n TbF_{n+4}$ (M = K, Rb, or Cs; $n \geq 2$) are known.

No numerical estimate has been given for the Tb^{IV}/Tb^{III} potential, but it must certainly be more positive than $+1.23$ V, since dissolution of any oxide containing Tb^{IV} gives only Tb^{III} in solution and oxygen is evolved.

19-13 Oxidation State II for Eu, Sm, and Yb

All of the lanthanides can give Ln^{2+} ions under unusual conditions, for example, trapped in place of Ca^{2+} in CaF_2, or by reduction of $NdCl_3$, $DyCl_3$, or $TmCl_3$ by Na naphthalenide in THF. Alkali metals, Tl and In, reduce the tribromides and triiodides of Sm, Dy, Tm, and Yb to give $M^I LnX_3$ crystalline compounds.[42] The colors and reduction potentials for these ions (in THF) are reported to be Tm^{2+} (emerald, -2.3 V), Dy^{2+} (brown, -2.5 V), and Nd^{2+} (cherry, -2.6 V). However, it is only the elements Eu, Yb, and Sm that have significant "normal" chemistry (i.e., based on true Ln^{2+} ions) in this oxidation state. We have already mentioned organometallic compounds of these same three Ln^{II} species in Sections 19-9 and 19-10. Properties of the Eu^{2+}, Yb^{2+}, and Sm^{2+} ions are given in Table 19-4.

The stabilities of the Eu^{2+}, Yb^{2+}, and Sm^{2+} ions correlate with the third ionization enthalpies of the atoms and the sublimation enthalpies of the metals. The Eu^{2+}(aq) ion is readily obtained by reducing Eu^{3+}(aq) with Zn or Mg, while preparation of the others requires use of Na/Hg or electrolysis. The aqueous Eu^{2+} solutions are easily handled, but those of Sm^{2+} and Yb^{2+} are rapidly oxidized by air and by water itself. The Ln^{2+} ions show many resemblences to Ba^{2+}, giving insoluble sulfates, for example, but soluble hydroxides. Europium can be easily separated from other lanthanides by Zn reduction followed by precipitation of the other Ln^{3+} hydroxides.

[42]G. Schilling and G. Meyer, *Z. anorg. allg. Chem.* **1996,** *622,* 759.

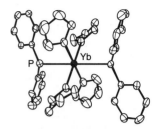

Figure 19-7 A drawing of the molecular structure of $Yb[PPh_2]_2(THF)_4$ (from Ref. 40).

Among the many simple Ln^{II} compounds that are well characterized in the solid state are the air- and moisture-sensitive halides, $LnCl_2$, $LnBr_2$, LnI_2 (which have 7- or 8-coordinated Ln^{2+} ions), the oxides (LnO), sulfides (LnS), carbonates, sulfates, and amides $(Ln[N(SiMe_3)_2](THF)_2)$. The sulfides have NaCl structures and exhibit metallic conductance. More recently the phosphides of Sm and Yb, $Ln(PPh_2)_2(THF)_4$, have been reported.[43] The preparative reaction as well as the structures (Fig. 19-7) are interesting:

$$LnI_3(THF)_3 + 3KPPh_2 \xrightarrow{THF} Ln(PPh_2)_2(THF)_4 + \tfrac{1}{2}Ph_2PPPh_2 + 3KI$$

SmI_2 finds many uses as a reducing agent and coupling agent in organic[44] and main-group[45] chemistry.

Ln^{II} compounds react with donor solvents such as THF to form solvates. SmI_2 forms a 6-coordinate, polymeric (I-bridged) solvate, $SmI_2(Me_3CCN)_2$, an 8-coordinate solvate, $SmI_2[O(CH_2CH_2OMe)_2]_2$, with triglyme, and a series of solvates all having pentagonal bipyramidal coordination with linear I—Sm—I units.[46]

Molecular complexes are also formed, examples being trispyrazolylborate complexes of Yb[47] and β-dike complexes of europium.[48]

19-14 Other Lower Oxidation State Compounds

For lanthanide metals other than Eu, Sm, and Yb, compounds in lower oxidation states generally involve metal—metal bonding rather than discrete Ln^{2+} ions. In general they are formed by heating LnX_3 with Ln in tantalum containers.

Sesquichlorides of a unique type are formed by Y, Gd, and Tb. In Gd_2Cl_3 or $[Gd_4^{6+}(Cl^-)_6]_n$, there are infinite chains of metal atoms in octahedra sharing opposite edges (Fig. 19-8); chlorine atoms are located over triangles formed by three Gd atoms. These halides can be further reduced at 800°C to GdCl and TbCl. The latter materials are graphitelike platelets. They have a layer structure like ZrCl and ScCl where close-packed double layers of metal atoms alternate with double layers of

[43]G. W. Rabe *et al.*, *Inorg. Chem.* **1995,** *34,* 4521.
[44]G. A. Molander, *Chem. Rev.* **1992,** *92,* 29.
[45]K. Mochida *et al.*, *J. Chem. Soc., Chem. Commun.* **1995,** 2275.
[46]W. J. Evans *et al.*, *J. Am. Chem. Soc.* **1995,** *117,* 8999.
[47]A. Sella, *J. Chem. Soc., Chem. Commun.* **1994,** 2689.
[48]W. J. Evans *et al.*, *Inorg. Chem.* **1994,** *33,* 6435.

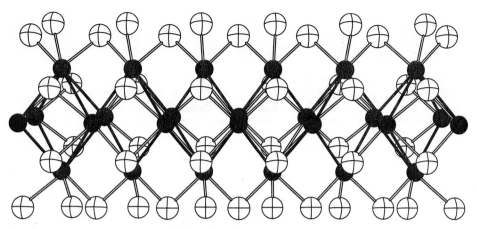

Figure 19-8   A portion of the $\frac{1}{\infty}[Gd_4Cl_6]$ chain in Gd_2Cl_3. Interchain bridging chlorine atoms at the top and bottom are duplicates.

halide atoms (XMMX···XMMX) and are related to the CdI_2 type (XMX···XMX) structure.

There are three other lanthanides that give well-characterized LnI_2 compounds (La, Ce, Pr, Gd), but these are actually Ln^{III} species having metallic conductance. This results from the fact that they are best described as $Ln^{3+}(I^-)_2e^-$, with delocalized electrons.

LaI, the only binary halide containing Ln^I, also appears to have metallic conductance.[49]

There are many other formally low-valent lanthanide compounds that doubtless display extensive metal—metal bonding or contain M_n cluster units, such as $[M_5C_2]$ units in the $M_5C_2Cl_9$ phases (M = La—Pr), where the M_5C_2 units are trigonal bipyramidal M_5 clusters with a C_2 unit inside on the 3-fold axis.[50]

Additional References

K. A. Gschneidner and L. Eyring, Eds., *Handbook on the Physics and Chemistry of Rare Earths*, Elsevier, Lausanne, An annual series, Vol. 189, 1994.

F. A. Hart, in *Comprehensive Coordination Chemistry*, Vol. 5, Pergamon Press, Oxford, 1987.

G. Meyer and L. R. Morss, Eds., *Synthesis of Lanthanide and Actinide Compounds*, Kluwer Academic Publishers, Dordrecht, 1991.

Progress in the Science and Technology of the Rare Earths, Vol. 1, 1964. (A continuing series with reviews on extraction, solution chemistry, magnetic properties, analysis, halides, oxides, and so on.)

S. P. Sinha, Ed., *Lanthanide and Actinide Research* (Journal, Vol. 1, 1986).

[49]J. D. Martin and J. D. Corbett, *Angew. Chem. Int. Ed. Engl.* **1995**, *34*, 233.

[50]G. Meyer *et al.*, *Z. anorg. allg. Chem.* **1995**, *621*, 1299.

Chapter 20

THE ACTINIDE ELEMENTS

20-1 Occurrence and General Properties

The actinide elements (which are all radioactive) up to $Z = 103$ are listed in Table 20-1 along with some of their properties. The principal isotopes that can be obtained in macroscopic amounts are listed in Table 20-2. The significance of the term "actinide" series is justified by the overall agreement of the chemistry of these elements with that concept, although it is not as clearly justified as is the lanthanide concept. Note in Table 20-1 that some irregular variation in the $5f/6d$ electron distribution does occur.

Only thorium, protactinium, and uranium occur in Nature in amounts sufficient for practical extraction. Thorium and uranium occur in enriched deposits, workable by normal mining procedures, and are available in ton quantities from deposits in Canada, USA, South Africa, Australia, and Namibia.

Actinium and protactinium are present in uranium minerals but recovery is not generally practiced. Neptunium (^{237}Np and ^{239}Np) and plutonium (^{239}Pu) are present in minute amounts in uranium ores because they result from reaction of neutrons with uranium isotopes, not due to survival from primordial formation. All other actinides are entirely synthetic. Methods of preparation for those up to Cf are given in Table 20-2; syntheses of the other elements are noted in Section 20-18.

The atomic spectra of the elements are very complex and it is difficult to identify levels in terms of quantum numbers and configurations. For chemical behavior the lowest configuration is of greatest importance and the competition between $5f^n7s^2$ and $5f^{n-1}6d7s^2$ is of interest.

For the elements in the first half of the f shell it appears that less energy is required for the promotion of $5f \rightarrow 6d$ than for the $4f \rightarrow 5d$ promotion in the lanthanides; there is thus a greater tendency to supply more bonding electrons with the corollary of higher valences in the actinides. The second half of the actinides more closely resembles the lanthanides.

Another difference is that the $5f$ orbitals have a greater spatial extension relative to the $7s$ and $7p$ orbitals than the $4f$ orbitals have relative to the $6s$ and $6p$ orbitals. The greater spatial extension of the $5f$ orbitals has been shown experimentally; the esr spectrum of UF_3 in a CaF_2 lattice shows structure attributable to the interaction of fluorine nuclei with the electron spin of the U^{3+} ion. This implies a small overlap of $5f$ orbitals with fluorine and constitutes an f covalent contribution to the ionic bonding. With the neodymium ion a similar effect is *not* observed. Because they occupy inner orbitals, the $4f$ electrons in the lanthanides are not accessible for

Table 20-1 The Actinide Elements and Some of Their Properties[a]

Z	Symbol	Name	Electronic structure	Metal mp (°C)	Metal Radius (Å)	$-E°$ (V) 0-III	$-E°$ (V) 0-II	Ionic radius (Å) M^{3+}	Ionic radius (Å) M^{4+}
89	Ac	Actinium	$6d^17s^2$	1100	1.898	2.13		1.26	
90	Th	Thorium	$6d^27s^2$	~1750	1.798	1.17			1.08
91	Pa	Protactinium	$5f^26d7s^2$ or $5f^16d^27s^2$	1572	1.642	1.49		1.18	1.04
92	U	Uranium	$5f^36d^17s^2$	1132	1.542	1.66		1.17	1.03
93	Np	Neptunium	$5f^57s^2$	639	1.503	1.79		1.14	1.00
94	Pu	Plutonium	$5f^67s^2$	640	1.523	2.00		1.15	1.01
95	Am	Americium	$5f^77s^2$	1173	1.730	2.07		1.12	0.99
96	Cm	Curium	$5f^76d^17s^2$	1350	1.743	2.06	2.0	1.11	0.99
97	Bk	Berkelium	$5f^97s^2$	986	1.704	1.97	0.9	1.10	0.97
98	Cf	Californium	$5f^{10}7s^2$	900	1.694	2.01	1.6	1.09	0.96
99	Es	Einsteinium	$5f^{11}7s^2$		(1.69)	1.98	2.2		
100	Fm	Fermium	$5f^{12}7s^2$		(1.94)	1.95	2.3		
101	Md	Mendelevium	$5f^{13}7s^2$		(1.94)	1.66	2.5		
102	No	Nobelium	$5f^{14}7s^2$		(1.94)	1.78	1.6		
103	Lr	Lawrencium	$5f^{14}6d^17s^2$		(1.71)	2.06			

[a]For 1st ionization energies see N. Trautman et al., *Angew. Chem. Int. Ed. Engl.* **1995**, *34*, 1713. For other extensive physical data see: J. Emsley, *The Elements*, Clarendon Press, Oxford, 3rd ed., 1997. The names proposed for elements above 103 are: 104, Rutherfordium, Rf; 105, Dubnium, Db; 106, Seaborgium, Sg; 107, Bohrium, Bh; 108, Hassium, Hs; 109, Meitnerium, Mt. *C&EN* **1997**, *September 7*, 10.

Table 20-2 Principal Actinide Isotopes Available in Macroscopic Amounts[a]

Isotope	Half-life	Source
^{227}Ac	21.7 yr	Natural $^{226}Ra(n\gamma)^{227}Ra \xrightarrow[41.2\ min]{\beta^-} {}^{227}Ac$
^{232}Th	1.39×10^{10} yr	Natural; 100% abundance
^{231}Pa	3.28×10^5 yr	Natural; 0.34 ppm of U in uranium ores
^{235}U	7.13×10^8 yr	Natural; 0.7204% abundance
^{238}U	4.50×10^9 yr	Natural; 99.2739% abundance
^{237}Np	2.20×10^6 yr	$^{235}U(n\gamma)^{236}U(n\gamma)^{237}U \xrightarrow[6.75\ days]{\beta^-} {}^{237}Np$ [and $^{238}U(n, 2n)^{237}U$]
^{238}Pu	86.4 yr	$^{237}Np(n\gamma)^{238}Np \xrightarrow{\beta^-} {}^{238}Pu$
^{239}Pu	24,360 yr	$^{238}U(n\gamma)^{239}U \xrightarrow[23.5\ min]{\beta^-} {}^{239}Np \xrightarrow[23.5\ days]{\beta^-} {}^{239}Pu$
^{242}Pu	3.79×10^5 yr	Successive $n\gamma$ in ^{239}Pu
^{244}Pu	8.28×10^7 yr	Successive $n\gamma$ in ^{239}Pu
^{241}Am	433 yr	$^{239}Pu(n\gamma)^{240}Pu(n\gamma)^{241}Pu \xrightarrow[13.2\ yr]{\beta^-} {}^{241}Am$
^{243}Am	7650 yr	Successive $n\gamma$ in ^{239}Pu
^{242}Cm	162.5 days	$^{241}Am(n\gamma)^{242m}Am \xrightarrow[16.0\ h]{\beta^-} {}^{242}Cm$
^{244}Cm	18.12 yr	$^{239}Pu(4n\gamma)^{243}Pu \xrightarrow[5.0\ h]{\beta^-} {}^{243}Am(n\gamma)^{244}Am \xrightarrow[26\ min]{\beta^-} {}^{244}Cm$
^{249}Bk	325 days	Successive $n\gamma$ on ^{239}Pu; $^{248}Cm(n\gamma) \xrightarrow{\beta^-} {}^{247}Bk$
^{252}Cf	2.57 yr	Successive $n\gamma$ in ^{242}Pu

[a] ^{237}Np and ^{239}Pu are available in multikilogram quantities; ^{238}Pu, ^{242}Pu, ^{241}Am, ^{243}Am, and ^{244}Cm in amounts of 100 g or above; ^{244}Pu, ^{252}Cf, ^{249}Bk, ^{248}Cm (4.7×10^5 yr), ^{253}Es (20 days), ^{254}Es (1.52 years for α) in milligram and ^{257}Fm (94 days) in microgram quantities. Other long-lived isotopes are known but can be obtained only in traces by use of accelerators [e.g., ^{247}Bk ($\sim 10^4$ yr), ^{252}Es, 471.7 days].

bonding purposes; and virtually no compound in which 4f orbitals are used can be said to exist.

In the actinide series, therefore, the energies of the 5f, 6d, 7s, and 7p orbitals are about comparable over a range of atomic numbers (especially U to Am), and since the orbitals also overlap spatially, bonding can involve any or all of them. In the chemistries this situation is indicated by the fact that the actinides are much more prone to complex formation than are the lanthanides, where the bonding is more ionic. The difference from lanthanide chemistry is usually attributed to the contribution of covalent hybrid bonding involving 5f electrons.

A further point is that since the energies of the 5f, 6d, 7s, and 7p levels are comparable, the energies involved in an electron shifting from one to another, say 5f to 6d, may lie *within* the range of chemical binding energies. Thus the electronic structure of the element in a given oxidation state may vary between compounds and in solution be dependent on the nature of the ligands. It is accordingly also often impossible to say *which* orbitals are being utilized in bonding or to decide meaningfully whether the bonding is covalent or ionic. Theoretical studies of the relativistic effects (Section 1-8) in these heavy elements indicate these to be more important than for the lanthanides, suggesting a pronounced "actinide contraction."[1]

[1]P. Schwerdtfeger *et al., J. Am. Chem. Soc.* **1995,** *117,* 6597.

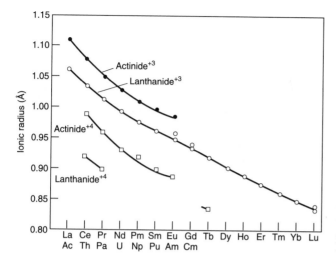

Figure 20-1 Radii of actinide and lanthanide ions. (Reproduced by permission from D. Brown, *Halides of the Lanthanides and Actinides*, Wiley-Interscience, New York, 1968.)

The ionic radii for the commonest oxidation states (Table 20-1) are compared with those of the lanthanides in Fig. 20-1. There is clearly an "actinide" contraction, and the similarities in radii of both series correspond to similarities in their chemical behavior for properties that depend on the ionic radius, such as hydrolysis of halides. It is also generally the case that similar compounds in the same oxidation state have similar crystal structures that differ only metrically.

The electronic absorption spectra of the actinide ions, like those of the lanthanides, are due to transitions within the $5f^n$ levels and consist of narrow bands, relatively uninfluenced by ligand fields; the intensities are generally ~10 times those of the lanthanide bands.

Spectra involving only one f electron are simple, consisting of only a single transition $^2F_{5/2} \rightarrow {}^2F_{7/2}$. For the f^7 configuration (Cm^{3+}; cf. Gd^{3+}) the lowest excited state lies about 4 eV above the ground level, so that these ions show only charge-transfer absorption in the ultraviolet. Most actinide species have complicated spectra.

The magnetic properties of the actinide ions are considerably harder to interpret than those of the lanthanide ions, although there are similarities. The experimental magnetic moments are usually lower than the values calculated by using Russell-Saunders coupling, and this appears to be due both to ligand field effects similar to those operating in the d transition series and to inadequacy of this coupling scheme. Since $5f$ orbitals can participate to some extent in covalent bonding, ligand effects are to be expected.

For the ions Pu^{3+} and Am^{3+}, the phenomenon noted for Sm^{3+} and Eu^{3+} is found; since the multiplet levels are comparable to kT, anomalous temperature dependence of the susceptibilities occurs.

GENERAL CHEMISTRY OF THE ACTINIDES

Given the close similarities in preparations and properties of actinide compounds in a given oxidation state, it is convenient to discuss some general features and to follow this by additional descriptions for the separate elements. Methods of chemical separations of the elements are also discussed. Just as Ln is often used as a generic symbol for lanthanide elements, An is so used for the actinides.

20-2 The Metals[2]

A general method for preparation of all An metals is by reduction of AnF_3 or AnF_4 with vapors of Li, Mg, Ca, or Ba at 1100 to 1400°C; the chlorides or oxides are sometimes used. There are some special methods such as the preparation of Th or Pa from their tetraiodides by the van Arkel-de Boer process, or the following reaction for the relatively volatile americium:

$$2La + Am_2O_3 \xrightarrow{1200°C} 2Am{\uparrow} + La_2O_3$$

Uranium

This is one of the densest metals (19.07 g cm^{-3} at 25°C) and has three crystalline modifications. It forms a wide range of intermetallic compounds (U_6Mn, U_6Ni, USn_3, etc.), but owing to its unique crystal structures, it cannot form extensive ranges of solid solutions. Uranium is chemically reactive and combines directly with most elements. In air the surface is rapidly converted into a yellow and subsequently a black nonprotective film. Powdered uranium is frequently pyrophoric. The reaction with water is complex; boiling water forms UO_2 and hydrogen, the latter reacting with the metal to form a hydride, which causes disintegration. Uranium dissolves rapidly in hydrochloric acid (a black residue often remains) and in nitric acid, but slowly in sulfuric, phosphoric, or hydrofluoric acid. It is unaffected by alkalis. An important reaction of uranium is that with hydrogen, forming the hydride which is a useful starting material for the synthesis of uranium compounds.

Thorium

This metal is white but tarnishes in air. It can be readily machined and forged. It is highly electropositive, resembling the lanthanide metals, and is pyrophoric when finely divided. It is attacked by boiling water, by oxygen at 250°C, and by N_2 at 800°C. Dilute HF, HNO_3, and H_2SO_4, and concentrated HCl or H_3PO_4, attack thorium only slowly, and concentrated nitric acid makes it passive. The attack of hot 12 M hydrochloric acid on thorium gives a black residue that appears to be a complex hydride approximating to $ThH(O)(Cl)(H_2O)$ that is very pyrophoric and appears to give ThO on heating in vacuum.

Plutonium

This metal is similar to U chemically; it is pyrophoric and must be handled with extreme care owing to the health hazard. Also, above a certain critical size the pure metal can initiate a nuclear explosion. The metal is unique in having at least six allotropic forms below its melting point, each with a different density, coefficient

[2]L. B. Asprey et al., Adv. Inorg. Chem. **1987,** 31, 1.

Table 20-3 Oxidation States of the Actinide Elements[a]

Ac	Th	Pa	U	Np	Pu	Am	Cm	Bk	Cf	Es	Fm	Md	No
						2			2	2	2	2	2
3		3	3	3	3	**3**	**3**	**3**	**3**	**3**	**3**	**3**	**3**
	4	**4**	**4**	4	4	4	4	4	4				
		5	5	**5**	5	5							
			6	6	**6**	6							
				7	7								

[a]Bold number signifies most stable state.

of expansion, and resistivity; curiously, if the phase expands on heating, the resistance decreases. Plutonium forms numerous alloys.

Protactinium is silvery and relatively unreactive, but *actinium* and *curium* are so radioactive that they glow. *Neptunium* is similar to uranium and plutonium in appearance. *Americium* is similar and is very electropositive, dissolving readily in acids to give the Am^{3+} ion.

The metals, Cm, Bk, and Cf, have been made by reduction of the oxides by lanthanum.[3]

20-3 Survey of Oxidation States

The known oxidation states of the actinides are given in Table 20-3. With the exception of Th and Pa, the common oxidation state, and for trans-americium elements the dominant oxidation state, is +3, and the behavior is similar to the +3 lanthanides. Thorium and the other elements in the +4 state show resemblances to Hf^{IV} or Ce^{IV}, whereas Pa and the elements in the +5 state show some resemblances to Ta^V. Exceptions to the latter statement are the *dioxo ions* MO_2^+; for U, Np, Pu, and Am that are related to the MO_2^{2+} ions in the +6 state. Redox potentials are given in Table 20-4.

Oxidation State +2

Americium and the heavier elements show the +2 state; the $M^{3+} + e = M^{2+}$ potentials are given in Table 20-5. The Md^{2+} ion is even more stable than Eu^{2+} or Yb^{2+}; these and the other +2 ions have properties resembling those of Ba^{2+} and the +2 lanthanide ions.

Oxidation State +3

This is the most common oxidation state, although for Th, Pa, and U it is of secondary importance. The general chemistry closely resembles that of the lanthanides. The halides MX_3 may be readily prepared and are easily hydrolyzed to MOX. The oxides M_2O_3 are known only for Ac, Pu, and heavier elements. In aqueous solution

[3]A. G. Seleznver *et al.*, *Radiokhimiya* **1989**, *31*, 20.

Table 20-4   Formal Reduction Potentials (V) of the Actinides for 1 M Perchloric Acid
Solutions at 25°C (brackets indicate estimate)

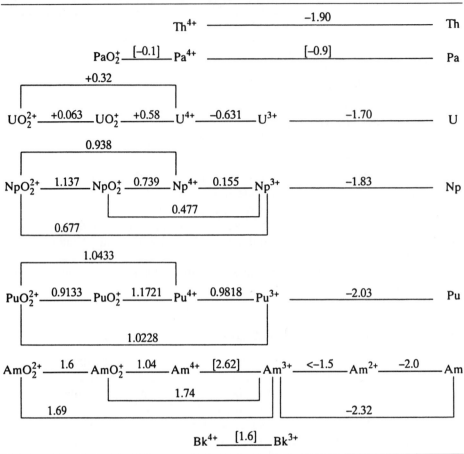

there are M^{3+} ions, and insoluble hydrated fluorides and oxalates can be precipitated.
Isomorphism of crystalline solids is common. Since Ln^{3+} and An^{3+} ions are similar
in size, the formation of complex ions and their stability constants are similar. Thus
it is difficult to separate actinide from lanthanide elements, though it can be done
by ion-exchange or solvent-extraction procedures.

Table 20-5   Actinide Potentials, $M^{3+} + e = M^{2+}$

Element	E^0 (V)	Element	E^0 (V)
Am	−2.3	Es	−1.3
Cm	−4.4	Fm	1.0
Bk	−2.8	Md	−0.15
Cf	−1.6	No	1.45

Oxidation State +4

This is the principal one for Th and has considerable importance for Pa, U, and Pu. For Am, Cm, Bk, and Cf it is much more easily reduced and stable only under special conditions of complexing. Points of importance concerning the +4 state are (a) the dioxides MO_2 from Th to Cf all have the fluorite lattice, (b) the tetrafluorides MF_4 are isostructural with lanthanide tetrafluorides, (c) the chlorides and bromides are known only for Th, Pa, U, and Np, presumably owing to the inability of the halogen to oxidize the heavier metals, and for iodides only those of Th, Pa, and U exist, (d) oxohalides MOX_2 can be made for Th–Np, for example, by the reaction

$$3MX_4 + Sb_2O_3 \xrightarrow{450\,°C} 3MOX_2 + 2SbX_3\uparrow$$

(e) hydrolysis, complexation, and disproportionation are important in aqueous solution, as discussed later.

Oxidation State +5

This is the principal state for Pa, whose chemistry resembles that of Nb^V and Ta^V. For U^V, Np^V, Pu^V, and Am^V, however, there is no particular resemblance to Nb^V and Ta^V because of the importance of the AnO_2^+ ions. Comparatively few solid An^V compounds, other than those of Pa^V and Pu^V are known; the most notable are the U^V halides, the pentafluorides PaF_5, NbF_5, and UF_5, fluoro anions, namely, AnF_6^-, AnF_7^{2-}, and AnF_8^{3-}, for Pa—Pu and the oxochlorides $PaOCl_3$ and $UOCl_3$.

Oxidation State +6

The most important examples are the AnO_2^{2+} ions of U, Np, Pu, and Am (to be discussed later). Otherwise, there are only a few compounds such as the hexafluorides of U, Np, and Pu, as well as UCl_6, various oxohalides and oxohalide complexes (UOF_4, UOF_5^-, and $UOCl_5^-$) and the uranium alkoxides, $U(OR)_6$.

Oxidation State +7

There are only a few marginally stable Np^{VII} and Pu^{VII} compounds, which formally contain the dioxo unit AnO_2^{3+} in alkaline solution.

20-4 The Dioxo Ions: AnO_2^+, AnO_2^{2+}, and AnO_2^{3+}

The dioxo ions are remarkably stable with respect to the strength of the M—O bond. Unlike some other oxo ions, they can persist through a variety of chemical changes, and they behave like cations whose properties are intermediate between those of M^+ or M^{2+} ions of similar size but greater charge. The MO_2 group even appears more or less as an "yl" group in certain oxide and oxo ion structures; furthermore, whereas MoO_2F_2 or WO_2F_2 are molecular halides, in UO_2F_2 there is a linear O—U—O group with F bridges. The stability of UO_2^{2+} and PuO_2^{2+} ions in aqueous solution is shown by the very long half-life for exchange with $H_2^{18}O$ ($>10^4$ h).

So far as is known, the AnO_2^{n+} ions are all linear. The An—O distances vary somewhat from compound to compound in the solid state and for UO_2^{2+} the frequency of the antisymmetric stretching mode (ν, in cm^{-1}), seen easily in the ir, and the U—O distance (D in Å), are related by the equation

$$D = 81.2\nu^{-2/3} + 0.85$$

The order of stabilities derived from Raman spectra appears to be $UO_2^{2+} > NpO_2^{2+} > PuO_2^{2+}$. For aqueous solutions of UO_2^{2+} there is a linear correlation between the frequency of the $O{=}U{=}O$ symmetric stretch (ν_1 cm^{-1}) and the average number of ligands so that the stoichiometries in solutions can be estimated. Complex formation leads to weakening of the UO_2^{2+} axial bonds in the order $OH^- > CO_3^{2-} > C_2O_4^{2-} > F^- > SO_4^{2-}$ and this order reflects the corresponding decrease in stability constants (log β).[4] The unstable UO_2^+ ion disproportionates as follows:

$$2UO_2^+ + 4H^+ = U^{4+} + UO_2^{2+} + 2H_2O \qquad K = 1.7 \times 10^6$$

Visible light photolysis of UO_2^{2+} ion in water gives a long lived, highly oxidizing excited state that is strongly luminescent and a strong oxidant ($E^0 = 2.6$ V). Since it is unaffected by O_2, it may be considered as a photooxygenation catalyst. Interaction of $*UO_2^{2+}$ with a reductant would give UO_2^+ which undergoes autoxidation forming UO_2^{2+} and H_2O_2 quantitatively;[5] the overall reaction is

$$2UO_2^+ + O_2 + 2H^+ \longrightarrow 2UO_2^{2+} + H_2O_2$$

20-5   Actinide Ions in Aqueous Solution

The formal reduction potentials in aqueous solution given in previous tables indicate that the electropositive character of the metals increases with increasing Z and that the stability of the higher oxidation states decreases. Table 20-6 gives the general properties, color, preparation, and reactions for the most important cations. Note that for short-lived isotopes that decay by α emission or spontaneous fission cause heating and consequent chemical effects in solids and solutions. Thus the heat outputs, Wg^{-1}, for three isotopes are reckoned to be as follows: ^{238}Pu 0.5, ^{241}Am 0.1, ^{242}Cm 122. Radiation-induced decomposition of water leads to H and OH radicals, H_2O_2 production, and so on, and in solution higher oxidation states such as Pu^V, Pu^{VI}, and Am^{IV,VI} are reduced. Chemical reactions observable with a short-lived isotope, for example, ^{242}Cm (163 days), may differ when a longer-lived isotope is used; thus Cm^{IV} can be observed only when ^{244}Cm (17.6 years) is employed.

The possibility of several cationic species introduces complexity into the aqueous chemistries, particularly of U, Np, Pu, and Am. Thus all four oxidation states of Pu can coexist in appreciable concentrations in a solution. The solution chemistries and the oxidation-reduction potentials are further complicated by the formation in the presence of ions other than perchlorate, of cationic, neutral, or anionic complexes. Furthermore, even in solutions of low pH, hydrolysis and the formation of polymeric ions occurs. Third, there is the additional complication of disproportionation of certain ions, which is particularly dependent on the pH.

[4]D. A. Palmer et al., Inorg. Chem. **1992**, 31, 5280.
[5]A. Bakac and J. H. Espenson, Inorg. Chem. **1995**, 34, 6034.

Table 20-6 The Principal Actinide Ions in Aqueous Solution[a]

Ion	Color[a]	Preparation	Stability
U^{3+}	Purple	Na or Zn/Hg on UO_2^{2+}	Slowly oxidized by H_2O, rapidly by air to U^{4+}
Np^{3+}	Purplish	$H_2(Pt)$ or electrolytic	Stable in water; oxidized by air to Np^{4+}
Pu^{3+}	Blue violet	SO_2, NH_2OH on higher states	Stable to water and air; easily oxidized to Pu^{4+}
Am^{3+}	Pink[b]	I^-, SO_2, etc., on higher states, or by dissolution of AmO_2 in conc. HNO_3	Stable; starting material for the higher-valence states
Cm^{3+}	Pale yellow		Stable; not oxidized chemically
U^{4+}	Green	Air or O_2 on U^{3+}	Stable; slowly oxidized by air to UO_2^{2+}
Np^{4+}	Green	SO_2 on NpO_2^+ in H_2SO_4	Stable; slowly oxidized by air to NpO_2^+
Pu^{4+}	Anion dependent[c]	SO_2 or NO_2 on PuO_2^{2+}	Stable for up to a year in $6M$ HCl (but not $HClO_4$ or HNO_3); disproportionates in low acid $\rightarrow Pu^{3+} + PuO_2^{2+}$
Am^{4+}	Pink red	$Am(OH)_4$ in $15\,M$ NH_4F	Stable in $15\,M$ NH_4F; reduced by I^-
Cm^{4+}	Pale yellow	CmF_4 in $15\,M$ CsF	Stable only 1 hr at 25°C
UO_2^+	Clear	Transient species	Stability greatest pH 2–4; disproportionates to U^{4+} and UO_2^{2+}; luminesces in the blue
NpO_2^+	Green	Np^{4+} and hot HNO_3	Stable; disproportionates only in strong acid
PuO_2^+	Pink	Hydroxylamine on PuO_2^{2+}, electrochemical synthesis at pH 3	Always disproportionates; most stable at low acidity; it can be stored at pH 7 for up to one week
AmO_2^+	Pale yellow	Am^{3+} with OCl^-, cold $S_2O_8^{2-}$	Disproportionates in strong acid; reduced (2%/hr) by products of own α radiation
UO_2^{2+}	Yellow	Oxidize U^{4+} with HNO_3, etc.	Very stable; difficult to reduce
NpO_2^{2+}	Purple	Oxidize lower states with Ce^{4+}, MnO_4^-, O_3, BrO_3^-, etc.	Stable; easily reduced
PuO_2^{2+}	Yellow brown		Stable; fairly easy to reduce
AmO_2^{2+}	Light purplish-red		Reduced (4%/hr) by products from own α radiation but is stable for ~18 hr when complexed by carbonate ions, giving $[AmO_2(CO_3)_3]^{4-}$.

[a] Ac^{3+}, Th^{4+}, Cm^{3+}, and ions of Pa are colorless.
[b] In dilute solutions; golden brown in concentrated solutions.
[c] Color is tan in $HClO_4$ solution $[Pu(H_2O)_9]^{4+}$, red-brown in HCl, and emerald green in HNO_3.

Since extrapolation to infinite dilution is impossible for most of the actinide ions, owing to hydrolysis—for example, Pu^{4+} cannot exist in solution below 0.05 M in acid—sometimes only approximate oxidation potentials can be given. The potentials are sensitive to the anions and other conditions.

20-6　Complexes and Stereochemistry

The actinides have a far greater tendency to form complexes than do the lanthanides. There are extensive series of complexes with oxo anions of all types (RCO_2^-, NO_3^-, SO_4^{2-}, CO_3^{2-}, $H_nPO_4^{-3+n}$, $C_2O_4^{2-}$), halide ions, BH_4^-, and other ligands, especially chelating ones.

A vast amount of data exists on complex ion formation in solution, since this has been of primary importance in connection with solvent extraction, ion-exchange behavior, and precipitation reactions involved in the technology of actinide elements. The general tendency to complex ion formation decreases in the direction controlled by factors such as ionic size and charge, so that the order is generally $M^{4+} > MO_2^{2+} > M^{3+} > MO_2^+$. For anions the order of complexing ability is generally: uninegative ions, $F^- > NO_3^- > Cl^- > ClO_4^-$; binegative ions, $CO_3^{2-} > ox^{2-} > SO_4^{2-}$.

There are a great many actinide nitrate complexes, many of which are important in separation procedures whereby the elements are extracted from aqueous nitric acid into nonpolar solvents. Of these, the UO_2^{2+} complexes are best characterized; they are typically 8-coordinate with two bidentate NO_3^- ions and two neutral ligands (H_2O, THF, DMSO, and R_3PO) forming a distorted equatorial hexagon.

Because of the enormous variety of ligands that form actinide complexes and the number of oxidation states, the stereochemistry found in complexes and compounds of the actinides is extraordinary. An overview is presented in Table 20-7. The relatively large sizes of the actinide ions coupled with the high electrostatic attraction due to formal charges of +3 to +6, along with the fact that a large number of valence shell orbitals are potentially available for bonding, all lead to the result that higher coordination numbers, especially 8 and 9, are very common.

For the +3 oxidation state, where there is much resemblance to the lanthanides, octahedral coordination is often found, but so also is coordination number 9. Eight-coordination is very common for the +4 oxidation state. In the +5 oxidation state the AnF_8^{3-} ions afford rare examples of discrete cubic coordination. In oxidation state +6, the hexafluorides are strictly octahedral.

The stereochemistry of the MO_2^{2+} complexes varies considerably. Thus for uranium the OUO unit can have 4–6 ligand atoms in, or close to, the equatorial plane, giving tetragonal, pentagonal, or hexagonal bipyrimidal coordination. The 6-coordinate species are least common and 8-coordinated species most numerous. The latter commonly have an essentially flat equatorial six-membered ring as in $Na[UO_2(\eta^2\text{-}O_2CMe)_3]$ or $UO_2(NO_3)_2(OPPh_3)_2$, but puckered rings are also formed.

CHEMISTRY OF ACTINIUM, THORIUM, AND PROTACTINIUM

20-7　Actinium

Actinium occurs in traces in uranium minerals, but it is made on a milligram scale by neutron capture in radium (Table 20-2). The actinium +3 ion is separated from

Table 20-7 Stereochemistry of Actinides

Oxidation state	Coordination number	Geometry[a]	Examples
+3	3	Pyramidal	$U[CH(SiMe_3)_2]_3$
	6	Octahedral	$An(acac)_3$, $[An(H_2O)_6]^{3+}$
	7	Pentagonal bipyramidal	$UI_3(THF)_4$
	8	Bicapped trigonal prism	$PuBr_3$, $[AmCl_2(H_2O)_6]^+$
	9	Tricapped trigonal prism	UCl_3, $AmCl_3$
	9	Irregular	$(NH_4)U(SO_4)_2(H_2O)_4$
+4	4	Distorted	$U(NPh_2)_4$
	5	Distorted *tbp*	$U_2(NEt_2)_8$
	6	Octahedral	UCl_6^{2-}, $UCl_4(HMPA)_2$
	8	Cubic (O_h)	$(Et_4N)_4[U(NCS)_8]$
		Dodecahedral	$[Th(ox)_4]^{4-}$, $Th(S_2CNEt_2)_4$
		Fluorite structure	ThO_2, UO_2
		Square antiprism	$ThI_4(s)$, $U(acac)_4$, $Cs_4[U(NCS)_8]$
	9	Capped square antiprism	$Th(trop)_4 \cdot H_2O$
	10	Bicapped square antiprism	$K_4Th(ox)_4 \cdot 4H_2O$
	10	?	$An(trop)_5^-$ (Th or U)
	12	Irregular icosahedral	$[Th(NO_3)_6]^{2-}$
		Distorted cubo-octahedral	$An(BH_4)_4$ (Np, Pu)
	14	Complex	$An(BH_4)_4$ (Th, Pa, U)
+5	6	Octahedral	UF_6^-, $\alpha\text{-}UF_5$ (infinite chain)
	7		$\beta\text{-}UF_5$
	8	Cubic (O_h)	Na_3AnF_8 (An = Pa, U, No)
	9	Complex	PaF_7^{2-} in K_2PaF_7
+6	6	Octahedral	AnF_6, Li_4UO_5(distorted), UCl_6
	3,4,5,6,7	See text	AnO_2^{2+} complexes
	8	?	$M_2^I UF_8$
+7	6	Octahedral	Li_5NpO_6, $Na_3O_4(OH)_2 \cdot nH_2O$

[a] For detailed discussion of crystal structures, many of which are more complicated, see A. F. Wells, *Structural Inorganic Chemistry*, 5th ed., Oxford University Press, Oxford, UK, 1984.

the excess of radium and isotopes of Th, Po, Bi, and Pb formed simultaneously by ion-exchange elution or by solvent extraction with thenoyltrifluoroacetone.

The general chemistry of Ac^{3+} in both solid compounds and solution, where known, is very similar to that of lanthanum, as would be expected from the similarity in position in the Periodic Table and in radii (Ac^{3+}, 1.10; La^{3+}, 1.06 Å) together with the noble gas structure of the ion. Thus actinium is a true member of Group 3, the only difference from lanthanum being in the expected increased basicity. The increased basic character is shown by the stronger absorption of the hydrated ion on cation-exchange resins, the poorer extraction of the ion from concentrated nitric acid solutions by tributyl phosphate, and the hydrolysis of the trihalides with water vapor at ~1000°C to the oxohalides AcOX; the lanthanum halides are hydrolyzed to oxide by water vapor at 1000°C.

The crystal structures of actinium compounds, where they have been studied, for example, in AcH_3, AcF_3, Ac_2S_3, and AcOCl, are the same as those of the analogous lanthanum compounds.

Table 20-8   Some Thorium Compounds

Compound	Form	mp (°C)	Properties
ThO_2	White, crystalline; fluorite structure	3220	Stable, refractory, soluble in $HF + HNO_3$
ThN	Refractory solid	2500	Slowly hydrolyzed by water
ThS_2	Purple solid	1905	Metal-like; soluble in acids
$ThCl_4$	Tetragonal white crystals	770	Soluble in and hydrolyzed by H_2O, Lewis acid
$Th(NO_3)_4 \cdot 5H_2O$	White crystals		Very soluble in H_2O, alcohols, ketones, ethers
$Th(IO_3)_4$	White crystals		Precipitated from 50% HNO_3, very insoluble
$Th(acac)_4$	White crystals	171	Sublimes in a vacuum 160°C
$Th(BH_4)_4$	White crystals	204	Sublimes in a vacuum about 40°C
$Th(C_2O_4)_2 \cdot 6H_2O$	White crystals		Precipitated from up to 2 M HNO_3

The study of even milligram amounts of actinium is difficult owing to the intense γ-radiation of its decay products that rapidly build up in the initially pure material.

20-8   Thorium

The principal ore is *monazite* found in India and other countries; thorium is extracted by a complex process culminating in solvent extraction of the nitrate.

Some representative thorium compounds are given in Table 20-8; only few compounds in oxidation states lower than +4 are known.

Binary Compounds

The *oxide* ThO_2 is formed on heating the nitrate or the hydroxide. Addition of H_2O_2 to aqueous solutions of salts gives an insoluble peroxide of variable composition.

The *halides* are obtained in high temperature reactions of the metal with halogens, HF, CCl_4, etc., as white polymeric solids that are commonly contaminated with oxide or other impurities due to exposure to traces of H_2O or O_2, reaction with glass or other vessels. *Oxohalides* $ThOX_2$ result from interaction of ThO_2 with ThX_4 at elevated temperature. The hydrated fluoride which can be precipitated from aqueous solutions by HF, can be dehydrated in a stream of HF.

Various borides, sulfides, carbides, nitrides, etc., have been obtained by direct interaction of the elements at elevated temperatures. Like other actinide and lanthanide metals, thorium also reacts at elevated temperatures with hydrogen. Products with a range of compositions can be obtained, but two definite phases, ThH_2 and Th_4H_{15}, have been characterized.

Complexes and Salts[6]

Nitrate and sulfate solutions in water on evaporation give $Th(NO_3)_4 \cdot 5H_2O$ and $Th(SO_4)_2 \cdot 8H_2O$; other salts can be obtained also. The nitrate is a common starting

[6]D. L. Clark *et al.*, *Chem. Rev.* **1995**, *95*, 25.

material and is soluble in alcohols. The addition of F^-, IO_4^-, $C_2O_4^{2-}$, and PO_4^{3-} to *strongly* acidic solutions precipitates insoluble salts. This allows the separation of Th from elements having +3 or +4 ions that also give similar precipitates in solutions of lower acidity but form soluble complex anions in strong acid. The Th^{4+} ion is larger and less readily hydrolyzed than other +4 cations but even so, above pH *ca.* 3, hydrolysis occurs giving monomeric $Th(OH)_n^{+4-n}$ and more complicated polymeric species. The high charge also facilitates the formation of ions such as $[Th(NO_3)(H_2O)_n]^{3+}$ or $[ThCl_2(H_2O)_n]^{2+}$; the unusual $[Th(CO_3)_5]^{6-}$ ion is also known.

Halide Complexes

Direct interaction of thorium metal turnings with Br_2 or I_2 in THF at 0°C gives $ThX_4(THF)_4$; these complexes are hydrocarbon soluble and useful starting materials for making compounds such as $Th(OAr)_4$, $Th(NR_2)_4(THF)_2$, pyrazolyl-borates, etc.[7]

Alkoxides

A variety of types are known. These may be homoleptic, such as $Th(2,6-Bu^tC_6H_3O)_4$, or mixed ligand as in $Th(OBu^t)_4py_2$ or $Th(OAr)[N(SiMe_3)_2]_3$. The homoleptic species show a propensity for giving dimers, tetramers, and more complicated species, especially for smaller alkoxides like OMe. Bulky alkoxides or aryloxides are likely to give monomers or dimers as in the equilibrium for $R = CHPr_2^i$:[8]

$$2Th(OR)_4 \rightleftharpoons Th_2(OR)_8$$

In the dimer the structure can be considered as two *tbp* ThO_5 units joined along a common axial-equatorial edge. The reaction of $ThBr_4(THF)_4$ with the potassium salt of an even bulkier alkoxo group $(2,6-Bu^tC_6H_3O)$ gives $ThBr_2(OAr)_2(THF)_2$ that can be alkylated to give $Th(OAr)_2(CH_2SiMe_3)_2$.[9] The latter compound reacts with H_2 to give $Th_3(\mu_3-H)_2(\mu_2-H)_4(OAr)_6$. Another hydrido cluster is $Th_6H_7Br_{14}$[10] that has an octahedral Th_6 core.

Amido Compounds

These often resemble the alkoxides both in the method of synthesis and in stoichiometries. Thus interaction of $ThBr_4(THF)_4$ with KNR_2 gives $Th(NPh_2)_4(THF)$ or $K[Th(NMePh)_5]$.[11] The amido species can be used to make alkoxides by interaction with alcohols.

[7]D. L. Clark *et al.*, *Inorg. Chem.* **1992**, *31*, 1628; R. A. Andersen *et al.*, *J. Alloys Compd.* **1994**, *213/214*, 11; J. A. McCleverty *et al.*, *Polyhedron* **1996**, *15*, 2023.
[8]D. L. Clark *et al.*, *Inorg. Chem.* **1995**, *34*, 5416.
[9]D. L. Clark *et al.*, *Organometallics* **1996**, *15*, 949.
[10]T. P. Braun and A. Simon, *Chem. Eur. J.* **1996**, *2*, 511.
[11]D. L. Clark *et al.*, *Inorg. Chem.* **1995**, *34*, 1695. For other amides see P. Scott and P. B. Hitchcock, *J. Chem. Soc., Dalton Trans.* **1995**, 603; D. L. Clark *et al.*, *Organometallics* **1995**, *14*, 2799.

Phosphorus Compounds

These are rare. The compounds[12] $Th[P(CH_2CH_2PMe_2)_2]_4$ and $ThCl_2[N(CH_2-CH_2PPr^i)_2]_2$ have, respectively, a bidentate P,P,P ligand with a pendant group and a similar chelating PNP ligand. Organometallic types, $Cp_2Th(PR_2)_2$[13] and $Cp_2Th(PR_2)Ni(CO)_2$, are also known.

Other Complexes

A variety of neutral complexes are formed with β-diketonate, dithiocarbamate, carbamate, and related ligands. The $Th(acac)_4$ is electrochemically reduced to $[Th^{III}(acac)_4]^-$ which loses $acac^-$ to give $Th(acac)_3$—one of the very few Th^{III} complexes known.[14]

While Th^{IV} compounds are diamagnetic, a paramagnetic species[15] with the sterically demanding ligand amino[ethyl(trimethylsilyl)amido] (tren') has a singly reduced 1,4-diazobutadiene ligand, i.e., $(Bu_2^tDAB)^{\cdot-}$.

Thorium(IV) complexes of some catecholate, hydroxypyridonate, and desferrioxamine ligands have been investigated and stability constants determined in connection with searches for sequestering agents for Pu^{4+}.[16]

20-9   Protactinium

While protactinium is present at less than the parts per million level in uranium ores, this is still the major source because it can be extracted from the residues accumulated in large-scale production of uranium. There are extreme technical difficulties because of colloid formation but more than 100 g have been isolated.

The main oxidation state is V, but there is no Pa^{5+} aquo ion because of hydrolysis to species such as $PaO(OH)^{2+}$ and $PaO(OH)_2^+$ or the formation of complexes with virtually all anions. The fluoro complexes, PaF_6^-, PaF_7^{2-}, and PaF_8^{3-} are of greatest importance and many solids containing them have been isolated. Other stable complexes are formed with $C_2O_4^{2-}$, SO_4^{2-}, citrate, tartrate, β-diketones, and so on.

In the solid state, an important Pa^V compound is the oxide (Pa_2O_5) obtained by ignition of most other compounds in air. On further heating it gives lower oxide phases and ultimately PaO_2.

The pentahalides are all known. The chloride sublimes at 160°C in vacuum even though it has a polymeric structure consisting of infinite chains of irregular pentagonal bipyramidal $PaCl_7$ units sharing edges. It is soluble in THF but readily hydrolyzes to oxo chlorides such as Pa_2OCl_8 or $Pa_2O_3Cl_4$.

The oxidation state IV can be obtained in aqueous solution by reduction of Pa^V with Cr^{2+} or Zn/Hg but Pa^{IV} is rapidly reoxidized by air. Solid compounds include PaO_2, PaF_4, and $PaCl_4$, and chloro, bromo, and other complexes have been characterized. The esr spectra suggest that Pa^V is a $5f^1$ rather than a $6d^1$ ion.

[12]P. G. Edwards *et al., Organometallics* **1995,** *14,* 3649; *J. Chem. Soc., Dalton Trans.* **1995,** 3401.
[13]C. J. Burns *et al., Organometallics* **1994,** *13,* 3491.
[14]A. Vallat *et al., J. Chem. Soc., Dalton Trans.* **1990,** 921.
[15]P. Scott and P. B. Hitchcock, *J. Chem. Soc., Chem. Commun.* **1995,** 579.
[16]K. N. Raymond *et al., Inorg. Chem.* **1996,** *35,* 4128.

THE CHEMISTRY OF URANIUM

The existence of the element was recognized by Klaproth in 1789, but only in 1841 did Peligot actually isolate the metal itself. However, it was not until the discovery of uranium fission by Meitner, Hahn, and Strassmann in 1939 that it became commercially important. Its most important ores are *uraninite* (usually called *pitchblend*, and approximating to UO_2) and uranium vanadates.

20-10 Binary Compounds

Solid state compounds of uranium have been much studied because of their unusual magnetic properties and their inertness. They are usually made by direct interaction with uranium metal.

The main *oxides* are UO_2, which is brown-black and highly non-stoichiometric, U_3O_8, greenish-black, and UO_3 orange yellow. The latter is best made by heating uranyl nitrate or "ammonium uranate" (see below) and the other oxides can be made from UO_3:

$$3UO_3 \xrightarrow{700°C} U_3O_8 + \tfrac{1}{2}O_2$$

$$UO_3 + CO \xrightarrow{350°C} UO_2 + CO_2$$

The oxides readily dissolve in HNO_3 to give the UO_2^{2+} ion. An air-sensitive hydrous oxide $UO_2 \cdot nH_2O$ is formed from UCl_4 on hydrolysis.[17]

The addition of H_2O_2 to solutions of UO_2^{2+} gives a pale yellow peroxide best described as $(UO_2^{2+})(O_2^{2-})(H_2O)_2$ while action of NaOH and H_2O_2 forms the stable salt $Na_4[UO_2(O_2)_3]$.

The metal reacts on heating with B, C, N, P, As, etc., to give semi-metallic solids: compounds of Si, S, Se, and Te such as US, U_7Te_{12}, and $CsCuUTe_3$ are also known.[18]

Uranates

The fusion of uranium oxides with alkali or alkaline earth carbonates, or thermal decomposition of salts of the uranyl acetate anion, gives orange or yellow materials generally referred to as uranates, for example,

$$2UO_3 + Li_2CO_3 \longrightarrow Li_2U_2O_7 + CO_2$$

$$Li_2U_2O_7 + Li_2CO_3 \longrightarrow 2Li_2UO_4 + CO_2$$

$$Li_2UO_4 + Li_2CO_3 \longrightarrow Li_4UO_5 + CO_2$$

Addition of aqueous NH_3 to $UO_2(NO_3)_2$ solutions affords the so-called ammonium diuranate. This is mainly a hydrated uranyl hydroxide containing NH_4^+. Below 580°C it gives UO_3 and above, U_3O_8.

[17]D. Rai *et al.*, *Inorg. Chem.* **1990**, *29*, 260.
[18]J. A. Cody and J. A. Ibers, *Inorg. Chem.* **1995**, *34*, 3165.

Table 20-9 Uranium Halides[a]

+3	+4	+5	+6
UF_3, green	UF_4, green	UF_5, white-blue	UF_6, colorless
UCl_3, red	UCl_4, green	U_2Cl_{10}, red-brown	UCl_6, green
UBr_3, red	UBr_4, brown	UBr_5, dark red	
UI_3, black	UI_4, black		

[a]Other halides are U_2F_9, U_3F_{13}, U_4F_{17}, U_5F_{22}. Mass spectra show the existence of UF_3Cl and UF_2Cl_2 in the gas phase (J. K. Gibson, *Radiochim. Acta* **1994**, 65, 227.)

Uranates do not contain discrete ions such as UO_4^{2-}. They have octahedral U^{VI} with unsymmetrical oxygen coordination such that two U—O bonds are short (~1.92 Å), constituting a sort of uranyl group, with other longer U—O bonds in the plane normal to this UO_2 axis linked into chains or layers. However, Na_4UO_5 has strings of UO_6 octahedra sharing opposite corners.

20-11 Uranium Halides and Their Adducts

The principal halides[19] are given in Table 20-9.

Fluorides

Uranium trifluoride, which may be obtained by reduction of UF_4 with H_2 or by reaction of UF_4 with UN at 950°C is a high-melting, nonvolatile solid. It resembles the lanthanide trifluorides in being insoluble in water or dilute acids.

The *tetrafluoride*, also insoluble in water, can be prepared in several ways, *inter alia*, the following:

$$UO_2 \xrightarrow[500\text{-}600°C]{C_2Cl_4F_2} UF_4$$

$$NH_4UF_5 \xrightarrow{500°C} UF_4$$

It can also be precipitated in hydrated form from aqueous solutions of U(IV).

The *pentafluoride*, which exists in two crystalline modifications, melts to give an electrically conducting liquid. It may be obtained by reduction of UF_6 (see later).

Uranium hexafluoride is the most important fluoride and is made on a large scale, since it is the compound used in gas diffusion plants for the separation of uranium isotopes. It is made by the reactions

$$UO_2 \xrightarrow{HF} UF_4 \xrightarrow{ClF_3} UF_6$$

[19]R. D. Hunt *et al., Inorg. Chem.* **1994**, 33, 388; N. P. Freestone and J. H. Holloway, *Topics f-Elem. Chem.* **1991**, 2, 67 (423 refs).

A laboratory preparation is the reaction of uranium metal and ClF_3.[20] The hexafluoride forms colorless crystals (mp 64.1°C) with a vapor pressure of 115 mm Hg at 25°C. It is a powerful fluorinating agent (e.g., it converts CS_2 to SF_4) and is rapidly hydrolyzed by water. The hexafluoride is also used to make UF_4 and UF_5 by the following reactions:

$$UF_6 + Me_3SiCl \longrightarrow Me_3SiF + \tfrac{1}{2}Cl_2 + UF_5$$

$$UF_6 + 2Me_3SiCl \longrightarrow 2Me_3SiF + Cl_2 + UF_4$$

Other Halides

The *trihalides* are insoluble polymers made by passing HCl or HBr over UH_3; UI_3 is formed by high temperature interaction of U with I_2 and Zn.[21]

The *tetrachloride* is readily obtained by refluxing UO_3 with hexachloropropene, probably *via* UCl_6 as intermediate. The bromide and iodide are similarly made, but UI_4 decomposes slowly at room temperature to UI_3 and I_2.

The *hexachloride* made by chlorination of U_3O_8 and C, or of UCl_4 at *ca.* 400°C forms green crystals that sublime in vacuum. In CH_2Cl_2 it slowly decomposes to the red-brown volatile crystalline pentachloride that has a structure of the M_2Cl_{10} (M = Ta, Mo) type.

Halide Adducts

Most of the halides are Lewis acids and form adducts with Lewis bases. Both UF_5 and UF_6 react with nitrogen bases to give molecular or ionic species, examples being $[(bipy)_2H]^+[UF_6]^-$, $UF_4(2\text{-}FC_5H_4N)$, and $U_2F_{12}(bipy)$.[22]

Although UCl_4 can be reduced to $UCl_3 \cdot nTHF$ by NaH and used for other syntheses, the best starting material if uranium turnings are available is the iodide $UI_3(THF)_4$ formed on reaction of the metal with I_2 in THF at 0°C.[23] It is very air-sensitive and has a pentagonal bipyramidal structure. Other adducts[24] of UI_3 as well as similar ones for UCl_3 and UBr_3 are known, e.g., with MeCN, dme, and py. The pale green $UCl_4(THF)_4$ and adducts with MeCN, RNC, $OPPh_3$, and similar ones from UBr_4 and UI_4 have been characterized.[25]

All the halides form *halogeno anions* by interaction of the halide with M^+X^- in solvents such as $SOCl_2$, in melts or, for fluorides, sometimes water. Examples are $[UF_6]^{2-}$, $[UF_7]^{3-}$, and $[UF_8]^{4-}$ for U^{IV}, $Rb[U^VF_6]$ and $Na_2[U^{VI}F_8]$. Chloro complexes such as $(Me_4N)_3[UCl_8]$ are known for U^V as well as *pseudohalide anions* such as $[U(NCS)_8]^{4-}$ from the reaction of UCl_4 and KNCS in MeCN.

[20]I. C. Tornieporth-Oetting and T. M. Klapötke, *Inorg. Chim. Acta* **1994**, *215*, 5.
[21]J. Collins *et al.*, *Inorg. Chim. Acta* **1994**, *222*, 91.
[22]R. Bougon *et al.*, *Inorg. Chem.* **1994**, *33*, 4510.
[23]A. P. Sattelberger *et al.*, *Inorg. Chem.* **1989**, *28*, 1771; **1994**, *33*, 2248.
[24]J. Drozdzyúski and J. G. H. du Preez, *Inorg. Chim. Acta* **1994**, *218*, 203.
[25]C. J. Burns *et al.*, *Inorg. Chem.* **1996**, *35*, 537; W. G. Van der Sluys *et al.*, *Inorg. Chim. Acta* **1993**, *204*, 251; J. G. H. du Preez *et al.*, *J. Chem. Soc., Dalton Trans.* **1991**, 2585; *Inorg. Chim. Acta* **1991**, *189*, 67.

20-12   Uranium Hydrides[26]

The important solid hydride UH_3 is often more suitable for the synthesis of compounds than is the bulk metal which reacts reacts rapidly and exothermically with hydrogen at 250 to 300°C to give a pyrophoric black powder. The reaction is reversible:

$$U + \tfrac{3}{2}H_2 = UH_3 \qquad \Delta H_f^0 = -129 \text{ kJ mol}^{-1}$$

The hydride decomposes at somewhat higher temperatures to give extremely reactive, finely divided metal. A study of the isostructural deuteride by X-ray and neutron diffraction shows that the deuterium atoms lie in a distorted tetrahedron equidistant from four uranium atoms; no U—U bonds appear to be present, and the U—D distance is 2.32 Å. The stoichiometric hydride UH_3 can be obtained, but the stability of the product with a slight deficiency of hydrogen is greater.

Some typical useful reactions are the following:

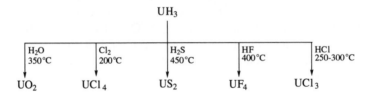

The reduction of UCl_4 with $LiBu^t$ in alkanes gives an ill-defined but reactive hydride species.

Borohydrides

Uranium in oxidation states III and IV has quite extensive chemistry with BH_4^- and also with $BH_3CH_3^-$. The compound $U(BH_4)_4$, which has full T_d symmetry (all BH_4^- are η^3) decomposes on heating in solution to give $U(BH_4)_3$. This can also be made by interaction of UH_3 and B_2H_6 in THF: it forms numerous adducts, such as $U(BH_4)_3(THF)_3$, $U(BH_4)_3(dmpe)_2$, and $U(BH_4)_3[Ph_2Ppy]_2$. $U(BH_4)_4$ reacts with ethers to give a structurally diverse series of products depending on the ether; examples are monomeric $U(BH_4)_4(THF)_2$ and $[U(BH_4)_4(Et_2O)]_\infty$. It is convenient to note here that borohydrides of Pa, Th, Np, and Pu have properties similar to those of $U(BH_4)_3$ but there are some structural differences. The adduct $UH(BH_4)_3(dme)$ has a U—H bond. There are also some alkylborohydrides[27] such as $U(H_3BMe)_4$ and some organometallic derivatives noted later.

20-13   Nitrogen and Phosphorus Compounds

Like most transition metals, uranium and the other actinides (cf. Th above) form complexes with a wide variety of N-ligands. Neptunium and plutonium often give similar species.

[26]J. M. Haschke in *Topics f-Elem. Chem.* **1991**, 2, 1 (78 refs).
[27]C. Villiers and M. Ephritikhine, *J. Chem. Soc., Chem. Commun.* **1995**, 979.

Dialkylamides[28] are known for U^{III}, U^{IV}, and U^V. Some examples are U[N(SiMe$_3$)$_2$]$_3$, [U(NEt$_2$)$_3$]BF$_4$, and U[N(C$_6$H$_5$)$_2$]$_4$. They can undergo reactions characteristic of amido compounds such as insertion reactions with CS$_2$ or CO$_2$ leading to dithiocarbamate or carbamate complexes respectively and addition reactions with donor ligands. The ligand [N(CH$_2$CH$_2$NSiMe$_3$)$_3$]$^{3-}$ (L) gives tripodal amino (triamido) complexes of U and Th of the type [U(μ-Cl)L]$_2$. The Cl atoms can be substituted by OBu^t, BH$_4$, or C$_5$R$_5$.[29]

Imido complexes[30] are few, examples being Me$_3$SiN=U^V[N(SiMe$_3$)$_2$]$_3$, Me$_3$SiN=U^{VI}F[N(SiMe$_3$)$_2$]$_3$, and Cp$_2^*$U(N-2,4,6-Bu^{t}C$_6$H$_2$).

A variety of *polypyrazolylborates*, mostly with HB(3,5-Me$_2$pz)$_3^-$ (L) have been made, examples being UX$_3$L, X = Cl, MeCO$_2$ and LUCl$_2$[CH(SiMe$_3$)$_2$].[31]

Finally *porphyrins* and *phthalocyanins* can give "sandwich" type compounds such as U^{IV}(porph)$_2$, U^{IV}(Pc)$_2$, and [U^{III}(porph)$_2$]$^-$.[32] Thorium gives similar complexes.

Phosphorus complexes are rather rare. The 8-coordinate complex U[P(CH$_2$CH$_2$PMe$_2$)$_2$]$_4$[33] is isostructural with its thorium analogue.

20-14 Oxygen and Sulfur Ligands

The most extensive range of compounds are the alkyl or aryloxides that are known for U^{III} to U^{VI}; the U^{IV} compounds are similar to those for Th^{IV}. They are usually made from dialkylamides and ROH, but isopropyls can be obtained by dissolving uranium and I$_2$ in Pr^{i}OH.[34] Typical compounds are U(OMe)$_6$, U$_2$(OEt)$_{10}$, U(OAr)$_4$, and U(OAr)$_3$ where Ar = 2,6-Bu^{t}C$_6$H$_3$.[35] There are also anionic complexes such as (Bu$_4^t$N)[U$_2$(OBu^t)$_9$] and [Li(THF)$_4$][U(OAr)$_5$].

Other O-ligand complexes include the 8-coordinate acetylacetonate U^{IV}(acac)$_4$, compounds of the O-tripod L = CpCo[P(O)(OEt)$_2$]$_3^-$ such as UCl$_3$L(THF), and UCl$_2$L$_2$.

For sulfur, dithiocarbamates and thiolates have long been known, some recent examples being U(SBu^n)$_4$, [U(SBu^n)$_6$]$^{2-}$, [U(SPh)$_6$]$^{2-}$, and [Li dme]$_4$[U(SCH$_2$-CH$_2$S)$_4$].[36] The homoleptic polychalcogenide K$_4$[U(Se$_2$)$_4$] was made by heating U with Se and K$_2$Se at 300°C; the anion has four η^2-Se$_2$ units in a distorted dodecahedral structure.[37]

[28]J. C. Berthet *et al.*, *J. Chem. Soc., Dalton Trans.* **1995**, 3019; **1996**, 947; D. L. Clark *et al.*, *Inorg. Chem.* **1994**, *33*, 2248; S. E. Turman and W. G. van der Sluys, *Polyhedron* **1992**, *11*, 3139.

[29]P. Scott and P. B. Hitchcock, *J. Chem. Soc., Dalton Trans.* **1995**, 603.

[30]C. J. Burns *et al.*, *J. Am. Chem. Soc.* **1995**, *117*, 9448; R. G. Denning *et al.*, *J. Chem. Soc., Chem. Commun.* **1994**, 2601.

[31]J. Takats *et al.*, *Inorg. Chim. Acta* **1995**, *229*, 315; M. P. C. Campello *et al.*, *J. Organomet. Chem.* **1995**, *484*, 37.

[32]K. M. Kadish *et al.*, *J. Am. Chem. Soc.* **1993**, *115*, 8153.

[33]P. G. Edwards *et al.*, *Organometallics* **1995**, *14*, 3649.

[34]N. N. Sauer *et al.*, *Inorg. Chem.* **1995**, *34*, 4862.

[35]W. G. Van der Sluys and A. P. Sattelberger, *Chem. Rev.* **1990**, *90*, 1027; C. J. Burns *et al.*, *Inorg. Chem.* **1994**, *33*, 4245; J. M. Berg *et al.*, *Inorg. Chem.* **1993**, *32*, 647.

[36]M. Ephritikhine *et al.*, *J. Chem. Soc., Dalton Trans.* **1994**, 3563; **1995**, 237; K. Tatsumi *et al.*, *Inorg. Chem.* **1990**, *29*, 4928.

[37]A. C. Sutorik and M. G. Kanatzidis, *J. Am. Chem. Soc.* **1991**, *113*, 7754.

20-15   Aqueous Chemistry; Uranyl Compounds

The aqueous chemistry of uranium species is quite complicated owing to extensive complexation as well as hydrolytic reactions which lead to polymeric ions under appropriate conditions. The formal potentials for 1 M $HClO_4$ have been given in Table 20-5. In the presence of other anions the values differ: thus for the U^{4+}/U^{3+} couple in 1 M $HClO_4$ the potential is -0.631 V, but in 1 M HCl it is -0.640 V. The simple ions and their properties are also listed in Table 20-6. Because of hydrolysis, aqueous solutions of uranium salts have an acid reaction that increases in the order $U^{3+} < UO_2^{2+} < U^{4+}$. The main hydrolyzed species of UO_2^{2+} at 25°C are UO_2OH^+, $(UO_2)_2(OH)_2^{2+}$, and $(UO_2)_3(OH)_5^+$, but the system is a complex one and the species present depend on the medium; at higher temperatures the monomer is most stable but the rate of hydrolysis to UO_3 of course increases. The solubility of large amounts of UO_3 in UO_2^{2+} solutions is also attributable to formation of UO_2OH^+ and polymerized hydroxo-bridged species. The U^{3+} ion, which is readily obtained in perchlorate or chloride solutions by reduction of UO_2^{2+} electrolytically or with zinc amalgam, is a powerful reducing agent. Its oxidation by I_2 or Br_2 occurs by an outer-sphere mechanism.

The solutions of U^{3+} in 1 M HCl are stable for days, but in more acid solution, spontaneous oxidation occurs more rapidly. The hydrate $UF_3 \cdot H_2O$ and $U_2(SO_4)_3 \cdot 5H_2O$ can be obtained from fluoride and sulfate solutions, respectively. The sulfate $(NH_4)U(SO_4)_2 \cdot (H_2O)_4$ is a crystalline salt that can be handled in air. The U^{III} is 9-coordinate but the arrangement does not correspond to any ideal polyhedron.

The U^{4+} ion is only slightly hydrolyzed in molar acid solutions:

$$U^{4+} + H_2O \rightleftharpoons U(OH)^{3+} + H^+ \quad K_{25°C} = 0.027 \ (1 \ M \ HClO_4, NaClO_4)$$

but it can also give polynuclear species in less acid solutions. The U^{4+} ion gives insoluble precipitates with F^-, PO_4^{3-}, and IO_3^- from acid solutions ($cf.$ Th^{4+}).

Although the UO_2^+ ion is extremely unstable towards disproportionation (Section 20-4), its aqueous chemistry has been much studied by stopped-flow methods. In DMSO it is more stable, both thermodynamically and kinetically (half-life, ~0.5 h).

The UO_2^{2+} ion in aqueous solution is stable and forms many complexes. It can be reduced to U^{IV}, for example by Cr^{2+}; in the reverse process, use of ^{18}O tracer has shown that some oxidants (PbO_2, H_2O_2, and MnO_2) give almost entirely $U(^{18}O)_2^{2+}$, while O_2 and O_3 give $U(^{16}O)(^{18}O)^{2+}$.

In acidic aqueous solutions UO_2^{2+} acts as a photocatalyst involving $^*UO_2^{2+}$ for oxidation of benzene by H_2O_2.[38]

Uranyl Compounds[39]

The most common uranium compounds are those containing the linear, symmetrical UO_2^{2+} ion. This ion readily adds 4–6 donor atoms in its equatorial plane to give

[38]Y. Mao and A. Bakac, *Inorg. Chem.* **1996**, *35*, 3925.
[39]R. G. Denning, *Struct. Bonding (Berlin)* **1992**, *79*, 215.

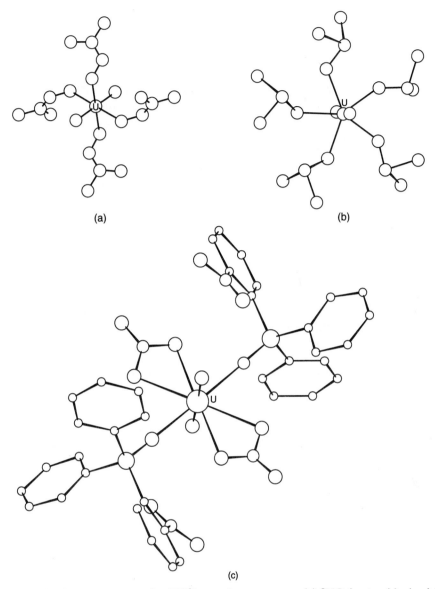

(a)

(b)

(c)

Figure 20-2 Some representative UO_2^{2+} complex structures. (a) $[UO_2(\text{acetamidoxime})_4]^{2+}$, (b) $[UO_2(DMSO)_4]^{2+}$, and (c) $UO_2(NO_3)_2(Ph_3PO)_2$.

structures such as those shown in Fig. 20-2. The number of known complexes of UO_2^{2+} is enormous, and virtually every kind of oxygen donor ligand as well as many nitrogen donors and even sulfur donors have been found in such complexes.

One of the most important uranyl complexes is the *nitrate,* which crystallizes with six, three, or two molecules of water, depending on whether it is obtained from dilute, concentrated, or fuming nitric acid. In each case, there are two bidentate nitrate ions and two water molecules coordinated equatorially. The anhydrous compound can be obtained by the reactions

$$U + N_2O_4(l) \xrightarrow{MeCN} UO_2(NO_3)_2 \cdot N_2O_4 \cdot 2MeCN \xrightarrow{163\,°C} UO_2(NO_3)_2$$

The nitrate is of major importance because the extraction of uranyl nitrate from aqueous nitric acid into nonpolar solvents is a classic method for separating and purifying the element. Typically, phosphate esters, such as $[(C_4H_9)(C_2H_5)\text{-}CHCH_2O]_2P(O)(OH)$, are used and the extracted species is $UO_2(NO_3)_2[(RO)_2PO_2]_2$.

Uranyl carbonato complexes are important and much studied since they occur in Nature and are of environmental concern: this is also the case for other actinide carbonate complexes.[40] The CO_3^{2-} ion is exceptionally strongly bound to UO_2^{2+} and similar actinide ions. There are several naturally occurring minerals such as UO_2CO_3 while the anion $[UO_2(CO_3)_3]^{4-}$ has importance in the extraction of U from ores and is responsible for the migration of UO_2^{2+} in ground waters. Interaction of the 4– ion with $HClO_4$ proceeds as follows:

$$3[UO_2(CO_3)_3]^{4-} + 6H^+ \rightleftharpoons [(UO_2)_3(CO_3)_6]^{6-} + 3CO_2 + 3H_2O$$

In the structures of the carbonates, the CO_3^{2-} group can be chelate as in $[UO_2(CO_3)_3]^{2-}$ or in more complex compounds, bridging.

Other uranyl salts are given by organic acids,[41] sulfate, halides, and so on; the water-soluble acetate in the presence of an excess of sodium acetate in dilute acetic acid gives a crystalline precipitate of $NaUO_2(OCOCH_3)_3$. There is considerable research on the use of cyclic hexadentate and related ligands to form strong complexes with the UO_2^{2+} ion.

One of the characteristic features of UO_2^{2+} compounds is fluorescence, and uranyl oxalate is used as an actinometer.

Oxohalides

These stable compounds are soluble in water and are made by reactions such as

$$UCl_4 + O_2 \xrightarrow{350\,°C} UO_2Cl_2 + Cl_2$$

$$UO_3 + 2HF \xrightarrow{400\,°C} UO_2F_2 + H_2O$$

The oxofluoride is polymeric with F bridges. Some dioxofluorouranates, e.g., $Na_2[UO_2F_4]$, are also known.[42]

Alkoxide and aryloxide complexes[43] such as the pentagonal bipyramidal $UO_2(o\text{-}2,6\text{-}Pr_2^iC_6H_3O)_2py_3$, $[Na\ THF_3][UO_2(o\text{-}2,6\text{-}Me_2C_6H_3O)_4]$ and alkoxides such as $UO_2(OBu^t)_2(OPPh_3)$ are known.

[40]D. L. Clark *et al.*, *Chem. Rev.* **1995**, *95*, 25; *Inorg. Chem.* **1995**, *34*, 4797.
[41]A. Bismondo *et al.*, *Inorg. Chim. Acta* **1994**, *223*, 151; J. Leciejewicz *et al.*, *Struct. Bonding (Berlin)* **1995**, *82*, 43 (uranyl carboxylates, 78 refs).
[42]M. K. Chaudhuri *et al.*, *J. Chem. Soc., Dalton Trans.* **1994**, 2693.
[43]C. J. Burns *et al.*, *Inorg. Chem.* **1995**, *34*, 4079.

Other Ligands

In addition to amides[44] such as $UO_2[N(SiMe_3)_2]_2(OPPh_3)_2$ there are numerous compounds of the type $[UO_2L]^{2+}$, where L is a macrocyclic N_6 ligand. Schiff base complexes[45] commonly show strong affinity for binding anions such as Cl^-, $H_2PO_4^{2-}$, and NO_2^- that can be very specific in aqueous solutions.[46]

In view of the vast quantities of uranium in sea water (4.5×10^{17} kg as $[UO_2(CO_3)_4]^{4-}$), complexes that can be used for solvent extraction have been studied. Tripodal ligands like tris[2-(2-carboxyphenoxy)ethyl]amine provide at least one H-bond donor for interaction with oxo groups as well as donor ligands for coordination to the metal center. These ligands are powerful extractants for UO_2^{2+} into $CHCl_3$ solutions of the ligands.[47] The extraction of uranyl nitrate into ethers and other oxygenated solvents has been used for UO_2^{2+} extraction and purification for many years. Since H-bonding is probably critical in complexing by "uranophiles," much data on H-bonding to UO_2^{2+} species has been collected.[48]

Some other species of interest include an imido analogue of UO_2^{2+} having a linear $O{=}U{=}NR$ group, $R = P(m\text{-}tol)_3$, made[49] by the reaction

$$[UOCl_5]^- + Me_3Si[NP(m\text{-}tol)_3] \longrightarrow Me_3SiCl + [UOCl_4NP(m\text{-}tol)_3]^-$$

The red Ph_4P^+ salt is stable in air and soluble in CH_2Cl_2 and MeCN. The nitrido oxo cation NUO^+ has been identified in the gas phase by mass spectra; it was obtained by ion-molecule reactions from U^+.

20-16 Organometallic Chemistry of the Actinides

The overwhelming majority of organoactinide compounds are formed by uranium together with a few by thorium. The first organoactinide compound, Cp_3UCl, was made in 1956 and the majority of those made since have Cp, Cp*, C_8H_8, or other ring systems as ligands.

Cyclopentadienyls[50]

These and compounds of substituted cyclopentadienyls such as Cp* or $C_5H_3(SiMe_3)_2$ constitute the major class. Examples are listed in Table 20-10. It may be noted that (a) there are compounds in all oxidation states from III to VI, (b) there is a wide variety of X groups extending from halogens to alkyls and aryls, amido, imido,

[44]C. J. Burns et al., Inorg. Chem. **1992**, 31, 3724.
[45]R. M. Maurya and R. C. Maurya, Rev. Inorg. Chem. **1995**, 15, 1 (346 refs); E. R. Dockal et al., Polyhedron **1996**, 15, 245.
[46]D. N. Reinhoudt et al., J. Am. Chem. Soc. **1994**, 116, 4341.
[47]K. N. Raymond et al., Inorg. Chim. Acta **1995**, 240, 593.
[48]K. N. Raymond et al., J. Am. Chem. Soc. **1992**, 114, 8138
[49]R. G. Denning et al., J. Chem. Soc., Chem. Commun. **1994**, 2601.
[50]M. F. Lappert et al., J. Chem. Soc., Dalton Trans. **1995**, 3335; B. D. Zwick et al., Transuranium Elem. Symp. **1990**, Am. Chem. Soc., Washington DC, 1992; M. Ephritikhine, New J. Chem. **1992**, 16, 451 (recent advances); B. E. Bursten and R. J. Strittmatter, Angew. Chem. Int. Ed. Engl. **1991**, 30, 1069; J. Am. Chem. Soc. **1991**, 113, 552 (bonding and electronic structures).

Table 20-10 Representative Cyclopentadienyl Uranium Compounds[a]

U[III]	U[IV]	U[V]	U[VI]
Cp$_3$U	Cp*UCl$_3$	Cp$_2^*$UO(OAr), Ar = 2,6-Pr^{i}C$_6$H$_3$	Cp$_2^*$U(NPh)$_2$
Cp$_2$UCl	Cp$_2$UX$_2$, X = Cl, NEt$_2$, BH$_4^-$, Me[b]	(C$_5$H$_4$Me)UNBu^t	
Cp*UX$_2$(THF), X = I, o-2,6- Pr^{i}C$_6$H$_3$	[Cp$_3$U(H$_2$O)$_2$]$^+$	[Cp*U(NMe$_2$)$_3$THF]$^+$	
	Cp$_4$U		

[a]Compounds of Th and other actinides are known for M[III] and M[IV], e.g., Cp$_4$Th, Cp$_4$Np, Cp$_3$Cf. For Th see D. L. Clark *et al.*, *Organometallics* **1996**, *15*, 1488 (Cp*, alkyl and aryloxide derivatives, e.g., Cp*Th(OAr)(CH$_2$SiMe$_3$)$_2$.
[b]For Cp$_2$U(NHR)$_2$, R = 2,6-Me$_2$C$_6$H$_3$, Me, Bu^t made by interaction of Cp$_2$UMe$_2$ with the amine. The aryl Cp$_2$U(NHAr)$_2$ loses ArNH$_2$ in solvents to give the imide Cp$_2$U = N(2,6-Me$_2$C$_6$H$_3$) while reactions of Cp$_2$U(NHR)$_2$ with acetylenes produces Cp$_2$U(C≡CR')$_2$, R' = Ph, Bu^t. See M. S. Eisen *et al.*, *J. Chem. Soc., Dalton Trans.* **1996**, 2541.

alkyl and aryl oxo, hydrido, MPh$_3$, M = Si, Ge, Sn, etc., and (c) many compounds are Lewis acids and give adducts with THF, N-donors, acetylenes, and so on. The compounds are made by standard reactions of NaCp or TlCp on halides in THF or dme, e.g.,

$$UCl_4 + 3TlCp \xrightarrow{dme} Cp_3UCl + 3TlCl$$

$$UCl_4 + 4KCp \xrightarrow{C_6H_6} Cp_4U + 4KCl$$

$$UCl_3 + 3KCp \xrightarrow[reflux]{C_6H_6} Cp_3U + 3KCl$$

$$U + 3CpH \longrightarrow Cp_3U + {}^3/_2H_2$$

Generally Cp* compounds are more stable than Cp ones and have a greater solubility in organic solvents. Some chemistry of the cyclopentadienyls[51] of U[IV] is shown in Fig. 20-3.

Other hydrides can be made for U and Th by the reaction

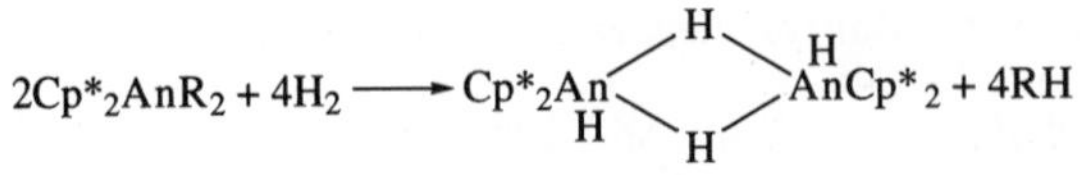

$$2Cp^*_2AnR_2 + 4H_2 \longrightarrow Cp^*_2An\underset{H}{\overset{H}{\underset{\diagdown}{\diagup}}}\underset{H}{\overset{H}{\underset{\diagup}{\diagdown}}}AnCp^*_2 + 4RH$$

The hydrogen in these molecules is rather hydridic and the compounds undergo rapid and interesting reactions with alcohols, ketones, and halocarbons. With CO, at low temperature, there is an insertion into the An—H bond to give η^2-formyls, but these are converted at higher temperatures into the remarkable bridged species,

[51]M. Ephritikhine *et al.*, *J. Chem. Soc., Dalton Trans.* **1992**, 1573.

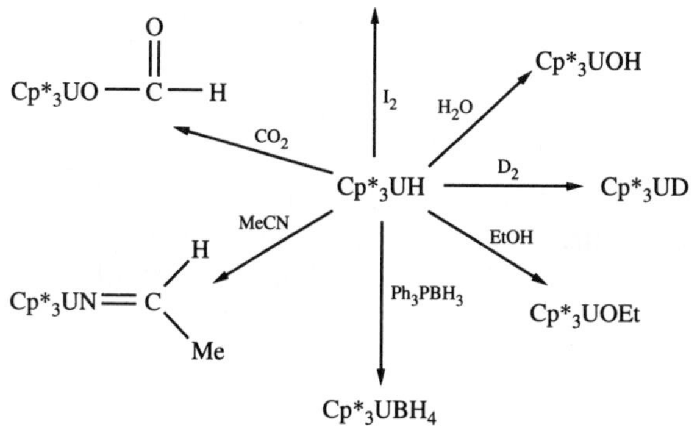

Figure 20-3 Some reactions of Cp_3^*UH, $Cp^* = C_5Me_5$, $C_5H_4Bu^t$, or $C_5H_4SiMe_3$.

as shown in the following reaction scheme (where Th carries two Cp*):

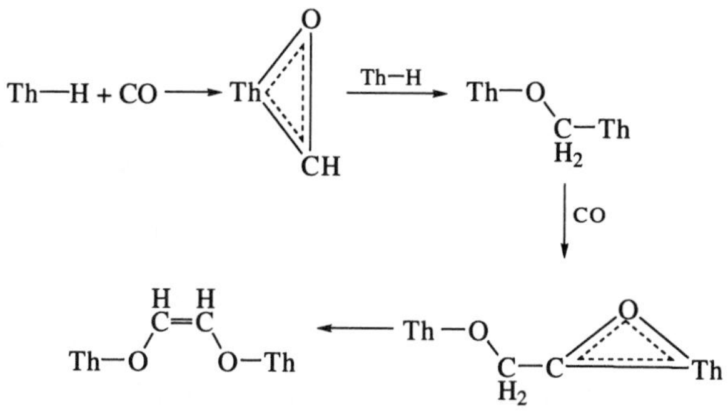

The most studied *alkyl and aryl compounds* are of the type Cp_3AnR, $An = Th$, U, Np, and their Cp* analogues. Unlike the *d*-block alkyls, decomposition by β-H transfer and alkene elimination is not a major route but both thermal and photochemical pathways are multiple and complex.

Such compounds undergo migratory insertion reactions with CO,[52] RNC, or SO_2, as shown in the following equations:

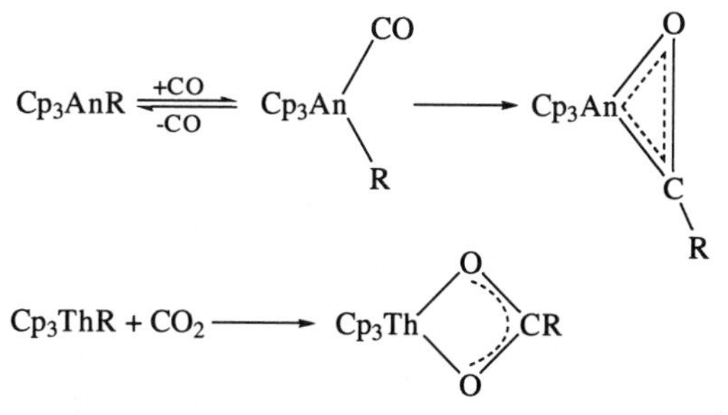

---

[52]C. Villiers and M. Ephritikhine, *J. Chem. Soc., Dalton Trans.* **1994,** 3397 (mechanism and aryl rearrangements).

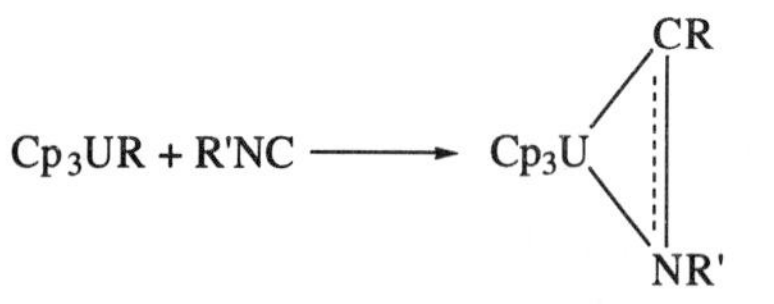

$$Cp_3UR + R'NC \longrightarrow Cp_3U\underset{NR'}{\overset{CR}{\diagup}}$$

The η^2-bonding of the CO and RNC insertion products is characteristic of these actinide compounds, but has some parallels in the organometallic chemistry of the d-block elements. There are similar insertions into An—NR_2 bonds to give products with η^2 ligands, namely,

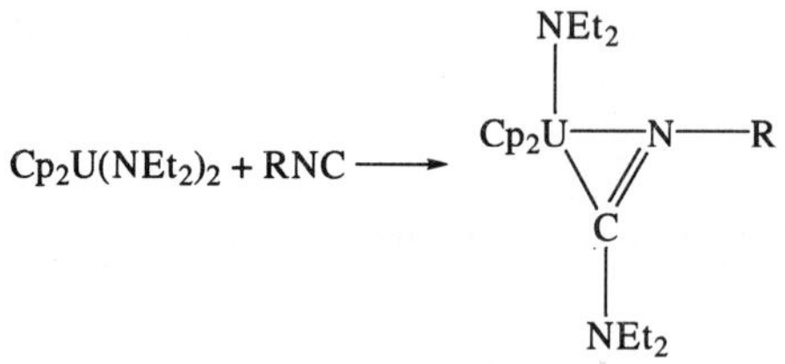

$$Cp_2U(NEt_2)_2 + RNC \longrightarrow \underset{\underset{NEt_2}{|}}{\overset{\overset{NEt_2}{|}}{Cp_2U}}\text{—N—R}$$

Alkyl Complexes

For Th and U the homoleptic allyls (or 2-methylallyls), which are unstable above 0°C, may be prepared by the reaction

$$AnCl_4 + 4(allyl)MgX \longrightarrow An(allyl)_4 + 4MgXCl$$

The allyl groups are bound in an η^3 fashion, and may be displaced by reactions with HX or ROH to give, for example, $U(C_3H_5)_3X$ or $[U(C_3H_5)_3(OR)]_2$. More stable An—allyl linkages are known in compounds that also contain η^5-Cp or η^5-Cp* ligands.

Annulene and Cycloheptatrienyl Compounds[53]

The green, pyrophoric compound $U(C_8H_8)_2$, known as *uranocene,* was first made in 1968 by the reaction

$$UCl_4 + 2K_2C_8H_8 \longrightarrow U(C_8H_8)_2 + 4KCl$$

but it can be made in other ways. The molecular structure (Fig. 20-4) is that of a sandwich (*cf.* Cp_2Fe) with planar rings and D_{8h} symmetry. The Th, Pa, Np, and Pu analogues are all known. Just as for Cp and arenes, mono COT compounds are known, e.g., $[\eta\text{-}C_8H_8U(\mu\text{-}SPr^i)_2]_2$, which is made by reacting the thiol with $U(COT)(BH_4)$.[54] The cycloheptatrienyls can be made by interaction of UCl_4 and C_7H_8 in THF to give the anion $[U(\eta\text{-}C_7H_7)_2]^-$; the C_7H_7 rings are planar.[55]

[53]A. Streitwieser *et al., Organometallics* **1991,** *10,* 1922; **1993,** *12,* 5023.
[54]C. J. Burns *et al., Organometallics* **1993,** *12,* 1497.
[55]M. Ephritikhine *et al., J. Chem. Soc., Chem. Commun.* **1995,** 183.

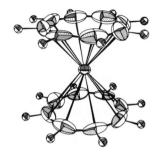

Figure 20-4 The molecular structure of An(C₈H₈)₂ compounds (An = U and Th).

Alkyls and Aryls

Some σ-alkyl and aryl compounds have been made with supporting ligands such as Cp, Cp*, bulky aryloxides such as o-2,6-Bu$_2^t$C$_6$H$_3$O,[56] and tertiary phosphines, as exemplified by Cp$_2^*$UCH(SiMe$_3$)$_2$[57] and UR$_4$(dmpe)$_2$. Homoleptic compounds have proved more difficult to make. There are, however, the compounds Th(CH$_2$Ph)$_4$, *ato* complexes [Li tmed]$_3$[MMe$_7$], M = Th, U, and the blue U[CH(SiMe$_3$)$_2$]$_3$ which is stable as the solid at −60°C though unstable in solution over a few hours.[58]

Actinide Carbonyls

The carbonyl U(CO)$_6$ can exist below 20 K and its CO stretch, 1961 cm^{-1}, implies appreciable π-bonding. The compound (Me$_3$SiC$_5$H$_4$)$_3$UCO is stable under CO in solution but cannot be isolated. However, the similar compound (C$_5$Me$_4$H)$_3$UCO, which has a CO stretch at 1880 cm^{-1} in the solid, is stable and has been structurally characterized.[59] The low CO stretch suggests U → CO back-bonding between U $5f\pi$ and CO π^* orbitals. Carbon monoxide also inserts into the U–C bond of (C$_5$H$_4$Me)$_3$UBu^t to give an acyl and insertion of ethylene also occurs.[60]

Finally a *carbene compound* with a U=C bond is Cp$_3$U=CHPMe$_3$.[61] Various alkyl and hydrido Cp species can catalyze oligomerization of terminal acetylenes, activate C–H bonds, hydrogenate alkenes, and so on.[62]

Carbollide Compounds

These are relatively scarce but can be obtained[63] by reactions such as

$$UBr_4(MeCN)_2 + 2Li_2C_2B_9H_{11} \xrightarrow{MeCN} Li_2[U^{IV}(\eta^5\text{-}C_2B_9H_{11})_2Br_2]$$

$$UI_3(THF)_4 + Li_2C_2B_9H_{11} \xrightarrow{tmeda} [Li\,tmeda][U(C_2B_9H_{11})_2I_2(THF)_2]$$

[56]D. L. Clark *et al.*, *Organometallics* **1996**, *15*, 949.
[57]I. L. Fragalà *et al.*, *Organometallics* **1996**, *15*, 205.
[58]A. P. Sattelberger *et al.*, *Organometallics* **1989**, *8*, 855.
[59]E. Carmona *et al.*, *J. Am. Chem. Soc.* **1995**, *117*, 2649.
[60]R. A. Andersen *et al.*, *Organometallics* **1995**, *14*, 3942.
[61]R. Bau *et al.*, *Organometallics* **1990**, *9*, 6494; R. E. Cramer *et al.*, *Organometallics* **1990**, *9*, 694.
[62]M. S. Eisen *et al.*, *J. Am. Chem. Soc.* **1995**, *117*, 6343.
[63]D. Rabinovitch *et al.*, *Organometallics* **1996**, *35*, 1425.

THE TRANSURANIUM ELEMENTS[64]

20-17 Neptunium, Plutonium, Americium

Though several isotopes of each of these elements are known, only a few (listed in Table 20-2) have been obtained in macroscopic amounts. Both ^{237}Np and ^{239}Pu are available on a large (~kilogram) scale from spent uranium fuel rods of reactors, but the technical problems in working with above-tracer amounts of these highly radioactive and (especially for Pu) toxic elements are formidable. Nevertheless, a number of effective separation methods have been developed. The principles upon which separation procedures are based are the following:

1. *Stabilities of Oxidation States.* The stabilities of the major ions involved are: $UO_2^{2+} > NpO_2^{2+} > PuO_2^{2+} > AmO_2^{2+}$; $Am^{3+} > Pu^{3+} \gg Np^{3+}$, U^{4+}. It is thus possible (see also Table 20-5) by choice of suitable oxidizing or reducing agents to obtain a solution containing the elements in different oxidation states; they can then be separated by precipitation or solvent extraction. For example, Pu can be oxidized to PuO_2^{2+} while Am remains as Am^{3+}. The former can then be removed by solvent extraction or the latter by precipitation of AmF_3.

2. *Extractability into Organic Solvents.* As noted previously, the MO_2^{2+} ions can be extracted from nitrate solutions into organic solvents. The M^{4+} ions can be extracted into tributyl phosphate in kerosene from 6 M nitric acid solutions; the M^{3+} ions can be similarly extracted from 10 to 16 M nitric acid; and neighboring actinides can be separated by a choice of conditions.

3. *Precipitation Reactions.* Only M^{3+} and M^{4+} give insoluble fluorides or phosphates from acid solutions; the higher oxidation states either give no precipitate or can be prevented from precipitation by complex formation with sulfate or other ions.

4. *Ion-Exchange Procedures.* Although ion-exchange procedures, both cationic and anionic, can be used to separate the actinide ions, they are best suited for small amounts of material.

To illustrate briefly the practical employment of these principles on a large scale, the following processes may be mentioned.

1. *The hexone [MeC(O)CH$_2$CHMe$_2$] method,* in which both U and Pu are oxidized to the AnO_2^{2+} ions, which are then extracted into an organic solvent by the hexone, leaving the fission products in the aqueous layer. Uranium is then separated from Pu by selectively (SO_2) reducing the latter to Pu^{4+} and solvent extracting only the UO_2^{2+}.

2. *The tributyl phosphate extraction method,* where control of the concentration of nitric acid from 6 to 16 M allows control of the transfer of various actinide ions into kerosene containing 30% $(C_4H_9O)_3PO$.

3. *The lanthanum fluoride cycle* in which hexone extraction along with LaF_3 as a carrier precipitate are used.

[64]G. T. Seaborg, *Acc. Chem Res.* **1995**, *28*, 257.

20-18 Compounds of Np, Pu, and Am

Oxides

The important oxides are the *dioxides,* obtained by heating nitrates or hydroxides in air. They are isostructural with UO_2, but PuO_2 is nonstoichiometric unless roasted in air at 1200°C. No AnO_3 compounds are known, though corresponding hydrated substances, for example, $NpO_3 \cdot 2H_2O$, can be made. Only americium gives a lanthanide-like oxide (Am_2O_3); the oxide AmO_2 is used in smoke detectors where the α-particles ionize air between electrodes causing a tiny current to flow. Smoke particles absorb the ions and stop the current flow, which sounds the alarm.

Halides

These are listed in Table 20-11. While their preparations and properties show much similarity to those of the uranium halides, there is a steady decrease in stability of the higher oxidation states with increasing atomic number (which continues in the succeeding elements). In short, the "actinide" concept becomes more and more valid.

The fluorides AnF_3 and AnF_4 can be precipitated from aqueous solutions in hydrated form. The hexafluorides have been greatly studied because they are volatile; the melting points and stabilities decrease in the order $U > Np > Pu$. Plutonium hexafluoride is so very much less stable than UF_6 that at equilibrium, the partial pressure of PuF_6 is only 0.004% of the fluorine pressure. Hence PuF_6 formed by fluorination of PuF_4 at 750°C must be quenched immediately by a liquid nitrogen probe. The compound also undergoes self-destruction by α-radiation damage, especially in the solid; it must also be handled with extreme care owing to the toxicity of Pu.

Neptunium hexafluoride has a $5f^1$ configuration according to esr and absorption spectra, the octahedral field splitting the sevenfold orbital degeneracy of the $5f$

Table 20-11 Halides of Np, Pu, and Am[a]

+2	+3	+4	+6
	NpF_3, purple-black	NpF_4, green	NpF_6, orange mp 55.1°C
	PuF_3, purple	PuF_4, brown	PuF_6, red-brown, mp 51.6°C
	AmF_3, pink	AmF_4, tan	
	$NpCl_3$, white	$NpCl_4$, red-brown	
	$PuCl_3$, emerald		
$AmCl_2$, black	$AmCl_3$, pink		
	$NpBr_3$, green	$NpBr_4$, red-brown	
	$PuBr_3$, green		
$AmBr_2$, black	$AmBr_3$, white		
	NpI_3, brown		
	PuI_3, green		
AmI_2, black	AmI_3, yellow		

[a]Certain oxohalides $M^{III}OX$, M^VOF_3, and $M^{VI}O_2F_2$ are known.

electron and leaving a ground state that has only spin degeneracy. This quenching of the orbital angular momentum is similar to that in the first-row d-transition group.

Other Compounds

A substantial number of compounds, particularly of plutonium, are known, and most of them closely resemble their uranium analogues. The hydride systems of Np, Pu, and Am are more like that of thorium than that of uranium and are complex. Thus nonstoichiometry up to $MH_{2.7}$ is found in addition to stoichiometric hydrides such as PuH_2 and AmH_2. As with uranium, many complex salts are known, such as Cs_2PuCl_6, $NaPuF_5$, $KPuO_2F_3$, $NaPu(SO_4)_2 \cdot 7H_2O$, and $Cs_2Np(NO_3)_6$.

Aqueous Chemistry

The formal reduction potentials of Np, Pu, and Am were given in Table 20-5 and information on the ions in Table 20-6. Solutions of the IV to VI oxidation states undergo spontaneous reduction because of their own α radiation. For Np (like U), the potentials of the four oxidation states are separated, but unlike the chemistry of U, the NpO_2^+ ion is comparatively stable. For Pu, however, the potentials are close and at 25°C in 1 M $HClO_4$ we have

$$K = \frac{[Pu^{VI}][Pu^{III}]}{[Pu^V][Pu^{IV}]} = 10.7$$

which indicates that measurable amounts of all four states can be present at once. The Am ions that can exist in appreciable concentrations are Am^{3+}, AmO_2^+ and AmO_2^{2+}, with Am^{3+} being the usual state in acid solution. However, for 1 M basic solution the $Am(OH)_4$–$Am(OH)_3$ couple has a value of +0.5 V, nearly 2 V less than for the Am^{4+}/Am^{3+} couple in acid solution. Thus pink $Am(OH)_3$ can be readily converted into black $Am(OH)_4$ or hydrous AmO_2 by the action of hypochlorite.

As with uranium, the solution chemistry is complicated, owing to hydrolysis and polynuclear ion formation, complex formation with anions other than perchlorate, and disproportionation reactions of some oxidation states. The tendency of ions to displace a proton from water increases with increasing charge and decreasing ion radius, so that the tendency to hydrolysis increases in the same order for each oxidation state, that is, Am > Pu > Np > U and $M^{4+} > MO_2^{2+} > M^{3+} > MO_2^+$; simple ions such as NpO_2OH^+ or $PuOH^{3+}$ are known in addition to polymeric species that in the case of plutonium can have molecular weights up to 10^{10}.

The complexing tendencies decrease, on the whole, in the same order as the hydrolytic tendencies. The formation of complexes shifts the oxidation potentials, sometimes influencing the relative stabilities of oxidation states; thus the formation of sulfate complexes of Np^{4+} and NpO_2^{2+} is strong enough to cause disproportionation of NpO_2^+.

As for the carbonate complexes of UO_2^{2+}, there has been similar study of NpO_2^+ in groundwater where CO_3^{2-} is in high concentration and is a strong complexing agent for actinides.[65] The main species characterized in solutions

[65]D. L. Clark et al., J. Am. Chem. Soc. 1996, 118, 2089.

having $[Bu_4^nN]^+$ as cation are $[NpO_2(CO_3)_3]^{5-}$, $[NpO_2(CO_3)_2(H_2O)_2]^{3-}$, and $[NpO_2(CO_3)(H_2O)_3]^-$.

Typical plutonium disproportionations at low acidities are:

$$3Pu^{4+} + 2H_2O \rightleftharpoons PuO_2^{2+} + 2Pu^{3+} + 4H^+$$

$$2Pu^{4+} + 2H_2O \rightleftharpoons PuO_2^+ + Pu^{3+} + 4H^+$$

$$PuO_2^+ + Pu^{4+} \rightleftharpoons PuO_2^{2+} + Pu^{3+}$$

EXAFS studies of Pu^{IV} in HNO_3 solutions have shown that the complexes $(Et_4N)_2$-$[Pu(NO_3)_6]$, $Pu(NO_3)_2^{2+}$, and $Pu(NO_3)_4$ have structures similar to the solid state structures of Th and Np nitrates with $\mu_2\text{-}NO_3$ ligands.[66] Similar studies of alkaline Np^{VII} solutions prepared by bubbling O_3 through a 2.5 M NaOH solution of $NpO_2(OH)_2$ show the presence of a trans dioxo ion,[67] formally NpO_2^{3+}, which must be coordinated by OH^- and possibly H_2O ligands.

In view of the extreme toxicity of Pu in particular, the removal of actinides from living organisms is an important goal. Various sequestering agents have been studied. A new class comprises tetracatecholate ligands with amine backbones that give an 8-coordinate cage for Pu^{IV} with ligating terephthalamido groups.[68]

20-19 The Heavier Elements[69]

The isotope ^{242}Cm was first isolated among the products of α-bombardment of ^{239}Pu, and its discovery actually preceded that of americium. Isotopes of other elements were first identified in products from the first hydrogen bomb explosion (1952) or in cyclotron bombardments. Although Cm, Bk, and Cf have been obtained in macro amounts (Table 20-2), much of the chemical information has been obtained on the tracer scale. For the later elements, i.e., those with $Z > 100$, identification of a few atoms of short lifetime has required the use of very rapid separation techniques and detection based on their nuclear properties.

For Cm, Bk, and Cf the correspondence of the actinide and the lanthanide series becomes most clearly revealed. The position of curium corresponds to that of gadolinium where the f shell is half-filled. For curium the +3 oxidation state is the normal state in solution, although unlike gadolinium, a solid tetrafluoride, CmF_4, has been obtained. Berkelium has +3 and +4 oxidation states, as would be expected from its position relative to terbium, but the +4 state of terbium does not exist in solution, whereas for Bk it does. Although CfF_4 and CfO_2 have been made, the remaining elements have only the +3 and +2 states.

Ion exchange has been the principal tool in the isolation of these elements. By detailed comparison with the elution of lanthanide ions and by extrapolating data

[66]D. K. Veirs *et al.*, *Inorg. Chem.* **1996**, *35*, 2841.
[67]D. L. Clark *et al.*, *J. Am. Chem. Soc.* **1997**, *119*, 5259.
[68]K. N. Raymond *et al.*, *Inorg. Chem.* **1992**, *31*, 4904; **1996**, *35*, 4128.
[69]D. C. Hoffman, *Radiochim. Acta* **1996**, *72*, 1; *C&EN* **1994**, *May 2*. Extensive reviews.

for the lighter actinides such as Np^{3+} and Pu^{3+}, the order of elution of the heavier actinides can be accurately forecast.

The main problems in the separations are (a) separation of the actinides as a group from the lanthanide ions (which are formed as fission fragments in the bombardments which produce the actinides) and (b) separation of the actinides from one another.

The first problem is solved by the use of concentrated HCl as eluting agent; since the actinide ions form chloride complexes more easily, they are desorbed first from a cation-exchange resin, thus effecting a *group* separation. Conversely, the actinides are more strongly adsorbed on anion-exchange resins. The actinide ions are effectively separated from each other by elution with citrate or similar eluant.

After separation by ion exchange, the actinides may be precipitated by fluoride or oxalate in macroscopic amounts or collected on an insoluble fluoride precipitate for trace quantities.

Curium

Solid curium compounds are known, e.g., CmF_3, CmF_4, $CmCl_3$, $CmBr_3$, white Cm_2O_3 (mp 2265°C), and black CmO_2. Where X-ray structural studies have been made—and these are difficult, since amounts of the order of 0.5 μg must be used to avoid fogging of the film by radioactivity and because of the destruction of the lattice by emitted particles—the compounds are isomorphous with other actinide compounds.

The potential of the Cm^{4+}/Cm^{3+} couple must be greater than that of the Am^{4+}/Am^{3+} couple, which is 2.6 to 2.9 V, so that solutions of Cm^{4+} must be unstable. When CmF_4, prepared by dry fluorination of CmF_3, is treated with 15 M CsF at 0°C, a pale yellow solution is obtained that appears to contain Cm^{4+} as a fluoro complex. The solution exists for only an hour or so at 10°C owing to reduction by the effects of α-radiation; its spectrum resembles that of the isoelectronic Am^{3+} ion.

The solution reactions of Cm^{3+} closely resemble those of the lanthanide and actinide +3 ions, and the fluoride, oxalate, phosphate, iodate, and hydroxide are insoluble. Complexes appear to be weaker than those of preceding elements.

Magnetic measurements on CmF_3 diluted in LaF_3 and also the close resemblance of the absorption spectra of CmF_3 and GdF_3 support the hypothesis that the ion has the $5f^7$ configuration.

Berkelium

As the analogue of Tb, this element has both +3 and +4 states in both solid compounds and in solutions. Berkelium(II) has been detected in aqueous media by absorption spectra[70] after pulse radiolysis of Bk^{III}. Many solid compounds have been structurally characterized, for example, BkO_2 (fluorite structure), Bk_2O_3, BkX_3 (F, Cl, Br, I), BkOX (Cl, Br, I), and $(Me_4N)_2[BkCl_6]$ which is isostructural with K_2PtCl_6.[71]

[70]J. C. Sullivan *et al., Inorg. Chem.* **1988,** *27,* 597.
[71]L. R. Morss *et al., J. Less Common Met.* **1991,** *169,* 1.

Californium

The compounds CfF_4 and CfO_2 are well characterized. In the $+3$ state there are halides, Cf_2O_3, the sulfide, oxohalides, and Cp_3Cf. For the $+2$ state $CfBr_2$ and CfI_2 are known.

Einsteinium

Although the initial discovery was based on only 17 atoms, several isotopes are known and a few solids characterized: EsO_2, $EsCl_3$, $EsOCl$, and EsI_2.

Elements 101–112

The characterization of the heavier elements becomes increasingly difficult. They are made by irradiation of thin targets by other atoms such as ^{18}O, the product recoiling from the target usually along with others. Fast chemical studies can be made in some cases, e.g., for ^{260}Lr ($t_{1/2}$ 3 m), ^{261}Rf ($t_{1/2}$ 65 s), and ^{262}Db ($t_{1/2}$ 34 s). Thus dubnium could be eluted from cation exchange resin columns by acid and the most stable oxidation state in solution shown to be $+5$. For the heaviest elements, e.g., $^{272}111$, a nickel target was bombarded by ^{207}Bi ions:[72]

$$^{64}Ni + {}^{207}Bi \longrightarrow {}^{272}111 + {}^{1}n$$

In this case only three nuclei ($t_{1/2} = 1.5$ milliseconds) were observed. Even more extreme was the detection of *one* atom of element 112 after bombardment of a lead target with high energy zinc atoms for two weeks.[73]

Additional References

B. Alexander, *Chem. Rev.* **1995**, *95*, 273. Macrocyclic ligand complexes of actinides and lanthanides (506 references).

K. A. Gschneidner Jr., L. R. Eyring, G. R. Chopin, and G. A. Lander, Eds., *Handbook of Physics and Chemistry of Rare Earths*, Vol. 18, Elsevier-North Holland, Amsterdam, 1994. Comparison of lanthanides with actinides.

A. Harper, *Insights Spec. Inorg. Chem.*, Roy. Soc. Chem., Cambridge, UK. Technical aspects of nuclear industry.

B. Jung, N. M. Edelstein, and G. T. Seaborg, *ACS Sym. Ser.* **1994**, *565*, 361. Hydrolysis of actinide cations (74 references).

J. J. Katz, G. T. Seaborg, and L. R. Morss, *The Chemistry of the Actinide Elements*, Vols. 1, 2, Chapman and Hall, London, 1986.

T. J. Marks and I. L. Fragala, *Fundamental and Technological Aspects of Organo f-Element Chemistry*, D. Reidel, Dordrecht, 1985.

G. Meyer and L. R. Morss, Eds., *Synthesis of Lanthanide and Actinide Compounds*, Kluwer, Dordrecht, 1991.

N. B. Mikheev and A. N. Kamenskaya, *Coord. Chem. Rev.* **1991**, *109*, 1. Complexes in low oxidation states (also of lanthanides).

[72]S. Hofmann *et al.*, *Z. Phys. A: Hadrons Nucl.* **1995**, *350*, 281; See also G. Herrman, *et al.*, *Angew. Chem. Int. Ed. Engl.* **1995**, *34*, 1713.
[73]*C&EN* **1996**, *Feb 26*, 6.

L. R. Morss, R. Lester, and J. Finger, Eds., *Transuranium Elements,* Symposium, ACS, Washington, DC, 1992.

G. T. Seaborg, *Modern Alchemy. Selected Papers of G. T. Seaborg,* World Scientific, London, 1994.

W. G. Van der Sluys and A. P. Sattelberger, *Chem. Rev.* **1990,** *90,* 1027. Alkoxides (82 refs).

F. Wastin, *et al., J. Alloys Compd.* **1995,** *219,* 232 (solid actinide compounds, 33 refs).

Part 4

THE ROLE OF ORGANOMETALLIC CHEMISTRY IN CATALYSIS

Chapter 21

FUNDAMENTAL REACTION STEPS OF TRANSITION METAL CATALYZED REACTIONS

Catalysts are substances that accelerate the rates of chemical reactions, facilitate the establishment of equilibria and are capable of greatly enhancing product selectivities; they allow chemical transformations to be performed with increased efficiency, minimal waste and reduced energy consumption. It is not surprising therefore that the vast majority of products of the chemical industry involve a catalyst at some stage in their manufacture. This applies to bulk chemicals ("commodities") produced on a large scale as the starting materials for innumerable end-products, such as alcohols, ketones, carboxylic acids, hydrocarbons such as olefins and dienes that can be polymerized to polyolefins (e.g., polyethylene, polypropylene, and rubbers), and also increasingly to fine chemicals, i.e., high added-value compounds produced on a smaller scale, as well as pharmaceuticals.

The term "catalyst" is often ambiguous and not clearly defined. Most catalysts are heterogeneous, for example, finely divided metal particles or a metal oxide. Here the term refers to substances that contain some catalytically active centers, of usually unknown structures and concentrations. On the other hand, homogeneous catalysts that operate in solution are usually derived from well-defined precursors. The simplest example of a homogeneous catalyst is H^+, for example, in the acid catalyzed esterification of carboxylic acids. In this chapter and the following chapter we will be concerned primarily with the role of transition metal complexes as catalysts. In these systems all of the metal species can be expected to be active. The reaction usually starts with a stable precursor complex which brings about a chemical reaction by entering a *catalytic cycle* involving a series of metal complexes linked to each other by consecutive reaction steps. In such a cycle no one species can therefore be said to be "the" catalyst, and it is sometimes possible to enter the cycle at several different points.

In the early part of this century, coal and coal tar products were the main source of bulk chemicals. Acetylene was the major feedstock, obtained by converting coal to calcium carbide followed by hydrolysis. As the petroleum and natural gas industries developed, ethylene and other products obtained by "cracking" hydrocarbons became increasingly important. The present-day chemical industry is almost exclusively olefin-based. The interaction of olefins or acetylenes with transition metals is therefore of key importance in catalytic reactions and forms the basis for the majority of processes described in this chapter.

The majority of catalyzed processes employ heterogeneous catalysts which have the obvious advantage of ease of separation from the product. However, where high selectivity and mild reaction conditions are required, homogeneous catalysts with their well-defined ligand systems and high chemical uniformity have the advantage. It is also possible to "heterogenize" a homogeneous catalyst, either by attachment to a solid support or by using two immiscible media, without altering the underlying chemistry of these catalysts (see Section 22-16).

The rise of homogeneous catalysis, as well as the understanding of the mechanistic principles of many heterogeneously catalyzed reactions, is inextricably linked to the development of organometallic chemistry.[1] Catalytic reactions can be understood on the basis of a limited number of basic reaction types. This chapter will consider the fundamental reaction steps involved in transition metal catalyzed reactions; the next chapter will deal with catalytic reaction types and processes.

Catalytic reaction cycles are based on a number of reaction principles, such as coordinative unsaturation/ligand substitution for substrate or reagent binding to the metal center, oxidative addition, migration (insertion) reactions to achieve suitable functionalization of the substrate, and reductive elimination. They provide the mechanistic framework for the understanding of catalytic processes that involve the making and breaking of $M-C$ bonds. There are numerous model reactions which indicate the wide-ranging synthetic potential of metal-mediated reactions; many of these offer promise but have not yet become the basis of effective catalysts.

21-1 Coordinative Unsaturation

Coordinatively saturated catalyst precursors become "activated" by ligand loss. Ligand dissociation is most commonly induced by heating and is assisted by the solvent and the steric requirements of the dissociating ligand, as well as its donor strength. For example, although triphenylphosphine is quite bulky (cone angle 145°), $RhCl(PPh_3)_3$ dissociates only to a small extent at 25°C:

$$RhCl(PPh_3)_3 \rightleftharpoons RhCl(PPh_3)_2 + PPh_3 \qquad K = 2.3 \times 10^{-7}\,M$$

$$\underset{16e}{} \qquad \underset{14e}{}$$

At higher temperature, the red mononuclear complex is converted to the orange halide-bridged species:

$$2RhCl(PPh_3)_3 \rightleftharpoons (Ph_3P)_2Rh(\mu\text{-}Cl)_2Rh(PPh_3)_2 + 2PPh_3 \qquad K = 3 \times 10^{-4}\,M$$

The degree of ligand dissociation is dependent on donor and π-acceptor strength, i.e., neither strongly basic trialkylphosphines nor ligands such as CO dissociate as readily as triarylphosphines in most circumstances. For this reason complexes of trialkylphosphines often provide isolable models for catalytically active species

[1]W. A. Herrmann and B. Cornils, *Angew. Chem. Int. Ed. Engl.* **1997**, *36*, 1047.

without being active themselves. Ligand basicity can also determine whether the binding of a substrate such as ethylene follows a dissociative process, e.g., to give the square 16-electron complex $RhCl(C_2H_4)(PPh_3)_2$:

$$RhCl(PPh_3)_3 + C_2H_4 \rightleftharpoons RhCl(C_2H_4)(PPh_3)_2 + PPh_3 \qquad K = 0.4$$

or whether ligand binding is associative, as in related iridium PEt_3 complexes where the 18-electron trigonal-bipyramidal adduct $IrCl(C_2H_4)_2(PEt_3)_2$ is favored over $IrCl(C_2H_4)(PEt_3)_2$.[2]

Ligand size determines coordination numbers as well as reactivity. Thus phosphine complexes of Pd^0, frequently used as precursors in palladium catalyzed reactions, may be 4-, 3-, or 2-coordinate, as in $Pd(PMe_3)_4$, $Pd(PPr^i_3)_3$, and $Pd(PPhBu^t_2)_2$, respectively. For nickel phosphine and phosphite complexes, the dissociation constant K_d for the equilibrium

$$NiL_4 \rightleftharpoons NiL_3 + L \qquad K_d \,(25\,°C, \text{benzene})$$

varies from 6×10^{-10} M for $P(p\text{-}OC_6H_4Me)_3$ with a cone angle of $128°$, to $K_d = 5 \times 10^{-2}$ M for $PMePh_2$ ($\theta = 136°$), while for PPh_3 ($\theta = 145°$) dissociation is complete and no $Ni(PPh_3)_4$ can be detected in solution. The reaction enthalpy of binding a ligand L' to a metal complex fragment ML_n is a sensitive measure of the combined steric and electronic effects and is indicative of the reactivity of ML_n in other reactions such as oxidative addition.[3]

The creation of vacant coordination sites by chemical reactions is exemplified by the generation of Pd^0 species from Pd^{II} precursors in the presence of alkylating agents *via* reductive elimination:

$$L_2PdX_2 \xrightarrow[-\,2MX]{RM} L_2PdR_2 \xrightarrow[-\,R_2]{} L_2Pd$$

In the case of the very bulky chelate complex (21-I), the reduction process generates a T-shaped intermediate whose ligand binding enthalpy decreases in the order $L = H_2 > N_2 > C_2H_4$, i.e., a trend which is the inverse of what would normally be expected.[4]

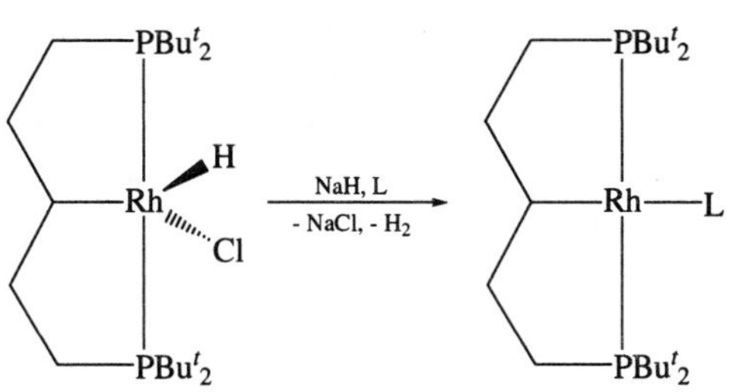

(21-I)

Sites may be made more readily available by the use of ligands with high *trans effect*. For example, the substitution of Cl^- ligand in $PtCl_4^{2-}$ by C_2H_4 is slow but is

[2]D. Milstein *et al.*, *Organometallics* **1996,** *15,* 4093.
[3]K. G. Caulton *et al.*, *Organometallics* **1996,** *15,* 4900.
[4]D. Milstein *et al.*, *Organometallics* **1996,** *15,* 1839.

accelerated by the addition of $SnCl_2$, which forms $[PtCl_3(SnCl_3)]^{2-}$ where the $SnCl_3$ ligand labilizes the Cl in *trans* position. Similarly, while PPh_3 loss from $RhCl(PPh_3)_3$ is unfavorable, the addition of H_2 to give $Rh(H)_2Cl(PPh_3)_3$ (21-IIa) results in facile dissociation of a PPh_3 ligand due to the *trans* effect of H, to give trigonal-bipyramidal (21-IIb):

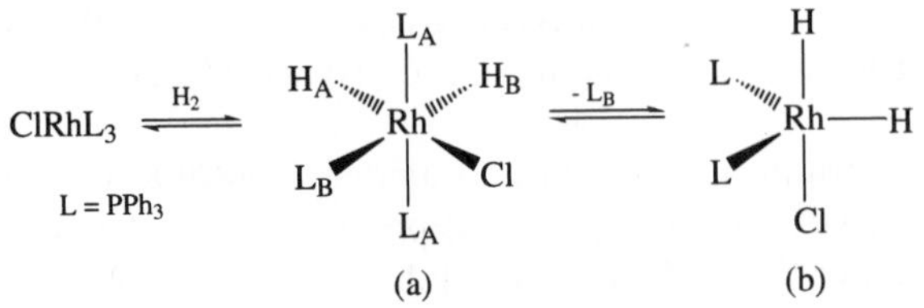

$$ClRhL_3 \rightleftharpoons$$

L = PPh_3

(a)   (b)

(21-II)

The fluxionality of ligands with potentially varying coordination modes can be a very important factor in stoichiometric and catalytic reactions. For example, indenyl complexes are often markedly more reactive than their cyclopentadienyl analogues; for example, $(\eta^5\text{-ind})Rh(CO)_2$ undergoes substitution of a CO ligand by PPh_3 *via* an S_N2 process 3.8×10^8 times faster than $(\eta^5\text{-Cp})Rh(CO)_2$.[5] This dramatic difference is the result of a "ring-slippage" of the indenyl ligand from η^5 to η^3 to accommodate the incoming ligand in an associative substitution reaction:

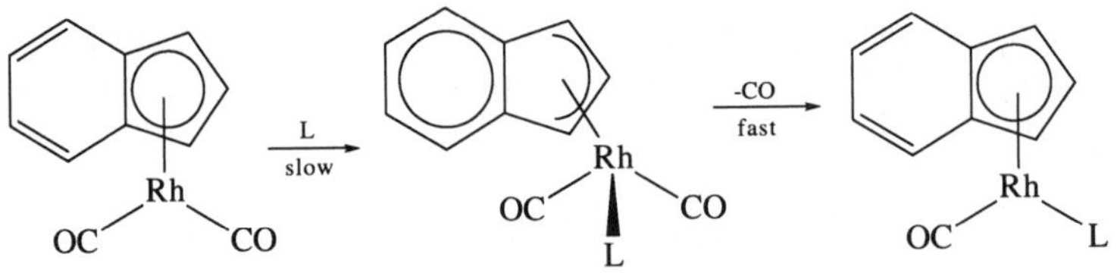

This pathway is energetically favored by the gain in aromaticity of the six-membered ring in the η^3-intermediate, a stabilization not available to Cp. It should be noted, however, that this "indenyl effect" is specific to an associative mechanism; substitution reactions of $RuCl(\eta^5\text{-ind})(PPh_3)_2$, which proceed *via* a dissociative pathway, are only one order of magnitude faster than those of $RuCl(Cp)(PPh_3)_2$.[6]

Thermal substitution reactions of 18-electron complexes are often sluggish. By contrast, 17-electron species are highly reactive, e.g., $V(CO)_6$ undergoes ligand substitution $\sim 10^{10}$ times faster than $Cr(CO)_6$ *via* an associative mechanism. Short-lived 17-electron intermediates are known to participate in substitution reactions of $Co_2(CO)_8$:

$$Co_2(CO)_8 \rightleftharpoons 2Co(CO)_4$$

$$Co(CO)_4 + L \rightleftharpoons Co(CO)_3L + CO$$

$$Co(CO)_3L + Co(CO)_4 \longrightarrow Co_2(CO)_7L$$

[5]J. C. Green *et al.*, *Organometallics* **1993,** *12,* 3688.
[6]J. Gimeno *et al.*, *Organometallics* **1996,** *15,* 302.

Similarly, substitution reactions may be redox-catalyzed:

$$(CO)_nML \xrightarrow{-e} [(CO)_nML]^+$$

$$[(CO)_nML]^+ + L' \longrightarrow [(CO)_nML']^+ + L$$

$$[(CO)_nML']^+ + (CO)_nML \longrightarrow (CO)_nML' + [(CO)_nML]^+$$

An example is the opening of a halide bridge in $Mo_2Cp_2(\mu\text{-SMe})_3(\mu\text{-Cl})$ on one-electron oxidation[7]:

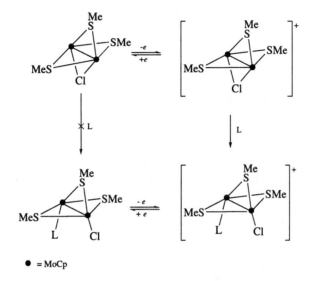

● = MoCp

Whereas thermally induced ligand substitution reactions with unreactive substrates such as $Fe(CO)_5$ require forcing conditions and tend to give low yields and mixtures of products, the addition of catalysts such as $CoCl_2$ gives single substitution products, $Fe(CO)_4(PR_3)$, in high yield.[8]

Photolysis of $L_nM(X)_2$ to give X_2 and coordinatively unsaturated fragments ML_n is rarely important in catalytic reactions but is an elegant way to probe the reactivity of ML_n species in mechanistic studies. An example is the reductive elimination of H_2 on flash photolysis of $H_2Ru(dmpm)_2$ to give $Ru(dmpm)_2$, a highly reactive species which gives η^2-complexes with ethylene and benzene and readily undergoes oxidative addition reactions.[9]

21-2 Oxidative Addition

Coordinatively unsaturated compounds, whether transition metal or main group element, are capable of accepting electron donor ligands, i.e., they act as Lewis

[7]J. Talarmin et al., J. Chem. Soc., Dalton Trans. **1996,** 3967.
[8]M. O. Albers et al., Inorg. Synth. **1989,** 26, 59.
[9]R. N. Perutz et al., Organometallics **1997,** 16, 1410.

acids. The Lewis acid character may be strong, as in electron-deficient early transition metal complexes and in compounds carrying electron-withdrawing ligands such as fluoride:

$$PF_5 + F^- \rightleftharpoons PF_6^-$$

$$TiCl_4 + 2L \rightleftharpoons TiCl_4L_2$$

while more electron-rich 16-electron compounds may bind ligands reversibly to give 18-electron species:

$$trans\text{-}IrCl(CO)(PPh_3)_2 + CO \rightleftharpoons IrCl(CO)_2(PPh_3)_2$$

$$PdCl_4^{2-} + Cl^- \rightleftharpoons PdCl_5^{3-}$$

In addition to this Lewis-acid behavior, 16- and 18-electron metal complexes can act as Lewis bases, i.e., they possess accessible electron pairs. This Lewis basic character depends strongly on the donor and acceptor strengths of the ligands and is very pronounced in complexes of strong donors such as trialkylphosphines. For example, whereas $CpCo(CO)_2$ shows little basic character and little tendency to react with electrophiles such as CH_3I, $CpCo(PMe_3)_2$ is a strong "metallic base".[10] Such compounds are particularly reactive towards oxidative addition reactions:

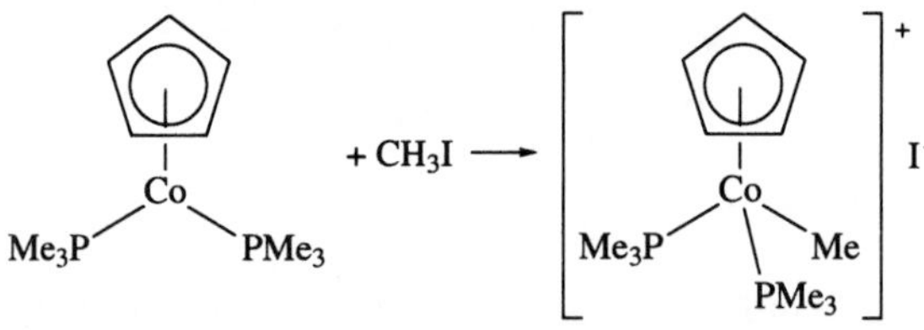

Metal basicity may be probed by protonation. While neutral species such as $Fe(CO)_5$ are only weak bases, metal carbonyl anions are readily protonated to give metal hydrides. The basicity is increased by the presence of ligands such as Cp, and Cp_2ReH for example has a basicity comparable to that of ammonia.

Examples are:

$$Fe(CO)_5 + H^+ \rightleftharpoons [HFe(CO)_5]^+$$

$$Ni\{P(OEt)_3\}_4 + H^+ \rightleftharpoons [HNi\{P(OEt)_3\}_4]^+$$

$$Cp_2ReH + H^+ \rightleftharpoons [Cp_2ReH_2]^+$$

$$[Mn(CO)_5]^- + H^+ \rightleftharpoons HMn(CO)_5$$

[10]H. Werner et al., Angew. Chem. Int. Ed. Engl. **1983**, 22, 927.

These protonations can formally be considered as oxidations, i.e., $[Mn^{-I}(CO)_5]^-$ is oxidized to $HMn^I(CO)_5$, and Re^{III} in Cp_2ReH to Re^V in the dihydrido cation.

Oxidative addition can occur when a complex behaves *simultaneously as a Lewis base* and a *Lewis acid*: the metal must be able to provide electrons for ligand binding, and must possess vacant coordination sites to accommodate two additional ligands. In general terms the reaction can be envisaged as

Very often this process can be understood on the basis of a concerted mechanism leading to a product in which X and Y are mutually *cis*. In many cases, however, the actual mechanism is extremely complicated and may involve electron transfer to give odd-electron intermediates, as well as subsequent rearrangements of the primary product to one or more stable stereoisomers; therefore, the nature of the final product does not necessarily give a guide to the course of the reaction. Some specific examples are given later.

For oxidative addition reactions to proceed, a number of conditions need to be fulfilled.

(i) The metal must possess a non-bonding electron pair.

(ii) The complex must be coordinatively unsaturated and possess two sites for accommodating X and Y.

(iii) Since the process involves the oxidation of the metal by two units, from M^n to M^{n+2}, the higher oxidation state must be energetically accessible and stable.

These conditions are typically fulfilled by transition metals with d^8 and d^{10} electron configurations, notably Fe^0, Ru^0, Os^0, Rh^I, Ir^I, Ni^0, Pd^0, Pt^{II}, and Pt^0 (see later for a discussion of Ni^{II}/Ni^{IV} and Pd^{II}/Pd^{IV}). The stability of the higher oxidation states increases within a transition metal triad, so that, for example, Ir^{III} is more stable than Rh^{III}. This trend is most pronounced for the nickel triad where there are numerous examples of the oxidation of Pt^{II} to Pt^{IV} but stable products of oxidative addition to Pd^{II} have only recently been discovered.

Oxidative addition reactions are not restricted to transition metal complexes, but occur also in main group chemistry, although they are not usually considered in those terms. Examples are:

$$(CH_3)_2S + I_2 \rightleftharpoons (CH_3)_2SI_2$$

$$PF_3 + F_2 \rightleftharpoons PF_5$$

$$SnCl_2 + Cl_2 \rightleftharpoons SnCl_4$$

General features of oxidative addition reactions may be summarized as follows:

1. Square-planar 16-electron complexes of d^8 metals possess two "vacant sites" and on oxidative addition give octahedral 18-electron products. This is

frequently the entry into a catalytic cycle. The reverse of this reaction is reductive elimination (see later) which regenerates the 16-electron species.

An especially well-studied square 16-electron complex is Vaska's compound, *trans*-IrCl(CO)(PPh$_3$)$_2$, which undergoes reactions such as:

$$trans\text{-}Ir^{I}Cl(CO)(PPh_3)_2 + HCl \rightleftharpoons Ir^{III}(H)(Cl)_2(CO)(PPh_3)_2$$

Similar addition reactions occur with X—Y = H$_2$, Cl$_2$, CH$_3$I, RSH, and many others, with cleavage of the X—Y bond (*cf.* Table 21-1).

2. In addition to substrates of type X—Y where on reaction the X—Y bond is broken, compounds with multiple bonds may add oxidatively *without* X—Y bond cleavage, to give new complexes that have three- membered rings. In effect, one component of the multiple bond is broken and employed

Table 21-1 Reagents for Oxidative Addition Reactions to Metal Complexes[a]

Atoms Separate	Atoms Remain Attached
X—X	
H$_2$, Cl$_2$, Br$_2$, I$_2$, (SCN)$_2$	O$_2$
RSSR, R$_2$B—BR$_2$[b]	SO$_2$
C—C	
C$_2$Ph$_6$, (CN)$_2$, Ph—CN,	CF$_2$=CF$_2$,
MeC(CN)$_3$	(NC)$_2$C=C(CN)$_2$, cyclopropanes
	perfluorocyclopropene[c]
H—X	
HCl, HBr, HI, C$_6$F$_5$OH, C$_6$F$_5$SH, H$_2$O,	RC≡CR′
H$_2$S, PH$_3$, ROH,	RN=C=S
NH$_3$, C$_6$F$_5$NH$_2$, C$_4$H$_4$NH, HC≡CR,	RN=C=O
C$_5$H$_6$, H—CH$_2$CN, HCN, HCO$_2$R,	RN=C=NR′
C$_6$H$_6$, C$_6$F$_5$H, HSiR$_3$, HSiCl$_3$,	
H—B$_5$H$_8$, CH$_4$, RCHO	R$_2$C=C=O
C—X	CS$_2$
CH$_3$I, C$_6$H$_5$Br, CH$_2$Cl$_2$, CCl$_4$, CH$_3$COCl,	
PhCOCl, PhCH$_2$COCl, CF$_3$COCl,	(CF$_3$)$_2$C=O, (CF$_3$)$_2$C=S,
RCO$_2$R′, ROR′, R$_2$S, RC≡C—SiR$_3$[d]	CF$_3$CN
	epoxides[e]
M—X	
Ph$_3$PAuCl, HgCl$_2$, MeHgCl, R$_3$SnCl, R$_3$SiCl, RGeCl$_3$,	
H$_8$B$_5$Br, Ph$_2$BX	
Ionic	
PhN$_2^+$BR$_4^-$, CPh$_3^+$BF$_4^-$	

[a]Oxidative-addition of CH bonds of arenes and alkanes is discussed in Section 21-4.
[b]T. B. Marder *et al.*, *Inorg. Chem.* **1994**, *33*, 4623.
[c]R. P. Hughes *et al.*, *J. Am. Chem. Soc.* **1989**, *111*, 8919.
[d]K. G. Caulton *et al.*, *Organometallics* **1997**, *16*, 292.
[e]D. Milstein *et al.*, *J. Am. Chem. Soc.* **1990**, *112*, 6411.

in the new ligand—metal bonds, thus reducing the X—Y bond order:

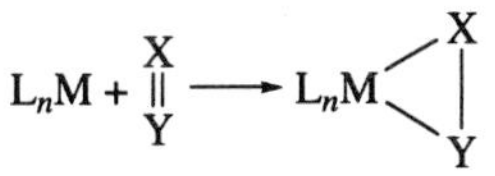

$$L_nM + \overset{X}{\underset{Y}{\|}} \longrightarrow L_nM \overset{X}{\underset{Y}{\diagdown}}$$

Whether or not a metallacyclic description is appropriate for such an adduct between the L_nM fragment and X=Y depends on the π-acceptor strength of X=Y. For example, since acetylenes are stronger π-acceptors than olefins, acetylene complexes have, with few exceptions, the character of metallacyclopropenes (21-III), with C—C—R angles in the region of 140° rather than 180° as in the free alkyne. For ethylene derivatives, electron-withdrawing substituents (R = F, CN)[11] lead to extensive back-bonding and a metallacyclopropane-type structure (21-IV), with R—C—R angles much smaller than 120° and C—C distances close to single-bond values (*ca.* 1.45–1.51 Å).

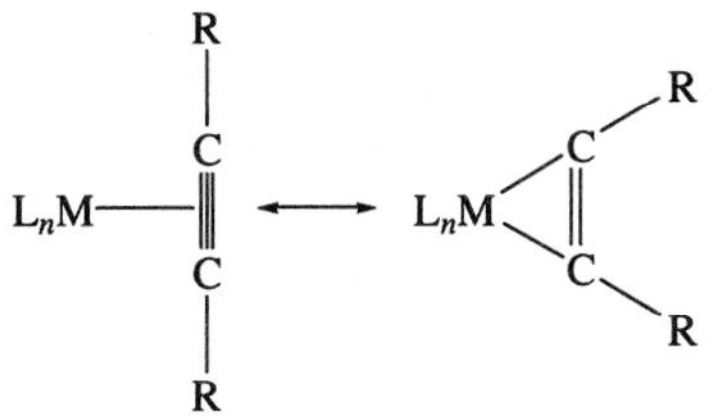

(21-III) 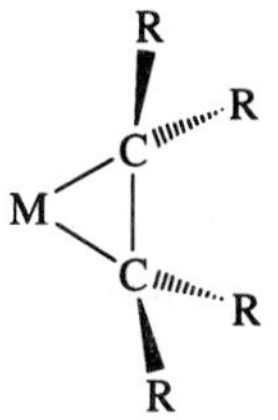   (21-IV)

On the other hand, for a given ligand the differing electron-donor characteristics of the complex fragment L_nM can determine whether or not the addition of an unsaturated ligand is best described as formation of a π-complex or as oxidative addition. For example, the geometry of the ethylene ligand in Zeise's salt $K[PtCl_3(C_2H_4)]$ differs little from that of free ethylene, with sp^2-hybridized C-atoms and a short C=C double bond of 1.354 Å (free ethylene: 1.337 Å). By contrast, the C—C bond of the ethylene ligand in $Os(C_2H_4)(CO)_4$ is elongated to 1.49 Å,[12] typical of a structure of type (21-IV) due to extensive back-bonding.

Early transition metals have a strong preference for the highest possible oxidation state. For example, the highly reactive species Cp_2Ti, generated by photolysis of Cp_2TiMe_2, reacts with acetylenes to form stable metallacyclopentadienes:

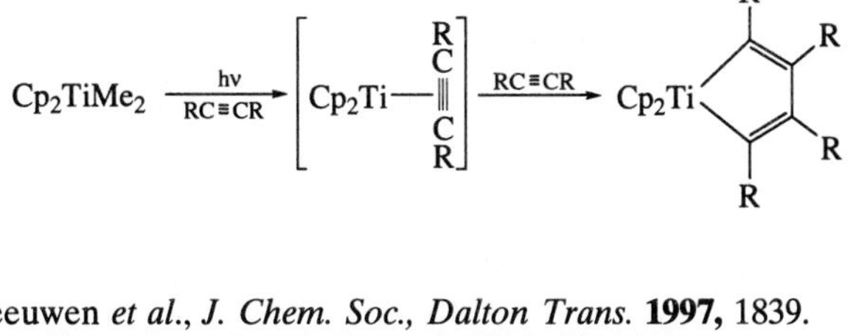

[11] P. W. N. M. van Leeuwen *et al.*, *J. Chem. Soc., Dalton Trans.* **1997,** 1839.
[12] N. Sheppard *et al.*, *J. Chem. Soc., Farad. Trans. I*, **1994,** *90*, 1449.

Electrophiles such as O_2 and ketones with electron-withdrawing substituents commonly form metallacyclic oxidative addition products[13]:

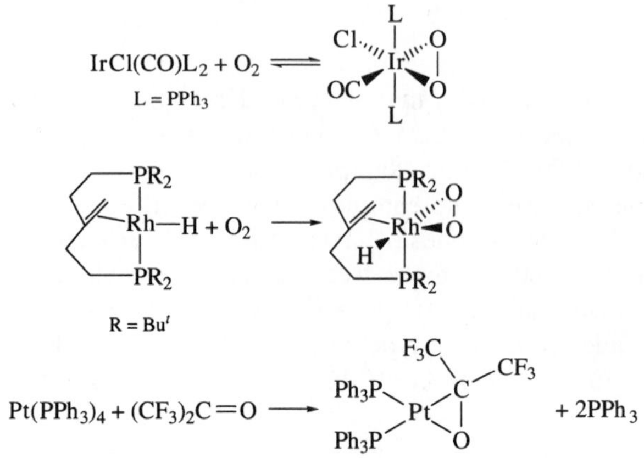

$$IrCl(CO)L_2 + O_2 \rightleftharpoons$$

L = PPh_3

R = Bu^t

$$Pt(PPh_3)_4 + (CF_3)_2C{=}O \longrightarrow \qquad + 2PPh_3$$

Heterocumulenes X=C=Y (CO_2, CS_2, O=C=NR, etc.) tend to give products of type (21-V) where one double bond adds to the metal center. On the other hand, electron-rich Mo^{II} and W^{II} complexes react with these substrates by cleaving one X=C double bond to give oxo, sulfido, or imido species:[14]

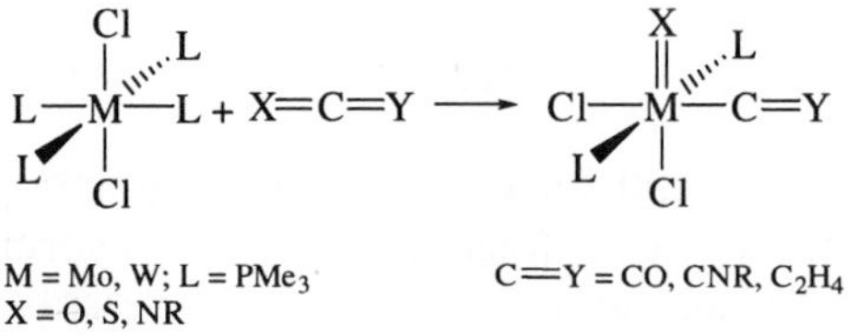

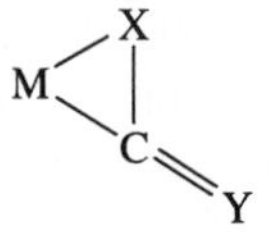

$$L{-}M{-}L + X{=}C{=}Y \longrightarrow Cl{-}M{-}C{=}Y$$

M = Mo, W; L = PMe_3 C=Y = CO, CNR, C_2H_4
X = O, S, NR

(21-V)

Highly electropositive elements such as metallic ytterbium undergo oxidative addition reactions even with imines to give azametallacyclopropanes[15]:

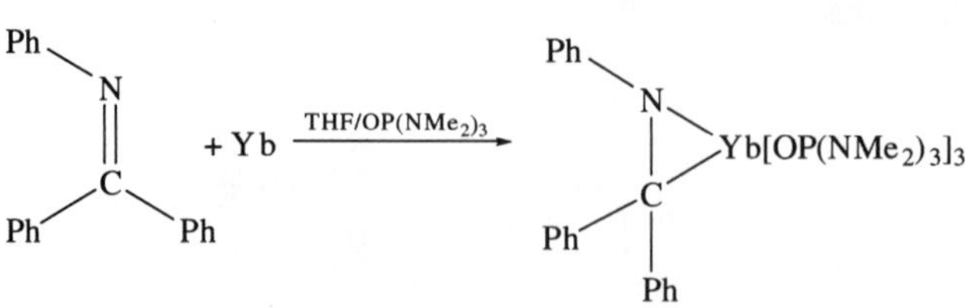

$$+ Yb \xrightarrow{\text{THF/OP(NMe}_2)_3} Yb[OP(NMe_2)_3]_3$$

[13]D. Milstein *et al., Chem. Commun.* **1996,** 1673; K. Vrieze *et al., Organometallics* **1997,** *16,* 979.
[14]K. A. Hall and J. M. Mayer, *J. Am. Chem. Soc.* **1992,** *114,* 10402.
[15]Y. Fujiwara *et al., Organometallics* **1996,** *15,* 5476.

3. Donor ligands exert a strong influence and favor the formation of the oxidative addition product. In the reaction of iridium(I) complexes with carboxylic acids, for example,

$$\text{trans-Ir}^{I}\text{X(CO)L}_2 + \text{RCO}_2\text{H} \rightleftharpoons \text{Ir}^{III}(\text{H})(\text{X})(\text{O}_2\text{CR})(\text{CO})\text{L}_2$$

the equilibria lie further to the Ir^{III} side in the order X = Cl < Br < I and L = PPh_3 < $PMePh_2$ < PMe_2Ph < PMe_3. There is, however, no direct correlation expected between the pK_a values of various acids HX, measured in water, and their propensity to add to metals. Similarly, the rates of oxidative addition of H_2 to $\text{Ru}^0(R_2PCH_2CH_2PR_2)_2$ strongly depends on the nature of R[16]:

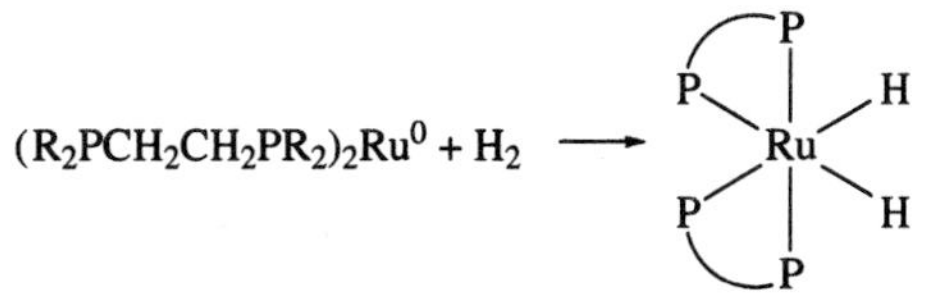

$$(R_2PCH_2CH_2PR_2)_2Ru^0 + H_2 \longrightarrow$$

R =	C_2F_5	Ph	C_2H_5	Me
relative value	1	120	2000	3400

Carbon—carbon coupling reactions of aryl halides are commonly catalyzed by palladium triarylphosphine complexes and proceed well for aryl bromides and iodides while aryl chlorides are generally unreactive. More basic chelating trialkylphosphines, however, render palladium sufficiently electrophilic to undergo rapid oxidative addition with chlorobenzene:

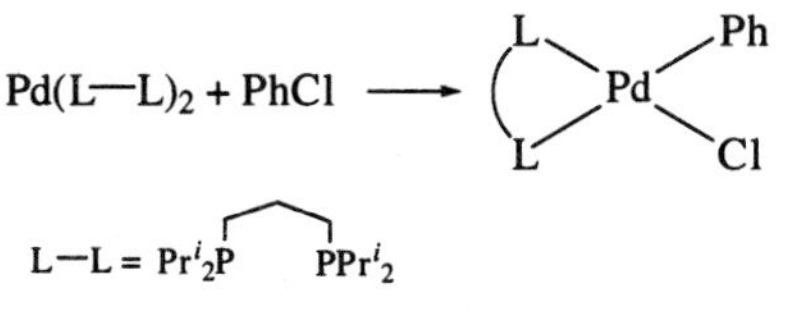

$$\text{Pd(L—L)}_2 + \text{PhCl} \longrightarrow$$

$$\text{L—L} = Pr^i_2P \quad PPr^i_2$$

In this way, chloroarenes can be converted to amides, esters, acids, and aldehydes or reduced to arenes.[17]

It was mentioned before that the oxidative addition of X—Y to L_nM may involve odd-electron intermediates. The steric and electronic influence of ligands may combine in certain circumstances to produce very stable odd-electron species, i.e., the oxidative addition is prevented from proceeding to completion. This is illustrated by the ligand-dependent behavior of tungsten(0) complexes. Whereas $W(CO)_5$ forms a simple adduct with disulfides, $W(CO)_3(PPr^i_3)_2$ gives a stable 17-electron W^I radical. By contrast, the related phen complex gives the W^{II} oxidative addition product.[18]

[16]R. N. Perutz et al., J. Am. Chem. Soc. **1995**, 117, 10047.
[17]M. Portnoy and D. Milstein, Organometallics **1993**, 12, 1665.
[18]G. J. Kubas et al., J. Am. Chem. Soc. **1994**, 116, 7917.

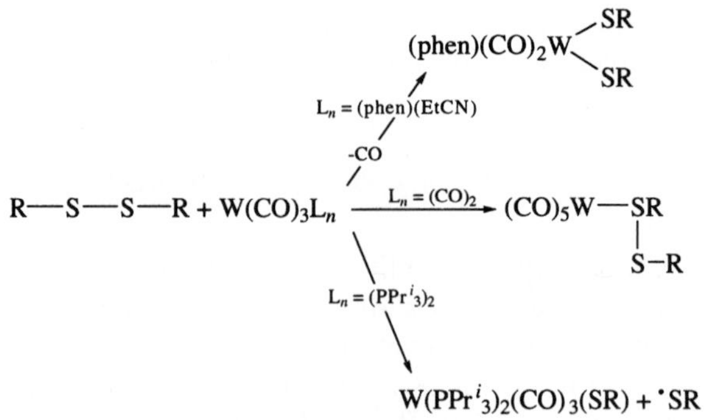

$$(phen)(CO)_2W\overset{\displaystyle SR}{\underset{\displaystyle SR}{<}}$$

$$L_n = (phen)(EtCN)$$

$$-CO$$

$$R-S-S-R + W(CO)_3L_n \xrightarrow{\ L_n = (CO)_2\ } (CO)_5W-SR$$
$$\underset{\displaystyle S-R}{}$$

$$L_n = (PPr^i_3)_2$$

$$W(PPr^i_3)_2(CO)_3(SR) + {}^{\bullet}SR$$

Equally important are steric influences in the product. For example, diaryl dichalcogenides ArE−EAr (E = S, Se, Te) have long been known to react with metallic mercury to give stable oxidative addition products $Hg(EAr)_2$:

$$Hg + ArE-EAr \longrightarrow Hg(EAr)_2$$

While this holds true for EAr = $TeC_6H_2Me_3$, the more crowded $TeC_6H_2Pr^i_3$ derivative is sufficiently labile for reductive elimination to be induced by very small energy changes, such as changes in solvent polarity or slight pressure.[19]

4. Compounds with an 18-electron configuration cannot undergo oxidative addition reactions without the expulsion of a ligand, as otherwise the maximum coordination number in the oxidized state would be exceeded, for example:[20]

$$Ru^0(CO)_4(PMe_3) + CH_3I \longrightarrow Ru^{II}(CH_3)(I)(CO)_3(PMe_3) + CO$$

$$Mo^0(CO)_4(bipy) + HgCl_2 \longrightarrow Mo^{II}Cl(HgCl)(CO)_3(bipy) + CO$$

5. Only mononuclear species have been considered so far. Oxidative addition reactions can, however, proceed at more than one metal center in bi- or polynuclear complexes. Compounds that have bridging groups, with or without a metal−metal bond, can undergo additions to one or both metal atoms, to give products such as

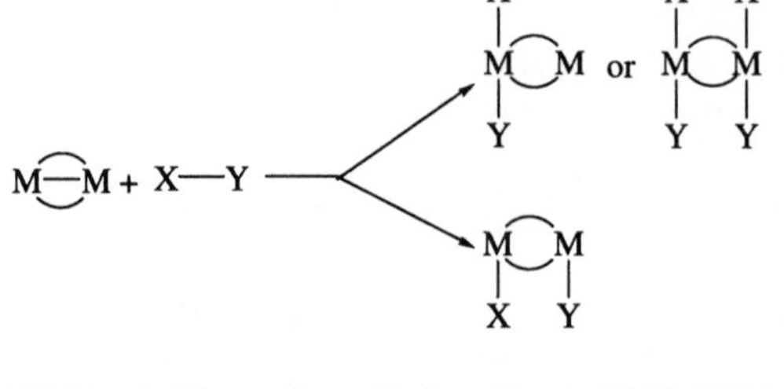

$$M-M + X-Y \longrightarrow$$

[19]M. Bochmann and K. J. Webb, *J. Chem. Soc., Dalton Trans.* **1991**, 2325.
[20]G. Cardaci et al., *Inorg. Chem.* **1993**, *32*, 554.

Addition to give M—X and M—Y results in formal one-electron oxidation of each metal:

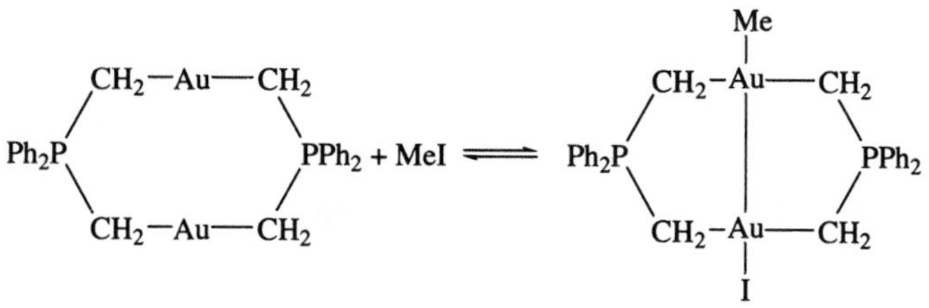

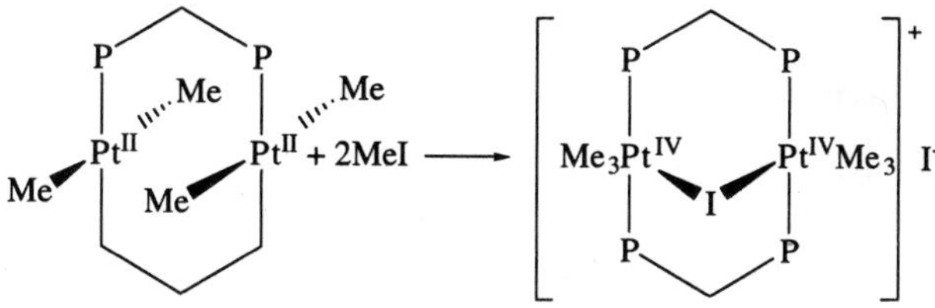

Multiple oxidative addition, for example, of RCX_3 (X = Cl, Br) to M^0 precursors, may lead to mixed-valence clusters[21]:

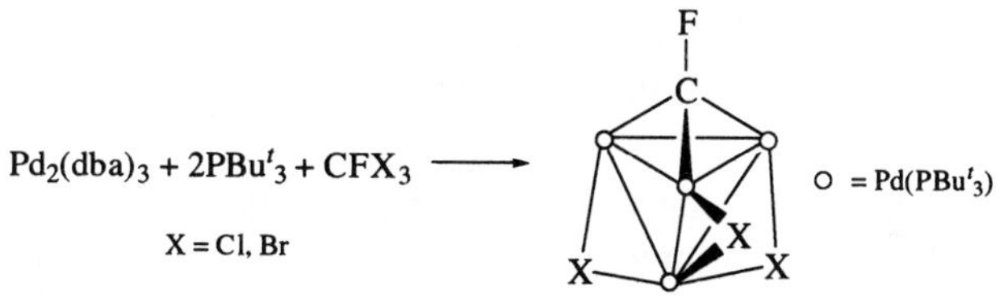

$$Pd_2(dba)_3 + 2PBu^t_3 + CFX_3 \longrightarrow$$

X = Cl, Br

O = Pd(PBu'$_3$)

Oxidative additions can also occur with metal—metal multiple bonds, for example[22]:

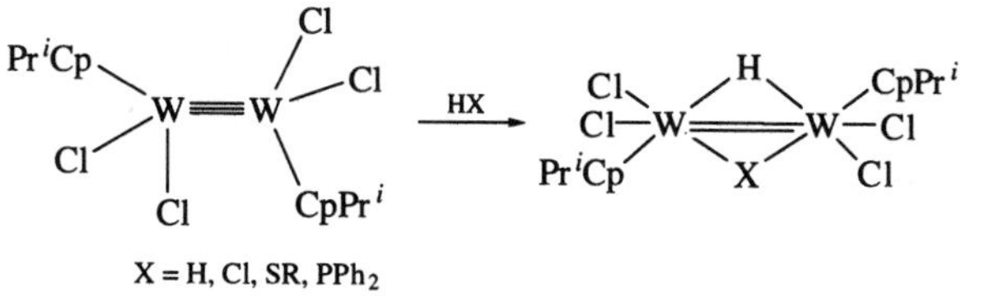

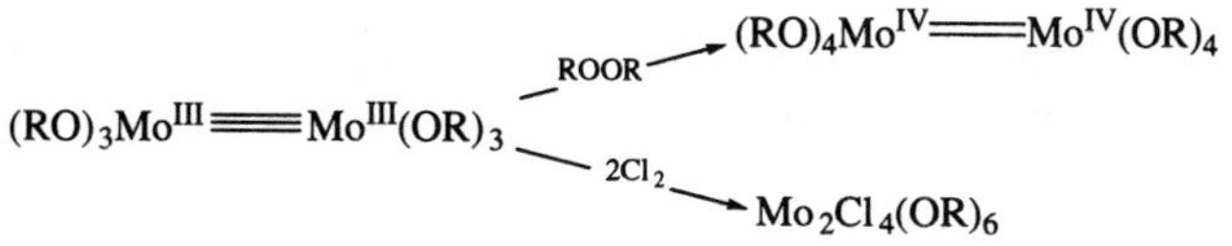

X = H, Cl, SR, PPh$_2$

$(RO)_3Mo^{III}\equiv\equiv Mo^{III}(OR)_3$

$\xrightarrow{ROOR} (RO)_4Mo^{IV}=\!\!=Mo^{IV}(OR)_4$

$\xrightarrow[-2Cl_2]{} Mo_2Cl_4(OR)_6$

[21]D. M. P. Mingos *et al., Chem. Commun.* **1997**, 285.
[22]F. A. Cotton *et al., Inorg. Chem.* **1992**, *31*, 5308; M. L. H. Green *et al., J. Chem. Soc., Dalton Trans.* **1991**, 1397.

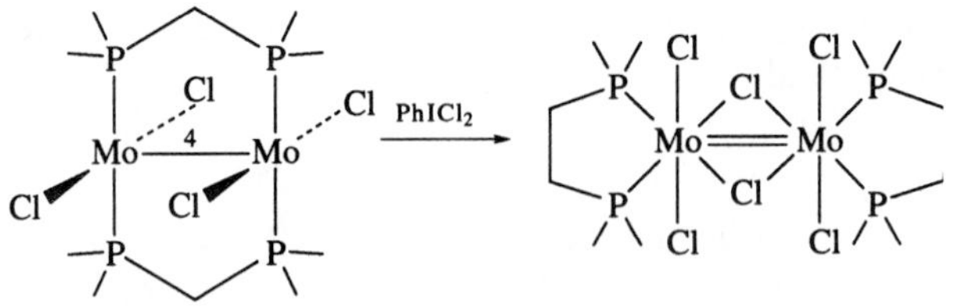

Whereas ketones typically give oxidative addition products in which the C—O bond is retained, the reaction of $R_2C=O$ with $W_2(OR)_6py_2$ ($W^{III}-W^{III}$) leads to reductive C—O cleavage to give $W_2^V(O)(\mu\text{-}CR_2)(OR)_6(py)$.[23]

Other examples of compounds known for their ability to oxidatively add to metals are given in Table 21-1.

Addition Reactions of Specific Molecules

Hydrogen Addition. In spite of the high H—H bond energy of 450 kJ mol^{-1} (stronger than C—H, C—C, and even most C—F bonds) and the lack of bond polarity, dihydrogen reacts with many transition metal complexes with great facility. Oxidative addition of H_2 occurs for electron-rich late transition metal complexes as well as early transition metal complexes in low oxidation states under very mild conditions (ambient temperature and pressure). The reaction is frequently reversible, the hydrogen being removed by sweeping with N_2 or argon or by evacuation, as in the case of Vaska's compound:

$$IrCl(CO)(PPh_3)_2 + H_2 \rightleftharpoons Ir(H)_2Cl(CO)(PPh_3)_2$$

Molecular hydrogen interacts with a metal center using its σ- and σ^*-orbitals as donor- and acceptor orbitals, as explained in Section 2-15. The approach of H_2 leads to the polarization, elongation, and finally cleavage of the H—H bond in a concerted, three-center addition process (21-VI).

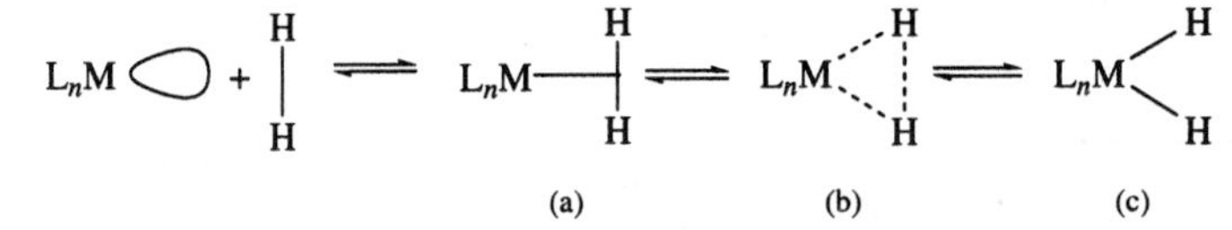

(21-VI)

In most cases the product will be the "classical" metal hydride (21-VIc), with structures like (21-VIa) or (21-VIb) as transition states. Increasingly often, however, it is observed that the reaction does not proceed to the oxidative addition product but gives complexes of dihydrogen (21-VIa) or "elongated dihydrogen" (21-VIb) instead.

The addition of H_2 to $IrCl(CO)L_2$ can occur in two ways, with H_2 oriented in either the Cl—Ir—CO or the L—Ir—L planes. Electron withdrawing ligands (Cl,

[23]M. H. Chisholm *et al.*, *Organometallics* **1990**, *9*, 602.

CO) stabilize the transition state and exert electronic control over the stereochemistry of H_2 oxidative addition[24]:

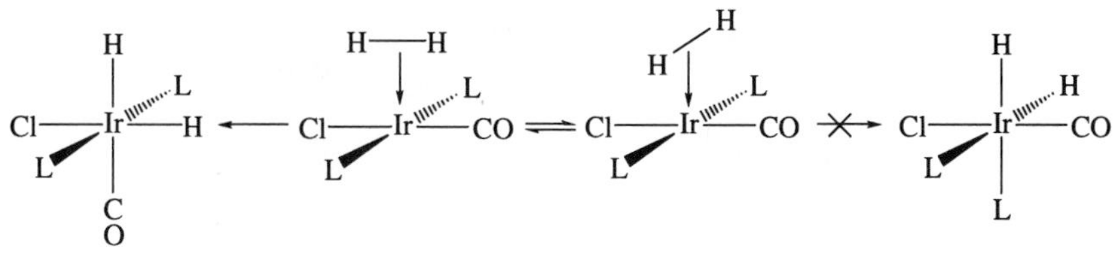

The rate of H_2 addition increases with the polarity of the solvent, indicative of a polar transition state.[25]

Whereas $IrCl(CO)(PPh_3)_3$ has a very rich oxidative addition chemistry, the reaction of H_2 with the analogous Rh compound does not give a stable product but leads to labilization of CO and formation of a binuclear complex. The reactivity of the Rh^I compounds decreases in the order $X = I > Br > Cl$.[26]

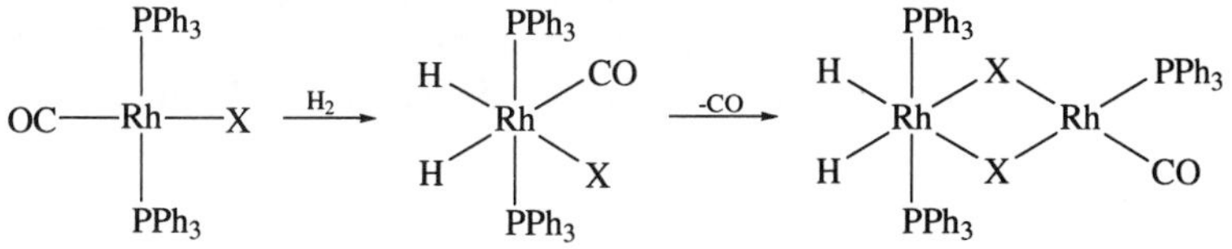

For H_2 oxidative addition to Rh^I or Ir^I to take place, a square-planar geometry is required. Where this is not the case, as with the *tbp* tris(pyrazolyl)borate complex (21-VII), dissociation of a pyrazolyl arm precedes the H_2 addition step[27]:

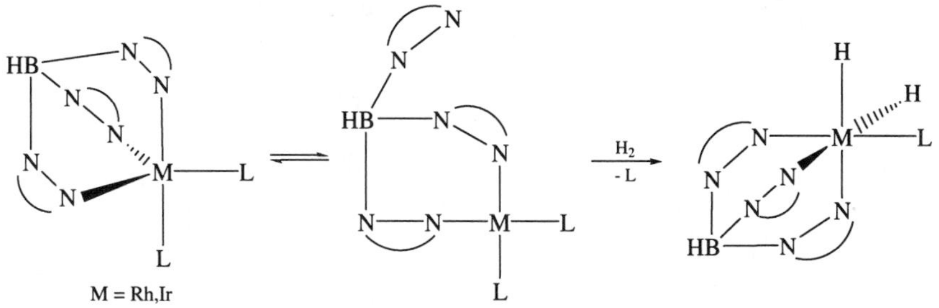

M = Rh,Ir

(21-VII)

Apart from the well-known oxidative additions to Ir^I, hydrogen can also be added to Ir^{III}. In the case of (21-VIII) the reaction proceeds at room temperature; the resulting Ir^V tetrahydride reductively eliminates H_2 only on heating above 130°C and is a highly active catalyst for the transfer dehydrogenation of cyclooctane.[28]

[24]M. B. Hall *et al.*, *Inorg. Chem.* **1992,** *31,* 317; *Coord. Chem. Rev.* **1994,** *135,* 845.
[25]J. D. Atwood *et al.*, *Organometallics* **1997,** *16,* 3371.
[26]S. B. Duckett and R. Eisenberg, *J. Am. Chem. Soc.* **1993,** *115,* 5292.
[27]W. J. Oldham and D. M. Heinekey, *Organometallics* **1997,** *16,* 467.
[28]W. C. Kaska *et al.*, *J. Am. Chem. Soc.* **1997,** *119,* 840.

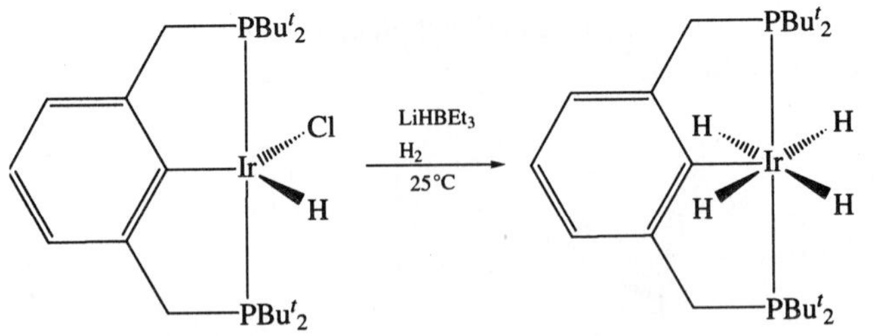

(21-VIII)

An example of H_2 oxidative addition to early transition metal complexes is the reaction with $WX_2(PMe_3)_4$ to give $W(H)_2X_2(PMe_3)_4$. The product is stable to reductive elimination if $X = Cl$ but H_2 addition is reversible if $X = I$.[29]

As mentioned before, the reaction with H_2 often gives η^2-dihydrogen complexes instead of oxidative addition products.[30] Neutron diffraction and spectroscopic data suggest that these can span a wide range of H—H bonds, from "normal" H_2 complexes with H—H bond lengths of ~ 0.85 Å of type (21-VIa) to elongated H_2 complexes, while typical dihydride complexes (21-VIc) $L_nM(H)_2$ have $r_{H-H} > 1.5$ Å.[31] Dihydrogen complexes may also arise by the reaction of hydrido complexes with HX if X = weakly coordinating anion, as in the case of the Nb^{III} and Fe^{II} examples below; note that the possible Nb^V and Fe^{IV} products are not formed.[32,33]

$$(C_5H_4SiMe_3)_2Nb(H)(CNR) \xrightarrow{CF_3COOH} [(C_5H_4SiMe_3)_2Nb(\eta^2\text{-}H_2)(CNR)]^+$$

$$[HFe(dppe)_2(CO)]^+ \overset{HOTf}{\rightleftharpoons} [Fe(\eta^2\text{-}H_2)(dppe)_2(CO)]^{2+}$$

Whether or not protonation leads to a classical polyhydride or an H_2 complex may depend on the nature of the ligand X *trans* to the hydride, with X = π-donor (Cl) stabilizing an η^2-H_2 complex while X = H does not[34]:

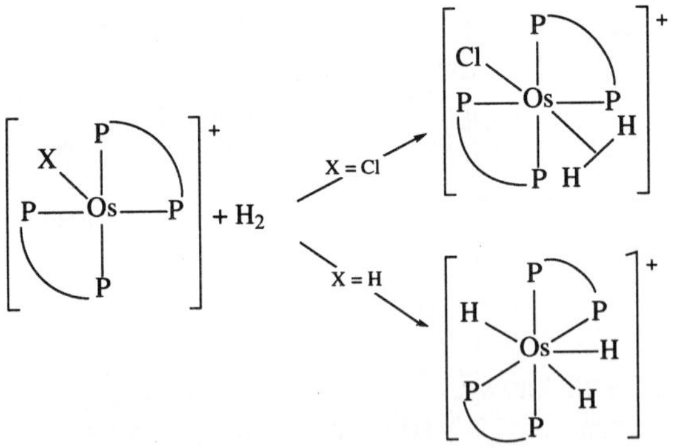

[29]D. Rabinovich and G. Parkin, *J. Am. Chem. Soc.* **1993**, *115*, 353.
[30]P. J. Jessop and R. H. Morris, *Coord. Chem. Rev.* **1992**, *121*, 155; D. M. Heinekey and W. J. Oldham, *Chem. Rev.* **1993**, *93*, 913.
[31]F. Maseras *et al.*, *Organometallics* **1996**, *15*, 2947.
[32]A. Otero *et al.*, *J. Am. Chem. Soc.* **1997**, *119*, 6107.
[33]R. H. Morris *et al.*, *J. Chem. Soc., Dalton Trans.* **1997**, 1663.
[34]A. Mezzetti *et al.*, *Inorg. Chem.* **1997**, *36*, 711.

HX Additions

The addition of HCl or HBr to *trans*-IrCl(CO)(PPh$_3$)$_2$ in non-polar solvents such as benzene leads to *cis*-addition products, suggesting a concerted mechanism. However, if wet or polar solvents (DMF or benzene-methanol) are used, *cis/trans* mixtures are obtained.

In polar solvents, where HX is dissociated, there are two possible reaction pathways—initial attack by H$^+$ or X$^-$, namely,

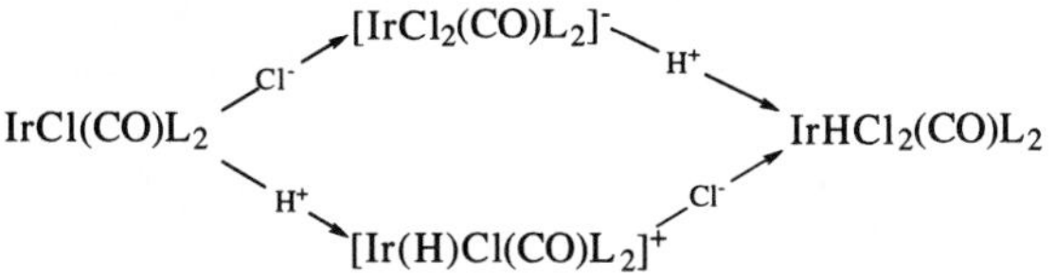

Kinetic studies on IrCl(CO)L$_2$ compounds in methanol have confirmed initial attack by Cl$^-$. However, protonation by CF$_3$SO$_3$H, where the anion is non-coordinating, gives [IrHCl(CO)L$_2$(MeOH)]$^+$. The more basic complex IrCl(COD)(PEtPh$_2$)$_2$ is also first attacked by Cl$^-$, followed by immediate protonation of the resulting complex anion. The 5-coordinate intermediates are fluxional, and hence the stereochemistry of the final product is determined by thermodynamic stability.

The *oxidative addition of water* is observed with electron-rich trialkylphosphine complexes of Ir^I and Pt^0, as well as with dialkyl complexes of Pd^{II} and Pt^{II}.[35] In the latter case, the initially formed M(H)(OH) complex is immediately hydrolyzed, to give products such as (21-IX(a)).[36] Note that for M = Pd these provide rare examples of thermally stable Pd^{IV} species.

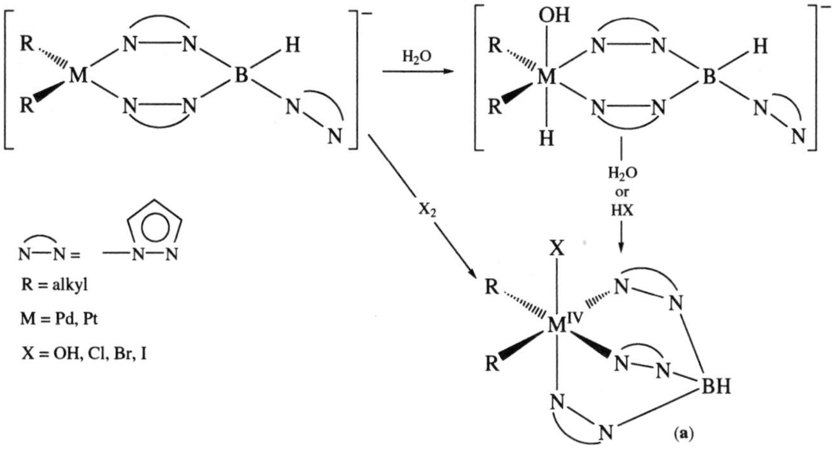

(21-IX)

Water also oxidatively adds to photochemically generated WCp$_2^*$ to give Cp$_2^*$W(H)(OH).[37]

[35] A. J. Canty and G. van Koten, *Acc. Chem. Res.* **1995**, *28*, 406; A. J. Canty *et al.*, *J. Organomet. Chem.* **1996**, *510*, 281.
[36] A. J. Canty *et al.*, *Organometallics* **1996**, *15*, 5713; **1997**, *16*, 2175.
[37] M. Yoon and D. R. Tyler, *Chem. Commun.* **1997**, 639.

Alcohols and phenols similarly give oxidative addition compounds; the structure of the product depends on the steric hindrance of the OAr ligands. For example, $Mo(PMe_3)_6$ reacts with C_6H_5OH to give *trans*-$Mo(OC_6H_5)_4(PMe_3)_2$, while $C_6H_2Me_3OH$ affords the expected hydrido aryloxide, $Mo(H)(OC_6H_2Me_3)(PMe_3)_4$.[38] Similarly, phenols add to $[Ir(COD)(PMe_3)_3]Cl$ to give *mer*-$(Me_3P)_3Ir(H)(OAr)Cl$.[39]

Thiols may react in analogous fashion, e.g., their addition to $Ru(CO)_2L_3$ gives $RuH(SR)(CO)_2L_2$ (L = PPh_3); a hydrido selenolate species is also known.[40]

N—H bonds may oxidatively add to a variety of metals. For example, heating $[Ir(COD)(PMe_3)_3]Cl$ with heterocyclic amines gives products of the type *mer*-$Ir(H)(NR_2)Cl(PMe_3)_3$.[41] The addition of aniline to $Ru(C_2H_4)(PMe_3)_4$ gives $(PMe_3)_4Ru(H)(NHPh)$, *via* the cyclometallated complex $(PMe_3)_3Ru(CH_2PMe_2)(H)$ as intermediate. The rate determining step is the opening of the metallacycle, and the process may be catalyzed by traces of water.[42]

In the reaction of anilines $H_2NC_6H_4X$ with $(silox)_3Ta^{III}$, C—N bond cleavage competes with N—H bond addition; for example, the former is favored if X = CF_3 ($silox$ = Bu_3^tSiO).[43]

$$(silox)_3Ta + H_2NC_6H_4X \xrightarrow{X=CF_3} \begin{cases} (silox)_3Ta\begin{smallmatrix} H \\ NHC_6H_4X \end{smallmatrix} \\ (silox)_3Ta\begin{smallmatrix} NH_2 \\ C_6H_4X \end{smallmatrix} \end{cases}$$

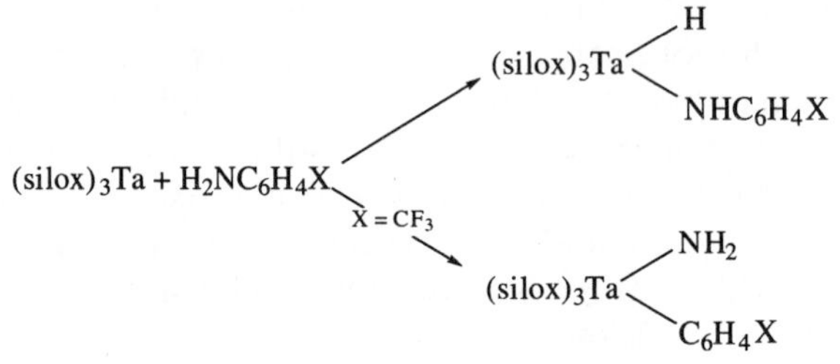

The P—H bonds of primary phosphines RPH_2 add to "zirconocene" to give unstable zirconium phosphido hydrides $CpZr(H)(PHR)$ from which C—H activation products such as fulvalene complexes $[CpZr(\mu\text{-}PHR)]_2(\eta^5:\eta^5\text{-}C_{10}H_8)$ can be isolated.[44] The addition of $Ph_2P(O)H$ to $M(PEt_3)_3$ (M = Pd, Pt) gives (21-X); the Pd derivative catalyzes the hydrophosphinylation of alkynes.[45]

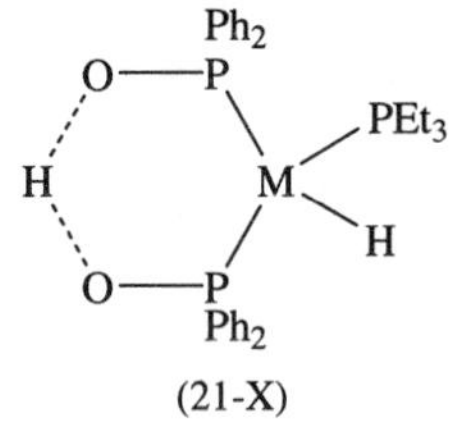

(21-X)

[38]G. Parkin *et al.*, *Organometallics* **1996**, *15*, 3910.
[39]J. S. Merola *et al.*, *Inorg. Chem.* **1993**, *32*, 1681.
[40]B. R. James *et al.*, *Inorg. Chem.* **1991**, *30*, 4617.
[41]F. T. Lapido and J. S. Merola, *Inorg. Chem.* **1990**, *29*, 4172.
[42]R. A. Andersen *et al.*, *Organometallics* **1991**, *10*, 1875.
[43]P. T. Wolczanski *et al.*, *J. Am. Chem. Soc.* **1996**, *118*, 5132.
[44]D. W. Stephan *et al.*, *Organometallics* **1993**, *12*, 3145.
[45]M. Tanaka *et al.*, *Organometallics* **1996**, *15*, 3259.

Addition of X_2

The oxidative reactions of metal complexes with halogens are too numerous to mention, and the formation of complexes with less common oxidation states such as Pd^{IV} by this route has already been pointed out. While in reactions of L_nM with X_2 usually both X atoms are bound to the metal, salt formation or attack of X on a coordinated ligand has also been observed, for example,[46]

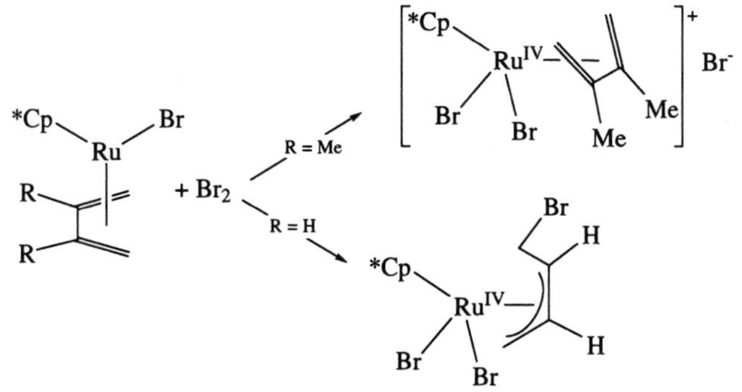

The oxidative addition of chalcogen-chalcogen bonds in RE—ER is widely applied in transition metal chemistry and is often a useful route for the preparation of thiolates, selenolates and, to a more limited extent, tellurolates. The oxidative addition of X_2 = HO—OH and $PhC(O)O—OC(O)Ph$ is also known and, in the case of Pt^{II} complexes, gives predominantly *trans* products:[47]

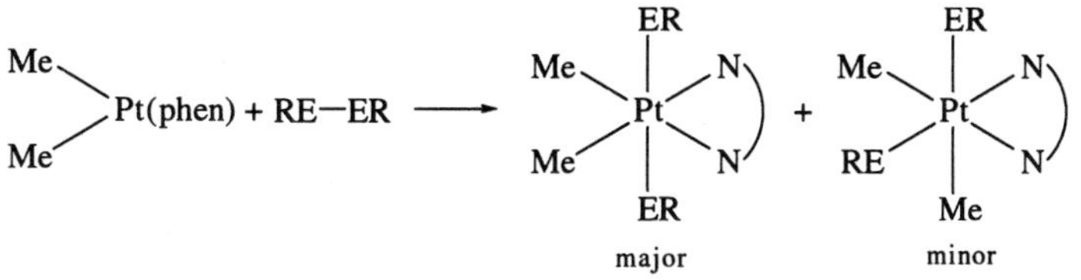

Comparatively recent is the isolation of stable metal-boryl complexes by the oxidative addition of B—B bonds, e.g., to Pt^0 or Rh^I:

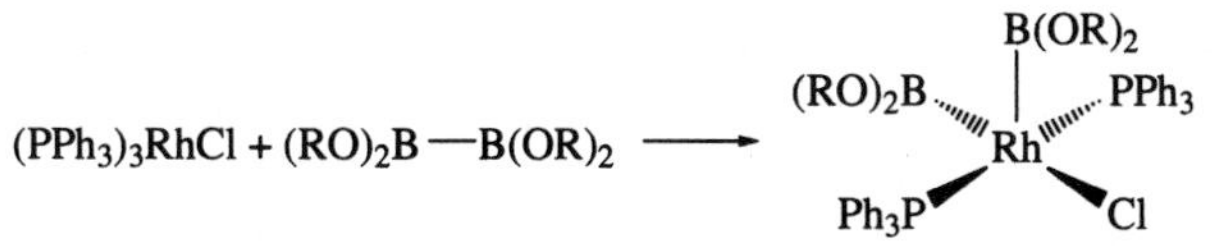

$$(PPh_3)_3RhCl + (RO)_2B—B(OR)_2 \longrightarrow$$

The reaction is the basis for the metal catalyzed diborylation of alkenes.[48]

[46]K. Kirchner et al., Organometallics, **1997**, 16, 427.
[47]R. J. Puddephatt et al., J. Chem. Soc., Dalton Trans. **1993**, 1835.
[48]T. B. Marder et al., Inorg. Chem. **1997**, 36, 272; S. Sakaki and T. Kikuno, Inorg. Chem. **1997**, 36, 226.

Organic Halides

The addition of organic halides to electrophilic metal centers can follow several different mechanisms. For a concerted *cis*-addition mechanism, a 3-center transition state of low polarity (21-XI) may be expected. Other possibilities include an S_N2 type mechanism, involving either inversion (structure 21-XIIa) or retention (structure 21-XIIb) at the C atom, followed by dissociation of X^- to give an ionic intermediate:

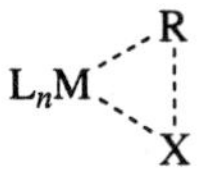

Alternatively, the addition may proceed *via* a radical "non-chain"

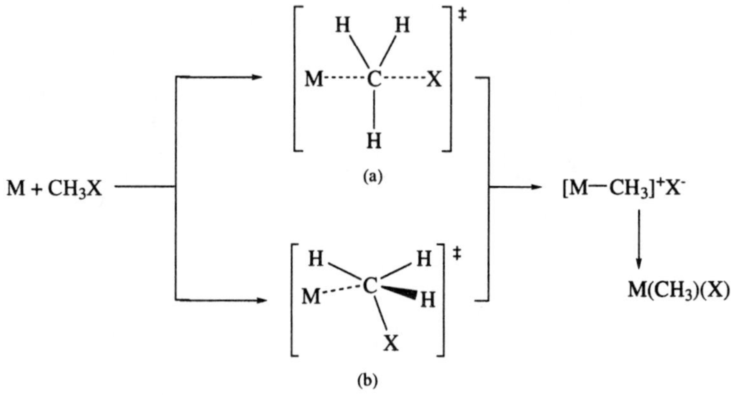

$$R—X + M^0 \longrightarrow [\overset{\frown}{R \smile X \smile M}]^\ddagger \longrightarrow [R^\bullet \ XM^\bullet] \xrightarrow{\text{fast}} R—M^{II}—X$$

or radical chain mechanism:[49]

$$RX + RM^0 \longrightarrow R—M^{II}-X + R^\bullet$$

$$R^\bullet + M^0 \longrightarrow RM^0, \text{ etc.}$$

The oxidative addition of CH_3I to $IrCl(CO)L_2$ involves a highly polar transition state of type (21-XIIa), unlike H_2 addition.[50] A similar linear transition state is preferred for the CH_3I addition to $[Rh(CO)_2I_2]^-$, a key step in the methanol carbonylation process (see Section 22-7). A transition state involving retention of configuration at carbon (21-XIIb) (which is distinct from a concerted structure 21-XI) may also be possible but is energetically disfavored.[51] By contrast, the addition of iodomethane to $Rh(Sacac)(CO)(PPh_3)$ apparently proceeds *via* a concerted three-center transition state, and the rates are little affected by changes in solvent polarity.[52]

[49]C. Amatore and F. Pflüger, *Organometallics* **1990**, *9*, 2276.
[50]A. Prock *et al.*, *Organometallics* **1993**, *12*, 2044.
[51]D. B. Cook *et al.*, *J. Am. Chem. Soc.* **1996**, *118*, 3029.
[52]R. van Eldik *et al.*, *Inorg. Chem.* **1991**, *30*, 2207.

Acyl iodides RCOI add to $[Rh(CO)_2I_2]^-$ to give $[RC(O)Rh(CO)_2I_3]^-$ which readily loses CO. The Pr^nCO and Pr^iCO complexes equilibrate *via* decarbonylation and isomerization steps.[53]

The reactions of aryl halides Ar—X with zerovalent metal species usually follow the trend of C—X bond strengths X = I > Br > Cl. Complexes such as $Ni(PEt_3)_4$ can give rise either to M^{II} products, $(Et_3P)_2Ni^{II}(Ar)(X)$, or Ni^I species, $NiX(PEt_3)_3$, often as competing products, depending on the halide and solvent polarity, arguably formed *via* a tight radical ion pair $\{Ni(PEt_3)_3^{\cdot+}ArX^{\cdot-}\}$ as common intermediate.[54] The oxidative addition of aryl halides to Pd^0 complexes is usually formulated as involving PdL_2 or PdL_3 compounds, though the intimate mechanism of this reaction is more complex. If the zerovalent palladium species are generated by reduction of $PdCl_2L_2$, anionic halide complexes may be involved, e.g. $[ClPdL_2]^-$, to give a pentacoordinate anionic aryl-Pd^{II} center as the initial product.[55] Similarly, reduction of Pd^{II} precursors with organometallic reagents or electrochemical reduction of $Pd^{II}(Ar)XL_2$ gives strongly nucleophilic $[ArPd^0L_2]^-$, which rapidly reacts with Ar'X to give an unstable *tbp* $[(Ar)(Ar')PdXL_2]^-$, which eliminates X^- within a few ms.[56]

Whereas tetrahedral chelate complexes $Pd(L—L)_2$ do not react with PhI, 1:1 mixtures of L—L with $Pd(dba)_2$ give (L—L)Pd(dba) and its dissociation product, 2-coordinate (L—L)Pd. Both 3- and 2-coordinate species oxidatively add iodobenzene, though the latter species are more reactive.[57] By contrast, the isolable 14-electron complex $Pd[P(o\text{-}tol)_3]_2$ reacts with $p\text{-}Bu^tC_6H_4Br$ *via* a dissociative pathway that produces a mono-phosphine intermediate:

$$L—Pd—L \underset{-ArBr}{\overset{ArBr}{\underset{+L}{\overset{-L}{\rightleftharpoons}}}} L—Pd—ArBr \overset{slow}{\longrightarrow} L—Pd\overset{Ar}{\underset{Br}{{\Big\langle}}} \overset{fast}{\longrightarrow} {}^{Ar}_{L}{\Big\rangle}Pd\overset{Br}{\underset{Br}{{\Big\langle}}}Pd{}^L_{Ar}$$

$$L = P\left(\underset{Me}{-{\Big\langle}{\bigcirc}{\Big\rangle}}\right)_3$$

There is evidence that PdL intermediates should give faster oxidative addition rates than PdL_2, although in most cases their concentration is too low to be kinetically significant. In the case of L = $P(o\text{-}tol)_3$, steric hindrance inhibits the reactivity of PdL_2 but enhances the concentration of PdL.[58]

Oxidative addition reactions of bromomethylbipyridine derivatives to Pt^{II} is an elegant route to polynuclear Pt^{IV} complexes with dendrimer structures.[59]

It must be stressed that in many cases it is not easy to be certain of the reaction pathway, and traces of oxygen, paramagnetic impurities, or light may have a profound effect. The donor and acceptor characteristics of ligands are important in directing the course of the reaction. For example, while $CpRh(CO)_2$ reacts

[53]L. A. Howe and E. E. Bunel, *Polyhedron* **1995**, *14*, 167.
[54]T. T. Tsou and J. K. Kochi, *J. Am. Chem. Soc.* **1979**, *101*, 6319.
[55]C. Amatore *et al.*, *J. Am. Chem. Soc.* **1993**, *115*, 9531.
[56]C. Amatore *et al.*, *Chem. Eur. J.* **1996**, *2*, 957.
[57]C. Amatore *et al.*, *J. Am. Chem. Soc.* **1997**, *119*, 5176.
[58]J. F. Hartwig and F. Paul, *J. Am. Chem. Soc.* **1995**, *117*, 5373.
[59]R. J. Puddephatt *et al.*, *Organometallics* **1996**, *15*, 43.

with primary and secondary perfluoroalkyl iodides with selective alkylation at the metal center to give CpRhCO(R$_F$)I, the reaction of CpRh(PMe$_3$)$_2$ leads to *exo*-fluoroalkylation of Cp:

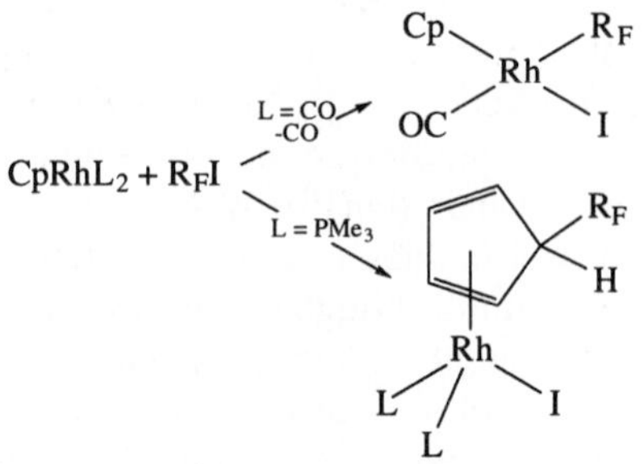

In view of the very similar reactivity of *prim-* and *sec*-R$_F$I, radical intermediates are unlikely to be involved.[60]

The addition of perhaloalkanes (usually CCl$_4$) to alkenes (Kharash reaction) is generally promoted by light, peroxides, and a range of metal complexes and is commonly thought to involve a radical chain mechanism. On the other hand, in the presence of Rh complexes, CCl$_4$ addition products RhX(Cl)(CCl$_3$)(CO)L$_2$ can be isolated, and an addition-alkene insertion reaction sequence may be feasible in some cases.[61]

Dichloromethane and chloroform add to Rh^I and Pd^0 compounds to give chloromethyl complexes.[62] Double oxidative addition may, however, also be found, for example,[63,64]

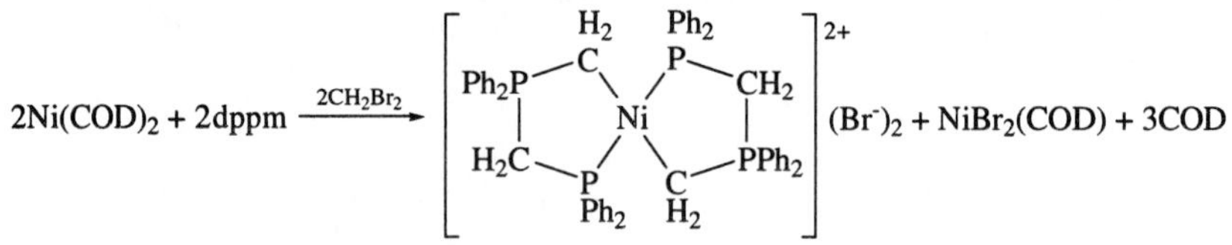

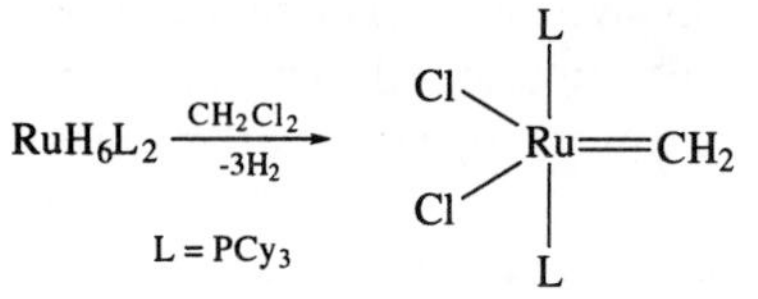

The Ru complex is an important olefin metathesis catalyst.

In view of the relative C—X bond energies, *C—F bond activation* can be expected to be strongly disfavored. Nevertheless, there is an increasing number of cases where C—F bond addition to electron-rich metal complexes is observed. For example, C$_6$F$_6$ oxidatively adds to the Cp*Rh(PMe$_3$) fragment, whereas C$_6$F$_5$H undergoes only C—H activation.[65] Hexafluorobenzene slowly adds to

[60]R. P. Hughes *et al.*, *Organometallics* **1997**, *16*, 5.
[61]C. White *et al.*, *J. Chem. Soc., Chem. Commun.* **1991**, 165.
[62]H. Nishiyama *et al.*, *Organometallics* **1991**, *10*, 2706; P. Leoni, *Organometallics* **1993**, *12*, 2432.
[63]C. P. Kubiak *et al.*, *Organometallics* **1992**, *11*, 1392.
[64]M. Oliván and K. G. Caulton, *Chem. Commun.* **1997**, 1733.
[65]R. N. Perutz *et al.*, *Organometallics* **1994**, *12*, 522.

$(Bu_2^tPCH_2CH_2PBu_2^t)(C_6H_6)Ni$ *via* an isolable η^2-complex:[66]

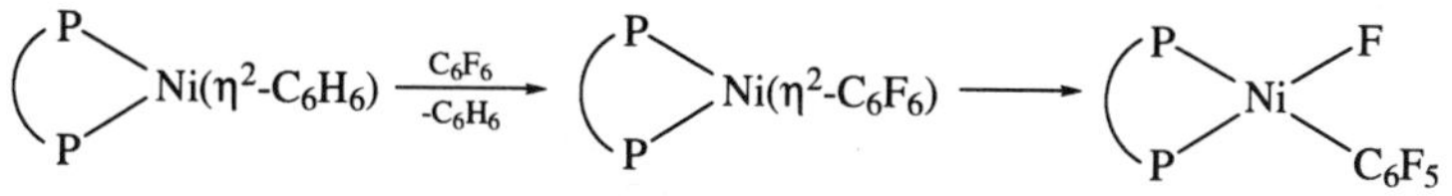

The addition of an *ortho*-C—F bond of pentafluoropyridine to Ni(PEt$_3$)$_3$ is significantly faster than C$_6$F$_6$ addition and occurs even with C$_5$HF$_4$N, in preference to C—H activation.[67]

The very electron-rich ruthenium hydride *cis*-RuH$_2$(dmpe)$_2$ (but *not* its dppe analogue) reacts with C$_6$F$_6$ under unusually mild conditions ($-78°$C) and with C$_6$F$_5$H gives exclusively C—F activation:

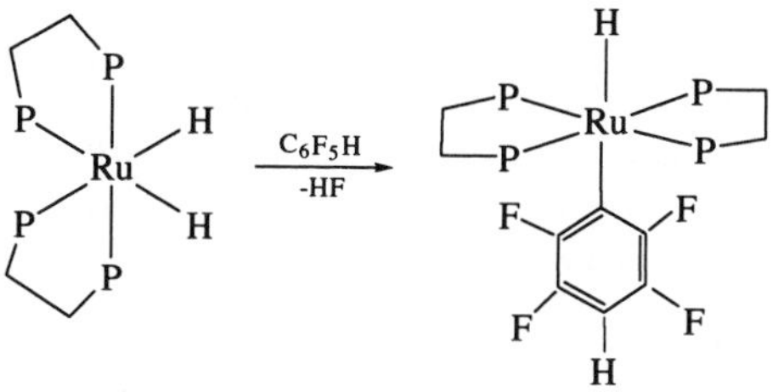

The data are consistent with an electron-transfer process involving a caged radical ion pair[68]:

$$RuH_2 + C_6F_6 \longrightarrow \{RuH_2^+C_6F_6^-\} \xrightarrow{-HF} \{RuH + C_6F_5\} \longrightarrow Ru(H)(C_6F_5)$$

Finally, oxidative addition of R—X may lead to *unusual oxidation states*, such as Ni^{IV} and Pd^{IV}. Thus the addition of alkyl, allyl, benzyl, or phenacyl bromides and iodides to Pd^{II} complexes with chelating N-donor ligands gives remarkably stable Pd^{IV} trialkyls,[69] and CH$_3$I addition to C—O chelate-stabilized nickel trimethylphosphine compounds affords rare examples of stable Ni^{IV} alkyls[70]:

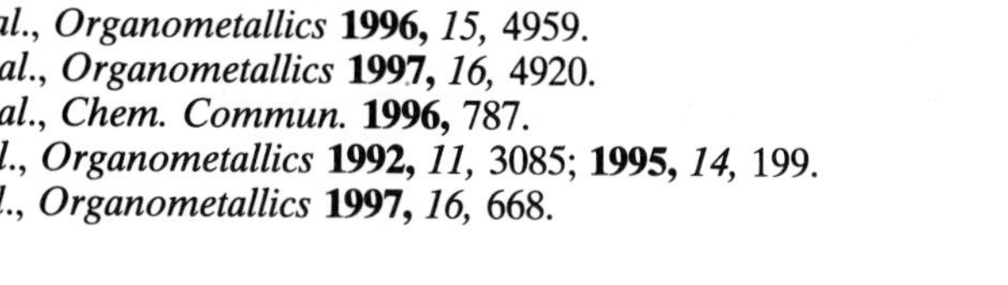

L = PMe$_3$

[66]R. Pörschke *et al.*, *Organometallics* **1996**, *15*, 4959.
[67]R. N. Perutz *et al.*, *Organometallics* **1997**, *16*, 4920.
[68]R. N. Perutz *et al.*, *Chem. Commun.* **1996**, 787.
[69]A. J. Canty *et al.*, *Organometallics* **1992**, *11*, 3085; **1995**, *14*, 199.
[70]H. F. Klein *et al.*, *Organometallics* **1997**, *16*, 668.

Addition Reactions of Si—H Bonds

The interaction of silanes with metal complexes strongly resembles the behavior of H_2, i.e., both η^2-silane complexes (21-XIII) and oxidative addition products (21-XIV) are observed. In binuclear complexes, silyl bridges of type (21-XV) may be found.

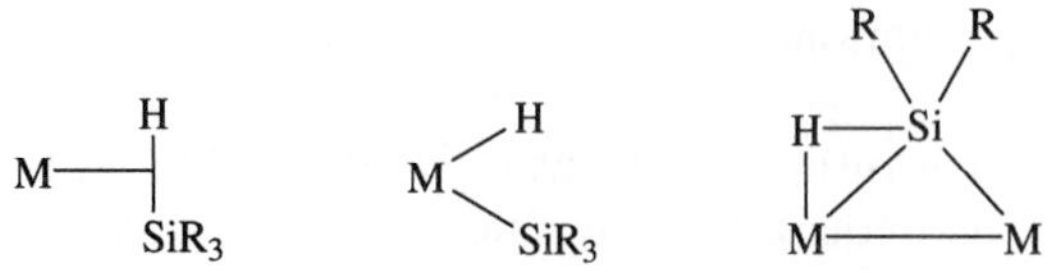

(21-XIII)   (21-XIV)   (21-XV)

The reactivity of silanes has been intensively investigated in connexion with hydrosilylation and silane dehydrocoupling reactions (see Sec. 22-3).

Electron-rich phosphine complexes as well as metal carbonyls have been found to add silanes; the latter often require photolysis to generate a reactive unsaturated metal fragment. Some examples are[71,72,73]

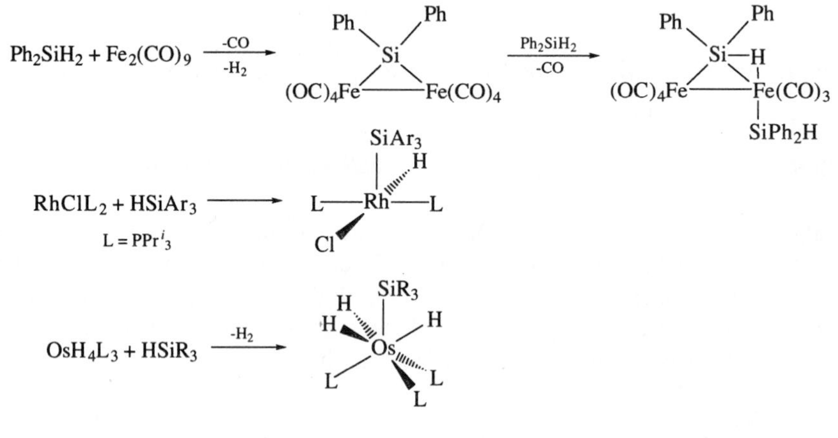

Complexes are frequently fluxional, leading to hydrogen scrambling *via* reversible oxidative addition/reductive elimination steps:[74]

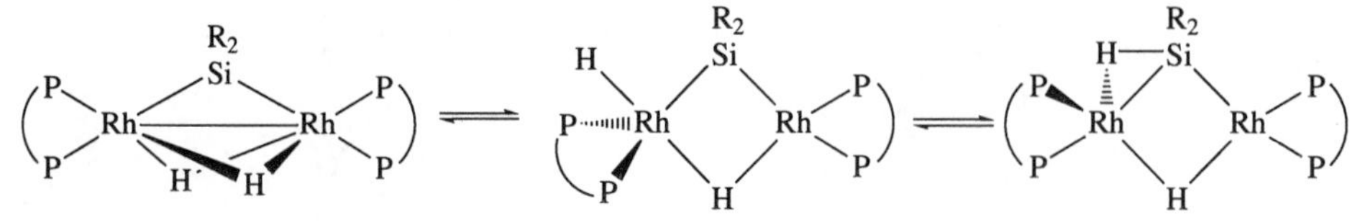

[71]R. S. Simons and C. A. Tessier, *Organometallics* **1996**, *15*, 2604.
[72]K. Osakada et al., *Organometallics* **1997**, *16*, 2063.
[73]W. R. Roper et al., *Chem. Eur. J.* **1997**, *3*, 1608.
[74]M. D. Fryzuk et al., *Organometallics* **1996**, *15*, 2871. For SiR_2H-bridged W_2 complexes see G. J. Kubas et al., *Inorg. Chem.* **1997**, *36*, 3341.

Silane addition to Rh[I] alkyls may be accompanied by alkane elimination to give rare examples of Rh[I] silyl species:[75]

$$\text{MeRhL}_4 + \text{HSiPh}_3 \xrightarrow[-\text{L}]{-\text{CH}_4} \text{Ph}_3\text{SiRhL}_3$$

L = PMe₃

Double silane addition to Rh[I] leads to Rh[V] products, e.g., (21-XVI).[74] Similarly, reaction of HSiCl₃ with (arene)Cr(CO)₃ under photolytic conditions gives the Cr[IV] complex (21-XVII).[76]

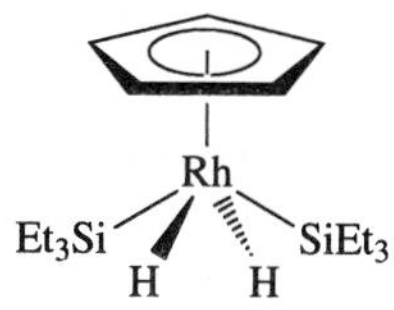

(21-XVI) 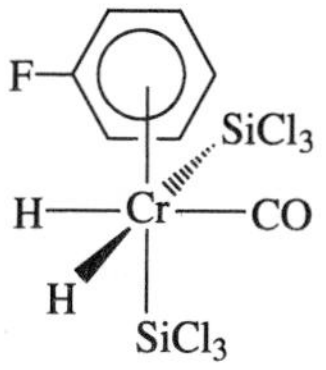   (21-XVII)

The silanols HR₂SiOH react with IrCl(C₂H₄)(PEt₃)₂ by Si—H rather than O—H addition.[77]

There are many cases where η^2-silane coordination is the ground state structure, as in the η^2-SiH₄ complex (21-XVIII) (R = Pr^i).[78] The η^2-silanes may also form chelate structures, as in (21-XIX) and (21-XX).[79,80]

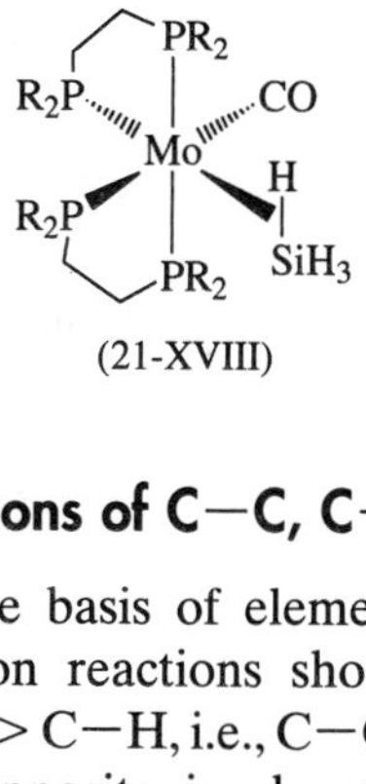

(21-XVIII) 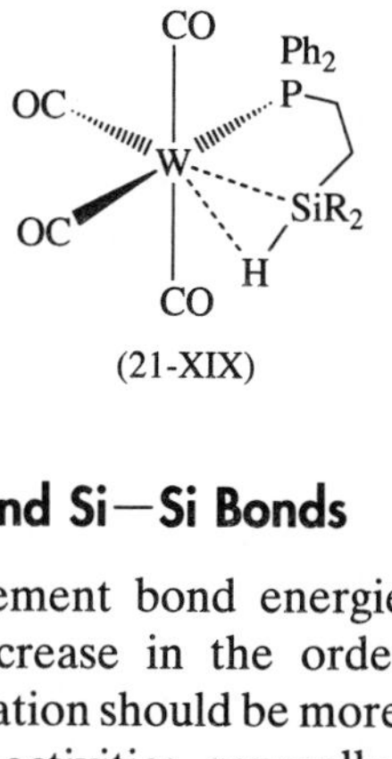   (21-XIX) 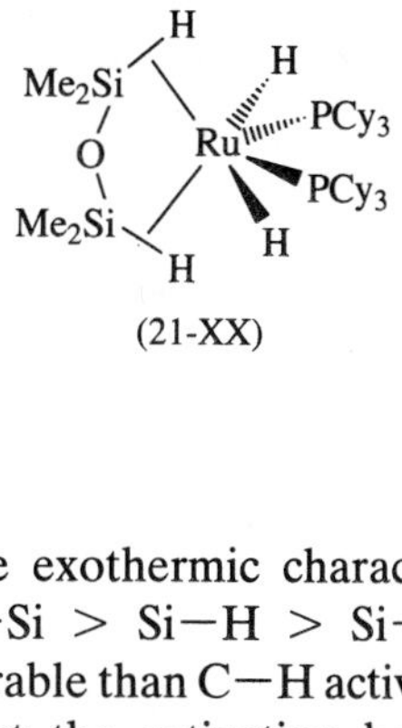   (21-XX)

Additions of C—C, C—Si, and Si—Si Bonds

On the basis of element—element bond energies, the exothermic character of addition reactions should decrease in the order Si—Si > Si—H > Si—C > C—C > C—H, i.e., C—C activation should be more favorable than C—H activation. The opposite is observed: reactivities generally reflect the activation barriers, Si—H < Si—Si < C—H ≈ Si—C << C—C.[81] Addition reactions of C—C bonds

[75]P. Hoffmann et al., J. Organomet. Chem. **1995**, 490, 51.
[76]K. J. Klabunde et al., Inorg. Chem. **1995**, 34, 278.
[77]D. Milstein et al., J. Am. Chem. Soc. **1995**, 117, 5865.
[78]G. L. Kubas et al., J. Am. Chem. Soc. **1995**, 117, 1159.
[79]U. Schubert and H. Gilges, Organometallics **1996**, 15, 2373.
[80]S. Sabo-Etienne et al., J. Am. Chem. Soc. **1997**, 119, 3167.
[81]U. Schubert, Angew. Chem. Int. Ed. Engl. **1994**, 33, 419.

to metals are therefore largely restricted to strained systems, for example:[82,83]

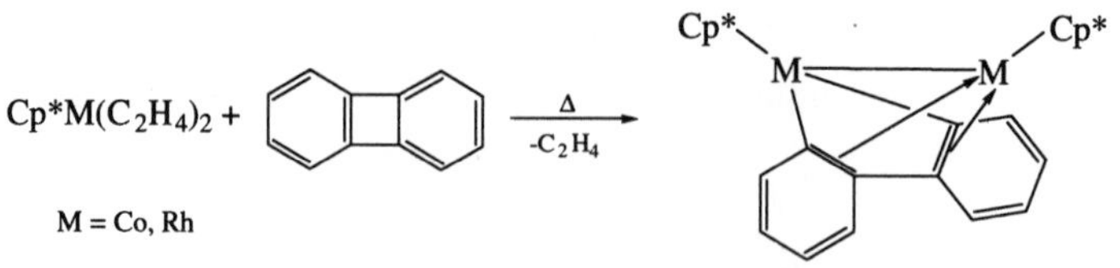

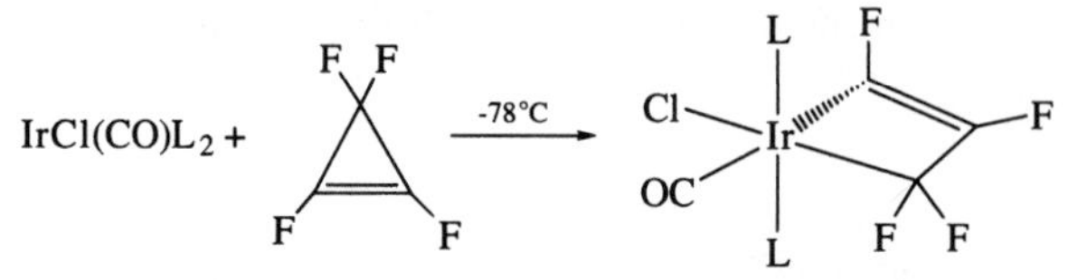

$$Cp^*M(C_2H_4)_2 +$$

M = Co, Rh

$$IrCl(CO)L_2 +$$

However C—C addition may be preceded by C—H activation, as is seen with the oxidative addition of cyclopropanes or in cases where strained metallacycles may be formed:[84]

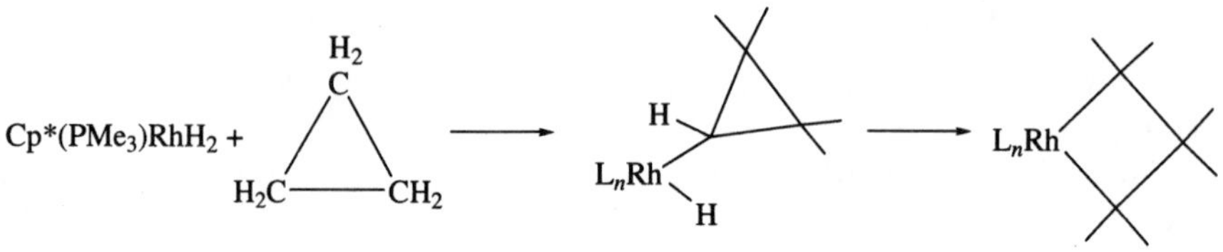

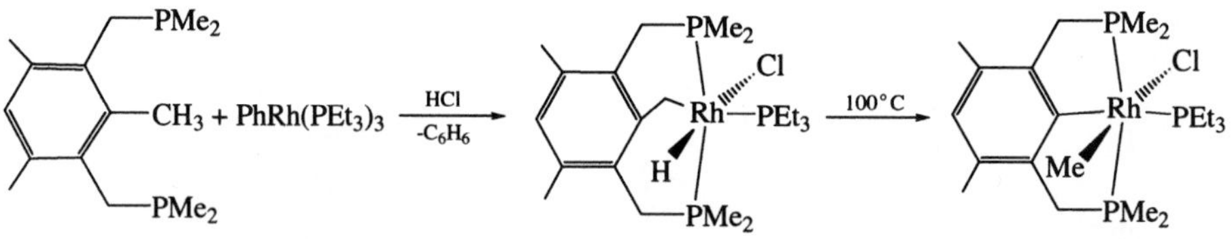

$$Cp^*(PMe_3)RhH_2 +$$

Si—C bonds are activated by (L—L)M chelate complexes with small bite angles:[85,86]

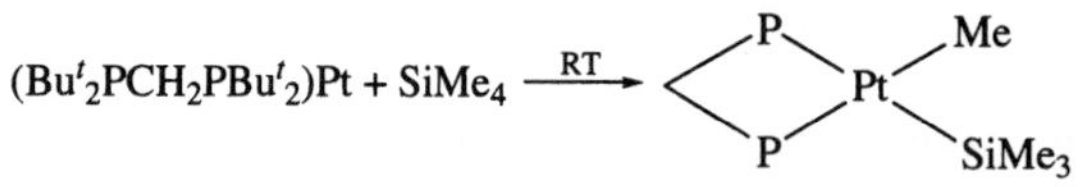

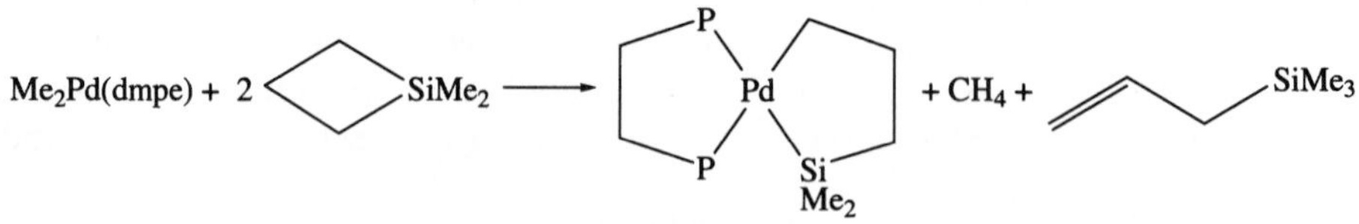

$$(Bu^t_2PCH_2PBu^t_2)Pt + SiMe_4 \xrightarrow{RT}$$

$$Me_2Pd(dmpe) + 2$$

[82] W. D. Jones *et al.*, *Organometallics* **1997**, *16*, 2016.
[83] R. P. Hughes *et al.*, *J. Am. Chem. Soc.* **1989**, *111*, 8919.
[84] D. Milstein *et al.*, *J. Am. Chem. Soc.* **1995**, *117*, 9774.
[85] P. Hoffmann et al., *Angew. Chem. Int. Ed. Engl.* **1990**, *29*, 880.
[86] M. Tanaka *et al.*, *Organometallics* **1997**, *16*, 3246.

Strained silyl-bridged ferrocenes add to Pt⁰. This is the key reaction in the Pt catalyzed ring-opening polymerization of these *ansa*-ferrocenes:[87]

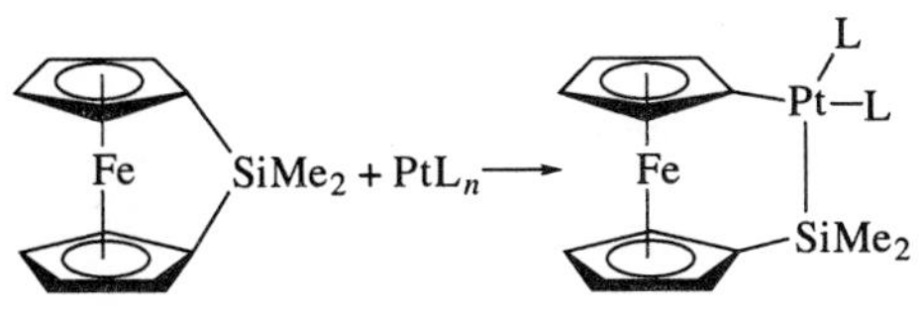

Alkynyl-Si and -Sn compounds readily add to metals to give acetylide complexes, e.g,[88,89]

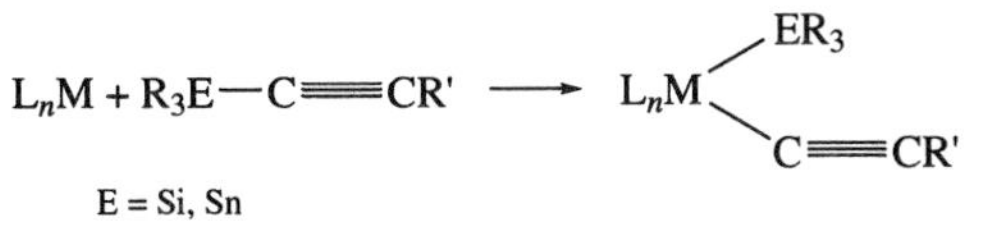

$$E = Si, Sn$$

Finally, disilanes give oxidative addition products with chelating Pd phosphine complexes which, unlike non-chelating complexes, are thermally stable.[81]

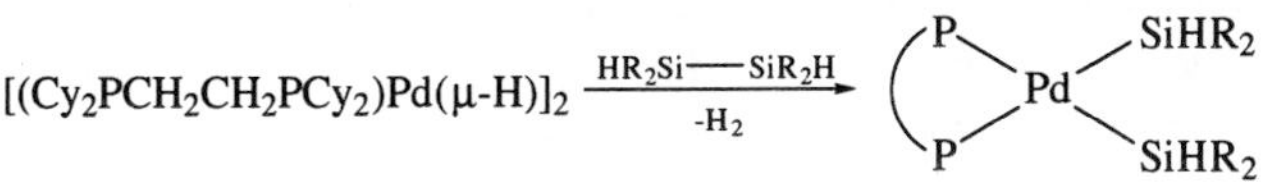

Oxidative Addition of Other Molecules

Aldehydes do not commonly undergo oxidative additions to give $L_nM(H)(COR)$ but are capable of doing so if stable chelate rings result, as in the reactions:

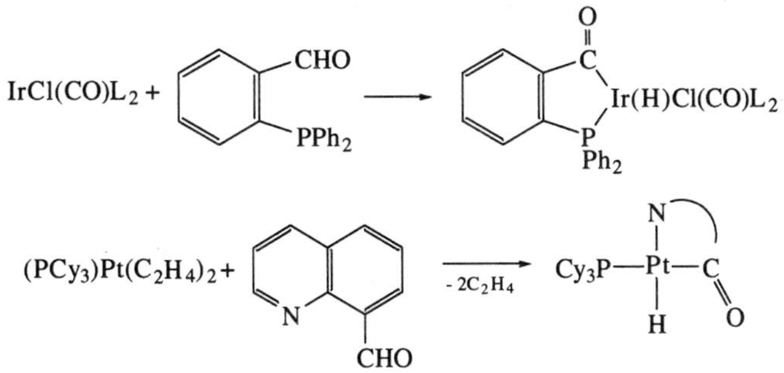

Addition of formaldehyde can occur, for example, with $IrCl(PMe_3)_3$ to give $Ir(H)(CHO)(PMe_3)_3$.

[87]J. B. Sheridan *et al.*, *J. Chem. Soc., Dalton Trans.* **1997**, 711.
[88]K. G. Caulton *et al.*, *Organometallics* **1997**, *16*, 292.
[89]H. Werner *et al.*, *Organometallics* **1997**, *16*, 803.

Epoxides react with electron-rich Rh or Ir complexes under $C-O$ activation, though the primary product rapidly rearranges to the hydrido β-oxoalkyl complex:[90]

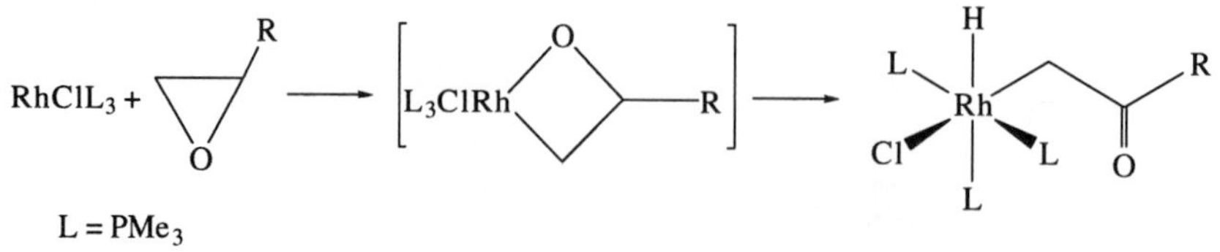

$L = PMe_3$

The chelate effect is important in the oxidative additions of $P-C$ bonds which, in the case of nickel, give $P-O$ and $P-N$ chelate complexes of the type used as ethylene oligomerization catalysts in the Shell higher olefin process (SHOP),[91] for example,

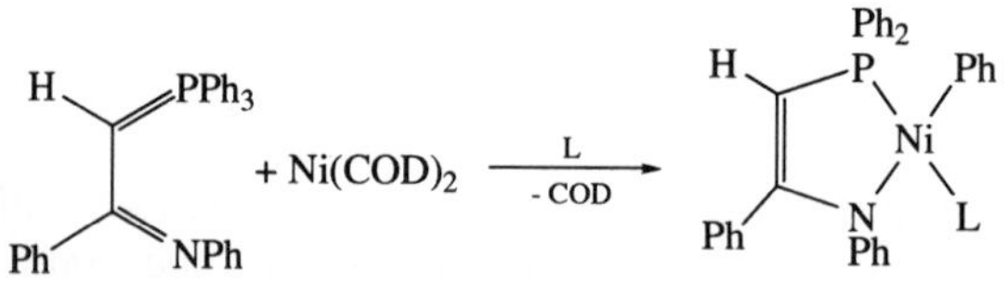

Platinum(0) complexes, PtL_4, react with halosilanes R_3SiX to give $L_2Pt(X)$-(SiR_3); reactivity decreases in the order $X = I > Br >> Cl$ (no reaction), and increases with electron-donor strength of $L = PPh_3$ (no reaction) $<<$ $PMe_2Ph < PMe_3 < PEt_3$.[92]

Formal oxidative addition products are also obtained in the reductive coupling of CO or CNR ligands of early transition metal complexes, as in some Nb and Ta compounds:[93]

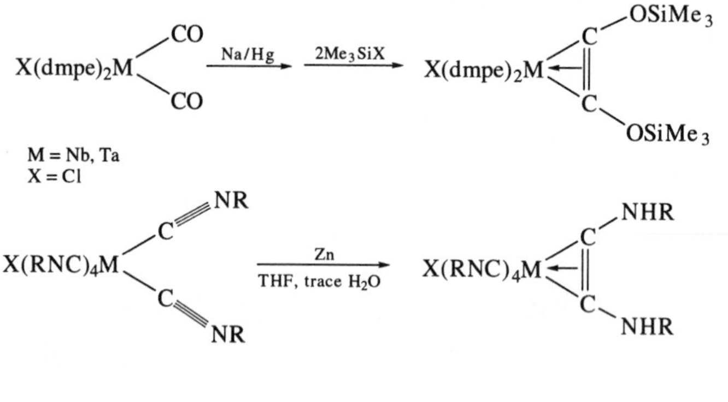

$M = Nb, Ta$
$X = Cl$

21-3 Elimination Reactions

Elimination reactions are important in catalytic cycles as terminating steps. In concerted reactions the principle of microscopic reversibility applies; oxidative addi-

[90]D. Milstein *et al.*, *J. Am. Chem. Soc.* **1990**, *112*, 6411.
[91]P. Braunstein *et al.*, *J. Chem. Soc., Dalton Trans.* **1996**, 3571.
[92]M. Tanaka *et al.*, *Organometallics* **1997**, *16*, 4696.
[93]E. M. Carnahan and S. J. Lippard, *J. Am. Chem. Soc.* **1992**, *114*, 4166.

tions should therefore be reversible, and reductive eliminations should follow the same mechanistic path and be intramolecular. This is unlikely, however, for compounds formed by radical additions.

The elimination reactions of dialkyl compounds of d^8 metals are particularly well studied. There are significant variations in thermal stability of metal dialkyls L_2MR_2, depending on M and L. Good donor ligands, such as bipy or trialkylphosphines, stabilize the dialkyl complex, and platinum(II) compounds are much more resistant to reductive elimination than Pd^{II} analogues.[94] For concerted intramolecular reductive elimination to occur, the two alkyl ligands must be *cis*, i.e., reductive elimination from *trans* complexes requires isomerization. The reaction is initiated by loss of a phosphine ligand L:

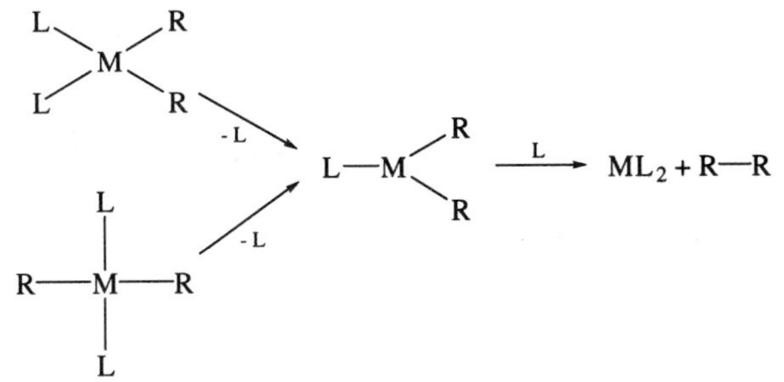

Elimination is often assisted and accelerated by the addition of π-acceptors such as acetylenes or olefins.[95]

Reductive elimination is of course not restricted to the coupling of alkyl groups; for example, heating $L_2Pd(Ph)NR_2$ gives amines $Ph-NR_2$, again *via* an intramolecular mechanism involving a 3-coordinate intermediate.[96]

Hydrido alkyl species $L_nM(H)(R)$ are particularly prone to elimination of $R-H$; this thermodynamically favored reaction is the reverse of $C-H$ activation (see Section 21-4) and explains why for a long time intermolecular $C-H$ activation remained elusive. For example, the protolysis of $(TMEDA)PtMe_2$ by HCl does not lead to the direct electrophilic attack of H^+ on the Pt—Me bond but gives thermally unstable hydrido alkyl $(TMEDA)Pt(H)ClMe_2$ which undergoes reductive elimination *via* a coordinatively unsaturated 5-coordinate intermediate:[97]

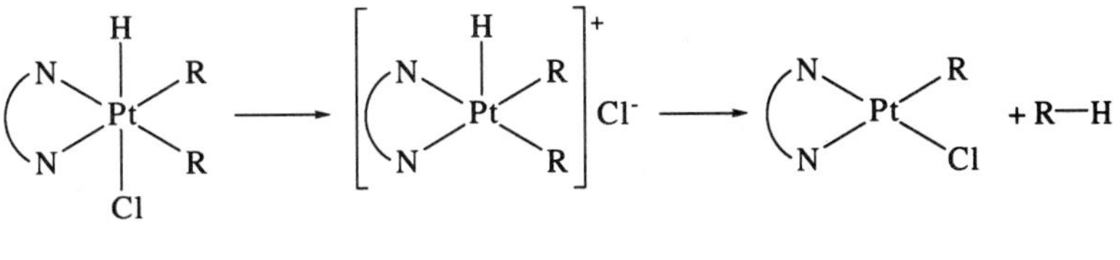

Reductive elimination is generally induced by heating or, less commonly, photolysis and employed to generate highly reactive complexes L_nM suitable for the activation of unreactive substrates such as hydrocarbons. Usually this involves

[94]J. M. Brown and N. A. Cooley, *Chem. Rev.* **1988**, *88*, 1031.
[95]F. Ozawa *et al.*, *J. Am. Chem. Soc.* **1994**, *116*, 2844.
[96]M. S. Driver and J. F. Hartwig, *J. Am. Chem. Soc.* **1995**, *117*, 4708.
[97]J. A. Labinger *et al.*, *J. Am. Chem. Soc.* **1996**, *118*, 5961.

18-electron complexes, such as $(PMe_3)_4Ru(CH_2Ph)(H)$ to give $Ru(PMe_3)_4$,[98] or $Cp^*(PMe_3)Rh(H)Ph$ to generate $Cp^*Rh(PMe_3)$ (*cf.* C—F activation).[99] On the other hand, the reductive elimination of ethane from $Cp^*(PPh_3)RhMe_2$ is accelerated by a factor of 3×10^9 by one-electron oxidation to the corresponding 17-electron radical cation.[100]

The reductive elimination product from alkyl halide complexes $L_nM(X)_yR_2$ is usually the alkane R—R. In some cases, however, C—X bond formation may compete with C—C elimination, as in $(dppe)Pt(I)Me_3$:[101]

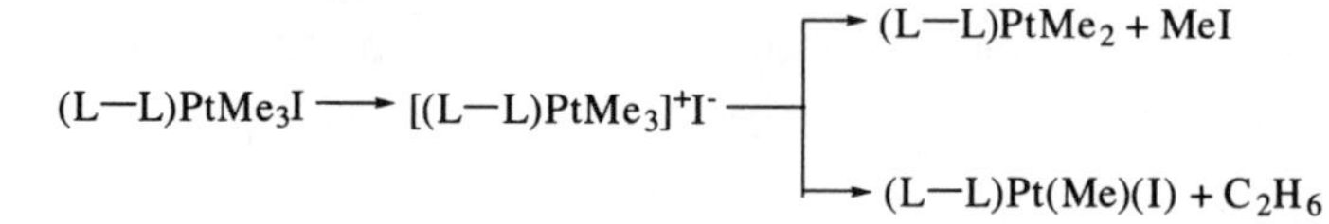

$$(L—L)PtMe_3I \longrightarrow [(L—L)PtMe_3]^+I^- \begin{cases} \longrightarrow (L—L)PtMe_2 + MeI \\ \\ \longrightarrow (L—L)Pt(Me)(I) + C_2H_6 \end{cases}$$

Other examples of reductive eliminations not involving C—C coupled products are the generation of "CpCp*Hf" from $CpCp^*HfH\{Si(SiMe_3)_3\}$,[102] the elimination of ROH from $Cp^*Ir(PPh_3)_3(H)(OR)$,[103] and the formation of an allenylphosphonium ligand as in (21-XXI(a)):[104]

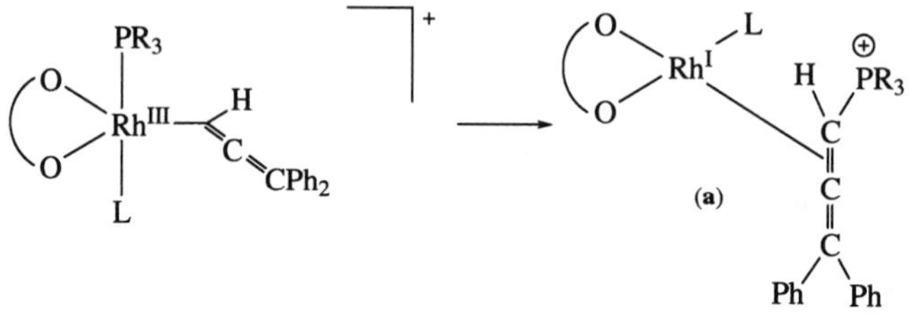

(21-XXI)

α- and β-Elimination

The decomposition of metal alkyls carrying at least one H atom in the β-position frequently occurs by β-H elimination followed by reductive elimination of the resulting M(H)(R') species:

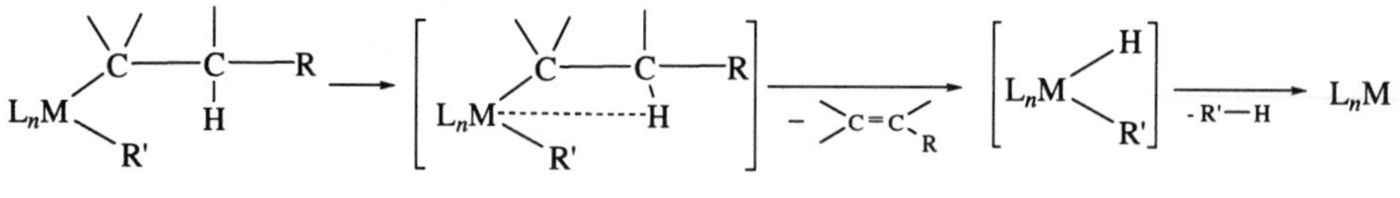

This decomposition pathway leads to equimolar amounts of alkenes and alkanes. Complexes with β-agostically bonded alkyl groups are likely intermediates; structures of this kind may sometimes be the ground state, as in $[(P—P)Pt(\eta^2\text{-}C_2H_5)]^+$.[105]

[98]R. A. Andersen *et al., J. Am. Chem. Soc.* **1991**, *113*, 6492.
[99]R. N. Perutz *et al., Organometallics* **1994**, *13*, 522.
[100]A. Pedersen and M. Tilset, *Organometallics* **1993**, *12*, 56.
[101]K. I. Goldberg *et al., J. Am. Chem. Soc.* **1994**, *116*, 1573.
[102]T. D. Tilley *et al., Organometallics* **1997**, *16*, 8.
[103]R. G. Bergman *et al., Organometallics* **1991**, *10*, 1462.
[104]L. A. Oro *et al., Organometallics* **1997**, *16*, 4572.
[105]J. L. Spencer *et al., J. Chem. Soc., Dalton Trans.* **1992**, 2653.

The alkene elimination product usually dissociates readily but remains coordinated in some cases, as in a compound of the type (21-XXII).[106] The β-H elimination is the predominant termination step in catalytic alkene oligomerizations and polymerizations.

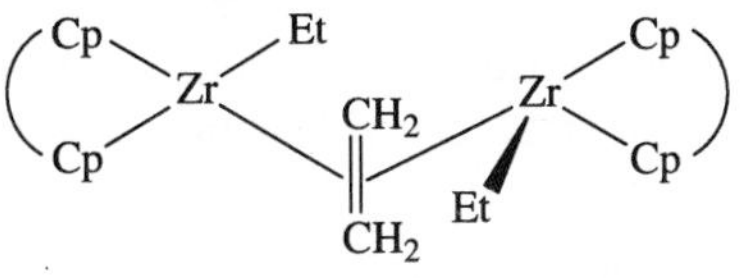

(21-XXII)

The β-H elimination may involve phenyl ligands, as in the decomposition of Cp(PMe₃)₂VPh₂ to the benzyne complex Cp(PMe₃)₂V(η^2-C₆H₄),[107] or metal amide ligands to give ketimines,[108]

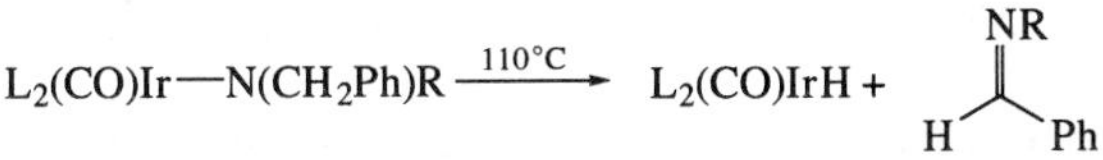

$$L_2(CO)Ir—N(CH_2Ph)R \xrightarrow{110°C} L_2(CO)IrH + \underset{H \quad Ph}{\overset{NR}{\diagup\diagdown}}$$

The β-methyl elimination reactions are rare by comparison, examples being the decomposition of zirconium neopentyl complexes:[109]

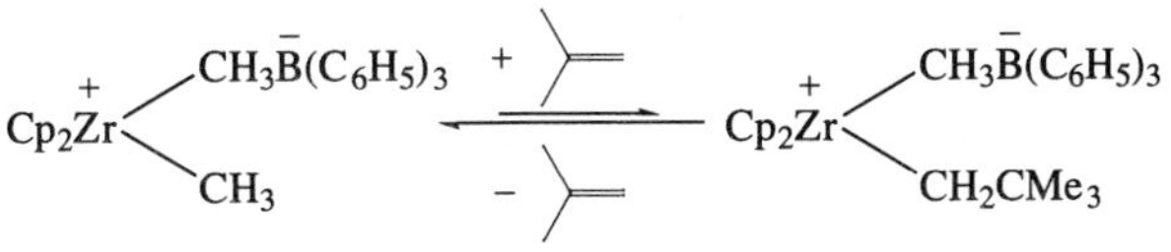

The elimination of an α-hydrogen atom is generally less facile, but common for Group 5 and 6 metals where it can lead to alkylidene complexes [e.g., $L_nM = Cp*W(NO)$]:[110]

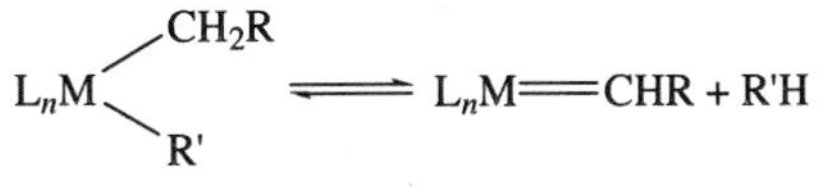

Similarly, α-methyl elimination is possible, as in (diimine)PtXMe₂(EMe₃) where the tendency to eliminate EMe₂ increases in the sequence E = Si < Ge < Sn.[111]

21-4 Cleavage of C—H Bonds; Alkane Activation; Cyclometallation Reactions[112,113]

The cleavage of C—H bonds represents an oxidative addition reaction in most cases, but is such an important area that we treat it separately. Activation of C—H

[106]P. Royo *et al.*, *Organometallics* **1997**, *16*, 1553.
[107]J. H. Teuben *et al.*, *Organometallics* **1996**, *15*, 2523.
[108]J. F. Hartwig, *J. Am. Chem. Soc.* **1996**, *118*, 7010.
[109]A. D. Horton, *Organometallics* **1996**, *15*, 2675.
[110]E. Tran and P. Legzdins, *J. Am. Chem. Soc.* **1997**, *119*, 5071.
[111]C. J. Levy and R. J. Puddephatt, *Organometallics* **1997**, *16*, 4115.
[112]J. R. Chipperfield: *C—H Bond Activation*, in: F. R. Hartley (Ed.) , *Chemistry of the Platinum Group Metals: Recent Developments*, Elsevier, Amsterdam, 1991, p. 147.
[113]R. G. Bergman *et al.*, *Acc. Chem. Res.* **1995**, *28*, 154.

bonds can occur *intramolecularly*, leading to metallacycles and ligand transformations within the metal coordination sphere, or *intermolecularly*, between reactive complexes and alkanes. The former has synthetic potential and is relevant for the understanding of some thermolysis reactions, while the latter is more challenging and the basis for catalytic alkane activation reactions. Examples of such catalytic reactions are included in this Section for convenience, as is the related C—H and C—S activation chemistry relevant to the *hydrodesulfurization* of heterocycles.

Intramolecular C—H Activation

Reactions leading to the activation of C—H bonds in α- or β-position have been mentioned in the previous section. An example for reversible α-H transfer is the equilibrium

$$Os_3(CO)_{10}(\mu\text{-}H)(\mu\text{-}CH_3) \rightleftharpoons Os_3(CO)_{10}(\mu\text{-}H)_2(\mu\text{-}CH_2)$$

The decomposition of metal dimethyl complexes may proceed *via* a 4-center transition state:

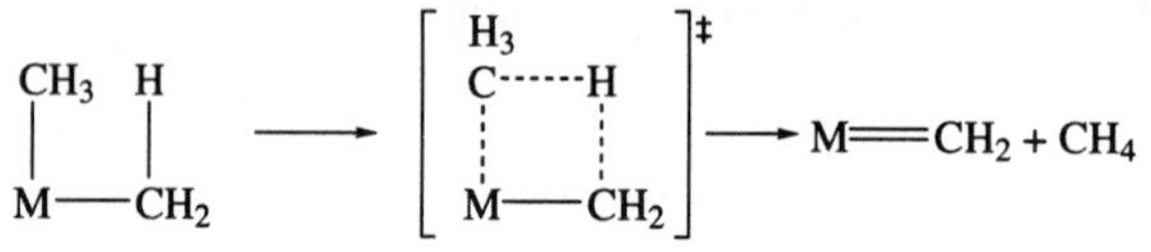

There are several types of cyclometallation reactions.

1. Reactions involving sp^3 hybridized carbon atoms of alkyl ligands. These may involve γ- or δ-C—H bonds; the C—H activation reaction may be followed by reductive elimination:

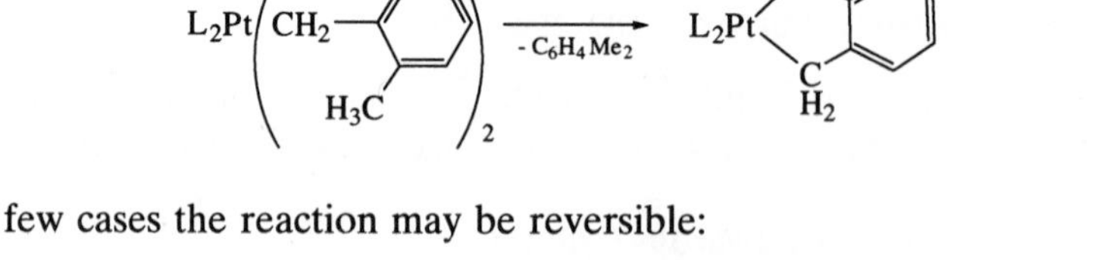

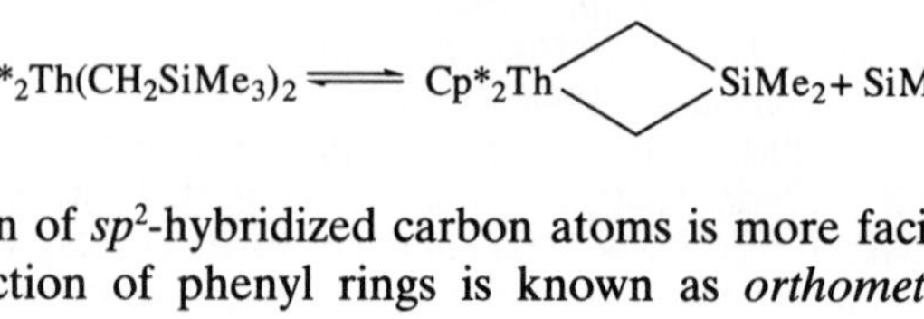

In a few cases the reaction may be reversible:

The metallation of sp^2-hybridized carbon atoms is more facile than that of sp^3-C; the reaction of phenyl rings is known as *orthometallation*. Some specific examples include the orthometallation of silyl ligands,[114] the formation of a metallacyclohexadiene,[115] and the reversible C—H activation of a Cp ligand:[116]

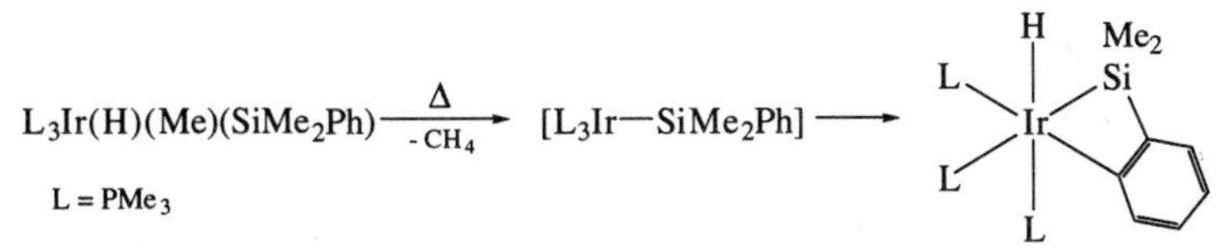

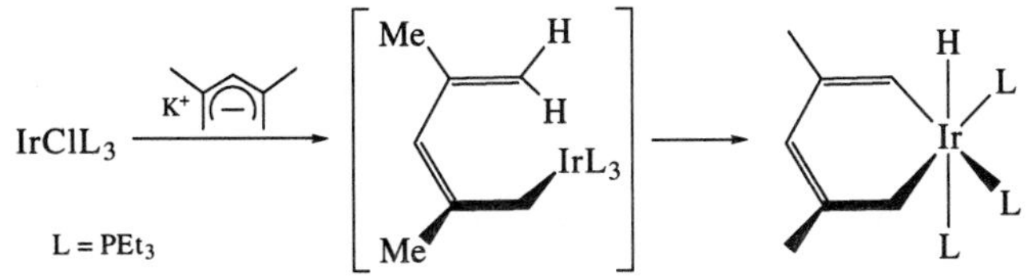

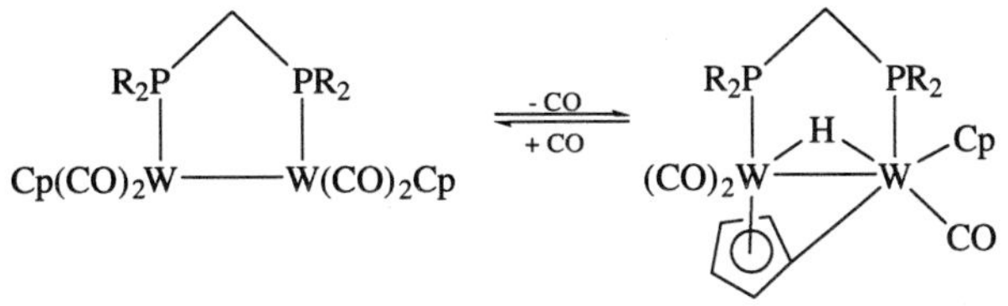

2. Reactions of C—H bonds attached to donor ligands. These may lead to the formation of rings of various sizes. The orthometallation of aryl substituents is particularly common; those involving PPh₃ may contribute to catalyst deactivation reactions. The metallation of ligands may be reversible, particularly where 3- and 4-membered rings are formed.[117]

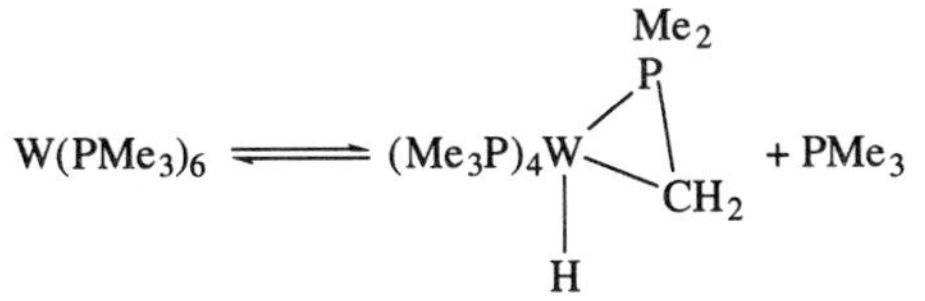

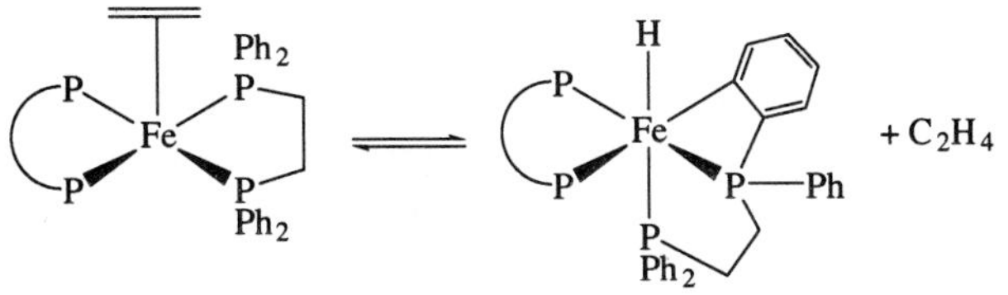

Orthometallation of PPh₃ is common and explains for example the ability of RuHCl(PPh₃)₃ to catalyze the exchange of the *ortho*—H atoms of PPh₃ for

[114]M. Aizenberg and D. Milstein, *Organometallics* **1996**, *15*, 3317.
[115]J. R. Bleeke et al., *Organometallics* **1997**, *16*, 606.
[116]M. A. Ruiz et al., *Organometallics* **1997**, *16*, 354.
[117]G. Parkin et al., *J. Am. Chem. Soc.* **1992**, *114*, 4611.

deuterium. Stable oxidative addition products of PPh$_3$ are found for iridium:

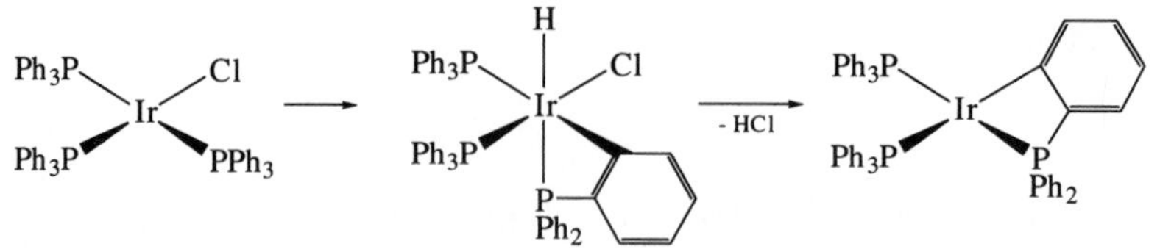

Palladium readily causes C—H activation in the *ortho*-position.

Five-membered metallacycles are favored over four- and six-membered rings:[112]

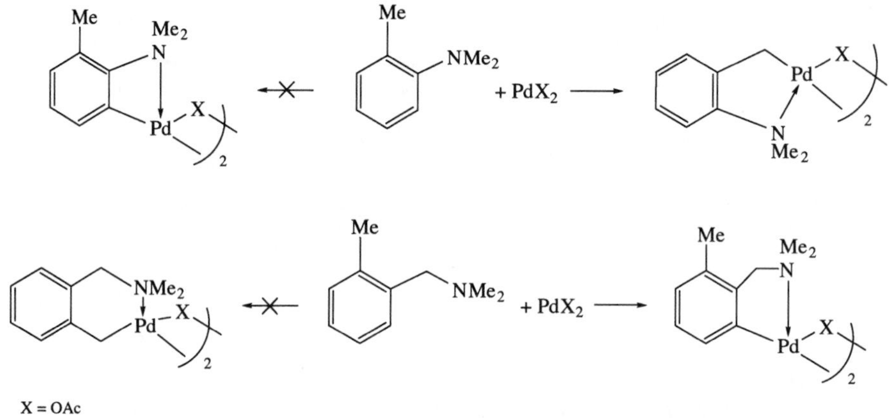

X = OAc

The intramolecular C—H activation of ethylene in complexes of the type CpIr(L)(C$_2$H$_4$) is induced by uv irradiation and is favored over reaction with the solvent; the process evidently involves a "cage complex" intermediate.[118]

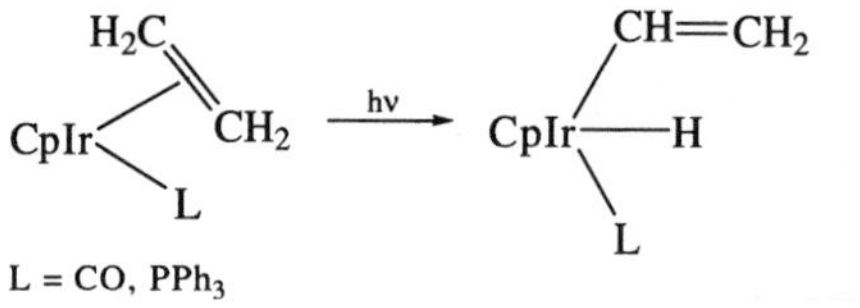

L = CO, PPh$_3$

An example of a reaction sequence involving both C—H and β-C—C activation reactions is the stepwise transformation of a neopentyl into a trimethylenemethyl ligand in (SiP$_3$)Ru complexes [SiP$_3$ = MeSi(CH$_2$PMe$_2$)$_3$]:[119]

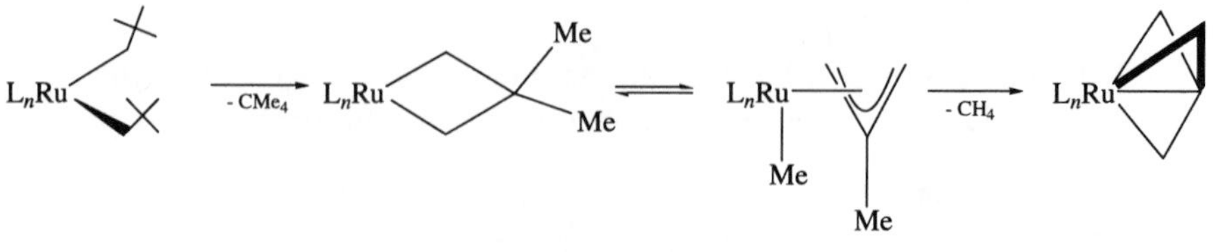

[118]R. N. Perutz *et al.*, *Organometallics* **1993,** *12*, 2933.
[119]R. A. Andersen *et al.*, *J. Am. Chem. Soc.* **1997,** *119*, 11244.

Intermolecular C—H Activation

Highly electrophilic early transition metal and lanthanide complexes as well as electron-rich coordinatively unsaturated complexes of late transition metals, notably Rh, Ir, Pd, and Pt, are able to activate C—H bonds of arenes and saturated hydrocarbons. There are remarkable mechanistic similarities between them.

Exchange reactions such as

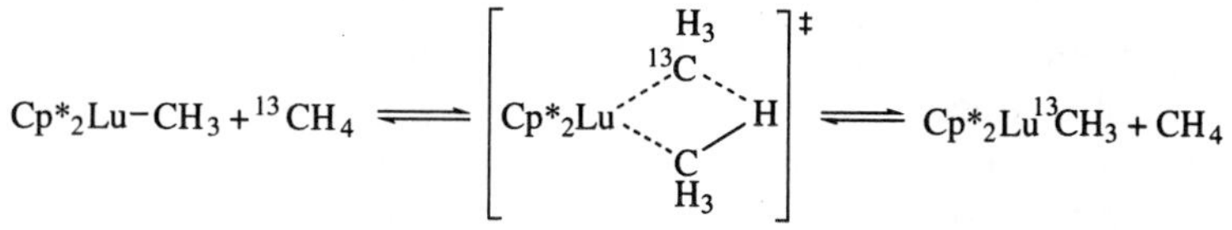

$$Cp*_2Lu-CH_3 + {}^{13}CH_4 \rightleftharpoons \left[Cp*_2Lu \begin{array}{c} {}^{13}C \\ H_3 \\ \cdots \\ C \\ H_3 \end{array} H \right]^{\ddagger} \rightleftharpoons Cp*_2Lu{}^{13}CH_3 + CH_4$$

proceed *via* a four-membered transition state. In view of the high oxidation state of the lanthanide or actinide elements in these complexes, oxidative addition is an unlikely process; exchange reactions of this type have been termed "*σ-bond metathesis.*" Such complexes typically catalyze the H/D exchange with deuterated solvents. Aromatic and benzylic C—H bonds are more reactive than alkanes; for example, $[Cp^*_2Y(\mu\text{-}H)]_2$ reacts with benzene and toluene to give Cp^*_2Y-Ph and $Cp^*_2Y-CH_2Ph$, respectively.[120] Functionalized substrates such as acetonitrile are readily C—H activated, but are also capable of insertion reactions, for example:[121]

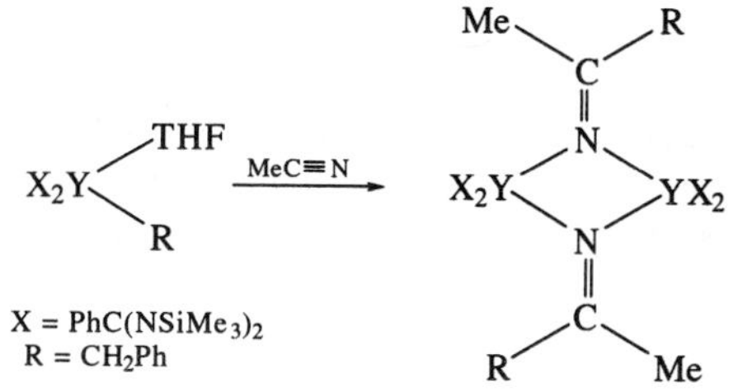

X = PhC(NSiMe₃)₂
R = CH₂Ph

Bulky titanium alkylamido complexes are capable of equilibrating with aliphatic and aromatic hydrocarbons *via* concerted 1,2-elimination/1,2-addition processes; a Ti=NR imido species is formed as a transient intermediate. Kinetic parameters suggest a four-center transition state with an almost linear N···H···R arrangement:[122]

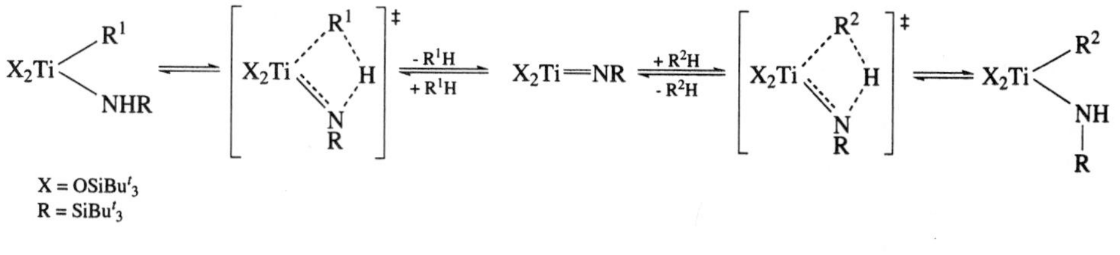

X = OSiBu'₃
R = SiBu'₃

[120] J. H. Teuben *et al.*, *Organometallics* **1993,** *12,* 3531.
[121] J. H. Teuben *et al.*, *Organometallics* **1996,** *15,* 2291.
[122] J. L. Bennett and P. T. Wolczanski, *J. Am. Chem. Soc.* **1997,** *119,* 10696.

Palladium(II) in highly acidic media is capable of attacking even methane, as in the reaction[123]

$$CH_4 + Pd^{2+} + CF_3COOH \longrightarrow CH_3O_2CCF_3 + Pd^0 + 2H^+$$

The reaction of $PdX_2(PPh_3)_2$ (X = Cl, Br, I) with hydrocarbons at 70–130°C is thought to lead to Pd^{IV} compounds, $Pd(H)(R)X_2L_2$, which readily lose HX.[124]

Coordinatively unsaturated 14- or 16-electron fragments L_nM, where M has a d^6, d^8, or d^{10} configuration, are capable of oxidatively adding C—H bonds of arenes and alkanes and have been studied in considerable detail. Calculations suggest that the reaction proceeds *via* an η^2-alkane complex.[125] More electron-rich as well as heavier transition metal centers, i.e., 3rd-row metals, facilitate C—H oxidative addition. In the case of C—H addition to Pd and Pt phosphine complexes ML_2, high activation barriers ($\sim$30 kcal mol^{-1}) have been calculated for monodentate phosphines, whereas chelating phosphines lead to values as low as $\sim$4 kcal mol^{-1} (M = Pt).[126]

The oxidative addition of alkanes to Rh^I and Ir^I species CpML is very facile. In the case of the tris(pyrazolyl)borate complex $Tp^*Rh(CO)_2$ [Tp^* = HB(2,4-Me$_2$pyr)$_3$], the lifetimes of the intermediates were determined by ultrafast time-resolved infrared spectroscopy. In this case de-chelation of one of the pyrazolyl arms was found to precede the C—H oxidative addition step. The proposed intermediates for R—H addition and their lifetimes are shown in Fig. 21-1.[127]

The addition of R—H to $Cp^*Ir(CO)_2$ can be carried out photochemically under supercritical conditions. The reaction is greatly enhanced by H_2, although with ethane in the presence of D_2 only $Cp^*(CO)Ir(H)(Et)$ and not $Cp^*Ir(CO)(D)(Et)$ is formed.[128]

Arenes add to metals much more readily than alkanes, due to their ability to form comparatively stable η^2-arene complexes prior to C—H activation. Arene coordination is favored by Cp compared to Cp*, and by Rh^I compared to Ir^I. The equilibrium

L = PMe₃

is strongly in favor of benzene activation.[129] The reaction of Ar—H with [Ir(COD)-(PMe$_3$)$_3$]Cl gives *mer*-Ir(H)ArClL$_3$; if ArH = pyridine the reaction is specific to the C—H bond α to the nitrogen atom.[130]

[123]A. Sen, *Acc. Chem. Res.* **1988,** *21,* 421. For C—H activation reactions of Pd see also V. V. Grushin, *Chem. Rev.* **1996,** *96,* 2011.
[124]A. N. Vedernikov et al., *J. Chem. Soc., Chem. Commun.* **1994,** 121.
[125]C. Hall and R. N. Perutz, *Chem. Rev.* **1996,** *96,* 3125.
[126]S. Sakaki et al., *J. Chem. Soc., Dalton Trans.* **1997,** 803.
[127]R. G. Bergman et al., *Science* **1997,** *278,* 260.
[128]M. Poliakoff et al., *Organometallics* **1996,** *15,* 1804.
[129]W. D. Jones et al., *J. Am. Chem. Soc.* **1993,** *115,* 7685.
[130]H. E. Selnau and J. S. Merola, *Organometallics* **1993,** *12,* 1583.

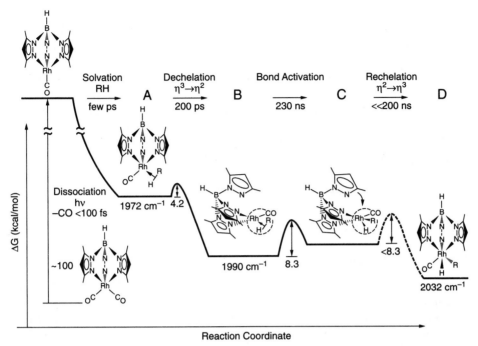

Figure 21-1   Proposed mechanism and energy diagram for the C—H activation reaction of Tp*Rh(CO)$_2$ in alkane solution. These energy differences are estimates from separate ultrafast and nanosecond experiments. The stabilities of the intermediates are shown relative to each other and are not intended to be absolute (from Ref. 127).

Terminal acetylenes readily add to low-valent metal centers. The addition of PhC≡CH to [Ir(RCN)(CO)(PPh$_3$)$_2$]$^+$ requires RCN dissociation to give an η^2-alkyne intermediate.[131]

The addition of alkanes to Ir^{III} species such as [CpIr(Me)L]$^+$ can proceed *via* two pathways, oxidative addition **(A)** or σ-bond metathesis akin to that in lanthanide complexes **(B)**.

$$
\begin{array}{c}
\text{CH}_3 \\
| \\
\text{M}\!\!-\!\!\text{H} \\
\end{array}
$$

[131]C. S. Chin *et al.*, *Inorg. Chem.* **1993**, *32*, 5901.

Calculations show that for $M = [CpIr^{III}(PH_3)(Me)]^+$ the oxidative addition mechanism **A** is the low-energy pathway, while Rh^{III} may adopt path **B**.[132] With complexes containing very labile ligands, such as the η^1-dichloromethane complex $[Cp^*(PMe_3)IrMe(ClCH_2Cl)]^+$, methane activation takes place under very mild conditions and at temperatures as low as 10°C, while benzene adds rapidly at $-30°C$.[133]

A very different type of alkane activation is the reaction of CH_4 with the Rh^{II} metalloradical (TMP)Rh (TMP = tetramesitylporphyrinato) which proceeds under 1-10 bar CH_4 even in benzene; there is no C_6H_6 activation. A trimolecular transition state $Rh\cdots CH_3H\cdots Rh$ appears to be involved.[134]

$$2(TMP)Rh^{\bullet} + CH_4 \rightleftharpoons (TMP)RhCH_3 + (TMP)RhH$$

$$K_{eq}\,(70°C) = 7300$$

Catalytic C—H Activation

Rhodium and iridium complexes of basic trialkylphosphines catalyze the dehydrogenation of alkanes to alkenes and H_2, or the transfer dehydrogenation of alkanes in the presence of H_2 receptors such as $CH_2=CHBu^t$. For example, $RhCl(CO)(PMe_3)_2$ becomes active after photolytic CO dissociation, while the compounds $RhClL_{2,3}$ (L = PMe_3, PPr_3^i or PCy_3) are thermally active. The initial step is the formation of a Rh^{III} dihydride:[135]

$$ClRhL_2 + \bigcirc \longrightarrow H_2RhClL_2 + \bigcirc$$

$$L = PPr^i_3, PCy_3$$

Phosphino enolate complexes such as (21-XXIII) catalyze the transfer dehydrogenation of cyclooctane with norbornene at 60–90°C under H_2 pressure.[136] Iridium(III) dihydrides with P—C—P chelate ligands of type (21-VIII) are thermally stable at 150–200°C and catalyze the dehydrogenation of cyclooctane at rates of up to 12 turnovers min^{-1} (200°C). The transfer-dehydrogenation of ethylcyclohexane with this catalyst gives ethylcyclohexenes, EtPh, and styrene.[137] Mechanistic studies of the transfer dehydrogenation of cyclooctane with $CH_2=CHBu^t$ in the presence of IrH_2ClL_2 indicate that cyclooctane coordination is required for the reductive

[132]M. B. Hall et al., J. Am. Chem. Soc. **1996**, 118, 6068; M. D. Su and S. Y. Chu, J. Am. Chem. Soc. **1997**, 119, 5373.
[133]B. A. Arndtsen and R. G. Bergman, Science **1995**, 270, 1970.
[134]A. E. Sherry and B. B. Wayland, J. Am. Chem. Soc. **1990**, 112, 1259.
[135]K. C. Shih and A.S. Goldman, Organometallics **1993**, 12, 3390.
[136]P. Braunstein et al., Organometallics **1996**, 15, 5551.
[137]C. M. Jensen et al., Chem. Commun. **1996**, 2083; **1997**, 461; For the dehydrogenation of alkanes without an H_2-acceptor see C. M. Jensen et al., Chem. Commun. **1997**, 2273.

elimination of $CH_3CH_2Bu^t$:[138]

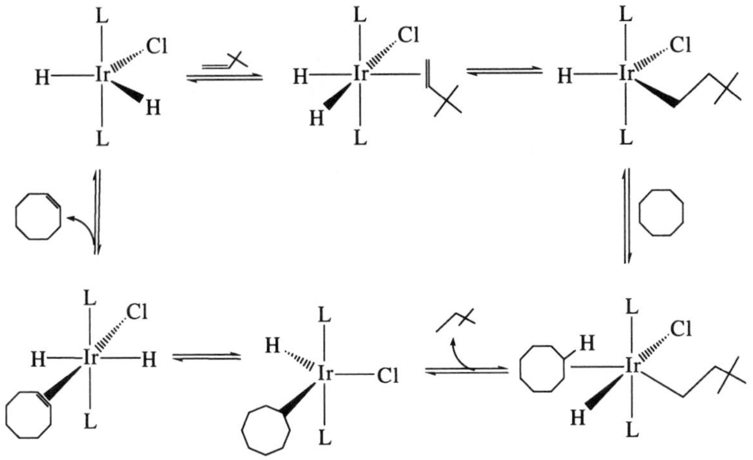

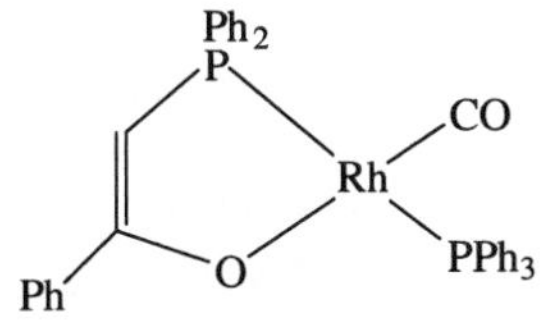

(21-XXIII)

Dibenzylideneacetone Pt^0 complexes catalyze the arylation of Si—H bonds *via* C—H bond activation of substituted arenes.[139]

Heterogeneous catalysts activate C—H bonds at significantly higher temperatures. For example, a Fe/Co modified Mo-supported acidic ZSM-5 zeolite catalyst dehydrogenates methane under non-oxidizing conditions at 700°C to a mixture of C_2-C_4 alkanes/alkenes and C_6-C_{12} aromatics such as benzene and naphthalene.[140]

Activation of Sulfur Heterocycles; Hydrodesulfurization

The removal of sulfur compounds from fossil fuels is of great commercial and environmental importance. This process is carried out by treatment with H_2 using heterogeneous catalysts (hydrodesulfurization, HDS) and generates H_2S; this in turn is converted to elemental sulfur by oxidation with SO_2 over aluminum oxide at 300°C (Claus process):[141]

$$2H_2S + SO_2 \xrightarrow[300°C]{Al_2O_3} {}^3/_8 S_8 + 2H_2O$$

Of the various sulfur-containing compounds found in crude fossil fuel, such as thiols, sulfides, disulfides, etc., thiophenes are the most difficult to remove.

[138]J. Belli and C. M. Jensen, *Organometallics* **1996**, *15*, 1532.
[139]M. Tanaka *et al.*, *Organometallics* **1993**, *12*, 2065.
[140]M. Ichikawa *et al.*, *Chem. Commun.* **1997**, 1455.
[141]A. Shaver *et al.*, *Angew. Chem. Int. Ed. Engl.* **1996**, *35*, 2362.

Homogeneous model complexes shed light on the reactivity of thiophene and benzothiophene. Thiophene is known to coordinate to metals in η^1- to η^5-fashion, as in (21-XXIV) to (21-XXVII). The η^1-S and η^2-C,C structures are believed to act as immediate precursors to C—H and C—S addition reactions. This was confirmed spectroscopically for the reaction of $(pp_3)Ru(N_2)$ with thiophene:

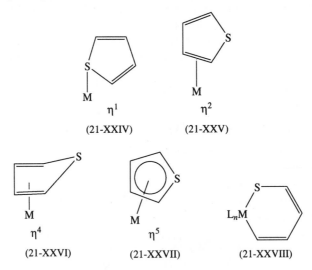

Formation of the unstable adduct $(pp_3)Ru(SC_4H_4)$ is followed by fast C—H activation to give $(pp_3)Ru(H)(2\text{-}C_4H_3S)$, whereas with 2,5-dimethylthiophene the reaction stops at the S-adduct stage.[142] The photochemical reaction with $RhCl(CO)(PMe_3)_2$, however, activates even dimethylthiophene, with C—H cleavage in the 3-position.[143] By contrast, the thermal reaction of $Cp^*Rh(H)(Ph)(PMe_3)$ with thiophene leads to a C—S addition product of type (21-XXVIII); the reaction is reversible.[144] Similar thiametallacycles are obtained for $L_nM = Pt(PEt_3)_2$ and dibenzothiophene; they release 2-phenylthiophenol on treatment with HCl.[145]

The C—S and the C—H addition products may equilibrate.[146]

[142]C. Bianchini *et al.*, *Organometallics* **1997**, *16*, 4611.
[143]L. D. Field *et al.*, *Organometallics* **1996**, *15*, 872.
[144]W. D. Jones and L. Dong, *J. Am. Chem. Soc.* **1991**, *113*, 559.
[145]J. J. García *et al.*, *Organometallics* **1997**, *16*, 3216; for MO calculations on these reactions see S. Harris *et al.*, *Organometallics* **1997**, *16*, 2448.
[146]M. Paneque *et al.*, *Organometallics* **1996**, *15*, 2678.

The cleavage of the C—S bond in η^4-thiophene complexes can be achieved by protonation:[147]

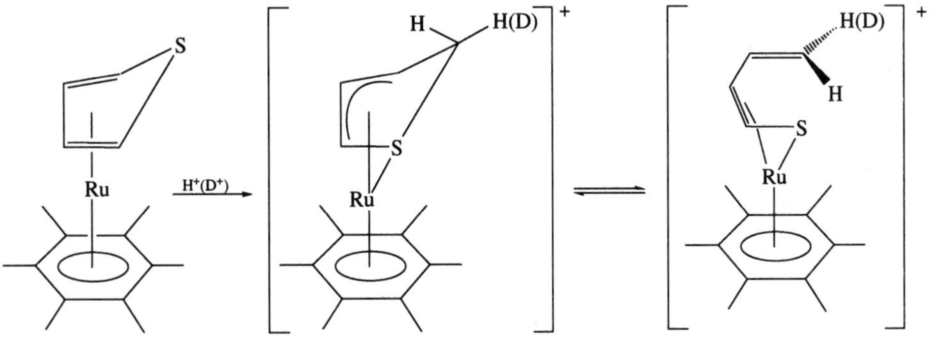

Removal of sulfur from these complexes remains challenging. Catalysts such as $[M(COD)(PPh_3)_2]^+$ (M = Rh, Ir) only hydrogenate the C=C bond of benzothiophene, whereas the hydrogenation with rhodium triphos complexes gives 2-ethylthiophenol.[148] The addition of $Co_4(CO)_{12}$ to (21-XXVIII, ML_n = Cp*Ir), however, leads to partial desulfurization, presumably with formation of cobalt sulfide.[149]

21-5 Migration (Insertion) Reactions

The concept of "insertion" is widely applicable in chemistry when defined as a reaction wherein any atom or group of atoms is inserted between two atoms initially bound together:

$$L_nM——X + YZ \longrightarrow L_nM—(YZ)—X$$

The term "insertion" carries no mechanistic implications. The reaction may proceed by the migration of X to a ligated YZ molecule coordinated *cis*, in the manner of an intramolecular nucleophilic attack on YZ — a common pathway — or it may be the result of direct attack of YZ on X.

Insertions may be termed 1,1, 1,2, 1,3, 1,4, and so on, depending on the atom to which the migrating group is transferred. Thus we have:

1,1-insertion: $M——H + CH_2N_2 \xrightarrow{-N_2} M——CH_3$

1,2-insertion: $M——H + CF_2=CF_2 \longrightarrow M——CF_2CF_2H$

$$M——R + O_2 \longrightarrow M——O——O——R$$

[147]T. B. Rauchfuss *et al.*, *J. Am. Chem. Soc.* **1993**, *115*, 4943.
[148]R. A. Sánchez-Delgado *et al.*, *Organometallics* **1997**, *16*, 2465; C. Bianchini *et al.*, *J. Am. Chem. Soc.* **1997**, *119*, 4945.
[149]R. J. Angelici *et al.*, *Organometallics* **1996**, *15*, 1414.

1,3-insertion: $M\!-\!H + N_2CHR \longrightarrow M\!-\!N\!=\!N\!-\!CH_2R$

1,4-insertion: $M\!-\!H + CH_2\!=\!CH\!-\!CH\!=\!CH_2 \longrightarrow M\!-\!CH_2CH\!=\!CHCH_3$

The insertion reactions of CO, alkenes, and alkynes are particularly important. An early example (Berthelot, 1869) is:

$$SbCl_5 + 2HC\!\equiv\!CH \longrightarrow Cl_3Sb(CH\!=\!CHCl)_2$$

Insertion reactions frequently involve the reactions of electrophiles (CO_2, SO_2) with M—H, M—C, M—N, and M—O bonds. Some representative examples are:

$$R_3SnNR'_2 + CO_2 \longrightarrow R_3SnOC(O)NR'_2$$

$$Ti(NR_2)_4 + 4CS_2 \longrightarrow Ti(S_2CNR_2)_4$$

$$R_3PbR' + SO_2 \longrightarrow R_3PbOS(O)R'$$

$$[(NH_3)_5RhH]^{2+} + O_2 \longrightarrow [(NH_3)_5Rh\!-\!OOH]^{2+}$$

$$(OC)_5MnCH_3 + CO \longrightarrow (OC)_5Mn\!-\!C(O)CH_3$$

Note that while oxidative addition reactions (Section 21-2) could be defined as insertions of metals into R—X bonds, the reactions under discussion here only deal with insertions into M—X bonds, without changes in the formal metal oxidation state.

Some representative examples of insertion or group transfer reactions are given in Table 21-2.

Insertion of Carbon Monoxide

The most intensely studied insertion reactions are those of CO into metal—carbon bonds to form metal acyls. These reactions are fundamental to industrially important catalytic reactions such as carbonylation and hydroformylations (Sections 22-5 and 22-6).

The 18-electron complexes $CH_3Mn(CO)_5$ and $CpFe(CO)_2Me$ have proved particularly amenable to studies of the CO insertion mechanism. With ^{13}CO and ^{14}CO as labeled reagents, it has been shown that

(a) the CO molecule that becomes the acyl ligand is *not* derived from external CO but is one already coordinated to the metal atom; i.e., the process involves a migration of an alkyl ligand to a coordinated CO:

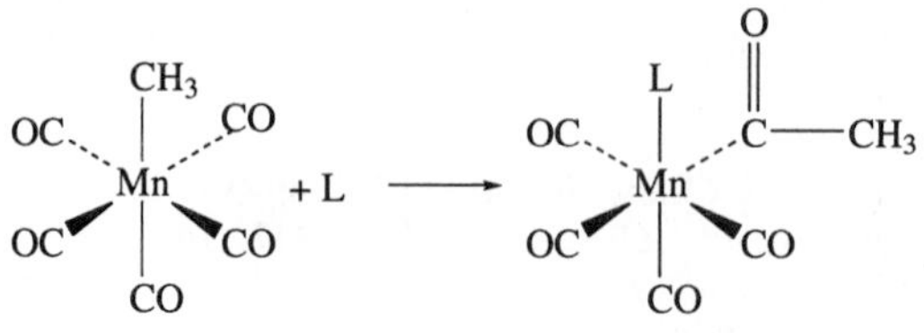

$L = {}^{14}CO$ or PPh_3

Table 21-2 Some Representative Insertion or Group
Transfer Reactions

"Inserted" molecule	Bond	Product
CO	M−CR₃	M−C(O)CR₃
	M−OH, M−OR	MC(O)OH, MC(O)(OR)
	M−NR₂	MC(O)NR₂
RNC	M−H	MCH=NR
	M−R′	MCR′=NR
RCN	M−R′	M−N=CRR′[a]
CO₂	M−H	MO₂CH[b]
	M−NR₂	MOC(O)NR₂
	M−OH, M−OR	MOCO₂H(R)[c]
	M−M′	M−CO₂M′[d]
CS₂	M−H	MS₂CH and MSC(S)H
	M−NR₂	MS₂CNR₂
	M−M′	MSC(S)M′
RN=C=NR	M−H	M(RNCHNR)
RNCO	M−H	M(RNCHO)
	M−NR′₂	M[RNC(O)NR′₂][e]
SO₂	M−C	MS(R)O₂ or MOS(O)R
	M−OR	MO₂SOR
SO₃	M−CH₃	MOSO₂CH₃
O₂	M−H	M−OOH
	M−CR₃	M−OOCR₃, MOCR′₃[f]
S₈	M−CR₃	MSCR₃
C₂H₄	M−H	MC₂H₅[g]
	M−R	M−CH₂CH₂R[h]
	M−C(O)R	M−C₂H₄C(O)R[i]
	M−OH	M−CH₂CH₂OH[j]
RC≡CR′	M−H	MC(R)=CHR′ *cis* or *trans*[k]
	M−R″	MC(R)=CR′R″
H₂C=C=CH₂	M−R	M(η³-CH₂CRCH₂)
CH₂=CHCN	M−PR₂	MCH(CN)CH₂PR′₂[l]
CH₂N₂	M−Cl	MCH₂Cl[m]
SnCl₂	M−Cl	M−SnCl₃
	M−M	MSn(Cl₂)M

[a] R. F. Jordan *et al.*, *Organometallics* **1991**, *10*, 1406.

[b] D. J. Darensbourg *et al.*, *J. Am. Chem. Soc.* **1990**, *112*, 9252.

[c] D. J. Darensbourg *et al.*, *Inorg. Chem.* **1991**, *30*, 2418; M. Orchin *et al.*, *Organometallics* **1993**, *12*, 1714.

[d] L. H. Gade *et al.*, *Chem. Commun.* **1996**, 1751.

[e] P. Legzdins *et al.*, *Organometallics* **1994**, *13*, 569.

[f] E. B. Coughlin and J. E. Bercaw, *Organometallics* **1992**, *11*, 465.

[g] T. Ziegler *et al.*, *Inorg. Chem.* **1990**, *29*, 4530.

[h] T. Ziegler *et al.*, *J. Am. Chem. Soc.* **1997**, *119*, 5939; P. E. M. Siegbahn *et al.*, *Organometallics* **1996**, 15, 5542.

[i] K. Vrieze *et al.*, *Organometallics* **1997**, *16*, 68.

[j] J. C. M. Ritter and R. G. Bergman, *J. Am. Chem. Soc.* **1997**, *119*, 2580.

[k] H. Berke *et al.*, *Organometallics* **1992**, *11*, 563.

[l] D. S. Glueck *et al.*, *J. Am. Chem. Soc.* **1997**, *119*, 5039.

[m] R. McCrindle and A. J. McAlees, *Organometallics* **1993**, *12*, 2445.

(b) the incoming CO is, as a rule, added *cis* to the acyl group;

(c) the CO "insertion" can be affected by the addition of ligands other than CO, as well as the solvent.

The migration of an alkyl ligand to a coordinated CO generates a coordinatively unsaturated species (21-XXIXa) which can be stabilized either by the solvent (S) or as a η^2-bound acyl (21-XXIXb) by coordination of the oxygen to the metal. In general, several mechanistic pathways are therefore possible:

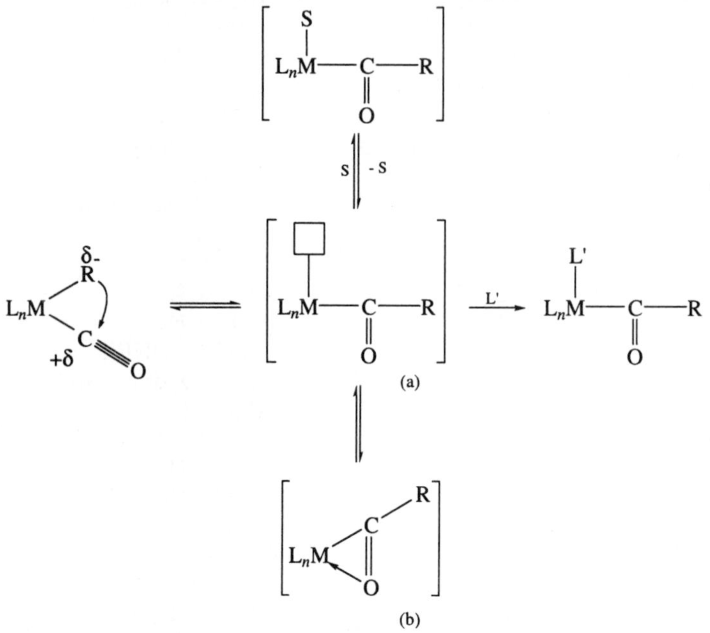

(21-XXIX)

Infrared and nmr studies have provided evidence for solvent participation, e.g., the formation of a solvated acyl complex $(CO)_4(S)MnCOR$ decreases for DMSO > DMF > py > MeCN.[150] On the other hand, flash photolysis experiments in non-coordinating hydrocarbon solvents suggest the formation of an η^2-acyl intermediate (21-XXIXb).[151]

While the incoming ligand L' is usually coordinated *cis* to the acyl ligand, the extent to which this rule applies depends on the stereochemical rigidity of the 5-coordinate intermediate $L_4MnC(O)R$. If ligand association is slow, it may be preceded by rearrangements, and *trans* addition may result, for example,[152]

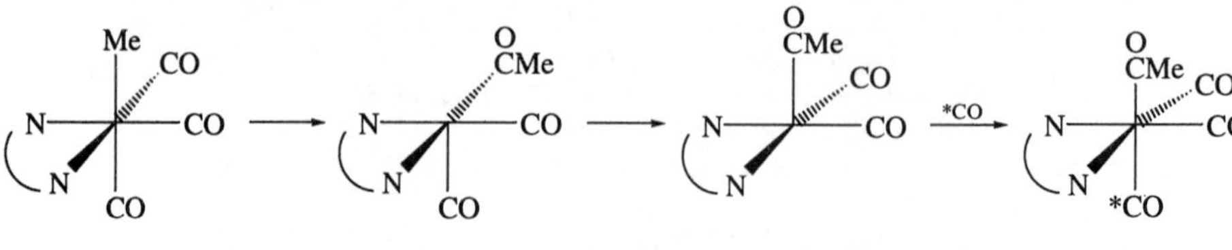

[150]T. L. Bent and J. D. Cotton, *Organometallics* **1991**, *10*, 3156.
[151]P. C. Ford *et al.*, *Organometallics* **1993**, *12*, 4739.
[152]V. Riera *et al.*, *J. Chem. Soc., Chem. Commun.* **1991**, 1058.

An activation energy for the methyl insertion step of ~70 kJ mol^{-1} has been estimated, although the influence of solvents is important.

The migration of an alkyl ligand to a coordinated CO ligand in *cis* position is assisted by the polarization of CO, and proceeds through a three-center transition state:

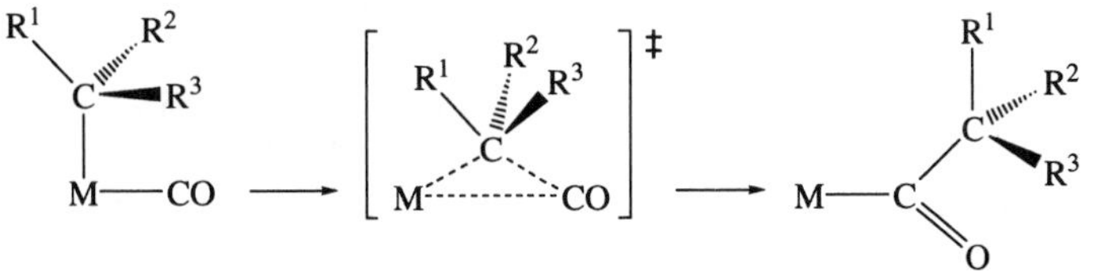

In studies with chiral ligands $CR^1R^2R^3$ it was found that alkyl transfer proceeds with *retention of configuration*. One example is the CO insertion into the Fe—C bond of $CpFe(CO)_2(CHDCHDCMe_3)$ on reaction with PPh_3.

While in manganese and other electronically saturated complexes an η^2-bonded acyl (21-XXIXb) is observed as an unstable intermediate, there are numerous examples of stable η^2-acyls of Group 4–6 metals.[153] For molybdenum a third, β-agostic bonding mode was observed which is in equilibrium with the alkyl-carbonyl isomer:[154]

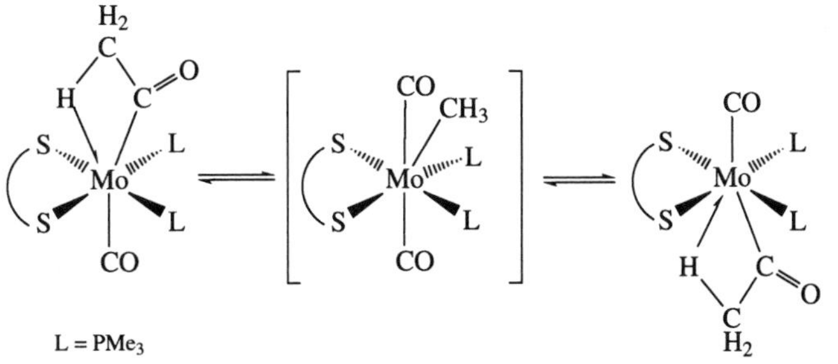

L = PMe$_3$

Carbonyl insertions may follow associative or dissociative mechanisms. For example, alkyl migration to CO in $LFe(CO)_2Me$ (L = Cp or η^5-ind) is associative and induced by phosphines, with rates increasing with decreasing cone angle and increasing basicity, apparently as the consequence of a weak complex between the two reactants; there is however no $\eta^5 \to \eta^3$ indenyl "ring slippage."[155] Similarly, aryl and methyl CO migrations in $Cp^*Rh(CO)(R)X$ are induced by phosphines and are first order in both [Rh] and [PR$_3$]; the rates are accelerated by electron-releasing aryl ligands and increasing electronegativity of X = I < Br < Cl.[156]

The rate of carbonyl insertion in chelating Pd complexes $(N-O)Pd(R)L$ decreases with increasing basicity of the phosphine ligand L and with L = PCy$_3$

[153]L. D. Durfee and I. P. Rothwell, *Chem. Rev.* **1988**, *88*, 1059.
[154]E. Carmona *et al.*, *J. Am. Chem. Soc.* **1991**, *113*, 4322; *J. Organomet. Chem.* **1995**, *500*, 61.
[155]M. Bassetti *et al.*, *J. Chem. Soc., Dalton Trans.* **1996**, 3527.
[156]P. M. Maitlis *et al.*, *J. Chem. Soc., Dalton Trans.* **1990**, 1799; *Organometallics* **1991**, *10*, 4015.

proceeds only after dissociation of the hemilabile N—ligand:[157]

$$\underset{L}{\overset{N}{\underset{|}{O-Pd-R}}} \quad \overset{CO, -L}{\underset{-CO, +L}{\rightleftharpoons}} \quad \underset{CO}{\overset{N}{\underset{|}{O-Pd-R}}} \quad \overset{+L}{\underset{-L}{\rightleftharpoons}} \quad \underset{R}{\overset{N \quad L}{\underset{|}{O-Pd-CO}}} \quad \rightleftharpoons \quad \underset{N}{\overset{L}{\underset{|}{O-Pd-COR}}}$$

Further evidence for a dissociative pathway is given by the rates of carbonylation of the Pd—Me bond in cationic chelate complexes [(P—P)PdMe(S)]$^+$ which are at least 10 times higher than those of the corresponding neutral chloride complexes.[158]

Promotion of Alkyl Migration

Alkyl migration to CO can be speeded up in various ways.

1. There can be solvent participation, as discussed above, which stabilizes intermediates. Polar solvents are particularly effective, and rate increases of up to 10^4 have been observed for CO insertions of MeMn(CO)$_5$.

2. Rate increases of orders of magnitude can be obtained by addition of Lewis acids like BF$_3$ or AlCl$_3$, or in the heterogeneous systems, by Lewis acid sites on alumina that coordinate to the acyl group and thus drive the reaction, probably *via* steps of the type,

$$\underset{M-C\equiv O}{\overset{Me}{|}} \quad \overset{AlX_3}{\longrightarrow} \quad \underset{M-C\equiv O \rightarrow AlX_3}{\overset{Me}{\underset{\delta+ \qquad \delta-}{|}}} \quad \longrightarrow \quad \overset{X}{\underset{Me}{M \diagup \diagdown AlX_2}} \quad \overset{CO}{\longrightarrow} \quad \underset{Me}{\overset{CO}{M-C}\diagdown \underset{}{O \rightarrow AlX_3}}$$

3. Migration can be accelerated by chemical or electrochemical oxidation of 18-electron complexes to 17-electron cations, present in catalytic concentration. For example, [CpFe^{III}(CO)(PPh$_3$)Me]$^+$ undergoes carbonylation at least 10^6 times faster than CpFe^{II}(CO)(PPh$_3$)Me. The reaction is independent of [CO] and accelerated by poorer electron donor phosphine ligands:[159]

$$L_nFe\diagdown^{Me}_{CO} \longrightarrow \left[L_nFe\diagdown^{R}_{CO}\right]^+ \overset{fast}{\longrightarrow} \left[L_nFe-\overset{O}{\overset{||}{C}}-R\right]^+ \overset{CO}{\longrightarrow} \left[L_nFe\diagdown^{CO}_{CR}\underset{O}{}\right]^+ \longrightarrow L_nFe\diagdown^{CO}_{CR}\underset{O}{}$$

4. Protonic acids may facilitate transfer but there is dependence on solvent. Thus CpFe(CO)$_2$Me is normally unreactive but with 1% HBF$_4$ in CH$_2$Cl$_2$ insertion of CO is much more rapid at 1 bar pressure.

5. Many catalytic reactions involving CO are promoted by halide ions, especially iodide (*cf.* methanol carbonylation, Section 22-7). For example, in

[157]K. J. Cavell *et al.*, *J. Chem. Soc., Dalton Trans.* **1996**, 2197; *Chem. Commun.* **1996**, 781.
[158]P. W. N. M. van Leewen *et al.*, *Organometallics* **1992**, *11*, 1598.
[159]A. Prock *et al.*, *Organometallics* **1991**, *10*, 3479.

anionic osmium clusters $[Os_3(CO)_{10}(\mu\text{-}CH_2)(\mu\text{-}X)]^-$ the rate for the insertion

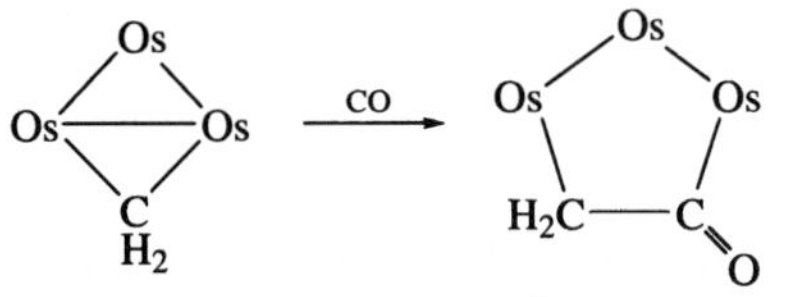

is $>10^2$ times faster than in $Os_3(CO)_{10}(\mu\text{-}CH_2)$. This may be attributed to the acyl group, a good electron acceptor, being better able than $\mu\text{-}CH_2$ to stabilize the electron density injected by X^-.

Insertion of CO into M—H Bonds

The insertion reactions of CO into M—H bonds could be expected to give formyl complexes,

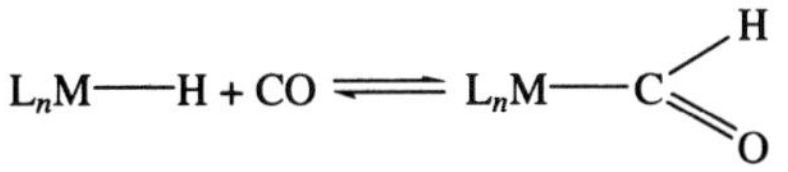

but for transition metals is *thermodynamically unfavorable*. Thus for early transition metals the reaction MH + CO is generally endothermic, whereas MR + CO is exothermic. The difference lies in the fact that M—H bonds are stronger than M—C bonds by up to $\sim$120 kJ mol^{-1}. For actinides the difference is only $\sim$60 kJ mol^{-1}, and for this reason one of the few direct insertions is observed,

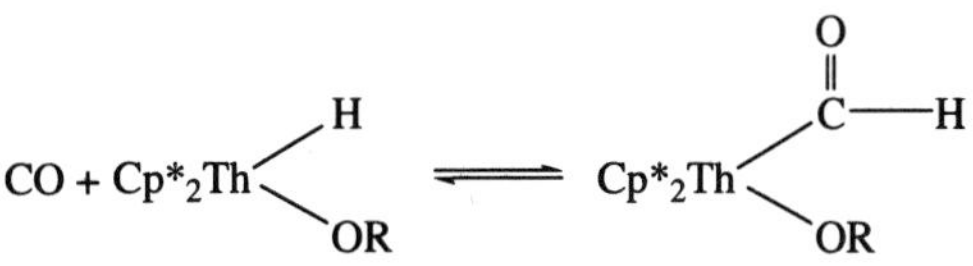

where the formyl, like the acyl in actinides, has carbene-like character.

The reaction

$$HMn(CO)_5(g) + CO(g) \rightleftharpoons (CO)_5MnCHO \qquad \Delta H \approx +20 \text{kJ mol}^{-1}$$

is endothermic, and even more unfavorable when the entropy contribution of CO is included. Calculations for the conversion of $[(CO)_4FeH]^-$ to $[(CO)_3Fe(CHO)]^-$ show that the hydride is more stable by 84 to 92 kJ mol^{-1}.

However, kinetic studies of $XM(CO)_n$ compounds indicate that when X = H, the rate of CO substitution reactions is much faster than when X = Cl or CH$_3$, so that it is possible to have *kinetically* significant amounts of a formyl intermediate. Rapid substitution reactions involving formyls occur only for first row transition elements and ruthenium:

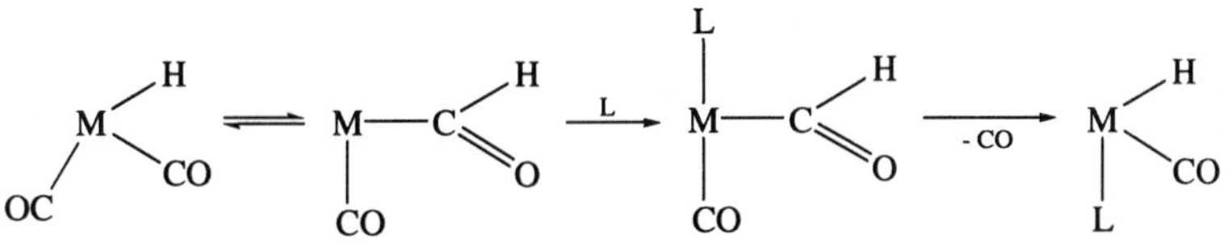

Attempts to speed up H migration to CO by use of Lewis acids, which is effective for CH_3 migrations, does not lead to formyls and instead gives reactions such as

$$(CO)_5MnH + BCl_3 \rightleftharpoons (CO)_5Mn\text{---}H\text{---}BCl_3$$

Formyl intermediates have been postulated as steps in the Fischer-Tropsch reduction of CO to hydrocarbons (Section 22-5). Coordinated CO can insert into the M—H bonds of metal hydrides, particularly when the reaction is driven by the oxophilic character of early transition metals. Such reactions may lead to C—C bond formation. Some examples are:

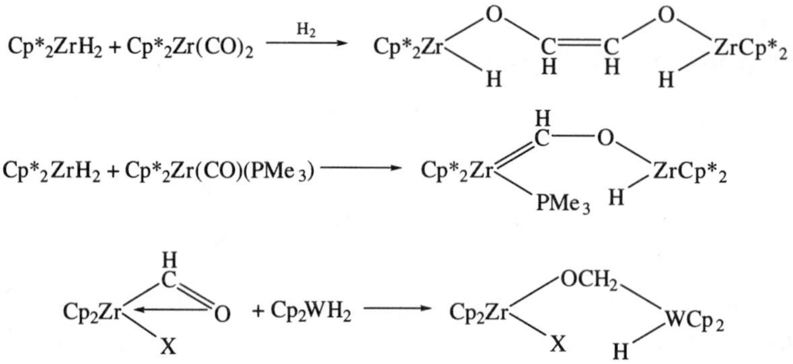

Other Aspects of CO Insertion Reactions

The double insertion of CO to give β-keto acyls,

is thermodynamically unfavorable. Compounds such as $(R_3P)_2ClPdC(O)C(O)R$ can be made by the oxidative addition of, e.g., pyruvyl chloride to Pd^0 but they readily extrude CO, and although reactions of the type

(Y = OH, NR$_2$)

are known to be catalyzed by $Co_2(CO)_8$ and palladium compounds, they do not involve two consecutive CO insertion steps. In the Pd catalyzed synthesis of β-keto amides the apparent "double carbonylation" product is formed by the reductive coupling of a RC(O) and a C(O)NR$_2$ ligand:[160]

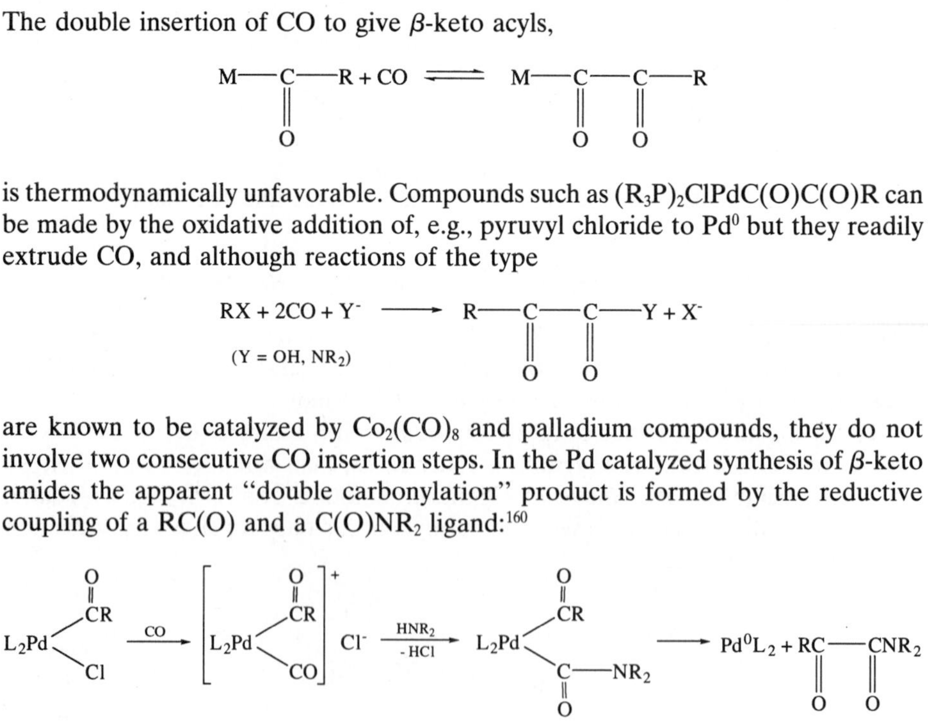

[160]A. Yamamoto et al., Pure Appl. Chem. **1991**, 63, 687.

Although acyl ligands are generally not prone to undergo further insertion reactions, those of palladium readily insert olefins. This reaction is the basis for the alternating copolymerization of CO and alkenes to give polyketones $[-C(O)CH_2CHR-]_n$ (see Sec. 22-9). Some products of successive CO and alkene insertions have proved isolable if stabilized with rigid nitrogen chelate ligands:[161]

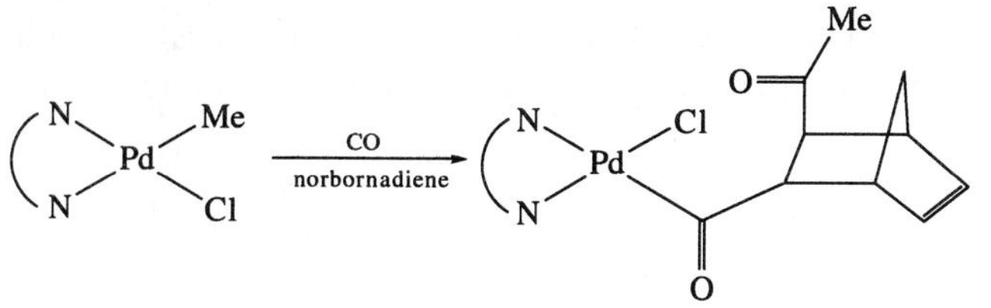

The successive insertion of CO and alkynes into M—R bonds is also known:[162]

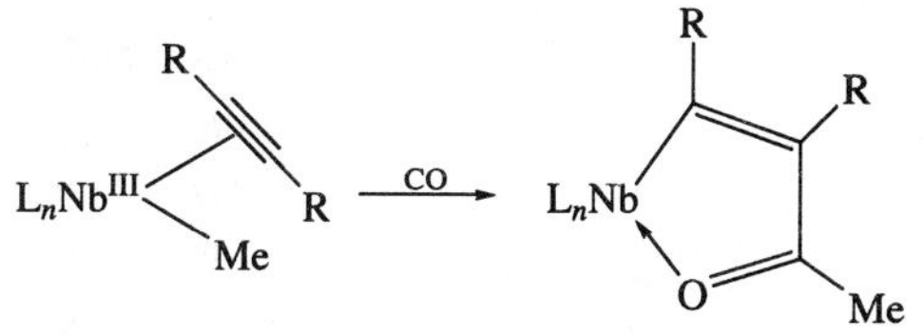

Carbon monoxide is also known to insert into metal alkoxides (M—OR), di-alkylamides (M—NR_2), and some hydroxymethyls (M—CH_2OH). For the reaction of the alkoxide $(PPh_3)_2Ir(CO)(OMe)$ with CO the intermediate $[Ir(CO)_3(PPh_3)_2]^+$ OMe^- has been identified, and the "insertion" therefore proceeds *via* external nucleophilic attack of OMe^- on Ir—CO rather than intramolecular OMe migration. True intramolecular transfer is, however, evident for (dppe)PtMe(OMe) where the rate of OMe migration is much faster than for Me migration, to give (dppe)PtMe(COOMe).

The reductive coupling of acyl ligands to give enediolato units (21-XXX) is seen for oxophilic early transition metals and actinides, for example,

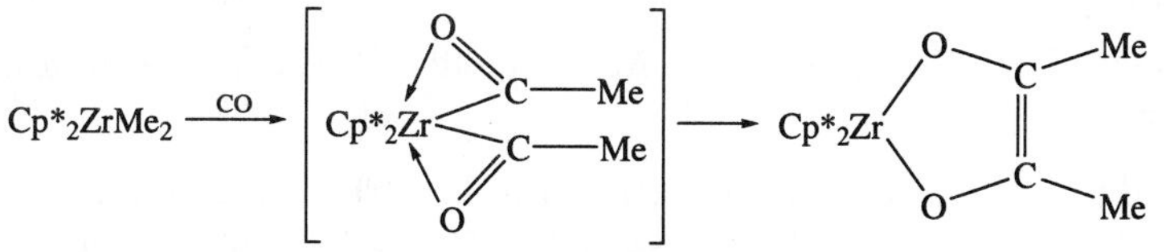

(21-XXX)

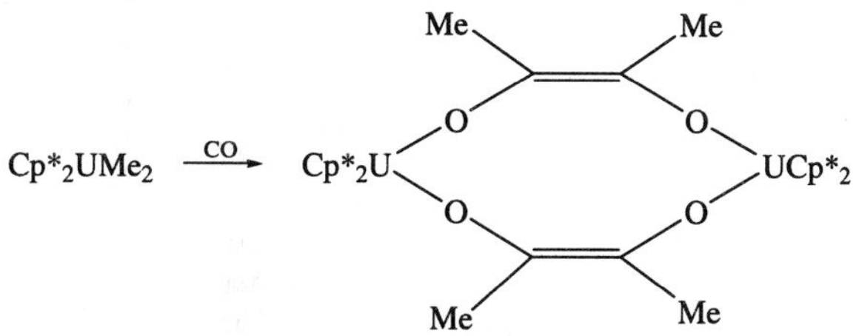

Products of type (21-XXX) are quite common for isocyanide insertions (see later).

[161]C. J. Elsevier *et al.*, *J. Chem. Soc., Chem. Commun.* **1993**, 1203.
[162]M. Etienne *et al.*, *Organometallics* **1993**, *12*, 4010.

With compounds containing several alkyl groups, CO insertions can lead to the reductive elimination of ketones, for example,

$$WMe_6 \xrightarrow{\text{CO}} \left[Me_5W(COMe) \xrightarrow[-Me_2CO]{} WMe_4 \right] \longrightarrow \longrightarrow W(CO)_6 + 3Me_2CO$$

Insertion of CO into M—Si bonds is not common but several examples are known. The migratory insertion of CS into a M—Si bond on CO addition has also been observed:[163]

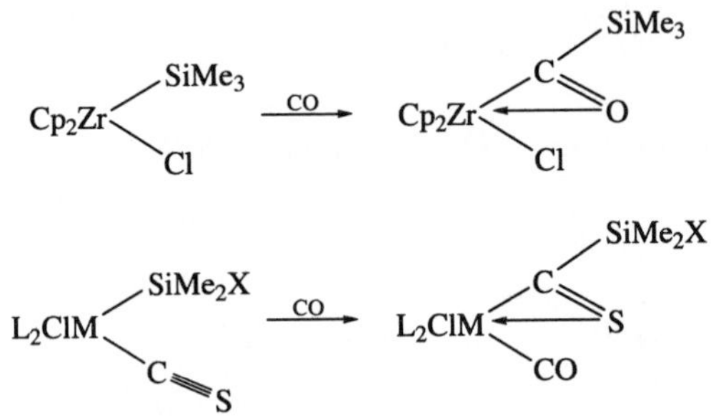

M = Ru, Os
X = OEt, Cl

Isocyanide Insertions

Since RNC is isoelectronic with CO, a similar chemistry could be expected. However, isocyanides differ from CO in several important aspects.

1. Isocyanides are more reactive than CO, and multiple insertions into M—C bonds are a common occurrence.
2. The nitrogen atom in iminoacyl insertion products MC(=NR)R′ is more basic than the oxygen atom in acyls, and η^2-bonding of iminoacyls is therefore common.
3. Unlike CO, RNC inserts into M—H bonds to give η^1- and η^2-formimidoyl compounds.

In addition to insertion products into M—CH$_3$ of types (21-XXXI) to (21-XXXIII), rearrangement to the 1-azaallyl (21-XXXIV) has also been observed, as has the involvement of iminoacyls in the formation of metallacycles (21-XXXV) if the insertion is conducted in polar solvents such as acetone or MeCN.[164]

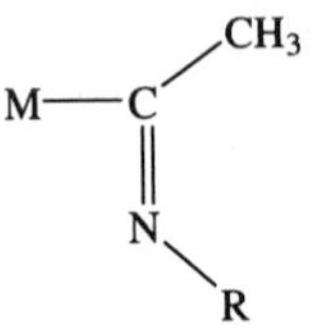

(21-XXXI) 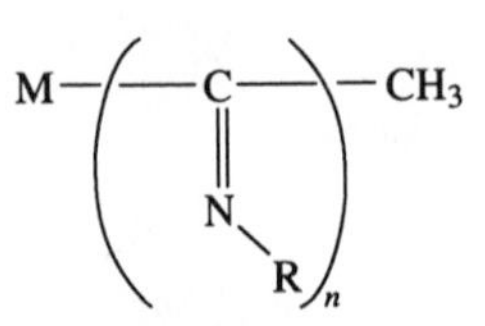   (21-XXXII) 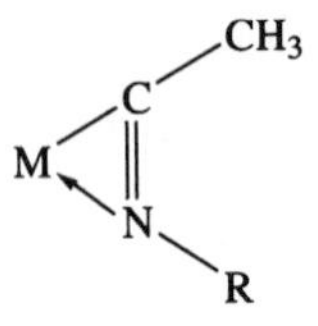   (21-XXXIII)

[163]W. R. Roper *et al.*, *Organometallics* **1992**, *11*, 3931.
[164]M. L. Poveda *et al.*, *Organometallics* **1997**, *16*, 2263.

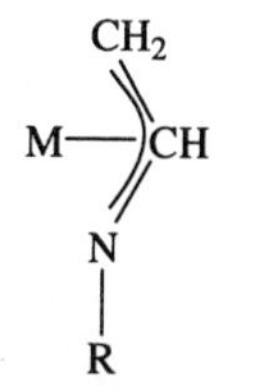

(21-XXXIV) 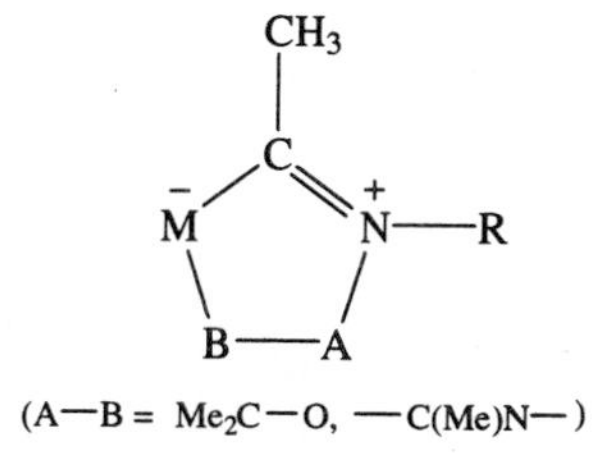   (21-XXXV)

$(A-B = Me_2C-O, -C(Me)N-)$

In complexes containing both CO and CNR ligands, formation of the iminoacyl is favored over acyl; the formation of η^2-C(NR)R' is however reversible:[165]

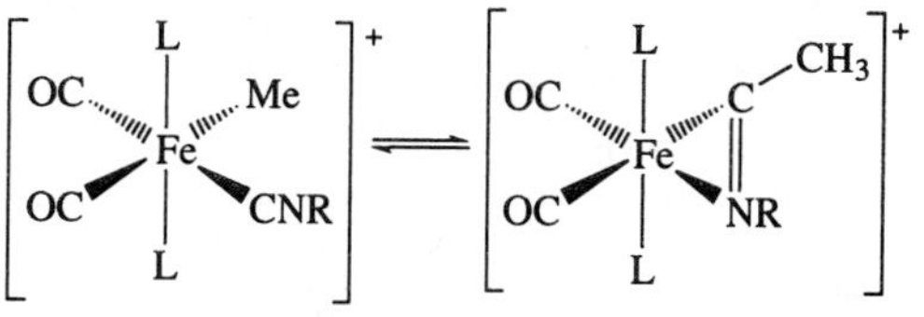

Exchange between acyls and iminoacyls is possible and in the case of Mo^{II} complexes proceeds *via* a 7-coordinate alkyl intermediate:[166]

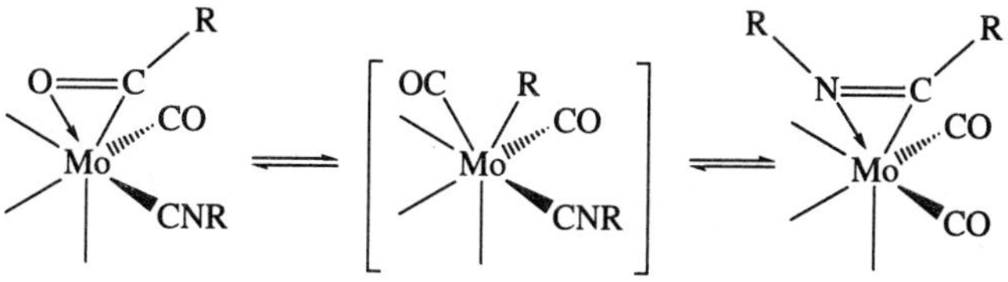

An unusual complex containing only iminoacyl ligands is obtained by the reaction[167]

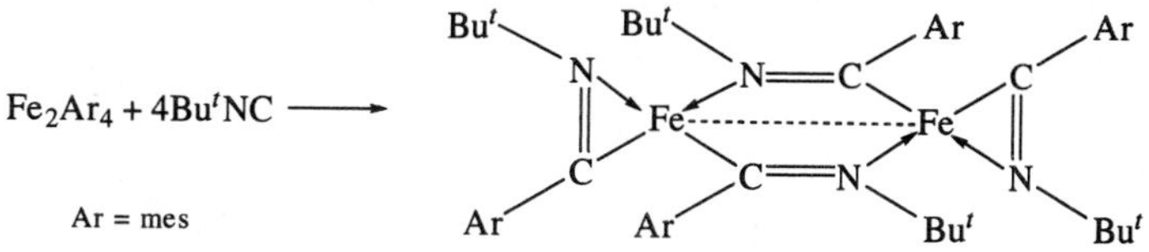

$Fe_2Ar_4 + 4Bu^tNC \longrightarrow$

Ar = mes

Both early and late transition metals form η^2-iminoacyls which may be converted to η^1-compounds on addition of donor ligands:[168,169]

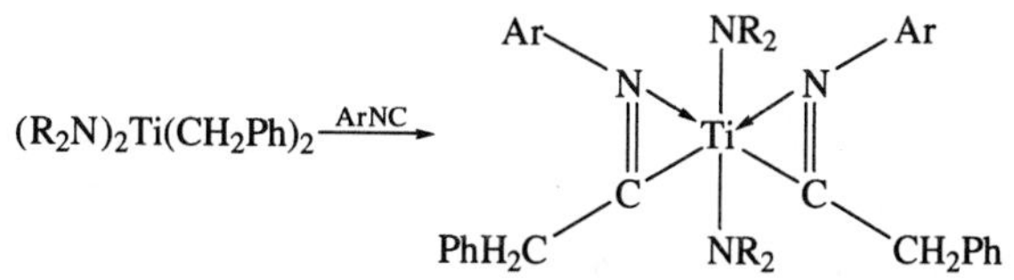

$(R_2N)_2Ti(CH_2Ph)_2 \xrightarrow{ArNC}$

[165]G. Cardaci *et al.*, *Inorg. Chem.* **1992**, *31*, 63.
[166]L. J. Sánchez *et al.*, *J. Chem. Soc., Dalton Trans.* **1996**, 3687.
[167]C. Floriani *et al.*, *Organometallics* **1993**, *12*, 2414.
[168]I. P. Rothwell *et al.*, *Chem. Commun.* **1997**, 1109.
[169]E. Carmona *et al.*, *Organometallics* **1990**, *9*, 583.

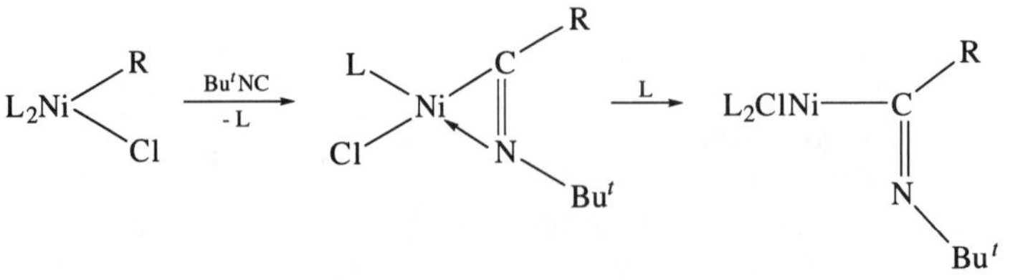

Coupling of η^2-iminoacyls is frequently observed, for example, on heating tungsten η^2-iminoacyls:[170]

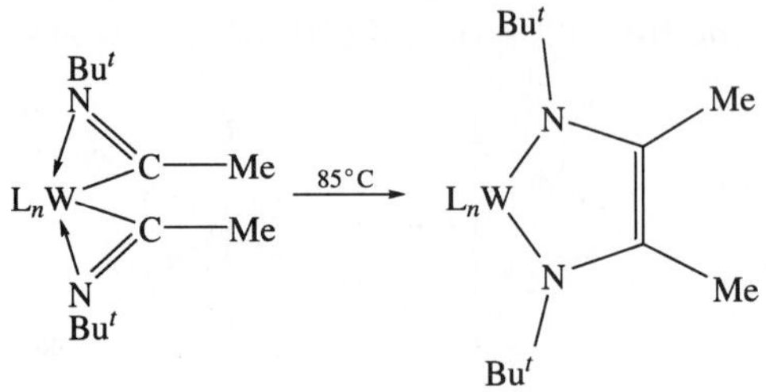

The mechanism of RNC insertion into Pd^{II}-R proceeds *via* a cationic RNC adduct; the methyl migration to coordinated RNC is rate-limiting:[171]

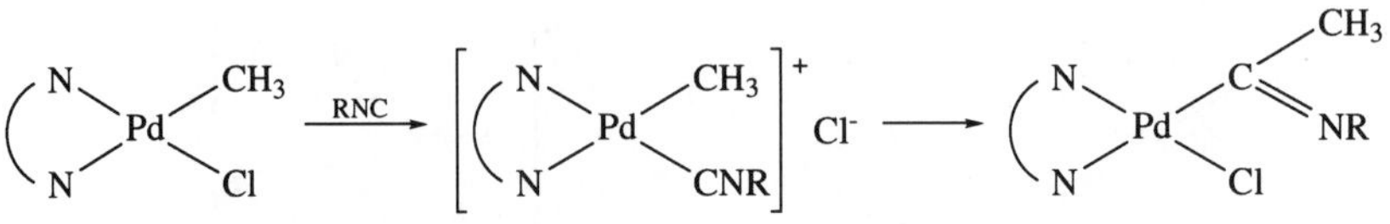

Isocyanides also insert into M—M' single bonds to give η^2-iminoacyl-like structures, with N coordinated to the more electropositive metal.[172]

Examples of multiple isocyanide insertions into Pd—C,[173] Pd—H,[174] and even Pd—S,[175] bonds are compounds (21-XXXVI) - (21-XXXVIII):

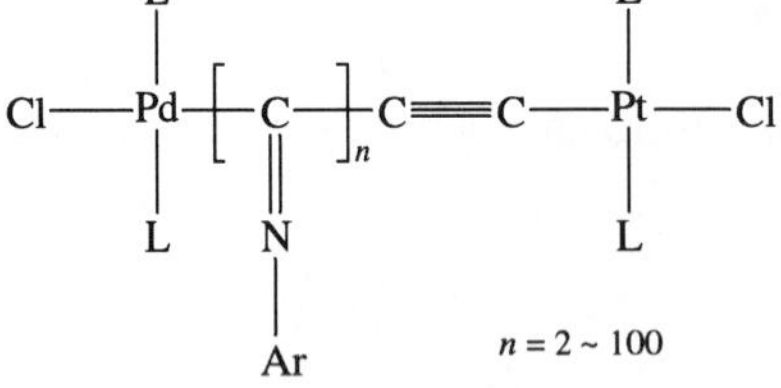

(21-XXXVI)

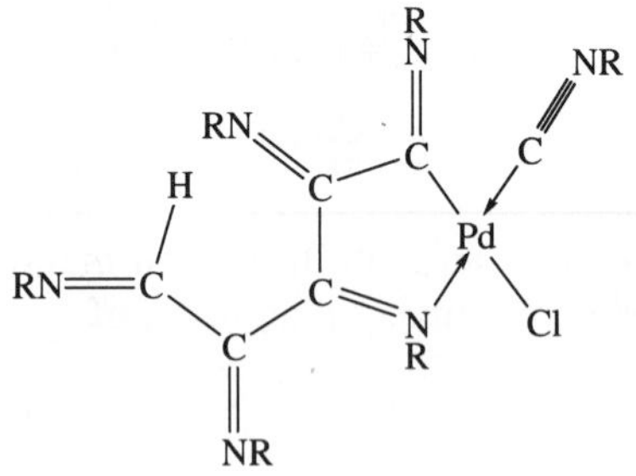

(21-XXXVII)

[170]J. M. Boncella *et al.*, *Organometallics* **1997**, *16*, 1779.
[171]K. Vrieze *et al.*, *Organometallics* **1997**, *16*, 2948.
[172]L. H. Gade *et al.*, *Angew. Chem. Int. Ed. Engl.* **1996**, *35*, 1338; *Chem. Commun.* **1996**, 219.
[173]S. Takahashi *et al.*, *Angew. Chem. Int. Ed. Engl.* **1992**, *31*, 851.
[174]T. Tanase *et al.*, *Organometallics* **1996**, *15*, 3404.
[175]H. Kuniyasu *et al.*, *J. Am. Chem. Soc.* **1997**, *119*, 4669.

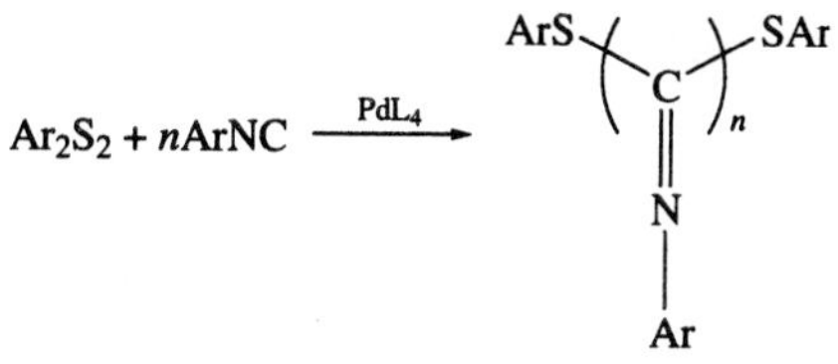

$$Ar_2S_2 + nArNC \xrightarrow{PdL_4} \left(\begin{array}{c} ArS \diagdown \diagup SAr \\ C \\ \| \\ N \\ | \\ Ar \end{array} \right)_n$$

(21-XXXVIII)

Products of complex stuctures may thus arise, as in the insertion of xyNC into Ir=NR bonds (21-XXXIX), and in the case of nickel complexes as side-products of the catalyzed polymerization of PhNC.[176,177]

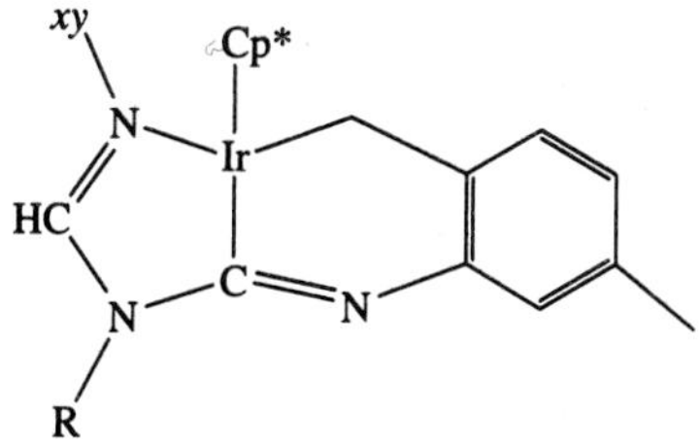

(21-XXXIX)

Nitric Oxide

The insertion of NO into metal—carbon bonds was first discovered by Frankland in his original studies on zinc alkyls. The course of the reaction depends on whether the alkyl complex is diamagnetic or paramagnetic. If it is diamagnetic, the reaction involves a radical intermediate (21-XL) which reacts rapidly with the radical NO to generate an N-methyl, N-nitrosohydroxylaminato group:

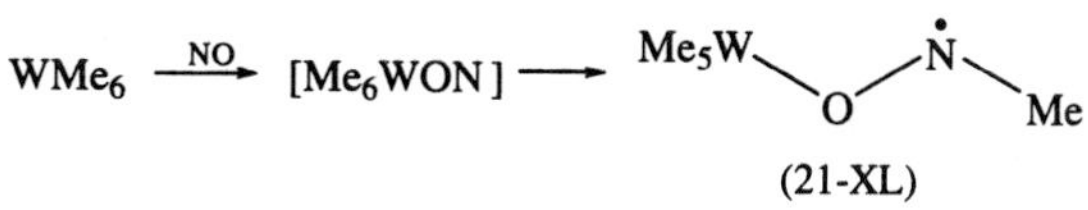

$$WMe_6 \xrightarrow{NO} [Me_6WON] \longrightarrow Me_5W \diagdown_O \diagup N \diagdown_{Me}$$

(21-XL)

$$Me_5WONMe + NO \longrightarrow Me_5W \begin{array}{c} O-N \diagdown Me \\ | \\ O \diagdown\diagup N \end{array}$$

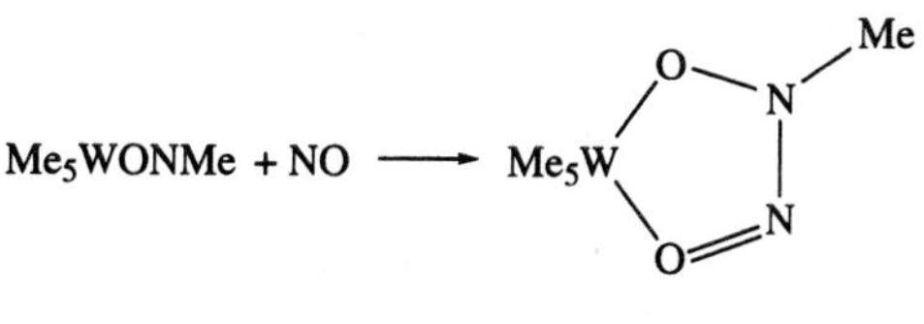

On the other hand, paramagnetic alkyls such as $Re(O)Me_4$ or Cp_2NbMe_2 cannot react in this way since the initial insertion product *must* be diamagnetic. This intermediate, which may have an η^2-ONMe group, decomposes with formation of an M=O bond, releasing MeN: which dimerizes to MeN=NMe. The nitrene MeN

[176]G. Wilkinson *et al.*, *J. Chem. Soc., Dalton Trans.* **1996,** 3771.
[177]W. B. Euler *et al.*, *Chem. Commun.* **1997,** 257.

may also be trapped by an olefin to give an aziridene, for example,

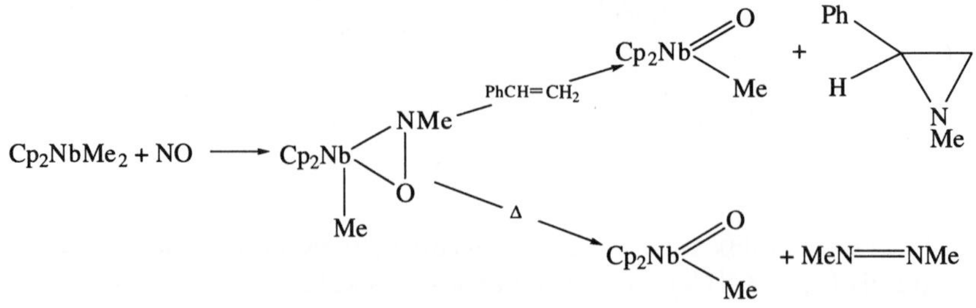

Isolable nitrosyl alkyl complexes undergo NO insertions on treatment with donor ligands:

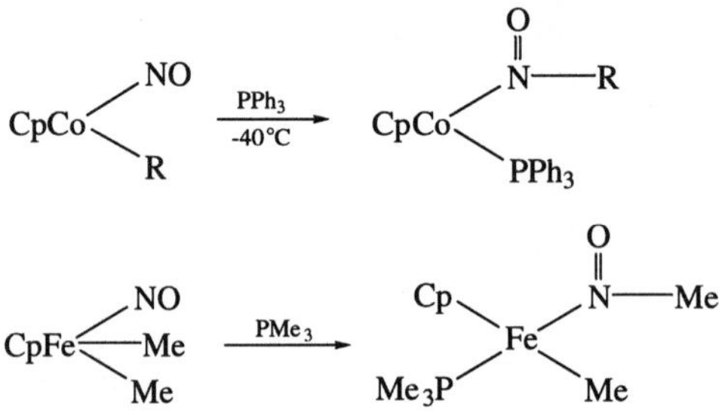

The ONR moiety may be deprotonated to an η^3-heteroallyl ligand:

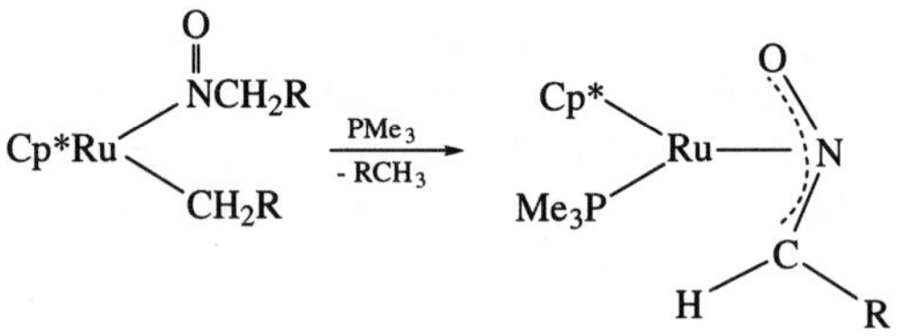

Insertions of NO^+ are also known, for example:

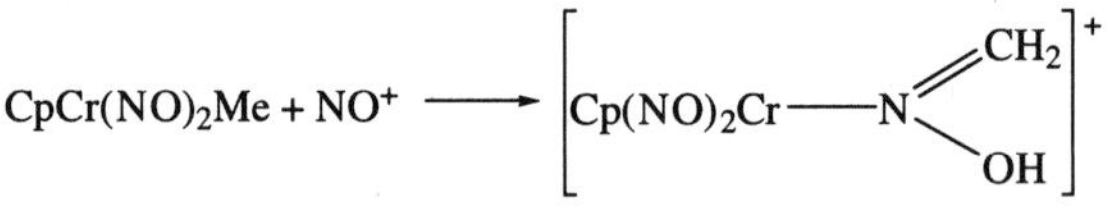

21-6   1,2-Insertions: C=C, C—C, and C=X

Alkene Insertions

The reaction of alkenes and other unsaturated substances with transition metal hydrido or alkyl complexes is of prime importance in catalytic reactions such as hydrogenation, hydroformylation, and polymerization (see Chapter 22). It is one of the major methods for synthesizing metal-to-carbon bonds. The reverse reaction, the β-hydride or β-alkyl transfer-alkene elimination reaction has already been discussed (Section 21-3).

The insertion of an alkene into a M—H bond is considered a concerted intramolecular process involving a planar cyclic transition state, resulting in a *cis* addition to the C=C bond:

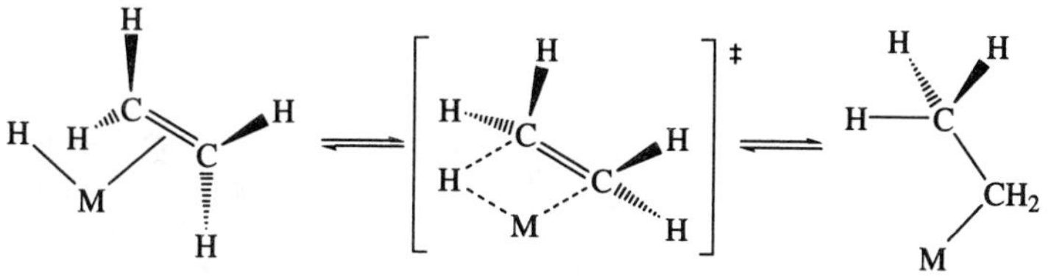

Alkene insertions into M—CR$_3$ bonds proceed *via* a similar transition state, with the CR$_3$ moiety tilting towards the terminal alkene-C, with retention of configuration of the CR$_3$ group. Note that in either case in the M(R)(alkene) complex two coordination sites are involved, whereas in the product only one is occupied. For this reason the stability of alkyls depends on blocking this second coordination site to prevent the reverse reaction.

The first step in the alkene insertion sequence is C=C coordination,

$$L_nM\!\!-\!\!H + RCH\!\!=\!\!CH_2 \;\rightleftharpoons\; L_nM(H)(RCH\!\!=\!\!CH_2)$$

but such equilibria are observed only rarely. Some hydrido alkene complexes are known; where the ligands are *trans*, isomerization precedes the insertion step, for example,

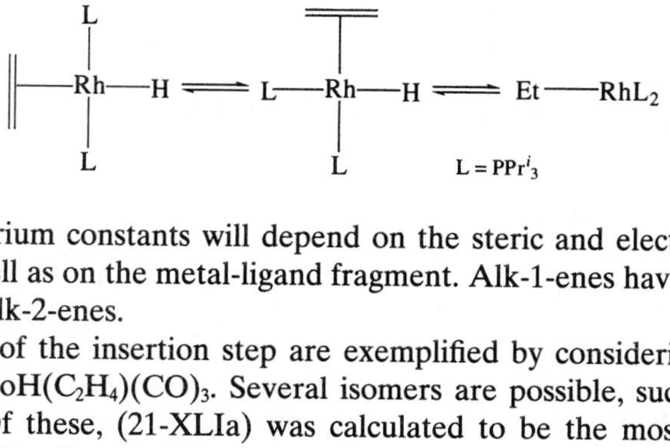

The equilibrium constants will depend on the steric and electronic nature of the alkene as well as on the metal-ligand fragment. Alk-1-enes have constants ~50 times those of alk-2-enes.

The details of the insertion step are exemplified by considering the possible geometries of CoH(C$_2$H$_4$)(CO)$_3$. Several isomers are possible, such as (21-XLIa) to (21-XLIc). Of these, (21-XLIa) was calculated to be the most stable due to optimized back-bonding to the ethylene ligand, but rearrangement to an isomer which has the C=C bond aligned for the H-transfer step is required for the reaction to proceed. The resulting insertion products may have structures (21-XLId) or (21-XLIe); the latter was found to be more stable and less strained. Note that these primary products are stabilized by β-agostic interactions.[178] Similarly, the insertion of ethylene into M—CH$_3$ bonds gives primary products stabilized by γ-agostic bonding.[179]

[178]T. Ziegler *et al.*, *Inorg. Chem.* **1990**, *29*, 4530.
[179]T. Yoshida *et al.*, *Organometallics* **1995**, *14*, 746; R. Ahlrichs *et al.*, *J. Am. Chem. Soc.* **1994**, *116*, 4919.

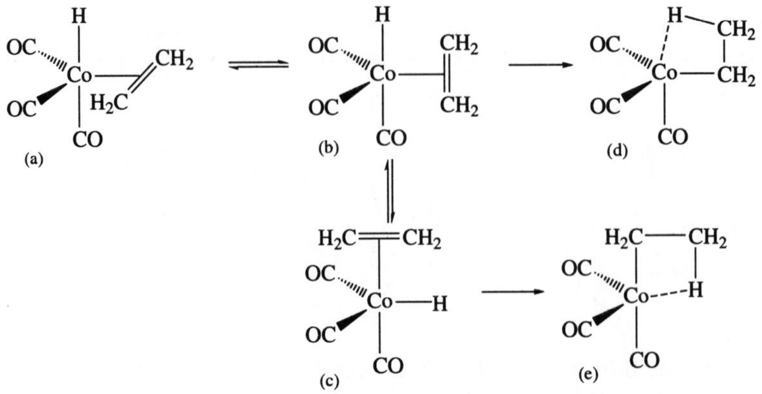

(21-XLI)

Specific examples of alkene insertions are the reactions of cationic hydrides $[Cp_2ZrH(L)]^+$ to give metal alkyls which may either be stabilized by agostic interactions, as in (21-XLII), or free of such interactions (21-XLIII), depending on L.[180] The insertion of isobutene into M—C bonds of electron-deficient zirconium complexes was found to be reversible.[109]

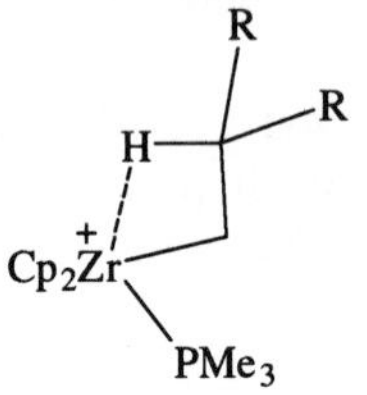

(21-XLII) 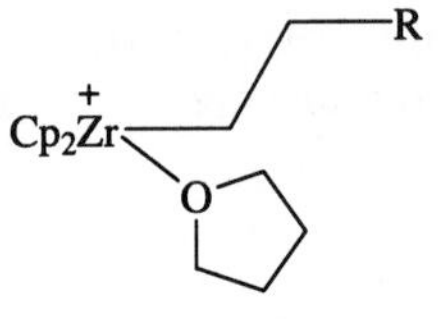   (21-XLIII)

A reaction that is frequently applied in synthesis is the hydrozirconation of alkenes:[181]

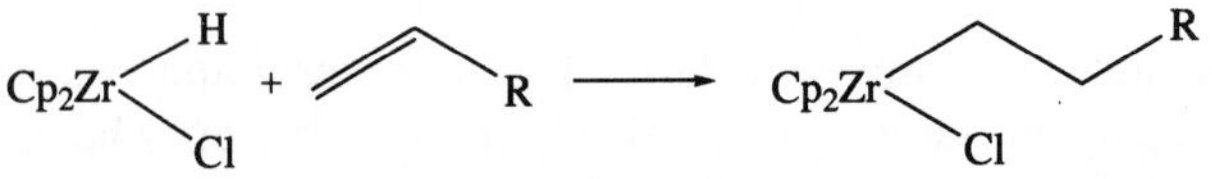

There are relatively few well-defined insertion products of alkenes into M—alkyl bonds, although in catalytic systems this reaction can be very fast. An example is the single insertion of propene into a Zr—benzyl bond:[182]

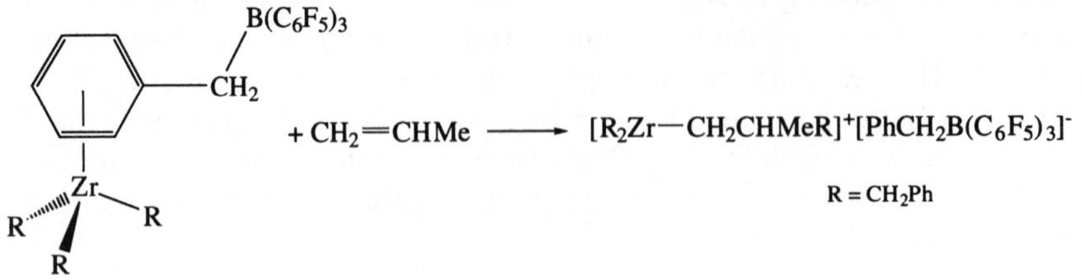

$+ CH_2{=}CHMe \longrightarrow [R_2Zr{-}CH_2CHMeR]^+[PhCH_2B(C_6F_5)_3]^-$

$R = CH_2Ph$

[180]R. F. Jordan et al., Organometallics **1994**, 13, 148, 1424.
[181]S. R. Stobart et al., Inorg. Chem. **1997**, 36, 3745.
[182]C. Pellecchia et al., Organometallics **1994**, 13, 298.

Of particular interest are complexes of the type $[(L-L)M(R)(alkene)]^+$ (M = Ni, Pd) which are the resting states of nickel or palladium catalysts for olefin oligomerizations, polymerizations, and alkene/carbon monoxide copolymerizations. The ligands $L-L$ are either chelating diphosphines or rigid nitrogen ligands. Some, like $[(phen)PdMe(C_2H_4)]^+$ oligomerize ethylene and react sufficiently slowly to allow the energetics of migratory alkene insertion reaction to be determined.[183]

The alternating copolymerization of CO with ethylene proceeds *via* the reaction steps

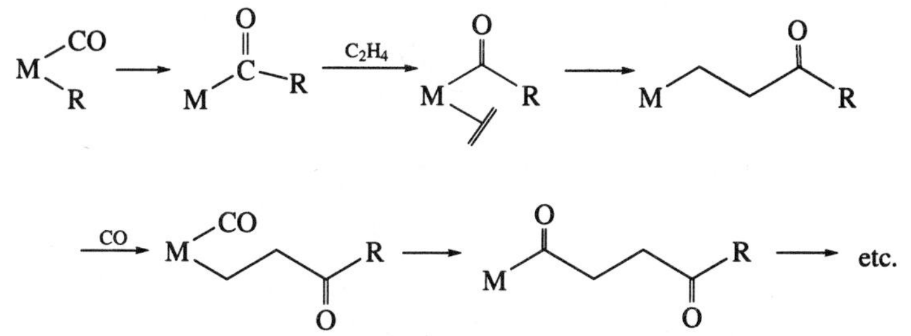

A model for such a reaction sequence is (21-XLIV).[184] In the case of Pd^{II} complexes of rigid bidentate nitrogen ligands, products of multiple successive insertions of alkenes and CO have proved isolable.[185] The insertion of an alkene into the Pd—acyl bond of a neutral acyl chloro complex leads to displacement of the halide ligand and formation of a chelate-stabilized product (21-XLV).[186]

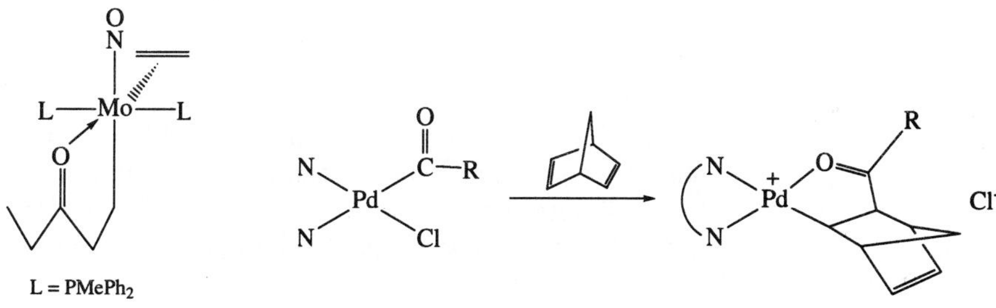

L = PMePh₂

(21-XLIV)   (21-XLV)

The insertion of ethylene into Pd^{II} allyl complexes with hemilabile P—O ligands gives $(1,2,5-\eta^3)$-pentenyl complexes.[187] The insertion of alkenes into M—C bonds of η^2-iminoacyl and η^2-pyridyl,[188] into Rh—B bonds of boryl complexes,[189] into M—Si,[190] M—P,[191] and M—O bonds is also known, an example for the latter being

[183]F. C. Rix and M. Brookhart, *J. Am. Chem. Soc.* **1995**, *117*, 1137.
[184]G. L. Hillhouse *et al.*, *Organometallics* **1997**, *16*, 2335.
[185]C. J. Elsevier *et al.*, *J. Am. Chem. Soc.* **1994**, *116*, 977.
[186]K. Vrieze *et al.*, *Organometallics* **1997**, *16*, 68.
[187]S. Mecking and W. Keim, *Organometallics* **1996**, *15*, 2650.
[188]R. F. Jordan *et al.*, *J. Org. Chem.* **1993**, *58*, 5595; *J. Am. Chem. Soc.* **1994**, *116*, 4491.
[189]R. T. Baker and J. C. Calabrese, *J. Am. Chem. Soc.* **1993**, *115*, 4367.
[190]T. D. Tilley *et al.*, *Organometallics* **1989**, *8*, 2284.
[191]D. S. Glueck *et al.*, *J. Am. Chem. Soc.* **1997**, *119*, 5039.

the reaction[192]

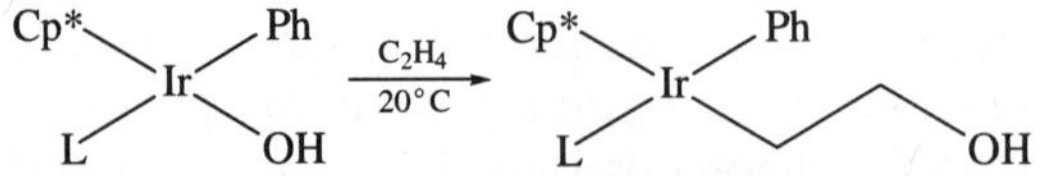

The insertion of olefins and hetero-alkenes and -alkynes (ketones, nitriles) into the M—C bonds of metallacyclopropenes, as present in metal alkyne and particularly benzyne complexes, leads to five-membered metallacycles.[193,194] This reaction has widespread synthetic applications:[195]

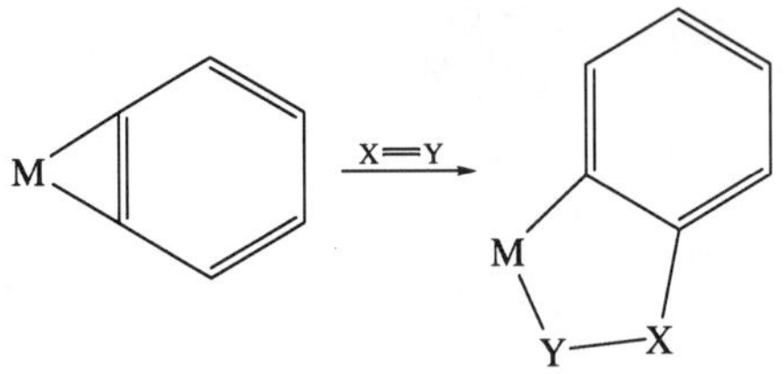

Alkyne Insertions

Alkynes insert into M—H and M—C bonds, typically to give *cis*-addition products,

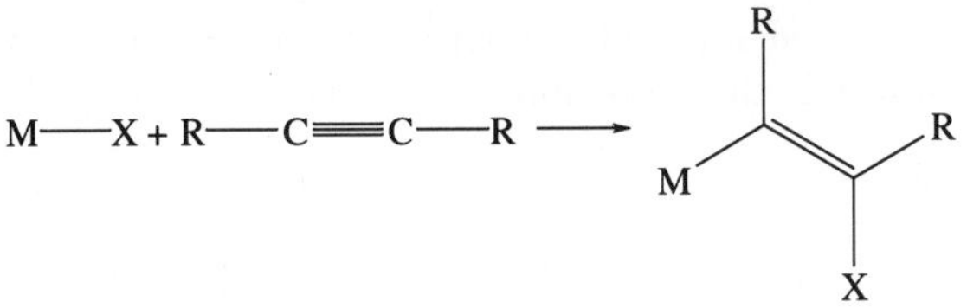

although under kinetic control mixtures of both (*E*)- and (*Z*)-vinyl complexes as well as stereospecific *trans*-insertion products may be found, as for example in the insertion of $RO_2C-C\equiv C-CO_2R$ into Nb—H and Ru—H bonds.[196] The double insertion of 2-butyne into a Zr—Me bond is accompanied by isomerization to give a pentadienyl complex:[197]

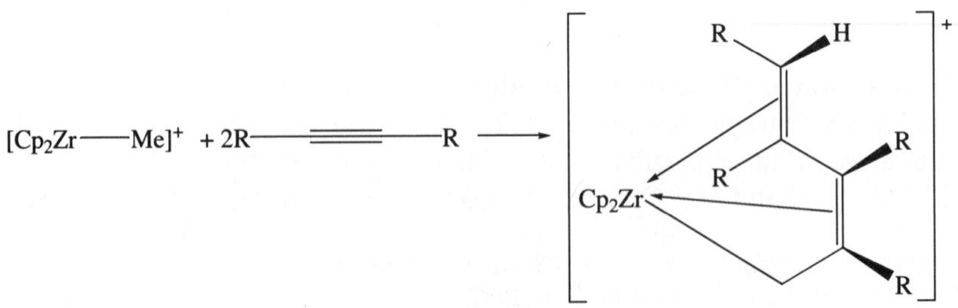

[192]J. C. M. Ritter and R. G. Bergman, *J. Am. Chem. Soc.* **1997**, *119*, 2580.

[193]U. Rosenthal *et al.*, *Organometallics* **1997**, *16*, 2886.

[194]R. G. Bergman *et al.*, *J. Am. Chem. Soc.* **1991**, *113*, 3404.

[195]F. Mathey *et al.*, *J. Am. Chem. Soc.* **1997**, *119*, 9417.

[196]A. Otero *et al.*, *Organometallics* **1996**, *15*, 5507; J. D. Vessey and R. J. Mawby, *J. Chem. Soc., Dalton Trans.* **1993**, 51.

[197]A. D. Horton and A. G. Orpen, *Organometallics* **1992**, *11*, 8.

The insertion of alkynes into W^{II}−C bonds gives η^2-vinyl complexes with carbenoid character:[198]

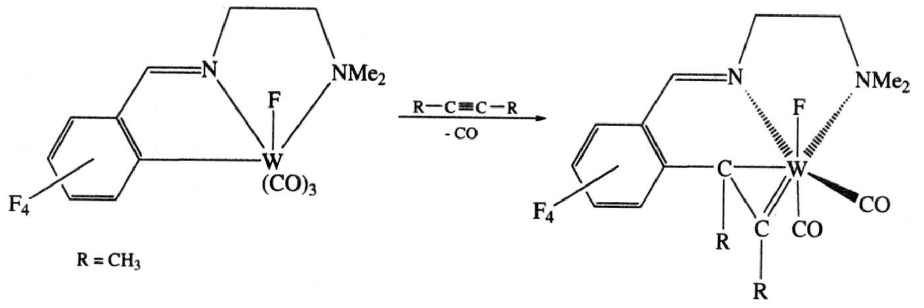

R = CH$_3$

Platinum diboryl complexes $(PPh_3)_2Pt(BCat)_2$ react with alkynes to give *cis*-diborylation products, *via* a Pt-vinyl intermediate.[199] Alkynes are also known to insert into M−Si, M−N, and M−Cl bonds.[200]

Insertions of Aldehydes, Ketones, and Nitriles

The insertion of formaldehyde into M−H bonds can in principle lead to M−OCH$_3$ or M−CH$_2$OH products, though for M = Co(CO)$_3$ the methoxo product is thermodynamically preferred.[201] Some examples of ketone[202] and nitrile[203] insertions are:

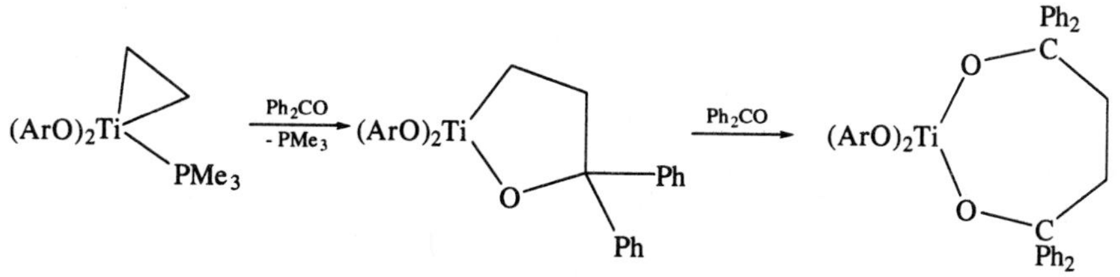

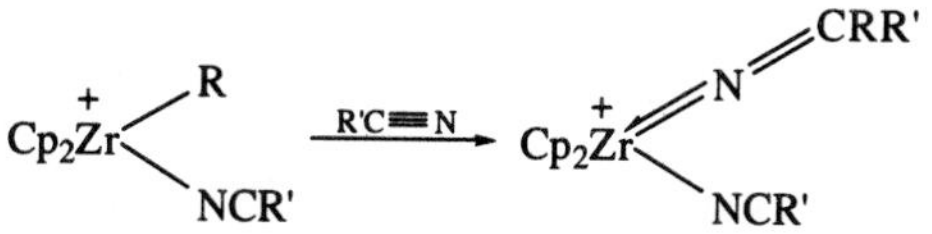

Insertions of CO$_2$ and Related Electrophilic Heterocumulenes

The insertion of carbon dioxide probably involves CO$_2$ complexes as intermediates and is known for alkyls, hydrides,[204] dialkylamides, dialkylphosphides, hydroxides,[205]

[198]T. G. Richmond *et al.*, *Organometallics* **1993**, *12*, 3382.
[199]C. N. Iverson and M. R. Smith III, *Organometallics* **1996**, *15*, 5155.
[200]W. R. Roper *et al.*, *Organometallics* **1996**, *15*, 1793; T. J. Marks *et al.*, *Organometallics* **1994**, *13*, 439; J. Dupont *et al.*, *Organometallics* **1997**, *16*, 2386.
[201]L. Versluis and T. Ziegler, *J. Am. Chem. Soc.* **1990**, *112*, 6763.
[202]I. P. Rothwell *et al.*, *J. Am. Chem. Soc.* **1997**, *119*, 8630.
[203]R. F. Jordan *et al.*, *Organometallics* **1991**, *10*, 1406.
[204]A. Otero *et al.*, *J. Chem. Soc., Dalton Trans.* **1995**, 3409.
[205]M. D. Roundhill *et al.*, *Inorg. Chem.* **1992**, *31*, 3831.

alkoxides,[206] and with ethylene complexes. Both η^1- and η^2-products may result, sometimes in equilibrium with each other:[207]

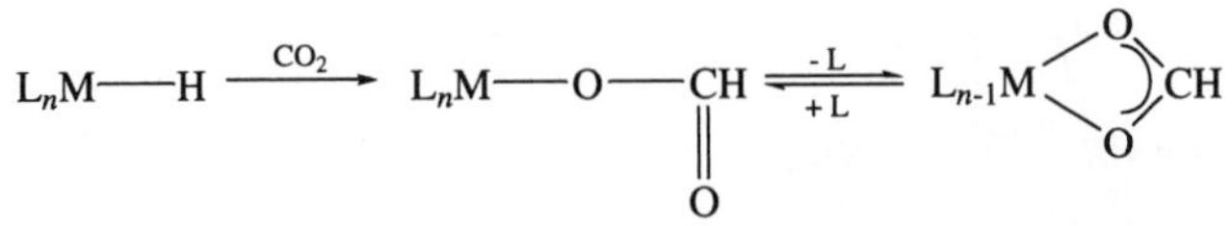

For late transition metals η^1-carboxylate insertion products are favored, whereas electron-deficient early transition metals give η^2-products. Carbon disulfide reacts similarly. Some examples are[208]

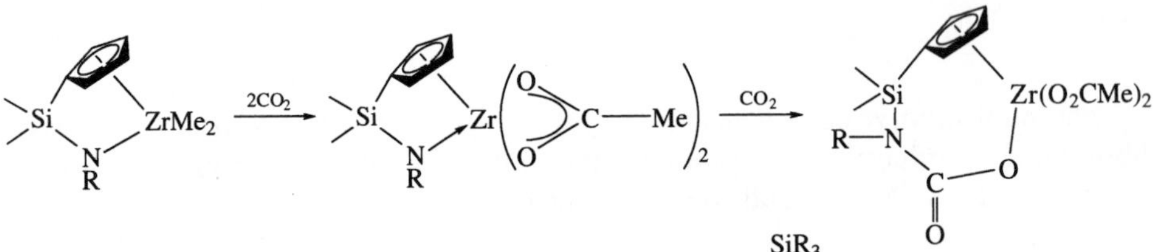

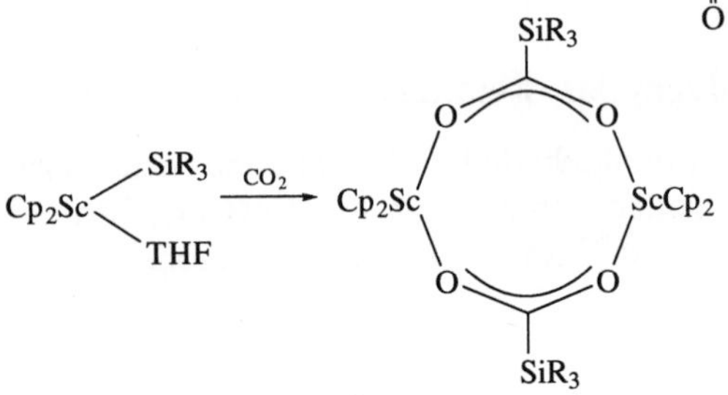

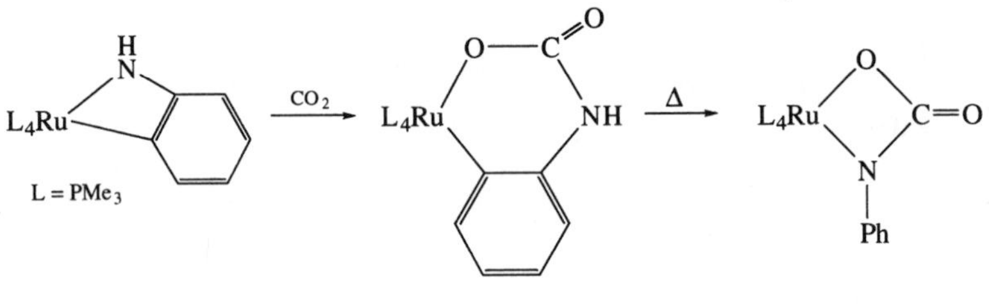

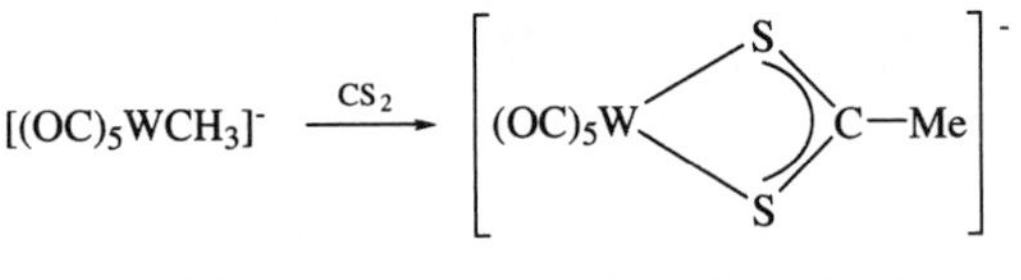

The reactions are reversible in some cases:[209]

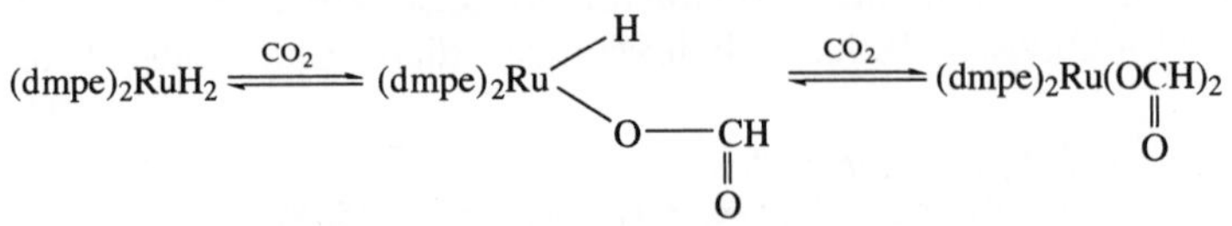

[206]M. Orchin *et al.*, *Organometallics* **1993**, *12*, 1714.
[207]D. J. Darensbourg *et al.*, *J. Am. Chem. Soc.* **1990**, *112*, 9252.
[208]L. Kloppenburg and J. L. Petersen, *Organometallics* **1996**, *15*, 7; T. D. Tilley *et al.*, *Inorg. Chem.* **1990**, *29*, 4355; R. G. Bergman *et al.*, *J. Am. Chem. Soc.* **1991**, *113*, 6499; D. J. Darensbourg *et al.*, *Organometallics* **1991**, *10*, 6.
[209]R. N. Perutz *et al.*, *Organometallics* **1996**, *15*, 5166.

For dialkylamides of early transition metals the insertion reactions are not intramolecular but are catalyzed by traces of amines, so that we have the sequence

$$Me_2NH + CO_2 \rightleftharpoons Me_2NCO_2H$$

$$L_nMNMe_2 + Me_2NCO_2H \rightleftharpoons L_nM(O_2CNMe_2) + HNMe_2$$

Cumulenes X=C=Y other than CO_2, such as C_3O_2, PhNCO, PhNCS, and $Ph_2C=C=O$ can also undergo insertion reactions, such as:

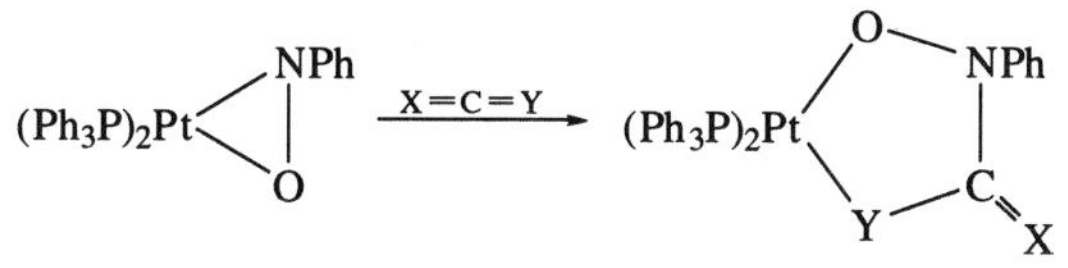

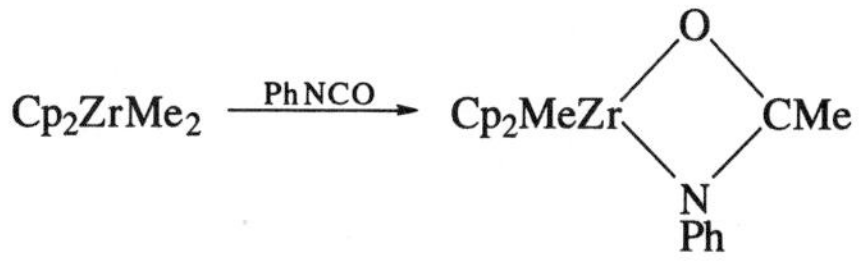

These reactions proceed by dipolar transition states, which enables the electrophilic carbon in the heterocumulene to directly attack the X ligand of the M—X bond. The reactions are therefore facilitated if X is a good donor; for example for a series of complexes Cp*(NO)(R)W—X (X = amide, alkoxide or alkyl), the reactivity of isocyanates was found to decrease in the order W—N > W—O > W—C.[210]

Sulfur dioxide can insert in various ways as shown in (21-XLVI). The usual product is the *S*-sulfinate (21-XLVIa), although the other isomers are known. The insertion of SO_2 into M—OR bonds, e.g., of $Ir(OR)(CO)L_2$, gives oxygen-coordinated sulfite complexes.[211]

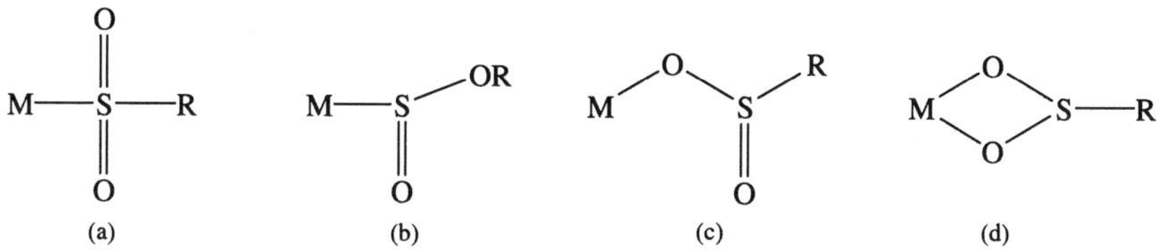

(21-XLVI)

The insertion of SO_2 appears to proceed by several different mechanisms, one of which involves *exo* S_E2 attack, as follows:

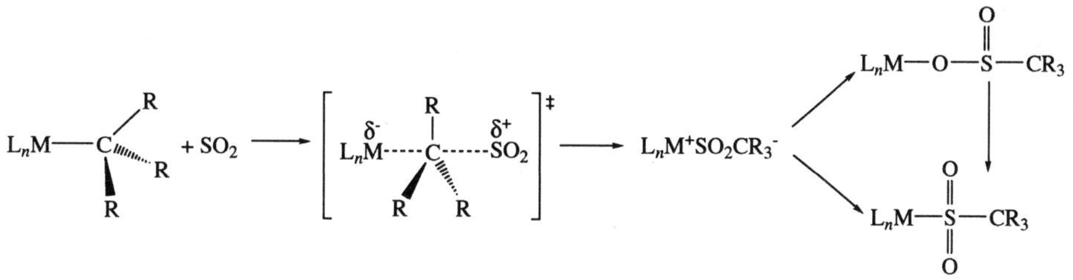

[210]P. Legzdins *et al.*, *Organometallics* **1994**, *13*, 569.
[211]M. R. Churchill *et al.*, *Organometallics* **1994**, *13*, 141.

The insertion of SO_2 into the Co—R group of cobaloximes is a radical chain reaction of the type

$$R\text{——}Co \rightleftharpoons R^{\cdot} + Co$$

$$Co + SO_2 \longrightarrow CoSO_2$$

$$CoSO_2 + RCo \longrightarrow CoSO_2R + Co$$

Chapter 22

HOMOGENEOUS CATALYSIS BY TRANSITION METAL COMPLEXES

Transition metal catalyzed reactions are the basis of many important industrial processes, such as hydrogenations, carbonylation reactions, and the low-pressure polymerization of ethylene and propene. Many of these processes use heterogeneous, solid-phase materials as catalysts. While such catalysts can be highly efficient, the difficulties in characterizing the nature of the active species in heterogeneous systems are formidable, and their development has therefore been largely empirical. Nevertheless, in many cases there can be no doubt that the chemistry of heterogeneous catalysts mirrors closely the reaction pathways established for homogeneous (solution phase) catalysts. In this chapter we will concentrate on transformations of organic molecules mediated by well-defined transition metal complexes as homogeneous catalysts.

The characteristics of homogeneous catalysis are

  (i) dispersion at the molecular level, i.e., the catalytically active species and the substrate molecules are in the same phase;
  (ii) the catalysts (or at least the catalyst precursor complexes) can be unequivocally characterized by spectroscopic means and synthesized reproducibly;
  (iii) each metal center is potentially a catalytically active site; all these sites show chemical uniformity.

In the majority of catalytic reactions discussed in this chapter it has been possible to rationalize the reaction mechanism on the basis of the spectroscopic or structural identification of reaction intermediates, kinetic studies, and model reactions. Most of the reactions involve steps already discussed in Chapter 21, such as oxidative addition, reductive elimination, and insertion reactions. One may note, however, that it is sometimes difficult to be sure that a reaction is indeed homogeneous and not catalyzed heterogeneously by a decomposition product, such as a metal colloid, or by the surface of the reaction vessel. Some tests have been devised, for example the addition of mercury would poison any catalysis by metallic platinum particles but would not affect platinum complexes in solution, and unsaturated polymers are hydrogenated only by homogeneous catalysts.

There have been numerous attempts to overcome the technical problems of product separation from homogeneously catalyzed systems by attaching the catalyst to a support or working under mixed-phase conditions, e.g., with immiscible mixtures of polar and non-polar solvents (see Section 22-16). These approaches still encounter difficulties but are now increasingly successful.

22-1   Hydrogenation Reactions

We will begin our discussion with the reactions of dihydrogen with unsaturated substrates. The ability of finely divided metallic nickel to catalyze hydrogenation reactions of alkenes and arenes was discovered by Sabatier at the turn of the century. The reaction is generally "structure insensitive," i.e., it is catalyzed by metallic particles, colloids, as well as mononuclear soluble complexes. Simple heterogeneous catalysts, such as Raney nickel and palladium on charcoal, are widely used. Metal complexes have, however, the advantage of substrate-, regio-, and, given suitable ligands, enantio-selectivity, and it is in the selectivity aspect where they have made a major impact.

The first rapid and practical system for the homogeneous reduction of alkenes, alkynes, and other unsaturated substrates at room temperature and 1 bar pressure used the complex $RhCl(PPh_3)_3$, also known as Wilkinson's catalyst. Since then many other tertiary phosphine complexes have been studied as hydrogenation catalysts, including neutral compounds like $RuHCl(PPh_3)_3$ and cationic complexes such as $[Rh(diene)(PPh_3)_2]^+$ and $[RuH(PMe_2Ph)_5]^+$. There are three main types of complexes:

(1) Those without a metal—hydride bond, for example $RhCl(PPh_3)_3$ and $[Rh(PR_3)_2(solvent)_2]^+$, which react with molecular hydrogen under oxidative addition.

(2) Complexes with an M—H bond, such as $RhH(CO)(PPh_3)_3$ and $RuHCl(PPh_3)_3$, where hydrogenation is not initiated by H_2.

(3) Early transition metal and f-block element hydrides, such as $(Cp_2^*LuH)_2$ where catalytic cycles do *not* involve oxidative addition reactions.

Examples of non-platinum metal hydrogenation catalysts include (arene)-chromiumtricarbonyls which will hydrogenate dienes, alkynes, and so on, while $ReH_7(PCy_3)_2$ will selectively hydrogenate acenaphthalene. Lanthanides and early transition metals are discussed later. Those catalysts operate *via* non-radical processes, but a few systems are known to involve radical reactions. The complex $[CoH(CN)_5]^{3-}$ is a water-soluble catalyst that is selective for the hydrogenation of α,β-unsaturated compounds.

Reversible *cis*-Dihydrido Catalysts

The dissociation of PPh_3 from the distorted square d^8 complex $RhCl(PPh_3)_3$ in benzene occurs only to a very small extent ($K = 2.3 \times 10^{-7}$ M at 25°C), and the binding constant for hydrogenation substrates such as alkenes is low, the constant for ethylene being the largest ($K = 0.4$, *cf.* Section 21-1). Under hydrogen, solutions of $RhCl(PPh_3)_3$ rapidly become yellow due to the oxidative addition of H_2 to give *cis*-$H_2RhCl(PPh_3)_3$. One PPh_3 ligand is labilized by the strong *trans* effect of H and rapidly dissociates at room temperature to give a five-coordinate fluxional rhodium(III) species (21-II). This species is then capable of coordinating alkenes, to give, for example, the cyclohexene complex (22-I) where a *cis* orientation of the phosphines is sterically favored and one of the PPh$_3$ ligands is labilized by a *trans*-hydride. In the transition state of the M—H to alkene transfer step the M—H and C=C bonds are coplanar, in a four-center arrangement, as in (22-II).

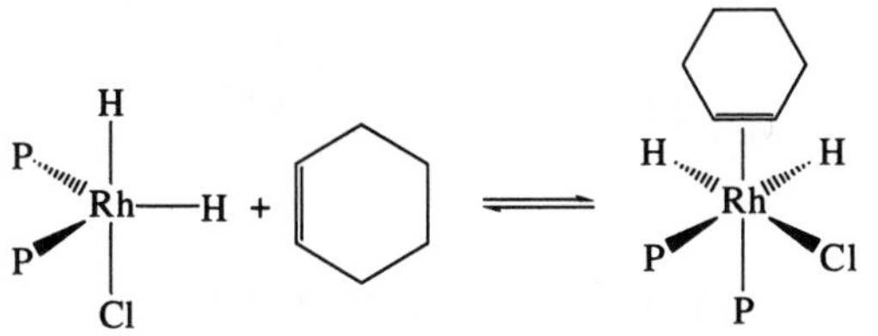

$$(22\text{-}I) \qquad\qquad\qquad (22\text{-}II)$$

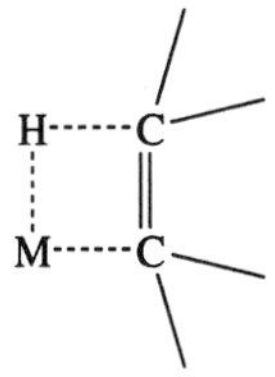

When a *cis* phosphine geometry is enforced by the use of a chelating diphosphine, hydride transfer to coordinated alkene is very rapid. Detailed kinetic studies of H-transfer in the system

$$trans\text{-}HRh(C_2H_4)(PPr_3)_2 \rightleftharpoons cis\text{-}HRh(C_2H_4)(PPr_3)_2 \rightleftharpoons EtRh(PPr_3)_2$$

indicate that the complex with *cis* phosphines is the active one. By contrast, a rhodium complex with a rigid *trans* diphosphine was found to give only slow hydrogen transfer to alkenes.

To complete the alkene hydrogenation reaction sequence, the first hydrogen transfer must be followed by a second, which results in the reductive elimination of the alkane product. This proceeds through a three-centered transition state. The catalytic cycle is shown in Fig. 22-1 but the process is actually more complicated since the equilibria are dependent on phosphine, alkene, rhodium concentrations, temperature, and pressure.

Rhodium triphenylphosphine catalysts are sensitive to steric influences of the alkene substrate; the rates of hydrogenation decrease with increasing alkene substi-

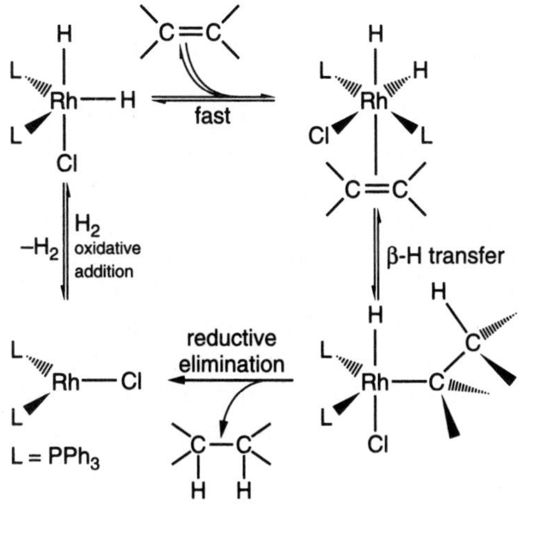

Figure 22-1 Simplified catalytic cycle for the hydrogenation of C=C bonds by species derived from RhClL$_3$ or from [(alkene)$_2$RhCl]$_2$ + L (L = triarylphosphine). Possible solvent coordination is disregarded. The cycle shows only the major intermediates involved for millimolar rhodium concentrations with large alkene concentrations under ambient conditions. Similar cycles operate for cationic complexes where the *cis*-dihydrido species is of the type [Rh(H)$_2$(PR$_3$)$_2$(S)$_2$]$^+$ (S = solvent).

tution in the sequence

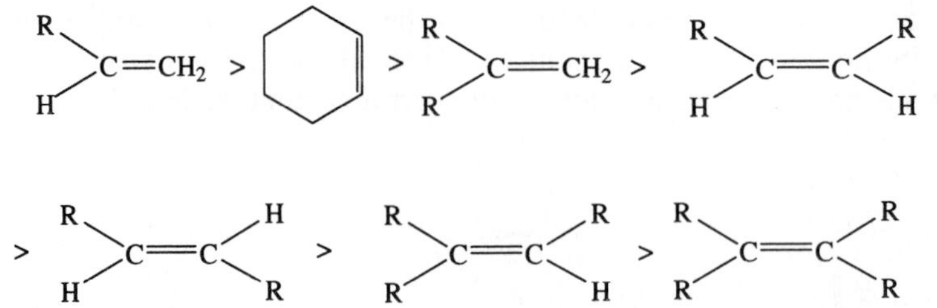

They are also selective for C=C over C=O. Highly regioselective hydrogenations are possible, for example,

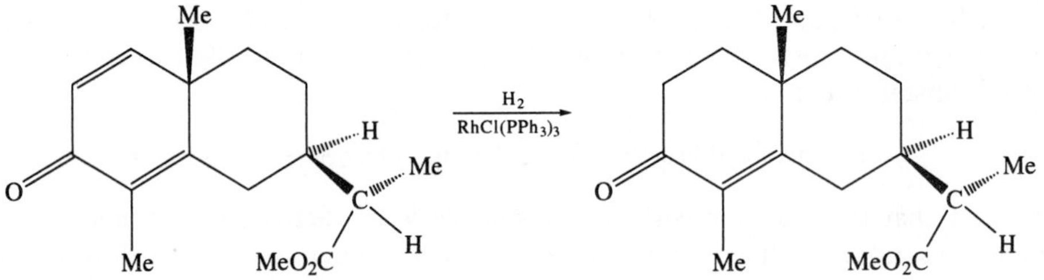

Complexes with trialkylphosphines which are more basic and sterically less demanding than triarylphosphines are generally less active due to a lower tendency to dissociation. Ligands, as well as the polarity of the reaction medium, may also influence the nature of the active species. For example, whereas in the case of triphenylphosphine the hydrogenation cycle is initiated by a Rh^{III} dihydride species, hydrogenations with the water-soluble analogue RhCl(tppms)$_3$ (tppms = Ph$_2$PC$_6$H$_4$SO$_3$Na) involve instead a Rh^I hydride formed by water-assisted reductive HCl elimination (*cf.* monohydride complexes discussed below):

$$RhCl(tppms)_3 + H_2 \rightleftharpoons HRh(tppms)_3 + H^+ + Cl^-$$

The reaction rates increase therefore with increasing water content in the solvent mixture.[1]

Iridium complexes such as IrCl(COD)(L-L), [Ir(COD)(L-L)]$^+$ and [Ir(COD)(py)(PR$_3$)$_2$]$^+$ are active catalysts for the hydrogenation of a variety of substrates, particularly tetrasubstituted alkenes, e.g. [Ir(COD)(py)(PCy$_3$)]$^+$, known as Crabtree's catalyst, hydrogenates tetramethylethylene. On the other hand, [Ir(COD)(py)$_2$]$^+$ is inactive, apparently since it does not add H$_2$. The hydrido cation [Ir(H)$_2$(COD)(η^2-Pr$^i_2$PCH$_2$CH$_2$OMe)]$^+$, with a chelating P—O ligand, selectively hydrogenates phenylacetylene to styrene, *via* Ir—CH=CHPh intermediates.[2]

The stereochemistry of hydrogen addition to C=C may be influenced by the polarity of the alkene substrate. For example, the palladium catalyzed H/D exchange and isomerization of *cis*-crotonic acid to the *trans* isomer is regio- and stereoselective

[1]F. Joó *et al.*, *J. Chem. Soc., Chem. Commun.* **1993**, 1602.
[2]L. A. Oro *et al.*, *Organometallics* **1993**, *12*, 1823.

as a result of the preferential orientation of the unsaturated substrate induced by both C=C and Pd−H polarization[3]:

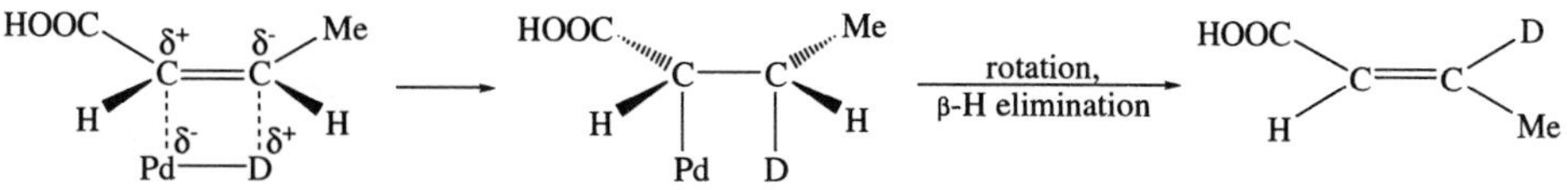

The hydrogenation of polynuclear aromatic heterocycles is important in the context of coal liquefaction and hydrodenitrogenation (for hydrodesulfurization see Section 21-4), but generally requires forcing conditions. A variety of Ru, Os, Rh, and Ir complexes are capable of hydrogenating nitrogen heterocycles such as quinoline (Q); for example, the complex $[Rh(COD)Q_2]^+$ has been isolated from catalytic runs and is one of the species involved in the cycle.[4] At higher temperatures and pressures Ru catalysts hydrogenate even esters and alcohols.[5]

Monohydride Complexes

Compounds such as $RuHCl(PPh_3)_3$, $RhH(CO)(PPh_3)_3$, and $IrH(CO)(PPh_3)_3$, i.e., species which do not require H_2 oxidative addition prior to olefin coordination, also act as hydrogenation catalysts. The hydride $RhH(PPh_3)_3$ is a much more active catalyst than $RhCl(PPh_3)_3$. Compounds such as $RuCl_2(PPh_3)_3$ or $RuCl_2(CO)(PPh_3)_2$ undergo hydrogenolysis to hydride species in the presence of HCl acceptors, for example:

$$RuCl_2(PPh_3)_3 + H_2 + NEt_3 \longrightarrow RuHCl(PPh_3)_3 + Et_3NHCl$$

In these cases the catalytic cycle differs from the one discussed previously in that the first hydrogen transfer to C=C takes place prior to H_2 addition. The cycle for $RhH(CO)(PPh_3)_3$ is shown in Fig. 22-2. The 18-electron precursor complex dissociates in benzene to a 16-electron square bis(phosphine) species which is capable of binding the olefin. In agreement with the assumption of a ligand dissociation step, catalysis is suppressed by the addition of excess PPh_3.

In contrast to the mild conditions required for the rhodium catalyst, $CoH(CO)(PPh_3)_3$ is active only at ~150°C and 150 bar hydrogen pressure, while $IrH(CO)(PPh_3)_3$ will hydrogenate alkenes at ~50°C.

Hydrogenations with $RhH(CO)(PPh_3)_3$ are highly selective for alk-1-enes compared to alk-2-enes. The alk-1-ene can react with the metal−hydrogen bond in either a Markovnikov or an anti-Markovnikov fashion, giving rise to a secondary branched or a primary straight-chain alkyl, respectively. An internal alkene can, of course, give only a branched alkyl. We can now explain the selectivity as follows. In the square species **A** and **B** of Fig. 22-2 the bulky triphenylphosphine groups are in *trans* positions, and the result is that a primary alkyl (i.e., Rh−CH$_2$CH$_2$R) will experience much less steric interaction than will a more bulky branched alkyl, e.g., Rh−CH(Me)R. This less stable secondary alkyl species can therefore be

[3]J. Yu and J. B. Spencer, *J. Am. Chem. Soc.* **1997**, *119*, 5257.
[4]R. A. Sánchez-Delgado *et al., Organometallics* **1993**, *12*, 4291; R. H. Fish *et al., J. Am. Chem. Soc.* **1992**, *114*, 5187.
[5]H. T. Teunissen and C. J. Elsevier, *Chem. Commun.* **1997**, 667.

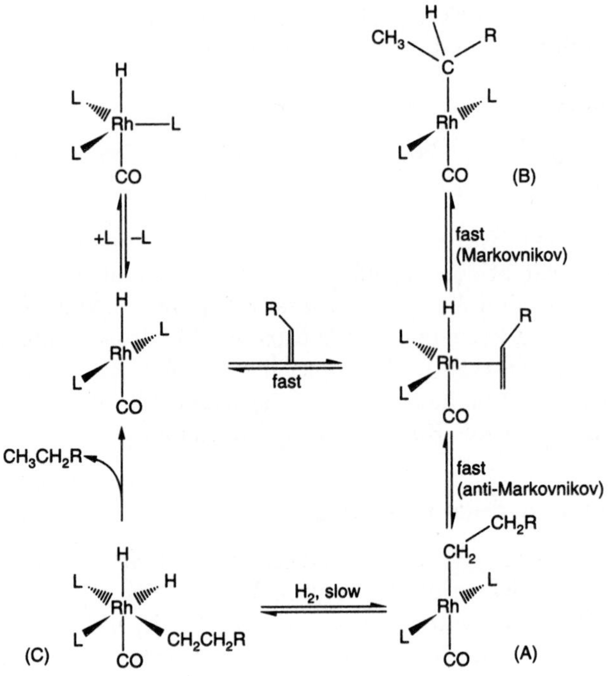

Figure 22-2 Catalytic cycle for the hydrogenation and isomerization of alk-1-enes by RhH(CO)L₃ (L = PPh₃) at 25°C and 1 bar pressure.

expected to undergo the reverse β-hydrogen transfer reaction to give olefin and Rh–H more easily and at a competitive or faster rate than the slow H_2 oxidative addition to give species **C**; this oxidative addition is the rate-limiting step in the catalytic cycle. That the secondary alkyl complex is formed in equilibrium with the primary alkyl complex is shown by the comparability of the rates of isomerization (alk-1-ene to alk-2-ene) and hydrogenation.

Similar steric factors introduced by bulky PPh₃ groups are also important in the hydroformylation reaction of alkenes that is catalyzed by the same system (Section 22-6) and which, in the presence of a large excess of PPh₃, gives high selectivity for linear n-aldehydes.

The phosphite cluster [HRh{P(OR)₃}₂]₃ catalyzes alkene hydrogenations, allegedly without fragmentation into mononuclear species, while the dimer [Rh₂(μ-H){P(OPrⁱ)₃}₂]₂ hydrogenates alkynes *via* μ-acetylene complexes:

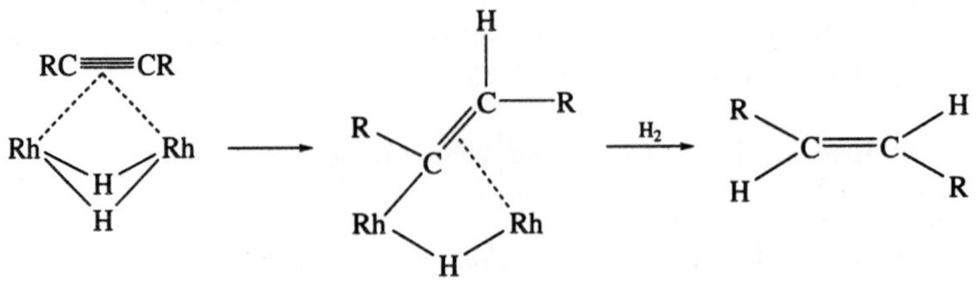

The ruthenium hydrogenation catalyst RuH{HB(pz)$_3$}(PPh$_3$)$_2$ is readily proton-ated to give a dihydrogen complex:

$$RuH\{HB(pz)_3\}(PPh_3)_2 \xrightleftharpoons[NEt_3]{H^+} [Ru(H_2)\{HB(pz)_3\}(PPh_3)_2]^+$$

This reaction explains the incorporation of deuterium into the product if hydrogena-tions are carried out in the presence of D$_2$O.[6]

Asymmetric Hydrogenation

The development of the catalytic hydrogenation system based on RhCl(PPh$_3$)$_3$ and methods for the resolution of optical isomers of tertiary phosphines occurred around the same time (1965), and this led to the possibility of asymmetric catalytic hydroge-nation of prochiral unsaturated substances with C=C, C=O, and C=N bonds using transition metal complexes with chiral phosphine ligands. Such tertiary phos-phines are of three types:

(i) those that have chiral phosphorus atoms; these were prepared early in the development of asymmetric catalysis but are now little used;

(ii) phosphines with chiral hydrocarbon substituents; and

(iii) chelating phosphines with a chiral bridge connecting the two phosphorus atoms. This bridge may contain chiral carbon atoms or an element related by a C_2 axis, as in 2,2′-binaphthyl units.

There are now hundreds of chiral ligands. Some representative examples are (o-anisyl)(cyclohexyl)methylphosphine, (+)- or (−)-2,3-O-isopropylidene-2,3-dihy-droxy-1,4- bis(diphenylphosphino)butane, usually abbreviated (+)- or (−)-DIOP (22-III), which is readily accessible from tartaric acid as a naturally occurring chiral precursor, neomenthyldiphenylphosphine (22-IV), and (+)-"CAMPHOS" (22-V). The ligand trans-1,2-bis(diphenylphosphinoxo)cyclopentane (22-VI) was found very useful for the hydrogenation of substrates without functional groups. The binaphthyl derivative (22-VII) is an example of a ligand with axial chirality which frequently gives exceptionally high optical yields; it is known as BINAP.[7,8] A more bulky derivative, MeO−BIPHEP (22-VIII), is even more enantioselective in a number of catalyzed reactions, including hydrogenations with Ru complexes and palladium catalyzed C−C bond-forming reactions.[9] Another recently developed example with a rigid chiral backbone is (22-IX).[10]

(22-III)   (22-IV)

[6]C. P. Lau et al., Organometallics 1997, 16, 34.
[7]R. Noyori and H. Takaya, Acc. Chem. Res. 1990, 23, 345.
[8]R. Noyori, Science 1990, 248, 1194.
[9]P. S. Pregosin et al., J. Am. Chem. Soc. 1997, 119, 6315.
[10]X. Zhang et al., J. Am. Chem. Soc. 1997, 119, 1799.

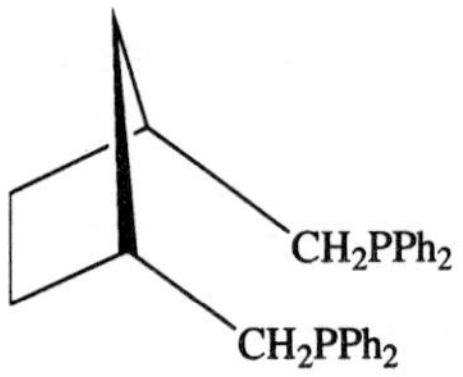

(22-V) 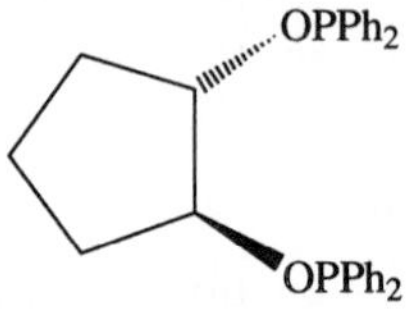   (22-VI) 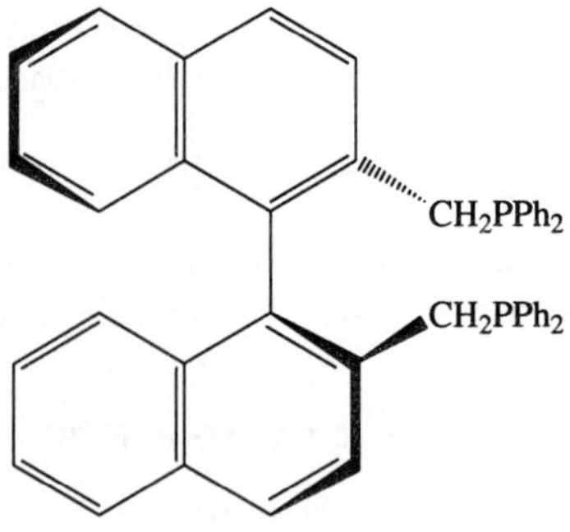   (22-VII)

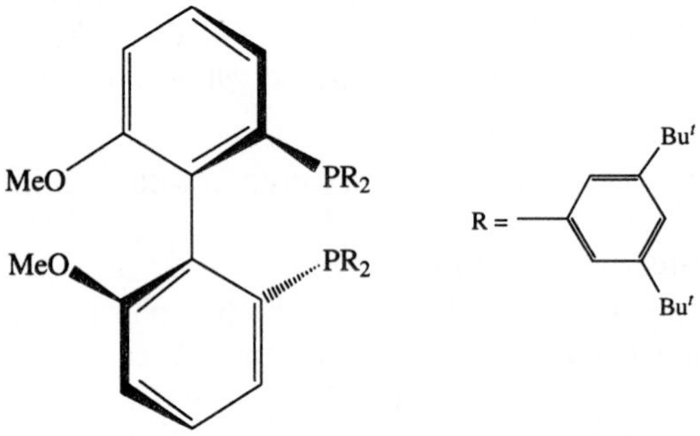

(22-VIII) 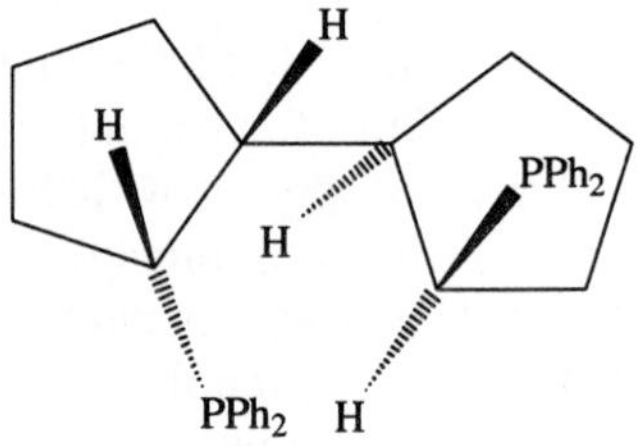   (22-IX)

In most of these ligands the chiral center is far removed from the site where the prochiral substrate coordinates to the transition metal, so that little diastereo-differentiation might be expected. However, the chirality of the backbone controls the conformation of the bulky diarylphosphine groups and hence generates a chiral pocket around the metal, with aryl groups in axial and equatorial positions. This imposes C_2 symmetry on the complex, as shown in (22-X). Viewed from the side, such a complex can schematically be divided into sterically hindered and open quadrants (22-XI). A prochiral substrate will then naturally bind with the π-face that leads to the product with the least steric repulsion.

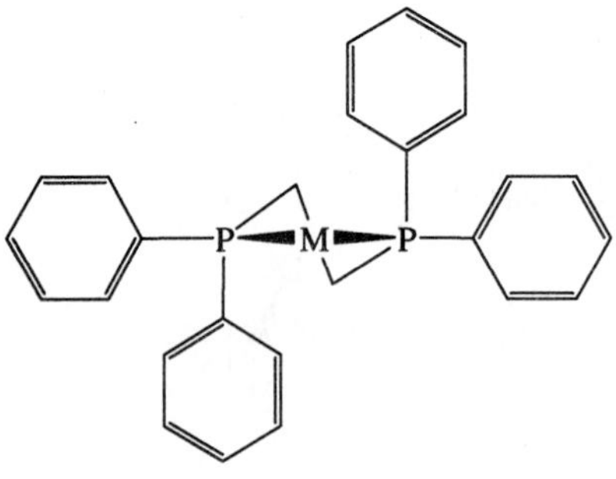

(22-X) 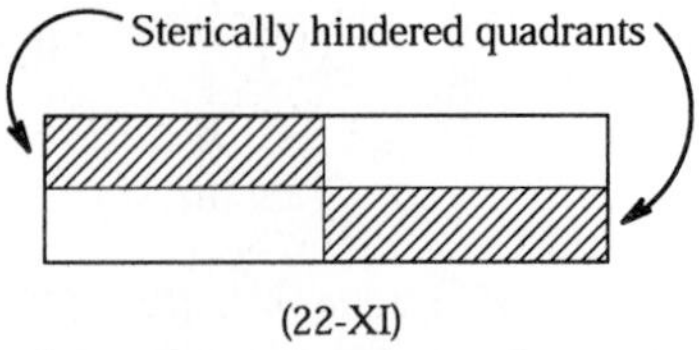   (22-XI)

A ligand with a more direct interaction between chiral centers and the substrate binding pocket is the bis(phospholane) (22-XII);[11] rhodium hydrogenation catalysts

[11]M. J. Burk et al., J. Am. Chem. Soc. **1993**, *115*, 10125.

containing this ligand give enantioselectivities of nearly 100% e.e. (e.e. = enantiomeric excess, i.e., the difference in concentration between the two possible product enantiomers).

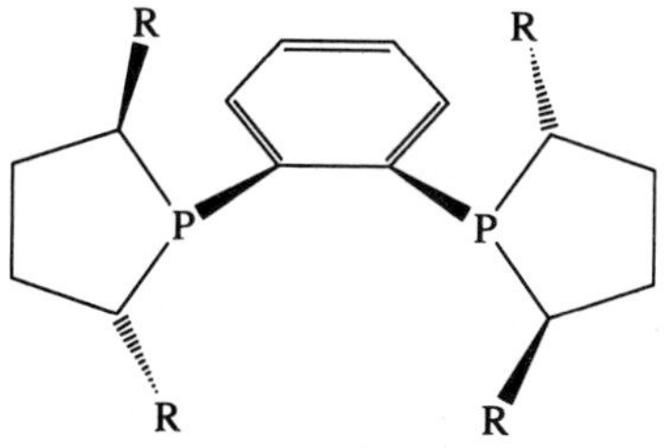

(22-XII)

Chiral catalysis is employed in several industrial processes leading to enantiomerically pure products, such as the amino acid L-dihydroxyphenylalanine (L-DOPA) required for the treatment of Parkinson's disease, L-menthol, and carbapenems.[7,8] The hydrogenation of prochiral functionalized alkenes, notably α-acetamidocinnamic esters, has attracted particular attention:

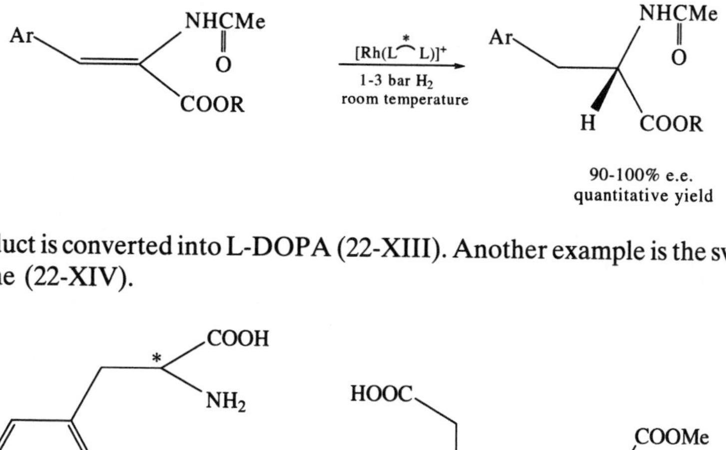

90-100% e.e.
quantitative yield

The product is converted into L-DOPA (22-XIII). Another example is the sweetener aspartame (22-XIV).

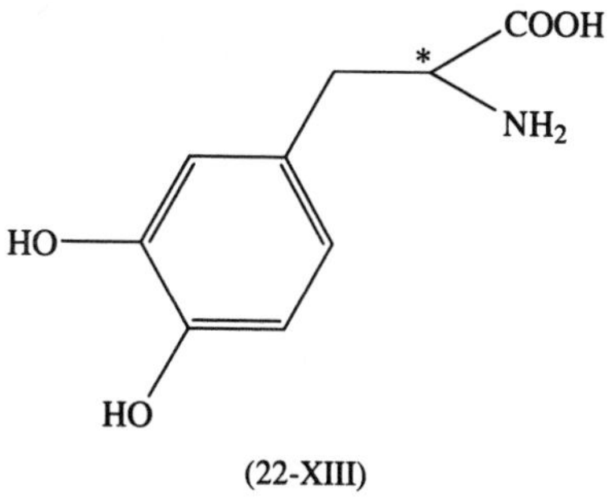

(22-XIII)

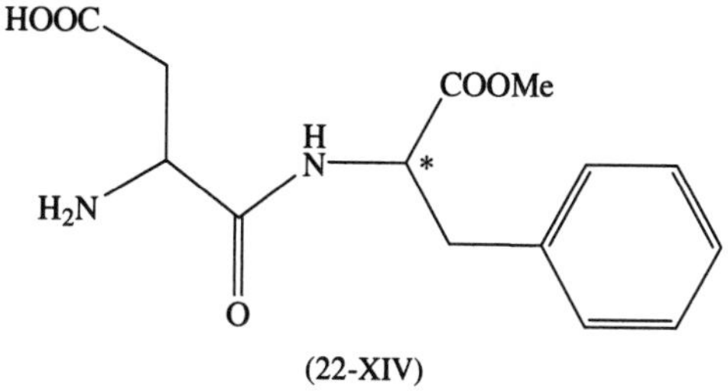

(22-XIV)

In asymmetric catalysis a prochiral substrate binds to an enantiomerically pure catalyst to generate a pair of diastereomeric intermediates. The energy difference and the rate of exchange between them controls the optical yield (e.e.) of the final product. In the case of α-aminocinnamic acid derivatives, the acyl auxiliary on the nitrogen is required to enable the substrate to form a chelate complex with rhodium.[12] The mechanism of this reaction is shown in Fig. 22-3; the ligand in this case is DIPAMP (22-XV).

[12]C. R. Landis et al., J. Am. Chem. Soc. **1993,** 115, 4040.

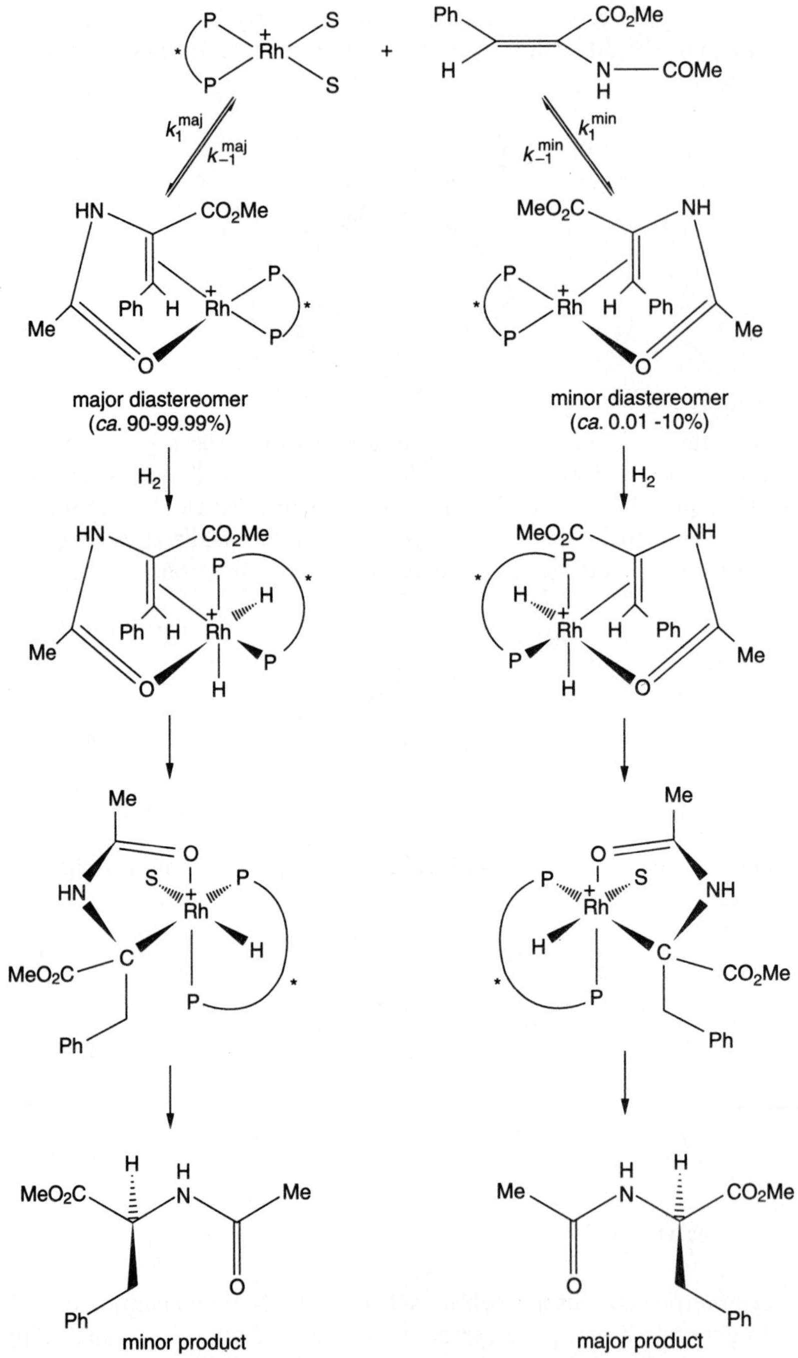

Figure 22-3 Mechanism of asymmetric hydrogenation of methyl (Z)-α-acetamidocinnamate using rhodium complexes with chiral diphosphine ligands. The minor complex diastereomer leads to the major (R) product, while the dominant complex in solution leads to the minor (S) isomer. See also C. R. Landis *et al.*, *J. Am. Chem. Soc.* **1993,** *115,* 4040.

(22-XV)

The enantioselective step is the oxidative addition of H_2 to the square diastereomeric substrate complexes that are in rapid dissociative equilibrium. *The major enantiomer of the product* arises from the *minor substrate-catalyst diastereomer*; this isomer cannot always be detected since it reacts much more rapidly with H_2 than the major diastereomer. Molecular modelling suggests that the principal enantiodifferentiating interactions are between the enamide ester function and the nearest arene substituent of the chiral diphosphine.[12] The large increase in reaction rate for the minor diastereomer arises from the increased stability of the corresponding dihydro intermediate, that is, the enantioselective step is under product control.

Asymmetric hydrogenation has been achieved for C=C bonds also using ruthenium catalysts, notably those containing the BINAP ligand.[13] Sulfonated analogues of BINAP give water soluble rhodium complexes which catalyze asymmetric hydrogenations in water, without reduction in optical yield.[14] Rhodium complexes of ligand (22-XII) are particularly effective for the hydrogenation of C=N bonds to chiral amines.[15] Cationic rhodium complexes in reversed micelles also catalyze C=N hydrogenations with high enantio-selectivities.[16] The asymmetric hydrogenation of cyclic ketones to chiral alcohols is achieved with Ru, Rh, or Ir compounds, e.g., $[Ir(COD)(BINAP)]^+$.[17]

Finally, metal colloids can adsorb chiral molecules; such surface-modified particles can catalyze hydrogenations in high optical yields. An example is platinum colloids treated with cinchona alkaloids.[18]

Hydrogenations with Early Transition Metal and Lanthanide Catalysts

A number of early transition metal compounds, e.g., Ti^{III} hydrides, have long been known as hydrogenation catalysts. Similarly, metallocene compounds of lanthanides and actinides can be extremely active, as a comparison of the turnover numbers of 1-hexene hydrogenations at 25°C (1 bar H_2) show: Cp_2^*LuH, 120,000; $[Ir(COD)-(py)(PCy_3)]PF_6$, 6400; $[Rh(COD)(PPh_3)_2]PF_6$, 4000; $RuHCl(PPh_3)_3$, 3000; $RhCl(PPh_3)_3$, 650.

It is clear that oxidative addition of H_2 to these high-valent compounds cannot occur; the catalytic cycle is most likely to involve four-membered transition states

[13]H. Takaya *et al., Organometallics* **1996,** *15,* 1521.
[14]K. Wan and M. E. Davis, *J. Chem. Soc., Chem. Commun.* **1993,** 1262.
[15]C. Bolm, *Angew. Chem. Int. Ed. Engl.* **1993,** *32,* 232.
[16]J. M. Buriak and J. A. Osborn, *Organometallics* **1996,** *15,* 3161.
[17]H. Takaya *et al., J. Am. Chem. Soc.* **1993,** *115,* 3318.
[18]H. Bönnemann and G. A. Braun, *Chem. Eur. J.* **1997,** *3,* 1200.

for the H-transfer and M—C hydrogenolysis steps:

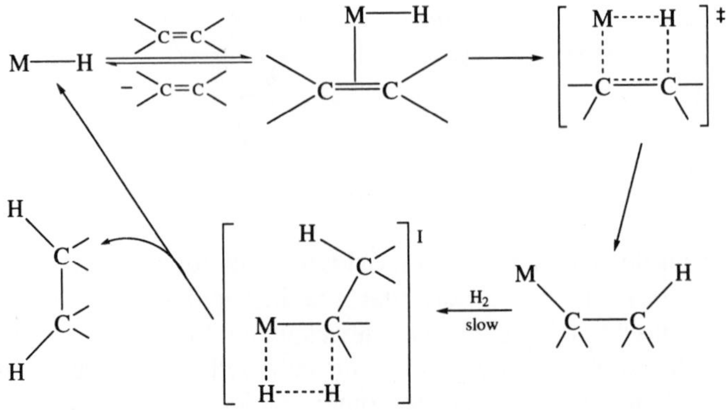

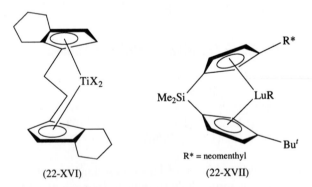

Alkenes without functional groups are difficult to hydrogenate enantioselectively with noble-metal catalysts. Titanium complexes with a C_2 symmetric chiral ligand framework, as in the ansa-titanocene (22-XVI), reduce aryl-substituted C=C bonds in very high isolated and optical yields:[19]

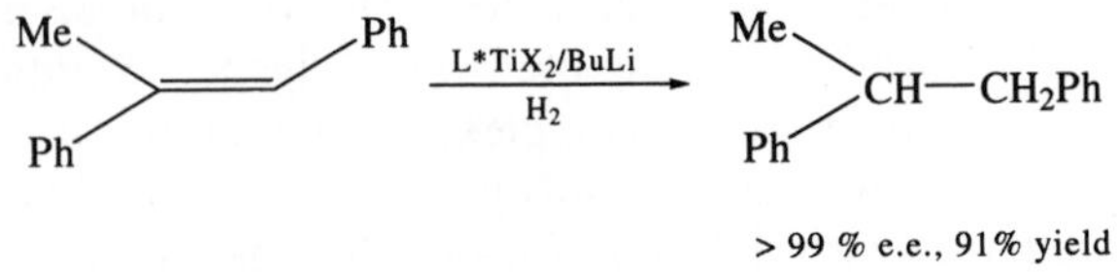

$$\underset{Ph}{\overset{Me}{\diagdown}}\!\!=\!\!\underset{}{\overset{Ph}{\diagup}} \xrightarrow[\text{H}_2]{\text{L*TiX}_2/\text{BuLi}} \underset{Ph}{\overset{Me}{\diagdown}}\!\!CH\!\!-\!\!CH_2Ph$$

> 99 % e.e., 91% yield

The catalysts are Ti[III] hydrides which are generated *in situ* by reduction of enantiomerically pure L*TiCl$_2$ with butyllithium and PhSiH$_3$; they are also highly effective in the asymmetric hydrogenation of imines and enamines.[20] The lutetium ansa-metallocene complex (22-XVII) catalyzes the deuteration of 1-pentene (63% e.e.)[21] Related samarium compounds hydrogenate imines.[22]

TiX$_2$

(22-XVI)

Me$_2$Si   LuR   R*   Bu^t

R* = neomenthyl

(22-XVII)

A different type of hydrogenation catalysts are aryloxo compounds of niobium and tantalum which are highly effective in the hydrogenation of arenes, including

[19]R. D. Broene and S. L. Buchwald, *J. Am. Chem. Soc.* **1993**, *115*, 12569.
[20]S. L. Buchwald *et al.*, *J. Am. Chem. Soc.* **1994**, *116*, 5985; 11714.
[21]T. J. Marks *et al.*, *Organometallics* **1997**, *16*, 4486.
[22]T. J. Marks *et al.*, *J. Am. Chem. Soc.* **1997**, *119*, 3745.

the aryl substituents of their ligands. The Ta compound is made by the reaction

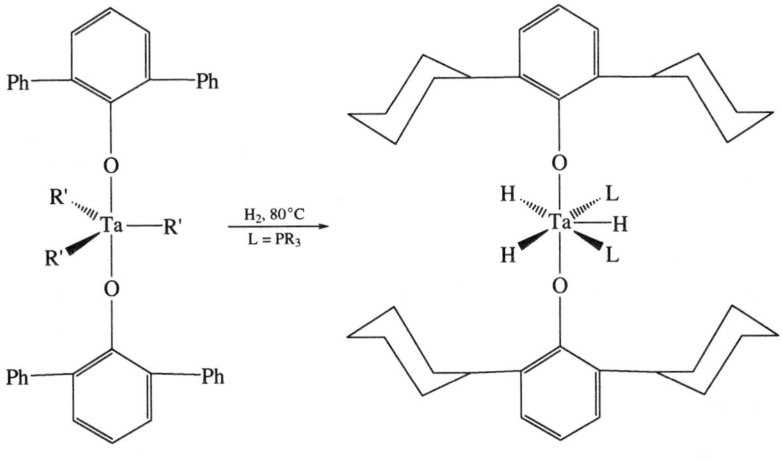

and hydrogenates arenes from one π-face only to give all-*cis* products:[23]

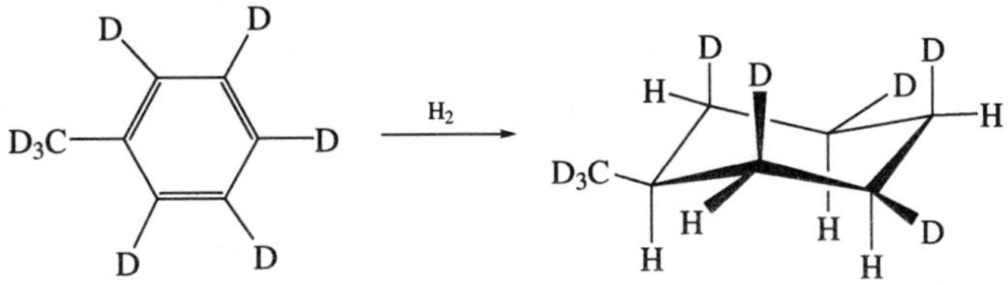

The related Nb^{III} 1,3-hexadiene compound $(ArO)_3Nb(\eta^4\text{-}C_6H_8)$ ($Ar = C_6H_3Pr^i_2$) catalyzes the hydrogenation of cyclohexadiene to cyclohexane, as well as its disproportionation to cyclohexene and benzene.[24]

Transfer Hydrogenations

In many reactions molecular hydrogen may be replaced by other H-sources such as methanol, isopropanol, or formic acid; even cyclic ethers such as dioxane or THF can be used in homogeneous transfer hydrogenation. The hydrogen donor coordinates to the metal and undergoes β-hydrogen transfer:

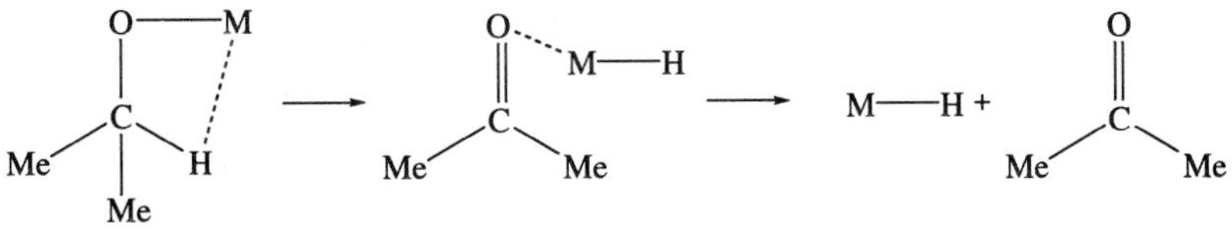

Examples are the reduction of imines by isopropanol catalyzed by $RuCl_2(PPh_3)_3$ in the presence of base,[25] to give secondary amines and acetone, and the asymmetric

[23] I. P. Rothwell, *Chem. Commun.* **1997**, 1331.
[24] I. P. Rothwell *et al., J. Am. Chem. Soc.* **1997**, *119*, 3490.
[25] G. Z. Wang and J. E. Bäckvall, *J. Chem. Soc., Chem. Commun.* **1992**, 980.

analogue of this reaction using formic acid/triethylamine in the presence of chiral ruthenium catalysts:[26]

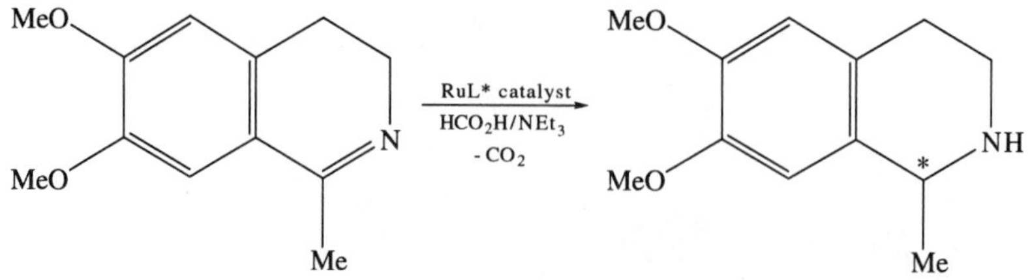

Formic acid is also effective in the Pd catalyzed asymmetric reduction of ester functions $R-OCO_2Me$ to $R-H$ (R = chiral allyl group).[27]

Hydrogenations of Other Unsaturated Molecules

The hydrogenation of aromatics has been a topic of interest since Sabatier's first synthesis of cyclohexane from benzene with metallic nickel. The role of Nb and Ta aryloxides as catalysts for this reaction was mentioned earlier. Another system that has been studied in detail comprises allyl and hydride complexes of cobalt, e.g., $(\eta^3$-$C_3H_5)Co[P(OMe)_3]_3$. Like the Nb and Ta compounds cobalt gives cyclohexane with *cis* stereoselectivity. The active species is probably the hydride, generated from (allyl)CoL$_3$ on hydrogenolysis, which reacts with arenes in a stepwise manner.

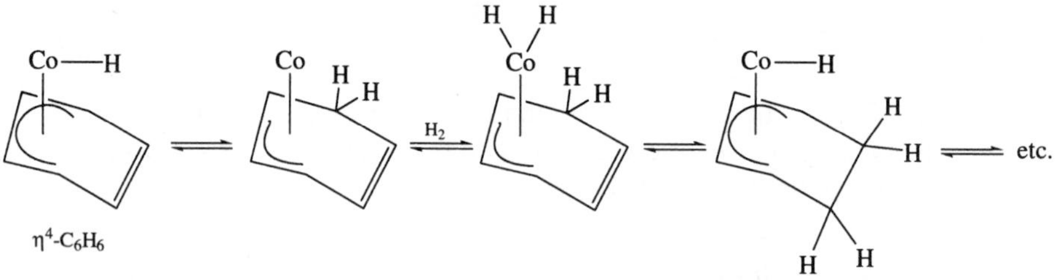

The reduction of CO_2 to formic acid is catalyzed by Rh and Ru complexes under basic conditions. Calculations suggest an η^2-CO_2 complex as intermediate in the sequence

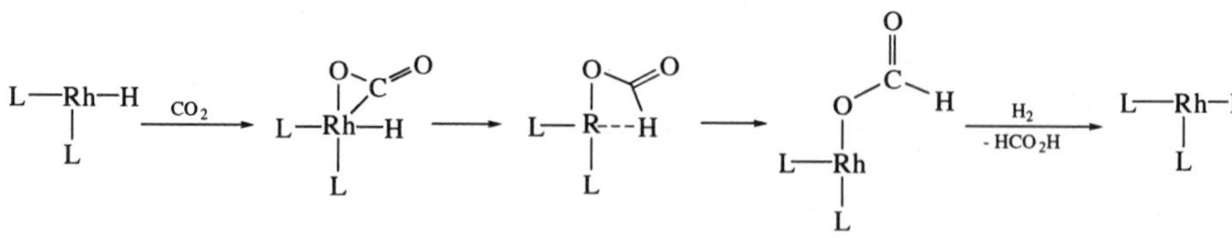

with a rate-limiting liberation of formic acid.[28] Anionic ruthenium complexes also catalyze the formation of formate from CO_2; a binuclear methoxide-bridged complex $[Ru_2(\mu$-OMe)(OMe)_4(CO)_6]^-$ is proposed as the active species.[29]

[26]R. Noyori *et al., J. Am. Chem. Soc.* **1996,** *118,* 4916.
[27]T. Hayashi *et al., Chem. Commun.* **1996,** 1767.
[28]W. Leitner *et al., J. Am. Chem. Soc.* **1997,** *119,* 4432.
[29]G. Süss-Fink *et al., Organometallics* **1996,** *15,* 3416.

22-2 Alkene Isomerization[30]

The isomerization of C=C double bonds is frequently an undesirable by-product in hydrogenation, hydroformylation, oligomerization, and other reactions, but is sometimes used as an essential part of a reaction sequence. Both homogeneous and heterogeneous catalysts are effective, and it is likely that similar mechanistic principles are followed in either case. An important application of C=C bond isomerization is part of the Shell Higher Olefin Process where the higher fraction ($C_{20}+$) of 1-alkenes, produced by Ni catalyzed ethylene oligomerization, are isomerized to internal alkenes before being cleaved in an olefin metathesis step (*cf.* Section 22-13).

Olefin isomerizations can follow two different mechanisms, depending on whether or not the metal species involved contains an M—H bond. Nickel and palladium complexes, but also iron and rhodium, can induce isomerization *via* a π-allyl mechanism:

$$H_2C\!=\!CH\!-\!CH_2R \;\rightleftharpoons\; H_2C\overset{\overset{\displaystyle H}{\underset{|}{C}}}{\diagdown}\underset{\underset{\displaystyle H}{M\diagdown}}{\overset{\displaystyle H}{\underset{|}{CR}}} \;\rightleftharpoons\; H_3C\!-\!CH\!=\!CHR$$

An example is the isomerization of 1,3- to 1,5-cyclooctadiene catalyzed by [RhCl(COD)]$_2$.

The second mechanism is a 1,2-H shift:

$$\begin{array}{c}
\text{RCH}_2\text{CH}=\text{CH}_2 \quad \text{RCH}_2\text{CH}=\text{CH}_2 \\
\text{M}\!-\!\text{H} \longrightarrow \qquad\qquad | \\
\uparrow \qquad\qquad\qquad \text{M}\!-\!\text{H} \\
\text{RCH}=\text{CHMe} \leftarrow \Big| \qquad\qquad\qquad | \\
\text{RCH}=\text{CH}\!-\!\text{CH}_3 \leftarrow \text{RCH}_2\text{CH}\!-\!\text{CH}_3 \\
| \qquad\qquad\qquad | \\
\text{M}\!-\!\text{H} \qquad\qquad \text{M}
\end{array}$$

This reaction frequently accompanies reactions catalyzed by metal hydride species, e.g., hydrides of Co, Rh, or Ni.

An elegant test for these two reaction pathways is the deuterium shift in the reaction

$$\underset{\text{Et}}{\overset{\text{Et}}{\diagdown}}\underset{\text{D}}{\overset{1}{C}}\text{—}\overset{2}{\text{CH}}=\overset{3}{\text{CH}_2} \quad\begin{array}{c}\xrightarrow{\;1,3\text{-shift}\;} \quad \underset{\text{Et}}{\overset{\text{Et}}{\diagdown}}C\!=\!CH\!-\!CH_2D \\[2em] \xrightarrow{\;1,2\text{-shift}\;} \quad \underset{\text{Et}}{\overset{\text{Et}}{\diagdown}}C\!=\!CD\!-\!CH_3\end{array}$$

[30]W. A. Herrmann, in: B. Cornils and W. A. Herrmann (Eds.), *Applied Homogeneous Catalysis with Organometallic Compounds*, VCH, Weinheim, 1996, p. 980.

Isomerizations have wide synthetic applications. For example, $[Ru(H_2O)_6]^{2+}$ catalyzes the isomerization of allylic alcohols to carbonyl compounds,[31] and propynylic alcohols are isomerized to α,β-unsaturated ketones:[32]

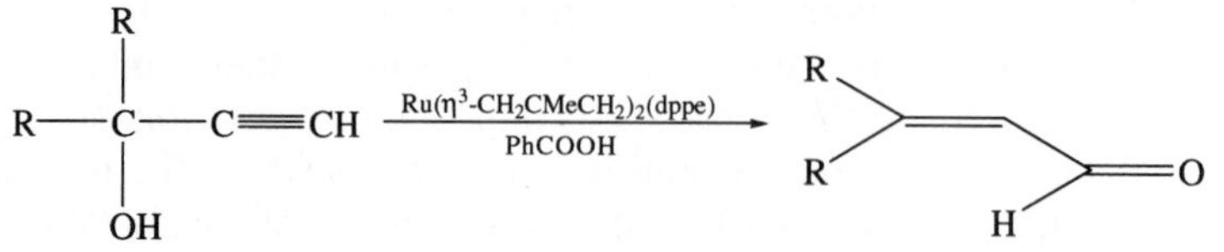

Rhodium BINAP complexes catalyze enantioselective 1,3-H shifts:[7]

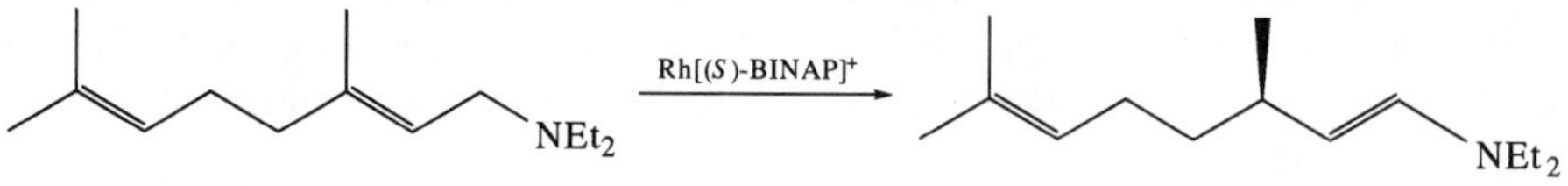

22-3   Hydrosilylation and Hydroboration Reactions

The hydrosilation or hydrosilylation reaction is similar to hydrogenation except that $H—SiR_3$ instead of H_2 is added to a C=C double bond, leading usually to linear products:

$$RCH{=}CH_2 + HSiR_3 \longrightarrow RCH_2CH_2SiR_3$$

The reaction was first discovered using chloroplatinic acid as a catalyst; H_2PtCl_6 in alcohols is commonly referred to as Speier's catalyst. It is effective in extremely low concentrations (10^{-5} to 10^{-8} mol of Pt/mol of reactant). The reaction is used in silicone technology in various ways; for example, hydrosilylation of acrylonitrile allows the incorporation of C_2H_4CN groups into rubbers for self-sealing fuel tanks and for cross-linking reactions in the curing of gums to elastomers. It appears that in reactions initiated by H_2PtCl_6 in isopropanol, platinum colloids may be involved which are very active. Other systems, like those catalyzed by well-defined Fe, Mn, Co, Ru, Rh, Ir, Pd, and Pt complexes, are clearly homogeneous.

A widely accepted cycle for olefin hydrosilylation is the so-called Chalk-Harrod mechanism, where it is assumed that H migration to the alkene is faster than silyl migration, followed by reductive elimination from a silyl-alkyl intermediate,

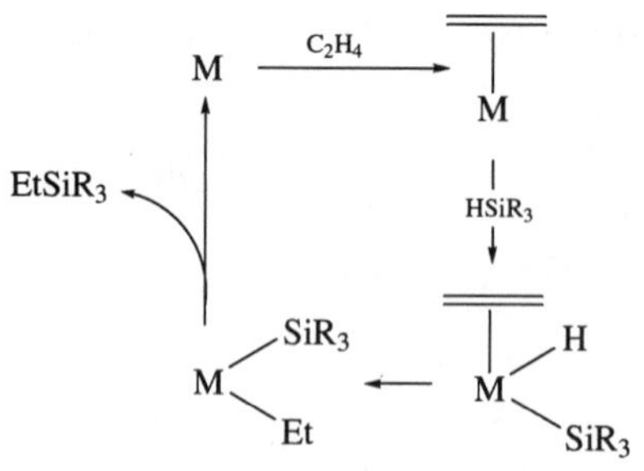

[31]D. V. McGrath and R. H. Grubbs, *Organometallics* **1994**, *13*, 224.
[32]P. H. Dixneuf *et al.*, *Chem. Commun.* **1997**, 1201.

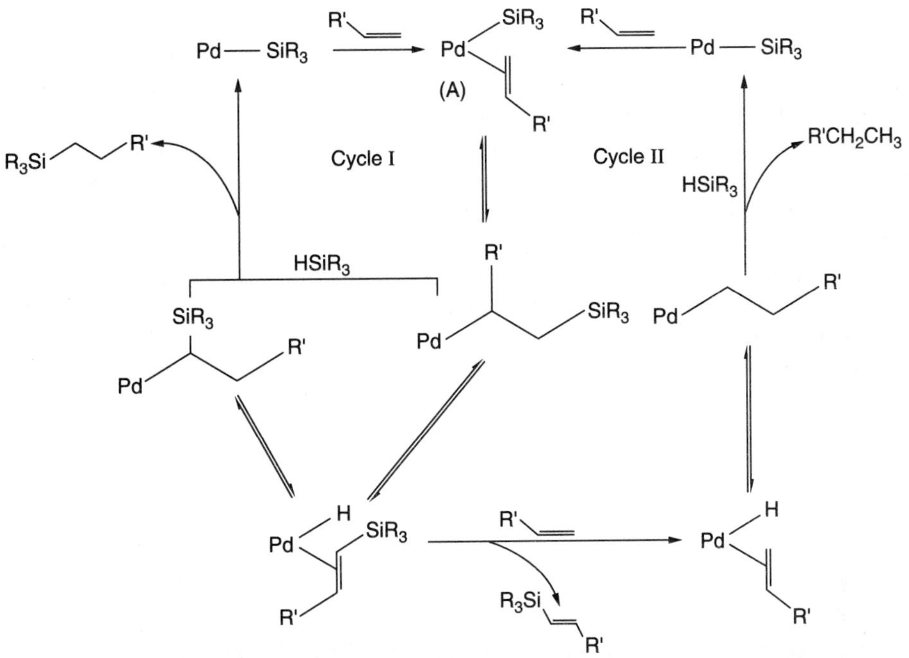

Figure 22-4 Mechanism for the hydrosilylation and dehydrogenative silylation of 1-alkenes catalyzed by cationic palladium complexes; Pd represents [(phen)Pd]$^+$. The palladium alkene complex A is the resting state of the cycle. Cycle I denotes the hydrosilylation cycle, Cycle II describes the dehydrogenative silylation reaction.

However, it is now known that silyl migration is a very facile process, and in some cases both β-silyl and β-H migrations were found to be operative even at low ($<-40°C$) temperatures.[33] The data for hydrosilylations catalyzed by CpRh(H)(SiR$_3$)(C$_2$H$_4$) are in agreement with facile silyl migration and a Rh^V intermediate:[34]

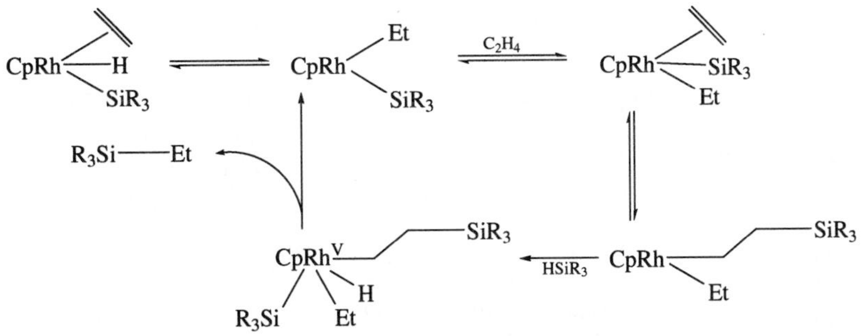

A related mechanism has been established for hydrosilylations catalyzed by cationic palladium complexes and is shown in Fig. 22-4. The mechanisms for both

[33]M. Brookhart *et al., J. Am. Chem. Soc.* **1997**, *119*, 906.
[34]S. B. Duckett and R. N. Perutz, *Organometallics* **1992**, *11*, 90.

the Rh and Pd catalysts also account for the dehydrogenative alkene silylations that are often found; such products arise where the reaction of the alkyl intermediate $M-CHR'CH_2SiR_3$ with $HSiR_3$ is slow compared to β-H elimination. This is typically the case with styrene, which gives $R_3SiCH=CHPh$ and PhEt.[32,35]

The hydrosilylation of $C=O$ and $C=N$ bonds can be achieved with many catalytic systems, for example, manganese where the efficiency of the catalyst precursor decreases in the order $(PPh_3)(CO)_4MnC(O)Me \gg (CO)_5MnC(O)Me > (CO)_5MnMe \gg Mn_2(CO)_{10}$.[36] With suitable chelating chiral phosphine or nitrogen ligands Ru, Rh, and Ir catalysts hydrosilylate ketones enantioselectively.[37] Chiral ansa-titanocene hydrides, generated from (22-XVI, $X = F$) and $PhSiH_3$, catalyze the asymmetric hydrosilylation of imines.[38]

We also mention here the dehydrocoupling of silanes which gives silane oligomers and polymers[39]:

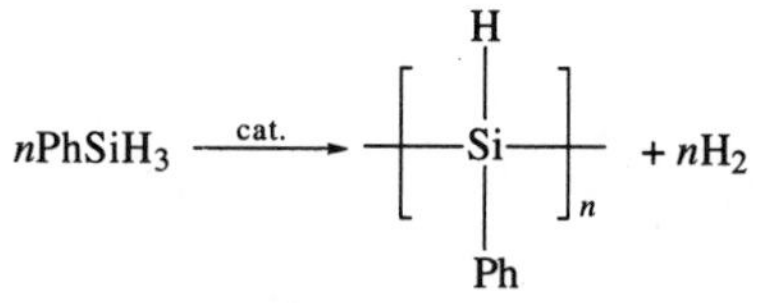

$$n\text{PhSiH}_3 \xrightarrow{\text{cat.}} \left[\begin{array}{c} H \\ | \\ -Si- \\ | \\ Ph \end{array} \right]_n + n\text{H}_2$$

The reaction is catalyzed by reduced Ti, Zr, or Hf metallocene complexes.[40]

Hydroboration and Diborylation Reactions

While there are numerous hydroborations of alkenes that proceed without catalysts, transition metal complexes are very effective in accelerating this reaction and in influencing product selectivity. Catecholborane is commonly used:

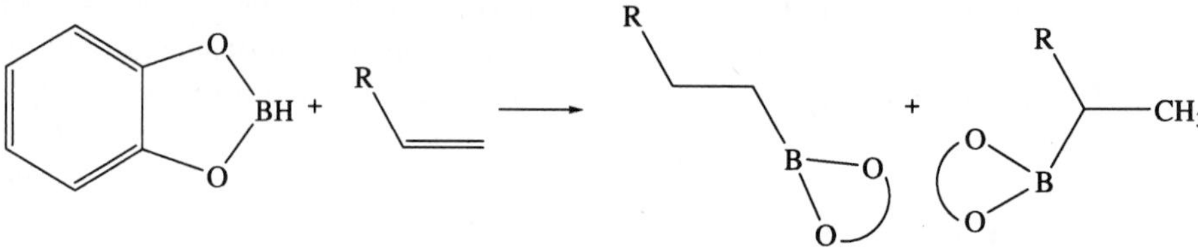

The reaction is catalyzed by lanthanide complexes Cp_2^*LnR,[41] although noble metal catalysts, notably rhodium, are most widely applied, particularly in *asymmetric hydroboration*.[42] The mechanism is likely to be similar to hydrosilylation. The products may be oxidized with H_2O_2 and converted to alcohols or amines.

[35]R. Takeuchi and H. Yasue, *Organometallics* **1996**, *15*, 2098.
[36]A. R. Cutler *et al.*, *Organometallics* **1996**, *15*, 2764.
[37]S. H. Bergens *et al.*, *Organometallics* **1996**, *15*, 3782; S. Uemura *et al.*, *Chem. Commun.* **1996**, 847; *Organometallics* **1996**, *15*, 370.
[38]S. L. Buchwald *et al.*, *J. Am. Chem. Soc.* **1996**, *118*, 6784.
[39]T. D. Tilley, *Acc. Chem. Res.* **1993**, *26*, 22.
[40]R. M. Shaltout and J. Y. Corey, *Organometallics* **1996**, *15*, 2866; M. Tanaka, *Organometallics* **1997**, *16*, 2765.
[41]K. N. Harrison and T. J. Marks, *J. Am. Chem. Soc.* **1992**, *114*, 9220.
[42]J. M. Brown *et al.*, *Chem. Commun.* **1997**, 173; A. Togni *et al.*, *Organometallics* **1997**, *16*, 255.

The titanium complex $Cp_2^*Ti(C_2H_4)$ catalyzes the dehydrogenative borylation of ethylene,[43]

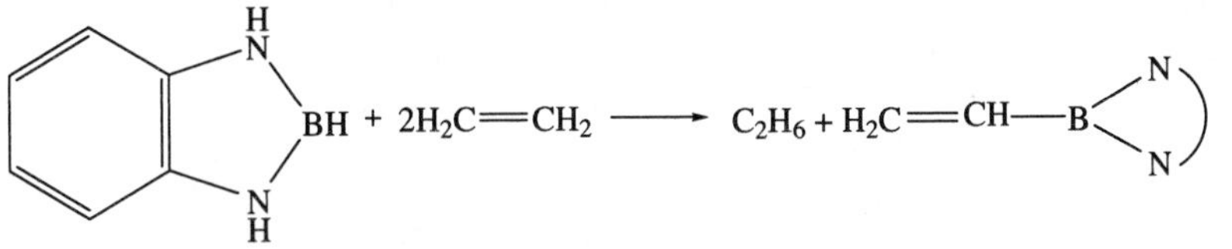

Similarly, palladium phosphite complexes catalyze the addition of B—Si bonds of silylboranes to alkynes; the reaction is a typical *cis* addition with high regioselectivity.[44]

Platinum(0) complexes such as $Pt(PPh_3)_4$ and $Pt(COD)_2$ catalyze the addition of diboranes CatB—BCat to alkenes and alkynes (Cat = catecholate $C_6H_4O_2$).

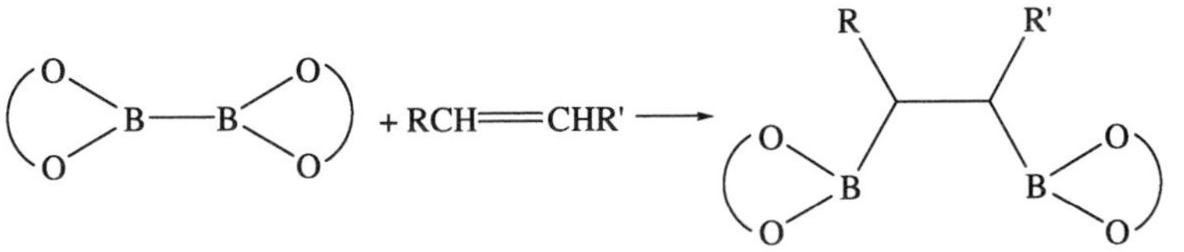

Phosphine dissociation from platinum boryl intermediates $(PPh_3)_2Pt(BCat)_2$ allows the coordination of an alkyne prior to the boryl transfer step. The diborylation of alkenes requires phosphine-free catalysts.[45,46] Boron—boron bonds react with α,β-unsaturated ketones with 1,4-addition.[47]

22-4 Alkene Hydrocyanation

Various complexes of Cu, Ni, and Pd, especially nickel phosphite complexes, such as $Ni[P(O\text{-}o\text{-}C_6H_4Me)_3]_4$, are active for the addition of HCN to alkenes and alkynes. Hydrogen cyanide will oxidatively add to low-valent metal complexes,

$$NiL_4 + HCN \underset{+L}{\overset{-L}{\rightleftharpoons}} NiH(CN)L_3 \underset{+L}{\overset{-L}{\rightleftharpoons}} NiH(CN)L_2$$

and while in the case of L = phosphine this would lead to unreactive nickel cyanide compounds, phosphite ligands, as better electron acceptors, stabilize nickel cyano hydride species. The hydride can then undergo alkene insertions.

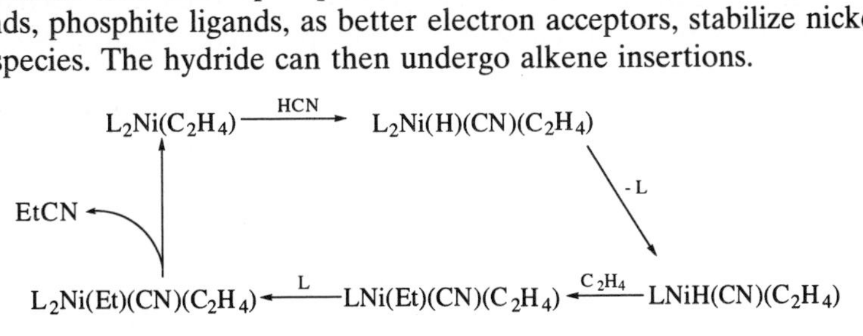

[43]M. R. Smith III *et al.*, *J. Am. Chem. Soc.* **1997**, *119*, 2743.
[44]M. Tanaka *et al.*, *Chem. Commun.* **1997**, 1229.
[45]T. B. Marder *et al.*, *Organometallics* **1996**, *15*, 5127; N. Miyaura *et al.*, *Organometallics* **1996**, *15*, 713.
[46]C. N. Iverson and M. R. Smith III, *Organometallics* **1997**, *16*, 2757; N. Miyaura *et al.*, *Chem. Commun.* **1997**, 689.
[47]T. B. Marder *et al.*, *Chem. Commun.* **1997**, 2051.

The reaction can be carried out asymmetrically, using nickel complexes of chiral phosphite ligands. Examples are the enantioselective hydrocyanation of norbornene using ligand (22-XVIII),[48] and of vinylnaphthalene derivatives with (22-XIX).[49] The latter is a precursor for the anti-inflammatory drug naproxen.

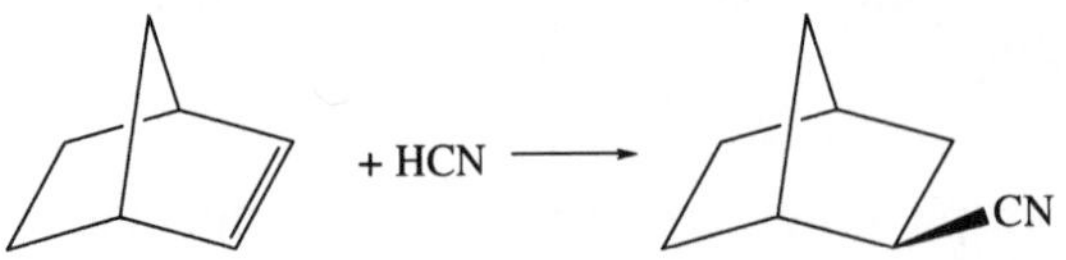

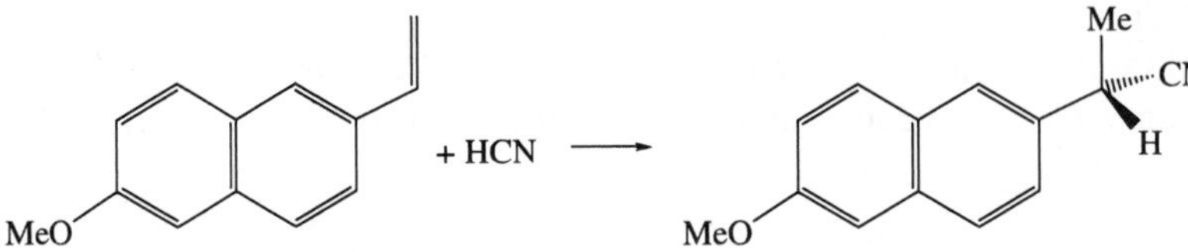

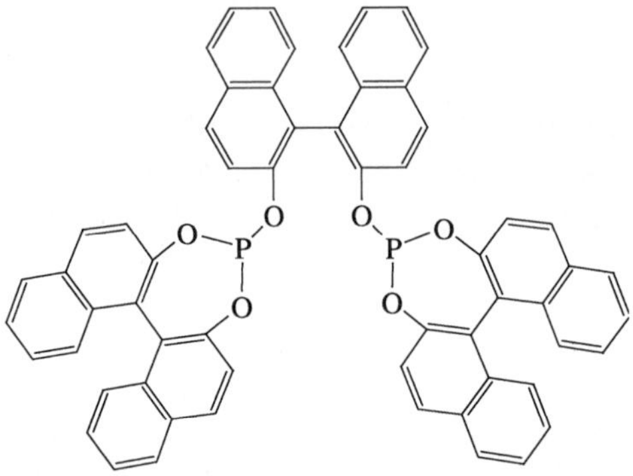

(22-XVIII) 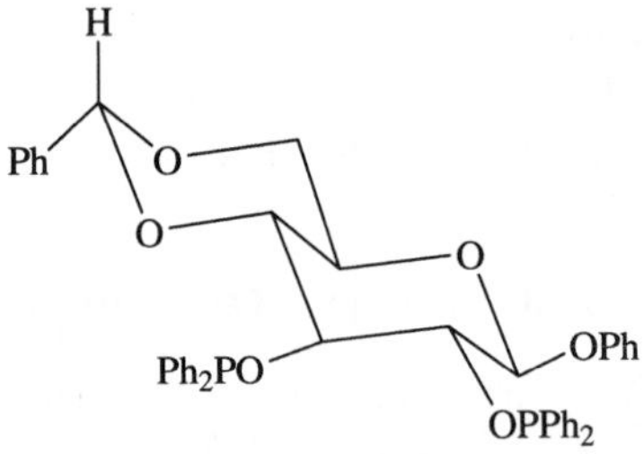   (22-XIX)

The most important use is the hydrocyanation of butadiene to adiponitrile, $NC-(CH_2)_4-CN$, a precursor to hexamethylenediamine for the synthesis of nylon. The process goes stepwise. The first addition of HCN involves nickel allyl intermediates and gives a mixture of linear and branched products in a ratio of ~70:30.

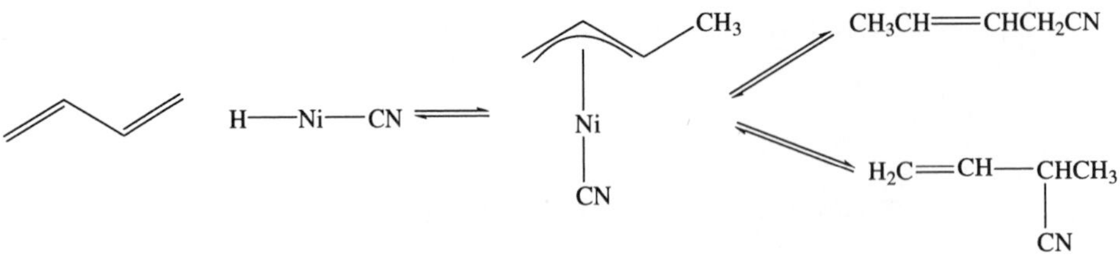

In the second stage the product is isomerized to 3- and 4-pentenenitriles using a nickel catalyst promoted by Lewis acids, followed by the addition of a second HCN

[48]M. J. Baker and P. G. Pringle, *J. Chem. Soc., Chem. Commun.* **1991,** 1292.
[49]T. V. RajanBabu and A. L. Casalnuovo, *Pure Appl. Chem.* **1994,** *66,* 1535.

to the C=C bond. The selectivities to Markovnikov and anti-Markovnikov addition products can be adjusted by suitable choice of ligand and the Lewis acid promoter, with BPh_3 giving the highest levels of linear product. The influence of the Lewis acid is probably steric since BPh_3 can bind to CN,

$$L_n HNi \longrightarrow CN + BPh_3 \rightleftharpoons L_n HNi \longrightarrow CN \longrightarrow BPh_3$$

Alkynes can also be hydrocyanated to α,β-unsaturated nitriles by heating alkyne, acetone cyanohydrin (as HCN source), and $Ni[P(OPh)_3]_4$ in refluxing toluene.

22-5 Reactions of Carbon Monoxide and Hydrogen

Mixtures of carbon monoxide and hydrogen, also known as "synthesis gas," are important for many industrial processes. Such gas mixtures can be generated independent of the carbon source, from coal, crude oil, or natural gas, and are a potentially important source for more complex chemicals, as well as for hydrogen. The system $CO/H_2/CO_2/H_2O$ will be discussed first.

The Water Gas Shift (WGS) Reaction

Mixtures of CO, H_2, and CO_2 may be obtained by (a) controlled oxidation or catalytic "steam reforming" of CH_4 or light petroleum fractions (naphtha, C_6 - C_8), and (b) gasification of coal with oxygen and steam at ~1500°C. Carbon dioxide can be removed from the product stream by scrubbing with aqueous base.

Carbon monoxide may also be recovered from the gaseous effluents of blast furnaces or air oxidation of coke by complexing with $CuAlCl_4$ in aromatic solvents.

A critical prerequisite for the use of synthesis gas or CO in most reactions catalyzed by transition metals is that sulfur compounds, such as H_2S, thiols, or COS, derived from the sulfur content of crude oil or coal must first be removed (hydrodesulfurization, see Section 21-4).

The water gas shift is an equilibrium reaction catalyzed by metals and an important source of industrial hydrogen:

$$CO(g) + H_2O(g) \rightleftharpoons CO_2(g) + H_2 \qquad \Delta H^0_{298°C} = -41.2 \text{ kJ mol}^{-1}$$
$$\Delta G^0_{298°C} = -28.5 \text{ kJ mol}^{-1}$$

If liquid water is present, however, the reaction is mildly endothermic,

$$CO(g) + H_2O(l) \rightleftharpoons CO_2 + H_2(g) \qquad \Delta H^0_{298°C} = +2.8 \text{ kJ mol}^{-1}$$
$$\Delta G^0_{298°C} = -19.9 \text{ kJ mol}^{-1}$$

but nevertheless highly favored because of the large positive entropy change.

The WGS reaction is usually catalyzed heterogeneously over iron-chromium and zinc-copper oxides; there are homogeneous model reactions and catalytic systems that operate in aqueous solution. The WGS mechanism is best illustrated by the reactions of $Fe(CO)_5$:

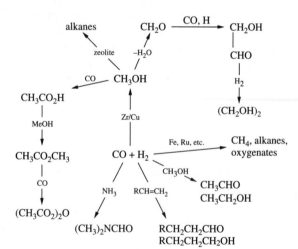

A key step is the formation of a metallacarboxylic acid M—COOH by attack of water or OH^- on coordinated CO. This species then undergoes β-H elimination to give CO_2 and a metal hydride. This reaction can act as the source of H_2 in a number of homogeneously catalyzed reactions, or alternatively as the source of CO, for example,[50]

$$CO_2 + H_2 + Me_2NH \longrightarrow Me_2NC(O)H + H_2O$$

An example of a homogeneous WGS catalyst is $Rh_4(CO)_{12}$ in aqueous pyridine.[51] Analogous reactions with CO and alcohols instead of water are also known.[52]

Some important reactions of CO, H_2, and methanol are shown in Fig. 22-5.

The Reduction of Carbon Monoxide by Hydrogen; the Fischer-Tropsch Reaction

The simplest reduction product of CO is methane, a reaction originally discovered by Sabatier using nickel catalysts and still used as part of industrial processes (methanation).[53] The names most readily connected with CO/H_2 chemistry are

Figure 22-5 Reaction products derived from synthesis gas and methanol.

[50]A. Baiker *et al.*, *Chem. Commun.* **1996**, 1497.
[51]G. Fachinetti *et al.*, *Inorg. Chem.* **1994**, *33*, 1719.
[52]R. G. Miller *et al.*, *Organometallics* **1993**, *12*, 1161.
[53]B. B. Pearce, M. V. Twigg, and C. Woodward, in: *Catalyst Handbook*, 2nd. ed., M. V. Twigg (ed.), Wolfe Publ. Ltd., London, 1989, p. 340.

Table 22-1 Thermodynamic Data for $CO-H_2$ Reactions (at 500 K)[a]

Reaction	ΔG kJ mol^{-1}	Log K_p
$CO + 3H_2 = CH_4 + H_2O$	-96.22	10.065
$CO + H_2 = HCHO$	50.62	-5.293
$CO + 2H_2 = CH_3OH$	21.23	-2.222
$2CO + 3H_2 = HOCH_2-CH_2OH$	65.92	-6.891

[a]From D. R. Stull, E. F. Westrum, Jr., and G. C. Sinke, *The Thermodynamics of Organic Compounds,* John Wiley, New York, 1969. Methanation of CO_2 is also feasible, $CO_2 + 4H_2 = CH_4 + 2H_2O(l)$, $\Delta H_{298} = -252.9$ kJ mol^{-1} as is the reduction to formates.

F. Fischer and H. Tropsch, who first described the conversion of synthesis gas into hydrocarbons and oxygen-containing compounds ("oxygenates") over heterogeneous transition metal catalysts such as iron/zinc oxide. This reaction was developed into a process for the conversion of coal into gasoline. At present such a process is economically feasible only where coal is plentiful and cheap while access to oil products is limited. Currently only South Africa operates plants using the Fischer-Tropsch process.

Some thermodynamic data for CO/H_2 reactions are given in Table 22-1. Only the reduction to methane is energetically favorable; nevertheless, the reduction of CO to give alcohols is feasible using appropriate catalysts.

Metals such as Fe, Co, Ni, or Ru on alumina or other oxide supports convert CO and H_2 to hydrocarbons. Using different catalysts and reaction conditions either CH_4, liquid hydrocarbons, high molecular weight paraffins, methanol, higher alcohols, olefins, and aromatics can be obtained, though rarely (with the exception of CH_4 and methanol) with high selectivity. Hydrocarbons typically exhibit a Schulz-Flory type molecular weight distribution.

There has been extensive speculation about the reaction mechanism, all the more so since there are no satisfactory homogeneous models or productive catalysts. The formation of hydrocarbons requires a condensation reaction of surface methylene units with surface-H and/or CH_3, to give "polymethylene" products:

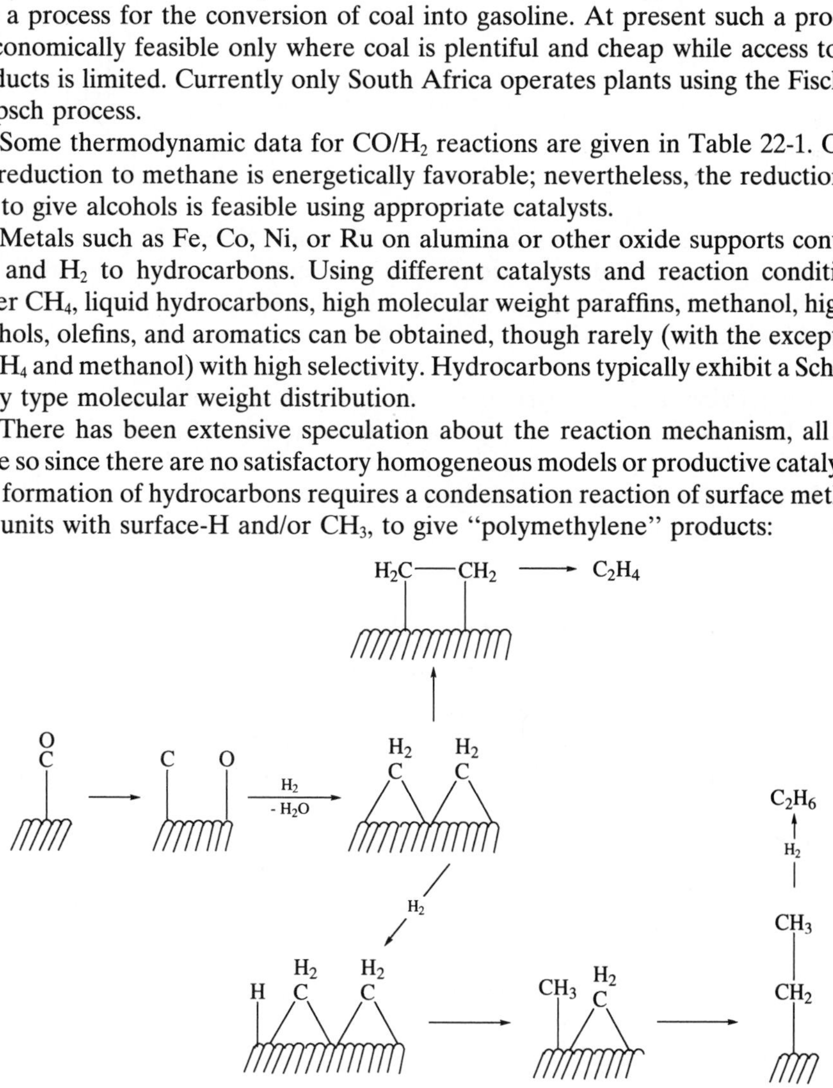

The possibility of such a condensation sequence has also been demonstrated for rhodium methylene complexes which on thermolysis give selectively methane and propene:[54]

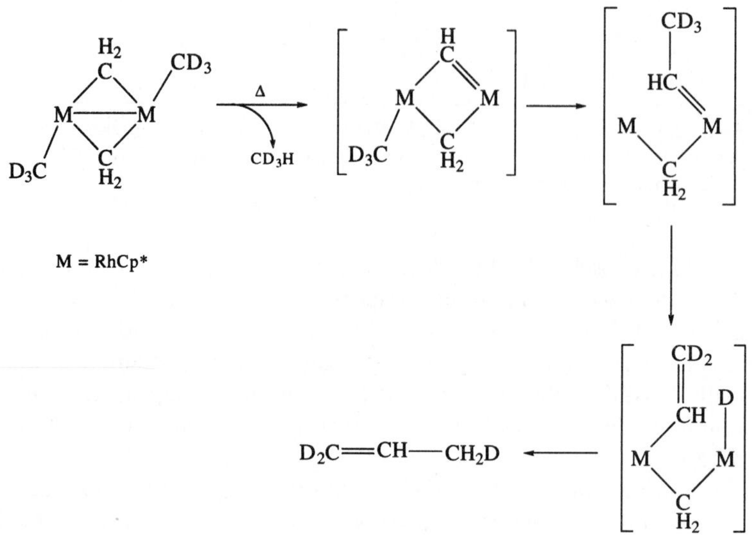

$$M = RhCp^*$$

A key step in the production of hydrocarbons from CO must be the dissociation of CO into surface carbido and oxo species, a reaction not readily reproduced in mononuclear CO complexes but found in clusters. Reactions of this type have focused interest on the analogy between surfaces and metal clusters.[55]

By far the most important synthesis gas reaction is its conversion into methanol, using copper/zinc oxide catalysts under relatively mild conditions (50 bar, 100-250°C). Methanol is further carbonylated to acetic acid (see Section 22-7), so that CH_3CO_2H, methyl acetate, and acetic anhydride can all be made from simple CO and H_2 feedstocks. Possible pathways to oxygenates in cobalt catalyzed reactions are shown in Fig. 22-6.

22-6   Hydroformylation of Unsaturated Compounds[56]

The hydroformylation of alkenes, discovered by Otto Roelen of Ruhrchemie in 1938, is one of the most important homogeneously catalyzed reactions. The name "hydroformylation" refers to the effective addition of a hydrogen atom and a formyl (CHO) group to a C=C double bond of an olefin; the reaction is also known under its older name of the *oxo reaction*.

The original reaction used cobalt as the catalyst, and the commercial process requires high temperatures (150 - 180°C) and pressures (>200 bar). It produces a mixture, roughly 3:1 of both linear and branched-chain aldehydes, as well as alcohols. These products are formally the consequence of anti-Markovnikov and Markovni-

[54]P. M. Maitlis *et al.*, *Chem. Eur. J.* **1995**, *1*, 549; *Chem. Commun.* **1996**, 1.
[55]B. C. Gates, *Angew. Chem. Int. Ed. Engl.* **1993**, *32*, 228.
[56]C. D. Frohning and C. W. Kohlpaintner in: B. Cornils and W. A. Herrmann (Eds.), *Applied Homogeneous Catalysis with Organometallic Compounds*, VCH, Weinheim, 1996, p. 29.

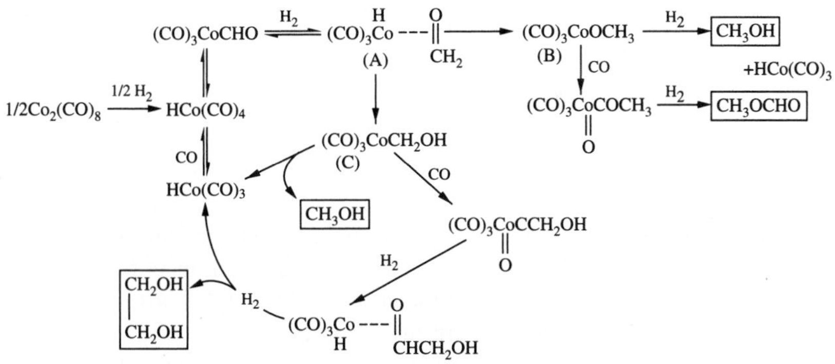

Figure 22-6 The C_1 and C_2 syntheses using $Co_2(CO)_8$ as precursor (adapted from H. M. Feder and J. M. Rathke, *Ann. N.Y. Acad. Sci.,* **1980,** 333, 45). Note that H transfer to coordinated formaldehyde A, can give a methoxide B, or a hydroxymethyl C, either of which can then be reduced by H_2 or can insert CO, the product of which is then reduced. The $HCo(CO)_3$ is recycled.

kov addition, respectively:

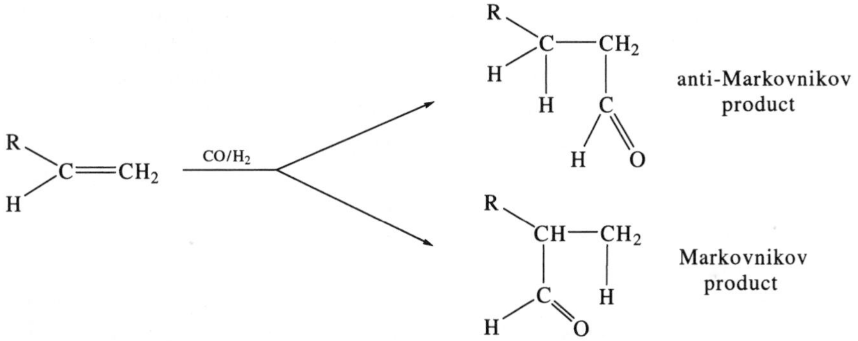

Under pressure of CO and H_2, the cobalt catalyst precursor is transformed into cobalt carbonyl hydride, $HCo(CO)_4$. The main steps of the reaction mechanism, first elucidated by D. S. Breslow and R. F. Heck, involve (a) β-hydrogen transfer to the coordinated olefin, (b) the insertion of CO to form an acyl intermediate, and (c) the hydrogenolysis of the acyl, with formation of the aldehyde product:

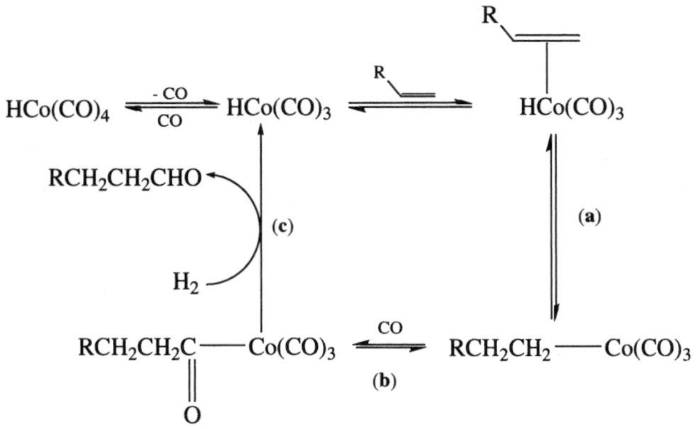

The last step could also be intermolecular,

$$\text{R}-\overset{\overset{\displaystyle O}{\|}}{\text{C}}-\text{Co(CO)}_3 + \text{HCo(CO)}_4 \xrightarrow{\text{CO}} \text{RCH} + \text{Co}_2\text{(CO)}_8$$

which is established in stoichiometric reactions and where kinetic data for the overall reaction indicate that the complexation of the alkene is the rate-limiting step. However, high pressure spectroscopic studies of catalytic systems suggest that external $HCo(CO)_4$ does not participate in the hydrogen cleavage step.

The initiating species $HCo(CO)_3$ is generated from trigonal-bipyramidal $HCo(CO)_4$ by loss of one equatorial CO ligand and hence has C_s symmetry.[57] Such a species is ideally suited to alkene binding. The importance of adequate concentrations of coordinatively unsaturated species for rapid turnover is illustrated by the fact that at very high pressures even "inert" gases such as nitrogen or argon are able to compete with alkene and H_2 for coordination sites and hence strongly retard the reactions.[58]

The cobalt process is difficult to operate, partly because of the high pressures involved, and partly because of the need to recycle volatile cobalt carbonyls. Another disadvantage is the loss of ~15% of the alkene due to hydrogenation; condensation and ketone by-products are also formed. A Shell modification adding trialkylphosphine to the cobalt catalyst increased selectivity for linear aldehydes and allowed lower reaction pressures, but gave lower activity and increased hydrogenation.

A number of metals catalyze the hydroformylation reaction, of which rhodium is by far the most active, Rh >> Co > Ir, Ru > Os > Pt. Platinum and ruthenium are mainly of academic interest, although $L_2PtCl(SnCl_3)$ complexes with chiral ligands find use in asymmetric alkene hydroformylations.[59] In most cases, and certainly in industrial processes, cobalt has now been replaced by rhodium.

The introduction of rhodium has allowed the development of processes which operate under much milder conditions and lower pressures, are highly selective, and avoid loss of alkene by hydrogenation. Although the catalyst is active at moderate temperature, plants are usually operated at 120°C to give a high n/iso (linear/branched) ratio. The key to selectivity is the use of triphenylphosphine in large excess which leads to >95% straight chain anti-Markovnikov product. The process is used for the hydroformylation of propene to n-butyraldehyde, allyl alcohol to butanediol, and maleic anhydride to 1,4-butanediol, tetrahydrofuran, and γ-butyrolactone.

The rhodium catalyst is based on $RhH(CO)(PPh_3)_3$, and a fairly detailed picture of the mechanism has now been obtained (Fig. 22-7). The initial step is the generation of a 16-electron square intermediate, **A**, from the 18-electron precursor. This is followed by alkene coordination and hydrogen transfer to give the alkyl species **B**. The latter undergoes CO addition (**C**) and insertion to give the acyl derivative **D**, which subsequently undergoes oxidative addition of molecular hydrogen to give the hydridoacyl complex **E**. This last step, which involves a change in the oxidation state of rhodium, is probably rate-determining. The final steps are another H transfer

[57]T. Ziegler et al., Organometallics **1993**, 12, 3586.
[58]F. Piacenti et al., Organometallics **1997**, 16, 4235.
[59]B. E. Hanson et al., Organometallics **1993**, 12, 848.

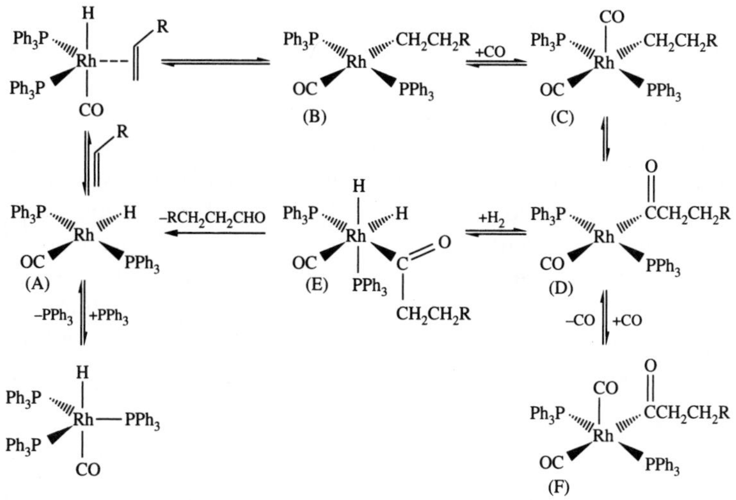

Figure 22-7 Simplified catalytic cycle for hydroformylation using rhodium complexes. Note that the configurations of complexes are not known with certainty and that five-coordinate species are fluxional. Rhodium can be added as $Rh(acac)(CO)(PPh_3)$, $HRh(CO)(PPh_3)_3$, or similar complexes. The solvent in C_2H_4 or $CH_3CH=CH_2$ hydroformylation is the aldehyde trimer which is in equilibrium with aldehyde.

to the carbon atom of the acyl group, i.e., the reductive elimination of aldehyde, followed by dissociation of the product and regeneration of **A**. An excess of CO over H_2 inhibits the hydroformylation reaction, probably through the formation of five-coordinate dicarbonyl acyl complexes of type **F** which cannot react with hydrogen.

The industrial process operates with high PPh_3 concentrations so that species of type **A** dominate, although the steric hindrance of the PPh_3 reduces the tendency for alkene binding. It has been argued that at lower $[PPh_3]$ another square species, $RhH(CO)_2(PPh_3)$, may be important where alkene binding is kinetically more favorable. The energy profile for such a hydroformylation sequence has been calculated and shows alkyl intermediates of type **C** to be particularly stable. The results also suggest that square intermediates experience strong stabilization through solvation.[60]

Very high phosphine concentrations increase the selectivity for the *n*-aldehyde product isomer, evidently by suppressing the formation of monophosphine species. The high selectivity arises since steric hindrance in the alkyl intermediate B is less for a primary alkyl than for a secondary one; the selectivity is therefore promoted by the phosphine cone angle. At low phosphine concentrations the straight and branched chain aldehydes are formed in a ~3:1 ratio, as in the unmodified cobalt system.

The intermediates shown in Fig. 22-7 are too unstable for isolation. However, detailed nmr studies have confirmed some of them while it has been possible, either by use of C_2F_4 which gives stable species such as $Rh(CF_2F_2H)(CO)(PPh_3)_2$, or by using the similar but more stable iridium compounds, to characterize analogues for most of the intermediates in the cycles.

[60]K. Morokuma *et al.*, *Organometallics* **1997,** *16*, 1065.

Rhodium catalyzed hydroformylations are strongly dependent on the nature of the ligands. Suitably bulky phosphites, for example (22-XX), combine high activity with high regioselectivity for linear aldehydes.[61] Others, such as (22-XXI), allow catalytic reactions to be carried out in organic media followed by extraction of the catalyst into aqueous acid by protonation of the amino substituents.[62] Complexes of the flexible phosphite (22-XXII) hydroformylate styrene with very high regioselectivities for the branched aldehyde.[63] By contrast, the binuclear complex *rac*-(22-XXIII) hydroformylates 1-hexene to give predominantly linear heptanal, e.g., *n/iso* = 96.5:3.5. There seems to be a cooperative effect between the two metal centers, with intramolecular H and CO transfer between the rhodium atoms. This cannot happen in the *meso* diastereomer, which in consequence gives much lower activity and regioselectivity.[64]

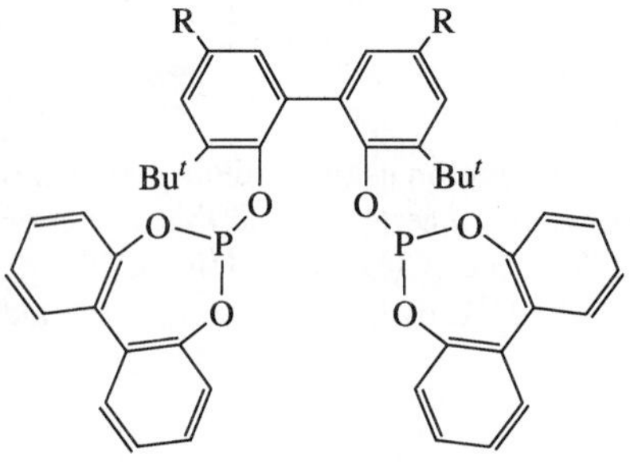

R = H, OMe

(22-XX)

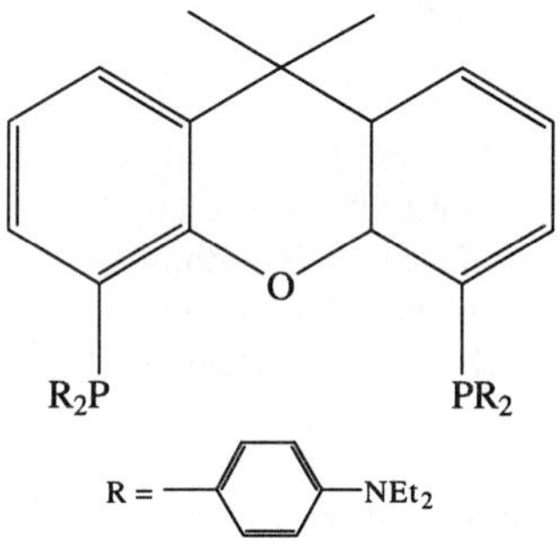

$R = $ —NEt$_2$

(22-XXI)

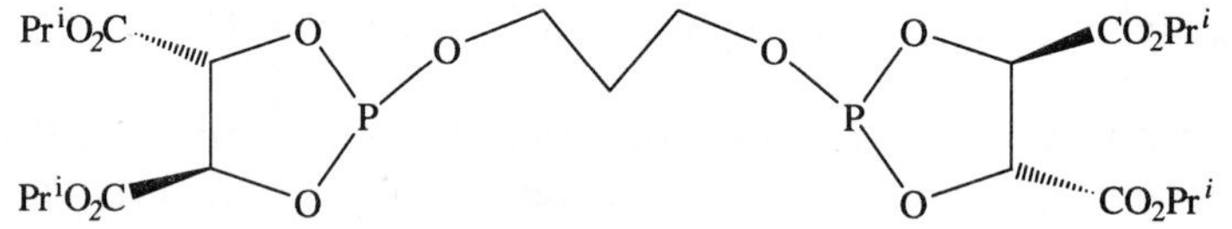

(22-XXII)

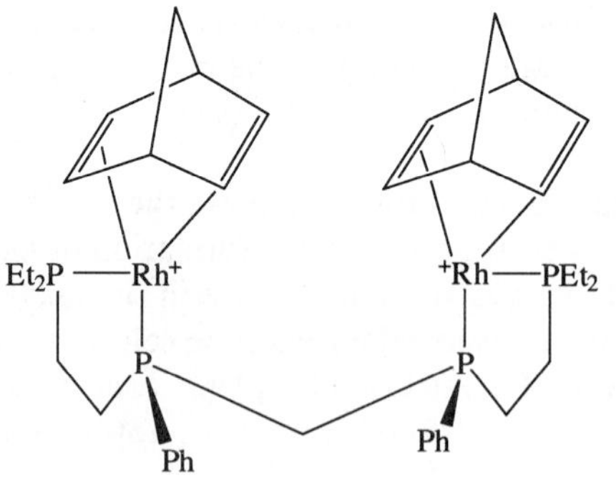

(22-XXIII)

[61]G. D. Cuny and S. L. Buchwald, *J. Am. Chem. Soc.* **1993,** *115,* 2066; P. W. N. M. van Leeuwen *et al., Organometallics* **1996,** *15,* 835.
[62]P. W. N. M. van Leeuwen *et al., Organometallics* **1997,** *16,* 3027.
[63]T. J. Kwok and D. J. Wink, *Organometallics* **1993,** *12,* 1954.
[64]G. G. Stanley *et al., Science* **1993,** *260,* 1784; G. Süss-Fink, *Angew. Chem. Int. Ed. Engl.* **1994,** *33,* 67.

Water soluble ligands, notably $P(C_6H_4SO_3)_3^{3-}$,[65] allow hydroformylations to be conducted in a two-phase system (see later). Such a process has been commercialized by Rhône-Poulenc and Ruhrchemie.

A wide variety of substances can be hydroformylated. Conjugated dienes may give a number of products including monohydrogenated monoaldehydes. The use of chelating phosphines allows selective hydroformylation of butadiene to pentanal. The mechanism differs from that for monoalkenes since addition of $M-H$ to conjugated dienes leads to allylic species, which may be present as η^1- or η^3-allyls:

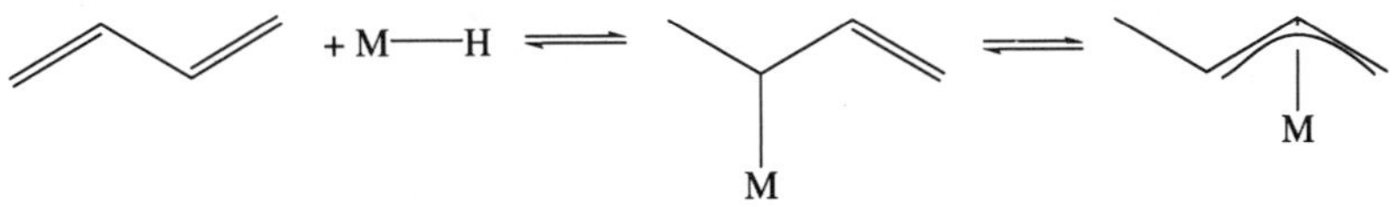

Hydroformylation can also be achieved using ruthenium complexes such as $Ru(CO)_3(PPh_3)_2$, by platinum-tin catalysts,[59] and by $PtH(Ph_2POH)(PPh_3)$ made from $Pt(COD)_2$, PPh_3, and Ph_2POH. The latter system yields ketones when under high ethylene pressure. Ketone formation can also be observed in other systems and occurs by the reaction sequence:

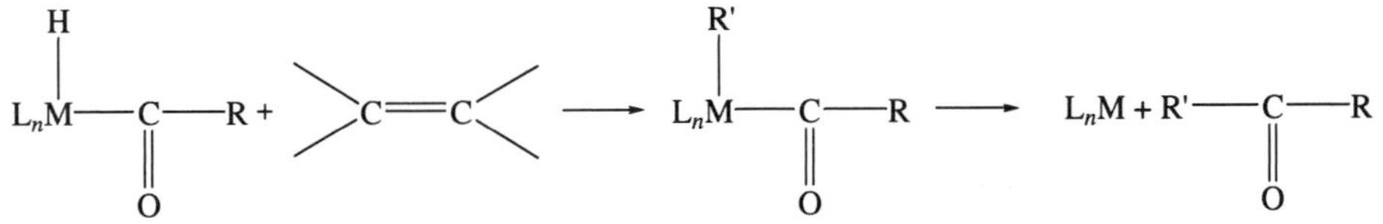

This will be favored by low CO and hydrogen concentration.

The hydroformylation of *formaldehyde*, a route to ethylene glycol *via* glycolaldehyde, may involve the steps

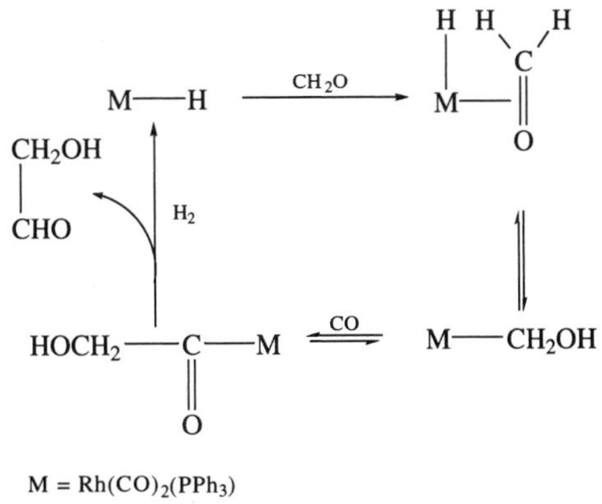

$M = Rh(CO)_2(PPh_3)$

The reaction is strongly affected by base which may reversibly deprotonate $HRh(CO)_2L_2$ to an anionic species which then reacts with formaldehyde and $[H\ base]^+$ to give the hydroxymethyl intermediate. Using iridium analogues as models, several postulated reaction intermediates could be isolated.[66]

[65]B. E. Hanson *et al., Inorg. Chim. Acta* **1995,** *229,* 329.
[66]A. S. C. Chan and H. S. Shieh, *Inorg. Chim. Acta* **1994,** *218,* 89.

The *silylhydroformylation* of unsaturated substrates including acetylenes[67] and aldehydes[68] is catalyzed by various rhodium complexes and is a synthetically versatile reaction, for example:

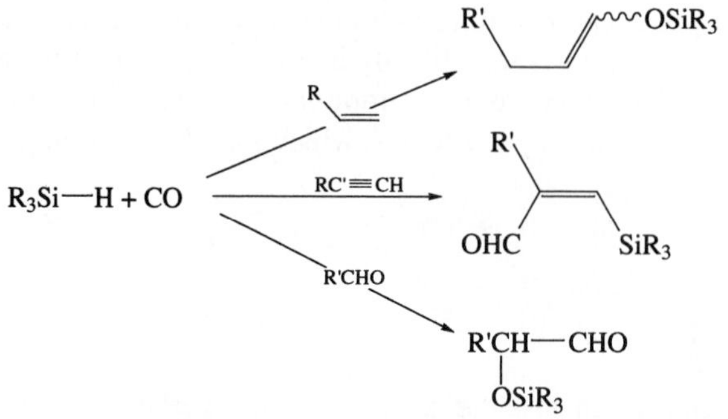

The *asymmetric hydroformylation* of alkenes is catalyzed by a number of Rh and Pt complexes with chiral chelating ligands.[69] A number of binaphthyl-based chiral phosphites[70] and mixed phosphine-phosphite ligand like (22-XXIV) ["*(R,S)*-BINAPHOS"] have been found particularly effective, e.g., in the asymmetric hydroformylation of styrene in >90% optical yield.[71]

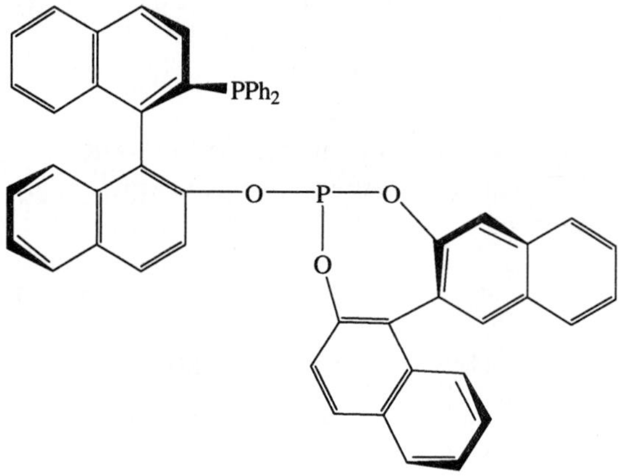

(22-XXIV)

22-7 Carbonylation Reactions

Reactions using CO in the presence of water (a synthesis gas equivalent *via* water-gas shift), alcohols, amines, and so on, are generally called carbonylations. Various syntheses are summarized in Fig. 22-8.

[67]J. Q. Zhou and H. Alper, *Organometallics* **1994,** *13,* 1586.
[68]M. E. Wright and B. B. Cochran, *Organometallics* **1996,** *15,* 317.
[69]F. Agbossou *et al., Chem. Rev.* **1995,** *95,* 2485.
[70]P. C. J. Kramer *et al., Organometallics* **1997,** *16,* 2929.
[71]K. Nozaki *et al., J. Am. Chem. Soc.* **1997,** *119,* 4413; *Organometallics* **1997,** *16,* 2981.

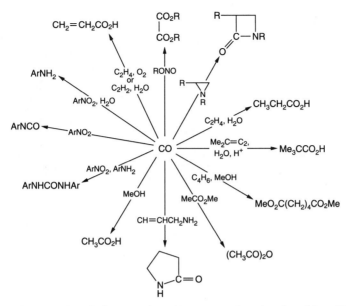

Figure 22-8 Some carbonylation reactions. Reppe reactions involve CO + H₂O or CO + ROH. Koch reactions involve CO + H₂O in strongly acidic solution. Oxidative carbonylations may involve O₂, NO, or ROOR.

Alcohol Carbonylation: The Acetic Acid Process[72]

The carbonylation of methanol to acetic acid and methyl acetate, and the carbonylation of the latter to acetic anhydride, was found by W. Reppe at BASF in the 1940s, using iodide-promoted cobalt salts as catalyst precursors. This process required very high pressure (600 bar) as well as high temperatures (230°C) and gave *ca.* 90% selectivity for acetic acid.

As is the case of hydroformylation, the use of rhodium allows much milder conditions to be used. Such a process was started by Monsanto in 1966; it operates at 30–60 bar and 150–200°C and is now the world's largest process for acetic acid production (>5 million tons per year). In view of the corrosive nature of the reagents, Hastalloy or zirconium reactors have to be used.

As was the case with cobalt, the key feature is the presence of iodide in the cycle. Some water is also present in the methanol feedstream, but is also produced by the reaction

$$CH_3COOH + CH_3OH \rightleftharpoons CH_3COOCH_3 + H_2O$$

Water levels are important for the efficiency of the cycle as well as for product selectivity; with very dry methanol carbonylation of MeCOOMe to acetic anhydride occurs.

[72]P. M. Maitlis *et al.*, *J. Chem. Soc., Dalton Trans.* **1996**, 2187.

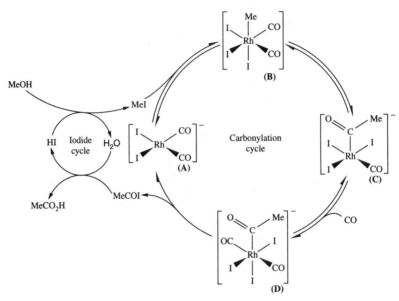

Figure 22-9 Simplified cycle for the rhodium catalyzed carbonylation of methanol. The oxidative addition of iodomethane to **A** is the rate-limiting step.

A simplified mechanistic scheme is shown in Fig. 22-9. Note that the process is the result of two cycles coupled to each other: (a) the iodide cycle, which converts methanol to iodomethane and acetyliodide to acetic acid, and (b) the main carbonylation cycle. The species $[Rh(CO)_2I_2]^-$ is actually connected to a third cycle, the water gas shift, which removes water from the system to give CO_2 and H_2; it also ties up some of the catalyst as $[Rh(CO)_2I_4]^-$.

The process is zero order in [CO] and [MeOH], an indication that the rate-limiting step is the oxidative addition of MeI to $[Rh(CO)_2I_2]^-$ (**A**) to give the methyl complex **B**. **B** is an unstable transient but could be detected by *in situ* infrared spectroscopy.[72] There is a fast methyl migration to CO to give the acyl **C**.

The monocarbonyl **C** decomposes by MeI rather than MeCOI elimination but the latter process is rapid in the presence of excess CO. Compound **C** was structurally characterized; it is an iodide-bridged dimer in the solid state. A free-energy profile for the sequence **A** → **B** → **C** has been constructed from kinetic data and is shown in Fig. 22-10.

Iridium has long been known as a carbonylation catalyst but might be expected to be less active in view of the greater stability of iridium alkyl species, and indeed it was shown that in the iridium analogue of the carbonylation cycle the rate-limiting step is no longer the oxidative addition of MeI but the methyl migration to CO to give the acyl. The carbonylation rates are, however, highly sensitive to the reaction conditions, particularly the concentration of methanol which has a strong enhancing effect, and on the presence of promoters such as Lewis acids or metal carbonyls (e.g., SnI_2 or $Ru(CO)_4I_2$) which can act as I^- acceptors, whereas the addition of iodide suppresses the reaction. The data are in agreement with a cycle involving

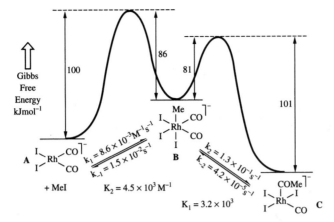

Figure 22-10 Free energy profile of part of the Rh catalyzed methanol carbonylation cycle derived from kinetic data for the interconversion of **A**, **B**, and **C** (cf. Fig. 22-9) at 35°C in CH$_2$Cl$_2$–MeI. All three species are in equilibrium with each other (adapted from P. M. Maitlis *et al.*, *J. Chem. Soc., Dalton Trans.* **1996**, 2187).

solvated and neutral carbonyl species:

This iridium reaction is the basis of BP's new "Cativa" process launched in 1996,[73] which offers significant operational efficiency gains and is expected to eventually replace the current rhodium catalysts.

Higher alcohols can also be carbonylated, although the rate decreases sharply, MeOH >> EtOH > PrOH. Alkene intermediates are involved, and propanol gives

[73]P. Layman, *C&EN,* 1 July **1996,** p. 7.

both *n*- and *iso*-butanoic acid:[72]

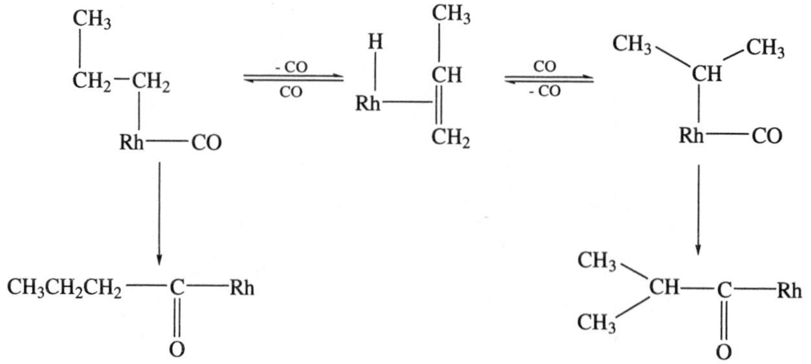

None of these carbonylation cycles requires donor ligands such as phosphines. Electron-rich trialkyl phosphines might however be expected to increase the rate of oxidative addition. At higher water concentrations $RhCl(CO)(PEt_3)_2$ is indeed significantly more active than $[RhCl(CO)_2]_2$,[74] although in prolonged reactions phosphonium salts will be formed. The double carbonylation of CH_2I_2 is also possible:[75]

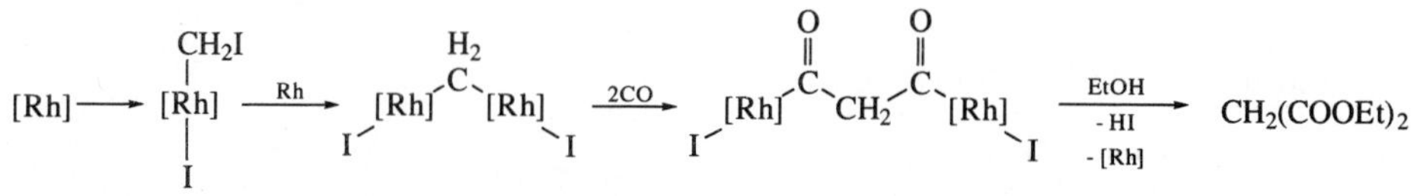

Heteropolyacids ion-changed with Group 8 metals, $M[W_{12}PO_{40}]$ on silica, catalyze the gas-phase carbonylation of methanol or dimethyl ether to methyl acetate.[76]

The homologation of acetic acid to higher carboxylic acids, especially to propionic acid, is also possible, e.g., with Ru/iodide catalysts. The reaction is thought to involve reduction and carbonylation steps, such as[77]

$$MeCOOH + HI \longrightarrow MeCOI + H_2O$$

$$MeCOI + H_2 \longrightarrow MeCHO + HI$$

$$MeCHO + H_2 \longrightarrow EtOH$$

$$MeCOOH + EtOH \longrightarrow MeCOOEt + H_2O$$

$$MeCOOEt + HI \longrightarrow MeCOOH + EtI$$

$$EtI + CO \longrightarrow EtCOI$$

$$EtCOI + H_2O \longrightarrow EtCOOH + HI$$

_____

$$MeCOOH + CO + 2H_2 \longrightarrow EtCOOH + H_2O$$

[74]D. J. Cole-Hamilton *et al., Chem. Commun.* **1997,** 1835.
[75]D. J. Cole-Hamilton *et al., J. Chem. Soc., Chem. Commun.* **1994,** 745.
[76]R. W. Wegman, *J. Chem. Soc., Chem. Commun.* **1994,** 947.
[77]G. Braca *et al.* in: *Oxygenates by Homologation or CO Hydrogenation with Metal Complexes,* G. Braca, Ed., Kluwer Academic, Dordrecht, **1994.**

Another approach to acetic acid is the dehydrogenation of methanol which is thermodynamically feasible:

$$2CH_3OH \longrightarrow CH_3COOH + 2H_2 \qquad \Delta G^0 = \text{-}51.7 \text{ kJ mol}^{-1}$$

This reaction is catalyzed by complexes such as $[Ru(SnCl_3)_5(PPh_3)]^{3-}$ and converts methanol to methyl acetate.[78]

Other Carbonylation Reactions

While the direct carbonylation of ethylene to acrolein $H_2C{=}CHCHO$ is mildly endothermic, it becomes exothermic on addition of triethylorthoformate:

$$C_2H_4 + CO + HC(OEt)_3 \longrightarrow CH_2{=\!=}CHC(OEt)_2 + HCOOEt$$

The reaction is catalyzed by $[Cp_2^*Zr(\mu\text{-S})_2Rh(CO)_2]^{-}$.[79]

The carbonylation of butadiene is catalyzed by $Co_2(CO)_8$ in the presence of pyridine at ~300 bar.

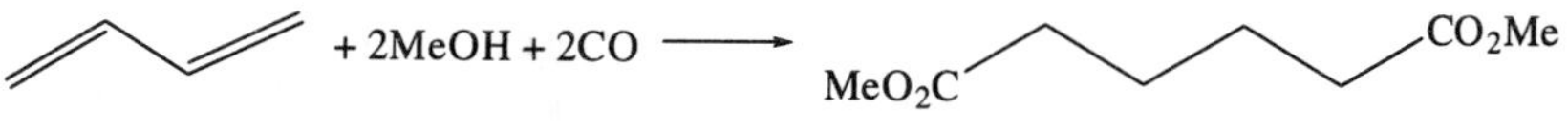

The ester can be hydrolyzed to adipic acid, a nylon-6 precursor. The reaction proceeds stepwise and involves $Co-COOMe$ species which react with butadiene under insertion:

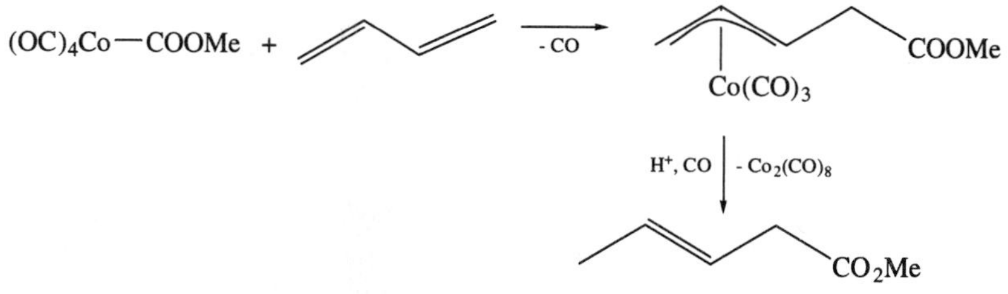

The *carbonylation of nitro arenes* can lead to amines, isocyanates, carbamates, azo compounds, or ureas.[80] Iron, ruthenium, rhodium, or palladium complexes have been used as catalysts. The carbonylation to give isocyanates, for example, involves the steps

[78]S. Shinoda *et al., Catalysis Surveys from Japan* **1997,** *1,* 25.
[79]P. Kalck *et al., Organometallics* **1993,** *12,* 1021.
[80]A. M. Tafesh and J. Weiguny, *Chem. Rev.* **1996,** *96,* 2035.

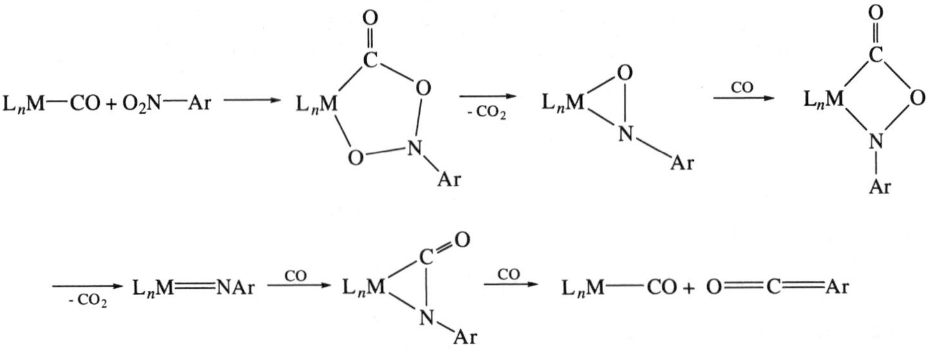

$$L_nM-CO + O_2N-Ar \longrightarrow L_nM \overset{O}{\underset{O-N}{\diagdown}} \xrightarrow[-CO_2]{} L_nM \overset{O}{\underset{N}{\diagdown}} \xrightarrow{CO} L_nM \overset{O}{\underset{N}{\diagdown}}$$

$$\xrightarrow[-CO_2]{} L_nM{=}NAr \xrightarrow{CO} L_nM \overset{C=O}{\underset{N}{\diagdown}} \xrightarrow{CO} L_nM-CO + O{=}C{=}Ar$$

The M=NAr intermediate may also be reduced to give anilines. Carbamates are produced by the reaction

$$ArNO_2 + 3CO + CH_3OH \longrightarrow ArNHC(O)OMe + 2CO_2$$

and complexes with $-C(O)OMe$ and $-C(O)NHAr$ ligands are involved as intermediates.[81]

In the presence of trapping agents such as aldehydes or acetylenes, the carbonylation of nitroarenes can be used to synthesize heterocycles, for example,

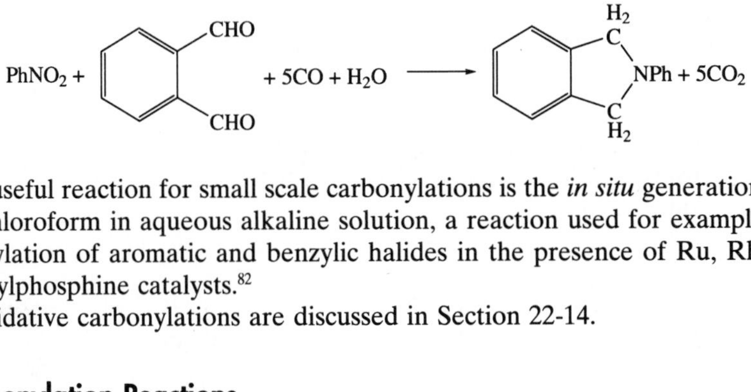

$$PhNO_2 + \overset{CHO}{\underset{CHO}{\bigcirc}} + 5CO + H_2O \longrightarrow \overset{H_2C}{\underset{H_2C}{\bigcirc}}NPh + 5CO_2$$

A useful reaction for small scale carbonylations is the *in situ* generation of CO from chloroform in aqueous alkaline solution, a reaction used for example in the carbonylation of aromatic and benzylic halides in the presence of Ru, Rh, or Pd triphenylphosphine catalysts.[82]

Oxidative carbonylations are discussed in Section 22-14.

Decarbonylation Reactions

The insertion of CO into M—C bonds to give acyls is reversible (Section 21-5). However, CO can be irreversibly removed from organic molecules; these reactions involve acyl intermediates. Thus aldehydes, acyl, and aroyl halides can be decarbonylated either stoichiometrically or catalytically by complexes such as $RhCl(PPh_3)_3$ or $RhCl(CO)(PPh_3)_2$. An example of a stoichiometric decarbonylation is

$$RhCl(PPh_3)_3 + RCOCl \xrightarrow{-PPh_3} RhCl_2(COR)(PPh_3)_2$$

$$RCl + RhCl(CO)(PPh_3)_2 \longleftarrow RhRCl_2(CO)(PPh_3)_2$$

[81]J. D. Gargulak and W. L. Gladfelter, *J. Am. Chem. Soc.* **1994**, *116*, 3792.
[82]V. V. Grushin and H. Alper, *Organometallics* **1993**, *12*, 3846.

With rhodium PPh$_3$ complexes the system becomes catalytic only >200°C, but with [Rh(dppp)$_2$]BF$_4$ efficient catalytic decarbonylation can be achieved.

It is convenient to note here that a common reaction of alcohols with transition metal halides in the presence of phosphine is the decarbonylation and sometimes dehydrogenation to give metal carbonyls. These reactions proceed *via* alkoxide intermediates which then undergo β-H transfer:

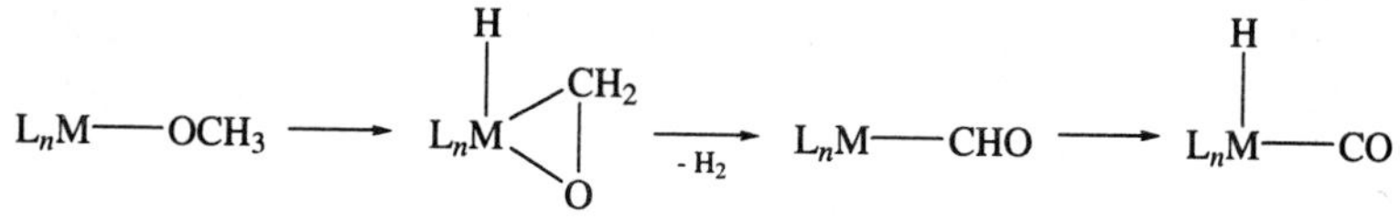

$$L_nM\!-\!OCH_3 \longrightarrow L_nM\!\overset{H}{\underset{O}{\diagdown CH_2}} \xrightarrow[-H_2]{} L_nM\!-\!CHO \longrightarrow L_n\overset{H}{M}\!-\!CO$$

22-8 C—C Cross Coupling and Related Reactions

The arylation of activated alkenes with aryl halides in the presence of base was discovered by R. F. Heck in 1971 and is now one of the standard methods for C—C bond formation. The catalysts are mostly palladium or nickel phosphine complexes, which react *via* a succession of oxidative addition and insertion reactions, as shown in the following simplified cycle:

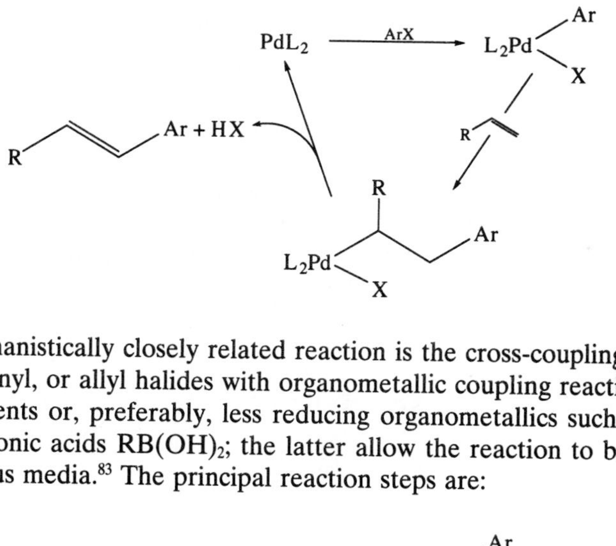

Another, mechanistically closely related reaction is the cross-coupling reaction of aryl, benzyl, vinyl, or allyl halides with organometallic coupling reactions, such as Grignard reagents or, preferably, less reducing organometallics such as tin compounds or boronic acids RB(OH)$_2$; the latter allow the reaction to be conducted even in aqueous media.[83] The principal reaction steps are:

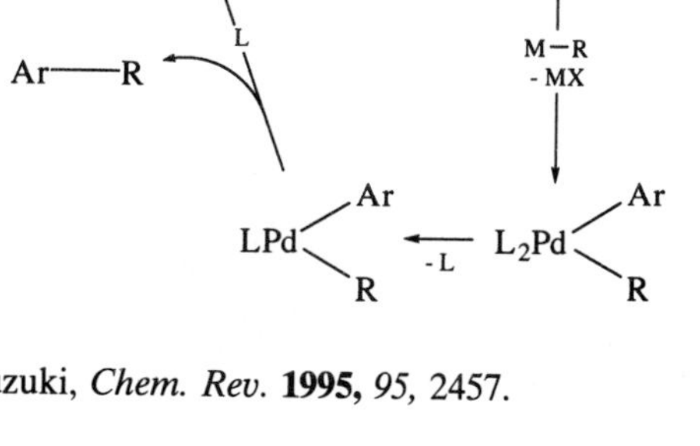

[83]N. Miyaura and A. Suzuki, *Chem. Rev.* **1995**, 95, 2457.

Highly active catalysts are obtained using P(o-tol)$_3$ which is sterically more hindered than PPh$_3$ and leads to two-coordinate species PdL$_2$. A thermally stable but highly active species is the palladacyclic compound (22-XXV), which catalyzes for example the arylation of *n*-butylacrylate with 4-bromobenzaldehyde with very high turnovers.[84] It was shown that under catalytic conditions the Pd^{II} metallacycle is converted into Pd^{0}L$_2$, e.g., in the coupling of aryl bromides with anilines to give *tert* arylamines.[85] Heterocyclic carbenes, until recently a hardly used ligand type in homogeneous catalysis, form complexes of type (22-XXVI) which are also highly active in the arylation of H$_2$C=CHCOOBu with chloro- and bromoarenes.[86]

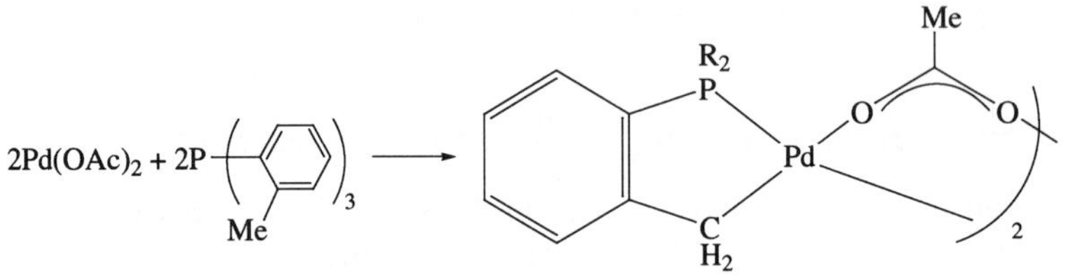

(22-XXV)

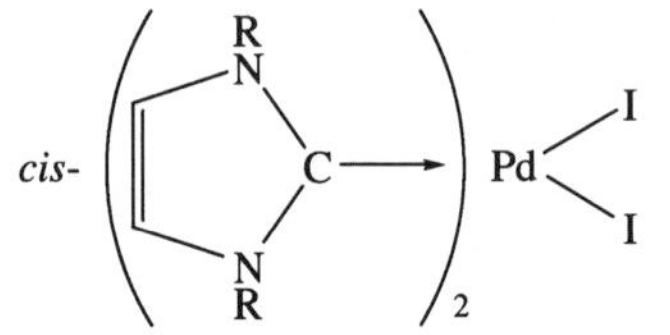

(22-XXVI)

Palladium deposited onto the mesoporous material MCM-41 by vacuum sublimation of CpPd(allyl) gives a heterogeneous catalyst for Heck arylations.[87]

There are very numerous synthetic applications of the Heck reaction, notably in the coupling of naphthyl bromide with ethylene to give a precursor to naproxen:[88]

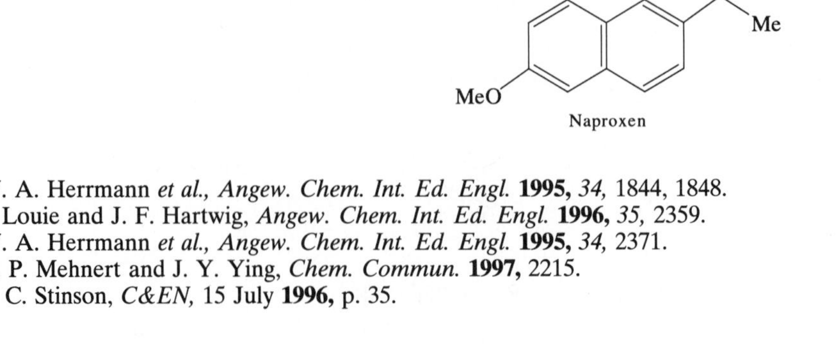

Naproxen

[84]W. A. Herrmann *et al., Angew. Chem. Int. Ed. Engl.* **1995,** *34,* 1844, 1848.
[85]J. Louie and J. F. Hartwig, *Angew. Chem. Int. Ed. Engl.* **1996,** *35,* 2359.
[86]W. A. Herrmann *et al., Angew. Chem. Int. Ed. Engl.* **1995,** *34,* 2371.
[87]C. P. Mehnert and J. Y. Ying, *Chem. Commun.* **1997,** 2215.
[88]S. C. Stinson, *C&EN,* 15 July **1996,** p. 35.

Other applications include the coupling of 1,4-diiodobenzene with 1,4-diethynylbenzene to give rigid-rod polymers:[89]

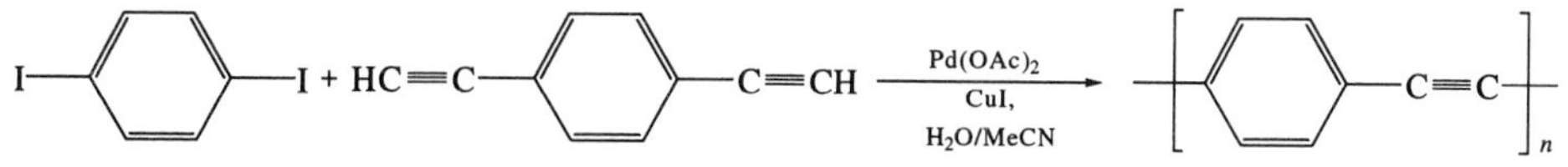

There are numerous coupling reactions between allylic halides or acetates and nucleophiles; these Ni or Pd catalyzed reactions involve π-allyl intermediates:[90]

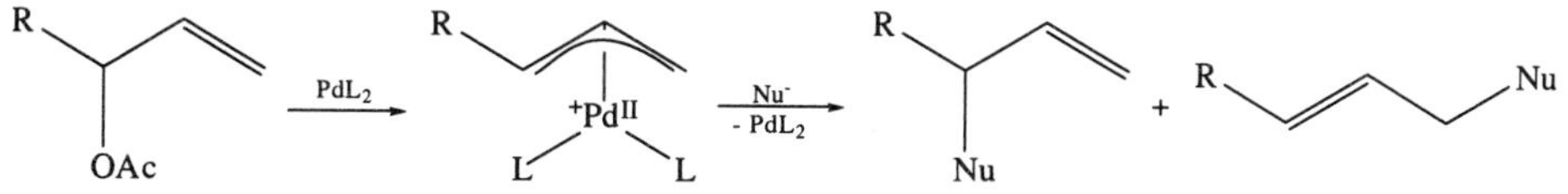

The attack of Nu⁻ on the coordinated π-allyl generates a Pd-olefin complex in the first instance which is often too unstable to be observable but can sometimes be identified.[91]

There are numerous applications of such allylic coupling reactions to enantioselective C—C bond formation, with often high optical yields.[92] A range of chiral ligands has been developed for such reactions, including for example (22-VIII).[9]

Successive Heck reactions, including carbonylation steps, have been used in cascade reactions which allow the simultaneous formation of five or more C—C bonds.[93]

Carbon-nitrogen bonds can be made using similar coupling reactions. With nickel complexes of chelating phosphines such as 1,1′-bis(diphenylphosphino)ferrocene, even the generally unreactive aryl chlorides can be converted to anilines.[94]

In *hydroacylation reactions* the C—H bond of an aldehyde is in effect added across a C=C bond:

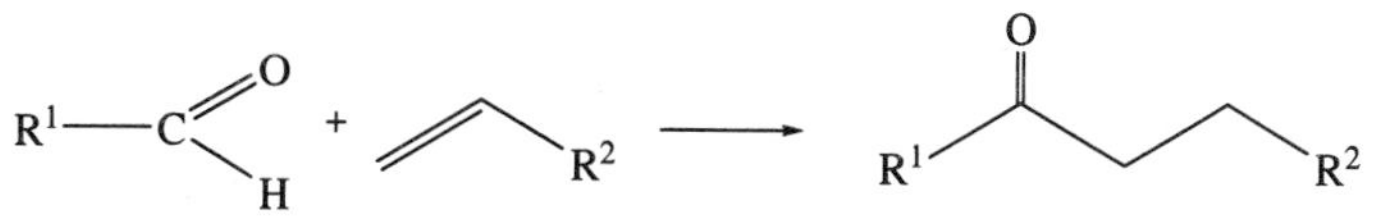

The mechanism of the intermolecular (as opposed to the better known intramolecular) version of this reaction is outlined below and involves the oxidative addition of an aldehyde C—H bond to Co¹:[95]

[89]C. J. Li *et al., Chem. Commun.* **1997,** 1569.
[90]B. M. Trost and D. L. van Vranken, *Chem. Rev.* **1996,** *96,* 395.
[91]M. Reggelin *et al., Angew. Chem. Int. Ed. Engl.* **1997,** *36,* 2108.
[92]T. Hayashi *et al., Chem. Commun.* **1997,** 561; M. Wills *et al., Chem. Commun.* **1997,** 1053; S. R. Gilbertson and C. W. T. Chang, *Chem. Commun.* **1997,** 975; B. M. Trost and R. C. Bunt, *Angew. Chem. Int. Ed. Engl.* **1996,** *35,* 99.
[93]E. Negishi *et al., Angew. Chem. Int. Ed. Engl.* **1996,** *35,* 2125.
[94]J. P. Wolfe and S. L. Buchwald, *J. Am. Chem. Soc.* **1997,** *119,* 6054. See also M. S. Driver and J. F. Hartwig, *J. Am. Chem. Soc.* **1996,** *118,* 7217.
[95]C. P. Lenges and M. Brookhart, *J. Am. Chem. Soc.* **1997,** *119,* 3165. See also B. Bosnich *et al., Chem. Commun.* **1997,** 589.

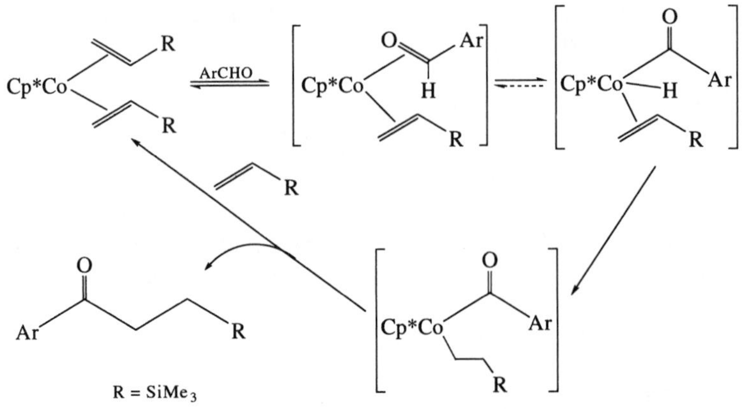

R = SiMe₃

An alternative reaction, the addition of an olefinic C—H bond to C=O, is catalyzed by Lewis acidic titanium aryloxide complexes; chiral binaphthol ligands give high optical yields.[96]

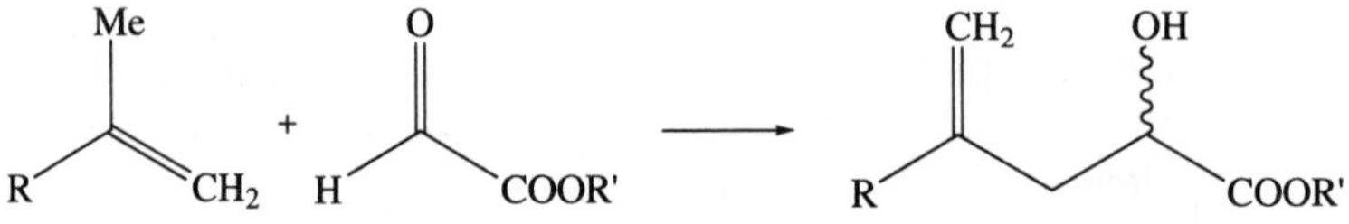

Organolanthanides catalyze the *hydroamination of alkenes and alkynes, via* the reaction sequence

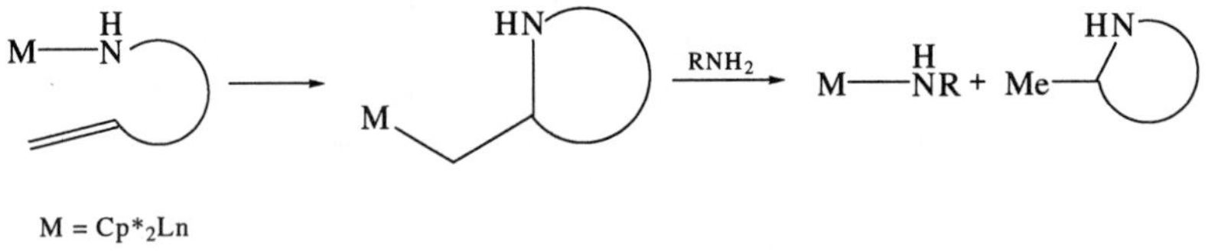

M = Cp*₂Ln

This reaction can also proceed intermolecularly; samarium and lutetium metallocene complexes catalyze the reaction of internal alkynes with primary amines to give enamines as precursors to ketimines.[97]

A number of catalysts are capable of adding carbenes or nitrenes to C=C bonds, to give cyclopropanes and aziridines, respectively:

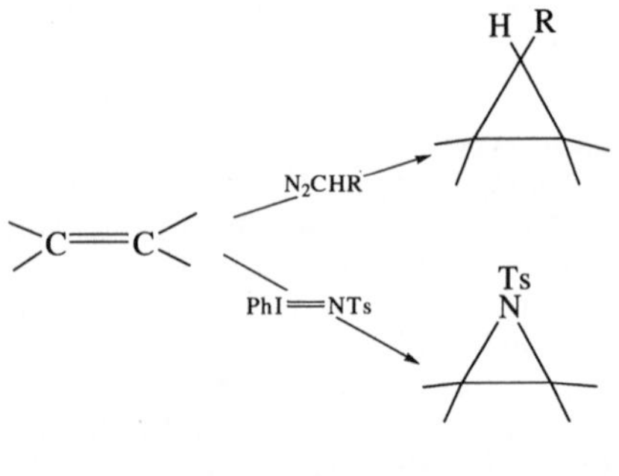

[96]K. Mikami *et al., Chem. Commun.* **1997,** 281.
[97]Y. Li and T. J. Marks, *Organometallics* **1996,** *15,* 3770.

Catalysts include copper pyrazolylborates $(Tp)Cu(C_2H_4)$, Ru^{II} porphyrins, and Rh^{II} carboxylates.[98] Metal carbene intermediates are likely; such species are also present in the ruthenium-catalyzed stereoselective coupling of α-diazocarbonyl compounds:[99]

$$2 \quad \underset{O}{\overset{N=N}{\diagdown}}\text{—R} \quad \xrightarrow[-2N_2]{CpRuCl(PPh_3)_2} \quad R\text{—}\underset{O \;\; O}{\diagup\diagdown}\text{—R}$$

The reaction of diazo compounds with aldehydes is catalyzed by methylrhenium trioxide and in the presence of PR_3 (R = Bu, Ph) as O-acceptor leads to the formation of alkenes:[100]

$$\underset{R^1}{\overset{O}{\diagdown}}\underset{H}{\diagup} + \underset{N_2}{\overset{R^2}{\diagdown}}\underset{}{\diagup R^3} + PR_3 \xrightarrow[-N_2]{CH_3ReO_3} R^1\text{∿CH}\!=\!\underset{R^3}{\overset{R^2}{C}} + O\!=\!PR_3$$

22-9 Alkene Oligomerizations and Polymerizations

Many transition metal complexes catalyze C—C bond formation between alkenes. The dimerization, oligomerization, and polymerization of non-functionalized olefins such as ethylene and propene are particularly valuable and have developed into major industrial processes. Whether a reaction leads to dimers, oligomers, or high-molecular weight polymers depends on the relative rates of chain propagation, k_p, and chain termination k_t, i.e., the reaction is kinetically controlled. Transforming a C=C double bond into two C—C single bonds is an exothermic process; polymers are therefore thermodynamically favored. The termination step in a polymerization reaction is (usually) β-H elimination or β-H transfer to monomer, which generates a terminal C=C bond and is usually slower than chain propagation; the metal hydride thus formed initiates a new polymer chain. Higher temperatures favor β-H elimination.

$$M\text{—}CH_2CH_2\text{—}\textcircled{P} \longrightarrow M\text{—}H + H_2C\!=\!CH\text{—}\textcircled{P}$$

$\textcircled{P}$ = polymer or oligomer chain

Polymerization reactions follow an insertion mechanism, that is, alkene coordination to a vacant site on the active metal species, followed by a migratory alkyl transfer step. The addition of donor molecules which can compete with the alkene for coordination sites is therefore a means of reducing the rate of propagation and allows β-H elimination to take place, so that a polymerization reaction might be converted to oligomerization or dimerization. On the other hand, metals which

[98]C. M. Che et al., Chem. Commun. 1997, 1205; M. M. Díaz-Requejo et al., Organometallics 1997, 16, 4399; P. Lahuerta et al., Organometallics 1997, 16, 880.
[99]P. Rigo et al., Chem. Commun. 1997, 2163.
[100]W. A. Herrmann and M. Wang, Angew. Chem. Int. Ed. Engl. 1991, 30, 1641.

give β-H eliminations very easily, notably nickel, can be induced to give polymers if ligands are added which make the chain termination step less favorable, as discussed later.

Zieger-Natta Polymerization

Three different catalyst types are recognized for alkene polymerizations: (a) Heterogeneous $TiCl_3$ based catalysts (Ziegler catalysts); (b) heterogeneous catalysts using chromium on silica, either by impregnation of SiO_2 with Cr_2O_3 (Phillips catalyst) or with Cp_2Cr (Union Carbide); (c) homogeneous catalysts based on metallocene complexes of Group 4 metals.

In addition, cationic alkyl complexes of nickel and palladium have been developed which follow mechanistic pathways very similar to those of metallocenes.

The discovery by K. Ziegler in 1953 that hydrocarbon solutions of $TiCl_4$ in the presence of $AlEt_3$ give heterogeneous suspensions of finely divided $TiCl_3$ that polymerize ethylene at only 1 bar pressure and moderate temperatures has led to an extremely diverse chemistry in which aluminum alkyls are used to generate transition metal alkyl species on the surface of the catalyst particles. G. Natta found that these catalysts also polymerized propene to give predominantly one stereoisomer, isotactic polypropene (see later). At about the same time the chromium based catalysts were developed, which are less active but are highly selective for the polymerization of ethylene to a highly linear polymer, with very few side branches. These are now major industrial processes, with an annual polyolefin production of >50 million tons (1993).

Since the polymers produced in these processes are highly regular they can pack into crystalline domains and give materials of comparatively high density ("high density polyethylene," HDPE, 0.94–0.96 g cm^{-3}). This contrasts with the low density polyethylene (LDPE) produced in the absence of a catalyst under very high pressures (1000-3000 bar) and temperatures (150–230°C) by a radical process which gives an extensively branched material consisting of chains of widely differing molecular weights and very different properties from the (compositionally identical) polymer obtained with metal catalysts.

In the Ziegler-Natta polymerization process $TiCl_4$ is reduced by $AlEt_3$ to polymeric $TiCl_3$. There are several $TiCl_3$ modifications, depending on their mode of synthesis; all have layer structures. Preformed $TiCl_3$ is also used and highly active catalysts are produced by ball-milling $TiCl_3$ with other metal halides, such as $MgCl_2$ which has a similar layer structure. The large number of crystal defects produced in this way is important for catalytic activity.

Polymerization takes place at the edges or corners of crystallites where metal atoms are necessarily coordinatively unsaturated. The reaction steps are those expected for a migratory alkyl transfer mechanism (Section 21-6) and has become known as the Cossee-Arlman mechanism:

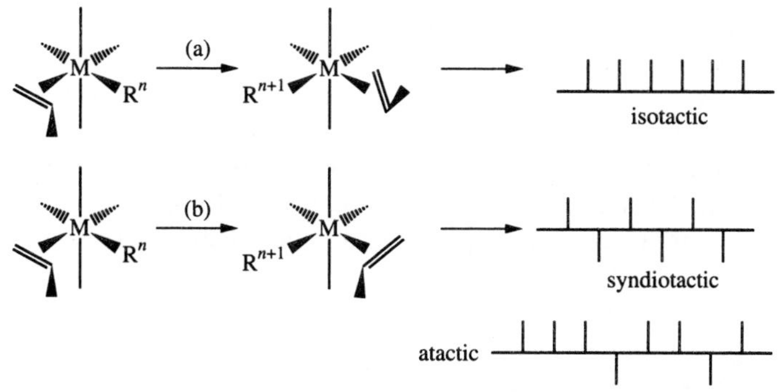

Figure 22-11 Stereochemistry of successive propane insertion steps into M—R bonds to give isotactic polypropene (a) and syndiotactic polypropene (b). In the absence of a stereocontrol mechanism atactic polypropene is formed. Note that in (a) the prochiral monomers coordinate to the metal with the same π-face, so that an isotactic polymer is formed. This stereochemistry is favored by C_2-symmetric ligands of type (22-XXVIII). In reaction (b) the second monomer coordinates with the opposite π-face to the first; this stereochemistry is enforced by metallocenes of C_s-symmetry (22-XXX, R = H).

The polymerization of propene (and other 1-alkenes) can lead to a variety of products depending on the different relative orientations of the methyl substituents. Three main types are distinguished,

(1) isotactic polypropene (i-PP), where the CH_3 groups are all aligned in the same way;

(2) syndiotactic polypropene (s-PP), where every second —CH(Me)— unit has the opposite stereochemistry to the first, and

(3) atactic polypropene (a-PP), where all methyl groups are randomly oriented.

Isotactic and s-PP arise as a consequence of the stereochemistries in two consecutive insertion steps, as shown in Fig. 22-11.

In heterogeneous catalysts this stereoselectivity to isotactic polypropene is achieved by control of the way the monomer is coordinated by the chiral β-carbon of the polymer chain ("chain end control"). Thus conformation **A** in (22-XXVII) is sterically favored over intermediate **B**.[101]

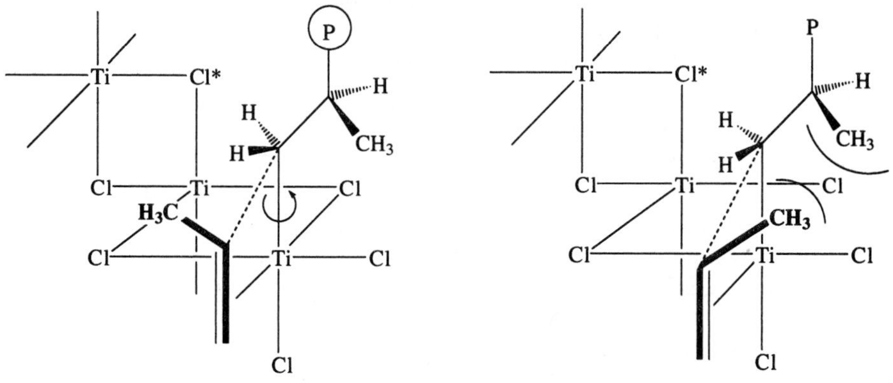

(22-XXVII)

[101]M. Bochmann, *J. Chem. Soc., Dalton Trans.* **1996**, 255.

Metallocene complexes of Ti, Zr, and Hf have attracted considerable attention in recent years because of their high activity and since their ligand framework can be tailored to a wide variety of polymerization requirements.[102] The active species is the 14-electron cationic alkyl $[Cp_2M—R]^+$, with a pseudo-tetrahedral structure and a vacant site suitable for forming a weak complex with ethylene. Calculations show that the alkyl transfer to ethylene is stabilized by an α-agostic interaction with the metal, with a very low (*ca.* 2 kJ mol^{-1}) activation barrier:[103]

The active species is most commonly generated by treating a metallocene dichloride with methylalumoxane (MAO), which acts both as an alkylating agent and as a Lewis acidic acceptor for a methyl ligand, probably leading to structures such as (6-XX).[104] Alternatively, the same cationic species can be generated by the addition of $Ph_3C^+B(C_6F_5)_4^-$ or $B(C_6F_5)_3$; the zwitterion formed in the latter case can dissociate into $[Cp_2ZrMe]^+[MeB(C_6F_5)_3]^-$:

$$Cp_2ZrMe_2 + MAO \rightleftharpoons [Cp_2ZrMe]^+[Me—MAO]^-$$

$$Cp_2ZrMe_2 + CPh_3^+ \longrightarrow [Cp_2ZrMe]^+ + Ph_3CMe$$

In the presence of MAO or AlMe$_3$ the 14-electron species is stabilized by reversible adduct formation, giving $[Cp_2Zr(\mu\text{-Me})_2AlMe_2]^+$.[101] Very weakly basic anions such as $B(C_6F_5)_4^-$ are required to stabilize the cationic metallocene species; anions have a strong effect on catalytic activity.[105]

The structure of ligands in metallocene complexes determines activity, stereoselectivity, and molecular weight of 1-alkene polymerizations, by controlling the preferential conformation of the growing polymer chain which in turn controls the stereochemistry of monomer coordination ("enantiomorphic site control"). The difference between this and the chain-end control mechanism mentioned earlier is that stereo errors due to misinsertions can be repaired.[101,106]

Examples of important metallocene catalysts are the C_2 symmetric compounds (22-XXVIII) and (22-XXIX), which are highly selective for isotactic polypropene,[107]

[102]W. Kaminsky and M. Arndt, *Adv. Polym. Sci.* **1997,** *127,* 143.
[103]T. Ziegler *et al., J. Organomet. Chem.* **1995,** *497,* 91.
[104]A. R. Barron *et al., J. Am. Chem. Soc.* **1995,** *117,* 6465.
[105]T. J. Marks *et al., J. Am. Chem. Soc.* **1997,** *119,* 2582; *Organometallics* **1997,** *16,* 842.
[106]H. H. Brintzinger *et al., Angew. Chem. Int. Ed. Engl.* **1995,** *34,* 1143.
[107]H. H. Brintzinger *et al., Organometallics* **1997,** *16,* 3413.

the zirconocenes (22-XXX) which give syndiotactic polypropene if R = H but isotactic polymers if R = Bu^t,[108] and the so-called "constrained geometry" Cp-amido complexes (22-XXXI), which are highly effective in the copolymerization of ethylene with higher 1-alkenes.

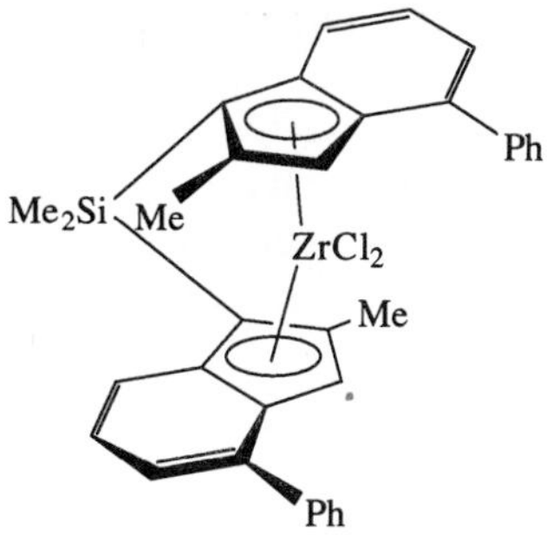

(22-XXVIII)

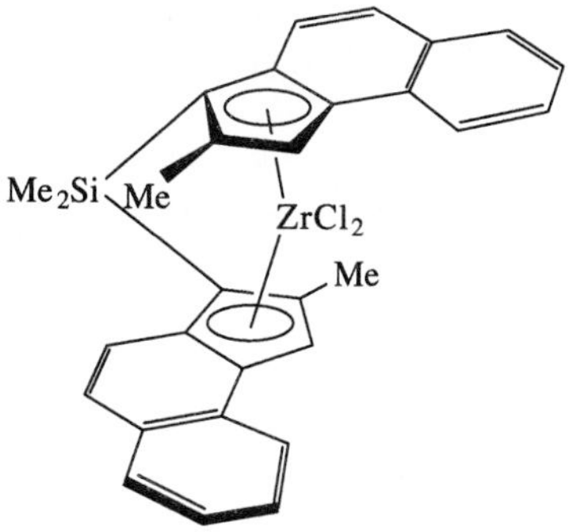

(22-XXIX)

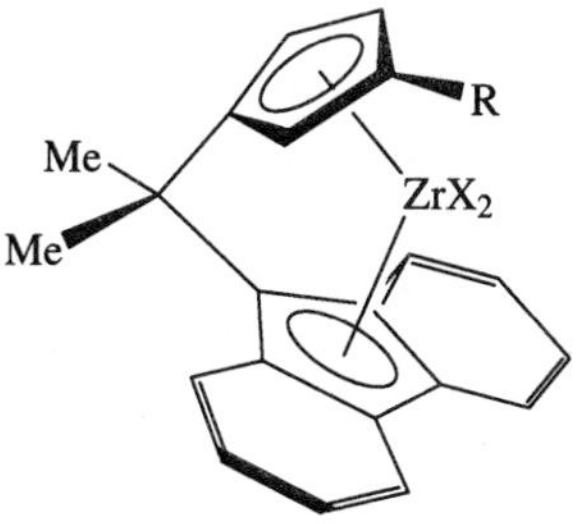

(22-XXX)

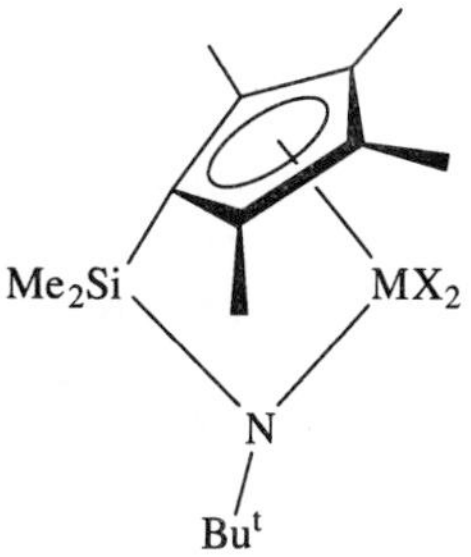

(22-XXXI)

The effect of the ligands is to sterically control the stereochemistry of monomer coordination prior to insertion, as indicated in Fig. 22-11. For C_2-symmetric complexes the coordination pocket may be represented by two hindered and two open quadrants (cf. scheme 22-XI). A prochiral monomer such as propene will adopt the orientation where repulsive interactions between the propene-methyl group, the ligand framework, and the growing polymer chain are minimized. With rac-bis(indenyl)metallocene complexes this mechanism necessarily results in an isotactic polymer.

The scope of ligand modification has been explored to a very considerable extent. For example, the bis(fluorenyl) complex Me$_2$Si(Flu)$_2$ZrCl$_2$ gives atactic polypropene of unusually high molecular weight that is elastomeric.[109] Half-sandwich complexes such as Cp*TiMe$_3$ activated with B(C$_6$F$_5$)$_3$ are highly active in a variety of polymerization reactions, including the cyclopolymerization of 1,5-hexadiene[110]

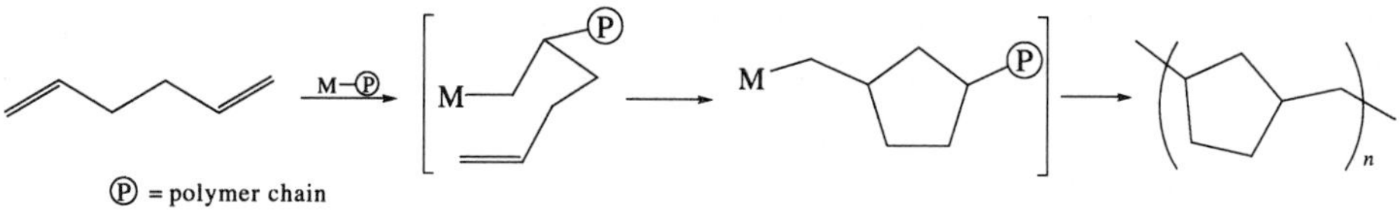

Ⓟ = polymer chain

[108]G. Fink et al., Chem. Eur. J. **1997**, 3, 585.
[109]L. Resconi et al., Organometallics **1996**, 15, 998; see also R. M. Waymouth et al., J. Am. Chem. **1997**, 119, 11174.
[110]M. C. Baird et al., J. Organomet. Chem. **1995**, 497, 143.

The Cp*TiX$_3$ system also catalyzes the syndiotactic polymerization of styrene. In contrast to all the metallocene catalysts discussed above which are d^0 systems, there is evidence that in the styrene catalyst the active species is Ti^{III}, [Cp*TiMe]$^+$.[111]

A number of Group 4 complexes without Cp ligands but based on chelating aryloxo[112] or amido ligands[113] are also known; some of the latter catalyze the "living" polymerization of 1-hexene to polymers with a very narrow molecular weight distribution.[114]

Metallocene complexes of lanthanides are also highly active for ethylene (but *not* higher 1-alkene) polymerizations. Hydrogenolysis of metal alkyl precursor complexes generates the hydrides Cp*$_2$LnH *in situ* which give rapid chain growth; they are also capable of the ring-opening polymerization of strained cyclic alkenes such as methylenecyclopropane:[115]

In contrast to Group 4 catalysts, lanthanide alkyl complexes tend to react with higher alkenes under σ-bond metathesis to give alkane and Ln-vinyl species, rather than insertion products.[116]

Group 5 complexes tend to be less active in alkene polymerizations, although vanadium-based Ziegler catalysts have long been known. The compounds Cp*Nb(η^4-diene)X$_2$ (X = CH$_3$ or Cl), activated by MAO, catalyze the living polymerization of ethylene at $-20°$C.[117]

Chromium Catalysts

There has been much debate about the nature of the catalytically active species in heterogeneous Phillips and Union Carbide catalysts.[118] Soluble Cr^{III} alkyl cations, such as [Cp*CrR(OEt$_2$)$_2$]$^+$ and the Cr analogue of (22-XXXI) (MX$_2$ = CrR) polymerize ethylene[119]; by contrast, alkyls of Cr^{II} are inactive.[120] On the other hand, treatment of chelating Cr^{II} siloxides with AlMe$_3$ gives active catalysts.[121]

[111]A. Grassi *et al., Organometallics* **1996,** *15,* 480.

[112]C. J. Schaverien *et al., J. Am. Chem. Soc.* **1995,** *117,* 3008; K. Morokuma *et al., J. Am. Chem. Soc.* **1997,** *119,* 7190.

[113]M. Bochmann *et al., J. Chem. Soc., Dalton Trans.* **1997,** 2487; A. D. Horton and J. de With, *Organometallics* **1997,** *16,* 5424.

[114]R. R. Schrock *et al., J. Am. Chem. Soc.* **1997,** *119,* 3830; J. D. Scollard and D. H. McConville, *J. Am. Chem. Soc.* **1996,** *118,* 10008.

[115]T. J. Marks *et al., J. Am. Chem. Soc.* **1996,** *118,* 7900.

[116]Slow 1-alkene polymerization has been found: E. B. Coughlin and J. E. Bercaw, *J. Am. Chem. Soc.,* **1992,** *114,* 7606.

[117]K. Mashima *et al., Organometallics* **1995,** *14,* 2633.

[118]K. H. Theopold, *Acc. Chem. Res.* **1990,** *23,* 263.

[119]K. H. Theopold *et al., Organometallics* **1996,** *15,* 5473; 5284; V. J. Jensen and K. J. Børve, *Organometallics* **1997,** *16,* 2514.

[120]K. H. Theopold *et al., Chem. Eur. J.* **1997,** *3,* 1668.

[121]A. C. Sullivan *et al., J. Chem. Soc., Chem. Commun.* **1993,** 1132.

Nickel and Palladium Catalysts

Cationic alkyl complexes of nickel and palladium of type (22-XXXII) stabilized by rigid chelating diazadiene (DAD) ligands polymerize 1-alkenes; the nickel complex is particularly active.[122] The activities are significantly higher than those of neutral nickel complexes.

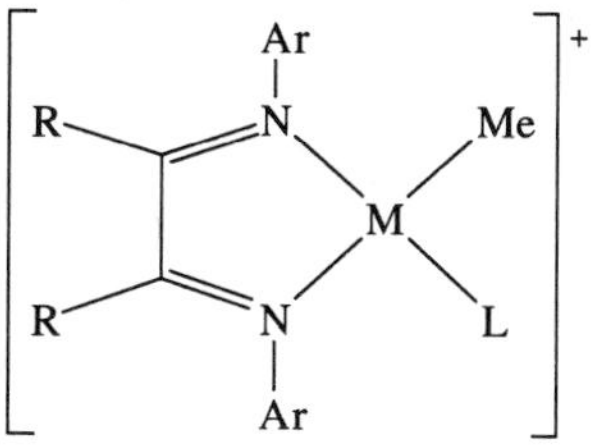

$$
\left[
\begin{array}{c}
\begin{array}{ccc}
 & Ar & \\
R- & N & Me \\
 & \diagdown & \\
 & M & \\
 & \diagup \diagdown & \\
R- & N & L \\
 & Ar &
\end{array}
\end{array}
\right]^{+}
$$

(22-XXXII)

In contrast to early transition metal catalysts which are poisoned by donor ligands and very sensitive to water, these compounds can be introduced into the system, for example, with L = OEt$_2$, or generated from (DAD)NiBr$_2$ and MAO. As was the case with metallocenes, non-coordinating counter anions, here B{C$_6$H$_3$(CF$_3$)$_2$}$_4^-$, are required. The steric requirements of the DAD ligands are important: high molecular weight polymers are obtained if Ar = 2,6-Pr$_2^i$C$_6$H$_3$, whereas with Ar = C$_6$H$_5$ ethylene is oligomerized to α-olefins.[123] Steric hindrance is an important factor in inhibiting the β-elimination step sufficiently to allow the build-up of long polymer chains. When β-H elimination does occur, re-insertion leads to branching with these catalysts:

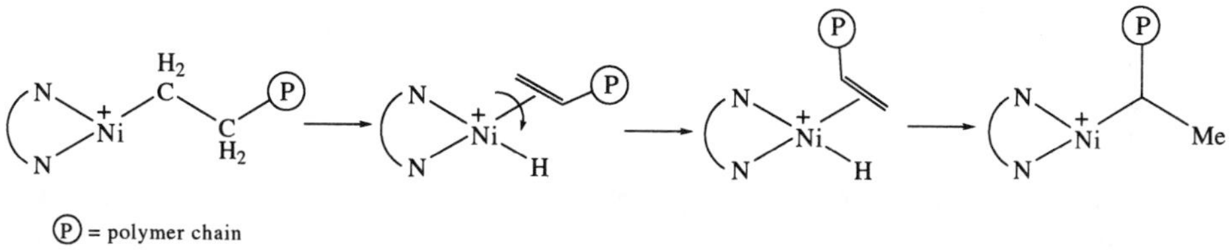

Ⓟ = polymer chain

Repetition of this step allows the metal to migrate down the polymer chain, giving rise to long-chain branched polymers.[124] These systems are now being developed by industry.

The *alternating co-polymerization of CO and ethylene*[125] is catalyzed by palladium and nickel complexes:

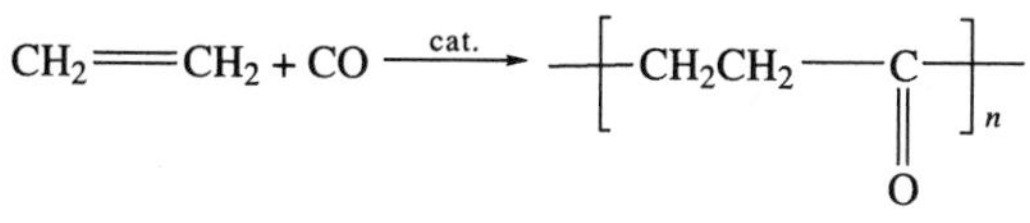

$$
CH_2{=}CH_2 + CO \xrightarrow{\text{cat.}} \left[CH_2CH_2 - \underset{\underset{O}{\|}}{C} \right]_n
$$

[122]M. Brookhart *et al., J. Am. Chem. Soc.* **1995,** *117,* 6414.
[123]M. Brookhart *et al., Organometallics* **1997,** *16,* 2005.
[124]T. Ziegler *et al., J. Am. Chem. Soc.* **1997,** *119,* 6177.
[125]E. Drent and P. H. M Budzelaar, *Chem. Rev.* **1996,** *96,* 663.

The reaction was first found by Reppe in the 1940s using $K_2Ni(CN)_4$ in water. Palladium complexes of chelating ligands are highly active, whereas complexes of monodentate phosphines react with ethylene and CO in the presence of methanol to give methyl propionate, i.e., the carbonylation proceeds without a chain propagation sequence. The reaction can be initiated with palladium alkyls, $[L_2PdR]^+$, as in the case of ethylene polymerizations, or by Pd^{2+} in the presence of methanol to give carbomethoxy species:

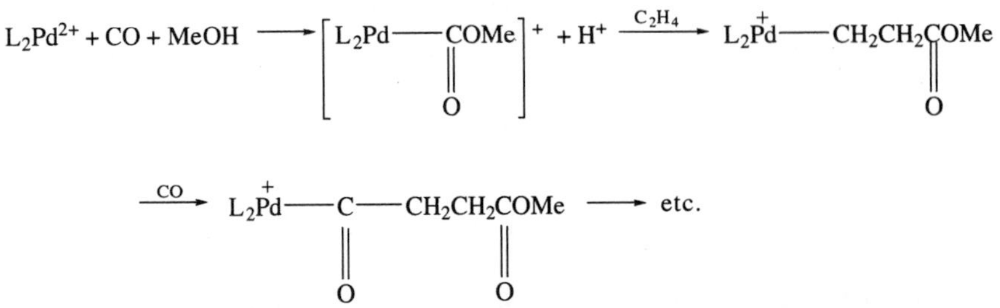

$$L_2Pd^{2+} + CO + MeOH \longrightarrow \left[L_2Pd\!-\!\underset{\underset{O}{\|}}{C}OMe \right]^+ + H^+ \xrightarrow{\ C_2H_4\ } L_2\overset{+}{Pd}\!-\!CH_2CH_2\underset{\underset{O}{\|}}{C}OMe$$

$$\xrightarrow{\ CO\ } L_2\overset{+}{Pd}\!-\!\underset{\underset{O}{\|}}{C}\!-\!CH_2CH_2\underset{\underset{O}{\|}}{C}OMe \longrightarrow \text{etc.}$$

The $Pd-CH_2CH_2C(O)R$ species is stabilized by chelate formation with the acyl oxygen [*cf.* structure (21-XLV)]; this bonding interaction makes the ethylene step essentially irreversible. The barriers of migration reactions in the system $[(phen)PdR(L)]^+$ increase in the order[126]

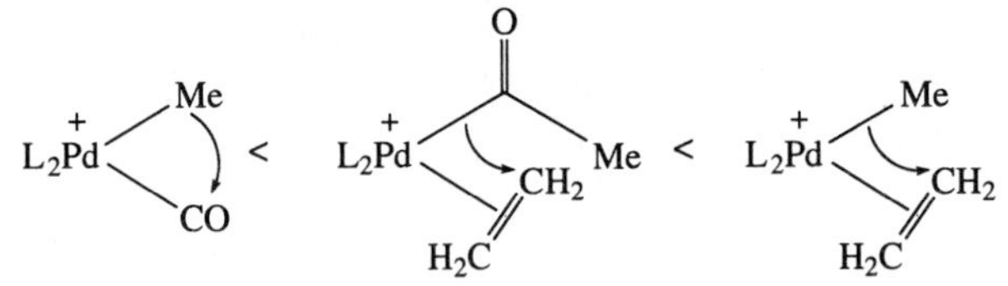

Higher alkenes are also copolymerized with CO, again in a strictly alternating sense.[127]

Unlike early transition metal polymerization catalysts which do not tolerate functional groups, cationic palladium complexes are able to copolymerize ethylene with methyl acrylate.[128]

The ethylene/CO copolymerization process is now being commercialized by Shell.

Alkene Dimerization and Oligomerizations[129]

Allyl complexes of nickel with monodentate phosphines, e.g., (22-XXXIII), are highly active catalysts for the dimerization of propene.[130] Nickel hydride species

[126]M. Brookhart *et al., J. Am. Chem. Soc.* **1996,** *118,* 4746. See also P. Margl and T. Ziegler, *Organometallics* **1996,** *15,* 5519; K. Morokuma *et al., Organometallics* **1996,** *15,* 5568.
[127]B. Rieger *et al., Macromol. Rapid Commun.* **1996,** *17,* 559.
[128]M. Brookhart *et al., J. Am. Chem. Soc.* **1996,** *118,* 267.
[129]G. Wilke, *Angew. Chem. Int. Ed. Engl.* **1988,** *27,* 185.
[130]J. Evans *et al., J. Chem. Soc., Dalton Trans.* **1994,** 1337.

are involved:

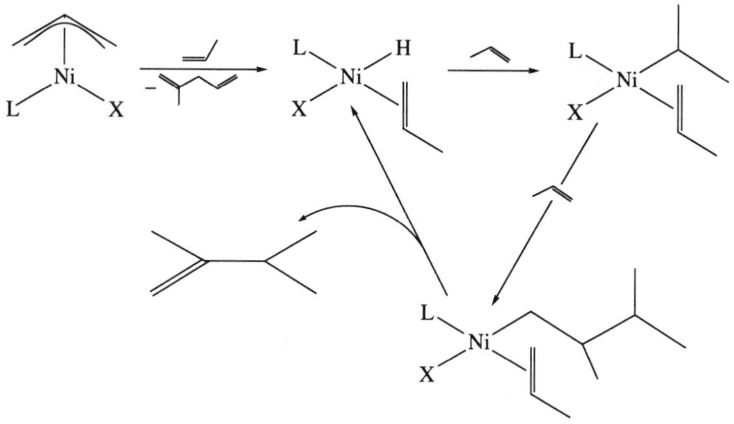

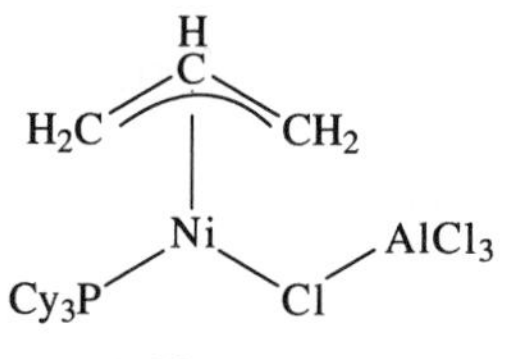

(22-XXXIII)

This is one of the most active homogeneous catalysts known.

The reaction is the basis of the "Dimersol" process for the production of octane enhancers.[131] The activity of these catalysts is strongly dependent on the steric properties of the phosphine ligands, $PEt_3 \ll PPh_3 < PCy_3 < PPr_2^iBu^t$.[129]

Nickel allyl complexes in the presence of chiral bidentate ligands catalyze the enantioselective codimerization of ethylene with norbornene and with styrene:[129]

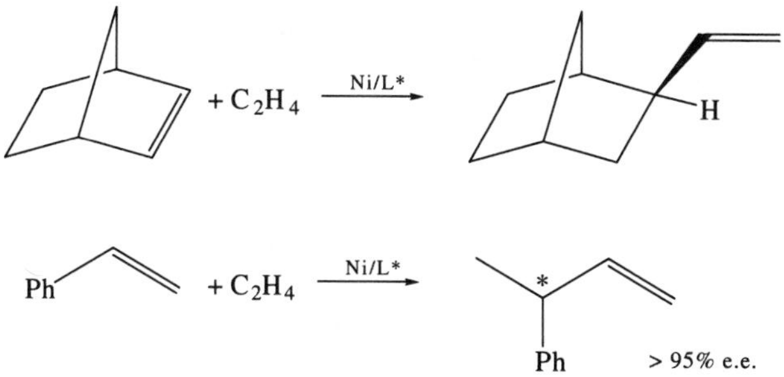

Both nickel and palladium monophosphine complexes catalyze the dimerization of methyl acrylate; the tail-to-tail dimer is a nylon-6,6 precursor. The cycle involves

[131]W. Keim, *Angew. Chem. Int. Ed. Engl.* **1990**, *29*, 235.

five- and seven-membered chelate rings:[132]

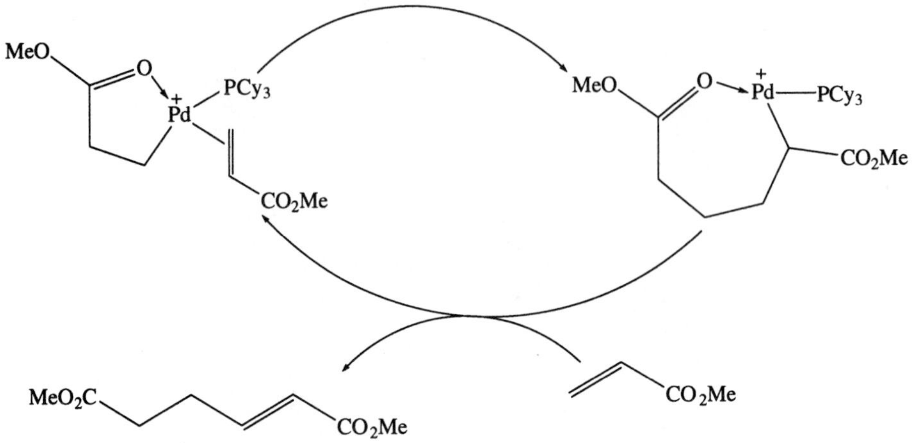

Less well defined catalysts, such as $RuCl_2(DMSO)_4$/carboxylic acid, catalyze the linear tail-to-tail dimerization of acrylonitrile.[133]

The oligomerization of ethylene to higher α-olefins is catalyzed by nickel(II) chelate complexes such as (22-XXXIV). The active species is a nickel hydride, generated *in situ*,

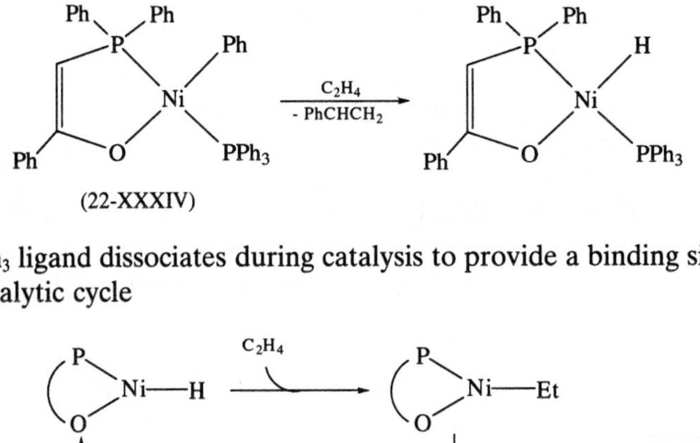

(22-XXXIV)

The PPh₃ ligand dissociates during catalysis to provide a binding site for ethylene. The catalytic cycle

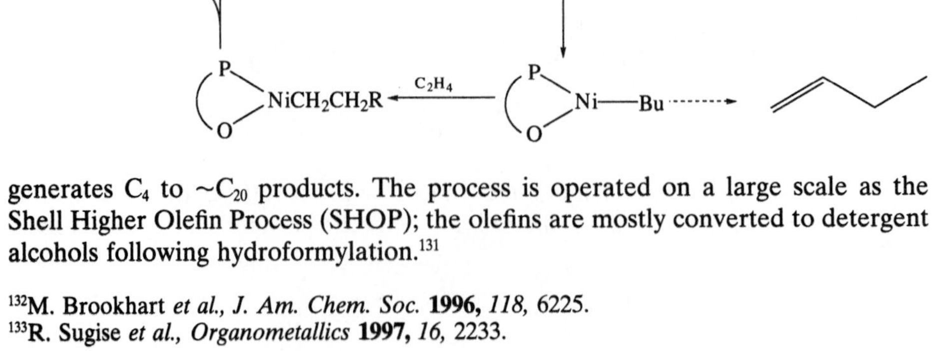

generates C_4 to $\sim C_{20}$ products. The process is operated on a large scale as the Shell Higher Olefin Process (SHOP); the olefins are mostly converted to detergent alcohols following hydroformylation.[131]

[132]M. Brookhart et al., J. Am. Chem. Soc. **1996**, 118, 6225.
[133]R. Sugise et al., Organometallics **1997**, 16, 2233.

22-10 Reactions of Conjugated Dienes

Nickel and palladium oligomerize butadiene *via* π-allyl complexes. Nickel compounds, either using "naked nickel" sources such as $Ni(COD)_2$ or $Ni(C_2H_4)_2$ or complexes with donor ligands, are particularly active in cyclooligomerizations and are well studied. The coupling of two C_4 units to a C_8 ligand corresponds to an oxidative addition reaction, giving a Ni^{II} bis(allyl) species which fluctuates between η^1 and η^3 bonding modes. Reductive elimination in the presence of phosphine or phosphite ligands L produces butadiene cyclodimers (divinylcyclobutane, vinyl cyclohexene, and 1,5-cyclooctadiene), with the selectivity being controlled by L. In the absence of donor ligands a third butadiene molecule is incorporated to give 1,5,9-cyclododecatriene.[129,131]

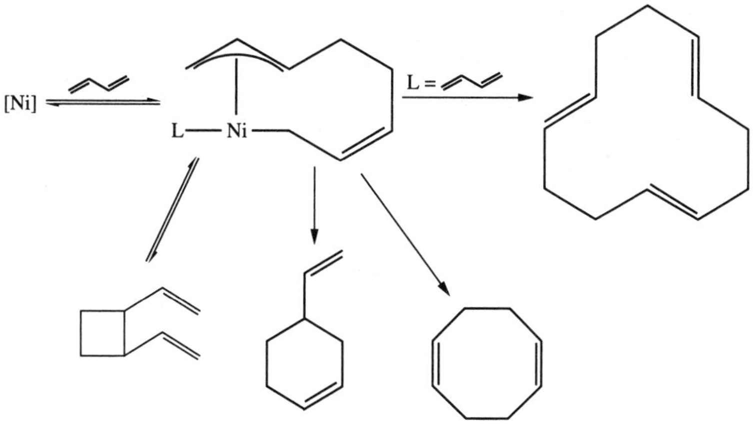

In the presence of HX = HO_2CR, HOR, HNR_2, or a C—H acidic compound, palladium complexes catalyze the *telomerization* of butadiene, a selective route to 2,7-octadienyl derivatives:

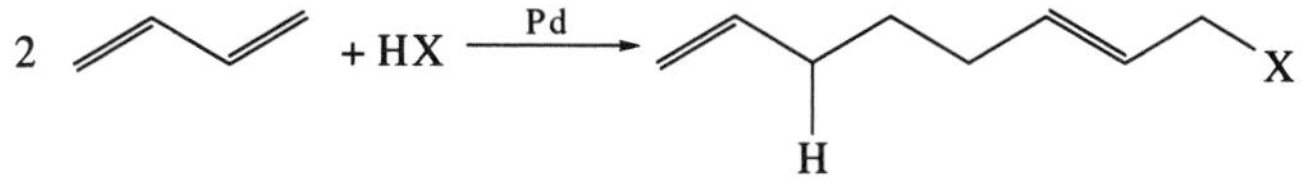

The dimerization and cyclotrimerization of butadiene are also catalyzed by titanium but are less well investigated. The Ti^{IV} η^1,η^3-octadienyl complex (22-XXXV) is isolable; it catalyzes the dimerization of butadiene to the thermodynamically favored product, vinylcyclohexene.[134]

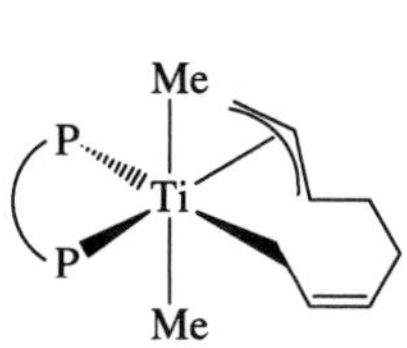

(22-XXXV)

[134]G. S. Girolami *et al., Organometallics* **1997,** *16,* 3055.

The polymerization of butadiene is catalyzed by nickel allyl complexes in the absence of donor ligands, $(\eta^3$-allyl)NiX, where X = weakly coordinating anion, such as CF_3CO_2, BF_4, or $B[C_6H_3(CF_3)_2]_4$. The polymerization proceeds stereospecifically to the 1,4-polymer.[135] Cationic complexes such as $[(\eta^3$-$RC_3H_4)Ni(butadiene)]^+$ have been isolated and are very active. Living *cis*-1,4-polybutadiene is produced with (allyl)$Ni(O_2CCF_3)$ complexes.[136]

22-11 Reactions of Alkynes

Alkynes can be selectively dimerized, cyclotrimerized, or polymerized with a large variety of transition metal and lanthanide catalysts; nickel also catalyzes the cyclotetramerization of $HC\equiv CH$ to cyclooctatetraene. Very electrophilic complexes such as $Cp*_2LnR$ and Group 4 compounds,[137] as well as 18-electron species such as $Cp*RuH_3(L)$ and $Ru(Tp)Cl(PPh_3)_2$, catalyze the linear dimerization of terminal alkynes:[138]

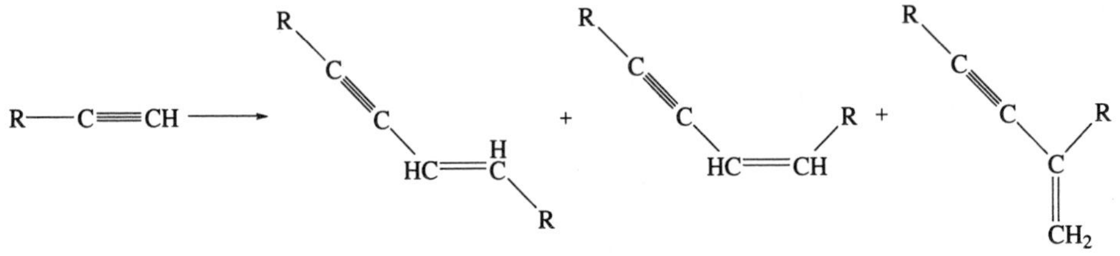

The product selectivity depends on the stereochemistry of the alkyne insertion step, for example:

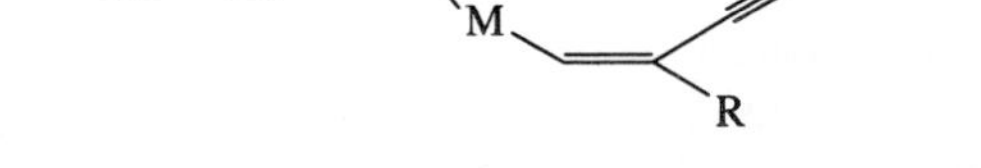

The cyclotrimerization to arenes is catalyzed by, for example, $NbCl_5$, $TaCl_5$, $Mo_2Cl_6(THT)_3$,[139] $CpCo(\eta^4$-$C_6Me_6)$, and CpRh(COD). The reaction proceeds stepwise and may involve metallacycloheptatriene, η^4-benzene, or Dewar benzene-like

[135]S. Tobisch *et al., Organometallics* **1996,** *15,* 3563.
[136]B. M. Novak *et al., J. Am. Chem Soc.* **1994,** *116,* 2366.
[137]J. H. Teuben *et al., Organometallics* **1991,** *10,* 1980; **1996,** *15,* 2291; M. Yoshida and R. F. Jordan, *Organometallics* **1997,** *16,* 4508.
[138]C. S. Yi and N. Liu, *Organometallics* **1996,** *15,* 3968; K. Kirchner *et al., Organometallics* **1996,** *15,* 5275.
[139]P. M. Boorman *et al., J. Chem. Soc., Dalton Trans.* **1996,** 4533.

intermediates:[140]

The reaction scheme shows metal-carbene/alkyne intermediates leading to η⁴-benzene, η²-benzene, and benzene.

Alternatively, with known metathesis catalysts such as $(PCy_3)_2Cl_2Ru=CHPh$, a metal carbene mechanism of alkyne trimerization has been proposed.[141]

Nitriles can be cotrimerized with alkynes to give pyridines. The reaction is catalyzed by $CpCo(C_2H_4)_2$ and similar labile Co complexes.

Many types of metal complexes polymerize alkynes, either, like WCl_6 or $(CO)_5W=C(Ph)OMe$, via metal-carbene intermediates,

$$M=CR_2 \xrightarrow{R-\!\!\!\equiv\!\!\!-R} M \xrightarrow{} M= \xrightarrow{etc.}$$

or by alkyne insertion into M-alkynyl bonds, as with $[(Tp)Rh(COD)]^+$ and related rhodium compounds.[142]

22-12 Valence Isomerization of Strained Hydrocarbons

A number of complexes, notably of Rh^I, Pd^0, and Ni^0 will induce valence isomerization of strained hydrocarbons. A typical example is the interconversion of norbornadiene and quadricyclane.

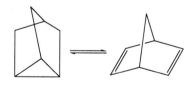

[140]C. Bianchini *et al., Organometallics* **1994,** *13,* 2010.
[141]J. U. Peters and S. Blechert, *Chem. Commun.* **1997,** 1983.
[142]F. Ozawa *et al., Organometallics* **1997,** *16,* 4497.

This reaction proceeds readily with metal ions such as Fe^I or Cu^I in the gas phase.[143] The forward reaction is catalyzed, for example, by $PdCl_2$(norbornadiene) while the norbornadiene $\rightarrow$ quadricyclane photoisomerization is sensitized by an Ir^{III} ortho-metallated complex, $[(bipy)_2Ir(2,2'\text{-bipyrid-3-yl-}C^3,N^1)]^{2+}$.

Most reactions of this type are considered to proceed by oxidative addition and cleavage of the C—C bond. For cyclopropane the reaction with $[RhCl(CO)_2]_2$ gives an isolable intermediate:

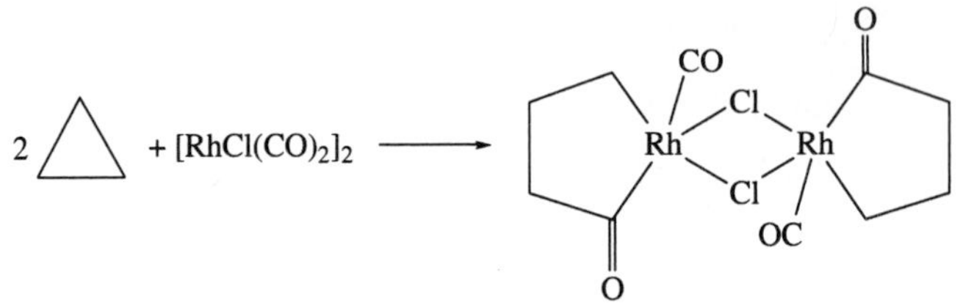

A few other intermediates have been isolated. With non-carbonyl complexes, e.g., $RhCl(PPh_3)_3$, the reaction is thought to proceed as

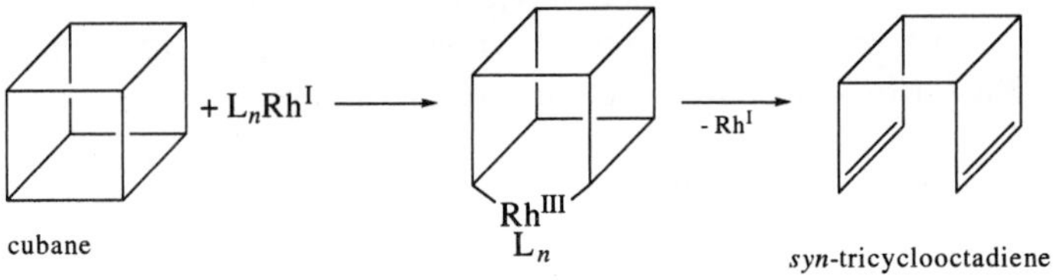

cubane Rh^{III} *syn*-tricyclooctadiene
                           L_n

The Ag^+ ion also catalyzes such isomerizations, but these proceed by electrophilic cleavage of C—C bonds followed by carbonium ion rearrangements:

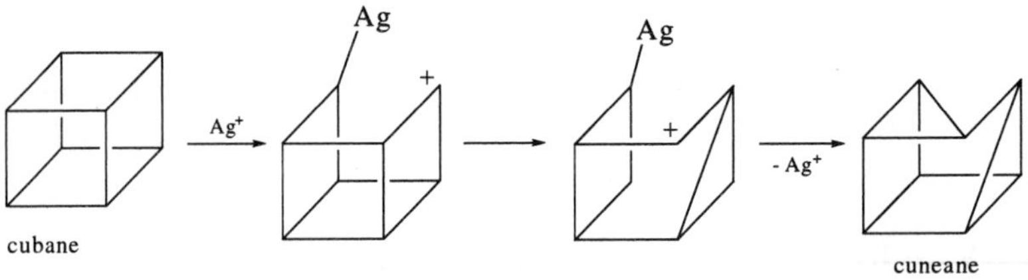

cubane   cuneane

22-13   Alkene and Alkyne Metathesis

Alkenes

The metathesis or dismutation reaction of alkenes is reversible and can be extremely rapid:

$$\begin{array}{ccc} CH_2{=\!=}CHR & CH_2 & CHR \\ + & \| \;\; + & \| \\ CH_2{=\!=}CHR & CH_2 & CHR \end{array}$$

[143]M. L. Gross *et al.*, *J. Am. Chem. Soc.* **1992**, *114*, 7801.

It was first discovered for heterogeneous systems where olefins were passed at 150 to 500°C over $Mo(CO)_6$ or $W(CO)_6$ and deposited on Al_2O_3, silica supported metals, and so on. Later a wide variety of homogeneous catalysts of Mo and W were used, typical examples being WCl_6 + $EtAlCl_2$ + EtOH, $Mo(py)_2(NO)_2Cl_2$ + AlR_3, and $MoCl_5$ + $AlCl_3$ in chlorobenzene. Rhenium oxide (Re_2O_7) on Al_2O_3 is also very effective for the metathesis of functionalized olefins.

There was much initial speculation on the mechanism but one involving metal alkylidenes, first suggested by the French chemists Herrisson and Chauvin, is now fully established by experiments involving direct reaction of alkenes with isolated alkylidene complexes.

There are several different types of catalysts but the mechanistic principles are the same. The metal alkylidene species (**A**) undergoes [2+2] cycloaddition with the alkene to give a metallacyclobutane **B** which then ring-opens to give a new alkylidene and the product alkene:

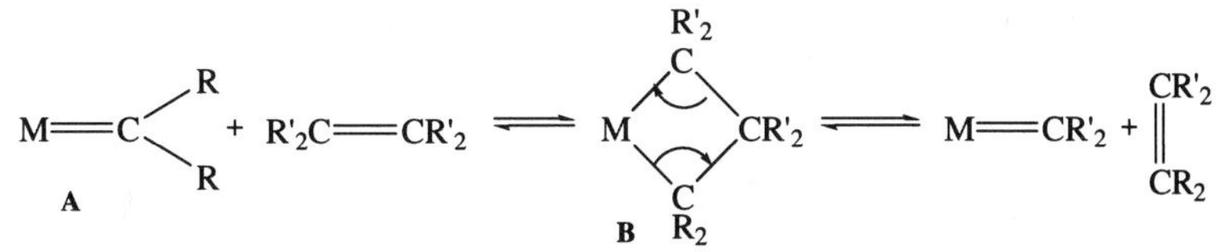

The initial $M=CR_2$ species is usually generated by an α-CH activation process from the metal alkyl.

There are several types of well-defined homogeneous metathesis catalysts:

1. Compounds which are stable as metallacyclobutanes but give alkylidene complexes as transient species:

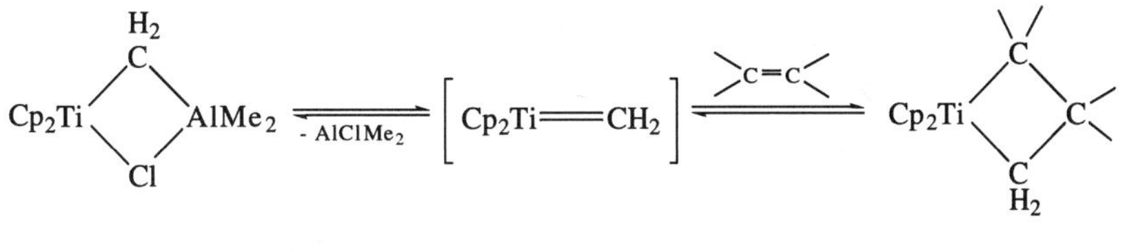

2. Complexes with alkylidene ligands. These are found for metals that typically give $M=O$ oxo species, such as Nb, Ta, Mo, W, and Re. These compounds are isolable and comparatively stable species, the best known examples being (22-XXXVI) (e.g., M = Mo or W, Ar = $C_6H_3Pr^i_2$, R = Bu^t). These complexes are also known as Schrock catalysts and are active without any activator. The activity can be tuned by the choice of R' in the alkoxide ligands, with R' = $CMe(CF_3)_2$ giving much higher activity than R' = Bu^t. Such complexes catalyze, for example, the ring-opening olefin metathesis (ROMP) of norbornene,

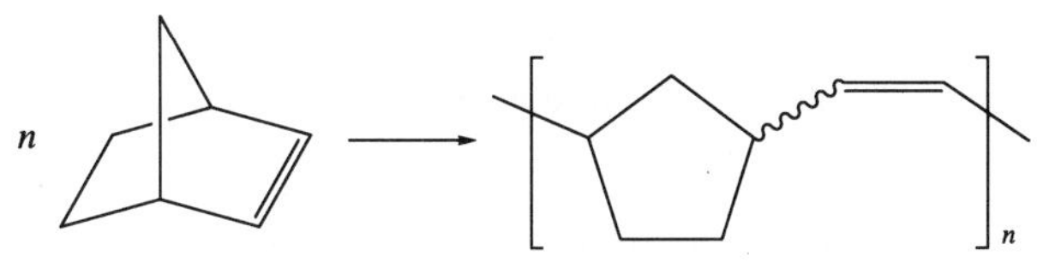

where the *cis/trans* content of the resulting polymer depends on the rate of rotational isomerization of the M=CHR unit.[144] This *syn/anti* isomerization

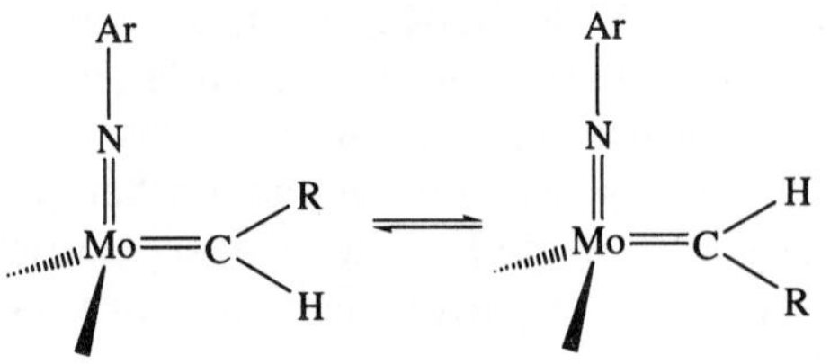

may in some cases be the rate-limiting step in a ROMP reaction.[145]

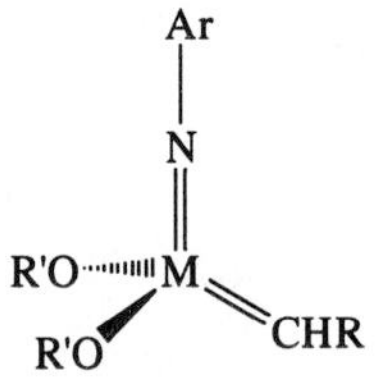

(22-XXXVI)

Related complexes require activation by a Lewis acid, for example:

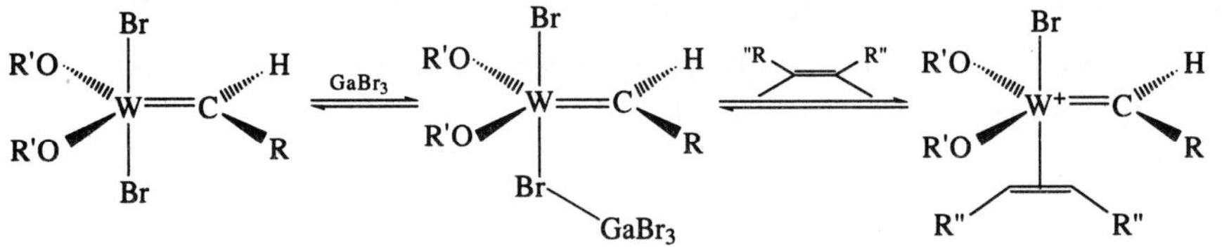

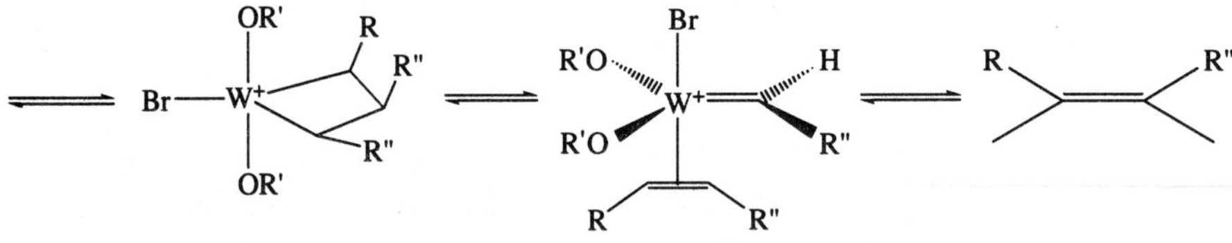

All these early transition metal catalysts are active for the metathesis of simple alkenes and the ROMP reaction of strained cycloalkenes but are sensitive to functional groups. However, catalysts tolerant of a variety of functional groups are formed by depositing CH_3ReO_3 on oxide supports; unlike heterogeneous catalysts such as Re_2O_7/Al_2O_3, the CH_3ReO_3 catalysts do not require metal alkyl activators.[146]

3. Noble metal catalysts. Ruthenium and other noble metals are also active, and while the active alkylidene species remained for a long time elusive,

[144]J. H. Oskam and R. R. Schrock, *J. Am. Chem. Soc.* **1993**, *115*, 11831.
[145]W. J. Feast *et al.*, *J. Chem. Soc., Chem. Commun.* **1994**, 1399.
[146]W. A. Herrmann *et al.*, *Angew. Chem. Int. Ed. Engl.* **1991**, *30*, 1636.

stable complexes of this kind can now be obtained:[147]

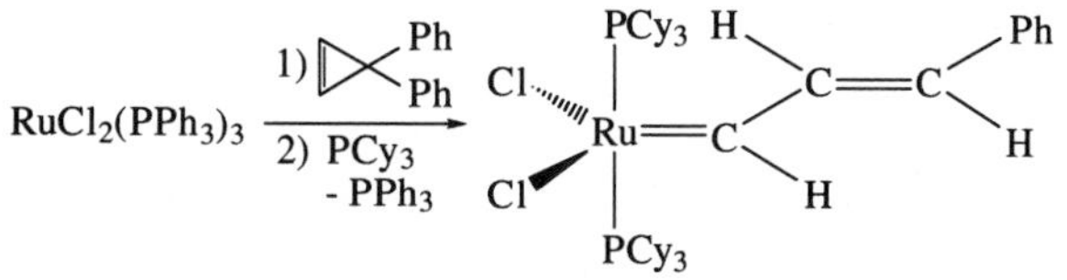

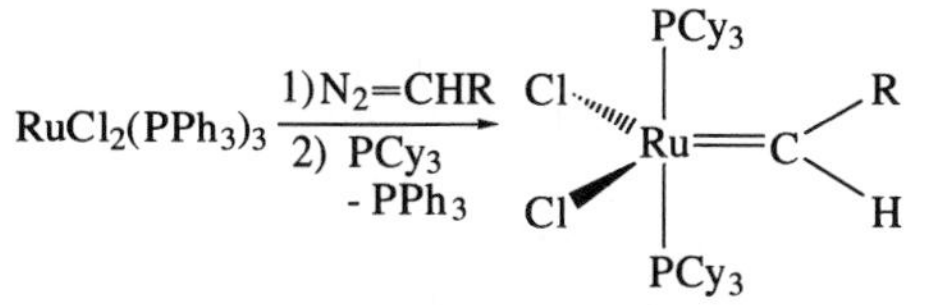

$$RuCl_2(PPh_3)_3 \xrightarrow[\substack{2) \; PCy_3 \\ - PPh_3}]{1) \; \text{(norbornadiene)} Ph} Cl_{\cdots}Ru=C\cdots \quad (\text{complex with } PCy_3, Cl, Cl, PCy_3) \quad C=C(Ph, H, H)$$

$$RuCl_2(PPh_3)_3 \xrightarrow[\substack{2) \; PCy_3 \\ - PPh_3}]{1) N_2=CHR} Cl_{\cdots}Ru=C(R, H) \quad (PCy_3, Cl, Cl, PCy_3)$$

These complexes are suitable for the ring-opening metathesis polymerization of functionalized norbornenes in water,[148] as well as ring-closure reactions *via* ethylene elimination.[149] The versatility of these catalysts has given rise to extensive applications in organic synthesis,[150,151] including, for example, amino acid derivatives.[152]

Catalysts without preformed alkylidene moieties may also be used. For example, the compounds (arene)MCl$_2$(L) (M = Ru, Os), where L = sterically demanding trialkylphosphine, are photochemically activated and polymerize norbornene.[153]

Alkyne Metathesis

Molybdenum, tungsten, and rhenium alkylidyne compounds similar to the complexes involved in alkene metathesis can metathesize alkynes *via* metallacyclobutadiene intermediates:

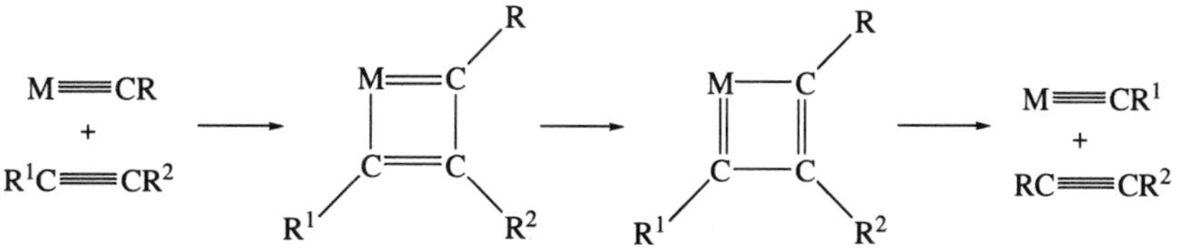

Tungstenocyclobutadiene complexes of the type proposed as intermediates have been characterized, for example, where Ar is the bulky aryl $C_6H_3Pr^i_2$-2,6:

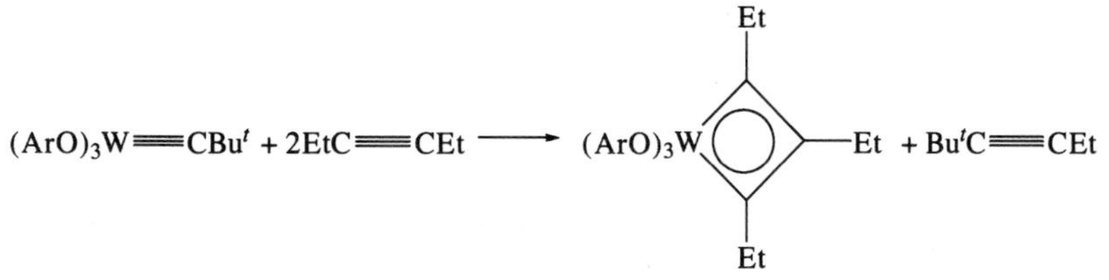

$$(ArO)_3W\equiv CBu^t + 2EtC\equiv CEt \longrightarrow (ArO)_3W(\text{cyclobutadiene with Et groups})-Et + Bu^tC\equiv CEt$$

[147]R. H. Grubbs *et al.*, *J. Am. Chem. Soc.* **1996**, *118*, 100.
[148]R. H. Grubbs *et al.*, *J. Am. Chem. Soc.* **1996**, *118*, 784; *Organometallics* **1996**, *15*, 4317.
[149]G. W. Coates and R. H. Grubbs, *J. Am. Chem. Soc.* **1996**, *118*, 229.
[150]M. Schuster and S. Blechert, *Angew. Chem. Int. Ed. Engl.* **1997**, *36*, 2036.
[151]R. H. Grubbs *et al.*, *J. Am. Chem. Soc.* **1997**, *119*, 3887.
[152]S. E. Gibson *et al.*, *Chem. Commun.* **1997**, 1107.
[153]A. Hafner *et al.*, *Angew. Chem. Int. Ed. Engl.* **1997**, *36*, 2121.

22-14 Oxidations

The catalytic oxidations of alkanes, alkenes, and other substances are of enormous technological and biological importance. In addition to classical oxidations of unsaturated substances like alkenes and arenes there is an increasing number of systems capable of catalyzing the selective oxidation of saturated hydrocarbons.

Oxidative Carbonylations

These are reactions involving CO, a substrate, and an oxidizing agent. Probably the most important reaction is that used for oxalate ester synthesis; the ester can be used as a precursor for ethylene glycol.

There are both homogeneous and heterogeneous processes. The latter uses palladium on carbon with NO as the oxygen carrier in a two-step gas-phase reaction at low pressure:

$$2NO + 2ROH + \frac{1}{2}O_2 \longrightarrow 2RONO + H_2O$$

$$2RONO + 2CO \longrightarrow (CO_2R)_2 + 2NO$$

The probable catalytic cycle involves RONO oxidative addition:

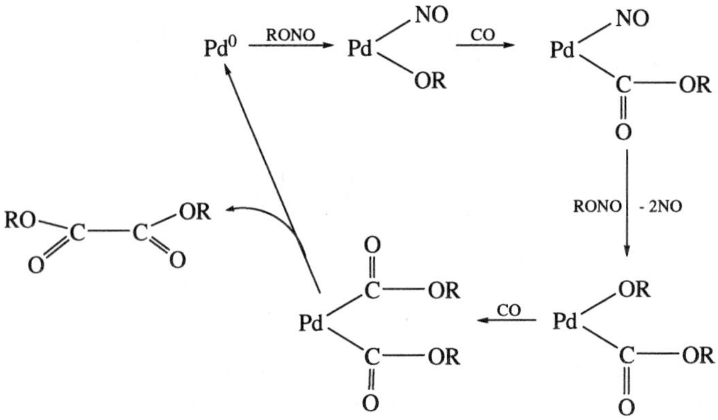

The oxidative carbonylation of aniline (Asahi Chemicals) produces alkylphenylcarbamate:

$$PhNH_2 + CO + EtOH + \frac{1}{2}O_2 \xrightarrow[83 \text{ bar}]{165°C} PhNHCO_2Et + H_2O$$

This reaction is catalyzed by palladium on carbon or $PdCl_2/CuCl_2$.

The oxidative carbonylation of methanol to dimethylcarbonate is catalyzed by Pd—Cu:

$$2MeOH + Bu^tO—OBu^t + CO \longrightarrow (MeO)_2C=O + 2Bu^tOH$$

Palladium Catalyzed Oxidation of Ethylene

The earliest large-scale catalytic use of palladium is the Wacker Process, which links together several well-known reactions:

$$C_2H_4 + PdCl_2 + H_2O \longrightarrow CH_3CHO + Pd + 2HCl$$

$$Pd + 2CuCl_2 \longrightarrow PdCl_2 + 2CuCl$$

$$2CuCl + 2HCl + \tfrac{1}{2}O_2 \longrightarrow 2CuCl_2 + H_2O$$

$$C_2H_4 + \tfrac{1}{2}O_2 \longrightarrow CH_3CHO$$

The process can be carried out either in one or in two stages; in the latter case the reoxidation by O_2 is done separately. The oxidation of higher alkenes gives ketones, e.g., acetone from propene.

When media other than water are used, related processes operate. Thus in acetic acid ethylene gives vinyl acetate, whereas vinyl ethers may be formed in alcohols. Both homogeneous and heterogeneous syntheses of vinyl acetate have been commercialized. The latter process (Hoechst) involves direct oxidation over a palladium-gold catalyst containing alkali acetate on a support:

$$C_2H_4 + CH_3COOH + \tfrac{1}{2}O_2 \longrightarrow CH_2 = CHO_2CCH_3 + H_2O$$

The mechanism of the Wacker oxidation has been much investigated. The following reactions are involved:

(1) $[PdCl_4]^{2-} + C_2H_4 \rightleftharpoons [PdCl_3(C_2H_4)]^- + Cl^-$

(2) $[PdCl_3(C_2H_4)]^- + H_2O \rightleftharpoons PdCl_2(H_2O)(C_2H_4) + Cl^-$

(3) $PdCl_2(H_2O)(C_2H_4) + H_2O \rightleftharpoons [Cl_2(H_2O)Pd-CH_2CH_2OH]^- + H^+$

(4) $[Cl_2(H_2O)Pd-CH_2CH_2OH]^- + H_2O \rightleftharpoons Cl(H_2O)Pd-CH_2CH_2OH + Cl^-$

There is nucleophilic attack by external water (reaction 3) at a neutral $Pd(C_2H_4)$ species to give a palladium hydroxyethyl intermediate. The following displacement of Cl^- by water is rate-limiting. The $Pd-CH_2CH_2OH$ complex decomposes by β-H elimination; the resulting vinyl alcohol isomerizes to acetaldehyde.

The oxidation of Pd^0 by Cu^{II} chloro complexes probably involves electron transfer *via* halide bridges.

A reaction related to the mechanism given above is the oxidation of Zeise's salt, $[Pt^{II}Cl_3(C_2H_4)]^-$, by $Pt^{IV}Cl_6^{2-}$. It was shown that this also involves nucleophilic attack by H_2O on the ethylene ligand of the Pt^{II} complex, followed by electron (not alkyl) transfer:[154]

$$[PtCl_3(C_2H_4)]^- + H_2O \rightleftharpoons [PtCl_3(CH_2CH_2OH)]^{2-} + H^+$$

$$[PtCl_3(CH_2CH_2OH)]^{2-} + PtCl_6^{2-} \longrightarrow [PtCl_5(CH_2CH_2OH)]^{2-} + PtCl_4^{2-}$$

Alkane Oxidations

Platinum(II), for example, K_2PtCl_4, in an aqueous medium in the presence of O_2, oxidizes ethane selectively to ethanol and ethylene glycol, whereas metallic Pt is required for the further oxidation of alcohols to carboxylic acids. Platinum metal also catalyzes the oxidation of alkenes to 1,2-diols. The presence of CO prevents further oxidation of the alcohol products.[155]

In the presence of Cl^- or I^-, $RhCl_3$ in $C_3F_7CO_2H/H_2O$ catalyzes the direct formation of methanol from methane, CO, and O_2. Ethane is oxidized more readily and gives mainly methanol, as well as C_2 products, but products due to C—C bond cleavage dominate over those generated by C–H activation. The alcohols produced are less reactive in this system than the alkane.[156] A bimetallic copper chloride/ metallic Pd system in CF_3CO_2H/H_2O in the presence of O_2 and CO at 150°C is highly selective for the oxidation of methane to methanol, with good catalytic activity.[157] The system $[NBu_4]VO_3$/pyrazine-2-carboxylic acid catalyzes the air oxidation of methane to give methyl hydroperoxide.[158]

Various alkane oxidations are catalyzed by iron complexes. Such reactions are important in view of the action of non-heme iron enzymes, such as methane monooxygenase, in hydrocarbon oxidations in biological systems. For example, the oxo-bridged complex $[Fe_2(TPA)_2(\mu\text{-}O)(\mu\text{-}OAc)]^{3+}$ [TPA = tris(2-pyridylmethyl)-amine] catalyzes the oxidation of cyclohexane with Bu^tOOH. Related complexes with an $Fe^{III}_2(\mu\text{-}O)(\mu\text{-}OAc)_2$ core oxidize cyclohexane or adamantane to give a mixture of alcohols and ketones.[159] Less well defined systems, e.g., $FeCl_3 \cdot 6H_2O$/ aldehyde/$AcOH$/O_2 are similarly active.[160]

[154]J. A. Labinger *et al.*, *Organometallics* **1994**, *13*, 755.

[155]A. Sen *et al.*, *J. Am. Chem. Soc.* **1992**, *114*, 6385; *J. Chem. Soc., Chem. Commun.* **1993**, 970.

[156]A. Sen *et al.*, *J. Am. Chem. Soc.* **1996**, *118*, 4574.

[157]A. Sen *et al.*, *J. Am. Chem. Soc.* **1997**, *119*, 6048.

[158]G. Süss-Fink *et al.*, *Chem. Commun.* **1997**, 397.

[159]L. Que *et al.*, *J. Am. Chem. Soc.* **1993**, *115*, 9524; M. Kodera *et al.*, *Chem. Commun.* **1996**, 1737.

[160] S. I. Murahashi *et al.*, *J. Am. Chem. Soc.* **1992**, *114*, 7913.

An iron(II) catecholate/hydroquinone/O_2 system oxidizes phenols to catechols. The reaction mimics the action of tyrosine hydroxylase, which gives dihydroxyphenylalanine.[161]

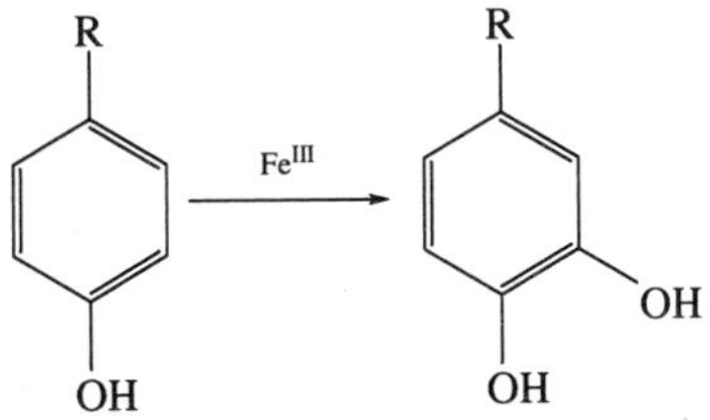

The oxo-bridged mixed-valent compound $[(phen)_2Mn^{III}(\mu\text{-}O)_2Mn^{IV}(phen)_2]^{3+}$, a model for part of the oxygen-evolving complex in photosystem II, oxidizes dihydroanthracene with high selectivity to anthracene.[162]

22-15 Oxygen Transfer Reactions from Peroxo and Oxo Complexes

An important commercial oxidation using *t*-butyl hydroperoxide is the epoxidation of propene catalyzed by molybdate compounds:

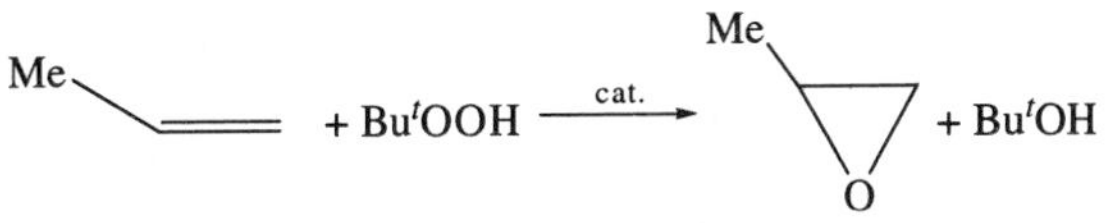

Peroxomolybdate species are involved, and the steps can be modelled using the well-characterized complex (22-XXXVII(a)), where L = $(Me_2N)_3PO$.

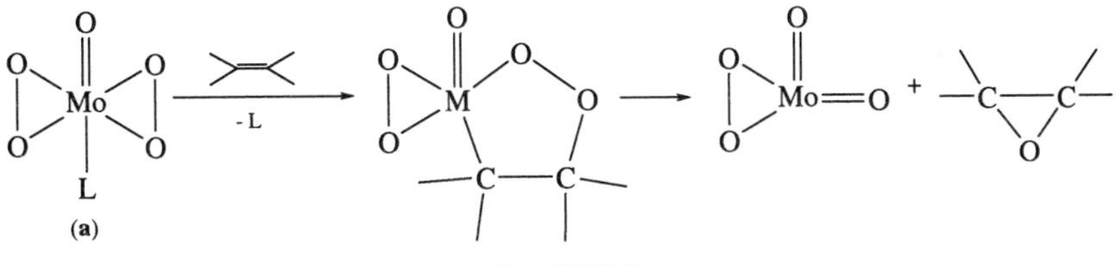

(22-XXXVII)

There are many synthetic applications of epoxide formation. Titanium alkoxides in the presence of diethyltartrate as a chiral ligand catalyze the epoxidation of allylic alcohols enantioselectively (Sharpless reaction). In the presence of singlet

[161]T. Funabiki *et al.*, *Chem. Commun.* **1997**, 151.
[162]K. Wang and J. M. Mayer, *J. Am. Chem. Soc.* **1997**, *119*, 1470.

oxygen this reaction converts alkenes into hydroxy-epoxides:[163]

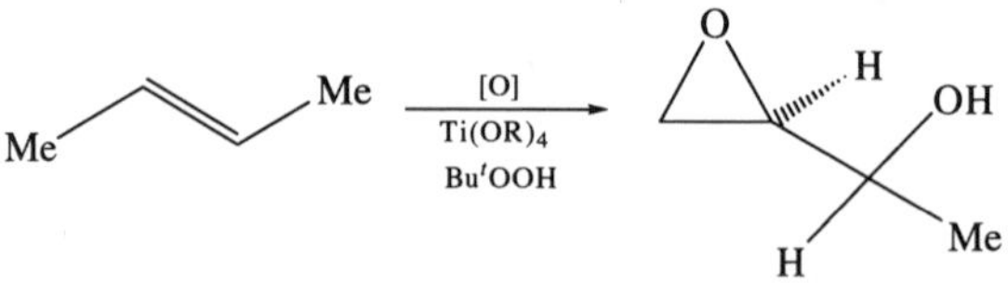

Manganese compounds of the type (22-XXXVIII) catalyze the epoxidation of alkenes with PhI=O, OCl⁻, and similar oxidants. There is a debate about the mechanism, though a metallaoxirane (22-XXXIX) intermediate is likely; with chiral Schiff base ligands the reaction is enantioselective.[164]

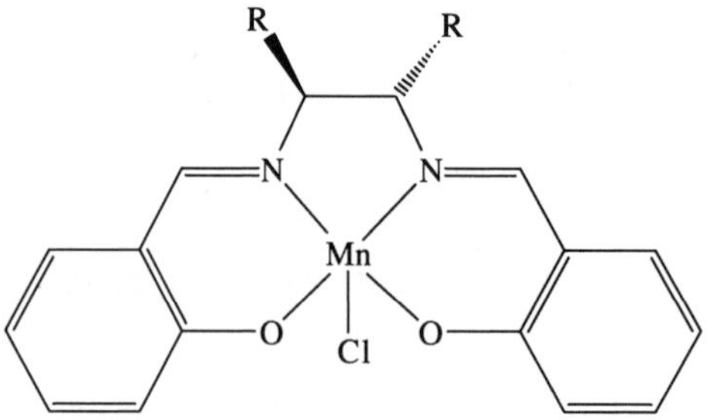

(22-XXXVIII) 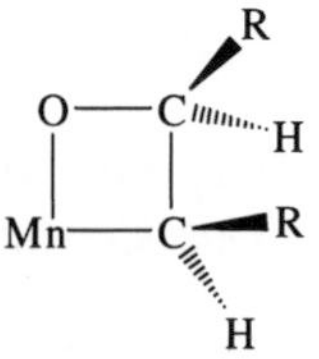   (22-XXXIX)

Methylrhenium trioxide is an effective catalyst for the epoxidation of olefins by aqueous H_2O_2. The reaction is accelerated by pyridine and efficiently oxidizes simple and functionalized 1-alkenes.[165]

The epoxidation of alkenes by O_2 is catalyzed by polydentate iron and by cobalt porphyrin complexes; the latter also operate in a perfluoroalkane/MeCN biphasic system.[166] Ruthenium complexes, such as (22-XL), catalyze alkene epoxidations with bis(acetoxy)iodobenzene.[167]

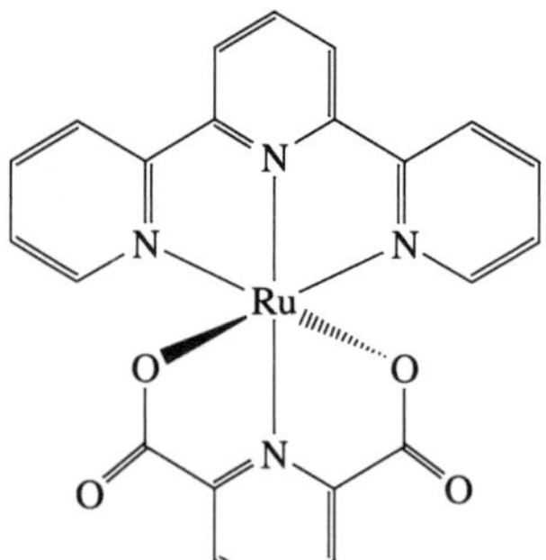

(22-XL)

[163]W. Adams and M. J. Richter, *Acc. Chem. Res.* **1994**, *27*, 57.
[164]T. Linker, *Angew. Chem. Int. Ed. Engl.* **1997**, *36*, 2060.
[165]W. A. Herrmann *et al.*, *Angew. Chem. Int. Ed. Engl.* **1991**, *30*, 1638; K. B. Sharpless *et al.*, *Chem. Commun.* **1997**, 1565.
[166]Y. Journaux *et al.*, *Chem. Commun.* **1997**, 2283; G. Pozzi, *Chem. Commun.* **1997**, 69.
[167]H. Nishiyama *et al.*, *Chem. Commun.* **1997**, 1863.

Osmium tetroxide reacts with alkenes to give 1,2-diols. Two mechanisms have been debated:

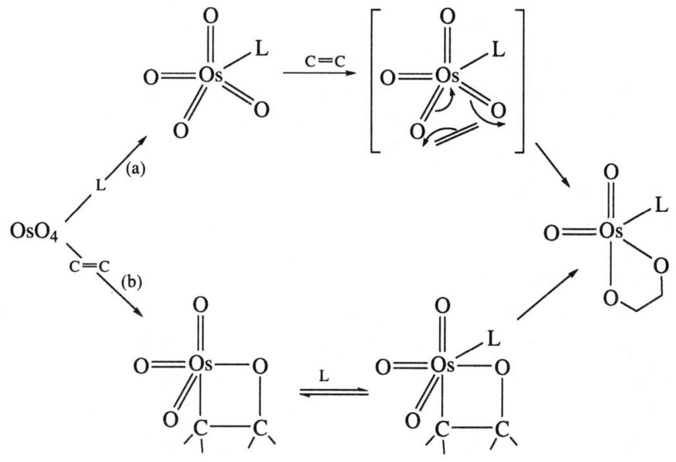

There is evidence that a concerted [2 + 3] step, *via* path (a), is favored over four-membered intermediates postulated for (b). With chiral ligands L the reaction proceeds enantioselectively.[168]

22-16 Supported Homogeneous and Phase Transfer Catalysis

A difficulty associated with homogeneous catalytic systems is the separation of the products from the reactants and the catalyst. Since homogeneous catalysts often display much greater selectivity and/or activity than heterogeneous systems, many attempts have been made to harness these benefits by "heterogenization." The main approaches that have been used are:

(1) Using a polymeric ligand, or attaching donor groups such as $-PPh_2$ to a polymer resin.

(2) Working in a biphasic system, usually a mixture of two immiscible liquids in the presence of an additive that facilitates molecule transfer between the two phases.

(3) Attaching soluble catalysts by covalent tethers or simple impregnation to a solid support such as a metal oxide.

An example for resins acting as ligands is polystyrene modified by attaching $-PPh_2$ groups to the phenyl rings. Such resins form complexes with, e.g., $RhCl(PPh_3)_3$ to give polymeric analogues of Wilkinson's catalyst and can be removed after the reaction by filtration. A disadvantage is that the coordination of the metal to the polymer is reversible, so that some catalyst is lost due to leaching.

A related approach is the attachment of a $-(CH_2)_2Si(OMe)_3$ group to a phosphine ligand. After formation of the metal complex the $-Si(OMe)_3$ groups can

[168]T. Ziegler *et al.*, *Organometallics* **1997**, *16*, 13; K. B. Sharpless *et al.*, *J. Am. Chem. Soc.* **1997**, *119*, 1840.

then be hydrolyzed to give a cross-linked polysiloxane network. These polymeric catalysts proved more active than their monomeric analogues.[169]

Water soluble polymers such as poly(enolato-co-vinyl alcohol-co-vinyl acetate) (PEVV) act as polymeric ligands to Rh in the biphasic hydroformylation of olefins.[170] Another type of polymeric ligands are phosphinated dendrimers, i.e., regularly branched polymers with an almost spherical morphology, which carry PPh$_2$ substituents on the surface, suitable, e.g., for attaching ruthenium complexes.[171]

Metal complexes can be attached to membranes. For example, Ru BINAP complexes (*cf.* Scheme 22-VII) and a Brønsted acid attached to a polydimethylsiloxane membrane act as effective heterogenized catalysts for asymmetric hydrogenations.[172]

Phase transfer catalysis involves typically an organic/aqueous biphasic system in the presence of a transfer agent such as a tetraalkylammonium salt which facilitates the exchange of the catalyst between the two phases, while the reactants and the products are usually retained in the organic layer. Almost all types of homogeneously catalyzed reactions can be carried out in this way.[173]

Water soluble ligands greatly facilitate catalyst separation. Examples are (22-XLI)[174] and (22-XLII),[175] which are used in hydroformylations and hydrogenation catalysis, respectively. Rhodium complexes of the sulfonated phosphine (22-XLI) are used in the production of butyraldehyde, a large-scale process developed by Ruhrchemie/Rhône-Poulenc.[174]

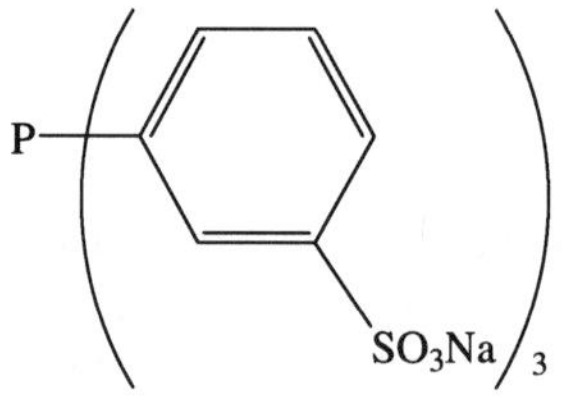

(22-XLI) 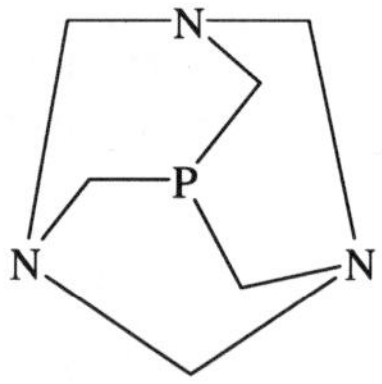   (22-XLII)

Such water soluble ligands are especially useful if the liquid/liquid biphasic system can be homogenized during the polymerization reaction to overcome mass transport limitations, but be separated into two phases thereafter for catalyst recycling. This can be achieved by adjusting the temperature, and/or by adding surfactants which generate reverse micelles. Separation is possible by lowering the temperature or changing the solvent polarity, and industrial hydroformylation catalysts have been developed on this basis.[176] Reversed micelles have also been used in imine hydrogenation.[16]

A combination of the phase-transfer and solid-support strategy was applied to asymmetric hydrogenations with complex (22-XLIII). Solutions of this complex in

[169]E. Lindner *et al., Chem. Eur. J.* **1997,** *3,* 1833.
[170]J. Chen and H. Alper, *J. Am. Chem. Soc.* **1997,** *119,* 893.
[171]J. P. Majoral *et al., Organometallics* **1997,** *16,* 3489.
[172]P. A. Jacobs *et al., Chem. Commun.* **1997,** 2323.
[173]E. S. Gore, *Platinum Metals Rev.* **1990,** *34,* 2.
[174]W. A. Herrmann and C. W. Kohlpaintner, *Angew. Chem. Int. Ed. Engl.* **1993,** *32,* 1524.
[175] D. J. Darensbourg *et al., Inorg. Chem.* **1994,** *33,* 200.
[176]J. Haggin, *C&EN* **1995,** 17 April, p. 25.

ethylene glycol adhere to controlled-pore glass particles as a surface film. The organic reactant is dissolved in a chloroform/cyclohexane phase. The system gives high enantioselectivity, high reaction rates due to the intimate mixing and the large surface area of the polar/non-polar phase boundary, and quantitative catalyst recovery.[177]

(22-XLIII)

Organic solvents are essentially immiscible with perfluoroalkyls at ambient temperature but give homogeneous phases on heating. This principle has been employed in rhodium catalyzed hydroformylations with $HRh(CO)[P\{CH_2CH_2-(CF_2)_5CF_3\}_3]_3$,[178] in Rh catalyzed hydroborations, and in cobalt and iridium catalyzed oxidations. A disadvantage is the need for fluorinated alkylphosphines. The oxidation reactions take advantage of the high solubility of O_2 in fluorocarbons.[179]

Operating under *supercritical conditions* allows the homogeneous mixing of various reactants at high concentrations. For example, palladium supported on polysiloxanes catalyzes the hydrogenation of cyclohexene in supercritical CO_2 significantly faster than in a common liquid phase.[180]

Metallocene polymerization catalysts can be supported on silica[181] or on mesoporous supports[182] with a coating of methylalumoxane (MAO) as activator. These catalysts tend to be less active than homogeneous analogues at lower temperatures but are comparable under higher operating temperatures.[181] Silica supported metallocene catalysts are used industrially in gas phase and slurry olefin polymerization processes where they have started to displace $TiCl_3$-based Ziegler catalysts. In gas phase operations the reaction presumably takes place in the MAO/alkene film on the particle surface, and the chemistry involved is identical to that in the homogeneous systems. Apart from acting as a means to heterogenize metallocene catalysts, the silica particles also control the morphology of the resulting polymer granules; this is important for subsequent processing. In view of the high activity of the catalysts, there is no need for catalyst recycling.

[177]K. T. Wan and M. E. Davis, *Nature* **1994**, *370*, 449.
[178]I. T. Horváth and J. Rábai, Science **1994**, *266*, 72.
[179]B. Cornils, *Angew. Chem. Int. Ed. Engl.* **1997**, *36*, 2057.
[180]M. G. Hitzler and M. Poliakoff, *Chem. Commun.* **1997**, 1667.
[181]R. Mülhaupt *et al.*, *J. Polym. Sci. Part A: Polym. Chem.* **1997**, *35*, 1.
[182]J. Tudor and D. O'Hare, *Chem. Commun.* **1997**, 603.

Additional References

H. Brunner and W. Zettlmeier, *Handbook of Enantioselective Catalysis with Transition Metal Compounds*, VCH, Weinheim, 1993.

H. M. Colquhoun, D. J. Thompson, and M. V. Twigg, *Carbonylation*, Plenum Press, New York, 1991.

B. Cornils and W. A. Herrmann, Eds., *Aqueous Phase Organometallic Catalysis,* Wiley-VCH, Weinheim, 1998.

B. Cornils and W.A. Herrmann, *Applied Homogeneous Catalysis with Organometallic Compounds*, VCH, Weinheim, 1996.

F. Diederich and P.J. Stang, Eds., *Metal-catalyzed Cross-coupling Reactions,* Wiley-VCH, Weinheim, 1998.

B. C. Gates, *Catalytic Chemistry*, Wiley, Chichester, 1992.

C. Girard and H. B. Kagan, Nonlinear Effects in Asymmetric Synthesis and Stereoselective Reactions, *Angew. Chem. Int. Ed.* **1998,** *37,* 2922.

F. R. Hartley, Ed., *Chemistry of the Platinum Metals*, Elsevier, Amsterdam, 1991.

J. F. Hartwig, Transition Metal Catalyzed Synthesis of Arylamines and Aryl Ethers from Aryl Halides and Triflates: Scope and Mechanism, *Angew. Chem. Int. Ed.* **1998,** *37,* 2047.

P. G. Jessop, T. Ikariya, and R. Noyori, Homogeneous Catalysis in Supercritical CO_2, *Chem. Rev.* **1995,** *95,* 259.

I. Ojima, Ed., *Catalytic Asymmetric Synthesis*, VCH, Weinheim, 1993.

G. W. Parshall and S. D. Ittel, *Homogeneous Catalysis*, 2nd ed., Wiley, New York, 1992.

A. Togni and L. M. Venanzi, Nitrogen Donors in Organometallic Chemistry and Homogeneous Catalysis, *Angew. Chem. Int. Ed. Engl.* **1994**, *33*, 497.

J. Tsuji, *Palladium Reagents and Catalysts,* John Wiley, New York, 1996.

Appendix 1

UNITS AND FUNDAMENTAL CONSTANTS

The most widely sanctioned set of units, the Système Internationale d'Unités or SI set, is summarized here. In this book (as elsewhere, and for good reason*) this system has been only partially adopted. Thus, we use joules (instead of calories) but retain other earlier units, most importantly, Angstroms, atmospheres, and degrees Celsius.

Presented in this appendix are a list of basic and derived SI units, some fundamental constants frequently required by inorganic chemists, and some useful conversion factors.

Basic SI Units

Physical Quantity	Name of Unit	Symbol for Unit
Length	meter	m
Mass	kilogram	kg
Time	second	s
Electric current	ampere	A
Thermodynamic temperature	kelvin	K
Luminous intensity	candela	cd
Amount of substance	mole	mol

*A. W. Adamson, *J. Chem. Educ.*, 1978, **55**, 634.

Derived SI Units

Physical quantity	Name of unit	Symbol for unit	Definition of unit
Energy	Joule	J	$kg\ m^2s^{-2}$
Force	Newton	N	$kg\ m\ s^{-2} = J\ m^{-1}$
Power	Watt	W	$kg\ m^2s^{-3} = J\ s^{-1}$
Pressure	Pascal	Pa	$kg\ m^{-1}s^{-2} = N\ m^{-2}$
Electric charge	Coulomb	C	$A\ s$
Electric potential difference	Volt	V	$kg\ m^2s^{-3}A^{-1} = J\ A^{-1}s^{-1}, J/C$
Electric resistance	Ohm	Ω	$kg\ m^2s^{-3}A^{-2} = V\ A^{-1}$
Electric capacitance	Farad	F	$A^2s^4kg^{-1}m^{-2} = A\ s\ V^{-1}$
Magnetic flux	Weber	Wb	$kg\ m^2s^{-2}A^{-1} = V\ s$
Inductance	Henry	H	$kg\ m^2s^{-2}A^{-2} = V\ s\ A^{-1}$
Magnetic flux density	Tesla	T	$kg\ s^{-2}A^{-1} = V\ s\ m^{-2}$
Frequency	Hertz	Hz	$Hz = s^{-1}$
Customary temperature, t	Degree Celsius	°C	$t[°C] = T[K] - 273.15$

Fundamental Constants

Quantity	Symbol	Value	SI unit
Speed of light in vacuum	c	$2.997\ 925 \times 10^8$	$m\ s^{-1}$
Elementary charge	e	$1.602\ 189 \times 10^{-19}$	C
Planck constant	h	$6.626\ 18 \times 10^{-34}$	J s
Avogadro constant	N_A	$6.022\ 04 \times 10^{23}$	mol^{-1}
Atomic mass unit	1u	$1.660\ 566 \times 10^{-27}$	kg
Electron rest mass	m_e	$0.910\ 953 \times 10^{-30}$	kg
Proton rest mass	m_p	$1.672\ 649 \times 10^{-27}$	kg
Neutron rest mass	m_n	$1.674\ 954 \times 10^{-27}$	kg
Faraday constant	F	$9.648\ 46 \times 10^4$	$C\ mol^{-1}$
Rydberg constant	R_∞	$1.097\ 373 \times 10^7$	m^{-1}
Bohr radius	a_0	$0.529\ 177 \times 10^{-10}$	m
Electron magnetic moment	μ_e	$9.284\ 83 \times 10^{-24}$	$J\ T^{-1}$
Proton magnetic moment	μ_p	$1.410\ 617 \times 10^{-26}$	$J\ T^{-1}$
Bohr magneton	μ_B	$9.274\ 08 \times 10^{-24}$	$J\ T^{-1}$
Nuclear magneton	μ_N	$5.050\ 82 \times 10^{-27}$	$J\ T^{-1}$
Molar gas constant	R	$8.314\ 41$	$J\ mol^{-1}\ K^{-1}$
Molar volume of ideal gas (stp.)	V_m	$0.022\ 413\ 8$	$m^3\ mol^{-1}$
Boltzmann constant	k	$1.380\ 662 \times 10^{-23}$	$J\ K^{-1}$

Conversion Factors

1 cal	$= 4.184$ joules (J)
1 eV/molecule	$= 96.485$ kJ mol^{-1}
	$= 23.061$ kcal mol^{-1}
1 kcal mol^{-1}	$= 349.76\ cm^{-1}$
	$= 0.0433$ eV
1 kJ mol^{-1}	$= 83.54\ cm^{-1}$
1 wavenumber (cm^{-1})	$= 2.8591 \times 10^{-3}$ kcal mol^{-1}
1 erg	$= 2.390 \times 10^{-11}$ kcal
1 centimeter (cm)	$= 10^8$ Å
	$= 10^7$ nm
1 picometer (pm)	$= 10^{-2}$ Å
1 nanometer (nm)	$= 10$ Å

IONIZATION ENTHALPIES OF THE ATOMS

Definition: The ionization enthalpies of an atom X are the enthalpies of the processes

$$X(g) = X^+(g) + e^-(g) \qquad \Delta H^0_{ion}(1)$$

$$X^+(g) = X^{2+}(g) + e^-(g) \qquad \Delta H^0_{ion}(2), \text{ and so on}$$

Older chemical literature commonly uses the term "ionization potential," which is $-\Delta H^0_{ion}$, usually expressed in electron-volts.

Values of $\Delta H^\bullet_{ion}(n)$ (kJ mol^{-1})

	First	Second	Third	Fourth
1 H	1311			
2 He	2372	5249		
3 Li	520.0	7297	11,810	
4 Be	899.1	1758	14,850	21,000
5 B	800.5	2428	2394	25,020
6 C	1086	2353	4618	6512
7 N	1403	2855	4577	7473
8 O	1410	3388	5297	7450
9 F	1681	3375	6045	8409
10 Ne	2080	3963	6130	9363
11 Na	495.8	4561	6913	9543
12 Mg	737.5	1450	7731	10,540
13 Al	577.5	1817	2745	11,580
14 Si	786.3	1577	3228	4355
15 P	1012	1903	2910	4955
16 S	999.3	2260	3380	4562
17 Cl	1255	2297	3850	5160
18 Ar	1520	2665	3950	5771
19 K	418.7	3069	4400	5876
20 Ca	589.6	1146	4942	6500
21 Sc	631	1235	2389	7130
22 Ti	656	1309	2650	4173
23 V	650	1414	2828	4600
24 Cr	652.5	1592	3056	4900

Values of $\Delta H°_{ion}$ (n) $(kJ\ mol^{-1})$ (*Continued*)

	First	Second	Third	Fourth
25 Mn	717.1	1509	3251	
26 Fe	762	1561	2956	
27 Co	758	1644	3231	
28 Ni	736.5	1752	3489	
29 Cu	745.2	1958	3545	
30 Zn	906.1	1734	3831	
31 Ga	579	1979	2962	6190
32 Ge	760	1537	3301	4410
33 As	947	1798	2735	4830
34 Se	941	2070	3090	4140
35 Br	1142	2080	3460	4560
36 Kr	1351	2370	3560	
37 Rb	402.9	2650	3900	
38 Sr	549.3	1064		5500
39 Y	616	1180	1979	
40 Zr	674.1	1268	2217	3313
41 Nb	664	1381	2416	3700
42 Mo	685	1558	2618	4480
43 Tc	703	1472	2850	
44 Ru	710.6	1617	2746	
45 Rh	720	1744	2996	
46 Pd	804	1874	3177	
47 Ag	730.8	2072	3360	
48 Cd	876.4	1630	3615	
49 In	558.1	1820	2705	5250
50 Sn	708.2	1411	2942	3928
51 Sb	833.5	1590	2440	4250
52 Te	869	1800	3000	3600
53 I	1191	1842		
54 Xe	1169	2050	3100	
55 Cs	375.5	2420		
56 Ba	502.5	964		
57 La	541	1103	1849	
72 Hf	760	1440	2250	3210
73 Ta	760	1560		
74 W	770	1710		
75 Re	759	1600		
76 Os	840	1640		
77 Ir	900			
78 Pt	870	1791		
79 Au	889	1980		
80 Hg	1007	1809	3300	
81 Tl	588.9	1970	2880	4890
82 Pb	715.3	1450	3080	4082
83 Bi	702.9	1609	2465	4370
84 Po	813			
86 Rn	1037			
88 Ra	509.1	978.6		
89 Ac	670	1170		

Appendix 3

ENTHALPIES OF ELECTRON ATTACHMENT (ELECTRON AFFINITIES) OF ATOMS

Definition: The enthalpy of electron attachment pertains to the process:

$$X(g) + e^- = X^-(g) \qquad \Delta H_{EA}^0$$

Older chemical literature commonly uses the term "electron affinity," defined as $-\Delta H_{EA}^0$.

Values of ΔH_{EA}^0 (kJ mol^{-1})^{a,b}

Z	Atom	$-\Delta H_{EA}^0$
1	H	72.77
3	Li	59.8(6)
5	B	27
6	C	122.3
7	N	*ca.* 0
8	O	141.1(3)
9	F	328.0(3)
11	Na	52.7(5)
13	Al	45
14	Si	132.2
15	P	71(1)
16	S	200.42(5)
17	Cl	348.8(4)
19	K	48.36(5)
33	As	77(5)
34	Se	194.96(3)
35	Br	324.6(4)
37	Rb	46.89(5)
51	Sb	101(5)
52	Te	190.15(3)
53	I	295.3(4)
55	Cs	45.5(2)
79	Au	222.8(1)

Values of ΔH_{EA}^0 (kJ mol^{-1})^{a,b} (*Continued*)

Z	Atom	$-\Delta H_{\text{EA}}^0$
83	Bi	97(2)
84	Po	183(30)
85	At	270(20)

[a] All taken from H. Hotop and W. C. Lineberger, *J. Phys. Chem. Reference Data*, 1975, **4**, 539, which should be consulted for sources and background.

[b] All alkaline earth and noble gas atoms have $\Delta H_{\text{EA}}^0 > 0$.

Appendix 4

IONIC RADII

One of the major factors in determining the structures of the substances that can be thought of, at least approximately, as made up of cations and anions packed together, is ionic size, especially the ratio of radii of the two or more ions present.

It is obvious from the nature of wave functions that no ion or atom has a precisely defined radius. The only way radii can be assigned is to determine how closely the centers of two atoms or ions actually approach each other in solid substances and then to assume that such a distance is equal or closely related to the sum of the radii of the two atoms or ions. Even this procedure is ambiguous. Further assumptions are required to get an empirically useful set of radii. Early efforts to do this were made by Goldschmidt and by Pauling. Their results satisfy the minimum requirements of being additive, internally consistent, and showing physically reasonable trends (e.g., increasing with Z for members of a group in the Periodic Table and decreasing with the degree of ionization for the cations of a given metallic element). However, it must be kept in mind that there is no absolute way to say where one ion ends and the neighboring one begins. Moreover, ions are undoubtedly somewhat elastic and their apparent radii can vary with environment, particularly with coordination number.

With the advent of relatively accurate experimental electron density maps based on high-precision X-ray diffraction methods, it became possible to see more realistically how electron density actually varies along the line between adjacent nuclei in an ionic crystal. An example is provided by LiF, as shown in Fig. A4-1. It can be seen that neither the Goldschmidt (G) nor, *a fortiori*, the Pauling (P) radii for Li^+ are at the minimum (M) of electron density. Thus these radii could be said to be "wrong" in an absolute sense, even though the complete sets to which they belong display internal consistency.

Recently, with the insight afforded by electron density maps and with an enormously larger base of data, new efforts to establish tables of ionic radii have been made, the most successful being those of Shannon and Prewitt. In Table A4-1 are listed a selection of these radii. For some additional values and detailed discussion of how these radii were derived, see footnotes to the table. There are many important trends and correlations to be found among

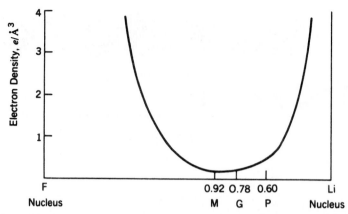

FIG. A4-1. A plot of experimentally measured electron density between adjacent F and Li nuclei in LiF. The Pauling and Goldschmidt radii of Li^+ are indicated by P and G, while M denotes the actual minimum (adapted from H. Krebs, *Fundamentals of Inorganic Crystal Chemistry*, McGraw-Hill, New York, 1968).

these results. Several of the principal ones are

1. For a cation that displays several coordination numbers (C.N.), the radius increases with increasing C.N.

2. For a given vertical group (e.g., the alkalis or alkaline earths) the radii increase with Z for a given coordination number.

3. For an isoelectronic series such as Na^+, Mg^{2+}, Al^{3+}, and Si^{4+}, the radii decrease with increasing charge.

4. For any element with several oxidation numbers, for example, Ti^{2+}, Ti^{3+}, and Ti^{4+}, the radius decreases with increasing charge.

5. For each of the various transition series, ions of the same charge show an overall trend towards decreasing radii with increasing Z. However, in the d block, these trends are not smooth, for reasons given in accounts of ligand field theory. The behavior of the lanthanide and actinide radii is discussed in Chapters 19 and 20, respectively.

6. For transition metals that can have both high- and low-spin states, for example, Fe^{2+}, the radius is larger for the high-spin ion.

TABLE A4-1
Ionic Radii[a,b] (Å)

Ion	C.N.[c]	Radius	Ion	C.N.[c]	Radius
A. *Alkali and Alkaline Earth Cations*					
Li^+	4	0.73	Fr^+	6	1.94
	6	0.90	Be^{2+}	4	0.41
	8	1.06		6	0.59
Na^+	4	1.13	Mg^{2+}	4	0.71
	6	1.16		6	0.86
	8	1.32		8	1.03
	12	1.53	Ca^{2+}	6	1.14

TABLE A4-1 (*Continued*)

Ion	C.N.^{c}	Radius	Ion	C.N.^{c}	Radius
K^+	4	1.51		8	1.26
	6	1.52		10	1.37
	8	1.65		12	1.48
	10	1.73	Sr^{2+}	6	1.32
	12	1.78		8	1.40
Rb^+	6	1.66		10	1.50
	8	1.75		12	1.58
	10	1.80	Ba^{2+}	6	1.49
	12	1.86		8	1.56
	14	1.97		10	1.66
Cs^+	6	1.81		12	1.75
	8	1.88	Ra^{2+}	8	1.62
	10	1.95		12	1.84
	12	2.02			

B. Group IB (11)

Ion	C.N.^{c}	Radius	Ion	C.N.^{c}	Radius
Cu^+	2	0.60	Ag^+	8	1.42
	4	0.74	Au^+	6	1.51
	6	0.91	Cu^{2+}	4	0.71
Ag^+	2	0.81		4 (sq)	0.71
	4	1.14		6	0.87
	4 (sq)	1.16	Au^{3+}	4 (sq)	0.82
	6	1.29		6	0.99

C. Group IIB (12)

Ion	C.N.^{c}	Radius	Ion	C.N.^{c}	Radius
Zn^{2+}	4	0.74	Cd^{2+}	12	1.45
	6	0.88	Hg^{2+}	2	0.83
	8	1.04		4	1.10
Cd^{2+}	4	0.92		6	1.16
	6	1.09		8	1.28
	8	1.24			

D. Other Non-Transition Metal Ions

Ion	C.N.^{c}	Radius	Ion	C.N.^{c}	Radius
NH_4^+	6	1.61 (?)	Tl^{3+}	8	1.12
Tl^+	6	1.64	Sb^{3+}	6	0.90
	8	1.73	Bi^{3+}	6	1.17
	12	1.84	Sc^{3+}	6	0.89
Pb^{2+}	6	1.33		8	1.01
	8	1.43	Y^{3+}	6	1.04
	10	1.54	C^{4+}	4	0.29
	12	1.63	Si^{4+}	4	0.40
B^{3+}	4	0.25		6	0.54
Al^{3+}	4	0.53	Ge^{4+}	4	0.53
	6	0.68		6	0.67
Ga^{3+}	4	0.61	Sn^{4+}	4	0.69
	6	0.76		6	0.83
In^{3+}	4	0.76		8	0.95
	6	0.94	Pb^{4+}	4	0.79
	8	1.06		6	0.92
Tl^{3+}	4	0.89		8	1.08
	6	1.03			

TABLE A4-1 (*Continued*)

Ion	C.N.[c]	Radius	Ion	C.N.[c]	Radius
E. *First Transition Series Metals*					
Ti^{2+}	6	1.00	Ni^{2+}	6	0.83
V^{2+}	6	0.93	Ti^{3+}	6	0.81
Cr^{2+}	6 (LS)	0.87	V^{3+}	6	0.78
	6 (HS)	0.94	Cr^{3+}	6	0.76
Mn^{2+}	4 (HS)	0.80	Mn^{3+}	6 (LS)	0.72
	6 (LS)	0.81		6 (HS)	0.79
	6 (HS)	0.97	Fe^{3+}	4 (HS)	0.63
Fe^{2+}	4 (HS)	0.77		6 (LS)	0.69
	6 (LS)	0.75		6 (HS)	0.79
	6 (HS)	0.92	Co^{3+}	6 (LS)	0.69
Co^{2+}	4 (HS)	0.72		6 (HS)	0.75
	6 (LS)	0.79	Ni^{3+}	6 (LS)	0.70
	6 (HS)	0.89		6 (HS)	0.74
Ni^{2+}	4	0.69	Ti^{4+}	6	0.75
	4 (sq)	0.63			
F. *Second Transition Series Elements*					
Pd^{2+}	4 (sq)	0.78	Rh^{3+}	6	0.81
	6	1.00	Nb^{4+}	6	0.82
Nb^{3+}	6	0.86	Mo^{4+}	6	0.79
Mo^{3+}	6	0.83	Ru^{4+}	6	0.76
Ru^{3+}	6	0.82	Rh^{4+}	6	0.74
G. *Third Transition Series Elements*					
Pt^{2+}	4 (sq)	0.74	W^{4+}	6	0.80
	6	0.94	Re^{4+}	6	0.77
Ta^{3+}	6	0.86	Os^{4+}	6	0.78
Ir^{3+}	6	0.82	Ir^{4+}	6	0.77
Hf^{4+}	6	0.85	Pt^{4+}	6	0.77
Ta^{4+}	6	0.82	Th^{4+}	6	1.08
H. *Anions*					
F^-	2	1.15	O^{2-}	8	1.28
	4	1.17	S^{2-}	6	1.70
	6	1.19	Se^{2-}	6	1.84
Cl^-	6	1.67	Te^{2-}	6	2.07
Br^-	6	1.82	OH^-	2	1.18
I^-	6	2.06		3	1.20
O^{2-}	2	1.21		4	1.21
	3	1.22		6	1.23
	4	1.24	N^{3-}	4	1.32
	6	1.26			

[a]Selected from R. D. Shannon, *Acta Crystallogr.*, 1976, **A32**, 751. This article gives radii for many other ions and also for other coordination numbers.

[b]For lanthanide and actinide radii, see Tables 19-1 and 20-1.

[c]Unannotated 6 means octahedral; no particular geometry is implied for other numbers unless stated [e.g., 4 (sq) means square]; LS and HS mean low spin and high spin, respectively.

BASIC CONCEPTS OF MOLECULAR SYMMETRY; CHARACTER TABLES

A5-1. Symmetry Operations and Elements

When we say that a molecule has *symmetry*, we mean that *certain parts of it can be interchanged with others without altering either the identity or the orientation of the molecule*. The interchangeable parts are said to be equivalent to one another by symmetry. Consider, for example, a trigonal bipyramidal molecule such as PF_5 (A5-I). The three equatorial P—F bonds, to F_1, F_2, and

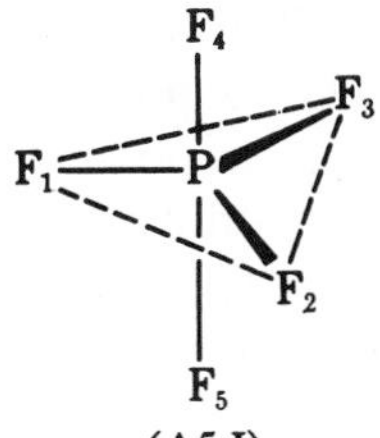

(A5-I)

F_3, are equivalent. They have the same length, the same strength, and the same type of spatial relation to the remainder of the molecule. Any permutation of these three bonds among themselves leads to a molecule indistinguishable from the original. Similarly, the axial P—F bonds, to F_4 and F_5, are equivalent. *But,* axial and equatorial bonds are different types (e.g., they have different lengths), and if one of each were to be interchanged, the molecule would be noticeably perturbed. These statements are probably self-evident, or at least readily acceptable, on an intuitive basis; but for systematic and detailed consideration of symmetry, certain formal tools are needed. The first set of tools is a set of *symmetry operations*.

Symmetry operations are geometrically defined ways of exchanging equivalent parts of a molecule. There are four kinds which are used conventionally and these are sufficient for all our purposes.

1. Simple rotation about an axis passing through the molecule by an angle $2\pi/n$. This operation is called a *proper rotation* and is symbolized C_n. If it is

repeated n times, of course the molecule comes all the way back to the original orientation.

2. Reflection of all atoms through a plane that passes through the molecule. This operation is called *reflection* and is symbolized σ.

3. Reflection of all atoms through a point in the molecule. This operation is called *inversion* and is symbolized **i**.

4. The combination, in either order, of rotating the molecule about an axis passing through it by $2\pi/n$ and reflecting all atoms through a plane that is perpendicular to the axis of rotation is called *improper rotation* and is symbolized S_n.

These operations are *symmetry operations if, and only if*, the appearance of the molecule is *exactly* the same after one of them is carried out as it was before. For instance, consider rotation of the molecule H_2S by $2\pi/2$ about an axis passing through S and bisecting the line between the H atoms. As shown in Fig. A5-1, this operation interchanges the H atoms and interchanges the S—H bonds. Since these atoms and bonds are equivalent, there is no physical (i.e., physically meaningful or detectable) difference after the operation. For HSD, however, the corresponding operation replaces the S—H bond by the S—D bond, and vice versa, and one can see that a change has occurred. Therefore, for H_2S, the operation C_2 is a symmetry operation; for HSD it is not.

These types of symmetry operation are graphically explained by the diagrams in Fig. 5A-2, where it is shown how an arbitrary point (0) in space is affected in each case. Filled dots represent points above the xy plane and open dots represent points below it. Let us examine first the action of proper rotations, illustrated here by the C_4 rotations, that is, rotations by $2\pi/4 = 90°$. The operation C_4 is seen to take the point 0 to the point 1. The application of C_4 twice, designated C_4^2, generates point 2. Operation C_4^3 gives point 3 and, of course, C_4^4, which is a rotation by $4 \times 2\pi/4 = 2\pi$, regenerates the original point. The set of four points, 0, 1, 2, 3 are permutable, cyclically, by repeated C_4 proper rotations and are equivalent points. It will be obvious

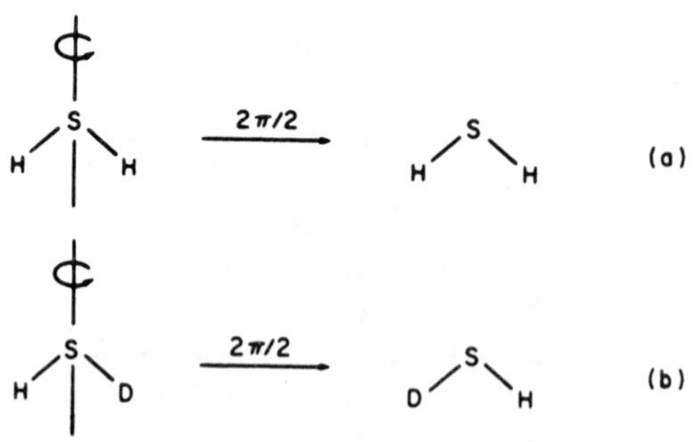

FIG. A5-1. The operation C_2 carries H_2S into an orientation indistinguishable from the original, but HSD goes into an observably different orientation.

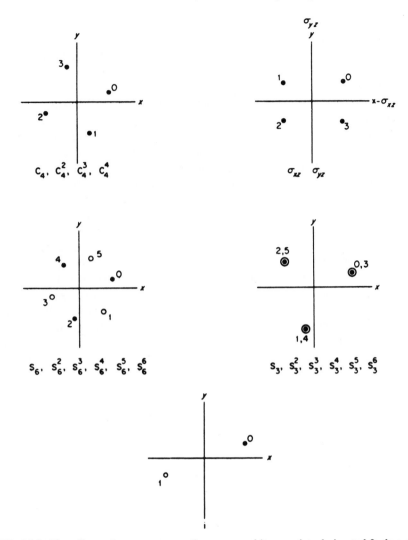

FIG. A5-2. The effects of symmetry operations on an arbitrary point, designated 0, thus generating sets of points.

that in general repetition of a C_n operation will generate a set of n equivalent points from an arbitrary initial point, provided that point lies off the axis of rotation.

The effect of reflection through symmetry planes perpendicular to the xy plane, specifically, σ_{xz} and σ_{yz} is also illustrated in Fig. A5-2. The point 0 is related to point 1 by the σ_{yz} operation and to the point 3 by the σ_{xz} operation. By reflecting either point 1 or point 3 through the second plane, point 2 is obtained.

The set of points generated by the repeated application of an improper rotation will vary in appearance depending on whether the order of the operation, S_n, is even or odd, order being the number n. A crown of n points, alternately up and down, is produced for n even, as illustrated for S_6. For n odd there is generated a set of $2n$ points which form a right n-sided prism, as shown for S_3.

Finally, the operation $\mathbf{i}$ is seen to generate from point 0 a second point, 1, lying on the opposite side of the origin.

Let us now illustrate the symmetry operations for various familiar molecules as examples. As this is done it will be convenient to employ also the concept of *symmetry elements*. A symmetry element is an *axis* (line), *plane*, or *point* about which symmetry operations are performed. The existence of a certain symmetry operation implies the existence of a corresponding symmetry element, and conversely, the presence of a symmetry element means that a certain symmetry operation or set of operations is possible.

Consider the ammonia molecule (Fig. A5-3). The three equivalent hydrogen atoms may be exchanged among themselves in two ways: by proper rotations, and by reflections. The molecule has an axis of threefold proper rotation; this is called a C_3 axis. It passes through the N atom and through the center of the equilateral triangle defined by the H atoms. When the molecule is rotated by $2\pi/3$ in a clockwise direction H_1 replaces H_2, H_2 replaces H_3, and H_3 replaces H_1. Since the three H atoms are physically indistinguishable, the numbering having no physical reality, the molecule after

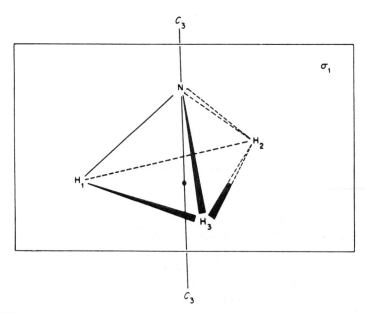

FIG. A5-3. The ammonia molecule, showing its threefold symmetry axis C_3, and one of its three planes of symmetry σ_1, which passes through H_1 and N and bisects the H_2—H_3 line.

rotation is indistinguishable from the molecule before rotation. This rotation, called a C_3 or threefold proper rotation, is a symmetry operation. Rotation by $2 \times 2\pi/3$ also produces a configuration different, but physically indistinguishable, from the original and is likewise a symmetry operation; it is designated C_3^2. Finally, rotation by $3 \times 2\pi/3$ carries each H atom all the way around and returns it to its initial position. This operation, C_3^3, has the same net effect as performing no operation at all, but for mathematical reasons it must be considered as an operation generated by the C_3 axis. This, and other operations which have no net effect, are called *identity* operations and are symbolized by **E**. Thus, we may write $C_3^3 = E$.

The interchange of hydrogen atoms in NH_3 by reflections may be carried out in three ways; that is, there are three planes of symmetry. Each plane passes through the N atom and one of the H atoms, and bisects the line connecting the other two H atoms. Reflection through the symmetry plane containing N and H_1 interchanges H_2 and H_3; the other two reflections interchange H_1 with H_3, and H_1 with H_2.

Inspection of the NH_3 molecule shows that no other symmetry operations besides these six (three rotations, C_3, C_3^2, $C_3^3 = E$, and three reflections, σ_1, σ_2, σ_3) are possible. Put another way, the only symmetry elements the molecule possesses are C_3 and the three planes that we may designate σ_1, σ_2, and σ_3. Specifically, it will be obvious that no sort of improper rotation is possible, nor is there a center of symmetry.

As a more complex example, in which all four types of symmetry operation and element are represented, let us take the $Re_2Cl_8^{2-}$ ion, which has the shape of a square parallepiped or right square prism (Fig. A5-4). This ion has altogether six axes of proper rotation, of four different kinds. First, the Re_1–Re_2 line is an axis of fourfold proper rotation, C_4, and four operations, C_4, C_4^2, C_4^3, $C_4^4 \equiv E$, may be carried out. This same line is also a C_2 axis, generating the operation C_2. It will be noted that the C_4^2 operation means rotation by $2 \times 2\pi/4$, which is equivalent to rotation by $2\pi/2$, that is, to the C_2 operation. Thus the C_2 axis and the C_2 operation are implied by, not independent of, the C_4 axis. There are, however, two other types of C_2 axis that exist independently. There are two of the type that passes through the centers of opposite vertical edges of the prism, C_2' axes, and two more that pass through the centers of opposite vertical faces of the prism, C_2'' axes.

The $Re_2Cl_8^{2-}$ ion has three different kinds of symmetry plane [see Fig. A5-4(b)]. There is a unique one that bisects the Re—Re bond and all the vertical edges of the prism. Since it is customary to define the direction of the highest proper axis of symmetry, C_4 in this case, as the vertical direction, this symmetry plane is horizontal and the subscript h is used to identify it, σ_h. There are then two types of vertical symmetry plane, namely, the two that contain opposite vertical edges, and two others that cut the centers of the opposite vertical faces. One of these two sets may be designated $\sigma_v^{(1)}$ and $\sigma_v^{(2)}$, the v implying that they are vertical. Since those of the second vertical set bisect

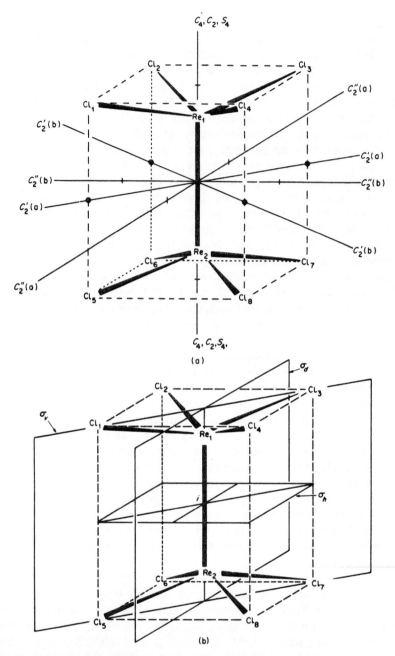

FIG. A5-4. The symmetry elements of the $Re_2Cl_8^{2-}$ ion. (a) The axes of symmetry. (b) One of each type of plane and the center of symmetry.

the dihedral angles between those of the first set, they are then designated $\sigma_d^{(1)}$ and $\sigma_d^{(2)}$, the d standing for dihedral. Both pairs of planes are vertical and it is actually arbitrary which are labeled σ_v and which σ_d.

Continuing with $Re_2Cl_8^{2-}$, we see that an axis of improper rotation is present. This is coincident with the C_4 axis and is an S_4 axis. The S_4 operation about this axis proceeds as follows. The rotational part, through an angle of $2\pi/4$, in the clockwise direction has the same effect as the C_4 operation. When this is coupled with a reflection in the horizontal plane, σ_h, the following shifts of atoms occur:

$$Re_1 \longrightarrow Re_2 \qquad Cl_1 \longrightarrow Cl_6 \qquad Cl_5 \longrightarrow Cl_2$$
$$Re_2 \longrightarrow Re_1 \qquad Cl_2 \longrightarrow Cl_7 \qquad Cl_6 \longrightarrow Cl_3$$
$$Cl_3 \longrightarrow Cl_8 \qquad Cl_7 \longrightarrow Cl_4$$
$$Cl_4 \longrightarrow Cl_5 \qquad Cl_8 \longrightarrow Cl_1$$

Finally, the $Re_2Cl_8^{2-}$ ion has a center of symmetry i and the inversion operation i can be performed.

In the case of $Re_2Cl_8^{2-}$ the improper axis S_4 might be considered as merely the inevitable consequence of the existence of the C_4 axis and the σ_h, and, indeed, this is a perfectly correct way to look at it. However, it is important to emphasize that there are cases in which an improper axis S_n exists without independent existence of either C_n or σ_h. Consider, for example, a tetrahedral molecule as depicted in (A5-II), where the $TiCl_4$ molecule is shown inscribed

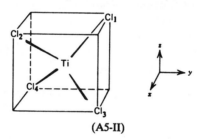

(A5-II)

in a cube and Cartesian axes, x, y, and z are indicated. Each of these axes is an S_4 axis. For example, rotation by $2\pi/4$ about z followed by reflection in the xy plane shifts the Cl atoms as follows:

$$Cl_1 \longrightarrow Cl_3 \qquad Cl_3 \longrightarrow Cl_2$$
$$Cl_2 \longrightarrow Cl_4 \qquad Cl_4 \longrightarrow Cl_1$$

Note, however, that the Cartesian axes are not C_4 axes (though they are C_2 axes) and the principal planes (namely, xy, xz, yz) are not symmetry planes. Thus we have here an example of the existence of the S_n axis without C_n or σ_h having any independent existence. The ethane molecule in its staggered configuration has an S_6 axis and provides another example.

A5-2. Symmetry Groups

The complete set of symmetry operations that can be performed on a molecule is called the *symmetry group* for that molecule. The word "group" is used here not as a mere synonym for "set" or "collection," but in a technical, mathematical sense, and this meaning must first be explained.

Introduction to Multiplying Symmetry Operations. We have already seen in passing that if a proper rotation C_n and a horizontal reflection σ_h can be performed, there is also an operation that results from the combination of the two which we call the improper rotation S_n. We may say that S_n is the product of C_n and σ_h. Noting also that the order in which we perform σ_h and C_n is immaterial,* we can write

$$C_n \times \sigma_h = \sigma_h \times C_n = S_n$$

This is an algebraic way of expressing the fact that successive application of the two operations shown has the same effect as applying the third one. For obvious reasons, it is convenient to speak of the third operation as being the product obtained by multiplication of the other two.

This example is not unusual. Quite generally, any two symmetry operations can be multiplied to give a third. For example, in Fig. A5-2 the effects of reflections in two mutually perpendicular symmetry planes are illustrated. It can be seen that one of the reflections carries point 0 to point 1. The other reflection carries point 1 to point 2. Point 0 can also be taken to point 2 by way of point 3 if the two reflection operations are performed in the opposite order. But a moment's thought will show that a direct transfer of point 0 to point 2 can be achieved by a C_2 operation about the axis defined by the line of intersection of the two planes. If we call the two reflections $\sigma(xz)$ and $\sigma(yz)$ and the rotation $C_2(z)$, we can write:

$$\sigma(xz) \times \sigma(yz) = \sigma(yz) \times \sigma(xz) = C_2(z)$$

It is also evident that

$$\sigma(yz) \times C_2(z) = C_2(z) \times \sigma(yz) = \sigma(xz)$$

and

$$\sigma(xz) \times C_2(z) = C_2(z) \times \sigma(xz) = \sigma(yz)$$

It is also worth noting that if any one of these three operations is applied twice in succession, we get no net result or, in other words, an identity operation, namely;

$$\sigma(xz) \times \sigma(xz) = E$$

$$\sigma(yz) \times \sigma(yz) = E$$

$$C_2(z) \times C_2(z) = E$$

---

*This is, however, a special case; in general, order of multiplication matters as noted later.

Introduction to a Group.   If we pause here and review what has just been done with the three operations $\sigma(xz)$, $\sigma(yz)$, and $C_2(z)$, we see that we have formed all the nine possible products. To summarize the results systematically, we can arrange them in the annexed tabular form. Note that we have added seven more multiplications, namely, all those in which the identity operation E is a factor. The results of these are trivial, since the product of any other, nontrivial operation with E must be just the nontrivial operation itself, as indicated.

	E	$C_2(z)$	$\sigma(xz)$	$\sigma(yz)$
E	E	$C_2(z)$	$\sigma(xz)$	$\sigma(yz)$
$C_2(z)$	$C_2(z)$	E	$\sigma(yz)$	$\sigma(xz)$
$\sigma(xz)$	$\sigma(xz)$	$\sigma(yz)$	E	$C_2(z)$
$\sigma(yz)$	$\sigma(yz)$	$\sigma(xz)$	$C_2(z)$	E

The set of operations E, $C_2(z)$, $\sigma(xz)$, and $\sigma(yz)$ evidently has the following four interesting properties:

1. There is one operation E, the identity, that is the trivial one of making no change. Its product with any other operation is simply the other operation.

2. There is a definition of how to multiply operations: we apply them successively. The product of any two is one of the remaining ones. In other words, this collection of operations is self-sufficient, all its possible products being already within itself. This is sometimes called the property of *closure*.

3. Each of the operations has an *inverse*, that is, an operation by which it may be multiplied to give E as the product. In this case, each operation is its own inverse, as shown by the occurrence of E in all diagonal positions of the table.

4. It can also be shown that if we form a triple product, this may be subdivided in any way we like without changing the result, thus

$$\sigma(xz) \ \times \ \sigma(yz) \ \times \ C_2(z)$$

$$= [\sigma(xz) \times \sigma(yz)] \times C_2(z) \ = \ C_2(z) \times C_2(z)$$

$$= \ \sigma(xz) \ \times \ [\sigma(yz) \times C_2(z)] = \sigma(xz) \times \sigma(xz)$$

$$= E$$

Products that have this property are said to obey the *associative law* of multiplication.

The four properties just enumerated are of fundamental importance. They are the properties—and the *only* properties—that any collection of symmetry operations must have to constitute a *mathematical group*. Groups consisting of symmetry operations are called *symmetry groups* or sometimes *point groups*. The latter term arises because all the operations leave the molecule fixed at a certain point in space. This is in contrast to other groups of symmetry operations, such as those that may be applied to crystal structures in which individual molecules move from one location to another.

The symmetry group we have just been examining is one of the simpler groups; but nonetheless, an important one. It is represented by the symbol C_{2v}; the origin of this and other symbols is discussed later. It is not an entirely representative group in that it has some properties that are *not* necessarily found in other groups. We have already called attention to one, namely, that each operation in this group is its own inverse; this is actually true of only three kinds of operation: reflections, twofold proper rotations, and inversion i. Another special property of the group C_{2v} is that all multiplications in it are *commutative;* that is, every multiplication is equal to the multiplication of the same two operations in the opposite order. It can be seen that the group multiplication table is symmetrical about its main diagonal, which is another way of saying that all possible multiplications commute. In general, multiplication of symmetry operations is *not* commutative, as subsequent discussion will illustrate.

For another simple, but more general, example of a symmetry group, let us recall our earlier examination of the ammonia molecule. We were able to discover six and only six symmetry operations that could be performed on this molecule. If this is indeed a complete list, they should constitute a group. The easiest way to see if they do is to attempt to write a multiplication table. This will contain 36 products, some of which we already know how to write. Thus we know the result of all multiplications involving **E**, and we know that

$$\mathbf{C_3} \times \mathbf{C_3} = \mathbf{C_3^2}$$

$$\mathbf{C_3} \times \mathbf{C_3^2} = \mathbf{C_3^2} \times \mathbf{C_3} = \mathbf{E}$$

It will be noted that the second of these statements means that $\mathbf{C_3}$ is the inverse of $\mathbf{C_3^2}$ and vice versa. We also know that **E** and each of the σ's is its own inverse. So all operations have inverses, thus satisfying requirement 3.

To continue, we may next consider the products when one σ_v is multiplied by another. A typical example is shown in Fig. A5-5(a). When point 0 is reflected first through $\sigma^{(1)}$ and then through $\sigma^{(2)}$, it becomes point 2. But point 2 can obviously also be reached by a clockwise rotation through $2\pi/3$, that is, by the operation $\mathbf{C_3}$. Thus we can write

$$\sigma^{(1)} \times \sigma^{(2)} = \mathbf{C_3}$$

If, however, we reflect first through $\sigma^{(2)}$ and then through $\sigma^{(1)}$, point 0 becomes point 4, which can be reached also by $\mathbf{C_3} \times \mathbf{C_3} = \mathbf{C_3^2}$. Thus we write

$$\sigma^{(2)} \times \sigma^{(1)} = \mathbf{C_3^2}$$

Clearly the reflections $\sigma^{(1)}$ and $\sigma^{(2)}$ do not commute. The reader should be able to make the obvious extension of the geometrical arguments just used

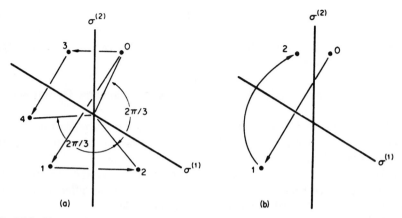

FIG. A5-5. The multiplication of symmetry operations (a) reflection times reflection and (b) reflection followed by C_3.

to obtain the following additional products:

$$\sigma^{(1)} \times \sigma^{(3)} = C_3^2$$

$$\sigma^{(3)} \times \sigma^{(1)} = C_3$$

$$\sigma^{(2)} \times \sigma^{(3)} = C_3$$

$$\sigma^{(3)} \times \sigma^{(2)} = C_3^2$$

There remain, now, the products of C_3 and C_3^2 with $\sigma^{(1)}$, $\sigma^{(2)}$, and $\sigma^{(3)}$. Figure A5-5(b) shows a type of geometric construction that yields these products. For example, we can see that the reflection $\sigma^{(1)}$ followed by the rotation C_3 carries point 0 to point 2, which could have been reached directly by the operation $\sigma^{(2)}$. By similar procedures all the remaining products can be easily determined. The complete multiplication table for this set of operations is given here.

	E	C_3	C_3^2	$\sigma^{(1)}$	$\sigma^{(2)}$	$\sigma^{(3)}$
E	E	C_3	C_3^2	$\sigma^{(1)}$	$\sigma^{(2)}$	$\sigma^{(3)}$
C_3	C_3	C_3^2	E	$\sigma^{(3)}$	$\sigma^{(1)}$	$\sigma^{(2)}$
C_3^2	C_3^2	E	C_3	$\sigma^{(2)}$	$\sigma^{(3)}$	$\sigma^{(1)}$
$\sigma^{(1)}$	$\sigma^{(1)}$	$\sigma^{(2)}$	$\sigma^{(3)}$	E	C_3	C_3^2
$\sigma^{(2)}$	$\sigma^{(2)}$	$\sigma^{(3)}$	$\sigma^{(1)}$	C_3^2	E	C_3
$\sigma^{(3)}$	$\sigma^{(3)}$	$\sigma^{(1)}$	$\sigma^{(2)}$	C_3	C_3^2	E

The successful construction of this table demonstrates that the set of six operations does indeed form a group. This group is represented by the symbol

C_{3v}. The table shows that its characteristics are more general than those of the group C_{2v}. Thus it contains some operations that are not, as well as some which are, their own inverse. It also involves a number of multiplications that are not commutative.

A5-3. Some General Rules for Multiplication of Symmetry Operations

In the preceding section several specific examples of multiplication of symmetry operations have been worked out. On the basis of this experience, the following general rules should not be difficult to accept:

1. The product of two proper rotations must be another proper rotation. Thus although rotations can be created by combining reflections [recall: $\sigma(xz) \times \sigma(yz) = C_2(z)$], the reverse is not possible.

2. The product of two reflections in planes meeting at an angle θ is a rotation by 2θ about the axis formed by the line of intersection of the planes (recall: $\sigma^{(1)} \times \sigma^{(2)} = C_3$ for the ammonia molecule).

3. When there is a rotation operation C_n and a reflection in a plane containing the axis, there must be altogether n such reflections in a set of n planes separated by angles of $2\pi/2n$, intersecting along the C_n axis (recall: $\sigma^{(1)} \times C_3 = \sigma^{(2)}$ for the ammonia molecule).

4. The product of two C_2 operations about axes that intersect at an angle θ is a rotation by 2θ about an axis perpendicular to the plane containing the two C_2 axes.

5. The following pairs of operations always commute:
   (*a*) Two rotations about the same axis.
   (*b*) Reflections through planes perpendicular to each other.
   (*c*) The inversion and any other operation.
   (*d*) Two C_2 operations about perpendicular axes.
   (*e*) C_n and σ_h, where the C_n axis is vertical.

A5-4. A Systematic Listing of Symmetry Groups, with Examples

The symmetry groups to which real molecules may belong are very numerous. However they may be systematically classified by considering how to build them up using increasingly more elaborate combinations of symmetry operations. The outline that follows, though neither unique in its approach nor rigorous in its procedure, affords a practical scheme for use by most chemists.

The simplest nontrivial groups are those of order 2, that is, those containing but one operation in addition to $\mathbf{E}$. The additional operation must be one that is its own inverse; thus the only groups of order 2 are

$$C_s: \mathbf{E}, \sigma$$

$$C_i: \mathbf{E}, \mathbf{i}$$

$$C_2: \mathbf{E}, \mathbf{C}_2$$

The symbols for these groups are rather arbitrary, except for C_2 which, we shall soon see, forms part of a pattern.

Molecules with C_s symmetry are fairly numerous. Examples are the thionyl halides and sulfoxides (A5-III), and secondary amines (A5-IV). Molecules having a center of symmetry as their *only* symmetry element are quite rare; two types are shown as (A5-V) and (A5-VI). The reader should find it very challenging, though not impossible, to think of others. Molecules of C_2 symmetry are fairly common, two examples being (A5-VII) and (A5-VIII).

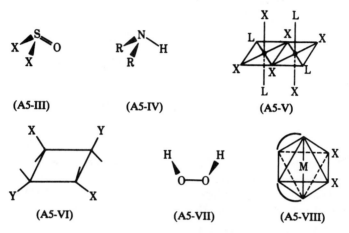

(A5-III)   (A5-IV)   (A5-V)

(A5-VI)   (A5-VII)   (A5-VIII)

The Uniaxial or C_n Groups.   These are the groups in which all operations are due to the presence of a proper axis as the sole symmetry element. The general symbol for such a group, and the operations in it, are

$$C_n: C_n, C_n^2, C_n^3, \cdots C_n^{n-1}, C_n^n \equiv E$$

A C_n group is thus of order n. We have already mentioned the group C_2. Molecules with pure axial symmetry other than C_2 are rare. Two examples of the group C_3 are shown in (A5-IX) and (A5-X).

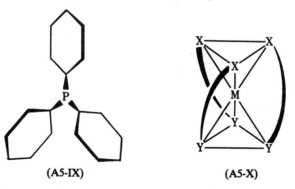

(A5-IX)   (A5-X)

The C_{nv} Groups.   If in addition to a proper axis of order n there is also a set of n vertical planes, we have a group of order $2n$, designated C_{nv}. This type of symmetry is found quite frequently and is illustrated in (A5-XI) to (A5-XV), where the values of n are 2 to 6.

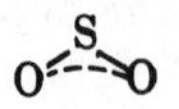

(A5-XI)    (A5-XII) 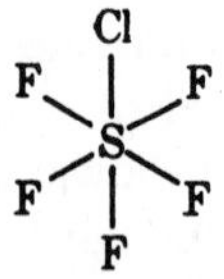   (A5-XIII)

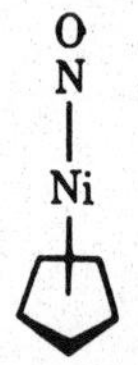

(A5-XIV) 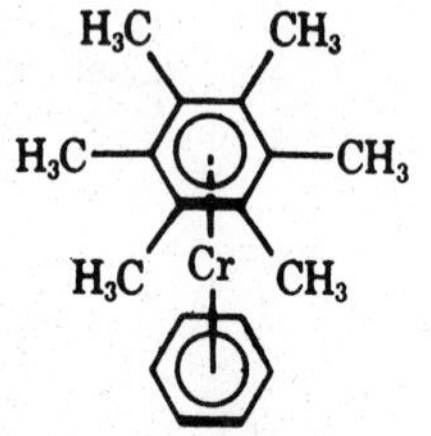   (A5-XV)

The C_{nh} Groups.   If in addition to a proper axis of order n there is also a horizontal plane of symmetry, we have a group of order $2n$, designated C_{nh}. The $2n$ operations include S_n^m operations that are products of C_n^m and σ_h for n odd, to make the total of $2n$. Thus for C_{3h} the operations are

$$C_3, \ C_3^2, \ C_3^3 \equiv E$$

$$\sigma_h$$

$$\sigma_h \times C_3 = C_3 \times \sigma_h = S_3$$

$$\sigma_h \times C_3^2 = C_3^2 \times \sigma_h = S_3^5$$

Molecules of C_{nh} symmetry with $n > 2$ are relatively rare; examples with $n = 2, 3$, and 4 are shown in (A5-XVI) to (A5-XVIII).

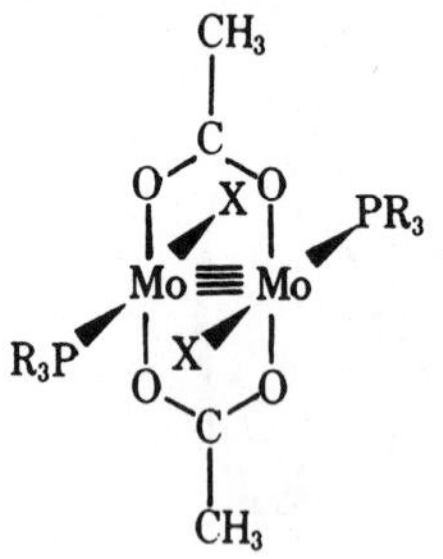

(A5-XVI) 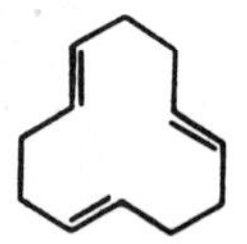   (A5-XVII) 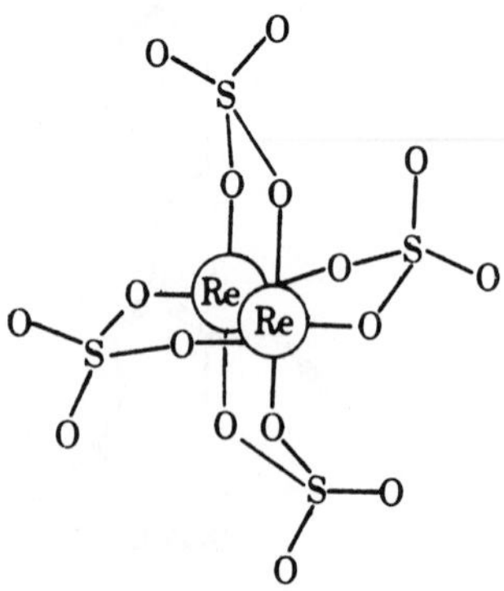   (A5-XVIII)

The D_n Groups.   When a vertical C_n axis is accompanied by a set of n C_2 axes perpendicular to it, the group is D_n. Molecules of D_n symmetry are, in general, rare, but there is one very important type, namely, the trischelates (A5-XIX) of D_3 symmetry.

(A5-XIX)

The D_{nh} Groups.   If to the operations making up a D_n group we add reflection in a horizontal plane of symmetry, the group D_{nh} is obtained. It should be noted that products of the type $C_2 \times \sigma_h$ will give rise to a set of reflections in vertical planes. These planes *contain* the C_2 axes; this point is important in regard to the distinction between D_{nh} and D_{nd}, mentioned next. The D_{nh} symmetry is found in a number of important molecules, a few of which are benzene (D_{6h}), ferrocene in an eclipsed configuration (D_{5h}), $Re_2Cl_8^{2-}$, which we examined previously, (D_{4h}), $PtCl_4^{2-}$ (D_{4h}), and the boron halides (D_{3h}) and PF_5(D_{3h}). All right prisms with regular polygons for bases as illustrated in (A5-XX) and (A5-XXI), and all bipyramids, as illustrated in (A5-XXII) and (A5-XXIII), have D_{nh}-type symmetry.

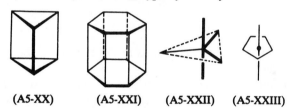

(A5-XX)   (A5-XXI)   (A5-XXII)   (A5-XXIII)

The D_{nd} Groups.   If to the operations making up a D_n group we add a set of vertical planes that bisect the angles between pairs of C_2 axes (note the distinction from the vertical planes in D_{nh}), we have a group called D_{nd}. The D_{nd} groups have no σ_h. Perhaps the most celebrated examples of D_{nd} symmetry are the D_{3d} and D_{5d} symmetries of $R_3W{\equiv}WR_3$ and ferrocene in their staggered configurations (A5-XXIV) and (A5-XXV).

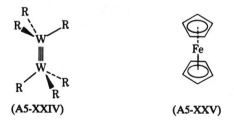

(A5-XXIV)   (A5-XXV)

Two comments about the scheme so far outlined may be helpful. The reader may have wondered why we did not consider the result of adding to the operations of C_n *both* a set of $n\sigma_v$'s and a σ_h. The answer is that this is simply another way of getting to D_{nh}, since a set of C_2 axes is formed along the lines of intersection of the σ_h with each of the σ_v's. By convention, and in accord with the symbols used to designate the groups, it is preferable to proceed as we did. Second, in dealing with the D_{nh}-type groups, if a horizontal plane is found, there must be only the n vertical planes *containing* the C_2 axis. If dihedral planes were also present, there would be, in all $2n$ planes and hence, as shown previously, a principal axis of order $2n$, thus vitiating the assumption of a D_n type of group.

The S_n Groups. Our scheme has, so far, overlooked one possibility, namely, that a molecule might contain an S_n axis as its only symmetry element (except for others that are directly subservient to it). It can be shown that for n odd, the groups of operations arising would actually be those forming the group C_{nh}. For example, take the operations generated by an S_3 axis:

$$S_3$$

$$S_3^2 = C_3^2$$

$$S_3^3 = \sigma_h$$

$$S_3^4 = C_3$$

$$S_3^5$$

$$S_3^6 = E$$

Comparison with the list of operations in the group C_{3h} shows that the two lists are identical.

It is only when n is an even number that new groups can arise that are not already in the scheme. For instance, consider the set of operations generated by an S_4 axis:

$$S_4$$

$$S_4^2 = C_2$$

$$S_4^3$$

$$S_4^4 = E$$

This set of operations satisfies the four requirements for a group and is not a set that can be obtained by any procedure previously described. Thus S_4, S_6, and so on are new groups. They are distinguished by the fact that they contain no operation that is not an S_n^m operation, even though it may be written in another way, as with $S_4^2 = C_2$ above.

Note that the group S_2 is not new. A little thought will show that the operation S_2 is identical with the operation i. Hence the group that could be called S_2 is the one we have already called C_i.

An example of a molecule with S_4 symmetry is shown in (A5-XXVI). Molecules with S_n symmetries are not very common.

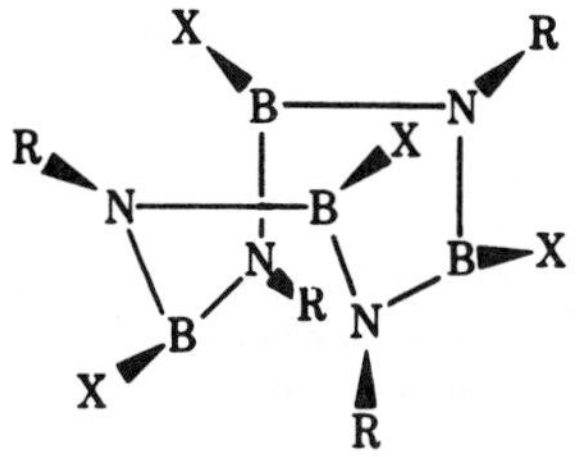

(A5-XXVI)

Linear Molecules. There are only two kinds of symmetry for linear molecules. There are those represented by (A5-XXVII), which have identical ends. Thus, in addition to an infinitefold rotation axis C_∞, coinciding with the molecular axis, and an infinite number of vertical symmetry planes, they have a horizontal plane of symmetry and an infinite number of C_2 axes perpendicular to C_∞. The group of these operations is $D_{\infty h}$. A linear molecule with different ends (A5-XXVIII), has only C_∞ and the σ_v's as symmetry elements. The group of operations generated by these is called $C_{\infty v}$.

$$A—B—C—B—A \qquad A—B—C—D$$
(A5-XXVII) (A5-XXVIII)

A5-5. The Groups of Very High Symmetry

The scheme followed in the preceding section has considered only cases in which there is a single axis of order equal to or > 3. It is possible to have symmetry groups in which there are several such axes. There are, in fact, seven such groups, and several of them are of paramount importance.

The Tetrahedron.   We consider first a regular tetrahedron. Figure A5-6 shows some of the symmetry elements of the tetrahedron, including at least one of each kind. From this it can be seen that the tetrahedron has altogether 24 symmetry operations, which are as follows:

There are three S_4 axes, each of which gives rise to the operations S_4, $S_4^2 \equiv C_2$, S_4^3, and $S_4^4 \equiv E$. Neglecting the S_4^4's, this makes $3 \times 3 = 9$.

There are four C_3 axes, each giving rise to C_3, C_3^2 and $C_3^3 \equiv E$. Again omitting the identity operations, this makes $4 \times 2 = 8$.

There are six reflection planes, only one of which is shown in Fig. A5-6, giving rise to six σ_d operations.

Thus there are $9 + 8 + 6 +$ one identity operation $= 24$ operations. This group is called T_d. It is worth emphasizing that despite the considerable amount of symmetry, there is no inversion center in T_d symmetry. There are,

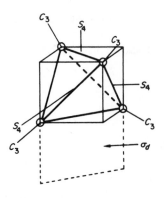

FIG. A5-6. The tetrahedron, showing some of its essential symmetry elements. All S_4 and C_3 axes are shown, but only one of the six dihedral planes σ_d.

of course, numerous molecules having full T_d symmetry, such as CH_4, SiF_4, ClO_4^-, $Ni(CO)_4$, and $Ir_4(CO)_{12}$, and many others where the symmetry is less but approximates to it.

If we remove from the T_d group the reflections, it turns out that the S_4 and S_4^3 operations are also lost. The remaining 12 operations (E, four C_3 operations, four C_3^2 operations and three C_2 operations) form a group, designated T. This group in itself has little importance, since it is very rarely, if ever, encountered in real molecules. However, if we then add to the operations in the group T a different set of reflections in the three planes defined so that each one contains two of the C_2 axes, and work out all products of operations, we get a new group of 24 operations (E, four C_3, four C_3^2, three C_2, three σ_h, i, four S_6, four S_6^5) denoted T_h. This, too, is rare, but it occurs in some "octahedral" complexes in which the ligands are planar and arranged as in (A5-XXIX). The important feature here is that each pair of ligands on

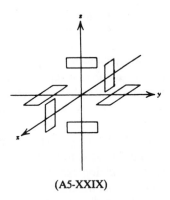

(A5-XXIX)

each of the Cartesian axes is in a different one of the three mutually perpendicular planes, xy, xz, and yz. Real cases are provided by $W(NMe_2)_6$ and several $M(NO_3)_6^{n-}$ ions in which the NO_3^- ions are bidentate.

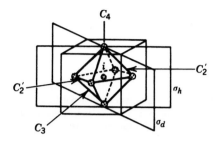

FIG. A5-7. The octahedron and the cube, showing one of each of their essential types of symmetry element.

The Octahedron and the Cube. These two bodies have the same elements, as shown in Fig. A5-7, where the octahedron is inscribed in a cube, and the centers of the six cube faces form the vertices of the octahedron. Conversely, the centers of the eight faces of the octahedron form the vertices of a cube. Figure A5-7 shows one of each of the types of symmetry element that these two polyhedra possess. The list of symmetry operations is as follows:

There are three C_4 axes, each generating C_4, $C_4^2 \equiv C_2$, C_4^3, $C_4^4 \equiv E$. Thus there are $3 \times 3 = 9$ rotations, excluding C_4^4's.

There are four C_3 axes giving four C_3's and four C_3^2's.

There are six C_2' axes bisecting opposite edges, giving six C_2'''s.

There are three planes of the type σ_h and six of the type σ_d, giving rise to nine reflection operations.

The C_4 axes are also S_4 axes and each of these generates the operations S_4, $S_4^2 \equiv C_2$ and S_4^3, the first and last of which are not yet listed, thus adding $3 \times 2 = 6$ more to the list.

The C_3 axes are also S_6 axes and each of these generates the new operations S_6, $S_6^3 \equiv i$, and S_6^5. The i counts only once, so there are then $(4 \times 2) + 1 = 9$ more new operations.

The entire group thus consists of the identity $+ 9 + 8 + 6 + 9 + 6 + 9 = 48$ operations. This group is denoted O_h. It is, of course, a very important type of symmetry since octahedral molecules (e.g. SF_6), octahedral complexes $[Co(NH_3)_6^{3+}$ and $IrCl_6^{3-}]$, and octahedral interstices in solid arrays are very common. There is a group O, which consists of only the 24 proper rotations from O_h, but this, like T, is rarely if ever encountered in Nature.

The Pentagonal Dodecahedron and the Icosahedron. These bodies (Fig. A5-8) are related to each other in the same way as are the octahedron and the cube, the vertices of one defining the face centers of the other, and vice versa. Both have the same symmetry operations, a total of 120! We shall not list them in detail but merely mention the basic symmetry elements: six C_5 axes; ten C_3 axes, fifteen C_2 axes, and fifteen planes of symmetry. The group of 120 operations is designated I_h and is often called the icosahedral group.

There is one known example of a molecule that is a pentagonal dodecahedron, viz., dodecahedrane, $C_{12}H_{12}$. The icosahedron is a key structural unit

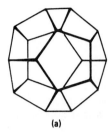

(a)   (b)

FIG. A5-8. The two regular polyhedra having I_h symmetry. (a) The pentagonal dodecahedron and (b) the icosahedron.

in boron chemistry, occurring in all forms of elemental boron as well as in the $B_{12}H_{12}^{2-}$ ion.

If the symmetry planes are omitted, a group called I consisting of only proper rotations remains. This is mentioned purely for the sake of completeness, since no example of its occurrence in Nature is known.

A5-6.   Molecular Dissymmetry and Optical Activity

Optical activity, that is, rotation of the plane of polarized light coupled with unequal absorption of the right- and left-circularly polarized components, is a property of a molecule (or an entire three-dimensional array of atoms or molecules) that is not superposable on its mirror image. When the number of molecules of one type exceeds the number of those that are their nonsuperposable mirror images, a net optical activity results. To predict when optical activity will be possible, it is necessary to have a criterion to determine when a molecule and its mirror image will not be identical, that is, superposable.

Molecules that are not superposable on their mirror images are called *dissymmetric.* This term is preferable to "asymmetric," which means "without symmetry," whereas dissymmetric molecules can and often do possess some symmetry, as will be seen.*

A compact statement of the relation between molecular symmetry properties and dissymmetric character is: *a molecule that has no axis of improper rotation is dissymmetric.*

This statement includes and extends the usual one to the effect that optical isomerism exists when a molecule has neither a plane nor a center of symmetry. It has already been noted that the inversion operation i is equivalent to the improper rotation S_2. Similarly, S_1 is a correct although unused way of representing σ, since it implies rotation by $2\pi/1$, equivalent to no net rotation, in conjunction with the reflection. Thus σ and i are simply special cases of improper rotations.

---

*Dissymmetry is sometimes called chirality, and dissymmetric chiral, from the Greek word χειρ for hand, in view of the left-hand/right-hand relation of molecules that are mirror images.

However, even when σ and i are absent, a molecule may still be identical with its mirror image if it possesses an S_n axis of some higher order. A good example of this is provided by the $(\text{—RNBX—})_4$ molecule shown in (5A-XXVI). This molecule has neither a plane nor a center of symmetry, but inspection shows that it can be superposed on its mirror image. As we have noted, it belongs to the symmetry group S_4.

Dissymmetric molecules either have no symmetry at all, or they belong to one of the groups consisting only of proper rotation operations, that is, the C_n or D_n groups. (Groups T, O, and I are, in practice, not encountered, though molecules in these groups must also be dissymmetric.) Important examples are the bischelate and trischelate octahedral complexes (A5-VIII), (A5-X), and (A5-XIX).

INDEX